Use these cards to...

- **order extra copies**
- **share your comments or suggestions with the editor**

...or call toll-free 1-800-877-GALE

...nc.

Please send me ____ copy(s) of **Inventing & Patenting Sourcebook / Inventor's Desktop Companion.**
To save 5% on future editions, I've checked the "standing order" space.

Copies		Enter as Standing Order (5% discount)
____ **Inventing & Patenting Sourcebook**	$75.00	____
____ **Inventor's Desktop Companion**	$24.95	____

Institution _____

Address _____

City, State & Zip _____

Attention _____
 Name   Title/Department

Phone (____) _____

Gale Research Inc.

Please send me ____ copy(s) of **Inventing & Patenting Sourcebook / Inventor's Desktop Companion.**
To save 5% on future editions, I've checked the "standing order" space.

Copies		Enter as Standing Order (5% discount)
____ **Inventing & Patenting Sourcebook**	$75.00	____
____ **Inventor's Desktop Companion**	$24.95	____

Institution _____

Address _____

City, State & Zip _____

Attention _____
 Name   Title/Department

Phone (____) _____

COMMENT CARD

Please use this postage paid card to make suggestions regarding the content, arrangement, indexing, or other features of **Inventing & Patenting Sourcebook.**

Name/Title _____

Institution _____

Address _____

City, State, & Zip _____

Phone (____) _____

BUSINESS REPLY MAIL
FIRST CLASS PERMIT NO. 17022 DETROIT, MI 48226

POSTAGE WILL BE PAID BY ADDRESSEE

**Order Department
Gale Research Inc.
P.O. Box 441914
Detroit, MI 48244-9980**

No postage
necessary if mailed
in the United States

These cards are for your convenience.

Use this card to order your own copy of Inventing & Patenting Sourcebook.

Make it a STANDING ORDER, and you will be sure of receiving new editions promptly... and at a 5% discount. (All editions come to you on a 30-day approval, and you may cancel your standing order at any time.)

BUSINESS REPLY MAIL
FIRST CLASS PERMIT NO. 17022 DETROIT, MI 48226

POSTAGE WILL BE PAID BY ADDRESSEE

**Order Department
Gale Research Inc.
P.O. Box 441914
Detroit, MI 48244-9980**

No postage
necessary if mailed
in the United States

...or call toll-free
1-800-877-GALE

BUSINESS REPLY MAIL
FIRST CLASS PERMIT NO. 17022 DETROIT, MI 48226

POSTAGE WILL BE PAID BY ADDRESSEE

**Inventing & Patenting Sourcebook
Gale Research Inc.
835 Penobscot Building
Detroit, MI 48226-9980**

No postage
necessary if mailed
in the United States

INVENTING
&
PATENTING
SOURCEBOOK

ISSN 1044-4742

INVENTING & PATENTING SOURCEBOOK

How to Sell and Protect Your Ideas

SECOND EDITION

RICHARD C. LEVY

Annette Piccirelli,

Editor

DETROIT • LONDON

Richard C. Levy

Editor: Annette Piccirelli
Associate Editors: Christopher Kasic, Camille Killens
Assistant Editors: Charles A. Beaubien, Tony Gerring,
Evelyn Sullen
Aided By: Cynthia E. Grayson
Senior Editor: Linda S. Hubbard

Supervisor of Systems and Programming: Theresa Rocklin
Programmer: Dan Bono
Typesetting Consultant: Brian Keith Partin

Research Manager: Victoria B. Cariappa
Editorial Associate: Lisa Lantz
Editorial Assistants: Daniel L. Day, Brian Escamilla,
Charles A. Jewell, Phyllis Shepherd

Production Manager: Mary Beth Trimper
Production Assistant: Shanna Heilveil

Art Director: Arthur Chartow
Keyliner: Nicholas A. Jakubiak

∞™ This book is printed on acid-free paper that meets the minimum requirements of American National Standard for Information Sciences— Permanence Paper for Printed Library Materials, ANSI Z39.48-1984.

♺ This book is printed on recycled paper that meets Environmental Protection Agency standards.

ISBN: 0-8103-7616-4
ISSN: 1044-4742

Printed in the United States of America

Published in the United States by Gale Research Inc.
Published simultaneously in the United Kingdom
by Gale Research International Limited
(an affiliated company of Gale Research Inc.)

To my wife, Sheryl, who brings to every one of our creative adventures a great intelligence, imagination, and spirit.

To my daughter, Bettie, who sets the pace, and has taught me more about the creative process than she may ever realize.

Best-selling author and inventor Richard C. Levy has developed and licensed more than 75 innovative concepts in the last decade. He is the author of **Secrets of Selling Inventions** and the creator of the popular game **Adverteasing**, which can be found on the shelves of department and toy stores nationwide.

"A rare and multi-talented individual . . ."—*Inventor's Digest*

Contents

Part One: Forms and Tables

Part Two: How to Protect and License Your Invention

Highlights

Inventing & Patenting Sourcebook (IPS) is the most comprehensive guide for helping inventors develop, protect, and license their inventions. It combines a thorough how-to section, a complete set of pertinent forms that can be photocopied, and detailed directory listings on several types of organizations and publications, including:

- ◆ National inventor associations
- ◆ Consultants and research firms
- ◆ Business incubators
- ◆ Venture capital sources
- ◆ Federal funding sources

- ◆ Online resources
- ◆ Patent attorneys and agents
- ◆ Youth innovation programs
- ◆ Patent and Trademark Depository Libraries
- ◆ Publications

IPS was conceptualized, compiled, and authored by professional inventor and marketeer Richard C. Levy, whose more than 75 licensed products include the best-selling game Adver*teasing*. Levy takes the inventor step-by-step through his personal strategies for successful licensing of inventions. Topics include patent, trademark, and copyright protection, lawyer selection, corporate presentations, contract negotiations, funding sources, and much more.

New in This Edition

This second edition also includes more than 2700 directory listings, hundreds of them appearing for the first time. Entries have been completely updated, with thousands of changes to names, addresses, telephone numbers, and descriptions.

These new features help make *IPS* more useful than ever:

- ◆ Approximately 370 federal laboratories offering inventors free technical assistance;
- ◆ Approximately 100 state venture capital programs;
- ◆ A complete listing of federal and state participants in the Small Business Innovation Research (SBIR) Program;
- ◆ A glossary of important inventing terms and phrases;
- ◆ Master Index now includes page references to terms and organizations in Levy's how-to essay.

Arrangement and Indexes

IPS is divided into four main parts:

- ◆ **Forms and Tables**, a collection of copyright-free forms and documents and convenient reference tables showing the distribution of U.S. patents;
- ◆ **How to Protect and License Your Invention**, an essay by Richard C. Levy;
- ◆ **Directory of Inventing and Patenting Information**, 15 sections covering associations, publications, state innovation programs, federal funding sources, and more;
- ◆ **Appendices and Master Index**, including an alphabetical guide to U.S. Patent Classifications, the Patent and Trademark Office Telephone Directory, the Copyright Office Telephone Directory, a glossary of inventing terms, and a comprehensive Master Index.

Acknowledgements

A book such as this depends heavily upon the cooperation and assistance of many people. During the course of my research, I contacted an extensive number of independent inventors, federal and state government officials, educators, patent attorneys, corporate executives, trade association directors, librarians, and members of inventor organizations. Their warm reception, hospitality, and interest made my assignment one that I will always fondly remember.

While many of those who have helped me are mentioned at appropriate places in the book, I do want to express special thanks to certain people.

Oscar Mastin, Public Affairs Officer at the Patent and Trademark Office, has been a friend for many years and an invaluable source of assistance. He provided me with much of the background material for the opening section of the book, and guided me through the PTO's bureaucracy to get the answers to my seemingly endless queries.

George P. Lewett, Chief, and Fred L. Hart, Operations Research Analyst, Office of Technology Evaluation and Assessment, National Institute of Standards and Technology, responded, as always, above and beyond the call of duty, and always with good cheer, to get me anything I required for my work. Even in the midst of an office relocation and a personal tragedy, Elizabeth M. Robertson, analyst for Clearinghouse for State and Local Initiatives on Productivity, Technology and Innovations, never missed a beat in her support of my requests for help. Her sense of responsibility, concern, and passion for her work will always stand out in my mind.

I owe a debt of appreciation to many people at the PTO, including: Donald J. Quigg, former Commissioner of Patent and Trademarks; Harry F. Manbeck, Jr., Commissioner of Patents and Trademarks; Norma Rose, administrative assistant to the Commissioner; Ed Kazenske, executive assistant to the Commissioner; Jim Lynch, Director, Office of Budget; John D. Hassett, director of the Office of General Services; Lou Massel, editor of *The Manual of Patent Examining Procedure*; Carole A. Shores, director of the Patenting and Trademark Depository Library Programs; Don Kelly, Group director of Group 320; Jeff Nase, supervisory petition examiner; J. Michael Thesz, special program examiner; Ruth Ann Nyblod, director of Project XL; Elizabeth Weimar, patent examiner, Group 180; Ann Faris, patent examiner, Group 260; Jeff Alderson, Office of Information Systems; Maureen Brown, computer systems analyst; Cameron Weiffenbach, director of Enrollment and Discipline; and Gerald Gerard Rogers, staff attorney, Office of the Assistant Commissioner for Trademarks.

At the Copyright Office, thanks to Dorothy Schrader, general counsel, and Boris Bohun-Chudyniv, analyst for the Federal Research Division; Janis Long, Majority Counsel, Senate Judicial Subcommittee on Patents, Copyrights, and Trademarks. Also of great help were: John Heizer, financial analyst for the Clearinghouse for State and Local Initiatives on Productivity, Technology and Innovations, Department of Commerce; Ray Barnes, acting director of Inventions and Innovation Program, Department of Energy; Howard Shapiro, press officer for the Federal Trade Commission; Maureen Wood, Office of Technology Evaluation and Assessment; Keith Gold, freedom of information officer, Federal Trade Commission; Marianne K. Clarke, senior policy analyst, Center for Policy Research and Analysis, National Governors' Association; Jeff Norris and Joyce Hamaty, public affairs specialists, National Science Foundation; Sheila Hooks, SBA Office of Communications; Howard A. MacCord, secretary of National Council of Intellectual Property Law Associations; Richard Sparks, program manager for Defense Technical Information Center; Gil Young, acting director of the Minnesota Department of Trade and Economic Development; Mark A. Spikell, Ed. D., professor, George Mason University; G. Thomas Cator, executive director of the Association of Small Business Development Centers; Herbert C. Wamsley, executive director of Intellectual Property Owners; Ada Winters, associate executive director of Intellectual Property Owners; Ginny Panholzer and Shannon S. Jamison, IPO; Michael Blommer, AIPLA; John T. Farady, president of Affiliated Inventors Foundation; Joanne M. Hayes, editor of *Inventor's Digest*; Jan Kosko, public affairs officer, NIST; Robert Faris, Esq., Nixon and Vanderhye;

Howard Doesher, Esq., Cushman, Darby and Cushman; John C. Sandefur, membership services director, Association of Small Business Development Centers; Dr. Martin J. Bernard III, Argonne National Laboratory; Bernard (Burt) Murphy, president of FAXPAT; and Robert MacCollum, a good friend and expert patent draftsman.

Special thanks to Michael Ross, a friend who is always there with a wise word on just about anything.

At Gale Research Inc., thanks to: Elizabeth A. Geiser for taking the phone call that started this adventure; Bob Elster and John Schmittroth, for their belief in my proposal and for being its steadfast champions; Bob Huffman, my first editor, who moonlights as a very talented game inventor; Beth Dempsey, the best PR and media pro I've ever worked with in the publishing industry; and Annette Piccirelli, editor of this edition and a person who has made invaluable contributions to it.

Introduction

Today's inventors often find themselves faced with a dilemma: while they usually have a surplus of creative ideas, they often don't have a good understanding of how to move their concepts into the marketplace. Those who do manage the long journey from concept to fully-marketed product find it can be a frustrating, time-consuming, and expensive effort. For example, simple errors filling out government forms or negligence in returning the forms can lead to delays, penalties, or an outright refusal by the government to consider the application to protect the invention. In addition, inventors are faced with a wide variety of law firms, invention marketing services, government agencies, and other organizations involved in protecting and marketing inventions, many of them charging hefty fees for their services. Consequently, inventors may easily spend a great deal of time and money trying to turn their ideas into reality.

Inventing & Patenting Sourcebook (IPS) was developed to give inventors practical advice and an extensive network of resources to get through the process of patenting and licensing their ideas. *IPS* provides inventors, innovators, and marketers of new products and inventions with a comprehensive how-to guide on developing, patenting, licensing, and marketing ideas and concepts. Richard C. Levy, inventor of the popular game Adver*teasing*, shares his own strategies and expertise throughout the how-to sections. Over the past 15 years, Levy has created and successfully licensed more than 75 products, which have in turn generated tens of millions of dollars.

In *IPS* Levy covers such topics as independent patent and trademark searching, laws protecting inventors from phony invention marketing companies, selecting companies to approach for licensing agreements, and procuring federal research and development funds. Complementing this essay are directory listings for thousands of agencies, programs, professionals, publications, and other resources that can provide invaluable assistance and information.

New Features in This Edition

Several new sections have been added to this edition of *IPS* to put new and veteran inventors in touch with more resources than ever before:

- ◆ A complete listing of federal and state participants in the Small Business Innovation Research (SBIR) Program;

- ◆ A complete listing of all district, regional, and branch offices of the U.S. Small Business Administration;

- ◆ Approximately 100 state-sponsored venture capital programs that can provide loans to innovative individuals and companies;

- ◆ Approximately 370 federal laboratories that provide free technical assistance and can often make arrangements to share facilities and lab space with inventors.

Other new features include:

- ◆ A glossary of commonly-used inventing and licensing terms and phrases, and

- ◆ An enhanced Master Index providing citations to organizations, individuals, and publications in Levy's how-to material as well as to the directory listings.

Method of Compilation

The database resources of Gale Research Inc. were used to supply information for various sections in Part Three: Directory of Inventing and Patenting Information. Other sections in Part Three were compiled from lists provided by various U.S. federal organizations, including the U.S. Patent and Trademark Office, the Small Business Administration, and the National Technical Information Service. For more details on sources used, consult the User's Guide immediately following this Introduction.

Acknowledgements

The editor would like to thank Richard Levy for his careful attention to deadlines and his constant enthusiasm throughout the course of this project. Special thanks are also extended to Charles Beaubien and Brian Partin who offered their computer expertise and devoted much of their time to this project.

Comments Welcome

Every effort was made to provide the most accurate, up-to-date information possible. Comments and suggestions for improvements are welcome, including areas of additional coverage. Please contact:

Editor
Inventing & Patenting Sourcebook
Gale Research, Inc.
835 Penobscot Bldg.
Detroit, MI 48226-4094
Phone: (313) 961-2242
Fax: (313) 961-6815
Toll-Free: 800-347-GALE

—Annette Piccirelli

User's Guide

Inventing & Patenting Sourcebook is divided into four parts:

Part One: Forms and Tables
Part Two: How to Protect and License Your Invention
Part Three: Directory of Inventing and Patenting Information
Part Four: Appendices and Master Index

Part One: Forms and Tables

This section contains 41 copyright-free forms that inventors can photocopy and file with the Patent and Trademark Office. Also included in this section are two tables showing the distribution of U.S. patents granted to U.S. residents and residents of foreign countries. All forms and tables are referenced in Richard C. Levy's essay.

Part Two: How to Protect and License Your Invention

This 114-page essay by Richard C. Levy takes you step-by-step through the process of getting your idea licensed and marketed. It includes how-to information on all aspects of inventing, including how to conduct patent, trademark, and copyright searches; how to complete patent applications; how to execute patent drawings and make prototypes; how to choose a patent attorney or agent; how to select a company to approach with your idea; and how to present your concept. Throughout the essay, Levy offers "insider" tips and personal anecdotes to help readers save time and money and avoid the common pitfalls associated with getting an idea patented and to the marketplace.

Part Three: Directory of Patenting and Inventing Information

Separate sections identify each category of resource, from governmental agencies to publications. Arrangement of entries within individual sections varies as appropriate to the material being presented. If a parent organization maintains one or more subunits, then those subunits will appear below their parent organizations, indented and in slightly smaller type. Except for the 13,700 registered patent attorneys and agents, all directory listings are cited in the Master Index.

Section 13: National and Regional Inventor Associations

- **Content:** Includes the name, address, telephone number, fax number (when available), contact name, and description (when available) of more than 90 U.S. inventor associations.
- **Arrangement:** Alphabetical by state, then alphabetical by association name.
- **Sources:** *Encyclopedia of Associations* (published by Gale Research, Inc.); supplemented by original research.

Section 14: Invention Consultants and Research Firms

- **Content:** Includes the name, address, telephone number, and fax number and description (when available) of more than 370 U.S. new product development firms, invention marketing organizations, and small business consultants.
- **Arrangement:** Alphabetical by state, then alphabetical by consulting organization name.

♦ **Sources:** *Consultants and Consulting Organizations Directory* and *Research Services Directory* (published by Gale Research, Inc.); supplemented by original research.

Section 15: University and Independent Innovation Research Centers

♦ **Content:** Includes the name, address, telephone number, fax number (when available), contact name, and description of nearly 300 university research parks, technology transfer programs, invention evaluation centers, small business development centers, and independent research centers.
♦ **Arrangement:** Alphabetical by state, then alphabetical by research center name.
♦ **Sources:** *Research Centers Directory* (published by Gale Research Inc.); supplemented by lists provided by the Association of Small Business Development Centers.

Section 16: Business Incubators

♦ **Content:** Includes the name, address, telephone number, and fax number and description (when available) of nearly 50 organizations established to encourage entrepreneurship and provide low-cost office and production space to innovative organizations.
♦ **Arrangement:** Alphabetical by state, then alphabetical by incubator name.
♦ **Sources:** *Small Business Sourcebook* (published by Gale Research Inc.).

Section 17: Private Sector Venture Capital Sources

♦ **Content:** Includes the name, address, telephone number, and fax number and description (when available) of more than 625 firms that provide loans to small businesses and innovative companies.
♦ **Arrangement:** Alphabetical by state, then alphabetical by venture capital firm name.
♦ **Sources:** *Small Business Sourcebook* (published by Gale Research Inc.).

Section 18: Public Sector Venture Capital Sources

♦ **Content:** Includes the name, address, telephone number, contact name, and description (when available) of nearly 100 state government organizations that provide loans to small businesses and innovative companies.
♦ **Arrangement:** Alphabetical by state, then alphabetical by venture capital source name.
♦ **Sources:** Lists provided by the U.S. Department of Commerce's Clearinghouse for State and Local Initiatives on Productivity, Technology and Innovations.

Section 19: Small Business Administration (SBA) Offices

♦ **Content:** Includes the name, address, telephone number, and description (when available) of more than 100 regional, district, and branch offices of the U.S. Small Business Administration.
♦ **Arrangement:** Alphabetical by state, then alphabetical by office name.
♦ **Sources:** Lists provided by the U.S. Small Business Administration's Office of Public Communications.

Section 20: Federal Funding Sources

♦ **Content:** Includes the name, address, telephone number, fax number (when available), contact name, and description of more than 100 federal agencies that participate in the Small Business Innovation Research Program and/or other granting and concept evaluation programs available to inventors.

◆ **Arrangement:** Alphabetical by federal agency, then alphabetical by organization name.
◆ **Sources:** Lists provided by the U.S. Department of Commerce's National Technical Information Service.

Section 21: Small Business Innovation Research (SBIR) Program

◆ **Content:** Includes the name, address, telephone number, and fax number, contact name, and description (when available) of all 12 federal agencies and 34 state organizations participating in the U.S. Small Business Administration's Small Business Innovation Research (SBIR) Program.
◆ **Arrangement:** The first group of listings in this section is arranged alphabetically by federal agency; the second group of listings in this section is arranged alphabetically by state, then alphabetically by program name.
◆ **Sources:** Lists provided by the U.S. Department of Commerce's Clearinghouse on State and Local Initiatives on Productivity, Technology and Innovations.

Section 22: Federal Laboratory Assistance for Inventors

◆ **Content:** Includes the name, address, telephone number, and contact name of nearly 370 federal laboratories offering free technical assistance and sometimes lab space to inventors.
◆ **Arrangement:** Alphabetical by federal agency, then alphabetical by laboratory name.
◆ **Sources:** Lists provided by the U.S. Department of Commerce's National Technical Information Service.

Section 23: Publications

◆ **Content:** Includes the name, address, telephone number, fax number (when available), and description (when available) of more than 150 directories, magazines, and newsletters on topics such as technology and innovation, research and development, entrepreneurship, intellectual property, and other subjects of interest to inventors.
◆ **Arrangement:** Alphabetical by publication name.
◆ **Sources:** *Directories in Print, Newsletters in Print*, and *Gale Directory of Publications and Broadcast Media* (all published by Gale Research, Inc.); supplemented by original research.

Section 24: Project XL: Youth Innovation Programs and Publications

◆ **Content:** Includes the name, address, telephone number (when available), contact name, and description of more than 30 programs and 130 publications designed to encourage creative thinking and develop problem-solving skills in students grades K-12.
◆ **Arrangement:** Arranged into two subsections, Programs and Publications. Each subsection is arranged alphabetically by program or publication name.
◆ **Sources:** Lists provided by the U.S. Patent and Trademark Office.

Section 25: Online Resources

◆ **Content:** Includes the name, address, telephone number, fax number (when available), and description of more than 100 U.S. and international databases on topics such as patents, trademarks, copyrights, intellectual property law, brand names, and new product development.
◆ **Arrangement:** Alphabetical by database name.
◆ **Sources:** *Computer-Readable Databases* (published by Gale Research Inc.).

Section 26: Patent and Trademark Depository Libraries

- ◆ **Content:** Includes the name, address, telephone number, and description (when available) of 70 Patent and Trademark Depository Libraries (PTDLs). *[Editor's Note: Three newly-approved PTDLs are listed on page 338.]*
- ◆ **Arrangement:** Alphabetical by state, then alphabetical by library name.
- ◆ **Sources:** Lists provided by the Office of Patent Depository Library Programs of the U.S. Patent and Trademark Office.

Section 27: Registered Patent Attorneys and Agents

- ◆ **Content:** Arranged into two separate lists, National Council of Intellectual Property Law Associations (NCIPLA) and Registered Patent Attorneys and Agents. The first list includes the name and address and contact name of 51 member groups within the NCIPLA. The second list includes the name, address, and telephone number of the more than 13,700 patent attorneys and agents registered to practice before the U.S. Patent and Trademark Office.
- ◆ **Arrangement:** The first list is arranged alphabetically by state, then alphabetically by NCIPLA member name. The second list is arranged alphabetically by state, then by city, and then by attorney or agent name.
- ◆ **Sources:** The NCIPLA list was provided by the National Council of Intellectual Property Law Associations. The list of registered patent attorneys and agents was provided by the U.S. Patent and Trademark Office, which no longer offers a print edition of the list, only a machine-readable format.

Part Four: Appendices and Master Index

Appendix A: U.S. Patent Classifications. Lists more than 300 classes and 95,000 subclasses used by the U.S. Patent and Trademark Office (PTO) to classify patents by subject. Information was obtained from the PTO's *Index to the U.S. Patent Classification.*

Appendix B: Patent and Trademark Office Telephone Directory. Provides the names, addresses, and telephone numbers for all PTO officials. Information was provided by the PTO.

Appendix C: Copyright Office Telephone Directory. Provides the names, addresses, and telephone numbers for all Copyright Office officials. Information was provided by the U.S. Library of Congress.

Appendix D: Patent and Trademark Office Fee Schedule. Lists all fees required for PTO patent, trademark, and general services, with prices effective as of December 16, 1991. Information was provided by the PTO.

Appendix E: Significant Inventors and Inventions in U.S. History. Lists brief descriptions of U.S. inventors and inventions that have had a significant impact on U.S. history. Information was compiled and provided by the National Council of Intellectual Property Law Associations.

Appendix F: Top 200 Corporations Receiving U.S. Patents in 1990. Information was compiled by Intellectual Property Owners (IPO) from data provided by the U.S. Patent and Trademark Office.

Appendix G: Glossary. This glossary defines more than 50 inventing terms and phrases. Terms are cross-referenced and arranged alphabetically. Information was compiled and provided by the U.S. Department of Energy and Argonne National Laboratory.

Master Index: Combines all organizations, agencies, programs, and publications into one comprehensive index. The index includes citations for directory listings (except for the 13,700 registered patent attorneys and agents) and, for the first time in this edition, includes citations for organizations, agencies, publications, and individuals listed in Richard C. Levy's how-to material.

Directory Citations. Directory citations appear with an entry number; a star (★) before an entry number indicates a name mentioned within the text of an entry. Citations for publications appear in italics. Names may also be cited under the keywords they contain.

Essay Citations. Index references for items in the how-to sections appear with an italicized page number preceded by *p.* or *pp.*

Examples:

Geographic Abbreviations for U.S. States and Territories

AL	Alabama		MT	Montana
AK	Alaska		NE	Nebraska
AS	American Samoa		NV	Nevada
AZ	Arizona		NH	New Hampshire
AR	Arkansas		NJ	New Jersey
CA	California		NM	New Mexico
CO	Colorado		NY	New York
CT	Connecticut		NC	North Carolina
DE	Delaware		ND	North Dakota
DC	District of Columbia		OH	Ohio
FL	Florida		OK	Oklahoma
GA	Georgia		OR	Oregon
GU	Guam		PA	Pennsylvania
HI	Hawaii		PR	Puerto Rico
ID	Idaho		RI	Rhode Island
IL	Illinois		SC	South Carolina
IN	Indiana		SD	South Dakota
IA	Iowa		TN	Tennessee
KS	Kansas		TX	Texas
KY	Kentucky		UT	Utah
LA	Louisiana		VT	Vermont
ME	Maine		VI	Virgin Islands
MD	Maryland		VA	Virginia
MA	Massachusetts		WA	Washington
MI	Michigan		WV	West Virginia
MN	Minnesota		WI	Wisconsin
MS	Mississippi		WY	Wyoming
MO	Missouri			

Table of Figures

Part One

Forms and Tables

Forms and Tables

(The page numbers listed in the right-hand column indicate where the forms and tables are mentioned in the text.)

Power of Attorney or Authorization of Agent, Not Accompanying Application

To The Commissioner of Patents and Trademarks:

The undersigned having, on or about the _____ day of _____, 19 _____, made application for

letters patent for an improvement in _____, Serial

Number _____, hereby appoints _____

of _____; State of _____, Registration

No. _____, his attorney (or agent), to prosecute said application, and to transact all

business in the Patent and Trademark Office connected therewith.

_____
(Signature)

Revocation of Power of Attorney or
Authorization of Agent

To the Commissioner of Patents and Trademarks:

The undersigned having, on or about the_____day of _____ , 19_____ ,

appointed _____ , of _____ ;

State of_____ , his attorney (or agent), to prosecute an application for letters patent

which application was filed on or about the _____ day of _____ , 19_____ , for

an improvement in _____ , Serial Number_____ ,

hereby revokes the power of attorney (or authorization of agent) then given.

_____

(Signature)

	ATTORNEY'S DOCKET NO.
# PATENT APPLICATION TRANSMITTAL LETTER	

TO THE COMMISSIONER OF PATENTS AND TRADEMARKS:

Transmitted herewith for filing is the patent application of _____

_____

for _____

_____

Enclosed are:

☐ _____ sheets of drawing.

☐ an assignment of the invention to _____

_____

☐ a certified copy of a _____ application.

☐ associate power of attorney.

☐ verified statement to establish small entity status under 37 CFR 1.9 and 1.27. ———

CLAIMS AS FILED

FOR	NO. FILED	NO. EXTRA		RATE	FEE	OR	RATE	FEE
			SMALL ENTITY				**OTHER THAN A SMALL ENTITY**	
BASIC FEE	▓▓▓▓	▓▓▓▓		▓▓▓▓	$	OR	▓▓▓▓	$
TOTAL CLAIMS	−20−	•		× $ 6 =	$	OR	× $ 12 =	$
INDEP CLAIMS	−3−	•		× $ 17 =	$	OR	× $ 34 =	$
MULTIPLE DEPENDENT CLAIM PRESENT				+ $ 55 =	$	OR	+ $ 110 −	$
				TOTAL	$	OR	TOTAL	$

* If the difference in col. 1 is less than zero, enter "0" in col. 2

☐ Please charge my Deposit Account No. _____ in the amount of $ _____
☐ A duplicate copy of this sheet is enclosed.

☐ A check in the amount of $ _____ to cover the filing fee is enclosed.

☐ The Commissioner is hereby authorized to charge payment of the following fees associated with this communication or credit any overpayment to Deposit Account No. _____ . A Duplicate copy of this sheet is enclosed.

   ☐ Any additional filing fees required under 37 CFR 1.16.

   ☐ Any patent application processing fees under 37 CFR 1.17

☐ The Commissioner is hereby authorized to charge payment of the following fees during the pendency of this application or credit any overpayment to Deposit Account No. _____ . A duplicate copy of this sheet is enclosed.

   ☐ Any filing fees under 37 CFR 1.16 for presentation of extra claims.

   ☐ Any patent application processing fees under 37 CFR 1.17.

   ☐ The issue fee set in 37 CFR 1.18 at or before mailing of the Notice of Allowance, pursuant to 37 CFR 1.311(b).

_____   _____
date   signature

Patent and Trademark Office - U.S. DEPARTMENT of COMMERCE

DECLARATION FOR PATENT APPLICATION

Docket No. _____

As a below named inventor, I hereby declare that:

My residence, post office address and citizenship are as stated below next to my name.

I believe I am the original, first and sole inventor (if only one name is listed below) or an original, first and joint inventor (if plural names are listed below) of the subject matter which is claimed and for which a patent is sought on the invention entitled _____, the specification of which

(check one) ☐ is attached hereto.
　　　　　　☐ was filed on _____ as
　　　　　　Application Serial No. _____
　　　　　　and was amended on _____ (if applicable).

I hereby state that I have reviewed and understand the contents of the above identified specification, including the claims, as amended by any amendment referred to above.

I acknowledge the duty to disclose information which is material to the examination of this application in accordance with Title 37, Code of Federal Regulations, §1.56(a).

I hereby claim foreign priority benefits under Title 35, United States Code, §119 of any foreign application(s) for patent or inventor's certificate listed below and have also identified below any foreign application for patent or inventor's certificate having a filing date before that of the application on which priority is claimed:

Prior Foreign Application(s)

Priority Claimed

(Number)	(Country)	(Day/Month/Year Filed)	Yes	No
(Number)	(Country)	(Day/Month/Year Filed)	Yes	No
(Number)	(Country)	(Day/Month/Year Filed)	Yes	No

I hereby claim the benefit under Title 35, United States Code, §120 of any United States application(s) listed below and, insofar as the subject matter of each of the claims of this application is not disclosed in the prior United States application in the manner provided by the first paragraph of Title 35, United States Code, §112, I acknowledge the duty to disclose material information as defined in Title 37, Code of Federal Regulations, §1.56(a) which occurred between the filing date of the prior application and the national or PCT international filing date of this application:

(Application Serial No.)	(Filing Date)	(Status—patented, pending, abandoned)
(Application Serial No.)	(Filing Date)	(Status—patented, pending, abandoned)

I hereby appoint the following attorney(s) and/or agent(s) to prosecute this application and to transact all business in the Patent and Trademark Office connected therewith:

Address all telephone calls to _____ at telephone no. _____.
Address all correspondence to _____
_____
_____
_____

I hereby declare that all statements made herein of my own knowledge are true and that all statements made on information and belief are believed to be true; and further that these statements were made with the knowledge that willful false statements and the like so made are punishable by fine or imprisonment, or both, under Section 1001 of Title 18 of the United States Code and that such willful false statements may jeopardize the validity of the application or any patent issued thereon.

Full name of sole or first inventor _____
Inventor's signature _____ Date _____
Residence _____ Citizenship _____
Post Office Address _____

Full name of second joint inventor, if any _____
Second Inventor's signature _____ Date _____
Residence _____ Citizenship _____
Post Office Address _____

(Supply similar information and signature for third and subsequent joint inventors.)

Form PTO-FB-A110 (8-83)

Form 4

VERIFIED STATEMENT (DECLARATION) BY A NON-INVENTOR
SUPPORTING A CLAIM BY ANOTHER FOR SMALL ENTITY STATUS

I hereby declare that I am making this verified statement to support a claim by _____ for small entity status for purposes of paying reduced fees under section 41(a) and (b) of Title 35, United States Code, with regard to the invention entitled _____ by inventor(s) _____

[ ] the specification filed herewith
[ ] application serial number _____, filed _____
[ ] patent number _____, issued _____ .

I hereby declare that I would qualify as an independent inventor as defined in 37 CFR 1.9(c) for purposes of paying fees under section 41(a) and (b) of Title 35, United States Code, if I had made the above identified invention.

I have not assigned, granted, conveyed or licensed and am under no obligation under contract or law to assign, grant, convey or license, any rights to the invention to any person who could not be classified as an independent inventor under 37 CFR 1.9(c) if that person had made the invention, or to any concern which would not qualify as a small business concern under 37 CFR 1.9(d) or a nonprofit organization under 37 CFR 1.9(a).

Each person, concern or organization to which I have assigned, granted, conveyed, or licensed or am under an obligation under contract or law to assign, grant, convey, or license any rights in the invention is listed below:

[ ] No such person, concern, or organization
[ ] Persons, concerns or organizations listed below*

* Note: Separate verified statements are required from each named person, concern or organization having rights to the invention averring to their status as small entities. (37 CFR 1.27)

NAME _____
ADDRESS _____
    [ ] INDIVIDUAL   [ ] SMALL BUSINESS CONCERN   [ ] NONPROFIT ORGANIZATION

NAME _____
ADDRESS _____ .
    [ ] INDIVIDUAL   [ ] SMALL BUSINESS CONCERN   [ ] NONPROFIT ORGANIZATION

NAME _____
ADDRESS _____
    [ ] INDIVIDUAL   [ ] SMALL BUSINESS CONCERN   [ ] NONPROFIT ORGANIZATION

      I acknowledge the duty To file, in this application or patent, notification of any change in status resulting in loss of entitlement to small entity status prior to paying, or at the time of paying, the earliest of the issue fee or any maintenance fee due after the date on which status as a small entity is no longer appropriate. (37 CFR 1.28(b))

      I hereby declare that all statements mad
    e herein of my own knowledge are true and that all statements made on information and belief are believed to be true; and further that these statements were wade with the knowledge that willful false statements and the like so made are punishable by fine or imprisonment, or both, under section 1001 of Title 18 of the United States Code, and that such willful false statements may jeopardize the validity of the application, any patent issuing thereon, or any patent to which this verified statement is directed.

NAME OF PERSON SIGNING _____
ADDRESS OF PERSON SIGNING _____
SIGNATURE _____ DATE _____

Rev. 11, Apr. 1989

Applicant or Patentee:_____ Attorney's
Serial or Patent No.:_____ _____ Docket No.:_____
Filed or Issued:_____ _____
For:_____ _____

VERIFIED STATEMENT (DECLARATION) CLAIMING SMALL ENTITY STATUS
(37 CFR 1.9(f) & 1.27(c)) - INDEPENDENT INVENTOR

As a below named inventor, I hereby declare that I qualify as an independent inventor as defined in 37 CFR 1.9(c) for purposes of paying reduced fees under section 41(a) and (b) of Title 35, United States Code, to the Patent and Trademark Office with regard to the invention entitled _____
described in

[ ] the specification filed herewith
[ ] application serial number _____, filed _____
[ ] patent number _____, issued _____ .

I have not assigned, granted, conveyed or licensed and am under no obligation under contract or law to assign, grant, convey or license, any rights to the invention to any person who could not be classified as an independent inventor under 37 CFR 1.9(c) if that person made the invention, or to any concern which would not qualify as a small business concern under 37 CFR 1.9(d) or a nonprofit organization under 37 CFR 1.9(a).

Each person, concern or organization to which I have assigned , granted, conveyed, or licensed or am under an obligation under contract or law to assign, grant, convey, or license any rights in the invention is listed below:

[ ] No such person, concern, or organization
[ ] Persons, concerns or organizations listed below*

* Note: Separate verified statements are required from each named person, concern or organization having rights to the invention averring to their status as small entities. (37 CFR 1.27)

NAME _____
ADDRESS _____
[ ] INDIVIDUAL   [ ] SMALL BUSINESS CONCERN   [ ] NONPROFIT ORGANIZATION

NAME _____
ADDRESS _____
[ ] INDIVIDUAL   [ ] SMALL BUSINESS CONCERN   [ ] NONPROFIT ORGANIZATION

NAME _____
ADDRESS _____
[ ] INDIVIDUAL   [ ] SMALL BUSINESS CONCERN   [ ] NONPROFIT ORGANIZATION

I acknowledge the duty To file, in this application or patent, notification of any change in status resulting in loss of entitlement to small entity status prior to paying, or at the time of paying, the earliest of the issue fee or any maintenance fee due after the date an which status as a small entity is no longer appropriate. (37 CFR 1.28(b))

I hereby declare that all statements made herein of my own knowledge are true and that all statements made on information and belief are believed to be true; and further that these statements were wade with the knowledge that willful false statements and the like so made are punishable by fine or imprisonment, or both, under section 1001 of Title 18 of the United States Code, and that such willful false statements may jeopardize the validity of the application, any patent issuing thereon, or any patent to which this verified statement is directed.

_____   _____   _____
NAME OF INVENTOR   NAME OF INVENTOR   NAME OF INVENTOR

_____   _____   _____
Signature of inventor   Signature of inventor   Signature of inventor

_____   _____   _____
Date   Date   Date

Rev. 11, Apr. 1989

Applicant or Patentee:_____ Attorney's
Serial or Patent No.:_____ Docket No.:_____ Filed or Is-
sued:_____
For:_____

VERIFIED STATEMENT (DECLARATION) CLAIMING SMALL ENTITY STATUS
(37 CFR 1.9(f) & 1.27(c)) - SMALL BUSINESS CONCERN

I hereby declare that I am

[ ] the owner of the small business concern identified below:
[ ] an official of the small business concern empowered to act on behalf of the concern identified below:
NAME OF CONCERN _____
ADDRESS OF CONCERN _____
_____

    I hereby declare that the above identified small business concern qualifies as a small business concern as defined in 13 CFR 121.12, and reproduced in 37 CFR 1.9(d), for purposes of paying reduced fees under section 41(a) and (b) of Title 35, United States Code, in that the number of employees of the concern, including those of its affiliates, does not exceed 500 persons. For purposes of this statement, (1) the number of employees of the business concern is the average over the previous fiscal year of the concern of the persons employed on a full-time, part-time or temporary basis during each of the pay periods of the fiscal year, and (2) concerns are affiliates of each other when either, directly or indirectly, one concern controls or has the power to control the other, or a third party or parties controls or has the power to control both.

    I hereby declare that rights under contract or law have been conveyed to and remain with the small business concern identified above with regard to the invention, entitled _____
_____by inventor(s)

described in

[ ] the specification filed herewith
[ ] application serial no._____ , filed _____
[ ] patent no._____ , issued _____
    If the rights held by the above identified small business concern are not exclusive, each individual, concern or organization having rights to the invention is listed below and no rights to the invention are held by any person, other than the inventor, who would not qualify as a small business concern under 37 CFR 1.9(d) or by any concern which would not qualify as a small business concern under 37 CFR 1.9(d) or a nonprofit organization under 37 CFR 1.9(e). NOTE: Separate verified statements are required from each named person, concern or organization having rights to the invention averring to their status as small entities. (37 CFR 1.27)

NAME _____
ADDRESS _____
    [ ] INDIVIDUAL   [ ] SMALL BUSINESS CONCERN   [ ] NONPROFIT ORGANIZATION

NAME _____
ADDRESS _____
    [ ] INDIVIDUAL   [ ] SMALL BUSINESS CONCERN   [ ] NONPROFIT ORGANIZATION

    I acknowledge the duty To file, in this application or patent, notification of any change in status resulting in loss of entitlement to small entity status prior to paying, or at the time of paying, the earliest of the issue fee or any maintenance fee due after the date on which status as a small entity is no longer appropriate. (37 CFR 1.28(b))

    I hereby declare that all statements made herein of my own knowledge are true and that all statements made on information and belief are believed to be true; and further that these statements were wade with the knowledge that willful false statements and the like so made are punishable by fine or imprisonment, or both, under section 1001 of Title 18 of the United States Code, and that such willful false statements may jeopardize the validity of the application, any patent issuing thereon, or any patent to which this verified statement is directed.

NAME OF PERSON SIGNING _____
TITLE OF PERSON OTHER THAN OWNER _____
ADDRESS OF PERSON SIGNING _____
_____

SIGNATURE _____ DATE _____ _

Rev. 11, Apr. 1989

Applicant or Patentee:_____ Attorney's
Serial or Patent No.:_____ Docket No.:_____
Filed or Issued:_____
For:_____

VERIFIED STATEMENT (DECLARATION) CLAIMING SMALL ENTITY STATUS
(37 CFR 1.9(f) & 1.27(c)) - NONPROFIT ORGANIZATION

I hereby declare that I am an official empowered to act on behalf of the nonprofit organization identified below:
   NAME OF ORGANIZATION _____
   ADDRESS OF ORGANIZATION _____
_____

TYPE OF ORGANIZATION
   [ ] UNIVERSITY OR OTHER INSTITUTION OF HIGHER EDUCATION
   [ ] TAX EXEMPT UNDER INTERNAL REVENUE SERVICE CODE (26 U.S.C. 501(a) and 501(c)(3))
   [ ] NONPROFIT SCIENTIFIC OR EDUCATIONAL UNDER STATUTE OF STATE OF THE UNITED STATES OF AMERICA
      (NAME OF STATE _____)
      (CITATION OF STATUTE _____)
   [ ] WOULD QUALIFY AS TAX EXEMPT UNDER INTERNAL REVENUE SERVICE CODE (26 U.S.C. 501(a) and 501(c) IF
LOCATED IN THE UNITED STATES OF AMERICA
   [ ] WOULD QUALIFY AS NONPROFIT SCIENTIFIC OR EDUCATIONAL UNDER STATUTE OF STATE OF THE UNITED
STATES OF AMERICA IF LOCATED IN THE UNITED STATES OF AMERICA
      (NAME OF STATE _____)
      (CITATION OF STATUTE _____)
    I hereby declare that the nonprofit organization identified above qualifies as a small business concern as defined in 37 CFR 1.9(e)
for purposes of paying reduced fees under section 41(a) and (b) of Title 35, United States Code with regard to the invention entitled
_____ by inventor(s)_____
described in
   [ ] the specification filed herewith
   [ ] application serial no._____ , filed _____
   [ ] patent no._____ , issued _____
    I hereby declare that rights under contract or law have conveyed to and remain with the nonprofit organization with regard to the above
identified invention.
    If the rights held by nonprofit organization are not exclusive, each individual, concern or organization having rights to the invention is
listed below* and no rights to the invention are held by any person, other than the inventor, who would not qualify as a small business concern
under 37 CFR 1.9(d) or by any concern which would not qualify as a small business concern under 37 CFR 1.9(d) or a nonprofit organization
under 37 CFR 1.9(e).
    *NOTE: Separate verified statements are required from each named person, concern or organization having rights to the invention
averring to their status as small entities. (37 CFR 1.27)

NAME _____
ADDRESS _____
   [ ] INDIVIDUAL   [ ] SMALL BUSINESS CONCERN   [ ] NONPROFIT ORGANIZATION

NAME _____
ADDRESS _____
   [ ] INDIVIDUAL   [ ] SMALL BUSINESS CONCERN   [ ] NONPROFIT ORGANIZATION

    I acknowledge the duty To file, in this application or patent, notification of any change in status resulting in loss of entitlement to small
entity status prior to paying, or at the time of paying, the earliest of the issue fee or any maintenance fee due after the date an which status as
a small entity is no longer appropriate. (37 CFR 1.28(b))

    I hereby declare that all statements made herein of my own knowledge are true and that all statements made on information and belief
are believed to be true; and further that these statements were wade with the knowledge that willful false statements and the like so made are
punishable by fine or imprisonment, or both, under section 1001 of Title 18 of the United States Code, and that such willful false statements
may jeopardize the validity of the application, any patent issuing thereon, or any patent to which this verified statement is directed.

NAME OF PERSON SIGNING _____
TITLE IN ORGANIZATION _____
ADDRESS OF PERSON SIGNING _____
_____

SIGNATURE _____ DATE _____

Rev. 11, Apr. 1989

INFORMATION DISCLOSURE CITATION

(Use several sheets if necessary)

ATTY. DOCKET NO.		SERIAL NO.	
APPLICANT			
FILING DATE		GROUP	

U.S. PATENT DOCUMENTS

*EXAMINER INITIAL		DOCUMENT NUMBER	DATE	NAME	CLASS	SUBCLASS	FILING DATE IF APPROPRIATE

FOREIGN PATENT DOCUMENTS

		DOCUMENT NUMBER	DATE	COUNTRY	CLASS	SUBCLASS	TRANSLATION YES	NO

OTHER DOCUMENTS *(Including Author, Title, Date, Pertinent Pages, Etc.)*

EXAMINER	DATE CONSIDERED

*EXAMINER: Initial if reference considered, whether or not citation is in conformance with MPEP 609; Draw line through citation if not in conformance and not considered. Include copy of this form with next communication to applicant.

Form PTO-FB-A820

(also form PTO-1449)

Patent and Trademark Office · U.S. DEPARTMENT of COMMERCE

AMENDMENT TRANSMITTAL LETTER	ATTORNEY'S DOCKET NO.

SERIAL NO.	FILING DATE	EXAMINER	GROUP ART UNIT

INVENTION

TO THE COMMISSIONER OF PATENTS AND TRADEMARKS:

Transmitted herewith is an amendment in the above-identified application.

Small entity status of this application under 37 CFR 1.27 has been established by a verified statement previously submitted.

A verified statement to establish small entity status under 37 CFR 1.9 and 1.27 is enclosed.

No additional fee is required.

The fee has been calculated as shown below:

	CLAIMS REMAINING AFTER AMENDMENT (1)		HIGHEST NO PREVIOUSLY PAID FOR (2)	PRESENT EXTRA (3)	SMALL ENTITY			OTHER THAN A SMALL ENTITY	
					RATE	ADDIT FEE	OR	RATE	ADDIT FEE
TOTAL	*	MINUS	**	-	x $6=	$		x $12=	$
INDEP	*	MINUS	***	-	x $17=	$		x $34=	$
FIRST PRESENTATION OF MULTIPLE DEP. CLAIM					+$55=	$		+$110=	$
					TOTAL ADDIT. FEE	$	OR	TOTAL	$

* If the entry in Col 1 is less than the entry in Col 2, write "0" in Col. 3.

** If the "Highest No Previously Paid For" IN THIS SPACE is less than 20, enter "20".

*** If the "Highest No Previously Paid For" IN THIS SPACE is less than 3, enter "3".

The "Highest No Previously Paid For" (Total or Indep.) is the highest number found in the appropriate box in Col 1

Please charge my Deposit Account No. _____ in the amount of $ _____ . A duplicate copy of this sheet is enclosed.

A check in the amount of $ _____ to cover the filing fee is enclosed.

The Commissioner is hereby authorized to charge payment of the following fees associated with this communication or credit any overpayment to Deposit Account No. _____ . A Duplicate copy of this sheet is enclosed.

Any additional filing fees required under 37 CFR 1.16.

Any patent application processing fees under 37 CFR 1.17

_____   _____
(date)   (signature)

Form PTO-FB-A520 (10-85)   Patent and Trademark Office · U.S. DEPARTMENT of COMMERCE
(also form PTO-1083)

Form 10

PETITION FOR REVIVAL OF AN APPLICATION FOR PATENT ABANDONED UNINTENTIONALLY UNDER 37 CFR 1.137(b), 37 CFR 1.155(c) OR 37 CFR 1.316(c)	Docket Number (Optional)

First named inventor:

Serial No.:   Group Art Unit:

Filed: Examiner:

Title:

Attention: Assistant Commissioner for Patents
Commissioner of Patents and Trademarks
Washington, D.C. 20231

NOTE: If information or assistance is needed in completing this form, please contact Petitions Information at (703)557-4282.

The above-identified application became abandoned for failure to file a timely and proper response to the Office action mailed on _____ , which set a _____ month/day period for response. The abandonment date of this application is _____ (i.e., the day after the expiration date of the period set for response plus any extensions of time obtained therefore).

APPLICANT HEREBY PETITIONS FOR REVIVAL OF THIS APPLICATION

NOTE: A grantable petition requires the following items:
    (1) Petition fee
    (2) Proposed response and/or issue fee
    (3) Verified statement that the abandonment was unintentional

1. Petition fee
  ☐ Small entity - fee $525.00
    ☐ Small entity statement enclosed herewith.
    ☐ Small entity statement previously filed.
  ☐ Other than small entity - fee $1,050.00

2. Proposed response and/or fee

  A. The proposed response and/or fee to the above-noted Office action in
    the form of _____ (identify type of response):
  ☐ has been filed previously on _____ .
  ☐ is enclosed herewith.

  B. The issue fee of $ _____
  ☐ has been paid previously on _____ .
  ☐ is enclosed herewith.

3. Verified statement that abandonment was unintentional

This application became abandoned unintentionally.

I hereby declare that all statements made herein of my own knowledge are true and that all statements made on information and belief are believed to be true; and further that these statements were made with the knowledge that willful false statements and the like so made are punishable by fine or imprisonment, or both, under Section 1001 of Title 18 of the United States Code, and that such willful false statements may jeopardize the validity of the application, any patent issuing thereon, or any patent to which this verified statement is directed.

_____
Date

_____
Signature

Telephone
Number: (____) _____

_____
Typed or printed name

_____
Address

_____

Enclosures: ☐ Response

☐ Fee Payment

☐ Small Entity Status Form

☐ _____

By completing the Certificate of Mailing, below, the date mailed will be considered the date this paper is filed.

CERTIFICATE OF MAILING [37 CFR 1.8(a)]

I hereby certify that this paper is being deposited with the United States Postal Service on the date shown below with sufficient postage as first class mail in an envelope addressed to the: Commissioner of Patents and Trademarks, Washington, DC 20231.

_____
Date

_____
Signature

_____
Typed or printed name

[Page 2 of 2]

Patent Worksheet

TRADE MARK (WORKING): _____

TAG LINE:_____

DESCRIPTION:_____

_____

_____

_____

PATENT NOTES:_____

_____

_____LODE #_____

WHO/DATE CONCEIVED:_____

WITNESSED:_____

SKETCH/PHOTO:

NOTES:_____

_____

_____

LICENSES/TIE-INS/SPIN-OFFS/ACCESSORIES/LINE CONCEPTS:___

_____

_____

_____

POTENTIAL MANUFACTURERS:_____

_____

_____

SEEN BY/DATE: _____ _____

_____ _____ _____

_____ _____ _____

Assignment of Patent

(No special form is prescribed for assignments, which may contain various provisions depending upon the agreement of the parties. The following form is a specimen of an assignment which has been used in some cases.)

Whereas, I_____, of _____, did obtain Letters Patent of the United States for an improvement in_____
No._____, dated_____; and whereas; I am now the sole owner of said patent; and,

Whereas,_____, of _____, whose post office address is_____, is desirous of acquiring the entire interest in the same; now, therefore, in consideration of the sum of _____dollars ($_____), the receipt of which is hereby acknowledged, and other good and valuable considerations, I, _____, by these presents do sell, assign, and transfer unto the said _____,the entire right, title, and interest in and to the said Letters Patent aforesaid; the same to be held and enjoyed by the said_____, for his own use and behoof, and for his legal representatives and assigns, to the full end of the term for said Letters Patent are granted, as fully and entirely as the same would have been held by me had this assignment and sale not been made.

Executed, this_____day of_____, 19____,

State of _____County of_____

at_____.

ss:

Before me personally appeared said_____and acknowledged the foregoing instruments to be his free act and deed this_____day of _____.

_____
(Notary Public)

[SEAL]

Form 13

Assignment of Application

(No special form is prescribed for assignments, which may contain various provisions depending upon the agreement of the parties. The following form is a specimen of an assignment which has been used in some cases.)

Whereas, I,_____ of_____ have invented certain new and useful improvements in_____, for which an application for United States Letters Patent was filed on _____, 19_____ Serial No._____, [if the application has been prepared but not yet filed, state "for which an application for United States Letters Patent was executed on_____," instead] and

Whereas, _____, of _____, whose post office address is_____, is desirous of acquiring the entire right, title and interest in the same;

Now, therefore, in consideration of the sum of _____dollars ($_____), the receipt whereof is hereby acknowledged, and other good and valuable consideration, I the said _____, by these presents do sell, assign and transfer unto said _____, the full and exclusive right to the said invention in the United States and the entire right, title, and interest in and to any and all Letters Patent which may be granted, therefore in the United States.

I hereby authorize and request the Commissioner of Patents and Trademarks to issue said Letters Patent to said _____, as the assignee of the entire right, title, and interest to the same, for his sole use and behoof; and for the use and behoof of his legal representatives, to the full end of the term for which said Letters Patent may be granted, as fully and entirely as the same would have been held by me had this assignment and sale not be made.

Executed this _____day of_____, 19_____, State of_____at

_____ County of_____.

ss:

Before me personally appeared said _____and acknowledged the foregoing instrument to be his free act and deed this _____day of_____, 19_____.

_____
(Notary Public)

[SEAL]

Design Patent Specification

Be it known that I (we), _____ and _____ have invented a new, original, and ornamental design for a _____, over which the following is a specification. Reference is made to the accompanying drawing, which forms a part hereof wherein:

FIGURE 1 is a perspective view of a _____ in accordance with the present design;

FIGURE 2 is a rear elevational view of said _____;

FIGURE 3 is a side elevational view of said _____, the side of the _____ not shown being a mirror image of the side shown in this view;

FIGURE 4 is a front elevational view of said _____.

What I (we) claim is:

The ornamental design for a _____ or similar article, as shown and described.

(Note: Attached to this application are the draftsman drawings.)

TRADEMARK/SERVICE MARK APPLICATION, PRINCIPAL REGISTER, WITH DECLARATION	MARK (Identify the mark)
	CLASS NO. (If known)

TO THE ASSISTANT SECRETARY AND COMMISSIONER OF PATENTS AND TRADEMARKS:

APPLICANT NAME:

APPLICANT BUSINESS ADDRESS:

APPLICANT ENTITY: (Check one and supply requested information)

☐ Individual - Citizenship: (Country) _____

☐ Partnership - Partnership Domicile: (State and Country) _____
Names and Citizenship (Country) of General Partners: _____
_____

☐ Corporation - State (Country, if appropriate) of Incorporation: _____

☐ Other: (Specify Nature of Entity and Domicile) _____
_____

GOODS AND/OR SERVICES:

Applicant requests registration of the above-identified trademark/service mark shown in the accompanying drawing in the United States Patent and Trademark Office on the Principal Register established by the Act of July 5, 1946 (15 U.S.C. 1051 et. seq., as amended.) for the following goods/services: _____
_____
_____
_____

BASIS FOR APPLICATION: (Check one or more, but NOT both the first AND second boxes, and supply requested information)

☐ Applicant is using the mark in commerce on or in connection with the above identified goods/services. (15 U.S.C. 1051(a), as amended.) Three specimens showing the mark as used in commerce are submitted with this application.
- Date of first use of the mark anywhere: _____
- Date of first use of the mark in commerce which the U.S. Congress may regulate: _____
- Specify the type of commerce: _____
   (e.g., interstate, between the U.S. and a specified foreign country)
- Specify manner or mode of use of mark on or in connection with the goods/services: _____
_____
   (e.g., trademark is applied to labels, service mark is used in advertisements)

☐ Applicant has a bona fide intention to use the mark in commerce on or in connection with the above identified goods/services. (15 U.S.C. 1051(b), as amended.)
- Specify intended manner or mode of use of mark on or in connection with the goods/services: _____
_____
   (e.g., trademark will be applied to labels, service mark will be used in advertisements)

☐ Applicant has a bona fide intention to use the mark in commerce on or in connection with the above identified goods/services, and asserts a claim of priority based upon a foreign application in accordance with 15 U.S.C. 1126(d), as amended.
- Country of foreign filing: _____ • Date of foreign filing: _____

☐ Applicant has a bona fide intention to use the mark in commerce on or in connection with the above identified goods/services and, accompanying this application, submits a certification or certified copy of a foreign registration in accordance with 15 U.S.C. 1126(e), as amended.
- Country of registration: _____ • Registration number: _____

Note: Declaration, on Reverse Side, MUST be Signed

DECLARATION

The undersigned being hereby warned that willful false statements and the like so made are punishable by fine or imprisonment, or both, under 18 U.S.C. 1001, and that such willful false statements may jeopardize the validity of the application or any resulting registration, declares that he/she is properly authorized to execute this application on behalf of the applicant; he/she believes the applicant to be the owner of the trademark/service mark sought to be registered, or, if the application is being filed under 15 U.S.C. 1051(b), he/she believes applicant to be entitled to use such mark in commerce; to the best of his/her knowledge and belief no other person, firm, corporation, or association has the right to use the above identified mark in commerce, either in the identical form thereof or in such near resemblance thereto as to be likely, when used on or in connection with the goods/services of such other person, to cause confusion, or to cause mistake, or to deceive; and that all statements made of his/her own knowledge are true and all statements made on information and belief are believed to be true.

_____   _____
Date   Signature

_____   _____
Telephone Number   Print or Type Name and Position

INSTRUCTIONS AND INFORMATION FOR APPLICANT

To receive a filing date, the application must be completed and **signed by the applicant** and submitted along with:

1. The prescribed fee for each class of goods/services listed in the application;
2. A drawing of the mark in conformance with 37 CFR 2.52;
3. If the application is based on use of the mark in commerce, three (3) specimens (evidence) of the mark as used in commerce for each class of goods/services listed in the application. All three specimens may be the same and may be in the nature of: (a) labels showing the mark which are placed on the goods; (b) a photograph of the mark as it appears on the goods, (c) brochures or advertisements showing the mark as used in connection with the services.

Verification of the application - The application must be signed in order for the application to receive a filing date. Only the following person may sign the verification (Declaration) for the application, depending on the applicant's legal entity: (1) the individual applicant; (b) an officer of the corporate applicant; (c) one general partner of a partnership applicant; (d) all joint applicants.

Additional information concerning the requirements for filing an application are available in a booklet entitled **Basic Facts about Trademarks**, which may be obtained by writing:

U.S. DEPARTMENT OF COMMERCE
Patent and Trademark Office
Washington, D.C. 20231

Or by calling: (703) 557-INFO

This form is estimated to take 15 minutes to complete. Time will vary depending upon the needs of the individual case. Any comments on the amount of time you require to complete this form should be sent to the Office of Management and Organization, U.S. Patent and Trademark Office, U.S. Department of Commerce, Washington D.C., 20231, and to the Office of Information and Regulatory Affairs, Office of Management and Budget, Washington, D.C. 20503.

<table>
<tr><td>AMENDMENT TO ALLEGE USE
UNDER 37 CFR 2.76, WITH
DECLARATION</td><td>MARK (Identify the mark)</td></tr>
<tr><td></td><td>SERIAL NO.</td></tr>
</table>

TO THE ASSISTANT SECRETARY AND COMMISSIONER OF PATENTS AND TRADEMARKS:

APPLICANT NAME:

Applicant requests registration of the above-identified trademark/service mark in the United States Patent and Trademark Office on the Principal Register established by the Act of July 5, 1946 (15 U.S.C. 1051 et. seq., as amended). Three specimens showing the mark as used in commerce are submitted with this amendment.

☐ Check here if Request to Divide under 37 CFR 2.87 is being submitted with this amendment.

Applicant is using the mark in commerce on or in connection with the following goods/services:

_____

_____

(NOTE: Goods/services listed above may not be broader than the goods/services identified in the application as filed)

Date of first use of mark anywhere: _____

Date of first use of mark in commerce
which the U.S. Congress may regulate: _____

Specify type of commerce: (e.g., interstate, between the U.S. and a specified foreign country) _____

_____

Specify manner or mode of use of mark on or in connection with the goods/services: (e.g., trademark is applied to labels, service mark is used in advertisements) _____

_____

The undersigned being hereby warned that willful false statements and the like so made are punishable by fine or imprisonment, or both, under 18 U.S.C. 1001, and that such willful false statements may jeopardize the validity of the application or any resulting registration, declares that he/she is properly authorized to execute this Amendment to Allege Use on behalf of the applicant; he/she believes the applicant to be the owner of the trademark/service mark sought to be registered; the trademark/ service mark is now in use in commerce; and all statements made of his/her own knowledge are true and all statements made on information and belief are believed to be true.

_____   _____
Date   Signature

_____   _____
Telephone Number   Print or Type Name and Position

INSTRUCTIONS AND INFORMATION FOR APPLICANT

In an application based upon a bona fide intention to use a mark in commerce, applicant must use its mark in commerce before a registration will be issued. After use begins, the applicant must submit, along with evidence of use (specimens) and the prescribed fee(s), **either**:

> (l) **an Amendment to Allege Use** under 37 CFR 2.76, or
> (2) a Statement of Use under 37 CFR 2.88.

The difference between these two filings is the timing of the filing. Applicant may file an Amendment to Allege Use before approval of the mark for publication for opposition in the **Official Gazette**, or, if a final refusal has been issued, prior to the expiration of the six month response period. Otherwise, applicant must file a Statement of Use after the Office issues a Notice of Allowance. The Notice of Allowance will issue after the opposition period is completed if no successful opposition is filed. Neither Amendment to Allege Use or Statement of Use papers will be accepted by the Office during the period of time between approval of the mark for publication for opposition in the **Official Gazette** and the issuance of the Notice of Allowance.

Applicant may call (703) 557-5249 to determine whether the mark has been approved for publication for opposition in the **Official Gazette.**

Before filing an Amendment to Allege Use or a Statement of Use, applicant must use the mark in commerce on or in connection with **all** of the goods/services for which applicant will seek registration, **unless** applicant submits with the papers, a request to divide out from the application the goods or services to which the Amendment to Allege Use or Statement of Use pertains. (See: 37 CFR 2.87, Dividing an application)

Applicant **must** submit with an Amendment to Allege Use or a Statement of Use:

> (l) the appropriate fee of $100 per class of goods/services listed in the Amendment to Allege Use or the Statement of Use, and

> (2) three (3) specimens or facsimiles of the mark as used in commerce for each class of goods/services asserted (e.g., photograph of mark as it appears on goods, label containing mark which is placed on goods, or brochure or advertisement showing mark as used in connection with services).

Cautions/Notes concerning completion of this Amendment to Allege Use form:

> (l) The goods/services identified in the Amendment to Allege Use must be within the scope of the goods/services identified in the application as filed. Applicant may delete goods/services. Deleted goods/services may not be reinstated in the application at a later time.

> (2) Applicant **may list dates** of use for only one item in each class of goods/services identified in the **Amendment to Allege Use.** However, applicant must have used the mark in commerce on all the goods/services in the class. Applicant must identify the particular item to which the dates apply.

> (3) Only the following person may sign the verification of the Amendment to Allege Use, depending on the applicant's legal entity: (a) the individual applicant; (b) an officer of corporate applicant; (c) one general partner of partnership applicant; (d) all joint applicants.

This form is estimated to take 15 minutes to complete. Time will vary depending upon the needs of the individual case. Any comments on the amount of time you require to complete this form should be sent to the Office of Management and Organization, U.S. Patent and Trademark Office, U.S. Department of Commerce, Washington D.C., 20231, and to the Office of Information and Regulatory Affairs, Office of Management and Budget, Washington, D.C. 20503.

STATEMENT OF USE UNDER 37 CFR 2.88, WITH DECLARATION	MARK (Identify the mark)
	SERIAL NO.

TO THE ASSISTANT SECRETARY AND COMMISSIONER OF PATENTS AND TRADEMARKS:

APPLICANT NAME:

NOTICE OF ALLOWANCE ISSUE DATE:

Applicant requests registration of the above-identified trademark/service mark in the United States Patent and Trademark Office on the Principal Register established by the Act of July 5, 1946 (15 U.S.C. 1051 et. seq., as amended). Three (3) specimens showing the mark as used in commerce are submitted with this statement.

☐ Check here only if a Request to Divide under 37 CFR 2.87 is being submitted with this Statement.

Applicant is using the mark in commerce on or in connection with the following goods/services: (Check One)

☐ Those goods/services identified in the Notice of Allowance in this application.

☐ Those goods/services identified in the Notice of Allowance in this application except: (Identify goods/services to be deleted from application) _____

_____

_____

Date of first use of mark anywhere: _____

Date of first use of mark in commerce which the U.S. Congress may regulate: _____

Specify type of commerce: (e.g., interstate, between the U.S. and a specified foreign country) _____

_____

Specify manner or mode of use of mark on or in connection with the goods/services: (e.g., trademark is applied to labels, service mark is used in advertisements) _____

_____

The undersigned being hereby warned that willful false statements and the like so made are punishable by fine or imprisonment, or both, under 18 U.S.C. 1001, and that such willful false statements may jeopardize the validity of the application or any resulting registration, declares that he/she is properly authorized to execute this Statement of Use on behalf of the applicant; he/she believes the applicant to be the owner of the trademark/service mark sought to be registered; the trademark/ service mark is now in use in commerce; and all statements made of his/her own knowledge are true and all statements made on information and belief are believed to be true.

_____
Date

_____
Signature

_____
Telephone Number

_____
Print or Type Name and Position

INSTRUCTIONS AND INFORMATION FOR APPLICANT

In an application based upon a bona fide intention to use a mark in commerce, applicant must use its mark in commerce before a registration will be issued. After use begins, the applicant must submit, along with evidence of use (specimens) and the prescribed fee(s), **either:**

(1) **an** Amendment to Allege Use under 37 CFR 2.76, or
(2) **a** Statement of Use under 37 CFR 2.88.

The difference between these two filings is the timing of the filing. Applicant may file an Amendment to Allege Use before approval of the mark for publication for opposition in the **Official Gazette**, or, if a final refusal has been issued, prior to the expiration of the six month response period. Otherwise, applicant must file a Statement of Use after the Office issues a Notice of Allowance. The Notice of Allowance will issue after the opposition period is completed if no successful opposition is filed. Neither Amendment to Allege Use or Statement of Use papers will be accepted by the Office during the period of time between approval of the mark for publication for opposition in the **Official Gazette** and the issuance of the Notice of Allowance.

Applicant may call (703) 557-5249 to determine whether the mark has been approved for publication for opposition in the **Official Gazette.**

Before filing an Amendment to Allege Use or a Statement of Use, applicant must use the mark in commerce on or in connection with **all** of the goods/services for which applicant will seek registration, **unless** applicant submits with the papers, a request to divide out from the application the goods or services to which the Amendment to Allege Use or Statement of Use pertains. (See: 37 CFR 2.87, Dividing an application)

Applicant **must** submit with an Amendment to Allege Use or a Statement of Use:

(1) the appropriate fee of $100 per class of goods/services listed in the Amendment to Allege Use or the Statement of Use, and

(2) three (3) specimens or facsimiles of the mark as used in commerce for each class of goods/services asserted (e.g., photograph of mark as it appears on goods, label containing mark which is placed on goods, or brochure or advertisement showing mark as used in connection with services).

Cautions/Notes concerning completion of this Statement of Use form:

(1) The goods/services identified in the Statement of Use must be identical to the goods/services identified in the Notice of Allowance. Applicant may delete goods/services. Deleted goods/services may not be reinstated in the application at a later time.

(2) Applicant **may list dates** of use for only one item in each class of goods/services identified in the **Statement of Use.** However, applicant must have used the mark in commerce on all the goods/services **in the class.** Applicant must identify the particular item to which the dates apply.

(3) Only the following person may sign the verification of the Statement of Use, depending on the applicant's legal entity: (a) the individual applicant; (b) an officer of corporate applicant; (c) one general partner of partnership applicant; (d) all joint applicants.

REQUEST FOR EXTENSION OF TIME UNDER 37 CFR 2.89 TO FILE A STATEMENT OF USE, WITH DECLARATION	MARK (Identify the mark)
	SERIAL NO.

TO THE ASSISTANT SECRETARY AND COMMISSIONER OF PATENTS AND TRADEMARKS:

APPLICANT NAME:

NOTICE OF ALLOWANCE MAILING DATE:

Applicant requests a six-month extension of time to file the Statement of Use under 37 CFR 2.88 in this application.

☐ Check here if a Request to Divide under 37 CFR 2.87 is being submitted with this request.

Applicant has a continued bona fide intention to use the mark in commerce in connection with the following goods/services: (Check one below)

☐ Those goods/services identified in the Notice of Allowance in this application.

☐ Those goods/services identified in the Notice of Allowance in this application except: (Identify goods/services to be **deleted** from application) _____

_____

This is the _____ request for an Extension of Time following mailing of the Notice of Allowance.
(Specify first - fifth)

If this is not the first request for an Extension of Time, check one box below. If the first box is checked, explain the circumstance(s) of the non-use in the space provided:

☐ Applicant has not used the mark in commerce yet on all goods/services specified in the Notice of Allowance; however, applicant has made the following ongoing efforts to use the mark in commerce on or in connection with each of the goods/services specified above:

_____

_____

If additional space is needed, please attach a separate sheet to this form

☐ Applicant believes that it has made valid use of the mark in commerce, as evidenced by the Statement of Use submitted with this request; however, if the Statement of Use is found by the Patent and Trademark Office to be fatally defective, applicant will need additional time in which to file a new statement.

The undersigned **being** hereby warned that willful false statements and the like so made are punishable **by fine or** imprisonment, or both, under 18 U.S.C. 1001, and that such willful false statements **may jeopardize the** validity of the application or any resulting registration, declares that he/she is **properly authorized** to execute this Request for Extension of Time to File a Statement of Use on behalf of the applicant; he/she believes the applicant to be the owner of the trademark/service mark sought to be registered; and all statements made of his/her own knowledge are true and all statements made on information and belief are believed to be true.

_____   _____
Date   Signature

_____   _____
Telephone Number   Print or Type Name and Position

INSTRUCTIONS AND INFORMATION FOR APPLICANT

Applicant must file a Statement of Use within six months after the mailing of the Notice of Allowance in an application based upon a bona fide intention to use a mark in commerce, UNLESS, within that same period, applicant submits a request for a six-month extension of time to file the Statement of Use. The request **must**:

> (1) be in writing,
> (2) include applicant's verified statement of continued bona fide intention to use the mark in commerce,
> (3) specify the goods/services to which the request pertains as they are identified in the Notice of Allowance, and
> (4) include a fee of $100 for each class of goods/services.

Applicant may request four further six-month extensions of time. No extension may extend beyond 36 months from the issue date of the Notice of Allowance. Each request must be filed within the previously granted six-month extension period and must include, in addition to the above requirements, a showing of **GOOD CAUSE**. This good cause showing must include:

> (1) applicant's statement that the mark has not been used in commerce yet on all the goods or services specified in the Notice of Allowance with which applicant has a continued bona fide intention to use the mark in commerce, **and**
>
> (2) applicant's statement of ongoing efforts to make such use, which may include the following: (a) product or service research or development, (b) market research, (c) promotional activities, (d) steps to acquire distributors, (e) steps to obtain required governmental approval, or (f) similar specified activity .

Applicant may submit one additional six-month extension request during the existing period in which applicant files the Statement of Use, unless the granting of this request would extend beyond 36 months from the issue date of the Notice of Allowance. As a showing of good cause, applicant should state its belief that applicant has made valid use of the mark in commerce, as evidenced by the submitted Statement of Use, but that if the Statement is found by the PTO to be defective, applicant will need additional time in which to file a new statement of use.

Only the following person may sign the verification of the Request for Extentsion of Time, depending on the applicant's legal entity: (a) the individual applicant; (b) an officer of corporate applicant; (c) one general partner of partnership applicant; (d) all joint applicants.

search request form

Copyright Office
Library of Congress
Washington, D.C. 20559

Reference & Bibliograpy
Section
(202) 707-6850
8:30 a.m.-5 p.m. Monday-Friday
(Eastern time)

Type of work:

☐ Book   ☐ Music   ☐ Motion Picture   ☐ Drama   ☐ Sound Recording
☐ Photograph/Artwork   ☐ Map   ☐ Periodical   ☐ Contribution   ☐ Architectural Work

Search information you require:

☐ Registration   ☐ Renewal   ☐ Assignment   ☐ Address

Specifics of work to be searched:

TITLE: _____
_____

AUTHOR: _____
_____

COPYRIGHT CLAIMANT (if known): _____
(name in © notice)

APPROXIMATE YEAR DATE OF PUBLICATION/CREATION: _____

REGISTRATION NUMBER (if known): _____

OTHER IDENTIFYING INFORMATION: _____
_____
_____

If you need more space please attach additional pages.

Estimates are based on the Copyright Office fee of $20.00 an hour or fraction of an hour consumed. The more information you furnish as a basis for the search the better service we can provide.

Names, titles, and short phrases are not copyrightable.

Please read Circular 22 for more information on copyright searches.

YOUR NAME: _____ DATE: _____

ADDRESS: _____
_____
_____

DAYTIME TELEPHONE NO. (_____) _____

Convey results of estimate/search by telephone   Fee enclosed? ☐ yes   Amount $ _____
☐ yes   ☐ no   ☐ no

Form 20

FORM TX

UNITED STATES COPYRIGHT OFFICE

REGISTRATION NUMBER

TX	TXU

EFFECTIVE DATE OF REGISTRATION

Month	Day	Year

DO NOT WRITE ABOVE THIS LINE. IF YOU NEED MORE SPACE, USE A SEPARATE CONTINUATION SHEET.

1

TITLE OF THIS WORK ▼

PREVIOUS OR ALTERNATIVE TITLES ▼

PUBLICATION AS A CONTRIBUTION If this work was published as a contribution to a periodical, serial, or collection, give information about the collective work in which the contribution appeared.   **Title of Collective Work ▼**

If published in a periodical or serial give: **Volume ▼**   **Number ▼**   **Issue Date ▼**   **On Pages ▼**

2

a

NAME OF AUTHOR ▼

DATES OF BIRTH AND DEATH
Year Born ▼   Year Died ▼

Was this contribution to the work a "work made for hire"?
☐ Yes
☐ No

AUTHOR'S NATIONALITY OR DOMICILE
Name of Country
OR { Citizen of ▶_____
     Domiciled in ▶_____

WAS THIS AUTHOR'S CONTRIBUTION TO THE WORK
Anonymous?   ☐ Yes ☐ No
Pseudonymous? ☐ Yes ☐ No

If the answer to either of these questions is "Yes," see detailed instructions.

NATURE OF AUTHORSHIP   Briefly describe nature of the material created by this author in which copyright is claimed. ▼

b

NAME OF AUTHOR ▼

DATES OF BIRTH AND DEATH
Year Born ▼   Year Died ▼

Was this contribution to the work a "work made for hire"?
☐ Yes
☐ No

AUTHOR'S NATIONALITY OR DOMICILE
Name of country
OR { Citizen of ▶_____
     Domiciled in ▶_____

WAS THIS AUTHOR'S CONTRIBUTION TO THE WORK
Anonymous?   ☐ Yes ☐ No
Pseudonymous? ☐ Yes ☐ No

If the answer to either of these questions is "Yes," see detailed instructions.

NATURE OF AUTHORSHIP   Briefly describe nature of the material created by this author in which copyright is claimed. ▼

c

NAME OF AUTHOR ▼

DATES OF BIRTH AND DEATH
Year Born ▼   Year Died ▼

Was this contribution to the work a "work made for hire"?
☐ Yes
☐ No

AUTHOR'S NATIONALITY OR DOMICILE
Name of Country
OR { Citizen of ▶_____
     Domiciled in ▶_____

WAS THIS AUTHOR'S CONTRIBUTION TO THE WORK
Anonymous?   ☐ Yes ☐ No
Pseudonymous? ☐ Yes ☐ No

If the answer to either of these questions is "Yes," see detailed instructions.

NATURE OF AUTHORSHIP   Briefly describe nature of the material created by this author in which copyright is claimed. ▼

NOTE

Under the law, the "author" of a "work made for hire" is generally the employer, not the employee (see instructions). For any part of this work that was "made for hire" check "Yes" in the space provided, give the employer (or other person for whom the work was prepared) as "Author" of that part, and leave the space for dates of birth and death blank.

3

YEAR IN WHICH CREATION OF THIS WORK WAS COMPLETED This information must be given in all cases.   ◄ Year

DATE AND NATION OF FIRST PUBLICATION OF THIS PARTICULAR WORK   Complete this information Month ▶ _____ Day ▶ _____ Year ▶ _____ ONLY if this work has been published.   ◄ Nation

4

See instructions before completing this space.

COPYRIGHT CLAIMANT(S) Name and address must be given even if the claimant is the same as the author given in space 2.▼

TRANSFER If the claimant(s) named here in space 4 are different from the author(s) named in space 2, give a brief statement of how the claimant(s) obtained ownership of the copyright.▼

DO NOT WRITE HERE OFFICE USE ONLY

APPLICATION RECEIVED

ONE DEPOSIT RECEIVED

TWO DEPOSITS RECEIVED

REMITTANCE NUMBER AND DATE

MORE ON BACK ▶   • Complete all applicable spaces (numbers 5-11) on the reverse side of this page.
• See detailed instructions.   • Sign the form at line 10.

DO NOT WRITE HERE

Page 1 of_____pages

Form 21

EXAMINED BY

CHECKED BY

☐ CORRESPONDENCE
Yes

☐ DEPOSIT ACCOUNT
FUNDS USED

FORM TX

FOR
COPYRIGHT
OFFICE
USE
ONLY

DO NOT WRITE ABOVE THIS LINE. IF YOU NEED MORE SPACE, USE A SEPARATE CONTINUATION SHEET.

PREVIOUS REGISTRATION Has registration for this work, or for an earlier version of this work, already been made in the Copyright Office?

☐ **Yes** ☐ **No** If your answer is "Yes," why is another registration being sought? (Check appropriate box) ▼

☐ This is the first published edition of a work previously registered in unpublished form.

☐ This is the first application submitted by this author as copyright claimant.

☐ This is a changed version of the work, as shown by space 6 on this application.

If your answer is "Yes," give: **Previous Registration Number** ▼   **Year of Registration** ▼

5

DERIVATIVE WORK OR COMPILATION Complete both space 6a & 6b for a derivative work; complete only 6b for a compilation.

a. Preexisting Material Identify any preexisting work or works that this work is based on or incorporates. ▼

b. Material Added to This Work Give a brief, general statement of the material that has been added to this work and in which copyright is claimed. ▼

See instructions
before completing
this space.

6

—space deleted—

7

REPRODUCTION FOR USE OF BLIND OR PHYSICALLY HANDICAPPED INDIVIDUALS   A signature on this form at space 10, and a check in one of the boxes here in space 8, constitutes a non-exclusive grant of permission to the Library of Congress to reproduce and distribute solely for the blind and physically handicapped and under the conditions and limitations prescribed by the regulations of the Copyright Office: (1) copies of the work identified in space 1 of this application in Braille (or similar tactile symbols); or (2) phonorecords embodying a fixation of a reading of that work; or (3) both.

a ☐ Copies and Phonorecords   b ☐ Copies Only   c ☐ Phonorecords Only

See instructions.

8

DEPOSIT ACCOUNT If the registration fee is to be charged to a Deposit Account established in the Copyright Office, give name and number of Account.

Name ▼   **Account Number** ▼

CORRESPONDENCE Give name and address to which correspondence about this application should be sent.   Name/Address/Apt/City/State/Zip ▼

Area Code & Telephone Number ▶

Be sure to
give your
daytime phone
◀ number.

9

CERTIFICATION* I, the undersigned, hereby certify that I am the

Check one ▶

☐ author
☐ other copyright claimant
☐ owner of exclusive right(s)
☐ authorized agent of _____

of the work identified in this application and that the statements made by me in this application are correct to the best of my knowledge.

Name of author or other copyright claimant, or owner of exclusive right(s) ▲

Typed or printed name and date ▼ If this is a published work, this date must be the same as or later than the date of publication given in space 3.

_____   date ▶ _____

☞ Handwritten signature (X) ▼

10

**MAIL
CERTIFI-
CATE TO**

**Certificate
will be
mailed in
window
envelope**

Name ▼

Number/Street/Apartment Number ▼

City/State/ZIP ▼

YOU MUST:
● Complete all necessary spaces
● Sign your application in space 10
**SEND ALL 3 ELEMENTS
IN THE SAME PACKAGE:**
1. Application form
2. Non-refundable $10 filing fee
   in check or money order
   payable to *Register of Copyrights*
3. Deposit material
MAIL TO:
Register of Copyrights
Library of Congress
Washington, D.C. 20559

11

☆ U.S. GOVERNMENT PRINTING OFFICE 1988—241-428/80,009

February 1989—65,000

Form 21 (cont.)

Filling Out Application Form TX

Detach and read these instructions before completing this form. Make sure all applicable spaces have been filled in before you return this form.

BASIC INFORMATION

When to Use This Form: Use Form TX for registration of published or unpublished non-dramatic literary works, excluding periodicals or serial issues. This class includes a wide variety of works: fiction, non-fiction, poetry, textbooks, reference works, directories, catalogs, advertising copy, compilations of information, and computer programs. For periodicals and serials, use Form SE.

Deposit to Accompany Application: An application for copyright registration must be accompanied by a deposit consisting of copies or phonorecords representing the entire work for which registration is to be made. The following are the general deposit requirements as set forth in the statute:

Unpublished Work: Deposit one complete copy (or phonorecord).

Published Work: Deposit two complete copies (or phonorecords) of the best edition.

Work First Published Outside the United States: Deposit one complete copy (or phonorecord) of the first foreign edition.

Contribution to a Collective Work: Deposit one complete copy (or phonorecord) of the best edition of the collective work.

The Copyright Notice: For published works, the law provides that a copyright notice in a specified form "shall be placed on all publicly distributed copies from which the work can be visually perceived." Use of the copyright notice is the responsibility of the copyright owner and does not require advance permission from the Copyright Office. The required form of the notice for copies generally consists of three elements: (1) the symbol "©", or the word "Copyright," or the abbreviation "Copr."; (2) the year of first publication; and (3) the name of the owner of copyright. For example: "© 1981 Constance Porter." The notice is to be affixed to the copies "in such manner and location as to give reasonable notice of the claim of copyright."

For further information about copyright registration, notice, or special questions relating to copyright problems, write:

Information and Publications Section, LM-455
Copyright Office
Library of Congress
Washington, D.C. 20559

LINE-BY-LINE INSTRUCTIONS

1 SPACE 1: Title

Title of This Work: Every work submitted for copyright registration must be given a title to identify that particular work. If the copies or phonorecords of the work bear a title (or an identifying phrase that could serve as a title), transcribe that wording *completely* and *exactly* on the application. Indexing of the registration and future identification of the work will depend on the information you give here.

Previous or Alternative Titles: Complete this space if there are any additional titles for the work under which someone searching for the registration might be likely to look, or under which a document pertaining to the work might be recorded.

Publication as a Contribution: If the work being registered is a contribution to a periodical, serial, or collection, give the title of the contribution in the "Title of this Work" space. Then, in the line headed "Publication as a Contribution," give information about the collective work in which the contribution appeared.

2 SPACE 2: Author(s)

General Instructions: After reading these instructions, decide who are the "authors" of this work for copyright purposes. Then, unless the work is a "collective work," give the requested information about every "author" who contributed any appreciable amount of copyrightable matter to this version of the work. If you need further space, request additional Continuation sheets. In the case of a collective work, such as an anthology, collection of essays, or encyclopedia, give information about the author of the collective work as a whole.

Name of Author: The fullest form of the author's name should be given. Unless the work was "made for hire," the individual who actually created the work is its "author." In the case of a work made for hire, the statute provides that "the employer or other person for whom the work was prepared is considered the author."

What is a "Work Made for Hire"? A "work made for hire" is defined as: (1) "a work prepared by an employee within the scope of his or her employment"; or (2) "a work specially ordered or commissioned for use as a contribution to a collective work, as a part of a motion picture or other audiovisual work, as a translation, as a supplementary work, as a compilation, as an instructional text, as a test, as answer material for a test, or as an atlas, if the parties expressly agree in a written instrument signed by them that the work shall be considered a work made for hire." If you have checked "Yes" to indicate that the work was "made for hire," you must give the full legal name of the employer (or other person for whom the work was prepared). You may also include the name of the employee along with the name of the employer (for example: "Elster Publishing Co., employer for hire of John Ferguson").

"Anonymous" or "Pseudonymous" Work: An author's contribution to a work is "anonymous" if that author is not identified on the copies or phonorecords of the work. An author's contribution to a work is "pseudonymous" if that author is identified on the copies or phonorecords under a fictitious name. If the work is "anonymous" you may: (1) leave the line blank; or (2) state "anonymous" on the line; or (3) reveal the author's identity. If the work is "pseudonymous" you may : (1) leave the line blank; or (2) give the pseudonym and identify it as such (for example: "Huntley Haverstock, pseudonym"); or (3) reveal the author's name, making clear which is the real name and which is the pseudonym (for example: "Judith Barton, whose pseudonym is Madeline Elster"). However, the citizenship or domicile of the author **must** be given in all cases.

Dates of Birth and Death: If the author is dead, the statute requires that the year of death be included in the application unless the work is anonymous or pseudonymous. The author's birth date is optional, but is useful as a form of identification. Leave this space blank if the author's contribution was a "work made for hire."

Author's Nationality or Domicile: Give the country of which the author is a citizen, or the country in which the author is domiciled. Nationality or domicile **must** be given in all cases.

Nature of Authorship: After the words "Nature of Authorship" give a brief general statement of the nature of this particular author's contribution to the work. Examples: "Entire text"; "Coauthor of entire text"; "Chapters 11-14"; "Editorial revisions"; "Compilation and English translation"; "New text."

3 SPACE 3: Creation and Publication

General Instructions: Do not confuse "creation" with "publication." Every application for copyright registration must state "the year in which creation of the work was completed." Give the date and nation of first publication only if the work has been published.

Creation: Under the statute, a work is "created" when it is fixed in a copy or phonorecord for the first time. Where a work has been prepared over a period of time, the part of the work existing in fixed form on a particular date constitutes the created work on that date. The date you give here should be the year in which the author completed the particular version for which registration is now being sought, even if other versions exist or if further changes or additions are planned.

Publication: The statute defines "publication" as "the distribution of copies or phonorecords of a work to the public by sale or other transfer of ownership, or by rental, lease, or lending"; a work is also "published" if there has been an "offering to distribute copies or phonorecords to a group of persons for purposes of further distribution, public performance, or public display." Give the full date (month, day, year) when, and the country where, publication first occurred. If first publication took place simultaneously in the United States and other countries, it is sufficient to state "U.S.A."

4 SPACE 4: Claimant(s)

Name(s) and Address(es) of Copyright Claimant(s): Give the name(s) and address(es) of the copyright claimant(s) in this work even if the claimant is the same as the author. Copyright in a work belongs initially to the author of the work (including, in the case of a work made for hire, the employer or other person for whom the work was prepared). The copyright claimant is either the author of the work or a person or organization to whom the copyright initially belonging to the author has been transferred.

Transfer: The statute provides that, if the copyright claimant is not the author, the application for registration must contain "a brief statement of how the claimant obtained ownership of the copyright." If any copyright claimant named in space 4 is not an author named in space 2, give a brief, general statement summarizing the means by which that claimant obtained ownership of the copyright. Examples: "By written contract"; "Transfer of all rights by author"; "Assignment"; "By will." Do not attach transfer documents or other attachments or riders.

5 SPACE 5: Previous Registration

General Instructions: The questions in space 5 are intended to find out whether an earlier registration has been made for this work and, if so, whether there is any basis for a new registration. As a general rule, only one basic copyright registration can be made for the same version of a particular work.

Same Version: If this version is substantially the same as the work covered by a previous registration, a second registration is not generally possible unless: (1) the work has been registered in unpublished form and a second registration is now being sought to cover this first published edition; or (2) someone other than the author is identified as copyright claimant in the earlier registration, and the author is now seeking registration in his or her own name. If either of these two exceptions apply, check the appropriate box and give the earlier registration number and date. Otherwise, do not submit Form TX; instead, write the Copyright Office for information about supplementary registration or recordation of transfers of copyright ownership.

Changed Version: If the work has been changed, and you are now seeking registration to cover the additions or revisions, check the last box in space 5, give the earlier registration number and date, and complete both parts of space 6 in accordance with the instructions below.

Previous Registration Number and Date: If more than one previous registration has been made for the work, give the number and date of the latest registration.

6 SPACE 6: Derivative Work or Compilation

General Instructions: Complete space 6 if this work is a "changed version," "compilation," or "derivative work," and if it incorporates one or more earlier works that have already been published or registered for copyright, or that have fallen into the public domain. A "compilation" is defined as "a work formed by the collection and assembling of preexisting materials or of data that are selected, coordinated, or arranged in such a way that the resulting work as a whole constitutes an original work of authorship." A "derivative work" is "a work based on one or more preexisting works." Examples of derivative works include translations, fictionalizations, abridgments, condensations, or "any other form in which a work may be recast, transformed, or adapted." Derivative works also include works "consisting of editorial revisions, annotations, or other modifications" if these changes, as a whole, represent an original work of authorship.

Preexisting Material (space 6a): For derivative works, complete this space and space 6b. In space 6a identify the preexisting work that has been recast, transformed, or adapted. An example of preexisting material might be: "Russian version of Goncharov's 'Oblomov'." Do not complete space 6a for compilations.

Material Added to This Work (space 6b): Give a brief, general statement of the new material covered by the copyright claim for which registration is sought. **Derivative work** examples include: "Foreword, editing, critical annotations"; "Translation"; "Chapters 11-17." If the work is a **compilation**, describe both the compilation itself and the material that has been compiled. Example: "Compilation of certain 1917 Speeches by Woodrow Wilson." A work may be both a derivative work and compilation, in which case a sample statement might be: "Compilation and additional new material."

7 SPACE 7: Manufacturing Provisions

Due to the expiration of the Manufacturing Clause of the copyright law on June 30, 1986, this space has been deleted.

8 SPACE 8: Reproduction for Use of Blind or Physically Handicapped Individuals

General Instructions: One of the major programs of the Library of Congress is to provide Braille editions and special recordings of works for the exclusive use of the blind and physically handicapped. In an effort to simplify and speed up the copyright licensing procedures that are a necessary part of this program, section 710 of the copyright statute provides for the establishment of a voluntary licensing system to be tied in with copyright registration. Copyright Office regulations provide that you may grant a license for such reproduction and distribution solely for the use of persons who are certified by competent authority as unable to read normal printed material as a result of physical limitations. The license is entirely voluntary, nonexclusive, and may be terminated upon 90 days notice.

How to Grant the License: If you wish to grant it, check one of the three boxes in space 8. Your check in one of these boxes, together with your signature in space 10, will mean that the Library of Congress can proceed to reproduce and distribute under the license without further paperwork. For further information, write for Circular R63.

9,10,11 SPACE 9, 10, 11: Fee, Correspondence, Certification, Return Address

Deposit Account: If you maintain a Deposit Account in the Copyright Office, identify it in space 9. Otherwise leave the space blank and send the fee of $10 with your application and deposit.

Correspondence (space 9): This space should contain the name, address, area code, and telephone number of the person to be consulted if correspondence about this application becomes necessary.

Certification (space 10): The application can not be accepted unless it bears the date and the **handwritten signature** of the author or other copyright claimant, or of the owner of exclusive right(s), or of the duly authorized agent of author, claimant, or owner of exclusive right(s).

Address for Return of Certificate (space 11): The address box must be completed legibly since the certificate will be returned in a window envelope.

FORM SE

UNITED STATES COPYRIGHT OFFICE

REGISTRATION NUMBER

U

EFFECTIVE DATE OF REGISTRATION

Month	Day	Year

DO NOT WRITE ABOVE THIS LINE. IF YOU NEED MORE SPACE, USE A SEPARATE CONTINUATION SHEET.

1

TITLE OF THIS SERIAL ▼

Volume ▼   Number ▼   Date on Copies ▼   Frequency of Publication ▼

PREVIOUS OR ALTERNATIVE TITLES ▼

2

NOTE

Under the law. the "author" of a "work made for hire" is generally the employer. not the employee (see instructions) For any part of this work that was "made for hire" check "Yes" in the space provided. give the employer (or other person for whom the work was prepared) as "Author" of that part. and leave the space for dates of birth and death blank

a

NAME OF AUTHOR ▼

DATES OF BIRTH AND DEATH
Year Born ▼   Year Died ▼

Was this contribution to the work a "work made for hire"?
☐ Yes
☐ No

AUTHOR'S NATIONALITY OR DOMICILE
Name of Country
OR { Citizen of ▶_____
Domiciled in ▶_____

WAS THIS AUTHOR'S CONTRIBUTION TO THE WORK
Anonymous?   ☐ Yes ☐ No
Pseudonymous?   ☐ Yes ☐ No

If the answer to either of these questions is "Yes." see detailed instructions

NATURE OF AUTHORSHIP   Briefly describe nature of the material created by this author in which copyright is claimed. ▼
☐ Collective Work   Other:

b

NAME OF AUTHOR ▼

DATES OF BIRTH AND DEATH
Year Born ▼   Year Died ▼

Was this contribution to the work a "work made for hire"?
☐ Yes
☐ No

AUTHOR'S NATIONALITY OR DOMICILE
Name of country
OR { Citizen of ▶_____
Domiciled in ▶_____

WAS THIS AUTHOR'S CONTRIBUTION TO THE WORK
Anonymous?   ☐ Yes ☐ No
Pseudonymous?   ☐ Yes ☐ No

If the answer to either of these questions is "Yes." see detailed instructions

NATURE OF AUTHORSHIP   Briefly describe nature of the material created by this author in which copyright is claimed. ▼
☐ Collective Work   Other:

c

NAME OF AUTHOR ▼

DATES OF BIRTH AND DEATH
Year Born ▼   Year Died ▼

Was this contribution to the work a "work made for hire"?
☐ Yes
☐ No

AUTHOR'S NATIONALITY OR DOMICILE
Name of Country
OR { Citizen of ▶_____
Domiciled in ▶_____

WAS THIS AUTHOR'S CONTRIBUTION TO THE WORK
Anonymous?   ☐ Yes ☐ No
Pseudonymous?   ☐ Yes ☐ No

If the answer to either of these questions is "Yes." see detailed instructions.

NATURE OF AUTHORSHIP   Briefly describe nature of the material created by this author in which copyright is claimed. ▼
☐ Collective Work   Other:

3

a YEAR IN WHICH CREATION OF THIS ISSUE WAS COMPLETED This information must be given in all cases.
_____ ◀ Year

b DATE AND NATION OF FIRST PUBLICATION OF THIS PARTICULAR ISSUE
Complete this information ONLY if this work has been published.
Month ▶_____ Day ▶_____ Year ▶_____
◀ Nation

4

See instructions before completing this space.

COPYRIGHT CLAIMANT(S) Name and address must be given even if the claimant is the same as the author given in space 2.▼

TRANSFER If the claimant(s) named here in space 4 are different from the author(s) named in space 2, give a brief statement of how the claimant(s) obtained ownership of the copyright.▼

DO NOT WRITE HERE OFFICE USE ONLY

APPLICATION RECEIVED

ONE DEPOSIT RECEIVED

TWO DEPOSITS RECEIVED

REMITTANCE NUMBER AND DATE

MORE ON BACK ▶
• Complete all applicable spaces (numbers 5-11) on the reverse side of this page.
• See detailed instructions.

Form 22

DO NOT WRITE HERE

Page 1 of_____pages

EXAMINED BY

CHECKED BY

☐ CORRESPONDENCE
Yes

FORM SE

FOR
COPYRIGHT
OFFICE
USE
ONLY

DO NOT WRITE ABOVE THIS LINE. IF YOU NEED MORE SPACE, USE A SEPARATE CONTINUATION SHEET.

PREVIOUS REGISTRATION   Has registration for this issue, or for an earlier version of this particular issue, already been made in the Copyright Office?

☐ Yes   ☐ No   If your answer is "Yes," why is another registration being sought? (Check appropriate box) ▼

a. ☐ This is the first published version of an issue previously registered in unpublished form.

b. ☐ This is the first application submitted by this author as copyright claimant.

c. ☐ This is a changed version of this issue, as shown by space 6 on this application.

If your answer is "Yes," give: **Previous Registration Number** ▼   **Year of Registration** ▼

5

DERIVATIVE WORK OR COMPILATION   Complete both space 6a & 6b for a derivative work; complete only 6b for a compilation.

a.   **Preexisting Material**   Identify any preexisting work or works that this work is based on or incorporates. ▼

b.   **Material Added to This Work**   Give a brief, general statement of the material that has been added to this work and in which copyright is claimed. ▼

6

See instructions
before completing
this space.

—space deleted—

7

REPRODUCTION FOR USE OF BLIND OR PHYSICALLY HANDICAPPED INDIVIDUALS   A signature on this form at space 10, and a check in one of the boxes here in space 8, constitutes a non-exclusive grant of permission to the Library of Congress to reproduce and distribute solely for the blind and physically handicapped and under the conditions and limitations prescribed by the regulations of the Copyright Office: (1) copies of the work identified in space 1 of this application in Braille (or similar tactile symbols); or (2) phonorecords embodying a fixation of a reading of that work; or (3) both.

a ☐ Copies and Phonorecords   b ☐ Copies Only   c ☐ Phonorecords Only

8

See instructions.

DEPOSIT ACCOUNT   If the registration fee is to be charged to a Deposit Account established in the Copyright Office, give name and number of Account.
Name ▼   **Account Number** ▼

9

CORRESPONDENCE   Give name and address to which correspondence about this application should be sent.   Name/Address/Apt/City/State/Zip ▼

Area Code & Telephone Number ▶

Be sure to
give your
daytime phone
◀ number.

CERTIFICATION*   I, the undersigned, hereby certify that I am the   ☐ author
☐ other copyright claimant
Check one ▶   ☐ owner of exclusive right(s)
of the work identified in this application and that the statements made   ☐ authorized agent of _____
by me in this application are correct to the best of my knowledge.   Name of author or other copyright claimant, or owner of exclusive right(s) ▲

10

Typed or printed name and date ▼ If this application gives a date of publication in space 3, do not sign and submit it before that date.

_____ date ▶ _____

☞   Handwritten signature (X) ▼

* 17 U S C § 506(e) Any person who knowingly makes a false representation of a material fact in the application for copyright registration provided for by section 409, or in any written statement filed in connection with the application, shall be fined not more than $2 500.

Form 22 (cont.)

SHORT FORM SE

UNITED STATES COPYRIGHT OFFICE

REGISTRATION NUMBER

EFFECTIVE DATE OF REGISTRATION
(Assigned by Copyright Office)

Month	Day	Year

APPLICATION RECEIVED

ONE DEPOSIT RECEIVED

TWO DEPOSITS RECEIVED

EXAMINED BY

CORRESPONDENCE ☐

DO NOT WRITE ABOVE THIS LINE.

1

TITLE OF THIS SERIAL AS IT APPEARS ON THE COPY

Volume▼   Number▼   Date on Copies▼   ISSN▼

2

NAME AND ADDRESS OF THE AUTHOR AND COPYRIGHT CLAIMANT IN THIS COLLECTIVE WORK MADE FOR HIRE

3

DATE OF PUBLICATION OF THIS PARTICULAR ISSUE
Month▼   Day▼   Year▼

YEAR IN WHICH CREATION OF
THIS ISSUE WAS COMPLETED
(IF EARLIER THAN THE YEAR OF
PUBLICATION):
Year▼

CERTIFICATION*: I, the undersigned, hereby certify that I am the copyright claimant or the authorized agent of the copyright claimant of the work identified in this application, that all the conditions specified in the instructions on the back of this form are met, and that the statements made by me in this application are correct to the best of my knowledge.

Handwritten signature (X) _____

Typed or printed name of signer _____

PERSON TO CONTACT FOR CORRESPONDENCE ABOUT THIS CLAIM

Name▶ _____

Daytime telephone number▶ _____

Address (if other than given below) ▶ _____

DEPOSIT ACCOUNT

Account number▶ _____

Name of account▶ _____

MAIL CERTIFI-CATE TO

Name▼

Number/Street/Apartment Number▼

City/State/ZIP▼

Certificate will be mailed in window envelope

*17 U.S.C. §506(e): Any person who knowingly makes a false representation of a material fact in the application for copyright registration provided for by section 409, or in any written statement filed in connection with the application, shall be fined not more than $2,500.

March 1990—50,000   **Form 23**   ☆U.S. GOVERNMENT PRINTING OFFICE: 1990–262-308/13

✪ Filling Out Short Form SE

BASIC INFORMATION
Read these instructions before completing this form. Make sure all applicable spaces have been filled in before you return this form.

When to Use This Form: All the following conditions must be met in order to use this form. If any one of the conditions does not apply, you must use Form SE. Incorrect use of this form will result in a delay in your registration.

The claim must be in a collective work.
The work must be essentially an all-new collective work or issue.
The author must be a citizen or domiciliary of the United States.
The work must be a work made for hire.
The author(s) and claimant(s) must be the same person(s) or organization(s).
The work must be first published in the United States.

Deposit to Accompany Application: An application for registration of a copyright claim in a serial issue first published in the United States must be accompanied by a deposit consisting of two copies (or phonorecords) of the best edition.

Fee: The filing fee of $10.00 must be sent for each issue to be registered. Do not send cash or currency.

Mailing Requirements: It is important that you send the application, the deposit copies, and the $10.00 fee together in the same envelope or package. Send to: Register of Copyrights, Library of Congress, Washington, D.C. 20559.

Reproduction for Use of Blind or Physically Handicapped Individuals: A signature on this form and a check in one of these boxes constitutes a nonexclusive grant of permission to the Library of Congress to reproduce and distribute solely for the blind and physically handicapped under the conditions and limitations prescribed by the regulations of the Copyright Office: (1) copies of the work identified in space 1 of this application in Braille (or similar tactile symbols); or (2) phonorecords embodying a fixation of a reading of that work; or (3) both.

☐ Copies only ☐ Phonorecords only ☐ Copies and phonorecords

Collective Work: The term "collective work" refers to a work , such as a serial issue, in which a number of contributions are assembled into a collective whole. A claim in the "collective work" extends to all copyrightable authorship created by employees of the author, as well as any independent contributions in which the claimant has acquired ownership of the copyright.

Publication: The statute defines "publication" as "The distribution of copies or phonorecords of a work to the public by sale or other transfer of ownership, or by rental, lease, or lending;" a work is also "published" if there has been an "offering to distribute copies or phonorecords to a group of persons for purposes of further distribution, public performance, or public display."

Creation: A work is "created" when it is fixed in a copy (or phonorecord) for the first time.

Work Made for Hire: A "work made for hire" is defined as: (1) a work prepared by an employee within the scope of his or her employment; or (2) a work specially ordered or commissioned for certain uses (including use as a contribution to a collective work), if the parties expressly agree in a written instrument signed by them that the work shall be considered a work made for hire. The employer is the author of a work made for hire.

The Copyright Notice: For works first published on or after March 1, 1989, the law provides that a copyright notice in a specified form "may be placed on all publicly distributed copies from which the work can be visually perceived." Use of the copyright notice is the responsibility of the copyright owner and does not require advance permission from the Copyright Office. The required form of the notice for copies generally consists of three elements: (1) the symbol "©", or the word "Copyright," or the abbreviation "Copr."; (2) the year of first publication; and (3) the name of the owner of copyright. For example: "©1989 Jane Cole." The notice is to be affixed to the copies "in such manner and location as to give reasonable notice of the claim of copyright." Works first published prior to March 1, 1989, **must** carry the notice or risk loss of copyright protection.

For information about notice requirements for works published before March 1, 1989, or other copyright information, write: Information Section, LM-401, Copyright Office, Library of Congress, Washington, D.C. 20559.

SPACE-BY-SPACE INSTRUCTIONS

1 SPACE 1: Title

Every work submitted for copyright registration must be given a title to identify that particular work. Give the complete title of the periodical, including the volume, number, issue date, or other indicia printed on the copies. If possible, give the International Standard Serial Number (ISSN).

2 SPACE 2: Author and Copyright Claimant

Give the fullest form of the author and claimant's name. If there are joint authors and owners, give the names of all the author/owners. (It is assumed that the authors and claimants are the same, that the work is made for hire, and that the claim is in the collective work).

3 SPACE 3: Date of Publication of This Particular Work

Give the exact date on which publication of this issue first took place. The full date, including month, day, and year must be given.

Year in Which Creation of This Issue Was Completed: Give the year in which this serial issue was fixed in a copy or phonorecord for the first time. If no year is given, it is assumed that the issue was created in the same year in which it was published. The date must be the same as or no later than the publication date.

Certification: The application cannot be accepted unless it bears the handwritten signature of the copyright claimant or the duly authorized agent of the copyright claimant.

Person to Contact for Correspondence About This Claim: Give the name and telephone number, including area code, of the person to whom any correspondence concerning this claim should be addressed. Give the address only if it is different from the address for mailing of the certificate.

Deposit Account: If the filing fee is to be charged against a Deposit Account in the Copyright Office, give the name and number of the account in this space. Otherwise, leave the space blank and forward the $10.00 filing fee with your application and deposit.

Mailing Address of Certificate: This address must be complete and legible since the certificate will be mailed in a window envelope.

FORM PA
UNITED STATES COPYRIGHT OFFICE

REGISTRATION NUMBER

PA	PAU

EFFECTIVE DATE OF REGISTRATION

Month _____ Day _____ Year _____

DO NOT WRITE ABOVE THIS LINE. IF YOU NEED MORE SPACE, USE A SEPARATE CONTINUATION SHEET.

1

TITLE OF THIS WORK ▼

PREVIOUS OR ALTERNATIVE TITLES ▼

NATURE OF THIS WORK ▼ See instructions

2

a

NAME OF AUTHOR ▼

DATES OF BIRTH AND DEATH
Year Born ▼   Year Died ▼

Was this contribution to the work a "work made for hire"?
☐ Yes
☐ No

AUTHOR'S NATIONALITY OR DOMICILE
Name of Country
OR { Citizen of ▶ _____
Domiciled in ▶ _____

WAS THIS AUTHOR'S CONTRIBUTION TO THE WORK
Anonymous?   ☐ Yes ☐ No
Pseudonymous? ☐ Yes ☐ No

If the answer to either of these questions is "Yes," see detailed instructions.

NATURE OF AUTHORSHIP Briefly describe nature of the material created by this author in which copyright is claimed. ▼

b

NAME OF AUTHOR ▼

DATES OF BIRTH AND DEATH
Year Born ▼   Year Died ▼

Was this contribution to the work a "work made for hire"?
☐ Yes
☐ No

AUTHOR'S NATIONALITY OR DOMICILE
Name of country
OR { Citizen of ▶ _____
Domiciled in ▶ _____

WAS THIS AUTHOR'S CONTRIBUTION TO THE WORK
Anonymous?   ☐ Yes ☐ No
Pseudonymous? ☐ Yes ☐ No

If the answer to either of these questions is "Yes," see detailed instructions.

NATURE OF AUTHORSHIP Briefly describe nature of the material created by this author in which copyright is claimed. ▼

c

NAME OF AUTHOR ▼

DATES OF BIRTH AND DEATH
Year Born ▼   Year Died ▼

Was this contribution to the work a "work made for hire"?
☐ Yes
☐ No

AUTHOR'S NATIONALITY OR DOMICILE
Name of Country
OR { Citizen of ▶ _____
Domiciled in ▶ _____

WAS THIS AUTHOR'S CONTRIBUTION TO THE WORK
Anonymous?   ☐ Yes ☐ No
Pseudonymous? ☐ Yes ☐ No

If the answer to either of these questions is "Yes," see detailed instructions.

NATURE OF AUTHORSHIP Briefly describe nature of the material created by this author in which copyright is claimed. ▼

NOTE

Under the law, the "author" of a "work made for hire" is generally the employer, not the employee (see instructions). For any part of this work that was "made for hire" check "Yes" in the space provided, give the employer (or other person for whom the work was prepared) as "Author" of that part, and leave the space for dates of birth and death blank.

3

YEAR IN WHICH CREATION OF THIS WORK WAS COMPLETED This information must be given in all cases.
◀ Year

DATE AND NATION OF FIRST PUBLICATION OF THIS PARTICULAR WORK
Complete this information ONLY if this work has been published.
Month ▶ _____ Day ▶ _____ Year ▶ _____
◀ Nation

4

See instructions before completing this space.

COPYRIGHT CLAIMANT(S) Name and address must be given even if the claimant is the same as the author given in space 2.▼

TRANSFER If the claimant(s) named here in space 4 are different from the author(s) named in space 2, give a brief statement of how the claimant(s) obtained ownership of the copyright.▼

MORE ON BACK ▶
- Complete all applicable spaces (numbers 5-9) on the reverse side of this page.
- See detailed instructions.
- Sign the form at line 8.

DO NOT WRITE HERE

Page 1 of _____ pages

Form 24

DO NOT WRITE ABOVE THIS LINE. IF YOU NEED MORE SPACE, USE A SEPARATE CONTINUATION SHEET.

PREVIOUS REGISTRATION Has registration for this work, or for an earlier version of this work, already been made in the Copyright Office?

☐ Yes ☐ No   If your answer is "Yes," why is another registration being sought? (Check appropriate box) ▼

☐ This is the first published edition of a work previously registered in unpublished form.

☐ This is the first application submitted by this author as copyright claimant.

☐ This is a changed version of the work, as shown by space 6 on this application.

If your answer is "Yes," give: **Previous Registration Number** ▼   **Year of Registration** ▼

5

DERIVATIVE WORK OR COMPILATION   Complete both space 6a & 6b for a derivative work; complete only 6b for a compilation.
a. **Preexisting Material**   Identify any preexisting work or works that this work is based on or incorporates. ▼

_____

_____

b. **Material Added to This Work**   Give a brief, general statement of the material that has been added to this work and in which copyright is claimed. ▼

_____

_____

6

See instructions
before completing
this space.

DEPOSIT ACCOUNT   If the registration fee is to be charged to a Deposit Account established in the Copyright Office, give name and number of Account.
Name ▼   **Account Number** ▼

7

CORRESPONDENCE   Give name and address to which correspondence about this application should be sent.   Name/Address/Apt/City/State/Zip ▼

_____

_____

Area Code & Telephone Number ▶

Be sure to
give your
daytime phone
◀ number.

CERTIFICATION*   I, the undersigned, hereby certify that I am the
Check only one ▼

☐ author

☐ other copyright claimant

☐ owner of exclusive right(s)

☐ authorized agent of_____
                      Name of author or other copyright claimant, or owner of exclusive right(s) ▲

of the work identified in this application and that the statements made
by me in this application are correct to the best of my knowledge.

8

Typed or printed name and date ▼ If this is a published work, this date must be the same as or later than the date of publication given in space 3.

_____   date ▶ _____

Handwritten signature (X) ▼

* 17 U.S.C. § 506(e): Any person who knowingly makes a false representation of a material fact in the application for copyright registration provided for by section 409, or in any written statement filed in connection with the application, shall be fined not more than $2,500.

Form 24 (cont.)

■March 1989—200.000   ☆U.S. GOVERNMENT PRINTING OFFICE: 1989—241-428/80,014

FORM VA
UNITED STATES COPYRIGHT OFFICE

REGISTRATION NUMBER

_____

VA   VAU

EFFECTIVE DATE OF REGISTRATION

_____

| Month | Day | Year |

DO NOT WRITE ABOVE THIS LINE. IF YOU NEED MORE SPACE, USE A SEPARATE CONTINUATION SHEET.

1

TITLE OF THIS WORK ▼   **NATURE OF THIS WORK ▼** See instructions

PREVIOUS OR ALTERNATIVE TITLES ▼

PUBLICATION AS A CONTRIBUTION If this work was published as a contribution to a periodical, serial, or collection, give information about the collective work in which the contribution appeared.   **Title of Collective Work ▼**

If published in a periodical or serial give: **Volume ▼**   **Number ▼**   **Issue Date ▼**   **On Pages ▼**

2

a

NAME OF AUTHOR ▼   **DATES OF BIRTH AND DEATH**
Year Born ▼   Year Died ▼

Was this contribution to the work a "work made for hire"?
☐ Yes
☐ No

AUTHOR'S NATIONALITY OR DOMICILE
Name of Country
OR { Citizen of ▶_____
    { Domiciled in ▶_____

WAS THIS AUTHOR'S CONTRIBUTION TO THE WORK
Anonymous?   ☐ Yes ☐ No
Pseudonymous?   ☐ Yes ☐ No
If the answer to either of these questions is "Yes," see detailed instructions.

NATURE OF AUTHORSHIP Briefly describe nature of the material created by this author in which copyright is claimed. ▼

NOTE

Under the law, the "author" of a "work made for hire" is generally the employer, not the employee (see instructions). For any part of this work that was "made for hire" check "Yes" in the space provided, give the employer (or other person for whom the work was prepared) as "Author" of that part, and leave the space for dates of birth and death blank.

b

NAME OF AUTHOR ▼   **DATES OF BIRTH AND DEATH**
Year Born ▼   Year Died ▼

Was this contribution to the work a "work made for hire"?
☐ Yes
☐ No

AUTHOR'S NATIONALITY OR DOMICILE
Name of country
OR { Citizen of ▶_____
    { Domiciled in ▶_____

WAS THIS AUTHOR'S CONTRIBUTION TO THE WORK
Anonymous?   ☐ Yes ☐ No
Pseudonymous?   ☐ Yes ☐ No
If the answer to either of these questions is "Yes," see detailed instructions.

NATURE OF AUTHORSHIP Briefly describe nature of the material created by this author in which copyright is claimed. ▼

c

NAME OF AUTHOR ▼   **DATES OF BIRTH AND DEATH**
Year Born ▼   Year Died ▼

Was this contribution to the work a "work made for hire"?
☐ Yes
☐ No

AUTHOR'S NATIONALITY OR DOMICILE
Name of Country
OR { Citizen of ▶_____
    { Domiciled in ▶_____

WAS THIS AUTHOR'S CONTRIBUTION TO THE WORK
Anonymous?   ☐ Yes ☐ No
Pseudonymous?   ☐ Yes ☐ No
If the answer to either of these questions is "Yes," see detailed instructions.

NATURE OF AUTHORSHIP Briefly describe nature of the material created by this author in which copyright is claimed. ▼

3

YEAR IN WHICH CREATION OF THIS WORK WAS COMPLETED This information must be given in all cases.   ◀ Year

DATE AND NATION OF FIRST PUBLICATION OF THIS PARTICULAR WORK
Complete this information ONLY if this work has been published. Month ▶ _____ Day ▶ _____ Year ▶ _____   ◀ Nation

4

See instructions before completing this space.

COPYRIGHT CLAIMANT(S) Name and address must be given even if the claimant is the same as the author given in space 2.▼

TRANSFER If the claimant(s) named here in space 4 are different from the author(s) named in space 2, give a brief statement of how the claimant(s) obtained ownership of the copyright.▼

DO NOT WRITE HERE OFFICE USE ONLY

APPLICATION RECEIVED

_____

ONE DEPOSIT RECEIVED

_____

TWO DEPOSITS RECEIVED

_____

REMITTANCE NUMBER AND DATE

MORE ON BACK ▶ • Complete all applicable spaces (numbers 5-9) on the reverse side of this page.
• See detailed instructions.   • Sign the form at line 8.

DO NOT WRITE HERE

Page 1 of_____pages

4

EXAMINED BY

FORM VA

CHECKED BY

☐ CORRESPONDENCE
Yes

☐ DEPOSIT ACCOUNT
FUNDS USED

FOR
COPYRIGHT
OFFICE
USE
ONLY

DO NOT WRITE ABOVE THIS LINE. IF YOU NEED MORE SPACE, USE A SEPARATE CONTINUATION SHEET.

PREVIOUS REGISTRATION Has registration for this work, or for an earlier version of this work, already been made in the Copyright Office?
☐ Yes ☐ No   If your answer is "Yes," why is another registration being sought? (Check appropriate box) ▼
☐ This is the first published edition of a work previously registered in unpublished form.
☐ This is the first application submitted by this author as copyright claimant.
☐ This is a changed version of the work, as shown by space 6 on this application.
If your answer is "Yes," give: **Previous Registration Number** ▼   **Year of Registration** ▼

5

DERIVATIVE WORK OR COMPILATION   Complete both space 6a & 6b for a derivative work; complete only 6b for a compilation.
a. **Preexisting Material**   Identify any preexisting work or works that this work is based on or incorporates. ▼

6

See instructions before completing this space.

b. **Material Added to This Work**   Give a brief, general statement of the material that has been added to this work and in which copyright is claimed.▼

DEPOSIT ACCOUNT   If the registration fee is to be charged to a Deposit Account established in the Copyright Office, give name and number of Account.
Name ▼   **Account Number** ▼

7

CORRESPONDENCE   Give name and address to which correspondence about this application should be sent.   Name/Address/Apt/City/State/Zip ▼

Be sure to give your daytime phone ◀ number.

Area Code & Telephone Number ▶

CERTIFICATION*   I, the undersigned, hereby certify that I am the
Check only one ▼
☐ author
☐ other copyright claimant
☐ owner of exclusive right(s)
☐ authorized agent of_____
Name of author or other copyright claimant, or owner of exclusive right(s) ▲

8

of the work identified in this application and that the statements made
by me in this application are correct to the best of my knowledge.

Typed or printed name and date ▼ If this is a published work, this date must be the same as or later than the date of publication given in space 3.

_____ date ▶ _____

👉 **Handwritten signature (X)** ▼

MAIL CERTIFI-CATE TO

Certificate will be mailed in window envelope

Name ▼

Number/Street/Apartment Number ▼

City/State/ZIP ▼

Have you:
• Completed all necessary spaces?
• Signed your application in space 8?
• Enclosed check or money order for $10 payable to *Register of Copyrights?*
• Enclosed your deposit material with the application and fee?
MAIL TO: Register of Copyrights, Library of Congress, Washington, D.C. 20559.

9

* 17 U.S.C. § 506(e): Any person who knowingly makes a false representation of a material fact in the application for copyright registration provided for by section 409, or in any written statement filed in connection with the application, shall be fined not more than $2,500.

:·U.S. GOVERNMENT PRINTING OFFICE: 1988—202-133·80,008   August 1988—60,000

Filling Out Application Form VA

Detach and read these instructions before completing this form. Make sure all applicable spaces have been filled in before you return this form.

BASIC INFORMATION

When to Use This Form: Use Form VA for copyright registration of published or unpublished works of the visual arts. This category consists of "pictorial, graphic, or sculptural works," including two-dimensional and three-dimensional works of fine, graphic, and applied art, photographs, prints and art reproductions, maps, globes, charts, technical drawings, diagrams, and models.

What Does Copyright Protect? Copyright in a work of the visual arts protects those pictorial, graphic, or sculptural elements that, either alone or in combination, represent an "original work of authorship." The statute declares: "In no case does copyright protection for an original work of authorship extend to any idea, procedure, process, system, method of operation, concept, principle, or discovery, regardless of the form in which it is described, explained, illustrated, or embodied in such work."

Works of Artistic Craftsmanship and Designs: "Works of artistic craftsmanship" are registrable on Form VA, but the statute makes clear that protection extends to "their form" and not to "their mechanical or utilitarian aspects." The "design of a useful article" is considered copyrightable "only if, and only to the extent that, such design incorporates pictorial, graphic, or sculptural features that can be identified separately from, and are capable of existing independently of, the utilitarian aspects of the article."

Labels and Advertisements: Works prepared for use in connection with the sale or advertisement of goods and services are registrable if they contain "original work of authorship." Use Form VA if the copyrightable material in the work you are registering is mainly pictorial or graphic; use Form TX if it consists mainly of text. **NOTE:** Words and short phrases such as names, titles, and slogans cannot be protected by copyright, and the same is true of standard symbols, emblems, and other commonly used graphic designs that are in the public domain. When used commercially, material of that sort can sometimes be protected under state laws of unfair competition or under the Federal trademark laws. For information about trademark registration, write to the Commissioner of Patents and Trademarks, Washington, D.C. 20231.

Deposit to Accompany Application: An application for copyright registration must be accompanied by a deposit consisting of copies representing the entire work for which registration is to be made.

Unpublished Work: Deposit one complete copy.

Published Work: Deposit two complete copies of the best edition.

Work First Published Outside the United States: Deposit one complete copy of the first foreign edition.

Contribution to a Collective Work: Deposit one complete copy of the best edition of the collective work.

The Copyright Notice: For published works, the law provides that a copyright notice in a specified form "shall be placed on all publicly distributed copies from which the work can be visually perceived." Use of the copyright notice is the responsibility of the copyright owner and does not require advance permission from the Copyright Office. The required form of the notice for copies generally consists of three elements: (1) the symbol "©", or the word "Copyright," or the abbreviation "Copr."; (2) the year of first publication; and (3) the name of the owner of copyright. For example: "© 1981 Constance Porter." The notice is to be affixed to the copies "in such manner and location as to give reasonable notice of the claim of copyright."

For further information about copyright registration, notice, or special questions relating to copyright problems, write:

Information and Publications Section, LM-455
Copyright Office, Library of Congress, Washington, D.C. 20559

LINE-BY-LINE INSTRUCTIONS

1 SPACE 1: Title

Title of This Work: Every work submitted for copyright registration must be given a title to identify that particular work. If the copies of the work bear a title (or an identifying phrase that could serve as a title), transcribe that wording *completely* and *exactly* on the application. Indexing of the registration and future identification of the work will depend on the information you give here.

Previous or Alternative Titles: Complete this space if there are any additional titles for the work under which someone searching for the registration might be likely to look, or under which a document pertaining to the work might be recorded.

Publication as a Contribution: If the work being registered is a contribution to a periodical, serial, or collection, give the title of the contribution in the "Title of This Work" space. Then, in the line headed "Publication as a Contribution," give information about the collective work in which the contribution appeared.

Nature of This Work: Briefly describe the general nature or character of the pictorial, graphic, or sculptural work being registered for copyright. Examples: "Oil Painting"; "Charcoal Drawing"; "Etching"; "Sculpture"; "Map"; "Photograph"; "Scale Model"; "Lithographic Print"; "Jewelry Design"; "Fabric Design."

2 SPACE 2: Author(s)

General Instructions: After reading these instructions, decide who are the "authors" of this work for copyright purposes. Then, unless the work is a "collective work," give the requested information about every "author" who contributed any appreciable amount of copyrightable matter to this version of the work. If you need further space, request additional Continuation Sheets. In the case of a collective work, such as a catalog of paintings or collection of cartoons by various authors, give information about the author of the collective work as a whole.

Name of Author: The fullest form of the author's name should be given. Unless the work was "made for hire," the individual who actually created the work is its "author." In the case of a work made for hire, the statute provides that "the employer or other person for whom the work was prepared is considered the author."

What is a "Work Made for Hire"? A "work made for hire" is defined as: (1) "a work prepared by an employee within the scope of his or her employment"; or (2) "a work specially ordered or commissioned for use as a contribution to a collective work, as a part of a motion picture or other audiovisual work, as a translation, as a supplementary work, as a compilation, as an instructional text, as a test, as answer material for a test, or as an atlas, if the parties expressly agree in a written instrument signed by them that the work shall be considered a work made for hire." If you have checked "Yes" to indicate that the work was "made for hire," you must give the full legal name of the employer (or other person for whom the work was prepared). You may also include the name of the employee along with the name of the employer (for example: "Elster Publishing Co., employer for hire of John Ferguson").

"Anonymous" or "Pseudonymous" Work: An author's contribution to a work is "anonymous" if that author is not identified on the copies or phonorecords of the work. An author's contribution to a work is "pseudonymous" if that author is identified on the copies or phonorecords under a fictitious name. If the work is "anonymous" you may: (1) leave the line blank; or (2) state "anonymous" on the line; or (3) reveal the author's identity. If the work is "pseudonymous" you may: (1) leave the line blank; or (2) give the pseudonym and identify it as such (for example: "Huntley Haverstock, pseudonym"); or (3) reveal the author's name, making clear which is the real name and which is the pseudonym (for example: "Henry Leek, whose pseudonym is Priam Farrel"). However, the citizenship or domicile of the author **must** be given in all cases.

Dates of Birth and Death: If the author is dead, the statute requires that the year of death be included in the application unless the work is anonymous or pseudonymous. The author's birth date is optional, but is useful as a form of identification. Leave this space blank if the author's contribution was a "work made for hire."

Author's Nationality or Domicile: Give the country of which the author is a citizen, or the country in which the author is domiciled. Nationality or domicile **must** be given in all cases.

Nature of Authorship: Give a brief general statement of the nature of this particular author's contribution to the work. Examples: "Painting"; "Photograph"; "Silk Screen Reproduction"; "Co-author of Cartographic Material"; "Technical Drawing"; "Text and Artwork."

3 SPACE 3: Creation and Publication

General Instructions: Do not confuse "creation" with "publication." Every application for copyright registration must state "the year in which creation of the work was completed." Give the date and nation of first publication only if the work has been published.

Creation: Under the statute, a work is "created" when it is fixed in a copy or phonorecord for the first time. Where a work has been prepared over a period of time, the part of the work existing in fixed form on a particular date constitutes the created work on that date. The date you give here should be the year in which the author completed the particular version for which registration is now being sought, even if other versions exist or if further changes or additions are planned.

Publication: The statute defines "publication" as "the distribution of copies or phonorecords of a work to the public by sale or other transfer of ownership, or by rental, lease, or lending"; a work is also "published" if there has been an "offering to distribute copies or phonorecords to a group of persons for purposes of further distribution, public performance, or public display." Give the full date (month, day, year) when, and the country where, publication first occurred. If first publication took place simultaneously in the United States and other countries, it is sufficient to state "U.S.A."

4 SPACE 4: Claimant(s)

Name(s) and Address(es) of Copyright Claimant(s): Give the name(s) and address(es) of the copyright claimant(s) in this work even if the claimant is the same as the author. Copyright in a work belongs initially to the author of the work (including, in the case of a work made for hire, the employer or other person for whom the work was prepared). The copyright claimant is either the author of the work or a person or organization to whom the copyright initially belonging to the author has been transferred.

Transfer: The statute provides that, if the copyright claimant is not the author, the application for registration must contain "a brief statement of how the claimant obtained ownership of the copyright." If any copyright claimant named in space 4 is not an author named in space 2, give a brief, general statement summarizing the means by which that claimant obtained ownership of the copyright. Examples: "By written contract"; "Transfer of all rights by author"; "Assignment"; "By will." Do not attach transfer documents or other attachments or riders.

5 SPACE 5: Previous Registration

General Instructions: The questions in space 5 are intended to find out whether an earlier registration has been made for this work and, if so, whether there is any basis for a new registration. As a rule, only one basic copyright registration can be made for the same version of a particular work.

Same Version: If this version is substantially the same as the work covered by a previous registration, a second registration is not generally possible unless: (1) the work has been registered in unpublished form and a second registration is now being sought to cover this first published edition; or (2) some-

one other than the author is identified as copyright claimant in the earlier registration, and the author is now seeking registration in his or her own name. If either of these two exceptions apply, check the appropriate box and give the earlier registration number and date. Otherwise, do not submit Form VA; instead, write the Copyright Office for information about supplementary registration or recordation of transfers of copyright ownership.

Changed Version: If the work has been changed, and you are now seeking registration to cover the additions or revisions, check the last box in space 5, give the earlier registration number and date, and complete both parts of space 6 in accordance with the instructions below.

Previous Registration Number and Date: If more than one previous registration has been made for the work, give the number and date of the latest registration.

6 SPACE 6: Derivative Work or Compilation

General Instructions: Complete space 6 if this work is a "changed version," "compilation," or "derivative work," and if it incorporates one or more earlier works that have already been published or registered for copyright, or that have fallen into the public domain. A "compilation" is defined as "a work formed by the collection and assembling of preexisting materials or of data that are selected, coordinated, or arranged in such a way that the resulting work as a whole constitutes an original work of authorship." A "derivative work" is "a work based on one or more preexisting works." Examples of derivative works include reproductions of works of art, sculptures based on drawings, lithographs based on paintings, maps based on previously published sources, or "any other form in which a work may be recast, transformed, or adapted." Derivative works also include works "consisting of editorial revisions, annotations, or other modifications" if these changes, as a whole, represent an original work of authorship.

Preexisting Material (space 6a): Complete this space **and** space 6b for derivative works. In this space identify the preexisting work that has been recast, transformed, or adapted. Examples of preexisting material might be "Grunewald Altarpiece"; or "19th century quilt design." Do not complete this space for compilations.

Material Added to This Work (space 6b): Give a brief, general statement of the **additional** new material covered by the copyright claim for which registration is sought. In the case of a derivative work, identify this new material. Examples: "Adaptation of design and additional artistic work"; "Reproduction of painting by photolithography"; "Additional cartographic material"; "Compilation of photographs." If the work is a compilation, give a brief, general statement describing both the material that has been compiled **and** the compilation itself. Example: "Compilation of 19th Century Political Cartoons."

7,8,9 SPACE 7, 8, 9: Fee, Correspondence, Certification, Return Address

Deposit Account: If you maintain a Deposit Account in the Copyright Office, identify it in space 7. Otherwise leave the space blank and send the fee of $10 with your application and deposit.

Correspondence (space 7): This space should contain the name, address, area code, and telephone number of the person to be consulted if correspondence about this application becomes necessary.

Certification (space 8): The application cannot be accepted unless it bears the date and the **handwritten signature** of the author or other copyright claimant, or of the owner of exclusive right(s), or of the duly authorized agent of the author, claimant, or owner of exclusive right(s).

Address for Return of Certificate (space 9): The address box must be completed legibly since the certificate will be returned in a window envelope.

MORE INFORMATION

Form of Deposit for Works of the Visual Arts

Exceptions to General Deposit Requirements: As explained on the reverse side of this page, the statutory deposit requirements (generally one copy for unpublished works and two copies for published works) will vary for particular kinds of works of the visual arts. The copyright law authorizes the Register of Copyrights to issue regulations specifying "the administrative classes into which works are to be placed for purposes of deposit and registration, and the nature of the copies or phonorecords to be deposited in the various classes specified." For particular classes, the regulations may require or permit "the deposit of identifying material instead of copies or phonorecords," or "the deposit of only one copy or phonorecord where two would normally be required."

What Should You Deposit? The detailed requirements with respect to the kind of deposit to accompany an application on Form VA are contained in the Copyright

Office Regulations. The following does not cover all of the deposit requirements, but is intended to give you some general guidance.

For an Unpublished Work, the material deposited should represent the entire copyrightable content of the work for which registration is being sought.

For a Published Work, the material deposited should generally consist of two complete copies of the best edition. Exceptions: (1) For certain types of works, one complete copy may be deposited instead of two. These include greeting cards, postcards, stationery, labels, advertisements, scientific drawings, and globes; (2) For most three-dimensional sculptural works, and for certain two-dimensional works, the Copyright Office Regulations require deposit of identifying material (photographs or drawings in a specified form) rather than copies; and (3) Under certain circumstances, for works published in five copies or less or in limited, numbered editions, the deposit may consist of one copy or of identifying reproductions.

FORM SR

UNITED STATES COPYRIGHT OFFICE

REGISTRATION NUMBER

_____

SR   SRU

_____

EFFECTIVE DATE OF REGISTRATION

_____

Month   Day   Year

DO NOT WRITE ABOVE THIS LINE. IF YOU NEED MORE SPACE, USE A SEPARATE CONTINUATION SHEET.

1

TITLE OF THIS WORK ▼

PREVIOUS OR ALTERNATIVE TITLES ▼

NATURE OF MATERIAL RECORDED ▼ See instructions.
- ☐ Musical   ☐ Musical-Dramatic
- ☐ Dramatic   ☐ Literary
- ☐ Other _____

2

a

NAME OF AUTHOR ▼

DATES OF BIRTH AND DEATH
Year Born ▼   Year Died ▼

Was this contribution to the work a "work made for hire"?
- ☐ Yes
- ☐ No

AUTHOR'S NATIONALITY OR DOMICILE
Name of Country
OR { Citizen of ▶ _____
Domiciled in ▶ _____

WAS THIS AUTHOR'S CONTRIBUTION TO THE WORK
Anonymous?   ☐ Yes ☐ No
Pseudonymous?   ☐ Yes ☐ No

If the answer to either of these questions is "Yes," see detailed instructions.

NATURE OF AUTHORSHIP Briefly describe nature of the material created by this author in which copyright is claimed. ▼

b

NAME OF AUTHOR ▼

DATES OF BIRTH AND DEATH
Year Born ▼   Year Died ▼

Was this contribution to the work a "work made for hire"?
- ☐ Yes
- ☐ No

AUTHOR'S NATIONALITY OR DOMICILE
Name of country
OR { Citizen of ▶ _____
Domiciled in ▶ _____

WAS THIS AUTHOR'S CONTRIBUTION TO THE WORK
Anonymous?   ☐ Yes ☐ No
Pseudonymous?   ☐ Yes ☐ No

If the answer to either of these questions is "Yes," see detailed instructions.

NATURE OF AUTHORSHIP Briefly describe nature of the material created by this author in which copyright is claimed. ▼

c

NAME OF AUTHOR ▼

DATES OF BIRTH AND DEATH
Year Born ▼   Year Died ▼

Was this contribution to the work a "work made for hire"?
- ☐ Yes
- ☐ No

AUTHOR'S NATIONALITY OR DOMICILE
Name of Country
OR { Citizen of ▶ _____
Domiciled in ▶ _____

WAS THIS AUTHOR'S CONTRIBUTION TO THE WORK
Anonymous?   ☐ Yes ☐ No
Pseudonymous?   ☐ Yes ☐ No

If the answer to either of these questions is "Yes," see detailed instructions.

NATURE OF AUTHORSHIP Briefly describe nature of the material created by this author in which copyright is claimed. ▼

NOTE

Under the law, the "author" of a "work made for hire" is generally the employer, not the employee (see instructions). For any part of this work that was "made for hire" check "Yes" in the space provided, give the employer (or other person for whom the work was prepared) as "Author" of that part, and leave the space for dates of birth and death blank.

3

YEAR IN WHICH CREATION OF THIS WORK WAS COMPLETED This information must be given in all cases.
◀ Year

DATE AND NATION OF FIRST PUBLICATION OF THIS PARTICULAR WORK
Complete this information ONLY if this work has been published.
Month ▶ _____ Day ▶ _____ Year ▶ _____
◀ Nation

4

COPYRIGHT CLAIMANT(S) Name and address must be given even if the claimant is the same as the author given in space 2.▼

See instructions before completing this space.

TRANSFER If the claimant(s) named here in space 4 are different from the author(s) named in space 2, give a brief statement of how the claimant(s) obtained ownership of the copyright.▼

DO NOT WRITE HERE
OFFICE USE ONLY

APPLICATION RECEIVED

ONE DEPOSIT RECEIVED

TWO DEPOSITS RECEIVED

REMITTANCE NUMBER AND DATE

MORE ON BACK ▶
- Complete all applicable spaces (numbers 5-9) on the reverse side of this page.
- See detailed instructions.
- Sign the form at line 8.

DO NOT WRITE HERE

Page 1 of _____ pages

Form 26

DO NOT WRITE ABOVE THIS LINE. IF YOU NEED MORE SPACE, USE A SEPARATE CONTINUATION SHEET.

PREVIOUS REGISTRATION   Has registration for this work, or for an earlier version of this work, already been made in the Copyright Office?

☐ **Yes**   ☐ **No**   If your answer is "Yes," why is another registration being sought? (Check appropriate box) ▼

☐ This is the first published edition of a work previously registered in unpublished form.

☐ This is the first application submitted by this author as copyright claimant.

☐ This is a changed version of the work, as shown by space 6 on this application.

If your answer is "Yes," give: **Previous Registration Number** ▼   **Year of Registration** ▼

5

DERIVATIVE WORK OR COMPILATION   Complete both space 6a & 6b for a derivative work; complete only 6b for a compilation.

a. Preexisting Material   Identify any preexisting work or works that this work is based on or incorporates. ▼

b. Material Added to This Work   Give a brief, general statement of the material that has been added to this work and in which copyright is claimed.▼

6

See instructions
before completing
this space.

DEPOSIT ACCOUNT   If the registration fee is to be charged to a Deposit Account established in the Copyright Office, give name and number of Account.
Name ▼   **Account Number** ▼

7

CORRESPONDENCE   Give name and address to which correspondence about this application should be sent.   Name/Address/Apt/City/State/Zip ▼

Area Code & Telephone Number ▶

Be sure to
give your
daytime phone
◀ number.

CERTIFICATION*   I, the undersigned, hereby certify that I am the

Check one ▼

☐ author

☐ other copyright claimant

☐ owner of exclusive right(s)

☐ authorized agent of _____
Name of author or other copyright claimant, or owner of exclusive right(s) ▲

8

of the work identified in this application and that the statements made
by me in this application are correct to the best of my knowledge.

Typed or printed name and date ▼ If this is a published work, this date must be the same as or later than the date of publication given in space 3.

_____ date ▶ _____

✍ Handwritten signature (X) ▼

**MAIL
CERTIFI-
CATE TO**

**Certificate
will be
mailed in
window
envelope**

Name ▼

Number/Street/Apartment Number ▼

City/State/ZIP ▼

9

Have you:
• Completed all necessary
  spaces?
• Signed your application in space
  8?
• Enclosed check or money order
  for $10 payable to *Register of
  Copyrights?*
• Enclosed your deposit material
  with the application and fee?

MAIL TO: Register of Copyrights,
Library of Congress, Washington,
D.C. 20559.

FORM RE

UNITED STATES COPYRIGHT OFFICE

REGISTRATION NUMBER

EFFECTIVE DATE OF RENEWAL REGISTRATION
. . . . (Month)   (Day)   (Year)

DO NOT WRITE ABOVE THIS LINE.   FOR COPYRIGHT OFFICE USE ONLY

---

① Renewal Claimant(s)

RENEWAL CLAIMANT(S), ADDRESS(ES), AND STATEMENT OF CLAIM: (See Instructions)

1
Name . . . .
Address . . . .
Claiming as . . . .
(Use appropriate statement from instructions)

2
Name . . . .
Address . . . .
Claiming as . . . .
(Use appropriate statement from instructions)

3
Name . . . .
Address . . . .
Claiming as . . . .
(Use appropriate statement from instructions)

---

② Work Renewed

TITLE OF WORK IN WHICH RENEWAL IS CLAIMED:

RENEWABLE MATTER:

CONTRIBUTION TO PERIODICAL OR COMPOSITE WORK:

Title of periodical or composite work: . . . .

If a periodical or other serial, give: Vol. . . . . No. . . . . Issue Date . . . .

---

③ Author(s)

AUTHOR(S) OF RENEWABLE MATTER:

---

④ Facts of Original Registration

ORIGINAL REGISTRATION NUMBER:	ORIGINAL COPYRIGHT CLAIMANT:
. . . .	

ORIGINAL DATE OF COPYRIGHT:

• If the original registration for this work was made in published form, give: } OR {   • If the original registration for this work was made in unpublished form, give:

DATE OF PUBLICATION: . . . .
(Month)   (Day)   (Year)

DATE OF REGISTRATION: . . . .
(Month)   (Day)   (Year)

Form 27

	EXAMINED BY:	RENEWAL APPLICATION RECEIVED:	FOR COPYRIGHT OFFICE USE ONLY
	CHECKED BY:		
	CORRESPONDENCE ☐ Yes	REMITTANCE NUMBER AND DATE:	
	DEPOSIT ACCOUNT FUNDS USED ☐		

DO NOT WRITE ABOVE THIS LINE. FOR COPYRIGHT OFFICE USE ONLY

RENEWAL FOR GROUP OF WORKS BY SAME AUTHOR: To make a single registration for a group of works by the same individual author published as contributions to periodicals (see instructions). give full information about each contribution. If more space is needed, request continuation sheet (Form RE/CON).

⑤ **Renewal for Group of Works**

1
Title of Contribution: . . . .
Title of Periodical: . . . . Vol. . . . . . . No. . . . . Issue Date . . . .
Date of Publication: . . . . Registration Number: . . . .
(Month) (Day) (Year)

2
Title of Contribution: . . . .
Title of Periodical: . . . . Vol. . . . . . . No. . . . . Issue Date . . . .
Date of Publication: . . . . Registration Number: . . . .
(Month) (Day) (Year)

3
Title of Contribution: . . . .
Title of Periodical: . . . . Vol. . . . . . . No. . . . . Issue Date . . . .
Date of Publication: . . . . Registration Number: . . . .
(Month) (Day) (Year)

4
Title of Contribution: . . . .
Title of Periodical: . . . . Vol. . . . . . . No. . . . . Issue Date . . . .
Date of Publication: . . . . Registration Number: . . . .
(Month) (Day) (Year)

5
Title of Contribution: . . . .
Title of Periodical: . . . . Vol. . . . . . . No. . . . . Issue Date . . . .
Date of Publication: . . . . Registration Number: . . . .
(Month) (Day) (Year)

6
Title of Contribution: . . . .
Title of Periodical: . . . . Vol. . . . . . . No. . . . . Issue Date . . . .
Date of Publication: . . . . Registration Number: . . . .
(Month) (Day) (Year)

7
Title of Contribution: . . . .
Title of Periodical: . . . . Vol. . . . . . . No. . . . . Issue Date . . . .
Date of Publication: . . . . Registration Number: . . . .
(Month) (Day) (Year)

DEPOSIT ACCOUNT: (If the registration fee is to be charged to a Deposit Account established in the Copyright Office, give name and number of Account.)

Name: . . . .
Account Number: . . . .

CORRESPONDENCE: (Give name and address to which correspondence about this application should be sent.)
Name: . . . .
Address: . . . . (Apt.)
. . . .
(City) (State) (ZIP)

⑥ **Fee and Correspondence**

CERTIFICATION: I, the undersigned, hereby certify that I am the: (Check one)
☐ renewal claimant   ☐ duly authorized agent of: . . . .
(Name of renewal claimant)
of the work identified in this application, and that the statements made by me in this application are correct to the best of my knowledge.
☞ Handwritten signature: (X) . . . .
Typed or printed name: . . . .
Date: . . . .

⑦ **Certification (Application must be signed)**

MAIL CERTIFICATE TO

. . . .
(Name)
. . . .
(Number, Street and Apartment Number)
. . . .
(City) (State) (ZIP code)

(Certificate will be mailed in window envelope)

⑧ **Address for Return of Certificate**

Form 27 (cont.)

APPLICATION FOR
Renewal Registration

HOW TO REGISTER A RENEWAL CLAIM:

- **First:** Study the information on this page and make sure you know the answers to two questions:

    (1) What are the renewal time limits in your case?

    (2) Who can claim the renewal?

- **Second:** Turn this page over and read through the specific instructions for filling out Form RE. Make sure, before starting to complete the form, that the copyright is now eligible for renewal, that you are authorized to file a renewal claim, and that you have all of the information about the copyright you will need.

- **Third:** Complete all applicable spaces on Form RE, following the line-by-line instructions on the back of this page. Use typewriter, or print the information in dark ink.

- **Fourth:** Detach this sheet and send your completed Form RE to: Register of Copyrights, Library of Congress, Washington, D.C. 20559. Unless you have a Deposit Account in the Copyright Office, your application must be accompanied by a check or money order for $6, payable to: *Register of Copyrights*. Do not send copies, phonorecords, or supporting documents with your renewal application.

WHAT IS RENEWAL OF COPYRIGHT? For works originally copyrighted between January 1, 1950 and December 31, 1977, the statute now in effect provides for a first term of copyright protection lasting for 28 years, with the possibility of renewal for a second term of 47 years. If a valid renewal registration is made for a work, its total copyright term is 75 years (a first term of 28 years, plus a renewal term of 47 years). Example: For a work copyrighted in 1960, the first term will expire in 1988, but if renewed at the proper time the copyright will last through the end of 2035.

SOME BASIC POINTS ABOUT RENEWAL:

(1) There are strict time limits and deadlines for renewing a copyright.

(2) Only certain persons who fall into specific categories named in the law can claim renewal.

(3) The new copyright law does away with renewal requirements for works first copyrighted after 1977. However, copyrights that were already in their first copyright term on January 1, 1978 (that is, works originally copyrighted between January 1, 1950 and December 31, 1977) **still have to be renewed** in order to be protected for a second term.

TIME LIMITS FOR RENEWAL REGISTRATION: The new copyright statute provides that, in order to renew a copyright, the renewal application and fee must be received in the Copyright Office "within one year prior to the expiration of the copyright." It also provides that all terms of copyright will run through the end of the year in which they would otherwise expire. Since all copyright terms will expire on December 31st of their last year, all periods for renewal registration will run from December 31st of the 27th year of the copyright, and will end on December 31st of the following year.

To determine the time limits for renewal in your case:

(1) First, find out the date of original copyright for the work. (In the case of works originally registered in unpublished form, the date of copyright is the date of registration; for published works, copyright begins on the date of first publication.)

(2) Then add 28 years to the year the work was originally copyrighted.

Your answer will be the calendar year during which the copyright will be eligible for renewal, and December 31st of that year will be the renewal deadline. Example: a work originally copyrighted on April 19, 1957, will be eligible for renewal between December 31, 1984, and December 31, 1985.

WHO MAY CLAIM RENEWAL: Renewal copyright may be claimed only by those persons specified in the law. Except in the case of four specific types of works, the law gives the right to claim renewal to the individual author of the work, regardless of who owned the copyright during the original term. If the author is dead, the statute gives the right to claim renewal to certain of the author's beneficiaries (widow and children, executors, or next of kin, depending on the circumstances). The present owner (proprietor) of the copyright is entitled to claim renewal only in four specified cases, as explained in more detail on the reverse of this page.

CAUTION: Renewal registration is possible only if an acceptable application and fee are **received** in the Copyright Office during the renewal period and before the renewal deadline. If an acceptable application and fee are not received before the renewal deadline, the work falls into the public domain and the copyright cannot be renewed. The Copyright Office has no discretion to extend the renewal time limits.

INSTRUCTIONS FOR COMPLETING FORM RE

SPACE 1: RENEWAL CLAIM(S)

• **General Instructions:** In order for this application to result in a valid renewal, space 1 must identify one or more of the persons who are entitled to renew the copyright under the statute. Give the full name and address of each claimant, with a statement of the basis of each claim, using the wording given in these instructions.

• **Persons Entitled to Renew:**

A. The following persons may claim renewal in all types of works except those enumerated in Paragraph B, below:

1. The author, if living. State the claim as: *the author.*

2. The widow, widower, and/or children of the author, if the author is not living. State the claim as: *the widow (widower) of the author* . . . .
(Name of author)

and/or *the child (children) of the deceased author* . . . .
(Name of author)

3. The author's executor(s), if the author left a will and if there is no surviving widow, widower, or child. State the claim as: *the executor(s) of the author*
. . . .
(Name of author)

4. The next of kin of the author, if the author left no will and if there is no surviving widow, widower, or child. State the claim as: *the next of kin of the deceased author* . . . . *there being no will.*
(Name of author)

B. In the case of the following four types of works, the proprietor (owner of the copyright at the time of renewal registration) may claim renewal:

1. Posthumous work (a work as to which no copyright assignment or other contract for exploitation has occurred during the author's lifetime). State the claim as: *proprietor of copyright in a posthumous work.*

2. Periodical, cyclopedic, or other composite work. State the claim as: *proprietor of copyright in a composite work.*

3. "Work copyrighted by a corporate body otherwise than as assignee or licensee of the individual author." State the claim as: *proprietor of copyright in a work copyrighted by a corporate body otherwise than as assignee or licensee of the individual author.* (This type of claim is considered appropriate in relatively few cases.)

4. Work copyrighted by an employer for whom such work was made for hire. State the claim as: *proprietor of copyright in a work made for hire.*

SPACE 2: WORK RENEWED

• **General Instructions:** This space is to identify the particular work being renewed. The information given here should agree with that appearing in the certificate of original registration.

• **Title:** Give the full title of the work, together with any subtitles or descriptive wording included with the title in the original registration. In the case of a musical composition, give the specific instrumentation of the work.

• **Renewable Matter:** Copyright in a new version of a previous work (such as an arrangement, translation, dramatization, compilation, or work republished with new matter) covers only the additions, changes, or other new material appearing for the first time in that version. If this work was a new version, state in general the new matter upon which copyright was claimed.

• **Contribution to Periodical, Serial, or other Composite Work:** Separate renewal registration is possible for a work published as a contribution to a periodical, serial, or other composite work, whether the contribution was copyrighted independently or as part of the larger work in which it appeared. Each contribution published in a separate issue ordinarily requires a separate renewal registration. However, the new law provides an alternative, permitting groups of periodical contributions by the same individual author to be combined under a single renewal application and fee in certain cases.

If this renewal application covers a single contribution, give all of the requested information in space 2. If you are seeking to renew a group of contributions, include a reference such as "See space 5" in space 2 and give the requested information about all of the contributions in space 5.

SPACE 3: AUTHOR(S)

• **General Instructions:** The copyright secured in a new version of a work is independent of any copyright protection in material published earlier. The only "authors" of a new version are those who contributed copyrightable matter to it. Thus, for renewal purposes, the person who wrote the original version on which the new work is based cannot be regarded as an "author" of the new version, unless that person also contributed to the new matter.

• **Authors of Renewable Matter:** Give the full names of all authors who contributed copyrightable matter to this particular version of the work.

SPACE 4: FACTS OF ORIGINAL REGISTRATION

• **General Instructions:** Each item in space 4 should agree with the information appearing in the original registration for the work. If the work being renewed is a single contribution to a periodical or composite work that was not separately registered, give information about the particular issue in which the contribution appeared. You may leave this space blank if you are completing space 5.

• **Original Registration Number:** Give the full registration number, which is a series of numerical digits, preceded by one or more letters. The registration number appears in the upper right hand corner of the certificate of registration.

• **Original Copyright Claimant:** Give the name in which ownership of the copyright was claimed in the original registration.

• **Date of Publication or Registration:** Give only one date. If the original registration gave a publication date, it should be transcribed here; otherwise the registration was for an unpublished work, and the date of registration should be given.

SPACE 5: GROUP RENEWALS

• **General Instructions:** A single renewal registration can be made for a group of works if **all** of the following statutory conditions are met: (1) all of the works were written by the same author, who is named in space 3 and who is or was an individual (not an employer for hire); (2) all of the works were first published as contributions to periodicals (including newspapers) and were copyrighted on their first publication; (3) the renewal claimant or claimants, and the basis of claim or claims, as stated in space 1, is the same for all of the works; (4) the renewal application and fee are "received not more than 28 or less than 27 years after the 31st day of December of the calendar year in which all of the works were first published"; and (5) the renewal application identifies each work separately, including the periodical containing it and the date of first publication.

Time Limits for Group Renewals: To be renewed as a group, all of the contributions must have been first published during the same calendar year. For example, suppose six contributions by the same author were published on April 1, 1960, July 1, 1960, November 1, 1960, February 1, 1961, July 1, 1961, and March 1, 1962. The three 1960 copyrights can be combined and renewed at any time during 1988, and the two 1961 copyrights can be renewed as a group during 1989, but the 1962 copyright must be renewed by itself, in 1990.

Identification of Each Work: Give all of the requested information for each contribution. The registration number should be that for the contribution itself if it was separately registered, and the registration number for the periodical issue if it was not.

SPACES 6, 7 AND 8: FEE, MAILING INSTRUCTIONS, AND CERTIFICATION

• **Deposit Account and Mailing Instructions (Space 6):** If you maintain a Deposit Account in the Copyright Office, identify it in space 6. Otherwise, you will need to send the renewal registration fee of $6 with your form. The space headed "Correspondence" should contain the name and address of the person to be consulted if correspondence about the form becomes necessary.

• **Certification (Space 7):** The renewal application is not acceptable unless it bears the handwritten signature of the renewal claimant or the duly authorized agent of the renewal claimant.

• **Address for Return of Certificate (Space 8):** The address box must be completed legibly, since the certificate will be returned in a window envelope.

REGISTRATION NUMBER

TX	TXU	PA	PAU	VA	VAU	SR	SRU	RE

Effective Date of Supplementary Registration

. . . .
        MONTH   DAY   YEAR

DO NOT WRITE ABOVE THIS LINE. FOR COPYRIGHT OFFICE USE ONLY

(A) **Basic Instructions**

TITLE OF WORK:

REGISTRATION NUMBER OF BASIC REGISTRATION:

YEAR OF BASIC REGISTRATION:

NAME(S) OF AUTHOR(S):

NAME(S) OF COPYRIGHT CLAIMANT(S):

(B) **Correction**

LOCATION AND NATURE OF INCORRECT INFORMATION IN BASIC REGISTRATION:

Line Number . . . . Line Heading or Description . . . .

INCORRECT INFORMATION AS IT APPEARS IN BASIC REGISTRATION:

CORRECTED INFORMATION:

EXPLANATION OF CORRECTION: (Optional)

(C) **Amplification**

LOCATION AND NATURE OF INFORMATION IN BASIC REGISTRATION TO BE AMPLIFIED:

Line Number . . . . Line Heading or Description . . . .

AMPLIFIED INFORMATION:

EXPLANATION OF AMPLIFIED INFORMATION: (Optional)

DO NOT WRITE ABOVE THIS LINE. FOR COPYRIGHT OFFICE USE ONLY

CONTINUATION OF: (Check which) ☐ PART B OR ☐ PART C

(D) Continuation

DEPOSIT ACCOUNT: If the registration fee is to be charged to a Deposit Account established in the Copyright Office, give name and number of Account.

Name . . . . Account Number . . . .

CORRESPONDENCE: Give name and address to which correspondence should be sent:

Name . . . . Apt. No. . . . .

Address . . . .
(Number and Street)   (City)   (State)   (ZIP Code)

(E) Deposit Account and Mailing Instructions

CERTIFICATION ✱ I, the undersigned, hereby certify that I am the: (Check one)

☐ author ☐ other copyright claimant ☐ owner of exclusive right(s) ☐ authorized agent of: . . . .
(Name of author or other copyright claimant, or owner of exclusive right(s))

of the work identified in this application and that the statements made by me in this application are correct to the best of my knowledge.

Handwritten signature: (X) . . . .

Typed or printed name: . . . .

Date: . . . .

✱ 17 USC §50b(e) FALSE REPRESENTATION – Any person who knowingly makes a false representation of a material fact in the application for copyright registration provided for by section 409 or in any written statement filed in connection with the application, shall be fined not more than $2,500.

(F) Certification (Application must be signed)

MAIL CERTIFICATE TO

(Certificate will be mailed in window envelope)

. . . .
(Name)

. . . .
(Number, Street and Apartment Number)

. . . .
(City)   (State)   (ZIP code)

(G) Address for Return of Certificate

U S GOVERNMENT PRINTING OFFICE 1988—202-133 60.014

April 1988—15,000

Form 28 (cont.)

USE THIS FORM WHEN:

- An earlier registration has been made in the Copyright Office; and

- Some of the facts given in that registration are incorrect or incomplete; and

- You want to place the correct or complete facts on record.

FORM CA

UNITED STATES COPYRIGHT OFFICE
LIBRARY OF CONGRESS
WASHINGTON, D.C. 20559

Application for
Supplementary Copyright Registration

To Correct or Amplify Information Given in the
Copyright Office Record of an Earlier Registration

What is "Supplementary Copyright Registration"? Supplementary registration is a special type of copyright registration provided for in section 408(d) of the copyright law.

Purpose of Supplementary Registration. As a rule, only one basic copyright registration can be made for the same work. To take care of cases where information in the basic registration turns out to be incorrect or incomplete, the law provides for "the filing of an application for supplementary registration, to correct an error in a copyright registration or to amplify the information given in a registration."

Earlier Registration Necessary. Supplementary registration can be made only if a basic copyright registration for the same work has already been completed.

Who May File. Once basic registration has been made for a work, any author or other copyright claimant, or owner of any exclusive right in the work, who wishes to correct or amplify the information given in the basic registration, may submit Form CA.

Please Note:

- Do not use Form CA to correct errors in statements on the copies or phonorecords of the work in question, or to reflect changes in the content of the work. If the work has been changed substantially, you should consider making an entirely new registration for the revised version to cover the additions or revisions.

- Do not use Form CA as a substitute for renewal registration. For works originally copyrighted between January 1, 1950 and December 31, 1977, registration of a renewal claim within strict time limits is necessary to extend the first 28-year copyright term to the full term of 75 years. This cannot be done by filing Form CA.

- Do not use Form CA as a substitute for recording a transfer of copyright or other document pertaining to rights under a copyright. Recording a document under section 205 of the statute gives all persons constructive notice of the facts stated in the document and may have other important consequences in cases of infringement or conflicting transfers. Supplementary registration does not have that legal effect.

How to Apply for Supplementary Registration:

First: Study the information on this page to make sure that filing an application on Form CA is the best procedure to follow in your case.

Second: Turn this page over and read through the specific instructions for filling out Form CA. Make sure, before starting to complete the form, that you have all of the detailed information about the basic registration you will need.

Third: Complete all applicable spaces on this form, following the line-by-line instructions on the back of this page. Use typewriter, or print the information in dark ink.

Fourth: Detach this sheet and send your completed Form CA to: Register of Copyrights, Library of Congress, Washington, D.C. 20559. Unless you have a Deposit Account in the Copyright Office, your application must be accompanied by a non-refundable filing fee in the form of a check or money order for $10 payable to: *Register of Copyrights.* Do not send copies, phonorecords, or supporting documents with your application, since they cannot be made part of the record of a supplementary registration.

What Happens When a Supplementary Registration is Made? When a supplementary registration is completed, the Copyright Office will assign it a new registration number in the appropriate registration category, and issue a certificate of supplementary registration under that number. The basic registration will not be expunged or cancelled, and the two registrations will both stand in the Copyright Office records. The supplementary registration will have the effect of calling the public's attention to a possible error or omission in the basic registration, and of placing the correct facts or the additional information on official record. Moreover, if the person on whose behalf Form CA is submitted is the same as the person identified as copyright claimant in the basic registration, the Copyright Office will place a note referring to the supplementary registration in its records of the basic registration.

PLEASE READ DETAILED INSTRUCTIONS ON REVERSE

Please read the following line-by-line instructions carefully and refer to them while completing Form CA.

INSTRUCTIONS
For Completing FORM CA (Supplementary Registration)

PART A: BASIC INSTRUCTIONS

• *General Instructions:* The information in this part identifies the basic registration to be corrected or amplified. Each item must agree exactly with the information as it already appears in the basic registration (even if the purpose of filing Form CA is to change one of these items).

• *Title of Work:* Give the title as it appears in the basic registration, including previous or alternative titles if they appear.

• *Registration Number:* This is a series of numerical digits, pre-ceded by one or more letters. The registration number appears in the upper right hand corner of the certificate of registration.

• *Registration Date:* Give the year when the basic registration was completed.

• *Name(s) of Author(s) and Name(s) of Copyright Claim-ant(s):* Give all of the names as they appear in the basic registra-tion.

PART B: CORRECTION

• *General Instructions:* Complete this part **only** if information in the basic registration was incorrect at the time that basic registration was made. Leave this part blank and complete Part C, instead, if your purpose is to add, update, or clarify information rather than to rectify an actual error.

• *Location and Nature of Incorrect Information:* Give the line number and the heading or description of the space in the basic registration where the error occurs (for example: "Line number 3 . . . Citizenship of author").

• *Incorrect Information as it Appears in Basic Registration:* Transcribe the erroneous statement exactly as it appears in the basic registration.

• *Corrected Information:* Give the statement as it should have ap-peared.

• *Explanation of Correction (Optional):* If you wish, you may add an explanation of the error or its correction.

PART C: AMPLIFICATION

• *General Instructions:* Complete this part if you want to provide any of the following: (1) additional information that could have been given but was omitted at the time of basic registration; (2) changes in facts, such as changes of title or address of claimant, that have oc-curred since the basic registration; or (3) explanations clarifying infor-mation in the basic registration.

• *Location and Nature of Information to be Amplified:* Give the line number and the heading or description of the space in the basic registration where the information to be amplified appears.

• *Amplified Information:* Give a statement of the added, updated, or explanatory information as clearly and succinctly as possible.

• *Explanation of Amplification (Optional):* If you wish, you may add an explanation of the amplification.

PARTS D, E, F, G: CONTINUATION, FEE, MAILING INSTRUCTIONS AND CERTIFICATION

• *Continuation (Part D):* Use this space if you do not have enough room in Parts B or C.

• *Deposit Account and Mailing Instructions (Part E):* If you main-tain a Deposit Account in the Copyright Office, identify it in Part E. Otherwise, you will need to send the non-refundable filing fee of $10 with your form. The space headed "Correspondence" should contain the name and address of the person to be consulted if correspondence about the form becomes necessary.

• *Certification (Part F):* The application is not acceptable unless it bears the handwritten signature of the author, or other copyright clai-mant, or of the owner of exclusive right(s), or of the duly authorized agent of such author, claimant, or owner.

• *Address for Return of Certificate (Part G):* The address box must be completed legibly, since the certificate will be returned in a window envelope.

Form 28 (cont.)

Confidentiality Agreement

(This agreement may be used between the inventor and a vendor quoting on the manufacture of the invention or the elements thereof.)

It is understood and agreed that all information relating to _____ is considered proprietary to _____ and that all rights therein are reserved. However, this information is revealed to the undersigned on strictly confidential basis to permit the undersigned and the undersigned's company to evaluate it. The undersigned agrees to respect the confidential nature of the disclosure, oral and/or written, and not to disclose such information to others, and that the undersigned may discuss such information only with personnel of the undersigned's company who agree to respect the confidential nature of this agreement. The undersigned also agrees not to use, or induce or permit others to use any of the information for any other purpose whatsoever.

The above obligation shall not apply or shall cease to apply, as appropriate, to the information disclosed when it can be established by the undersigned that:

    (1) the information was in the public domain at the time of the disclosure;

    (2) such information was known to the undersigned prior to this disclosure; or

    (3) such information has become known to the undersigned from an independent source which did not receive it from _____ under an obligation of confidence.

The undersigned also warrants that he has authority to bind the undersigned's company to the terms of this confidentiality agreement.

By:_____

Title:_____

For:_____
        (Company)

Date:_____

Idea Submission Agreement I

(This agreement is representative of the type of submission paper used by manufacturers who entertain outside submissions.)

While _____(Company) wishes to take every opportunity to improve its products and add profitable ones to its line, it has found certain precautions necessary in accepting disclosures from persons not in its employ. For an idea to be considered, this form must be completed in full, signed, and returned with any disclosures of an idea or invention.

Date: _____, 19_____

To: _____

I am submitting to you, for your evaluation and permanent record, copies of certain ideas, suggestions, or other materials having to do with _____
_____
_____
_____
_____
_____

The information I am submitting to you consists of the following (please check the appropriate blanks):

       _____ (1) Description
       _____ (2) Drawing or sketches
       _____ (3) Samples
       _____ (4) Copy of a patent application(s)
       _____ (5) Other

In doing so, I agree to the conditions listed below and further agree that such conditions shall apply to any additional disclosures made incidental to the original material submitted.

(Signature) _____

(Name Printed) _____

(Address) _____

Conditions of Submission:

(1)   All submissions or disclosures of ideas are voluntary on the part of the submitter. No confidential relationship is established by submission or implied from receipt of the submitted material.

(2)   Patented ideas and ideas covered by pending applications for patent are considered only with the understanding that the submitter agrees to rely for his/her protection solely on such rights as he/she may have under the patent laws of the United States.

(3)   Ideas that have not been covered by a patent or pending application for patent are considered only with the understanding that the use to be made of such ideas and the compensation, if any, to be paid for them are matters resting solely on the discretion of the Company.

(4)   If the subject matter offered the Company is a proposed trademark, advertising slogan, or merchandising plan, susceptible to trademark or copyright protection, the Company will examine it only under the terms set forth in this Agreement. The submitter shall rely for his/her protection solely on such rights as he/she may have under the copyright and trademark laws of the United States.

(5)   The foregoing conditions may not be modified or waived.

Idea Submission Agreement II

(This agreement is representative of the type of submission paper used by manufacturers who entertain outside submissions)

Agreement made and entered into this _____ day of _____, 19 _____, by and between _____, (Discloser's full name and address) and _____, a corporation organized under the laws of the State of _____, with offices located at _____, and

Whereas, Discloser is a developer of, or has licensing rights to, concepts for _____, and

Whereas, Discloser represents that he/she has developed a certain concept, device or other proprietary subject matter more specifically described at the end of this Agreement and on the attachment hereto (hereinafter referred to as the "Item"), and

Whereas, _____ desires to evaluate the commercial utility of the item, and

Whereas, in order to make this evaluation possible, it will be necessary for Discloser to disclose confidential information concerning the Item to _____.

Now, therefore, in consideration of the mutual promises hereinafter contained, and for other good and valuable consideration, the parties agree as follows:

1) Discloser shall make full disclosure with respect to the Item to employees of_____ or one of its affiliates (collectively, _____) and shall submit to _____ all relevant data in connection therewith. The disclosure by Discloser to _____ is solely to enable_____ to evaluate the Item in order to determine its commercial utility. _____ is under no obligation to market or produce the Item, unless and until a formal written agreement is entered into, and the obligations of _____ shall be only those which are set forth in any such agreement.

2) Discloser hereby represents to _____ that the Item is his own individual creation and wholly and solely the property of Discloser and that Discloser has not assigned, sold, licensed, mortgaged, pledged, or otherwise transferred or encumbered the Item or entered into any agreement to do any of the foregoing with respect to the Item. The execution and performance of this Agreement by Discloser does not violate any contract, agreement or other restriction to which Discloser is a party or by which it is bound or any rights of any third party.

3) The disclosure of the Item and all information incidental thereto is confidential and shall be received by _____ in confidence. _____ shall not disclose such confidential information to others and shall take reasonable steps to prevent such disclosure. _____ agrees to use the same degree of care in protecting and safeguarding the confidentiality of the concepts and information disclosed hereunder as it uses for

its own information of like importance. _____shall not be liable for inadvertent disclosure or use of the Item by persons who are or have been in its employ, unless_____fails to exercise the degree of care set forth above.

4) It is understood that _____s willingness to evaluate the Item is not to be construed as an admission of the Item's novelty, priority, or originality. Discloser understands that _____ may have rights to the Item or particular elements thereof, due to prior access to information similar to the Item or elements thereof, including, by way of illustration and not limitation, prior patents, prior publication, prior submissions to _____ _____ by others, prior development by_____s personnel or representatives, prior use by_____, prior knowledge, or prior sale. Accordingly, consideration of the Item by _____shall not deprive _____ of its existing rights, if any, with respect to the Item or any element thereof.

5) Without limiting the generality of the provisions of paragraph 4 hereof, the obligations of _____ hereunder are not applicable to such information which:

    a)   prior to disclosure by Discloser, was already known to _____ as evidenced by records kept in the ordinary course of business of _____ _____ or by proof of actual use by _____.

    b)   was known to the public or generally available to the public prior to the date of disclosure.

    c)   becomes known to the public or is generally available to the public subsequent to the date of said disclosure through no act of _____ contrary to the obligations imposed by this Agreement.

    d)   is disclosed by Discloser to an unrelated third party without restriction.

    e)   is approved for public release by Discloser.

    f)   is rightfully received from a third party without similar restriction and without breach of this Agreement.

    g)   is independently developed by _____ without breach of this Agreement.

    h)   is required to be disclosed by judicial or government action.

    i)   is disclosed in judicial or governmental proceeding subject to a protective order.

_____shall be free of any obligations restricting disclosure and use of the information provided by Discloser hereunder, subject to Discloser's patent rights, if any of the provisions of a) through i) of this paragraph 5 are applicable to the information disclosed.

6) Upon submission of the Item to _____,_____shall consider the Item and as promptly as practicable advise Discloser of _____s interest or lack of interest therein, all subject to the terms, conditions and provisions of this Agreement.

7) _____ shall not be obligated to take any action with regard to the Item other than pursuant to paragraphs 3 and 6 hereof.

8) _____ will, upon request, return any letters, drawings, descriptions, specifications, or other materials submitted to it in connection with the Item.

9) The provisions of this Agreement shall apply to any additional or supplemental information pertaining to the Item provided by Discloser to_____.

10) This writing reflects the entire agreement between the parties concerning the Item, and no modification, amendment, waiver or cancellation of this Agreement or any provision hereof shall have any validity or effect whatsoever unless in writing and signed by both parties hereto. Without limiting the generality of the foregoing, no agreement relating to the purchase or use of the Item by

_____ or any of its affiliates, or relating to the terms of or consideration of such purchase or use, or relating to any compensation to, or reimbursement or any expenses of, Discloser, shall be binding upon either party hereto unless in writing and signed by both parties hereto.

11) This Agreement shall be governed by, construed and enforced in accordance with the internal laws of the State of _____, without reference to principles or conflict of laws.

12) This Agreement shall be binding upon, and inure the benefit of, the Discloser and _____ (and the affiliates of_____) and their respective heirs, executors, administrators, successors, and assigns.

In witness whereof, the parties have signed this Agreement on the respective dates hereinafter written.

The Item is generally described as follows:

SEE ATTACHED

_____
(Company)

By:_____

_____
(Discloser)

Date:_____

Agreement to Hold Secret and Confidential

(This agreement may be used between the inventor and a manufacturer to whom an invention is being submitted for possible acquisition/licensing.)

The below described invention, idea, or concept (hereinafter referred to as INVENTION) is being submitted to _____(hereinafter referred to as COMPANY) by_____of _____on _____, 19___ (hereinafter referred to as INVENTOR) who is the inventor of record. The undersigned, in consideration of examining said INVENTION, with a purpose to opening negotiations to obtain a license to manufacture and sell said INVENTION, hereby agrees on behalf of himself/herself and said COMPANY that he/she represents, that:

(1) He/she (during or after the termination of employment with said COMPANY, will keep said INVENTION and any information pertaining to it, in confidence.

(2) He/she will not disclose said INVENTION or data related thereto to anyone save for employees of said COMPANY, sufficient information about said INVENTION to enable said COMPANY to continue with negotiations for said license, and that anyone in said COMPANY to whom said INVENTION is revealed, shall be informed of the confidential nature of the disclosure and shall agree to hold secret and confidential the information, and be bound by the terms thereof, to the same extent as if they had signed this Agreement.

(3) Neither he/she nor said COMPANY shall use any of the information provide to produce said INVENTION until agreement is reached with INVENTOR.

(4) He/she has the authority to make this Agreement on behalf of said COMPANY.

It is understood, nevertheless, that the undersigned and said COMPANY shall not be prevented by the Agreement from selling any product heretofore sold by said COMPANY, or any product in the development or panning stage, as of the date first above written, or any product disclosed in any heretofore issued U.S. Letters Patent or otherwise known to the general public.

The terms of the preceding section releasing, under certain conditions, the obligation to hold the disclosure in confidence does not, however, constitute a waiver of any patent, copyright, or other rights which said INVENTOR or any licensee thereof may have against the undersigned or said COMPANY.

INVENTOR: _____

COMPANY: _____

DATE: _____

OMB NUMBER: 0693-0009
APPROVAL EXPIRES: 8-31-1991

NIST-1262
(11-90)
DAO 203-26

U.S. DEPARTMENT OF COMMERCE
NATIONAL INSTITUTE OF STANDARDS AND TECHNOLOGY

SINGLE BUSINESS APPLICATION
ADVANCED TECHNOLOGY PROGRAM PROPOSAL COVER SHEET

Public reporting burden for this collection of information is estimated to average two (2) hours per response, including the time for reviewing instructions, searching existing data sources, gathering and maintaining the data needed, and completing and reviewing the collection of information. Send comments regarding this burden estimate or any other aspect of this collection of information, including suggestions for reducing this burden, to George Uriano, Director, Advanced Technology Program, Technology Building, Room B110, Gaithersburg, Maryland 20899; and to the Office of Management and Budget, Paperwork Reduction Project 0693-0009, Washington, D.C. 20503.

1. ANNOUNCEMENT NUMBER	2. CATALOG NUMBER **11.612**	3. PROPOSAL NUMBER	4. DATE RECEIVED
5. DATE SUBMITTED	6. TECHNOLOGY AREA		

7. NAME AND ADDRESS OF SUBMITTING ORGANIZATION

8. TYPE OF APPLICATION

☐ NEW ☐ REVISED

DATE PREVIOUSLY SUBMITTED _____

9. SUBMITTING ORGANIZATION

☐ PROFIT ☐ NONPROFIT

☐ SMALL BUSINESS ☐ INDEPENDENT RESEARCH INSTITUTION

10. NAME OF TECHNICAL CONTACT
(Address required, if different than submitting organization)

11. NAME OF BUSINESS CONTACT
(Address required, if different than submitting organization)

TELEPHONE NUMBER:

FAX NUMBER:

E-MAIL ADDRESS:

TELEPHONE NUMBER:

FAX NUMBER:

E-MAIL ADDRESS:

12. PROPOSAL TITLE

13. PROPOSAL ABSTRACT

14. PROPOSAL KEYWORDS

15. PROPOSED START DATE	16. DURATION YEARS/MONTHS	17. SUBMITTING ORGANIZATION'S CONGRESSIONAL DISTRICT(S)

ELECTRONIC FORM

Form 33

FIRST YEAR SPENDING PLAN

18. PERSONNEL DIRECT COSTS	19. WORK YEARS	20. AMOUNT
A. TECHNICAL PERSONNEL SALARIES/WAGES		
B. TECHNICAL PERSONNEL FRINGE BENEFITS		
C. ADMINISTRATIVE SUPPORT SALARIES/WAGES		
D. ADMINISTRATIVE SUPPORT FRINGE BENEFITS		
E. SUBTOTAL PERSONNEL COSTS (LINES 18A THRU 18D)		

21. OTHER DIRECT COSTS	
A. EQUIPMENT	
B. MATERIALS/SUPPLIES	
C. SUBCONTRACTS	
D. TRAVEL	
E. OTHER	
F. SUBTOTAL OTHER DIRECT COSTS (LINES 21A THRU 21E)	
22. TOTAL DIRECT COSTS (LINES 18E AND 21F)	
23. TOTAL DIRECT COSTS REQUESTED FROM ATP	
24. TOTAL INDIRECT COSTS (TO BE ABSORBED BY APPLICANT)	
25. TOTAL FIRST YEAR COSTS (LINES 22 AND 24)	

PROPOSED BUDGET SUMMARY FOR PROJECT DURATION

26. SOURCE OF SUPPORT	27. YEAR ONE	28. YEAR TWO	29. YEAR THREE	30. TOTAL
A. ATP (DIRECT COSTS ONLY)				
B. APPLICANT (INCLUDE INDIRECT COSTS)				
C. OTHER FEDERAL				
D. STATE/LOCAL				
E. OTHER PRIVATE				
F. TOTAL (LINES 26A THRU 26E)				

31. REMARKS

32. CERTIFICATION: BY SIGNING AND SUBMITTING THIS PROPOSAL, I CERTIFY THAT NO INDIRECT COSTS WILL BE CHARGED TO THE ADVANCED TECHNOLOGY PROGRAM, SHOULD THE AWARD BE RECEIVED. TO THE BEST OF MY KNOWLEDGE AND BELIEF, ALL DATA IN THIS APPLICATION ARE TRUE AND CORRECT. AND THE FOLLOWING QUESTIONS HAVE BEEN TRUTHFULLY ANSWERED.

	YES	NO
A. IS THE ORGANIZATION DELINQUENT ON ANY FEDERAL DEBT? (If yes, provide explanation)		
B. DOES PROPOSAL CONTAIN PROPRIETARY INFORMATION?		
C. IS CERTIFICATE CONCERNING NON-USE OF FEDERAL FUNDS FOR LOBBYING ACTIVITIES ATTACHED?		
D. IS DRUG-FREE WORK-PLACE CERTIFICATE ATTACHED?		
E. WAS PROPOSAL OR A VERY SIMILAR PROPOSAL SUBMITTED TO ANOTHER FEDERAL AGENCY? (If yes, identify agency)		

33. AUTHORIZED ORGANIZATIONAL REPRESENTATIVE (NAME AND TITLE TYPED)	34. DATE
35. SIGNATURE	36. TELEPHONE NUMBER

NIST-1262 (11-90) (BACK)
ELECTRONIC FORM

Form 33 (cont.)

1.   Enter the ATP Solicitation number (see cover of application).

2.   For ATP use (Catalog Of Federal Assistance Number).

3.   For ATP use.

4.   For ATP use.

5.   Enter the date that proposal was submitted to the ATP.

6.   For ATP use.

7.   Enter the submitting organization's name, street address, city, state's two-letter abbreviation, and zip code.

8.   Check "new" if this is a first time submission. Check "revised" if this proposal has been previously submitted to the ATP and enter the date of the previous submission.

9.   Check all blocks that apply to submitting organization.

10.   Enter the name, telephone number, Fax number, and Electronic Mail address of the individual to be contacted about the technical aspects of the proposal. Include mailing address if different from submitting organization.

11.   Enter the name of and data for the individual to be contacted about the business aspects of the proposal (enter "same" if the individual is the same as entry 10).

12.   Enter the title of the proposal (90 character limit).

13.   Enter a "non-proprietary" abstract of the proposed work (maximum ten (10) lines).

14.   Enter technical keywords (90 character limit).

15.   Enter proposed starting date.

16.   Enter total duration of the proposed work in years/months. Note: Maximum duration allowed is three (3) years.

17.   List the congressional district(s) where the proposed work is to be completed. This information is available at the nearest public library or may be obtained by contacting the district offices of the members of the U. S. House of Representatives.

Single Business First Year Spending Plan
and
Proposed Budget Summary For Project Duration

18.   List in this section the direct personnel costs for the first year.

18 A.   Enter the scientific, engineering, other professional, and technician work years and salaries.

18 B.   Enter the costs of fringe benefits, i.e., health insurance, retirement, leave/vacation, life insurance, social security, etc. associated with the technical staff.

18 C.   Enter the direct managerial and administrative support work years and salaries.

18 D.   Enter the costs of fringe benefits (see 18 B. above) associated with the managerial and administrative support staff.

18 E.   Subtotal all personnel work years and costs for the first year of the proposed program (lines 18A thru 18D).

19.   Enter in this column the first year's staff time in terms of work years. Example one individual working full time all year is represented by 1.0 work year. An individual who works full time for six months is represented as 0.5 work year.

20.   Enter in this column the direct costs (salaries and related costs) associated with the work years.

21.   List in this section the direct costs of other objects (non personnel costs) associated with the first year of the proposed program.

21 A.   Enter the estimated amount of funds to be spent on equipment. In block 31 "Remarks" list those items that cost $10,000 or more and a justification of need (continue on separate sheet if necessary).

21 B.   Enter the amount needed for expendable materials and supplies.

21 C.   Enter the amount set aside for subcontracts. List the proposed subcontractor(s), the amount, and specific scope of work in Section 5.f.

21 D.   Enter estimated amount for travel associated with this program.

21 E.   Enter any other items that do not fit into the categories above. Please provide detailed description in block 31 "Remarks".

21 F.   Enter subtotal of Other Direct Costs (lines 21A thru 21E).

22.   Enter Total direct Costs (lines 18E and 21F).

23.   Enter the amount of direct costs to be funded by the ATP. In most cases this amount will equal the amount shown on line 22. however, some single businesses may have other sources of support or may wish to go beyond the required minimum (absorbing the indirect costs) and absorb part of the direct costs.

24.   Enter the amount of Indirect Costs to be absorbed by the applicant (see ATP Rule Sec. 295.2(c) for definition of indirect costs).

25. Enter the Total First Year Costs (lines 22 and 24) for the proposed program.

26. Enter the sources and amount of support for each year of the proposed program (columns 27 thru 29 and column 30 Total all years).

26 A. Enter amount requested from the ATP. Note the amount shown in column 27 (first year) should equal the amount shown on line 23 above.

26 B. Enter the amount to be absorbed by the applicant. Note for column 27 the amount should **at least** be equal to the amount shown on line 24 above.

26 C. Enter any funds that the applicant is receiving or expecting to receive from other Federal agencies for this program.

26 D. Enter any funding received from State or Local government agencies in support of this proposed program.

26 E. Enter any other private sources of funding for this proposed program. List the funding organization(s) and amount(s) in block 31. "Remarks".

26 F. Enter the total for each year of the proposed program. Note that the amount shown in column 27 should equal the amount shown on line 25 above.

32. Answer the questions concerning Federal debt, proprietary information, lobbying, drug-free workplace, and proposal being submitted to other Federal agencies, attach additional documentation as needed. This form must be signed by an official of the submitting organization who has delegated contractual and fiduciary authority.

OMB NUMBER: 0693-0009
APPROVAL EXPIRES: 8-31-1991

NIST-1263
(11-90)
DAO 203-26

U.S. DEPARTMENT OF COMMERCE
NATIONAL INSTITUTE OF STANDARDS AND TECHNOLOGY

JOINT VENTURE BUSINESS APPLICATION
ADVANCED TECHNOLOGY PROGRAM PROPOSAL COVER SHEET

Public reporting burden for this collection of information is estimated to average two (2) hours per response, including the time for reviewing instructions, searching existing data sources, gathering and maintaining the data needed, and completing and reviewing the collection of information. Send comments regarding this burden estimate or any other aspect of this collection of information, including suggestions for reducing this burden, to George Uriano, Director, Advanced Technology Program, Technology Building, Room B110, Gaithersburg, Maryland 20899; and to the Office of Management and Budget, Paperwork Reduction Project 0693-0009, Washington, D.C. 20503.

1. ANNOUNCEMENT NUMBER	2. CATALOG NUMBER **11.612**	3. PROPOSAL NUMBER	4. DATE RECEIVED
5. DATE SUBMITTED	6. TECHNOLOGY AREA		

7. NAME AND ADDRESS OF SUBMITTING ORGANIZATION

8. TYPE OF APPLICATION

☐ NEW   ☐ REVISED

DATE PREVIOUSLY SUBMITTED _____

9. SUBMITTING ORGANIZATION

☐ PROFIT   ☐ NONPROFIT

☐ SMALL BUSINESS ☐ INDEPENDENT RESEARCH INSTITUTION

10. NAME OF TECHNICAL CONTACT
(Address required, if different than submitting organization)

TELEPHONE NUMBER:
FAX NUMBER:
E-MAIL ADDRESS:

11. NAME OF BUSINESS CONTACT
(Address required, if different than submitting organization)

TELEPHONE NUMBER:
FAX NUMBER:
E-MAIL ADDRESS:

12. PROPOSAL TITLE

13. PROPOSAL ABSTRACT

14. PROPOSAL KEYWORDS

15. PROPOSED START DATE	16. DURATION YEARS/MONTHS	17. SUBMITTING ORGANIZATION'S CONGRESSIONAL DISTRICT(S)

ELECTRONIC FORM

Form 34

FIRST YEAR SPENDING PLAN FOR JOINT VENTURE

18. PERSONNEL DIRECT COSTS	19. WORK YEARS	20. AMOUNT
A. TECHNICAL PERSONNEL SALARIES/WAGES		
B. TECHNICAL PERSONNEL FRINGE BENEFITS		
C. ADMINISTRATIVE SUPPORT SALARIES/WAGES		
D. ADMINISTRATIVE SUPPORT FRINGE BENEFITS		
E. SUBTOTAL PERSONNEL COSTS (LINES 18A THRU 18D)		

21. OTHER DIRECT COSTS	
A. EQUIPMENT	
B. MATERIALS/SUPPLIES	
C. SUBCONTRACTS	
D. TRAVEL	
E. OTHER	
F. SUBTOTAL OTHER DIRECT COSTS (LINES 21A THRU 21E)	
22. TOTAL DIRECT COSTS (LINES 18E AND 21F)	
23. TOTAL INDIRECT COSTS	
24. TOTAL FIRST YEAR COSTS (LINES 22 AND 23)	
25. TOTAL FIRST YEAR MATCHING FUNDS	
26. TOTAL FIRST YEAR ATP REQUEST	

PROPOSED BUDGET SUMMARY FOR PROJECT DURATION

27. SOURCE OF SUPPORT	28. YEAR ONE	29. YEAR TWO	30. YEAR THREE	31. YEAR FOUR	32. YEAR FIVE	33. TOTAL
A. ATP						
B. R&D PARTICIPANTS						
C. STATE/LOCAL						
D. OTHER FEDERAL						
E. OTHER						
F. TOTALS (LINES 27A THRU 27E)						

34. REMARKS

35. CERTIFICATION: BY SIGNING AND SUBMITTING THIS PROPOSAL, I CERTIFY THAT THE VENTURE PARTICIPANTS SHALL CONTRIBUTE MORE THAN 50 PERCENT OF EACH YEAR'S COSTS IN THE FORM OF MATCHING FUNDS. TO THE BEST OF MY KNOWLEDGE AND BELIEF, ALL DATA IN THIS APPLICATION ARE TRUE AND CORRECT. AND THE FOLLOWING QUESTIONS HAVE BEEN TRUTHFULLY ANSWERED.

	YES	NO
A. IS THE ORGANIZATION DELINQUENT ON ANY FEDERAL DEBT? (If yes, provide explanation)		
B. DOES PROPOSAL CONTAIN PROPRIETARY INFORMATION?		
C. IS CERTIFICATE CONCERNING NON-USE OF FEDERAL FUNDS FOR LOBBYING ACTIVITIES ATTACHED?		
D. IS DRUG-FREE WORK-PLACE CERTIFICATE ATTACHED?		
E. WAS PROPOSAL OR A VERY SIMILAR PROPOSAL SUBMITTED TO ANOTHER FEDERAL AGENCY? (If yes, identify agency)		

36. AUTHORIZED ORGANIZATIONAL REPRESENTATIVE (NAME AND TITLE TYPED)	37. DATE
38. SIGNATURE	39. TELEPHONE NUMBER

NIST-1263 (11-90) (PAGE 2)
ELECTRONIC FORM

Form 34 (cont.)

JOINT VENTURE RESEARCH AND DEVELOPMENT PARTICIPANTS SUPPLEMENTAL SHEET

1. NAME OF ORGANIZATION

2. CONTACT

3. ADDRESS (INCLUDE STREET, CITY, STATE, ZIP CODE)

4. TELEPHONE NUMBER

5. FAX NUMBER

6. E-MAIL ADDRESS

7. TYPE OF ORGANIZATION

☐ PROFIT ☐ NON-PROFIT ☐ SMALL BUSINESS ☐ INDEPENDENT RESEARCH INSTITUTE ☐ UNIVERSITY ☐ GOVERNMENT

8. CONGRESSIONAL DISTRICT

---

1. NAME OF ORGANIZATION

2. CONTACT

3. ADDRESS (INCLUDE STREET, CITY, STATE, ZIP CODE)

4. TELEPHONE NUMBER

5. FAX NUMBER

6. E-MAIL ADDRESS

7. TYPE OF ORGANIZATION

☐ PROFIT ☐ NON-PROFIT ☐ SMALL BUSINESS ☐ INDEPENDENT RESEARCH INSTITUTE ☐ UNIVERSITY ☐ GOVERNMENT

8. CONGRESSIONAL DISTRICT

---

1. NAME OF ORGANIZATION

2. CONTACT

3. ADDRESS (INCLUDE STREET, CITY, STATE, ZIP CODE)

4. TELEPHONE NUMBER

5. FAX NUMBER

6. E-MAIL ADDRESS

7. TYPE OF ORGANIZATION

☐ PROFIT ☐ NON-PROFIT ☐ SMALL BUSINESS ☐ INDEPENDENT RESEARCH INSTITUTE ☐ UNIVERSITY ☐ GOVERNMENT

8. CONGRESSIONAL DISTRICT

Joint Venture Business Application
Advanced Technology Program Proposal Cover Sheet

1. Enter the ATP Solicitation number (see cover of application).

2. For ATP use (Catalog Of Federal Assistance Number).

3. For ATP use.

4. For ATP use.

5. Enter the date that proposal was submitted to the ATP.

6. For ATP use.

7. Enter the submitting organization's name, street address, city, state's two-letter abbreviation, and zip code. Use the attached Joint Venture Research and Development Participants Supplemental Sheet to provide information related to the participating organizations of the Joint Venture. This supplemental sheet may be copied as needed.

8. Check "new" if this is a first time submission. Check "revised" if this proposal has been previously submitted to the ATP and enter the date of the previous submission.

9. Check all blocks that apply to submitting organization.

10. Enter the name, telephone number, fax number, and electronic mail address of the individual to be contacted about the technical aspects of the proposal. Include mailing address if different from submitting organization.

11. Enter the name of and data for the individual to be contacted about the business aspects of the proposal (enter "same" if the individual is the same as entry 10).

12. Enter the title of the proposal (90 character limit).

13. Enter a "non-proprietary" abstract of the proposed work (maximum ten (10) lines).

14. Enter technical keywords (90 character limit).

15. Enter proposed starting date.

16. Enter duration of the proposed work in years/months.
    Note: Maximum duration allowed is five (5) years.

17. List the congressional district(s) where the proposed work is to be completed. This information is available at the nearest public library or may be obtained by contacting the district offices of the members of the U. S. House of Representatives.

Joint Venture Business First Year Spending Plan
and
Proposed Budget Summary For Project Duration

18.   List in this section the direct personnel costs for the first year.

18 A.   Enter the scientific, engineering, other professional, and technician work years and salaries.

18 B.   Enter the costs of fringe benefits, i.e., health insurance, retirement, leave/vacation, life insurance, social security, etc. associated with the technical staff.

18 C.   Enter the direct managerial and administrative support work years and salaries.

18 D.   Enter the costs of fringe benefits (see 18B above) associated with the managerial and administrative support staff.

18 E.   Subtotal all personnel work years and costs for the first year of the proposed program (lines 18A thru 18D).

19.   Enter in this column the first year's staff time in terms of work years. Example one individual working full time all year is represented by 1.0 work year. An individual who works full time for six months is represented as 0.5 work year.

20.   Enter in this column the direct costs (salaries and related costs) associated with the work years.

21.   List in this section the direct costs of other objects (non personnel costs) associated with the first year of the proposed program.

21 A.   Enter the estimated amount of funds to be spent on equipment. In block 34 "Remarks" list those items that cost $10,000 or more and a justification of need (continue on separate sheet if necessary).

21 B.   Enter the amount needed for expendable materials and supplies.

21 C.   Enter the amount set aside for subcontracts. List the proposed subcontractor(s), the amount, and specific scope of work (Section 5.f).

21 D.   Enter estimate amount for travel associated with this program.

21 E.   Enter any other items that do not fit into the categories above. Please provide detailed description in block 34 "Remarks".

21 F.   Enter subtotal of Other Direct Costs (lines 21A thru 21E).

22.   Enter Total Direct Costs (lines 18E and 21F).

23.   Enter Total Indirect Costs to be charged to the proposed program.

24.   Enter the Total First Year Costs (lines 22 and 23).

25.   Enter the amount of the first year's matching funds. Note this amount must be greater than 50% of the amount shown on line 24.

26.   Enter the amount request from the ATP. Note this amount must be less than 50% of the amount shown on line 24.

27.   Enter the sources and amount of support for each year of the proposed program (columns 28 thru 32 and column 33 total for all years).

27 A.   Enter amount request from the ATP for each year. Note that the amount shown on line 28 (first year) should equal the amount shown on line 26 above. Note that the amounts shown in columns 29 thru 32 must be less than 50% of the amounts shown on line 27F.

27 B.   Enter the amount to be provided by the R&D participants for each year.

27 C.   Enter any fund contributed by State or local government agencies.

27 D.   Enter any funds received from other Federal agencies supporting the proposed work. List the Federal agency and contact person in block 34. "Remarks".

27 E.   Enter amounts contributed by any other private source(s) for this proposed work list each funding organization and the amount contributed in Block 34. "Remarks".

27 F.   Enter the total amount for each year of the proposed program. Note that the amount shown in column 28 should be equal to or greater than the amount shown on line 25.

35.   Answer the questions concerning Federal debt, proprietary information, lobbying, drug-free workplace, and proposal being submitted to other Federal agencies, attach additional documentation as needed. This form must be signed by an official of the submitting organization who has delegated contractual and fiduciary authority.

JOINT VENTURE MATCHING FUNDS ANALYSIS FOR FIRST YEAR

1. SOURCE OF SUPPORT	2. QUALIFYING MATCHING FUNDS CATEGORIES						3. NON-MATCHING OTHER FUNDS	4. TOTAL FIRST YEAR FUNDS/SOURCE
	A. CASH	B. LABOR		C. NEW EQUIPMENT	D. OWNED EQUIPMENT	E. TOTAL QUALIFYING MATCHING FUNDS		
		(1) FULL-TIME	(2) PART-TIME					
5. RESEARCH AND DEVELOPMENT PARTICIPANTS								
A.								
B.								
C.								
D.								
E.								
F.								
G.								
H.								
I.								
J.								
K.								
L.								
M. SUBTOTAL R&D PARTICIPANTS MATCHING FUNDS (LINES 5A THRU 5L)								
6. OTHER SOURCES OF SUPPORT								
A. STATE/LOCAL								
B. OTHER FEDERAL								
C. OTHER								
D.								
E.								
F.								
G. SUBTOTAL OTHER SOURCES OF SUPPORT (LINES 5A THRU 6F)								
7. TOTAL NON-ADVANCED TECHNOLOGY PROGRAM SUPPORT (LINES 5M AND 6G)								
8. ADVANCED TECHNOLOGY PROGRAM AMOUNT REQUESTED								
9. TOTAL SUPPORT ALL SOURCES FOR FIRST YEAR (LINES 7 AND 8)								
10. FINAL TOTAL (PAGES _____ THRU _____)								

NIST-1263 (11-90) (PAGE 4)
ELECTRONIC FORM

Form 34 (cont.)

Joint Venture Matching Funds Analysis For First Year

1.   List in this section the organizations that are contributing to the proposed work.

2.   Listed in section are the various matching funds categories. For the definitions of these matching funds categories see the ATP Rule section 295.2(e).

2 A.   Enter the amount of cash contributed as part of the total Joint Venture fund.

2 B(1). Enter the cash value of the in-kind "full time" staff.

2 B(2). Enter the cash value of the in-kind "part time" staff. Note: The ATP Rule requires that the individuals must be contributing at least 50% of their time and that the total value of the in-kind part time contribution may not exceed 20% of the total annual matching funds amount.

2 C.   Enter the purchase price of in-kind new equipment deemed necessary to carry out the proposed work.

2 D.   Enter the current depreciated value of owned equipment deemed necessary to carry out the proposed work using the participant's preestablished depreciation accounting methods.

2 E.   Enter the total amount of qualifying matching funds (columns 2A thru 2D).

3.   Enter any additional non-qualifying funds that are being contributed to the proposed work.

4.   Enter the Total First Year Funds (columns 2E and 3).

5.   List in this section the individual R&D participants, those organizations which are performing the research described in the Research and Development Plan. Use additional copies of this form if needed.

5 M.   Enter subtotal R&D participants matching funds (lines 5A thru 5L or additional entries as required).

6.   List in this section other individual organizations contributing to the proposed work.

6 A.   Enter the amount of state/local funds supporting the proposed work in columns 2A, 2E, and 4.

6 B.   Enter the amount of Federal funds (other than ATP) supporting the proposed work in columns 3 an 4. Note: Federal funds **can not** count toward the matching funds.

6 C. thru 6 F. Enter other private organizations' contribution to the proposed work in columns 2A, 2E, and 4.

6 G.   Enter the subtotals for all other support of the proposed work.

7.   Enter the total non-ATP support for the proposed work (lines 5M and 6G).

8.   Enter the amount of ATP funds requested for the first year in column 4. Note: this amount should be the same as the amount shown in line 26 and line 27A/column 28 of the First

2

Year spending Plan and the Proposed Budget Summary For Project Duration; and must be less than 50% of the amount shown in line 7/column 4.

9. Enter the total support from all sources for the first year of the proposed work.

10. Enter Grand Totals from additional pages if necessary.

INFORMATION REQUEST FORM

CLEARINGHOUSE FOR STATE AND LOCAL INITIATIVES ON PRODUCTIVITY, TECHNOLOGY, AND INNOVATION

To get the names and phone numbers of people in your state's assistance programs who will help you move your ideas into the marketplace, you may mail or fax this information request form to:

The Clearinghouse for State and Local Initiatives on Productivity, Technology, and Innovation
8001 Forbes Place, Room 304
Springfield, VA 22151

Fax: (703) 321-8199

Name: _____

Company: _____

Street or P.O. Box: _____

City: _____ State: _____ Zip Code: _____

Telephone: (___) _____.   Date: _____

Fax Number: (___) _____   Profession: _____

   I need to know _____
_____
_____
_____
_____
_____
_____
_____
_____
_____

FORM A 11/91

NIST-1019
(REV. 4-90)

OMB APPROVAL NUMBER 0693-0002

APPROVAL EXPIRES MARCH 31, 1993

U.S. DEPARTMENT OF COMMERCE
NATIONAL INSTITUTE OF STANDARDS AND TECHNOLOGY

ENERGY-RELATED INVENTION EVALUATION REQUEST

PUBLIC REPORTING BURDEN FOR THIS COLLECTION OF INFORMATION IS ESTIMATED TO AVERAGE 0.1 HOURS PER RESPONSE INCLUDING THE TIME TO READ THE INSTRUCTIONS. SEND COMMENTS REGARDING THIS BURDEN ESTIMATE OR ANY OTHER ASPECT OF THIS COLLECTION OF INFORMATION, INCLUDING SUGGESTIONS FOR REDUCING THIS BURDEN TO GEORGE P. LEWETT, CHIEF, OFFICE OF ENERGY RELATED INVENTIONS, BUILDING TRF411, ROOM A115, GAITHERSBURG, MARYLAND 20899; AND TO THE OFFICE OF INFORMATION AND REGULATORY AFFAIRS, OFFICE OF MANAGEMENT AND BUDGET, WASHINGTON, D.C. 20503.

NAME AND ADDRESS OF INVENTOR

TELEPHONE NUMBER

NAME AND ADDRESS OF OWNER, IF DIFFERENT FROM ABOVE

TELEPHONE NUMBER

REQUEST IS BEING SUBMITTED BY (check which)

☐ INVENTOR   ☐ OWNER   ☐ OTHER

NAME AND ADDRESS OF SUBMITTER, IF NOT INVENTOR OR OWNER

SIZE OF COMPANY INVOLVED
(Write number of employees; $ gross last year; N/N if none)
NUMBER OF EMPLOYEES | $ GROSS LAST YEAR

THIS BOX IS FOR OFFICE USE ONLY	
DATE	ER NUMBER
CLASSIFICATION	
TECHNICAL CATEGORY	
ANALYST	DATE
HOW DID YOU LEARN OF THIS PROGRAM?	
OTHER (identify)	

NAME OR TITLE OF THIS INVENTION

STATUS OF INVENTION DEVELOPMENT (check to indicate both the steps completed and the current status; highest number checked will indicate current status)

0 ☐ Concept Definition   1 ☐ Concept Development   2 ☐ Laboratory Test   3 ☐ Engineering Design   4 ☐ Working Model

5 ☐ Prototype Development   6 ☐ Prototype Test   7 ☐ Production Engineering   8 ☐ Limited Production/ Marketing   9 ☐ Production and Marketing

PATENT STATUS (check one)

0 ☐ Not patentable   1 ☐ Not applied for   2 ☐ Disclosure Document Program   3 ☐ Patent applied for   4 ☐ Patent granted

Patent Numbers

CHECK THE ITEM BELOW THAT MOST NEARLY DESCRIBES WHY YOU ARE REQUESTING EVALUATION

☐ 1. I wish the U.S. Government to provide funds to support development of the invention or new concept. Support is first needed for (write in):

_____

☐ 2. Development is complete. I need assistance to bring my invention or product into full utilization. Assistance is needed in (check whichever applies):

☐ General Marketing   ☐ Selling to the Government   ☐ Business Management   ☐ Other _____

☐ 3. I only desire an opinion that the disclosure describes a technically valid invention. This information is for:

☐ Use in obtaining private development support   ☐ Other (specify in disclosure)

☐ 4. The Small Business Administration suggested I request evaluation from NIST in connection with a loan application.

☐ 5. Other (specify) _____

☐ YES   ☐ NO   Has the invention been described to other agencies of the Government? (If yes, discuss in disclosure.)

☐ YES   ☐ NO   Has the invention been disclosed to any private companies, patent attorneys, etc.? (If yes, discuss in disclosure.)

SUBMITTER'S COPY - KEEP THIS COPY AND RETURN OERI COPY WITH DISCLOSURE.

ELECTRONIC FORM

Form 36

TEAR OFF ALONG PERFORATION AND RETURN ONLY ORIGINAL OERI COPY (PAGES 3 AND 4) WITH THE DISCLOSURE.

MEMORANDUM OF UNDERSTANDING

I have read the Program Description and Statement of Policy on pages 1 and 2 of this form. As the owner, or with the authority from the owner who is listed on page 3, I have attached (or previously submitted) a disclosure of the identified invention for the purpose of evaluation by the National Institute of Standards and Technology pursuant to Section 14 of Public Law 93-577.

I understand that to help protect property rights in an unpatented invention, an appropriate statement or notation should be applied to the title page or first page of the invention description, and that if the description is so marked, the Government will consider all information that is in fact (a) trade secret or (b) commercial or financial information that is privileged or confidential, as coming within the exemption set out in 5 U.S.C. 552(b) (4). Accordingly, I have checked directly below, the box which is applicable to this invention.

	YES	NO
An appropriate statement has been applied to the information I have submitted.	☐	☐
Please apply an appropriate statement to all material I have submitted describing the invention to which this request pertains. (Example: This material contains commercial or financial information which is confidential.)	☐	☐
No statement is required because the information submitted is not confidential.	☐	

I also understand that NIST will evaluate the invention described in the invention disclosure on the following conditions:

1. The Government will, in the evaluation process, restrict access to the description to those persons, within or without the Government, who have a need for purposes of administration or evaluation and will restrict their use of this information to such purposes.

2. The information submitted will not be returned and may be retained as a Government record.

3. The Government may make additional copies of the material submitted if required to facilitate the review process.

4. The acceptance of the information for evaluation does not, in itself, imply a promise to pay, a recognition of novelty or originality, or a contractual relationship such as would render the Government liable to pay for use of the information submitted.

5. The provisions of this Memorandum of Understanding shall also apply to additions to the disclosure made by me incidental to the evaluation of the invention.

_____
Date

_____
Signature

_____
Status
(Owner, Business or Company Representative, Patent Attorney, Interested Party, etc.)

_____
Printed or Typed Name

U.S. DEPARTMENT OF COMMERCE
NATIONAL INSTITUTE OF STANDARDS AND TECHNOLOGY

OFFICE OF ENERGY-RELATED INVENTIONS
ENERGY-RELATED INVENTION EVALUATION REQUEST

INSTRUCTIONS FOR SUBMISSION OF INVENTION DISCLOSURES AND SUBSTANTIATING MATERIAL FOR EVALUATION.

After reading this page and the following page, complete page 3 of both OERI and Submitter copies. Check appropriate box on page 4 and sign, date, and complete the Memorandum of Understanding. Retain the Submitter's Copy for your records. Detach the OERI copy (pages 3 and 4) and send with your invention disclosure to:

> Office of Energy-Related Inventions
> National Institute of Standards and Technology
> Gaithersburg, Maryland 20899

A written disclosure of your invention, in the English language, must be attached to the OERI copy of this form. This disclosure should include an outline, a complete description of the invention and information to substantiate any claims for performance. Drawings or patents, where appropriate, should be included.

The quality of the evaluation will depend upon the quality of your submission. It should include or cover the following:

1. PURPOSE of the invention. Discuss, if appropriate, where it can be used to best advantage: By industry? By individuals? By the Government? Emphasize the energy conservation or energy production potential.

2. The EXISTING METHOD(S), if any, of performing the function of the invention. Disadvantages of the existing method(s).

3. The NEW METHOD, using your invention. Details of the operation of the invention, identifying specific features which are new. If the invention is conceptual in nature, discuss typical applications.

4. CONSTRUCTION of the invention, showing changes, deletions, improvement over the old method(s).

5. DATA AND CALCULATIONS. If tests have been conducted, detail the test conditions, controls, and results. Energy savings or efficiency estimates should be documented by calculations and data, if available. Theoretical analyses should include the pertinent equations, definitions of terminology, and references.

6. STATUS OF DEVELOPMENT. Include information on stage of research, development, preproduction or production. Discuss proprietary nature, circumstances of public disclosure, instances of disclosure to government agencies, etc.

7. DIFFICULTIES encountered or to be expected in exploiting your invention. Reasons why it has not been patented, manufactured, used, or accepted. What needs to be done to bring the invention closer to use?

PROGRAM DESCRIPTION AND STATEMENT OF POLICY

The Federal Nonnuclear Energy Research and Development Act of 1974 (Public Law 93-577) recognized the importance of encouraging invention and innovation in a national energy program. Section 14 of the Act directs the National Institute of Standards and Technology (NIST) (formerly the National Bureau of Standards) to give particular attention to the evaluation of promising energy-related inventions, particularly those from individual inventors and small companies. The Office of Energy-Related Inventions (OERI) was established at NIST to carry out the provisions of Section 14. Its duties include conducting analyses of submitted inventions to determine their technical and commercial feasibility for saving or producing energy, and bringing noteworthy concepts to the attention of the Department of Energy (DOE).

The principal objective of the OERI effort is to assist DOE in in identifying inventions that are ready to be moved into the private sector, but may require business management assistance, or inventions that require further research and development (R&D), prototype fabrication, or laboratory tests in order to bring them to the point where they can compete with other DOE projects for program R&D funds. The evaluation of inventions submitted will, therefore, be performed principally as a service to DOE. Thus, the outcome of an evaluation will be either a recommendation for action by DOE in connection with the invention, or notification to the inventor that his invention is not being so recommended. It should be noted that a recommendation by OERI is no guarantee that DOE will provide assistance in developing a given invention.

A decision not to recommend action by DOE does not necessarily mean that the invention is considered scientifically unsound or without practical value. Also, a favorable evaluation by OERI should not be construed as being a ruling as to the patentability of any feature of an invention. The inventor should apply for a patent whenever such action is thought to be appropriate. OERI will provide no assistance in filing or prosecuting patent applications. Inventors interested in patent protection should discuss the matter with a registered patent attorney or agent.

To safeguard such proprietary rights as may exist in a submission, OERI will restrict access to invention disclosures to those persons having a need for purposes of administration or evaluation. Accordingly, in accepting invention disclosures for evaluation, an explicit statement is required (see page 4) that the information does or does not come within one of the exemptions of the Freedom of Information Act. If, for example, the disclosure contains information that is (a) a trade secret or (b) commercial or financial information that is privileged or confidential, such information falls within the exemption that is set out in 5 U.S.C. 552(b) (4). Thus, if the disclosure is protectable, the following or a similar statement should be applied to the title page or first page of the disclosure: "The disclosure contains information which is (a) a trade secret or (b) commercial or financial information that is privileged or confidential."

THE PRIVACY ACT OF 1974 (Public Law 93-579) 5 U.S.C. 552(a) requires that you be provided with certain information in connection with this form. You should know that:

1. The authority for collecting this data is the Federal Nonnuclear Energy Research and Development Act of 1974 (Public Law 93-577).

2. The furnishing of the information is entirely voluntary on your part.

3. The principal purpose for which the data will be used is to conduct an evaluation of your invention to determine its technical validity and potential for saving or producing energy.

4. The routine uses which may be made of the information submitted in this form are as follows:

   a. Disclosure to those employees of the Office of Energy-Related Inventions or other Federal agencies having need for the information, either to perform evaluations or administer the evaluation program.

   b. Disclosure to a contractor of the National Institute of Standards and Technology having need for the information in the performance of a contract to perform evaluations of inventions and having agreed to hold the information in confidence.

   c. Disclosure to a Member of Congress submitting a request involving your invention, when you have requested his assistance.

   d. Disclosure to any persons with your written authorization.

NIST Agreement of Nondisclosure

I agree to handle Invention Disclosures received by me from the Office of Energy-Related Inventions pursuant to Section 14 of the Federal Nonnuclear Energy Research and Development Act of 1974.

I further agree that I shall hold in confidence for NIST any such Invention Disclosures provided to me by the Office of Energy-Related Inventions, and shall not disclose any Invention Disclosure or any portion thereof to anyone without the written authorization of the Contracting Officer.

My obligations under the Agreement of Nondisclosure shall not extend to any information or technical data

> a) which is now available or which later becomes available to the general public, other than by any breach of this agreement;
>
> b) which is obtained from any source other than the Officer of Energy-Related Inventions by proper means and without notice of any obligation to hold such information or technical data in confidence; or
>
> c) which is developed without the use of any Invention Disclosure provided to me by the Office of Energy-Related Inventions.

I further agree not to make, have made, or permit to be made, any copies of any Invention Disclosure or portions thereof, except with the written permission of Mr. George P. Lewett, Chief, Office of Energy-Related Inventions. Upon completion of the task called for in the letter, I shall return the Invention Disclosure and any copies thereof to Mr. Lewett.

If, upon examination of an Invention Disclosure, I feel that I have any financial interest or any relation with a third party which might be deemed likely to affect the integrity and impartiality of the performance of the task specified in the letter, I shall provide Mr. Lewett with a complete written disclosure of such interest or relationship prior to undertaking the task and shall not proceed with the task without the written authorization of Mr. Lewett.

If any invention or discovery is conceived or first actually reduced to practice by me in the course of or under this task, I shall promptly furnish Mr. Lewett with complete information thereon; and NIST shall have the sole power to determine whether or not and where a patent application shall be filed, and to determine the disposition of the title and rights in and to any invention or discovery and any patent application or patent that may result. The judgment of NIST on these matters shall be accepted as final and I agree to execute all documents and do all things necessary or proper to carry out the judgment of NIST.

_____
(signature)

_____
(typed name)

_____
(date)

U.S. DEPARTMENT OF DEFENSE
SMALL BUSINESS INNOVATION RESEARCH (SBIR) PROGRAM
PROPOSAL COVER SHEET
Failure to fill in all appropriate spaces may cause your proposal to be disqualified.

TOPIC NUMBER: _____

PROPOSAL TITLE: _____

_____

FIRM NAME: _____

MAIL ADDRESS: _____

_____

CITY: _____ STATE: _____ ZIP: _____

PROPOSED COST: _____ PHASE I OR II: _____ PROPOSED DURATION: _____
PROPOSAL IN MONTHS

BUSINESS CERTIFICATION: YES   NO

▶ Are you a small business as described in paragraph 2.2?   ☐   ☐

▶ Are you a minority or small disadvantaged business as defined in paragraph 2.3? ☐   ☐

▶ Are you a woman-owned small business as described in paragraph 2.4?   ☐   ☐

▶ Will you permit the government to disclose the information on Appendix B, if your proposal does not result   ☐   ☐
in an award, to any party that may be interested in contacting you for further information or possible
investment?

▶ Has this proposal been submitted to other US government agency/agencies; or DoD components, or other   ☐   ☐
SBIR Activity? If yes, list the name(s) of the agency, DoD component or other SBIR office in the spaces to
the left below. If it has been submitted to another SBIR activity list the Topic Numbers in the spaces to the
right below:

_____   _____

_____   _____

▶ Number of employees including all affiliates (average for preceding 12 months)   _____

PROJECT MANAGER/PRINCIPAL INVESTIGATOR   CORPORATE OFFICIAL (BUSINESS)

NAME: _____   NAME: _____

TITLE: _____   TITLE: _____

TELEPHONE: _____   TELEPHONE: _____

For any purpose other than to evaluate the proposal, this data except Appendix A and B shall not be disclosed outside the Government
and shall not be duplicated, used or disclosed in whole or in part, provided that if a contract is awarded to this proposer as a result of or in
connection with the submission of this data, the Government shall have the right to duplicate, use or disclose the data to the extent
provided in the funding agreement. This restriction does not limit the Government's right to use information contained in the data if it is
obtained from another source without restriction. The data subject to this restriction is contained on the pages of the proposal listed on the
line below.

PROPRIETARY INFORMATION: _____

DISCLOSURE PERMISSION STATEMENTS: All data on Appendix A are releasable. All data on Appendix B, of an awarded contract, are
also releasable.

_____ _____ _____ _____
SIGNATURE OF PRINCIPAL INVESTIGATOR   DATE   SIGNATURE OF CORPORATE BUSINESS OFFICIAL   DATE

Nothing on this page is classified or proprietary information/data.

Form 38

U.S. DEPARTMENT OF DEFENSE

DEFENSE SMALL BUSINESS INNOVATION RESEARCH (SBIR) PROGRAM
PHASE I—FY 19__
COST PROPOSAL

Background:

The following items, as appropriate, should be included in proposals responsive to the DOD Solicitation Brochure.

Cost Breakdown Items (in this order, as appropriate):

1. Name of offeror
2. Home office address
3. Location where work will be performed
4. Title of proposed effort
5. Topic number and topic title from DOD Solicitation Brochure
6. Total Dollar amount of the proposal (dollars)
7. Direct material costs
   a. Purchased parts (dollars)
   b. Subcontracted items (dollars)
   c. Other
      (1) Raw material (dollars)
      (2) Your standard commercial items (dollars)
      (3) Interdivisional transfers (at other than cost) (dollars)
   d. Total direct material (dollars)
8. Material overhead (rate _____%) × total direct material = dollars
9. Direct labor (specify)
   a. Type of labor, estimated hours, rate per hour and dollar cost for each type.
   b. Total estimated direct labor (dollars)
10. Labor overhead
    a. Identify overhead rate, the hour base and dollar cost.
    b. Total estimated labor overhead (dollars)
11. Special testing (include field work at Government installations)
    a. Provide dollar cost for each item of special testing
    b. Estimated total special testing (dollars)
12. Special equipment
    a. If direct charge, specify each item and cost of each
    b. Estimated total special equipment (dollars)
13. Travel (if direct charge)
    a. Transportation (detailed breakdown and dollars)
    b. Per Diem or subsistence (details and dollars)
    c. Estimated total travel (dollars)
14. Consultants
    a. Identify each, with purpose, and dollar rates
    b. Total estimated consultants costs (dollars)
15. Other direct costs (specify)
    a. Total estimated direct cost and overhead (dollars)
16. General and administrative expense
    a. Percentage rate applied
    b. Total estimated cost of G&A expense (dollars)
17. Royalties (specify)
    a. Estimated cost (dollars)
18. Fee or profit (dollars)
19. Total estimate cost and fee or profit (dollars)
20. The cost breakdown portion of a proposal must be signed by a responsible official, and the person signing must have typed name and title and date of signature must be indicated.
21. On the following items offeror must provide a yes or no answer to each question.
    a. Has any executive agency of the United States Government performed any review of your accounts or records in connection with any other government prime contract or subcontract within the past twelve months? If yes, provide the name and address of the reviewing office, name of the individual and telephone/extension.
    b. Will you require the use of any government property in the performance of this proposal? If yes, identify.
    c. Do you require government contract financing to perform this proposed contract? If yes, then specify type as advanced payments or progress payments.
22. Type of contract proposed, either cost-plus-fixed-free or firm-fixed price.

SMALL BUSINESS INNOVATION RESEARCH (SBIR) PROGRAM
PROJECT SUMMARY

TOPIC NUMBER: _____

PROPOSAL TITLE: _____

_____

FIRM NAME: _____

PHASE I or II PROPOSAL: _____

Technical Abstract (Limit your abstract to 200 words with no classified or proprietary information/data.)

Anticipated Benefits/Potential Commercial Applications of the Research or Development

List a maximum of 8 Key Words that describe the Project.

_____   _____

_____   _____

_____   _____

_____   _____

Nomination Form

Donald J. Quigg
Excellence in Education Award

Name of Nominee(s): _____
Address: _____

_____

Phone Number: _____
Position: _____
Business Address (if appropriate): _____

_____

Name of Nominator: _____
Address: _____

_____

Phone Number: _____
Position: _____

Category (Circle One):   Educator   Parent   Professional Society
                               Student   Business   Government

---

General Directions

The descriptions requested below should be completed on separate sheets of paper, typewritten, double-spaced, on 8 1/2" x 11" paper, and accompanied by this nomination form. Five copies of the nomination are required.

Selection Criteria:

1. **Accomplishments (or professional achievement):** A description, in essay form, of specific accomplishments in promoting higher order thinking skills, resulting in improvement to a program, class, or student. Include in this section a description of the nominee's outstanding abilities as a communicator and leader. (Major emphasis in judging placed on essay.)

2. **Community Involvement:** A description of any community activities of the nominee outside the professional sphere in which he or she has participated in pursuit of higher order thinking skills.

3. **Awards and Publications:** A list of any professional awards received and a list of any professional publications germane to the nominee's pursuit of higher order thinking skills.

Mail nomination to:

Project XL
U.S. Patent and Trademark Office
Crystal Park 1, Suite 507
Washington, DC 20231

Nominations for the 1992 Award must be received in the Patent and Trademark Office by close of business June 30, 1992.
For further information contact: Ruth Nyblod, Administrator for Project X L, Phone: (703) 557-1610.

THE DONALD J. QUIGG
EXCELLENCE IN EDUCATION AWARD

Established by the Patent and Trademark Office Society

Purpose:

To recognize the efforts of an individual (or group) to promote the teaching of creative and inventive thinking skills at all levels of the curricula, in conjunction with the Patent and Trademark Office's Project XL. Project XL was initiated by Donald J. Quigg, Assistant Secretary and Commissioner of Patents and Trademarks 1985-1989, to foster national competitiveness worldwide by encouraging analytical thinking and problem solving among America's youth. The goals of Project XL are:

• To generate awareness among educators, parents, businesses, governments, educational associations, and professional societies of the importance of applied thinking skills;

• to motivate educators to use the inventive process as a vehicle through which students apply the skills of inquiry and creative and critical thinking to real life problem-solving experiences;

• to identify and, where possible, provide the tools needed to accomplish these educational goals; and

• to establish a network of information, communication and support for persons interested in participating in the Project XL vision.

Naming the award for Commissioner Quigg adds prestige to the honor because of his well-known support of educators throughout the country who promote innovation in the classroom, in the community, and across the Nation.

Eligibility of Nominees:

The person (or group) selected for the Donald J. Quigg Excellence in Education Award shall be a leader in the quest for education excellence from one of six categories: Educator, Student, Parent, Business, Professional Society, or Government.

The awardee shall be one who has contributed significantly and measurably to the advancement of innovation through the teaching of higher order thinking skills. The award may be given for a specific single but lasting contribution, or for a series of activities which, by advancing the promotion of innovation at the elementary, secondary, or college level, has benefited the Nation's youth as a whole. In either case, the awardee should embody qualities of integrity, dedication and performance which are marks of professionalism in his or her field.

Award Selection and Presentation:

The winner(s) of the award will be chosen by a panel of distinguished experts in the fields relating to the Patent and Trademark Office and Project XL during a Roundtable discussion. The panel will consist of a representative from each of the following five categories: (1) Patent and Trademark Office Society; (2) National Inventive Thinking Association; (3) Patent and Trademark Office/Project XL; (4) Educator; and (5) Inventor Representative.

The 1991 award will be presented by the President of the Patent and Trademark Office Society on or about October 18, 1991, at a special awards dinner in Washington, D.C. The award winner(s) will receive a suitably engraved plaque and an all-expense-paid trip with one guest to Washington, D.C. (based on U.S. Government per diem rates effective at the time of the award presentation).

Judges decision will be final on August 31, 1991. The winner(s) will be notified the first week of September 1991.

Part Two

How to Protect and License Your Invention

Section 2

Preface

"A chief event of life is the day in which we have encountered a mind that startled us."—Ralph Waldo Emerson

"Heavier-than-air flying machines are impossible," the celebrated British mathematician and physicist William Kelvin assured everyone back in 1895.

"...man can never tap the power of the atom," said Nobel Laureate and physicist Robert Andrews Millikan, credited with being first to isolate the electron and measure its charge.

"Everything that can be invented has been invented," offered another man of vision, Charles H. Duell, Director of the U.S. Patent Office in 1899.

TRW reminded us of these profound observations in a 1985 *The Wall Street Journal* ad which was tagged, "There's no future in believing something can't be done. The future is in making it happen."

What the above gentlemen did not know, quite obviously, is what inventors from fully equipped labs to basement workshops across America are proving everyday: there is no future in the word impossible. Results first. Theory second.

We are privileged! We live in a land of opportunity. Nowhere in the world do people have more freedom and encouragement to innovate, to be different and individual than in America.

We are always looking to challenge the previous and reach new levels of interest and involvement by doing things in novel ways. Our history is replete with examples of Yankee ingenuity, with independent and courageous individuals who succeeded by doing things differently, and with dreamers who believed in themselves and their ideas.

"It takes a special kind of independence to invent something. You put yourself and your ideas on the line. And maybe people will say that you're crazy or that you're impractical," said President George Bush when, as vice president, he spoke at the National Museum of American History to commemorate the 150th anniversary of the Patent Act of 1836.

"But for over more than two centuries, millions of Americans have ignored the ridicule," he continued. "They've worked on ideas. From those ideas, they've started businesses. And many of those

1

businesses have grown and are today our great industrial companies—companies like Xerox, Ford Motor Company, American Telephone and Telegraph and Apple Computer. Think of what America would be like if the skeptics had silenced the inventors."

We have been on the leading edge of uncertainty, experimentation, and exploration since the first pilgrims set out for the New World and came ashore at Plymouth, Massachusetts in 1620. In 1850, British subject James Nasmyth, best known for inventing the steam hammer, said about inventive Americans, "There is not a working boy of average ability in the New England States...who has not an idea of some mechanical invention or improvement...by which he hopes to better his position, or rise to fortune."

Alexis de Tocqueville, the French writer who visited America in 1831, wrote, "They [Americans] all have a lively faith in the perfectibility of man, and judge that the diffusion of knowledge must necessarily be advantageous, and the consequences of ignorance fatal. They [Americans] consider society as a body in a state of improvement, humanity as a changing scene, in which nothing is, or ought to be, permanent; and what appears to them today to be good, may be superseded by something better tomorrow."

Our heritage is rich with examples of American inventors, tinkerers, daydreamers, and gadgeteers, working from basements and garages, who dared to be different and refused to trade incentive for security. Many of their names have become celebrated: George Westinghouse created the first air brake; W.H. Carrier gave us air conditioning; Benjamin Franklin put us on bicycles; Edwin H. Land developed the 60-second camera; Charles Goodyear first vulcanized rubber; George Pullman designed double-deck sleeping cars on trains; Robert Goddard started the race to space; King S. Gillette made the first safety razor; and William S. Burroughs patented the first mechanical adding machine.

Other Americans are not so well known, but their inventions are: W.H. Carothers created nylon; Whitcomb L. Judson patented the zipper; Bette Graham invented the quick-drying paint that covers typing mistakes; Luther Childs Crowell made the square bottom paper bag; Walter Hunt stuck us with the safety pin; James Ritty rang up tremendous sales with the first cash register; Mary Anderson made motoring safer with the first windshield wipers; Garett A. Morgan changed our driving habits and city streets with traffic signals and is also credited with a pioneering gas mask; and Stephanie Kwolek, one of Dupont's leading chemists, discovered the "miracle fiber" Kevlar.

Margaret Knight, remembered as "the female Edison," received some 26 patents for such diverse items as a window frame and sash, machinery for cutting shoe soles, and improvements to internal combustion engines.

Not all inventions are so conventional. Two Harvard University researchers, Philip Leder and Timothy A. Stewart, made history on April 12, 1988 when they were awarded a patent for genetically engineered mice for cancer research—the first patent on animals ever granted.

As diverse a group of individuals as this appears to be, these people share many things in common, things that you too will require to fulfill your aspirations and see your inventions patented, licensed, manufactured, and marketed.

On an intellectual level, they have what Emerson called that "gleam of light that flashes across the mind from within." They did not dismiss their own thoughts without notice. They abided by their own spontaneous impression. Successful inventors permit nothing to affect the integrity of their minds. Successful inventors do not procrastinate because procrastination is the thief of time.

These people shared the challenges every inventor faces when trying to protect and exploit a new product. Whether it is a wheel mounting device, a method for manufacturing semiconductors, an adjustable hoe attachment for a rake, fluidic rotational speed sensors, a U-joint mount, or a better mousetrap, the same basic steps apply.

This book is not based on theory. I've personally developed or co-developed and licensed more than 75 original products and concepts to companies ranging from Fortune 500 diversified manufacturers to small independent businesses. Taken together, these products have generated tens of millions of dollars. This book provides straightforward "how-to" information and serves as a guide to the enormous amount of information and services available to the independent inventor.

In addition to the formal information on how to protect your ideas through U.S. government devices such as patents, copyrights, and trademarks, the *Inventing & Patenting Sourcebook* is crammed full of suggestions, modus operandi, and "insider" tips and pointers on invention marketing, prototyping, corporate licensing, and funding research and development. This is information every independent inventor needs, but only a fortunate few possess. My objective is to guide the inventor down the frequently murky road leading from concept to market fulfillment.

Educational Background is Not Everything

My university degree is in television and film with a minor in English. I have never had a course in marketing. I am a totally self-taught inventor, designer, and marketeer. I know, therefore, better than most, what the independent inventor requires in terms of information. I also know the limited resources and time restraints the independent inventor faces.

I have done my best to make this book as comprehensive as possible, designing it to deliver empirical information on what to do and how to do it. It is not just another directory that tells you where but not how.

Levy's Golden Rules

To be successful in the exciting business of product development it takes much more than a good idea and a strong patent. Understanding and practicing the following six points is critical to the inventor who attempts to beat what often appear to be insurmountable odds.

1) **Don't take yourself too seriously.** Don't take your idea too seriously either. The world will probably survive without your idea. Industry will probably survive without your idea. You might need it to survive, but no one else does.

2) **You can't do it all yourself.** Remember the words of John Donne, "No man is an island, entire of itself; every man is a piece of the continent, a part of the main."

The success I have experienced is the result of unselfish, highly talented, and creative associates willing to face the frustrations, rejections, and seemingly open-ended time frames that are inherent in any product development and marketing exercise. I have also been lucky to have met and worked with very creative, understanding, and courageous corporate executives willing to believe in me and gamble on my ideas.

It is the cross-pollination and subsequent synergism of entrepreneurs and intrapreneurs (executives) that results in success, success in which all parties share. For if any link in this often complex and serpentine chain breaks, an entire project could flag.

3) **Keep egos under control.** Unchecked egocentricity can be the source for major failure in the development and licensing of new concepts. Arrogance has no place in the process. I have always found that my concepts are enhanced by the right touch. Working together or in competition, other people contribute time and time again to making an idea more useful or marketable.

4) **Learn to take rejection.** Rejection can be positive if it is turned in to constructive growth. Don't let it shake you or your confidence.

3

I have rarely licensed a product to the first manufacturer who sees it. And for every product I've licensed there are at least five that were prototyped and did not make it.

5) **Don't do it for the financial rewards** alone. You should be motivated by the gamesmanship. It may sound trite, but people who do things just for the money usually come up shortchanged.

6) **Be patient.** It's going to take time. Except for baked goods and newspapers, nothing is made overnight. Even though the market is big, the competition is ferocious. Murphy's first corollary is: Nothing is as easy as it looks. Murphy's second corollary is: Everything takes longer than you think.

The Beatles worked for ten years in the clubs of Liverpool before getting a break. Michael Jackson had made millions of dollars by age twenty-two, but let us not forget he began performing professionally at age five. Closer to home, it took me three years of pitching the idea for this book before Gale Research Inc. decided to publish it. Ideas generally don't diminish with time, but grow and find their proper environment.

Longfellow said it best in the work, "The Ladder of St. Augustine": "The heights by great men reached and kept were not attained by sudden flight. But they, while their companions slept, were toiling upward in the night."

Information is power, but only when properly understood and utilized. This volume is packed with power and the instructions to help you use it effectively. If you carefully apply this information to an innovative and novel concept, putting your own English on the lessons implied, you'll have that added edge that could mean success.

In closing, I'll again quote President Bush. "Let us rededicate ourselves to ensuring that America is always a land in which, for those who dare to create new technologies and new businesses, the air is clear and the sky is open and the energies of man are free, where the enthusiasm for invention and the spirit of enterprise is always part of the American spirit, where men and women who want to build and dream can always find a home...Today, tomorrow, and for all time to come."

Section 3

Patents and the PTO

"Creativity is contagious, pass it on!"—Albert Einstein

It is smart to initiate every form of appropriate invention protection available under the law. First, this sends a signal to those to whom you disclose a concept, whether they be potential licensees or investors, that you are serious, committed, and willing to go that extra mile. Second, it tends to keep honest people honest, much in the same way locks do on doors. Keep in mind, though, that locks have never stood in the way of a good second-story man. A dishonest person or company bent on misappropriating a piece of intellectual property will do it regardless of the protective steps you may take. In most cases, strong protection is a prerequisite to any licensing agreement. You'll find that companies prefer to license protected products over nonprotected products as extra insurance against competition.

The U.S. government offers three ways to protect your ideas: patents, copyrights, and trademarks. As each of these can be important to the independent inventor, in this section I will attempt to cut through "bureaucratese" and explain each in simple and understandable terms. I will address the questions most frequently asked and then some.

This section is not intended to be a comprehensive text or detailed legal guide on patent, copyright, and trademark law. The source material for this writing weighed more than 15 pounds! My purpose is to enlighten you, make you comfortable with the material and processes, share some personal experiences, and, in the end, better equip you to handle matters, make decisions, save some money, and otherwise defend yourself and your innovative concepts in what can be a very rough-and-tumble marketplace. Protecting your invention can be a tortuous journey—the steps outlined here should remove some of the tougher obstacles.

People can confuse patents, copyrights, and trademarks. Although these three kinds of intellectual property share certain characteristics, they serve different purposes. Let's begin with patents.

5

Patents

What is a Patent?

A patent for an invention is a grant of property right by the U.S. government to the inventor (or his heirs or assigns), acting through the Patent and Trademark Office (PTO).

The right conferred by the patent grant is, in the language of the statute and of the grant itself, "the right to exclude others from making, using, or selling" the invention. A patent does **not** grant the right to make, use, or sell an invention; instead it grants the **right to exclude others** from making, using, or selling an invention.

There are three types of patents:

1) **Utility patents,** which are granted for new, useful, and nonobvious processes, machines, manufactured articles, compositions, or improvements in any of the above. Examples of utility patents range from the well-known artificial heart valve (#4,490,859) to the microwave clothes dryer (#4,490,923).

See **Fig. 1 on pages 7-14 for an example of a utility patent illustration and complete specification.**

2) **Design patents,** which are available for the invention of new, original, and ornamental designs for articles of manufacture. A design patent only protects the appearance of an article and not its structure or utilitarian features. Examples of design patents range from the ordinary combined toothbrush holder/tumbler (#278,586) to a seaweed plate garnish for a sushi dish (#278,565).

See **Fig. 2 on page 15 for an example of a design patent illustration.**

3) **Plant patents,** which are provided to anyone who has invented or discovered and asexually reproduced any distinct and new variety of plant, including cultivated spores, mutants, hybrids, and newly found seedlings, other than a tuber-propagated plant or a plant found in an uncultivated state. Examples of plant patents include a new and distinct cultivar of African violet (#5,383), a variety of almond tree (#5,382), and a chrysanthemum plant named Organdy (#5,278). Illustrations for plant patents are usually in color and therefore it is not practical to reproduce a drawing.

What Can Be Patented?

The patent law specifies the general field of subject matter that can be patented.

In the language of the statute, a person who "invents or discovers any new and useful process, machine, manufacture, or composition of matter or any new and useful improvements thereof, may obtain a [utility] patent," subject to the conditions and requirements of the law. The word **"process"** as used in the statute refers to a method, usually an industrial or technical method. The term **"machine"** used in the statute needs no explanation. The term **"manufacture"** refers to articles which are made, and includes all manufactured articles. The term **"composition of matter"** relates to chemical compositions and may include mixtures of ingredients as well as new chemical compounds. These classes of subject matter taken together include practically everything which is made by man and the process for making them.

United States Patent [19]

DeLay, Jr.

[11] Patent Number: 4,557,395

[45] Date of Patent: Dec. 10, 1985

[54] PORTABLE CONTAINER WITH INTERLOCKING FUNNEL

[75] Inventor: Victor A. DeLay, Jr., Largo, Fla.

[73] Assignee: E-Z Out Container Corp., Clearwater, Fla.

[21] Appl. No.: 717,439

[22] Filed: Mar. 28, 1985

[51] Int. Cl.⁴ B65D 3/04
[52] U.S. Cl. .... 220/86 R; 220/85 F; 220/1 C; 141/98
[58] Field of Search .... 220/86 R, 85 F, 1 C; 141/98, 331; 220/360

[56] References Cited

U.S. PATENT DOCUMENTS

1,554,589	9/1925	Long	220/1 C X
3,410,438	11/1968	Bartz	220/1 C
4,010,863	3/1977	Ebel	220/1 C
4,149,575	4/1979	Fisher	220/85 F X
4,162,020	7/1979	Kirkland	220/1 C X
4,296,838	10/1981	Cohen	220/1 C X
4,301,841	11/1981	Sandow	220/1 C X

Primary Examiner—Steven M. Pollard
Attorney, Agent, or Firm—Stanley M. Miller

[57] ABSTRACT

A portable, vented container for dirty oil, of the type having a small fill spout and having increased utility when used in conjunction with a funnel. A vent closure member and a funnel securing latch are integral with the funnel so that when the funnel is inverted and positioned in surmounting relation to the container, the vent closure member closes the vent and the securing latch is engaged by a fill spout cap which engagement secures the funnel against movement and hence maintains the vent closure as well. Removal of the fill spout cap releases the funnel, and positioning the funnel into its operative position relative to an automotive oil drain plug separates the vent closure portion of the funnel from the vent. An elongate extension member having a flexible medial portion is further provided.

20 Claims, 12 Drawing Figures

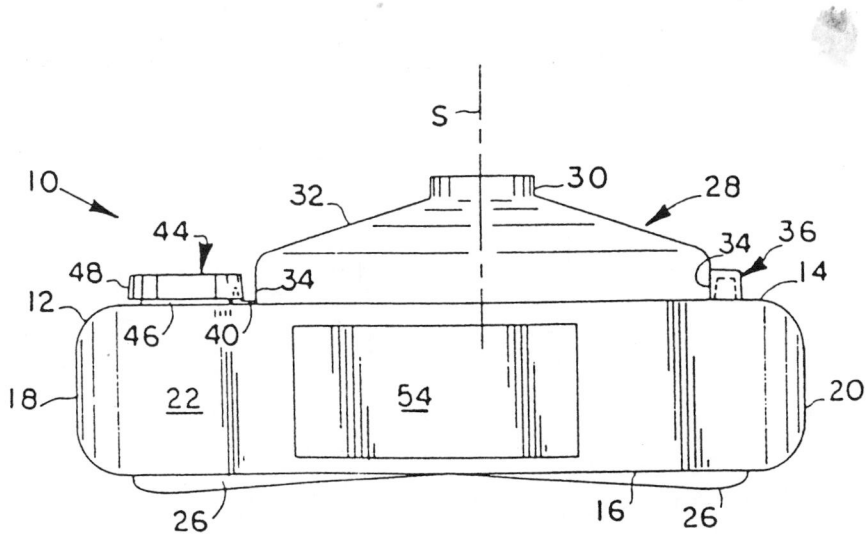

Figure 1

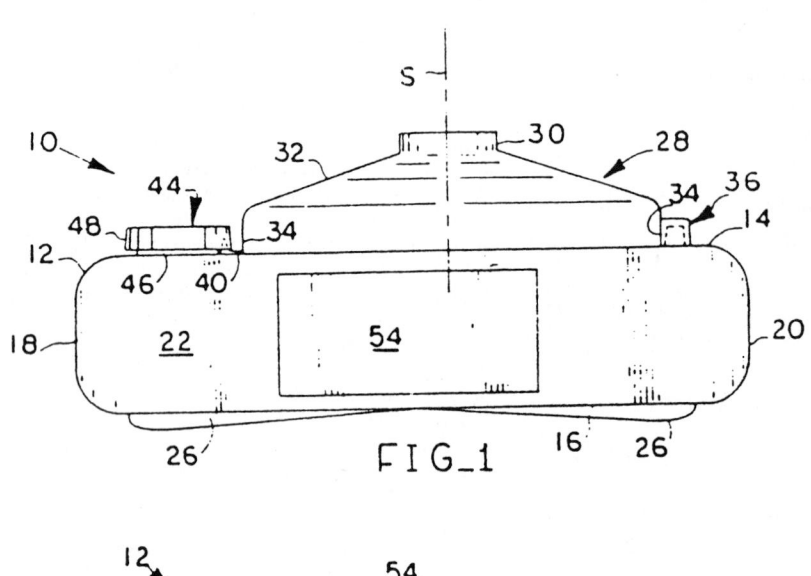

FIG_1

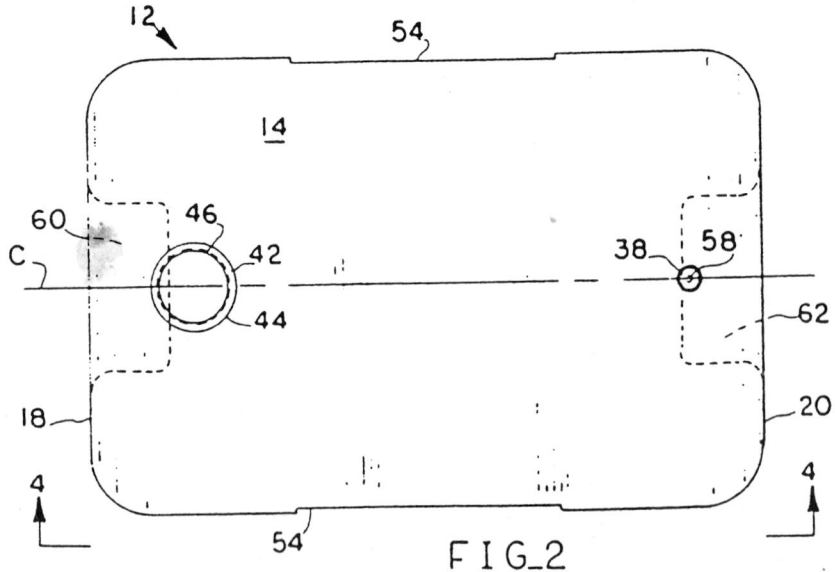

FIG_2

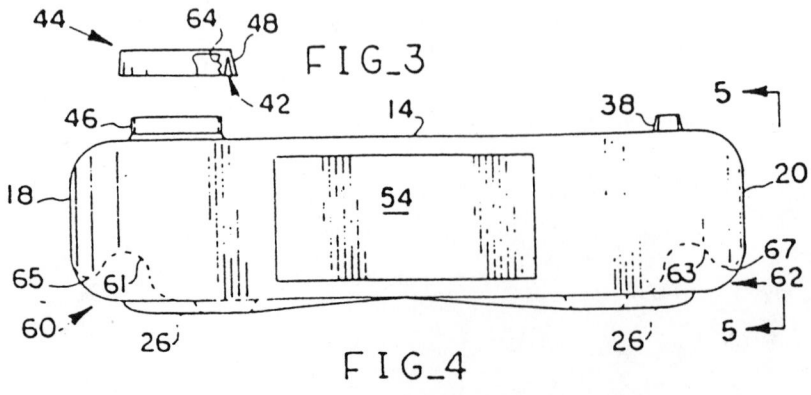

FIG_3

FIG_4

Figure 1 (cont.)

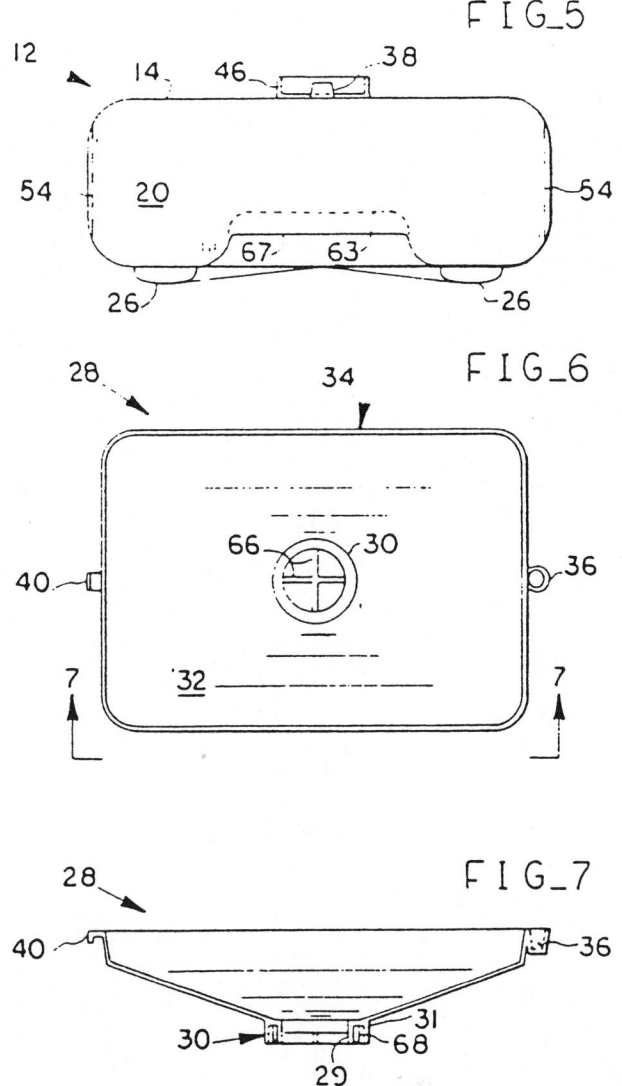

FIG_5

FIG_6

FIG_7

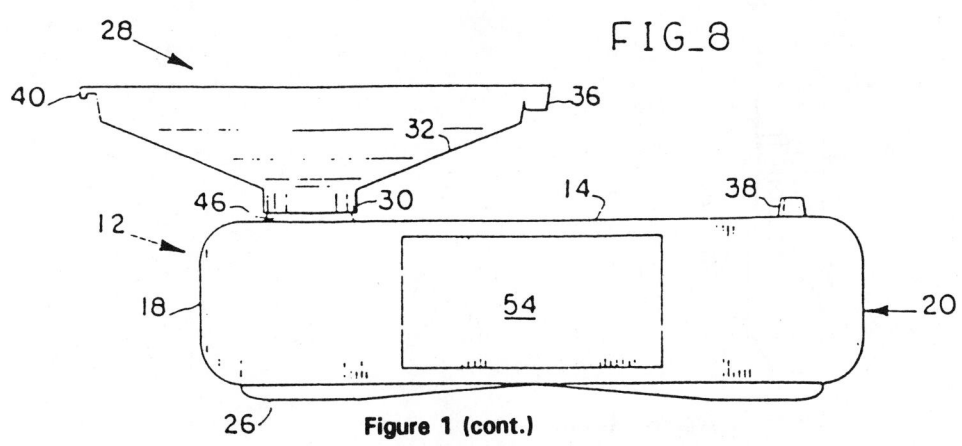

FIG_8

Figure 1 (cont.)

U.S. Patent   Dec. 10, 1985   4,557,395

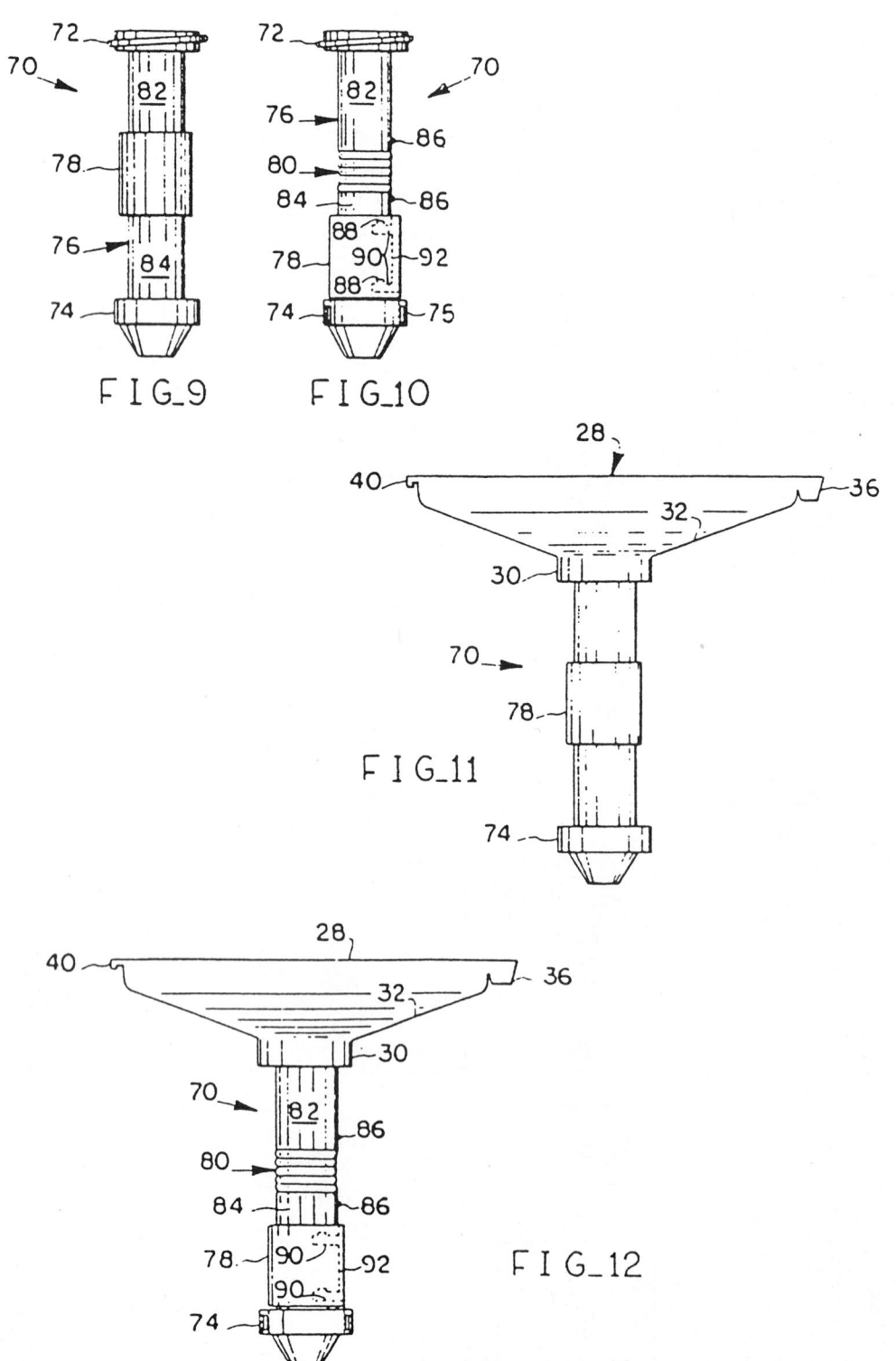

FIG_9   FIG_10

FIG_11

FIG_12

Figure 1 (cont.)

4,557,395

PORTABLE CONTAINER WITH INTERLOCKING FUNNEL

BACKGROUND OF THE INVENTION

1. Field of the Invention

This invention relates generally to containers having small fill spouts, and more particularly this invention relates to a vented container the vent of which is closed when the funnel is stored in latching engagement with the container body.

2. Description of the Prior Art

A thorough description of the prior art in the the field to which this invention pertains may be found in my co-pending application having a filing date of Sept. 14, 1983, Ser. No. 06/531,948. Moreover, the most pertinent prior art is believed to be the container for dirty oil disclosed in said application.

Other patents of interest are: U.S. Pat. Nos. 4,403,692 to Pollacco (1983); 822,854 to Cosgrave (1906); 2,576,154 to Trautvetter (1951); 4,098,393 to Meyers (1978); 4,217,940 to Wheeler and others (1980); and 4,301,841 to Sandow (1981).

Of the known containers, only the container provided by the present inventor and disclosed in the above-identified patent application contains a means whereby the funnel of the container can be conveniently stored when not in use.

Containers having small fill spouts are normally vented to allow the air inside the container to escape as liquid fluids are charged thereinto. Typically, the vent is provided in the form of an upstanding coupling which is provided with a closure member in the form of a cap which may or may not be attached to the coupling itself. Where the cap is attached to the coupling, its loss is safeguarded against but still the user of the container must remember to open and close the vent as needed. Vent caps that are not attached to their couplings are usually lost.

There is a need, therefore, for a vent cap that is safeguarded against loss, and which also opens and closes the vent as needed without requiring the user thereof to remember to open and close such vent.

Another common problem with small-mouthed containers is that the funnels which must be used therewith are often lost. Pollacco solves this problem by permanently securing his funnel to his container. This storage expedient is unsatisfactory because it is important to maintain funnels of the type used to fill automotive crankcases in a substantially clean condition as the introduction of dirt into a crankcase can damage engine parts.

Therefore, there is a need for a funnel storage apparatus capable of storing a funnel in an inverted position when it is not in use. The storage apparatus that is needed would also safeguard against the loss of the funnel.

The art has heretofore developed elongate funnel extension members of the type disclosed by Cosgrave, Trautvetter, and the present inventor, but the same are inflexible and thus inadequate and lacking in utility in certain specific environments.

SUMMARY OF THE INVENTION

The longstanding but heretofore unfulfilled need for a portable container for dirty oil having the desireable features of a self-opening and self-closing vent, a funnel that is storable in an inverted position and which is also secured against loss, is now fulfilled by the invention disclosed hereinafter and summarized as follows.

The container is of parallelepiped form and has finger-receiving recesses formed in its opposite ends, on the underside thereof, which recesses are grasped by an individual when transporting the container.

The top of the container includes a large, imperforate medial portion against which the rim of the funnel is seated when the funnel is in its storage position.

A fill spout of small diameter projects upwardly from the top of the container, and is disposed near the periphery of the container so the medial portion of the container can receive the stored funnel, as aforesaid.

A sleeve member which defines a vent opening projects upwardly from the top of the container as well, but is disposed in longitudinally spaced relation to the fill spout so that it is near the periphery of the container opposite from the fill spout.

The longitudinal axis of symmetry of the container bisects the finger-receiving recesses or handles, the fill spout, the vent-defining sleeve, and the funnel when the latter is in its stored position. In this manner, the container is stable when transported.

The funnel has an integral vent closure member that projects outwardly from the rim of the funnel, in radial relation to the funnel's axis of symmetry. A latch member used to secure the stored funnel against movement is also formed integral to the funnel, extends radially with respect to said axis from the rim thereof, and is positioned in opposition to the vent closure member.

The funnel's size and the amount of space between the fill spout and the vent opening are selected so that when the funnel is inverted and placed in the center of the medial portion of the top wall of the container, and properly rotated about its axis of symmetry, the vent closure member will align with and seal the vent opening and the latch which is opposed to the vent closure member will be positioned in close proximity to the fill spout.

A novel fill spout closure member in the form of a double-walled cap, when brought into screw threaded engagement with the fill spout, will seal the spout and simultaneously overlie the funnel latch to secure the funnel against displacement.

The novel cap's first wall is internally threaded and thus adapted for screw threaded engagement with the externally threaded fill spout. It outer wall defines an annular recess having an open bottom, which recess surrounds the first wall and which recess receives the funnel latch therewithin. The annular configuration of the recess eliminates any need for aligning the cap with respect to the latch.

In this manner, the act of inverting the funnel and placing it in its storage position on the top wall of the funnel will close the vent if the proper alignment is made. Once the vent has been closed, no further alignment is required as the sealing of the fill spout by the novel cap will also secure the funnel as desired.

Thus, when the funnel is deployed into its operative configuration, the user of the invention need only remove the fill spout cap, as such will release the funnel from its stored position. The act of placing the funnel's spout into the container's fill spout then serves to open the vent.

A funnel extension member having a flexible medial portion is also disclosed hereinafter. A slideably mounted rigid sleeve member serves to delete the flexi-

Figure 1 (cont.)

4,557,395

3

bility function of the extension member when desired when such sleeve member is positioned in registration with the flexible portion of the member. However, the flexibility of the member is restored upon slidingly displacement of the sleeve away from the flexible medial portion.

An important object of this invention, therefore, is to provide a container for dirty oil that includes a funnel as an attachment to the container so that the funnel is not easily misplaced.

Another object is to provide an attachment means that protects the sloping inside walls of the funnel contamination when the funnel is stored.

Another object of this invention is to provide a means whereby the vent of a container can be automatically opened and closed at the time the container's funnel is placed into its operative position and its storage position, respectively.

Other objects will become apparent as this description proceeds.

The invention accordingly comprises the features of construction, combination of elements and arrangement of parts that will be exemplified in the construction hereinafter set forth, and the scope of the invention will be indicated in the claims.

BRIEF DESCRIPTION OF THE DRAWINGS

For a fuller understanding of the nature and objects of the invention, reference should be made to the following detailed description, taken in connection with the accompanying drawings, in which:

FIG. 1 is a side elevational view of the container with the funnel stored in its inverted position thereatop;

FIG. 2 is a top plan view of the container body member;

FIG. 3 is a partially cut away side elevational view of the novel fill spout closure means;

FIG. 4 is a side elevational view taken along line 4—4 of FIG. 2;

FIG. 5 is an end view taken along line 5—5 of FIG. 4;

FIG. 6 is a top plan view of the novel funnel member;

FIG. 7 is a side elevational view of the funnel member taken along line 7—7 of FIG. 6;

FIG. 8 is a side elevational view, like that of FIG. 4, which shows the funnel member engaging the fill spout of the container body;

FIG. 9 is a side elevational view of the novel funnel downspout extension member with the rigid sleeve in its locked position;

FIG. 10 is a side elevational view of the funnel downspout extension member with the rigid sleeve in its unlocked position;

FIG. 11 is a side elevational view showing the extension member operatively coupled to the funnel member with the sleeve in its locked position; and

FIG. 12 is a side elevational view showing the extension member operatively coupled to the funnel member with the sleeve in its unlocked position.

Similar reference numerals refer to similar parts throughout the several views of the drawings.

DETAILED DESCRIPTION OF THE PREFERRED EMBODIMENT

Referring now to FIG. 1, it will there be seen that an illustrative embodiment of the invention is designated by the reference numeral 10 as a whole. The container body 12 has a parallelepiped construction when seen in

4

perspective. Visible in FIG. 1 are the container's top wall 14, bottom wall 16, its left and right end walls 18, 20, a side wall 22, and support members collectively designated 26.

The novel funnel is indicated generally by the numeral 28. Funnel 28 includes downspout 30, sloping or converging walls 32, and an annular rim 34.

A vent closure member 36 is integrally formed with the rim 34 and extends therefrom as shown. The closure member 36 overlies a vent shroud 38 which is shown in phantom lines in FIG. 1.

A latch 40 is also integrally formed with the funnel rim 34 and is on the opposite side thereof relative to the vent closure member 36. The latch 40 has an "L" shape as shown. The horizontal leg of the latch abuts the top wall 14 of the container 12 and extends radially with respect to the axis of symmetry S of the funnel 28. It terminates in an upstanding leg (shown in phantom lines in FIG. 1) that extends into a cavity 42, which cavity 42 is an annular recess as shown in FIG. 2.

Referring again to FIG. 1, fill spout cap 44 is internally threaded to mate with the external threads of the fill spout 46. The annular latch-receiving recess 42 is formed by the provision of annular wall 48 that surrounds the spout 46, said annular wall depending to the periphery of the top wall of cap 44. The diameter of the top wall of cap 44 is greater than the diameter of the fill spout 46 by an amount substantially equal to the width of the latch-receiving recess 42.

The placement of the upstanding portion of latch 40 in the annular cavity 42 maintains the funnel 28 in its inverted, stored position until the cap 44 is removed.

The space designated 54 in FIG. 1 is a display space and accommodates a label which may have imprinted thereon the trademark of the device and other information.

Returning now to FIG. 2, it will there be seen that the longitudinal axis of symmetry of the device 10 is indicated by the centerline C. It bisects the vent 58 which is formed in the top wall 14 of the container 10 and which is surrounded by vent shroud 38, the fill spout 46, and the longitudinally spaced handles 60, 62 of the invention. The width of the handles 60, 62 is sufficient to accommodate four fingers of a human hand. Both of the label-accommodating recesses 54, 54 mentioned in connection with the description of FIG. 1 are shown in FIG. 2 as well.

The vent closure member 36 slideably and snugly engages the outer walls of the shroud 38, thereby closing the vent opening 58, when funnel 28 is in the inverted storaage position, as aforesaid.

FIG. 3 shows the internal threads 64 on the cap 44 and the annular wall 48 that depends to the periphery of the cap top wall to define the annular cavity 42 into which the upstanding portion of latch 40 extends.

The externally threaded fill spout 46 is shown in FIG. 4, which FIG. shows the container 12 with funnel 28 and cap 44 separated therefrom.

The handles 60, 62 include concave surfaces 61, 63, respectively, and convex surfaces 65, 67, the former of which are abutted by fingertips when the container is carried and the latter of which provide a comfortable rounded weight bearing surface.

An end view of the container 12 is provided in FIG. 5.

A top view of the novel funnel 28 appears in FIG. 6. A strainer 66 formed by a pair of cross bars is formed where the downwardly sloping walls 32 of the funnel 28

Figure 1 (cont.)

4,557,395

5

merge with the funnel's downspout. The generally rectangular planform of the funnel 28 conforms to the planform of the container body 12 as shown in FIG. 2, but the corresponding dimensions of the funnel are smaller.

The downspout 30 of funnel 28 is internally threaded as indicated by the reference numeral 68 appearing in FIG. 7, and is thus adapted for screw threaded engagement with the externally threaded fill spout 46. Accordingly, the downspout 30 of the funnel 28 is coupled to fill spout 46 when it is desired to charge the container with dirty oil. This operative positioning of the funnel 28 and fill spout 46 is depicted in FIG. 8. A comparison of FIGS. 1 and 8 indicates that the removal of cap 44 from spout 46 releases latch 40 so that funnel 28 can be separated from its engagement with top wall 14 of container 12, restored to its upright configuration, and coupled with the spot 46. The separation of the funnel 28 and the container body top wall 14 also separates the vent closure member 36 from vent shroud 38, which separation exposes vent 58 (FIG. 2) to ambient. The internal threads 68 of downspout 30 are formed in outer wall 31 thereof. An inner wall 29 is spaced radially inwardly of outer wall 31, and is concentric therewith. Accordingly, dirty oil contacts inner wall 29 only.

The truncate downspout 30 of funnel 28 is provided because some vehicle are built close to the ground. However, other vehicles are built higher from the ground and the use of a downspout extension member becomes advisable.

An improved downspout extension member is shown in FIGS. 9–12, and is designated 70 as a whole. It includes an externally threaded adapter 72 which is coupled to the internally threaded downspout 30 of funnel 28 when in use, as shown in FIGS. 11 and 12. Another adapter 74 at the lower end of the extension member 70 is internally threaded as at 75 (FIG. 10) to mate with the external threads of the fill spout 46. An elongate medial portion 76 interconnects the upper and lower adapters 72 and 74.

A slideably mounted rigid sleeve member 78 is shown mid-length of the medial portion 76 in FIG. 9. When the sleeve member 78 is locked into this position by means disclosed hereinafter, the novel extension member 70 can be used in the same manner as conventional downspout extension members, which use is depicted in FIG. 11.

However, when the sleeve 78 is unlocked and slideably displaced to its lowermost position, which position is depicted in FIG. 10, such displacement frees a flexible member 80 from confinement so that it is free to bend. More specifically, upper portion 82 of the downspout extension member medial portion 76 and lower portion 84 thereof may be displaced from their axial alignment with each other, i.e., their respective axes of longitudinal symmetry may be made oblique to one another. As shown in FIG. 12, when the flexible member 80 is free, funnel 28 can be moved in any direction relative to lower coupling 74, or vice versa.

FIGS. 10 and 12 both show the means employed to lock and unlock sleeve 78 as desired. A pair of vertically spaced beads, collectively designated 86, are formed on upper and lower portions 82, 84 of the extension member medial portion 76. A pair of vertically spaced bead-receiving cavities, collectively designated 88, are formed internally of sleeve member 78, so that the sleeve 78 is locked into overlying relation to the flexible member 80 when beads 86 are disposed therein.

6

To unlock the sleeve 78, the user of the inventive apparatus grasps sleeve 78 and slides it upwardly by a distance equal to the depth of the bead-receiving cavities 88. Each bead 86 will then be positioned in channels 90 which are also formed internally of sleeve 78. The user of the device then rotates the sleeve 78 until the beads 86 have traveled the length of the arcuate channels 90, which length could be a quarter of an inch, for example. This rotation of sleeve 78 will bring the beads 86 into registration with a vertically extending channel 92 so that the sleeve 78 can be moved to the position shown in FIGS. 10 and 12.

It will thus be seen that the objects set forth above, and those made apparent from the foregoing description, are effectively attained and since certain changes may be made in the above construction without departing from the scope of the invention, it is intended that all matters contained in the foregoing description or shown in the accompanying drawings shall be interpreted as illustrative and not in a limiting sense.

It is also to be understood that the following claims are intended to cover all of the generic and specific features of the invention herein described, and all statements of the scope of the invention which, as a matter of language, might be said to fall therebetween.

Now that the invention has been described,
What is claimed is:

1. A container of the type having a small fill spout and having increased utility when used in conjunction with a funnel, comprising:
   a container body member of generally parallelepiped configuration,
   a fill spout formed in a top wall of said container body member and projecting upwardly therefrom,
   a vent means in the form of an aperture formed in said top wall,
   a funnel member having a rim, converging sidewalls, and a downspout,
   said fill spout and funnel downspout adapted for releasable engagement with one another,
   a vent closure member secured to said funnel rim and projecting outwardly therefrom,
   said vent closure member closing said vent when brought into registration therewith.

2. The container of claim 1, further comprising,
   a fill spout closure means in the form of a cap member,
   a latch member secured to and projecting outwardly from said funnel rim,
   said cap member adapted to releasably engage said latch member when said funnel member is inverted and disposed atop said container top wall and when said cap member is releasably engaged to said fill spout.

3. The container of claim 2, wherein said vent closure member and said latch member are secured to said rim in opposed relation to each other.

4. The container of claim 3, further comprising,
   a sleeve-shaped shroud member disposed in surrounding relation to said aperture and projecting upwardly from said container top wall,
   said vent closure member adapted to engage said shroud member when said funnel is inverted and said vent closure member is brought into releasable engagement with said shroud member.

5. The container of claim 4, further comprising,
   a first handle means formed in said container body member at a first end thereof,

Figure 1 (cont.)

4,557,395

7

a second handle means formed in said container body member at a second end thereof which is longitudinally spaced from said first end,

each of said first and second handle means defined by a concavity formed in the bottom wall of said container body member and by a convexity contiguous thereto and continuous therewith, said convexity merging with an end wall of said container body member.

6. The container of claim 5, wherein the depth of the concavity forming a handle means is greater than the height of the convexity contiguous thereto.

7. The container of claim 5, wherein said first and second handle means are disposed transverse to and are bisected by the longitudinal axis of symmetry of said container body member.

8. The container of claim 3, wherein said cap member has a top wall having a diameter greater than the outer diameter of said fill spout, wherein an annular wall depends to the periphery of said cap top wall, wherein an annular cavity is defined between said fill spout and said depending wall, and wherein said latch member is specifically configured to enter into said annular cavity when brought into registration therewith.

9. The container of claim 8, wherein said latch member has a generally L-shaped configuration.

10. The container of claim 3, wherein said fill spout, said vent and said funnel member, latch member and vent closer member are collectively aligned with the longitudinal axis of symmetry of said container body member when said funnel member is inverted, when said vent closure member is disposed in engaging relation to said vent, and when said latch member is disposed in engaging relation to said fill spout cap.

11. The container of claim 1, wherein a strainer means is positioned within said funnel member at the juncture of said converging sidewalls and said downspout.

12. The container of claim 1, wherein said funnel member has a generally rectangular configuration when seen in plan view, and wherein said latch member and vent closure member are disposed mid-length of the opposite truncate sidewalls of said funnel member.

13. The container of claim 1, wherein said fill spout is externally threaded and wherein said funnel member downspout is internally threaded.

8

14. The container of claim 1, further comprising, an elongate funnel downspout extension member having a first end adapted to releasably engage said funnel downspout and a second end adapted to releasably engage said fill spout, and said downspout extension member having a flexible medial portion.

15. The container of claim 14, further comprising, a rigid sleeve-shaped locking member, having a length greater than the length of said flexible medial portion and having an inside diameter slightly greater than the outside diameter of said downspout extension member, disposed in ensleeving relation to said flexible medial portion and restricting said downspout extension member from flexing at said medial portion.

16. The container of claim 15, further comprising, means for selectively locking and unlocking said sleeve member into and out of its restricting engagement with said medial portion, respectively.

17. The container of claim 16, wherein said means for selectively locking and unlocking said sleeve member includes a pair of vertically spaced bead members formed on said downspout extension member, one of which is positioned above said flexible medial portion and one of which is positioned below said flexible medial portion, and wherein said sleeve member has a pair of cooperatively spaced bead-receiving cavities formed therein, which cavities are interconnected by a vertical slot and which cavities are formed at the end of associated channels orthogonal to said vertical slot.

18. The container of claim 13, wherein said funnel member downspout further comprises a cylindrical outer wall within which said internal threads are formed, and a cylindrical inner wall spaced radially inwardly of said outer wall so that dirty oil contacts only said inner wall when the container is used.

19. The container of claim 18, wherein said downspout inner wall is concentric with said downspout outer wall.

20. The container of claim 19, wherein the spacing between said downspout outer and inner walls is sufficient to receive therebetween said externally threaded fill spout.

* * * * *

Figure 1 (cont.)

United States Patent

Des. 237,427
Patented Nov. 4, 1975

237,427

SHOE

William H. Thornberry, Newtown, Conn., assignor to
Uniroyal, Inc.

Filed July 26, 1974, Ser. No. 492,307

Term of patent 14 years

Int. Cl. D2—*04*

U.S. Cl. D2—310

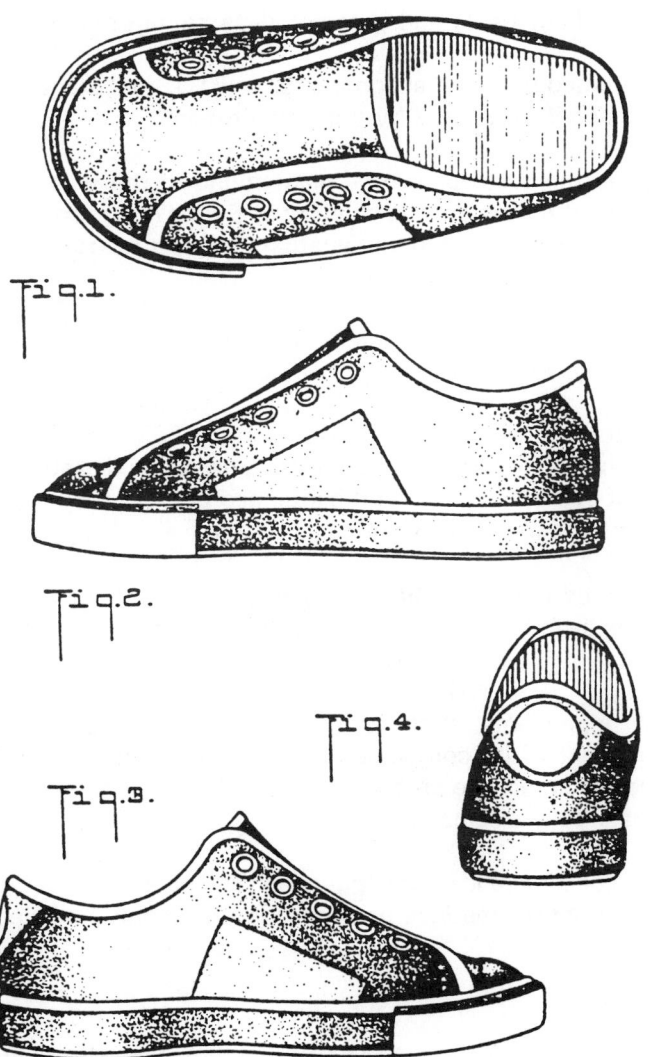

Fig.1.

Fig.2.

Fig.4.

Fig.3.

FIG. 1 is a plan view of a shoe embodying my new design;

FIG. 2 is a side elevational view of the FIG. 1 article;

FIG. 3 is a side elevational view of the FIG. 1 article; and

FIG. 4 is an end elevational view of the FIG. 1 article.

I claim:

The ornamental design for a shoe, substantially as shown and described.

References Cited

UNITED STATES PATENTS

D. 118,131	12/1939	Pick	D2—313
D. 173,699	12/1954	Hosker	D2—310
D. 226,461	3/1973	Nelson	D2—309

LOIS S. LANIER, Primary Examiner

Figure 2

In a 1980 Supreme Court decision the Court indicated that the PTO could not refuse a patent simply because it covered living subject matter. The PTO subsequently announced on April 17, 1987, that it would consider patents on animals after an agency appellate board ruled that oysters are patentable subject matter. This patent, however, did not issue. But it wasn't long until one did.

On April 12, 1988, patent history was made when the PTO issued the first patent covering an animal, specifically genetically engineered mice for cancer research.You might also find it interesting that each week an increasing number of patents are issued on computer software. If software design is your specialty, you may want to consider applying for a patent. You can apply for it simultaneously with copyright protection.

What Cannot Be Patented?

Many things are not open to patent protection.

The laws of nature, physical phenomena, and abstract ideas are not patentable subject matter.

Human beings cannot be patented because, according to the PTO, the "grant of a limited, but exclusive property right in a human being is prohibited by the Constitution."

A new mineral discovered in the earth or a new plant found in the wild is not patentable subject matter. Likewise, Einstein could not patent his celebrated $E = mc2$; nor could Newton have patented the law of gravity. Such discoveries are manifestations of nature, free to all men and reserved exclusively for none.

The Atomic Energy Act of 1954 excludes the patenting of inventions useful solely in the utilization of special nuclear material or atomic energy for atomic weapons.

The patent law specifies that the subject matter must be "useful." The term "useful" in this connection refers to the condition that the subject matter has a useful purpose and is able to function as intended, i.e. a machine which will not operate to perform the intended purpose would not be called useful, and therefore would not be granted patent protection.

Interpretations of the statute by the courts have defined the limits of the field of subject matter that can be patented. Thus it has been held that printed matter and methods of doing business cannot be patented.

In the case of mixtures of ingredients, such as medicines, a patent cannot be granted unless the mixture is more than the effect of its components. It is of interest to note that so-called "patent medicines" generally are not patented; the phrase "patent medicine" in this connection means only that the ingredients have not been fully disclosed.

A patent cannot be obtained on a mere idea or suggestion. The patent is granted for the new machine, manufacture, etc., itself, and not for the idea or suggestion for a new machine.

Do Patents Matter?

You bet they do! Just ask 64-year-old Robert W. Kearns, inventor of the intermittent windshield wipers now standard equipment on millions of vehicles worldwide. In a recent settlement, Ford Motor Company agreed to pay 10.2 million for infringing on Kearns' patent. Kearns is using the proceeds from this win to move ahead and sue Chrysler Corp., Honda Motor Co., Nissan Motor Co., and other automakers for similar infringement.

Jerome Lemelson of Princeton, New Jersey, a man who claims to have been awarded more than 400 patents, took Mattel Toys to court for infringing his 1969 patent for a flexible plastic track. The self-

described "tinkerer" claims he was financially damaged by sales of the track on which Mattel's popular Hot Wheels cars run. Between 1972 and 1986, Mattel reportedly sold $545 million worth of Hot Wheels, six percent of which (32.7 million) was for the sale of track.

On November 7, 1989, a federal court in Chicago awarded Lemelson $24.8 million in damages. If the damages are tripled and assessed interest, the final penalty could earn the inventor more than $100 million.

Expected to go to trial during the autumn of 1991 is the suit brought by 73-year-old inventor Charles Petrosky against the famous shoemaker Nike. Petrosky claims that Nike swiped his invention for air conditioned soles when it introduced the Air Jordan athletic shoe.

"There's no doubt Nike knew about Mr. Petrosky and his idea for an air cushion sole before the introduction of the Air Max and Air Jordan," Joel M. Freed, Petrosky's attorney, told *The Washington Post*. Freed said his client patented an air cushion sole in 1978 and had attempted to interest Nike in his innovation long before the company marketed the Air Jordan in the 1980s.

Independent inventors are not the only ones to rely on patents to protect their innovations; corporations do, too. Since 1986, lawyers representing Texas Instruments Inc. have squeezed more than $650 million out of competitors who marketed products that infringed its patents.

In October, 1990 a federal judge ordered Eastman Kodak Co. to pay Polaroid Corp. $909 million, the highest patent infringement award ever issued. This was for infringements of Polaroid's patents on its instant cameras.

As of this writing, a major fight is underway between Apple Computer, Inc. and Microsoft Corp. over rights to certain graphic features that make personal computers easier to operate and are a critical element in the aura around the MacIntosh computer.

Even smaller companies aggressively enforce the rights given to them through patents. In September, 1990, a federal court jury found that Advanced Micro Devices Inc. had copied a microchip design owned by a much smaller San Diego firm. AMD has been ordered to cough up $26 million. It plans to appeal.

It is wise to proceed under the assumption that patents are important. In fact, few companies will license an idea unless it has been or can be protected by patent.

"If there is no patent, what are we licensing?" is a standard query when an inventor has not made application for patent protection.

According to *The Washington Post*, in the past decade, the number of intellectual property suits filed in federal courts has quadrupled, far outpacing the overall growth of litigation. Congress, in the same time frame, passed over a dozen laws to strengthen intellectual property.

Whether it's Walt Disney Company protecting Snow White or Xerox guarding the proprietary copier, almost every week a property rights case is brought to court. Chief Judge Howard T. Markey of the U.S. Court of Appeals for Washington, DC Federal Circuit, a patent's court of last resort, is upholding patents 80 percent of the time.

Typically, the stronger the patent protection, the better the contract you will be able to negotiate. This is because a meaningful patent should keep any competition from manufacturing, using or selling your idea for at least a limited time. If nothing else, your manufacturer will have a head start.

Japan's Sumitomo Electric Industries was closed down by Corning Glass Works in 1987 when Corning won a patent infringement suit over its design of optical fibers.

As Harry F. Manbeck, Jr., Assistant Secretary of Commerce and Commissioner of Patents and Trademarks, impresses upon new graduates of the Patent Academy, "The only way our innovators can maintain their technological lead is to give them a strong, high quality patent system. That's our job!"

Patent Laws

The U.S. Constitution gives Congress the power to enact laws relating to patents, in Article I, section 8, which reads, "Congress shall have power...to promote the progress of science and useful arts, by securing for limited times to authors and inventors the exclusive right to their respective writings and discoveries." Under this power Congress has from time to time enacted various laws relating to patents. The first patent law was enacted in 1790. The current law is a general revision which was enacted July 19, 1952, and which went into effect January 1, 1953. It is codified in Title 35, United States Code.

The patent law specifies the subject matter for which a patent may be obtained and the conditions of patentability. The law establishes the Patent and Trademark Office (PTO) for administering the law relating to the granting of patents, and contains various other provisions relating to patents.

The Patent and Trademark Office

"The Patent Office is a curious record of the fertility of the mind of man when left to its own resources..." wrote an English lady who visited America in 1828.

The Patent and Trademark Office (PTO) administers the patent laws as they relate to the granting of patents for inventions, and performs other duties vis-a-vis patents. It examines applications for patents to determine if the applicants are entitled to patents under the law, and grants patents when they are so entitled; it publishes issued patents; records assignments of patents; maintains a search room for the use of the public to examine issued patents; and supplies copies of records and other papers. **The PTO has no jurisdiction over questions of infringement and the enforcement of patents, nor over matters relating to the promotion or utilization of patents or inventions.**

Today the PTO has 3,898 employees. More than 164,000 patent applications were received in fiscal year 1991, and next year the PTO estimates it will see approximately 180,000 applications.

More than 4.5 million patents have been awarded since the Office was established in 1802. In fiscal year 1990, 96,727 patents were issued.

According to the PTO, patents issued to individuals in the U.S. rose 32 percent in the past five years, from 13,000 in 1986 to 18,300 in 1990.

PTO Office Hours: The PTO's working hours are 8:30 a.m. to 5:00 p.m., Monday through Friday, excluding federal holidays in the District of Columbia.

Terms of U.S. Patents

The term of a **utility patent** is 17 years from the date of issue, subject to the payment of maintenance fees. The right conferred by a utility patent grant extends throughout the U.S. and its territories and possessions.

The term of a **design patent** is 14 years from the date of issue, and is not subject to the payment of maintenance fees. The right conferred by a design patent grant extends throughout the U.S. and its territories.

The term of a **plant patent** is 17 years and is not subject to maintenance fees.

Cost of a U.S. Patent

See Appendix D, PTO Fee Schedule, on page 961 for the current fees effective as of December 16, 1991. This schedule notwithstanding, it is wise to call the PTO or your nearest Patent and Trademark Depository Library and double check that the fees have not changed at the time you make your application. If you send in the wrong fee, it can cause all kinds of bureaucratic headaches, not to mention penalty charges.

The least you'll pay to file a patent application is the filing fee, and if a patent issues, an issue fee. Once the patent issues you'll have a 17-year window of protection as long as you pay the required maintenance fees. While the fees may change, the payment cycles will probably remain as follows: Payment must be made between three and three and a half years after the date of issue; seven and seven and a half years after date of issues; and eleven and eleven and a half years after date of issue. At this time, you can figure about $3,340 to maintain a patent over its lifetime.

Who May Apply for a U.S. Patent?

According to the law, only the inventor may apply for a patent, with certain exceptions. If you are a co-inventor, you may make a joint application. Financial contributors are not considered joint inventors, and cannot be joined in the application as an inventor. If you make an innocent mistake of erroneously omitting or naming an inventor, the application can be corrected. If a partisan who is not the inventor should apply for a patent, the patent, if awarded, would be invalid. Any person who falsely applies as the inventor is subject to criminal penalties.

If the inventor is dead, the application may be made by legal representatives, i.e. the administrator or executor of the estate. If the inventor is not mentally competent, the application for patent may be made by a guardian. If an inventor refuses to apply for a patent or cannot be found, a joint inventor or a person having a proprietary interest in the invention may apply on behalf of the missing inventor.

Officers and employees of the PTO are prohibited by law from applying for a patent or acquiring, directly or indirectly, except by inheritance or bequest, any patent or any right or interest in any patent.

Can Foreigners Apply For U.S. Patents?

Yes. The patent laws of the United States do not discriminate with respect to the citizenship of the inventor. Any inventor, regardless of citizenship, may apply for a patent on the same basis as a U.S. citizen. In fact, the number of foreigners obtaining U.S. patents continues to increase.

"The world is waking up as far as the value of intellectual property is concerned, and ours is the richest of markets," commented PTO Commissioner Donald J. Quigg, referring to the growth of foreign patenting in the United States.

In 1987, foreign inventors were issued 48 percent of the U.S. utility patents, compared with 37 percent in 1976. In 1987, to no one's surprise, Japan had five companies among the top 12 corporations receiving U.S. utility patents.

See Appendix F, Top 200 Corporations Receiving U.S. Patents in 1990, on page 977 for more information.

Conditions for Obtaining a Patent

In order for an invention to be patentable it must be new as defined in the patent law, which provides that an invention **cannot** be patented if:

"a) The invention was known or used by others in this country, or patented or described in a printed publication in this or a foreign country, before the invention thereof by the applicant for a patent; or

"b) The invention was patented or described in printed publication in this or a foreign country or in a public use or on sale in this country more than one year prior to the application for patent in the United States..."

Even if your exact idea is not shown by the prior art, i.e., patents on file, and involves one or more differences over the most nearly similar thing already known, a patent still may be refused if the differences wouldn't be obvious. The subject matter you seek to patent must be sufficiently different from what has been used or described before. For example, the substitution of one material for another, or changes in size, are ordinarily not patentable.

Care must be taken not to permit any public disclosure of a patentable idea unless you are prepared to make the application within one year from the date of disclosure. In other words, exhibiting a new invention at an invention exposition or new product fair, writing a magazine piece about a new idea, or permitting someone else to report in print on your idea will make it ineligible for patent protection if a patent application is not filed within one year from the time of its first public exposure.

Inventors should be very careful, therefore, about disclosing their inventions prior to making patent application. Many inventors who do not have the money to patent their products think that by exposing them at expos and fairs, and in the mass media, investors and/or licensees might see them, make an offer, and cover the patent with advance monies. Maybe so. But if you take this dangerous route before making a patent application, be aware that the clock starts ticking when you go public. On top of this, you have exposed an unprotected idea to the world, a dangerous and questionable move under any circumstances. Knock-off artists frequent such events.

While I support public events such as expos and fairs as a method to promote and publicize inventions, I do not recommend them as a place to display unprotected products that you plan to patent one day.

Models are Not Generally Required

Models of inventions were once required as part of invention applications, and these models became part of the patent record. Such models are no longer generally required (the description of the invention in the specifications, and the drawings, must be sufficiently full and complete, and capable of being understood, to disclose the invention without aid of a model), and will not be accepted unless specifically called for by the examiner.

When the invention relates to a composition of matter, the applicant may be asked to furnish specimens of the composition, or of its ingredients or intermediates, for the purpose of inspection or experiments.

Patent Application Time

The time to process a patent application, from the time the application is filed up until the time of issue or abandonment, is called the **patent pendency time**. In fiscal year 1990, the average patent pendency time for utility, plant, and reissue patents was 18.3 months; in fiscal year 1990, the average patent pendency time for design patents was 31.2 months.

The process may go more rapidly if you qualify for a Special Status Patent.

What are Special Status Patents?

The PTO allows special cases to qualify for accelerated processing. No special forms are required. All you need to do if you qualify is attach a letter of explanation to your application.

Certain pre-set categories automatically permit special status:

1) **Age (65).** If you are 65 years of age or older, your application may be made special by sending along a copy of your birth certificate. No fee is required with such a petition.

2) **Illness (Terminal).** If the state of health of the applicant is such that he or she might not be available to assist in the normal prosecution of the application, the application can be made special by sending along a physician's letter attesting to that state of health. No fee is required with such a petition.

3) **Infringer (Actual).** Under certain conditions, an application may be made special because of actual infringement (but not for prospective infringement) upon payment of a fee and the filing of a petition alleging facts under oath or declaration to show, or indicating why it is not possible to show (1) that there is an infringing device or product actually on the market or method in use; (2) when the device, product, or method alleged to infringe was first discovered to exist; supplemented by an affidavit or declaration from your attorney or agent to show; (3) that a rigid comparison of the alleged infringing device, product, or method with the claims of the application has been made; (4) that, in the applicant's opinion, some of the claims are unquestionably infringed; (5) that the applicant has made or caused to be made a careful and thorough search of the prior art or has good knowledge of the salient prior art; and, (6) that the applicant believes all of the claims in their application are allowable.

Models or specimens of the infringing product or your product should not be submitted unless requested.

4) **Manufacture.** If you have a manufacturer ready to go, this also may qualify your application for special handling. I have requested and received this on numerous occasions. You pay a fee and allege under oath or declaration that: (a) the licensee has the capital available (indicate the approximate amount) and the facilities (briefly describe) to manufacture your invention in quantity or that sufficient capital and facilities will be made available if a patent is granted; (b) the prospective licensee will not manufacture, or will not increase present manufacture, unless certain that the patent will be granted; and (c) you and/or your licensee stand obligated to produce the product immediately if the patent issues which protects the investment of capital and facilities.

Furthermore, you or your attorney must file an affidavit or declaration to show that: (1) you have searched or caused to be searched the prior art; and (2) you believe all the claims in your application to be allowable.

5) **Energy.** No fee is required to make special an application for an invention that will materially contribute to: (1) the discovery or development of energy resources; or (2) the more efficient use and conservation of energy resources. Examples of inventions in the first category would be developments in fossil fuels (natural gas, coal, and petroleum), nuclear energy, and solar energy. Examples in the second category would include inventions relating to the reduction of energy consumption in combustion systems, industrial equipment, household appliances, etc.

6) **Superconductors.** In accordance with the President's proposal directing the   PTO to accelerate the processing of patent applications and adjudication of disputes involving superconductivity technologies when requested by the applicant to do so, the PTO will comply accordingly. No fee is required.

Examples of such inventions would include those directed to the superconductive materials themselves as well as to their manufacture and application.

7) **Environment.** If the invention will materially enhance the quality of the environment by contributing to the restoration or maintenance of the basic life-sustaining natural elements—air, water, and soil—you can request to make it special. No fee is required for such a petition.

8) **DNA.** In recent years revolutionary genetic research has been conducted involving recombinant deoxyribonucleic acid (recombinant DNA). Recombinant DNA appears to have extraordinary potential benefit for humans. It has been suggested, for example, that research in this field may lead to ways of controlling or treating cancer and hereditary defects.

A petition to make special in this category must contain an affidavit or declaration in writing to the effect that your invention is directly related to safety of research in the field of recombinant DNA. A fee must be paid.

General PTO Information and Correspondence

The Patent and Trademark Office (PTO) is a huge bureaucracy. Without a map and compass, so to speak, it can be frustrating or near impossible for a beginner (especially one living outside the National Capital area), to find out who handles what and how to directly contact any particular person. **To help you quickly reach the most appropriate person for your needs, see Appendix B, Patent and Trademark Office Telephone Directory, on page 943.**

If you wish to mail a letter to someone at the Patent and Trademark Office, address your letter to the addressee care of: Patent and Trademark Office, Washington, DC 20231.

Only Express Mail sent to a specific street address and office will arrive. The PTO cannot receive express mail nor can it pick up Post Office-to-Post Office Express Mail.

Special PTO mail box numbers should be used to allow forwarding of particular types of mail to the appropriate areas as quickly as possible. Mail will be forwarded directly to the appropriate area without being opened. If a document other than the specified type identified for each box is addressed to that box, expect delays in correct delivery.

Special PTO Post Office Boxes

Commissioner of Patents and Trademarks, Washington, DC 20231

Box 3: Mail for the Office of Personnel from NFC.

Box 4: Mail for the Assistant Commissioner for External Affairs, or the Office of Legislation and International Inquiries.

Box 5: Non-fee mail related to trademarks.

Box 6: Mail for the Office of Procurement.

Box 7: Reissue applications for patents involved in litigation.

Box 8: All papers for the Office of Solicitor except letters relating to pending litigation.

Box 9: Coupon orders for U.S. patents and trademarks.

Box 10: Orders for certified copies of patents and trademark applications.

Box 11: Electronic Ordering Services (EOS).

Box AF: Expedited procedure for processing amendments and other responses after final rejection.

Box FWC: Requests for File Wrapper Continuation Applications.

Box DAC: Petition to revive abandoned patents.

Box Issue Fee: Issue Fee Transmittals and associated fees and corrected drawings.

Box M Fee: Letters related to interferences and applications and patents involved in interference.

Box Non Fee: Non-fee amendments to patent applications.

Amendment

Box Pat. Ext.: Applications for patent term extensions.

Box PCT: Mail related to applications filed under the Patent Cooperation Treaty.

Box reexam: Mail related to re-examination.

Patent

Application: New patent applications and associated papers and fees.

Trademark

Application: New trademark applications and associated papers and fees.

Separate letters (but not necessarily in separate envelopes) should be written in relation to each distinct subject of inquiry, such as assignments, payments, orders for printed copies of patents, orders for copies of records, and requests for other services. None of these should be included with letters responding to Office actions in applications.

If your letter concerns a patent application, include the serial number, filing date, and Group Art Unit number. When a letter concerns a patent, include the name of the patentee, the title of the invention, the patent number, and date of issue.

If ordering a copy of an assignment, provide the book and page or microfiche reel and frame record, as well as the name of the inventor; otherwise, the PTO will assess an additional charge to cover the time consumed in making the search for the assignment.

The PTO will not send or show you applications for patents. They are not open to the public, and no information concerning patent applications is released except on written authority of the applicant, assignee, or designated attorney, or when necessary to the conduct of the PTO's business. You can write for and receive, however, records of any decisions, the records of assignments other than those relating to assignment of patent applications; books; or other records and papers in the PTO that are open to the public.

The PTO will not respond to inquiries concerning the novelty and patentability of an invention before an application is filed; give advice as to possible infringement of a patent; advise of the propriety of filing an application; respond to inquiries as to whether or to whom any alleged invention has been

patented; or act as an expounder of the patent law or as a counselor for individuals, except in deciding questions arising before it in regularly filed cases.

Keep copies of everything you send the PTO. Don't forget to put your name and return address on all papers and envelopes.

In the event of any major interruption in the country's mail service, the PTO has established a contingency plan for filing any paper or paying any fee in the Office.

If the Commissioner of Patents and Trademarks determines that such an emergency exists, the Commissioner will cause to be printed a notice of the plan in *The Wall Street Journal* and make it available by telephone at (703) 557-3158. Also, certain publications, patent bar groups, and other organizations closely associated with the patent system will be notified. Termination of the emergency program will be similarly announced. Where the postal emergency is not nationwide, the Commissioner will designate the areas of the country in which the procedures outlined will be in effect.

The plan calls for the U.S. Department of Commerce District Offices to be designated as emergency receiving stations for filing papers and paying fees in the PTO.

PTO Buildings

In speaking with PTO officials, you'll find that they'll often refer to buildings by acronyms (e.g. "His office is at CP34.") The designations for all PTO offices are listed below to help you decipher the building codes. Every building except HCHB and Newington is located in Arlington, VA.

Important: These acronyms are for reference use only. Do not send your PTO correspondence to these addresses or it will be delayed by the U.S. Postal Service. All PTO correspondence should be sent to Washington, DC 20231.

CM1	Crystal Mall 1, 1911 Jefferson Davis Hwy.
CM2	Crystal Mall 2, 1921 Jefferson Davis Hwy.
CP1	Crystal Plaza 1, 2001 Jefferson Davis Hwy.
CP2	Crystal Plaza 2, 2011 Jefferson Davis Hwy.
CP3	Crystal Plaza 3, 2021 Jefferson Davis Hwy.
CP4	Crystal Plaza 4, 2201 Jefferson Davis Hwy.
CP6	Crystal Plaza 6, 2221 Jefferson Davis Hwy.
CP34	Crystal Plaza 34, 2021 Jefferson Davis Hwy.
CS5	Crystal Square 5, 1755 Jefferson Davis wy.
CG2	Crystal Gate 2, 1225 Jefferson Davis Hwy.
PK1	Crystal Park 1, 2011 Crystal Dr.
PK2	Crystal Park 2, 2121 Crystal Dr.
ST	South Tower, 2809 Jefferson Davis Hwy.

NT	North Tower, 2805 Jefferson Davis Hwy.
EADS	1232 S. Eads St.
FERN	1411 S. Fern St.
HCHB	Herbert C. Hoover Bldg., 14th and Constitution Ave. NW, Washington, DC
NEWINGTON	7401 Fullerton Rd., Springfield, VA

PTO Disclosure Document Program

If you are not ready or don't care to apply for a patent yet but want to officially evidence and register the conception date of an invention, the PTO offers its **Disclosure Document Program**. For a fee of $6, the Office will preserve your idea on file for a period of two years. This inexpensive recognition will strengthen your case if any conflict arises as to the date of your conception, but is not meant as a replacement of an inventor's notebook or actual patent.

The requirements are simple. Send the PTO a paper disclosing the invention. Although there are no stipulations as to content, and claims are not required, the benefits afforded by the Disclosure Document Program will depend directly upon the adequacy of your disclosure. Therefore, it is strongly recommended that the document contain a clear and complete explanation of the manner and process of making and using the invention in sufficient detail to enable a person having ordinary knowledge in the field of the invention to make and use the invention. When the nature of the invention permits, a drawing or sketch should be included. The use or utility of the invention should be described, especially in chemical inventions.

All disclosure must be on paper having dimensions or being folded to dimensions not to exceed 8.5 x 13 inches (21.6 by 33 cm). Do not submit written matter or drawings on media such as linen or plastic drafting material. Photographs are also acceptable. Number each page, and make sure that the text and drawings are of such quality as to permit reproduction.

In addition to the $6 fee, the Disclosure Document must be accompanied by a stamped, self-addressed envelope (SASE) and a duplicate request, signed by you as the inventor. These papers will be stamped by the PTO upon receipt, and the duplicate request will be returned to you in the SASE together with a notice indicating that the Disclosure Document may be relied upon only as evidence that a patent application should be diligently filed if patent protection is desired.

Your request may take the following form:

"The undersigned, being the inventor of the disclosed invention, requests that the enclosed papers be accepted under the Disclosure Document Program, and that they be preserved for a period of two years."

Warning: The two-year retention period should not be considered to be a "grace period" during which you can wait to file a patent application without possible loss of benefits. It must be recognized that in establishing priority of invention an affidavit or testimony referring to the Disclosure Document must usually also establish diligence in completing the invention or in filing the patent application since the filing of the Disclosure Document.

Also be reminded that any public use or sale in the U.S. or publication of the invention anywhere in the world more than one year prior to the filing of your patent application on the invention disclosed will prohibit the granting of a patent on it.

The Disclosure Document is not a patent application, and the date of its receipt in the Patent and Trademark Office will not become the effective filing date of any patent application subsequently filed. It will be retained for two years and then be destroyed unless referred to in a separate letter in a related application within two years.

The program does not diminish the value of the conventional witnessed and notarized records as evidence of conception of your invention, but it should provide a more credible form of evidence than that provided by the popular practice of mailing a disclosure to oneself or another person by registered mail.

Section 4

How to Conduct a Patent Search

"If you hear hoofbeats, expect a horse."—Anonymous

Why Make a Patent Search?

Before making an application for patent protection, it is advisable to see whether or not any prior art exists, i.e. if the same concept has been patented by someone else. This is done through a **patent search**.

A thorough search is recommended for numerous reasons. First, making a search involves less expense than trying to obtain a patent and having it rejected on the basis of prior art. If the search reveals that the invention cannot be protected as engineered, the cost of preparing and filing an application, as well as significant time and energy, will be saved.

Further, even if none of the earlier patents show all the details of the invention, they may point out important features or better ways of doing the job. If this is the case, you may not want to get patent protection on an invention that could encounter commercial difficulty.

If nothing is found in the search that would prevent or delay application, the information gathered during the search will prove helpful, acquainting you with the details of patents related to the invention.

In addition, you may wish to search a patent file to:

1) learn about existing U.S. patents to avoid possible infringement actions;
2) know the state of the art and monitor development in a specific technology in order to be aware of the latest development and to update one's knowledge or to locate relevant documents;
3) assessing novelty and patentability of own developments with a view of applying for a U.S. patent or foreign industrial property right;
4) judge possible actions, such as opposition proceedings concerning the validity of existing U.S. patents;
5) evaluate a specific technology and identify possible licensors;
6) identify alternate technology and its sources;

27

7) locate sources of know-how in a specific field;

8) improve an existing product or process;

9) develop new products or processes;

10) judge an alleged innovation whether it qualifies for development, production, or financial assistance such as venture capital;

11) solve a specific technical problem;

12) assess a particular technical approach whether it had not been tried before and might be worth pursuing or it would lead to wasteful duplication of research effort;

13) monitor the activities of competitors both within the country and abroad;

14) survey the market in order to identify a gap or to discover new trends or entrepreneurial opportunities at an early stage, and to forecast major changes in both competitors and technology; and

15) supplement school curricula and aid teachers involved in programs to stimulate creativity in young children such as Invent America! and the Patent and Trademark Office's Project XL.

Who Conducts a Patent Search?

There are three ways to approach the search. You can have a patent attorney do it for you; you can engage the services of a professional patent search firm or individual; or you can conduct the search yourself.

Law Firm Initiated Search

Nothing is wrong with having a law firm handle the search. Be aware that lawyers typically do not conduct the searches themselves; instead they hire a search firm or individual and mark-up the fees they are charged. Law firms may step on search fees anywhere from 40 to 100 percent or more depending upon the firm and city.

If you decide to retain a lawyer for this task, remember to insist upon an estimate of the costs in advance. The fee will be based upon how far back and encompassing you want the search.

See Section 27, Registered Patent Attorneys and Agents, on page 345 for information on how to hire a patent lawyer or agent.

Direct Hire Professional Search

If you want to save yourself the lawyer's mark-up, consider hiring a patent searcher. Typically, they are listed in the Yellow Pages under "Patent Searchers." Be careful to contact a firm or person who specializes in this work and not an invention marketing organization listed under a toll-free number. Many such firms use the search as a hook to sucker unsuspecting inventors into their rapacious grasp. Get all the facts before committing yourself. **Always ask for an estimate of charges in advance.**

In Washington, DC, searchers currently charge between $30.00 and $60.00 per hour. This range depends a great deal upon whether it is an independent or large firm. A simple search can run between six to seven hours for a mechanical patent. Chemical and electrical patents average eight to ten hours. Fixed price deals for $180 are available, providing three hours and six references.

There will be a fee for making copies as well. Some searchers mark this up while some go at cost. I've received estimates from 40 cents per page to $3.00 per patent (no matter how many pages).

If you live in a city with one of the PTO's 70 Patent Depository Libraries (PTDLs), a librarian can usually help you find a list of those who make a living searching patents. Librarians are not encouraged to personally recommend any patent searcher or search organization.

See Section 26, Patent and Trademark Depository Libraries, on page 337 for more information.

The better the lawyer and/or searcher understands the invention, the better the search. Highlight the novel features of your invention. This explanation can be made through drawings or sketches, models, written description, a discussion, or a combination of these. Take the time; it'll be worth it.

Correspondence concerning the search should be kept in a safe place, as you may need it to prove dates and other facts about the invention later.

Do-it-Yourself Search

You may opt to do the patent search yourself. Several methods are available.

1) The PTO runs a Patent Public Search Room located in Arlington, Virginia. Here, every U.S. patent granted since 1836 (over four million) may be searched and examined. Patents are arranged according to a classification system of more than 400 classes and 115,000 subclasses. **See Appendix A, U.S. Patent Classifications, on page 693 for a complete list.**

The Patent Public Search Room is really something to behold. You can touch and feel original documents, including everything from Abraham Lincoln's 1849 patent (# 6,469) for a device to buoy vessels over shoals to Auguste Bartholdi's design patent on a statue entitled "Liberty Enlightening The World" (a/k/a The Statue of Liberty).

For more information on important historical patents see Appendix E, Significant Inventions and Inventors in U.S. History on page 967.

For more information about the PTO's Patent Public Search Room, call (703) 557-4357.

Many inventors like to make at least one pilgrimage to the PTO's Crystal Plaza facility. It is located less than five minutes from National Airport by taxi and Metro Rail (Blue and Yellow Lines - Crystal City Station). Many fine hotels are within walking distance.

When you arrive at the facility the PTO will issue you, at no cost, a non-transferable User Pass for the day. It is wise to double check the hours of operation by calling (703) 308-0595. Depending where in the facility you want to search, the Patent Public Search Room is typically open from 8 a.m. to 8 p.m. When you call for information, tell the PTO staff member what specifically you want to search. The staff member will be able to tell you the hours that particular search room is open.

2) Near the Patent Public Search Room is **The Scientific Library of the Patent and Trademark Office.** The Scientific Library makes publicly available over 120,000 volumes of scientific and technical books in various languages, about 90,000 bound volumes of periodicals devoted to science and technology, the official journals of 77 foreign patent organizations, and over 12 million foreign patents. The hours are from 8:45 a.m. to 4:45 p.m.

3) If you cannot make the trip to Washington, DC, you may inspect copies of the patents at a **Patent and Trademark Depository Library (PTDL).** The Patent and Trademark Depository Library System, a nationwide network of 70 prestigious academic, research, and public libraries, continues to be one of the PTO's most effective mechanisms for publicly disseminating patent information. PTDLs receive current issues of U.S. patents and maintain collections of earlier issued patents. The scope of these collections varies from library to library, ranging from patents of only recent years to all or most of the patents issued since 1790. Due to the variations in the scope of patent collections among the PTDLs

and their hours of service to the public, you should call first to find out when it is open, to avoid any possible inconvenience.

How to Conduct a Patent Search

Whether you hire someone to conduct your patent search or do it yourself, certain steps must be taken. Here is a brief guide to manual searches for U.S. patents.

Step 1: If you know the PATENT NUMBER, go to the *Official Gazette* to read a summary of the patent. This publication is available at any of the above-mentioned patent search facilities and in many public library reference rooms.

Step 2: If you know the PATENTEE or ASSIGNEE, look at the *Patent/Assignee Index* to locate the PATENT NUMBER. This is available at any of the patent search facilities. In Crystal City, it is on microfiche and in card catalogues.

Step 3: If you know the SUBJECT, start with the *Index to the U.S. Patent Classification.* **The most recent Index is included in this book as Appendix A on page 693**. This will help you prepare for a search and familiarize you with the PTO's indexing system. You'll save time by doing some homework before you begin searching, i.e. familiarizing yourself with the *Index* and locating the class and subclass numbers for terms that pertain to your invention.

Step 4: Once you have jotted down the class(es) and subclass(es) out of the *Index*, refer to the *Manual of Classification* and check this information vis-a-vis the hierarchy to see if they are close to what you need. The *Manual of Classification* is available at all patent search facilities.

Step 5: Using the class/subclass numbers you have found, look at the *U.S. Patent Classification Subclass and Numeric Listing* and copy the patent numbers of patents assigned to the selected class/subclass. If you are at the Crystal City facility, take the class/subclass numbers into the stacks of patents and begin "pulling shoes". To pull shoes is to physically remove patent groupings from the open shelves.

Step 6: Next, using the *Official Gazette* again, look at the patent summaries for the patents you have pulled. At the Crystal City facility, you will not have to refer back to this publication since the actual patents are there.

Step 7: Upon locating the relevant patents, examine the complete patent in person or on microfilm, depending upon where you conduct the search.

If you do not locate the patent you are seeking, try again using another subject class and subclass.

Searching Patents Via Telefax

There are many patent search services that now offer same-day service via telefax. For a price, some will give service within the hour. These services have people on duty at the PTO's Patent Public Search Room in Crystal City, VA. Typically, they will make copies of the original patents and then rush them to a nearby fax machine.

This is not an inexpensive service. Faxpat of Arlington, VA, for example, charges the following rates. I include this specific data as a guideline and not an endorsement. I encourage you to compare prices from several services before engaging one.

U.S. Patents:

Sent same day: $.90/page, $6 minimum each

Sent next day: $6 each

Sent in 3-5 days: $4 each

Foreign Patents and Literature:

Same day, next day, etc.: $.90/page, $10 minimum each

Trademarks:

Sent same or next day: $6 each

Sent in 3-5 days: $4 each

U.S. Patent or Trademark Files: $.90/page, $35 minimum each

Cited references are processed as same-day patents unless instructed otherwise.

Files may be adhesive-bound with clear plastic covers, or hole-punched and metal clipped. However, files will be hole-punched unless adhesive binding is requested.

Faxing rate:

Document cost plus $1.50/page, $15.00 minimum each. Nonlocal taxes add $.26/page.

Online Patent Searches

The Automated Patent Searching System (APS) provides patent examiners and public searchers with a full-text search tool. The software used for APS text searching is called Messenger.

Currently, five APS text terminals are located in the Patent Public Search Room for public use.

APS Text Search gives you online access to two files:

1) USPAT: U.S. patents spanning from January 1971 to date, updated weekly; largest search patent file on APS approximately one million U.S. patents with more than 5 billion words.

2) JPOABS: Japanese patent abstracts file. English language of unexamined Japanese patent applications; data is supplied to the PTO by the Japanese Patent Office. Updated semiannually.

Public access to online APS Text Search is provided in the PTO's Patent Public Search Room, Monday-Friday between 8 a.m. and 5 p.m., except federal holidays when the room is closed.

Untrained public searchers can request an assisted while-you-wait search. An assisted search is conducted by helpful PTO staff for a fee.

APS text search training is offered to the public for free. There are two levels of training: novice and expert. The application for text search training is available at the Search Room. I would have included it in this book, but it is frequently changed. I recommend you call Patti Young, technical information

specialist at the Search Room, for details on how to sign up and when courses are scheduled. Her direct line is (703) 308-0595.

Costs:

APS Text Search connect time:	$40.00 per hour
Printed page:	$.10 per page
Assisted Search fee:	$25.00 per hour

You may pay with cash, a personal check, PTO or APS Deposit Accounts, or Master Card/Visa.

PTDLs Offering APS

13 of the 70 PTDLs have APS terminals. They are located in Auburn University, AL; Boston, MA; Cleveland, OH; Dallas, TX; Houston, TX; Jackson, MS; Los Angeles, CA; Madison, WI; Minneapolis, MN; Newark, NJ; Providence, RI; Raleigh, NC; and Salt Lake City, UT. **See Section 26, Patent and Trademark Depository Libraries, on page 337 for addresses and telephone numbers.**

CASSIS

A less sophisticated but still invaluable tool for searching patents is the Classification and Search Support Information System (CASSIS). It is available at Patent and Trademark Depository Libraries free of charge.

CASSIS is in CD-ROM format and allows keyword searching on three years of patent abstracts and on utility patent titles back to 1969. Plant and design patent titles are included back to 1977. Searches can be modified using date, assignee, and state or country information. CASSIS retrieves complete subclass listings for individual or merged subclasses, and displays current classifications of all U.S. patents issued since 1790. The *Manual of Classification* can also be searched by keyword or displayed by class/subclass.

CASSIS receives high marks from inventors. George L. Dwight of Pueblo, CO, writes, "Today I went to Denver and used your system. Man what a fine thing it is!"

Other electronic products available at PTDLs:

ASIST is a database on CD-ROM which has a number of files that complement CASSIS. The ASIST disc includes the *Index to U.S. Patent Classification*, the *IPC-USPC Concordance*, and the attorney roster information from the *Attorneys and Agents Registered to Practice Before the U.S. PTO*. Also included are the Patentee-Assignee File; the Classification Orders Index File, which is a comprehensive list of subclass numbers abolished and established since 1976 with corresponding Classification Order number and effective date; and the Reassignment File, which shows assignments at date-of-issue and subsequent reassignments after issue for certain dates.

Other CD-ROM products that have been distributed to the PTDLs on a pilot basis include trademark information (Trademarks), Patent Cooperation Treaty (PCT) data (PraCTis), Patent Image-Full Text for the Weekly Issue of December, 1988, and Patent Abstracts of Japan.

The CASSIS and ASIST CD-ROM products are available for purchase by the public. They may be ordered from the U.S. PTO's Office of Electronic Information Products and Services, Crystal Mall 2, Room 304, Washington, DC 20231.

More Help With Your Patent Search—If you Have the Time

If you want help after referring to the Index to the U.S. Patent Classification, you may write to the Commissioner of Patents and Trademarks for free assistance.

Sample Letter

Date:

Attn. Commissioner of Patents and Trademarks
Patent Search Division
Washington, DC 20231

Dear Commissioner:

Please let me know what subject area or class(es) and subclass(es) cover my idea. The enclosed sketches on the back of this sheet, with each part labeled, show my intended invention.

My idea has certain features of structure, mode of operation, and intended uses which I have defined below.

I understand that there is no charge for this information.

Thank you for your help.

Sincerely,

_____
(Signature)

(1) Features of structure, or how it is constructed.

(2) Mode of operation, or how it works.

(3) Intended uses, or purpose of idea.

(4) Rough sketches of idea, viewed from all sides, with labels to identify each part. (If necessary use extra sheets of paper either plain or lined.) These sketches may be made in pencil and need not be drawn to scale.

How to Order Copies of Searched Patents

You may order copies of original patents or cross-referenced patents contained in subclasses comprising the field of search from the Patent and Trademark Office. Mail your requests to: Patent and Trademark Office, Box 9, Washington, DC 20231.

Payment may be made by check, coupons, or money. Postage is free. Expect to wait up to four weeks for your copies.

For the convenience of attorneys, agents, and the general public in paying any due fees, deposit accounts may be established with the PTO with a minimum deposit of $50. For information on this service, call (703) 308-0902.

What Do You Do with Your Search Results?

Study the results of the patent search. You're out of luck if any of the patented inventions are exactly like yours; your invention may even be infringing on another invention. On the other hand, one or more patents may describe inventions that are intended for the same purpose as yours, but are different in various ways. Look these over and decide whether it is worthwhile to proceed.

If the features that make your invention different from the prior art provide important advantages, you should discuss the situation with your attorney to determine whether there is a fair chance of obtaining a patent covering these features.

I have found from experience that a good patent attorney can often get some claim(s) to issue, albeit not always strong ones. A patent for patent's sake is usually possible. Whether it will be worth the paper it is printed on is another matter. Do not take the decision to patent your idea lightly because the patent process is not inexpensive. The average utility patent will cost about $2,500.

Section 5

How to Apply for a Patent

"Success depends upon staying power. The reason for failure in most cases is lack of perseverance."—James Russell Miller

It is a good idea to log all of your correspondence in and out of your office. This will help you keep track of deadlines as well as give you a record of paper flow. Losing paperwork or missing deadlines can be costly and time consuming.

Your patent application must be directed to the Commissioner of Patents and Trademarks and must include:

1) A written document which comprises a specification (description and claims), and an oath or declaration. Two forms are used in association with patent applications: Power of Attorney or Authorization of Agent, Not Accompanying Application; and Revocation of Power of Attorney or Authorization of Patent Agent.

Power of Attorney Form: If you file your application through an attorney or agent, no power of attorney form is required (it will be written into the oath or declaration). If, however, you decide to appoint an attorney or agent after submitting your application, you may use **Form 1 in Part One: Forms and Tables.**

Revocation of Power of Attorney Form: If after submitting your application you decide to change attorney or agent use **Form 2 in Part One: Forms and Tables.**

2) A drawing in those cases in which a drawing is required; and

3) The correct filing fee.

The specification and oath or declaration must be legibly written or printed in permanent ink on one side of the paper only. The PTO prefers typewriting on legal size paper, 8 to 8.5 x 10.5 to 13 inches, 1.5 or double spaced, with margins of one inch on the left-hand side and at the top. If the filed papers are not correctly, legibly, or clearly written, the Office may require typewritten or printed papers.

The application for patent will not be accepted for examination until all of the required parts, complying with the rules of presentation, are received. If the papers and parts are incomplete or so defective they

35

cannot be accepted as a complete application for examination, you will be notified about the found deficiencies and given a time period in which to remedy them.

To help make sure everything is together, a copy of a **Patent Application Transmittal Letter is shown as Form 3 in Part One: Forms and Tables.** This will serve as your cover document.

The four formal responses used by the PTO Application Division to notify inventors of defects in their applications are reproduced as **Fig. 3, Notice of Informal Application, on page 37; Fig. 4, Notice of Incomplete Application, on page 38; Fig. 5, Notice to File Missing Parts of an Application—No Filing Date, on page 39; and Fig. 6, Notice to File Missing Parts of an Application—Filing Date Granted, on page 40.** It's a good idea to look them over because anyone who does patent work usually receives one at some point.

Upon receiving one of these notices, a surcharge may be required. If you do not respond within the prescribed time period, your application will be returned or discarded. The filing fee may be refunded if the application is refused as incomplete; however, you'll probably be charged a handling fee.

File all the application documents together; otherwise, each part must be signed and accompanied by a letter accurately and clearly connecting it with the other parts of your application. This can be a nightmare, so send everything together from the start and save yourself headaches.

Every application received by the PTO is numbered in serial order, and you are informed of the serial number and filing date of the application by a filing receipt. **The filing receipt does not mean that a patent has been awarded, only that your application is at the PTO.**

Keep copies of everything you submit. If you ask the PTO for copies of what you've submitted, they will charge you a service fee.

Patent applications are not open to the public, and no information concerning them is released except by written permission from the applicant, assignee, or designated attorney, or when the information is necessary to the conduct of PTO business.

Oath or Declaration

By law the inventor must file an oath or declaration. The inventor must make an oath or declare that he/she is believed to be the original and first inventor of the subject matter of the application. Additionally, the inventor must make various other allegations required by law and by PTO rules.

If you opt for an oath, it must be sworn before a notary public or other officer authorized to administer oaths.

I favor using the declaration in lieu of an oath, making it part of the original application. A declaration does not require notarization.

A sample blank application, which includes the declaration, is shown as Form 4 in Part One: Forms and Tables. The PTO does not supply blank forms.

Reference to Drawings

If you have drawings, the PTO requires a brief description of several views of the drawing and the detailed description of the invention shall refer to the different views by specifying the numbers of the figures and to the different parts by use of reference letters or numerals.

UNITED STATES DEPARTMENT OF COMMERCE
Patent and Trademark Office
Address: COMMISSIONER OF PATENTS AND TRADEMARKS
Washington, D.C. 20231

SERIAL NUMBER	FILING DATE	FIRST NAMED APPLICANT	ATTY. DOCKET NO.

DATE MAILED:

NOTICE OF INFORMAL APPLICATION
(Attachment to Office Action)

This application does not conform with the rules governing applications for the reason(s) checked below. The period within which to correct these requirements and avoid abandonment is set in the accompanying Office action.

A. A new oath or declaration, identifying this application by the serial number and filing date is required. The oath or declaration does not comply with 37 CFR 1.63 in that it:

1. ☐ was not executed in accordance with either 37 CFR 1.66 or 1.68.

2. ☐ does not identify the city and state or foreign country of residence of each inventor.

3. ☐ does not identify the citizenship of each inventor.

4. ☐ does not state whether the inventor is a sole or joint inventor.

5. ☐ does not state that the person making the oath or declaration:

    a. ☐ has reviewed and understands the contents of the specification, including the claims, as amended by any amendment specifically referred to in the oath or declaration.

    b. ☐ believes the named inventor or inventors to be the original and first inventor or inventors of the subject matter which is claimed and for which a patent is sought.

    c. ☐ acknowledges the duty to disclose information which is material to the examination of the application in accordance with 37 CFR 1.56(a).

6. ☐ does not identify the foreign application for patent or inventor's certificate on which priority is claimed pursuant to 37 CFR 1.55, and any foreign application having a filing date before that of the application on which priority is claimed, by specifying the application serial number, country, day, month, and year of its filing.

7. ☐ does not state that the person making the oath or declaration acknowledges the duty to disclose material information as defined in 37 CFR 1.56(a) which occurred between the filing date of the prior application and filing date of the continuation-in-part application which discloses and claims subject matter in addition to that disclosed in the prior application (37 CFR 1.63(d)).

8. ☐ does not include the date of execution.

9. ☐ does not use permanent ink, or its equivalent in quality, as required under 37 CFR 1.52(a) for the: ☐ signature ☐ oath/declaration.

10. ☐ contains non-initialed alterations (See 37 CFR 1.52(c) and 1.56).

11. ☐ does not contain the clause regarding "willful false statements..." as required by 37 CFR 1.68.

12. ☐ Other:

B. Applicant is required to provide:

1. ☐ A statement signed by applicant giving his or her complete name. A full name must include at least one given name without abbreviation as required by 37 CFR 1.41(a).

2. ☐ Proof of authority of the legal representative under 37 CFR 1.44.

3. ☐ An abstract in compliance with 37 CFR 1.72(b).

4. ☐ A statement signed by applicant giving his or her complete post office address (37 CFR 1.33(a)).

5. ☐ A copy of the specification written, typed, or printed in permanent ink, or its equivalent in quality as required by 37 CFR 1.52(a).

6. ☐ Other:

FORM PTO-152 (REV. 4-87)

Figure 3

UNITED STATES DEPARTMENT OF COMMERCE
Patent and Trademark Office
Address: COMMISSIONER OF PATENTS AND TRADEMARKS
Washington, D.C. 20231

SERIAL NUMBER	FILING DATE	FIRST NAMED APPLICANT	ATTY DOCKET NO

[   ]

DATE MAILED:

Notice of Incomplete Application

A filing date has NOT been assigned to the above identified application papers for the reason(s) shown below.

1. ☐ The specification (description and claims):
   a. ☐ is missing
   b. ☐ has pages _____ missing.
   c. ☐ does not include a written description of the invention.
   d. ☐ does not include at least one claim in compliance with 35 U.S.C. 112.

A complete specification in compliance with 35 U.S.C. 112 is required.

2. ☐ A drawing of Figure(s) _____ described in the specification is required in compliance with 35 U.S.C. 111.

3. ☐ A drawing of applicant's invention is required since it is necessary for the understanding of the subject matter of the invention in compliance with 35 U.S.C. 113.

4. ☐ The inventor's name(s) is missing. The full names of all inventors are required in compliance with 37 CFR 1.41.

5. ☐ Other items missing but not required for a filing date:

All of the above-noted omissions, unless otherwise indicated, must be submitted within TWO MONTHS of the date of this notice or the application will be returned or otherwise disposed of. Any fee which has been submitted will be refunded less a $50.00 handling fee. See 37 CFR 1.53(c).

The filing date will be the date of receipt of all the items required above, unless otherwise indicated. Any assertions that the items required above were submitted, or are not necessary for a filing date, must be by way of a petition directed to the attention of the Office of the Assistant Commissioner for Patents accompanied by the $140.00 petition fee (37 CFR 1.17(h)). If the petition alleges that no defect exists, a request for refund of the petition fee may be included in the petition.

Direct the response to, and questions about, this notice to the undersigned, Attention: Application Branch, and include the above Serial Number and Receipt Date.

Enclosed:
   ☐ "General Information Concerning Patents". See page _____.
   ☐ Copy of a patent to assist applicant in making corrections.
   ☐ "Notice to File Missing Parts of Application", Form PTO-1532.
   ☐ Other: _____

For: Manager, Application Branch
(703) 557-_____

FORM PTO-1123 (REV. 4-87)

Figure 4

UNITED STATES DEPARTMENT OF COMMERCE
Patent and Trademark Office
Address: COMMISSIONER OF PATENTS AND TRADEMARKS
Washington, D.C. 20231

SERIAL NUMBER	FILING DATE	FIRST NAMED APPLICANT	ATTY DOCKET NO.

DATE MAILED:

NOTICE TO FILE MISSING PARTS OF APPLICATION— NO FILING DATE

(Attachment to Form PTO-1123)

In order to avoid payment by applicant of the surcharge required if items 1 and 3-6 are filed after the filing date the following items are also brought to applicant's attention at this time.

If all missing parts of this form and on the "Notice of Incomplete Application" are filed together, the total amount owed by applicant as a ☐ large entity ☐ small entity (verified statement filed) is $ _____.

1. ☐ The statutory basic filing fee is: ☐ missing ☐ insufficient. Applicant as a ☐ large entity ☐ small entity must submit $_____ to complete the basic filing fee and MUST ALSO SUBMIT THE SURCHARGE, IF REQUIRED, AS INDICATED BELOW.

2. ☐ Additional claim fees of $_____ as a ☐ large entity, ☐ small entity, including any required multiple dependent claim fee, are required. Applicant must submit the additional claim fees or cancel the additional claims for which fees are due. NO SURCHARGE IS REQUIRED FOR THIS ITEM.

3. ☐ The oath or declaration:
   ☐ is missing.
   ☐ does not cover items required on the "Notice of Incomplete Application".
   An oath or declaration in compliance with 37 CFR 1.63, referring to the above Serial Number and Receipt Date is required. A SURCHARGE, IF REQUIRED, MUST ALSO BE SUBMITTED AS INDICATED BELOW.

4. ☐ The oath or declaration does not identify the application to which it applies. An oath or declaration in compliance with 37 CFR 1.63, identifying the application by the above Serial Number and Receipt Date is required. A SURCHARGE, IF REQUIRED, MUST ALSO BE SUBMITTED AS INDICATED BELOW.

5. ☐ The signature to the oath or declaration is: ☐ missing; ☐ a reproduction; ☐ by a person other than the inventor or a person qualified under 37 CFR 1.42, 1.43, or 1.47. A properly signed oath or declaration in compliance with 37 CFR 1.63, referring to the above Serial Number and Receipt Date is required. A SURCHARGE, MUST ALSO BE SUBMITTED AS INDICATED BELOW.

6. ☐ The signature of the following joint inventor(s) is missing from the oath or declaration:
   _____. Applicant(s) should provide, if possible, an oath or declaration signed by the omitted inventor(s), identifying this application by the above Serial Number and Receipt Date. A SURCHARGE, IF REQUIRED, MUST ALSO BE SUBMITTED AS INDICATED BELOW.

7. ☐ A $20.00 processing fee is required for returned checks. (37 CFR 1.21(m)).

8. ☐ Other:

Required items 1-7 above SHOULD be filed, if possible, with any items required on the "Notice of Incomplete Application" enclosed with this form. If concurrent filing of all required items is not possible, items 1-7 above must be filed no later than two months from the filing date of this application. The filing date will be the date of receipt of the items required on the "Notice of Incomplete Application." If items 1 and 3-6 above are submitted after the filing date, THE PAYMENT OF A SURCHARGE OF $110.00 for large entities, or $55.00 for small entities who have filed a verified statement claiming such status, is required. (37 CFR 1.16(e)).

Applicant must file all the required items 1-7 indicated above within two months from any filing date granted to avoid abandonment. Extensions of time may be obtained by filing a petition accompanied by the extension fee under the provisions of 37 CFR 1.136(a).

Direct the response to, and any questions about, this notice to the undersigned, Attention: Application Branch.

A copy of this notice MUST be returned with response.

_____

For: Manager, Application Branch
(703) 557-3254
FORM PTO-1532 (REV. 7-87)

For Office Use Only	
☐ 102	☐ 202
☐ 103	☐ 203
☐ 104	☐ 204
☐ 105	☐ 205

OFFICE COPY

Figure 5

UNITED STATES DEPARTMENT OF COMMERCE
Patent and Trademark Office
Address: COMMISSIONER OF PATENTS AND TRADEMARKS
Washington, D.C. 20231

SERIAL NUMBER	FILING DATE	FIRST NAMED APPLICANT	ATTY DOCKET NO

DATE MAILED:

NOTICE TO FILE MISSING PARTS OF APPLICATION—
FILING DATE GRANTED

A filing date has been granted to this application. However, the following parts are missing.

If all missing parts are filed within the period set below, the total amount owed by applicant as a
☐ large entity, ☐ small entity (verified statement filed), is $ _____.

1. ☐ The statutory basic filing fee is: ☐ missing. ☐ insufficient. Applicant as a ☐ large entity, ☐ small entity, must submit $ _____ to complete the basic filing fee and MUST ALSO SUBMIT THE SURCHARGE AS INDICATED BELOW.

2. ☐ Additional claim fees of $ _____ as a ☐ large entity, ☐ small entity, including any required multiple dependent claim fee, are required. Applicant must submit the additional claim fees or cancel the additional claims for which fees are due. NO SURCHARGE IS REQUIRED FOR THIS ITEM.

3. ☐ The oath or declaration:
   ☐ is missing.
   ☐ does not cover items omitted at the time of execution.

   An oath or declaration in compliance with 37 CFR 1.63, identifying the application by the above Serial Number and Filing Date is required. A SURCHARGE MUST ALSO BE SUBMITTED AS INDICATED BELOW.

4. ☐ The oath or declaration does not identify the application to which it applies. An oath or declaration in compliance with 37 CFR 1.63 identifying the application by the above Serial Number and Filing Date is required. A SURCHARGE MUST ALSO BE SUBMITTED AS INDICATED BELOW.

5. ☐ The signature to the oath or declaration is: ☐ missing; ☐ a reproduction; ☐ by a person other than the inventor or a person qualified under 37 CFR 1.42, 1.43, or 1.47. A properly signed oath or declaration in compliance with 37 CFR 1.63, identifying the application by the above Serial Number and Filing Date is required. A SURCHARGE MUST ALSO BE SUBMITTED AS INDICATED BELOW.

6. ☐ The signature of the following joint inventor(s) is missing from the oath or declaration:
   _____. Applicant(s) should provide, if possible an oath or declaration signed by the omitted inventor(s), identifying this application by the above Serial Number and Filing Date. A SURCHARGE MUST ALSO BE SUBMITTED AS INDICATED BELOW.

7. ☐ The application was filed in a language other than English. Applicant must file a verified English translation of the application and a fee of $26.00 under 37 CFR 1.17(k), unless this fee has been paid NO SURCHARGE UNDER 37 CFR 1.16(e) IS REQUIRED FOR THIS ITEM.

8. ☐ A $20.00 processing fee is required for returned checks. (37 CFR 1.21(m)).

9. ☐ Your filing receipt was mailed in error because check was returned.

10. ☐ Other:

A Serial Number and Filing Date have been assigned to this application. However, to avoid abandonment under 37 CFR 1.53(d), the missing parts and fees identified above in items 1 and 3-6 must be timely provided ALONG WITH THE PAYMENT OF A SURCHARGE OF $110.00 for large entities or $55.00 for small entities who have filed a verified statement claiming such status. The surcharge is set forth in 37 CFR 1.16(e). Applicant is given ONE MONTH FROM THE DATE OF THIS LETTER, OR TWO MONTHS FROM THE FILING DATE of this application, WHICHEVER IS LATER, within which to file all missing parts and pay any fees. Extensions of time may be obtained by filing a petition accompanied by the extension fee under the provisions of 37 CFR 1.136(a).

Direct the response to, and any questions about, this notice to the undersigned, Attention: Application Branch.

A copy of this notice __MUST__ be returned with response.

_____
For: Manager, Application Branch
(703) 557-3254

FORM PTO-15 (REV 7-87)

For Office Use Only	
☐ 102	☐ 202
☐ 103	☐ 203
☐ 104	☐ 204
☐ 105	☐ 205

OFFICE COPY

Figure 6

A sample United States Patent, including drawings and references, is shown as Fig. 1 on pages 7-14.

Filing Fees

Caution: Fees may change from time to time, especially when the fiscal year begins on October 1st of each year. People outside of the PTO information loop are not alerted to price hikes. It is not exactly front-page news. There are penalties for making incorrect payments, so double check the required amount with the PTO before sending in a check. **The latest PTO fee information is available by calling (703) 557-4357. The PTO Fees Schedule, printed as Appendix D on page 961, is current per publication, but still should be confirmed.**

Small Entity Status Reduces Fees by Half

On August 27, 1982, Public Law 97-247 provided that effective October 1, 1982, funds would be made available to the PTO to reduce by 50 percent the payment of fees by independent inventors, small business concerns, and nonprofit organizations.

The reduced fees include those for patent application, extension of time, revival, appeal, patent issues, statutory disclaimer, and maintenance on patents based on applications filed since August 27, 1982.

Fees that are not reduced include petition and processing (other than revival), document supply, certificate of correction, request for reexamination, international application fees, and certain maintenance fees. **Four PTO forms are available for claiming small entity status. See Forms 5, 6, 7, and 8 in Part One: Forms and Tables.**

What is an Independent Inventor?

The PTO considers an inventor as independent if the inventor (1) has not assigned, granted, conveyed, or licensed, and (2) is under no obligation under contract or law to assign, grant, convey, or license any rights in the invention to any person who could not likewise be classified as an independent inventor if that person had made the invention, or to any concern which would not qualify as a small concern or a nonprofit organization.

What is a Small Business Concern?

The PTO defines a small business as one whose number of employees, including those of its affiliates, does not exceed 500. The definition also requires a small business, for this purpose, to be one which has not assigned, granted, conveyed, or licensed, and is under no obligation under contract or law to assign, grant, convey, or license, any rights in the invention to any person who could not be classified as an independent inventor if that person had made the invention, or to any concern which would not qualify as a small business concern or a nonprofit organization.

What is a Nonprofit Organization?

To be recognized as nonprofit, an organization must be so accredited by a nationally recognized accrediting agency or association or of the type described in Section 501(c)(3) of the IRS Code of 1954 (26 U.S.C. 501(c)(3)) and which is exempt from taxation under 26 U.S.C. 501(a).

How to Fax Your Patent Application to the PTO

Effective November 1, 1988, certain papers to be filed in national patent applications and re-examination proceedings for consideration by the Office of the Assistant Commissioner for Patents,

the Office of the Deputy Assistant Commissioner for Patents, and the Patent Examining Groups (Patent Examining Corps) may be submitted to the PTO by fax.

The provision of 37 CFR 1.33(a), requiring signatures on amendments and other papers filed in applications, is hereby waived to the extent that a faxed signature is acceptable. The paper that is used as the original for the fax must have an original signature, and should be retained by the applicant or the representative as evidence of the content of the faxed page. No special format, addressing information or written ratification is required for faxes. However, the paper size must be 8.5 x 14 inches or smaller to be accepted.

A facsimile center has been established in the Patent Examining Corps to receive and process submissions. The filing date accorded the submission will be the date the complete transmission is received by the PTO facsimile unit unless that date is a Saturday, Sunday or federal holiday within the District of Columbia, in which case the official date of receipt will be the next business day.

Each transmission session must be limited to papers to be filed in a single national patent application or reexamination proceeding. It is recommended that the serial number of the application or control number of the reexamination be entered as part of the sender's identification, if possible. It is also recommended that the sending fax machine generate a report confirming transmission for each transmission session. The transmitting activity report should be retained along with the original.

The papers, including authorizations to charge deposit accounts, which may be submitted using this procedure, are limited to those which may be filed in national patent applications and reexamination proceedings and which are to be considered by the PTO organizations named above. Examples of such papers are amendments, responses to restriction requirements, requests for reconsideration before the examiner, petitions, terminal disclaimers, powers of attorney, notices of appeal, and appeal briefs.

New or continuing patent applications of any type, assignments, issue fee payments, maintenance fee payments, declarations or oaths under 37 CFR 1.63 or 1.67, and formal drawings are excluded, as are all papers relating to international patent applications. Papers to be filed in applications that are subject to a secrecy order under 37 CFR 5.1-5.8, and directly related to the secrecy order content of the application, are also excluded. Informal communications between applicant and the examiner, such as proposed claims for interview purposes, are permissible and are encouraged. Informal communications from applicants will not be made of record in the application or reexamination and must be clearly identified as informal such as by including the word "DRAFT" on each paper. To facilitate informal communications from examiners, applicants are encouraged to supply their fax numbers on communications to the Office.

The facsimile submissions may include a certificate for each paper stating the date of transmission. A copy of the facsimile submission with a certificate attached thereto will be evidence of transmission of the paper should the original be misplaced. The person signing the certificate should have a reasonable basis to expect that the paper would be faxed on the date indicated. An example of a preferred certificate is:

Certification of Facsimile Transmission

I hereby certify that this paper is being facsimile transmitted to the Patent and Trademark Office on the date shown below.

_____

Type or print name of person signing certification

_____   _____

Signature   Date

When possible, the certification should appear on a portion of the paper being transmitted. If the certification is presented on a separate paper, it must identify the application to which it relates, and the type of paper being transmitted, e.g., amendment, notice of appeal, etc.

In the event that the facsimile submission is misplaced or lost in the PTO, the submission will be considered filed as of the date of the transmission, if the party who transmitted the paper:

1) Informs the PTO of the previous facsimile transmission promptly after becoming aware that the submission has been misplaced or lost;

2) Supplies another copy of the previously transmitted submission with the Certification of the Transmission; and

3) Supplies a copy of the sending unit's report confirming transmission of the submission. In the event that a copy of the report is not available, the party who transmitted the paper may file a declaration under 37 CFR 1.68 which attests on a personal knowledge basis or to the satisfaction of the Commissioner to the previous timely transmission.

If all criteria above cannot be met, the PTO will require applicant to submit a verified showing of facts. You must indicate to the satisfaction of the Commissioner the date the PTO received the submission.

The facsimile center will have five facsimile units and will be staffed during the business hours of 8:30 a.m. and 5:00 p.m., Monday through Friday, excluding holidays. Although the units may normally be accessed at all times, including non-business hours, there may be times when reception is not possible due to equipment failure or maintenance requirements. Accordingly, applicants are cautioned not to rely on the availability of this service at the end of response periods.

The telephone numbers for accessing the fax machines are (703) 308-3719, (703) 308-3720, and (703) 308-3721. The facsimile center staff can be reached at telephone number (703) 308-1353 during normal business hours.

Specification (Description and Claims - Utility Patents)

The specification is a written description of the invention and of the manner and process of making and using it, and is required to be in such full, clear, concise, and exact language as to enable any person skilled in the technological area to which the invention pertains, or with which it is most nearly connected, to make and use it. Even after all these years and a familiarity with patent descriptions, I still cannot write them.

The specification must set forth the precise invention for which a patent is solicited, in such a manner as to distinguish it from other inventions and from what is old. It must describe completely a specific embodiment of the process, machine, manufacture, composition of matter, or improvement invented, and must explain the method of operation or principle whenever applicable. The best way you see to carry out the invention also must be described.

In the case of an improvement, your specification must particularly point out the part(s) of the process, machine, manufacture, or composition of matter to which the improvement relates, and the description should be confined to the specific improvement and to such parts as necessarily cooperate with it or as may be required to complete understanding or description of it.

Title and Abstract

The title of your invention, which should be as short and specific as possible, should be a heading on the first page of the specification, if it does not otherwise appear at the beginning of the application.

A brief abstract of the technical disclosure in the specification must be set forth in a separate page immediately following the claims in a separate paragraph under the heading "Abstract of the Disclosure." The purpose of the abstract is to enable the PTO and the public generally to determine quickly from a cursory inspection the nature and gist of the technical disclosure. It is not used for interpreting the scope of the claims. I have included the following as an example of an abstract:

An electronic toy doll including electronic circuitry for selectively generating a number of simulation sounds typically associated with a mystic or science-fantasy character. In respective operational modes, the sound of wind, the sound of breathing, an eerie pseudo-random sequence of musical notes and sounds representing the operation of a weapon are selectively generated. In a further operational mode, a random one of a predetermined number of responses are provided upon generation of an actuation signal. The random response may be considered to be an answer to inquiries. Preferred circuitry for generating the simulation sounds is described; accessories, adapted for removable interconnection with the circuitry are also described.

Summary of the Invention

A short summary of your invention indicating its nature and substance, which may include a statement of the object of the invention, should precede the detailed description. Your summary should, when set forth, be commensurate with the invention you claim and any object recited should be that of the invention as claimed. I have included the following as an example of a summary:

The present invention provides a relatively inexpensive, rugged electronic doll. The doll selectively simulates the sounds typically associated with a mystic or science-fantasy character. The sounds of wind, breathing, weapons operation, and an eerie tune of pseudo-random musical notes are selectively produced. Additionally, the future is "foretold" through generation of a random one of a predetermined number of gong sounds generated in response to an actuation signal. The operational mode is controlled by a central function select logic circuit in cooperation with remote switches. The remote switches may be located in the doll body, or may be disposed in removably interconnectable remote accessories.

Claims

I would recommend that you hire a lawyer to prepare and prosecute a utility patent **(see Section 27, Registered Patent Attorneys and Agents, on page 345)**. The preparation of a patent application is a highly complex business and not to be taken lightly.

If you are bent, however, on writing your own description and claims, it is best to study many patents for style and content. Then locate an updated copy of the PTO's Statutory Requirements of Claims at the nearest Patent and trademark Depository Library. You can photocopy them from the Manual of Patent Examining Procedure which is available at each PTDL.

For an example of Claims, see below; also Fig. 1, Utility Patent Illustration with Specifications on pages 7-14.

In the meantime, here are some pointers to get you going:

a) The specification section of the patent must conclude with a claim specifically pointing out and distinctly claiming the subject matter which the application regards as your invention or discovery.

b) The claims are brief descriptions of the subject matter of your invention, eliminating unnecessary details and reciting all essential features necessary to distinguish the invention from what is old. The claims are the operative part of the patent. Novelty and patentability are

judged by the claims, and, when a patent is granted, questions of infringement are judged by the courts on the basis of the claims.

c) When more than one claim is presented, they may be placed in dependent form in which a claim may refer back to and further restrict one or more preceding claims.

d) A claim in multiple dependent form shall contain a reference, in the alternative only, to more than one claim previously set forth and then specify a further limitation of the subject matter claimed. A multiple dependent claim shall not serve as a basis for any other multiple dependent claim. A multiple dependent claim shall be construed to incorporate by reference all the limitations of the particular claim in relation to which it is being considered.

e) The claims or claims must conform to the invention as set forth in the remainder of your specification and the terms and phrases used in the claims must find clear support or antecedent basis in the description so that the meaning of the terms in the claims may be ascertainable by reference to the description.

A sample of Claims is shown, in part, below:

1) An electronic doll comprising:

*a body;

*function select means, disposed on said body, for selectively generating respective mode control signals;

*electronic signal generator means, disposed in said body and responsive to said mode control signals, including:

  *means for generating a wind simulation signal representative of the sounds of wind;

  *means for generating a breathing simulation signal representative of the sounds of breathing;

  *means for generating a weapons simulation signal representative of weapons fire;

  *means for generating pseudo-random musical notes; and

  *means for generating one of a predetermined number of responses to an actuation signal, said one response being determined in a random manner;

*transducer means for generating audible output signals representative of electrical input signals applied thereto; and

*means, responsive to said mode control signals, for selectively applying said respective electrical simulation signals as input signals to said transducer means.

2) The electronic doll of claim 1 wherein said means for generating a wind simulation signal comprises:

*pseudo-random signal generator means, responsive to clock signals applied thereto for producing a pseudo-random signal, said pseudo-random signal having a digital value changing in a pseudo-random manner at a rate in accordance with said clock signal;

*oscillator means for generating said clock signal at a frequency in accordance with a frequency control signal applied thereto; and

*irregular signal generator means, responsive to said pseudo-random signal, for generating an irregular signal which alternatively rises or falls in amplitude in an irregular manner, said irregular signal being applied to said oscillator means as said frequency control signal.

3) The electronic doll of claim 2 wherein said pseudo-random signal generator means comprises:...[etc.].

Arrangement of Application Elements

The elements of the application should appear in the following order:

1) Title of the invention; or an introductory portion stating the inventor(s) name(s), citizenship, and residence of the applicant, and the title of the invention may be used.

2) Cross-references to related applications, if any.

3) Brief summary of the invention.

4) Brief description of the several views of the drawing, if there are drawings.

5) Detailed description.

6) Claim(s).

7) Abstract of the disclosure.

You are also encouraged to file an **Information Disclosure Citation (Form 9)** at the time of filing your patent application or within three months after the filing date of the application or two months after a filing receipt is received. If filed separately, be sure to tag it with the Group Art Unit to which the action is assigned as indicated on the filing receipt. This form will enable you to provide the PTO with a uniform listing of citations.

While the filing of information disclosure statements is voluntary, the form must be accompanied by an explanation of relevance of each listed item, a copy of each listed patent or publication of other item of information and a translation of the pertinent portions of foreign documents (only if an existing translation is available to you).

Examiners will consider all citations submitted and place their initials adjacent the citations in the boxes provided on the form.

Patent Drawings

When you apply for a patent, it is required by law in most cases that you furnish a drawing of the invention together with your application. This includes practically all inventions except compositions of matter or processes, but a drawing may also be useful in the case of many processes.

The drawing must show every feature of the invention specified in your claims and is required by the PTO rules to be in a particular form. The Office specifies the size of the sheet on which the drawing is made, the type of paper, the margins, and other details relating to the making of the drawing. The reason for specifying the standards in detail is that the drawings are printed and published in a uniform

style when the patent issues, and the drawings must also be such that they can be readily understood by persons using the patent descriptions.

No names or other identification will be permitted within the "sight" of the drawing, and applicants are expected to use the space above and between the hole locations to identify each sheet of drawings. This identification may consist of the attorney's name and docket number or the inventor's name and case number and may include the sheet number and the total number of sheets filed (for example, "sheet 2 of 4").

Drawings in colors other than black and white are not acceptable unless the drawing requirements are waived. Only the Deputy Assistant Commissioner for Patents can make the decision.

Color drawings are permitted for plant patents where color is a distinctive feature.

See Fig. 1, Utility Patent with Specifications, on pages 7-14, for an example of format used in patent drawings.

How To Get Patent Drawings Done

You have several options. You can do-it-yourself. To this end standards for drawings are included below to guide your efforts and keep you to the letter of the required standards. I have never attempted to do the drawings for a utility patent, but I have been successful at doing a few uncomplicated design patent drawings.

You could ask that your patent attorney arrange for the drawings. This is all right, but understand that just as in the search business, the lawyer or law firm will be adding a premium of typically no less than 40 percent to the fees charged by the draftsman. If you go with the law firm, get a guarantee that the drawings will pass PTO muster. If they are rejected for any technical reason whatsoever, you do not want to be double billed, i.e. pay for the draftsman's mistakes.

On more than one occasion I have had to have extra charges taken off my bills for the reworking of unacceptable drawings.

The surest and least expensive way, if you are not a draftsman, is to take bids and hire your own draftsman. Find the fair market price by calling around. Look in the telephone directory or ask at a regional patent library for candidates. Get the same guarantee for PTO acceptance from the draftsman if you contract the work yourself.

Standards for Patent Drawings

a) **Paper and ink.** Drawings must be made upon paper which is flexible, strong, white, smooth, nonshiny, and durable. Two-ply or three-ply bristol board is preferred. The surface of the paper should be calendered and of a quality which will permit erasure and correction with India ink. India ink, or its equivalent in quality, is preferred for pen drawings to secure perfectly black solid lines. The use of white pigment to cover lines is not acceptable.

b) **Size of sheet and margins.** The size of the sheets on which drawings are made may either be exactly 8.5 x 14 inches (21.6 x 35.6 cm) or exactly 21.0 x 29.7 cm (DIN size A4). All drawing sheets in a particular application must be the same size. One of the shorter sides of the sheet is regarded as its top.

1) On 8.5 x 14 inch drawing sheets, the drawing must include a top margin of 2 inches (5.1 cm) and bottom and side margins of 0.25 inch (6.4 mm) from the edges, thereby leaving a "sight" precisely 8 x 11.75 inches (20.3 x 29.8 cm). Margin border lines are not permitted. All work must be included within the

"sight." The sheets may be provided with two 0.25 inch (6.4 mm) diameter holes having their centerlines spaced 11/16 inch (17.5 mm) below the top edge and 2.75 inches (7.0 cm) apart, said holes being equally spaced from the respective side edges.

2) On 21.0 x 29.7 cm drawing sheets, the drawing must include a top margin of at least 2.5 cm, a left side margin of 2.5 cm, a right side margin of 1.5 cm, and a bottom margin of 1.0 cm. Margin border lines are not permitted. All work must be contained within a sight size not to exceed 17 x 26.2 cm.

c) **Character of lines.** All drawings must be made with drafting instruments or by a process which will give them satisfactory reproduction characteristics. Every line and letter must be durable, black, sufficiently dense and dark, uniformly thick and well defined; the weight of all lines and letters must be heavy enough to permit adequate reproduction. This direction applies to all lines however fine, to shading, and to lines representing cut surfaces in sectional views. All lines must be clean, sharp, and solid. Fine or crowded lines should be avoided. Solid black should not be used for sectional or surface shading. Freehand work should be avoided wherever it is possible to do so.

d) **Hatching and shading.** Hatching should be made by oblique parallel lines spaced sufficiently apart to enable the lines to be distinguished without difficulty. Heavy lines on the shade side of objects should preferably be used except where they tend to thicken the work and obscure reference characters. The light should come from the upper left-hand corner at a 45-degree angle. Surface delineations should preferably be shown by proper-shading, which should be open.

e) **Scale.** The scale to which a drawing is made ought to be large enough to show the mechanism without crowding when the drawing is reduced to two-thirds in reproduction, and views of portions of the mechanism on a larger scale should be used when necessary to show details clearly; two or more sheets should be used if one does not give sufficient room to accomplish this end, but the number of sheets should not be more than is necessary.

f) **Reference characters.** The different views should be consecutively numbered figures. Reference numerals (and letters, but numerals are preferred) must be plain, legible, and carefully formed, and not be encircled. They should, if possible, measure at least one-eighth of an inch (3.2 mm) in height so that they may bear reduction to one twenty-fourth of an inch (1.1 mm); and they may be slightly larger when there is sufficient room. They should not be so placed in the close and complex parts of the drawing as to interfere with a thorough comprehension of the same, and therefore should rarely cross or mingle with the lines. When necessarily grouped around a certain part, they should be placed at a little distance, at the closest point where there is available space, and connected by lines with the parts to which they refer. They should not be placed upon hatched or shaded surfaces but when necessary, a blank space may be left in the hatching or shading where the character occurs so that it shall appear perfectly distinct and separate from the work. The same part of an invention appearing in more than one view of the drawing must always be designated by the same character, and the same character must never be used to designate different parts. Reference signs not mentioned in the description shall not appear in the drawing and vice versa.

g) **Symbols, legends.** Graphic drawing symbols and other labeled representations may be used for conventional elements when appropriate, subject to approval by the Patent and Trademark Office. The elements for which such symbols and labeled representations are used must be adequately identified in the specification. While descriptive matter on drawings is not permitted, suitable legends may be used, or may be required, in proper cases, as in diagrammatic views and flow sheets or to show materials or where labeled representations are employed to illustrate conventional elements. Arrows may be required, in proper cases, to

show direction of movement. The lettering should be as large as, or larger than, the reference characters.

h) **[Reserved]** (Note: The government keeps this space available for future specifications to avoid re-lettering each item in the list.)

i) **Views.** The drawing must contain as many figures as may be necessary to show the invention; the figures should be consecutively numbered if possible in the order in which they appear. The figure may be plain, elevation, section, or perspective views, and detail views of portions of elements, on a larger scale if necessary, may also be used. Exploded views, with the separated parts of the same figure embraced by a bracket, to show the relationship or order of assembly of various parts are permissible. When necessary, a view of a large machine or device in its entirety may be broken and extended over several sheets if there is no loss in facility of understanding the view. Where figures on two or more sheets form in effect a single complete figure, the figures on the several sheets should be so arranged that the complete figure can be understood by laying the drawing sheets adjacent to one another. The arrangement should be such that no part of any of the figures appearing on the various sheets are concealed and that the complete figure can be understood even though spaces will occur in the complete figure because of the margins on the drawing sheets. The plane upon which a sectional view is taken should be indicated on the general view by a broken line, the ends of which should be designated by numerals corresponding to the figure number of the sectional view and have arrows applied to indicate the direction in which the view is taken. A moved position may be shown by a broken line superimposed upon a suitable figure if this can be done without crowding, otherwise a separate figure must be used for this purpose. Modified forms of construction can only be shown in separate figures. Views should not be connected by projection lines nor should centerlines be used.

j) **Arrangement of views.** All views on the same sheet should stand in the same direction and, if possible, stand so that they can be read with the sheet held in an upright position. If views longer than the width of the sheet are necessary for the clearest illustration of the invention, the sheet may be turned on its side so that the top of the sheet with the appropriate top margin is on the right-hand side. One figure must not be placed upon another or within the outline of another.

k) **Figure for Official Gazette.** The drawing should, as far as possible, be so planned that one of the views will be suitable for publication in the *Official Gazette* as the illustration of the invention.

l) **Extraneous matter.** Identifying indicia (such as the attorney's docket number, inventor's name, number of sheets, etc.) not to exceed 2.75 inches (7.0 cm) in width may be placed in a centered location between the side edges within three-fourths inch (19.1 mm) of the top edge. Authorized security markings may be placed on the drawings provided they be outside the illustrations and are removed when the material is declassified. Other extraneous matter will not be permitted upon the face of a drawing.

m) **Transmission of drawings.** Do not send original patent drawings with your application. Under new regulations in effect since January, 1989, a clean photocopy will suffice as long as the lines are uniformly thick, black, and solid. Should the PTO request a change in the drawing, you may submit a revised copy of the original.

Drawings transmitted to the PTO must never be folded. They can be sent flat, protected by a sheet of heavy binder's board, or may be rolled in a suitable mailing tube. If received creased or mutilated, new drawings will be required.

How to Effect Drawing Changes

Under regulations effective January 1, 1991, no one is allowed to remove and change original patent drawings once they have been submitted to the PTO. Prior to this date, original drawings could be picked up at the PTO by bonded draftsmen, reworked, and resubmitted.

Therefore, keep your original drawings on file and submit only copies of the drawings with your application. Then should you receive a request from the PTO for changes (Form 948), you can make them or have them made for you from the originals.

Make sure that the art unit number, serial number, and number of drawing sheets is written on the reverse side of the drawings.

You may delay filing the new drawings until receipt of the "Notice of Allowability." If delayed, the new drawings must be filed within the three month shortened statutory period set for response in the before-mentioned form. You may request an extension. Drawings should be filed as a separate paper, with a transmittal letter addressed to the PTO official draftsman or examiner.

In any case, once you have submitted drawings, no changes will be permitted, other than correction of informalities, unless the examiner has approved the proposed changes.

If you want to discuss the particulars of how to make drawing changes or anything else related to drawing changes, you may contact the Chief Draftsman at (703) 557-6404.

Applications Filed Without Drawings

Not all applications require drawings. It has been a long approved procedure, for example, to accept a process case (i.e. a case having only process or method claims) which is filed without drawings.

Other situations where drawings are usually not considered essential for a filing date are:

1) **Coated articles or products.** If the invention resides only in coating or impregnating a conventional sheet, e.g. paper or cloth, or an article of known and conventional character with particular composition.

2) **Articles made from a particular material or composition.** If the invention consists in making an article of a particular material or composition, unless significant details of structure or arrangement are involved in the claims.

3) **Laminated structures.** If the invention involves only laminations of sheets (and coatings) of specified material, unless significant details of structure or arrangement (other than the mere order of layers) are involved in the claims.

4) **Articles, apparatus, or systems where sole distinguishing feature is presence of a particular material.** If the invention resides solely in the use of particular material in an otherwise old article, apparatus, or system recited broadly in its claim.

Photographs are not normally considered to be proper drawings. Photographs are acceptable for a filing date and are generally considered to be informal drawings. Photographs are only acceptable where they come within special categories. Photolithographs are never acceptable.

The PTO is willing to accept black and white photographs or photomicrographs (not photolithographs or other reproductions of photographs made using screens) printed on sensitized paper in lieu of India ink drawings, to illustrate inventions which are incapable of being accurately or adequately depicted

by India ink drawings restricted to the following categories: crystalline structures, metallurgical microstructures, textile fabrics, grain structures, and ornamental effects.

Models are generally not required as part of an application or patent. They have not been for many, many years. And save for cases involving perpetual motion, or a composition of matter, you will probably never have to submit one.

If your invention is, however, a microbiological invention, a deposit of the microorganism is required.

Section 6

How the PTO Processes Your Application

"I do not view the process with any misgivings...Like the Mississippi it just keeps rolling along."—Winston Churchill

If your application passes initial muster, it will be assigned to the appropriate examining group, and then to an examiner. Applications are handled in the order received.

The application examination inspects for compliance with the legal requirements and includes a search through U.S. patents, prior foreign patent documents which are available in the PTO, and available literature to ensure that the invention is new. A decision is reached by the examiner in light of the study and the result of the search.

First Office Action

You or your attorney will be notified of the examiner's decision by what the PTO refers to as an **action**. An action is actually a letter that gives the reasons for any adverse action or any objection or requirement. Noted will be any appropriate references or information that you'll find useful in making the decision to continue the prosecution of the application or to drop it.

If the invention is not considered patentable subject matter, the claims will be rejected. If the examiner finds that the invention is not new, the claims will be rejected; but the claims may also be rejected if they differ somewhat from what is found to be obvious. It is not uncommon for some or all of the claims to be rejected on the examiner's first action; very few applications sail through as first submitted.

A sample Examiner's Action is shown as Fig. 7 on page 54.

Your First Response

Let's say the examiner gives you the thumbs down on all or some of your claims. Your next move, if you wish to continue prosecuting the patent, is to respond, specifically pointing out the supposed errors in the examiner's action. Patent examiners have a lot on their plates and their units are typically understaffed for the amount of work they handle. For example, the PTO reported in November of 1988 that 6,500 biotechnology patents were awaiting a decision or about 70 per government examiner.

UNITED STATES DEPARTMENT OF COMMERCE
Patent and Trademark Office

Address : COMMISSIONER OF PATENTS AND TRADEMARKS
Washington, D.C. 20231

☐ This application has been examined   ☐ Responsive to communication filed on _____   ☐ This action is made final.

A shortened statutory period for response to this action is set to expire_____ month(s), _____ days from the date of this letter.
Failure to respond within the period for response will cause the application to become abandoned.   35 U.S.C. 133

Part I   THE FOLLOWING ATTACHMENT(S) ARE PART OF THIS ACTION:

1. ☐ Notice of References Cited by Examiner, PTO-892.   2. ☐ Notice re Patent Drawing, PTO-948.
3. ☐ Notice of Art Cited by Applicant, PTO-1449.   4. ☐ Notice of Informal Patent Application, Form PTO-152.
5. ☐ Information on How to Effect Drawing Changes, PTO-1474.   6. ☐ _____

Part II   SUMMARY OF ACTION

1. ☐ Claims _____ are pending in the application.

        Of the above, claims _____ are withdrawn from consideration.

2. ☐ Claims _____ have been cancelled.

3. ☐ Claims _____ are allowed.

4. ☐ Claims _____ are rejected.

5. ☐ Claims _____ are objected to.

6. ☐ Claims _____ are subject to restriction or election requirement.

7. ☐ This application has been filed with informal drawings under 37 C.F.R. 1.85 which are acceptable for examination purposes.

8. ☐ Formal drawings are required in response to this Office action.

9. ☐ The corrected or substitute drawings have been received on _____ . Under 37 C.F.R. 1.84 these drawings are ☐ acceptable. ☐ not acceptable (see explanation or Notice re Patent Drawing, PTO-948).

10. ☐ The proposed additional or substitute sheet(s) of drawings, filed on _____ has (have) been ☐ approved by the examiner. ☐ disapproved by the examiner (see explanation).

11. ☐ The proposed drawing correction, filed on _____, has been ☐ approved. ☐ disapproved (see explanation).

12. ☐ Acknowledgment is made of the claim for priority under U.S.C. 119. The certified copy has ☐ been received ☐ not been received ☐ been filed in parent application, serial no. _____ ; filed on _____

13. ☐ Since this application appears to be in condition for allowance except for formal matters, prosecution as to the merits is closed in accordance with the practice under Ex parte Quayle, 1935 C.D. 11; 453 O.G. 213.

14. ☐ Other

PTOL-326 (Rev. 6-88)   **EXAMINER'S ACTION**

Figure 7

In this area alone the PTO predicts that the number of applications will grow by 12 percent a year through the early 1990s.

Examiners must process a specific number of patents to be considered productive by their superiors for periodic job performance ratings. The bottom line is that as careful as they try to be, they make mistakes that can be reversed with careful argument.

Your response should address every ground of objection and/or rejection. Show where the examiner is wrong. The mere allegation that the examiner has erred is not enough.

Your response will cause the examiner to reconsider, and you'll be notified if the claims are rejected, or objections or requirements made, in the same manner as after the first examination. This second Office action usually will be made final.

Feel free to call your examiner up on the telephone to discuss your case. I have always found them to be hospitable and helpful. His or her telephone number will appear at the end of the Office action or you can look it up in **Appendix B, Patent and Trademark Office Telephone Directory, on page 943.**

Depending upon how serious the matter, you or your attorney might wish to make an appointment to personally visit the examiner. Don't just drop in unannounced. It is to your benefit that the examiner has the time to prepare for your visit and get up to speed on the case. Remember that personal interviews do not remove the necessity for response to Office actions within the required time, and the action of the Office is based solely on the written record.

If you feel that you can handle the matter alone, try it and save the attorney's fees. Patent examiners will meet with inventors. I've always found it to be a rewarding experience.

Final Rejection

On the second or later consideration, the rejection of claims may be made final. Your response is then limited to appeal and further amendment is restricted. You may petition the Commissioner in the case of objections or requirements not involved in the rejection of any claim. Response to a final rejection must include cancellation of, or appeal from the rejection of each claim so rejected and, if any claim stands allowed, compliance with any requirement or objection as to its form.

In determining such final rejection, your examiner will repeat or state all grounds of rejection then considered applicable to your claims as stated in the application.

The odds? As in the case of the examination by the Office, patents are granted in the case of about two out of every three applications filed.

Making Amendments to Your Application

The preceding section referred to amendments to an application. Following are some details concerning amendments:

1) The applicant may amend before or after the first examination and action as specified in the rules, or when and as specifically required by the examiner.

2) After final rejection or action, amendments may be made canceling claims or complying with any requirement of form which has been made but the admission of any such amendment or its refusal, and any proceedings relative thereto, shall not operate to relieve the application from its condition as subject to appeal or to save it from abandonment.

3) If amendments touching the merits of the application are presented after final rejection, or after appeal has been taken, or when such amendment might not otherwise be proper, they may be admitted upon a showing of good and sufficient reasons why they are necessary and were not earlier presented.

4) No amendment can be made as a matter of right in appealed cases. After decision on appeal, amendments can only be made as provided in the rules.

5) The specifications, claims, and drawing must be amended and revised when required, to correct inaccuracies of description and definition of unnecessary words, and to secure correspondence between the claims, the description, and the drawing.

All amendments of the drawings or specifications, and all additions thereto, must conform to at least one of them as it was at the time of the filing of the application. Matter not found in either, involving a departure from or an addition to the original disclosure, cannot be added to the application even though supported by a supplemental oath or declaration, and can be shown or claimed only in a separate application.

The claims may be amended by canceling particular claims, by presenting new claims, or by amending the language of particular claims (such amended claims being in effect new claims). In presenting new or amended claims, the applicant must point out how they avoid any reference or ground rejection of record which may be pertinent.

Erasures, additions, insertions, or alterations of the papers and records must not be made by the applicant. Amendments are made by filing a paper, directing or requesting that specified changes or additions be made. The exact word or words to be stricken out or inserted in the application must be specified and the precise point indicated where the deletion or insertion is to be made.

Amendments are "entered" by the Office through the making of proposed deletions by drawing a line in red ink through the word or words canceled and by making the proposed substitutions or insertions in red ink, small insertions being written in at the designated place and larger insertions being indicated by reference.

No change in the drawing may be made except by permission of the Office. Permissible changes in the construction shown in any drawing may be made only by the bonded draftsmen. A sketch in permanent ink showing proposed changes, to become part of the record, must be filed for approval by the Office before the corrections are made. The paper requesting amendments to the drawing should be separate from other papers.

If the number or nature of the amendments render it difficult to consider the case, or to arrange the papers for printing or copying, the examiner may require the entire specification or claims, or any part thereof, to be rewritten.

The original numbering of the claims must be preserved throughout the prosecution. When claims are canceled, the remaining claims must not be renumbered. When claims are added by amendment or substituted for canceled claims, they must be numbered by the applicant consecutively beginning with the number next following the highest numbered claim previously presented. When the application is ready for allowance, the examiner, if necessary, will renumber the claims consecutively in the order in which they appear or in such order as may have been requested by applicant.

To better organize your request for amendment, a copy of the PTO's Amendment Transmittal Letter format appears as Form 3 in Part One: Forms and Tables.

Time for Response and Abandonment

The maximum period given for response is six months, but the Commissioner has the right to shorten the period to no less than thirty days. The typical response time allowed to an Office action is three months. If you want a longer time, you usually have to pay some extra money for an extension. The amount of the fee depends upon the response time desired. If you miss any target date, your application will be abandoned by the PTO and made no longer pending. However, if you can show whereby your failure to prosecute was unavoidable or unintentional, the application can be revived by the Commissioner.

The revival requires a petition to the Commissioner and a fee for petition, which should be filed without delay. As of this writing the fee required to have your case reviewed is $50. **See Form 11, the official PTO form used to petition for revival of an application for a patent abandoned unintentionally, in Part One: Forms and Tables.**

Abandonment of Patents

There are two kinds of abandonment: **intentional** (it's your fault) and **unintentional** (due to circumstances beyond your control). If the reason for abandonment is your fault, let's say you lost track of the dates and missed the deadline, then you must pay the due fee (e.g. $185 for a small entity design patent, $525 for a small entity utility patent) plus a penalty for your mistake, an additional $525. A costly error!

If your reason is unintentional, for example, you never received the notice from the PTO, then you must pay $50 to have your petition considered plus the required fee (e.g. $185 for small entity design patents, $525 for small entity utility patents). You may wish to add a notarized letter to this form explaining your story in detail. If you're going to blame the PTO, you'd better have your facts in order and in writing because it is tough to prove a PTO error.

Once I was involved once in a case in which the PTO typed the wrong zip code on our paperwork, an error which caused a month delay in delivery of our paperwork, and ultimately resulted in our getting slapped with abandonment papers.

There was no question the PTO had erred. I was able to cure this with a phone call to a senior PTO official. It was an open and shut case, as far as he was concerned. Upon seeing the typo, he personally caused the abandonment to be withdrawn. We never even had to pay the petition fees. There was no question about who was wrong.

See Form 11, the official PTO form used to petition for revival of an application for a patent abandoned unintentionally, in Part One: Forms and Tables.

If you have any questions about how to handle a petition, do not hesitate to call (703) 557-4282 for the latest information. I have had occasion to revive patents and the folks who answer this line are extremely helpful.

Mail petitions for revival to: Commissioner, Patent and Trademark Office, Box DAC, Washington, DC 20231.

Appeal to the Board of Patent Appeals and Interferences and to the Courts

If the examiner circles his or her wagons and begins to stonewall, there is a higher court. Rejections that have been made final may be appealed to the Board of Patent Appeals and Interferences. This august body consists of the Commissioner of Patents and Trademarks, the Deputy Commissioner, the Assistant Commissioners, and the examiners-in-chief, but typically each appeal is heard by only three

members. An appeal fee is required and you must file a brief in support of your position. You can even get an oral hearing if you pay enough.

An alternative to appeal in situations where you wish consideration of different claims or further evidence is to file a new continuation application. This requires a filing fee and the claims and evidence for which consideration is desired.

If the Board goes against you, there is yet a higher court, the Court of Appeals for the Federal Circuit. Or you might file a civil action against the Commissioner in the U.S. District Court for the District of Columbia. He won't take it personally. It goes with the territory. The Court of Appeals for the Federal Circuit will review the record made in the Office and may affirm or reverse the Office's action. In a civil action, you may present testimony in the court, and the court will make a decision.

Channels of Ex Parte Review are illustrated as Fig. 8 on page 59.

Interferences

Parallel development is a phenomenon that should not be discounted. On numerous occasions a company executive has said to me, "I've seen that concept twice in the last month," or something to this effect. At times, two or more applications may be filed by different inventors claiming substantially the same patentable invention. A patent can only be granted to one of them, and a proceeding known as an **interference** is instituted by the Office to determine who is the first inventor and entitled to the patent. About one percent of all applications filed become engaged in an interference proceeding.

Interference proceedings may also be instituted between an application and a patent already issued, if the patent has not been issued for more than one year prior to the filing of the conflicting application, and if that the conflicting application is not barred from being patentable for some other reason.

The priority question is determined by a board of three examiners-in-chief on the evidence submitted. From the decision of the Board of Patent Appeals and Interferences, the losing party may appeal to the Court of Appeals for the Federal Circuit or file a civil action against the winning party in the appropriate U.S. district court.

The terms "conception of the invention" and "reduction to practice" are encountered in connection with priority questions. "Conception of the invention" refers to the completion of the devising of the means for accomplishing the result. "Reduction to practice" refers to the actual construction of the invention in physical form. In the case of a machine it includes the actual building of the machine. In the case of an article or composition it includes the actual carrying out of the steps in the process; actual operation, demonstration, or testing for the intended use is usually required. The filing of a regular application for patent completely disclosing the invention is treated as equivalent to reduction to practice. The inventor who proves to be the first to conceive the invention and the first to reduce it to practice will be held to be the prior inventor, but more complicated situations cannot be stated this simply.

Here is a case when it is important to be able to have evidence that proves when you first had an idea and when the prototype was made. This is why you should keep careful and accurate records throughout the development of an idea. The Disclosure Document Program was established by the PTO for this purpose.

Form 12, Patent Worksheet, in Part One: Forms and Tables, is designed to help inventors keep track of when ideas originated; for information on the Disclosure Document Program, see pages 25 and 26.

Channels of Ex Parte Review

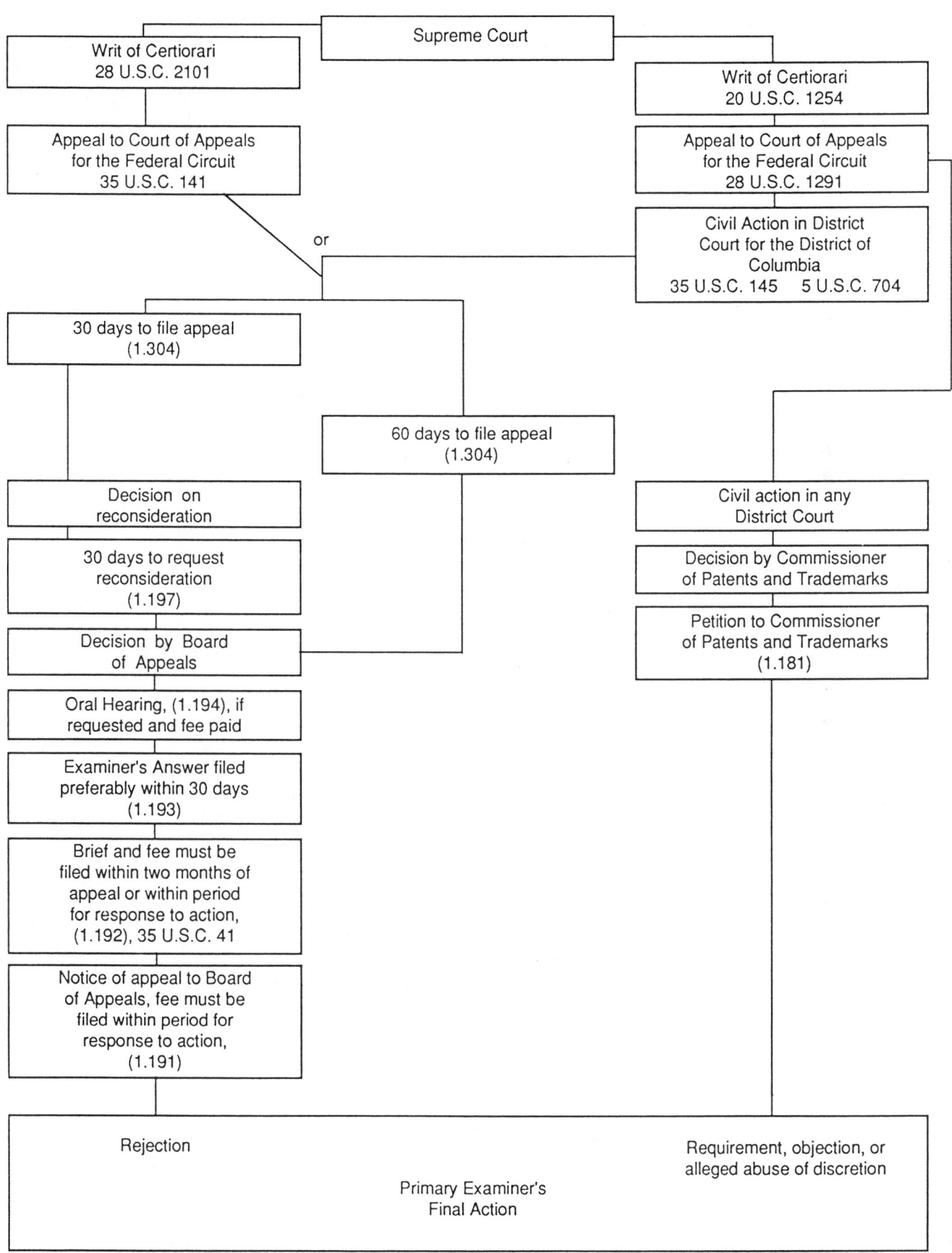

Figure 8

Section 7

Allowance and Issue of Patents

"A person who never made a mistake never tried anything new."—Albert Einstein

If your utility patent is found to be allowable, a **notice of allowance** will be sent to you or your attorney. Within three months from the date of the notice you must pay an issue fee.

What Rights Does a Patent Give You?

It's a pretty exciting moment when you get your first patent. It comes bound inside a beautiful oyster white folder that has the United States Constitution screened in blue as its background. The large official gold seal of the Patent and Trademark Office is embossed thereon with two red ribbons furcate as a tail.

Between the covers of this folder is your patent, a grant that gives you the inventor(s) "the right to exclude others from making, using or selling the invention throughout the United States" and its territories and possessions for a period of 17 years (14 for a design patent) subject to the payment of maintenance fees as provided by law. The patent does not give you the right to make, use, or sell the invention. Any person is ordinarily free to make, use, or sell anything he or she pleases, and a grant from Uncle Sam is not required.

If you receive a patent for a new soda pop and the marketing of said beverage is prohibited by law, the patent will not help you. Nor may you market said soda pop if by doing so you infringe the prior rights of others.

Since the essence of the right granted by a patent is the right to exclude others from commercial exploitation of the invention, the patentee is the only one who may make, use, or sell the invention. Others may not do so without your authorization. You may assign your rights in the invention to another person or company.

After a patent has expired, anyone may make, use, or sell the invention without permission of the patentee, provided the matter covered by other unexpired patents is not used. The terms may not be extended save for by a special act of Congress.

Maintenance Fees

All utility patents which issue from applications filed on or after December 12, 1980 are subject to the payment of maintenance fees which must be paid to keep the patent in force. These fees are due at 3.5, 7.5, and 11.5 years from the date the patent is granted and can be paid without a surcharge during the six-month period preceding each due date, e.g. three years to three years and six months, etc. The amounts of the maintenance fees are subject to change every three years.

The PTO does not mail notices to patent owners advising them that a maintenance fee is due.

If you have a patent attorney tracking your business, he or she will let you know when the money is due. An attorney gets paid every time your business moves across his or her desk. But if you are doing it by yourself and you miss a payment, it may result in the expiration of the patent. A six month grace period is provided when the maintenance fee may be paid with a surcharge.

Patents relating to some pharmaceutical inventions may be extended by the Commissioner for up to five years to compensate for marketing delays due to federal FDA pre-marketing regulatory procedures. Patents relating to all other types of inventions can only be extended by congressional legislation.

Can Two People Own A Patent?

Yes. Two or more people may jointly own patents as either inventors, investors, or licensees. Most of my patents are joint ownerships. Anyone who shares in the ownership of a patent, no matter how small a part they might own, has the right to make, use, or sell it for his or her own profit, unless prohibited from doing so by prior agreement. It is accordingly dangerous to assign part interest in a patent of yours without having a definite agreement hammered out vis-a-vis respective rights and obligations to each other.

Can You Sell Your Patent?

Yes. The patent law provides for the transfer or sale of a patent, or of an application for patent, by a contract. When assigned the patent, the assignee becomes the owner of the patent and has rights identical to those of the original patentee.

Assignment of Patents Applications

Should you wish to assign your patent or patent application to a third party (manufacturer, investor, university, employer, etc.), this is possible by filing the appropriate form, **Assignment of Patent, or Assignment of Patent Application, Forms 13 and 14, located in Part One: Forms and Tables.**

You can sell all or part interest in a patent. If you prefer, you could even sell it by geographic region.

I consider patents as valuable properties personal assets. Never assume that because you have been unsuccessful in selling a patent that it has no value. You might sell it eventually or find someone infringing it, thus turning it to positive account.

Infringement of Patents

Infringement of a patent consists in the unauthorized making, using, or selling of the patented invention within the territory of the U.S. during the term of the patent. If your patent is infringed, it is your right to seek relief in the appropriate federal court.

When I see an apparent infringement of a patent of ours, as has occurred occasionally over the years, the first thing do is call the company and set a meeting. I am not litigious. Things can often be worked out between parties. Thus far, I have always been able to do this. Court battles over patents can be long and expensive affairs. And, if you want to continue working in your particular field, it is wise to not make too many corporate enemies.

I have a close friend who beat a large manufacturer in court to the tune of a million plus dollars. After splitting the award with his partner and attorney, and paying taxes and costs, he came away with about $250,000. It sounded good at first. But he soon realized that every door in his industry was closed to him. None of the major companies wanted to see his concepts. They still don't. It was a high price to pay because the right idea with the right company can be worth millions.

Several years ago I saw an infringement of a patent we hold. One call to the company's president, and a quick fax of our patent, brought immediate relief in the form of a settlement on all items made to date and a royalty on all future sales of the item. It was an early Christmas. Further, I was invited to submit new ideas for licensing consideration.

If your friendly approach is turned away, and you are sure of your position, then the next step is to get a lawyer and decide if a **Temporary Restraining Order (TRO)** is appropriate. A TRO is an injunction to prevent the continuation of the infringement. You may also ask the court for an award of damages because of the infringement. In such an infringement suit, the defendant may raise the question of the validity of the patent, which is then decided by the court. The defendant may also aver that what is being done does not constitute infringement.

Infringement is determined primarily by the language of the claims of the patent and, if what the defendant is making does not fall within the language of any of the claims of the patent, there is no infringement.

The PTO has no jurisdiction over questions relating to infringement of patents. In examining applications for patent, no determination is made as to whether the patent-seeking invention infringes any prior patent.

To Sue or Not to Sue

If you catch someone infringing on your patent, you may be faced with the prospect of having to sue for damages. This can be a costly exercise. According to Stephen R. May, Manager at Intellectual Property Services Department at Pacific Northwest Laboratory, "a full-blown patent lawsuit that actually goes to trial will probably cost a minimum of $75,000 to $100,000, although a very simple case could cost less." In most instances, Mr. May reports, the cost can be $250,000 and up.

His advice to inventors is: "If you believe someone is infringing your patent, an attorney can draft a cease and desist letter for as little as a few hundred dollars. This might resolve the matter if the infringer ceases, but in many cases it does not."

The expensive part of any lawsuit is discovery, in which the parties exchange documents and take depositions of potential witnesses. The photocopying bill alone can run into the thousands of dollars

and the process can last anywhere from six months to several years. Trials tend to run from one to six weeks with decisions rendered in a matter of days in the case of a jury, or as long as several months if the verdict is by a judge. If you lose, appeals take more time and money.

What Does "Patent Pending" Signify?

These words are put on a product by a manufacturer to inform the public (and competition) that an application for patent on that item is on file at the PTO. It means stay away. The law imposes a fine on those who use these words falsely to deceive the public.

Section 8

Specific Kinds of Patents

"The more we observe and are aware of, the more mental connections we can make that will result in new and relevant ideas."—Melissa Strickland

If you want an inexpensive patent that will give you little actual protection because they are easy to end-run, but still meet the requirement that a patent issue, maybe a design patent is for you. It appears to be popular with lots of folks. Over 6,000 new designs were patented last year.

Inventors get design patents on just about anything, e.g. baby bibs, sweatbands, tissue box holders, dishes, ammo boxes, game boards, vending machines, telephones, pencils and pens, and even internal combustion engines.

Rubbermaid has one for a cereal container. Totes protects its umbrella handle designs. The Parker Pen Company takes them out for writing instruments. And the Ford Motor Company has them on car parts such as automobile quarter panels.

For little expense and effort a design patent is another way to stake out a claim. It permits you to legally post a no trespass sign in the form of **Patent Pending, Patent Applied For,** or **Patent** and the number of the patent.

The range of ornamental appearances that have been patented during the 147-year history of design patents is most impressive. Over 270,000 designs have received design patent protection since the first one was granted to George Bruce for "Printing Type" on November 9, 1842.

If you've invented any new, original, and ornamental design for an article of manufacture, a design patent may be appropriate. A design patent protects only the appearance of an article, and not its structure or utilitarian features. The proceedings relating to granting of design patents are the same as those relating to other patents with few differences.

In design cases as in "mechanical" cases, novelty and unobviousness are necessary prerequisites to the grant of a patent. In the case of designs, the inventive novelty resides in the shape or configuration or ornamentation as determining the appearance or visual aspect of the object or article of manufacture in contradistinction to the structure of a machine, article of manufacture, or the constitution of a

composition of matter. Simply put, it is the appearance presented by the object that creates an impression upon the mind of the observer.

A utility patent and a design patent may be based upon the same object matter; however, there must be a clearly patentable distinction between them.

Are Patent Attorneys Required for Design Patents?

This is a personal decision, but I never engage a patent attorney's services for design patent applications. Unlike the complicated business of utility patents, the design patent application process is very easy and uncomplicated. I have found that paying a lawyer to do my design patents is like tossing money out the window because the application form is simple, esoteric language is not required to draft the claims, and searches are actually so pleasant an experience that I will often do them myself.

There is not much to read in a design patent search. You're just looking at a lot of line drawings, many of which are fascinating.

To give you an idea how they are presented, refer to Fig. 2, Design Patent Illustration, on page 15.

It cost me $500 to learn that lawyers were not necessary in such matters. At the time I had been using a fine law firm for utility patents, and I naturally released our first design patent to them. I knew no better.

My lawyer said that he would "write up the specifications." I was asked to provide his draftsman with a prototype of the design. It was a tricycle with a mainframe shaped like a toothpaste tube. It became Procter & Gamble's Crest Fluorider. When the design patent arrived, I saw the simplicity of the specifications (**see Design Patent Specification below and Form 15 in Part One: Forms and Tables.**) Design patents are obvious cash cows for patent attorneys. I have removed the product name from Form 15 so that you will be able to use its format for your design patents by inserting a title. This is a sample only and, unlike other Forms in this book, should not be returned to the Patent Office. The petition, specification, and claim must be typed on legal-sized paper.

Design Patent Title

The title is of great importance in a design patent application. It serves to identify the article in which the design is embodied and which is shown in the drawing, by a name generally used by the public. The title should be to a specific definite article. Thus a stove would be called a "Stove" and not "Heating Device." The same title is used in the preamble to the specification, in the description of the drawing, and in the claim.

To allow latitude of construction it is permissible to add to the title "or similar article." The title must be in the singular.

Design Patent Specification

See Form 15, a sample application for design specification, in Part One: Forms and Tables. There is nothing complicated about it. There is no need to pay a lawyer for this work. Unlike the utility patent, this is one you can do yourself and save the expense of legal fees.

Copy the form and insert your name and the name of your design where indicated in the first paragraph. If you've invented a bath toy, for example, type "child's bath toy" for the design's name.

Your patent draftsman can tell you which views to designate in the section that follows (figure 1, figure 2, etc.) The views should be a figure description line for each view the draftsman gives you.

In the last section, repeat the name of your item in the space indicated.

Attach to this form a declaration claiming small business status to get the discounted fee (if appropriate) and your check in the correct amount. **See Form 7 in Part One: Forms and Tables for a copy of the Declaration for Claiming Small Entity Status—Small Business Concern. It will qualify you for the discounted small business fee. See Appendix D, PTO Fee Schedule, on page 961 for the fee and double-check it with the PTO office at (703) 305-8000 to make sure it has not changed since this printing.**

Design Patent Drawings

Unless you are capable of doing this work, hire a competent, experienced patent draftsman to make your drawings. The requirements for drawings are strictly enforced. Professional draftsmen will stand behind their work and guarantee revisions if requested by the Office due to inconsistencies in the drawings.

The claim of the design patent determines its classification, i.e. the appropriate class and subclass into which the design patent will be placed. This classification is designated as the "original classification" of the patent. Copies of a patent may be placed in other subclasses for the convenience of the examiner as an aid in searching during the examination process. Additional copies are designated "cross reference."

The PTO changes the Design Patent Classification System as required to provide an appropriate area for each patented design. Areas which show great activity are expanded while other areas, used infrequently as activity fades, are compressed into other subclasses.

A copy of the operative Design Patent Classification System may be found in Appendix A, U.S. Patent Classifications, on page 693.

Plant Patents

The law provides for the granting of a patent to anyone who has invented or discovered and asexually reproduced any distinct and new variety of plant, including cultivated sports, mutants, hybrids, and newly found seedlings, other than a tuber-propagated plant or a plant found in an uncultivated state.

Asexually propagated plants are those that are reproduced by means other than from seeds, such as by the rooting of cuttings, by layering, budding, grafting, and inarching.

With reference to tuber-propagated plants, for which a plant patent cannot be obtained, the term "tuber" is used in its narrow horticultural sense as meaning a short, thickened portion of an underground branch. The only plants covered by the term "tuber-propagated" are the Irish potato and the Jerusalem artichoke.

Plant Variety Protection Act

The Plant Variety Protection Act (Public Law 91-557), approved December 24, 1970, provides for a system of protection for sexually reproduced varieties, for which protection was not previously provided, under the administration of a Plant Variety Protection Office within the Department of Agriculture. Requests for information regarding the protection of sexually reproduced varieties should be addressed to Commissioner, Plant Variety Protection Office, Consumer and Marketing Service, Grain Division, 6525 Bellcrest Rd., Hyattsville, Maryland 20782.

Elements of a Plant Application

An application for a plant patent consists of the same parts as other applications and must be filed in duplicate, but only one need be signed and executed; the second copy may be a legible carbon copy of the original. Two copies of color drawings must be submitted.

The reason for submitting two copies is that one, the photocopy, must go to the Department of Agriculture for an advisory report on plant variety. The original is retained at the PTO.

Applications for a plant patent which fail to include two copies of the specification and two copies of the drawing when in color, will be accepted for filing only. The Application Division will notify you immediately if something is missing from the filing. You'll be given one month to rectify the situation. Failure to do so will result in loss of the filing date and the fee paid.

Plant Patent Specification and Claim

The specification should include a complete, detailed description of the plant and its characteristics that distinguish the plant over related known varieties and its antecedents. The specification must be expressed in botanical terms (in the general form followed in standard botanical text books or publications dealing with the varieties of the kind of plant involved), rather than the non-botanical characterizations commonly found in nursery or seed catalogs. The specification should also include the origin or parentage of the plant variety sought to be patented and must particularly point out where and in what manner the variety of plant has been asexually reproduced.

When color is a distinctive feature of the plant, it should be positively identified in the specification by reference to a designated color as given by a recognized color dictionary, for example, cherry red blooms. Where the plant variety originated as a newly found seedling, the specification must fully describe the conditions (cultivation, environment, etc.) under which the seedling was found growing to establish that it was not found in an uncultivated state.

A plant patent is granted on the entire plant. It follows that only one claim is necessary and only one is permitted.

Plant Patent Oath or Declaration

Your oath or declaration as inventor, in addition to the statements required for other applications, must include the statement that you asexually reproduced the new plant variety. Where the plant is a newly found plant the oath or declaration must also state that it was found in a cultivated area.

Plant Patent Drawings

Plant patent drawings are not mechanical drawings and should be artistically and competently executed. The drawing must disclose all the distinctive characteristics of the plant capable of visual representation. When color is a distinguishing characteristic of the new variety, the drawing must be in color. Two duplicate copies of color drawings must be submitted. Color drawings may be made either in permanent water color or oil, or in lieu thereof may be photographs made by color photography or properly colored on sensitized paper. The paper in any case must correspond in size, weight, and quality to the paper required for other drawings. Mounted photographs are acceptable.

All color drawings should be mounted so as to provide a two-inch margin at the top for Office markings when the patent is printed.

Plant Specimens

Specimens of the plant variety, its flower, or fruit, should not be submitted unless specifically called for by the examiner. If the PTO wants to inspect a plant that cannot be physically submitted, it will send an examiner to the growing site.

Fees and Correspondence

All inquiries relating to plant patents and pending plant patent applications should be directed to the Patent and Trademark Office and not to the Department of Agriculture.

For more information on fees, see Appendix D, PTO Fee Schedule, on page 961.

Treaties and Foreign Patents

The rights granted by a U.S. patent extend only throughout the territory of the U.S. and have no effect in a foreign country. Therefore, to receive patent protection in other countries you'll have to make separate application(s) in each of the other countries or in regional patent offices. Almost every country has its own patent law.

The laws in many countries differ from our own. In most foreign countries, publication of the invention before the date of the application will bar the right to a patent. Most foreign countries require maintenance fees and that the patented invention be manufactured in that country within a certain period, usually about three years. If no manufacturing occurs within that period, the patent may be subject to the grant of compulsory licenses to any person who may apply for a license.

The Paris Convention

There is a treaty relating to patents which is followed by 93 countries, including the United States, and is known as the Paris Convention for the Protection of Industrial Property. It provides that each country guarantees to the citizens of the other countries the same rights in patent and trademark matters that it gives to its own citizens. The Paris Convention is administered by the World Intellectual Property Organization (WIPO) in Geneva, Switzerland.

The treaty also provides for the right of priority in the case of patents, design patents, and trademarks. This right means that, on the basis of a regular first application filed in one of the member countries, the applicant may, within a certain period of time, apply for protection in all the other member countries. These later applications will than be regarded as if they had been filed on the same day as the first application. Thus, these later applications will have priority over applications for the same invention which may have been filed during the same period by other persons. Moreover, these later applications, being based on the first application, will not be invalidated by any acts accomplished in the interval, such as, for example, publication or exploitation of the invention, sale of copies of the design, or use of the trademark.

The time frame allowed for subsequent applications in other member countries is 12 months in the case of utility patents and six months in the case of design patents and trademarks.

Patent Cooperation Treaty

Negotiated at a diplomatic conference in Washington, DC, in June of 1970, the Patent Cooperation Treaty (PCT) came into force on January 24, 1978, and is presently adhered to by the 39 countries listed alphabetically below.

List of Patent Cooperation Treaty Member States

1) Austria
2) Australia
3) Barbados
4) Belgium
5) Brazil
6) Bulgaria
7) Cameroon
8) Central Africa Republic
9) Chad
10) Congo
11) Democratic People's Republic of Korea (North Korea)
12) Denmark
13) Finland
14) France
15) Gabon
16) Germany, Federal Republic of
17) Hungry
18) Italy
19) Japan
20) Liechtenstein
21) Luxembourg
22) Madagascar
23) Malawi
24) Mali
25) Mauritania
26) Monaco
27) Netherlands
28) Norway
29) Republic of Korea (South Korea)
30) Romania
31) Senegal
32) Soviet Union
33) Sri Lanka
34) Sudan
35) Sweden
36) Switzerland
37) Togo
38) United Kingdom
39) United States of America

The PCT facilitates the filing of applications for patent on the same invention in member countries by providing, among other benefits, for centralized filing procedures and a standardized application format.

Under U.S. law it is necessary, in the case of inventions made in that country, to obtain a license from the Commissioner of Patents and Trademarks before applying for a patent in a foreign country. Such a license is required if the foreign application is to be filed before an application is filed in the United States or before the expiration of six months from the filing of an application in the U.S.

If the invention has been ordered to be kept secret, the consent to the filing abroad must be obtained from the Commissioner of Patents and Trademarks during the period the secret is in effect.

There is a basic fee for the first thirty pages, a basic supplemental fee for each page over thirty, a designation fee per member country or region, etc. **PCT fees are listed in Appendix D, PTO Fee Schedule, on page 961.**

Foreign Applications for U.S. Patents

Any person of any nationality may make application for a U.S. patent so long as that person is the inventor of record. The inventor must sign the same oath and declaration (with certain exceptions).

No U.S. patent can be obtained if the invention was patented abroad more than one year before filing in the U.S. Six months are allowed in the case of a design patent.

An application for a patent filed in the U.S. by any person who has previously regularly filed an application for a patent for the same invention in a foreign country (which affords similar privileges to citizens of the U.S.) shall have the same force and effect for the purpose of overcoming intervening acts of others. The requirement is that it be filed in the U.S. on the date on which the application for a patent (for the same invention) was first filed in the foreign country, provided that the application in the U.S. is filed within 12 months (six months in the case of a design patent) from the earliest date on which any such foreign application was filed. A copy of the foreign application certified by the patent office of the country in which it was filed is required to secure this right of priority.

If any application for patent has been filed in any foreign country prior to application in the U.S., the applicant must, in the oath or declaration accompanying the application, state the country in which the earliest such application has been filed, giving the date of filing the application. All applications filed more than a year before the filing in the U.S. must also be recited in the oath or declaration.

An oath or declaration must be made with respect to each and every application. When the applicant is in a foreign country, the oath or affirmation may be before any diplomatic or consular officer of the U.S. It may also be made before any officer having an official seal and authorized to administer oaths in the foreign country, whose authority shall be proved by a certificate of a diplomatic or consular officer of the U.S. In all cases, the oath is to be attested by the proper official seal of the officer before whom the oath is made.

When the oath is taken before an officer in the country foreign to the U.S., all the application papers (except the drawings) must be attached together and a ribbon passed one or more times through all the sheets. The ends of the ribbons are to be brought together under the seal before the latter is affixed and impressed, or each sheet must be impressed with the official seal of the officer before whom the oath was taken.

If the application is filed by the legal representative (executive, administrator, etc.) of a deceased inventor, the legal representative must make the oath or declaration.

When a declaration is used, the ribboning procedure is not necessary, nor is it necessary to appear before an official in connection with the making of a declaration.

A foreign applicant may be represented by any patent attorney or agent who is registered to practice before the U.S. Patent and Trademark Office.

For a comparison of patents issued to U.S. residents and patents filed by residents of foreign countries, see Tables 1 and 2 in Part One: Forms and Tables.

Section 9

Trademarks and Copyrights

*"When you can do the common things in life in an uncommon way,
you will command the attention of the
world."—George Washington Carver*

Coke. Greyhound. Scrabble. Crest. NBC's peacock. Mr. Goodwrench. The Campbell Kids. Morris the Cat. Paramount's mountain and stars. The MGM "pussy cat". They are all well known trademarks and are very important to their owners as tools to sell products and services.

Trademarks can be very important to inventors as a tool for helping to sell an invention or new concept. A good trademark creates fast product identification and can help tell the product's story. I always spend a great deal of time creating an appropriate trademark for a product that requires one as part of its release package. For example, my suggestion to Procter & Gamble that it license our patented tricycle for Crest to use as a premium would not have had the same sex appeal without the trademark, Crest Fluorider.

When told the name of our proposed original ride-on, the product's brand manager immediately responded, "Wow! That sounds exciting, a Crest Fluorider. When can we see a prototype?"

Trademark Office

Trademarks are handled by the Trademark Office of the Patent and Trademark Office, under the directorship of an Assistant Commissioner for Trademarks. It employs a total of 110 examining attorneys.

In 1990, the Trademark Office handled over 127,000 trademark applications. The federal government has issued over one million trademarks since the passage of the Trademark Act of 1905. Today, over 620,000 have active status. **Trademark Examination Activities, Fig. 9 on page 74, will give you some idea of how applications are routed and the time period needed to get them on the official trademark register.**

73

TRADEMARK EXAMINATION ACTIVITIES*

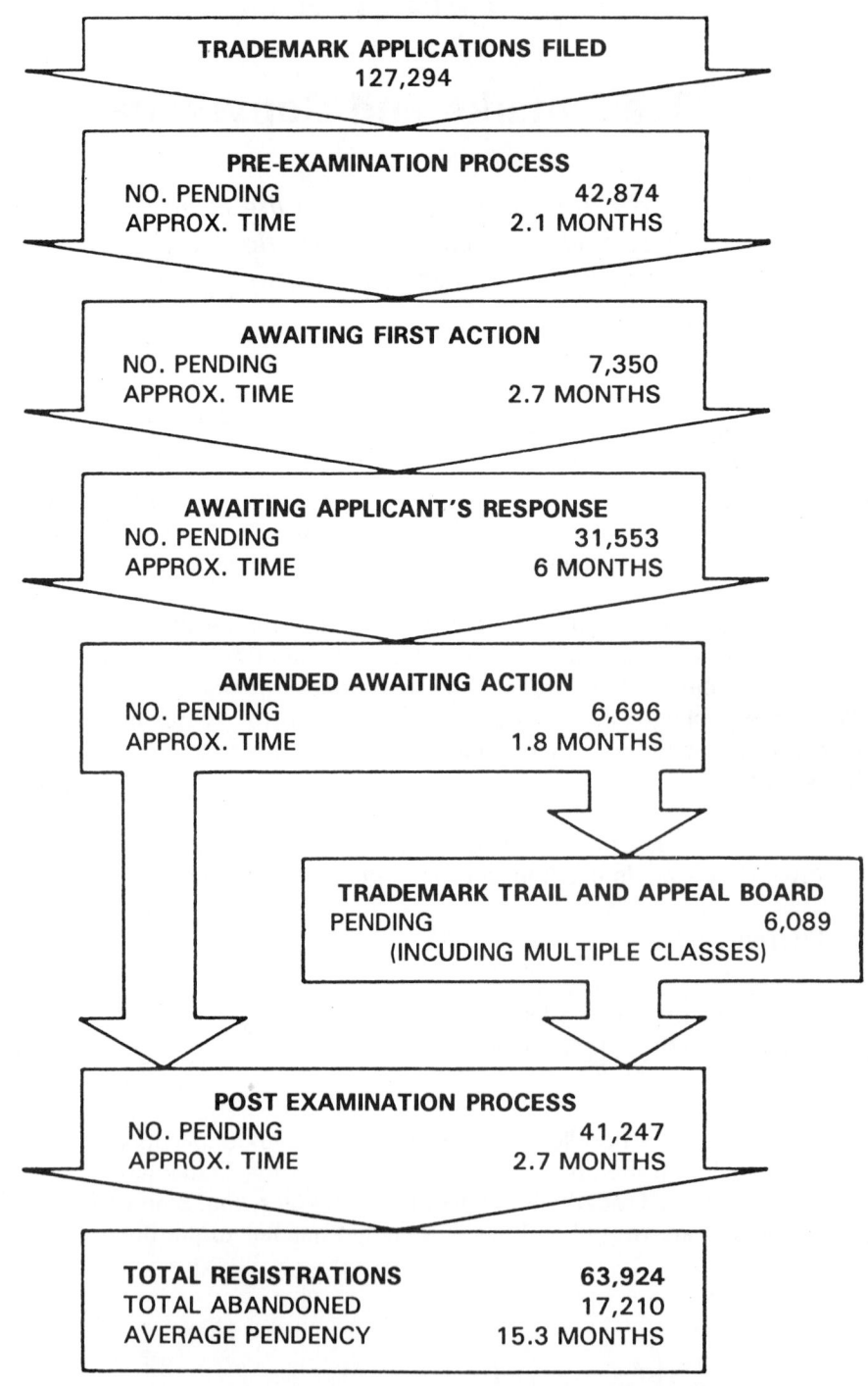

TRADEMARK APPLICATIONS FILED
127,294

PRE-EXAMINATION PROCESS
NO. PENDING   42,874
APPROX. TIME   2.1 MONTHS

AWAITING FIRST ACTION
NO. PENDING   7,350
APPROX. TIME   2.7 MONTHS

AWAITING APPLICANT'S RESPONSE
NO. PENDING   31,553
APPROX. TIME   6 MONTHS

AMENDED AWAITING ACTION
NO. PENDING   6,696
APPROX. TIME   1.8 MONTHS

TRADEMARK TRAIL AND APPEAL BOARD
PENDING   6,089
(INCUDING MULTIPLE CLASSES)

POST EXAMINATION PROCESS
NO. PENDING   41,247
APPROX. TIME   2.7 MONTHS

TOTAL REGISTRATIONS   63,924
TOTAL ABANDONED   17,210
AVERAGE PENDENCY   15.3 MONTHS

*This figure represents a simplified work-flow diagram with statistics on cases at various stages of processing at the end of FY 1987.

Figure 9

Trademark Correspondence and Information

All correspondence should be addressed to: Commissioner of Patents and Trademarks, Washington, DC 20231, unless you have the name of a particular official.

For information on a particular trademark, call (703) 557-5249.

For names of officials and their direct dial phone numbers, refer to Appendix B, Patent and Trademark Office Telephone Directory, on page 943.

What Are Trademarks?

Trademark. A "trademark," as defined in section 45 of the 1946 Trademark Act (Lanham Act) "includes any word, name, symbol, or device, or any combination thereof adopted and used by a manufacturer or merchant to identify his goods and distinguish them from those manufactured or sold by others." Examples of trademarks include Coca-Cola, Barbie, Ford, and Adver*teasing*.

Service Mark. A mark used in the sale or advertising of services to identify the services of one person and distinguish them from the services of others. Titles, character names, and other distinctive features of radio or television programs may be registered as service marks notwithstanding that they, or the programs, may advertise the goods of the sponsor. Examples of service marks include American Express, Mr. Goodwrench, and The Bill Cosby Show.

Certification Mark. A mark used upon or in connection with the products or services of one or more persons other than the owner of the mark to certify regional or other origin, material, mode of manufacture, quality, accuracy, or other characteristics of such goods or services, or that the work or labor on the goods or services was performed by members of a union or other organization. Examples of certification marks of quality include The UL symbol of Underwriters Laboratories and 100% Pure Florida's Seal of Approval. Examples of certification marks of service include AAA Approved Auto Repair and the MPAA's motion picture ratings.

Collective Mark. Trademark or service mark used by the members of a cooperative, an association, or other collective group or organization. Marks used to indicate membership in a union, an association, or other organization may be registered as Collective Membership Marks. Examples of collective marks include the NCAA, the National Rifle Association of American Member, AAA, and Sigma Delta Chi.

Trade and Commercial Name. Marks differ from trade and commercial names that are used by manufacturers, industrialists, merchants, agriculturists, and others to identify their businesses, vocations, or occupations, or other names or titles lawfully adopted by persons, firms, associations, companies, unions, and other organizations. The latter are not subject to registration unless actually used as trademarks. Examples of trade and commercial names include Coca-Cola Company, Gale Research Inc., and Sony Corp. of America.

To apply for a trademark or service mark, see Form 16 in Part One: Forms and Tables.

Function of Trademarks

The primary function of a trademark is to indicate the origin of a product or service; however, trademarks also serve to guarantee the quality of the goods bearing the mark and, through advertising, serve to create and maintain a demand for the product. Trademark rights are acquired only through use of the trademark; this use must continue if the rights you can acquire are to be preserved. Registration of a trademark in the Patent and Trademark Office does not in itself create or establish any exclusive rights, but it is recognition by the government of your rights to use the mark in commerce to distinguish your goods from those of others.

Intent-to-Use

Since November 16, 1989, the PTO began accepting trademark applications for intent-to-use, i.e. a bona fide intention to use a mark in commerce. If you make such an application, understand that you must make use of the mark in commerce before a registration will be issued. After use begins, you must submit, along with specimens evidencing use and a fee of $100 per class of goods or services in applications, either 1) an Amendment to Allege Use or 2) a Statement of Use. **(See Forms 17 and 18 in Part One: Forms and Tables.)** The difference between the two filings is the time of the filing. **(Also see Form 19, Request for Extension of Time to File a Statement of Use in Part One: Forms and Tables.)**

At the end of 1989, there were approximately 100,000 pending intent-to-use trademark applications. By the end of 1990, there were 140,000 pending applications! This trend is expected to continue.

This year, 48 percent of all trademark applications were filed under intent-to-use.

Do You Need a Federal Trademark Registration?

While federal registration is not necessary for trademark protection, registration on the Principal Register does provide certain advantages:

1) A constructive date of first use of the mark in commerce. This gives registrant nationwide priority as of that date, except as to certain prior users or prior applicants;
2) The right to sue in federal court for trademark infringement;
3) Recovery of profits, damages, and costs in a federal court infringement action and the possibility of treble damages and attorneys' fees;
4) Constructive notice of a claim of ownership. This eliminates a good faith defense for a party adopting the trademark subsequent to the registrant's date of registration;
5) The right to deposit the registration with Customs in order to stop the importation of goods bearing an infringing mark;
6) Prima facie evidence of the validity of the registration, registrant's ownership of the mark, and of registrant's exclusive right to use the mark in commerce in connection with the use of goods or services specified in the certificate;
7) The possibility of incontestability, in which case the registration constitutes conclusive evidence of the registrant's exclusive right, with certain limited exceptions, to use the registered mark in commerce;
8) Limited grounds for attacking a registration once it is five years old;
9) Availability of criminal penalties and treble damages in an action for counterfeiting a registered trademark; and
10) A basis for filing trademark applications in foreign countries.

Marks Not Subject to Registration

A trademark cannot be registered if it:

a) Consists of or comprises immoral, deceptive, or scandalous matter or matter that may disparage or falsely suggest a connection with persons, living or dead, institutions, beliefs, or national symbols, or bring them into contempt or disrepute;

b) Consists of or comprises the flag or coat of arms or other insignia of the United States, or of any state or municipality, or of any foreign nation, or any simulation thereof;

c) Consists of or comprises a name, portrait, or signature identifying a particular living individual except by his written consent, or the name, signature, or portrait of a deceased

president of the United States during the life of his widow, if any, except by the written consent of the widow;

d) Consists of or comprises a mark which so resembles a mark registered in the Patent and Trademark Office or a mark or trade name previously used in the United States by another and not abandoned, as to be likely when applied to the goods of another person, to cause confusion, or to cause mistake, or to deceive.

Registrable Marks

Principal Register: The trademark, if otherwise eligible, may be registered on the *Principal Register* unless it consists of a mark which, 1) when applied to the goods/service of the applicant is merely descriptive or deceptively misdescriptive of them, except as indications of regional origin, or 2) is primarily merely a surname.

Such marks, however, may be registered on the *Principal Register*, provided they have become distinctive as applied to the applicant's goods in commerce. The Commissioner may accept as prima facie evidence that the mark has become distinctive as applied to applicant's goods/services in commerce, proof of substantially exclusive and continuous use thereof as a mark by the applicant in commerce for the five years next preceding the date of filing of the application for registration.

Supplemental Register: All marks capable of distinguishing your goods and not registrable on the *Principal Register*, which have been in lawful use in commerce for the year preceding your filing for registration, may be registered on the *Supplemental Register*. A mark on this register may consist of any trademark, symbol, label, package, configuration of goods, name, word, slogan, phrase, surname, geographical name, numeral, or device, or any combination of the foregoing.

Trademark Search

Just like patent applications, it is advisable for you to make a search of registered marks before filing an application.

There are three ways available. Have your patent attorney do it; engage the services of a professional trademark search organization; or do it yourself.

Law Firm Trademark Search

A patent attorney will gladly handle a trademark search. The attorney will not do it personally, but instead will rely on the services of a professional trademark search firm. Law firms may step on search fees 40 percent to 100 percent or more depending upon what the market will bear.

If you decide to hire a lawyer for this job, ask for an estimate of costs up front. Lawyers most often use computer-aided search methods, which can take a few minutes. There may be a minimum charge. To avoid surprises, also find out in advance how much the lawyer will charge for copies.

Lawyers and professional searchers often charge more for logo searches. In my opinion, both take an equal amount of effort, yet a firm may charge $75 for trademarks and $150 for logos.

Professional Trademark Search

If you want to save yourself the lawyer's mark-up, consider hiring a trademark searcher yourself. They are listed in the Yellow Pages under "Trademark Searchers" or "Patent Searchers." Patent searchers will often handle trademark searches as well.

Get an estimate. An average price is $75 per hour with a one hour minimum. It should not take anywhere near one hour to do a single mark. Significant portions of the trademark operation are already fully automated. For over five years, the public has been able to access the PTO's electronic files concerning registrations and applications through the Trademark Reporting and Application Monitoring system (TRAM). Via this computer-aided search, it can take a few minutes to access bibliographic, status, and location data. Access to TRAM often obviates the need to consult the actual file.

Add to this price about $3 per copy of each trademark copied.

Do-it-Yourself Trademark Search

You can do the search utilizing two methods:

1) The Trademark Office has a Patent Public Search Room, located in Crystal Plaza 2, 2nd Fl., 2011 Jefferson Davis Hwy., Arlington, VA 22202. Once you learn the layout of the Search Room (which takes about five minutes), you can breeze through a manual search in 20 minutes.

2) If a visit to Arlington is inconvenient, see if there is a Patent and Trademark Depository Library (PTDL) in your area. It will offer numerous trademark reference books, e.g. Gale Research Inc.'s *Tradenames Dictionary*, Compu-Mark's *Directory of U.S. Trademarks*, etc., as well as an online computer search capability (See TRAM reference above).

Depending upon how extensive the search, costs can run anywhere from $10 and up for computer time. Often appointments must be made to access the computer services.

According to Commissioner Manbeck, the next generation trademark search system, dubbed X-Search, should be brought online in 1993. This system will allow: 1) mark searching; 2) access to applications and registration status and location; 3) access to assignment records; 4) access to the *Trademark Manual of Examining Procedure*; and 5) access to the *Design Code Search Manual*.

For PTDL listing, See Section 26, Patent and Trademark Depository Libraries, on page 337.

T-Search

T-Search is the PTO's online trademark search system to research word marks and design elements of trademarks. The PTO currently offers an experimental public use program on this system.

Terms of a Trademark

Unlike a copyright or patent, trademark rights can last indefinitely if the mark continues to perform a source-indicating function. The term of the federal trademark registration is 10 years, with 10 year renewal terms. However, between the fifth and sixth year after the date of registration, the registrant must file an affidavit stating the mark is currently in use in commerce. If no affidavit is filed, the registration will be canceled.

How to Register for a Trademark

A trademark application package consists of four items:

1) A written application;
2) A drawing of the mark;
3) Five specimens or facsimiles of the mark;
4) The required filing fee.

Trademark Application

Due to the recent changes in the trademark law, the PTO was unable to provide the updated forms for this publication. Trademark forms are available from the PTO, (703) 557-5249 or from Department of Commerce offices.

Trademark Pendency

At the end of fiscal year, the average time between the filing of an application and the PTO's mailing of the examining attorney's initial action on the application was 4.8 months. For applications based upon the use of a mark in commerce, the average time between filing and registration or abandonment was 15.3 months.

Trademark Classification

In the upper right hand corner of the application form, write the classification number that refers to your product. An application in which a single fee is submitted must be limited to the goods or to the services comprised in a single class.

The International Schedule of Goods and Services is reproduced as Fig. 10 on page 80.

Trademark Drawings

Your drawing must be a fairly exact representation of the mark as actually used in connection with the invention. You do not have to draw a service mark if it is not capable of representation by a drawing. If your application is for registration of a word, letter or numeral, or any combination thereof, not depicted in special form, the drawing may be simply the mark typed in upper case letters on paper, otherwise complying with the requirements.

Here are the exact specifications required by the PTO for trademark drawings:

Paper and Ink. The drawing must be made upon pure white durable paper, the surface of which must be calendered and smooth. A good grade of bond paper is suitable. India ink alone must be used for pen drawings to secure perfectly black solid lines. The use of white pigment to cover lines is not acceptable.

Size of Sheet and Margins. The size of the sheet on which a drawing is made must be 8.5 x 11 inches. One of the shorter sides of the sheet should be regarded as its top. When the figure is longer than the width of the sheet, the sheet should be turned on its side with the top at the right. The size of the mark must be such as to leave a margin of at least one inch on the sides and bottom of the paper and at least one inch between it and the heading.

Heading. Across the top of the drawing, beginning one inch from the top edge and not exceeding one-fourth of the sheet, there must be placed a heading, listing in separate lines, applicant's name, applicant's post office address, date of first use, date of first use in commerce, and the goods or services recited in the application (or typical items of goods or services if a number is recited in the application). This heading may be typewritten.

Character of Lines. All drawings, except as otherwise provided, must be made with the pen or by a process which will give them satisfactory reproduction characteristics. Every line and letter, names included, must be black. This direction applies to all lines, however fine, and to shading. All lines must be clean, sharp, and solid, and they must not be too fine or crowded. Surface shading, when used, should be open. A photolithographic reproduction or printer's proof copy may be used if otherwise suitable. Photocopies are not acceptable.

International schedule of classes of goods and services

Goods

1 — Chemicals used in industry, science, photography, as well as in agriculture, horticulture, and forestry; unprocessed artificial resins; unprocessed plastics; manures; fire extinguishing compositions; tempering and soldering preparations; chemical substances for preserving foodstuffs; tanning substances; adhesives used in industry.

2 — Paints, varnishes, lacquers; preservatives against rust and against deterioration of wood; colourants; mordants; raw natural resins; metals in foil and powder form for painters, decorators, printers and artists.

3 — Bleaching preparations and other substances for laundry use; cleaning, polishing, scouring and abrasive preparations; soaps; perfumery, essential oils, cosmetics, hair lotions; dentifrices.

4 — Industrial oils and greases; lubricants; dust absorbing, wetting and binding compositions; fuels (including motor spirit) and illuminants; candles, wicks.

5 — Pharmaceutical, veterinary, and sanitary preparations; dietetic substances adapted for medical use, food for babies; plasters, materials for dressings material for stopping teeth, dental wax, disinfectants; preparations for destroying vermin; fungicides, herbicides.

6 — Common metals and their alloys; metal building materials; transportable buildings of metal; materials of metal for railway tracks; non-electric cables and wires of common metal; ironmongery, small items of metal hardware; pipes and tubes of metal; safes; goods of common metal not included in other classes; ores

7 — Machines and machine tools; motors (except for land vehicles); machine coupling and belting (except for land vehicles); agricultural implements; incubators for eggs.

8 — Hand tools and implements (hand operated); cutlery; side arms; razors.

9 — Scientific, nautical, surveying, electric, photographic, cinematographic, optical, weighing, measuring, signalling, checking (supervision), life-saving and teaching apparatus and instruments; apparatus for recording transmission or reproduction of sound or images; magnetic data carriers, recording discs; automatic vending machines and mechanisms for coin-operated apparatus; cash registers, calculating machines, data processing equipment and computers; fire-extinguishing apparatus.

10 — Surgical, medical, dental, and veterinary apparatus and instruments, artificial limbs, eyes and teeth; orthopedic articles; suture materials.

11 — Apparatus for lighting, heating, steam generating, cooking, refrigerating, drying, ventilating, water supply, and sanitary purposes.

12 — Vehicles; apparatus for locomotion by land, air or water.

13 — Firearms; ammunition and projectiles; explosives; fireworks.

14 — Precious metals and their alloys and goods in precious metals or coated therewith, not included in other classes; jewelry, precious stones; horological and other chronometric instruments.

15 — Musical instruments.

16 — Paper and cardboard and goods made from these materials, not included in other classes; printed matter; bookbinding material; photographs; stationery; adhesives for stationery or household purposes; artists' materials; paint brushes; typewriters and office requisites (except furniture); instructional and teaching material (except apparatus); plastic materials for packaging (not included on other classes); playing cards; printers' type; printing blocks.

17 — Rubber, gutta-percha, gum, asbestos, mica and goods made from these materials and not included in other classes; plastics in extruded form for use in manufacture; packing, stopping and insulating materials; flexible pipes, not of metal.

18 — Leather and imitations of leather, and goods made from these materials and not included in other classes; animal skins, hides; trunks and travelling bags; umbrellas, parasols and walking sticks; whips, harness and saddlery.

19 — Building materials (non-metallic); non-metallic rigid pipes for building; asphalt, pitch and bitumen; non-metallic transportable buildings; monuments, not of metal.

20 — Furniture, mirrors, picture frames; goods (not included in other classes) of wood, cork, reed, cane, wicker, horn, bone, ivory, whalebone, shell, amber, mother-of-pearl, meerschaum and substitutes for all these materials, or of plastics.

21 — Household or kitchen utensils and containers (not of precious metal or coated therewith); combs and sponges; brushes (except paint brushes); brush-making materials; articles for cleaning purposes; steel wool; unworked or semi-worked glass (except glass used in building); glassware, porcelain and earthenware, not included in other classes.

22 — Ropes, string, nets, tents, awnings, tarpaulins, sails, sacks; and bags (not included other classes); padding and stuffing materials (except of rubber or plastics); raw fibrous textile materials.

23 — Yarns and threads, for textile use.

24 — Textile and textile goods, not included in other classes; bed and table covers.

25 — Clothing, footwear, headgear.

26 — Lace and embroidery, ribbons and braid; buttons, hooks and eyes, pins and needles; artificial flowers.

27 — Carpets, rugs, mats and matting; linoleum and other materials for covering existing floors; wall hangings (non-textile).

28 — Games and playthings; gymnastic and sporting articles not included in other classes; decorations for Christmas trees.

29 — Meats, fish, poultry and game; meat extracts; preserved, dried and cooked fruits and vegetables; jellies, jams; eggs, milk and milk products; edible oils and fats; salad dressings; preserves.

30 — Coffee, tea, cocoa, sugar, rice, tapioca, sago, artificial coffee; flour, and preparations made from cereals, bread, pastry and confectionery, ices; honey, treacle; yeast, baking-powder; salt, mustard, vinegar, sauces, (except salad dressings) spices; ice.

31 — Agricultural, horticultural and forestry products and grains not included in other classes; living animals; fresh fruits and vegetables; seeds, natural plants and flowers; foodstuffs for animals, malt.

32 — Beers; mineral and aerated waters and other non-alcoholic drinks; fruit drinks and fruit juices; syrups and other preparations for making beverages.

33 — Alcholic beverages (except beers).

34 — Tobacco; smokers' articles; matches.

Services

35 — Advertising and business.

36 — Insurance and financial.

37 — Construction and repair.

38 — Communication.

39 — Transportation and storage.

40 — Material treatment.

41 — Education and entertainment.

42 — Miscellaneous.

Figure 10

Extraneous Matter. Extraneous matter must not appear upon the face of the drawing.

Linings for Color. Where color is a feature of the mark, the color or colors may be designated in the drawing by means of conventional linings as shown in the following chart:

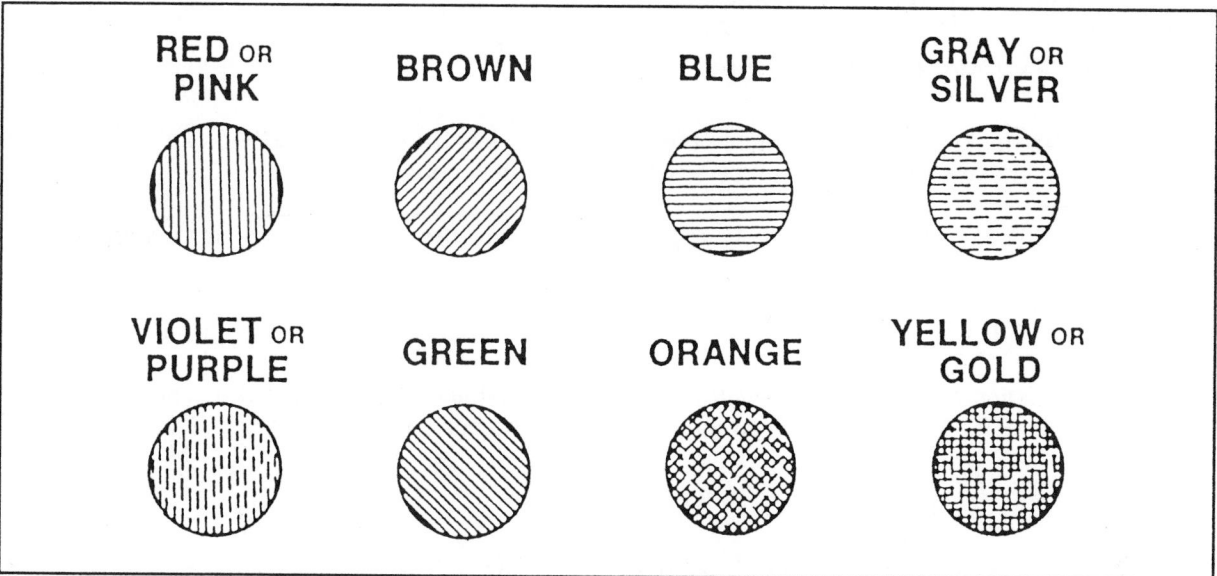

Transmission of Drawings. Other than typed drawings, you should send your artwork to the PTO flat and well protected by a sheet of heavy binder's board, or rolled and posted in a suitable mailing tube.

Informal Drawings. A drawing that does not conform with these requirements may be accepted for the purpose of examination only, but it must be ultimately corrected or a new one furnished, as required, before the application will be allowed or the mark can be published.

Trademark Specimens/Facsimiles

Trademark Specimens. A trademark may be placed in any manner on the goods, or their containers, tags, labels, or displays. The five specimens submitted must be duplicates of the actual labels, tags, containers, or displays or portions thereof, when made of suitable material and capable of being arranged flat and of a size no larger than 8.5 x 13 inches. Three-dimensional or bulky material submitted as specimens will not be accepted, and may delay your filing date.

Trademark Facsimiles. If you must submit facsimiles of your trademark, send in five copies of a suitable photograph or other acceptable reproduction, not larger than 8.5 x 13 inches and clearly and legibly showing the mark and all matter used in connection with it.

Service Mark Specimens or Facsimiles. In the case of a service mark, specimens or facsimiles of the mark as used in the sale or advertising of the services must be furnished unless impossible from the nature of the mark or the manner in which it is used, in which event some other representation must be submitted.

If your service mark is not used in a printed or written form, only three recordings will be required.

Trademark Filing Fee

The fee is $175 for each class of goods or services for which an application is made. At least $175 must be submitted for the application to be given a filing date. Personal or business checks may be submitted. Application fees are nonrefundable.

Trademark Examination Procedure

Applications are docketed and examined in the order of receipt. If the examiner finds any reason that the proposed mark will not pass, you'll be advised of the objections.

You are given six months to respond. If you fail to respond, the application for trademark will be abandoned.

In fiscal year 1988, the wait for a trademark to be awarded was an average of 13.3 months.

For more information see Fig. 9, Trademark Examination Activities, on page 74. For a schedule of fees, see Appendix D, PTO Fee Schedule. You may also check the current fees by calling (703) 557-INFO.

Copyrights

What do Hal David's song "Do You Know the Way to San Jose?," Cadaco's hit game "Adver*teasing*," *Baby Talk* magazine, "The Graduate," and this book have in common? If you said they are all protected by copyright, you are correct.

Copyrights are very different than patents and trademarks. A patent primarily prevents inventions, discoveries, or advancements of useful processes from being manufactured, used, or marketed. A trademark is a word, name, or symbol to indicate origin, and in so doing distinguish the products and services of one company from those of another. Copyrights protect the form of expression rather than the subject matter of the writing. They protect the original works of authors and other creative people against copying and unauthorized public performance.

One of the most aggressive enforcers of copyrights is the Walt Disney Company, Burbank, CA. The owner of popular properties such as Mickey Mouse, Donald Duck, Roger Rabbit, Dick Tracy, Goofy, The Little Mermaid, and so on, Disney estimates that 25 percent of its 800 lawsuits and regulatory cases each year involve copyright and trademark issues. This is not surprising when you consider that the entertainment company's consumer products division earned more than $742 million in 1990.

Robert Ogden, vice president and counsel for Disney's consumer products division, says that illegally manufactured merchandise brings in more than $1 billion each year to bootleggers and pirates.

Officials report the United States loses some $40 billion per year worldwide from copyright violations.

The Copyright Office

Copyrights are not handled by the PTO. For this discussion we move across the Potomac River from the PTO's Crystal City, Virginia headquarters, up Independence Avenue, and on to Capitol Hill to the Library of Congress which is primarily responsible for administering copyright law.

What is a Copyright?

Copyright is a form of protection provided by the laws of the United States (Title 17, U.S. Code) to the authors of "original works of authorship" including literary, dramatic, musical, artistic, and certain

other intellectual works. The U.S. Constitution authorized Congress to establish the copyright legislation. The first federal copyright act was passed in 1790. The most recent change in copyright law took effect on March 1, 1989, which states under the Berne Convention treaty that the created works of U.S. citizens are now afforded copyright protection in 79 foreign countries. For more detailed information on the Berne Law, call (202) 783-3238.

What Can Be Copyrighted?

This protection is available to both published and unpublished works. I slap copyright notices on everything I create. Copyrights can be as important to an inventor as to an author, which is why I have included a discussion of them in this book. Instructions and other written instruments such as background papers, concept papers, drawings, photographs, and the like that relate to inventions are all protected under U.S. copyright laws.

How to Investigate the Copyright Status of a Work

There are several ways to investigate whether a work is under copyright. Here are the main ones:

1) Examine a copy of the work (or, if the work is a sound recording, examine the disk, tape cartridge, or cassette in which the recorded sound is fixed, or the album cover, sleeve, or container in which the recording is sold) for such elements as a copyright notice, place and date of publication, author and publisher.

2) Make a personal search of the Copyright Office catalogs and other records.

3) Have the Copyright Office make a search for you.

Individual Searches of Copyright Records

The Copyright Office is located in the Library of Congress, James Madison Memorial Bldg., 101 Independence Ave. SE, Washington, DC.

Most records of the Copyright Office are open to public inspection and searching from 8:30 a.m. to 5 p.m. Monday through Friday (except federal holidays). The various records available to the public include an extensive card catalog, an automated catalog containing records of assignments and related documents. Other records, including correspondence files and deposit copies, are not open to the public. However, they may be inspected upon request and payment of a $20-per-hour search fee.

If you wish to do your own searching in the Copyright Office, you will be given assistance in locating the records you need and in learning searching procedures. If the Copyright Office staff does the search, a search fee of $20-per-hour will be charged. The search will not be done while you wait.

To save you time, I have included a copy of the Copyright Office's Search Request form. **See Form 20 in Part One: Forms and Tables.** Make as many copies of this form as you need and fill them out before you go to the Library of Congress.

How to Secure a Copyright

The way in which copyright protection is secured under the present law is frequently misunderstood. In the past, one was required to fill out special forms and send them to the Library of Congress together with a check and copies of the original work. Under the new law, no publication or registration or other action in the Copyright Office is required to secure copyright.

Under present law, copyright is secured "automatically" when the work is created, and the work is "created" when it is fixed in a copy or phonographically recorded for the first time. In general "copies" are material objects from which a work can be read or visually perceived either directly or with the aid of a machine or device, such as books, manuscripts, sheet music, film, videotape, or microfilm.

However, it is still prudent to make a formal application with the Library of Congress. This is to establish a "public record" of your claim; receive a certificate of registration (required if you ever have to go into a court of law over infringement); and if you receive a copyright within five years of publication, it will be considered prima facie evidence in a court of law.

If you would like to get specific information on the process, call a Copyright Public Information Specialist at (202) 479-0700 weekdays between the hours of 8:30 a.m. and 5 p.m. These folks are a wealth of information. You can also write for information to Information and Publications Section, LM-455, Copyright Office, Library of Congress, Washington, DC 20559.

Notice of Copyright

Before you publicly show or distribute your work, notice of copyright is required. The use of the copyright notice is your responsibility and does not need any special advance permission from, or registration with, the Copyright Office.

The notice for visually perceptible copies should contain all the following three elements:

1) **The copyright symbol** © (the letter C in a circle), or the word "Copyright," or the abbreviation "Copr."

2) **The year of first publication of said work.** In the case of complications or derivative works incorporating previously published material, the year of first publication of the compilation or derivative work is enough. The year may be omitted where a pictorial, graphic, or sculptural work, with accompanying text (if any) is reproduced in or on greeting cards, postcards, stationary, jewelry, dolls, toys, or any useful articles.

3) **The name of the owner of copyright** in the work, or an abbreviation by which the name can be recognized, or a generally known alternative of the owner's name.

Example: © 1992 Richard C. Levy

You should affix the notice in such a way as to give it "reasonable notice of the claim of copyright."

How Long Copyright Protection Endures

A work that was created on or after January 1, 1978, is automatically protected from the moment of its creation, and is usually given a term enduring for the author's life, plus an extra fifty years after the author's death. In the case of a "joint work prepared by two or more authors who did not work for hire," the term lasts for fifty years after the last surviving author's death. For works made for hire, and for anonymous and pseudonymous works (unless the author's identity is revealed in Copyright Office records), the duration of copyright will be 75 years from publication or one hundred years from creation, whichever is shorter.

Works that were created before the present law came into effect, but had neither been published nor registered for copyright before January 1, 1978, have been automatically brought under the statute and are now given federal copyright protection. The duration of copyright in these works will generally be computed in the same way as for works created on or after January 1, 1978 (the life-plus-50 or 75/100-year terms apply to them as well). However, all works in this category are guaranteed at least 25 five years of statutory protection.

Under the law in effect before 1978, copyright was secured either on the date a work was published, or on the date of registration if the work was registered in unpublished form. In either case, the copyright endured for a first term of 28 years from the date secured. During the last (28th) year of the first term, the copyright was eligible for renewal. The new copyright law has extended the renewal term from 28 to 47 years for copyrights that were subsisting on January 1, 1978, making these works eligible for a total term of protection of 75 years. However, the copyright must be timely renewed to receive the 47 year period of added protection.

Deposit: The new law requires that you deposit two copies of the published work in the Copyright Office for use by the Library of Congress. If the work is unpublished, one copy is required. This should be done within three months of publication of notice of copyright. A failure to do so may mean a fine. Even so, such an omission would not affect your copyright protection.

What Copyrights Do Not Protect

Ideas cannot be copyrighted. The same is true of the name or title given to a product and the method or methods for doing something.

Copyright protects only the particular manner in which you express yourself in a literary, artistic, or musical form. Copyright protection does not extend to ideas, systems, devices, or trademark material involved in the development, merchandising, or usage of a product.

Application Forms

The forms you will probably need include:

For Original Registration

Form TX: For published and unpublished non-dramatic literary works.

Form SE: For serials, works issued or intended to be issued in successive parts bearing numerical or chronological designations and intended to be continued indefinitely (periodicals, newspapers, magazines, newsletters, annuals, journals, etc).

Short Form/SE and Form SE/Group: Specialized SE forms for use when certain requirements are met.

Form PA: For published and unpublished works of the performing arts (musical and dramatic works, pantomimes and choreographic works, motion pictures and other audiovisual works).

Form VA: For published and unpublished works of the visual arts (pictorial, graphic, and sculptural works, including architectural works).

Form SR: For published and unpublished sound recordings.

For Renewal Registration

Form RE: For claims to renewal copyright in works copyrighted under the law in effect through December 31, 1977 (1909 Copyright Act).

For Corrections and Amplifications

Form CA: For application for supplementary Copyright Registration. Use when an earlier registration has been made in the Copyright Office; and some of the facts given in that

registration are incorrect or incomplete; and you want to place the correct or complete facts on record.

See Forms 21-28 in Part One: Forms and Tables.

The cost is only $10 per registration. This fee has been the same since 1978, and, unlike PTO fees, an act of Congress is needed to increase it. **If you desire original forms to register copyrights, call the forms hotline at (202) 707-9100 and leave your request as a recorded message.** The Copyright Office has established this hotline to speed up service. Requests made on the hotline are handled promptly. The recorders are connected 24 hours a day, seven days a week.

Section 10

How to License Your Invention

"There are certain things that our age needs. It needs, above all, courageous hope and the impulse to creativeness."—Bertrand Russell

Competition drives American industry and creates a fertile market for new inventions. No scarcity exists of progressive manufacturers willing to consider appropriate, new, and innovative products from independent inventors. Many companies in every field strive to produce better and more competitive products on a continuing basis, and often rely upon the independent inventor.

What Victor Hugo wrote in 1852 has never rung truer than today: "Greater than the tread of mighty armies is an idea whose time has come." The difficulty is bridging the supply and demand gap between the creators of new ideas and those who possess the capabilities and facilities to manufacture and market them successfully.

America constantly demands new products. We are, after all, a "throw away" society. We purchase something, use it for a while, and upgrade to the next generation of the product as soon as it's affordable. Generally, we do not take good care of things, nor do we spend much money fixing them when they break.

The biggest problem inventors have is how to license their innovations. While many people have the ability to dream and innovate, few of them also have the ability to sell. Many great ideas never come to market because of this.

"I knew it was a good idea, but I didn't have the knowledge or experience...," 78-year-old West Palm Beach, Florida inventor Emma DeSarro told *The Washington Post*, referring to her idea for the in-line roller skate.

In the late 1950s, Emma DeSarro created a skate design that would give birth to a $250-million-a-year industry, in-line roller skates (a/k/a Rollerblades). Her proof is U.S. patent No. 3,387,852: "Detachable and Removable Roller Skates," based upon "a single longitudinal line of wheels mounted on special frames."

She reportedly spent about $900 developing a prototype and getting the patent. Then after being rejected by the Chicago Roller Skate Company, in desperation she contacted an invention marketing company that took her money and didn't deliver.

87

There are two ways to get your invention manufactured and marketed. One is to license it to a firm that specializes in manufacturing and selling products. The other is to raise venture capital and become the manufacturer. Both have pluses and minuses. Neither is risk free. There are no guarantees either way, just different opportunities. Both are in the best and proudest traditions of American entrepreneurism.

I subscribe to the school of thought that says it is best to license inventions to companies run by executives who know more than you do about how to cost-effectively develop, manufacture, package, market, and promote products to specifically targeted users. I have no desire to go through the learning curve to acquire an expertise in all that's needed to do what it is a manufacturer does. If you are thinking of doing this, remember Murphy. Murphy's first corollary is: Nothing is as easy as it looks. Murphy's second corollary is: Everything takes longer than you think.

My aim is to create and maintain an atmosphere where I can be my most creative and productive, a situation free from supervisors, meetings, suits, power lunches, time-tested methods, office politics, and assorted "administrivia." You don't do this by raising venture capital and starting a manufacturing company. You do this by accepting the risks inherent in independent product development and licensing your creative yield.

You have three methods to address the business of invention marketing and licensing. The first is the do-it-yourself approach. I prefer this method. Over the years I have found that putting together my own team and making my own deals works best. I prefer to take my chances, not someone else's. It is more fun, I have better control, and the rewards are far greater.

Another way is to take your inventions to an inventor's exposition and run it up the flag pole and see if someone salutes. Under no circumstances would I recommend this avenue for an unprotected concept—or any concept for that matter. Expos can be good for publicity but not much else.

Finally, you can engage the services of an agent or broker. Good agents are few and far between. The best agents become so involved that they become virtual co-developers. I speak from personal experience, based on occasionally representing products other than my own. Nothing less will do. Jon W. Bayless, a partner in Sevin Rosen Bayless Venture Fund, says "A third party who is not committing his future to the business has no part to play in the process. Third party involvement is frequently a reason for a project to lose the consideration it may deserve."

Let's first look at this third possibility because as soon as the patent issues, you are likely to hear from one species of agent: the infamous invention marketing service.

Invention Marketing Services

How Do They Get Your Name?

Don't be surprised if you first learn of your patent award from an "invention marketing" company. When your patent issues, a notice of it will be automatically carried in the *Official Gazette* of the United States Patent and Trademark Office. This publication, issued weekly since 1872, comes out each Tuesday. It publicly records the following information about each new patent:

1) the name, city, and state of residence of the applicant, with the Post Office address in the case of unassigned patents;
2) the same data for the assignee, if any;
3) the filing date;
4) the serial number of the application;
5) the patent number;

6) the title of invention;
7) the number of claims;
8) the U.S. classification by class and subclass;
9) a selected figure of the drawing, if any, except in the case of a plant patent;
10) a claim or claims;
11) international classification;
12) U.S. patent application data, if any; and,
13) foreign priority application data, if any.

In the case of a reissue patent they publish the additional data of the number and date of the original patent and original application.

Invention marketing companies subscribe to the *Official Gazette* as an inexpensive way to obtain a qualified mailing list of inventors recently granted patent protection. They typically approach the independent inventors rather than their assignee corporations.

The "Pros" and the "Cons"

Caveat emptor! Separating the good invention marketing companies from the bad ones is not easy. The invention marketing business is rife with nonperforming, paracreative slugs who prey on and take advantage of unsuspecting, inexperienced, frustrated, and hungry independent inventors. I have never known of one invention marketing organization that advertises via direct mail, classified ads, radio, or television spots to be honest or effective. They are often called by a very appropriate appellation: front money frauds.

Alan A. Tratner, president of Inventors Workshop International, says about such parasites, "We view the unsavory practices of many of these firms and individuals as a 'cancer' in the inventing community that needs to be eradicated immediately. Too often inventors lose large amounts of money and are derailed by the unfulfilled promises and come-ons of these companies."

Florida inventor Frederick L. Jones, a member of the Palm Beach Society of American Inventors, agrees about invention marketing companies. "They prey on people. They'll take all the money they can," says Jones.

Frank Smirne, president of F. Smirne Plastic Co., Allentown, Pennsylvania, suggests that inventors get advice from other inventors before dealing with invention marketeers.

Al Lawrence Smith of the Patent and Trademark Office cautions, "Beware of these people." He suggests that inventors only permit marketeers to have their patented products on a contingency, non-exclusive basis.

"Pure and simple garbage," is how Fred L. Hart, Outreach Inventor Liaison at NIST's Office of Energy-Related Inventions, describes the work most invention marketing services do for inventors. In a meeting with Hart at his Gaithersburg, MD office he showed me two submitted invention assessment reports that a Pittsburgh invention marketing firm had prepared for two different inventors and their inventions. The first was for a frost pump. The second was for a magnetic engine (a/k/a perpetual motion machine). Both reports were 34 pages in length. Both were identical, word-for-word, except for the name of the invention and the respective product descriptions.

The inventors paid the invention marketing company for the same boilerplate. It was obvious that a standard text had been designed and was being used for every product. In the case of the perpetual motion machine, Hart said the company had written and submitted three separate but identical reports on behalf of three different inventors who felt they had discovered the way around gas engines. The

reports were all signed by the same man who tagged his name with the initials P.E. (Professional Engineer).

According to charges made in 1991 by the Federal Trade Commission (FTC), American Idea Management (AIM) of Stoneham, Massachusetts, has misrepresented the nature, quality, and success rate of invention promotion services they have sold to inventors since 1984. Also named in the charges are AIM's two successor corporations, Idea Management and Patent Assistance Corp. of Stoneham and Technology Licensing Consultants, Inc. of Pittsburgh, PA; and corporate owners Suzanne Kameese of Lynfield, MA, and Anita and Lowell French of Pittsburgh, PA. The FTC proposed a $570,000 (plus interest) fine for consumer redress.

Among the false claims, the FTC complaints allege, are representations that:

*The defendants' clients have realized financial gains when, in fact, no inventor has earned more money than he or she paid the defendants for their services.
*The free "initial review" and "research report" are honest, competent and objective evaluations of merit or marketability of the invention when, in fact, they were not.
*The defendants perform substantial services to develop or refine clients' ideas when, in most instances, they did not.
*They have no special access to, or have been retained by, manufacturers seeking new product ideas, though, in fact, neither claim was true.

The Federal Trade Commission's proposed consent decrees, which require the approval of the courts in which they were filed, would permanently ban the defendants from making the kinds of misrepresentations alleged in the complaints, and from misrepresenting any material aspect of their services.

It can almost be guaranteed that if you opt to engage the services of an invention marketing company the odds are that the only ones who'll get rich will be the invention marketeers on your payroll. Gerald Udell, a professor at Southwestern University and owner of the Innovation Institute, told *The Wall Street Journal* that he estimates invention marketing companies rack up more than $100 million a year.

I tested one of these companies myself. I got it through a toll-free number heard on late night radio. We submitted an idea for an absurd device that kept individual strands of spaghetti fresh. The concept was to insert spaghetti strands into clear plastic extrusions capped off at both ends. "Terrific. Lots of potential," I was told by telephone. All I had to do was pay about $400 for an assessment.

Watch out for them. They are easily recognizable. You'll find their ads running on all-night radio talk shows, at 3 a.m. sponsoring Million Dollar Movie reruns, and in the classified ads sections of newspapers and magazines that appeal to do-it-yourselfers, e.g. *Popular Mechanics, Popular Science, Income Opportunities,* etc.

The direct mail pieces they send might be in the form of a postal card. Here is an example:

IMPORTANT NOTICE

To: Owner of U.S. Patent

We have located six companies that produce, market, or sell products in a field to which your invention might apply. You may wish to contact these manufacturers if you are interested in one or more of the following:

A. Licensing your patent.

B. Finding a company to produce your invention.

C. Securing marketing, distribution, and/or sales help.

D. Hiring design or technical support.

Then comes the kicker. You are asked to remit a fee of $75 plus $1 for each company named.

Other direct mail offers come in #10 envelopes overflowing with all kinds of slick fliers and official-looking confidential disclosure agreements with diploma-like borders. Rarely will you be able to make out the signature of the "authorized agent" who signed off on the standard "Dear Inventor" cover letter. If they feature any products which have been successfully licensed to industry, it is doubtful you will recognize any of them. The company's toll-free number will be something like 1-800-SUCCESS.

They use all kinds of subtle techniques to compensate for the fact that they have no successful products to show off. For example, I recall seeing one promotional flier that depicted a beautiful, modern building with the caption "4th Floor, Chicago, Illinois." The company wanted the reader to assume that this was the base of its corporate headquarters. However, no reference was made to the picture or office building anywhere in the flier's copy.

The more sophisticated invention marketing companies wrap themselves in Old Glory and apple pie. Their names incorporate words like American, Federal, and National and they have Washington, DC postal addresses.

How to Identify the Good Guys

One of the acid tests I use to separate the honest and serious professional invention marketing companies from those involved in unfair business practices is whether or not they want any advance payments from the inventor for anything whatsoever.

"The firm we dealt with appears to be interested in marketing our idea for us," a woman from Virginia wrote me, "but they are quoting us a price of from $3,000-$9,000 for this service. Also, they want a slice of any profits (up to 60 percent)."

The most reputable invention marketing services will not require front money from the inventor. You invested your ingenuity, time, and money in creating the product. The marketer must invest ingenuity, time, and money in causing the product to be licensed. What is more fair than that? The moment you pay for marketing services, the carrot is removed. There is no risk. With nothing to lose, and a gain already in the bank, the incentive for your marketing partner is reduced.

Charles F. Mullen, an inventor from Houston, Texas, works part time as sales manager for national accounts at an advertising agency, and says he spends the rest of his time inventing and helping others market their products. "We work straight percentage. We don't charge them anything up front," he says. "That's the trouble with the industry. Most of the people get them up front." Mr. Mullen claims to have licensed 14 inventions all in different fields.

A reputable and active marketing company will be able to demonstrate a track record of successful products and satisfied inventor clients. References should be available. Ask for a list of clients and their telephone numbers. The products they have marketed should be available somewhere other than their office waiting rooms.

Read the small print. I saw one flier that showed numerous products leaving the impression that they had been licensed by the invention marketing group. Then I saw the disclaimer. "These products are not intended to represent success for inventors who have worked with our firm." They should be able to substantiate licensing agreements claimed and royalty earnings. Do they make the majority of their

earnings from royalties or from inventor service fees? If it is not from royalties, you're in the wrong place.

Always check the name of the invention marketing company and/or individual marketer with the local Better Business Bureau, state agency for consumer fraud, and your Attorney General's office. See if any complaints have ever been registered against the company or its officers and representatives.

Ask fellow inventors. Word about bad apples spreads like wild fire. Many inventor organizations maintain files on the worst invention marketing organizations and frequently write them up in their newsletters.

See Section 13, National and Regional Inventor Organizations, on page 117, and Section 23, Publications, on page 291 for more information on publications.

Inventors' Digest Litmus Test

Amateur inventors, according to Joanne M. Hayes, editor of *Inventors' Digest*, a publication of Affiliated Inventors Foundation, Inc. (AIF), are full of questions, not the least of which is, "How can I find somebody to help me with my invention who won't rip me off?"

AIF says that the following questions should help you to identify practices which are warning signals and typify activities operating with doubtful honesty. "If you check any of the non-bold responses, be warned that there is a strong possibility you are dealing with crooks. We recommend that you avoid doing business with any company that does not answer the following questions satisfactorily," advises Hayes.

How to Rate an Invention Marketing Company:

Do they imply that the Disclosure Document Program (DDP) of the U.S. Patent Office offers protection or that it is a viable substitute for a U.S. Patent?
> yes **no**

Do they offer a meaningful evaluation of inventions at a reasonable price (less than $200)?
> **yes** no

Does the evaluation report point out weaknesses as well as strengths of your invention?
> **yes** no

Does the evaluation report make specific recommendations concerning such important areas as patenting, model making, marketing, and protection?
> **yes** no

Do they recommend a patent search before recommending expensive marketing services?
> **yes** no

If they offer to perform a patent search, do they state specifically that the search will include an opinion of patentability by a registered patent attorney or agent?
> **yes** no

If you purchased a patent search, did they tell you the name of the individual responsible for the search and the name and registration number of the individual providing the opinion of patentability?
> **yes** no

Do they ask for exorbitant front-end cash fees running to hundreds or thousands of dollars for their "best efforts" to sell or present your invention to industry?

      yes   **no**

Do they tell you that in most cases to sell or license the rights to your invention you need a working model?

      yes   **no**

Do they tell you anything in conversation, such as the "company is all lined up to buy your invention," which does not end up in the written contract?

      yes   **no**

If they think your invention is so promising, are they willing to offer their marketing services on a no-cash, straight-commission basis?

      yes   no

Do they suggest trying to market your potentially patentable invention before it reaches the patent pending stage?

      yes   **no**

Do they recommend filing for a patent application before beginning marketing efforts?

      yes   no

Do they quote a firm, fixed price for preparation and filing a patent application (except for patent drawings, which are priced by the sheet)?

      yes   no

Do they tell you about the patent application costs after filing, such as the normally required amendments, the final issue fee, maintenance fees, etc.?

      yes   no

Are they willing to provide you with names and addresses of inventors who have used their services?

      yes   no

Is the company willing to provide you with a copy of its latest financial statement indicating its net profit?

      yes   no

If you checked even one non-bold response, you should investigate the company in greater detail before proceeding. It is also wise to check invention companies out with your State Attorney General, and even the Federal Trade Commission.

Federal Trade Commission

The Federal Trade Commission (FTC) has received so many complaints about invention marketing companies that it has a complaint category for them.

If you wish to register a complaint against an invention marketing company, write to: Stanley Ciurczak, Chief, Complaint Division, Room 692, Federal Trade Commission, Washington, DC 20580.

Based upon a case decided in 1978, the FTC has determined that the following practices used in the advertising and marketing of idea or invention promotion or development services are unfair or deceptive trade practices and are unlawful under Section 5 (a) (1) of the FTC Act:

1) For a seller of idea or invention promotion or development services to misrepresent, directly or indirectly, that potential purchasers will be provided with evaluations or appraisals of the patentability, merit, or marketability of ideas or inventions.

2) To represent, directly or indirectly, that the seller of idea or invention promotion or development services, or its officers, agents, representatives, or employees are registered patent attorneys or patent agents, or are qualified to practice before the U.S. Patent and Trademark Office, unless such is a fact.

3) To misrepresent, directly or indirectly, the scope, nature, or quality of the services performed to develop or refine ideas or inventions.

4) To misrepresent, directly or indirectly, the scope, nature, or quality of the services performed to introduce or promote ideas or inventions to industry.

5) To represent, directly or indirectly, that a seller of idea or invention development service has special access to manufacturers or has been retained to locate new product ideas, unless such is a fact.

6) To misrepresent, directly or indirectly, that a person, partnership, corporation, government agency, or other entity endorses or uses the services of a seller or provider of services.

7) For sellers of idea or invention promotion or development services to fail to disclose, when price information is provided to potential purchasers, all significant fees or charges that may be incurred by purchasers in connection with such services.

8) To misrepresent, directly or indirectly, the background, qualifications, experience, or expertise of a seller or provider of services.

9) For a seller of idea or invention promotion or development services to induce through misleading or deceptive representations the purchase of services that have little or no inherent value, or to offer to provide services that grossly exceed the value of the services actually provided. It is also an unfair or deceptive act or practice to retain money from the sale of such services.

If you want to find out whether the FTC has any ongoing investigation into a particular invention marketing company, use a Freedom of Information Act (FOIA) request. The FTC will not divulge any particulars about an active investigation, legal or otherwise, or even about complaints received; however, if a consent order has been issued against the invention marketing company, you'll find out.

You may also read between the lines. If under an FOIA request you ask for copies of any and all FOIA requests for information about a specific company(s), you just may get in return copies of requests sent to the FTC by law firms representing said company(s). The law firm would not be making a request about its client unless something was amiss.

For example, I once made an FOIA request to find out whether anyone else was interested in a specific invention marketing firm. The FTC sent me copies of requests, although denied, from a television news department, and a law firm. The fact that the FTC had denied their requests told me that said invention marketing firm was being looked at for some reason. **Send your Freedom of Information Requests to:**

Keith Golden, FOIA Officer, Federal Trade Commission, 6th Street and Pennsylvania Avenue, N.W., Washington, DC 20580. The phone number for the FTC is (202) 326-2222; fax: (202) 326-2050.

Protective Legislation

The invention marketing business went unregulated for years. Now, however, some states such as Minnesota, Oklahoma, Virginia, Washington, and Tennessee have enacted protective legislation on behalf of resident inventors.

Let's take a quick look at some of the issues that the states address. This will give you a good idea of what to watch for and what to ask prospective invention marketing companies. If you want a full text of the legislation—which I recommend you acquire—contact the state legislature and it will be sent to you free of charge.

If your state does not have protective legislation for inventors yet, get the word to the appropriate elected officials. Every state should have some sort of regulation on its books that sets fair standards for invention marketing companies.

Oklahoma State legislator Tim Pope heard his inventor's requests for legislative assistance and subsequently helped push through a law regulating marketeers in his state. "These companies are simply ripping off inventors," Pope told *The Wall Street Journal*.

Let's take a look at the laws of Minnesota and Virginia. I shall highlight some important points they make.

Minnesota (Invention Services 325A.02)

1) Notwithstanding any contractual provision to the contrary, inventors have the unconditional right to cancel a contract for invention development services for any reason at anytime before midnight of the third business day following the date the inventor gets a fully executed copy.

2) A contract for invention development services shall be set in no less than 10-point type.

3) An invention developer who is not a lawyer may not give you legal advice with respect to patents, copyrights, or trademarks.

4) The invention marketer must tell you (1) the total number of customers who have contracted with him up to the last thirty days; (2) the number of customers who have received, by virtue of the invention marketer's performance, an amount of money in excess of the amount of money paid by such customers to the invention marketer pursuant to a contract for invention development services.

5) The contract shall state the expected date of completion of invention marketing services.

6) Every invention marketer rendering invention development services must maintain a bond issued by a surety admitted to do business in the state, and equal to either ten percent of the marketer's gross income from the invention development business during the preceding fiscal year, or $25,000, whichever is larger.

Virginia, Chapter 18,59.1-209

1) No invention developer may acquire any interest, partial or whole, in the title to the inventor's invention or patent rights, unless the invention developer contracts to manufacture

the invention and acquires such interest for this purpose at or about the time the contract for manufacture is executed.

2) The developer must tell you if they intend to spend more for their services than the cash fee you will have to pay.

3) The Attorney General has the mandate to enforce the provisions of this chapter, and recover civil penalties.

Marketing Via Inventor Expos, Trade Shows, Conferences, and Workshops

Expos

While some inventor expositions are worthwhile and well-meaning, such as the annual National Inventors Day Exhibit sponsored by the Patent and Trademark Office or regional shows sponsored by inventor organizations, many slick operators disguise their motives by organizing inventor expositions and fairs. Know thy promoters and their motives.

Some wolves in sheep's clothing invite inventors suffering from "sellitus" to display prototypes, even drawings or photographs of their inventions. They charge for booth exhibition space and advertising in publications that are released in conjunction with the event. The general public is charged an admission fee to see the inventions. Some even charge exhibitors a broker or commission fee should an invention be licensed through its exposure at the show. Another scheme is to offer the inventor a large cash buy out should a product sell, and the promoter walks off with the royalty points.

When all is said and done, my experience confirms that few if any meaningful contacts ever come out of these shows from the inventor's standpoint, while the promoters make money and receive publicity.

IMPORTANT! Should you be contemplating such a show, keep one very important thing in mind. If you intend to apply for patent protection on an invention, its display may foul your chances. Exhibiting a new invention in public for sale or otherwise will make it ineligible for patent protection if a patent application is not filed within one year from the time of the invention's first public exposure.

Trade Shows

Every industry takes part in trade fairs, including the butchers, the bakers, and the candlestick makers. According to the Trade Show Bureau, in 1990, 9,000 trade shows took place in the United States, with 3,289 of them in excess of 10,000 square feet or 100 booths.

The proliferation of trade shows in America has created a trade show industry, complete with its own newsletters, magazines, and professional organizations.

National, regional, and local events promote the sale of almost anything you can imagine. There are trade shows for everything from hardware, consumer electronics, apparel, and aircraft to nuclear medicine, dental equipment, toys, comic books, and musical instruments. If it is manufactured and sold, you can be sure there is a trade show somewhere, sometime.

The Patent and Trademark Office (PTO) even co-sponsors an annual trade show. The National Inventors Expo is held in February at the PTO's search room in Crystal City, Virginia. It features exhibits by independent inventors of their patented inventions, and large and small businesses of interest to the inventor. For current information, call (703) 557-3341. Admission is free and the event is open to the public.

In fact, most trade shows are admission free "to the trade." All you usually need is a business card to enter the exhibition area.

While I do not recommend trade shows as the best place to present or license patented inventions, they are a "must" for getting the beat on any particular market and its dynamics. It's all there for you to see. Competitors line up side by side for the all important buyers to compare the manufacturers' products and pricing.

If you go to the shows, remember that companies have paid many thousands of dollars to participate. Their primary reason to do a trade fair is to ring up sales. They are not there to license concepts.

The sales force does not exist to review new concepts. It is responsible to sell. It is both fruitless and dangerous to impose on and expose inventions to sales people. They are, however, excellent sources of information and normally delighted to chat about their products, the state of the market, and industry particulars. Information on the market can be gathered in a very convenient and time-effective manner.

There are exceptions, of course. In some industries, R&D executives attend trade shows to get a feel for the competition as well as host "invited" outside inventors with products to show. Presentations are typically conducted in hotel suites away from the exhibition site. It is best to call the corporate headquarters in advance of the trade show and check policy.

I attend the shows to scout for new product introductions, pick up information handouts and samples, and make personal contacts. No better or more cost-effective way exists to acquire product literature than at a trade show. Manufacturers publish fliers and information kits just for trade show distribution. And most come with price lists!

I take empty flight bags with me in which to transport home everything collected. Sturdy ones. I never rely on the paper or plastic bags some of the companies supply. The material is too valuable (and heavy!). Thanks to trade shows over the past decade, I now have a comprehensive reference library.

Trade shows are also an excellent place to meet and network with executives to whom you would otherwise not have access. They rarely take their "bodyguards" to trade shows; it is too expensive and, after all, they also go to meet new people. They even make it easier by wearing name tags!

I have made super contacts in convention hotel elevators, lobby queues, and taxi-shares to and from the exhibition centers.

The best kinds of shows at which to meet senior executives are the national or international ones. The smaller regional or local trade shows are typically staffed by sales people alone. Nevertheless, such shows provide a less hectic atmosphere and many of the same resource materials.

Many exhibitors host receptions and special events for buyers. If you get lucky, you may be invited to attend.

There are several ways to find out about where and when trade shows for any particular industry will be taking place. Gale Research Inc. annually publishes the *Trade Shows Worldwide*, a comprehensive listing of over 11,800 national and international shows. Check your local library for a copy.

Other ways to track down trade shows:

1) Ask a manufacturer or distributor in your field of invention. The sales and marketing people will have such information readily available.

2) Contact the trade association that covers your field of invention. About 3,600 trade associations operate on the national level in America. A great start is through Gale Research Inc.'s latest

Encyclopedia of Associations, available at most public libraries. The 1992 edition features detailed entries describing over 22,300 active associations, organizations, clubs and other nonprofit membership groups in virtually every field of human endeavor.

Conferences and Meetings

There are perhaps more conferences and meetings going on than trade shows. It doesn't take much to have either. Technically all you require is a number of experts sitting around a table discussing a field of interest.

The biggest difference between trade shows and conferences and meetings is that you almost always pay to attend conferences. This is because the primary reason for a conference is to hear experts speak, pick their brains, share your ideas, and network.

Conferences are excellent places to get to know the people behind the products. Socializing is encouraged and the atmosphere is calmer than at trade shows. There is no pressure to sell or buy. The object is to brainstorm and trade ideas. Participants can increase their "know-how" and "know-who" at the same time.

Many trade fairs have conferences or seminars as part of the program. And many conferences offer simultaneous resource fairs.

There are several ways to find out about where and when conferences for any particular industry will be taking place.

1) Ask a manufacturer or distributor in your field of invention. Many larger manufacturers have training departments which can provide helpful information.

2) Contact the trade association that covers your field of invention. There are about 3,600 trade associations operating on the national level in America. Again, a great start is through Gale Research Inc.'s latest *Encyclopedia of Associations* available at most public libraries. If the particular association holds a conference, it will usually be listed in its entry.

3) Ask department heads and professors at a nearby university where your field of interest is taught. Universities, especially those teaching engineering and related technical fields, will have a current schedule of conferences on hand.

National Innovation Workshops

The Department of Energy and the National Institute of Standards and Technology's Office of Technology Evaluation Assessment support National Innovation Workshops each year in different regions of the country to bring inventors in contact with local innovation sources. This program provides a network for innovation and is recognized as successful throughout the inventor community.

The National Innovation Workshop series was initiated in the spring of 1980. Its format is a standardized two-day seminar. There are two addresses each day (keynote and luncheon) by nationally known speakers; a wide variety of free, how-to and technical publications available from federal, state and local government agencies and other organizations; and daily periods of eight to ten concurrent workshops, totaling 48 to 60 individual sessions.

There are seven cosponsors at a national level: The National Congress of Inventor Organizations; American Intellectual Property Law Association; Licensing Executives Society; Association of Small Business Development Centers; National Society of Professional Engineers; U.S. Department of Commerce; and the U.S. Department of Energy.

Workshop topics include patenting and protection; estimating the worth of an invention; licensing; marketing; new business start-up; the business plan; R&D and venture financing; the DOE-NIST Energy-Related Inventions Program; and SBIR Programs. The cost of attending is about $85.

National Innovation Workshops for fiscal year 1992 will take place in the following cities:

Baltimore, MD	March 1992
Honolulu, HI	April 1992
Toledo, OH	May 1992
Syracuse, NY	June 1992
Albuquerque, NM	September 1992
Little Rock, AR	October 1992
Portland, OR	November 1992

Do-It-Yourself Marketing

This book is a guide to locating information and doing it yourself. All the information required in this regard is provided in the sections that follow. But remember this rule: never get so wrapped up in the marketing of the invention that you forget to market yourself.

First Impressions

Inventors are always selling two things: the concept and themselves. Personal credibility is often more important than any single creative concept.

It is critical that a corporate executive buy the inventor as much as the invention. You may be capable of dreaming up numerous innovative products for a company to consider, and will want to be invited back again and again. Without respect from corporate executives, your products will never be taken seriously.

PTO Can Help You Get the Word Out

The Patent and Trademark Office cannot help develop or market your invention, but it will publish, at the request of the patent owner, a notice in its *Official Gazette* that the patent is available for licensing or sale. The fee for this is a very reasonable $7.

Organizations Offer a Helping Hand

Inventor and business organizations, as well as state and university assistance programs can be of great value to the inventor preparing presentation packages for prospective licensees. **For more information, see Section 13, National and Regional Inventor Associations, on page 117, and Section 15, University and Independent Innovation Research Centers, on page 145.**

Marketing is a Science

Marketing inventions is a science; no less a discipline than electronics, mechanical engineering, or any other field of study. It can be learned. Of course, as with any talent, some natural abilities may make it easier for some, but nothing is out of reach if you take the time.

Section 11

How to Select Which Company to Approach

"One of the reasons mature people stop learning is that they become less willing to risk failure."—John Gardner

Remember the old saw that says if you build a better mousetrap, the world will beat a path to your door? Maybe so back in 1889 when Ralph Waldo Emerson suggested this, but today the reverse is true for inventors without track records. Amateur inventors spend a lot of time beating paths to the doors of companies looking to license better mouse traps.

Selecting which companies to approach as possible licensees for your invention is a matter that requires a great deal of thought. It is not something to be taken lightly. You don't want to make the same mistake Rick Blaine (Humphrey Bogart) made in the movie Casablanca. "I came to Casablanca for the waters," he said. When told that he was in the middle of a desert, he adds, "I was misinformed."

You must be well informed. Your first approach to a company requires the kind of detailed analysis, imagination, and forethought associated with championship chess. In fact, your approach should resemble the game of chess insomuch as it consists of a planned attack and defense, and has winning as its goal.

Just as much effort is required in dealing with a small company as with a large one. Don't be afraid to approach the major players. Remember, 80 percent of the business in any particular industry is typically done by 20 percent of the companies. You'll want a company capable of engineering and then following through on a success. The larger a company is, the better and faster it pays.

Put time into your decision on which manufacturer to approach. Don't make the error of insufficient options. Some rejection is to be expected, so you'll want as many targets as possible.

Study corporate product lines. Do store checks. Get new product catalogues. Companies do what they do, and bringing them something out of their discipline is usually a lost effort. You would not, for example, go to Black & Decker with a new type of record player. Black & Decker manufactures tools and labor saving devices. Round pegs don't fit square holes.

Attempting to get a company to purchase an inappropriate product is not only a waste of everyone's time, but it does nothing for your reputation, which is more precious then your invention.

Corporations exist to make money. Executives, especially those in lucrative profit-sharing plans and incentive programs, want their corporations to be successful, but at what risk? I have found that most senior executives will listen to any scheme that rings of potential profit. That profit can come in the form of a new product or a labor-saving device. But it is rare that an executive will rock the proverbial boat for untested, unfamiliar products, especially those that fall outside the company's expertise or channels of distribution.

Many large firms are guided by numbers as much as products. Lawyers and accountants tend to become CEOs before R&D executives do. These bean counters see products in terms of SKUs (stock keeping units) only. They like to keep the pipeline filled with line extensions of already proven and profitable products. Their attitude is that a major breakthrough in medicine, for example, would be nice, but let's keep the mouthwashes and toothpastes coming. These types would rather take a popular pudding and put it on a stick than gamble on creating a new novelty food. They are so preoccupied with statistics that they forget companies can create new products.

One company licensed a product of ours and was going to produce it until a consumer study showed that its popularity would upset existing business. The company opted for the status quo and dropped the item. "Why should we erode our market share in an industry we almost totally control?" reasoned the company president. "If we bring your product out, the consumers will obviously love it, but we'll just point to opportunities our competition does not realize exist."

On the other end of the spectrum, a senior research and development executive from a $1 billion plus company told me once, "It's my responsibility to waste $2 million per year on long shots. I would not be doing my job if I didn't." Unfortunately, there are few corporate executives with this entrepreneurial attitude.

The first move is yours. It is not an easy one. Going to the wrong company with the wrong product can cost valuable time, do nothing to enhance your contacts, and even bring grief. You want your idea in the right hands because, as advertising legend Bill Burnbach said, "An idea can turn to dust or magic depending on the talent that rubs against it."

NIH Syndrome

NIH, as used here, is not the acronym for the government's National Institutes of Health. NIH is corporate jargon for "Not Invented Here," a syndrome from which many companies suffer. It means that such companies have in-house research and development staffs and do not entertain outside patent submissions.

It is hard to tell which companies fall into the NIH category. This learning curve is part of every selection exercise.

There are two sides to the NIH issue, and corporate policy can change with different administrations. Many executives feel that no insulated group of salaried product development people, no matter how brilliant, can come up with winning products day in and day out. There are companies that cannot afford to pay engineers and designers to sit around and dream up ideas all day long. These kinds of companies are always worth approaching.

Other firms are against outside licensing of patented ideas because they would rather see the millions of dollars paid in royalties kept inside for its own research and development activities. They are not typically structured to interact with independent inventors.

I like to tell executives who believe they can do it all in-house that it is the spirit of the independent inventor that built America. To completely shut out the independent inventor is to severely limit one's opportunities and horizons. Then I remind them of these stories:

Kodak, America's largest manufacturer of photographic products, should have developed the instant camera. It didn't. The 60-second camera was invented and produced by Edwin Land, a maverick inventor. And when Kodak ultimately decided to imitate Land's invention, it was stopped in its tracks by the courts.

IBM, a name synonymous with computer innovation, completely missed the hand-held calculator market. The Japanese captured the lucrative market and never gave an inch.

The Swiss, the once undisputed world leaders in watchmaking, got their chimes rung when the Japanese came out with cheap, plastic timepieces utilizing microchips, capturing billions of consumer dollars in the process. Only in recent years, with the plastic Swatch, have the Swiss begun to rebound.

And the U.S. television networks, with worldwide news-gathering operations, passively watched as CNN got started because they thought no one would pay to see news 24 hours a day. CNN has become such a success that the networks have all entered the cable market, albeit late. It's doubtful they'll ever recover their market share from CNN.

Breaking the Code

Many companies that work with outside developers do not encourage "unknowns" and return inquires with a letter like this:

"Our advertising, research, marketing, and new product planning staffs are primarily responsible for creativity and development. Corporate policy precludes us from either encouraging or accepting unsolicited ideas from persons outside the Company. While an idea may seem feasible to the submitter, there are usually a number of factors that would make it impractical for us to implement. Moreover, many of the unsolicited ideas that we receive from both nonprofessional and professional sources have previously been submitted in one form or another."

Reading between the lines, this letter is not as negative as it appears to be at a glance. The first clue, that the company does not do everything internally, is that its internal staffs are not "exclusively responsible," but rather "primarily responsible" for creativity and development. This means that outside people support the company's research and development.

The next good news comes when the letter states that the company cannot encourage or accept "unsolicited ideas from persons outside the Company." I think this means that the company solicits ideas from a trusted base of outside creative sources.

Companies that send back letters similar to this are worth a second look.

Public Companies vs. Private Companies

I have no preference for either one. My decision is based upon what I am able to find out about a company and whether it is best for my product. I maintain, at all costs, the mental frame of mind that I am evaluating the company rather than an inferior position of being considered by the company.

Public Companies

The best place I know to obtain deep and detailed information on a publicly traded company is the Securities and Exchange Commission (SEC), Washington, DC. This independent, bipartisan, quasi-

judicial federal agency was created July 2, 1934 by act of Congress. It requires a public disclosure of financial and other data about companies whose securities are offered for public sale. There are 11,000 companies registered.

A company must register under the act of 1934 if: 1) its securities are registered on a national securities exchange or 2) its assets exceed $3 million with a class of equity securities held by 500 or more investors.

The SEC operates a public reference room at 450 5th Street, N.W., Washington, DC 20549. It is open from 9:00 a.m. to 5:00 p.m. daily. The telephone number is (202) 272-7450. Consumer Telecommunications for the Deaf-TTY-Voice is (202) 272-7065.

This specially staffed and equipped facility provides all of the publicly available records of the Commission. These include corporate registration statements, periodic company reports, annual reports to shareholders, tender offers and acquisition reports, and much more.

SEC reference libraries located in New York City and Chicago are open to the public. They are located at 26 Federal Plaza, (212) 264-1207 and 1204 S. Dearborn St., (312) 353-7433, respectively.

Requests by Mail

If you find it inconvenient to visit one of the Public Reference Rooms, the Commission will, upon written request, send copies of any document or information. Send a written request stating the documents or information needed, and indicate you will pay the copying and shipping charges. Also include a daytime telephone number. Address all correspondence to: Securities and Exchange Commission, Public Reference Branch, Stop 1-2, 450 5th St., N.W., Washington, DC 20549.

Bechtel Information Services also provides prompt and low-cost research and copying services. It is located at 15740 Shady Grove Road, Gaithersburg, MD 20877-1454. In Maryland, the telephone number is (301) 258-4300; outside Maryland, phone toll-free: 1-800-231-DATA.

Corporate annual reports are on file with the SEC, or can be obtained directly from the company. There is no charge for annual reports that are ordered from the company. The SEC copies would require copying. Contact the executive in charge of Investor Relations or the Senior Vice President and Chief Financial Officer at the particular company that you are researching.

Another way to obtain annual reports and SEC filings is to tell a stockbroker that you are interested in purchasing stock in a particular company and need to see the company's annual report.

Somewhere within an annual report it will say something like this: A copy of the company's annual report on Form 10-K, as filed with the SEC, will be furnished without charge upon written request to the Office of the Corporate Secretary. The 10-K is research pay dirt!

Form 10-K

I find this annual report to be the most useful of all SEC filings. In summary, it will tell you the registrant's state of business. This form is filed within 90 days after the end of the company's fiscal year. The SEC retains 10-Ks for ten years.

Part I of the 10-K reveals, among other things:

a) **When the company was organized and incorporated.** You will want to know how long the company has been in business to gauge its experience. What you would expect from an established company may vary from what you would tolerate at a start-up firm.

b) **What the company produces, percentages of sales any one item may be, seasonal/nonseasonal, etc.** It is critical to have a complete picture of company's product lines, their strengths and markets, any seasonality or other restrictions to the appropriateness of your item, and if and how your product could be positioned.

c) **How the company markets, for instance via independent sales representatives or its own regional staff offices.** It is important to know how a company gets something onto the market and where sales staff loyalty is, i.e. company employees typically have more loyalty than independent sales reps who handle more than one company's line.

d) **Whether or not it pays royalties and how much per year.** You can often see how much work the company does with outside developers and whether it licenses anything at all. An example of such wording is this from one corporate 10-K: "We review several thousand ideas from professionals outside the Company each year."

I recall another 10-K that read, "The Company is actively planning to expand its business base as a licenser of its products." Statements such as these show that door is open.

e) **The amount of money the company spends to advertise and promote its products.** If your product will require heavy promotion, and the company does not promote its lines, you are at the wrong place.

f) **Details on design and development.** You should know before approaching a company whether an internal design and development group exists and how strong it is. I found one 10-K in which a company stated, "Management believes that expansion of its R&D department will reduce expenses associated with the use of independent designers and engineers and enable the Company to exert greater control over the design and quality of its products." It could not be more obvious that outside inventors were not wanted.

g) **Significant background on production capabilities.** Often it is valuable to know in advance what the company's in-house production capabilities are, and what its out-sourcing experiences are in your field of invention. It's no use taking a technology to a company that does not have the experience to produce it.

h) **Terms of long-term leases.** It can be important to know whether a company owns or rents its facilities as a measurement of its strength and capabilities. An inventory of real estate can also give you an excellent overview of warehouses, plants, offices, etc.

i) **If the company is involved in any legal proceedings, law suits, for instance.** You may not want to go with a manufacturer that is being sued right and left. Maybe it has just risen from a bankruptcy and is still not strong financially. All of this kind of information is an excellent indicator of corporate health.

j) **The security ownership of certain beneficial owners and management.** This is vital to understanding the pecking order and power structure. Here is where you'll see who owns how much stock (including family members), and what percentage of the company this represents. The ages and years with the company are also shown.

l) **Competition.** This section will give you a frank assessment of the company's competition and its ability to compete. One 10-K I read once admitted, "The Company competes with many larger, better capitalized companies in design and development..." It is unlawful to paint a rosy picture when it doesn't exist. The 10-K is one of the few places you can get an accurate

picture. Would you want to license a product to a company that states, for example, "...most of the Company's competitors have financial resources, manufacturing capability, volume and marketing expertise which the Company does not have." This tells me to check out the competition.

m) **Exhibits.** On occasion, a company will attach to its 10-K exhibits such as employee stock option plans, licensing agreements, executive employment contracts, leases, letters of credit, etc. Information from any one of such documents could be important in a future negotiation. If a company denies your request for a certain contractual term, it would be beneficial to know there is corporate precedent for your receiving said stipulation.

I once read the employment agreement between the executive with whom I was negotiating and his company. This helped me to estimate his worth to the company and what he could and could not make happen on my behalf. I even knew the type of car he leased.

By combining the information available in 10-Ks with your observations, you will be much better prepared to do business.

Form 10-Q

The Form 10-Q is a report filed quarterly by most registered companies. It includes unaudited financial statements and provides a continuing view of the company's financial position during the year.

The 10-K and 10-Q are the two filings I find most valuable. However, the SEC has many other reports available. The best way to view a full inventory of records on file is to contact the nearest SEC regional office and pick up or have sent to you one of the Commission's booklets.

Private Companies

It is not as easy to find detailed information on privately held companies. No regulations require that they fill out the kinds of revealing reports public companies must. Nevertheless, it is important to gather as much background information as possible.

Here are some questions I ask before approaching a private company. The answers come to me from a combination of sources ranging from state incorporation records to interviews with competition, suppliers, retailers (as appropriate), and the owners themselves.

1) **Is the company a corporation, partnership, or sole proprietorship?** This can have legal ramifications from the standpoint of liabilities the licensee assumes. A lawyer can advise you on the pluses and minuses of each situation.

2) **When was the company organized or incorporated?** If a corporation, in which state is it registered? When a company was organized will give you some idea as to its experience. The more years in business, the more tracks in the sand are left. The state it is registered in will tell you where you may have to go to sue it.

3) **Who are the company's owners, partners, or officers?** Always know with whom you are going into business. In the end companies are people, not just faceless institutions.

4) **What are the company's bank and credit references?** How a company pays its bills is important for obvious reasons, and its capital base is worth assessing.

5) **Is the manufacturer the source for raw material?** Does it do the fabrication? Such information will help estimate a company's capabilities for bringing your invention to the marketplace.

6) **How many plants does the company own (lease), and what is the total square footage?** Does it warehouse? This kind of information will help complete the corporate picture.

7) **What products are currently being manufactured or distributed?** You don't want to waste time pitching companies that do not manufacture your type of invention. Maybe a company you thought to be a manufacturer is really only a distributor.

8) **How does the company distribute?** Find out about the direct sales force. Make inquiries concerning outside sales representatives and number of jobbers. Does the company use mail order, house-to-house, mass marketing, or some other form of distribution? This information will quickly reveal how a company delivers its product and whether its system is appropriate to your product. With a mass market item, it would be foolish to approach a firm that markets door-to-door, regardless of its success.

9) **What kind of marketing and advertising support can you expect your product to receive?** Specify whether or not your product will receive national, regional, or local promotion. Will announcements be carried on television, radio, or in print? Does the company provide point of purchase promotion? It makes no sense to license a product requiring promotion to a firm that does not promote its lines.

If You Need Help In Locating A Company

The *Thomas Register*

One of the best sources for product and corporate profiles is the *Thomas Register.* Available in most public library reference rooms, "Thomcat", as it is known, contains information on more than 120,000 U.S. companies in alphabetical order, including addresses and phone numbers, asset ratings, company executives, the location of sales offices, distributors, plants, and service and engineering offices. If you know a brand name, you can locate it in the *Thomcat Brand Names Index.*

If you wish to purchase your own set of books, the *Thomas Register* costs $249 (plus $15.80 shipping and handling) for the 26 volumes (1992). Like an encyclopedia, it requires considerable shelf space. To order call: 1-800-222-7900.

S&P's *Register*

Another excellent source is Standard & Poor's *Register of Corporations, Directors and Executives,* available at public libraries. Consisting of three volumes, it carries data on more than 45,000 corporations including their zip codes; telephone numbers; names, titles, and functions of approximately 400,000 officers, directors, and executives. A separate volume selects 70,000 key executives for special biographical sketches. The last volume contains a classified industrial index.

The S&P *Register of Corporations, Directors and Executives* (1992) costs $525 (no shipping and handling). To order call: 1-800-221-5277.

After you have read all the literature and investigated the company, you must ask yourself if the company can deliver and if you will be comfortable working with its people. The abbreviation "Inc."

after a company's name is not significant. Nice offices, a few secretaries, a fax and copying machine do not a successful licensee make.

Section 12

How to Present a Concept

"An important idea not communicated persuasively is like having no idea at all."—William Bernbach

Avoid Cold Calls

Presentations are best when carefully choreographed and staged. Nothing is usually gained ambushing executives outside of their offices. As Agesilaus II, King of Sparta, said, "It is circumstance and proper timing that give action its character and make it either good or bad."

For every rule about when to sell, another rule proves it wrong. The best rule on when to sell ideas is whenever possible. Timing is everything. I operate under the principle that when you have something hot, burn it. When it gets cold, sell it for ice.

I licensed our first electronic toy, StarBird, to Milton Bradley after Thanksgiving, and the manufacturer premiered it only weeks later at American International Toy Fair. Milton Bradley put its resources on the line to make it happen in spite of the toy's complicated structure.

I created the game, Adver*teasing*, which was licensed in June to Cadaco and manufactured and in stores by mid-September of the same year. By early December it had racked up sales of over 250,000 units.

While these examples are exceptions to the rule, had I waited until the "best" time to make the pitch, opportunity might have been delayed or the product may never have been licensed. The marketplace is temperamental and erratic.

Another rule to consider is this: The faster a product is made available, the faster it begins to generate income.

Curtain Up. Light the Lights.

Once you have been extended an invitation to display your concept to a manufacturer, it's major show time. And if you thought inventing was tough, you haven't experienced hardship. The moment you walk into a company's conference room with your invention at the ready, you pass into an eerie

109

twilight zone. You are the hero in a Nintendo video game and the corporate executives comprise an array of varied personalities. Some pray for you to succeed. Others are gremlins out to gobble you up, shoot you down, and otherwise obliterate your ideas faster than Octorok can toss stones at Link.

Thomas Alva Edison once observed, "Society is never prepared to receive any inventions. Every new thing is resisted, and it takes years for the inventor to get people to listen to him before it can be introduced."

"Creation is a stone thrown uphill against the downward rush of habit," an anonymous inventor said.

World history is rich in stories about people's resistance to new ideas and change. Some seem unbelievable in retrospect. I find it helpful and comforting to recall some of these stories before I begin a presentation.

Impressionist art that sells for millions of dollars today was met with onslaughts of condemnation and cavil when first introduced. The innovative Stravinsky was once labeled "cynically hell-bent to destroy music as an art form." Today he is regarded as a genius.

When railroads were established, farmers protested that the "iron horse" would scare their cattle to death and stop hens from laying eggs. The British Association for the Advancement of Science insisted the automobile would fail because a human driver "has not the advantage of the intelligence of the horse in shaping his path."

Back in the 1930s, Charles B. Darrow invented a game called Monopoly. It was rejected by six or seven companies, including Parker Brothers. Parker Brothers eventually saw the light and published Monopoly, which annually sells more than three million units. Darrow became the first millionaire game inventor.

Remember that inflexibility is inherent in the system.

In order for companies to operate efficiently, they must have rules and controls. Loose cannons do not last long in corporate environments. Any organization must understandably have some established routine to survive. Therefore, executives, especially in larger companies, often fall into predictable and ordered routines. Your job is to break this routine, make people believe in you, and interest the company in buying your concept.

Preparing Your Proposal

I always back up my verbal presentations with written proposals. Do not depend on the company to copy your proposals. Bring enough originals with you for the meeting and then some. You cannot afford to have someone inadvertently lose a critical page.

All of my proposals begin with a concept summary. This is nothing elaborate, just a solid paragraph that paints a picture of the product and its overall concept and objective.

This is followed by a technical section that addresses cost and manufacturing. You are well advised to have a rough estimate of what the invention will cost to manufacture. Don't expect the company to know.

I always try to provide the following data with every item (as appropriate):

> 1) A sheet listing all components with respective prices from various sources. Pricing from three different sources is good. When possible, a mix of domestic and offshore numbers is best. Do not forget to include the volume the quotes are based upon, plus vendor contacts.

2) Note the type of material(s) desirable, e.g., polyethylene, wood, board, etc. Provide substitutions and options for consideration.

3) When you calculate the item's cost, do not forget to consider the price of assembly (if any). The quoting vendors will be helpful here.

Every written proposal should also contain:

1) **Detailed operating instructions:** Take nothing for granted. The worst thing that can happen is a client's inability to use your item after you have departed. Illustrate with pictures if required. No item is too easy.

2) **A marketing plan:** Highlight your item's advantages over existing product(s) and what makes it unique. Define its appeal and target audience. Suggest follow-ups, including second generations and line extensions of your product (as appropriate). Manufacturers like products that have a future, especially if they will be required to spend lots of start-up dollars in the development and launch phases.

3) **Trademarks:** Offer possible trademarks. If a trademark search has been done, include the results of your search or status of any applications. The right mark can go a long way in securing a sale.

4) **Patents:** Include an update on any patent searches, PTO actions, etc. If a patent has been issued, attach a copy of it. The more detailed and comprehensive your work, the quicker and easier the corporate evaluation of your product will be.

5) **Ad Campaign/Copy:** Suggest advertising direction, artwork, and slogans whenever appropriate. This type of work makes presentations even more persuasive and polished.

6) **Test results:** Report the results of any formal testing or focus group sessions. Include photographs and videotapes whenever appropriate.

7) **Inventor's Background/Capabilities:** If you are unknown to the company, provide personal background and describe yourself and your capabilities to support further development and manufacture of the invention under consideration. After all, who understands an invention better than its inventor? Your experience may save the manufacturer a great deal of money.

The aim here is to show the manufacturer that your front line is strong and the bench is deep. The more confident the company feels about you the better.

Watch Your Language

You cannot be too specific in proposals. Spell out everything. Take nothing for granted. Do not assume words and terms have the same meaning to every reader, especially when you're making presentations to potential foreign licensees.

Although much is known about how new products are developed, there is no consensus on the meaning of key terms and definitions.

"With the increased emphasis in developing new commercial products and processes to help the U.S. become more competitive internationally, the need for a common language in the innovation process is becoming even more important," says George Lewett, chief of the National Bureau of Energy-Related Inventions at National Institute of Standards and Technology (NIST).

It is of interest to note that the NIST, the Department of Energy, and the National Society of Professional Engineers are collaborating to develop a consensus language for use in describing the innovation process.

So far, seven organizations have nominated members to a task group. They are: American Institute of Chemical Engineers, American Society for Engineering Education, American Society of Mechanical Engineers, Commercial Development Association, Industrial Research Institute, Institute of Industrial Engineers, and NSPE.

I hope to incorporate their glossary in a future edition of this book.

Manufacturers Don't License Ideas

It is imperative that you develop a prototype prior to disclosing the invention to potential licensees. A prototype is defined as an original model on which something is patterned. If you do not have the time, money, or commitment to build a prototype, the odds of licensing the idea are nil.

The most effective kind of prototype is what is called a "looks like-works like" version. There are no short cuts. Manufacturers don't license "ideas". They react to physical matter. Don't count on people being able to "imagine" what your product will look like or how it will operate. Even if they could, busy executives do not usually have the time or interest to engage in such typically futile exercises.

Executives love to touch and feel prototypes. Kick the tires, so to speak. Knowing this, do your best to have prototypes that most resemble and operate like a production model. Go that extra mile to ensure the prototypes are solid and have perceived value. Prototypes must be solid because often they take quite a beating at the hands of potential licensees. Don't be surprised to get them back broken. It happens even at the best of companies. This comes with the territory.

Don't take any of this lightly. You must be as sophisticated and slick in your presentation to a potential licensee as it will have to be in its pitch to the trade and/or the consumer. While getting a product known is relatively easy, marketing a need is something else. And that's your ultimate goal.

Note: It has been over 110 years since the PTO has required prototypes or mock-ups of inventions to accompany patent applications.

Making Prototypes

If you cannot make the prototype, plenty of places will provide assistance. In some cities prototype makers can be found in the telephone directory. Universities and engineering schools often have workshops where connections can be made to get something done. Local inventor groups are a wonderful source of information as well. **See Section 13 on page 117, National and Regional Inventor Associations and Section 15 on page 145, University and Independent Innovation Research Centers for more information.**

Many invention marketing companies offer prototype making services. Be careful. Know what you are getting into when contracting to have a product prototyped. I would not engage anyone to do any prototype work without inspecting their shop and checking references.

It is prudent to require that any prototype maker sign a Confidentiality Agreement before you show or discuss your product. **A sample agreement is shown as Form 29 in Part One: Forms and Tables.**

Multiple Submissions

If you have more than one prototype, and the situation is appropriate, you may wish to consider making submissions to more than one manufacturer at the same time. I have no set rule about this and take it case by case, guided by experience.

If a company asks to hold off further presentations until it has an opportunity to review the item more in depth, try to set guidelines. In fairness to everyone, some products require a reasonable number of days to be properly considered. However, if you feel the company is asking for an unreasonable period of time, seek some holding money to hold the product out of circulation. The amount of time and money is negotiable. Also insist that the product not be shown to anyone outside the company. **To keep track of who has seen what and when, a Patent Worksheet appears as Form 12 in Part One: Forms and Tables.**

Mutual Dependency

You need the company or you would not be there. Show yourself as being independently creative, while at the same time taking the "we approach" and not the "I approach."

In order for your product to sustain itself through the review and development process, it will **need a champion.** Typically this standard-bearer will come from among those attending your first meeting. Get others involved. Turn "your idea" into "our idea."

If You Sign My Paper, I'll Sign Yours

Many companies ask that an agreement be signed for the submission of outside ideas. This may surface when you first approach the company or happen on the day you appear to make the formal presentation. I have never had a problem with such requests. I always know with whom I am dealing and feel confident in the relationship. If I did not, I wouldn't be there in the first place.

A suspicious attitude may seriously inhibit your progress. Put your time and energies into creating concepts versus overprotecting them.

Idea Submission by Outsiders

Outside submission agreements take many forms. **Form 30, Idea Submission Agreement I and Form 31, Idea Submission Agreement II are commonly used formats (see Part One: Forms and Tables).** Most companies use variations of these forms. Use these forms to familiarize yourself with the types of document, and as examples, in case you want to use one to protect yourself before looking at another individual's ideas.

Agreement to Hold Secret and Confidential

It is appropriate, in some instances for the inventor to have the company sign an agreement to hold a product secret and confidential. **A sample agreement is shown as Form 32 in Forms and Tables.**

A Fish Story

The art of product presentation is not unlike the sport of fly fishing. It takes time and patience. In casting, the lure is presented and then pulled back. The lure's movement, aided by twitches, pauses, and jerks of the rod by the angler, entices the fish to strike. In both cases the goal is to hook a big one!

Part Three

Directory of Inventing and Patenting Information

Section 13

National and Regional Inventor Associations

"When the world seems large and complex, we need to remember that great world ideas all begin in some home neighborhood."—Konrad Adenaur

Inventors are not known for their business acumen or marketing expertise. Getting from the workshop drawing board to the corporate conference room can be a long, tough, and lonely haul. Operating on the frontiers of an emerging idea is difficult enough. Spearheading the development, licensing, and manufacture of an invention for commercial utilization can leave even the most experienced players nonplused.

Inventors can receive guidance and support from myriad sources, such as patent lawyers, government agencies, universities, reference books, business and technical resource centers, and so forth. But often the most sincere information, albeit not always the most useful, comes from colleague inventors, peers who understand and also pursue the "quest for fire."

Inventors are no different than any group of people with a common interest. They love to get together and explore professional issues and share empirical experiences—success stories, and heartbreaks, "insider" information, personal contacts, methods, techniques, and dreams.

Inventor groups often take root where inventors practice their trade. Seeking to build relationships among themselves, independent inventors have formed organizations throughout the country that provide professional and social forums. These assemblies offer members product development support and guidance and resource information. A common objective is to stimulate self-fulfillment, creativity, and problem-solving.

Some inventor groups are more sophisticated and active than others, some more organized, but most have something beneficial to offer. This section covers a wide range of support organizations available to independent inventors. I encourage you to be selective in your association with any organization and make sure that those charging fees provide the appropriate kinds of services and data.

There are two basic types of inventor groups, local clubs and national organizations.

Local Inventor Clubs

Local clubs are real grass roots. They typically encompass only a limited geographic area, for example Albuquerque Invention Club or the Inventor's Council of Hawaii. Often they adopt names that are not so obvious. For example, one Cincinnati, Ohio inventors group calls itself The Salmon Club. They say they chose this name because "inventors always swim upstream."

The local clubs can be great fun and informative, not to mention wonderful places to strike up friendships. They range in size from several members to several hundred members. Rarely do they have formal offices. Meetings are held at hotels, restaurants, or members' homes.

The Salmon Club, for example, is large enough to run occasional seminars for its members and guests. One of its recent seminars was a two-day event with speakers presenting oral and written information on topics including invention evaluation, licensing, financing, basic patent law, incubators, and marketing.

Such clubs don't spend their money on slick newsletters and staff salaries. They are what you would expect from smallish, local organizations. Volunteers do their best with what they have at hand.

Local clubs do not promise to deliver anything more than you would expect, either. They are usually not the best place to pick up fast-breaking information on intellectual property legislation. But what I particularly like about these kinds of clubs is their lack of pretension, and the warmth, hospitality, community, and comradeship they engender.

Frederick L. Jones, a member of the Palm Beach Society of American Inventors, says his club has approximately 150 members. Asked how many of his colleagues make a successful living from inventing, he responded, "Practically none. We're all dreamers. Inventors are all dreamers. If we weren't we wouldn't be inventors."

Typical of most independent inventors in his club, Fred Jones holds nine patents and has yet to license one. "I've been too busy inventing," he says.

Chuck Mullen, who bills himself as an inventor and a consultant for new idea development, is chairman of the board of advisors of the Houston Inventors Association, which has reported around 220 active members. How many have been successful at licensing their inventions? "Not many," he says, "Maybe five or six. Most join the Association to learn."

Alexander T. Marinaccio, president of Inventors Clubs of America, says the average inventor's club lasts only two years. "Everybody joins, wants to get rich and then they all drop out."

National Inventor Organizations

National inventor organizations are often businesses owned and operated by business people, many strictly for profit. They run the gamut from reputable, high-powered lobbying organizations based in the nation's capital, to "nonprofit" societies that make money selling magazine subscriptions, official-looking certificates, book clubs, group insurance policies, discounted car rentals, and kindred fare.

When you join a national organization your expectations and dues are much higher than at the local level; so is the organization's overhead required to support their high-paid, ever-traveling executives and toll-free telephone numbers. In return for the steeper membership dues, you'll want the kind of information unavailable via local clubs, for example, access to major corporate executives, state-of-the-art marketing and patenting, trademark and copyright advice, quality product evaluations, and so forth. Very few national organizations can deliver all of this under one roof.

Ask questions before you join. Compare membership offers. Even the best group can disappoint when seen in the light of unrealistic expectations. Don't be misled by slick brochures, certificates, directories, travel and insurance discounts, nonprofit status, press release services, and book clubs. The bottom line is what the organization can do for you.

Ask for member and business references. You want to know which companies rely upon the organization for input. Talk to corporate R&D executives who express faith in the organization. Money can shotgun press releases and newsletters out to a range of diversified manufacturers; it does not guarantee results.

Over and above dues, find out whether the organization expects a piece of your invention in return for its assistance. Make sure it is not just an invention marketing company in disguise. **Ask for the salary of full-time organization executives and relevant budget information. As a potential member, this kind of information should be made available to you.** Beware if they won't give out this information. You want to know how they'll be spending your dues.

Selected Organization Profiles

Affiliated Inventors Foundation, Inc. (AIF)
2132 E. Bijou St.
Colorado Springs, CO 80909-5950   Phone: (719) 635-1234
John T. Farady, Exec. Dir.

The Foundation was established in 1975 as a national organization with honorary members in 41 states. Membership is by invitation only, and the majority of its members are holders of U.S. patents. It was founded by John Farady and the late Tom Wille, an inventor, electrical engineer, and an avid student of Thomas Edison's rival Nickola Tesla.

AIF has very been busy in recent years. From 1986 through 1990, the company claims to have performed more than 11,410 preliminary appraisals of inventions. Of these, 3,291 were rejected during the appraisal stage. According to Foundation executive director, John Farady, about 30 percent of all inventions they see have "fatal" flaws and are thus rejected.

Since it opened for business the Foundation has reportedly evaluated more than 23,000 inventions.

"Our educational goal is to provide independent inventors with sufficient information about each phase of invention development to help them make better decisions in their own best interests," says Farady.

"We do not require any advance cash fees," he continues. "And our contract negotiation services are on a no-fee, straight commission basis should we be successful."

The Foundation states that "unlike the front-end-cash-fee companies, it is not the aim of the Foundation to offer marketing services to so many inventors that it cannot focus on anything properly. This organization keeps the membership controlled in such a way that it can be most effective."

The Foundation provides the following services to member inventors: free educational materials; free consultations; free preliminary appraisals; invention evaluations; low-cost patent and trademark services; funding opportunities; marketing assistance services; commission-based negotiation services (five to 15 percent range); and an invention publicity program, for example, features member inventions in the bimonthly magazine, *Inventors' Digest*, which is mailed to thousands of manufacturers.

According to the magazine's editor, Joanne Hayes, "*Inventors' Digest* has been published since 1985 and our main goal has been to provide inventors with informative articles which will guide them through the invention development process and will, hopefully, help them make informed decisions.

We also have a section called Invention Mart which lets manufacturing companies know about patented and patent pending inventions that are available for sale or licensing.

I have always liked AIF and what co-founder John Farady is trying to do for his membership. The Foundation is one national organization that has much to offer independent inventors and its publication is very informative.

Write or call for an information package. Check out the Foundation's references with any of the following organizations: Colorado Springs Chamber of Commerce, 100 Chase Stone Center, Colorado Springs, CO 80902; Dun & Bradstreet (any local office) Report No. DUNS-06-062-9748; and Central Bank of Colorado Springs, 2308 E. Pikes Peak Ave., Colorado Springs, CO 80909.

Intellectual Property Owners, Inc.
1255 23rd St., N.W., Ste. 850
Washington, DC 20037   Phone:(202) 466-2396
Herbert C. Wamsley, Exec. Dir.   Telecopier: (202) 833-3636
Ada B. Winter, Assoc. Exec. Dir.   Telex: 248959

Intellectual Property Owners (IPO) is a nonprofit association representing people who own patents, trademarks, and copyrights. It was founded in 1972 by a group of individuals who were concerned about the lack of understanding of intellectual property rights in the United States. Members include about 100 large and medium sized corporations and a growing number of small businesses, universities, patent attorneys, and independent inventors.

IPO's corporate members include the likes of Monsanto, Standard Oil, Westinghouse Electric, United Technologies, Procter & Gamble, IBM, Ciba-Geigy, Upjohn, AT&T, Amoco, and Union Carbide. IPO also has memberships for small businesses as well as individuals. Herb Wamsley, IPO executive director, says that he is interested in expanding his organization's membership base to include more individual members such as independent inventors, investors, and attorneys.

IPO conducts an active government and public relations program in Washington, DC. One of its best known programs is the Inventor of the Year Award, given each spring to an inventor whose invention was either patented or first made commercially available during the previous year. In the spring, IPO sponsors a conference for small inventors and entrepreneurs in cooperation with the PTO.

In my opinion, IPO is outstanding and well-respected, and the flagship among intellectual property organizations. It offers its members the most current information on patent, trademark, and copyright issues, as well as a strong voice on Capitol Hill and at the PTO.

For example, during 1990, IPO mailed about 20 editions of its *Washington Briefs*, usually within a few hours after significant intellectual property events occurred. More detailed information is published in its printed newsletter, *IPO News*. Much of this information was available from no other source.

Section Arrangement

The following entries are arranged geographically by state, then alphabetically by association name. All association names in this section are referenced in the Master Index.

Alaska

★1★ Alaska Inventor Association
PO Box 241801   Phone: (907) 273-5473
Anchorage, AK 99524
Robin Zerbil, Pres.

★2★ Inventors Institute of Alaska
PO Box 871327   Phone: (907) 376-7555
Wasilla, AK 99687
Jay James, Pres.

Arkansas

★3★ Arkansas Inventors Congress, Inc.
PO Box 411   Phone: (501) 229-4515
Dardanelle, AR 72834
Garland E. Bull, President

★4★ Inventors Clubs of America, Arkansas Chapter
Rte. 64, Box 135
Chester, AR 72934
Vernon Zellers, Pres.

California

★5★ Asian Inventors and Entrepreneurs Co-op
876 Zeiler Ave.   Phone: (818) 892-9731
Arleta, CA 91331
Mel Afareo, Contact

★6★ California Engineering Foundation
913 1/2 K St., Ste. A   Phone: (916) 442-6369
Sacramento, CA 95814
Robert J. Kuntz, Director

★7★ Community Entrepreneurs Organization (CEO)
PO Box 2781   Phone: (415) 435-4461
San Rafael, CA 94912
Dr. R. Crandall, Exec.Dir.

★8★ High Technology Entrepreneurial Council
6200 Antioch St.   Phone: (415) 339-3895
Oakland, CA 94611
Bruce Methven, President

Persons in Oakland, CA interested in high technology computer software and hardware. Provides software developers and high technology entrepreneurs with a forum for developing marketing strategies, building business relationships, and exchanging experience and information. **Members:** 200.

★9★ Inventors Clubs of America, California Chapter
600 St. Francis Blvd.   Phone: (415) 992-2425
Daly City, CA 94015
Robert Hagman, Exec. Officer

★10★ Inventors Workshop International
3201 Corte Malpaso, Ste. 304   Phone: (805) 484-9786
Camarillo, CA 93012
Alan A. Tratner, Pres.

★11★ National Inventors Foundation, Inc. (NIF)
345 W. Cypress St.   Phone: (818) 246-6540
Glendale, CA 91204
Ted DeBoer, Exec.Dir.

★12★ Spirit of the Future Creative Institute (SFCI)
3308 1/2 Mission St., Ste. 300
San Francisco, CA 94110
Gary Marchi, Founder & Dir.

Colorado

★13★ Affiliated Inventors Foundation, Inc. (AIF)
2132 E. Bijou St.   Phone: (719) 635-1234
Colorado Springs, CO 80909-5950
John T. Farady, Pres.

★14★ Colorado Inventors Council
PO Box 88   Phone: (303) 854-3851
Holyoke, CO 80734
Michael R. McKensie, Director

★15★ National Inventors Cooperative Association
PO Box 6585   Phone: (303) 797-7540
Denver, CO 80206
Morton J. Levand, Contact

★16★ Rocky Mountain Inventors and Entrepreneurs Congress
PO Box 4365   Phone: (303) 443-6200
Denver, CO 80204-0365
Bill Van Arsdale, Director

Connecticut

★17★ Connecticut Innovations, Inc.
845 Brook St.   Phone: (203) 258-4305
Rocky Hill, CT 06067
David Driver, Director

★18★ Inventors Association of Connecticut
PO Box 3325   Phone: (203) 259-1174
Westport, CT 06880
Bruce Mims, President

District of Columbia

★19★ American Association of Engineering Societies
1111 19th St. NW, Ste. 608   Phone: (202) 296-2237
Washington, DC 20036
Mitchell H. Bradley, President

★20★ Intellectual Property Owners, Inc.
1255 23rd St. NW, Ste. 850   Phone: (202) 466-2396
Washington, DC 20037
Herbert C. Wamsley, Executive Director

Founded: 1972. People who own patents, trademarks, and copyrights.
Publications: IPO News and Washington Briefs. Offers Inventor of the Year Award. For a further discussion of this group's activities, see the introduction at the beginning of this chapter.

Florida

★21★ The Inventors Club
WSRE-TV   Phone: (904) 484-1200
1000 College Blvd.
Pensacola, FL 32504
William R. Bowman, Contact

★22★ Inventors Council of Central Florida
822 E. Wallace St.   Phone: (407) 857-8242
Orlando, FL 32809
David Flinchbaugh, Pres.

★23★ Inventors Society of South Florida
PO Box 4306   Phone: (407) 736-6594
Boynton Beach, FL 33424
Betty White, Contact

★24★ Sunrise Inventors Association
6402 Gamble Dr.
Orlando, FL 32818
Steve Chandler, Pres.

★25★ Tampa Bay Inventor's Council
13543 Periwinkle Ave.   Phone: (813) 391-0315
Seminole, FL 34646
Ray Purdy, Pres.

Georgia

★26★ Inventor Associates of Georgia, Inc.
241 Freyer Dr. NE   Phone: (404) 427-8024
Marietta, GA 30060
Hal Stribling, Pres.

★27★ Inventors Clubs of America (ICA)
3166 Maple Dr.   Phone: (404) 938-5089
Atlanta, GA 30309
Don M. Koening, CEO

Hawaii

★28★ Inventors Council of Hawaii
PO Box 27844   Phone: (808) 595-4296
Honolulu, HI 96827
George K.C. Lee, Pres.

Illinois

★29★ American Inventors Council
624 Cedar St.   Phone: (815) 968-1040
PO Box 4304
Rockford, IL 61110
Nicholas G. Parnello, Pres.

★30★ Inventors Council of Illinois
53 W. Jackson, Ste. 1643   Phone: (312) 939-3329
Chicago, IL 60604
Don Moyer, President

Indiana

★31★ Indiana Inventors Association, Inc. (IIA)
PO Box 2388   Phone: (317) 745-5597
Indianapolis, IN 46206
Randall Redelman, Pres.

★32★ Inventor & Entrepreneur Society of Indiana
Purdue University   Phone: (219) 989-2354
PO Box 2224
Hammond, IN 46323
Prof. Daniel J. Yovich, Exec. Dir.

Iowa

★33★ Triple I Inventors
c/o 7-Cities Sod Co.   Phone: (319) 391-1663
12554 210th St.
Davenport, IA 52804
Norman Fry, Contact

Kansas

★34★ Association of Collegiate Entrepreneurs
The Wichita State University   Phone: (316) 689-3000
1845 Fairmount   Fax: (316) 689-3687
Wichita, KS 67208-1595
Sharon Ryan, Director

★35★ Kansas Association of Inventors, Inc. (KAI)
2015 Lakin   Phone: (316) 792-1375
Great Bend, KS 67530   Fax: (316) 792-3406
Shirley Ridgel, Contact

Massachusetts

★36★ Inventors Association of New England (IANE)
PO Box 335   Phone: (617) 862-5008
Lexington, MA 02173
Dr. Donald D. Job, V.Pres.

★37★ Worcester Area Inventors USA
42 Shattuck St.   Phone: (508) 835-6435
Worcester, MA 01605
Edith Morgan, Contact

Michigan

★38★ Alliance of Women Entrepreneurs
PO Box 6731   Phone: (616) 455-2424
Grand Rapids, MI 49506
Jane Ross, President

★39★ Inventors Association of Metro Detroit
4405 Gratiot   Phone: (313) 772-7888
East Detroit, MI 48021
Peter D. Keefe, President

★40★ Inventors Council of Michigan (INCOM)
2727 2nd Ave.   Phone: (313) 963-0616
Detroit, MI 48201
Barbara Bach Eldersveld, Exec.Dir.

★41★ Michigan Biotechnology Institute
3900 Collins Rd.   Phone: (517) 337-3181
Lansing, MI 48910

Promotes the commercialization of biotechnology in the state. Provides in-house research and provides technology transfer.

★42★ Michigan Energy and Resource Research Association (MERRA)
328 Executive Plaza   Phone: (313) 964-5030
1200 6th St.
Detroit, MI 48226
John Mogk, Director

Serves as a statewide partnership between industry, university, and government, promoting energy-resource-technology research and working to bring public and private research and development grants and contracts into the state. Provides information about the Small Business Innovation Research (SBIR) program.

Minnesota

★43★ Inventors and Designers Education Association (IDEA)
PO Box 268   Phone: (612) 430-1116
Stillwater, MN 55082
Bill Baker, Director

★44★ Inventors and Designers Network
PO Box 8
Pillager, MN 56473
Bob Gillson, Contact

★45★ Inventors Education Network
P.O. Box 14775   Phone: (612) 379-7387
Minneapolis, MN 55414
Marge Braddock, Contact

★46★ Minnesota Project Innovation, Inc.
1200 Washington Ave. S., Ste. M100   Phone: (612) 338-3280
Minneapolis, MN 55415
James S. Hayes, Executive Director

★47★ Society of Minnesota Inventors (MIS)
PO Box 335   Phone: (612) 253-2537
St. Cloud, MN 56302
Helen Saatzer, Sec.-Treas.

Mississippi

★48★ American Association of Entrepreneurial Dentists
c/o C.E. Moore Co.   Phone: (601) 842-1036
420 Magazine St.
Tupelo, MS 38801
Dr. Charles E. Moore, Pres.

★49★ Mississippi Research and Development Center
3825 Ridgewood Rd.   Phone: (601) 982-6425
Jackson, MS 39211-6453
R.W. Parkin, Marketing Consultant

★50★ Mississippi Society of Scientists
508 Cindy Ln.
Pearl, MS 39208

★51★ Society of Mississippi Inventors
PO Box 5111   Phone: (601) 984-6047
Jackson, MS 39296
Eric R. Rommerdale, Pres.

Missouri

★52★ Inventors Association of St. Louis
PO Box 16544   Phone: (314) 432-1291
St. Louis, MO 63105
Roberta Toole, Director

★53★ Missouri Inventors Association
c/o American Marketing Group, Inc.   Phone: (314) 725-8078
11182 Towne Sq.
St. Louis, MO 62123
Richard T. Pisani, Contact

★54★ United Inventors Association of the United States of America
6611 Clayton Rd.   Phone: (314) 721-3842
St. Louis, MO 63117
Roberta Toole, Director

Montana

★55★ Montana Inventors Association
RR 1, Box 37   Phone: (406) 733-5031
Highwood, MT 59450
Fred E. Davison, Pres.

★56★ Yellowstone Inventors Association
PO Box 23306   Phone: (406) 259-9110
Billings, MT 59104
W.T. George, Pres.

Nebraska

★57★ Kearney Inventors Association
c/o Kearney Development Council   Phone: (308) 237-3101
2001 Ave. A, Box 607
Kearney, NE 68847
Ron Tillery, President

★58★ Lincoln Inventors Association
PO Box 94666   Phone: (402) 471-3782
Lincoln, NE 68509
Steve Williams, Contact

Nevada

★59★ High Technology Entrepreneurs Council
2620 S. Maryland Pkwy., Unit 153   Phone: (702) 736-6775
Las Vegas, NV 89109
George S. Sanders, Secretary-Treasurer

★60★ Nevada Inventors Association
PO Box 9905   Phone: (702) 322-9636
Reno, NV 89507
Don Costar, Contact

New Jersey

★61★ Inventrepreneurs' Forum (IF)
813 Columbus Dr.   Phone: (201) 833-2461
Teaneck, NJ 07666
Dr. Thomas M. Noone, Pres.

New Mexico

★62★ Albuquerque Invention Club
Box 30062   Phone: (505) 266-3541
Albuquerque, NM 87190
Dr. Albert Goodman, President

New York

★63★ Inventors of Greece in the USA
Pan Hellenic Society   Phone: (516) 223-5958
2053 Narwood Ave.
South Merrick, NY 11566
Dr. Kimon M. Louvaris, Director

★64★ Inventors Workshop International Education Foundation, Syracuse Chapter (IWIEF-S)
205 S. Central Ave.   Phone: (315) 656-9210
Minoa, NY 13116
Ray B. Di Pietro, Chm.

★65★ Investors Society of Western New York
PO Box 23654
Rochester, NY 14693
Jenny C. Servo, Director

★66★ New York Society of Professional Inventors
116 Stuart Ave.   Phone: (516) 598-3228
Amityville, NY 11701
Philip Knapp, Pres.

North Dakota

★67★ North Dakota Inventors Congress
Box 1530   Phone: (701) 252-4830
Jamestown, ND 58401
Arvid Brockman, Chairman

Ohio

★68★ Akron/Youngstown Inventors Organization
1225 W. Market St.   Phone: (216) 864-5550
Akron, OH 44313
Ned Oldham, Contact

★69★ Inventors Connection of Greater Cleveland
PO Box 46-254   Phone: (216) 581-4546
Bedford, OH 44146
Robert Abernathy, President

★70★ Inventors Council of Dayton, Inc.
140 E. Monument Ave.   Phone: (513) 224-8513
Dayton, OH 45402
George Pierce, Pres.

★71★ Inventors Council of Greater Lorain County
1005 N. Abbe Rd.   Phone: (216) 365-7771
Elyria, OH 44035
Rich Schoenberger, Pres.

★72★ Inventors Council of Ohio (Columbus)
3539 Lacon Rd.
Hilliard, OH 45459
Charles R. Morrison, Contact

★73★ Inventors Network of Columbus
1445 Summit St.   Phone: (614) 291-7900
Columbus, OH 43201
Paul Cover, Pres.

★74★ Ohio Inventors Association
9855 Sandridge Rd.   Phone: (614) 797-4434
Millfield, OH 45761
Ron Docie, Pres.

★75★ Yankee Ingenuity Program
623 Grant St.
Kent, OH 44240
Charles Clark, Pres.

Oklahoma

★76★ Invention Development Society, Inc.
8230 SW 8th St.   Phone: (405) 376-2362
Oklahoma City, OK 73128
William L. Enter, Sr., President

★77★ Oklahoma Inventors Congress
PO Box 54625-1625   Phone: (405) 848-1991
Oklahoma City, OK 73154-1625
Albert N. Janco, Pres.

Pennsylvania

★78★ American Society of Inventors (ASI)
PO Box 58426   Phone: (215) 622-4060
Philadelphia, PA 19102
Jay Cohen, Contact

★79★ Inventors League
403 Longfield Rd.
Philadelphia, PA 19118

★80★ Northwestern Inventors Council
Gannon University   Phone: (814) 871-7619
University Sq.
Erie, PA 16541
Robert K. Jordan, Pres.

South Carolina

★81★ Inventors Association of South Carolina
222 Terrace Way
Columbia, SC 29205
Olivia Mixon, Contact

Tennessee

★82★ Tennessee Inventors Association
PO Box 11225   Phone: (615) 376-6894
Knoxville, TN 37939-1225
George Wrenn, Pres.

Texas

★83★ Houston Inventors Association
2916 W. T.C. Jester Blvd., Ste. 108   Phone: (713) 686-7676
Houston, TX 77018
Kenneth A. Roddy, Pres.

★84★ Innovex Inc.
4144 N. Central Expy., Ste. 1125   Phone: (214) 265-1540
Dallas, TX 75204
Eloyd Murphy, President

★85★ Inventors Clubs of America, Texas Chapter
PO Box 927
Abilene, TX 79604
Scottie D. Williams, Exec. Officer

★86★ Network of American Inventors and Entrepreneurs
11371 Walters Rd.   Phone: (713) 537-8277
Houston, TX 77067
Wessie Cramer, Exec. Dir.

★87★ Texas Inventors Association
4000 Rock Creek Dr., #100   Phone: (214) 528-8050
Dallas, TX 75204
Tom E. Workman, President

Utah

★88★ Intermountain Society of Inventors
9888 S. Darin Dr.   Phone: (801) 571-2617
Sandy, UT 84070
John Winder, Contact

Virginia

★89★ Association for Science, Technology and Innovation (ASTI)
PO Box 1242   Phone: (703) 759-5104
Arlington, VA 22210
Dr. Mary Ellen Mogee, Pres.

★90★ Invent America!
510 King St., Ste. 420   Phone: (703) 684-1836
Alexandria, VA 22314
Nancy Metz, Exec.Dir.

Washington

★91★ Inventors Association of Washington
PO Box 1725   Phone: (206) 455-5520
Bellevue, WA 98009
David V. Sires, Contact

★92★ Northwest Inventors Association
723 E. Highland Dr.   Phone: (206) 947-4307
Arlington, WA 98223
Henry B. Ehrhardt, Contact

Wisconsin

★93★ Midwest Inventors Group, Inc.
PO Box 518
Chippewa Falls, WI 54729
Steve Henry, Contact

Phone: (715) 723-5061

Section 14

Invention Consultants and Research Firms

"An idea can turn to dust or magic, depending on the talent that rubs against it."—William Bernbach

What is consulting? I like the definition offered by England's Institute of Management Consultants:

> "The service provided by an independent and qualified person or persons in identifying and investigating problems concerned with policy, organization, procedures and methods; recommending appropriate action and helping to implement these recommendations."

I was introduced to the product development business when I was hired as a marketing consultant by an entrepreneur who had an idea for a child's computer. As you can imagine, I believe in using consultants whenever their expertise can contribute to the progress of a project.

There are as many types of consultants as there are problems to solve. Consultants can bring new techniques and approaches to bear on an inventor's work. Their contribution can range from helping to bridge a technological gap to the special knowledge and talent required to successfully license or market a particular innovation.

In the case of the child's first computer project, an electronics wizard originated the basic product with help from several consultants. A former Texas Instruments engineer, for example, was consulted regarding the best microprocessor to use. My services were engaged to provide information on how to best position the item within the marketplace.

Advisors can provide impartial points-of-view by seeing challenges in a fresh light. They operate outside existing frameworks and free from existing beliefs, politics, problems, and procedures inherent in many organizations or situations.

Most consultants operate on the basis of an hourly rate plus expenses. Inventors, however, by the nature of their work, are often able to make equity deals whereby in return for their advice, consultants are given participation in any profits the invention might generate. Inventors should think long and hard before doing something like this because it is often less expensive to risk the cash and hold aa many points as possible in-house.

One example of an inventor who took in consulting partners to make his project materialize is Seattle waiter, Robert Angel, inventor of the popular game "Pictionary." Angel knew that he had a terrific concept, but needed some graphic support. He asked artist Gary Everson to help him design the game board in return for points in the venture. Everson agreed. The then 26-year-old inventor also needed assistance in making business decisions, so he gave points, as well, to an accountant named Terry Langston. "Pictionary" went on to be the best-selling game of 1987, grossing more than $52 million at retail. In 1988 the product's sales soared to an astounding $120 million. To date more than 14 million units of the quick draw game have been sold. The three partners have become wealthy, receiving royalties on every game sold.

Do not think that consultants and research organizations have all the answers. They do not. Consulting is hard work and not everything can be solved as quickly as one would like. Do not look for miracle solutions. Shop around. Get references on any consultant or research organization you are considering. Don't be impressed by a consultant's or organization's professional association alone (for example, if they are part of a university). Their success rate in fields related to yours is what matters. Their "know-who" is as important as their "know-how"—that is, how much they can accomplish with a single phone call. Results are what you want, not just paper reports.

Section Arrangement

The following entries are arranged geographically by state, then alphabetically by consulting organization name. All organization names listed in this section are referenced in the Master Index.

Arizona

★94★ CCT Corporation
4354 N. Cotton Ln.   Phone: (602) 853-0077
Litchfield Park, AZ 85340

★95★ Integrated Circuit Engineering Corporation
15022 N. 75 St.   Phone: (602) 998-9780
Scottsdale, AZ 85260   Fax: (602) 948-1925
Glen Madland, Chairman

Founded: 1964. Data collection and analysis, market and opinion research, technological forecasting, testing and analysis, and failure and construction analysis in the field of microelectronics. Provides technical consulting to users, manufacturers, and suppliers to the semiconductor industry; legal consulting on patents; and expert witness testimony. Long-range project involves the development of a failure analysis laboratory in Bordeaux, France. **Publications:** Status Report (annually); Icecap (monthly); Report on the Integrated Circuit Industry (annually).

★96★ Trans Energy Corporation
2026 W. Campus Dr.   Phone: (602) 438-8005
Tempe, AZ 85282

★97★ Turoff Consulting Service
P.O. Box 5740   Phone: (602) 948-3956
Scottsdale, AZ 85261

★98★ Van Cleve Associates
6932 E. Second St.   Phone: (602) 296-2587
Tucson, AZ 85710

California

★99★ Advanced Technology Resources
P.O. Box 80   Phone: (916) 626-4104
El Dorado, CA 95623

The ATR Service includes patent and technology monitoring activities as well as the ATR Database, a unique collection of patent and scientific information concerning developments in imaging materials and technology. The ATR patent monitoring service serves engineers and scientists who wish to keep abreast of competitive activities.

★100★ AML Information Services
Box 405   Phone: (415) 927-0340
Corte Madera, CA 94925   Fax: (415) 927-7250
Anne Morgens, Owner

Founded: 1982. Patent monitoring and ordering in the areas of drugs, vaccines, biochemistry, genetic engineering, molecular biology, microbiology, antibody technology, clinical diagnostics, biomedical devices, and diagnostic imaging. **Publications:** Patent Alert (monthly newsletter).

★101★ Andrew Allison, Consultant
27360 Natoma Rd.   Phone: (415) 941-6065
Los Altos Hills, CA 94022

★102★ Cheskin Plus Masten
835 Page Mill Rd.   Phone: (415) 856-8300
Palo Alto, CA 94304

★103★ Cole/Green/Associates, Inc.
25835 Narbonne Ave., Ste. 250   Phone: (310) 534-2000
Los Angeles, CA 90717

★104★ Communication Technologies
140 2nd St., Ste. 600   Phone: (415) 541-9551
San Francisco, CA 94105

★105★ Computer Modules, Inc.
2348C Walsh Ave.   Phone: (408) 496-1881
Santa Clara, CA 95051

★106★ Consulting Industries International
Worldway Center, No. 91435   Phone: (310) 670-1177
Los Angeles, CA 90009

★107★ Creative Strategies Ventures Corp.
20074 Beatty Ridge Rd.   Phone: (408) 354-4774
Los Gatos, CA 95030

★108★ David Kessler & Associates
1306 N. Wilton Pl.   Phone: (213) 462-6043
Hollywood, CA 90028

★109★ The Devoir Group
8720 6th Ave.   Phone: (310) 672-3248
Inglewood, CA 90305-2412

★110★ Disk/Trend, Inc.
1925 Landings Dr.   Phone: (415) 961-6209
Mountain View, CA 94043

★111★ EOS Technologies, Inc.
606 Wilshire Blvd., Ste. 700   Phone: (213) 458-1791
Santa Monica, CA 90401   Fax: (213) 394-0034
Bryan Gabbard, President

Provides research, analysis, and consulting on systems engineering and concept definitions as applied to national defense, energy resources, and commercial technology transfer.

★112★ ETEL, Inc.
12 Highgate Ct.   Phone: (510) 527-3734
Berkeley, CA 94707

★113★ Ford Bubala & Associates
16400 Pacific Coast Hwy., Ste. 211   Phone: (310) 592-4581
Huntington Beach, CA 92649

★114★ Fred Rohe, Organic Marketing
11270 Clayton Creek Rd.   Phone: (707) 994-7744
Lower Lake, CA 95457

★115★ H.B. Hindin Associates, Inc.
P.O. Box 2035   Phone: (916) 265-8424
Grass Valley, CA 95945

Offers new product evaluation, legal services including consultation and expert witness, patent evaluation and investigation, market research and analysis, product development, market surveys and economic viability.

★116★ Herrgott Associates
1673 Kamsack Dr.   Phone: (408) 730-2209
Sunnyvale, CA 94087

★117★ Innovation of California
854 Fremont St.   Phone: (415) 321-9394
Menlo Park, CA 94025

★118★ Innovative Technology Associates
3639 E. Harbor Blvd., Ste. 203E   Phone: (805) 650-9353
Ventura, CA 93001

★119★ James M. Degen & Company, Inc.
P.O. Box 1169   Phone: (805) 963-9808
Santa Barbara, CA 93102

★120★ Keck & Company Business Consultants
410 Walsh Rd.   Phone: (415) 854-9588
Atherton, CA 94025

★121★ Koeth & Associates, Inc.
9 Rincon   Phone: (714) 731-6738
Irvine, CA 92720

★122★ Laird Durham Company
P.O. Box 26470   Phone: (415) 981-3820
San Francisco, CA 94126

★123★ Lawrence M. Liggett
1856 Piedras Cir.   Phone: (510) 820-9304
Danville, CA 94526

★124★ Lindquist Consultants
P.O. Box 8355   Phone: (510) 524-6685
Berkeley, CA 94707

★125★ Lofts and Associates
31291 Holly Dr.   Phone: (714) 499-2147
South Laguna, CA 92677

★126★ Malcolm E.A. Kaufman & Associates
2400 Broadway, Ste. 500   Phone: (310) 828-1500
Santa Monica, CA 90404

★127★ Margiloff & Associates
382 E. California Blvd., Ste. 301   Phone: (818) 793-7829
Pasadena, CA 91106

★128★ The Market Connection
4020 Birch St., Ste. 203   Phone: (714) 731-6273
Newport Beach, CA 92660

★129★ McCorkle Research Services
3200 Rimhill Rd.   Phone: (818) 248-5311
La Crescenta, CA 91214

★130★ Mertec
1524 Hacienda Pl.   Phone: (714) 622-3972
Pomona, CA 91768

★131★ Microglobe Corp.
3333 Bowers Ave., Ste. 130   Phone: (408) 986-0375
PO Box 3019
Santa Clara, CA 95055

★132★ Newport Computer Consultants
980 Paulerino   Phone: (714) 546-2868
Costa Mesa, CA 92626

★133★ Nicholson Associates
3218 Oakes Dr.   Phone: (510) 526-2546
Hayward, CA 94542

★134★ PAK Associates
P.O. Box 904   Phone: (510) 522-6137
El Cerrito, CA 94530

Biotechnology experts with services offered in corporate intelligence, telecommunications, and trademarks. Additional services include current awareness, document delivery, manual searching, online searching, and product tracking.

★135★ Palo Alto Management Group, Inc.
2685 Marine Way   Phone: (415) 968-4374
Mountain View, CA 94043

★136★ Product Intro West, Inc.
The Product Introduction Company
24879 Mulholland Hwy.   Phone: (818) 887-5668
Calabasas, CA 91302

★137★ RCS Associates
1603 Danbury Dr.   Phone: (714) 624-1801
Claremont, CA 91711

★138★ Reference Desk
Box 12122   Phone: (619) 459-1513
La Jolla, CA 92037
Ellen Slotoroff Zyroff, Ph.D., Chief Operating Officer

Founded: 1983. Provides computerized database searching, library research, consulting on information sources, bibliography compilation, article photocopy and document retrieval, company profiles, product and industry surveys, patent searches, and current awareness updates in the areas of business and industry, engineering, psychology, medicine, education, law, humanities, history, ancient studies, classics, linguistics, and current events.

★139★ Richard Siedlecki, Marketing
2674 E. Main St., Ste. C-170   Phone: (805) 658-7000
Ventura, CA 93003

★140★ Richard W. Hynes & Associates
Sky Harbor   Phone: (714) 886-8261
495 W. Edgerton Dr.
San Bernardino, CA 92405

★141★ Roger Johnson and Associates, Inc.
1725 N. Santa Anita Ave.   Phone: (818) 355-1443
Arcadia, CA 91006

★142★ Rosenau Consulting Company
1003 Wilshire Blvd., Ste. 202   Phone: (310) 394-0790
Santa Monica, CA 90401

★143★ Rothchild Consultants, Inc.
256 Laguna Honda Blvd.   Phone: (415) 681-3700
San Francisco, CA 94116-1496

★144★ The Rowland Company/West
5700 Wilshire Blvd., Ste. 475   Phone: (310) 479-3363
Los Angeles, CA 90036

★145★ Royalco International Inc.
749 N. Leonard St.   Phone: (213) 888-2006
Montebello, CA 90640

★146★ Sherman Kingsbury
25 Locust Ave.   Phone: (415) 388-4408
Mill Valley, CA 94941

★147★ **Soyfoods Consulting Services**
P.O. Box 234   Phone: (510) 283-2991
Lafayette, CA 94549

★148★ **Struman & Associates, Inc.**
5655 Lindero Canyon Rd., Ste. 407   Phone: (818) 889-9666
Westlake Village, CA 91362

★149★ **Thomas W. Vaughan**
98 Flynn Ave., Ste. B   Phone: (415) 968-0527
Mountain View, CA 94043

★150★ **Van Norman & Co.**
PO Box 66218   Phone: (310) 397-7536
Los Angeles, CA 90066

★151★ **Wiedenmann & Associates**
3915 Olive Ave.   Phone: (310) 426-7681
Long Beach, CA 90807

★152★ **William G. Garner**
10820 Amigo Ave.   Phone: (818) 368-5971
Northridge, CA 91326

★153★ **Wine Investor**
3284 Barham Blvd.   Phone: (213) 876-7590
Los Angeles, CA 90068-1454

★154★ **Ychem International Corporation**
72 Herringbone Ct.   Phone: (805) 498-5902
Newbury Park, CA 91320

Colorado

★155★ **Al Bergamo Consultant**
812 N. Hancock Ave.   Phone: (719) 634-6775
Colorado Springs, CO 80903

★156★ **DeWitt Marketing Resources**
2318 Eastwood Dr.   Phone: (303) 493-5200
Fort Collins, CO 80525

★157★ **Hyland/Associates**
7027 S. Chapparal Cir.   Phone: (303) 877-0417
Aurora, CO 80016

★158★ **Information Brokers of Colorado**
2888 Bluff St., Ste. 152   Phone: (303) 449-8896
Boulder, CO 80301

★159★ **Jeffery M. Ferguson & Associates**
6180 Lehman Dr., Ste. B-203   Phone: (719) 598-4913
Colorado Springs, CO 80918

★160★ **Ken Garrett & Associates**
4232 Quince Ct.   Phone: (303) 447-9798
Boulder, CO 80302

Connecticut

★161★ **Business Development Associates, Inc.**
46 Main St.   Phone: (203) 972-3812
New Canaan, CT 06840

★162★ **Calle' and Company**
132 Round Hill Rd.   Phone: (203) 661-4889
Greenwich, CT 06831-3745

★163★ **Eastern Marketing Corporation**
15 Innis Ln.   Phone: (203) 637-3999
Old Greenwich, CT 06870

★164★ **Edward Hart Associates**
23 Fox Hollow Dr.   Phone: (203) 623-0033
Windsor Locks, CT 06096

★165★ **Gediman Research Group, Inc.**
26 6th St.   Phone: (203) 348-0009
Stamford, CT 06905

★166★ **Glendinning Associates**
1 Glendinning Pl.   Phone: (203) 226-4711
Westport, CT 06880

★167★ **The Greenfield Consulting Group, Inc.**
274 Riverside Ave.   Phone: (203) 221-0411
Westport, CT 06880

★168★ **Innovators & Associates, Inc.**
82 N. Main St.   Phone: (203) 668-6669
Suffield, CT 06087-2102

★169★ **International Resource Development Inc.**
P.O. Box 1716   Phone: (203) 966-2525
New Canaan, CT 06840

★170★ **JHHMarket Directions**
7 Canton Rd.   Phone: (203) 651-9753
West Simsbury, CT 06092

★171★ **Leferman Associates, Inc.**
22 Knapp St.   Phone: (203) 329-2043
Stamford, CT 06907

★172★ **New Directions Group**
145 Witch Ln.   Phone: (203) 853-6907
Norwalk, CT 06853

★173★ **New England Consulting Group**
55 Greens Farms Rd.   Phone: (203) 226-9200
Westport, CT 06880

★174★ **New Product Dynamics**
240 N. Main St.   Phone: (203) 233-0900
West Hartford, CT 06107

★175★ **Prince Associates**
83 Hoyt St.   Phone: (203) 327-2097
Darien, CT 06820

★176★ **Ridgefield Marketing Services**
24 Forest Glen Dr.   Phone: (203) 663-3088
Killingworth, CT 06417

★177★ **Robert Talmage**
163 Bowery Rd.   Phone: (203) 966-2082
New Canaan, CT 06840

★178★ **Smith, Stanley & Co., Inc.**
P.O. Box 1651   Phone: (203) 655-7664
Darien, CT 06820

★179★ Spectrum Consulting Services
1 RR Fox Rd.   Phone: (203) 928-0491
Putnam, CT 06260

★180★ Stockman and Associates, Inc.
215 Clapboard Ridge Rd.   Phone: (203) 661-3338
Greenwich, CT 06831-3305

★181★ Strategic Alternatives Inc.
194 Main St.   Phone: (203) 849-1217
Norwalk, CT 06851

★182★ Technology Management Group
25 Science Park   Phone: (203) 786-5445
New Haven, CT 06511

★183★ Walter C. Beard, Inc.
P.O. Box 340   Phone: (203) 758-2194
Middlebury, CT 06762

★184★ Youth Research (Division of Consumer Sciences Inc.)
Brookfield Commons B-22   Phone: (203) 797-0666
246 Federal Rd.
Brookfield, CT 06804

Delaware

★185★ Seehusen & Associates
239 Broadhaven Rd.   Phone: (302) 239-4690
Hockessin, DE 19707

District of Columbia

★186★ Bruce W. McGee and Associates
7826 Eastern Ave., NW, Ste. 306   Phone: (202) 726-7272
Washington, DC 20012

★187★ Commercial Associates Incorporated
1001 Connecticut Ave.   Phone: (202) 331-1363
Washington, DC 20036

★188★ International Science and Technology Institute, Inc.
1129 20th St. NW, Ste. 800   Phone: (202) 785-0831
Washington, DC 20036   Fax: (202) 223-3865
Mr. Nihal Goonewardene, Chairman & C.E.O.

Founded: 1977. Technology transfer, especially energy applications for developing countries. Provides information on technologies available, appraises technological needs, conducts feasibility studies, and offers conference management services. Research results published in professional journals, books, magazines, and project reports.

Florida

★189★ Aaron Cohen Marketing Services (ACMS), Inc.
P.O. Box 810815   Phone: (407) 488-4631
Boca Raton, FL 33481

★190★ Chandler Associates
1890 14th St., Ste. 311   Phone: (904) 277-4488
Amelia Island, FL 32034

★191★ Delmar W. Karger
506 Circle Dr.   Phone: (904) 892-7297
DeFuniak Springs, FL 32433

★192★ Dr. Dvorkovitz and Associates
Box 1748   Phone: (904) 677-7033
Ormond Beach, FL 32075   Fax: (904) 677-7113
Dr. Vladimir Dvorkovitz, President

Founded: 1961. Provides new product and process development. Investigates technological developments with objective of obtaining licenses for clients. Provides data, prototypes, samples, assistance in arranging meetings, and negotiation consultation.

★193★ Food Fitness International, Inc.
5975 Sunset Dr., Ste. 601   Phone: (305) 854-1065
Miami, FL 33143

★194★ Harry Prebluda
4101 Pine Tree Dr., Ste. 803   Phone: (305) 531-6927
Miami Beach, FL 33140

★195★ The Hawksbill Company
20533 Biscayne Blvd., Ste. 4-418   Phone: (305) 933-4969
Aventura, FL 33180

★196★ Inovatek Advisors, Inc.
Tarpon Towers, Ste. 200   Phone: (813) 937-7387
905 E. Lake St.
Tarpon Springs, FL 34689-4827

★197★ The Landis Group, Inc.
1551 Forum Pl., Ste. 500 A   Phone: (407) 684-3636
West Palm Beach, FL 33401

★198★ Martin J. Friedman
1710 Daytonia Rd.   Phone: (305) 861-8342
Miami Beach, FL 33141

★199★ W.H. Reed Company
5734 Oakview Ln.   Phone: (813) 637-0043
Punta Gorda, FL 33950

★200★ Zanes Communications Consultants
1833 Halstead Blvd., Ste. 1512   Phone: (904) 668-4191
Tallahassee, FL 32308

Georgia

★201★ Abbey Information Systems
1002 Citadel Dr.   Phone: (404) 633-7446
Atlanta, GA 30324

★202★ Business Ventures Corporation
6030 Dawson Blvd., Ste. E   Phone: (404) 729-8000
Norcross, GA 30093

★203★ Carpax Associates, Inc.
5755 Dupree Dr., Ste. 240   Phone: (404) 859-9075
Atlanta, GA 30327

★204★ Darden Research Corporation
1534 N. Decatur Rd., NE   Phone: (404) 377-9294
Atlanta, GA 30307

★205★ **Fry Consultants, Inc.**
1 Park Place, Ste. 450   Phone: (404) 352-2293
1900 Emery Steet, N.W.   Fax: (404) 352-2299
Atlanta, GA 30318
Dr. Kenneth Bernhardt, Research Director
Founded: 1942. Conducts merger and acquisition studies, industrial marketing research, consumer research, long-range planning, competitive situation analysis, new venture research, strategic analysis of markets and industries, analysis of the buying decision process, and technology transfer studies. Research results published in client reports.

★206★ **Omni-Facts Information Services**
3500 Tritt Springs Circle, Ste. 100   Phone: (404) 565-4455
Marietta, GA 30062

★207★ **Scotti Marketing Research**
1800 Peachtree Rd., NW, Ste. 500   Phone: (404) 352-0686
Atlanta, GA 30309

Illinois

★208★ **Amos & Montgomery Food Business Group**
627 N. York Rd.   Phone: (708) 941-3320
Elmhurst, IL 60126

★209★ **Arthur Fakes**
600 Thames Dr.   Phone: (708) 372-8050
Schaumburg, IL 60193-4345

★210★ **Clinton Associates, Inc.**
102 Wilmot Rd., Ste. 395   Phone: (708) 945-4075
Deerfield, IL 60015

★211★ **Corporate Marketing Management, Inc.**
188 Industrial Dr., Ste. 12   Phone: (708) 941-9300
Elmhurst, IL 60126

★212★ **Creative Research Associates, Inc.**
500 N. Michigan, 12th Fl.   Phone: (312) 828-9200
Chicago, IL 60611

★213★ **Dechert-Hampe & Company**
200 W. Adams, Ste. 2004   Phone: (312) 407-7240
Chicago, IL 60606

★214★ **FMS Consultants**
5801 N. Sheridan Rd., Ste. 3D   Phone: (312) 561-7362
Chicago, IL 60660

★215★ **G-A-M Associates**
616 Laramie   Phone: (708) 256-2016
Wilmette, IL 60091

★216★ **Glen Consulting Group**
770 N. Halsted St.   Phone: (312) 733-4800
Chicago, IL 60622

★217★ **Greenwood-Sutz and Associates Inc.**
146 James Ct.   Phone: (708) 998-5520
Glenview, IL 60025

★218★ **Herbst LaZar Bell Inc.**
345 N. Canal   Phone: (312) 454-1116
Chicago, IL 60606

★219★ **Industrial Electronics Consultants**
P.O. Box 68653   Phone: (708) 980-9595
Schaumburg, IL 60168-0653

★220★ **Information Resources, Inc.**
150 N. Clinton St.   Phone: (312) 726-1221
Chicago, IL 60606

★221★ **Intercon Research Associates, Ltd.**
6865 Lincoln Ave.   Phone: (708) 982-1100
Lincolnwood, IL 60646

★222★ **Ken-Quest Limited**
425 S. Chester Ave.   Phone: (708) 692-2509
Park Ridge, IL 60068

★223★ **Kenneth L. Rubel and Associates**
Box 46245   Phone: (312) 929-0191
Chicago, IL 60646-0245

★224★ **Kubba Consultants, Inc.**
O'Hare Office Center N.   Phone: (708) 296-1224
2720 River Rd.
Des Plaines, IL 60018

★225★ **Kuczmarski and Associates**
7 E. Chestnut St., Ste. 401   Phone: (312) 642-2982
Chicago, IL 60611

★226★ **The Learning Curve**
247 E. Chestnut, Ste. 2304   Phone: (312) 649-0970
Chicago, IL 60611

★227★ **Management Planning Associates, Inc.**
2275 Half Day Rd., Ste. 350   Phone: (708) 945-2421
Bannockburn, IL 60015

★228★ **Marketing Management Pro Temp.**
2107 Pioneer Ln.   Phone: (708) 251-8498
P.O. Box 72
Wilmette, IL 60091

★229★ **Marketron Associates**
P.O. Box 822   Phone: (708) 665-3400
Wheaton, IL 60189

★230★ **Mesa Associates**
P.O. Box 363   Phone: (708) 255-5983
Arlington Heights, IL 60006

★231★ **Porter/Matjasich & Associates**
154 W. Hubbard, Ste. 504   Phone: (312) 670-4355
Chicago, IL 60610

★232★ **SEARCHLINE**
4914 Columbia Ave.   Phone: (708) 964-0127
P.O. Box F
Lisle, IL 60532-0188
Ronald J. Scheidelman, Ph.D., Director
Founded: 1968. Information and literature research, market research, technological forecasting, and product development in the areas of science, engineering, biomedicine and law, especially regarding new products, product liability, patents, and health. Current project focuses on the origin, distribution, fate, and chemical speciation of copper in the environment.

★233★ **Technology Marketing Group**
950 Lee St., Ste. 206   Phone: (708) 297-1404
Des Plaines, IL 60016

★234★ Technology Search International, Inc.
500 E. Higgins Rd.   Phone: (708) 593-2111
Elk Grove Village, IL 60007-1437

★235★ Transcience Associates, Inc.
1112 Hinman Ave.   Phone: (708) 475-1125
Evanston, IL 60202
Harold Kantner, President

Founded: 1979. Computer and biomedical sciences; control systems; hydraulic device development, including photovoltaic-controlled agricultural systems; electrochemistry and electronics; mathematical sciences; physics and solar energy; and technology transfer. Offers development, promotion, brokerage, and licensing support services. Research results published in project reports.

★236★ Xpand Inc.
P.O. Box 188
Glen Ellyn, IL 60138   Phone: (708) 231-5770

Indiana

★237★ Gluys and Assoc.
P.O. Box 399   Phone: (317) 462-4168
Greenfield, IN 46140

★238★ O A Laboratories and Research, Inc.
410 W. 10th St.   Phone: (317) 639-2626
Indianapolis, IN 46202

Kansas

★239★ Charles, Charles & Associates Inc.
8676 W. 96th St., Ste. A   Phone: (913) 341-1354
Overland Park, KS 66212

★240★ Ed Greif and Company, Inc.
Box 40247   Phone: (913) 236-6083
Overland Park, KS 66204

Kentucky

★241★ HLPM, Inc.
10503 Timberwood Cir., Ste. 220   Phone: (502) 429-0015
Louisville, KY 40223

Maine

★242★ Dan Fisher
Rte. 1, Box 122   Phone: (207) 646-6660
Wells, ME 04090

Manufacturing management consultant specializing in new product introduction.

★243★ G. Thomas Cartier & Associates
311 Middle Rd.   Phone: (207) 781-3813
Falmouth, ME 04105

★244★ Gorham International Incorporated
P.O. Box 8   Phone: (207) 892-2216
Gorham, ME 04038

Maryland

★245★ Charles S. Lerner Associates
7201 Marbury Rd.   Phone: (301) 320-2938
Bethesda, MD 20817

★246★ Dahlstrom Associates, Inc.
Rte. 8, Box 154   Phone: (301) 733-1900
Hagerstown, MD 21740

★247★ Mankind Research Unlimited, Inc.
1315 Apple Ave.   Phone: (301) 587-8686
Silver Spring, MD 20910
Dr. Carl Schleicher, President

Founded: 1972. Human systems design, alternate energy sources, research and development for innovative technology, and bionic, biocybernetics, biomedical, and high speed learning systems. Research results published in project reports.

★248★ Marketing Consultants International, Incorporated
645 E. 1st St.   Phone: (301) 791-0290
Hagerstown, MD 21740

Provides marketing and sales consulting services. Specializes in new "high technology" systems requiring pioneering. Specific product areas include speech recognition, machine generated voice output, voice and data communications protection technologies, and factory and office automation. Activities include market research, sales planning, application engineering, distribution method development, organization of sales operations; development of sales and marketing business plans, new product introduction to marketplace, public relations and product promotion campaigns, assistance in implementing management by objectives (MBO), development of product literature and brochures, salesmen training and professional selling skills, assistance in developing professional presentations (sales or otherwise), client representation as public speaker on key issues or sales presentations, manufacturer representative, special tasks that require a sales and marketing orientation, and establishment of export sales and distribution.

★249★ Trademarkers Ltd.
PO Box 30535   Phone: (301) 229-7777
Bethesda, MD 20814

Massachusetts

★250★ Alliance Engineering & Research Corporation
57 Kilby St.   Phone: (617) 935-7400
P.O. Box 532
Woburn, MA 01801

★251★ Anthony Fowler and Company
20 Walnut St.   Phone: (617) 237-4201
Wellesley Hills, MA 02181

★252★ Billings Group
184 Walden St.   Phone: (508) 369-2790
Concord, MA 01742

★253★ Bismark Corporation
P.O. Box 879   Phone: (508) 385-6889
East Dennis, MA 02641

★254★ Center for Product Development (Division of Arthur D. Little, Inc.)
25 Acorn Park   Phone: (617) 864-5770
Cambridge, MA 02140

★255★ Concord Associates, Inc.
34 Monument Sq.   Phone: (508) 369-1087
Concord, MA 01742

★256★ Consulting Resources Corporation
6 Northbrook Park   Phone: (617) 863-1222
Lexington, MA 02173

★257★ Enbede Company
24 Oxbow Rd.   Phone: (617) 862-5008
Lexington, MA 02173

★258★ Francis A. McLoughlin, Consultant
P.O. Box 182   Phone: (617) 237-5262
Wellesley Hills, MA 02181

Overseas business consultant focusing on the technology transfer of patents and ideas, natural resources, and real estate and other business investment ventures. Active in the Far East and Japan.

★259★ George Freedman
Five Brook Trail Rd.   Phone: (508) 358-2350
Wayland, MA 01778

★260★ H.S. Holappa and Associates
9 Hart Rd.   Phone: (617) 334-5716
Box 249
Lynnfield, MA 01940

★261★ Hartwood Associates, Inc.
2 Forest St.   Phone: (617) 862-3363
Lexington, MA 02173

★262★ Jeffrey D. Marshall
21 Central St.   Phone: (508) 887-9788
Topsfield, MA 01983

★263★ John T. Donoghue
16 Colby Ct.   Phone: (201) 287-3691
North Andover, MA 01845

★264★ Kalba International, Inc.
23 Sandy Pond Rd.   Phone: (617) 259-9589
Lincoln, MA 01773

★265★ Kazmaier Associates, Inc.
676 Elm St.   Phone: (508) 371-1732
Concord, MA 01742

★266★ Michael Gigliotti and Associates Incorporated
42 Rogers St.   Phone: (508) 281-3310
Box 591
Gloucester, MA 01931-0591

★267★ Qube Resources
124 Mt. Auburn St., Ste. 200   Phone: (617) 576-5837
Cambridge, MA 01930

★268★ Stewart A. Washburn, Consultant to Sales & Marketing Management
46R Old Main St.   Phone: (508) 947-8607
Lakeville, MA 02347

★269★ Strategic Marketing & Communications, Inc.
Ten High St., Ste. 630   Phone: (617) 426-8646
Boston, MA 02110

★270★ Telematics Resource Group
1209 Windsor Dr.   Phone: (508) 877-3453
Framingham, MA 01701

★271★ Thomson & Thomson
500 Victory Rd.   Phone: (617) 479-1600
North Quincy, MA 02171   Fax: (617) 786-8273
Robert Becker, President

Founded: 1920. Conducts trademark, company name, design, and ancillary research. Publications: Trademark Alert (periodical listing of new trade names). Contact: Toll-free 800-692-8833; toll-free in Canada 800-338-1867; telex 6971430.

★272★ Venditti & Marshall
21 Central St.   Phone: (508) 887-9788
Topsfield, MA 01983

★273★ Venture Development Corporation
1 Apple Hill   Phone: (508) 653-9000
Natick, MA 01760
Lewis Solomon, President

Founded: 1971. Provides customized market research, market planning, strategic planning, and acquisition and new venture research in the areas of computers, communication systems, instrumentation, office equipment, electronic components, and consumer electronics. Offers the VENTURESEARCH quick response service, including database access and consultation regarding specific market questions. Publications: Industry Reports (identifying key technology trends, competitive strategies, and areas of risk and opportunity); The Uniterruptible Power Supply Industry: 1987-1991; The Intelligent Copier/Printer Industry; Venturecasts (annual compilation).

★274★ Victoria A. Mularski
68 Orchard St.   Phone: (413) 743-0754
Adams, MA 01220

Michigan

★275★ Brand Consulting Group
17117 W. Nine Mile Rd., Ste. 1020   Phone: (313) 559-2100
Southfield, MI 48075

★276★ DAVCO Solar Corporation
P.O. Box 335   Phone: (517) 238-2155
Coldwater, MI 49036

★277★ PRS Consulting Group
2301 W. Big Beaver, Ste. 620   Phone: (313) 649-7110
Troy, MI 48084

Minnesota

★278★ Bio-Metric Systems, Inc.
9924 W. 74th St.   Phone: (612) 829-2700
Eden Prairie, MN 55344   Fax: (612) 829-2743
Patrick Guire, R&D Director

Founded: 1979. Biotechnology and sensor systems for detection of biological materials, toxins, drugs, and other biologically active components in the environment. Provides patent technology in biomembrane chemical sensors, biocompatibility control technology, detection/diagnostic kits, and proprietary data and reagents. Technical strengths include bio-organics, bioanalysis, protein and immunochemistry, immunoassay technology, microbiology, enzymology, cell biology, and immunology.

★279★ Cliff Johnson Marketing Consulting
4932 Newton Ave., S.   Phone: (612) 922-4319
Minneapolis, MN 55409

★280★ Daniel H. Talbot and Associates
418 Woodlawn Ave.   Phone: (612) 699-2828
Saint Paul, MN 55105

★281★ Ideas To Go, Inc.
1 Main St., SE, Ste. 504   Phone: (612) 331-1570
Minneapolis, MN 55414

★282★ James H. Owens Associates, Inc.
1221 Nicollet Mall, Ste. 320   Phone: (612) 338-3785
Minneapolis, MN 55403

★283★ Minnesota Cooperation Office for Small Business & Job Creation, Inc. (MCO)
5001 W. 80th St., Ste. 1005   Phone: (612) 830-1230
Minneapolis, MN 55437

★284★ Proud Ind., Inc.
1217 Laurel Ave.   Phone: (612) 646-9428
P.O. Box 4515
Saint Paul, MN 55104

★285★ Strategy Associates Inc.
3120 W. 29th St.   Phone: (612) 927-5044
Minneapolis, MN 55416

Montana

★286★ Kevin C. McGuire
P.O. Box 684   Phone: (406) 777-3833
Stevensville, MT 59870

Consultant in new business ventures, venture capital, new products, and investment capital.

Nebraska

★287★ Chandler & Associates Ltd.
4220 S. Pratt St.   Phone: (402) 453-4560
Omaha, NE 68111

Nevada

★288★ Scott-Brown Engineers
575 Ethan Ln.   Phone: (702) 849-1830
Carson City, NV 89704-9773

New Hampshire

★289★ Intermarketing Group, Management Consultants
29 Holt Rd.   Phone: (603) 672-0499
Amherst, NH 03031

★290★ Paul Hansen Associates
234 Parker Rd.   Phone: (603) 497-2854
Goffstown, NH 03045

★291★ Strategic Business Development Services/ International
Bog Rd., Box 657   Phone: (914) 241-4945
Hillsboro, NH 03244

★292★ Wolff Consultants
1 Buck Rd.   Phone: (603) 643-6015
P.O. Box 1003
Hanover, NH 03755

New Jersey

★293★ Applied Research Techniques, Inc.
1200 Rte. 46   Phone: (201) 263-0880
Parsippany, NJ 07054

★294★ Challenger Industries, Ltd.
1552 Palisade Ave.   Phone: (201) 592-9090
Fort Lee, NJ 07024
P.C. Torigian, Contact

Firm providing marketing research and development, including product development, quality control monitoring, product marketing consulting, patent development, and development of injectable medicaments.

★295★ CHI Research/Computer Horizons, Inc.
10 White Horse Pike   Phone: (609) 779-0911
Haddon Heights, NJ 08035   Fax: (609) 546-9633
Dr. Francis Narin, President

Founded: 1968. Quantitative measurement of research and development activity, including assessment of papers, patents, citations, domains, subfields, laboratories, programs, companies, institutions, agencies, economic sectors, and countries. Specializes in patent and literature citation analysis.

★296★ Cross Gates Consultants
P.O. Box 822   Phone: (908) 238-3381
East Brunswick, NJ 08816

★297★ Cuisine Crafts, Inc.
P.O. Box 1141   Phone: (201) 670-7371
Ridgewood, NJ 07451
Steven Kingsley, Owner

Founded: 1981. Food processing and biochemical engineering, including product design and development, laboratory experimentation, information research, editorial services, patent application assistance, and other evaluation and consulting services. Research results published in trade journals.

★298★ Donna Aughey Ely & Associates, Inc.
14 Pine St.   Phone: (201) 267-0859
Morristown, NJ 07960

★299★ Fluoramics Inc.
18 Industrial Ave.   Phone: (201) 825-8110
Mahwah, NJ 07430

★300★ Food Business Institute, Inc.
P.O. Box 709   Phone: (201) 429-7737
Bloomfield, NJ 07003

Consultants specializing in food industry related business projects and problems, including marketing strategy and planning, advertising and sales promotion, marketing research, new product introductions, test markets, and packaging. Provides both business expertise with technical

experience support, offering a comprehensive and in-depth approach to each assignment.

★301★ Gaylord Associates
28 Newcomb Dr.   Phone: (908) 665-1255
New Providence, NJ 07974

★302★ Glickman Research Associates, Inc.
160 Paris Ave.   Phone: (201) 767-8888
Northvale, NJ 07647

★303★ Irwin P. Sharpe and Associates
50 Greenwood Ave.   Phone: (201) 731-7800
West Orange, NJ 07052

★304★ JBM Associates
PO Box 216   Phone: (201) 267-1871
Convent Station, NJ 07961

Offers new product development, marketing and distribution, and expertise in the acquisition and manufacturing of these products. Serves the automotive, OEM, aftermarket and consumer electronics industries, and government agencies. Active in the United States. Small business firm.

★305★ Kahn/Larsen/Walsh, Inc.
350 W. Passaic St.   Phone: (201) 368-2260
Rochelle Park, NJ 07662

★306★ Lee Slurzberg Research, Inc.
158 Linwood Plaza   Phone: (201) 461-6100
Fort Lee, NJ 07024

★307★ Marketing Perceptions, Inc.
15 Engle St.   Phone: (201) 569-6088
Englewood, NJ 07631

★308★ Matryxx Corporation
PO Box 1201   Phone: (908) 752-8235
Piscataway, NJ 08854
Darryl L. North, Contact

Technical services for the chemical processing industry with specialization in scrubbers/towers, aerators and oxidation pond units, hoods and ducts, oil separators, plate settlers, and plating solution evaporator systems. Specific services include data collection and analysis, site reviews, process surveys, hazardous waste permit review assistance, patent services, and chemical process equipment design reviews. Also recommends, sizes, and designs equipment for process or pollution control, especially fiberglass and thermoplastics. **Publications:** Newsbreak (newsletter).

★309★ May Bender Design Associates
Princeton Corporate Plaza   Phone: (908) 329-8388
7 Deer Park Dr., Ste. M
Monmouth Junction, NJ 08852

★310★ Meadowlands Consumer Center
The Plaza at the Meadows   Phone: (201) 865-4900
700 Plaza Dr.
Secaucus, NJ 07094

Offers expertise in marketing, advertising, new products, marketing research, and management consulting. Serves the packaged goods, financial services, leisure and travel, imported products, household products, health and beauty aids, and photographic industries.

★311★ Noone Associates
813 Columbus Dr.   Phone: (201) 833-2461
Teaneck, NJ 07666

★312★ Princeton Polymer Laboratories, Inc.
501 Plainsboro Rd.   Phone: (609) 799-2060
Plainsboro, NJ 08536   Fax: (609) 779-2558
Dr. Peter Wachtel, President

Founded: 1969. Research and development on new products and processes, market and technological studies, economic evaluation, and consulting in the area of polymers. Research and development capabilities include monomer and polymer synthesis and modification, compounding, and polymer blending, grafting, characterization, testing, and end use evaluation. Studies high-temperature, liquid crystal polymers, ethylene-carbon monoxide copolymers, photodegradable polymers, and polyethyene waxes. Also provides patent confirmation, research corroboration, and economic feasibility studies in the areas of plastics, adhesives, coatings, films, thermoplastics, fibers, foams, additives, and elastomers.

★313★ Profit Planning Associates
P.O. Box 1265   Phone: (201) 746-9225
Montclair, NJ 07042

★314★ R.C. Ropp
138 Mountain Ave.   Phone: (201) 647-7353
Warren, NJ 07060

★315★ Ralph J. Sigona, Associates
1575 Center Ave.   Phone: (201) 461-3067
Fort Lee, NJ 07024

★316★ Robert Auerbach Associates
157 D Wilson Dr.   Phone: (609) 667-7934
Maple Shade, NJ 08052

★317★ Rowin Group, Inc.
101 Cedar Ln.   Phone: (201) 836-0762
Teaneck, NJ 07666

★318★ S Plus H New Products, Inc.
377 Windsor Rd.   Phone: (201) 265-3130
River Edge, NJ 07661

★319★ Thomas M. Poole & Associates
52 Mason Dr.   Phone: (609) 924-2271
Princeton, NJ 08540

★320★ William B. Hardy
260 Metape S.   Phone: (908) 356-7093
Bound Brook, NJ 08805

★321★ Wolff Associates, Inc.
165 Laurel Hill Rd.   Phone: (201) 263-1409
Mountain Lakes, NJ 07046

New Mexico

★322★ Marketing Counselors International
PO Box 9870   Phone: (505) 982-9792
Santa Fe, NM 87504-9870

New York

★323★ ACC - Market Research and Information
55 5th Ave., 17th Fl.   Phone: (212) 627-4100
New York, NY 10003

Provides a wide range of marketing research services, both to consumer and business-to-business, as well as to government agencies.

Specialties include new product development, market studies, psychologically probing qualitative research and trend spotting.

★324★ Andrews Workshops in Creativity
61 Jane St.   Phone: (212) 675-0846
New York, NY 10014

★325★ Ann Scheib Associates
12 Cochrane Ave.   Phone: (914) 693-2157
Dobbs Ferry, NY 10522

★326★ Associated Industrial Designers, Incorporated
75 Livingston St.   Phone: (718) 624-0034
Brooklyn Heights, NY 11201

★327★ Bobrow Consulting Group, Inc.
1025 Fifth Ave., Ste. 3FN   Phone: (212) 249-7001
New York, NY 10028-0134

★328★ Boice Dunham Group
437 Madison Ave.   Phone: (212) 752-5550
New York, NY 10022

★329★ BRX/Global Inc.
169 Rue de Ville   Phone: (716) 442-0590
Rochester, NY 14618

★330★ Carsch Consultant Company
1255 North Ave.   Phone: (914) 633-7292
New Rochelle, NY 10804

★331★ Charles Davidoff
5 Secatoag Ave.   Phone: (516) 883-3700
Port Washington, NY 11050

★332★ COMART-KLP
21 Penn Plaza   Phone: (212) 714-2550
New York, NY 10001

★333★ Contact - Europe
18 E. 68th St.   Phone: (212) 288-9999
New York, NY 10021

★334★ Copywriter's Council of America (Freelance)
7 Putter Ln., Bldg. 102   Phone: (516) 924-8555
Middle Island, NY 11953-0102

★335★ Creative Marketing Group, Inc.
166 E. 61st St.   Phone: (212) 755-0505
New York, NY 10021

★336★ Data Development Corp.
120 5th Ave.   Phone: (212) 633-1100
New York, NY 10011

★337★ David Shepard Associates, Inc.
Two Micole Ct.   Phone: (516) 271-5567
Dix Hills, NY 11746

★338★ Deskey Associates, Inc.
45 W. 36th St.   Phone: (212) 967-3360
New York, NY 10018

★339★ Destiny Kinal Marketing Consultancy
105 Chemung St.   Phone: (607) 565-8317
Waverly, NY 14892

★340★ Dolgins Associates
170 E. 83rd St.   Phone: (212) 719-1300
New York, NY 10028

★341★ E. Janet Berry
274 Madison Ave.   Phone: (212) 679-0580
New York, NY 10016

★342★ E. N. Rysso & Associates
21 Jordan Rd.   Phone: (315) 732-2206
New Hartford, NY 13413-2311

★343★ Embryonics New Product Workshop, Inc.
Taghkonic Rd.   Phone: (518) 537-4605
Elizaville, NY 12523

★344★ Emerson Marketing Agency, Inc.
17 Battery Pl.   Phone: (212) 269-0344
New York, NY 10004

★345★ Evelyn Konrad Associates
136 E. 64th St., Penthouse C   Phone: (212) 371-0024
New York, NY 10021

★346★ Guideline Research Corporation
3 W. 35th St.   Phone: (212) 947-5140
New York, NY 10001

Full service market research company offering strategic research and corporate image studies, product evaluation, development research for new or existing products, sensory testing, product optimization, advertising and concept research, "executive" and industrial research, media research, and legal research and claim substantiation.

★347★ The Hanover Consultancy
141 E. 63rd St.   Phone: (212) 223-4392
New York, NY 10021

★348★ I. Krone Associates
777 3rd Ave.   Phone: (212) 935-0990
New York, NY 10017

★349★ INFO-MARKET, Inc.
P.O. Box 25500   Phone: (716) 377-2050
Rochester, NY 14625

★350★ Isidore & Paulson Inc.
810 7th Ave.   Phone: (212) 586-9300
New York, NY 10022

★351★ J & H Enterprises
P.O. Box 48   Phone: (914) 628-7868
Mahopac Falls, NY 10542

Consulting firm offering marketing research services including strategic planning, product development, business management, and advertising.

★352★ Jelling & Associates
21 Spring Hill Rd.   Phone: (516) 621-0060
Roslyn Heights, NY 11577

★353★ Jennifer B. Freeman
600 W. 111th St., Ste. 7C1   Phone: (212) 662-2775
New York, NY 10025

★354★ Jim Cotton Graphic Communications
417 Canal St., 8th Fl.   Phone: (212) 431-1923
New York, NY 10013

★355★ John Gartner & Co., Technical Marketing Consultants
26 Chasewood Ln.   Phone: (716) 688-7876
East Amherst, NY 14051-1813

★356★ **Kathryn Alexander Enterprises, Inc.**
215 W. 95th St., Ste. 14L   Phone: (212) 222-0216
New York, NY 10025

★357★ **Klein Biomedical Consultants, Inc.**
215 W. 90th St.   Phone: (212) 362-0579
New York, NY 10024

★358★ **KV Marketing Incorporated**
12 Huntsville Rd.   Phone: (914) 232-7566
Box K
Katonah, NY 10536

★359★ **Leon Fleischer, International Marketing**
245 E. 25th St.   Phone: (212) 685-3624
New York, NY 10010

★360★ **Linick Marketing Research Organization
(Division of the Linick Group, Inc.)**
Linick Bldg.   Phone: (516) 924-3888
Middle Island, NY 11953-0102

★361★ **Lynn Greenberg Associates**
36 Country Ridge Rd.   Phone: (914) 723-3121
Scarsdale, NY 10583

Marketing research firm offering strategic planning services and research involving both qualitative and quantitative methods. Experienced with focus groups, interviewing, new product development and marketing, packaging, and advertising. The children's market is a specialty.

★362★ **Management by Design Associates, Inc.**
12 Skyline Dr.   Phone: (516) 673-5111
Huntington, NY 11743-7114

★363★ **Management Practice, Inc.**
342 Madison Ave.   Phone: (212) 867-7948
New York, NY 10173

★364★ **Marilyn Landis Hauser**
5 Peter Cooper Rd.   Phone: (212) 228-7395
New York, NY 10010

★365★ **Marketing Decisions Co.**
165 W. End Ave.   Phone: (212) 496-5464
New York, NY 10023

★366★ **Marketing World, Ltd.**
115 Henry St.   Phone: (516) 546-3441
Freeport, NY 11520

★367★ **Northstar Research Associates, Inc.**
545 8th Ave.   Phone: (212) 594-1117
New York, NY 10018

★368★ **Olewine Company, Inc.**
645 Madison Ave.   Phone: (212) 246-3197
New York, NY 10022

★369★ **The O'Neil Group, Ltd.**
216 Mitchell Ave.   Phone: (315) 426-8455
Syracuse, NY 13207

★370★ **Opatow Associates, Incorporated**
919 3rd Ave.   Phone: (212) 421-4837
New York, NY 10022

★371★ **Packaged Facts**
581 Avenue of the Americas   Phone: (212) 627-3228
New York, NY 10011
David A. Weiss, President

Founded: 1962. Trademark research, syndicated market studies, legal research, back-dated clipping, and advertisement tracking of a particular product, company, or industry. Provides consumer market studies in such areas as the food and beverage industry, personal care products, household products, and advertising specialties. Specific projects include prospective client searches, aquisition/merger research, and historical research for presentations and promotional materials. Research results published in reports. **Publications:** Consumer Market Studies.

★372★ **Peterson & Blyth Associates, Inc.**
216 E. 45th St.   Phone: (212) 557-5566
New York, NY 10017

★373★ **The Qume Group, Ltd.**
141 E. 44th St., Ste. 607   Phone: (212) 286-9048
New York, NY 10017

★374★ **Rain Hill Group, Inc.**
90 Broad St.   Phone: (212) 483-9162
New York, NY 10004   Fax: (212) 514-6217
Rajappan Valagopal, Contact

Founded: 1977. Identifies technology and/or market-based opportunities to augment diversification and growth strategies. Services include screening and review licensing, joint venture research, marketing and distribution studies, minority equity investment research, and acquisition opportunity forecasting in business and technology.

★375★ **Rex Associates, Inc.**
157 E. 57th St.   Phone: (212) 935-0100
New York, NY 10022

★376★ **Ross Research Center, Inc.**
103 Mt. Airy Rd.   Phone: (914) 271-8701
Croton-on-Hudson, NY 10520

★377★ **S. Liebmann & Associates, Inc.**
509 Madison Ave.   Phone: (212) 751-7821
New York, NY 10022

★378★ **Saul Soloway**
180 Broadview Ave.   Phone: (914) 636-6970
New Rochelle, NY 10804

★379★ **Schoonmaker Associates**
P.O. Drawer M   Phone: (516) 473-8741
Coram, NY 11727

★380★ **Spier Corp.**
50 Park Ave.   Phone: (212) 679-4180
New York, NY 10016

★381★ **Stanley Lomas Associates, Inc.**
Major Lockwood Rd.   Phone: (914) 764-5195
Pound Ridge, NY 10576

★382★ **T. H. Land Research Group, Incorporated**
94 Smith Ave.   Phone: (914) 666-2904
P.O. Box 9
Mount Kisco, NY 10549

★383★ **Timbertech Incorporated**
161 Altamont Ave.   Phone: (914) 631-2922
Tarrytown, NY 10591

★384★ Total Concept Company
7580 Northfield Ln.   Phone: (315) 637-5157
Manlius, NY 13104

★385★ Trademark Research Corp.
300 Park Ave., S.   Phone: (212) 228-4084
New York, NY 10010

Supplier of U.S. federal, state, and common law trademark searches. Also has an international trademark watching service, and publishes U.S. trademark office files (full text and images) on CD-ROM. Serves legal and business communities (public and private sectors). Active worldwide, with most efforts in the United States.

★386★ Venture Planning Group, Inc.
350 5th Ave., Ste. 3308   Phone: (212) 564-2838
New York, NY 10118

★387★ Willis S. Steinitz
3229 Skillman Ave.   Phone: (516) 764-3320
Oceanside, NY 11572

North Carolina

★388★ Robertson & Associates, Inc.
P.O. Drawer B   Phone: (919) 827-4881
Pinetops, NC 27864

★389★ Spencer & Associates
417 East Boulevard, Ste. 106   Phone: (704) 376-3682
Charlotte, NC 28203

Ohio

★390★ Brand Research
1600 Keith Bldg.   Phone: (216) 696-4550
Cleveland, OH 44115

★391★ Burke Consulting & Analytical Services
(Division of Burke Marketing Research Inc.)
800 Broadway   Phone: (513) 381-8898
Cincinnati, OH 45202

★392★ Cory Dillon Associates
111 Schreyer Pl., E.   Phone: (614) 262-8211
Columbus, OH 43214

★393★ D. Gentile Associates
1 Dolphin Dr.   Phone: (216) 945-4540
Akron, OH 44313

★394★ Information Consulting, Incorporated
2074 Arlington Ave., Ste. C   Phone: (614) 486-7755
P.O. Box 21865
Columbus, OH 43221

★395★ Robert A. Westman & Associates
359 Quarry Ln.   Phone: (216) 856-4149
Warren, OH 44483

★396★ Strategic Research Center
1 Corporate Exchange   Phone: (216) 831-2410
25825 Science Park Dr.
Cleveland, OH 44122

Oregon

★397★ Search Northwest Associates
Ports O'Call - Swan Island   Phone: (503) 285-6560
4505 N. Channel
Portland, OR 97217

Pennsylvania

★398★ Bob Barry
PO Box H   Phone: (215) 353-7333
Newtown Square, PA 19073

★399★ Comquest, Inc.
512 W. Lancaster Ave.   Phone: (215) 688-3288
Wayne, PA 19087
Rosemary H. Driehaus, President

Founded: 1982. Business research and information, including computerized literature searches and analysis, marketing and opinion reports, patent and trademark information, and executive and expert identification. Research results published in reports to clients. **Publications:** Update (monthly newsletter).

★400★ The Consumer Network, Inc.
3624 Science Center   Phone: (215) 386-5890
Philadelphia, PA 19104

★401★ Creative Design Service
Norman Hills Rd.   Phone: (717) 296-7416
Rte. 1, Box 115
Milford, PA 18337

★402★ Cresheim Company, Inc.
PO Box 27785   Phone: (215) 836-1400
Philadelphia, PA 19118   Fax: (215) 836-1403
James E. Barrett, Managing Director

Founded: 1968. Business negotiating and strategy planning, product and program development and management, organization assessment, opinion research, market and economic research, venture management, industrial marketing, telemarketing, and forecasting, specializing in the areas of construction, distribution, insurance, and transportation. Research results published in reports to clients and through the National Technical Information Service of the U.S. Department of Commerce.

★403★ Domestic & International Technology
115 West Ave.   Phone: (215) 885-7670
Jenkintown, PA 19046

★404★ Gellman Research Associates
115 West Ave., Ste. 201   Phone: (215) 884-7500
Jenkintown, PA 19046   Fax: (215) 884-1385
Frank Berardino, President

Founded: 1972. Data collection and analysis, information/literature research, economic impact studies, cost allocation, benefit-cost analysis, new product feasibility, technology transfer, and microeconomic analysis of technology, transportation, and public policy. Research results published in technical journals.

★405★ George W. K. King Associates Consulting Engineers
1050 Eagle Rd.   Phone: (215) 968-4483
Newtown, PA 18940

★406★ **GRADTECH**
415 S. 19th St.　　Phone: (215) 893-4165
Philadelphia, PA 19146　　Fax: (215) 893-4178
Graham Campbell, Executive Director

Founded: 1985. Identification, research and development, and commercialization of inventions and other new technologies that address the health care concerns of society. Services include market reserch, product evaluation and feasibility studies, intellectual property protection, prototype research and development, and licensing and marketing negotiation. Acts as a general partner whose purpose is to manage and provide seed funding for product development projects.

★407★ **HDR Group**
635 Mallard Rd.　　Phone: (215) 964-8555
Wayne, PA 19087

★408★ **IMS America Ltd., Market Diagnostics**
660 W. Germantown Pike　　Phone: (215) 834-5000
Plymouth Meeting, PA 19462

★409★ **Info/Consult**
Box 204　　Phone: (215) 667-0266
Bala Cynwyd, PA 19004
Gabrielle S. Revesz, Contact

Founded: 1980. Reviews new developments in the fields of chemistry, biomedicine, pharmaceuticals, engineering, patents, and technical sciences. Research results published in confidential reports to clients.

★410★ **Invention Submission Corp.**
903 Liberty Ave.　　Phone: (412) 288-1300
Pittsburgh, PA 15222

★411★ **Joseph W. Prane**
213 Church Rd.　　Phone: (215) 635-2008
Elkins Park, PA 19117

★412★ **Lewis B. Weisfeld**
1 Franklin Town Blvd., Ste. 1204　　Phone: (215) 567-7235
Philadelphia, PA 19103

★413★ **Marilyn A. Lashner, Media Analysis & Communications Research**
PO Box 3165-G　　Phone: (215) 884-4331
Meadowbrook, PA 19046

★414★ **Marketing Network**
1455 Brentwood Ct.　　Phone: (215) 395-1400
Allentown, PA 18104

★415★ **PrimeLife Marketing (Division of The Data Group)**
2260 Butler Pike, Ste. 150　　Phone: (215) 834-3003
Plymouth Meeting, PA 19462

★416★ **Response Innovations, Inc.**
45 Banbury Rd.　　Phone: (717) 566-3849
Hummelstown, PA 17036

★417★ **S. Pullano Associates**
310 State Rd.　　Phone: (215) 649-4176
Gladwyne, PA 19035

★418★ **Target Group, Incorporated**
1000 Harston Ln.　　Phone: (215) 233-4083
Philadelphia, PA 19118

★419★ **Techni Research Associates, Inc.**
Willow Grove Plaza　　Phone: (215) 657-1753
York & Davisville Rds.
Willow Grove, PA 19090

★420★ **Thomas G. Brown Associates**
209 Fox Ln.　　Phone: (215) 565-4393
Wallingford, PA 19086

★421★ **Verna Engineering, Inc.**
416 E. Main St.　　Phone: (412) 276-7611
Carnegie, PA 15106

★422★ **Vision Systems International**
3 Milton Dr.　　Phone: (215) 736-0994
Yardley, PA 19067

Rhode Island

★423★ **Advantage Research Corp.**
16 Shewatuck Bldg.　　Phone: (401) 294-6640
610 Ten Rod Rd.
North Kingstown, RI 02852

South Carolina

★424★ **Ludwig Consultants, Inc.**
Rte. 1, Box 35　　Phone: (803) 423-2452
Marion, SC 29571

Automotive and aerospace consultant active in product development and improvement. Services include patent evaluation, preparation and review; and design, tooling and testing compatible with automated fabrication, assembly and test machinery. Offers expertise in hydraulics, servo systems, pneumatics, vacuum, structures, emission control systems, and aircraft.

★425★ **Semaphore, Inc.**
828 Woodrow St.　　Phone: (803) 799-6464
Columbia, SC 29205

Tennessee

★426★ **Business Consulting Associates**
424 Church St.　　Phone: (615) 259-1400
Nashville, TN 37219

★427★ **McGlinchey Enterprises, Inc.**
310 N. River Oaks Rd.　　Phone: (901) 685-7870
Memphis, TN 38119

★428★ **Venture Marketing Associates, Inc.**
P.O. Box 171392　　Phone: (901) 795-6720
Memphis, TN 38187

Texas

★429★ **Arnold P. Hanson, Jr.**
1528 Harrington Dr.　　Phone: (214) 578-9436
Plano, TX 75075

★430★ Belden Associates
3102 Oaklawn Ave., Ste. 500
Dallas, TX 75219
Deanne Ternimi, President
Phone: (214) 522-8630
Fax: (214) 522-0926

Founded: 1940. Offers product, opinion, advertising, communications, market and sales, and legal research services. Tests and develops products. Conducts new product acceptance studies, prototype research, package and label tests, brand name tests, brand share-of-market surveys, consumer characteristics analysis, consumer behavior surveys, readership studies, audience and circulation analysis, media selection studies, readability tests, political polls, image studies, public relations surveys, shareholder relations surveys, employee attitude surveys, community relations surveys, company name identification studies, health care surveys, trademark confusion studies, packaging studies, and product design research.

★431★ Curly Baca & Associates
16702 Rugby Ct.
Spring, TX 77379-7543
Phone: (713) 370-7283

★432★ Diversified Chemistry Inc.
2501 Gulf Freeway, No. 5
Dickinson, TX 77539
Phone: (713) 337-5745

★433★ Energy-Environmental Research and Development Company
4638 Adkins Dr.
Corpus Christi, TX 47811
Frank D. Kodras, P.E., Chief Operating Officer
Phone: (512) 857-2158

Founded: 1980. Offers product planning, invention searches, and marketing in the areas of chemical processes, cryogenics, nuclear power, engineering, gold and silver ore leaching, management information systems, and patents. Services include market planning, product development, survey taking, patent trend analysis, technology transfer, and development of environmental control systems. Research focuses on coal bacterial desulfurization, flue gas cleaning systems, chemical and natural gas engineering, pollution control, injection additives, coal/water/oil boiler fuels, hazardous wastes, liquid fertilizers, and byproduct utilization and recycling. Research results published in technical society journals and company reports and by the Environmental Protection Agency and the U.S. Government Printing Office in confidential research reports. **Publications:** Innovations & Inventions Newsletter (quarterly).

★434★ Gelb Consulting Group Inc.
3701 Kirby Dr.
Houston, TX 77098
Phone: (713) 526-5711

★435★ Marketing Plans, Programs and Services
1630 Harold
Houston, TX 77006
Phone: (713) 526-7681

★436★ Phillip Townsend Associates, Inc.
P.O. Box 90327
Houston, TX 77290
Phone: (713) 873-8733

★437★ Probe Research, Inc.
2723 Valley View Ln.
Dallas, TX 75234
Phone: (214) 241-6696

★438★ PROTEC
P.O. Box 590011
Houston, TX 77259
Phone: (713) 486-1587

★439★ Rolnick & Associates, Inc.
4615 SW Fwy.
Houston, TX 77027
Phone: (713) 623-8500

★440★ Tool Tron Industries
RFD 4, Ste. 4967
Boerne, TX 78006
Phone: (512) 755-2295

Utah

★441★ Communication Certification Laboratory
1940 W. Alexander St.
Salt Lake City, UT 84119
Thomas C. Jackson, President
Phone: (801) 972-6146
Fax: (801) 972-8432

Founded: 1971. Offers certification, evaluation, quality assurance, and technical documentation services. Prepares quality assurance procedure manuals for federal and state regulatory commissions. Conducts product evaluations and on-site inspections of manufacturing procedures and evaluates designs for compliance with copyright law standards. Assists in filing applications with the Federal Communications Commission and provides type acceptance and certification testing to meet regulations. Also performs testing and filing for certification of telecommunications products to meet Canadian and Japanese government regulations. Research results published in reports to the Federal Communications Commission and clients. **Publications:** Quarterly newsletter.

★442★ Linkage Genetics
1515 W. 2200 S., Ste. C
Salt Lake City, UT 84119
Mark Walton, General Manager
Phone: (801) 975-1188
Fax: (801) 975-1244

Founded: 1973. Commercialization of technology via services and technology licenses in plant biotechnology, including molecular biology, cell and tissue biology, microbiology, phytochemistry, plant breeding and genetics, and applied ecology.

★443★ Materials Research, Inc.
790 E. 700 S.
Centerville, UT 84014
Dr. R. Natesh, President
Phone: (801) 298-4000
Fax: (801) 298-1714

Founded: 1973. Product and process development and testing in the materials industry. Provides characterization of metals and materials, analysis of ultrafine particles, analysis of air and waterborne pollutants, inspection of microelectric devices, failure analysis, corrosion studies, investigation of product contamination, examination of welds and diffusion bonding problems, technical feasibility studies, forensic investigations and accident reconstruction, patent infringement and product liability studies, geothermal kinetic studies, and legal testimony. Also studies problems related to caking, porosity, and discoloration; production, fabrication, and tooling; and lubrication, viscosity, and adhesion. Long-range projects include analysis of defect structure in silicon and chemical reduction of nuclear steam generator crevice deposits. **Publications:** Analysis of Defect Structures in Silicon (series of 20); Chemical Reduction of Nuclear Steam Generator Crevice Deposites (series of 2); and Boron Nitride Coating for High Performance Electric Heaters.

Virginia

★444★ Clark Engineering
916 W. 25th St.
Norfolk, VA 23517
Stephen E. Clark, President
Phone: (804) 625-1140
Fax: (804) 625-5917

Founded: 1979. Energy, specifically low temperature thermal energy conversion; environmental sciences, including pollution control device development and testing; engineering, economic, and marketing studies; computer simulation and analysis; patent evaluation; and prototype performance testing. Research results published in project reports.

★445★ Derwent, Inc.
1313 Dolly Madison Blvd., Ste. 401
McLean, VA 22101
Ginger Roberts, Online Training Executive
Phone: (703) 790-0400
Fax: (703) 790-1426

Founded: 1981. Analyzes, classifies, indexes, abstracts, and codes patent documents from 29 patent issuing authorities. The resulting information is available in printed material, microform, or online.

Publications: On-Line News; Patent News; literature services include Ringdoc, pharmaceutical abstract service; Chemical Reations Documentation Service; Biotechnology Abstracts; and Vetdoc and Pestdoc, providing abstracts in the fields of veterinary products and pest control.

★446★ **First Washington Associates**
1501 Lee Hwy., Ste. 302   Phone: (703) 525-0966
Arlington, VA 22209

★447★ **Government Liaison Services, Inc.**
3030 Clarendon Blvd., Ste. 209   Phone: (703) 524-8200
Arlington, VA 22201

★448★ **HCI Corporation**
P.O. Box 936   Phone: (703) 463-1095
Lexington, VA 24450

★449★ **Hoeft Associates**
P.O. Box 2323   Phone: (703) 415-0366
2001 Jefferson Davis Highway
Arlington, VA 22202
Pamela A. Hoeft, President
Founded: 1978. Data collection and analysis, information and literature research, market research, product liability research, and patent, trademark, and copyright services in the areas of chemistry, electrical and electronics engineering, and physics.

★450★ **Interdevelopment, Inc.**
2001 S. Jefferson Davis Hwy., Ste. 307   Phone: (703) 415-1020
Arlington, VA 22202

★451★ **International Defense Consultant Services, Inc.**
7315 Hooking Rd.   Phone: (703) 827-9032
McLean, VA 22101

★452★ **Marketing Department**
P.O. Box 3525   Phone: (703) 824-8787
Alexandria, VA 22302

★453★ **Nathan Associates**
2 Colonial Pl.   Phone: (703) 516-7700
2101 Wilson Blvd., Ste. 1200
Arlington, VA 22201

★454★ **Science & Engineering Consultants, Inc.**
Reston Town Center   Phone: (703) 471-1600
1820 Discovery St., Ste. 300
Reston, VA 22090

★455★ **SEARCH Corporation**
655 Mine Ridge Rd.   Phone: (703) 759-3560
Great Falls, VA 22066   Fax: (703) 759-9778
Jean Tibbetts, President
Founded: 1973. Specializes in finding available new products, processes, and technology for industry. Conducts technology, market, information, patent, acquisition, and photograph searches. Monitors activity in biotechnology, medical devices, electronics, and any other field of technology.

★456★ **Southeastern Institute of Research, Inc.**
2325 W. Broad St.   Phone: (804) 358-8981
Richmond, VA 23220

★457★ **Technology Research Corporation**
Springfield Professional Park   Phone: (703) 451-8830
8328-A Traford Ln.
Springfield, VA 22152
V. Daniel Hunt, President
Founded: 1979. Collects and analyzes data, performs experimental review and synthesis, conducts information and literature research, and provides systems design and technological forecasting in such fields as defense systems, advanced manufacturing technology, applied expert systems, computer-aided design and manufacturing, computer-integrated manufacturing, applied arctificial intelligence, robotics, machine vision, and expert systems. Related services include technical and management consulting, requirements analysis, program planning and evaluation, technology assessment, risk assessment, and technolgy transfer. Long-range projects focus on superconductivity technology assessment. **Publications:** Mechatronics: Japan's Newest Threat.

★458★ **Venture Services Group**
P.O. Box 1609   Phone: (703) 836-9290
Alexandria, VA 22313

Washington

★459★ **ARA Corporation**
2844 Cascadia Ave.   Phone: (206) 723-4600
Seattle, WA 98144

★460★ **Gulbranson & Associates**
15903 NE 40th Way
Redmond, WA 98052
Management consultant specializing in new idea or item development and automotive wholesale and retail services.

★461★ **Jerome K. Miller**
P.O. Box 1460-A   Phone: (206) 356-7590
Friday Harbor, WA 98250-1460

★462★ **Northwest Trade Adjustment Assistance Center**
900 4th Ave., Ste. 2430   Phone: (206) 622-2730
Seattle, WA 98164

★463★ **RFR Corporation**
P.O. Box 6517
Bellevue, WA 98008

★464★ **Sound Marketing Services, Inc.**
P.O. Box 22443   Phone: (206) 328-0239
Seattle, WA 98122-0443

★465★ **Technical Game Services Inc.**
133 Winslow Way, Ste. 3   Phone: (206) 842-5104
Bainbridge Island, WA 98110

Wisconsin

★466★ **Food Evaluation Center**
2730 W. Silver Spring Dr.   Phone: (414) 461-7760
Milwaukee, WI 53209

★467★ **Philip Adam & Associates**
8223 N. Lake Dr.   Phone: (414) 352-4426
Milwaukee, WI 53217

Wyoming

★468★ Centennial Engineering & Research, Inc.

237 N. Main, Ste. 1   Phone: (307) 672-1711
Sheridan, WY 82801   Fax: (307) 674-5014
Thomas L. Barker, President

Founded: 1976. Information collection, systems design, product development, patent evaluation, and economic analysis of electronic and mechanical systems, remote water systems, control systems, photo cell control equipment, and microprocessors. Also offers water, soils, concrete, and asphalt testing.

Section 15

University and Independent Innovation Research Centers

"Doing easily what others find difficult is talent; doing what is impossible for talent is genius."—Amiel

Many universities have formal programs specifically designed to assist inventors with ideas or patents. Nearly every state has at least one college or university that can provide research and development facilities to technically oriented companies or individuals.

State governments are heavily involved with university research through cooperative programs and support of research parks and technology transfer programs. Known at some universities as "Advanced Technology Centers" or "Centers of Excellence," these programs are designed to increase cooperation between academic institutions and state-based industries. They assist in the creation of new firms through the development of new technology, attracting new business to the state, and increasing competition.

Specifically, technology transfer programs facilitate the transmission of new technologies from the laboratory to the private sector. Such technologies can become the impetus for the creation of new businesses, new products, or revitalization of industries.

Research parks are planned groupings of technology companies, often near universities, that encourage university/private partnerships. They draw industry to a particular location and provide incubator services.

Other kinds of services that are available at many university research centers include innovation evaluation; invention testing; counseling on the start-up of new high technology businesses; assistance in identifying potential research funding from the federal government under programs such as the Small Business Innovation Research (SBIR) Grants and the Energy-Related Inventions Program (ERIP); counseling regarding patents, trademarks, and copyrights; and assistance in identifying sources of government contracts for goods or services needed. Many centers also have excellent libraries of technical information and computer literature searching capabilities.

Whenever I require specialized engineering skills on a project, especially to build complicated prototypes, one of the first places I check is the talent pool available at the local university research

center. The costs are reasonable and there is no better place to get fresh, young minds working on a resolution to a particular problem.

Universities do have specialties, so it is best to check in advance before going out to the campus. For example, some universities concentrate on biomedical and life sciences research, others handle aerospace engineering, electronic systems, mechanics, physical sciences, fabrication technologies, and so forth.

Idea evaluation is the first major step after a concrete, detailed idea has been developed. This is a critical phase since every subsequent phase requires the investment of more time and money. University research centers are an excellent place to have an invention evaluated to determine its overall technical and commercial feasibility.

The university research centers listed on the following pages may be useful in these evaluation activities. They have people to help determine whether the new invention is a marked improvement over its competition; whether it is likely to be commercially viable; what the probable demand for it will be; who will produce it; and how it will be distributed. These specialists can help an inventor arrive at the decision to go ahead to the commercialization stage, to redesign the invention, or to kill the project altogether.

The failure rate of technological innovators is estimated by one Michigan study to be one quarter that of all entrepreneurs during the first five years of business. One way to decrease the chances of failure is to ensure that all the resources are available when required. A good university research center program can help toward this end.

Listed on the following pages are principal university research institutes, technology transfer centers, and research parks whose purpose is to promote innovation, invention, and product development. **See Section 16, Business Incubators, on page 175 for additional information on related services.**

Section Arrangement

The following entries are arranged geographically by state, then alphabetically by research center name. If the research center is an integral part of a university, then the university name will be listed above the research center name. Both the research center name and the university name are listed in the Master Index.

Alabama

★469★ Auburn University
Center for International Commerce
205 Thach Hall
Auburn, AL 36849
Dr. William Boulton, Director
Phone: (205) 844-2352

Investigates technology transfer activities in U.S., Japan, and European countries.

★470★ Economic Development Institute
3354 Haley Center
Auburn, AL 36849
Bettye Burkhalter, Director
Phone: (205) 826-4704

Facilitates technology transfer activities through interdisciplinary technology assistance teams and projects. Serves as a resource for generating economic development projects through supporting faculty and graduate research opportunities, administering community economic development plans, and assisting contract and grant development. Offers assistance to faculty, graduate students, professionals, business and industry and local and state governments in proposal and project development.

★471★ Cummings Research Park
Chamber of Commerce
PO Box 408
Huntsville, AL 35804-0408
Steve Golinveaux, Director
Phone: (205) 535-2018

4,000-acre site linking University of Alabama in Huntsville research and educational resources with Park tenants, particularly in the areas of aerospace, missile research and development, applied optics, artificial intelligence, software development, and data communications. Houses the state's supercomputer, accessible by universities and industries throughout Alabama.

★472★ Jacksonville State University
Center for Economic Development and Business Research
Rm. 114, Merrill Hall
Jacksonville, AL 36265
Pat W. Shaddix, Director
Phone: (205) 231-5781

Industrial needs analysis for cities and counties and business research, including retail and service studies. Special projects include economic development strategies for Cleburne County, marketing plan considerations for Calhoun County, and economic best fit studies for Oxford, Alabama. **Publications:** Monographs.

★473★ University of Alabama at Birmingham
Alabama Small Business Development Center
Medical Towers Bldg.
1717 11th Ave., Ste. 419
Birmingham, AL 35294
Jeff D. Gibbs, State Director
Phone: (205) 934-7260
Fax: (205) 934-7645

Member of the Association of Small Business Development Centers.

★474★ University of Alabama in Huntsville
Center for the Management of Science and Technology
Ste. 126
Administrative Science Bldg.
Huntsville, AL 35899
Dr. William E. Souder, Director
Phone: (205) 895-6407

Management of high technology projects and new product innovation processes, including research and development management, organization design, group dynamics, and strategic management. **Publications:** Annual Report.

Alaska

★475★ University of Alaska
Alaska Small Business Development Center
430 W. 7th Ave.
Anchorage, AK 99501
Jan Fredericks, State Director
Phone: (907) 274-7232
Fax: (907) 274-9524

Member of the Association of Small Business Development Centers.

Arizona

★476★ Arizona Small Business Development Center
108 N. 40th St., #148
Phoenix, AZ 85034
David E. Smith, State Director
Phone: (602) 392-5224
Fax: (602) 392-5300

Member of the Association of Small Business Development Centers.

★477★ Arizona State University Research Park
2049 E. ASU Circle
Tempe, AZ 85284
LeEtta Overmyer, Executive Director
Phone: (602) 752-1000

323-acre university-related research park, affiliated with Arizona State University. Seeks to link technological results of university and individual research with private industry. The Park provides tenants with leased land for construction of research and development facilities, laboratories, offices, pilot plants, facilities for production or assembly of prototype products, and University and government research facilities.

★478★ Association of University Related Research Parks
4500 S. Lakeshore Dr.
Ste. 475
Tempe, AZ 85282-7055
Chris Boettcher, Executive Director
Phone: (602) 752-2002

Serves as forum for the exchange of information on planning, construction, marketing, and managing of university-related research parks, particularly information on university-industry relations, innovation, and technology transfer to the private sector. Monitors legislative and regulatory actions affecting the development and operation of research parks. Acts as a clearinghouse for career opportunities. Maintains a database, Best North America, which provides members with information on research activities at select research universities on the continent. **Publications:** The Research Park Forum (quarterly newsletter).

★479★ Research Corporation Technologies
6840 E. Broadway Blvd.   Phone: (602) 296-6400
Tucson, AZ 85710-2815
Dr. Gary M. Munsinger, President

Evaluates, protects, develops, and commercializes inventions from colleges, universities, medical research organizations, and other nonprofit laboratories. Among the variety of inventions that have been developed and marketed are vitamins, genetic engineering techniques, the maser-laser concept, hybrid seed corn, cortisone, anticancer drugs and other pharmaceuticals, laboratory and medical instruments, and chemical, electronic, and mechanical equipment. Provides incentives for invention disclosure, funds applied research, new business formation, and joint ventures. **Publications:** Annual Report; Newsletters (occasionally); Brochures.

Arkansas

★480★ University of Arkansas
Arkansas Center for Technology Transfer
131 Engineering Research Center   Phone: (501) 575-3747
Fayetteville, AR 72701
William H. Rader, Director

Provides research and technical facilities to analyze factors such as production, workforce motivation and training, and technical problems. Facilitates research and technology transfer in the areas of manufacturing technology, interactive technology, productivity and industrial efficiency, and business incubation and analysis. Specific projects include bar-code reading of trucks at weigh stations, scales to weigh-in-motion, tactile sensing, and advanced brakes, clutches, and transmissions.

★481★ University of Arkansas at Little Rock
Arkansas Small Business Development Center
100 S. Main, Ste. 401   Phone: (501) 324-9043
Little Rock, AR 72201   Fax: (501) 324-9049
Paul E. McGinnis, State Director

Member of the Association of Small Business Development Centers.

California

★482★ Babson College
Center for Entrepreneurial Studies
Babson Park, CA 02157   Phone: (617) 239-5288
Prof. Neil C. Churchill, Director

Entrepreneurship and new and growing businesses, including studies in venture capital, starting and financing new value-creating ventures, family businesses, and harvesting enterprises through IPO's, merger or sale, and family succession.

★483★ California Small Business Development Center
801 K St., 16th Fl., Ste. 1600   Phone: (916) 324-9234
Sacramento, CA 95814   Fax: (916) 322-3524
Edward Kawahara, State Director

Member of the Association of Small Business Development Centers.

★484★ Inventors Workshop International Education Foundation
HQ, Inventor Center USA   Phone: (805) 484-9786
3201 Corte Malpaso, #304-A
Camarillo, CA 93012
Alan Arthur Tratner, President

New product development and market research focusing on technology-oriented inventions. Also studies patent classifications. Maintains a database of U.S. inventors/members and invention development resources. **Publications:** Invent! (magazine).

★485★ Research Institute for the Management of Technology
215 N. Marengo, Third Fl.   Phone: (818) 584-9139
Pasadena, CA 91101

Management issues, technology policy, and the international competitiveness of technology industries. Develops programs facilitating the commercialization of environmental and other technology from national research facilities, particularly the National Aeronautics and Space Administration (NASA) and its Jet Propulsion Laboratory (JPL), operated by the California Institute of Technology. Maintains a database of technology companies in southern California.

★486★ SRI International
333 Ravenswood Ave.   Phone: (415) 326-6200
Menlo Park, CA 94025-3493
Dr. William F. Miller, President

Conducts research in more than 100 disciplines in the physical and life sciences, engineering, industrial management, social sciences, and public policy. Engages in 2,000 projects per year involving scientists, engineers, industry experts, and management consultants from hundreds of fields. Areas of research and consulting include economics, energy, engineering and development, environment, health, industry consulting, information sciences, management consulting, public policy, national security, and physical, life, and social sciences. **Publications:** Annual Report; SRI Journal.

★487★ Stanford University
Center for Economic Policy Research
100 Encina Commons   Phone: (415) 725-1874
Stanford, CA 94305
Dr. John B. Shoven, Director

Funds public policy research at the University in the economics of regulation, innovation and contribution to economic growth of high technology industries, macroeconomics, economics of energy and natural resources, and fiscal and monetary policy, including the federal budget, social security, tax treatment of the family, and alternative tax systems. Research and conference programs range from analytical and conceptual matters, such as ways of measuring the benefits of public services and costs of government programs, to empirical studies of trends in savings, capital formation, regulation, prices, interest rates, productivity, technological change, and the distribution of income and wealth in the U.S. **Publications:** Discussion Paper Series; Newsletter (semiannually).

★488★ Office of Technology Licensing
857 Serra St., Second Fl.   Phone: (415) 723-0651
Stanford, CA 94305-6225
Niels Reimers, Director

Licenses technology in the fields of scientific and medical instruments, pharmaceuticals, chemicals, computer software and databases, integrated circuit technology, optics, and microbiology. Evaluates, markets, and negotiates licensing agreements with industry.

★489★ Stanford Research Park
857 Serra St.   Phone: (415) 725-6886
Stanford, CA 94305-6225
Zera Murphy, Managing Director

655-acre site linking research resources of the University with Park tenants, particularly in the areas of electronics, space, publishing, pharmaceutics, and chemistry. Activities between park tenants and the University community include cooperative research ventures, instruction, and consulting.

★490★ University of California, Berkeley
Institute of Transportation Studies

109 McLaughlin Hall   Phone: (415) 642-3585
Berkeley, CA 94720
Prof. Adib Kanafani, Director

Transportation planning, management of traffic and transportation systems, transit finance, new systems and technology development and transfer, policy and economics, regional and national planning, and energy conservation. Studies all modes of transportation. Also stimulates research in transportation engineering at the University's College of Engineering. **Publications:** ITS Review; Research Reports; Working Papers: Dissertation Series; Annual Reports.

★491★ University of California, Irvine
Office of University/Industry Research & Technology

345 Administrative Bldg.   Phone: (714) 856-7295
Irvine, CA 92717
David G. Schetter, Director

University and industry partnership and cooperative research development, including industrial contract development, review, and approval; science and technology research identification; consortia participation; new consortia initiation; and federal and state science and technology center participation.

★492★ University of California, Los Angeles
Entrepreneurial Studies Center

John E. Anderson Graduate School of   Phone: (213) 825-2985
   Management
405 Hilgard Ave., Ste. 4284
Los Angeles, CA 90024-1481
Alfred E. Osborne, Jr., Ph.D., Director

Entrepreneurship, venture initiation, and interdisciplinary studies of business development, including finance, operations, information systems, organization, and behavior. Maintains oral histories of entrepreneurs. **Publications:** IMPACT (newsletter).

★493★ University of Southern California
Office of Patent & Copyright Administration

3716 S. Hope St., Ste. 313   Phone: (213) 743-4926
Los Angeles, CA 90007
Rosanne Dutton, Director

Transfers technology from the University to the private sector, including the areas of medicine, engineering, pharmacy, gerontology, and architecture.

Colorado

★494★ Center for Advanced Technology

PO Box 483   Phone: (303) 482-2916
601 S. Howes St.
Fort Collins, CO 80522
Kathleen Byington, Contact

235-acre research park that facilitates the exchange of University research resources with Park tenants.

★495★ Center for the New West

600 World Trade Center   Phone: (303) 592-5310
1625 Broadway
Denver, CO 80202
Philip M. Burgess, President

Research institute focusing on trends in demographics, enterprise formation, and job creations in 19 western states, including Alaska and Hawaii. Seeks to improve the quality and usefulness of information on the New Economy, which is characterized by expanding global competition, dramatic demographic shifts, technological change, rapidly changing consumer tastes, entrepreneurship, and the growing impact of innovation on enterprise formation. Focuses on national and international trends and their impact on businesses and communities in the western region of the U.S. Emphasizes new enterprise development and expansion of existing enterprises through innovation; capital formation, including the formation, availability, and cooperative use of public and private capital; technology; expanded international trade, particularly in the Asian Pacific Rim region; human capital, including demographic trends and their impact on economic growth and productivity; and area and regional development, including infrastructure and innovative institutional arrangements. **Publications:** Points West (biennial newsletter); Profile of the West (biennial compendium of 200 demographic, economic, and social indicators on the western states); Occasional Reports.

★496★ Colorado Small Business Development Center

Colorado Dept. of Economics   Phone: (303) 892-3809
   Development   Fax: (303) 892-3848
1625 Broadway, Ste. 1710
Denver, CO 80202
Rick Garcia, Director

Member of the Association of Small Business Development Centers.

★497★ Colorado State University
Manufacturing Excellence Center

Dept. of Mechanical Engineering   Phone: (303) 491-6558
Ft. Collins, CO 80523
Dr. C. Byron Winn, Director

Research and technology transfer.

★498★ University of Colorado
University of Colorado Foundation Inc.

Box 1140   Phone: (303) 492-8134
Boulder, CO 80306
John P. Holloway, Patent Officer

Facilitates research and technology transfer activity at the University in the areas of medicine, optoelectronics, molecular biology, pharmacy, and chemistry. **Publications:** Patent Administration and Technology Transfer at the University of Colorado.

★499★ University of Denver
Denver Research Institute

2050 E. Iliff   Phone: (303) 871-2628
Denver, CO 80208
Larry Brown, Director

Industrial product and process development, including studies of energy sources and technologies, materials processing, composites and nonmetallic materials, and minerals and materials extraction and processing; economics and management, including studies of regional economics and growth management, energy and resource economics, business planning and management, and technology management; defense technology, including studies of ordnance, terminal ballistics, vulnerability assessment, blast effects, electronic systems, and management and policy; environmental technology, including studies of particulate control, wastewater, atmospheric sciences, and geothermal energy; service systems, including studies on human service delivery, health care, institutional, management, program, and experimental evaluation, information transfer, and legal and judicial programs; instructional/training and learning systems, including analysis, evaluation, simulation and design, and development of educational and training activities; instrumentation and measurement, including research and development of industrial, defense, photographic, photogrammetric, and biomedical technologies; and international development, including studies of industrial development, technology transfer, management development, institutional development, information systems, and resource development. Provides data processing facilities, programming assistance, and technical service in support of research in various department of the University.

Connecticut

★500★ Connecticut Innovations, Inc.
Technology Assistance Center   Phone: (203) 258-4305
845 Brook St.
Rocky Hill, CT 06067-3405
Eric C. Ott, Director

Stimulates technological innovation and economic growth in Connecticut in the areas of aerospace, biotechnology, computer applications, energy systems, materials technology, medical technology, and telecommunications. Provides risk capital for new product, process, and service development, and to launch and market new products. Finances start-up of high-technology companies with seed capital and small technology-based firms doing federal research and development leading to commerical products.

★501★ NERAC, Inc.
One Technology Dr.   Phone: (203) 872-7000
Tolland, CT 06084
Dr. Daniel U. Wilde, President

Examines results of government and other research and development activities and applies these results to industrial and nonindustrial sectors of the economy. Also studies ramifications of transfer and utilization of scientific and technological developments. Maintains close ties with worldwide information sources, including specialized data centers, libraries, governmental agencies, universities, and research institutes.

★502★ Science Park
Five Science Park   Phone: (203) 786-5000
New Haven, CT 06511
William W. Ginsberg, President

Research park established by Yale University, state of Connecticut, city of New Haven, and Olin corporation. 80-acre technology and light industrial site provides scientific facilities, services, and assistance for tenants to interface with University research resources. Houses the New Enterprise Center, an incubator facility. **Publications:** Science Park Business-to-Business Directory Update; Acess.

★503★ University of Connecticut
Connecticut Small Business Development Center
368 Fairfield Rd., U-41, Rm. 422   Phone: (203) 486-4135
Storrs, CT 06269-2041   Fax: (203) 486-1576
John P. O'Connor, State Director

Member of the Association of Small Business Development Centers.

★504★ Yale University
Economic Growth Center
Box 1987, Yale Sta.   Phone: (203) 432-3620
27 Hillhouse Ave.
New Haven, CT 06520
T. Paul Schultz, Director

Processes of economic growth and economic relations between the developing and economically advanced countries. Program emphasizes cross-sectional and intertemporal studies, policy analysis, and applied microeconomic theory in the study of the behavior of households and firms. Projects include technology choice and transfer, household consumption, investment and demographic behavior, agricultural research and productivity growth, interrelated factor markets, labor markets and the returns to education and migration, income distribution, and international economic relations, including monetary and trade policies. Geographic spread is worldwide with particular emphasis on Asia, Latin America, and Africa. **Publications:** Economic Growth Center Series; Reprints; Discussion Papers.

★505★ Office of Cooperative Research
246 Church St., Ste. 401   Phone: (203) 432-7240
New Haven, CT 06510
Dr. Robert K. Bickerton, Director

Facilites the transfer of technology from the University to business and industry. Patents and licenses University inventions in the areas of physical sciences and engineering, medicine and biotechnology, and computer sciences. Administers industrial liasion programs in computer science.

Delaware

★506★ University of Delaware
Delaware Small Business Development Center
Purnell Hall, Ste. 005   Phone: (302) 451-2747
Newark, DE 19716   Fax: (302) 451-6750
Linda L. Fayeweather, State Director

Member of the Association of Small Business Development Centers.

★507★ University of Delaware Research Foundation
Univ. of Delaware   Phone: (302) 451-2136
210 Hullihen Hall
Newark, DE 19716
Howard E. Simmons, Jr., President

Supports University research by awarding grants to faculty members, also supports invention and patent activities for faculty members. Facilitates technology transfer between the University and industry.

District of Columbia

★508★ Corporation for Enterprise Development
777 N. Capitol St., NE, Ste. 801   Phone: (202) 408-9788
Washington, DC 20002
Doug Ross, President

Research, development, and dissemination of entrepreneurial policy initiatives at the local, state, and federal levels. Specific studies include economic climate of the states, transfer payment investment, pension funds, job creation, seed capital assessment, international exchange, flexible manufacturing networks, and state capital market analysis. Emphasizes economic empowerment of economically disadvantaged populations. **Publications:** The Entrepreneurial Economy Review (quarterly); Making the Grade: The Development Report Card for the States (annually); State Enterprise Development Implementation Packets (paper series); State Strategy Memoranda (paper series); Investing In.

★509★ Howard University
District of Columbia Small Business Development Center
6th & Fairmont St. NW, Rm. 128   Phone: (202) 806-1550
Washington, DC 20059   Fax: (202) 797-6393
Nancy A. Flake, State Director

Member of the Association of Small Business Development Centers.

★510★ Industrial Research Institute, Inc.
1550 M St. NW   Phone: (202) 872-6350
Washington,, DC 20005
Charles F. Larson, Executive Director

Independent, nonprofit research organization, formed as a consortium of approximately 270 industrial companies with technical research departments. Promotes, through cooperative efforts of its members, economical and effective techniques of organization, adminstration, and operation of industrial research. Seeks to generate industrial and academic research collaboration as well as industry/government cooperation in matters related to research. Also endeavors to stimulate

and develop an understanding of research as a force in economic, industrial, and social activities and encourages high standards in the field of industrial research. **Publications:** Research Technology Management (bimonthly).

★511★ **Knowledge Transfer Institute**

1308 4th St. SW   Phone: (202) 554-9434
Washington, DC 20024
Dr. Ronald G. Havelock, Director

Knowledge dissemination, transfer, and use, including scientific knowledge and technology transfer and use in education, medicine, and other areas. Conducts studies of networks involving schools and universities, the role of external innovations agents in dissemination and use of educational innovations, transfer of new cancer technologies into routine patient care, the effects of the mandating of legislation on use of new knowledge, and utilization of new technologies by managers.

Florida

★512★ **Central Florida Research Park**

12424 Research Pkwy.   Phone: (407) 282-3944
Ste. 100
Orlando, FL 32826
Joe Wallace, Director of Marketing

1,027-acre site zoned for commercial and light manufacturing adjacent to the University of Central Florida. Established to create an environment which promotes and fosters relationships between industry and the University. **Publications:** Central Florida Research Park Update (newsletter).

★513★ **First Coast Technology Park**

4567 St. John's Bluff Rd. S.   Phone: (904) 646-2710
Jacksonville, FL 32216
Curtis D. Bullock, Executive Director

275-acre research park affiliated with University of North Florida. Facilitates cooperative research and development activities between Park tenants and the University community.

★514★ **Florida Atlantic Research Park**

Office of Academic Affairs   Phone: (407) 367-3068
PO Box 3091
Boca Raton, FL 33431-0991
Dr. J.S. Tennant, Executive Director

Proposed 52-acre research and development park on the campus of Florida Atlantic University. Serves as a bridge between the research interests of the tenant companies and research activities of the University community. The Park features an innovation center to aid in transferring technology and house University functions such as the Small Business Development Center and NASA/Southern Technology Application Center.

★515★ **Florida State University**
Florida Economic Development Center

335 College of Business   Phone: (904) 644-1044
Tallahassee, FL 32306-1007
Roy Thompson, Director

Business planning, venture and seed capital, insurance, purchasing, community development, and management. Conducts target industry studies, area business analysis studies, and downtown business surveys. **Publications:** Special Reports/Studies; Small Business News and Views (weekly column); Florida Venture Capital Handbook; Special Sources of Credit; Business Planning Guide.

★516★ **Innovation Park (Tallahassee, FL)**

1673 W. Dirac Dr.   Phone: (904) 575-6381
Tallahassee, FL 32304
1977 Linda Mizerski, Contact

Joint research park of Florida A&M University, Florida State University, Leon County Research and Development Authority, and the City of Tallahassee. Fosters research partnerships between the Universities and industry tenants.

★517★ **Progress Center: University of Florida**
        Research and Technology Park

One Progress Blvd., Box 10   Phone: (904) 462-4040
Alachua, FL 32615
Gordon Carlisle, Director

200-acre research and technology park open to both public and private research and manufacturing organizations. High-technology development is emphasized, including the areas of electronics, biotechnology, advanced materials, pharmacology, and agriculture. Provides a link between University researchers and industry and is designed to transfer new technologies from the laboratory to the marketplace. Mainframe computer resources available. **Publications:** Progress Report (quarterly).

★518★ **Tampa Bay Area Research & Development**
        Authority

Administration 275   Phone: (813) 974-2890
Univ. of South Florida
Tampa, FL 33620
John Hennessey, Executive Director

University-affiliated research and development park leasing land to industries needing facilities for research and related scientific manufacturing in medicine, engineering, and natural sciences. Tenants share University research facilities and services, use faculty as consultants, and utilize graduate students as part-time work force. Oversees development of university-related research parks in the Tampa Bay area, particularly University Technology Center.

★519★ **University of Florida**
Biotechnology Institute for Technology Transfer

One Progress Blvd., Box 26   Phone: (904) 462-3904
Alachua, FL 32615
Lenie Breeze, Interim Director

Identifies and markets commercially-significant biotechnology patents and discoveries of the University. Collaborates with industry.

★520★ **University of Miami**
Innovation and Entrepreneurship Institute

PO Box 249117   Phone: (305) 284-4692
Coral Gables, FL 33124
Carl McKeney, Executive Director

Entrepreneurship and innovation in Florida, including studies on high-technology ventures and African-American and Cuban-American entrepreneurs. Promotes interaction of entrepreneurs and capital and service providers. **Publications:** Research Report Series; Friends of the Forum Directory.

★521★ **University of South Florida**
Florida Small Business Development Center

College of Business Administration   Phone: (813) 974-4274
Tampa, FL 33620
Bill Manck, Director

Small business operations, entrepreneurship, and success and failure factors for small business development and management, including developing business plans, marketing strategy, and loan packages. **Publications:** Speaking of Small Business (monthly newsletter).

★522★ **University of West Florida**
Florida Small Business Development Center
Bldg. 38, Rm. 107   Phone: (904) 474-3016
Pensacola, FL 32514   Fax: (904) 474-2030
Jerry G. Cartwright, State Coordinator

Member of the Association of Small Business Development Centers.

★523★ **University Technology Center**
c/o VRS Realty Services   Phone: (813) 281-0601
7650 Courtney Campbell Causeway,
  Ste. 1100
Tampa, FL 33607-1432
Daniel Woodward, Project Manager

Park provides an interface between research and development tenants and University of South Florida resources, particularly in the areas of medical technology and engineering sciences. Tenants access University research facilities and services, faculty consulting, and graduate student assistance in research and development and prototype assembly activities.

Georgia

★524★ **Georgia Institute of Technology**
Advanced Technology Development Center
430 Tenth St., NW, Ste. N-116   Phone: (404) 894-3575
Atlanta, GA 30318
H. Wayne Hodges, Director

Component of the University of Georgia system, located at Georgia Institute of Technology. Serves as a conduit to the University system's research programs, faculty, and facilities for developing technology businesses, particularly in the areas of aerospace vehicles and equipment, biotechnology products, telecommunications equipment, computers and peripheral devices, computer software, electronic equipment, medical devices, instrumentation and test equipment, pharmaceuticals, new materials, and robotics. Operates Technology Business Center to provide office, research and development, laboratory, and manufacturing space to early-stage, high-technology companies.

★525★ **Economic Development Laboratory**
Georgia Tech Research Institute   Phone: (404) 894-3841
Atlanta, GA 30332
Dr. David S. Clifton, Jr., Director

Supports 12 regional offices throughout Georgia to assist local industries in the areas of agricultural technology, industrial energy conservation, hazardous waste management, safety engineering, industrial hygiene, asbestos abatement, market planning and research, target industry analysis, cost-benefit analysis, energy modelling, technology transfer, industrial training, productivity improvement, small business assistance, analytical chemistry, and indoor air pollution. **Publications:** Engineering Reviews (occasionally); Technical Briefs (occasionally); Industrial Advisor (quarterly newsletter); Industrial Energy Conserver (quarterly newsletter); AMTC Quarterly; Environmental Spectrum: Poultry Tech (quarterly newsletter); Research Focus (quarterly newsletter).

★526★ **Technology Policy and Assessment Center**
Office of Interdisciplinary Programs Phone: (404) 894-2330
Atlanta, GA 30332
Dr. Alan L. Porter, Director

Policy and societal aspects of science and technology, both domestic and international. Studies include technology and impact assessment, technological innovation and diffusion of innovations, cost, benefit analysis, socioeconomic development, research and development policy and management, energy and environmental policy, and the interdisciplinary research process.

★527★ **Georgia State University**
International Center for Entrepreneurship
College of Business Administration   Phone: (404) 651-3782
Univ. Plaza
Atlanta, GA 30303
Dr. Francis W. Rushing, Director

Conducts and facilitates studies in entrepreneurship. Serves as a resource center.

★528★ **University of Georgia**
Center for East-West Trade Policy
204 Baldwin Hall   Phone: (404) 542-2985
Athens, GA 30602
Prof. Gary K. Bertsch, Co-Director

Monitors and evaluates U.S. and western alliance policies governing East-West trade, particularly the role of governments, export controls, and technology transfer.

★529★ **Georgia Small Business Development Center**
Chicopee Complex-1180 E. Broad St. Phone: (404) 542-5760
Athens, GA 30602
Hank Logan, State Director

Member of the Association of Small Business Development Centers.

★530★ **University of Georgia Research Foundation, Inc.**
Boyd Graduate Studies Research   Phone: (404) 542-5969
  Center
Ste. 609
Athens, GA 30602
Dr. Joe L. Key, Executive Vice President

Primarily life sciences, agriculture, veterinary medicine, pharmacy, chemistry, physics, and forestry resources. Subcontracts research projects to the University, handles all University technology transfer, and intellectual property issues.

Hawaii

★531★ **Maui Research and Technology Park**
Maui Economic Development Board, Inc. Phone: (808) 871-6802
PO Box 187
Kahului Maui, HI 96732
Donald G. Malcolm, President

300-acre research park fostering research activities between the academic sector and Park tenants, particularly in the areas of optic systems, electronics design and assembly, information systems and telecommunications, biotechnology, and alternate energy. **Publications:** Newsletter; Proceedings of Symposium.

★532★ **Mililani Technology Park**
300 Kahelu Ave. #35   Phone: (808) 548-8996
Mililani, HI 96789
William M. Bass, Executive Director

547-acre research and development park, affiliated with the University of Hawaii at Manoa. Facilitates cooperative research and development activities between Park tenants and University community, particularly in the areas of marine microbiology, oceanography, and alternative energy production and other forms of ocean-related high technology. Operates a pumping system capable of delivering approximately 13,300 gallons per minute of nutrient rich, pathogen-free, cold seawater from 2,100 feet below the ocean surface. **Publications:** Hawaii High Tech Journal (quarterly); Hawaii High Technology Business Directory (biennially).

★533★ University of Hawaii at Hilo
Hawaii Small Business Development Center
523 W. Lanikaula St.   Phone: (808) 933-3515
Hilo, HI 96720   Fax: (808) 933-3683
Janet M. Nye, State Director
Member of the Association of Small Business Development Centers.

Idaho

★534★ Boise State University
Idaho Small Business Development Center
1910 University Dr.   Phone: (208) 385-1640
Boise, ID 83725
Ronald R. Hall, Director
Member of the Association of Small Business Development Centers.

★535★ Idaho State University Research Park
Box 8044   Phone: (208) 236-2430
Pocatello, ID 83209-8044
Ron E. Millick, Director
77-acre research park affiliated with Idaho State University. Facilitates the exchange of University research community with Park tenants, particularly in the areas of health professions, life sciences, pharmacy, engineering, electronics, and business.

Illinois

★536★ Argonne National Laboratory Office of the Director
9700 S. Cass Ave.   Phone: (708) 972-7229
Argonne, IL 60439
Norman Peterson, Director
Facilitates the exchange of Argonne research resources and inventions with industry on the state and national level and develops industrywide partnerships for the Laboratory. Collaborates with Argonne/University of Chicago Development Corporation to oversee patenting and licensing activities. **Publications:** Patents Available for Licensing; Newsletter.

★537★ Bradley University
Technology Commercialization Center
Lovelace Technology Center   Phone: (309) 677-2263
Peoria, IL 61625
Dr. William M. Hammond, Director
Multidisciplinary organization assisting in the development and commercialization of new products and in the transfer of new technologies and manufacturing processes. Activities include product and material testing, prototype development, microelectronic design layout, medical instrumentation, accelerated corrosive testing, printed circuit board layout, sports medicine instrumentation, optical electronics investigations, and microprocessor control system design.

★538★ Chicago Technology Park
2201 W. Campbell Park Dr.   Phone: (312) 829-7252
Chicago, IL 60612
Nina M. Klarich, President
A 56-acre joint research and development park for companies in biotechnology, pharmaceuticals, engineering, computers, and other areas of science. Seeks to coordinate industry, university, and government partnerships to stimulate the formation of science-based companies and economic development in the Chicago area. Provides access to university and hospital resources, offers assistance in the creation of new venture companies, and provides space in an incubator building.

★539★ College of DuPage
Technology Commercialization Center
Business & Professional Institute   Phone: (708) 858-2800
22nd & Lambert Rd.
Glen Ellyn, IL 60137-6599
John J. Sygielski, Program Manager
Links high technology businesses to College and other resources to assist in the production and commercialization of new ideas and products and to enhance the transfer of technologies from College laboratories into the marketplace. **Publications:** Cutting Edge (monthly newsletter).

★540★ Illinois Institute of Technology
Technology Commercialization Center
10 W. 35th St.   Phone: (312) 567-5115
Research Tower, room 18F4-2
Chicago, IL 60616
Thomas M. Jacobius, Director
Provides assistance and support to small businesses, entrepreneurs, and inventors through technology transfer, including technical and business feasibility studies.

★541★ Illinois Small Business Development Center
Dept. of Commerce & Community Affairs Phone: (217) 524-5856
620 E. Adams St., 6th Fl.   Fax: (217) 785-6328
Springfield, IL 62701
Jeff Mitchell, State Director
Member of the Association of Small Business Development Centers.

★542★ Illinois State University
Technology Commercialization Center
Rm. 215   Phone: (309) 438-7127
Media Services Bldg.
Normal, IL 61761
Jerry W. Abner, Director
General technology commercialization. Provides innovation evaluation service for new product and business proposals, including complete research and development assistance for those proposals which meet the economic development objectives of the I-TEC Illinois program. **Publications:** Annual Report; Quarterly Report (both distributed to I-TEC).

★543★ Northern Illinois University
Technology Commercialization Center
DeKalb, IL 60115-2874   Phone: (815) 753-1238
Dr. Larry Sill, Director
Assists inventors, entrepreneurs, and small businesses with product research and development, construction of prototypes, technical and commercial assessments, patent applications, licensing, and locating funding sources.

★544★ Northwestern University
Technology Innovation Center
1840 Oak Ave.   Phone: (708) 866-1818
Evanston, IL 60201
Richard Holbrook, Director
Matches University resources with the needs of state business by developing small business innovation research programs, linking businesses to share technologies, developing international technology cooperatives, commercializing University technology, and providing business planning activities for entrepreneurs. Research includes entrepreneurship, technology transfer, technology commercialization, economic development, and business strategy. **Publications:** Technology Transfer in Third World.

★545★ Northwestern University/Evanston Research Park

1710 Orrington Ave.   Phone: (708) 475-7170
Evanston, IL 60201
Ron Kysiak, Executive Director

A 24-acre research park which encourages the exchange of research activities between the University and Park tenants and transfer technological advances to basic industry. Current tenants include the Basic Industry Research Laboratory of Northwestern University, which focuses on manufacturing and applied materials research, the Institute for Learning Sciences, an artificial intelligence center; and the Computer-Integrated Manufacturing Demonstration Center and Kraft General Foods bio-tech research division. The Park provides a small-business incubator system to support newly-developing, high-technology companies, offers technical assistance, and administers a seed capital fund.

★546★ Sangamon State University
Center for Entrepreneurship and Enterprise Development

Springfield, IL 62794-9243   Phone: (217) 786-6571
Richard J. Judd, Ph.D., Director

Performs marketing research (including customer/client surveys), economic analyses, and organizational analyses and development for private and nonprofit organizations. **Publications:** Economic Business Review (quarterly).

★547★ Southern Illinois University at Carbondale
Technology Center

Southern Illinois Small Business   Phone: (618) 536-7551
    Incubator
Carbondale, IL 62901-6706
Martha Cropper, Director

Facilitates the production, transfer, and commercialization of new technology in agriculture, forestry, engineering, materials, mining, and biomedicine by investigating technical feasibility, commercial development, marketing, and cash flow. **Publications:** Connections (six times per year).

★548★ Southern Illinois University at Edwardsville
Technology Commercialization Center

Box 1108   Phone: (618) 692-2166
Edwardsville, IL 62026-1108
James W. Mager, Director

Provides assistance to inventors, entrepreneurs, and businesses in the commercialization of ideas and products, particularly in the areas of robotics, CAD/CAM, optical coatings, laser communications, automatic inspection, inferfacing microcomputers to instrumentation, computer simulation, analysis of variance, statistical quality control, statistical process control, plant operations, materials planning, linear and nonlinear programming, radioactive biochemicals, human resources, and marketing.

★549★ University of Illinois
Bureau of Economic and Business Research

428 Commerce W.   Phone: (217) 333-2330
1206 S. Sixth St.
Champaign, IL 61820
William R. Bryan, Director

Economics and business, including studies in business expectations, health economics, forecasting and planning, public utilities, innovation, entrepreneurship, consumer behavior, poverty problems, small business operations and problems, investment and growth, productivity, and research methodology. **Publications:** Project Reports; Monographs; Bulletins; Illinois Business Review (four times a year); Illinois Economic Outlook (annually); Illinois Statistical Abstract (annually); Quarterly Review of Economics and Business; Research Projects and Publications (annual report of the faculty of the College of Commerce and Business Administration).

★550★ Business Development Service

109 Coble Hall   Phone: (217) 333-8357
801 S. Wright St.
Champaign, IL 61820
Arthur H. Perkins, Director

Business development aimed at University entrepreneurs assists in producing and commercializing new ideas and products and promoting technology transfer from laboratories into the marketplace.

★551★ Office of the Vice Chancellor for Research

417 Swanlund Bldg.   Phone: (217) 333-7862
601 E. John St.
Champaign, IL 61820
Dillon Mapother, Associate Vice Chancellor

Transfers technology from the University to business and industry, particularly in the areas of engineering and computing. Activities include negotiating research contracts and commercial testing agreements, licensing University-owned software and patents, and consulting on University policies on classified research and academic integrity.

★552★ University of Illinois at Chicago
Technology Commercialization Program

PO Box 4348   Phone: (312) 996-9131
M/C 345
Chicago, IL 60680
G.B. Van Wagenen, Director

Produces and commercializes new ideas and products and promotes technology transfer in the fields of biotechnology, robotics, mechanical devices, engineering software, and data processing hardware components. Arranges assistance for qualified clients' requirements for testing, prototyping, demonstration, market surveys, and business planning. Maintains a database of Chicago venture firms, patent attorneys, and neighborhood development organizations. **Publications:** Workshops on Business Plans; Workshops on Technology Transfer (semiannually).

★553★ Western Illinois University
Center for Business and Economic Research

College of Business   Phone: (309) 298-1594
525 Stipes Hall
Macomb, IL 61455
Dr. Richard E. Hattwick, Director

Business and economics, with emphasis on labor markets, occupational education, regional economic analysis, and entrepreneurial case studies. **Publications:** Monographs; Working Papers; Journal of Behavioral Economics; Illinois Business Leader (newspaper); American Business Leader (newspaper).

★554★ Technology Commercialization Center

Rm. 212 Seal   Phone: (309) 298-2211
Macomb, IL 61455
Daniel Voorhis, Director

Commercializes new ideas and products. Promotes the transfer of technology from the laboratory to the marketplace. Provides assistance to develop technology-based products and businesses through technical and commercial assessments and design assistance.

Indiana

★555★ Ball State University
Center for Entrepreneurial Resources
Carmichael Hall, Rm. 201   Phone: (317) 285-1588
Muncie, IN 47306
Dr. B.J. Bischoff, Director

Entrepreneurship and development of existing businesses, including corporate training, executive development, intrapreneurship, adult literacy, technology transfer, strategic planning, human resource development, needs assessment, creativity, program evaluation, consultant selection, computer software, new product development, sales training, marketing strategy development, hospitality training, and customer service strategy development. **Publications:** Practical Guides for Professions (annually).

★556★ Indiana Institute for New Business Ventures
One N. Capital Ave., Ste. 1275   Phone: (317) 264-2820
Indianapolis, IN 46204
David C. Clegg, Contact

Assists small business ventures by determining ways to upgrade management skills. Promotes small business innovation research.

★557★ Indiana Small Business Development Center
Economic Development Council   Phone: (317) 634-1690
1 N. Capitol, Ste. 200   Fax: (317) 264-6855
Indianapolis, IN 46204
Stephen G. Thrash, State Director

Member of the Association of Small Business Development Centers.

★558★ Indianapolis Center for Advanced Research
611 N. Capitol Ave.   Phone: (317) 262-5000
Indianapolis, IN 46204
Sid Johnson, President

Diagnostic and therapeutic applications of ultrasound, urban technology, technology transfer, computer engineering, software development for engineering applications, medical instrumentation, automated manufacturing, and advanced electronics. Operates NASA-supported industrial application center for technology transfer to the private sector. The center is located on campus of Indiana University-Purdue University at Indianapolis. **Publications:** Newsletter (quarterly).

★559★ Purdue Industrial Research Park
Purdue Research Foundation/Division of   Phone: (317) 494-1727
   Real Estate
1220 Potter Dr.
West Lafayette, IN 47906
Stan M. Mithoefer, Director of Real Estate

Provides facilities and University research and technical support services to industrial tenants.

★560★ Purdue University
Electrical Engineering Industrial Institute
School of Electrical Engineering   Phone: (317) 494-3538
Electrical Engineering Bldg.
West Lafayette, IN 47907
Douglas B. Morrison, Manager, Industrial Relations

Operates as a limited industrial membership organization where member industrial companies and technical organizations pay an annual fee of $10,000 for various Center privileges and services. Coordinates cooperative research programs between members and interested faculty. Acts as an intermediary to acquaint engineers in member organizations with research projects in the School and provides technical assistance. **Publications:** Annual Research Summary.

Iowa

★561★ Iowa State University
Center for Industrial Research and Service
500 Research Park   Phone: (515) 294-3420
Ames, IA 50011
Lloyd E. Anderson, Interim Director

Problem areas of business, manufacturing, technology transfer, productivity, new product design, manufacturing processes, marketing, and related topics. Acts as a problem-handling facility and a clearinghouse for efforts to help Iowa's industry grow through studies highlighting not only production and management problems but also markets, marketing, and profit potential of possible new developments.

★562★ Iowa Small Business Development Center
137 Lynn Ave.   Phone: (515) 292-6351
Ames, IA 50010   Fax: (515) 292-0020
Ronald A. Manning, State Director

Member of the Association of Small Business Development Centers.

★563★ Iowa State University Research Park
125 Beardshear Hall   Phone: (515) 294-5121
Ames, IA 50011
Leonard C. Goldman, President

195-acre site on the University's South Campus facilitating interaction between corporate research laboratories and the University research community.

Kansas

★564★ Kansas Technology Enterprise Corporation
112 W. 6th St., Ste. 400   Phone: (913) 296-5272
Topeka, KS 66603
William G. Brundage, Ph.D., President

Stimulates the commercialization of new technologies and promotes the growth of Kansas enterprises. KTEC finances Centers of Excellence at universities throughout Kansas, and awards matching grants for collaborative research, provides seed capital, offers technical referral services, and funds Small Business Innovation Research Grants and the Invention Development Assist Program (IDAP). Maintains a database on Kansas Technology Resource (KTR). **Publications:** Innovator (quarterly); Investments in Kansas; Annual Report.

★565★ Pittsburg State University
Center for Technology Transfer
School of Technology & Applied Science   Phone: (316) 235-4114
Pittsburg, KS 66762
Harvey Dean, Director

Development, introduction, and transfer of technology to Kansas industries, particularly to wood and plastics industries. Emphasizes design, testing, and development of products and processing methods, including the applications of computer-aided design, computer numerical control, and robotics to manufacturing. Maintains National Wood Technology Center and dimensional metrology and metallurgy laboratories.

★566★ University of Kansas
Space Technology Center
Raymond Nichols Hall   Phone: (913) 864-4775
Lawrence, KS 66045
Prof. B.G. Barr, Director

Supports development of new knowledge, concepts, and technology for surveying earth resources and evaluating environmental quality, including remote sensing, technology transfer, flight research, economic and business research, microprocessor control, computer integrated manufacturing, mineral resource surveys, geochronology, and energy

research. Staff also works with industry and government in applying newly developed technology. The Center facilitates the activities of the Augmented Telerobotic Laboratory, Computer Aided Systems Engineering Laboratory, Flight Research Laboratory, Isotope Geochemistry Laboratory, Kansas Applied Remote Sensing Program, Kansas Biological Survey, Radiation Physics Laboratory, Radar Systems and Remote Sensing Laboratory, and Telecommunications and Information Systems Laboratory.

★567★ University of Kansas Center for Research, Inc.
2291 Irving Hill Dr.   Phone: (913) 864-3441
Lawrence, KS 66045
Dr. Carl E. Locke, Ph.D., Director

Remote sensing technology (development, analysis, and applications), energy, environmental quality, aircraft performance improvement, stress analysis, communications systems, technology transfer, microprocessing, transportation, biomedical engineering, isotope geochemistry, and radiation physics. **Publications:** Annual Report.

★568★ Wichita State University
Center for Entrepreneurship
008 Devlin Hall—Campus Box 147   Phone: (316) 689-3000
Wichita, KS 67208
Prof. Gerald Graham, Director

Entrepreneurship, particularly the effect entrepreneurial education has on business start-ups and success rates, and profile studies of entrepreneurs. **Publications:** Center Report (twice yearly); The Business Heritage Series; The Complete Information Book for Entrepreneurs and Small Business Managers; Entrepreneurship: Your Future in Business.

★569★ Kansas Small Business Development Center
Campus Box 148   Phone: (316) 689-3193
Wichita, KS 67208   Fax: (316) 689-3647
Thomas H. Hull, State Director

Member of the Association of Small Business Development Centers.

Kentucky

★570★ NKU Foundation Research/Technology Park
Highland Heights, KY 41076   Phone: (606) 572-5126
Kenneth R. Lucas, Contact

75-acre research park operated by the Northern Kentucky University Foundation, Inc. Facilitates the exchange of research resources between the University and Park tenants.

★571★ University of Kentucky
Kentucky Small Business Development Center
465 E. High St., Ste. 201   Phone: (606) 257-7668
Lexington, KY 40507   Fax: (606) 258-1907
Janet S. Holloway, State Drirector

Member of the Association of Small Business Development Centers.

Louisiana

★572★ Louisiana State University
Office of Technology Transfer
3990 W. Lakeshore Dr.   Phone: (504) 388-6941
Baton Rouge, LA 70808
Donald W. Pennington, Director

Identifies, protects, and transfers technology that originates from University research activities. Activities include patent work, finding state or national licensees, and starting new entrepreneurial projects. Projects include work in such areas as computer chips, enzymes, instrumentation, genetic engineering, drugs, and aquaculture. **Publications:** Newsletter (occasionally).

★573★ Northeast Louisiana University
Louisiana Small Business Development Center
College of Business Administration   Phone: (318) 342-5506
700 University Ave.   Fax: (318) 342-5510
Monroe, LA 71209
Dr. John P. Baker, Director

Member of the Association of Small Business Development Centers.

Maine

★574★ Center for Technology Transfer
59 Exeter St.   Phone: (207) 780-4616
Portland, ME 04102
Robert Dalton, Director

Partnership between the Maine Science and Technology Commission, the University of Maine, the University of Southern Maine, the Maine Technical College System, and the metals and electronic industries of the state. Facilitates technology transfer between university research and Maine metals and electronics manufacturing industries. **Publications:** CTT Newsletter (quarterly).

★575★ University of Southern Maine
Center for Business and Economic Research
96 Falmouth St.   Phone: (207) 780-4020
Portland, ME 04103
Richard J. Clarey, Dean of SBEM

Business expansion in Maine, including management and small business development. **Publications:** Maine Business Indicators (quarterly); Reports.

★576★ Maine Small Business Development Center
96 Falmouth St.   Phone: (207) 780-4420
Portlamd, ME 04103   Fax: (207) 780-4810
Robert H. Hird, State Director

Member of the Association of Small Business Development Centers.

★577★ Office of Sponsored Research
96 Falmouth St.   Phone: (207) 780-4411
Portland, ME 04103
Dr. Robert J. Goettel, Director

Economic development, the formation and management of business enterprises, health and human services, education, marine resources, state and local governance, science and technology, and organized camping. The Center is organized into five cooperating institutes: Marine Law Institute, Human Services Development Institute, The New Enterprise Institute, Health Policy Unit, and National Child Welfare Resource Center. **Publications:** Network News (a runaway suicide prevention newsletter); Territorial Sea (legal developments in the management of interjurisdictional resources); Journal (quarterly); Annotated Bibliography (annually); Grants/Contracts (monthly).

Maryland

★578★ Johns Hopkins University
Bayview Research Campus
Dome Corp.   Phone: (301) 955-7724
1629 Thames St., Ste. 301
Baltimore, MD 21231
David Hash, Director, Property Development

Biomedical research park established to provide government academia and industry with space for offices and laboratory facilities.

★579★ Maryland Small Business Development Center

Dept. of Economic & Employment Development
217 E. Redwood St., 10th Fl.
Baltimore, MD 21202
Michael Long, Director

Phone: (301) 333-6995
Fax: (301) 333-6608

Member of the Association of Small Business Development Centers.

★580★ R&D Village

Montgomery County
Office of Economic Development
101 Monroe St., Ste. 1500
Rockville, MD 20850
Dyan Brasington, Director

Phone: (301) 217-2345

1,200-acre site linking biomedical research and development activities between government, park tenants, and academia. Houses the Center for Advanced Research in Biotechnology, a joint research venture between the National Institute of Standards and Technology, University of Maryland, and Montgomery county government; also houses a Johns Hopkins University facility focusing on advanced study programs in computer science, electrical engineering, and technical management. The Village also includes the Shady Grove Life Sciences Center, Shady Grove Executive Center, Decoverly, and The Washingtonian. **Publications:** Economic Focus Newsletter.

★581★ University of Maryland
Center for Public Issues in Biotechnology

2200 Seymors Hall
College Park, MD 20742
Dr. Darrell Hueth, Director

Phone: (301) 405-1290

Legal, policy, and ethical issues related to biotechnology and its applications. Interests are divided into five main areas: 1) environmental applications and risk assessment, including regulation of risks associated with the release of genetically altered organisms; 2) human and animal applications, including definitions and analyses of creation science and evolution, use of human body parts and cells for research and treatment, availability and cost of biotechnical experimental treatments, and human gene therapy and eugenics; 3) social, economic, and political impacts of the biotechnology industry; 4) business issues, including intellectual property rights, patenting of life forms, and industry/university ties; and 5) international issues such as the conversion of medical advances to biological warfare and foreign investment in biotechnology.

★582★ Engineering Research Center

College of Engineering
Wind Tunnel Bldg.
Rm. 2104
College Park, MD 20742
Dr. David Barbe, Executive Director

Phone: (301) 405-3906

Engineering and scientific disciplines. Activities include a Technology Extension Service (TES), which offers technical assistance to the Maryland business community; a Technology Advancement Program (TAP), an incubator for start-up companies engaging in the development of technically oriented products and services; a Technology Initiatives Program (TIP) to support technological capabilities within the University; and the Maryland Industrial Partnerships (MIPS) program, a matching fund for industry sponsored research. **Publications:** ERC Update (quarterly newsletter).

★583★ Michael D. Dingman Center for Entrepreneurship

College of Business Management
College Park, MD 20742
Dr. Charles O. Heller, Director

Phone: (301) 405-2144

Entrepreneurship, new venture creation, and venture capital.

★584★ University of Maryland Science and Technology Center

7505 Greenway Center Dr., 001
Greenbelt, MD 20770
Daniel Colton, Vice President

Phone: (301) 982-9400

466-acre research and development park focusing in the areas of math, physics, computer sciences, electrical engineering, and mechanical engineering. Researchers are provided access on a contract basis to University of Maryland faculty and equipment. The Supercomputing Research Center is housed at the Center.

Massachusetts

★585★ Harvard University
Office for Patents, Copyrights & Licensing

Ste. 256, Univ. Pl.
124 Mt. Auburn St.
Cambridge, MA 02138
Joyce Brinton, Director

Phone: (617) 495-3067

Facilitates University-industry relations through technology licensing in the areas of applied sciences, recombinant DNA, hybridoma technology, software and courseware, chemistry, therapeutics, vaccines, bioprocesses, and medical, veterinary, and agricultural diagnostics.

★586★ Massachusetts Biotechnology Research Park

373 Plantation St.
Worcester, MA 01605
Raymond L. Quinlan, Executive Director

Phone: (508) 755-2230

75-acre research park located adjacent to the University of Massachusetts Medical Center; affiliated with Clark University, Worcester Polytechnic Institute, Worcester Foundation for Experimental Biology, and Tufts University School of Medicine. Research and development space designed for biotechnology. Fosters the exchange between the research communities of the affiliated institutions and park tenants.

★587★ Massachusetts Institute of Technology
MIT Research Program on the Management of Technology

50 Memorial Dr.
E52-535
Cambridge, MA 02139
Prof. Edward B. Roberts, Chairman

Phone: (617) 253-4934

Managerial research on technology-based innovation, with emphasis on industry and government. Studies focus on organizations in the United States, Europe, Asia, and Latin America.

★588★ Technology and Development Program

1 Amherst St.
Cambridge, MA 02139
Prof. Fred Moavenzadeh, Director

Phone: (617) 253-7227

Technological problems facing developing countries in the fields of energy, public works, manufacturing, and socioeconomics. Coordinates interdisciplinary research projects with academic institutions in developing countries to promote science, technology, and development. **Publications:** TDP Publication Series.

★589★ Technology Licensing Office

Rm. E32-300
77 Massachusetts Ave.
Cambridge, MA 02139
John Preston, Director

Phone: (617) 253-6966

Commercializes technology from the Institute in the areas of biotechnology, biomedicine, ceramics, chemistry, computers, electrooptics, integrated circuits, and polymers. Markets inventions and software developed at Lincoln Laboratory.

Massachusetts Institute of Technology (Cont.)

★590★ University Park

77 Massachusetts Ave.   Phone: (617) 253-5278
Cambridge, MA 02139
Ron Suduiko, Contact

27-acre research park which fosters interaction between the Institute's research community and Park tenants.

★591★ University of Massachusetts at Amherst
Massachusetts Small Business Development Center

School of Management, Rm. 205   Phone: (413) 545-6301
Amherst, MA 01003   Fax: (413) 545-1273
John F. Ciccarelli, State Director

Member of the Association of Small Business Development Centers.

★592★ Worcester Polytechnic Institute
Management of Advanced Automation Technology Center

Dept. of Management   Phone: (508) 831-5000
Worcester, MA 01609
Arthur Gerstenfeld, Director

Management of technological innovations in the areas of computer-aided manufacturing, decision support systems, office automation, group technology, manufacturing communication networks, automation assembly process, application of bar coding, user friendly shop floor control, and manufacturing strategy. Conducts applied research for the specific technological needs of industrial clients. **Publications:** Newsletter (quarterly).

Michigan

★593★ Eastern Michigan University
Center for Entrepreneurship

121 Pearl St.   Phone: (313) 487-0225
Ypsilanti, MI 48197
Dr. Patricia B. Weber, Director

Applied research toward the development of entrepreneurship and growth management. Focuses on the vital transition from start-up to sustained long-term growth and stability. A major project involves following the longitudinal progress of 150 companies. **Publications:** Working Paper Series.

★594★ Industrial Technology Institute

PO Box 1485   Phone: (313) 769-4000
2901 Hubbard Rd.
Ann Arbor, MI 48106
George Kuper, President

Design and application of advanced manufacturing technologies and processes and the inherent social consequences of new technologies. Specific areas include computer-aided design and engineering, computer-aided manufacturing (flexible machining, assembly, inspection, and materials handling), manufacturing information systems, artificial intelligence, computer communication networks, computer-based information and technology transfer products and services, and economic, organizational, and social impacts. **Publications:** Gateway: The MAP/TOP Reporter (bimonthly newsletter); Modern Michigan (quarterly manufacturing journal).

★595★ Michigan State University
Center for the Redevelopment of Industrialized States

403 Olds Hall   Phone: (517) 353-3255
College of Social Science
East Lansing, MI 48824-1047
Dr. Jack H. Knott, Director

Economic and social changes affecting the state of Michigan and ways to diversify and improve the state's business climate. Addresses issues such as technological innovation and strategic human resources management, modeling municipal expenditure patterns in Michigan, determinants of state industrial policy expenditures, implementing new technologies, organizational training practices and the facilitation of technological change, models of plant and investment decisions, and implications of changes in world automobile production for local communities. Maintains the Michigan Data Archive, a database of over 3,000 variables on every Michigan county, available for free to the public. **Publications:** Newsletter; Reprints.

★596★ Technology Transfer Center

51 Kellogg Center   Phone: (517) 355-1660
East Lansing, MI 48824
John L. Pearson, Director

Transfers technology from the University to business and industrial sectors.

★597★ Michigan Technological University
Bureau of Industrial Development

Houghton, MI 49931   Phone: (906) 487-2470
Richard E. Tieder, Director

Small business, business analysis and operations research, natural resource economics, and technology transfer; provides a broad base of knowledge for developing the resources, industries, markets, and communities of Michigan's Upper Peninsula through research, service, and academic activities.

★598★ Transportation Technology Transfer Center

Dept. of Civil Engineering   Phone: (906) 487-2102
Houghton, MI 49931
Dr. Bernard D. Alkire, Director

Transfers technology in transportation engineering to state county and municipal officials. **Publications:** Newsletter.

★599★ Oakland Technology Park

Schostak Brothers & Co., Inc.   Phone: (313) 262-1000
First Center Office Plaza
26913 Northwestern Hwy.
Southfield, MI 48034
Philip J. Houdek, Vice President

1,800-acre joint research, development, and incubator site of Oakland University, Oakland Community College, and industry. Facilitates technology transfer between the University and College and Park tenants, enhancing the research strength of the affiliated institutions. Facilities are oriented toward robotics, engineering, automation, computer technology, and advanced manufacturing applications.

★600★ University of Michigan
Center for Management and Technology Transfer

Business & Industrial Assistance   Phone: (313) 998-6201
   Division
506 E. Liberty, 3rd Fl.
Ann Arbor, MI 48104-2210
Lawrence R. Crockett, Director

Transfers technology from the University to businesses and industry, particularly in the area of automated engineering. Emphasizes common problems of manufacturers, manufacturing facilities at risk of closing, and revitalization of closed industrial facilities.

★601★ Wayne State University
Michigan Small Business Development Center

2727 2nd Ave., MCHT   Phone: (313) 577-4848
Detroit, MI 48202   Fax: (313) 577-4222
Norman J. Schlafmann, State Director

Member of the Association of Small Business Development Centers.

Wayne State University (Cont.)

★602★ Technology Transfer Center
3rd Fl. FAB Bldg.   Phone: (313) 577-2788
Detroit, MI 48202
Nate R. Borofsky, Director

Assists manufacturers, entrepreneurs, and inventors in gaining access to technical resources at the University.

★603★ Western Michigan University
Institute for Technological Studies
Kalamazoo, MI 49008   Phone: (616) 387-4017
Molly Williams, Director

Provides administrative support for internal and external research of the University. Also links industry with University resources by providing technical assessments, information, referrals, testing, and instruction.

★604★ Technology Transfer Center
350 E. Michigan Ave., Ste. 29   Phone: (616) 387-2714
Kalamazoo, MI 49007
William H. Cotton, Director

Transfers technology from the University to the business and industrial sectors.

Minnesota

★605★ Midwest Technology Development Institute
Ste. 815   Phone: (612) 297-6300
245 E. 6th St.
St. Paul, MN 55101-1940
Wm. C. Norris, President/Chairman

Independent, nonprofit corporation established by the states of Minnesota, Illinois, Indiana, Kansas, Nebraska, Ohio, North Dakota and South Dakota. Seeks to enhance competitiveness and economic growth in the Midwest through the establishment of cooperative research and development ventures involving industry, universities, and government. Institute ventures include the Rural Enterprise Partnership, which conducts case studies of midwestern farmers successfully using new tillage techniques and crops and livestock management systems; and the Advanced Ceramics and Composites Partnership, a coalition to promote start-up programs in materials. Also seeks to facilitate the equitable transfer of technology among industry, universities, and domestic and foreign governments. **Publications:** Technology Advance.

★606★ Minnesota Small Business Development Center
Dept. of Trade & Economic Development Phone: (612) 297-5770
900 American Center Bldg.   Fax: (612) 296-1290
150 E. Kellogg Blvd.
St. Paul, MN 55101
Randall Olsen, State Director

Member of the Association of Small Business Development Centers.

★607★ Minnesota Technology Corridor
Midland Sq. Bldg.   Phone: (612) 348-7140
331 Second Ave. S.
Ste. 700
Minneapolis, MN 55401
Judy Cedar, Project Coordinator

128-acre site established to foster technology transfer, research and development, and prototype manufacturing activities between the University of Minnesota and related businesses.

★608★ University of Minnesota
Center for the Development of Technological Leadership
107 Lind Hall   Phone: (612) 624-5747
Institute of Technology
Minneapolis, MN 55455
Y. Shulman, Director

Technology transfer and technology leadership, including technology transfer methods and pilot experiments.

★609★ Office of Patents and Licensing
1919 Univ. Ave.   Phone: (612) 624-0550
St. Paul, MN 55104
John Thuente, Director

Transfers technology from the University to business and industry in the areas of health sciences, engineering, and agriculture.

★610★ Office of Research and Technology Transfer Administration
1919 Univ. Ave.   Phone: (612) 624-1648
St. Paul, MN 55104
A.R. Potami, Associate Vice President

Serves as the research support unit for University of Minnesota faculty members by administering non-programmatic aspects of all research, training, and public service projects funded by external sources. Reviews and processes all proposals and awards for research, training, and public service projects and is responsible for financial management, cash receipt, and financial reporting of project funds. Works with faculty to stimulate the disclosure of patentable discoveries resulting from University research. Assists faculty in locating potential sources of support for research, training, and public service programs.

Mississippi

★611★ Institute for Technology Development
700 N. State St.   Phone: (601) 960-3600
Ste. 500
Jackson, MS 39202
Leonard R. Vernamonti, President

Identifies the state's technological needs in the areas of advanced living systems, microelectronics, and space remote sensing and coordinates the commercialization of research and development from the state's university, government and small business laboratories. **Publications:** Innovations (newsletter).

★612★ Mississippi Research and Technology Park
PO Box 2720   Phone: (601) 323-5783
Starkville, MS 39759
Gray Swoope, Executive Director

218-acre site facilitating the transfer of technology to research, industrial, and economic development use, operating under the Oktibbeha County Economic Development Authority in cooperation with Mississippi State University, Industrial Development Foundation, Tennessee Valley Authority, and city of Starkville. Operates research laboratories, incubator and multitenant facilities, professional and business services, and office space to foster high technology research collaboration between the University and industry tenants. **Publications:** Mississippi Research and Technology Park (newsletter).

★613★ Southern Rural Development Center
Box 5446   Phone: (601) 325-3207
Mississippi State, MS 39762
Dr. H. Doss Brodnax, Jr., Director

Jointly sponsored extension and experiment station support service of Mississippi State University and Alcorn State University. Offers information and assistance to extension services and experiment stations in 29 land-grant universities in 13 southern states, Puerto Rico, and the

Virgin Islands, facilitates interaction between research and extension personnel, and provides staff training at regional and subregional levels. Sponsors rural and community development programs, especially studies on use/value assessment of agricultural lands in the South, rural crime, and rural banking and credit. Develops and tests models for computer technology transfer, community resource development, delivery of community services, database development, rural transportation, and water policy. **Publications:** Annual Reports; Series Reports; Capsules (monthly newsletter).

★614★ University of Mississippi
Mississippi Small Business Development Center
School of Business Administration   Phone: (601) 232-5001
Old Chemistry Bldg., Rm. 216
University, MS 38677
Larry Martin, Director

Small business management, including feasibility studies, business law, venture capital, government contracting, and financial, production, and personnel management.

Missouri

★615★ St. Louis University
Jefferson Smurfit Center for Entrepreneurial Studies
School of Business & Administration   Phone: (314) 534-7204
3674 Lindell Blvd.
St. Louis, MO 63108
Dr. Robert Brockhaus, Director

Venture capital and endowed positions in entrepreneurship.

★616★ University of Missouri
Missouri Research Park
215 Univ. Hall   Phone: (314) 882-3397
Columbia, MO 65211
Rick Finholt, Executive Director

A 750-acre research park designed to be a link between academia and industry as a center for research and development in such fields as advanced manufacturing and toxic waste technology.

★617★ Missouri Small Business Development Center
300 University Pl.   Phone: (314) 882-0344
Columbia, MO 65211   Fax: (314) 884-4297
Max E. Summers, State Director

Member of the Association of Small Business Development Centers.

★618★ University of Missouri—Columbia
Geographic Resources Center
Dept. of Geography   Phone: (314) 882-1404
Stewart Hall, Rm. 217
Columbia, MO 65211
Dr. Christopher Salter, Director

Remote sensing, spatial analysis, geographic information systems, classification systems, and data compression. Projects include multi-resource geographic/pictorial databases, environmental assessment, crop production mapping, land use inventory, wildlife habitat evaluation, automated radiotelemetry studies, automated digitizing, and remote sensing/information systems technology transfer. **Publications:** Annual Report.

★619★ University of Missouri—Rolla
Center for Technology Transfer and Economic Development
212 Mechanical Engineering Annex   Phone: (314) 341-4555
Rolla, MO 65401
H. Dean Keith, Director

Technological innovation.

★620★ Washington University
Industrial Contracts and Licensing
1 Brookings Dr.   Phone: (314) 889-5889
St. Louis, MO 63130
H.S. Leahy, Director

Transfers technology from the University to business and industry in the area of biotechnology.

Montana

★621★ Montana Small Business Development Center
Montana Dept. of Commerce   Phone: (406) 444-4780
1424 9th Ave.
Helena, MT 59620
Evan McKinney, State Director

Member of the Association of Small Business Development Centers.

★622★ Montana State University
Technology Park
1711 W. College   Phone: (406) 587-4480
Bozeman, MT 59717
Ron Flair, Director

90-acre research park at Montana State University. Provides access for park tenants to the research community and facilities at the University. **Publications:** Montana State University Resources Catalog.

Nebraska

★623★ University of Nebraska at Omaha
Nebraska Business Development Center
60th & Dodge Sts., CBA Rm. 407   Phone: (402) 554-2521
Omaha, NE 68182   Fax: (402) 554-3747
Robert E. Bernier, State Director

Member of the Association of Small Business Development Centers.

Nevada

★624★ Dandini Research Park
DRI Research Foundation   Phone: (702) 673-7315
PO Box 60220
Reno, NV 89506
Dale F. Schulke, Contact

470-acre site that links the research and development activities of Park tenants with the Desert Research Institute's technological equipment, personnel, laboratories, and training programs. Instrument design, environmental testing, and research and development services may be done in cooperation with DRI staff. The Desert Research Institute is a nonprofit, separately incorporated research organization affiliated with the University of Nevada System.

★625★ University of Nevada—Reno
Nevada Small Business Development Center
College of Business Administration, Rm.   Phone: (702) 784-1717
   411   Fax: (702) 784-4337
Reno, NV 89557
Sam Males, State Director

Member of the Association of Small Business Development Centers.

New Hampshire

★626★ PTC Research Foundation
Franklin Pierce Law Center   Phone: (603) 228-1541
2 White St.
Concord, NH 03301
Robert Shaw, Director

Patents, trademarks, copyrights, invention, and legal and practical systems for dealing with industrial and intellectual property, both in the U.S. and worldwide. **Publications:** IDEA (a journal of law and technology); Monographs; Project Reports.

★627★ University of New Hampshire
Center for Venture Research
Whitmore School of Business & Finance   Phone: (603) 862-3369
Durham, NH 03824
Dr. William Wetzel, Jr., Director

Research concerned with financing new technology-based firms. **Publications:** IIVCRN Newsletter.

★628★ Industrial Research and Consulting Center
Thompson Hall   Phone: (603) 862-3750
Durham, NH 03824
Dr. Donald C. Sundberg, Executive Director

Promotes research and development relationships between the private sector and the University. Organizes problem solving teams and makes available University instrumentation and computer facilities and research and development laboratories. Assists business and industry with product development, process development, long-range research and planning, modeling, software development, technical troubleshooting, feasibility studies, development of laboratory testing procedures, market analysis, risk analysis, and planning and educational programs. Develops patents and licenses technology and other intellectual property of the University.

★629★ New Hampshire Small Business Development Center
108 McConnell Hall   Phone: (603) 862-2200
Durham, NH 03824   Fax: (603) 862-4468
Helen M. Goodman, State Director

Member of the Association of Small Business Development Centers.

New Jersey

★630★ Innovation and Productivity Strategies Research Program
606 Hill Hall   Phone: (201) 648-5837
360 Dr. Martin Luther King, Jr. Blvd.
Rutgers Univ.
Newark, NJ 07102
Frank Hall, Codirector

Innovation and productivity improvement strategies in U.S. and international industrial establishments. Studies include productivity strategies for invention and management of new product commercialization from an international perspective, strategic partnerships between large firms and small hi-tech firms, entrepreneurship strategies of small high-tech industries, intrapreneurship strategies for corporations, employee attitudes and behaviors, quality management techniques, and joint ventures and technology transfer strategies. Collaborates with other universities in the U.S. and overseas to carry out international projects.

★631★ John P. Caufield Technology Extension Center for Investigational Cancer Treatment
1 Bruce St.   Phone: (201) 456-4600
Newark, NJ 07103
David M. Goldenberg, Sc.D., M.D., Director

Develops and provides new and more effective technologies for the early detection, diagnosis, and treatment of cancer; facilitates the transfer of proven and investigational methods to New Jersey practitioners and hospitals; and serves as a resource for the New Jersey pharmaceutical industry by supporting clinical trials necessary to discover and develop new products.

★632★ New Jersey Institute of Technology Center for Information Age Technology
Newark, NJ 07102   Phone: (201) 596-3035
William R. Kennedy, Executive Director

Serves as a bridge between the Institute and industry/public sector to transfer new and existing computer and information technologies from university research laboratories to state and local governments, nonprofit organizations, and small- and medium-sized businesses.

★633★ Industry Consortium for CAD/CAM Robotics
323 Martin Luther King, Jr. Blvd.   Phone: (201) 596-3647
Newark, NJ 07102
Dr. Kevin McDermott, Director

Factory automation, specifically in computer-aided design/computer-aided manufacturing (CAD/CAM) robotics and technology transfer problems. Studies include control systems, artificial intelligence, knowledge-based automated systems, robot controllers, ergonomics in robot/human interface, computer graphics in industrial systems, numerical control (NC/CNC/DNC) machines, robot time and motion, product quality, solid mechanics and finite element analysis, and cost/benefit analyses.

★634★ Princeton University Princeton Forrestal Center
105 College Rd. E.   Phone: (609) 258-7720
Princeton, NJ 08540
David H. Knights, Director of Marketing

1,750-acre research park of Princeton University. The Center is designed as a planned multiuse development area creating an interdependent mix of academia and business enterprise in the Princeton area.

★635★ Rutgers University Fisheries and Aquaculture Technology Extension Center
PO Box 231   Phone: (201) 932-8959
New Brunswick, NJ 08903
Dr. Richard Lutz, Director

Fisheries and aquaculture. Research and technology transfer activities include enhancing hard clam production through the development and transfer of new and existing technologies; developing site-selection criteria for hard clam and finfish aquaculture; utilizing existing, genetically-selected, disease-resistant stocks of oysters for repopulation of Delaware Bay; developing and transfering technologies to reduce oyster mortalities due to parasites; and seafood marketing research. Conducts mission-oriented projects to enhance industrial productivity and economic growth in the state.

★636★ New Jersey Small Business Development Center
180 University Ave.   Phone: (201) 648-5950
Newark, NJ 07102   Fax: (201) 648-1110
Brenda Hopper, State Director

Member of the Association of Small Business Development Centers.

Rutgers University *(Cont.)*

★637★ Technology Management Research Center

Graduate School of Management   Phone: (201) 648-1077
92 New St.
Newark, NJ 07102
Prof. George F. Farris, Director

Applied research in technology management, especially technology development, implementation and adoption of new technology, trends in technology, and strategic management of technology. **Publications:** Working Paper Series.

New Mexico

★638★ New Mexico Research Park

New Mexico Institute of Mining &   Phone: (505) 835-5600
   Technology
Socorro, NM 87801
Laurance H. Lattman, Vice President

Research park affiliated with New Mexico Institute of Mining and Technology. Facilitates the exchange of research resources between the Institute's research community and Park tenants.

★639★ New Mexico State University
Arrowhead Research Park

Box 30001-Dept. 3RED   Phone: (505) 646-2022
Las Cruces, NM 88003
Dr. Averett S. Tombes, Vice Pres./Research & Econ. De

365-acre research park at New Mexico State University. Fosters opportunities for exchange between Park tenants and the University community. Multidisciplinary activities include heavy metal recovery from waste streams, endoscope research, solar energy research, language translations, and waste water treatment research, computer telemetry data, fiber optic imaging. Maintains access to a geothermal energy source.

★640★ Rio Grande Research Corridor

Pinion Bldg., Ste. 358   Phone: (505) 827-5886
1220 St. Francis Dr.
Sante Fe, NM 87501
Ponziano Ferarracio, Acting Director

A 340-mile research corridor located in central New Mexico operating as a center of high technology research and development under the auspices of the New Mexico Science and Technology Commercialization Commission with funding administered by the New Mexico Research and Development Institute; affiliated with New Mexico Institute of Mining and Technology, University of New Mexico, and New Mexico State University. Links University research resources with government laboratories and private research facilities, particularly in the areas of non-invasive diagnosis, materials, explosives technology, plant genetic engineering, computer research applications, and commercial product development. **Publications:** R&D Forum (monthly reports).

★641★ Santa Fe Community College
New Mexico Small Business Development Center

PO Box 4187   Phone: (505) 438-1362
Santa Fe, NM 87502-4187   Fax: (505) 438-1237
Randy W. Grissom, State Director

Member of the Association of Small Business Development Centers.

★642★ University of New Mexico
Technology Application Center

Albuquerque, NM 87131   Phone: (505) 277-3622
Dr. Stanley A. Morain, Director

Retrieves, processes, and analyzes satellite and aerial data for earth resources and develops geographic information systems (GIS). Image processing and GIS activities include mineral exploration, cover type mapping, habitat mapping and modeling, and surveys of archeological locations. Photo search and retrieval services include satellite images, aerial photos, maps, digital data, LANDSAT data, and photos from Gemini, Apollo, Apollo-Soyuz, Skylab, and Space Shuttle missions. Designated as a NASA Industrial Applications Center to assist in commercializing the national program for space especially in the areas of remote sensing and image dissemination. Offers national and international visiting scientist programs providing customized technical assistance and training in remote sensing and image processing. **Publications:** Remote Sensing of Natural Resources: A Quarterly Literature Review.

New York

★643★ Columbia University
Center for Studies in Innovation and Entrepreneurship

719 Uris Hall   Phone: (212) 854-2830
New York, NY 10027
Michael Tushman, Director

Designs for innovation, entrepreneurship, and technology and organizations.

★644★ Office of Science and Technology Development

411 Low Memorial Library   Phone: (212) 854-8444
116th St. & Broadway
New York, NY 10027
Jack M. Granowitz, Director

Transfers technology from the University to business and industry, especially in the area of biotechnology.

★645★ Cornell University
Cornell Business and Technology Park

20 Thornwood Dr., Ste. 103   Phone: (607) 255-5341
Ithaca, NY 14850
Richard E. deVito, Marketing Manager

200-acre research park serving as a conduit between Cornell University and business, especially in electronics, computer manufacturing, and biotechnology, and a home for the independent development of technologies resulting from efforts by Cornell researchers. Leases space for business incubator activities and start-up companies. **Publications:** The Park Light.

★646★ Cornell Research Foundation, Inc.

Cornell Business & Technology Park   Phone: (607) 257-1081
20 Thornwood Dr., Ste. 105
Ithaca, NY 14850
H. Walter Haeussler, President

Patent and technology marketing arm of the University. Activities include technology transfer, evaluating legal and business contracts, and assisting with patenting and licensing. Holds title to patents and promotes technology transfer to industry. **Publications:** Selected Technology Available for Licensing (semiannually).

★647★ MetroTech

333 Jay St.   Phone: (718) 260-3665
Brooklyn, NY 11201
Dr. Seymour Scher, President

10-block, 16-acre urban research park linking Polytechnic University research resources with information technology industries.

★648★ New York University
Center for Entrepreneurial Studies

Leonard N. Stern School of Business   Phone: (212) 285-6150
90 Trinity Pl., Rm. 421
New York, NY 10006
Prof. William D. Guth, Director

Conducts and funds research on the factors that promote entrepreneurship and lead to the creation of new wealth and business

revenues and on business venturing within established firms. Topics include the major pitfalls and obstacles to start-ups, securing of venture capital, psychology and sociology of entrepreneurship, management of new ventures, innovation and innovative problem-solving, organizing for corporate venturing, control and reward systems, and cross-cultural environments that stimulate entrepreneurship. Also operates a program focusing on entrepreneurship in nonprofit corporations. **Publications:** CES Reports (semiannually); Journal of Business Venturing (copublished with Wharton, quarterly); INE Reports (for nonprofit corporations).

★649★ Initiative for Nonprofit Entrepreneurship

Stern School of Business
90 Trinity Pl., Rm. 421 Phone: (212) 285-6548
New York, NY 10006
Laura Landy, Director

Works with those who are seeking entrepreneurial solutions to the problem of revenue generation. Promotes approaches that enhance success and reduce the risk of venturing, including the correlation between planning, investment, management, size, and venture success in nonprofit corporations and competition between nonprofit and for-profit corporations. **Publications:** INE Reports (semiannual newsletter); Research Reports.

★650★ Northern Advanced Technologies Corporation

PO Box 72 Phone: (315) 265-2194
State Univ. College at Potsdam
Potsdam, NY 13676
Steven C. Hychkano, Executive Director

Provides support for technology transfer and research and development in the areas of nondestructive testing, software development, computer applications, materials processing, and high technology. Clients are offered building space, general assistance, and seed money for a variety of projects to commercialize new technology in the Potsdam area. Prepares marketing studies. Nonprofit corporation governed by a board of directors consisting of members from Clarkston University, State University College at Potsdam, Saint Lawrence County Industrial Development Agency, Village of Potsdam, and Town of Potsdam.

★651★ Rensselaer Polytechnic Institute
Center for Science and Technology Policy

School of Management Phone: (518) 276-6836
Troy, NY 12180-3590
Randy Norsworthy, Director

Factors influencing technical change and its conversion to use. Programs and interests include academic structures and values in university-industry interactions; technically-based regional economic development; comparative study of technical structures in selected countries outside the European community; worker behavior and its effects on product quality, productivity, and costs; institutions, origins, and growth of computer science; impact of technical alliances on strategic technical planning in the semiconductor, pharmaceutical, agrichemical, and other science-based industries; impact of technical change on service industries; factors affecting the introduction of technical change into manufacturing processes; and role of professional organizations in the evolution of industrial research. **Publications:** Series on Technology Policy and Economic Growth.

★652★ George M. Low Center for Industrial Innovation

Troy, NY 12180 Phone: (518) 276-6023
Dr. Christopher W. LeMaistre, Director

Consists of six component centers-Rensselaer Design Research Center, Center for Manufacturing Productivity and Technology Transfer, Center for Integrated Electronics, Decision Sciences and Engineering Systems Department, Scientific Computation Research Center, and Automation and Robotics Center. Coordinates research related to high technology business and industrial productivity and quality, including solid geometric modeling, computer-integrated manufacturing, vision systems, and computer-aided design of very-large-scale-integrated circuits.

★653★ Rensselaer Technology Park

100 Jordan Rd. Phone: (518) 283-7102
Troy, NY 12180
Michael Wacholder, Director

Two-square mile project of Rensselaer Polytechnic Institute zoned for research, development, and light manufacturing activities of technological businesses. Serves as a conduit for joint research activities, consultancies, refresher studies, associate programs, and human interactions between Park tenants and the Institute.

★654★ State University of New York at Albany
New York State Small Business Development Center

Suny Plaza, Rm. S. 523 Phone: (518) 443-5398
Albany, NY 12246 Fax: (518) 465-4992
James L. King, State Director

Member of the Association of Small Business Development Centers.

★655★ State University of New York at Binghamton
Office of Research and Sponsored Programs

Administration Bldg., Rm. 242 Phone: (607) 777-6136
PO Box 6000
Binghamton, NY 13902-6000
Stephen A. Gilje, Assoc. V. Provost for Research

Coordinates preparation and submission of grant and contract proposals for sponsored programs. Assists faculty and administrators in developing preliminary and formal proposals, locating potential sponsors, processing and transmitting applications, and negotiating budgets, grants, and contract awards. Administers ongoing grant activities, including personnel, purchasing, fiscal reporting, property control, subcontracting and project management, technology transfer, and grant modifications. **Publications:** Graduate Studies Research News (monthly).

★656★ State University of New York at Stony Brook
Center for Industrial Cooperation

Rm. 127 Engineering Phone: (516) 689-6000
Stony Brook, NY 11794
Randolph H. Cope, Director

Assists in establishing applied research on topics of interest to local industry by keeping businesses aware of ongoing research at the University and providing consulting arrangements with faculty through the Stony Brook Research Foundation. Also assists in providing access to technical information sources.

★657★ State University of New York Health Science
Center at Brooklyn
Office of Research Adminstration

450 Clarkson Ave. Phone: (718) 270-1128
Box 69
Brooklyn, NY 11203
Anthony Selvadvrai, Director

Administers sponsors programs, extramural funding, and contracts for research in medical and basic sciences. Also transfers technology from the University to business and industry.

★658★ Syracuse University
Institute for Energy Research

103 College Pl. Phone: (315) 423-3353
Syracuse, NY 13244-4010
Dr. Walter Meyer, Director

Application of advanced industrial processes, forecasting of technological trends, planning and evaluation of research and technology transfer programs, technology assessment of direct coal liquifaction, low-level radioactive waste disposal, development of corrosion-resistant thin films to reduce radiation build-up in nuclear reactors, development of pattern recognition artificial intelligence-based electric and gas load forecasting systems, nondestructive analysis for hydrogen in steel and other materials, and energy-related computer-based information systems. Maintains a database of utility rate payers, New York manufacturers, and national R&D laboratories. **Publications:** Annual Report.

Syracuse University *(Cont.)*

★659★ Technology and Information Policy Program
L.C. Smith College of Engineering   Phone: (315) 443-1890
329 Link Hall
Syracuse, NY 13244-1120
Barry Bozeman, Director

Nexus of technology, information, and public policy, particularly computer-based technological forecasting and assessment, management of scientific and technical information flows in organizations, technological innovation studies, and Pacific Rim science policy. **Publications:** T.

North Carolina

★660★ Research Triangle Institute
PO Box 12194   Phone: (919) 541-6000
3040 Cornwallis Rd.
Research Triangle Park, NC 27709-2194
F. Thomas Wooten, President

Chemical analysis, chemical synthesis, polymer science, life sciences and toxicology, environmental sciences and engineering, physics and aerosol technology, earth sciences, policy analysis and public sector management, economics, social sciences, statistical science and survey research, computer sciences, engineering, health care and pharmaceuticals, medical devices, public health research, health effects, environmental protection, social issues, public services, digital systems and computer architecture, semiconductors, energy, and industrial processes. Specific interests include aerospace technology, clean rooms, speech processors, international development, efficient delivery of services, AIDS, substance abuse, environmental measurements and quality assurance, risk assessment, and hazardous materials. **Publications:** Hypotenuse (bimonthly); Annual Report (January).

★661★ Research Triangle Park
2 Hanes Dr.   Phone: (919) 549-8181
PO Box 12255
Research Triangle Park, NC 27709
James O. Roberson, President

6,800-acre tract centrally located between North Carolina State University, University of North Carolina at Chapel Hill, and Duke University, owned and administered by Research Triangle Foundation. Facilitates interaction between industrial and governmental research and development organizations with the research communities of the three universities. **Publications:** Newsletter.

★662★ University of North Carolina
North Carolina Small Business Development Center
4509 Creedmoor Rd., Ste. 201   Phone: (919) 571-4154
Raleigh, NC 27612   Fax: (919) 571-4161
Scott R. Daugherty, State Director

Member of the Association of Small Business Development Centers.

★663★ University Research Park (Charlotte, NC)
Ste. 1980, Two First Union Plaza   Phone: (704) 375-6220
Charlotte, NC 28282
Seddon Goode, Jr., President

2,700-acre site providing companies located in the park with research and educational interaction with University of North Carolina at Charlotte.

North Dakota

★664★ Center for Innovation and Business Development
Box 8103   Phone: (701) 777-3132
Univ. Sta.
Grand Forks, ND 58202
Bruce Gjovig, Director

Works with applied research personnel at the University and other institutions to facilitate the commercialization of new technologies by providing research support services to entrepreneurs, inventors, and small manufacturers in the areas of invention evaluation, technology commercialization, SBIR applications, technical development, licensing, and business development, including market feasibility studies and marketing and business plans. **Publications:** Annual Report; Entrepreneur Kit; Business Plan for Start-ups; Marketing Plan: Step by Step.

★665★ University of North Dakota
North Dakota Small Business Development Center
University Sta. Box 7308   Phone: (701) 777-3700
Grand Fork, ND 58202   Fax: (701) 777-3650
Walter Kearns, State Director

Member of the Association of Small Business Development Centers.

Ohio

★666★ Battelle Memorial Institute
505 King Ave.   Phone: (614) 424-6424
Columbus, OH 43201
Dr. Douglas E. Olesen, President and Chief Executive

Independent, nonprofit international research and innovative organization serving industry and government in the generation, application, and commercialization of technology in the areas of advanced materials, electronics, defense, health and environment, engineering and manufacturing technology, transportation, and information systems. Conducts marine research at three coastal locations: Florida Marine Research Facility (Daytona Beach), Northwest Marine Research Laboratory (Sequim, Washington), and the Ocean Sciences and Technology Laboratory (Duxbury, Massachusetts). Specialized facilities and units include: Center for High-Speed Commercial Flight, Aviation Safety Reporting System Project Office, Advanced Materials Center for the Commercial Development of Space, Copper Data Center, Magnesium Research Center, National Center for Biomedical Infrared Spectroscopy, Health and Population Center, Human Factors and Organizational Effectiveness Research Center, Technology and Society Research Center, Office of Low-Level Waste Technology, Battelle Pressure Vessel and Piping Test Facility, and Battelle Center for Materials Fabrication. **Publications:** Battelle Today (four times per year); Annual Report; Published Papers and Articles (annually).

★667★ Lorain County Community College
Advanced Technologies Center
1005 N. Abbe Rd.   Phone: (216) 366-6618
Elyria, OH 44035
Dr. Ted H. Gasper, Jr., Director

Serves as an application center in the transfer of technological information to the design, manufacture, and marketing of systems in robotics, flexible manufacturing, microelectronics, computer system maintenance and repair, computer-aided design, injection molding simulation, and computer numerical control products.

★668★ Miami Valley Research Institute

1850 Kettering Tower
Dayton, OH 45423
John F. Torley, President

Phone: (513) 228-7987

Facilitates the transfer of basic and applied scientific technological research from Miami Valley Research Park tenants and member institutions to production, manufacture, and marketing of materials and services. Solicits public and private grants for research personnel and facilities of member institutions and recruits tenants for the Park.

★669★ Ohio Small Business Development Center

30 E. Broad St., 23rd Fl.
PO Box 1001
Columbus, OH 43226
Jack Brown, State Director

Phone: (614) 466-2711
Fax: (614) 466-0829

Member of the Association of Small Business Development Centers.

★670★ Ohio State University
Research Park

108 Bricker Hall
190 N. Oval Mall
Columbus, OH 43210
James C. Dowell, Special Assistant

Phone: (614) 292-9250

200-acre park offering research-oriented companies a site for their administrative, research, and development facilities to foster exchange between the University and industry. Seeks to enhance the University's teaching and research capabilities through stimulating the exchange of ideas and sharing of resources between University and tenant researchers. Coordinates University and tenant resources in developing commercial applications for new discoveries and technologies.

★671★ Ohio University
Innovation Center and Research Park

One President St.
Athens, OH 45701-2979
John Jarvis, Manager

Phone: (614) 593-1818

A 232-acre research park and 3,500-foot business incubator at Ohio University created to foster entrepreneurial activities and to provide technical and business assistance to new and expanding companies. Facilitates consulting, product testing, and technical assistance between tenants, the University, and the community at large.

★672★ University Technology Incorporated

Research Bldg. #1
11000 Cedar Ave.
Cleveland, OH 44106
Brian O'Riordan, President

Phone: (216) 368-5514

Established to strengthen regional business and industry, stimulate research, and provide a channel for the commercialization of Case Western Reserve University campus technologies. Identifies, evaluates, and implements development strategies for technologies, including intellectual property and protection; patent strategy; defining business and marketing opportunities; designing business structures; and forming business, commercial, and financial relationships. Licenses proprietary rights to companies prepared to invest in further on-campus research and development of technology, including established businesses and newly formed ventures.

Oklahoma

★673★ Research and Development Institute of the United States

PO Box 700270
Tulsa, OK 74170-0270
Daniel B. McDevitt, President

Phone: (918) 627-1181

Promotes and facilitates the formation of ad hoc single discipline and interdisciplinary research and development teams between academic institutions and private industry and organizations. Emphasizes the application of communication, computer, and control technology; the development of expert learning, diagnostic, training, and operating systems; and dynamic models and simulation of physical and operating complexes. Maintains on online catalog of human and physical resources available at 200 member institutions for research and development activities. **Publications:** MID-SOUTH R&D.

★674★ Southeastern Oklahoma State University
Oklahoma Small Business Development Center

PO Box 2584, Sta. A
Durant, OK 74701
Grady L. Pennington, State Director

Phone: (405) 924-0277
Fax: (405) 924-8531

Member of the Association of Small Business Development Centers.

★675★ University of Oklahoma
University Research Park

1700 Lexington Ave.
Norman, OK 73069
George Hargett, Airpark Administrator

Phone: (405) 325-7233

Provides facilities and services on a 1,500-acre tract for research administered by the University's Office of Research Administration, as well as U.S. and state government, industry, and University of Oklahoma laboratories located in the Park.

Oregon

★676★ Advanced Science & Technology Institute

Univ. of Oregon
318 Hendricks Hall
Eugene, OR 97403
Dr. Robert S. McQuate, Executive Director

Phone: (503) 346-3189

Bridges university research activities and resources with the corporate community. Encourages cooperative research projects with industry through its Industrial Associates Program. Serves as a broker for patenting and licensing arrangements between industry and parent universities. **Publications:** Newsletter; Directory of Oregon State University and University of Oregon Researchers.

★677★ Oregon Graduate Center Science Park

1600 NW Compton Dr.
Ste. 300
Beaverton, OR 97006
Bert Gredvig, President/CEO

Phone: (503) 690-1025

A 200-acre research park adjacent to Oregon Graduate Institute of Science and Technology which serves as a site for interaction between tenants and OGI faculty, students, and facilities, including joint scientific research ventures and internship opportunities for Center students.

★678★ Oregon Small Business Development Center

Lane Community College
99 W. 10th Ave., Ste. 216
Eugene, OR 97401
Edward Cutler, Ph.D., State Director

Phone: (503) 726-2250
Fax: (503) 345-6006

Member of the Association of Small Business Development Centers.

★679★ Oregon State University
Office of Vice President for Research, Graduate Studies, and International Programs

Corvallis, OR 97331
Dr. George H. Keller, Vice President

Phone: (503) 737-3467

Coordinates research activities of the University, including individual projects in various academic schools and special research organizations. Also administers the Technology Transfer Program and international activities of the University.

★680★ Riverfront Research Park

Univ. Planning Office   Phone: (503) 346-5566
1295 Franklin Blvd.
Eugene, OR 97403
Diane K. Wiley, Representative

67-acre site adjacent to Willamette River and the University of Oregon. Provides opportunities for research interaction between the University and Park tenants engaged in such activities as industrial research and development, biotechnology, materials science, environmental technology, advanced computer and software development, and business, educational, and governmental research and consulting services. Seeks to assist in the diversification of the economic base of the state, particularly the Eugene-Springfield metropolitan area, and provide additional financial opportunities for the University and its community.

★681★ Southern Willamette Research Corridor

408 SW Monroe   Phone: (503) 757-6851
Corvallis, OR 97333
Dr. John Byrne, Chairman

Joint research center of Oregon State University, University of Oregon, Lane Community College, Linn-Benton Community College, and cities of Albany, Corvalis, Eugene, Junction City, and Springfield. 40-mile research, development, and specialized manufacturing park facilitating cooperative ventures between participating Colleges and Universities, industry, and local government. **Publications:** Southern Willamette Research Corridor Briefing; Annual Report; Annual Legislative Briefing; Newsletter.

★682★ Sunset Research Park

PO Box 809   Phone: (503) 929-2477
Corvallis, OR 97339
B. Bond Starker, General Manager

85-acre research park affiliated with and located near Oregon State University. Facilitates the exchange of research resources and expertise between Park tenants and faculty at the University; especially in the areas of biotechnology, engineering, instrumentation, superconductivity, food science, and natural resources.

Pennsylvania

★683★ Ben Franklin Technology Center of Central and Northern Pennsylvania, Inc.

Fifth Fl. Rider Bldg.   Phone: (814) 863-4558
120 S. Burrowes St.
University Park, PA 16801
John E. Werner, Jr., President

Seeks to develop technologies to create new processes and products for technology transfer to industries in Pennsylvania in the areas of food, forestry, and agricultural sciences, advanced materials, coal and minerals, manufacturing, management and control systems, and biotechnology.

★684★ California University of Pennsylvania Mon Valley Renaissance

Box 62   Phone: (412) 938-5938
California, PA 15419
Richard H. Webb, Executive Director

Multiprogram consortium with a focus on applied research and economic development services to business and industry.

★685★ Carnegie Mellon University Biotechnology Center

4400 Fifth Ave.   Phone: (412) 268-3188
Mellon Institute
Pittsburgh, PA 15213-2683
Dr. Edwin Minkley, Director

Transfers molecular biology technology into viable commercial processes. Activities emphasize large-sale production of recombinantly engineeered proteins, from cost-efficient expression systems through in-house commercial production.

★686★ Center for Entrepreneurial Development

120 S. Whitfield St.   Phone: (412) 361-5000
Pittsburgh, PA 15206
Prof. Dwight M. Baumann, Executive Director

Entrepreneurial activities, including small business and industrial projects as a vehicle for research and experimentation in teaching, advancement of technology and management sources, innovation and entrepreneurship, and transfer of university-based technology in these fields. Aids potential businesspersons in overcoming barriers in moving projects from their conception to realization within the business community, assists professionals in starting their own businesses, and provides technical advice and introductions to lending institutions, potential clients, and suppliers.

★687★ Drexel University Center for Research in Technology and Strategy

Management Dept.   Phone: (215) 895-2148
32nd & Chestnut St.
Philadelphia, PA 19104
Dr. Alok K. Chakarabarti, Director

Interdisciplinary studies in technology and innovation, including mergers and acquisitions, productivity, corporate strategies, national security, and human resource management.

★688★ Lehigh University Center for Innovation Management Studies

Rauch Business Center #37   Phone: (215) 758-4819
Bethlehem, PA 18015
Alden S. Bean, Director

University/industry cooperative research center at Lehigh University focusing on management of technological innovation. Seeks to understand the reasons for the success or failure of technological innovation emphasizing the role of management in improving industrial innovation. Specific projects include comparison of management of external versus internal technological ventures, analysis of the response of the stock market to corporate financial decisions concerning research and development, examination of the productivity and creativity of research and development teams, success and failure of high-tech innovations, termination decisions in monitoring research and development projects, technological cycles and industrial innovation, leadership and productivity in innovation activities, cooperative ventures in large and small firms, boundary management in innovation groups, and creation of a database on technological entrepreneurship. **Publications:** The Network (quarterly newsletter).

★689★ Office of Research and Sponsored Programs

526 Brodhead Ave.   Phone: (215) 758-3020
Bethlehem, PA 18015
Mr. John M. Cheezum, Jr., Acting Director

Administers and coordinates, as administrative agency of the University, sponsored and cooperative research supported by government agencies, industry, and technical associations, including studies in physical, natural, social, and engineering sciences, and the humanities. Assists faculty and students in unsponsored research and scholarly efforts. **Publications:** Research Notes; Research Perspective.

Lehigh University *(Cont.)*

★690★ Small Business Development Center
Rauch Business Center   Phone: (215) 758-3980
Bethlehem, PA 18015
Dr. John W. Bonge, Director

Problems faced by small businesses, the impact of the general economy on the formation and operation of small business, and characteristics on entrepreneurs. **Publications:** Lehigh University Small Business Reporter (semiannually); Financing Guide for Northampton, Lehigh and Berks County; Market Planning Guide; Export Planning Guide; Lehigh Valley Business Support Services; Financing Your Business.

★691★ Pennsylvania State University
Center for Enterprise Creation and Growth
501 Business Administration Bldg.   Phone: (814) 865-1866
University Park, PA 16802
Dr. David N. Allen, Director

Entrepreneurship, focusing on structure, policy, services, and performance in business incubators; value-added contributions of Pennsylvania's business incubators; building-oriented business assistance programs; and faculty commercialization of research. **Publications:** Enterprise (quarterly).

★692★ Center for Regional Business Analysis
College of Business Administration   Phone: (814) 865-7669
Research Division
108 Business Administration Bldg.,
  No. 2
University Park, PA 16802
Rodney Erickson, Director

Industrial analysis and economic forecasting and impacts. **Publications:** The Pennsylvania Business Survey (monthly newsletter).

★693★ Research and Technology Transfer
114 Kern Graduate Bldg.   Phone: (814) 865-6331
University Park, PA 16802
Dr. K. Jack Yost, Associate Vice President

Transfers technology from the University to business and industry in the areas of electronics, materials, structural ceramics, polymers, and manufacturing.

★694★ University of Pennsylvania
Office of Corporate Programs and Technology
133 S. 36th St., Ste. 419   Phone: (215) 898-9585
Philadelphia, PA 19104-3246
Dr. George C. Farnbach, Director

Transfers technology from the Univeristy to business and industry in the fields of biotechnology, chemistry and chemical engineering, dental medicine, diagnostics, electrical engineering, laboratory devices and reagents, medical devices, mechanical engineering, materials science, pharmaceuticals, robotics factory automation, software, and veterinary medicine.

★695★ Office of Sponsored Programs
133 S. 36th St.   Phone: (215) 898-7293
Philadelphia, PA 19104-3246
Anthony Merritt, Director

Administers extramurally sponsored research for all departments and research units of the University and handles processing of research applications. Responsible for licensing of inventions and other technology transfer activities.

★696★ Pennsylvania Small Business Development
Center
Vance Hall, 4th Fl.   Phone: (215) 898-1219
Philadelphia, PA 19104   Fax: (215) 898-1299
Gregory L. Higgins, State Director

Member of the Association of Small Business Development Centers.

★697★ Sol C. Snider Entrepreneurial Center
3733 Spruce St.   Phone: (215) 898-4856
Vance Hall, Ste. 400
Philadelphia, PA 19104-6374
Dr. Ian C. MacMillan, Director

Entrepreneurship in private sector, public sector, and nonprofit organizations, including venture capital, corporate venturing, development of internal corporate entrepreneurship, technology transfer, and university spin-offs.

★698★ University of Pittsburgh
NASA Industrial Applications Center
823 William Pitt Union   Phone: (412) 648-7000
Pittsburgh, PA 15260
Lani Hummel, Ph.D., Executive Director

Transfers technology developed by the National Aeronautics and Space Administration (NASA) and other government agencies to business and industrial clients. Activities center on new product identification and testing, engineering analyses, corporate resource redeployment initiatives, literature searching and document procurement, database system design and electronic publishing, and science curriculum enhancements for grades K-12. The Center is hard-wired to the NASA Recon database and it maintains direct link with researchers in NASA field centers, federal research laboratories, and academic centers. **Publications:** United States Political Science Documents (annually); Federal Laboratories Database (annually).

Puerto Rico

★699★ University of Puerto Rico
Puerto Rico Small Business Development Center
PO Box 5253 College Sta.   Phone: (809) 834-3590
Mayaguez, PR 00709   Fax: (809) 834-3790
Jose M. Romaguera, Executive Director

Member of the Association of Small Business Development Centers.

Rhode Island

★700★ Bryant College
Rhode Island Small Business Development Center
1150 Douglas Pike   Phone: (401) 232-6111
Smithfield, RI 02917   Fax: (401) 232-6319
Douglas H. Jobling, State Director

Member of the Association of Small Business Development Centers.

South Carolina

★701★   Clemson University
Emerging Technology Development and Marketing Center

338 Univ. Sq.   Phone: (803) 656-4237
PO Box 5703
Clemson, SC 29634-5703
Dr. B.E. Gilliland, Director

Provides technical and marketing support to businesses and University faculty, staff, and students to facilitate the commercialization, production, and transfer of new products in South Carolina.

★702★   South Carolina Research Authority

PO Box 12025   Phone: (803) 799-4070
Columbia, SC 29211
Robert E. Henderson, Ph.D., Director

Independent, nonprofit organization that manages the activities of the 266-acre Carolina Research Park at University of South Carolina at Columbia, 11-acre Charleston Research Park at Clemson University, and the Trident Research Center, an incubator. Promotes cooperative research and development activities between the Universities and park tenants. The Authority serves as the contractor and manager of the Rapid Acquisition of Manufactured Parts (RAMP) program, a flexible computer integrated manufacturing system being developed for the U.S. Navy; the Product Data Exchange Specification (PDES) program, the development of an international standard for the exchange of engineering drawings or blueprints in a computer-interpretable format; and the Plasma Arc program, a process to improve extracting ferrochrome and other materials from their ores.

★703★   University of South Carolina
South Carolina Small Business Development Center

Columbia, SC 29208   Phone: (803) 777-4907
                         Fax: (803) 777-4403
John M. Lenti, State Director

Member of the Association of Small Business Development Centers.

★704★   Southeast Manufacturing Technology Center

College of Engineering   Phone: (803) 777-9595
Columbia, SC 29201-9980
Prof. William F. Ranson, Director

Cooperative effort between the University of South Carolina College of Engineering, the University of South Carolina Small Business Development Center, and South Carolina technical colleges. Goal is to introduce new manufacturing technology processes to small-and medium-sized industries. Technology targeted for transfer is developed in various national research laboratories, universities, or vendor research and development facilities. Primary emphasis is on the automation of the manufacturing process, specifically, computer-aided design, computer-aided manufacturing, computer-aided engineering, process and resource planning, automated inspection, materials handling, quality control, and communications. Projects have included development of a robotics system for a forklift manufacturer and the redesign of a 30-foot tall tree stump crusher.

South Dakota

★705★   University of South Dakota
South Dakota Small Business Development Center

414 E. Clark St.   Phone: (605) 677-5272
Vermillion, SD 57069   Fax: (605) 677-5427
Donald D. Greenfield, State Director

Member of the Association of Small Business Development Centers.

Tennessee

★706★   Memphis State University
Tennessee Small Business Development Center

Memphis, TN 38138   Phone: (901) 678-2500
                         Fax: (901) 678-4072
Kenneth J. Burns, State Director

Member of the Association of Small Business Development Centers.

★707★   Tennessee Center for Research and Development

10521 Research Dr., Ste. 400   Phone: (615) 675-9505
Knoxville, TN 37932
B. James George, President

Research, educational, and public service organization headquartered in the Tennessee Technology Corridor and affiliated with the University of Tennessee at Knoxville, Oak Ridge National Laboratory, and Tennessee Valley Authority. Promotes economic development in Tennessee through two components: a research division to conduct basic and applied studies for industrial development and to perform technology transfer activities, and a for-profit subsidiary to provide management support services and temporary laboratory facilities to new companies. Serves as a resource center for public information on the effects of science and technology on the environment.

★708★   Tennessee Technology Foundation

Oak Ridge/Knoxville Technology   Phone: (615) 694-6772
  Corridor
10915 Hardin Rd.
PO Box 23184
Knoxville, TN 37933-1184
Dr. David A. Patterson, President

Research park linking the research resources of the University of Tennessee at Knoxville, Oak Ridge National Laboratory, and Tennessee Valley Authority with tenants, particularly in the areas of advanced materials, biotechnology, information sciences, waste technology, measurements and control, and electrotechnology. Fosters the transfer of technology from research and development resources in the Corridor and in Tennessee. Maintains power quality testing laboratory, and laser training center.

★709★   University of Tennessee Research Corporation

415 Communications Bldg.   Phone: (615) 974-1882
Knoxville, TN 37996-0344
Dr. K.W. Heathington, President

Supports research at various campuses of the University of Tennessee by investigating the patentability and marketability of faculty inventions and creative works, procuring patent protection and/or copyright registration, and commercializing new technology through licensing, sales, and the formation of new ventures.

Texas

★710★   Baylor University
Center for Private Enterprise

PO Box 98003   Phone: (817) 755-3766
Hankamer School of Business
Waco, TX 76798-8003
Dr. W. James Truitt, Director

Entrepreneurship and private enterprise, including studies on characteristics of entrepreneurs, student and teacher attitudes toward the private enterprise system, business taxation, and women entrepreneurs.

Baylor University *(Cont.)*

★711★ John F. Baugh Center for Entrepreneurship

Hankamer School of Business   Phone: (817) 755-2265
PO Box 98011
Waco, TX 76798-8011
D. Ray Bagby, Ph.D., Director

Entrepreneurship, including identifying and evaluating new ventures, acquiring capital and other resources, and starting, developing, and divesting new business interests. **Publications:** Entrepreneurship: Theory and Practice (quarterly journal); News Venture (newsletter).

★712★ North Texas Small Business Development Center

1402 Corinth St.   Phone: (214) 565-5831
Dallas, TX 75215   Fax: (214) 565-5815
Norbert R. Dettmann, Regional Director

Member of the Association of Small Business Development Centers.

★713★ Northwest Texas Small Business Development Center

Texas Tech University   Phone: (806) 745-3973
2579 S. Loop 289, Ste. 114   Fax: (806) 745-6207
Lubbock, TX 79423
Craig Bean, Director

Member of the Association of Small Business Development Centers.

★714★ Pan American University
Center for Entrepreneurship and Economic Development

School of Business Bldg., Rm. 124   Phone: (512) 381-3361
1201 W. Univ. Dr.
Edinburg, TX 78539-2999
Dr. J. Michael Patrick, Director

South Texas business assistance and economic research, focusing on business plans, economic area profiles, economic impact studies, economic development planning, market feasibility studies, international trade, urban and rural commercial revitalization, industrial park development and feasibility, and land use surveys.

★715★ RiverBend

2501 Gravel Dr.   Phone: (817) 284-5555
Fort Worth, TX 76118-6999
David Newell, Contact

1,100-acre research park which fosters interaction between the University of Texas at Arlington research community and industry, particularly in the areas of robotics and automation.

★716★ Sam Houston State University
Center for Business and Economic Research

Huntsville, TX 77341-2056   Phone: (409) 294-1518
Jo Ann Duffy, Director

Designed to support faculty research activities in area business and industry. **Publications:** Journal of Business Strategies (semiannually).

★717★ Southern Methodist University
Caruth Institute of Owner-Managed Business

Cox School of Business, Box 333   Phone: (214) 692-3326
Dallas, TX 75275-0333
Prof. Jerry White, Director

Entrepreneurs, including successful characteristics and entrepreneurs as managers. **Publications:** Newsletter.

★718★ Texas A&M University
Technology Business Development

Texas Engineering Experiment Sta., Ste.   Phone: (409) 845-0538
310
Wisenbaker Engineering Research
   Center
College Station, TX 77843-3369
Helen Baca Dorsey, Ph.D., Director

Statewide agency to facilitate the commercialization of innovative technologies developed in Texas. Cofacilitates the start-up of a limited number of commercial operations each year involving university research in engineering and assists in licensing university-developed technology. Provides assistance to engineering researchers, Texas entrepreneurs, and small companies in technical research and development, marketing and pricing studies, technical evaluations, and production and financial planning of unique ideas. Performs strategic planning and economic feasibility studies for Texas municipalities through its Municipal Assistance Program. Maintains the Texas Innovation Information Network System. Provides access to databases of the Southern Technology Applications Center for Texas clients. **Publications:** Diversification Report (quarterly newsletter).

★719★ Texas A&M University Research Park

College Station, TX 77843-1120   Phone: (409) 845-7275
Dr. Mark L. Money, Vice Chancellor

Located on 434 acres west of main campus, the Park serves to assist private utilization of University resources, to promote closer ties between industry engaged in research and the University, to improve the quality and productivity of University research activities, and to accelerate the dissemination of new knowledge and the transfer of new technologies.

★720★ Texas Engineering Experiment Station

College Station, TX 77843   Phone: (409) 845-1321
Dr. Herbert H. Richardson, Director

Engineering and related sciences, including traditional engineering disciplines; multidisciplinary centers in space research, new materials, biochemical engineering, food protein research and development, knowledge systems, electrochemical systems, hydrogen, and thermodynamics; and research divisions with technology-related activities in the System's colleges of architecture, business, education, geosciences, liberal arts, science, and veterinary medicine. Major research thrusts include biotechnical engineering, electrooptics and telecommunications, artificial intelligence and expert systems, electrochemical engineering, environmental engineering, and manufacturing systems. Assists technology transfer, commercialization, and economic development of University research through the Technology Business Development Division. **Publications:** Windows (four per year); Research Report; Annual Report; Engineering Issues.

★721★ Texas Research and Technology Foundation

14785 Omicron Dr.   Phone: (512) 677-6000
San Antonio, TX 78245
John F. D'Aprix, CEO and President

Supports basic and applied science and new advanced technology enterprises in the San Antonio area by providing research and development facilities, equipment, endowment, venture capital, and business and scientific expertise. Manages the 1,500-acre Texas Research Park providing industry with access to the resources of several educational and research organizations. Activities include creation of research centers conducting studies in areas such as human performance, neuroscience, toxicology, and clinical trials. Also conducts inventory and analyses of scientific and technical resources in San Antonio. **Publications:** Foundation Forum; TRTFOCUS.

★722★ Texas Research Park

Texas Research & Technology   Phone: (512) 342-6063
  Foundation
14785 Omicron Dr.
San Antonio, TX 78245
Chris Harness, Vice President

Supports technology development efforts of private industry by providing facilities for basic and applied research and access to the resources of local educational and research organizations. 1,300 acres of the Park are available to private firms through sale or long-term lease; 200 acres are reserved for nonprofit research and development. The Institute of Biotechnology, supporting basic and applied research in the biosciences, the Institute of Applied Sciences, providing technology transfer to industry and government, and the Invention and Investment Institute, offering business and technical services in support of advanced technology business ventures, are collaborative programs based at the Park.

★723★ University of Houston
Small Business Development Center

601 Jefferson, Ste. 2330   Phone: (713) 752-8444
Houston, TX 77002   Fax: (713) 752-8484
Betsy J. Gatewood, Regional Director

Member of the Association of Small Business Development Centers.

★724★ University of North Texas
University Center for Economic Development and
  Research

PO Box 12988   Phone: (817) 565-3437
Denton, TX 76203-2988
Dr. Weinstein, Head

Performs labor and manpower analysis, economic planning and forecasting, market analysis, and location and feasibility studies to promote economic development. Transfers technology from the University to business and industry.

★725★ University of Texas
Center for Technology Development and Transfer

Ernest Cockrell Jr., Hall 10.340   Phone: (512) 471-3700
College of Engineering
Austin, TX 78712-1080
Dr. Dale E. Klein, Director

Links university researchers with industrial needs and facilitates the commercialization of academic research, university-generated technology, scientific information, and other intellectual property. Activities range from serving as a clearinghouse for technology to product development and the formation of businesses. Also serves as the hub for a statewide technology development and transfer network.

★726★ University of Texas at Austin
IC2 Institute

2815 San Gabriel   Phone: (512) 478-4081
Austin, TX 78705
Dr. George Kozmetsky, Director

Management of technology, creative and innovative management, measurement of the state of society, dynamic business development and entrepreneurship, new methods of economic analysis, and the evaluation of attitudes, concerns, and opinions on key issues. **Publications:** Newsletter (quarterly).

★727★ University of Texas at El Paso
Institute for Manufacturing and Materials Management

901 Education Bldg.   Phone: (915) 747-5336
500 W. Univ. Ave.
El Paso, TX 79968
Donald A. Michie, Ph.D., Director

Materials management, focusing on extraction of raw materials, physical metallurgy studies, technology transfer, and production of commercial products. Coordinates binational resources between the U.S. and Mexico, emphasizing customs procedures, bilateral negotiations, trade area analyses, and technology transfer. Initiates infrastructure studies involving southern border ports of entry, transborder transportation, Maquiladora economic impact, Maquiladora supplier sources, Texas ad valorem tax rulings, and regional/community economic development.

★728★ University of Texas at San Antonio
South Texas Border Small Business Development
  Center

San Antonio, TX 78285   Phone: (512) 224-0791
    Fax: (512) 222-9834
Robert M. McKinley, Regional Director

Member of the Association of Small Business Development Centers.

Utah

★729★ University of Utah
Office of Technology Transfer

295 Chipeta Way, Ste. 280   Phone: (801) 581-7792
Salt Lake City, UT 84108
Dr. Norman Brown, Director

Technology transfer and licensing; new business start-up based on university-developed technologies; and intellectual property management. Maintains a technology targeting database. **Publications:** Innovations (semiannual).

★730★ University of Utah Research Park

505 Wakara Way   Phone: (801) 581-8133
Salt Lake City, UT 84108
Charles A. Evans, Director

300-acre tract designed to facilitate scientific research projects between government agencies, industrial organizations, and the University. Facilities include general and technical libraries, scientific and research equipment, and computer center. The Park is the site of 25 buildings.

★731★ Utah Small Business Development Center

102 W. 500 S., Ste. 315   Phone: (801) 581-7905
Salt Lake City, UT 84101   Fax: (801) 581-7814
David A. Nimkin, Executive Director

Member of the Association of Small Business Development Centers.

★732★ Utah State University
Utah State University Research and Technology Park

1780 N. Research Pkwy.   Phone: (801) 750-6924
Ste. 108
North Logan, UT 84321
Wayne Watkins, Director

30-acre research park at Utah State University that fosters interaction between the University research community and Park tenants. Administers a technology transfer program and provides incubator services to start-up companies.

Vermont

★733★ University of Vermont
Vermont Small Business Development Center

Extension Service-Morrill Hall   Phone: (802) 656-4479
Burlington, VT 05405   Fax: (802) 656-8642
Norris A. Elliott, State Director

Member of the Association of Small Business Development Centers.

Virgin Islands

★734★ University of the Virgin Islands
Virgin Islands Small Business Development Center
PO Box 1087   Phone: (809) 776-3206
St. Thomas, VI 00804   Fax: (809) 775-3756
Soloman S. Kabuka, State Director

Member of the Association of Small Business Development Centers.

Virginia

★735★ Center for the Productive Use of Technology
1100 S. Washington St.   Phone: (703) 549-3611
Alexandria, VA 22314
David S. Bushnell, Director

Technology transfer, work group dynamics and motivation, production measurement, human factors and sociotechnical systems design, and training systems design and development. Assists public and private sector organizations in planning, implementing, and evaluating alternative strategies for quality management and productivity improvement.

★736★ George Mason University
Institute for Advanced Study in the Integrative
** Sciences**
Thompson Hall, Rm. 219   Phone: (703) 764-6427
4400 Univ. Dr.
Fairfax, VA 22030
Dr. John N. Warfield, Director

Acts to stimulate on-campus research activity by encouraging cooperation between the University and high-technology industry. Principal investigators specialize in systems research, communications research, systems design, and computer science.

★737★ James Madison University
Center for Entrepreneurship
Harrisonburg, VA 22807   Phone: (703) 568-6334
Dr. Roger H. Ford, Director

New business startup, small business growth and assistance, and rural economic development. Develops models and systems for university and community cooperation for economic development. Maintains a regional listing of entrepreneurs and small businesses. **Publications:** JMU Entrepreneurs (semiannual newsletter); Entrepreneurship and Small Business Research Series.

★738★ Virginia Center for Innovative Technology
CIT Bldg., Ste. 600   Phone: (703) 689-3000
2214 Rock Hill Rd.
Herndon, VA 22070
Linwood Holton, President

Facilitates the transfer of technology for commonwealth universities to industry in biotechnology, computer-aided engineering, information technology, materials science engineering, fiber optics, coal, space, semicustom integrated systems, wood and renewable resources, power electronics, electrochemical processes, magnetic bearings, advanced ceramics, and command, control, communications and intelligence (C3I). **Publications:** CIT Brochure.

★739★ Virginia Polytechnic Institute and State
** University**
Virginia Tech Corporate Research Park
1800 Kraft Dr.   Phone: (703) 231-3600
Blacksburg, VA 24060
Fred R. Meade, Director

120 acres adjacent to the main campus and university airport provides building sites and buildings for lease to companies interested in developing or expanding a research relationship with the University. Houses an innovation center and other buildings to provide facilities for start-up companies requiring support of University programs.

★740★ Virginia Small Business Development Center
Dept. of Economic Development   Phone: (804) 371-8258
1021 E. Cary St., 11th Fl.   Fax: (804) 371-8185
Richmond, VA 23206
Robert D. Smith, State Coordinator

Member of the Association of Small Business Development Centers.

Washington

★741★ University of Washington
Northwest Policy Center
327 Parrington Hall, DC-14   Phone: (206) 543-7900
Seattle, WA 98195
David S. Harrison, Executive Director

Workforce, workplace, and entrepreneurial policy, rural-urban interdependence, K-12 and higher education, environmental quality, capital availability for small businesses, and flexible manufacturing networks, with emphasis on strategies for economic vitality in the Northwest states of Alaska, Idaho, Montana, Oregon, and Washington. **Publications:** Changing Northwest (bimonthly newsletter).

★742★ Office of Technology Transfer
Ste. 301   Phone: (206) 543-5900
4225 Roosevelt Way NE, XD-40
Seattle, WA 98105
Donald Baldwin, Director

Transfers technology from the University to business and industry, particularly in the areas of bioengineering, biotechnology, engineering, and medicine. **Publications:** Annual Report; Listing of Available Technologies and Information Pieces.

★743★ Washington State University
Research and Technology Park
NE 1615 Eastgate Blvd.   Phone: (509) 335-5526
Pullman, WA 99164-1802
Larry M. Simonsmeier, Director

Encourages research interaction between Washington State University and industry. Seeks to promote economic development in southeastern Washington. The Park leases land to industries engaged in research, development, and light manufacturing. Industrial tenants share University research facilities and services, use faculty as consultants, and utilize graduate students as a part-time work force. Major research areas include agriculture, forestry, mines, veterinary medicine, plant and animal biotechnology, and engineering. **Publications:** Quarterly Newsletter.

★744★ Washington Small Business Development Center
245 Todd Hall   Phone: (509) 335-1576
Pullman, WA 99164-4740   Fax: (509) 335-0949
Lyle M. Anderson, State Director

Member of the Association of Small Business Development Centers.

West Virginia

★745★ Center for Education and Research with
** Industry**
1050 4th Ave.   Phone: (304) 696-3093
Huntington, WV 25755
Larry Kyle, Director

Facilitates joint ventures between academia, business, and government by linking campus resources and faculty expertise with technology

transfer activities around the state. Administers joint research programs between academia and industry.

★746★ Center for Entrepreneurial Studies and Development, Inc.

West Virginia Univ.   Phone: (304) 293-5551
College of Engineering
P.O. Box 6101
Morgantown, WV 26506
Dr. Jack Byrd, Executive Director

Business operations improvement, employee relations, management, and systems. Operations improvement studies focus on materials handling systems, cost reduction, work standards development, training manual development, facilities utilization and planning, work methods, inventory control systems, production automation, computer applications, quality control, and facility relocation models. Employee relations studies focus on job incentive systems, wage payments, job enrichment, employee motivation programs, job evaluation, labor/management relations, and employee staffing. Management studies focus on development programs, organization development, management incentives, succession planning, policy manual development, and small business organizations. Systems studies focus on financial planning, computer systems, sales forecasting, strategic planning, office systems improvements, competition, insurance policies, and business plan development. Operates Discovery Lab, an environment supporting the development of new manufacturing technologies in West Virginia by inventors and University faculty, students, and staff; Howard Purdum, technical director.

★747★ Morgantown Industrial/Research Park

1000 DuPont Rd.   Phone: (304) 292-9453
Bldg. 510
Morgantown, WV 26505-9654
John R. Snider, Executive Vice President

600-acre research and development site, affiliated with West Virginia University, zoned for commercial use. Links University research resources with industrial research and development activities. The 200-acre Industrial Park provides pilot plants for industrial research tenants to test findings for entry into commercial markets and further generate research in the 400-acre Research Park.

★748★ West Virginia Small Business Development Center

1115 Virginia St. E.   Phone: (304) 348-2960
Charleston, WV 25301   Fax: (304) 348-0127
O. Eloise Jack, State Director

Member of the Association of Small Business Development Centers.

★749★ Wheeling Jesuit College
National Technology Transfer Center

316 Washington Ave.   Phone: (304) 243-2455
Wheeling, WV 26003
Thomas Acker, Director

Transfers federally sponsored research and technology to U.S. business and industry to increase industrial competitiveness.

Wisconsin

★750★ Fox Valley Technical College
Technical Research Innovation Park (TRI-Park)

1825 N. Bluemound Dr.   Phone: (414) 735-5600
P.O. Box 2277
Appleton, WI 54913-2277
Stanley Spanbauer, President

Seeks to attrct and encourage liaisons of highly technical businesses and industries with the College. Encourages marketing ventures, supports inventors and entrepreneurs, fosters research and development for the paper and printing industries, and promotes new technology industries as tenants in the Park. Operates the D.J. Bordini Technological Innovation Center of the College that includes a technical library, quality and productivity improvement center, communications network area, product development service center, flexography laboratory, high technology demonstration laboratories, and facilities for conferences and classes.

★751★ Marquette University
Center for the Study of Entrepreneurship

College of Business Administration   Phone: (414) 288-5578
Milwaukee, WI 53233
Dr. Paul D. Reynolds, Director

New business formation, including studies on nature of the entrepreneur; background and training, opportunities for independent venture, social, economic, and legal climates conducive to new firm formation, business failures, and incidence of new firm formation by region, industry type, and growth of industry type.

★752★ Medical College of Wisconsin
MCW Research Foundation

8701 Watertown Plank Rd.   Phone: (414) 257-8219
Milwaukee, WI 53226
D.H. Westermann, Executive Vice President

Creates intellectual property from ideas and processes of College researchers for license of products and services to the marketplace primarily in the areas of biophysics, biomedical engineering, biochemistry, microbiology, pharmacology, medical instrumentation, medical imaging, magnetics, computer software design, and genetic engineering, and immunology.

★753★ University of Wisconsin
Wisconsin Small Business Development Center

432 N. Lake St., Rm. 423   Phone: (608) 263-7794
Madison, WI 53706   Fax: (608) 262-3878
Member of the Association of Small Business Development Centers.

★754★ University of Wisconsin—Madison
University-Industry Research Program

1215 WARF Bldg.   Phone: (608) 263-2840
610 Walnut St.
Madison, WI 53705
Robert M. Bock, Director

Identifies scientific resources of faculty, information, and facilities at the University for application to needs of industry, commerce, and government. Facilitates technology transfer activities with industry and provides advisory services to faculty on patents and consortia development. Maintains a database on University research activities and resources. **Publications:** Touchstone (quarterly); Activity Report (annually).

★755★ University of Wisconsin—Milwaukee
Center for Urban Transportation Studies

PO Box 784   Phone: (414) 229-5787
Milwaukee, WI 53201
Dr. Edward Beimborn, Director

Transportation engineering and planning, mass transit, outreach, and technology transfer. Studies include alternative methods of organizing transportation planning, transit sensitive land use design, highway projects effects on land use, investigations of transport improvement, evaluation, and methodologies through private financing, development and improvement of disaggregate mode choice models, ridesharing alternatives for major trip generators, trip tour modeling, the use of microcomputers for transportation planning, improved methods of transport management, rural transit systems, and the dissemination of transportation research results. **Publications:** Research Report.

University of Wisconsin—Milwaukee *(Cont.)*

★756★ Office of Industrial Research and Technology Transfer

PO Box 340   Phone: (414) 229-5000
Milwaukee, WI 53201
Irving D. Ross, Jr., Director

Serves as a catalyst in developing university/industry research programs and facilitates the transfer of technology between the University and industry. Handles patents, copyrights, licensing and proprietary agreements.

★757★ Urban Research Center

Rm. 450, Physics Bldg.   Phone: (414) 229-5916
PO Box 413
Milwaukee, WI 53201
Dr. Sammis B. White, Director

Public policy, emphasizing metropolitan economic revitalization, urban education, technological innovation and transfer, state and metropolitan public policy, health care organization and policy, gerontology, employment and manpower, and housing and urban development. Also studies medical technology adoption decisions in hospitals and gangs in midwestern cities. **Publications:** Occasional Papers; Reports (bimonthly).

★758★ University of Wisconsin—Stout
Center for Innovation and Development

Menomonie, WI 54751   Phone: (715) 232-1385
Dr. Glenn Gehring, Manager

Manufacturing productivity, electronics, and process evaluation. Evaluates inventions and new products, develops prototype models of innovative products, and evaluates materials for fabrication of new products.

★759★ University Research Park (Madison, WI)

Univ. of Wisconsin—Madison   Phone: (608) 262-3677
1265 Wisconsin Alumni Research
   Foundation Bldg.
610 Walnut St.
Madison, WI 53705
Wayne McGown, Director

The 296-acre Park facilitates technology transfer between the research produced on the University of Wisconsin—Madison campus and applied research of industry and provides a long-term endowment income to the University. Leases land to private companies and agencies and offers laboratory and office services to start-up companies. Provides access to the University Industry Research Program for matching research interests.

★760★ Wisconsin for Research, Inc.

565 Science Dr., Ste. A   Phone: (608) 238-3031
Madison, WI 53711
Reed Coleman, President

Seeks to encourage economic development by promoting the transfer of University of Wisconsin research and technology to businesses in the state. Operates the Madison Business Incubator to generate economic development in Dane County by providing office space, support services, and business assistance. Administers a seed capital fund.

Wyoming

★761★ Wyoming Small Business Development Center

Casper College   Phone: (307) 235-4825
111 W. 2nd St., Ste. 416   Fax: (307) 473-7243
Casper, WY 82601
Member of the Association of Small Business Development Centers.

Section 16

Business Incubators

"Success depends upon staying power. The reason for failure in most cases is lack of perseverance."—James Russell Miller

Business incubators provide start-up companies with below-market rates for office and lab space. They are designed to encourage entrepreneurship and minimize obstacles to new business formation and growth, particularly for high-technology firms. In addition, these facilities offer shared support for clerical, reception, and computer services.

Business incubators may also provide programs that assist in the development of business plans and marketing strategies, advise firms on personnel, accounting, and legal matters, and identify sources of financing. Professionals may also evaluate product lines and manufacturing processes, assist in the use of state-of-the-art design and manufacturing tools, and identify special expertise at universities and other research centers. Many incubators are located on or near universities or research/technology parks. **See Section 15, University and Independent Innovation Research Centers, on page 145 for more information.**

Section Arrangement

The following entries are arranged geographically by state, then alphabetically by incubator name. All incubator names in this section are referenced in the Master Index.

Arkansas

★762★ East Arkansas Business Incubator System, Inc.
5501 Krueger Dr.   Phone: (501) 935-8365
Jonesboro, AR 72401   Fax: (501) 932-1982

Leases office, lab, and manufacturing space to new technology-based businesses. Provides technical and management advice.

California

★763★ Victor Valley College
Small Business Incubator Project
18422 Bear Valley Rd.   Phone: (619) 245-4271
Victorville, CA 92392-9699   Fax: (619) 245-9744

Colorado

★764★ Canon City Business Incubator
402 Valley Rd.   Phone: (719) 275-8601
Canon City, CO 81212   Fax: (719) 275-4400

A small business incubator that assists entrepreneurs in the business start-up process and gives aid to new businesses to help ensure their survival.

★765★ Denver Enterprise Center
3003 Arapahoe St.   Phone: (303) 296-9400
Denver, CO 80205   Fax: (303) 296-5542

A small business incubator.

★766★ Jefferson County Business Innovation Center
1667 Cole Blvd., Ste. 400   Phone: (303) 238-0913
Golden, CO 80401   Fax: (303) 237-1103

A small business incubator.

★767★ Mesa County Small Business Incubator
304 W. Main St.   Phone: (303) 243-5242
PO Box 3080   Fax: (303) 241-0771
Grand Junction, CO 81505

Assists entrepreneurs in the business start-up process and in managing new businesses.

★768★ Pueblo Business Technology Center
301 N. Main St.   Phone: (719) 546-1133
Pueblo, CO 81003

A small business incubator that assists entrepreneurs in the business start-up process and gives aid to new businesses to help ensure their survival.

Idaho

★769★ Business Center for Innovation and Development
11100 Airport Dr.   Phone: (208) 772-0584
Hayden Lake, ID 83835   Fax: (208) 772-6196

A small business incubator that assists entrepreneurs in the business start-up process and gives aid to new businesses to help ensure their survival.

★770★ Idaho Innovation Center
2300 N. Yellowstone   Phone: (208) 523-1026
Idaho Falls, ID 83401   Fax: (208) 523-1049
A small business incubator.

Illinois

★771★ Bradley Industrial Incubator Program
c/o Area Jobs Development Association   Phone: (815) 933-2537
PO Box 845   Fax: (815) 933-5519
Kankakee, IL 60901

★772★ The Decatur Industrial Incubator
2121 U.S. Rte. 51, S.   Phone: (217) 423-2832
Decatur, IL 62521   Fax: (217) 423-7214

★773★ Des Plaines River Valley Enterprise Zone Incubator Project
Will County Office Bldg.   Phone: (815) 726-0028
302 N. Chicago St.
Joliet, IL 60432

Louisiana

★774★ Northeast Louisiana Incubator Center
State Rte. 594   Phone: (318) 343-2262
Swartz School Rd.
Monroe, LA 71203

Michigan

★775★ Delta Market Street Incubator
470 Market St., SW   Phone: (616) 451-2561
Grand Rapids, MI 49503   Fax: (616) 243-1013

Incubator is operated by Delta Properties, 1300 4 Mile Rd., NW, Grand Rapids, MI 49504; (616)451-2561.

★776★ Grand Rapids Terminal Warehouse
446 Granville Ave., SW Phone: (616) 451-2561
Grand Rapids, MI 49503 Fax: (616) 243-1013
Incubator is operated by Delta Properties, 1300 4 Mile Rd., NW, Grand Rapids, MI 49504; (616)451-2561.

★777★ Jackson Industrial Incubator
Jackson Business Development Center
414 N. Jackson St. Phone: (517) 787-0442
Jackson, MI 49201
Offers training, consultation, and other support services.

★778★ Metropolitan Center for High Technology
2727 2nd Ave. Phone: (313) 963-0616
Detroit, MI 48201 Fax: (313) 963-7606
Conducts, coordinates, and facilitates research and development of new technology. Activities include collaborating with client businesses and universities in the areas of automated vehicles, biotechnology, biomedicine, rehabilitation engineering, toxicology, information systems, robotics, and computer engineering. Research components include Information Technology Institute and Institute of Chemical Toxicology at Wayne State University. Databases: Biomedical technology corporations. Research results are published in papers and presentations. Public Services: Acts as a technology business incubator, providing low-cost office space and start-up services to include professional and management assistance to high technology companies.

Minnesota

★779★ College of St. Thomas
Entrepreneurial Enterprise Center
1107 Hazeltine Blvd. Phone: (612) 448-8800
Chaska, MN 55318 Fax: (612) 448-8895
Incubator facility with space for forty to fifty tenants. Provides flexible space at competitive rents with short-term leases, shared office services, access to local network of experienced business experts, and access to seed capital.

★780★ St. Paul Small Business Incubator
Department of Planning & Economic Phone: (612) 228-3301
    Development
25 W. 4th St.
1000 City Hall Annex
St. Paul, MN 55102

Mississippi

★781★ Jackson Enterprise Center
Battlefield Park Industrial Complex Phone: (601) 352-0957
931 Hwy. 80, W. Fax: (601) 948-3250
Jackson, MS 39204
Act as a business incubator to create jobs and reduce small business failures by providing services to businesses and entrepreneurs during the criticial early stages of development.

New York

★782★ Business Incubator (Brooklyn)
Local Development Corp. of East New Phone: (718) 385-6700
    York Fax: (718) 385-7505
116 Williams Ave.
Brooklyn, NY 11207

★783★ Greater Syracuse Business Incubator Center
1201 E. Fayette St. Phone: (315) 475-0844
Syracuse, NY 13210 Fax: (315) 475-8460

★784★ Incubator Industries Building (Buffalo)
Buffalo Urban Renewal Agency Phone: (716) 851-5035
920 City Hall
Buffalo, NY 14202

★785★ Syracuse Incubator
Central New York Technology Phone: (315) 470-1350
    Development Organization
c/o Knowledge Systems & Research,
    Inc.
500 S. Salina St., Ste. 230
Syracuse, NY 13202

★786★ Troy Incubator Program
Incubator Center Phone: (518) 276-6658
1223 Peoples Ave.
Troy, NY 12180

North Carolina

★787★ North Carolina Technological Development Authority
PO Box 13169 Phone: (919) 733-7022
Durham, NC 27709-3169
Provides grants to establish incubator facilties for small firms and stimulates the development of new and existing small businesses. Also oversees the Innovation Research Fund.

★788★ Regional Small Business Incubator Facility
Roanoke-Chowan Technical College Phone: (919) 332-4042
Rte. 2, Box 46-A
Ahoskie, NC 27910

Ohio

★789★ Akron Industrial Incubator
58 W. Center St. Phone: (216) 375-2173
Akron, OH 44308

★790★ Barberton Incubator
Barberton Community Development Phone: (216) 745-3141
    Corp.
503 W. Park Ave.
Barberton, OH 44203

Oklahoma

★791★ Atoka Industrial Incubator
Atoka Kiamichi Area Vo-Tech
School-Atoka
PO Box 220
Atoka, OK 74525-0220
Phone: (405) 889-7321

★792★ Bryan County Skill Center Incubator
810 Waldron Dr.
Durant, OK 74701
Phone: (405) 924-7081

★793★ Canadian County Commercial Incubator
3901 Valley Park Dr.
El Reno, OK 73036
Phone: (405) 262-7197

★794★ Central Oklahoma Business and Job Development Corp. Incubator
3 Court Cir.
Drumright, OK 74030
Phone: (918) 352-2551

★795★ Entrepot Incubator
10810 E. 45th St., Ste. 400
Tulsa, OK 74146
Phone: (918) 622-3999

★796★ Hugo Industrial Incubator
Kiamichi Area Vo-Tech School-Hugo
107 S. 15th
Hugo, OK 74743
Phone: (405) 326-6491

★797★ Kiamichi Area Vo-Tech-Idabel Incubator
Rte. 3, Box 177
Idabel, OK 74745

★798★ Kiamichi Area Vo-Tech-Stigler Incubator
PO Box 488
Stigler, OK 74462
Phone: (918) 967-2493

★799★ McAlester Economic Development Service, Inc. Incuabtor
1628 S. George Nigh Expy.
McAlester, OK 74501
Phone: (918) 423-5735

★800★ McAlester Industrial Incubator
Kiamichi Area Vo-Tech
School-McAlester
PO Box 308
McAlester, OK 74502
Phone: (918) 426-0940

★801★ New Ventures, Inc. Incubator
Sullivan Industrial Park
1900 Ray Davis Blvd.
Seminole, OK 74868
Phone: (405) 382-7450

★802★ Oklahoma City Innovation Center Ltd. Incubator
101 Park Ave., Ste. 500
Oklahoma City, OK 73102
Phone: (405) 235-3127

★803★ Rural Enterprises Incubator
10 Waldron Dr.
Durant, OK 74701
Phone: (405) 924-5094

★804★ Tri-County Business Assistance Center Incubator
PO Box 3428
Bartlesville, OK 74006
Phone: (918) 333-3422

★805★ Tulsa Innovation Center Incubator
1216 N. Lansing Ave.
Tulsa, OK 74106
Phone: (918) 596-2600

Pennsylvania

★806★ Altoona Business Incubator
6th Ave. & 45th St.
Altoona, PA 16602
Phone: (814) 949-2030

★807★ East Liberty Incubator
Center for Entreprenuerial Development, Inc.
120 S. Whitfield St.
Pittsburgh, PA 15206
Phone: (412) 361-5000

★808★ Greenville Business Incubator
12 N. Diamond St.
Greenville, PA 16125
Phone: (412) 588-1161

★809★ Meadville Industrial Incubator
Meadville Area Industrial Commission
RD-2 Dunham Rd.
Meadville, PA 16335
Phone: (814) 337-8200

Vermont

★810★ North Bennington Business Incubator
Bennington County Industrial Corp. (BCIC)
PO Box 357
North Bennington, VT 05257
Phone: (802) 442-8975

Washington

★811★ Spokane Business Incubator
S. 3707 Godfrey Blvd.
Spokane, WA 99204-5753
Phone: (509) 458-6340
Fax: (509) 623-4276

Section 17

Private Sector Venture Capital Sources

*"Never bring up the artillery until you bring up the
ammunition."—Anonymous*

Venture capitalists and inventors have much in common. High rollers and risk takers in the finest American tradition, they live on the edge, pushing their respective envelopes in the hopes of generating technological advancement coupled with great financial reward. It is this blend of business acumen and inventive thinking that will help keep the U.S. a nation of suppliers rather than buyers of new technologies.

Banks are creditors. They're interested in the immediate future, yet most heavily influenced by the past. Loan officers examine the product and market position of the company for assurance that the invention can return a steady flow of sales and generate enough cash to repay the loan and interest.

Venture capital companies are owners. They gamble on the future. Venture capitalists can provide the money and management skills that permit inventors to bring their inventions to fruition and market. Inventors, in turn, give the venture capitalists an opportunity to get what they seek, typically a three to five times return on their investment within five to seven years.

Because their investment is unprotected in the event of failure, most venture capital firms set rigorous policies for venture proposal size, maturity of the seeking company, and requirements and evaluation procedures to reduce risks.

Projects requiring under $250,000 are of limited interest because of the high cost of investigation and administration; however, some venture firms will consider smaller proposals, if the investment is intriguing enough. Most venture capitalists live in the $250,000 to $1.5 million atmosphere.

Experienced in putting together "marriages" between good ideas and good money, these wheeling, deal-makers have relationships with a wide variety of potential investors such as corporations, insurance companies, union pension funds, university endowments, and wealthy individuals. Yet, unlike passive investors and traditional lenders, venture capitalists take a hands-on proactive role in managing the companies in which they invest. In the end, investors invest in the venture capitalist as much as in any invention they may represent.

While the inventor works on product development, the venture capitalist gets the money to keep the pump primed and prepares long-term corporate business programs and marketing and personnel recruitment plans. Most inventors are not skilled at these activities and welcome the assistance.

The inventor's motivation most often involves more than money, i.e. things like ego and pride. The venture capitalist is in it typically for money alone, the invention being a vehicle to this end. For the relationship to work the parties must be highly compatible. The chemistry between inventor and venture capitalist is, therefore, a major factor in how well their business will develop. Inventors should look at venture capitalists as working partners and not just investors. The relationship begins with the ritual of "contact" which leads to "courtship" and then to "investigation."

According to the Small Business Administration (SBA), the typical venture capital company receives over 1,000 proposals a year. Probably 90 percent will be rejected quickly because they don't fit the established geographical, technical, or market area policies of the firm, or because they have been poorly presented.

The remaining 10 percent are investigated carefully. The venture capitalist may spend between $2,000 and $3,000 per company to hire consultants that will make preliminary investigations. According to the SBA, these result in 10 to 15 proposals of interest. Then, second investigations, more thorough and more expensive than the first, reduce the number of companies under active consideration to only three or four. Eventually the firm invests in one or two of these.

The entire process can take anywhere from three to six months. Before any attempts are made to raise money, the venture capitalist and inventor had best become very familiar with each other. Annulments are rare. Divorces can be messy.

The best way to check out the capabilities of venture capitalists is to speak with the management of other companies the venture capitalist has set-up. And don't look only into successful enterprises, but find people who have experienced total flops to see how the venture capitalist handles both success and failure. Any seasoned venture capitalist will have had a share of good and bad deals.

An inventor would be well advised to engage the services of an attorney and or CPA before signing on with any venture capitalist. The money you spend will be money well spent if it gets you a good deal or saves you from a bad one.

Section Arrangement

The following entries are arranged geographically by state, then alphabetically by venture capital source name. All entries in this section are are referenced in the Master Index.

Alabama

★812★ Alabama Capital Corp.
16 Midtown Pk., E.   Phone: (205) 476-0700
Mobile, AL 36606   Fax: (205) 476-0026

A minority enterprise small business investment corporation. No industry preference.

★813★ First SBIC of Alabama
16 Midtown Pk., E.   Phone: (205) 476-0700
Mobile, AL 36606   Fax: (205) 476-0026

A small business investment corporation. No industry preference.

★814★ Hickory Venture Capital Corp.
200 Westside Sq., Ste. 100   Phone: (205) 539-1931
Huntsville, AL 35801-4816   Fax: (205) 539-5130

A small business investment corporation. No industry preference.

Alaska

★815★ Calista Business Investment Corp.
601 W. 5th Ave., Ste. 200   Phone: (907) 279-5516
Anchorage, AK 99501   Fax: (907) 272-5060

A minority enterprise small business investment corporation. No industry preference.

Arizona

★816★ Kayenta Regional Business Development Office
PO Box 545   Phone: (602) 697-3572
Kayenta, AZ 86033   Fax: (602) 697-3206

★817★ Rocky Mountain Equity Corp.
2525 E. Camelback Rd., Ste. 275   Phone: (602) 955-6100
Phoenix, AZ 85016   Fax: (602) 956-5909

A small business investment corporation. No industry preference.

★818★ Valley National Investors, Inc.
201 N. Central Ave., A-835   Phone: (602) 261-1577
Phoenix, AZ 85004   Fax: (602) 261-1734

A small business investment corporation. Industry preferences include manufacturing and proprietary technology with middle market acquisitions.

Arkansas

★819★ Capital Management Services, Inc.
1910 N. Grant St., Ste. 200   Phone: (501) 664-8613
Little Rock, AR 72207

A minority enterprise small business investment corporation. No industry preference.

★820★ Power Ventures, Inc.
829 Hwy. 270, N.   Phone: (501) 332-3695
Malvern, AR 72104   Fax: (501) 337-4393

A minority enterprise small business investment corporation. No industry preference.

★821★ Small Business Investment Capital, Inc.
PO Box 3627   Phone: (501) 455-3590
Little Rock, AR 72203

A small business investment corporation. Retail grocery industry preferred.

California

★822★ ABC Capital Corp.
PO Box 300   Phone: (818) 355-3577
Arcadia, CA 91066   Fax: (818) 355-5577

A minority enterprise small business investment corporation. No industry preference.

★823★ Accel Partners (San Francisco)
1 Embarcadero Center, Ste. 3820   Phone: (415) 989-5656
San Francisco, CA 94111   Fax: (415) 989-5554

Private venture capital firm. Prefers to invest in telecommunications, software, or biotechnology/medical products industries.

★824★ Adler and Co. (San Bruno)
950 Elm Ave., Ste. 330   Phone: (415) 589-8320
San Bruno, CA 94066   Fax: (415) 589-3921

Private venture capital supplier. Provides all stages of financing.

★825★ Advanced Technology Ventures (Menlo Park)
1000 El Camino Real, Ste. 360   Phone: (415) 321-8601
Menlo Park, CA 94025-4327

★826★ Alan Patricof Associates, Inc. (Palo Alto)
1 Embarcadero Pl.   Phone: (415) 494-9944
2100 Geng Rd., Ste. 220   Fax: (415) 494-6751
Palo Alto, CA 94303

Prefers diversified equity investments in both technology and nontechnology.

★827★ Allied Business Investors, Inc.
428 S. Atlantic Blvd., Ste. 201   Phone: (818) 289-0186
Monterey Park, CA 91754   Fax: (818) 289-2369

A minority enterprise small business investment corporation. No industry preference.

★828★ Ally Finance Corp.
9100 Wilshire Blvd., Ste. 408   Phone: (213) 550-8100
Beverly Hills, CA 90212   Fax: (213) 550-6136

A small business investment corporation. No industry preference.

★829★ Apple Computer, Inc.
20525 Mariani Ave., M/S 38-G   Phone: (408) 974-5686
Cupertino, CA 95014

★830★ Asian American Capital Corp.
1251 W. Tennyson Rd., Ste. 4   Phone: (415) 887-6888
Hayward, CA 94544

A minority enterprise small business investment corporation. Diversified industry preferences.

★831★ Asset Management Co.
2275 E. Bayshore, Ste. 150   Phone: (415) 494-7400
Palo Alto, CA 94303

Venture capital firm. High-technology industries preferred.

★832★ Associates Venture Capital Corp.
300 Montgomery St., Ste. 421   Phone: (415) 956-1444
San Francisco, CA 94104

No industry preference.

★833★ Bain Capital
1 Embarcadero Center, Ste. 3400   Phone: (415) 434-1022
San Francisco, CA 94111-3723

★834★ Bay Partners
10600 N. De Anza Blvd., Ste. 100   Phone: (408) 725-2444
Cupertino, CA 95014   Fax: (408) 446-4502

Venture capital supplier. Provides start-up financing primarily to West Coast technology companies that have highly qualified management teams. Initial investments range from $100,000 to $800,000. Where large investments are required, the company will act as lead investor to bring in additional qualified venture investors.

★835★ Bay Venture Group
1 Sansome St., Ste. 2000   Phone: (415) 951-4674
San Francisco, CA 94104-4433

A small business investment corporation. High-technology industries preferred; geographic region limited to the San Francisco Bay area.

★836★ Bessemer Venture Partners (Menlo Park)
3000 Sand Hill Rd., Bldg. 3, Ste. 225   Phone: (415) 854-2200
Menlo Park, CA 94025   Fax: (415) 854-7415

★837★ Brentwood Associates (Irvine)
1920 Main St., Ste. 820   Phone: (714) 251-1010
Irvine, CA 92714   Fax: (714) 251-1011

★838★ Brentwood Associates (Los Angeles)
11150 Santa Monica Blvd., Ste. 1200   Phone: (213) 477-6611
Los Angeles, CA 90025-3314   Fax: (213) 477-1011

Venture capital supplier. Provides start-up and expansion financing to technology-based enterprises specializing in computing and data processing, electronics, communications, materials, energy, industrial automation, and bioengineering and medical equipment. Investments generally range from $1 to $3 million.

★839★ Brentwood Associates (Menlo Park)
3000 Sand Hill Road, Bldg. 3, Ste. 260   Phone: (415) 854-7691
Menlo Park, CA 94025   Fax: (415) 854-9513

★840★ Bryan and Edwards (Menlo Park)
3000 Sand Hill Rd., Bldg. 1, Ste. 190   Phone: (415) 854-1555
Menlo Park, CA 94025   Fax: (415) 854-5015

★841★ Bryan and Edwards (San Francisco)
600 Montgomery St., 35th Fl.   Phone: (415) 421-9990
San Francisco, CA 94111   Fax: (415) 421-0471

A small business investment corporation. No industry preference.

★842★ Burr, Egan, Deleage, and Co. (Costa Mesa)
650 Town Center Dr., Ste. 580   Phone: (714) 557-9250
Costa Mesa, CA 92626   Fax: (714) 557-1231

★843★ Burr, Egan, Deleage, and Co. (San Francisco)
1 Embarcadero Center, 40th Fl.   Phone: (415) 362-4022
San Francisco, CA 94111   Fax: (415) 362-6178

Private venture capital supplier. Invests start-up, expansion, and acquisitions capital nationwide. Principal concerns are strength of the management team; large, rapidly expanding markets; and unique products for services. Past investments have been made in the fields of biotechnology and pharmaceuticals, CATV, communications, software, computer systems and peripherals, distributorships, RCCs, electronics and electrical components, environmental control, health services, medical devices and instrumentation, and radio and cellular telecommunications.

★844★ Cable and Howse Ventures (Palo Alto)
435 Tasso St., Ste. 115   Phone: (415) 322-8400
Palo Alto, CA 94301   Fax: (415) 322-6487

Venture capital supplier. Provides start-up and early stage financing to enterprises in the western United States, although a national perspective is maintained. Interested in proprietary or patentable technology. Investments range from $500,000 to $2 million.

★845★ Campbell Venture Management
375 Forest Ave.   Phone: (415) 853-0766
Palo Alto, CA 94301-2521

★846★ Canaan Venture Partners (Menlo Park)
2884 Sand Hill Rd., Ste. 115   Phone: (415) 854-8092
Menlo Park, CA 94025-7022

Venture capital supplier. Will consider all stages of investment (seed, development, and expansion) and all industry sectors.

★847★ Charterway Investment Corp.
222 S. Hill St., Ste. 800   Phone: (213) 687-8539
Los Angeles, CA 90012   Fax: (213) 626-8238

A minority enterprise small business investment corporation. No industry preference.

★848★ Comdisco, Inc. Venture Leasing (Menlo Park)
3000 Sand Hill Rd., Bldg. 2, Ste. 210   Phone: (415) 854-9484
Menlo Park, CA 94025   Fax: (415) 854-4026

Prefers start-up businesses in fields of semiconductors, computer hardware and software, computer services and systems, telecommunications, and medical biotechnology. Investments range from $500,000 to $5 million.

★849★ Comdisco, Inc. Venture Leasing (San Francisco)
101 California St., 38th Fl.   Phone: (415) 421-1800
San Francisco, CA 94111   Fax: (415) 421-1800

Prefers start-up businesses in fields of semiconductors, computer hardware and software, computer services and systems, telecommunications, and medical biotechnology. Investments range from $500,000 to $5 million.

★850★ Concord Partners (Palo Alto)
435 Tasso St., Ste. 305   Phone: (415) 327-2600
Palo Alto, CA 94301

Venture capital supplier. Diversified in terms of stage of development, industry classification, and geographic location. Areas of special interest include computers, telecommunications, health care and medical products, energy, and leveraged buy outs. Preferred investments range from $2 to $3 million, with a $1 million minimum and $10 million maximum.

★851★ Continental Investors, Inc.
8781 Seaspray Dr.   Phone: (714) 964-5207
Huntington Beach, CA 92646

A minority enterprise small business investment corporation. No industry preference.

★852★ Crosspoint Investment Corp.
1 1st St., Ste. 2   Phone: (415) 948-8300
Los Altos, CA 94022

A small business investment corporation. No industry preference.

★853★ Crosspoint Venture Partners (Irvine)
18552 MacArthur, No. 400   Phone: (714) 852-1611
Irvine, CA 92715   Fax: (714) 852-9804

★854★ Crosspoint Venture Partners (Los Altos)
1 1st St., Ste. 2   Phone: (415) 948-8300
Los Altos, CA 94022

Venture capital partnership. Seeks to invest start-up capital in unique products, services, and/or market opportunities located in the western United States. Primary interests lie in communication devices and systems; biotechnology; medical products, instruments, and equipment; computer software and productivity tools; computers and computer peripherals; industrial automation and controls; semiconductor devices and equipment; instrumentation; and related service and distribution businesses. Investments range from $50,000 to $3 million.

★855★ Developers Equity Capital Corp.
1880 Century Pk., E., Ste. 311   Phone: (213) 277-0330
Los Angeles, CA 90067

A small business investment corporation. Real estate preferred.

★856★ Dougery, Wilder and Howard (San Mateo)
155 Bovet Rd., Ste. 650   Phone: (415) 358-8701
San Mateo, CA 94402-3113   Fax: (415) 358-8706

Venture capital supplier. Prefers to invest in small to medium-sized companies headquartered or primarily operating in the west and southwest. The typical company sought is in the computers/communications or medical/biotechnology industries, is privately owned, has an experienced management team, serves growth markets, and can achieve at least $30 million in sales. Initial investment ranges from $250,000 to $2.5 million; and may be expanded by including other venture capital firms.

★857★ DSV Partners (Newport Beach)
620 Newport Center Dr., Ste. 990   Phone: (714) 759-5657
Newport Beach, CA 92660

Prefers investing in early-stage technology companies.

★858★ El Dorado Technology Partners
2 N. Lake Ave., Ste. 480   Phone: (818) 793-1936
Pasadena, CA 91101   Fax: (818) 793-2613

Private venture capital firm. Prefers to invest in the communications, electronics, industrial products and services, and medical/health care industries.

★859★ El Dorado Ventures (Cupertino)
20300 Stevens Creek Blvd., Ste. 395   Phone: (408) 725-2474
Cupertino, CA 95014   Fax: (408) 252-2762

★860★ El Dorado Ventures (Pasadena)
2 N. Lake Ave., Ste. 480   Phone: (818) 793-1936
Pasadena, CA 91101

Private venture capital firm. Prefers to invest in the communications, electronics, industrial products and services, and medical/health care industries.

★861★ Enterprise Partners
5000 Birch St., Ste. 6200   Phone: (714) 833-3650
Newport Beach, CA 92660   Fax: (714) 955-2508

Venture capital fund. Prefers electronics, medical technology, and health care ventures in southern California.

★862★ Equitable Capital Corp.
855 Sansome St., 2nd Fl.   Phone: (415) 434-4114
San Francisco, CA 94111   Fax: (415) 434-0479

A minority enterprise small business investment corporation. No industry preference.

★863★ Fairfield Venture Partners (Costa Mesa)
650 Town Center Dr., Ste. 810   Phone: (714) 754-5717
Costa Mesa, CA 92626   Fax: (714) 754-6802

★864★ First American Capital Funding, Inc.
38 Corporate Pk., Ste. B   Phone: (714) 660-9288
Irvine, CA 92714   Fax: (714) 660-0119

A minority enterprise small business investment corporation. No industry preference.

★865★ First Century Partners (Menlo Park)
3000 Sand Hill Rd., Bldg. 2, Ste. 220   Phone: (415) 854-4025
Menlo Park, CA 94025   Fax: (415) 854-4063

★866★ First Century Partners (San Francisco)
350 California St., 21st Fl.   Phone: (415) 955-1612
San Francisco, CA 94104

Private venture capital firm. High-technology, medical, and specialty retail industries are preferred; has no interest in real estate or entertainment investments. Minimum investment is $1 million.

★867★ First SBIC of California (Costa Mesa)
650 Town Center Dr., 17th Fl.   Phone: (714) 556-1964
Costa Mesa, CA 92626   Fax: (714) 546-8021

A small business investment corporation and venture capital company. No industry preference.

★868★ First SBIC of California (Palo Alto)
Security Pacific Capital   Phone: (415) 424-8011
2400 Sand Hill Rd., Ste. 100   Fax: (415) 324-6830
Menlo Park, CA 94025

A small business investment corporation. No industry preference.

★869★ First SBIC of California (Pasadena)
2 N. Lake St., Ste. 940   Phone: (818) 304-3451
Pasadena, CA 91101   Fax: (818) 440-9931

A small business investment corporation. No industry preference.

★870★ G C and H Partners
1 Maritime Plaza, 20th Fl.   Phone: (415) 981-5252
San Francisco, CA 94111-3580   Fax: (415) 951-3699

A small business investment corporation. No industry preference.

★871★ Glenwood Management
3000 Sand Hill Rd., Bldg. 4, Ste. 230   Phone: (415) 854-8070
Menlo Park, CA 94025   Fax: (415) 854-4961

Venture capital supplier. Provides start-up and strategic financing to high-technology companies.

★872★ Grace Ventures Corp./Horn Venture Partners
20300 Stevens Creek Blvd., Ste. 330   Phone: (408) 725-0774
Cupertino, CA 95014   Fax: (408) 725-0327

★873★ Greylock Management Corp. (Palo Alto)
755 Page Mill Rd., Ste. B-140   Phone: (415) 493-5525
Palo Alto, CA 94304-1007

★874★ Hall, Morris & Drufva
5000 Birch St., Ste. 10100   Phone: (714) 253-4360
Newport Beach, CA 92660

A small business investment corporation. No industry preference. Provides capital for small and medium-sized companies through participation in private placements of subordinated debt and preferred

and common stock. Offers growth-acquisition and later-stage venture capital.

★875★ **Hambrecht and Quist Co.**
1 Bush St.   Phone: (415) 986-5500
San Francisco, CA 94104   Fax: (415) 576-3624
A small business investment corporation. High-technology industries preferred.

★876★ **Hambrecht and Quist Venture Partners**
1 Bush St.   Phone: (415) 576-3300
San Francisco, CA 94104   Fax: (415) 576-3624
Venture capital firm. Prefers start-up investments in high-technology and biotechnology industries. Investments range from $500,000 to $5 million.

★877★ **Harvest Ventures, Inc. (Cupertino)**
19200 Stevens Creek Blvd., Ste. 220   Phone: (408) 996-3200
Cupertino, CA 95014   Fax: (408) 996-1765
Private venture capital supplier. Prefers to invest in high-technology, growth-oriented companies with proprietary technology, large market potential, and strong management teams. Specific areas of interest include computers, computer peripherals, semiconductors, telecommunications, factory automation, military electronics, medical products, and health care services. Provides seed capital of up to $250,000; investments in special high growth and leveraged buy out situations range from $500,000 to $2 million. Will serve as lead investor and arrange financing in excess of $2 million in association with other groups.

★878★ **Helio Capital, Inc.**
624 S. Grand Ave., Ste. 2700   Phone: (213) 721-8053
Los Angeles, CA 90017-3328   Fax: (213) 622-3582
A minority enterprise small business investment corporation. No industry preference.

★879★ **Hewlett-Packard Co., Corporate Investments**
3000 Hanover St.   Phone: (415) 857-2314
Palo Alto, CA 94304   Fax: (415) 857-7113

★880★ **HMS Group**
170 Middlefield Rd., Ste. 150   Phone: (415) 324-4672
Menlo Park, CA 94025   Fax: (415) 324-4684
Prefers communications industries.

★881★ **Idanta Partners**
10975 Torreyana Rd., Ste. 304   Phone: (619) 452-9690
San Diego, CA 92121-1114
Venture capital partnership. Provides start-up and second-stage financing; will also invest in special situations such as leveraged buy outs.

★882★ **Imperial Ventures, Inc.**
PO Box 92991   Phone: (213) 417-5830
Los Angeles, CA 90009
A small business investment corporation. Prefers low to medium technology companies located in California earning a minimal pre-tax profit of $500,000.

★883★ **Inman and Bowman**
4 Orinda Way, Bldg. D, Ste. 150   Phone: (415) 253-1611
Orinda, CA 94563

★884★ **Institutional Venture Partners**
3000 Sand Hill Rd., Bldg. 2, Ste. 290   Phone: (415) 854-0132
Menlo Park, CA 94025   Fax: (415) 854-5762
Venture capital fund. Invests in early-stage ventures with significant market potential in the computer, information sciences, communications, and life sciences fields.

★885★ **Interscope Investments, Inc.**
10900 Wilshire Blvd., Ste. 1400   Phone: (213) 208-8525
Los Angeles, CA 90024   Fax: (213) 208-1764
Venture capital firm. No industry preference.

★886★ **InterVen Partners**
800 S. Figueroa St., Ste. 760   Phone: (213) 622-1922
Los Angeles, CA 90017   Fax: (213) 622-9035
Venture capital fund. Diversified industry preferences; West Coast only.

★887★ **Interwest Partners (Menlo Park)**
3000 Sand Hill Rd., Bldg. 3, Ste. 255   Phone: (415) 854-8585
Menlo Park, CA 94025
Venture capital fund. Both high-tech and low- or non-technology companies are considered. No oil, gas, real estate, or construction projects.

★888★ **Ivanhoe Venture Capital Ltd.**
737 Pearl St., Ste. 201   Phone: (619) 454-8881
La Jolla, CA 92037   Fax: (619) 454-8880
A small business investment corporation. No industry preference.

★889★ **J. H. Whitney and Co. (Menlo Park)**
3000 Sand Hill Rd., No. 1-270   Phone: (415) 854-0500
Menlo Park, CA 94025   Fax: (415) 854-5447

★890★ **Jafco America Ventures, Inc. (San Francisco)**
555 California St., Ste. 2450   Phone: (415) 788-0706
San Francisco, CA 94104   Fax: (415) 788-0709

★891★ **Julian, Cole and Stein**
11777 San Vicente Blvd., Ste. 604   Phone: (213) 826-8002
Los Angeles, CA 90049   Fax: (213) 820-5805

★892★ **Kleiner, Perkins, Caufield, and Byers (Palo Alto)**
2 Embarcadero Pl.   Phone: (415) 424-1660
2200 Geng Rd., Ste. 205   Fax: (415) 856-2760
Palo Alto, CA 94303

★893★ **Kleiner, Perkins, Caufield, and Byers (San Francisco)**
4 Embarcadero Center, Ste. 3520   Phone: (415) 421-3110
San Francisco, CA 94111   Fax: (415) 421-3128
Provides seed, start-up, second- and third-round, and bridge financing to companies on the West Coast. Past investments have been made in the following fields: computers and computer peripherals, software, office equipment, medical products and instruments, microbiology, genetic engineering, telecommunications, instrumentation, semiconductors, lasers and optics, and unique consumer products and services. Areas avoided for investment include real estate, motion pictures, solar energy, hotels and motels, restaurants, resort areas, oil and gas, construction, and metallurgy. Investments range from $1 to $5 million.

★894★ **Marwit Capital Corp.**
180 Newport Center Dr., Ste. 200   Phone: (714) 640-6234
Newport Beach, CA 92660   Fax: (714) 759-1363
A small business investment corporation. Varied industry preferences, but no start-ups or retail.

★895★ **Matrix Partners**
2500 Sand Hill Rd., Ste. 113   Phone: (415) 854-3131
Menlo Park, CA 94025   Fax: (415) 854-3296
Private venture capital partnership. Investments range from $500,000 to $1 million.

★896★ Mayfield Fund
2200 Sand Hill Rd., Ste. 200
Menlo Park, CA 94025
Phone: (415) 854-5560
Fax: (415) 854-5712

Venture capital partnership. Prefers high technology and biomedical fields.

★897★ MBW Management, Inc. (Los Altos)
350 2nd St., Ste. 7
Los Altos, CA 94022
Phone: (415) 941-2392
Fax: (415) 941-2865

★898★ McCown De Leeuw and Co. (Menlo Park)
3000 Sand Hill Rd., Bldg. 3, Ste. 290
Menlo Park, CA 94025
Phone: (415) 854-6000
Fax: (415) 854-0853

★899★ Menlo Ventures
3000 Sand Hill Rd., Bldg. 4, Ste. 100
Menlo Park, CA 94025
Phone: (415) 854-8540

Venture capital supplier. Provides start-up and expansion financing to companies with experienced management teams, distinctive product lines, and large growing markets. Primary interest is in technology-oriented, service, consumer products, and distribution companies. Investments range from $500,000 to $3 million; also provides capital for leveraged buy outs.

★900★ Merrill, Pickard, Anderson, and Eyre (Menlo Park)
2480 Sand Hill Rd., Ste. 200
Menlo Park, CA 94025
Phone: (415) 856-8880
Fax: (415) 856-2571

Private venture capital partnership. Provides start-up and early stage financing to companies with experienced management teams, with the ability to grow to $50 to $100 in annual revenues within four-to-six years and with distinctive product lines. High-technology industries are preferred. The company is not interested in financial, real estate, or consulting companies. Investments range from $750,000 to $2.5 million.

★901★ MK Global Ventures
2471 E. Bayshore Rd., Ste. 520
Palo Alto, CA 94303
Phone: (415) 424-0151
Fax: (415) 494-2753

★902★ Mohr, Davidow Ventures
3000 Sand Hill Rd., Bldg. 1, Ste. 240
Menlo Park, CA 94025
Phone: (415) 854-7236

★903★ Montgomery Securities
600 Montgomery St.
San Francisco, CA 94111
Phone: (415) 627-2454
Fax: (415) 627-2909

Private venture capital and investment banking firm. Diversified, but will not invest in real estate or energy-related industries. Involved in both start-up and later-stage financing.

★904★ Morgan Stanley Venture Partners L.P. (San Francisco)
555 California St.
San Francisco, CA 94104
Phone: (415) 576-2351

★905★ Myriad Capital, Inc.
328 S. Atlantic Blvd., Ste. 200-A
Monterey, CA 91754
Phone: (818) 570-4548
Fax: (818) 570-9570

A minority enterprise small business investment corporation. Prefers investing in production and manufacturing industries.

★906★ New Enterprise Associates (Menlo Park)
3000 Sand Hill Rd., Bldg. 4, Ste. 235
Menlo Park, CA 94025
Phone: (415) 854-2215
Fax: (415) 854-8083

Venture capital supplier. Concentrates on technology-based industries that have the potential for product innovation, rapid growth, and high profit margins. Investments range from $250,000 to $2 million. Past investments have been made in the following industries: computer software, medical and life sciences, computers and peripherals, communications, semiconductors, specialty retailing, energy and alternative energy, defense electronics, materials, and specialty chemicals. Management must demonstrate intimate knowledge of its marketplace and have a well-defined strategy for achieving strong market penetration.

★907★ New Enterprise Associates (San Francisco)
Russ Bldg., Ste. 1025
235 Montgomery St.
San Francisco, CA 94104
Phone: (415) 956-1579
Fax: (415) 981-4168

Venture capital supplier. Concentrates in technology-based industries that have the potential for product innovation, rapid growth, and high profit margins. Investments range from $250,000 to $2 million. Past investments have been made in the following industries: computer software, medical and life sciences, computers and computer peripherals, communications, semiconductors, specialty training, energy and alternative energy, defense electronics, materials, and specialty chemicals. Management must demonstrate intimate knowledge of its marketplace and have a well-defined strategy for achieving strong market penetration.

★908★ New Kukje Investment Co.
3670 Wilshire St., Ste. 418
Los Angeles, CA 90010
Phone: (213) 389-8679

A minority enterprise small business investment corporation. No industry preference.

★909★ New West Ventures (San Diego)
4350 Executive Dr., Ste. 206
San Diego, CA 92121
Phone: (619) 457-0722
Fax: (619) 457-0829

A small business investment corporation. No industry preference.

★910★ Oak Investment Partners (Menlo Park)
3000 Sand Hill Rd., Bldg. 3, Ste. 240
Menlo Park, CA 94025
Phone: (415) 854-8825
Fax: (415) 854-5259

Private venture capital firm. Prefers computer, biotechnology, and communications industries.

★911★ Olympic Venture Partners II (San Francisco)
101 California St., Ste. 4035
San Francisco, CA 94111
Phone: (415) 362-4433

Prefers companies located on the West Coast involved in high technology fields such as micro-electronics, biotechnology, communications, software, and medical instrumentation.

★912★ Opportunity Capital Corp.
39650 Liberty St., Ste. 425
Fremont, CA 94538
Phone: (415) 651-4412
Fax: (415) 651-4415

A minority enterprise small business investment corporation. No industry preference.

★913★ Oxford Partners (Santa Monica)
233 Wilshire Blvd., Ste. 830
Santa Monica, CA 90401
Phone: (213) 458-3135
Fax: (213) 394-0189

Independent venture capital partnership. Prefers to invest in high-technology industries, including biotechnology, health-related, computer-related, communications, medical, and electronic components and instrumentation.

★914★ Pacific Capital Fund, Inc.
675 Mariners' Island Blvd., Ste. 103
San Mateo, CA 94404
Phone: (415) 574-4747

A minority enterprise small business investment corporation. No industry preference.

★915★ Paragon Partners
3000 Sand Hill Rd., Bldg. 2, Ste. 190
Menlo Park, CA 94025
Phone: (415) 854-8000
Fax: (415) 854-7260

★916★ Pathfinder Venture Capital Funds (Menlo Park)
3000 Sand Hill Rd., Bldg. 3, Ste. 255   Phone: (415) 854-0650
Menlo Park, CA 94025

Venture capital supplier. Provides start-up and early-stage financing to emerging companies in the medical, computer, pharmaceuticals, and data communications industries. Emphasis is on companies with proprietary technology or market positions and with substantial potential for revenue growth.

★917★ Paul Capital Partners
2200 Sand Hill Rd., Ste. 240   Phone: (415) 854-4653
Menlo Park, CA 94025   Fax: (415) 854-8939

Venture capital firm. Prefers medical, biotechnology, and computer-related electronics industries.

★918★ PBC Venture Capital, Inc.
PO Box 6008   Phone: (805) 395-3555
Bakersfield, CA 93386   Fax: (805) 395-3443

A small business investment corporation. No industry preference. Area limited to California.

★919★ PCF Venture Capital Corp.
675 Mariners' Island Blvd., Ste. 103   Phone: (415) 574-4747
San Mateo, CA 94404

A small business investment corporation. No industry preference.

★920★ Positive Enterprises, Inc.
1166 Post St., Ste. 200   Phone: (415) 885-6600
San Francisco, CA 94109   Fax: (415) 928-6363

A minority enterprise small business investment corporation. No industry preference.

★921★ Quest Ventures (Menlo Park)
3000 Sand Hill Rd., Ste. 3-255   Phone: (415) 989-2020
Menlo Park, CA 94025

★922★ Quest Ventures (San Francisco)
555 California St., Ste. 2840   Phone: (415) 989-2020
San Francisco, CA 94104   Fax: (415) 394-9291

Independent venture capital partnership.

★923★ R and D Funding Corp.
3945 Freedom Cir., Ste. 800   Phone: (408) 980-0990
Santa Clara, CA 95054   Fax: (408) 980-1702

Venture capital firm. Invests in high-growth businesses. Direct investment in research and development.

★924★ Robertson-Stephens, Co.
1 Embarcadero, Ste. 3100   Phone: (415) 781-9700
San Francisco, CA 94111   Fax: (415) 781-0278

Investment banking firm. Considers investments in any attractive merging-growth area, including product and service companies. Key preferences include health care, hazardous waste services and technology, biotechnology, software, and information services. Maximum investment is $5 million.

★925★ Round Table Capital Corp.
655 Montgomery St., Ste. 700   Phone: (415) 392-7500
San Francisco, CA 94111   Fax: (415) 362-7967

A small business investment corporation. No industry preference.

★926★ RSC Financial Corp.
501 Ojai Ave.   Phone: (805) 646-2925
Ojai, CA 93023

A minority enterprise small business investment corporation. No industry preference.

★927★ San Joaquin Capital Corp.
PO Box 2538   Phone: (805) 323-7581
Bakersfield, CA 93303

A small business investment corporation. No industry preference. Dealings occur in the $100,000-$200,000 range.

★928★ SAS Associates
515 S. Figueroa St., Ste. 600   Phone: (213) 624-4232
Los Angeles, CA 90071-3396   Fax: (213) 688-1431

★929★ Security Pacific Capital Corp. (Costa Mesa)
650 Town Center Dr., 17th Fl.   Phone: (714) 556-1964
Costa Mesa, CA 92626   Fax: (714) 546-8021

A small business investment corporation. No industry preference.

★930★ Security Pacific Venture Capital Group
2 N. Lake Ave., Ste. 940   Phone: (818) 304-3451
Pasadena, CA 91101   Fax: (818) 440-9931

A small business investment corporation. Does not invest in real estate, construction, entertainment, or publishing ventures.

★931★ Sequoia Capital
3000 Sand Hill Rd., Bldg. 4, Ste. 280   Phone: (415) 854-3927
Menlo Park, CA 94025   Fax: (415) 854-2977

Private venture capital partnership with $300 million under management. Provides financing for all stages of development of well-managed companies with exceptional growth prospects in fast-growth industries. Past investments have been made in computers and peripherals, communications, health care, biotechnology, and medical instruments and devices. Investments range from $350,000 for early stage companies to $4 million for late stage accelerates.

★932★ Sierra Ventures
3000 Sand Hill Rd., Bldg. 1, Ste. 280   Phone: (415) 854-1000
Menlo Park, CA 94025   Fax: (415) 854-5593

Venture capital partnership. Diversified, with interests in early-stage financing. Prior investment experience in technology-related service businesses, environmental technology, telecommunications, semiconductors, and computer companies.

★933★ Sigma Partners
2884 Sand Hill Rd., Ste. 121   Phone: (415) 854-1300
Menlo Park, CA 94025-7022   Fax: (415) 854-1323

★934★ Southern California Ventures (Irvine)
19800 MacArthur Blvd., Ste. 830   Phone: (714) 752-9341
Irvine, CA 92715

Prefers investing with biomedical, high-technology, and computer telecommunications companies.

★935★ Spectra Enterprise Associates
200 N. Westlake Blvd., Ste. 215   Phone: (805) 373-8537
Westlake Village, CA 91362

Venture capital partnership. Early-stage investors, with initial investments ranging from $500,000 to $2 million.

★936★ Sprout Group (Menlo Park)
3000 Sand Hill Rd., Bldg. 3, Ste. 245   Phone: (415) 854-2300
Menlo Park, CA 94025   Fax: (415) 854-1025

★937★ Summit Partners L.P.
4675 MacArthur Ct.   Phone: (714) 476-2700
Newport Beach, CA 92660   Fax: (714) 476-5074

★938★ Sunwestern Investment Group (Rancho Santa Fe)
Box 1506   Phone: (619) 756-1469
Rancho Santa Fe, CA 92067   Fax: (619) 259-0470

★939★ Sutter Hill Ventures
755 Page Mill Rd., Ste. A-200 Phone: (415) 493-5600
Palo Alto, CA 94304 Fax: (415) 858-1854
Venture capital partnership. Provides seed early-stage financing to high-tech companies.

★940★ TA Associates (Palo Alto)
435 Tasso St. Phone: (415) 328-1210
Palo Alto, CA 94301 Fax: (415) 326-4933
Private venture capital firm. Prefers technology companies and leveraged buy outs. Provides from $1 to $20 million in investments.

★941★ Taylor and Turner
220 Montgomery St., Penthouse No. 10 Phone: (415) 398-6821
San Francisco, CA 94104 Fax: (415) 398-3220

★942★ TBM Associates, Inc.
Westerly Pl. Phone: (714) 752-7811
1500 Quail St., Ste. 540 Fax: (714) 752-8519
Newport Beach, CA 92660
A small business investment corporation. Provides all stages of financing to technology-based and oil and gas service industries; no geographic preference. Experience of management team and high growth market potential are emphasized. Investments range from $500,000 to $5 million.

★943★ Technology Funding, Inc.
2000 Alameda de las Pulgas, Ste. 250 Phone: (415) 345-2200
San Mateo, CA 94403 Fax: (415) 345-1797
Private venture capital supplier. Provides primarily late first-stage and early second-stage equity financing. Also offers secured debt with equity participation to venture capital backed companies. Investments range from $500,000 to $1 million.

★944★ Technology Venture Investors
3000 Sand Hill Rd., Bldg. 4, Ste. 210 Phone: (415) 854-7472
Menlo Park, CA 94025 Fax: (415) 854-4187
Private venture capital partnership. Primary interest is in technology companies, a with minimum investment of $1 million.

★945★ 3i Ventures (Menlo Park)
3000 Sand Hill Rd., Bldg. 3, Ste. 105 Phone: (415) 854-3330
Menlo Park, CA 94025 Fax: (415) 854-4044
Venture capital supplier. Provides start-up and early-stage financing to companies in high-growth fields such as microelectronics, computers, telecommunications, biotechnology, health sciences, and industrial automation. Investments generally range from $500,000 to $2 million.

★946★ Trinity Ventures Ltd.
155 Bovet Rd., Ste. 700 Phone: (415) 358-9700
San Mateo, CA 94402 Fax: (415) 358-9785
Venture capital fund. Provides up to $3 million in seed capital to third-round financing for West Coast companies. Emphasis is on high technology, with flexibility to enter all areas at all stages of development. Will not consider real estate, oil and gas, or construction ventures.

★947★ Union Venture Corp.
445 S. Figueroa St. Phone: (213) 236-4092
Los Angeles, CA 90071 Fax: (213) 688-0101

★948★ U.S. Venture Partners
2180 Sand Hill Rd, Ste. 300 Phone: (415) 854-9080
Menlo Park, CA 94025 Fax: (415) 854-3018
Venture capital partnership. Prefers the specialty retail, consumer products, technology, and biomedical industries.

★949★ USVP-Schlein Marketing Fund
2180 Sand Hill Rd., Ste. 300 Phone: (415) 854-9080
Menlo Park, CA 94025 Fax: (415) 854-3018
Venture capital fund. Prefers specialty retailing/consumer products companies.

★950★ Venrock Associates
755 Page Mill, A-230 Phone: (415) 493-5577
Palo Alto, CA 94304 Fax: (415) 493-6443
Private venture capital supplier. Prefers high-technology start-up equity investments.

★951★ Ventana Growth Fund L.P. (Irvine)
2301 Dupont Dr., Ste. 430 Phone: (714) 476-2204
Irvine, CA 92715 Fax: (714) 752-0223

★952★ Ventana Growth Fund L.P. (San Diego)
Rio Vista Towers, Ste. 500 Phone: (619) 291-2757
8880 Rio Diego Dr. Fax: (619) 295-0189
San Diego, CA 92108
Venture capital firm that engages in business with biomedical, environmental, health services, and solid state materials and electronics operations.

★953★ Vista Capital Corp.
7716 Lookout Dr. Phone: (619) 453-0780
La Jolla, CA 92037
A small business investment corporation. No industry preference.

★954★ Volpe, Welty and Co.
1 Maritime Plaza, 11th Fl. Phone: (415) 956-8120
San Francisco, CA 94111 Fax: (415) 986-6754

★955★ Walden Capital Partners
750 Battery St., 7th Fl. Phone: (415) 391-7225
San Francisco, CA 94111

★956★ Weiss, Peck and Greer Venture Partners L.P. (San Francisco)
555 California St., Ste. 4760 Phone: (415) 622-6864
San Francisco, CA 94104 Fax: (415) 989-5108

★957★ Wolfensohn Associates, L.P. (Palo Alto)
900 Welch Rd., Ste. 210 Phone: (415) 322-9966
Palo Alto, CA 94304-1803 Fax: (415) 322-8017

Colorado

★958★ Capital Health Management (Denver)
2084 S. Milwaukee St. Phone: (303) 692-8600
Denver, CO 80210 Fax: (303) 692-9656

★959★ Centennial Business Development Fund Ltd.
1999 Broadway, Ste. 2100 Phone: (303) 298-9066
Denver, CO 80202 Fax: (303) 292-3519
Venture capital fund. Prefers to invest in later-stage companies.

★960★ The Centennial Funds
1999 Broadway, Ste. 2100 Phone: (303) 298-9066
Denver, CO 80202
Venture capital fund. Prefers to invest in early-stage companies in the Rocky Mountain region.

★961★ Columbine Ventures
6312 S. Fiddlers Green Cir., No. 260N Phone: (303) 694-3222
Englewood, CO 80111 Fax: (303) 694-9007

★962★   Hill, Carman, Kirby and Washing
885 Arapahoe   Phone: (303) 442-5151
Boulder, CO 80302   Fax: (303) 442-8525

★963★   Weiss, Peck and Greer Venture Partners L.P. (Boulder)
1113 Spruce St., Ste. 300   Phone: (303) 443-1023
Boulder, CO 80302   Fax: (303) 938-8278

★964★   William Blair and Co. (Denver)
1225 17th St., Ste. 2440   Phone: (303) 825-1600
Denver, CO 80202   Fax: (303) 296-2337
Investment banker and venture capital supplier. Provides all stages of financing to growth companies; also deals in leveraged acquisitions.

Connecticut

★965★   Bank of New Haven
PO Box 1874   Phone: (203) 865-4500
New Haven, CT 06508-1874   Fax: (203) 624-9467
A small business investment corporation. No industry preference.

★966★   Canaan Venture Partners (Rowayton)
105 Rowayton Ave.   Phone: (203) 855-0400
Rowayton, CT 06853
Venture capital supplier. Primary concern is strong entrepreneurial management. There are no geographic or industry-specific constraints.

★967★   Capital Impact Corp.
961 Main St.   Phone: (203) 384-5984
Bridgeport, CT 06601
A small business investment corporation. No industry preference.

★968★   Capital Resource Co. of Connecticut, L.P.
699 Bloomfield Ave.   Phone: (203) 243-1114
Bloomfield, CT 06002
A small business investment corporation. No industry preference.

★969★   Consumer Venture Partners
3 Pickwick Plaza   Phone: (203) 629-8800
Greenwich, CT 06830   Fax: (203) 629-2019
Invests in consumer products and services industries.

★970★   DCS Growth Fund
PO Box 740   Phone: (203) 637-1704
Old Greenwich, CT 06870

★971★   Fairfield Venture Partners (Stamford)
1275 Summer St.   Phone: (203) 358-0255
Stamford, CT 06905
Venture capital firm. Diversified. Prefers early-stage financing. Minimum investment is $500,000.

★972★   Financial Opportunities, Inc.
174 South Rd.   Phone: (203) 741-9727
Enfield, CT 06082
A small business investment corporation. Prefers full franchise convenience stores.

★973★   First Connecticut SBIC
1000 Lafayette Blvd.   Phone: (203) 366-4726
Bridgeport, CT 06604
A small business investment corporation.

★974★   Grayrock Capital Inc.
36 Grove St.   Phone: (203) 966-8392
New Canaan, CT 06840

★975★   International Technology Ventures, Inc.
Prime Capital   Phone: (203) 964-9869
1177 Summer St.   Fax: (203) 964-0862
Stamford, CT 06905

★976★   James B. Kobak and Co.
774 Hollow Tree Ridge Rd.   Phone: (203) 655-8764
Darien, CT 06820   Fax: (203) 655-2905
Venture capital supplier and consultant. Provides assistance to new ventures in the communications field through conceptualization, planning, organization, raising money, and control of actual operations. Special interest is in magazine publishing.

★977★   Marcon Capital Corp.
49 Riverside Ave.   Phone: (203) 226-6893
Westport, CT 06880
A small business investment corporation. Media preferred.

★978★   Marketcorp Venture Associates
285 Riverside Ave.   Phone: (203) 222-1000
Westport, CT 06880
Venture capital firm. Prefers to invest in consumer-market businesses, including the packaged goods, specialty retailing, communications, and consumer electronics industries.

★979★   Oak Investment Partners (Westport)
1 Gorham Island   Phone: (203) 226-8346
Westport, CT 06880   Fax: (203) 227-0372

★980★   Oxford Partners (Stamford)
Soundview Plaza   Phone: (203) 964-0592
1266 Main St.   Fax: (203) 964-3192
Stamford, CT 06902
Independent venture capital partnership. Prefers to invest in high-technology industries. Initial investments range from $500,000 to $1.5 million; up to $3 million over several later rounds of financing.

★981★   Prince Ventures (Westport)
1 Gorham Island, 2nd Fl.   Phone: (203) 227-8332
Westport, CT 06880   Fax: (203) 226-5302
Provides early-stage financing for medical and life sciences ventures.

★982★   Regional Financial Enterprises
36 Grove St., 3rd Fl.   Phone: (203) 966-2800
New Canaan, CT 06840
A small business investment corporation. Prefers to invest in high-technology companies or nontechnology companies with significant growth prospects, located anywhere in the United States. Investments range from $1 to $5 million.

★983★   Saugatuck Capital Co.
1 Canterbury Green   Phone: (203) 348-6669
Stamford, CT 06901   Fax: (203) 324-6995
Private investment partnership. Seeks to invest in building products, transportation, health care products and services, energy services and products, process control instrumentation, industrial and automation equipment, test and measurement instrumentation, communications, fasteners, filtration equipment and filters, and valves and pumps. Prefers leveraged buy out situations, but will consider start-up financing. Investments range from $1 to $5 million.

★984★   The SBIC of Connecticut, Inc.
965 White Plains Rd.   Phone: (203) 261-0011
Trumbull, CT 06611   Fax: (203) 452-9699
A small business investment corporation. No industry preference.

★985★ The Vista Group (New Canaan)
36 Grove St.   Phone: (203) 972-3400
New Canaan, CT 06840

Venture capital supplier. Provides start-up and second-stage financing to technology-related businesses that seek to become major participants in high-growth markets of at least $100 million in annual sales. Areas of investment interest include information systems, communications, computer peripherals, medical products and services, retailing, agrigenetics, biotechnology, low technology, no technology, instrumentation, and genetic engineering.

★986★ Xerox Venture Capital (Stamford)
Headquarters   Phone: (203) 968-3000
800 Long Ridge Rd.
Stamford, CT 06904

Venture capital subsidiary of operating company. Prefers to invest in document processing industries.

Delaware

★987★ Morgan Investment Corp.
902 Market St.   Phone: (302) 651-3808
Wilmington, DE 19801   Fax: (302) 655-1710

District of Columbia

★988★ Allied Investment Corp.
1666 K St., NW, Ste. 901   Phone: (202) 331-1112
Washington, DC 20006   Fax: (202) 659-2053

A small business investment corporation. No industry preference.

★989★ Allied Venture Partnership
1666 K St., NW, Ste. 901   Phone: (202) 331-1112
Washington, DC 20006   Fax: (202) 659-2053

Venture capital fund. No industry preference.

★990★ American Security Capital Corp., Inc.
730 15th St., NW, Ste. A-5   Phone: (202) 624-4843
Washington, DC 20005   Fax: (202) 628-2670

A small business investment corporation. No industry preference.

★991★ Broadcast Capital, Inc.
1771 N St., NW, 4th Fl.   Phone: (202) 429-5393
Washington, DC 20036   Fax: (202) 775-2991

A minority enterprise small business investment corporation. Invests only in radio and TV stations. Investments lie in the $300,000-$400,000 range.

★992★ Consumers United Capital Corp.
2100 M St., NW   Phone: (202) 872-5390
Washington, DC 20037   Fax: (202) 872-5494

A minority enterprise small business investment corporation. No industry preference.

★993★ DC Bancorp Venture Capital Co.
1801 K St., NW   Phone: (202) 955-6970
Washington, DC 20006   Fax: (202) 955-8981

A small business investment corporation. No industry preference. Preferred geographic area includes the District of Columbia, Maryland, and Virginia.

★994★ Fulcrum Venture Capital Corp.
1030 15th St., NW, Ste. 203   Phone: (202) 785-4253
Washington, DC 20005   Fax: (202) 789-1062

A minority enterprise small business investment corporation. No industry preference.

★995★ Minority Broadcast Investment Corp.
1200 18th St., NW, Ste. 705   Phone: (202) 293-1166
Washington, DC 20036   Fax: (202) 293-1181

A minority enterprise small business investment corporation. Communications and media industry preferred.

★996★ Syncom Capital Corp.
1030 15th St., NW, Ste. 203   Phone: (202) 293-9428
Washington, DC 20005   Fax: (202) 789-1062

A minority enterprise small business investment corporation. Telecommunications media industry only.

Florida

★997★ Allied North American Co.
111 E. Las Olas Blvd.   Phone: (305) 763-8484
Fort Lauderdale, FL 33301   Fax: (305) 527-0904

A small business investment corporation. No industry preference.

★998★ Gold Coast Capital Corp.
3550 Biscayne Blvd., Rm. 601   Phone: (305) 576-2012
Miami, FL 33137

A small business investment corporation. No industry preference.

★999★ Ideal Financial Corp.
780 NW 42nd Ave., Ste. 501   Phone: (305) 442-4665
Miami, FL 33126   Fax: (305) 446-0602

A minority enterprise small business investment corporation. No industry preference.

★1000★ J and D Capital Corp.
12747 Biscayne Blvd.   Phone: (305) 893-0303
North Miami, FL 33181   Fax: (305) 891-2338

A small business investment corporation. No industry preference.

★1001★ Market Capital Corp.
1102 N. 28th St.   Phone: (813) 247-1357
PO Box 31667
Tampa, FL 33631

A small business investment corporation. Grocery industry preferred.

★1002★ South Atlantic Capital Corp.
614 W. Bay St., Ste. 200   Phone: (813) 253-2500
Tampa, FL 33606-2704

Venture capital supplier. Provides long-term working capital for privately owned, rapidly growing companies located in the southeast.

★1003★ Southeast Venture Capital Limited 1
3250 Miami Center   Phone: (305) 375-6470
201 S. Biscayne Blvd.
Miami, FL 33131

A small business investment corporation. No industry preference, but prefers to invest in firms located in the southwest.

★1004★ Universal Financial Services, Inc.
225 NE 35th St.   Phone: (305) 573-1496
Miami, FL 33137

A minority enterprise small business investment corporation. No industry preference.

★1005★ Venture Group, Inc.
5433 Buffalo Ave.
Jacksonville, FL 32208
Phone: (904) 353-7313
Fax: (904) 353-3032

A minority enterprise small business investment corporation. Automotive industry preferred.

Georgia

★1006★ Bellsouth Ventures Corp.
1100 Peachtree St., NE, Rm. 7F02
Atlanta, GA 30309
Phone: (404) 249-4550
Fax: (404) 249-4768

★1007★ Investor's Equity, Inc.
PO Box 18859
Atlanta, GA 30326
Phone: (404) 266-8300

A small business investment corporation. Invests in manufacturing companies.

★1008★ Noro-Moseley Partners
4200 Northside Pkwy., Bldg. 9
Atlanta, GA 30327
Phone: (404) 233-1966
Fax: (404) 239-9280

Venture capital partnership. Prefers to invest in private, diversified small and medium-sized growth companies located in the southeastern United States.

★1009★ North Riverside Capital Corp.
50 Technology Park/Atlanta
Norcross, GA 30092
Phone: (404) 446-5556
Fax: (404) 446-8627

A small business investment corporation. No industry preference.

Hawaii

★1010★ Bancorp Hawaii SBIC, Inc.
130 Merchant St., Ste. 22T
PO Box 2900
Honolulu, HI 96846
Phone: (808) 537-8557
Fax: (808) 521-7602

A small business investment corporation. No industry preference.

★1011★ Pacific Venture Capital Ltd.
222 S. Vineyard St., No. PH-1
Honolulu, HI 96813-2445
Phone: (808) 521-6502

A minority enterprise small business investment corporation.

Illinois

★1012★ Allstate Venture Capital
Allstate Plaza, S. Bldg. G5D
Northbrook, IL 60062
Phone: (708) 402-5681
Fax: (708) 402-0880

Venture capital supplier. Investments are not limited to particular industries or geographical locations. Interest is in unique products or services that address large potential markets and offer great economic benefits; strength of management team is also important. Investments range from $500,000 to $5 million.

★1013★ Alpha Capital Venture Partners L.P.
3 1st National Plaza, 14th Fl.
Chicago, IL 60602
Phone: (312) 372-1556

A small business investment corporation. No industry preference; however, no real estate, oil and gas, or start-up ventures are considered. All investments are structured to provide equity participation in the business; no straight loans are considered.

★1014★ Ameritech Development Corp.
10 S. Wacker Dr., 21st Fl.
Chicago, IL 60606
Phone: (312) 609-6000

Venture capital supplier. Prefers business and technology industries that complement their core telecommunications business.

★1015★ Amoco Venture Capital Co.
200 E. Randolph Dr.
Chicago, IL 60601
Phone: (312) 856-6523

A minority enterprise small business investment corporation. Interested in technical industries or those that relate to Amoco Corporation needs.

★1016★ Batterson, Johnson and Wang Venture Partners
303 W. Madison St., Ste. 1110
Chicago, IL 60606-3300
Phone: (312) 269-0300
Fax: (312) 269-0021

★1017★ Brinson Partners, Inc.
70 W. Madison Ave., 9th Fl.
Chicago, IL 60602-4298
Phone: (312) 220-7100
Fax: (312) 220-7199

★1018★ Business Ventures, Inc.
20 N. Wacker Dr., Ste. 1741
Chicago, IL 60606
Phone: (312) 346-1580

A small business investment corporation. No industry preference; considers only ventures in the Chicago area.

★1019★ Capital Health Management (Chicago)
122 S. Michigan, Ste. 1915
Chicago, IL 60603
Phone: (312) 427-1227
Fax: (312) 427-1247

★1020★ The Capital Strategy Group, Inc.
20 N. Wacker Dr.
Chicago, IL 60606
Phone: (312) 444-1170

Investment banker and venture capital supplier. Provides financing to start-up and early-stage companies, located in the Midwest, in the manufacturing and service industries.

★1021★ Caterpillar Venture Capital, Inc.
100 NE Adams St.
Peoria, IL 61602
Phone: (309) 675-5503

Venture capital subsidiary of operating firm. Prefers to invest in industrial electronics, advanced materials, and environmental-related industries.

★1022★ Chicago Community Ventures, Inc.
25 E. Washington St., Ste. 2015
Chicago, IL 60602-1809
Phone: (312) 726-6084
Fax: (312) 726-0167

A minority enterprise small business investment corporation. No industry preference.

★1023★ Cilcorp Ventures, Inc.
300 Liberty St.
Peoria, IL 61602
Phone: (309) 672-5158
Fax: (309) 672-5000

★1024★ Comdisco Venture Group
6111 N. River Rd.
Rosemont, IL 60018
Phone: (708) 698-3000
Fax: (708) 518-5465

Venture capital subsidiary of operating firm.

★1025★ Continental Illinois Venture Corp.
231 S. LaSalle St.
Chicago, IL 60697
Phone: (312) 828-8021
Fax: (312) 987-0887

A small business investment corporation. Provides start-up and early-stage financing to growth-oriented companies with capable management teams, proprietary products, and expanding markets.

★1026★ **Essex Venture Partners**
190 S. LaSalle, Ste. 2800
Chicago, IL 60603
Phone: (312) 444-6040
Fax: (312) 444-6034
Prefer investing in health care companies.

★1027★ **First Analysis Corp.**
233 S. Wacker Dr., Ste. 9600
Chicago, IL 60606
Phone: (312) 258-1400

★1028★ **First Capital Corp. of Chicago**
3 1st National Plaza
Chicago, IL 60670
Phone: (312) 732-5414
A small business investment corporation. No industry preference.

★1029★ **First Chicago Venture Capital**
3 1st National Plaza, Ste. 1330
Chicago, IL 60670-0501
Phone: (312) 732-5400
Venture capital supplier. Invests a minimum of $1 million in early-stage situations to a maximum of $25 million in mature growth or buy out situations. Emphasis is placed on a strong management team and unique market opportunity.

★1030★ **Frontenac Capital Corp.**
208 S. LaSalle St., Rm. 1900
Chicago, IL 60604
Phone: (312) 368-0047
Fax: (312) 368-9520
A small business investment corporation. No industry preference.

★1031★ **Golder, Thoma and Cressey**
120 S. LaSalle St., Ste. 630
Chicago, IL 60603
Phone: (312) 853-3322
Fax: (312) 853-3354
Private equity investors. Provides financing for start-up and leveraged buy out situations. No geographic or industry limitations, with the exception of real estate. Past investments have been made in the health care field and in information services. Investments range from $2 to $10 million.

★1032★ **IEG Venture Management, Inc.**
10 S. Riverside Plaza
Chicago, IL 60606
Phone: (312) 993-7500
Fax: (312) 454-0369
Venture capital supplier. Provides start-up financing primarily to technology-based companies located in the Midwest.

★1033★ **Marquette Venture Partners**
1751 Lake Cook Rd., Ste. 550
Deerfield, IL 60015
Phone: (708) 940-1700
Fax: (708) 940-1724

★1034★ **Mesirow Venture Capital**
350 N. Clark St.
Chicago, IL 60610
Phone: (312) 670-6092
A small business investment corporation. No industry preference.

★1035★ **Prince Ventures (Chicago)**
10 S. Wacker Dr., Ste. 2575
Chicago, IL 60606
Phone: (312) 454-1408
Fax: (312) 454-9125

★1036★ **Sears Investment Management Co.**
Xerox Center
55 W. Monroe St., 32nd Fl.
Chicago, IL 60603
Phone: (312) 875-0463
Fax: (312) 875-7529

★1037★ **Seidman, Fisher and Co.**
233 N. Michigan Ave., Ste. 1812
Chicago, IL 60601
Phone: (312) 856-1812
Private venture capital supplier. Provides early-stage and growth-equity financing to companies with proprietary or patented products or services that deal with large and rapidly growing industrial markets; limited interest in consumer markets. Leveraged buy outs and turn-around of mature companies are considered under certain circumstances. Investments range from $200,000 to $2 million.

★1038★ **Tower Ventures, Inc.**
Sears Tower, BSC 11-11
Chicago, IL 60684
Phone: (312) 875-0571
Fax: (312) 906-0164
A minority enterprise small business investment corporation.

★1039★ **Walnut Capital Corp. (Chicago)**
2 N. LaSalle, Ste. 2410
Chicago, IL 60602
Phone: (312) 346-2033
A small business investment corporation. No industry preference.

★1040★ **William Blair and Co. (Chicago)**
135 S. LaSalle St.
Chicago, IL 60603
Phone: (312) 236-1600
A small business investment corporation. Provides all stages of financing to growth companies; also deals in leveraged acquisitions.

★1041★ **Wind Point Partners (Chicago)**
321 N. Clark St., Ste. 3010
Chicago, IL 60610
Phone: (312) 245-4949
Fax: (312) 245-4940

Indiana

★1042★ **Circle Ventures, Inc.**
3228 E. 10th St.
Indianapolis, IN 46201
Phone: (317) 636-7242
Fax: (317) 637-7581
A small business investment corporation. Prefers second-stage, leveraged buy out, and growth financings.

★1043★ **First Source Capital Corp.**
PO Box 1602
South Bend, IN 46634
Phone: (219) 236-2180
Fax: (219) 236-2719
A small business investment corporation. No industry preference.

★1044★ **Heritage Venture Partners**
135 N. Pennsylvania St., Ste. 2380
Indianapolis, IN 46204
Phone: (317) 635-5696
Venture capital fund. Prefers communications industries, especially broadcasting.

★1045★ **Mount Vernon Venture Capital Co.**
PO Box 40177
Indianapolis, IN 46240
Phone: (317) 469-0400
A small business investment corporation. No industry preference.

★1046★ **White River Capital Corp.**
PO Box 929
Columbus, IN 47202
Phone: (812) 376-1759
Fax: (812) 376-1709
A small business investment corporation. No industry preference.

Iowa

★1047★ **Allsop Venture Partners (Cedar Rapids)**
2750 1st Ave., NE, Ste. 210
Cedar Rapids, IA 52402
Phone: (319) 363-8971
Fax: (319) 363-9519

★1048★ **InvestAmerica Venture Group, Inc. (Cedar Rapids)**
101 2nd St., SE, Ste. 800
Cedar Rapids, IA 52401
Phone: (319) 363-8249
Venture capital fund management company. Prefers investing in manufacturing, service, and retail with minimum investments of $300,000 to $500,000.

Kansas

★1049★   Allsop Venture Partners (Overland Park)
7400 College Blvd., Ste. 302   Phone: (913) 338-0820
Overland Park, KS 66210   Fax: (913) 338-1019

★1050★   Kansas Venture Capital, Inc.
6700 Antioch Rd., Ste. 460   Phone: (913) 233-1368
Overland Park, KS 66204-1200   Fax: (913) 262-3509
A small business investment corporation. No industry preference.

Kentucky

★1051★   Equal Opportunity Finance, Inc.
420 Hurstbourne Ln., Ste. 201   Phone: (502) 423-1943
Louisville, KY 40222   Fax: (502) 423-1945
A minority enterprise small business investment corporation. No industry preference; geographic areas limited to Indiana, Kentucky, Ohio, and West Virginia.

★1052★   Mountain Ventures, Inc.
PO Box 1738   Phone: (606) 864-5175
London, KY 40743-1738
A small business investment corporation. No industry preference.

Louisiana

★1053★   Capital for Terrebonne, Inc.
27 Austin Dr.   Phone: (504) 868-3933
Houma, LA 70360
A small business investment corporation. No industry preference.

★1054★   Dixie Business Investment Co.
401-1/2 Lake St.   Phone: (318) 559-1558
PO Box 588
Lake Providence, LA 71254
A small business investment corporation. No industry preference.

★1055★   Premier Venture Capital Corp.
PO Box 1511   Phone: (504) 389-4421
Baton Rouge, LA 70821
A small business investment corporation. No industry preference.

★1056★   SCDF Investment Corp.
1006 Surrey St.   Phone: (318) 232-3769
PO Box 3005
Lafayette, LA 70502
A minority enterprise small business investment corporation. No industry preference.

Maine

★1057★   Maine Capital Corp.
70 Center St.   Phone: (207) 772-1001
Portland, ME 04101
A small business investment corporation. No industry preference.

Maryland

★1058★   ABS Ventures Limited Partnerships (Baltimore)
10 N. Calvert St., Ste. 735   Phone: (301) 727-1700
Baltimore, MD 21202
Invests in the computer software, health care, and biotechnology industries.

★1059★   Broventure Capital Management
16 W. Madison St.   Phone: (301) 727-4520
Baltimore, MD 21201   Fax: (301) 727-1436
Venture capital partnership. Provides start-up capital to early-stage companies, expansion capital to companies experiencing rapid growth, and capital for acquisitions. Initial investments range from $400,000 to $750,000.

★1060★   Greater Washington Investments, Inc.
5454 Wisconsin Ave.   Phone: (301) 656-0626
Chevy Chase, MD 20815   Fax: (301) 656-4053
A small business investment corporation. Provides financing to small developing companies primarily in the health care and manufacturing industries. Does not invest in start-ups.

★1061★   Grotech Partners, L.P. (Timonium)
9690 Deereco Rd.   Phone: (301) 560-2000
Timonium, MD 21093   Fax: (301) 560-1910

★1062★   New Enterprise Associates (Baltimore)
1119 St. Paul St.   Phone: (301) 244-0115
Baltimore, MD 21202   Fax: (301) 752-7721
Private free-standing venture capital firm. Prefers high-technology and medical-related industries.

★1063★   Security Financial and Investment Corp.
7720 Wisconsin Ave., Ste. 207   Phone: (301) 951-4288
Bethesda, MD 20814
A minority enterprise small business investment corporation. No industry preference.

★1064★   T. Rowe Price
100 E. Pratt St.   Phone: (301) 547-2000
Baltimore, MD 21202
Venture capital supplier. Offers specialized investment services to meet the needs of companies in various stages of growth.

★1065★   Triad Investor's Corp.
1629 Thames St., Ste. 301   Phone: (301) 955-4652
Baltimore, MD 21231-3430   Fax: (301) 955-7755

★1066★   Washington Resources Group, Inc.
11130 Rockville Pke., Ste. 1203   Phone: (301) 816-7955
Rockville, MD 20852   Fax: (301) 816-7961

Massachusetts

★1067★   ABS Ventures Limited Partnerships (Woburn)
400 Unicon Park Dr.   Phone: (617) 935-0240
Woburn, MA 01801   Fax: (617) 937-0477

★1068★   Advanced Technology Ventures (Boston)
10 Post Office Sq.   Phone: (617) 423-4050
Boston, MA 02109   Fax: (617) 423-4573
Private venture capital firm. Prefers early-stage financing in high-technology industries.

★1069★ **Advent Atlantic Capital Co. L.P.**
75 State St.   Phone: (617) 345-7200
Boston, MA 02109   Fax: (617) 345-7201

A small business investment corporation. Communications industry preferred.

★1070★ **Advent Industrial Capital Co. L.P.**
75 State St.   Phone: (617) 345-7200
Boston, MA 02109   Fax: (617) 345-7201

A small business investment corporation. Communications industry preferred.

★1071★ **Advent International Corp.**
101 Federal St.   Phone: (617) 951-9420
Boston, MA 02110   Fax: (617) 951-0567

Venture capital firm. Specializes in working with companies that need assistance in accessing international markets and major corporations.

★1072★ **Advent IV Capital Co.**
75 State St.   Phone: (617) 345-7200
Boston, MA 02109   Fax: (617) 345-7201

A small business investment corporation. Communications industry preferred.

★1073★ **Advent V Capital Co.**
75 State St.   Phone: (617) 345-7200
Boston, MA 02109   Fax: (617) 345-7201

A small business investment corporation. Communications industry preferred.

★1074★ **Aegis Venture Funds**
1 Cranberry Hill   Phone: (617) 862-0200
Lexington, MA 02173   Fax: (617) 862-6156

★1075★ **Aeneas Venture Corp.**
600 Atlantic Ave.   Phone: (617) 523-4400
Boston, MA 02210   Fax: (617) 523-1063

Diversified venture capital firm. Minimum investment is $1 million.

★1076★ **American Research and Development**
45 Milk St.   Phone: (617) 423-7500
Boston, MA 02109   Fax: (617) 423-9655

Independent private venture capital partnership. All stages of financing; no minimum or maximum investment.

★1077★ **Ampersand Ventures**
55 William St., Ste. 240   Phone: (617) 239-0700
Wellesley, MA 02081   Fax: (617) 239-0824

Venture capital supplier. Provides start-up and early-stage financing to technology-based companies. Investments range from $500,000 to $1 million.

★1078★ **Analog Devices Enterprises**
1 Technology Way   Phone: (617) 329-4700
PO Box 9106   Fax: (617) 326-8703
Norwood, MA 02062-9106

Venture capital division of Analog Devices, Inc., a supplier of integrated analog circuits. Prefers to invest in industries involved in analog devices.

★1079★ **Applied Technology Partners**
1 Cranberry Hill   Phone: (617) 862-8622
Lexington, MA 02173-7397   Fax: (617) 862-8367

Prefers investing with companies involved in semiconductors, electronics, computers, software, telecommunications, and information services.

★1080★ **Atlas Venture**
1 Cambridge Center   Phone: (617) 621-1600
Cambridge, MA 02142   Fax: (617) 621-1638

★1081★ **Bain Capital Fund**
2 Copley Pl.   Phone: (617) 572-3000
Boston, MA 02116   Fax: (617) 572-3274

Private venture capital firm. No industry preference, but avoids investing in high-tech industries. Minimum investment is $500,000.

★1082★ **BancBoston Ventures, Inc.**
100 Federal St.   Phone: (617) 434-2442
Boston, MA 02110   Fax: (617) 434-1383

A small business investment corporation. Provides start-up and first-and second-round financing to companies in the communications, computer hardware and software, electronic components and instrumentation, industrial products, and health care industries. Investments range from $500,000 to $2 million.

★1083★ **Battery Ventures (Boston)**
200 Portland St.   Phone: (617) 367-1011
Boston, MA 02114   Fax: (617) 367-1070

★1084★ **Bessemer Venture Partners (Wellesley Hills)**
83 Walnut St.   Phone: (617) 237-6050
Wellesley Hills, MA 02181   Fax: (617) 235-7068

★1085★ **Bever Capital Corp.**
TBM Associates, Inc.   Phone: (617) 951-9920
1 International Plaza, 23rd Fl. Fax: (617) 951-0569
Boston, MA 02110-2600

A small business investment corporation. Provides all stages of financing to technology-based and oil and gas service industries; no geographic preference. Experience of management team and high-growth market potential are emphasized. Investments range from $200,000 to $400,000.

★1086★ **BMW Technologies, Inc.**
655 N. Shore Rd.
Riviera, MA 02151

★1087★ **Boston Capital Ventures L.P.**
Old City Hall   Phone: (617) 227-6550
45 School St.   Fax: (617) 227-3847
Boston, MA 02108

Venture capital firm. Prefers health care and high-technology industries.

★1088★ **Burr, Egan, Deleage, and Co. (Boston)**
1 Post Office Sq., Ste. 3800   Phone: (617) 482-8020
Boston, MA 02109   Fax: (617) 482-1944

Private venture capital supplier. Invests start-up, expansion, and acquisitions capital nationwide. Principal concerns are strength of the management team; large, rapidly expanding markets; and unique products or services. Past investments have been made in the fields of electronics, health, and communications. Investments range from $750,000 to $5 million.

★1089★ **Business Achievement Corp.**
1172 Beacon St.   Phone: (617) 965-0550
Newton Centre, MA 02161   Fax: (617) 969-0181

A small business investment corporation. No industry preference.

★1090★ **Capital Formation Service**
Boston College
96 College Rd., Rahner House   Phone: (617) 552-4091
Chestnut Hill, MA 02167

Provides assistance to clients requiring financing from nonconventional sources, such as quasi-public financing programs; state, federal, and local programs; venture capital; and private investors.

★1091★ Charles River Ventures
67 Battery March St., Ste. 600   Phone: (617) 439-0477
Boston, MA 02110   Fax: (617) 439-0084

Venture capital partnership. Diversified, with preferences in high technology (software and communications), environmental, financial, and health care technology.

★1092★ Chestnut Capital International II L.P.
75 State St.   Phone: (617) 345-7200
Boston, MA 02109

A small business investment corporation. Communications industry preferred.

★1093★ Claflin Capital Management, Inc.
77 Franklin St.   Phone: (617) 426-6505
Boston, MA 02110   Fax: (617) 482-0016

Private venture capital firm investing its own capital. No industry preference but prefers early stage companies.

★1094★ Comdisco, Inc. Venture Leasing (Boston)
101 Federal St., 27th Fl.   Phone: (617) 330-9011
Boston, MA 02110   Fax: (617) 330-9018

Prefers start-up businesses in fields of semiconductors, computer hardware and software, computer services and systems, telecommunications, and medical and biotechnology. Investments range from $500,000 to $5 million.

★1095★ Copley Venture Partners
600 Atlantic Ave., 13th Fl.   Phone: (617) 722-6030
Boston, MA 02210   Fax: (617) 523-7739

★1096★ Eastech Management Co.
260 Franklin St., Ste. 530   Phone: (617) 439-6130
Boston, MA 02110

Private venture capital supplier. Provides start-up and first- and second-stage financing to companies in the following industries: communications, computer-related electronic components and instrumentation, and industrial products and equipment. Will not consider real estate, agriculture, forestry, fishing, finance and insurance, transportation, oil and gas, publishing, entertainment, natural resources, or retail. Investments range from $400,000 to $900,000, with a minimum of $250,000. Eastech prefers that portfolio companies be located in New England or on the East Coast, within two hours of the office.

★1097★ Fidelity Venture Associates, Inc.
Fidelity Investments   Phone: (617) 728-6488
82 Devonshire St.   Fax: (617) 439-0439
Boston, MA 02109

★1098★ First United SBIC, Inc.
135 Will Dr.   Phone: (617) 828-6150
Canton, MA 02021   Fax: (617) 821-2682

A small business investment corporation. Prefers grocery distribution investments.

★1099★ Fleet Venture Partners II
1740 MAS Ave.   Phone: (508) 263-0177
Boxborough, MA 01719

Venture capital partnership. No industry preference.

★1100★ Greylock Management Corp. (Boston)
1 Federal St.   Phone: (617) 423-5525
Boston, MA 02110   Fax: (617) 482-0059

Private venture capital partnership. Minimum investment of $250,000; preferred investment size of over $1 million. Will function either as deal originator or investor in deals created by others.

★1101★ Hambrecht and Quist, Inc.
50 Rowes Wharf   Phone: (617) 574-0500
Boston, MA 02110   Fax: (617) 574-0547

★1102★ Hambro International Venture Fund (Boston)
160 State St.   Phone: (617) 523-7767
Boston, MA 02109   Fax: (617) 523-8394

Private venture firm. Seeks to invest in mature companies as well as in high-technology areas, from start-ups to leveraged buy outs. Investments range from $500,000 to $3 million, with initial investments ranging from $800,000 to $1 million.

★1103★ Highland Capital Partners
1 International Pl.   Phone: (617) 330-8765
Boston, MA 02110   Fax: (617) 330-8768

Industry preferences include health care, software, and telecommunications.

★1104★ John Hancock Venture Capital Management, Inc.
1 Financial Center, 44th Fl.   Phone: (617) 348-3707
Boston, MA 02111   Fax: (617) 350-0305

Venture capital supplier. Diversified investments.

★1105★ Massachusetts Business Development Corp.
1 Liberty Sq.   Phone: (617) 350-8877
Boston, MA 02109

Provides loans to private for-profit and nonprofit firms for the purchase or construction of fixed business assets and for working capital. There are six types of loans: working capital, leveraged buyout, second mortgages, government guaranteed loans, SBA(504) loans, and long-term loans.

★1106★ Massachusetts Community Development Finance Corp.
131 State St., Ste. 600   Phone: (617) 742-0366
Boston, MA 02109

Provides financing for small businesses and for commercial, industrial, and residential business developments through community development corporations (CDCs) in depressed areas of Massachusetts. Three investment programs are offered: the Venture Capital Investment Program, the Community Development Program, and the Small Loan Guarantee Program.

★1107★ Matrix Partners III
1 International Pl., Ste. 3250   Phone: (617) 345-6740
Boston, MA 02110-2600   Fax: (617) 345-6773

Private venture capital partnership. Investments range from $500,000 to $1 million.

★1108★ MDT Advisers, Inc.
25 Acorn Pk.   Phone: (617) 864-5770
Cambridge, MA 02140   Fax: (617) 547-1376

★1109★ Merrill, Pickard, Anderson, and Eyre (Waltham)
1000 Winter St., Ste. 1080   Phone: (617) 890-0670
Waltham, MA 02154   Fax: (617) 890-5830

★1110★ Morgan, Holland Ventures Corp.
1 Liberty Sq.   Phone: (617) 423-1765
Boston, MA 02109   Fax: (617) 338-4362

Venture capital partnership. Provides start-up, early-stage, and expansion financing to companies that are pioneering applications of proven technology; also will consider nontechnology-based companies with strong management teams and plans for expansion. Investments range from $500,000 to $1 million, with a $6 million maximum.

★1111★ New England Capital Corp.
1 Washington Mall, 7th Fl.   Phone: (617) 573-6400
Boston, MA 02108   Fax: (617) 573-7575

A small business investment corporation. Invests in high- or low-technology manufacturing, service, and distribution industries, primarily in the northeastern states.

★1112★ New England MESBIC, Inc.
530 Turnpike St.   Phone: (508) 688-4326
North Andover, MA 01845   Fax: (508) 689-2261

A minority enterprise small business investment corporation. No industry preference.

★1113★ Northeast Small Business Investment Corp.
16 Cumberland St.   Phone: (617) 267-3983
Boston, MA 02115

A small business investment corporation. No industry preference.

★1114★ Orange Nassau Capital Corp.
TBM Associates, Inc.   Phone: (617) 951-9920
1 International Plaza, 23rd Fl.
Boston, MA 02110-2600

A small business investment corporation. Provides all stages of financing to technology-based and oil and gas service industries; no geographic preference. Experience of management team and high-growth market potential are emphasized. Investments range from $200,000 to $400,000.

★1115★ P. R. Venture Partners, L.P.
40 Rowes Wharf   Phone: (617) 439-6700
Boston, MA 02110   Fax: (617) 439-0594

★1116★ Palmer Partners L.P.
300 Unicorn Park Dr.   Phone: (617) 933-5445
Woburn, MA 01801   Fax: (617) 933-0698

Venture capital partnership. Provides financing to new enterprises and existing growth companies in high growth situations. No industry preference, but does not invest in real estate.

★1117★ Plant Resources Venture Funds
75 State St., Ste. 2200   Phone: (617) 345-9440
Boston, MA 02109   Fax: (617) 345-9878

★1118★ Security Pacific Capital Corp. (Boston)
101 Federal St., 19th Fl.   Phone: (617) 542-7601
Boston, MA 02110   Fax: (617) 451-2589

★1119★ Sprout Group (Boston)
1 Center Plaza   Phone: (617) 570-8720
Boston, MA 02108

Venture capital affiliate of Donaldson, Lufkin, and Jenrette. Provides early-stage financing to companies specializing in retailing, financial services, and high-technology products. Will also invest in selected later-stage, emerging/growing companies and in management buy outs of mature companies. Investments range from $1 to $2 million, with a minimum of $500,000.

★1120★ Summit Ventures
1 Boston Pl., Ste. 3420   Phone: (617) 742-5500
Boston, MA 02108   Fax: (617) 742-6138

Venture capital firm. Prefers to invest in emerging, profitable, growth companies in the electronic technology, environmental services, and health care industries. Investments range from $1 to $4 million.

★1121★ TA Associates (Boston)
45 Milk St.   Phone: (617) 338-0800
Boston, MA 02109   Fax: (617) 574-6728

Private venture capital partnership. Technology companies, media communications companies, and leveraged buy outs preferred. Will provide from $1 to $20 million in investments.

★1122★ TBM II Capital Corp.
TBM Associates, Inc.   Phone: (617) 951-9920
1 International Plaza, 23rd Fl.   Fax: (617) 951-0569
Boston, MA 02110-2600

A small business investment corporation. Provides all stages of financing to technology-based and oil and gas service industries; no geographic

preference. Experience of management team and high-growth market potential are emphasized. Investments range from $200,000 to $400,000.

★1123★ 3i Capital
99 High St., Ste. 1530   Phone: (617) 542-8560
Boston, MA 02110   Fax: (617) 542-0394

Venture capital supplier. Provides capital for growth, acquisition, share repurchase, or leveraged buy out to companies in communications, computer, consumer, distribution, electronics, industrial products, manufacturing equipment, medical/health, and retail industries. Investments range from $500,000 to $5 million.

★1124★ 3i Ventures (Boston)
99 High St., Ste. 1530   Phone: (617) 542-8560
Boston, MA 02110   Fax: (617) 542-0394

Venture capital supplier. Provides start-up and early-stage financing to companies in high-growth industries such as biotechnology, communications, electronics, and health care.

★1125★ Transportation Capital Corp.
45 Newbury St., Rm. 207   Phone: (617) 536-0344
Boston, MA 02116

A minority enterprise small business investment corporation. Specializes in taxicabs and taxicab medallion loans.

★1126★ Vadus Capital Corp.
TBM Associates, Inc.   Phone: (617) 951-9920
1 International Plaza, 23rd Fl.
Boston, MA 02110-2600

A small business investment corporation. Provides all stages of financing to technology-based and oil and gas service industries; no geographic preference. Experience of management team and high-growth market potential are emphasized. Investments range from $200,000 to $400,000.

★1127★ Venture Capital Fund of New England II
160 Federal St., 23rd Fl.   Phone: (617) 439-4646
Boston, MA 02110   Fax: (617) 439-4652

Venture capital fund. Prefers New England high-technology companies that have a commercial prototype or initial product sales. Will provide up to $500,000 in first-round financing.

★1128★ Venture Founders Corp.
1 Cranberry Hill   Phone: (617) 863-0900
Lexington, MA 02173   Fax: (617) 862-1945

Venture capital fund. CAD/CAM and robotics, materials components, information technology, biotechnology, and medical industries preferred. Will provide $250,000 to $1 million in start-up and first-round financing. Preferred initial investment size is between $100,000 and $750,000.

★1129★ Vimac Corp.
12 Arlington St.   Phone: (617) 267-2785
Boston, MA 02116   Fax: (617) 267-5836

Venture capital supplier. Provides early-stage financing to businesses in the computer industry, and in information technologies.

★1130★ Zero Stage Capital Co., Inc. (Cambridge)
1 Broadway   Phone: (617) 876-5355
Kendall Sq.   Fax: (617) 876-1248
Cambridge, MA 02142

Michigan

★1131★ Battery Ventures (Ann Arbor)
425 N. Main St.   Phone: (313) 663-1666
Ann Arbor, MI 48104-1133   Fax: (313) 663-7358

★1132★ Dearborn Capital Corp.
PO Box 1729   Phone: (313) 337-8577
Dearborn, MI 48121-1729   Fax: (313) 845-6124

A minority enterprise small business investment corporation. Loans to minority-owned, operated, and controlled suppliers to Ford Motor Company, Dearborn Capital Corporation's parent.

★1133★ Demery Seed Capital Fund
3307 W. Maple Rd., Ste. 101   Phone: (313) 433-1722
Birmingham, MI 48010

Seed capital fund. Diversified, but interested in food processing. Invests in start-up companies in Michigan.

★1134★ The Edward Lowe Group, Inc.
PO Box 385   Phone: (616) 445-3881
58220 Decatur Rd.   Fax: (616) 445-2648
Cassopolis, MI 49031

★1135★ Federated Capital Corp.
30955 Northwestern Hwy., Ste. 300   Phone: (313) 737-1300
Farmington Hills, MI 48334   Fax: (313) 626-1544

A small business investment corporation. No industry preference.

★1136★ MBW Management, Inc. (Ann Arbor)
4251 Plymouth Rd.   Phone: (313) 747-9401
PO Box 986   Fax: (313) 747-9704
Ann Arbor, MI 48106

Manages a small business investment corporation and two venture capital funds. Prefers high-tech industries and leveraged buy outs for manufacturing companies. No geographic limitations.

★1137★ Metro-Detroit Investment Co.
30777 Northwestern Hwy., Ste. 300   Phone: (313) 851-6300
Farmington Hills, MI 48334-2549   Fax: (313) 851-9551

A minority enterprise small business investment corporation. Food store industry preferred.

★1138★ Motor Enterprises, Inc.
3044 W. Grand Blvd.   Phone: (313) 556-4273
Detroit, MI 48202   Fax: (313) 564-9909

A minority enterprise small business investment corporation. Prefers manufacturing.

Minnesota

★1139★ Capital Dimensions Ventures Fund, Inc.
2 Apple Tree Sq., Ste. 335   Phone: (612) 854-3506
Minneapolis, MN 55425-1637

A minority enterprise small business investment corporation. No industry preference.

★1140★ Cherry Tree Ventures
1400 Northland Plaza   Phone: (612) 893-9012
3800 W. 80th St.   Fax: (612) 893-9036
Minneapolis, MN 55431

Venture capital supplier. Provides start-up and early-stage financing. Fields of interest include communications, medical devices, health care services, applications and systems software, and microprocessor-based systems for the office and factory. There are no minimum or maximum investment limitations.

★1141★ FBS Venture Capital Co.
1st Bank Pl., E.   Phone: (612) 370-4764
120 S. 6th St.   Fax: (612) 370-3853
Minneapolis, MN 55480

A small business investment corporation. Generally invests in high-technology companies, although they are not necessarily preferred.

★1142★ ITASCA Growth Fund, Inc.
19 NE 3rd St.   Phone: (218) 326-0754
Grand Rapids, MN 55744   Fax: (218) 327-2242

A small business investment corporation. No industry preference.

★1143★ Medical Innovation Capital, Inc.
Opus Center, Ste. 421   Phone: (612) 931-0154
9900 Bren Rd., E.   Fax: (612) 931-0003
Minneapolis, MN 55343

★1144★ North Star Ventures, Inc.
3434 Norwest   Phone: (612) 936-4500
Minneapolis, MN 55402   Fax: (612) 333-5425

A small business investment corporation. Invests start-up and early-stage capital in all industry segments excluding real estate, with a preference toward companies in high-technology, electronics, and/or medical industries. Investments range from $400,000 to $1 million, with a preferred limit of $1 million.

★1145★ Northland Capital Corp.
613 Missabe Bldg.   Phone: (218) 722-0545
Duluth, MN 55802   Fax: (218) 722-7241

A small business investment corporation. No industry preference.

★1146★ Norwest Venture Capital Management, Inc. (Minneapolis)
2800 Piper Jaffray Tower   Phone: (612) 667-1650
222 S. 9th St.   Fax: (612) 667-1660
Minneapolis, MN 55402-3388

Small business investment company. Invests in all industries except real estate.

★1147★ Oak Investment Partners (Minneapolis)
4550 Norwest Center   Phone: (612) 339-9322
90 S. 7th St.
Minneapolis, MN 55402

Prefers investing in retail industries.

★1148★ Pathfinder Venture Capital Funds (Minneapolis)
7300 Metro Blvd., Ste. 585   Phone: (612) 835-1121
Minneapolis, MN 55439   Fax: (612) 835-8389

Venture capital supplier. Provides start-up and early-stage financing to emerging companies in the medical, computer, pharmaceuticals, and data communications industries. Emphasis is on companies with proprietary technology or market positions and with substantial potential for revenue growth.

★1149★ Piper Jaffray Ventures, Inc.
Piper Jaffray Tower   Phone: (612) 342-6310
222 S. 9th St.   Fax: (612) 342-6979
Minneapolis, MN 55440

★1150★ Threshold Ventures, Inc.
15500 Wayzata Blvd.   Phone: (612) 473-2051
12 Oak Center, Ste. 819
Minnetonka, MN 55391

A small business investment corporation. No industry preference.

Mississippi

★1151★ Sun-Delta Capital Access Center, Inc.
819 Main St. Phone: (601) 335-5291
Greenville, MS 38701

A minority enterprise small business investment corporation. No industry preference.

★1152★ Vicksburg SBIC
PO Box 821568 Phone: (601) 636-4762
Vicksburg, MS 39182-1568 Fax: (601) 636-9476

A small business investment corporation. No industry preference.

Missouri

★1153★ Allsop Venture Partners (St. Louis)
55 W. Port Plaza, Ste. 575 Phone: (314) 434-1688
St. Louis, MO 63146 Fax: (314) 434-6560

★1154★ Bankers Capital Corp.
3100 Gillham Rd. Phone: (816) 531-1600
Kansas City, MO 64109

A small business investment corporation. No industry preference.

★1155★ Capital for Business, Inc. (Kansas City)
1000 Walnut, 18th Fl. Phone: (816) 234-2357
Kansas City, MO 64106 Fax: (816) 234-2333

A small business investment corporation. No industry preference.

★1156★ Capital for Business, Inc. (St. Louis)
11 S. Meramec, Ste. 800 Phone: (314) 854-7427
St. Louis, MO 63105 Fax: (314) 746-8739

A small business investment corporation. Focuses primarily on later-stage expansion and acquisition in the manufacturing and distribution industries.

★1157★ Gateway Associates L.P.
8000 Maryland Ave., Ste. 1190 Phone: (314) 721-5707
St. Louis, MO 63105 Fax: (314) 721-5135

★1158★ Intercapco, Inc.
7800 Bonhomme Ave. Phone: (314) 863-0600
St. Louis, MO 63105 Fax: (314) 863-8179

A small business investment corporation. No industry preference.

★1159★ InvestAmerica Venture Group, Inc. (Kansas City)
Commerce Tower Bldg. Phone: (816) 842-0114
911 Main St., Ste. 2724A Fax: (816) 471-7339
Kansas City, MO 64105

A small business investment corporation. No industry preference.

★1160★ MBI Venture Capital Investors, Inc.
850 Main St. Phone: (816) 889-1700
Kansas City, MO 64105

A small business investment corporation. No industry preference.

★1161★ United Missouri Capital Corp.
1010 Grand Ave. Phone: (816) 556-7333
Kansas City, MO 64106 Fax: (816) 860-7143

A small business investment corporation. No industry preference.

Nebraska

★1162★ United Financial Resources Corp.
PO Box 1131 Phone: (402) 734-1250
Omaha, NE 68101

A small business investment corporation. Engages in the grocery industry in the Midwest only.

Nevada

★1163★ Enterprise Finance Capital Development Corp.
1 E. 1st St., Ste. 1102 Phone: (702) 329-7797
PO Box 3597
Reno, NV 89501

A small business investment corporation. No industry preference.

New Jersey

★1164★ Accel Partners (Princeton)
1 Palmer Sq. Phone: (609) 683-4500
Princeton, NJ 08542 Fax: (609) 683-0384

Venture capital firm. Telecommunications, software, and health care industries preferred.

★1165★ AMEV Venture Management, Inc.
333 Thornall St., 2nd Fl. Phone: (908) 603-8500
Edison, NJ 08837

Manages two limited partnerships and one small business investment corporation. Prefers leveraged buy outs and later-stage financing. Investments range from $1 to $3 million.

★1166★ BCI Advisors, Inc.
Glenpointe Center, W. Phone: (201) 836-3900
Teaneck, NJ 07666 Fax: (201) 836-6368

Venture capital firm. Prefers later-stage financing.

★1167★ Bradford Associates
22 Chambers St., 4th Fl. Phone: (609) 921-3880
Princeton, NJ 08540

Venture capital firm. No industry preference.

★1168★ Bridge Capital Investors
Glenpoint Center, W. Phone: (201) 836-3900
Teaneck, NJ 07666 Fax: (201) 836-6368

Venture capital partnership. Investments range from $3 to $5 million.

★1169★ Carnegie Hill Co.
202 Carnegie Center, 2nd Fl. Phone: (609) 520-0500
Princeton, NJ 08540 Fax: (609) 520-1160

★1170★ Domain Associates
1 Palmer Sq. Phone: (609) 683-5656
Princeton, NJ 08542 Fax: (609) 683-9789

★1171★ DSV Partners (Princeton)
221 Nassau St. Phone: (609) 924-6420
Princeton, NJ 08542 Fax: (609) 683-0174

Private venture capital supplier. Provides seed, research and development, start-up, and first- and second-stage financing to companies specializing in communications, computers, electronic

components and instrumentation, energy/natural resources, genetic engineering, industrial products and equipment, and medical-related products and services. Will not consider real estate. Preferred investment is $750,000 or more, with a minimum of $250,000.

★1172★ Edelson Technology Partners
Park 80, W., Plaza 2   Phone: (201) 843-4474
Saddle Brook, NJ 07662   Fax: (201) 843-5479
Venture capital partnership. Prefers high-tech industries.

★1173★ Edison Venture Fund
997 Lenox Dr., Ste. 3   Phone: (609) 896-1900
Lawrenceville, NJ 08648   Fax: (609) 896-0066
Private venture capital firm. No industry preference.

★1174★ Eslo Capital Corp.
212 Wright St.   Phone: (201) 242-4488
Newark, NJ 07114
A small business investment corporation. No industry preference.

★1175★ First Princeton Capital Corp.
5 Garret Mt. Plaza   Phone: (201) 278-8111
West Patterson, NJ 07424
A small business investment corporation. No industry preference.

★1176★ InnoVen Group
Park 80, W., Plaza 1   Phone: (201) 845-4900
Saddle Brook, NJ 07662   Fax: (201) 845-3388
Venture capital firm. Prefers to invest in high-tech industries. Also prefers second- and third-round financing.

★1177★ Johnston Associates, Inc.
181 Cherry Valley Rd.   Phone: (609) 924-3131
Princeton, NJ 08540
Venture capital supplier. Seeks high-technology, medical, and biologically oriented concepts in order to initiate and guide the establishment of a company by providing seed and start-up financing.

★1178★ MBW Management, Inc. (Morristown)
365 South St., 2nd Fl.   Phone: (201) 285-5533
Morristown, NJ 07960   Fax: (201) 285-5108

★1179★ Monmouth Capital Corp.
125 Wyckoff Rd.   Phone: (908) 542-4927
PO Box 335
Eatontown, NJ 07724
A small business investment corporation. No industry preference.

★1180★ Rutgers Minority Investment Co.
180 University   Phone: (201) 648-5627
Newark, NJ 07102   Fax: (201) 648-1100
A minority enterprise small business investment corporation. No industry preference.

★1181★ Taroco Capital Corp.
716 Jersey Ave.   Phone: (201) 798-5000
Jersey City, NJ 07302
A minority enterprise small business investment corporation. Focuses on Chinese-Americans.

★1182★ Unicorn Ventures II LP
6 Commerce Dr.   Phone: (908) 276-7880
Cranford, NJ 07016   Fax: (908) 276-5635
A small business investment corporation. No industry preference.

★1183★ Unicorn Ventures Ltd.
6 Commerce Dr.   Phone: (908) 276-7880
Cranford, NJ 07016   Fax: (908) 276-5635
A small business investment corporation. No industry preference.

New Mexico

★1184★ Ads Capital Corp.
524 Camino del Monte Sol   Phone: (505) 983-1769
Santa Fe, NM 87501
Venture capital supplier. Prefers to invest in manufacturing or distribution companies.

★1185★ Albuquerque SBIC
501 Tijeras Ave., NW   Phone: (505) 247-0145
PO Box 487
Albuquerque, NM 87102
A small business investment corporation. No industry preference.

★1186★ Associated Southwest Investors, Inc.
6400 Uptown Blvd., NE, Ste. 580W   Phone: (505) 881-0066
Albuquerque, NM 87110   Fax: (505) 881-0118
A minority enterprise small business investment corporation. No industry preference.

★1187★ Industrial Development Corp. of Lea County
PO Box 1376   Phone: (505) 397-2039
Hobbs, NM 88240   Fax: (505) 392-2300
Certified development company.

★1188★ New Mexico Business Development Corp.
6001 Marble, NE, No. 6   Phone: (505) 268-1316
Albuquerque, NM 87110
Venture capital firm.

★1189★ Roswell Community Development Co.
PO Drawer 70   Phone: (505) 623-5695
Roswell, NM 88202-0070   Fax: (505) 624-6870
Small business investment company. Industry preferences include manufacturing, electronics, aviation, and food processing.

★1190★ Sage Management Partners
1650 University Blvd., NE, Ste. 317   Phone: (505) 242-3755
Albuquerque, NM 87102   Fax: (505) 242-8307

★1191★ United Mercantile Capital Corp.
PO Box 37487   Phone: (505) 883-8201
Albuquerque, NM 87176
A small business investment corporation. Manufacturing and distribution preferred.

New York

★1192★ Adler and Co. (New York City)
250 Park Ave., 6th Fl.   Phone: (212) 682-3800
New York, NY 10177
Private venture capital supplier. Provides all stages of financing.

★1193★ Alan Patricof Associates, Inc. (New York)
545 Madison Ave., 15th Fl.
New York, NY 10022
Phone: (212) 753-6300
Fax: (212) 319-6155
Venture capital firm. No industry preference, but avoids real estate. Interested in all stages of financing. Also handles turnarounds and leveraged buy outs.

★1194★ AMEV Capital Corp.
1 World Trade Center, Ste. 5001
New York, NY 10048-0024
Phone: (212) 775-9100
Fax: (212) 775-9773
Venture capital supplier. Diversified with respect to industry, stage of development, and geographic location. Preferred minimum investment is $1 million; able to lead or participate in syndications of up to $10 million or more.

★1195★ ASEA-Harvest Partners II
767 3rd Ave.
New York, NY 10017
Phone: (212) 838-7776
A small business investment corporation. No industry preference.

★1196★ Avdon Capital Corp.
1413 Ave. J
Brooklyn, NY 11230
Phone: (718) 692-0950
Fax: (718) 253-0146
A minority enterprise small business investment corporation. No industry preference.

★1197★ Bessemer Venture Partners (New York)
630 5th Ave.
New York, NY 10111
Phone: (212) 708-9303
Fax: (212) 265-5826
Venture capital partnership. No industry preference.

★1198★ Bohlen Capital Corp.
767 3rd Ave.
New York, NY 10017
Phone: (212) 838-7776
A small business investment corporation. No industry preference.

★1199★ BT Capital Corp.
280 Park Ave.
New York, NY 10017
Phone: (212) 454-1916
Fax: (212) 454-2421
A small business investment corporation. No industry preference.

★1200★ The Business Loan Center
704 Broadway, 2nd Fl.
New York, NY 10003
Phone: (212) 979-6688
A small business loan company.

★1201★ Capital Investors and Management Corp.
210 Canal St., Ste. 611
New York, NY 10013
Phone: (212) 964-2480
Fax: (212) 349-9160
A minority enterprise small business investment corporation. No industry preference.

★1202★ Chase Manhattan Capital Corp.
1 Chase Manhattan Plaza, 7th Fl.
New York, NY 10081
Phone: (212) 552-7138
A small business investment corporation. No industry preference.

★1203★ Chemical Venture Capital Associates
885 3rd Ave., Ste. 810
New York, NY 10028
Phone: (212) 230-2280
A small business investment corporation. No industry preference.

★1204★ Citicorp Venture Capital Ltd. (New York City)
Citicorp Center
399 Park Ave., 6th Fl.
New York, NY 10043
Phone: (212) 559-1127
Fax: (212) 888-2940
A small business investment corporation. Invests in the fields of information processing and telecommunications, transportation and energy, and health care; provides financing to companies in all stages of development. Also provides capital for leveraged buy out situations.

★1205★ Clinton Capital Corp.
477 Madison Ave., Ste. 707
New York, NY 10022
Phone: (212) 888-7440
Fax: (212) 486-2935
A small business investment corporation. No industry preference.

★1206★ CMNY Capital Co., Inc.
135 E. 57th St., Ste. 27
New York, NY 10022-2101
Phone: (212) 909-8400
A small business investment corporation. No industry preference.

★1207★ Concord Partners (New York City)
535 Madison Ave.
New York, NY 10022
Phone: (212) 906-7000
Fax: (212) 308-5107
Venture capital partnership. Diversified in terms of stage of development, industry classification, and geographic location. Areas of special interest include computer software, electronics, environmental services, biopharmaceuticals, health care, and oil and gas.

★1208★ CVC Capital Corp.
131 E. 62nd St.
New York, NY 10021
Phone: (212) 319-7210
A minority enterprise small business investment corporation. Radio and television industries preferred.

★1209★ CW Group, Inc.
1041 3rd Ave.
New York, NY 10021
Phone: (212) 308-5266
Fax: (212) 644-0354
Venture capital supplier. Interest is in the health care field, including diagnostic and therapeutic products, services, and biotechnology. Invests in companies at developing and early stages.

★1210★ Demuth, Folger and Terhune
1 Exchange Plaza at 55 Broadway
New York, NY 10006
Phone: (212) 509-5580
Fax: (212) 363-7965
Venture capital firm with preferences for technology, medical, and health care investments.

★1211★ E. M. Warburg, Pincus and Co., Inc.
466 Lexington Ave.
New York, NY 10017
Phone: (212) 878-0600
Fax: (212) 878-9351
Privately owned financial services organization engaged in venture banking and investment counseling.

★1212★ Edwards Capital Co.
2 Park Ave.
New York, NY 10016
Phone: (212) 686-2568
Fax: (212) 213-6234
A small business investment corporation. Transportation industry preferred.

★1213★ Elf Aquitainers, Inc.
280 Park Ave., 36th Fl., W.
New York, NY 10017-1216
Phone: (212) 922-3000
Fax: (212) 922-3001

★1214★ Elk Associates Funding Corp.
600 3rd Ave., 38th Fl.
New York, NY 10016
Phone: (212) 972-8550
A minority enterprise small business investment corporation. Transportation industry preferred.

★1215★ Elron Technologies, Inc.
850 3rd Ave., 10th Fl.
New York, NY 10022
Phone: (212) 935-3110
Fax: (212) 935-3882
Venture capital supplier. Provides incubation and start-up financing to high-technology companies.

★1216★ Equico Capital Management Corp.
1221 Avenue of the Americas, 32nd Fl.   Phone: (212) 382-8000
New York, NY 10020   Fax: (212) 382-8118

A minority enterprise small business investment corporation. No industry preference.

★1217★ Euclid Partners Corp.
50 Rockefeller Plaza, Ste. 1022   Phone: (212) 489-1770
New York, NY 10020   Fax: (212) 757-1686

Venture capital firm. Prefers medical, biotechnology, computer, and information processing industries.

★1218★ European Development Corp.
767 3rd Ave., 7th Fl.   Phone: (212) 838-9721
New York, NY 10017

A small business investment corporation. No industry preference.

★1219★ Everlast Capital Corp.
350 5th Ave., Ste. 2805   Phone: (212) 695-3910
New York, NY 10118

A minority enterprise small business investment corporation. No industry preference.

★1220★ Exim Capital Corp.
9 E. 40th St., 15th Fl.   Phone: (212) 683-3375
New York, NY 10016   Fax: (212) 689-4118

A minority enterprise small business investment corporation. No industry preference.

★1221★ Fair Capital Corp.
212 Canal St., Ste. 611   Phone: (212) 964-2480
New York, NY 10013

A minority enterprise small business investment corporation. No industry preference.

★1222★ Ferranti High Technology, Inc.
501 Madison Ave.   Phone: (212) 688-9828
New York, NY 10022   Fax: (212) 688-9710

A small business investment corporation. No industry preference.

★1223★ First Boston Corp.
55 E. 52nd St.   Phone: (212) 909-2000
New York, NY 10055   Fax: (212) 308-9151

Investment banker. Provides financing to the oil and gas pipeline, hydroelectric, medical technology, consumer products, electronics, aerospace, and telecommunications industries. Supplies capital for leveraged buy outs.

★1224★ First Century Partnership
1345 Avenue of the Americas, 47th Fl.   Phone: (212) 698-6382
New York, NY 10105   Fax: (212) 698-5517

Private venture capital firm. Diversified investment policy; minimum investment is $1.5 million.

★1225★ First New York Management Co.
2 Metrotech Center   Phone: (718) 797-5990
Brooklyn, NY 11201   Fax: (718) 797-5986

A small business investment corporation. No industry preference.

★1226★ Fleet Venture Partners (New York)
60 E. 42nd St., 3rd Fl.   Phone: (212) 907-5000
New York, NY 10017   Fax: (212) 907-5615

★1227★ Fortieth Street Venture Partners (New York City)
545 Madison Ave., Ste. 800   Phone: (212) 421-0045
New York, NY 10022   Fax: (212) 421-0067

Venture capital firm. No industry preference.

★1228★ Foster Management Co.
437 Madison Ave.   Phone: (212) 753-4810
New York, NY 10022

Private venture capital supplier. Not restricted to specific industries or geographic locations; diversified with investments in the health care, transportation, broadcasting, communications, energy, and home furnishings industries. Investments range from $2 to $15 million.

★1229★ Franklin Corp.
767 5th Ave.   Phone: (212) 486-2323
New York, NY 10153   Fax: (212) 755-5451

A small business investment corporation. No industry preference; no start-ups.

★1230★ Fresh Start Venture Capital Corp.
313 W. 53rd St., 3rd Fl.   Phone: (212) 265-2249
New York, NY 10019

A minority enterprise small business investment corporation. No industry preference.

★1231★ Fundex Capital Corp.
525 Northern Blvd.   Phone: (516) 466-8551
Great Neck, NY 11021

A small business investment corporation. No industry preference.

★1232★ Hambro International Venture Fund (New York)
650 Madison Ave.   Phone: (212) 223-7400
New York, NY 10022   Fax: (212) 223-0305

Venture capital supplier. Seeks to invest in mature companies as well as in high-technology areas from start-ups to leveraged buy outs.

★1233★ Hanover Capital Corp.
315 E. 62nd St., 6th Fl.   Phone: (212) 980-9670
New York, NY 10021

A small business investment corporation. No industry preference.

★1234★ Harvest Partners Ltd.
767 3rd Ave.   Phone: (212) 838-7776
New York, NY 10017

A small business investment corporation. No industry preference.

★1235★ Harvest Ventures, Inc. (New York)
767 3rd Ave.   Phone: (212) 838-7776
New York, NY 10017

Private venture capital supplier. Prefers to invest in high-technology, growth-oriented companies with proprietary technology, large market potential, and strong management teams.

★1236★ Holding Capital Management Corp.
685 5th Ave., 14th Fl.   Phone: (212) 486-6670
New York, NY 10022

A small business investment corporation. No industry preference.

★1237★ Hycliff Partners
6 E. 43rd St., 28th Fl.   Phone: (212) 986-7500
New York, NY 10017   Fax: (212) 986-8178

★1238★ Ibero-American Investors Corp.
104 Scio St.   Phone: (716) 262-3440
Rochester, NY 14604   Fax: (716) 262-3441

A minority enterprise small business investment corporation. No industry preference.

★1239★ Inco Venture Capital Management
1 New York Plaza, 37th Fl.   Phone: (212) 612-5620
New York, NY 10004   Fax: (212) 612-5617

★1240★ **Instoria, Inc., and Providentia Ltd.**
15 W. 54th St., 2nd Fl.   Phone: (212) 957-3232
New York, NY 10019   Fax: (212) 957-9866

★1241★ **Intercontinental Capital Funding Corp.**
432 Park Ave., S., Ste. 1307   Phone: (212) 689-2484
New York, NY 10016

A minority enterprise small business investment corporation. No industry preference.

★1242★ **Interstate Capital Co., Inc.**
380 Lexington Ave.   Phone: (212) 972-3445
New York, NY 10017

A small business investment corporation. No industry preference.

★1243★ **J. H. Whitney and Co. (New York)**
630 5th Ave., Rm. 3200   Phone: (212) 757-0500
New York, NY 10111   Fax: (212) 247-3146

★1244★ **Jafco America Ventures, Inc. (New York)**
180 Maiden Ln., 21st Fl.   Phone: (212) 269-8900
New York, NY 10038   Fax: (212) 269-3212

★1245★ **Josephberg, Grosz and Co., Inc.**
538 Madison Ave., 5th Fl.   Phone: (212) 935-1050
New York, NY 10022

Venture capital firm. No industry preference.

★1246★ **Kwiat Capital Corp.**
576 5th Ave.   Phone: (212) 391-2461
New York, NY 10036

A small business investment corporation. No industry preference.

★1247★ **Lambda Funds Management**
41 E. 57th St., 31st Fl.   Phone: (212) 838-0005
New York, NY 10022

Venture capital partnership. Prefers to invest in manufacturers, service companies, and leveraged buy outs.

★1248★ **Lawrence, Tyrrell, Ortale, and Smith (New York)**
515 Madison Ave., 29th Fl.   Phone: (212) 826-9080
New York, NY 10022   Fax: (212) 759-2561

Venture capital firm. No industry preference.

★1249★ **M and T Capital Corp.**
1 M & T Plaza   Phone: (716) 842-5881
Buffalo, NY 14240   Fax: (716) 842-4436

A small business investment corporation. No industry preference.

★1250★ **Manufacturers Hanover Venture Capital Corp.**
270 Park Ave.   Phone: (212) 270-3220
New York, NY 10017   Fax: (212) 983-0626

Venture capital and leveraged buy out firm. Invests in leveraged buy outs and growth equity.

★1251★ **McCown De Leeuw and Co. (New York)**
900 3rd Ave., 28th Fl.   Phone: (212) 418-6539
New York, NY 10022   Fax: (212) 418-6584

★1252★ **Medallion Funding Corp.**
205 E. 42nd St., Ste. 2020   Phone: (212) 682-3300
New York, NY 10017   Fax: (212) 983-0351

A minority enterprise small business investment corporation. Transportation industry preferred.

★1253★ **Michael Fredericks and Co.**
2 Wall St., 4th Fl.   Phone: (212) 732-1600
New York, NY 10005   Fax: (212) 732-1872

Private venture capital supplier. Provides start-up and early-stage financing, and supplies capital for buy outs and acquisitions.

★1254★ **Minority Equity Capital Co., Inc.**
42 W. 38th St., Ste. 604   Phone: (212) 768-4240
New York, NY 10018   Fax: (212) 768-4246

A minority enterprise small business investment corporation. No industry preference.

★1255★ **ML Technology Ventures, L.P.**
World Financial Center
North Tower
New York, NY 10281-1325

★1256★ **Monsey Capital Corp.**
125 Rte. 59   Phone: (914) 425-2229
Monsey, NY 19052   Fax: (914) 425-9419

A minority enterprise small business investment corporation. No industry preference.

★1257★ **Morgan Stanley Venture Partners L.P. (New York)**
1251 Avenue of the Americas   Phone: (212) 703-4000
New York, NY 10020   Fax: (212) 703-6503

★1258★ **N.A.B. Nordic Investors Ltd.**
c/o DnC Capital Corp.   Phone: (212) 315-6500
600 5th Ave.   Fax: (212) 307-1589
New York, NY 10020

★1259★ **Nazem and Co.**
600 Madison Ave., 14th Fl.   Phone: (212) 644-6433
New York, NY 10022   Fax: (212) 751-2731

Venture capital fund. Electronics and medical industries preferred. Will provide seed and first- and second-round financing.

★1260★ **New Enterprise Associates (New York City)**
119 E. 55th St.   Phone: (212) 371-8210
New York, NY 10022   Fax: (212) 486-4227

Venture capital supplier. Concentrates on technology-based industries that have the potential for product innovation, rapid growth, and high profit margins. Investments range from $250,000 to $2 million. Past investments have been made in the following industries: computer software, medical and life sciences, computers and peripherals, communications, semiconductors, specialty retailing, energy and alternative energy, defense electronics, and materials and specialty chemicals. Management must demonstrate intimate knowledge of its marketplace, and have a well-defined strategy for achieving strong market penetration.

★1261★ **New York Job Development Authority**
605 3rd Ave.   Phone: (212) 818-1700
New York, NY 10158   Fax: (212) 682-1476

Assists companies wishing to expand or build new facilities, thereby retaining existing jobs or creating new employment opportunities. Provides loans and loan guarantees.

★1262★ **Noro Capital Ltd.**
767 3rd Ave., 7th Fl.   Phone: (212) 838-9720
New York, NY 10017

A small business investment corporation. No industry preference.

★1263★ Norstar Venture Capital
1 Norstar Plaza   Phone: (518) 447-4043
Albany, NY 12203

Venture capital supplier. No industry preference. Typical investment is between $500,000 and $1 million.

★1264★ North American Funding Corp.
177 Canal St.   Phone: (212) 226-0080
New York, NY 10013   Fax: (212) 219-1379

A minority enterprise small business investment corporation. No industry preference.

★1265★ Northwood Ventures (New York)
485 Underhill Blvd., Ste. 205   Phone: (516) 364-5544
Syosset, NY 11791   Fax: (516) 364-0879

★1266★ Novatech Resource Corp.
375 Park Ave., Ste. 3401   Phone: (212) 832-1988
New York, NY 10152   Fax: (212) 826-1497

★1267★ NYBDC Capital Corp.
PO Box 738   Phone: (518) 463-2268
Albany, NY 12201-0738   Fax: (518) 463-0240

A small business investment corporation. No industry preference.

★1268★ Pan Pac Capital Corp.
121 E. Industry Ct.   Phone: (516) 586-7653
Deer Park, NY 11729

A minority enterprise small business investment corporation. No industry preference.

★1269★ Pierre Funding Corp.
605 3rd Ave.   Phone: (212) 490-9540
New York, NY 10158   Fax: (212) 573-6329

A minority enterprise small business investment corporation. No industry preference.

★1270★ Prospect Group, Inc.
667 Madison Ave., 25th Fl.   Phone: (212) 758-8500
New York, NY 10021   Fax: (212) 593-6127

Venture capital supplier. Investments focus on computer communications and software products, fiber optics, genetic engineering, biotechnology, health care management, and solar energy.

★1271★ Prudential Venture Capital
717 5th Ave., 11th Fl.   Phone: (212) 753-0901
New York, NY 10022   Fax: (212) 826-6798

Venture capital fund. Specialty retailing, medical and health services, communications, and technology companies preferred. Will provide $3 to $7 million in equity financing for later-stage growth companies.

★1272★ R and R Financial Corp.
1451 Broadway   Phone: (212) 790-1400
New York, NY 10036   Fax: (212) 790-0900

A small business investment corporation. No industry preference.

★1273★ Rain Hill Group, Inc.
90 Broad St.   Phone: (212) 483-9162
New York, NY 10004   Fax: (212) 514-6217

Prefers to invest in manufacturing endeavors, including chemical, pharmaceutical, food, electronics, and environmental, but not in distribution, financial services, or retail.

★1274★ Rand SBIC, Inc.
1300 Rand Bldg.   Phone: (716) 853-0802
Buffalo, NY 14203   Fax: (716) 854-8480

A small business investment corporation. No industry preference.

★1275★ Revere Fund, Inc.
575 5th Ave., 18th Fl.   Phone: (212) 661-5290
New York, NY 10017   Fax: (212) 661-5294

Business development company. Obtains current income and capital gains through the purchase of private placements of corporate debt obligations with equity features or preferred stock. Invests in companies with sales over $10 million. Maximum investment in a new portfolio company is approximately $1 million. Does not invest in oil and gas and real estate.

★1276★ Rothschild Ventures, Inc.
1 Rockefeller Plaza   Phone: (212) 757-6000
New York, NY 10020   Fax: (212) 977-3150

Private venture capital firm. Prefers seed and all later-stage financing.

★1277★ S and S Venture Associates Ltd.
80 Cuttermill Rd.   Phone: (516) 773-4000
Great Neck, NY 11021

★1278★ Salomon Brothers Venture Capital
7 World Trade Center, 31st Fl.   Phone: (212) 783-5928
New York, NY 10048   Fax: (212) 783-3350

★1279★ Schroder Ventures
787 7th Ave., 29th Fl.   Phone: (212) 841-3880
New York, NY 10019   Fax: (212) 582-1405

★1280★ 767 Limited Partnership
767 3rd Ave.   Phone: (212) 838-7776
New York, NY 10017

A small business investment corporation. No industry preference.

★1281★ Situation Ventures Corp.
56-20 59th St.   Phone: (718) 894-2000
Maspeth, NY 11378

A minority enterprise small business investment company.

★1282★ Sprout Group (New York City)
140 Broadway, 42nd Fl.   Phone: (212) 504-3600
New York, NY 10005   Fax: (212) 504-3444

Venture capital supplier.

★1283★ TCW Special Placements Fund I
200 Park Ave., Ste. 2200   Phone: (212) 972-1440
New York, NY 10166   Fax: (212) 297-4025

Venture capital fund. Companies with sales of $25 to $100 million preferred. Will provide up to $20 million in later-stage financing for recapitalizations, restructuring management buy outs, and general corporate purposes.

★1284★ Telesciences Capital Corp.
8625 5th Ave.   Phone: (718) 748-7213
Brooklyn, NY 11209

A small business investment corporation. No industry preference.

★1285★ Tessler and Cloherty, Inc.
155 Main St.   Phone: (914) 265-4244
Cold Spring, NY 10516

A small business investment corporation. No industry preference.

★1286★ TLC Funding Corp.
660 White Plains Rd.   Phone: (914) 332-5200
Tarrytown, NY 10591   Fax: (914) 332-5660

A small business investment corporation. No industry preference.

★1287★ Triad Capital Corp. of New York
960 Southern Blvd. Phone: (212) 589-5000
Bronx, NY 10459 Fax: (212) 589-5101

A minority enterprise small business investment corporation. No industry preference.

★1288★ Vega Capital Corp.
720 White Plains Rd. Phone: (914) 472-8550
Scarsdale, NY 10583

A small business investment corporation. Diversified industry preferences.

★1289★ Venture Capital Fund of America, Inc.
509 Madison Ave., Ste. 812 Phone: (212) 838-5577
New York, NY 10022 Fax: (212) 838-7614

★1290★ Venture Opportunities Corp.
110 E. 59th St., 29th Fl. Phone: (212) 832-3737
New York, NY 10022

A minority enterprise small business investment corporation. No industry preference. Second- or third-stage for expansion, mergers, or acquisitions. No start-up or seed capital investments.

★1291★ Venture SBIC, Inc.
249-12 Jericho Tpke. Phone: (516) 352-0068
Floral Park, NY 11001 Fax: (516) 775-5707

A small business investment corporation. No industry preference.

★1292★ Warburg Pincus Ventures, Inc. (New York)
466 Lexington Ave. Phone: (212) 878-0600
New York, NY 10017 Fax: (212) 878-9351

Private venture capital firm. No industry preference.

★1293★ Weiss, Peck and Greer Venture Partners L.P. (New York)
1 New York Plaza Phone: (212) 908-9500
New York, NY 10004 Fax: (212) 908-9652

★1294★ Welsh, Carson, Anderson, and Stowe
200 Liberty, Ste. 3601 Phone: (212) 945-2000
New York, NY 10281 Fax: (212) 945-2016

Venture capital partnership. High-technology industries preferred. Also interested in leveraged buy outs. Minimum investment is $5 million.

★1295★ Winfield Capital Corp.
237 Mamaroneck Ave. Phone: (914) 949-2600
White Plains, NY 10605

A small business investment corporation. No industry preference.

★1296★ Wolfensohn Associates, L.P. (New York)
599 Lexington Ave. Phone: (212) 909-8100
New York, NY 10022 Fax: (212) 909-8158

★1297★ Wood River Capital Corp.
667 Madison Ave., 15th Fl. Phone: (212) 758-8500
New York, NY 10021

A small business investment corporation. No industry preference.

★1298★ Yusa Capital Corp.
622 Broadway, 2nd Fl. Phone: (212) 420-1350
New York, NY 10012 Fax: (212) 420-1355

A minority enterprise small business investment corporation. No industry preference.

North Carolina

★1299★ Delta Capital, Inc.
227 N. Tryon St., Ste. 201 Phone: (704) 372-1410
Charlotte, NC 28202

A small business investment corporation. Real estate industry preferred.

★1300★ Kitty Hawk Capital Ltd.
1640 Independence Center Phone: (704) 333-3777
Charlotte, NC 28246

No industry preference. Does not engage in real estate, single store retailing, or natural resources investments.

★1301★ NCNB SBIC Corp.
1 NCNB Plaza (T05-2) Phone: (704) 374-5583
Charlotte, NC 28255

A small business investment corporation. No industry preference. Involved with later-stage, expansion, and buy out investments.

★1302★ NCNB Venture Co.
1 NCNB Plaza (T38) Phone: (704) 374-5723
Charlotte, NC 28255

A small business investment corporation. No industry preference.

★1303★ Southgate Venture Partners (Charlotte)
227 N. Tryon St., Ste. 201 Phone: (704) 372-1410
Charlotte, NC 28202 Fax: (704) 342-4478

Private venture capital firm. Diversified.

Ohio

★1304★ A. T. Capital Corp.
900 Euclid Ave., T-18 Phone: (216) 737-4970
Cleveland, OH 44101 Fax: (216) 737-3177

A small business investment corporation. No industry preference.

★1305★ Banc One Capital Corp.
100 E. Broad St. Phone: (614) 248-5800
Columbus, OH 43271-0251

A small business investment corporation. No industry preference.

★1306★ Brantley Venture Partners, L.P.
20600 Chagrin Blvd., Ste. 520, Tower E. Phone: (216) 283-4800
Cleveland, OH 44122 Fax: (216) 283-5324

★1307★ Cardinal Development Capital Fund I
155 E. Broad Phone: (614) 464-5557
Columbus, OH 43215 Fax: (614) 464-8706

Private venture capital firm. Provides expansion capital to manufacturing and service firms, particularly in Ohio and other parts of the Midwest. Avoids investments in real estate and resource recovery systems.

★1308★ Center City MESBIC, Inc.
40 S. Main St., Ste. 762 Phone: (513) 461-6164
Dayton, OH 45402

A minority enterprise small business investment corporation. Diversified industries.

★1309★ Clarion Capital Corp.
1801 E. 9th, Ste. 1520 Phone: (216) 687-1096
Cleveland, OH 44114 Fax: (216) 694-3545

A small business investment corporation. Interested in specialty chemicals, instrumentation, and health care.

★1310★ First Ohio Capital Corp.
PO Box 1868   Phone: (419) 259-7141
Toledo, OH 43603

A small business investment corporation. No industry preference.

★1311★ Fortieth Street Venture Partners (Cincinnati)
3712 Carew Tower   Phone: (513) 579-0101
Cincinnati, OH 45202   Fax: (513) 579-0971

Venture capital firm. No industry preference.

★1312★ Gries Investment Corp.
Statler Office Tower, Ste. 1500   Phone: (216) 861-1146
1127 Euclid Ave.   Fax: (216) 861-0106
Cleveland, OH 44115

A small business investment corporation. No industry preference.

★1313★ Lubrizol Business Development Co.
29400 Lakeland Blvd.   Phone: (216) 943-4200
Wickliffe, OH 44092   Fax: (216) 944-8112

Venture capital supplier. Provides seed capital and later-stage expansion financing to emerging companies in the biological, chemical, and material sciences whose technology is applicable to and related to the production and marketing of specialty and fine chemicals.

★1314★ Miami Valley Capital, Inc.
Talbott Tower, Ste. 315   Phone: (513) 222-7222
131 N. Ludlow St.   Fax: (513) 222-7448
Dayton, OH 45402

A small business investment corporation. No industry preference.

★1315★ Morgenthaler Ventures
700 National City Bank Bldg.   Phone: (216) 621-3070
Cleveland, OH 44114   Fax: (216) 621-2817

Private venture capital supplier. Provides start-up and later-stage financing to all types of business in North America; prefers not to invest in real estate and mining. Investments range from $500,000 to $3 million.

★1316★ National City Capital Corp.
1965 E. 6th St., No. 400   Phone: (216) 575-2491
Cleveland, OH 44114   Fax: (216) 575-3355

A small business investment corporation. Prefers providing equity for expansion programs, recapitalizations, acquisitions, and management buy outs.

★1317★ Primus Venture Partners
1 Cleveland Center, Ste. 2700   Phone: (216) 621-2185
1375 E. 9th St.   Fax: (216) 621-4543
Cleveland, OH 44114

Venture capital partnership. Provides seed, early-stage, and expansion financing to companies located in Ohio and the Midwest. Does not engage in gas, oil, or real estate investments.

★1318★ River Capital Corp. (Cleveland)
796 Huntington Bldg.   Phone: (216) 781-3655
Cleveland, OH 44115

A small business investment corporation. No industry preference.

★1319★ Rubber City Capital Corp.
1144 E. Market St.   Phone: (216) 796-9167
Akron, OH 44316

A minority enterprise small business investment corporation. Prefers investing with tire and automotive supply companies.

★1320★ Scientific Advances, Inc.
601 W. 5th Ave.   Phone: (614) 424-7005
Columbus, OH 43201   Fax: (614) 424-4874

Venture capital partnership. Seeks small to medium-sized specialty chemicals or natural gas companies with innovative, proven technologies.

★1321★ Seed One
Park Pl.   Phone: (216) 650-2338
10 W. Streetsboro St.
Hudson, OH 44236

Private venture capital firm. No industry preference. Equity financing only.

★1322★ Society Venture Capital Corp.
800 Superior Ave., 14th Fl.   Phone: (216) 689-5776
Cleveland, OH 44114   Fax: (216) 689-3204

A small business investment corporation. Diversified industry preferences.

★1323★ Tomlinson Capital Corp.
13700 Broadway   Phone: (216) 587-3400
Garfield Heights, OH 44125

A small business investment corporation. Miniature supermarket industry preferred.

Oklahoma

★1324★ Alliance Business Investment Co. (Tulsa)
1 Williams Center, Ste. 2000   Phone: (918) 584-3581
Tulsa, OK 74172   Fax: (918) 582-3403

A small business investment corporation. No industry preference.

★1325★ Davis Venture Partners (Tulsa)
1 Williams Center, Ste. 2000   Phone: (918) 584-7272
Tulsa, OK 74172   Fax: (918) 582-3403

★1326★ Rubottom, Dudash and Associates, Inc.
4870 S. Lewis, Ste. 180   Phone: (918) 742-3031
Tulsa, OK 74105

Management and investment consultants. Emphasis on retail, wholesale, and light fabrication.

Oregon

★1327★ InterVen II L.P.
227 SW Pine St., Ste. 200   Phone: (503) 223-4334
Portland, OR 97204

A small business investment corporation. No industry preference.

★1328★ Olympic Venture Partners II (Portland)
10300 SW Greenburg Rd., Ste. 440   Phone: (503) 245-5900
Portland, OR 97223   Fax: (503) 245-1083

★1329★ Trendwest Capital Corp.
803 Main St., Ste. 404   Phone: (503) 882-8059
Klamath Falls, OR 97601

A small business investment corporation. No industry preference.

Pennsylvania

★1330★ Alan Patricof Associates, Inc. (Radnor)
5 Radnor Corp. Center, Ste. 470   Phone: (215) 687-3030
Radnor, PA 19087   Fax: (215) 687-8520

★1331★ Alliance Enterprise Corp.
1801 Market St., 3rd Fl.   Phone: (215) 977-3925
Philadelphia, PA 19103   Fax: (215) 977-6579

A minority enterprise small business investment corporation. Prefers broadcasting, manufacturing, and fast-food investments.

★1332★ Capital Corp. of America
225 S. 15th St., Ste. 920   Phone: (215) 732-1666
Philadelphia, PA 19102

A small business investment corporation. No industry preference.

★1333★ Core States Enterprise Fund
PO Box 7618, F.C. 1-8-7-67   Phone: (215) 973-6519
Philadelphia, PA 19101-7618   Fax: (215) 973-6900

Venture capital supplier. Invests with any industry except real estate or construction. Minimum investment is $1 million.

★1334★ Enterprise Venture Capital Corp. of Pennsylvania
551 Main St., Ste. 303   Phone: (814) 535-7597
Johnstown, PA 15901   Fax: (814) 535-8677

A small business investment corporation. No industry preference.

★1335★ Erie SBIC
32 W. 8th St., Ste. 615   Phone: (814) 453-7964
Erie, PA 16501

A small business investment corporation. No industry preference.

★1336★ First SBIC of California (Washington)
Security Pacific Capital Corp.   Phone: (412) 223-0707
PO Box 512
Washington, PA 15301

A small business investment corporation. Backs proven general management in any industry at any stage. Investment size ranges from $200,000 to $10,000,000.

★1337★ First Valley Capital Corp.
640 Hamilton Mall   Phone: (215) 776-6766
Allentown, PA 18101

A small business investment corporation. No industry preference.

★1338★ Fostin Capital Corp.
681 Andersen Dr.   Phone: (412) 928-1400
Pittsburgh, PA 15220

★1339★ Genesis Seed Management Co.
c/o Howard, Lawson & Co.   Phone: (215) 988-0010
2 Penn Center Plaza
Philadelphia, PA 19102

★1340★ Greater Philadelphia Venture Capital Corp., Inc.
225 S. 15th St., Ste. 920   Phone: (215) 732-3415
Philadelphia, PA 19102

A minority enterprise small business investment corporation. No industry preference.

★1341★ Hillman Medical Ventures, Inc. (Berwyn)
1235 Westlakes Dr., No. 395   Phone: (215) 251-0600
Berwyn, PA 19312   Fax: (215) 251-0606

★1342★ Keystone Venture Capital Management Co.
121 S. Broad St., Ste. 310   Phone: (215) 985-5519
Philadelphia, PA 19107

Private venture capital partnership. Provides equity-based expansion financing to companies in the computer, communications, automated equipment, medical, and other industries. Also provides financing for small leveraged buy outs.

★1343★ Loyalhanna Venture Fund
PO Box 36   Phone: (412) 928-1440
Ligonier, PA 15658   Fax: (412) 928-0108

Venture capital firm. No industry preference.

★1344★ Meridian Capital Corp.
Horsham Business Center   Phone: (215) 957-7500
455 Business Center Dr., Ste. 200   Fax: (215) 957-7521
Horsham, PA 19044

A small business investment corporation. No industry preference.

★1345★ Meridian Venture Partners
The Fidelity Court Bldg., Ste. 220   Phone: (215) 254-2999
259 Radnor-Chester Rd.   Fax: (215) 254-2996
Radnor, PA 19087

★1346★ Philadelphia Ventures
200 S. Broad St., 8th Fl.   Phone: (215) 732-4445
Philadelphia, PA 19102   Fax: (215) 732-4644

Venture capital partnership. Provides start-up and early-stage financing to companies offering products or services based on technology or other proprietary capabilities. Industries of particular interest are information processing equipment and services, medical products and services, data communications, and industrial automation. Initial investments range from $500,000 to $2 million. Also supplies capital for buy out situations.

★1347★ PNC Venture Capital Group
Pittsburgh National Bank Bldg.   Phone: (412) 762-2248
5th Ave. & Wood, 19th Fl.
Pittsburgh, PA 15222

A small business investment corporation. Prefers to invest in later-stage and leveraged buy out situations. Requires that sales range from $5-$50 million; and EBIT from $1-$5 million.

★1348★ S. R. One Ltd.
Fidelity Court Bldg., Ste. 190   Phone: (215) 254-2944
259 Radnor Chester Rd.   Fax: (215) 254-2940
Radnor, PA 19087

Investments limited to life sciences and health care, including companies involved in human diagnostics or therapeutics, pharmaceuticals, or biologicals.

★1349★ Salween Financial Services, Inc.
228 N. Pottstown Pke.   Phone: (215) 524-1880
Exton, PA 19341

A minority enterprise small business investment corporation. No industry preference.

★1350★ Security Pacific Capital Corp. (Washington)
PO Box 512   Phone: (412) 223-0707
Washington, PA 15301

Wholly-owned bank subsidiary. Seeks proven CEOs in any industry at any stage for $250,000 to $10,000,000 direct investment.

★1351★ VenWest, Inc.
Westinghouse Credit Corp.
1 Oxford Center, 7th Fl.   Phone: (412) 393-3162
301 Grant St.   Fax: (412) 393-3158
Pittsburgh, PA 15219

★1352★ Zero Stage Capital Co., Inc. (State College)
1346 S. Atherton St.   Phone: (814) 231-1330
State College, PA 16801   Fax: (814) 231-1333
Venture capital firm. Industry preferences include high-technology start-up companies located in central and northern Pennsylvania.

Puerto Rico

★1353★ North America Investment Corp.
PO Box 1831   Phone: (809) 754-6177
Hato Rey, PR 00919   Fax: (809) 754-6181
A minority enterprise small business investment corporation. Diversified industry preferences.

Rhode Island

★1354★ Domestic Capital Corp.
815 Reservoir Ave.   Phone: (401) 946-3310
Cranston, RI 02910
A small business investment corporation. No industry preference.

★1355★ Fleet Venture Partners (Providence)
111 Westminster St.   Phone: (401) 278-6770
Providence, RI 02903   Fax: (401) 751-1274
Venture capital fund. Managed leveraged buy outs.

★1356★ Monarch Narragansett Venture, Inc.
Fleet Center, 9th Fl.   Phone: (401) 751-1000
50 Kennedy Plaza
Providence, RI 02903
A small business investment corporation. No industry preference.

★1357★ Moneta Capital Corp.
99 Wayland Ave.   Phone: (401) 454-7500
Providence, RI 02906-4314   Fax: (401) 455-3636
A small business investment corporation. No industry preference.

★1358★ Old Stone Capital Corp.
1 Old Stone Sq.   Phone: (401) 278-2559
Providence, RI 02903
A small business investment corporation. Real estate industry preferred.

South Carolina

★1359★ Charleston Capital Corp.
111 Church St.   Phone: (803) 723-6464
PO Box 328
Charleston, SC 29402
A small business investment corporation. Prefers secured investments. Does not engage in pure equity transactions.

★1360★ Floco Investment Co., Inc.
PO Box 919   Phone: (803) 389-2731
Lake City, SC 29560
A small business investment corporation. Invests only in retail grocery stores (supermarkets).

★1361★ Lowcountry Investment Corp.
4401 Daley St.   Phone: (803) 554-9880
PO Box 10447   Fax: (803) 745-2735
Charleston, SC 29411
A small business investment corporation. Grocery industry preferred.

★1362★ Reedy River Ventures, Inc.
PO Box 17526   Phone: (803) 232-6198
Greenville, SC 29606   Fax: (803) 271-8374
A small business investment corporation. No industry preference, but does not invest with retail, high-technology, or early-stage companies.

Tennessee

★1363★ American Health Capital Associates, L.P.
278 Franklin Rd., Ste. 240   Phone: (615) 377-0416
Brentwood, TN 37027   Fax: (615) 377-0416

★1364★ Chickasaw Capital Corp.
67 Madison Ave.   Phone: (901) 523-6404
PO Box 387   Fax: (901) 578-2939
Memphis, TN 38147
A minority enterprise small business investment corporation. No industry preference.

★1365★ Financial Resources, Inc.
200 Jefferson Ave., Ste. 750   Phone: (901) 527-9411
Memphis, TN 38103   Fax: (901) 523-9212
A small business investment corporation. No industry preference.

★1366★ International Paper Capital Formation, Inc.
Tower 2, 4th Fl.   Phone: (901) 763-6282
6400 Poplar Ave., 4-061   Fax: (901) 763-7278
Memphis, TN 38197
A minority enterprise small business investment corporation. Focuses on small and disadvantaged businesses in the paper industry. Involvement includes expansion, refinancing, and acquisitions, but no start-up projects. Requires a minimum investment of $50,000 to $300,000.

★1367★ Lawrence, Tyrrell, Ortale, and Smith (Nashville)
3100 W. End Ave., Ste. 500   Phone: (615) 383-0982
Nashville, TN 37203
Private venture capital firm. Prefers to invest in health care industries.

★1368★ Massey Burch Investment Group
310 25th Ave., N., Ste. 103   Phone: (615) 329-9448
Nashville, TN 37203   Fax: (615) 329-9237
Venture capital firm. No industry preference. Investments range from $1 to $3 million.

★1369★ Tennessee Venture Capital Corp.
PO Box 3001   Phone: (615) 244-6935
Nashville, TN 37219
A minority enterprise small business investment corporation. No industry preference.

★1370★ Valley Capital Corp.
Krystal Bldg., Ste. 212   Phone: (615) 265-1557
100 W. Martin Luther King Blvd.
Chattanooga, TN 37402
A minority enterprise small business investment corporation. Diversified industry preferences. Prefers business dealings in the southeast within a four-hour driving radius of the firm.

Texas

★1371★ Acorn Ventures, Inc.
520 Post Oak Blvd., Ste. 130   Phone: (713) 622-9595
Houston, TX 77027
No industry preference.

★1372★ Alliance Business Investment Co. (Houston)
910 Louisiana   Phone: (713) 224-8224
3990 1 Shell Plaza
Houston, TX 77002
A small business investment corporation. No industry preference.

★1373★ Austin Ventures L.P.
1300 Norwood Tower   Phone: (512) 479-0055
114 W. 7th St.   Fax: (512) 476-3952
Austin, TX 78701
Administers investments through two funds, Austin Ventures L.P. and Rust Ventures L.P. Prefers to invest in start-up/emerging growth companies located in the southwest, and in special situations such as buy outs, acquisitions, and mature companies. No geographic limitations are placed on later-stage investments. Past investments have been made in broadcasting, cable television, data communications, factory automation, lodging, telecommunications, and general manufacturing. Investments range from $1 to $2 million for start-up financing, and from $2 to $5 million for special situation companies.

★1374★ Banc One Capital Partners Corp.
300 Crescent Ct., Ste. 1600   Phone: (214) 979-4361
Dallas, TX 75201   Fax: (214) 979-4355
A small business investment corporation. Specializes in later-stage investments and buy outs. Prefers basic manufacturing, wholesale, distribution, and health care with a minimum investment of $1 million and minimum revenues of $15 million.

★1375★ BCM Technologies, Inc.
1709 Dryden, Ste. 901   Phone: (713) 795-0105
Houston, TX 77030   Fax: (713) 795-4602

★1376★ Brittany Capital Co.
7557 Rambler Rd., Ste. 818   Phone: (214) 363-1541
Dallas, TX 75231-4165   Fax: (214) 363-2670
A small business investment corporation. No industry preference.

★1377★ Capital Marketing Corp.
100 Nat Gibbs Dr.   Phone: (817) 431-7309
PO Box 1000
Keller, TX 76248
A small business investment corporation.

★1378★ Capital Southwest Corp.
12900 Preston Rd., Ste. 700   Phone: (214) 233-8242
Dallas, TX 75230
A small business investment corporation engaged in manufacturing and retail investments. Prefers later-stage and management buy out opportunites.

★1379★ Charter Venture Group, Inc.
2600 Citadel Plaza Dr., Ste. 600   Phone: (713) 863-0704
Houston, TX 77008   Fax: (713) 691-7566
A small business investment corporation. No industry preference.

★1380★ Chen's Financial Group, Inc.
6671 SW Fwy., Ste. 505   Phone: (713) 772-8868
Houston, TX 77074-2213   Fax: (713) 772-1271
A minority enterprise small business investment corporation. No industry preference.

★1381★ Citicorp Venture Capital Ltd. (Dallas)
2001 Ross, Ste. 3050   Phone: (214) 880-9670
Dallas, TX 75201   Fax: (214) 953-1495
A small business investment corporation. No industry preference.

★1382★ Criterion Investments
1000 Louisiana St., Ste. 6200   Phone: (713) 751-2400
Houston, TX 77002   Fax: (713) 751-2626
Venture capital fund. Raises venture capital. Interested in companies headquartered in the Sunbelt region.

★1383★ Cureton & Co., Inc.
2050 Houston Natural Gas Bldg.   Phone: (713) 658-9806
Houston, TX 77002   Fax: (713) 658-0476
Prefers oilfield service and environmental manufacturing, service, and distribution in Texas, Oklahoma, and Louisiana. Requires $1 million minimum financing.

★1384★ Davis Venture Partners (Dallas)
2121 San Jacinto St., Ste. 975   Phone: (214) 954-1822
Dallas, TX 75201   Fax: (214) 969-0256

★1385★ Dougery, Wilder and Howard (Dallas)
17950 Preston Rd., Ste. 990   Phone: (214) 250-0909
Dallas, TX 75252
Venture capital supplier. Prefers to invest in small to medium-sized companies headquartered or primarily operating in the west and southwest. The typical company sought is in the computers/communications or medical/biotechnological industries.

★1386★ Energy Capital Corp.
808 Travis St.   Phone: (713) 236-0006
Houston, TX 77002
A small business investment corporation. Specializes in oil and gas energy industries.

★1387★ Enterprise Capital Corp.
515 Post Oak Blvd., Ste. 310   Phone: (713) 621-9444
Houston, TX 77027
A small business investment corporation. No industry preference.

★1388★ FCA Investment Co.
5847 San Felipe, Ste. 850   Phone: (713) 781-2857
Houston, TX 77057   Fax: (713) 781-7195
A small business investment corporation. No industry preference; no start-ups.

★1389★ Interwest Partners (Dallas)
1 Galleria Tower   Phone: (214) 392-7279
13355 Noel Rd., Ste. 1375/LB 65   Fax: (214) 490-6348
Dallas, TX 75240

★1390★ Mapleleaf Capital Corp.
55 Waugh, Ste. 710   Phone: (713) 880-4494
Houston, TX 77007   Fax: (713) 880-4494
A small business investment corporation. No industry preference.

★1391★ May Financial Corp.
8333 Douglass Ave., Ste. 400   Phone: (214) 987-5200
Lock Box 82   Fax: (214) 987-1994
Dallas, TX 75225
Brokerage firm working with a venture capital firm. Prefers food, oil and gas, and electronics industries.

★1392★ MESBIC Financial Corp. of Houston
811 Rusk, Ste. 201   Phone: (713) 228-8321
Houston, TX 77002   Fax: (713) 546-2229
A minority enterprise small business investment corporation. Prefers fast-growth, viable, minority-owned companies.

★1393★ MESBIC Ventures, Inc.
12655 N. Central Expy., Ste. 710   Phone: (214) 991-1597
Dallas, TX 75243

A minority enterprise small business investment corporation. No industry preference.

★1394★ Minority Enterprise Funding, Inc.
17300 El Camino Real, Ste. 107B   Phone: (713) 488-4919
Houston, TX 77058   Fax: (713) 488-3786

A minority enterprise small business investment corporation. No industry preference.

★1395★ MSI Capital Investments
6500 Greenville Ave., Ste. 720   Phone: (214) 265-1801
Dallas, TX 75206-1012   Fax: (214) 265-1804

★1396★ Phillips-Smith Specialty Retail Group
15110 Dallas Pkwy., Ste. 310   Phone: (214) 387-0725
Dallas, TX 75248

Prefers retail industry investments.

★1397★ Red River Ventures, Inc.
400 Chisholm Pl., Ste. 402   Phone: (214) 422-4999
Plano, TX 75075   Fax: (214) 423-3116

★1398★ Retzloff Capital Corp.
PO Box 41250   Phone: (713) 466-4690
Houston, TX 77240-1250   Fax: (713) 466-3238

A small business investment corporation. No industry preference.

★1399★ Rotan Mosle Technology, Inc.
NCNB Bldg., Ste. 3800   Phone: (713) 236-3180
700 Louisiana St.   Fax: (713) 223-3815
Houston, TX 77002

★1400★ San Antonio Venture Group, Inc.
2300 W. Commerce St., Ste. 300   Phone: (512) 978-0513
San Antonio, TX 78207   Fax: (512) 978-0540

A small business investment corporation. No industry preference.

★1401★ SBI Capital Corp.
PO Box 570368   Phone: (713) 975-1188
Houston, TX 77257-0368

A small business investment corporation. No industry preference; Texas businesses only.

★1402★ South Texas SBIC
First Capital Group of Texas
PO Box 15616   Phone: (512) 573-5151
San Antonio, TX 78212   Fax: (512) 736-5449

A small business investment corporation. No industry preference.

★1403★ Southern Orient Capital Corp.
2419 Fannin, Ste. 200   Phone: (713) 225-3369
Houston, TX 77002

A minority enterprise small business investment corporation. No industry preference.

★1404★ Southwest Enterprise Associates
2 Lincoln Centre, Ste. 1266   Phone: (214) 991-1620
5420 LBJ Fwy.   Fax: (214) 490-4051
Dallas, TX 75240

Venture capital supplier. Concentrates on technology-based industries that have the potential for product innovation, rapid growth, and high profit margins. Investments range from $250,000 to $1.5 million. Past investments have been made in the following industries: computer software, medical and life sciences, computers and peripherals, communications, semiconductors, and defense electronics. Management

must demonstrate intimate knowledge of its marketplace and have a well-defined strategy for achieving strong market penetration.

★1405★ Southwest Venture Partnerships
300 Convent, Ste. 1400   Phone: (512) 227-1010
San Antonio, TX 78205   Fax: (512) 227-1343

Venture capital partnership. Invests in maturing companies located primarily in the southwest. Invests in $1 million range.

★1406★ SRB Partners Fund Ltd.
Sevin Rosen Management Co.   Phone: (214) 702-1100
2 Galleria Tower, Ste. 1670   Fax: (214) 960-1749
Dallas, TX 75240

Venture capital firm. Prefers financing the electronic and information sciences.

★1407★ Sunwestern Capital Corp.
3 Forest Plaza   Phone: (214) 239-5650
12221 Merit Dr., Ste. 1300   Fax: (214) 701-0024
Dallas, TX 75251

A small business investment corporation. No industry preference.

★1408★ Tenneco Ventures, Inc.
PO Box 2511   Phone: (713) 757-8229
Houston, TX 77252   Fax: (713) 651-1666

Venture capital supplier. Provides financing to small, early-stage growth companies. Areas of interest include energy-related technologies, factory automation, biotechnology, and health care services. Prefers to invest in Texas-based companies, but will consider investments elsewhere within the United States. Investments range from $250,000 to $1 million; will commit additional funds over several rounds of financing, and will work with other investors to provide larger financing.

★1409★ Texas Commerce Investment Co.
PO Box 2558   Phone: (713) 236-4553
Houston, TX 77252-8032   Fax: (713) 236-5803

A small business investment corporation. No industry preference.

★1410★ United Oriental Capital Corp.
908 Town & Country Blvd., Ste. 310 Phone: (713) 461-3909
Houston, TX 77024

A minority enterprise small business investment corporation. No industry preference.

★1411★ Ventex Partners Ltd.
1000 Louisiana St., Ste. 1110   Phone: (713) 659-7860
Houston, TX 77002

A small business investment corporation. Prefers manufacturing and processing, retail, wholesale, medical and health services, communications, transportation, and computer-related investments.

★1412★ Western Financial Capital Corp.
17772 Preston Rd.   Phone: (214) 380-0044
Dallas, TX 75252   Fax: (214) 380-1371

A small business investment corporation. Medical industry preferred.

Utah

★1413★ Deseret Certified Development Corp.
(Midvale)
7050 Union Park Center, No. 570   Phone: (801) 566-1163
Midvale, UT 84047

Maintains an SBA(504) loan program, designed for community development and job creation, and an intermediary loan program, through Farmer's Home Administration.

Virginia

★1414★ Deseret Certified Development Corp. (Provo)
2696 N. University, Ste. 240
Provo, UT 84604
Phone: (801) 374-1025
Fax: (801) 374-1051

Maintains an SBA(504) loan program, designed for community development and job creation, and an intermediary loan program, through Farmer's Home Administration.

★1415★ Utah Ventures
419 Wakara Way
Salt Lake City, UT 84108
Phone: (801) 583-5922
Fax: (801) 583-4105

Virginia

★1416★ Atlantic Venture Partners (Alexandria)
101 N. Union St., Ste. 220
Alexandria, VA 22314-3217
Phone: (703) 548-6026
Fax: (703) 683-5348

Private venture capital partnership. Provides all stages of financing primarily to high-technology businesses. Initial investments generally range from $250,000 to $750,000. Under certain circumstances, an investment of less than $100,000 or more than $1 million will be undertaken. May provide financing of $1 to $10 million (or more) in participation with other venture capital firms.

★1417★ Atlantic Venture Partners (Richmond)
PO Box 1493
Richmond, VA 23212
Phone: (804) 644-5496

Private venture capital partnership. Provides all stages of financing primarily to high-technology businesses. Initial investments generally range from $250,000 to $750,000. Under certain circumstances, an investment of less than $100,000 or more than $1 million will be undertaken. May provide financing of $1 to $10 million (or more) in participation with other venture capital firms.

★1418★ Basic Investment Corp.
6723 Whittier Ave.
McLean, VA 22101
Phone: (703) 356-4300

A minority enterprise small business investment corporation. No industry preference.

★1419★ Crestar Capital
PO Box 1776
Richmond, VA 23214
Phone: (804) 643-7358
Fax: (804) 648-3313

A small business investment corporation. No industry preference.

★1420★ Hampton Roads SBIC (HRSBIC)
420 Bank St.
Norfolk, VA 23510
Phone: (804) 622-2312
Fax: (804) 622-5563

A small business investment corporation. Prefers investing in small businesses in the immediate geographical area of Virginia.

★1421★ James River Capital Associates
PO Box 1776
Richmond, VA 23214
Phone: (804) 643-7323
Fax: (804) 648-3313

A small business investment corporation. No industry preference.

★1422★ Metropolitan Capital Corp.
2550 Huntington Ave.
Alexandria, VA 22303
Phone: (703) 960-4698
Fax: (703) 329-4623

A small business investment corporation. Equity or loans with equity features. No retail or real estate.

★1423★ Quest Tech Capital Corp.
9990 Lee Hwy.
Fairfax, VA 22030-1720
Phone: (212) 922-9320

A small business investment corporation. No industry preference.

★1424★ River Capital Corp. (Falls Church)
2830 Graham Rd.
Falls Church, VA 22042-1638
Phone: (703) 207-9460
Fax: (703) 560-7329

A small business investment corporation. No industry preference.

★1425★ Sovran Funding Corp.
Sovran Center, 3rd Fl.
1 Commercial Pl.
Norfolk, VA 23510
Phone: (804) 441-4041
Fax: (804) 441-4725

A small business investment corporation. No industry preference.

★1426★ Walnut Capital Corp. (Vienna)
8300 Boone Blvd., Ste. 780
Vienna, VA 22182
Phone: (703) 448-3771
Fax: (703) 448-7751

A small business investment corporation. No industry preferences.

Washington

★1427★ Cable and Howse Ventures (Bellevue)
777 108th Ave., Ste. 2300
Bellevue, WA 98004
Phone: (206) 646-3030
Fax: (206) 646-3041

Venture capital investor. Provides start-up and early-stage financing to enterprises in the western United States, although a national perspective is maintained. Interests lie in proprietary or patentable technology. Investments range from $50,000 to $2 million.

★1428★ Capital Resource Corp.
1001 Logan Bldg.
Seattle, WA 98101
Phone: (206) 623-6550
Fax: (206) 623-1816

A small business investment corporation. Prefers investments in the Pacific northwest.

★1429★ Norwest Venture Capital Management, Inc. (Bellevue)
777 108th Ave., NE, Ste. 2460
Bellevue, WA 98004-5117
Phone: (206) 646-3444

A small business investment corporation. No industry preference.

★1430★ Olympic Venture Partners II (Kirkland)
2420 Carillon Pt.
Kirkland, WA 98033
Phone: (206) 889-9192

Prefers funding for early-stage, technology-oriented companies.

★1431★ Palms and Co., Inc.
515 Lake St., S., Ste. 103
Kirkland, WA 98033-6437
Phone: (206) 828-6774
Fax: (206) 827-5528

Private venture capital supplier and investment banker. Provides all stages of financing to companies located anywhere in the United States and Canada.

★1432★ The Phoenix Partners
1000 2nd Ave., Ste. 3600
Seattle, WA 98104
Phone: (206) 624-8968
Fax: (206) 624-1907

★1433★ Pierce Nordquist Associates L.P.
5350 Carillon Pte.
Kirkland, WA 98033
Phone: (206) 822-4100
Fax: (206) 827-4086

Prefers investing in advanced materials and related technologies.

★1434★ Washington Trust Equity Corp.
Washington Trust Financial Center
PO Box 2127
Spokane, WA 99210
Phone: (509) 455-3821
Fax: (509) 353-4125

A small business investment corporation. No industry preference.

Wisconsin

★1435★ Bando McGlocklin Capital Corp.

13555 Bishops Ct., Ste. 205   Phone: (414) 784-9010
Brookfield, WI 53005   Fax: (414) 784-3426

A small business investment corporation. Fixed assets-based lender.

★1436★ Bank One Venture

111 E. Wisconsin Ave.   Phone: (414) 765-2274
Milwaukee, WI 53202   Fax: (414) 765-2235

A small business investment corporation. No industry preference.

★1437★ Capital Investments, Inc.

744 N. 4th St., Ste. 540   Phone: (414) 273-6560
Milwaukee, WI 53203

A small business investment corporation. Investments are limited to later-stage companies located in the Midwest, involved in manufacturing, distribution, and service.

★1438★ Future Value Ventures, Inc.

622 N. Water St., Ste. 500   Phone: (414) 278-0377
Milwaukee, WI 53202   Fax: (414) 278-0377

A small business investment corporation. Provides financing to companies owned by socially and economically disadvantaged persons. Managers should have proven management ability. Prefers businesses with potential to create jobs. Flexible regarding industry preference.

★1439★ InvestAmerica Venture Group, Inc.
(Milwaukee)

600 E. Mason St.   Phone: (414) 276-3839
Milwaukee, WI 53202   Fax: (414) 276-1885

A small business investment corporation. Prefers later-stage and acquisition financings of $750,000 to $2,000,000 with equity participation. Will not consider real estate investments.

★1440★ Lubar and Co., Inc.

3380 1st Wisconsin Center   Phone: (414) 291-9000
Milwaukee, WI 53202

Private investment and management firm. Interests are in energy-related products and services, manufacturers of industrial products, and unique or niche businesses.

★1441★ M and I Ventures Corp.

770 N. Water St.   Phone: (414) 765-7910
Milwaukee, WI 53202   Fax: (414) 765-7850

A small business investment corporation. Prefers manufacturing, distribution, electronics, and technology-related industries.

★1442★ Venture Investors of Wisconsin, Inc.

565 Science Dr.   Phone: (608) 233-3070
Madison, WI 53711   Fax: (608) 238-5120

Venture capital firm. No industry preference.

★1443★ Wind Point Partners (Racine)

1525 Howe St.   Phone: (414) 631-4030
Racine, WI 53403   Fax: (414) 631-4975

Section 18

Public Sector Venture Capital Sources

"It takes courage to be creative, just as soon as you have a new idea, you're in the minority of one."—E. Paul Torrance

In addition to private sector venture capital sources, there are also pools of public sector venture capital. California, Connecticut, Massachusetts, and Ohio are just a few of the states that have venture capital funds.

During the 1980s, U.S. states and some cities developed a range of policies and programs to stimulate economic development. The promotion of scientific and technological innovation has been a particular focus of many of these programs.

The following list contains descriptions of 100 such programs along with relevant contacts.

Don't expect state venture capital to be any easier to secure than private sector venture capital. It is simply another opportunity available.

Section Arrangement

The following entries are arranged geographically by state, then alphabetically by venture capital source name. All entries in this section are referenced in the Master Index.

Alaska

★1444★ Alaska Science and Technology Foundation
550 W. 7th Ave., Ste. 360   Phone: (907) 562-5818
Anchorage, AK 99501-3555
John Sibert, Executive Director

Arkansas

★1445★ Arkansas Science and Technology Authority
Seed Capital Investment Program
100 Main St., Ste. 450   Phone: (501) 371-3554
Little Rock, AR 72201
Dr. John W. Ahlen, President

California

★1446★ California Department of Commerce
Innovation Development Loan Program
1211 L St., Ste. 600
Sacramento, CA 95814

★1447★ California Energy Commission
Energy Technology Advancement Program
MS-43   Phone: (916) 324-3553
1516 9th St.
Sacramento, CA 95814
George Simons, Program Manager

★1448★ California Office of Small Business
New Product Development Program
California Department of Commerce   Phone: (916) 445-6545
1121 L. St., Ste. 501
Sacramento, CA 95814
Richard Nelson, Executive Director

Qualified affiliated regional corporations are authorized to make royalty-based investments in small firms to bring to market new products and processes that are beyond the theoretical development stage.

Colorado

★1449★ Colorado Advanced Technology Institute
Colorado Institute for Artificial Intelligence
1625 Broadway, Ste. 700   Phone: (303) 620-4777
Denver, CO 80202
Dr. J. Jeffrey Richardson, Director

Connecticut

★1450★ Connecticut Innovations, Inc.
845 Brook St.   Phone: (203) 258-4305
Rocky Hill, CT 06067-3405
David C. Driver

Provides seed capital for promising new technology-based firms. Managed by Connecticut Innovations Incorporated through its Ely Whitney Fund.

★1451★ High Technology Start-Up Financing Program
845 Brook St.
Rocky Hill, CT 06067-3405
David C. Driver, Executive Director

★1452★ Connecticut Office of Business Development
Technology Investment Fund
770 Chapel St.   Phone: (203) 787-8031
New Haven, CT 06510
Salvatore J. Brancati, Director

Manages the state-sponsored seed capital funds.

★1453★ Connecticut Seed Ventures Group
200 Fisher Dr.   Phone: (203) 677-0183
Avon, CT 06001
Sam McKay, General Partner

★1454★ Technology Assistance Center
SBIR Assistance Grant Program
845 Brook St.   Phone: (203) 258-4305
Rocky Hill, CT 06067-4305
Eric Ott

Delaware

★1455★ University of Delaware
Delaware Research Partnership
Office of Research & Patents   Phone: (302) 451-2136
201 Hullihen Hall
Newark, DE 19716
Robert D. Varrin, Assoc. Provost for Research

Hawaii

★1456★ Hawaii Department of Business and Economic Development
Hawaii Innovation Development Program
Business Services   Phone: (808) 548-4608
PO Box 2359
Honolulu, HI 96804
Thomas J. Smyth, Division Head

★1457★ Hawaii High Technology Development Corporation
Hawaii SBIR Matching Grant Program
300 Kahelu Ave., #35   Phone: (808) 625-5293
Mililani, HI 96789
William M. Bass, Jr., Executive Director

★1458★ Hawaii Strategic Fund Corporation
Hawaii Strategic Fund
Hawaii Dept. of Business & Economic   Phone: (808) 548-7743
   Development
737 Bishop St.
PO Box 2359
Honolulu, HI 96804
Carl Swanholm, Science & Tech. Officer

Illinois

★1459★ Frontenac Venture Company
Illinois Venture Fund
208 S. LaSalle St., Ste. 1900   Phone: (312) 368-0044
Chicago, IL 60604
Joan Fortune, General Partner

★1460★ Illinois Department of Commerce & Community Affairs
Business Innovation Fund
620 E. Adams St.   Phone: (217) 782-3891
Springfield, IL 62701
Dick LeGrand

★1461★ Technology Venture Investment Program
100 W. Randolph St.   Phone: (312) 814-2387
Ste. 3-400
Chicago, IL 60601
Grant Skeens, Manager

★1462★ Illinois Office of Technology Advancement and Development
Technology Challenge Grant Fund
100 W. Randolph St., Ste. 3-400   Phone: (312) 814-3982
Chicago, IL 60601
John Straus, Director

★1463★ Northwestern University/Evanston Research Park
1710 Orrington Ave., Ste. 200   Phone: (708) 475-7170
Evanston, IL 60201
Ronald C. Kysiak, Executive Director

Indiana

★1464★ Indiana Corporation for Innovation Development
Venture Capital Program
One N. Capitol Ave., Ste. 250   Phone: (317) 635-7325
Indianapolis, IN 46204
Mr. Marion Dietrich, President

★1465★ Indiana Corporation for Science and Technology
One N. Capitol Ave., Ste. 925   Phone: (317) 635-3058
Indianapolis, IN 46204
Delbert J. Schuh, II, Acting President

★1466★ SBIR Bridge Grant Funding Pilot
One N. Capitol Ave., Ste. 925
Indianapolis, IN 46204-2242
Delbert J. Schuh, II, Acting President

★1467★ Indiana Institute for New Business Ventures
Enterprise Advisory Service
One N. Capitol Ave., Ste. 420   Phone: (317) 264-2820
Indianapolis, IN 46204
Bruce K. Kidd, Dir. of Business Assistance

Iowa

★1468★ Invest America
Iowa Venture Capital Fund
300 Mor America Bldg.
Cedar Rapids, IA 52401
David Schroder

★1469★ Iowa Department of Economic Development
Iowa Product Development Corporation
200 E. Grand Ave.   Phone: (515) 281-3704
Des Moines, IA 50309
Jude T. Conway, Acting Director

★1470★ SBIR Proposal Grants - Iowa Innovation Program
200 E. Grand Ave.   Phone: (515) 281-3036
Des Moines, IA 50309
Doug Getter

★1471★ Wallace Technology Transfer Foundation
First Interstate Bank Bldg.   Phone: (515) 243-1487
317 6th Ave., Ste. 840
Des Moines, IA 50309
Daniel K. Dittemore, Deputy Director

Kansas

★1472★ Kansas Department of Commerce
Kansas Venture and Seed Capital Tax Credit Program
Division of Existing Industry   Phone: (913) 296-5298
   Development
400 SW 8th, 5th Fl.
Topeka, KS 66603-3957
Mr. R.S. Montgomery, Director

Allows an income tax credit of 25% on investments in certified local seed capital pools. The certified pools must have a minimum private cash investment of $200,000. The Kansas Venture Capital Company Act allows the transfer and acquisition of state income tax credits for investors in certified Kansas venture capital companies. The state also provides a tax credit for certain research and development activities.

★1473★ Kansas Technology Enterprise Corp.
SBIR Bridge Grant Program
112 W. 6th St., Ste. 400   Phone: (913) 296-5272
Topeka, KS 66603
Dr. William Brundage, President

★1474★ Kansas Technology Enterprise Corporation
Ad Astra Fund, L.P.
112 W. 6th St., Ste. 400   Phone: (913) 296-5272
Topeka, KS 66603
Dr. William Brundage, President

  ★1475★ Applied Research Matching Grant Program
  112 W. 6th St., Ste. 400   Phone: (913) 296-5272
  Topeka, KS 66603
  Dr. William Brundage, President

Kentucky

★1476★ Kentucky Cabinet for Economic Development
Commonwealth Venture Fund
Dept. of Finance   Phone: (502) 564-7670
12400 Capital Plaza Tower
Frankfort, KY 40601
Dennis Fleming, Executive Director

  ★1477★ Kentucky Office of Business and Technology
  Capital Plaza Tower, 24th Fl.
  Frankfort, KY 40601
  Debbie Kimbrough, Executive Director

Louisiana

★1478★ Louisiana Economic Development
  Corporation
Louisiana Venture Capital Incentive Program
PO Box 94185   Phone: (504) 342-5675
Baton Rouge, LA 70804-9185
Mike Williams, Executive Director

★1479★ Louisiana Seed Capital Corporation
Louisiana Seed Capital Fund, L.P.
PO Box 3435   Phone: (504) 383-1508
Baton Rouge, LA 70821
Kerry Mitchell, Director

Maine

★1480★ Maine Science and Technology Commission
Centers of Innovation
State House Sta. 147
Augusta, ME 04333   Phone: (207) 289-3703
Robert Kidd, Director

Maryland

★1481★ Maryland Office of Technology Development
Challenge Grant Program
Maryland Dept. of Economic &   Phone: (301) 333-6990
   Employment Development
217 E. Redwood St., 12th Fl.
Baltimore, MD 21202
Selig Solomon, Director

Massachusetts

★1482★ Massachusetts Centers of Excellence
  Corporation
9 Park St.   Phone: (617) 727-7430
Boston, MA 02108
Megan Jones, Executive Director

★1483★ Massachusetts Product Development
  Corporation
55 Union St.   Phone: (617) 727-1133
Boston, MA 02108-2409
Samuel Leiken, President

★1484★ Massachusetts Technology Development
  Corporation
131 State St., Ste. 215   Phone: (617) 723-4920
Boston, MA 02109
John F. Hodgman, President

Michigan

★1485★ Michigan Department of Commerce
Michigan State Research Fund
106 W. Allegan   Phone: (517) 373-7550
Lansing, MI 48913

  ★1486★ Michigan Strategic Fund
  PO Box 30234
  Lansing, MI 48909
  Peter Plastrik, Director

  ★1487★ Michigan Strategic Fund
  Michigan Seed Capital Program
  PO Box 30234
  Lansing, MI 48909
  Peter Plastrik, President

★1488★ Michigan State Retirement System
Venture Capital Division
PO Box 15128   Phone: (517) 373-4330
Lansing, MI 48901
Paul Rice, Administrator

★1489★ Michigan Strategic Fund
Product Development Program
PO Box 30234   Phone: (514) 373-7550
Lansing, MI 48909
Peter Plastrik, President

Minnesota

★1490★ Greater Minnesota Corporation
1250 International Center II   Phone: (612) 338-6666
920 Second Ave., S.
Minneapolis, MN 55402
Jacques Koppel, President

★1491★ IAI Venture Capital Group
Superior Ventures
1100 Dain Tower   Phone: (612) 371-7935
PO Box 357
Minneapolis, MN 55440
Mitchell Dann, Managing Partner

★1492★ Minnesota Higher Education Coordinating
Board
Enterprise Development Partnership
400 Capital Sq.   Phone: (612) 296-9586
550 Cedar St.
St. Paul, MN 55101
Dorothy Dahlenburg, Program Coordinator

Missouri

★1493★ Missouri Department of Economic
Development
Innovation Center Seed Capital Program
PO Box 118   Phone: (314) 751-3906
Jefferson City, MO 65102
Lisa Kane, Manager

★1494★ Missouri IncuTech Foundation
Missouri Enterprise Business Assistance Center
800 W. 14th St.
Rolla, MO 65401
Dennis Roedemeier, Executive Director

★1495★ St. Louis Technology Center, Incorporated
10143 Paget Dr.   Phone: (314) 432-4204
St. Louis, MO 63132
Gene J. Boesch, Managing Director

Montana

★1496★ Montana Science and Technology Alliance
Research and Development Fund
46 N. Last Chance Gulch, Ste. 2B   Phone: (406) 449-2778
Helena, MT 59620
Carl E. Russell, Executive Director

★1497★ Seed Capital Investment Fund
46 W. Last Chanch Gulch, Ste. 28
Helena, MT 59620
Carl E. Russell, Executive Director

Nebraska

★1498★ Nebraska Research and Development
Authority
NBC Center, Ste. 780   Phone: (402) 475-5109
Lincoln, NE 68508
Dr. Jack L. Bishop, Jr., President

★1499★ Nebraska Enterprise Fund
1248 O St., 780 NCV Center
Lincoln, NE 68508-1424
Harvey Schwartz, Director

Makes equity investments in start-up businesses which export at least 80% of their product or service outside Nebraska. The Authority leverages other public and private financial resources for individual projects as appropriate. Long-term venture capital or conventional financing can also be arranged through the Authority.

New Jersey

★1500★ New Jersey Commission on Science and
Technology
SBIR Bridge Grants Program
20 W. State St., CN832   Phone: (609) 633-2740
Trenton, NJ 08625
David Hochman, Interim Executive Director

New Mexico

★1501★ New Mexico Research and Development
Institute
Applied Research Program
1220 S. St. Francis Dr., Ste. 358   Phone: (505) 827-5886
Santa Fe, NM 87501
Ponziano Ferraraccio, Acting Director

★1502★ Entrepreneurial Capital Program
1220 S. St. Francis Dr., Ste. 358   Phone: (505) 827-5886
Santa Fe, NM 87501
Ponziano Ferraraccio, Acting Director

New York

★1503★ New York Department of Economic Development
Industrial Effectiveness Program
1 Commerce Plaza, Rm. 920   Phone: (518) 474-1131
Albany, NY 12245
Larry Barker, Director

★1504★ New York Office of the State Comptroller
New York Business Venture Partnership
Albany, NY 12210   Phone: (518) 474-4003
John Hull, Director of Investments
Invests in state start-up businesses.

★1505★ New York Science and Technology Foundation
Corporation for Innovation Development
99 Washington Ave., Ste. 1730   Phone: (518) 474-4349
Albany, NY 12210
H. Graham Jones, Executive Director

★1506★ Technology and Disabilities Program
99 Washington Ave., Ste. 1730   Phone: (518) 473-9746
Albany, NY 12210
Mark S. Tebanno, Manager

★1507★ New York State Department of Agriculture and Markets
Agricultural Research and Development Grants Program
Division of Agriculture & Support   Phone: (518) 457-7076
 Services
1 Winners Circle, Capital Plaza
Albany, NY 12235-0001
William Kimball, Program Manager

★1508★ New York State Energy Research and Development Authority
Energy Efficiency Assistance Program
2 Rockefeller Plaza   Phone: (518) 465-6251
Albany, NY 12223
Robert Callender, Manager

Assists existing state industries and attracts new industry to the state by providing technical and financial assistance to firms to reduce their energy use and related costs. The program offers financial and technical support for feasibility studies of innovative and energy-efficient process modifications; and demonstration of innovative energy-efficient process technology.

★1509★ New York State Science and Technology Foundation
Small Business Innovation Research Promotion Program
99 Washington Ave., Ste. 1731   Phone: (518) 474-4349
Albany, NY 12210
H. Graham Jones, Executive Director

North Carolina

★1510★ North Carolina Biotechnology Center
79 Alexander Dr., Bldg. 4501   Phone: (919) 541-9366
PO Box 13547
Research Triangle Park, NC 27709
Dr. Charles E. Hammer, President

★1511★ North Carolina Board of Science and Technology
116 W. Jones St.   Phone: (919) 733-6500
Raleigh, NC 27611
Dr. Earl R. MacCormac, Executive Director

★1512★ North Carolina Department of Commerce
North Carolina Technological Development Authority
430 N. Salisbury St.   Phone: (919) 733-7022
Raleigh, NC 27611
Brent Lane, Executive Director

★1513★ North Carolina Public Employee Pension Fund
Venture Capital Investments Program
325 N. Salisbury St.   Phone: (919) 733-9307
Raleigh, NC 27611
C. Douglas Chappell, State Treasurer

★1514★ North Carolina Technological Development Authority
Innovation Research Fund
430 N. Salisbury St.   Phone: (919) 733-7022
Raleigh, NC 27611
Brent Lane, Executive Director

★1515★ New Industry Fund
A component of the Innovation Research Fund seed capital investment strategy. It directly leverages private capital in the commercialization of successful projects. Provides an additional source of equity capital as a catalyst for private sector investments.

Ohio

★1516★ Ohio Business Development Division
Ohio Steel Futures Program
77 S. High, 28th Fl.   Phone: (614) 466-2317
Columbus, OH 43266
Paul Leonard, Lt. Governor, Acting Dir.

★1517★ Ohio Public Employees Retirement Fund
Pension Fund Venture Set-Aside Program
277 E. Town St.   Phone: (614) 466-7320
Columbus, OH 43215
Robert McLaughlin, Investment Officer

★1518★ Ohio's Thomas Edison Program
Edison Seed Development Fund
77 S. High St., 26th Fl.   Phone: (614) 466-3086
Columbus, OH 43266
Chris Coburn, Executive Director

★1519★ Ohio SBIR Program
77 S. High St., 26th Fl.   Phone: (614) 466-5867
Columbus, OH 43266
G. Mark Skinner, Manager

Oklahoma

★1520★ Oklahoma Center for the Advancement of Science and Technology
Oklahoma Health Research Program
205 NW 63rd, Ste. 305   Phone: (405) 848-2633
Oklahoma City, OK 73116-8209
Dr. Carolyn Smith, Executive Director

Funds research conducted by health scientists working at Oklahoma institutions of higher education, nonprofit research organizations, and private business facilities. Projects are funded for $25,000 to $100,000 per year. The Program seeks to improve health care availability and efficiency.

★1521★ SBIR Phase I Incentive Funding and Matching Funds Program
205 NW 63rd St., Ste. 305
Oklahoma City, OK 73116-8209
Dr. Carolyn Smith, Executive Director

★1522★ Venture Capital Tax Credit Program
204 NW 63rd, Ste. 305
Oklahoma City, OK
Dr. Carolyn Smith, Executive Director

Oregon

★1523★ Oregon Pension System
Pension Fund Venture Set-Aside Program
Oregon State Treasury   Phone: (503) 378-4111
159 State Capitol Bldg.
Salem, OR 97310
John B. Fewel, Investment Officer

★1524★ Oregon Resource and Technology Development Corporation
Applied Research Contract Fund
1934 NE Broadway   Phone: (503) 282-4462
Portland, OR 97232-1502
John Beaulieu, President

★1525★ Seed Capital Fund
1934 NE Broadway
Portland, OR 97223
John Beaulieu, President

Pennsylvania

★1526★ Ben Franklin Partnership Program
Seed Venture Capital Fund Program
352 Forum Bldg.   Phone: (717) 787-4147
Harrisburg, PA 17120
Robert Coy, Jr., Director

★1527★ Small Business Research Seed Grant Program
352 Forum Bldg.   Phone: (717) 787-4147
Harrisburg, PA 17120
Robert Coy, Jr., Director

★1528★ Public Schools Employees Retirement Board
Pension Fund Investments Program
SERS   Phone: (717) 787-8540
PO Box 125
Harrisburg, PA 17108

Rhode Island

★1529★ Rhode Island Partnership for Science and Technology
Rhode Island SBIR Support Program
7 Jackson Walkway   Phone: (401) 277-2601
Providence, RI 02903
Bruce Lang, Executive Director

South Carolina

★1530★ Palmetto Seed Capital Fund
1330 Lady St., Ste. 607   Phone: (808) 779-5759
Columbia, SC 29201
Richard F. Bannon, President and CEO

South Dakota

★1531★ Division of Enterprise Initiation
Future Fund Program
Governor's Office of Economic   Phone: (800) 872-6190
  Development
711 Wells Ave.
Pierre, SD 57501-3369
Kenneth Schaak, Director

Tennessee

★1532★ Tennessee Department of Economic and Community Development
Tennessee Growth Fund Program
Rachel Jackson Bldg., 7th Fl.   Phone: (615) 741-6201
Nashville, TN 37219-5308
Philip Trauernicht

Texas

★1533★ Texas Department of Commerce
Office of Advanced Technology
PO Box 12728   Phone: (512) 320-9407
Austin, TX 78711
Keren Ware Cummins, Program Manager

★1534★ Texas Office of Advanced Technology
Product Commercialization Fund
Texas Department of Commerce   Phone: (512) 320-9407
PO Box 12728
Austin, TX 78711
Keren Ware Cummins, Program Manager

★1535★ Texas Office of Advanced Technology
Product Development Fund
Texas Department of Commerce   Phone: (512) 320-9407
PO Box 12728
Austin, TX 78711
Keren Ware Cummins, Program Manager

★1536★ Texas State Pension Fund Organizations
Texas Growth Fund
Permanent University Fund   Phone: (512) 499-4337
210 W. Sixth St.
Austin, TX 78701
Michael E. Patrick, Exec. Vice-Chancellor

Utah

★1537★ Utah Technology Finance Corporation
185 S. State St., Ste. 208   Phone: (801) 364-4346
Salt Lake City, UT 84111
Dr. Richard E. Turley, Executive Director

Virginia

★1538★ Virginia's Center for Innovative Technology
Centers for Entrepreneurship and Business
   Incubators
2214 Rock Hill Rd.   Phone: (703) 689-3025
CIT Tower, Ste. 600
Herndon, VA 22070
David Miller, Project Director

Washington

★1539★ Washington State Investment Board
421 S. Capital Way   Phone: (206) 753-6810
Mail Stop FR-31
Olympia, WA 98504-0916
Basil Schwan, Executive Director

The Board allows its investment managers broad authority to diversify pension fund investments and seek higher rates of return. There are no geographic limitations on investments.

West Virginia

★1540★ West Virginia Economic Development
   Authority
West Virginia Capital Company Credit Program
Capital Complex, Bldg. 6, Rm. 525   Phone: (304) 348-2234
Charleston, WV 25305
Joseph W. Valis, Executive Director

Wisconsin

★1541★ Wisconsin Bureau of Development Finance
SBIR Bridge Financing Program
Wisconsin Department of Development   Phone: (608) 267-9383
123 W. Washington Ave.
PO Box 7970
Madison, WI 53707
Caroline Garber, Director

Provides winners of federal Small Business Innovation Research Phase I awards with loans of up to $40,000 while between Phase I and Phase II.

25% of the loan amount must be matched by the award recipient. If the research project does not lead to product commercialization, the loan is treated as a grant. The state also operates and advisory proposal review program which provides a critique of pre-submission SBIR applicants by a two-person team consisting of a university scientist and a representative from a successful SBIR company.

Wyoming

★1542★ Wyoming Science, Technology and Energy
   Authority
University Sta.   Phone: (307) 721-2345
PO Box 3985
Laramie, WY 82071
Dr. Shelby Gerking, Director

Section 19

Small Business Administration (SBA) Offices

"Ideas, like young wine, should be put in storage and taken up again
only after they have been allowed to ferment
and ripen."—Richard Strauss

The U.S. Small Business Administration (SBA) was created by Congress in 1953 to help America's entrepreneurship form successful small enterprises.

Small businesses are the backbone of the American economy. They create two of every three new jobs, produce 40 percent of the gross national product, and invent more than one half the nation's technological innovations. Today 20 million small companies provide opportunities for all Americans.

Inventors are small businesses, whether they manufacture their concepts themselves or license them to larger manufacturers. As small business owners and operators, inventors must know how to manage and finance their enterprises. This is where SBA can be of assistance.

Through workshops, individual counseling, publications, and videotapes, the SBA helps entrepreneurs understand and meet challenges of operating businesses, such as financing, marketing, and management.

The SBA offers a full range of specialized financing, including the following programs:

*International Trade Loan Guarantees offer financing to U.S. based facilities or equipment for producing goods or services for export.

*Export Revolving Line of Credit Guarantees help firms penetrate foreign markets.

*Small Loan Guarantees help businesses needing capital of $50,000 or less.

*Handicapped Assistance Loans help businesses owned by physically disabled persons and private nonprofit organizations that employ disabled persons and operate in their interest.

*Loans to Vietnam Era Veterans help start, operate, and expand small businesses.

SBA provides small businesses with long-term loans and venture capital by licensing, regulating, and investing in privately owned and managed Small Business Investment Companies.

The SBA works with three partner organizations to help with management:

*Service Corps of Retired Executives (SCORE) provides training and one-on-one counseling free of charge.

*Small Business Development Centers (SBDC) provide training, counseling, research, and other specialized assistance at more than 600 locations nationwide. **See Section 15 on page 145 for a listing of the 50 members of the ASBDC.**

*Small Business Institutes (SBI) are located at more than 500 universities and free management studies performed by advanced business students under faculty direction.

The SBA has business development specialists available in 115 field offices nationwide. The listing follows.

Small Business Answer Desk: SBA toll-free number: 1-800-827-5722.

Section Arrangement

The following entries are arranged geographically by state, then alphabetically by office name. All offices are referenced in the Master Index.

Alabama

★1543★ Birmingham District Office
2121 8th Ave. N., Ste. 200   Phone: (205) 731-1344
Birmingham, AL 35203-2398

Alaska

★1544★ Anchorage District Office
222 W. 8th Ave., Rm. A36   Phone: (907) 271-4022
Anchorage, AK 99501

Arizona

★1545★ Phoenix District Office
2005 N. Central Ave., 5th Fl.   Phone: (602) 379-3737
Phoenix, AZ 85004

★1546★ Tucson Post of Duty
300 W. Congress St., Rm. 3V   Phone: (602) 629-6715
Tucson, AZ 85701

Arkansas

★1547★ Little Rock District Office
320 W. Capitol Ave., Rm. 601   Phone: (501) 378-5871
Little Rock, AR 72201

California

★1548★ Fresno District Office
2719 N. Air Fresno Dr.   Phone: (209) 487-5189
Fresno, CA 93727-1547

★1549★ Glendale District Office
330 N. Grand Blvd.   Phone: (213) 894-2956
Glendale, CA 91203

★1550★ Sacramento Branch Office
660 J St., Rm. 215   Phone: (916) 551-1426
Sacramento, CA 95814-2413

★1551★ Sacramento Disaster Area Office 4
1825 Bell St., Suite 208   Phone: (916) 978-4578
Sacramento, CA 95825

★1552★ San Diego District Office
880 Front St., Ste. 4-S-29   Phone: (619) 557-5440
San Diego, CA 92188

★1553★ San Francisco District Office
211 Main St., 4th Fl.   Phone: (415) 974-0649
San Francisco, CA 94105-1988

★1554★ San Francisco Regional Office
450 Golden Gate Ave.   Phone: (415) 556-7489
San Francisco, CA 94102

Covers activities in Arizona, California, Guam, Hawaii, and Nevada.

★1555★ Santa Ana District Office
901 W. Civic Ctr. Dr., Rm. 160   Phone: (714) 836-2494
Santa Ana, CA 92703

★1556★ Ventura Post of Duty
6477 Telephone Rd., Ste. 10   Phone: (805) 642-1866
Ventura, CA 93003-4459

Colorado

★1557★ Denver District Office
721 19th St., Rm. 407   Phone: (303) 844-6501
Denver, CO 80201-0660

★1558★ Denver Regional Office
999 18th St., Ste. 701   Phone: (303) 294-7001
Denver, CO 80202

Covers activities in Colorado, Montana, North Dakota, South Dakota, Utah, and Wyoming.

Connecticut

★1559★ Hartford District Office
330 Main St., 2nd Fl.   Phone: (203) 240-4700
Hartford, CT 06106

Delaware

★1560★ Wilmington Branch Office
920 N. King St., Rm. 412   Phone: (302) 573-6295
Wilmington, DE 19801

District of Columbia

★1561★ Washington District Office
1111 18th St., NW, 6th Fl.   Phone: (202) 634-1500
Washington, DC 20036

Florida

★1562★ Coral Gables District Office
1320 S. Dixie Hwy., Ste. 501   Phone: (305) 536-5521
Coral Gables, FL 33146

★1563★ Jacksonville District Office
7825 Baymeadows Way, Ste. 100-B   Phone: (904) 443-1900
Jacksonville, FL 32256-7504

★1564★ Tampa Post of Duty
700 Twiggs St., Rm. 607   Phone: (813) 228-2594
Tampa, FL 33602

★1565★ West Palm Beach Post of Duty
5601 Corporate Way, Ste. 402   Phone: (407) 689-3922
West Palm Beach, FL 33407

Georgia

★1566★ Atlanta Disaster Area Office 2
120 Ralph McGill Blvd., 14th Fl.   Phone: (404) 347-3771
Atlanta, GA 30308

★1567★ Atlanta District Office
1720 Peachtree St., NW, 6th Fl.   Phone: (404) 347-2441
Atlanta, GA 30367

★1568★ Atlanta Regional Office
1375 Peachtree St., NE, 5th Fl.   Phone: (404) 347-2797
Atlanta, GA 30367-8102

Covers activities in Alabama, Florida, Georgia, Kentucky, Mississippi, North Carolina, South Carolina, and Tennessee.

★1569★ Statesboro Post of Duty
52 N. Main St., Rm. 225   Phone: (912) 489-8719
Statesboro, GA 30458

Guam

★1570★ Agana Branch Office
Pacific Daily News Bldg., Rm. 508   Phone: (671) 472-7277
Agana, GU 96910

Hawaii

★1571★ Honolulu District Office
300 Ala Moana Blvd., Rm. 2213   Phone: (808) 541-2990
Honolulu, HI 96850

Idaho

★1572★ Boise District Office
1020 Main St., Ste. 290   Phone: (208) 334-1696
Boise, ID 83702

Illinois

★1573★ Chicago District Office
219 S. Dearborn St., Rm. 437   Phone: (312) 353-4528
Chicago, IL 60604-1779

★1574★ Chicago Regional Office
230 S. Dearborn St., Rm. 510   Phone: (312) 353-0359
Chicago, IL 60604-1593

Covers activities in Illinois, Indiana, Michigan, Minnesota, Ohio, and Wisconsin.

★1575★ Springfield, Illinois Branch Office
511 W. Capitol St., Ste. 302   Phone: (217) 492-4416
Springfield, IL 62704

Indiana

★1576★ Indianapolis District Office
575 N. Pennsylvania St., Rm. 578   Phone: (317) 226-7272
Indianapolis, IN 46204-1584

Iowa

★1577★ Cedar Rapids District Office
373 Collins Rd. NE, Rm. 100   Phone: (319) 393-8630
Cedar Rapids, IA 52402-3118

★1578★ Des Moines District Office
210 Walnut St., Rm. 749   Phone: (515) 284-4422
Des Moines, IA 50309

Kansas

★1579★ Wichita District Office
110 E. Waterman St., 1st Fl.   Phone: (316) 269-6571
Wichita, KS 67202

Kentucky

★1580★ Louisville District Office
600 M.L. King, Jr. Pl., Rm. 188   Phone: (502) 582-5976
Louisville, KY 40202

Louisiana

★1581★ New Orleans District Office
1661 Canal St., Ste. 2000   Phone: (504) 589-6685
New Orleans, LA 70112

★1582★ Shreveport Post of Duty
500 Fannin St., Rm. 8A-08   Phone: (318) 226-5196
Shreveport, LA 71101

Maine

★1583★ **Augusta District Office**
40 Western Ave., Rm. 512   Phone: (207) 622-8378
Augusta, ME 04330

Maryland

★1584★ **Baltimore District Office**
10 N. Calvert St., 3rd Fl.   Phone: (301) 962-4392
Baltimore, MD 21202

Massachusetts

★1585★ **Boston District Office**
10 Causeway St., Rm. 265   Phone: (617) 565-5590
Boston, MA 02222-1093

★1586★ **Boston Regional Office**
155 Federal St., 9th Fl.   Phone: (617) 451-2023
Boston, MA 02110
Covers activities in Connecticut, Maine, Massachusetts, New Hampshire, Rhode Island, and Vermont.

★1587★ **Springfield, Massachusetts Branch Office**
1550 Main St., Rm. 212   Phone: (413) 785-0268
Springfield, MA 01103

Michigan

★1588★ **Detroit District Office**
477 Michigan Ave., Rm. 515   Phone: (313) 226-6075
Detroit, MI 48226

★1589★ **Marquette Branch Office**
300 S. Front St.   Phone: (906) 225-1108
Marquette, MI 49885

Minnesota

★1590★ **Minneapolis District Office**
100 N. 6th St., Ste. 610   Phone: (612) 370-2324
Minneapolis, MN 55403-1563

Mississippi

★1591★ **Gulfport Branch Office**
One Hancock Plaza, Ste. 1001   Phone: (601) 863-4449
Gulfport, MS 39501-7758

★1592★ **Jackson District Office**
101 W. Capitol St., Ste. 400   Phone: (601) 965-4378
Jackson, MS 39201

Missouri

★1593★ **Kansas City District Office**
1103 Grand Ave., 6th Fl.   Phone: (816) 374-6708
Kansas City, MO 64106

★1594★ **Kansas City Regional Office**
911 Walnut St., 13th Fl.   Phone: (816) 426-2989
Kansas City, MO 64106
Covers activities in Iowa, Kansas, Missouri, and Nebraska.

★1595★ **St. Louis District Office**
815 Olive St., Rm. 242   Phone: (314) 539-6600
St. Louis, MO 63101

★1596★ **Springfield, Missouri Branch Office**
620 S. Glenstone St., Ste. 110   Phone: (417) 864-7670
Springfield, MO 65802-3200

Montana

★1597★ **Helena District Office**
301 S. Park, Rm. 528   Phone: (406) 449-5381
Helena, MT 59626

Nebraska

★1598★ **Omaha District Office**
11145 Mill Valley Rd.   Phone: (402) 221-4691
Omaha, NE 68154

Nevada

★1599★ **Las Vegas District Office**
301 E. Stewart St., Rm. 301   Phone: (702) 388-6611
Las Vegas, NV 89125

★1600★ **Reno Post of Duty**
50 S. Virginia St., Rm. 238   Phone: (702) 784-5268
Reno, NV 89505

New Hampshire

★1601★ **Concord District Office**
55 Pleasant St., Rm. 210   Phone: (603) 225-1400
Concord, NH 03302-1257

New Jersey

★1602★ **Camden Post of Duty**
2600 Mt. Ephrain Ave.   Phone: (609) 757-5183
Camden, NJ 08104

U.S. Small Business Administration (Cont.)

★1603★ **Newark District Office**
60 Park Place, 4th Fl.
Newark, NJ 07102
Phone: (201) 645-2434

New Mexico

★1604★ **Albuquerque District Office**
625 Silver Ave., S.W., Ste. 320
Albuquerque, NM 87102
Phone: (505) 755-1868

New York

★1605★ **Albany Post of Duty**
445 Broadway, Rm. 222
Albany, NY 12207
Phone: (518) 472-6300

★1606★ **Buffalo Branch Office**
111 W. Huron St., Rm. 1311
Buffalo, NY 14202
Phone: (716) 846-4301

★1607★ **Elmira Branch Office**
333 E. Water St., 4th Fl.
Elmira, NY 14901
Phone: (607) 734-8130

★1608★ **Melville Branch Office**
35 Pinelawn Rd., Rm. 102E
Melville, NY 11747
Phone: (516) 454-0750

★1609★ **New York District Office**
26 Federal Plaza, Rm. 3100
New York, NY 10278
Phone: (212) 264-4355

★1610★ **New York Regional Office**
26 Federal Plaza, Rm. 31-08
New York, NY 10278
Phone: (212) 264-7772

Covers activities in New Jersey, New York, Puerto Rico, and the Virgin Islands.

★1611★ **Niagara Falls Disaster Area Office 1**
360 Rainbow Blvd. S., 3rd Fl.
Niagara Falls, NY 14303
Phone: (716) 282-4612

★1612★ **Rochester Post of Duty**
100 State St., Rm. 601
Rochester, NY 14614
Phone: (716) 263-6700

★1613★ **Syracuse District Office**
100 S. Clinton St., Rm. 1071
Syracuse, NY 13260
Phone: (315) 423-5383

North Carolina

★1614★ **Charlotte District Office**
200 N. College St., Ste. A-2015
Charlotte, NC 28202-2173
Phone: (704) 344-6563

North Dakota

★1615★ **Fargo District Office**
657 2nd Ave. N., Rm. 218
Fargo, ND 58108-3086
Phone: (701) 239-5131

Ohio

★1616★ **Cincinnati Branch Office**
550 Main St., Rm. 5028
Cincinnati, OH 45202
Phone: (513) 684-2814

★1617★ **Cleveland District Office**
1240 E. 9th St., Rm. 317
Cleveland, OH 44199
Phone: (216) 522-4180

★1618★ **Columbus District Office**
85 Marconi Blvd., Rm. 512
Columbus, OH 43215
Phone: (614) 469-6860

Oklahoma

★1619★ **Oklahoma City District Office**
200 N.W. 5th St., Ste. 670
Oklahoma City, OK 73102
Phone: (405) 231-4301

Oregon

★1620★ **Portland District Office**
222 SW Columbia, Ste. 500
Portland, OR 97201-6605
Phone: (503) 326-2682

Pennsylvania

★1621★ **Harrisburg Branch Office**
100 Chestnut St., Ste. 309
Harrisburg, PA 17101
Phone: (717) 782-3840

★1622★ **King of Prussia District Office**
475 Allendale Rd., Ste. 201
King of Prussia, PA 19406
Phone: (215) 962-3846

★1623★ **King of Prussia Regional Office**
475 Allendale Rd., Ste. 201
King of Prussia, PA 19406
Phone: (215) 962-3700

Covers activities in Delaware, the District of Columbia, Maryland, Pennsylvania, Virginia, and West Virginia.

★1624★ **Pittsburgh District Office**
960 Penn Ave., 5th Fl.
Pittsburgh, PA 15222
Phone: (412) 644-2780

★1625★ **Wilkes-Barre Branch Office**
20 N. Pennsylvania Ave., Rm. 2327
Wilkes-Barre, PA 18701
Phone: (717) 826-6497

Puerto Rico

★1626★ Hato Rey District Office
Carlos Chardon Ave., Rm. 691
Hato Rey, PR 00918
Phone: (809) 766-4002

Rhode Island

★1627★ Providence District Office
380 Westminister Mall, 5th Fl.
Providence, RI 02903
Phone: (401) 528-4561

South Carolina

★1628★ Columbia District Office
1835 Assembly St., Rm. 358
Columbia, SC 29202
Phone: (803) 765-5376

South Dakota

★1629★ Sioux Falls District Office
101 S. Main Ave., Ste. 101
Sioux Falls, SD 57102-0527
Phone: (605) 336-4231

Tennessee

★1630★ Nashville District Office
50 Vantage Way, Ste. 201
Nashville, TN 37228-1504
Phone: (615) 736-5850

Texas

★1631★ Austin Post of Duty
300 E. 8th St., Rm. 520
Austin, TX 78701
Phone: (512) 482-5288

★1632★ Corpus Christi Branch Office
400 Mann St., Ste. 403
Corpus Christi, TX 78401
Phone: (512) 888-3331

★1633★ Dallas District Office
100 Commerce St., Rm. 3C-36
Dallas, TX 75242
Phone: (214) 767-0605

★1634★ Dallas Regional Office
8625 King George Dr., Bldg. C
Dallas, TX 75235-3391
Phone: (214) 767-7643
Covers activities in Arkansas, Louisiana, New Mexico, Oklahoma, and Texas.

★1635★ El Paso District Office
10737 Gateway W., Ste. 320
El Paso, TX 79935
Phone: (915) 541-7586

★1636★ Ft. Worth Branch Office
819 Taylor St., Rm. 10A27
Ft. Worth, TX 76102
Phone: (817) 334-3777

★1637★ Ft. Worth Disaster Area Office 3
4400 Amon Carter Blvd., Ste. 102
Ft. Worth, TX 76155
Phone: (817) 267-1888

★1638★ Harlingen District Office
222 E. Van Buren St., Rm. 500
Harlingen, TX 78550
Phone: (512) 427-8533

★1639★ Houston District Office
2525 Murworth, Ste. 112
Houston, TX 77054
Phone: (713) 660-4401

★1640★ Lubbock District Office
1611 Tenth St., Ste. 200
Lubbock, TX 79401
Phone: (806) 743-7462

★1641★ Marshall Post of Duty
505 E. Travis, Rm. 103
Marshall, TX 75670
Phone: (214) 935-5257

★1642★ San Antonio District Office
7400 Blanco Rd., Ste. 200
San Antonio, TX 78216
Phone: (512) 229-4535

Utah

★1643★ Salt Lake City District Office
125 S. State St., Rm. 2237
Salt Lake City, UT 84138-1195
Phone: (801) 524-5800

Vermont

★1644★ Montpelier District Office
87 State St., Rm. 205
Montpelier, VT 05602
Phone: (802) 828-4474

Virgin Islands

★1645★ St. Croix Post of Duty
4C & 4D Este Sion Frm., Rm. 7
St. Croix, VI 00820
Phone: (809) 778-5380

★1646★ St. Thomas Post of Duty
Veterans Dr., Rm. 283
St. Thomas, VI 00801
Phone: (809) 774-8530

Virginia

★1647★ Richmond District Office
400 N. 8th St., Rm. 3015
Richmond, VA 23240
Phone: (804) 771-2617

Washington

★1648★ Seattle District Office
915 Second Ave., Rm. 1792   Phone: (206) 442-5534
Seattle, WA 98174-1088

★1649★ Seattle Regional Office
2615 4th Ave., Rm. 440   Phone: (206) 442-8544
Seattle, WA 98121

Covers activities in Alaska, Idaho, Oregon, and Washington.

★1650★ Spokane District Office
W. 601 First Ave., 10th Fl. E.   Phone: (509) 353-2807
Spokane, WA 99204

West Virginia

★1651★ Charleston Branch Office
550 Eagan St., Ste. 309   Phone: (304) 347-5220
Charleston, WV 25301

★1652★ Clarksburg District Office
168 W. Main St., 5th Fl.   Phone: (304) 623-5631
Clarksburg, WV 26301

Wisconsin

★1653★ Eau Claire Post of Duty
500 S. Barstow Common, Rm. 37   Phone: (715) 834-9012
Eau Claire, WI 54701

★1654★ Madison District Office
212 E. Washington Ave., Rm. 213   Phone: (608) 264-5261
Madison, WI 53703

★1655★ Milwaukee Branch Office
310 W. Wisconsin Ave., Ste. 400   Phone: (414) 291-3941
Milwaukee, WI 53203

Wyoming

★1656★ Casper District Office
100 East B. St., Rm. 4001   Phone: (307) 261-5761
Casper, WY 82602-2839

Section 20

Federal Funding Sources

*"Myth: Complex problems require complex solutions arrived at
through complex thinking." — Eric M. Bienstock*

Uncle Sam wants your better ideas!

The U.S. federal government funds almost one-half of all research and development in our country. The 700 federal laboratories receive a budget of $72 billion. More than one billion dollars each year goes to small companies for R&D through direct awards. But pursuing it is not for everyone. Tapping the pot of green at the end of Uncle Sam's red, white, and blue rainbow takes a major commitment of both time and money. And unless you are willing to tolerate a cumbersome and inordinate amount of paper work, red tape, and costly bureaucratic excesses and delays, this exercise may not be for you.

On the other hand, there appears to be something for everyone. Washington's interest ranges from innovations in astronomy, earth, environmental, and marine sciences to chemistry, physics, engineering, and materials sciences. It awards money for breakthroughs in robotics, genetics, plastics, adhesives, fusion, optics,

Commerce Business Daily

In order to alert potential sources to emerging government R&D interests, agencies must by law publish advanced notice of R&D opportunities in the *Commerce Business Daily (CBD)*.

Published Monday through Saturday (except on federal holidays), copies of *CBD* may be obtained at the Department of Commerce field offices and at most major urban and university libraries. Subscriptions to *CBD* may be obtained from The Superintendent of Documents, Government Printing Office, Washington, DC 20402. The GPO takes Visa and Mastercard.

First class mail subscriptions to *CBD* cost $261 per year; $130 for six months. Second class delivery costs $208 per year; $104 for six months.

Because much of the material contained in *CBD* is time sensitive, a first class subscription is recommended.

As in any marketing situation, it's "know how" combined with "know who" that makes the difference. I recommend personal contact as the best way to get information on R&D opportunities. As helpful as the *Commerce Business Daily* is, there's no substitute for personal visits and phone follow-ups.

National Technical Information Service

While the federal government provides opportunities for inventors to sell and/or receive R&D grants, it also shares its expertise and equipment to aid inventors in their research. It does this through its agencies, laboratories, and engineering centers.

The National Technical Information Service (NTIS), part of the U.S. Department of Commerce and the largest government clearinghouse after the Library of Congress, publishes and sells a wide variety of bulletins, journals, catalogs, and directories that could be of interest to many inventors. It is through these information products that the federal government announces more than 150,000 summaries of U.S. and foreign government sponsored R&D and engineering activities annually.

NTIS also manages the Federal Computer Products Center which provides access to software, data files, and databases produced by federal agencies.

The nearly 2 million reports in NTIS archives represent hundreds of billions of dollars of taxpayer-supported research, development, and analysis. More than 1.5 million of those reports are referenced in NTIS's online bibliographic database to help make this national resource easily accessible to the private sector.

"Imagine an annual catalog which describes U.S. government-owned inventions available for licensing, often on an exclusive basis; a directory describing expertise, equipment for sharing, and special services available from federal laboratories; a catalog presenting some of the best new R&D and engineering results produced by government laboratories and contractors," says Edward Lehman, director of NTIS Office of Applied Technology.

NTIS generates annual revenues of approximately $30 million through information collection, codification, dissemination, and archiving arrangements with more than 765 federal organizations.

Of particular interest to inventors are the following products and services:

Government Inventions for Licensing Abstract Newsletter (weekly). Describes each invention and identifies supporting material.

Catalog of Government Inventions Available for Licensing (annual). Contains more than 1,000 inventions, divided into 43 subject areas.

Tech Notes (monthly). Contains more than 100 fact sheets arranged by subject.

Federal Laboratory Technology Catalog (annual). Describes more than 1,000 processes, instruments, materials, equipment, software services, and techniques.

Directory of Federal Laboratory and Technology Resources (annual). Lists hundreds of federal laboratories willing to share their expertise, equipment, and sometimes their facilities. **See Section 22, Federal Laboratory Assistance, beginning on page 269 for the listing.**

To receive the most recent NTIS Products and Services catalog call: 1-800-336-4700. This catalog is free.

An eight minute video tape entitled *NTIS—The Competitive Edge* is also available. The tape gives an overview of NTIS and its activities. To obtain your free copy call: (703) 487-4650 and ask for PR-858/827.

The National Technical Information Service's Center for the Utilization of Federal Technology is located at 5285 Port Royal Rd., Springfield, VA 22161.

Key NTIS Telephone Numbers

Subscription order:	(703) 487-4630
Reports and computer product orders	(703) 487-4650
Rush orders (reports and computer products)	1-800-336-4700
Rush orders in Virginia	(703) 487-4700
Fax	(703) 321-8547
Telex	89-9405
Assistance identifying a title	(703) 487-4780
Assistance tracing an order	(703) 487-4660
Assistance with NTIS deposit accounts	(703) 487-4770
Federal Computer Products Center	(703) 487-4763

NTIS Annual Conference

Each year NTIS holds a two-day conference designed to bring leaders in business, academia, government, and the information sciences the latest news about what government information is available, and how to find and use it.

For more information contact: NTIS Annual Conference, Rm. 203F, Springfield, VA 22161; phone (703) 487-4079.

National Institute of Standards and Technology

Originally established by Congress as the National Bureau of Standards in 1901, the organization was renamed the National Institute of Standards and Technology in 1988 and given a broader mandate "to assist industry in the development of technology...needed to improve product quality, to modernize manufacturing processes, to ensure product reliability...and to facilitate the rapid commercialization...of products based on new scientific discoveries." This explicit mission builds on long-standing agency responsibility to develop, maintain, and retain custody of the national standards of measurement.

The 3,000 scientists, engineers, technicians, and support personnel at NIST's laboratories in Gaithersburg, MD and Boulder, CO conduct basic and applied research in the physical sciences and engineering, developing generic technology, measurement techniques, and related services. The Institute also contributes to public health and safety.

NIST advances technology through world-class science and engineering research. Measurements, testing procedures, quality assurance, and innovations developed at NIST have helped build the technical infrastructure upon which much of the U.S. economy rests. NIST's laboratories perform research across a broad spectrum of disciplines, affecting virtually every industry. Primary fields of NIST research include chemical science and technology, physics, material science and engineering, electronics and electrical engineering, manufacturing engineering, computer systems, building technology, fire safety, computing, and applied mathematics.

Reflecting its role as the only federal laboratory exclusively dedicated to serving the needs of U.S. industry, NIST offers more than 300 types of calibration testing, 1,000 standard reference materials

for calibrating instruments and evaluating test methods, 24 standard reference data centers, laboratory accreditation programs, and free evaluations of energy-related inventions.

Under the Omnibus Trade and Competitiveness Act of 1988, the U.S. Department of Commerce reorganized the National Institute of Standards and Technology and established a Technology Administration headed by an Under Secretary reporting to the Secretary of Commerce. In April of 1990, Commerce Secretary Mosbacher announced the appointment of Robert M. White as Under Secretary of Commerce for Technology to head the new Technology Administration.

NIST developed three major programs that may be of interest to independent inventors. Its mandate is to enhance the country's technological competitiveness through the fast and effective transfer of new technologies to U.S. industries. The programs are:

1) Regional Centers for the Transfer of Manufacturing Technology;

2) State Technology Extension Program; and

3) Advanced Technology Program.

A Clearinghouse for State and Local Initiatives on Productivity, Technology and Innovations has also been established as part of the Under Secretary's office to develop and provide information on programs state-by-state.

If you would like to visit the NIST, directions to its Gaithersburg, MD campus are located on the next pages.

What are Regional Centers for the Transfer of Manufacturing Technology?

These regional centers are intended to provide direct support to small- and medium-sized manufacturing firms in automating and modernizing their facilities. If you are an inventor who has his/her own manufacturing facility, NIST could help you develop technically and financially sound plans for modernizing production runs.

Five centers now funded through the program are:

*The Great Lakes Manufacturing Technology Center in Cleveland, OH, operated by the Cleveland Advanced Manufacturing Program;
*The Northeast Manufacturing Technology Center in Troy, NY, operated by the Rensselaer Polytechnic Institute.
*The Southeast Manufacturing Technology Center at the University of South Carolina in Columbia, SC.
*The Midwest Manufacturing Technology Center at the Industrial Technology Institute in Ann Arbor, MI.
*The Mid-America Manufacturing Technology Center at the Kansas Technology Enterprise Corporation of Topeka, KS.

NIST is creating these centers in partnership with nonprofit organizations established by state and local governments, universities, or companies. NIST is authorized to provide up to 50 percent of the operating funds for these centers for their first three years, reducing thereafter to zero after six years. For updated information on the Regional Centers, contact Dr. Philip Nanzetta, Program Director,

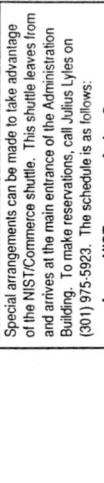

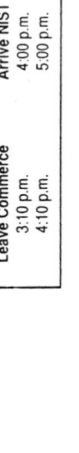

NIST

Getting To
Getting From
Getting Around

The National Institute of Standards and Technology is located near Gaithersburg, Maryland, just off Interstate Route 270, about 25 miles (40 kilometers) from the center of Washington, DC. From Washington (northbound on Rt. 270), visitors to NIST should enter the Collector/ Distributor lanes at Exit 9 and take Exit 10, Rt. 117 West. Turn left at the first traffic light and proceed through the main gate (Gate A). If coming southbound on Rt. 270, take Exit 11a, Rt. 124 West (Quince Orchard Rd.). Visitors should go to the receptionist in the main lobby of the Administration Building for directions or for shuttle service to other buildings on the site. The Institute provides shuttle service for official visitors and staff to the Shady Grove Metro Station and back. The shuttle leaves NIST on the hour and half-hour starting at 8:00 a.m., with the final trip at 5:30 p.m. Visitors using Metro can meet the NIST shuttle at the west side "Kiss & Ride" bus kiosk at the Shady Grove Station. A Metrorail map, area road map, and NIST site map are provided for your convenience.

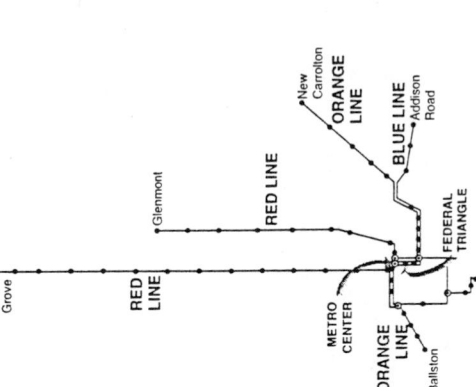

Special arrangements can be made to take advantage of the NIST/Commerce shuttle. This shuttle leaves from and arrives at the main entrance of the Administration Building. To make reservations, call Julius Lyles on (301) 975-5923. The schedule is as follows:

Leave NIST	Arrive Commerce
9:10 a.m.	10:00 a.m.
12:30 p.m.	1:20 p.m.
Leave Commerce	**Arrive NIST**
3:10 p.m.	4:00 p.m.
4:10 p.m.	5:00 p.m.

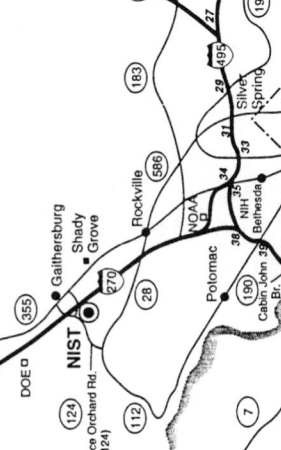

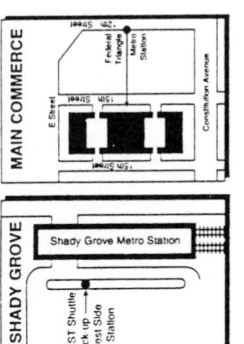

Figure 11

NIST Site Map

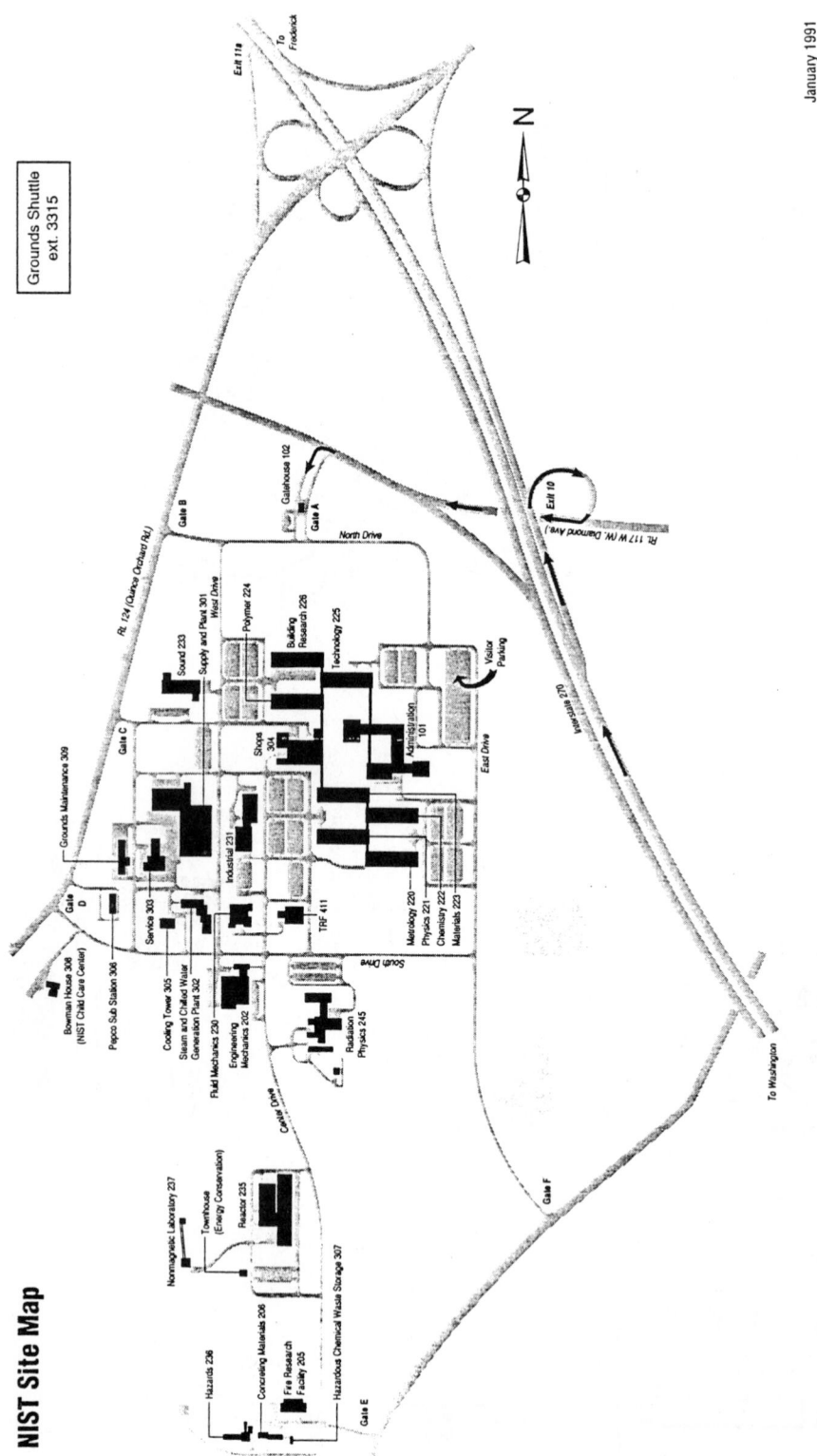

Grounds Shuttle
ext. 3315

N

January 1991

Figure 12

Manufacturing Technology Centers, NIST, B124 Metrology Bldg., Gaithersburg, MD 20899. Telephone: (301) 975-3414; fax: (301) 963-6556.

What is the State Technology Extension Program?

The State Technology Extension Program is designed to work with state and local technology outreach and economic development groups. The program staff will provide support to state and local groups who assist small- and medium-size business in applying technology to solve technical problems.

For detailed and updated information on the program contact: Joseph Berke, Chief, Research and Technology Applications Program, NIST, A343, Physics Bldg., Gaithersburg, MD 20899. Telephone: (301) 975-5017.

What is the Advanced Technology Program?

The Advanced Technology Program (ATP) is a new program to assist U.S. businesses in carrying out research and development on precompetitive, generic technologies. The ATP emphasizes enabling technologies that underlie a wide range of potential applications which would offer significant benefits to the nation's economy.

What is precompetitive, generic technology? Precompetitive means R&D activities up to the stage where technical uncertainties are sufficiently reduced to allow preliminary assessment of commercial potential, and prior to development of application-specific commercial prototypes.

Generic technology means concepts, components, processes, or scientific investigations that potentially could be applied to a broad range of products or processes.

ATP can provide either technology development or planning grants to U.S. companies or consortia of firms. The program can sponsor cooperative research between private industry and federal laboratories such as the National Institute of Standards and Technology.

Any U.S. business, independent R&D organization, or industrial consortium may apply for an ATP grant. Any given company or joint venture may submit more than one proposal. **Application forms for single businesses and joint ventures, respectively, and instructions on how to fill them out are included in Part I: Forms and Tables. See Forms 33 and 34.**

No direct funding will be provided to universities or government organizations, but they may participate as members of a joint venture.

Awards to individual companies are limited to $2 million over three years. Awards to joint ventures are for up to five years and are limited only by available funds. The ATP was funded at over $30 million during fiscal year 1991.

Who will own the rights to developments under the ATP? In general, grant recipients may elect to patent inventions or copyright software developed under an ATP grant. In most cases, the government will retain a nonexclusive license for the use of technologies or materials developed under the ATP. The government will also be entitled to a share of any licensing fees or royalties paid on technologies developed through the ATP, in proportion to the government's share of the cost in developing the technology.

For additional information on the Advanced Technology Program, write or call: Advanced Technology Program, A430, Administration Bldg., NIST, Gaithersburg, MD 20899. Telephone: (301) 975-2636.

The ATP has established a telephone hotline with a periodically updated status report on proposal applications. This number is (301) 975-2273.

Commerce Clearinghouse: An Information Bonanza

Finding information about local, state, and federal technology development and commercialization programs used to be like finding the proverbial needle in a haystack. That was, at least, until October 1990, when the U.S. Department of Commerce formally inaugurated its Clearinghouse for State and Local Initiatives on Productivity, Technology and Innovations. Now information and expert guidance is just a phone call away.

Established by Congress under the Omnibus Trade and Competitiveness Act of 1988, the Clearinghouse's primary mission is to "help share the wealth," says Under Secretary for Technology Robert M. White, referring to the information it gathers and disseminates.

"State and local governments are in the forefront of efforts to improve productivity, technology, and innovation," he adds. "Centers of excellence, seed and venture capital programs, and manufacturing technology extension services are only a sampling of the variety of approaches which states and localities have undertaken to stimulate technological innovation.

The Clearinghouse has a database with information from more than 700 local, state, and federal technology programs and from more than 400 individuals working in technology programs and policy.

Technology programs are classified by the Clearinghouse into 12 categories: research center, university research grant, technology transfer, university research park, tax policy, management assistance, technical assistance, incubator, business capital, manufacturing extension, training, and policy development.

"Business is great!" says Elizabeth M. Robertson, an analyst at the Clearinghouse. She reports that almost half of the calls are from individuals, e.g. inventors, entrepreneurs, and businesspersons.

"I tend to give people who call at least three or four numbers and tell them that what they are doing is getting into a network," Ms. Robertson explains. While it can be very frustrating trying to get information from a large government agency, she sees her job, in part, as helping "to keep the frustration factor way down" for people who call for assistance.

There are three main groups that call the Clearinghouse for assistance. They are state and local officials; the private sector; and federal officials.

Queries are wide ranging, but here are some examples of the kinds of questions the Clearinghouse responds to:

*The state of New Hampshire wanted to set up a research grant program and sought to learn about how other states had done it. The Clearinghouse was able to provide a list of people who had experience in designing and initiating such programs.

*An engineer called to obtain a list of who in the United States could help him with opto-electronic research. The Clearinghouse gave him a list of laboratories conducting R&D in the field.

*Some independent inventors call to get something as simple as the telephone number for the nearest inventor organization.

*At the federal level, the Assistant Secretary for Intergovernmental Affairs might call to find out who's who in a specific field of technology for the purpose of organizing a conference.

No task is too large or too small for the Clearinghouse team. It is also free of charge to anyone who calls or writes. **(See Form 35, Information Request Form in Part One: Forms and Tables.)** It is one of

the most valuable services the federal government provides to those involved in invention and technology transfer.

"It is indeed rare, but refreshing, to have one's needs met in such a timely manner. You are to be commended," wrote Myrta A. Mason of the Association for Manufacturing Technology, referring to her experience with John Heizer of the Clearinghouse.

Marilyn Ross, vice president of About Books, Inc. praised the work of Elizabeth Robertson in a letter to Robert White. "I've contacted dozens of government agencies and talked with scores of people. Elizabeth stands out as being absolutely the most informed and cooperative."

Ms. Robertson is knowledgeable, patient, and eager to assist anyone who calls for information. Her direct phone number is (703) 487-4968. Also available to answer questions at the same number is John Heizer, a financial economist and a 15-year veteran at the Department of Commerce. Joseph P. Allen manages the Office of Technology Commercialization in which the Clearinghouse is situated.

There are three booklets available from the Clearinghouse free of charge.

A Guide to Professional Program Developers and Administrators. This 40-page booklet describes 21 national clearinghouses that provide information on community and economic development programs, state and local government management, or technology and technology transfer.

A Guide to Professional Program Developers and Administrators. This 48-page booklet is broken down into four parts and offers a master list, by organization, with separate sections for the federal government, state and local institutions and organizations, universities, and nongovernmental organizations.

A Guide to Directories. This 28-page booklet summarizes 19 directories of programs and initiatives undertaken by federal, state, and local governments to stimulate business competitiveness. The directories are grouped into two broad categories: Technology and R&D; general business and economic development.

Ordering information: The Clearinghouse for State and Local Initiatives on Productivity, Technology and Innovations, Technology Administration, U.S. Department of Commerce, RM H-4418, Washington, DC 20230. Telephone: (202) 377-8100; fax: (202) 377-0253.

What is Going on in Nonenergy-Related Invention Development?

The most exciting part of this activity for the independent inventor will be the technical evaluation of promising inventions that are not energy related—a program modeled after NIST's successful 12-year-old program for energy-related inventions managed by NIST for the Department of Energy. Please see below for information on the Office of Energy-Related Inventions and its evaluation and funding program.

This part of the program still has no funding. But it will and for updates, contact: George P. Lewett, Chief, Office of Energy-Related Inventions, NIST, Gaithersburg, MD 20899; phone: (301) 975-5500.

Energy-Related Inventions Program (ERIP)

"The NIH (Not Invented Here) does not apply here," says George Lewett, Director of Office of Technology Evaluation and Assessment (OTEA) at the National Institute of Standards and Technology.

The National Institute of Standards and Technology (NIST) and the Department of Energy (DOE), under provisions of the Federal Non-nuclear Energy Research and Development Act of 1974 (P.L. 93-577), have combined to offer a marvelous opportunity to inventors of energy-related concepts, devices, products, materials, or industrial processes. Called the Energy-Related Inventions Program (ERIP), it is designed as a process to discover and assist the development of worthy inventions which might otherwise never be commercialized.

From an NIST facility in Gaithersburg, MD, just outside Washington, DC, off I-270, the program provides a chance for independent inventors and small businesses with promising energy-related inventions to obtain federal assistance in the development and commercialization of their inventions.

Since its inception OTEA has received more than 28,000 evaluation requests, accepted about half of these for evaluation, and recommended 550 for U.S. government support. Approximately $24 million in grants or contracts have been awarded to date on some 350 recommended inventions; the remainder are either in process (94) or have received other than financial assistance.

OTEA is under the experienced leadership of George P. Lewett. "We currently review about 200 inventions a month for evaluation and we recommend, fairly steadily, 3-4 inventions per month," he said. And the system works.

Ask A.S. Richardson, inventor of a power line galloping prevention device and galloping indicator device. He submitted his invention to OTEA in 1985 and in 1990 was awarded $147,950 by the DOE.

Ask Leon Lazare, inventor of an improved refrigeration system. He submitted the invention to OTEA in 1986 and by 1987 had landed a DOE award in the amount of $92,500 and additional funding from the state and private sources amounting to over $300,000.

Ask Michael C. Catapano, inventor of an improved device for testing heat exchanger tubes used in power plants and process industries. He submitted his invention to OTEA in 1986 and in 1990 received an award of $80,230 from the DOE.

Ask Carol Balzar, inventor of a mobile biomass processing unit that manufactures burnable logs from wood waste residue. The inventor received $92,000 through the Department of Energy.

Ask Joyce A. Kostura, inventor of an oil recovery process modified by steam drive employing high velocity noncondensible gas. The inventor was awarded $79,500 to extend the scale model work to collect oil production data.

Ask Michael Gondouin, inventor of a process for converting natural gas into liquid hydrocarbons. Granted $79,000 to build an engineering prototype and establish operating variables.

Inventions Covered by OTEA

Invention disclosures evaluated by OTEA are categorized as follows:

    Fossil Fuel Production
    Direct Solar
    Other Natural Sources
    Combustion Engines and Components
    Transportation Systems, Vehicles, and Components
    Building, Structures, and Components
    Industrial Processes
    Miscellaneous

You Keep the Rights. Uncle Sam Keeps Your Secret

It is the government's policy that inventors retain patent rights to their inventions, and that these rights are not compromised by the OTEA evaluation.

OTEA maintains the confidentiality of invention disclosures submitted by limiting access to them. All federal government personnel involved in the program sign a statement advising them of penalties for disclosure or misuse of information under 18 U.S.C. 1905. Nongovernment consultants also sign an agreement of nondisclosure which includes agreement not to undertake invention reviews that may subject the consultant to a conflict of interest.

OTEA Submission Process

Submissions are handled by the Office of Energy-Related Inventions, U.S. Department of Commerce, Gaithersburg, MD 20899; phone: (301) 975-5500.

The disclosure of an invention should include information required by NIST Form 1019 **(See Form 36, Energy-Related Invention Evaluation Request, in Part One: Forms and Tables)**, but the format may vary widely depending on a number of factors. Here are some suggestions that you should consider in preparing the description of the invention to be submitted with the Evaluation Request Form 1019.

Make a Complete Disclosure. The principal requirement in submitting a request is a thorough and complete invention disclosure which describes the invention in detail. It is most important to submit all information which is available even if the method of presentation and organization is not professional in nature. Test data and information on how tests were conducted are particularly critical, when applicable, since no testing will be done by NIST as part of its evaluation.

Emphasize the Energy Relation. The program is interested in all energy-related inventions including both those that involve energy conservation and those that involve alternative sources of energy. This means everything from new methods of recovery to drill bits. Your disclosure should emphasize and document, to the extent possible, the amount of energy saved or made available through an alternate source.

It is only after the invention reaches the commercialization stage that its ultimate contribution to the solution of our energy problem can be realized. It is not necessary to calculate energy savings exactly, but the potential should be indicated.

Describe Your Competition. Make an effort to find out if there are any similar products on the market. Detail the known competition and document why your innovation is superior technically or from an energy or economic standpoint.

Give the Status of Your Invention. Address the question of what needs to be done to bring your invention closer to fruition. Spell out what you would expect from the U.S. government in the event of a favorable evaluation.

Highlight the Innovation. As you can imagine, NIST has seen all kinds of concepts involving common devices, such as windmills, wave machines, furnaces, carburetors, internal combustion engines, and space heaters. And like so many ideas, many of those submitted are neither new nor innovative. Be sure, therefore, to point out and highlight novel principles or innovations that make yours different, particularly if it falls into a common device class.

Be Factual, Realistic. Your proposal will be evaluated by savvy technical and business-oriented professionals. Prepare your disclosure with that in mind and do not make outrageous claims that cannot be justified or substantiated by data or information in your disclosure.

Time to Process Your Request. Do not expect an immediate response to your request for evaluation. The evaluation process is time-consuming and there are a large number of submittals to consider. Remember that submission of an invention to NIST for evaluation is no guarantee that it will be recommended to the Department of Energy (DOE), and a recommendation is no guarantee that your invention will be accepted for a DOE grant award.

The OTEA Evaluation Process

The OTEA evaluation process, under the direction of Howard Robb, is the formal procedure that will determine, in an objective way, whether the invention you have submitted is appropriate for recommendation to the Department of Energy (DOE) for support. OTEA considers an invention appropriate for DOE consideration if it is technically feasible, will offer a favorable impact on the energy situation and holds potential for commercial success. The focus is on your technology; your credentials or capabilities are not a factor in OTEA decision making.

Each disclosure is logged in and tracked through a computerized system that records the number of days spent in each stage of the five-step evaluation process and the number of days required to obtain the opinions of government consultants.

1) **Disclosure Acknowledgment.** OTEA will immediately acknowledge by letter the receipt of each disclosure received. Every disclosure must be accompanied by a signed evaluation request, NIST Form 1019 **(See Form 37 in Part One: Forms and Tables).**

2) **Disclosure Review & Analysis.** OTEA will review the invention disclosures to determine whether they are complete, readable, technically sufficient, and within the scope of its program. Inventors are notified by letter whether their disclosure is accepted for first-stage evaluation or is rejected. This is, in effect, a preliminary screening.

3) **First-Stage Evaluation.** In this initial screening process, OTEA will determine whether an invention is promising enough to warrant an in-depth analysis. OTEA evaluators typically seek brief expert opinions (invention reviews) independently from two consultants before deciding whether to place the invention into second-stage evaluation or to reject it.

The consultants are asked to comply with the following rules:

> a) Do not transmit information on the invention disclosure to anyone who has not signed an NIST Agreement of Nondisclosure **(Form 37, located in Part One: Forms and Tables).**
>
> b) Do not copy the disclosure.
>
> c) Do not write on or otherwise mark the disclosure.
>
> d) Return the complete disclosure with the invention report to OTEA by certified mail or other secure transmittal method.

OTEA evaluators and their consultants are expected to be liberal and to highlight the positive aspects of an invention rather than seek reasons for rejection. The inventor is notified by letter of said decision. The actual evaluation paper is not made available, but a summary of its finding is communicated to the inventor.

4) **Second-Stage Evaluation.** OTEA will determine whether to recommend an invention to DOE for support. OTEA evaluators obtain a deep analysis of the invention from a consultant before deciding whether to recommend the invention to DOE or to reject it. The consultant, to the extent possible, will try to provide information and references to pertinent literature that may

be of assistance to the OTEA evaluator and the inventor. The inventor is notified by letter of this decision too, in either case, and is sent a copy of the second-stage consultant's analysis.

5) **Recommendation to DOE.** OTEA will prepare a transmittal document which includes all pertinent material submitted by the inventor, the second-stage consultant's in-depth analysis, any other pertinent material obtained or developed by OTEA during the evaluation, and an OTEA evaluator's report outlining the reason for recommending the invention and the appropriate next step to DOE. Inventors are provided a copy of this document and the name of an OTEA coordinator whom they may contact.

At each stage in the evaluation process, you may present additional information, which is fully considered before further decisions are made. If your invention is rejected at any stage, you may submit additional information addressing the problems raised by OTEA and request re-evaluation. This "Open Appeal" feature is considered an intrinsic part of the evaluation process.

"We never close a case," says Fred Hunt of OTEA. "15 to 20 percent of the inventions we recommend to DOE have been turned down at some point in the evaluation process."

The evaluation process is primarily a selection process which does not necessarily include analysis of every aspect of an invention. **OTEA will provide you with any useful information it obtains or develops during the evaluation process of your innovation.** The brief first-stage invention reviews made by consultants are not released, but information contained in them may be included in OTEA's correspondence with you.

Department of Energy

Upon receipt of a recommendation from OTEA, an Invention Coordinator from DOE's Inventions and Innovation Programs Division analyzes it and opens negotiations on the type and amount of support (if any) to be provided. A letter is written to the inventor. The inventor's response to this letter is treated as a preliminary proposal by DOE.

The OTEA recommendation, the inventor's preliminary proposal, and other information available to the Invention Coordinator (such as information concerning other DOE projects in related fields) are considered. Also taken into account by the official are the inventor's business/management capabilities and the resources available to the inventor. The Invention Coordinator then decides whether federal support is warranted and, if so, what type of support should be provided.

DOE support under this program may consist of a grant award, a contract award, testing service at a DOE facility, guidance in obtaining private venture capital, or other types of assistance.

In practice, each case is handled on an individual basis. There is no standard operational procedure. Each case is unique.

Inventors, especially those in the early stages of development, often require an amount of money that exceeds DOE's authority and guidelines. Grants have ranged from $20,000 to $200,000, averaging about $85,000, with most granted to support technical research, scientific testing, or business planning. DOE cannot support marketing efforts, although it can help inventors obtain market information. As the program has evolved, the DOE staff, under the direction of Ray Barnes, has sought to fine-tune financial support to the needs and abilities of individual inventors and their technologies, and to expand the range of non-financial services.

Among the many benefits the Program provides to independent inventors is credibility. The NIST evaluation has served many inventors as a kind of "Good Housekeeping Seal of Approval" that they converted into a bankable asset. It was reported that one inventor parlayed the NIST evaluation into

an endorsement by, and financial support from, a state agency, then went on to build a business that sold $32 million worth of his product. Another testified that he "couldn't get a venture capitalist to talk to him" until he had the NIST evaluation and DOE support as credentials.

When asked about the benefits of this NIST/DOE combined program, many inventors cited the gain in "credibility" as important as the money awards.

Inquiries may be directed care of U.S. Department of Energy, CE-521, 5E052, Washington, DC 20585; phone: (202) 586-1692.

Department of Defense

In my opinion, Department of Defense (DoD) presents one of the best opportunities for an individual or small firm to land federal R&D funds and/or sell innovative technologies. The Pentagon has an active and wide ranging Small Business Innovation Research Program (SBIR) which makes special efforts to identify creative individuals, small businesses, and small disadvantaged businesses with R&D capabilities and/or better ideas. DoD not only has an interest in soliciting business, but it also entertains outside queries and submissions.

Individuals and firms with strong R&D capabilities in science and engineering and other disciplines are encouraged to participate. DoD's SBIR Program goals include stimulating technological innovation in the private sector, strengthening the role of small business in meeting DoD research and development needs, fostering and encouraging participation by minority and disadvantaged persons in technological innovation, and increasing the commercial application of DoD-supported research and development results.

How Can You Get Current DoD R&D Information?

There are several ways to approach DoD under the SBIR Program.

a) You can go directly to the different departments to present your concepts and/or to receive information on what they need in terms of outside assistance. The names and addresses of same are provided in this Section.

b) Each year the Department of Defense, Office of Small Business Innovation Research, conducts seminars around the country at which it presents "wish lists" and explains the SBIR program. To get general information on the where and when of the seminars contact:

Bob Wrenn, SBIR Coordinator
OSD/SADBU
U.S. Department of Defense
The Pentagon - Rm. 2A340
Washington, DC 20301-3061   Phone: (202)697-9383

"The amazing thing about these SBIR meetings is that I sit there and meet this endless stream of awesomely bright people," says Richard Sparks, program manager for the Defense Technical Information Center (DTIC). Nonetheless, he adds that the presentations accepted by the military in 1983 would never make the final cut in 1991. "The level of sophistication required has risen dramatically because the topics are so much more sophisticated."

There is no charge to register for your SBIR program and to receive DoD program solicitations. Since fiscal year 1990, there have been two solicitations per year, a 60-day solicitation on May 1st and a 90-day solicitation on October 1st. They typically cover over 450 different topics. "Once a solicitation period begins," Sparks explains, "we handle between 200 and 300 phone calls per day."

There are several ways to place your name on the list to receive solicitations. You can go to a Small Business Administration Office (see page 219 for complete list), phone (202) 205-7777, or contact:

Defense Technical Information Center,
Att: DTIC-SBIR, Alexandria, VA 22304-6145   Phone: (703) 224-6902

c) In September, 1991, DoD announced the availability of a two-volume video tape with a running time of two-and-a-half hours in which officials from various branches and agencies explain how to answer their solicitations. The second half of the tape features successful winners of SBIR grants telling how they won the business.

For registered users of DTIC, the tapes cost $27. If you are not a DTIC subscriber, call National Technical Information Services at (703) 487-4600 to find out the current cost to you.

d) The Defense Technical Information Center (DTIC) is a great place to learn what the Pentagon is looking for in terms of technologies and gadgets. "We exist to help the independent inventors and small businesses cut through the red tape and bureaucracy," says a most helpful and hospitable Richard Sparks, program manager for DTIC and the pointman on the front line with outside queries. And there are no flacks to block your access, he answers his own phone. Mr. Sparks can be reached at (703) 274-6902 or via 1-800-225-3842.

DTIC is the central storehouse of scientific and technical information resulting from and describing R&D projects that are funded by the Department of Defense. It has 1.8 million technical reports on file.

DTIC users include university libraries, special libraries, DoD agencies, other U.S. government agencies ranging from the U.S. Coast Guard to the CIA, commercial companies like Union Carbide, General Electric, and IBM, independent inventors, and students.

During the fiscal year 1991 DTIC provided 3,837 small business requesters approximately 36,720 technical information packages and 16,600 technical reports on the 1,102 DoD solicitation topics. Additionally, DTIC distributed 7,183 copies of the solicitation documents to small R&D businesses.

URI Program

DTIC information is also available via University Research Initiative (URI), a basic DoD research program. Eighty-six universities throughout the U.S. currently participate in the program which provides free access to DTIC products and services.

Terminal Access to DTIC

You may gain DTIC access via the Defense RDT&E Online System (DROLS) by using one of a wide variety of available terminals. DROLS can communicate with any terminal (CRT or typewriter) which employs the standard AASCII asynchronous protocol. Terminal communications speeds are 300, 1200, or 2400 baud (30, 120, or 240 characters per second) in even parity. Access may be also gained by using TYMNET commercial data communications network or direct.

This service is limited to unclassified access only. Users requiring classified access will be required to use specialized UNIVAC terminals with dedicated telephone lines.

* There is a charge of $30.00 per connect hour or proportionate share.

* Subscribers to this service must have a deposit account with the National Technical Information Service.

*Users must have a DTIC user code which is assigned when contract is registered at DTIC.

* Users will not be charged for time they input technical data into the DTIC databases.

If you are interested in this service, contact:

Agnes Taylor
Defense Technical Information Center
ATTN.: Online Support Office (DTIC-BLD)
Bldg. 5, Cameron Station   Phone: (202) 274-7709
Alexandria, VA 22304-6145   Autovon: 284-7709

What Do DTIC Services Cost?

Most of DTIC services are free to registered users. The only charges are for paper copy documents, which cost $5 for 1-100 pages and $.07 for each additional page over 100, and microfiche copies, which are $.95 per demand document and $.35 per document supplied under the Automatic Document Distribution.

How Can You Become a Registered DTIC User?

You can receive an information packet that describes DTIC's services and products and instructions on registration if you contact:

Defense Technical Information Center
ATTN.: Registration and Services Section (DTIC-FDRB)
Bldg. 5, Cameron Station   Phone: (703) 274-6871
Alexandria, VA 22304-6145   Autovon: 284-6871

Mr. Sparks says that each year near the end of August, military departments begin to publish SBIR Requests for Proposals (RFPs) in the *Commerce Business Daily* and mail out formal Program Solicitations. Then beginning around October 1st for the next 13 weeks, he and his Defense Technical Information Centers are "open for business" (see list below) to assist people wishing to compete for research and development grants or present concepts.

While "officially" he only has the budget to handle outside queries during this time period, Mr. Sparks is a very warm, knowledgeable, and helpful individual who will never turn away an inventor in need of guidance. "You never know what you'll find," he says. "One that fascinated me most was a three mile-long magnetic gun for putting things into orbit at low cost without expending fuel. These guys were a couple of wiz kids!"

You may call or visit at the following locations:

Defense Technical Information Center
ATTN.: DTIC-SBIR
Bldg. 5, Cameron Station
Alexandria, VA 22304-6145
800-225-3842 (Toll-free)
(703) 274-6902 (Commercial for VA, AK, HI)

DTIC Boston On-Line Service Facility
DTIC-BOS
Bldg. 1103, Hanscom AFB
Bedford, MA 01731-5000
(617) 377-2413

DTIC Albuquerque Regional Office
AFWL/SUL Bldg. 419
Kirtland AFB, NM 87117-6008
(505) 846-6797

DTIC Los Angeles On-Line Service Facility
Defense Contract Administration Services Region
222 N. Sepulveda Blvd.
El Segundo, CA 90245-4320
(213) 355-4170

Guide to Unsolicited Proposals

You do not have to follow a particular format for the submission of proposals. However, a proposal should, at a minimum, cover the points set forth below in the order indicated. Elaborate proposals or presentations are neither necessary nor desirable. Two copies of your proposal should be submitted.

The DoD follows established competitive procurement procedures for awarding contracts. In other cases it may issue statements of interest or similar notices to announce various program opportunities in such publications as the *Commerce Business Daily*.

DoD also awards contracts based upon unsolicited proposals.

Competitive Proposals

Normally, competitive procurements will be initiated by the issuance of a formal request for proposal (RFP). The RFP will contain instructions telling you how to prepare the proposal.

Unsolicited Proposals

A formal unsolicited proposal should be in the form of a detailed document signed by the inventor. This document forms the basis for further technical evaluations and for possible contract negotiations. Make sure each proposal includes the following elements as appropriate:

1) Cover Sheet
2) Abstract
3) Narrative
4) Cost Proposal (should be a document that is separate from original proposal)

Cover Sheet. Each proposal must have a cover sheet providing the information set forth in the sample form entitled **Small Business Innovation Research (SBIR) Program Proposal Cover Sheet shown as Form 38 in Part One: Forms and Tables.**

Abstract. Each proposal should start with a narrative in which you describe the relevance of your invention to the DoD mission. Your personal qualifications should also be stated. **A sample Small Business Innovation Research (SBIR) Program Project Summary is offered for your use in Form 39 located in Part One: Forms and Tables.**

The Department of Defense is primarily interested in technical competence. The people who'll evaluate your proposals want to know that you understand what the project involves and that you can perform the work required. Once your technical competence has been established, DoD will consider costs.

Be sure to specifically address these points:

1) **Purpose and Objective.** State briefly the primary purpose, general objective, and expected results of the proposal, such as:

> a) State the problem or problems your invention will contribute to solving, and the anticipated contribution of any research to that solution.

> b) Enumerate the specific objective of any additional research and specify the questions that your research will attempt to answer. This will be important if you hope for DoD moneys to finance further research.

> c) State the expected consequences of successful completion of your research, including potential economic and other benefits.

> d) Discuss the existing interest by potential beneficiaries or users.

2) **Previous or Ongoing Related Work.** Document any knowledge you may have of other related activities with appropriate references to the literature or currently ongoing R&D. In particular, show how your proposed project relates to these activities and how your invention/research will extend the level of knowledge in the field.

3. **Statement of Work.** If your proposal involves additional R&D, give full and complete technical details of the procedures that will be followed throughout the scope of your work. Outline it phase by phase. Give time frames and objectives.

4. **Organization, Facilities and Qualifications.** Fully describe who you are, if individual, or what your company does.

5. **Patents.** If you believe you are presenting a patentable idea, I suggest that you file, as a protection to yourself and to the government, necessary patent applications with the Patent and Trademark Office, or otherwise identify your intention to do this. Mark your proposal as containing proprietary data.

Cost Estimate

Give a detailed estimate of all costs. To help you formalize this, I am including a **Defense Small Business Innovation Research (SBIR) Program Cost Proposal (Form 40 located in Part One: Forms and Tables)**. Take from it those parts that are most appropriate to your project and situation.

Proprietary Data

If you do not want your invention or idea disclosed to the public or used by the government for any other purpose than proposal evaluation, clearly label it as containing proprietary data. Don't just do it on the cover sheet, but rather on every page that holds such information.

Note that this restriction does not limit the government's right to use or disclose any data contained in your proposal if it is obtainable from another source (or from yourself previously) without restriction.

Notification of DoD Action

It may take up to six months for you to get an answer! Be patient. Your proposal will be acknowledged upon its receipt.

Security

Always submit unclassified material when possible. When this is not possible, make sure you properly label documents as being classified. Ask the Pentagon for a copy of its *Industrial Security Manual for Safeguarding Classified Information* (DoD 5220.22-M). Paragraph 11 of this publication will tell you how to correctly classify your submissions.

Around the Federal Triangle

I have conscientiously read through all federal R&D activity descriptions and selected for this section only those agencies and departments that are potential funding sources to independent inventors. I have omitted those agencies and departments whose interest is more in reports and studies than in patentable inventions and innovative concepts. Many federal R&D budgets are applied towards scientific research programs and policy research and analysis studies and the like and, thus, inappropriate.

Because locating and obtaining federal R&D funding requires a significant commitment of time and resources, the advice of NASA Headquarters Small Business Specialist Mark Kilkenny should be remembered. He recommends that you concentrate on the important few and not waste time and paperwork on the unimportant many.

IMPORTANT!! You should critically evaluate yourself before you submit a proposal for a government contract. And you should be equally critical in evaluating any idea you submit. The federal government is interested in only the most experienced people with pertinent novel ideas. Don't waste its time or your own on half-baked ideas or proposals for contracts that you are not qualified to carry out.

If you feel your invention would be of interest to a particular agency or that an agency appears to be funding research and development in your field of expertise, get in touch.

If the system starts to frustrate you either in the finding of appropriate government officials or the way in which you are being handled by the government officials you find, don't hesitate a moment to enlist the assistance of your elected representatives on Capitol Hill. You'll be amazed at how quickly a call from a U.S. Senator's or Representative's office can make things happen.

Section Arrangement

The following entries are arranged alphabetically by federal department, followed by an alphabetic listing of programs. Organizations and programs in this section are referenced in the Master Index.

National Aeronautics and Space Administration

★1657★ Ames Research Center (ARC)
Moffett Field, CA 94035   Phone: (415) 604-5000
                        Fax: (415) 604-4003
Lawrence Milov, External Relations Officer

Research and development programs in aeronautics, life sciences, space science, space technology, and flight research. Other efforts are aimed at human factors in space, earth resources study, thermal protection systems for atmosphere entry vehicles, computational physics and chemistry, and artificial intelligence and autonomous system. **Publications:** NASA Technical Reports.

★1658★ Dryden Flight Research Facility
PO Box 273   Phone: (805) 258-3311
Edwards Air Force Base, CA 93523   Fax: (805) 258-3567
Leonie Boarden, Small Business Specialist

Manned flight research, including systems, configuration, and technology developments for high performance aircraft. **Publications:** X-Press Magazine (biweekly).

★1659★ George C. Marshall Space Flight Center
Huntsville, AL 35812   Phone: (205) 544-2121
Conrad Walker, Small Business Specialist

Serves as a primary center for the design and development of space transportation systems, elements of the Space Station, scientific applications payloads, and other systems for present and future space exploration. Operates the Program Development Directorate, which generates plans for promising new programs; Institutional and Program Support Directorate, which provides various supporting services; and Science and Engineering Directorate, which provides technical support and furnishes a research base.

★1660★ Goddard Space Flight Center (GSFC)
Procurement Analysis Branch   Phone: (301) 286-6575
Code 263   Fax: (301) 286-2575
Greenbelt, MD 20771
Janet Jew, Small Business Specialist

Research interests include orbital spacecraft development, tracking and data acquisition systems, space physics and astronomy payloads, upper atmospheric research, weather and climate, earth dynamics and resources, information systems, sounding rocket and payload development, planetary science, and sensors and experiments in environmental monitoring and ocean dynamics. Maintains an automated source system comprised of approximately 4,000 firms. The Center encourages businesses to submit a Bidders Mailing List Application (SF129) to be added to this system.

★1661★ Headquarters Contracts and Grants Division
L'entant Plaza, Ste. 425   Phone: (202) 453-2090
Washington, DC 20546   Fax: (202) 755-1020
Darleen Druyun, Assistant Administrator

Responsible for planning, negotiating, awarding, and administering contracts based on procurement requirements, including system engineering services, reliability studies, basic and applied research and development, mobile lecture-demonstration units, exhibits, motion picture services, management analysis surveys and automatic data processing equipment, and software services. The Division also has agency-wide responsibility for negotiating and executing NASA contracts with foreign governments and commercial organizations. **Contact:** Mark E. Kilkenny, Procurement System Branch Chief (202) 453-1840; Eugene D. Rosen, Director, Small and Disadvantaged Business Utilization (202) 453-2088; James T. Rose, Assistant Administrator, Commercial Programs Office (202) 453-1123.

★1662★ Jet Propulsion Laboratory (JPL)
California Institute of Technology   Phone: (818) 354-5722
4800 Oak Grove Dr.
Pasadena, CA 91109
Margo Kuhn, Small Business Specialist

Has the primary responsibility of exploration of the solar system with unmanned spacecraft. Research and development activities include aeronautics and aerospace, communications, computer science and mathematics, earth and space sciences, electronics, and physics. **Publications:** (1) Annual Report, Highlights; (2) Closeup. **Contact:** Mr. T. May, Minority Business Specialist (818) 354-2121.

★1663★ John F. Kennedy Space Center
Kennedy Space Center, FL 32899   Phone: (407) 867-7110
                    Fax: (407) 867-2217
Ann Watson, Small Business Specialist

Responsible for assembly, checkout, servicing, launch, recovery, and operational support of space transportation system elements. Principle research activity is the design and development of launch and landing facilities. Technology transfer programs are administered by the Center's Advanced Projects and Technology Office. **Publications:** NASA Tech Briefs.

★1664★ Langley Research Center
Hampton, VA 23665   Phone: (804) 864-1000
Joe Foglia, Small Business Specialist

Research and development in aeronautics and space technology, including acoustics and noise reduction, aerodynamics, aerospace vehicle structures and materials, aerothermodynamics, avionics technology, environmental quality monitoring technology, sensor and data acquisition technology, long-haul aircraft, military support, and advanced space vehicle configuration. Activities also include technology transfer and providing development support to government agencies, industry, and other NASA centers.

★1665★ Lewis Research Center
21000 Brookpark Rd.   Phone: (216) 433-4000
Cleveland, OH 44135
Steven Fedor, Small Business Officer

Responsible for managing and design and development of the power generation, storage, and distribution system for the U.S. space station. Areas of research include air breathing and space propulsion systems; turbomachinery thermodynamics and aerodynamics; fuels and combustion; aero and space propulsion systems; power transmission; tribology; internal engine computational fluid dynamics; high temperature engine instrumentation, space communications; and space and terrestrial energy processes, systems technology, and applications.

★1666★ Lyndon B. Johnson Space Center (JSC)
Houston, TX 77058   Phone: (713) 483-4511
     Fax: (713) 483-6200
Mr. R.L. Duppstadt, Small Business Specialist

Primary mission is the development of spacecraft for manned space flight programs and the conduct of manned flight operations; selection and training of astronauts; providing management in systems engineering and integration and business and operations management for the Space Shuttle Program; conducting investigations of earth resources technology; and conducting space life sciences research.

★1667★ National Space Technology Laboratories (NSTL)
John C. Stennis Space Center   Phone: (601) 688-3680
NASA Procurement Office, Bldg. 1100
Stennis Space Center, MS 39529-8000
Becky Dubisson, Small Business Specialist

Primary installation for static test firing of large rocket engines and propulsion systems. Beside NASA, resident agencies include NOAA, and departments of Army, Navy, and Interior; U.S. Coast Guard; as well as groups from various universities. These groups are involved in oceanographic, meteorological, and environmental research and, together with NASA, form a scientific and technical community, each pursuing its own programmatic objectives but collectively producing a scientific base for technology exchange. **Contact:** Mr. P.B. Higdon, Contracting Officer, Earth Resources Laboratory (601) 688-1641.

★1668★ Office of Space Science and Applications (OSSA)
OSSA Steering Committee   Phone: (202) 453-1409
Code SPS   Fax: (202) 755-3370
Washington, DC 20546
Leonard A. Fisk, Associate Administrator

Operationally responsible for scientific research into the nature and origin of the universe and applying space systems and techniques to solve problems of Earth.

★1669★ Small Business Innovation Research Program
Washington, DC 20546   Phone: (202) 271-5650
     Fax: (202) 271-5566
Harry Johnson, Manager

Areas of soliciting proposals are based on the needs of NASA's programs and missions as described by the roles of each NASA Center.

National Science Foundation

★1670★ Division of Policy Research and Analysis
1800 G St. NW, Rm. 1233   Phone: (202) 357-9689
Washington, DC 20550   Fax: (202) 357-7745
Dr. Peter House, Division Director

Funds policy research, including business firms specializing in science and innovation policy and technology and resource policy.

★1671★ Industry/University Cooperative Research Centers
1800 G St. NW, Rm. 1121   Phone: (202) 357-7636
Washington, DC 20550
Alex Schwaizjopt, Program Manager

Fosters university-industry interaction at university centers by supporting a limited number of research projects. **Contact:** Alex Schwarzkopf, Manager.

★1672★ Office of Small Business Research and Development
1800 G St. NW, Rm. V-502   Phone: (202) 653-5202
Washington, DC 20550   Fax: (202) 653-7699
Dr. Donald Senich, Director

Provides information and serves as a referral point for small businesses interested in the Foundation's research or procurement opportunities, or how to submit proposals. Also compiles and publishes information on research grants and contracts awarded by NSF to small business. **Contact:** Linda L. Boutchyard, Program Specialist (202) 357-7464.

★1673★ Small Business Innovation Research Program
1800 G St. NW, Rm. V-502   Phone: (202) 653-5002
Washington, DC 20550
Roland Tibbetts, Program Manager

Acts as the principal NSF research opportunity organization for small businesses. Research topics include physics; chemistry; mathematical sciences; computer sciences; materials research; electrical, communications, and systems engineering; chemical, biochemical, and thermal engineering; mechanics; structures, and materials engineering; cellular biosciences; molecular biosciences; biotic systems and resources; behavioral and neural sciences; social and economic sciences; information sciences; astronomy; atmospheric sciences, earth sciences; ocean sciences; design, manufacturing, and computer engineering; and advanced scientific computing. **Contact:** Mr. Ritchie Coryell, Program Manager (202) 357-7527.

U.S. Department of Agriculture

★1674★ Agricultural Research Service
Bldg. 005, BARC-W   Phone: (301) 344-3191
Beltsville, MD 20705
Edward Knipling, Deputy Admin.

Research interests include physical, biological, and chemical engineering; clothing, housing, and household economics; industrial and food products and processing methods for agricultural commodities; and soil and water, crop, animal husbandry, entomology, agricultural engineering, and energy. Published solicitations are not made by the office, but proposals are received at any time.

★1675★ Office of Grants and Program Systems (OGPS)
USDA-CSRS   Phone: (202) 401-1761
Aerospace Bldg., Rm. 328   Fax: (202) 401-6869
Washington, DC 20250
Dr. William D. Carlson

Maintains Competitive Research Grants Office, which publishes its solicitations in the Federal Register. Grants are awarded on a competitive basis. Proposals are requested in the areas of biological nitrogen fixation, biological stress on plants, and human nutrition.

★1676★ Small Business Innovation Research Program
14th & Independence Ave. SW, Rm. 127   Phone: (202) 720-7117
West   Fax: (202) 720-3001
Washington, DC 20250
Jim House

U.S. Department of Commerce

★1677★ National Institute of Standards and Technology (NIST)
Bldg. 301, Rm. B128   Phone: (301) 963-7723
Gaithersburg, MD 20899
Keith Chandler, Small Business Specialist

NIST is organized into three technical laboratories as follows: (1) National Engineering Laboratory (NEL) furnishes technology and technical services to public and private sectors to address national needs and to solve problems in the public interest. NEL conducts research in engineering and applied science; builds and maintains competence in the necessary disciplines required to carry out research; and develops engineering data and measurement capabilities. (2) National Measurement Laboratory (NML) provides the national system of physical, chemical, and materials measurement; coordinates the system with measurement systems of other nations, and furnishes services leading to uniform measurement; and conducts research leading to methods of measurement, standards, and data on the properties of materials. (3) National Computer and Telecommunication Laboratory (NCTL), which develops computer standards, conducts research, and provides scientific and technical services in the selection, acquisition, application, and use of computer technology; manages a government-wide program for standards development and use, including management of federal participation in voluntary standardization.

★1678★ National Institute of Standards and Technology (OERI)
Office of Energy-Related Inventions
Bldg. 411, Rm. A-115   Phone: (301) 975-5500
Gaithersburg, MD 20899   Fax: (301) 975-3839
George L. Lewett

Invention disclosures evaluated by OERI are categorized as follows: fossil fuel production; direct solar; other natural resources; combustion engines and components; transportation systems, vehicles, and components; building, structures, and components; industrial processes, and miscellaneous. (For further information, see introduction to this chapter.)

★1679★ National Oceanic and Atmospheric Administration (NOAA)
Silver Spring Metro Center Two   Phone: (301) 427-2829
SSMC2 Sta. 4301   Fax: (301) 427-2806
1325 F1W Hwy.
Silver Springs, MD 20910
Scott Cook, Small Business Specialist

Research and development interests in environmental monitoring and prediction include automation of meteorological observations, analysis, and communication; remote sensing; oceanic monitoring; stratospheric measurement; satellite technology and development of ground equipment; weather modification; and modeling of atmospheric and oceanic processes, including climate variation. **Contact:** Edward Tiernan, SBIR Program Manager Chief, NOAA/NESDIS, Orta FB4, Room 3316, Suitland, MD 20233 (202) 763-2418.

U.S. Department of Defense

★1680★ Defense Nuclear Agency (DNA)
Office of Contracts   Phone: (703) 325-6626
6801 Telegraph Rd.   Fax: (703) 325-2955
Alexandria, VA 22310
Mrs. Patricia Brooks, Director, Business Utilization

Seeks research and development from small business firms with strong capabilities in the nuclear weapons effects, including simulation, instrumentation, directed energy, nuclear hardening and survivabilty, security of nuclear weapons, and operational planning. **Contact:** Small Business Innovation Research Program, Attn: OAAM/SBIR, Jim Gerding (202) 325-7018.

U.S. Department of Energy

★1681★ Assistant Secretary for Conservation and Renewable Energy
1000 Independence Ave. SW   Phone: (202) 586-9260
Washington, DC 20585
Dr. Robert L. SanMartin

Programs in conservation and renewable energy support research that attracts limited or no venture capital because the risks are too high or the payoff too long-range or unpredictable. Renewable energy programs involve research, development, and proof-of-concept experiments to enable industry to bring resource and technology options into the marketplace. Research activities are carried out through the award of grants and contracts.

★1682★ Assistant Secretary for Defense Programs
1000 Independence Ave. SW   Phone: (202) 586-2177
Washington, DC 20585   Fax: (202) 586-1567
Richard Claytor

Directs the DOE's programs for nuclear weapons research, development, testing, production, and surveillance; manages programs for the production of the special materials used by the weapons program within the Department; and manages the defense nuclear waste and by-products program. In addition, awards grants and contracts to universities and other research groups for some defense programs.

★1683★ Assistant Secretary for Nuclear Energy
1000 Independence Ave. SW   Phone: (202) 586-6450
Washington, DC 20585   Fax: (202) 586-8353
William H. Young

Responsible for programs and projects for nuclear power generation and fuel technology; evaluation of alternative reactor fuel cycle concepts; development of naval nuclear propulsion plants and cores; and nuclear waste technology and remedial action programs. Unsolicited proposals for projects in nuclear energy should be sent to the Unsolicited Proposal Coordinator, Reports and Analysis Branch, Procurement and Assistance Management Directorate, Department of Energy, Washington, DC 20585.

★1684★ Morgantown Energy Technology Center (METC)
3610 Collins Ferry Rd.   Phone: (304) 291-4764
PO Box 880
Morgantown, WV 26507-0880
Thomas Bechtel, Director

Fossil fuel research and development laboratory responsible for projects in unconventional gas recovery, fluidized bed combustion, gas stream cleanup, fuel cells, heat engines, component development for coal conversion and utilization devices, surface coal gasification,

instrumentation and control, oil shale technology, tar sands, underground goal gasification, and arctic and offshore technologies. **Telecommunication Access:** Alternate telephone (304) 599-4511.

★1685★ Office of Energy Research (OER)
1000 Independence Ave. SW
Washington, DC 20585
Robert Hunter, Jr., Director-Designate
Phone: (202) 586-5430
Fax: (202) 586-4120

Manages research programs in the basic energy sciences, high energy physics, and fusion energy; administers DOE programs supporting university researchers; funds research in mathematical and computational sciences critical to the use and development of supercomputers; and administers a financial support program for research and development not funded elsewhere in the Department. The Office also manages the Small Business Innovation Research Program for the Department. **Contact:** Jean Marrow (301) 353-5544.

★1686★ Pittsburgh Energy Technology Center (PETC)
PO Box 10940
Pittsburgh, PA 15236
Dr. Sun W. Chun, Director
Phone: (412) 892-6400
Fax: (412) 892-6127

Primary mission is to promote the use of coal and its derived synthetic fuels in an environmentally sound manner. Research and development programs cover coal liquefaction, alternate fuels, coal preparation, flue gas cleanup, combustion technology, magnetohydrodynamics, solids transport, and peat and anthracite. The Center is also responsible for handling all unsolicited research proposals pertaining to any part of the DOE's fossil energy program. **Publications:** (1) PETC Quarterly Technical Progress Report; (2) D.P. Databeat; and (3) Energizer (all quarterly). **Telecommunication Access:** Alternate telephone (412) 892-6128.

★1687★ Small Business Innovation Research Program (SBIR)
Mail Stop ER-64, GTN
Washington, DC 20585
Jean Marrow, Contact
Phone: (301) 353-5544
Fax: (301) 353-3884

Stimulates technological innovation; uses small business to meet federal research and development needs; fosters and encourages participation by minority and disadvantaged persons in technological innovation; and increases private sector commercializition of innovations derived from federal research and development. **Telecommunication Access:** Alternate telephone (301) 353-5867.

U.S. Department of Health and Human Services

★1688★ National Institute of Allergy and Infectious Diseases
Control Data Bldg., Rm. 3C07
6003 Executive Blvd.
Bethesda, MD 20205
Lew Pollack, Chief of Contract Mgt. Branch
Phone: (301) 496-0611
Fax: (301) 402-0972

Supports research on causes, diagnosis, treatment, and prevention of infections, allergic, and other immunologically mediated diseases. Contract programs are concerned with bacterial and viral vaccines, antiviral substances, transplantation, and viral and allergen reagents.

★1689★ National Institute of Arthritis and Musculoskeletal and Skin Diseases (NIAMS)
5333 Westbard Ave., 9A35
Westwood Bldg., Rm. 602
Bethesda, MD 20892
Patrick Sullivan
Phone: (301) 496-7111
Fax: (301) 496-9721

Conducts and supports basic and clinical research on arthritis, osteoarthritis, lupus, muscle disease, psoriasis, acne, ichthyosis, and vitiligo. Offers investigator-initiated grants, reseasrch center grants, individual and institutional research training research awards, career development awards, and contracts to public and private research institutions and organizations. **Contact:** Lawrence E. Shulman, Director (301) 496-4353.

★1690★ National Institute of Dental Research (NIDR)
Contract Management Section
Westwood Bldg., Rm. 521
Bethesda, MD 20892
Ms. Marion Blevins
Phone: (301) 496-7311
Fax: (301) 402-1261

Supports investigations into the causes, means of prevention, diagnosis, and treatment of oral diseases, including dental caries, periodontal diseases, lesions of soft and hard tissue, and oral-facial abnormalities. Also supports biomaterials research, improvements in anesthesia and analgesia, basic epidemiological research, and clinical trials.

★1691★ National Institute of General Medical Sciences (NIGMS)
Research Contract Branch
Division of Contracts & Grants
Bldg. 31, Rm. 1B44
9000 Rockville Pike
Bethesda, MD 20892
David Snight
Phone: (301) 496-4487
Fax: (301) 402-0178

Supports research in the sciences basic to medicine, behavioral sciences, and clinical disciplines. The Institute fosters multidisciplinary approaches to research and employs the full range of support mechanisms, including research project grants, program project grants, research center grants, career development awards, awards to new investigators, and institutional and individual fellowships.

★1692★ National Institute of Neurological and Communicative Disorders and Stroke (NINCDS)
Contract Management Branch
Federal Bldg., Rm. 901
7550 Wisconsin Ave.
Bethesda, MD 20892
Lawrence Fitzgerald
Phone: (301) 496-9203
Fax: (301) 402-2211

Conducts, coordinates, and supports research concerned with the cause, development, diagnosis, therapy, and prevention of disorders and diseases of the central nervous system and the communicative and sensory systems. Research includes neurological and mental development, infectious diseases, multiple sclerosis, epilepsy, head injury and stroke, biomedical engineering and instrumentation, and neural prostheses.

★1693★ National Institutes of Health (NCI)
National Cancer Institute
9000 Rockvillw Pike
Bldg. 31, Rm. 11A48
Bethesda, MD 20892
Samuel Broder, Director
Phone: (301) 496-5615
Fax: (301) 402-0338

Plans, directs, conducts, and coordinates a national research program on the detection, diagnosis, cause, prevention, treatment, and palliation of cancers. Conducts and directs research performed in its own laboratories

and through contracts; supports construction of facilities necessary for research on cancer; supports demonstration projects on cancer control; and collaborates with cancer research with industrial concerns. **Contact:** Louis P. Greenberg, Acting Chief, Research and Contracts Branch.

★1694★ National Institutes of Health
National Eye Institute

Research Contract Branch   Phone: (301) 496-4487
Division of Contracts & Grants   Fax: (301) 402-0178
Bldg. 31, Rm. 1B44
9000 Rockville Pike
Bethesda, MD 20892
John Becenvo

Supports research on the cause, natural history, prevention, diagnosis, and treatment of disorders of the eye and the visual system, especially glaucoma, retinal disease, corneal diseases, cataract, and sensory-motor disorders. Support is provided for basic studies, clinical trials, and the development of animal models for vision disorders. **Publications:** Planning Report (every five years). **Meetings:** Grand Rounds (weekly) and Neuro-Ophthalmology Seminar (monthly).

★1695★ National Institutes of Health (NHLBI)
National Heart, Lung, and Blood Institute

Contracts Operations Branch   Phone: (301) 496-7033
Westwood Bldg., Rm. 654
Bethesda, MD 20892
Robert Carlsen, Chief

Supports basic and clinical research, development, and other activities related to the prevention, diagnosis, and treatment of cardiovascular, lung, and blood diseases. The Institute plans and directs research in the development, trial, and evaluation of drugs and devices relating to the prevention such diseases.

★1696★ Office of Human Development Services
(HDS)

200 Independence Ave. SW   Phone: (202) 245-1787
HHH Bldg., Rm. 338-F   Fax: (202) 245-6699
Washington, DC 20201
Cynthia Haile Selassie, Small Business Specialist

Oversees programs that contribute to the social and economic well-being of a number of vulnerable populations, including the elderly, children, youth and families, the developmentally disabled, and Native Americans.

U.S. Department of the Air Force

★1697★ Air Force Arnold Engineering Development
Center (AEDC)

AEDC/PKP   Phone: (615) 454-4407
Arnold Air Force Base, TN 37389-5000   Fax: (615) 454-7330
William Lamb, Small Business Specialist

Conducts development, certification, and qualification testing on aircraft, missile, and space systems. Also conducts research and technology programs to develop advanced testing techniques and instrumentation and to support the design of new test facilities. **Publications:** AEDC Test Highlights.

★1698★ Air Force Flight Test Center (AFFTC)

AFTC/BC   Phone: (805) 277-2619
Edwards Air Force Base, CA 93523 Fax: (805) 277-3784
Judy Vaughn

Flight testing of aerospace vehicles. Specific activities include: (1) conduct and support of aircraft systems tests; (2) flight evaluation and recovery of aerospace research vehicles and development testing of aerodynamic decelerators; (3) management and operation of USAF Test Pilot School; (4) management and operation of the Utah Test and Training Range and the Edwards Flight Test Range; and (5) support for and participation in agency, foreign, and contractor test and evaluation programs. **Telecommunication Access:** Alternate telephone (617) 861-4974.

★1699★ Air Force Systems Command

AFSC/PKU   Phone: (301) 981-6107
HQ Bldg., Rm. 204   Fax: (301) 981-9660
Andrews Air Force Base
Washington, DC 20334
Carol Bebel, Executive for Small Business

Command includes project divisions that acquire aircraft, missiles, space systems, electronic systems, and conventional weapons; test centers; laboratories that conduct research in the mathematical, physical, engineering, and environmental sciences; and divisions that are involved in medical research and technology, identification of foreign technology, and contract management.

★1700★ Air Force Systems Command
Aeronautical Systems Division (Eglin Air Force Base)

AFDTC/BC   Phone: (904) 882-2843
Eglin Air Force Base, FL 32542-5000   Fax: (904) 537-5372
Ralph K. Frangioni, Jr.

Armament systems research and development, testing, and procurement, including total program management responsibility for Air Force non-nuclear munitions, including guided bombs, mines, fuzes, flares, bomb racks, missiles, aerial targets, and munition handling and transportation equipment.

★1701★ Air Force Systems Command
Aeronautical Systems Division (Wright-Patterson Air
Force Base)

ASD/BC   Phone: (513) 255-5322
Wright-Patterson Air Force Base   Fax: (513) 476-4037
Dayton, OH 45433-6503
Norman Hayes, Small Business Specialist

Development and acquisition of aeronautical systems, including aircraft engines, aircraft wheels and brakes, airborne communication systems, aircraft bombing and navigation systems, and aircraft instruments, and aeronautical reconnaissance systems, and mobile land-based tactical information processing and interpretation facilities. **Contact:** Jim Beach, Small and Disadvantaged Business Utilization Specialist.

★1702★ Air Force Systems Command
Space Division

Los Angeles Air Force Sta.   Phone: (213) 363-2855
PO Box 92960   Fax: (213) 363-2044
Worldway Postal Center
Los Angeles, CA 90009-2960
Chuck Willeme, Deputy for Small Business

Acquisition and evaluation of space systems and equipment; management of research, tracking, telemetry, and recovery; feasibility studies; acquisition, production, quality assurance, and installation of assigned space and missile systems; and quality assurance and installation of boosters, and aerospace ground equipment to support launch control and recovery.

★1703★ Air Force Weapons Laboratory (AFWL)

Phillips Lab (AFSC)
Civilian Personnel Liaison Office
Kirtland Air Force Base, NM 87117-6008

Phone: (505) 846-4709
Fax: (505) 846-1724

Responsible for research efforts in the areas of survivable systems and directed energy weapons, including particle beam technology, advanced weapons technology, high power microwave technology, and space nuclear power.

★1704★ Ballistic Missile Office

Attn: BMO/BC
Norton Air Force Base
San Bernardino, CA 92409
Terry Carey, Chief, Small Business Office

Phone: (714) 382-4304
Fax: (714) 382-3568

Responsible for intercontinental ballistic missile development. Office integrates activities of contractors, retaining overall engineering responsibility for a particular system.

★1705★ Defense Advanced Research Projects Agency (DARPA)

3701 N. Fairfax Dr.
Arlington, VA 22203-1714
Stephen L. Squires, Director

Phone: (703) 696-2414
Fax: (703) 696-2200

Pursues imaginative and innovative research ideas and concepts offering significant military utility. Role in basic research is to develop selected new ideas from conception to hardware prototype for transfer to development agencies. Programs are conducted through contracts with industrial, university, and nonprofit organizations, focusing on improved strategic, conventional, rapid-development, and sea-power forces; and on scientific investigation into technologies for the future. Unsolicited proposals will be accepted if they are related to any of the current technical programs assigned to DARPA. DARPA invites small businesses to submit proposals under its SBIR program, which has as its objectives stimulating technological innovation in the private sector, strengthening the needs, fostering and encouraging participation by minority and disadvantaged persons in technological innovation, and increasing commercial application of Department of Defense-supported research and development. **Contact:** Bud Durand, SBIR Progam Manager (202) 694-1626.

★1706★ Directorate of Research and Development Procurement

Aeronautical Systems Division
Wright-Patterson Air Force Base
Dayton, OH 45433
Barbara Cooke, Small Business Specialist

Phone: (513) 255-3825

Procurement of exploratory and advanced research and development in the areas of air breathing, electric, and advanced propulsion; fuels and lubricants; and power generation, including molecular electronics, bionics, lasers, vehicle environment, photo materials, position and motion sensing devices, navigation, communications, flight dynamics, materials science, and life support. **Contact:** Dorothy Muhlhauser, Small and Disadvantaged Business Utilization Specialist.

★1707★ Eastern Space and Missile Center (ESMC)

ASMC/BC
Patrick Air Force Base, FL 32925-5432
Carolyn-Lee Thompson, Small Business Specialist

Phone: (407) 494-2207
Fax: (407) 494-2983

Launches the Titan III space booster and represents Department of Defense interests in Space Shuttle operations. Research, development, and procurement program is confined primarily to range test instrumentation, including radar, telemetry, electro-optics, impact locations, data reduction, range/mission safety and control, weather timing and firing, and frequency control and analysis. **Telecommunication Access:** Alternate telephone (305) 494-2208.

★1708★ Electronic Systems Division (ESD)

ESD/BC
Hanscom Air Force Base, MA
  01731-5000
Al Hart

Phone: (617) 377-4973

Development, acquisition, and delivery of electronic systems and equipment for the command, control, and communications function of aerospace forces. **Telecommunication Access:** Alternate telephone (617) 861-4973/4.

★1709★ Office of Scientific Research (AFOSR)

Bldg. 410
Bolling Air Force Base
Washington, DC 20332
Louise Harrison, Small Business Specialist

Phone: (202) 767-4943
Fax: (202) 767-0466

Considers basic research proposals from industrial and small business concerns and nonprofit organizations. Accepts unsolicited proposals. Any scientific investigator may make a preliminary inquiry to obtain advice on the degree of interst in a project, or may submit a specific research proposal. Proposals should indicate the field of the investigation and objectives sought, describing previous work and related grants and contract held, if any; in addition, it should outline the approach planned for the research and should include estimates of the time and cost requirements. The principal investigator should be named and an outline of professional background included. Brochures describing research ares of interest and procedures for submitting proposals may be obtained.

★1710★ Operational Test and Evaluation Center (AFOTEC)

PL/BC
Bldg. 499, Rm. 100
Kirtland Air Force Base, NM 87117-5023
Larry Jones, Chief

Phone: (505) 846-8515
Fax: (505) 844-3819

Independent agency responsible for testing and evaluation of systems being developed for Air Force and joint service use. Center's focus is on the operation effectiveness and suitability of the Air Force's future weapons and supporting equipment, as well as identifying deficiencies requiring corrective action. **Contact:** Nancy Lindquist, SADBU Specialist.

★1711★ Rome Air Development Center (RADC)

Rome Lab
Attn: BC
Griffis Air Force Base, NY 13441
Richard Smith, Small Business Specialist

Phone: (315) 330-4256

Provides a technology base for projects that pertain to command, control, communications, and intelligence research and development, including radar, airborne and space-based reconnaissance and sensing systems, vast arrays of communications systems, and computers. Has also been designated as a lead laboratory for research and development of the technology of photonics.

★1712★ Space Technology Center

Kirtland Air Force Base, NM 87117-6008 Phone: (505) 846-4701

Manages three Air Force laboratories, integrating space technology efforts in order to explore military space capabilities and the needs of future space systems. Specific areas of interest include directed energy research, nuclear weapons effects, survivability issues, rocket propulsion, and the earth and space environment.

★1713★ Strategic Defense Initiative Organization (SDIO)
SDIO/SBIR
The Pentagon, Rm. 1E-167
Washington, DC 20301-7100
Carl Nelson, SBIR Coordinator
Phone: (703) 693-1671
Fax: (703) 693-1695

Supports programs involved with kinetic energy weapons, and surveillance satellites, and directed energy weapons research. **Contact:** Submission of unsolicited proposals should be directed to Col. Kluter, SDIO/OSD, Washington, DC 20301-7100 (202) 653-0034.

U.S. Department of the Army

★1714★ Aberdeen Proving Ground Installation Support Activity
Attn: STEAP-SB
Bldg., 310 Wing 3A
Aberdeen Proving Ground, MD
   21005-5001
Bonnie Maliszewski, Small Business Advisor
Phone: (301) 278-3878
Fax: (301) 278-2001

Research and development, production, and post-production testing of components and complete items of weapons, systems, ammunition, and combat and support vehicles. Also tests items of individual equipment in use throughout the Army.

★1715★ Army Armament Research, Development, and Engineering Center (ARDC)
ARDC
ATTN: SMCAR-SB
Dover, NJ 07801
Ed Smith, Small Business Advisor
Phone: (201) 724-4104
Fax: (201) 793-4803

Responsible for management of research, development, life-cycle engineering, and initial acquisition of weapon systems and support equipment.

★1716★ Army Aviation Systems Command (AVSCOM)
ATTN: AMSAV-V
4300 Goodfellow Blvd., Bldg. 102E
St. Louis, MO 63120-1798
Angelina F. Amato, Small Business Advisor
Phone: (314) 263-2200

Aviation design, research, development, maintenance, engineering, stock and supply control, and technical assistance to users of all Army aviation and aerial delivery equipment. Evaluates prototype hardware for fueling and defueling equipment for use in combat areas and in solving fuel contamination problems.

★1717★ Army Ballistic Research Laboratory (BRL)
CRDEC
ATTN: AMSMC/SBA (A)
Bldg. E4415
Aberdeen Proving Ground, MD
   21010-5423
John Rasmussen, Small Business Specialist
Phone: (301) 671-4726

Provides basic and applied research in mathematics, physics, chemistry, biophysics, and the engineering sciences related to the solution of problems in ballistics and vulnerability technology.

★1718★ Army Chemical Research, Development, and Engineering Center
CRDEC
ATTN: AMSMC-SBA (A)
Bldg. E4415
Aberdeen Proving Ground, MD
   21010-5423
John Rasmussen, Contact
Phone: (301) 671-2309

Research and development in the fields of chemical, smoke, and flame weapons. Activities also include studies in pollution abatement and environmental control technology.

★1719★ Army Cold Regions Research and Engineering Laboratory
72 Lyme Rd.
Hanover, NH 03755-1290
Charles Zdunczyk, Small Business Specialist
Phone: (603) 646-4323
Fax: (603) 646-4488

Conducts research pertaining to characteristics and events unique to cold regions, especially winter conditions, including design of facilities, structures and equipment, and refining methods for building, traveling, living, and working in cold environments.

★1720★ Army Communications-Electronics Command (CECOM)
HQ CECOM
ATTN: DRSEL-SB AMSEL-SB
Fort Monmouth, NJ 07703
John Meschler
Phone: (908) 532-4511

Covers full spectrum of services to the U.S. soldier in the field of communications and electronics. The first steps in converting concepts into new military materiel are taken in the CECOM Research and Development Center and its laboratories. These laboratories are responsible for research and development of communications and electronics equipment and systems.

★1721★ Army Construction Engineering Research Laboratory (CERL)
Procurement/Supply
PO Box 4005
Champaign, IL 61820-1305
Julie Flowers
Phone: (217) 352-6511
Fax: (217) 373-7222

Mission is to provide research and development to support Army programs in facility construction, operation, and maintenance. Efforts focus on vertical construction as applied to buildings and structures rather than heavy construction. Laboratory has four main divisions: Engineering and Materials Division, Energy Systems Division, Environmental Division, and Facilities Systems Division. Center emphasizes communication and interchange of information with academic, engineering, and construction activities within the Department of Defense, other governmental agencies, and the private sector. **Publications:** CERL Reports (quarterly).

★1722★ Army Engineer Topographic Laboratories (USAETL)
Fort Belvoir, VA 22060-5546
Mrs. M.L. Williams, Small Business Specialist
Phone: (703) 355-2608
Fax: (703) 355-5176

Research and development in the topographic sciences, including mapping, charting, terrain analysis, geodesy, remote sensing, point positioning, surveying, and land navigation. Provides scientific and technical advisory services to meet environmental design criteria requirements of military materiel developers and in support of geographic intelligence and land environmental resources inventory requirements. Composed of the following laboratories and operational centers:

Geographic Sciences Laboratory, Research Institute, Space Programs Laboratory, Terrain Analysis Center, and Topographic Developments Laboratory.

★1723★ **Army Engineer Waterways Experiment Station (WES)**
3909 Halls Ferry Rd.   Phone: (601) 636-3111
Vicksburg, MS 39180   Fax: (601) 634-4238
Mr. A.J. Breithaupt, Utilization Specialist

Principle research, testing, and development facility of the Corps of Engineers. Operates six laboratories, including Coastal Engineering Research Center, Environmental Laboratory, Geotechnical Laboratory, Hydraulics Laboratory, Information Technology Laboratory, and Structures Laboratory, which conduct research in soil mechanics, concrete, engineering geology, pavements, weapons effects, protective structures, water quality, and dredge materials. **Publications:** List of Publications of the Waterways Experiment Station. **Meetings:** Annual Aquatic Plant Control Research and Operations Review (November), attendance is open to all.

★1724★ **Army Materials Technology Laboratory**
ATTN: SLCMT-PR   Phone: (617) 923-5005
Watertown, MA 02172   Fax: (617) 923-5328
Henry Mahler, Small Business Advisor Dir.

Manages Army's research and development programs relating to materials technology, solid mechanics, lightweight technology, and manufacturing testing technology. Is a center for excellence in the area of corrosion protection. **Meetings:** Sponsors Sagamore Army Materials Research Conference (annually, in August for invited participants and Military Handbook-17 Coordinating Group Meeting (semiannually, in May/November).

★1725★ **Army Medical Research and Development Command**
ATTN: SGRD-SADBU   Phone: (301) 663-2744
Fort Detrick   Fax: (301) 663-2982
Frederick, MD 21701-5012
Audrey L. Wolfe, Assistant Director, SADBU

Research involves assessment, prevention, diagnosis, and treatment of infectious diseases that would hamper military operations; disease vector surveillance; combat casualty care; health hazards of military materiel; factors limiting soldier effectiveness; prevention of oral disease; and dental materials. Operates research and development laboratories, including Army Aeromedical Research Laboratory, Army Biomedical Research and Development Laboratory, Army Institute of Dental Research, Army Institute of Surgical Research, Army Medical Research Institute of Chemical Defense, Army Medical Research Institute of Infectious Diseases, Army Research Institute of Environmental Medicine, and Army Walter Reed Institute of Research. Also supports medical research through contract research awards. Announcements of specific proposals solicited are published in Commerce Business Daily.

★1726★ **Army Missile Command (MICOM)**
Redstone Arsenal   Phone: (205) 876-2561
ATTN: AMSMI-SB   Fax: (205) 842-2622
Huntsville, AL 35809
Rex Conners, Small Bus. Advisor Cmdr.

Responsible for design and development; product, production, and maintenance engineering; and new equipment design of training devices in the areas of rockets, guided missiles, targets, air defense, fire control equipment, test equipment, missile launching and ground support equipment, and metrology and calibration equipment. **Telecommunication Access:** Alternate telephone (205) 876-5441.

★1727★ **Army Tank-Automotive Command (TACOM)**
ATTN: AMSTA-CB   Phone: (313) 574-5388
Warren, MI 48297-5000   Fax: (313) 574-5011
Louise Lather, Small Business Officer

Responsible for research, design, development, engineering, test management, modification, product assurance, integrated logistics support, acquisition, and deployment of wheeled and tracked vehicles and associated automotive equipment. Research and development are the functional responsibility of TACOM's Tank-Automotive Research, Development, and Engineering Center. **Contact:** Tom Clynes, Small Business Office (313) 574-5406.

★1728★ **Ballistic Missile Defense Systems Command**
PO Box 1500   Phone: (205) 955-3410
Huntsville, AL 35807   Fax: (205) 955-3958
Virginia B. Wright

Research and development in the fields of radar, interceptors, optics, discrimination, and data processing applicable to ballistic missile defense, including analysis of new and novel applications of science and engineering seeking revolutionary approaches to ballistic missile defense.

★1729★ **Belvoir Research, Development and Engineering Center**
ATTN: STRBE-U   Phone: (703) 664-5134
Fort Belvoir, VA 22060-5606   Fax: (703) 353-3739
Cindy Cherry, Small Business Advisor

Research, development, engineering, and initial production buys in the areas of mobility/countermobility, survivability, energy, and logistics. Research is carried out by the Combat Engineering Directorate, Logistics Support Directorate, and the Materials, Fuels, and Lubricants Directorate. Related support services are provided by the Advanced Systems Concepts Directorate, Information Management Directorate, Product Assurance and Engineering Directorate, and Resource Management Directorate. **Telecommunication Access:** Alternate telephone (703) 644-2482.

★1730★ **Benet Weapons Laboratory**
ATTN: SMCWV-TPA   Phone: (518) 266-5765
Watervliet, NY 12180-5000   Fax: (518) 266-4682
Betty Brownell, Small Business Advisor

Research, development, engineering, and design of mortars, recoilless rifles, and cannons for tanks and towed and self-propelled artillery. Laboratory is a division of U.S. Army Close Combat Armament Center (Picatinny Arsenal, New Jersey 07806-5000), which publishes patent disclosures for inventions.

★1731★ **Dugway Proving Ground (DPG)**
ATTN: STEDP-DOC   Phone: (801) 522-2102
Dugway, UT 84022-5202   Fax: (801) 831-2085
Robert O. Andrus, Small Business Advisor

Conducts field and laboratory tests to evaluate chemical and radiological weapons and defense systems and materiel. Also conducts biological defense research.

★1732★ **Jefferson Proving Ground (JPG)**
STEJP-EH-C   Phone: (812) 273-7226
Madison, IN 47250-5100   Fax: (812) 273-7294
Ms. Mary N. Greenhill, Small Business Advisor

Processing, assembling, and acceptance testing of ammunition and ammunition components. Receives, stores, maintains and issues assigned industrial stocks, including calibrated components.

★1733★ Laboratory Command (LABCOM)
ATTN: DRDEL-SB   Phone: (301) 394-2963
2800 Powder Mill Rd.
Adelphi, MD 20783-1145
Tom Rogers, Small Business Advisor

Research, development, engineering, and initial procurement of assigned items in the areas of electronic signal intelligence, electric warfare, atmospheric sciences, target acquisition and combat surveillance, electronic fuzing, radars, sensors, night vision, radar frequency and optical devices, nuclear weapons effects, instrumentation and simulation, and fluidics. Has direct control and management of seven Army laboratories, including Atmospheric Sciences Laboratory, Ballistic Research Laboratory, Electronics Technology and Devices Laboratory, Human Engineering Laboratory, Materials Technology Laboratory, and Army Vulnerability Assessment Laboratory.

★1734★ Materiel Command (AMC)
ATTN: AMCSB   Phone: (703) 274-8185
5001 Eisenhower Ave.   Fax: (703) 274-5451
Alexandria, VA 22333-0001
Kurt Wussow, Disadvantaged Utilization

Responsible for the life cycle of U.S. Army hardware, including research, development, procurement, production, supply, and maintenance. Provides supervisory, planning, and budgetary direction to installations where contracts are executed and administered. **Contact:** John Flakenham, Doris Agnew, Janet Tull. **Telecommunication Access:** Alternate phone number (202) 274-8186.

★1735★ Natick Research, Development, and Engineering Center
ATTN: STRNC-2SB   Phone: (508) 651-4995
Natick, MA 01760-5008   Fax: (508) 651-5286
Phillip Haddad, Small Business Advisor

Research, development, and engineering in advanced systems concepts, aero-mechanical engineering, food engineering, individual protection, and science and advanced technology.

★1736★ Research and Development Directorate
Attn: CEDB   Phone: (202) 272-0725
20 Massachusetts Ave. NW, Rm. 4117   Fax: (202) 504-4609
Washington, DC 20314-1000
Diane Sisson, Dir., Small Bus. Utilization

Serves as scientific advisor to the Chief of Engineers for Research and Development; directs the activities of all Corps laboratories; exercises responsibility for Army technology-based programs in environmental quality; and maintains responsibility for the planning and budgeting of the Corps research and development program and the management of all resources, including military research, development, test, and evaluation; civil works research and development appropriations; and mission support funding.

★1737★ Small Business Innovation Research Program (SBIR)
Headquarters Laboratory   Phone: (301) 394-4602
Code AMSLC-TP-TS   Fax: (301) 394-4795
2800 Powder Mill Rd.
Adelphi, MD 20783
Joseph Forry, SBIR Coordinator

Will provide funding for nearly 500 topics in fiscal year 1991.

★1738★ Test and Evaluation Command (TECOM)
ATTN: AMSTE-PR   Phone: (301) 278-5184
Aberdeen Proving Ground, MD   Fax: (301) 278-8602
   21005-5055
Debbie Alt, Small Business Advisor

Directs research activities, proving grounds, installations, boards, and facilities required to test equipment, weapons, and materiel systems intended for use by U.S. Army.

★1739★ Water Resources Support Center
Humphreys Engineering Center   Phone: (703) 355-2153
Kingman Bldg.   Fax: (703) 355-3181
Telegraph & Leaf Rds.
Fort Belvoir, VA 22060-5580
John Carpenter

Research and development in the topographic sciences; implementation of emergency operations to provide the capabilities to transfer data between civil organizations; and remote sensing, water control data systems, and telecommunications research planning. Administers the Waterborne Commerce Statistics Center, which collects and reports on traffic and tonnage.

★1740★ White Sands Missile Range (WSMR)
Attn: STEWS-SA-B   Phone: (505) 678-1401
White Sands Missile Range, NM   Fax: (505) 678-4975
   88002-5031
Luis E. Sosa, Small Business Advisor

Conducts testing and evaluation of Army missiles, rockets, warheads, and special weapons. Major directorates include National Range Operations, Army Materiel Test and Evaluation, Instrumentation Directorate, and Directed Energy Directorate. Major tenants are Naval Ordnance Missile Test Station, Air Fore Range Operations Office, U.S. Army Vulnerability Assessment Laboratory, LABCOM, U.S. Army Atmospheric Sciences Laboratory, and U.S. Army TRADOC Analysis Command.

★1741★ Yuma Proving Ground (YPG)
ATTN: STEYP-CR-S   Phone: (602) 328-6171
Yuma Proving Ground, AZ 85364-1530   Fax: (602) 343-1966
Steve Foster, Small Business Advisor

Conducts research and development, production, and post-production testing of components and complete items of weapons, systems, ammunition, and combat and support vehicles in desert environments.

U.S. Department of the Interior

★1742★ Bureau of Mines
2401 E St. NW   Phone: (202) 634-1303
Washington, DC 20241   Fax: (202) 634-4659
Mr. T.S. Ary, Acting Director

Conducts research and collects, interprets, and analyzes information involving mineral reserves and the production, consumption, and recycling of mineral materials. **Publications:** (1) Guide for the Submission of Unsolicited Research and Development; (2) Reports of Investigations Information Circulars; (3) Minerals Yearbook; and (4) Mineral Facts and Problems. **Contact:** Chief Mining Engineer (202) 634-1303; Division of Procurement (202) 634-4704.

★1743★ U.S. Geological Survey
Geologic Division
National Center   Phone: (703) 648-6600
12201 Sunrise Valley Dr.
Mail Stop 911
Reston, VA 22092
Dr. Benjamin A. Morgan, Chief Geologist

Conducts programs to assess energy and mineral resources, identify and predict geologic hazards, and investigate the effects of climate.

U.S. Department of the Navy

★1744★ David Taylor Research Center (DTRC)
Code 3000   Phone: (301) 227-1220
Bethesda, MD 20084-5000   Fax: (301) 227-1847
David Tychnan, Small Business Specialist

Navy's principal research, development, test, and evaluation center for naval vehicles. Research information may be obtained from the National Technical Information Service, 5285 Port Royal Road, Springfield, VA 22161.

★1745★ Marine Corps Headquarters
Code L-2   Phone: (703) 696-1022
U.S. Marine Corps (LS)
Washington, DC 20380
Mrs. Sheila D'Agostino, Small Business Representative

Electronic equipment, specialized vehicles, and equipment peculiar to the Marine Corps.

★1746★ Naval Air Development Center (NADC)
Code 094   Phone: (215) 441-2456
Street Rd.   Fax: (215) 441-1955
Warminster, PA 18974-5000
John Scott, Deputy for Small Business

Navy's principal center for research, development, testing, and evaluation of naval aircraft systems, including simulation, personal safety, aircraft stuctures, aircraft hydraulics and fluids, magnetics, air to air strike, antisubmarine warfare systems, computers, and software. Organized into the following departments: Antisubmarine Warfare Systems Department, Battle Force Systems Department, Tactical Air Systems Department, Air Vehicles and Crew Systems Technology Department, Communication and Navigation Technology Department, Mission Avionics Technology Department, and Systems and Software Technology Department.

★1747★ Naval Air Engineering Center
Code 00M   Phone: (908) 323-2064
Bldg. 129   Fax: (908) 323-2165
Lakehurst, NJ 08733-5133
Francis J. Galbraith, Small Business Specialist

Research, development, test, system evaluation, and engineering in support of aircraft and shipboard interface systems, including launch and recovery systems, support equipment, and visual landing aids.

★1748★ Naval Air Propulsion Center (NAPC)
Box 7176   Phone: (609) 538-6863
Trenton, NJ 08628-0176   Fax: (609) 538-6528
Jim Thaler, System Development Mgr.

Technical and engineering support for air breathing propulsion systems, including their accessories and components, fuels, and lubricants. Also provides technical and advisory services and consulting services on matters relating to the development, evaluation, and support of new air breathing propulsion systems.

★1749★ Naval Air Systems Command (NAVAIR)
Jefferson Plaza, Rm. 424   Phone: (703) 692-0934
Washington, DC 20361-0001   Fax: (703) 746-6868
Sarah Cross, Small Business Representative

Design, development, testing, and evaluation of aircraft, airborne weapon systems, avionics, related photographic and meteorological equipment, ranges, and targets. Principal research and engineering components include the Naval Air Engineering Center, Naval Air Propulsion Center, Naval Air Test Center, Naval Training Systems Center, Naval Weapons Evaluation Facility, and Pacific Missile Test Center. **Contact:** Barbara Williams, Small Business Representative (202)692-0936; Mr. J. A. Johnson, Technical Representative, (AIR 303M), Room 424, Washington, DC 20361.

★1750★ Naval Avionics Center
6000 E. 21st St.   Phone: (317) 353-7009
Indianapolis, IN 46218   Fax: (317) 353-3066
D. Middleton, Small Business Specialist

Research, development, pilot, and limited manufacturing and depot maintenance on avionics and related equipment.

★1751★ Naval Civil Engineering Laboratory (NCEL)
Naval Construction Battalion Center   Phone: (805) 982-5992
 (Code 641)   Fax: (805) 982-4361
Port Hueneme, CA 93043-5000
Mary Lorenzana, Small Business Representative

Principal Navy research, development, test, and evaluation center for shore and seafloor facilities.

★1752★ Naval Coastal Systems Center (NCSC)
Panama City, FL 32407   Phone: (904) 234-4347
Johnny L. Peace, Small Business Specialist

Research, development, test, and evaluation center for mine and undersea countermeasures, special warfare, amphibious warfare, diving, and other naval mission that take place primarily in the coastal regions. Hosts several tenant activities, including the Navy Experimental Diving Unit and the Naval Diving and Salvage Training Center. **Contact:** Mr. W.R. Donaldson (904) 234-4862.

★1753★ Naval Explosive Ordnance Disposal Technology Center
Attn: Code 50   Phone: (301) 743-6850
Navel EOD Technology Center Fax: (301) 743-6947
Indian Head, MD 20640-5070
Claude Manley, Head R&D Dept.

Research and development of specialized equipment, tool, techniques, and procedures required to support operational explosive ordnance disposal units in the location, neutralization, and disposal of surface and underwater explosive ordnance. Also provides support toward the demilitarization of chemical weapons. **Telecommunication Access:** Alternate telephone (301) 743-4841.

★1754★ Naval Medical Research and Development Command (NMRDC)
National Naval Medical Center   Phone: (301) 295-6490
Bldg. 54   Fax: (301) 295-0286
Bethesda, MD 20889
Sandy Shepard, Small Business Representative

Research, development, test, and evaluation programs in submarine and diving medicine, aviation medicine, fleet occupational health, human performance, combat casualty care, infectious disease, oral and dental health, and electromagnetic radiation. Major components include Naval Aerospace Medical Research Laboratory, Naval Dental Research Institute, Naval Health Research Center, Naval Medical Research Institute, and Naval Submarine Medical Research Laboratory.

★1755★ Naval Ocean Systems Center (NOSC)
271 Catalina Blvd.   Phone: (619) 225-2707
San Diego, CA 92152-5000
Forrest L. Hodges, Small Business Representative

Research, development, test, and evaluation for command, control, and communications; ocean surveillance, surface- and air-launched weapons systems; and submarine arctic warfare. Also involved in development of technologies that support these activities, including ocean engineering, environmental sciences, marine bioscience, electronics, computer sciences, and atmospheric sciences. **Publications:** (1) NOSC Technical Manuals; (2) Technical Documents; and (3) Technical Reports.

★1756★ Naval Oceanographic Office (NAVOCEANO)
Bldg. 2105   Phone: (601) 688-4166
Stennis Space Center, MS 39522-5001   Fax: (601) 688-5776
Don Hutchison

Collects, analyzes, and displays oceanographic data in support of Naval operations and establishment. Activities include oceanographic, hydrographic, magnetic, gravity, navigational, and acoustic surveys in support of mapping, charting, and geodesy. **Publications:** (1) NAVOCEANO's Reference Publications; (2) Special Publications; and (3) Technical Reports.

★1757★ Naval Ordnance Missile Test Station (NOMTS)
Bldg. N-103   Phone: (505) 678-6115
White Sands Missile Range, NM 88002   Fax: (505) 678-2499
Les Gibbs, Technical Representative

Supports research, development, test, and evaluation programs in flight testing and guided missiles, rocket gun, and directed energy programs.

★1758★ Naval Ordnance Station
Code 114D   Phone: (301) 743-6611
Naval Ordnance Sta.   Fax: (301) 743-6685
Indian Head, MD 20640-5000
Sue Pulliam, Small Business Specialist

Research, development, test, and evaluation of ammunition, pyrotechnics, and solid propellant used in missiles, rockets, and guns.

★1759★ Naval Research Laboratory (NRL)
Code 3413   Phone: (202) 767-2914
4555 Overlook Ave. SW   Fax: (202) 767-0523
Washington, DC 20375
George Smith, Small Business Representative

Multidisciplinary scientific research and technological development directed toward materials, equipment, techniques, systems, and related operational procedures for the Navy. **Publications:** (1) NRL Review; (2) Fact Book (annually).

★1760★ Naval Sea Systems Command (NAVSEA)
Code 02K   Phone: (703) 746-3183
Washington, DC 20362-5101   Fax: (703) 746-3110

Provides material support to the Navy and Marine Corps for ships, submarines, and other sea platforms; shipboard combat systems and components; other surface and undersea warfare and weapons systems; and ordnance expendables. **Contact:** George W. Gatling, Jr., SBIR Coordinator, Crystal Plaza 6, Room 850, Crystal City, VA 20362.

★1761★ Naval Surface Warfare Center (NSWC)
Attn: CG   Phone: (703) 663-8391
Dahlgren, VA 22448-5000   Fax: (703) 663-7169
Jim Howard

Principal research, development, test, and evaluation center for surface ship weapons systems, ordnance, mines, and strategic systems support. Current areas of interest include low observables technology, applications of artificial intelligence to naval systems, mission and weapons analysis in support of the Navy's use of space systems, and development of technology for advanced autonomous weapons. **Publications:** (1) NSWC Briefs; (2) On the Surface (both weekly).

★1762★ Naval Training Systems Center
Code NTC-092   Phone: (407) 646-5448
Orlando, FL 32813-7100   Fax: (407) 646-5506
Helen Falch, Small Business Specialist

Responsible for procurement of training aids and equipment for the Army, Navy, Marine Corps, Air Force, and other government activities, including research and development in simulation, training psychology, human factors and human engineering, and design and engineering of training equipment. **Contact:** Wiley V. Dykes, Technical Representitive, Code N-731 (305) 646-4629/5464.

★1763★ Naval Underwater Systems Center (NUSC)
Code 45, Bldg. 1170   Phone: (401) 841-2675
Newport, RI 02841   Fax: (401) 841-4243
James M. Woodside, Technical Representative

Research, development, test, and evaluation for submarine warfare and weapons systems, including sonar, electromagnetics combat control, weapon systems, launchers, undersea ranges, and combat systems analysis. **Contact:** Marvin Berger, Technical Representative (New London Laboratory), New London, CT 06320 (203) 440-4811.

★1764★ Naval Weapons Station
Code 11A   Phone: (804) 887-4644
Yorktown, VA 23691   Fax: (804) 887-9402
LCDR J. Dickey, Small Business Representative

Development of explosives processing and explosives loading methods for Navy weapons. **Telecommunication Access:** Alternate telephone (804) 887-4744.

★1765★ Naval Weapons Support Center
Code SB, Bldg. 64   Phone: (812) 854-1542
Crane, IN 47522   Fax: (812) 854-3465
Mrs. Reva Swango, Small Business Representative

Design, engineering, and inservice engineering, evaluation, and analysis programs required in providng support for ships and craft components, shipboard weapons systems, and assigned expendable and non-expendable ordnance items. **Contact:** James F. Short. Jr., Technical Representative, Code 505, (812) 854-1625/1626.

★1766★ Naval Weapons Systems Center (NWC)

Code 004
China Lake, CA 93555-6001
Lois Herrington, Small Business Specialist

Phone: (619) 939-2712
Fax: (619) 939-2582

Research, development, test, and evaluation of air warfare and missile systems, including missile propulsion, warheads, fuzes, avionics and fire control, and missile guidance. Also acts as national range facility for parachute test and evaluation. **Publications:** (1) Naval Weapons Center Technical Publications; (2) Current Technical Events (irregular newsletter). **Contact:** George F. Linsteadt, Technical Representative, Code O1T3 (619) 939-2305.

★1767★ Office of Naval Research (ONR)

Code 1111MA
800 N. Quincy St.
Arlington, VA 22217-5000
Neal Gerr, Dir. of Mathematical Sciences

Phone: (301) 696-4313
Fax: (703) 696-3945

Mission is to encourage and promote naval research. Offers broad programs to encourage and assist young scientists, including (1) ONR Naval Science Awards Program, which recognizes achievements of high school science students; (2) ONR Young Investigators Program, which is intended to attract the best young academic researchers at U.S. universities; (3) ONR Graduate Fellowship Program which supports 135 students at about 80 universities in nine fields of science and engineering; (4) ONR University Research Initiatives Block Research Program, which offers interaction research between university and naval laboratory personnel; (5) ONR Summer Faculty Research Program, which allows university researchers to spend the summer working at a naval research activity; and (6) ONR Instrumentation Program, which allows for the purchase of major, high-cost university research equipment. **Contact:** Joseph C. Ely (202) 696-6525.

★1768★ Pacific Missile Test Center (PMTC)

Code 6330
Point Mugu, CA 93042
Norman Nix, Small Business Specialist

Phone: (805) 989-8432
Fax: (805) 488-1694

Primary test and evaluation facility for air-launched weapons and airborne electronic warfare systems.

★1769★ Space and Naval Warfare Systems Command (SPAWAR)

Code OOK
2451 Crystal Park Dr.
Arlington, VA 22202
Mrs. Betty Geesey, Asst. Dir. for Small Business

Phone: (703) 602-6092
Fax: (703) 602-6309

Research, development, test, and evaluation for command, control, and communications; undersea and space surveillance; electronic test equipment; and electronic materials, components, and devices. **Contact:** Robert H. Branner, Head of Operational ASW Systems Branch (SPAWAR 661), (202) 433-4729.

U.S. Department of Transportation

★1770★ Coast Guard Office of Engineering and Development

S.W. G-E
2100 2nd St. NW
Washington, DC 20593-0001
P.A. Bunch, Chief

Phone: (202) 267-1844
Fax: (202) 267-4136

Expects to spend $19-million in 1989 to support reseach and maintain and improve search and rescue systems, environmental protection, marine safety, and aids to navigation. Majority of the research and development tasks are conducted by private business on contract, and almost all research is applied, aimed at developing improved systems for Coast Guard use.

★1771★ Federal Aviation Administration

Small Business Coordinator
800 Independence Ave. SW
Washington, DC 20591
Inez C. Willams

Phone: (202) 267-8881
Fax: (202) 267-5071

Expects to spend $165-million in 1989 to conduct engineering and development programs in air traffic control, advanced computer applications, navigation, weather, aviation medicine, aircraft safety, and environment. For procurement information on the Federal Aviation Administration Technical Center contact: Procurement Officer, ANA-51B, FAA Technical Center, Atlantic City International Airport, NJ 08405 (609) 484-4000.

★1772★ Federal Highway Administration (FHWA)

Office of Contracts & Procurement
400 7th St. SW
Washington, DC 20509
Frank J. Waltos, Administrator

Phone: (202) 366-4242
Fax: (202) 366-3705

Research on civil, mechanical, geotechnical, chemical, hydraulic, electrical, environmental, and human factors engineering, including direct and contract research and development relating to traffic operations, new construction techniques, and social and environmental aspects of highways and programs. **Contact:** George H. Duffy, Disadvantaged Business Enterprise Division (202) 366-1586.

★1773★ Federal Railroad Administration

Office of Procurement
400 7th St. SW, Rm. 8222
Washington, DC 20590
Joseph Kerner, Contact

Phone: (202) 366-0565
Fax: (202) 366-3055

Will obligate $9-million to continue emphasis on safety of train operations, including testing and evaluation, computer modeling, systems engineering, and safety hazard analysis.

★1774★ National Highway Traffic Safety Adminstration

Office of Contracts & Procurement
400 7th St. SW
Washington, DC 20590
Thomas Stafford, Director

Phone: (202) 366-0607
Fax: (202) 366-9555

Plans to spend $30-million for motor vehicle research, traffic safety research and demonstration, and other statistical and analytical studies in 1989. Enters into contracts with private industry, educational institutions, nonprofit organizations, and state and local governments for defects investigations, crashworthiness programs, alcohol traffic safety programs, systems operations, emergency medical services, safety manpower development, driver/vehicle interaction, experimental vehicles, occupant packaging testing, biomechanics, passive restraint tests, computer support, management studies, and data acquisition.

★1775★ Procurement Operations Division

Office of the Secretary of Transportation
Washington, DC 20590

Phone: (202) 366-4952
Fax: (202) 366-7510

Anticipates spending $7-million on broad-based policy research on domestic and international transportation issues of importance to the nation. Procurement Operations Division contracts for studies covering social, economic, environmental, safety and policy-oriented transportation research. Such studies include data collection, modeling, economic, and financial studies and projects to support national transportation policy development and evaluation.

★1776★ Transportation Systems Center (TSC)

DOT SBIR Program Office, DTS-22   Phone: (617) 494-2051
Kendall Sq.   Fax: (617) 494-2497
Cambridge, MA 02142
Dr. George Kovatch

Industrial-funded research and analysis organization that applies technical skills to national transportation and logistics problems. Expertise includes radionavigation, human factors, telecommunications, qtructural analysis, railroad inspection technology, industry analysis, emergency and readiness planning, air traffic automation, explosives detection, security and surveillance, and information systems development. Manages the Department of Transportation Small Business Innovation Research Program. **Publications:** (1) National Transportation Statistics; (2) Transportation Safety Information (annual); and (3) Project Directory.

Section 21

Small Business Innovation Research (SBIR) Program

*"Ideas are such funny things; they never work unless you
do."—Herbert V. Prochnow*

In 1982 Congress passed the Small Business Innovation Development Act creating the federal Small
Business Innovation Research (SBIR) Program. The purpose of the program is to increase the
opportunity for small firms to participate in federal research and development. In addition to
encouraging small business participation, the program is designed to stimulate the conversion of
research findings into commercial application.

The law designated the Small Business Administration (SBA) as the agency to run the program, govern
its policy, monitor its progress, and then analyze its result.

The SBIR grant program has given out about $1 billion this past decade, according to a recent survey
of 1,457 SBIR grantees by the General Accounting Office.

U.S. House Small Business Committee Chairperson John LaFalce calls it the most effective technology
program in the history of our country. "The public investment in small business innovation has been
productive beyond anybody's expectations," LaFalce said during an October 3, 1991 oversight hearing.

How SBIR Works

Under the SBIR Program, the participating federal agencies request highly competitive proposals from
small businesses in response to solicitations outlining their R&D requirements. After evaluating the
proposals, each agency awards funding agreements for determining the technical feasibility of the
research and development concepts proposed.

These awards are as follows:

Phase I: Awards up to $50,000 that are made for research projects to evaluate the scientific and
technical merit and feasibility of an idea.

Let's say that you have an idea for a device that could, if successful, solve a problem posed by one of the SBIR agencies. There just might be $50,000 in that agency's budget to help you prove the concept.

Phase II: The Phase I projects with the most potential are funded to further develop the proposed idea for one or two years. Phase II awards can be as high as $500,000.

If you are successful in realizing the first stage of your R&D effort, and the sponsoring agency thinks you are onto something, you just might qualify for another half million dollars.

Phase III: Once you get into the final stage, or the commercialization process, there are no more federal SBIR funds available.

At this point, the federal government encourages you to seek private sector investment and support. While the government may extend follow-up production contracts for your technology, it no longer wants to be your partner. Hopefully the federal seed money has been enough to get you off the ground.

State-Supported SBIR Programs

State governments, anxious to build their own industrial bases, have actively supported the SBIR Program by: 1) promoting SBIR to small businesses; 2) providing information and technical assistance to SBIR applicants; 3) providing matching funds to SBIR Phase I and II recipients; and 4) helping firms to obtain Phase III funding from both private and state sources.

Why do the states do this? States see the independent inventor and small businessperson as a good investment because the chances are that technologies developed in a particular state will stay there once commercialized. Innovation leads to hard goods, goods create jobs, jobs employ people, people pay taxes, and so forth.

How to Contact SBIR Agencies

What I attempt to do in this section is provide you with two options for approaching SBIR agencies, learning about their needs, and seeing if your work qualifies for funding.

The most direct way is to refer to the listings on page 262, and tackle the agency head-on. Here you will find all the information needed to get to the source with a phone call or letter. Write down the name, telephone number, and address of the contact person at a particular agency and either shop an invention or technology you have created or get a copy of that agency's most recent "wish list" of problems it wants solved.

If, on the other hand, you want some assistance on how to approach a particular SBIR agency, chances are that your state has an SBIR office at which a professional can guide you along, step by step, and help you identify a home for your invention or technology. To this end, **please refer to pages 263, where all the state SBIR offices and key personnel are listed.**

A sample of one agency's SBIR forms may be found as Forms 38-40 located in Part One: Forms and Tables.

For specific information on SBIR programs, call 1-800-827-5722. This number will connect you with the SBA's Small Business answer desk.

Small Business Administration

Each agency makes its own awards using contracts, grants, or co-operative agreements. But the Small Business Administration (SBA) is charged with the formulation and issues policy direction for government-wide SBIR programs. For further information contact:

Raymond Marchakitus
Small Business Administration
SBIR
1441 L Street, N.W.
Premier Bldg. #414
Washington, DC 20416

There are SBA Regional Offices that can also provide you with information on the SBIR program and what's happening at any time throughout the participating agencies and departments. **For a complete listing of local SBA offices, see Section 19 beginning on page 219.**

Section Arrangement

The following entries are arranged alphabetically by federal department, then alphabetically by program name. All program names are referenced in the Master Index.

★1777★ Environmental Protection Agency
SBIR Program
c/o Office of Research & Development   Phone: (202) 382-5744
401 M St., SW
Washington, DC 20460

★1778★ National Aeronautics and Space Administration
SBIR Program
SBIR Office—Code RB   Phone: (202) 453-2848
600 Independence Ave., SW
Washington, DC 20546

★1779★ National Science Foundation
SBIR Program
c/o SBIR Program Manager   Phone: (202) 357-7527
1800 G. St., NW
Washington, DC 20550

★1780★ Nuclear Regulatory Commission
SBIR Program
c/o Office of Nuclear Regulatory   Phone: (301) 427-4250
Research
Washington, DC 20460

★1781★ U.S. Department of Agriculture
SBIR Program
c/o Director, Office of Grant & Program   Phone: (202) 475-5022
Systems
West Auditors Bldg., Rm. 112
15th & Independence Ave., SW
Washington, DC 20251

★1782★ U.S. Department of Commerce
SBIR Program
c/o Director, Office of Small &   Phone: (202) 377-1472
Disadvantaged Business Utilization
Rm. 6411
14th & Constitution Ave., NW
Washington, DC 20230

★1783★ U.S. Department of Defense
SBIR Program
c/o Director, Small Business & Economic   Phone: (703) 614-1151
Utilization
Office of Secretary of Defense
The Pentagon, Rm. 2A340
Washington, DC 20301

★1784★ U.S. Department of Education
SBIR Program
c/o SBIR Program Coordinator   Phone: (202) 254-8247
Office of Education & Improvement
Mail Stop 40
Washington, DC 20208

★1785★ U.S. Department of Energy
SBIR Program
c/o SBIR Program Manager   Phone: (301) 353-5867
Dept. of Energy
Washington, DC 20545

★1786★ U.S. Department of Health and Human Services
SBIR Program
c/o Director, Office of Small &   Phone: (202) 245-7300
Disadvantaged Business Utilization
Rm. 513D
200 Independence Ave., SW
Washington, DC 20201

★1787★ U.S. Department of the Interior
SBIR Program
c/o Chief Scientist   Phone: (202) 643-1305
Bureau of Mines
2401 E. St., NW
Washington, DC 20241

★1788★ U.S. Department of Transportation
SBIR Program
c/o SBIR Program Manager   Phone: (617) 494-2051
Transportation Systems Center
Kendall Sq.
Cambridge, MA 02142

Section Arrangement

The following entries are arranged geographically by state, then alphabetically by program name. All programs are referenced in the Master Index.

Arizona

★1789★ Arizona Department of Commerce
Office of Business Retention and Expansion
3800 N. Central Ave., Ste. 1500   Phone: (602) 280-1335
Phoenix, AZ 85012
Gervaise Tompkin, Director

Offers business assistance and counseling to existing basic business companies based in the state. Utilizes a computer network to link various businesses. Provides staff support for the Governor's Advisory Council on science and technology. Operates a program to assist communities to establish a Retention and Expansion (R&E) Task Force and conduct R&E programs.

Arkansas

★1790★ Arkansas Science and Technology Authority
SBIR Program Assistance
100 Main St., Ste. 450   Phone: (501) 371-3554
Little Rock, AR 72201
Dr. John W. Ahlen, Director

Assists small businesses in obtaining federal Small Business Innovation Research funding and aides eligible businesses in targeting and submitting grant proposals.

Colorado

★1791★ Colorado University Business Advancement
Centers
4700 Walnut, Ste. 101   Phone: (303) 444-5723
Boulder, CO 80301
Karen Eye, Director

A statewide program operated by the University of Colorado to provide technology transfer services to state businesses. Offers business management consulting for technology-based firms and small manufacturing firms. Provides information, training, and consulting assistance to firms with the federal Small Business Innovation Research program and other federal research and development contract opportunities. Operates the Business Research and Information Network (BRAIN) to offer expert advice and maintain a database of faculty experts available from the University of Colorado at Boulder. Cosponsors an annual technology transfer conference with federal laboratories, universities, and businesses. Centers are maintained in Boulder, Grand Junction, Durango, Colorado Springs, Trinidad, and Burlington.

Connecticut

★1792★ Connecticut SBIR Assistance Grant Program
845 Brook St.   Phone: (203) 258-4305
Rocky Hill, CT 06067-3405
Eric Ott, Director

Encourages participation in the federal Small Business Innovation Research (SBIR) grant program and provides financial assistance to companies who have received SBIR Phase I awards and have applied for Phase II awards. Eligible companies employ no more than 250 people and are located or conduct research in the state. Also encourages industry-acdemia collaboration in developing projects.

Delaware

★1793★ University of Delaware
Delaware Small Business Development Center
99 Kings Hwy.   Phone: (302) 739-4271
PO Box 1401
Dover, DE 19903
Gary Smith, Director

Provides federal Small Business Innovation Research matching grants; will match the amount of a Phase I award. Offers management counseling, business training, and information. Helps businesses develop plans and offers financial advice and consulting services. Also receives state funds.

Florida

★1794★ Florida High Technology & Industry Council
SBIR/University Researcher Program
Colling Bldg.. Rm. 501-A   Phone: (904) 487-3134
107 W. Gaines St.
Tallahassee, FL 32399
Ray Iannucci, Director

Provides opportunities for university researchers to form a business, apply for federal Small Business Innovation Research (SBIR) grants, and produce products for sale. Offers workshops on incorporation, market identification and assessment, advertising strategies, intellectual property law, and SBIR applications. Serves as a clearinghouse of university and industrial research projects.

★1795★ NASA-Southern Technology Applications
Center
One Progress Blvd., Box 24   Phone: (904) 462-3913
Alachua, FL 32615
J. Ronald Thornton, Director

University-based technology transfer and economic development organization. Services include market and technology feasibility studies, federal Small Business Innovation Research and entrepreneurial assistance, workshops, special research projects, and access to over 1,700 electronic databases internationally.

Hawaii

★1796★ Hawaii High Technology Development Corporation
Hawaii SBIR Matching Grant Program
300 Kahelu Ave., #35   Phone: (808) 625-5293
Miliani, HI 96789
William M. Bass, Jr., Director

Provides up to 50% of the federal Phase I Small Business Innovation Research award up to $25,000 as a matching grant to eligible applicants.

Indiana

★1797★ Ball State University
SBIR Proposal Assistance
Office of Research   Phone: (317) 285-1600
1825 Riverside Ave.
Muncie, IN 47306
James L. Pyle, Director

Assists small firms or individuals in submitting proposals to perform research in the federal Small Business Innovation Research program. Develops proposal strategies and provides links with available University research resources.

★1798★ Indiana Corporation for Science and Technology
SBIR Bridge Grant Funding Pilot
One N. Capitol Ave., Ste. 925   Phone: (317) 635-3058
Indianapolis, IN 46204-2242
Delbert J. Schuh, II, Director

Makes conditional loans to companies which are Small Business Innovation Research Phase I recipients. Awards are limited to a maximum of $50,000.

★1799★ Indiana Institute for New Business Ventures
Enterprise Advisory Service
One N. Capitol, Ste. 420   Phone: (317) 264-2820
Indianapolis, IN 46204
Bruce K. Kidd, Director

Encourages and supports the development of growth-oriented enterprises throughout the state. Serves as a catalyst to link entrepreneurs with the management, technical, and financial resources to start and operate high-growth companies. Conducts conferences and workshops on managing and financing growing business enterprises. Operates a $3,000,000 incubator fund for development throughout the state.

Iowa

★1800★ Iowa Department of Economic Development
SBIR Proposal Grants - Iowa Innovation Program
200 E. Grand Ave.   Phone: (515) 281-3036
Des Moines, IA 50309
Douglas Getter, Director

Provides grants to underwrite up to 50% of the federal Small Business Innovation Research proposal preparation cost. Also provides interim Phase II financial support.

Kansas

★1801★ Kansas Technology Enterprise Corp.
SBIR Bridge Grant Program
112 W. 6th. St., Ste. 400   Phone: (913) 296-5272
Topeka, KS 66603
William G. Brundage, Director

Provides matching funds to small businesses for preparations of proposals under the federal Small Business Innovation Research (SBIR) program. Firms are eligible for up to one-half of the cost of preparing Phase I and II proposals, up to a maximum of $5,000. Offers a support network for SBIR concept evaluation, identification of appropriate solicitation topics, federal agency contact, and technical assistance.

Kentucky

★1802★ Kentucky Cabinet for Economic Development
Kentucky Office of Business and Technology
Capital Plaza Tower, 24th Fl.   Phone: (502) 564-7670
Frankfort, KY 40601
Debbie Kimbrough, Director

Provides a focus on the technology-based businesses in the commonwealth. Serves as a link between businesses and the technological resources and research capabilities of the private and public sectors. Provides and manages incentive grants to facilitate economic development and support local technology assistance programs, including Small Business Innovation Research bridge grants; up to 60% of the Phase I award is available to firms.

Louisiana

★1803★ Louisiana Economic Development Corporation
Louisiana Venture Capital Incentive Program
PO Box 94185   Phone: (504) 342-5675
Baton Rouge, LA 70804-9185
Mike Williams, Director

Operates a matching grant program of up to $50,000 for federal Small Business Innovation Research Phase I winners. Also operates three programs to encourage availability of seed and venture capital: the Louisiana Venture Capital Co-investment Program, Louisiana Venture Capital Match Program, and the Louisiana Minority Venture Capital Match Program.

Maine

★1804★ Centers of Innovation
State House Sta. 147   Phone: (207) 289-3703
Augusta, ME 04333
Robert Kidd, Director

Grants up $5,000 for federal Small Business Innovation Research proposal development by the Maine Science and Technology Commission. Supports Centers of Innovation in aquaculture, biomedical technology, manufacturing productivity, and metals and electronics manufacturing.

Michigan

★1805★ Michigan Department of Commerce
Michigan State Research Fund
106 W. Allegan   Phone: (517) 335-2139
Lansing, MI 48913
Sharon Woollard, Contact

Offers interim grants for successful Phase I federal Small Business Innovation Research projects which are expecting Phase II funding. Provides grants to organizations with commercially feasible research and development projects. Additional funding is available for projects with potential for prototype development.

★1806★ Michigan Energy and Resource Research
Association (MERRA)
Executive Plaza Bldg.   Phone: (313) 964-5030
1200 Sixth St., Ste. 328
Detroit, MI 48226
John Mogk, Director

Consortium of state government, university, and private businesses that assist with the procurement of research and development funding and transfer of research results and technological innovations. Supports the operation of the federal Small Business Innovation Research Program, the Michigan State Research Fund, and various university and industry research and technology development initiatives.

Minnesota

★1807★ Minnesota Project Innovation, Inc.
Hazeltine Gates Office Bldg.   Phone: (612) 448-8826
1107 Hazeltine Blvd.
Chaska, MN 55318
James W. Swiderski, Director

Partnership of government, business, and others which seeks to promote technological innovation and technology transfer and create private sector financing opportunities and new jobs in the state. Assists small business in obtaining federal Small Business Innovation Research (SBIR) grants. Disseminates information about SBIR programs and refers applicants for technical assistance on proposal development. Operates SBIR-SEARCH, an electronic database to match research/product capabilities with interests of funding agencies. Assistance includes topic search, proposal preparation information and materials, agency contacts, and research center profiles. Charges $75.00 per search.

Missouri

★1808★ Missouri Small Business Innovation
Research Program
PO Box 118   Phone: (314) 751-3906
Jefferson City, MO 65102
Jeffrey J. Kormann, Director

Matches current research solicitations from 12 federal agencies to capabilities and interests in the state by identifying university researchers who can assist in the preparation of quality research proposals. Also serves as a liaison between small businesses and federal agencies.

Montana

★1809★ University Technical Assistance Program
(UTAP)
402 Roberts Hall   Phone: (406) 994-3812
Bozeman, MT 59717
William R. Taylor, Director

Provides technical and managerial assistance to manufacturing companies in the state that are unable to hire consultants. As a NASA Industrial Application Center Affiliate, activities include publicizing its database, locating technical experts in U.S. laboratories, performing computer searches, facilitating searches for companies, and assisting in publicity and preparation of federal Small Business Innovation Research proposals. Additional assistance includes on-site analysis and consultation, technical training, technical publications, short courses and seminars, and copies of military and federal standards and specifications.

Nebraska

★1810★ Nebraska SBIR Assistance Program
W191 Nebraska Hall   Phone: (402) 472-5600
University of Nebraska
Lincoln, NE 68588
Thomas W. Spilker, Director

Assists state firms and researchers in locating and applying for appropriate federal Small Business Innovation Research (SBIR) grants.

New Jersey

★1811★ New Jersey Commission on Science and
Technology
SBIR Bridge Grants Program
20 W. State St., CN832   Phone: (609) 633-2740
Trenton, NJ 08625
Joseph Montemarano, Director

Awards bridge grants of up to $40,000 to small companies that have received Phase I federal Small Business Innovation Research grants. Grants are awarded on a competitive basis with preference given to companies doing research and development in an area designated as important to the state's economic growth.

New York

★1812★ New York State Science and Technology
Foundation
Small Business Innovation Research Promotion
Program
99 Washington Ave., Ste. 1731   Phone: (518) 474-4349
Albany, NY 12210
H. Graham Jones, Director

Gives matching research funds of up to $50,000 to firms receiving Phase I federal Small Business Innovation Research grants to develop research into a commercial phase.

North Dakota

★1813★ Center for Innovation and Business Development

PO Box 8103   Phone: (701) 777-3132
University Sta.
Grand Forks, ND 58202
Bruce Gjovig, Director

Assists small businesses in obtaining grants from the federal Small Business Innovation (SBIR) Prgram and provides up to $1,000 to firms for SBIR proposal preparation. Provides technical and business support services to entrepreneurs, inventors, and small manufacturers in northwest Minnesota and North Dakota. Assists in the formation of start-up ventures or expansion and diversification of existing companies. Provides technical information and business and marketing plans to assist startup manufacturing firms. Operates a rural manufacturing initiative to provide business and technical support. Qualified applicants are eligible to receive up to $15,000 in product development, manufacturing process development, plant and equipment needs evaluation, and market research. Additional services available on a contract basis and small equity positions are requested to generate ongoing funds.

Ohio

★1814★ Ohio Department of Development (TIE-IN)
Technology Information Exchange-Innovation Network

PO Box 1001   Phone: (614) 466-2115
Columbus, OH 43266
Keith Ewald, Director

A comprehensive inventory of research and development activity in the state, including the Ohio Patents, Venture Opportunities, and Corporate Research and Development files. Serves as a link between entrepreneurs and businessess, and experts and investors. Maintains a current list of topics for the 12 federal agencies participating in the Small Business Innovation Research program.

★1815★ Ohio's Thomas Edison Program
Ohio SBIR Program

77 S. High St., 26th Fl.   Phone: (614) 466-5867
Columbus, OH 43266
G. Mark Skinner, Director

Assists small businesses in securing federal Small Business Innovation Research (SBIR) funding. Operates seven regional Technical Assistance Centers to provide local companies with direct assistance in identifying topics and preparing proposals. Companies which have secured federal Phase I awards and have the potential to commercialze any SBIR technologies in Ohio may apply for an Ohio SBIR Bridge Grant. Grants are up to $35,000 to maintain continuity between the two phases of federal funding. Administers the Winners' Support System to provide liaison services to federal agencies, guidance in securing Phase II funding, and assistance in locating follow-on funding sources.

Oklahoma

★1816★ Oklahoma Center for the Advancement of Science and Technology
SBIR Phase I Incentive Funding and Matching Funds Program

205 N.W. 63rd St., Ste. 305   Phone: (405) 848-2633
Oklahoma City, OK 73116-8209
Dr. Carolyn Wendel Smith, Director

Provides eligible Oklahoma small businesses with matching awards of 50% of the cost of preparing a technical Phase I federal Small Business Innovation Research proposal up to $3,000.

Rhode Island

★1817★ Rhode Island Partnership for Science and Technology
Rhode Island SBIR Support Program

7 Jackson Walkway   Phone: (401) 277-2601
Providence, RI 02903
Bruce R. Lang, Director

Established to increase federal Small Business Innovation Research (SBIR) projects in the state. Activities focus on proposal consulting service, Phase I Proposal Preparation Grants ($1,000), Phase I University/College Consultant Grants (50% or up to $2,500), or Phase II Bridge Grants (50% of Phase I grant or up to $2,500).

South Dakota

★1818★ South Dakota Future Fund Program

Governor's Office of Economic   Phone: (800) 872-6190
   Development
711 Wells Ave.
Pierre, SD 57501-3369
Kenneth Schaak, Director

Provides grants for research projects to link state university faculty with private industry.

★1819★ South Dakota SBIR Center

Dakota State University   Phone: (605) 256-5555
East Hall
Madison, SD 57042
Melvin Ustad, Director

Disseminates information from various federal agencies which participate in the Small Business Innovation Research (SBIR) program. Develops partnerships between university faculty and private companies to increase the number of SBIR grants awarded to state firms. Conducts seminars on the SBIR program.

Texas

★1820★ Texas A&M University
Texas Technology Business Development

310 Wisenbaker Engineering Research   Phone: (409) 845-0538
   Center
College Station, TX 77843-3369
Dr. Helen Baca Dorsey, Director

Provides research commercialization assistance, technical assistance to entrepreneurs, economic development assistance, and access to

technology and market databases, publications, conferences, and seminars. Serves as a liaison for TechCom, a technology transfer company at the University. Identifies and brokers research concepts to promote commercialization of University research. Works to license intellectual property and form new ventures.

Virginia

★1821★ Center for Entrepreneurship and Business Incubators

CIT Tower, Ste, 600   Phone: (703) 689-3025
2214 Rock Hill Rd.
Herndon, VA 22070
David Miller, Director

Provides commercializaton assistance to federal Phase II Small Business Innovation Research winners and assistance to SBIR applicants. Offers outreach services to assist new businesses and provides educational programs in entrepreneurship. Administers incubators to provide physical space for start-up businesses to design, construct, and test prototypes with the assistance of university experts. Offers business counseling.

Wisconsin

★1822★ Wisconsin Bureau of Development Finance
Wisconsin SBIR Bridge Financing Program

Bureau of Development Finance   Phone: (608) 267-9383
123 W. Washington Ave.
PO Box 7970
Madison, WI 53707
Caroline Garber, Director

Provides winners of the federal Small Business Innovation Research (SBIR) program Phase I awards with loans of up to $40,00 while between Phase I and Phase II (25% of the loan amount must be matched by the award recipient). Operates an advisory proposal review program to critique presubmission SBIR applications.

Section 22

Federal Laboratory Assistance for Inventors

"The era of the "intelligent man/woman" is almost over and a new one is emerging—the era of the "creative man/woman."—Pinchas Noy

Currently there are more than 700 U.S. federal government research and development facilities, representing approximately 85 percent of the government's in-house R&D efforts. These facilities employ more than 100,000 scientists and engineers, one-sixth of the nation's total, and expend approximately $72 billion annually. The laboratories cover a full spectrum of science and technology, including material sciences, biotechnology, medicine, energy, electronics, optics, and environment, to name a few.

The federal laboratories seek to connect themselves with private industry and are willing to share their expertise, sometimes their facilities, and often special services. At a December meeting Energy Secretary James D. Watkins told private sector technical specialists, "We in government must get a lot smarter and more aggressive in working with you [the private sector]."

Even as a toy inventor, I contacted federal laboratories on many occasions for assistance with my work. Since most toys are no more than a miniaturization of real world things, I frequently find myself looking for advice on how to transfer some costly and esoteric technology into an affordable, promotable feature for a toy.

I have always found the scientists and engineers on the other end of the line to be more than willing to answer my questions and even, at times, to conduct experiments for me or fax critical background data. Two of the mandates under which federal labs operate are to expand secondary use of product and process technology, and to assist small companies through incubators.

Not every contact listed is able to handle public inquiries. However, all contacts are interested in working to further the transfer of technology.

Section Arrangement

The following entries are arranged alphabetically by federal department, then alphabetically by laboratory name. All laboratory names are referenced in the Master Index.

Environmental Protection Agency

★1823★ Atmospheric Environment Engineering Research Laboratory
Research Triangle Park, NC 27711   Phone: (919) 541-2157
Blair Martin

★1824★ Atmospheric Research and Exposure Assessment Laboratory
Research Triangle Park, NC 27711   Phone: (919) 541-3779
Ronald Patterson

★1825★ Environmental Monitoring Systems Laboratory (Cincinnati)
26 W. Martin Luther King St.   Phone: (513) 569-7364
Cincinnati, OH 45268
Robert Booth

★1826★ Environmental Monitoring Systems Laboratory (Las Vegas)
PO Box 15027   Phone: (702) 798-2530
Las Vegas, NV 89114
Donald Wruble

★1827★ Environmental Research Laboratory (Athens)
USEPA   Phone: (404) 546-3128
College Sta. Rd.
Athens, GA 30613
Robert R. Swank

★1828★ Environmental Research Laboratory (Corvallis)
USEPA/ERL   Phone: (503) 757-4601
200 SW 35th St.
Corvallis, OR 97333
Tom Murphy

★1829★ Environmental Research Laboratory (Gulf Breeze)
USEPA/ERL   Phone: (904) 934-9299
Sabine Island
Gulf Breeze, FL 35261
Raymond G. Wilhour

★1830★ Environmental Research Laboratory (Narragansett)
USEPA/ERL   Phone: (401) 782-6030
S. Ferry Rd.
Narragansett, RI 02882
Allan D. Beck

★1831★ Health Effects Research Laboratory
Research Triangle Park, NC 27711   Phone: (919) 541-2541
Joseph Elder

★1832★ Risk Reduction Environmental Laboratory
26 W. Martin Luther King St.   Phone: (513) 569-7276
Cincinnati, OH 45268
John Convery

★1833★ Robert S. Kerr Environmental Research Laboratory
PO Box 1198   Phone: (405) 332-8800
Ada, OK 74820
Marion R. Scalf

Federal Aviation Administration

★1834★ Civil Aeromedical Institute
Aviation Toxicology Laboratory
PO Box 25082   Phone: (405) 686-4806
Oklahoma City, OK 73125
William E. Collins

National Aeronautics and Space Administration

★1835★ Ames Research Center
Mail Stop 204-2   Phone: (415) 694-6370
Moffett Field, CA 94035
Maylene Duenas

★1836★ George C. Marshall Space Flight Center
Mail Code AT01   Phone: (205) 554-0962
Huntsville, AL 35812
Ismail Akbay

★1837★ Goddard Space Flight Center
Technology Utilization/Transfer Office
Mail Code 702   Phone: (301) 286-6242
Greenbelt, MD 20771
Donald S. Friedman

★1838★ Jet Propulsion Laboratory
4800 Oak Grove Dr.   Phone: (818) 354-8300
Mail Stop 180-801
Pasadena, CA 91109
Gordon S. Chapman

★1839★ **John C. Stennis Space Center**
Bldg. 1103   Phone: (601) 688-1929
Mail Code HA-00
Stennis Space Center, MS 39529
Robert Barlow

★1840★ **John F. Kennedy Space Center**
Mail Code PT-PMO-A   Phone: (305) 867-3017
Kennedy Space Center, FL 32899
Thomas M. Hammond

★1841★ **Langley Research Center**
Mail Stop 139A   Phone: (804) 864-2484
Hampton, VA 23665-5225
John Samos

★1842★ **Lewis Research Center**
Mail Stop 7-3   Phone: (216) 433-5568
21000 Brookpark Rd.
Cleveland, OH 44135
Daniel G. Soltis

★1843★ **Lyndon B. Johnson Space Center**
Mail Code IC4   Phone: (713) 483-3809
Houston, TX 77058
Dean C. Glenn

National Science Foundation

★1844★ **National Astronomy and Ionosphere Center**
Cornell University   Phone: (607) 256-3734
Space Sciences Bldg.
Ithaca, NY 14853
Eugene Bartell

★1845★ **National Center for Atmospheric Research**
PO Box 3000   Phone: (303) 484-5151
Boulder, CO 80307
Larry Anderson

★1846★ **National Optical Astronomy Observatory**
PO Box 26732   Phone: (602) 325-9361
Tucson, AZ 85726
L. Daggert

★1847★ **National Radio Astronomy Observatory**
Edgemont Rd.   Phone: (804) 296-0211
Charlottesville, VA 22903
James L. Desmond

★1848★ **National Solar Observatory**
Sunspot, NM 88349   Phone: (505) 434-1390
Frank A. Hegwer

Smithsonian Institution

★1849★ **National Museum of Natural History**
NHB 401   Phone: (202) 357-2661
10th & Constitution Ave. NW
Washington, DC 20560
Stanwyn Shetler

★1850★ **National Zoological Park & Conservation Center**
Washington, DC 20008   Phone: (202) 673-4705
Devro G. Kleinman

★1851★ **Smithsonian Astrophysical Observatory**
60 Garden St.   Phone: (617) 495-7100
Cambridge, MA 02138
Irwin Shapiro

★1852★ **Smithsonian Environmental Research Center**
PO Box 28   Phone: (301) 798-4424
Edgewater, MD 20137
David Correll

★1853★ **Smithsonian Tropical Research Institute**
Balboa, Canal Zone, APO
Miami, FL 34002
Frank Morris

Tennessee Valley Authority

★1854★ **Tennessee Valley Authority**
400 W. Summit Hill Dr.   Phone: (615) 632-6435
2D46 OCH
Knoxville, TN 37902
H. Brown Wright

U.S. Department of Agriculture

★1855★ **Agricultural Research Service**
Poisonous Plant Research Laboratory
1150 E. 1400 N.   Phone: (801) 752-2941
Logan, UT 84321
Lynn F. James

★1856★ **Animal Parasite Research Unit**
PO Box 952   Phone: (205) 826-4382
Auburn, AL 36830
Phillip H. Klesius

★1857★ **Appalachian Fruit Research Station**
USDA-ARS-NER   Phone: (304) 725-3451
Rte. 2, Box 45
Kearneysville, WV 25430
Bill A. Butt

★1858★ Appalachian Soil and Water Conservation Research Laboratory
PO Box 867
Beckley, WV 25801
Phone: (304) 252-6426

★1859★ Aquatic Plant Management Laboratory
USDA-ARS-SR-SAA
3205 College Ave.
Ft. Lauderdale, FL 33314
Kerry K. Steward
Phone: (305) 475-0541

★1860★ Arid Watershed Management Research Unit
2000 E. Allen Rd.
Tucson, AZ 85719
Frasier Renard
Phone: (602) 629-6381

★1861★ Athropod-Borne Animal Diseases Research Laboratory
USDA-S&E-ARS
University Sta.
PO Box 3965
Laramie, WY 82071
Thomas E. Walton
Phone: (307) 721-0304

★1862★ Beltsville Agricultural Research Center Technology Transfer Office
Bldg. 005, Rm. 404
Beltsville, MD 20705
James Hall
Phone: (301) 344-4045

★1863★ Beltsville Human Nutrition Research Center
USDA-ARS-NER
Bldg. 308, Rm. 223
BARC-E
Beltsville, MD 20705
Helene N. Guttman
Phone: (301) 344-1627

★1864★ Beneficial Insects Research Laboratory
501 S. Chapel St.
Newark, DE 19713
R.J. Dysart
Phone: (302) 731-7330

★1865★ Biological Control of Insects Research
USDA-S&E-ARS-NCR
PO Box 7629
Research Park
Columbia, MO 65205
Carlo M. Ignoffo
Phone: (314) 875-5361

★1866★ Boll Weevil Research Unit
USDA-ARS
PO Box 5364
Mississippi State, MS 39762
James Smith
Phone: (601) 323-2230

★1867★ Carl Hayden Bee Research Center
2000 E. Allen Rd.
Tucson, AZ 85719
Eric H. Erickson
Phone: (602) 629-6380

★1868★ Cereal Crops Research Unit
ARS
501 N. Walnut St.
Madison, WI 53705
David M. Peterson
Phone: (608) 262-4482

★1869★ Cereal Rust Laboratory
University of Minnesota
1551 Lindig Ave.
St. Paul, MN 55108
K.J. Leonard
Phone: (612) 376-4647

★1870★ Children's Nutrition Research Center
Baylor College of Medicine
1709 Dryden, Ste. 601
Houston, TX 77030
Buford L. Nichols
Phone: (713) 799-6006

★1871★ Citrus and Sub-Tropical Products Laboratory (Colorado)
NTIA/ITS.D1
325 Broadway
Boulder, CO 80303-3328
Val O'Day
Phone: (303) 497-3484

★1872★ Citrus and Sub-Tropical Products Laboratory (Florida)
USDA-S&E-ARS-SR
PO Box 1909
Winter Haven, FL 33880
Robert E. Berry
Phone: (813) 293-4133

★1873★ Coastal Plain Experiment Station
USDA-ARS-SR
PO Box 748
Tifton, GA 31793
James L. Butler
Phone: (912) 386-3585

★1874★ Conservation and Production Research Laboratory
USDA-ARS
PO Drawer Box 10
Bushland, TX 79012
B.A. Stewart

★1875★ Corn Insects Research Laboratory
RR 3, PO Box 45B
Akeny, IA 50021
Wilbur D. Guthrie
Phone: (515) 964-6664

★1876★ Cotton Production Research Unit
USDA-ARS   Phone: (803) 669-6664
PO Box 2131
Florence, SC 29503
Raymond F. Moore

★1877★ Cotton Quality Research Station
USDA-ARS   Phone: (803) 656-2488
PO Box 792
Clemson, SC 29631
C. Kenneth Bragg

★1878★ Crop Science Research Laboratory
USDA-ARS-SR   Phone: (601) 323-2230
PO Box 5367
Mississippi State, MS 39762
Johnie N. Jenkins

★1879★ Cropping Systems and Water Quality Research Unit
207 Business Loop 70 E.   Phone: (314) 875-5331
Columbia, MO 65203
A.T. Hjelmfelt

★1880★ Cropping Systems Research Laboratory
Rte. 3, Box 215   Phone: (806) 746-5353
Lubbock, TX 79401
Jerry E. Quisenberry

★1881★ Crops Research Laboratory
USDA-ARS-SAA   Phone: (919) 693-5151
PO Box 1555
Oxford, NC 27565-1555
Harvey W. Spurr, Jr.

★1882★ Crops Research Laboratory
Sugar Beet Production Unit
Colorado State University   Phone: (303) 482-7717
Ft. Collins, CO 80523
Richard J. Hecker

★1883★ Dairy Forage Research Center
University of Wisconsin   Phone: (608) 263-2030
1925 Linden Dr. W.
Madison, WI 53706
Larry D. Satter

★1884★ Eastern Regional Research Center
USDA-S&E-ARS-NAA   Phone: (215) 233-6595
600 E. Mermaid Ln.
Philadelphia, PA 19118
John P. Cherry

★1885★ Field Crops Mechanization Research Unit
PO Box 36   Phone: (601) 686-2385
Stoneville, MS 38776
J. Ray Williford

★1886★ Forage and Livestock Research Laboratory
USDA-ARS   Phone: (405) 262-5291
PO Box 1199EI
Reno, OK 73036
William A. Phillips

★1887★ Foreign Disease-Weed Science Research Laboratory
USDA-ARS   Phone: (301) 663-7344
Fort Detrick, Bldg. 1301
Frederick, MD 21701
William M. Dowler

★1888★ Forest Products Laboratory
1 Gifford Pinchot Dr.   Phone: (608) 264-5636
Madison, WI 53705
Rodney G. Larson

★1889★ Fruit and Vegetable Chemistry Laboratory
USDA-S&E-ARS   Phone: (213) 796-0239
262 S. Chester Ave.
Pasadena, CA 91106
Vincent P. Maier

★1890★ Georgia Coastal Plain Experiment Station
Nematodes, Weeds and Crops Research
Tifton, GA 31793   Phone: (912) 386-3372
Alva W. Johnson

★1891★ Grand Forks Human Nutrition Research Center
2420 2nd Ave. N.   Phone: (701) 795-8353
PO Box 7166, University Sta.
Grand Forks, ND 58202
Forrest H. Nielsen

★1892★ Grassland, Soil and Water Research Laboratory
USDA-ARS   Phone: (817) 770-6500
808 E. Blackland Rd.
Temple, TX 76502
Clarence W. Richardson

★1893★ Honey Bee Breeding, Genetics, and Physiology Research Laboratory
1157 Ben Hur Rd.   Phone: (504) 766-6064
Baton Rouge, LA 70820
Thomas E. Rinderer

★1894★ Horticultural Crops Research Laboratory (Corvallis, OR)
3420 NW Orchard Ave.   Phone: (503) 757-4544
Corvallis, OR 97330
Robert G. Linderman

★1895★ Horticultural Crops Research Laboratory (Fresno, CA)
ARS   Phone: (209) 453-3000
2021 S. Peach Ave.
Fresno, CA 93727
Patrick V. Vail

★1896★ Human Nutrition Research Center on Aging
Tufts University   Phone: (617) 956-0310
711 Washington St.
Boston, MA 02111
Donald Therriault

★1897★ Insect Attractants, Behavior, and Basic Biology Research Laboratory
USDA-ARS   Phone: (904) 374-5701
1700 SW 23rd Dr.
PO Box 14565
Gainesville, FL 32604
Herbert Oberlander

★1898★ Insect Biology and Population Management Research Laboratory
USDA-ARS   Phone: (912) 382-6904
PO Box 784
Tifton, GA 31793-0748
Charlie E. Rogers

★1899★ Insects Affecting Man and Animals Research Laboratory
USDA-S&E-ARS-SR   Phone: (904) 374-5900
1600 SW 23rd Dr.
PO Box 14565
Gainesville, FL 32604
Gary A. Mount

★1900★ Intermountain Research Station
324 25th St.   Phone: (801) 625-5412
Ogden, UT 84401
Carter B. Gibbs, Deputy Director

★1901★ Jamie Whitten Delta States Research Center Cotton Ginning Research Unit
PO Box 256   Phone: (601) 686-2385
Stoneville, MS 38776
W. Stanley Anthony

★1902★ Jamie Whitten Delta States Research Center Cotton Physiology and Genetics Research Laboratory
PO Box 345   Phone: (601) 686-2311
Stoneville, MS 38776
William R. Meredith

★1903★ Livestock Insects Research Laboratory
University of Nebraska   Phone: (402) 471-5267
East Campus
305A Plant Industry Bldg.
Lincoln, NE 68583
G.D. Thomas

★1904★ National Animal Disease Center
USDA-ARS-CPA   Phone: (515) 239-8201
PO Box 70
Ames, IA 50010
Phillip A. O'Berry

★1905★ National Clonal Germplasm Repository
33447 Peoria Rd.   Phone: (503) 757-4448
Corvallis, OR 97333
H.B. Lagerstedt

★1906★ National Monitoring and Residue Analysis Laboratory
USDA APHIS S&T   Phone: (601) 863-8124
PO Box 3209
Gulfport, MS 39505-3209
Joseph H. Ford

★1907★ National Sedimentation Laboratory
PO Box 1157   Phone: (601) 232-2900
Oxford, MS 38655
C.K. Mutchler

★1908★ National Seed Storage Laboratory
USDA-ARS   Phone: (303) 484-0402
Colorado State University
Ft. Collins, CO 80523
Loren Wiesner

★1909★ National Soil Dynamics Laboratory
PO Box 792   Phone: (205) 887-8596
Auburn, AL 36831-0792
A.C. Burt

★1910★ National Soil Erosion Laboratory
USDA-S&E-ARS   Phone: (317) 494-8673
Purdue University
Soil Bldg.
West Lafayette, IN 47907
John M. Laflen

★1911★ New England Plant, Soil and Water Lab
University of Maine   Phone: (207) 581-3266
Orono, ME 04469
William Clapham

★1912★ New York State Agricultural Experiment Station Germplasm Resources Laboratory
USDA-ARS   Phone: (315) 787-2244
Geneva, NY 14456-0462
Stephen Kresovich

★1913★ North Appalachian Experimental Watershed
PO Box 478   Phone: (614) 545-6349
Coshocton, OH 43812
L.B. Owens, Research Leader

★1914★ **North Central Forest Experiment Station**
1992 Folwell Ave.   Phone: (612) 642-5252
St. Paul, MN 55108
Nancy R. Walters

★1915★ **North Central Soil Conservation Research Laboratory**
USDA-ARS   Phone: (612) 589-3411
N. Iowa Ave.
Morris, MN 56267
Charles A. Onstad

★1916★ **Northeast Watershed Research Laboratory**
111 Research Bldg. A   Phone: (814) 865-2048
University Park, PA 16802
Harry B. Pionke

★1917★ **Northeastern Forest Experiment Station**
370 Reed Rd.   Phone: (215) 690-3048
Broomall, PA 19008
Robert Lewis

★1918★ **Northern Grain Insects Research Laboratory**
USDA-S&E-ARS-NCR   Phone: (605) 693-3241
RR 3
Brookings, SD 57006
Gerald R. Sutter

★1919★ **Northern Great Plains Research Center**
USDA-S&E-ARS-NPA   Phone: (701) 663-6445
PO Box 459
Mandan, ND 58554
Albert B. Frank

★1920★ **Northern Regional Research Center**
USDA-S&E-ARS, MWA   Phone: (309) 685-4011
1815 N. University St.
Peoria, IL 61604
L.H. Princen

★1921★ **Nursery Crops Research Laboratory**
359 Main Rd.   Phone: (614) 363-1129
Delaware, OH 43015
Lawrence Schreiber

★1922★ **Ohio Agricultural Research and Development Center**
1680 Madison Ave.   Phone: (216) 263-3777
Wooster, OH 44691-4096
Robert Furbee

★1923★ **Pacific Northwest Forest and Range Experiment Station**
PO Box 3890   Phone: (503) 294-5645
Portland, OR 97208
Richard O. Woodfin

★1924★ **Pacific Northwest Laboratory**
PO Box 999   Phone: (509) 375-2789
Richland, WA 99352
Marv Clement

★1925★ **Pacific Southwest Forest Experiment Station**
PO Box 245   Phone: (415) 486-3286
Berkeley, CA 94701
Richard Hubberd

★1926★ **Plant Physiology and Photosynthesis Research Unit**
North Carolina State University Phone: (919) 737-2661
Box 7620
Raleigh, NC 27695
Donald E. Moreland

★1927★ **Plant Science and Water Conservation Division Plant Science Research Laboratory**
1301 N. Western   Phone: (405) 624-4126
Stillwater, OK 74075
Robert L. Burton

★1928★ **Plant Science Research Laboratory**
University of Minnesota   Phone: (612) 373-1679
Agronomy Dept.
1509 Gortner Ave.
St. Paul, MN 55108
Gordon C. Marten

★1929★ **Plum Island Animal Disease Center**
USDA-ARS   Phone: (516) 323-2500
PO Box 848
Greenport, NY 11944
Roger G. Breeze

★1930★ **Red River Valley Agricultural Research Center**
USDA-ARS   Phone: (701) 239-1370
PO Box 5677
University Sta.
Fargo, ND 58105
Don C. Zimmerman

★1931★ **Red River Valley Potato Research Laboratory**
311 5th Ave. NE   Phone: (218) 773-2473
PO Box 113
East Grand Forks, MN 56721
Paul H. Orr

★1932★ **Regional Poultry Research Center**
USDA-ARS-MWA   Phone: (517) 337-6828
3606 E. Mt. Hope Rd.
East Lansing, MI 48823
Richard L. Witter

★1933★ Rice Research Unit
Rte. 7, Box 999   Phone: (409) 752-2741
Beaumont, TX 77706
Charles N. Bollich

★1934★ Richard B. Russell Agriculture Research Center
USDA-ARS-SAA   Phone: (404) 546-3541
PO Box 5677
Athens, GA 30613
David Zimmer

★1935★ Rocky Mountain Forest and Range Experimental Station
240 W. Prospect Rd.   Phone: (303) 498-1282
Ft. Colins, CO 80526
Robert Hamre

★1936★ Roman L. Hruska U.S. Meat Animal Research Center
USDA-ARS-CPA   Phone: (402) 762-3241
State Spur 18D
PO Box 166
Clay Center, NE 68933
John D. Crouse

★1937★ Snake River Conservation Research Center
USDA-ARS   Phone: (208) 423-5582
Rte. 1
3793 North, 3600 E.
Kimberly, ID 83341
David L. Carter

★1938★ Soil and Water Conservation Research Unit
Darlington Hwy.   Phone: (803) 669-5203
PO Box 3039
Florence, SC 29502
Patrick G. Hunt

★1939★ South Central Poultry Research Laboratory
PO Box 5367   Phone: (601) 323-2230
Mississippi State, MS 39762
James W. Deaton

★1940★ Southeast Poultry Research Laboratory
USDA-ARS-SAA   Phone: (404) 546-3433
934 College Sta. Rd.
PO Box 5657
Athens, GA 30605
Charles W. Beard

★1941★ Southeast Watershed Research Laboratory
PO Box 946   Phone: (912) 386-3462
Tifton, GA 31793
Walter G. Knisel

★1942★ Southeastern Forest Experiment Station
Forest Service   Phone: (303) 258-2850
PO Box 2680
Ashville, NC 28802
Gordon Lewis

★1943★ Southeastern Fruit and Tree Nut Laboratory
USDA-ARS-SAA   Phone: (912) 956-5656
PO Box 87
Byron, GA 31008
James W. Snow

★1944★ Southern Forest Experiment Station
P&A AD, T-10210   Phone: (504) 589-6712
Loyola Ave., Bldg. 701
New Orleans, LA 70113
James Bell

★1945★ Southern Piedmont Conservation Laboratory
USDA-S&E-ARS-SAA   Phone: (404) 769-5631
Hwy. 53
PO Box 555
Watkinsville, GA 30677
Maurice H. Frere

★1946★ Southern Regional Research Center
USDA-S&E-ARS-SRRC   Phone: (504) 286-4212
PO Box 19687
1100 Robert E. Lee Blvd.
New Orleans, LA 70179
John A. Barkate

★1947★ Southern Weed Science Laboratory
USDA-ARS   Phone: (601) 686-2311
PO Box 350
Stoneville, MS 38776
Stephen O. Duke

★1948★ Southwestern Cotton Ginning Research Laboratory
PO Box 578   Phone: (505) 526-6381
Mesilla Park, NM 88047
Sidney E. Hughs

★1949★ Soybean and Nitrogen Fixation Research Unit
North Carolina State University   Phone: (919) 737-3171
PO Box 7620
4114 Williams Hall
Raleigh, NC 27695-7620
Richard F. Wilson

★1950★ Soybean Production Research Unit
PO Box 196   Phone: (601) 686-9311
Stoneville, MS 38776
Thomas C. Kilen

★1951★ Stored-Product Insects Research and
Development Laboratory
USDA-ARS-SAA   Phone: (912) 233-7981
PO Box 22909
Savannah, GA 31403
Robert Davis

★1952★ Subtropical Agricultural Research Laboratory
PO Box 267   Phone: (512) 565-2423
Weslaco, TX 78596
W.G. Hart

★1953★ Subtropical Horticultural Research Unit
13601 Old Cutler Rd.   Phone: (305) 238-9321
Miami, FL 33158
Jennifer L. Sharp

★1954★ Sugarbeet Production Research Unit
USDA-ARS   Phone: (408) 443-2253
1636 E. Alisal St.
Salinas, CA 93915
James E. Duffus

★1955★ Tobacco and Forage Research
Animal Pathology Bldg., Rm. 107-A   Phone: (606) 257-4770
Lexington, KY 40546-0076
Steven J. Crafts-Brandner

★1956★ Tree Fruit Research Laboratory
USDA-ARS   Phone: (509) 664-2280
1104 N. Western Ave.
Wenatchee, WA 98801
Max W. Williams

★1957★ Tropical Crops and Germplasm Research Unit
PO Box 70   Phone: (809) 834-2435
Mayaguez, PR 00709
Antonio Sotomayor-Rios

★1958★ Tropical Fruit and Vegetable Research
Laboratory
USDA-ARS-PWA   Phone: (808) 988-2158
2727 Woodlawn Dr.
PO Box 2280
Honolulu, HI 96804
J.E. Gilmore

★1959★ U.S. Grain Marketing Research Laboratory
USDA-ARS-NPA   Phone: (913) 776-2701
1515 College Ave.
Manhattan, KS 66502

★1960★ U.S. Horticultural Laboratory
USDA-ARS-SR   Phone: (407) 897-7300
2120 Camden Rd.
Orlando, FL 32803
Richard T. Mayer

★1961★ U.S. Livestock Insects Laboratory
USDA-ARS-SPA   Phone: (512) 257-3566
PO Box 232
Kerrville, TX 78029-0232
S.E. Kunz

★1962★ U.S. Plant, Soil and Nutrition Laboratory
USDA-S&E-ARS   Phone: (607) 255-8002
Tower Rd.
Ithaca, NY 14853
D.R. Van Campen

★1963★ U.S. Regional Pasture Research Laboratory
Curtin Rd.   Phone: (814) 863-0939
University Park, PA 16802
R.R. Hill

★1964★ U.S. Sugarcane Field Laboratory
USDA-ARS-MSA   Phone: (504) 872-5042
PO Box 470
Houma, LA 70361
Rex W. Millhollon

★1965★ U.S. Vegetable Laboratory
USDA-ARS-SAA   Phone: (803) 556-0840
2875 Savannah Hwy.
Charleston, SC 29414
George Fassuliotis

★1966★ U.S. Water Conservation Laboratory
USDA-ARS-PWA   Phone: (602) 261-4356
4331 E. Broadway Rd.
Phoenix, AZ 85040
Herman Bouwer

★1967★ Veterinary Toxicology and Entomology
Research Laboratory
USDA-S&E-ARS   Phone: (409) 260-9371
PO Drawer GE
College Station, TX 77841
Donald A. Witzel

★1968★ Water Quality and Watershed Research
Laboratory
PO Box 1430   Phone: (405) 924-5066
Durant, OK 74702
S.J. Smith

★1969★ Western Cotton Research Laboratory
4135 E. Broadway   Phone: (602) 261-3524
Phoenix, AZ 85040
Thomas J. Henneberry

★1970★ Western Regional Research Center
USDA-ARS-PWA   Phone: (415) 559-5600
800 Buchanan St.
Albany, CA 94710
Martin H. Rogoff

★1971★ Yakima Agricultural Research Laboratory
USDA-ARS　　　　　　　　　Phone: (509) 575-5877
Pacific W. Area
3706 W. Nob Hill Blvd.
Yakima, WA 98902
J.L. Krysan

U.S. Department of Commerce

★1972★ Environmental Research Laboratories
Aeronomy Laboratory
325 Broadway　　　　　　　　Phone: (303) 797-3218
Boulder, CO 80303
Daniel L. Albritten

★1973★ Environmental Research Laboratories
Air Resources Laboratory
8060 13th St., Rm. 927　　　　Phone: (301) 427-7684
Silver Spring, MD 20910
Lester Machta

★1974★ Environmental Research Laboratories
Atlantic Oceanographic and Meteorological Laboratories
4301 Rickenbacker Causeway　　Phone: (305) 361-4300
Miami, FL 33149
Hugo F. Bezdek

★1975★ Environmental Research Laboratories
Environmental Science Group
RE2　　　　　　　　　　　　Phone: (303) 497-6378
3100 Marine St.
Boulder, CO 80303
William Hooke

★1976★ Environmental Research Laboratories
Geophysical Fluid Dynamics Laboratory
PO Box 308　　　　　　　　Phone: (609) 452-6502
Princeton, NJ 08540
Jerry D. Mahlman

★1977★ Environmental Research Laboratories
Great Lakes Environmental Research Laboratory
2205 Commonwealth Blvd.　　Phone: (313) 668-2244
Ann Arbor, MI 48105-1593
Alfred M. Beeton

★1978★ Environmental Research Laboratories
National Severe Storms Laboratory
1313 Halley Circle　　　　　Phone: (405) 366-0429
Norman, OK 73069
Robert A. Maddox

★1979★ Environmental Research Laboratories
Office of Oceanic and Atmospheric Research
R/E5X2　　　　　　　　　Phone: (303) 497-6914
325 Broadway
Boulder, CO 80303
B.L. Trotter

★1980★ Environmental Research Laboratories
Pacific Marine Environmental Laboratory
7600 Sand Point Way NE
Seattle, WA 98115　　　　　Phone: (206) 526-6239
Eddie L. Bernard

★1981★ Environmental Research Laboratories
Space Environment Laboratory
325 Broadway　　　　　　　Phone: (303) 497-3311
Boulder, CO 80303
Ernest Hildner

★1982★ Environmental Research Laboratories
Wave Propogation Laboratory
R/E/WP　　　　　　　　　Phone: (303) 497-6261
325 Broadway
Boulder, CO 80303
Steven Clifford

★1983★ National Climatic Data Center
NESDIS　　　　　　　　　Phone: (704) 259-0476
Federal Bldg.
Asheville, NC 28801
Kenneth Hadeen

★1984★ National Institute of Standards and Technology
Office for Technology Utilization
Bldg. 101, Rm. 537　　　　Phone: (301) 975-3087
Gaithersburg, MD 20899
David Edgerly

★1985★ National Institute of Standards and Technology
Office of Research and Technology Applications
Bldg. 101, Rm. 537　　　　Phone: (301) 975-3088
Gaithersburg, MD 20899
James M. Wyckoff

★1986★ National Institute of Standards and Technology
State Extension Activities Program
Bldg. 101, Rm. 537　　　　Phone: (301) 975-3084
Gaithersburg, MD 20899
Joseph G. Berke

★1987★ National Institute of Standards and Technology
Technology Development and Small Business Program
Bldg. 101, Rm. 537　　　　Phone: (301) 975-3084
Gaithersburg, MD 20899
Bruce E. Mattson

★1988★ National Marine Fisheries Service
Atlantic Environmental Group
RR 7A, Box 522A　　　　　Phone: (401) 789-9326
Narragansett, RI 02882
Merton C. Ingham

★1989★ National Marine Fisheries Service
Auke Bay Laboratory
PO Box 210155   Phone: (907) 789-7231
Auke Bay, AK 99821
George R. Snyder

★1990★ National Marine Fisheries Service
Beaufort Laboratory
Beaufort, NC 28516-9722   Phone: (919) 728-4595
Ford Cross

★1991★ National Marine Fisheries Service
Charleston Laboratory
217 Ft. Johnson Rd.   Phone: (803) 762-1200
PO Box 12067
Charleston, SC 29412
Robert Kifer

★1992★ National Marine Fisheries Service
Galveston Laboratory
4700 Ave. U   Phone: (409) 766-3500
Galveston, TX 77550
Edward F. Klima

★1993★ National Marine Fisheries Service
Gloucester Laboratory
Emerson Ave.   Phone: (617) 281-3600
Gloucester, MA 01930
Robert J. Learson

★1994★ National Marine Fisheries Service
Honolulu Laboratory
PO Box 3830   Phone: (808) 943-1221
Honolulu, HI 96812
Richard Shomura

★1995★ National Marine Fisheries Service
Milford Laboratory
Milford, CT 06460   Phone: (203) 878-2459
Anthony Calabrese

★1996★ National Marine Fisheries Service
Mississippi Laboratory
3209 Frederick St.   Phone: (601) 688-3650
PO Drawer 1207
Pascagoula, MS 39567
Andrew J. Kemmerer

★1997★ National Marine Fisheries Service
Narragansett Laboratory
S. Ferry Rd.   Phone: (401) 789-9326
Narragansett, RI 02882
Kenneth Sherman

★1998★ National Marine Fisheries Service
National Systematics Laboratory
Smithsonian Institution   Phone: (202) 357-2550
Washington, DC 20560
Bruce B. Collette

★1999★ National Marine Fisheries Service
Northeast Fisheries Center
Woods Hole, MA 02543   Phone: (617) 548-5123
Allen Peterson

★2000★ National Marine Fisheries Service
Northwest and Alaska Fisheries Center
7600 Sand Point Way NE   Phone: (206) 526-4000
Seattle, WA 98115-0070
William Aron

★2001★ National Marine Fisheries Service
Oxford Laboratory
Oxford, MD 21654   Phone: (301) 226-5193
Aaron Rosenfield

★2002★ National Marine Fisheries Service
Pacific Environmental Group
PO Box 831   Phone: (408) 646-3311
Monterey, CA 93942
Andrew Bakun

★2003★ National Marine Fisheries Service
Panama City Laboratory
Panama City, FL 32407-7499   Phone: (904) 234-6541
Eugene L. Nakamura

★2004★ National Marine Fisheries Service
Sandy Hook Laboratory
Highlands, NJ 07732   Phone: (201) 872-0200
Ann Studholme

★2005★ National Marine Fisheries Service
Southeast Fisheries Center
75 Virginia Beach Dr.   Phone: (305) 361-4286
Miami, FL 33149
Richard Berry

★2006★ National Marine Fisheries Service
Southwest Fisheries Center
Box 271   Phone: (619) 453-2820
La Jolla, CA 92038
Izadore Barrett

★2007★ National Marine Fisheries Service
Tiburon Laboratory
3150 Paradise Dr.   Phone: (415) 435-3149
Tiburon, CA 94920
Norman J. Abramson

★2008★ National Weather Service
Climate Analysis Center
Rm. 606, WWB
Washington, DC 20233   Phone: (301) 763-8167
David R. Rodenhuis

★2009★ National Weather Service
Equipment Test and Evaluation Branch
RD 1, Box 105
Sterling, VA 22170   Phone: (703) 471-5302
Richard Stone

★2010★ National Weather Service
Hydrologic Research Laboratory
3060 13th St., Rm. 530
Silver Spring, MD 20910   Phone: (301) 427-7619
Edward Johnson

★2011★ National Weather Service
Integrated Systems Laboratory
8060 13th St., Rm. 530
Silver Spring, MD 20910   Phone: (301) 427-7809
Richard Waters

★2012★ National Weather Service
National Data Buoy Center
NSTL Station, MS 39529
Phone: (601) 688-2800
Jerry McCall

★2013★ National Weather Service
National Meteorological Center
Development Division   Phone: (301) 763-8005
WWB, Rm. 204
Washington, DC 20233
John A. Brown

★2014★ National Weather Service
Techniques Development Laboratory
Gramax Bldg., Rm. 825   Phone: (301) 427-7768
3060 13th St.
Silver Spring, MD 20910
Harry R. Glahn

★2015★ NOAA Office of Climatic and Atmospheric
   Research
Office of Oceanic and Atmospheric Research
WSC5, Rm. 825   Phone: (301) 443-8415
Rockville, MD 20852
Michael Hall

★2016★ Satellite Applications Laboratory
NESDIS   Phone: (301) 763-8282
Rm. 601, WWB
Washington, DC 20233
Donald B. Miller

★2017★ Satellite Research Laboratory
NESDIS   Phone: (301) 763-8078
Rm. 712, WWB
Washington, DC 20233
George Ohring

U.S. Department of Defense

★2018★ Armed Forces Radiobiology Research Institute
Bethesda, MD 20814   Phone: (301) 295-1210
Capt. Bumgamer

U.S. Department of Energy

★2019★ Ames Laboratory
Iowa State University
Office & Laboratory Bldg., Rm. 119   Phone: (515) 294-2635
Ames, IA 50011
Daniel E. Williams

★2020★ Argonne National Laboratory Technology
   Transfer Center
Bldg. 207   Phone: (312) 972-4929
9700 S. Cass Ave.
Argonne, IL 60439
Brian Frost

★2021★ Bettis Atomic Power Laboratory
PO Box 1449   Phone: (412) 476-6111
West Mifflin, PA 15122
Nick Masterson

★2022★ Brookhaven National Laboratory
Office of Research and Technology Applications
Bldg. 475   Phone: (516) 282-2103
Upton, NY 11973
William Marcuse

★2023★ EG&G Mound Applied Technologies
Technology Exchange
PO Box 3000   Phone: (513) 865-3829
Miamisburg, OH 45343-0987
Japnell D. Braun

★2024★ Energy Technology Engineering Center
PO Box 1449   Phone: (818) 700-5532
Canoga Park, CA 91304
Guy Ervin

★2025★ Environmental Measurement Laboratory
376 Hudson St.   Phone: (212) 620-3616
New York, NY 10014-3621
Gail dePlanque

★2026★ **Fermi National Accelerator Laboratory**
Fermilab - M.S. 208   Phone: (312) 840-3333
PO Box 500
Batavia, IL 60510
Richard A. Carrigan

★2027★ **Hanford Engineering Development Laboratory**
PO Box 1970   Phone: (509) 376-4063
MS L5-55
Richland, WA 99352
Fred Reich

★2028★ **Idaho National Engineering Laboratory**
INEL-ORTA   Phone: (208) 526-8318
PO Box 1625
MS 3402
Idaho Falls, ID 83415
Jane M. Welch

★2029★ **Knolls Atomic Power Laboratory**
PO Box 1072
Schenectady, NY 12301
Gerald Sabian

★2030★ **Lawrence Berkeley Laboratory**
50A-4112   Phone: (415) 486-6502
University of California
Berkeley, CA 94720
Pepi Ross

★2031★ **Lawrence Livermore National Laboratory**
Technology Transfer Initiative Program
PO Box 808 L-795   Phone: (415) 422-6416
Livermore, CA 94550
Gordon Longerbeam

★2032★ **Los Alamos National Laboratory**
Industrial Initiative Office
M899   Phone: (505) 667-3839
Los Alamos, NM 87545
Ron Barks

★2033★ **Lovelace Biomedical and Environmental**
  Research Institute
Inhalation Toxicology Research Institute
PO Box 5890   Phone: (505) 844-4542
Albuquerque, NM 87185
Teri Coons

★2034★ **Morgantown Energy Technology Center**
Collins Ferry Rd.   Phone: (304) 291-4620
PO Box 880
Morgantown, WV 26507-0880
Wennona Brown

★2035★ **MSU-DOE Plant Research Laboratory**
Michigan State University   Phone: (517) 353-2270
East Lansing, MI 48824
Gary Watson

★2036★ **New Brunswick Laboratory**
9800 S. Cass Ave., Bldg. 350   Phone: (312) 972-2446
Argonne, IL 60439
Carleton D. Bingham

★2037★ **Notre Dame Radiation Laboratory**
University of Notre Dame   Phone: (219) 283-7502
Notre Dame, IN 45556
John Bentley

★2038★ **Oak Ridge Associated Universities**
PO Box 117   Phone: (615) 576-3365
Oak Ridge, TN 37831-0117
Wanda Penland

★2039★ **Oak Ridge National Laboratory**
PO Box 2008   Phone: (615) 574-4192
4500 N, MS 257
Oak Ridge, TN 37831-6257
Donald W. Jared

★2040★ **Oak Ridge National Laboratory**
Energy Division
PO Box X   Phone: (615) 574-5953
Oak Ridge, TN 37830
Jon Soderstrom

★2041★ **Pittsburgh Energy Technology Center**
PO Box 10940   Phone: (412) 675-4756
Pittsburgh, PA 15236
Kay Downey

★2042★ **Princeton Plasma Physics Laboratory**
Princeton University   Phone: (609) 243-3009
PO Box 451
Princeton, NJ 08543
Joseph File

★2043★ **Rocky Flats Plant**
Bldg. 060   Phone: (303) 966-2416
PO Box 464
Golden, CO 80402-2416
Frederick J. Fraikor

★2044★ **Sandia National Laboratories**
Technology Transfer Department
Dept. 6010   Phone: (505) 846-0387
Albuquerque, NM 87185
Daniel E. Arvizu

★2045★ Savannah River Ecology Laboratory
PO Drawer E   Phone: (803) 725-2472
Aiken, SC 29801
Robert I. Nestor

★2046★ Savannah River Laboratory
Dupont Co.   Phone: (803) 725-3020
Aiken, SC 29808
John Corey

★2047★ Solar Energy Research Institute
1617 Cole Blvd.   Phone: (303) 231-7115
Golden, CO 80401
H. Dana Moran

★2048★ Stanford Linear Accelerator Center
Stanford University   Phone: (415) 926-2130
PO Box 4349
Stanford, CA 94309
Herman H. Murphy

★2049★ William H. Bates Linear Accelerator Center
Massachusetts Institute of   Phone: (617) 245-6600
  Technology
PO Box 846
Middleton, MA 01949
Ernest J. Moniz

U.S. Department of Health and Human Services

★2050★ Alcohol, Drug Abuse and Mental Health Administration
5600 Fishers Ln.   Phone: (301) 443-4797
Rm. 12-105
Rockville, MD 20857
Fredrick K. Goodwin

★2051★ Center for Devices and Radiological Health
Office of Science and Technology
ATTN: HFZ-150   Phone: (301) 443-7003
12200 Wilkins Ave.
Rockville, MD 20852
Edward Mueller

★2052★ Centers for Disease Control
1600 Clifton Rd. NE   Phone: (404) 329-1900
Executive Park, Bldg. 24
MS E-20
Atlanta, GA 30333
Carl H. Blank

★2053★ Food and Drug Administration
National Center for Toxicological Research
HFT-2   Phone: (501) 541-4516
Jefferson, AR 72079
Arthur R. Noris

★2054★ Laboratory of Biomedical and Environmental Science
UCLA   Phone: (213) 825-9431
900 Veteran Ave.
Los Angeles, CA 90024-1786
William J. Moffitt

★2055★ Laboratory of Radiobiology and Environmental Health
University of California   Phone: (415) 666-1636
School of Medicine
San Francisco, CA 94143
Sheldon Wolff

★2056★ National Cancer Institute
Frederick Cancer Research Facility
Frederick, MD 21701-1013   Phone: (301) 698-1108
Henry J. Hearn

★2057★ National Eye Institute
Intramural Research Program
NIH Bldg. 41, 6A04   Phone: (301) 496-3552
9000 Rockville Pike
Bethesda, MD 20205
Jin H. Kinoshita

★2058★ National Institute of Allergy and Infectious Diseases
NIH Bldg. 10, Rm. 11C103   Phone: (301) 496-3006
Bethesda, MD 20892
Gordon D. Wallace

★2059★ National Institute of Child Health and Human Development
Office of Planning and Evaluation
NIH Bldg. 31, Rm. 2A10   Phone: (301) 496-1877
Bethesda, MD 20892
James G. Hill

★2060★ National Institute of Dental Research
NIH Bldg. 31, Rm. 522   Phone: (301) 496-2883
Bethesda, MD 20892
Alice M. Horowitz

★2061★ National Institute of Diabetes
NIH Bldg. 31, Rm. 9A03   Phone: (301) 496-4955
Bethesda, MD 20892
Benjamin Burton

★2062★ National Institute of Environmental Health Science
Box 12233   Phone: (919) 541-3205
Research Triangle Park, NC 27709
Martin Rodbell

★2063★ **National Institute of Mental Health**
Division of Communications and Education
5600 Fishers Lane, Rm. 15-105   Phone: (301) 443-3600
Rockville, MD 20857
Julius Segal

★2064★ **National Institute of Mental Health**
Intramural Research Program
WAW-Rm. 536   Phone: (202) 373-7571
St. Elizabeths Hospital
Washington, DC 20032
Richard Wyatt

★2065★ **National Institute of Mental Health**
Office of State and Community Liaison
5600 Fishhers Lane, Rm. 11C-26   Phone: (301) 443-3606
Rockville, MD 20857
James W. Stockdill

★2066★ **National Institute of Neurological and**
Communicative Disorders and Stroke
NIH Bldg. 31, Rm. 8A03   Phone: (301) 496-9271
Bethesda, MD 20892
Zekin Shakhashiri

★2067★ **National Institute on Aging**
Gerontology Research Center
4940 Eastern Ave.   Phone: (301) 550-1707
Baltimore, MD 21244
Daniel Rogers

★2068★ **National Institute on Drug Abuse**
Addiction Research Center
PO Box 5180   Phone: (301) 955-7594
Baltimore, MD 21224
Jerome Jaffe

★2069★ **National Institute on Drug Abuse**
Division of Prevention and Communication
5600 Fishers Lane, Rm. 53   Phone: (301) 443-6780
Rockville, MD 20857
Barry S. Brown

★2070★ **National Institutes of Health**
Division of Computer Research and Technology
NIH Bldg. 12A, Rm. 3023   Phone: (301) 496-5206
Bethesda, MD 20892
L. Lee Manuel

★2071★ **National Institutes of Health**
National Cancer Institute
NIH Bldg. 32, Rm. 11A23   Phone: (301) 496-1148
Bethesda, MD 20892
Elliott Stonehill

★2072★ **National Institutes of Health**
National Heart, Lung, and Blood Institute
NIH Bldg. 31, Rm. 4A21   Phone: (301) 496-4236
Bethesda, MD 20205
Larry Blaser

★2073★ **Office for Substance Abuse Prevention**
Division of Communication Programs
5600 Fishers Lane, Rm. 13A-54   Phone: (301) 443-0373
Rockville, MD 20857
Robert W. Denniston

★2074★ **Radiation Biology Laboratory**
RBL   Phone: (301) 443-2329
12441 Parklawn Dr.
Rockville, MD 20852
William H. Klein

U.S. Department of Justice

★2075★ **Federal Bureau of Investigation Laboratory**
Forensic Science Research and Training Unit
Quantico, VA 22135   Phone: (703) 640-6131
Cecil Yates

U.S. Department of the Air Force

★2076★ **Air Force Armament Laboratory**
AFATL/DOIR   Phone: (904) 882-4013
Eglin AFB, FL 32542-5434
James E. Krug

★2077★ **Air Force Arnold Engineering Development**
Center
Arnold Air Force Station, TN   Phone: (615) 454-7621
  37389-5000
Dale F. Vosika

★2078★ **Air Force Astronautics Laboratory**
AFAL/TSTR   Phone: (805) 275-5014
Edwards AFB, CA 93523-5000
Chris Degnan

★2079★ **Air Force Engineering and Services Center**
AFESC/RD   Phone: (904) 283-6494
Tyndall AFB, FL 32403-6001
Bob Van Orman

★2080★ **Air Force Flight Test Center**
Edwards AFB, CA 93523   Phone: (805) 277-2410
William T. Twinting

★2081★ **Air Force Geophysics Laboratory**
AFGL/XO   Phone: (617) 861-3606
Hanscom AFB, MA 01731-5000
Rene Cormier

★2082★ **Air Force Human Resources Laboratory**
AFHRL/PRT   Phone: (512) 536-3426
IR&D
Brooks AFB, TX 78235-5601
Douglas Blair

★2083★ **Air Force Rocket Propulsion Laboratory**
STINFO Office   Phone: (805) 277-5677
AFRPL/TSPR
Edwards AFB, CA 93523
Rosemary Degnan

★2084★ **Air Force Weapons Laboratory**
AFWL/CA   Phone: (505) 844-9856
Kirtlan AFB, NM 87117-6008
Patrick Rodriguez

★2085★ **Air Force Wright Aeronautical Laboratories**
ORTA, AFWAL/XO   Phone: (513) 255-2788
Rm. 219
Wright Patterson AFB, OH
    45433-6523
Cindy Ingalls

★2086★ **Frank J. Seiler Research Laboratory**
FJSRL, NA   Phone: (303) 472-3120
USAF Academy, CO 80840
Ken Siegenthaler, Chief Scientist

★2087★ **Human Systems Division**
HQ HSD/SORX   Phone: (512) 536-3687
Brooks AFB, TX 78235-5000
Edward J. Wright, Tech. Plans & Analysis Ofc.

★2088★ **Rome Air Development Center**
Code RADC-XP   Phone: (315) 330-3705
Griffiss AFB, NY 13441-5700
Billy G. Oaks

U.S. Department of the Army

★2089★ **Air Force School of Aerospace Medicine**
USAFSAM/TSZ   Phone: (512) 536-3836
Brooks AFB, TX 78235
James T. Harrison

★2090★ **Army Aeromedical Research Laboratory**
ATTN: SGRD-UAX-SI   Phone: (205) 255-6907
PO Box 577
Ft. Rucker, AL 36362-5292
Diana Hemphill

★2091★ **Army Armament Research, Development, and Engineering Center**
ATTN: SMCAR-AST   Phone: (201) 724-7954
Bldg. 1
Picatinny Arsenal, NJ 07806-5000
Marie Saunders

★2092★ **Army Atmospheric Sciences Laboratory**
ATTN: SLCAS-DP-P   Phone: (505) 678-4917
White Sands, NM 88002-5501
Richard Himebrook

★2093★ **Army Aviation Systems Command**
ATTN: AMSAV-NR   Phone: (314) 263-1082
4300 Goodfellow Blvd.
St. Louis, MO 63120-1798
Roy J. Warhover

★2094★ **Army Ballistic Research Laboratory**
ATTN: SLCBR-D   Phone: (301) 278-6955
Bldg. 328
Aberdeen Proving Ground, MD
    21005-5066
Ronald Hinkle

★2095★ **Army Biomedical Research and Development Laboratory**
ATTN: SGRD-UBZ-C   Phone: (301) 663-2024
Bldg. 568
Fort Detrick
Frederick, MD 21701-5010
Lee Merrell

★2096★ **Army Center for Signals Warfare**
ATTN: AMSEL-RD-SW-OS   Phone: (703) 347-6464
Vint Hill Farms Sta.
Warrenton, VA 22186-5100
Royal H. Burkhardt

★2097★ **Army Chemical Research, Development, and Engineering Center**
ATTN: SMCCR-OPP   Phone: (301) 671-2031
Aberdeen Proving Ground, MD
    21010-5423
Richard Dimmick

★2098★ **Army Cold Regions Research and Engineering Laboratory**
ATTN: CRREL-CS   Phone: (603) 646-4237
72 Lyme Rd.
Hanover, NH 03755-1920
Andrew Assur

★2099★ **Army Combat Systems Test Activity**
ATTN: STECS   Phone: (301) 274-4102
Bldg. 400
Aberdeen Proving Ground, MD
    21005-5059
Palmer Paules

★2100★ Army Communications-Electronics Command
ATTN: AMSEL-RD-TPPO-L   Phone: (201) 544-2239
Ft. Monmouth, NJ 07703-5205
Albert J. Feddeler

★2101★ Army Construction Engineering Research Laboratory
ATTN: CERER-TAO   Phone: (217) 373-6789
PO Box 4005
Champaign, IL 61824-4005
Rob Gorham

★2102★ Army Electronics Technology and Devices Laboratory
SLCET-DT   Phone: (201) 544-4666
Fort Monmouth, NJ 07703-5000
Richard Stern

★2103★ Army Engineer Topographic Laboratories
ATTN: CEETL-RT   Phone: (703) 355-2629
Ft. Belvoir, VA 22060-5546
George N. Simcox

★2104★ Army Engineer Waterways Experiment Station
ATTN: CEWES-FV   Phone: (601) 634-4113
PO Box 631
Vicksburg, MS 39181-0631
Philip Stewart

★2105★ Army Human Engineering Laboratory
ATTN: SLCHE-SS-IR   Phone: (301) 278-5817
Aberdeen Proving Ground, MD
   21005-5001
Dean Westerman

★2106★ Army Institute of Surgical Research
ATTN: SGRD-USA   Phone: (512) 221-2340
Ft. Sam Houston
San Antonio, TX 78234-6200
David Howard

★2107★ Army Materials Technology Laboratory
ATTN: SLCMT-DA   Phone: (617) 923-5527
Watertown, MA 02172-0001
George R. Thomas

★2108★ Army Medical Research Institute of Chemical Defense
ATTN: SGRD-UV-R   Phone: (301) 671-2363
Aberdeen Proving Ground, MD
   21010-5425
Lloyd Roberts

★2109★ Army Medical Research Institute of Infectious Diseases
ATTN: SGRD-UIZ-D   Phone: (301) 663-2227
Fort Detrick, Bldg. 1425
Frederick, MD 21701-5011
Michael A. Chirigos, Deputy for Science

★2110★ Army Missile Command
ATTN: AMSMI-RD-TI   Phone: (205) 876-5449
Redstone Arsenal, AL 35898-5243
Steven Smith

★2111★ Army Natick RD&E Center
ATTN: STRNC-EMP   Phone: (508) 651-5296
Natick, MA 01760-5014
Robert Rosenkrans

★2112★ Army Night Vision and Electro-Optics Laboratory
ATTN: AMSEL-RD-NV-D   Phone: (703) 664-5308
Ft. Belvoir, VA 22060-5677
Clarence Johnson, Public Affairs Officer

★2113★ Army Research Institute for Behavioral and Social Sciences
ATTN: PERI-PO   Phone: (202) 274-8029
5001 Eisenhower Ave.
Alexandria, VA 22333-5600

★2114★ Army Research Institute of Environmental Medicine
Bldg. 42   Phone: (617) 651-4891
ATTN: SGRD-UE-RSI
Natick, MA 01760-5007
Carol Joriman

★2115★ Army Research Office
ATTN: SLCRO-TS   Phone: (919) 549-0641
PO Box 12211
Research Triangle Park, NC
   27709-2211
David W. Seitz

★2116★ Army Tank-Automotive Research, Development, and Engineering Center
ATTN: AMSTA-CK   Phone: (313) 574-5270
Warren, MI 48397-5000
Robert Hostetler

★2117★ Army Vulnerability Assessment Laboratory
ATTN: SLCVA-DPC   Phone: (505) 678-2650
White Sands Missile Range, NM
   88002-5513
Tom Reader

★2118★ Army Walter Reed Institute of Research
ATTN: SGRD-UWZ-I   Phone: (202) 576-2274
Washington, DC 20307-5100
B. Nolan Dale

★2119★ Army Walter Reed Medical Center
Army Institute of Dental Research
ATTN: SGRD-UDR   Phone: (202) 576-3254
Bldg. 40
Washington, DC 20307-5300
Gino C. Battistone

★2120★ Aviation Applied Technology Directorate
USAARTA (AVSCOM)   Phone: (804) 878-2000
Ft. Eustis, VA 23604-5577
John L. Shipley

★2121★ Belvoir Research, Development and
Engineering Center
ATTN: STRBE-IL   Phone: (703) 664-1068
Fort Belvoir, VA 22060-5606
Connie Harrison

★2122★ Dugway Proving Ground
ATTN: STEDP-SD   Phone: (801) 522-3314
Dugway, UT 84022
Lothar L. Salomon

★2123★ Harry Diamond Laboratories
ATTN: SLCHD-TT   Phone: (202) 394-4210
2800 Powder Mill Rd.
Adelphi, MD 20783-1197
Clifford E. Lanham

★2124★ Letterman Army Institute of Research
SGRD-ULZ-IR   Phone: (415) 561-2641
Presidio of San Francisco, CA
   94129-6800
Jack Keller

★2125★ Picatinny Arsenal
Benet Laboratory
Watervliet Arsenal   Phone: (518) 266-5807
Watervliet, NY 12189
Francis A. Heiser

U.S. Department of the Interior

★2126★ Alaska Fish and Wildlife Research
   Center—Anchorage
1011 E. Tudor Rd.
Anchorage, AK 99503   Phone: (907) 786-3512
A.W. Palmisano

★2127★ Albany Research Center
1450 Queen Ave. SW   Phone: (503) 967-5896
PO Box 70
Albany, OR 97321
William D. Riley

★2128★ Bureau of Mines
Twin Cities Research Center
5629 Minnehaha Ave. S.   Phone: (612) 725-4562
Minneapolis, MN 55417
Richard Dick

★2129★ Denver Research Center
Denver Federal Center   Phone: (303) 236-0747
Bldg. #20
Denver, CO 80225
Guy Johnson

★2130★ Denver Wildlife Research Center
APHIS   Phone: (303) 236-7820
Federal Center, Bldg. 16
PO Box 25266
Denver, CO 80225-0266
Russell Reidinger

★2131★ Environmental Research Laboratory (Duluth)
USEPA/ERL   Phone: (218) 720-5548
6201 Congdon Blvd.
Duluth, MN 55804
Nelson Thomas

★2132★ Fish and Wildlife Service
National Ecology Research Center—Ft. Collins
Creekside One   Phone: (303) 226-9398
2627 Redwing Rd.
Ft. Collins, CO 80526

★2133★ Fish and Wildlife Service
National Wetlands Research Center
1010 Gause Blvd.   Phone: (504) 646-7564
Slidell, LA 70458
Robert E. Stewart

★2134★ Geological Survey—Central Region
Mail Stop 507   Phone: (303) 236-5825
Box 25046
Denver Federal Center
Denver, CO 80225
Lee Aggers

★2135★ Geological Survey—National Center
12201 Sunrise Valley Dr.   Phone: (703) 648-4443
Mail Stop 407
Reston, VA 22092
Ethan T. Smith

★2136★ Geological Survey—Western Region
345 Middlefield Rd.　　　　Phone: (415) 329-4002
MS 144
Menlo Park, CA 94025
George Gryc

★2137★ National Fisheries Contaminant Research Center
Rte. 2　　　　　　　　Phone: (314) 875-5399
4200 New Haven Rd.
Columbia, MO 65201
Ell-Piret Multer

★2138★ National Fisheries Research Center—Gainesville
7920 NW 71st St.
Gainesville, FL 32606
James McCann

★2139★ National Fisheries Research Center—Great Lakes
1451 Green Rd.　　　　　Phone: (313) 994-3331
Ann Arbor, MI 48105
Jon G. Stanley

★2140★ National Fisheries Research Center—La Crosse
2630 Fanta Reed Rd.　　　Phone: (608) 783-6451
PO Box 818
La Crosse, WI 54602
Fred Meyer

★2141★ National Fisheries Research Center—Leetown
Box 700　　　　　　　Phone: (304) 725-8461
Kerneysville, WV 25430
Jan E. Riffe

★2142★ National Fisheries Research Center—Seattle
Bldg. 204, Naval Sta.　　　Phone: (206) 526-6282
Seattle, WA 98115
Alfred C. Fox

★2143★ National Wildlife Health Research Center
6006 Schroeder Rd.　　　Phone: (608) 271-4640
Madison, WI 53711
Wallace R. Hansen

★2144★ Northern Prairie Wildlife Research Center
PO Box 2096　　　　　Phone: (701) 252-5363
Jamestown, ND 58402
Ray Stendell

★2145★ Patuxent Wildlife Research Center
Laurel, MD 20708　　　Phone: (301) 498-0239
Nancy C. Coon

★2146★ Rolla Research Center
1300 Biship Ave.　　　　Phone: (314) 364-3169
PO Box 280
Rolla, MO 65401
Jim Stephenson

★2147★ Salt Lake City Research Center
729 Arapeen Dr.　　　　Phone: (801) 524-6113
Salt Lake City, UT 84108-1283
Bill Nissen

★2148★ Spokane Research Center
315 Montgomery Ave.　　Phone: (509) 484-1610
Spokane, WA 99207-2291
Robert Bates

★2149★ Tuscaloosa Research Center
PO Box L, University　　　Phone: (205) 758-0491
Tuscaloosa, AL 35486
Ron Church

U.S. Department of the Navy

★2150★ David Taylor Research Center
Code 0117　　　　　　Phone: (202) 227-1037
Bethesda, MD 20084-5000
Basil Nakonechny

★2151★ Marine Corps Development and Education Command
Tehchnology Transfer Office
Quantico, VA 22134　　　Phone: (703) 640-3133
R.E. Ouellette

★2152★ Naval Aerospace Medical Research Laboratory
Code 00B2　　　　　　Phone: (904) 452-3286
Naval Air Sta.
Pensacola, FL 32508-5700
Kathleen Mayer

★2153★ Naval Air Development Center
Code 024　　　　　　Phone: (215) 441-2033
Warminster, PA 18974-5000
Jerome S. Bortman, Technology Transfer Coord.

★2154★ Naval Air Engineering Center
Code 5331　　　　　　Phone: (201) 323-2574
Lakehurst, NJ 08733-5000
Eileen Foy

★2155★ Naval Air Propulsion Center
S&T Department (PE3)　　Phone: (609) 896-5713
Trenton, NJ 08628
Albert Martino

★2156★ Naval Air Test Center
CT23   Phone: (301) 863-1134
Patuxent River, MD 20670
A.M. Rossetti

★2157★ Naval Avionics Center
Office of Research and Technology Application
Code 802   Phone: (317) 353-7075
6000 E. 21st St.
Indianapolis, IN 46219-2189
Larry Halbig

★2158★ Naval Biosciences Laboratory
Naval Supply Center   Phone: (415) 466-5955
Code 82
Oakland, CA 95624
Richard C. Hedstrom

★2159★ Naval Blood Research Laboratory
Boston University   Phone: (617) 247-6700
615 Albany Ave.
Boston, MA 02118

★2160★ Naval Civil Engineering Laboratory
LO3C   Phone: (805) 982-4070
Port Hueneme, CA 93043
Jerry Dummer

★2161★ Naval Coastal Systems Center
Code 710T   Phone: (904) 235-5275
Panama City, FL 32407-5000
J.D. Wright

★2162★ Naval Dental Research Institute
Naval Training Center   Phone: (312) 688-5647
Bldg. 1-H
Great Lakes, IL 60088
Myra J. Portis

★2163★ Naval Explosive Ordnance Disposal Technology
Center
Indian Head, MD 20640-5070
Bert Stevenson   Phone: (301) 743-4430

★2164★ Naval Health Research Center
PO Box 85122   Phone: (619) 524-4516
San Diego, CA 92138-9174

★2165★ Naval Medical Research Institute
Technology Transfer Office
Bethesda, MD 20014
L. Kiesow   Phone: (202) 295-1310

★2166★ Naval Ocean R&D Activity
Code 115T   Phone: (601) 688-5211
NSTL Sta.
Bay St. Louis, MS 39529
George E. Stanford

★2167★ Naval Ocean Systems Center
Code 0141   Phone: (619) 553-2103
San Diego, CA 92152-5000
Richard November

★2168★ Naval Oceanographic Office
Code 0010   Phone: (601) 688-4594
NSTL Sta.
Bay St. Louis, MS 39522-5001
Richard Hess

★2169★ Naval Ordnance Missile Test Station
White Sands Missile Range, NM   Phone: (505) 678-2101
   88002
A.F. Schreader

★2170★ Naval Research Laboratory
Code 1003.2   Phone: (202) 767-3744
Washington, DC 20375-5000
Richard Fulper, Jr.

★2171★ Naval Submarine Medical Research Laboratory
Box 900   Phone: (203) 449-3398
Naval Submarine Base
Groton, CT 06349-5900
S.M. Luria

★2172★ Naval Surface Weapons Center
White Oak, Code D211   Phone: (301) 394-1505
Silver Spring, MD 20903-5000
Ramsey D. Johnson

★2173★ Naval Underwater Systems Center
ORTA, Code 102   Phone: (203) 440-4116
Bldg. 80
New London, CT 06320
John F. Griffin

★2174★ Naval Weapons Center
Industrial and Governmental Liason Office
Code 374   Phone: (619) 939-1074
China Lake, CA 93555-6001
George F. Linsteadt

★2175★ Naval Weapons Station Earle
Code 70   Phone: (201) 577-2145
Colts Neck, NJ 07722
M. Gray

★2176★ Navy Clothing and Textile Research Facility
21 Strathmore Rd.   Phone: (508) 651-4680
Natick, MA 01760-2490
John Mylotte

★2177★ Navy Personnel R&D Center
Technical Information Office
Code 232   Phone: (619) 553-9308
San Diego, CA 92152-6800
Robert Turney

★2178★ Pacific Missile Test Center
Code 1032   Phone: (805) 989-7124
Point Mugu, CA 93042
Dan Kimsey, Technology Transfer Officer

★2179★ Polaris Missile Facility, Atlantic
Code SPC 053   Phone: (803) 764-7775
Charleston, SC 29408
Edwin J. Durr

U.S. Department of Transportation

★2180★ FAA Aeronautical Center
Protection and Survival Laboratory
PO Box 25082   Phone: (405) 686-4851
Oklahoma City, OK 73125
Richard F. Chandler

★2181★ Federal Aviation Administration Technical
Center
ATTN: ACL-1   Phone: (609) 484-6689
Atlantic City Airport
Atlantic City, NJ 08405
A.A. Lupinetti

★2182★ Federal Highway Administration
Turner-Fairbanks Research Center
HRT-1   Phone: (703) 285-2035
6300 Georgetown Pike
McLean, VA 22101
Stanley R. Byington

★2183★ Transportation Systems Center
Office of Technology Sharing
Kendall Sq.   Phone: (617) 494-2486
Code DTS-31
Cambridge, MA 02142
R.V. Giangrande

★2184★ Transportation Test Center
PO Box 11130   Phone: (719) 584-0546
Pueblo, CO 81001
David A. Watts

★2185★ U.S. Coast Guard Research Development Center
Avery Point   Phone: (203) 441-2604
Groton, CT 06340
Samuel F. Powell

★2186★ Vehicle Research and Test Center
PO Box 37   Phone: (513) 666-4511
East Liberty, OH 43319
James E. Hofferberth

U.S. Department of Veterans Affairs

★2187★ Hines VA Hospital
Rehabilitation R&D Center
Code 151L, Box 20   Phone: (312) 343-7200
Hines, IL 60141
John Trimble

★2188★ VA Medical Center
Rehabilitation Research and Development Center
3801 Miranda Ave.   Phone: (415) 858-3991
Palo Alto, CA 94304
Alvin H. Sacks

★2189★ VA Medical Center
Rehabilitation Research and Development Unit
1670 Clairmont Rd.   Phone: (404) 321-5828
Decatur, GA 30033
Franklyn K. Coombs

★2190★ VA Prosthesics R&D Center
Rehabilitation R&D Evaluation Unit
103 S. Gay St.   Phone: (301) 962-2333
Baltimore, MD 21202
Mercia C. Decker

Section 23

Publications

"The man with a new idea is a crank until the idea succeeds."—Mark Twain

Publications can provide a wide variety of information to inventors. I cannot emphasize enough their value and the time you should dedicate to researching and reading them.

In the United States, an estimated 12,300 magazines are listed as commercial publications. If you add on smaller publications and newsletters, the total jumps to no less than 25,000.

I love prospecting through publications. Often when I need creative stimulation, I will go to a nearby library and spend hours leafing through a variety of publications. If nothing else, this exercise tends to focus my direction.

It is a sure bet that even the most general circulation magazines will run a few stories each year on inventions and creativity, and maybe even something on your particular field of interest. I keep up with what is being covered through the *Reader's Guide to Periodical Literature* (The H.W. Wilson Co.) at my local library. The *Reader's Guide* keeps track of hundreds of publications day-to-day.

There are some publications I read on a regular basis, however. I mention them in hopes that it might help you organize your reading list.

Newspapers and News Magazines

The Wall Street Journal. The *Journal* provides the most current, fast-breaking information on industry. It was through a piece in this newspaper that I learned what was going on at Proctor & Gamble's Crest brand, information that led to my selling P&G what was to become a $12-million premium program (Crest Fluorider).

The Wall Street Journal, Forbes, Fortune, Business Week. These publications are the best way to track the ins and outs of senior executives and the ups and downs of corporate earnings, trends, and competition. *Time and Newsweek* are also must reads.

Commerce Business Daily. In order to alert potential sources to emerging government R&D interests, agencies must, by law, publish advance notice of R&D opportunities in the *Commerce Business Daily*. Published Monday through Saturday (except on federal holidays), copies of *CBD* may be obtained at the Department of Commerce field offices and at most major urban and university libraries. If you want personal copies, write or call the Government Printing Office (GPO). Subscriptions to *CBD* may be obtained from The Superintendent of Documents, Government Printing Office, Washington, DC 20402; (202) 275-3054. The GPO takes Visa and Mastercard.

Trade Journals and Newsletters

Although the news is not as fresh as you'll find in a daily newspaper, industry trade magazines and newsletters are excellent sources for in-depth information. And they carefully track and report a very broad range of executive assignments, not just the upper-most echelon.

It was through a toy trade magazine, for example, that I obtained the name of the Milton Bradley senior vice president of research and development who licensed our first toy, STARBIRD.

These kinds of esoteric business trade publications are found in many major libraries that offer comprehensive business reference rooms. University technical libraries also usually have a wide range of these kinds of magazines.

Advertising Age and Ad Week. There are no better sources than these weekly publications for the latest information on Fortune 500 new introductions, and who's spending how much to promote and introduce which products.

The following are some controlled circulation publications that I find of great use, overall interest, and inspiration. Contact the publishers to see if you qualify for a free subscription. There are hundreds of such magazines available to inventors and product developers in almost every field.

These magazines make their money through the sale of advertising, not the sale of subscriptions. One of the best things about them is the abundance of trade advertisements.

Tech Briefs. The National Aeronautics and Space Administration (NASA) sponsors the publication of *Tech Briefs*, a monthly high-gloss magazine dedicated to the transfer of technology from the space program to industry. Its technical section covers new product ideas, electronic components and circuits, materials, machinery, computer programs, fabrication technology and life sciences.

For information about how to receive *Tech Briefs*, contact Associated Business Publications, 41 E. 42nd St., Ste. 921, New York, NY 10017-5391; phone (212) 490-3999; fax (212) 986-7864.

NASA Spinoffs. If you have applied NASA technology to your products and processes, you can receive free publicity in *NASA Spinoffs*, an annual publication designed to tell consumers how NASA technologies are being applied by industry.

To find out if you qualify, contact Linda Watts at (301) 621-0241.

Product Design & Development. Chilton's *PD&D* is billed as a product news magazine for design engineers. Its dedicated coverage ranges from printers/plotters to powered metals and motion control devices. If you have an interest in valves and switches, there is no better source.

For information about how to receive *PD&D*, contact Chilton, Chilton Way, Radnor, PA 19089; phone (215) 964-4000; fax (215) 964-4100.

Designfax. *Designfax* is dedicated to product design engineering. It has a fine section on new materials, for example, long-fiber composites, epoxy potting compounds, ceramic coatings, and optical release films. It has high advertisement content and provides reader service cards for more information.

For information about how to receive *Designfax*, contact Huebcore Communications, Inc., 6521 Davis Industrial Parkway, Solon, OH 44139; phone (216) 248-1125.

Electronics. This is a monthly magazine that reports on a wide range of subjects from corporate reorganizations and news analysis to new product and technology introductions. It includes executive profiles, reviews, and news of upcoming meetings.

For information about how to receive *Electronics*, contact VNU Business Publications, Inc., 10 Holland Dr., Hasbrouck Heights, NJ 07604; phone (201) 393-6000; fax (201) 393-6388.

Electronic Engineering Times. Billing itself as the "Industry Newspaper for Engineers and Technical Management," *Electronic Engineering Times* is one of the better free publications in its field. It serves up the latest news on technologies, business, and new products. Its reporters cover major trade shows.

For information about how to receive *Electronic Engineering Times*, contact CMP Publications, 600 Community Dr., Manhasset, NY 11030; phone (516) 562-5882.

Inventor-Assistance Program News. In 1987, the Department of Energy launched the *Inventor-Assistance Program News*, the main information transfer mechanism of the States Inventors Initiative (SII). SII is conducted by the Pacific Northwest Laboratory located in Richland, VA.

For information on this publication contact: Ray Watts, Editor, K6-545, Pacific Northwest Laboratory, PO Box 999, Richland, VA 99352; phone (509) 376-4348; fax (509) 376-8054.

Inventors may also find general science magazines such as *Popular Mechanics*, *Popular Science*, *Discover*, *Omni*, and so forth of interest.

In addition to the publications listed in this chapter, there is a large section on publications for the young inventor in Section 24, Project XL: Youth Innovation Programs and Publications, beginning on page 307.)

As valuable as these publications are, it is important to remember that there is no substitute for personal contacts. It's "know-how" combined with "know-who" that makes the difference. I recommend a phone call to authors and editors as the best way to get information on R&D dollar opportunities out of many publications.

Section Arrangement

The following entries are arranged alphabetically by publication title. All titles in this section are referenced in the Master Index.

★2191★ A-Z of U.K. Brands
Euromonitor Publications Ltd.
87-88 Turnmill St.
London EC1M 5QU, England

Over 500 manufacturers of leading (by sales) brand name products in the United Kingdom. **Frequency:** Irregular; previous edition 1984; latest edition 1990. **Price:** 75 British pounds.

★2192★ Advanced Manufacturing Technology
Technical Insights, Inc.   Phone: (201) 568-4744
PO Box 1304   Fax: (201) 568-8247
Fort Lee, NJ 07024-9967

Reports on technological advances that are contributing to robotics, computer graphics, flexible automation, computer-integrated manufacturing, and new techniques for machining. Covers techniques for cutting costs, improving product quality, and increasing productivity. Includes supplements on current techniques and available systems. **Frequency:** Monthly. **Price:** $470/yr.

★2193★ Advanced Science & Technology Institute—CONNECTIONS
Advanced Science & Technology   Phone: (503) 346-3189
   Institute   Fax: (503) 346-2040
University of Oregon
318 Hendricks Hall
Eugene, OR 97403

Reports on developments in research and technology to promote economic development and technology transfer. **Frequency:** Quarterly. **Price:** Free.

★2194★ Aerosol Science & Technology
Elsevier Science Publishing Co., Inc.   Phone: (212) 989-5800
655 Avenue of the Americas
New York, NY 10010

Science journal. **Frequency:** 8x/yr. **Subscription:** $396 institutions; $425 other countries.

★2195★ Aerospace & Defense Science
PO Drawer 033619   Phone: (407) 773-5711
Indialantic, FL 32903-0619

Magazine covering all aspects of U.S. defense policy, technology and science. **Frequency:** Quarterly. **Subscription:** $36 U.S. and Canada; $48 other countries.

★2196★ American Association of Small Research Companies—Members Directory
American Association of Small Research Phone: (617) 491-7906
   Companies
Northern Ventures Corp.
222 3rd St., Ste. 3150
Cambridge, MA 02142

About 400 small research and development companies covering most scientific disciplines.

★2197★ American Innovations
Publishing & Business Consultants   Phone: (213) 732-3477
951 S. Oxford, Ste. 109
Los Angeles, CA 90006

Magazine featuring breakthroughs in science and high technology. **Frequency:** Quarterly. **Subscription:** $23.99; $30.99 other countries.

★2198★ American Investors Market Finance Directory
American Investors Market
561 Pershing Dr.
Walnut Creek, CA 94596

Small Business Investment Companies, (SBICs), financial companies, and institutions lending money for business investment purposes such as venture capital, equity financing, mortgages, real estate development, and equipment and machinery capital purchases. **Frequency:** Biennial, in even years.

★2199★ American Scientist
Sigma Xi, The Scientific Research   Phone: (919) 549-0097
   Society
PO Box 13975
Research Triangle Park, NC 27709-3975

Scientific magazine. **Frequency:** 6x/yr. **Subscription:** $28. $4.75 single issue.

★2200★ Annals of Science
Taylor & Francis   Phone: (215) 785-5800
1900 Frost Rd., Ste. 101
Bristol, PA 19007

Independent review of the development of science since the Renaissance. **Frequency:** 6x/yr. **Subscription:** $346 institutions.

★2201★ Applied Science & Technology Index
The H.W. Wilson Co.   Phone: (212) 588-8400
950 University Ave.
Bronx, NY 10452

Index of science and technology periodicals. **Frequency:** Monthly (not issued July).

★2202★ ASTI Newsletter
Association for Science, Technology, &   Phone: (703) 241-2850
   Innovation (ASTI)
PO Box 1242
Arlington, VA 22210

Publishes news of ASTI, which promotes the development, demonstration, and application of policies, standards, and techniques for improving management of innovation, and news of other organizations with similar goals. Recurring features include book reviews. **Frequency:** Bimonthly. **Price:** Included in membership.

★2203★ The Authority Report

Arkansas Science & Technology Authority
100 Main St., Ste. 450
Little Rock, AR 72201

Phone: (501) 324-9006
Fax: (501) 324-9012

Concentrates on issues related to the development of Arkansas' scientific and technological resources. Contains notices of publications available and related conferences as well as news of research, including research and grants funded by the Authority. Recurring features include interviews, reports of meetings, and news of educational opportunities. **Frequency:** Quarterly. **Price:** Free.

★2204★ B.O.S.S. Directory of Research and Developments Laboratories/Facilities in Canada

Business Opportunity Sourcing Section (BOSS)
Industry, Science, & Technology
235 Queens St.
Ottawa, ON, Canada K1A 0H5

Phone: (613) 954-5031
Fax: (613) 954-1894

Federally controlled and privately owned laboratories in Canada. **Frequency:** Annual, November. **Price:** Free.

★2205★ Biocatalysis

Harwood Academic Publishers
PO Box 786, Cooper Sta.
New York, NY 10276

Phone: (212) 206-8900

Science journal. **Frequency:** 4x/yr.

★2206★ Biographical Dictionary of Scientists: Engineers and Inventors

Peter Bedrick Books
2112 Broadway, Rm. 318
New York, NY 10023

Phone: (212) 496-0751

About 200 engineers and inventors. **Price:** $28.00.

★2207★ Biotechnology Law Report

Mary Ann Liebert, Inc.
20 W. 3rd St., 2NDFL
Media, PA 19063-2824

Phone: (215) 892-9580
Fax: (215) 892-9577

Covers legal developments affecting the fields of biotechnology and genetic engineering. Discusses patent law, product liability, biomedical law, contract and licensing law, and international law. Describes pertinent legislation, regulatory actions, personnel and company changes, litigation resolution, and international developments. Publishes complete texts of significant court decisions, regulations, and legislation. Recurring features include book reviews and a calendar of events. **Frequency:** Bimonthly. **Price:** $375/yr.

★2208★ BioWorld Magazine

Io Publishing Inc.
217 S. B St.
San Mateo, CA 94401

Phone: (415) 696-6555

Magazine covering the biotechnology industry. **Frequency:** 10x/yr. **Subscription:** Free to qualified subscribers; $75 other countries

★2209★ BNA's Patent, Trademark, & Copyright Journal

Bureau of National Affairs, Inc. (BNA)
1231 25th St. NW
Washington, DC 20037

Phone: (202) 452-4500

Monitors developments in the intellectual property field, including patents, trademarks, and copyrights. Covers proposed and enacted legislation, litigation, Patent and Trademark Office decisions, Copyright Office practices, activities of professional associations, government contracting, and international developments. **Frequency:** Weekly. **Price:** $784/yr.

★2210★ Brand Names: Who Owns What

Facts on File, Inc.
460 Park Ave. S.
New York, NY 10016

Phone: (212) 683-2244

Over 750 firms and their 15,000 brand names. **Frequency:** Irregular; previous edition June 1986; latest edition 1988. **Price:** $65.00.

★2211★ Business Capital Sources

IWS, Inc.
24 Canterbury Rd.
Rockville Centre, NY 11570

Phone: (516) 766-5850

About 1,500 banks, insurance and mortgage companies, commercial finance, leasing, and venture capital firms that lend money for business investment. **Frequency:** Annual, January. **Price:** $15.00.

★2212★ BUSINESS (Database on Trade Opportunities and Business Contacts)

BUSINESS Datenbanken GmbH
Poststrasse 42
W-6900 Heidelberg, Germany

About 25,000 manufacturers, importers and exporters, research establishments, chambers of commerce, trade promotion agencies, and investment companies offering or seeking business opportunities worldwide.

★2213★ Business Organizations, Agencies, and Publications Directory

Gale Research, Inc.
835 Penobscot Bldg.
Detroit, MI 48226-4094

Phone: (313) 961-2242

Approximately 24,000 organizations and publications of all kinds that are helpful in business, including trade, business, commercial, and labor associations; government agencies and advisory organizations; commodity and stock exchanges; United States and foreign diplomatic offices; regional planning and development agencies; convention, fair, and trade organizations; franchise companies; banks and savings and loans; hotel/motel systems; publishers; newspapers; information centers; computer information services; research centers; graduate schools of business; special libraries, periodicals, directories, etc. **Frequency:** Biennial, in even years. **Price:** $315.00.

★2214★ Business Week—R&D Scoreboard Issue

McGraw-Hill, Inc.
1221 Avenue of the Americas
New York, NY 10020

Phone: (212) 512-2000

★2215★ Buyouts Directory of LBO Financing Sources
Venture Economics, Inc.   Phone: (617) 449-2100
75 2nd Ave., Ste. 700
Needham, MA 02194

Over 600 sources of acquisition financing, including banks, asset-based lenders, small business investment companies, insurance companies, and venture capital firms. **Price:** $145.00, plus $5.00 shipping.

★2216★ Canadian Business—The Small Business Issue
CB Media Ltd.   Phone: (416) 364-4266
70 The Esplanade, 2nd Fl.
Toronto, ON, Canada M5E 1R2

List of government agencies offering training programs and other forms of assistance for small businesses; venture capital sources; firms offering cash for equity in small businesses; computerized databases concerned with business and management, especially small business concerns. **Frequency:** Annual, in October. **Price:** 3 Canadian dollars (current and 1988 editions).

★2217★ Catalog of Government Inventions Available for Licensing to U.S. Businesses
Center for the Utilization of Federal   Phone: (703) 487-4650
   Technology
National Technical Information Service
   (NTIS)
U.S. Dept. of Commerce
5285 Port Royal Rd.
Springfield, VA 22161

About 1,000 federal inventions, developed during the previous year, that are available for licensing by U.S. companies. **Frequency:** Annual, in Spring. **Price:** $54.00, plus $3.00 shipping (NTIS number PB91-100206).

★2218★ CDA Reporter
Commercial Development Association   Phone: (202) 659-0060
   (CDA)
1330 Conneticut Ave. NW, Ste. 300
Washington, DC 20036-1702

Covers the activities of the Association, whose members are engaged in "identifying, evaluating, and establishing profitable new businesses." Provides limited coverage of related industry events. Recurring features include news of members, a calendar of events, and the column titled President's Message. **Frequency:** Quarterly. **Price:** Included in membership.

★2219★ Chartered Institute of Patent Agents— Register of Patent Agents
Chartered Institute of Patent Agents
Staple Inn Bldgs.
London WC1V 7PZ, England

Member patent agents in the United Kingdom. **Frequency:** Annual, in March. **Price:** 2 British pounds.

★2220★ China Exchange Newsletter
Committee on Scholarly Communication   Phone: (202) 334-2718
   with the People's Republic of China
National Academy of Sciences
2101 Constitution Ave. NW
Washington, DC 20418

Covers recent developments in science, technology, and scholarship in the People's Republic of China. Discusses exchange activities with the Committee and worldwide exchange activities of China, often listing participants. Includes an extensive bibliography of recent publications on China. **Frequency:** Quarterly. **Price:** Free.

★2221★ CLEAR News
The Council of State Governments   Phone: (606) 231-1892
Iron Works Pike   Fax: (606) 231-1943
PO Box 11910
Lexington, KY 40578-1910

Concentrates on professional licensing. Addresses such issues as reciprocity, alternatives to individual licensure, sunset, antitrust, and administrative rule-making. Recurring features include news of members, conferences, committees, programs, and legislation, and a column titled State Lines. **Frequency:** Quarterly. **Price:** Included in membership; $30/yr.

★2222★ Companies and Their Brands
Gale Research Inc.   Phone: (313) 961-2242
835 Penobscot Bldg.
Detroit, MI 48226-4094

Over 42,000 companies that manufacture, distribute, import, or otherwise market consumer-oriented products. **Frequency:** Annual, in January. **Price:** $330.00.

★2223★ Computer Industry Litigation Reporter
Andrews Publications, Inc.   Phone: (215) 399-6600
PO Box 1000   Fax: (216) 399-6610
Westtown, PA 19395

Tracks significant computer law developments relating to copyright, patent and trademark infringement, alleged misappropriation of trade secrets, antitrust allegations, and user-vendor breach of contracts. Publishes complete texts of key decisions in addition to "unbiased news summaries of cases around the country. **Frequency:** Semimonthly. **Price:** $825/yr.

★2224★ Computer Software Protection Law
The Bureau of National Affairs, Inc.   Phone: (202) 452-4200
1231 25th St. NW
Washington, DC 20037

Treatise provides analysis of copyright law, trade secret, and patent protection of computer programs and databases. **Frequency:** Annual. **Subscription:** $550.

★2225★ Conference Papers Index
Cambridge Scientific Abstracts   Phone: (301) 961-6750
7200 Wisconsin Ave.
Bethesda, MD 20814

Journal covering worldwide scientific and technical conferences. **Frequency:** 6x/yr. plus annual index. **Subscription:** $935; $975 other countries.

★2226★ Contract Research Organizations in the EEC
UNIPUB   Phone: (301) 459-7666
4611-F Assembly Dr.
Lanham, MD 20706-4391

Contract research organizations in the European Economic Community. **Price:** $55.00.

★2227★ Copyright Clearance Center—Report

Copyright Clearance Center
27 Congress St.
Salem, MA 01970

Phone: (508) 744-3350
Fax: (508) 741-2318

Publicizes Center services and activities and reports general news of the copyright community. **Frequency:** Quarterly. **Price:** Included in membership; $10/yr.

★2228★ Copyright Law Reports

Commerce Clearing House, Inc.
4025 W. Peterson Ave.
Chicago, IL 60646

Phone: (312) 583-8500

Copyright law publication. **Frequency:** Monthly. **Subscription:** $430.

★2229★ Corporate Venturing News

Venture Economics, Inc.
75 2nd Ave., Ste. 700
Needham, MA 02158

Phone: (617) 449-2100

Concerned with corporate development strategies, particularly strategic alliances and business relationships such as joint ventures, research and development contracts, and licensing agreements. Lists recent deals by industry, profiles companies, and provides articles on strategic partnering, internal venturing, corporate venture capital, university-industry technology transfer, business incubators, and industry consortia. **Frequency:** 18/yr. **Price:** $345/yr.

★2230★ C3I News

Washington Defense Reports, Inc.
3918 Prosperity Ave., No. 318
Fairfax, VA 22031-3334

Phone: (703) 573-1600
Fax: (703) 573-1604

Focuses on the area of command, control, and communications intelligence (C3I). Provides information on government contracts, research and development programs, effectiveness of new systems, and related political and legislative issues. Recurring features include news of research and reports of meetings. **Frequency:** Monthly. **Price:** $128/yr.

★2231★ Decisions & Developments

Decisions & Developments Publishing
    Co.
PO Box 342
Wayland, MA 01778

Phone: (617) 890-5678

Reports on court decisions relating to intellectual property matters. **Frequency:** Bimonthly. **Price:** $84/yr.

★2232★ Development Directory

Omnigraphics, Inc.
Penobscot Bldg.
645 Griswold, 24th Fl.
Detroit, MI 48226

Phone: (313) 961-1340

Organizations, foundations, academic institutions, government agencies, professional associations, and firms pursuing research and development interests worldwide. **Price:** $110.00.

★2233★ Directory of American Research and Technology

R.R. Bowker Co.
121 Chanlon Rd.
New Providence, NJ 07974

Phone: (908) 464-6800

Over 11,000 publicly and privately owned research facilities in the U.S. and Canada. **Frequency:** Annual, in December. **Price:** $259.95 (ISSN 0886-0076).

★2234★ Directory of Associations of Inventors

World Intellectual Property Organization
34, chemin des Colombettes
CH-1211 Geneva, Switzerland

About 120 associations of inventors, worldwide, primarily national groups; includes about 35 state and national organizations in the United States. **Frequency:** Biennial, odd years. **Price:** Free.

★2235★ Directory of Intellectual Property Lawyers and Patent Agents

Clark Boardman Co. Ltd.
375 Hudson St.
New York, NY 10014

Phone: (212) 929-7500

More than 7,000 patent agents, lawyers, and law firms specializing in intellectual property law, including patents, trademarks, copyrights, unfair trade, and trade secrets. **Frequency:** Latest edition, 1989. **Price:** $145.00, cloth; $125.00, paper.

★2236★ Directory of Operating Small Business Investment Companies

Small Business Administration
1441 L St. NW, Rm. 808
Washington, DC 20416

Phone: (202) 653-6672

About 570 operating small business investment companies holding regular licenses and licenses under the section of the Small Business Investment Act covering minority enterprise SBICs. **Frequency:** Semiannually, June and December. **Price:** Free.

★2237★ Directory of Professional Representatives

European Patent Office
Erhardstrasse 27
W-8000 Munich, Germany

Representatives of European national patent offices. **Frequency:** Annual.

★2238★ Discover

Family Media, Inc.
3 Park Ave.
New York, NY 10016

Phone: (212) 779-6200

Magazine focusing on science, from astronomy to zoology. **Frequency:** Monthly. **Subscription:** $24. $2.95 single issue.

★2239★ Diversification Report

Technology Business Development
Texas Engineering Experiment Station,
    Ste. 310 WERC
College Station, TX 77843-3369

Phone: (409) 845-0538

Communicates economic development and technology transfer. **Frequency:** Quarterly. **Price:** Free.

★2240★ Drug and Device Product Approval List

National Technical Information Service
U.S. Dept. of Commerce
5285 Port Royal Rd.
Springfield, VA 22161

Phone: (703) 487-4630

Lists the most recent new drug approvals, new animal and medical devices, and licensees issued for biological products. **Frequency:** Monthly. **Price:** $100/yr.

★2241★ East/West Technology Digest

Welt Publishing Co.
1413 K St. NW, Ste. 800
Washington, DC 20005

Phone: (202) 371-0555
Fax: (202) 408-9369

Supplies information on new technology and license offers, with addresses for obtaining further information. Recurring features include news of research. **Frequency:** Monthly. **Price:** $99/yr.

★2242★ Edison Entrepreneur

Thomas Edison Program
Ohio State Dept. of Development
77 S. High St., 26th Fl.
Columbus, OH 43215

Phone: (614) 466-3086
Fax: (614) 644-5758

Covers a variety of subjects dealing with research and development throughout the state of Ohio. Recurring features include interviews and news of research. **Frequency:** Quarterly. **Price:** Free.

★2243★ Electro Manufacturing

Worldwide Videotex
Babson Pk.
PO Box 138
Boston, MA 02157

Phone: (617) 449-1603

Covers research and development of products and services in manufacturing. Includes company profiles, studies in productivity, and new technology. **Frequency:** Monthly.

★2244★ Endeavour

Pergamon Press, Inc.
Maxwell House
Fairview Park
Elmsford, NY 10523

Phone: (914) 592-7700

Magazine serving as a review of science and technology. **Frequency:** 4x/yr. **Subscription:** $90

★2245★ The Entrepreneurial Economy Review

Corporation for Enterprise Development   Phone: (202) 408-9788
777 N. Capitol NE, Ste. 801
Washington, DC 20002

Reviews enterprise support, economic development, and economic opportunity policies and programs. Recurring features include letters to the editor, news of research, book reviews, and columns titled Perspective, Initiative, In Sight, and Of Note. **Frequency:** 4/yr. **Price:** $78/yr.

★2246★ Entrepreneurial Manager's Newsletter

Center for Entrepreneurial Management,   Phone: (212) 633-0060
  Inc.   Fax: (202) 627-9247
180 Varick St., Penthouse
New York, NY 10014

"Designed to provide accurate and authoritative information relative to subjects of concern to entrepreneurial managers." Covers management, taxes, finance, marketing, information sources, and educational programs. Recurring features include news of seminars, book reviews, news of research and survey results, and columns titled Entrepreneurs's Hall of Fame, Mind Your Own Business, Resources, Accounting, Personal, Miscellaneous, and The Business Exchange. **Frequency:** Monthly. **Price:** Included in membership; $71/yr.

★2247★ Eureka! The Canadian Inventor's Newsletter

Canadian Industrial Innovation   Phone: (519) 885-5870
  Centre/Waterloo   Fax: (519) 885-5729
156 Columbia St. W.
Waterloo, ON, Canada N2L 3L3

Serves as a forum for Canadian inventors and innovators. **Frequency:** Quarterly. **Price:** Free.

★2248★ European Sources of Scientific and Technical Information

Longman Group Ltd.
Westgate House
The High
Harlow, Essex CM20 1YR, England

Over 1,500 patents and standards offices, national offices of information, and organizations active in scientific fields in Europe, including Soviet bloc nations. Provides English-language version of foreign terminology. **Frequency:** Biennial, in even years. **Price:** $217.00.

★2249★ European Venture Capital Association— Yearbook

Peat Marwick McLintock
1 Puddle Dock
Blackfriars
London EC4V 3PD, England

About 175 member companies that provide venture capital funding and services; 10 national venture capital associations in Europe. **Frequency:** Annual, May/June. **Price:** 20 British pounds.

★2250★ Factory Automation News

Market Intelligence Research Co.
2525 Charleston Rd.
Mountain View, CA 94043

Phone: (415) 961-9000
Fax: (415) 961-5042

Reports on activities in the industrial automation field. Concentrates on market strategies, international joint ventures, licensing, and news of research and development. Includes company profiles. **Frequency:** Monthly.

★2251★ Federal Research in Progress (FEDRIP)

Office of Product Management   Phone: (703) 487-4929
National Technical Information Service
  (NTIS)
U.S. Dept. of Commerce
5285 Port Royal Rd.
Springfield, VA 22161

Database of more than 138,000 federally-funded research projects in progress in the physical, engineering, health, agriculture, and life science areas.

★2252★ Federal Research Report

Business Publishers, Inc.   Phone: (301) 587-6300
951 Pershing Dr.
Silver Spring, MD 20910

Provides information on research and development funds available from federal agencies and bureaus or associations that provide support money for research and development. Lists items in categories including environment/energy, transportation, medicine/health, education, and social sciences. **Frequency:** Weekly. **Price:** $160.

★2253★ Foreign Technology: An Abstract Newsletter

National Technical Information Service (NTIS)
U.S. Dept. of Commerce
5285 Port Royal Rd.
Springfield, VA 22161

Phone: (703) 487-4630

Carries abstracts of reports on the field of foreign technology. Covers biomedical technology; civil, construction, structural, and building engineering; communications; computer, electro, and optical technology; energy, manufacturing, and industrial engineering; and physical and materials sciences. Recurring features include notices of publications available and a form for ordering reports from NTIS. **Frequency:** Weekly. **Price:** $135/yr.

★2254★ Gorman's New Product News

Gorman Publishing Co.
8750 W. Bryn Mawr Ave.
Chicago, IL 60631

Phone: (312) 693-3200
Fax: (312) 693-0568

Reports on new consumer product introductions in the U.S. and abroad that will be sold in food and drug stores. Gives a brief description of each product and lists the manufacturer. **Frequency:** Monthly. **Price:** $345/yr.

★2255★ Government Inventions for Licensing: An Abstract Newsletter

National Technical Information Service (NTIS)
U.S. Dept. of Commerce
5285 Port Royal Rd.
Springfield, VA 22161

Phone: (703) 487-4630

Abstracts reports on mechanical devices and equipment and other government inventions in chemistry, nuclear technology, biology and medicine, metallurgy, and electrotechnology, as well as optics and lasers, patent applications, and miscellaneous instruments. Recurring features include a form for ordering reports from NTIS. **Frequency:** Weekly. **Price:** $235/yr.

★2256★ Handbook of Business Finance and Capital Sources

AMACOM Books
American Management Association
135 W. 50th St.
New York, NY 10020

Phone: (212) 903-8089

Lists of several hundred private sources of funds for business, including investment companies, venture capital firms, leasing companies, insurance firms, commercial and industrial (Morris Plan) banks, real estate investment trusts, savings and loan associations, and about 160 federal and state government sources. **Frequency:** Irregular; latest edition, 1985. **Price:** $95.00.

★2257★ Healthcare Technology & Business Opportunities

Biomedical Business International, Inc.
1524 Brookhollow Dr.
Santa Ana, CA 92706

Phone: (714) 755-5757
Fax: (714) 755-5704

Provides information on new technologies available for license or transfer, distributors seeking products, and suppliers seeking distributors. Reports on joint ventures, acquisitions sought, strategic partnering, product and corporate divestitures, and product and business development services. Recurring features include listings of U.S., European, and Japanese patents and thier applications and columns titled New Technology, Business Opportunities, Patents, and Information Resources. **Frequency:** Monthly. **Price:** $325/yr.

★2258★ Hi-Tech Alert

Communication Research Associates, Inc.
10606 Mantz Rd.
Silver Spring, MD 20903

Phone: (301) 445-3230

Provides "fresh, timely, usable news of hi-tech developments in plain English." Monitors the areas of electronic mail, office automation, desktop publishing, computer-aided research, online information services, and personal computer applications. Recurring features include notices of publications available and news of educational opportunities. **Frequency:** Monthly. **Price:** $98/yr.

★2259★ High-Tech Materials Alert

Technical Insights, Inc.
PO Box 1304
Ft. Lee, NJ 07024-9967

Phone: (201) 568-4744
Fax: (201) 568-8247

Provides details of developments in high-performance materials, ranging from alloys and metallic whiskers and ceramic and graphite fibers, their fabrication, and industrial applications. Recurring features include news of research, listings of new patents, and a calendar of events. **Frequency:** Monthly. **Price:** $467/yr.

★2260★ High Technology Business

Infotechnology Publishing Corp.
9990 Lee Hwy., No. 301
Fairfax, VA 22030-1720

Magazine containing information on current trends and developments in the full range of high technology industries. Material is directed toward business people in the high technology field. **Frequency:** Monthly. **Subscription:** Free to qualified subscribers; $60. $5 single issue.

★2261★ InKnowVation Newsletter

Innovation Development Institute
45 Beach Bluff Ave., Ste. 300
Swampscott, MA 01907

Phone: (617) 595-2920

Description: Focuses on SBIR (Small Business Innovative Research) program opportunities for small businesses entering early-stage, high-risk research and development ventures. Discusses how SBIR programs can be used to obtain initial funding or to test promising ideas.

★2262★ Innovation

A.G. Publications Ltd.
PO Box 7422
31070 Haifa, Israel

★2263★ Innovation News

Aremco Products, Inc.
PO Box 429
Ossining, NY 10562-0429

Phone: (914) 762-0685
Fax: (914) 762-1663

Features technical research, updates on process equipment, and photographs of ceramic materials. Covers high temperature design, including process information and case histories; and describes equipment used in ceramic processing. **Frequency:** Biennially. **Price:** Free.

★2264★ Innovation Showcase

Canadian Industrial Innovation Centre/Waterloo
156 Columbia St. W.
Waterloo, ON, Canada N2L 3L3

Phone: (519) 885-5870
Fax: (519) 885-5279

Focuses on corporate development activities, successful inventors, and other topics of interest. **Frequency:** Quarterly. **Price:** Free.

★2265★ International Invention Register

Catalyst   Phone: (619) 723-8064
Box 547
Fallbrook, CA 92028

An advertising newspaper devoted to classified ads describing inventions for sale or license; published approximately quarterly, but may vary to coincide with certain international trade shows; a typical issue may include 2,000 patents. **Frequency:** Latest edition, April 1990. **Price:** $18.00 per year.

★2266★ International New Product Newsletter

Transcommunications International, Inc.   Phone: (617) 887-0100
Box 1146   Fax: (617) 639-2623
Marblehead, MA 01945

Provides "advance news of new products and processes, primarily from sources outside the U.S." Emphasizes new products which can cut costs and improve efficiency. Recurring features include the column Special Licensing Opportunities which lists new products and processes that are available for manufacture under license, or are for sale or import. **Frequency:** Monthly. **Price:** $175/yr.

★2267★ International Product Alert

Marketing Intelligence Service, Ltd.   Phone: (716) 374-6326
33 Academy St.   Fax: (716) 374-5217
Naples, NY 14512

Provides concise reports on new products in 18 countries outside of the U.S. and Canada. Covers foods, beverages, non-prescription drugs, cosmetics, toiletries, pet products, and miscellaneous household items. Also lists products that are extensions of existing lines and lists packaging changes. Recurring features include occasional copies of advertising. **Frequency:** Semimonthly. **Price:** $600/yr.

★2268★ Invent!

Mindsight Publishing   Phone: (805) 388-3097
3201 Corte Malpaso, Ste. 304
Camarillo, CA 93010

★2269★ Inventor Entrepreneur Network

Edward Zimmer
6175 Jackson Rd.
Ann Arbor, MI 48103

Publishes stories and ideas for inventions. Features essayss on raising money, selling, and benefits of licensing. Provides information on contacts for technical and business services and new product brokers. Includes news of conferences, a calendar of events, and a list of new patents. **Frequency:** Monthly. **Price:** Free.

★2270★ Inventors' Digest

Affiliated Inventors Foundation, Inc.   Phone: (719) 635-1234
2132 E. Bijou St.
Colorado Springs, CO 80909-5950

Magazine covering the invention process including development and marketing. **Frequency:** 6x/yr. **Subscription:** $20/yr. U.S.; $25/yr. Canada; $38/yr. overseas.

★2271★ Inventor's Gazette

Inventors Association of America   Phone: (714) 980-6446
PO Box 1531
Rancho Cucamonga, CA 91730

Provides information on inventions, inventors, and patents. Recurring features include letters to the editor, news of members, and humorous quotes. **Frequency:** Monthly. **Price:** $24/yr.

★2272★ Inventors News

Inventors Clubs of America, Inc.   Phone: (404) 938-5089
PO Box 450261
Atlanta, GA 30345

Designed to keep inventors abreast with the latest information on patents and trademarks. Carries news of the Club and information on exhibits and inventions. Recurring features include editorials, news of research, letters to the editor, news of members, a calendar of events, and a column titled Patents for Sale. **Frequency:** Bimonthly. **Price:** Included in membership.

★2273★ IVCI Directory of Venture Capital Clubs

International Venture Capital Institute,   Phone: (203) 323-3143
   Inc. (IVCI)
Baxter Associates, Inc.
PO Box 1333
Stamford, CT 06904

About 115 venture capital clubs; international coverage. **Frequency:** Semiannual, spring and fall. **Price:** $9.95 per issue.

★2274★ IVCI Venture Capital Digest

International Venture Capital Institute   Phone: (203) 323-3143
   (IVCI)
Box 1333
Stamford, CT 06904

Lists notices of individuals or firms seeking joint venture deals or with products or services of interest to those who are. **Frequency:** Quarterly. **Price:** $40.00.

★2275★ Japan Consumer Electronics Scan

Kyodo News International, Inc.   Phone: (212) 397-3723
50 Rockefeller Plaza, Ste. 815   Fax: (212) 307-1532
New York, NY 10020

Reports on new consumer products made available by Japanese companies. Includes information on price, target markets, and availability. **Frequency:** Weekly.

★2276★ The Journal of Proprietary Rights

Law & Business, Inc.   Phone: (201) 894-8538
Prentice-Hall, Inc.
910 Sylvan Ave.
Englewood Cliffs, NJ 07632

Covers trends involving patent, trade secret, trademark, and intellectual property protection issues, including practical solutions. **Frequency:** Monthly. **Price:** $275/yr.

★2277★ Journal of the Copyright Society of the U.S.A.

Copyright Society of the U.S.A.   Phone: (212) 998-6194
c/o New York University Law School
40 Washington Sq. S.
New York, NY 10012

Journal focusing on copyright law. **Frequency:** Quarterly. **Subscription:** $125.

★2278★ Knowledge in Policy

Transaction Periodicals Consortium   Phone: (201) 932-2280
Rutgers - The State University
Dept. 3089
New Brunswick, NJ 08903

International journal devoted to the development of an interdisciplinary science of knowledge transfer. **Frequency:** Quarterly. **Subscription:** $30; $60 institutions; $45 other countries.

★2279★ Les Nouvelles

Licensing Executives Society   Phone: (216) 771-2600
    International
1444 W. 10th St., No. 403
Cleveland, OH 44113

Concerned with licensing and related subjects. Covers technology, patents, trade marks, and licensing "know-how" world-wide. **Frequency:** Quarterly. **Price:** Included in membership.

★2280★ Licensing Law and Business Report

Clark Boardman Company, Ltd.   Phone: (212) 929-7500
375 Hudson St.   Fax: (212) 924-0460
New York, NY 10014

Focuses on a specific area within the licensing field in each issue. Provides analysis of court cases and recent developments affecting the design of licensing agreements worldwide, including such topics as antitrust law, tax considerations, and technology management consulting. Recurring features include an annual table of cases. **Frequency:** 6/yr. **Price:** $125/yr.

★2281★ The Licensing Letter

EPM Communications Inc.   Phone: (718) 469-9330
488 E. 18th St.   Fax: (718) 469-7124
Brooklyn, NY 11226

Concerned with all aspects of licensed merchandising, "the business of associating someone's name, likeness or creation with someone else's product or service, for a consideration." Recurring features include statistics, research, events, mechanics, available properties, and identification of licensors, licensing accounts, and licensees. **Frequency:** Monthly. **Price:** $235/yr.

★2282★ Manufacturing Technology: An Abstract Newsletter

National Technical Information Service   Phone: (703) 487-4630
    (NTIS)
U.S. Dept. of Commerce
5285 Port Royal Rd.
Springfield, VA 22161

Reports on Computer Aided Design, Computer Aided Manufacturing, technology transfer, and other matters related to manufacturing technology. Also provided information on subjects such as planning, marketing and economics, and research program administration. **Frequency:** Weekly. **Price:** $135/yr.

★2283★ Martindale-Hubbell Law Directory

Martindale-Hubbell, Inc.   Phone: (201) 464-6800
Reed Publishing (USA) Inc.
Box 1001
Summit, NJ 07902-1001

Lawyers and law firms in the United States and its possessions, Canada, and abroad; includes a biographical section by firm, and a separate list of patent lawyers and attorneys in government service. **Frequency:** Annual, in January. **Price:** $305.00, plus $32.00 shipping.

★2284★ Mass High Tech

Mass Tech Times, Inc.   Phone: (617) 924-5100
755 Mt. Auburn St.
Watertown, MA 02172

High technology business newspaper. **Frequency:** 2x/wk. **Subscription:** $28; $50 two years.

★2285★ Medical Device Patents Letter

Washington Business Information, Inc.   Phone: (703) 247-3424
1117 N. 19th St., Ste. 200   Fax: (703) 247-3421
Arlington, VA 22209-1798

Monitors patent status of new medical devices in the U.S. and worldwide. Reports new licenses granted and new product options. **Frequency:** Monthly. **Price:** $537.

★2286★ Merchant & Gould Computer Law Newsletter

Merchant, Gould, Smith, Edell, Welter &   Phone: (612) 336-4623
    Schmidt   Fax: (612) 332-9081
3100 Norwest Center
90 S. 7th St.
Minneapolis, MN 55402

Covers developments and provides information on computer law and intellectual property. **Frequency:** Semiannually. **Price:** Free.

★2287★ National Venture Capital Association— Membership Directory

National Venture Capital Association   Phone: (703) 528-4370
1655 N. Fort Myer Dr., Ste. 700
Arlington, VA 22209

Over 220 venture capital firms, including subsidiaries of banks and insurance companies. **Frequency:** Annual, in November. **Price:** Free; send self-addressed, business envelope, stamped with $1.85 postage.

★2288★ NCPLA Newsletter

National Council of Patent Law   Phone: (617) 227-7400
    Associations (NCPLA)
Lahive & Cockfield
60 State St., Ste. 510
Boston, MA 02109

Provides legal information concerning the patent, trademark, and copyright fields and news of the member state and local associations. **Frequency:** Quarterly. **Price:** $50/yr.

★2289★ NCST Quarterly Briefing

National Coalition for Science &   Phone: (202) 833-2322
    Technology (NCST)
1155 21st St. NW, Ste. 1000
Washington, DC 20036

Concentrates on the political activities of the National Coalition for Science & Technology regarding items such as animal rights, technology transfer, technology innovation, and other areas in which science policy affects society. Recurring features include book reviews, a calendar of events, and columns titled The Chairman's Column, The Director's Column, and Nest Action Alert. **Frequency:** Quarterly. **Price:** Included in membership; $30/yr.

★2290★ New From Europe

Prestwick Publications, Inc.   Phone: (305) 427-2924
390 N. Federal Hwy., No. 401
Deerfield Beach, FL 33441

Contains market forecasts, trends, and descriptions of new products and technologies from Europe. Descriptions include the developer's name and address and an explanation of why the new product is superior to existing products or processes. Provides an overview of the European economy, its research and new product emphasis, and governmental actions that will affect future market activity. Recurring features include news of research. **Frequency:** Monthly. **Price:** $275/yr.

★2291★   New From Japan

Prestwick Publications, Inc.   Phone: (305) 427-2924
390 N. Federal Hwy., No. 401
Deerfield Beach, FL 33441

Describes new Japanese products and technologies and explains why they are superior to existing products or processes. Covers consumer products, energy conserving processes and products, manufacturing methods, and electronic products. Recurring features include news of research. **Frequency:** Monthly. **Price:** $275/yr.

★2292★   New From U.S.

Prestwick Publications, Inc.   Phone: (305) 427-2924
390 N. Federal Hwy., No. 401
Deerfield Beach, FL 33441

Describes new products and technologies researched and developed in the U.S. Examines a single product and its use and applications in depth in each issue. **Frequency:** Monthly. **Price:** $275/yr.

★2293★   New Mexico R&D Forum

New Mexico Research & Development   Phone: (505) 277-3661
    Institute   Fax: (505) 277-8463
Research & Development
    Communications Office
University of New Mexico
457 Washington Ave. SE, Ste. M
Albuquerque, NM 87108

Concerned with developments relating to technology in New Mexico. Reports on new projects being funded by the Institute, profiles technology firms and laboratories in New Mexico, and contains news briefs on pertinent legislation. Recurring features include information on workshops and useful publications, news of research, book reviews, and a calendar of events. **Frequency:** Monthly. **Price:** Free.

★2294★   New Product Development

Point Publishing Company, Inc.   Phone: (201) 295-8258
PO Box 1309
Point Pleasant, NJ 08742

Concentrates on issues relating to new product research and development, idea generation, marketing, distribution, design, and other aspects of product development within national and international companies. Recurring features include interviews, news of research, reports of meetings, book reviews, and a calendar of events. **Frequency:** Monthly. **Price:** $75/yr.

★2295★   New Technology Week

King Communications Group, Inc.   Phone: (202) 638-4260
627 National Press Bldg.
Washington, DC 20045

Carries news on evolving technologies, especially those in defense-related fields. Follows legislation and government agency action affecting defense and high-tech industries. Lists recipients of foundation and research grants in the U.S. Recurring features include a calendar of events and news of employment opportunities. **Frequency:** Weekly. **Price:** $495/yr.

★2296★   Nurse-Entrepreneur's Exchange

David Norris   Phone: (707) 763-6021
47 6th St.   Fax: (707) 763-1706
Petaluma, CA 94952

Provides news and information of interest to nurse entrepreneurs. **Frequency:** Bimonthly. **Price:** $45/yr.

★2297★   Official Gazette of the United States Patent and Trademark Office: Patents

U.S. Patent & Trademark Office   Phone: (703) 557-3158
U.S. Dept. of Commerce   Fax: (202) 275-0019
Washington, DC 20231

Persons and corporations granted patents, as part of the description of newly issued patents; also includes patents available for license or sale. **Frequency:** Weekly. **Price:** $28.00 per issue; $593.00 per year (S/N 703-033-00000-8; ISSN 0098-1133). **Send Orders To:** Superintendent of Documents, U.S. Government Printing Office, Washington, DC 20402 (202-783-3238).

★2298★   Official Gazette of the United States Patent and Trademark Office: Trademarks

U.S. Patent & Trademark Office   Phone: (703) 557-3158
U.S. Dept. of Commerce
Washington, DC 20231

List of persons and corporations recently granted trademarks. **Frequency:** Weekly. **Price:** $9.00 per issue; $312.00 per year (S/N 703-034-00000-4). **Send Orders To:** Superintendent of Documents, U.S. Government Printing Office, Washington, DC 20402 (202-783-3238).

★2299★   Online Patents and Tradenames Databases

ASLIB (Association for Information Management)
Information House
20-24 Old St.
London EC1V 9AP, England

More than 35 databases concerned with patents information in 15 countries, and about 20 hosts and producers. **Frequency:** Biennial, even years. **Price:** 21 British pounds.

★2300★   Patenting and Marketing Your Invention

Van Nostrand Reinhold Company, Inc.   Phone: (212) 254-3232
115 5th Ave.   Fax: (212) 254-9499
New York, NY 10003

List of patent libraries and United States government bookstores with information on patents. **Frequency:** Published 1986. **Price:** $39.95.

★2301★   Patents

Washington, DC 20231   Phone: (703) 557-3341

Publication reporting on patents. **Frequency:** Weekly. **Subscription:** $347; $433.75 other countries.

★2302★   Product Alert

Marketing Intelligence Service, Ltd.   Phone: (716) 374-6326
33 Academy St.   Fax: (716) 374-5217
Naples, NY 14512

Reports on new consumer goods launched in American retailing, including foods and beverages, non-prescription drugs, cosmetics and toiletries, and miscellaneous household items. Lists products that are an extension of an existing product line, package changes, and marketing plans. Recurring features include pictures as well as descriptions of the products. **Frequency:** Weekly. **Price:** $600/yr.

★2303★   Research and Development—R&D Magazine

Cahners Publishing Co.   Phone: (708) 635-8800
1350 E. Touhy Ave.
PO Box 5080
Des Plaines, IL 60017-5080

Magazine serving research scientists, engineers, and technical managers. Reports significant advances, problems, and trends that affect

the performance, funding, and administration of research. **Frequency:** Monthly. **Subscription:** Controlled; $45 others.

★2304★ **Research Centers Directory**
Gale Research Inc.   Phone: (313) 961-2242
835 Penobscot Bldg.   Fax: (313) 961-6083
Detroit, MI 48226-4094

Over 12,000 university-related and other nonprofit research organizations which are established on a permanent basis and carry on continuing research programs in all areas of study; includes research institutes, laboratories, experiment stations, computing centers, research parks, technology transfer centers, and other facilities and activities; coverage includes Canada. **Frequency:** Annual. **Price:** Base edition, $400.00; supplement, $260.00.

★2305★ **Research Money**
Evert Communications, Ltd.   Phone: (613) 728-4621
982 Wellington St.   Fax: (613) 728-0385
Ottawa, ON, Canada K1Y 2X8

Supplies reports and analyses of the forces driving science and technology investment in Canada, with special emphasis on government policies, granting programs, and other incentives for industry and universities. Tracks major expenditures on research and development and highlights areas where research monies are available. Recurring features include interviews, news of research, reports of meetings, news of educational opportunities, and a calendar of events. **Frequency:** 20/yr. **Price:** $275/yr.

★2306★ **Research Services Directory**
Gale Research Inc.   Phone: (313) 961-2242
835 Penobscot Bldg.   Fax: (313) 961-6083
Detroit, MI 48226

Approximately 4,170 commercial laboratories, consultants, firms, data collection and analysis centers, individuals, and facilities in the private sector that conduct contract research in all areas of business, government, humanities, social science, and science and technology. **Price:** $290.00.

★2307★ **Research-Technology Management**
Industrial Research Institute, Inc.   Phone: (202) 872-6350
1550 M. St. NW
Washington, DC 20005

Magazine for managers of research, development, and technology implementation. **Frequency:** 6x/yr. **Subscription:** $40; $70 institutions, libraries, and corporations. $15 single issue.

★2308★ **SBIC Directory and Handbook of Small Business Finance**
International Wealth Success, Inc.   Phone: (516) 766-5850
24 Canterbury Rd.
Rockville Centre, NY 11570

Over 400 small business investment companies (SBIC's) that lend money for periods from 5 to 20 years to small businesses. **Frequency:** Annual, January. **Price:** $15.00, payment with order.

★2309★ **Science & Technology Serial Reports: Europe and Latin America**
Joint Publications Research Service   Phone: (703) 487-4650
  (JPRS)
National Technical Information Service
  (NTIS)
Springfield, VA 22162

★2310★ **Science & Technology Serial Reports: Japan**
Joint Publications Research Service   Phone: (703) 487-4650
  (JPRS)
National Technical Information Service
  (NTIS)
Springfield, VA 22162

★2311★ **Small Business Guide to Federal R&D Funding Opportunities**
Office of Small Business Research &   Phone: (202) 357-7464
  Development
National Science Foundation
1800 G St. NW, Rm. 1250
Washington, DC 20550

Federal agencies and their major components with significant research and development programs. **Price:** Free (S/N 038-000-00522-7). **Send Orders To:** Superintendent of Documents, U.S. Government Printing Office, Washington, DC 20402 (202-783-3238).

★2312★ **The Software Law Bulletin**
7721 Kalohelani Pl.
Honolulu, HI 96825

Provides coverage of software-related legal issues. Includes review and analysis of the different aspects of software law, including new cases (both reported and unreported), law review articles, state regulations and statuses, federal government documents, and international issues. **Frequency:** Monthly.

★2313★ **Southeast Asia High Tech Review**
Mead Ventures, Inc.   Phone: (602) 234-0044
PO Box 44952
Phoenix, AZ 85064

Covers investment in high technology in southeast Asian countries. Focuses on news of developments in the electronic products industry in Taiwan, Indonesia, Hong Kong, Thailand, Singapore, Phillipines, and Malaysia. **Frequency:** Monthly.

★2314★ **Soviet Technology Alert**
Elsevier Science Publishing Company, Inc.
655 Avenue of the Americas
New York, NY 10010

Covers news of research and development in technology in Eastern Europe, particularly the Soviet Union. **Frequency:** Monthly.

★2315★ **Status Report of the Energy-Related Inventions Program**
Office of Energy-Related Inventions   Phone: (301) 975-5500
National Institute of Standards &
  Technology
U.S. Dept. of Commerce
Gaithersburg, MD 20899

Inventors of items recommended for possible Department of Energy support. **Frequency:** Annual, in October. **Send Orders To:** National Technical Information Service, Springfield, VA 22161 (703-487-4780).

★2316★ *Superconductor Week*
Atlantic Information Services, Inc.   Phone: (202) 775-9008
1050 17th St. NW, No. 480   Fax: (202) 331-9542
Washington, DC 20036

Tracks news of the semiconductor industry. Covers companies involved in research and new product development, funding and financial support, and innovations in use. **Frequency:** Weekly.

★2317★ *Tech Notes*
National Technical Information Service   Phone: (703) 487-4630
   (NTIS)
U.S. Dept. of Commerce
5285 Port Royal Rd.
Springfield, VA 22161

Presents fact sheets on recently developed U.S. government laboratory technology, selected as having potential commercial or practical application. **Frequency:** Monthly. **Price:** $175/yr.

★2318★ *Technology Access Report*
University R&D Opportunities   Phone: (415) 507-0190
7 Mt. Lassen Dr., Ste. D251   Fax: (415) 507-0661
San Rafael, CA 94903

Covers news and analysis of research and development, resources, contracts and funding, joint ventures, licensing, and other information regarding technology, transfer and commercialization processes, and management. **Price:** $447/yr.

★2319★ *Technology Forecasts and Technology Surveys*
PWG Publications   Phone: (213) 273-3486
205 S. Beverly Dr., Ste. 208
Beverly Hills, CA 90212

Covers new developments in advanced technology and predicts future trends in areas such as sales volumes, consumer demand, new technological advances, and developments in the methodology for forecasting future trends. Concerned with a range of technologies, including electronics, computers, medical technology, chemicals, pulp and paper, food, and materials. **Frequency:** Monthly. **Price:** $144/yr.

★2320★ *Technology Transfer Highlights*
Argonne National Laboratory   Phone: (708) 972-6393
   Technology Transfer Center   Fax: (708) 972-5230
9700 S. Cass Ave.
Argonne, IL 60439

Provides information on federal technology; state and regional technology transfer; and small business transfer and commercialization. **Frequency:** Quarterly. **Price:** Free.

★2321★ *Technology Transfer Society—Newsletter*
Technology Transfer Society
611 N. Capitol Ave.
Indianapolis, IN 46204

Serves the communication and information needs of technology transfer professionals and policy makers. Focuses on issues and methodologies critical to the effective transfer of technology. **Frequency:** 12/yr. **Price:** Included in membership.

★2322★ *Technology Update*
Predicasts, Inc.   Phone: (216) 795-3000
11001 Cedar Ave.
Cleveland, OH 44106

Compiles "technology news abstracted from more than 1000 industry and trade journals, government reports, research studies and other documents." Covers technical management, agriculture, chemistry, health and medicine, energy, engineering, transportation, communications, environment, lifestyle and leisure, and education. **Frequency:** Weekly. **Price:** $200/yr.

★2323★ *TOWERS Club, U.S.A.—Newsletter*
TOWERS Club, U.S.A.   Phone: (206) 574-3084
PO Box 2038
Vancouver, WA 98668-2038

Intended for freelance writers, publishers, entrepreneurs, and those engaged in marketing their own creative efforts. Provides an exchange of news, quotes, and clippings. Recurring features include columns titled News, Tips, and Sources, Readin' Jerry's Mail, and M/O Mini Clinic. **Frequency:** Monthly, except May and December. **Price:** $60/yr.

★2324★ *Trademark Design Register*
Trademark Register   Phone: (202) 662-1233
National Press Bldg., No. 1297
Washington, DC 20045

Owners of over 15,000 registered logos, symbols, and design trademarks. **Frequency:** Annual, in May. **Price:** $267.00; payment must accompany order.

★2325★ *Trademark Report*
Tobacco Merchants Association of the   Phone: (609) 275-4900
   U.S., Inc. (TMA)   Fax: (609) 275-8379
PO Box 8019
Princeton, NJ 08543

Tracks tobacco product and tobacco accessory trademarks and brand names from test markets through registration. Covers renewals and cancellations. **Frequency:** Monthly.

★2326★ *The Trademark Reporter*
U.S. Trademark Assn.   Phone: (212) 986-5880
6 E. 45th St.
New York, NY 10017

Legal journal reporting on trademarks. **Frequency:** 6x/yr. **Subscription:** $80 members.

★2327★ *Trademark World—Directory Section*
Intellectual Property Publishing Ltd.
2 Parkside
Ravenscourt Park
London, England

List of international trademark services and professionals in private practice and industry. **Frequency:** Ten issues per year. **Price:** $375.00 annual subscription, postpaid.

★2328★ *Trademarks*
U.S. Patent & Trademark Office   Phone: (703) 557-3341
Dept. of Commerce
Washington, DC 20231

Publication reporting on trademarks. **Frequency:** Weekly. **Subscription:** $302; $377.50 other countries.

★2329★ TRADEMARKSCAN—FEDERAL
Thomson & Thomson   Phone: (617) 479-1600
International Thomson Information, Inc.   Fax: (617) 786-8273
500 Victory Rd.
North Quincy, MA 02171

Database of more than 1,000,000 federal trademark registrations and applications filed with the U.S. Patent and Trademark Office.

★2330★ TRADEMARKSCAN—STATE
Thomson & Thomson   Phone: (617) 479-1600
International Thomson Information, Inc.   Fax: (617) 786-8273
500 Victory Rd.
North Quincy, MA 02171

Database of more than 800,000 trademarks registered with the Secretaries of State in the U.S. and Puerto Rico.

★2331★ United States Patents Quarterly
Bureau of National Affairs, Inc. (BNA)   Phone: (202) 452-4200
1231 25th St. NW   Fax: (202) 822-8092
Washington, DC 20037

Reports decisions dealing with patents, trademarks, copyrights, unfair competition, trade secrets, and computer chip protection. **Frequency:** Weekly. **Price:** $928/yr.

★2332★ University R&D
University R&D, Inc.   Phone: (512) 346-2372
PO Box 9802-677
Austin, TX 78766

Contains abstracts of university research needing sponsorship and technology for license. **Price:** $125/yr.

★2333★ USSR Technology Update
Delphic Associates   Phone: (703) 556-0278
c/o Mary Heslin
7700 Leesburg Pike, No. 250
Falls Church, VA 22043

Provides current information on Soviet activities in trade and technology. Presents articles on topics "ranging from fifth generation computer research to industrial automation." Compiles information from Soviet scientific and technical journals in areas including lasers, energy technology, fiber optics, low temperature physics, and computer technology. Also lists U.S. patents granted to Soviet and East European countries. **Frequency:** Biweekly. **Price:** $400/yr.

★2334★ Venture Capital Directory
Forum Publishing Co.   Phone: (516) 754-5000
383 E. Main St.
Centerport, NY 11721

Over 400 members of the Small Business Administration and the Small Business Investment Company that provide funding for small and minority businesses. **Frequency:** Annual. **Price:** $9.95.

★2335★ Venture Capital: Where to Find It
National Association of Small Business   Phone: (202) 833-8230
   Investment Companies (NASBIC)
1156 15th St. NW, Ste. 1101
Washington, DC 20005

About 400 member firms licensed as small business investment companies (SBICs) under the Small Business Investment Act of 1958; associate and sustaining members who are non-SBIC investors in small

businesses or suppliers of services are included. **Frequency:** Biennial. **Price:** $5.00.

★2336★ Venture—Venture Capital 100 Issue
Venture Magazine, Inc.   Phone: (212) 682-7373
801 2nd Ave., 14th Fl.
New York, NY 10017

List of 100 largest venture capital investors ranked by dollars invested directly in new ventures; includes venture capital private partnerships, business development companies, small business investment companies (SBICs), family trust managers, and venture capital divisions of corporations and institutions. **Frequency:** Annual, summer. **Price:** $3.00.

★2337★ Who's Who in Venture Capital
John Wiley & Sons, Inc.   Phone: (212) 850-6331
605 3rd Ave.
New York, NY 10158

Companies employing about 650 individuals involved in investment and venture capital. **Price:** $29.95.

★2338★ World Directory of Sources of Patent Information
World Intellectual Property Organization
34, chemin des Colombettes
CH-1211 Geneva, Switzerland

About 50 information centers with collections of patent documents, periodicals specializing in patent related matters, and offering other information services concerning technological, legal, and economic data on industrial property licensing and technology transfer. **Frequency:** Latest edition, February 1991. **Price:** 100 Swiss francs.

★2339★ World Electronic Developments
Prestwick Publications, Inc.   Phone: (305) 427-2924
390 N. Federal Hwy., No. 401
Deerfield Beach, FL 33441

Contains forecasts, trends, and descriptions of new developments in electronics in western Europe, Japan, and the United States. Covers automation and robotics, communications and information processing, computer processing and computer-aided design, and circuits and electronics. Recurring features include statistics and news of research. **Frequency:** Monthly. **Price:** $195/yr.

★2340★ World Intellectual Property Report
BNA International, Inc.
The Bureau of National Affairs, Inc.
17 Dartmouth St.
London SW1H 9BL, England

★2341★ World Patent Information
Pergamon Press, Inc.   Phone: (914) 592-7700
Maxwell House
Fairview Park
Elmsford, NY 10523

Journal serving as worldwide forum for the exchange of information among professionals in the patent information and documentation field. **Frequency:** 4x/yr. **Subscription:** $185.

★2342★ *World Technology/Patent Licensing Gazette*
Techni Research Associates, Inc.   Phone: (215) 657-1753
Willow Grove Plaza
Willow Grove, PA 19090

Leading firms, private and government research laboratories, universities, inventors, consultants, and others that have new products, new process developments, and new technologies available for license or acquisition; also lists related seminars and meetings. **Frequency:** Bimonthly. **Price:** $136.00 per year.

Section 24

Project XL: Youth Innovation
Programs and Publications

*"If the world didn't have inventors, everything would be an unsolved
mystery."—Bettie Levy, Sixth Grader*

It is a widely shared opinion that Americans are not as imaginative and inventive as they once were. Such opinions can be debated, but virtually everyone would agree that more and more we are becoming a nation of buyers rather than suppliers of new technology. How we are to survive in the future is likely to depend upon America's most precious natural resource: our schoolchildren.

Project XL is an outreach program of the Patent and Trademark Office (PTO) and an integral part of the U.S. Department of Commerce's Private Sector Initiative Program. Designed to encourage the development of inventive thinking and problem-solving skills among America's youth, the Project's principal focus is promoting educational programs that teach critical and creative thinking, and fostering national proliferation of such programs. The overall objective of Project XL, which was initiated by Donald J. Quigg, Assistant Secretary, Commissioner of Patents and Trademarks, is to ensure the nation's position as a world technological leader in the next century and to guarantee that Americans will have the innovative skills to meet the challenges of an increasingly competitive world.

"I believe that the schools of this great nation are filled with Edisons, Wrights, Marconis, Whitneys, and Bells, along with other potential thinkers who can change the world," says Commissioner Quigg. "The very least we can do is to help them realize their potential—to nurture those young people who will inherit and build the future."

Here are some examples of young people who made a difference:

*At age 14, one schoolboy invented a rotary brush device to remove husks from wheat in the flour mill run by his father's friend. The young inventor was Alexander Graham Bell.

*At age 16, another one of America's junior achievers saved his pennies to buy materials for his chemistry experiments. While still a teenager, he set his mind on developing a commercially viable aluminum refining process. By age 25, Charles Hall received a patent on his revolutionary electrolytic process.

*While only 19 years old, another imaginative youth designed and built his first helicopter. In the summer of 1909, it nearly took flight. Years later, Igor Sikorsky perfected his design and saw his early dreams change aviation history.

*Robert Fulton was 14 when he invented the manually operated paddle wheel, as was the father of television, Philo T. Farnsworth, when he conceived his optical scanning idea.

One of the many benefits of Project XL is that young people and their parents and teachers will gain an increased awareness of new technology's importance to advancing society and strengthening the domestic economy. Project XL aims to instill an increased appreciation of the contributions inventors make to our way of life and recapture the spirit of those golden years at the turn of the century when inventors were heralded as true American heroes.

Project XL is comprised of the following components:

*National coordination of efforts to teach inventive thinking and problem-solving skills at every level of public and private education throughout the country;

*Presentation of national and regional conferences to promote the teaching of critical and creative thinking skills and the inventive process;

*Establishment of an Education Roundtable, an open forum and national discussion network, drawing upon the talents and resources of public and private sector leaders to develop and promote programs;

*Development of a broad-based speakers bureau on the topics of invention, problem-solving, creativity, thinking skills, and related topics;

*Dissemination of an informational guide called the Inventive Thinking Project, designed to channel students in grades K-12 into the inventive thinking process through the creation of their own unique inventions or innovations;

*Creation of an educator's resource guide to include programs, materials, literature, organizations, and other sources that promote thinking across all disciplines;

*Curriculum development for special teaching materials designed to stress problem-solving, the value of creative thinking, and the importance of American inventors;

*Identification of government programs and resources that focus on the development of future problem-solvers in all fields;

*Establishment of an Inventive Thinking Center, a collection of literature, videotapes, and other curriculum materials;

*Bestowment of the annual Donald J. Quigg Excellence in Education Award (**see Form 41 in Part One: Forms and Tables**).

The following directory includes listings for Project XL programs and publications for the teaching of creativity, critical thinking, problem-solving, and invention.

Section Arrangement

The following entries are arranged into two subsections: Programs and Publications. Each subsection is arranged alphabetically by program or publication. All programs and publications in this section are referenced in the Master Index.

Programs

★2343★ Connecticut Invention Convention
104 Washington Circle   Phone: (203) 523-8005
West Hartford, CT 06119
Michelle Munson, Chairperson

Competition for Connecticut students who have created original inventions. Designed to encourage and share creative productivity among today's youth. Age range: grades K-12.

★2344★ Creativity and Innovation
New York State Education Department   Phone: (518) 474-3954
Technology Education
1 Commerce Plaza, Rm. 1619
Albany, NY 12234
John Fabozzi, Bureau Chief

Study course designed to enhance 9-12 grade students' creative thinking, problem-solving skills, and production of innovative products and/or processes.

★2345★ Design and Research Exhibition (DARE)
New Jersey Department of Education   Phone: (609) 771-2068
Center for Technology Education
Armstrong Hall
Trenton State College CN4700
Trenton, NJ 08650-4700
Dr. Patricia Hutchinson, Project Director

Designed to enhance problem-solving skills and technological literacy among public, private, and vocational high school students through a noncompetitive design and invention program. Outstanding designs are exhibited in an annual state exhibition.

★2346★ Duracell NSTA Scholarship Competition
National Science Teachers Association   Phone: (202) 328-5800
1742 Connecticut Ave., N.W.
Washington, DC 20009
Kathleen A. Rapp, Special Programs Coordinator

The contest's objective is to encourage children in grades 9-12 to invent a working device powered by Duracell batteries that can make life easier, educate, entertain, serve as a warning device, provide light, make sounds, etc. Competition offers $30,000 in college scholarships and cash awards to 41 winners annually. Entries due in late January.

★2347★ Envisioneering
Belfast Central School   Phone: (716) 365-2646
Belfast, NY 14711
Michael J. Doyle

Also known as Creativity and Innovation: Beyond Paper Solutions, this half-unit course is designed to enhance creative thinking and problem solving in students in grades 9-12.

★2348★ Futuremakers Inventor/Mentor Program
Saturday Academy, Oregon Graduate   Phone: (503) 690-1190
  Center
19600 N.W. Von Neumann Dr.
Beaverton, OR 97006
Gail Whitney, Project Director

The program encourages the development of reasoning and problem-solving skills in students in grades 9-12 through the process of invention.

★2349★ Imagination Celebration/Invention Convention
Imagination Celebration Patent Office   Phone: (518) 473-0823
Cultural Education Center, Rm. 9B38
Empire State Plaza
Albany, NY 12230
Dr. Vivianne Anderson, Coordinator

Awards certificates to students in grades 6-9 in New York who submit descriptions of their inventions. Top inventors receive an Imagination Celebration patent and their inventions are displayed at the State Museum.

★2350★ Invent America!
U.S. Patent Model Foundation   Phone: (703) 684-1836
501 King St., Ste. 420
Alexandria, VA 22314
Kevin O'Brien, Vice President

Encourages creativity and productivity by developing the problem-solving and analytical skills of K-8 grade students. Includes state, regional, and national invention competition with grants and awards for students, teachers, and schools. Distributes educational material to elementary schools and sponsors a national, annual conference to encourage classroom instruction on invention and creativity.

★2351★ Invent, Iowa!
c/o U.S. Senator Tom Harkin   Phone: (515) 284-4574
Rm. 733, Federal Bldg.
Des Moines, IA 50309
Dianne Liepa, Deputy Administrator

Statewide program encourages creativity and productivity by developing problem-solving and analytical skills in children grades K-8. Sponsors annual convention.

★2352★ Invention Program
Midland Public Schools   Phone: (517) 835-7128
1305 E. Sugnet
Midland, MI 48640
Ms. Jody Pagel

Lesson program which studies work and machines, concentrating on critical thinking and problem solving. Age range: grades 7-9.

★2353★ Inventors Association of New England Youth Education Program

Youth Education Committee   Phone: (617) 244-4679
P.O. Box 335
Lexington, MA 02173
Samuel C. Smith, Chairman

Offers technical assistance to elementary and middle schools whose curricula include invention education. Compiles information on school invention programs and contests.

★2354★ Inventors Association of St. Louis Youth Programs

P.O. Box 16544   Phone: (314) 534-2677
St. Louis, MO 63105
Roberta Toole, Executive Director

Sponsors programs on innovation and invention for children in grades 2-9, gifted children, and adults. Is developing a program to use inventions and innovation to study science.

★2355★ InVenture

Inventors Council of Dayton, Ohio   Phone: (513) 439-4497
c/o Ronald T. Dodge Co.
P.O. Box 9488
Dayton, OH 45409
Dr. Ronald J. Versic, Chairman

Classroom program designed to promote innovation by encouraging middle and high school students to create a product, process, or art form and take the first step to protect the invention. Publishes teachers manual, guide for students, and a videotape of lessons.

★2356★ KIDSCON

National Foundation for Gifted & Talented Children
16400 Pacific Coast Hwy., #217
Huntington Beach, CA 92649
Ms. Jacqui Jeffrey, Executive Director

Creative KIDSCONference is a three-day program offering approximately 100 workshops for gifted children in grades 1-8 and their parents. Three of the workshops in 1990 were concerned with inventing. Toll-free number: 1-800-GIFTED-1.

★2357★ Mentor/FACETS Connection Invention Convention

Springfield Public Schools   Phone: (413) 787-7015
195 State St., PO Box 1410
Springfield, MA 01102-1410
Linda Tammi, Supervisor

Program for children grades 5 and 6 stresses use of higher-level thinking skills and encourages creativity in identifying a problem and solving it with an invention.

★2358★ Midland Public Schools Invention Program

Northeast Intermediate School   Phone: (517) 835-7128
1305 E. Sugnet
Midland, MI 48640
Jody Pagel, Contact

Uses inventions as a way of teaching critical thinking and problem solving and encourages students to design products to solve every day problems. Designed for grades 7-9; also taught to special education students at all levels and gifted students in elementary schools.

★2359★ Mini-Invention Innovation Team Contest

New Jersey Department of Education   Phone: (201) 290-1900
Division of Vocational Education
225 W. State St.
Trenton, NJ 08625
Sylvia M. Kaplan, Director

Sponsors competitions to inspire children in grades K-9 to think creatively. Competition winners within school districts progress to the regional competition and top contestants enter statewide finals. Conducts teacher in-service workshops and provides technical assistance to students.

★2360★ Minnesota Student Inventors Congress

South Central ECSU   Phone: (507) 389-5101
1610 Commerce Dr.
North Mankato, MN 56001
Paul Olsonr, Coordinator

Statewide competition and exhibition for winners of student (grades K-12) invention fairs.

★2361★ New Hampshire Young Inventors Program

Academy of Applied Science   Phone: (603) 228-4530
98 Washington St.
Concord, NH 03301
Dr. Mary Stuart Gile, Vice President

The Program provides teacher training, an instruction booklet, and a statewide Young Inventors' Celebration for students grades K-12.

★2362★ New Jersey Invention Convention

Silver Burdett & Ginn   Phone: (201) 285-7740
250 James St.
Morristown, NJ 07960
Andrew Socha, Product Manager

Encourages students to apply basic science skills in a creative and productive manner through classroom, school, or district-wide events. Sponsors international convention where student inventions are exhibited, judged, and awards are presented. Publishes guidelines for teachers and procedures for students.

★2363★ New Mexico Invention Convention

113 W. Yucca
Clovis, NM 88101
Ms. Pat Thomas

Offers a 10-week lesson plan on inventing, which culminates with an invention convention.

★2364★ New York Invention Convention

Imagination Celebration/Council on the   Phone: (518) 563-5222
    Arts for Clinton County
P.O. Box 451
64 Margaret St.
Plattsburgh, NY 12901
JoAnn Perry, Contact

Exhibits and judges inventions of Clinton County K-12 grade students at the annual Imagination Celebration Community Showcase Festival.

★2365★ New York Young Inventors Program

New York City Teacher Centers Consortium
Phone: (212) 475-3737
260 Park Ave. S.
New York, NY 10010
Myrna Cooper, Director

Trains K-6 grade teachers in New York City to foster inventive and creative thinking in heterogeneous settings. Encourages students to learn how to investigate real problems using scientific processes of inquiry, understand technology, and develop new products and designs as a result of these experiences. Sponsors competitions and bestows awards at each grade level.

★2366★ San Diego Invention Program

San Diego Unified School District
Phone: (619) 293-8552
Education Center, Rm. 2005
4100 Normal St.
San Diego, CA 92103-2682
Jo Anne Schaper, Contact

Encourages strong involvement from K-12 students in the national INVENT AMERICA! contest.

★2367★ SBG Invention Convention

Silver Burdett & Ginn
Phone: (201) 285-7740
250 James St.
Morristown, NJ 07960
Andrew Socha

The Convention is designed to encourage students in grades 1-9 to apply basic science skills in a creative and productive manner. The grand prize is a computer and an all-expenses-paid trip to the National Science Teachers Association meeting.

★2368★ Texas Invention Convention

Richardson Independent School District
Phone: (214) 470-5202
400 S. Greenville Ave.
Richardson, TX 75080
Dr. Leonard Molotsky, Contact

Designed to stimulate K-12 grade students' imaginations and promote problem-solving and creative thinking skills through invention conception, development, and marketing. Sponsors school competitions; winners compete at the district Invention Convention.

★2369★ Toledo Public Schools Invention Convention

Toledo Public Schools
Phone: (419) 729-8315
Manhattan Blvd. & Elm St.
Toledo, OH 43608
Robert Frisch, Contact

Designed to promote cognitive and creative thinking skills in 3-6 grade students through the development and exhibition of inventions, display boards, and tributes to inventors.

★2370★ Tualatin Invention Convention

Tualatin Elementary School
Phone: (503) 684-2359
19945 S.W. Boones Ferry Rd.
Tualatin, OR 97062
Evelyn Andrews, Contact

Trains teachers to work with students in grades K-6 to develop inventions that are judged at the Invention Convention.

★2371★ U.S. Cognetics

700 Hollydell Ct.
Phone: (609) 582-7000
Sewell, NJ 08080
Dr. Theodore R. Gourley

The program's objective is to teach creative problem solving to students in grades K-12 through fact gathering, problem analysis, idea generation, and solution selection in a group setting. Since its purpose is to achieve a goal as opposed to competing against others, certificates are given instead of prizes.

★2372★ Weekly Reader National Invention Contest

Weekly Reader
Phone: (203) 638-2752
245 Long Hill Rd.
Middletown, CT 06457
Dr. Irwin Siegelman, Editorial Director

Presents awards to K-8 students nationwide for inventions or innovations that are judged on originality, usefulness in addressing real needs, workability, and clarity of presentation. Awards a grand prize for elementary school students and for middle school students and bestows awards for specific grade levels.

★2373★ Western New York Invention Program

Buffalo Public Schools
Phone: (716) 842-3693
419 City Hall
Buffalo, NY 14202
Marge Korzelius, Contact

Draws innovation and invention entries from pre-kindergarten through grade 12 students and displays the top invention from each school or grade at the Buffalo Science Museum. Sponsors a teacher in-service workshop and publishes instructional materials.

★2374★ Young Inventors Fair

Metro Educational Cooperative Service Unit
Phone: (612) 490-0058
3499 Lexington Ave. N.
Arden Hills, MN 55126
Janet Robb, Chair

A year-long program for the Twin Cities consisting of workshops and events that teach and encourage students grades 4-9 to invent. Also offers training for teachers and holds special classes.

Publications

★2375★ ABCs of Books and Thinking Skills K-8

Book Lures, Inc.
Phone: (314) 272-4242
P.O. Box 9450
O'Fallon, MO 63366

Applications of thinking skills for K-8 grade students. **Author/Editor:** Nancy Polette.

★2376★ ABCs of Reading, Thinking and Literacy 7-12

Book Lures, Inc.
Phone: (314) 272-4242
P.O. Box 9450
O'Fallon, MO 63366

Step-by-step directions for multi-level activities covering 50 separate thinking skills for grades 7-12. **Author/Editor:** Nancy Polette and Gloria Levine.

★2377★ Acting, Creating and Thinking (ACT)
Sundance Publishers & Distributors, Inc. Phone: (508) 486-9201
Newton Rd.
Littleton, MA 01460

Each ACT packet contains 10 activity sheets used to reinforce oral language and critical-creative thinking skills with elementary and middle school students. ACT is designed for use with Take Part books which are children's stories in play form. **Author/Editor:** Dr. Olive Stafford Niles and Audrey A. Friedman.

★2378★ Analytical Reading and Reasoning
Innovative Sciences, Inc. Phone: (203) 359-1311
300 Broad St.
Stamford, CT 06901

Text book designed to enhance reading comprehension of high school students by improving the thinking and reasoning process that support higher-order comprehension. **Author/Editor:** Dr. Arthur Whimbey.

★2379★ Analyze
Kolbe Concepts, Inc. Phone: (602) 840-9770
P.O. Box 15050
Phoeniz, AZ 85060

Activity book containing problem solving situations that stimulate the use of analysis.

★2380★ Apple Shines
Good Apple, Inc. Phone: (217) 357-3981
1204 Buchanan St.
P.O. Box 299
Carthrage, IL 62321

"Think-and-then-write" activities to stimulate elementary and middle school students to use original ideas as a source. Each activity includes an introduction, instruction, and follow-up. **Author/Editor:** Bob Eberle.

★2381★ Art and Perception - The Flexible Line
Foxtail Press Phone: (214) 552-3922
P.O. Box 2996
La Habra, CA 90632

Information and activity-based book focusing on line. **Author/Editor:** Betty Lewis and Marge Tezak.

★2382★ The Art of Perceiving Problems
Foxtail Press Phone: (214) 522-3922
P.O. Box 2996
La Habra, CA 90632

Booklet to help intermediate students become aware of problem situations, understand mental blocks, and learn techniques of dealing with problems in unique and creative ways. **Author/Editor:** Eileen Babcock and Marilyn Brown.

★2383★ The Art of Resolving Problems
Foxtail Press Phone: (214) 522-3922
P.O. Box 2996
La Habra, CA 90632

Booklet presenting middle school teachers and students with sequential methods for creative problem solving. **Author/Editor:** Marilyn Brown and Eileen Babcock.

★2384★ Basic Thinking Skills
Society for Visual Education Phone: (312) 525-1500
1345 W. Diversey Pkwy.
Chicago, IL 60614

Four sound filmstrips designed to teach middle school students to think: Finding the Main Idea, Coming to Conclusions, Developing Ideas, and Deciding on the Facts. Includes teacher's guide.

★2385★ Be an Inventor
Harcourt Brace Jovanovich Phone: (212) 614-3000
111 Fifth Ave.
New York, NY 10003

Shows children how to turn their ideas into inventions by leading them step-by-step through the inventive process. Designed for teachers, students, and parents. **Author/Editor:** Barbara Taylor.

★2386★ The Book for Women Who Invent or Want To
Women Inventors Project
P.O. Box 689
Waterloo, ON, Canada N2J 4B8

Discusses the creative process from the idea to product stage, including the patent process and marketing. Covers how to create a network of women inventors. **Author/Editor:** Elizabeth Wallace.

★2387★ Brain Muscle Builders - Games to Increase Your Natural Intelligence
Trillium Press, Inc. Phone: (914) 783-2999
Box 921, Madison Sq. Sta.
New York, NY 10159

Book encouraging game playing as a means of improving thinking skills. **Author/Editor:** Marco Meirovitz and Paul I. Jacobs.

★2388★ Brain Scratchers
Kolbe Concepts, Inc. Phone: (602) 840-9770
P.O. Box 15050
Phoenix, AZ 85060

Instructions for designing crossword, hidden puzzle, and maze games.

★2389★ A Catalog of Programs for Teaching Thinking
Research For Better Schools Phone: (215) 574-9300
444 N. Third St.
Philadelphia, PA 19123

Catalog for teachers providing concise summaries of some major commercial published programs that teach thinking. **Author/Editor:** Janice Kruse and Barbara Z. Presseisen.

★2390★ Catalog of Publications and Services
Center for Creative Learning
P.O. Box 619
Honeoye, NY 14471

Catalog listing publications related to thinking skills and creativity as well as programs and workshops offered at the Center.

★2391★ Challenge
Dale Seymour Publications Phone: (415) 324-2800
P.O. Box 10888
Palo Alto, CA 94303

Magazine for students in grades 5-8. **Author/Editor:** Carole Greenes, George Immerzeel, Linda Schulman, and Rita Spungin.

★2392★ Challenge Boxes
Dale Seymour Publications   Phone: (415) 324-2800
P.O. Box 10888
Palo Alto, CA 94303

Book of 50 activities, each with a creative thinking theme, that can be used with middle school students as a learning center or enrichment challenge. **Author/Editor:** Katherine Valentino.

★2393★ The Challenge of the Unknown
W.W. Norton Company, Inc.   Phone: (212) 354-5500
500 Fifth Ave.
New York, NY 10110

Videotape with teacher's manual for a seven-part film series on mathematics and problem solving for grades 5-9. **Author/Editor:** Hillary C. Maddus.

★2394★ Challenges for Children
NL Associates, Inc.
P.O. Box 1199
Highstown, NJ 08250

Contains activities to stimulate creative thinking and challenge elementary school children to solve problems.

★2395★ Chrysalis - Nurturing Creative and Independent Thought in Children
Zephr Press
430 S. Essex Ln.
Tucson, AZ 85711

Designed to provide middle and high school teachers with a framework for educational experiences for developing creativity, self-reliance, and an independent approach to learning. Seven units cover thinking and feeling, health, scientific phenomena, esthetics, fine art, world problems, exploring the future, and a seqenced plan for independent learning. **Author/Editor:** Micki McKisson.

★2396★ Circles of Creativity
Point Publishing Co.   Phone: (201) 295-8258
P.O. Box 1309
Point Pleasant Beach, NJ 08742

Brainstorming aid consisting of 276 idea stimulator words and phrases located on rotating disks. The disks are used to prompt ideas through the use of random stimuli. **Author/Editor:** Arthur B. VanGundy and James Betts.

★2397★ Citizenship Decision-Making
Addison-Wesley Publishing Co.   Phone: (617) 944-3700
South St.
Reading, MA 01867

Set of 25 social studies supplementary lessons for elementary and junior high school teachers that focus on building students' skills with the tasks of making, judging, and influencing decisions. **Author/Editor:** Roger La Raus and Richard C. Remy.

★2398★ Cognetics
The Talent Network   Phone: (609) 582-7000
Educational Information & Resource
   Center
700 Hollydell Ct.
Sewell, NJ 08080

Annual school program to stimulate creativity and creative problem solving. Subscription includes newsletters and other supplemental material and three manuals providing a background in creativity and creative problem solving, six problems to be solved, and a calendar, forms, and procedures for the current program year. **Author/Editor:** Theodore J. Gourley and Judith Burr.

★2399★ The Cort Thinking Program
Pergamon Press, Inc.   Phone: (914) 592-7700
Maxwell House, Fairview Park
Elmsford, NY 10523

Program of six units with each unit containing 10 lessons that teach separate thinking skills. Includes teacher's handbook and 10 student workcards for each lesson in the first five units; unit six includes a teacher's handbook and student textbook. **Price:** $185.00. **Author/Editor:** Edward de Bono.

★2400★ Creative Encounters with Creative People
Good Apple, Inc.   Phone: (217) 357-3981
1204 Buchanan St.
P.O. Box 299
Carthrage, IL 62321

Contains analysis of creative personalities such as Henry Ford, Walt Disney, and Joni Eareckson. Each unit includes a biographical sketch, creative encounters, and work sheets. **Author/Editor:** Janice Gudeman.

★2401★ Creative Kids
GCT Inc.   Phone: (205) 478-4700
P.O. Box 6448
Mobile, AL 36660

Magazine containing stories, poetry, music, artwork, plays, experiments, and other features contributed by children aged five to eighteen. The work represents children's ideas, questions, fears, concerns, and pleasures. **Author/Editor:** Fay L. Gold.

★2402★ Creative Problem Solving: A Guide for Trainers and Management
Quorum Books   Phone: (203) 226-3571
88 Post Rd. W.
Westport, CT 06881

Presents a six-step problem solving model designed to deal with problems that require creative solutions. Includes guidelines for teachers. **Author/Editor:** Dr. Arthur B. VanGundy.

★2403★ Creative Problem Solving Techniques
Trillium Press   Phone: (914) 783-2999
Box 921, Madison Sq. Sta.
New York, NY 10159

Handbook of creative problem solving procedures for middle and high school students. **Author/Editor:** Julie L. Ellis.

★2404★ Creative Writing in Action
Good Apple, Inc.   Phone: (217) 357-3981
1204 Buchanan St.
P.O. Box 299
Carthrage, IL 62321

Book designed to enhance writing and speaking skills. Student activities focus on characterizations, dialogue development, setting, and plot. **Author/Editor:** Elizabeth Marten and Nina Crosby.

★2405★ *Creativity 1,2,3*
Trillium Press   Phone: (914) 738-2999
Box 921, Madison Sq. Sta.
New York, NY 10159

Book providing ideas on how to make creativity an integral part of the elementary classroom. Explains ideas for incorporating creativity and encourages teachers to increase their own creativity. **Author/Editor:** Susan Ellis Baum and Martha Cray Andrews.

★2406★ *Critical Thinking and Thinking Skills: State of the Art Definitions and Practice in Public Schools*
Research for Better Schools   Phone: (215) 574-9300
444 N. Third St.
Philadelphia, PA 19123

Book providing an overview of the history of critical thinking from 1938 to the 1980s. Contains a reference list and appendix listing thinking skill meetings and conferences in the United States. **Author/Editor:** Barbara Z. Pressisen.

★2407★ *Critical Thinking: How to Evaluate Information and Draw Conclusions*
The Center for Humanities, Inc.   Phone: (914) 666-4100
Communications Park, Box 3000
Mt. Kisco, NY 10549-9989

Videotape uses scenarios with teenagers to explain concept evaluation and establish critical thinking skills. Includes teacher's manual. **Price:** $197.00.

★2408★ *Daughters of Invention: An Invention Workshop for Girls*
Women Inventors Project
P.O. Box 689
Waterloo, ON, Canada N2J 4B8

Handbook suggests guidelines for planning invention workshops for girls. **Author/Editor:** Rachelle Beauchamp and Lisa Avedon.

★2409★ *Delta Science Modules*
Delta Education   Phone: (603) 889-8899
P.O. Box M
Nashua, NH 03061-6012

Activity-based science series consisting of science topic modules for grades K-6. Modules cover concepts in life, earth, and phyical science and are accompanied by a classroom kit of materials for activities.

★2410★ *Detecting and Deducing - Preparing for Logical Thinking*
Foxtail Press   Phone: (214) 522-3922
P.O. Box 2996
La Habra, CA 90632

Booklet to assist intermediate school students with logic, become aware of assumptions, use facts to reach valid conclusions, avoid generalizations, consider evidence, and recognize and use the syllogistic pattern for logic. **Author/Editor:** Eileen Babcock, Marilyn Brown, and Betty Lewis.

★2411★ *The Discovery*
Foxtail Press   Phone: (214) 522-3922
P.O. Box 2996
La Habra, CA 90632

Describes a creative problem solving process for teachers and students integrating all areas of study. **Author/Editor:** Ken Mittan.

★2412★ *Do-It-Yourself Critical and Creative Thinking*
Kolbe Concepts, Inc.   Phone: (602) 840-9770
P.O. Box 15050
Phoeniz, AZ 85060

Explains thinking skills and provides confidence-building activities.

★2413★ *Education and Learning to Think*
National Academy Press   Phone: (202) 334-3318
2102 Constitution Ave., N.W.
Washington, DC 20418

Addresses the question of what American educators can do to teach higher order skills more effectively. **Author/Editor:** Lauren B. Resnick.

★2414★ *Effective Questions to Strengthen Thinking*
Foxtail Press   Phone: (214) 522-3922
P.O. Box 2996
La Habra, CA 90632

Booklet addresses logical, critical, and creative thinking through questioning. Questions are for all grade levels; most material is for middle and high school students. **Author/Editor:** Marilyn Brown.

★2415★ *Elementary Science Study* (ESS)
Delta Education   Phone: (603) 889-8899
P.O. Box M
Nashua, NH 03061-6012

Designed to teach children to think critically about the scientific world through hands-on experience. Incorporates different instructional strategies and styles; includes activity kits for scientific investigation with materials for eight students.

★2416★ *Evaluate*
Kolbe Concepts, Inc.   Phone: (602) 840-9770
P.O. Box 15050
Phoenix, AZ 85060

More than 50 activities designed to develop levels of logical thinking and evaluation with elementary and middle school children.

★2417★ *Expanding Creative Imagination*
NL Associates, Inc.
P.O. Box 1199
Highstown, NJ 08250

Student workbook containing over 100 exercises and activities to develop major aspects of creative thinking. Teacher's guide contains information and directions to accompany student book; activities require no teacher preparation.

★2418★ *Exploring the Lives of Gifted People in the Arts*
Good Apple, Inc.   Phone: (217) 357-3981
1204 Buchanan St.
P.O. Box 299
Carthrage, IL 62321

Profiles contemporary heroes such as Bart Connor, Dr. Norman Vincent Peale, and Peter Strauss. Encourages students to become involved in creative problem solving and decision making processes and to expand career awareness. **Author/Editor:** Kathy Balsamo.

★2419★ Exploring the Lives of Gifted People in the Sciences

Good Apple, Inc.　　　　　　Phone: (217) 357-3981
1204 Buchanan St.
P.O. Box 1204
Carthrage, IL 62321

Six units containing an interview with contemporary heroes including Dr. Sally Ride, Barbara Jordan, and Dr. Isaac Asimov, followed by over 40 activities. **Author/Editor:** Kathy Balsamo.

★2420★ Eye Cue Puzzles Sets

Dale Seymour Publications　　　Phone: (415) 324-2800
P.O. Box 10888
Palo Alto, CA 94303

Set of 40 challenges designed to promote visual thinking skills.

★2421★ Fact Fantasy and Folklore

Good Apple, Inc.　　　　　　Phone: (217) 357-3981
1204 Buchanan St.
P.O. Box 299
Carthrage, IL 62321

Stimulates critical thinking using 11 well-known fairy tales through role play, creative activities, and open-ended questions. Includes background clues for the teacher and lesson plans. **Author/Editor:** Greta B. Lipson and Baxter Morrison.

★2422★ Focus on Thinking

Focus on Thinking Foundation
Box 430
Ivermere, BC, Canada V0A 1K0

Focus on Thinking I - An Introduction to Thinking Skills contains interviews with many of Canada's top thinking skills consultants. Focus on Thinking II is a television program of interviews of top consultants from Canada and the United States.

★2423★ Future Options - Unlimited

Foxtail Press　　　　　　　Phone: (214) 522-3922
P.O. Box 2996
La Habra, CA 90632

Booklet on students' future needs as adults. Includes exercises requiring research and recordkeeping, vocabulary needed for jobs and careers of the future, background materials, news items, and study units. **Author/Editor:** Dr. Eldon Meyler and Donald David Zielinski.

★2424★ Game Puzzles for the Joy of Thinking

Kadon Enterprises, Inc.　　　Phone: (301) 437-2163
1227 Lorene Dr., Ste. 16
Pasadena, MD 21122

Catalog of original games and puzzles based on mathematical principles and geometric sets. **Author/Editor:** Kathy Jones.

★2425★ The Gifted Child Today

GCT Inc.　　　　　　　　　Phone: (205) 478-4700
P.O. Box 6448
Mobile, AL 36660

Magazine designed to meet the needs of parents and teachers in working with gifted, creative, and talented children. **Author/Editor:** Marvin J. Gold.

★2426★ Gifted Children Monthly

Gifted & Talented Publications, Inc.　　Phone: (609) 582-0277
213 Hollydell Dr.
Sewell, NJ 08080

Provides columns on a variety of topics. **Subscription:** $24.00.

★2427★ The Great Bridge Lowering

Trillium Press　　　　　　　Phone: (914) 783-2999
Box 921, Madison Sq. Sta.
New York, NY 10159

Guide to creative thinking for elementary and middle school students. Encourages development of the ability to generate ideas, break mind sets, create new ideas, and elaborate on ideas. **Author/Editor:** Sandra Warren.

★2428★ How Can We Teach Intelligence?

Research For Better Schools　　Phone: (215) 574-9300
444 N. Third St.
Philadelphia, PA 19123

Report on three programs that train aspects of intelligence. Includes references and bibliography. **Author/Editor:** Robert J. Sternberg.

★2429★ How do You Figure That? Ways of Problem Solving

The Center for Humanities, Inc.　　Phone: (914) 666-4100
Communications Park, Box 3000
Mt. Kisco, NY 10549-9989

Set of four filmstrips/cassettes covering a problem to be systematically solved using a five-step procedure by students in grades 3-6. Includes teacher's guides. **Price:** $139.00.

★2430★ How to Invent

IFI/Plenum Data Corp.　　　　Phone: (703) 683-1085
302 Swann Ave.
Alexandria, VA 22301

Introduces basic principles, methods, and tools of the inventing process. Includes drawings from famous patents, information on protecting inventions, and a history of inventions in the United States. **Author/Editor:** B. Edward Shelesinger, Jr.

★2431★ The Idea Generator

Experience in Software, Inc.　　Phone: (415) 644-0694
2039 Shattuck Ave., Ste. 401
Berkeley, CA 94704

Offers middle and high school students a step-by-step approach to solving problems. **Author/Editor:** Gerald I. Nierenberg.

★2432★ Imagine That

Rainbow Planet
5110 Comwell Dr.
Gig Harbor, WA 98335

Album of children's songs that paint a picture of children's fantasies. **Author/Editor:** Jim Valley.

★2433★ Improving the Quality of Student Thinking
Association for Supervision & Curriculum Phone: (703) 549-9110
Development
Order Processing Department
125 N. West St.
Alexandria, VA 22314-2798

Videotape created for staff development. Contains explanations of methods for improving the quality of student thinking.

★2434★ Invention Convention Procedural Manual
Richardson Independent School District Phone: (214) 238-8111
400 S. Greenville Ave.
Richardson, TX 75081

Covers administrative aspects of organizing an invention convention at the classroom, building, or district level. A companion book to Inventive Thinking: A Teacher-Student Handbook.

★2435★ Inventioneering
Good Apple, Inc. Phone: (217) 357-3981
1204 Buchanan St.
P.O. Box 299
Carthrage, IL 62321

Provides invention instructions, content, process and product objectives, and photographs of student inventions. **Author/Editor:** Bob Stanish and Carol Singletary.

★2436★ The Inventive Child
Encyclopedia Britannica Educational Phone: (312) 347-7400
Corp.
425 N. Michigan Ave.
Chicago, IL 60611

Series of 20 film programs providing a basis for creative thinking and problem solving activities in science, social studies, communication arts, industrial arts, home economics, art, music, and interdisciplinary humanities programs. **Author/Editor:** Grace N. Lacey.

★2437★ The Inventive Imagination to Illumination
Foxtail Press Phone: (214) 522-3922
P.O. Box 2996
La Habra, CA 90632

Activity booklet of exercises to enhance inventive thinking. Based on the work of Guilford, Torrance, and Parnes. **Author/Editor:** Marilyn Brown.

★2438★ The Inventive Innovation to Ingenuity
Foxtail Press Phone: (214) 522-3922
P.O. Box 2996
La Habra, CA 90632

Booklet containing background information, teacher directions, and student activities on invention, imagination, resourceful thinking, and the invention process. **Author/Editor:** Marilyn Brown.

★2439★ Inventive Thinking: A Teacher-Student Handbook
Richardson Independent School District Phone: (214) 238-8111
400 S. Greenville Ave.
Richardson, TX 75081

Handbook for teachers and students provides activities on creativity and a four step method for teaching the inventive process; includes bibliography.

★2440★ The Inventor's Guide
Haley Publications
Box 335
Lexington, MA 02173

Guide to the inventing process, documenting the process, inventing on and off the job, assessing market and financial worth, protecting an invention, building a business, and producing a product. **Author/Editor:** Dr. Donald D. Job.

★2441★ The Journal of Creative Behavior
Creative Education Foundation, Inc. Phone: (716) 884-2744
437 Franklin St.
Buffalo, NY 14202

Journal containing articles which tend to be theoretical on the fields of creativity and problem solving. **Frequency:** Quarterly.

★2442★ Just Think Program Series and Stretch Think Program Series
Thomas Geal Publications, Inc.
P.O. Box 370540
Montara, CA 94037

Curricula to teach thinking from the pre-school through grade 8. Each full year thinking program includes lessons, instructions, worksheets, and a matrix of curricular content objectives. **Author/Editor:** Sydney Billig Tyler.

★2443★ Keep Them Thinking
Illinois Renewal Institute, Inc. Phone: (312) 991-6300
200 E. Wood St., Ste. 250
Palatine, IL 60067

Three books present lesson designs to teach, rehearse, practice, and transfer explicit thinking skills: Level I: Primary (K-4), Level II: Intermediate (5-8), and Level III: Advanced (9-12). **Author/Editor:** Jim Bellanca, Kay Opeka, and Robin Fogarty.

★2444★ The Lateral Thinking Machine
Gemini Group
RD #2, Box 117
Bedford, NY 10506

IBM personal computer-based program for problem solving. **Author/Editor:** Ken Finn.

★2445★ Lessons in Logic: Unravelling Common Complexities
Kolbe Concepts, Inc. Phone: (602) 840-9770
P.O. Box 15050
Phoenix, AZ 85060

Course teaching creative problem solving methodologies. Provides information needed for specific problem solving approaches and includes practice activities.

★2446★ Life After School
Simon & Schuster Phone: (212) 698-7000
Julian Mesmer Division
1230 Avenue of the Americas
New York, NY 10020

Self-help career development book for 11th and 12th graders which enables students to identify and determine their creative skills and possible creative careers.

★2447★ Literature is for Thinking (LIFT)
Sundance Publishers & Distributors, Inc.   Phone: (508) 486-9201
Newton Rd.
Littleton, MA 01460

Series of supplementary materials to be used with literary books to stimulate high school students' development of critical and creative thinking and language skills. Includes rational for the program, teacher's guide, and 14-18 reproducable activity sheets for students' use. **Author/Editor:** Katherine Paterson.

★2448★ Logic Number Problems
Dale Seymour Publications   Phone: (415) 324-2800
P.O. Box 10888
Palo Alto, CA 94303

Collection of 50 number puzzles designed to provide experience in problem solving and thinking skills for middle and high school mathematics students. **Author/Editor:** Wade H. Sherard III.

★2449★ Looking Glass Logic: Problems and Solutions
The Perfection Form Co.   Phone: (712) 644-2831
1000 N. Second Ave.
Logan, IA 51546

Workbook uses illustrations and quotations from Lewis Carroll's classics to demostrate the principles of logic. **Author/Editor:** Kristin Kalsem.

★2450★ Managing Group Creativity
American Management Associations   Phone: (212) 903-8089
135 W. 50th St.
New York, NY 10020

Provides teachers with information on group dynamics, effectiveness, and problem solving. Includes references to research literature and questionnaires for evaluating problem-solving techniques. **Author/Editor:** Dr. Arthur B. VanGundy.

★2451★ Mastering Reading Through Reading
Innovative Sciences, Inc.   Phone: (203) 359-1311
300 Broad St.
Stamford, CT 06901

Secondary level text book geared to improving the thinking and reasoning processes that support higher-order comprehension. Includes teacher's guide. **Author/Editor:** Dr. Arthur Whimbey.

★2452★ Mathematics Pentathlon
Pentathlon Institute   Phone: (317) 782-1553
P.O. Box 20590
Indianapolis, IN 46220

Text paragraph Series of 20 instructional games which foster the development of active problem solving. Includes instruction and tournament activities for primary and secondary students.

★2453★ Max Think
Max Think, Inc.
230 Crocker Ave.
Piedmont, CA 94610

IBM-based computer software for processing ideas. Includes instruction manual. **Author/Editor:** Neil Larson.

★2454★ Mental Menus
Illinois Renewal Institute, Inc.   Phone: (312) 991-6300
200 E. Wood St., Ste. 250
Palatine, IL 60067

Lesson plans for 24 critical and creative thinking skills including inferencing, contrasting, determining bias, analyzing assumptions, and hypothesizing. **Price:** $14.95. **Author/Editor:** Rubin Fogarty and James Bellanca.

★2455★ The Million Dollar Idea
New Product Development   Phone: (201) 295-8258
P.O. Box 1309
Point Pleasant, NJ 08742

Paperback book on new product ideas and developing them. **Author/Editor:** Jim Betts.

★2456★ Mind Games: Puzzles and Logic
The Perfection Form Co.   Phone: (712) 644-2831
1000 N. Second Ave.
Logan, IA 51546

Explains three types of logic problems and instructs students how to break problems into smaller units, examine each part, and identify relationships between parts. **Author/Editor:** Kristin Kalsem.

★2457★ Mind Joggers
NL Associates, Inc.
P.O. Box 1199
Highstown, NJ 08250

Contains activities for elementary and middle school students covering thinking and reasoning, math, language and writing, and listening and remembering. **Author/Editor:** Susan S. Petreshene.

★2458★ Mind Movers - Creative Homework Assignments
Addison-Wesley Publishing Co.   Phone: (617) 944-3700
South St.
Reading, MA 01867

Weekly homework assignments, most of which contian higher level thinking skills, for students in grades 3-6. **Author/Editor:** Diane Hart and Margaret Rechif.

★2459★ New Frontiers/Nuevas Fronteras
Pergamon Press, Inc.   Phone: (914) 592-7700
Maxwell House, Fairview Park
Elmsford, NY 10523

Oral, early learning congitive curriculum program with a strong emphasis on mathematics and science. Designed to encourage young Spanish-speaking children to learn English and develop thinking skills and problem solving strategies. **Author/Editor:** Barbara Coffingan Cox, Janet McCaulay, and Manual Ramierz III.

★2460★ A New Way to Use Your Bean - Developing Thinking Skills in Children
Trillium Press   Phone: (914) 783-2999
Box 921, Madison Sq. Sta.
New York, NY 10159

Book containing 30 cooking activities that introduce critical thinking, creative thinking, logic, and problem solving. Includes descriptions of thinking processes utilized, materials and equipment needed, precooking activities, cooking suggestions, and suggestions for additional activities. **Author/Editor:** Darlene Freeman.

★2461★ Oceanography
Engine-Uity, Ltd.   Phone: (602) 997-7144
P.O. Box 9610
Phoeniz, AZ 85068

Writing activities to help children learn how to think creatively and to use critical thinking skills.

★2462★ Odyssey of the Mind: Problems to Develop Creativity
Creative Competitions, Inc.   Phone: (609) 881-1603
P.O. Box 27
Glassboro, NJ 08028

Book of 57 long-term and spontaneous problems used in Odyssey of the Mind competitions. **Price:** $12.50. **Author/Editor:** C. Samuel Micklus.

★2463★ OM-AHA! Problems to Develop Creative Thinking Skills
Creative Competitions, Inc.   Phone: (609) 881-1603
P.O. Box 27
Glassboro, NJ 08028

Book of problems designed to allow students to use their imaginations. Includes nine long-term problems, more than 40 spontaneous problems, and nine Odyssey of the Mind World Finals warm-up problems. **Price:** $15.50. **Author/Editor:** C. Samuel Micklus.

★2464★ Patent Pending
Social Studies School Service   Phone: (213) 839-2436
P.O. Box 802
Culver City, CA 90232-0802

Film providing an overview of the U.S. patent system and highlights inventions in use today.

★2465★ Perception, Inc.
Illinois Renewal Institute, Inc.   Phone: (312) 991-6300
200 E. Wood St., Ste. 250
Palatine, IL 60067

Software program designed to provide early elementary students with practice in perception, attributing, and classification. Includes student activity book and teacher's guide.

★2466★ Planning for Thinking
Illinois Renewal Institute, Inc.   Phone: (312) 991-6300
200 W. Wood St., Ste. 250
Palatine, IL 60067

Book to guide instructional leaders in designing and implementing an effective thinking curriculum for their districts. **Author/Editor:** James Bellanca and Robin Fogarty.

★2467★ Problem Solving and Comprehension
Lawrence Erlbaum Associates, Inc.   Phone: (201) 666-4110
365 Broadway
Hillsdale, NJ 07642

Addresses how to increase secondary students' problem solving and test taking skills and learning comprehension. **Author/Editor:** Arthur Whimbey and Jack Lockhead.

★2468★ Problem Solving in Science
Curriculum Associates, Inc.   Phone: (508) 667-8000
5 Esquire Rd.
North Billerica, MA 01862

Activity-based program that builds reading comprehension and creative problem solving in the content area of science. **Author/Editor:** Louis James Taris and James Robert Taris.

★2469★ Problems! Problems! Problems! Discussions and Activities Designed to Enhance Creativity
Creative Competitions, Inc.   Phone: (609) 881-1603
P.O. Box 27
Glassboro, NJ 08028

Features a discussion on creativity and over 60 creative long-term and spontaneous problems. **Price:** $10.95. **Author/Editor:** C. Samuel Micklus and T. Gourley.

★2470★ Product Improvement Checklist
Point Publishing Co.   Phone: (201) 295-8258
P.O. Box 1309
Point Pleasant Beach, NJ 08742

Contains 576 stimulator words and phrases to be used as a brainstorming aid to prompt ideas by provoking associations between unrelated stimuli and aspects of a problem. **Price:** $10.00. **Author/Editor:** Arthur B. VanGundy.

★2471★ Project: Problem Solving
Kolbe Concepts, Inc.   Phone: (602) 840-9770
P.O. Box 15050
Phoenix, AZ 85060

Kit for do-it-yourself patterns and constructions for bending light, designing mazes, creating film strip viewers, discovering variances in air rockets, and other activities.

★2472★ Project Success Enrichment
Creative Child Concepts
Station 111, PO Box 61100
Seattle, WA 98121

Nationally-validated (NDN) language arts and art curriculum appropriate for grades 2-8 that uses cooperative learning strategies to address critical and creative thinking processes, self-management, and social skills. **Author/Editor:** Carolyn G. Bronson, Sally Maryatt, and Karlene George.

★2473★ Prolific Thinkers Guide
Dale Seymour Publications   Phone: (415) 324-2800
P.O. Box 10888
Palo Alto, CA 94303

Guide to teaching critical and creative thinking using a hands-on approach. **Author/Editor:** Gary A. Carnow and Constance Gibson.

★2474★ Quizzles - Logic Problem Puzzles
Dale Seymour Publications   Phone: (415) 324-2800
P.O. Box 10888
Palo Alto, CA 94303

Booklet of 38 problems designed as student worksheets that require inventive thinking. Also available is More Quizzles - Logic Problem Puzzles containing 48 puzzles of increasing difficulty. **Author/Editor:** Wayne Williams.

★2475★ Risk Taking
Kolbe Concepts, Inc.   Phone: (602) 840-9770
P.O. Box 15050
Phoeniz, AZ 85060

Activities designed to help elementary and middle school students practice decision making skills.

★2476★ Science Curriculum Improvement Study
Delta Education   Phone: (603) 889-8899
P.O. Box M
Nashua, NH 03061-6012

Elementary science series using hands-on activities to develop thinking skills. Through each topic, students proceed in discover, concept, introduction, and concept application activities.

★2477★ Secrets and Surprises
Good Apple, Inc.   Phone: (217) 357-3981
1204 Buchanan St.
P.O. Box 299
Carthrage, IL 62321

Contains 183 activities focusing on creative thinking to encourage language development for elementary and middle school children. **Author/Editor:** Joe Wayman and Lorraine Plum.

★2478★ Simulations
Knolbe Concepts, Inc.
P.O. Box 15050
Phoeniz, AZ 85060

Presents 20 realistic situations that involve exercising judgement, wighing alternatives, and considering consequences.

★2479★ Solve - Action Problem Solving
Curriculum Associates, Inc.   Phone: (508) 667-8000
5 Esquire Rd.
North Billerica, MA 01862

Five step program for students in grades 4-9 that integrates strategies with a problem solving procedure. Problems are presented through short stories and are illustrated by action photos. Student books and teacher's guides are available for: Book I - Whole Numbers; Book II - Fractions; Book III - Decimals and Percents. **Price:** $3.95 for student books; $5.95 for teacher's guides. **Author/Editor:** Brian E. Enright.

★2480★ Sound Ideas
Foxtail Press   Phone: (214) 522-3922
P.O. Box 2996
La Habra, CA 90632

Activity booklets with teacher directions and background information on critical thinking: Critical Thinking Grades 1-2, Critical Thinking Grades 4-5, Logical Thinking Grades 7-8, and Creative Thinking Grades 7-8.

★2481★ Springboards to Creative Thinking
NL Associates, Inc.
P.O. Box 1199
Highstown, NJ 08250

Contains 101 activites to stimulate creative thinking in students in grades 3-6. **Author/Editor:** Patricia Tyler Muncy.

★2482★ Start Them Thinking
Illinois Renewal Institute, Inc.   Phone: (312) 991-6300
200 E. Wood St., Ste. 250
Palatine, IL 60067

Handbook of 37 classroom strategies designed to promote thinking and stimulate active participation in the K-4 classroom. **Author/Editor:** Robin Fogarty and Kay Opeka.

★2483★ Stories to Stretch Minds
NL Associates, Inc.
P.O. Box 1199
Highstown, NJ 08250

Presents short story situations designed to be read aloud. Listeners are to ask yes or no questions to decide how or why situations occured. Available in four volumes. **Author/Editor:** Nathan Levy.

★2484★ Strategic Reasoning
Innovative Sciences, Inc.   Phone: (203) 359-1311
300 Broad St.
Stamford, CT 06901

Systematic approach to improving the thinking process. **Author/Editor:** Dr. Arthur Whimbey.

★2485★ Teachers Teaching Thinking
Illinois Renewal Institute, Inc.   Phone: (312) 991-6300
200 E. Wood St., Ste. 250
Palatine, IL 60067

Videotape series for teachers showing thinking skills being taught in a classroom within the context of a subject. Each of four tapes addresses a different age group and subject: Structured Interaction with Thinking, Teaching Explicit Skills of Thinking, Setting the Climate for Thinking, and Metacognitive Processing About Thinking. **Price:** $79.00 each or $295.00 for set of 4. **Author/Editor:** Robin Fogarty and Jim Bellanca.

★2486★ Teaching and Learning Mathematical Problem Solving: Multiple Research Perspectives
Lawrence Erlbaum Associates, Inc.   Phone: (201) 666-4110
365 Broadway
Hillsdale, NJ 07642

Compilation of 24 papers presented at San Diego State University in 1983 on learning mathematical problem solving. **Author/Editor:** Edward A. Silver.

★2487★ Techniques of Structured Problem Solving
Van Nostrand Reinhold Co., Inc.   Phone: (212) 254-3232
115 Fifth Ave.
New York, NY 10003

Booklet containing descriptions and evaluations of over 100 techniques for analyzing and redefining problems, and generating, evaluating, selecting, and implementing ideas. Includes bibliography and index. **Author/Editor:** Dr. Arthur B. VanGundy.

★2488★ Technology, Innovation & Entrepreneurship for Students Magazine (TIES Magazine)
Drexel University   Phone: (215) 895-2386
College of Design Arts
Philadelphia, PA 19104

Drawn upon the information resources of industry, business, government, and education to assist teachers interested in helping students increase technological literacy and capability. **Frequency:** 6 per year. **Subscription:** Free to teachers and administrators; $15 for others.

★2489★ *Think and Reason*

Weekly Reader   Phone: (203) 638-2400
245 Long Hill Rd.
Middletown, CT 06457

Paperback activity books designed to teach students thinking skills through problem-oriented activities and puzzles. **Author/Editor:** Donald Barnes and Arlene Burgdorf.

★2490★ *A Thinker's Log*

Illinois Renewal Institute, Inc.   Phone: (312) 991-6300
200 E. Wood St., Ste. 250
Palatine, IL 60067

Booklet designed to provide high school students with a recording device for reflections on concepts encountered in current studies and class interactions. **Author/Editor:** Robin Fogarty.

★2491★ *The Thinker's Toolbox*

Dale Seymour Publications   Phone: (415) 324-2800
P.O. Box 10888
Palo Alto, CA 94303

Book introduces ways to teach and strengthen divergent thinking skills. Presents 16 tools used in problem solving and contains more than 50 problems that allow for a choice of tools to create alternative solutions. **Author/Editor:** Pamela and David Thornbury.

★2492★ *The Thinking Log*

Illinois Renewal Institute, Inc.   Phone: (312) 991-6300
200 E. Wood St., Ste. 250
Palatine, IL 60067

A 24-page workbook for elementary school students to be used to respond to ideas presented in class. **Price:** $52.50 for 30. **Author/Editor:** Robin Fogarty.

★2493★ *Thinking Posters: Keys to Critical Thinking*

Sundance Publishers & Distributors, Inc. Phone: (508) 468-9201
Newton Rd.
Littleton, MA 01460

Series of lessons designed to improve students' (grades 5-8) critical thinking by teaching them to distinguish good from bad critical thinking. **Author/Editor:** D.N. Perkins.

★2494★ *Thinking Skills: Meanings, Models, and Materials*

Research For Better Schools   Phone: (215) 574-9300
444 N. Third St.
Philadelphia, PA 19123

Brief synopsis, models, and discussions of available thinking programs. **Author/Editor:** Barbara Z. Presseisen.

★2495★ *Thinking Skills Set*

Kolbe Concepts, Inc.   Phone: (602) 840-9770
P.O. Box 15050
Phoenix, AZ 85060

Presents examples of critical thinking and eight creative behaviors, with practical applications, to enhance student understanding of the thinking process.

★2496★ *Thinking to Write - A Work Journal Program*

Curriculum Development Associates,   Phone: (202) 293-1760
Inc.
1211 Connecticut Ave., N.W., Ste. 414
Washington, DC 20036

Journals intended to provide students with an opportunity to assume control of their thinking and to reflect on their thinking process and problem solving abilities. Includes a teacher's edition. **Author/Editor:** Frances R. Link and Shannon Almquist.

★2497★ *Thinking Visually*

Dale Seymour Publications   Phone: (415) 324-2800
P.O. Box 10888
Palo Alto, CA 94303

Strategy manual that describes approaches to problem solving through visual thinking. **Author/Editor:** Robert McKim.

★2498★ *The Thoughtwave Curriculum - Applying Creative Thinking to Problem Solving*

Foxtail Press   Phone: (214) 522-3922
P.O. Box 2996
La Habra, CA 90632

Model for primary-level teachers interested in using a computer with students to develop higher-order thinking skills.

★2499★ *The Unconventional Invention Book*

Good Apple, Inc.   Phone: (217) 357-3981
1204 Buchanan St.
P.O. Box 299
Carthrage, IL 62321

Collection of reproducible activities to stimulate inventive thinking and creating inventions. **Author/Editor:** Bob Standish.

★2500★ *Wake Up Your Creative Genius*

William Kaufmann, Inc.   Phone: (415) 948-5810
95 First St.
Los Altos, CA 94022

Sourcebook of ideas providing techniques and practical examples that can be applied in realistic settings. Topics cover ways to increase and nurture creativity and methods for having ideas accepted and protected. **Author/Editor:** Kurt Hanks and Jay Parry.

★2501★ *Warm-up to Creativity*

Good Apple, Inc.   Phone: (217) 357-3981
1204 Buchanan St.
P.O. Box 299
Carthrage, IL 62321

Games drawn from language arts and social studies skills to provide a conscious and deliberate crossover to divergent ways of thinking. **Author/Editor:** Bob Eberle.

★2502★ *What to Do?*

Trillium Press   Phone: (914) 783-2999
Box 921, Madison Sq. Sta.
New York, NY 10159

Book of problem solving ideas with reproducible worksheets to assist students with fact collecting, defining the problem, and arriving at a solution. **Author/Editor:** Kathryn T. Hegeman.

★2503★ *What's Next*

Kolbe Concepts, Inc.   Phone: (602) 840-9770
P.O. Box 15050
Phoeniz, AZ 85060

Set of 38 drawings and a cassette tape designed to encourage young children to develop observation skills by identifying discrepant events and predicting outcomes.

★2504★ *Wise Owl*

Sundance Publishers & Distributors, Inc.   Phone: (508) 486-9201
Newton Rd.
Littleton, MA 01460

Educational package for teachers of grades 1-3. Designed to develop critical and creative thinking skills. Includes classic children's paperback book, reproducible activity cards with philosopical discussion questions, and teacher's guide for directing activities. **Author/Editor:** Lenore Carlisle and Gareth B. Matthews.

Section 25

Online Resources

"The realization that there are other points of view is the beginning of wisdom."—Charles M. Campbell

BIZ. ORBIT. CompuServe. DIALOG. BYTE. NEXIS. These words look and sound like ITT cable addresses, but represent just six of the major players in an industry encompassing thousands of online databases. Such an abundance of information is available electronically that choosing the appropriate database can be a complicated business.

The Gale Research Inc. publication *Computer-Readable Databases: A Directory and Data Sourcebook* reports on more than 6,000 databases representing a 57 percent increase from 1987 through 1991.

Years ago, electronic databases were used and understood only by specially trained librarians. Not any more. Today online databases are available to anyone, and no prior knowledge of computers is required. The learning curve is short. A few online databases can be learned with less than 20 minutes of practice.

Online databases are not the end-all by any means. A personal touch is often required to iron out the finer points and make sure that some original material has not been omitted. The data are, after all, input by human beings. But, online databases provide more information, faster and easier and more cost-effectively than manual searches.

The inventor is required to stay informed about myriad fields of research and development, many outside one's specialty. Before the advent of online databases this was a difficult, time-consuming task. But with the present network of electronic tracking systems, the inventor is able to monitor fast-breaking events as they occur. All one requires is a PC connected to a phone line, and entry to numerous online databases. And, of course, you can now maintain select databases on your PC utilizing CD-ROM technology.

"Eighty to ninety percent of all the scientists that have ever lived are alive now," historian D.J. Price pointed out in his 1961 book Science Since Babylon. Thus, it is a good guess to venture that eighty to ninety percent of man's written communication has been put to paper in our lifetime.

Many of the publications cited in the chapter entitled Publications for the Inventor, and many more, are available via online databases. Your access to online databases may be as close as your own

computer terminal or as far away as your nearest public library. For more information on databases for use by inventors, see the subheading Online Patent Searches located in Section 4, How to Conduct a Patent Search, on page 31.

Section Arrangement

The following entries are arranged alphabetically by database name. All names in this section are referenced in the Master Index.

★2505★ ADVERTISE

Deutscher Sparkassenverlag GmbH
Am Wallgrassen 115
Postfach 800330
D-7000 Stuttgart 80, Germany

Covers offers and requests of companies for worldwide cooperative business ventures. **Type of Database:** Directory. **Record Items:** Company banking firm; indication of request or offer; type of cooperative venture; company name; company address; contact name, telephone number, fax number; company type; company description; language (English, French, German, or Italian); country; industry code; Standard Industrial Classification (SIC) code; ZIP Code; sales in thousand DM; number of employees; input date; expiry date.

★2506★ APIPAT

American Petroleum Institute (API)   Phone: (212) 366-4040
Central Abstracting & Information
   Services
275 7th Ave., 9th Fl.
New York, NY 10001

Provides worldwide coverage of patents related to petroleum refining, the petrochemical industry, and synthetic fuels. **Type of Database:** Bibliographic. **Record Items:** Document number; patent title; patent assignee; other source; patent codes; priority codes; patent family codes; designated states; abstract (for records since 1980); International Patent Classification (IPC) codes. **Availability:** Online through DIALOG Information Services, Inc.

★2507★ Automated Patent Searching (APS)

MicroPatent   Phone: (203) 786-5500
25 Science Park   Fax: (203) 786-5499
New Haven, CT 06511

Contains complete patent data for all patents issued by the U. S. Patent and Trademark Office (PTO). **Type of Database:** Full-text. **Record Items:** Patent number; application number; filing date; issue date; patent assignee; inventor name; state/country; patent classification codes; title; U.S. references abstract. **Availability:** CD-ROM through the producer.

★2508★ BNA Patent, Trademark, & Copyright Daily

The Bureau of National Affairs, Inc.   Phone: (202) 452-4132
   (BNA)
BNA ONLINE
1231 25th St. NW
Washington, DC 20037

Provides details on legal, regulatory, and legislative developments affecting patent, copyright, trademark, and unfair competition laws. **Type of Database:** Full-text. **Availability:** Online through LEXIS.

★2509★ BNA's Patent, Trademark, & Copyright Journal

The Bureau of National Affairs, Inc.   Phone: (202) 452-4132
   (BNA)   Fax: (203) 822-8092
BNA ONLINE
1231 25th St. NW
Washington, DC 20037

Interprets and analyzes developments in intellectual property issues. **Type of Database:** Full-text.

★2510★ BRANDY

Toyo Information Systems Co., Ltd.
Shinbashi-Sanwa-Toyo Bldg.
1-11-7, Shinbashi, Minato-ku
Tokyo 105, Japan

Contains trademarks registered in Japan. **Type of Database:** Full-text. **Availability:** Online through Toyo Information Systems Co., Ltd. **Contact:** Mr. N. Naito, Manager, Information and Database Service Department. Telex: 252 2817 TISTOK J. Fax: 03 35754355. Phone: 03 35754300.

★2511★ BREV

Belgium Ministere des Affaires Economiques
Office de la Propriete Industrielle (OPRI)
24-26, rue J.A. de Mot
B-1040 Brussels, Belgium

Contains references and abstracts of Belgian patents and other European patents brought into force in Belgium. **Type of Database:** Bibliographic. **Record Items:** International Patent Classification codes; English-language translation; descriptive text. **Availability:** Online through BELINDIS. **Contact:** Dr. Francois-Dominique Declerck, Data Base Manager. Telex: 20627 COMHAN. Fax: 02 2310256. Phone: 02 2336426.

★2512★ British Library Catalogue: Science Reference and Information Service

British Library
Science Reference & Information Service (SRIS)
25 Southampton Bldgs.
Chancery Ln.
London WC2A 1AW, England

Contains full bibliographic descriptions of books and periodicals published in Britain and elsewhere in the areas of science, technology, and business. **Type of Database:** Bibliographic. **Record Items:** Title; author; publisher; series title; publication date; classmarks; subject headings; SRIS building locations and shelfmarks. **Contact:** Alan Gomersall, SRIS Director. Telex: 266959 SCIREF G. Fax: 071-323 7930. Phone: 071-323 7494.

★2513★ BYTE

McGraw-Hill, Inc.   Phone: (603) 924-9281
One Phoenix Mill Ln.
Peterborough, NH 03458

Provides information of interest to computer programmers, including new products, product reviews and comparisons, and highlights of technological developments in the computer industry. **Type of Database:** Full-text. **Record Items:** Article title; journal title; publication date; page numbers; journal title code; ISSN; section heading; byline; article text; special features (accompanying captions in graphs, tables, illustrations, and photographs).

★2514★ BYTE Information Exchange (BIX)
McGraw-Hill, Inc.   Phone: (603) 924-7681
One Phoenix Mill Ln.
Peterborough, NH 03458

An online electronic conferencing system offering a forum for the discussion and exchange of information on a variety of topics of interest to personal computer users. **Type of Database:** Bulletin board; full-text; software. **Contact:** Toll-free: 800-227-2983 (outside New Hampshire). For BIX registration only, call toll-free 800-225-4129.

★2515★ Canadian Patent Index
University of British Columbia Library   Phone: (604) 822-5404
1956 Main Hall   Fax: (604) 822-3893
Vancouver, BC, Canada V6T 1Y3

Provides citations to Canadian patents. **Type of Database:** Bibliographic.

★2516★ Canadian Patent Reporter (CPR)
Canada Law Book Inc.   Phone: (416) 841-6472
240 Edward St.   Fax: (416) 841-5085
Aurora, ON, Canada L4G 3S9

Provides the complete text and headnotes of decisions on copyright, patent, trademarks, industrial design, and intellectual property. **Type of Database:** Full-text. **Record Items:** Date; style of cause; citation; jurisdiction; judges/dates; summary; counsel; decision; publication date. **Contact:** Colleen DuHart, Manager. Toll-free: 800-263-2036; 800-263-3269 (in Ontario and Quebec).

★2517★ Canadian Trade Marks (TMRK)
STM Systems Corp.   Phone: (613) 737-7373
WISDOM Information Services   Fax: (613) 737-9479
2300 St. Laurent Blvd.
Ottawa, ON, Canada K1G 4K1

Provides information on registered and pending marks in Canada. **Type of Database:** Directory. **Record Items:** Registration number; registration date; application number; application date; priority date; renewal date; registered owner; agent or representative; trademark (word and/or design); disclaimer; products or services for which the trademark is registered; basis of claim; associated marks; footnotes. Formatted fields include: wordmark, design, certification mark, distinguishing guise marks, date the trademark was first used, and filing, priority, and registration dates. **Contact:** Scott Flewelling, Market Analyst.

★2518★ CD-Namesearch
Control Data Canada, Ltd.   Phone: (613) 598-0217
Information Services Group   Fax: (613) 563-1716
130 Albert St., 11th Fl.
Ottawa, ON, Canada K1P 5G4

Contains the complete official U.S., U.K., and Canadian trademark databases. Includes images of design marks. **Type of Database:** Full-text; graphic.

★2519★ CD-ROM Electronic News
Ellis Enterprises, Inc. (EEI)   Phone: (405) 749-0273
4205 McAuley Blvd., Ste. 385   Fax: (405) 751-5168
Oklahoma City, OK 73120

Provides information on CD-ROM technology, products, and software currently in use or available for purchase. **Type of Database:** Full-text; bulletin board; directory. **Availability:** Online through producer. Computer is directly accessible by dialing (405) 236-3750.

★2520★ Chinese Patent Abstracts in English Data Base
International Patent Documentation Center (INPADOC)
Mollwaldplatz 4
A-1040 Vienna, Austria

Contains citations and English-language abstracts of all patents published in the People's Republic of China since the opening of the Chinese Patent Office on April 1, 1985. **Type of Database:** Bibliographic. **Record Items:** Title of initial publication for a given invention in China (called a patent basic); subsequent Chinese patent document titles concerning the same invention; application country; application date; application number; authors (inventors); International Patent Classification (IPC) code; number of patents; patent assignee; patent country code; publication date.

★2521★ CIB
Institut National de la Propriete Industrielle (INPI)
26 bis, rue de Leningrad
F-75800 Paris Cedex 8, France

Contains patent classification codes from the International Patent Classification (IPC). **Type of Database:** Full-text.

★2522★ CLAIMS/CITATION
IFI/Plenum Data Corp.   Phone: (703) 683-1085
302 Swann Ave.   Fax: (703) 683-0246
Alexandria, VA 22301

Provides more than 5,000,000 patent references cited during the patent examination process against each United States patent, and references to the patents in which it has subsequently been cited. **Type of Database:** Bibliographic. **Record Items:** Patent number; number of later patent citations; patent citation numbers.

★2523★ CLAIMS/CLASS
IFI/Plenum Data Corp.   Phone: (703) 683-1085
302 Swann Ave.   Fax: (703) 683-0264
Alexandria, VA 22301

Provides a classification code and title dictionary for all classes and subclasses of the U.S. Patent Classification System. **Type of Database:** Full-text. **Record Items:** Uniterm code/text; USC code; general or compound term code; CDB fragment code/text; molecular formula.

★2524★ CLAIMS/Comprehensive Data Base
IFI/Plenum Data Corp.   Phone: (703) 683-1085
302 Swann Ave.   Fax: (703) 683-0246
Alexandria, VA 22301

Provides bibliographic and major claim information on chemical patents issued by the United States Patent and Trademark Office since 1950; general, electrical, and mechanical patents since 1963; and design patents since 1980; abstracts are included since 1971. **Type of Database:** Bibliographic. **Record Items:** Abstract; claim text; CAS Registry Numbers; application country/date/number; controlled term; document type; family member country/date/number; field availability; file segment; fragment code; IPC code; inventors; patent assignee/country/number; publication date; role indicator; title; uniterm code; UPC code.

★2525★ CLAIMS/Reassignment & Reexamination
IFI/Plenum Data Corp.   Phone: (703) 683-1085
302 Swann Ave.   Fax: (703) 683-0246
Alexandria, VA 22301

Provides information on patents whose ownership has been reassigned from the original assignee to another company or individual since 1980, and patents reexamined since 1981 by the U.S. Patent and Trademark Office at the request of a second party who has raised substantial new

questions regarding the patentability of a patent's claims. Also includes patents which have expired due to non-payment. **Type of Database:** Bibliographic. **Record Items:** Reexamination request number/date; requestor and requestor location; reexamination certificate date/number/text; reassignment date/type; new patent assignee; expiration date.

★2526★ CLAIMS/UNITERM

IFI/Plenum Data Corp. Phone: (703) 683-1085
302 Swann Ave. Fax: (703) 683-0246
Alexandria, VA 22301

Provides bibliographic and major claim information on chemical patents issued by the United States Patent and Trademark Office since 1950; general, electrical, and mechanical patents since 1963; and design patents since 1980. **Type of Database:** Bibliographic. **Record Items:** Title; abstract; claim text; country code; application country/date/number; inventor name/country; class code; CAS Registry Number; document type; International Patent Classification (IPC) code; patent assignee; patent assignee country/code; issue date; patent number; publication year.

★2527★ CLAIMS/U.S. Patent Abstracts

IFI/Plenum Data Corp. Phone: (703) 683-1085
302 Swann Ave. Fax: (703) 683-0246
Alexandria, VA 22301

Provides bibliographic and major claim information on chemical patents issued by the United States Patent and Trademark Office since 1950; general, electrical, and mechanical patents since 1963; design patents since 1980; abstracts are included since 1971. **Type of Database:** Bibliographic. **Record Items:** Title; abstract; claim text; country code; application country/date/number; inventor name/country; class code; CA reference number; document type; International Patent Classification (IPC) code; patent assignee; patent assignee country/code; issue date; patent number; publication year; CAS Registry Number.

★2528★ Classification and Search Support Information System (CASSIS)

U.S. Patent & Trademark Office Phone: (703) 557-6154
Office of Electronic Data Conversion & Fax: (703) 557-0668
 Dissemination
Office of the Director
Crystal Park 2, Ste. 1100B
Washington, DC 20231

Contains patent classification information, bibliographic information, and the Manual of Classification for patents. **Type of Database:** Full-text; bibliographic. **Record Items:** Patent; year of issue; state/country of the inventor's residence; assignee at time of issue; title and status; current patent abstracts (up to three years).

★2529★ CLINPAT

Registro de la Propiedad Industrial de Espana
Departamento de Informacion Tecnologica
Calle Panama, 1
28071 Madrid, Spain

Contains the complete text of the current (5th) edition of the International Patent Classification (IPC), with the terms translated into Spanish. **Type of Database:** Full-text. **Availability:** Online through producer.

★2530★ COMLINE

COMLINE International Corp.
Shugetsu Bldg.
3-12-7, Kita-Aoyama, Minato-ku
Tokyo 107, Japan

Covers Japanese industry, with emphasis on new products, new technologies, ongoing research, and marketing strategies. **Type of**

Database: Bibliographic. **Contact:** In the United States, COMLINE International Corp., 10601 S. De Anza Blvd., Suite 216, Cupertino, CA 95014. Telephone: (408) 257-9956. Facsimile: (408) 257-0695.

★2531★ Compu-Mark U.K. On-Line

Compu-Mark (UK) Ltd.
New Premier House, Ste. 3
150 Southampton Row
London WC1B 5AL, England

Contains more than 250,000 registrations, applications, and pending applications filed with the British Patent Office. **Type of Database:** Full-text. **Record Items:** Trademark; registration of application number; Part B marks; classes of goods and/or services; status: pending unpublished applications, active published applications/registrations, inactive marks (abandoned, cancelled, or expired); WHO, INN's, or ISO pesticide names; year and page of publication in Trade Marks Journal; owner; codes indicating changes to any aspect of the trademark's status. **Contact:** David I. Sheppard, Director.

★2532★ Compu-Mark U.S. On-Line

Compu-Mark U.S.
500 Victory Rd.
North Quincy, MA 02171-1545

Contains all trademarks contained in the U.S. federal and state registers. Includes 1 million U.S. and 300,000 state trademarks. **Type of Database:** Full-text. **Record Items:** Trademark; owner; registration of application numbers, with code indicating federal or state origins; classes of goods or services by U.S. and international classification; status: pending applications, active registrations, inactive (abandoned, cancelled, or expired), state registration, affidavits; publication date in official gazette; texts to provide information on changes of ownership, assignment, other.

★2533★ Computerized Administration of Patent Documents Reclassified According to the IPC (CAPRI)

International Patent Documentation Center (INPADOC)
Mollwaldplatz 4
A-1040 Vienna, Austria

Contains references to worldwide patent documentation issued before 1973 which have been or are being reclassified according to the International Patent Classification. **Type of Database:** Bibliographic. **Record Items:** Country of publication; document type; document number; International Patent Classification (IPC) symbols; edition(s) in which the given IPC symbol is valid (IPC 1, 2, 3...); reclassifying organization. **Contact:** Dipl.-Kfm. Norbert Fux, Director, Sales Department.

★2534★ Current Patents Evaluation

Current Patents Ltd.
34-42 Cleveland St.
London W1P 5FB, England

Provides evaluations and merit ratings of current therapeutic patents registered worldwide covering cardiovascular, central nervous system (CNS), and antimicrobial. **Type of Database:** Full-text. **Record Items:** Patent title; annotated title; journal title; patent assignee; inventor; designated states; publication date; priority application number and date; filing date; section code; International Patent Classification (IPC) codes; key references; abstract; evaluation; merit rating. **Contact:** Amanda Stembridge, Director, Online Services.

★2535★ Current Patents Fast-Alert
Current Patents Ltd.
34-42 Cleveland St.
London W1P 5FB, England

Provides information on pharmaceutical patents and patent applications registered internationally and in the United States. **Type of Database:** Directory. **Record Items:** Patent title; patent number; company name; drug code; abstract. **Availability:** Online through Data-Star. **Contact:** Amanda Stembridge, Director, Online Services.

★2536★ Database on Legal Precedents Regarding Intellectual Property Rights
Kinki University
Industrial & Law Information Institute
Kowakae 3-4-1, Higashiosaka-shi
Osaka 577, Japan

Contains the complete text of legal precedents for laws concerning intellectual property rights. **Type of Database:** Full-text. **Availability:** Online through Heiwa Information Center Company, Ltd.

★2537★ Deutsche Patent Datenbank (PATDPA)
Deutsches Patentamt
Zweibrueckenstr. 12
D-8000 Munich 2, Germany

Contains references, abstracts, and illustrations from patent documents published by the Deutsches Patentamt. **Type of Database:** Bibliographic; graphic. **Record Items:** Title of invention; inventor; system number; application country, date, kind, number, and type; document type; entry date and week; family member country, publication date, kind, number, and publication type; International Patent Classification; language; patent country; publication date. **Contact:** Peter Holtkemeier, Dipl. Ing. Informations-Service.

★2538★ DMARK
Compu-Mark
St. Pietersvliet 7
B-2000 Antwerp, Belgium

Contains all verbal and verbal/numerical components of trademarks currently in force in West Germany. **Type of Database:** Bibliographic. **Record Items:** Trademark name; registration number; international class; legal status.

★2539★ DYNIS
Control Data Canada, Ltd.   Phone: (613) 598-0217
Information Services Group   Fax: (613) 563-1716
130 Albert St., 11th Fl.
Ottawa, ON, Canada K1P 5G4

Contains the complete text of registered and pending trademark applications in Canada. **Type of Database:** Full-text. **Contact:** Mr. Jean Millette, Sales Administrator.

★2540★ ECLATX
Institut National de la Propriete Industrielle (INPI)
26 bis, rue de Leningrad
F-75800 Paris Cedex 8, France

Contains the full text of the international classification scheme of the European Patent Office (EPO), which is built according to the hierarchical structure of the International Patent Classification (IPC). **Type of Database:** Full-text. **Availability:** Online throught Questel. **Contact:** Catherine Pagis, Marketing Manager.

★2541★ EDOC
Institut National de la Propriete Industrielle (INPI)
26 bis, rue de Leningrad
F-75800 Paris Cedex 8, France

Provides cross-referenced numbers to patents issued by different countries for the same invention. **Type of Database:** Directory. **Availability:** Online through Questel. **Contact:** Catherine Pagis, Marketing Manager.

★2542★ The Electronic Whole Earth Catalog
Broderbund Software   Phone: (415) 492-3200
17 Paul Dr.
San Rafael, CA 94903-2101

Provides expert evaluations of innovative new products, services, and technologies. **Type of Database:** Directory; graphic; audio; full-text.

★2543★ EPAT
Institut National de la Propriete Industrielle (INPI)
26 bis, rue de Leningrad
F-75800 Paris Cedex 8, France

Lists patents applied for and published in the European Patent Office's printed [European Patent Bulletin]. **Type of Database:** Full-text. **Record Items:** Depositor; title; technical section; country of origin; place of publication; date of filing; publication date; publication and registration numbers; applicant name; inventor name; representative priority rights; title; technical sections according to the International Patent Classification (IPC); opposition; designated states; original titles in French, English, and German. **Contact:** Catherine Pagis, Marketing Manager.

★2544★ ESPACE EP-A
European Patent Office (EPO)
Erhardtstr. 27
D-8000 Munich 2, Germany

Contains the complete text of European patent applications, including illustrations and claims. **Type of Database:** Full-text; graphic. **Record Items:** Inventor; applicant; agent; application title (in English, French, and German); International Patent Classification (IPC) code; country of origin; countries designated; application date; application number; opponents; dates of opposition.

★2545★ ESPACE FIRST
European Patent Office (EPO)
Erhardtstr. 27
D-8000 Munich 2, Germany

Provides the complete text images of the first pages of all European patent applications and international patent applications filed in accordance with the Patent Cooperation Treaty (PCT). **Type of Database:** Full-text. **Record Items:** Inventor; applicant; agent; application title (in English, French, and German); International Patent Classification (IPC) code; country of origin; countries designated; application date; application number; opponents; dates of opposition.

★2546★ ESPACE WORLD
World Intellectual Property Organization (WIPO)
34, Chemin des Colombettes
CH-1211 Geneva 20, Switzerland

Contains full images of European patent applications, including illustrations and text. **Type of Database:** Graphic; full-text. **Record Items:** Inventor; applicant; agent; application title (in English, French, and German); International Patent Classification (IPC) code; country of origin; countries designated; application date; application number; opponents; dates of opposition.

★2547★ European Patent Searching (EPS)

MicroPatent
25 Science Park
New Haven, CT 06511

Phone: (203) 786-5500
Fax: (203) 786-5499

Contains citations and abstracts for European Patent Office (EPO) patent applications. **Type of Database:** Full-text. **Record Items:** Patent number; issue year; patent assignee; state/country; patent classification codes; title; abstract. **Availability:** CD-ROM through producer.

★2548★ EXPORT

VVO Licensintorg
Oy LISTECH Ltd.
Hameentie 103
PO Box 16
00550 Helsinki, Finland

Contains information on technological products manufactured in the Soviet Union. **Type of Database:** Full-text. **Availability:** Online through HELECON. **Contact:** Hans Lindqvist.

★2549★ Exposition

International Center for Scientific & Technical Information (ICSTI)
Ul. Kuusinena 21B
125252 Moscow, Union of Soviet Socialist Republics

Contains information on new products and developments from countries participating in the Council for Mutual Economic Assistance (COMECON). **Type of Database:** Directory. **Record Items:** Product; production program; industry; brand name; exhibiting company; production partner; company with documentation. **Contact:** GBI Help Desk.

★2550★ Financial Times Business Reports: Fintech

Financial Times Business Information (FTBI)
Financial Times Electronic Publishing
One Southwark Bridge
London SE1 9LS, England

Contains the complete text of articles appearing in Financial Times newsletters covering developments in the business aspects of new technology in six key areas-telecom markets, the electronic office, personal computer markets, factory automation, software markets, and computer product updates. **Type of Database:** Full-text. **Record Items:** Article title; source; publication date; page numbers; publication year; named companies; length of article; abstract.

★2551★ FMARK

Institut National de la Propriete Industrielle (INPI)
26 bis, rue de Leningrad
F-75800 Paris Cedex 8, France

Covers all applied French trademarks. **Type of Database:** Full-text **Record Items:** Trademark name; designated products; classification code; agent; applicant; filing number and date; publication number; renewal information; design and color. **Availability:** Online through Questel. **Contact:** Catherine Pagis, Marketing Manager.

★2552★ FPAT

Institut National de la Propriete Industrielle (INPI)
26 bis, rue de Leningrad
F-75800 Paris Cedex 8, France

Lists French patents applied for and published in the printed Bulletin Officiel de la Propriete Industrielle. **Type of Database:** Full-text. **Record Items:** Article title; author; journal title; volume and issue numbers; page numbers; publication date; representative priority rights; technical sections according to the International Patent Classification (IPC). **Contact:** Catherine Pagis, Marketing Manager.

★2553★ Friday Memo

Information Industry Association (IIA)
555 New Jersey Ave. NW, Ste. 800
Washington, DC 20001

Phone: (202) 639-8262
Fax: (202) 638-4403

Reports on developments in the information industry and within the Information Industry Association and its divisions and service councils. **Type of Database:** Full-text.

★2554★ FullText

MicroPatent
25 Science Park
New Haven, CT 06511

Phone: (203) 786-5500
Fax: (203) 786-5499

Contains the complete text of all patents issued by the U.S. Patent and Trademark Office (PTO), including descriptions of drawings, background and summary of the invention, examples, and claims. **Type of Database:** Full-text; graphic. **Record Items:** Patent number; issue year; patent assignee; state/country; patent classification codes; title; description of drawings; description of invention; examples; claims. **Availability:** CD-ROM through the producer.

★2555★ Government-Industry Data Exchange Program (GIDEP)

U.S. Navy
Naval Warfare Assessment Center
GIDEP Operations Center
Corona, CA 91720

Phone: (714) 273-4677
Fax: (714) 273-5200

Facilitates the exchange of technical data between government and private industry on parts, components, materials, and processes. **Type of Database:** Numeric; full-text. **Availability:** Online through producer. **Contact:** Edwin T. Richards, Program Manager for Reliability.

★2556★ GRAFMAR

Registro de la Propiedad Industrial de Espana
Departamento de Informacion Tecnologica
Calle Panama, 1
28071 Madrid, Spain

Contains approximately 190,000 graphic images associated with trademarks, and includes bibliographic citations. **Type of Database:** Bibliographic; graphic. **Contact:** Miguel Angel Gutierrez Carbajal, Director, Department of Information Technology.

★2557★ IMARK

Compu-Mark
St. Pietersvliet 7
B-2000 Antwerp, Belgium

Contains all valid trademarks in Italy, including unpublished pending applications filed in the Provincial offices of Commerce, unpublished registrations, registrations published in the Official Bulletin as well as Italian-French and French-Italian trademarks. **Type of Database:** Full-text. **Record Items:** Trademark number; application date; trademark assignee; assignee address; product description. **Availability:** Online through Questel.

★2558★ IMSMARQ

Imsmarq AG
Industriestr. 9
CH-6301 Zug, Switzerland

Covers trademarks currently in force and applied for all classes of goods and services in Denmark, Finland, Germany, Italy, Norway, Sweden, the United Kingdom, the United States, and the International Register (WIPO). **Type of Database:** Directory. **Availability:** Online through producer. **Contact:** In the United States, Patricia Penyak, 660 W.

Germantown Pike, Plymouth Meeting, PA 19462-1048. Telephone: (215) 834-5089. Facsimile: (215) 834-5100.

★2559★ INPADOC Data Base (IDB)

International Patent Documentation Center (INPADOC)
Mollwaldplatz 4
A-1040 Vienna, Austria

Contains references to worldwide patent documentation for 56 countries, which accounts for 96 percent of the world's currently published patent documents. **Type of Database:** Bibliographic. **Record Items:** Country of publication; document type; document publication date; country of priority; number of application (serves as the basis of priority); priority date; International Patent Classification (IPC) symbol (if present); inventor name; name of owner; applicant name; invention title; national classification symbol; other legally related domestic application. **Contact:** Dipl.-Kfm. Norbert Fux, Director, Sales Department.

★2560★ INPAMAR

Registro de la Propiedad Industrial de Espana
Departamento de Informacion Tecnologica
Calle Panama, 1
28071 Madrid, Spain

Contains details of trademarks, trade names, business signs, and international trademarks registered in Spain. **Type of Database:** Full-text. **Contact:** Miguel Angel Gutierrez Carbajal, Director, Department of Information Technology.

★2561★ INSPEC

Institution of Electrical Engineers (IEE)
Michael Faraday House
Six Hills Way
Stevenage SG1 2AY, England

Contains citations and abstracts of the world's technical literature dealing with physics, electrical engineering, electronics and telecommunications, control technology, computers and computing, and information technology. **Type of Database:** Bibliographic. **Record Items:** Abstract number; article title; author; author affiliation name and address; journal title abbreviation; country of publication; volume and issue numbers; page numbers; number of references; publication date; CODEN; ISSN; CCCC; document type; language; treatment codes; abstract. **Contact:** In North and South America, INSPEC, IEEE Service Center, 445 Hoes Lane, P.O. Box 1331, Piscataway, NJ 08855-1331. Telephone: (908) 562-5549. Telex: 833233 IEEE PWAY. Facsimile: (908) 981-0027.

★2562★ ISC Data Bank

Invention Submission Corp. (ISC)   Phone: (412) 288-1300
903 Liberty Ave.   Fax: (412) 288-1354
Pittsburgh, PA 15222

Contains information on new product ideas and interested manufacturers. **Type of Database:** Directory. **Contact:** Peter Geiringer, Sales Director.

★2563★ Japan High Tech Review

Kyodo News International, Inc. (KNI)   Phone: (212) 586-0152
50 Rockefeller Plaza, Rm. 803
New York, NY 10020

Contains news and analyses of Japan's high technology industries. **Type of Database:** Full-text. **Contact:** Toshiaki Mitsudome, Director, Online Services. Phone: (212) 397-3723. Telex: 201975 KYODO USA. Fax: (212) 397-3721.

★2564★ Japio

Japan Patent Information Organization (Japio)
5-16-1, Toranomon
Minato-ku
Tokyo 105, Japan

Contains abstracts and drawings of Japanese patent and utility model documents and covers design and trademark information as well. **Type of Database:** Bibliographic; graphic. **Record Items:** For Japanese documents: application number and date; publication number; publication date; registration number and date; priority country name, number, and date; International Patent Classification code; Japanese Patent Classification code; applicant name; inventor name; title of invention; file history; code of attorney; abstract; illustration; keywords. **Contact:** Yasushi Furukawa, Manager, International Affairs Section.

★2565★ JURINPI

Institut National de la Propriete Industrielle (INPI)
26 bis, rue de Leningrad
F-75800 Paris Cedex 8, France

Contains references and abstracts of French legal decisions concerning patents issued and trademarks registered since 1952. **Type of Database:** Bibliographic. **Record Items:** Jurisdiction; date of decision; parties involved; patent number and title; abstract; precedent cases; bibliographic reference. **Contact:** Catherine Pagis, Marketing Manager.

★2566★ KPTN

Korea Institute of Industry & Technology Information (KINITI)
PO Box 205, Cheongryangri
Seoul, Republic of Korea

Provides citations and abstracts of all Korean patent specifications. **Type of Database:** Bibliographic. **Record Items:** Applied year and month; granted year and month; patent title; patentee; inventor; patent number and date (applied, unexamined, examined, registered); priority; abstract. **Contact:** Mr. Jae Bok Kim, Director, International Cooperation Division.

★2567★ KUMO

Korea Institute of Industry & Technology Information (KINITI)
PO Box 205, Cheongryangri
Seoul, Republic of Korea

Indexes all examined specifications of Korean utility models. **Type of Database:** Bibliographic. **Record Items:** Granted year and month; applied year and month; patent title; patentee; inventor; International Patent Classification (IPC) codes; patent number and date (applied, unexamined, examined, registered); priority. **Contact:** Mr. Jae Bok Kim, Director, International Cooperation Division.

★2568★ LEXIS Federal Patent, Trademark, & Copyright Library

Mead Data Central, Inc. (MDC)   Phone: (513) 865-6800
LEXIS
9393 Springboro Pike
PO Box 933
Dayton, OH 45401

Provides the complete text of U.S. Court decisions and publications related to patent, trademark, and copyright laws. **Type of Database:** Full-text. **Availability:** Online through LEXIS.

★2569★ LEXPAT

Mead Data Central, Inc. (MDC) Phone: (513) 865-6800
9393 Springboro Pike
PO Box 933
Dayton, OH 45401

Provides the complete text of utility patents issued by the U.S. Patent and Trademark Office (PTO). **Type of Database:** Full-text. **Availability:** Online through NEXIS.

★2570★ LINCUP News Database

LINC Resources, Inc. Phone: (614) 885-5599
4820 Indianola Ave. Fax: (614) 433-0852
Columbus, OH 43214

Facilitates the exchange of information related to the distribution and dissemination of special education products. **Type of Database:** Directory; bulletin board. **Contact:** Chuck Lynd, Director, Information Services.

★2571★ LitAlert

Research Publications, Inc. Phone: (703) 920-5050
Rapid Patent Service Fax: (703) 685-3987
1921 Jefferson Davis Hwy., Ste. 1821-D
Arlington, VA 22202

Provides information on unpublished, unresolved, and current U.S. patent and trademark litigation. **Type of Database:** Full-text; bibliographic. **Record Items:** Patent or trademark number and title; patent publication date; document type; inventor; patent assignee; other patent or trademark numbers; trademark status date; trademark; court location; defendant; plaintiff; docket number; filing date; action date; description of the action; note; names of law firms or attorneys. **Contact:** Eleanor Roberts, Telemarketing Representative.

★2572★ Local Government Information Network (LOGIN)

Login Services Corp. Phone: (612) 297-6900
245 E. 6th St., Ste. 809 Fax: (612) 297-6333
St. Paul, MN 55101-9006

Contains summaries of experiences and expertise from local government jurisdictions. **Type of Database:** Directory; full-text. **Record Items:** Summary; contact organization; contact person. **Contact:** Joan Coleman, Manager, Customer Services.

★2573★ McGraw-Hill Publications Online

McGraw-Hill Information Services Co. Phone: (609) 426-5000
Princeton-Hightstown Rd. Fax: (609) 426-7352
Hightstown, NJ 08520

Contains the complete text of the following 34 industry-specific McGraw-Hill magazines and newsletters that serve as a source of background information on companies and industry, the economy and international markets, labor and management, government, and technology: bltA Aerospace Daily (1989 to the present). **Type of Database:** Full-text. **Record Items:** Article title; journal title; publication date; volume and issue numbers; page numbers; journal title code; ISSN; section heading; byline; article text; special features (accompanying captions in graphs, tables, illustrations, and photographs). **Availability:** Online through DIALOG Information Services, Inc.

★2574★ MIC/Info

Management Information Corp. (MIC) Phone: (609) 428-1020
401 E. Rte. 70 Fax: (609) 428-1683
PO Box 5062
Cherry Hill, NJ 08034

Covers news and developments in the voice and data communications industry, including new product announcements and company and financial changes. **Type of Database:** Full-text. **Availability:** Online through the producer. **Contact:** Carol Bell, Circulation Manager.

★2575★ NUANS

Control Data Canada, Ltd. Phone: (613) 598-0217
Information Services Group Fax: (613) 563-1716
130 Albert St., 11th Fl.
Ottawa, ON, Canada K1P 5G4

Contains corporate names and trademarks in Canada currently in use. **Type of Database:** Full-text. **Record Items:** Name; evaluation of corporate name based on phonetics, letter content, root word, coined words, and synonyms; distinctive versus descriptive terms; line of business; geographical proximity. **Contact:** Mr. Jean Millette, Sales Administrator.

★2576★ OG/PLUS

Research Publications, Inc. Phone: (703) 920-5050
Rapid Patent Service Fax: (703) 685-3987
1921 Jefferson Davis Hwy., Ste. 1821-D
Arlington, VA 22202

Covers patents granted by the United States Patent and Trademark Office (PTO). **Type of Database:** Full-text; bibliographic; graphic. **Record Items:** Patent number; original number (if reissue); title; issue date; inventor name; inventor state/country; assignee name; assignee state/country; filing date; serial number; related patents/applications/priority country; priority date; priority application number; IPC class; U.S. class; number of claims. **Contact:** Eleanor Roberts, Telemarketing Representative.

★2577★ ORBPAT

ORBIT Search Service Phone: (703) 442-0900
8000 Westpark Dr. Fax: (703) 893-4632
McLean, VA 22102

Assists the user in crossfile searching of patents databases accessible through the ORBIT Search Service. **Type of Database:** Directory.

★2578★ PAPERCHEM

Institute of Paper Science & Technology Phone: (404) 853-9500
575 14th St. NW Fax: (404) 853-9510
Atlanta, GA 30318

Provides citations and abstracts to international patent and journal literature dealing with pulp and paper technology. **Type of Database:** Bibliographic. **Record Items:** Author; editor; corporate author; article title; journal title; publication date; volume and issue numbers; page numbers; abstract; language. **Contact:** D. Gail Stahl, Database Manager.

★2579★ PATDATA

BRS Information Technologies Phone: (703) 442-0900
8000 Westpark Dr. Fax: (703) 889-4632
McLean, VA 22102

Contains detailed information and abstracts for all utility patents issued by the U.S. Patent and Trademark Office since 1971, and all reissue patents which have been issued since July 1, 1975. **Type of Database:** Bibliographic. **Record Items:** Patent number; title; inventor name and address; patent assignee; patent application date; foreign priority; original U.S. patent classification codes; cross-reference U.S. patent

classification codes; international patent classification codes; U.S. patents cited; foreign patents cited; abstract.

★2580★ Patent History

Research Publications, Inc.   Phone: (703) 920-5050
Rapid Patent Service   Fax: (703) 685-3987
1921 Jefferson Davis Hwy., Ste. 1821-D
Arlington, VA 22202

Contains the complete history of all active U.S. patents for the past 17 years, including a complete record of all patent status changes and litigation history for all patent suit cases reported to the U.S. Patent and Trademark Office (excluding those cases already covered in the Official Gazette of the U.S. PTO earlier in the period). **Type of Database:** Full-text; bibliographic. **Record Items:** Patent number; title of invention; state/country; assignee code; classification; issue year; status history; issue date; inventor name; assignee name; class title; plaintiffs; defendants; court; docket number; filing date; date of latest action; notes; other affected patents. **Contact:** Eleanor Roberts, Telemarketing Representative.

★2581★ Patent Status File

Research Publications, Inc.   Phone: (703) 920-5050
Rapid Patent Service   Fax: (703) 685-3987
1921 Jefferson Davis Hwy., Ste. 1821-D
Arlington, VA 22202

Provides information on status changes to U.S. patents over the last 17 years, and tracks 20 patent status change actions as reported by the U.S. Patent and Trademark Office (PTO). **Type of Database:** Full-text. **Record Items:** Patent number; reissue number; document type; change code; change of action; Official Gazette (or withdrawal notice) date; notes; subject heading. **Contact:** Eleanor Roberts, Telemarketing Representative.

★2582★ PatentImages

MicroPatent   Phone: (203) 786-5500
25 Science Park   Fax: (203) 786-5499
New Haven, CT 06511

Contains the complete text plus images of all U.S. patents. **Type of Database:** Bibliographic; full-text; graphic. **Record Items:** Patent number; issue year; patent assignee; state/country; patent classification codes; title; text; image. **Availability:** CD-ROM through producer.

★2583★ PATGRAPH

Deutsches Patentamt
Zweibrueckenstr. 12
D-8000 Munich 2, Germany

Contains graphics and chemical formulas derived from patents. **Type of Database:** Numeric; graphic. **Availability:** Online through STN International. **Contact:** Peter Holtkemeier, Dipl. Ing. Informations-Service.

★2584★ PATOS European Legal Status and Alterations

Wila-Verlag Wilhelm Lampl KG
Landsberger Str. 191a
D-8000 Munich 21, Germany

Contains citations to the legal status of European patents, published patent applications, and EURO-PCT applications published by the European Patent Office. **Type of Database:** Bibliographic. **Record Items:** Patent and application number; document type; procedure; language; designated states; international patent classification; hybrid codes; title; application type; name and address of inventor; assignee and agent; application and publication date; priority information. **Contact:** Lydia Rapp, Information Consultant.

★2585★ PATOS European Patent Applications

Wila-Verlag Wilhelm Lampl KG
Landsberger Str. 191a
D-8000 Munich 21, Germany

Contains citations to European patent applications submitted by the German Patent Office. **Type of Database:** Bibliographic. **Record Items:** Patent and application number; document type; procedure; language; designated states; international patent classification; hybrid codes; title; application type; name and address of inventor; assignee and agent; application and publication date; priority information.

★2586★ PATOS European Patents

Wila-Verlag Wilhelm Lampl KG
Landsberger Str. 191a
D-8000 Munich 21, Germany

Contains citations to European patents granted by the European Patent Office. **Type of Database:** Bibliographic. **Record Items:** Patent and application number; document type; procedure; language; designated states; international patent classification; hybrid codes; title; name and residence of inventor; assignee and agent; application and publication dates; examination information with references; priority information.

★2587★ PATOS German Patent Applications and Utility Models

Wila-Verlag Wilhelm Lampl KG
Landsberger Str. 191a
D-8000 Munich 21, Germany

Contains citations to first patent applications and utility models published by the German Patent Office. **Type of Database:** Bibliographic; full-text. **Record Items:** Patent number; document type; international patent classification; title; name and address of inventor; assignee and legal representative; dates of filing; publication and grant; examination information; priority information.

★2588★ PATOS German Patents

Wila-Verlag Wilhelm Lampl KG
Landsberger Str. 191a
D-8000 Munich 21, Germany

Contains citations to German patents granted by the German Patent Office. **Type of Database:** Bibliographic. **Record Items:** Patent number; document type; international patent classification; title; name and address of inventor; assignee and legal representative; dates of filing; publication and grant; examination information; priority information.

★2589★ PATOS PCT Applications

Bertelsmann InformationsService GmbH
Landsberger Str. 191a
D-8000 Munich 21, Germany

Contains citations to international patent applications submitted to the Patent Cooperation Treaty (PCT). **Type of Database:** Bibliographic. **Record Items:** Patent number; document type; international patent classification; title; name and address of inventor; assignee and legal representative; dates of filing; publication and grant; examination information; priority information. **Availability:** Online through Bertelsmann InformationsService GmbH.

★2590★ PHARMSEARCH

Institut National de la Propriete Industrielle (INPI)
26 bis, rue de Leningrad
F-75800 Paris Cedex 8, France

Contains information on pharmaceutical patents published in France and the United States, and by the European Patent Office (EPO). **Type of Database:** Properties; bibliographic; graphic. **Record Items:** Chemical

names preparation processes; described therapeutic activities; English abstract; administration information relative to the patent application; publication; applicant; and International Patent Classification (IPC). **Availability:** Online through Questel. **Contact:** Catherine Pagis, Marketing Manager.

★2591★ PTS New Product Announcements/Plus (NPA/Plus)

Predicasts
11001 Cedar Ave.
Cleveland, OH 44106

Phone: (216) 795-3000
Fax: (216) 229-9944

Contains the complete text of news releases that announce business events, including new product introductions and modifications, new technologies and processes, license agreements, new facilities and expansions, mergers and acquisitions, contract awards, litigation, and corporate financial results as issued by the developing company or its marketing agent. **Type of Database:** Full-text. **Record Items:** Dateline; release date; title; text; company name; D-U-N-S number; ticker symbol; CUSIP; event name and code; product name and code; use name and code; geographic name and code; price; performance specification; trade name. **Contact:** Emily Melton, Director, Marketing, or Customer Service Department.

★2592★ PTS PROMT

Predicasts
11001 Cedar Ave.
Cleveland, OH 44106

Phone: (216) 795-3000
Fax: (216) 229-9944

Provides abstracts and the complete text of journal articles, newspaper articles, and other sources of worldwide market and technology information relating to more than 200,000 companies, new products, plant capacities and expansions, market spread shares, sales and consumption, government regulations, mergers and acquisitions, and other topics. **Type of Database:** Bibliographic; full-text. **Record Items:** Article title; journal title; publication date; page numbers; abstract; country code and name; product code and name; event code and description; company name; D-U-N-S number; ticker symbol; CUSIP number.

★2593★ Rapra Trade Names (RAPTN)

Rapra Technology Ltd.
Information Centre
Shawbury
Shrewsbury SY4 4NR, England

Contains citations and abstracts of worldwide literature on tradenames of particular interest to companies producing rubber and plastics materials and products, as well as to suppliers to the industry and users of its products. **Type of Database:** Bibliographic. **Record Items:** Trade name; company name; product category; product description; graphic design; source citation. **Contact:** Paul Cantrill, Controller, Information Services, or Elaine Davison.

★2594★ Search Master Intellectual Property Library

Matthew Bender & Company, Inc.
1275 Broadway
Albany, NY 12204-2694

Phone: (518) 487-3000

Covers U.S. intellectual property rights. **Type of Database:** Full-text. **Availability:** CD-ROM through producer.

★2595★ Technology Assessment and Forecast Reports Data Base

U.S. Patent & Trademark Office
Technology Assessment & Forecast
  Program (TAF)
CM2-304
Washington, DC 20231

Phone: (703) 557-5652
Fax: (703) 557-0668

Contains statistical and other data relating to patents granted by the U.S. Patent and Trademark Office (PTO). **Type of Database:** Directory; statistical. **Record Items:** Patent number; assignee for specific corporation or government; inventor's residence as state or country; date of patent application; independent inventors (street address, city, state and ZIP Code); abstract. **Contact:** Jane S. Myers, Manager.

★2596★ Thomas New Industrial Products Database

Thomas Publishing Co.
Thomas Online
One Penn Plaza
250 W. 34th St.
New York, NY 10119

Phone: (212) 290-7291
Fax: (212) 290-7362

Provides technical information on new industrial products and systems introduced by American and foreign manufacturers and distributors. **Type of Database:** Directory. **Record Items:** Product name; Standard Industrial Classification (SIC) code; model number; price; synonymous names for the product; product features; attributes; performance specifications; press release publication date; trade names; manufacturer/seller name; address; telephone number. **Contact:** Gary Craig, Manager, Thomas Online.

★2597★ Thomas Register

Thomas Publishing Co.
Thomas Online
One Penn Plaza
250 W. 34th St.
New York, NY 10119

Phone: (212) 290-7291
Fax: (212) 290-7362

Provides information on products made in the United States and Canada, the companies that make them, and where they are made. **Type of Database:** Directory. **Record Items:** Company name; address; telephone number; facsimile number; telex; assets; number of employees; parent company name; officers; foreign trade status; products; Standard Industrial Classification (SIC) codes; trade names.

★2598★ TMA Trademark Report

Tobacco Merchants Association of the
  United States (TMA)
231 Clarkville Rd.
PO Box 8019
Princeton, NJ 08543-8019

Phone: (609) 275-4900
Fax: (609) 275-8379

Contains a listing of tobacco-related trademark activity as reported in the U.S. Patent Office Official Gazette. **Type of Database:** Full-text. **Contact:** Dr. Thomas C. Slane, Vice President.

★2599★ TMINT

Institut National de la Propriete Industrielle (INPI)
26 bis, rue de Leningrad
F-75800 Paris Cedex 8, France

Covers all international trademarks in force, filed, and renewed with the World Intellectual Property Organization (WIPO) under the Madrid Agreement. **Type of Database:** Full-text. **Record Items:** Trademark name; designated products; classification code; registered owner; registration number, date, and duration of the mark; country of origin of registration; countries for which protection is claimed or refused; colors claimed and classification of figurative elements of device marks. **Contact:** Catherine Pagis, Marketing Manager.

★2600★ Trade Marks Opposition Board (TMOB)

STM Systems Corp. Phone: (613) 737-7373
WISDOM Information Services Fax: (613) 737-9479
2300 St. Laurent Blvd.
Ottawa, ON, Canada K1G 4K1

Contains the complete text of decisions and rulings issued by the Canadian Trade Marks Opposition Board. **Type of Database:** Full-text. **Record Items:** Trademark; application number; type of opposition; style of cause; ruling officer; date of decision; text. **Contact:** Scott Flewelling, Market Analyst.

★2601★ Trade Names Database (TND)

Gale Research Inc. Phone: (313) 961-2242
835 Penobscot Bldg. Fax: (313) 961-6815
Detroit, MI 48226-4094

Provides information on more than 250,000 trade names, trademarks, and brand names of consumer-oriented products and their 50,000 manufacturers, importers, marketers, or distributors worldwide. **Type of Database:** Directory. **Record Items:** Trade name; brief description of the product; name of the company which manufactures, imports, or distributes the item; address. **Contact:** Kathleen Young Marcaccio, Database Specialist.

★2602★ Trademark Access

Trademark Research Corp. Phone: (212) 228-4084
300 Park Ave. S. Fax: (212) 228-5090
New York, NY 10010

Contains information on all active registered and pending trademarks on file in the U.S. Patent and Trademark Office. **Type of Database:** Full-text; graphic. **Record Items:** Trademark text and enhancements; serial number; U.S. class number; status date; date of first use of mark; goods/services description; mark type; owner name; registration number; series code; status text and code; permuted trademark text and enhancements. **Availability:** CD-ROM through producer. **Contact:** Peter Skrobela.

★2603★ Trademark Weekly Text File

U.S. Patent & Trademark Office Phone: (703) 557-6154
Office of Electronic Data Conversion &
  Dissemination
Office of the Director
Crystal Park 2, Ste. 1100B
Washington, DC 20231

Provides details on registered and pending U.S. Federal trademarks as published in the U.S. Patent and Trademark Office's Official Gazette. **Type of Database:** Full-text.

★2604★ Trademarkscan-Federal

Thomson & Thomson Phone: (617) 479-1600
500 Victory Rd. Fax: (617) 786-8273
North Quincy, MA 02171-1545

Provides information on all active registered and pending trademarks on file in the U.S. Patent and Trademark Office, and all inactive trademarks since January 1984. **Type of Database:** Full-text; graphic. **Record Items:** Trademark text and enhancements; serial number; U.S. class number; status date; date of first use of mark; goods/services description; mark type; owner name; registration number; series code; status text and code; permuted trademark text and enhancements. **Availability:** Online through DIALOG Information Services, Inc. **Contact:** Anthea P. Gotto, Product Manager.

★2605★ Trademarkscan-State

Thomson & Thomson Phone: (617) 479-1600
500 Victory Rd. Fax: (617) 786-8273
North Quincy, MA 02171-1545

Provides information on trademarks registered with the Secretaries of State of all 50 U.S. states and in Puerto Rico. **Type of Database:** Full-text; graphic. **Record Items:** Trademark text and enhancements; rotated trademark; design type; state of registration; U.S. class number; international class; goods/service description; mark type; registration number; status; date of registration; date of renewal; date of cancellation; date of first use; current owner name. Corporate name records are not included. Trade names, assumed names, and fictitious names are also not generally included, but may be identified for some states. **Contact:** Anthea P. Gotto, Product Manager.

★2606★ Trademarkscan-U.K.

Thomson & Thomson Phone: (617) 479-1600
500 Victory Rd. Fax: (617) 786-8273
North Quincy, MA 02171-1545

Provides information on all active registered and pending trademarks on file with the United Kingdom Patent Office. **Type of Database:** Full-text. **Record Items:** Trademark text and enhancements; serial number; International Patent Classification (IPC) number; status date; date of first use of mark; goods/services description; mark type; owner name; registration number; series code; status text and code; permuted trademark text and enhancements. **Availability:** Online through DIALOG Information Services, Inc.

★2607★ TRANSIN

Transinove International
c/o INPI
26 bis, rue de Leningrad
F-75800 Paris Cedex 8, France

Indexes offers and requests for patented technologies, new products and inventions, and innovative ideas in need of development from the private and public sectors worldwide. **Type of Database:** Directory. **Record Items:** International Patent Classification (IPC) code; type of opportunity code; development stage; descriptive title; technology level and descriptors; description (in French); technical skill descriptors; agency or company source; contact person name and address; publication reference number; file input date. **Availability:** Online through Questel. **Contact:** Sylvie Couillard, Assistant Manager.

★2608★ UK Trade Marks

Great Britain Patent Office
State House
66-71 High Holborn
London WC1R 4TP, England

Contains information on nearly 600,000 British trademarks that are active or pending. **Type of Database:** Full-text. **Record Items:** Trademark; number; application date; status; owner name and address; class; goods description; usage limitations; agents and service address; owner country code. **Availability:** Online through PFDS Online. **Contact:** Mr. W. Preacher.

★2609★ U.S. Copyrights

DIALOG Information Services, Inc. Phone: (415) 858-3810
3460 Hillview Ave. Fax: (415) 858-7069
Palo Alto, CA 94304

Provides registration details for all active copyright and mask-work registrations on file at the U.S. Patent and Trademark Office. **Type of Database:** Full-text. **Record Items:** Title; class; U.S. Library of Congress Retrieval Code; status; registration number; date registered; date of creation; publication date; author; application author; owner; contents; notes; limitation on new matter; imprint; registration deposit. **Availability:**

Online through DIALOG Information Services, Inc. **Contact:** DIALOG Customer Service.

★2610★ U.S. Patent Classification (USPC) System

U.S. Patent & Trademark Office   Phone: (703) 557-6154
Office of Electronic Data Conversion &   Fax: (703) 557-0668
  Dissemination
Office of the Director
Crystal Park 2, Ste. 1100B
Washington, DC 20231

Constitutes 120,000 patent document classifications to assist patent examiners and others in the field in assessing the novelty, intrinsic merit, and utility of inventions deposited at the U.S. Patent and Trademark Office. **Type of Database:** Full-text. **Contact:** Edward J. Earls, Director, Office of Documentation. Phone: (703) 557-0400.

★2611★ U.S. Patents

Derwent, Inc.   Phone: (703) 790-0400
1313 Dolley Madison Blvd., Ste. 303   Fax: (703) 790-1426
McLean, VA 22101

Contains patent information for all inventions patented in the United States, including bibliographic information, the complete text of all claims, and complete front page information. **Type of Database:** Bibliographic; full-text. **Record Items:** Document title; inventor; patent assignee; publication date; application details; priority number; notes; patent classification numbers; International Patent Classification (IPC) code; examiner's field of search; document type; file segment; citations; abstract; text of main claim; text of subsequent claims. **Availability:** Online through ORBIT Search Service.

★2612★ USCLASS

Derwent, Inc.   Phone: (703) 790-0400
1313 Dolley Madison Blvd., Ste. 303   Fax: (703) 790-1426
McLean, VA 22101

Contains classification information for nearly 5 million patents issued by the U.S. Patent and Trademark Office. **Type of Database:** Bibliographic. **Availability:** Online through ORBIT Search Service.

★2613★ WESTLAW Intellectual Property Library

West Publishing Co.   Phone: (612) 228-2500
50 W. Kellogg Blvd.
PO Box 64526
St. Paul, MN 55164-0526

Contains the complete text of U.S. federal court decisions, statutes and regulations, specilized files, and texts and periodicals dealing with intellectual property law. **Type of Database:** Full-text. **Availability:** Online through WESTLAW.

★2614★ WESTLAW Texts and Periodicals Library

West Publishing Co.   Phone: (612) 228-2500
50 W. Kellogg Blvd.
PO Box 64526
St. Paul, MN 55164-0526

Contains the complete text of selected articles from more than 350 textbooks, law reviews, and bar association journals. **Type of Database:** Full-text. **Availability:** Online through WESTLAW.

★2615★ World Bank of Technology

Dr. Dvorkovitz & Associates (DDA)   Phone: (904) 677-7033
PO Box 1748   Fax: (904) 677-7113
Ormond Beach, FL 32074

Covers licensable technology around the world. **Type of Database:** Directory. **Record Items:** Title; description; subject category; individual. **Contact:** Barbara Witkowski, Data Processing Manager.

★2616★ World Patents Index (WPI)

Derwent Publications Ltd.
Rochdale House
128 Theobalds Rd.
London WC1X 8RP, England

Supplies titles and other details of general, mechanical, electrical, and chemical patents, covering the patent literature of leading industrial countries and using IPC codes for areas such as human necessities, performing operations, transporting, chemistry, textiles, building, construction, mechanics, lighting, heating, instruments, nuclear science, and electricity. **Type of Database:** Bibliographic. **Record Items:** Article title; author; author affiliation; journal title; publication date; publisher; patent number; abstract; patent information; equivalents which are patent numbers in other countries and priorities (filing date and serial number for first filing). **Contact:** Dr. Brian Gore, Marketing Manager, Literature Services. In North America, Derwent Inc., 1313 Dolley Madison Blvd., Suite 303, McLean, VA 22101. Telephone: (703) 790-0400. Toll-free: 800-451-3451. Facsimile: (703) 790-1426.

★2617★ WPI/APIPAT

Derwent Publications Ltd.
Rochdale House
128 Theobalds Rd.
London WC1X 8RP, England

Contains patents covering petroleum refining, the petrochemical industry, and synthetic fuels. **Type of Database:** Bibliographic. **Record Items:** Document number; patent title; patent assignee; company code; inventor; patent codes; priority codes; patent family codes; patent country; designated states; abstract; International Patent Classification (IPC) codes. **Availability:** Online through ORBIT Search Service. **Contact:** Dr. Brian Gore, Marketing Manager, Literature Services. •

Section 26

Patent and Trademark Depository Libraries

"A man's judgment cannot be better than the information on which he has based it."—Arthur Hays Sulzberger

The Patent and Trademark Depository Libraries (PTDLs) are a nationwide network of 70 academic, research, and public libraries in which you may, among other things, inspect copies of patents and trademarks.

PTDLs are an invaluable source of information and support to the independent inventor. In my opinion, they are one of the best uses of PTO funds and an absolute must to visit for inventors living outside the Washington, DC area.

If you have an invention to patent, you face a long, tough trek across a landscape littered with wasted money, abandoned ideas, and failures. Your trip is more likely to be successful and cost-effective if it begins at a PTDL.

I think every inventor should do at least one hands-on patent search to fully understand and appreciate the process, as well as get a comprehensive idea of what prior art is out there is a specific area. Obviously, not everyone can visit the PTO's Patent Public Search Room in Arlington, Virginia. But, thanks to an ever-expanding network of PTDLs, help is never far away. The PTDLs are your link to the PTO. "My goal is to open four new PTDLs a year," says Carole A. Shores, director of the PTO's Office of Patent and Trademark Depository Library Programs. "I want to see one in every state and U.S. territory," she hastens to add. Ms. Shores opened her first PTDL in 1977.

"Our mission is a very simple one," she continues. "It is to bring more information and more help to all the people out there who need it and can't afford to pay big money to get it. What makes our program special is that we really listen to their [inventors] needs and when they bring requests to us, we try very hard to give them what they want."

Milwaukee inventor James Pinske, creator of B'zarts flying toys, says of the PTDL in his neighborhood, "If I got into a snag, I just went to the librarian and got complete information." Don Coster, president of the Nevada Inventor's Association, agrees. "I think without those Patent and Trademark Depository Libraries we'd [independent inventors] be floating in limbo."

Formerly known as Patent Depository Libraries (PDLs), the Patent and Trademark Depository Libraries continue to be one of the PTO's most effective mechanisms for publicly disseminating patent and trademark information. PTDLs receive current issues of U.S. patents and maintain collections of previously issued patents. The scope of these collections varies from library to library, ranging from only the most recent years (minimum 20-year back file collection on microfilm) to all or most of the patents awarded since 1970. Due to the variations in the scope of information available at each PTDL, and their different hours of operation, you should call first to make sure the nearest one to you has what you require, to avoid any possible inconvenience. A complete list of PTDLs together with their addresses and telephone numbers begins on the next page.

In addition to patents and trademarks, PTDLs have publications such as *The Manual of Classification, Index to the U.S. Patent Classification, Official Gazette of the United States Patent and Trademark Office,* etc.

Patent collections at PTDLs are open to the general public and I have always found the librarians most willing to take the time and interest to help newcomers gain effective access to the information they require. Such librarians are given special ongoing training by the PTO in the use and understanding of the U.S. Patent Classification System.

"Something is always going on at the PTDLs," boasts Carole A. Shores. To illustrate this, she showed me a recent copy of *Ad Libs*, the official newsletter of the PTDLs. It was wall-to-wall with news of PTDL outreach programs.

One of the most useful programs put on by the PTDLs is a one-day seminar on accessing patent and trademark information. This seminar is free of charge and, in the best of budget times, the PTDL conducts five to seven of these programs per year. Jay D. James, president of Inventor's Institute of Alaska, wrote to the Commissioner of Patents and Trademarks after attending a PTDL seminar, that it was, "the finest, most complete information seminar I have ever attended."

"My husband and I left with our heads and briefcases packed with valuable information," was how Pauline J. Eatherly, president of Corely Concepts, Cerritos, CA, described her experience at a PTDL seminar.

For information on PTDL seminars or to learn more about library programs, you may call the Office of Patent Depository Library Programs (PDLP) at (703) 557-9686 or fax it at (703) 557-7275. If Carole Shores is not available, a member of her staff will be more than willing to assist you. Working along with Carole in the PDLP office are technical information specialists Martha Crocket-Sneed, Amanda Putnam, and James Arshem.

The address for the Office of Patent Depository Library Programs is: Crystal Mall, Bldg. 2, Rm. 306, PTO, Washington, DC 20231.

[Editor's note: Three new PTDLs were approved as this book went to press, bringing the total number of PTDLS from 70 to 73. The three new PTDLs are: Siegesmund Engineering Library, Potter Center, Purdue University, West Lafayette, IN 47907; Abigail S. Timme Library, Ferris State University, 901 State St., Big Rapids, MI 49307; and Evansdale Library, West Virginia University, PO Box 6105, Morgantown, WV 26506-6105.]

Section Arrangement

The following entries are arranged geographically by state, then alphabetically by library name. All names in this section are referenced in the Master Index.

Alabama

★2618★ **Auburn University Libraries**
Science and Technology Department
Auburn University, AL 36849-5606   Phone: (205) 844-1747

★2619★ **Birmingham Public Library**
Government Documents Department
2100 Park Pl.   Phone: (205) 226-3680
Birmingham, AL 35203
Online Access: DIALOG Information Services, U.S. Patent Classification System.

Alaska

★2620★ **Anchorage Municipal Libraries**
Z.J. Loussac Public Library
3600 Denali St.   Phone: (907) 261-2916
Anchorage, AK 99503-6093

Arizona

★2621★ **Arizona State University**
Noble Science and Engineering Library
Tempe, AZ 85287   Phone: (602) 965-7607
Online Access: DIALOG Information Services, Pergamon ORBIT InfoLine, Inc., BRS Information Technologies, NASA/RECON, Integrated Technical Information System (ITIS), U.S. Patent Classification System, OCLC, STN International, WILSONLINE. Performs searches on fee basis.

Arkansas

★2622★ **Arkansas State Library**
One Capitol Mall   Phone: (501) 682-2053
Little Rock, AR 72201-1081

California

★2623★ **California State Library**
Government Publications Section
Library-Courts Bldg.   Phone: (916) 322-4572
P.O. Box 942837
Sacramento, CA 94237-0001

★2624★ **Los Angeles Public Library**
Science, Technology and Patents Department
630 W. Fifth St.   Phone: (213) 612-3273
Los Angeles, CA 90071-2097
Online Access: DIALOG Information Services, LEXIS, NEXIS, U.S. Patent Classification System, Bookline, LEXPAT; internal database; OnTyme Electronic Message Network Service (electronic mail service). Performs searches on fee basis.

★2625★ **San Diego Public Library**
Science Section
820 E St.   Phone: (619) 236-5813
San Diego, CA 92101
Online Access: DIALOG Information Services, U.S. Patent Classification System. Performs searches free of charge.

★2626★ **Sunnyvale Patent Clearinghouse**
1500 Partridge Ave., Bldg. 7   Phone: (408) 730-7290
Sunnyvale, CA 94087

Colorado

★2627★ **Denver Public Library**
Business, Science and Government Publications
    Department
1357 Broadway   Phone: (303) 640-8847
Denver, CO 80203
Online Access: DIALOG Information Services, Pergamon ORBIT InfoLine, Inc., BRS Information Technologies, U.S. Patent Classification System. Performs searches on fee basis.

Connecticut

★2628★ **Patent Library**
25 Science Park   Phone: (203) 786-5447
New Haven, CT 06511

Delaware

★2629★ **University of Delaware Library**
Reference Department
Newark, DE 19717-5267   Phone: (302) 451-2965

District of Columbia

★2630★ **Howard University Libraries**
500 Howard Pl., N.W.   Phone: (202) 806-7252
Washington, DC 20059

Florida

★2631★ Broward County Main Library
Government Documents Department
100 S. Andrews Ave.   Phone: (305) 357-7444
Fort Lauderdale, FL 33301

★2632★ Miami-Dade Public Library
Business and Science Department
101 W. Flagler St.   Phone: (305) 375-2665
Miami, FL 33130-2585

Online Access: DIALOG Information Services, BRS Information Technologies, U.S. Patent Classification System, LOGIN.

★2633★ University of Central Florida Library
PO Box 25,000   Phone: (407) 823-2562
Orlando, FL 32816-0666

★2634★ University of South Florida
Patent Library
Library LIB122   Phone: (813) 974-2726
4202 E. Fowler Ave.
Tampa, FL 33620-5400

Georgia

★2635★ Georgia Institute of Technology
Price Gilbert Memorial Library
Department of Microforms
225 North Ave.   Phone: (404) 894-4508
Atlanta, GA 30332-0900

Online Access: DIALOG Information Services, Pergamon ORBIT InfoLine, Inc., BRS Information Technologies, Wharton Econometric; BITNET, ALANET (electronic mail services). Performs searches on fee basis.

Hawaii

★2636★ Hawaii State Public Library
634 Pensacola St.   Phone: (808) 586-3477
Honolulu, HI 96814

Idaho

★2637★ University of Idaho Library
Science Department
Moscow, ID 83843
                    Phone: (208) 885-6235

Illinois

★2638★ Chicago Public Library
Science and Technology Information Center
400 N. Franklin St.   Phone: (312) 269-2865
Chicago, IL 60610

Online Access: DIALOG Information Services, Pergamon ORBIT InfoLine, Inc., BRS Information Technologies, Dow Jones News/

Retrieval, Mead Data Central, OCLC, VU/TEXT Information Services, Info Globe, LEGI-SLATE, DataTimes.

★2639★ Illinois State Library
300 S. 2nd St.   Phone: (217) 782-5659
Springfield, IL 62701-1796

Online Access: BRS Information Technologies, DIALOG Information Services, LEXIS, CAS ONLINE, OCLC, U.S. Patent Classification System, TechCentral; CLSI (internal database); ALANET (electronic mail service).

Indiana

★2640★ Indianapolis-Marion County Public Library
Business, Science and Technology Division
PO Box 211   Phone: (317) 269-1741
Indianapolis, IN 46206

Online Access: Access to DIALOG Information Services, NEXIS, U.S. Patent Classification System, Statistical Information System (STATIS).

Iowa

★2641★ State Library of Iowa
Information Services and Patent Depository
East 12th & Grand   Phone: (515) 281-4118
Des Moines, IA 50319

Kansas

★2642★ Wichita State University
Ablah Library
Campus Box 68   Phone: (316) 689-3155
Wichita, KS 67208-1595

Kentucky

★2643★ Louisville Free Public Library
Reference and Adult Services
301 York St.   Phone: (502) 561-8617
Louisville, KY 40203-2257

Louisiana

★2644★ Louisiana State University
Business Administration/Government Documents
    Department
Troy H. Middleton Library   Phone: (504) 388-2570
Baton Rouge, LA 70803

Online Access: DIALOG Information Services, Pergamon ORBIT InfoLine, Inc., BRS Information Technologies, U.S. Patent Classification System; DIALMAIL (electronic mail service). Performs searches on fee basis.

Maryland

★2645★ University of Maryland
Engineering and Physical Sciences Library
Reference Services
College Park, MD 20742   Phone: (301) 405-9157
Online Access: DIALOG Information Services, BRS Information Technologies, DTIC, U.S. Patent Classification System, OCLC.

Massachusetts

★2646★ Boston Public Library
PO Box 286   Phone: (617) 536-5400
Boston, MA 02117
Online Access: DIALOG Information Services.

★2647★ University of Massachusetts
Physical Sciences Library
Graduate Research Center   Phone: (413) 545-1370
Amherst, MA 01003

Michigan

★2648★ Detroit Public Library
Technology and Science Department
5201 Woodward Ave.   Phone: (313) 833-1450
Detroit, MI 48202
Online Access: NEXIS, U.S. Patent Office Classification System.

★2649★ University of Michigan
Engineering Transportation Library
312 UGL   Phone: (313) 764-7494
Ann Arbor, MI 48109-1185

Minnesota

★2650★ Minneapolis Public Library and Information
Center
Technology and Science Department
300 Nicollet Mall   Phone: (612) 372-6570
Minneapolis, MN 55401
Online Access: DIALOG Information Services, Pergamon ORBIT InfoLine, Inc., BRS Information Technologies, U.S. Patent Classification System; internal database. Performs searches free of charge.

Mississippi

★2651★ Mississippi Library Commission
PO Box 10700
Jackson, MS 39289-0700

Missouri

★2652★ Linda Hall Library
5109 Cherry St.   Phone: (816) 363-4600
Kansas City, MO 64110

★2653★ St. Louis Public Library
Applied Sciences Department
1301 Olive St.   Phone: (314) 214-2288
St. Louis, MO 63103

Montana

★2654★ Montana College of Mineral Science and
Technology Library
Patent Center
W. Park St.   Phone: (406) 496-4281
Butte, MT 59701
Online Access: DIALOG Information Services, Pergamon ORBIT InfoLine, Inc., BRS Information Technologies, STN International; FAPRS (internal database). Performs searches on fee basis.

Nebraska

★2655★ University of Nebraska, Lincoln
Engineering Library
Nebraska Hall, 2nd Fl. W.   Phone: (402) 472-3411
Lincoln, NE 68588-0410
Online Access: DIALOG Information Services. Performs searches on fee basis.

Nevada

★2656★ University of Nevada
Government Publications Department
University Library   Phone: (702) 784-6579
Reno, NV 89557-0044
Online Access: DIALOG Information Services, BRS Information Technologies, RLIN, U.S. Patent Classification System, LIBS 100 System; Nevada Documents Online (internal database). Performs searches on fee basis.

New Hampshire

★2657★ University of New Hampshire
University Library
Patent Collection
Durham, NH 03824   Phone: (603) 862-1777

New Jersey

★2658★ Newark Public Library
Pure, Applied, and Social Sciences Division
5 Washington St.   Phone: (201) 733-7782
PO Box 630
Newark, NJ 07102-3175

Online Access: DIALOG Information Services, BRS Information Technologies, NEXIS, OCLC, Pergamon ORBIT InfoLine, Inc., VU/TEXT Information Services, WILSONLINE, U.S. Patent Classification System. Performs searches on fee basis.

★2659★ Rutgers University
Library of Science and Medicine
Government Documents Department
PO Box 1029   Phone: (908) 932-2895
Piscataway, NJ 08855-1029

New Mexico

★2660★ University of New Mexico
Centennial Science and Engineering Library
100 S. Andrews Ave.   Phone: (505) 277-4412
Albuquerque, NM 87131

New York

★2661★ Buffalo and Erie County Public Library
Science and Technology Department
Lafayette Sq.   Phone: (716) 858-7101
Buffalo, NY 14203

Online Access: DIALOG Information Services, U.S. Patent Classification System, Chemical Information Systems, Inc. (CIS). Performs searches on fee basis.

★2662★ New York Public Library
521 W. 43rd St.   Phone: (212) 714-8529
New York, NY 10036-4396

Online Access: U.S. Patent Classification System.

★2663★ New York State Library
Reference Services
Cultural Education Center   Phone: (518) 473-4636
Albany, NY 12230
Christine A. Bain, Associate Librarian

Online Access: OCLC, U.S. Patent Classification System, BRS Information Technologies, DIALOG Information Services, NLM, Mead Data Central, LEXIS, NEXIS, STN International, WILSONLINE. Performs searches on fee basis.

North Carolina

★2664★ North Carolina State University
D.H. Hill Library
Documents Department
Box 7111   Phone: (919) 515-3280
Raleigh, NC 27695-7111

Online Access: U.S. Patent Classification System, DIALOG Information Services. Performs searches on fee basis.

North Dakota

★2665★ University of North Dakota
Chester Fritz Library
University Sta.   Phone: (701) 777-4888
Grand Forks, ND 58202

Ohio

★2666★ Cleveland Public Library
Documents Collection
325 Superior Ave.   Phone: (216) 623-2870
Cleveland, OH 44114-1271

Online Access: OCLC, DIALOG Information Services, BRS Information Technologies, OhioPI (Ohio Public Information Utility), U.S. Patent Classification System, Hannah Legislative Service, Pergamon ORBIT InfoLine, Inc.

★2667★ Ohio State University Libraries
Information Services Department
1858 Neil Ave. Mall   Phone: (614) 292-6175
Columbus, OH 43210

★2668★ Public Library of Cincinnati and Hamilton County
Science and Technology Department
800 Vine St.   Phone: (513) 369-6936
Cincinnati, OH 45202-2071

Online Access: DIALOG Information Services, WILSONLINE, BRS Information Technologies, NEXIS, U.S. Patent Classification System. Performs U.S. Patent searches free of charge; other searches on fee basis.

★2669★ Toledo-Lucas County Public Library
Science and Technology Department
325 Michigan St.   Phone: (419) 259-5212
Toledo, OH 43624

Online Access: DIALOG Information Services, BRS Information Technologies, Dow Jones News/Retrieval, CAS ONLINE, U.S. Patent Classification System; TLM (internal database).

Oklahoma

★2670★ Oklahoma State University Library
Stillwater, OK 74078-0375   Phone: (405) 744-7086

Oregon

★2671★ Oregon State Library
State Library Bldg.   Phone: (503) 378-4239
Summer & Court Sts.
Salem, OR 97310

Online Access: DIALOG Information Services, LEXIS, ISIS, LEGISNET, BRS Information Technologies, NLM, EROS Data Center, Oregon Legislative Information System; ALANET, TYMNET, OCLC, Oregon Public Access Catalog (OPAC) (electronic mail services).

Pennsylvania

★2672★ Carnegie Library of Pittsburgh
Science and Technology Department
4400 Forbes Ave.   Phone: (412) 622-3138
Pittsburgh, PA 15213

Online Access: DIALOG Information Services, OCLC, Pergamon ORBIT InfoLine, Inc., U.S. Patent Classification System, STN International. Performs searches on fee basis.

★2673★ Free Library of Philadelphia
Government Publications Department
Logan Sq.   Phone: (215) 686-5331
Philadelphia, PA 19103

★2674★ Pennsylvania State University Libraries
Documents Section
C207 Pattee Library   Phone: (814) 865-4861
University Park, PA 16802

Rhode Island

★2675★ Providence Public Library
225 Washington St.   Phone: (401) 455-8027
Providence, RI 02903

Online Access: DIALOG Information Services, OCLC. Performs searches on fee basis.

South Carolina

★2676★ Medical University of South Carolina Library
171 Ashley Ave.   Phone: (803) 792-2372
Charleston, SC 29425

Tennessee

★2677★ Memphis and Shelby County Public Library
and Information Center
Business/Science Department
1850 Peabody Ave.   Phone: (901) 725-8876
Memphis, TN 38104

★2678★ Vanderbilt University
Stevenson Science Library
419 21st Ave., S.   Phone: (615) 322-2775
Nashville, TN 37240-0007

Online Access: DIALOG Information Services, OCLC, BRS Information Technologies, CAS ONLINE, U.S. Patent Classific ation System. Performs searches on fee basis.

Texas

★2679★ Dallas Public Library
J. Erik Jonsson Central Library
Government Publications Division
1515 Young St.   Phone: (214) 670-1468
Dallas, TX 75201

Online Access: DIALOG Information Services, U.S. Patent Classification System, BRS Information Technologies, Pergamon ORBIT InfoLine, Inc. Performs searches on fee basis.

★2680★ Rice University
Fondren Library
Division of Government Publications and Special
Resources
Box 1892   Phone: (713) 527-8101
Houston, TX 77251-1892

Online Access: U.S. Patent Classification System, LEGI-SLATE. Performs searches free of charge.

★2681★ Texas A&M University
Evans Library
Documents Division
College Station, TX 77843-5000   Phone: (409) 845-2551

★2682★ University of Texas at Austin
McKinney Engineering Library
Rm. 1.3 ECJ   Phone: (512) 471-1610
Austin, TX 78712

Utah

★2683★ University of Utah
Marriott Library
Salt Lake City, UT 84112   Phone: (801) 581-8394

Online Access: DIALOG Information Services, Pergamon ORBIT InfoLine, Inc., BRS Information Technologies.

Virginia

★2684★ Virginia Commonwealth University
University Library Services
Documents and Interlibrary Loan
Box 2033 - 901 Park Ave.   Phone: (804) 367-1104
Richmond, VA 23284-2033

Washington

★2685★ University of Washington
Engineering Library
Engineering Library Bldg., FH-15   Phone: (206) 543-0740
Seattle, WA 98195

Online Access: DIALOG Information Services, STN International, Pergamon ORBIT InfoLine, Inc., BRS Information Technologies, OCLC.

Wisconsin

★2686★ Milwaukee Public Library
814 W. Wisconsin Ave.   Phone: (414) 278-3247
Milwaukee, WI 53233

Online Access: OCLC, U.S. Patent Classification System.

★2687★ University of Wisconsin—Madison
Kurt F. Wendt Library
215 N. Randall Ave.   Phone: (608) 262-6845
Madison, WI 53706

Section 27

Registered Patent Attorneys and Agents

Caveat Emptor!—Proverb

Do you need a patent attorney?

The answer is probably yes. It is perfectly legal to prepare your own patent application. You can conduct your own proceedings in the Patent and Trademark Office (PTO). But, unless you are familiar with such matters and have studied them in detail, you could experience considerable difficulty and extreme frustration.

Preparing an application for a **utility patent** and conducting proceedings at the PTO requires a knowledge of patent law and PTO practice, and a knowledge of the scientific or technical matters involved with the particular invention. In Part Two: How to Protect and License Your Invention, I provide a comprehensive, step-by-step explanation of the entire patenting process. It will help you understand the various steps your lawyer will take, and identify those steps you might feel comfortable handling yourself.

While a patent may be obtained in many cases by persons not skilled in such esoteric work, there would be no assurance that the patent awarded would adequately protect the particular invention. I, therefore, highly recommend that a qualified patent attorney be retained for your utility patent work. Most inventors employ the services of registered patent attorneys or patent agents.

The PTO is allowed by law to make rules and regulations governing patent attorney conduct and the recognition of patent attorneys and agents to practice before the PTO. Agents and attorneys who are not recognized by the PTO for this practice are not permitted by law to represent inventors in their patent actions.

The PTO has the power to disbar, or suspend from practicing before it, persons guilty of gross misconduct, but this can only be done after a full hearing with the presentation of clear and convincing evidence concerning the misconduct.

The PTO will receive and, in appropriate cases, act upon complaints against attorneys and agents. If you wish to register a complaint, contact Cameron Weiffenbach, Director, Office of Enrollment, PO Box OED, U.S. Patent and Trademark Office, Washington, DC 20231; phone: (703) 557-2012.

How Much do Legal Services Cost?

It is not inexpensive to retain a patent attorney. The amount of time a patent attorney will have to put into any particular matter will depend a great deal upon the complexity of the invention. Below are some guidelines for costs, arranged according to the ordered sequence of steps a lawyer will take.

Patent Search. The first thing a patent lawyer will rightfully suggest is that you authorize a patent search to see what, if any, prior art exists. The cost of a patent search will depend upon the scope of your patent. Rarely does a lawyer do the search. Lawyers normally engage the services of a professional patent searcher or patent search organization. A premium will be added to whatever the lawyer is charged by the patent searcher. This premium can range from 40 to 100 percent or more. Some inventors avoid this added cost by either independently doing the patent search or by hiring a patent searcher directly. The results are then handed over to an attorney for an opinion.

If you are quoted too low a figure from a lawyer, it will probably not be much of a search. Searches done fast may not worth much. Done correctly, a search takes some time.

I have found some search firms that have $150 specials. Typically, you get six pieces of prior art, copies included, for the package price. As in most things in life, you get what you pay for. If the searcher stops at three and there are six pieces of prior art, you may have lost $150. To be useful, a search must be comprehensive.

A good manual search can take up to seven hours. Searchers work at hourly rates. In Washington, DC, fees range from $30 to $60 per hour. Generally it is less expensive to employ an independent searcher than a large firm. Some firms require minimums of up to $230. Independent searchers often do not insist upon minimums.

Chemical and electrical patent searches typically cost more than mechanical patent searches because, according to the searchers, they take more time (an average of eight to ten hours).

Some searchers will augment their work through the use of computer databases. **See the introductory essay heading Online Patent Searches on page 31 and Section 25, Online Resources, on page 323 for details on how to conduct a patent search yourself.**

You will also have to pay a fee to have the search results photocopied. Some independent searchers charge their cost or $.40 per page; larger firms get $3 per copy no matter how few or how many pages.

Patent Drawings. Lawyers do not personally do patent drawings either. They employ the services of a draftsman skilled in such matters. Sometimes the draftsmen are in-house staffers. Smaller practitioners use free-lancers, whose fees they step on as much as 40 percent. In Washington, DC, the average price charged by a bonded draftsman is $35 per hour. It takes an average of three hours to do one sheet, so you can expect to pay around $100 per sheet. Patents comprise numerous sheets of drawings depending upon the complexity of the invention.

Patent Application. To prepare a patent application, attorneys charge anywhere from $2,500 to $6,000 depending on the complexity of the invention. The more complex the patent, the more you can expect to pay. On the other hand, simple inventions can be done for as little as $1,000.

It is a buyers market. And just like anything else, it usually pays to shop around. Patent attorneys are able to give pretty close estimates of expected charges once they see the scope of an invention and its claims.

When you have an acceptable estimate, get your attorney to agree in writing to that price and cap it off. If you do not do this, you may find yourself caught in what I call fee creep. Make a package deal. A price cap discourages the lawyer from extending the case unnecessarily.

Get a handle on the photocopying charges for the application work in advance. Patent work generates a great deal of paper and it can be a gold mine for law firms. I found one large firm charging $.50 per photocopied page and sending three copies of each document, one for each of my partners. I quickly put a stop to this abuse, requesting one copy of each document. I can copy documents for a couple cents per page. The little things add up.

The fees charged by patent attorneys and agents for their professional services are not subject to regulation by the PTO. Solid evidence of overcharging may afford a basis for PTO action, but the Office rarely intervenes in disputes concerning fees.

Legal Forms. There are five forms which lawyers charge to prepare. Copies of these are found in the Forms and Tables.

1. Declaration for a Patent Application (**Form 4**)
2. Power of Attorney or Authorization of Agent, Not Accompanying Application (**Form 1**)
3. Revocation of Power of Attorney or Authorization of Agent (**Form 2**)
4. Assignment of Patent (**Form 13**)
5. Assignment of Patent Application (**Form 14**)

What to Look for in a Patent Attorney

You should look for an attorney who specializes in your field of invention. Just as you would not hire a dermatologist to do the job of a cardiologist, even though both are licensed physicians, you would not want to hire a patent attorney with a background in mechanical engineering to do an electronic patent.

Larger firms offer quite an array of patent specialists from which to choose. But the smaller firms and many independent practitioners will often take anything that comes along. Do not be timid about requesting the technical qualifications of any attorney you are considering.

If your attorney cannot read your schematics, engineering drawings, or similar technical specifications, chances are your patent will not be nearly as complete and strong as it could be. Furthermore, you will be paying a premium to educate the attorney.

Bigger firms are not always better. At very large patent firms you will be small potatoes compared to lucrative corporate retainers. In such cases, your account may be assigned to a "spear carrier," who may be very good, but will seldom get the time to concentrate on your work. The primary responsibility of junior associates is to carry the workload for senior partners, the so-called "rainmakers," who bring in the accounts that pay lunch, health club, limo, and kindred perks.

Big patent law firms make their real money defending the patents of corporate clients in court, not from two and three thousand dollar application jobs on behalf of independents. Experienced patent attorneys know that the chances of an entrepreneur inventor being worth much more than what is earned on the patent application are slim. The majority of independents are dead ends financially. Big corporate clients can be "cash cows."

I am most comfortable with the independent specialist, even if this approach means a different attorney every time I change disciplines.

347

Do not rely on any patent attorney for prototyping advice, manufacturing processes, or insights into the day-to-day complexities of marketing. These are not their fields of expertise.

National Council of Intellectual Property Law Associations

Patent attorneys who are active in the National Council of Intellectual Property Law Associations (NCIPLA) should be particularly up-to-date on PTO matters. The Council is a 35-year-old organization consisting of some 40 local and regional patent law associations.

In addition to promoting exchange of information and lobbying, NCIPLA keeps its members current on legislative and executive actions which might affect the nation's intellectual property system.

For more information on NCIPLA, contact your nearest member association (listed on the following pages), or contact:

National Council of Intellectual Property Law Associations
c/o Howard A. MacCord, Jr., NCIPLA Secretary
Rhodes, Coats, and Bennett
1600 1st Tower
Greensboro, NC 27401   Phone: (919) 273-4422

Attorneys and Agents Registered to Practice before the U.S. Patent and Trademark Office

I cannot overemphasize that the preparation and prosecution of an application that will adequately protect your invention is an undertaking that requires knowledge of patent law and PTO practices. It also requires a knowledge of technical aspects of the invention. It is for this reason that I highly recommend that you hire only a registered patent attorney or agent to do your patent work.

To be admitted to this select register, a person must comply with the regulations prescribed by the PTO, which require that a person be of good moral character and repute and have the legal and scientific credentials necessary to render a valuable service. Some of these qualifications must be demonstrated by passing an examination. Those admitted to the examination must have a college degree in engineering or physical science or the equivalent of such a degree.

The PTO registers attorneys at law, referred to as "patent attorneys", as well as persons who are not members of the bar, referred to as "patent agents". Insofar as the work of preparing an application for patent and conducting the prosecution in the PTO is concerned, patent agents are typically as well qualified as patent attorneys, although patent agents cannot conduct patent litigation in the courts or perform various services which the local jurisdiction considers as practicing law. For example, a patent agent would not be allowed to draw up a contract relating to a patent, such as an assignment or a license, if the agent resides in a state that considers such contracts as practicing law.

Some individuals and organizations that are not registered with the PTO advertise their services in the fields of patent searching and invention marketing and development. Such individuals and organizations cannot represent inventors before the PTO. Caveat emptor! They are not subject to PTO discipline, and the Office cannot assist you in dealing with them.

While calling to acquire information on organizations that I felt might be helpful to inventors, I came across a nifty scam used by some lawyers to attract patent, trademark, and copyright clients. What appeared in telephone directories and source material to be professional councils specific to patents, trademarks, and copyrights were actually general law offices. In one case the phone numbers for three different councils led to the same law firm. The person listed as president of the three respective organizations was a senior partner in the law firm.

The secretary who answered the phone pitched me on the fees charged for various services. This is a highbrow bait and switch that should be avoided.

The PTO cannot recommend any particular attorney or agent, or aid in the selection of an attorney or agent, as by stating in response to inquiry that a named patent attorney, agent, or firm is "reliable" or "capable."

The most current names and addresses of 13,700 individuals authorized to represent inventors before the PTO begin on page 350. This listing reprints information compiled and provided by the PTO.

In the listings, a number sign (#) appears beside the name of each registrant who is a patent agent. An asterisk (*) appears in the listings besides the name of each registrant who is an officer or employee of the U.S. government. Registrants who are officers or employees of the U.S. government cannot, otherwise than in the discharge of their official duties, represent inventors before the PTO.

Section Arrangement

The following entries are arranged geographically by state, then alphabetically by association name. All associations are referenced in the Master Index.

California

★2688★ Los Angeles Patent Law Association
c/o William J. Robinson
Graham & James
725 S. Figueroa St., 34th Fl.
Los Angeles, CA 90017-5434

★2689★ Orange County Patent Law Association
c/o Bruce B. Brunda
Stetina and Brunda
24221 Calle De La Louisa, Ste. 401
Laguna Hills, CA 92653

★2690★ Peninsula Intellectual Property Law Association
c/o Edith A. Rice
Raychem Corporation
300 Constitution Dr.
Menlo Park, CA 94025

★2691★ San Francisco Patent and Trademark Law Association
c/o Neil A. Smith
Limbach, Limbach & Sutton
2001 Ferry Bldg.
San Francisco, CA 94111

Colorado

★2692★ Colorado Bar Association Patent, Trademark, and Copyright Section
c/o Gregory W. O'Connor
Samsonite Corporation
11200 E. 45th Ave.
Denver, CO 80239

Connecticut

★2693★ Connecticut Bar Association Intellectual Property Law Section
c/o Karen A. Molitor, Esq.
Shipman & Goodwin
1 American Row
Hartford, CT 06103-2819

★2694★ Connecticut Patent Law Association
c/o Ted Carvis
St. Onge, Steward, Johnson & Reens
986 Bedford St.
Stamford, CT 06905

District of Columbia

★2695★ Bar Association of D.C.
Susan J. Mack, Esq.
Sughrue, Mion, Zinn, MacPeak & Seas
2100 Pennsylvania Ave. NW
Washington, DC 20037

★2696★ Patent, Trademark, and Copyright Section
c/o Robert G. Weilacher
Beveridge, DeGrandi & Weilacher
Federal Bar Bldg. W.
1819 H St. NW, Ste. 1100
Washington, DC 20006

★2697★ Maryland Patent Law Association
c/o Frank Robbins
Venable, Baetjer, Howard & Civiletti
1201 New York Ave. NW, Ste. 1000
Washington, DC 20005-3917

Florida

★2698★ South Florida Patent Law Association
c/o Henry W. Collins
Patent Counsel
Cordis Corporation
PO Box 025700
Miami, FL 33102-5700

Georgia

★2699★ State Bar of Georgia Patent, Trademark, and Copyright Section
c/o Todd Deaveau
Hurt, Richardson, Garner, Todd & Cadenhead
1400 Peachtree Place Tower
999 Peachtree St. NE
Atlanta, GA 30309-3999

Illinois

★2700★ Intellectual Property Law Association of Chicago
c/o Timothy T. Patula
Patula & Associates
116 S. Michigan Ave., 14th Fl.
Chicago, IL 60603

Indiana

**★2701★ Indiana State Bar Association
Patent, Trademark, and Copyright Section**
c/o Bobby B. Gillenwater
ISBA PTC Section
1 Summit Square, Ste. 600
Fort Wayne, IN 46802

Iowa

★2702★ Iowa Intellectual Property Law Association
c/o Lee Murrah
Rockwell International
Avionics Group
400 Collins Rd. NE
Cedar Rapids, IA 52402

Massachusetts

★2703★ Boston Patent Law Association
c/o David J. Powsner
Lahive & Cockfield
60 State St., #510
Boston, MA 02109

Michigan

★2704★ Michigan Patent Law Association
c/o Ernest R. Helms
General Motors Corp., Patent Section
3031 W. Grand Blvd.
PO Box 33114
Detroit, MI 48232

★2705★ Saginaw Valley Patent Law Association
c/o Paula Ruhr
Dow Chemical Co.
Patent Dept., Bldg. 1776
Midland, MI 48674

**★2706★ State Bar of Michigan
Intellectual Property Law Section**
c/o Charles E. Burpee
Warner, Norcross & Judd
900 Old Kent Bank Bldg.
Grand Rapids, MI 49503-2489

Minnesota

**★2707★ Minnesota Intellectual Property Law
Association**
c/o Philip M. Goldman
Office of Patent Counsel
3M Company
3M Center, PO Box 33427
St. Paul, MN 55133-3427

Missouri

**★2708★ Bar Association of Metropolitan St. Louis
Patent, Trademark, and Copyright Section**
c/o John W. Kepler, III
Heller & Kepler
721 Emerson, Ste. 569
St. Louis, MO 63141

New Jersey

**★2709★ American Association of Registered Patent
Attorneys and Agents**
c/o Albert L. Gazzola
22 Main St.
Montvale, NJ 07645

**★2710★ National Intellectual Property Law
Association**
c/o Samuel R. Williamson
AT&T Bell Laboratories
Rm. 4P-354
200 Laurel Ave.
Middletown, NJ 07748

★2711★ New Jersey Patent Law Association
c/o David A. Draegert
The BOC Group, Inc.
100 Mountain Ave.
Murray Hill, NJ 07974

**★2712★ New Jersey State Bar
Patent, Trademark, and Copyright Section**
c/o Arthur J. Plantamura
10 Butterworth Dr.
Morristown, NJ 07960

New York

★2713★ Central New York Patent Law Association
c/o H. Walter Haeussler
Cornell Research Foundation, Inc.
20 Thornwood Dr., Ste. 105
Ithaca, NY 14850

★2714★ Eastern New York Patent Law Association
c/o William A. Teoli
General Electric Co.
Corporate Research & Development
PO Box 8, Bldg. K1, Rm. 3A68
Schenectady, NY 12302

★2715★ New York Patent, Trademark, and Copyright Law Association
c/o Frank F. Scheck
1155 Avenue of the Americas
New York, NY 10036

★2716★ Niagara Frontier Patent Law Association
c/o Laird F. Miller
National Gypsum Co.
1650 Military Rd.
Buffalo, NY 14217-1198

★2717★ Rochester Intellectual Property Law Association
c/o Paul F. Morgan
Xerox Corporation
Xerox Sq. - 020
Rochester, NY 14644

North Carolina

★2718★ Carolina Patent, Trademark, and Copyright Law Association
c/o Charles B. Elderkin
Bell, Seltzer, Park & Gibson
PO Box 34009
Charlotte, NC 28234

Ohio

★2719★ Allegheny County Bar Association Intellectual Property Section
c/o John W. Jordan, IV
Grigsby, Gaca & Davies
1 Gateway Center, 10th Fl.
Pittsburgh, OH 15222

★2720★ Cincinnati Intellectual Property Law Association
c/o Eric W. Guttag
Proctor & Gamble Co.
6090 Center Hill Rd.
Cincinnati, OH 45224

★2721★ Cleveland Patent Law Association
c/o Edwin W. (Ned) Oldham
Oldham, Oldham, Webber Co., L.P.A.
Twin Oaks Estate
1225 W. Market St.
Akron, OH 44313

★2722★ Columbus Patent Law Association
c/o Robert B. Watkins
Watkins, Dunbar & Pollick
2941 Kenny Rd., Ste. 260
Columbus, OH 43221

★2723★ Dayton Intellectual Property Law Association
c/o Joseph J. Nauman
2102 W. Dorothy Lane
Dayton, OH 45439

★2724★ Ohio State Bar Association Patent, Trademark, and Copyright Section
c/o Ralph Jocke
Parker Hannifin Corporation
17325 Euclid Ave.
Cleveland, OH 44112

★2725★ Oklahoma Bar Association Patent, Trademark, and Copyright Section
c/o Tom Weaver
Halliburton Services
PO Drawer 1431
Duncan, OH 73536

★2726★ Toledo Patent Law Association
c/o David D. Murray
William, Brinks, Olds, Hofer, Gilson & Lione
1130 Edison Plaza
Toledo, OH 43604-1537

Oregon

★2727★ Oregon Patent Law Association
c/o David Petersen
Klarquist, Sparkman, et al.
121 SW Salmon, Ste. 1600
Portland, OR 97204

Pennsylvania

★2728★ Philadelphia Patent Law Association
c/o John C. Dorfman, Esq.
Dann, Dorfman, Herrell & Skillman
3 Mellon Bank Ctr., Ste. 900
15th St. and S. Penn Sq.
Philadelphia, PA 19102-2440

★2729★ Pittsburgh Intellectual Property Law Association
c/o Kenneth J. Stachel
PPG Industries, Inc.
1 PPG Place
Pittsburgh, PA 15272

Texas

★2730★ American Intellectual Property Law Association
c/o David A. Rose
Conley, Rose & Tayon
600 Travis, Ste. 1850
Houston, TX 77002

★2731★ **Austin Intellectual Property Law Association**
c/o John Fisher
3M Corporation
PO Box 2963
Austin, TX 78769

★2732★ **Dallas-Fort Worth Patent Law Association**
c/o Robert A. Felsman
Felsman, Bradley, Gunter & Kelly
2850 Continental Plaza
777 Main St.
Ft. Worth, TX 76102

★2733★ **Houston Intellectual Property Law Association**
c/o Frank S. Vaden, III
Vaden, Eickenroht, Thompson & Boulware
1 Riverway, Ste. 1100
Houston, TX 77056-1903

★2734★ **State Bar of Texas Intellectual Property Law Section**
c/o William L. LaFuze
Vinson & Elkins
1001 Fannin, Ste. 2912
Houston, TX 77002

Utah

★2735★ **State Bar of Utah Patent, Trademark, and Copyright Section**
c/o Allen R. Jensen
Workman, Nydegger & Jensen
1000 Eagle Gate Tower
60 E. South Temple
Salt Lake City, UT 84111

Virginia

★2736★ **Virginia State Bar Intellectual Property Law Section**
c/o James H. Laughlin, Jr.
Benoit, Smith & Laughlin
2001 Jefferson Davis Hwy.
Arlington, VA 20006

Washington

★2737★ **Washington Patent Law Association**
c/o John Mason
Microsoft Corporation
1 Microsoft Corporation
Bldg. 8, 2 South
Redmond, WA 98052

Wisconsin

★2738★ **Wisconsin Intellectual Property Law Association**
c/o Gary A. Essmann
Andrus, Sceales, Starke & Sawall
100 E. Wisconsin Ave.
Milwaukee, WI 53202

Section Arrangement

The following entries are arranged geographically by state, then alphabetically by city and name of agent or attorney. The cross-hatch (#) after the name indicates that the person is a patent agent; the asterisk (*) after the name indicates that the person is employed by the U.S. Government and not available to accept clients. Names in this section are not listed in the Master Index.

Alabama

Kenneth L. Cleveland
Cleveland & Cleveland
2330 Highland Avenue South
Birmingham, AL 35205   (205) 322-1811

William B. Hairston, III
Engel, Hairston & Johanson
4th Floor
109 N. 20th Street
Birmingham, AL 35203   (205) 328-4600

Donald H. Jones
1425 21st Street, South
Suite 200
Birmingham, AL 35205   (205) 933-2525

James D. Long
P.O. Box 590052
Suite 100
No. 6 Office Park Circle
Birmingham, AL 35259   (205) 871-1443

Thad G. Long
Bradley, Arant, Rose & White
1400 Park Place Tower
Birmingham, AL 35203   (205) 252-4500

Wm. Randall May
Griffin, Allison, May, Alvis & Fuhrmeister
4513 Valleydale Rd.
Birmingham, AL 35242   (205) 991-6367

Roy Leon Mims
3712 Woodvale Road
Birmingham, AL 35223   (404) 659-3859

J. David Pugh
Beadley, Arant, Rose & White
1400 Park Place Tower
Birmingham, AL 35203   (205) 521-8000

Theodore T. Robin, Jr.
4524 Pine Mountain Road
Birmingham, AL 35213   (205) 870-7268

Martin C. Ruegsegger
Bellsouth Services Inc.
South E9D1
3535 Colonnade Parkway
Birmingham, AL 35243   (205) 977-3524

Woodford R. Thompson, Jr.
Jennings, Carter, Thompson & Veal
2001 Park Place North, Ste. 525
Birmingham, AL 35203   (205) 324-1524

Robert J. Veal
Jennings, Carter, Thompson & Veal
2001 Park Place North, Ste. 525
Birmingham, AL 35203   (205) 324-1524

George L. Williamson
P.O. Box 2351
Daphne, AL 36526   (205) 661-1888

Arthur B. Beindorff#
2812 Burning Tree Mountain Rd.
Decatur, AL 35603   (205) 350-1256

Raymond H. Quist
516 E. Juniper Pl. Dr.
Foley, AL 36535   (205) 943-5162

James F. Quinn*
U.S. Army
DSJA
Attn: ATZN-JA
Fort Mc Clellan, AL 36205   (205) 848-5435

James Everette Staudt#
406 Audubon Street
Hartselle, AL 35640   (205) 773-6292

L. Frederick Hilbers
P.O. Box 19393
Homewood, AL 35219   (205) 871-1939

Frank M. Caprio
Bradley, Arant, Rose & White
200 Clinton Ave., W., Ste. 900
Huntsville, AL 35801   (205) 517-5100

Steven M. Clodfelter#
Phillips & Beumer, P.C.
1100 Jordan Lane, Ste. K
Huntsville, AL 35816   (205) 536-8261

Leonard Flank
1502 Elmwood Dr., S.E.
Huntsville, AL 35801   (205) 539-9704

John Calder Garvin, Jr.
117 Jefferson Street, North
Huntsville, AL 35801   (205) 536-3264

Harold W. Hilton#
7233 Statton Drive
Huntsville, AL 35802   (205) 882-2431

Charles A. Phillips
Phillips & Beumer, P.C.
1100 Jordan Lane, Ste. K
Huntsville, AL 35816   (205) 536-8261

George J. Porter
P.O. Box 4123
Huntsville, AL 35815   (205) 883-9212

Wayland H. Riggins
8802 Louis Drive
Huntsville, AL 35802   (205) 881-1428

Gary L. Rigney
Rigney, Garvin & Webster, P.C.
117 Jefferson St. North
Huntsville, AL 35801   (205) 536-3264

Floyd E. Roberts, III
NASA
Marshall Space Flight Ctr.
Mail Code EH-34 Bldg. 4612
Room 1101
Huntsville, AL 35812   (205) 544-1967

Jack Wendel Voigt, Sr.#
2601 Vista Drive
Huntsville, AL 35803   (205) 881-1594

Joseph H. Beumer*
NASA
Off. of Chief Coun.
Marshall Space Flight Ctr., AL 35812
(205) 544-0013

Robert L. Broad, Jr.*
NASA
Marshall Space Flight Center
Off. of Chief Counsel
Attn: CC01
Marshall Space Flight Ctr., AL 35812
(205) 881-1594

Jerry L. Seemann*
NASA
CC01/Off. of Chief Coun.
Marshall Space Flight Ctr., AL 35812
(205) 544-0026

Steve M. Tesney*
NASA/MSFC
P.O. Box 9104
Marshall Space Flight Ctr., AL 35812
(205) 544-3400

James A. Berneburg
Dravo Natural Resources Company
61 St. Joseph Street
P.O. Box 1685
Mobile, AL 36633   (205) 438-3531

Gregory M. Friedlander
11 South Florida Street
Mobile, AL 36606   (205) 470-0303

L. Daniel Morris, Jr.
Blount Inc.
P.O. Box 949
Montgomery, AL 36101   (205) 244-4341

Robert A. Petrusek#*
Tennessee Valley Authority
Off. of General Coun.
Natl. Fertilizer Dev.
1F/108C
Muscle Shoals, AL 35660   (205) 386-2363

Donald W. Phillion, Sr.
203 Carlisle Way
Rainbow City, AL 35901   (205) 442-1676

Freddie M. Bush#*
U.S. Army Missile Comm.
Attn: AMSMI-GC-IP
Redstone Arsenal, AL 35898   (205) 876-5107

Hay Kyung Chang*
U.S. Missile Command
AMSMI-GC-IP
Intellectual Property Law Div.
Redstone Arsenal, AL 35898   (205) 876-5106

James T. Deaton#*
U.S. Army Missile Comm.
Attn: AMSMI-GC-IP
Redstone Arsenal, AL 35898   (205) 876-5106

J. Keith Fowler#*
U.S. Army Missile Command
Attn. AMSMI-GC-IP
Redstone Arsenal, AL 35898   (205) 876-1121

Howard G. Garner, Jr.*
Dept. of the Army
Army Missile Command
Attn. AMSMI-GC-IP
Redstone Arsenal, AL 35898   (205) 876-1121

Jack M. Glandon*
U.S. Army Missile Command
AMSMI-GC-IP
Redstone Arsenal, AL 35898   (205) 876-1121

Hugh P. Nicholson*
U.S. Army Missile Cmd.
Legal Office
AMSMI-GC-IP
Nicholson
Redstone Arsenal, AL 35898   (205) 876-1121

Isaac Pugh Espy
Gray Espy & Nettles
P.O. Box 2786
Tuscaloosa, AL 35403   (205) 758-5591

Alaska

Lloyd V. Anderson, Jr.
Birch, Horton, Bittner, Pestinger & Anderson
1127 W. 7th Ave.
Anchorage, AK 99501   (907) 276-1550

Daniel Anthony Gerety
Delaney, Wiles, Hayes, Reitman & Brubaker,
   Inc.
1007 W. 3rd Ave.
Anchorage, AK 99501   (907) 279-3581

Kenneth P. Jacobus
Hughes, Thorsness, Gantz, Powell & Brundin
509 W. 3rd Ave.
Anchorage, AK 99501   (907) 274-7522

Michael J. Tavella#
6900 Rovena Street
Anchorage, AK 99518   (907) 349-2495

Terrance A. Turner
Owens & Turner, P.C.
1500 West 33rd Ave., Ste. 200
Anchorage, AK 99503   (907) 276-3963

Richard L. Blackmer#
P.O. Box 80286
College, AK 99708   (907) 276-3963

Arizona

Charles R. Lewis
P.O. Box 947
8929 E. Lazywood Place
Carefree, AZ 85377   (602) 488-3408

Wayne D. House#
W.L. Gore & Associates, Inc.
1500 N. Fourth St.
Flagstaff, AZ 86004   (602) 526-3030

Edward C. Threedy
15928 East Echo Hills Drive
Fountain Hills, AZ 85268   (602) 837-8637

Lockwood D. Burton
4424 W. Keating Circle
Glendale, AZ 85308   (602) 843-4825

Joseph S. Failla#
4201 W. Angela Drive
Glendale, AZ 85308   (602) 978-8750

Arthur S. Stewart
6540 W. Butler Dr., Ste. 3-85
Glendale, AZ 85302   (602) 939-0569

George Aichele
21300 S. Heather Ridge Cir.
Green Valley, AZ 85614   (602) 648-8241

John C. L. Cowen
1093 S. Paseo Del Prado
Green Valley, AZ 85614   (602) 625-0565

John M. Johnson#
110 W. Paseo Tesoro
Green Valley, AZ 85614   (602) 625-3190

Karl E. Sager#
696 W. Rio Altar
Green Valley, AZ 85614   (602) 648-1706

Joseph R. Black, Jr.
1346 W. Nopal Avenue
Mesa, AZ 85202   (602) 231-3939

Robert A. Farley#
Motorola, Inc.
2200 W. Broadway Road
(Mail Drop M285)
Mesa, AZ 85282   (602) 962-2310

Richard W. Gurtler
1304 W. Mountain View Drive
Mesa, AZ 85201   (602) 969-2504

James H. Gray#
210 East Chateau Circle
Payson, AZ 85541   (602) 474-5015

Noel G. Artman
17300 N. 88th Ave., #337
Peoria, AZ 85381   (602) 977-2431

Robert D. Atkins#
Motorola, Inc.
4250 E. Camelback Rd.
Suite 300K
Phoenix, AZ 85018   (602) 952-4704

Anthony J. Baca#
Ag Communication Systems Corp.
2500 West Utopia Road
Bldg. 3
PO Box 52179
Phoenix, AZ 85072   (602) 581-4681

Joe E. Barbee
Motorola, Inc.
4250 E. Camelback Road
Suite 300K
Phoenix, AZ 85018   (602) 952-4704

Frank Timothy Barber
Greyhound Corp.
Greyhound Tower
Phoenix, AZ 85077   (602) 248-5676

Michael D. Bingham
Motorola, Inc.
Pat. Dept. Suite 300K
4250 E. Camelback Rd.
Phoenix, AZ 85018   (602) 952-4703

Frank J. Bogacz
Motorola, Inc.
4250 E. Camelback Road
Suite 300-K
Phoenix, AZ 85018   (602) 952-4701

Bradley J. Botsch#
Motorola Inc.
4250 E. Camelback Rd.
Suite 300K
Phoenix, AZ 85018   (602) 952-4702

Kenneth R. Bowers, Jr.
2346 East Orangewood Avenue
Phoenix, AZ 85020   (602) 870-9818

William C. Cahill
Cahill, Sutton & Thomas
1400 Valley Bank Center
Phoenix, AZ 85073   (602) 258-8008

Charles E. Cates
Cates & Phillips
Suite 1210
2700 N. Central Ave.
Phoenix, AZ 85004   (602) 248-0982

Vincent F. Chiappetta
Meyer, Hendricks, Victor, Osborn & Maledon
2700 N. 3rd St.
Phoenix, AZ 85004   (602) 263-8700

Lowell E. Clark#
5901 E. Calle Del Sud
Phoenix, AZ 85018   (602) 945-5818

Thomas William De Mond
2936 West Larkspur
Phoenix, AZ 85029   (602) 866-9724

356

James F. Duffy
13430 N. 2nd Street
Phoenix, AZ 85022   (602) 942-8615

Don J. Flickinger#
1817 N. 3rd St., Ste. 106
Phoenix, AZ 85004   (602) 271-0092

William John Foley
Cahill, Sutton & Thomas
1400 Valley Bank Ctr.
Phoenix, AZ 85073   (602) 258-8008

Donald P. Gabrielson
3335 E. Garfield
Phoenix, AZ 85008   (602) 275-8351

Edward R. J Glady, Jr.
Fennemore, Craig, Von Ammon, Udall &
     Powers
2 N. Central Ave., Ste. 2200
Phoenix, AZ 85004   (602) 251-2527

Marvin A. Glazer
Cahill, Sutton & Thomas
155 Park One
2141 E. Highland Avenue
Phoenix, AZ 85016   (602) 956-7000

Hugh P. Gortler#
Allied-Signal, Inc.
111 South 34th Street
P.O. Box 5217
Phoenix, AZ 85010   (602) 231-1000

Lowell W. Gresham
1817 North Third Street
Suite 106
Phoenix, AZ 85004   (602) 256-7041

Robert M. Handy
Motorola, Inc.
Pat. Dept. - Suite 300K
4350 E. Camelback Rd.
Phoenix, AZ 85018   (602) 952-4704

Richard Grant Harrer
Cates & Phillips
Suite 1210
2700 North Central Ave.
Phoenix, AZ 85004   (602) 248-0982

Herbert E. Haynes, Jr.#
Karsten Manufacturing Corp.
2201 W. Desert Cove
Phoenix, AZ 85029   (602) 277-1300

Gregory G. Hendricks
GTE Service Corp.
2500 West Utopia Road
Phoenix, AZ 85027   (602) 581-4136

Robert A. Hirschfeld
4723 N. 44th Street
Phoenix, AZ 85018   (602) 840-0342

Charles R. Hoffman
Cahill, Sutton & Thomas
2141 E. Highland Avenue
155 Park One
Phoenix, AZ 85016   (602) 956-7000

Jerry J. Holden
Allied-Signal Aerospace Co.
P.O. Box 5217
Phoenix, AZ 85010   (602) 231-3333

William W. Holloway, Jr.
626 E. Orangewood Avenue
Phoenix, AZ 85020   (602) 995-2221

William P. Hovell
Robbins & Green, P.A.
3300 N. Central
Suite 1800
Phoenix, AZ 85012   (602) 248-7999

Bernard L. Howard
Greyhound Corporation
Greyhound Tower
Phoenix, AZ 85077   (602) 248-5776

Grant L. Hubbard
Nelson, Hubbard & Roediger
2623 North Seventh Street
Phoenix, AZ 85006   (602) 263-8782

Miriam Jackson#
Motorola, Inc.
4250 East Camelback Road
Suite 300K
Phoenix, AZ 85018   (602) 952-4704

Dale E. Jepsen
Honeywell Inc., M/S DV9L
P.O. Box 21111
Phoenix, AZ 85036   (602) 869-1336

Charles W. Jirauch
Streich, Lang, Weeks & Cardon
PO Box 471
Phoenix, AZ 85001   (602) 229-5200

Michael K. Kelly
100 West Washington, #2100
Phoenix, AZ 85003   (602) 229-5545

William E. Koch
Motorola Inc.
Suite 300K
4250 E. Camelback Rd.
Phoenix, AZ 85018   (602) 952-4703

Elliott Kurzman
5115 E. Windsor Avenue
Phoenix, AZ 85008   (602) 952-9035

Stuart T. Langley#
Motorola Inc.
4250 E. Cambelback Road
Phoenix, AZ 85018   (602) 952-4704

Michael A. Lechter
100 West Washington
Suite 2100
Phoenix, AZ 85003   (602) 229-5200

Warren F. B. Lindsley, Sr.
Camel Square, Suite 200E
4350 E. Camelback Rd.
Phoenix, AZ 85018   (602) 840-7310

Donald Julius Lisa
Lisa & Kubida, P.C.
2700 N. Central Avenue
Suite 1225
Phoenix, AZ 85004   (602) 222-5771

James W. Mc Farland
Allied-Signal Aerospace Co.
Patent Dept.
111 S. 34th Street
Phoenix, AZ 85034   (602) 231-1882

Jordan M. Meschkow
1817 North Third Street
Suite 106
Phoenix, AZ 85004   (602) 256-6996

A. Donald Messenheimer
Lisa & Lisa
2700 N. Central Avenue
Suite 1225
Phoenix, AZ 85004   (602) 222-5771

John K. Mickevicius
Suite 106
1817 North 3rd Street
Phoenix, AZ 85004   (602) 941-2555

Terry L. Miller
Allied-Signal Aerospace Co.
111 S. 34th Street
P.O. Box 5217
Phoenix, AZ 85010   (602) 231-1881

Anthony Miologos
GTE Service Corp.
2500 W. Utopia Road
Phoenix, AZ 85027   (602) 581-4314

Victor Myer
12014 S. Tomi Drive
Phoenix, AZ 85044   (602) 893-3621

Gregory J. Nelson
Nelson & Roediger
2623 North Seventh Street
Phoenix, AZ 85006   (602) 263-8782

Walter W. Nielsen
Motorola Inc.
Pat. Dept. - 300K
4250 E. Camelback Rd.
Phoenix, AZ 85018   (602) 952-4700

Tod R. Nissle
4041 N. Central Avenue
Suite 415
Phoenix, AZ 85012   (602) 279-5564

Eugene Arthur Parsons
Motorola Inc.
4250 E. Cambelback Rd.
Suite 300K
Phoenix, AZ 85018   (602) 952-4700

Robert A. Parsons#
1817 N. Third St.
Suite 106
Phoenix, AZ 85004   (602) 271-0092

James Harold Phillips
Bull Horn Information Sys. Inc.
13430 N. Black Canyon Hwy.
Phoenix, AZ 85029   (602) 862-6542

Jordan C. Powell
Motorola Inc.
4250 E. Camelback Road
Suite 300K
Phoenix, AZ 85018   (602) 952-4701

Edward C. Rapp
Maricopa County Superior Court
201 West Jefferson
Phoenix, AZ 85003   (602) 269-4404

Janelle F. Raupp
Cahill, Sutton & Thomas
155 Park One
2141 E. Highland Avenue
Phoenix, AZ 85016   (602) 956-7000

Joseph H. Roediger
Nelson & Roediger
2623 North Seventh Street
Phoenix, AZ 85006   (602) 263-8782

Arthur A. Sapelli
Honeywell Inc.
M/S DV9L
P.O. Box 21111
Phoenix, AZ 85036   (602) 869-1425

M. David Shapiro
Cahill, Sutton & Thomas
155 Park One
2141 E. Highland Avenue
Phoenix, AZ 85016   (602) 956-7000

H. Gordon Shields
Blake & Shields
7830 N. 23rd Ave.
Phoenix, AZ 85021   (602) 995-0490

Martin Lee Stoneman
525 East Cheery Lynn Road
Phoenix, AZ 85012   (602) 263-9200

Samuel J. Sutton, Jr.
Cahill, Sutton & Thomas
2141 E. Highland, #155
Phoenix, AZ 85016   (602) 956-7000

Joel E. Thompson
4500 North 32nd St.
Suite 100
Phoenix, AZ 85018   (602) 956-8191

Carl R. Von Hellens
Cahill, Sutton & Thomas
2141 E. Highland Ave.
Suite 155
Phoenix, AZ 85016   (602) 956-7000

Robert A. Walsh
Allied-Signal Aerospace Co.
Patent Dept. G4 301-1RA
111 S. 34th Street
Phoenix, AZ 85034   (602) 231-2510

Thomas G. Watkins, III
Cahill, Sutton & Thomas
2141 E. Highland Avenue
Suite 155
Phoenix, AZ 85016   (602) 956-7000

Paul F. Wille
Motorola Inc.
4250 E. Camelback Rd.
Suite 300K
Phoenix, AZ 85018   (602) 952-4700

Harry A. Wolin
Motorola, Inc.
4250 E. Cambelback
Suite 300K
Phoenix, AZ 85018   (602) 952-4702

Richard W. Toelken#
Box 27124
4581 Parent Road
Prescott Valley, AZ 86312   (602) 772-0295

Donald R. Bentz
P.O. Box 6434
Scottsdale, AZ 85261   (602) 481-9708

J. Stanley Edwards
Edwards & Edwards
10505 North 69th Street
Suite 800
Scottsdale, AZ 85253   (602) 991-1938

Lannas S. Henderson, Jr.
6801 E. Camelback Road
Apt. P-103
Scottsdale, AZ 85251   (602) 990-0684

Edward W. Hughes
6451 E. Cholla St.
Scottsdale, AZ 85254   (602) 948-3356

John S. Lieb
8010 E. Morgan Trail
Suite 10
Scottsdale, AZ 85258   (602) 483-1285

H. Barry Moyerman
7141 E. Mc Donald Drive
Scottsdale, AZ 85253   (602) 991-3556

Ruth Moyerman
7141 E. Mc Donald Dr.
Scottsdale, AZ 85253   (602) 991-3556

Richard R. Mybeck
8010 E. Morgan Trail
Suite 10
Scottsdale, AZ 85258   (602) 483-1285

Lavalle D. Ptak
4420 N. Saddlebag Trail
Suite 102
Scottsdale, AZ 85251   (602) 994-1003

David G. Rosenbaum
Rosenbaum & Schwartz, P.A.
6991 East Camelback Road
Suite B-360
Scottsdale, AZ 85251   (602) 481-9575

Kevin S. Taylor
13001 North 49 Place
Scottsdale, AZ 85254   (602) 996-4487

Harry M. Weiss
4204 N. Brown Ave.
Scottsdale, AZ 85251   (602) 994-8888

Burton Ralph Turner
130 Siesta Way
Sedona, AZ 86336   (602) 282-2602

J. Edwin Coates
9414 Cedar Hill Circle
Sun City, AZ 85351   (602) 974-0284

Harry P. Eichin#
10023 Lancaster Dr.
Sun City, AZ 85351   (602) 977-8019

James A. Hauer
10221 West Edgewood Drive
Sun City, AZ 85351   (602) 933-4942

Mary B. Moshier
16807 103rd Avenue
Sun City, AZ 85351   (602) 974-5676

George C. Nebesar
10327 Prairie Hills Circle
Sun City, AZ 85351   (602) 972-8285

John W. Overman
10909 Palmeras Drive
Sun City, AZ 85373   (602) 974-5697

Frank Cristiano, Jr.#
17430 Conquistador Dr.
Sun City West, AZ 85375   (602) 584-5274

Norman C. Fulmer
20443 135th Avenue
Sun City West, AZ 85375   (602) 584-7470

Thomas A. Seeman
14224 Franciscan
Sun City West, AZ 85375   (602) 584-6610

Frederick Burton Sellers#
12637 Rampart Drive
Sun City West, AZ 85375   (602) 975-1760

Harold C. Weston
17814 Conquistador Dr.
Sun City West, AZ 85375   (602) 584-9857

Ronald M. Halvorsen
10310 East Silvertree Court
Sun Lakes, AZ 85248   (602) 895-7389

Nedwin Berger
2645 E. Southern
Apt. A-229
Tempe, AZ 85282   (602) 831-3449

Louise S. Heim#
2720 S. Rita Avenue
Tempe, AZ 85282   (602) 829-0386

Charles P. Padgett, Jr.
1634 E. Del Rio Drive
Tempe, AZ 85282   (602) 820-2158

Floyd E. Anderson#
IBM
9000 S. Rita Road
90A/061-2
Tucson, AZ 85744   (602) 799-2544

H. Walter Clum#
3940 N. Romero Road
No. 11
Tucson, AZ 85705   (602) 293-1394

Antonio R. Durando
33 North Stone, Suite 850
Tucson, AZ 85701   (602) 792-3223

Samuel W. Engle
38231 S. Samaniego Dr.
Tucson, AZ 85737   (602) 825-1537

Victor Flores
3721 W. Goret Rd.
Tucson, AZ 85745   (602) 743-0715

Eric A. Gifford#
2664 N. Alvernon Way, #412
Tucson, AZ 85712   (602) 743-0715

Franklin L. Gubernick#
380 E. Ft. Lowell Road
Suite 226
Tucson, AZ 85705   (602) 743-0715

John H. Holcombe
IBM Corp.
Intellectual Property Law Dept.
90A/301
Tucson, AZ 85744   (602) 629-4102

James M. Mc Clanahan
7473 E. Broadway
Tucson, AZ 85710   (602) 296-7373

Joseph Daniel Odenweller
6222 E. Placita Aspecto
Tucson, AZ 85715   (602) 296-7373

Mark E. Ogram
780 S. Freeman Road
Tucson, AZ 85748   (602) 298-1210

David G. Perry#
6235 N. Windemere St.
Tucson, AZ 85704   (602) 742-0982

James A. Pershon
IBM Corporation
Dept. 90A - 061-2
Patent Operations
Tucson, AZ 85744   (602) 799-2537

Timothy J. Reckart
Research Corp. Technologies, Inc.
6840 E. Broadway Blvd.
Tucson, AZ 85710   (602) 296-6400

Manny W. Schecter
IBM Corporation
IPLAW 90A/061-2
Tucson, AZ 85744   (602) 799-2550

Herbert F. Somermeyer
8421 East Fernhill Drive
Tucson, AZ 85715   (602) 296-5815

Horace St. Julian
IBM Corporation
9000 S. Rita Road
90A/061
Tucson, AZ 85744   (602) 799-2800

Jerome M. Teplitz
Research Corporation
6840 East Broadway Blvd.
Tucson, AZ 85710   (602) 296-6400

David A. Wiersma#
Research Corp. Technologies, Inc.
6840 E. Broadway Blvd.
Tucson, AZ 85710   (602) 296-6400

Herbert L. Martin
P.O. Box 3506
W. Sedona, AZ 86340   (602) 282-3688

Harold H. Card, Jr.*
P.O. Box 98
Winslow, AZ 86047   (602) 289-4539

Arkansas

Boyd D. Cox
26 E. Center Street
Fayetteville, AR 72701   (501) 521-2052

Robert Raymond Keegan
Suite G, Sherwood Building
130 North College Avenue
Fayetteville, AR 72701   (501) 521-4412

R. Donald Pitts#
410 Skyline Dr.
Harrison, AR 72601   (501) 741-3914

Stephen D. Carver
Suite 800 - Pleasant Valley Corp. Ctr.
2024 Arkansas Valley Drive
Little Rock, AR 72212   (501) 224-1500

Ray F. Cox, Jr.
Wright, Lindsey & Jennings
2200 Worthen Bank Bldg.
Little Rock, AR 72203   (501) 371-0808

Hermann Ivester
Ivester, Henry, Skinner & Camp
111 Center St., Ste. 1200
Little Rock, AR 72201   (501) 376-7788

Joseph A. Strode
Bridges, Young, Matthews, Holmes & Drake
315 E. 8th Ave.
PO Box 7808
Pine Bluff, AR 71611   (501) 534-5532

Frank P. Troseth
1406 Dutchmans Drive
Rogers, AR 72756   (501) 925-2174

Jerry L. Mahurin#
Route 1, Box 320A
Traskwood, AR 72167   (501) 778-3872

California

Jack Calvin Munro
5210 Lewis Road, Unit 10
Agoura Hills, CA 91301   (818) 991-1687

Taylor M. Belt
1825 Shoreline Drive
Apt. 305
Alameda, CA 94501   (415) 523-1284

Laurence C. Bonar#
Univ. of California
1320 Harbor Bay Parkway
Suite 150
Alameda, CA 94501   (415) 748-6600

Albert A. Jecminek
Triton Biosciences Inc.
1501 Harbor Bay Pkwy.
Alameda, CA 94501   (415) 769-5203

Joan V. Mc Cormack
University of California
Pat., Trdemk. Copyright Off.
1320 Harbor Bay Parkway
Suite 150
Alameda, CA 94501   (415) 748-6600

Howard B. Scheckman
1448 E. Shore Drive
Alameda, CA 94501   (415) 769-1990

Palmer Martin Simpson, Jr.
University of California
1320 Harbor Bay Pkwy.
Suite 150
Alameda, CA 94501   (415) 748-6600

Louis J. Strom
University of California
1320 Harbor Bay Parkway
Suite 150
Alameda, CA 94501   (415) 748-6600

Margaret A. Connor*
U.S. Dept. of Agriculture
Western Regional Res. Center
800 Buchanan St.
Albany, CA 94710   (415) 486-3208

Matthias L. Tam
248 E. Main St., #201
Alhambra, CA 91801   (818) 289-9616

Cecilia L. Yu
801 South Garfield Ave.
Suite 212
Alhambra, CA 91801   (818) 281-0280

Gabriel N. Hanover, II
3488 Rubio Crest Drive
Altadena, CA 91001   (818) 791-7377

David J. Arthur
Rockwell International Corp.
Pats., Mail Code HA33
3370 Miraloma Avenue
Anaheim, CA 92803   (714) 762-3848

Harold Burg
P.O. Box 18776
Anaheim, CA 92807   (714) 637-8691

Thomas M. Deforest
Fujitsu Bus. Communications of America, Inc.
3190 Miraloma Ave.
Anaheim, CA 92806   (714) 630-7721

Allen A. Dicke, Jr.#
224 Mall Way
Anaheim, CA 92804   (714) 527-3766

H. Frederick Hamann
Rockwell International Corp.
3370 Miraloma Avenue
HA 52
Anaheim, CA 92803   (714) 762-1663

Wallace B. King
Morris & King
600 N. Euclid Ave.
Suite 202
Anaheim, CA 92801   (714) 956-1600

George A. Montanye
Rockwell International Corp.
3370 Miraloma Avenue
Anaheim, CA 92803   (714) 762-6300

Arthur E. Oaks
7662 Northfield Ave.
Anaheim, CA 92807   (714) 970-0575

Gregory D. Ogrod
Rockwell International Corp.
3370 Miraloma Avenue
HA33
Anaheim, CA 92803   (714) 762-3848

William F. Porter, Jr.
Cal Comp Inc.
2411 West La Palma Avenue
Anaheim, CA 92803   (714) 821-2523

Tom Streeter
Rockwell International Corp.
Patent Department - HA52
3370 Miraloma Avenue
P.O. Box 3105
Anaheim, CA 92803   (714) 762-5662

James R. Thornton#
U.S. Borax Research Corp.
412 Crescent Way
Anaheim, CA 92801   (714) 490-6000

James D. Leimbach,#
8430 E. Saratoga
Anaheim Hills, CA 92808   (714) 921-1739

William Hinman Hooper
1300 Summit Lake Drive
Angwin, CA 94508   (707) 965-2163

Edward F. Jaros
P.O. Box 829
Arcadia, CA 91066   (818) 447-7783

Kenneth Watson Mateer
13153 Tonopah Street
Arleta, CA 91331   (818) 768-9770

William H. Dana
10 Camino Por Los Arboles
Atherton, CA 94027   (415) 321-0888

Chester Martin Mc Closkey
Norac Co., Inc.
405 S. Motor Avenue
P.O. Box F
Azusa, CA 91702   (213) 334-2908

Janis E. Kerber
3501 Sterling Road
Bakersfield, CA 93306   (805) 871-2027

Peter J. Szabo
Alliance Fund Distributors
810 Oxford Way
Benicia, CA 94510   (707) 747-5104

Bertram Bradley
Miles Laboratories, Inc.
4th and Parker Sts.
Berkeley, CA 94710   (415) 420-4326

Roy L. Brown
2832 College Ave.
Berkeley, CA 94705   (415) 486-1981

David T. Emerson#
2739 Regent St.
Berkeley, CA 94705   (415) 540-8495

Elizabeth F. Enayati
Cutter Biological, Miles Inc.
4th & Parker Streets
P.O. Box 1986
Berkeley, CA 94701   (415) 420-5000

Valentin D. Fikovsky#
University of California
Patent Office
Berkeley, CA 94720   (415) 642-5000

James A. Giblin
Miles Inc.
4th & Parker Sts.
Berkeley, CA 94710   (415) 420-5511

George C. Gorman
952 the Alameda
Berkeley, CA 94707   (415) 524-9520

Marcus Lothrop
Lothrop & West
726 Euclid Avenue
Berkeley, CA 94708   (415) 986-5833

Paul R. Martin
Lawrence Berkeley Lab.
1 Cyclotron Road
Berkeley, CA 94720   (415) 486-7025

Kathleen S. Moss
Lawrence Berkeley Laboratory
1 Cyclotron Road
Berkeley, CA 94720   (415) 486-7025

Alvah Levern Snow
1806 San Antonio Ave.
Berkeley, CA 94707   (415) 525-6647

William Takacs
886 Arlington Ave.
Berkeley, CA 94707   (415) 524-0257

Berthold J. Weis#
Lawrence Berkeley Laboratory
One Cyclotron Road
Off. of Laboratory Coun.
Berkeley, CA 94720   (415) 486-7025

Judith A. Woods
Xoma Corp.
2910 Seventh St.
Berkeley, CA 94710   (415) 644-1170

Steve W. Ackerman
Cooper, Epstein & Hurewitz
345 North Maple Drive
Suite 200
Beverly Hills, CA 90210   (213) 278-1111

Sanford Astor
9401 Wilshire Blvd.
Suite 1105
Beverly Hills, CA 90212   (213) 274-6122

Robert Boyd Block
P.O. Box 10717
Beverly Hills, CA 90213   (213) 550-8725

Warner W. Clements#
13435 Java Drive
Beverly Hills, CA 90210   (213) 276-7918

I. Morley Drucker
Drucker & Sommers
9465 Wilshire Blvd.
Suite 328
Beverly Hills, CA 90212   (213) 278-6852

Joseph R. Evanns
Evanns & Walsh
119 N. San Vicente Blvd.
Suite 206
Beverly Hills, CA 90211   (213) 273-0938

Charles L. Hartman
Drucker & Sommers
9465 Wilshire Blvd.
Suite 328
Beverly Hills, CA 90212   (213) 278-6852

Harvey Sander Hertz
9777 Wilshire Blvd.
Suite 500
Beverly Hills, CA 90212   (213) 278-9673

Booker T. Hogan
9595 Wilshire Blvd.
Suite 900
Beverly Hills, CA 90212   (213) 278-0966

Daniel R. Kimbell
Drucker & Sommers
9465 Wilshire Blvd.
Suite 328
Beverly Hills, CA 90212   (213) 278-6852

Robert H. Lentz
Litton Industries, Inc.
360 N. Crescent Dr.
Beverly Hills, CA 90210   (213) 859-5153

Michael F. Mc Entee
9595 Wilshire Blvd.
Suite 900
Beverly Hills, CA 90212   (213) 273-3342

Richard Morganstern
9595 Wilshire Blvd.
Suite 800
Beverly Hills, CA 90212   (213) 858-7796

Cyrus S. Nownejad
9935 Santa Monica Blvd.
Second Floor
Beverly Hills, CA 90212   (213) 557-9005

Michael A. Painter
Cooper, Epstein & Hurewitz
345 North Maple Drive
Suite 200
Beverly Hills, CA 90210 (213) 278-1111

Howard N. Sommers
Drucker & Sommers
9465 Wilshire Blvd.
Suite 328
Beverly Hills, CA 90212 (213) 278-6852

Paul D. Supnik
Nemschoff & Supnik
9601 Wilshire Blvd.
Suite 735
Beverly Hills, CA 90210 (213) 274-8281

Thomas A. Turner
9100 Wilshire Blvd.
Suite 800
Beverly Hills, CA 90212 (213) 273-1870

Edward C. Walsh
Evanns & Walsh
119 N. San Vicente Blvd.
Suite 206
Beverly Hills, CA 90211 (213) 273-0938

Nita J. Almquist
P.O. Box 864
Bishop, CA 93515 (619) 873-4211

Fred Flam
P.O. Box 833
Bonsall, CA 92003 (619) 945-7777

Henry M. Stanley
165 E. Hilton Drive
Boulder Creek, CA 95006 (408) 338-9561

Timothy H. Briggs
P.O. Box 401
Brea, CA 92622 (714) 733-9868

William M. Dooley
Unocal Science & Tech. Div.
Unocal Corporation
376 S. Valencia Avenue
P.O. Box 76
Brea, CA 92621 (714) 528-7201

Daniel R. Farrell
Unocal Corporation
376 S. Valencia Avenue
P.O. Box 76
Brea, CA 92621 (714) 528-7201

Yale S. Finkle
Unocal Corporation
376 S. Valencia Avenue
Science & Tech. Div.
Brea, CA 92621 (714) 528-7201

Shlomo R. Frieman
Union Oil Co. of Calif.
P.O. Box 76
Brea, CA 92621 (714) 528-7201

William O. Jacobson
Union Oil Company of Calif.
DBA Unocal
P.O. Box 76
Brea, CA 92621 (714) 528-7201

John A. Kane
Unocal Corporation
376 S. Valencia Avenue
P.O. Box 76
Brea, CA 92621 (714) 528-7201

Michael Albert Kondzella
Union Oil Company of Calif.
Patent Dept.
P.O. Box 76
Brea, CA 92621 (714) 528-7201

Michael H. Laird
Union Oil Co. of Calif.
376 Valencia
P.O. Box 76
Brea, CA 92621 (714) 528-7201

Dorothy B. Mc Kenzie-Wardell
Unocal Corporation
376 S. Valencia Avenue
P.O. Box 76
Brea, CA 92621 (714) 528-7201

Robert A. Schruhl
Carroll & Gilbert
711 South Brea Ave.
Brea, CA 92621-000 (714) 671-9963

John F. Sicotte
Gary Steven Findley
& Associates
2700 E. Imperial Hwy.
Suite J
Brea, CA 92621 (714) 524-5210

Gregory F. Wirzbicki
Union Oil Co. of Calif.
P.O. Box 7600
Brea, CA 92622 (714) 528-7201

Louis L. Dachs
Lockheed-California Company
P.O. Box 551
Burbank, CA 91520 (818) 847-5291

Philip Hoffman
4421 Riverside Dr., #101
Burbank, CA 91505 (818) 955-9770

Harry P. Levin
Electro Energy Corp.
120 Elm Court
Burbank, CA 91502 (213) 845-2666

William J. Schneider, Jr.#
1220 East Providencia Avenue
Burbank, CA 91501 (818) 409-0153

Laurence Coit
1369 Vancouver Avenue
Burlingame, CA 94010 (415) 347-7149

Milton F. Custer#
4885 North Point
Byron, CA 94514 (415) 634-5322

Billy G. Corber
Lockheed Corp.
4500 Park Granada Blvd.
Calabasas, CA 91399 (818) 712-2390

Henry Kolin
26124 Roymor Drive
Calabasas, CA 91302 (818) 880-4253

David O Reilly
23603 Park Sorrento
Suite 103
Calabasas, CA 91302 (818) 883-3600

Frederic Paul Smith
22900 Wrencrest Drive
Calabasas, CA 91302 (818) 992-8090

Fay I. Konzem
American Patent Institute
2707 N. Los Pinos Cir.
Camarillo, CA 93010 (805) 987-1880

Joseph M. St. Amand, Jr.
255 Camarillo Drive
Camarillo, CA 93010 (805) 482-6629

William B. Anderson
1158 Harriet Avenue
Campbell, CA 95008 (408) 379-5329

James Bartholomew#
7312 Independence Ave
Apt. 8
Canoga Park, CA 91303 (818) 704-1622

John J. Casparro#
23326 Sandalwood St.
Canoga Park, CA 91307 (818) 992-4067

Robert P. Egermeier
22354 Malden St.
Canoga Park, CA 91304 (818) 348-1670

David C. Faulkner
Rockwell International Corp.
Rocketdyne Patent Dept. F B18
6633 Canoga Avenue
Canoga Park, CA 91304 (818) 700-4616

Harry Bruce Field
Rockwell International Corp.
6633 Canoga Avenue
Canoga Park, CA 91303 (818) 700-4616

Lawrence N. Ginsberg
Rockwell International Corp.
6633 Canoga Avenue
Canoga Park, CA 91304 (818) 700-3629

Steven E. Kahm
Rockwell International Corp.
Rocketdyne Division
6633 Canoga Avenue
Canoga Park, CA 91303 (818) 700-4619

Gilbert Kivenson
22030 Wyandotte St.
Canoga Park, CA 91303 (213) 883-5707

Thomas R. Waite
8609 De Sota Avenue
Suite 147
Canoga Park, CA 91304 (213) 341-4474

Richard E. Cummins
870 Park Avenue
Apt. 116
Capitola, CA 95010 (408) 462-4325

Duane C. Bowen
2551 State Street
Carlsbad, CA 92008   (714) 729-8446

David J. Harshman
7824 Gaviota Circle
Carlsbad, CA 92009   (619) 942-9084

John J. Murphey
Pacific Center One
Suite 260
701 Palomar Airport Road
Carlsbad, CA 92009   (619) 431-0091

Van Wesley Smart
26123 Atherton Dr.
Carmel, CA 93923   (408) 624-5598

Joseph E. Gerber
5441 Fair Oaks Blvd.
Suite B-1
Carmichael, CA 95608   (916) 971-3725

Charles G. Miller
20014 S. Camba Ave.
Carson, CA 90746   (213) 604-1494

Alfred Fafarman#
Applied Physics Consultants
P.O. Box 2994
Castro Valley, CA 94546   (415) 530-2326

Steven P. Brown#
Optical Disc Corporation
17517-H Fabrica Way
Cerritos, CA 90701   (714) 522-2370

Gilbert P. Hyatt#
P.O. Box 3357
Cerritos, CA 90703   (714) 995-1087

Howard A. Kenyon
19620 S. Jeffrey Circle
Cerritos, CA 90701   (213) 860-4770

Herbert Eckerling
20336 Coraline Circle
Chatsworth, CA 91311   (213) 998-9037

Don A. Hollingsworth#
10511 Keokuk Ave.
Chatsworth, CA 91311   (818) 998-3465

William A. Kemmel, Jr.
Syncor International Corp.
20001 Prairie Street
Chatsworth, CA 91311   (818) 886-7400

Stephen J. Church*
Naval Weapons Center
Off. of Pat. Coun.
Code 006
China Lake, CA 93555   (619) 939-3733

Donald E. Lincoln*
Dept. of the Navy
Naval Weapons Center
Off. of Coun., Code 006
China Lake, CA 93555   (619) 939-1839

Stuart H. Nissim*
Department of the Navy
Off. of Gen. Counsel
Naval Weapons Center
China Lake, CA 93555   (619) 939-1839

Melvin J. Sliwka, Jr.*
Naval Weapons Center
Code 006 Building #00100
China Lake, CA 93555   (619) 939-1816

James H. Meadows#
P.O. Box 122143
Chula Vista, CA 92012   (619) 939-1816

Patrick J. Schlesinger
Rohr Inds., Inc.
P.O. Box 878
Chula Vista, CA 92012   (619) 691-2555

Doris Drucker#
636 Wellesley Dr.
Claremont, CA 91711   (714) 626-3172

Harold S. Gault
2549 N. Mountain Avenue
Claremont, CA 91711   (714) 626-5500

Edward B. Johnson#
754 West Ninth Street
Claremont, CA 91711   (714) 626-8147

Harry Loberman
876 Occidental Drive
Claremont, CA 91711   (714) 624-4198

Bernard V. Ousley
140 West Foothill Blvd.
Claremont, CA 91711   (714) 621-7949

Robert Nathan Schlesinger#
RNS Research Institute
P.O. Box 1117
Claremont, CA 91711   (714) 980-9540

Paul E. Calrow
211 Round House Place
Clayton, CA 94517   (415) 672-7901

Thomas J. Murphy
Box 432
7200 Amoloc Lane
Coloma, CA 95613   (916) 626-4162

Edward J. Holzrichter#
Whittaker Coatings Research Center
1231 S. Lincoln Avenue
P.O. Box 825
Colton, CA 92324   (714) 825-6292

James W. Lucas
Technical Enterprises
1401 Bonnie Doone
Corona Del Mar, CA 92625   (714) 644-9500

Ronald James Clark
Discovision Associates
P.O. Box 6600
Costa Mesa, CA 92628   (714) 957-3000

Roy A. Ekstrand
125 E. Baker Street
Suite 240
Costa Mesa, CA 92626   (714) 662-7733

Ronald C. Hudgens
Discovision Associates
2183 Fairview Road
Suite 211
Costa Mesa, CA 92627   (714) 957-3000

William J. Kearns
3350 California Street
Costa Mesa, CA 92626   (714) 641-9435

John H. Lynn
Lynn & Lynn
Suite 1160
3200 Park Center Drive
Costa Mesa, CA 92626   (714) 641-4712

Michael K. O Neill
650 Town Center Drive
Suite 740
Costa Mesa, CA 92626   (714) 540-8700

John Paul Scherlacher
Spensley Horn Jubas & Lubitz
650 Town Center Drive
Center Tower
Suite 1930
Costa Mesa, CA 92626   (714) 557-2047

Robert M. Taylor, Jr.
Lyon and Lyon
3200 Park Center Drive
Suite 1170
Costa Mesa, CA 92626   (714) 751-6606

Eugene H. Valet
Archive Corporation
Legal Dept.
1650 Sunflower Avenue
Costa Mesa, CA 92626   (714) 966-4741

Michael A. Shimokaji
One Mahogany Run
Coto De Caza, CA 92679   (714) 768-6788

James R. Eckel#
5104 Copperfield Lane
Culver City, CA 90230   (213) 839-0108

Richard K. Ehrlich
4901 S. Overland Avenue
Culver City, CA 90230   (213) 559-7415

Robert Jacobs
Beehler & Pavitt
100 Corporate Pointe
Suite 330
Culver City, CA 90230   (213) 215-3183

John H. Kusmiss
12200 Allin Street
Culver City, CA 90230   (213) 313-0926

John T. Matlago
100 Corporate Pointe
Suite 330
Culver City, CA 90230   (213) 215-3275

Ralf H. Siegemund
100 Corporate Pointe
Suite 330
Culver City, CA 90230   (213) 215-3183

Mark A. Aaker
Apple Computer, Inc.
20525 Mariani Ave.
M.S. 38-1
Cupertino, CA 95014   (408) 974-4347

Hal Jay Bohner
Measurex Corp.
One Results Way
Cupertino, CA 95014   (408) 255-1500

Timothy Daniel Casey
Apple Computer, Inc.
10431 North De Anza Blvd.
MS: 38-I
Cupertino, CA 95014   (408) 996-1010

Lawrence W. Granatelli
Tandem Computers Inc.
10435 N. Tantau Ave.
Loc. 200-16
Cupertino, CA 95014   (408) 996-1010

Alan H. Haggard
Hewlett-Packard Company
10900 Wolfe Road
Cupertino, CA 95014   (408) 447-0077

Douglas A. Kundrat
Hewlett-Packard Co.
Legal Dept. M/S 41AJ
10900 Wolfe Road
Cupertino, CA 95014   (408) 447-0289

Robert T. Martin#
Apple Computer, Inc.
20525 Mariani Avenue
M/S 38-I
Cupertino, CA 95014   (408) 974-4700

John B. Mc Gowan
Hewlett-Packard Company
U.S. Field Operations
19320 Pruneridge Avenue
Cupertino, CA 95014   (408) 865-6878

Theodore S. Park
Tandem Computers Inc.
10435 N. Tantau Ave.
Cupertino, CA 95014   (408) 865-4692

Jonathan B. Penn
Apple Computer, Inc.
10431 N. De Anza Blvd.
M.S. 38-1
Cupertino, CA 95014   (408) 982-7673

Franklyn Charles Weiss
Apple Computer, Inc.
10431 N. De Anza Blvd.
M.S. 38-I
Cupertino, CA 95014   (408) 974-1210

John Helms Warden
6666 Vinalhaven Court
Cypress, CA 90630   (714) 893-8443

Stephen L. Hurst
142 Wyandotte Avenue
Daly City, CA 94014   (415) 992-7473

Robert Edward Havranek
130 Virginia Court
Danville, CA 94526   (415) 820-8082

Sharon R. Kantor#
Firenza Group, Ltd.
893 Richard Lane
Danville, CA 94526   (415) 820-8082

Pauline M. Naillon
17 Diamond Dr.
Danville, CA 94526   (415) 837-5848

Michael D. Nelson
665 S. Hartz Avenue
Apt. 210
Danville, CA 94526   (415) 837-8019

Martin G. Reiffin
5439 Blackhawk Dr.
Danville, CA 94506   (415) 736-3799

Elizabeth Lassen
Calgene, Inc.
1920 Fifth Street
Davis, CA 95616   (916) 753-6313

Donna E. Scherer#
Calgene, Inc.
1920 Fifth St.
Davis, CA 95616   (916) 753-6313

Howard Weinberg
1008 Clark Court
Davis, CA 95616   (916) 753-7611

John R. Ross, Jr.
P.O. Box 2138
13020 Long Boat Way
Del Mar, CA 92014   (916) 753-7611

Leonard Zalman
526 Stratford Ct., #E
Del Mar, CA 92014   (619) 755-1726

William S. Bernheim
Whitaker & Bernheim
255 North Lincoln Street
Dixon, CA 95620   (916) 678-4447

John C. Grant#
7429 Firestone Pl., No. 3
Downey, CA 90241   (213) 927-0805

Ronald L. Juniper
9025 Florence Ave.
Suite A
Downey, CA 90240   (213) 861-0796

Kenneth T. Theodore
9012 Suva Street
Downey, CA 90240   (213) 420-7017

James T. English#
1134 Calle Adra
Duarte, CA 91010   (818) 303-6229

Glenn H. Lenzen, Jr.
Hexcel Corp.
11555 Dublin Blvd.
Dublin, CA 94568   (415) 828-4209

John F. O Flaherty
Hexcel Corp.
11555 Dublin Blvd.
Dublin, CA 94568   (415) 828-4200

Sidney Sternick
7567 Amador Valley Blvd.
Suite 106
Dublin, CA 94568   (415) 829-9270

Ervin Frederic Johnston
1904 Ventana Way
El Cajon, CA 92020   (619) 448-2228

William B. Wong
South Coast Air Quality
Management District
Legal Division
9150 Flair Drive
El Monte, CA 91731   (818) 572-2045

Anthony T. Cascio
Aura Systems, Inc.
2335 Alaska Ave.
El Segundo, CA 90245   (213) 643-5300

Robert E. Cunha
Xerox Corp.
701 S. Aviation Blvd.
ESAE- 335
El Segundo, CA 90245   (213) 333-7292

Terrell P. Lewis
Rockwell International Corp.
2230 East Imperial Hwy.
El Segundo, CA 90245   (213) 647-5225

Daniel J. O'Neill
Xerox Corporation
1960 East Grand Ave.
El Segundo, CA 90245   (213) 333-9802

William W. Propp
Xerox Corporation
101 Continental Blvd.
Room 550
El Segundo, CA 90245   (213) 333-8642

Sheldon F. Raizes
Xerox Corp.
101 Continental Blvd.
ESC 1-801
El Segundo, CA 90245   (213) 333-3670

Charles T. Silberberg
Rockwell International Corp.
2230 East Imperial Hwy.
El Segundo, CA 90732   (213) 647-5224

Barry P. Smith
Xerox Corporation
1960 E. Grand Avenue
El Segundo, CA 90245   (213) 333-8635

Gregory J. Giotta#
Cetus Corp.
1400 53rd Street
Emeryville, CA 94608   (415) 420-3152

Grant D. Green
Chiron Corp.
4560 Horton St.
Emeryville, CA 94608   (415) 655-8730

Barbara G. Mc Clung
Chiron Corp.
Intell. Prop. Dept.
4560 Horton St.
Emeryville, CA 94608   (415) 655-8730

Philip L. Mc Garrigle
Cetus Corporation
1400 53rd St.
Emeryville, CA 94608   (415) 420-3217

Stacey R. Sias, Codon#
Cetus Corp.
1400 53rd St.
Emeryville, CA 94608   (415) 420-3197

Lewis E. Massie
Abaris Corp.
2218 13th Street
Encinitas, CA 92024   (714) 436-0061

Lewis Anten
16830 Ventura Blvd.
Encino, CA 91436   (818) 501-3535

Joseph A. Compton#
4847 Hayvenhurst Ave.
Encino, CA 91436   (818) 501-3535

Albert O. Cota#
5460 White Oak Avenue
Suite A-331
Encino, CA 91316   (818) 905-0848

Donald Diamond
16133 Ventura Blvd.
7th Floor
Encino, CA 91436   (213) 986-2221

Robert Louis Finkel
16055 Ventura Blvd.
Suie 915
Encino, CA 91436   (213) 986-5000

Thomas Gunzler#
16929 Escalon Dr.
Encino, CA 91436   (818) 988-1040

Thomas D. Linton, Jr.
3404 Colville Place
Encino, CA 91436   (818) 784-3303

Seymour Rosenberg
15915 Ventura Blvd.
Suite 303
Encino, CA 91436   (213) 905-6888

Ralph Eugene Walters
1420 Westwood Place
Escondido, CA 92026   (619) 741-0135

Loyal M. Hanson
P.O. Box 430
Fallbrook, CA 92028   (619) 723-0620

Richard W. Keefe
1439 Knoll Park Lane
Fallbrook, CA 92028   (619) 728-3779

Leonard D. Schappert
P.O. Box 1366
Ferndale, CA 95536   (707) 786-4253

Aaron Passman
54 Monserrat Avenue
Foothill Ranch, CA 92610   (714) 588-7438

Kenneth A. Cox
704 Somerset Lane
Forest City, CA 94404   (415) 574-0179

Albert J. Miller
P.O. Box 219
Fort Bragg, CA 95437   (707) 937-1855

Joseph H. Smith#
Applied Biosystems, Inc.
850 Lincoln Centre Drive
Foster City, CA 94404   (415) 358-8114

T. R. Zegree#
421 Beach Park Blvd.
Foster City, CA 94404   (415) 574-2639

Harland L. Burge, Jr.
Putman, Strid & Burge
Suite 350
17330 Brookhurst
Fountain Valley, CA 92708   (714) 842-4484

Wilfred G. Caldwell
18242 Cabrillo Court
Fountain Valley, CA 92708   (714) 762-4517

Donald L. Royer
18765 Santa Isadora St.
Fountain Valley, CA 92708   (714) 968-9156

Harvey Citrin
2358 Castillejo Way
Freemont, CA 94539   (714) 968-9156

James P. Hillman
45010 Pawnee Dr.
Fremont, CA 94539   (415) 651-4144

Richard B. Main
47000 Warm Springs Blvd.
Suite 427
Fremont, CA 94539   (415) 438-9310

Mark D. Miller
Kimble, Mac Michael & Upton
P.O. Box 9489
Fresno, CA 93792   (209) 435-5500

Victor Sepulveda
1345 Bulldog Lane
Fresno, CA 93710   (209) 225-7666

Rodney K. Worrel
Worrel & Worrel
Civic Center Square
2444 Main St., Suite 130
Fresno, CA 93721   (209) 486-4526

Dale E. Bennett#
801 N. Mountain View Pl.
Fullerton, CA 92631   (714) 871-8577

Richard P. Burgoon, Jr.
Beckman Instruments, Inc.
2500 Harbor Blvd.
Fullerton, CA 92635   (714) 773-7610

Norman E. Carte#
924 White Water Drive
Fullerton, CA 92633   (714) 992-2354

William C. Daubenspeck*
Hughes Aircraft Company
P.O. Box 3310
Bldg. 618, M.S. E425
Fullerton, CA 92634   (714) 732-8097

Gary T. Hampson
Beckman Instruments, Inc.
2500 Harbor Blvd.
Fullerton, CA 92634   (714) 773-6922

Paul R. Harder
Beckman Instruments, Inc.
2500 Harbor Blvd.
Fullerton, CA 92634   (714) 773-6909

Wen Liu
Beckman Instruments, Inc.
2500 Harbor Blvd.
Fullerton, CA 92634   (714) 773-6929

Ben E. Lofstedt
Christian Law Center
2701 E. Chapman Avenue
Suite 112
Fullerton, CA 92634   (714) 738-8822

William H. May
Beckman Instruments, Inc.
2500 Harbor Blvd.
Fullerton, CA 92634   (714) 773-6973

Robert T. Spaulding#
550 Elinor Drive
Fullerton, CA 92635   (714) 879-5000

Robert Jay Steinmeyer
609 Lemon Hill Terrace
Fullerton, CA 92632   (714) 526-5229

Cleveland R. Williams
1506 N. Sycamore
Fullerton, CA 92631   (714) 773-0748

Tai S. Cho#
9681 Garden Grove Blvd.
Suite 200
Garden Grove, CA 92644   (714) 636-0882

Thomas L. Venezia
9301 Shannon Avenue
Garden Grove, CA 92641   (714) 638-8011

Howard L. Johnson
16010 Crenshaw Blvd.
Gardena, CA 90249   (213) 323-6396

Chris Papageorge
14625 S. Vermont
Apt. 5
Gardena, CA 90247   (213) 324-9890

Michael Aguilar
3508 Angelus Avenue
Glendale, CA 91208   (213) 249-9898

James E. Brunton
Suite 500
225 W. Broadway
Glendale, CA 91204   (213) 956-7154

Ted De Boer#
National Inventors Foundation, Inc.
345 W. Cypress St.
Glendale, CA 91204   (213) 246-6540

Richard S. Erbe
Walt Disney Imagineering
1401 Flower Street
Glendale, CA 91221   (818) 544-7626

Theodore Hawley Lassagne
1627 Sheridan Road
Glendale, CA 91206   (818) 246-1256

Clarence J. Morrissey
P.O. Box 11036
Glendale, CA 91206   (818) 354-6834

Robert Catlett Smith
Wagner & Middlebrook
3541 Ocean View Blvd.
Glendale, CA 91208   (818) 957-3340

John Emery Wagner
Wagner & Middlebrook
3541 Ocean View Blvd.
Glendale, CA 91208   (818) 957-3340

Eugene Carl Ziehm#
Nestle USA, Inc.
800 N. Brand Blvd.
Glendale, CA 91203   (818) 549-6221

Alan H. Thompson
630 N. Wildwood Avenue
Glendora, CA 91740   (818) 335-3045

William C. Schubert
Santa Barbara Research Center
75 Coromar Drive
Goleta, CA 93117   (805) 562-2108

Keith L. Johnson
17211 Chatsworth St., #30
Granada Hills, CA 91344   (818) 831-0554

Thomas J. Clough
Ensci, Inc.
125 Sharon Lane
Grover City, CA 93433   (703) 704-1884

Mary Ann Dillahunty#
107 San Pedro Road
Half Moon Bay, CA 94019   (415) 726-6599

Eustace B. Franklin#
Eufra Marketing
11435 Oxford Ave.
Hawthorne, CA 90250   (213) 671-9801

Karl J. Hoch, Jr.
Northrop Corp.
One Northrop Ave.
Mail Stop 30/110/99
Hawthorne, CA 90250   (213) 332-5052

Linval B. Castle
22693 Hesperian Blvd.
Suite 270
Hayward, CA 94541   (415) 887-1346

James R. Shay
Nellcor Inc.
25495 Whitesell St.
Hayward, CA 94545   (415) 887-1346

Sanford S. Wadler
Bio-Rad Laboratories
1000 Alfred Nobel Drive
Hercules, CA 94547   (415) 724-7000

Thomas D. Kiley
986 Baileyana Road
Hillsborough, CA 94010   (415) 348-6974

Barry A. Bisson
17111 Beach Blvd.
Suite 207
Huntington Beach, CA 92646   (714) 848-0479

Michael R. Collins
Beech & Collins
20422 Beach Blvd.
Suite 325
Huntington Beach, CA 92648   (714) 960-2501

Elbert D. Craft#
17311 Almelo Lane
Huntington Beach, CA 92649   (714) 846-5317

Sharon N. Fennessy#
9811 La Cresta Circle
Huntington Beach, CA 92646   (714) 846-5317

Walter C. Glowski
Mc Donnell Douglas Corp.
5301 Bolsa Ave.
Huntington Beach, CA 92647   (714) 896-3713

Charles S. Gumpel
9601 Onset Circle
Huntington Beach, CA 92646   (714) 964-3361

Merrill G. Hinton, Jr.
16161 Ballantine Lane
Huntington Beach, CA 92647   (714) 846-4887

James G. O Neill
325 - 21st Street
Huntington Beach, CA 92648   (714) 960-3436

Thomas A. Schenach#
6531 Meath Circle
Huntington Beach, CA 92647   (714) 892-8886

Paul H. Ware
8910 2nd Avenue
Inglewood, CA 90305   (213) 751-7680

Joseph C. Andras
Price, Gess & Ubell
2100 SE Main Street
Suite 250
Irvine, CA 92714   (714) 261-8433

Leslie Badin, Jr.
4882 Basswood Lane
Irvine, CA 92715   (714) 786-1113

June M. Bostich#
Baxter Healthcare, Inc.
2132 Michelson Dr.
Irvine, CA 92715   (714) 474-6400

Bruce M. Canter
Baxter Healthcare Corp.
2132 Michelson Dr.
Irvine, CA 92715   (714) 474-6419

Debra D. Condino
Baxter Healthcare Corp.
2132 Michelson Drive
Irvine, CA 92715   (714) 474-6473

Woodie D. English, III
Del Mar Avionics
1601 Alton Avenue
Irvine, CA 92714   (714) 250-3200

Albin H. Gess
Price, Gess & Ubell
2100 S.E. Main Street
Suite 250
Irvine, CA 92714   (714) 261-8433

Howard J. Klein
Klein & Szekeres
4199 Campus Drive
Suite 700
Irvine, CA 92715   (714) 854-5502

Donald J. Koprowski
Kawasaki Motors Corp.
9950 Jeronimo Road
Irvine, CA 92718   (714) 458-5604

Howard R. Lambert
Allergan, Inc.
2525 Dupont Drive
Irvine, CA 92715   (714) 752-4749

Daniel C. Mallery
Price, Gess & Ubell
2100 S.E. Main St., Suite 250
Irvine, CA 92714   (714) 261-8433

Richard L. Myers
Baxter Healthcare Corp.
2132 Michelson Drive
Irvine, CA 92715   (714) 474-6450

Edward D. O Brian
Price, Gess & Ubell, P.C.
2100 S.E. Main Street
Suite 250
Irvine, CA 92714   (714) 261-8433

Thomas J. Plante
Plante, Strauss, Vanderburgh
& Connors
11 Solana
Irvine, CA 92715   (714) 854-4466

Joseph W. Price, Jr.
Price, Gess & Ubell
2100 S.E. Main St.
Suite 250
Irvine, CA 92714   (714) 261-8433

Michael C. Schiffer
Baxter Healthcare Corporation
2132 Michelson Drive
Irvine, CA 92715   (714) 474-6447

Sandra S. Schultz
Medtronic, Inc.
Law Dept.
18011 S. Mitchell Ave.
Irvine, CA 92714   (714) 496-6297

Janice A. Sharp#
Christie, Parker & Hale
5 Park Plaza, Suite 1440
Irvine, CA 92714   (714) 476-0757

Kenneth L. Sherman
Price, Gess & Ubell
2100 S.E. Main St., Ste. 250
Irvine, CA 92714   (714) 261-8433

John R. Shewmaker
Pfizer Inc.
17600 Gillette Ave.
Irvine, CA 92714   (714) 250-8222

Joseph E. Szabo
18852 Via Messina
Irvine, CA 92715   (714) 854-1038

Gabor L. Szekeres
Klein & Szekeres
4199 Campus Drive
Suite 700
Irvine, CA 92715 (714) 854-5502

Roger C. Turner
17600 Gillette Avenue
Irvine, CA 92714 (714) 250-8378

Franklin D. Ubell
Price Gess & Ubell
2100 S.E. Main St.
Suite 250
Irvine, CA 92714 (714) 261-8433

Martin Andries Voet
Allergan, Inc.
2525 Dupont Dr.
Irvine, CA 92713 (714) 724-5984

Randall G. Wick
AST Research, Inc.
16215 Alton Parkway
P.O. Box 19658
Irvine, CA 92713 (714) 727-7777

Leo J. Young
Western Digital Corp.
8105 Irvine Center Drive
Suite 370
Irvine, CA 92718 (714) 932-5628

Corwin R. Horton
Suite 2001, The Park
25 Mann Drive
Kentfield, CA 94904 (415) 453-5443

William T. O Neil
2142 La Canada Crest Dr. No. 3
La Canada, CA 91011 (818) 248-3252

Louis J. Bachand, Jr.
P.O. Box 12330
La Crescenta, CA 91214 (818) 352-8841

Christine M. Bellas
Fitch, Even, Tabin
& Flannery
Regents Square II, Ste. 510
4250 Executive Square
La Jolla, CA 92037 (619) 552-1311

Selwyn S. Berg
7730 A Hershel Avenue
La Jolla, CA 92037 (619) 459-2374

Douglas A. Bingham
Scripps Research Institute
Office of Patent Counsel
10666 North Torrey Pines Rd.
TPC8
La Jolla, CA 92037 (619) 554-2937

Nancy K. Dahl
Lyon & Lyon
4250 Executive Square
Suite 660
La Jolla, CA 92037 (619) 552-8400

Thomas Fitting
Scripps Research Institute
Office of Patent Counsel
10666 N. Torrey Pines Road
TPC8
La Jolla, CA 92037 (619) 554-2937

Bryant R. Gold
Fitch, Even, Tabin
& Flannery
4250 Executive Square
Suite 510
La Jolla, CA 92037 (619) 552-1311

Andrew F. Hillhouse, Jr.
P.O. Box 12708
La Jolla, CA 92037 (619) 457-4173

Natalie Jensen#
The Salk Institute
10010 N. Torrey Pines Road
La Jolla, CA 92037 (714) 453-4100

Karol J. Mysels
8327 La Jolla Scenic Dr.
La Jolla, CA 92037 (714) 453-6988

Frederick W. Pepper
Spensley Horn Jubas & Lubitz
4225 Executive Square
Suite 1400
La Jolla, CA 92037 (619) 455-5100

Ernest Arthur Polin
5810 Caminito Cardelina
La Jolla, CA 92037 (619) 459-5334

Richard J. Reilly
1556 Virginia Way
La Jolla, CA 92037 (614) 454-4439

Dennis P. Ritz
505 Colima Street
La Jolla, CA 92037 (619) 456-2129

James J. Schumann
Fitch, Even, Tabin
& Flannery
Regents Square II.
4250 Executive Sq., Ste. 510
La Jolla, CA 92037 (619) 552-1311

Stephanie L. Seidman
Fitch, Even, Tabin
& Flannery
4250 Executive Square
Suite 510
La Jolla, CA 92038 (619) 552-1311

Ronni L. Sherman
Stratagene
11099 N. Torrey Pines Rd.
La Jolla, CA 92037 (619) 535-5431

Richard J. Warburg
Lyon & Lyon
4250 Executive Square
Suite 660
La Jolla, CA 92037 (619) 552-0159

John R. Wetherell, Jr.
Spensley Horn Jubas
& Lubitz
4225 Executive Square
Suite 1400
La Jolla, CA 92037 (619) 455-5100

I. Louis Wolk
939 Coast Blvd.
La Jolla, CA 92037 (619) 454-5958

Kenneth J. Woolcott
IDEC Pharmaceuticals Corp.
11099 N. Torrey Pines Road
La Jolla, CA 92037 (619) 458-0600

Kazuyuki Yamasaki
Spensley Horn Jubas & Lubitz
4225 Executive Square
Suite 1400
La Jolla, CA 92037 (619) 455-5100

Charles Chalmers Logan, II
7851 University Avenue
Suite 102
La Mesa, CA 92041 (619) 463-7344

Richard M. Stanley
P.O. Box 2005
La Mesa, CA 92044 (619) 460-5878

Joseph M. Hageman
18815 Elizondo St.
La Puente, CA 91744 (818) 912-4661

Donald Edward Nist
2403 Sloan Drive
La Verne, CA 91750 (714) 596-6659

David J. Aston
3315 Beechwood Drive
Lafayette, CA 94549 (415) 283-5889

Roger G. Ditzel#
1790 Ivanhoe Ave.
Lafayette, CA 94549 (415) 946-1379

Robert J. Henry
3812-C Happy Valley Rd.
Lafayette, CA 94549 (415) 283-5146

Donald J. Mc Rae
1050 Via Roble
Lafayette, CA 94549 (415) 284-4760

John Stephen Rhoades
613 Huntleigh Drive
Lafayette, CA 94549 (415) 284-1172

George W. Wasson
3123 Indian Way
Lafayette, CA 94549 (415) 283-4420

Derrick M. Reid
26 La Costa Court
Laguna Beach, CA 92651 (714) 497-8800

Herb Boswell#
La Paz Office Plaza
25283 Cabot Road
Suite 209
Laguna Hills, CA 92653 (714) 380-4890

Bruce B. Brunda
Stetina & Brunda
24221 Calle De La Louisa
Suite 401
Laguna Hills, CA 92653   (714) 855-1246

Robert D. Buyan
Stetina & Brunda
24221 Calle De La Louisa
Suite 401
Laguna Hills, CA 92653   (714) 855-1246

Mark B. Garred
Stetina & Brunda
24221 Calle Dela Louisa
Suite 401
Laguna Hills, CA 92653   (714) 855-1246

Hisako Muramatsu
27362 Lost Colt Dr.
Laguna Hills, CA 92653   (714) 643-1227

Gordon Lloyd Peterson
Weissenberger & Peterson
24012 Calle De La Plata
Suite 470
Laguna Hills, CA 92653   (714) 380-4046

Howard E. Sandler
23046 Avenida De La Carlota
Suite 600
Laguna Hills, CA 92653   (714) 588-5755

Kit M. Stetina
Stetina & Brunda
24221 Calle De La Louisa
Suite 401
Laguna Hills, CA 92653   (714) 855-1246

Frank J. Uxa, Jr.
Weissenberger & Peterson
24012 Calle De La Plata
Suite 470
Laguna Hills, CA 92653   (714) 380-4046

Harry G. Weissenberger
Weissenberger, Peterson,
Uxa & Myers
24012 Calle De La Plata
Suite 470
Laguna Hills, CA 92653   (714) 380-4046

Robert J. Baran
29131 Ridgeview Drive
Laguna Niguel, CA 92677   (714) 831-5247

Gerald L. Floyd
28921 Curlew Lane
Laguna Niguel, CA 92677   (714) 831-6679

John L. Hummer
P.O. Box 6427
Laguna Niguel, CA 92677   (714) 485-8272

Francis X. Lo Jacono, Sr.
14 Hyannis Lane
Laguna Niguel, CA 92677   (714) 661-8882

Allan D. Mockabee
42740 Montello Drive
Lake Elizabeth, CA 93532   (805) 724-1865

David F. O Brien
2102 E. Tern Bay Ln.
Lakewood, CA 90712   (213) 616-3336

Delmas R. Buckley, Jr.
411 South L St.
Livermore, CA 94550   (415) 940-8400

Lafayette E. Carnahan#*
U.S. Dept. of Energy
Off. of Pat. Coun.
P.O. Box 808, L-376
Livermore, CA 94550   (415) 422-1430

Clifton E. Clouse, Jr.#*
U.S. Dept. of Energy
Off. of Pat. Coun.
P.O. Box 808, L-376
Livermore, CA 94550   (415) 422-1429

Martin I. Finston#
Lawrence Livermore National Laboratory
Classification Office, L-302
P.O. Box 808
Livermore, CA 94550   (415) 423-3055

Robert R. Fredlund, Jr.*
U.S. Dept. of Energy
S.F. Operations Office
Lawrence Livermore Natl. Lab.
P.O. Box 808/L-368
Livermore, CA 94550   (415) 423-3022

Roger Sherwin Gaither*
U.S. Dept. of Energy
Off. of Pat. Coun.
P.O. Box 808, L-376
Livermore, CA 94550   (415) 422-4367

Nora A. Hackett#
Lawrence Livermore Nat'L. Lab.
Patent Group L-703
P.O. Box 808
7000 East Avenue
Livermore, CA 94550   (415) 422-7820

Michael B.K. Lee
LLNL
P.O. Box 808 L-703
Livermore, CA 94550   (415) 423-8051

Shyamala Rajender
Lawrence Livermore National Lab.
L-701, Box 808
Livermore, CA 94550   (415) 422-7275

Frederick A. Robertson
704 Wimbledon Lane
Livermore, CA 94550   (415) 447-6787

Henry P. Sartorio#
Univ. of Calif.
Lawrence Livermore Lab.
7000 East Ave.
Livermore, CA 94550   (415) 422-7816

Miguel A. Valdes*
U.S. Dept. of Energy
Off. of Patent Coun.
Livermore Office
P.O. Box 808, L-376
Livermore, CA 94550   (415) 422-1427

Basil B. Travis
1820 W. Hwy. 12
P.O. Box 287
Lodi, CA 95241   (209) 333-8379

Douglas M. Thompson
25313 Doria Ave.
Lomita, CA 90717   (213) 325-4981

Walter John Adam
4109 E. Ocean Blvd., Apt. 2
Long Beach, CA 90803   (213) 434-2925

Gary M. Anderson
Fulwider, Patton, Lee &
Utecht
11 Golden Shore
5th Floor
Long Beach, CA 90802   (213) 432-0453

I. Michael Bak-Boychuk
400 Oceangate Plaza
Suite 325
Long Beach, CA 90802   (213) 432-8419

Herbert A. Birenbaum
Atlantic Richfield Co.
300 Oceangate
Suite 1597
Long Beach, CA 90802   (213) 590-4421

Donald L. Carlson
6301 Bixby Hill Rd.
Long Beach, CA 90815   (213) 431-0644

Gregory A. Cone
Mc Donnell Douglas Corp.
C1-H009 (122-23)
3855 Lakewood Blvd.
Long Beach, CA 90846   (213) 593-6812

Marcia A. Devon
One World Trade Center
Suite 1850
Long Beach, CA 90831   (213) 495-4000

George Walter Finch
Mc Donnell Douglas Corp.
MS 78-81
3855 Lakewood Blvd.
Long Beach, CA 90846   (213) 593-6834

Benjamin Hudson
Mc Donnell Douglas Corp.
C1-H009 (78-81)
3855 Lakewood Blvd.
Long Beach, CA 90846   (213) 593-7579

L. Lee Humphries
7821 Tibana St.
Long Beach, CA 90808   (213) 596-6962

Paul T. Loef
Mc Donnell Douglas Corp.
3855 Lakewood Blvd.
Long Beach, CA 90846   (213) 593-7563

George R. Loftis
P. O. Box 1602
Long Beach, CA 90801   (213) 593-7563

William D. Mooney
Sohio Supply Company
401 E. Ocean Blvd.
Suite 705
Long Beach, CA 90802   (213) 436-4868

Thomas A. Runk
Fulwider, Patton, Lee & Utecht
11 Golden Shore
Suite 510
Long Beach, CA 90801   (213) 432-0453

John P. Scholl
Mc Donnell Douglas Corp.
3855 Lakewood Blvd.
C1-H009
Long Beach, CA 90846   (213) 593-8189

Vern D. Schooley
Fulwider, Patton, Lee
& Utecht
11 Golden Shore
Long Beach, CA 90802   (213) 432-0453

James M. Skorich
Mc Donnell Douglas Corp.
3855 Lakewood Blvd.
Mail Code 78-81
Long Beach, CA 90846   (213) 593-7198

Donald E. Stout
Mc Donnell Douglas Corp.
3855 Lakewood Blvd.
C1-H009 M.S. 78-81
Long Beach, CA 90846   (213) 593-6823

Paul M. Stull
Fulwider, Patton, Lee & Utecht
11 Golden Shore
Long Beach, CA 90801   (212) 432-0543

Charles H. Thomas, Jr.
Cislo & Thomas
4201 Long Beach Blvd.
Suite 405
Long Beach, CA 90807   (213) 595-8422

Francis A. Utecht
Fulwider, Patton, Lee
& Utecht
11 Golden Shore
Suite 510
Long Beach, CA 90802   (213) 432-0453

Georges A. Maxwell
11362 Wallingsford Road
Los Alamitos, CA 90720   (213) 431-6255

Victor R. Beckman
900 N. San Antonio Road
Suite 205
Los Altos, CA 94022   (415) 949-3103

Mervin Halstead
1131 Hillslope Pl.
Los Altos, CA 94022   (415) 948-8324

Leon F. Herbert
610 Twelve Acres Drive
Los Altos, CA 94022   (415) 948-8653

Paul L. Hickman
P.O. Box Q
Los Altos, CA 94023   (415) 949-1846

Baylor G. Riddell
1032 Dartmouth Lane
Los Altos, CA 94024   (415) 968-8062

Bruce D. Riter
101 First Street
Suite 208
Los Altos, CA 94022   (415) 948-6235

David Stewart Romney
978 Altos Oaks Drive
Los Altos, CA 94024   (415) 965-0367

Michael L. Sherrard
Patent Law
72 Doud Drive
Los Altos, CA 94022   (415) 949-2316

Edwin A. Sloane#
1040 Covington Road
Los Altos, CA 94024   (415) 968-2916

H. Donald Volk
374 Benvenue Ave.
Los Altos, CA 94022   (408) 742-0691

Edward Y. Wong
P.O. Box 1161
Los Altos, CA 94023   (415) 857-3873

Arthur J. Deex
25396 La Loma Dr.
Los Altos Hills, CA 94022   (415) 949-1830

John C. Oberlin
26140 Robb Rd.
Los Altos Hills, CA 94022   (415) 948-8157

David B. Abel#
Graham & James
725 South Figueroa St.
34th Floor
Los Angeles, CA 90017   (213) 624-2500

Colin P. Abrahams
3600 Wilshire Blvd.
Suite 1520
Los Angeles, CA 90010   (213) 385-4281

Carolyn R. Adler
Romney Golant Martin
Seldon & Ashen
10920 Wilshire Blvd.
Suite 1000
Los Angeles, CA 90024   (213) 208-1100

Naveed Alam
453 S. Spring
Suite 1017
Los Angeles, CA 90013   (213) 489-3131

Leonard A. Alkov
Hughes Aircraft Co.
Corp. Pats. & Licensing
P.O. Box 45066
7200 Hughes Terrace Bldg.
Los Angeles, CA 90045   (213) 568-6081

Roy L. Anderson
Lyon & Lyon
611 West Sixth Street
34th Floor
Los Angeles, CA 90017   (213) 489-1600

William L. Androlia
Koda & Androlia
1880 Century Park East
Los Angeles, CA 90067   (213) 277-1391

Gene Wesley Arant
Arant, Kleinberg & Lerner
2049 Century Park East
Los Angeles, CA 90067   (213) 557-1511

Erik M. Arnhem
4113 Beverly Blvd.
Los Angeles, CA 90004   (213) 660-5067

Robert M. Ashen
Romney Golant Martin
Seldon & Ashen
Suite 1000
10920 Wilshire Blvd.
Los Angeles, CA 90024   (213) 208-1100

Craig B. Bailey
Fulwider, Patton, Lee
& Utecht
10877 Wilshire Blvd.
Tenth Floor
Los Angeles, CA 90024   (213) 824-5555

Paul D. Bangor, Jr.
Spensley Horn Jubas
& Lubitz
1880 Century Park East
Fifth Floor
Los Angeles, CA 90067   (213) 553-5050

Michael Barclay
Spensley Horn Jubas
& Lubitz
1880 Century Park East
5th Floor
Los Angeles, CA 90067   (213) 553-5050

Richard A. Bardin
Fulwider, Patton, Lee
& Utecht
10877 Wilshire Blvd.
Tenth Floor
Los Angeles, CA 90024   (213) 824-5555

Vernon D. Beehler
Beehler, Pavitt, Siegemund,
Jagger, Martella & Dawes
3435 Wilshire Blvd.
Suite 1100
Los Angeles, CA 90010   (213) 385-7087

Vincent J. Belusko
Spensley Horn Jubas
& Lubitz
1880 Century Park East
Suite 500
Los Angeles, CA 90067   (213) 553-5050

John M. Benassi
Lyon & Lyon
611 W. 6th Street
34th Floor
Los Angeles, CA 90017   (213) 489-1600

William J. Benman, Jr.
10850 Wilshire Blvd.
Suite 800
Los Angeles, CA 90024   (213) 475-3112

Luc P. Benoit
Benoit Law Corp.
2551 Colorado Blvd.
Los Angeles, CA 90041   (213) 255-0000

Richard P. Berg
Ladas & Parry
3600 Wilshire Blvd.
Suite 1520
Los Angeles, CA 90010   (213) 385-4281

Brian M. Berliner#
Poms, Smith, Lande & Rose
2121 Avenue of the Stars
Suite 1400
Los Angeles, CA 90067   (213) 277-8141

Robert Berliner
Nilsson, Robbins, Dalgarn,
Berliner, Carson & Wurst
201 North Figueroa Street
5th Floor
Los Angeles, CA 90012   (213) 977-1001

Rod S. Berman
Spensley Horn Jubas
& Lubitz
1880 Century Park East
Suite 500
Los Angeles, CA 90067   (213) 553-5050

Suzanne L. Biggs
Lyon & Lyon
611 West Sixth Street
Los Angeles, CA 90017   (213) 955-1600

Henry Martyn Bissell
6820 La Tijera Blvd.
Suite 106
Los Angeles, CA 90045   (213) 776-3122

Roger W. Blakely, Jr.
Blakely, Sokoloff, Taylor
& Zafman
12400 Wilshire Blvd.
7th Floor
Los Angeles, CA 90025   (213) 207-3800

Thomas L. Blasdell
Crawford Blasdell & Reimann
11755 Wilshire Blvd.
Suite 1500
Los Angeles, CA 90025   (213) 478-7442

Jeffrey J. Blatt
Blakely, Sokoloff, Taylor
& Zafman
12400 Wilshire Blvd.
7th Floor
Los Angeles, CA 90025   (213) 207-3800

Coe A. Bloomberg
Lyon & Lyon
611 West Sixth Street
34th Floor
Los Angeles, CA 90017   (212) 489-1600

Nathan Boatner
445 S. Figueroa St.
Suite 2600
Los Angeles, CA 90071   (213) 489-6871

Breton A. Bocchieri
Blakely, Sokoloff, Taylor
& Zafman
12400 Wilshire Blvd.
7th Floor
Los Angeles, CA 90025   (213) 207-3800

Raymond A. Bogucki
Merchant, Gould, Smith, Edell,
Welter & Schmidt
11100 Santa Monica Blvd.
Suite 1700
Los Angeles, CA 90025   (213) 445-1140

Louis J. Bovasso
Poms, Smith, Lande,
& Rose
2121 Avenue of the Stars
Suite 1400
Los Angeles, CA 90067   (213) 277-8141

Ralph M. Braunstein
Reagin & King
12400 Wilshire Blvd.
Suite 500
Los Angeles, CA 90025   (213) 820-5864

E. Lawrence Brevik
B Two M Industries, Inc.
12923 S. Spring St.
Los Angeles, CA 90061   (213) 770-1871

Patrick Francis Bright
Bright & Lorig
515 South Figueroa St.
Suite 1200
Los Angeles, CA 90071   (213) 627-7774

James C. Brooks
Lyon & Lyon
611 West Sixth Street
34th Floor
Los Angeles, CA 90017   (213) 489-1600

Charles D. Brown
Hughes Aircraft Co.
Pats. & Licensing
Bldg. C1, Mail Station A126
P.O. Box 45066
Los Angeles, CA 90045   (213) 568-6084

Marc E. Brown
Fleishman & Damon
1901 Avenue of the Stars
Suite 931
Los Angeles, CA 90067   (213) 277-3338

Clarence L. Browning
Whann & Connors
315 West Ninth St.
Los Angeles, CA 90015   (213) 622-7163

James R. Brueggemann
Pretty Schroeder Brueggemann
& Clark
444 S. Flower Street
Suite 2000
Los Angeles, CA 90071   (213) 489-4442

William J. Burke
Aerospace Corporation
P.O. Box 92957
Mail Stop M1/040
Los Angeles, CA 90009   (213) 336-6708

Deanna C. Bushendorf
Pretty, Schroeder, Bruggemann
& Clark
444 S. Flower St.
Suite 2000
Los Angeles, CA 90071   (213) 489-4442

Richard E. Campbell
Poms, Smith, Lande & Rose
2121 Avenue of the Stars
14th Floor
Los Angeles, CA 90067   (213) 788-5000

M. Michael Carpenter
Poms, Smith, Lande & Rose
2121 Avenue of the Stars
Suite 1400
Los Angeles, CA 90067   (213) 277-8141

M. John Carson, III
Nilsson, Robbins, Dalgarn,
Berliner, Carson & Wurst
201 North Figueroa Street
5th Floor
Los Angeles, CA 90012   (213) 977-1001

Nathan Cass
Unisys Corporation
Law Dept.
10920 Wilshire Blvd.
4th Floor
Los Angeles, CA 90024   (213) 443-5693

David Charness
612 North Sepulveda Blvd.
Los Angeles, CA 90049   (213) 472-8012

John Stirling Christopher
AVCO Bldg., Suite 800
10850 Wilshire Blvd.
Los Angeles, CA 90024   (213) 470-9258

Eric T.S. Chung
Chung & Stein
2029 Century Park East
Suite 480
Los Angeles, CA 90067   (213) 556-2104

Gary Alan Clark
Pretty, Schroeder, Bruegggemann
& Clark
444 S. Flower
20th Floor
Los Angeles, CA 90071   (213) 489-4442

Vincent W. Cleary*
U.S. Air Force
SD/JA
Box 92960
L.A. Air Force Station
Los Angeles, CA 90009   (213) 643-0916

Lisa A. Clifford
Poms, Smith, Lande & Rose
2121 Ave. of the Stars
Suite 1400
Los Angeles, CA 90067   (213) 277-8141

Paul M. Coble
Hughes Aircraft Co.
7200 Hughes Terrace
P.O. Box 45066
Bldg. C01, M.S. A-126
Los Angeles, CA 90045   (213) 568-7824

Lawrence S. Cohen
Freilich, Hornbaker & Rosen
10960 Wilshire Blvd.
Suite 1434
Los Angeles, CA 90024   (213) 477-0578

David W. Collins
Romney Golant Martin
Seldon & Ashen
10920 Wilshire Blvd.
Suite 1000
Los Angeles, CA 90024   (213) 208-1100

John H. Colter
Baker & Mc Kenzie
Citicorp Plaza
36th Floor
725 S. Figueroa Street
Los Angeles, CA 90017   (213) 629-3000

Mary S. Consalvi
Lyon & Lyon
611 West Sixth Street
34th Floor
Los Angeles, CA 90017   (213) 489-1600

Lewis M. Dalgarn
Nilsson, Robbins, Dalgarn,
Berliner, Carson & Wurst
201 North Figueroa Street
5th Floor
Los Angeles, CA 90012   (213) 977-1001

Christopher Darrow
Poms, Smith, Lande
& Rose
2121 Avenue of the Stars
Suite 1400
Los Angeles, CA 90067   (213) 277-8141

Daniel L. Dawes#
Beehler & Pavitt
P.O. Box 92400
Los Angeles, CA 90009   (213) 215-3183

Wanda K. Denson-Low
Hughes Aircraft Company
7200 Hughes Terrace
C1, A126
Los Angeles, CA 90045   (213) 568-6972

Robert W. Dickerson, Jr.
Lyon & Lyon
611 West Sixth Street
34th Floor
Los Angeles, CA 90017   (213) 489-1600

Arthur V. Doble
12301 Wilshire Blvd.
Suite 512
Los Angeles, CA 90025   (213) 826-5505

Bernard Philip Drachlis
5012 Eagle Rock Blvd.
Los Angeles, CA 90041   (213) 255-1918

William H. Drummond
Fulwider Patton Lee
& Utecht
3435 Wilshire Blvd.
Suite 2400
Los Angeles, CA 90010   (213) 380-6800

Bradford J. Duft
Lyon & Lyon
611 West Sixth Street
34th Floor
Los Angeles, CA 90017   (213) 489-1600

Vijayalakshmi D. Duraiswamy
Hughes Aircraft Company
Pats. & Licensing
7200 Hughes Terrace
Bldg. C1, M.Station A126
Los Angeles, CA 90045   (213) 568-6076

Michael S. Elkind
Nilsson, Robbins, Dalgarn,
Berliner, Carson & Wurst
210 North Figueroa Street
5th Floor
Los Angeles, CA 90012   (213) 977-1001

Natan Epstein
Beehler & Pavitt
P.O. Box 92400
Los Angeles, CA 90009   (213) 215-3183

Saul Epstein
Suite 500
1880 Century Park E.
Los Angeles, CA 90067   (213) 553-2223

David L. Fehrman#
Graham & James
725 S. Figueroa St.
34th Floor
Los Angeles, CA 90017   (213) 624-2500

John F. Feldsted
Rogers & Wells
201 North Figueroa Street
16th Floor
Los Angeles, CA 90012   (213) 580-1236

A. M. Fernandez
Freilich, Hornbaker, Rosen
& Fernandez
10960 Wilshire Blvd.
Suite 1434
Los Angeles, CA 90024   (213) 477-0578

Andra M. Finkel
10920 Wilshire Blvd.
Suite 1000
Los Angeles, CA 90024   (213) 208-4900

Don Berry Finkelstein
Ladas & Parry
Suite 1520
3600 Wilshire Blvd.
Los Angeles, CA 90010   (213) 385-4281

Hallie A. Finucane
Merchant & Gould
333 S. Grand Avenue
Suite 1650
Los Angeles, CA 90071   (213) 485-0100

Thomas A. Fournie
1880 Century Park East
Suite 1018
Los Angeles, CA 90067   (213) 879-1401

Arthur Freilich
Freilich, Hornbaker, Rosen,
& Fernandez, P.C.
10960 Wilshire Blvd.
Suite 1434
Los Angeles, CA 90024   (213) 477-0578

Frank Frisenda, Jr.
Frisenda, Morris & Nicholson
11755 Wilshire Blvd.
10th Floor
Los Angeles, CA 90025   (213) 478-4540

Mavis S. Gallenson
Ladas & Parry
3600 Wilshire Blvd.
Suite 1520
Los Angeles, CA 90010   (213) 385-4281

Bernard R. Gans
Poms, Smith, Lande,
& Rose
2121 Avenue of the Stars
Suite 1400
Los Angeles, CA 90067   (213) 277-8141

Paul Lawrence Gardner
Spensley, Horn, Jubas & Lubitz
1880 Century Park East
Suite 500
Los Angeles, CA 90067   (213) 553-5050

Mark E. Garscia
Nilsson, Robbins, Dalgarn, Berliner,
Carson & Wurst
201 N. Figueroa Street
Fifth Floor
Los Angeles, CA 90012   (213) 977-1001

George Henry Gates, III
Merchant, Gould, Smith, Edell,
Welter & Schmidt
11100 Santa Monica Blvd.
Suite 1700
Los Angeles, CA 90025   (213) 445-1140

James W. Geriak
Lyon and Lyon
611 West Sixth Street
34th Floor
Los Angeles, CA 90017   (213) 489-1600

Rabindra N. Ghose
8167 Mulholland Terrace
Los Angeles, CA 90046   (818) 880-4533

William West Glenny
2121 Avenue of the Stars
Suite 1400
Los Angeles, CA 90067   (213) 557-0382

Joseph H. Golant
Ashen, Golant, Martin
& Selden
10920 Wilshire Blvd.
Suite 1000
Los Angeles, CA 90024   (213) 208-1100

Sol L. Goldstein
TRW Inc.
One Space Park
E2/7073
Los Angeles, CA 90278   (213) 812-1516

Robert A. Green
Nilsson, Robbins, Dalgarn,
Berliner, Carson & Wurst
201 North Figueroa Street
5th Floor
Los Angeles, CA 90012   (213) 977-1001

William P. Green
201 North Figueroa Street
Fifth Floor
Los Angeles, CA 90012   (213) 977-1001

J. Nicholas Gross
Spensley Horn Jubas & Lubitz
1880 Century Park East
Fifth Floor
Los Angeles, CA 90067   (213) 553-5050

Stephen D. Gross
Blakely, Sokoloff, Taylor
& Zafman
12400 Wilshire Blvd.
7th Floor
Los Angeles, CA 90025   (213) 207-3800

Georgann S. Grunebach
Hughes Aircraft Co.
7200 Hughes Terrace
Building CO 1, M.S. A126
P. O. Box 45066
Los Angeles, CA 90045   (213) 568-7200

David S. Guttman
Poms, Smith, Lande & Rose
2121 Ave. of the Stars
Suite 1400
Los Angeles, CA 90067   (213) 277-8141

Deborah L. Haake
Lyon & Lyon
611 W. 6th St.
34th Floor
Los Angeles, CA 90017   (213) 489-1600

Michael M. Hachigian
4250 Wilshire Blvd.
Second Floor
Los Angeles, CA 90010   (213) 933-5743

David A. Hall
Pretty, Schroeder, Brueggemann
& Clark
444 S. Flower Street
Suite 2000
Los Angeles, CA 90071   (213) 489-4442

John J. Hall
1631 Beverly Blvd.
Los Angeles, CA 90026   (213) 250-1145

Gunther O. Hanke
Fulwider Patton Lee
& Utecht
3435 Wilshire Blvd.
Suite 2400
Los Angeles, CA 90010   (213) 432-0453

John D. Harriman, II
Blakely, Sokoloff, Taylor
& Zafman
12400 Wilshire Blvd.
7th Floor
Los Angeles, CA 90025   (213) 207-3800

Michael D. Harris
Poms, Smith, Lande & Rose
2121 Avenue of the Stars
Suite 1400
Los Angeles, CA 90067   (213) 277-8141

Charles S. Haughey
Hughes Aircraft Co.
Bldg. C1, M.S. A126
P.O. Box 45066
Los Angeles, CA 90045   (213) 568-7025

Gary A. Hecker
Blakely, Sokoloff, Taylor
& Zafman
12400 Wilshire Blvd.
7th Floor
Los Angeles, CA 90025   (213) 207-3800

Steven D. Hemminger
Lyon & Lyon
611 W. 6th Street
34th Floor
Los Angeles, CA 90017   (213) 489-1600

James A. Henricks
Pretty, Schroeder, Brueggemann
& Clark
444 S. Flower Street
Suite 2000
Los Angeles, CA 90071   (213) 489-4442

Janis C. Henry
Poms, Smith, Lande & Rose
2121 Ave. of the Stars
Los Angeles, CA 90067   (213) 277-8141

David L. Henty
Graham & James
725 South Figueroa St.
34th Floor
Los Angeles, CA 90017   (213) 624-2500

Henry L. Herold
1412 Butler Ave., Apt. 24
Los Angeles, CA 90025   (213) 390-8673

Kenjiro Hidaka#
2040 Pelham Avenue
Los Angeles, CA 90025   (213) 474-2668

Howard L. Hoffenberg
1409 Midvale Avenue
Apt. 320
Los Angeles, CA 90024   (213) 478-3328

Jon E. Hokanson
Morgan, Lewis & Bockius
801 S. Grand Avenue
Twenty-Second Floor
Los Angeles, CA 90017   (213) 612-1392

George W. Hoover, III
Blakely, Sokoloff, Taylor
& Zafman
12400 Wilshire Blvd.
7th Floor
Los Angeles, CA 90025   (213) 207-3800

Martin R. Horn
Spensley, Horn, Jubas
& Lubitz
1880 Century Park E.
Los Angeles, CA 90067   (213) 553-5050

Robert David Hornbaker
Freilich, Hornbaker, Rosen
& Fernandez
Suite 1434
10960 Wilshire Blvd.
Los Angeles, CA 90024   (213) 477-4039

Harlan P. Huebner
900 Wilshire Blvd.
Suite 1000
Los Angeles, CA 90017   (213) 626-7766

Michael Hurey
Blakely, Sokoloff, Taylor
& Zafman
12400 Wilshire Blvd.
Seventh Floor
Los Angeles, CA 90025   (213) 207-3800

Eric S. Hyman
Blakely, Sokoloff, Taylor
& Zafman
12400 Wilshire Blvd.
7th Floor
Los Angeles, CA 90025   (213) 207-3800

Keiichiro Imai
3264 Granville Avenue
Los Angeles, CA 90066   (213) 398-8444

Richard Thomas Ito
Fulwider, Patton, Lee
& Utecht
10877 Wilshire Blvd.
Tenth Floor
Los Angeles, CA 90024   (213) 824-5555

Bruce A. Jagger
Beehler & Pavitt
P.O. Box 92400
Los Angeles, CA 90009   (213) 215-3183

Joseph T. Jakubek
Pretty, Schroeder, Brueggemann
& Clark
444 S. Flower St., Ste. 2000
Los Angeles, CA 90071   (213) 489-4442

Allan W. Jansen
Lyon & Lyon
611 W. Sixth Street
Suite 3400
Los Angeles, CA 90017   (213) 489-1600

Matthew F. Jodziewicz
3447 Mandeville Canyon Road
Los Angeles, CA 90049   (213) 476-2129

Walter E. Johansen, III
11661 San Vicente Blvd.
Los Angeles, CA 90049   (213) 207-1795

Anthony W. Karambelas
Hughes Aircraft Co.
P.O. Box 45066
Bldg. C1/M.S. A121
7200 Hughes Terrace
Los Angeles, CA 90045   (213) 568-7233

Brian W. Kasell
Poms, Smith, Lande & Rose
2121 Ave. of the Stars
Suite 1400
Los Angeles, CA 90067   (213) 277-8141

Elwood S. Kendrick
550 S. Flower Street
7th Floor
Los Angeles, CA 90017   (213) 683-5214

Evan M. Kent
Spensley Horn Jubas & Lubitz
1880 Century Park East
Fifth Floor
Los Angeles, CA 90067   (213) 553-5731

Wellesley R. Kime
8745 Appian Way
Los Angeles, CA 90046   (213) 650-0635

Stephen Lowell King
Blakely, Sokoloff, Taylor
& Zafman
12400 Wilshire Blvd.
Suite 700
Los Angeles, CA 90025   (213) 207-3800

Marvin H. Kleinberg
Arant, Kleinberg & Lerner
2049 Century Park East
Suite 1080
Los Angeles, CA 90067   (213) 557-1511

H. Henry Koda
Koda & Androlia
1880 Century Park East
Suite 519
Los Angeles, CA 90067   (213) 277-1391

William K. Konrad
Spensley Horn Jubas
& Lubitz
1880 Century Pk. E.
Suite 500
Los Angeles, CA 90067   (213) 553-5050

Gerald A. Koris
2717 Bottlebrush Drive
Los Angeles, CA 90077   (213) 475-2475

Alexander L. Kormos#
Focusvision International
12688 Dewey Street
Los Angeles, CA 90066   (213) 391-4275

Stephen S. Korniczky
Lyon & Lyon
611 W. 6th St.
Suite 3400
Los Angeles, CA 90017   (213) 489-1600

Gilbert Gerald Kovelman
Fulwider, Patton, Lee
& Utecht
10877 Wilshire Blvd.,
Tenth Flr.
Los Angeles, CA 90024   (213) 824-5555

Bruce D. Kuyper
Irell & Manella
333 S. Hope St., Ste. 3300
Los Angeles, CA 90071   (213) 620-1555

Reena Kuyper#
Nilsson, Robbins, Dalgarn,
Berliner, Carson & Wurst
201 N. Figueroa St., 5th Flr.
Los Angeles, CA 90012   (213) 977-1001

Mary E. Lachman#
Hughes Aircraft Co.
P.O. Box 45066
Bldg. C1, M/S A126
Los Angeles, CA 90045   (213) 334-9118

Michael B. Lachuk
Poms, Smith, Lande & Rose
2121 Avenue of the Stars
Suite 1400
Los Angeles, CA 90067   (213) 277-8141

Gary E. Lande
Poms, Smith, Lande & Rose
2121 Avenue of the Stars
Suite 1400
Los Angeles, CA 90067   (213) 277-8141

Robert C. Laurenson
Lyon & Lyon
611 W. Sixth St., 34th Flr.
Los Angeles, CA 90017   (213) 489-1600

Don C. Lawrence
3406 Granville Av.
Los Angeles, CA 90066   (213) 398-5656

Marshall A. Lerner
Arant, Kleinberg & Lerner
2049 Century Park East
Suite 1080
Los Angeles, CA 90067   (213) 557-1511

Donald G. Lewis
Freilich, Hornbaker, Rosen,
& Fernandez
10960 Wilshre Blvd.
Suite 1434
Los Angeles, CA 90024   (213) 477-4030

Peter I Lippman
Ashen, Golant, Martin
& Seldon
10920 Wilshire Blvd.
Suite 1000
Los Angeles, CA 90024   (213) 208-1100

Don F. Livornese
Spensley, Horn, Jubas
& Lubitz
1880 Century Park East
Suite 500
Los Angeles, CA 90067   (213) 553-5050

William E. Lloyd, Jr.
11601 Wilshire Blvd.
Suite 1830
Los Angeles, CA 90025   (213) 477-1200

Kam C. Louie
Sun, Louie & Hehmann
3550 Wilshire Blvd.
Suite 1250
Los Angeles, CA 90010   (213) 625-0327

Stuart Lubitz
Spensley, Horn, Jubas
& Lubitz
1880 Century Pk. E.
5th Floor
Los Angeles, CA 90067   (213) 553-5050

Edward Joseph Lynch
Crosby, Heafey, Roach & May
700 South Flower , Suite 2400
Los Angeles, CA 90017   (213) 628-6733

Richard E. Lyon, Jr.
Lyon & Lyon
611 West Sixth Street
Los Angeles, CA 90017   (213) 489-1600

Robert Edward Lyon
Lyon and Lyon
611 W. Sixth Street
34th Floor
Los Angeles, CA 90017   (213) 489-1600

Michael J Mac Dermott
Harris, Kern, Wallen
& Tinsley
650 S. Grand Avenue
Top Floor
Los Angeles, CA 90017   (213) 626-5251

Kurt A. Mac Lean
Poms, Smith, Lande
& Rose
2121 Avenue of the Stars
Suite 1400
Los Angeles, CA 90067   (213) 277-8141

Thomas H. Majcher
Fulwider, Patton, Lee
& Utecht
10877 Wilshire Blvd.
Tenth Floor
Los Angeles, CA 90024   (213) 824-5555

David N. Makous
Fulwider, Patton, Lee
& Utecht
10877 Wilshire Blvd.
Tenth Floor
Los Angeles, CA 90024   (213) 824-5555

Gary D. Mann
Spensley, Horn, Jubas & Lubitz
1880 Century Park East
5th Floor
Los Angeles, CA 90067   (213) 553-5050

Mario A. Martella
Beehler & Pavitt
P.O. Box 92400
Los Angeles, CA 90009   (213) 215-3183

John L. Maxin#
Merchant, Gould, Smith,
Edell, Welter & Schmidt
11100 Santa Monica Blvd.
Suite 1700
Los Angeles, CA 90025   (213) 445-1140

William H. Maxwell, Jr.
12301 Wilshire Blvd.
Los Angeles, CA 90025   (213) 820-3911

J. Donald Mc Carthy
Lyon & Lyon
611 W. 6th Street
34th Floor
Los Angeles, CA 90017   (213) 489-1600

John D. Mc Conaghy
Lyon & Lyon
611 W Sixth Street
34th Floor
Los Angeles, CA 90017   (213) 489-1600

Maria E. Mc Cormack
Blakely, Sokoloff, Taylor
& Zafman
12400 Wilshire Blvd.
Suite 700
Los Angeles, CA 90025   (213) 207-3800

Paul H. Meier
Lyon & Lyon
611 West Sixth Street
34th Floor
Los Angeles, CA 90017   (213) 489-1600

Hope E. Melville
Lyon & Lyon
611 W. Sixth St.
34th Floor
Los Angeles, CA 90017   (213) 489-1600

Stuart Lowell Merkadeau
Graham & James
725 South Figueroa Street
34th Floor
Los Angeles, CA 90017   (213) 689-5561

David J. Meyer
Bright & Lorig
515 S. Figueroa Street
Los Angeles, CA 90071   (213) 627-7774

Scott R. Miller
Riordan & Mc Kinzie
300 S. Grand Avenue
29th Floor
Los Angeles, CA 90071   (213) 629-4824

Wendell S. Miller
1341 Comstock Avenue
Los Angeles, CA 90024   (213) 274-1205

Steven M. Mitchell
Hughes Aircraft Company
7200 Hughes Terrace
Los Angeles, CA 90045   (213) 568-6091

Louis Mok
Spensley, Horn, Jubas
& Lubitz
1880 Century Park East
Fifth Floor
Los Angeles, CA 90067   (213) 553-5050

Robert G. Moll
Spensley Horn Jubas & Lubitz
1880 Century Park East
Suite 500
Los Angeles, CA 90067   (213) 553-5050

Frederick E. Mueller
650 S. Grand Avenue
14th Floor
Los Angeles, CA 90017   (213) 626-6061

Joseph E. Mueth
700 S. Flower Street
Suite 2200
Los Angeles, CA 90017   (213) 688-7407

David B. Murphy
Lyon & Lyon
611 W. Sixth Street
34th Floor
Los Angeles, CA 90017   (213) 489-1600

John S. Nagy
Fulwider, Patton, Lee
& Utecht
10877 Wilshire Blvd.
Tenth Floor
Los Angeles, CA 90024   (213) 824-5555

Deborah A. Neville
Sedgwick, Detert, Moran
& Arnold
3701 Wilshire Blvd.
Suite 900
Los Angeles, CA 90010   (213) 386-2833

Byard G. Nilsson
Nilsson, Robbins, Dalgarn,
Berliner, Carson & Wurst
201 North Figueroa Street
5th Floor
Los Angeles, CA 90012   (213) 977-1001

Earnest F. Oberheim
Hughes Aircraft Co.
Bldg. C1, MS A126
P.O. Box 45066
Los Angeles, CA 90045   (714) 679-1877

Kenneth H. Ohriner
Lyon & Lyon
611 West Sixth Street
34th Floor
Los Angeles, CA 90017   (213) 489-1600

David J. Oldenkamp
Poms, Smith, Lande & Rose
2121 Avenue of the Stars
Suite 1400
Los Angeles, CA 90067   (213) 277-8141

Douglas E. Olson
Lyon and Lyon
611 West 6th
34th Floor
Los Angeles, CA 90017   (213) 489-1600

Jeffrey M. Olson
Lyon & Lyon
611 W. Sixth Street
34th Floor
Los Angeles, CA 90017   (213) 489-1600

Mildred Oncken
455 S. Berendo Street
Apt. 301
Los Angeles, CA 90020   (213) 385-0975

Deidre A. Oppenheimer
3939 Veselich Avenue
Apt. 123
Los Angeles, CA 90039   (213) 665-8222

David G. Parkhurst
Fulwider, Patton, Lee
& Utecht
10877 Wilshire Blvd.
Tenth Floor
Los Angeles, CA 90024   (213) 824-5555

Sherman O. Parrett
Irell & Manella
Suite 3300
333 S. Hope Street
Los Angeles, CA 90071   (213) 620-1555

James W. Paul
Fulwider, Patton, Lee
& Utecht
10877 Wilshire Blvd.
Tenth Floor
Los Angeles, CA 90024   (213) 824-5555

William H. Pavitt, Jr.
Beehler & Pavitt
P.O. Box 92400
Los Angeles, CA 90009   (213) 215-3183

Karol M. Pessin
Lyon & Lyon
611 W. 6th St.
34th Floor
Los Angeles, CA 90017   (213) 489-1600

Leo J. Peters
Fulwider, Patton, Lee
& Utecht
10877 Wilshire Blvd.
10th Floor
Los Angeles, CA 90024   (213) 489-1600

William Poms
Poms, Smith, Lande
& Rose
2121 Avenue of the Stars
Suite 1400
Los Angeles, CA 90067   (213) 277-8141

Edward G. Poplawski
Pretty, Schroeder, Brueggemann
& Clark
444 S. Flower St
Suite 2000
Los Angeles, CA 90071   (213) 489-4442

Laurence H. Pretty
Pretty, Schroeder, Brueggemann
& Clark
444 S. Flower Street
Suite 2000
Los Angeles, CA 90071   (213) 489-4442

John A. Rafter, Jr.
Lyon & Lyon
611 W. Sixth Street
Suite 3400
Los Angeles, CA 90017   (213) 489-1600

Ronald W. Reagin
Blakely, Sokoloff, Taylor
& Zafman
12400 Wilshire Blvd.
Suite 700
Los Angeles, CA 90025   (213) 207-3800

Jerrold B. Reilly
Lyon & Lyon
611 West Sixth Street
34th Floor
Los Angeles, CA 90017   (213) 489-1600

Thomas C. Reynolds
Spensley Horn Jubas
& Lubitz
1880 Century Park East
Fifth Floor
Los Angeles, CA 90067   (213) 553-5050

David B. Ritchie
Lyon & Lyon
611 West Sixth Street
Suite 3400
Los Angeles, CA 90017   (213) 489-1600

Billy A. Robbins
Nilsson, Robbins, Dalgarn,
Berliner, Carson & Wurst
201 North Figueroa Street
5th Floor
Los Angeles, CA 90012   (213) 977-1001

William J. Robinson
Poms, Smith, Lande
& Rose
2121 Avenue of the Stars
Suite 1400
Los Angeles, CA 90067   (213) 277-8141

Jesse D. Roggen#
Poms, Smith, Lande
& Rose, P.C.
2121 Avenue of the Stars
Suite 1400
Los Angeles, CA 90067   (213) 788-5000

Alan C. Rose
Poms, Smith, Lande
& Rose, P.C.
2121 Avenue of the Stars
Suite 1400
Los Angeles, CA 90067   (213) 277-8141

Leon David Rosen
Freilich, Hornbaker,
& Rosen
10960 Wilshire Blvd.
Suite 1434
Los Angeles, CA 90024   (213) 477-0578

Charles Rosenberg
Poms, Smith, Lande & Rose
2121 Avenue of the Stars
Suite 1400
Los Angeles, CA 90067   (213) 277-8141

L. Kenneth Rosenthal#
Strategic Innovation Services
10306 Rossbury Place
Los Angeles, CA 90064   (213) 558-3760

Ellsworth R. Roston
Roston & Schwartz
5900 Wilshire Blvd.
Suite 1430
Los Angeles, CA 90036   (213) 938-3657

Gregory L. Roth
Pretty, Schroeder, Brueggemann
& Clark
Suite 2000
444 South Flower Street
Los Angeles, CA 90071   (213) 489-4442

W. Norman Roth
Roth & Goldman
Pacific Mutual Bldg.
523 W. Sixth Street
Suite 840
Los Angeles, CA 90014   (213) 688-1143

Francis G. Rushford
Poms, Smith, Lande &
Rose, P.C.
2121 Avenue of the Stars
Suite 1400
Los Angeles, CA 90067   (213) 277-8141

Michael W. Sales
Hughes Aircraft Company
7200 Hughes Terrace
Los Angeles, CA 90045   (213) 568-7028

Carol K. Samek
Stroock & Stroock & Lavan
2029 Century Park East
Suite 1800
Los Angeles, CA 90067   (213) 556-5970

Robert Jay Schaap
6820 La Tijera Blvd.
Suite 107
Los Angeles, CA 90045   (213) 645-6460

Edward C. Schewe
Lyon & Lyon
611 W. Sixth St., 34th Flr.
Los Angeles, CA 90017   (213) 489-1600

Carol A. Schneider
Lyon & Lyon
611 W. Sixth St.
Suite 3400
Los Angeles, CA 90017   (213) 489-1600

Seymour A. Scholnick
1906 S. Roxbury Dr.
Los Angeles, CA 90035   (213) 870-0066

Robert A. Schroeder
Pretty, Schroeder, Brueggemann
& Clark
444 S. Flower Street
Suite 2000
Los Angeles, CA 90071   (213) 489-4442

Charles H. Schwartz
Roston & Schwartz.
5900 Wilshire Blvd.
Suite 1430
Los Angeles, CA 90036   (213) 938-3657

John A. Scillieri
Ashen, Golant, Martin
& Seldon
10920 Wilshire Blvd.
Suite 1000
Los Angeles, CA 90024   (213) 208-1100

Michael D. Scott
Graham & James
725 S. Figueroa Street
Suite 1200
Los Angeles, CA 90017   (213) 624-2500

Robert A. Seldon
Ashen Golant Martin & Seldon
10920 Wilshire Blvd.
Suite 1000
Los Angeles, CA 90024   (213) 208-1100

Todd B. Serota
Poms, Smith, Lande
& Rose
2121 Avenue of the Stars
Suite 1400
Los Angeles, CA 90067   (213) 277-8141

James H. Shalek
Lyon & Lyon
611 W. 6th Street
34th Floor
Los Angeles, CA 90017   (213) 489-1600

Edmond F. Shanahan
725 S. Norton Ave.
Los Angeles, CA 90005   (213) 385-1331

James J. Short
Lyon and Lyon
611 West 6th Street
34th Floor
Los Angeles, CA 90017   (213) 489-1600

Ira M. Siegel
Blakely, Sokoloff, Taylor
& Zafman
12400 Wilshire Blvd.
Suite 700
Los Angeles, CA 90025   (213) 207-3800

David M. Simon
Spensley Horn Jubas
& Lubitz
1880 Century Park East
Suite 500
Los Angeles, CA 90067   (213) 553-5050

Marvin O. Sleven#
13257 Chalon Road
Los Angeles, CA 90049   (213) 476-1063

Jeffrey L. Slusher#
Poms, Smith, Lande
& Rose
2121 Ave. of the Stars
Suite 1400
Los Angeles, CA 90067   (213) 277-8141

Thomas M. Small
Baker & Mckenzie
725 S. Figueroa St.
36th Floor
Los Angeles, CA 90017   (213) 629-7181

Charles A. Smiley, Jr.
7461 Beverly Blvd.
Suite 303
Los Angeles, CA 90036   (213) 933-5604

Guy Porter Smith
Poms, Smith, Lande,
& Rose
2121 Avenue of the Stars
Suite 1400
Los Angeles, CA 90067   (213) 277-8141

Roland N. Smoot
Lyon and Lyon
611 W. Sixth Street
34th Floor
Los Angeles, CA 90017　　　(213) 489-1600

Stanley William Sokoloff
Blakley, Sokoloff, Taylor
& Zafman
12400 Wilshire Blvd.
Suite 700
Los Angeles, CA 90025　　　(213) 207-3800

Conrad R. Solum, Jr.
Lyon and Lyon
611 W. Sixth Street
34th Floor
Los Angeles, CA 90017　　　(213) 489-1600

William C. Steffin
Lyon & Lyon
611 W. Sixth Street
34th Floor
Los Angeles, CA 90017　　　(213) 489-1600

Robert Steinberg
Irell & Manella
333 S. Hope Street
Suite 3300
Los Angeles, CA 90071　　　(213) 620-1555

Lewis B. Sternfels
3100 Inglewood Blvd.
Los Angeles, CA 90066　　　(213) 391-6665

Samuel Beckner Stone
Lyon and Lyon
611 W. Sixth Street
34th Floor
Los Angeles, CA 90017　　　(213) 489-1600

William J. Streeter
Hughes Aircraft Company
Bldg. C1, M.S. A126
7200 Hughes Terrace
P.O. Box 45066
Los Angeles, CA 90045　　　(213) 334-9108

Daniel F. Sullivan
Poms, Smith, Lande
& Rose
2121 Avenue of the Stars
Suite 1400
Los Angeles, CA 90067　　　(213) 277-8141

Craig S. Summers
Pretty, Schroeder, Bruggemann
& Clark
444 S. Flower Street
Suite 2000
Los Angeles, CA 90071　　　(213) 489-4442

Philip C. Swain
Kirkland & Ellis
300 S. Grand Avenue
Suite 3000
Los Angeles, CA 90071　　　(213) 680-8400

Steven A. Swernofsky
Lyon & Lyon
611 West Sixth Street
Suite 3400
Los Angeles, CA 90017　　　(213) 489-1600

William Edward Thomson, Jr.
Lyon and Lyon
611 West Sixth Street
34th Floor
Los Angeles, CA 90017　　　(213) 489-1600

Robert R. Thornton
Carlsmith, Ball, Wichman, Murray,
Case, Mukai & Ichiki
555 S. Flower St.
25th Floor
Los Angeles, CA 90071　　　(213) 955-1200

Walton Eugene Tinsley
Harris, Kern, Wallen & Tinsley
Quinby Bldg.
650 S. Grand Ave.
Los Angeles, CA 90017　　　(213) 626-5251

Norton R. Townsley#
3525 Saint Susan Pl.
Los Angeles, CA 90066　　　(213) 390-2425

R. Joseph Trojan
900 Wilshire Blvd.
Suite 1000
Los Angeles, CA 90017　　　(213) 626-7766

Joan S. Trygstad
10467 Raybet Road
Los Angeles, CA 90077　　　(213) 470-4575

Robert M. Unruh
991 Montecito Dr.
Los Angeles, CA 90031　　　(213) 221-8144

Michael L. Wachtell
Buchalter, Nemer, Fields
& Younger
601 S. Figueroa St.
Los Angeles, CA 90017　　　(213) 891-0700

Jeannette M. Walder
Hughes Aircraft Co.
Corporate Patents & Licensing
Bldg. C1, Mail Station A126
P.O. Box 45066
Los Angeles, CA 90045　　　(714) 759-6460

Les J. Weinstein
Pepper, Hamilton & Scheetz
444 S. Flower St.
Suite 1900
Los Angeles, CA 90071　　　(213) 688-5608

David Weiss
2551 Colorado Blvd.
Los Angeles, CA 90041　　　(213) 254-5020

Robert Charles Weiss
Lyon and Lyon
611 W. 6th Street
Suite 3400
Los Angeles, CA 90017　　　(213) 489-1600

John T. Wiedemann
Pretty, Schroeder, Brueggemann
& Clark, P.C.
444 S. Flower Street
Suite 2000
Los Angeles, CA 90071　　　(213) 489-4442

Wayne E. Willenberg
Spensley, Horn, Jubas & Lubitz
1880 Century Park East
5th Floor
Los Angeles, CA 90067　　　(213) 553-6052

Charles E. Wills
725 S. Figueroa Street
34th Floor
Los Angeles, CA 90017　　　(213) 689-5123

Garth A. Winn
Spensley, Horn, Jubas
& Lubitz
1880 Century Park East
Fifth Floor
Los Angeles, CA 90067　　　(213) 553-5050

Michael J. Wise
Lyon & Lyon
611 W. 6th St.
Suite 3400
Los Angeles, CA 90017　　　(213) 489-1600

Roger R. Wise
Spensley, Horn, Jubas
& Lubitz
1880 Century Park East
Suite 500
Los Angeles, CA 90067　　　(213) 553-5050

Harold E. Wurst
Nilsson, Robbins, Dalgarn,
Berliner, Carson & Wurst
201 North Figueroa Street
5th Floor
Los Angeles, CA 90012　　　(213) 977-1001

Joseph A. Yanny
1925 Century Park East
Suite 1260
Los Angeles, CA 90067　　　(213) 551-2966

Ben J. Yorks
Blakely, Sokoloff, Taylor
& Zafman
12400 Wilshire Blvd.
7th Floor
Los Angeles, CA 90025　　　(714) 557-3800

Norman Zafman
Blakely, Sokoloff, Taylor
& Zafman
12400 Wilshire Blvd.
7th Floor
Los Angeles, CA 90025　　　(213) 207-3800

Richard H. Zaitlen
Spensley, Horn, Jubas
& Lubitz
1880 Century Park East
Suite 500
Los Angeles, CA 90067　　　(213) 553-5050

Harris Zeitzew
Meyers, Bianchi & Mc Connell
11859 Wilshire Blvd.
Fourth Floor
Los Angeles, CA 90025　　　(213) 312-0772

William W. Burns
15720 Winchester Blvd.
Los Gatos, CA 95030　　　(408) 395-5111

Kim Effron#
117 Calle Nivel
Los Gatos, CA 95030   (408) 374-8144

Owen Lester Lamb
P.O. Box 2254
Los Gatos, CA 95031   (408) 395-2321

Henry E. Otto, Jr.
969 Cherrystone Drive
Los Gatos, CA 95032   (408) 356-5378

Arthur R. Sorkin#
Tesuji, Inc
12340 Indian Trail Rd.
Los Gatos, CA 95030   (408) 867-5830

James E. Mc Taggart
21470 Rambla Vista
Malibu, CA 90265   (213) 456-8854

Gordon A. Shifrin#
7145 Fernhill Drive
Malibu, CA 90265   (213) 457-2317

Rodolfo Aquirre
16 Monterey Court
Manhattan Beach, CA 90266   (213) 545-3577

Randall M. Heald
1613 2nd Street
Manhattan Beach, CA 90266   (213) 568-6084

Melvin Arthur Klein
8 San Miguel Court
Manhattan Beach, CA 90266   (213) 546-1270

Newell C. Rodewald#
P.O. Box 1061
Manhattan Beach, CA 90266   (805) 245-2275

Joseph Shulsinger
Suite 117, Box J
Manhattan Beach, CA 90266   (213) 838-1444

Frank Wattles
P.O. Box 3514
Manhattan Beach, CA 90266   (213) 372-0454

Robert H. Fraser
3812 Via Dolce
Marina Del Rey, CA 91292   (213) 821-1242

William H. Benz
Ciotti & Murashige
545 Middlefield Road
Suite 200
Menlo Park, CA 94025   (415) 327-7250

Robert P. Blackburn
Ciotti & Murashige
545 Middlefield Road
Suite 200
Menlo Park, CA 94025   (415) 327-7250

Karl Bozicevic
Irell & Manella
545 Middlefield Road
Suite 200
Menlo Park, CA 94025   (415) 327-7250

Herbert G. Burkard
Raychem Corp.
300 Constitution Dr.
Menlo Park, CA 94025   (415) 361-3338

Yuan Chao
Raychem Corp.
Mail Stop 120/6600
300 Constitution Drive
Menlo Park, CA 94025   (415) 361-5979

Paula N. Chavez
901 1/2 Olive St.
Menlo Park, CA 94025   (415) 329-0773

John Y. Chen#
SRI International
333 Ravenswood Avenue
Menlo Park, CA 94025   (415) 859-2446

Thomas Edward Ciotti
Ciotti & Murashige
545 Middlefield Road
Suite 200
Menlo Park, CA 94025   (415) 327-7250

Troy G. Dillahunty
Raychem Corporation
300 Constitution Drive
Menlo Park, CA 94025   (415) 361-5106

David S. Dolberg
Irell & Manella
545 Middlefield Rd.
Suite 200
Menlo Park, CA 94025   (415) 327-7250

Janet E. Farrant#
Membrane Technology & Res. Inc.
1360 Willow Road
Menlo Park, CA 94025   (415) 328-2228

Urban Hart Faubion
SRI International
333 Ravenswood Ave.
Menlo Park, CA 94025   (415) 859-4550

Ralph L. Freeland, Jr.
Burns, Doane, Swecker
& Mathis
3000 Sand Hill Road
Building 4, Suite 160
Menlo Park, CA 94025   (415) 854-7400

Marguerite E. Gerstner#
Raychem Corporation
300 Constitution Drive
Menlo Park, CA 94025   (415) 361-5857

Kenneth M. Goldman
Irell & Manella
545 Middlefield Rd.
Suite 200
Menlo Park, CA 94025   (415) 327-7252

Dennis E. Kovach
Raychem Corporation
300 Constitution Drive
Menlo Park, CA 94025   (415) 324-6545

Robert E.D. Krebs
Burns, Doane, Swecker
& Mathis
3000 Sand Hill Rd.
Bldg. 4, Suite 160
Menlo Park, CA 94025   (415) 854-7400

Ronald S. Laurie
Ciotti & Murashige, Irell
& Manella
545 Middlefield Road
Suite 200
Menlo Park, CA 94025   (415) 327-7250

Bernard J. Lyons#
Raychem Corporation
300 Constitution Drive
Menlo Park, CA 94025   (415) 361-3960

Gladys H. Monroy
Ciotti & Murashige, Irell
& Manella
545 Middlefield Road
Suite 200
Menlo Park, CA 94025   (415) 327-7250

Kate H. Murashige
Ciotti & Murashige, Irell
& Manella
545 Middlefield Road
Suite 200
Menlo Park, CA 94025   (415) 327-7250

William Thomas Nye
Commtech International
Management Corp.
545 Middlefield Road
Suite 180
Menlo Park, CA 94025   (415) 328-0190

James W. Peterson
Burns, Doane, Swecker
& Mathis
3000 San Hill Road
Bldg. 4, Suite 160
Menlo Park, CA 94025   (415) 854-7400

Dianne E. Reed
Irell & Manella
545 Middlefield Rd.
Suite 200
Menlo Park, CA 94025   (415) 327-7250

Norman E. Reitz
P.O. Box 2630
Menlo Park, CA 94026   (415) 732-9940

Edith A. Rice
Raychem Corp.
300 Constitution Dr.
Menlo Park, CA 94025   (415) 361-3331

Timothy H.P. Richardson#
Raychem Corp.
Pat. Dept.
300 Constitution Dr.
Menlo Park, CA 94025   (415) 361-2004

Roberta L. Robins
Irell & Manella
545 Middlefield Road
Suite 200
Menlo Park, CA 94025   (415) 327-7250

Debra A. Shetka
Irell & Manella
545 Middlefield Road
Suite 200
Menlo Park, CA 94025   (415) 327-7250

John P. Spitals
Irell & Manella
545 Middlefield Road
Suite 200
Menlo Park, CA 94026   (415) 327-7250

Carol A. Stratford#
Neurex Corp.
3760 Haven Ave.
Menlo Park, CA 94025   (415) 853-1500

James E. Toomey
Burns, Doane, Swecker
& Mathis
3000 Sand Hill Rd., Bldg. 4
Suite 160
Menlo Park, CA 94025   (415) 854-7400

A. Stephen Zavell
Raychem Corporation
300 Constitution Drive
Menlo Park, CA 94025   (415) 361-2990

Robert E. Wickersham
1 Glen Drive
Mill Valley, CA 94941   (415) 361-2990

David B. Harrison
Quantum Corp.
500 Mc Carthy Blvd.
Milpitas, CA 95035   (408) 894-4287

Richard M. Ladden
1404 Acadia Avenue
Milpitas, CA 95035   (408) 262-5993

Elizabeth A. Lawler
1609 Greenwood Way
Milpitas, CA 95035   (408) 263-0780

Michael D. Rostoker
LSI Logic Corp.
1551 Mc Carthy Blvd.
Milpitas, CA 95035   (408) 433-7380

Thomas H. Williams
Greyhawk Systems, Inc.
1557 Centre Pointe Drive
Milpitas, CA 95035   (408) 945-1776

Fred N. Schwend
456 Ivy Glen Dr.
Mira Loma, CA 91752   (714) 685-0327

Ronald B. Blanchard#
24481 Dardania
Mission Viejo, CA 92691   (714) 770-5161

Clifford B. Boehmer
25266 Pacifica
Mission Viejo, CA 92691   (714) 896-1914

Grover A. Frater
P.O. Box 3545
Mission Viejo, CA 92690   (714) 495-8777

Bradley Lionel Jacobs#
26816 La Sierra Dr.
Mission Viejo, CA 92691   (714) 582-1554

Jonathan B. Orlick
27111 South Ridge Drive
Mission Viejo, CA 92692   (714) 957-3000

Darrell Gene Brekke*
NASA-Ames Research Center
Mail Stop 200-11
Moffett Field, CA 94035   (415) 694-5104

J. Carl Cooper#
Pixel Instruments Corp.
15288 Via Pinto
Monte Sereno, CA 95030   (408) 986-9090

F. David La Riviere
Schroeder, Davis & Orliss
P.O. Box 3080
Monterey, CA 93942   (408) 649-1122

James E. Noble
1176 Harrison St.
Monterey, CA 93940   (408) 373-2269

Marvin Jabin
Jabin & Jabin
701 S. Atlantic Blvd.
Monterey Park, CA 91754   (213) 570-1117

Edmund W. Rusche, Jr.
Litton Systems, Inc.
Aero Products Division
6101 Condor Drive
Moorpark, CA 93021   (805) 378-2913

John R. Murtha
1253 Larch Avenue
Moraga, CA 94556   (415) 376-1498

Ronald C. Fish
16590 Oak View Circle
Morgan Hill, CA 95037   (415) 376-1498

Robert J. Grassi
66 Mulberry Court
Morgan Hill, CA 95037   (408) 779-3110

Daniel C. Moyles
1470 Bluebonnet Way
Morgan Hill, CA 95037   (408) 778-2000

Daniel C. Mc Kown
355 Fairview Avenue
Morro Bay, CA 93442   (805) 541-5148

J. L. Bohan
P.O. Box 4720
Mountain View, CA 94040   (408) 272-2688

John H. Grate#
Catalytica, Inc.
430 Ferguson Drive
Bldg. 3
Mountain View, CA 94043   (415) 960-3000

John F. Lawler
P.O. Box 638
Mountain View, CA 94042   (415) 948-8608

Kenneth J. Nussbacher
Daisy Systems Corporation
700 E. Middlefield Road
Mountain View, CA 94039   (415) 960-7182

Thomas Henry Olson
4309 Collins Court
Apt. 6
Mountain View, CA 94040   (415) 941-2391

Karen S. Perkins
1166 Nilda Avenue
Mountain View, CA 94040   (415) 961-1166

Paul F. Schenck#
1934 Miramonte Avenue
Mountain View, CA 94040   (415) 964-7902

Peter R. Shearer
California Biotechnology Inc.
2450 Bayshore Parkway
Mountain View, CA 94043   (415) 962-5860

Harry G. Thibault
Abbott Critical Care Sys.
1212 Terra Bella Avenue
Mountain View, CA 94043   (415) 940-7465

Frank M. Weyer
American Innovative
Products, Inc.
P.O. Box 1477
Mountain View, CA 94041   (415) 965-1421

Eugene T. Wheelock
Catalytica Inc.
430 Ferguson Drive, Bldg. 3
Mountain View, CA 94043   (415) 960-3000

Daniel E. Altman
Knobbe, Martens, Olson & Bear
620 Newport Center Drive
16th Floor
Newport Beach, CA 92660   (714) 760-0404

Lowell Anderson
Knobbe, Martens, Olson
& Bear
620 Newport Center Drive
Sixteenth Floor
Newport Beach, CA 92660   (714) 760-0404

James Barth Bear
Knobbe, Martens, Olson,
& Bear
620 Newport Center Drive
Sixteenth Floor
Newport Beach, CA 92660   (714) 760-0404

Richard E. Bee
519 Ventaja
Newport Beach, CA 92660   (714) 644-8618

George Frazier Bethel
Beehler, Pavitt, & Bethel
Suite 200
610 Newport Center Drive
Newport Beach, CA 92660   (714) 640-0900

Patience K. Bethel#
Beehler, Pavitt & Bethel
Suite 200
610 Newport Center Drive
Newport Beach, CA 92660   (714) 640-0900

William B. Bunker
Knobbe, Martens, Olson,
& Bear
620 Newport Center Drive
Sixteenth Floor
Newport Beach, CA 92660   (714) 760-0404

377

Robert R. Deveza#
Knobbe, Martens, Olson
& Bear
620 Newport Center Drive
16th Floor
Newport Beach, CA 92660   (714) 760-0404

Dennis H. Epperson
Knobbe, Martens, Olson
& Bear
620 Newport Center Drive
16th Floor
Newport Beach, CA 92660   (714) 760-0404

Morland C. Fischer
4650 Von Karman Avenue
Newport Beach, CA 92660   (714) 476-0600

Kenneth W. Float
Hughes Aircraft Co.
500 Superior Avenue
Newport Beach, CA 92658   (714) 759-7335

Jeffery D. Frazier#
Harness, Dickey & Pierce
610 Newport Center Drive
Suite 1520
Newport Beach, CA 92660   (714) 760-6233

Michael T. Gabrik
Harness, Dickey & Pierce
280 Newport Center Drive
Suite 130
Newport Beach, CA 92660   (714) 760-6233

Curtis L. Harrington
Hawes & Fischer
660 Newport Center Dr.
Suite 460
Newport Beach, CA 92660   (714) 759-6601

James E. Hawes
660 Newport Center Drive
Suite 460
Newport Beach, CA 92660   (714) 759-6601

L. Harmon Hook
660 Newport Center Drive
Suite 460
Newport Beach, CA 92660   (714) 759-6601

Scott Hunter
1300 Bristol Street, N.
Suite 180
Newport Beach, CA 92660   (714) 833-9922

Louis J. Knobbe
Knobbe, Martens, Olson,
& Bear
620 Newport Center Drive
16th Floor
Newport Beach, CA 92660   (714) 760-0404

William Gregory Lane
Harness, Dickey & Pierce
280 Newport Center Drive
Suite 130
Newport Beach, CA 92660   (714) 760-6233

James Francis Lesniak
Knobbe, Martens, Olson
& Bear
620 Newport Center Drive
Sixteenth Floor
Newport Beach, CA 92660   (714) 760-0404

Thomas P. Mahoney
Mahoney & Schick
4000 Mac Arthur Blvd.
Suite 6200
Newport Beach, CA 92660   (714) 851-8081

Don W. Martens
Knobbe, Martens, Olson,
& Bear
620 Newport Center Drive
Sixteenth Floor
Newport Beach, CA 92660   (714) 760-0404

William H. Nieman
Knobbe, Martens, Olson,
& Bear
620 Newport Center Drive
Sixteenth Floor
Newport Beach, CA 92660   (714) 760-0404

Harvey Charles Nienow
1300 Dove Street
Suite 200
Newport Beach, CA 92660   (714) 851-8585

Darrell L. Olson
Knobbe, Martens, Olson,
& Bear
620 Newport Center Drive
Suite 1600
Newport Beach, CA 92660   (714) 760-0404

Gordon H. Olson
Knobbe, Martens, Olson,
& Bear
620 Newport Center Drive
Sixteenth Floor
Newport Beach, CA 92660   (714) 760-0404

Manuel Quioque
Roberts & Quiogue
660 Newport Center Drive
Suite 1400
P.O. Box 8569
Newport Beach, CA 92658   (714) 640-6200

Joseph R. Re#
Knobbe, Martens, Olson
& Bear
620 Newport Center Drive
16th Floor
Newport Beach, CA 92660   (714) 760-0404

Larry K. Roberts
Roberts & Quiogue
660 Newport Center Drive
Suite 1400
P.O. Box 8569
Newport Beach, CA 92658   (714) 640-6200

William C. Rooklidge
Knobbe, Martens, Olson
& Bear
620 Newport Center Drive
16th Floor
Newport Beach, CA 92660   (716) 760-0404

Arthur S. Rose
Knobbe, Martens, Olson,
& Bear
620 Newport Center Drive
Sixteenth Floor
Newport Beach, CA 92660   (714) 760-0404

Edward A. Schlatter
Knobbe Martens Olson
& Bear
620 Newport Center Drive
Sixteenth Floor
Newport Beach, CA 92660   (714) 760-0404

Jerry T. Sewell
Knobbe, Martens, Olson
& Bear
620 Newport Center Drive
Sixteenth Floor
Newport Beach, CA 92660   (714) 760-0404

John B. Sganga, Jr.
Knobbe, Martens, Olson
& Bear
620 Newport Center Drive
Sixteenth Floor
Newport Beach, CA 92660   (714) 760-0404

Andrew H. Simpson#
Knobbe, Martens, Olson
& Bear
620 Newport Center Drive
Suite 1600
Newport Beach, CA 92660   (714) 760-0404

Milton R. Spielman
220 Vin Eboli
Newport Beach, CA 92663   (714) 675-7667

Leonard Tachner
3990 Westerly Place
Suite 295
Newport Beach, CA 92660   (714) 752-8525

W. Gerard Von Hoffmann, III
Knobbe, Martens, Olson
& Bear
620 Newport Center Drive
Suite 1600
Newport Beach, CA 92660   (714) 760-0404

Frank L. Zugelter
10221 Riverside Drive
Suite 207
North Hollywood, CA 91602   (213) 769-3411

William H. Fleeson#
11019 Nestle Ave.
Northridge, CA 91324   (213) 363-1465

Andrew S. Jordan#
17900 Schoenborn St., #21
Northridge, CA 91325   (818) 881-3141

David A. Kemper
10913 Reseda Blvd.
Northridge, CA 91326   (213) 885-1101

Joseph Kriensky
10214 Topeka Drive
Northridge, CA 91324   (213) 368-1326

378

Alexander Linger
8935 Garden Grove Avenue
Northridge, CA 91325   (818) 886-3305

Roy J. Mankovitz
18842 Kilfinan Street
Northridge, CA 91326   (818) 363-1383

Francis R. Reilly
19000 Merridy Street
Northridge, CA 91324   (213) 886-3065

Perry E. Turner
P.O. Drawer E
Northridge, CA 91328   (818) 360-6485

Bruce E. Francone*
U.S. Dept. of Air Force
HQBMO/JA
Norton A.F.B., CA 92409   (714) 382-6433

R. Russel Austin#
215 Butterfield Dr.
Novato, CA 94945   (415) 898-9950

Donald J. De Geller
2725 Topaz Drive
Novato, CA 94947   (415) 892-2147

Andrew E. Barlay
100 Oak Street
Suite 5
Oakland, CA 94607   (510) 836-0514

Donald L. Beeson
Beilock, Collins & Beeson
One Kaiser Plaza
Suite 2360
Oakland, CA 94612   (510) 832-8700

Stephen L. Berger#
766-B Kingston Avenue
Oakland, CA 94611   (415) 658-2256

H. Michael Brucker
166 Santa Clara Ave.
Oakland, CA 94601   (415) 658-2500

Howard S. Cohen#
1330 Broadway
Suite 1150
Oakland, CA 94612   (415) 465-0828

James Robert Cypher
405 - 14th Street
Suite 1607
Oakland, CA 94612   (415) 832-4111

Glen R. Grunewald
Grunewald & Lampe
166 Santa Clara Avenue
Oakland, CA 94610   (415) 658-2500

James M. Hanley*
U.S. Dept. of Energy
1333 Broadway
Oakland, CA 94612   (415) 273-6240

Henry G. Hardy
6500 Chabot Road
Oakland, CA 94618   (415) 658-2244

Joel J. Hayashida
Clorox Company
P.O. Box 24305
Oakland, CA 94623   (415) 271-7847

Henry H. Johnson#
494 58th Street
Oakland, CA 94609   (415) 655-4464

Paul Maria Klein, Jr.
719 54th St.
Oakland, CA 94609   (415) 776-7426

Elliott B. Lieberman
Morinelli & Lieberman
6009 Buenaventura Avenue
Oakland, CA 94605   (415) 430-9366

Michael J. Mazza
Clorox Company
Legal Services
1221 Broadway
Oakland, CA 94612   (415) 271-7521

Ernest H. Mc Coy
Bruce and Mc Coy
One Kaiser Plaza
Suite 2360
Oakland, CA 94612   (415) 836-2400

Jane R. Mc Laughlin
Cetus Corporation
511 Florence Avenue
Oakland, CA 94618   (415) 420-3223

Malcolm Caven Mc Quarrie
National Refractories
& Minerals Corporation
One Kaiser Plaza
Suite 650
Oakland, CA 94612   (415) 452-8742

Clinton H. Neagley
DNA Plant Technology Corp.
6701 San Pablo Avenue
Oakland, CA 94608   (415) 547-2395

Thomas S. O Dwyer*
U.S. Dept. of Energy
Off. of Chief Counsel
1333 Broadway
Oakland, CA 94612   (415) 273-6428

Harry A. Pacini#
Clorox Company
Legal Division
P.O. Box 24305
Oakland, CA 94623   (415) 271-7416

Kathleen A. Skinner
One Kaiser Plaza
Suite 2360
Oakland, CA 94612   (510) 832-8700

Richard M. Snead#
335 Warwick Ave.
Oakland, CA 94610   (415) 452-1441

Wayne P. Sobon#
400 Perkins Street, #402
Oakland, CA 94610   (415) 444-7514

Stanley M. Teigland
James River Corp.
300 Lakeside Drive
Oakland, CA 94612   (415) 874-3631

Robert R. Tipton
2101 Webster Street Tower
Suite 1500
Oakland, CA 94612   (510) 465-9330

Allen B. Wagner
Univ. of Calif.
Off. of Gen. Coun.
300 Lakeside Dr., 7th Flr.
Oakland, CA 94612   (415) 987-9800

Joseph Gregory Walsh
10732 Fallbrook Way
Oakland, CA 94605   (415) 568-6959

Stephen M. Westbrook
Clorox Co.
1221 Broadway
Oakland, CA 94612   (510) 271-7296

Malcolm B. Wittenberg
Crosby, Heafey, Roach
& May
1999 Harrison St.
Oakland, CA 94612   (415) 763-2000

Harris Zimmerman
1330 Broadway
Suite 1150
Oakland, CA 94612   (415) 465-0828

Milton C. Hansen
1072 Turnstone Way
Oceanside, CA 92056   (619) 722-4980

Claron N. White
3571 Papaya Way
Oceanside, CA 92054   (619) 439-4020

William G. Anderson
Sunkist Growers, Inc.
Patent Law Dept.
760 East Sunkist Street
Ontario, CA 91761   (714) 983-9811

Stephen J. Koundakjian
Sunkist Growers, Inc.
760 E. Sunkist Street
Ontario, CA 91761   (714) 983-9811

Richard F. Carr
Gausewitz, Carr and Rothenberg
One City Blvd. West
Suite 830
Orange, CA 92668   (714) 634-4003

William L. Chapin
2410 W. Palm Avenue
Orange, CA 92668   (714) 978-6130

Margie R. Dickinson
505 City Parkway West
Suite 1000
Orange, CA 92668   (714) 634-4540

Richard L. Gausewitz
Gausewitz, Carr & Rothenberg
1 City Blvd. W.
Suite 830
Orange, CA 92668   (714) 634-4003

Willie Krawitz
3001 Chapel Hill Rd.
Orange, CA 92667   (714) 974-1190

Walter F. Krstulja
196 S. Amberwood Street
Orange, CA 92669   (714) 639-2748

Allan Rothenberg
Poms, Smith, Lande & Rose
One City Blvd. West
Suite 830
Orange, CA 92668   (714) 634-4003

G. Donald Weber, Jr.
505 City Parkway West
Suite 1000
Orange, CA 92668   (714) 634-4540

Steven W. Wilcox#
9520 Beacon Avenue
Orangevale, CA 95662   (916) 988-2179

Merwyn G. Brosler#
1 Meadow Court
Orinda, CA 94563   (415) 254-5593

David J. Goren#
16 Sanborn Road
Orinda, CA 94563   (415) 254-5593

Armand G. Guibert#
61 Avenida De Orinda
#29A
Orinda, CA 94563   (415) 254-4334

David Lowell Hagmann
464 Camino Sobrante
Orinda, CA 94563   (415) 254-3146

John H. Mc Carthy
17 Camino Lenada
Orinda, CA 94563   (415) 254-4954

John B. Miller, Jr.
Lawler, Bonham & Walsh
300 Esplanade Dr., Ste. 1900
Oxnard, CA 93031   (805) 485-8921

Frank A. Campbell
P.O. Box 25
Pacific Palisades, CA 90272   (213) 333-5552

Thomas Long Flattery
439 Via De La Paz
Pacific Palisades, CA 90272   (213) 454-3768

Robert D. Kummel
14225 Sunset Blvd.
Pacific Palisades, CA 90272   (213) 454-4548

Robert E. Malm
16624 Pequeno Pl.
Pacific Palisades, CA 90272   (213) 459-3992

William K. Rieber
15200 Sunset Blvd.
Suite 214
Pacific Palisades, CA 90272   (213) 454-5770

Paul Richter Wylie, Jr.
15200 Sunset Blvd.
Suite 210
Pacific Palisades, CA 90272   (213) 459-8439

Roy H. Davies
679 Parkview Court
Pacifica, CA 94044   (415) 355-1344

Allan R. Fowler
P.O. Box 3305
Palm Desert, CA 92261   (619) 346-8377

Serge Abend
Xerox Corp.
Pat. Dept.
3333 Coyote Hill Rd.
Palo Alto, CA 94304   (415) 494-4262

Kenneth R. Allen
Townsend & Townsend
379 Lytton Avenue
Palo Alto, CA 94301   (415) 326-2400

Hunter L. Auyang
Wilson, Sonsini, Goodrich
& Rosati
Two Palo Alto Square
Suite 900
Palo Alto, CA 94306   (415) 493-9300

Robert A. Barr
3787 Redwood Circle
Palo Alto, CA 94306   (415) 493-9300

Patrick J. Barrett
Hewlett-Packard Co.
Legal Dept., M/S 20b0
3000 Hanover St.
Palo Alto, CA 94304   (415) 857-3489

James T. Beran
Xerox Corporation
3333 Coyote Hill Road
Palo Alto, CA 94304   (415) 494-4253

Edward H. Berkowitz
Varian Associates
611 Hansen Way
Palo Alto, CA 94303   (415) 424-5403

Gerard Alan Blaufarb
Syntex U.S.A. Inc.
3401 Hillview Ave.
P.O. Box 10850
Palo Alto, CA 94304   (415) 855-6176

William J. Bohler
Townsend & Townsend
379 Lytton Avenue
Palo Alto, CA 94301   (415) 326-2400

Roger S. Borovoy
Brown & Bain
600 Hansen Way
Suite 100
Palo Alto, CA 94306   (415) 856-9411

Charles Lee Botsford
P.O. Box 1116
Palo Alto, CA 94302   (415) 277-1561

Howard R. Boyle#
Hewlett-Packard Co.
Legal Department
M.S. 20b0
P.O. Box 10301
Palo Alto, CA 94303   (415) 857-4316

Julian Caplan
Flehr, Hohbach, Test,
Albritton & Herbert
850 Hansen Way, Ste. 200
Palo Alto, CA 94304   (415) 494-8700

Steven F. Caserza
Cooley Godward Castro
Huddleson & Tatum
5 Palo Alto Square
4th Floor
Palo Alto, CA 94306   (415) 494-7622

Jean C. Chognard
P. O. Box 406
Palo Alto, CA 94302   (415) 857-2541

Y. Ping Chow#
Syntex Corp.
3401 Hillview Avenue
Palo Alto, CA 94304   (415) 852-1356

Pauline A. Clarke#
Syntex U.S.A. Inc.
Pat. Dept. A2-200
3401 Hillview Avenue
Palo Alto, CA 94303   (415) 852-1355

Stanley Z. Cole
Varian Associates, Inc.
611 Hansen Way
Palo Alto, CA 94303   (415) 424-5408

Peter J. Dehlinger
Dressler, Goldsmith, Shore,
Sutker & Milnamow, Ltd.
350 Cambridge Avenue
Suite 100
Palo Alto, CA 94306   (415) 323-8302

Peter N. Detkin
Wilson, Sonsini, Goodrich
& Rosati
Five Palo Alto Square, Ste. 900
Palo Alto, CA 94306   (415) 493-9300

John A. Dhuey
Syntex Corp.
3401 Hillview Ave.
Palo Alto, CA 94303   (415) 855-6118

Terrence Enroth Dooher
Varian Associates
611 Hansen Way
Palo Alto, CA 94303   (414) 424-5167

Karen B. Dow
Townsend & Townsend
379 Lytton Ave.
Palo Alto, CA 94301   (415) 623-1683

Barry W. Elledge#
Cooley, Godward, Castro,
Huddleson & Tatum
5 Palo Alto Square
4th Flr.
Palo Alto, CA 94306   (415) 494-7622

Gary R. Fabian#
350 Cambridge Ave.
Suite 100
Palo Alto, CA 94306 (415) 323-8302

Dennis S. Fernandez
Fenwick & West
2 Palo Alto Square
Suite 500
Palo Alto, CA 94306 (415) 694-0600

John S. Ferrell#
Fenwick & West
Two Palo Alto Square
Suite 500
Palo Alto, CA 94306 (415) 858-2148

Gerald M. Fisher
Varian Associates
611 Hansen Way
Palo Alto, CA 94301 (415) 424-5407

Stephen P. Fox
Hewlett-Packard Co.
3000 Hanover Street
MS 20b0
Palo Alto, CA 94304 (415) 847-3510

Derek P. Freyberg
Syntex (U.S.A.) Inc.
3401 Hillview Avenue
P.O. Box 10850
Palo Alto, CA 94303 (415) 855-6166

Willliam S. Galliani
Flehr, Hohbach, Test,
Albritton & Herbert
200 Page Mill Road
Suite 200
Palo Alto, CA 94306 (415) 324-8888

Thomas N. Giaccherini
Anglin & Giaccherini
360 Forest Ave.
Suite 201
Palo Alto, CA 94301 (415) 321-4455

Joanne M. Giesser
Sandoz Crop. Protection Corp.
975 California Ave.
Palo Alto, CA 94304 (415) 354-3588

Nancy J. Gracey#
Cooley, Godward, Castro,
Huddleson & Tatum
Five Palo Alt Sq., Ste. 400
Palo Alto, CA 94306 (415) 494-7622

William E. Green
550 Hamilton Avenue
Suite 301
Palo Alto, CA 94301 (415) 321-9992

Edward B. Gregg
Flehr, Hohbach, Test,
Albritton & Herbert
200 Page Mill Road
Suite 200
Palo Alto, CA 94306 (415) 324-8888

Marshall C. Gregory
2211 Park Blvd.
Palo Alto, CA 94306 (415) 321-5030

Roland I. Griffin
Hewlett-Packard Co.
Mail Stop 20B-0
3000 Hanover St.
Palo Alto, CA 94304 (415) 857-2805

Saundra S. Hand#
2450 West Bayshore Rd.
No. 7
Palo Alto, CA 94303 (415) 424-9335

Keith L. Hargrove
Townsend & Townsend
379 Lytton Ave.
Palo Alto, CA 94301 (415) 326-2400

Paul C. Haughey
Townsend & Townsend
379 Lytton Avenue
Palo Alto, CA 94301 (415) 326-2400

David Lloyd Hayes
Fenwick & West
Two Palo Alto Square
Palo Alto, CA 94306 (415) 494-0600

Paul M. Hentzel
441 Nevada Avenue
Palo Alto, CA 94301 (415) 326-8254

James M. Heslin
Townsend & Townsend
379 Lytton Avenue
Palo Alto, CA 94301 (415) 326-9800

Willis Edward Higgins
Flehr, Hohbach, Test,
Albritton & Herbert
200 Page Mill Road
Suite 200
Palo Alto, CA 94306 (415) 324-8888

Joseph I. Hirsch
Syntex U.S.A. Inc.
3401 Hillview Avenue
Palo Alto, CA 94304 (415) 855-6186

James E. Hite, III
842 Wintergreen Way
Palo Alto, CA 94303 (415) 856-8848

Irene Y. Hu#
Cooley, Godward, Castro,
Huddleson & Tatum
5 Palo Alto Square
4th Floor
Palo Alto, CA 94306 (415) 494-7622

Lester E. Johnson
Syntex U.S.A. Inc.
Pat. Law Dept.
3401 Hillview Avenue
P.O. Box 10850
Palo Alto, CA 94303 (415) 855-6593

Allston L. Jones
Cohen & Ostler
525 University Ave.
Suite 410
Palo Alto, CA 94301 (415) 321-3835

Alan M. Krubiner
Syntex (U.S.A.) Inc.
3401 Hillview Ave.
P.O. Box 10850
Palo Alto, CA 94304 (415) 855-6133

Katharine Ku
1130 Waverley Street
Palo Alto, CA 94301 (415) 723-0651

Jacqueline S. Larson
Alza Corporation
950 Page Mill Road
Post Office Box 10950
Palo Alto, CA 94303 (415) 494-5639

David J. Larwood
Brown & Bain
600 Hansen Way
Suite 100
Palo Alto, CA 94306 (415) 856-9411

Theodore J. Leitereg*
Syntex U.S.A. Inc.
3401 Hillview Avenue
Palo Alto, CA 94304 (415) 852-1091

David P. Lentini#
350 Cambridge Avenue
Suite 100
Palo Alto, CA 94306 (415) 323-8302

Brian Lewis
Syntex U.S.A. Inc.
3401 Hillview Avenue
P.O. Box 10850
Palo Alto, CA 94303 (415) 852-3097

Francis H. Lewis, Jr.
Fenwick & West
Two Palo Alto Sq.
Palo Alto, CA 94306 (415) 494-0600

David A. Lowin
Syntex U.S.A. Inc.
Pat. Dept.
3401 Hillview Avenue
Palo Alto, CA 94303 (415) 855-6167

Stephen C. Macevicz
Dnax Research Institute
Of Molecular & Cellular Biology
901 California Avenue
Palo Alto, CA 94304 (415) 496-1204

Edward H. Maker
Hewlett-Packard Company
Intellectual Property Section
Legal Dept.
P.O. Box 10301, M/S 20b0
Palo Alto, CA 94303 (415) 857-5143

Edward Lewis Mandell
Alza Corp.
950 Page Mill Rd.
Palo Alto, CA 94304 (408) 867-5311

Elizabeth Manning
Syntex U.S.A. Inc.
3401 Hillview Avenue
P.O. Box 10850
Palo Alto, CA 94303 (415) 855-5986

Julie Y. Mar-Spinola
Heller, Ehrman, White
& Mc Auliffe
525 University Ave., 11th Flr.
Palo Alto, CA 94301   (415) 324-7600

Thomas J. Mc Naughton
Syntex (U.S.A.) Inc.
3401 Hillview Avenue
P.O. Box 10850
Palo Alto, CA 94303   (415) 852-1560

Michelle M. Mc Spadden
Syntex U.S.A. Inc.
3401 Hillview Avenue
P.O. Box 10850
Palo Alto, CA 94303   (415) 855-6184

Stuart P. Meyer
Fenwick, Davis & West
2 Palo Alto Square
Palo Alto, CA 94306   (494) 060-0

D. Byron Miller, Jr.
Alza Corporation
P.O. Box 10950
950 Page Mill Road
Palo Alto, CA 94303   (415) 496-8150

Harold J. Milstein
Syntex U.S.A. Inc.
3401 Hillview Avenue
Palo Alto, CA 94303   (415) 855-5575

Wayne W. Montgomery
Syntex U.S.A. Inc.
3401 Hillview Ave.
M.S. A2-200
Palo Alto, CA 94303   (415) 855-5564

Annette M. Moore
Syntex (U.S.A.) Inc.
3401 Hillview Ave.
Palo Alto, CA 94304   (415) 855-5564

Tom M. Moran
Syntex (U.S.A.) Inc.
3401 Hillview Ave.
Palo Alto, CA 94304   (415) 855-6137

Raymond R. Moser, Jr.#
Flehr, Hohbach, Test,
Albritton & Herbert
850 Hansen Way
Suite 200
Palo Alto, CA 94304   (415) 494-8700

Richard L. Neeley
Cooley, Godward, Castro,
Huddleson & Tatum
Five Palo Alto Square
4th Floor
Palo Alto, CA 94306   (415) 494-7622

Richard B. Nelson
Varian Associates
Pat. Dept.
611 Hansen Way
Palo Alto, CA 94303   (415) 424-5593

Allen E. Norris#
Sandoz Crop Protection
975 California Avenue
Palo Alto, CA 94304   (415) 354-3592

Vernon A. Norviel
Townsend & Townsend
379 Lytton Avenue
Palo Alto, CA 94301   (415) 326-2400

Sheri M. Novack
Varian Associates, Inc.
3100 Hansen Way
Palo Alto, CA 94304   (415) 424-6604

Linda J. Nyari
Syntex U.S.A. Inc.
Pat. Law Dept. A2-200
3401 Hillview Avenue
Palo Alto, CA 94304   (415) 852-1309

Andrew Pickholtz
Wilson, Sonsini, Goodrich
& Rosati
Two Palo Alto Square
Palo Alto, CA 94306   (415) 493-9300

Shelley G. Precivale
Syntex (U.S.A.) Inc.
3401 Hillview Avenue
P.O. Box 10850
Palo Alto, CA 94303   (415) 493-9300

Michael S. Rabson
Wilson, Sonsini, Goodrich
& Rosati
2 Palo Alto Square
Suite 900
Palo Alto, CA 94306   (415) 493-9300

Edward J. Radlo
Fenwick, Davis & West
2 Palo Alto Square
Palo Alto, CA 94306   (415) 494-0600

Barbara Rae-Venter
Cooley Godward Castro
Huddleson & Tatum
5 Palo Alto Square
4th Floor
Palo Alto, CA 94306   (415) 494-7622

Matthew C. Rainey
419 Margarita Avenue
Palo Alto, CA 94306   (415) 494-2346

Irving Shale Rappaport
1500 Edgewood Drive
Palo Alto, CA 94303   (415) 321-7024

Carol J. Roth
Syntex U.S.A., Inc.
Pat. Law Dept., A2-200
3401 Hillview Ave.
P.O. Box 10850
Palo Alto, CA 94303   (415) 852-1698

Norman P. Rousseau
Syntex Corp.
3401 Hillview Ave.
P.O. Box 10850
Palo Alto, CA 94304   (415) 855-6446

Bertram I. Rowland
Cooley, Godwand, Castro,
Huddleson & Tatum
5 Palo Alto Sq., 4th Flr.
Palo Alto, CA 94306   (415) 494-7622

Robert P. Sabath
Hewlett-Packard Co.
M.S. 20B0
3000 Hanover St.
Palo Alto, CA 94303   (415) 857-3864

Paul L. Sabatine
Alza Corp.
950 Page Mill Rd.
Palo Alto, CA 94303   (415) 494-5224

John F. Schipper
3133 Flowers
Palo Alto, CA 94306   (415) 321-7449

Eberhard G.H. Schmoller
Consolidated Freightways, Inc.
3350 W. Bayshore Road
Palo Alto, CA 94303   (415) 855-9100

David Schnapf
Varian Associates Inc.
611 Hansen Way
Palo Alto, CA 94303   (415) 424-5406

Herbert R. Schulze
Hewlett-Packard Co.
3000 Hanover Street
Stop 20B0
Palo Alto, CA 94304   (415) 857-4377

Peter J. Sgarbossa
Varian Associates, Inc.
611 Hansen Way
Palo Alto, CA 94303   (415) 424-5402

David N. Slone
Townsend & Townsend
379 Lytton Avenue
Palo Alto, CA 94301   (415) 326-2400

Jonathan A. Small
Xerox Corporation
3333 Coyote Hill Road
Palo Alto, CA 94304   (415) 494-4268

John A. Smart
Townsend & Townsend
379 Lytton Ave.
Palo Alto, CA 94301   (415) 326-2400

Albert C. Smith
Fenwick, Davis & West
2 Palo Alto Square
Suite 800
Palo Alto, CA 94306   (415) 494-0600

Robert K. Stoddard#
560 Oxford Avenue
Suite 8
Palo Alto, CA 94306   (415) 856-3344

Steven Faraday Stone
Alza Corp.
950 Page Mill Road
Palo Alto, CA 94304   (415) 494-5283

John Stoner, Jr.
101 Alma Street,
No. 608
Palo Alto, CA 94301   (415) 323-2306

Greg T. Sueoka
Fenwick & West
Two Palo Alto Square
Suite 500
Palo Alto, CA 94306   (415) 494-0600

Gerald F. Swiss
Dehlinger & Swiss
350 Cambridge Ave.
Suite 100
Palo Alto, CA 94306   (415) 323-8302

Herwig Von Morze#
Syntex (U.S.A.) Inc.
3401 Hillview Avenue
Palo Alto, CA 94304   (415) 855-5160

Thomas M. Webster#
Xerox Corporation
3333 Coyote Hill Road
Palo Alto, CA 94304   (415) 494-4266

Paul J. Weiner
469 Grant Avenue, Apt. J
Palo Alto, CA 94306   (415) 326-4312

Eric H. Willgohs
Townsend & Townsend
379 Lytton Ave.
Palo Alto, CA 94301   (415) 326-2400

James M. Williams
Hewlett-Packard Company
3000 Hanover
M.S. 20B0
Palo Alto, CA 94303   (415) 857-5949

James J. Wong
Syntex, Inc.
3401 Hillview Avenue
P.O. Box 10850
Palo Alto, CA 94303   (415) 855-5918

Henry Kissinger Woodward
Townsend & Townsend
379 Lytton Avenue
Palo Alto, CA 94301   (415) 326-2400

Jack H. Wu
Hewlett Packard Company
Legal Department
Mail Stop 20B0
3000 Hanover Street
Palo Alto, CA 94304   (415) 857-2177

John C. Yakes
320 Palo Alto Avenue
Suite A-1
Palo Alto, CA 94301   (415) 323-2300

Lisa M. Yamonaco#
Xerox Corporation
3333 Coyote Hill Road
Palo Alto, CA 94304   (415) 494-4298

C. Michael Zimmerman
Flehr, Hohbach, Test,
Albritton & Herbert
850 Hansen Way, Ste. 200
Palo Alto, CA 94304   (415) 494-8700

Terry J. Anderson
1353 Via Zumaya
Palos Verdes Estates, CA 90274
    (213) 544-3222

Lois C. Babcock#
2516 Via Tejon
Suite 316
Palos Verdes Estates, CA 90274
    (213) 373-6119

Noel F. Heal
2516 Via Tejon
Suite 316
Palos Verdes Estates, CA 90274
    (213) 373-1311

John J. Mc Cormack
2101 Thorley Road
Palos Verdes Estates, CA 90274
    (213) 373-7018

Denton L. Anderson
Sheldon & Mak
201 S. Lake Avenue
Suite 800
Pasadena, CA 91101   (818) 796-4000

Philip J. Anderson
Christie, Parker & Hale
350 W. Colorado Blvd.
Pasadena, CA 91109   (818) 795-5843

Edward O. Ansell
California Institute of Tech.
1201 E. California Blvd.
Pasadena, CA 91125   (818) 356-4567

Andrew J. Belansky
Christie, Parker and Hale
350 W. Colorado Blvd.
P.O. Box 7068
Pasadena, CA 91109   (818) 795-5843

Charles Berman
Sheldon & Mak
201 S. Lake Avenue
Suite 800
Pasadena, CA 91101   (818) 796-4000

Earl C. Briggs#
606 Michigan Blvd.
Pasadena, CA 91107   (213) 796-4939

Norman E. Brunell
300 N. Lake Avenue
Suite 425
Pasadena, CA 91101   (818) 793-3000

Hayden A. Carney
Christie, Parker and Hale
P.O. Box 7068
350 W. Colorado Blvd.
Pasadena, CA 91105   (213) 795-5843

John D. Carpenter
Christie, Parker & Hale
350 W. Colorado Blvd.
Suite 500
Pasadena, CA 91109   (818) 795-5843

Norman L. Chalfin#
Calif. Institute of Tech.
4800 Oak Grove Drive
Pasadena, CA 91109   (213) 354-6833

William P. Christie
Christie, Parker & Hale
350 West Colorado Blvd.
Pasadena, CA 91105   (818) 795-5843

Teresa P. Clark
Sheldon & Mak
201 S. Lake
Suite 800
Pasadena, CA 91101   (213) 681-9000

Edwin Roderick Cline
Christie, Parker and Hale
P.O. Box 7068
350 W. Colorado Blvd.
Pasadena, CA 91109   (213) 681-5637

Thomas J. Daly
Christie, Parker & Hale
P.O. Box 7068
Pasadena, CA 91109   (818) 795-5843

Edward Joseph Darin
301 E. Colorado Blvd.
Suite 518
Pasadena, CA 91101   (213) 793-0689

David A. Dillard
Christie, Parker & Hale
P.O. Box 7068
Pasadena, CA 91109   (213) 795-5843

Stephen Donovan
Sheldon & Mak
201 South Lake Avenue
Suite 800
Pasadena, CA 91101   (818) 796-4000

Elgin C. Edwards
Calif. Inst. of Technology
305-6
1201 E. California Blvd.
Pasadena, CA 91103   (818) 356-4567

Thomas L. Ewing#
California Institute of
Technology
4800 Oak Grove Dr.
MS 156-200
Pasadena, CA 91109   (818) 354-5161

Michael B. Farber
Sheldon & Mak
201 S. Lake Avenue
Suite 800
Pasadena, CA 91101   (818) 796-4000

Felix L. Fischer#
Christie, Parker & Hale
350 West Colorado Blvd.
Suite 500
Pasadena, CA 91109   (818) 795-5843

Robert D. Fish#
330 South Mentor, #134
Pasadena, CA 91106   (213) 489-1600

Vincent Gerard Gioia
Christie, Parker & Hale
350 W. Colorado Blvd.
Suite 500
Pasadena, CA 91105 (818) 795-5843

Frederick Gotha
80 South Lake Avenue
Suite 823
Pasadena, CA 91101 (818) 796-1849

William Jacquet Gribble
1421 Glengarry Road
Pasadena, CA 91105 (213) 254-4142

John Peter Grinnell
Christie, Parker and Hale
P.O. Box 7068
Pasadena, CA 91105 (818) 795-5843

William W. Haefliger
Suite 512
201 S. Lake Ave.
Pasadena, CA 91101 (213) 684-2707

C. Russell Hale
Christie, Parker and Hale
350 W. Colorado Blvd.
P.O. Box 7068
Pasadena, CA 91109 (213) 795-5843

Edwin L. Hartz
Christie, Parker and Hale
350 West Colorado Blvd.
5th Floor
Pasadena, CA 91109 (818) 795-5843

Casey Heeg
Sheldon & Mak
201 S. Lake Ave.
Suite 800
Pasadena, CA 91101 (818) 796-4000

David L. Hoffman
Christie, Parker & Hale
350 W. Colorado Blvd.
Suite 500
Pasadena, CA 91105 (818) 795-5843

Rowland William Johnston
Christie, Parker and Hale
P.O. Box 7068
Pasadena, CA 91109 (213) 795-5843

J. Leslie Jones, Sr.#
P.O. Box 233
Pasadena, CA 91102 (213) 792-7280

Thomas H. Jones*
NASA
Jet Propulsion Lab.
4800 Oak Grove Dr.
Pasadena, CA 91109 (818) 354-5179

Richard L. Klein
California Institute of Tech.
Jet Propulsion Lab.
Off. of Pat. Coun.
4800 Oak Grove Drive
Pasadena, CA 91109 (818) 354-1867

Carl Kustin, Jr.
Christie Parker & Hale
350 Colorado Blvd.
Pasadena, CA 91109 (213) 681-5637

Gordon R. Lindeen, III
Christie, Parker & Hale
P.O. Box 7068
350 W. Colorado Blvd.
Suite 500
Pasadena, CA 91109 (818) 795-5843

Danton K. Mak
Sheldon & Mak
201 South Lake Avenue
Suite 800
Pasadena, CA 91101 (818) 796-4000

Sara L. Mandel
Sheldon & Mak
201 S. Lake Avenue
Suite 800
Pasadena, CA 91101 (213) 312-9900

Charles O. Marshall, Jr.
Sheldon & Mak
201 South Lake Avenue
Suite 800
Pasadena, CA 91101 (818) 796-4000

Walter G. Maxwell
Christie, Parker and Hale
350 W. Colorado Blvd.
Pasadena, CA 91009 (213) 681-5637

John M. May
Christie, Parker & Hale
P.O. Box 7068
Pasadena, CA 91109 (818) 795-5843

Paul F. Mc Caul*
NASA
NASA Resident Off.-J PL
4800 Oak Grove Dr.
Pasadena, CA 91109 (213) 354-2734

Donald D. Mon
101 S. Madison Ave.
Pasadena, CA 91101 (818) 793-9173

George J. Netter
215 N. Marengo Avenue
Pasadena, CA 91101 (818) 578-0703

Russell R. Palmer, Jr.
Christie, Parker & Hale
P.O. Box 7068
350 W. Colorado Blvd.
Suite 500
Pasadena, CA 91109 (818) 795-5843

D. Bruce Prout
Christie, Parker and Hale
350 W. Colorado Blvd.
Suite 500
Pasadena, CA 91105 (818) 795-5843

L. T. Rahn
Christie, Parker and Hale
350 W. Colorado Blvd.
Suite 500
Pasadena, CA 91109 (818) 795-5843

Michael J. Ram
Sheldon & Mak
201 S. Lake Ave.
Suite 800
Pasadena, CA 91101 (818) 796-4000

Richard Dorland Seibel
Christie, Parker and Hale
P.O. Box 7068
350 W. Colorado Blvd.
Pasadena, CA 91109 (818) 795-5843

William Douglas Sellers
Sellers and Brace
430 South Virginia Ave.
Pasadena, CA 91107 (818) 578-0430

Jeffrey G. Sheldon
Sheldon & Mak
201 S. Lake Avenue
Suite 800
Pasadena, CA 91101 (818) 796-4000

Luther Price Speck, II
California Institute of Tech.
Jet Propulsion Lab.
4800 Oak Grove Dr.
Pasadena, CA 91109 (818) 354-6060

Henry P. Stevens#
Calif. Institute of Technology
4800 Oak Grove Drive
Pasadena, CA 91103 (818) 354-3203

Richard Joseph Ward, Jr.
Christie, Parker & Hale
350 W. Colorado Blvd.
P.O. Box 7068
Pasadena, CA 91109 (818) 795-5843

Clarence W. Martin
18499 Fair Oaks Drive
Penn Valley, CA 95946 (916) 432-4442

Joseph E. Machamer#
526 Acadia Dr.
Petaluma, CA 94954 (707) 763-4708

William H.F. Howard
928 Kingston Avenue
Suite 703
Piedmont, CA 94611 (415) 654-8636

George K. De Brucky
101 S. Kraemer Blvd.
Suite 202
Placentia, CA 92670 (714) 993-3387

William D. Mc Cann
Hallgrimson, Mc Nichols,
Mc Cann & Inderbitzen
5000 Hopyard Rd., Ste. 400
Pleasanton, CA 94588 (415) 460-3700

Ron Billi*
Dept. of the Navy
Off. of Pat. Coun.
Code 0061, Bldg. 36, Rm. 1082
Pacific Missle Test Center
Point Mugu, CA 93042 (805) 982-1601

David S. Kalmbaugh*
U.S. Dept. of Navy
Code 0061
Office of Counsel
Point Mugu, CA 93042   (805) 989-8266

Leo R. Carroll
General Dynamics Corporation
Pomona/Valley Sys. Divs.
P.O. Box 2507
MZ 1-25
Pomona, CA 91769   (714) 868-1033

Edward Ronald Grant
1623 Juniper Ridge
Pomona, CA 91766   (714) 622-7008

Michael A. Glenn
115 Lake Road
Portola Valley, CA 94028   (415) 851-4709

Robert Oswald Webster
1255 Westridge Dr.
Portola Valley, CA 94028   (415) 851-0454

Gregory O. Garmong
13126 Silver Saddle Lane
Poway, CA 92064   (619) 451-0660

Emmette Rudolph Holman#
28401 Ridgethorne Ct.
Rancho Palos Verdes, CA 90274
(213) 541-2141

Arnold W. Lieman
26622 Fond Du Lac Rd.
Rancho Palos Verdes, CA 90274
(213) 378-5086

Irving B. Osofsky
28327 San Nicolas Dr.
Rancho Palos Verdes, CA 90274
(213) 377-5829

John A. Sarjeant
26832 Indian Peak Road
Rancho Palos Verdes, CA 90274
(213) 373-7846

Lee W. Tower
19 Saddle Road
Rancho Palos Verdes, CA 90274
(213) 548-3709

Alfons Valukonis#
6760 Los Verdes Dr.
Rancho Palos Verdes, CA 90274
(213) 541-3549

Leo Francis Costello
P.O. Box 7000-343
Redondo Beach, CA 90277   (213) 377-3146

James E. Crawford#
357 Avenue E
Redondo Beach, CA 90277   (213) 540-2930

Benjamin De Witt
TRW Inc.
One Space Park
Bldg. E2/Room 7073
Redondo Beach, CA 90278   (213) 535-1243

John Holtrichter, Jr.
P.O. Box 227
Redondo Beach, CA 90277   (213) 544-3033

Robert W. Keller
TRW, Inc.
Bldg. E2-7073
One Space Park
Redondo Beach, CA 90278   (213) 812-1520

Monty Koslover
145 Via Monte Doro
Redondo Beach, CA 90277   (213) 378-7498

William B. Leach
TRW Inc.
Space & Defense Sector E2/7062
One Space Park
Redondo Beach, CA 90278   (213) 812-1509

Ronald L. Taylor
TRW Inc.
One Space Park
M.S. E2-6051b
Redondo Beach, CA 90278   (213) 812-1521

Harold H. Wilson#
Patent Prosecution Service
2105 Rockefeller Lane
Apt. 7
Redondo Beach, CA 90278   (213) 372-6273

George B. Almeida#
Ampex Corporation
401 Broadway
Redwood City, CA 94063   (415) 367-3331

Jeffrey S. Gananian
Ropers, Majeski, Kohn, Bentley,
Wagner & Kane
1001 Marshall St.
Redwood City, CA 94063   (415) 364-8200

Douglas M. Gilbert
AMPEX Corp.
401 Broadway M.S. 3-35
Redwood City, CA 94063   (415) 367-3336

Richard P. Lange
Ampex Corporation
401 Broadway
MS 3-35
Redwood City, CA 94063   (415) 397-3338

Ralph Leonard Mossino
Ampex Corp.
401 Broadway
M.S. 3-35
Redwood City, CA 94063   (415) 367-3333

Richard J. Roddy
Ampex Corporation
401 Broadway
M.S. 3-35
Redwood City, CA 94063   (415) 367-3338

Joel D. Talcott
Ampex Corporation
401 Broadway
MS 3-36
Redwood City, CA 94063   (415) 367-3330

Richard Paul Alberi
191 W. Linden Street
Reedley, CA 93654   (209) 638-6059

Ellis A. Pangborn#
1481 Santa Rosa Circle
Reedley, CA 93654   (209) 637-2292

Donald Richard Nyhagen
7540 Corbin Avenue
Apt. 4
Reseda, CA 91335   (818) 886-4910

Candace P. Olsen
19119 Sherman Way, #205
Reseda, CA 91335   (818) 705-4264

Joel G. Ackerman
Stauffer Chemical Company
1200 South 47th Street
Richmond, CA 94804   (415) 231-1194

Edwin Hale Baker
Stauffer Chemical Co.
1200 S. 47th St.
Richmond, CA 94804   (415) 231-1193

Michael Joseph Bradley
Stauffer Chemical Co.
1200 S. 47th St.
Richmond, CA 94804   (415) 231-1017

Lynn Marcus-Wyner
ICI Americas, Inc.
1200 S. 47th St.
Box 4023
Richmond, CA 94804   (415) 231-1200

Witta O. Priester#
Chevron Res. & Tech. Co.
100 Chevron Way
Richmond, CA 94802   (415) 620-3019

Gerald Franklin Baker
Baker & Houston
114 S. China Lake Blvd.
Suite A
Ridgecrest, CA 93555   (714) 375-1618

Harvey A. Gilbert
625 La Paloma
Ridgecrest, CA 93555   (619) 371-2567

Kenneth G. Pritchard
P.O. Box 306
Ridgecrest, CA 92556   (619) 375-4020

William Thomas Skeer*
515 W. Weiman Avenue
Ridgecrest, CA 93555   (619) 446-4359

Frank R. Lafontaine
2213 Raintree Lane
Riverbank, CA 95367   (209) 869-4413

William G. Becker
Bourns, Inc.
1200 Columbia Ave.
Riverside, CA 92507   (714) 781-5138

John Hance Crowe#
4296 Orange St.
Riverside, CA 92501   (714) 684-5833

Herbert E. Kidder
4376 Maplewood Place
Riverside, CA 92506   (714) 683-7854

Fritz B. Peterson#
3290 Monroe St.
Riverside, CA 92504   (714) 688-5621

Willard M. Graham#
55 Cypress Way
Rolling Hills Estates, CA 90274
    (213) 541-3045

Walter J. Jason
49 Cypress Way
Rolling Hills Estates, CA 90274
    (213) 541-7525

John S. Bell
Aerojet-General Corp.
P.O. Box 13618
Sacramento, CA 95853   (916) 355-3788

Robert D. Eastham
724 35th St.
Sacramento, CA 95816   (916) 355-3788

Carey B. Huscroft
801 12th Street
Suite 600
Sacramento, CA 95814   (916) 444-9600

Mark C. Jacobs
3033 El Camino Avenue
Sacramento, CA 95821   (916) 485-5000

Bernhard Kreten
77 Cadillac Drive
Suite 245
Sacramento, CA 95825   (916) 921-6181

John P. O Banion
9288 Linda Rio Drive
Sacramento, CA 95826   (916) 363-7272

Robert G. West
Lothrop and West
555 Capitol Mall
Suite 1525
Sacramento, CA 95814   (916) 444-5412

Robert M. West
Lothrop & West
555 Capitol Mall
Suite 1525
Sacramento, CA 95814   (916) 444-5412

Gary L. Jordan
2636 El Camino Real North
Salinas, CA 93907   (408) 647-4148

Michael A. Kaufman
107 Suffield Avenue
San Anselmo, CA 94960   (415) 457-2797

C. Douglas De Freytas
Gillette, Loof, Langton
& Hagner
1848 Commercenter East
P.O. Box 1390
San Bernardino, CA 92402   (714) 884-1247

Stephen R. Seccombe
Sheldon & Mak
290 North D Street
Suite 700
San Bernardino, CA 92401   (714) 889-3649

Carole F. Barrett
890 Cabot Court
San Carlos, CA 94070   (415) 593-5312

Albert L. Gabriel
209 Avenida Del Mar
Suite 202
San Clemente, CA 92672   (714) 498-9041

Cathryn A. Campbell
Pretty, Schroeder,
Brueggemann & Clark
4370 La Jolla Village Dr.
Suite 700
San Diegeo, CA 92122   (619) 535-9001

David L. Baker
P.O. Box 927419
San Diego, CA 92192   (619) 546-9312

Freling E. Baker
Baker, Maxham, Jester
& Meador
Symphony Towers
750 B Street, Suite 2770
San Diego, CA 92101   (619) 233-9004

Laurance E. Banghart
3864 Mt. Ainsworth Avenue
San Diego, CA 92111   (619) 277-4125

Stanley Alan Becker
Dressler, Goldsmith, Shore,
Sutker & Milnamow, Ltd.
11300 Sorrento Valley Road
Suite 200
San Diego, CA 92121   (619) 546-1555

Ralph S. Branscomb
Charmasson, Branscomb & Holz
1200 3rd Avenue
Suite 1200
San Diego, CA 92101   (714) 236-9500

Carl R. Brown
Brown, Martin, Haller
& Meador
110 West C St.
13th Floor
San Diego, CA 92101   (714) 238-0999

Rodney F. Brown
Nydegger & Associates
4350 La Jolla Village Dr.
Suite 950
San Diego, CA 92122   (619) 455-5700

Theresa A. Brown#
Pretty, Schroeder,
Brueggemann & Clark
4370 La Jolla Village Dr.
Suite 700
San Diego, CA 92122   (619) 535-9001

Edward William Callan
3033 Science Park Road
San Diego, CA 92121   (619) 457-2340

Donald W. Canady
Dressler, Goldsmith, Shore,
Suter & Milnamow, Ltd.
11300 Sorrento Valley Road
Suite 200
San Diego, CA 92121   (619) 546-1555

John M. Carson
Knobbe, Martens, Olson & Bear
501 W. Broadway
Suite 1700
San Diego, CA 92101   (619) 235-8550

Henri J. A. Charmasson
1545 Hotel Circle South
Suite 150
San Diego, CA 92108   (619) 294-2922

Morris Cohen
444 W. C Street
Suite 220
San Diego, CA 92101   (619) 233-3650

K. David Crockett#
1572 Thomas Ave.
San Diego, CA 92109   (619) 581-0738

Ginger R. Dreger#
Knobbe, Martens, Olson
& Bear
501 West Broadway
Suite 1700
San Diego, CA 92101   (619) 235-8550

Walter W. Duft
Baker, Maxham, Jester
& Meador
Symphony Towers
750 B Street, Suite 2770
San Diego, CA 92101   (619) 233-9004

John Robert Duncan, Jr.
General Dynamics Corp.
Convair Div.-MZ 103-10
P.O. Box 85357
San Diego, CA 92138   (714) 547-3542

Charles J. Fassbender
Unisys Corporation
10850 Via Frontera
San Diego, CA 92127   (714) 451-4507

Harvey Fendelman*
Dept. of the Navy
Naval Ocean System Center
Code 0012
San Diego, CA 92152   (619) 553-3001

William C. Fuess
Dressler, Goldsmith, Shore,
Sutker & Milnamow, Ltd.
11300 Sorrento Valley Road
Suite 200
San Diego, CA 92121   (619) 546-1555

Frank Donald Gilliam
4565 Ruffner Street
Suite 104
San Diego, CA 92111   (619) 292-0901

Stephen A. Gratton
Nydegger & Associates
4350 La Jolla Village Dr.
Suite 950
San Diego, CA 92122   (619) 455-5700

Jeffrey W. Guise#
13255 Denara Road
San Diego, CA 92130   (619) 792-0323

John L. Haller
Brown, Martin, Haller
& Meador
110 West C Street
13th Floor
San Diego, CA 92101   (619) 238-0999

Drew S. Hamilton
Knobbe, Martens, Olson
& Bear
501 W. Broadway
Suite 1700
San Diego, CA 92101   (619) 235-8550

Robert D. Harder*
U.S. Dept. of Navy
Naval Ocean Systems Center
Legal Counsel for Patents
Code 0012
San Diego, CA 92152   (619) 553-6364

Andrew B. Hellewell#
Benson Resource Mgt., Inc.
2425 San Diego Avenue
Suite 107
San Diego, CA 92110   (619) 260-1494

Dinah N. Hill
Spensley, Horn, Jubas
& Lubitz
101 W. Broadway, Suite 1980
San Diego, CA 92101   (619) 696-0567

Stacy L. Howells
Brown, Martin, Haller
& Mc Clain
110 West C Street
13th Floor
San Diego, CA 92101   (619) 238-0999

Ned A. Israelsen
Knobbe, Martens, Olson
& Bear
501 W. Broadway
Suite 1700
San Diego, CA 92101   (619) 235-8550

Michael H. Jester
Baker, Maxham, Jester &
Meador
Symphony Towers
750 B St., Suite 2770
San Diego, CA 92101   (619) 233-9004

Michael A. Kagan*
Naval Ocean Systems Center
Off. of Legal Coun.
For Patents
San Diego, CA 92152   (619) 553-3001

Thomas Glenn Keough*
Dept. of Navy
Naval Ocean Sys. Ctr.
San Diego, CA 92152   (714) 553-3001

Anita M. Kirkpatrick
Knobbe, Martens, Olson
& Bear
501 W. Broadway
Suite 1700
San Diego, CA 92101   (619) 235-8550

Bernard L. Kleinke
Laff, Whitesel, Conte
& Saret
101 W. Broadway
Suite 1580
San Diego, CA 92101   (619) 232-6060

Henry G. Kohlmann
Cipher Data Products, Inc.
9715 Business Park Avenue
San Diego, CA 92131   (619) 693-7252

Alfred Waldemar Kozak
Unisys Corporation
Law Dept. MS 1000
10850 Via Frontera
San Diego, CA 92127   (619) 451-4531

Peter L. Lagus
S-Cubed
3398 Carmel Mtn. Rd.
San Diego, CA 92121   (714) 453-0060

John C. Lambertsen
Knobbe, Martens, Olson,
& Bear
501 W. Broadway
Suite 1700
San Diego, CA 92101   (619) 235-8550

John F. Land
Spensley Horn Jubas
& Lubitz
101 West Broadway
San Diego, CA 92101   (619) 696-0567

Kamwah Li
Dain & Li
555 W. Beech St.
Suite 222
San Diego, CA 92101   (619) 234-4576

John R. Lindsay
Brown, Martin, Haller
& Mc Clain
110 West C St.
13th Floor
San Diego, CA 92101   (714) 238-0999

Peter Adams Lipovsky*
U.S. Navy
Legal Counsel for Pats.
Code 0012
Naval Ocean Systems Center
San Diego, CA 92152   (619) 553-3001

April C. Logan
Knobbe, Martens, Olson & Bear
501 W. Broadway
Suite 1700
San Diego, CA 92101   (619) 235-8550

Neil F. Martin
Brown, Martin, Haller
& Mc Clain
110 West C Street
13th Floor
San Diego, CA 92101   (619) 238-0999

Robert J. Mawhinney
5335 Westknoll Drive
San Diego, CA 92109   (714) 270-2705

Lawrence A. Maxham
Baker, Maxham, Jester
& Meador
Symphony Towers
750 B Street, Suite 2770
San Diego, CA 92101   (714) 233-9004

James Mc Cafferty
Van Camp Seafood Co., Inc.
11555 Sorrento Valley Road
San Diego, CA 92121   (619) 481-7734

James W. Mc Clain
Brown, Martin, Haller
& Mc Clain
110 West C Street
Suite 1300
San Diego, CA 92101   (619) 238-0999

Joseph F. Mc Lellan
1322 Scott Street
Suite 202
San Diego, CA 92106   (619) 222-6800

Terrance A. Meador*
Baker, Maxham, Jester
& Meador
Symphony Towers
750 B Street, Suite 2770
San Diego, CA 92101   (619) 233-9044

Russell Ben Miller
Qualcomm, Inc.
10555 Sorrento Valley Road
San Diego, CA 92121   (619) 587-1121

Hugo F. Mohrlock
2867 Grandview St.
San Diego, CA 92110   (714) 276-4456

Linda Rae Neyenesch
1200 3rd Avenue
Suite 1200
San Diego, CA 92101   (619) 236-9500

Neil K. Nydegger*
Nydegger & Harshman
4350 La Jolla Village Drive
Suite 950
San Diego, CA 92122   (619) 455-5700

G.T. Parsons#
General Dynamics Corp.
Convair Div.
MZ 12-1031
P.O.Box 85357
San Diego, CA 92186   (619) 547-3402

George Edward Pearson
973 Manor Way
San Diego, CA 92106   (619) 222-7025

David G. Perryman
Pretty, Schroeder, Brueggemann
& Clark
4370 La Jolla Village Drive
Suite 700
San Diego, CA 92122   (619) 535-9001

Thomas M. Phillips
5227 Middleton Rd.
San Diego, CA 92109   (619) 274-6686

Jerry R. Potts
Laff, Whitesel, Conte & Saret
101 W. Broadway
Suite 1580
San Diego, CA 92101   (619) 232-6060

Katherine Proctor#
Brown, Martin, Haller
& Mc Clain
110 West C Street
13th Floor
San Diego, CA 92101   (619) 238-0999

Stephen E. Reiter
Pretty, Schroeder,
Brueggeman & Clark
4370 La Jolla Village Dr.
Suite 700
San Diego, CA 92122   (619) 535-9001

William L. Respess
Ligand Pharmaceuticals Inc.
9393 Towne Centre Drive
Suite 100
San Diego, CA 92121   (619) 535-3900

J. Ronald Richbourg
Unisys Corp.
10850 Via Frontera
MS 1000
San Diego, CA 92127   (619) 451-4506

Daniel Robbins
550 West C Street
Suite 1100
San Diego, CA 92101   (619) 233-6868

John L. Rogitz#
Nydegger & Harshman
9350 La Jolla Village Dr.
Suite 950
San Diego, CA 92122   (619) 955-5700

George J. Rubens
2117 Blackmore Ct.
San Diego, CA 92109   (619) 272-2260

Joseph Carl Schwalbach
110 West C Street
Suite 1212
San Diego, CA 92101   (619) 234-1002

Peter P. Scott#
119 W. Walnut Avenue
Apt. 6
San Diego, CA 92103   (619) 298-4875

Jerry Roland Seiler
Knobbe, Martens, Olson & Bear
501 W. Broadway
Suite 1700
San Diego, CA 92101   (619) 235-8550

Tom Sherrard
2285 Comstock Road
San Diego, CA 92111   (619) 541-7852

Paul C. Steinhardt
Hybritech Incorporated
11095 Torreyana Road
San Diego, CA 92121   (619) 535-8407

Dennis G. Stenstrom
Unisys Corporation
10850 Via Frontera
MS 1000
San Diego, CA 92127   (619) 535-8407

Curtis Wayne Stephens
15603 Caldas De Reyes
San Diego, CA 92128   (619) 592-9152

Vincent E. Sullivan
110 West C Street
Suite 2202
San Diego, CA 92101   (619) 235-0550

John P. Sumner
Harness, Dickey & Pierce
450 B Street, Suite 890
San Diego, CA 92101   (619) 238-4122

James F. Sweeney
Office of the District Attorney
P.O. Box X-1011
San Diego, CA 92112   (619) 238-4122

Stephen P. Swinton
Brobeck, Phleger & Harrison
550 W. C Street
Suite 1300
San Diego, CA 92101   (619) 234-1966

Howard C. Tarr
5995 Eldergardens Street
San Diego, CA 92120   (619) 265-7877

Calif Kip Tervo
6387 Caminito Lazaro
San Diego, CA 92111   (619) 234-4034

Thomas J. Tighe
5080 Shoreham Place
Suite 202
San Diego, CA 92122   (619) 450-1881

Bruno J. Verbeck
110 West C Street
Suite 1212
San Diego, CA 92101   (714) 234-1130

James A. Ward
Baker, Maxham, Jester & Meador
Symphony Towers
750 B Street, Suite 2770
San Diego, CA 92101   (619) 233-9004

William P. Waters
Laff, Whitesel, Conte
& Saret
101 West Broadway
Suite 1580
San Diego, CA 92101   (619) 232-6060

Adam Cochran
Vestar, Inc.
650 Cliffside Drive
San Dimas, CA 91773   (619) 232-6060

Raul V. Aguilar
444 Markert St.
Suite 2750
San Francisco, CA 94111   (415) 392-5100

Gary T. Aka
Townsend & Townsend
One Market Plaza
Stewart Street Tower
San Francisco, CA 94105   (415) 493-2590

Elmer S. Albritton
Flehr, Hohbach, Test,
Albritton & Herbert
3400 Four Embarcadero Center
San Francisco, CA 94111   (415) 781-1989

James W. Ambrosius
Chevron Corporation
Law Dept.
P.O. Box 7141
San Francisco, CA 94120   (415) 894-4712

Ernest M. Anderson
Eckhoff, Hoppe, Slick,
Mitchell & Anderson
Four Embarcadero Center
Suite 760
San Francisco, CA 94111   (415) 391-7160

William L. Anthony, Jr.
Townsend & Townsend
Steuart Street Tower
One Market Plaza
San Francisco, CA 94105   (415) 543-9600

Elliot B. Aronson
Townsend & Townsend
One Market Plaza
2000 Steuart Tower
San Francisco, CA 94105   (415) 543-9600

Richard Elliott Backus
Flehr, Hohbach, Test,
Albritton & Herbert
Four Embarcadero Center
Suite 3400
San Francisco, CA 94111   (415) 781-1989

Arthur B. Bakalar
2730 Lyon St.
San Francisco, CA 94123   (415) 346-7119

Stephen E. Baldwin
Heller, Ehrman, White
& Mc Auliffe
333 Bush St.
San Francisco, CA 94104   (415) 324-7192

Donald L. Bartels
Mc Cubbrey, Bartels, Meyer
& Ward
One Post Street
Suite 2700
San Francisco, CA 94104   (415) 391-6665

Thomas Richard Baruch
3954 Clay Street
San Francisco, CA 94118   (415) 386-4668

Erwin J. Basinski
1456 Jones St.
Suite 37
San Francisco, CA 94109 (415) 441-5188

Kevin L. Bastian
Townsend & Townsend
One Market Plaza
20th Floor
Steuart St. Tower
San Francisco, CA 94105 (415) 543-9600

W. Patrick Bengtsson
Limbach, Limbach & Sutton
2001 Ferry Building
San Francisco, CA 94111 (415) 433-4150

Robert J. Bennett
Townsend & Townsend
Steuart St. Tower
One Market Plaza
San Francisco, CA 94105 (415) 543-9600

Michael N. Berg
Pillsbury, Madison & Sutro
225 Bush Street
P.O. Box 7880
San Francisco, CA 94120 (415) 983-1895

Lowell C. Bergstedt
601 Van Ness Avenue
Box E3519
San Francisco, CA 94102 (415) 928-4274

Janis John Otto Biksa
Owen, Wickersham & Erickson
455 Market Street
19th Floor
San Francisco, CA 94105 (415) 882-3200

Scott R. Bortner#
Fliesler, Dubb, Meyer
& Lovejoy
Suite 400
4 Embarcadero Center
San Francisco, CA 94111 (415) 362-3800

Michael K. Bosworth
Chevron Corporation
P.O. Box 7141
San Francisco, CA 94120 (415) 894-9575

David Jay Brezner
Flehr, Hohbach, Test,
Albritton & Herbert
Suite 3400
Four Embarcadero Center
San Francisco, CA 94111 (415) 781-1989

Theodore G. Brown
Townsend & Townsend
One Market Plaza
Steuart Street Tower, #2000
San Francisco, CA 94105 (415) 543-9600

J. A. Buchanan, Jr.
100 Pine Street
Suite 250
San Francisco, CA 94111 (415) 788-7799

Henry C. Bunsow
Townsend & Townsend
One Market Plaza
Steuart St. Tower
20th Floor
San Francisco, CA 94105 (415) 543-9600

William H. Callaway
Austgen Biojet
500 Sansome Street
Suite 500
San Francisco, CA 94111 (415) 989-8333

Richard K. Cannon
Cartwright, Slobodin, Bokelman,
Borowsky, Wartnick, Moore & Harris
101 California Street
Suite 2600
San Francisco, CA 94111 (415) 433-0440

Charles M. Carman, Jr.#
1648 Great Highway
San Francisco, CA 94122 (415) 665-8399

Claude J. Caroli
Chevron Corporation
Law Dept. Pat. Tdmk. & Contract Div.
555 Market Street
San Francisco, CA 94105 (415) 894-5863

Peter G. Carroll
Limbach, Limbach & Sutton
2001 Ferry Bldg.
San Francisco, CA 94111 (415) 433-4150

Matt W. Carson
Chevron Corp.
P.O. Box 7141
San Francisco, CA 94120 (415) 894-5765

Anthony J. Castro
2035 9th Avenue
San Francisco, CA 94116 (415) 753-8672

Vincent J. Cavalieri, Jr.
Chevron Corporation
Law Dept., Pat. Div.
P.O. Box 7141
San Francisco, CA 94120 (415) 894-5435

Guy W. Chambers
Townsend & Townsend
Steuart Street Tower, 20th Floor
One Market Plaza
San Francisco, CA 94105 (415) 543-9600

Robert Boyd Chickering
Flehr, Hohbach, Test,
Albritton & Herbert
Four Embarcadero Center
Suite 3400
San Francisco, CA 94111 (415) 781-1989

Robert C. Colwell
Townsend & Townsend
Steuart Street Tower
One Market Plaza
San Francisco, CA 94105 (415) 543-9600

Roger L. Cook
Townsend & Townsend
One Market Plaza
Steuart Street Tower
San Francisco, CA 94105 (415) 543-9600

Raymond Cranfill#
Limbach, Limbach & Sutton
2001 Ferry Bldg.
San Francisco, CA 94110 (415) 433-4150

Luann Cserr
Flehr, Hobbach, Test,
Albritton & Herbert
Four Embarcadero Center
Suite 3400
San Francisco, CA 94111 (415) 781-1989

Mark A. Dalla Valle#
Limbach, Limbach & Sutton
2001 Ferry Bldg.
San Francisco, CA 94111 (415) 433-4150

Philip A. Dalton, Jr.
Flehr, Hohbach, Test,
Albritton & Herbert
Suite 3400
Four Embarcadero Center
San Francisco, CA 94111 (415) 781-1989

Herbert Davis
Chevron Corporation
555 Market Street
San Francisco, CA 94105 (415) 894-2021

Mark L. Davis
Chevron Corp.
555 Market St.
Room 4031
San Francisco, CA 94105 (415) 894-3867

Joel J. De Young
Chevron Res. Co.
P.O. Box 7141
San Francisco, CA 94120 (415) 894-5574

Thomas George Dejonghe
Chevron Research Co.
Law Dept.
555 Market Street
San Francisco, CA 94105 (415) 894-3546

James A. Deland#
Townsend & Townsend
One Market Plaza
Steuart Street Tower
20th Floor
San Francisco, CA 94105 (415) 543-9600

Samuel D. Delich
Graham & James
Suite 300
One Maritime Plaza
San Francisco, CA 94111 (415) 954-0220

Michael E. Dergosits
Limbach, Limbach & Sutton
2001 Ferry Bldg.
San Francisco, CA 94111 (415) 433-4150

Veronica C. Devitt
Limbach, Limbach & Sutton
2001 Ferry Bldg.
San Francisco, CA 94111 (415) 433-4150

Mark K. Dickson
Townsend & Townsend
Steuart Street Tower
One Market Plaza
San Francisco, CA 94105 (415) 543-9600

Anthony Bernard Diepenbrock
Townsend and Townsend
One Market Plaza
Steuart St. Tower
San Francisco, CA 94105   (415) 543-9600

Gerald P. Dodson
Townsend & Townsend
Steuart Street Tower
20th Floor
One Market Plaza
San Francisco, CA 94105   (415) 543-9600

Hana H. Dolezalova
Phillips, Moore, Lempio
& Finley
177 Post St., Ste. 800
San Francisco, CA 94108   (415) 421-2674

Walter H. Dreger
Flehr, Hohbach, Test,
Albritton, & Herbert
Four Embarcadearo Center
Suite 3400
San Francisco, CA 94111   (415) 781-1989

Hubert E. Dubb
Fliesler, Dubb, Meyer
& Lovejoy
Four Embarcadero Center
Suite 400
San Francisco, CA 94111   (415) 362-3800

Stephen C. Durant
Flehr, Hohbach, Test,
Albritton & Herbert
4 Embarcadero Center
Suite 3400
San Francisco, CA 94111   . (415) 781-1989

Robert S. Dweck#
Four Embarcadero Center
Suite 3390
San Francisco, CA 94111   (415) 986-8833

William J. Egan, III
Heller, Ehrman, White
& Mc Auliffe
333 Bush St.
San Francisco, CA 94104   (415) 772-6000

Alfred A. Equitz
Limbach, Limbach & Sutton
2001 Ferry Bldg.
San Francisco, CA 94111   (415) 433-4150

Roger W. Erickson
Owen, Wickersham & Erickson, P.C.
455 Market St., 19th Fl.
San Francisco, CA 94105   (415) 882-3200

Noemi C. Espinosa
Townsend & Townsend
One Market Plaza
Steuart Street Tower-20th Floor
San Francisco, CA 94105   (415) 543-9600

Stephen M. Everett
Limbach, Limbach & Sutton
2001 Ferry Bldg.
San Francisco, CA 94111   (415) 433-4150

Hugh D. Finley
Phillips, Moore, Lempio
& Finley
Suite 800
177 Post Street
San Francisco, CA 94108   (415) 421-2674

Paul D. Flehr
Flehr, Hohbach, Test,
Albritton & Herbert
Suite 3400
Four Embarcadero Center
San Francisco, CA 94111   (415) 781-1989

Martin C. Fliesler
Fliesler, Dubb, Meyer
& Lovejoy
Four Embarcadero Center
Suite 400
San Francisco, CA 94111   (415) 362-3800

Dirks B. Foster
Townsend & Townsend
Steuart St. Tower, 20th Floor
One Market Plaza
San Francisco, CA 94105   (415) 543-9600

Thomas M. Freiburger
650 California Street
29th Floor
San Francisco, CA 94108   (415) 781-0310

Richard C. Gaffney, Sr.
Chevron Corporation
555 Market Street
San Francisco, CA 94105   (415) 781-0310

Thomas Allen Gallagher
100 Green Street
Third Floor
San Francisco, CA 94111   (415) 989-8080

Gideon Gimlan
Fliesler, Dubb, Meyer & Lovejoy
Four Embarcadero Center
Suite 400
San Francisco, CA 94111   (415) 362-3800

Philip A. Girard
Limbach, Limbach & Sutton
2001 Ferry Bldg.
San Francisco, CA 94111   (415) 433-4150

Neil D. Greenstein
Pillsbury, Madison & Sutro
235 Montgomery Street
P.O. Box 7880
San Francisco, CA 94120   (415) 983-6430

Albert Price Halluin
Fliesler, Dubb, Meyer & Lovejoy
Four Embarcadero Center
Suite 400
San Francisco, CA 94111   (415) 362-3800

James F. Hann
Townsend & Townsend
2000 Steuart Street Tower
One Market Plaza
San Francisco, CA 94105   (415) 543-9600

Ian Hardcastle
100 Green Street
Third Floor
San Francisco, CA 94111   (415) 989-4327

Thomas B. Haverstock
Limbach, Limbach & Sutton
2001 Ferry Bldg.
San Francisco, CA 94111   (415) 433-4150

M. Henry Heines
Townsend & Townsend
Steuart St. Tower, 20th Flr.
One Market Plaza
San Francisco, CA 94105   (415) 543-9600

Alvin E. Hendricson
500 Sutter Street
Suite 604
San Francisco, CA 94102   (415) 981-4463

Thomas Oliver Herbert
Flehr, Hohbach, Test,
Albritton & Herbert
4 Embarcadero Center
Suite 3400
San Francisco, CA 94111   (415) 781-1989

Walter D. Herrick
512 Wisconsin Street
San Francisco, CA 94107   (415) 282-3393

Robert Charles Hill
235 Montgomery Street
Suite 1741
San Francisco, CA 94104   (415) 421-2080

Albert J. Hillman
Townsend and Townsend
Steuart Street Tower
One Market Plaza
San Francisco, CA 94105   (415) 543-9600

Harold C. Hohbach
Flehr, Hohbach, Test,
Albritton & Herbert.
Four Embarcadero Center
Suite 3400
San Francisco, CA 94111   (415) 781-1989

James S. Hsue
Majestic, Parsons, Siebert
& Hsue
Four Embarcadero Center
Suite 1450
San Francisco, CA 94111   (415) 362-5556

William Michael Hynes
Townsend and Townsend
Steuart Street Tower
One Market Plaza
San Francisco, CA 94105   (415) 543-9600

Donald B. Jarvis
530 Dewey Blvd.
San Francisco, CA 94116   (415) 557-3601

Charles H. Jew
Majestic, Parsons, Siebert & Hsue
4 Embarcadero Center
Suite 1450
San Francisco, CA 94111   (415) 362-5556

Bruce H. Johnsonbaugh
Eckhoff, Hoppe, Slick,
Mitchell & Anderson
Four Embarcadero Center
Suite 760
San Francisco, CA 94111   (415) 391-7160

Frank E. Johnston
Limbach, Limbach & Sutton
2001 Ferry Bldg.
San Francisco, CA 94111   (415) 433-4150

Kenneth M. Kaslow
Limbach, Limbach & Sutton
2001 Ferry Bldg.
San Francisco, CA 94111   (415) 433-4150

Kevin R. Kaster
1334 Chestnut St.
San Francisco, CA 94123   (415) 673-6191

Edward J. Keeling
Chevron Corporation
Law Dept.
555 Market Street
Room 425
San Francisco, CA 94105   (415) 894-2420

Steven P. Koda
Townsend & Townsend
Steuart Street Tower
One Market Plaza
San Francisco, CA 94105   (415) 543-9600

Victoria S. Kolakowski
Mc Cubbrey, Bartels, Meyer
& Ward
One Post Street
Suite 2700
San Francisco, CA 94104   (415) 391-6665

Warren J. Krauss
Sedgwick, Detert,
Moran & Arnold
One Embarcadero Center
16th Floor
San Francisco, CA 94111   (415) 781-7900

Charles E. Krueger
Townsend & Townsend
20th Floor Steuart Street Tower
One Market Plaza
San Francisco, CA 94105   (415) 543-9600

Warren P. Kujawa
Townsend & Townsend
One Market Plaza
Steuart St. Tower, 20th Floor
San Francisco, CA 94105   (415) 543-9600

Laura L. Kulhanjian
Flehr, Hohbach, Test,
Albritton & Herbert
Four Embarcadero Center
Suite 3400
San Francisco, CA 94111   (415) 781-1989

S. Russell La Paglia
Chevron Research Co.
P.O. Box 7141
San Francisco, CA 94120   (415) 894-2220

David N. Lathrop
100 Green Street
San Francisco, CA 94111   (415) 989-8080

Leona L. Lauder
One Market Plaza
Steuart Street Tower
18th Floor
San Francisco, CA 94105   (415) 777-9257

Paul S. Lempio
Phillips, Moore, Lempio
& Finley
177 Post Street
Suite 800
San Francisco, CA 94108   (415) 421-2674

George C. Limbach
Limbach, Limbach and Sutton
2001 Ferry Bldg.
San Francisco, CA 94111   (415) 433-4150

Karl A. Limbach
Limbach, Limbach and Sutton
2001 Ferry Bldg.
San Francisco, CA 94111   (415) 433-4150

David Eugene Lovejoy
Fliesler, Dubb, Meyer
& Lovejoy
4 Embarcadero Center
Suite 400
San Francisco, CA 94111   (415) 362-3800

Donald N. Mac Intosh
Flehr, Hohbach, Test,
Albritton & Herbert
Four Embarcadero Center
Suite 3400
San Francisco, CA 94111   (415) 781-1989

Harry J. Macey
Townsend & Townsend
One Market Plaza
Steuart Street Tower
20th Fl.
San Francisco, CA 94105   (415) 543-9600

Martin F. Majestic
Majestic, Parsons, Siebert
& Hsue
Four Embarcadero Center
Suite 1450
San Francisco, CA 94111   (415) 362-5556

Miriam E. Majofis
Flehr, Hohbach, Test, Albritton
& Herbert
4 Embarcadero Center
Suite 3400
San Francisco, CA 94111   (415) 781-1989

C. Woodworth Marsh
3530 Wawona Street
San Francisco, CA 94116   (415) 564-5855

Duane H. Mathiowetz
Leboeuf, Lamb, Leiby
& Mac Rae
One Embarcadero Center
Suite 611
San Francisco, CA 94111   (415) 951-1100

J. Thomas Mc Carthy
Univ. of San Francisco
2130 Fulton Street
San Francisco, CA 94117   (415) 666-6517

James Bruce Mc Cubbrey
Mc Cubbrey, Bartels, Meyer
& Ward
100 Bush Street
26th Floor
San Francisco, CA 94104   (415) 391-6665

John L. Mc Gannon
Townshend and Townshend
One Market Plaza
Steuart Street Tower
San Francisco, CA 94105   (415) 326-2400

Annette M. Mc Garry#
Chevron Corporation
555 Market Street
4th Floor
San Francisco, CA 94120   (415) 894-7700

Virginia S. Medlen
Limbach, Limbach & Sutton
2001 Ferry Bldg.
San Francisco, CA 94111   (415) 433-4150

Mark J. Meltzer
335 Prentiss Street
San Francisco, CA 94110   (415) 864-3586

Sheldon R. Meyer
Fliesler, Dubb, Meyer
& Lovejoy
4 Embarcadero Center
Suite 400
San Francisco, CA 94111   (415) 362-3800

Virginia H. Meyer
Mc Cubbrey, Bartels, Meyer
& Ward
One Post Steet
Suite 2700
San Francisco, CA 94104   (415) 391-6665

Bernard H. Meyers
First Interstate Bank of Ca.
Law Division - Sta. Floor
345 Calif. St.
San Francisco, CA 94104   (415) 773-7895

Dennis D. Miller
Franzel & Share
101 Market Street
Suite 610
San Francisco, CA 94105   (415) 543-1171

Mark E. Miller
Fliesler, Dubb, Meyer
& Lovejoy
4 Embarcadero Center
Suite 400
San Francisco, CA 94111   (415) 362-3800

Warren E. Miller
Stoney & Youngberg
1 California Street
San Francisco, CA 94123   (415) 981-1314

James F. Mitchell
Eckhoff, Hoppe, Slick,
Mitchell & Anderson
Four Embarcadero Center
Suite 760
San Francisco, CA 94111   (415) 391-7160

Mark B. Mondry
Townsend & Townsend
Steuart Street Tower
One Market Plaza
San Francisco, CA 94105   (415) 543-9600

Carlisle M. Moore
Phillips, Moore, Lempio
& Finley
177 Post Street
Suite 800
San Francisco, CA 94108   (415) 421-2674

Richard A. Nebb#
Limbach, Limbach & Sutton
2001 Ferry Building
San Francisco, CA 94111   (415) 433-4150

Keiichi Nishimura
Heller, Ehrman, White
& Mc Auliffe
333 Bush St.
San Francisco, CA 94104   (415) 772-6533

Melville Owen
Owen, Wickersham & Erickson, P.C.
455 Market St.
19th Fl.
San Francisco, CA 94105   (415) 882-3200

Raymond Owyang
20 Wood Street
San Francisco, CA 94118   (415) 668-5531

Gerald P. Parsons
Majestic, Parsons, Siebert
& Hsue
4 Embarcadero Center
Suite 1450
San Francisco, CA 94111   (415) 362-5556

Howard M. Peters
Phillips, Moore, Lempio
& Finley
177 Post Street
Suite 800
San Francisco, CA 94108   (415) 421-2674

Clarence R. Pfeiffer
554 Rockdale Drive
San Francisco, CA 94127   (415) 334-1790

Michael J. Pollock
Limbach, Limbach & Sutton
2001 Ferry Bldg.
San Francisco, CA 94111   (415) 433-4150

David J. Power#
Chevron Corporation
555 Market Street
Room 419
San Francisco, CA 94105   (415) 894-1224

David Roy Pressman
1237 Chestnut St., #4
San Francisco, CA 94109   (415) 776-3960

Alfons Puishes
1095 Market St.
Grant Bldg.
Suite 805
San Francisco, CA 94103   (415) 863-3911

Cathy E. Rincon
Chevron Corp.
555 Market Street
Suite 4037
San Francisco, CA 94105   (415) 894-1172

Jill L. Robinson
Fliesler, Dubb, Meyer
& Lovejoy
4 Embarcadero Center
Suite 400
San Francisco, CA 94111   (415) 362-3800

Gerald B. Rosenberg
Fliesler, Dubb, Meyer
& Lovejoy
4 Embarcadero Center
Suite 400
San Francisco, CA 94111   (415) 362-3800

Steven H. Roth#
Chevron Corporation
555 Market Street
Room 447
San Francisco, CA 94105   (415) 894-9309

Charles P. Sammut
Visa International
P.O. Box 8999
San Francisco, CA 94124   (415) 570-3166

Ernest Arthur Schaal
Chevron Corp.
555 Market Street
San Francisco, CA 94120   (415) 894-3695

Milton W. Schlemmer
345 California Street
Suite 2525
San Francisco, CA 94104   (415) 986-8833

John W. Schlicher
Townsend & Townsend
Steuart Street Tower, 20th Flr.
One Market Plaza
San Francisco, CA 94105   (415) 543-9600

Bruce W. Schwab
Townsend & Townsend
One Market Plaza
Steuart Street Tower
San Francisco, CA 94105   (415) 543-9600

George M. Schwab
Townsend & Townsend
1 Market Plaza
Steuart Tower
20th Floor
San Francisco, CA 94105   (415) 543-9600

Johann G. Seka
Townsend & Townsend
Steuart Street Tower
One Market Plaza
San Francisco, CA 94105   (415) 543-9600

Gerald T. Sekimura
Limbach, Limbach & Sutton
2001 Ferry Bldg.
San Francisco, CA 94111   (415) 433-4150

Harold Shain
254 Edgewood Ave.
San Francisco, CA 94117   (415) 731-6712

Laurence J. Shaw#
338 Third Ave., #1
San Francisco, CA 94118   (415) 751-2523

Philip M. Shaw, Jr.
Limbach, Limbach & Sutton
2001 Ferry Bldg.
San Francisco, CA 94111   (415) 433-4150

Stephen C. Shear
Flehr, Hohbach, Test,
Albritton & Herbert
Suite 3400
Four Embarcadero Center
San Francisco, CA 94111   (415) 781-1989

James A. Sheridan
Flehr, Hohbach, Test,
Albritton & Herbert
Four Embarcadero Center
Suite 3400
San Francisco, CA 94111   (415) 781-1989

Richard J. Sheridan
Chevron Corporation
Law Dept., Pat. Unit
P.O. Box 7141
San Francisco, CA 94120   (415) 894-3171

Craig E. Shinners
Townsend & Townsend
One Market Plaza
Steuart Street Tower
20th Floor
San Francisco, CA 94105   (415) 543-9600

J. Suzanne Siebert
Majestic, Parsons, Siebert
& Hsue
Four Embarcadero Center
Suite 1450
San Francisco, CA 94111   (415) 362-5556

Thomas F. Smegal, Jr.
Townsend and Townsend
Steuart St. Tower
One Market Plaza
20th Floor
San Francisco, CA 94105   (415) 543-9600

Karen S. Smith
Flehr, Hohbach, Test,
Albritton & Herbert
4 Embarcadero Center
Suite 3400
San Francisco, CA 94111   (415) 781-1989

Neil Arthur Smith
Limbach, Limbach & Sutton
2001 Ferry Bldg.
San Francisco, CA 94111   (415) 433-4150

William M. Smith
Towsend & Townsend
Steuart Street Tower
One Market Plaza
San Francisco, CA 94105   (415) 326-2400

Lawrence S. Squires
Chevron Corporation
Law Department
55 Market Street
San Francisco, CA 94105   (415) 894-4986

Michael A. Stallman
Limbach, Limbach & Sutton
2001 Ferry Bldg.
San Francisco, CA 94111   (415) 433-4150

John L. Stavert
AT&T
795 Folsom St.
Suite 690
San Francisco, CA 94107   (415) 442-3452

Joseph L. Strabala
One Market Plaza
Suite 4100
Spear Street Tower
San Francisco, CA 94105   (415) 981-8083

Jeanne C. Suchodolski
Townsend & Townsend
One Market Plaza
Steuart St. Tower
20th Floor
San Francisco, CA 94105   (415) 543-9600

John P. Sutton
Limbach, Limbach & Sutton
2001 Ferry Bldg.
San Francisco, CA 94111   (415) 433-4150

Reginald J. Suyat
Heller, Ehrman, White
& Mc Auliffe
333 Bush St.
San Francisco, CA 94104   (415) 772-6432

Aldo J. Test
Flehr, Hobbach, Test,
Albritton & Herbert
Four Embarcadero Center
Suite 3400
San Francisco, CA 94111   (415) 781-1989

M. Patricia Thayer
Howard, Rice, Nemerovski, Canady,
Robertson & Falk
3 Embarcadero Center
7th Floor
San Francisco, CA 94111   (415) 434-1600

Thomas C. Tokos
Sherman & Sterling
4 Embarcadero Center
San Francisco, CA 94111   (415) 981-5500

Robert D. Touslee
Chevron Corporation
555 Market St., 4th Flr.
San Francisco, CA 94105   (415) 894-4628

Charles E. Townsend, Jr.
Townsend and Townsend
One Market Plaza
20th Floor
San Francisco, CA 94105   (415) 543-9600

Richard F. Trecartin
Flehr, Hohbach, Test,
Albritton & Herbert
Four Embarcadero Center
Suite 3400
San Francisco, CA 94111   (415) 781-1989

William K. Turner
Chevron Corp.
P.O. Box 7141
San Francisco, CA 94120   (415) 894-3530

John Klaas Uilkema
Limbach, Limbach & Sutton
2001 Ferry Bldg.
San Francisco, CA 94111   (415) 433-4150

Allen H. Uzzell
Chevron Corp.
Law Dept.
555 Market St.
Suite 400
San Francisco, CA 94105   (415) 894-4743

Pierre C. Van Rysselberghe
Limbach, Limbach & Sutton
2001 Ferry Building
San Francisco, CA 94111   (415) 433-4150

Paul W. Vapnek
Townsend and Townsend
One Market Plaza
Steuart St. Tower
San Francisco, CA 94105   (415) 543-9600

Larry E. Vierra
Fliesler, Dubb, Meyer &
Lovejoy
4 Embarcadero Center
Suite 400
San Francisco, CA 94111   (415) 362-3800

Paul M. Vuksich
773 - 15th Avenue
San Francisco, CA 94118   (415) 751-7039

Carrie L. Walthour
Limbach, Limbach & Sutton
2001 Ferry Bldg.
San Francisco, CA 94111   (415) 433-4150

Calvin B. Ward
Mc Cubbrey, Bartels, Meyer
& Ward
One Post Street, Suite 2700
San Francisco, CA 94104   (415) 391-6665

Jeffrey K. Weaver#
Townsend & Townsend
One Market Plaza
Steuart Street Tower
San Francisco, CA 94105   (415) 326-2400

Ellen Lauver Weber
Townsend & Townsend
One Market Plaza
Steuart Street Tower
San Francisco, CA 94127   (415) 543-9600

Kenneth A. Weber
Townsend & Townsend
20th Floor Steuart Street Tower
One Market Plaza
San Francisco, CA 94105   (415) 543-9600

James C. Weseman
Limbach, Limbach & Sutton
2001 Ferry Bldg.
San Francisco, CA 94111   (415) 433-4150

Douglas E. White
Acronational
101 California Street
Suite 980
San Francisco, CA 94111   (415) 421-1546

J. William Wigert, Jr.
Limbach, Limbach & Sutton
2001 Ferry Bldg.
San Francisco, CA 94111   (415) 433-4150

Gary S. Williams
Flehr, Hohbach, Test,
Albritton & Herbert
Four Embarcadero Center
Suite 3400
San Francisco, CA 94111   (415) 494-8700

Warren S. Wolfeld
Fliesler, Dubb, Meyer
& Lovejoy
Four Embarcadero Center
Suite 400
San Francisco, CA 94111   (415) 362-3800

Michael E. Woods
Townsend & Townsend
Steuart Street Tower, 20th Flr.
One Market Plaza
San Francisco, CA 94105   (415) 543-9600

Edward S. Wright
Flehr, Hohbach, Test,
Albritton & Herbert
Four Embarcadero Center
Suite 3400
San Francisco, CA 94111   (415) 781-1989

Jerry G. Wright
Flehr, Hohbach, Test,
Albritton & Herbert
Four Embarcadero Center
Suite 3400
San Francisco, CA 94111   (415) 781-1989

Philip Yau#
Majestic, Parsons, Siebert
& Hsue
Four Embarcadero Center
Suite 1450
San Francisco, CA 94114   (415) 362-5556

Ronald Loh - Hwa Yin
Limbach, Limbach & Sutton
2001 Ferry Bldg.
San Francisco, CA 94111   (415) 433-4150

Steven J. Adamson
7061 Via Serena
San Jose, CA 95139   (415) 433-4150

Warren Michael Becker
Fliesler, Dubb, Meyer
& Lovejoy
60 South Market Street
Suite 1570
San Jose, CA 95113   (408) 287-8278

Thomas R. Berthold
IBM Corporation
951-029
5600 Cottle Road
San Jose, CA 95193   (408) 997-4793

Patrick T. Bever#
Skjerven, Morrill, Mac Pherson,
Franklin & Friel
25 Metro Plaza
Suite 700
San Jose, CA 95110   (408) 283-1222

Robert Bruce Brodie
IBM Corp.
5600 Cottle Rd.
San Jose, CA 9519?   (408) 997-4223

W. Douglas Carothers, Jr.
S-Mos Systems, Inc.
2460 North First Street
San Jose, CA 95131   (408) 954-9055

David H. Carroll
Skjerven, Morrill, Mac Pherson,
Franklin & Friel
25 Metro Drive, Suite 700
San Jose, CA 95110   (408) 283-1222

Robert H. Chen
Acer America
401 Charcot Avenue
San Jose, CA 95131   (408) 922-0333

Kenneth D'Alessandro
Lyon & Lyon
111 West St. John Street
Suite 400
San Jose, CA 95113   (408) 993-1555

Joseph T. Downey
IBM Corp.
Intellectual Property Law Dept.
Almaden Research Center (KO2/802)
650 Harry Road
San Jose, CA 95120   (408) 927-3360

Jacques M. Dulin
111 North Market Street
Suite 900
San Jose, CA 95113   (408) 286-0700

Jack Warren Edwards
95 South Market Street
Suite 300
San Jose, CA 95113   (408) 298-8886

Charles K. Epps
VMX, Inc.
2115 O'Nel Drive
San Jose, CA 95131   (408) 441-1144

Thomas C. Feix
111 N. Market St.
Suite 900
San Jose, CA 95113   (408) 286-0700

Richard Karl Franklin
Skjerven, Morrill, Macpherson,
Franklin & Friel
25 Metro Drive
Suite 700
San Jose, CA 95110   (408) 283-1222

John A. Frazzini
Hickman & Frazzini
Ten Almaden Blvd.
Suite 1100
San Jose, CA 95113   (408) 288-6500

Bradley A. Greenwald#
Skjerven, Morrill, Mac Pherson,
Franklin & Friel
25 Metro Dr., Suite 700
San Jose, CA 95110   (408) 283-1222

Robert O. Guillot
Rosenblum, Parish & Bacigalupi, P.C.
160 W. Santa Clara Street
Fifteenth Floor
San Jose, CA 95113   (408) 977-0120

Forrest E. Gunnison
Skjerven, Morrill, Mac Pherson,
Franklin & Freil
25 Metro Drive
Suite 700
San Jose, CA 95110   (408) 283-1222

Claude A.S. Hamrick
Rosenblum, Parish & Bacigalupi, P.C.
160 W. Santa Clara Street
Fifteenth Floor
San Jose, CA 95113   (408) 977-0120

Michael L. Harrison
Harrison & Kaylor
4320 Stevens Creek Blvd.
Suite 150
San Jose, CA 95129   (408) 241-2220

Mark A. Haynes
Fliesler, Dubb, Meyer
& Lovejoy
60 S. Market Street
Suite 1570
San Jose, CA 95113   (408) 287-8278

David W. Heid
Skjerven, Morrill, Mac Pherson,
Franklin & Friel
25 Metro Drive
Suite 700
San Jose, CA 95110   (408) 283-1222

Michael Hetherington
Lyon & Lyon
99 Almaden Blvd.
Suite 700
San Jose, CA 95113   (408) 993-1555

David H. Jaffer
Rosenblum, Parish & Bacigalupi, P.C.
160 W. Santa Clara Street
Fifteenth Floor
San Jose, CA 95113   (408) 977-0120

Ivor J. James, Jr.
2120 Briarwood Drive
San Jose, CA 95125   (408) 266-7346

David J. Kappos
6225 Mahan Drive
San Jose, CA 95123   (408) 226-1854

Yoshio Katayama
P.O. Box 1525
San Jose, CA 95109   (408) 286-0333

Patrick T. King
Hickman & Frazzini
Ten Almaden Blvd.
Suite 1100
San Jose, CA 95113   (408) 288-6500

B. Noel Kivlin#
Skjerven, Morrill, Mac Herson,
Franklin & Friel
25 Metro Drive
Suite 700
San Jose, CA 95110   (408) 283-1222

Keith E. Kline
111 N. Market Street
Suite 325
San Jose, CA 95113   (408) 947-7122

Norman R. Klivans, Jr.
Skjerven, Morrill, Mac Pherson,
Franklin & Friel
25 Metro Drive
Suite 700
San Jose, CA 95110   (408) 283-1222

Edward C. Kwok
Skjerven, Morrill, Mac Pherson,
Franklin & Friel
25 Metro Drive
Suite 700
San Jose, CA 95110   (408) 283-1222

Peter R. Leal
IBM Corporation
5600 Cottle Road
Dept. 951/029
San Jose, CA 95193   (408) 997-4175

John James Leavitt
777 North First Street
Suite 610
San Jose, CA 95112   (408) 286-2262

Kenneth E. Leeds
Skjerven, Morrill, Mac Pherson,
Franklin & Friel
25 Metro Drive
Suite 700
San Jose, CA 95110   (408) 283-1222

Elizabeth E. Leitereg
Skjerven, Morrill, Mac Pherson,
Franklin & Friel
25 Metro Drive
Suite 700
San Jose, CA 95110   (408) 283-1222

Thomas Scott Mac Donald
Skjerven, Morrill, Mac Pherson,
Franklin & Friel
25 Metro Drive
Suite 700
San Jose, CA 95110   (408) 283-1222

Alan H. Mac Pherson
Skjerven, Morrill, Mac Pherson
Franklin & Friel
25 Metro Drive
Suite 700
San Jose, CA 95110   (408) 283-1222

Walter J. Madden, Jr.
Skjerven, Morrill, Mac Pherson,
Franklin & Friel
25 Metro Drive
Suite 700
San Jose, CA 95110   (408) 283-1222

Stuart J. Madsen
Rosenblum, Parish & Bacigalupi
160 W. Santa Clara St.
15th Floor
San Jose, CA 95113   (408) 977-0120

Robert B. Martin
IBM
Almaden Res. Ctr., KO2/802
650 Harry Road
San Jose, CA 95120   (408) 927-3362

Terry A. Mc Hugh
111 West S. John Street
Suite 620
P.O. Box 2-E
San Jose, CA 95109   (408) 297-9733

I. Robert Mednick
888 N. First Street
San Jose, CA 95112   (408) 293-8565

Douglas R. Millett
1262 Silverado Drive
San Jose, CA 95120   (408) 268-1765

Leslie G. Murray
IBM Corporation
Intell. Prop. Law, 951/029
5600 Cottle Road
San Jose, CA 95193   (408) 997-4070

Brian D. Ogonowsky
Skjerven, Morrill, Mac Pherson,
Franklin & Friel
25 Metro Drive
Suite 700
San Jose, CA 95110   (408) 283-1222

Donald J. Pagel, Jr.
675 North First Street
Suite 1304
San Jose, CA 95112   (408) 292-0550

Bradley A. Perkins
VLSI Technology, Inc.
1109 Mc Kay Drive
San Jose, CA 95131   (408) 434-3159

Mark J. Protsik#
111 West St. John Street
Suite 620
P.O. Box 2-E
San Jose, CA 95109   (408) 297-9733

Louis J. Quick
3560 Parkland Avenue
San Jose, CA 95117   (408) 243-3552

William David Reese*
Internal Revenue Service
District Counsel
55 S. Market
Suite 505
San Jose, CA 95113   (408) 291-4674

Jerald E. Rosenblum
Rosenblum, Parish & Bacigalupi, P.C.
160 W. Santa Clara Street
15th Floor
San Jose, CA 95113   (408) 977-0120

Joseph A. Sawyer, Jr.
Chips & Technologies, Inc.
3050 Zanker Road
San Jose, CA 95134   (408) 434-0600

Otto Schmid, Jr.
IBM Corp.
Intellectual Property Law Dept.
5600 Cottle Road
San Jose, CA 95193   (408) 997-4182

Thomas Schneck, Jr.
111 West St. John Street
Suite 620
P.O. Box 2-E
San Jose, CA 95109   (408) 297-9733

Robert R. Schroeder
General Electric Company
175 Curtner Avenue
M/C 822
San Jose, CA 95125   (408) 925-2707

Michael Shenker
Skjerven, Morrill, Mac Pherson,
Franklin & Friel
25 Metro Drive, Suite 700
San Jose, CA 95110   (408) 283-1222

Guy W. Shoup
Skjerven, Morrill, Mac Pherson,
Franklin & Friel
25 Metro Drive
Suite 700
San Jose, CA 95110   (408) 432-9432

Melvyn D. Silver
Suden & Silver
111 N. Market Street
Suite 715
San Jose, CA 95113   (408) 298-9755

Raymond G. Simkins
3311 Beacon Lane
San Jose, CA 95118   (408) 266-8036

Robert S. Smith#
Engineering Enterprise
1263 Emory Street
San Jose, CA 95126   (408) 287-1894

David E. Steuber
Skjerven, Morrill, Mac Pherson,
Franklin & Friel
25 Metro Drive
Suite 700
San Jose, CA 95110   (408) 283-1222

Edward M. Suden
Suden & Silver
111 N. Market Street
Suite 715
San Jose, CA 95113   (408) 298-9755

John P. Taylor
658 New Dorset Court
San Jose, CA 95136   (408) 224-5050

Laura M. Terlizzi
Skjerven, Morrill, Mac Pherson,
Franklin & Friel
25 Metro Drive, Ste. 700
San Jose, CA 95110   (408) 283-1222

Liza B.K. Toth
Hopkins & Carley
150 Almaden Blvd.
15th Floor
San Jose, CA 95113   (408) 286-9800

Samuel Edward Turner
1450 Glenwood Avenue
San Jose, CA 95125   (408) 297-4039

William B. Walker, Sr.
2021 the Alameda
Suite 110
San Jose, CA 95126   (408) 249-7750

Raymond J. Werner
S-MOS
2460 North First St.
San Jose, CA 95131   (408) 922-0200

Paul John Winters
Skjerven, Morrill, Mac Pherson,
Franklin & Friel
25 Metro Drive
Suite 700
San Jose, CA 95110   (408) 283-1222

Jack M. Wiseman
12 South First Street
Suite 911
San Jose, CA 95113   (408) 294-6824

Edel M. Young#
XILINX, Inc.
2100 Logic Drive
San Jose, CA 95124   (408) 879-4969

Egbert Walter Mark
1516 140th Ave.
San Leandro, CA 94578   (415) 357-8442

Jerry N. Lulejian
Carsel & Lulejian
1118 Palm Street
San Luis Obispo, CA 93401   (805) 544-8510

Mitchell Saffian#
Guidance Software Corp.
793 Higuera St.
Suite 15
San Luis Obispo, CA 93401   (805) 545-9630

John A. Bucher
241 North San Mateo Drive
San Mateo, CA 94401   (415) 347-8871

Janet K. Castaneda
Harrison & Eakin
1700 S. El Camino Real
Suite 405
San Mateo, CA 94402   (415) 571-7500

James E. Eakin
Harrison, Harrison & Eakin
1700 S. El Camino Real
Suite 405
San Mateo, CA 94402   (415) 571-7500

Donald C. Feix
241 North San Mateo Drive
San Mateo, CA 94401   (415) 342-4508

Frank A. Neal
4237 Bettina
San Mateo, CA 94403   (415) 345-2477

David E. Newhouse
2855 Campus Drive
Suite 225
San Mateo, CA 94403   (415) 345-4930

Elizabeth E. Strnad#
554 Edgewood Road
San Mateo, CA 94402   (415) 343-8484

Ralph C. Grove#
1149 Crestwood Street
San Pedro, CA 90732   (213) 831-3480

Arnold Thomas Bertolli
8 Bradcliff Ct.
San Rafael, CA 94901   (415) 453-2763

Larry D. Jonnson
185 N. Redwood Drive
Suite 130
San Rafael, CA 94903   (415) 499-8822

Richard P. Maloney
Arundel Associates Inc.
P.O. Box 12716
San Rafael, CA 94913   (415) 472-3050

Melvin Robert Stidham
1050 Northgate Drive
Suite 100
San Rafael, CA 94903   (415) 472-3164

Draper B. Gregory
111 Deerwood Place
Suite 370
San Ramon, CA 94583   (415) 820-7323

Dix A. Newell
Chevron Chemical Company
6001 Bollinger Canyon Road
San Ramon, CA 94583   (415) 842-5745

Raymond J. Suberlak
ADP Automotive Claims
Services, Inc.
2010 Crow Canyon Place
San Ramon, CA 94583   (415) 866-1100

David J. Althoen
18641 Silver Maple Way
Santa Ana, CA 92705   (714) 639-1733

Gary Appel
1170 W. Civic Center Dr.
Santa Ana, CA 92703   (714) 558-0366

John J. Connors
Plante, Strauss, Vanderburgh
& Connors
1020 N. Broadway
Suite 305
Santa Ana, CA 92701   (714) 667-1570

Arthur W. Fuzak
Kimstock, Inc.
2200 S. Yale St.
Santa Ana, CA 92704   (714) 546-6850

Abe Goldstein
Goldstein, Block & Block
11577 Forum Way
Suite A
Santa Ana, CA 92705   (714) 832-2518

Arnold Grant
1522 Martingale Place
Santa Ana, CA 92705   (714) 832-2518

Robert H. Himes
1282 Landfair Circle
Santa Ana, CA 92705   (714) 544-3660

Harold C. Horwitz
2127 North Main Street
Santa Ana, CA 92706   (714) 542-1102

Robert B. C. Newcomb
19052 Chadbourne Lane
Santa Ana, CA 92705   (714) 838-4706

Richard J. Otto, Jr.
Space Applications Corp.
200 E. Sandpointe Ave.
Suite 200
Santa Ana, CA 92707   (714) 662-2855

Thomas L. Peterson
ITT Corporation
ITT Electromechanical Com. Worldwide
1851 E. Deere Avenue
P.O. Box 35000
Santa Ana, CA 92705   (714) 757-8444

Frank C. Price#
13812 Sandhurst Place
Santa Ana, CA 92705   (714) 544-7907

Lawrence D. Sassone#
900 N. Broadway
Suite 725
Santa Ana, CA 92701   (714) 547-5611

Robert L. Sassone
900 North Broadway
Suite 725
Santa Ana, CA 92701   (714) 547-5611

Robert E. Strauss
Plante Strauss Vanderburgh
& Connors
1020 North Broadway
Suite 305
Santa Ana, CA 92701   (714) 667-1570

James D. Thackrey#
13852 Dall Lane
Santa Ana, CA 92705   (714) 731-0705

John E. Vanderburgh
Plante, Strauss, Vanderburgh
& Connors
1020 N. Broadway
Suite 305
Santa Ana, CA 92701   (714) 667-1570

Walter A. Hackler
2372 S.E. Bristol, Suite B
Santa Ana Heights, CA 92707
    (714) 851-5010

Catherine J. Bos
1114 State Street, Apt. 200
Santa Barbara, CA 93105   (805) 963-1347

Harry W. Brelsford
233 East Carillo Street
Suite C
Santa Barbara, CA 93101   (805) 966-2281

Daniel J. Meaney, Jr.
P.O. Box 22307
Santa Barbara, CA 93121   (805) 687-6909

Michael G. Petit#
841 Weldon Road
Santa Barbara, CA 93109   (805) 965-4452

Edward E. Sachs
53 Tierra Cielo Lane
Santa Barbara, CA 93105   (805) 966-3833

Allen H. Sochel
P.O. Box 5548
1485 E. Valley Road
Santa Barbara, CA 93150   (805) 562-7539

Philip J. Wyatt#
Wyatt Technology Corporation
802 East Cota Street
Santa Barbara, CA 93103   (805) 963-5904

Eugene T. Battjer
Unisys Corporation
San Tomas At Central Expressway
M/S 12-33
Santa Clara, CA 95052   (408) 987-3000

Paul Davis
3COM Corp.
5400 Bayfront Plaza
P.O. Box 58145
Santa Clara, CA 95052   (408) 562-6400

Lawrence Edelman
Intel Corp.
Intellectual Property
MS GR1-21
3535 Garrett Dr.
Santa Clara, CA 95052   (408) 765-1852

Lloyd B. Guernsey#
FMC Corp.
Mach. Pat., Tdmk & Lic. Dept.
900 Lafayette Street
Suite 608
Santa Clara, CA 95050   (408) 241-8320

Michael J. Hughes
Hughes, Bedolla & Diener, Inc.
2350 Mission College Blvd.
Suite 1150
Santa Clara, CA 95054   (408) 727-9991

William J. Iseman
Intel Corporation
MS GR1-21
3535 Garrett Dr.
Santa Clara, CA 95052   (408) 765-1877

Ronald C. Kamp
FMC Corp.
900 Lafayette
Suite 608
Santa Clara, CA 95050   (408) 241-6728

Richard C. Liu#
3564 Londonderry Dr.
Santa Clara, CA 95050   (408) 248-9074

Richard D. Lowe
P.O. Box 296
Santa Clara, CA 95052   (408) 984-1555

Alan Johnston Moore
FMC Corp.
900 Lafayette Street
Suite 608
Santa Clara, CA 95050   (408) 241-7538

Lee Patch#
National Semiconductor Corp.
2900 Semiconductor Drive
P.O. Box 58090
M.S. 16-135
Santa Clara, CA 95052   (408) 721-7195

L. A. Pursglove#
Box 4418
Santa Clara, CA 95054   (408) 244-1667

Charles E. Quarton
Intel Corporation
3535 Garrett Drive
Santa Clara, CA 95052   (408) 765-1239

James W. Rose
National Semiconductor Corp.
2900 Semiconductor Dr.
P.O. Box 58052
M/S 16-135
Santa Clara, CA 95052   (408) 721-5365

Steven W. Roth#
ROLM
4900 Old Ironsides Dr.
Santa Clara, CA 95123   (408) 492-7664

Thomas E. Schatzel
3211 Scott Blvd.
Suite 201
Santa Clara, CA 95054   (408) 727-7077

Cheryl L. Shavers#
3036 Carleton Place
Santa Clara, CA 95051   (408) 246-8310

Carl L. Silverman
Intel Corporation
3065 Bowers Avenue
Santa Clara, CA 95051   (408) 765-1142

Robert J. Stern
Applied Materials, Inc.
P.O. Box 58039
Santa Clara, CA 95052   (408) 727-5555

Douglas L. Weller
836 Fremont St.
Santa Clara, CA 95050   (408) 985-0642

Gail W. Woodward#
2166 San Rafael Avenue
Santa Clara, CA 95051   (408) 246-4853

Jeffrey A. Hall
212 Clinton Street
Santa Cruz, CA 95062   (408) 423-1365

Stanley M. Weir#
227 Fern Street
Santa Cruz, CA 95060   (408) 425-5767

George H. Nicholson
1094 Clubhouse Drive
Santa Maria, CA 93455   (805) 937-0816

Keith D. Beecher
Jessup, Becher & Slehofer
2001 Wilshire Blvd.
Suite 500
Santa Monica, CA 90403   (213) 829-4525

Daniel M. Cislo
Cislo & Thomas
233 Wilshire Blvd.
Suite 900
Santa Monica, CA 90401   (213) 451-0647

Donald M. Cislo
Cislo & Thomas
233 Wilshire Blvd.
Suite 900
Santa Monica, CA 90401   (213) 451-0647

Albert M. Herzig
Haight, Brown & Bonesteel
201 Santa Monica
Santa Monica, CA 90406   (213) 458-1000

Laura A. Pitta#
1308 San Vicente Blvd.
Santa Monica, CA 90402   (213) 451-5280

John D. Raiford
1448 Fifteenth Street
Suite 107
Santa Monica, CA 90404   (213) 459-6910

Brian F. Rowe#
1223 Wilshire Blvd., #263
Santa Monica, CA 90403-0   (213) 453-6216

Lyle J. Schlyer#
USA Petroleum Corp.
2701 Ocean Park Blvd.
Suite 210
Santa Monica, CA 90405   (213) 399-3054

Harold Steingold#
407 16th Street
Santa Monica, CA 90402   (213) 393-2335

Stanley R. Steingold#
407 16th St.
Santa Monica, CA 90402   (213) 393-2335

Walter Richard Thiel
322 18th St.
Santa Monica, CA 90402   (213) 394-7597

Gary F. Grafel
Keegan & Coppin, Inc.
1355 N. Dutlon Avenue
Santa Rosa, CA 95401   (707) 528-1400

William C. Milks, III
Hewlett-Packard Company
1412 Fountain Grove Pkwy.
M/S 2US-Y
Santa Rosa, CA 95403   (707) 577-4744

Jane Bieberman De Nuzzo#
4335 Woodstock Road
Santa Ynez, CA 93460   (805) 688-3092

Daniel Strugar
9336 Nalini Court
Santee, CA 92071   (619) 258-9750

David A. Boone
P.O. Box 2125
Saratoga, CA 95070   (408) 291-6000

Richard B. Catto#
P.O. Box 243
14711 Bohlman Rd.
Saratoga, CA 95071   (408) 867-4059

Craig W. Hartsell
12291 Kosich Court
Saratoga, CA 95070   (408) 252-0780

Nathan N. Kallman
20900 Sarahills Drive
Saratoga, CA 95070   (408) 867-9289

Robert S. Kelly
19191 Portos Pl.
Saratoga, CA 95070   (408) 867-5648

Joseph Mednick
13424 Beaumont Ave.
Saratoga, CA 95070   (408) 867-0186

Gerald L. Moore
14590 Horseshoe Drive
Saratoga, CA 95070   (408) 867-9031

George Edgar Roush
14660 B Big Basin Way
Saratoga, CA 95070   (408) 741-1876

J.B. Mc Guire
P.O. Box 2230
Sausalito, CA 94966   (415) 435-6478

Kyle Eppele
Seagate Technology, Inc.
Building 14
920 Disc Drive
Scotts Valley, CA 95066   (408) 439-7319

Edward P. Heller, III
Seagate Technology, Inc.
920 Disc Drive
P.O. Box 66360
Scotts Valley, CA 95067   (408) 439-7291

Andrew A. Steiner
7777 Bodega Ave., L-2
Sebastopol, CA 95472   (707) 823-8134

Edward M. Bayer
16601 Nordhoff Street
Sepulveda, CA 91343   (818) 892-3913

William M. Harris
15736 Tuba Street
Sepulveda, CA 91343   (213) 892-8706

Harold R. Beck#
3970 Cody Road
Sherman Oaks, CA 91403   (818) 789-2484

Ralph Deutsch#
Deutsch Res. Labs., Ltd.
3647 Scadlock Lane
Sherman Oaks, CA 91403   (213) 789-2779

Roger A. Marrs
15233 Ventura Blvd.
Suite 820
Sherman Oaks, CA 91403   (213) 788-4115

William K. Quarles, Jr.
Sunkist Growers, Inc.
14130 Riverside Drive
Sherman Oaks, CA 91423   (818) 986-4800

Thomas I. Rozsa
15303 Ventura Blvd.
Suite 800
Sherman Oaks, CA 91403   (818) 783-0990

Allan M. Shapiro
Linden Court
15315 Magnolia Blvd.
Suite 105
Sherman Oaks, CA 91403   (818) 784-4228

Benjamin Franklin Spencer
175 North Mountain Trail Avenue
Sierra Madre, CA 91024   (818) 355-9233

Manuel J. Rodriguez, Jr.
2040 St. Louis Ave.
Signal Hill, CA 90806   (213) 498-0177

Harry C. Burgess
5410 Ball Drive
Soquel, CA 95073   (408) 475-6307

Paul Bernhard Fihe, II#
P.O. Box 126
Soquel, CA 95073   (408) 462-5079

Melanie A. Calvert
630 A Orange Grove Avenue
South Pasadena, CA 91030   (818) 441-2596

Paul A. Weilein
P.O. Box 3801
South Pasadena, CA 91030   (818) 799-1564

Robert M. Mc Manigal
1701 Camden Parkway
South Pasedena, CA 91030   (818) 799-1571

Walter Eugene Buting
Genentech, Inc.
460 Point San Bruno Blvd.
South San Francisco, CA 94080
          (415) 266-1715

Edwin Paul Ching
Berlex Biosciences
213 East Grand Avenue
South San Francisco, CA 94080
          (415) 266-7476

Debra J. Glaister#
Genentech Inc.
460 Pt. San Bruno Blvd.
South San Francisco, CA 94080
          (415) 266-3177

Janet E. Hasak
Genetech, Inc.
460 Point San Bruno Blvd.
South San Francisco, CA 94080
          (415) 266-1000

Darlene W. Hayes#
CODON
213 E. Grand Ave.
South San Francisco, CA 94080
          (415) 952-7070

Max D. Hensley
Genentech, Inc.
460 Point San Bruno Blvd.
South San Francisco, CA 94080
          (415) 266-1994

Margaret A. Horn
Genencor International, Inc.
180 Kimball Way
South San Francisco, CA 94080
          (415) 742-7536

Dennis G. Kleid
Genentech, Inc.
460 Point San Bruno Blvd.
South San Francisco, CA 94080
          (415) 266-1713

Richard B. Love
Genentech, Inc.
460 Point San Bruno Blvd.
South San Francisco, CA 94080
          (415) 266-1000

Lisabeth Feix Murphy
Athena Neurosciences, Inc.
800f Gateway Blvd.
South San Francisco, CA 94080
          (415) 877-0900

Nancy A. Oleski#
Genentech, Inc.
460 Point San Bruno Blvd.
South San Francisco, CA 94080
          (415) 266-2910

James G. Passe
Genencor, Inc.
180 Kimball Way
South San Francisco, CA 94080
          (415) 742-7595

Stephen Raines
Genentech, Inc.
460 Point San Bruno Blvd.
South San Francisco, CA 94080
          (415) 266-1705

Daryl B. Winter
Genentech, Inc.
460 Point San Bruno Blvd.
South San Francisco, CA 94080
          (415) 266-1249

Albert Macovski
Stanford University
Dept. of Elec. Eng.
Durand 109
Stanford, CA 94305   (415) 723-2708

Herman H. Murphy
Stanford Linear Accelerator Center
P.O. Box 4349
Stanford, CA 94309   (415) 926-2213

Max Geldin
3386 Canton Way
Studio City, CA 91604   (818) 763-5803

Lawrence V. Link, Jr.
3739 Mound View Avenue
Studio City, CA 91604   (818) 980-6258

Robert T. Merrick
10572 Mahoney Drive
Sunland, CA 91040   (818) 353-2639

Rodger N. Alleman
Lockheed Missiles & Space Co.
P.O. Box 3504
1111 Lockheed Way
Sunnyvale, CA 94088   (408) 742-0691

Clifton L. Anderson
333 Cobalt Way
Suite 107
Sunnyvale, CA 94086   (408) 245-0820

Keith G. Askoff#
Blakely, Sokoloff, Taylor
& Zafman
1245 Oakmead Parkway
Suite 101
Sunnyvale, CA 94086   (408) 720-8598

Bradley J. Bereznak
Blakely Sokoloff Taylor
& Zafman
1245 Oakmead Parkway
Suite 101
Sunnyvale, CA 94086   (408) 720-8598

Richard A. Brown
Jackson, Brown & Efting
465 S. Mathilda Ave.
Suite 304
Sunnyvale, CA 94086   (408) 732-3114

Richard Harry Bryer
Lockheed Missiles & Space Co., Inc.
1111 Lockheed Way
Sunnyvale, CA 94088   (408) 742-0691

Russell Adams Cannon#
961 Harney Way
Sunnyvale, CA 94087   (408) 739-6612

Charles D.B. Curry*
Dept. of the Navy
Office of Counsel
Strategic Systems Program Off.
1111 Lockheed Way, Bldg. 181-B
Sunnyvale, CA 94089   (408) 756-3662

Vernon R. Gard#
Blakely, Sokoloff, Taylor
& Zafman
1245 Oakmead Pkwy.
Suite 101
Sunnyvale, CA 94086   (408) 720-8598

Wayne O. Hadland*
U.S. Dept. of the Navy
Strategic Sys. Program Off.
Code SPLE-52
1111 Lockheed Way, Bldg. 181-B
Sunnyvale, CA 94089   (408) 756-3662

David R. Halvorson
Blakely, Sokoloff, Taylor
& Zafman
1245 Oakmead Parkway
Sunnyvale, CA 94086   (408) 720-8598

Tracy L. Hurt
Blakely, Sokoloff, Taylor
& Zafman
1245 Oakmead Pkwy., Ste. 101
Sunnyvale, CA 94086   (408) 720-8598

Yemmanur Jayachandra#
1713 Chitamook Court
Sunnyvale, CA 94087   (408) 737-9762

Bo-In Lin#
IBM
1299 Orleans Drive
Bldg. 4
Sunnyvale, CA 94089   (408) 235-2448

James D. Mc Farland
Blakely, Sokoloff,
Taylor & Zafman
1245 Oakmead Parkway
Sunnyvale, CA 94086   (408) 720-8598

Ronald J. Meetin
Signetics Company, A Div. of
North American Phillips Corp.
Pat. Dept., M/S 54
811 E. Arques Avenue
Sunnyvale, CA 94088   (408) 991-2046

John J. Morrissey, Jr.
Lockheed Missiles & Space Co., Inc.
Dept. 26-02, Bldg. 101
1111 Lockheed Way
Sunnyvale, CA 94088   (415) 742-0691

James E. Parsons#
1065 Greco Ave., #A212
Sunnyvale, CA 94087   (408) 746-0644

James M. Pershon#
AT&T Network Systems
1090 E. Duane Ave.
Sunnyvale, CA 94086   (408) 522-4751

James C. Scheller, Jr.
Blakely, Sokoloff, Taylor
& Zafman
1245 Oakmead Parkway
Suite 101
Sunnyvale, CA 94086   (408) 720-8598

Dudley W. Swedberg#
Amdahl Corporation
M.S. 323
1250 East Arques Ave.
P.O. Box 3470
Sunnyvale, CA 94088   (408) 746-6000

Edwin H. Taylor
Blakely, Sokoloff, Taylor
& Zafman
1245 Oakmead Pkwy.
Suite 101
Sunnyvale, CA 94086   (408) 720-8598

J. Vincent Tortolano
Advanced Micro Devices, Inc.
901 Thompson Place
P.O. Box 3453, M.S. 68
Sunnyvale, CA 94088   (408) 982-6037

Lester J. Vincent
Blakely, Sokoloff, Taylor
& Zafman
1245 Oakmead Pkwy.
Suite 101
Sunnyvale, CA 94086   (408) 720-8598

Kenneth L. Warsh*
Department of the Navy
Strategic Systems Program Office
Code SPLE5
1233 N. Matilda Ave., Bldg. 181-B
Sunnyvale, CA 94088   (408) 756-3665

Leslie S. Miller
Siemens-Pacesetter, Inc.
12884 Bradley Avenue
Sylmar, CA 91342   (818) 362-6822

Malcolm J. Romano
Siemens-Pacesetter, Inc.
12884 Bradley Ave.
Sylmar, CA 91342   (818) 362-6822

Lisa Prechter Weinberg#
Siemens-Pacesetter, Inc.
12884 Bradley Avenue
Sylmar, CA 91342   (818) 362-6822

Anthony T. Zachary
19601 Greenbriar Drive
Tarzana, CA 91356   (213) 342-4021

Julia E. Abers
AMGEN Inc.
1900 Oak Terrace Lane
Thousand Oaks, CA 91320   (805) 499-5725

Louise P. Anderson#
711 Woodbine Court
Thousand Oaks, CA 91360   (805) 492-6760

Stanley E. Anderson, Jr.
711 Woodbine Court
Thousand Oaks, CA 91360   (805) 492-6760

Thomas E. Byrne
AMGEN Inc.
1900 Oak Terrace Lane
Thousand Oaks, CA 91320   (805) 499-5725

Daniel M. Chambers#
AMGEN, Inc.
Amgen Center
1840 Dehavilland Dr.
Thousand Oaks, CA 91320   (805) 499-5725

John J. Deinken
Rockwell International Corp.
P.O. Box 1085
Thousand Oaks, CA 91358   (805) 373-4556

Robert E. Geauque
167 Windsong Street
Thousand Oaks, CA 91360   (805) 492-5331

Richard J. Mazza
AMGEN Inc., Amgen Center
1840 Dehavilland Drive
Thousand Oaks, CA 91320   (805) 499-5725

John C. Mc Farren
Rockwell International Corp.
Pat. Dept. M/S 083-A15
P.O. Box 1085
Thousand Oaks, CA 91360   (805) 373-4566

Henry P. Nowak
AMGEN Center
Legal Dept.
1900 Oak Terrace Lane
Thousand Oaks, CA 91360   (805) 499-5725

Steven M. Odre
AMGEN, Inc.
1900 Oak Terrace Lane
Thousand Oaks, CA 91360   (805) 499-5725

Robert D. Weist
AMGEN, Inc.
1900 Oak Terrace Lane
Thousand Oaks, CA 91320   (805) 499-5725

Robert B. Winter#
AMGEN Inc.
Amgen Center
1840 Dehavilland Dr.
Thousand Oaks, CA 91320   (805) 499-5725

G. Joseph Buck
Suite 300 - 15
3868 Carson Street
Torrance, CA 90503   (213) 540-8840

Ronald M. Goldman
Roth & Goldman
21515 Hawthorne Blvd.
Union Bank Tower - Suite 214
Torrance, CA 90503   (213) 316-5399

Terje Gudmestad
Hughes Aircraft Company
3100 West Lomita Blvd.
Torrance, CA 90509   (213) 517-5742

John E. Halamka
Halamka & Halamka
Suite 590
21515 Hawthorne Blvd.
Torrance, CA 90503   (213) 316-6100

Irving Keschner
21515 Hawthorne Blvd.
Suite 940
Torrance, CA 90503   (213) 543-5200

Michael M. Schuster#
Hi - Shear Corp.
2600 Skypark Dr.
Torrance, CA 90509   (213) 784-4003

Irwin Shuldiner
2231 W. 235th Street
Torrance, CA 90501   (213) 325-5649

Gerald Singer
Singer & Singer
3142 Pacific Coast Hwy.
Suite 208
Torrance, CA 90505   (213) 530-2202

Edward A. Sokolski
3868 Carson Street
Apt. 105
Torrance, CA 90503   (213) 540-5631

Gilbert E. Moody
Moody, Johnson & Mach
250 W. Main Street
Turlock, CA 95380   (209) 632-1086

Gordon K. Anderson#
14632 Pacific Street
Tustin, CA 92680   (714) 730-3460

Noel B. Hammond#
1311 Mitchell Ave.
Tustin, CA 92680   (714) 731-6949

Harold Leo Jackson
Jackson & Jones, P.C.
17592 Irvine Blvd.
Tustin, CA 92680   (714) 832-2080

Stanley R. Jones
Jackson & Jones
17592 Irvine Blvd.
Tustin, CA 92680   (714) 832-2080

Stuart W. Knight
145 W. Main St., #210
Tustin, CA 92680   (714) 730-4808

Forrest E. Logan
275 Centennial Way
Suite 205
Tustin, CA 92680   (714) 730-5553

Steven R. Markl
14902 Bridgeport
Tustin, CA 92680   (714) 838-7339

Suzanne T. Michel#
33 Union Square
Apartment 1415
Union City, CA 94587   (415) 471-1954

Robert W. Cramer
Knapp Petersen & Clarke
70 Universal City Plaza
Suite 400
Universal City, CA 91608   (213) 508-5000

Joseph A. Nicassio
25743 N. Hogan Dr., #F7
Valencia, CA 91355   (213) 508-5000

Harold Dale Messner
1021 Nebraska Street
Vallejo, CA 94590   (707) 577-8709

Bruce D. Jimerson
27375 Coolwater Ranch Rd.
Valley Center, CA 92082   (714) 749-1991

Gerald L. Price
Calvert Engineering Inc.
7051 Hayvenhurst Ave.
Van Nuys, CA 91406   (213) 781-6029

Julius L. Rubinstein
Central Valley Professional
Center
Lower Level
6454 Van Nuys Blvd.
Van Nuys, CA 91401   (818) 781-9922

Albert S. Sheppard#
13610 Valerio Street
Van Nuys, CA 91405   (818) 902-1130

Walter Unterberg#
5709 Burnet Avenue
Van Nuys, CA 91411   (818) 780-6333

Lawrence D. Weber
14416 Hamlin Street
Suite 209
Van Nuys, CA 91401   (818) 988-3248

James H. Griffith
2674 E. Main Street
Suite C-230
Ventura, CA 93003   (805) 643-5617

Kenneth Jennings Hovet
3585 Maple Street
Suite 140
Ventura, CA 93003   (805) 642-2101

Marvin E. Jacobs
Koppel & Jacobs
2151 Alessandro Drive
Ventura, CA 93001   (805) 648-5194

Dennis Blaine Haase
P.O. Box 1587
Visalia, CA 93291   (209) 733-1844

George W. Mc Laughlin
Borton, Petrini & Conron
206 South Mooney Blvd.
Visalia, CA 93291   (209) 627-5600

Joseph R. Dwyer
P.O. Box 3183
Vista, CA 92083   (619) 945-0211

Theodore J. Bielen, Jr.
Bielen, Peterson & Lampe
1990 N. California Blvd.
Suite 720
Walnut Creek, CA 94596   (415) 937-1515

Edward Brosler
3100 Tice Creek Drive
Apt. 2
Walnut Creek, CA 94595   (415) 932-7693

Janet Pauline Clark
Dow Chemical Co.
2800 Mitchell Dr.
Bldg. 340
Walnut Creek, CA 94598   (415) 944-2054

Murray K. Hatch
1511 Treat Blvd.
Suite 500
Walnut Creek, CA 94598   (415) 930-0777

Robert T. Kloeppel
2673 Velvet Way
Walnut Creek, CA 94596   (415) 933-5274

Thomas Raymond Lampe
Bielen, Peterson & Lampe
1990 N. California Blvd.
Suite 720
Walnut Creek, CA 94596   (415) 937-1515

Rankin Allen Milliken
Schapp & Hatch
1511 Treat Blvd.
Suite 500
Walnut Creek, CA 94598   (415) 930-0777

D. Wendell Osborne
Dow Chemical Company
2800 Mitchell Drive
Walnut Creek, CA 94598   (415) 944-2041

John A. Perona#
31 Hanson Lane
Walnut Creek, CA 94596   (415) 935-0758

Richard E. Peterson
Bielen, Peterson & Lampe
1990 N. California Blvd.
Suite 720
Walnut Creek, CA 94596   (415) 937-1515

Robert R. Stringham#
Dow Chemical Co.
2800 Mitchell Dr.
Walnut Creek, CA 94598   (415) 944-2042

Boniard I. Brown
1500 West Covina Parkway
Suite 113
West Covina, CA 91790   (213) 338-0100

Don A. Hart
Honeywell Inc.
T&CSD
1200 East San Bernardino Rd.
West Covina, CA 91790   (818) 915-9137

D. James Schwedler
1129 E. Walnut Creek Pkwy.
West Covina, CA 91790   (818) 919-3816

David G. Alexander#
Koppel & Jacobs
Suite 302
31255 Cedar Valley Drive
Westlake Village, CA 91362   (818) 707-1911

Donald C. Glynn
1019 Barrow Court
Westlake Village, CA 91361   (805) 497-4488

Warren T. Jessup
Jessup Beecher & Slehofer
875 Westlake Blvd.
Suite 205
Westlake Village, CA 91361   (818) 991-7062

Richard S. Koppel
Koppel & Jacobs
31255 Cedar Valley Drive
Suite 302
Westlake Village, CA 91362   (818) 707-1911

Richard D. Slehofer
Jessup, Beecher & Slehofer
875 Westlake Blvd.
Suite 205
Westlake Village, CA 91361   (818) 991-7062

Edgar Waite Averill, Jr.
Averill & Varn
8244 Painter Ave.
Whittier, CA 90602   (213) 698-8039

Al A. Canzoneri#
15912 Silvergrove Drive
Whittier, CA 90604   (213) 943-4924

Matthew P. Lynch
7433 Quakertown Avenue
Winnetka, CA 91306   (818) 898-0020

Gregory B. Wood
Rehwald, Rameson, Lewis
& Wood
5855 Topanga Canyon Blvd.
Suite 310
Woodland Hills, CA 91367   (818) 703-7500

John D. Bauersfeld
Kelly, Bauersfeld & Lowry
6320 Canoga Avenue
Suite 1650
Woodland Hills, CA 91367   (818) 347-7900

Gerald L. Cline
Litton Industries, Inc.
5500 Canoga Avenue
M/S 30
Woodland Hills, CA 91367   (818) 716-3139

Marc S. Colen
6355 Topanga Canyon Blvd.
Suite 331
Woodland Hills, CA 91367   (818) 716-2891

Clark E. De Larvin
22920 Cass Avenue
Woodland Hills, CA 91364   (818) 716-2891

Donald J. Ellingsberg#
23547 Burbank Blvd.
Woodland Hills, CA 91367   (213) 346-9750

Harold E. Gillmann
Litton Industries, Inc.
MS-30
5500 Canoga Ave.
Woodland Hills, CA 91367   (818) 716-3138

Scott W. Kelley
Kelly, Bauersfeld & Lowry
6320 Canoga Avenue
Suite 1650
Woodland Hills, CA 91367   (818) 347-7900

John E. Kelly
Kelly, Bauersfeld & Lowry
6320 Canoga Avenue
Suite 1650
Woodland Hills, CA 91367   (818) 347-7900

James F. Kirk
Litton Industries
5500 Canoga Avenue
Woodland Hills, CA 91367   (818) 712-7220

Sebron Koster
23411 Berdon Street
Woodland Hills, CA 91367   (818) 883-5972

Elliott N. Kramsky
5850 Canoga Avenue
Suite 400
Woodland Hills, CA 91367   (818) 992-5221

Stuart O. Lowry
Kelly, Bauersfeld & Lowry
6320 Canoga Avenue
Suite 1650
Woodland Hills, CA 91367   (818) 347-7900

Milton Madoff
6200 1/2 Nita Ave.
Woodland Hills, CA 91367   (213) 884-0505

Richard J. Mc Mullen
5155 Llano Drive
Woodland Hills, CA 91364   (213) 347-5434

Janine Rickman Novatt
Kelly, Bauersfeld & Lowry
6320 Canoga Ave., Suite 1650
Woodland Hills, CA 91367   (818) 347-7900

John Joseph Posta, Jr.
5850 Canoga Avenue
Suite 400
Woodland Hills, CA 91367   (818) 348-1088

L. David Rish
Litton Systems, Inc.
5500 Canoga Avenue
M/S 30
Woodland Hills, CA 91367   (818) 712-7212

Robert Malcolm Sperry, Sr.
23390 Ostronic Drive
Woodland Hills, CA 91367   (818) 340-4629

James M. Steinberger
4384 Charlemont Avenue
Woodland Hills, CA 91364   (818) 346-7828

A. Donald Stolzy
6301 Glade Avenue
Apt. K-110
Woodland Hills, CA 91367   (818) 716-0881

Timothy Thut Tyson
4600 Willens Avenue
Woodland Hills, CA 91364   (818) 992-5528

Robert P. Whipple
5850 Canoga Ave.
Woodland Hills, CA 91367   (818) 883-9882

Douglas A. Chaikin
2995 Woodside Road
Suite 400-382
Woodside, CA 94062   (415) 851-8118

Harvey Gunther Lowhurst
P.O. Box 620241
Woodside, CA 94062   (415) 366-2874

Robert R. Meads
5472 Club View Dr.
Yorba Linda, CA 92686   (714) 777-5534

Colorado

Keith E. Archer#
3151 S. Vaugh Way
Suite 300
Aurora, CO 80014   (886) 271-3225

Christine A. Peterson#
14214 E. First Dr. - #C2
Aurora, CO 80011   (303) 340-8647

Fred A. Winans
5405 S. Jasper Way
Aurora, CO 80015   (303) 693-5351

Jean M. Macheledt
1401 Saint Andrews Drive
Bloomfield, CO 80020   (303) 466-0589

Harold A. Burdick
5305 Spine Road
Suite B-East
Boulder, CO 80301   (303) 444-5205

Roland P. Campbell
P.O. Box 3171
Boulder, CO 80307   (303) 444-5205

Jennie M. Caruthers#
Greenlee & Associates
5370 Manhattan Circle
Boulder, CO 80303   (303) 499-8080

Bruce E. Dahl
Chrisman, Bynum & Johnson
1401 Walnut St.
Suite 500
Boulder, CO 80302   (303) 444-4820

Donald M. Duft
5078 Cottonwood Drive
Boulder, CO 80301   (303) 530-0456

Donna M. Ferber#
Greenlee & Associates, P.C.
5370 Manhattan Circle
Suite 201
Boulder, CO 80303   (303) 499-8080

John R. Flanagan
P. O. Box 4309
Boulder, CO 80306   (303) 449-0884

Lorance L. Greenlee
Greenlee & Associates, P.C.
5370 Manhattan Circle
Suite 201
Boulder, CO 80303   (303) 499-8080

Barbara A. Gyure#
Greenlee & Associates, P.C.
5370 Manhattan Circle
Suite 201
Boulder, CO 80303   (303) 499-8080

Earl Clark Hancock
3445 Penrose Place
Suite 210
Boulder, CO 80301   (303) 447-2060

Robert E. Harris
5305 Spine Road
Suite B-East
Boulder, CO 80301   (303) 444-5205

Robert W. Jackson#
Chrisman, Bynum & Johnson, P.C.
1401 Walnut Street
Suite 500
Boulder, CO 80302   (303) 444-4820

Donald W. Margolis
3405 Penrose Place
Suite 104
Boulder, CO 80301   (303) 443-3818

Diane H. Mc Clearn#
Greenlee & Associates
5370 Manhattan Circle
Suite 201
Boulder, CO 80303   (303) 499-8080

Thomas W. O Rourke
7212 Old Post Road
Boulder, CO 80301   (303) 530-3566

Louis G. Puls#
Associated Patent Services, Inc.
P.O. Box 3264
Boulder, CO 80307   (303) 444-7937

Charles Edwin Rohrer
6971 Hunter Place
Boulder, CO 80301   (303) 924-9721

Jon Sheldon Saxe
Synergen, Inc.
1885 33rd Street
Boulder, CO 80301   (303) 938-6278

Charles L. Sharp, Jr.
1919 14th Street
Suite 330
Boulder, CO 80302   (303) 444-2456

Martin Layne Shively#
Micro Motion, Inc.
7070 Winchester Circle
Boulder, CO 80301   (303) 530-8401

Francis A. Sirr
3445 Penrose Place
Suite 210
Boulder, CO 80303   (303) 447-2060

Sally A. Sullivan#
Greenlee & Associates
5370 Manhattan Circle
Suite 201
Boulder, CO 80303   (303) 499-8080

Margaret M. Wall
Greenlee & Associates
5370 Manhattan Circle
Suite 201
Boulder, CO 80303   (303) 499-8080

Ellen P. Winner
Greenlee & Associates, P.C.
5370 Manhattan Circle
Suite 201
Boulder, CO 80303   (303) 499-8080

Carl M. Wright
646 Furman Way
Boulder, CO 80303   (303) 499-8080

James Ralph Young
Chrisman, Bynum & Johnson, P.C.
Continental Building
1401 Walnut St.
Suite 500
Boulder, CO 80302   (303) 444-4820

Harold K. Johnston
7050 W. 120th Ave.
Suite 50
Broomfield, CO 80020   (303) 466-1787

Frank H. Thomson
Ball Corporation
10 Longs Peak Dr.
Broomfield, CO 80021   (303) 460-2580

Gary Boone#
6547 N. Academy Blvd.
Suite 531
Colorado Springs, CO 80918   (719) 599-3229

Lawrence Stephen Galka
108 E. Cheyenne Road
Colorado Springs, CO 80906   (303) 633-4444

Linda N.F. Gould
Berniger, Berg, Rioth
& Diver
P.O. Box 1716
Colorado Springs, CO 80901   (303) 475-9900

Richard W. Hanes
1670 N. Newport Rd.
Suite 300
Colorado Springs, CO 80916   (719) 637-4330

William J. Kubida
Digital Equipment Corporation
1110 Chapel Hills Drive
CXN2-31
Colorado Springs, CO 80920   (719) 260-3022

Anthony W. Raskob, Jr.#
823 East High Street
Colorado Springs, CO 80903   (719) 520-5330

Phillip A. Rein
102 S. Tejon
Suite 1100
Colorado Springs, CO 80903   (719) 635-9993

Roger L. Schneider
Atmel Corporation
1150 E. Cheyenne Mtn. Blvd.
Colorado Springs, CO 80906   (719) 540-1837

Gregg I. Anderson
Holland & Hart
555 17th Street
Suite 2900
Denver, CO 80202   (303) 295-8000

Martin G. Anderson
Martin Marietta Denver Aerospace
P.O. Box 179
MS 1400
Denver, CO 80201   (303) 977-6474

Robert D. Anderson#
Sheridan, Ross & Mc Intosh
1700 Lincoln Street
35th Floor
Denver, CO 80203   (303) 863-9700

Glenn K. Beaton
4582 S. Ulster Street Parkway
Suite 403
Denver, CO 80237   (303) 850-9900

Todd P. Blakely
Sheridan, Ross & Mc Intosh
One United Bank Center
1700 Lincoln Street
35th Floor
Denver, CO 80203   (303) 863-9700

Phillips V. Bradford#
Colorado Advanced Technology
Institute
1625 Broadway
Suite 700
Denver, CO 80202   (303) 620-4777

T. D. Bratschun
Holme Roberts & Owen
1700 Broadway
Suite 1800
Denver, CO 80290   (303) 861-7000

Duane C. Burton
1100 Writer Commons
1720 South Bellaire Street
Denver, CO 80222   (303) 691-9119

William Scott Carson
Dorr, Carson, Sloan
& Peterson
3010 E. 6th Avenue
Denver, CO 80206   (303) 333-3010

Curtis H. Castleman, Jr.
Gates Corporation
900 S. Broadway
Denver, CO 80209   (303) 744-4685

Alan B. Clay#
Sheridan, Ross & Mc Intosh
One United Bank Center
1700 Lincoln Street
35th Floor
Denver, CO 80203   (303) 863-9700

Charles C. Corbin
1050 Lafayette
Apt. 304
Denver, CO 80218   (303) 831-7323

Edwin H. Crabtree
3773 Cherry Creek
North Drive
Suite 900
Denver, CO 80206   (303) 322-7460

Ralph F. Crandell
Holland & Hart
555 17th Street
Suite 2900
P.O. Box 8749
Denver, CO 80201   (303) 295-8390

Robert G. Crouch
Holland & Hart
555 17th St.
Suite 2900
Denver, CO 80202   (303) 295-8266

Phillip L. De Arment
Martin Marietta Group
Mail Stop DC- 2400
P.O. Box 179
Denver, CO 80201   (303) 971-1190

David F. Dockery#
Sheridan, Ross &
Mc Intosh, P.C.
1700 Lincoln St., 35th Flr.
Denver, CO 80203   (303) 863-9700

Robert C. Dorr
Dorr, Carson, Sloan
& Peterson
3010 E. 6th Avenue
Denver, CO 80206   (303) 333-3010

William Griffith Edwards
1700 Broadway
Suite 1200
Denver, CO 80290   (303) 861-1456

Martha E. Ely
P.O. Box 6459
Denver, CO 80206   (719) 577-8420

Frank A. Elzi
1638 S. Jasmine St.
Denver, CO 80224   (303) 756-4656

Gary D. Fields
Fields, Lewis, Pittenger
& Rost
1720 Bellaire Street
Suite 1100
Denver, CO 80222   (303) 758-8400

Craig C. Groseth
Sheridan, Ross & Mc Intosh
One United Bank Center
1700 Lincoln Street
35th Floor
Denver, CO 80203   (303) 863-9700

Joseph C. Herring
Herring & Associates
3257 S. Steele Street
Denver, CO 80210   (303) 756-2372

Conrad B. Houser
Mobil Oil Corp.
Box 17772
Denver, CO 80217   (303) 293-6100

James L. Johnson
Sheridan, Ross & Mc Intosh
One United Bank Center
1700 Lincoln St.
35th Flr.
Denver, CO 80203   (303) 863-9700

Joseph J. Kelly
Klaas & Law
738 Pearl Street
Denver, CO 80203   (303) 837-1616

Bruce G. Klaas
Klaas & Law
738 Pearl St.
Denver, CO 80203   (303) 837-1616

Joseph E. Kovarik
Sheridan, Ross & Mc Intosh, P.C.
1700 Lincoln St., Ste 3500
Denver, CO 80203   (303) 863-9700

Christopher Jon Kulish
Sheridan, Ross & Mc Intosh
One United Bank Center
1700 Lincoln Street
35th Floor
Denver, CO 80203   (303) 863-9700

Ancel W. Lewis, Jr.
Fields, Lewis, Pittenger
& Rost
1720 S. Bellaire Street
Suite 1100
Denver, CO 80222   (303) 758-8400

John R. Ley
Holland & Hart
555 Seventeenth Street
No. 2900 - P.O. Box 8749
Denver, CO 80201   (303) 295-8392

John David Lister
Manville Corporation
P.O. Box 5108
Denver, CO 80217   (303) 978-2159

Thomas R. Marsh
Sheridan, Ross & Mc Intosh
One United Bank Center
1700 Lincoln Street
35th Floor
Denver, CO 80203   (303) 863-9700

C.E. Martine, III
Rothgerber, Appel, Powers
& Johnson
One Tabor Center, Suite 3000
1200 17th Street
Denver, CO 80202   (303) 623-9000

Michael D. Mc Intosh
Sheridan, Ross & Mc Intosh
One United Bank Center
1700 Lincoln Street
35th Floor
Denver, CO 80203   (303) 863-9700

John A. Mc Kinney
Holme Roberts & Owen
1700 Lincoln Street
Suite 4100
Denver, CO 80203   (303) 861-7000

John C. Moran
At&T Bell Laboratories
11900 N. Pecos Street
Denver, CO 80234   (303) 538-3940

James E. Nelson
Schafer, Rooke & Nelson
738 Pearl Street
Denver, CO 80203   (303) 861-4040

Loren D. Nelson
Ophir Corporation
Suite 100
3190 S. Wadsworth Blvd.
Denver, CO 80227   (303) 986-1512

Gregory W. O Connor
Samsonite Corporation
11200 East 45th Avenue
Denver, CO 80239   (303) 373-6165

William P. O Meara
Klaas & Law
738 Pearl Street
Denver, CO 80203   (303) 837-1616

H. W. Oberg, Jr.
The Gates Corporation
900 S. Broadway
Denver, CO 80209   (303) 744-4743

Nellie Clare Olszanski#
Klaas & Law
738 Pearl St.
Denver, CO 80203   (303) 837-1616

Michael J. Pfister
8225 E. Lehigh Ave.
Denver, CO 80237   (303) 771-5420

James E. Pittenger
Field, Lewis, Pittenger
& Rost
1100 Writer Commons Bldg.
1720 S. Bellaire St., Ste. 1100
Denver, CO 80222   (303) 758-8400

Gary M. Polumbus
Holland & Hart
555 17th Street
Suite 2900
Denver, CO 80202   (303) 295-8391

John Edward Reilly
1554 Emerson Street
Denver, CO 80218   (303) 830-2014

Kyle W. Rost
Fields, Lewis, Pittenger
& Rost
1720 S. Bellaire St.
Suite 1100
Denver, CO 80222   (303) 758-8400

Philip H. Sheridan
Sheridan, Ross & Mc Intosh
One United Bank Center
1700 Lincoln Street
35th Floor
Denver, CO 80203   (303) 863-9700

David A. Shifrin
Sheridan, Ross & Mc Intosh
One United Bank Center
1700 Lincoln St., 35th Flr.
Denver, CO 80203   (303) 863-9700

403

Jack C. Sloan
Dorr, Carson, Sloan
& Peterson
3010 E. 6th Avenue
Denver, CO 80206   (303) 333-3010

Brian D. Smith
Fields, Lewis, Pittenger
& Rost
1720 S. Bellaire Street
Suite 1100
Denver, CO 80222   (303) 758-8400

Brenda L. Speer
Sheridan, Ross & Mc Intosh
One United Bank Center
1700 Lincoln St., 35th Flr.
Denver, CO 80203   (303) 863-9700

Sabrina Crowley Stavish
Sheridan, Ross & Mc Intosh
One United Bank Center
1700 Lincoln Street
35th Floor
Denver, CO 80203   (303) 863-9700

Barry J. Swanson
Beaton & Swanson, P.C.
4582 S. Ulster St. Pkwy.
Suite 403
Denver, CO 80237   (303) 850-9900

Mike L. Tompkins
Sheridan, Ross & Mc Intosh
One United Bank Center
1700 Lincoln Street
35th Floor
Denver, CO 80203   (303) 836-9700

Roy K. Uenishi#
1255 - 19th Street
Suite 311
Denver, CO 80202   (303) 295-2655

H. B. Van Valkenburgh, III#
738 Pearl St.
Denver, CO 80203   (303) 837-1616

Douglass F. Vincent
Fields, Lewis, Pittenger
& Rost
1720 S. Bellaire St.
Suite 1100
Denver, CO 80222   (303) 758-8400

David Volejnicek
AT&T Bell Laboratories
11900 N. Pecos Street
Room 2U52
Denver, CO 80234   (303) 538-4154

Glenn Lowell Webb#
Dorr, Carson, Sloan
& Peterson
3010 East 6th Avenue
Denver, CO 80206   (303) 333-3010

David R. Weiss*
U.S. Securities & Exchange Comm.
410 17th Street
Suite 700
Denver, CO 80202   (303) 844-2071

Paul J. White
2212 Osceola Street
Denver, CO 80212   (303) 458-7316

Bruce R. Winsor
Holland & Hart
555 17th St.
Suite 2900
Denver, CO 80202   (303) 295-8055

Lesley S. Witt
Sheridan, Ross & Mc Intosh
One United Bank Center
1700 Lincoln Street
35th Floor
Denver, CO 80203   (303) 863-9700

Max L. Wymore
3441 South Ivy Way
Denver, CO 80222   (303) 756-9162

Mervyn L. Young
675 S. University Blvd., #501
Denver, CO 80209   (303) 733-5687

Thomas H. Young
Rothgerber, Appel, Powers
& Johnson
1200 17th St., Suite 3000
Denver, CO 80202   (303) 623-9000

Lee W. Zieroth
Cohen Brame & Smith
1700 Lincoln Street
Suite 1800
Denver, CO 80203   (303) 837-8800

David F. Zinger
Sheridan, Ross & Mc Intosh
One United Bank Center
1700 Lincoln Street
35th Floor
Denver, CO 80203   (303) 863-9700

Kenton L. Freudenberg
P.O. Box 841
Durango, CO 81302   (303) 259-2765

Maxwell C. Freudenberg
P.O. Box 841
Durango, CO 81301   (303) 259-2765

Jerry T. Kearns
P.O. Box 4596
Englewood, CO 80155   (303) 773-9222

Ken Koestner
Telectronics & Cordiss Pacing Sys.
7400 S. Tucson Way
Englewood, CO 80112   (303) 790-8000

Robert M. Krone
3271 Cherryridge Road
Englewood, CO 80110   (303) 761-5657

John R. Linton
Re/Max International, Inc.
P.O. Box 3907
Englewood, CO 80111   (303) 770-5531

John S. Mc Guire
Purcell & Mc Guire
5350 South Dtc Parkway
Englewood, CO 80111   (303) 773-9222

Robert E. Purcell#
Purcell & Mc Guire
5350 S. Dtc Parkway
Englewood, CO 80111   (303) 773-9222

Timothy R. Schulte
U.S. West, Inc.
7800 E. Orchard Rd.
Suite 480
Englewood, CO 80111   (303) 793-6566

Carl C. Batz
2207 Charolais Drive
Fort Collins, CO 80526   (303) 484-9023

Karl E. Bring
Hewlett-Packard Company
Legal Dept.
3404 E. Harmony Road
M.S. 79
Fort Collins, CO 80525   (303) 229-3800

William W. Cochran, II
Hewlett-Packard Company
Legal Dept.
3404 E. Harmony Road
M.S. 79
Fort Collins, CO 80525   (303) 229-3800

Hugh H. Drake
P.O. Box 727
Fort Collins, CO 80522   (303) 493-0123

Dean P. Edmundson
Stuart Professional Park
Suite 2160
1136 E. Stuart Street
Fort Collins, CO 80525   (303) 224-9502

Jeffery B. Fromm
Hewlett-Packard Company
Legal Dept.
3404 E. Harmony Road
M.S. 79
Fort Collins, CO 80525   (303) 229-7247

Guy J. Kelley
Hewlett-Packard Co.
Legal Dept.
3404 E. Harmony Rd.
M.S. 79
Fort Collins, CO 80525   (303) 229-3800

Edward L. Miller
Hewlett-Packard Co.
Legal Dept.
3404 E. Harmony Rd.
M.S. 79
Fort Collins, CO 80525   (303) 229-3800

Luke Santangelo
315 West Oak Street
7th Floor
Fort Collins, CO 80521   (303) 224-3100

Augustus W. Winfield#
Hewlett-Packard Co.
Legal Dept., M.S. 79
3404 E. Harmony Rd.
Fort Collins, CO 80525   (303) 229-3142

John P. Bailey
240 Old Y Road
Golden, CO 80401   (303) 526-0099

John Howard Barney#
Tosco Corporation
18200 W. Highway 72
Golden, CO 80403   (303) 425-6021

Thomas S. Birney
Bradley, Campbell & Carney, P.C.
1717 Wash. Ave.
Golden, CO 80401   (303) 278-3300

Jerald J. Devitt
Devitt & Weiszmann
1301 Arapahoe St.
Suite 300
Golden, CO 80401   (303) 279-3344

Michael A. Goodwin#
Adolph Coors Company
MS 120
Golden, CO 80401   (303) 277-2463

Kenneth L. Richardson
Solar Energy Research Inst.
1617 Cole Blvd.
Golden, CO 80401   (303) 231-7724

Steven G. Austin#
G. E. Shields, P.C.
12600 W. Colfax
Suite C460
Lakewood, CO 80215-5000   (303) 233-9911

Clarence O. Babcock#
5 South Flower St.
Lakewood, CO 80226   (303) 238-0912

Raymond Fink
44 Union Blvd.
Suite 620
Lakewood, CO 80228   (303) 988-1568

Timothy J. Martin
44 Union Blvd.
Suite 620
Lakewood, CO 80228   (303) 988-0800

Dana S. Rewoldt
44 Union Street
Suite 620
Lakewood, CO 80228   (303) 080-0

E. Michael Byorick
5908 S. Clayton St.
Littleton, CO 80121   (303) 797-3692

C. Kelley Crossman#
P. O. Box 3129
Littleton, CO 80161   (303) 740-9452

Jack E. Ebel
Marathon Oil Company
P.O. Box 269
Littleton, CO 80160   (303) 794-2601

Monte L. Gleason
P.O. Box 2034
Littleton, CO 80161   (303) 295-5586

Jack L. Hummel
Marathon Oil Co.
7400 S. Broadway
P.O. Box 269
Littleton, CO 80160   (303) 794-2601

John L. Isaac
5545 S. Lee Street
Littleton, CO 80127   (303) 756-1311

Richard A. Kulp
6861 S. Clayton Way
Littleton, CO 80122   (303) 741-3969

Edna M. O Connor
4701 Homestead Drive
Littleton, CO 80123   (303) 797-9057

Cornelius P. Quinn
7254 S. Chase Way
Littleton, CO 80123   (303) 978-0477

Norvell Edward Von Behren
7463 S. Marion Street
Littleton, CO 80122   (303) 794-9516

Christopher J. Byrne
Storage Technology Corp.
Legal Dept.
2270 South 88th St.
Louisville, CO 80028   (303) 794-9516

William E. Hein
P.O. Box 335
Loveland, CO 80539   (303) 667-6741

Donald W. Erickson
P.O. Box 482
Road W35
Norwood, CO 81423   (303) 327-4803

James M. Graziano
4662 Weld Co. Rd. 34
Platteville, CO 80651   (303) 535-4743

James Robert Young
1333 W. 120th Ave.
Suite 302
Westminster, CO 80234   (303) 457-4975

David S. Woronoff
33876 Cliff Road
P.O. Box 823
Windsor, CO 80550   (303) 686-5458

Connecticut

Joseph R. Spalla
17 Oxbow Drive
Avon, CT 06001   (203) 673-9201

Frederick J. Mc Carthy, Jr.
54 Palmer Road
Bantam, CT 06750   (203) 567-9698

Joseph R. Carvalko, Jr.
34 Redwood Drive
Bethel, CT 06801   (203) 798-9310

Robert R. Hubbard
13 Fawn Road
Bethel, CT 06801   (203) 744-3527

Lawrence A. Cox#
CIGNA Corporation
900 Cottage Grove Road
Bloomfield, CT 06002   (203) 726-8930

George B. Yntema
61 Vernon Road
Bolton, CT 06043   (203) 643-9358

Paul J. Lerner
225 Montowese Street
Branford, CT 06405   (203) 481-6550

Terrence Mac Laren
7 Howd Avenue
Branford, CT 06405   (203) 488-2597

John R. Doherty
4380 Main Street
Bridgeport, CT 06606   (203) 371-7130

James E. Espe
General Elec. Co.
International Pat. Operation
1285 Boston Avenue
Bridgeport, CT 06601   (203) 382-3515

Arthur M. King
General Electric Co.
1285 Boston Avenue
Bridgeport, CT 06601   (203) 382-3508

Philip G. Luckhardt
General Electric Co.
1285 Boston Ave.
Bldg. 30EE
Bridgeport, CT 06601   (203) 382-2625

James W. Mitchell
General Electric Co.
International Pat. Operation
1285 Boston Ave.
23 CW
Bridgeport, CT 06601   (203) 382-2768

Guido Moeller
General Electric Company
International Pat. Operation
1285 Boston Avenue
Bridgeport, CT 06601   (203) 382-3870

Bernard Snyder
General Electric Co.
International Pat. Operation
1285 Boston Ave.
Bldg., 23CW
Bridgeport, CT 06601   (203) 382-4225

Henry Irwin Steckler
General Electric Co.
1285 Boston Avenue
Bldg. 23CW
Bridgeport, CT 06601   (203) 382-2000

Howard F. Van Denburgh
General Electric Co.
1285 Boston Ave.
Bldg. 23 LW
Bridgeport, CT 06601   (203) 382-3285

Edward Herbert#
Route 44
Canton, CT 06019   (203) 693-2204

Dale Lynn Carlson
Olin Corporation
350 Knotter Drive
Cheshire, CT 06410   (203) 271-4059

Ralph D Alessandro
Olin Research Ctr.
350 Knotter Drive
Cheshire, CT 06410   (203) 271-4055

James B. Haglind#
Olin Corporation
350 Knotter Drive
Cheshire, CT 06410   (203) 271-4057

H. Samuel Kieser
Olin Corporation
350 Knotter Drive
Cheshire, CT 06410   (203) 271-4000

Donald M. Papuga
Olin Corporation
350 Knotter Drive
P.O. Box 350
Cheshire, CT 06410   (203) 271-4037

Gregory S. Rosenblatt
Olin Corporation
350 Knotter Drive
P.O. Box 586
Cheshire, CT 06410   (203) 271-4049

William A. Simons
Olin Corporation
350 Knotter Drive
P.O. Box 586
Cheshire, CT 06410   (203) 271-4063

John R. Wahl
Olin Corporation
Olin Research Center
350 Knotter Dr.
P.O. Box 586
Cheshire, CT 06410   (203) 271-4373

Paul Weinstein
Olin Corp.
350 Knotter Drive
P.O. Box 586
Cheshire, CT 06410   (203) 271-4052

Donald H. Winslow#
134 Tow Path Lane
Cheshire, CT 06410   (203) 272-9205

Charles G. Nessler
P.O. Box H
Chester, CT 06412   (203) 526-9149

Edward L. Kochey, Jr
Box 66
Smith Hill Road
Colebrook, CT 06021   (203) 379-5359

Curtis W. Carlson
277 Cognewaugh Road
Cos Cob, CT 06807   (203) 661-2210

Norman L. Balmer
Union Carbide Corporation
E-1266
39 Old Ridgebury Road
Danbury, CT 06817   (203) 794-6343

Sharon A. Blinkoff
Ethan Allen Inc.
Off. of Gen. Coun.
Ethan Allen Dr.
P.O. Box 1966
Danbury, CT 06813   (203) 743-8509

Paul L. Bollo
57 North Street
Suite 210
Danbury, CT 06810   (203) 798-8360

Robert C. Brown
Union Carbide Corp.
Law Dept. E-3268
39 Old Ridgebury Road
Danbury, CT 06817   (203) 794-6207

Shirley L. Church
Union Carbide Corporation
Law Dept. E2-258
39 Old Ridgebury Road
Danbury, CT 06817   (203) 794-6130

Gerald L. Coon
Union Carbide Corp.
Law Dept.-Pat. Sect.
39 Old Ridgebury Road
Danbury, CT 06817   (203) 794-6221

Aldo John Cozzi
Union Carbide Corp.
Law Dept.-Pat. Sect.
Old Ridgebury Rd.
Danbury, CT 06817   (203) 794-6197

Janet R. Elliott#
Advanced Technology Materials, Inc.
7 Commerce Drive
Danbury, CT 06810   (203) 794-1100

Clyde V. Erwin, Jr.
Union Carbide Corp.
Law Dept. E2
39 Old Ridgebury Road
Danbury, CT 06817   (203) 794-6350

J. Hart Evans
Union Carbide Corp.
39 Old Ridgebury Road
E3-267
Danbury, CT 06817   (203) 794-6201

Reynold Joseph Finnegan
Union Carbide Corp.
Old Ridgebury Road
Danbury, CT 06817   (203) 794-6336

Alvin H. Fritschler
Union Carbide Corp.
Law Dept.-Pat. Sect.
39 Old Ridgeway Rd.
Danbury, CT 06817   (203) 794-6116

William F. Gray
Union Carbide Chemicals &
Plastics Co., Inc.
39 Old Ridgebury Road
Danbury, CT 06817   (203) 794-2512

Robert A. Hays
Hughes Danbury Optical Sys., Inc.
100 Wooster Heights Rd.
M.S. 811
Danbury, CT 06810   (203) 797-6045

Sharon H. Hegedus
Union Carbide Corp.
Law Dept., E2
39 Old Ridgebury Rd.
Danbury, CT 06817   (203) 797-6045

Lawrence G. Kastriner
Union Carbide Corp.
39 Old Ridgebury Rd.
Law Dept.
Danbury, CT 06817   (203) 794-6120

Marylin Klosty
Union Carbide Corp.
39 Old Ridgebury Road
Danbury, CT 06817   (203) 794-6237

Stanley Ktorides
Union Carbide Corp.
39 Old Ridgebury Rd.
Danbury, CT 06817   (203) 794-6122

Joseph F. Leightner
Union Carbide Chemicals &
Plastics Co., Inc.
D2-243
39 Old Ridgebury Rd.
Danbury, CT 06817   (203) 794-2815

Paul W. Leuzzi, III
Union Carbide Corp.
39 Old Ridgebury Road
Danbury, CT 06817   (203) 794-6346

Jean B. Mauro
Union Carbide Corp.
Old Ridgebury Road
Danbury, CT 06817   (203) 794-6394

Cornelius Francis O Brien
Union Carbide Corp.
39 Old Ridgebury Road
Danbury, CT 06817   (203) 794-6160

Thomas I. O Brien
Union Carbide Corp.
39 Old Ridgebury Road
Danbury, CT 06817   (203) 794-6194

Chung K. Pak
Union Carbide Industrial
Gas Inc.
Law Dept.
39 Old Ridgebury Road
Danbury, CT 06817   (203) 794-6122

John S. Piscitello
Union Carbide Corp.
39 Old Ridgebury Road
Danbury, CT 06817   (203) 794-6241

Leo A. Plum
Union Carbide Chemicals
& Plastics Co., Inc.
39 Old Ridgebury Road
Danbury, CT 06817   (203) 794-6280

Dominic J. Terminello
Union Carbide Chemicals
& Plastics Co., Inc.
39 Old Ridgebury Rd.
Danbury, CT 06817 (203) 794-5015

Eugene C. Trautlein
Union Carbide Chemicals
& Plastics Co., Inc.
39 Old Ridgebury Rd.
Danbury, CT 06817 (203) 794-6303

Clement J. Vicari
Union Carbide Chemical
& Plastics Co.
Old Ridgebury Rd.
Danbury, CT 06817 (203) 794-6224

Gary L. Wamer
First Brands Corporation
83 Wooster Heights Rd.
P.O. Box 1911
Danbury, CT 06813 (203) 731-2304

S. A. Giarratana
4 Short Lane
Darien, CT 06820 (203) 655-7161

Albert Willis Scribner
6 Country Club Road
Darien, CT 06820 (203) 322-2770

Kenneth C. Baran#
Pratt & Whitney
400 Main Street
East Hartford, CT 06108 (203) 565-3727

Francis J. Maguire, Jr.
33 Colby Drive
East Hartford, CT 06108 (203) 289-5023

Eric W. Petraske
United Technologies Corp.
One Riverview Square
7th Floor
East Hartford, CT 06108 (203) 548-2537

Terrance J. Radke
United Technologies Corp.
Pat. Dept.
One Riverview Square
East Hartford, CT 06101 (203) 548-2503

Eric A. Shevchenko#
Pratt & Whitney
400 Main Street
M/S 118-27
East Hartford, CT 06108 (203) 565-2169

Loren P. Stolp
United Technologies Corp.
400 Main Street
East Hartford, CT 06108 (203) 565-6452

Raymond E. Stone, Jr.
United Technologies Corp.
Pratt & Whitney Group
400 Main Street
East Hartford, CT 06108 (203) 565-2847

George C. Butenkoff
169 Wells Rd.
East Windsor, CT 06088 (203) 623-0441

Benton Blair
Rose Lane
East Woodstock, CT 06244 (203) 928-2334

H. Gibner Lehmann
Five Kent Road
Easton, CT 06612 (203) 372-7695

K. Gibner Lehmann
Five Kent Road
Easton, CT 06612 (203) 372-7695

Barry R. Lipsitz
55 Sweetbriar Trail
Easton, CT 06612 (203) 268-4440

Herman A. Michelson
12 Princess Pine Lane
Easton, CT 06612 (203) 374-7346

Robert L. Olson
36 St. Thomas Street
Enfield, CT 06082 (203) 745-2307

Erwin F. Berrier, Jr.
General Electric Co.
3135 Easton Turnpike
Fairfield, CT 06431 (203) 373-2446

Clarence Arthur Green
Perman & Green
425 Post Road
Fairfield, CT 06430 (203) 259-1800

Mark F. Harrington
Perman & Green
425 Post Road
Fairfield, CT 06430 (203) 259-1800

Maurice M. Klee
1951 Burr Street
Fairfield, CT 06430 (203) 255-1400

David N. Koffsky
Perman & Green
425 Post Road
Fairfield, CT 06430 (203) 259-1800

Peter W. Krehbiel
Babcock Industries Inc.
425 Post Road
Fairfield, CT 06430 (203) 255-7158

Charles R. Miranda
136 Misty Wood Lane
Fairfield, CT 06430 (203) 259-5318

Martey Robert Perman
Perman & Green
425 Post Road
Fairfield, CT 06430 (203) 259-1800

David Enoch Pitchenik
113 Wagon Hill Rd.
Fairfield, CT 06430 (203) 351-7500

George Richard Powers
General Electric Co.
W3-D1
3135 Easton Turnpike
Fairfield, CT 06431 (203) 373-2835

Arthur V. Puccini
General Electric Co.
3135 Easton Turnpike - W2E
Fairfield, CT 06431 (203) 373-3374

Stephen J. Rudy
245 Unquowa Road
Suite 135
Fairfield, CT 06430 (203) 254-3735

Melvin I. Stoltz
Mattern, Ware, Stoltz
& Fressola
34 Sherman Court
P.O. Box 783
Fairfield, CT 06430 (203) 255-8881

Michael J. Tully
Perman & Green
425 Post Road
Fairfield, CT 06430 (203) 259-1800

Thomas L. Tully
Perman & Green
425 Post Road
Fairfield, CT 06430 (203) 259-1800

Patrick J. Walsh
2713 Congress St.
Fairfield, CT 06430 (203) 259-1800

Mark R. Warefield
General Electric Co.
3135 Easton Turnpike
W3C
Fairfield, CT 06431 (203) 373-2448

Breffni X. Baggot
United Technologies Corp.
4 Farm Springs
Farmington, CT 06032 (203) 676-5755

Robert P. Hayter
Otis Elevator Company
Patent Dept.
Four Farm Springs
Farmington, CT 06032 (203) 676-5760

Charles Morgan Hussey
Emhart Corp.
426 Colt Hwy.
Farmington, CT 06032 (203) 678-3226

Clifford L. Tager
United Technologies Corp.
Otis Elevator Co.
4 Farm Springs
Farmington, CT 06032 (203) 676-5758

Robert T. Casey
71 Frank Street
Forestville, CT 06010 (203) 583-1515

Jeffrey F. Kushin
97 Deerfield Drive
Glastonbury, CT 06033 (203) 633-3066

Paul R. Audet
American Can Packaging Inc.
American Lane
Greenwich, CT 06830 (203) 552-3260

Ernestine C. Bartlett
American Can Packaging Inc.
American Lane
P.O. Box 2600
Greenwich, CT 06836   (203) 552-3261

Michael A. Ciomek
Amax Inc
55 Railroad Avenue
Greenwich, CT 06836   (203) 629-6119

James Benton Grant
Cummings & Lockwood
Two Greenwich Plaza
Greenwich, CT 06830   (203) 869-1200

Norton Steele Johnson
26-B Putnam Green
Greenwich, CT 06860   (203) 531-1728

James Joseph Mc Keever#
Suite 101
78 East Putnam Avenue
Greenwich, CT 06830   (203) 622-6550

Alfred E. Miller
150-152 Mason Street
Greenwich, CT 06830   (203) 661-1900

Gregg C. Benson
Pfizer Incorporated
Eastern Point Road
Groton, CT 06340   (203) 441-4901

Mervin E. Brokke
Pfizer, Inc.
Patent Dept.
Eastern Point Rd.
Groton, CT 06340   (203) 441-5920

William Carter Everett
General Dynamics Corp.
Electric Boat Div.
Eastern Point Rd.
Groton, CT 06340   (203) 441-8252

Gregory E. Gardiner#
Pfizer Inc.
Eastern Point Rd.
Groton, CT 06340   (203) 441-3176

John T. Lumb
Pfizer Inc.
Eastern Point Road
Groton, CT 06340   (203) 441-4902

Donald S. Mc Farlin
Pfizer Inc.
Patent Dept.
Eastern Point Road
Groton, CT 06340   (203) 441-5820

James M. Mc Manus
Pfizer Inc.
Eastern Point Rd.
Groton, CT 06340   (203) 441-4903

John S. Oki
Pfizer Incorporated
Eastern Point Road
Groton, CT 06340   (203) 441-3500

Arthur Dean Olson
Pfizer Inc.
Patent Dept.
Eastern Point Road
Groton, CT 06340   (203) 441-4904

Paul D. Thomas#
271 Plant Street
Groton, CT 06340   (203) 445-6476

James S. Rose#
1820 Durham Road
Guilford, CT 06437   (203) 453-3120

Thomas P. O Day
66 Charlton Hill
Hamden, CT 06518   (203) 248-1483

Robert A. Seemann#
89 Earl Avenue
Hamden, CT 06514   (203) 288-2122

Marie F. Zuckerman#
2911 Dixwell Avenue
Suite 305B
Hamden, CT 06518   (203) 248-3907

James E. Alix
Chilton, Alix & Van Kirk
750 Main Street
Hartford, CT 06103   (203) 527-9211

Dominic J. Chiantera
United Technologies Corp.
Pat. Dept.
Hartford, CT 06101   (203) 548-2526

Ralph H. Chilton
Chilton, Alix & Van Kirk
750 Main Street
Hartford, CT 06103   (203) 527-9211

Alan C. Cohen
United Technologies Corporation
United Technologies Bldg.
Pat. Dept.
Hartford, CT 06101   (203) 548-2505

Peter Louis Costas
Costas, Montgomery & Dorman, P.C.
3 Lewis Street
Hartford, CT 06103   (203) 278-9892

Diane F. Covello
Chilton, Alix & Van Kirk
750 Main Street
Hartford, CT 06103   (203) 527-9211

Lloyd D. Doigan
United Technologies Corporation
United Technologies Bldg.
Hartford, CT 06101   (203) 548-2567

Joseph A. Fischetti
Mc Cormick, Paulding & Huber
266 Pearl Street
Hartford, CT 06103   (203) 549-5290

Gene D. Fleischhauer
United Technologies Corp.
United Tech. Bldg.
Pat. Dept.
Hartford, CT 06101   (203) 548-2514

Norman Friedland
United Technologies Corp.
One Financial Plaza
Hartford, CT 06101   (203) 548-2534

Robert J. Galiette
Le Boeuf, Lamb,
Leiby & Mac Rae
6 Central Row
Hartford, CT 06103   (203) 549-6960

Robert E. Greenstien
United Technologies Corp.
Hartford, CT 06101   (203) 548-2536

Michael Grillo#
United Technologies Corp.
United Technologies Bldg.
Patent Dept.
Hartford, CT 06101   (203) 548-2534

James K. Grogan
Mc Cormick, Paulding & Huber
266 Pearl Street
Hartford, CT 06103   (203) 549-5290

Harry J. Gwinnell
United Technologies Corp.
Patent Dept.
Hartford, CT 06101   (203) 548-2508

William W. Habelt
Patent Dept.
United Technologies Corp.
Hartford, CT 06101   (203) 548-2562

Frederick J. Haesche
Mccormick, Paulding & Huber
266 Pearl St.
Hartford, CT 06103   (203) 549-5290

Vernon F. Hauschild
Mc Cormick, Paulding & Huber
266 Pearl Street
Hartford, CT 06103   (203) 548-2527

Christopher T. Hayes
United Technologies Corp.
United Technologies Building
Hartford, CT 06101   (203) 548-2562

Donald Joseph Hayes
Hayes & Reinsmith
Cityplace
Hartford, CT 06103   (203) 727-9956

John C. Hilton
Mccormick, Paulding and Huber
266 Pearl St.
Hartford, CT 06103   (203) 549-5290

Hubert F. Howson*
Dept. of Veterans Affairs
450 Main Street
Hartford, CT 06103   (203) 240-3316

Donald Keith Huber
Mccormick, Paulding and Huber
266 Pearl St.
Hartford, CT 06103   (203) 549-5290

Joseph Kentoffio
Mc Cormick, Paulding & Huber
266 Pearl Street
Hartford, CT 06103   (203) 549-5290

Richard H. Kosakowski
United Technologies Corp.
1 Financial Plaza
Hartford, CT 06101   (203) 548-2569

John C. Linderman
Mc Cormick, Paulding & Huber
266 Pearl St.
Hartford, CT 06103   (203) 549-5290

Robert E. Lucia
Emhart Corp.
P.O. Box 2730
Hartford, CT 06101   (203) 678-3371

Willajeanne F. Mc Lean
University of Connecticut
Law School
55 Elizabeth Street
Hartford, CT 06105   (203) 241-4638

Pamela J. Mercier#
United Technologies Corp.
United Technologies Bldg.
Patent Dept.
Hartford, CT 06101   (203) 548-2538

Eugene F. Miller
Loctite Corporation
Hartford Square North
Ten Columbus Blvd.
Hartford, CT 06106   (203) 520-5021

Donald W. Muirhead
United Technologies Corp.
United Technologies Bldg.
Hartford, CT 06101   (203) 548-2578

Jack Michael Pasquale
Mc Cormick, Paulding & Huber
266 Pearl Street
Hartford, CT 06103   (203) 549-5290

Theodore Roy Paulding
Mc Cormick, Paulding & Huber
266 Pearl St.
Hartford, CT 06103   (203) 549-5290

James M. Rashid
United Technologies Corp.
Patent Dept.
Hartford, CT 06101   (203) 548-2539

R. William Reinsmith
Hayes & Reinsmith
185 Asylum Street
City Place
34th Floor
Hartford, CT 06103   (203) 727-9956

Stephen E. Revis
United Technologies Corp.
Patent Dept.
Hartford, CT 06101   (203) 548-2533

Brian L. Ribando
Litton Industries, Inc.
100 Constitution Plaza
Suite 1614
Hartford, CT 06103   (203) 727-0023

L. James Ristas
Chilton, Alix & Van Kirk
750 Main Street
Hartford, CT 06103   (203) 527-9211

George Joseph Romanik
United Technologies Corp.
United Technologies Bldg.
Hartford, CT 06101   (203) 548-2508

Peter R. Ruzek
United Technologies Corp.
United Technologies Bldg.
Hartford, CT 06101   (203) 548-2535

Stephen A. Schneeberger
United Technologies Corporation
United Technologies Bldg.
Hartford, CT 06101   (203) 548-2509

Robert S. Smith
57 Pratt Street
Suite 513
Hartford, CT 06103   (203) 249-1857

Troxell K. Snyder
United Technologies Corporation
Patent Dept.
United Technologies Building
Hartford, CT 06101   (203) 548-2571

Charles E. Sohl
United Technologies Corp.
Pat. Dept.
One Financial Plaza
Hartford, CT 06101   (203) 548-2506

John M. Swiatocha
United Technologies Corp.
United Technologies Building
Hartford, CT 06101   (203) 548-2513

Roger A. Van Kirk
Chilton, Alix & Van Kirk
750 Main Street
Hartford, CT 06103   (203) 527-9211

Robert C. Walker
United Technologies Corp.
UTC Bldg.
Hartford, CT 06101   (203) 728-7810

Melvin Pearson Williams
United Technologies Corp.
One Financial Plaza
Hartford, CT 06101   (203) 728-7940

Guy D. Yale
Chilton, Alilx & Van Kirk
750 Main St.
Hartford, CT 06103   (203) 527-9211

Paul D. Greeley
12 Brookfield Drive
Huntington, CT 06484   (203) 929-6403

E. Seward Stevens
Guion and Stevens
93 West Street
Box 338
Litchfield, CT 06759   (203) 567-0821

Eldon Harmon Luther
491 Joshuatown Road
Lyme, CT 06371   (203) 526-2910

Radford W. Luther
492 Joshuatown Road
Lyme, CT 06371   (203) 576-4080

Ronald G. Cummings
308 Race Hill Road
Madison, CT 06443   (203) 421-4864

William W. Jones
6 Juniper Lane
Madison, CT 06443   (203) 245-2418

Walter J. Mc Murray
23 Windward Lane
Madison, CT 06443   (203) 785-4253

Peter F. Willig
22 Wilshire Road
Madison, CT 06443   (203) 245-3539

Donald Francis Bradley
65 Ludlow Road
Manchester, CT 06040   (203) 649-7951

Ira S. Dorman
Watkins Centre
935 Main Street
Manchester, CT 06040   (203) 649-1862

Glenn E. Karta
Uniroyal Chemical Co., Inc.
Law Dept.
World Headquarters
Middlebury, CT 06749   (203) 573-4321

Ira Jay Krakower
Uniroyal Chemical Co., Inc.
World Headquarters
Benson Road
Middlebury, CT 06749   (203) 573-4335

John A. Shedden#
Uniroyal Chemical Co., Inc.
World Headquarters
Law Dept.
Middlebury, CT 06749   (203) 573-4388

Raymond D. Thompson
Uniroyal Chemical Co., Inc.
World Headquarters
Middlebury, CT 06749   (203) 573-4385

Alan E. Steele
60 Morgan Street
Middletown, CT 06457   (203) 346-4768

Marc A. Block
IBM Corp.
472 Wheelers Farms Rd.
43A05
Milford, CT 06460   (203) 783-7000

John L. Peterson
18 Alden Place
Milford, CT 06460   (203) 878-3247

Harold M. Snyder
Dorr-Oliver Incorporated
612 Wheeler'S Farm Road
P.O. Box 3819
Milford, CT 06460   (203) 876-5419

Kenneth B. Adolphson
Ware, Fressola, Van Der Sluys
& Adolphson
Bradford Green, Bldg. Five
755 Main St., P.O. Box 224
Monroe, CT 06468   (203) 876-5419

Dennis J. Delaney
450 Monroe Turnpike
Monroe, CT 06468   (203) 261-6115

Alfred A. Fressola
Ware, Fressola, Van Der
Sluys & Adolphson
Bradford Green, Bldg. Five
755 Main St., P.O. Box 224
Monroe, CT 06468   (203) 261-1234

Stephen T. Keohane#
Ware, Fressola, Van Der Sluys
& Adolphson
755 Main Street
P.O. Box 224
Monroe, CT 06468   (203) 261-1234

John Herbert Mulholland
71 Old Zoar Road
Monroe, CT 06468   (203) 285-9117

Peter C. Van Der Sluys
Ware, Fressola, Van Der Sluys
& Adolphson
Bradford Green, Bldg. 5
755 Main St., P.O. Box 224
Monroe, CT 06468   (203) 261-1234

Robert H. Ware
Ware, Fressola, Van Der Sluys, & Adolphson
Bradford Green, Bldg. 5
755 Main St.
PO Box 224
Monroe, CT 06468   (203) 261-1234

Robert A. Hlavacek#
6 Fox Hill Road
Naugatuck, CT 06770   (203) 723-2079

Charles Joseph Fickey
7 Siwanoy Lane
New Canaan, CT 06840   (203) 966-8020

Spencer Everett Olson
15 Cobblers Green
New Canaan, CT 06840   (203) 966-1770

James J. Salerno
236 S. Bald Hill Road
New Canaan, CT 06840   (203) 966-7091

Joseph Church Sweet, Jr.
45 Wee Burn Drive
New Canaan, CT 06840   (203) 966-0909

Vincent J. Vasta, Jr.
41 Pond View Lane
New Canaan, CT 06840   (203) 966-7416

Alvin Joseph Riddles
Candlewood Isle
Box 34
New Fairfield, CT 06812   (203) 746-3470

Robert H. Bachman
Bachman & La Pointe, P.C.
Suite 1201
900 Chapel Street
New Haven, CT 06510   (203) 777-6628

Raymond Bower
Southern New England Telephone
300 George Street
New Haven, CT 06510   (203) 771-7340

Bradford J. Chaucer
109 Church Street
Suite 510
New Haven, CT 06510   (203) 785-0076

Anthony P. Delio, II
121 Whitney Avenue
New Haven, CT 06510   (203) 787-0595

Barry L. Kelmachter
Bachman & Lapointe, P.C.
900 Chapel Street
Suite 1201
New Haven, CT 06510   (203) 777-6628

Gregory P. La Pointe
Bachman & La Pointe, P.C.
900 Chapel Street
Suite 1201
New Haven, CT 06510   (203) 777-6628

Ron K. Levy
284 W. Elm Street
New Haven, CT 06515   (203) 387-2944

Robert H. Montgomery
Costas & Montgomery, P.C.
246 Church Street
New Haven, CT 06510   (203) 787-2708

Peter W. Peterson
Delio & Associates
121 Whitney Ave.
New Haven, CT 06510   (203) 787-0595

Richard S. Strickler
Bachman & La Pointe, P.C.
Suite 1201
900 Chapel Street
New Haven, CT 06510   (203) 777-6628

Bruce Edward Hosmer
158 1/2 Prospect Hill
New Milford, CT 06776   (203) 355-0869

Haynes N. Johnson
P.O. Box 420
New Milford, CT 06776   (203) 355-1184

Murray J. Kessler
Cohen & Kessler
62 Bridge Street
New Milford, CT 06776   (305) 354-4488

Edward K. Welch, II
Loctite Corporation
705 N. Mountain Road
Newington, CT 06111   (203) 280-3597

Albert William Hilburger
3 Laurel Hill Drive S.
Niantic, CT 06357   (203) 739-0202

Charles F. Blaich#
CTC & Associates
7 Corporate Drive, #118
North Haven, CT 06473   (203) 234-9636

John C. Andres
United States Surgical Corp.
150 Glover Avenue
Norwalk, CT 06856   (203) 866-5050

Rasma B. Balodis#
R. T. Vanderbilt Co., Inc.
30 Winfield St.
P.O. Box 5150
Norwalk, CT 06856   (203) 853-1400

Thomas R. Bremer
United States Surgical Corp.
150 Glover Avenue
Norwalk, CT 06856   (203) 866-5050

Jay Lionel Chaskin
80 Washington Street
Norwalk, CT 06854   (203) 838-8589

Richard L. Croiter
Continental Can Company, Inc.
800 Connecticut Avenue
Norwalk, CT 06856   (203) 855-5963

Ernest Fanwick
Burndy Corp.
Richards Ave.
Norwalk, CT 06856   (203) 852-8440

Mark Farber
United States Surgical Corp.
150 Glover Avenue
Norwalk, CT 06850   (203) 845-4585

Richard J. Gallagher
James River Corporation
P.O. Box 6000
Norwalk, CT 06856   (203) 854-2511

Neil D. Gershon
United States Surgical Corp.
150 Glover Avenue
Norwalk, CT 06856   (203) 845-1480

Edwin T. Grimes
Perkin-Elmer Corp.
Pat. Law Dept.
761 Main Avenue
Norwalk, CT 06856   (203) 762-6803

Burtsell J. Kearns
Burndy Corporation
Richards Avenue
Norwalk, CT 06856   (203) 852-8343

Samuel Kriegel
Purdue Frederick Co.
100 Connecticut Avenue
Norwalk, CT 06856   (203) 853-0123

Thomas Patrick Murphy
Perkin-Elmer Corporation
761 Main Avenue
Norwalk, CT 06859   (203) 762-4304

Jerry M. Presson
Hubbell Incorporated
584 Derby Milford Road
P.O. Box 549
Orange, CT 06477   (203) 799-4256

Walter C. Bernkopf
General Elec. Co
41 Woodford Avenue
Plainville, CT 06062   (203) 747-7135

Richard A. Menelly
General Electric Co.
41 Woodford Avenue
Plainville, CT 06062   (203) 747-7153

David G. Coker#
Schlumberger Doll Research
Old Quarry Road
Ridgefield, CT 06877   (203) 431-5209

Bernard F. Crowe
17 Langstroth Drive
Ridgefield, CT 06877   (203) 792-1063

Charles T. Emmons#
327 Old Branchville Road
Ridgefield, CT 06877   (203) 544-8523

Marc D. Foodman
Schlumberger-Doll Res. Ctr.
Legal Dept.
Old Quarry Road
Ridgefield, CT 06877   (203) 431-5550

David E. Frankhouser
Boehringer Ingelheim Corp.
90 East Ridge
P.O. Box 368
Ridgefield, CT 06877   (203) 431-5920

Louis F. Heeb
24 Strawberry Ridge Road
Ridgefield, CT 06877   (203) 438-6713

Michael G. Kroposki
Boehringer Ingelheim Corp.
90 East Ridge
Ridgefield, CT 06877   (203) 431-5858

Leonard W. Pojunas, Jr.
Schlumberger - Doll Research
Pat. Dept.
Old Quarry Road
Ridgefield, CT 06877   (203) 431-5507

Daniel Reitenbach
Boehringer Ingelheim Corp.
90 East Ridge
P.O. Box 368
Ridgefield, CT 06877   (203) 431-5844

Alan R. Stempel
Boehringer Ingelheim
Pharmaceuticals, Inc.
90 East Ridge
Ridgefield, CT 06877   (203) 431-5873

Mary - Ellen M. Timbers
Boehringer Ingelheim Corporation
Legal - Pat. Dept.
90 East Ridge
P.O. Box 368
Ridgefield, CT 06877   (203) 431-5916

Seymour Polansky#
72 Florence Road
Riverside, CT 06878   (203) 637-4020

Joanne T. Rauchfuss#
41 Terrace Ave.
Riverside, CT 06878   (203) 637-9391

Willard Weber Roberts
4 Rainbow Drive
Riverside, CT 06878   (203) 637-0513

William D. Soltow, Jr.
P.O. Box 655
Riverside, CT 06878   (203) 637-3937

Marshall E. Rosenberg#
2B Kingsley Court
Rocky Hill, CT 06067   (203) 637-3937

Lewis Clifford Brown
P.O. Box 286
Rowayton, CT 06853   (203) 637-3937

John K. Conant
23 Flicker Ln.
Rowayton, CT 06853   (203) 866-1700

Andrew K. Mc Colpin
61 Bluff Avenue
Rowayton, CT 06853   (203) 838-8815

Prescott W. May
Altschuler, May & Stanek, P.C.
12 Bank Street
Seymour, CT 06483   (203) 888-4144

Satya P. Asija
7 Woonsocket Ave.
Shelton, CT 06484   (203) 736-0774

David K. Dabbiere
Richardson-Vicks, Inc.
One Far Mill Crossing
Shelton, CT 06484   (203) 925-7660

Barry E. Deutsch
Black & Decker U.S.A. Inc.
6 Armstrong Road
Shelton, CT 06484   (203) 926-3016

Anthony D. Sabatelli#
Richardson-Vicks, Inc.
1 Far Mill Crossing
Shelton, CT 06484   (203) 925-7907

Paul A. Sobel
444 Wooded Lane
Shelton, CT 06484   (203) 336-1811

Lloyd P. Stauder
William Carter Co.
1000 Bridgeport Ave.
Shelton, CT 06484   (203) 926-5006

Charles E. Mc Tiernan
RR #2 Box 553
Sherman, CT 06784   (203) 355-0829

Victor E. Libert
965 Hopmeadow Street
P.O. Box 538
Simsbury, CT 06070   (203) 651-9321

Frederick A. Spaeth
965 Hopmeadow Street
P.O. Box 538
Simsbury, CT 06070   (203) 651-9321

Anthony H. Handal
Handal & Morofsky
80 Washington Street
South Norwalk, CT 06854   (203) 838-8589

James D. Mackinnon
360 Beelzebub Road
South Windsor, CT 06074   (203) 644-0816

Roger B. Mc Cormick
Gerber Scientific, Inc.
83 Gerber Road West
South Windsor, CT 06074   (203) 644-1551

Olin T. Sessions
Gerber Scientific, Inc.
83 Gerber Road West
South Windsor, CT 06074   (203) 644-1551

George A. Dalin#
197A Heritage Village
Southbury, CT 06488   (203) 264-9418

Lawrence Hager
Southbury Plaza Prof. Bldg. 201
Southbury, CT 06488   (203) 264-3515

Allen D. Brufsky
181 Old Post Road
P.O. Box 59
Southport, CT 06490   (203) 255-8900

Armand Cifelli
Kramer, Brufsky & Cifelli, P.C.
181 Old Post Road
P.O. Box 59
Southport, CT 06490   (203) 255-8900

John F. Cullen
86 Fawn Ridge Lane
Southport, CT 06490   (203) 255-3236

Robert J. Eck
245 Daybreak Road
Southport, CT 06490   (203) 259-1128

Arthur T. Fattibene
2480 Post Road
Southport, CT 06490   (203) 255-4400

Paul A. Fattibene
2480 Post Road
Southport, CT 06490   (203) 255-4400

Julia D. Hart
Kramer, Brufsky & Cifelli, P.C.
181 Old Post Road
Southport, CT 06490   (203) 255-8900

Barry Kramer
Kramer, Brufsky & Cifelli, P.C.
181 Old Post Road
P.O. Box 59
Southport, CT 06490   (203) 255-8900

Gerald E. Linden
Kramer, Brufsky & Cifelli
181 Old Post Road
P.O. Box 59
Southport, CT 06490   (203) 255-8900

Duane C. Basch#
Xerox Corp.
800 Long Ridge Road
P.O. Box 1600
Stamford, CT 06904   (203) 968-3118

Gregory J. Battersby
Grimes & Battersby
8 Stamford Forum
301 Tresser Blvd.
Stamford, CT 06901   (203) 324-2828

Howard M. Bollinger
Parmelee, Bollinger & Bramblett
460 Summer Street
Stamford, CT 06901   (203) 327-2650

Garold E. Bramblett, Jr.
Parmelee, Bollinger & Bramblett
460 Summer Street
Stamford, CT 06901   (203) 327-2650

Alice Comins Brennan
American Cyanamid Company
1937 W. Main Street
Stamford, CT 06904   (203) 348-7331

R. Bruce Brooks#
27 Northill Street
Apt. 1F
Stamford, CT 06907   (203) 324-5253

Harvey M. Brownrout
Xerox Corp.
P.O. Box 1600
Stamford, CT 06904   (203) 968-3106

William H. Calnan, IV
American Cyanamid Co.
Pat. Law Dept.
P.O. Box 60
Stamford, CT 06904   (203) 348-7331

Francis Noel Carten
17 Mill Stone Circle
Stamford, CT 06903   (203) 329-1771

James R. Cartiglia
St. Onge, Steward, Johnston
& Reens
986 Bedford Street
Stamford, CT 06905   (203) 324-6155

Thaddius J. Carvis, Jr.
St. Onge Steward Johnston
& Reens
986 Bedford Street
Stamford, CT 06905   (203) 324-6155

Lawrence A. Cavanaugh
20 Victoria Lane
Stamford, CT 06902   (203) 229-8502

John H. Chapman*
Chapman & Moran
One Landmark Square
Stamford, CT 06901   (203) 358-9390

Edward A. Conroy, Jr.
American Cyanamid Co.
1937 W. Main Street
Stamford, CT 06904   (203) 348-7331

Eugene Stephen Cooper
CBS, Inc.
227 High Ridge Road
Stamford, CT 06905   (203) 327-2000

Charles F. Costello, Jr.
American Cyanamid Corp.
1937 W. Main St.
P.O. Box 60
Stamford, CT 06904   (203) 348-7331

Samuel S. Cross
Kelley, Drye & Warren
Six Stamford Forum
Stamford, CT 06901   (203) 324-1400

F. Eugene Davis, IV
P.O. Box 8206
Stamford, CT 06905   (203) 329-8117

Douglas E. Denninger
American Cyanamid Co.
1937 West Main Street
P.O. Box 60
Stamford, CT 06904   (203) 321-2255

Kenneth J. Dow
American Cyanamid Company
Pat. Law Dept.
1937 W. Main Street
P.O. Box 60
Stamford, CT 06904   (203) 321-2659

David Fink
6 Todd Lane
Stamford, CT 06905   (203) 322-9130

Denis Arthur Firth#
St. Onge Steward Johnston
& Reens
986 Bedford Street
Stamford, CT 06905   (203) 324-6155

Hensley M. Flash
112 Lawn Avenue
Stamford, CT 06902   (203) 348-0289

Joseph M. Fowler
American Actuator Corporation
P.O. Box 384
Stamford, CT 06904   (203) 324-6334

James R. Frederick
Parmelee, Bollinger & Bramblett
460 Summer Street
Stamford, CT 06901   (203) 327-2650

Philip M. French
51 Bank Street
Stamford, CT 06901   (201) 357-1332

Henry Z. Friedlander
85 Riverside Ave.
Stamford, CT 06905   (203) 357-9277

Alan M. Gordon
American Cyanamid Company
1937 W. Main Street
P.O. Box 60
Stamford, CT 06904   (203) 348-7331

David P. Gordon
65 Woods End Road
Stamford, CT 06905   (203) 329-1160

Charles W. Grimes
Grimes & Battersby
Eight Stamford Forum
5th Floor
Stamford, CT 06901   (203) 324-2828

John Joseph Hagan
American Cyanamid Co.
1937 W. Main St.
Stamford, CT 06904   (203) 348-7331

James K. Hammond
St. Onge Steward Johnston
& Reens
986 Bedford
Stamford, CT 06905   (203) 324-6155

Gordon L. Hart
77 Rocky Rapids Road
Stamford, CT 06903   (203) 322-0006

John W. Hogan, Jr.
American Cyanamid Co.
Pat. Law Dept.
1937 W. Main Street
P.O. Box 60
Stamford, CT 06904   (203) 348-7331

Herbert Girard Jackson
Amer. Cyanamid Co.
Pat. Dept.
1937 W. Main St.
Stamford, CT 06904   (203) 348-7331

Albert C. Johnston
St. Onge Steward Johnston
& Reens
986 Bedford Street
Stamford, CT 06905   (212) 324-6155

Joseph J. Kaliko
One Landmark Square
Stamford, CT 06901   (203) 359-4370

Michael J. Kelly
American Cyanamid Co.
1937 W. Main Street
Stamford, CT 06904   (203) 348-7331

Barry Jay Kesselman
Xerox Corporation
800 Long Ridge Road
Stamford, CT 06904   (203) 968-3931

Milton E. Kleinman
General Signal Corp.
2 High Ridge Park
Stamford, CT 06904   (203) 357-8800

Peter K. Kontler
One Landmark Square
Suite 400
Stamford, CT 06901   (203) 348-6719

Mary M. Krinsky
St. Onge Steward Johnston
& Reens
986 Bedford Street
Stamford, CT 06905   (203) 324-6155

Warren Kunz
83 Chestnut Hill Lane
Stamford, CT 06903   (203) 329-9351

Nathaniel Levin
Pitney Bowes
Mail Drop 50-01
Stamford, CT 06926   (203) 351-6024

Joseph Levinson
Parmelee, Bollinger & Bramblett
460 Summer Street
Stamford, CT 06901   (203) 327-2650

Eugene Lieberstein
2151 Long Ridge Road
Stamford, CT 06903   (203) 794-6127

Karen A. Lowney
American Cyanamid Co.
Patent Law. Dept.
1937 West Main St.
Stamford, CT 06904   (203) 321-2200

Charles R. Malandra, Jr.
Pitney Bowes
World Headquarters
One Elmcroft Road
Stamford, CT 06926   (203) 351-7501

Stephen P. Mc Namara
St. Onge Steward Johnston
& Reens
986 Bedford Street
Stamford, CT 06905   (203) 324-6155

Robert E. Meyer
Pitney Bowes, Inc.
World Headquarters
One Elmcroft Road
Stamford, CT 06926   (203) 351-6237

Eric Y. Munson
460 Summer St.
Stamford, CT 06901   (203) 325-1361

Alphonse R. Noe
American Cyanamid Co.
1937 W. Main St.
Stamford, CT 06904   (203) 348-7331

Carmella A. O Gorman
American Cyanamid Co.
1937 W. Main Street
Stamford, CT 06904   (203) 321-2701

Robert T. Orner
Gte Service Corporation
One Stamford Forum
Stamford, CT 06904   (203) 965-3045

John F. Osterndorf
248 Thornwood Road
Stamford, CT 06903   (203) 322-3982

Charles G. Parks, Jr.
Pitney Bowes Inc.
World Headuarters
One Elmcroft Rd.
Stamford, CT 06926   (203) 351-6236

G. Kendall Parmelee
Parmelee, Bollinger & Bramblett
460 Summer St.
Stamford, CT 06901   (203) 327-2650

David C. Petre
Xerox Corp.
800 Long Ridge Rd.
P.O. Box 1600
Stamford, CT 06904   (203) 968-3231

Maria M. Potter
Walden Book Compay, Inc.
201 High Ridge Road
Stamford, CT 06904   (203) 352-2737

George W. Rauchfuss, Jr.
Parmelee, Bollinger &
Bramblett
460 Summer Street
Stamford, CT 06901   (203) 327-2650

Robert P. Raymond
American Cyanamid Co.
1937 W. Main St.
Stamford, CT 06904   (203) 321-2672

Louis H. Reens
St. Onge Steward Johnston
& Reens
986 Bedford Street
Stamford, CT 06905   (203) 324-6155

Jack W. Richards
American Cyanamid Co.
1937 W. Main St.
Stamford, CT 06904   (203) 321-2413

Stuart N. Roth
Olin Corp.
120 Long Ridge Road
Stamford, CT 06904   (203) 356-2480

Charles N.J. Ruggiero
Grimes & Battersby
8 Stamford Forum
5th Floor
Stamford, CT 06901   (203) 324-2828

Karin A. Russo#
Pitney Bowes
One Elmcroft
Stamford, CT 06926   (203) 849-3074

Melvin J. Scolnick
Pitney Bowes, Inc.
World Headquarters
Stamford, CT 06926   (203) 351-7513

James M. Serafino
206 Skymeadow Drive
Stamford, CT 06903   (203) 322-1855

Lawrence E. Sklar
Pitney Bowes, Inc.
1 Elmcroft Rd.
Stamford, CT 06926   (203) 351-7573

Raymond E. Smiley
General Signal Corp.
Patent Dept.
High Ridge Park
P.O. Box 10010
Stamford, CT 06904   (203) 329-4289

Evelyn M. Sommer
Champion Internatl. Corp.
1 Champion Plaza
Stamford, CT 06921   (203) 358-7680

William J. Speranza
St. Onge Steward Johnston
& Reens
986 Bedford Street
Stamford, CT 06905   (203) 324-6155

Ronald J. St. Onge
St. Onge, Steward, Johnston
& Reens
986 Bedford Street
Stamford, CT 06905   (203) 324-6155

Merrill Franklin Steward
Stonge, Steward, Johnson
& Reens
986 Bedford St.
Stamford, CT 06905   (203) 324-6155

Mark Paul Stone
945 Summer Street
Stamford, CT 06905   (203) 967-2507

Thomas V. Sullivan
41 Malvern Road
Stamford, CT 06905   (203) 322-4601

Thomas S. Szatkowski
American Cyanamid Company
Pat. Law Div.
1937 W. Main Street
P.O. Box 60
Stamford, CT 06904   (203) 321-2249

Frank J. Thompson
111 Prospect Street
Stamford, CT 06901   (203) 348-9987

Estelle J. Tsevdos
American Cyanamid
1937 West Main Street
P.O. Box 60
Stamford, CT 06904   (203) 321-2756

Frank M. Van Riet
American Cyanamid Co.
1937 W. Main Street
Stamford, CT 06904   (203) 321-2614

Peter Vrahotes
Pitney Bowes Inc.
World Headquarters 50-01
Stamford, CT 06926   (203) 351-7566

Donald P. Walker
Pitney Bowes, Inc.
World Headquarters
One Elmcroft Road
Stamford, CT 06926   (203) 351-7510

David A. Warmbold
American Cyanamid Co.
1937 W. Main St.
P.O. Box 60
Stamford, CT 06904   (203) 321-2466

Robert H. Whisker
Pitney Bowes Inc.
One Elmcroft Rd.
Stamford, CT 06926   (203) 351-7471

Wesley W. Whitmyer, Jr.
St. Onge Steward Johnston
& Reens
986 Bedford Street
Stamford, CT 06905   (203) 324-6155

Gene S. Winter
St. Onge Steward Johnston
& Reens
986 Bedford Street
Stamford, CT 06905   (203) 324-6155

Martin David Wittstein
Pitney-Bowes, Inc.
One Elmcroft
Stamford, CT 06926   (203) 351-7239

Ronald Zibelli
Xerox Corp.
P.O. Box 1600
800 Long Ridge Road
Stamford, CT 06904   (203) 698-3451

Michael G. Marinangeli
GTE Corporation
One Stanford Forum
Stamford, CT 06904   (203) 965-2278

Brian A. Collins#
United Technologies Sikorsky
Aircraft Division
6900 North Main Street
Stratford, CT 06601   (203) 386-4234

Istrate Ionescu
1435 North Avenue
Stratford, CT 06497   (203) 381-9400

Peter Kent
Avco Lycoming Div.
550 Main St.
Stratford, CT 06497   (203) 385-2352

Robert E. Kline
Sikorsky Aircraft
6900 Main Street
Stratford, CT 06601   (203) 386-6046

Allen A. Meyer
122-A Cayuga Lane
Stratford, CT 06497   (203) 377-7131

Milton Zucker
273A Agawam Dr.
Stratford, CT 06497   (203) 377-5835

William C. Nealon
40 Crane Hill Road
Suffield, CT 06078   (203) 668-0226

John H. Crozier
1934 Huntington Turnpike
Trumbull, CT 06611   (203) 582-9561

Raghunath V. Date#
45 Gibson Avenue
P.O. Box 194
Trumbull, CT 06611   (203) 268-7188

F. A. Iskander
14 Stirrup Drive
Trumbull, CT 06611   (203) 268-0256

Charles Patrick Martin
45 North Street
Trumbull, CT 06611   (203) 386-7504

Nicholas Skovran, Sr.
6 Havermill Road
Trumbull, CT 06611   (203) 261-2321

Harry F. Smith
P.O. Box 815
Trumbull, CT 06611   (203) 426-7186

George M. Yahwak
25 Skytop Drive
Trumbull, CT 06611   (203) 268-1951

Charles A. Warren
P.O. Box 2416
Vernon, CT 06066   (203) 749-6406

Aldo A. Algieri#
Bristol-Myers Company
5 Research Parkway
Wallingford, CT 06492   (203) 284-6024

John J. Balser
Bristol-Myers Co.
Pharmaceutical R&D Div.
5 Research Parkway
Wallingford, CT 06492   (203) 284-6121

Michelle A. Cepeda
Bristol-Myers Company
5 Research Parkway
P.O. Box 5100
Wallingford, CT 06492   (203) 284-6944

William T. Han#
Bristol-Myers Squibb Co.
5 Research Pkwy.
P.O. Box 5100
Wallingford, CT 06492   (203) 284-6025

David M. Morse
Bristol-Myers Co.
5 Research Parkway
Wallingford, CT 06492   (203) 284-6997

Richard P. Ryan#
Bristol-Myers Squibb Company
Legal Div., Pat. Dept.
5 Research Parkway
P.O. Box 5100
Wallingford, CT 06492   (203) 284-6075

Mollie M. Yang#
Bristol-Myers Company
5 Research Parkway
Wallingford, CT 06492   (203) 284-6758

Neophytos Ganiaris#
Nettleton Hollow Road
Washington, CT 06793   (203) 868-2718

William C. Crutcher
Gager, Henry & Narkis
One Exchange Place
P.O. Box 2480
Waterbury, CT 06722   (203) 597-5116

Dallett Hoopes
21 Church Street
Waterbury, CT 06702   (203) 575-1773

Deborah J. Barnett
50 Brainard Rd.
West Hartford, CT 06117   (203) 236-7665

Richard A. Dornon
Colt Industries Inc.
Charter Oak Blvd.
West Hartford, CT 06110   (203) 236-0651

Abraham Friedman
25 Pinecrest Road
West Hartford, CT 06117   (203) 236-5072

John H. Midney
140 Wood Pond Road
West Hartford, CT 06107   (203) 521-5719

Timothy F. Mills
Corporate Center West
433 S. Main St.
Suite 217
West Hartford, CT 06110   (203) 561-5410

Howard Scott Reiter
Colt Industries Inc.
Charter Oak Boulevard
P.O. Box 330651
West Hartford, CT 06133   (203) 523-2250

Edward A. Siegel
65 Flagg Road
West Hartford, CT 06117   (203) 285-4188

John Blevney Willard
111 Hunter Drive
West Hartford, CT 06107   (203) 521-7702

Jae H. Kim
Miles Inc.
Patent - Legal Dept.
400 Morgan Lane
West Haven, CT 06516   (203) 937-2340

Barbara A. Shimei
Miles Inc.
400 Morgan Lane
West Haven, CT 06516   (203) 937-2786

Pamela A. Simonton
Miles Inc.
400 Morgan Lane
West Haven, CT 06516   (203) 937-2501

Dieter J. Schaefer#
53 Godfrey Road West
Weston, CT 06883   (203) 937-2501

Alfred I. Wirtenberg
15 Wilson Road
Weston, CT 06883   (203) 544-9270

William A. Aguele
12 Gault Park Drive
Westport, CT 06880   (203) 226-1563

A. Sidney Alpert
University Patents, Inc.
P.O. Box 901
Westport, CT 06881   (203) 255-6044

Donald C. Caulfield
11 Catamount Road
Westport, CT 06880   (203) 226-0324

Francis M. Fazio
26 Pequot Trail
Westport, CT 06880   (203) 227-5059

William Kaufman
53 C Compo Beach Rd.
Westport, CT 06880   (203) 221-1803

Lloyd L. Mahone
270 Hillspoint Road
Westport, CT 06880   (203) 226-4682

Martin M. Novack
1465 Post Road East
P.O. Box 901
Westport, CT 06881   (203) 255-4373

Andrew M. Riddles
26 Oak Street
Westport, CT 06880   (203) 454-0442

Russell M. Lipes, Jr.
74 Boulter Road
Wethersfield, CT 06109   (203) 529-0724

Steven H. Flynn
244 Thayer Pond Road
Wilton, CT 06897   (203) 762-5980

Andrew L. Gaboriault
296 Chestnut Hill Road
Wilton, CT 06897   (203) 762-9030

Ernest Gergely
94 Wild Duck Road
Wilton, CT 06897   (203) 834-9906

Alfred H. Hemingway, Jr.
221 Linden Tree Road
Wilton, CT 06897   (203) 762-3826

Edward R. Hyde
261 Danbury Road
P.O. Box 494
Wilton, CT 06897   (203) 762-5444

Keith E. Mullenger
44 Ryders Lane
Wilton, CT 06897   (203) 762-9378

Richard H. Berneike
Combustion Engineering, Inc.
1000 Prospect Hill Rd.
Windsor, CT 06095   (203) 285-9106

Michael Alan Cantor
Fishman, Dionne & Cantor
88 Day Hill Road
Windsor, CT 06095   (203) 688-4470

Arthur F. Dionne
Fishman, Dionne & Cantor
88 Day Hill Road
Windsor, CT 06095   (203) 688-4470

David S. Fishman
Fishman, Dionne & Cantor
88 Day Hill Road
Windsor, CT 06095   (203) 688-4470

Arthur Edmond Fournier, Jr.
Combustion Eng., Inc.
1000 Prospect Hill Rd.
Windsor, CT 06095   (203) 285-9112

Robert J. Hoffberg
AIW-Alton Iron Works, Inc.
P.O. Box 20
Windsor, CT 06095   (203) 683-0731

Kevin E. Mc Veigh
Fishman, Dionne & Cantor
88 Dayhill Rd.
Windsor, CT 06095   (203) 688-4470

Morris N. Reinisch
4 Woods Way
Woodbury, CT 06798   (203) 688-4470

Gurdon R. Abell#
Perrin Road
RR 1, Box 202
Woodstock, CT 06281   (203) 974-2442

Delaware

Earl Christensen
3014 Wrangle Hill Road
Bear, DE 19701   (302) 834-7341

John C. Andrade
Parkowski, Noble & Guerke
116 West Water Street
P.O. Box 598
Dover, DE 19903   (302) 678-3262

Anthony P. Mentis
44 Candlewicke
Dover, DE 19901   (302) 697-3474

Floyd D. Higel#
Three H Search Systems, Ltd.
R.D. 3, Box 43
Frankford, DE 19945   (302) 732-9274

Roger Arnold Hines
3929 Heather Drive
Greenville, DE 19807   (302) 655-2265

Edward J. Kaliski
P.O. Box 3661
Greenville, DE 19807   (302) 652-5666

Don A. Erlandson
11 Guenever Drive
New Castle, DE 19720   (302) 322-3028

Stanislav Antolin#
Lanxide Corp.
P.O. Box 6077
1300 Marrows Rd.
Newark, DE 19714   (302) 456-6565

Bill N. Baron#
University of Delaware
Institute of Energy Conversion
Newark, DE 19716   (302) 451-6229

John S. Campbell#
W. L Gore & Assoc. Incs
555 Paper Mill Rd.
P.O. Box 9206
Newark, DE 19714   (302) 738-4880

Joseph P. Klimowicz
E.I. Du Pont De Nemours & Co.
Engineering Dept.
P.O. Box 6090
Newark, DE 19714   (302) 366-2613

Carol A. Lewis
Lanxide Corporation
Tralee Industrial Park
Newark, DE 19711   (302) 454-6420

Thomas J. Lundy
82 E. Park Place
Newark, DE 19711   (302) 999-2842

Michael T. Mc Cartin#
22 Merriman Rd.
Newark, DE 19713   (302) 737-2648

William E. Mc Shane
Lanxide Corporation
Tralee Industrial Park
P.O. Box 6077
Newark, DE 19714   (302) 456-6561

Mark G. Mortenson
Lanxide Corporation
One Tralee Industrial Park
Newark, DE 19711   (302) 454-6420

John A. Parkins#
14 Vassar Drive
Newark, DE 19711   (302) 731-5698

Jeffrey R. Ramberg#
Lanxide Corp.
1300 Marrows Road
P.O. Box 6077
Newark, DE 19714   (302) 456-6200

James H. Ryan
Office of Research & Patents
University of Delaware
Hulliken Hall
Newark, DE 19716   (302) 451-2136

Gary Allyn Samuels
W.L. Gore & Assoc., Inc.
551 Paper Mill Road
Newark, DE 19814   (302) 738-4880

Robert B. Stevenson
23 Lamatan Road
Newark, DE 19711   (302) 774-5306

Dena Meyer Weker
W.L. Gore & Associates
551 Paper Mill Rd.
P.O. Box 9206
Newark, DE 19714   (302) 292-4130

John G. Abramo
Abramo & Abramo
105 W. 9th Street
P.O. Box 668
Wilmington, DE 19899   (302) 652-3504

Samuel V. Abramo
Abramo & Abramo
105 West Ninth Street
P.O. Box 668
Wilmington, DE 19899   (302) 652-3504

Zarah Ainbinder#
1300 Grinnell Road
Wilmington, DE 19803   (302) 478-6314

William Stanley Alexander
Hercules, Inc.
Hercules Plaza
Wilmington, DE 19894   (302) 594-6942

William K. Baggott
E. I. Du Pont De Nemours and Co.
1007 Market St.
Wilmington, DE 19898   (302) 992-3202

Nathan Bakalar
Connolly and Hutz
1220 Market Street
P.O. Box 2207
Wilmington, DE 19801   (302) 658-9141

Claude L. Beaudoin
508 Whitby Dr.
Wilmington, DE 19803   (302) 478-2449

Richard M. Beck
Connolly and Hutz
1220 Market Street
P.O. Box 2207
Wilmington, DE 19899   (302) 658-9141

Robert W. Black
E.I. Dupont De Nemours and Co.
Du Pont Bldg. Room 8131
1007 Market St.
Wilmington, DE 19898   (302) 992-3216

Samuel S. Blight
E. I. Du Pont De Nemours
& Company
Legal Dept., Pat. Div.
1007 Market Street
Wilmington, DE 19898   (302) 992-4922

Charles L. Board, Sr.
404 Stafford Road
Wilmington, DE 19803   (302) 655-6643

Paul U. Bockrath#
1403 Stoneleigh Road
Wilmington, DE 19803   (302) 478-7533

Robert R. Bonczek
E.I. Du Pont De Nemours and Co.
1007 Market Street
D-1028
Wilmington, DE 19898   (302) 594-3603

Mary Webb Bourke
Connolly, Bove, Lodge
& Hutz
P.O. Box 2207
1220 Market Street
Wilmington, DE 19899   (302) 658-9141

Alanson Gray Bowen, Jr.
E.I. Du Pont De Nemours
& Company
Patent Division
Legal Dept.
Wilmington, DE 19898   (302) 992-3227

Albert Frank Bower
Connolly, Bove, Lodge
& Hutz
1220 Market Street
10th Floor
Wilmington, DE 19801   (302) 658-9141

Michael K. Boyer
E.I. Du Pont De Nemours & Co.
Legal Dept. - Pat. Div.
Du Pont Bldg., D-7069
1007 Market St.
Wilmington, DE 19898   (302) 773-0504

Donald L. Bruton
820 N. French St.
Wilmington, DE 19801   (302) 571-3847

Thomas J. Bucknum
E. I. Du Pont De Nemours & Co.
Legal Dept.
Wilmington, DE 19898   (302) 992-4334

Richard Hunt Burgess
Du Pont Co.
1007 Market St.
Legal, M-2704
Wilmington, DE 19898   (302) 774-5325

Lawton Arthur Burrows, Jr.
E. I. Du Pont De Nemours
& Co.
1007 Market St.
Wilmington, DE 19898   (302) 774-2729

Frederick Frank Butzi
1102 Dardel Drive
Wilmington, DE 19803   (302) 478-3357

Mark J. Caplan
E.I. Du Pont De Nemours & Co.
Legal Dept.
1007 Market St.
Wilmington, DE 19898   (302) 773-4111

Sydney R. Chirlin
Young, Conaway, Stargatt
& Taylor
P.O. Box 391
Wilmington, DE 19899   (302) 571-6622

Lynne M. Christenbury
E.I. Du Pont De Nemours & Co.
Legal Dept. P-17-1110
1007 Market Street
Wilmington, DE 19898   (302) 992-5481

Joel D. Citron#
Central Res. & Dev. Dept.
E.I. Du Pont De Nemours & Co., Inc.
Experimental Station
P.O. Box 80301
Wilmington, DE 19880   (302) 695-3642

Harvey L. Cohen
Himont Inc.
Three Little Falls Centre
2801 Centerville Rd.
P.O. Box 15439
Wilmington, DE 19850   (302) 996-6233

Robert Collat#
49 Bancroft Mills
Apt. P-14
Wilmington, DE 19806   (302) 654-1632

Michael Conner
E.I. Du Pont
Legal Dept.
Barley Mill Plaza
Wilmington, DE 19898   (302) 992-4924

Arthur G. Connolly, Sr.
Connolly, Bove, Lodge
& Hutz
1220 Market Bldg.
P.O. Box 2207
Wilmington, DE 19899   (302) 658-9141

James T. Corle
E. I. Du Pont De Nemours and Co.
Legal Dept., B-11208
1007 Market St.
Wilmington, DE 19898   (302) 774-8536

James Anthony Costello
E.I. Du Pont De Nemours
& Co., Inc.
1007 Market Street
Wilmington, DE 19898   (302) 992-4926

Richard F. B. Cox#
2009 Woodbrook Drive
Wilmington, DE 19810   (302) 475-5288

Jeffrey F. Craft
Connolly & Hutz
1220 Market Street
Wilmington, DE 19801   (302) 658-9141

Paul E. Crawford
Connolly & Hutz
P.O. Box 2207
Wilmington, DE 19899   (302) 658-9141

Harry Cress, Jr.#
2516 Foulk Woods Road
Wilmington, DE 19810   (302) 475-3157

John E. Crowe
Hercules Inc.
Hercules Plaza
Market Street
Wilmington, DE 19899   (302) 594-6949

Francis J. Crowley
514 Kerfoot Farm Road
Wilmington, DE 19803   (302) 658-2028

Joseph P. Daniszewski#
E. I. Dupont De Nemours & Co.
Central Res. & Dev. Dept.
P14-1126 Barley Mill Plaza
Wilmington, DE 19898   (302) 992-2611

Gerald E. Deitch
E. I. Du Pont De Nemours & Co.
Legal Dept.
10th & Market Streets
Wilmington, DE 19898   (301) 774-1000

Louis Del Vechio
E.I. Du Pont De Nemours and Co.
Legal Dept.
1007 Market St.
Wilmington, DE 19898   (302) 992-

Hazel L. Deming#
207 Old Mill Lane
Wilmington, DE 19803   (302) 764-5616

Thomas J. Des Rosier
E.I. Du Pont De Nemours & Co.
Legal Dept. - Pat. Division
1007 Market St.
Wilmington, DE 19898   (302) 774-2575

William E. Dickheiser
ICI Americas Inc.
Intellectual Property Section
New Murphy Rd. & Concord Pike
Wilmington, DE 19897   (302) 575-3731

Roseanne R. Duffy
E.I. Du Pont De Nemours & Co.
Barley Mill Plaza
Bldg. 17, Room 1166
Wilmington, DE 19898   (302) 992-2908

John Edward Dull
E.I. Du Pont De Nemours and Co.
Legal Dept.
Wilmington, DE 19898   (302) 774-6183

David Edwards
Hercules Incorporated
Hercules Plaza
1313 N. Market Street
Wilmington, DE 19894   (302) 594-6952

Barry Estrin
E.I. Du Pont De Nemours
& Company
Legal Dept.
Wilmington, DE 19898   (302) 992-3230

John D. Fairchild
Connolly and Hutz
1220 Market Bldg.
P.O. Box 2207
Wilmington, DE 19899   (302) 658-9141

Charles E. Feeny
E.I. Du Pont De
Nemours & Co.
Legal Dept.
1007 Market St.
Wilmington, DE 19898   (302) 774-9445

Blair Q. Ferguson#
E. I. Du Pont De Nemours
& Co.
Experimental Station
P.O. Box 80400, E400/2427
Wilmington, DE 19880   (302) 695-7244

Lynn N. Fisher
E. I. Du Pont De Nemours
& Co., Inc.
Legal Dept.
1007 Market Street
Wilmington, DE 19898   (302) 992-3221

James A. Fitzgerald#
E.I. Du Pont De Nemours
& Co., Inc.
P.O. Box 80302
Wilmington, DE 19880   (302) 695-3663

Linda A. Floyd
E.I. Du Pont De Nemours & Co.
Legal Dept.
1007 Market St.
Wilmington, DE 19898   (302) 992-4929

James J. Flynn
E. I. Du Pont De Nemours and Co.
1007 Market St.
Wilmington, DE 19898   (302) 774-4668

James A. Forstner
E. I. Du Pont De Nemours
& Company
Legal Dept.
Wilmington, DE 19898   (302) 992-3232

Richard T. Foster
Connolly & Hutz
1220 Market Bldg.
P.O. Box 2207
Wilmington, DE 19899   (302) 658-9141

George A. Frank
E. I. Du Pont De Nemours
& Company
1007 Market Street
Wilmington, DE 19898   (302) 992-3218

Hilmar L. Fricke
E.I. Du Pont De Nemours and Co.
Du Pont Bldg., Rm. 7131
1007 Market St.
Wilmington, DE 19898   (302) 774-7892

Ralph T. Gallegos
Du Pont Co.
Legal Dept.
Wilmington, DE 19898   (302) 774-7892

Richard G. Gantt#
E.I. Du Pont De Nemours
& Co., Inc.
Experimental Station E357
Wilmington, DE 19898   (302) 695-2870

Sharon A. Gibson#
Himont, Inc.
P.O. Box 15439
Wilmington, DE 19850   (302) 996-5025

Mark Goldberg
Hercules Incorporated
Patent Section
Hercules Plaza
Wilmington, DE 19894   (302) 594-6974

Kevin W. Goldstein
Duane, Morris & Heckscher
1201 Market St.
Suite 1500
Wilmington, DE 19801   (302) 571-5550

Andrew G. Golian
E.I. Du Pont De Nemours & Co.
Barley Mill Plaza
Wilmington, DE 19805   (302) 992-3228

David J. Gould
E.I. Du Pont De Nemours and Co.
Legal Dept.
1007 Market St.
Wilmington, DE 19898   (302) 774-2487

Theodore C. Gregory
E.I. Du Pont De Nemours and Co.
Legal Dept.
1007 Market Street
Wilmington, DE 19898   (302) 992-4925

John E. Griffiths
E.I. Du Pont De Nemours
& Co.
Legal Dept.
Pat. Div.
Wilmington, DE 19898   (302) 992-4941

John Ellsworth Griffiths
2705 Marklyn Drive
Wilmington, DE 19810   (302) 475-7961

William H. Hamby
E.I. Du Pont De Nemours
& Company
1007 Market Street
Legal Dept.
Wilmington, DE 19898   (302) 774-2575

Earl L. Handley
E. I. Du Pont De Nemours and Co.
1007 Market Street
Wilmington, DE 19898   (302) 773-4325

Gerald A. Hapka
E. I. Du Pont De Nemours & Co.
Legal Dept.
1007 Market St.
Wilmington, DE 19898   (302) 774-9466

William P. Hauser#
E.I. Dupont De Nemours and Co.
Electronics Dept.
Barley Mill Plaza 21-2354
Wilmington, DE 19898   (302) 992-3433

Selma Hayman
Berg, Bifferato, Tighe
& Cottrell, P.A.
First Federal Plaza
Suite 300, P.O. Box 33
Wilmington, DE 19899   (602) 571-8600

David E. Heiser
E. I. Du Pont De Nemours
& Company
Legal Department
1007 Market Street
Wilmington, DE 19898   (302) 892-1926

John S. Hendrickson
E.I. Du Pont De Nemours & Co.
Legal Dept., Pat. Div.
M-2710
Wilmington, DE 19898   (302) 773-3214

C. Harold Herr
112 Rockingham Drive
Wilmington, DE 19803   (302) 478-3871

W. Victor Higgs
E.I. Du Pont De Nemours and Co.
1007 Market St.
Wilmington, DE 19898   (302) 774-6278

Frank C. Hilberg, Jr.
E. I. Du Pont De
Nemours & Co.
1007 Market Street
Wilmington, DE 19898   (302) 774-2012

Donald Allen Hoes
211 Sorrel Drive
Surrey Park
Wilmington, DE 19803   (302) 478-5853

Joseph Lee Hollowell#
P.O. Box 5447
Wilmington, DE 19808   (302) 478-5853

George H. Hopkins
217 Hitching Post Dr.
Wilmington, DE 19803   (302) 478-0831

Joanne L. Horn
Hinmont Incorporated
P.O. Box 15439
Wilmington, DE 19850   (302) 996-6154

Edward Leigh Hunt, Sr.
1809 Wyckwood Court
Wilmington, DE 19803   (302) 656-4189

Frederick D. Hunter
E. I. Du Pont De Nemours
& Co., Inc.
1007 Market St.
Wilmington, DE 19898   (302) 774-2383

Donald Wayne Huntley
E.I. Dupont De Nemours and Co.
Legal Dept.
1007 Market St.
Wilmington, DE 19898   (302) 774-3811

Rudolf Edward Hutz
Connolly, Bove, Lodge
& Hutz
1220 Market Street
Wilmington, DE 19801   (302) 658-9141

Martha S. Imbalzano#
E.I. Du Pont De Nemours & Co.
Barley Mill Plaza 21
Wilmington, DE 19898   (301) 992-3435

Lawrence Isakoff#
E.I. Dupont De Nemours Co.
Legal Dept.
M-2714
10th & Market Sts.
Wilmington, DE 19898   (302) 774-5302

Roy V. Jackson#
Hercules Incorporated
Hercules Plaza
Wilmington, DE 19894   (302) 594-6958

Thomas E. Jackson#
ICI Americas Inc.
Law Dept. - Intellectual Prop. Section
Concord Pike & New Murphy Rd.
Wilmington, DE 19897   (302) 575-2884

Robert Jacobs
Jacobs & Crumplar, P.A.
2 East 7th Street
P.O. Box 1271
Wilmington, DE 19899   (302) 656-5445

James L. Jersild
E.I. Du Pont De Nemours and Co.
1007 Market St.
Wilmington, DE 19898   (302) 774-8618

James T. Jones, III
ICI Americas, Inc.
Murphy Road & Concord Pike
Wilmington, DE 19897   (302) 575-3759

Karin J. Karel#
E.I. Du Pont De Nemours, Inc.
E 262/234-B
Wilmington, DE 19880   (302) 695-2508

Elliott A. Katz
E. I. Du Pont De Nemours & Co.
1007 Market St.
Wilmington, DE 19898   (302) 774-5330

Michael B. Keehan
Hercules Incorporated
Pat. Sect./Law Dept.
1313 N. Market Street
Wilmington, DE 19894   (302) 594-6954

Don M. Kerr
E. I. Du Pont De Nemours
& Company, Inc.
Legal Dept., BMP-17-1168
P.O. Box 80017
Wilmington, DE 19880   (302) 992-3215

Karen Kay King
E.I. Du Pont De Nemours
& Company
Legal Department
1007 Market Street
Wilmington, DE 19898   (302) 992-5909

Robert C. Kline
E.I. Du Pont De Nemours and Co.
1007 Market St.
Wilmington, DE 19898   (302) 774-5876

John J. Klocko, III
E.I. Du Pont De Nemours & Co.
Legal Department
1007 Market Street
Wilmington, DE 19898   (312) 774-2408

Charles S. Knothe
14 the Commons
3516 Silverside Road
Wilmington, DE 19810   (302) 478-8800

Chris P. Konkol
E.I. Du Pont De Nemours & Co.
Legal Dept., Pat. Div.
B-11204
1007 Market St.
Wilmington, DE 19898   (302) 478-8800

Costas S. Krikelis
P.O. Box 7228
Wilmington, DE 19803   (302) 992-3437

Charles E. Krukiel
E. I. Du Pont De Nemours
& Co.
Legal Dept.
1007 Market Street
Wilmington, DE 19898   (302) 774-7186

Mark D. Kuller
Hercules Inc.
Hercules Plaza
Wilmington, DE 19894   (302) 594-6923

James K. Luchs
32 Hayloft Circle
Wilmington, DE 19808   (302) 239-7020

David M. Lukoff
1510 Turkey Run Road
Wilmington, DE 19803   (302) 475-0110

Stanley Lukoff
1023 Parkside Drive
Wilmington, DE 19803   (302) 658-7815

John M. Lynn
E.I. Du Pont
Legal Dept.
1007 Market Street
B-11204
Wilmington, DE 19898   (302) 773-6879

Thomas H. Magee
E.I. Du Pont De
Nemours & Co.
1007 Market Street
Wilmington, DE 19898   (302) 992-5093

Andrea V. Malinowski
Du Pont Company
Legal Dept.
M-2614A
Wilmington, DE 19898   (302) 793-6177

Steven H. Markowitz
Hercules Inc.
Hercules Plaza
Wilmington, DE 19894   (302) 594-5000

Joshua W. Martin, III
Diamond State Telephone Co.
300 Delaware Ave.
Suite 714
Wilmington, DE 19801   (302) 427-7715

Prisca C. Marvin
Connolly & Hutz
1220 Market Bldg.
P.O. Box 2207
Wilmington, DE 19899   (302) 658-9141

Nancy S. Mayer
E.I. Du Pont De Nemours
& Company, Inc.
Legal Dept., Pat. Div.
1007 Market Street
Wilmington, DE 19899   (302) 992-3257

J.R. Mc Grath
100 Rue Mandaleine
Wilmington, DE 19807   (302) 652-0161

Kathleen H. Mc Keown#
E. I. Du Pont De Nemours
& Co.
101 Beech St.
P.O. Box 80840
Wilmington, DE 19880   (302) 695-0492

Robert G. Mc Morrow, Jr.
Connelly, Bove, Lodge
& Hutz
1220 Market Bldg.
P.O. Box 2207
Wilmington, DE 19899   (302) 658-9141

George M. Medwick
E.I. Dupont De Nemours
& Co., Inc.
BMP-17-1158
1007 Market St.
Wilmington, DE 19898   (302) 992-3220

Pamela Meitner
E. I. Dupont De Nemours
& Co.
1007 Market St.
Room D7015A
Wilmington, DE 19898   (302) 774-8720

Thomas Michael Meshbesher
Connolly, Bove, Lodge,
& Hutz
P.O. Box 2207
Wilmington, DE 19899   (302) 658-9141

Clinton F. Miller
800 Greenwood Road
Wilmington, DE 19807   (302) 652-7435

Philip E. Miller#
E.I. Du Pont De Nemours
& Co., Inc.
Chestnut Run Plaza
Laurel Run Bldg., G1C13
Wilmington, DE 19880   (302) 999-2825

Bruce M. Monroe#
E.I. Du Pont De Nemours
& Co., Inc.
BMP21-2364
Wilmington, DE 19880   (302) 992-3442

James Robert Morrison
95 Colorado Avenue
Wilmington, DE 19803   (302) 764-0362

Bruce W. Morrissey
E.I. Dupont De Nemours & Co.
Legal Dept.
1007 Market Street
Wilmington, DE 19898   (302) 992-4927

Carl W. Mortenson
Mortenson & Uebler, P.A.
Lindell Square, Suite 4
1601 Milltown Road
Wilmington, DE 19808   (302) 998-9400

James L. Newsom, Jr.
E. I. Du Pont De Nemours
& Company
101 Beech Street
P.O. Box 80840
Wilmington, DE 19880   (302) 695-0796

Jane E. Obee
5 Peirce Road
Wilmington, DE 19803   (302) 695-0796

Nicholas E. Oglesby, Jr.
16 Perth Drive
Wilmington, DE 19803   (302) 478-3291

Charlotte R. Otto#
E.I. Du Pont De Nemours
& Company
Barley Mill Plaza, 21-2358
P.O. Box 80021
Wilmington, DE 19880   (302) 992-3436

John A. Parrish
E.I. Du Pont De
Nemours & Co., Inc.
Pat. Div., Legal Dept.
3728A Montchanin Bldg.
Wilmington, DE 19898   (302) 774-2012

Donald F. Parsons, Jr.
Morris, Nichols, Arsht
& Tunnell
1201 N. Market Street
P.O. Box 1347
Wilmington, DE 19899   (302) 658-9200

Joanne W. Patterson#
Hercules Incorporated
Hercules Plaza
Wilmington, DE 19894   (302) 594-6961

Harold Pezzner
Connolly, Bove, Lodge
& Hutz
1220 Market Bldg.
Box 2207
Wilmington, DE 19899   (302) 658-9141

Robert Joseph Reichert
4660 Malden Drive
Wilmington, DE 19803   (302) 764-2966

Norbert Frederick Reinert
E.I. Du Pont De Nemours & Co.
1007 Market Street
Wilmington, DE 19898   (302) 774-4706

Dean R. Rexford#
2323 W. 16th St.
Wilmington, DE 19806   (302) 652-8893

Vernon R. Rice
E.I. Du Pont De Nemours
& Co., Inc.
1007 Market Street
Du Pont Bldg. 7016
Wilmington, DE 19898   (302) 774-2806

Annette L. Richter
E.I. Du Pont De Nemours & Co.
Legal Dept., P17-1174
Wilmington, DE 19898   (302) 992-3217

Louis H. Rombach
201 North Pembrey Drive
Wilmington, DE 19803   (302) 478-3027

David J. Roper
ICI Americas Inc.
Law Dept., Intell. Prop. Sect.
Concord Pike & New Murphy Rd.
Wilmington, DE 19897   (302) 886-3000

Richard A. Rowe#
ICI Americas, Inc
Concord Pike & New Murphy Road
Wilmington, DE 19897   (302) 886-3727

Howard J. Rudge
E.I. Du Pont De Nemours & Co.
Legal Dept.
Dupont Bldg.
Wilmington, DE 19898   (302) 774-7202

Jay Willis Sanner
ICI America Inc.
Wilmington, DE 19897   (302) 886-3723

George N. Sausen#
E. I. Du Pont De Nemours & Co.
Central Res. & Dev. Dept.
Barley Mill Plaza 14/2108
Wilmington, DE 19898   (302) 992-2722

Andrew L. Schaeffer
E.I. Du Pont De Nemours
& Co., Inc.
Legal Department
1007 Market Street
Wilmington, DE 19898   (302) 774-1000

John F. Schmutz
E. I. Du Pont De Nemours and Co.
1007 Market Street
7038 Du Pont Bldg.
Wilmington, DE 19898   (302) 774-7202

Patricia A. Schreck
E.I. Du Pont De Nemours
& Co.
1007 Market Street
Wilmington, DE 19898   (302) 774-7018

Ralph C. Schreyer#
2522 Deepwood Drive
Wilmington, DE 19810   (302) 475-3531

Sol Schwartz
E.I. Du Pont De Nemours
& Company
1007 Market Street
Wilmington, DE 19898   (302) 774-3177

Patricia M. Scott
E.I. Du Pont De Nemours
& Co., Inc.
1007 Market Street
BMP-17-1128
Wilmington, DE 19898   (302) 992-2236

Robert J. Shafer
E.I. Du Pont De Nemours & Co.
Legal Dept.
1007 Market St.
Wilmington, DE 19898   (302) 774-5783

John M. Sheehan
ICI Americas Inc.
Law Dept.
Wilmington, DE 19897   (302) 886-3721

James E. Shipley
E.I. Du Pont De Nemours
& Co.
Wilmington, DE 19898   (302) 773-3006

Charles J. Shoaf
E.I. Dupont De Nemours
& Co.
Legal Dept.
1007 Market St.
Wilmington, DE 19898   (302) 774-6745

David M. Shold#
E.I. Du Pont De Nemours & Co.
1007 Market Street
Wilmington, DE 19898   (302) 773-3264

Suzanne E. Shue#
E.I. Du Pont De Nemours
& Co.
1007 Market St.
Wilmington, DE 19898   (302) 773-5897

Barbara C. Siegell
E.I. Du Pont De Nemours
& Company
Legal Dept.
BMP-17/2286
Wilmington, DE 19898   (302) 992-4931

Faith L. K. Silver#
E.I. Du Pont De Nemours & Co.
Experimental Station
Bldg. 357, Engineering Dept.
Wilmington, DE 19880   (302) 695-2870

William J. Sipio#
E.I. Du Pont De Nemours
& Co., Inc.
Experimental Station
Med. Prods. Dept. E400/2427
Wilmington, DE 19880   (302) 695-7071

Robert J. Smat#
E.I. Du Pont De Nemours
& Co.
Electronics Dept.
Barley Mill Plaza, 30-1236
Wilmington, DE 19880   (302) 892-1921

Marion Cole Staves
Hercules Incorporated
Law Dept.
Pat. Section
Hercules Plaza
Wilmington, DE 19894   (302) 594-6953

Paul R. Steyermark
E.I. Du Pont De Nemours & Co.
Legal Dept.
M-2600
Wilmington, DE 19898   (302) 774-8966

Carol K. Stouffer#
E.I. Dupont De Nemours & Co., Inc.
Cr & D Dept.
1007 Market St.
Wilmington, DE 19898   (302) 695-4087

Daniel W. Sullivan
E.I. Du Pont De Nemours
& Co.
1007 Market St.
Wilmington, DE 19898   (302) 892-7915

Robert Francis Sullivan#
E. I. Du Pont De Nemours
& Company
Montchanin Bldg., Room 3724
10th & Shipley Sts.
Wilmington, DE 19898   (302) 374-2250

Ivan Gabor Szanto
Hercules Incorporated
Law Dept., Pat. Sect.
Hercules Plaza
Wilmington, DE 19894   (302) 594-6955

Wilkin Evans Thomas, Jr.
E.I. Du Pont De Nemours & Co.
Legal Dept.
Barley Mill Plaza
P.O. Box 80017
Wilmington, DE 19880   (302) 992-3210

Melford F. Tietze
712 Greenhill Ave.
Wilmington, DE 19805   (302) 654-6606

Roslyn T. Tobe
Hercules Incorporated
Hercules Plaza
1313 Markeat Street
Eighth Floor
Wilmington, DE 19894   (302) 594-6927

Edwin Tocker
E. I. Dupont De Nemours and Co.
Legal Dept.
10th and Market Sts.
Wilmington, DE 19898   (302) 774-2383

Maria P. Tungol
Connolly Bove, Lodge
& Hutz
P.O. Box 2207
1220 Market Bldg.
Wilmington, DE 19899   (302) 658-9141

S. Maynard Turk
Hercules Inc.
Hercules Plaza
Wilmington, DE 19894   (302) 594-7034

Ernest A. Uebler
Mortenson & Uebler, P.A.
Suite 4
1601 Milltown Road
Wilmington, DE 19808   (302) 998-9400

Roger O. Uhler#
Du Pont Co.
BMP 21-2126
P.O. Box 80021
Wilmington, DE 19880   (302) 992-3124

Eustathios Vassiliou#
E.I. Du Pont De Neours & Co.
Barley Mill Plaza, P30-1234
P.O. Box 80030
Wilmington, DE 19880   (302) 992-3166

Daniel J. Vrencur#
Du Pont Company
P.O. Box 80301
Experimental Station
301/311
Wilmington, DE 19880   (302) 695-4306

Frank R. Waite#
16 Sorrel Drive
Wilmington, DE 19803   (302) 478-4083

P. Michael Walker
E.I. Du Pont De Nemours & Co.
1007 Market St.
Wilmington, DE 19898   (302) 773-0167

Weston B. Wardell, Jr.
E. I. Du Pont De Nemours and Co.
1007 Market St.
Wilmington, DE 19898   (302) 992-3231

Charles A. Weigel
E.I. Dupont De Nemours
& Co., Inc.
1007 Market St.
Wilmington, DE 19898   (302) 992-3219

Howard P. West, Jr.
E.I. Du Pont De Nemours and Co.
1007 Market St.
Wilmington, DE 19898   (302) 892-7906

Robert W. Whetzel
Richards, Layton & Finger
One Rodney Square
P.O. Box 551
Wilmington, DE 19899   (302) 658-6541

Douglas E. Whitney
Morris, Nichols, Arsht
& Tunnell
P.O. Box 1347
1201 N. Market St.
Wilmington, DE 19899   (302) 575-7282

Don O. Winslow
Du Pont Company
Legal Dept.
Barley Mill Plaza, P17
Wilmington, DE 19805   (302) 992-3229

Herbert M. Wolfson
1213 Brook Drive
Wilmington, DE 19803   (302) 762-1476

Caroline J. Yun
E.I. Du Pont De Nemours
& Co.
1007 Market St.
Du Pont Bldg., Rm. 7056-A
Wilmington, DE 19898   (302) 778-2443

Patricia J. Smink Rogowski
Connolly & Hutz
P.O. Box 2207
1220 Market Bldg.
Wilmington, DE 19899   (302) 778-2443

District of Columbia

Frank L. Abbott
3805 Kanawha St.
Washington, DC 20015   (202) 362-0165

David S. Abrams
Roylance, Abrams, Berdo
& Goodman
Suite 204
1225 Connecticut Ave., N.W.
Washington, DC 20036   (202) 659-9076

Martin Abramson
Pollock, Vande Sande & Priddy
Suite 800
1990 M St., N.W.
Washington, DC 20036   (202) 331-7111

Harold W. Adams*
NASA
Off. of Assoc. Gen. Coun.
Intellectual Property
Mail Code GP
Washington, DC 20546   (202) 453-2417

Wilsie H. Adams, Jr.
Mc Kenna, Conner & Cuneo
1575 Eye St., N.W.
Suite 800
Washington, DC 20005   (202) 789-7652

Reid G. Adler
Finnegan, Henderson, Farabow,
Garrett & Dunner
1775 K Street, N.W.
Washington, DC 20006   (202) 293-6850

Stephen G. Adrian
Armstrong, Nikaido, Marmelstein,
Kubovcik & Murray
1725 K St., N.W., Ste. 1000
Washington, DC 20006   (202) 659-2930

Charles D. Ahearn, Jr.
Davis, Graham & Stubbs
1200 19th Street, N.W.
Suite 500
Washington, DC 20036   (202) 822-8660

Irwin Morton Aisenberg
Berman, & Aisenberg
1730 R. I. Ave., N.W.
Washington, DC 20036   (202) 293-1404

Richard L. Aitken
Lane & Aitken
Watergate Office Bldg.
2600 Virginia Ave., N.W.
Washington, DC 20037   (202) 337-5556

Stanislaus Aksman
Willian Brinks Olds Hofer
Gilson & Lione
2000 K Street, N.W.
Suite 200
Washington, DC 20006   (202) 429-0625

Jennifer A. Albert
Arent, Fox, Kintner,
Plotkin & Kahn
Washington Square
1050 Connecticut Ave., N.W.
Washington, DC 20036   (202) 857-6171

Louis Allahut*
Dept. of the Navy
Off. of Coun.
Naval Air Systems Comm.
Air-OOC5A
Washington, DC 20361   (703) 692-3418

Robert F. Allnutt
Pharmaceutical Manufacturers Assoc.
1100 15th Street, N.W.
Washington, DC 20005   (202) 835-3430

Scott M. Alter#
Saidman, Sterne, Kessler
& Goldstein
1225 Connecticut Ave., N.W.
Washington, DC 20036   (202) 833-7533

John C. Altmiller
Kenyon & Kenyon
1025 Connecticut Avenue
Washington, DC 20036   (202) 429-1776

Burton A. Amernick
Pollock, Vande Sande & Priddy
Suite 800
1990 M St., N.W.
Washington, DC 20036   (202) 331-7111

Walter D. Ames
Watson, Cole, Grindle
& Watson
1400 K Street, N.W.
Washington, DC 20005   (202) 628-0088

Steven M. Amundson*
U.S. Dept. of Justice
Room 410
550 Eleventh St., N.W.
Washington, DC 20530   (202) 724-7364

Larry N. Anagnos#
Antonelli, Terry & Wands
Suite 600
1919 Pennsylvania Ave., N.W.
Washington, DC 20006   (202) 828-0300

Richard Alan Anderson
Cushman, Darby & Cushman
1615 L St., N.W.
11th Floor
Washington, DC 20036   (202) 861-3060

Roger B. Andewelt*
U. S. Dept. of Justice
Antitrust Div.
10th St. & Pa. Ave., N.W.
Washington, DC 20530   (202) 633-2562

Donald R. Antonelli
Antonelli, Terry & Wands
Suite 600
1919 Pennsylvania Ave., N.W.
Washington, DC 20006   (202) 828-0300

Steven Anzalone
Finnegan, Henderson, Farabow,
Garrett & Dunner
1300 I St., N.W.
Washington, DC 20005   (202) 408-4000

Andres M. Arismendi, Jr.
Banner, Birch, Mc Kie
& Beckett
One Thomas Circle, N.W.
Suite 600
Washington, DC 20005   (202) 296-5500

James E. Armstrong, III
Armstrong, Nikaido, Marmelstein
& Kubovcik
1725 K St., N.W.
Suite 912
Washington, DC 20006   (202) 659-2930

James B. Arpin
Sutherland Asbill & Brennan
1275 Pennsylvania Ave., N.W.
Washington, DC 20004   (202) 383-0100

Thomas P. Athridge, Jr.*
U.S. Federal Trade Commission
601 Penna. Ave., N.W.
Sute 2311
Washington, DC 20580   (202) 326-2824

Jeffrey I. Auerbach
Weil, Gotshal & Manges
1615 L Street, N.W.
Suite 700
Washington, DC 20036   (202) 682-7033

Ormand R. Austin
MCI Communications Corp.
1133 19th Street, N.W.
Washington, DC 20036   (202) 887-2525

Aslan Baghdadi
Kirkland & Ellis
655 15th St., N.W.
Suite 1200
Washington, DC 20005   (202) 879-5227

Matthew T. Bailey#
Finnegan, Henderson, Farabow
Garrett & Dunner
1300 I Street, N.W.
Washington, DC 20005   (202) 408-4000

Robert D. Bajefsky
Finnegan, Henderson, Farabow,
Garrett & Dunner
1300 I St., N.W.
Washington, DC 20005   (202) 408-4000

Hollie L. Baker
Hale & Dorr
1455 Pennsylvania Ave., N.W.
Suite 1000
Washington, DC 20004   (202) 393-0800

Raymond N. Baker
Shanley & Baker
2233 Wisconsin Avenue, N.W.
Washington, DC 20007   (202) 333-5800

Thomas W. Banks
Finnegan, Henderson, Farabow,
Garrett & Dunner
1300 I Street, N.W.
Washington, DC 20005   (202) 408-4000

Donald W. Banner
Banner, Birch, Mc Kie
& Beckett
One Thomas Circle, N.W.
Washington, DC 20005   (202) 296-5500

Pamela I. Banner
Sughrue, Mion, Zinn,
Macpeak & Seas
2100 Pennsylvania Ave., N.W.
Suite 800
Washington, DC 20037   (202) 293-7060

Teresa J. Banta
3133 Connecticut Ave.
Apt. 311
Washington, DC 20008   (202) 232-6571

M. Paul Barker#
Finnegan, Henderson, Farabow,
Garrett & Dunner.
1300 I Street, N.W.
Washington, DC 20005   (202) 408-4000

Harry E. Barlow*
U.S. Dept. of Navy
Naval Air Systems Command
AIR- 00C5B
Washington, DC 20361   (202) 692-3455

Richard P. Bauer
Fitzpatrick, Cella, Harper
& Scinto
1001 Pennsylvania Ave., N.W.
Washington, DC 20004   (202) 347-8100

Robert M. Bauer#
4415 Q Street, N.W.
Washington, DC 20007   (202) 337-7234

James R. Beckers
Staas & Halsey
1825 K Street, N.W.
Washington, DC 20006   (202) 872-0123

William Wade Beckett
Banner, Birch, Mc Kie
& Beckett
One Thomas Circle, N.W.
Washington, DC 20005   (202) 296-5500

Michael D. Bednarek
Marks, Murase & White
Suite 750
2001 L Street, N.W.
Washington, DC 20036   (202) 955-4900

John W. Behringer
Sutherland, Asbill & Brennan
1275 Penn. Ave., N.W.
Washington, DC 20004   (202) 383-0125

Michael E. Belk
Shanley & Baker
2233 Wisconsin Avenue N.W.
Washington, DC 20007   (202) 333-5800

Edward L. Bell
Banner, Birch, McKie
& Beckett
One Thomas Circle, N.W.
Washington, DC 20005   (201) 296-5500

Robert P. Bell#
Banner, Birch, Mc Kie
& Beckett
1001 G St., N.W.
Washington, DC 20001   (202) 508-9100

Townsend Mikell Belser, Jr.
Pollock, Vande, Sande & Priddy
1990 M Street, N.W.
Suite 800
Washington, DC 20036   (202) 331-7111

John M. Belz
Leydig, Voit & Mayer
Suite 300
700 Thirteenth Street, N.W.
Washington, DC 20005   (202) 737-6770

Karen G. Bender
Morgan, Lewis & Bockus
1800 M Street
Washington, DC 20036   (202) 872-5000

Robert H. Berdo, Sr.
Roylance, Abrams, Berdo & Farley
Suite 315
1225 Conn. Ave., N.W.
Washington, DC 20036   (202) 659-9076

Joseph W. Berenato, III
Intellectual Propery Law
3421 M St., N.W.
Suite 1739
Washington, DC 20007   (301) 469-7291

Melvin G. Berger
Brand & Leckie
1730 K Street, N.W.
Suite 1000
Washington, DC 20006   (202) 347-7002

Herbert Berl*
U.S. Dept. of Justice
Comm. Litigation Branch
Civil Div.
Washington, DC 20530   (202) 724-7283

Stanford Warner Berman
Berman & Aisenberg
1730 R. I. Ave., N.W.
Washington, DC 20036   (202) 293-1404

Eugene L. Bernard
2500 Virginia Ave., N.W.
Washington, DC 20037   (202) 337-5954

Bruce G. Bernstein#
Finnegan, Henderson, Farabow,
Garrett & Dunner
1300 I St., N.W.
Suite 700
Washington, DC 20005   (202) 408-4346

Frank L. Bernstein
Sughrue, Mion, Zinn,
Mac Peak & Seas
2100 Pennsylvania Ave., N.W.
Washington, DC 20037   (202) 293-7060

Howard L. Bernstein
Sughrue Mion Zinn
Mac Peak & Seas
2100 Pennsylvania Ave., N.W.
Washington, DC 20037   (202) 293-7060

James D. Berquist
Cushman, Darby & Cushman
1615 L St., N.W.
11th Floor
Washington, DC 20036   (202) 861-3767

John R. Berres#
Communications Satellite Corp.
950 L Enfant Plaza, S.W.
Washington, DC 20024   (202) 863-6181

Mark S. Bicks
Roylance, Abrams, Berdo
& Goodman
1225 Connecticut Ave., N.W.
Washington, DC 20036   (202) 659-9076

W. Mark Bielawski
Fisher, Christen & Sabol
2000 M Street, N.W.
Suite 590
Washington, DC 20036   (202) 659-2000

Waddell Alexander Biggart, II
Sughrue, Mion, Zinn,
Macpeak & Seas
2100 Pennsylvania Ave., N.W.
Washington, DC 20037   (202) 293-7060

Harold J. Birch
Banner, Birch Mc Kie
& Beckett
One Thomas Circle, N.W.
Washington, DC 20005   (202) 296-5500

Donald J. Bird
Cushman, Darby & Cushman
1615 L Street, N.W.
11th Floor
Washington, DC 20036   (202) 861-3000

Scott Harris Blackman*
U.S. Court of Appeals
Federal Circuit
717 Madison Pl., N.W.
Suite 915
Washington, DC 20439   (202) 633-5866

Melvin Blecher
Rothwell, Figg, Ernst
& Kurz
1700 K St., N.W.
Suite 800
Washington, DC 20006   (202) 833-5740

Matthew P. Blischak
Cleary, Gottlieb, Steen
& Hamilton
1752 N Street, N.W.
Washington, DC 20036   (202) 728-2700

Mark Boland
Sughrue, Mion, Zinn,
Macpeak & Seas
2100 Pennsylvania Ave., N.W.
Washington, DC 20037   (202) 293-7060

Michael J. Boland
4424 Lingan Road, N.W.
Washington, DC 20007   (202) 337-0402

Thomas R. Boland
Vorys, Sater, Seymour
& Pease
Suite 1111
1828 L St., N.W.
Washington, DC 20036   (202) 296-2929

David B. Bonham#
Kenyon & Kenyon
1025 Conn. Ave., N.W.
Washington, DC 20036   (212) 429-1776

Donald D. Bosben
Fleit, Jacobson, Cohn, Price
Holman & Stern
400 Seventh Street, N.W.
Washington, DC 20004   (202) 638-6666

Mary J. Boswell
Finnegan, Henderson, Farabow,
Garrett & Dunner
1300 I Street, N.W.
Washington, DC 20005   (202) 693-6850

Walter Y. Boyd
Finnegan, Henderson, Farabow,
Garrett & Dunner
1775 K Street, N.W.
Washington, DC 20006   (202) 293-6850

James W. Brady, Jr.
Dickstein, Shapiro & Morin
2101 L Street, N.W.
Washington, DC 20037   (202) 785-9700

William D. Breneman
Breneman & Georges
Suite 290 - International Square
1850 K Street, N.W.
Washington, DC 20006   (202) 467-5800

James W. Brennan*
National Oceanic
& Atmospheric Admin.
14th & Constitution Ave., N.W.
Herbert C. Hoover Bldg., Rm. 5814
Washington, DC 20230   (202) 377-3043

Barry E. Bretschneider
Fish & Richardson
601 Thirteenth St., N.W.
Washington, DC 20005   (202) 783-5070

David William Brinkman
Cushman, Darby & Cushman
1615 L Street, N.W.
11th Floor
Washington, DC 20036   (202) 861-3033

William Lee Brooks
Howrey & Simon
1730 Pennsylvania Ave.
Washington, DC 20006   (120) 238-3662

Alvin Browdy
Browdy and Neimark
417 Seventh Street, N.W.
Washington, DC 20004   (202) 628-5197

Roger L. Browdy
Browdy and Neimark
419 Seventh Street, N.W.
Washington, DC 20004   (202) 628-5197

Norman V. Brown*
U.S. Navy
Asst. Secretary
S&L-CAG
Washington, DC 20360   (202) 692-5632

Scott A. Brown
Kenyon & Kenyon
1025 Connecticut Ave., N.W.
Washington, DC 20036   (202) 429-1776

Terrence L.B. Brown
P. O. Box 33772
Washington, DC 20033   (202) 722-4889

James T. Bruce, III
Wiley, Rein & Fielding
1776 K Street, N.W.
Washington, DC 20006   (202) 828-7552

Carl I. Brundidge
Antonelli, Terry & Wands
1919 Pennsylvania Ave., N.W.
Suite 600
Washington, DC 20006   (202) 828-0300

Brian Garrett Brunsvold
Finnegan, Henderson, Farabow,
Garrett & Dunner
1300 I St., N.W.
Washington, DC 20005   (202) 408-4000

B. Frederick Buchan, Jr.*
U.S. Dept. of Justice
Civil Division
Commercial Litigation Branch
550 11th Street, N.W.
Washington, DC 20530   (202) 724-7276

Mary-Elizabeth Buckles
Porter, Wright, Morris
& Arthur
1233 20th Street, N.W.
4th Floor
Washington, DC 20036   (202) 778-3000

George Joseph Budock*
Dept. of the Air Force
Off. of the Judge Advocate
1900 Half St., S.W.
Washington, DC 20324   (202) 475-1386

William T. Bullinger
Cushman, Darby & Cushman
1615 L Street, N.W.
Suite 1100
Washington, DC 20036   (202) 861-3000

John P.D. Bundock, Jr.
Leydig, Voit & Mayer
Suite 300
700 Thirteenth Street, N.W.
Washington, DC 20005   (202) 737-6770

William J. Bundren
Leydig, Voit & Mayer
Suite 300
700 Thirteenth Street, N.W.
Washington, DC 20005   (202) 737-6770

Kenneth J. Burchfiel
Sughrue, Mion, Zinn,
Macpeak & Seas
2100 Pennsylvania Ave., N.W.
Washington, DC 20037   (202) 293-7060

Richard V. Burgujian
Finnegan, Henderson, Farabow,
Garrett & Dunner
1300 I Street, N.W.
Washington, DC 20005   (202) 408-4000

Walter W. Burns, Jr.
Barnes & Thornburg
1815 H Street, N.W.
Washington, DC 20006   (202) 955-4500

Matthew J. Bussan#
Staas & Halsey
1825 K Street, N.W.
Washington, DC 20006   (202) 872-0123

Roy W. Butrum
Watson, Cole, Grindle
& Watson
1400 K Street, N.W.
Washington, DC 20005   (202) 628-0088

John J. Byrne
Baker & McKenzie
815 Connecticut Ave., N.W.
Washington, DC 20006   (202) 298-8290

Thomas Joseph Byrnes*
Dept. of Justice
Commercial Litigation Branch
Civil Division
Washington, DC 20530   (202) 724-7221

Frank E. Caffoe#
Finnegan, Henderson, Farabow,
Garrett & Dunner
1300 I Street, N.W.
Washington, DC 20005   (202) 408-4000

Kenneth L. Cage*
Willian Brinks Olds Hofer
Gilson & Lione
2000 K Street, N.W.
Washington, DC 20006   (202) 429-0625

Maurice U. Cahn
Leydig, Voit & Mayer
Suite 300
700 Thirteenth Street, N.W.
Washington, DC 20005   (202) 737-6770

John W. Caldwell, Sr.#
3344 5th Street, S.E.
Washington, DC 20032   (202) 562-1087

Charles W. Calkins
Kenyon & Kenyon
1025 Connecticut Avenue, N.W.
Washington, DC 20036   (202) 429-1776

Terry S. Callaghan#
Finnegan, Henderson, Farabow,
Garrett & Dunner
1300 I St., N.W.
Washington, DC 20005   (202) 408-4276

Celine T. Callahan
Finnegan, Henderson, Farabow,
Garrett & Dunner
1300 I St., N.W.
Washington, DC 20005   (202) 408-4000

John T. Callahan
Ostrolenk, Faber, Gerb
& Soffen
1725 K Street
Suite 1108
Washington, DC 20006   (202) 457-7785

Frederick F. Calvetti
Morgan & Finnegan
1627 I Street, N.W.
Washington, DC 20006   (202) 857-7887

John J. Camby
4405 Burlington Place, N.W.
Washington, DC 20016   (202) 362-6463

Alan I. Cantor
Banner, Birch McKie
& Beckett
One Thomas Cicle, N.W.
Washington, DC 20005   (202) 296-5500

Herbert I. Cantor
Wegner & Bretschneider
P.O. Box 18218
Washington, DC 20036   (202) 887-0400

Jay M. Cantor
Spencer & Frank
1111 Nineteenth Street, N.W.
Washington, DC 20036   (202) 828-8000

Mary A. Capria
Jones, Day, Reavis
& Pogue
1450 G St., N.W.
Suite 700
Washington, DC 20005   (202) 879-7604

Virginia B. Caress#*
U.S. Dept. of Energy
Off. of Asst. Gen. Coun.
For Intell. Prop., GC-42 Forstl
1000 Independence Ave., S.W.
Washington, DC 20585   (202) 586-2802

Dean E. Carlson
Pollock, Vande Sande & Priddy
Suite 800
1990 M St., N.W.
Washington, DC 20036   (202) 331-7111

John F. Carney
Banner, Birch, McKie & Beckett
One Thomas Circle
Suite 600
Washington, DC 20005   (202) 296-5500

Bruce G. Chapman
Finnegan, Henderson, Farabow,
Garrett & Dunner
1300 I Street, N.W.
Washington, DC 20005   (202) 293-6850

Alex Chartove
Spensley, Horn, Jubas
& Lubitz
Suite 1212
1050 17th Street, N.W.
Washington, DC 20036   (202) 223-5700

Warren M. Cheek, Jr.
Wenderoth, Lind & Ponack
Southern Building
Suite 700
805 Fifteenth St., N.W.
Washington, DC 20005   (202) 371-8850

Chun-I Chiang*
U.S. Dept. of Justice
550 11th Street, N.W.
Washington, DC 20530   (202) 724-7364

Gay Chin
Suite 724
1800 K Street, N.W.
Washington, DC 20006   (301) 897-6301

Elzbieta M. Chlopecka
Pollock, Vande Sande & Priddy
1990 M Street, N.W.
Suite 800
Washington, DC 20036   (202) 331-7111

Melissa L. Chun#*
EPA
401 M St., S.W.
Mail Code H7505C
Washington, DC 20460   (703) 557-7700

Michele A. Cimbala#
Saidman, Sterne, Kessler
& Goldstein
1225 Connecticut Ave., N.W.
Suite 300
Washington, DC 20036   (202) 466-0800

Conrad J. Clark
Dickinson, Wright, Moon,
Van Dusen & Freeman
1901 L Street N.W.
Suite 800
Washington, DC 20036   (202) 457-0160

Ronald D. Cohn
Fleit, Jacobson, Cohn, Price
Holman & Stern
400 Seventh Street, N.W.
Washington, DC 20004   (202) 638-6666

Joseph V. Colaianni
Pennie & Edmonds
1730 Pennsylvania Ave., N.W.
Suite 1000
Washington, DC 20006   (202) 393-0177

Lawrence L. Colbert
5139 33rd Steet, N.W.
Washington, DC 20008   (202) 363-2309

Amy L. Tsui Collins
Finnegan, Henderson, Farabow,
Garrett & Dunner
1300 I Street, N.W.
Washington, DC 20005   (202) 408-4000

Kendrew H. Colton
Cushman, Darby & Cushman
1615 L Street, N.W.
Suite 1100
Washington, DC 20036   (202) 861-3000

Chris Comuntzis
Cushman, Darby & Cushman
1615 L Street, N.W.
Washington, DC 20036   (202) 861-3623

Edward J. Connors, Jr.*
U.S. Department of Navy
Naval Sea Systems Command
Sea 00l53
Washington, DC 20362   (703) 602-3730

Richard E. Constant, Sr.*
U.S. Dept. of Energy
1000 Independence Ave., S.W.
Washington, DC 20585   (202) 586-2802

Robert E. Converse, Jr.
Finnegan, Henderson, Farabow,
Garrett & Dunner
1300 I Street, N.W.
Washington, DC 20005   (202) 408-4000

Iver P. Cooper
Browdy and Neimark
419 Seventh Street, N.W.
Suite 300
Washington, DC 20004   (202) 628-5197

Ruffin B. Cordell#
Baker & Mc Kenzie
815 Connecticut Ave., N.W.
Suite 1100
Washington, DC 20006   (202) 452-7000

Monique L. Cordray
Cushman, Darby & Cushman
1615 L St., N.W.
Suite 1100
Washington, DC 20036   (202) 861-3699

Cornell D.M.J. Cornish
Suite 301
1101 New Hampshire Ave., N.W.
Washington, DC 20037   (202) 429-9705

David K. Cornwell
Saidman, Sterne, Kessler
& Goldstein
1225 Connecticut Avenue
Suite 300
Washington, DC 20036   (202) 833-7533

Patrick J. Coyne
Collier, Shannon, Rill
& Scott
Suite 308
1055 Thomas Jeff. Street, N.W.
Washington, DC 20007   (202) 342-8606

Paul M. Craig, Jr.
Barnes & Thornburg
1815 H Street, N.W.
Washington, DC 20006   (202) 955-4500

Vincent M. Creedon
Wenderoth, Lind & Ponack
805 15th Street, N.W.
Suite 700
Washington, DC 20005   (202) 371-8850

Rae E. Cronmiller
NRECA
1800 Mass. Ave., N.W.
Washington, DC 20036   (202) 857-9593

Charles A. Cross
Rothwell, Figg, Ernst
& Kurz
1700 K Street, N.W.
Suite 800
Washington, DC 20006   (202) 833-5740

David J. Cushing
Sughrue, Mion, Zinn,
Macpeak & Seas
2100 Pennsylvania Ave., N.W.
Washington, DC 20037   (202) 293-7060

James H. Czerwonky*
National Oceanic
& Atmospheric Admin. -NMFS
1825 Connecticut Avenue, N.W.
Washington, DC 20235   (202) 673-5464

Christopher J. Daley
Banner, Birch, Mc Kie
1001 G St., N.W.
Eleventh Floor
Washington, DC 20001   (202) 508-9100

Scott M. Daniels
Armstrong, Nikaido, Marmelstein
& Kubovcik
1725 K Street, N.W.
Washington, DC 20006   (202) 659-2930

John S. Davidson
Cushman Darby & Cushman
1615 L St., N.W.
Eleventh Floor
Washington, DC 20036   (202) 861-3660

Akin Thornwall Davis
Cushman, Darby & Cushman
1615 L Street, N.W.
Washington, DC 20036   (202) 861-3047

Garrett V. Davis#
Fisher, Christen & Sabol
2000 M Street, N.W.
Suite 590
Washington, DC 20036   (202) 659-2000

James F. Davis
Howrey & Simon
Suite 900
1730 Pennsylvania Ave., N.W.
Washington, DC 20006   (202) 383-6589

Michael Rhodes Davis
Wenderoth, Lind & Ponack
805 Fifteenth Street, N.W.
Suite 700
Washington, DC 20005   (202) 371-8850

Carl M. De Franco, Jr.
Beveridge, De Grandi & Weilacher
1819 H Street, N.W.
Suite 1100
Washington, DC 20006   (202) 659-2811

Joseph A. De Grandi
Beveridge, De Grandi & Weilacher
1819 H Street, N.W.
Washington, DC 20006   (202) 659-2811

John P. De Luca
Venable, Baetjer, Howard
& Civiletti
1201 New York Avenue, N.W.
Suite 1000
Washington, DC 20005   (202) 962-4866

Vincent M. De Luca
Bernard, Rothwell & Brown
1700 K Street, N.W.
Washington, DC 20006   (202) 833-5740

Donald B. Deaver
Cushman, Darby & Cushman
1615 L Street, N.W.
11th Floor
Washington, DC 20036   (202) 861-3000

Gregg A. Delaporta
Finnegan, Henderson, Farabow,
Garrett & Dunner
1300 I St., N.W.
Washington, DC 20005   (202) 408-4000

Gita A. Delsing#
Wegner & Bretschneider
1233 20th St., N.W.
Washington, DC 20036   (202) 887-0400

Arthur P. Demers
Pennie & Edmonds
1730 Pennsylvania Avenue, N.W.
Washington, DC 20006   (202) 393-0177

Caroline D. Dennison
Roylance, Abrams, Berdo
& Goodman
1225 Connecticut Ave., N.W.
Suite 315
Washington, DC 20036   (202) 659-9076

Paul Devinsky
Marks Murase & White
Suite 750
2001 L Street, N.W.
Washington, DC 20036   (202) 955-4900

Vito J. Di Pietro*
U.S. Dept. of Justice
Commercial Litigation Branch
Civil Division
550 11th Street, N.W.
Washington, DC 20530   (202) 724-7223

Richard R. Diefendorf#
Fleit, Jacobson, Cohn, Price
Holman & Stern
400 Seventh Street, N.W.
Washington, DC 20004   (202) 638-6666

Bryan C. Diner
Finnegan, Henderson, Farabow,
Garrett & Dunner
1300 I Street, N.W.
Washington, DC 20005   (202) 408-4000

Clarence M. Ditlow, III
Center for Auto Safety
2001 S Street, N.W.
Suite 410
Washington, DC 20009   (202) 328-7700

Kenneth W. Dobyns*
U.S. Navy
Naval Sea Systems Comm.
SEA-00l5
Washington, DC 20362   (202) 692-7077

Howard D. Doescher
Cushman, Darby & Cushman
1615 L Street, N.W.
Washington, DC 20036   (202) 861-3000

John Donofrio#
Kirkland & Ellis
655 Fifteenth Street, N.W.
Suite 1200
Washington, DC 20005   (202) 879-5000

Charles Richard Donohoe
Samsung Electronics Co., Ltd.
Suite 250, One Westin Center
2445 M St., N.W.
Washington, DC 20037   (202) 296-0227

Patrick A. Doody#
Banner, Birch, Mc Kie
& Beckett
1001 G St., N.W.
Washington, DC 20001   (202) 508-9100

James Lawrence Dooley
Cushman, Darby & Cushman
1615 L Street, N.W.
11th Floor
Washington, DC 20036   (202) 861-3000

James N. Dresser
Antonelli, Terry & Wands
1919 Pennsylvania Ave., N.W.
Suite 600
Washington, DC 20006   (202) 828-0300

Norman A. Drezin*
Federal Trade Commission
601 Pennsylvania Ave., N.W.
Suite 3033
Washington, DC 20580   (202) 326-2903

Patricia M. Drost
Fitzpatrick, Cella, Harper
& Scinto
1001 Pennsylvania Ave., N.W.
Washington, DC 20004   (202) 347-8100

William Anthony Drucker
3rd Floor
1111 19th Street, N.W.
Washington, DC 20036   (202) 828-8048

Folsom E. Drummond
5415 Conn. Avenue, N.W.
Apt. 320
Washington, DC 20015   (202) 363-6420

Jean K. Dudek
Fitzpatrick, Cella,
Harper & Scinto
1001 Pennsylvania Ave., N.W.
Suite 650N
Washington, DC 20004   (202) 347-8100

Donald R. Dunner
Finnegan, Henderson, Farabow,
Garrett & Dunner
1300 I St., N.W.
Washington, DC 20005   (202) 408-4000

Tracy-Gene G. Durkin#
Sterne, Kessler, Goldstein
& Fox
1225 Connecticut Ave., N.W.
Washington, DC 20036   (202) 466-0800

Kathleen G. Dussault
Finnegan, Henderson, Farabow,
Garrett & Dunner
1300 I Street, N.W.
Washington, DC 20005   (202) 408-4000

Frank J. Dynda*
U.S. Dept. of the Navy
Naval Air Systems Command
AIR- 00C5
Washington, DC 20361   (202) 692-7810

Edward E. Dyson
Baker & Mckenzie
815 Conn. Ave., N.W.
Washington, DC 20006   (202) 298-8290

Barry A. Edelberg*
U.S. Dept. of Navy
Naval Research Lab.
Code 3008.2
4555 Overlook Ave., S.W.
Washington, DC 20375   (202) 298-8290

G. Paul Edgell
Cushman, Darby & Cushman
Eleventh Floor
1615 L Street, N.W.
Washington, DC 20036   (202) 822-0944

James W. Edmondson
Finnegan, Henderson, Farabow,
Garrett & Dunner
1300 I Street, N.W.
Washington, DC 20005   (202) 293-6850

Gary R. Edwards
Evenson, Wands, Edwards,
Lenahan & McKeown
1825 K Street, N.W.
Washington, DC 20006   (202) 457-9090

Carol P. Einaudi
Finnegan, Henderson, Farabow,
Garrett & Dunner
1300 I Street, N.W.
Washington, DC 20005   (202) 408-4000

Daniel L. Ellis#
Fidelman & Wolffe
1233 20th Street, N.W.
Suite 300
Washington, DC 20036   (202) 833-8801

Mary Anne Ellis
Howrey & Simon
1730 Pennsylvania Ave., N.W.
Washington, DC 20006   (202) 783-0800

Michael C. Elmer
Finnegan, Henderson, Farabow,
Garrett & Dunner
1775 K St., N.W.
Suite 600
Washington, DC 20006   (202) 293-6850

Alvin J. Englert#*
U.S. Dept. of Commerce
Off. of Gen. Coun.
Room 4610
14th & Constitution Ave., N.W.
Washington, DC 20230   (201) 377-5394

William T. Enos
Bernard Rothwell & Brown, P.C.
1700 K Street, N.W.
Suite 800
Washington, DC 20006   (202) 833-5740

Randall G. Erdley
Fleit, Jacobson, Cohn, Price,
Holman & Stern
400 Seventh Street, N.W.
Washington, DC 20004   (202) 638-6666

Barbara G. Ernst
Bernard, Rothwell & Brown, P.C.
1700 K Street, N.W.
Washington, DC 20006   (202) 833-5740

Robert W. Esmond#
Saidman, Sterne, Kessler
& Goldstein
Suite 300
1225 Connecticut Ave., N.W.
Washington, DC 20036   (202) 833-7533

Mitchell S. Ettinger
Dunnells, Duvall, Bennett
& Porter
2100 Pennsylvania Ave., N.W.
Washington, DC 20037   (202) 861-1400

Hubert E. Evans
Armstrong, Nikaido, Marmelstein
& Kubovcik
1725 K Street, N.W.
Washington, DC 20006   (202) 659-2930

Linda S. Evans#
Wegner & Bretschneider
1233 20th St., N.W.
3rd Floor
Washington, DC 20036   (202) 887-0400

Donald D. Evenson
Barnes & Thornburg
1815 H Street, N.W.
Suite 800
Washington, DC 20006   (202) 955-4500

John D. Fado*
U.S. Dept. of Agriculture
Res. & Operations Div., Rm. 23215
Off. of General Counsel
Washington, DC 20250   (202) 447-2421

Ford F. Farabow, Jr.
Finnegan, Henderson, Farabow,
Garrett & Dunner
1300 I St., N.W.
Washington, DC 20005   (202) 408-4000

Horacio A. Farach
Fleit, Jacobson, Cohn,
Price, Holman & Stern
400 Seventh St., N.W.
Washington, DC 20004   (202) 638-6666

John J. Fargo*
U.S. Dept. of Justice
Civil Div.
Commercial Litigation Branch
Washington, DC 20530   (202) 724-7415

Raymond E. Farrell*
Dept. of the Navy
Off. of Coun.
Naval Air Systems Comm.
Washington, DC 20361   (703) 692-7810

Donald J. Featherstone#
Saidman, Sterne, Kessler
& Goldstein
1225 Connecticut Ave., N.W.
Suite 300
Washington, DC 20036   (202) 466-0800

John Theodore Fedigan
Wenderoth, Lind, and Ponack
Southern Bldg., Suite 700
805 Fifteenth Street, N.W.
Washington, DC 20005   (202) 371-8850

Morris Fidelman
Fidelman & Wolffe
1233 20th Street, N.W.
Suite 300
Washington, DC 20036   (202) 833-8801

Edward A. Figg
Bernard, Rothwell & Brown, P.C.
1700 K Street, N.W.
Washington, DC 20006   (202) 833-5740

Jay M. Finkelstein
Spensley, Horn, Jubas
& Lubitz
1050 17th Street, N.W.
Washington, DC 20036   (202) 828-8000

William J. Fisher
Banner, Birch, Mc Kie
& Beckett
1001 G Street, N.W.
11th Floor
Washington, DC 20001   (202) 508-9100

Joseph T. Fitzgerald#
Spensley Horn Jubas & Lubitz
2000 M Street, N.W.
Suite 720
Washington, DC 20036   (202) 223-5700

Martin Fleit
Fleit, Jacobson, Cohn, Price
Holman & Stern
400 Seventh Street, N.W.
Washington, DC 20004   (202) 638-6666

Mark L. Fleshner#
Spencer & Frank
1111 19th St., N.W.
Suite 1200
Washington, DC 20036   (202) 828-8000

Raz E. Fleshner#
Finnegan, Henderson, Farabow,
Garrett & Dunner
1300 I St., N.W.
Washington, DC 20005   (502) 408-4266

D. Andrew Floam#
Fleit, Jacobson, Cohn,
Price, Holman & Stern
Jenifer Bldg.
400 7th St., N.W.
Washington, DC 20004   (202) 638-6666

Karl W. Flocks
3110 Chain Bridge Road, N.W.
Washington, DC 20016   (202) 363-3862

Howard M. Flournoy
P.O. Box 6113
Washington, DC 20044   (703) 455-7594

Kerry A. Flynn
Finnegan, Henderson, Farabow,
Garrett & Dunner
1300 I St., N.W.
Washington, DC 20005   (202) 408-4129

Christopher P. Foley
Finnegan, Henderson, Farabow,
Garrett & Dunner
1300 I Street, N.W.
Washington, DC 20005   (202) 408-4000

Jean Burke Fordis
Finnegan, Henderson, Farabow,
Garrett & Dunner
1300 I Street, N.W.
Washington, DC 20005   (202) 408-4000

David S. Forman
Finnegan, Henderson, Farabow,
Garrett & Dunner
1300 I Street, N.W.
Washington, DC 20005   (202) 293-6850

John L. Forrest, Jr.*
U.S. Dept. of Navy
Naval Air Systems Comm.
AIR- 00C-5-5870
Washington, DC 20361   (202) 692-3455

John D. Foster
Leydig, Voit & Mayer
700 Thirteenth St., N.W.
Suite 300
Washington, DC 20005   (202) 737-6770

Kevin J. Fournier#
Sughrue, Mion, Zinn,
Mac Peak & Seas
2100 Pennsylvania Ave.
Washington, DC 20037   (202) 293-7060

Samuel L. Fox
Saidman, Sterne, Kessler
& Goldstein
1225 Connecticut Avenue
Suite 300
Washington, DC 20036   (202) 833-7533

Robert J. Frank
Spencer & Frank
1111 19th St., N.W.
Washington, DC 20036   (202) 828-8000

Martin I. Fuchs#
Finnegan, Henderson, Farabow,
Garrett & Dunner
1300 I St., N.W.
Washington, DC 20005   (202) 408-4000

Arthur Joseph Gajarsa
Joseph, Gajarsa, Mc Dermott
& Reiner, P.C.
1300 19th Street, N.W.
Suite 400
Washington, DC 20036   (202) 331-1955

Arthur Sellers Garrett
Finnegan, Henderson, Farabow,
Garrett & Dunner
1300 I St., N.W.
Washington, DC 20005   (202) 408-4000

John C. Garvey
Staas & Halsey
1825 K Street, N.W.
Suite 816
Washington, DC 20006   (202) 872-0123

James G. Gatto
Marks Murase & White
Suite 750
2001 L Street, N.W.
Washington, DC 20036   (202) 955-4919

Robert James Gaybrick
Finnegan, Henderson, Farabow,
Garrett & Dunner
1300 I Street, N.W.
Washington, DC 20005   (202) 408-4000

Arthur P. Gershman
1003 K Street N.W.
Suite 820
Washington, DC 20001   (202) 408-4000

James E. Gilchrist
American Mining Congress
1920 N Street, N.W.
Suite 300
Washington, DC 20036   (202) 861-2876

Jim W. Gipple
Gipple & Hale
P.O. Box 40513
Washington, DC 20016   (703) 448-1770

Jeremiah Glassman*
U.S. Dept. of Justice
Civil Rights Division
10th & Pennsylvania Ave., N.W.
Washington, DC 20530   (202) 633-4092

Stephen C. Glazier
Ginsburg, Feldman & Bress
1250 Connecticut Ave., N.W.
Washington, DC 20036   (202) 637-9000

Raymond C. Glenny
Mason, Fenwick & Lawrence
1225 Eye Street, N.W.
Suite 1000
Washington, DC 20005   (202) 289-1200

Gregory J. Glover
Covington & Burling
1201 Pennsylvania Ave., N.W.
P.O. Box 7566
Washington, DC 20044   (202) 662-5566

Ricky S. Goldman
Suite 901
1725 K Street, N.W.
Washington, DC 20006   (202) 659-2366

Jorge A. Goldstein
Saidman, Sterne, Kessler
& Goldstein
1225 Connecticut Avenue
Suite 300
Washington, DC 20036   (202) 833-7533

Richard A. Gollhofer
Staas & Halsey
1825 K Street
Washington, DC 20006   (202) 872-0123

Michael A. Gollin
Sive, Paget & Riesel, P.C.
1275 Penn. Ave., N.W.
Suite 202
Washington, DC 20004   (202) 783-8720

Michael J. Gonet*
U.S. Navy
NAVAIR
AIR-OOC-5-5870
Patents Division
Washington, DC 20361   (202) 692-3456

Alfred N. Goodman
Roylance, Abrams, Berdo
& Goodman
1225 Connecticut Ave., N.W.
Suite 204
Washington, DC 20036   (202) 659-9076

Marian R. Gordon*
Federal Communications Comm.
1919 M Street, N.W.
Washington, DC 20554   (202) 632-6363

Thomas W. Gorman
Howrey & Simon
1730 Pennsylvania Ave., N.W.
Washington, DC 20006   (202) 783-0800

Mary E. Gormley
Armstrong, Nikaido, Marmelstein,
Kubovcik & Murray
1725 K Street, N.W.
Suite 1000
Washington, DC 20006   (202) 659-2930

Corinne R. Gorski#
Finnegan, Henderson, Farabow,
Garrett & Dunner
1300 I St., N.W.
Washington, DC 20005   (202) 408-4000

Lawrence Jay Gotts
Kirkland & Ellis
655 15th Street, N.W.
Washington, DC 20005   (202) 879-5000

James M. Gould*
U.S. International Trade Comm.
500 E Street, S.W.
Suite 401
Washington, DC 20436   (202) 252-1578

Peter W. Gowdey
Cushman, Darby & Cushman
1615 L Street, N.W.
11th Floor
Washington, DC 20036   (202) 861-3078

Barry W. Graham
Finnegan, Henderson, Farabow,
Garrett & Dunner
1300 I Street, N.W.
Washington, DC 20005   (202) 408-4000

Patricia D. Granados*
U.S. Court of Appeals
Federal Circuit
717 Madison Pl., N.W.
Suite 210
Washington, DC 20439   (202) 633-6583

Paul Grandinetti
Levy, Zito & Grandinetti
1511 K Street, N.W.
Suite 425
Washington, DC 20005   (202) 429-4560

Stanley Barry Green
Pollock, Vande Sande & Priddy
Suite 800
1990 M St., N.W.
Washington, DC 20036   (202) 331-7111

Susan H. Griffen
Finnegan, Henderson, Farabow,
Garrett & Dunner
1300 I Street, N.W.
Washington, DC 20005   (202) 408-4000

Alan M. Grimaldi
Howrey & Simon
1730 Penn. Ave., N.W.
Washington, DC 20006   (202) 383-6989

Barry L. Grossman
Banner, Birch, Mc Kie
& Beckett
One Thomas Circle
Washington, DC 20005   (202) 296-5500

Francis W. Guay
1713 18th Street, N.W.
Washington, DC 20009   (202) 234-3546

Louis Gubinsky
Sughrue, Mion, Zinn,
Macpeak & Seas
2100 Pennsylvania Ave., N.W.
Washington, DC 20037   (202) 293-7060

Mark J. Guttag
Spencer & Frank
1111 19th Street, N.W.
Washington, DC 20036   (202) 293-7060

H. Sanders Gwin, Jr.
Sughrue, Mion, Zinn,
Mac Peak & Seas
2100 Pennsylvania Ave., N.W.
Washington, DC 20037   (202) 293-7060

Thomas S. Hahn
Lane & Mittendorf
919 18th St., N.W.
Suite 800
Washington, DC 20006   (202) 785-4949

Charles S. Hall
Finnegan, Henderson, Farabow,
Garrett & Dunner
Suite 600
1775 K St., N.W.
Washington, DC 20006   (202) 293-6850

Melvin L. Halpern*
U.S. Court of Appeals
Federal Circuit
717 Madison Place, N.W.
Washington, DC 20433   (202) 633-6567

James Dillard Halsey, Jr.
Staas & Halsey
1825 K St., N.W.
Washington, DC 20006   (202) 872-0123

Philip G. Hampton, II
Kenyon & Kenyon
1025 Connecticut Ave., N.W.
Washington, DC 20036   (202) 429-1776

John A. Hankins
Evenson, Wands, Edwards,
Lenahan & Mc Keown
1825 K St., N.W.
Suite 919
Washington, DC 20006   (202) 457-9090

Brian W. Hannon
Sughrue, Mion, Zinn,
Macpeak & Seas
2100 Pennsylvania Ave., N.W.
Washington, DC 20037   (202) 293-7060

Donald W. Hanson
Huff & Hanson
Suite 725
1400 K Street, N.W.
Washington, DC 20005   (202) 466-6386

Alisa A. Harbin#
Armstrong, Nikaido, Marmelstein,
Kubovick & Murray
1725 K St., N.W.
Suite 1000
Washington, DC 20006   (202) 659-2930

Larry Harbin
Cushman, Darby & Cushman
1615 L Street, N.W.
Suite 1100
Washington, DC 20036   (202) 861-3000

Carol R. Harney
Finnegan, Henderson, Farabow,
Garrett & Dunner
1300 I Street, N.W.
Washington, DC 20005   (202) 408-4000

Charles H. Harris#
Fulbright & Jaworski
801 Pennsylvania Ave., N.W.
4th Floor
Washington, DC 20004   (202) 622-4746

Scott C. Harris
Cushman, Darby & Cushman
1615 L Street, N.W.
Washington, DC 20036   (202) 641-0727

Paul C. Hashim#
Robbins & Laramie
2100 Pennsylvania Ave., N.W.
Suite 600
Washington, DC 20037   (202) 887-5050

Fred W. Hathaway
Robbins & Laramie
2100 Pennsylvania Ave, N.W.
Suite 600
Washington, DC 20037   (202) 887-5050

Ken-Ichi Hattori
Armstrong, Nikaido, Marmelstein
& Kubovcik
1725 K Street, N.W.
Washington, DC 20006   (202) 659-2930

Gary L. Hausken*
Dept of Justice
Civil Division, Commercial
Litigation Branch
550 11th St., N.W.
Washington, DC 20530   (202) 659-2930

Donald E. Hayes, Jr.#
Robbins & Laramie
2100 Pennsylvania Ave., N.W.
Washington, DC 20037   (202) 887-5050

Lorenzo B. Hayes#
4729 Queens Chapel Terr., N.E.
Washington, DC 20017   (202) 526-7880

Laurence R. Hefter
Finnegan, Henderson, Farabow,
Garrett & Dunner
1300 I St., N.W.
Washington, DC 20005   (202) 408-4000

Robert E. Helfrich
Bishop, Cook, Purcell
& Reynolds
1400 L Street, N.W.
Washington, DC 20005   (202) 371-5709

Douglas B. Henderson
Finnegan, Henderson, Farabow,
Garrett & Dunner
1300 I Street, N.W.
Washington, DC 20005   (202) 408-4000

William R. Henderson*
U.S. Dept. of Navy
Naval Sea Systems Comm.
Off. of Gen. Counsel
Washington, DC 20362   (202) 692-7077

Bruce J. Hendricks
Bernard, Rothwell & Brown, P.C.
1700 K Street, N.W.
Suite 800
Washington, DC 20006   (202) 833-5740

Charles E. Hepner
Kenyon & Kenyon
1025 Connecticut Ave., N.W.
Washington, DC 20036   (202) 429-1775

Toni-Junell Herbert
Saidman, Sterne, Kessler
& Goldstein
1225 Connecticut Ave., N.W.
Suite 300
Washington, DC 20036   (202) 466-0878

William F. Herbert
Staas & Halsey
1825 K Street, N.W.
Washington, DC 20006   (202) 872-0123

Ronald S. Hermenau#
2628 Woodley Pl., N.W.
Washington, DC 20008   (202) 234-9708

Wayne W. Herrington*
U.S. International Trade Comm.
Off. of Gen. Counsel
500 E Street, S.W.
Washington, DC 20436   (202) 252-1092

David A. Hey#
Antonelli, Terry & Wands
1919 Pennsylvania Ave., N.W.
Suite 600
Washington, DC 20006   (202) 828-0300

David W. Highet
Venable, Baetjer, Howard
& Civiletti
1201 New York Ave., N.W.
Suite 1100
Washington, DC 20005   (202) 962-4800

Judson R. Hightower*
U.S. Dept. of Energy
1000 Independence Ave., S.W.
Washington, DC 20585   (202) 586-3499

David W. Hill
Finnegan, Henderson, Farabow,
Garrett & Dunner
1300 I St., N.W.
Washington, DC 20005   (202) 408-4000

Joseph A. Hill*
Dept. of Justice
Comm. Litigation Br.
Civil Div.
Washington, DC 20530   (202) 724-7275

Ramon R. Hoch#
Sughrue, Mion, Zinn,
Mac Peak & Seas
2100 Pennsylvania Ave., N.W.
Suite 800
Washington, DC 20037   (202) 293-7060

Michael P. Hoffman*
U.S. Dept. of Energy
Washington, DC 20585   (202) 586-3441

Mark L. Hogge
Fisher, Christen & Sabol
2000 M St., N.W.
Suite 590
Washington, DC 20036   (202) 659-2000

Dale Curtis Hogue, Sr.
Pennie & Edmonds
1730 Pennsylvania Ave., N.W.
Suite 1000
Washington, DC 20006   (202) 393-0177

Barry I. Hollander
Fisher, Christen & Sabol
2000 M Street, N.W.
Suite 590
Washington, DC 20036   (202) 659-2000

John C. Holman
Fleit, Jacobson, Cohn, Price,
Holman & Stern
400 Seventh Street, N.W.
Washington, DC 20004   (202) 638-6666

John E. Holmes
Robbins & Laramie
2100 Pennsylvania Ave., N.W.
Suite 600
Washington, DC 20037   (202) 887-5050

Alvin D. Hooper
Akin, Gump, Strauss, Hauer
& Feld
1333 New Hampshire Ave., N.W.
Suite 400
Washington, DC 20036   (202) 887-4455

Thomas M. Horgan
Banner, Birch, Mc Kie
& Beckett
1001 G St., N.W.
11th Floor
Washington, DC 20001   (202) 508-9100

Scott A. Horstemeyer
Saidman, Sterne, Kessler
& Goldstein
1225 Connecticut Ave., N.W.
Suite 300
Washington, DC 20036   (202) 466-0800

Dale H. Hoscheit
Banner, Birch, Mc Kie
& Beckett
One Thomas Circle, N.W.
Washington, DC 20005   (202) 296-5500

Alfred F. Hoyte, Jr.#
7734 16th St., N.W.
Washington, DC 20012   (202) 723-6589

Stuart Tin Fah Huang
Cushman, Darby & Cushman
1615 L Street, N.W.
Washington, DC 20036   (202) 861-3708

Frank L. Huband*
Natl. Science Foundation
Room 1134
1800 G St., N.W.
Washington, DC 20550   (202) 357-9545

David F. Hubbuch#
U.S. Information Agency
301 4th Street, S.W.
Washington, DC 20547   (202) 485-7767

John W. Huckert
7400 Benjamin Franklin Station
Washington, DC 20044   (301) 258-9418

Donald N. Huff
Huff & Hanson
1400 K Street, N.W.
Washington, DC 20005   (202) 466-6386

Richard L. Huff
Keil & Weinkauf
1101 Conn. Ave., N.W.
Suite 620
Washington, DC 20036   (202) 659-0100

James A. Hughes, Jr.*
U.S. Dept. of the Army
Off. of General Counsel
Pentagon, Room 2E725
SAGC
Washington, DC 20310   (202) 697-4349

Patrick H. Hume
3830 Macomb Street, N.W.
Washington, DC 20016   (202) 244-2772

Nathaniel A. Humphries
Mason, Fenwick & Lawrence
Suite 1000
1225 Eye Street, N.W.
Washington, DC 20005   (202) 289-1200

Lawrence A. Hymo
Cushman, Darby & Cushman
1615 L Street, N.W.
11th Floor
Washington, DC 20036   (202) 861-3000

Joseph A. Hynds#
Rothwell, Figg, Ernst & Kurz
1700 K St., N.W.
Suite 800
Washington, DC 20006   (202) 833-5740

Jeffrey L. Ihnen
Venable, Baetjer, Howard
& Civiletti
1201 New York Ave., N.W.
Suite 1000
Washington, DC 20005   (202) 962-4810

John R. Inge
Sughrue, Mion, Zinn,
Macpeak & Seas
2100 Pennsylvania Ave., N.W.
Washington, DC 20037   (202) 293-7060

Edward S. Irons
919 18th Street, N.W.
Suite 800
Washington, DC 20006   (202) 785-6938

Thomas L. Irving
Finnegan, Henderson, Farabow,
Garrett & Dunner
1300 I St., N.W.
Washington, DC 20005   (202) 408-4000

Christopher P. Isaac
Finnegan Henderson Farabow
Garrett & Dunner
1300 I Street, N.W.
Washington, DC 20005   (202) 408-4000

Yoji Ito#
Nippondenso Company, Ltd.
Cushman, Darby & Cushman
1615 L Street, N.W.
Washington, DC 20036   (202) 861-3627

Jean H. Jackson*
U.S. International Trade Comm.
500 E Street S.W.
Suite 707
Washington, DC 20436   (202) 252-1104

Jerome D. Jackson
Finnegan, Henderson, Farabow,
Garrett & Dunner
1300 I Street, N.W.
Washington, DC 20005   (202) 408-4021

Thomas H. Jackson
Banner, Birch, Mc Kie
& Beckett
1001 G St., N.W.
Eleventh Floor
Washington, DC 20001   (202) 508-9100

Matthew M. Jacob
Wenderoth, Lind and Ponack
Southern Bldg., Suite 700
805 Fifteenth Street, N.W.
Washington, DC 20005   (202) 371-8850

Gary M. Jacobs
Fitzpatrick, Cella, Harper
& Scinto
1001 Pennsylvania Ave., N.W.
Suite 650
Washington, DC 20004   (202) 347-8100

Harvey B. Jacobson, Jr.
Fleit, Jacobson, Cohn, Price
Holman & Stern
400 Seventh Street, N.W.
Washington, DC 20004   (202) 638-6666

Harvey B. Jacobson, Sr.
Fleit, Jacobson, Cohn
Price, Holman & Stern
400 7th St., N.W.
Washington, DC 20004   (202) 638-6666

Mishrilal L. Jain
Fish & Richardson
601 Thirteenth St., N.W.
Washington, DC 20005   (202) 783-5070

John M. Jakes
Finnegan, Henderson, Farabow,
Garrett & Dunner
1300 I Street, N.W.
Washington, DC 20005   (202) 408-4000

David A. Jakopin
Cushman, Darby & Cushman
1615 L Street, N.W.
11th Floor
Washington, DC 20036   (202) 861-3000

George Jameson*
U.S. Dept. of the Navy
Patents, Code 1208.2
Naval Research Laboratory
Washington, DC 20375   (202) 767-3428

Linda T. Jaron#
Lalos & Keegan
900 17th Street, N.W.
Suite 900
Washington, DC 20006   (202) 887-5555

Edmund Michael Jaskiewicz
1730 M Street, N.W.
Suite 501
Washington, DC 20036   (202) 296-2295

Jeremy M. Jay#
Leydig, Voit & Mayer
700 Thirteenth St., N.W.
Suite 300
Washington, DC 20005   (202) 737-6770

Thomas H. Jenkins
Finnegan, Henderson, Farabow,
Garrett & Dunner
1300 I Street, N.W.
Washington, DC 20005   (202) 408-4000

Tipton D. Jennings
Finnegan, Henderson Farabow,
Garrett & Dunner
1300 I Street, N.W.
Washington, DC 20005   (202) 408-4000

Henry R. Jiles#
Wenderoth, Lind & Ponack
Southern Bldg.
805 15th Street, N.W.
Washington, DC 20005   (202) 371-8850

Cynthia P. Johnson#*
U.S. International Trade Comm.
Office of General Counsel
500 E St., S.W.
Washington, DC 20436   (202) 252-1098

Doris J. Johnson#
Finnegan, Henderson, Farabow,
Garrett & Dunner
1300 I St., N.W.
Washington, DC 20005   (202) 408-4244

James W. Johnson, Jr.
1631 Jonquil Street, N.W.
Washington, DC 20012   (202) 944-6853

Lance G. Johnson
Banner, Birch, Mc Kie
& Beckett
One Thomas Circle, N.W.
Suite 600
Washington, DC 20005   (202) 296-5500

Lori-Ann Johnson#
Finnegan, Henderson, Farabow,
Garrett & Dunner
1300 I St. N.W.
Washington, DC 20005   (202) 408-4270

Robert J. Jondle
Venable, Baetjer, Howard
& Civiletti
1201 New York Ave., N.W.
Ste. 1000
Washington, DC 20005   (202) 962-4800

Herman O. Jones#
169 Chesapeake St., S.W.
Washington, DC 20032   (202) 562-8652

Raymond C. Jones#
Finnegan, Henderson, Farabow,
Garrett & Dunner
1300 I St., N.W.
Washington, DC 20005   (502) 408-4000

Raymond N. Jones#
1426 Leegate Road, N.W.
Washington, DC 20012   (202) 726-0730

Wayne A. Jones
Cushman, Darby & Cushman
Suite 1100
1615 L Street, N.W.
Washington, DC 20036-6000   (202) 861-3685

Mervin L. Jordan*
U.S. Air Force
JACP
1900 Half Street
Room 5160
Washington, DC 20324   (202) 475-1386

Richard D. Jordan#
Sterne, Kessler, Goldstein
& Fox
1225 Connecticut Ave., N.W.
Suite 300
Washington, DC 20036   (202) 466-0800

Kevin E. Joyce
Cushman, Darby & Cushman
1615 L Street, N.W.
11th Floor
Washington, DC 20036   (202) 861-3000

Daniel W. Juffernbruch
Staas & Halsey
1825 K Street, N.W.
Washington, DC 20006   (202) 872-0123

Richard A. Kaba
Shea & Gardner
1800 Massachusetts Ave., N.W.
Washington, DC 20036   (202) 775-3013

Sarah Anne Kagan
Banner, Birch, Mc Kie
& Beckett
1001 G St., N.W.
Eleventh Floor
Washington, DC 20001   (202) 508-9100

Daniel Kalish*
Dept. of the Navy
Naval Research Lab.
4555 Overlook Ave., S.W.
Code 3008.2
Washington, DC 20375   (202) 767-2246

Michael D. Kaminski
Finnegan, Henderson, Farabow
Garrett & Dunner
1300 I Street, N.W.
Washington, DC 20005   (202) 408-4000

Ronald P. Kananen
Marks Murase & White
Suite 750
2001 L Street, N.W.
Washington, DC 20036   (202) 955-4900

Lawrence M. Kaplan*
U.S. Court of Appeals
For the Federal Circuit
National Courts Bldg.
717 Madison Place, N.W.
Washington, DC 20439   (202) 955-4900

Albert Anthony Kashinski*
U.S. Dept. of Interior
Office of Solicitor
18th & E Sts., N.W.
Washington, DC 20240   (202) 343-5207

Alan J. Kasper
Sughrue, Mion, Zinn,
Macpeake & Seas
2100 Pennsylvania Avenue, N.W.
Washington, DC 20037   (202) 293-7060

Andrew B. Katz
Elliott, Mannino &
Flaherty, P.C.
1225 I Street, N.W.
Suite 400
Washington, DC 20005   (202) 408-1600

Mitchell A. Katz#
Finnegan, Henderson, Farabow,
Garrett & Dunner
1300 I Street, N.W.
Washington, DC 20005   (202) 408-4000

Craig R. Kaufman#
Cushman, Darby & Cushman
1615 L St., N.W.
11th Floor
Washington, DC 20036   (202) 861-3600

Daniel C. Kaufman
Ritts, Brickfield & Kaufman
600 New Hampshire Ave., N.W.
Washington, DC 20037   (202) 342-0800

Thomas F. Kaufman
Willkie Farr & Gallagher
1155 21st Street, N.W.
Suite 600
Washington, DC 20036   (202) 328-8000

Donald Allen Kaul
Brownstein Zeidman & Schomer
1401 New York Avenue N.W.
Suite 900
Washington, DC 20005   (202) 879-5755

Harvey Kaye
Cohen & Burg, P.C.
2555 M Street, N.W.
Washington, DC 20037   (202) 785-0773

Irving Kayton
2000 Pennsylvania Ave., N.W.
Suite 3450
Washington, DC 20006   (202) 223-1177

Francis A. Keegan
Lalos & Keegan
900 17th Street
Suite 900
Washington, DC 20006   (202) 887-5555

Herbert B. Keil
Keil & Weinkauf
1101 Conn. Ave., N.W.
Washington, DC 20036   (202) 659-0100

Gabor J. Kelemen
Spencer & Frank
1111 19th St., N.W.
Washington, DC 20036   (202) 828-8000

Julie Ring Keller
Venable, Baetjer, Howard
& Civiletti
1201 New York Ave., N.W.
Suite 1000
Washington, DC 20005   (202) 962-4800

Michael L. Keller
Squire, Sanders & Dempsey
1201 Pennsylvania Avenue, N.W.
P.O. Box 407
Washington, DC 20044   (202) 626-6600

Kevin J. Kelley
Bryan, Cave, Mc Pheeters
& Mc Roberts
700 13th St., N.W.
Washington, DC 20005   (202) 508-6117

David M. Kelly
Finnegan, Henderson, Farabow,
Garrett & Dunner
1300 I Street, N.W.
Washington, DC 20005   (202) 408-4000

Michael R. Kelly#
Finnegan, Henderson, Farabow,
Garrett & Dunner
1300 I St., N.W.
Washington, DC 20007   (202) 408-4000

Robert F. Kempf*
NASA
400 Maryland Avenue, S.W.
Washington, DC 20546   (202) 453-2424

Alan J. Kennedy*
NASA
Off. of the Associate General
Counsel (Intell. Property)
Mail Code GP
Washington, DC 20546   (202) 453-2421

Edward J. Kessler
Saidman, Sterne, Kessler
& Goldstein
1225 Connecticut Avenue
Washington, DC 20036   (202) 466-0800

Bradford E. Kile
Baker & Mc Kenzie
815 Conn. Ave., N.W.
Washington, DC 20006   (202) 452-7000

Luke A. Kilyk
Sughrue, Mion, Zinn,
Macpeak & Seas
2100 Pennsylvania Ave., N.W.
Washington, DC 20037   (202) 293-7060

Lyle K. Kimms
Pennie & Edmonds
1730 Pennsylvania Ave., N.W.
Washington, DC 20006   (202) 393-0177

Allen Kirkpatrick, III
Cushman, Darby & Cushman
1615 L Street, N.W.
11th Floor
Washington, DC 20036   (202) 861-3500

Gordon Kit
Sughrue, Mion, Zinn,
Macpeak & Seas
2100 Pennsylvania Ave., N.W.
Washington, DC 20037   (202) 293-7060

Alan P. Klein*
Dept. of Navy
Code SPAWAR-00C41
Space & Naval Warfare Sys. Comm.
Washington, DC 20363   (202) 692-2893

Peter L. Klempay
1003 K Street, N.W.
Suite 820
Washington, DC 20001   (202) 737-6610

Timothy J. Klima
Cushman, Darby & Cushman
1615 L Street N.W.
Washington, DC 20036   (202) 861-3662

Richard G. Kline
Kline, Rommel & Colbert
One Farragut Square South
Seventh Floor
Washington, DC 20006   (202) 393-5200

Thomas M. Kline#*
U.S. Navy
Naval Sea Systems Command
Washington, DC 20362   (202) 393-5200

G. Lloyd Knight, Jr.
Cushman, Darby & Cushman
1615 L Street, N.W.
11th Floor
Washington, DC 20036   (202) 861-3000

Robert J. Koch
Fulbright & Jaworski
801 Pennsylvania Ave., N.W.
Washington, DC 20004   (202) 662-0200

James L. Kohnen
Nms
1775 Pennsylvania Ave., N.W.
Washington, DC 20006   (202) 662-0200

Paul N. Kokulis
Cushman, Darby & Cushman
1615 L Street, N.W.
11th Floor
Washington, DC 20036   (202) 861-3000

Epaminondas Philip Koltos*
U.S. Dept. of the Interior
Office of the Solicitor
18th and C Streets, N.W.
Washington, DC 20240   (202) 343-4471

Anne M. Kornbau
Browdy & Neimark
419 7th Street, N.W.
Washington, DC 20004   (202) 628-5197

Judy K. Kosovich*
Dept. of Energy
Off. of Gen. Coun.
GC-42, 6f067
Washington, DC 20585   (202) 586-2802

Vincent P. Kovalick
Finnegan, Henderson, Farabow,
Garrett & Dunner
1300 I Street, N.W.
Washington, DC 20005   (202) 408-4000

William Kovensky
Roylance, Abrams, Berdo
& Goodman
1225 Connecticut Ave., N.W.
Suite 315
Washington, DC 20036   (202) 659-9076

Frank J. Kowalski
Fleit, Jacobson, Cohn, Price
Holman & Stern
400 Seventh Street, N.W.
Washington, DC 20004   (202) 626-3260

Bruce E. Kramer
Sughrue, Mion, Zinn,
Macpeak & Seas
2100 Pennsylvania Ave., N.W.
Washington, DC 20037   (202) 293-7060

Melvin Kraus
Antonelli, Terry & Wands
Suite 600
1919 Pennsylvania Ave., N.W.
Washington, DC 20006   (202) 828-0300

Louis F. Kreek
4636 Verplanck Pl. N.W.
Washington, DC 20016   (202) EM3-5991

Carla Magda Krivak#
Staas & Halsey
1825 K Street, N.W.
Suite 816
Washington, DC 20006   (202) 872-0123

Karen I. Krupen#
Cushman, Darby & Cushman
1615 L St., N.W.
Eleventh Floor
Washington, DC 20036   (202) 861-3690

Ronald J. Kubovcik
Armstrong, Nikaido, Marmelstein
& Kubovcik
1725 K St., N.W.
Washington, DC 20006   (202) 659-2930

James J. Kulbaski#
3101 Hawthorne Dr., N.E.
Washington, DC 20017   (202) 667-8711

Norman N. Kunitz
Spencer & Frank
1111 19th St., N.W.
Washington, DC 20036   (202) 828-8000

Raymond A. La Forge*
Federal Communications Comm.
1919 M Street, N.W.
Washington, DC 20554   (202) 653-8117

David L. Ladd
Wiley, Rein & Fielding
1776 K Street, N.W.
Washington, DC 20006   (202) 429-7030

Jack L. Lahr
Foley & Lardner
1775 Pennsylvania Ave., N.W.
Washington, DC 20006   (202) 862-5336

Peter N. Lalos
Lalos & Keegan
900 17th Street, N.W.
Suite 900
Washington, DC 20006   (202) 887-5555

Richard A. Lambert*
U.S. Dept. of Energy
1000 Independence Ave., S.W.
Washington, DC 20585   (202) 586-2807

John Thomas Lanahan
Ward, Lazarus & Grow
1711 N Street, N.W.
Washington, DC 20036   (202) 331-8160

Sheldon I. Landsman
Sughrue, Mion, Zinn,
Macpeak & Seas
2100 Pennsylvania Ave., N.W.
Suite 800
Washington, DC 20037   (202) 663-7933

Joseph M. Lane
Lane & Aitken
Suite 600 Watergate Off. Bldg.
2600 Virginia Ave., N.W.
Washington, DC 20037   (202) 337-5556

Thomas Joseph Lannon
105 Whittier St., N.W.
Washington, DC 20012   (202) 726-2882

John C. Laprade
1511 K Street, N.W.
Suite 738-740
Washington, DC 20005   (202) 328-3350

James R. Laramie
Robbins & Laramie
2100 Pennsylvania Ave., N.W.
Suite 600
Washington, DC 20037   (202) 887-5050

Douglas N. Larson
Banner, Birch, Mc Kie
& Beckett
One Thomas Circle, N.W.
Suite 600
Washington, DC 20005   (202) 466-9122

Robert J. Lasker
Watson, Cole, Grindle
& Watson
Suite 725
1400 K Street, N.W.
Washington, DC 20005   (202) 628-0088

Maryann S. Lastova*
Court of Appeals for the
Federal Circuit
717 Madison Place, N.W.
Washington, DC 20439   (202) 633-6533

Joseph P. Lavelle
Howrey & Simon
1730 Pennsylvania Ave., N.W.
Washington, DC 20006   (202) 783-0800

Lawrence M. Lavin
Finnegan, Henderson, Farabow,
Garrett & Dunner
1300 I Street, N.W.
Washington, DC 20005   (202) 408-4000

Irvin A. Lavine
Mason, Fenwick & Lawrence
1225 Eye Street, N.W.
Suite 1000
Washington, DC 20005   (202) 289-1200

Nina M.S. Lawrence*
NASA
400 Maryland Ave., S.W.
Washington, DC 20546   (202) 453-2417

Dale S. Lazar
Cushman, Darby & Cushman
1615 L Street, N.W.
11th Floor
Washington, DC 20036   (202) 861-3527

Robert M. Lazo*
U.S. Nuclear Regulatory Comm.
Atomic Safety & Licensing
Board Panel
Washington, DC 20555   (301) 492-7842

Samuel Lebowitz
5736 26th St., N.W.
Washington, DC 20015   (202) 537-0248

Frederick J. Lees
George Washington Univ.
Government Contracts Program
2020 K Street, N.W.
Washington, DC 20052   (202) 994-5272

Charles B. Lefkoff
1120 Lamont Street N.W.
Washington, DC 20010   (202) 797-7782

Allen M. Lenchek
1400 K Street, N.W.
Suite 725
Washington, DC 20005   (202) 408-7522

Michelle N. Lester
Cushman, Darby & Cushman
1615 L Street, N.W.
Suite 1100
Washington, DC 20036   (202) 861-3693

Jack Q. Lever, Jr.
Willian Brinks Olds Hofer
Gilson & Lione
2000 K Street, N.W.
Suite 200
Washington, DC 20006   (202) 429-0625

Sherman Levy
808 Investment Bldg.
1511 K Street, N.W.
Washington, DC 20005   (202) 628-7625

Basil J. Lewris
Finnegan, Henderson, Farabow,
Garrett & Dunner
1300 I St., N.W.
Washington, DC 20005   (202) 408-4000

Thomas H. Liddle, III*
U.S. Dept. of Justice
Pennsylvania Ave.
Washington, DC 20530   (202) 724-8312

Nancy J. Linck
Cushman, Darby & Cushman
1615 L Street, N.W.
Suite 1100
Washington, DC 20036   (202) 861-3658

Jeffrey A. Lindeman#
Finnegan, Henderson, Farabow,
Garrett & Dunner
1300 I St., N.W.
Washington, DC 20005   (202) 408-4000

Richard Linn
Marks Murase & White
2001 L Street, N.W.
Suite 750
Washington, DC 20036   (202) 955-4900

Steven E. Lipman
Fish & Richardson
601 Thirteenth St., N.W.
Washington, DC 20005   (202) 783-5070

Raymond F. Lippitt
Cushman, Darby & Cushman
1615 L Street, N.W.
Washington, DC 20036   (202) 861-3512

Charles E. Lipsey
Finnegan, Henderson, Farabow,
Garrett & Dunner
1300 I St., N.W.
Washington, DC 20005   (202) 408-4000

Morris Liss
Pollock, Vandesande & Priddy
Suite 800
1990 M Street, N.W.
Washington, DC 20036   (202) 331-7111

Shmuel Livnat#
Browdy and Neimark
419 Seventh St., N.W.
Suite 300
Washington, DC 20004   (202) 628-5197

James R. Longacre#
Squire, Sanders & Dempsey
1201 Pennsylvania Ave., N.W.
P.O. Box 407
Washington, DC 20044   (202) 626-6600

Stacey J. Longanecker#
Venable Baetjer Howard
& Civiletti
1201 New York Ave., N.W.
Washington, DC 20005   (202) 962-4800

Phillip G. Lookadoo
Ballard, Spahr, Andrews
& Ingersoll
555 13th St., N.W.
Suite 900 E.
Washington, DC 20004   (202) 383-8820

Robert V. Lottmann*
NASA
400 Maryland Avenue, S.W.
Washington, DC 20546   (703) 487-7406

Alan D. Lourie*
Court of Appeals
Federal Circuit
National Courts Building
717 Madison Place, N.W.
Washington, DC 20439   (202) 633-5851

Carl George Love
Cushman, Darby & Cushman
Suite 1100
1615 L Street, N.W.
Washington, DC 20036   (202) 861-3000

Ethel G. Love
4513 17th Street, N.E.
Washington, DC 20017   (202) 529-8857

Paul T. Lubeck*
U.S. Dept. of Justice
Antitrust Division
555 4th Street, N.W.
Room 10-435
Washington, DC 20530   (202) 724-7966

Robert Bennett Lubic
2033 M Street, N. W.
Washington, DC 20036   (202) 452-8200

John C. Luce
Cushman, Darby & Cushman
1615 L Street, N.W.
11th Fl.
Washington, DC 20036   (202) 861-3000

Paul J. Luckern*
U.S. International Trade Comm.
Off. of Admin. Law Judges
500 E Street, N.W.
Room 213
Washington, DC 20436   (202) 252-1697

Raphael V. Lupo
Willian Brinks Olds Hofer
Gilson & Lione
2000 K Street, N.W.
Suite 200
Washington, DC 20006   (202) 429-0625

Harry Lupuloff*
NASA
Code GP
Washington, DC 20546   (202) 453-2420

James C. Lydon
Armstrong, Nikaido, Marmelstein
Kubovcik & Murray
1725 K St., N.W.
Suite 1000
Washington, DC 20006   (202) 659-2930

Christopher H. Lynt
Spencer & Frank
1111 19th Street, N.W.
Washington, DC 20036   (202) 768-0956

John A. Mac Evoy
Winston & Strawn
1400 L St., N.W.
Washington, DC 20005   (202) 371-5769

Susan J. Mack
Sughrue, Mion, Zinn,
Macpeak & Seas
2100 Pennsylvania Ave., N.W.
Washington, DC 20037   (202) 293-7060

Thomas J. Macpeak
Sughrue, Mion, Zinn,
Macpeak & Seas
2100 Pennsylvania Ave., N.W.
Washington, DC 20037   (202) 293-7060

Laura A. Majerus
Finnegan, Henderson, Farabow,
Garrett & Dunner
1300 I St., N.W.
Washington, DC 20005   (202) 408-4012

Theodore Major*
U.S. Postal Service
Room 6427
475 L'Enfant Plaza, S.W.
Washington, DC 20260   (202) 268-3018

Michael A. Makuch
Beveridge, De Grandi & Weilacher
1819 H Street, N.W.
Suite 1100
Washington, DC 20006   (202) 659-2811

Donald M. Malone*
U.S. Dept. of Commerce
Off. of Gen. Counsel
Room 5876
Herbert C. Hoover Bldg.
Washington, DC 20230   (202) 377-8843

Scott D. Malpede
Fitzpatrick, Cella, Harper
& Scinto
1001 Pennsylvania Ave., N.W.
Suite 650
Washington, DC 20004   (202) 347-8100

William H. Mandir
Sughrue, Mion, Zinn,
Mac Peak & Seas
2100 Pennsylvania Avenue, N.W.
Washington, DC 20037   (202) 663-7959

John Raymond Manning*
NASA
Office of General Counsel
400 Maryland Ave., S.W.
Washington, DC 20546   (202) 453-2416

John G. Mannix*
NASA
400 Maryland Ave.
Mail Code ADB-1
Washington, DC 20546   (202) 453-1033

Robert J. Marchick*
U.S. Dept. of Energy
1000 Independence Ave., S.W.
Washington, DC 20585   (202) 586-4792

George T. Marcou
Marks Murase & White
Suite 750
2001 L Street, N.W.
Washington, DC 20036   (202) 955-4900

Michael S. Marcus
Morgan & Finnegan
1627 I Street, N.W.
Washington, DC 20006   (202) 857-7887

Charles M. Marmelstein
Armstrong, Nikaido, Marmelstein
& Kubovcik
1725 K Street, N.W.
Suite 912
Washington, DC 20006   (202) 659-2930

Juan Carlos A. Marquez#
Armstrong, Nikaido, Marmelstein,
Kubovcik & Murray
1725 K Street, N.W.
Suite 1000
Washington, DC 20006   (202) 659-2930

James H. Marsh, Jr.
Staas & Halsey
1825 K Street, N.W.
Washington, DC 20006   (202) 872-0123

Virgil Homer Marsh
Fisher, Christen & Sabol
2000 M Street, N.W.
Suite 590
Washington, DC 20036   (202) 659-2000

Edgar H. Martin
Cushman, Darby & Cushman
1615 L Street, N.W.
Washington, DC 20036   (202) 861-3521

James Thomas Martin
1521 44th Street, N.W.
Washington, DC 20007   (202) 296-3650

William D. Martin, Jr.
4404 Sheriff Road, N.E.
Washington, DC 20019   (202) 399-8097

Peter B. Martine
Finnegan, Henderson, Farabow,
Garrett & Dunner
1300 I Street, N.W.
Washington, DC 20005   (202) 408-4000

Kevin P. Mc Andrews#
Finnegan, Henderson, Farabow,
Garrett & Dunner
1300 I Street N.W.
Washington, DC 20005   (202) 408-4267

Clifton E. Mc Cann
Lane & Aitken
Suite 600
2600 Virginia Ave., N.W.
Washington, DC 20037   (202) 337-5556

Helen M. Mc Carthy
Beveridge, De Grandi & Weilacher
Federal Bar Bldg.
1819 H Street, N.W.
Washington, DC 20006   (202) 659-2811

William Thomas Mc Clain
Finnegan, Henderson, Farabow,
Garrett & Dunner
1300 I Street, N.W.
Washington, DC 20005   (202) 408-4000

Garland Thomas Mc Coy*
NASA
400 Maryland Ave., S.W.
Washington, DC 20546   (202) 453-2412

Barbara C. Mc Curdy
Finnegan, Henderson, Farabow
Garrett & Dunner
1300 I Street, N.W.
Washington, DC 20005   (202) 408-0000

Thomas E. Mc Donnell*
Dept. of the Navy
Naval Research Laboratory
4555 Overlook Ave., S.W.
Washington, DC 20375   (202) 767-3427

Sean M. Mc Ginn
Sughrue, Mion, Zinn,
Mac Peak & Seas
2100 Pennsylvania Ave., N.W.
Washington, DC 20037   (202) 293-7060

Harold C. Mc Gurk, IV
Staas & Halsey
1825 K St., N.W.
Suite 816
Washington, DC 20006   (202) 872-0123

Michael R. Mc Gurk
Finnegan, Henderson, Farabow,
Garrett & Dunner
1300 I Street, N.W.
Washington, DC 20005   (202) 408-4000

Christopher L. Mc Kee
Banner, Birch, Mc Kie
1001 G Street, N.W.
11th Floor
Washington, DC 20001   (202) 508-9100

Francis G. Mc Kenna
Anderson & Pendleton, C.A.
1000 Connecticut Ave., N.W.
Washington, DC 20036   (202) 659-2334

James F. Mc Keown
Evenson, Wands, Edwards,
Lenahan & Mc Keown
1825 K Street, N.W.
Suite 919
Washington, DC 20005   (202) 457-9090

Edward F. Mc Kie, Jr.
Banner, Birch, Mc Kie
& Beckett
One Thomas Circle, N.W.
Washington, DC 20005   (202) 296-5500

Le-Nhung Mc Leland
Armstrong, Nikaido,
Marmelstein & Kubovcik
1725 K Street, N.W.
Suite 912
Washington, DC 20006   (202) 659-2930

Robert G. Mc Morrow
Sughrue, Mion, Zinn,
Macpeak & Seas
2100 Pennsylvania Ave., N.W.
Washington, DC 20037   (202) 293-7060

William J. Mc Nichol, Jr.
Reid & Priest
Market Square
701 Pennsylvania Ave., N.W.
Washington, DC 20004   (202) 508-4000

Paul F. Mc Quade
Pillsbury, Madison & Sutro
1667 K St., N.W., Suite 1100
Washington, DC 20006   (202) 463-2316

John J. Mc Veigh
Fisher, Wayland, Cooper & Leader
1255 23rd St. NW, Ste. 800
Washington, DC 20037   (202) 775-3544

Bernard A. Meany, Sr.
Ginsburg, Feldman & Bress
1250 Conn. Ave., N.W.
Washington, DC 20036   (202) 637-9000

Nina L. Medlock
Banner, Birch, Mc Kie
& Beckett
1001 G Street NW, 11th Fl.
Washington, DC 20001   (202) 508-9100

Joseph T. Melillo*
U. S. Dept. of Justice
Antitrust Div.
Litigation II
555 4th St. NW
Washington, DC 20001   (202) 724-7984

Allen S. Melser
Mason, Fenwick & Lawrence
1225 Eye Street, N.W.
Suite 1000
Washington, DC 20005   (202) 289-1200

Evelyn K. Merker
Beveridge, De Grandi & Weilacher
1819 H Street, N.W.
Suite 1100
Washington, DC 20006   (202) 659-2811

Darryl Mexic
Sughrue, Mion, Zinn,
Macpeak & Seas
2100 Pennsylvania Ave., N.W.
Washington, DC 20037   (202) 293-7060

Richard S. Meyer
Howrey & Simon
1730 Penn. Ave., N.W.
Washington, DC 20006   (202) 783-0800

Kenneth John Meyers
Finnegan, Henderson, Farabow,
Garrett & Dunner
1300 I Street, N.W.
Washington, DC 20005   (202) 408-4000

Edward F. Miles*
U.S. Dept. of the Navy
Code 1208.2
Naval Research Laboratory
Washington, DC 20375   (202) 767-3427

John T. Miller
Wenderoth, Lind and Ponack
805 Fifteenth Street, N.W.
Suite 700
Washington, DC 20005   (202) 371-8850

Robert C. Millonig#
Finnegan, Henderson, Farabow,
Garrett & Dunner
1300 I Street N.W.
Washington, DC 20005   (202) 408-4264

Leo Millstein*
INTELSET
3400 International Drive, N.W.
Washington, DC 20008   (202) 944-6800

Herbert H. Mintz
Finnegan, Henderson, Farabow,
Garrett & Dunner
1300 I Street, N.W.
Washington, DC 20005   (202) 408-4000

John H. Mion
Sughrue, Mion, Zinn,
Macpeak & Seas
2100 Pennsylvania Avenue, N.W.
Washington, DC 20037   (202) 293-7060

Ralph A. Mittelberger
Fish & Richardson
601 Thirteenth St., N.W.
Washington, DC 20005   (202) 783-5070

George T. Mobille
Cushman, Darby & Cushman
1615 L Street, N.W.
11th Floor
Washington, DC 20036   (202) 861-3000

Robert A. Molan
Cushman, Darby & Cushman
Eleventh Floor
1615 L Street, N.W.
Washington, DC 20036   (202) 822-0944

James B. Monroe#
Finnegan, Henderson, Farabow,
Garrett & Dunner
1300 I Street, N.W.
Washington, DC 20005   (202) 408-0000

Gregory E. Montone
Antonelli, Terry & Wands
1919 Pennsylvania Ave., N.W.
Suite 600
Washington, DC 20006   (202) 828-0300

John P. Moran
Staas & Halsey
1825 K Street, N.W.
Washington, DC 20006   (202) 872-0123

Ruth N. Mordouch
Berman & Aisenberg
1730 R. I. Ave., N.W.
Washington, DC 20036   (202) 293-1404

Ethel L. Morgan*
Federal Energy Regulatory Comm.
Washington, DC 20426   (202) 357-8540

Thomas J. Morgan
Lyon & Lyon
1225 Eye Street, N.W.
Suite 1150
Washington, DC 20005   (202) 845-8400

Roy L. Morris
Alc Communications Corp.
1990 M Street, N.W.
Suite 500
Washington, DC 20036   (202) 293-0593

Robert H. Morse
Galland, Kharasch, Morse
& Garfinkle, P.C.
1054 31st Street, N.W.
Washington, DC 20007   (202) 342-5260

William R. Moser*
U.S. Dept. of Energy
1000 Independence Ave., S.W.
Washington, DC 20585   (202) 586-2802

Gerald J. Mossinghoff
Pharmaceutical Manufacturers Assn.
1100 15th Street, N.W.
Suite 900
Washington, DC 20005   (202) 835-3420

Richard J. Moura
Banner, Birch, Mc Kie
& Beckett
1001 G Street, N.W.
Washington, DC 20001   (202) 508-9100

Douglas P. Mueller
Wegner & Bretschneider
P.O. Box 18218
Washington, DC 20036   (202) 887-0400

Clair Xavier Mullen, Jr.
Finnegan, Henderson, Farabow,
Garrett & Dunner
1300 I Street, N.W.
Washington, DC 20005   (202) 408-4000

John Joseph Murphy*
Dept. of Justice
Litigation II Section
Antitrust Division
555 Fourth St., N.W.
Washington, DC 20001   (202) 307-0931

Robert B. Murray, Jr.
Armstrong, Nikaido, Marmelstein
& Kubovcik
1725 K Street, N.W.
Washington, DC 20006   (202) 659-2930

Barry J. Nace
Paulson, Nace, Norwind
& Sellinger
1814 N Street, N.W.
Washington, DC 20036   (202) 463-1999

Dean H. Nakamura#
Sughrue, Mion, Zinn, Macpeak
& Seas
2100 Pennsylvania Ave., N.W.
Suite 800
Washington, DC 20037   (202) 663-7429

Ronald F. Naughton
Armstrong, Nikaido, Marmelstein
& Kubovcik
1725 K Street, N.W.
Suite 912
Washington, DC 20006   (202) 659-2930

Paul F. Neils
Sughrue, Mion, Zinn,
Macpeak & Seas
2100 Pennsylvania Ave., N.W.
Washington, DC 20037   (202) 293-7060

Sheridan L. Neimark
Browdy and Neimark
419 Seventh St., N.W.
Washington, DC 20004   (202) 628-5197

Richard David Nevius
4000 Cathedral Ave., N.W.
Washington, DC 20016   (202) 333-8652

Bart G. Newland
Bernard, Rothwell & Brown, P.C.
1700 K Street, N.W.
Washington, DC 20006   (202) 833-5740

Mark M. Newman
Barnes & Thornburg
Suite 800
1815 H Street, N.W.
Washington, DC 20006   (202) 955-4500

James A. Niegowski
Banner, Birch, Mc Kie
& Beckett
One Thomas Circle, N.W.
Washington, DC 20005   (202) 296-5500

David T. Nikaido
Armstrong, Nikaido, Marmelstein
& Kubovcik
1725 K St., N.W.
Suite 912
Washington, DC 20006   (202) 659-2930

Jeffrey Nolton
Wenderoth, Lind and Ponack
805 15th Street, N.W.
Suite 700
Washington, DC 20005   (202) 371-8850

Lawrence Geoffrey Norris
Bernard Rothwell & Brown
1700 K Street, N.W.
Washington, DC 20006   (202) 833-5740

William M. Nugent
1730 K Street, N.W.
Suite 304
Washington, DC 20024   (202) 863-5462

Kevin M. O Brien
Baker & Mc Kenzie
815 Connecticut Avenue, N.W.
Washington, DC 20006   (202) 452-7032

Patrick J. O Connell
Finnegan, Henderson,
Farabow, Garrett & Dunner
1300 I St., N.W.
Washington, DC 20005   (202) 408-4000

Thomas J. O Connell
Fitzpatrick, Cella, Harper
& Scinto
1001 Pennsylvania Ave., N.W.
Washington, DC 20004   (202) 347-8100

Shawn W. O Dowd#
Kenyon & Kenyon
1025 Connecticut Ave., N.W.
Washington, DC 20036   (202) 429-1776

Dennis P. O Reilley
Finnegan, Henderson, Farabow,
Garrett & Dunner
1300 I St., N.W.
Washington, DC 20005   (202) 408-4000

Charles L. O Rourke
Finnegan, Henderson, Farabow,
Garrett & Dunner
1300 I Street, N.W.
Washington, DC 20005   (202) 408-4000

Joseph H. O Toole
Armstrong, Nikaido, Marmelstein
& Kubovcik
1725 K Street, N.W.
Suite 912
Washington, DC 20006   (202) 659-2930

William L. Oen#
Sughrue, Mion, Zinn,
Mac Peak & Seas
2100 Pennsylvania Ave., N.W.
Suite 800
Washington, DC 20037   (202) 663-7435

Peter D. Olexy
Sughrue, Mion, Zinn,
Macpeak & Seas
2100 Pennsylvania Ave., N.W.
Washington, DC 20037   (202) 293-7060

Warren E. Olsen
Fitzpatrick, Cella, Harper
& Scinto
1001 Pennsylvania N.W.
Suite 650
Washington, DC 20004   (202) 347-8100

George E. Oram, Jr.
Armstrong, Nikaido, Marmelstein
& Kubovcik
1725 K Street, N.W.
Suite 912
Washington, DC 20006   (202) 659-2930

Karen Lee Orzechowski
1730 K Street, N.W.
Washington, DC 20006   (202) 466-0995

J. Frank Osha
Sughrue, Mion, Zinn,
Macpeak & Seas
2100 Pennsylvania Ave., N.W.
Washington, DC 20037   (202) 293-7060

Jonathan P. Osha#
Wegner & Bretschneider
1233 20th Street
Washington, DC 20036   (202) 887-0400

Perry Palan
Barnes & Thornburg
1815 H Street, N.W.
Suite 800
Washington, DC 20006   (202) 955-4500

Gerson S. Panitch
Finnegan, Henderson, Farabow,
Garrett & Dunner
1300 I Street, N.W.
Washington, DC 20005   (202) 408-4000

Michael D. Parker#
Levy, Bushnell, Zito
& Grandinetti
1511 K Street, N.W.
Suite 425
Washington, DC 20005   (202) 429-4560

Scott F. Partridge
Banner, Birch, Mc Kie
& Beckett
One Thomas Circle, N.W.
Washington, DC 20005   (202) 296-5500

Herbert W. Patterson
Finnegan, Henderson, Farabow,
Garrett & Dunner
1300 I Street, N.W.
Washington, DC 20005   (202) 408-4000

Jerome A. Patterson*
U.S. Dept. of State
Agency for Internatl. Dev.
Washington, DC 20523   (202) 647-8874

John C. Paul
Finnegan, Henderson, Farabow,
Garrett & Dunner
1300 I Street, N.W.
Washington, DC 20005   (202) 408-4000

Mark G. Paulson#
Cushman, Darby & Cushman
1615 L Street, N.W.
Washington, DC 20036   (202) 861-3000

Kenneth E. Payne
Finnegan, Henderson, Farabow,
Garrett & Dunner
1300 I St., N.W.
Washington, DC 20005   (202) 408-4000

Leslie J. Payne#
Vorys, Sater, Seymour & Pease
1828 L Street, N.W.
Suite 1111
Washington, DC 20036   (202) 822-8200

Donald G Peck*
U.S. Dept. of Navy
Naval Sea Systems Command
Off. of Coun., Code 00l5
Washington, DC 20362   (202) 602-7077

Peter Peckarsky
950 25th St., N.W.
Suite 402 North
Washington, DC 20037   (202) 342-0675

Nils E. Pedersen
Wenderoth, Lind & Ponack
805 15th Street, N.W.
Suite 700
Washington, DC 20005   (202) 371-8850

Don J. Pelto
Finnegan, Henderson, Farabow,
Garrett & Dunner
1300 I Street, N.W.
Washington, DC 20005   (202) 408-4000

Stephen A. Pendorf
Sughrue, Mion, Zinn,
Macpeak & Seas
2100 Pennsylvania Ave., N.W.
Washington, DC 20037   (202) 663-7952

Edward A. Pennington
Staas & Halsey
1825 K St., N.W.
Washington, DC 20006   (202) 872-0123

Glenn J. Perry
Cushman, Darby & Cushman
1615 L Street, N.W.
11th Floor
Washington, DC 20036   (202) 861-3070

Stephen L. Peterson
Finnegan, Henderson, Farabow
Garrett & Dunner
1300 I St., N.W.
Washington, DC 20005   (202) 408-4000

Thomas L. Peterson
Banner, Birch, McKie
& Beckett
One Thomas Circle, N.W.
Suite 600
Washington, DC 20005   (202) 296-5500

Henry A. Petri*
U. S. Court of Appeals
For the Federal Circuit
717 Madison Place, N.W.
Washington, DC 20439   (202) 633-5859

John M. Petruncio*
Department of Air Force
Office of Judge Advocate Gen.
1900 Half Street
Washington, DC 20324   (202) 475-1386

George R. Pettit
Pollock, Vande Sande & Priddy
1990 M Street, N.W.
#800
P.O. Box 19088
Washington, DC 20036   (202) 331-7111

Scott Kevin Pickens
Staas & Halsey
1825 K St., N.W.
Washington, DC 20006   (202) 872-0123

Frank V. Pietrantonio
Kenyon & Kenyon
1025 Connecticut Ave., N.W.
Washington, DC 20036   (202) 429-1776

David M. Pitcher
Staas & Halsey
1825 K Street, N.W.
Suite 816
Washington, DC 20016   (202) 872-0123

Michael Terry Platt
Dickinson, Wright, Moon,
Van Dusen & Freeman
1901 L St., N.W.
Washington, DC 20036   (202) 497-0160

Robert C. Platt
1250 24th St., N.W.
Suite 600
Washington, DC 20037   (202) 675-6322

William E. Player
Wegner & Bretschneider
1233 20th Street, N.W.
Washington, DC 20036   (202) 887-0400

Robert H. Plotkin*
U.S. Dept. of Justice
P.O. Box 561
Ben Franklin Station
Washington, DC 20044   (202) 272-4694

Elliott I. Pollock
Pollock, Vande Sande & Priddy
1990 M St., N.W.
Suite 800
Washington, DC 20036   (202) 331-7111

Laurence H. Posorske#
Banner, Birch, McKie
& Beckett
1001 G St., N.W.
Eleventh Floor
Washington, DC 20001   (202) 508-9100

Martin S. Postman
Armstrong, Nikaido, Marmelstein,
Kubovcik & Murray
Suite 1000
1725 K Street, N.W.
Washington, DC 20006   (202) 659-2930

Joseph M. Potenza
Banner, Birch, Mc Kie
& Beckett
One Thomas Circle, N.W.
Suite 600
Washington, DC 20005   (202) 296-5500

Jane E.R. Potter
Finnegan, Henderson, Farabow,
Garrett & Dunner
1300 I St., N.W.
Washington, DC 20005   (202) 408-4122

D. Douglas Price
Fleit, Jacobson, Cohn, Price
Holman & Stern
400 Seventh Street, N.W.
Washington, DC 20004   (202) 638-6666

Robert R. Priddy
Pollock, Vande Sande & Priddy
1990 M St., N.W.
Suite 800
P.O. Box 19088
Washington, DC 20036   (202) 331-7111

Edward M. Prince
Cushman, Darby & Cushman
1615 L Street, N.W.
Suite 1100
Washington, DC 20036   (202) 861-3044

Daniel M. Pritchett, III
4200 19th Street, N.E.
Washington, DC 20018   (202) 526-1228

James Prizant
Kenyon & Kenyon
1025 Connecticut Ave., N.W.
Washington, DC 20036   (202) 775-2838

Paul A. Pumpian*
U.S. Small Business Admin.
1441 L Street, N.W.
Washington, DC 20416   (202) 653-6170

William H. Punter#
2400 Virginia Ave., N.W.
Apt. 1117
Washington, DC 20037   (202) 653-6170

Melanio R. Quintos
Armstrong, Nikaido, Marmelstein,
Kubovcik & Murray
1725 K Street
Suite 1000
Washington, DC 20006   (202) 659-2930

Steven M. Rabin
Spencer & Frank
1111 19th Street, N.W.
Suite 1200
Washington, DC 20036   (202) 828-8000

Richard B. Racine
Finnegan, Henderson, Farabow,
Garrett & Dunner
1300 I Street, N.W.
Washington, DC 20005   (202) 408-4000

Lawrence R. Radanovic
Watson, Cole, Grindle
& Watson
1400 K Street, N.W.
Washington, DC 20005   (202) 628-0088

W. Scott Railton
Reed Smith Shaw & Mc Clay
1200 18th Street N.W.
Washington, DC 20036   (202) 457-6100

William F. Rauchholz#
Beveridge, De Grandi & Weilacher
1819 H St., N.W., Ste. 1100
Washington, DC 20006   (202) 659-2811

Leonard Rawicz
Skadden, Arps, Slate,
Meagher & Flom
1440 New York Ave.
Washington, DC 20005   (202) 371-7001

Andrew E. Rawlins#
Finnegan, Henderson, Farabow,
Garrett & Dunner
1300 I St., N.W.
Washington, DC 20005   (202) 408-4344

Michael B. Ray#
Saidman, Sterne, Kessler
& Goldstein
1225 Connecticut Ave.
Suite 300
Washington, DC 20036   (202) 466-0800

Marvin Reich
12th Floor
1111 Nineteenth Street, N.W.
Washington, DC 20036   (202) 828-8048

George R. Repper
Bernard, Rothwell & Brown, P.C.
1700 K Street, N.W.
Suite 800
Washington, DC 20006   (202) 833-5740

Richard E. Rice
U. S. Dept. of Justice
Civil Division
550 11th St., N.W.
Room 8008
Washington, DC 20530   (202) 307-0380

Gordon B. Richman
Hogan & Hartson
Columbia Square
555 13th St., N.W.
Washington, DC 20004   (202) 637-5600

Ted R. Rittmaster#
Spensley, Horn, Jubas
& Lubitz
2000 M Street, N.W.
Suite 720
Washington, DC 20036   (202) 223-5700

Beatrice N. Robbins#
Venable, Baetjer, Howard
& Civiletti
1201 New York Avenue N.W.
Washington, DC 20005   (202) 962-4800

Frank Edward Robbins
Venable, Baetjer, Howard
& Civiletti
1201 New York Avenue, N.W.
Washington, DC 20005   (202) 962-4800

John C. Robbins#
Venable, Baetjer, Howard
& Civiletti
1201 New York Ave., N.W.
Suite 1000
Washington, DC 20005   (202) 962-4871

John Tyssowski Roberts
Beveridge, De Grandi & Weilacher
1819 H Street, Suite 1100
Washington, DC 20006   (202) 659-2811

Jon L. Roberts
Arter & Hadden
1801 K St., N.W.
Suite 400k
Washington, DC 20006   (202) 775-7980

Dennis C. Rodgers
Beveridge, De Grandi & Walacher
1819 H Street, N.W.
Washington, DC 20006   (202) 659-2811

Elizabeth Miller Roesel
Kirkland & Ellis
655 15th St., N.W.
Suite 1200
Washington, DC 20005   (202) 879-5044

John M. Romary
Finnegan, Henderson, Farabow,
Garrett & Dunner
1300 I St., N.W.
Washington, DC 20005   (202) 408-4000

John Marshall Rommel
Kenyon & Kenyon
Suite 600
1025 Connecticut Ave., N.W.
Washington, DC 20036   (202) 775-2868

Lawrence A. Root*
U.S. Dept. of Navy
Naval Research Laboratory
Off. of Assoc. Counsel (Pats).
Code 3008.2
Washington, DC 20375   (202) 767-3427

Herbert C. Rose
Leydig, Voit & Mayer
Suite 300
700 Thirteenth Street, N.W.
Washington, DC 20005   (202) 737-6770

Stuart W. Rose#
Finnegan, Henderson, Farabow,
Garrett & Dunner
1300 I Street, N.W.
Washington, DC 20005   (202) 408-4000

Stephen J. Rosenman
Finnegan, Henderson, Farabow,
Garrett & Dunner
1300 I Street, N.W.
Washington, DC 20005   (202) 408-4000

Abraham J. Rosner
Sughrue, Mion, Zinn,
Macpeak & Seas
2100 Pennsylvania Ave., N.W.
Washington, DC 20037   (202) 293-7060

Larry L. Rothenberg
Fulbright & Jaworski
801 Pennsylvania Ave., N.W.
Washington, DC 20004   (202) 662-0200

Gideon Franklin Rothwell, IV
Bernard, Rothwell & Brown
1700 K St., N.W.
Suite 800
Washington, DC 20006   (202) 833-5740

Stephen M. Roylance#
Venable, Baetjer, Howard
& Civilletti
1201 New York Ave., N.W.
Washington, DC 20005   (202) 926-4800

Joseph J. Ruch, Jr.
Sughrue, Mion, Zinn,
Macpeak & Seas
2100 Pennsylvania Ave., N.W.
Washington, DC 20037   (202) 663-7931

Peter T. Rutkowski
Banner, Birch, Mc Kie
& Beckett
One Thomas Circle N.W.
Suite 600
Washington, DC 20005   (202) 296-5500

Eugene Sabol
Fisher, Christen & Sabol
2000 M Street, N.W.
Suite 590
Washington, DC 20036   (202) 659-2000

Perry J. Saidman
Saidman Design Law Group
1201 Connecticut Ave.
Suite 550
Washington, DC 20036   (202) 223-0800

Frederick N. Samuels
Leydig, Voit & Mayer
700 13th St., N.W.
Suite 300
Washington, DC 20005   (202) 737-6770

Jeffrey D. Sanok#
Kenyon & Kenyon
1025 Connecticut Ave., N.W.
Suite 600
Washington, DC 20036   (202) 429-1716

Albert J. Santorelli
Finnegan, Henderson, Farabow,
Garrett & Dunner
1300 I St., N.W.
Washington, DC 20005   (202) 408-4000

Michael G. Savage
Kirkland & Ellis
655 - 15th Street, N.W.
Suite 1200
Washington, DC 20005   (202) 879-5914

Steven P. Schad
Banner, Birch, McKie
& Beckett
1001 G Street, 11th Fl.
Washington, DC 20001   (202) 508-9100

Michele M. Schafer#
Finnegan, Henderson, Farabow,
Garrett & Dunner
1300 I St., N.W.
Suite 700
Washington, DC 20005   (202) 408-4000

Walter A. Scheel
503 H St., S.W.
Washington, DC 20024   (202) 554-5908

Daniel B. Schein#
Mason, Fenwick & Lawrence
1225 I St., N.W.
Suite 1000
Washington, DC 20005   (202) 289-1200

Jonathan L. Scherer
Fleit, Jacobson, Cohn, Price
Holman & Stern
400 Seventh Street, N.W.
Jenifer Bldg.
Washington, DC 20004   (202) 638-6666

Alan E. Schiavelli#
Antonelli, Terry & Wands
1919 Pennsylvania Ave., N.W.
Suite 600
Washington, DC 20006   (202) 828-0300

Charles F. Schill
Adduci, Mastriani, Meeks & Schill
1140 Connecticut Ave., N.W.
Suite 250
Washington, DC 20036   (202) 467-6300

Stanley A. Schlitter
Kirkland & Ellis
655 Fifteenth St., N.W.
Suite 1200
Washington, DC 20005   (202) 879-5920

Jerome Schnall#
2801 Quebec St., N.W.
#735
Washington, DC 20008   (202) 362-8872

Jerold I. Schneider
Spencer & Frank
1111 19th Street, N.W.
Washington, DC 20036   (202) 828-8000

John W. Schneller
Spencer & Frank
12th Floor
1111 - 19th Street, N.W.
Washington, DC 20036   (202) 828-8000

William E. Schuyler, Jr.
Suite 400
4801 Massachusetts Ave., N.W.
Washington, DC 20016   (202) 895-1540

Jackie J. Schwartz#
Fleit, Jacobson, Cohn,
Price, Holman & Stern
400 7th St., N.W.
Washington, DC 20004   (202) 638-6666

Nigel L. Scott
Scott & Yallery - Arthur
7603 Georgia Avenue, N.W.
Suite 200
Washington, DC 20012   (202) 882-5770

Thomas J. Scott, Jr.
Howrey & Simon
1730 Pennsylvania Ave., N.W.
Washington, DC 20006   (202) 882-5770

Watson T. Scott
Cushman, Darby & Cushman
1615 L Street, N.W.
11th Floor
Washington, DC 20036   (202) 861-3067

Joseph Scovronek
5930 14th St., N.W., Apt. A1
Washington, DC 20011   (202) 723-0538

Mary Helen Sears
Ginsburg, Feldman & Bress
1250 Connecticut Avenue, N.W.
Washington, DC 20036   (202) 637-9181

Robert John Seas, Jr.
Sughrue, Mion, Zinn,
Macpeak & Seas
2100 Pennsylvania Ave., N.W.
Suite 800
Washington, DC 20037   (202) 293-7060

David W. Selesnick#
Fleit, Jacobson, Cohn, Price,
Holman & Stern
400 Seventh Street, N.W.
Washington, DC 20004   (202) 638-6666

David Hopkins Semmes
Pierson, Semmes & Finley
1054 31st Street, N.W.
Washington, DC 20007   (202) 333-4000

John Gibson Semmes
3286 M St., N.W.
P.O. Box 3559
Washington, DC 20007   (202) 965-1234

Forest Charles Sexton
Waldron & Associates
2120 L Street N.W.
Suite 200
Washington, DC 20037   (202) 955-5999

Robin E. Shaffer
Spencer & Frank
1111 19th St., N.W.
Washington, DC 20036   (202) 828-8000

Kurt Shaffert*
U.S. Dept. of Justice
Antitrust Div.
Judiciary Center Bldg.
555 4th St., N.W., Rm. 9846
Washington, DC 20001   (202) 307-0987

Mark R. Shanks
Finnegan, Henderson, Farabow,
Garretet & Dunner
1300 I Street, N.W.
Washington, DC 20005   (202) 408-4000

John P. Shannon, Jr.
Lane & Aitken
Watergate Office Bldg.
2600 Virginia Ave., N.W.
Washington, DC 20037   (202) 337-5556

Bruce Steven Shapiro
Banner, Birch, McKie
& Beckett
1001 G St., N.W.
Eleventh Floor
Washington, DC 20001   (202) 337-5556

Linda J. Shapiro
Mason, Fenwick & Lawrence
Suite 1000
1225 Eye Street, N.W.
Washington, DC 20005   (202) 289-1200

F. Barry Shay
2808 Northampton Street, N.W.
Washington, DC 20015   (202) 363-2457

Michael J. Shea#
Banner, Birch, McKie
& Beckett
1001 G St., N.W.
Eleventh Floor
Washington, DC 20001   (202) 508-9100

Sol Sheinbein*
U.S. Dept. of Navy
Space & Naval Warfare
Systems Command
Washington, DC 20363   (202) 602-2893

Ronald J. Shore
Antonelli, Terry, Stout
& Kraus
1919 Pennsylvania Ave., N.W.
Suite 600
Washington, DC 20006   (202) 828-0300

Darle M. Short
Roylance, Abrams, Berdo
& Goodman
1225 Conn. Ave., N.W.
Suite 315
Washington, DC 20036   (202) 659-9076

Neil B. Siegel
Sughrue, Mion, Zinn,
Macpeak & Seas
2100 Pennsylvania Ave., N.W.
Washington, DC 20037   (202) 293-7060

M. Howard Silverstein*
U.S. Dept. of Agriculture
Off. of Gen. Coun.
Washington, DC 20250   (202) 447-5474

Jeffrey A. Simenauer*
Cushman, Darby & Cushman
Eleventh Floor
1615 L Street, N.W.
Washington, DC 20036   (202) 861-3085

Donald James Singer*
Hq. USAF
AF/JACP
1900 Half St., S.W.
Room 5160
Washington, DC 20324   (202) 475-1386

George M. Sirilla
Cushman, Darby & Cushman
1615 L Street, N.W.
11th Floor
Washington, DC 20036   (202) 861-3000

Joseph M. Skerpon
Banner, Birch, McKie
& Beckett
1001 G St., N.W., 11th Flr.
Washington, DC 20001   (202) 508-9100

Paul J. Skwierawski
Sughrue, Mion, Zinn,
Macpeak & Seas
2100 Pennsylvania Ave., N.W.
Washington, DC 20037   (202) 293-7060

Robert Vincent Sloan
Sughrue, Mion, Zinn,
Macpeak & Seas
2100 Pennsylvania Ave., N.W.
Washington, DC 20037   (202) 293-7060

Michael R. Slobasky
Fleit, Jacobson, Cohn, Price
Holman & Stern
400 Seventh Street, N.W.
Washington, DC 20004   (202) 638-6666

Dean D. Small#
Sughrue, Mion, Zinn,
Macpeak & Seas
2100 Pennsylvania Ave., N.W.
Suite 800
Washington, DC 20037   (202) 293-7060

Homer Ashby Smith#
Fleit, Jacobson, Cohn,
Price, Holman & Stern
400 Seventh Street, N.W.
Jenifer Bldg., Ste. 600
Washington, DC 20004   (202) 638-6666

John G. Smith#
Finnegan, Henderson, Farabow,
Garrett & Dunner
1300 I St., N.W.
Washington, DC 20005   (202) 408-4092

Randolph A. Smith
Fish & Richardson
601 13th St. NW
5th Fl. N.
Washington, DC 20005   (202) 783-5070

Richard Harold Smith
Finnegan, Henderson, Farabow,
Garrett & Dunner
1300 I St. N.W.
Washington, DC 20005   (202) 408-4000

Samuel B. Smith, Jr.*
U.S. Air Force
JACP
1900 Half Street
Washington, DC 20324   (202) 475-1386

Ronald Ralph Snider
1607 31st. St. N.W.
Washington, DC 20007   (202) 337-3274

John Pennington Snyder#
Watson, Cole, Grindle
& Watson
1400 K Street, N.W.
Washington, DC 20005   (202) 628-0088

Stephen A. Soffen
Ostrolenk, Faber, Gerb
& Soffen
1725 K Street, N.W.
Washington, DC 20006   (202) 457-7785

Allen M. Sokal
Finnegan, Henderson, Farabow
Garrett & Dunner
1300 I Street, N.W.
Washington, DC 20005   (202) 408-4000

William I. Solomon
Antonelli, Terry, Stout & Kraus
1919 Pennsylvania Ave., N.W.
Suite 600
Washington, DC 20006   (202) 828-0300

David L. Soltz#
Finnegan, Henderson, Farabow,
Garrett & Dunner
1300 I St., N.W.
Washington, DC 20005   (202) 408-4345

Andrew C. Sonu
Finnegan, Henderson, Farabow,
Garret & Dunner
1300 I Street, N.W.
Washington, DC 20005   (202) 408-4000

Zosan S. Soong
Finnegan, Henderson, Farabow,
Garrett & Dunner
1300 I Street, N.W.
Washington, DC 20005   (202) 408-4000

John A. Sopp
Keil & Weinkauf
Suite 620
1101 Connecticut Ave., N.W.
Washington, DC 20036   (202) 659-0100

George H. Spencer
Spencer & Frank
1111 19th St., N.W.
Washington, DC 20036   (202) 828-8000

W. Murray Spruill#
Saidman, Sterne, Kessler
& Goldstein
1225 Connecticut Ave., N.W.
Washington, DC 20036   (202) 466-0800

Harry John Staas
Staas & Halsey
1825 K St., N.W.
Suite 816
Washington, DC 20006   (202) 872-0123

Norman L. Stack, Jr.*
U.S. Navy
Navsea Code 56W32
Naval Sea Systems Command
Washington, DC 20362   (703) 602-8017

Alfred A. Stadnicki
International Telecommunicatios
Satellite Organization
3400 International Drive, N.W.
Washington, DC 20008   (202) 944-6855

Lawrence A. Stahl
Fitzpatrick, Cella, Harper
& Scinto
1001 Pennsylvania Ave., N.W.
Suite 650
Washington, DC 20004   (202) 347-8100

Wayland W. Stallard#
4407 Garrison Street, N.W.
Washington, DC 20016   (202) 363-1122

A. Fred Starobin
419 7th St., N.W.
Suite 300
Washington, DC 20004   (202) 393-2717

Marvin R. Stern
Fleit, Jacobson, Cohn, Price,
Holman & Stern
400 Seventh Street, N.W.
Jenifer Building
Washington, DC 20004   (202) 638-6666

Richard H. Stern
1300 19th St., N.W.
Suite 300
Washington, DC 20036   (202) 659-1385

Robert Greene Sterne
Saidman, Sterne, Kessler
& Goldstein
1225 Conn. Ave., Ste. 300
Washington, DC 20036   (202) 466-0800

Gene W. Stockman
Staas & Halsey
1825 K Street, N.W.
Washington, DC 20006   (202) 872-0123

Charles J. Stockstill*
Dept. of the Navy
Naval Res. Lab., Code 3008.20
Off. of Gen. Coun.
4555 Overlook Ave., S.W.
Washington, DC 20375   (202) 767-3427

Leonard Francis Stoll*
U.S. Dept. of Air Force
AF/JACP
1900 Half St., S.W.
Washington, DC 20324   (202) 475-1386

Donald R. Stone
Mc Kenna & Cuneo
1575 Eye Street, N.W.
Washington, DC 20005   (202) 789-7500

Michael Stone
Wenderoth, Lind & Ponack
805 Fifteenth Street, N.W.
Southern Bldg., Suite 700
Washington, DC 20005   (202) 371-8850

Donald E. Stout
Antonelli, Terry, Stout
& Kraus
1919 Pennsylvania Ave., N.W.
Suite 600
Washington, DC 20006   (202) 828-0300

George B. Stover
Beveridge, De Grandi
& Weilacher
1819 H St., N.W.
Suite 1100
Washington, DC 20006   (202) 659-2811

John H. Stowe*
U.S. Dept. of Navy
Naval Air Systems Command
Code 00C5
Washington, DC 20361   (703) 692-7810

Lloyd Joseph Street
Cushman, Darby & Cushman
1615 L Street, N.W.
11th Floor
Washington, DC 20036   (201) 861-3000

Jur Strobos*
U.S. Court of Appeals
For the Federal Circuit
717 Madison Place
Room 809
Washington, DC 20439   (202) 633-5909

Richard L. Stroup
Finnegan, Henderson, Farabow,
Garrett & Dunner
1300 I St., N.W.
Washington, DC 20005   (202) 408-4000

Charles L. Sturtevant
3231 Reservoir Road, N.W.
Washington, DC 20007   (202) 337-5723

Stephen T. Sullivan
Finnegan, Henderson, Farabow,
Garrett & Dunner
1300 I Street, N.W.
Washington, DC 20005   (202) 408-4000

Marcia H. Sundeen
Cushman, Darby & Cushman
1615 L Street, N.W.
Suite 1100
Washington, DC 20036   (202) 861-3000

Gary N. Sundick*
U.S. Securities & Exchange Comm.
450 5th Street, N.W.
Washington, DC 20549   (202) 272-3871

Wilson D. Swayze, Jr.
Antonelli, Terry, Stout
& Kraus
1919 Penn. Ave., N.W.
Suite 600
Washington, DC 20006   (202) 828-0300

George W. Swenson#
Vorys, Sater, Seymour
& Pease
1828 L Street, N.W.
Suite 1111
Washington, DC 20036   (202) 822-8200

Brett S. Sylvester
Sughrue, Mion, Zinn, Macpeak
& Seas
2100 Pennsylvania Ave., N.W.
Washington, DC 20037   (202) 293-7060

C. Scott Talbot
Kenyon & Kenyon
1025 Connecticut Ave., N.W.
Washington, DC 20036   (202) 775-2818

W. Warren Taltavall, III
Cushman, Darby & Cushman
1615 L Street, N.W.
11th Floor
Washington, DC 20036   (202) 861-3000

David L. Tarnoff#
Roylance, Abrams, Berdo
& Goodman
1225 Connecticut Ave., N.W
Suite 315
Washington, DC 20036   (202) 659-9076

Bruce A. Tassan
Dickinson & Wright
1901 L St., N.W.
Suite 800
Washington, DC 20036   (202) 457-0160

Rodger L. Tate
Banner, Birch, Mc Kie
& Beckett
1001 G St., N.W, 11th Flr.
Washington, DC 20001   (202) 508-9100

Cheri M. Taylor
Finnegan, Henderson, Farabow,
Garrett & Dunner
1300 I Street, N.W.
Washington, DC 20005   (202) 408-4031

Hosea E. Taylor#
1749 N. Portal Drive, N.W.
Washington, DC 20012   (301) 829-1323

Roger D. Taylor
Finnegan, Henderson, Farabow,
Garrett & Dunner
1300 I Street, N.W.
Suite 700
Washington, DC 20005   (202) 408-4000

Jennifer A. Tegfeldt
Fitzpatrick, Cella, Harper
& Scinto
1001 Pennsylvania Ave., N.W.
Suite 650
Washington, DC 20004   (202) 347-8100

Rene D. Tegtmeyer
Fish & Richardson
601 Thirteenth St., N.W.
Washington, DC 20005   (202) 783-5070

David T. Terry
Antonelli, Terry, Stout
& Kraus
1919 Pennsylvania Ave., N.W.
Suite 600
Washington, DC 20006   (202) 828-0300

Melinda B. Thaler
Crowell & Morning
1001 Pennsylvania Avenue, N.W.
Washington, DC 20004   (202) 624-2500

C. Douglas Thomas#
Sughrue, Mion, Zinn,
Mac Peak & Seas
2100 Penn. Ave., N.W.
Washington, DC 20037   (202) 293-7060

Dirk D. Thomas
Finnegan, Henderson, Farabow,
Garrett & Dunner
1300 I Street, N.W.
Washington, DC 20005   (202) 408-4000

John E. Thomas#
Wegner, Cantor, Mueller & Player
1233 20th Street, N.W.
Suite 300
P.O. Box 18218
Washington, DC 20036   (202) 887-0400

Marian T. Thomson
Fleit, Jacobson, Cohn, Price
Holman & Stern
400 Seventh Street, N.W.
Jenifer Building
Washington, DC 20004   (202) 638-6666

Mark J. Thronson
Dickstein, Shapiro & Morin
2101 L Street, N.W.
Washington, DC 20037   (202) 785-9700

Michael H. Tobias#
Leydig, Voit & Mayer
700 13th St., N.W., Ste. 300
Washington, DC 20005   (202) 737-6770

Albert Tockman
Armstrong, Nikaido, Marmelstein
& Kubovcik
1725 K Street, N.W.
Suite 1000
Washington, DC 20006   (202) 659-2930

Allan J. Topol
Covington & Burling
1201 Pennsylvania Ave., N.W.
Washington, DC 20044   (202) 662-5402

Richard Lee Torczon, Jr.#*
U.S. Dept. of Commerce
Office of Gen. Counsel
Room H-4610
Washington, DC 20230   (202) 377-5394

Oscar A. Towler, III
Dickstein, Shapiro & Morin
2101 L Street, N.W.
Washington, DC 20037   (202) 785-9700

Guy K. Townsend#
Sughrue, Mion, Zinn,
Mac Peak & Seas
2100 Pennsylvania Ave., N.W.
Washington, DC 20037   (202) 663-7958

William Cecil Townsend*
Dept. of the Navy
Naval Sea Systems Command
Code Sea 00l5
Washington, DC 20362   (703) 602-7077

Francis G. Toye
1817 Sudbury Rd., N.W.
Washington, DC 20012   (202) 829-9007

Dalton L. Truluck#
Antonelli, Terry, Stout
& Kraus
1919 Pennsylvania Ave., N.W.
Suite 600
Washington, DC 20006   (202) 828-0300

Peter K. Trzyna
Sutherland, Asbill & Brennan
1275 Pennsylvania Avenue, N.W.
Suite 800
Washington, DC 20004   (202) 383-0100

Leon R. Turkevich#
Cushman, Darby & Cushman
1615 L Street, N.W.
Eleventh Floor
Washington, DC 20036   (202) 861-3671

John B. Turner#
Finnegan, Henderson, Farabow,
Garrett & Dunner
1300 I Street, N.W.
Washington, DC 20005   (202) 408-4198

Richard C. Turner
Sughrue, Mion, Zinn,
Macpeak & Seas
2100 Pennsylvania Ave., N.W.
Suite 800
Washington, DC 20037   (202) 293-7060

Richard H. Tushin
Watson, Cole, Grindle
& Watson
1400 K Street, N.W.
Suite 725
Washington, DC 20005   (202) 628-0088

Jon F. Tuttle
Dorsey & Whitney
1330 Connecticut Ave., N.W.
Suite 200
Washington, DC 20036   (202) 857-0700

Archie W. Umphlett
Phillip Petroleum Co.
1776 Eye Street, N.W.
Suite 700
Washington, DC 20006   (202) 833-0900

Mark E. Ungerman
Fulbright & Jaworski
801 Pennsylvania Ave., N.W.
Washington, DC 20004   (202) 662-0200

Thomas B. Van Poole, Jr.
Mason, Fenwick & Lawrence
1225 Eye Street, N.W.
Suite 1000
Washington, DC 20005   (202) 289-1200

George Vande Sande
Pollock, Vande Sande & Priddy
1990 M Street, N.W.
Suite 800
Washington, DC 20036   (202) 331-7111

J. Derek Vandenburgh#*
U.S. Court of Appeals
For the Federal Circuit
717 Madison Pl., N.W.
Suite 816
Washington, DC 20439   (202) 633-6579

Lynn Vandenburgh
Wegner, Cantor, Mueller & Player
1233 20th St., N.W.
Suite 300
Washington, DC 20036   (202) 887-0400

David Edwards Varner
Cushman, Darby & Cushman
1615 L Street, N.W.
11th Floor
Washington, DC 20036   (202) 861-3539

Robert Stanley Vermut*
U.S. Dept. of Transportation
Federal Railroad Admin.
400 Seventh Street, S.W.
Washington, DC 20590   (202) 366-0618

James M. Verna
Fleit, Jacobson, Cohn, Price,
Holman & Steton
Jenifer Bldg.
400 7th St., N.W.
Washington, DC 20004   (202) 638-6666

Jerry Dean Voight
Finnegan, Henderson, Farabow,
Garrett & Dunner
1300 I St., N.W.
Suite 700
Washington, DC 20005   (202) 408-4000

Adam C. Volentine#
Wenderoth, Lind & Ponack
805 15th Street, N.W.
Suite 700
Washington, DC 20005   (202) 371-8850

Catherine M. Voorhees#
Spencer & Frank
1111 19th St., N.W.
Washington, DC 20036   (202) 828-8000

Stanley A. Wal#
Antonelli, Terry & Wands
Suite 600
1919 Pennsylvania Ave., N.W.
Washington, DC 20006   (202) 828-0326

James S. Waldron
2120 L St., N.W., Ste. 200
Washington, DC 20037   (202) 955-5999

James H. Wallace, Jr.
Wiley, Rein & Fielding
1776 K St., N.W.
Washington, DC 20006   (202) 429-7240

T. Lester Wallace#
Cushman, Darby & Cushman
1615 L St., 11th Floor
Washington, DC 20036   (202) 861-3642

Wallace G. Walter
Cushman, Darby & Cushman
1615 L Street, N.W.
Suite 1100
Washington, DC 20036   (202) 861-3000

William M. Wannisky
Fitzpatrick Cella, Harper
& Scinto
1001 Pennsylvania Ave., N.W.
Suite 650
Washington, DC 20004   (202) 347-8100

Allen W. Wark
Marks Murase & White
2001 L Street, N.W.
Suite 750
Washington, DC 20036   (202) 955-4900

Steven E. Warner
Fitzpatrick, Cella, Harper
& Scinto
1001 Pennsylvania Ave.
Suite 650N
Washington, DC 20004   (202) 347-8100

Charles R. Watts
Wenderoth, Lind & Ponack
Southern Building, Suite 700
805 Fifteenth Street, N.W.
Washington, DC 20005   (202) 371-8850

Lynn F. Watts
Venable, Baetjer, Howard
& Civiletti
1201 New York Ave., N.W.
Suite 1000
Washington, DC 20005   (202) 962-4800

Ralph T. Webb
Roylance, Abrams, Berdo
& Goodman
1225 Connecticut Ave., N.W.
Washington, DC 20036   (202) 659-9076

Tamara L. Weber*
U.S. Navy
Navsea 00l54
2531 Jefferson Davis Hwy.
6S10
Washington, DC 20362   (703) 602-3730

Harold C. Wegner
Wegner, Cantor, Mueller
& Player
1233 20th St., N.W.
Suite 300
Washington, DC 20036   (202) 887-0400

Helmuth A. Wegner
Wegner, Cantor, Mueller
& Player
1233 20th St., N.W.
P.O. Box 18218
Washington, DC 20036   (202) 887-0400

Robert G. Weilacher
Beveridge, De Grandi & Weilacher
1819 H Street, N.W.
Washington, DC 20006   (202) 659-2811

Wendi L. Weinstein
Banner, Birch, Mc Kie
& Beckett
1001 G St., N.W.
Eleventh Floor
Washington, DC 20001   (202) 508-9100

James Price Welch
Armstrong, Nikaido, Marmelstein
& Kubovcik
1725 K Street, N.W.
Suite 1000
Washington, DC 20006   (202) 659-2930

Ashley J. Wells
Spencer & Frank
1111 - 19th Street, N.W.
Washington, DC 20036   (202) 828-8000

William K. Wells, Jr.
Kenyon & Kenyon
1025 Connecticut Ave., N.W.
Washington, DC 20036   (202) 429-1776

Gerald H. Werfel
Arent, Fox, Kintner,
Plotkin & Kahn
1050 Connecticut Avenue, N.W.
Washington, DC 20036   (202) 857-6165

William K. West, Jr.
Cushman, Darby & Cushman
1615 L Street, N.W.
Suite 1100
Washington, DC 20036   (202) 861-3542

Robert A. Westerlund, Jr.
Samsung Electronics Co., Ltd.
Suite 250, One Westin Center
2445 M Street, N.W.
Washington, DC 20037   (202) 296-0227

William F. Westerman
Armstrong, Nikaido, Marmelstein,
Kubovcik & Murray
1725 K Street, N.W.
Suite 1000
Washington, DC 20006   (202) 659-2930

John T. Whelan
Fitzpatrick, Cella, Harper
& Scinto
1001 Pennsylvania Ave., N.W.
Suite 650N
Washington, DC 20004   (202) 347-8100

Paul E. White, Jr.
Cushman, Darby & Cushman
1615 L Street, N.W.
Washington, DC 20036   (202) 861-3000

Stephen R. Whitt#
Finnegan, Henderson, Farabow,
Garrett & Dunner
1300 I St., N.W.
Washington, DC 20005   (202) 408-4000

Bruce T. Wieder*
U.S. Court of Appeals
Federal Circuit
717 Madison Place, N.W.
Washington, DC 20439   (202) 633-5869

Richard Wiener
Pollock, Vande Sande & Priddy
1990 M St., N.W.
Washington, DC 20036   (202) 331-7111

Otto M. Wildensteiner*
U.S Dept. of Transportation
C-15
400 7th St., S.W.
Washington, DC 20590   (202) 366-9161

Harlan B. Williams, Jr.#
Cushman, Darby & Cushman
1615 L St., N.W.
Eleventh Floor
Washington, DC 20036   (202) 861-3628

Harry S. Williams#
1845 47th Place N.W.
Washington, DC 20007   (202) 342-2687

Mark A. Williamson
Fitzpatrick, Cella, Harper
& Scinto
1001 Pennsylvania Ave., N.W.
Suite 650
Washington, DC 20004   (202) 347-8100

Mary J. Wilson#
Cushman, Darby & Cushman
1615 L Street, N.W.
11th Floor
Washington, DC 20036   (202) 861-3688

Thomas W. Winland
Finnegan, Henderson, Farabow,
Garrett & Dunner
1300 I Street, N.W.
Suite 700
Washington, DC 20005   (202) 408-4085

Morris Wiseman*
U.S. Air Force
Hq. USAF/JACP
1900 Half St., S.W.
Washington, DC 20324   (301) 475-1387

Wilford L. Wisner
Finnegan, Henderson, Farabow,
Garrett & Dunner
1300 I Street, N.W.
Washington, DC 20005   (202) 408-4075

John F. Witherspoon
1627 I Street, N.W.
Suite 1200
Washington, DC 20006   (202) 835-3700

Henry N. Wixon
Sterne, Kessler, Goldstein
& Fox
1225 Connecticut Ave., N.W.
Washington, DC 20036   (202) 833-7533

Franklin David Wolffe
Wegner, Cantor, Mueller & Player
1233 20th St., N.W.
Suite 300
P.O. Box 18218
Washington, DC 20036   (202) 887-0400

Susan A. Wolffe
Finnegan, Henderson, Farabow,
Garrett & Dunner
1300 I Street, N.W.
Washington, DC 20005   (202) 408-4000

Louis Woo
Pollock, Vande Sande & Priddy
1990 M Street, N.W.
Washington, DC 20036   (202) 331-7111

L. Allen Wood, Jr.
Spencer & Frank
Suite 1200
1111 19th Street, N.W.
Washington, DC 20036   (202) 828-8000

George N. Woodruff#
1000 6th Street St., S.W.
Suite 807
Washington, DC 20024   (202) 554-1228

Alesia M. Woodworth
Cushman, Darby & Cushman
Eleventh Floor
1615 L Street, N.W.
Washington, DC 20036   (202) 861-3000

Laurence Arthur Wright*
U.S. Air Force
Pat. Division
1900 Half Street, S.W.
Washington, DC 20324   (202) 475-1386

William H. Wright
Henderson & Sturm
1901 Pennsylvania Ave., N.W.
Suite 807
Washington, DC 20006   (202) 296-3854

Christopher P. Wrist
Fitzpatrick, Cella, Harper
& Scinto
1001 Pennsylvania Ave., N.W.
Suite 650N
Washington, DC 20004   (202) 347-8100

Jeffrey A. Wyand
Leydig, Voit & Mayer
Suite 300
700 Thirteenth Street, N.W.
Washington, DC 20005   (202) 737-6770

Edward R. Yoches
Finnegan, Henderson, Farabow,
Garrett & Dunner
1300 I Street, N.W.
Washington, DC 20005   (202) 408-4000

Richard G. Young
Beveridge, De Grandi & Weilacher
1819 H Street, N.W.
Suite 1100
Washington, DC 20006   (202) 659-2811

Thomas Zack*
U.S. Dept. of Energy
Off. of Gen. Coun.
1000 Independence Ave., S.W.
Room 6F067
Washington, DC 20585   (202) 586-0343

Albert J. Zervas
Marks Murase & White
2001 L Street, N.W.
Suite 750
Washington, DC 20036   (202) 955-4900

Robert F. Ziems
Finnegan, Henderson, Farabow,
Garrett & Dunner
1300 I Street, N.W.
Washington, DC 20005   (202) 408-4000

Donald E. Zinn, Sr.
Sughrue Mion Zinn Macpeak
& Seas
2100 Pennsylvania Avenue, N.W.
Washington, DC 20037   (202) 293-7060

Jonathan D. Zischkau
Crowell & Moring
1001 Pennsylvania Ave., N.W.
Washington, DC 20004   (202) 624-2608

Joseph J. Zito
Cintelli Legal
1919 Penn. Ave., N.W.
Suite 300
Washington, DC 20006   (202) 429-4560

Bruce C. Zotter
Finnegan, Henderson, Farabow
& Garrett
1300 I St., N.W.
Washington, DC 20005   (202) 408-4000

Henry Michael Zykorie
Wenderoth, Lind & Ponack
Southern Bldg.
805 Fifteenth Street, N.W.
Suite 700
Washington, DC 20005   (202) 371-8850

Florida

Elsie T. Apthorp
201 Park Place
Suite 106
Altamonte Springs, FL 32701
                                 (407) 260-6676

Harry R. Dumont
415 Montgomery Road
Suite 175
Altamonte Springs, FL 32714
                                 (305) 788-2788

Edward T. Mc Cabe
6547 Beachwood Rd.
Amelia Island, FL 32034   (904) 277-2738

Randall K. Pegg#
5201 First Coast Hwy.
Amelia Island, FL 32034   (919) 261-4640

Jack Hensel
8319 Roxboro Drive
Bayonet Point, FL 34667   (813) 868-1955

Harry G. Martin, Jr.
204 W. Sugarmaple Lane
Beverly Hills, FL 32665   (904) 746-4865

Leonidas Vlachos#
P.O. Box 1529
Big Pine Key, FL 33043   (305) 872-3390

Joseph P. Abate
IBM
Property Law Dept.
Internal Zip 4318
951 N.W. 51st St.
Boca Raton, FL 33431   (407) 982-6137

Herbert K. Anspach
2760 N.W. 29th Drive
Boca Raton, FL 33434   (305) 483-2403

Harry W. Barron
8221 Glades Road
Suite 202
Boca Raton, FL 33434   (305) 488-3000

Burton P. Beatty, Sr.
410 SW 7th Way
Boca Raton, FL 33486   (305) 368-1584

John C. Black
940 S.W. 15th St.
Boca Raton, FL 33432   (305) 391-7715

Bernard David Bogdon
IBM Corp.
Intell. Prop. Law. Dept.
Zip 4318
951 N.W. 51st St.
Boca Raton, FL 33432   (407) 443-2000

Charles H. Brown
6300 N.W. 2nd Ave.
Apt. 308
Boca Raton, FL 33432   (305) 994-1100

Winfield J. Brown, Jr.
IBM
Internal Zip 4318
951 N.W. 51st St.
Boca Raton, FL 33432   (407) 443-2928

Anibal Jose Cortina
2805 N.W. 29th Drive
Boca Raton, FL 33434   (407) 982-3302

Don Daniel Doty
6065 S. Verde Trail
Apt. G-211
Boca Raton, FL 33433   (305) 488-2362

John H. Faro
1800 Corporate Blvd.
Suite 202
Boca Raton, FL 33431   (407) 241-7444

Leonard Forman
6680 Burning Wood Dr.
Apt. 264
Boca Raton, FL 33433-6808   (406) 368-8406

Olney M. Gardiner
701 Marine Drive
Boca Raton, FL 33487   (305) 391-4199

George E. Grosser
IBM Corporation
951 NW 51st Street
Area 4318
Boca Raton, FL 33431   (305) 998-0430

Edward Halle
4001 N. Ocean Blvd.
Apt. 603-B
Boca Raton, FL 33431   (305) 394-5384

Rubin Hoffman
7615 Sierra Drive West
Boca Raton, FL 33433   (305) 392-2784

Paul T. Kashimba
IBM Corporation
Intellectual Property Law 4318
P.O. Box 1328
951 N.W. 51st Street
Boca Raton, FL 33431   (305) 443-7361

Robert Lieber
2762 Timber Creek Circle
Boca Raton, FL 33431   (407) 483-8780

H. Geoffrey Lynfield
7050 N.E. 7th Avenue
Boca Raton, FL 33487   (305) 997-5825

Daniel E. Mc Connell
IBM
Intellectual Property Law Dept.
Internal Zip 4318
951 N.W. 51st. Street
Boca Raton, FL 33342   (407) 982-4126

Martin J. Mc Kinley
IBM Corporation
Intellectual Property Law Dept.
Internal Zip 4318
951 N.W. 51st Street
Boca Raton, FL 33431   (407) 443-4708

Philip T. Mintz
198 N.W. 67th Street
Apt. 508
Boca Raton, FL 33487   (407) 994-2078

Raymond R. Skolnick#
10658 180th Court South
Boca Raton, FL 33434   (305) 483-6186

David J. Skrabec
Eltech Systems Corp.
6100 Glades Rd.
Suite 305
Boca Raton, FL 33434   (407) 487-3642

John C. Smith
IBM
P.O. Box 1328
Boca Raton, FL 33429   (407) 443-0625

Harold Hugh Sweeney, Jr.
1054 S. W. 7th Street
Boca Raton, FL 33486   (407) 392-4284

Stephen A. Terrile
IBM
Internal Zip 4318
951 N.W. 51st Street
Boca Raton, FL 33432   (407) 982-3181

Richard A. Tomlin
IBM Corporation
951 NW 51st Street
Boca Raton, FL 33431   (407) 443-9786

John T. Winburn
Leydig, Voit & Mayer, Ltd.
1900 Glades Road, Suite 350
Boca Raton, FL 33431   (407) 392-5332

Shaler G. Smith, Jr.#
27718 King'S Kew
Bonita Springs, FL 33923   (813) 992-3969

Daniel R. Collopy
Motorola, Inc.
1500 N.W. 22nd Ave.
Boynton Beach, FL 33426   (407) 364-2860

Michael J. De Luca#
Motorola, Inc.
1500 NW 22 Avenue
Boynton Beach, FL 33426   (407) 738-2860

James F. Higgins
13 Afton Place
Boynton Beach, FL 33462   (305) 965-3251

Vincent B. Ingrassia
Motorola Inc.
1500 NW 22 Avenue
Boynton Beach, FL 33426   (407) 738-2860

Philip P. Macnak
Motorola, Inc.
Pat. Law Dept.
1500 N.W. 22 Avenue
Boynton Beach, FL 33426   (407) 738-2860

Arthur T. Mc Keon
2086 S.W. 13th Way
Boynton Beach, FL 33436   (305) 737-6057

Warren D. Mc Phee
17 Fairway Drive
Boynton Beach, FL 33436   (407) 734-5366

Gregg E. Rasor#
Motorola, Inc.
1500 N.W. 22nd Ave.
M/S E2053
Boynton Beach, FL 33426   (407) 364-2952

Victor F. Volk#
646 Snug Harbor Dr.
Boynton Beach, FL 33435   (305) 732-4934

William E. Zitelli
Motorola, Inc.
Pat. Dept.
1500 N.W. 22nd Avenue
Boynton Beach, FL 33426   (407) 364-2862

Sidney Alfred Ochs
6404 21st Ave. W.
Bradenton, FL 34209   (813) 792-3121

Clement J. Paznokas
10424 Spoonbill Rd., W.
Bradenton, FL 34209   (813) 792-3796

Frederick Shapoe
6501-17th Avenue West
W-318
Bradenton, FL 34209   (813) 792-2555

Stuart A. White
6804 Ninth Avenue West
Brandenton, FL 34209   (813) 798-3279

Lewis P. Elbinger
121 North Osceola Avenue
Clearwater, FL 34617   (813) 449-0255

Daniel J. Hanlon, Jr.
2351 Irish Lane, Apt. 45
Clearwater, FL 33515   (813) 796-1250

John L. Harris
470 Palm Island, N.E.
Clearwater, FL 34630   (813) 446-1953

Charles E. Lykes, Jr.
300 Turner Street
Clearwater, FL 34616   (813) 441-8308

Joseph C. Mason, Jr.
Mason, Mason & Associates
19337 U.S. 19 N.
Suite 102
Clearwater, FL 34624   (813) 538-3800

Ronald E. Smith
Mason, Mason & Associates
19337 U.S. 19 N.
Suite 102
Clearwater, FL 34624 (813) 538-3800

Jacob H. Steinberg
1502 Cayman Way, #J2
Coconut Creek, FL 33066 (305) 977-8073

Edward Stern
2614 Nassau Bend
Apt. B-1
Coconut Creek, FL 33066 (305) 972-5118

Allen B. Curtis#
3204 Portofino Point, Apt. H2
Coconut Creek., FL 33066 (305) 979-2098

David K. Friedland
Leslie J. Lott &
& Associates, P.A.
338 Minorca Ave.
Coral Gables, FL 33134 (305) 448-7089

Meredith P. Sparks
5129 Granada Blvd.
Coral Gables, FL 33146 (305) 661-5756

J. P. J. Violette
UM P.O. Box 9148
Coral Gables, FL 33124 (305) 371-5261

Ronald V. Davidge#
4160 NW 113 Ave.
Coral Springs, FL 33065 (305) 341-0051

Leroy Greenspan
10844 NW 7th Street
Coral Springs, FL 33071 (305) 755-3262

Mark P. Kahler
10932 N.W. 13th Court
Coral Springs, FL 33071 (305) 751-345

Kenneth E. Merklen
11151 N.W. 15th St.
Coral Springs, FL 33071 (305) 753-7426

Marden S. Gordon#
P.O. Box 1097
Crystal Beach, FL 34681 (813) 787-5105

Charles W. Lanham, Jr.
330 Cornell Dr.
Daytona Beach, FL 32018 (904) 677-0272

James C. Wood
209 Pleasant Valley Drive
Daytona Beach, FL 32114 (904) 255-1445

James R. Hagen#
107 Amigos Road
De Bary, FL 32713 (407) 668-8342

Arthur J. Greif
Brunswick Corporation
2000 Brunswick Lane
Deland, FL 32724 (904) 736-1700

Theodore A. Te Grotenhuis
1420 N.W. 20th Ave.
CB 116
Delray Beach, FL 33445 (407) 243-8939

Roger L. Martin#
908 Sylvia Drive
Deltona, FL 32725 (305) 532-5935

Merle William Goodwin
2135 Lagoon Drive
Dunedin, FL 34698 (813) 733-4212

Stanley M. Miller
748 Broadway
Dunedin, FL 34698 (813) 733-8825

Wayne A. Warner*
U. S. Air Force
AFDDTC/JAN
Eglin Air Force Base, FL 32542
(904) 882-5335

Raymond G. Brodahl
1029 Grant Street
Englewood, FL 34224 (813) 475-7643

William A. Mikesell, Jr.
1000 Lee Street
Englewood, FL 34224 (813) 475-5228

Juliana Agon
Motorola Inc.
8000 West Sunrise Blvd.
Fort Lauderdale, FL 33322 (305) 475-6449

Thomas G. Berry
Motorola, Inc.
8000 W. Sunrise Blvd.
Fort Lauderdale, FL 33322 (305) 475-6449

Alvin S. Blum#
2350 Del Mar Place
Fort Lauderdale, FL 33301 (305) 462-5006

Michael J. Buchenhorner
Motorola Inc.
Pat. Law Dept.
8000 W. Sunrise Blvd.
Fort Lauderdale, FL 33322 (305) 475-3827

Kevin P. Crosby#
Malin, Haley & Mc Hale, P.A.
One East Broward Blvd.
Suite 1609
Fort Lauderdale, FL 33301 (305) 763-3303

Robert B. Czirr*
Internal Revenue Service
299 E. Broward Blvd.
Fort Lauderdale, FL 33301 (305) 527-7472

Dale P. Di Maggio
Malin, Haley & Mc Hale,
Di Maggio & Crosby
One East Broward Blvd.
Suite 1609
Fort Lauderdale, FL 33301 (305) 763-3303

William Joseph Flynn
Oltman & Flynn
915 Middle River
Suite 415
Fort Lauderdale, FL 33304 (305) 563-4814

Barry L. Haley
Malin, Haley, Mc Hale,
Di Maggio & Crosby
One East Broward Blvd.
Suite 1609
Fort Lauderdale, FL 33301 (305) 763-3303

Herman I. Hersh
3900 Galt Ocean Drive
Fort Lauderdale, FL 33308 (305) 566-0333

Adam A. Jorgensen#
915 Middle River Drive
Suite 415
Fort Lauderdale, FL 33304 (305) 563-4814

Pablo Meles
Motorola, Inc.
8000 W. Sunrise Blvd.
Suite 1159
Fort Lauderdale, FL 33322 (305) 475-6449

Alfred Musumeci
Argotec Inc.
3750 Hacienda Blvd.
Fort Lauderdale, FL 33314 (305) 584-7900

Daniel K. Nichols
Motorola, Inc.
Pat. Dept.
8000 W. Sunrise Blvd.
Fort Lauderdale, FL 33322 (305) 475-6449

John Harold Oltman
Oltman & Flynn
915 Middle River Drive
Suite 415
Fort Lauderdale, FL 33304 (305) 563-4814

Daniel S. Polley
Malin, Haley, Mc Hale,
Di Maggio & Crosby, P.A.
One East Broward Blvd.
Suite 1609
Fort Lauderdale, FL 33301 (305) 763-3303

Melvin K. Silverman
305 South Andrews Avenue
Suite 503
Fort Lauderdale, FL 33301 (305) 764-5155

Michael A. Slavin
Malin, Haley, Mc Hale,
Di Maggio & Crosby, P.A.
Suite 1609 1 E. Broward Blvd.
Fort Lauderdale, FL 33301 (305) 374-3311

Carl V. Wisner, Jr.
2709 N. E. 26th Terrace
Fort Lauderdale, FL 33306 (305) 564-2137

Joseph Zallen
2455 E. Sunrise Blvd.
Suite 802
Fort Lauderdale, FL 33304 (305) 565-9506

Frank P. Cyr
1004 La Paloma Blvd.
Fort Myers, FL 33903 (813) 731-1583

Robert H. Heise
994 N. Waterway Drive
Fort Myers, FL 33919 (813) 482-0613

William E. Noonan
P.O. Box 07338
Fort Myers, FL 33919　　　(813) 481-0900

David G. Thompson#
1017 El. Rio Ave.
Fort Myers, FL 33919　　　(813) 433-7798

Alfred E. Wilson
P.O. Box 50853
Fort Myers, FL 33905　　　(813) 694-1281

Charles A. Bevelacqua
1806 N.W. 93rd Drive
Gainesville, FL 32607　　　(904) 377-2806

Frank T. Johmann
8632 S.W. 42nd Place
Gainesville, FL 32608　　　(904) 335-7147

Thomas A. Redding
3501 W. University Avenue
Suite D-1
Gainesville, FL 32607　　　(904) 371-7341

David R. Saliwanchik
Saliwanchik & Saliwanchik
2421 N.W. 41st Street
Suite A-1
Gainesville, FL 32606　　　(904) 375-8100

Roman Saliwanchik
Saliwanchik & Saliwanchik
2421 N.W. 41st Street
Suite A-1
Gainesville, FL 32606　　　(904) 375-8100

Merrill Wilcox#
2911 N.W. 30th Terr.
Gainesville, FL 32605　　　(904) 376-1174

Kelly O. Corley
P.O. Box 273
Gonzalez, FL 32560　　　(904) 477-3041

Homer James Bridger#
300 Diplomat Parkway
Suite 608
Hallandale, FL 33009　　　(305) 456-4732

Carl Fissell, Jr.#
Coulter Electronics, Inc.
590 Coulter Way
Hialeah, FL 34983　　　(305) 995-0131

Gerald R. Hibnick
Coulter Electronics, Inc.
590 W. 20th St.
Hialeah, FL 33010　　　(305) 885-0131

Alvin Engelstein
Lerner & Greenberg, P.A.
P.O. Box 2480
Hollywood, FL 33022　　　(305) 925-1100

Laurence A. Greenberg
Lerner & Greenberg, P.A.
P.O. Box 2480
Hollywood, FL 33022　　　(305) 925-1100

Herbert L. Lerner
Lerner & Greenberg, P.A.
P.O. Box 2480
Hollywood, FL 33022　　　(305) 925-1100

Werner H. Stemer#
Lerner & Greenberg
1200 S. Fed. Highway
P.O. Box 2480
Hollywood, FL 33020　　　(305) 925-1100

Peter F. Hilder
624 Hampshire Lane
Holmes Beach, FL 33510　　　(813) 778-3903

Joel I. Rosenblatt
445 11th Avenue
Indialantic, FL 32903　　　(305) 727-7626

Donna Brooks
P.O. Box 11296
Jacksonville, FL 32211　　　(904) 721-1986

Richard E. Klein
Livermore Klein & Lott, P.A.
1750 Gulf Life Tower
Jacksonville, FL 32207　　　(904) 399-0500

Nathaniel L. Leek
6000 San Jose Blvd.
Jacksonville, FL 32217　　　(904) 731-8412

Thomas R. Madden#
Reichhold Chemicals Inc.
8540 Baycenter Rd.
Jacksonville, FL 32245　　　(904) 739-2170

Douglas E. Ringel#
4478 Craven Rd. W.
Jacksonville, FL 32257　　　(904) 733-2628

Thomas C. Saitta
6821 Southpoint Drive North
Suite 203
Jacksonville, FL 32216　　　(904) 737-5825

Steven R. Scott
Gabel, Taylor & Dees
1600 American Heritage Tower
76 S. Laura St.
Jacksonville, FL 32202　　　(904) 353-7329

Earl L. Tyner
112 W. Adams St.
Suite 1305
Jacksonville, FL 32202　　　(904) 355-9631

Arthur G. Yeager
1305 Barnett Bank Bldg.
112 W. Adams St.
Jacksonville, FL 32202　　　(904) 355-9631

Harrie M. Humphreys
2309 Costa Verde Blvd.
Jacksonville Beach, FL 32250
　　　　　　　　　　(904) 246-8376

Alexander Raymond Field
376 River Edge Road
Jupiter, FL 33477　　　(305) 744-9140

Herbert W. Mylius
United Technologies Corp.
17900 Beeline Hwy.
Jupiter, FL 33478　　　(407) 796-7698

James O. Harrell*
NASA
Mail Code PT-PAT
Kennedy Space Center, FL 32899
　　　　　　　　　　(305) 867-2544

William J. Sheehan, III*
NASA
John F. Kennedy Space Ctr.
Patent Counsel/ PT-PAT
Kennedy Space Center, FL 32899
　　　　　　　　　　(407) 867-2544

Leon Robbin
1111 Crandon Blvd.
A503
Key Biscayne, FL 33149　　　(305) 361-5067

Joseph B. Allen, III#
617 Whitehead St.
Key West, FL 33040　　　(305) 296-5031

Harris Cade Lockwood
4101-11 Northgate Drive
Kissimmee, FL 32741　　　(407) 846-4851

Henry M. Fendrich
Stromberg-Carlson Corporation
400 Rinehart Road
Lake Mary, FL 32746　　　(305) 849-3055

John S. Brown#
1510 W. Ariana Street
Box 462
Lakeland, FL 33803　　　(813) 686-7548

David D. Centola
125 Hypoluxo Road
Lantana, FL 33462　　　(305) 588-8821

James D. Haynes
655 Ulmerton Road
Bldg. 11
Largo, FL 34641　　　(813) 584-6100

James R. Hulen
Concept, Inc.
11311 Concept Blvd.
Largo, FL 34643　　　(813) 392-6464

Herbert William Larson, Sr.
7381 114th Avenue, North
Suite 406
Largo, FL 34643　　　(813) 546-0660

Harold D. Shall
1101 Belcher Rd., South
Suite B
Largo, FL 34641　　　(813) 536-2711

James Alfred Smith
35625 Cedar Lane
Leesburg, FL 34788　　　(904) 589-6703

Robert T. Ruff
1125 Gulf of Mexico Dr.
Longboat Key, FL 34228　　　(813) 383-5824

Ronald Fredric Weiszmann
Box 8084
Madeira Beach, FL 33738　　　(813) 397-6995

Joseph E. Kerwin
1890 Casade Court
Marco Island, FL 33937　　　(813) 394-0432

Arthur E. Wilfond
1857 Dogwood Drive
Marco Island, FL 33937   (813) 394-9447

Dennis L. Cook*
Harris Corporation
GASD Div. Counsel
P.O. Box 94000
Melbourne, FL 32902   (305) 727-4127

John L. De Angelis, Jr.
Harris Corporation
Electronic Systems Sector
P.O. Box 37
Building 2, Room 104
Melbourne, FL 32902   (407) 729-3353

Harry Martin Fleck, Jr.
Harris Corp.
1025 W. NASA Blvd.
Melbourne, FL 32919   (305) 727-9155

Leslie J. Hart
Harris Corp.
P.O. Box 883
Melbourne, FL 32901   (305) 724-2580

Charles C. Krawczyk
Harris Corp.
1025 W. NASA Blvd.
Melbourne, FL 32919   (305) 727-9156

Frederick W. Neitzke
2220 S. Front Street
Apt. 203
Melbourne, FL 32901   (407) 724-1924

Ferdinand M. Romano
Harris Corporation
Mail Stop: CB 2-10
1301 Woody Burke Rd.
Melbourne, FL 32901   (407) 724-3279

W. Joseph Shanley, Jr.
Harris Corporation
1025 W. NASA Blvd.
Melbourne, FL 32919   (407) 727-9124

William A. Troner
Harris Corporation
1025 W. NASA Blvd.
Melbourne, FL 32919   (407) 727-9247

Bernard H. Breymann
1925 Brickell Avenue
Penthouse 9
Miami, FL 33129   (305) 856-3948

Curtis D. Carlson#
Fowler, White, Burnett, Hurley,
Banick & Strickroot, P.A.
25 W. Flagler St.
Miami, FL 33130   (305) 358-6550

Michael C. Cesarano
Cesarano & Kain, P.A.
One Bayfront Plaza
100 S. Biscayne Blvd.
Miami, FL 33131   (305) 530-9100

Henry W. Collins
Cordis Corp.
P.O. Box 025700
Miami, FL 33102   (305) 824-2707

Robert M. Downey
Malloy, Downey & Malloy, P.A.
2 South Biscayne Blvd.
Suite 3760
Miami, FL 33131   (305) 374-8418

James A. Gale
Morgan, Lewis & Bockius
5300 Southeast Financial Center
200 S. Biscayne Blvd.
Miami, FL 33131   (305) 579-0414

Robert C. Kain, Jr.
Cesarano & Kain, P.A.
One Bayfront Plaza
100 S. Biscayne Blvd.
Miami, FL 33131   (305) 530-9100

John Cyril Malloy
Malloy, Downey & Malloy, P.A.
2 South Biscayne Blvd.
Suite 3760
Miami, FL 33131   (305) 374-8418

Sybil Meloy
Ruden, Barnett, Mc Closky,
Smith, Schuster & Russell, P.A.
701 Brickell Avenue
Suite 1900
Miami, FL 33131   (305) 789-2714

Raymond L. Robinson
801 Brickell Avenue
Suite 1200
Miami, FL 33130   (305) 374-4192

Jesus Sanchelima
235 S.W. Le Jeune Road
Miami, FL 33134   (305) 447-1617

Robert M. Schwartz
169 East Flagler St.
Suite 1122
Miami, FL 33131   (305) 447-1617

Cynthia G. Tymeson#
Baxter Diagnostics, Inc.
1851 Delaware Pkwy.
Mail Station 426
Miami, FL 33125   (305) 633-6461

Erwin Myles Barnett
7960 Hawthorne Ave.
Miami Beach, FL 33141   (305) 865-6996

Milton H. Gross
2862 Fairgreen Dr.
Miami Beach, FL 33140   (305) 532-5617

Phillip H. Pohl
1020 Meridian Ave.
Apt. 916
Miami Beach, FL 33139   (305) 531-7102

Jack Edward Dominik
Dominik, Stein, Saccocio
& Reese
Suite 225
6175 N.W. 153rd Street
Miami Lakes, FL 33014   (305) 556-9889

Francis L. Kubler
Dominik, Stein, Saccocio, Reese,
Colitz & Van Der Wall, P.A.
6175 NW 153rd Street
Suite 225
Miami Lakes, FL 33014   (305) 556-7000

Richard M. Saccocio
Dominik, Stein, Saccocio,
Reese, Colitz & Van Der Wall
Suite 225
6175 N.W. 153rd Street
Miami Lakes, FL 33014   (305) 556-7000

Robert J. Van Der Wall
Dominik, Stein, Saccocio, Reese,
Colitz & Van Der Wall
6175 N.W. 153rd Street
Suite 225
Miami Lakes, FL 33014   (305) 556-7000

Daniel P. Worth
3137 Nautilus Road
Middleburg, FL 32068   (904) 264-1907

Charles M. Kaplan
1331 Olympia Avenue
Route 1
Mount Dora, FL 32757   (904) 735-2922

James N. Buckner
5409 Foxhound Drive
Naples, FL 33942   (813) 643-0955

Myron E. Click
P.O. Box 10879
Naples, FL 33941   (813) 263-7675

Curtis R. Davis#
4522 Shearwater Lane
Naples, FL 33999   (813) 591-4197

Jack W. Heberling, Jr.#
422 Cypress Way East
Naples, FL 33942   (813) 598-2489

Merrill Nels Johnson
800 Harbour Dr.
Naples, FL 33940   (813) 262-8502

Doonan Dwight Mc Graw
245 St. James Way
Naples, FL 33942   (813) 353-3372

Clyde Christian Metzger
5890 Via Lugano
Naples, FL 33963   (813) 566-2550

Liber J. Montone
9242 Vanderbilt Drive
Naples, FL 33963   (813) 597-8781

John P. Murphy
14100 E. Tamiami Tr. 49
Naples, FL 33961   (315) 685-6608

Ernest H. Schmidt
5721 20th Avenue S.W.
Naples, FL 33999   (813) 455-3846

Lloyd F. Seebach#
18 Stymie Lane
New Smyrna Beach, FL 32168
(902) 427-6966

Tobias E. Levow
1700 N.E. 191st., Apt. 109
North Miami Beach, FL 33179
(305) 949-8690

Richard S. Ross
13899 Biscayne Blvd.
Suite 108
North Miami Beach, FL 33181
(305) 945-7444

William Thomas Clarke
2935 SW 32nd Avenue
Ocala, FL 32674
(904) 622-7127

Stanley C. Felton
611 S.E. 9th Avenue
Apt 34
Ocala, FL 32671
(904) 622-7947

Henry E. Millson, Jr.
2111 S.E. 84th St.
Ocala, FL 32676
(904) 351-5418

George J. Seligsohn
9367-E Southwest 83rd Avenue
Ocala, FL 32676
(904) 854-1252

Leon Chasan
490 Hickorynut Avenue
Oldsmar, FL 34677
(813) 785-6179

Robert W. Adams*
Dept. of Navy
NTSC, Code 004PC
12350 Research Parkway
Orlando, FL 32826
(407) 380-8204

Herbert L. Allen, Jr.
Duckworth, Allen & Dyer, P.A.
1 South Orange Avenue
Orlando, FL 32801
(305) 841-2330

James H. Beusse
Hobby & Beusse
1327 N. Mills Avenue
Orlando, FL 32803
(407) 896-5995

Robert W. Duckworth
Duckworth, Allen & Dyer, P.A.
One South Orange Avenue
Suite 600
Orlando, FL 32801
(305) 841-2330

Warren L. Franz
5401 Kirkman Road
Suite 477
Orlando, FL 32819
(407) 363-1982

Michael K. Gray
3218 TCU Blvd.
Orlando, FL 32817
(407) 657-0565

William M. Hobby, III
Hobby, Wiggins & Beusse
1327 N. Mills Avenue
Orlando, FL 32803
(305) 896-5995

John F. Miller
2532 Overlake Ave.
Orlando, FL 32806
(305) 859-0749

David Olsen
3602 Country Lakes Drive
Orlando, FL 32812
(407) 859-0630

Joseph Edward Papin
Hobby & Beusse
1327 North Mills Ave.
Orlando, FL 32803
(407) 896-5995

Arthur Richard Parker#
3291 El Primo Way
Orlando, FL 32808
(407) 292-6075

James L. Simon
Holland & Knight
P.O. Box 1526
800 N. Magnolia
Suite PA
Orlando, FL 32802
(305) 425-8500

Michael L. Slonecker
Martin Marietta Corp.
P.O. Box 555837
Mail Point - 186
Orlando, FL 32855
(407) 356-3405

Steven C. Stewart
Duckworth, Allen, Dyer
& Doppelt
Post Office Box 3791
Orlando, FL 32802
(407) 841-2330

Joseph V. Truhe, Sr.#
8764 Granada Blvd.
Orlando, FL 32819
(305) 876-4509

Lewis J. Lamm
P.O. Box 2817
Ormond Beach, FL 32075
(704) 441-5962

Macdonald J. Wiggins
P.O. Box 459
853 Mimosa Trail
Oviedo, FL 32765
(407) 366-1750

Jerrell P. Hollaway#
Consumer Engineering, Inc.
1330 Meadowbrook Road N.E.
Palm Bay, FL 32905
(305) 727-7625

James C. Pintner
Evenson, Wands, Edwards,
Lenahan & Mc Keown
5240 Babcock St., N.E.
Suite 306
Palm Bay, FL 32905
(407) 725-4760

Robert I. Smith
1701 Lamont St., N.W.
Palm Bay, FL 32907
(407) 768-0631

Charles E. Wands
Evenson, Wands, Edwards,
Lenahan & Mc Keown
5240 Babcock Street, N.E.
Suite 206
Palm Bay, FL 32905
(407) 725-4760

Theodore Bishoff
3546 S. Ocean Blvd.
Apt. 826
Palm Beach, FL 33480
(305) 588-8192

Harry S. Colburn, II
Alley, Maass, Rogers,
Lindsay & Chauncey
321 Royal Poinciana Plaza, S.
P.O. Box 431
Palm Beach, FL 33480
(305) 659-1770

Alfred H. Rosen
3546 South Ocean Blvd.
Suite 311
Palm Beach, FL 33480
(407) 582-0223

George A. Teacherson#
P.O. Box 762
Palm Beach, FL 33480
(407) 439-7005

Fred Ornstein
11811 Avenue of the PGA
Apt. 7-3F
Palm Beach Gardens, FL 33418
(407) 626-1709

Kevin P. Redmond
6960 SW Gator Trail
Palm City, FL 34990
(407) 283-1507

Richard A. Zambo
598 S.W. Hidden River Ave.
Palm City, FL 34990
(407) 220-9163

Oistein J. Bratlie
P.O. Box 2889
Palm Coast, FL 32037
(904) 445-0910

Robert B. Gerhardt
56 Beaver Dam Lane
Palm Coast, FL 32037
(201) 670-0607

Raymond F. Kramer
38 Cooper Lane
Palm Coast, FL 32137
(904) 446-3470

David A. Kiewit#
2420 Seneca Ct.
Palm Harbor, FL 34683
(813) 789-1915

Mary S. King
3338 Pattie Place
Palm Harbor, FL 34685
(813) 786-2122

John E. Becker*
U. S. Navy
Naval Coastal Systems Center
Code 051P
Panama City, FL 32407
(904) 235-5811

Harvey A. David
2808 Canal Drive
Panama City, FL 32405
(904) 763-8506

Thomas Y. Awalt, Jr.
Deep Seven Company
14260 Inncravity Pt. Road
Pensacola, FL 32507
(904) 492-0250

Alexander Kozel
5920 San Gabriel Dr.
Pensacola, FL 32504
(904) 477-3461

Thomas N. Wallin
Monsanto Company
P.O. Box 12830
Pensacola, FL 32575
(904) 968-8266

John William Whisler
Monsanto Co.
P.O. Box 12830
Pensacola, FL 32575   (904) 968-8272

Richard A. Wahl
3655 91st Avenue N.
Pinellas Park, FL 34666   (813) 576-4222

Robert S. Babayi#
Motorola Inc.
8000 W. Sunrise Blvd.
Plantation, FL 33321   (305) 475-6449

Walter A. Modance
808 Cypress Blvd.
Apt. 304
Pompano Beach, FL 33069   (305) 971-2362

William D. O Connor
651 S.W. 6th Street
CT- 609
Pompano Beach, FL 33060   (305) 781-2418

Stanley W. Sokolowski
2605 E. Atlantic Blvd.
Pompano Beach, FL 33062   (305) 782-6539

Aaron Mack Scharf
P.O. Box 1583
Ponte Vedra, FL 32004   (904) 285-5110

Oscar B. Waddell
11050 Elderberry Drive
Port Richey, FL 33568   (813) 863-0356

Thomas H. Buffton#
1838 S.E. Westmoreland Blvd.
Port St. Lucie, FL 33452   (305) 335-3785

William G. H. Finch#
3025 Morningside Blvd.
Port St. Lucie, FL 33452   (305) 335-5147

Harry T. Berriman#
2730 Luna Court
Punta Gorda, FL 33950   (813) 637-1841

Edward J. Brenner
4 Ocean Drive
Punta Gorda, FL 33950   (813) 637-1075

Frank A. Lukasik
2517 Rio Tiber Drive
Punta Gorda, FL 33950   (813) 639-3374

Robert J. Norton
Norton & Marryott
Suite 408
126 E. Olympia Avenue
Punta Gorda, FL 33950   (813) 639-0311

Harvey W. Rockwell#
701 Agui Esta Dr., #44
Punta Gorda, FL 33950   (813) 575-4318

Charles E. Vautrain, Jr.#
23465 Harborview Road
Apt. 724
Punta Gorda, FL 33980   (813) 629-6940

Robert K. Blackwood#
17900 Gulf Blvd., #13c
Redington Shores, FL 33708   (814) 397-0846

Jon L. Liljequist
5770 Pine Tree Drive
Sanibel, FL 33957   (813) 472-0805

John F. Ahern
6342 Midnight Pass Road
Apt. 467
Sarasota, FL 34242   (813) 346-1635

Marshall J. Breen
3345 Spring Mill Circle
Sarasota, FL 33579   (813) 924-8142

Richard G. Bremer#
5636 Pipers Waite
Sarasota, FL 33580   (813) 377-8595

William Genther Lambrecht
Williams, Parker, Harrison,
Dietz & Getzen
1550 Ringling Blvd.
Sarasota, FL 34236   (813) 366-4800

Willard R. Matthews, Jr.
1602 Stickney Point Road
Apt. 405
Sarasota, FL 34231   (813) 921-5459

William W. Mc Dowell, Jr.
4428 Deer Trail Blvd.
Sarasota, FL 34238   (813) 921-5459

David C. Noller
1040 Sylvan Drive
Sarasota, FL 34234   (813) 365-1935

Charles J. Prescott
2033 Wood Street
Suite 115
Sarasota, FL 34237   (813) 957-4208

William Julius Van Loo, Jr.#
1727 Bahia Vista Street
Sarasota, FL 34239   (813) 955-3887

Frederick P. Weidner, Jr.
8491 Woodbriar Dr.
Sarasota, FL 34238   (813) 921-1617

Billy J. Wilhite#
36 S. Washington Drive
Sarasota, FL 34236   (813) 388-3501

Jack N. Mc Carthy
655 Bimini Road
Satellite Beach, FL 32937   (305) 773-2081

Ruth M. Rife#
220 Hedgecock Court
Satellite Beach, FL 32937   (407) 777-1732

George B. Oujevolk
P. O. Box 273
Sebring, FL 33871   (813) 494-8209

Jerome F. Kramer
11554 Woodbridge Blvd.
Seminole, FL 34642   (813) 392-5257

Harry V. Strampel
10114 133rd Street North
Seminole, FL 34646   (813) 593-1308

Leon Gilden
933 Oleander Way
Suite 6
South Pasadena, FL 33707   (813) 345-9133

Alexander Skopetz
380 Holt Lane
Springhill, FL 34606   (904) 688-7603

Jack W. Wicks
9425 Blind Pass Rd., #1007
St Petersburg Beach, FL 33706
                                 (813) 360-5019

Walter J. Monacelli
720 36th Ave. North
St. Petersburg, FL 33704   (813) 525-5759

Hugh E. Smith
Camper's Gear
1935 1st Avenue S.
St. Petersburg, FL 33712   (813) 822-7592

Wilbur J. Kupfrian
1706 N.W. Fork Rd.
Stuart, FL 34994   (305) 692-1922

David M. Schiller
24 W. High Point Rd.
Stuart, FL 34996   (407) 287-2509

A. James Valliere
79 S. River Road
Stuart, FL 34996   (407) 287-4275

Charles P. Boberg
1218 N. Pebble Beach Blvd.
Sun City Center, FL 33573   (813) 634-7776

Martin J. Carroll
1206 Beach Blvd.
Sun City Center, FL 33570   (813) 634-7010

Arthur L. Morsell, Jr.
1212 Fordham Drive
Sun City Center, FL 33570   (813) 634-6062

Leslie G. Noller
623 Allegheny Drive
Sun City Center, FL 33570   (813) 634-3158

Louis Robertson
1411 Nashua Circle
Sun City Center, FL 33573   (813) 634-6105

Anna E. Mack
Racal Corp.
Law Dept.
1601 N. Harrison Pkwy.
Sunrise, FL 33323   (305) 846-5188

Jerry A. Miller
8571 NW 21st Court
Sunrise, FL 33322   (305) 742-0972

William A. Newton
Racal-Milgo
1601 N. Harrison Pkwy.
Sunrise, FL 33323   (305) 476-5159

Horace Schow, II
2816 Roscommon Dr.
Tallahassee, FL 32308   (904) 893-4028

Roy C. Lipton
8627 N.W. 83rd St.
Tamaral, FL 33321   (305) 720-8371

Donald Ruh Bahr
Questor Corporation
5750 A North Hoover Blvd.
Tampa, FL 33630   (813) 887-5274

Michael J. Colitz, Jr.
Dominik, Stein, Et Al.
3030 N. Rocky Point Dr., West
Suite 400
Tampa, FL 33607   (813) 289-2966

Franklin J. Cona#
Dominik, Stein, Saccocio
& Reese
312 East Harrison St.
Tampa, FL 33602   (813) 229-2122

Arthur W. Fisher, III
5553 West Waters Ave.
Suite 316
Tampa, FL 33634   (813) 885-2006

Robert Frank Frijouf
Frijouf, Rust & Pyle, P.A.
201 East Davis Blvd.
Tampa, FL 33606   (813) 254-5100

James Wm. Grace
Walter Industries, Inc.
P.O. Box 31601
Tampa, FL 33631   (813) 871-4456

Charles A. Mc Clure
243 Bayshore Blvd.
Tampa, FL 33606   (813) 251-1443

C. Douglas Mc Donald, Jr.
Pettis & Mc Donald, P.A.
P.O. Box 1528
Tampa, FL 33601   (813) 229-8176

John Joseph Mc Laughlin
Wagner, Cunningham, Vaughan
& Mc Laughlin, P.A.
708 Jackson Street
Tampa, FL 33602   (813) 223-7421

David L. Partlow
Gibbons, Smith, Cohn
& Arnett, P.A.
501 East Kennedy Blvd.
Ste. 906
Tampa, FL 33602   (813) 221-3730

David W. Pettis, Jr.
Pettis & Mc Donald, P.A.
P.O. Box 1528
Tampa, FL 33601   (813) 229-8176

Robert G. Pollock
Dme Systems, Inc.
3018 U.S. Hwy 301 North
Suite 100
Tampa, FL 33619   (813) 621-9624

Kenneth George Preston, Jr.
University of S. Florida
Div. of Technology Development
& Transfer
4202 East Fowler Ave., FAO 126
Tampa, FL 33620   (813) 974-2897

Ray S. Pyle
201 East Davis Blvd.
Tampa, FL 33606   (813) 254-5100

Benjamin P. Reese, II
Dominik, Stein, Saccocio,
Reese, Colitz & Van Der Wall
3030 N. Rocky Point Dr., W.
Suite 400
Tampa, FL 33607   (813) 289-2967

Peter H. Rehm#
14245 Les Palms Circle #2
Tampa, FL 33613   (813) 971-6594

Stefan V. Stein
3030 N. Rocky Point Dr., W.
Tampa, FL 33607   (813) 289-2966

William S. Van Royen
Pettis & Mc Donald
501 E. Kennedy Blvd.
Suite 700
Tampa, FL 33602   (813) 832-2965

Charles W. Vanecek#
Celotex Corp.
One Metro Center
4010 Boy Scout Blvd.
Tampa, FL 33607   (813) 873-4193

Edward T. Connors
521 W. Venice Ave.
Apt. 21
Venice, FL 33595   (813) 488-0391

Anton O. Oechsle#
1444 Trune Way
Venice, FL 34292   (813) 488-6045

Donald R. Andersen
P.O. Box 5230
Vero Beach, FL 32961   (407) 231-4064

Carroll F. Palmer
860 20th Place
Vero Beach, FL 32960   (407) 562-6222

John Webb Routh
1001 Bay Road
Unit 204A
Vero Beach, FL 32963   (407) 231-3179

Martin Kalikow
2260 Sunderland Avenue
Wellington, FL 33414   (407) 798-6128

Joseph W. Bain#
Eckert, Seamans, Cherin &
Mellott
515 N. Flagler Drive
400 Northbridge Tower
West Palm Beach, FL 33401   (407) 659-3900

James E. Jacobson, Jr.
Eckert, Seamans Cherin & Mellott
400 Northbridge Tower
515 North Flagler Dr.
West Palm Beach, FL 33401   (407) 659-3900

Gregory A. Nelson
Eckert, Seamans, Cherin
& Mellott
515 N. Flagler Drive
West Palm Beach, FL 32401   (407) 659-3900

J. Rodman Steele, Jr.
Eckert, Seamans, Cherin
& Mellott
515 N. Flagler Dr.
400 Northbridge Centre
West Palm Beach, FL 33401   (407) 659-3900

Edward M. Livingston
1455 W. Fairbanks Avenue
P.O. Box 2894
Winter Park, FL 32790   (407) 629-4545

Julian Carroll Renfro
1850 Lee Road
Suite 130
Winter Park, FL 32789   (305) 628-3600

Nathan Carl Schwartz
1915 S. Lakemont Ave.
Winter Park, FL 32792   (404) 629-5433

Georgia

Eugene S. Zimmer
Jones, Askew & Lunsford
P.O. Drawer 56527
Altanta, GA 30343   (404) 688-7500

William R. Alford
265 Cedar Springs Drive
Athens, GA 30605   (404) 353-7744

Vincent J. La Terza
University of Georgia
Research Foundation, Inc.
Boyd Graduate Studies Res. Ctr.
6th Flr., D.W. Brooks Drive
Athens, GA 30602   (404) 542-5944

Albert S. Anderson
Jones, Askew & Lunsford
P.O. Drawer 56527
Atlanta, GA 30343   (404) 688-7500

Anthony Bartholomew Askew
Jones, Askew & Lunsford
P.O. Drawer 56527
Atlanta, GA 30343   (404) 688-7500

Jason A. Bernstein
5281 Seaton Drive
Atlanta, GA 30338   (404) 457-3664

Thomas A. Boshinski
Mead Corporation
Legal Dept. Tower
127 Peachtree Street
Candler Bldg., Suite 700
Atlanta, GA 30303   (404) 658-8480

Thomas R. Boston
Coca-Cola Company
P.O. Drawer 1734
Atlanta, GA 30301   (404) 676-6682

James K. Boudreau
Coca-Cola Co.
P.O. Drawer 1734
Tower 2140
Atlanta, GA 30301   (404) 898-4872

William D. Brooks
Coca-Cola Co.
P.O. Drawer 1734
Atlanta, GA 30301   (404) 676-2103

James J. Cannon, Jr.
GTE Service Corp.
245 Perimeter Center Pkwy.
OLGL
Atlanta, GA 30346   (404) 391-8063

Eduardo M. Carreras
Coca-Cola Company
One Coca-Cola Plaza N.W.
Atlanta, GA 30313   (404) 676-3272

Laurence P. Colton
Hurt, Richardson, Garner,
Todd & Cadenhead
1400 Peachtree Place Tower
999 Peachtree Street, N.E.
Atlanta, GA 30309   (404) 870-6457

Carl M. Davis, II
Jones, Askew & Lunsford
P.O. Drawer 56527
Atlanta, GA 30343   (404) 688-7500

Todd Deveau
Newton, Hopkins & Ormsby
1010 the Equitable Bldg.
100 Peachtree St.
Atlanta, GA 30303   (404) 688-1788

Erwin Doerr#
Mead Corp.
Mead Packaging Div.
1040 W. Marietta St., N.W.
Atlanta, GA 30302   (404) 897-6752

Michael V. Drew
568 Fourteenth Street, N.W
Atlanta, GA 30318   (404) 876-4636

James L. Ewing, IV
Kilpatrick & Cody
3100 Equitable Bldg.
100 Peachtree Street
Atlanta, GA 30043   (404) 572-6500

Robert O. Fox
Jones, Askew & Lunsford
230 Peachtree St., N.E.
Suite 2000
Atlanta, GA 30303   (404) 688-7500

Roger T. Frost
Jones, Askew & Lunsford
P.O. Drawer 56527
Atlanta, GA 30343   (404) 688-7500

Arthur A. Gardner
Thomas & Kennedy
100 Galleria Parkway
Suite 1550
Atlanta, GA 30339   (404) 951-0931

Dale V. Gaudier
Neptune International
4360 Chamblee Dunwoody Road
Suite 410
Atlanta, GA 30341   (404) 458-1212

Michael J. Gilroy
Coca-Colo Co.
P.O. Drawer 1734
Atlanta, GA 30301   (404) 676-3207

Joel S. Goldman
Troutman, Sanders, Lockerman
& Ashmore
One Ravinia Drive
Suite 1600
Atlanta, GA 30346   (404) 658-8427

Jamie L. Greene
Kilpatrick & Cody
3100 Equitable Bldg.
100 Peachtree Street
Atlanta, GA 30303   (404) 681-7869

Gregory T. Gronholm
Jones, Askew & Lunsford
P.O. Drawer 56527
Atlanta, GA 30343   (404) 688-7500

John R. Harris
Jones, Askew & Lunsford
P.O. Drawer 56527
Atlanta, GA 30343   (404) 688-7500

Patrick F. Henry
2601 1st National Bank Bldg.
2 Peachtree St.
Atlanta, GA 30383   (404) 658-1754

William George Hervey
Needle & Rosenberg, P.C.
133 Carnegie Way
Suite 400
Atlanta, GA 30303   (404) 688-0770

Thomas Arthur Hodge
Jones, Askew & Lunsford
P.O. Drawer 56527
Atlanta, GA 30343   (404) 688-7500

George M. Hopkins
Hopkins & Thomas
100 Galleria Pkwy., N.W.
Suite 1550
Atlanta, GA 30339   (404) 951-0931

Louis T. Isaf
5770 Powers Ferry Road N.W.
Suite 202
Atlanta, GA 30327   (404) 951-2623

Elizabeth C. Jacobs
Jones, Askew & Lundsford
230 Peachtree St.
Suite 2000
Atlanta, GA 30303   (404) 688-7500

James D. Johnson
CYT RX Corporation
150 Technology Parkway
Technology Park Atlanta
Atlanta, GA 30092   (404) 368-9500

Harold D. Jones, Jr.
Jones, Askew & Lunsford
Suite 2000
230 Peachtree Street
Atlanta, GA 30303   (404) 255-9629

Brij M. Kapoor
Suite 1515
230 Peachtree St., N.W.
Atlanta, GA 30303   (404) 522-1122

James W. Kayden
Thomas & Kennedy
100 Galleria Parkway, N.W.
Suite 1550
Atlanta, GA 30339   (404) 951-0931

David P. Kelley
3822 Courtyard Drive
Atlanta, GA 30339   (404) 333-9159

Robert Bruce Kennedy
Kennedy & Kennedy
Suite 1550
1050 Crown Pointe Pkwy.
Atlanta, GA 30338   (404) 396-2244

Steven D. Kerr
Thomas & Kennedy
100 Galleria Parkway
Suite 590
Atlanta, GA 30339   (404) 951-0931

Andrew T. Knowles
P.O. Box 16245
Atlanta, GA 30321   (404) 951-0931

Jeffrey R. Kuester#
6190 Powers Ferry Rd., N.W.
Suite 550
Atlanta, GA 30339   (404) 951-2623

Nicholas N. Leach
Troutman, Sanders, Lockerman
& Ashmore
One Ravinia Drive
Suite 1600
Atlanta, GA 30346   (404) 658-8007

William C. Lee, III
The Coca-Cola Company
P.O. Drawer 1734
Atlanta, GA 30301   (404) 676-3209

Harry I. Leon
924 Bowen Street, N.W.
Atlanta, GA 30318   (404) 352-3882

Robert A. Lester
Coca Cola Co.
1 Coca Cola Plaza
P.O. Drawer 1734
Atlanta, GA 30301   (404) 676-2530

Dale Lischer
Jones, Askew & Lunsford
P.O. Drawer 56527
Atlanta, GA 30343   (404) 688-7500

J. Rodgers Lunsford, III
Jones, Askew & Lunsford
P.O. Drawer 56527
Atlanta, GA 30343   (404) 527-6568

Kathleen L. Maher
Jones, Askew & Lunsford
230 Peachtree St.
Suite 2000
Atlanta, GA 30303   (404) 688-7500

Harold L. Marquis
Mead Corporation
127 Peachtree St., N.E.
Chandler Bldg., Suite 700
Atlanta, GA 30303   (404) 658-8477

John R. Martin
The Coca-Cola Co.
One Coca-Cola Plaza
Atlanta, GA 30313   (404) 676-2102

Virginia M. Mc Guffey
Powell, Goldstein, Frazer
& Murphy
35 Broad Street
Atlanta, GA 30335   (404) 572-6652

Marvin L. Moore#
700 Lynn Cir. S.W.
Atlanta, GA 30311   (404) 691-0430

William H. Needle
Needle & Rosenberg, P.C.
133 Carnegie Way N.W.
Suite 400
Atlanta, GA 30303   (404) 688-0770

Patrea L. Pabst
Kilpatrick & Cody
100 Peachtree Street
Atlanta, GA 30303   (404) 572-6508

Peter G. Pappas
Jones, Askew & Lunsford
230 Peachtree St. N.W.
Suite 2000
Atlanta, GA 30303   (404) 688-7500

Billy J. Powell
1447 Peachtree St., N.E.
Suite 710
Atlanta, GA 30309   (404) 892-8046

Frederick W. Powers, III
Scientific-Atlanta, Inc.
One Technology Park
P.O. Box 105600
Atlanta, GA 30348   (404) 441-4872

John S. Pratt
Kilpatrick & Cody
100 Peachtree Street
Suite 3100
Atlanta, GA 30303   (404) 572-6367

Robert E. Richards
Jones, Askew & Lunsford
230 Peachtreet Street
Suite 2000
Atlanta, GA 30303   (404) 688-7500

Joy Lynne Richardson#
Coca-Cola Company
P.O. Drawer 1734
Atlanta, GA 30301   (401) 676-5616

Larry A. Roberts
Jones, Askew & Lunsford
P.O. Drawer 56527
Atlanta, GA 30343   (404) 688-7500

Walter A. Rodgers
1730 Gas Light Tower
Peachtree Center
Atlanta, GA 30303   (404) 523-6059

Walter M. Rodgers, Jr.
1730 Gas Light Tower
Peachtree Center
Atlanta, GA 30303   (404) 523-6059

Sumner C. Rosenberg
Needle & Rosenberg, P.C.
133 Carnegie Way N.W.
Suite 400
Atlanta, GA 30303   (404) 688-0770

Dean W. Russell
Kilpatrick & Cody
100 Peachtree Street
Suite 3100
Atlanta, GA 30303   (404) 681-7758

Stephen M. Schaetzel
Jones, Askew & Lunsford
P.O. Box 56526
Atlanta, GA 30343   (404) 688-7500

Mary L. Severson
Needle & Rosenberg, P.C.
133 Carnegie Way, N.W.
Suite 400
Atlanta, GA 30303   (404) 688-0770

Vivian L. Steadman#
924 Bowen Street, N.W.
Atlanta, GA 30318   (404) 352-3882

Larry W. Stults#
Kilpatrick & Cody
100 Peachtree St.
Suite 3100
Atlanta, GA 30303   (404) 572-6508

David S. Sudderth
Thomas & Kennedy
100 Galleria Parkway
Suite 1550
Atlanta, GA 30339   (404) 951-0931

George Marshall Thomas
Thomas, Kerr & Kayden
100 Galleria Parkway
Suite 1550
Atlanta, GA 30339   (404) 951-0931

James M. Thomson#
GTE Corporation
245 Perimeter Center Pkwy.
Atlanta, GA 30346   (404) 391-8064

James F. Vaughan
Hopkins & Thomas
100 Galleria Pkwy., N.W.
Suite 1550
Atlanta, GA 30339   (404) 951-0931

Charles L. Warner, II
Jones, Askew & Lunsford
P.O. Drawer 56527
Atlanta, GA 30343   (404) 688-7500

Daniel J. Warren
Lord, Bissell & Brook
1201 W. Peachtree St.
Suite 3700
Atlanta, GA 30309   (404) 870-4600

Walter Lewis Williamson
Bell South, Suite 1800
1155 Peachtree Street, N.E.
Atlanta, GA 30367   (404) 249-2603

Jeffrey E. Young
Jones, Askew & Lunsford
P.O. Drawer 56527
Atlanta, GA 30343   (404) 688-7500

Cheryl K. Zalesky
Kilpatrick & Cody
100 Peachtree Street
Suite 3100
Atlanta, GA 30303   (404) 572-6439

V. Lee Ringler
Suite 408, 500 Bldg.
501 Greene Street
Augusta, GA 30901   (404) 724-4000

Stanley L. Tate
Southwire Company
P.O. Box 1000
Carrollton, GA 30119   (404) 832-5375

James W. Wallis, Jr.
Southwire Company
One Southwire Drive
Carrollton, GA 30119   (404) 832-5375

Donald C. Studley
P.O. Box 55
Cleveland, GA 30528   (404) 865-2343

James B. Middleton
P.O. Box 1968
Decatur, GA 30031   (404) 377-5327

Robert L. Jay#
6934 Clear Lake Court
Doraville, GA 30360   (404) 395-1015

Sanford J. Asman
570 Vinington Court
Dunwoody, GA 30350   (404) 391-0215

Henry S. Jaudon#
108 Forest Avenue
Elberton, GA 30635   (404) 283-5332

William V. Adams*
U.S. Army
Off. of JAG
Attn. ATZB-JA
Fort Benning, GA 31905   (404) 544-4261

Marla J. Church
Elan Pharmaceutical Res. Corp.
1300 Gould Drive
Gainesville, GA 30501   (404) 534-8239

Frank Abbott Steinhilper
3361 Jean Marie Lane
Gainesville, GA 30506   (404) 535-0317

Michael C. Smith
P.O. Box 2222
La Grange, GA 30241   (404) 882-6664

James Allan Hinkle
Hinkle & Bull
175 Gwinnett Drive
Suite 300
Lawrenceville, GA 30245   (404) 995-8877

Robert Paul Barton
Lockheed-Georgia Co.
86 South Cobb Drive
Marietta, GA 30063   (404) 424-3388

Eric R. Katz
Lockheed Aeronautical Systems Co.
Dept. 91-10, Zone 0230
86 South Cobb Dr.
Marietta, GA 30063   (404) 494-3143

John J. Sullivan
5362 River Mill Circle S.E.
Marietta, GA 30068   (404) 992-0985

John J. Timar
IBM Corporation
Intellectual Prop. Law Dept.
3200 Wildwood Plaza, WG 16A
Marietta, GA 30067   (404) 835-8210

Joseph F. Villella, Jr.
IBM Corporation
3200 Wildwood Plaza
WG 16A
Marietta, GA 30067   (404) 835-8209

Frank Madonia
Miller Industries, Inc.
5555 Triangle Pkwy.
Suite 340
Norcross, GA 30092   (404) 368-8122

Charles L. Schwab
KHD Deutz of Amer. Corp.
Legal Services
3220 Pointe Parkway
Suite 1000
Norcross, GA 30092   (404) 246-3313

Edward Walter Somers
AT&T
2000 Northeast Expressway
Norcross, GA 30071   (404) 447-2040

W. Ferrel Bentley, Jr.#*
Federal Communications Comm.
P.O. Box 65
Powder Springs, GA 30073   (404) 943-6425

William B. Boykin
Kimberly-Clark Corp.
1400 Holcomb Bridge Rd.
Roswell, GA 30076   (404) 587-7293

Sherri V. Bull
215 Shadow Ledge Lane
Roswell, GA 30076   (404) 552-8892

Don Porter Bush
395 Saddle Horn Circle
Roswell, GA 30076   (404) 998-5834

Jeremiah J. Duggan
Kimberly-Clark Corporation
1400 Holcomb Bridge Road
Roswell, GA 30076   (404) 587-8626

Robert W. Hampton
1015 Windsor Trail
Roswell, GA 30076   (404) 594-7834

William D. Herrick
Kimberly-Clark Corp.
P.O. Box 103002
1400 Holcomb Bridge Road
Roswell, GA 30076   (404) 587-8096

William E. Maycock
Kimberly-Clark Corporation
1400 Holcomb Bridge Road
B200/1
Roswell, GA 30076   (404) 587-8621

Karl V. Sidor
Kimberly-Clark Corporation
1400 Holcomb Bridge Road
Bldg. 200/1
Roswell, GA 30076   (404) 587-7253

Patrick C. Wilson
Kimberly-Clark Corporation
1400 Holcomb Bridge Road
Roswell, GA 30076   (404) 587-7214

William S. Mc Curry, Jr.
242 Wiley Bottom Rd.
Savannah, GA 31411   (912) 598-0079

Albert M. Heiter
807-E Mallory Street
St. Simons Island, GA 31522   (912) 638-6193

David H. Robertson
P.O. Box 522
Stone Mountain, GA 30086   (404) 978-8310

Carla J. Dolce
1290 Drayton Woods Dr.
Tucker, GA 30084   (404) 493-6836

Earl D. Harris
P.O. Box 498
Watkinsville, GA 30677   (404) 769-7717

Hawaii

Robert C. Godbey
Gilbert & Jeynes
Suite 2424 Pauahi Tower
1001 Bishop St.
Honolulu, HI 96813   (808) 523-8894

Mark S. Holmes
315 Sand Island Road
Honolulu, HI 96819   (808) 842-5555

Martin E. Hsia
Cades Schutte Fleming
& Wright
1000 Bishop Street
Honolulu, HI 96813   (808) 521-0200

Kazuo Kiyonaga
1840-A Ninth Avenue
Honolulu, HI 96816   (808) 732-1912

George W. T. Loo
755 Mc Neill St., Suite B202
Honolulu, HI 96817   (808) 847-1056

Seth M. Reiss
Watanabe, Ing & Kawashima
745 Fort Street
Hawaii Tower, 5th Floor
Honolulu, HI 96813   (808) 544-8300

Donald A. Streck
191 Oko Street
Unit #1
Kailua Oahu, HI 96734   (808) 254-5696

James E. Smith
67 Karsten Drive
Wahiawa, HI 96786   (808) 621-5985

Idaho

William Johnson Bethurum, III
350 N. Ninth St.
Suite 304
Boise, ID 83702   (208) 345-7700

Gerard J. Carlson#
P.O. Box 4422
Boise, ID 83711   (208) 362-1316

Susan B. Collier#
Micron Technology, Inc.
2805 East Columbia Rd.
Boise, ID 83706   (208) 368-4514

Albert M. Crowder, Jr.
Micron Technology, Inc.
Legal Dept., M.S. 507
2805 E. Columbia Road
Boise, ID 83706   (208) 368-4504

Frank J. Dykas
First Interstate Center
877 W. Main St.
Suite 603
Boise, ID 83702   (208) 345-1122

Angus C. Fox, III
Micron Technology Inc.
2805 E. Columbia Road
Boise, ID 83706   (208) 389-4500

Paul F. Horton
1700 Vista Ave.
P.O. Box 5388
Boise, ID 83705   (208) 345-0241

Craig M. Korfanta
First Interstate Center
877 W. Main St.
Suite 603
Boise, ID 83702　　　(208) 345-1122

William H. Mac Allister, Jr.
Hewlett-Packard Company
Legal Dept., M.S.-374
11413 Chinden Blvd.
Boise, ID 83714　　　(208) 323-4038

John D. Merris
2324 Scyene Way
Boise, ID 83712　　　(208) 343-2311

David J. Paul#
Micron Technology, Inc.
2805 East Columbia Rd.
Boise, ID 83706　　　(208) 368-4515

Stanley N. Protigal
Micron Technology, Inc.
2805 East Columbia Road
Boise, ID 83706　　　(208) 389-4500

Jon P. Busack#
P.O. Box 69
Eagle, ID 83616　　　(208) 867-1796

Terrence L. J. Clausen
Lucero Computer Products
1320 Lincoln Road
Idaho Falls, ID 83401　　　(208) 524-0891

Robert A. De Groot#
Hopkins, French, Crockett, Springer & Hoopes
428 Park Ave.
PO Box 51219
Idaho Falls, ID 83405　　　(208) 523-4445

Alan D. Kirsch
EG&G Idaho, Inc.
P.O. Box 1625
1955 Fremont Ave.
Idaho Falls, ID 83415　　　(208) 526-1371

Ignacio Resendez*
U.S. DOE
785 Doe Place
Idaho Falls, ID 83402　　　(208) 526-1633

Warren Charles Porter
19 College Avenue
Rexburg, ID 83440　　　(208) 356-4616

Nicholas T. Bokides
P.O. Box 28
Weiser, ID 83672　　　(208) 549-0611

Illinois

Thomas D. Brainard
Abbott Laboratories
D-377/A P 6D-2
Routes 43 & 137
Abbott Park, IL 60064　　　(312) 937-6367

Thomas M. Breininger
Abbott Laboratories
One Abbott Park Road
Abbott Park, IL 60064　　　(708) 937-6367

Daniel W. Collins
Abbott Laboratories
D-377/AP6D-2
One Abbott Park Road
Abbott Park, IL 60064　　　(708) 937-6367

Steven R. Crowley#
Abbott Laboratories
D441 /AP 6D
Abbott Park, IL 60064　　　(312) 937-2335

Daniel R. Curry
Abbott Laboratories
Abbott Park, IL 60064　　　(312) 937-0890

Joan D. Eggert#
Abbott Laboratories
D. 9RK/AP6C-2
Abbott Park, IL 60064　　　(312) 937-8462

George A. Foster, Jr.#
Abbott Laboratories
D441, AP6D
One Abbot Pk. Rd.
Abbott Park, IL 60064　　　(708) 937-6555

Jerry F. Janssen
Abbott Laboratories
Dept. 377
1 Abbott Park Road
Abbott Park, IL 60064　　　(312) 937-4558

James D. Mc Neil
Abbott Laboratories
Abbott Park, IL 60064　　　(312) 937-8360

Lawrence S. Pope
Abbott Laboratories
D-377, AP6D-2
One Abbott Park Rd.
Abbott Park, IL 60064　　　(708) 937-6365

Priscilla E. Porembski
Abbott Laboratories
One Abbott Park Road
Abbott Park, IL 60064　　　(708) 937-4884

Richard D. Schmidt
Abbott Laboratories
D-377, AP6D-2
One Abbott Park Road
Abbott Park, IL 60064　　　(312) 937-6365

Frank S. Ungemach
Abbott Laboratories
D-377
AP-6D
Abbott Park, IL 60064　　　(708) 688-8727

Madhavi K. Vadnere#
Abbott Laboratories
PPD-0441, AP6D
One Abbott Park Road
Abbott Park, IL 60064　　　(708) 937-0147

Andrea C. Walsh#
Abbott Laboratories
One Abbott Park Rd.
AP6D
Abbott Park, IL 60064　　　(708) 937-9675

Steven F. Weinstock
Abbott Laboratories
One Abbott Park Road
Abbott Park, IL 60064　　　(708) 937-2341

Robert E. Wexler
Abbott Laboratories
Dept. 377 AP6D7
One Abbott Park Rd.
Abbott Park, IL 60064　　　(708) 937-6366

James L. Wilcox
Abbott Laboratories
D-377/AP6D
One Abbott Park Road
Abbott Park, IL 60064　　　(708) 937-8360

Wean Khing Wong
Abbott Laboratories
Dept. 377, AP6D
One Abbott Park Rd.
Abbott Park, IL 60064　　　(708) 937-9396

Charles W.B. Connors
Magneco/Metrel, Inc.
223 Interstate Road
Addison, IL 60101　　　(708) 543-6660

Mathew R.P. Perrone, Jr.
204 S. Main St.
Algonquin, IL 60102　　　(708) 658-5140

Powell L. Sprunger
20 Woodview Lane
Algonquin, IL 60102　　　(708) 658-6067

Russel P. Steele
P. O. Box 5
Alpha, IL 61412　　　(309) 529-4701

John M. Albrecht*
U.S. Dept. of Energy
Off. of Pat. Coun.
9800 South Cass Avenue
Argonne, IL 60439　　　(312) 972-2179

Thomas G. Anderson*
U.S. Depy of Energy
9800 S. Cass Avenue
Argonne, IL 60439　　　(312) 972-3216

Arthur Alan Churm
Argonne National Laboratory
9700 South Cass Avenue, Bldg. 201
Argonne, IL 60439　　　(312) 972-5951

Helen S. Cordell*
U.S. Dept. of Energy
Off. of Patent Coun.
9800 S. Cass Avenue
Argonne, IL 60439　　　(312) 972-2308

Tyrone Davis*
U.S. Dept of Energy
9800 S. Cass Ave.
Argonne, IL 60439　　　(708) 972-2177

Mark P. Dvorscak*
U. S. Department of Energy
Chicago Operations Office
9800 S. Cass Avenue
Argonne, IL 60439   (708) 972-2177

Robert J. Fisher*
U.S. Atomic Energy Comm.
Chicago Oper. Off.
9800 S. Cass Ave.
Argonne, IL 60439   (312) 972-2176

Hugh Walker Glenn, Jr.*
U. S. Dept. of Energy
9800 S. Cass Ave.
Argonne, IL 60439   (312) 972-2178

Paul A. Gottlieb*
U.S. Dept. of Energy
9800 S. Cass Avenue
Argonne, IL 60439   (312) 972-2169

William Lohff
Argonne National Laboratory
9700 S. Cass Avenue
Argonne, IL 60439   (312) 972-6408

Bradley W. Smith*
U.S. Dept. of Energy
9800 S. Cass Ave.
Argonne, IL 60439   (708) 972-2160

James W. Weinberger*
U.S. Dept. of Energy
9800 S. Cass Ave.
Argonne, IL 60439   (312) 972-2173

William C. Clarke
1264 North Race Avenue
Arlington Heights, IL 60004   (312) 577-1678

Michael J. Femal
Square D Company
812 N. Belmont Avenue
Arlington Heights, IL 60004   (312) 397-2600

Marvin Adolph Henrickson
710 E. Crabtree Drive
Arlington Heights, IL 60004   (312) 255-2435

Ronald J. La Porte
1722 Stratford Rd.
Arlington Heights, IL 60004   (312) 392-1030

Sang Ki Lee
Motorola, Inc.
Cellular Infrastructure Div.
Radio Telephone Systems Group
1501 W. Shure Drive
Arlington Heights, IL 60004   (312) 632-5597

E. Jerome Maas
1716 E. Hawthorne
Arlington Heights, IL 60004   (312) 392-1850

Jos. F. Shekleton
710 E. Waverly Dr.
Arlington Heights, IL 60004   (312) 259-4890

Robert A. Stenzel
516 E. Fairview
Arlington Heights, IL 60005   (708) 255-6868

Richard L. Dornfeld#
Walker Process Corp.
840 N. Russell Ave.
Aurora, IL 60506   (312) 892-7921

Frederick W. Neill
600 Fifth Street
Aurora, IL 60505   (312) 851-2306

Roger W. Nolan, Jr.
OSI Industries, Inc.
1225 Corporate Blvd.
Aurora, IL 60507   (312) 851-6600

Robert J. Lewis
LEAF, Inc.
2355 Waukegan Road
Bannockburn, IL 60015   (708) 940-7500

Bradford S. Allen
240 Castle Court
Barrington, IL 60010   (312) 381-5845

Morris P. Burkwall, Jr.#
Quaker Oats Company
617 W. Main Street
Barrington, IL 60010   (312) 381-1980

Jacque Louis Meister
111 North Avenue
Barrington, IL 60010   (312) 381-7747

Charles F. Meroni, Jr.
Meroni & Meroni
509 W. Main Street
Barrington, IL 60010   (312) 304-1500

Hideo Tomomatsu#
Quaker Oats Co.
Research Labs.
617 W. Main
Barrington, IL 60010   (708) 381-1980

Beverly A. Vandenburgh
400 E. Main St.
Barrington, IL 60010   (708) 381-1831

Thomas Raymond Vigil
836 South Northwest Hwy.
Barrington, IL 60010   (312) 382-6500

Stephen A. Litchfield*
8N160 Naperville Road
Bartlett, IL 60103   (312) 830-7179

Cornelius J O Connor
628 W. McLean Court
Bensenville, IL 60106   (312) 766-2811

Stephen F. Skala#
3839 S. Wenonah Ave.
Berwyn, IL 60402   (312) 788-5021

Rolland R. Hackbart#
730 Thompson Court
Buffalo, IL 60089   (312) 634-0779

Henry A. Weber
671 Aberdeen Lane
Buffalo Grove, IL 60089   (312) 537-6740

Stacy Y. Daniels
11445 79th St.
Burr Ridge, IL 60525   (312) 634-0779

Mary L. Rouhandeh
Feirich, Schoen, Mager & Green
2001 W. Main Street
P.O. Box 1570
Carbondale, IL 62903   (618) 529-3000

Kajane Mc Manus
120 Pueblo Road
Carpentersville, IL 60110   (312) 551-3323

Joseph P. O Halloran
131 W. James
Cary, IL 60013   (312) 639-5345

Raymond R. Kimpel
Summers, Watson & Kimpel
404 W. Church St.
Champaign, IL 61820   (217) 352-7629

Delphine Kranz
University of Illinois
417 Swanlund Administration Bldg.
601 E. John
Champaign, IL 61820   (217) 333-6807

Donald G. Flaynik, Jr.
223 E. Eames
Channahon, IL 60410   (815) 467-4333

Hugh A. Abrams
Neuman, Williams, Anderson
& Olson
77 W. Washington Street
Chicago, IL 60602   (312) 346-1200

Roger Aceto
Viskase Corporation
6855 West 65th Street
Chicago, IL 60638   (312) 496-4732

Paul L. Ahern
Leydig, Voit & Mayer, Ltd.
180 N. Stetson Ave.
Two Prudential Plaza
Suite 4900
Chicago, IL 60601   (312) 616-5600

Samuel L. Alberstadt
Niro, Scavone, Haller
& Niro, Ltd.
200 W. Madison Street
Suite 3500
Chicago, IL 60606   (312) 236-0733

John L. Alex
Lockwood, Alex, Fitz Gibbon
& Cummings
Three First Natl. Plaza
Suite 1700
Chicago, IL 60602   (312) 782-4860

Richard Elmont Alexander
Alexander, Unikel, Zalewa
& Tenenbaum
55 W. Monroe Street
Chicago, IL 60603   (312) RA6-7800

D. Dennis Allegretti
Allegretti & Witcoff, Ltd.
10 South Wacker Drive
Chicago, IL 60606   (312) 715-1000

Caliste Jay Alster
Trexler, Bushnell, Giangiorgi
& Blackstone, Ltd.
141 W. Jackson Blvd.
Suite 3440
Chicago, IL 60604   (312) 427-8082

Irwin C. Alter
Alter & Weiss
105 W. Adams St.
Suite 2700
Chicago, IL 60603   (312) 337-2100

Louis Altman
Laff, Whitesel, Conte
& Saret
401 N. Michigan Avenue
Chicago, IL 60611   (312) 661-2100

James M. Amend
Kirkland & Ellis
200 E. Randolph
Chicago, IL 60601   (312) 861-2154

Dalton L. Anderson
P.O. Box 912
Chicago, IL 60690   (312) 856-3611

David Albert Anderson
Willian Brinks Olds Hofer
Gilson & Lione, Ltd.
Nbc Tower, Suite 3600
455 N. Cityfront Plaza Dr.
Chicago, IL 60611   (312) 321-4210

Richard H. Anderson
Marshall, O'Toole, Gerstein,
Murray & Bicknell
Suite 2100 Two First Natl. Plaza
20 S. Clark Street
Chicago, IL 60603   (312) 346-5750

Theodore W. Anderson, Jr.
Leydig, Voit & Mayer, Ltd.
Two Prudential Plaza
Suite 4900
Chicago, IL 60601   (312) 616-5600

Stephen R. Arnold
Wallenstein, Wagner
& Hattis, Ltd.
311 S. Wacker Drive
53rd Floor
Chicago, IL 60606   (312) 554-3300

Leo Jhel Aubel
Wallenstein, Wagner, Hattis,
Strampel & Aubel
100 S. Wacker Drive
Chicago, IL 60606   (312) 641-1570

F. David Aubuchon
Navistar International
Transportation Corp.
455 Cityfront Plaza Dr.
Chicago, IL 60611   (312) 836-2320

John M. Augustyn
Leydig, Voit & Mayer, Ltd.
Two Prudential Plaza
Suite 4900
Chicago, IL 60601   (312) 616-5600

Jack Axelrood
Morton Chemical
Div. of Morton-Norwich Prods. Inc.
333 West Wacker Drive
Chicago, IL 60606   (312) 807-3189

Karen E. Ayd
The Quaker Oats Co.
345 Merchandise Mart Plaza
Chicago, IL 60654   (312) 222-7803

Y. Judd Azulay
Azulay & Azulay, P.C.
205 W. Wacker Dr.
Suite 1600
Chicago, IL 60606   (312) 236-6965

Mark C. Bach#
Trexler, Bushnell, Giangiorgi
& Blackstone, Ltd.
105 W. Adams St.
36th Floor
Chicago, IL 60603   (312) 704-1890

Michael H. Baniak
Willian Brinks Olds Hofer
Gilson & Lione Ltd.
Nbc Tower
455 N. Cityfront Plaza Dr.
Chicago, IL 60611   (312) 321-4200

Mark T. Banner
Allegretti & Witcoff, Ltd.
10 South Wacker Drive
Chicago, IL 60606   (312) 715-1000

Aaron A. Barlow
Roper & Quigg
200 S. Michigan Ave.
Suite 1000
Chicago, IL 60604   (312) 408-0855

James Patrick Barr
900 N. Lake Shore Drive
Suite 2114
Chicago, IL 60611   (312) 280-1652

Robert M. Barrett#
Hill, Van Santen, Steadman
& Simpson
70th Floor
Sears Tower
Chicago, IL 60606   (312) 876-0200

Alan L. Barry
Wallenstein, Wagner
& Hattis, Ltd.
311 S. Wacker Drive
53rd Floor
Chicago, IL 60606   (312) 554-3300

Steven R. Bartholomew
Fitch, Even, Tabin
& Flannery
135 S. La Salle St.
Suite 900
Chicago, IL 60603   (312) 372-7842

Sarah E. Bates
Kirkland & Ellis
200 E. Randolph Drive
Chicago, IL 60601   (312) 861-2266

Robert W. Beart
Illinois Tool Works Inc.
Patent Dept.
8501 W. Higgins Rd.
Chicago, IL 60631   (312) 693-3040

Richard P. Beem
Leydig, Voit & Mayer, Ltd.
Two Prudential Plaza
Suite 4900
Chicago, IL 60601   (312) 616-5600

Gregory Buckingham Beggs
Neuman, Williams, Anderson
& Olson
77 W. Washington Street
Chicago, IL 60602   (312) 346-1200

Robert S. Beiser
29 S. La Salle Street
Suite 325
Chicago, IL 60603   (312) 558-1900

Stanley Belsky
175 E. Delaware Place
Apt. 8608
Chicago, IL 60611   (312) 440-1638

Glen P. Belvis
Willian Brinks Olds Hofer
Gilson & Lione Ltd.
Nbc Tower, Suite 3600
455 N. Cityfront Plaza Dr.
Chicago, IL 60611   (312) 321-4200

Marvin N. Benn
Hamman & Benn
25 E. Washington Street
Suite 600
Chicago, IL 60602   (312) 372-2920

Joel W. Benson
Willian Brinks Olds Hofer
Gilson & Lione, Ltd.
Nbc Tower, Suite 3600
455 N. Cityfront Plaza Dr.
Chicago, IL 60611   (312) 321-4219

Martin J. Benson#
4107 N. Spaulding Ave.
Chicago, IL 60618   (312) 267-8241

Robert H. Benson
Leydig, Voit & Mayer Ltd.
One IBM Plaza
Suite 4600
Chicago, IL 60611   (312) 822-9666

Jack Charles Berenzweig
William Brinks Olds Hofer
Gilson & Lione Ltd.
Nbc Tower
455 N. Cityfront Plaza Dr.
Chicago, IL 60611   (312) 321-4212

Paul H. Berghoff
Allegretti & Witcoff, Ltd.
10 South Wacker Drive
Chicago, IL 60606   (312) 715-1000

Louis Bernat
135 South La Salle
Suite 1135
Chicago, IL 60604   (312) 346-3798

Denis A. Berntsen
Allegretti & Witcoff, Ltd.
10 South Wacker Drive
Chicago, IL 60606　　　(312) 715-1000

Albert W. Bicknell
Marshall, O'Toole, Gerstein
Murray & Bicknell
Two First National Plaza
Chicago, IL 60603　　　(312) 346-5750

William Joseph Birmingham
Leydig, Voit & Mayer, Ltd.
Two Prudential Plaza
Suite 4900
Chicago, IL 60601　　　(312) 616-5600

Raiford A. Blackstone, Jr.
Trexler, Bushnell, Giangiorgi
& Blackstone, Ltd.
105 W. Adams
36th Floor
Chicago, IL 60603　　　(312) 704-1890

James Bernard Blanchard
Willian Brinks Olds Hofer
Gilson & Lione Ltd.
Nbc Tower
455 N. Cityfront Plaza Dr.
Chicago, IL 60611　　　(312) 321-4207

Robert Emory Blankenbaker
Amoco Corporation
200 East Randolph Drive
Chicago, IL 60601　　　(312) 856-2054

Gunar John Blumberg
Amoco Corporation
200 E. Randolph Dr.
Mail Code 1906
Chicago, IL 60601　　　(312) 856-5967

Joel H. Bock
Kinzer, Plyer, Dorn,
& Mc Eachran
55 East Monroe Street
Chicago, IL 60603　　　(312) 726-4421

Douglas A. Boehm
Welsh & Katz, Ltd.
135 South La Salle Street
Suite 1625
Chicago, IL 60603　　　(312) 781-9470

Daniel A. Boehnen
Allegretti & Witcoff, Ltd.
10 South Wacker Drive
Chicago, IL 60606　　　(312) 715-1000

Jeffrey S. Boone
Akzo America Inc.
300 S. Riverside Plaza
Chicago, IL 60606　　　(312) 906-7590

Thomas S. Borecki
Roper & Quigg
200 S. Michigan Ave.
Suite 1000
Chicago, IL 60604　　　(312) 408-0855

Michael F. Borun
Marshall, O'Toole, Gerstein,
Murray & Bicknell
Two First National Plaza
Chicago, IL 60603　　　(312) 346-5750

George S. Bosy
Roper & Quigg
200 S. Michigan Ave.
Suite 1000
Chicago, IL 60604　　　(312) 408-0855

Richard J. Botos
Dressler, Goldsmith, Shore,
Sutker & Milnamow, Ltd.
Two Prudential Plaza
180 N. Stetson Ave., Ste. 4700
Chicago, IL 60601　　　(312) 616-5400

Charles Earhart Bouton
Amsted Industries, Inc.
205 N. Michigan Avenue
Boulevard Towers South - 44th Floor
Chicago, IL 60601　　　(312) 819-8481

Roland K. Bowler, II#
Dvorak & Traub
53 West Jackson Blvd.
Suite 1616
Chicago, IL 60604　　　(312) 922-6262

Gregory L. Bradley#
474 N. Lake Shore Drive
Apartment 2504
Chicago, IL 60611　　　(312) 464-3804

David C. Brezina
Lee, Mann, Smith, Mc Williams
& Sweeney
105 W. Adams St., Suite 300
Chicago, IL 60603　　　(312) 368-1300

Henry L. Brinks
Willian Brinks Olds Hofer
Gilson & Lione Ltd.
Nbc Tower
455 N. Cityfront Plaza Dr.
Chicago, IL 60611　　　(312) 321-4000

Edward J. Brosius
Amsted Industries Inc.
205 N. Michigan Ave.
44th Floor
Chicago, IL 60601　　　(312) 819-8482

Donald J. Brott
Marshall, O'Toole, Gerstein,
Murray & Bicknell
Two First National Plaza
Chicago, IL 60603　　　(312) 346-5750

Paul L. Brown
Emrich & Dithmar
Suite 3000
150 N. Wacker Drive
Chicago, IL 60606　　　(312) 368-8575

Robert E. Browne, Sr.
Wallenstein, Wagner, Hattis,
Strampel & Aubel, Ltd.
100 S. Wacker Drive
Suite 2100
Chicago, IL 60606　　　(312) 641-1570

Albert J. Brunett
Silverman, Cass, Singer
& Winburn, Ltd.
105 W. Adams Street
Suite 2700
Chicago, IL 60603　　　(312) 726-6006

Angelo J. Bufalino
Lockwood, Alex, Fitz Gibbon
& Cummings
Three First Natl. Plaza
Suite 1700
Chicago, IL 60602　　　(312) 782-4860

George Edward Bullwinkel
Burditt, Bowles & Radzius, Ltd.
333 W. Wacker Drive
Chicago, IL 60606　　　(312) 781-6667

Mark J. Buonaiuto
Mc Dermott, Will & Emery
227 West Monroe Street
31st Floor
Chicago, IL 60606　　　(312) 372-2000

Patrick N. Burkhart#
Hill, Van Santen, Steadman
& Simpson
70th Floor
Sears Tower
Chicago, IL 60606　　　(312) 876-0200

Marshall A. Burmeister
Burmeister, York, Palmatier,
Hamby & Jones
135 S. Lasalle St.
Chicago, IL 60603　　　(312) 782-6663

Mary Spalding Burns
Jones, Day, Reavis
& Pogue
225 W. Washington Street
Chicago, IL 60606　　　(312) 269-4291

Patrick G. Burns
Welch & Katz, Ltd.
135 S. La Salle Street
Suite 1625
Chicago, IL 60603　　　(312) 781-9470

Richard Bushnell
Trexler, Bushnell, Grangiorgi
& Blackstone Ltd.
105 W. Adams
36th Floor
Chicago, IL 60603　　　(312) 704-1890

Paul W. Busse
Marshall, O Toole, Gerstein,
Murray & Bicknell
Two First National Plaza
20 S. Clark St., Suite 2100
Chicago, IL 60603　　　(312) 346-5750

Joseph Peter Calabrese
Neuman, Williams, Anderson
& Olson
77 W. Washington
Chicago, IL 60602　　　(312) 346-1200

Charles G. Call
Allegretti, Newitt, Witcoff
& Mc Andrews, Ltd.
125 S. Wacker Dr.
Chicago, IL 60606   (312) 372-2161

James V. Callahan
Allegretti & Witcoff, Ltd.
10 South Wacker Drive
Chicago, IL 60606   (312) 715-1000

Myron C. Cass
Silverman, Cass, Singer
& Winburn, Ltd.
105 W Adams St.
27th Floor
Chicago, IL 60603   (312) 726-6006

John Jerome Cavanaugh
Neuman, Williams, Anderson
& Olson
77 West Washington St.
Suite 2000
Chicago, IL 60602   (312) 346-1200

Richard A. Cederoth
Neuman, Williams, Anderson
& Olson
77 W. Washington Street
Chicago, IL 60602   (312) 346-1200

Talivaldis Cepuritis
Dressler, Goldsmith, Shore,
Sutker & Milnamow, Ltd.
Two Prudential Plaza
180 N. Stetson Ave., Ste. 4700
Chicago, IL 60601   (312) 616-5400

Edward J. Chalfie
Keck, Mahin & Cate
8300 Sears Tower
233 S. Wacker Drive
Chicago, IL 60606   (312) 876-6167

Daniel R. Cherry
Welsh & Kratz, Ltd.
135 S. La Salle Street
Suite 1625
Chicago, IL 60603   (312) 781-9470

Michael J. Cherskov
Mason, Kolehmainen,
Rathburn & Wyss
20 N. Wacker Dr.
Suite 4200
Chicago, IL 60606   (312) 621-1300

Ernest Cheslow
Dressler, Goldsmith, Shore,
Sutker & Milnamow, Ltd.
Two Prudential Plaza
180 N. Stetson Ave., Ste. 4700
Chicago, IL 60601   (312) 616-5400

John William Chestnut
Tilton, Fallon, Lungmus
& Chestnut
100 S. Wacker Drive
Suite 960
Chicago, IL 60606   (312) 263-1841

Davis Chin
105 West Madison Street
Suite 1707
Chicago, IL 60602   (312) 726-6448

Daniel N. Christus
Wallenstein, Wagner
& Hattis, Ltd.
311 S. Wacker Drive
53rd Floor
Chicago, IL 60606   (312) 554-3300

John J. Chrystal
Ladas & Parry
224 S. Michigan Avenue
Chicago, IL 60604   (312) 236-9021

Jeffrey L. Clark
Wood, Dalton, Phillips,
Mason & Rowe
500 W. Madison Street
Suite 3800
Chicago, IL 60606   (312) 876-1800

Richard A. Clegg
Allegretti & Witcoff, Ltd.
Ten South Wacker Drive
Chicago, IL 60606   (312) 715-1000

James W. Clement
Clement & Ryan
150 N. Michigan Ave.
Suite 1250
Chicago, IL 60601   (312) 663-1200

John L. Cline
Willian Brinks Olds Hofer
Gilson & Lione Ltd.
NBC Tower
455 N. Cityfront Plaza Dr.
Chicago, IL 60611   (312) 321-4213

Eric C. Cohen
Welsh & Katz, Ltd.
135 S. La Salle Street
Suite 1625
Chicago, IL 60603   (312) 781-9470

John B. Conklin
Leydig, Voit & Mayer, Ltd.
180 N. Stetson Ave.
Two Prudential Plaza
Suite 4900
Chicago, IL 60601   (312) 616-5600

James J. Conlon
Conlon & Kerstein
Suite 1220
205 W. Randolph Street
Chicago, IL 60606   (312) 726-0545

P. Phillips Connor
Hill, Van Santen, Chiara
& Simpson
70th Floor, Sears Tower
233 S. Wacker Dr.
Chicago, IL 60601   (312) 876-0200

Robert F. I. Conte
Laff, Whitesel, Conte
& Saret
401 N. Michigan Avenue
Suite 2000
Chicago, IL 60611   (312) 661-2100

Granger Cook, Jr.
Cook, Egan, Mc Farron
& Manzo
135 S. La Sale St.
Suite 4100
Chicago, IL 60603   (312) 236-8500

Ronald B. Coolley
Arnold, White & Durkee
Quaker Tower, Suite 800
321 North Clark Street
Chicago, IL 60610   (312) 744-0090

Gordon Russell Coons
Leydig, Voit & Mayer, Ltd.
180 N. Stetson Ave.
Two Prudential Plaza
Suite 4900
Chicago, IL 60601   (312) 616-5600

Marc S. Cooperman
Allegretti & Witcoff, Ltd.
10 S. Wacker Drive
Chicago, IL 60606   (312) 715-1000

John W. Cornell
Lockwood, Alex, Fitzgibbon
& Cummings
Three First Natl. Plaza
Suite 1700
Chicago, IL 60602   (312) 782-4860

Terrance J. Coughlin
39 S. La Salle
Room 820
Chicago, IL 60603   (312) 332-7374

Lawrence J. Crain
Welsh & Kratz, Ltd.
135 S. La Salle Street
Suite 1625
Chicago, IL 60603   (312) 781-9470

Robert J. Crawford
Arnold, White & Durkee
321 N. Clark
Chicago, IL 60120   (312) 744-0090

Philip J. Crihfield
Sidley & Austin
One First National Plaza
Chicago, IL 60603   (312) 853-7650

John A. Crook, III
Willian Brinks Olds Hofer
Gilson & Lione Ltd.
NBC Tower - Suite 3600
455 N. Cityfront Plaza Dr.
Chicago, IL 60611   (618) 321-4200

Wannell M. Crook
Willian, Brinks, Olds, Hofer,
Gilson, Lione Ltd
NBC Tower
455 N. Cityfront Plaza Dr.
Chicago, IL 60611   (312) 321-4200

John R. Crossan
William, Brinks, Olds, Hofer,
Gilson & Lione, P.C.
455 N. Cityfront Plaza Dr.
Suite 3600
Chicago, IL 60611   (312) 321-4234

Eugene M. Cummings
Lockwood, Alex, Fitz Gibbon
& Cummings
Three First Natl. Plaza
Suite 1700
Chicago, IL 60602   (312) 782-4860

Robert P. Cummins
Bickel & Brewer
One First National Plaza
Suite 3300
Chicago, IL 60603   (312) 630-5300

Gary J. Cunningham
Amoco Corporation
200 E. Randolph Drive
P. O. Box 87703
Chicago, IL 60680   (312) 856-5965

Andreas M. Danckers
Laff, Whitesel, Conte
& Saret
401 N. Michigan Avenue
Suite 2000
Chicago, IL 60611   (312) 661-2106

Rodney A. Daniel
William, Brinks, Olds, Hofer,
Gilson & Lione Ltd.
NBC Tower, Suite 3600
455 N. Cityfront Plaza Dr.
Chicago, IL 60611   (312) 321-4238

Howard Helseth Darbo
Lee, Mann, Smith, Mc Williams
& Sweeney
105 W. Admas Street
Suite 300
Chicago, IL 60603   (312) 368-1300

Timothy Q. Delaney
William, Brinks, Olds, Hofer,
Gilson & Lione
NBC Tower
455 N. Cityfront Plaza Dr.
Chicago, IL 60611   (312) 321-4200

Daniel J. Deneufbourg
Emrich & Dithmar
150 N. Wacker Dr.
Suite 3000
Chicago, IL 60606   (312) 368-8575

Jo Anne M. Denison
333 W. Wacker Dr.
Suite 2600
Chicago, IL 60606   (312) 781-6609

John D. Dewey
Lockwood, Alex, Fitzgibbon
& Cummings
Three First Natl. Plaza
Suite 1700
Chicago, IL 60602   (312) 782-4860

Richard E. Dick
Dick & Harris
181 W. Madison Street
Suite 3800
Chicago, IL 60602   (312) 726-4000

Robert M. Didrick
Morton International, Inc.
Pat. & Tdmk. Dept.
100 N. Riverside Plaza
Chicago, IL 60606   (312) 807-2182

Melinda Lois Dierstein
1337 W. Fargo Ave., Apt. 11B
Chicago, IL 60626   (312) 465-7376

Alvin B. Dodek
Allegretti & Witcoff, Ltd.
10 S. Wacker Dr.
Suite 3000
Chicago, IL 60606   (312) 715-1000

Thomas J. Donovan
Laff, Whitesel, Conte
& Saret
401 N. Michigan Ave.
Suite 2000
Chicago, IL 60611   (312) 661-2100

Thomas E. Dorn
Kinzer, Plyer, Dorn
& Mc Eachran
55 E. Monroe St.
Chicago, IL 60603   (312) 236-1112

Vasilios D. Dossas
Tilton, Fallon, Lungmus
& Chestnut
100 S. Wacker Drive
Suite 960
Chicago, IL 60606   (312) 263-1841

Walter M. Douglas
Silverman, Cass & Singer
27th Floor
105 West Adams Street
Chicago, IL 60603   (312) 726-6006

James R. Dowdall
Neuman, Williams, Anderson
& Olson
77 W. Washington St.
Chicago, IL 60602   (312) 346-1200

James S. Downes
1864 North Orchard
Chicago, IL 60614   (312) 664-2439

Max Dressler
Dressler, Goldsmith, Shore,
Sutker & Milnamow, Ltd.
1800 Prudential
Chicago, IL 60601   (312) 527-4025

Grantland G. Drutchas
William, Brinks, Olds, Hofer,
Gilson & Lione
One IBM Plaza
Suite 4100
Chicago, IL 60611   (312) 822-9800

Christine A. Dudzik
Marshall, O Toole, Gerstein,
Murray & Bicknell
20 South Clark
Suite 2100
Chicago, IL 60603   (312) 346-5750

Jeffery M. Duncan
William, Brinks, Olds, Hofer,
Gilson & Lione Ltd.
One IBM Plaza
Suite 4100
Chicago, IL 60611   (312) 822-9800

Margaret M. Duncan
Amoco Corporation
P.O. Box 87703
Mail Code 1904
Chicago, IL 60680   (312) 856-2287

Jennifer A. Dunner
Laff, Whitesel, Conte
& Saret
401 N. Michigan Ave.
Chicago, IL 60614   (312) 661-2100

George F. Dvorak
Balogh, Osann, Kramer, Dvorak,
Genov & Traub
53 W. Jackson Blvd.
Chicago, IL 60604   (312) 922-6262

Thomas E. Earle#
Amoco Corporation
Pats. & Licensing Dept.
Mail Code 1907A
200 E. Randolph Drive
Chicago, IL 60601   (312) 856-5932

Vangelis Economou
6349 North Western Avenue
Suite 208
Chicago, IL 60659   (312) 743-1048

Donald E. Egan
Cook, Egan, McFarron & Manzo
135 S. La Salle St.
Suite 4100
Chicago, IL 60603   (312) 236-8500

C. Lyman Emrich, Jr.
Emrich & Dithmar
150 N. Wacker Drive
Chicago, IL 60606   (312) 368-8575

Ronald L. Engel#
440 S. La Salle
Suite 1931
Chicago, IL 60605   (312) 362-3414

Randall T. Erickson
Hill, Van Santen, Steadman
& Simpson, P.C.
70th Floor - Sears Tower
Chicago, IL 60606   (312) 876-0200

Patrick D. Ertel
Wood, Dalton, Phillips,
Mason & Rowe
500 W. Madison Street
Suite 3800
Chicago, IL 60606   (312) 876-1800

Francis A. Even
Fitch, Even, Tabin
& Flannery
135 South Lasalle St.
Suite 900
Chicago, IL 60603   (312) 372-7842

Irving Faber
150 S. Wacker Drive
Suite 500
Chicago, IL 60606   (312) 782-2833

Jody L. Factor
Dick & Harris
181 W. Madison St.
Ste. 3800
Chicago, IL 60606   (312) 726-4000

Martin Faier
Three First National Plaza
Suite 725
70 W. Madison Street
Chicago, IL 60602   (312) 332-2060

Thomas A. Fairhall#
Allegretti & Witcoff, Ltd.
10 S. Wacker Dr.
Chicago, IL 60606   (312) 715-1000

Jerome F. Fallon
Tilton, Fallon, Lungmus
& Chestnut
100 South Wacker Drive
Hartford Plaza
Chicago, IL 60606   (312) 263-1841

Scott B. Feder
Fitch, Even, Tabin
& Flannery
135 S. La Salle Street
Suite 900
Chicago, IL 60603   (312) 372-7842

Mark I. Feldman
Rudnick & Wolfe
203 N. La Salle Street
Chicago, IL 60601   (312) 368-7084

Robert W. Fieseler
Neuman, Williams, Anderson
& Olson
77 W. Washington Street
Chicago, IL 60602   (312) 346-1200

Thomas J. Filarski
Willian Brinks Olds Hofer
Gilson & Lione Ltd.
NBC Tower
455 N. Cityfront Plaza Dr.
Chicago, IL 60611   (312) 321-4200

Karl R. Fink
Dressler, Goldsmith, Shore,
Sutker & Milnamow, Ltd.
Two Prudential Plaza
180 N. Stetson Ave., Ste. 4700
Chicago, IL 60601   (312) 616-5400

Morgan L. Fitch
South Chicago Savings Bank
9200 S. Commercial Avenue
Chicago, IL 60617   (312) 768-1400

Morgan Lewis Fitch, Jr.
Fitch, Even, Tabin
& Flannery
135 S. Lasalle St.
Chicago, IL 60603   (312) 372-7842

James T. Fitz Gibbon
Lockwood, Alex, Fitz Gibbon
& Cummings
Three First Natl. Plaza
Suite 1700
Chicago, IL 60602   (312) 346-6540

John Francis Flannery
Fitch, Even, Tabin
& Flannery
135 S. Lasalle St.
Chicago, IL 60603   (312) 372-7842

Clarence J. Fleming
Jones, Day, Reavis
& Pogue
225 W. Washington St.
Suite 2600
Chicago, IL 60606   (312) 269-4105

F. Frederick Fondriest
Amoco Corporation
200 E. Randolph Dr.
Chicago, IL 60601   (312) 856-3966

Robert J. Fox
Fitch, Even, Tabin & Flannery
135 S. La Salle Street
Suite 900
Chicago, IL 60603   (312) 372-7842

John A. Franczyk
Butler, Rubin, Newcomer,
Saltarelli & Boyd
Three First Natl. Plaza
Suite 1505
Chicago, IL 60602   (312) 444-9660

William H. Frankel
Neuman, Williams, Anderson
& Olson
77 W. Wash. St.
Suite 2000
Chicago, IL 60602   (312) 346-1200

John C. Freeman#
Willian Brinks Olds Hofer
Gilson & Lione
455 North Cityfront Plaza Dr.
NBC Tower, Suite 3600
Chicago, IL 60611   (312) 321-4200

Roger J. French
Trexler, Bushnell, Giangiorgi
& Blackstone, Ltd.
105 W. Adams
36th Floor
Chicago, IL 60603   (312) 704-1890

Eugene F. Friedman
Monadnock Bldg.
53 W. Jackson Blvd.
Suite 1633
Chicago, IL 60604   (312) 922-8882

David M. Frischkorn
Allegretti & Witcoff, Ltd.
10 South Wacker Drive
Chicago, IL 60606   (312) 715-1000

Joseph A. Fuchs
Wallenstein, Wagner
& Hattis, Ltd.
311 South Wacker Drive
53rd Floor
Chicago, IL 60606   (312) 554-3300

Howard S. Fuhrman
Marshall, O Toole, Gerstein,
Murray & Bicknell
Two First National Plaza
20 S. Clark
Chicago, IL 60603   (312) 346-5750

James A. Gabala
Amoco Corporation
200 E. Randolph Dr.
P.O. Box 5910A
Chicago, IL 60680   (312) 856-3110

Ralph J. Gabric
Willian Brinks Olds Hofer
Gilson & Lione
3600 NBC Tower
455 N. Cityfront Plaza Dr.
Chicago, IL 60611   (312) 321-4253

Bruce M. Gagala
Leydig, Voit & Mayer
180 North Stetson Ave.
Two Prudential Plaza
Suite 4900
Chicago, IL 60601   (312) 616-5600

Priscilla F. Gallagher
Mc Andrews, Held & Malloy, Ltd.
Northwestern Atrium Center
500 W. Madison St.
Suite 3400
Chicago, IL 60661   (312) 707-8889

Norval B. Galloway, III
Amoco Corporation
200 E. Randolph Dr.
P.O. Box 87703
Mail Code 1907
Chicago, IL 60680   (312) 856-7180

Edward P. Gamson
Dressler, Goldsmith, Shore,
Sutker & Milnamow, Ltd.
Two Prudential Plaza
180 N. Stetson Ave., Ste. 4700
Chicago, IL 60601   (312) 616-5400

Lisa A. Garono
Rosenthal & Schanfield
55 East Monroe St.
46th Floor
Chicago, IL 60603   (312) 899-5511

John R. Garrett
Hill, Van Santen, Steadman
& Simpson, P.C.
70th Floor - Sears Tower
Chicago, IL 60606   (312) 876-0200

Stephen D. Geimer
Dressler, Goldsmith, Shore,
Sutker & Milnamow, Ltd.
Two Prudential Plaza
180 N. Stetson Ave., Ste. 4700
Chicago, IL 60601   (312) 616-5400

Henry J. Gens
8350 W. Addison
Suite 104
Chicago, IL 60634   (312) 589-2065

Timothy H. Gens
Olson & Hierl
20 N. Wacker Drive
Suite 3000
Chicago, IL 60606   (312) 580-1180

Gerald S. Geren
Hill, Van Santen, Steadman
& Simpson
70th Floor - Sears Tower
Chicago, IL 60606   (312) 876-0200

Allen H. Gerstein
Marshall, O'Toole, Gerstein,
Murray & Bicknell
Two First National Plaza
Suite 2100
Chicago, IL 60603   (212) 346-5750

Milton S. Gerstein#
6629 N. Francisco Avenue
Chicago, IL 60645   (312) 372-2926

Robert M. Gerstein
Marshall, O'Toole, Gerstein,
Murray & Bicknell
20 S. Clark St., Ste. 2100
Two First National Plaza
Chicago, IL 60603   (312) 346-5750

George Henry Gerstman
Pigott & Gerstman, Ltd.
2 N. La Salle Street
Chicago, IL 60602   (312) 263-4350

Richard Allen Giangiorgi
Trexler, Bushnell, Giangiorgi
& Blackstone, Ltd.
105 W. Adams Street
36th Floor
Chicago, IL 60603   (312) 704-1890

Edward D. Gilhooly
53 W. Jackson Blvd.
Suite 1516
Chicago, IL 60604   (312) 294-0350

Arthur Gwyer Gilkes
Leydig, Voit & Mayer, Ltd.
One IBM Plaza
Suite 4600
Chicago, IL 60611   (312) 822-9666

Hugh M. Gilroy
6313 North Wayne Avenue
Chicago, IL 60660   (312) 294-0350

Murray A. Gleeson
Mc Caleb, Lucas & Brugman
230 W. Monroe Street
Suite 2040
Chicago, IL 60606   (312) 236-4711

Jerome Goldberg
175 W. Jackson Blvd.
Suite 1629
Chicago, IL 60604   (312) 922-5031

Esther E. Goldman
Wood, Phillips, Mason,
Recktenwald & Van Santen
500 W. Madison, Ste. 3800
Chicago, IL 60606   (312) 876-1800

Stuart I. Graff
Schiff Hardin & Waite
7200 Sears Tower
233 S. Wacker Dr.
Chicago, IL 60606   (312) 876-1000

Jeffrey R. Gray
Lee, Mann, Smith, Mc Williams
& Sweeney
105 West Adams Street
Suite 300
Chicago, IL 60603   (312) 368-1300

Richard O. Gray, Jr.
Foley & Lardner
Suite 4950
70 West Madison Street
Chicago, IL 60602   (312) 444-9500

Raymond William Green
William, Brinks, Olds, Hofer,
Gilson & Lione, Ltd.
NBC Tower, Suite 3600
455 N. Cityfront Plaza Dr.
Chicago, IL 60611   (312) 321-4222

Robert F. Green
Leydig, Voit & Mayer, Ltd.
180 N. Stetson Ave.
Two Prudential Plaza
Suite 4900
Chicago, IL 60601   (312) 616-5600

Martin R. Greenstein
Baker & Mckenzie
2800 Prudential Plaza
Chicago, IL 60601   (312) 861-2770

Roger D. Greer
Welsh & Katz, Ltd.
135 S. La Salle Street
Suite 1625
Chicago, IL 60603   (312) 781-9470

Christopher T. Griffith
Leydig, Voit & Mayer, Ltd.
180 N. Stetson Ave.
Two Prudential Plaza
Suite 4900
Chicago, IL 60601   (312) 616-5600

Dennis A. Gross
Hill, Van Santen, Steadman,
& Simpson
233 S. Wacker Drive
70th Floor
Chicago, IL 60606   (312) 876-0200

J. Arthur Gross
Hill, Van Santen, Steadman,
& Simpson
70th Flr., Sears Tower
Chicago, IL 60606   (312) 876-0200

Lewis S. Gruber
400 E. Randolph
Apt. No. 3911
Chicago, IL 60601   (312) 938-8740

James C. Gumina
Allegretti & Witcoff, Ltd.
10 South Wacker Drive
Chicago, IL 60606   (312) 715-1000

Daniel M. Gurfinkel
Gerstman & Elllis, Ltd.
Two North La Salle Street
Suite 2010
Chicago, IL 60602   (312) 263-4350

Kevin W. Guynn
Hill, Van Santen, Steadman
& Simpson, P.C.
70th Floor Sear Tower
233 S. Wacker Drive
Chicago, IL 60606   (312) 876-0200

Edward A. Haight
Haight & Hofeldt
224 S. Michigan Avenue
Suite 600
Chicago, IL 60604   (312) 939-7909

Timothy J. Haller
Niro, Scavone, Haller
& Niro
181 W. Madison
Suite 4600
Chicago, IL 60602   (312) 236-0733

Jack Roger Halvorsen
Laff, Whitesel, Conte
& Saret
401 N. Michigan Avenue
Suite 2000
Chicago, IL 60611   (312) 661-2100

James John Hamill
Fitch, Even, Tabin
& Flannery
135 S La Salle St.
Suite 900
Chicago, IL 60603   (312) 372-7842

Steven J. Hampton
Mc Andrews, Held & Malloy, Ltd.
Northwestern Atrium Center
500 W. Madison St.
Suite 3400
Chicago, IL 60661   (312) 707-8889

John W. Harbst
Dressler, Goldsmith, Shore,
Sutker & Milnamow, Ltd.
Two Prudential Plaza
180 N. Stetson Ave., Ste. 4700
Chicago, IL 60601   (312) 616-5400

Robert L. Harmon
Willian Brinks Olds Hofer
Gilson & Lione Ltd.
Nbc Tower
455 N. Cityfront Plaza Dr.
Chicago, IL 60611   (312) 321-4208

Richard D. Harris
Dick & Harris
181 West Madison Street
Suite 3800
Chicago, IL 60602   (312) 726-4000

461

Wayne H. Harrold, Jr.
Ladas & Parry
224 S. Michigan Ave.
Suite 1200
Chicago, IL 60604   (312) 427-1300

Herbert D. Hart, III
Neuman, Williams, Anderson
& Olson
77 W. Wash. St.
Chicago, IL 60602   (312) 346-1200

H. Michael Hartmann
Leydig, Voit & Mayer, Ltd.
180 N. Stetson Ave.
Two Prudential Plaza
Suite 4900
Chicago, IL 60601   (312) 616-5600

Russell E. Hattis
Wallenstein, Wagner
& Hattis, Ltd.
311 S. Wacker Drive
53rd Floor
Chicago, IL 60606   (312) 554-3300

Barbara A. Heaphy
Allegretti & Witcoff, Ltd.
10 S. Wacker
Chicago, IL 60606   (312) 715-3418

John J. Held, Jr.
Mc Andrews, Held & Malloy, Ltd.
Northwestern Atrium Center
500 W. Madison Street
Suite 3400
Chicago, IL 60661   (312) 707-8889

Stephen B. Heller
Cook, Egan, Mc Farron
& Manzo, Ltd.
135 S. La Salle Street
Suite 4100
Chicago, IL 60603   (312) 236-8500

James R. Henes
Amoco Corporation
200 E. Randolph Drive
Chicago, IL 60601   (312) 856-2808

Stephen L. Hensley
Amoco Corporation
200 East Randolph Drive
Chicago, IL 60601   (312) 856-2764

Gary L. Hermanson
William, Brinks, Olds, Hofer,
Gilson & Lione
NBC Tower, Suite 3600
455 N. Cityfront Plaza Dr.
Chicago, IL 60611   (312) 321-4200

Brett A. Hesterberg
Leydig, Voit & Mayer
180 N. Stetson Ave.
Two Prudential Plaza
Suite 4900
Chicago, IL 60601   (312) 616-5600

Michael A. Hierl
Olson & Hierl
20 North Wacker Drive
Suite 3000
Chicago, IL 60606   (312) 580-1180

James Joseph Hill
Emrich & Dithmar
150 N. Wacker Drive
Suite 3000
Chicago, IL 60606   (312) 368-8575

Thomas E. Hill
Emrich & Dithmar
150 N. Wacker Drive
Suite 3000
Chicago, IL 60606   (312) 368-8575

Martin J. Hirsch
Marshall, O'Toole, Gerstein,
Murray & Bicknell
20 South Clark Street
Suite 2100
Chicago, IL 60603   (312) 346-5750

James Darwin Hobart
Hill, Van Santen, Steadman,
Chiara & Simpson
70th Flr., Sears Tower
233 S. Wacker Dr.
Chicago, IL 60606   (312) 876-0200

Douglass C. Hochstetler
Marshall, O'Toole, Gerstein,
Murray & Bicknell
Two First Natl. Plaza
Suite 2100
Chicago, IL 60603   (312) 346-5750

James M. Hoey
Clausen, Miller, Gorman,
Caffrey & Witous, P.C.
10 S. La Salle Street
Chicago, IL 60603   (312) 855-1010

Roy E. Hofer
William, Brinks, Olds, Hofer,
Gilson & Lione, Ltd.
NBC Tower
455 N. Cityfront Plaza Dr.
Chicago, IL 60611   (312) 321-4204

John R. Hoffman
Wood, Dalton, Phillips,
Mason & Rowe
500 W. Madison Street
Suite 3800
Chicago, IL 60606   (312) 876-1800

Richard B. Hoffman
Tilton, Fallon, Lungmus
& Chestnut
100 S. Wacker Dr.
Suite 960
Chicago, IL 60606   (312) 263-1841

Thomas J. Hoffmann
53 W. Jackson Blvd.
Chicago, IL 60604   (312) 663-1800

Kevin D. Hogg
Marshall, O'Toole, Gerstein,
Murray & Bicknell
Two First National Plaza
Suite 2100
Chicago, IL 60603   (312) 346-5750

Matthew R. Hooper
Amoco Comporation
Pats. & Lic. Dept.
200 E. Randolph Drive
Mail Code 1907
Chicago, IL 60601   (312) 856-5911

Allen Joseph Hoover
Dressler, Goldsmith, Shore,
Sutker & Milnamow, Ltd.
Two Prudential Plaza
180 N. Stetson Ave., Ste. 4700
Chicago, IL 60601   (312) 616-5400

Gerald D. Hosier
Hosier & Sufrin
100 South Wacker Drive
Suite 224
Chicago, IL 60606   (312) 726-1762

Joseph N. Hosteny, III
Niro, Scavone, Haller,
& Niro, Ltd.
181 W. Madison
Suite 4600
Chicago, IL 60602   (312) 236-0733

Clemens Hufmann
Mason, Kolehmainen, Rathburn
& Wyss
20 N. Wacker Dr.
Suite 4200
Chicago, IL 60606   (312) 621-1300

A. Blair Hughes#
Allegretti & Witcoff Ltd.
10 S. Wacker Drive
Suite 3000
Chicago, IL 60606   (312) 715-1000

Bradley J. Hulbert
Allegretti & Witcoff, Ltd.
10 South Wacker Drive
Chicago, IL 60606   (312) 715-1000

James Pickrell Hume
Willian, Brinks, Olds, Hofer,
Gilson & Lione Ltd.
One IBM Plaza
Chicago, IL 60611   (312) 822-9800

Kareem M. Irfan#
Arnold, White & Durkee
321 N. Clark Street
Suite 800
Chicago, IL 60610   (312) 744-0090

John P. Isacson, Jr.
Neuman, Williams, Anderson
& Olson
77 W. Washington St.
Chicago, IL 60602   (312) 346-1200

Linda Moncys Isacson
Roper & Quigg
200 S. Michigan Ave.
Suite 1000
Chicago, IL 60604   (312) 408-0855

Jerold A. Jacover
William, Brinks, Olds, Hofer,
Gilson & Lione, Ltd.
NBC Tower, Suite 3600
455 N. Cityfront Plaza Dr.
Chicago, IL 60611   (312) 321-4214

Melvin F. Jager
William, Brinks, Olds, Hofer,
Gilson & Lione Ltd.
One IBM Plaza
Suite 4100
Chicago, IL 60611   (312) 822-9800

James Joseph Jagoda
Wallenstein, Wagner
& Hattis, Ltd.
311 S. Wacker Drive
53rd Floor
Chicago, IL 60606   (312) 554-3300

Robert V. Jambor
Kinzer, Plyer, Dorn,
Mc Eachran & Jambor
55 E. Monroe St.
Suite 3905
Chicago, IL 60603   (312) 726-4421

John C. Janka
Niro, Scavone, Haller,
& Niro
181 W. Madison
Suite 4600
Chicago, IL 60602   (312) 236-0733

Michael J. Jaro
William, Brinks, Olds, Hofer,
Gilson & Lione
455 N. Cityfront Plaza Dr.
NBC Tower, Ste. 3600
Chicago, IL 60611   (312) 321-4200

L. Michael Jarvis
Mc Andrews, Held & Malloy, Ltd.
Northwestern Atrium Center
500 W. Madison Street
34th Floor
Chicago, IL 60606   (312) 707-9155

James Joseph Jennings, Jr.
Borg-Warner Corp.
200 S. Michigan Ave.
Chicago, IL 60604   (312) 322-8575

Everett A. Johnson
333 N. Michigan Ave.
Chicago, IL 60601   (312) 782-4010

Harold V. Johnson
William, Brinks, Olds, Hofer,
Gilson & Lione Ltd.
One IBM Plaza
Suite 4100
Chicago, IL 60611   (312) 822-9800

Lars S. Johnson
Quaker Oats Co.
Quaker Tower
Mail 25-7
321 N. Clark St.
Chicago, IL 60610   (312) 222-7111

Richard M. Johnson
Leydig, Voit & Mayer
One IBM Plaza
Suite 4600
Chicago, IL 60611   (312) 822-9666

Huw R. Jones
Allegretti & Witcoff, Ltd.
10 S. Wacker Drive
Suite 3000
Chicago, IL 60606   (312) 715-1000

Robert Bruce Jones
Fitch, Even, Tabin
& Flannery
135 S. Lasalle St.
Suite 900
Chicago, IL 60603   (312) 372-7842

David R. Josephs#
Niro, Scavone, Haller & Niro
181 W. Madison Street
Suite 4600
Chicago, IL 60602   (312) 236-0733

Paul G. Juettner
Juettner, Pyle, Lloyd
& Verbeck
221 North La Salle Street
Suite 850
Chicago, IL 60601   (312) 236-8123

Thomas R. Juettner
Juettner, Pyle, Lloyd
& Verbeck
221 North La Salle Street
Suite 850
Chicago, IL 60601   (312) 236-8123

Natalie D. Kadievitch
William, Brinks, Olds, Hofer,
Gilson & Lione
455 N. Cityfront Plaza Dr.
Suite 3600
Chicago, IL 60611   (312) 321-4200

Leonard J. Kalinowski
Emrich & Dithmar
150 North Wacker Drive
Suite 3000
Chicago, IL 60606   (312) 368-8575

Henry S. Kaplan
Dressler, Goldsmith, Shore,
Sutker & Milnamow, Ltd.
Two Prudential Plaza
180 N. Stetson Ave., Ste. 4700
Chicago, IL 60601   (312) 616-5400

Richard A. Kaplan
William, Brinks, Olds, Hofer,
Gilson & Lione
455 N. Cityfront Plaza Drive
Nbc Tower, Suite 3600
Chicago, IL 60611   (312) 321-4227

Kyle K. Kappes
Allegretti & Witcoff, Ltd.
10 South Wacker Drive
Suite 3000
Chicago, IL 60606   (312) 715-1000

Richard J. Karas
American National Can Co.
8770 W. Bryn Mawr Avenue
Chicago, IL 60631   (312) 399-3908

A. Sidney Katz
Welsh & Katz, Ltd.
135 S. La Salle Street
Suite 1625
Chicago, IL 60603   (312) 781-9470

Martin Lewis Katz
Dressler, Goldsmith, Shore,
Sutker & Milnamow, Ltd.
Two Prudential Plaza
180 N. Stetson, Ste. 4700
Chicago, IL 60601   (312) 616-5400

David D. Kaufman
1800 Prudential Plaza
Chicago, IL 60601   (312) 527-4025

Edward M. Keating
Kinzer, Plyer, Dorn
Mceachran & Jambor
55 E. Monroe St.
Chicago, IL 60603   (312) 726-4421

Esther O. Kegan
Kegan & Kegan, Ltd.
79 West Monroe Street
Chicago, IL 60603   (312) 782-6495

Michael L. Kenaga#
Ladas & Parry
224 S. Michigan Ave.
Suite 1200
Chicago, IL 60604   (312) 427-1300

Dolores T. Kenney
Olson & Hierl
20 North Wacker Drive
Suite 3000
Chicago, IL 60606   (312) 580-1180

Wayne A. Keown#
Allegretti & Witcoff, Ltd.
Ten S. Wacker Drive
30th Floor
Chicago, IL 60606   (312) 715-1000

Paul A. Kerstein
150 North Wacker Dr.
Suite 1717
Chicago, IL 60606   (312) 641-5441

John Kilyk, Jr.
Leydig, Voit & Mayer, Ltd.
180 N. Stetson Ave.
Two Prudential Plaza
Room 4900
Chicago, IL 60601   (312) 616-5600

Charles C. Kinne
Allegretti & Witcoff, Ltd.
Ten South Wacker Drive
Suite 3000
Chicago, IL 60606   (312) 715-1000

Layton F. Kinney
Sherwin-Williams
10909 S. Cottage Grove
Chicago, IL 60628   (312) 821-2219

James B. Kinzer
Kinzer, Plyer, Dorn
Mceachran & Jambor
55 E. Monroe Street
Suite 3905
Chicago, IL 60603   (312) 236-1112

John W. Klooster
Olson & Hierl
20 North Wacker Drive
Suite 3000
Chicago, IL 60606   (312) 580-1180

Jerome B. Klose
Marshall, O Toole, Gerstein,
Murray & Bicknell
20 South Clark Street
Suite 2100
Chicago, IL 60603   (312) 346-5750

Thomas A. Kmiotek
2243 N. Mango Avenue
Chicago, IL 60639   (312) 923-6871

Jeffrey G. Knoll#
Wood, Phillips, Mason,
Recktenwald & Van Santen
500 W. Madison, Ste. 3800
Chicago, IL 60606   (312) 876-1800

Philip M. Kolehmainen
Mason, Kolehmainen, Rathburn
& Wyss
20 N. Wacker Drive
Chicago, IL 60606   (312) 621-1300

Ludwig E. Kolman
Phelan, Pope & John, Ltd.
311 South Wacker Drive
Suite 4200
Chicago, IL 60606   (312) 362-0200

Andrew G. Kolomayets
Cook, Egan, Mc Farron
& Manzo, Ltd.
135 S. La Salle St.
Suite 4100
Chicago, IL 60603   (312) 236-8500

Paul J. Korniczky
Leydig, Voit & Mayer
180 N. Stetson Ave.
Two Prudential Plaza
Suite 4900
Chicago, IL 60601   (312) 616-5600

Nick C. Kottis
Amoco Corporation
P.O. Box 87703
200 E. Randolph Drive
Chicago, IL 60680   (312) 856-5171

Frank J. Kozak
Willian Brinks Olds Hofer
Gilson & Lione, P.C.
455 North Cityfront Plaza Drive
Nbc Tower, Suite 3600
Chicago, IL 60611   (312) 321-4200

John W. Kozak
Leydig, Voit & Mayer, Ltd.
180 N. Stetson Ave.
Two Prudential Plaza
Suite 4900
Chicago, IL 60601   (312) 616-5600

Nathan N. Kraus
6649 N. Maplewood Ave.
Chicago, IL 60645   (312) 262-5192

Richard A. Kretchmer
Amoco Corporation
200 E. Randolph Dr.
Chicago, IL 60601   (312) 856-5921

Daniel J. Krieger
Neuman, Williams, Anderson
& Olson
77 West Washington St.
Chicago, IL 60602   (312) 346-1200

Joseph Krieger
Mason, Kolehmainen, Rathburn
& Wyss
20 N. Wacker Dr.
Suite 4200
Chicago, IL 60606   (312) 621-1300

Robert G. Krupka
Kirkland & Ellis
200 E. Randolph Dr.
Chicago, IL 60601   (312) 861-2156

Linda A. Kuczma
Wallenstein, Wagner &
Hattis, Ltd.
311 S. Wacker Drive
53rd Floor
Chicago, IL 60606   (312) 554-3300

Jean D. Kuelper
Mc Andrews, Held & Malloy, Ltd
Northwetern Atrium Center
500 W. Madison Street
Suite 3400
Chicago, IL 60661   (312) 707-8889

Richard M. La Barge
Marshall, O'Toole, Gerstein,
Murray & Bicknell
2 First National Plaza
Suite 2100
Chicago, IL 60603   (312) 346-5750

Robert G. Ladd#
Amoco Corporation
200 E. Randolph Drive
Chicago, IL 60601   (312) 856-7847

Charles A. Laff
Laff, Whitesel, Conte
& Saret
401 N. Michigan Avenue
Suite 2000
Chicago, IL 60611   (312) 661-2100

Micheal D. Lake
Wallenstein, Wagner
& Hattis, Ltd.
311 S. Wacker Dr.
53rd Floor
Chicago, IL 60606   (312) 554-3300

Bradley G. Lane
William, Brinks, Olds, Hofer,
Gilson & Lione
455 N. Cityfront Plaza Drive
Suite 3600
Chicago, IL 60611   (312) 321-4200

Jay Calvin Langston, Jr.
Keil & Weinkauf
Suite 1216
135 S. La Salle Street
Chicago, IL 60603   (312) 265-5030

James V. Lapacek
S&C Electric Company
6601 N. Ridge Blvd.
Chicago, IL 60626   (312) 338-1000

Ronald E. Larson
Allegretti & Witcoff, Ltd.
10 S. Wacker Drive
Chicago, IL 60606   (312) 715-1000

John Curtis Le Fever
Viskase Corporation
6855 W. 65th Street
Chicago, IL 60638   (312) 496-4735

Jon B. Leaheey
Neuman, Williams, Anderson
& Olson
77 W. Washington St.
Chicago, IL 60602   (312) 346-1200

William M. Lee, Jr.
Lee, Mann, Smith, Mc Williams
& Sweeney
105 W. Adams Street
Suite 300
Chicago, IL 60603   (312) 368-1300

Wm. Marshall Lee
Lee, Mann, Smith, Mc Williams
& Sweeney
105 W. Adams Street
Suite 300
Chicago, IL 60603   (312) 368-1300

Edward Alan Lehman
Hill, Van Santen, Steadman,
Chiara & Simpson
70th Flr., Sears Tower
233 S. Wacker Dr.
Chicago, IL 60606   (312) 876-0200

David Lesht
Lockwood, Alex, Fitz Gibbon
& Cummings
Three First Natl. Plaza
Suite 1700
Chicago, IL 60602   (312) 782-4860

Norman Lettvin
208 S. La Salle Street
Chicago, IL 60604   (312) 782-8862

Russell E. Levine
Kirkland & Ellis
200 E. Randolph Drive
Chicago, IL 60601   (312) 861-2466

Seymour Levine
5515 N. Virginia Avenue
Chicago, IL 60625   (312) 878-0595

Timothy E. Levstik
Fitch, Even, Tabin
& Flannery
135 S. La Salle Street
Suite 900
Chicago, IL 60603   (312) 372-7842

Harry M. Levy
Emrich & Dithmar
150 N. Wacker Drive
Chicago, IL 60606   (312) 368-8575

C. Frederick Leydig, Jr.
Leydig, Voit & Mayer, Ltd.
180 N. Stetson Ave.
Two Prudential Plaza
Suite 4900
Chicago, IL 60601   (312) 616-5600

Henry Lilienheim
Ladas & Parry
224 S. Michigan Ave.
Chicago, IL 60604   (312) 236-9021

Robert L. Lindgren
Jones, Day, Reavis
& Pogue
225 W. Washington Street
Chicago, IL 60606   (312) 269-4107

Thomas B. Lindgren
Jones, Day, Reavis
& Pogue
225 W. Washington Street
Chicago, IL 60606   (312) 269-4340

Richard G. Lione
William, Brinks, Olds, Hofer,
Gilson & Lione Ltd.
NBC Tower
455 N. Cityfront Plaza Dr.
Chicago, IL 60611   (312) 321-4218

Steven G. Lisa
Hosier & Sufrin, Ltd.
100 S. Wacker Drive
Suite 224
Chicago, IL 60606   (312) 726-1762

Carl S. Lloyd
Kirkland & Ellis
200 E. Randolph Drive
Suite 5700
Chicago, IL 60601   (312) 861-2090

Robert A. Lloyd
Juettner, Pyle, Lloyd
& Verbeck
221 North La Salle Street
Suite 850
Chicago, IL 60601   (312) 236-8123

Fred Stark Lockwood
Lockwood, Alex, Fitz Gibbon
& Cummings
Three First Natl. Plaza
Suite 1700
Chicago, IL 60602   (312) 782-4860

Valerie E. Looper
Allegretti & Witcoff, Ltd.
10 South Wacker Dr.
Chicago, IL 60606   (312) 715-1000

Alan R. Loudermilk
1550 N. Lake Shore Dr.
Suite 16B
Chicago, IL 60610   (312) 280-5486

John K. Lucas
Williab, Brinks, Olds, Hofer,
Gilson & Lione Ltd.
NBC Tower
455 N. Cityfront Plaza Dr.
Chicago, IL 60611   (312) 321-4220

William E. Lucas
Mccaleb, Lucas and Brugman
230 W. Monroe St.
Suite 2040
Chicago, IL 60606   (312) 236-4711

Peter S. Lucyshyn
A.B. Dick Co.
5700 W. Touhy Ave.
Chicago, IL 60648   (312) 763-1900

Van Metre Lund
Neuman, Williams, Anderson
& Olson
77 W. Washington St.
Chicago, IL 60602   (312) 346-1200

John B. Lungmus
Tilton, Fallon, Lungmus
& Chestnut
100 S. Wacker Drive
Chicago, IL 60606   (312) 263-1841

Barbara J. Luther
William, Brinks, Olds, Hofer,
Gilson & Lione
455 N. Cityfront Plaza Drive
Suite 3600
Chicago, IL 60611   (312) 321-4200

Kathleen A. Lyons
Rocky & Rifkin
30 N. La Salle Street
Suite 2202
Chicago, IL 60602   (312) 704-5600

William Howard Magidson
Amoco Corporation
200 E. Randolph Drive
Chicago, IL 60680   (312) 856-3967

David W. Maher
Sonnenschein Carlin Nath
& Rosenthal
8000 Sears Tower
Chicago, IL 60606   (312) 876-8055

Timothy J. Malloy
Mc Andrews, Held & Malloy, Ltd.
Northwestern Atrium Center
500 W. Madison Street
Suite 3400
Chicago, IL 60606   (312) 707-8889

Dale A. Malone
Allegretti & Witcoff, Ltd.
Ten South Wacker Drive
Chicago, IL 60606   (312) 482-3800

Stephen J. Manich
Wood, Phillips, Mason,
Recktenwald & Van Santen
Northwestern Atrium Center
500 W. Madison Street, Ste. 3800
Chicago, IL 60606   (312) 876-1800

Basil P. Mann
Marshall, O'Toole, Gerstein,
Murray & Bicknell
20 South Clark Street
Chicago, IL 60603   (312) 346-5750

John Mcgregor Mann, Sr.
Lee, Mann, Smith, McWilliams
& Sweeney
105 W. Adams Street
Suite 300
Chicago, IL 60603   (312) 368-1300

Edward Manzo
Cook, Egan, Mc Farron, & Manzo
135 S. La Salle Street
Suite 4100
Chicago, IL 60603   (312) 236-8500

Joseph Robert Marcus
Welsh & Katz, Ltd.
135 S. La Salle Street
Suite 1625
Chicago, IL 60603   (312) 781-9470

David J. Marr
Trexler, Bushnell, Giangiorgi
& Blackstone, Ltd.
105 W. Adams Street
36th Floor
Chicago, IL 60603   (312) 704-1890

William A. Marvin
Welsh & Kratz, Ltd.
135 S. La Salle Street
Suite 1625
Chicago, IL 60603   (312) 781-9470

Lloyd W. Mason
Wood, Dalton, Phillips,
Mason & Rowe
500 W. Madison Street
Suite 3800
Chicago, IL 60606   (312) 876-1800

Mart C. Matthews
Quaker Oats Company
Law Dept., 25-6
P.O. Box 9001
Chicago, IL 60604   (312) 222-7574

Brian M. Mattson#
Hill, Van Santen, Steadman
& Simpson
Sears Tower
70th Floor
Chicago, IL 60606   (312) 876-0200

Paul D. Matukaitis
G.D. Searle & Company
P.O. Box 5110
Chicago, IL 60680   (708) 470-6300

Phillip H. Mayer
Leydig, Voit & Mayer, Ltd.
One IBM Plaza
Suite 4600
Chicago, IL 60611   (312) 822-9666

Michael P. Mazza
Niro, Scavone, Haller
& Niro, Ltd.
181 W. Madison St.
Suite 4600
Chicago, IL 60602   (312) 236-0733

George P. Mc Andrews
Mc Andrews, Held & Malloy, Ltd.
Northwestern Atrium Center
500 W. Madison Street
Suite 3400
Chicago, IL 60661   (312) 707-8889

William E. Mc Cracken
Marshall, O'Toole, Gerstein,
Murray & Bicknell
20 South Clark Street
Suite 2100
Chicago, IL 60603   (312) 346-5750

Scott P. Mc Donald
Jenner & Block
One IBM Plaza
Chicago, IL 60611   (312) 222-9350

John J. Mc Donnell
Allegretti & Witcoff, Ltd.
10 South Wacker Drive
Chicago, IL 60606   (312) 715-1000

Thomas C. Mc Donough
Vedder, Price, Kaufman
& Kammholz
222 N. La Salle Street
Suite 2600
Chicago, IL 60601   (312) 609-7853

Daniel C. Mc Eachran
Kinzer, Plyer, Dorn,
Mc Eachran & Jambor
55 E. Monroe St.
Chicago, IL 60603   (312) 726-4421

Gary W. Mc Farron
Cook, Egan, Mc Farron & Manzo
135 S. La Salle St.
Suite 4100
Chicago, IL 60603   (312) 236-8500

Frank J. Mc Gue
Olson & Hierl
20 North Wacker Drive
Suite 3000
Chicago, IL 60606   (312) 580-1180

Kirk M. Mc Inerney
Helene Curtis, Inc.
325 N. Wells Street
Chicago, IL 60610   (312) 661-0222

Michael R. Mc Kenna
500 West Maidson St.
Suite 3800
Chicago, IL 60661   (312) 321-0123

Janice M. Mc Lain
Amoco Corporation
200 E. Randolph Drive
P.O. Box 87703
Chicago, IL 60615   (312) 856-4138

F. William Mc Laughlin
Wood, Dalton, Phillips,
Mason & Rowe
500 W. Madison Street
Suite 3800
Chicago, IL 60606   (312) 876-1800

Terrence W. Mc Millin
Gerstmann & Ellis, Ltd.
Two North La Salle Street
Suite 2010
Chicago, IL 60602   (312) 263-4350

Michael B. Mc Murray
Jenner & Block
One IBM Plaza
Chicago, IL 60611   (312) 222-9350

Janet M. Mc Nicholas
Allegretti & Witcoff, Ltd.
10 S. Wacker Drive
Suite 3000
Chicago, IL 60606   (312) 715-1000

Dennis Michael Mc Williams
Lee, Mann, Smith, Mc Williams
& Sweeney
105 W. Adams Street
Suite 300
Chicago, IL 60603   (312) 368-1300

Ralph C. Medhurst
Amoco Corporation
200 East Randolph Drive
P.O. Box 87703
Mail Code 1905A
Chicago, IL 60601   (312) 856-5954

Richard B. Megley
FMC Corp.
200 E. Randolph Drive
Chicago, IL 60601   (312) 861-6650

Raymond M. Mehler
Lockwood, Alex, Fitz Gibbon
& Cummings
Three First Natl. Plaza
Suite 1700
Chicago, IL 60602   (312) 782-4860

Maureen C. Meinert
Marshall, O'Toole, Gerstein,
Murray & Bicknell
Two First Natl. Plaza
20 S. Clark St., Suite 2100
Chicago, IL 60603   (312) 346-5750

Irving H. Melnick
Gerlach & O'Brien
29 South La Salle Street
Chicago, IL 60603   (312) 332-6930

Alejandro Menchaca
Mc Andrews, Held & Malloy, Ltd.
Northwestern Atrium Center
500 West Madison St.
Suite 3400
Chicago, IL 60661   (312) 707-8889

David R. Metzger
Hill, Van Santen, Steadman
& Simpson
70th Floor - Sears Tower
Chicago, IL 60606   (312) 876-0200

Gerson E. Meyers
Dressler, Goldsmith, Shore,
Sutker & Milnamow, Ltd.
Two Prudential Plaza
180 N. Stetson Ave., Ste. 4700
Chicago, IL 60601   (312) 615-5400

Stephen G. Mican
Neuman, Williams, Anderson
& Olson
77 West Washington Street
Chicago, IL 60602   (312) 346-1200

Reuben Miller
9027 S. Luella Ave.
P.O. Box 17-558
Chicago, IL 60617   (312) 731-5049

John P. Milnamow
Dressler, Goldsmith, Shore,
Sutker & Milnamow, Ltd.
Two Prudential Plaza,
180 N. Stetson Ave., Ste. 4700
Chicago, IL 60601   (312) 616-5400

Michael E. Milz
William, Brinks, Olds, Hofer,
Gilson & Lione, P.C.
NBC Tower, Suite 3600
455 N. Cityfront Plaza Dr.
Chicago, IL 60611   (312) 321-4200

Dean A. Monco
Cook, Egan, Mc Farron
& Manzo, Ltd.
Suite 4100
135 S. La Salle St.
Chicago, IL 60603   (312) 236-8500

Donald D. Mondul
Foley & Lardner
3 First Natl. Plaza
Suite 4950
70 W. Madison Street
Chicago, IL 60602   (312) 444-9500

Marvin Moody
Hill, Van Santen, Steadman,
Chiara & Simpson
70th Flr., Sears Tower
233 S. Wacker Dr.
Chicago, IL 60606   (312) 876-0200

Carl E. Moore, Jr.
Marshall, O'Toole, Gerstein,
Murray & Bicknell
Two First National Plaza
Suite 2100
Chicago, IL 60603   (312) 346-5750

Lynn B. Morreale
Neuman, Williams, Anderson
& Olson
77 West Washington St.
Suite 2000
Chicago, IL 60602   (312) 346-1200

William Joseph Morris
Morris & Stella
200 W. Adams Street
Suite 1200
Chicago, IL 60606   (312) 782-2345

John S. Mortimer
Wood, Dalton, Phillips,
Mason & Rowe
500 W. Madison Street
Suite 3800
Chicago, IL 60606   (312) 876-1800

Charles H. Mottier
Leydig, Voit & Mayer, Ltd.
180 N. Stetson Ave.
Two Prudential Plaza
Suite 4900
Chicago, IL 60601   (312) 616-5600

Russell N. Muehleman
Hill, Van Santen, Steadman
& Simpson
70th Floor, Sears Tower
Chicago, IL 60606   (312) 876-0200

Wesley O. Mueller
Leydig, Voit & Mayer, Ltd.
Two Prudential Plaza
Suite 4900
Chicago, IL 60601   (312) 616-5600

Kurt Mullerheim
Keil & Witherspoon
135 S. La Salle Street
Suite 1216
Chicago, IL 60068   (312) 236-5030

Mark J. Murphy
Cook, Egan, Mc Farron &
Manzo
135 S. La Salle Street
Suite 4100
Chicago, IL 60603   (312) 236-8500

Owen Joseph Murray
Marshall, O'Toole, Gerstein,
Murray & Bicknell
2 First Natl. Plaza
Chicago, IL 60603   (312) 346-5750

Robert L. Murray#
7035 N. Ionia
Chicago, IL 60646   (312) 774-0410

William E. Murray
Amoco Corporation
200 E. Randolph Drive
P.O. Box 87703
Chicago, IL 60680   (312) 856-5990

James B. Muskal
Leydig, Voit & Mayer, Ltd.
180 N. Stetson Ave.
Two Prudential Plaza
Suite 4900
Chicago, IL 60601   (312) 616-5600

Richard John Myers
Myers & Ehrlich, Ltd.
53 W. Jackson Blvd.
Chicago, IL 60604   (312) 294-0350

James J. Myrick
Fitch, Even, Tabin
& Flannery
135 S. La Salle Street
Chicago, IL 60603   (312) 372-7842

Wayne E. Nacker
Morton Thiokol, Inc.
110 N. Wacker
Chicago, IL 60606   (312) 807-2146

James J. Napoli
Marshall, O'Toole, Gerstein,
Murray & Bicknell
Suite 2100, Two First Natl. Plaza
20 S. Clark Street
Chicago, IL 60603   (312) 346-5750

James P. Naughton
Roper & Quigg
200 S. Michigan Ave.
Suite 1000
Chicago, IL 60604   (312) 408-0855

Steven T. Naumann
Commonwealth Edison Co.
Engineering Staff-35 FNE
P.O. Box 767
Chicago, IL 60690   (312) 294-4367

Jon O. Nelson
Allegretti & Witcoff, Ltd.
10 S. Wacker Drive
Suie 3000
Chicago, IL 60606   (312) 715-1000

Thomas E. Nemo
Amoco Corporation
200 E. Randolph Drive
Chicago, IL 60680   (312) 856-4972

Sidney Neuman
Neuman, Williams, Anderson
& Olson
77 W. Washington St.
Chicago, IL 60602   (312) 346-1200

George B. Newitt
Allegretti & Witcoff, Ltd.
10 South Wacker Drive
Chicago, IL 60606   (312) 715-1000

Gerald M. Newman
Schoenberg, Fisher & Newman, Ltd.
222 S. Riverside Plaza
Suite 2700
Chicago, IL 60606   (312) 648-2300

Frank C. Nicholas
William, Brinks, Olds, Hoffer,
Gilson & Lione,
NBC Tower, 455 N. Cityfront Plaza Dr.
Suite 3600
Chicago, IL 60611   (312) 321-4279

G. Peter Nichols
William, Brinks, Olds, Hofer,
Gilson & Lione
455 N. Cityfront Plaza Dr.
Suite 3600
Chicago, IL 60611   (312) 321-4200

Mary Nicolaides
233 East Erie Street
Suite 1804
Chicago, IL 60611   (312) 337-1835

Keith K. Nicolls
Mc Caleb, Lucas & Brugman
230 W. Monroe St.
Chicago, IL 60606   (312) 236-4711

J. Richard Nighswander
Collins & Collins
332 S. Michigan Ave., #958
Chicago, IL 60604   (312) 663-4200

Anthony Nimmo
Marshall, O Toole, Gerstein,
Murray & Bicknell
Two First National Plaza
Chicago, IL 60603   (312) 346-5750

Raymond N. Nimrod
Roper & Quigg
200 S. Michigan Ave.
Suite 1000
Chicago, IL 60604   (312) 408-0855

Raymond P. Niro
Niro, Scavone, Haller,
& Niro
181 W. Madison
Suite 4600
Chicago, IL 60602   (312) 236-0733

Steven H. Noll
Hill, Van Santen, Steadman,
Chiara & Simpson, P.C.
70th Floor
Sears Tower
Chicago, IL 60606   (312) 876-0200

Joan I. Norek
Suite 1601
180 N. La Salle
Chicago, IL 60601   (312) 606-0000

Thomas E. Northrup
Dressler, Goldsmith, Shore,
Sutker & Milnamow, Ltd.
Two Prudential Plaza
180 N. Stetson Ave., Ste. 4700
Chicago, IL 60601   (312) 616-4500

Geoffrey M. Novelli
Trexler, Bushnell, Giangiorgi,
& Blackstone, Ltd.
105 W. Adams
36th Floor
Chicago, IL 60603   (312) 704-1890

John S. O Brien
Gerlach & O'Brien
29 S. La Salle Street
Suite 635
Chicago, IL 60603   (312) 332-6930

Edward M. O Toole
Marshall, O'Toole, Gerstein,
Murray & Bicknell
Two First National Plaza
Suite 2100
Chicago, IL 60603   (312) 346-5750

467

William P. Oberhardt
Roper & Quigg
200 S. Michigan Ave.
Suite 1000
Chicago, IL 60604   (312) 408-0855

Helen A. Odar
Kirkland & Ellis
200 East Randolph Drive
Chicago, IL 60601   (312) 861-2362

Thomas J. Odar
Jenner & Block
One IBM Plaza
Chicago, IL 60611   (312) 222-9350

Paul M. Odell
Dressler, Goldsmith, Shore,
Sutker & Milnamow, Ltd.
Two Prudential Plaza
18o N. Stetson Ave., Ste. 4700
Chicago, IL 60601   (312) 616-5400

Glenn W. Ohlson
Lee, Mann, Smith, Mc Williams
& Sweeney
105 W. Adams Street
Suite 300
Chicago, IL 60603   (312) 368-1300

Wallace L. Oliver, Jr.
Amoco Corporation
200 E. Randolph Drive
Chicago, IL 60680   (312) 856-5543

A. Andrew Olson, III
Paul R. Ray & Co., Inc.
200 S. Wacker Drive
Suite 3820
Chicago, IL 60606   (312) 876-0730

Arne M. Olson
Olson & Hierl
20 N. Wacker Drive
Suite 3000
Chicago, IL 60606   (312) 580-1180

Arthur Andrew Olson, Jr.
Leydig, Voit & Mayer
Two Prudential Plaza
Suite 4900
Chicago, IL 60601   (312) 616-5600

Keith H. Orum
Balogh, Osann, Kramer, Dvorak,
Genova & Traub
53 W. Jackson Blvd.
Suite 1616
Chicago, IL 60604   (312) 922-6262

Edward W. Osann, Jr.
29 S. La Salle Street
Suite 420
Chicago, IL 60603   (312) 782-5937

Charles S. Oslakovic
Leydig, Voit & Mayer
180 N. Stetson Ave.
Two Prudential Plaza
Suite 4900
Chicago, IL 60601   (312) 616-5600

John S. Pacocha
Dick & Harris
181 W. Madison St.
Suite 3800
Chicago, IL 60602   (312) 726-4000

Francois Newell Palmatier
Burmeister, York, Palmatier
& Zummer
135 South La Salle St.
Suite 1046
Chicago, IL 60603   (312) 782-6663

John S. Paniaguas
Mason, Kolehmainen, Rathburn
& Wyss
20 North Wacker Drive
Suite 4200
Chicago, IL 60606   (312) 621-1300

Joseph H. Paquin, Jr.
Mc Dermott, Will & Emery
227 West Monroe Street
Chicago, IL 60606   (312) 372-2000

J. Kevin Parker
Jones, Day, Reavis & Pogue
225 W. Washington St.
Chicago, IL 60606   (312) 269-4339

John L. Parker
39 S. La Salle Street
Chicago, IL 60603   (312) 263-6560

Todd S. Parkhurst
Schiff Hardin & Waite
7200 Sears Tower
Chicago, IL 60606   (312) 876-1000

Raymond E. Parks#
Fmc Corp.
200 E. Randolph Drive
Chicago, IL 60601   (312) 861-6652

Stanley M. Parmerter
Rockey & Rifkin
30 N. La Salle Street
Chicago, IL 60602   (312) 704-5600

Keith D. Parr
Lord, Bissell & Brook
115 S. La Salle Street
Chicago, IL 60603   (312) 443-0497

Daniel R. Pastirik
Lockwood, Alex, Fitz Gibbon
& Cummings
Three First Natl. Plaza
Suite 1700
Chicago, IL 60602   (312) 782-4860

Timothy T. Patula
Welsh & Katz, Ltd.
Suite 1625
135 South La Salle St.
Chicago, IL 60603   (312) 781-9470

Thomas D. Paulius
Lockwood, Alex, Fitz Gibbon
& Cummings
Three First Natl. Plaza
Chicago, IL 60602   (312) 782-4860

John J. Pavlak
Willian Brinks Olds Hofer
Gilson & Lione Ltd.
NBC Tower
455 N. Cityfront Plaza Dr.
Chicago, IL 60611   (312) 321-4215

Joan Pennington
Mason, Kolehmainen, Rathburn
& Wyss
20 N. Wacker Drive
Suite 4200
Chicago, IL 60606   (312) 621-1300

R. Jonathan Peters
333 West Wacker Drive
Suite 2500
Chicago, IL 60606   (312) 427-9626

Steven P. Petersen
Leydig, Voit & Mayer, Ltd.
Two Prudential Plaza
Suite 4900
Chicago, IL 60601   (312) 616-5600

Donald A. Peterson
Neuman, Williams, Anderson
& Olson
77 W. Washington Street
Suite 2000
Chicago, IL 60602   (312) 346-1200

Philip C. Peterson
Mason, Kolehmainen, Rathburn
& Wyss
Suite 4200
20 N. Wacker Drive
Chicago, IL 60606   (312) 621-1300

Thomas F. Peterson
Ladas & Parry
224 S. Michigan Avenue
Suite 1200
Chicago, IL 60604   (312) 427-1300

Maxwell J. Petesen
William, Brinks, Olds, Hofer,
Gilson & Lione, Ltd.
NBC Tower
455 N. Cityfront Plaza Dr., Ste. 3600
Chicago, IL 60611   (312) 321-4200

Ronald C. Petri
Amoco Corp.
200 E. Randolph Dr.
Chicago, IL 60601   (312) 856-4426

Philip T. Petti
Neuman, Williams, Anderson
& Olson
77 West Washington Street
Chicago, IL 60602   (312) 346-0850

Mark E. Phelps
Leydig, Voit & Mayer, Ltd.
180 N. Stetson Ave.
Two Prudential Plaza
Suite 4900
Chicago, IL 60601   (312) 616-5600

Richard S. Phillips
Wood, Phillips, Mason,
Recktenwald & Van Santen
500 W. Madison Street
Northwestern Atrium Ctr., Ste. 3800
Chicago, IL 60661   (312) 876-1800

Charles Francis Pigott, Jr.
Allegretti & Witcoff, Ltd.
10 South Wacker Drive
Suite 3000
Chicago, IL 60606   (312) 715-1000

R. Steven Pinkstaff#
Fitch, Even, Tabin
& Flannery
135 S. La Salle Street
Suite 900
Chicago, IL 60603   (312) 372-7842

Daniel C. Pinkus
3500 N. Lake Shore Drive
Chicago, IL 60657   (312) 528-5947

Michael Piontek
Juettner, Pyle, Lloyd
& Verbeck
221 N. La Salle St
Suite 850
Chicago, IL 60602   (312) 236-8123

Alfred H. Plyer, Jr.
Kinzer, Plyer, Dorn
Mc Eachran & Jambor
55 E. Monroe St.
Chicago, IL 60603   (312) 726-4421

Donald J. Pochopien
Mc Andrews, Held & Malloy, Ltd.
Northwestern Atrium Center
500 W. Madison St.
Suite 3400
Chicago, IL 60661   (312) 707-8889

Robert B. Polit#
Mc Andrews, Held & Malloy, Ltd.
Northwestern Atrium Center
500 West Madison St.
Suite 3400
Chicago, IL 60661   (312) 707-8889

William Patrick Porcelli
9920 S. Seeley Ave.
Chicago, IL 60643   (312) 239-5990

James W. Potthast
Potthast & Ring
Harbor House, Suite 100
3200 N. Lake Shore Drive
Chicago, IL 60657   (312) 929-9727

Nicholas A. Poulos
Roper & Quigg
200 S. Michigan Ave.
Suite 1000
Chicago, IL 60604   (312) 408-0855

William F. Prendergast
Willian Brinks Olds Hofer
Gilson & Lione
NBC Tower, Ste. 3600
455 N. Cityfront Plaza Dr.
Chicago, IL 60611   (312) 321-4200

Phyllis Y. Price
Kirkland & Ellis
200 East Randolph Dr.
Suite 6100
Chicago, IL 60601   (312) 861-2000

Russell Weston Pyle
Juettner Pyle Lloyd
& Verbeck
221 North La Salle Street
Suite 850
Chicago, IL 60601   (312) 236-8123

Christopher J. Reckamp#
1764 N. Clark St., Apt. #2
Chicago, IL 60614   (312) 236-8123

William E. Recktenwald
Wood, Phillips, Mason,
Recktenwald & Van Santen
Northwestern Atrium Ctr., Ste. 3800
500 West Madison St.
Chicago, IL 60661   (312) 876-1800

Edward W. Remus
Allegretti & Witcoff
10 South Wacker Drive
30th Floor
Chicago, IL 60606   (312) 715-1000

Christopher J. Renk
Allegretti & Witcoff
10 S. Wacker Drive
Chicago, IL 60606   (312) 715-1000

Robert H. Resis
Allegretti & Witcoff, Ltd.
10 South Wacker Drive
Suite 3000
Chicago, IL 60606   (312) 715-1000

Jonathan E. Retsky
William, Brinks, Olds, Hofer,
Gilson & Lione
NBC Tower, Ste. 3600
455 North Cityfront Plaza Dr.
Chicago, IL 60611   (312) 321-4200

Donald P. Reynolds
Mc Andrews, Held & Malloy, Ltd.
Northwestern Atrium Center
500 West Madison St.
Suite 3400
Chicago, IL 60661   (312) 707-8889

Jerry A. Riedinger
Allegretti & Witcoff, Ltd.
10 South Wacker Drive
Chicago, IL 60606   (312) 715-1000

Daniel M. Riess
Lockwood, Alex, Fitz Gibbon
& Cummings
Three First Natl. Plaza
Suite 1700
Chicago, IL 60602   (312) 782-4860

William T. Rifkin
Rockey & Rifkin
30 N. La Salle Street
Suite 2700
Chicago, IL 60602   (312) 704-5600

Reed F. Riley
Amoco Corp.
200 E. Randolph Dr.
Mail Code 1904
Chicago, IL 60601   (312) 856-5947

Thomas J. Ring
Potthast & Ring
3200 N. Lake Shore Drive
Suite 100
Chicago, IL 60657   (312) 929-9727

Ralph F. Risse#
Hill, Van Santen, Steadman,
& Simpson
233 S. Wacker Drive
Suite 7000
Chicago, IL 60606   (312) 876-0200

W. William Ritt, Jr.#
FMC Corp.
200 E. Randolph Drive
Chicago, IL 60601   (312) 861-6657

Richard A. Robbins
1482-H W. Summerdale Avenue
Chicago, IL 60640   (312) 878-0605

Charles L. Roberts
William, Brinks, Olds, Hofer,
Gilson & Lione
NBC Tower
455 N. Cityfront Plaza Dr., Ste. 3600
Chicago, IL 60611   (312) 321-4254

Melvin A. Robinson
Hill, Van Santen, Steadman
& Simpson
70th Floor - Sears Tower
Chicago, IL 60606   (312) 876-0200

Robert H. Robinson
USQ Corporation
101 S. Wacker Dr.
Dept. 157
Chicago, IL 60606   (312) 606-5802

David I. Roche
Baker & Mc Kenzie
One Prudential Plaza
130 E. Randolph Dr.
Chicago, IL 60601   (312) 861-8608

Keith Von Rockey
Rockey & Rifkin
30 N. La Salle Street
Suite 2700
Chicago, IL 60602   (312) 704-5600

Howard B. Rockman
Laff, Whitesel, Conte
& Saret
401 N. Michigan Avenue
Suite 2000
Chicago, IL 60611   (312) 661-2100

Donna M. Rogers
William, Brinks, Olds,
Hofer, Gilson & Lione
455 N. Cityfront Plaza Dr.
NBC Tower, Suite 3600
Chicago, IL 60611   (312) 321-4200

Harry J. Roper
Roper & Quigg
200 S. Michigan Ave.
Suite 1000
Chicago, IL 60604   (312) 408-0855

Gary M. Ropski
William, Brinks, Olds, Hofer,
Gilson & Lione Ltd.
NBC Tower - Suite 3600
455 N. Cityfront Plaza Dr.
Chicago, IL 60611   (312) 321-4200

John E. Rosenquist
Leydig, Voit & Mayer, Ltd.
180 N. Stetson Ave.
Two Prudential Plaza
Suite 4900
Chicago, IL 60601   (312) 616-5600

Thomas I. Ross
Hill, Van Santen, Steadman,
& Simpson
70th Floor, Sears Tower
233 S. Wacker Dr.
Chicago, IL 60606   (312) 876-0200

Seymour Rothstein
Allegretti & Witcoff, Ltd.
10 South Wacker Drive
Chicago, IL 60606   (312) 715-1000

Donnie Rudd
Schain, Firsel & Burney, Ltd.
222 N. La Salle
Suite 1910
Chicago, IL 60601   (312) 332-0200

Stephen G. Rudisill
Arnold, White & Durkee
Quaker Tower, Suite 800
321 N. Clark Street
Chicago, IL 60610   (312) 744-0090

Douglas W. Rudy
FMC Corporation
200 East Randolph Drive
Chicago, IL 60601   (312) 861-6654

Deborah S. Ruff
Keck, Mahin & Cate
8300 Sears Tower
233 S. Wacker Drive
Chicago, IL 60606   (312) 876-3400

Charles W. Rummler
Tilton, Fallon Lungmus
& Chestnut
100 S. Wacker Drive
Hartford Plaza, Suite 960
Chicago, IL 60606   (312) 236-3418

Donald W. Rupert
Roper & Quigg
200 S. Michigan Ave.
Suite 1000
Chicago, IL 60604   (312) 408-0855

Pamela J. Ruschau
Leydig, Voit & Mayer
Two Prudential Plaza
Suite 4900
Chicago, IL 60601   (312) 616-5600

Charles W. Ryan
Clement & Ryan
Suite 2150
150 N. Michigan Avenue
Chicago, IL 60601   (312) 781-2070

Robert C. Ryan
Mc Andrews, Held & Malloy, Ltd.
Northwestern Atrium Center
500 W. Madison Street
Suite 3400
Chicago, IL 60661   (312) 707-8889

William A. Ryan
Continental Materials Corp.
325 N. Wells Street
Chicago, IL 60610   (312) 661-7215

James D. Ryndak
Jenner & Block
One IBM Plaza
Chicago, IL 60611   (312) 222-9350

James P. Ryther
Jones, Day, Reavis
& Pogue
225 West Washington Street
Chicago, IL 60606   (312) 269-4102

Alan B. Samlan
Fox & Grove
Sears Tower, Suite 7818
233 S. Wacker Drive
Chicago, IL 60606   (312) 876-0500

Kenneth H. Samples
Fitch, Even, Tabin & Flannery
135 S. La Salle St.
Suite 900
Chicago, IL 60603   (312) 372-7842

Ronald A. Sandler
Jones, Day, Reavis
& Pogue
225 W. Washington Street
Chicago, IL 60606   (312) 782-3939

Larry L. Saret
Laff, Whitesel, Conte
& Saret
401 N. Michigan Avenue
Suite 2000
Chicago, IL 60611   (312) 661-2100

Steven J. Sarussi#
Allegretti & Witcoff
10 S. Wacker Street
Suite 3000
Chicago, IL 60606   (312) 715-1000

Lester J. Savit
Jones, Day, Reavis & Pogue
Suite 2600
225 West Washington
Chicago, IL 60606   (312) 782-3939

Nate Frank Scarpelli
Marshall, O'Toole, Gerstein,
Murray & Bicknell
Two First Natl. Plaza
20 S. Clark St., Ste. 2100
Chicago, IL 60603   (312) 346-5750

Thomas G. Scavone
Niro, Scavone, Haller
& Niro
181 W. Madison
Suite 4600
Chicago, IL 60602   (312) 236-0733

Julian Schachner
5733 N. Sheridan Rd.
Suite 10-C
Chicago, IL 60660   (312) 878-8619

James A. Scheer
Welsh & Katz
135 S. La Salle St.
Suite 1625
Chicago, IL 60603   (312) 781-9470

Dennis R. Schlemmer
Leydig, Voit & Mayer, Ltd.
180 North Stetson Ave.
Two Prudential Plaza
Suite 4900
Chicago, IL 60601   (312) 616-5600

James S. Schlifke
25 E. Washington St.
Suite 600
Chicago, IL 60602   (312) 372-7762

Richard J. Schlott
Amoco Corporation
200 E. Randolph Drive
Mail Code 1904
Chicago, IL 60680   (312) 856-5071

Jerold B. Schnayer
Welsh & Katz, Ltd.
135 S. La Salle Street
Suite 1625
Chicago, IL 60603   (312) 781-9470

Homer J. Schneider
Leydig, Voit & Mayer, Ltd.
180 N. Stetson Ave.
Two Prudential Plaza
Suite 4900
Chicago, IL 60601   (312) 616-5600

Robert J. Schneider
Mc Dermott, Will & Emery
227 W. Monroe St.
Chicago, IL 60606   (312) 372-2000

Richard A. Schnurr
Marshall, O'Toole, Gerstein,
Murray & Bicknell
Two First National Plaza
Suite 2100
Chicago, IL 60603   (312) 346-5750

Ekkehard Schoettle
Amoco Corp.
200 E. Randolph Drive
P.O. Box 87703
Chicago, IL 60601   (312) 856-5637

Donald E. Schreiber
Niro, Scavone, Haller & Niro
181 W. Madison - Suite 4600
Chicago, IL 60602   (312) 236-0733

Jerry A. Schulman
Seyfarth, Shaw, Fairweather
& Geraldson
55 E. Monroe Street
Chicago, IL 60603 (312) 346-8000

Robert K. Schumacher
Fitch, Even, Tabin
& Flannery
135 S. Lasalle St.
Suite 900
Chicago, IL 60603 (312) 372-7842

Gerald S. Schur
Welsh & Katz, Ltd.
135 S. La Salle Street
Suite 1625
Chicago, IL 60603 (312) 781-9470

Richard J. Schwarz
Hill, Van Santen, Steadman
& Simpson
70th Floor
Sears Tower
Chicago, IL 60606 (312) 876-0200

Robert J. Schwarz
Lockwood, Alex, Fitzgibbon
& Cummings
Three First Natl. Plaza
Suite 1700
Chicago, IL 60602 (312) 782-4860

Theodore R. Scott
Jones, Day, Reavis & Pogue
225 W. Washington Street
Suite 2600
Chicago, IL 60606 (312) 269-4103

Max Shaftal
Dick and Harris
181 W. Madison St.
Suite 3800
Chicago, IL 60602 (312) 726-4000

Steven M. Shape
Niro, Scavone, Haller,
& Niro
181 W. Madison
Suite 4600
Chicago, IL 60602 (312) 236-0733

Jeffrey S. Sharp
Marshall, O'Toole, Gerstein,
Murray & Bicknall
Two First National Plaza
Suite 2100
Chicago, IL 60603 (312) 346-5750

Gerald T. Shekleton
Welsh & Katz, Ltd.
135 S. La Salle Street
Suite 1625
Chicago, IL 60603 (312) 781-9470

Berton Scott Sheppard
Leydig, Voit & Mayer, Ltd.
180 N. Stetson Ave.
Two Prudential Plaza
Suite 4900
Chicago, IL 60601 (312) 616-5600

Stephen F. Sherry
Allegretti & Witcoff, Ltd.
10 South Wacker Drive
Suite 3000
Chicago, IL 60606 (312) 715-1000

Charles W. Shifley
Allegretti & Witcoff, Ltd.
10 South Wacker Drive
Suite 3000
Chicago, IL 60606 (312) 715-1000

Joseph E. Shipley
Fitch, Even, Tabin
& Flannery
135 S. La Salle Street
Suite 900
Chicago, IL 60603 (312) 372-7842

Jack Shore
Dressler, Goldsmith, Shore,
Sutker & Milnamow, Ltd.
Two Prudential Plaza
180 N. Stetson Ave., Ste. 4700
Chicago, IL 60601 (312) 616-5400

Tony T. Shu
Lee & Shu
208 S. La Salle Street
Suite 1400
Chicago, IL 60604 (312) 641-3303

Alvin D. Shulman
Marshall, O'Toole, Gerstein,
Murray & Bicknell
Two First National Plaza
20 S. Clark St., Ste. 2100
Chicago, IL 60603 (312) 346-5750

John H. Shurtleff
140 S. Dearborn St.
Suite 411
Chicago, IL 60603 (312) 236-5032

Steven P. Shurtz
William, Brinks, Olds, Hofer,
Gilson & Lione Ltd.
NBC Tower, Suite 3600
455 N. Cityfront Plaza Dr.
Chicago, IL 60611 (312) 321-4230

Douglas H. Siegel
Marshall, O Toole, Gerstein,
Murray & Bicknell
20 S. Clark St.
Chicago, IL 60603 (312) 346-5750

Joel E. Siegel
Dressler, Goldsmith, Shore,
Sutker & Milnamow, Ltd.
Two Prudential Plaza
180 N. Stetson Ave., Ste. 4700
Chicago, IL 60601 (312) 616-5400

Gustavo Siller, Jr.*
William, Brinks, Olds, Hofer,
Gilson & Lione Ltd.
455 N. Cityfront Plaza Dr.
NBC Bldg., 36th Floor
Chicago, IL 60611 (312) 321-4200

Howard E. Silverman
Dick & Harris
181 W. Madison St.
Suite 3800
Chicago, IL 60602 (312) 726-4000

John D. Simpson
Hill, Van Santen, Steadman,
& Simpson
70th Flr., Sears Tower
233 S. Wacker Dr.
Chicago, IL 60606 (312) 876-0200

Herbert Jay Singer
Silverman, Cass & Singer, Ltd.
105 W. Adams St.
Suite 2700
Chicago, IL 60603 (312) 726-6006

Robert E. Sloat
Amoco Corporation
200 E. Randolph Drive
P.O. Box 87703
Chicago, IL 60680 (312) 856-5275

Jamie S. Smith
Allegretti & Witcoff, Ltd.
10 South Wacker Drive
Chicago, IL 60606 (312) 715-1000

Jeffry W. Smith
Marshall, O'Toole, Gerstein,
Murray & Bicknell
Two First Nat'L Plaza
Suite 2100
Chicago, IL 60603 (312) 346-5750

Noel Irving Smith
Leydig, Voit & Mayer, Ltd.
Two Prudential Plaza
Suite 4900
Chicago, IL 60601 (312) 616-5600

Thomas Eugene Smith
Lee, Mann, Smith, Mc Williams
& Sweeney
105 W. Adams Street
Suite 300
Chicago, IL 60603 (312) 368-1300

James R. Sobieraj
William, Brinks, Olds, Hofer,
Gilson & Lione, Ltd.
NBC Tower
455 N. Cityfront Plaza Dr.
Chicago, IL 60611 (312) 321-4200

Julius L. Solomon#
2400 Lakeview
Suite 1508
Chicago, IL 60614 (312) 525-0425

Steven J. Soucar
Wood, Phillips, Mason,
Recktenwald & Van Santen
Northwestern Atrium Center
500 W. Madison St., Ste. 3800
Chicago, IL 60606 (312) 876-1800

George S. Spindler
Amoco Corporation
200 E. Randolph Drive
Chicago, IL 60601 (312) 856-5420

James Alexander Sprowl
Fitch, Even, Tabin
& Flannery
135 S. La Salle Street
Suite 900
Chicago, IL 60603   (312) 372-7842

Frank J. Sroka
Amoco Corp.
200 East Randolph Dr.
P.O. Box 87703
Chicago, IL 60680   (312) 856-5939

James G. Staples
Baker & Mc Kenzie
One Prudential Plaza
130 E. Randolph Dr.
Chicago, IL 60601   (312) 861-2766

Lewis T. Steadman
Hill, Van Santen, Steadman,
& Simpson
70th Fl., Sears Tower
233 S. Wacker Dr.
Chicago, IL 60606   (312) 876-0200

Edward J. Steeve#
7122 N. Odell
Chicago, IL 60631   (312) 631-1781

Roger H. Stein
Wallenstein, Wagner & Hattis, Ltd.
311 S. Wacker Drive
Suite 5300
Chicago, IL 60606   (312) 554-3300

Martin L. Stern
Laff, Whitesel, Conte
& Saret
401 N. Michigan Avenue
Suite 2000
Chicago, IL 60611   (312) 661-2100

Allan J. Sternstein
William, Brinks, Olds, Hofer,
Gilson & Lione Ltd.
NBC Tower, Ste. 3600
455 N. Cityfront Plaza Dr.
Chicago, IL 60611   (312) 321-4200

Robert W. Stevenson
William, Brinks, Olds, Hofer,
Gilson & Lione
NBC Tower, Suite 3600
455 North Cityfront Plaza Drive
Chicago, IL 60611   (312) 321-4200

Thomas K. Stine
Wallenstein, Wagner
& Hattis, Ltd.
311 S. Wacker Dr.
Suite 5300
Chicago, IL 60606   (312) 554-3300

Harold Victor Stotland
Emrich & Dithmar
Suite 3000
150 N. Wacker Drive
Chicago, IL 60606   (312) 372-2552

J. Terry Stratman
Emrich & Dithmar
Suite 3000
150 N. Wacker Drive
Chicago, IL 60606   (312) 372-2552

William A. Streff, Jr.
Kirkland & Ellis
200 E. Randolph Drive
59th Floor
Chicago, IL 60601   (312) 861-2126

Richard J. Streit
Ladas & Parry
224 S. Michigan Avenue
Suite 1200
Chicago, IL 60604   (312) 427-1300

William Creighton Stueber
Hill, Van Santen, Steadman,
Chiara & Simpson
70th Fl., Sears Tower
233 S. Wacker Dr.
Chicago, IL 60606   (312) 876-0200

Rae K. Stuhlmacher
Amoco Corp.
200 E. Randolph
P.O. Box 87703
Chicago, IL 60680   (312) 856-4763

Barry W. Sufrin
Laff, Whitesel, Conte
& Saret
401 N. Michigan Ave.
Suite 2000
Chicago, IL 60611   (312) 661-2100

Dennis K. Sullivan
Navistar International
Transportation Corp.
455 Cityfront Plaza Dr.
Law Offices - 13
Chicago, IL 60611   (312) 836-2311

Marshall W. Sutker
Dressler, Goldsmith, Shore,
Sutker & Milnamow, Ltd.
Two Prudential Plaza
180 N. Stetson Ave., Ste. 4700
Chicago, IL 60601   (312) 616-5400

James R. Sweeney
Lee, Mann, Smith, Mc Williams
& Sweeney
105 W. Adams Street
Suite 300
Chicago, IL 60603   (312) 368-1300

Slawomir Z. Szczepanski
William, Brinks, Olds, Hofer,
Gilson & Lione Ltd.
NBC Tower, Suite 3600
455 N. Cityfront Plaza Dr.
Chicago, IL 60611   (312) 321-4217

Julius Tabin
Fitch, Even, Tabin
& Flannery
135 S. Lasalle St.
Suite 900
Chicago, IL 60603   (312) 372-7842

Jill A. Tarzian
University of Illinois
1737 W. Polk
Suite 405
M.C. 225
Chicago, IL 60612   (312) 996-7762

Reginald K. Taylor#
Amoco Corporation
200 E. Randolph Dr.
P.O. Box 87703
Chicago, IL 60680   (312) 856-7847

Douglas B. Teaney
Dick & Harris
181 West Madison St.
Suite 3800
Chicago, IL 60602   (312) 726-4000

Frank R. Thienpont
230 West Monroe Street
Chicago, IL 60606   (312) 236-4711

Timothy L. Tilton
Tilton, Fallon, Lungmus
& Chestnut
100 S. Wacker Drive
Suite 960
Chicago, IL 60606   (312) 263-1841

Stanley J. Tomsa
Mason, Kolehmainen, Rathburn
& Wyss
20 North Wacker Drive
Suite 4200
Chicago, IL 60606   (312) 621-1300

James Earl Tracy
3407 West 83rd St.
Chicago, IL 60652   (312) 776-5447

Albert G. Tramposch
John Marshall Law School
Center for Intellectual
Property Law
315 S. Plymouth Court
Chicago, IL 60604-0001   (312) 987-1422

Steven R. Trybus
Roper & Quigg
200 S. Michigan Ave.
Suite 1000
Chicago, IL 60604   (312) 408-0855

Alan L. Unikel
Seyfarth, Shaw, Fairweather
& Geraldson
55 E. Monroe Street
Suite 4200
Chicago, IL 60603   (312) 346-8000

Charles C. Valauskas
Lockwood, Alex, Fitzgibbon
& Cummings
Three First National Plaza
Suite 1700
Chicago, IL 60602   (312) 782-4860

Brett A. Valiquet
Hill, Van Santen, Steadman,
& Simpson
70th Floor Sears Tower
Chicago, IL 60606   (312) 876-0200

James Van Santen
Hill, Van Santen, Steadman,
& Simpson, P.C.
70th Flr., Sears Tower
233 S. Wacker Dr.
Chicago, IL 60606 (312) 876-0200

William A. Van Santen, Jr.
Wood, Phillips, Mason,
Recktenwald & Van Santen
500 W. Madison Street
Northwestern Atrium Ctr., Ste. 3800
Chicago, IL 60661 (312) 876-1800

Paul M. Vargo
Dressler, Goldsmith, Shore,
Sutker & Milnamow Ltd.
Two Prudential Plaza
180 N. Stetson Ave., Ste. 4700
Chicago, IL 60478 (312) 616-5412

Timothy J. Vezeau#
Marshall, O Toole, Gerstein,
Murray & Bicknell
Two First National Plaza
20 S. Clark St., Ste. 2100
Chicago, IL 60603 (312) 346-5750

Karl A. Vick
William, Brinks, Olds, Hofer,
& Gilson & Lione
455 N. Cityfront Plaza Dr.
NBC Tower, Suite 3600
Chicago, IL 60611 (312) 321-4200

Robert A. Vitale
Niro, Scavone, Haller
& Niro, Ltd.
181 W. Madison Street
Suite 4600
Chicago, IL 60602 (312) 236-0733

Daniel W. Vittum, Jr.
Kirkland & Ellis
200 East Randolph Drive
Suite 6100
Chicago, IL 60601 (312) 861-2160

Gregory J. Vogler
Mc Andrews, Held & Malloy, Ltd.
Northwestern Atrium Center
500 W. Madison Street
Suite 3400
Chicago, IL 60661 (312) 707-8889

Robert Edward Wagner
Wallenstein, Wagner
& Hattis, Ltd.
311 S. Wacker Dr.
53rd Flr.
Chicago, IL 60606 (312) 554-3300

Robert J. Wagner
Amoco Corporation
200 E. Randolph Dr.
Chicago, IL 60601 (312) 856-5941

Richard B. Wakely
Fitch, Even, Tabin
& Flannery
135 S. La Salle Street
Suite 900
Chicago, IL 60603 (312) 372-7842

Bruce A. Walker
Amoco Corporation
200 E. Randolph
Chicago, IL 60601 (312) 856-3643

Sidney Wallenstein
Wallenstein, Wagner & Hattis, Ltd.
311 S. Wacker Dr.
53rd Flr.
Chicago, IL 60606 (312) 554-3300

Gomer Winston Walters
Wood, Phillips, Mason,
Recktenwald & Van Santen
500 W. Madison St.
Suite 3800
Chicago, IL 60606 (312) 876-1800

Ronald Lee Wanke
Jenner & Block
One IBM Plaza
44th Floor
Chicago, IL 60611 (312) 222-9350

Jeffrey S. Ward
Leydig, Voit & Mayer
180 N. Stetson Ave.
Two Prudential Plaza
Suite 4900
Chicago, IL 60601 (312) 616-5600

Robert M. Ward
Allegretti & Witcoff, Ltd.
10 South Wacker Drive
Suite 3000
Chicago, IL 60606 (312) 715-1000

Michael O. Warnecke
Neuman, Williams,
Anderson & Olson
77 W. Wash. St.
Chicago, IL 60602 (312) 346-1200

Damian G. Wasserbauer
Mason, Kolehmainen,
Rathburn & Wyss
20 N. Wacker Dr.
Suite 4200
Chicago, IL 60606 (312) 621-1300

Phillip H. Watt
Fitch, Even, Tabin
& Flannery
135 S. La Salle Street
Suite 900
Chicago, IL 60603 (312) 372-7842

William A. Webb
William, Brinks, Olds, Hofer,
Gilson & Lione, Ltd.
NBC Tower, Suite 3600
455 N. Cityfront Plaza Dr.
Chicago, IL 60611 (312) 321-4218

Eric H. Weimers
Jenner & Block
One IBM Plaza
Chicago, IL 60611 (312) 222-9350

Sandra B. Weiss
Jones, Day, Reavis & Pogue
225 W. Washington Street
Suite 2600
Chicago, IL 60606 (312) 782-3939

Donald L. Welsh
Welsh & Katz, Ltd.
135 S. La Salle Street
Suite 1625
Chicago, IL 60603 (312) 781-9470

David J. Werner#
641 W. Aldine #305
Chicago, IL 60657 (312) 935-6762

William M. Wesley
Neuman, Williams,
Anderson & Olson
77 W. Washington St.
Suite 2000
Chicago, IL 60602 (312) 346-1200

James Michael Wetzel
333 W. Wacker Drive
Suite 2600
Chicago, IL 60606 (312) 781-6668

George F. Wheeler
Neuman, Williams, Anderson
& Olson
77 W. Washington Street
Suite 2000
Chicago, IL 60602 (312) 346-1200

Gerald K. White
Morton International, Inc.
100 North Riverside Plaza
Morton International Bldg.
Chicago, IL 60606 (312) 807-2186

Kathleen A. White
Welsh & Katz, Ltd.
135 S. La Salle St.
Suite 1625
Chicago, IL 60603 (312) 781-9470

J. Warren Whitesel
Laff, Whitesel, Conte
& Saret
401 N. Michigan Avenue
Suite 2000
Chicago, IL 60611 (312) 649-2100

Michael D. Wiggins#
Wood, Phillips, Mason,
Recktenwald & Van Santen
500 W. Madison St.
Suite 3800
Chicago, IL 60661 (312) 876-1800

Keith B. Willhelm
Leydig, Voit & Mayer, Ltd.
180 N. Stetson Ave.
Two Prudential Plaza
Suite 4900
Chicago, IL 60601 (312) 616-5600

Roger A. Williams
G. D. Searle & Co.
P.O. Box 5110
Chicago, IL 60680 (312) 470-6500

Dennis J. Williamson#
Rockey & Rifkin
30 N. La Salle Street
Chicago, IL 60602 (312) 704-5600

Clyde F. Willian
William, Brinks, Olds, Hofer,
Gilson & Lione Ltd.
455 N. Cityfront Plaza
Chicago, IL 60611   (312) 321-4200

John B. Wilson
P.O. Box 2108
Chicago, IL 60690   (312) 580-3129

Leslie B. Wilson
720 S. Dearborn St., #503
Chicago, IL 60605   (312) 294-3854

Glen P. Winton
Welsh & Katz, Ltd.
135 S. La Salle Street
Suite 1625
Chicago, IL 60603   (312) 781-9470

David L. Witcoff
Kirkland & Ellis
200 E. Randolph Drive
Suite 6125
Chicago, IL 60601   (312) 861-2444

Sheldon W. Witcoff
Allegretti & Witcoff, Ltd.
10 South Wacker Drive
Suite 3000
Chicago, IL 60606   (312) 715-1000

Richard L. Wood
Welsh & Katz, Ltd.
135 S. La Salle Street
Rm. 1625
Chicago, IL 60603   (312) 781-9470

John S. Wrona
13351 Baltimore Avenue
Chicago, IL 60633   (312) 646-0022

Thomas A. Yassen
Amoco Corporation
200 E. Randolph Dr.
Chicago, IL 60601   (312) 856-6062

Foster York
Burmeister, York, Palmatier
& Zummer
135 S. La Salle St.
Chicago, IL 60603   (312) 782-6663

Richard W. Young
Gardner, Carton & Douglas
321 North Clark Street
Suite 3400
Chicago, IL 60610   (312) 245-8726

Richard A. Zachar
Pope, Ballard, Shepard
& Fowle, Ltd.
69 W. Washington St.
Suite 3300
Chicago, IL 60602   (312) 630-4225

Phiilip J. Zadeik
Baker & Mc Kenzie
2800 Prudential Plaza
130 East Randolph Dr.
Chicago, IL 60601   (312) 861-2852

James D. Zalewa
Rosenthal and Schanfield
55 East Monroe Street
Suite 4620
Chicago, IL 60603   (312) 236-5622

Mari - Kathleen F. Zaraza
5634 N. Kerbs Ave.
Chicago, IL 60646   (312) 777-5598

James P. Zeller
Marshall O'Toole, Gerstein,
Murray & Bicknell
Two First Natl. Plaza
Suite 2100
Chicago, IL 60603   (312) 346-5750

Lloyd L. Zickert
79 West Monroe Street
Chicago, IL 60603   (312) 236-1888

Anthony S. Zummer
135 S. La Salle St.
Suite 1046
Chicago, IL 60603   (312) 782-6663

Don W. Weber
212 E. Main Street
Collinsville, IL 62234   (618) 345-8424

Wayne Morris Russell
344 Division Street
Crete, IL 60417   (312) 672-5568

Floyd Boberg Harman
406 McHenry Ave.
Crystal Lake, IL 60014   (815) 459-0153

Russel D. Acton
1219 Robinson St.
Danville, IL 61832   (217) 446-1945

Charles A. Minne
2223 Denmark Road
Danville, IL 61832   (217) 443-6403

Augustus G. Douvas
1606 Coachmans Road
Darien, IL 60559   (312) 985-6122

A. Samuel Oddi
Northern Illinois University
College of Law
De Kalb, IL 60115   (815) 753-1980

Philip L. Bateman
Samuels, Miller, Schroeder,
Jackson & Sly
406 Citizens Bldg.
Decatur, IL 62523   (217) 429-4325

Michael F. Campbell#
A.E. Staley Manufacturing Co.
2200 East Eldorado St.
Decatur, IL 62525   (217) 421-2613

John F. Dunn
352 Millikin Court
Decatur, IL 62523   (217) 429-4000

Lori D. Tolly#
2282 Valley View Place
Decatur, IL 62522   (217) 425-1702

John Daniel Wood
A.E. Staley Manufacturing Co.
2200 East Eldorado Street
Decatur, IL 62521   (217) 421-2609

Craig M. Bell
Nutra Sweet Company
Box 730
1751 Lake Cook Road
Deerfield, IL 60015   (217) 421-2609

Robert A. Benziger
Travenol Laboratories, Inc.
One Baxter Parkway
Deerfield, IL 60015   (312) 291-4184

Michael P. Bucklo
Baxter Travenol Laboratories, Inc.
One Baxter Parkway
Deerfield, IL 60015   (312) 948-2422

Barry L. Clark
1759 We-Go Trail
Deerfield, IL 60015   (312) 945-1932

W. Dennis Drehkoff
Fujisawa U.S.A., Inc.
Parkway North Center
Three Parkway North
Deerfield, IL 60015   (708) 317-8839

Susan B. Fentress
Baxter Health Care Corp.
1 Baxter Parkway
Deerfield, IL 60015   (312) 948-3149

Paul C. Flattery
Baxter Health Care Corp.
One Baxter Pkwy.
Deerfield, IL 60015   (312) 948-4940

William B. Graham
Baxter Health Care Corporation
One Baxter Pkwy.
Deerfield, IL 60015   (312) 948-3000

Robert E. Hartenberger
Baxter Health Care Corp.
One Baxter Pkwy.
Deerfield, IL 60015   (312) 948-3779

Jeffrey M. Hoster
Nutra Sweet Company
Box 730
1751 Lake Look Road
Deerfield, IL 60015   (312) 940-9800

Marjorie Decou Hunter
Baxter Travenol Labs., Inc.
One Baxter Pkwy.
Deerfield, IL 60015   (312) 948-4934

Gregory J. Mancuso
Premark International Inc.
1717 Deerfield Road
Deerfield, IL 60015   (312) 405-6242

Charles R. Mattenson
Baxter International Inc.
DF2-2E
One Baxter Parkway
Deerfield, IL 60015   (312) 948-4928

Louise S. Pearson
Baxter Diagnostics Inc.
One Baxter Parkway
Bldg. 3-3E
Deerfield, IL 60015 (708) 948-4934

Kay H. Pierce
Baxter Health Care Corp.
One Baxter Parkway
Deerfield, IL 60015 (708) 948-3636

Bradford R.L. Price
Baxter International Inc.
Law Dept., Bldg. 2-2E
One Baxter Parkway
Deerfield, IL 60015 (312) 948-4948

Amy L.H. Rockwell
Baxter Healthcare Corporation
One Baxter Parkway
Deerfield, IL 60015 (312) 948-4942

John M. Sanders
Nutra Sweet Company
1751 Lake Cook Road
Deerfield, IL 60015 (708) 405-6898

Paul E. Schaafsma
Baxter International Inc.
One Baxter Pkwy., DF2-2E
Deerfield, IL 60015 (708) 948-3041

Thomas R. Schuman
Baxter Health Care Corp.
One Baxter Pkwy.
Deerfield, IL 60015 (708) 948-4946

Andrew Martin Solomon
Nutrasweet Company
Box 730
1751 Lake Cook Rd.
Deerfield, IL 60015 (708) 405-6634

Leigh Bannister Taylor
Premark International, Inc.
1717 Deerfield Road
Deerfield, IL 60015 (312) 405-6238

Maynard L. Youngs
Baxter Healthcare Corp.
One Baxter Pkwy.
Deerfield, IL 60015 (312) 948-4916

Harold W. Bergendorf
UOP Inc.
Algonquin & Mt. Prospect Rds.
Box 5017
Des Plaines, IL 60017 (312) 391-2518

Mary Jo Boldingh#
Allied-Signal Inc.
50 East Algonquin Rd.
Box 5016
Des Plaines, IL 60017 (708) 391-3383

Richard E. Conser
UOP
25 E. Algonquin Rd.
Des Plaines, IL 60017 (708) 391-2670

John Glenn Cutts, Jr.
UOP, Inc.
Box 5017
Des Plaines, IL 60017 (312) 391-2035

J. H. Hall
Allied Signal-Uop, Inc.
25 E. Algonquin Rd
Des Plaines, IL 60016 (312) 391-2033

Ronald H. Hausch
UOP Incorporated
25 E. Algonquin Road
Des Plaines, IL 60017 (312) 391-2516

Thomas Kerr Mc Bride
UOP Inc.
25 E. Algonquin Road
Des Plaines, IL 60017 (312) 391-2018

Richard G. Miller
25 East Algonquin Road
Des Plaines, IL 60017 (203) 794-6131

Frank S. Molinaro
UOP
Pat. Dept.
25 E. Algonquin Road
Des Plaines, IL 60017 (312) 391-2031

Richard R. Morris
UOP Inc.
Law Dept.
25 E. Algonquin Road
Des Plaines, IL 60017 (312) 391-2508

Raymond Harry Nelson
905 Center Street
Apt. 406
Des Plaines, IL 60016 (312) 824-8772

Dietmar H. Olesch
Sandoz Crop Protection Corp.
1300 East Touhy Avenue
Des Plaines, IL 60018 (312) 390-3728

William E. Parry
UOP
25 E. Algonquin Rd.
Des Plaines, IL 60017 (708) 391-2521

Eugene I. Snyder
UOP Group
25 E. Algonquin Rd.
Des Plaines, IL 60017 (312) 391-2061

John F. Spears, Jr.
UOP
Patent Dept.
25 E. Algonquin Road
Des Plaines, IL 60017 (312) 391-2037

John G. Tolomei
UOP
25 E. Algonquin Rd.
Des Plaines, IL 60017 (708) 391-2027

Warren K. Volles
UOP
25 E. Algonquin Rd.
Des Plaines, IL 60017 (708) 391-2023

Harold N. Wells, Jr.
Allied-Signal Inc.
50 E. Relgonquin Road
Des Plaines, IL 60017 (708) 391-3794

Thomas R. Wills
UOP
25 E. Algonquin Rd.
Des Plaines, IL 60017 (708) 391-2516

Anthony M. Berardi
4140 Main Street
Downers Grove, IL 60515 (312) 964-8820

George R. Clark
1501 Almond Court
Downers Grove, IL 60515 (312) 852-7754

Richard T. Lauterbach
Swift Adhesives
3100 Woodcreek Drive
Downers Grove, IL 60515 (312) 971-6791

Carmen B. Patti
Rockwell Internatl. Corp.
1431 Opus Place
Downers Grove, IL 60515 (312) 960-8055

Neil M. Rose
Sunbeam Corp.
1333 Butterfield Rd.
Downers Grove, IL 60515 (312) 719-4860

Bruce E. Burdick
Olin Corporation
T-189
Shamrock Street
East Alton, IL 62024 (618) 258-2362

Phillip J. Kardis
Madison County Courthouse
Third Judicial Circuit
Edwardsville, IL 62025 (618) 692-6200

William C. Grabarek
443 W. Pierce Street
P.O. Drawer G
Elburn, IL 60119 (312) 365-5404

Douglas J. Scheflow
Scheflow, Rydell, Travis
& Scheflow
63 Douglas Avenue
Suite 200
Elgin, IL 60120 (708) 695-2800

Edward C. Vandenburgh, III
6988 S. Pleasant Hill Rd.
Elizabeth, IL 61028 (815) 591-3351

Dillis V. Allen
1080 Nerge Rd.
Suite 205
Elk Grove Village, IL 60007 (708) 894-9100

Stanton Thomas Hadley
Ansco Photo-Optical
Products Corp.
1801 Touhy Ave.
Elk Grove Village, IL 60007 (708) 593-7404

Stephen A. Kozich
Videojet Systems International
2200 Arthur Ave.
Elk Grove Village, IL 60007 (312) 593-8800

Robert F. Van Epps
192 North York Road
Elmhurst, IL 60126 (708) 832-2333

Paul H. Gallagher
2530 Crawford Avenue
Evanston, IL 60201   (312) 475-0099

Richard R. Trexler
9509 Lawndale Ave.
Evanston, IL 60203   (708) 674-5866

Jacques John Filliung#
Sloan Valve Company
10500 Seymour Avenue
Franklin Park, IL 60131   (312) 671-4300

Willis J. Jensen
Duo-Fast Corp.
3702 N. River Road
Franklin Park, IL 60131   (312) 678-0100

Robert D. Teichert
Ekco Housewares Co.
9234 W. Belmont Ave.
Franklin Park, IL 60131   (312) 678-8600

Philip Jerome Zrimsek
Micro Switch
11 W. Spring St.
Freeport, IL 61032   (815) 235-5619

Allen L. Landmeier
Smith & Landmeier, P.C.
15 N. Second St.
P.O. Box 127
Geneva, IL 60134   (312) 232-2880

John A. Schaerli
Miner Enterprises Inc.
1200 E. State St.
Box 471
Geneva, IL 60134   (708) 232-3016

John L. Schmitt
312 W. State Street
Suite 101
P.O. Box 656
Geneva, IL 60134   (708) 232-1244

J. Leo Astrup#
21 West 604 Monticello
Glen Ellyn, IL 60137   (312) 469-3055

James P. Oxenham
64 Joyce Court
Glen Ellyn, IL 60137   (708) 858-8356

Walter L. Rees
366 Turner Avenue
Glen Ellyn, IL 60137   (312) 469-6124

Norman M. Shapiro
1198 Royal Glen Drive
Unit 221C
Glen Ellyn, IL 60137   (312) 932-0148

Michael G. Berkman
1701 E. Lake Avenue
Glenview, IL 60025   (708) 729-2222

Thomas W. Buckman
Illinois Tool Works, Inc.
Patent Dept.
3600 W. Lake Ave.
Glenview, IL 60025   (708) 657-4075

Donald W. Carlin
Kraft, Inc.
Kraft Court
Glenview, IL 60025   (312) 998-2488

Ralph E. Clarke, Jr.#
Zenith Electronics Corp.
1000 N. Milwaukee Avenue
Glenview, IL 60025   (312) 391-8099

John Harding Coult
Zenith Electronics Corporation
1000 Milwaukee Ave.
Glenview, IL 60025   (312) 391-8015

James Alan Geppert
120 Washington Street
Glenview, IL 60025   (312) 724-7679

Neal C. Johnson
Illinois Tool Works, Inc.
Patent Dept.
3650 W. Lake Avenue
Glenview, IL 60025   (312) 657-4073

Jack Kail
Zenith Electronics Corp.
Pat. Dept.
1000 Milwaukee Ave.
Glenview, IL 60025   (312) 391-8011

Francis C. Kowalik#
411 Crabtree Lane
Glenview, IL 60025   (708) 724-8388

John J. Kowalik
411 Crabtree Lane
Glenview, IL 60025   (312) 724-8388

Ronald H. Kullick
Kraft, Inc.
Kraft Court
Glenview, IL 60025   (312) 998-2493

Charles W. Mac Kinnon
Zenith Electronics Corp.
2134 Phillips Drive
Glenview, IL 60025   (312) 391-8052

Roland W. Norris
Zenith Electronics Corp.
Patent Dept.
1000 Milwaukee Ave.
Glenview, IL 60025   (708) 391-7000

John P. O Brien
Illinois Tool Works, Inc.
Pat. Dept.
3600 West Lake Avenue
Glenview, IL 60025   (708) 657-4075

John J. Pederson
Zenith Electronics Corp.
1000 Milwaukee Ave.
Glenview, IL 60025   (708) 391-7995

Benjamin Schlosser
330 Michael Manor
Glenview, IL 60025   (312) 966-0062

Charles J. Sindelar
Zenith Electronics Corporation
1000 Milwaukee Avenue
Glenview, IL 60025   (708) 671-2000

Edward Leonard Benno
17960 W. Hwy 120
Grays Lake, IL 60030   (312) 223-4906

Thomas W. Tolpin
444 Dell Lane
Highland Park, IL 60035   (312) 433-2556

Barbara R. Greenberg
1409 Burr Oak Rd., #308a
Hinsdale, IL 60521   (708) 323-1807

James Ramsey Hoatson, Jr.
Six Godair Park
Hinsdale, IL 60521   (312) 323-2624

Algird R. Ostis
201 E. Ogden Avenue
Suite 18-2
Hinsdale, IL 60521   (312) 325-3157

Edward Ptacek
17 W. 367 S. Frontage Rd.
Hinsdale, IL 60521   (312) 655-0977

Jack L. Uretsky
206 N. Grant
Hinsdale, IL 60521   (312) 323-5990

Mark E. Fejer
Speckman & Pauley, P.C.
2800 W. Higgins Road
Suite 365
Hoffman Estates, IL 60195   (708) 490-1400

Douglas H. Pauley
Speckman & Pauley, P.C.
2800 W. Higgins Road
Suite 365
Hoffman Estates, IL 60195   (708) 490-1400

Robert W. Slater
2300 N. Barrington Road
Suite 400
Hoffman Estates, IL 60195   (708) 490-5387

Thomas William Speckman
Speckman & Pauley, P.C.
2800 West Higgins Road
Suite 365
Hoffman Estates, IL 60195   (708) 490-1400

Arthur H. Bransky
18205 Hart Drive
Homewood, IL 60430   (312) 799-4028

John Vander Weit, Jr.
17924 S. Halsted
Suite 3N.E.
Homewood, IL 60430   (708) 957-7200

L. A. Combs
6508 Blackhawk Trail
Indian Head Park, IL 60545   (312) 246-3293

Ernest S. Kettelson
Barr Professional Bldg.
1520 N. Rock Run Drive
Joliet, IL 60435   (815) 741-1100

Paul Y. Feng
P.O. Box 424
Kenilworth, IL 60043   (708) 251-2021

Harold R. Schwappach
717 Terry Lane
La Grange, IL 60525   (708) 352-5325

Kent Barta
110 Indian Road
Lake Bluff, IL 60044   (708) 352-5325

Arthur N. Trausch, III
210 E. Witchwood Ln.
Lake Bluff, IL 60044   (708) 234-6366

Morando Berrettini
510 Broadsmoore Dr.
Lake Forest, IL 60045   (312) 853-3700

David L. Biek
466 Frost Place
Lake Forest, IL 60045   (312) 853-3700

John R. Diver
868 Larchmont Lane
Lake Forest, IL 60045   (312) 234-8314

Melburn E. Laundry
410 Circle Lane
Lake Forest, IL 60045   (312) 295-7177

Robert E. O Neill*
O'Neill & Bockelman, P.C.
250 East Illinois Road
Lake Forest, IL 60045   (312) 234-4422

Dorothy R. Thumler#
551 S. Beverly Pl.
Lake Forest, IL 60045   (708) 295-2551

Robert J. Zellner
648 Highview Terrace
Lake Forest, IL 60045   (312) 295-4785

Philip Hill
3256 Ridge Road
P.O. Box 187
Lansing, IL 60438   (312) 895-4404

Joseph W. Holloway
Rt 3 Box 175
Liberty, IL 62347   (217) 656-4355

Paul David Burgauer#
1110 Woodview Drive
Libertyville, IL 60048   (312) 362-0034

William Elliott Dominick
1260 Lake Street
Libertyville, IL 60048   (312) 362-8612

Bernard G. Donner#
907 N. Milwaukee Ave., 2-D
Libertyville, IL 60048   (708) 918-0721

Gildo E. Fato
515 Ash St.
Libertyville, IL 60048   (312) 362-0567

Donald L. Waller
1680 Young Drive
Libertyville, IL 60048   (708) 816-0374

Kenneth E. Roberts
31 Cumberland Drive
Lincolnshire, IL 60069   (708) 945-8938

Joy Ann G. Serauskas#
1 Fairfax Lane
Lincolnshire, IL 60069   (312) 945-2169

Thomas E. Torphy
Tenneco Automotive
Suite 300
100 Tri State International
Lincolnshire, IL 60069   (708) 940-6037

Gerald M. Walsh
529 Northgate
Lindenhurst, IL 60046   (708) 356-6522

Charles S. Cohen
Molex Inc.
2222 Wellington Court
Lisle, IL 60532   (708) 969-4747

Louis A. Hecht
Molex Inc.
2222 Wellington Court
Lisle, IL 60532   (312) 969-4550

Russell L. Mc Ilwain
1830 Middleton Avenue
Lisle, IL 60532   (312) 405-6243

William J. Michals
5121 Hawthorn Lane
Lisle, IL 60532   (708) 971-1649

Algis A. Tirva
Molex Incorp.
2222 Wellington Court
Lisle, IL 60532   (708) 969-4747

Stephen Z. Weiss
Molex Inc.
2222 Wellington Ct.
Lisle, IL 60532   (708) 969-4747

Marvin M. Chaban
1067 Apple Lane
Lombard, IL 60148   (312) 629-8126

Patrick J. O'Shea
916 E. St. Charles Road
Lombard, IL 60148   (312) 620-8551

Cedric M. Richeson
P.O. Box 1206
Lombard, IL 60148   (708) 932-9888

Emily A. Richeson#
Ashburn & Richeson
P.O. Box 1206
Lombard, IL 60148   (708) 932-9888

Jon Carl Gealow
2903 N. Bay View Lane
Mc Henry, IL 60050   (815) 385-2329

Terry R. Mohr
Mohr, Reilly, Prather
& Graham
420 N. Front Street
Mc Henry, IL 60050   (815) 385-1313

Mark A. Appleton
Laff, Whitesel, Conte
& Saret
P.O. Box 250
Moline, IL 61265   (309) 762-8568

Joel S. Carter
Deere & Co.
John Deere Road
Moline, IL 61265   (309) 752-6221

Duane A. Coordes
Deere & Co
John Deere Rd.
Moline, IL 61265   (309) 752-4383

Charles L. Dennis, II
Deere & Company
John Deere Road
Moline, IL 61265   (309) 765-5615

John O. Hayes
Deere & Co.
John Deere Rd.
Moline, IL 61265   (309) 765-4967

Michael C. Hlavaty
Deere & Co.
John Deere Rd.
Moline, IL 61265   (309) 765-4232

Raymond L. Hollister
Deere and Co.
John Deere Rd.
Moline, IL 61265   (309) 765-4451

Michael W. Mihm
Deere & Co.
John Deere Road
Moline, IL 61265   (309) 765-4856

Kevin J. Moriarty
Deere & Company
John Deere Road
Moline, IL 61265   (309) 765-4048

William A. Murray
3205 15th Street A
Moline, IL 61265   (309) 764-4972

John M. Nolan
Deere and Co.
Administrative Center
John Deere Rd.
Moline, IL 61265   (309) 752-4371

Jimmie Ralph Oaks
Deere and Co.
John Deere Road
Moline, IL 61265   (309) 765-4392

Kenneth J. Pedersen
Hyatt Legal Services
4575 16th St.
Suite 3
Moline, IL 61265   (309) 797-8188

Andrew J. Bootz
504 S. Albert
Mt. Prospect, IL 60056   (312) 255-6280

Charles F. Lind
350 W. Kensington
Suite 112
Mt. Prospect, IL 60056   (312) 392-0324

Ray Edward Snyder
200 E. Evergreen
Mt. Prospect, IL 60056   (312) 398-1525

Robert William Welch
22 South William
Mt. Prospect, IL 60056   (312) 259-3071

Joseph B. Barrett
Nalco Chemical Co.
One Nalco Center
Naperville, IL 60563   (708) 305-1000

Donald J. Breh
24 W581 Ohio Street
Naperville, IL 60540   (708) 357-3664

Aubrey L. Burgess
1191 Banbury Circle
Naperville, IL 60540   (312) 961-2470

Donald G. Epple
Nalco Chemical Company
1 Nalco Center
Naperville, IL 60566   (312) 961-9500

David L. Hurewitz
AT&T
Room 2A-401
2000 N. Naperville Rd.
P. O. Box 3033
Naperville, IL 60566   (708) 979-4637

Frederick S. Jerome
Amoco Corporation
P.O. Box 400
Warrenvile Rd. & Mill St.
Naperville, IL 60566   (312) 420-5456

Mathew L. Kalinowski
734 S. Sleight St.
Naperville, IL 60540   (312) 355-1504

James F. Lambe
Nalco Chemical Co.
One Nalco Center
Naperville, IL 60566   (312) 961-9500

Louis H. Le Mieux
Nalco Chemical Company
One Nalco Center
Naperville, IL 60566   (312) 983-2837

Bruce R. Mansfield
400 Olesen Drive
Naperville, IL 60540   (312) 420-1608

Carole A. Mickelson
Amoco Corporation
Mail Station H-7
P.O. Box 400
Naperville, IL 60566   (312) 420-4966

Robert A. Miller
Nalco Chemical Co.
One Nalco Center
Naperville, IL 60566   (312) 961-9500

Edward J. Mooney, Jr.
Nalco Chemical Co.
One Nalco Center
Naperville, IL 60555-5000   (312) 983-2812

Werner Ulrich
AT&T Bell Laboratories
2000 North Naperville Road
Naperville, IL 60566   (708) 979-3255

George E. Verhage
221 E. 14th Ave.
Naperville, IL 60563   (708) 355-3789

Ross T. Watland
AT&T Bell Laboratories
2000 N. Naperville Rd.
Naperville, IL 60566   (708) 979-2003

Herman Wissenberg#
Amoco Corp.
Amoco Research Center
P.O. Box 3011
Naperville, IL 60566   (708) 420-4955

Fred Gordon Thelander
209 April Ln.
North Aurora, IL 60542   (708) 896-8306

David C. Hannum
Abbott Laboratories
1400 Sheridan Rd.
Dept. 55B, Bldg. A3
North Chicago, IL 60064   (708) 937-4686

Peter Andress, Jr.
IMC Fertilizer Group, Inc.
2100 Sanders Road
Northbrook, IL 60062   (312) 272-9200

Robert A. Brown
2530 Shannon Road
P.O. Box 2127
Northbrook, IL 60065   (312) 272-3182

Lawrence William Brugman
2625 Techny Road
Apt. 712
Northbrook, IL 60062   (312) 272-3182

John A. Doninger
Premark International Inc.
2211 Sanders Road
Northbrook, IL 60062   (312) 498-8486

Thomas L. Farquer
International Minerals
& Chemical Corporation
2315 Sanders Road
Northbrook, IL 60062   (312) 205-2268

Sidney Norman Fox
555 Skokie Blvd.
Northbrook, IL 60062   (312) 498-3322

S. David Hoffman
Underwriters Labs., Inc.
333 Pfingsten Road
Northbrook, IL 60062   (312) 272-8800

Robert M. Mason
1943 Greenview
Northbrook, IL 60062   (708) 272-1464

Joyce R. Niblack
Niblack & Niblack, P.C.
555 Skokie Blvd.
Suite 205
Northbrook, IL 60062   (312) 291-9900

Robert L. Niblack
Niblack & Niblack, P.C.
555 Skokie Blvd.
Suite 205
Northbrook, IL 60067   (312) 291-9900

Howard E. Post
Internatl. Minn. & Chems. Corp.
2315 Sanders Road
Northbrook, IL 60062   (708) 564-8600

James B. Raden
151 Wellington Road
Northbrook, IL 60062   (708) 291-1901

Mary J. Schnurr
2923 White Pine Drive
Northbrook, IL 60062   (708) 291-1901

Marvin Smollar
170 Fairview
Northbrook, IL 60062   (708) 394-3737

William Garrettson Ellis
635 Woodland Lane
Northfield, IL 60093   (312) 446-7234

Norman R. Smith
Stepan Company
22 W. Frontage Road
Northfield, IL 60093   (708) 501-2240

Andrew Frank Zikas
Stepan Company
22 W. Frontage Road
Northfield, IL 60093   (708) 501-2287

John C. Brezina
Brezina & Buckingham
1000 Jorie Blvd.
Oak Brook, IL 60521   (708) 571-3838

Joseph M. Gartner
5 Baybrook Court
Oak Brook, IL 60521   (312) 654-0826

Edmond T. Patnaude
Patnaude, Batz & Videbeck
1S376 Summit Avenue
Court C
Oak Brook Terrace, IL 60181   (312) 627-4552

James N. Videbeck
Patnaude, Batz & Videbeck
1 S. 376 Summit Avenue
Court C
Oak Brook Terrace, IL 60181   (312) 627-4552

Robert L. Chandler#
9704 S. Kenneth Avenue
Oak Lawn, IL 60453   (312) 423-3386

Basil Emanuel Demeur
Knechtel & Demeur
132 S. Oak Park Ave.
Oak Park, IL 60302   (708) 848-0111

Charles A. Doktycz#
1023 N. Kenilworth
Oak Park, IL 60302   (312) 383-3384

Philip H. Kier
321 Home Avenue
Oak Park, IL 60302   (312) 386-5981

Robert George Petrinec#
617 S. East Ave.
Oak Park, IL 60304   (312) 848-8955

Carey A. Czarnik
8502 Tee Brook Drive
Orland Park, IL 60462   (312) 460-4937

M. Russell Bramwell
800 E. Northwest Hwy.
Suite 326
Palatine, IL 60067   (312) 359-5404

Anthony L. Cupoli
37 Peppertree Drive
Palatine, IL 60067   (312) 961-9500

Clifford A. Dean
719 Greenwood Drive
Palatine, IL 60067   (312) 359-1816

Larry I. Golden
Square D Company
Executive Plaza
Palatine, IL 60067   (312) 397-2600

Richard T. Guttman
Square D Co.
Executive Plaza
1415 S. Roselle Rd.
Palatine, IL 60067   (312) 397-2600

Jose W. Jimenez
Square D Company
Executive Plaza
Palatine, IL 60067   (708) 397-2600

Melvin A. Schechtman
Motorola, Inc.
335 Kensington Ct.
Palatine, IL 60067   (708) 705-1282

David R. Stacey
Square D Company
Legal Dept.
1415 S. Roselle Rd.
Palatine, IL 60067   (708) 397-2610

Robert James Black#
1400 Renaissance Drive
Suite 205
Park Ridge, IL 60068   (312) 635-6371

Richard W. Carpenter
1400 Renaissance Drive
Suite 205
Park Ridge, IL 60068   (312) 635-6357

Robert W. Dudley
634 North Overhill
Park Ridge, IL 60068   (312) 825-4633

Frank B. Hall
855 N. Northwest Hwy.
Park Ridge, IL 60068   (312) 825-2501

Jack Larsen
O419 Meacham Avenue
Park Ridge, IL 60068   (312) 823-1087

Adrienne B. Naumann
Otto & Blumenthal
3 S. Prospect
Suite 206
Park Ridge, IL 60068   (708) 698-1160

Frederick J. Otto
Otto & Blumenthal
3 South Prospect
Suite 6
Park Ridge, IL 60068   (312) 698-1160

Harold J. Rathbun
1892 De Cook Ave.
Park Ridge, IL 60068   (312) 825-1267

Robert Donald Silver
125 E. Kathleen Dr.
Park Ridge, IL 60068   (312) 825-5262

J. Wesley Blumenshine
Caterpillar Inc.
100 N.E. Adams St.
Peoria, IL 61629   (309) 675-6060

J. W. Burrows#
Caterpillar Inc.
100 N.E. Adams St.
Peoria, IL 61629   (309) 675-5676

Larry G. Cain#
Caterpillar Inc.
100 N.E. Adams Street
Peoria, IL 61629   (309) 675-4922

Randall E. Deck#*
USDA-ARS-MWA
Northern Regional Res. Center
1815 N. University Street
Peoria, IL 61604   (309) 685-4011

Clavin E. Glastetter#
Caterpillar Inc.
100 N.E. Adams Street
Peoria, IL 61629   (309) 675-5124

Eugene C. Goodale
Caterpillar Tractor Co.
100 N.E. Adams St.
Peoria, IL 61629   (309) 675-5089

John W. Grant#
Caterpillar Tractor Co.
100 N.E. Adams St.
Peoria, IL 61629   (309) 675-5613

Frank L. Hart
Caterpillar Inc.
100 N.E. Adams St.
Peoria, IL 61629   (309) 675-5313

William B. Heming
Caterpillar Inc.
100 NE Adams St.
Peoria, IL 61629   (309) 675-5509

Alan J. Hickman#
Caterpillar Inc.
100 N.E. Adams
Peoria, IL 61629   (309) 675-4517

Wei Wei Jeang#
Caterpillar Inc.
100 N.E. Adams Street
Peoria, IL 61629   (309) 675-4012

Joseph W. Keen
Caterpillar Inc.
100 N.E. Adams St.
Peoria, IL 61629   (309) 675-5753

Charles E. Lanchantin, Jr.
Caterpillar Inc.
100 N.E. Adams St.
Peoria, IL 61629   (309) 675-4013

David M. Masterson#
Caterpillar, Inc.
100 N.E. Adams St.
Peoria, IL 61629   (309) 675-4012

Robert A. Mc Fall
Caterpillar Inc.
Pat. Dept. AB6490
100 N.E. Adams St.
Peoria, IL 61629   (309) 675-4610

Robert E. Muir
Caterpillar Inc.
100 N.E. Adams Street
Peoria, IL 61629   (309) 675-4073

Stephen L. Noe
Caterpillar Inc.
100 N.E. Adams Street
Peoria, IL 61619   (309) 689-5589

Oscar G. Pence
Caterpillar Inc.
Patent Dept.
AB6490
100 N.E. Adams Street
Peoria, IL 61629   (309) 675-4460

William C. Perry#
Caterpillar Tractor Co.
100 N.E. Adams St.
Peoria, IL 61629   (309) 675-5083

Kenneth A. Rhoads#
Caterpillar Inc.
100 N.E. Adams St.
AB6490
Peoria, IL 61629   (309) 675-4015

Curtis P. Ribando#*
USDA-ARS
Northern Regional Res. Ctr.
1815 N. University St.
Peoria, IL 61604   (309) 685-4011

William Scott Thompson
Caterpillar Tractor Co.
100 N.E. Adams Street
Peoria, IL 61629   (309) 675-4452

Kirk A. Vander Leest#
Caterpillar
100 N.E. Adams
Peoria, IL 61629   (309) 675-6528

Loyal O. Watts#
Caterpillar, Inc.
100 N.E. Adams St.
Peoria, IL 61629   (309) 675-4923

Claude F. White#
Caterpillar Inc.
100 N.E. Adams St.
Patent Dept., AB 6490
Peoria, IL 61629   (309) 675-6008

Anthony N. Woloch
Caterpillar Inc.
Pat. Dept. AB6490
100 N.E. Adams St.
Peoria, IL 61629
(309) 675-5210

James R. Yee#
Caterpillar Inc.
Pat. Dept., AB 6490
100 N.E. Adams Street
Peoria, IL 61629
(309) 675-1517

Evan D. Roberts
122 N. Second St.
Peotone, IL 60468
(708) 258-6318

Brenda J. Ehrhardt
710 Country Club Heights Rd.
Apt. 102
Quincy, IL 62301
(217) 228-0614

D. James Bader
3677 Sauk Trail
Richton Park, IL 60471
(312) 481-3100

Nicholas Anthony Camasto
545 Thatcher
River Forest, IL 60305
(312) 366-0604

Frederick J. Krubel
289 Shenstone Road
Riverside, IL 60546
(708) 447-7838

Francis J, Lidd
247 Lawton Rd.
Riverside, IL 60546
(312) 447-2476

Glenn C. Sechen
424 Selborne Road
Riverside, IL 60546
(312) 447-7271

Claudio Giori
2975 Orange Brace Rd.
Riverwoods, IL 60015
(312) 447-7271

Thomas E. Currier
Connolly, Oliver, Coplan,
Close & Worden
124 N. Water Street
Suite 300
Rockford, IL 61104
(815) 968-7591

Robert M. Hammes, Jr.
Barber-Colman Co.
P.O. Box 7040
Rockford, IL 61125
(815) 397-7400

David H. Hitt
Sundstrand Corporation
4949 Harrison Avenue
P.O. Box 7003
Rockford, IL 61125
(815) 226-2852

Trevor B. Joike
Sundstrand Corporation
4949 Harrison Avenue
Rockford, IL 61125
(815) 226-6220

Ted E. Killingsworth, Jr.
Sundstrand Corporation
4949 Harrison Avenue
Rockford, IL 61125
(815) 226-6307

William D. Lanyi
Sundstrand Corporation
4949 Harrison Avenue
Rockford, IL 61125
(815) 226-7462

Leroy W. Mitchell
Leydig, Voit & Mayer, Ltd.
815 N. Church Street
Rockford, IL 61103
(815) 963-7661

Michael C. Payden
Leydig, Voit & Mayer, Ltd.
815 North Church Street
Rockford, IL 61103
(815) 963-7661

Vernon J. Pillote
310 Seventh Street
Rockford, IL 61104
(815) 964-9312

James A. Wanner
Sundstrand Corporation
4949 Harrison Avenue
Rockford, IL 61125
(815) 226-7912

Harold A. Williamson
Sundstrand Corp.
4949 Harrison Ave.
Rockford, IL 61108
(815) 226-7407

Michael S. Yatsko
Sundstrand Corporation
Dept. 912
4949 Harrison Avenue
Rockford, IL 61125
(815) 226-6348

Jeffrey M. Morris
Two Crossroads of Commerce
Suite 310
Rolling Meadows, IL 60008
(312) 394-3040

Gregory A. Conley
COVIA
9700 W. Higgins Road
Suite 400
Rosemont, IL 60018
(312) 518-4800

Paul W. Fish
Comdisco, Inc.
6111 N. River Road
Rosemont, IL 60018
(312) 698-3000

Paul W. Grauer
1300 Woodfield Drive
Suite 100
Schaumberg, IL 60173
(312) 240-9010

Lester N. Arnold
1409 Wright Blvd.
Schaumburg, IL 60193
(312) 893-1620

Kenneth W. Bolvin
Motorola, Inc.
1303 E. Algonquin Rd.
Schaumburg, IL 60196
(708) 576-5212

Frank J. Cerny, Jr.#
Motorola Incorporated
1301 E. Algonquin Road
Room 2918
Schaumburg, IL 60196
(312) 576-2443

Jon P. Christensen
Motorola, Inc.
1303 E. Algonquin Rd.
Schaumburg, IL 60196
(708) 576-3533

Shawn B. Dempster#
Motorola, Inc.
1303 E. Algonquin Rd.
3rd Flr.
Schaumburg, IL 60196
(708) 576-0053

Wayne Jay Egan
Motorola, Inc.
Pat. Dept.
1303 E. Algonquin Road
Schaumburg, IL 60196
(708) 576-0860

Douglas D. Fekete
Motorola, Inc.
1303 E. Algonquin Rd.
Schaumburg, IL 60196
(708) 538-2447

John A. Fisher
Motorola, Inc.
Intellectual Property Dept.
1303 E. Algonquin Rd.
Schaumburg, IL 60196
(708) 397-5000

James W. Gillman
Motorola, Inc.
1303 E. Algonquin Rd.
Schaumburg, IL 60196
(312) 576-5223

John W. Hayes#
Motorola, Inc.
1303 E. Algonquin Road
Schaumburg, IL 60196
(708) 576-5066

Val Jean F. Hillman
Motorola, Inc.
1303 E. Algonquin Rd.
3rd Floor
Schaumburg, IL 60196
(708) 576-5209

Raymond A. Jenski#
Motorola, Inc.
1303 E. Algonquin Road
Schaumburg, IL 60196
(312) 576-5223

Robert H. Kelly#
Motorola Inc.
1303 E. Algonquin Rd.
Schaumburg, IL 60173
(708) 576-7787

Robert Edward Knechtel
Knecthel & Demeur
600 N. Meacham Road
Suite 301
Schaumburg, IL 60173
(708) 517-7766

Timothy W. Markison#
Motorola Inc.
1303 E. Algonquin Rd.
Schaumburg, IL 60196
(312) 576-0256

Phillip H. Melamed
Motorola, Inc.
1303 E. Algonquin Rd.
Schaumburg, IL 60196
(312) 576-5218

John H. Moore
Motorola, Inc.
1303 E. Algonquin Road
Schaumburg, IL 60196
(312) 576-5218

F. John Motsinger
Motorola, Inc.
Motorola Center
1303 E. Algonquin Road
Schaumburg, IL 60196   (312) 576-5213

Steven G. Parmelee
Motorola, Inc.
1303 E. Algonquin Road
Schaumburg, IL 60196   (312) 576-0658

James S. Pristelski
Motorola, Inc.
1303 E. Algonquin Rd.
Schaumburg, IL 60196   (708) 576-5216

Vincent Joseph Rauner
Motorola Inc.
1303 E. Algonquin Road
Schaumburg, IL 60196   (312) 576-5220

Edward Milton Roney, III
Motorola Inc.
1303 E. Algonquin Rd.
Schaumburg, IL 60010   (312) 576-5222

Anthony J. Sarli, Jr.
Motorola, Inc.
1303 E. Algonquin Road
3rd Floor
Schaumburg, IL 60196   (708) 576-6364

Donald B. Southard
Motorola, Inc.
1303 East Algonquin Rd.
Schaumburg, IL 60196   (708) 576-5214

Darleen J. Stockley
Motorola Inc.
1303 E. Algonquin Rd.
Schaumburg, IL 60196   (708) 576-0659

Charles L. Warren
Motorola, Inc.
1303 E. Algonquin Road
Schaumburg, IL 60196   (708) 576-6364

Raymond J. Warren
Motorola Inc.
1303 E. Algonquin Rd.
Schaumburg, IL 60196   (708) 538-2725

Douglas B. White
600 Woodfield Drive
Suite 1160
Schaumburg, IL 60173   (708) 240-1616

Donald L. Barbeau#
Biomega
8707 Skokie Blvd.
Suite 107
Skokie, IL 60077   (708) 982-1400

Roberta L. Hastreiter
Searle
5200 Old Orchard Rd.
Skokie, IL 60077   (708) 982-7000

Mary J. Kanady
G.D. Searle & Company
5200 Old Orchard Road
Skokie, IL 60077   (312) 470-6501

William Gresham Lawler, Jr.
Brunswick Corporation
One Brunswick Plaza
Skokie, IL 60077   (312) 470-4321

Stuart J. Millman#
10101 Cherry Parkway
Suite 208
Skokie, IL 60076   (708) 677-6460

Raymond C. Nordhaus
8301 Karlov
Skokie, IL 60076   (312) 674-5739

Boris Parad
4711 Golf Rd.
Suite 700
Skokie, IL 60076   (708) 674-1620

David H. Sitrick
Sitrick & Sitrick
8340 N. Lincoln Avenue
Suite 104
Skokie, IL 60077   (708) 677-4411

Wayne Golomb
901 S. Second Street
Springfield, IL 62704   (217) 522-1581

John C. Albrecht
1044 N. Second Avenue
St. Charles, IL 60174   (312) 377-2415

Edward F. Jurow
1044 N. 6th Avenue
St. Charles, IL 60174   (312) 377-0484

Mark D. Hilliard
Panduit Corporation
17301 Ridgeland Avenue
Tinley Park, IL 60477   (312) 532-1800

Charles R. Wentzel
Panduit Corp.
17301 Ridgeland Ave.
Tinley Park, IL 60477   (312) 532-1800

Roger M. Fitz - Gerald
1104 S. Orchard Street
Urbana, IL 61801   (217) 344-5242

Mark F. La Marre
1437 Clairmont Court
Vernon Hills, IL 60061   (708) 680-8267

Michael T. Murphy
Rust-Oleum Corporation
11 Hawthorn Parkway
Vernon Hills, IL 60061   (312) 816-2220

R. Warren Comstock
Outboard Marine Corp.
100 Sea-Horse Dr.
Waukegan, IL 60085   (312) 689-5229

Mark W. Croll
Outboard Marine Corporation
100 Sea-Horse Drive
Waukegan, IL 60085   (312) 689-6187

Robert Kingsley Gerling
Outboard Marine Corp.
100 Sea Horse Drive
Waukegan, IL 60085   (312) 689-5247

James P. Hanrath
415 W. Washington Street
Suite 216
Waukegan, IL 60085   (312) 249-1420

Janet K. Hochstetler
414 Julian Street
Waukegan, IL 60085   (312) 662-0717

Walter Lothar Schlegel, Jr.
1 S. 311 Edgewood Walk
West Chicago, IL 60185   (708) 231-0821

Francis V. Cunningham#
5316 Central Avenue
Western Springs, IL 60558   (312) 246-5158

John G. Premo
110 51st Place
Western Springs, IL 60558   (708) 246-6528

William K. Serp
4027 Harvey Avenue
Western Springs, IL 60558   (708) 246-7856

Michael J. Higgins*
1603 Mayo Street
Wheaton, IL 60187   (312) 690-2098

Joseph P. Krause
1495 Castlewood Drive
Wheaton, IL 60187   (312) 690-5357

John M. Lorenzen
1329 Brighton Drive
Wheaton, IL 60187   (312) 690-8099

Robert L. Marsh
605 E. Roosevelt Road
Wheaton, IL 60187   (708) 665-8900

Margaret M. Parker#
1302 Scott Street
Wheaton, IL 60187   (312) 668-6550

Ronald G. Wesoloski#
1013 Shady Tree Ln.
Wheeling, IL 60090   (708) 459-6756

John Max Brown#
811 Lacrosse Ave.
Wilmette, IL 60091   (312) 256-2760

Sheldon Lee Epstein
P.O. Box 400
Wilmette, IL 60091   (708) 948-9292

Walter Christoph Ramm
Suite 302
444 Skokie Road
Wilmette, IL 60091   (312) 256-5425

William Brandt Ross
1334 Isabella
Wilmette, IL 60091   (708) 251-1890

Raymond A. Andrew
2 Kent Road
Winnetka, IL 60093   (312) 446-7868

Britton A. Davis
285 Linden Street
Winnetka, IL 60093   (312) 446-7868

Robert Gottschalk
P.O. Box 8436
Winnetka, IL 60093   (312) 446-5230

William H. Page, II
786 Foxdale Avenue
Winnetka, IL 60093   (708) 446-1458

John C. Shepard
575 Sunset Road
Winnetka, IL 60093   (708) 965-8660

Peter R. Bahn
RR 1, Box 261
Woodlawn, IL 62898   (618) 735-2897

Sterling R. Booth, Jr.#
RR 1, Box 50
Yates City, IL 61572   (309) 675-5136

Indiana

F. Kristen Koepcke
Hillenbrand Ind., Inc.
Rt. 46
Batesville, IN 47006   (812) 934-7361

Brian J. Leitten
Hillenbrand Industries, Inc.
Highway 46
Batesville, IN 47006   (812) 934-7261

Steve M. Mc Lary
Hillenbrand Industries, Inc.
Highway 46
Batesville, IN 47006   (812) 934-8747

Peter Peck-Koh Ho
1228 Ironwood Drive W.
Carmel, IN 46032   (317) 276-4460

Gary M. Gron
Cummins Engine Co., Inc.
Box 3005
Columbus, IN 47202   (812) 377-3554

Thomas J. Page#
P.O. Box 29242
Cumberland, IN 46229   (812) 377-3554

Mary G. Boguslaski
Miles Laboratories, Inc.
1127 Myrtle St.
P.O. Box 40
Elkhart, IN 46515   (219) 264-8384

R. Norman Coe
Miles Laboratories, Inc.
1127 Myrtle Street
Elkhart, IN 46515   (219) 262-7937

Louis E. Davidson
Miles Laboratories, Inc.
P.O. Box 40
1127 Myrtle Street
Elkhart, IN 46515   (219) 264-8393

John J. Gaydos
Gaydos & Ryan
118 South Second Street
Elkhart, IN 46516   (219) 294-1516

Jerome L. Jeffers
Miles Inc.
P.O. Box 40
Elkhart, IN 46515   (219) 264-8394

Andrew L. Klawitter
Miles Inc.
1127 Myrtle St.
P.O. Box 40
Elkhart, IN 46515   (219) 262-7148

Harry Thomas Stephenson
3201 Eastlake Drive North
Elkhart, IN 46514   (219) 262-2060

Albert W. Watkins#
CTS Corporation
905 West Blvd. N.
Elkhart, IN 46514   (219) 293-7511

Richard W. Winchell
Miles Inc.
1127 Myrtle St.
Elkhart, IN 46514   (219) 262-7748

Robert E. Carnahan#
1327 Timberlake Road
Evansville, IN 47710   (812) 426-2807

Warren Dale Flackbert
401 N. Weinbach Avenue
Suite D
Evansville, IN 47711   (812) 477-2434

George H. Morgan
401 Tyler Avenue
Evansville, IN 47715   (812) 476-4065

Robert H. Uloth#
Bristol-Myers Squibb Co.
2400 W. Lloyd Expressway
Evansville, IN 47721   (812) 429-8918

David L. Ahlersmeyer
Jeffers, Hoffman & Niewyk
1500 Anthony Wayne Bank Bldg.
Fort Wayne, IN 46807   (219) 426-1700

Edward P. Armstrong
Central Soya Co., Inc.
1300 Ft. Wayne Natl. Bk. Bldg.
Fort Wayne, IN 46801   (219) 425-5477

Michael T. Bates
Jeffers, Hoffman & Niewyk
444 East Main St.
Fort Wayne, IN 46802   (219) 426-1700

Eugene G. Botz#
8234 Ravinia Dr.
Fort Wayne, IN 46825   (219) 489-3055

George R. Caruso
8405 Lima Road
Fort Wayne, IN 46818   (219) 489-6233

Anthony J. Criso
Essex Group, Inc.
1601 Wall St.
Fort Wayne, IN 46802   (219) 461-4495

Kevin R. Erdman
Jeffers, Hoffman & Niewyk
444 East Main Street
Fort Wayne, IN 46815   (219) 426-1700

Bobby B. Gillenwater
Barnes & Thornburg
600 One Summit Square
Fort Wayne, IN 46802   (219) 423-9440

George A. Gust
909 Crestway Drive
Fort Wayne, IN 46819   (219) 747-9540

Joseph T. Hepp#
10614 Alderwood Lane
Fort Wayne, IN 46825   (219) 747-9540

John F. Hoffman
Jeffers, Hoffman & Niewyk
1500 Anthony Wayne Bank Bldg.
Fort Wayne, IN 46802   (219) 426-1700

Albert L. Jeffers
Jeffers, Hoffman & Niewyk
1500 Anthony Wayne Bank Bldg.
Fort Wayne, IN 46802   (219) 426-1700

Douglas E. Johnston
Tourkow, Crell, Rosenblatt
& Johnston
814 Anthony Wayne Bldg.
Fort Wayne, IN 46802   (219) 426-0545

Randall James Knuth
Jeffers, Hoffman & Niewyk
444 East Main St.
Fort Wayne, IN 46802   (219) 426-1700

Ralph E. Krisher, Jr.
3409 Rosewood Dr.
Fort Wayne, IN 46804   (219) 428-3283

David A. Lundy
Lundy & Walker
1020 Anthony Wayne Bk. Bldg.
Fort Wayne, IN 46802   (219) 422-1534

Anthony Niewyk
Jeffers, Hoffman & Niewyk
1500 Anthony Wayne Bank Bldg.
Fort Wayne, IN 46802   (219) 426-1700

George Pappas
701 S. Clinton Street
Suite 324
Fort Wayne, IN 46802   (219) 426-2340

Roger Mark Rickert
7500 Amber Road
Fort Wayne, IN 46804   (219) 672-2121

Richard Thompson Seeger
1067 Delaware Avenue
Fort Wayne, IN 46805   (219) 422-9829

Mark F. Smith
6911 Pointe Inverness
Fort Wayne, IN 46804   (219) 432-8194

Robert D. Sommer#
1022 Catalina Avenue
Fort Wayne, IN 46825   (219) 483-8919

Arlyce R. Stearns
1406 Tecumseh St.
Fort Wayne, IN 46805   (219) 428-0122

Lawrence A. Steward
Jeffers, Hoffman & Niewyk
1500 Anthony Wayne Bldg.
Fort Wayne, IN 46802   (219) 426-1700

George Hyman, Jr.
R.R. 2
Galveston, IN 46932   (219) 859-4622

Dugal S. Sickert
Dow/Elanco Res. Lab.
2001 W. Main Street
P.O. Box 708
Drop Code GL 18
Greenfield, IN 46140   (317) 277-4504

James L. Wilson
3534 43rd Street
Highland, IN 46322   (219) 924-4468

Roger M. Miller
1995 White Eagle Drive
Huntington, IN 46750   (219) 356-8027

Marilyn L. Amick#
2736 Lakeshire Lane
Indianapolis, IN 46268   (317) 875-8911

Edward P. Archer
University of Indiana
Indianapolis Law School
735 W. New York St.
Indianapolis, IN 46202   (317) 264-4998

Charles W. Ashbrook
Eli Lilly & Co.
Lilly Corporate Center
Indianapolis, IN 46285   (317) 276-6015

David H. Badger
William, Brinks, Olds, Hofer,
Gilson & Lione
One Indiana Square
Suite 2425
Indianapolis, IN 46204   (317) 636-0886

William F. Bahret
Woodard, Emhardt, Naughton,
Moriarty & Mc Nett
Bank One Center Tower
111 Monument Circle, Ste. 3700
Indianapolis, IN 46204   (317) 634-3456

Bruce J Barclay
Eli Lilly & Company
Lilly Corporate Center
307 E. Mc Carty Street
Indianapolis, IN 46285   (317) 276-3474

Michael D. Beck
Woodard, Emhardt, Naughton,
Moriarty & Mc Nett
Bank One Center Tower
111 Monument Circle, Ste. 3700
Indianapolis, IN 46204   (317) 634-3456

Spiro Bereveskos
Woodard, Emhardt, Naughton,
Moriarty & Mc Nett
Bank One Center Tower
111 Monument Circle, Ste. 3700
Indianapolis, IN 46204   (317) 634-3456

David E. Boone#
Eli Lilly & Company
Patent Department
Lilly Corporate Center
Indianapolis, IN 46285   (317) 276-3881

Daniel L. Boots
William, Brinks, Olds, Hofer,
Gilson & Lione
One Indiana Square
Suite 2425
Indianapolis, IN 46204   (317) 636-0886

Kyle S. Brant
Woodard, Emhardt, Naughton,
Moriarty & Mc Nett
Bank One Center Tower
111 Monument Circle, Ste. 3700
Indianapolis, IN 46204   (317) 634-3456

Alice A. Brewer
Dow Elanco
9002 Purdue Road
Indianapolis, IN 46268   (317) 870-7108

Clifford W. Browning
Woodard, Emhardt, Naughton,
Moriarty & Mc Nett
Bank One Center Tower
111 Monument Circle, Ste. 3700
Indianapolis, IN 46204   (317) 634-3456

John H. Calhoun, Jr.
6100 N. Keystone Avenue
Suite 333
Indianapolis, IN 46220   (317) 255-3438

Francis E. Cislak#
5331 N. Kenwood Ave.
Indianapolis, IN 46208   (317) 255-7115

Richard R. Clapp#
Barnes & Thornburg
1313 Merchants Bank Bldg.
11 South Meridian
Indianapolis, IN 46204   (317) 231-7461

William R. Coffey
Barnes & Thornburg
11 S. Meridian Street
Suite 1313
Indianapolis, IN 46204   (317) 231-7280

James A. Coles
Barnes & Thornburg
Suite 1313
11 South Meridan Street
Indianapolis, IN 46204   (317) 638-1313

Paul S. Collignon
6710 N. Riley
Indianapolis, IN 46220   (317) 251-7659

Richard D. Conard
Barnes & Thornburg
11 South Meridan Street
Suite 1313
Indianapolis, IN 46204   (317) 231-7285

Robert A. Conrad
Eli Lilly & Company
Lilly Corporate Center
Indianapolis, IN 46285   (317) 276-6013

Gerald V. Dahling
Eli Lilly & Co.
Corporate Center
Indianapolis, IN 46285   (317) 276-2965

John C. Demeter
Eli Lilly & Company
Pat. Dept.
Lilly Corporate Center
Indianapolis, IN 46285   (317) 276-2965

James M. Durlacher
Woodard, Emhardt, Naughton,
Moriarty & Mc Nett
Bank One Center Tower
111 Monument Circle, Ste. 3700
Indianapolis, IN 46204   (317) 634-3456

Richard J. Egan
Dow Consumer Products Inc.
9550 N. Zionsville Road
Indianapolis, IN 46268   (317) 873-7286

C. David Emhardt
Woodard, Emhardt, Naughton,
Moriarty & Mc Nett
Bank One Center Tower
111 Monument Circle, Ste. 3700
Indianapolis, IN 46204   (317) 634-3456

Steven A. Fontana#
Eli Lilly & Co.
Lilly Corporate Center
Indianapolis, IN 46285   (317) 276-2000

Carl A. Forest
7215 Normandy Way
Indianapolis, IN 46278   (317) 276-2000

R. Randall Frisk
Woodard, Emhardt, Naughton
Moriarty & Mc Nett
Bank One Center Tower
111 Monument Circle, Ste. 3700
Indianapolis, IN 46204   (317) 634-3456

Roland A. Fuller, III
Ice, Miller, Donadio & Ryan
One American Square
Box 82001
Indianapolis, IN 46282   (317) 236-2185

Kenneth A. Gandy
Woodard, Emhardt, Naughton,
Moriarty & Mc Nett
Bank One Center Tower
111 Monument Circle, Ste. 3700
Indianapolis, IN 46204   (317) 634-3456

Gerald Howard Glanzman
Willian Brinks Olds Hofer
Gilson & Lione
One Indiana Square
Suite 2425
Indianapolis, IN 46204   (317) 636-0886

Amy E. Hamilton
Eli Lilly & Co.
Lilly Corporate Center
Indianapolis, IN 46285   (317) 276-3169

Ronald S. Hansell
Amax Coal Co. Regulatory Affairs
Beechbank
1205 W. 64th St.
Indianapolis, IN 46260   (317) 253-9624

Nancy J. Harrison
Eli Lilly & Company
Lilly Corporate Center
Dept. MC 529
Indianapolis, IN 46285   (317) 276-2308

Thomas Q. Henry
Woodard, Emhardt, Naughton,
Moriarty & Mc Nett
Bank One Center Tower
111 Monument Circle, Ste. 3700
Indianapolis, IN 46204   (317) 634-3456

Charles W. Hoffmann
Emhart Industries, Inc.
3029 E. Wash. St.
Indianapolis, IN 46206   (317) 261-1417

Jerry E. Hyland
Barnes & Thornburg
11 South Meridian Street
1313 Merchants Bank Bldg.
Indianapolis, IN 46204   (317) 231-7288

Ettore V. Indiano
Ice Miller Donadio & Ryan
One American Square, Box 82001
Indianapolis, IN 46282   (317) 236-2290

Mark D. Janis
Barnes & Thornburg
11 S. Meridian St.
Suite 1313
Indianapolis, IN 46204   (317) 231-7473

Thomas P. Jenkins
Barnes & Thornburg
11 South Meridian Street
Indianapolis, IN 46204   (317) 231-7260

Joseph A. Jones
Eli Lilly & Co.
Patent Division
Lilly Corporate Center
Indianapolis, IN 46285   (317) 276-5183

Max J. Kenemore
Boehringer-Mannheim Corp.
Legal Dept.
9115 Hague Road
Indianapolis, IN 46250   (317) 576-7464

Dilip A. Kulkarni
Thomson Consumer Electronics, Inc.
600 N. Sherman Drive
Indianapolis, IN 46201   (317) 267-6633

Steven R. Lammert
Barnes & Thornburg
1313 Merchants Bank Bldg.
Indianapolis, IN 46204   (317) 231-7258

Robert E. Lee, Jr.
Eli Lillly & Company
Lillly Corporate Center
Indianapolis, IN 46285   (317) 276-2719

Daniel J. Lueders
Woodard, Emhardt, Naughton,
Moriarty & Mc Nett
Bank One Center Tower
111 Monument Circle, Ste. 3700
Indianapolis, IN 46204   (317) 634-3456

Lawrence M. Lunn
6125 U.S. Highway 31 South
Indianapolis, IN 46227   (317) 788-4000

William C. Martens, Jr.
Eli Lilly & Co.
Lilly Corporate Center
Indianapolis, IN 46285   (317) 276-2573

John C. Mc Nett
Woodard, Emhardt, Naughton,
Moriarty & Mc Nett
Bank One Center Tower
111 Monument Circle, Ste. 3700
Indianapolis, IN 46204   (317) 634-3456

Craig E. Mixan
Dow Chemical Company
4040 Vincennes Circle
Suite 601
Indianapolis, IN 46268   (317) 876-5838

John V. Moriarty
Woodard, Emhardt, Naughton,
Moriarty & Mc Nett
Bank One Center Tower
111 Monument Circle, Ste. 3700
Indianapolis, IN 46204   (317) 634-3456

Michael A. Morra
Bell Telephone Labs., Inc.
6612 E. 75th Street
Box 1008
Indianapolis, IN 46206   (317) 845-6012

Dwight Edward Morrison
250 Williams Drive
Indianapolis, IN 46260   (317) 251-0909

Joseph A. Naughton, Jr.
Woodard, Emhardt, Naughton,
Moriarty & Mc Nett
Bank One Center Tower
111 Monument Circle, Ste. 3700
Indianapolis, IN 46204   (317) 634-3456

Timothy E. Niednagel
Barnes & Thornburg
1313 Merchants Bank Bldg.
11 S. Meridan Street
Indianapolis, IN 46204   (317) 231-7551

Douglas K. Norman
Eli Lilly & Company
Lilly Corporate Center
Indianapolis, IN 46285   (317) 276-2958

Paul Overhauser
Baker & Daniels
300 N. Meridian Street
Suite 2700
Indianapolis, IN 46204   (317) 237-0300

Kathleen R.S. Page
Eli Lilly & Co.
Lilly Corporate Center
Indianapolis, IN 46285   (317) 277-4518

Sidney Persley
Eli Lilly & Co.
Patent Division
DC 1119
Lilly Corporate Ctr.
Indianapolis, IN 46285   (317) 276-2719

John F. Prescott, Jr.
Ice, Miller, Donadio & Ryan
One American Square
Box 82001
Indianapolis, IN 46282   (317) 236-2398

David B. Quick
Hughes & Hughes
10401 N. Meridian Street
Indianapolis, IN 46290   (317) 875-7524

Charles R. Reeves
Woodard, Emhardt, Naughton,
Moriarty & Mc Nett
Bank One Center Tower
111 Monument Circle, Ste. 3700
Indianapolis, IN 46204   (317) 634-3456

Richard A. Rezek
Barnes & Thornburg
11 South Meridan St.
1313 Merchants Bank Bldg.
Indianapolis, IN 46204   (317) 231-7283

Andrew J. Richardson
Locke, Reynolds, Boyd & Weisell
1000 Capital Center South
201 N. Illinois St.
Indianapolis, IN 46204   (317) 846-6522

James L. Rowe
Woodard, Emhardt, Naughton,
Moriarty & Mc Nett
Bank One Center Tower
111 Monument Circle, Ste. 3700
Indianapolis, IN 46204   (317) 634-3456

James J. Sales
Eli Lilly & Co.
Lilly Corporate Center
Indianapolis, IN 46285   (317) 276-3474

William B. Scanlon
Eli Lilly & Co.
Lily Corporate Center
Indianapolis, IN 46285   (317) 261-3159

Jack Schuman
Henderson, Daily, Withrow
& Devoe
2600 One Indiana Square
Indianapolis, IN 46204   (317) 639-4121

Everet F. Smith
Barnes & Thornburg
1313 Merchants Bank Bldg.
Indianapolis, IN 46204   (317) 231-7260

Gerald F. Smith
825 Queenswood Ct.
Indianapolis, IN 46217   (317) 888-2952

Robert A. Spray
7114 East 71st Street
Indianapolis, IN 46256   (317) 841-0113

Scott J. Stevens
Thomson Consumer Electronics
600 N. Sherman Drive
Mail Stop 6-209
Indianapolis, IN 46201   (317) 267-6631

Donald R. Stuart
Dow/Elanco Pat. Dept.
9002 Purdue Road
Indianapolis, IN 46268   (317) 871-8479

Houston L. Swenson
Eli Lilly and Co.
Lilly Corporate Center
Indianapolis, IN 46285   (317) 276-2923

Douglas J. Taylor
Eli Lilly & Company
Lilly Corporate Center
Indianapolis, IN 46285   (317) 276-3370

Jay G. Taylor
Ice, Miller, Donadio & Ryan
One American Square
Box 82001
Indianapolis, IN 46282   (317) 236-2150

Vincent O. Wagner
Woodard, Emhardt, Naughton,
Moriarty & Mc Nett
Bank One Center Tower
111 Monument Circle, Ste. 3700
Indianapolis, IN 46204   (317) 634-3456

Richard H. Weber#
3216 West 46th Street
Indianapolis, IN 46208   (317) 291-0354

Arthur Richard Whale
Baker & Daniels
Suite 2700
300 N. Meridian
Indianapolis, IN 46204   (317) 237-1403

Leroy Whitaker
Eli Lilly and Co.
Lilly Corporate Center
Indianapolis, IN 46285   (317) 276-2719

Craig A. Wood
22 East Washington Street
Suite 316
Indianapolis, IN 46204   (317) 637-5245

Harold Raymond Woodard
Woodard, Emhardt, Naughton,
Moriarty & Mc Nett
3700 Bank One Center Tower
Indianapolis, IN 46204   (317) 634-3456

D. Michael Young
Boehringer Mannheim Corp.
9115 Hague Road
P.O. Box 50100
Indianapolis, IN 46250   (317) 576-7340

Stephen E. Zlatos
Woodard, Emhardt, Naughton,
Moriarty & Mc Nett
Bank One Center Tower
111 Monument Circle, Ste. 3700
Indianapolis, IN 46204   (317) 634-3456

Robert S. Linne
Haynes International, Inc.
Patent Department
1020 W. Park Avenue
Kokomo, IN 46902   (317) 456-6115

Joseph J. Phillips#
513 Holly Lane
Kokomo, IN 46902   (317) 453-6132

Richard G. Kinney
Suite 425 South Tower
1000 East 80th Place
Merrillville, IN 46410   (219) 736-2110

Martin B. Barancik
General Elec. Corp.
Highway 69 South
Mt. Vernon, IN 47620   (812) 838-7966

Gilbert E. Alberding
Ball Corp.
345 S. High St.
Muncie, IN 47302   (317) 747-6422

Perry G. Cross
Cross, Marshall, Schuck,
De Weese, Cross & Feick
200 E. Wash. St.
Muncie, IN 47305   (317) 289-6151

Frank A. Steldt
1346 Pebble Brook Drive
Noblesville, IN 46060   (317) 896-5560

Ronald Perry Shipman
106 W. Benton St.
Oxford, IN 47971   (317) 385-2170

Florian S. Gregorczyk
2374 Saturn St.
Portage, IN 46368   (317) 385-2170

Marmaduke A. Hobbs
P.O. Box 367
105 N. Shelby Street
Salem, IN 47167   (812) 883-6145

William Nicholas Antonis
The Bendix Corp.
401 North Bendix Drive
South Bend, IN 46634   (219) 237-2450

Ken C. Decker
Allied Corporation
401 N. Bendix Dr.
South Bend, IN 46634   (219) 237-2455

Thomas J. Dodd
Oltsch, Knoblock & Hall
625 JMS Bldg.
South Bend, IN 46601   (219) 234-6091

Ryan M. Fountain
Barnes & Thornburg
Suite 600
100 N. Michigan
South Bend, IN 46601   (219) 233-1171

James D. Hall
Oltsch, Knoblock & Hall
625 JMS Bldg.
South Bend, IN 46601   (219) 234-6091

Leo H. Mc Cormick, Jr.
Allied-Signal Inc.
401 N. Bendix Drive
South Bend, IN 46628   (219) 237-2452

David R. Melton
Barnes & Thornburg
600 1st Source Bank Center
100 North Michigan St.
South Bend, IN 46601   (219) 233-1171

Craig E. Miller
Barnes & Thornburg
600 1st Source Bank Ctr.
100 North Michigan Street
South Bend, IN 46601   (219) 233-1171

Larry J. Palguta
Allied-Signal Inc.
Midwestern Regional Patent Dept.
P.O. Box 4001
401 N. Bendix Dr.
South Bend, IN 46634   (219) 237-2451

Charles V. Sweeney
Barnes & Thornburg
100 N. Michigan St.
South Bend, IN 46635   (219) 237-1139

David W. Van Story#
1837 Rockne Drive
South Bend, IN 46617   (219) 234-3410

Robert F. Meyer
6116 W. 29th Place
Speedway, IN 46224   (317) 291-3073

H. J. Barnett
2901 Ohio Blvd.
Room 150
Terre Haute, IN 47803   (812) 232-6362

Robert Hastings Dewey#
405 South 34th Street
Terre Haute, IN 47803   (812) 234-6976

Wendell R. Guffey
International Mineral
And Chemicals Corporation
1401 South Third Street
Terre Haute, IN 47802   (812) 232-0121

Kent R. Fase
2756 Hearthstone
Valparaiso, IN 46383   (219) 464-9676

Edward S. Sherman#
1300 Winding Ridge Lane
D-7
Valparaiso, IN 46383   (219) 462-3360

Todd A. Dawson
Zimmer
P.O. Box 708
Warsaw, IN 46580   (219) 267-6131

Margaret L. Geringer#
Zimmer, Inc.
P.O. Box 708
Warsaw, IN 46581   (219) 372-4275

Wendell E. Miller
1907 Crescent Dr.
Warsaw, IN 46580   (219) 267-2729

Paul D. Schoenle
Zimmer
P.O. Box 708
Warsaw, IN 46580   (219) 267-6131

Richard J. Godlewski
Med Institute
P.O. Box 2402
1220 Potter Drive
West Lafayette, IN 47906   (317) 463-7537

Iowa

Spencer Lorraine Blaylock
Research Foundation, Inc.
Iowa State Univ.
315 Beardsher Hall
Ames, IA 50011   (515) 294-4741

Robert O. Richardson
1445 - 14th Street
Bettendorf, IA 52722   (319) 359-0626

William Topping Metz
Jackson & Metz
306 Tama Bldg.
Burlington, IA 52601   (319) 752-2241

Meewon H. Kimm#
1810 Grand Blvd.
Cedar Falls, IA 50613   (319) 752-2241

Allan L. Harms
Wenzel, Piersall, Riccolo
& Harms, P.C.
4080 First Avenue, N.E.
Cedar Rapids, IA 52402   (319) 393-8900

John J. Horn
Rockwell International Corp.
Pat. Dept., M/S 175-100
400 Collins Rd., N.E.
Cedar Rapids, IA 52498   (319) 395-8280

Macka L. Murrah
Rockwell International Corp.
Mail Station 175-100
400 Collins Rd., N.E.
Cedar Rapids, IA 52498   (319) 395-8208

James C. Nemmers
Shuttleworth & Ingersoll, P.C.
P.O. Box 2107
500 Merchants Natl. Bank Bldg.
Cedar Rapids, IA 52406   (319) 365-9461

Winfrid O.E. Schellin
Simmons, Perrine, Albright
& Ellwood
1200 Merchants National Bk. Bldg.
Cedar Rapids, IA 52401   (319) 366-7641

John Howland Sherman
Norand Corporation
550 Second Street S.E.
Cedar Rapids, IA 52401   (319) 369-3661

Wayne H. Stoppelmoor, Jr.#
Square D Company
3700 6th St., S.W.
P.O. Box 3069
Cedar Rapids, IA 52406   (319) 365-4631

Gregory G. Williams
Rockwell International Corp.
400 Collins Road N.E.
M.S. 175-100
Cedar Rapids, IA 52498   (319) 395-2348

Stephen W. Southwick
Iowa Southern Utilities Co.
300 Sheridan Ave.
Centerville, IA 52544   (515) 437-4400

Albert E. Arnold, Jr.
1916 W. 38th St.
Davenport, IA 52806   (515) 437-4400

John E. Cepican
Henderson & Sturm
101 W. 2nd Street
Suite 204
Davenport, IA 52801   (319) 323-9731

H. Vincent Harsha
Henderson & Sturm
101 W. Second Street
Suite 204
Davenport, IA 52801   (319) 323-9731

Harold M. Knoth
Henderson & Sturm
101 W. 2nd Street
Davenport, IA 52801   (319) 323-9731

Morton S. Adler
Adler, Brennan & Joyce
317 Sixth Avenue
Suite 420
Des Moines, IA 50309   (515) 244-1391

James D. Birkenholz
974 - 73rd Street
Suite 10
Des Moines, IA 50312   (515) 223-1335

Robert Lee Farris
Massey-Ferguson Inc.
P.O. Box 1813
Des Moines, IA 50306   (515) 247-2100

Richard L. Fix
Henderson & Sturm
1213 Midland Financial Bldg.
206 Sixth Avenue
Des Moines, IA 50309   (515) 288-9589

Mark D. Frederiksen
Zarley, Mc Kee, Thomte,
Voorhees & Sease
801 Grand Ave.
Suite 3200
Des Moines, IA 50309   (402) 392-2280

Mark Davig Hansing
Zarley, Mc Kee, Thomte,
Voorhees & Sease
Suite 3200
801 Grand Ave.
Des Moines, IA 50309   (515) 288-3667

Kirk M. Hartung
Zarley, Mc Kee, Thomte,
Voorhees & Sease
Suite 3200
801 Grand Avenue
Des Moines, IA 50309   (515) 288-3667

H. Robert Henderson
Henderson & Sturm
1213 Midland Financial Bldg.
206 Sixth Avenue
Des Moines, IA 50309   (515) 288-9589

Kent A. Herink
Davis, Hockenberg, Wine,
Brown, Koehn & Shors
2300 Financial Center
666 Walnut Street
Des Moines, IA 50309   (515) 243-2300

Michael R. Hoffmann
Hoffmann & Udelhofen
Breakwater Building
3708 75th St.
Des Moines, IA 50322   (515) 270-8899

Brian J. Laurenzo
Davis, Hockenberg, Wine,
Brown, Koehn & Shors, P.C.
2300 Financial Center
Des Moines, IA 50309   (515) 243-2300

Rudolph L. Lowell
Davis, Hockenberg, Wine,
Brown, Koehn & Shors
2300 Financial Center
Des Moines, IA 50309   (515) 243-2300

Bruce Welcher Mc Kee
Zarley, Mc Kee, Thomte,
Voorhees & Sease
Suite 3200
801 Grand Ave.
Des Moines, IA 50309   (515) 288-3667

Michael J. Roth
Pioneer Hi-Bred
International, Inc.
700 Capital Square
400 Locust Street
Des Moines, IA 50309   (515) 245-3500

Edmund John Sease
Zarley, Mc Kee, Thomte,
Voorhees & Sease
Suite 3200, 801 Grand Ave.
Des Moines, IA 50309   (515) 288-3667

Michael O. Sturm
Henderson & Sturm
1213 Midland Financial Bldg.
206 Sixth Avenue
Des Moines, IA 50309   (515) 288-9589

Patricia A. Sweeney
Pioneer Hi-Bred International, Inc.
700 Capital Square
400 Locust Street
Des Moines, IA 50309   (515) 245-3500

Michael G. Voorhees
Zarley, Mc Kee, Thomte,
Voorhees & Sease
801 Grand Avenue
Suite 3200
Des Moines, IA 50309   (515) 288-3667

Donald H. Zarley
Zarley, Mc Kee, Thomte,
Voorhees & Sease
801 Grand Ave.
Suite 3200
Des Moines, IA 50309   (515) 288-3667

Charles D. Wingate
Marcus & Courtade, P.C.
107 S. Main St.
Fairfield, IA 52556   (515) 472-5945

Thomas E. Frantz#
Sheaffer Eaton Inc.
301 Avenue H
Fort Madison, IA 52627   (319) 372-3300

Lucas J. De Koster
1106 Main Street
Hull, IA 51239   (712) 439-2511

Ray V. Bailey
Millers Bay
RR2- Box 190
Milford, IA 51351   (712) 337-3571

Frank B. Hill
Bandag, Inc.
Bandag Center
Muscatine, IA 52761   (319) 262-1373

Allan P. Orsund
Maytag Co.
One Dependability Square
Newton, IA 50208   (515) 792-7000

Richard L. Ward#
The Maytag Co.
One Dependability Square
Newton, IA 50208   (515) 791-8851

David C. Larson
P.O. Box 246
Highways 71 & 9
Spirit Lake, IA 51360   (712) 336-4210

G. Brian Pingel
Shearer, Templer & Pingel, P.C.
3737 Woodland Avenue
437 Colony Park Bldg.
West Des Moines, IA 50265   (515) 225-3737

Kansas

Steven R. Janda#
217 E. 10th St.
Ellsworth, KS 67439   (913) 472-3769

Milton J. Chamberlain
RR 1, Box 135
Holton, KS 66436   (913) 472-3769

Thomas M. Scofield
4901 College Blvd.
Suite 104
Leawood, KS 66211   (913) 491-6474

John O. Mingle
P.O. Box 131
Manhattan, KS 66502   (913) 537-0838

Paul H. Harder#
4 Hickory Court
Newton, KS 67114   (316) 283-8262

Charles L. Johnson, Jr.
P.O. Box 545
201 E. Loula
Suite 203
Olathe, KS 66061   (913) 764-8773

D.A.N. Chase
Chase & Yakimo
4400 College Blvd.
Suite 130
Overland Park, KS 66211   (913) 339-9666

Chung L. Feng#
4005 W. 104th Terrace
Overland Park, KS 66207   (913) 649-9212

Joan O. Herman
Linde, Thompson, Fairchild,
Langworthy, Kohn & Van Dyke, P.C.
9300 Metcalf, Suite 1000
One Glenwood Place
Overland Park, KS 66212   (913) 649-4900

Kenneth W. Iles
9903 W. 129th St.
Overland Park, KS 66213   (913) 897-6738

Claude W. Lowe
10051 Roe Avenue
Overland Park, KS 66207   (913) 642-6212

Michael Yakimo, Jr.
Chase & Yakimo
4400 College Boulevard
Suite 130
Overland Park, KS 66211   (913) 339-9666

Keith Dillon Moore
P.O. Box 3563
Shawnee Mission, KS 66203   (913) 631-8420

Bruce J. Clark
Davis, Wright, Unrein, Hummer
& Mc Callister
3715 S.W. 29th Street
Topeka, KS 66614   (913) 273-4220

Stephen Adnan Jones
Kansas Board of Tax Appeals
915 South Harrison
Fourth Floor
Topeka, KS 66612   (913) 296-2388

Edward Linus Brown, Jr.
110 N. Market Street
Suite 420
Wichita, KS 67202   (316) 263-6400

John W. Carpenter
Widdowson & Carpenter
107 N. Market St.
Suite 401
Wichita, KS 67202   (316) 267-8381

Thomas Emil Harrison, Jr.
Boeing Military Airplanes.
P.O. Box 7730
M.S. K11-60
3801 S. Oliver Street
Wichita, KS 67277   (316) 526-2449

Lee W. Huffman
Boeing Military Airplane Company
3801 South Oliver
Wichita, KS 67210   (316) 526-7618

Harold J. Pfountz
Coleman Co., Inc.
P.O. Box 1762
Wichita, KS 67201   (316) 261-3197

John H. Widdowson
Widdowson & Carpenter
401 Bitting Bldg.
Wichita, KS 67202   (316) 267-8381

Kentucky

Theresa F. Camoriano
11508 Arbor Drive East
Anchorage, KY 40223   (502) 244-2705

Michael Ross Dowling
433 16th St.
P.O. Box 1689
Ashland, KY 41101   (606) 325-7682

Stanley M. Welsh
Ashland Oil, Inc.
2000 Ashland Dr.
P.O. Box 391
Ashland, KY 41114   (606) 329-5931

Richard Coale Willson
2000 Ashland Drive
Box 391 BL2
Ashland, KY 41114   (606) 329-4153

Philip R. Cloutier
906 Wrenwood
Bowling Green, KY 42101   (502) 782-3560

Bryan W. Le Sieur
P.O. Box 57
104 E. Jackson Street
Brownsville, KY 42210   (502) 597-2132

P. Joseph Clarke, Jr.
Clarke & Clarke
120 North Third Street
Box 297
Danville, KY 40422   (606) 236-2240

Kenneth F. Pearce
631 Denmark Drive
Danville, KY 40422   (606) 236-6401

Carl E. Knochelmann, Jr.
98 Garvey Avenue
Elsmere, KY 41018   (606) 342-9029

John Arthur Brady
Lexmark International, Inc.
Intellectual Prop. Law
Dept. 827A / Bldg. 035-2
740 New Circle Road, N.W.
Lexington, KY 40511   (606) 232-4785

D. Kendall Cooper
782 Cindy Blair Way
Lexington, KY 40503   (606) 224-4007

William J. Dick
IBM Corp.
740 New Circle Rd.
952/035-2
Lexington, KY 40511   (606) 232-5292

Steven Dale Gilliam
608 Autumn Lane
Lexington, KY 40502   (606) 266-7843

John Warren Girvin, Jr.
IBM Corporation
Dept. 952, Bldg. 035-2
740 New Circle Road, N.W.
Lexington, KY 40511   (606) 232-5292

J. Ralph King
King & Schickli
3070 Harrodsburg Road
Lexington, KY 40503   (606) 223-4050

Frank C. Leach, Jr.
P.O. Box 22455
Lexington, KY 40522   (606) 254-1395

Laurence R. Letson
2468 Heather Court
Lexington, KY 40503   (606) 278-1216

John J. Mc Ardle, Jr.
Lexmark International, Inc.
Intellectual Property Law
Dept. 827 A/ Bldg. 035-2
740 New Circle Rd., N.W.
Lexington, KY 40511   (606) 232-3939

Paul E. O Donnell, Jr.*
Social Security Administration
Off. of Hearings & Appeals
Suite 230, Bakhaus Bldg.
1500 W. Main Street
Lexington, KY 40505   (606) 233-2653

Warren D. Schickli
King & Schickli
3070 Harrodsburg Road
Suite 210
Lexington, KY 40503   (606) 223-4050

Jack E. Toliver
Landrum & Shouse
106 W. Vine Street
9th Floor
Lexington, KY 40507   (606) 255-2424

Richard A. Walker
Kuhlman Corporation
P.O. Box 6000
Lexington, KY 40544   (606) 224-4300

Ralph B. Brick
Polster, Polster & Lucchesi
2303 Tuckaho Road
Louisville, KY 40207   (502) 895-4672

Donald L. Cox
Lynch, Cox, Gilman
& Mahan, P.S.C.
500 Meidinger Tower
Louisville, KY 40202   (502) 589-4215

Scott R. Cox
Lynch, Cox, Gilman
& Mahan P.S.C.
500 Meidinger Tower
Louisville, KY 40202   (502) 588-4215

James P. Dowd#
8820 Tranquil Valley Lane
Louisville, KY 40299   (502) 588-4215

James C. Eaves, Jr.#
Middleton & Reutlinger
2500 Brown & Williamson Tower
Louisville, KY 40202   (502) 584-1135

Robert W. Fletcher
Hilliard-Lyons Pat. Management, Inc.
10509 Timberwood Circle
Louisville, KY 40223   (502) 429-0015

Frank P. Giacalone#
4623 Fox Run Road
Louisville, KY 40207   (502) 896-1539

James R. Higgins, Jr.
Middleton & Reutlinger
2500 Brown & Williamson Tower
Louisville, KY 40202   (504) 584-1135

Harold N. Houser
General Electric Co.
Appliance Park
AP 2-225
Louisville, KY 40225   (502) 452-4653

Charles G. Lamb
507 Oak Branch Road
Louisville, KY 40223   (502) 245-9607

William J. Mason
Premier Brands Ltd.
1303 Clear Springs Trace
Louisville, KY 40223   (502) 429-5207

Maurice L. Miller, Jr.
Robert, Miller, Thomas
& Dodson
Suite 101
200 Whittington Pkwy.
Louisville, KY 40222   (502) 425-2802

Thomas R. Payne
3504 Sorrento
Louisville, KY 40241   (502) 425-4207

Herbert Peter Price
Lynch, Cox, Gilman
& Mahan, P.S.C.
500 Meidinger Tower
Louisville, KY 40202   (502) 589-4215

Radford Monroe Reams, III
General Electric Co.
Appliance Park 2-225
Louisville, KY 40225   (502) 452-3331

Charles I. Sherman
Brown & Williamson
Tobacco Corp.
1500 B&W Tower
P.O. Box 35090
Louisville, KY 40232   (502) 425-3220

Vance A. Smith
10507 Timberwood Circle
Suite 208C
Louisville, KY 40223   (502) 423-9850

Edward Miller Steutermann
1332 South 2nd Street
Louisville, KY 40208   (502) 636-0467

Jon C. Winger
Winger & Steutermann
Portland Federal Bldg.
200 West Broadway
Suite 612
Louisville, KY 40202   (502) 583-7336

Bruce A. Yungman
10415 Edgewater Road
Louisville, KY 40223   (502) 245-6565

Nathan J. Cornfeld
2139 Griffith Ave.
Owensboro, KY 42301   (502) 684-2668

John J. Davis
P.O. Box 512
Pikeville, KY 41501   (606) 437-4026

Louisiana

Thomas E. Balhoff
Mathews, Atkinson, Guglielmo
Marks & Day
P.O. Box 3177
Baton Rouge, LA 70821   (504) 387-6966

Paul M. Bork
6113 Belle Grove
Baton Rouge, LA 70820   (504) 767-8377

David M. Bunnell
Ethyl Corporation
Pat. & Tdmk. Div.
451 Florida Blvd.
Baton Rouge, LA 70801   (504) 388-7098

Allen D. Darden
P.O. Box 4412
Baton Rouge, LA 70821   (504) 346-0285

Patricia J. Hogan#
Ethyl Corporation
451 Florida Blvd.
Baton Rouge, LA 70801   (504) 388-7023

Donald Lewis Johnson
5120 East Bluebell Drive
Baton Rouge, LA 70808   (504) 924-0703

William David Kiesel
Roy, Kiesel, Aaron,
Tucker & Zwick
2355 Drusilla Lane
Baton Rouge, LA 70809   (504) 927-9908

David E. La Rose
Ethyl Corp.
451 Florida Blvd.
Baton Rouge, LA 70801   (504) 388-7526

Paul H. Leonard, III
10639 Rondo Avenue
Baton Rouge, LA 70815   (504) 927-6991

Willard G. Montgomery
Ethyl Corporation
451 Florida Blvd.
Baton Rouge, LA 70801   (504) 388-7937

Henry E. Naylor
Exxon Res. & Eng. Company
P.O. Box 2226
Baton Rouge, LA 70817   (504) 359-7674

Herbert E. O Niell
Exxon Company, U.S.A.
P.O. Box 551
Baton Rouge, LA 70821   (504) 359-4290

J. Bradley Overton
Michael, Roy, Fogler & Assoc.
P.O. Box 1487
Baton Rouge, LA 70821   (504) 383-2222

Joel R. Penton
P.O. Box 16420B
Baton Rouge, LA 70893   (504) 778-3609

Phillip M. Pippenger
Ethyl Corporation
451 Florida Blvd.
Baton Rouge, LA 70801   (504) 388-7097

Penny L. Prater
Exxon Res. & Eng.
P.O. Box 2226
Baton Rouge, LA 70821   (504) 359-4487

Llewellyn Allen Proctor
11481 Sheraton Dr.
Baton Rouge, LA 70815   (504) 275-8689

David L. Ray
2051 Silverside Drive
Suite 205
Baton Rouge, LA 70808   (504) 343-8813

Reginald F. Roberts, Jr.#
P.O. Box 515
Baton Rouge, LA 70821   (504) 343-8500

John H. Runnels
Taylor, Porter, Brooks
& Phillips
P.O. Box 2471
Baton Rouge, LA 70821   (504) 387-3221

John F. Sieberth
Ethyl Corp.
Pat. & Tdmk. Division
451 Florida Blvd.
Baton Rouge, LA 70801   (504) 388-7925

Edgar E. Spielman, Jr.
Ethyl Corporation
Pat. & Tdmk. Div.
451 Florida Street
Baton Rouge, LA 70801   (504) 388-7604

Doris M. Thompson#
Ethyl Corp.
451 Florida Blvd.
Baton Rouge, LA 70801   (504) 388-7224

Robert C. Tucker
Roy, Kiesel, Aaron
& Tucker
2355 Drusilla Lane
Baton Rouge, LA 70809   (504) 927-9908

Rodney Bryant Jordan#
P.O. Box 39
US 171 N.
Florien, LA 71429   (318) 586-4212

James T. Cronvich
625 Posey Avenue
Harahan, LA 70123   (504) 737-0363

James C. Kesterson
Laitram Corporation
220 Laitram Lane
Harahan, LA 70123   (504) 733-6000

Joseph L. Lemoine, Jr.
Onebane, Donohoe, Bernard, Torian,
Diaz, Mc Namara & Abell
P.O. Drawer 3507
Lafayette, LA 70502   (318) 237-2660

William W. Stagg
Durio, Mc Goffin & Stagg
P.O. Box 51308
220 Heymann Blvd.
Lafayette, LA 70505   (318) 233-0300

Donald J. Griffith, Jr.#
Mc Neese State University
Dept. of Chem. & Elec. Engineering
P.O. Box 91735
Lake Charles, LA 70609   (318) 475-5858

Raymond G. Areaux
Milling, Benson, Woodward,
Hillyer, Pierson & Miller
909 Poydras Street
Suite 2300
New Orleans, LA 70112   (504) 569-7000

George A. Bode
2314 Broadway
New Orleans, LA 70125   (504) 861-8288

Len R. Brignac
Milling, Benson, Woodward,
Hillyer, Pierson & Miller
909 Poydras St.
Suite 2300
New Orleans, LA 70112   (504) 569-7000

Michael D. Carbo
Didriksen & Carbo
1122 Whitney Bank Bldg.
228 St. Charles Avenue
New Orleans, LA 70130   (504) 586-1600

Stephen R. Doody
Sessions, Fishman, Rosenson,
Boisfontaine, Nathan & Winn
201 St. Clarles Avenue
36th Floor
New Orleans, LA 70170   (504) 582-1530

Robert John Edwards
Mc Dermott International Inc.
1010 Common St.
New Orleans, LA 70112   (504) 587-5722

Raul V. Fonte
Freeport-McMoran Inc.
1615 Poydras Street
New Orleans, LA 70112   (504) 582-4234

Charles C. Garvey, Jr.
Pravel, Gambrell, Hewitt,
Kimball & Krieger
1515 Poydras Street
Suite 2300
New Orleans, LA 70112   (504) 524-7207

Michael L. Hoelter
Mc Dermott, Inc.
1010 Common Street
P.O. Box 60035
New Orleans, LA 70160   (504) 587-5709

Thomas St. Paul Keaty
Keaty & Keaty
2140 World Trade Center
No. 2 Canal Street
New Orleans, LA 70130   (504) 524-2100

Daniel N. La Haye
Mc Dermott, Inc.
1010 Common P.O. Box 60035
New Orleans, LA 70160   (504) 587-5719

Edward D. Markle
Adams & Reese
4500 One Shell Square
New Orleans, LA 70139   (504) 581-3234

Seth M. Nehrbass
Pravel, Gambrell, Hewitt,
Kimball & Krieger
1515 Poydras Street
Suite 2300
New Orleans, LA 70112   (504) 524-7207

Alexander H. Plache
Jones, Walker, Waechter, Poitevent,
Carrere & Denegre
201 St. Charles Ave.
New Orleans, LA 70170   (504) 582-8000

C. Emmett Pugh
Pugh & Associates
757 St. Charles Avenue
Suite 204
New Orleans, LA 70130   (504) 587-0000

Wylie R. Lynette
Mc Glinchey, Stafford, Cellini
& Lang, P.C.
643 Magazine Street
New Orleans, LA 70130   (504) 586-1200

Joseph T. Regard, III
Pugh / Associates
4917 St. Charles Ave.
New Orleans, LA 70115   (504) 587-0000

Lloyd N. Shields
Simon, Peragine, Smith
& Redfearn
3000 Energy Centre
1100 Poydras St., 30th Flr.
New Orleans, LA 70163   (504) 569-2030

Gregory C. Smith
Pravel, Gambrell, Hewitt,
Kimball & Krieger
1515 Poydras Center
Suite 2300
New Orleans, LA 70112   (504) 524-7207

Raymond C. Von Bodungen
1009 Opelousas Avenue
New Orleans, LA 70114   (504) 361-4247

William B. Miller
Dow Chemical Company
Patent Dept.
P.O. Box 400
Plaquemine, LA 70765   (504) 389-8914

James M. Pelton
Dow Chemical Company
Patent Dept., Bldg. 2507
P.O. Box 400
Plaquemine, LA 70765   (504) 389-1807

John M. Harrison
2139 E. Bert Kouns-Industrial Loop
Shreveport, LA 71105   (318) 797-3062

Robert C. Mai
1038 St. Peter Drive
Slidell, LA 70460   (504) 643-8064

John D. Jeter#
1403 Teche Drive
St. Martinville, LA 70582   (318) 394-5017

Maine

Charles A. Cutting
Upton Rd.
Andover, ME 04216   (207) 392-3741

David Francis Gould
Mc Cabe Nursery
2220 Ohio St.
Bangor, ME 04401   (207) 947-7822

Daniel H. Kane, Jr.
Fenton, Chapman, Fenton,
Smith & Kane, P.A.
109 Main St.
P.O. Box B
Bar Harbor, ME 04609   (207) 288-9393

William Rowsell Hulbert
Fish & Richardson
P.O. Box 90
Lincolnville, ME 04849   (207) 236-3508

Daniel L. Peabody
Thompson & Peabody, P.A.
P.O. Box 617
Route 302
Lakes Region Professional Ctr.
Naples, ME 04055   (207) 693-3030

Thomas L. Bohan
371 Fore Street
Portland, ME 04101   (207) 773-3132

Martin J. Robles
Pierce, Atwood, Scribner, Allen,
Smith & Lancaster
One Monument Square
Portland, ME 04101   (207) 773-6411

Laforest S. Saulsbury
519 Congress Street
Portland, ME 04101   (207) 773-8463

Joseph Shortill
Shortill & Shortill
6 Washington Street
P.O. Box 1069
Sanford, ME 04073   (207) 324-8070

Harold W. Lockhart
HCR #64
Box 122
South Bristol, ME 04568   (207) 644-8888

Wolfgang G. Fasse
Indian Pond Lane
P.O. Box K
St. Albans, ME 04971   (207) 938-4422

James W. Bock
Box 356
Swans Island, ME 04685   (207) 526-4368

Abbott Spear
Main Street
Warren, ME 04864   (207) 273-2768

Frederick R. Cantor
P.O. Box 981
Wells, ME 04090   (207) 646-6436

Francis M. Di Biase#
Scott Paper Co.
89 Cumberland St.
Westbrook, ME 04092   (207) 856-6911

Maryland

Nancy Ann Coleman
420 Hillcrest Drive
Aberdeen, MD 21001   (301) 272-3157

Richard C. Reed
P.O. Box 233
Accokeek, MD 20607   (301) 292-5618

Saul Elbaum*
U.S. Army
2900 Powder Mill Road
Adelphi, MD 20783   (202) 394-3790

Freda L. Krosnick*
U.S. Dept. of Army
U.S. Laboratory Command
2800 Powder Mill Road
Adelphi, MD 20783   (202) 394-1105

Paul E. Maslousky
2512 Hughes Road
Adelphi, MD 20783   (301) 439-4987

Thomas E. Mc Donald*
U.S. Dept. of the Army
Intellectual Property Division
2800 Powder Mill Road
Adelphi, MD 20783   (202) 394-3790

Guy M. Miller*
Department of the Army
Hdqrs. U.S. Army Labcom
2800 Powder Mill Road
Adelphi, MD 20783   (202) 394-3790

Muzio B. Roberto*
U.S. Army Laboratory Command
SLC1S-CC-1P
2800 Powder Mill Rd.
Aldelphi, MD 20783   (301) 394-3790

Egil Angeid#
Intersearch Inc.
855 Rudder Way
Annapolis, MD 21401   (301) 266-6771

Albert Harrison Helvestine
1729 Fairlop Trail
Epping Forest
Annapolis, MD 21401   (301) 849-8467

Harry A. Herbert, Jr.
1821 Manor Green Ct.
Annapolis, MD 21401   (301) 849-8467

Fendall Marbury
9 Neal Street
Annapolis, MD 21401   (301) 266-8254

Frank R. Ortolani, Sr.
1924 Harwood Road
Annapolis, MD 21401   (301) 757-8959

Charles William Helzer
P.O. Box 309
694 White Swan Drive
Arnold, MD 21012   (301) 261-1242

Robert E. Bushnell
200 N. Rolling Rd.
Baltimore, MD 21228   (202) 429-4560

John P. Carter#
Nova Pharmaceutical Corp.
6200 Freeport Centre
Baltimore, MD 21224   (301) 563-6168

James B. Eisel
Martin Marietta Laboratories
1450 S. Rolling Road
Baltimore, MD 21227   (301) 247-0700

Vincent L. Fabiano
Nova Pharmaceutical Corp.
6200 Freeport Centre
Baltimore, MD 21224   (301) 522-7000

Walter G. Finch
Fidelity Bldg., Suite 1501-03
206 N. Charles Street
Baltimore, MD 21201   (301) 539-8170

Peter Gibson#
2110 N. Charles St.
Baltimore, MD 21218   (301) 625-1196

Thomas Hoxie
Semmes, Bowen & Semmes
250 West Pratt Street
Baltimore, MD 21201   (301) 576-4781

Leonard J. Kerpelman
2403 W. Rogers Ave.
Baltimore, MD 21209   (301) 367-8855

Robert Thomas Killman
107 Midhurst Road
Baltimore, MD 21212   (301) 377-8116

Edmond P. Lazarus#*
U.S. Army
Corps of Engineers
P.O. Box 1715
31 Hopkins Plaza
Baltimore, MD 21203   (301) 962-4447

Thomas A. Lupica
801 South Luzerne Avenue
Baltimore, MD 21224   (301) 732-7604

Warren H. Mc Inteer
Price Waterhouse
7 St. Paul Street
Baltimore, MD 21230   (301) 685-0542

Frank H. Mc Kenzie, Jr.#
76 S. Morley St.
Baltimore, MD 21229   (301) 945-3930

Jeffrey R. Melnikoff
3507 Pinkney Road
Apt. 1H
Baltimore, MD 21215   (301) 358-9516

Charles F. Obrecht, Jr.
Obrecht & Obrecht
Suite 310
520 W. Fayette Street
Baltimore, MD 21201   (301) 685-6938

Bernard Joseph Ohlendorf
2924 Christopher Ave.
Baltimore, MD 21214   (301) 426-0216

Max S. Oppenheimer
1800 Mercantile Bldg.
2 Hopkins Plaza
Baltimore, MD 21201   (301) 244-7455

Robin J. Pecora
204 E. Preston St.
Baltimore, MD 21202   (301) 539-1990

Leslie Rajkay#
7617 L Hirondelle Club Road
Baltimore, MD 21204   (301) 828-5579

Brooke Schumm, III
Semmes, Bowen & Semmes
250 W. Pratt Street
Baltimore, MD 21201   (301) 539-5040

Bradley S. Thomas#
9332 Ramblebrook Road
Baltimore, MD 21236   (301) 529-3881

Alan G. Towner#
Martin Marietta Laboratories
1450 S. Rolling Road
Baltimore, MD 21227   (301) 247-0700

David Bruce Williams
W. R. Grace & Co.
10 East Baltimore St.
Baltimore, MD 21202   (301) 659-9059

Janelle S. Graeter#*
USDA-ARS-OCI
Building 005, Room 402
BARC-WEST
Beltsville, MD 20705   (301) 344-3676

Robert Halper
3118 Calverton Blvd.
Beltsville, MD 20705   (301) 572-4719

Beverly K. Johnson*
U.S. Dept. of Agriculture
ARS, Bldg. 5, Rm. 415
Beltsville, MD 20705   (301) 344-4032

Joseph A. Lipovsky#*
U.S. Dept. of Agriculture
Off. of Cooperative Interactions
Bldg. 005, Room 408
BARC-WEST
Beltsville, MD 20705   (301) 344-1003

David Robert Sadowski#*
USDA-ARS-OCI
Room 415 Bldg. 005
BARC-WEST
Beltsville, MD 20705   (301) 344-4302

Stephen C. Wieder#*
U.S. Dept. of Agriculture
Agricultural Res. Service
Room 402, Bldg. 005, BARC-WEST
Beltsville, MD 20705   (301) 344-3629

Jesse L. Abzug
IBM Corporation
Mail Stop 201
6600 Rockledge Drive
Bethesda, MD 20817   (301) 493-1214

Edward S. Ammeen#*
U.S. Dept. of Navy
David Taylor Res. Ctr.
Carderock Laboratory
Bethesda, MD 20084   (301) 227-1336

Robert Benson#*
National Institutes of Health
(DHHS, PHS) Patent Branch
Building 31, Rm. 2B-62
9000 Rockville Pike
Bethesda, MD 20892   (301) 496-7056

William M. Blackstone#
5225 Pooks Hill Road
Bethesda, MD 20814   (301) 493-6733

Robert K. Carpenter
3 Bethesda Metro Center
Suite 750
Bethesda, MD 20814   (301) 493-6733

Sidney Carter
5000 Battery Lane
Suite 403
Bethesda, MD 20814   (301) 907-6913

Arthur Edward Dowell, III
5121 Scarsdale Road
Bethesda, MD 20816   (301) 229-1815

Harleigh P. Ewell*
U.S. Consumer Product
Safety Commission
Off. of Gen. Coun.
5401 Westbard Avenue, Room 200
Bethesda, MD 20207   (301) 492-6980

Thomas G. Ferris*
Dept. of Health & Human Services
Patent Branch, OGC, DHHS
5A-03 Westwood Bldg.
Bethesda, MD 20892   (301) 496-7056

Gerald M. Forlenza
6401 Rockhurst Road
Bethesda, MD 20817   (301) 530-8518

Bernice W. Freundel#
9212 Bardon Rd.
Bethesda, MD 20014   (301) 897-5380

William T. Fryer, III
7507 Clarendon Rd.
Bethesda, MD 20814   (301) 656-9479

Claude Funkhouser
8808 Ridge Rd.
Bethesda, MD 20817   (301) 365-4137

Paul Louis Gomory
5609 Ogden Rd
Bethesda, MD 20816   (301) 320-4327

Jackson T. Hawkins,*
U.S. Navy
DTNSRDC
Code 1740.2
Bethesda, MD 20084   (202) 227-3767

John E. Hoel
IBM Corporation
Intellectual Property Law
Mail Stop 201
6600 Rockledge Drive
Bethesda, MD 20817   (301) 493-1216

Kenneth C. Hutchison#*
Naval School of Health Sciences
Bldg. 141, BC C, Room B25
Bethesda, MD 20814   (202) 295-6089

Milford A. Juten
4008 61st St. N.W.
Bethesda, MD 20816   (301) 229-2876

Howard Kaiser
U.S. Dept. of Navy
Off. of Gen. Coun.
Code 0205
David Taylor Research Center
Bethesda, MD 20084   (301) 227-1834

Jessie W. Karsted
4920 Sentinel Drive
Apt. 101
Bethesda, MD 20816   (301) 229-4603

Arthur D. Kellogg#
Shlesinger & Meyers
6550 Rock Spring Drive
Bethesda, MD 20817   (301) 365-8000

Jeffrey S. La Baw
IBM Corp.
Mail Stop 201
6600 Rockledge Drive
Bethesda, MD 20817   (301) 493-1218

Hadd S. Lane
4925 River Road
Bethesda, MD 20816   (301) 652-3875

Norman J. Latker
5112 Edgemore Lane
Bethesda, MD 20814   (301) 951-0375

James M. Leas#
5718 Wilson Lane
Bethesda, MD 20817   (301) 320-3329

David Lewis
Kawior & Feldman
6900 Wisconsin Avenue
Suite 406
Bethesda, MD 20815   (301) 654-8460

Leonard L. Lourie
7500 Westfield Drive
Bethesda, MD 20817   (301) 320-5411

Charles R. Macedo#
6100 Highboro Drive
Bethesda, MD 20817   (301) 229-0408

Herbert Magil
6708 Whittier Blvd.
Bethesda, MD 20817   (301) 229-0670

Alfred C. Marmor
7604 Quintana Court
Bethesda, MD 20817   (301) 365-2075

Luther A. Marsh*
U.S. Navy
Office of Pat. Counsel
David Taylor Naval Ship R&D Ctr.
Bethesda, MD 20084   (301) 227-1834

Thomas D. Mays#*
Nat. Institutes of Health
Off. of Techn. Dev.
Building 31, Room 4a51
9000 Rockville Pike
Bethesda, MD 20892   (301) 496-0477

Wilmer Mechlin
4733 Bethesda Avenue
Suite 350
Bethesda, MD 20814   (301) 652-6580

Robert W. Michell
7932 Maryknoll Avenue
Bethesda, MD 20817   (301) 229-4979

John A. Pekar#
9307 Adelaide Drive
Bethesda, MD 20817   (301) 530-8547

James M. Ready#*
U.S. Dept. of Navy
David Taylor Res. Center
Bethesda, MD 20084   (301) 227-1797

Joseph C. Redmond, Jr.
IBM Corp.
6600 Rockledge Drive
Mail Stop 201
Bethesda, MD 20817   (301) 493-1212

Gloria H. Richmond*
Dept. of Health & Human Services
Patent Branch, NIH Bldg. 31
Room 2B-62
9000 Rockville Pike,
Bethesda, MD 20892   (301) 496-7056

Howard L. Rose
Shlesinger & Myers
6550 Rock Spring Drive
Suite 240
Bethesda, MD 20817   (301) 365-8000

William H. Schultz#
4920 Redford Road
Bethesda, MD 20816   (301) 656-7334

Sheldon J. Singer
Singer & Recht, Chtd.
7315 Wisconsin Avenue
Suite 242W
Bethesda, MD 20814   (301) 654-6505

Avrom David Spevack*
U.S. Dept. of Navy
Naval Medical Research &
Development Command
National Naval Medical Center
Bethesda, MD 20814   (301) 295-6759

Jack Spiegel#*
Dept. Health & Human Services
NIH Patent Branch
Bldg. 31, 2b62
Bethesda, MD 20892   (301) 496-7056

Bernard Stickney
6021 Berkshire Dr.
Bethesda, MD 20814   (301) 530-1504

Walter Stolwein#
5211 Roosevelt Street
Bethesda, MD 20814   (301) 530-0921

John E. Tarcza#*
National Institutes of Health
9000 Rockville Pike
Bldg. 31, 2b62
Bethesda, MD 20892   (301) 496-7056

Norman G. Torchin
5225 Pooks Hill Road
Apt. 218N
Bethesda, MD 20814   (301) 493-6012

Lazar D. Wechsler#
10105 Ashburton Lane
Bethesda, MD 20817   (301) 530-8693

Lorraine A. Weinberger
5404 Linden Court
Bethesda, MD 20814   (301) 530-0429

Bruce M. Winchell
Martin Marietta Corp.
6801 Rockledge Drive
Bethesda, MD 20817   (301) 897-6356

Frederic K. Wine
9116 Friars Road
Bethesda, MD 20817   (301) 897-8843

Mark A. Wurm
IBM Corporation
6600 Rockledge Drive
Mail Stop 201
Bethesda, MD 20817   (301) 493-1217

Leon Zitver
6502 East Halbert Road
Bethesda, MD 20817   (301) 229-6725

Donald J. Arnold#
2811 Sudberry Lane
Bowie, MD 20715   (301) 464-1452

Francis E. Blake
6403 S. Homestake Drive
Bowie, MD 20715   (301) 262-0549

Morris Kaplan
3014 Tyson Lane
Bowie, MD 20715   (301) 464-8970

F. Richard Malzone
12509 Swirl Lane
Bowie, MD 20715   (301) 262-1082

Mayer Weinblatt#
3326 Memphis Lane
Bowie, MD 20715   (301) 262-3264

John K. Donaghy
6108 Mc Kay Drive
Brandywine, MD 20613   (301) 372-8685

John Root Hopkins, Jr.
Rte. 4, Box 289A
Dark Road
Cambridge, MD 21613   (301) 228-6380

Ronald J. Sasiela#
Coldwater Seafood Corporation
904 Woods Road
Cambridge, MD 21613   (301) 228-7500

Roland Alfred Anderson
3810 Club Dr.
Chevy Chase, MD 20015   (301) 656-4175

Ernst-Theodore Arndt
6420 Elmwood Road
Chevy Chase, MD 20815   (301) 656-2828

Harold S. Block#
6412 Ruffin Road
Chevy Chase, MD 20815   (301) 652-8915

George W. Boys
4811 Wellington Dr.
Chevy Chase, MD 20815   (301) 652-5972

Rupert Joseph Brady
Brady, O Boyle and Gates
920 Chevy Chase Bldg.
5530 Wisconsin Ave., N.W.
Chevy Chase, MD 20815   (301) 656-3326

Peter T. Dracopoulos
4515 Willard Avenue
Apt. 904-S
Chevy Chase, MD 20815   (301) 652-3330

Lawrence I. Field
3214 Pauline Dr.
Chevy Chase, MD 20815   (301) 656-3903

William L. Gates
Brady, O Boyle and Gates
920 Chevy Chase Bldg.
5530 Wisconsin Ave., N.W.
Chevy Chase, MD 20815   (301) 656-3326

Harold L. Jenkins
4407 Walsh St.
Chevy Chase, MD 20815   (301) 657-2471

Alfred Bernard Levine
P.O. Box 15968
Chevy Chase, MD 20815   (301) 588-2618

Sam Meerkreebs
5509 Greystone St.
Town of Somerset
Chevy Chase, MD 20815   (301) 652-7960

James G. O Boyle
714 Chevy Chase Bldg.
5530 Wisconsin Avenue
Chevy Chase, MD 20815   (301) 656-3326

John Petrakes#
9005 Levelle Drive
Chevy Chase, MD 20815   (301) 652-2595

Michael W. Werth
14 Grafton Street
Chevy Chase, MD 20815   (301) 986-0793

James A. Wong
714 Chevy Chase Bldg.
5530 Wisconsin Avenue
Chevy Chase, MD 20815   (301) 652-4372

Charles L. Harness
6705 Whitegate Rd.
Clarksville, MD 21029   (301) 531-6189

William G. Christoforo
10609 Blue Bell Way
Cockeysville, MD 21030   (301) 666-0824

Boyce C. Dent
Ward Machinery Co.
10615 Beaver Dam Road
Cockeysville, MD 21030   (301) 584-7700

Gerald Lee Lett
Litton Industries
5115 Calvert Road
College Park, MD 20740   (301) 454-9965

Vanessa L. Appleby#
W.R. Grace & Co.-Conn.
7379 Route 32
Columbia, MD 21044   (301) 531-4120

James P. Barr
W.R. Grace & Company
7379 Route 32
Columbia, MD 21044   (301) 531-4518

Edward J. Cabic
W.R. Grace & Co.
7379 Route 32
Columbia, MD 21044   (301) 531-4512

Steven Capella#
W.R. Grace & Co.-Conn.
7379 Route 32
Columbia, MD 21044   (301) 531-4514

Joseph P. Curtin#
7222 Single Wheel Path
Columbia, MD 21046   (301) 854-0766

Alfred Gluecksmann#
W.R. Grace & Co.
7379 Route 32
Columbia, MD 21044   (301) 531-4660

Jill H. Krafte
W.R. Grace & Co.-Conn.
Washington Research Center
7379 Route 32
Columbia, MD 21044   (301) 531-4509

William S. Ramsey#
5253 Even Star Place
Columbia, MD 21044   (301) 730-9467

Arthur Paul Savage
W.R. Grace and Co.
Research Division
7379 Route 32
Columbia, MD 21044   (301) 531-4511

Steven T. Trinker
W.R. Grace & Company
Pat. Dept.
7379 Route 32
Columbia, MD 21044   (301) 531-4508

Howard J. Troffkin
W.R. Grace & Co.
Legal Dept.
7379 Route 32
Columbia, MD 21044   (301) 531-4516

Robert Kinberg
Power International, Inc.
2127 Espey Ct.
Suite 210
Crofton, MD 21114   (301) 585-0333

James A. Kunkle
Box 548
Deale, MD 20751   (301) 261-9246

Harold A. Dixon
16204 Deer Lake Road
Derwood, MD 20855   (301) 926-1739

J. D. Miller
10031 Kaylorite Street
Dunkirk, MD 20754   (301) 855-8891

J. H. Fielding Jukes
22 Lynnbrook Terrace
Easton, MD 21601   (301) 822-9122

Donald E. Bullock#
1513 Shore Drive
Edgewater, MD 21037   (301) 261-7049

Fred L. Witherspoon, Jr.
1117 Paca Drive
Edgewater, MD 21037   (301) 261-4088

Frank W. Lane
Red Bird Farm
162 Russell Road
Elkton, MD 21921   (301) 398-0724

David I. Klein
Rosenberg, Maleson & Bilker
3444 Ellicott Center Drive
Suite 105
Ellicott City, MD 21043   (301) 465-6678

Morton J. Rosenberg
Rosenberg, Maleson & Bilker
3444 Ellicott Center Drive
Suite 105
Ellicott City, MD 21043   (301) 465-6678

Haven Ely Simmons
P.O. Box 1399
Ellicott City, MD 21043   (301) 750-3979

Dean T. Fisher
1901 Norwood Court - North
Fallston, MD 21047   (301) 347-9090

Joseph A. Finlayson, Jr.
8410 Indian Head Highway
Fort Washington, MD 20744   (301) 567-7230

Frank W. Miga
8007 Carey Branch Drive
Fort Washington, MD 20744   (301) 567-6840

Harry E. Thomason
609 Cedar Avenue
Fort Washington, MD 20744   (301) 292-5122

Thomas O. Maser*
National Security Agency
9800 Savage Road
Ft. George G. Meade, MD 20755
          (301) 859-6647

John R. Utermohle*
National Security Agency
Attn: R (P A)
Ft. George G. Meade, MD 20755
          (301) 859-6647

Fredric D. Abramson
21155 Burnham Rd.
Gaithersburg, MD 20882   (301) 840-9733

Isaac A. Angres#
6 War Admiral Court
Gaithersburg, MD 20878   (301) 926-2742

Mark S. Berninger#
Life Technologies Inc.
8717 Grovemont Circle
Gaithersburg, MD 20877   (301) 258-8205

Alvin Guttag
415 Russell Avenue
Apartment 108
Gaithersburg, MD 20877   (301) 330-4541

John P. Halvonik
845 Quince Orchard Blvd.
Suite E
Gaithersburg, MD 20878   (301) 990-9393

Florina B. Hoffer#*
National Institute of
Standards & Technology
Physics Building
Room A343
Gaithersburg, MD 20899   (301) 975-3084

Nathan Kaufman
427 Christopher Ave.
Apt. 22
Gaithersburg, MD 20879   (301) 948-8174

Tonya S. Lamb#
9717 Ambergate Ct.
Gaithersburg, MD 20882   (301) 253-6987

Robert D. Larrabee#*
National Institure of
Standards & Technology
Bldg. 225, Room A-347
Gaithersburg, MD 20899   (301) 975-2298

Catherine L. Mills#
8904 Clewiston Place
Gaithersburg, MD 20879   (301) 948-7342

Walter O. Ottesen
12704 Split Creek Court
Gaithersburg, MD 20878   (301) 869-8950

Michael W. York
5508 Griffith Rd.
Gaithersburg, MD 20882   (301) 253-4217

Edward D. C. Bartlett#
14026 Burntwoods Road
Glenelg, MD 21737   (301) 442-2203

Paul S. Clohan, Jr.*
NASA
Goddard Space Flight Center
Off. of Pat. Coun., Code 204
Greenbelt, MD 20771   (301) 442-2203

Robert D. Marchant*
NASA
Goddard Space Flight Center
Off. of Pat. Coun., Code 204
Greenbelt, MD 20771   (301) 286-7351

Harvey Ostrow*
NASA
Code 925
Goddard Space Flight Ctr.
Greenbelt, MD 20771   (301) 286-7087

Elizabeth J. Pawlak-Byczkowska#
7534 Mandan Road
Greenbelt, MD 20770   (301) 474-3391

Ronald F. Sandler*
NASA
Goddard Space Flight Ctr.
Code 204, Off. of Pat. Counsel
Greenbelt, MD 20771   (301) 344-9275

William D. Johnston, III
3458 Holland Cliff Road
Huntingtown, MD 20639   (301) 535-3015

James R. Garrett#
4300 College Hgts., Dr.
Hyattsville, MD 20782   (301) 779-1384

Joseph A. Genovese
7313 Adelphi Road
Hyattsville, MD 20783   (301) 779-1384

John H. Mack
7208 Hitching Post Lane
Hyattsville, MD 20783   (301) 422-8898

Richard C. Pinkham
5606 - 37th Avenue
Hyattsville, MD 20782   (301) 277-2068

Thomas H. Webb#
2226 Beechwood Road
Hyattsville, MD 20783   (301) 422-7324

Edward M. Woodberry#
7401 N. Hampshire Ave.
Apt. 903
Hyattsville, MD 20783   (301) 434-2609

Jo Anne S. Beery
211 Kearney Drive
Joppa, MD 21085   (301) 679-2010

Emory Lowell Groff, Jr.
Groff & O'Brien
3514 Plyers Mill Road
Kensington, MD 20895   (301) 949-0228

Vance Y. Hum#
3401 Bexhill Place
Kensington, MD 20895   (301) 585-4800

Anthony A. O Brien
Groff & O'Brien
3514 Plyers Mill Road
Kensington, MD 20895   (301) 949-0228

George T. Ozaki
4905 Bangor Drive
Kensington, MD 20895   (301) 949-8014

Charles B. Parker
9918 Kensington Pkwy.
Kensington, MD 20895   (301) 942-3350

David B. Newman, Jr.
Centennial Square
P.O. Box 2728
La Plata, MD 20646   (301) 934-6100

Donald L. Walton#
8502 Woodside Court
Lanham, MD 20706   (301) 552-2620

Robert Edmund Archibald
John Hopkins Univ.
Applied Physics Lab.
John Hopkins Rd.
Laurel, MD 20707   (301) 953-5632

Leander F. Aulisio#
13203 Claxton Drive
Laurel, MD 20708   (301) 776-7905

Mary L. Beall
The Johns Hopkins University
Applied Physics Lab.
John Hopkins Road
Laurel, MD 20707   (301) 953-5641

Howard W. Califano
Johns Hopkins University
Applied Physics Laboratory
John Hopkins Road
Laurel, MD 20707   (301) 953-5641

Francis A. Cooch, IV
Johns Hopkins University
Applied Physics Lab.
John Hopkins Road
Laurel, MD 20707   (301) 953-5632

Gordon C. Fell#
8716 Granite Lane
Laurel, MD 20708   (301) 953-7392

Larry Chauncey Hall
Westvaco Corporation
Johns Hopkins Rd.
Laurel, MD 20707   (301) 792-9100

James O. Olfson
General Elevator Co., Inc.
601 Nursery Road
Linthicum, MD 21090   (301) 789-0200

Leland R. Jorgensen
P.O. Box 850
215 W. Main Street
Middletown, MD 21769   (301) 293-2212

Stanley T. Krawczewicz
6103 87th Avenue
New Carrolton, MD 20784   (301) 577-7352

O.A. Neumann
6821 Old Stage Road
North Bethesda, MD 20852   (301) 881-4053

Michael K. Kirschner
13400 Moran Dr.
North Potomac, MD 20878   (301) 881-4053

Charles Albert Haase
14001 Sinepuxent Ave.
Cor. 140th Street
Ocean City, MD 21842   (301) 250-1015

Roland T. Bryan
P.O. Box 304
Oxford, MD 21654   (301) 226-5520

Gene Meredith Garner, II
6917 Jarrett Avenue
Oxon Hill, MD 20745   (301) 226-5520

Richard M. Mc Mahon
Scally, Scally & Mc Mahon, P.A.
8901 Harford Rd.
Parkville, MD 21234   (301) 661-8590

Edwin T. Yates, Jr.#
19 Club View Lane
Phoenix, MD 21131   (301) 667-4977

Rodney D. Bennett, Jr.
10609 Crossing Creek Road
Potomac, MD 20854   (301) 299-4204

Peter Feldman#
8802 Liberty Lane
Potomac, MD 20854   (301) 299-4617

William L. Freeh
2508 Chilham Place
Potomac, MD 20854   (301) 424-8618

Alan M. Friedman#
8510 Wilkesboro Lane
Potomac, MD 20854   (301) 983-2211

S. Rolfe Gregory#
11603 Milbern Drive
Potomac, MD 20854   (301) 299-6368

William D. Hall
10850 Stanmore Drive
Potomac, MD 20854   (301) 299-4053

Thomas P. Liniak
Shlesinger & Myers
10220 River Road
Suite 200
Potomac, MD 20854   (301) 365-8000

Geoffrey R. Myers
Hall, Myers & Rose
10220 River Road
Potomac, MD 20854   (301) 365-8000

Frank Louis Neuhauser
10030 Chapel Road
Potomac, MD 20854   (301) 299-9024

Gordon S. Parker
8901 Falls Rd.
Potomac, MD 20854   (301) 299-5659

Leroy Bruce Randall
11708 Rosalinda Drive
Potomac, MD 20854   (301) 299-6708

Joel Stearman
8506 Wild Olive Drive
Potomac, MD 20854   (301) 424-2877

Tak Ki Sung
10017 Chartwell Manor Court
Potomac, MD 20854   (301) 469-6812

Francis K. Zugel
10810 Old Coach Road
Potomac, MD 20854   (301) 299-7433

Jacob Block#
14112 Parkvale Road
Rockville, MD 20853   (301) 460-9799

Donna I. Bobrowicz
Organon Teknika Corporation
1330-A Piccard Drive
Rockville, MD 20850   (301) 258-5200

John K. Corbin#
307 Lorraine Drive
Rockville, MD 20852   (301) 881-7098

Michael M. De Angeli
1901 Research Blvd.
Suite 220
Rockville, MD 20850   (301) 424-3640

Ira Charles Edell
Epstein, Edell & Retzer
1901 Research Blvd.
Suite 220
Rockville, MD 20850   (301) 424-3640

Robert Howard Epstein
Epstein, Edell & Retzer
1901 Research Blvd.
Suite 220
Rockville, MD 20850   (301) 424-3640

Roy D. Frazier#
14109 Bauer Drive
Rockville, MD 20853   (301) 871-7324

Morton J. Frome
Katz, Frome, Slan
& Bleecker, P.A.
6116 Executive Blvd.
Suite 200
Rockville, MD 20852   (301) 230-5800

Karen M. Gerken#
Epstein, Edell & Retzer
1901 Research Blvd.
Suite 220
Rockville, MD 20850   (301) 424-3640

James C. Haight*
NIH Office of Tech. Transfer
6003 Executive Blvd.
Suite 307
Rockville, MD 20852   (301) 496-0750

Bernard Helfin#
13918 Marianna Drive
Rockville, MD 20853   (301) 871-7157

Margaret R. Howlett
12903 Atlantic Avenue
Rockville, MD 20851   (301) 881-7556

Leo J. Jennings#
11407 Stonewood Lane
Rockville, MD 20852   (301) 468-0103

Daniel M. Kennedy
10105 Burton Glen Dr.
Rockville, MD 20850   (301) 762-1963

Donald A. Kettlestrings
414 Hungerford Drive
Suite 211
Rockville, MD 20850   (301) 279-7577

Gregory E. Mc Neill
5004 Norbeck Road
Rockville, MD 20853   (301) 460-6789

Albert T. Meyers
17008 Cashell Road
Rockville, MD 20853   (301) 924-3466

Gordon K. Milestone
7708 Mary Cassatt Drive
Rockville, MD 20854   (301) 299-4146

Norma S. Milestone
7708 Mary Cassatt Drive
Rockville, MD 20854   (301) 299-4146

Aaron B. Retzer
Epstein, Edell & Retzer
1901 Research Blvd.
Suite 220
Rockville, MD 20850   (301) 424-3640

Mitchell J. Shein#*
U.S. Food & Drug Administration
1390 Piccard Drive
Rockville, MD 20850   (301) 427-1018

Rex Logan Sturm
Brown & Sturm
260 E. Jefferson St.
Rockville, MD 20850   (301) 762-2555

Marvin S. Towsend
8 Grovepoint Court
Rockville, MD 20854   (301) 279-0660

Christopher N. Sears#
5184 Spring Avenue
Shady Side, MD 20764   (301) 867-0016

William Britton Moore
531 Little John
Sherwood Forest, MD 21405   (301) 849-8303

Jay I. Alexander
8107 Eastern Ave.
Apt. D-404
Silver Spring, MD 20910   (301) 589-5176

Edward C. Allen
1903 Gatewood Place
Silver Spring, MD 20903   (301) 434-3575

Harold Ansher
11703 Fulham Street
Silver Spring, MD 20902   (301) 649-3752

Norton Ansher
13102 Middlevale Lane
Silver Spring, MD 20906   (301) 946-4642

Jack E. Armore
11817 Mentone Road
Silver Spring, MD 20906   (301) 946-1360

Hyland Bizot
2702 Harmon Road
Silver Spring, MD 20902   (301) 933-2199

Benjamin Herman Bochenek
1322 Xaveria Drive
Silver Spring, MD 20903   (301) 439-4711

Milton Buchler
9505 Ocala St.
Silver Spring, MD 20901   (301) 587-6673

Alice Lee Chen#
3013 Birchtree Lane
Silver Spring, MD 20906   (301) 460-8921

W. B. Childs
17 Piping Rock Road
Silver Spring, MD 20904   (301) 384-9343

Ervin M. Combs
9013 Sudbury Rd.
Silver Spring, MD 20901   (301) 589-1317

Arthur P. Cyr
11424 Maple View Drive
Silver Spring, MD 20902   (301) 949-1006

Justin P. Dunlavey
2027 Forest Hill Drive
Silver Spring, MD 20903   (301) 439-3354

Nathan Edelberg
11012 Lombardy Road
Silver Spring, MD 20901   (301) 593-2040

Ellen Marcie Emas
11701 Stonington Place
Silver Spring, MD 20902   (301) 593-8658

William Feldman
932 Schindler Drive
Silver Spring, MD 20903   (301) 593-3253

Daniel E. Fisher
2708 Shanandale Drive
Silver Spring, MD 20904   (301) 572-5326

Reuben Friedman
1821 Reedic Drive
Silver Spring, MD 20902   (301) 649-3916

Bernard A. Gelak#
115 Delford Ave.
Silver Spring, MD 20904   (301) 622-2676

Harold A. Gell, Jr.
13720 Lockdale Rd.
Silver Spring, MD 20906   (301) 460-0756

Lawrence Glassman
2203 Quinton Rd.
Silver Spring, MD 20910   (301) 358-9278

Benjamin J. Goldfarb
1001 Playford Lane
Silver Spring, MD 20901   (301) 593-3162

Herman Lewis Gordon
306 Ellsworth Dr.
Silver Spring, MD 20910   (301) 588-8968

Quinton E. Hodges
12508 White Dr.
Silver Spring, MD 20904   (301) 622-0167

Harold G. Jarcho
10113 Green Forest Drive
Silver Spring, MD 20903   (301) 434-8357

Roger D. Johnson*
U.S. Dept. of Navy
Off. of Pat. Coun. Code C72W
Naval Surface Warfare Center
10901 New Hampshire Avenue
Silver Spring, MD 20903   (202) 394-2174

Murray Katz
11435 Monterrey Drive
Silver Spring, MD 20902   (301) 949-3878

Bernard Konick
11617 Fulham Street
Silver Spring, MD 20902   (301) 649-3180

Roy Lake
1221 Burton Street
Silver Spring, MD 20910   (301) 588-7602

Joseph R. Liberman
609 Gilmoure Dr.
Silver Spring, MD 20901   (301) 593-1062

Eli Lieberman
1712 Overlook Drive
Silver Spring, MD 20903   (301) 434-2827

Hung C. Lin
8 Schindler Court
Silver Spring, MD 20903   (301) 454-6853

Sidney Marantz
835 Gist Avenue
Silver Spring, MD 20910   (301) 588-9265

Raymond E. Martin
423 St. Lawrence Dr.
Silver Spring, MD 20901   (301) 593-0555

Alex Mazel
15036 Candover Court
Silver Spring, MD 20906   (301) 598-5888

Aldrich F. Medbery
209 Hillmoor Drive
Silver Spring, MD 20901   (301) 593-7189

William Misiek
405 Royalton Rd.
Silver Spring, MD 20901   (301) 593-3953

Elbert L. Roberts#
13103 Brittany Drive
Silver Spring, MD 20904   (301) 384-3255

Samuel B. Rothberg#
1121 University Blvd., West
Apt. 811
Silver Spring, MD 20902   (301) 649-5030

Irving J. Rotkin
10202 Lariston Lane
Silver Spring, MD 20903   (301) 434-8882

Gersten Sadowsky
12400 Conn. Ave.
Silver Spring, MD 20906   (301) 949-7095

David S. Scrivener
1714 Overlook Drive
Silver Spring, MD 20903   (301) 434-7722

Lionel M. Shapiro
915 Burnt Crest Lane
Silver Spring, MD 20903   (301) 434-8494

Jacob Shuster*
U.S. Navy
N.S.W.C. - White Oak Lab.
Off. of Counsel (Patents)
Bldg. 25, Rm. 210
Silver Spring, MD 20903   (202) 394-2174

Lewis A. Thaxton#
520 Beaumont Road
Silver Spring, MD 20904   (301) 384-8224

Donald E. Townsend
117 Bluff Terrace
Silver Spring, MD 20902   (301) 681-5948

Kenneth E. Walden*
U.S. Dept. of Navy
Naval Surfaces Warfare Ctr.
Code C72W
Silver Spring, MD 20903   (202) 394-2182

James H. Warner
11865 Old Columbia Pike
Silver Spring, MD 20904   (301) 622-9576

William T. Webb#
14420 Cantrell Road
Silver Spring, MD 20905   (301) 384-6398

Steven R. Wegman
640 Concerto Lane
Silver Spring, MD 20901   (301) 681-8887

Stephen F.K. Yee
8605 Second Avenue
Silver Spring, MD 20910   (301) 589-1210

Charles Sukalo
4806 Silver Hill Rd.
Suitland, MD 20746   (301) 420-7707

Louis A. Scholz
850 State Route 32
Sykesville, MD 21784   (301) 795-9100

Nicholas S. Bromer#
6812 Westmoreland Ave.
Takoma Park, MD 20912   (301) 270-0575

Anthony V. Ciarlante#
4310 Brinkley Road
Temple Hills, MD 20748   (301) 423-4148

Elizabeth A. King#
7101 Berkshire Drive
Temple Hills, MD 20748   (301) 449-1649

Leonard Bloom
401 Washington Ave.
Suite 803
Towson, MD 21204   (301) 337-2295

Royal W. Craig
401 Washington Ave.
Towson, MD 21204   (301) 337-2295

Dennis A. Dearing
Black & Decker (U.S.) Inc.
701 East Joppa Road
Towson, MD 21204   (301) 583-3503

John David Del Ponti
Black & Decker Corp.
701 E. Joppa Road
Towson, MD 21204   (301) 583-2747

James R. Gaffey
401 Washington Avenue
Towson, MD 21204   (301) 337-2295

Robert M. Gamson
401 Washington Avenue
Towson, MD 21204   (301) 337-2295

Joseph Bruce Hoofnagle, Jr.
The Black & Decker Corporation
701 E. Joppa Road
Towson, MD 21204   (301) 583-2706

Bruce L. Lamb
401 Washington Avenue
Towson, MD 21204   (301) 337-2295

Edward D. Murphy
Black and Decker Mfg. Co.
701 E. Joppa Rd.
Towson, MD 21204   (301) 583-2867

Reginald F. Pippin, Jr.
7806 Ruxway Rd.
Towson, MD 21204   (301) 628-3229

Spencer T. Smith
Black & Decker Corp.
701 E. Joppa Road
Towson, MD 21204   (301) 583-3119

Harold Weinstein
Black & Decker Corp.
701 E. Joppa Rd.
Towson, MD 21204   (301) 583-2886

Charles E. Yocum
Black & Decker Corporation
701 E. Joppa Road
Towson, MD 21204   (301) 583-2956

Donald B. Moyer#
12107 Blaketon St.
Upper Marlboro, MD 20772   (301) 249-9064

Alfred C. Perham#
8409 Thornberry Drive, West
Upper Marlboro, MD 20772   (301) 627-2234

Dolph H. Torrence#
10904 Sutton Drive
Upper Marlboro, MD 20772   (301) 336-6854

Eugene F. Osborne, Sr.
3000 Old Taneytown Road
P.O. Box 423
Westminster, MD 21157   (301) 848-0861

Herbert S. Cockeram#
3804 Delano Street
Wheaton, MD 20902   (301) 933-9319

James Kee Chi#
1716 Arbor View Road
Wheaton, MD 20902   (301) 949-3974

Robert E. Sandt
42 Cliff Drive
Kinnaird'S Point
Worton, MD 21678   (301) 778-2307

Massachusetts

Barry Raymond Blaker#
4 Algonquin Rd.
Acton, MA 01720   (617) 263-3440

Arthur W. Fisher
Digital Equipment Corp.
100 Nagog Park
Acton, MA 01720   (617) 264-6805

Dewitt C. Seward, III#
P.O. Box 261
Acton, MA 01720   (617) 263-7704

William J. Driscoll
90 Main Street
Andover, MA 01810   (617) 475-6371

Edward A. Gordon
P.O. Box 2113
90 Main Street
Andover, MA 01810   (508) 475-6371

Kenneth A. Green#
19 Burton Farm Drive
Andover, MA 01810   (617) 475-8423

Richard F. Schuette
Hewlett-Packard Company
3000 Minuteman Rd.
Andover, MA 01810   (508) 687-1501

Herbert Warren Arnold
151 Mystic Street
Apt. M-5
Arlington, MA 02174   (617) 646-7210

Joseph J. Gano#
31 Davis Ave.
Arlington, MA 02174   (607) 643-9319

John D. Karagounis
687 Tiffany Street
Attleboro, MA 02703   (617) 222-0290

James Patrick Mc Andrews
Texas Instruments Inc.
34 Forest St.
Attleboro, MA 02703   (617) 699-3245

Peter Dulchinos
Raytheon
Hartwell Rd.
Bedford, MA 01730   (617) 274-2043

Robert T. Dunn
4 Cedar Ridge Drive
Bedford, MA 01730   (617) 275-6146

Andrew T. Karnakis
Millipore Corporation
80 Ashby Road
Bedford, MA 01730   (617) 275-9200

William J. O Brien
49 Glenridge Drive
Bedford, MA 01730   (617) 275-6289

Stuart B. Zigun
33 Lido Lane
Bedford, MA 01730   (617) 275-6921

William N. Anastos
28 Longmeadow Road
Belmont, MA 01930   (617) 281-0440

Robert T. Conway, Jr.
3 Long Avenue
Belmont, MA 02178   (617) 484-3943

James L. Diamond
22 Pine Street
Belmont, MA 02178   (617) 684-6148

Aubrey C. Brine
244 E. Lothrop St.
Beverly, MA 01915   (617) 927-4200

Elton T. Barrett
21 Hattie Lane
Billerica, MA 01821   (617) 667-6064

Gerald J. Cechony
Bull Horn Information Systems Inc.
Technology Park
Mail Station 412N
Billerica, MA 01821   (508) 294-6129

Lawrence A. Chaletsky
Cabot Corp.
Concord Road
Billerica, MA 01821   (617) 663-3455

Gary D. Clapp
Bull Horn Information Sys. Inc.
Technology Park
Mail Station 412N
Billerica, MA 01821   (508) 294-6742

Faith F. Driscoll
Bull Horn Information Systems Inc.
Technology Park
Mail Station 412N
Billerica, MA 01821   (508) 294-6165

Michelle B. La Roche#
Cabot Corp.
Billerica Technical Center
157 Concord Road
Billerica, MA 01821   (508) 670-6198

George M. Medeiros
574 Boston Road
Suite 13
Billerica, MA 01821   (617) 663-3467

Samuel Shiber#
446 Boston Road
Suite 333
Billerica, MA 01821   (617) 270-7035

John S. Solakian
Bull Horn Information Sys. Inc.
Technology Park MA02-412N
Billerica, MA 01821   (508) 294-6140

Gerald Altman
Morse, Altman, Dacey & Benson
73 Tremont Street
Boston, MA 02108   (617) 523-3515

Robert M. Asher
Dike, Bronstein, Roberts,
Cushman & Pfund
130 Water Street
Boston, MA 02109   (617) 542-8492

Michael E. Attaya
Cesari & Mc Kenna
30 Rowes Wharf
Boston, MA 02110   (617) 261-6800

Barbara A. Barakat
Lorusso & Loud
440 Commercial St.
Boston, MA 02109   (617) 227-0700

Charles J. Barbas
Cesari & Mc Kenna
30 Rowes Wharf
Boston, MA 02110   (617) 261-6800

Steven M. Bauer
Allegretti & Witcoff, Ltd.
75 State Street
Suite 2300
Boston, MA 02109   (617) 345-9100

Doris M. Bennett
Stone & Webster
245 Summer Street
Boston, MA 02107   (617) 589-8600

David B. Bernstein
Wolf, Greenfield & Sacks, P.C.
600 Atlantic Avenue
Federal Reserve Plaza
Boston, MA 02210   (617) 720-3500

Michael J. Bevilacqua
Hale & Dorr
60 State Street
Boston, MA 02109   (617) 742-9100

Richard J. Birch
Thompson, Birch, Gauthier
& Samuels
225 Franklin Street
Suite 3300
Boston, MA 02110   (617) 426-8989

Arthur Z. Bookstein
Wolf, Greenfield & Sacks, P.C.
Federal Reserve Plaza
600 Atlantic Avenue
Boston, MA 02210   (617) 720-3500

William E. Booth
Fish & Richardson
225 Franklin St.
Boston, MA 02110   (617) 542-5070

Joseph H. Born
Cesari & Mc Kenna
30 Rowes Wharf
Boston, MA 02110   (617) 261-6800

Damon Joseph Borrelli
Lahive & Cockfield
60 State Street
Suite 510
Boston, MA 02109   (617) 227-7400

Sewall P. Bronstein
Dike, Bronstein, Roberts,
Cushman & Pfund
130 Water St.
Boston, MA 02109   (617) 542-8492

Donald Brown
Dike, Bronstein, Roberts,
Cushman & Pfund
130 Water St.
Boston, MA 02109   (617) 542-8492

Linda M. Buckley
Dike, Bronstein, Roberts,
Cushman & Pfund
130 Water Street
Boston, MA 02109   (617) 542-8492

Paula A. Campbell
Testa, Hurwitz & Thiebeault
Exchange Place
53 State Street
Boston, MA 02109   (617) 367-7500

Katherine A. Caso#
Aldrich, Eastman & Waltch, Inc.
265 Franklin Street
Boston, MA 02110   (617) 439-9000

Adrienne M. Catanese
Gaston Snow & Ely Bartlett
One Federal Street
Boston, MA 02110   (617) 426-4600

D. Robert Cervera*
U.S. Securities and
Exchange Commission
J.W. Mc Cormack Post Office
And Courthouse
Boston, MA 02109   (617) 223-9900

Robert A. Cesari
Cesari & Mc Kenna
30 Rowes Wharf
Boston, MA 02110   (617) 261-6800

Franklin H. Chasen
Suite 710
44 School St.
Boston, MA 02108   (617) 367-9943

Frank S. Chow
Allegretti & Witcoff, Ltd.
75 State Street
Boston, MA 02109   (617) 345-9100

Stephen Y. Chow
Cesari & Mc Kenna
30 Rowes Wharf
Boston, MA 02110   (617) 261-6800

Francis J. Clark
Kendall Company
One Federal Street
Boston, MA 02110   (617) 574-7915

Paul T. Clark
Fish & Richardson
225 Franklin Street
Boston, MA 02110   (617) 542-5070

James E. Cockfield
Lahive & Cockfield
60 State Street
Boston, MA 02109   (617) 227-7400

Jerry Cohen
Cohen & Burg, P.C.
33 Broad Street
Boston, MA 02109   (617) 742-7840

David G. Conlin
Dike, Bronstein, Roberts,
Cushman & Pfund
130 Water St.
Boston, MA 02109   (617) 542-8492

Matthew E. Connors#
Samuels, Gauthier & Stevens
225 Franklin St.
Boston, MA 02210   (617) 426-9180

Peter F. Corless
Dike, Bronstein, Roberts,
& Cushman
130 Water Street
Boston, MA 02109   (617) 523-3400

Anne I. Craig
Lorusso & Loud
440 Commercial Street
Boston, MA 02109   (617) 227-0700

G. Eugene Dacey
Morse, Altman, Dacey
& Benson
73 Tremont Street
Boston, MA 02108   (617) 523-3515

Peter J. Devlin
Fish & Richardson
225 Franklin St.
Boston, MA 02110   (617) 542-5070

Maureen E. Dey
476 Commonwealth Ave.
No. 4
Boston, MA 02215   (617) 267-8993

Paul I. Douglas
The Gillette Company
800 Boylston St.
Prudential Tower Bldg.
38th Floor
Boston, MA 02199   (617) 421-7887

David Markham Driscoll
Wolf, Greenfield & Sacks, P.C.
Federal Reserve Plaza
600 Atlantic Avenue
Boston, MA 02210   (617) 720-3500

Albert P. Durigon
Weingarten, Schurgin, Gagnebin
& Hayes
Ten Post Office Square
Boston, MA 02109   (617) 542-2290

Ronald I. Eisenstein
Dike, Bronstein, Roberts,
Cushman & Pfund
130 Water Street
Boston, MA 02109   (617) 523-3400

Kristofer E. Elbing#
Fish & Richardson
225 Franklin St.
Boston, MA 02110   (617) 542-5070

Thomas J. Engellenner
Lahive & Cockfield
60 State Street
Boston, MA 02109   (617) 227-7400

Willis Marion Ertman
Fish & Richardson
225 Franklin Street
Boston, MA 02110   (617) 542-5070

Michael I. Falkoff
Lahive & Cockfield
60 State Street
Boston, MA 02109   (617) 227-7400

J. Peter Fasse
Fish & Richardson
225 Franklin Street
Boston, MA 02110   (617) 542-5070

Eugene A. Feher
Weingarten, Schurgin, Gagnebin
& Hayes
Ten Post Office Square
Boston, MA 02109   (617) 542-2290

David L. Feigenbaum
Fish & Richardson
225 Franklin Street
Boston, MA 02110   (617) 542-5070

James J. Foster
Wolf, Greenfield & Sacks, P.C.
600 Atlantic Avenue
Boston, MA 02210   (617) 720-3500

Steven J. Frank
Cesari and Mc Kenna
30 Rowes Wharf
Boston, MA 02110   (617) 261-6800

Janis K. Fraser
Fish & Richardson
225 Franklin St.
Boston, MA 02110   (617) 542-5070

John W. Freeman
Fish & Richardson
225 Franklin Street
Boston, MA 02110   (617) 542-5070

Timothy A. French
Fish & Richardson
225 Franklin Street
Boston, MA 02110   (617) 542-5070

R. W. Furlong
Fish & Richardson
225 Franklin Street
Boston, MA 02110   (617) 542-5070

John J. Gagel#
Fish & Richardson
225 Franklin Street
Boston, MA 02110   (617) 542-5070

Charles L. Gagnebin, III
Weingarten, Schurgin, Gagnebin
& Hayes
Ten Post Office Square
Boston, MA 02109   (617) 542-2290

Robert Trafton Gammons
Dike, Bronstein, Roberts,
Cushman & Pfund
130 Water St.
Boston, MA 02109   (617) 542-8492

Edward R. Gates
Wolf, Greenfield & Sacks, P.C.
600 Atlantic Avenue
Boston, MA 02210   (617) 720-3500

Herbert L. Gatewood
265 Franklin Street
16th Floor
Boston, MA 02110   (617) 737-9995

Maurice Edward Gauthier
Samuels, Gauthier, Stevens
& Kehoe
225 Franklin St.
Boston, MA 02110   (617) 426-9180

William C. Geary, III
Lahive & Cockfield
60 State Street
Boston, MA 02109   (617) 227-7400

Lawrence Gilbert
Boston University
881 Commonwealth Ave.
5th Floor
Boston, MA 02215   (617) 353-2212

Milton Edwin Gilbert
Wolf, Greenfield & Sacks, P.C.
Federal Reserve Plaza
600 Atlantic Avenue
Boston, MA 02210   (617) 720-3500

Susan G.L. Glovsky
One Bowdoin Square
Boston, MA 02114   (617) 227-8539

Lawrence M. Green
Wolf, Greenfield & Sacks, P.C.
Federal Reserve Plaza
600 Atlantic Avenue
Boston, MA 02210   (617) 720-3500

George L. Greenfield
Wolf, Greenfield & Sacks, P.C.
600 Atlantic Avenue
Boston, MA 02210   (617) 720-3500

Marvin Curtis Guthrie
Massachusetts General Hospital
Office of Tech. Affairs
75 Blossom Court
Boston, MA 02114   (617) 726-8608

Ann Lamport Hammitte
Lahive & Cockfield
60 State St.
Suite 510
Boston, MA 02109   (617) 227-7400

Paul J. Hayes
Weingarten, Schurgin, Gagnebin
& Hayes
Ten Post Office Square
Boston, MA 02109   (617) 542-2290

Mark J. Hebert
Fish & Richardson
225 Franklin Street
Boston, MA 02110   (617) 542-5070

Holliday C. Heine#
Weingarten, Schurgin, Gagnebin
& Hayes
Ten Post Office Square
Boston, MA 02109   (617) 542-2290

Therese A. Hendricks
Wolf, Greenfield & Sacks, P.C.
Federal Reserve Plaza
600 Atlantic Avenue
Boston, MA 02210   (617) 720-3500

Gilbert H. Hennessey, III
Fish & Richardson
225 Franklin Street
Boston, MA 02110   (617) 542-5070

Steven J. Henry
Wolf, Greenfield & Sacks, P.C.
600 Atlantic Avenue
Boston, MA 02215   (617) 720-3500

Charles Hieken
Fish & Richardson
225 Franklin St.
Boston, MA 02110   (617) 542-5070

Robert Eliot Hillman
Fish & Richardson
225 Franklin Street
Boston, MA 02110   (617) 542-5070

Beverly E. Hjorth
Weingarten, Schurgin, Gagnebin
& Hayes
Ten Post Office Square
Boston, MA 02109   (617) 542-2290

Jason M. Honeyman
Gaston & Snow
One Federal Street
Boston, MA 02110   (617) 426-4600

David E. Hoppe
294 Beacon St.
Boston, MA 02116   (617) 262-4509

Thomas W. Humphrey#
Fish & Richardson
225 Franklin Street
Boston, MA 02110   (617) 542-5070

John P. Iwanicki
Allegretti & Witcoff, Ltd.
75 State Street
Suite 2300
Boston, MA 02109   (617) 345-9100

Anthony J. Janiuk
Wolf, Greenfield
& Sacks, P.C.
Federal Reserve Plaza
600 Atlantic Ave.
Boston, MA 02210   (617) 720-3500

Bruce D. Jobse
Wolf, Greenfield & Sacks, P.C.
600 Atlantic Avenue
Boston, MA 02210   (617) 720-3500

Alfred Russell Johnson
Weingarten, Schurgin, Gagnebin
& Hayes
Ten Post Office Square
Boston, MA 02109   (617) 542-2290

Richard A. Jordan
Cesari & Mc Kenna
312 East Union Wharf
Boston, MA 02109   (617) 523-8100

Warren A. Kaplan
Wolf, Greenfield & Sacks, P.C.
600 Atlantic Ave.
Federal Reserve Bldg.
17th Floor
Boston, MA 02210   (617) 720-3500

Robin Doria Kelley#
Testa, Hurwitz & Thibeault
Exchange Place
53 State Street
Boston, MA 02109   (617) 367-4477

Bill Kennedy
Fish & Richardson
One Financial Center
Suite 2500
Boston, MA 02111   (617) 542-5070

Ann-Louise Kerner#
Lahive & Cockfield
60 State Street
Suite 510
Boston, MA 02109   (617) 227-7400

Celia H. Ketley#
Fish & Richardson
225 Franklin St.
Boston, MA 02110   (617) 542-5070

Daniel Kim
Bromberg, Sunstein & Casselman
Ten West Street
Seventh Floor
Boston, MA 02111   (617) 426-6464

Philip G. Koenig
Wolf, Greenfield & Sacks
600 Atlantic Avenue
Boston, MA 02210   (617) 720-3500

Ronald Joseph Kransdorf
Wolf, Greenfield & Sacks, P.C.
600 Atlantic Avenue
Federal Reserve Plaza
Boston, MA 02210   (617) 720-3500

Paul E. Kudirka
Wolf, Greenfield & Sacks, P.C.
600 Atlantic Avenue
Boston, MA 02210   (617) 720-3500

Toby H. Kusmer
Gaston & Snow
One Federal St.
Boston, MA 02110   (617) 426-4600

John A. Lahive, Jr.
Lahive & Cockfield
60 State Street
Boston, MA 02109   (617) 227-7400

James B. Lampert
Hale & Dorr
60 State Street
Boston, MA 02109   (617) 742-9100

Peter C. Lando#
Wolf, Greenfield & Sacks, P.C.
600 Atlantic Avenue
Boston, MA 02210   (617) 720-3500

Mark G. Lappin
Lahive & Cockfield
60 State Street
Boston, MA 02109   (617) 227-7400

Victor B. Lebovici
Weingarten, Schurgin Gagnebin
& Hayes
Ten Post Office Square
5th Floor
Boston, MA 02109   (617) 542-2290

G. Roger Lee
Fish & Richardson
225 Franklin Street
Boston, MA 02110   (617) 542-5070

John L. Lee#
Bromberg & Sunstein
10 West Street
Boston, MA 02111   (617) 426-6464

William H. Lee
Wolf, Greenfield & Sacks, P.C.
Federal Reserve Plaza
600 Atlantic Avenue
Boston, MA 02210   (617) 720-3500

Craig K. Leon
Six Whittier Place
Apt. 4-0
Boston, MA 02114   (617) 367-9266

W. Hugo Liepmann
Lahive & Cockfield
60 State Street
Suite 510
Boston, MA 02109   (617) 227-7400

Ernest V. Linek
Dike, Bronstein, Roberts
& Cushman
130 Water Street
Boston, MA 02109   (617) 523-3400

Willilam A. Loginov
Wolf, Greenfield & Sacks, P.C.
600 Atlantic Ave.
Boston, MA 02210   (617) 720-3500

Ralph A. Loren
Lahive & Cockfield
60 State Street
Boston, MA 02109   (617) 227-7400

Anthony M. Lorusso
440 Commercial Street
Boston, MA 02109   (617) 227-0700

Jeremiah Lynch
Lahive & Cockfield
60 State Street
Boston, MA 02109   (617) 227-7400

Kathleen A. Madden#
Fish & Richardson
One Financial Center
Suite 2500
Boston, MA 02111   (617) 542-5070

Gregory A. Madera
Fish & Richardson
225 Franklin St.
Boston, MA 02110   (617) 542-5070

Peter J. Manus
Wolf, Greenfield & Sacks, P.C.
Federal Reserve Plaza
600 Atlantic Avenue
Boston, MA 02210   (617) 720-3500

James E. Maslow
Lahive & Cockfield
Sixty State Street
Boston, MA 02109   (617) 227-7400

Stephen G. Matzuk
Weingarten, Schurgin, Gagnebin
& Hayes
Ten Post Office Square
Boston, MA 02109   (617) 542-2290

Lowell H. Mc Carter
Genzyme Corporation
75 Kneeland Street
Boston, MA 02111   (617) 451-1923

William R. Mc Clellan
Wolf, Greenfield & Sacks, P.C.
600 Atlantic Avenue
Boston, MA 02210   (617) 720-3500

Peter D. Mc Dermott
Allegretti & Witcoff, Ltd.
75 State Street
Suite 2300
Boston, MA 02109   (617) 345-9100

John P. Mc Gonagle
Mc Gonagle & Mc Gowan
11 Beacon Street
Boston, MA 02108   (617) 227-7755

John F. Mc Kenna
Cesari & Mc Kenna
30 Rowes Wharf
Boston, MA 02110   (617) 261-6800

Brian Michaelis
Weingarten, Schurgin,
Gagnebin & Hayes
Ten Post Office Square
Boston, MA 02109   (617) 542-2290

John B. Miller
Gadsby & Hannah
One Post Office Square
Boston, MA 02109   (617) 357-8700

Anthony J. Mirabito
Wolf, Greenfield & Sacks, P.C.
600 Atlantic Avenue
Boston, MA 02210   (617) 720-3500

James H. Morris#
Wolf, Greenfield & Sacks, P.C.
600 Atlantic Avenue
Boston, MA 01887   (617) 720-3500

James E. Mrose
Thomson & Mrose
468 Park Drive
Boston, MA 02215   (617) 262-6452

James Emil Mrose, Jr.
Fish & Richardson
225 Franklin St.
Boston, MA 02110   (617) 542-5070

Charles Francis Murphy
Massachusetts General Hospital
Off. of Tech. Affairs
Fruit Street
Boston, MA 02114   (617) 726-8608

Timothy M. Murphy
Bromberg & Sunstein
10 West Street
Boston, MA 02111   (617) 426-6464

Robert C. Nabinger
Fish & Richardson
225 Franklin Street
Boston, MA 02110   (617) 542-5070

Theodore Naccarella
Wolf, Greenfield & Sacks, P.C.
Federal Reserve Plaza
600 Atlantic Avenue
Boston, MA 02210   (617) 720-3500

James Lewis Neal
Hay & Dailey
One Center Plaza
Boston, MA 02108   (617) 227-3544

George W. Neuner
Dike, Bronstein, Roberts
& Cushman
130 Water St.
Boston, MA 02109   (617) 542-3400

Robert F. O Connell
Dike, Bronstein, Roberts,
Cushman & Pfund
130 Water St.
Boston, MA 02109   (617) 523-3400

Martin J. O Donnell
Cesari & Mc Kenna
30 Rowes Wharf
Boston, MA 02110   (617) 261-6800

Thomas C. O Konski
Cesari & Mc Kenna
30 Rowes Wharf
Boston, MA 02110   (617) 261-6800

Michael L. Oliverio
Wolf, Greenfield & Sacks
600 Atlantic Avenue
Boston, MA 02210   (617) 843-3612

Louis Litman Orenbuch
Wolf, Greenfield & Sacks
600 Atlantic Avenue
Boston, MA 02210   (617) 720-3500

Henry D. Pahl, Jr.
Dike, Bronstein, Roberts
& Cushman
130 Water Street
Boston, MA 02109   (617) 523-3400

Sam Pasternack
Choate, Hall & Stewart
Exchange Place
53 State Street
Boston, MA 02109   (617) 227-5020

Edward F. Perlman
Wolf, Greenfield & Sacks, P.C.
Federal Reserve Plaza
600 Atlantic Avenue
Boston, MA 02210   (617) 720-3500

Charles E. Pfund
Wolf, Greenfield & Sacks, P.C.
Federal Reserve Plaza
600 Atlantic Avenue
Boston, MA 02210   (617) 720-3500

Edmund R. Pitcher
Lahive & Cockfield
Sixty State Street
Suite 510
Boston, MA 02109   (617) 227-7400

Frank P. Porcelli
Fish & Richardson
225 Franklin Street
Boston, MA 02110   (617) 542-5070

David J. Powsner
Lahive & Cockfield
60 State Street
5th Floor
Boston, MA 02109   (617) 227-7400

Eric L. Prahl
Fish & Richardson
225 Franklin Street
Boston, MA 02110   (617) 542-5070

David S. Resnick
Dike, Bronstein, Roberts
& Cushman
130 Water Street
Boston, MA 02109   (617) 523-3400

Douglas W. Robinson
Wolf, Greenfield & Sacks
600 Atlantic Ave.
Federal Reserve Bank Bldg.
12th Floor
Boston, MA 02210   (617) 720-3500

Stephen J. Roe#
Lorusso & Loud
440 Commercial Street
Boston, MA 02109   (617) 227-0700

Gary E. Ross
Cesari & Mc Kenna
30 Rowes Wharf
Boston, MA 02110   (617) 261-6800

Stanley Sacks
Wolf, Greenfield & Sacks, P.C.
600 Atlantic Avenue
Boston, MA 02210   (617) 720-3500

I. Stephen Samuels
Samuels, Gauthier, & Stevens
225 Franklin St.
Suite 3300
Boston, MA 02110   (617) 426-9180

Thomas M. Saunders
Lorusso & Loud
440 Commercial Street
Boston, MA 02109   (617) 227-0700

John R. Schiffhauer
Fish & Richardson
225 Franklin Street
Boston, MA 02110   (617) 542-5070

Robert J. Schiller
Gaston & Snow
One Federal Street
Boston, MA 02110   (617) 426-4600

Stanley M. Schurgin
Weingarten, Schurgin, Gagnebin
& Hayes
Ten Post Office Square
Boston, MA 02109   (617) 542-2290

Edward R. Schwartz
Wolf, Greenfield & Sacks, P.C.
600 Atlantic Avenue
Boston, MA 02210   (617) 720-3500

Patricia A. Sheehan
Cesari & Mc Kenna
30 Rowes Wharf
Boston, MA 02110   (617) 261-6800

David Silverstein
Gaston & Snow
One Federal Street
Boston, MA 02110   (617) 426-4600

John M. Skenyon
Fish & Richardson
225 Franklin Street
Boston, MA 02110   (617) 542-5070

Mandel E. Slater
Gillette Co.
Prudential Tower Bldg.
Boston, MA 02199   (617) 421-7886

Arthur A. Smith, Jr.
Lorusso & Loud
440 Commercial St.
Boston, MA 02109   (617) 227-0700

Thomas V. Smurzynski
Lahive & Cockfield
60 State Street
Boston, MA 02109   (617) 227-7400

Neil A. Steinberg
Allegretti & Witcoff, Ltd.
75 State Street
Boston, MA 02109   (617) 345-9100

Richard L. Stevens
Samuels, Gauthier & Stevens
225 Franklin St.
Suite 3300
Boston, MA 02110   (617) 426-9180

Bruce D. Sunstein
Bromberg & Sunstein
10 West Street
7th Floor
Boston, MA 02111   (617) 426-6464

C. Hall Swaim
Hale & Dorr
60 State St.
Boston, MA 02109   (617) 742-9100

Robert Kanof Tendler
10 West Street
Boston, MA 02111   (617) 723-7268

David J. Thibodeau, Jr.
Cesari & Mc Kenna
30 Rowes Wharf
Boston, MA 02110   (617) 261-6800

Donal B. Tobin
Gillette Co.
Prudential Tower Bldg.
Boston, MA 02199   (617) 421-7889

Yuan-Kai Tsao
Fish & Richardson
225 Franklin Street
Boston, MA 02110   (617) 542-5070

Gary A. Walpert
Hale & Dorr
60 State Street
Boston, MA 02109   (617) 742-9100

Stanton D. Weinstein
Wolf, Greenfield & Sacks, P.C.
Federal Reserve Plaza
600 Atlantic Avenue
Boston, MA 02210   (617) 720-3500

Steven J. Weissburg
Hale & Dorr
60 State Street
Boston, MA 02109   (617) 742-9100

John L. Welch
Wolf, Greenfield & Sacks, P.C.
600 Atlantic Avenue
Boston, MA 02210   (617) 720-3500

Dorothy P. Whelan
Fish & Richardson
225 Franklin Street
Boston, MA 02110   (617) 542-5070

Gregory D. Williams
Dike, Bronstein, Roberts
& Cushman
130 Water Street
Boston, MA 02109   (617) 523-3400

John Noel Williams
Fish & Richardson
225 Franklin Street
Boston, MA 02110   (617) 542-5070

Stephen P. Williams
Ares-Serono, Inc.
Exchange Place
37th Floor
Boston, MA 02109   (617) 723-1300

Charles C. Winchester, Jr.
Fish & Richardson
225 Franklin Street
Boston, MA 02110   (617) 542-5070

David Wolf
Wolf, Greenfield & Sacks, P.C.
Federal Reserve Plaza
600 Atlantic Avenue
Boston, MA 02210   (617) 720-3500

Jerry D. Lentz#
Digital Equipment Co.
85 Swanson Rd.
Boxboro, MA 01719   (508) 264-5334

Joseph A. Cameron
43 Sheffield Rd.
Boxford, MA 01921   (508) 887-6471

William Hugh Mc Neill
13 Silver Brook Rd.
Boxford, MA 01921   (617) 887-9623

Steven N. Fox#
251 Chatham West Drive
Brockton, MA 02401   (617) 587-0106

Joseph H. Killion
4 Cypress St.
Brookline, MA 02146   (617) 232-0951

David Prashker
417 Washington Street
Brookline, MA 02146   (617) 232-7509

Wilfred J. Baranick
Badger Co., Inc.
One Bdwy.
Cambridge, MA 02142   (617) 494-7245

Philip P. Berestecki
Badger Co., Inc.
One Broadway
Cambridge, MA 02142   (617) 494-7203

Robert L. Berger
Polaroid Corp.
549 Technology Square
Cambridge, MA 02139   (617) 577-2202

David L. Berstein
BASF Bioresearch Corp.
195 Albany Street
Cambridge, MA 02139   (617) 868-5700

Sylvia Lyle Boyd
P.O. Box 1050
Cambridge, MA 02238   (617) 924-4380

Gregory B. Butler#
Cambridge Neuroscience
One Kendall Square
Building 700
Cambridge, MA 02139   (617) 924-4380

Sybil A. Campbell#
Polaroid Corp.
549 Technology Square
Cambridge, MA 02139   (617) 577-2576

Francis J. Caufield
Polaroid Corp.
Pat. Dept.
549 Technology Square
Cambridge, MA 02139   (617) 577-3532

Stacey L. Channing
Immulogic Pharmaceutical Corp.
One Kendall Square
Building 600
Cambridge, MA 02139   (617) 494-0060

David J. Cole#
Polaroid Corp.
Patent Department
549 Technology Square
Cambridge, MA 02139   (617) 494-0060

Alfred Ernest Corrigan
Polaroid Corp.
549 Technology Sq.
Cambridge, MA 02139   (617) 577-2974

Mark Douma
308 Pearl Street
Cambridge, MA 02139   (617) 491-8796

Bruce M. Eisen
Genetics Institute, Inc.
87 Cambridgepark Drive
Cambridge, MA 02140   (617) 876-1170

Robert M. Ford
Polaroid Corp.
549 Technology Square
Cambridge, MA 02139   (617) 577-2215

Karl Hormann
86 Sparks Street
Cambridge, MA 02138   (617) 491-8867

Bruce F. Jacobs
124 Mt. Auburn Street
Suite 200
Cambridge, MA 02138   (617) 576-5766

David A. Jacobs
M.I.T.
Box 91
Cambridge, MA 02139   (617) 272-7400

Ellen J. Kapinos
Genetics Institute, Inc.
87 Cambridge Park Drive
Cambridge, MA 02140   (617) 876-1170

Herbert S. Kassman
International Law Collaborative
One Kendall Square
Suite 2200
Cambridge, MA 02139   (617) 868-2285

John J. Kelleher
Polaroid Corp.
545 Technology Square - 6A
Cambridge, MA 02139   (617) 577-3372

Philip George Kiely
Polaroid Corp.
549 Technology Square
Cambridge, MA 02139   (617) 577-3691

Bernard Joseph Lacomis
20 Acorn Park
Cambridge, MA 01741   (617) 864-5770

Gaetano D. Maccarone
Polaroid Corp.
Pat. Dept.
549 Tech. Square
Cambridge, MA 02139   (617) 577-4592

Herbert Malsky
C.S. Draper Laboratory Inc.
555 Technology Square
Cambridge, MA 02139   (617) 258-4943

Janet K. Martinez
Westgate Apts., No. C-4
Cambridge, MA 02139   (617) 258-4943

Patricia A. Mc Daniels
Genetics Institute
Legal Department
87 Cambridge Park Dr.
Cambridge, MA 02140   (617) 876-5851

Stanley H. Mervis
Polaroid Corp.
Pat. Dept.
549 Technology Square
Cambridge, MA 02139   (617) 577-2281

Joseph Dale Michaels
BASF Bioresearch Corp.
195 Albany Street
Cambridge, MA 02139   (617) 868-5700

John Joseph Moss
17 Everett Street
Cambridge, MA 02138   (617) 864-1870

Kittie A. Murray
840 Memorial Drive
Cambridge, MA 02139   (617) 576-8611

Ronald J. Paglierani
Open Software Foundation, Inc.
11 Cambridge Center
Cambridge, MA 02142   (617) 621-8700

Mark J. Pandiscio
Schiller, Pandiscio & Kusmer
125 Cambridgepark Drive
Cambridge, MA 02140   (617) 499-2770

Nicholas A. Pandiscio
Schiller, Pandiscio & Kusmer
125 Cambridgepark Drive
Cambridge, MA 02140   (617) 499-2770

Frank R. Perillo
Hewlett Packard Co.
59 Dana Street
Cambridge, MA 02138   (508) 687-1501

Lulu Arline Pickering#
T Cell Sciences, Inc.
38 Sidney St.
Cambridge, MA 02139   (617) 621-1400

Li-Hsien Rin-Laures#
10 Chester Street
Cambridge, MA 02140   (617) 868-7910

Edward S. Roman
Polaroid Corporation
Patent Dept.
575 Technology Square, 3-A
Cambridge, MA 02139   (617) 577-2518

Sheldon W. Rothstein
Polaroid Corp.
575 Technology Square
3rd Floor
Cambridge, MA 02139   (617) 577-2793

H. Eugene Stubbs#
P.O. Box 1050
Cambridge, MA 02238   (617) 924-4380

David A. Tucker
Pandiscio & Pandiscio
125 Cambridgepark Dr.
Cambridge, MA 02140   (617) 499-2770

John S. Vale
Polaroid Corp.
Pat. Dept.
549 Tech. Square
Cambridge, MA 02139   (617) 577-3013

Claire M. Vasios#
Repligen Corp.
One Kendall Square
Bldg. 700
Cambridge, MA 02139   (617) 225-6000

Louis George Xiarhos
Polaroid Corp.
Pat. Dept.
549 Technology Square
Cambridge, MA 02139   (617) 577-4314

Leon R. Yankwich
Biogen, Inc.
14 Cambridge Center
Cambridge, MA 02142   (617) 252-9810

Sherman Gilbert Davis#
39 Emerson Way
Centerville, MA 02632   (617) 771-2241

Howard Burger#
255 North Road
Unit 62
Chelmsford, MA 01824   (508) 250-0245

Michael N. Raisbeck
85 High Street
Chelmsford, MA 01824   (508) 250-1236

Howard Paul Terry
37 Don Emerson Road
P.O. Box 61
Chesterfield, MA 01012   (413) 296-4703

William Nitkin
850 Boylston Street
Suite 424
Chestnut Hill, MA 02167   (617) 232-1854

Arsen Tashjian
82 Boylston St., #3
Chestnut Hill, MA 02167   (617) 964-3848

E. Mackay Fraser
LFE Corp.
55 Green Street
Clinton, MA 01510   (617) 835-1011

Ralph L. Cadwallader
30 Raymond Road
Concord, MA 01742   (617) 369-2906

Kevin S. Lemack
747 Main Street
Suite 200
Concord, MA 01742   (508) 369-0230

Henry Cooper Nields
747 Main Street
Suite 200
Concord, MA 01742   (617) 369-0230

Carlo S. Bessone#
GTE Service Corp.
100 Endicott Street
Danvers, MA 01923   (617) 777-1900

Robert F. Clark#
GTE Products Corp.
100 Endicott St.
Danvers, MA 01923   (617) 750-2393

Edward Joseph Coleman
GTE Service Corp.
100 Endicott St.
Danvers, MA 01923   (617) 777-1900

Martha A. Finnegan
GTE Products Corporation
100 Endicott Street
Danvers, MA 01923   (617) 750-2311

Carol A. Karolow
CAK Inc.
P.O. Box 181
Danvers, MA 01923   (508) 777-5022

David J. Koris
GTE Service Corp.
100 Endicott Street
Danvers, MA 01923   (508) 750-2943

Elizabeth A. Levy#
GTE Products Corp.
100 Endicott Street
Danvers, MA 01923   (508) 750-2275

William E. Meyer
GTE Products Corporation
100 Endicott Street
Danvers, MA 01923   (617) 750-2384

Kathryn A. Piffat
56 Ledgewood Drive
Danvers, MA 01923   (617) 750-2384

Joseph S. Romanow
GTE Products Corp.
100 Endicott Street
Danvers, MA 01923   (508) 777-1900

James Theodosopoulos
GTE Service Corporation
100 Endicott Street
Danvers, MA 01923   (508) 750-2308

Robert E. Walter
GTE Service Corporation
100 Endicott St.
Danvers, MA 01923   (508) 750-2509

Scott R. Foster
P.O. Box 136
Dover, MA 02030   (617) 421-7889

Raymond J. De Vellis
900 Mayflower Street
Duxbury, MA 02332   (617) 585-3377

Robert Evan Meyer
14 Colony Drive
P.O. Box 815
East Orleans, MA 02642   (508) 255-4531

Peter Lorillard Tailer
P.O. Box 1327
Edgartown, MA 02539   (617) 693-3658

Paul J. Murphy
48 Mc Kinley Street
Everett, MA 02149   (617) 387-3444

Hartley Hoskins#
42 Haynes Avenue
Falmouth, MA 02540   (508) 548-0179

Frederick M. Murdock
7 Saconesset Rd.
Falmouth, MA 02540   (617) 548-6070

Jules Jay Morris
Foxboro Company
M.D. B52-1J
33 Commercial Street
Foxboro, MA 02035   (508) 543-8750

Christine Rinik
Foxboro Company
33 Commercial Street
(B52-1K)
Foxboro, MA 02035   (508) 549-6160

William Gerald Gosz
Genzyme Corp.
One Mountain Rd.
Framingham, MA 01701   (508) 872-8400

Mark A. Hofer
Integrated Genetics, Inc.
31 New York Avenue
Framingham, MA 01701   (617) 872-8400

Esther A. H. Hopkins
1550 Worcester Rd., Apt. 524
Framingham, MA 01701   (508) 872-8148

Edward M. Kriegsman
Kriegsman & Kriegsman
883 Edgell Road
Suite 100
Framingham, MA 01701   (508) 877-8588

Irving Martin Kriegsman
883 Edgell Road
Framingham, MA 01701   (617) 877-8588

John T. Meaney
341 Pleasant Street
Framingham, MA 01701   (508) 872-7227

Arthur B. Moore
Dennison Mfg. Co.
300 Howard Street
Framingham, MA 01701   (617) 879-0511

John E. Toupal
116 Concord Street
Framingham, MA 01701   (617) 872-3781

Leonard J. Janowski
12 Northgate Road
Franklin, MA 02038   (617) 528-9967

Donald J. Shade
Gloucester Engineering Co., Inc.
P.O. Box 900
Gloucester, MA 01930   (508) 281-1800

Henry S. Miller, Jr.
P.O. Box 922
Groton, MA 01450   (508) 448-3130

Julian Lee Siegel*
U.S. Air Force
ESD/JAN
Hanscom A F B, MA 01731   (617) 377-4076

John M. Brandt
60 Thaxter St.
Hingham, MA 02043   (617) 749-2889

Irwin A. Shaw
John S. Cheever Co.
105 Research Road
Hingham, MA 02043   (617) 749-8110

Rufus M. Franklin
54 Centerwood Drive
Holden, MA 01520   (617) 829-5766

Robert D. Donley
3 Thayer Street
Hopedale, MA 01747   (617) 473-5313

R. Bruce Blance
Monsanto Company
730 Worcester St.
Indian Orchard, MA 01151   (413) 730-2827

Michael John Murphy
Monsanto Company
730 Worcester Street
Indian Orchard, MA 01151   (413) 730-2091

Herbert Edward Farmer
95 Whitcomb Ave.
Jamaica Plain, MA 02130   (617) 522-7464

William Lee Baker
W. R. Grace Co.
55 Hayden Avenue
Lexington, MA 02173   (617) 861-6600

David Edward Brook
Hamilton, Brook, Smith
& Reynolds
Two Militia Dr.
Lexington, MA 02173   (617) 861-6240

Alice Olek Carroll
Hamilton, Brook, Smith
& Reynolds, P.C.
Two Militia Drive
Lexington, MA 02173   (617) 861-6240

Chester Cekala
W.R. Grace & Co.
Patent Dept.
55 Hayden Ave.
Lexington, MA 02173   (617) 861-6600

William R. Clark
Raytheon Company
141 Spring Street
Lexington, MA 02173   (617) 862-4845

Philip L. Conrad#
Hamilton, Brook, Smith
& Reynolds, P.C.
2 Militia Drive
Lexington, MA 02173   (617) 861-6240

Walter F. Dawson
Raytheon Co.
141 Spring St.
Lexington, MA 02173   (617) 862-6600

Giulio A. De Conti
Hamilton, Brook, Smith
& Reynolds
Two Militia Drive
Lexington, MA 02173   (617) 861-6240

Alan T. Faber
Hamilton, Brook, Smith
& Reynolds
Two Militia Drive
Lexington, MA 02173   (617) 861-6240

Fred Fisher
7 Springdale Road
Lexington, MA 02173   (617) 466-3523

Michael P. Gilday#
497 Massachusetts Ave.
Lexington, MA 02173   (617) 862-2501

Patricia Granahan
Hamilton, Brook, Smith
& Reynolds
Two Militia Drive
Lexington, MA 02173   (617) 861-6240

George Grayson
9 Suzanne Road
Lexington, MA 02173   (617) 862-2517

Elizabeth A. Hanley#
Hamilton, Brook, Smith
& Reynolds, P.C.
Two Militia Drive
Lexington, MA 02173   (617) 861-6240

Thomas O. Hoover
Hamilton, Brook, Smith
& Reynolds
2 Militia Drive
Lexington, MA 02173   (617) 861-6240

John D. Hubbard
W.R. Grace & Co.
55 Hayden Avenue
Lexington, MA 02173   (617) 861-6600

Christopher L. Maginniss
Raytheon Company
141 Spring Street
Lexington, MA 02173   (617) 860-4848

Donald E. Mahoney
Raytheon Company
141 Spring Street
Lexington, MA 02173   (617) 862-4830

Denis G. Maloney
Raytheon Company
141 Spring Street
Lexington, MA 02173   (617) 860-4887

Margit Maus
Kendall Co.
17 Hartwell Ave.
Lexington, MA 02173   (617) 861-3232

Donald F. Mofford
Raytheon Company
141 Spring Street
Lexington, MA 02173   (617) 860-4866

Nelson Scott Pierce
Hamilton, Brook, Smith
& Reynolds
2 Militia Dr.
Lexington, MA 02173   (617) 861-6240

Leo R. Reynolds
Hamilton, Brook, Smith
& Reynolds
Two Militia Dr.
Lexington, MA 02173   (617) 861-6240

Richard M. Sharkansy
Raytheon Company
141 Spring Street
Lexington, MA 02173   (617) 860-4827

James M. Smith
Hamilton, Brook, Smith
& Reynolds
Two Militia Drive
Lexington, MA 02173   (617) 861-6240

Richard B. Smith#
Hamilton, Brook, Smith
& Reynolds, P.C.
Two Militia Dr.
Lexington, MA 02173   (617) 861-6240

Richard W. Wagner#
Hamilton, Brook, Smith
& Reynolds, P.C.
Two Militia Drive
Lexington, MA 02173   (617) 861-6240

Mary L. Wakimura
Hamilton, Brook, Smith
& Reynolds
Two Militia Drive
Lexington, MA 02173   (617) 861-6240

Edmund J. Walsh#
Raytheon Company
Pat. Dept.
141 Spring Street
Lexington, MA 02173   (617) 860-4886

John J. Wasatonic
W. R. Grace & Company
55 Hayden Avenue
Lexington, MA 02173   (617) 861-6600

Richard A. Wise
Hamilton, Brook, Smith
& Reynolds
2 Militia Drive
Lexington, MA 02173   (617) 861-6240

Nicholas Prasinos
Autumn Lane
Lincoln, MA 01773   (617) 259-9702

Martin Michael Santa
7 Orchard Lane
Lincoln, MA 01773   (617) 259-9628

Mary R. Bonzagni
Longmeadow Professional Park
171 Dwight Road
Longmeadow, MA 01106   (413) 567-2076

Malcolm J. Chisholm, Jr.
Longmeadow Professional Park
171 Dwight Road
Longmeadow, MA 01106   (413) 567-2076

Donald S. Holland
Longmeadow Professional Park
171 Dwight Road
Longmeadow, MA 01106   (413) 567-2076

Joseph S. Nelson
1 Druid Circle
Longmeadow, MA 01106   (413) 567-3418

Allegra A. Genest#
32 Brookside Street
Lowell, MA 01854   (413) 567-3418

Kenneth L. Milik
Wang Laboratories, Inc.
One Industrial Avenue
M/S 014-B7D
Lowell, MA 01851   (617) 967-3128

Gordon E. Nelson
Wang Laboratories
One Industrial Avenue
Lowell, MA 01851   (508) 967-2064

John H. Pearson, Jr.
Pearson & Pearson
12 Hurd Street
Lowell, MA 01852   (617) 452-1971

Sally L. Pearson#
Pearson & Pearson
12 Hurd Street
Lowell, MA 01852   (617) 452-1971

Scott K. Peterson
Wang Laboratories, Inc.
One Industrial Ave.
M/S 014-B7D
Lowell, MA 01851   (508) 967-5001

Michael H. Shanahan
Wang Laboratories, Inc.
One Industrial Ave.
Lowell, MA 01851   (617) 967-6020

Nathan D. Herkamp
GE Aircraft Engines
1000 Western Avenue
Lynn, MA 01910   (617) 594-2701

Derek P. Lawrence
General Electric Co.
GE Aircraft Engines
Maildrop 16002
Lynn, MA 01910   (617) 594-4627

Philip J. Mc Farland
10 Bancroft Street
Lynnfield, MA 01940   (617) 334-4342

Donald R. Castle
Harington Way
Manchester, MA 01944   (617) 526-7329

Paul John Cook
8 Washington Street
Manchester, MA 01944   (508) 526-7149

Donald N. Halgren
35 Central Street
Manchester, MA 01944   (617) 526-8000

George A. Herbster
27 Skytop Drive
Manchester, MA 01944   (617) 526-7033

Alvin Isaacs
Kendall Company
15 Hampshire St.
Mansfield, MA 02048   (508) 261-8412

Edward J. Scahill, Jr.
Kendall Co.
15 Hampshire St.
Mansfield, MA 02048   (508) 261-8544

Kenneth Wright Brown
75 Harbor Ave.
Marblehead, MA 01945   (617) 631-9135

George W. Crowley
74 Atlantic Avenue
Marblehead, MA 01945   (617) 639-1111

Herbert P. Kenway
Kenway & Crowley
210 Humphrey Street
Suite 105
Marblehead, MA 01945   (617) 639-1111

Martin Kirkpatrick
P.O. Box 1109
Marblehead, MA 01945   (617) 631-1334

John M. Gunther
Digital Equipment Corp.
2 Results Way
Marlboro, MA 01752   (617) 467-7083

Albert P. Cefalo
Digital Equipment Corp.
Legal Dept.
111 Powdermill Rd.
Maynard, MA 01754   (617) 493-8571

William C. Cray
Digital Equipment Corporation
Law Dept.
111 Powdermill Road
Maynard, MA 01754   (617) 493-2469

Robert J. Feltovic
Digital Equipment Corp.
111 Powdermill Road
Maynard, MA 01754   (508) 493-6734

Stephen J. Holmes#
Digital Equipment Corp.
111 Powdermill Road
M.S. 02-3/G1
Maynard, MA 01754   (508) 493-6064

Albert S. Johnston
Digital Equipment Corp.
Law Dept.
MSO/C10
111 Powdermill Road
Maynard, MA 01754   (508) 493-6526

Richard M. Kotulak
Digital Equipment Corp.
111 Powder Mill Rd.
Maynard, MA 01754   (508) 493-9763

Gerald E. Lester
Digital Equipment Corporation
Law Dept. MSO/C5
111 Powdermill Road
Maynard, MA 01754   (508) 493-6571

Robert C. Mayes
Digital Equipment Corporation
111 Powdermill Road
MS O/C5
Maynard, MA 01754   (617) 493-6984

John George Mesaros
Digital Equipment Corp.
111 Powdermill Road - MSO/D11
Maynard, MA 01754   (508) 493-4111

Ronald E. Myrick
Digital Equipment Corporation
111 Powdermill Road
Maynard, MA 01754   (508) 493-2132

Rama B. Nath
Digital Equipment Corporation
Patent Services Law Group
111 Powdermill Road
Maynard, MA 01754   (508) 493-2132

Richard J. Paciulan
Digital Equipment Corp.
111 Powdermill Road
MSO/C5
Maynard, MA 01754   (617) 493-8426

John V. Pezdek
Digital Equipment Corp.
111 Powdermill Road
MSO1/C5
Maynard, MA 01754   (508) 493-6502

Vincenzo D. Pitruzzella
Digital Equipment Corp.
111 Powdermill Road
Maynard, MA 01754   (617) 493-6604

Timothy C. Pledger
Digital Equipment Corp.
111 Powdermill Road
MSO/C5
Maynard, MA 01754   (508) 493-6355

Vincent J. Ranucci
Digital Equipment Corp.
111 Powdermill Road MSO/C5
Maynard, MA 01754   (508) 493-6984

Ronald T. Reiling
Digital Equipment Corp.
111 Powdermill Road
Maynard, MA 01754   (617) 493-2991

Thomas C. Siekman
Digital Equipment Corp.
Legal Dept.
111 Powder Mill Rd.
Maynard, MA 01754   (508) 493-4422

William P. Skladony
Digital Equipment Corp.
111 Powdermill Road
MSO/D11
Maynard, MA 01754   (508) 493-6335

Penelope A. Smith
Digital Equipment Corp.
111 Powdermill Road
MSO/C10
Maynard, MA 01754   (617) 493-4293

Maureen L. Stretch
Digital Equipment Corp.
111 Powdermill Road
Maynard, MA 01754   (508) 493-6983

William F. White
Digital Equipment Corp.
111 Powdermill Road
Maynard, MA 01754   (508) 493-7531

Barry N. Young
Digital Equipment Corp.
111 Powdermill Road
MSO/C10
Maynard, MA 01754   (508) 493-5479

Arthur S. Morgenstern
Ciba Corning Diagnostics Corp.
63 North Street
Medfield, MA 02052   (508) 359-3836

Nicholas I. Slepchuk, Jr.
Ciba Corning Diagnostics Corp.
63 North Street
Medfield, MA 02052   (508) 359-3867

John P. Morley
320 Upham St.
Melrose, MA 02176   (617) 665-7755

Edwin H. Paul, Jr.
34 Maple Street
Milford, MA 01757   (617) 478-2000

William Robert Sherman
15 Gay Street
Nantucket, MA 02554   (508) 228-3880

Richard L. Ballantyne
Prime Computer Inc.
Prime Park
Natick, MA 01760   (617) 655-8000

Richard J. Donahue*
Dept. of the Army
U.S. Army Natick Rd&E Center
Attn: STRNC-ZSL
Natick, MA 01760   (508) 651-4322

Anthony N. Fiore
Prime Computer, Inc.
Legal Dept.
Prime Park
MS 15-36
Natick, MA 01760   (617) 655-8000

Lawrence E. Labadini*
U.S. Army
RD&E Center
Kansas St.
Natick, MA 01760   (617) 651-4510

Ronald S. Cornell
Duracell Inc.
37 A Street
Needham, MA 02194   (617) 449-7600

James B. Mc Veigh, Jr.
Duracell Inc.
37 A Street
Needham, MA 02194   (617) 449-7600

Richard T. Oakes
Gould Inc.
374 Merrimac St.
Newburyport, MA 01950   (617) 465-4243

Frederick H. Brustman
48 Lantern Lane
Newton, MA 02159   (617) 527-6146

Robert Louis Goldberg
Dike, Bronstein, Roberts,
Cushman & Pfund
2345 Washington Street
Newton, MA 02162   (617) 244-4990

Daniel J. Klein
104 Woodchester Drive
Newton, MA 02167   (617) 332-9824

Paul G. Lunn
6 Kippy Drive
Newton, MA 02168   (508) 435-9001

Barbara Z. Terris#
269 Franklin Street
Newton, MA 02158   (617) 965-4940

Walter Joseph Kreske, Sr.
50 Tyler Terrace
Newton Centre, MA 02159   (617) 527-8198

Walter Juda#
Prototech Company
70 Jaconnet St.
Newton Highlands, MA 02161
    (617) 965-2720

Vincent Hilary Sweeney#
Sprague Electric Co.
96 Marshall Street
Research Center
North Adams, MA 01247   (413) 664-4524

John A. Haug
11 Ryder Circle
North Attleboro, MA 02760   (617) 699-3314

Carole M. Calnan
77 Ellis Ave.
Norwood, MA 02062   (617) 769-3240

David G. Rasmussen
Analog Devices, Inc.
Three Technology Way
Norwood, MA 02062   (617) 461-4073

Frank M. Ward, III
Allendale Mutual Insurance Co.
220 Norwood Park South
Norwood, MA 02062   (617) 762-0200

Richard Paul Crowley
901 Main Street
Osterville, MA 02655   (617) 428-4000

Thomas A. Kahrl
Wianno Place
901 Main Street
Osterville, MA 02655   (508) 428-4002

Abraham Ogman
9 Downing Road
Peabody, MA 01960   (508) 531-7532

John R. Castiglione
201 Wendell Avenue
Pittsfield, MA 01201   (413) 494-4531

Sidney Greenberg
18 Glenn Drive
Pittsfield, MA 01201   (413) 448-8336

William F. Mufatti
General Electric Co.
One Plastics Avenue
Pittsfield, MA 01201   (413) 448-4707

Kent I. Patashnick
674 North Street
P.O. Box 724
Pittsfield, MA 01202   (413) 499-2344

Gordon Needleman
72 Washington Street
Box 2019
Quincy, MA 02269   (617) 471-5632

Ellen C. Childress#
105 Green St.
Reading, MA 01867   (617) 471-5632

Edmund F. Chojnowski
P.O. Box 1188
S. Lancaster, MA 01561   (617) 368-8826

Owen J. Meegan
24 North Street
Salem, MA 01970   (617) 741-4135

Donald A. Teare
339 First Parish Road
P.O. Box 599
Scituate, MA 02066   (617) 545-7500

James H. Grover
63 Norwood Street
P.O. Box 296
Sharon, MA 02067   (617) 784-2635

Charles E. Cullen, Jr.
27 Main Circle
Shrewsbury, MA 01545   (617) 845-6510

Nicholas J. Cafarelli
60 Audubon Street
Springfield, MA 01108   (413) 739-5045

John J. Dempsey
Chapin, Neal and Dempsey, P.C.
1331 Main Street
Springfield, MA 01103   (413) 736-5401

William J. Farrington
Monsanto Co.
730 Worcester Street
Springfield, MA 01151   (413) 730-2811

Chester Edwin Flavin
Ross, Ross & Flavin
120 Maple Street
Springfield, MA 01103   (413) 733-3194

William Darrell Fosdick#
P.O. Box 80545
Springfield, MA 01138   (413) 733-3194

Leonard S. Michelman
Michelman & Feinstein
1333 East Columbus Avenue
P.O. Box 2992
Springfield, MA 01105   (413) 737-1166

Kenwood Ross
Ross, Ross & Flavin
120 Maple Street
Room 207
Springfield, MA 01103   (413) 733-3194

Michael J. Rye
Doherty, Wallace, Pillsbury
& Murphy, P.C.
1 Monarch Pl., 19th Flr.
1414 Main St.
Springfield, MA 01144   (413) 733-3194

Charles R. Fay
194 Justice Hill Rd.
Sterling, MA 01564   (617) 422-7146

Milton D. Bartlett#
566 Boston Post Road
Sudbury, MA 01776   (617) 443-2125

Maura K. Moran
250 Hudson Road
Sudbury, MA 01776   (508) 443-4558

Francis L. Conte
75 Burrill Street
Swampscott, MA 01907   (617) 592-9077

Jeffery D. Marshall
21 Central
Topsfield, MA 01983   (617) 887-9788

Robert S. Sanborn
Box 4248
Vineyard Haven, MA 02568   (617) 693-8527

William G. Auton*
U.S.A.F.
AFJACPB
Bldg. 104
424 Trapelo Road
Waltham, MA 02154   (617) 377-4072

Stanton E. Collier*
USAF
AFJACPB- Bldg. 104
424 Trapelo Road
Waltham, MA 02154   (617) 377-4072

Richard A. Covel
Foster-Miller, Inc.
350 Second Avenue
Waltham, MA 02154   (617) 890-3200

Frances P. Craig#
GTE Service Corporation
40 Sylvan Road
Waltham, MA 02254   (617) 466-4017

Brian M. Dingman
260 Bear Hill Road
Waltham, MA 02154   (617) 890-5678

Jacob N. Erlich*
U.S. Air Force
424 Trapelo Rd.
Waltham, MA 02154   (617) 377-4075

Irwin P. Garfinkle*
U.S. Dept. of Air Force
AFJACPB, Bldg. 104
424 Trapelo Road
Waltham, MA 02154   (617) 377-4072

Joseph S. Iandiorio
260 Bear Hill Road
Waltham, MA 02154   (617) 890-5678

Victor F. Lohmann, III#
GTE Service Corp.
40 Sylvan Road
Waltham, MA 02254   (617) 466-4018

Herbert E. Messenger
Thermo Electron Corp.
101 1st Ave.
Waltham, MA 02254   (617) 622-1173

Lawrence E. Monks
GTE Laboratories Inc.
40 Sylvan Road - M.S. #31
Waltham, MA 02254   (617) 890-8460

Robert Lang Nathans*
U.S. Air Force
AFJACB - Bldg. 104
424 Trapelo Road
Waltham, MA 02154   (617) 377-4072

Carl F. Ruoff
GTE Service Corp.
40 Sylvan Road
Waltham, MA 02254   (617) 466-4016

Joseph Stecewycz#
260 Bear Hill Road
Suite 207
Waltham, MA 02154   (617) 890-5678

William Stepanishen*
U.S. Air Force
AFJACPB
424 Trapelo Rd.
Bldg. 104
Waltham, MA 02154   (617) 377-4074

Thomas C. Stover, Jr.*
AFJACPB
Bldg. 104, 2nd Floor
424 Trapelo Road
Waltham, MA 02154   (617) 377-4072

Francis I. Sullivan, Jr.
GTE Government Systems Corp.
100 First Avenue
Waltham, MA 02254   (617) 466-3311

Robert Eugene Walrath
GTE Service Corp.
40 Sylvan Road
Waltham, MA 02254   (617) 466-4010

John S. Yeo
GTE Service Corporation
40 Sylvan Road
Waltham, MA 02254   (617) 466-4014

Jack M. Young
92-B Charles River Road
Waltham, MA 02154   (617) 893-3827

Herbert Epstein
199 Coolidge Ave.
Unit 103
Watertown, MA 02172   (617) 926-2603

Norman E. Saliba
Ionics, Inc.
65 Grove St.
Watertown, MA 02172   (617) 926-2500

Edward J. Collins
62 Claypit Hill Road
Wayland, MA 01778   (617) 358-5645

Herbert L. Bello
57 River Street
Wellesley, MA 02181   (617) 235-4457

Robert P. Cogan
EG & G, Inc.
45 William Street
Wellesley, MA 02181   (617) 237-5100

Leo M. Kelly
E.G. & G., Inc.
45 William St.
Wellesley, MA 02181   (617) 237-5100

Moisey M. Lerner
Tomin Corporation
P.O. Box 82-206
Wellesley, MA 02181   (617) 653-7860

William D. Roberson
19 Swarthmore Road
Wellesley, MA 02181   (617) 235-9256

Walter Fred#
385 Prospect Street
West Boylston, MA 01583   (617) 835-3296

James F. Baird
South Main Street
West Brookfield, MA 01585   (617) 867-2441

Thomas W. Underhill
P.O. Box 553
West Chatham, MA 02669   (617) 945-1427

Simon L. Cohen
Data General Corp.
4400 Computer Dr.
Westboro, MA 01580   (617) 870-7768

Robert L. Dulaney
Data General Corp.
4400 Computer Drive
Westboro, MA 01580   (617) 366-8911

Jacob Frank
Data General Corp.
4400 Computer Dr.
Westboro, MA 01580   (617) 820-7781

Joel Wall
Data General Corp.
4400 Computer Drive
Westboro, MA 01580   (508) 870-7777

Philip Colman
384 Glen Rd.
Weston, MA 02193   (617) 820-7781

Richard F. Benway
265 Washington Street
Westwood, MA 02090   (617) 329-6611

David R. Thornton
165 Fensview Drive
Westwood, MA 02090   (617) 762-9487

William M. Anderson
51 Broad Reach
Apt. T-102a
Weymouth, MA 02191   (617) 331-5293

Arthur K. Hooks#
1341 Green River Rd.
Williamstown, MA 01267   (413) 458-4622

Moonray Kojima
Box 627
Williamstown, MA 01267   (413) 458-2880

I. David Blumenfeld
General Electric Co.
Aircraft Instrument Dept.
50 Fordham Rd.
Wilmington, MA 01887   (617) 937-4727

David N. Caracappa
91 Hapgood Road
Winchendon, MA 01475   (508) 297-4722

Francis A. Di Luna
Roche, Carens & De Giacomo
Country Club Professional Bldg.
304 Cambridge Road
Woburn, MA 01801   (617) 933-5505

Natalie Trousof
15 Jacques Loeb Road
Woods Hole, MA 02543   (617) 933-5505

David Bennett
Norton Company
1 New Bond Street
Box Number 15008
Worcester, MA 01615   (508) 795-5000

Gerry A. Blodgett
Blodgett & Blodgett, P.C.
43 Highland St.
Worcester, MA 01608   (617) 753-5533

Norman S. Blodgett
Blodgett & Blodgett, P.C.
43 Highland St.
Worcester, MA 01608   (617) 753-5533

Thomas C. Blodgett#
Blodgett & Blodgett, P.C.
43 Highland Street
Worcester, MA 01609   (617) 753-5533

Stephen L. Borst
Norton Company
1 New Bond Street
Box Number 15008
Worcester, MA 01615   (508) 795-5000

Wilfred F. Des Rosiers, Sr.#
Blodgett & Blodgett, P.C.
43 Highland St
Worcester, MA 01608   (617) 753-5533

Arthur A. Loiselle, Jr.
Norton Company
Pat. Dept.
1 New Bond St.
Box 15008
Worcester, MA 01615   (550) 879-5255

Allan R. Redrow
Apt. 6
12 Brookside Avenue
Worcester, MA 01602   (508) 756-9299

Volker R. Ulbrich
Norton Company
1 New Bond Street
Box Number 15008
Worcester, MA 01615   (508) 795-5564

Michigan

Michael A. Mohr
Amway Corporation
7575 East Fulton Road
Ada, MI 49355   (616) 676-5416

Marion Duane Ford
Wacker Silicones Corporation
Patent Dept.
3301 Sutton Road
Adrian, MI 49221   (517) 263-5711

Elizabeth M. Anderson#
Warner Lambert Company
2800 Plymouth Road
Ann Arbor, MI 48105   (313) 996-7304

Jonathan A. Barney#
395 Village Green Blvd., #205
Ann Arbor, MI 48105   (313) 996-7304

Ross W. Campbell
Washtenaw County Courthouse
Suite 228
P.O. Box 8645
Ann Arbor, MI 48107   (313) 994-2552

James L. Cox, II
1014 Duncan St.
Ann Arbor, MI 48103   (313) 663-9366

Ronald A. Daignault
Warner-Lambert Company
2800 Plymouth Road
Ann Arbor, MI 48105   (313) 996-7530

James H. Dautremont
2124 Brockman
Ann Arbor, MI 48104   (313) 663-0058

James M. Deimen
325 East Eisenhower Pkwy.
Suite 2
Ann Arbor, MI 48104   (313) 994-5947

Ronald Feldbaum#
P.O. Box 3170
Ann Arbor, MI 48106   (313) 995-9872

Charles T. Graham
Harness, Dickey & Pierce
301 E. Liberty
Suite 555
Ann Arbor, MI 48104   (313) 662-5653

Kirk A. Kuhfeldt
Krass & Young
2001 Commonwealth Blvd.
Suite 301
Ann Arbor, MI 48105   (313) 662-0270

Ruth Hattan Newtson
Warner-Lambert Company
2800 Plymouth Road
Ann Arbor, MI 48105   (313) 996-7000

Steven L. Oberholtzer
Harness, Dickey & Pierce
301 East Liberty Street
Suite 555
Ann Arbor, MI 48104   (313) 662-5653

Vett Parsigian
788 Greenhills
Ann Arbor, MI 48105   (313) 665-6411

Eric J. Sosenko
Harness, Dickey & Pierce
301 E Liberty St., Suite 555
Ann Arbor, MI 48104   (313) 662-5653

James E. Stephenson
Harness, Dickey & Pierce
301 E. Liberty
Suite 555
Ann Arbor, MI 48104   (313) 662-5653

Joan V. Thierstein
Warner-Lambert Company
Legal Dept.
2800 Plymouth Road
Ann Arbor, MI 48105   (313) 996-7190

Francis J. Tinney
Warner-Lambert Company
2800 Plymouth Road
Ann Arbor, MI 48105   (313) 996-7295

Jason J. Young#
Krass & Young
2001 Commonwealth Blvd.
Suite 301
Ann Arbor, MI 48105   (313) 662-0270

George L. Boller
Siemens Automotive
2400 Excutive Hills Dr.
P.O. Box 217017
Auburn Hills, MI 48321   (313) 253-2664

J. Gordon Lewis
ITT Automotive, Inc.
3000 University Drive
Auburn Hills, MI 48321   (313) 340-3000

Michael J. Schmidt
GKN - Automotive Inc.
3300 University Drive
Auburn Hills, MI 48057   (313) 377-1267

Robert P. Seitter
ITT Automotive, Inc.
3000 University Drive
P.O. Box 57016
Auburn Hills, MI 48057   (313) 340-3447

Russel C. Wells
Siemens Corporation
2400 Executive Hills Dr.
P.O. Box 217017
Auburn Hills, MI 48321   (313) 340-3447

Franklin Clyde Harter
Whirlpool Corp.
2000 M-63
Benton Harbor, MI 49022   (616) 926-5020

Gene A. Heth
Whirlpool Corp.
2000 M-63 North
Benton Harbor, MI 49022   (616) 926-5600

Robert L. Judd
Whirlpool Corp.
2000 M-63
Benton Harbor, MI 49022   (616) 926-3511

Stephen D. Krefman
Whirlpool Corporation
2000 M-63
Benton Harbor, MI 49022   (616) 926-5013

James Stanley Nettleton
894 Sierra Drive
Benton Harbor, MI 49022   (616) 927-3894

Robert O. Rice
Whirlpool Corp.
2000 M-63
Benton Harbor, MI 49022   (616) 926-3870

Thomas J. Roth
Whirlpool Corporation
2000 M-63
Benton Harbor, MI 49022   (616) 926-5604

Thomas A. Schwyn#
Whirlpool Corporation
2000 M-63 North
Benton Harbor, MI 49022   (616) 926-5303

Thomas E. Turcotte
Whirlpool Corp.
2000 M-63, North
Benton Harbor, MI 49022   (616) 926-5021

Alex Rhodes
30100 Telegraph Road
Suite 460
Bingham Farms, MI 48025   (313) 646-4400

Thomas E. Anderson
Gifford, Groh, Van Ophem, Sheridan,
Sprinkle & Dolgorukov
280 N. Woodward
Suite 210
Birmingham, MI 48011   (313) 647-6000

John R. Benefiel
360 Birmingham Place
401 S. Woodward Avenue
Birmingham, MI 48011   (313) 644-1455

Barbara M. Burns
Gifford, Groh, Sprinkle,
Patmore & Anderson, P.C.
280 N. Woodward Ave., Ste. 400
Birmingham, MI 48009   (313) 647-6000

Lynn E. Cargill
30700 Telegraph Road
Suite 4550
Birmingham, MI 48010   (313) 646-2828

Susan M. Cornwall#
30700 Telegraph Road
Suite 4550
Birmingham, MI 48010   (313) 646-2828

T. I. Davenport
1055 Larchlea
Birmingham, MI 48009   (313) 643-3645

Paul Fitzpatrick
851 S. Glenhurst
Birmingham, MI 48009   (313) 644-1022

George Edward Frost
291 Hupp Cross Rd.
Birmingham, MI 48010   (313) 647-5508

Ernest I. Gifford
Gifford, Groh, Sherridan,
Sprinkle & Dolgorukov, P.C.
280 N. Woodward
Suite 210
Birmingham, MI 48011   (313) 647-6000

Irvin L. Groh
Gifford, Groh, Van Ophem, Sheridan
Sprinkle & Dolgorukov
280 N. Woodward
Suite 210
Birmingham, MI 48011   (313) 647-6000

Douglas J. Mc Evoy
Gifford, Groh, Sprinkle,
Patmore & Anderson
280 N. Woodward, Suite 400
Birmingham, MI 48009   (313) 647-6000

James P. Meloche
401 S. Woodward Avenue
Suite 360
Birmingham, MI 48009   (313) 644-2114

Thomas Traian Moga
Gifford, Groh, Sprinkle,
Patmore & Anderson
280 N. Woodward
Suite 400
Birmingham, MI 48009   (313) 647-6000

Theresa A. Orr
Gifford, Groh, Sprinkle, Patmore
& Anderson, P.C.
280 N. Woodward, Suite 400
Birmingham, MI 48009   (313) 647-6000

Alfred Lawrence Patmore, Jr.
Gifford, Groh, Sprinkle,
Patmore & Anderson
280 N. Woodward Avenue
Suite 400
Birmingham, MI 48009   (313) 647-6000

Anne G. Sabourin
Cargill & Associates
30700 Telegraph Rd.
Suite 4550
Birmingham, MI 48010   (313) 646-2828

John Lyons Shortley
5890 Snowshoe Circle
Birmingham, MI 48010   (313) 642-2791

Cass L. Singer
Cargill & Associates
30700 Telegraph Rd.
Suite 4550
Birmingham, MI 48010   (313) 642-2791

Douglas W. Sprinkle
Gifford Groh Sprinkle Patmore
& Anderson, P.C.
280 N. Woodward
Suite 400
Birmingham, MI 48009   (313) 647-6000

Marjory G. Basile
Miller, Canfield, Paddock
& Stone
P.O. Box 2014
1400 N. Woodward, Suite 100
Bloomfield Hills, MI 48303   (313) 645-5000

James H. Bower
2138 Randall Lane
Bloomfield Hills, MI 48013   (313) 642-2171

Gary L. Newtson
Harness, Dickey & Pierce
P.O. Box 828
Bloomfield Hills, MI 48013   (313) 641-1600

Theodore Van Meter
577 Cambridge Way
Bloomfield Hills, MI 48304   (313) 253-0436

Charles Frank Voytech
08494 4750 St.
Bloomingdale, MI 49026   (616) 521-3138

Bertram F. Claeboe
6307 Baldwin Circle
Brighton, MI 48116   (313) 227-2416

Paul D Arc Garty#
P.O. Box 121
Brooklyn, MI 49230   (517) 592-8765

Kenneth C. Witt
401 W. Front Street
Buchanan, MI 49107   (616) 695-3178

Lon H. Romanski
Suite 5, Crandell Bldg.
210 1/2 N. Micthell Avenue
Cadillac, MI 49601   (616) 775-0171

Fred Paul Kostka
7062 North Maple Drive
Coloma, MI 49038   (616) 468-5312

Douglas E. Mark
2152 Chevychase Drive
Davison, MI 48423   (313) 653-6470

Peter Abolins
Ford Motor Co.
Parklane Towers E., Suite 911
0ne Parklane Blvd.
Dearborn, MI 48126   (313) 337-3341

Glenn S. Arendsen
Ford Motor Co.
Office of Gen. Coun.
WHQ 1010
Dearborn, MI 48121   (313) 322-4898

Raymond L. Coppiellie
Ford Motor Company
Suite 911 Parklane Towers E.
One Parklane Blvd.
Dearborn, MI 48126   (313) 337-1069

Richard Dean Dixon
Ford Motor Co., Electronics Div.
Regent Court, Suite 804
P.O. Box 6200
Dearborn, MI 48126   (313) 337-1069

Jerome R. Drouillard
Ford Motor Company
Parklane Towers - 911 East
Dearborn, MI 48126   (313) 845-5101

Roger E. Erickson
Ford Motor Co.
Off. of Gen. Coun.
1078 WHQ
P.O. Box 1899
Dearborn, MI 48121   (313) 337-5462

Paul K. Godwin, Jr
Ford Motor Co.
911 E. Parklane Towers
Dearborn, MI 48126   (313) 337-8718

William Edwin Johnson
Ford Motor Co.
911 Parklane Towers East
Dearborn, MI 48126   (313) 323-2023

Anthony T. Lesnick
5021 Horger
Dearborn, MI 48126   (313) 584-3166

Dwight A. Lewis
23646 Rockford
Dearborn, MI 48124   (313) 562-8030

Allan J. Lippa
Ford Motor Company
Off. of General Counsel
911 E. Parklane Towers
Dearborn, MI 48126   (313) 594-1145

Joseph William Malleck
Ford Motor Company
Parklane Towers E., Suite 911
One Parklane Blvd.
Dearborn, MI 48126   (313) 323-8130

Adolph Gustav Martin
14327 Michigan Avenue
Dearborn, MI 48126   (313) 581-4444

Roger L. May
Ford Motor Co.
Off. of Gen. Coun.
Suite 911, Parkland Towers East
1 Parklane Blvd.
Dearborn, MI 48126   (313) 323-1903

Robert Ellsworth Mc Collum
Ford Motor Co.
911 Parklane Towers East
Dearborn, MI 48126   (313) 323-1904

Frank G. Mc Kenzie
Ford Motor Co.
Parklane Towers E., Suite 911
One Parklane Blvd.
Dearborn, MI 48126   (313) 323-0903

Lorraine S. Melotik
Ford Motor Co.
911 E. Parklane Towers
1 Parklane Blvd.
Dearborn, MI 48126   (313) 337-1069

Mark L. Mollon
Ford Motor Company
911 East Parklane Towers
Dearborn, MI 48126   (313) 845-5371

Damian Porcari
Ford Motor Company
One Parklane Blvd.
Suite 911 East
Dearborn, MI 48126   (313) 845-5101

Clifford Lincoln Sadler
Ford Motor Co.
911 Parklane Towers East
Dearborn, MI 48126   (313) 323-1823

Daniel M. Stock
Ford Motor Co.
Suite 911-Parklane Tower E.
One Parklane Blvd.
Dearborn, MI 48126   (313) 323-1289

Jay C. Taylor
1525 Belmont
Dearborn, MI 48128   (313) 274-3829

Keith L. Zerschling
Ford Motor Co.
Suite 911 - Parklane Towers East
One Parklane Blvd.
Dearborn, MI 48126   (313) 322-7725

Martin J. Adelman
Wayne State University
468 West Ferry
Detroit, MI 48202   (313) 577-3943

K. F. Barr, Jr.
General Motors Corporation
Patent Section
New Center One Bldg.
3031 W. Grand Blvd.
Detroit, MI 48202   (313) 974-1339

Gordon F. Belcher
General Motors Coporation
Patent Section
New Center One Bldg., Suite 450
3031 W. Grand Blvd.
Detroit, MI 48202   (313) 974-0075

Edward Joseph Biskup
General Motors Corp.
New Center One Bldg.
3031 W. Grand Blvd.
Detroit, MI 48202　　　(313) 974-1307

Cary W. Brooks
General Motors Corp.
Patent Section
P.O. Box 33114
Detroit, MI 48232　　　(313) 974-1351

Raymond I. Bruttomesso, Jr.
General Motors Corp.
New Center One Bldg..
3031 W. Grand Blvd.
P.O. Box 33114
Detroit, MI 48232　　　(313) 974-1335

Robert A. Choate
Barnes, Kisselle, Raisch Choate,
Whittemore & Hulbert, P.C.
3500 Penobscot Bldg.
Detroit, MI 48226　　　(313) 962-4790

John Gerard Chupa
Dykema Gossett
35th Floor
400 Renaissance Center
Detroit, MI 48243　　　(313) 568-6701

Robert C. Collins
Barnes, Kisselle, Raisch Choate,
Whittemore & Hulbert, P.C.
3500 Penobscot Bldg.
Detroit, MI 48226　　　(313) 962-4790

Howard N. Conkey
General Motors Corp.
Pat. Sect.-New Center One Bldg.
3031 W. Grand Blvd.
P.O. Box 33114
Detroit, MI 48232　　　(313) 974-1340

Joseph V. Coppola, Sr.
Dykema Gossett
35th Floor
400 Renaissance Center
Detroit, MI 48243　　　(313) 568-6671

Alfonse J. D Amico
Barnes, Kisselle, Raisch Choate
Whittemore & Hulbert, P.C.
1520 Ford Bldg.
Detroit, MI 48226　　　(313) 962-4790

Chester L. Davis, Jr.
Barnes, Kisselle, Raisch Choate,
Whittemore & Hulbert, P.C.
3500 Penobscot Bldg.
Detroit, MI 48226　　　(313) 962-4790

Linda M. Deschere
Barnes, Kisselle, Raisch,
Choate, Whittemore & Hulbert, P.C.
3500 Penobscot Building
Detroit, MI 48226　　　(313) 962-4790

D. Edward Dolgorukov
Dykema Gossett
35th Floor
400 Renaissance Center
Detroit, MI 48243　　　(313) 568-6701

Albert F. Duke
General Motors Corp.
3031 W. Grand Blvd.
Suite 450
P.O. Box 33114
Detroit, MI 48202　　　(313) 974-1358

Dean L. Ellis
General Motors Corp.
Pat. Section
P.O. Box 33114
3031 W. Grand Blvd.
Detroit, MI 48232　　　(313) 974-1323

B. Lynn Enderby
General Motors Corp.
1344 W. Grand Blvd.
Detroit, MI 48202　　　(313) 974-1594

Joseph W. Farley
710 Buhl Bldg.
Detroit, MI 48226　　　(313) 961-5190

Warren E. Finken
General Motors Corp.
Pat. Sect. Suite 450
3031 W. Grand Blvd.
P.O. Box 33114
Detroit, MI 48232　　　(313) 974-1304

Francis James Fodale
General Motors Corp.
Pat. Sect.-New Center One Bldg.
3031 W. Grand Blvd.
Detroit, MI 48232　　　(313) 974-1362

Basil C. Foussianes
Barnes, Kisselle, Raisch Choate,
Whittemore & Hulbert, P.C.
3500 Penobscot Bldg.
Detroit, MI 48226　　　(313) 962-4790

William H. Francis
Barnes, Kisselle, Raisch Choate
Whittemore & Hulbert
3500 Penobscot Bldg.
Detroit, MI 48226　　　(313) 962-4790

Jimmy L. Funke
General Motors Corp.
P. O. Box 33114
Detroit, MI 48232　　　(313) 974-1344

Herbert Furman
General Motors Corp.
Pat. Sect.-New Center One Bldg.
3031 W. Grand Blvd.
P.O. Box 33114
Detroit, MI 48232　　　(313) 974-1336

Lawrence J. Goffney, Jr.
Dykema Gossett
35th Floor
400 Renaissance Center
Detroit, MI 48243　　　(313) 568-6800

Richard D. Grauer
Dykema Gossett
35th Floor
400 Renaissance Center
Detroit, MI 48243　　　(313) 568-6701

Patrick M. Griffin
General Motors Corp.
Pat. Sect.-New Center One Bldg.
3031 W. Grand Blvd.
P.O. Box 33114
Detroit, MI 48232　　　(313) 974-1330

William H. Griffith
Barnes, Kisselle, Raisch, Choate,
Whittemore & Hulbert, P.C.
3500 Penobscot Bldg.
Detroit, MI 48226　　　(313) 962-4790

George Arthur Grove
General Motors Corp.
Patent Section
P.O. Box 33114
Detroit, MI 48232　　　(313) 974-1322

George H. Hathaway
Detroit Edison
2000 2nd Avenue
Detroit, MI 48226　　　(313) 237-8958

Lewis R. Hellman
ANR Pipeline Company
500 Renaissance Center
Detroit, MI 48243　　　(313) 496-3773

Ernest E. Helms
General Motors Corporation
Patent Section
P.O. Box 33114
Detroit, MI 48232　　　(313) 974-1346

William H. Honaker
Dykema Gossett
35th Floor
400 Renaissance Center
Detroit, MI 48243　　　(313) 568-6701

Tim G. Jaeger
General Motors Corp.
New Center One Bldg. - Suite 450
3031 W. Grand Blvd.
Pat. Section
Detroit, MI 48232　　　(313) 974-1329

Robert L. Kelly
Dykema Gossett
35th Floor
400 Renaissance Center
Detroit, MI 48243　　　(313) 568-6800

Charles E. Leahy
General Motors Corp.
Patent Section
P.O. Box 33114
Detroit, MI 48232　　　(313) 974-1369

Robert E. Luetje
Kolene Corporation
12890 Westwood Avenue
Detroit, MI 48223　　　(313) 273-9220

Kenneth H. Mac Lean, Jr.
Chrysler Motors Corp.
P.O. Box 1118
Detroit, MI 48288　　　(313) 956-4282

Creighton Roland Meland, Sr.
General Motors Corp.
Pat. Sect. New Center One Bldg.
3031 W. Grand Blvd.
P.O. Box 33114
Detroit, MI 48232   (313) 974-1327

Mark A. Navarre
General Motors Corp.
Pat. Section
Suite 450 - New Center One Bldg.
Detroit, MI 48202   (313) 974-1349

Theodore W. Olds, III
Dykema Gossett
35th Floor
400 Renaissance Center
Detroit, MI 48243   (313) 568-6800

Robert John Outland
General Motors Corp.
Patent Section, New Center 1 Bldg.
3031 W. Grand Blvd.
P.O. Box 33114
Detroit, MI 48232   (313) 974-1361

Ronald Lloyd Phillips
General Motors Corp.
Pat. Sect.-New Center One Bldg.
Suite 450
P.O. Box 33114
Detroit, MI 48232   (313) 974-1355

Lawrence Bruce Plant
General Motors Corp.
Pat. Sect.-New Center One Bldg.
3031 W. Grand Blvd.
P.O. Box 33114
Detroit, MI 48202   (313) 974-1350

Ralph T. Rader
Dykema Gossett
35th Floor
400 Renaissance Center
Detroit, MI 48243   (313) 568-6703

Charles R. Rutherford
Dykema Gossett
35th Floor
400 Renaissance Center
Detroit, MI 48243   (313) 568-5445

Donald F. Scherer
General Motors Corp.
Patent Section
3031 W. Grand Blvd.
P.O. Box 33114
Detroit, MI 48232   (313) 974-1353

William Adolph Schuetz
General Motors Corp.
Pat. Sect. New Center 1 Bldg.
3031 W. Grand Blvd.
P.O. Box 33114
Detroit, MI 48232   (313) 974-1306

Lee A. Schutzman
General Motors Corporation
3031 W. Grand Blvd.
P.O. Box 33122
Detroit, MI 48202   (313) 556-4417

Saul Schwartz
General Motors Corp.
Pat. Sect.-New Center One
Suite 450
P.O. Box 33114
Detroit, MI 48232   (313) 974-1317

Raymond E. Scott
Dykema Gossett
35th Floor
400 Renaissance Center
Detroit, MI 48243   (313) 568-6688

Lawrence J. Shurupoff
Federal - Mogul Corporation
P.O. Box 1966
Detroit, MI 48235   (313) 354-9439

Robert M. Sigler, Jr.
General Motors Corp.
3031 W. Grand Blvd.
Suite 450
Detroit, MI 48232   (313) 974-1359

Anthony Luke Simon
General Motors Corp.
Pat. Section
New Center One Bldg.
3031 W. Grand Blvd., Ste. 450
Detroit, MI 48202   (313) 974-0075

Robert A. Sloman
Dykema Gossett
35th Floor
400 Renaissance Center
Detroit, MI 48243   (313) 568-5364

A. Michael Tucker
General Motors Corporation
Pat. Section
P.O. Box 33114
Detroit, MI 48232   (313) 974-1366

Randy W. Tung
General Motors Corp.
Pat. Section
3031 W. Grand Blvd.
Detroit, MI 48202   (313) 974-1326

Charles Kenneth Veenstra
General Motors Corp.
3031 West Grand Blvd.
Suite 450
Detroit, MI 48202   (313) 974-1333

Robert James Wallace
General Motors Corp.
Pat. Section
P.O. Box 33114
Detroit, MI 48232   (313) 974-1309

William J. Waugaman
Barnes, Kisselle, Raisch Choate,
Whittemore & Hulbert, P.C.
3500 Penobscot Bldg.
Detroit, MI 48226   (313) 962-4790

James L. Wolfe#
Barnes, Kisselle, Raisch,
Choate, Whittemore & Hulbert
3500 Penobscot Bldg.
Detroit, MI 48226   (313) 962-4790

Thomas K. Ziegler
Dykema Gossett
35th Floor
400 Renaissance Center
Detroit, MI 48243   (313) 540-0868

Peter D. Keefe
Keefe & Associates, P.C.
24405 Gratiot Avenue
East Detroit, MI 48021   (313) 775-5680

Gerald R. Hershberger
519 Tawas Street
East Tawas, MI 48730   (517) 362-6293

Harold F. Mensing
6394 Peninsula Road
Erie, MI 48133   (313) 723-7435

Raymond J. Eifler, Sr.
30679 S. Wendybrook Ct.
Farmington Hills, MI 48334   (313) 723-7435

Gerald Kenneth Flagg
34405 W. 12 Mile Road
Farmington Hills, MI 48331   (313) 553-8010

Charles D. Lacina
27362 Skye Drive West
Farmington Hills, MI 48018   (313) 424-9021

John E. Mc Rae
28059 Kendallwood
Farmington Hills, MI 48018   (313) 553-4180

Robert D. Sanborn
32621 Biddestone Lane
Farmington Hills, MI 48018   (313) 477-2268

Donald C. Bolger
P.O. Box 500
Flint, MI 48501   (313) 733-0200

Donald G. Rockwell
Nill, Kirby & Rockwell, P.C
G-4413 Corunna Rd.
P.O. Box 4145
Flint, MI 48504   (313) 732-3320

John R. Faulkner
26460 W. 14 Mile Rd.
Box 601
Franklin, MI 48025   (313) 626-3640

Joel E. Bair
Varnum, Riddering, Schmidt
& Howlett
Suite 800
171 Monroe Avenue, N.W.
Grand Rapids, MI 49503   (616) 459-4186

James E. Bartek
Price, Heneveld, Cooper,
De Witt & Litton
P.O. Box 2567
Grand Rapids, MI 49501   (616) 949-9610

Frederick S. Burkhart
Price, Heneveld, Cooper,
De Witt & Litton
P.O. Box 2567
Grand Rapids, MI 49501   (616) 949-9610

Charles E. Burpee
Warner, Norcross & Judd
900 Old Kent Bldg.
Grand Rapids, MI 49503 (616) 459-6121

Robert J. Carrier#
Price, Heneveld, Cooper,
De Witt & Litton
P.O. Box 2567
Grand Rapids, MI 49501 (616) 949-9610

Carl S. Clark
Price, Heneveld, Cooper
De Witt & Litton
P.O. Box 2567
Grand Rapids, MI 49501 (616) 949-9610

Richard C. Cooper
Price, Heneveld, Cooper,
De Witt & Litton
P.O. Box 2567
Grand Rapids, MI 49501 (616) 949-9610

James D. Darnley, Jr.
Price, Heneveld, Cooper,
De Witt & Cooper
P.O. Box 2567
Grand Rapids, MI 49501 (616) 949-9610

William W. De Witt
Price, Heneveld, Cooper,
De Witt & Litton
P.O. Box 2567
Grand Rapids, MI 49501 (616) 949-9610

Timothy E. Eagle
Varnum, Riddering, Schmidt
& Howlett
171 Monroe Avenue, N.W.
Suite 800
Grand Rapids, MI 49503 (616) 459-4186

Michael D. Fishman
Varnum, Riddering, Schmidt
& Howlett
171 Monroe
Suite 800
Grand Rapids, MI 49503 (616) 459-4186

Richard A. Gaffin
Miller, Canfield, Paddock
& Stone
1200 Campau Square Plaza
99 Monroe Ave., N.W.
Grand Rapids, MI 49503 (616) 454-8656

Donald S. Gardner
Price, Heneveld, Cooper,
De Witt & Litton
P.O. Box 2567
Grand Rapids, MI 49501 (616) 949-9610

John A. Gazewood
Price, Heneveld, Cooper,
De Witt & Litton
P.O. Box 2567
Grand Rapids, MI 49501 (616) 949-9610

Daniel L. Girdwood
Price, Heneveld, Cooper,
De Witt & Litton
695 Kenmoor, S.E.
P.O. Box 2567
Grand Rapids, MI 49501 (616) 949-9610

Christopher D. Harrington
Knape & Vogt Manufacturing Co.
2700 Oak Industrial Dr., N.E.
Grand Rapids, MI 49505 (616) 459-3311

David M. Hecht
Dickinson, Wright, Moon,
Van Dusen & Freeman
650 Frey Building
300 Ottawa Avenue, N.W.
Grand Rapids, MI 49503 (616) 458-1300

Lloyd A. Heneveld
Price, Heneveld, Cooper,
De Witt & Litton
P.O. Box 2567
Grand Rapids, MI 49501 (616) 949-9610

Richard A. Kay
Varnum, Riddering, Schmidt
& Howlett
171 Monroe Avenue, N.W.
Suite 800
Grand Rapids, MI 49503 (616) 459-4186

Edward A. Ketterer, III
Steelcase Inc.
P. O. Box 1967
Grand Rapids, MI 49501 (616) 459-4186

Terence J. Linn
Price, Heneveld, Cooper,
De Witt & Litton
P.O. Box 2567
Grand Rapids, MI 49501 (616) 949-9610

Randall G. Litton
Price, Heneveld, Cooper,
De Witt & Litton
P.O. Box 2567
Grand Rapids, MI 49501 (616) 949-9610

Tom L. Lockhart
Varnum, Riddering, Schmidt
& Howlett
171 Monroe, N.W.
Suite 800
Grand Rapids, MI 49503 (616) 459-4186

John E. Mc Garry
Varnum, Riddering, Schmidt
& Howlett
Suite 800
171 Monroe Ave., N.W.
Grand Rapids, MI 49503 (616) 459-4186

Thomas M. Mc Kinley
Price, Heneveld, Cooper,
De Witt & Litton
P.O. Box 2567
Grand Rapids, MI 49501 (616) 949-9610

James A. Mitchell
Price, Heneveld, Cooper,
De Witt & Litton
P.O. Box 2567
Grand Rapids, MI 49501 (616) 949-9610

Glenn B. Morse
Waters, Morse & Harrington, P.C.
940 Calder Plaza Bldg.
Grand Rapids, MI 49503 (616) 458-7535

Kevin R. Peterson
Price, Heneveld, Cooper,
De Witt & Litton
P.O. Box 2567
Grand Rapids, MI 49501 (616) 949-9610

Peter P. Price
Price, Heneveld, Cooper,
De Witt & Litton
P.O. Box 2567
Grand Rapids, MI 49501 (616) 949-9610

Harold W. Reick
Price, Heneveld, Cooper,
De Witt & Litton
P.O. Box 2567
Grand Rapids, MI 49501 (616) 949-9610

Wilfred O. Schmidt
140 Monroe Centre
Suite 620
Grand Rapids, MI 49503 (616) 459-6232

Frank M. Scutch, III#
Price, Heneveld, Cooper,
De Witt & Litton
695 Kenmoor SE
P.O. Box 2567
Grand Rapids, MI 49501 (616) 949-9610

H. Lawrence Smith
Varnum, Riddering, Schmidt
& Howlett
Suite 800
171 Monroe Avenue, N.W.
Grand Rapids, MI 49503 (616) 459-4186

Randall S. Vaas#
Price, Heneveld, Cooper, De Witt
& Litton
695 Kenmoor S.E.
P. O. Box 2567
Grand Rapids, MI 49501 (616) 949-9610

Daniel Van Dyke
Price, Heneveld, Cooper,
De Witt & Litton
695 Kenmoor, S.E.
P.O. Box 2567
Grand Rapids, MI 49501 (616) 949-9610

Peter Visserman
Varnum, Riddering, Schmidt
& Howlett
171 Monroe Ave., N.W.
Suite 800
Grand Rapids, MI 49503 (616) 459-4186

John A. Waters
Waters & Morse, P.C.
125 Ottawa, N.W.
Suite 430
Grand Rapids, MI 49503 (616) 458-7535

A. Lewis Worthem, Jr.
Warner, Norcross & Judd
900 Old Kent Building
111 Lyon Street, N.W.
Grand Rapids, MI 49503 (616) 459-6121

William G. Conger
8415 Bridge Road
Grosse Ile, MI 48138 (313) 676-3894

513

Bernhard R. Swick
8035 Wood Drive
Grosse Ile, MI 48138   (313) 676-7238

Thomas W. Baumgarten, Jr.
St. Clair Intellectual
Property Consultants, Inc.
10 Stratford Place
Grosse Pointe, MI 48230   (313) 885-6814

Edmund M. Chung#
St. Clair Intellectual
Property Consultants, Inc.
10 Stratford Place
Grosse Pointe, MI 48230   (313) 885-6814

Charles E. Neff
384 McKinley
Grosse Pointe Farms, MI 48236
                    (313) 881-1409

Roger R. Kline
1342 Berkshire
Grosse Pointe Park, MI 48230
                    (313) 885-3168

Jean L. Carpenter
105 L Arbre Croche
Harbor Springs, MI 48033   (616) 347-3582

Mark P. Calcaterra
Chrysler Motors
12000 Chrysler Drive
Highland Park, MI 48288   (313) 956-2934

Edward A. Craig
Chrysler Motors
12000 Chrysler Drive
Highland Park, MI 48288   (313) 956-4182

Gerald P. Dundas
Chrysler Motors
12000 Chrysler Drive
Highland Park, MI 48288   (313) 956-2618

Wendell K. Fredericks
Chrysler Motors Corp.
12000 Chrysler Drive
Highland Park, MI 48288   (313) 956-4283

Christopher A. Taravella
Chrysler Corporation
12000 Chrysler Drive
Highland Park, MI 48288   (313) 956-2063

Michael L. Bauchan
Bauchan Law Offices, P.C.
P.O. Box 879
4611 W. Houghton Lake Dr.
Houghton Lake, MI 48629   (517) 366-5361

William D. Suomi
115 West Street
Howell, MI 48843   (517) 546-0741

Duncan F. Beaman
Beaman and Beaman
700 Harris Bldg.
Jackson, MI 49201   (517) 787-4511

Jerry King Harness
5022 Harriett
Jackson, MI 49203   (517) 788-5967

Christopher J. Kokoczka
600 S. Bowen Street
Jackson, MI 49203   (517) 781-3911

Joseph K. Andonian
The Upjohn Co.
Corp. Licensing
Kalamazoo, MI 49001   (616) 385-7327

Robert A. Armitage
Upjohn Co.
7000 Portage Road
Kalamazoo, MI 49001   (616) 385-7345

David G. Boutell
Flynn, Thiel, Boutell
& Tanis, P.C.
2026 Rambling Rd.
Kalamazoo, MI 49008   (616) 381-1156

Terryence F. Chapman
Flynn, Thiel, Boutell
& Tanis, P.C.
2026 Rambling Road
Kalamazoo, MI 49003   (616) 381-5465

Donald L. Corneglio, Jr.
Upjohn Company
Corp. Pat. Dept.
301 Henrietta Street
Kalamazoo, MI 49001   (616) 385-7200

Mark De Luca
Upjohn Company
301 Henrietta Street
Kalamazoo, MI 49001   (616) 385-5210

L. Martin Flynn
Flynn, Thiel, Boutell
& Tanis, P.C.
2026 Rambling Rd.
Kalamazoo, MI 49008   (616) 381-1156

William E. Heyd#
8152 Brookwood Drive
Kalamazoo, MI 49002   (616) 323-3124

Gordon W. Hueschen
715 The "H" Buildng
310 East Michigan Avenue
Kalamazoo, MI 49007   (616) 382-0030

William G. Jameson
Upjohn Company
301 Henrietta
Kalamazoo, MI 49001   (616) 385-7561

George T. Johannesen, Sr.
715 Amer. Natl. Bk. Bldg.
136 E. Michigan Avenue
Kalamazoo, MI 49006   (616) 382-0033

John Kekich
755 Clubview
Kalamazoo, MI 49009   (616) 375-2234

John J. Killinger
2404 South Park
Kalamazoo, MI 49001   (616) 343-3958

Paul J. Koivuniemi
Upjohn Company
Corporate Pats. & Tdmks.
301 Henrietta Street
Kalamazoo, MI 49001   (616) 385-7222

Samuel Kurlandsky
5124 Maple Ridge Drive
Kalamazoo, MI 49008   (616) 381-8515

Edward F. Rehberg
Upjohn Co.
Molecular Biology Research
301 Henrietta St.
Kalamazoo, MI 49001   (616) 385-5567

John Thomas Reynolds
2409 Cumberland St.
Kalamazoo, MI 49007   (616) 344-2746

Thomas M. Smith
Flynn, Thiel, Boutell
& Tanis, P.C.
2026 Rambling Rd.
Kalamazoo, MI 49008   (616) 381-1156

Earl C. Spaeth
3011 Bramble Drive
Kalamazoo, MI 49009   (616) 375-4851

Scott B. Stahl
Flynn, Thiel, Boutell
& Tanis, P.C.
2026 Rambling Rd.
Kalamazoo, MI 49008   (616) 381-1156

Gregory W. Steele
Upjohn Comany
301 Henrietta Street, 1920-32-1
Kalamazoo, MI 49001   (616) 385-7280

Bruce Stein
Upjohn Company
7000 Portage Road
Kalamazoo, MI 49001   (616) 385-7127

Ronald J. Tanis
Flynn, Thiel, Boutell
& Tanis
2026 Rambling Rd.
Kalamazoo, MI 49008   (616) 381-1156

Dale H. Thiel
Flynn, Thiel, Boutell
& Tanis, P.C.
2026 Rambling Rd.
Kalamazoo, MI 49008   (616) 381-1156

Lawrence T. Welch
Upjohn Co.
Pat. Law Dept.
301 Henrietta St.
Kalamazoo, MI 49001   (616) 385-7237

Sidney B. Williams, Jr.
Upjohn Company
Pat. Law Dept.
301 Henrietta St.
Kalamazoo, MI 49001   (616) 385-7236

Thomas A. Wootton
Upjohn Company
7000 Portage Rd.
Kalamazoo, MI 49001   (616) 385-7914

Debbie K. Wright
Upjohn Company
301 Henrietta St.
Kalamazoo, MI 49001   (616) 385-7569

Walter Potoroka, Sr.
3505 Adams Rd.
Lake Orion, MI 48035   (313) 693-1928

Winston E. Miller
Miller, Morriss & Pappas
219 S. Grand Ave.
Lansing, MI 48933   (517) 487-3791

William J. Morriss, Jr.
Miller, Morriss & Pappas
219 S. Grand
Lansing, MI 48933   (517) 487-3791

George P. Pappas
Miller, Morriss & Pappas
219 South Grand Ave.
Lansing, MI 48933   (517) 487-3791

James C. Mc Laughlin
Mc Laughlin & Mc Laughlin
1432 Duffield Road
Lennon, MI 48449   (313) 230-9080

Charles W. Chandler
33150 Schoolcraft
Livonia, MI 48150   (313) 522-0920

Thomas D. Baldwin
Route 1- Box 1922
Manistique, MI 49854   (906) 341-6430

Florian Harley Jabas
MJR Industries, Inc.
5600 - 13th Street
Menominee, MI 49858   (906) 863-4401

Mona Anand
Dow Chemical Co.
1776 Bldg.
Patent Dept.
Midland, MI 48640   (517) 636-0716

James B. Bieber, Jr.
Dow Chemical Company
Patent Dept.
P.O. Box 1967
Midland, MI 48641   (517) 636-8169

James E. Bittell
Dow Corning Corporation
Patent Dept.
Mail No. C01232
2200 W. Salzburg Road
Midland, MI 48640   (517) 496-5882

Carl Kenneth Bjork, Sr.#
The Dow Chemical Co.
Pat. Dept.
1776 Bldg.
P.O. Box 1967
Midland, MI 48640   (517) 636-2860

William F. Boley#
Dow Corning Corporation
2200 W. Salzburg Road
Midland, MI 48686   (517) 496-8119

Roger H. Borrousch#
Dow Corning Corp.
2200 W. Salzburg
P.O. Box 0994
Midland, MI 48686   (517) 496-4075

Andrea T. Borucki
Dow Chemical Co.
1776 Building
Midland, MI 48674   (517) 636-2820

Ronald Glenn Brookens
Dow Chemical Co.
Pat. Dept.
P.O. Box 1967
Midland, MI 48640   (517) 636-2174

Mildred S. Bruhn#
Dow Chemical Company
Pat. Dept.
P.O. Box 1967
Midland, MI 48641   (517) 636-0397

Margaret M. Brumm
Dow Chemical Co.
Patent Dept.
P.O. Box 1967
Midland, MI 48641   (517) 636-0903

Clifford Claborne Carter
4610 Washington St.
Midland, MI 48642   (517) 496-6464

Jerri J. Chase#
Dow Chemical Company
Pat. Dept.
P.O. Box 1967
Midland, MI 48641   (517) 636-0511

John L. Chiatalas
Dow Corning Corp.
Legal Dept.
Mail C01242
Midland, MI 48686   (517) 496-4217

Tai-Sam Choo
Dow Chemical Company
9008 Bldg.
Midland, MI 48640   (517) 636-2383

Vernon Dean Clausen#
Dow Chemical Co.
Pat. Dept.
P.O. Box 1967
Midland, MI 48641   (517) 636-0229

Gary C. Cohn
Dow Chemical Company
P.O. Box 1967
Midland, MI 48641   (517) 636-2464

Byron R. Crary
Dow Chemical Company
2030 Willard H. Dow Center
Midland, MI 48674   (517) 636-2638

Nemia C. Damocles#
1601 Swede Road
Midland, MI 48640   (517) 832-0924

Merlin Bruce Davey#
3031 N. Sturgeon Road
R.R. # 11
Midland, MI 48640   (517) 631-1542

Jim L. De Cesare
Dow Corning Corporation
Pat. Dept.
Mail C01232
Midland, MI 48686   (517) 496-4235

John M. De Meester
2013 Mapleleaf Drive
Midland, MI 48640   (517) 631-2927

J. Robert Dean, Jr.
Dow Chemical Company
Patent Department
1776 Bldg.
Midland, MI 48674   (517) 636-9215

Douglas N. Deline
Dow Chemical Co.
Midland, MI 48640   (517) 636-2938

Charles H. Ellerbrock
Dow Chemical Company
Pat. Dept.
1776 Bldg.
Washington St.
Midland, MI 48640   (517) 636-5984

Edward C. Elliott#
Dow Corning Corp.
Pat. Dept. - C01232
Midland, MI 48686   (517) 496-4161

Charles J. Enright
Dow Chemical Company
Patent Dept.
P.O. Box 1967
Midland, MI 48640   (517) 636-2604

Michael S. Feider#
Dow Chemical Co.
Pat. Dept.
1776 Bldg.
Midland, MI 48674   (517) 636-2798

David H. Fifield
Dow Chemical Co.
Pat. Dept.
P.O. Box 1967
Midland, MI 48641   (517) 636-9535

Robert Franklin Fleming, Jr.#
3909 Valley Drive
Midland, MI 48640   (517) 835-8775

Ann K. Galbraith
Dow Chemical Co.
Patent Department
1776 Bldg.
Midland, MI 48674   (517) 636-6916

Michael L. Glenn
P. O. Box 1965
Midland, MI 48641   (517) 636-6916

Roger E. Gobrogge
Dow Corning Corp.
Mail Co 1232
Midland, MI 48686   (517) 496-6306

Stephen Scott Grace
Dow Chemical Co.
Pat. Dept.
1776 Bldg.
Midland, MI 48640   (517) 636-3052

George A. Grindahl#
Dow Corning Corp.
2200 Salzburg Road
Midland, MI 48686   (517) 496-5784

James B. Guffey
Dow Chemical Company
Pat. Dept.
P.O. Box 1967
Midland, MI 48641   (517) 636-5984

Burke M. Halldorson
Dow Chemical Co.
Patent Dept.
1776 Bldg.
Midland, MI 48640   (517) 636-1330

J. William Hedelund#
1227 Holyrood St.
Midland, MI 48640   (517) 835-5423

Sharon S. Heins
Dow Chemical Company
2020 Willard H. Dow Center
Midland, MI 48674   (517) 636-3023

Howard W. Hermann
Dow Corning Corp.
Legal Dept.
Midland, MI 48686   (517) 496-4084

Lloyd E. Hessenaur, Jr.
3916 Woodside Drive
Midland, MI 48640   (517) 631-1872

Dan R. Howard
Dow Chemical Co.
1776 Bldg. Pruit Res. Center
Washington Street
Midland, MI 48674   (517) 636-2468

Arne R. Jarnholm
Dow Corning Corporation
Patent Dept. - Mail C01232
Midland, MI 48686   (517) 496-8763

Michael Stephen Jenkins#
Dow Chemical Company
Washington Street
P.O. Box 1967
Midland, MI 48640   (517) 636-2996

Paula A. Jonas#
Dow Chemical Company
Pat. Dept.
P.O. Box 1967
Midland, MI 48641   (517) 636-2749

S. Preston Jones#
Dow Chemical Company
Pat. Dept.
1776 Bldg.
Washington Avenue
Midland, MI 48674   (517) 636-2901

Lloyd S. Jowanovitz
The Dow Chemical Co.
1776 Bldg.
Pat. Dept.
Midland, MI 48674   (517) 636-2040

Nis H. Juhl#
Dow Chemical Company
Pat. Dept.
1776 Bldg., Pruitt Res. Center
Washington Street
Midland, MI 48674   (517) 636-0333

Bruce M. Kanuch
Dow Chemical Co.
Patent Dept.
P.O. Box 1967
Midland, MI 48640   (517) 636-2115

Karen L. Kimble
Dow Chemical Company
Pat. Dept.
1776 Bldg.
Washington Street
Midland, MI 48674   (517) 636-1687

Thomas A. Ladd
Dow Chemical Company
1776 Bldg.
Midland, MI 48674   (517) 636-2063

Raymond B. Ledlie
The Dow Chemical Co.
Pat. Dept.
1776 Bldg.
Midland, MI 48674   (517) 636-3378

Norman E. Lewis
Dow Corning Corp.
Pat. Dept.
Co-1232
Midland, MI 48686   (517) 496-4080

James Roger Lochhead
Dow Chemical Co.
2020 Dow Center
Midland, MI 48674   (517) 636-1103

Kenneth L. Loertscher
Dow Chemical Company
P.O. Box 1967
Midland, MI 48641   (517) 636-1304

Roderick B. Macleod
Dow Chemical Company
P.O. Box 1967
Midland, MI 48641   (517) 636-0308

Charles J. Maurer
Dow Chemcal Company
Pat. Dept.
1776 Bldg.
Washington Street
Midland, MI 48674   (517) 636-0829

Warren M. Mc Guire
Dow Chemical Company
Legal Dept.
2030 Willard H. Dow Center
Midland, MI 48674   (517) 636-9185

Robert L. Mc Kellar
Dow Corning Corporation
Mail Stop Co 1232
Midland, MI 48686   (517) 496-5026

Jack E. Moermond
2913 Dina St.
Midland, MI 48640   (517) 631-2163

Jonathan W. Morse
Dow Chem. Co.
P.O. Box 1967
Midland, MI 48641   (517) 636-8298

Ira Arthur Murphy
1804 Rapanos Drive
Midland, MI 48640   (517) 835-8949

William R. Norris
The Dow Chemical Co.
P.O. Box 1967
Midland, MI 48640   (517) 636-2554

Irene M. Nyquist
Dow Chem. Co.
Pat. Deppt.
1776 Bldg.
Midland, MI 48674   (517) 636-1707

Joe R. Prieto
Dow Chemical Company
Patent Dept.
P.O. Box 1967
1776 Bldg.
Midland, MI 48640   (517) 636-9361

Jack I. Pulley
Dow Corning Corporation
Mail Co1242
Midland, MI 48686   (517) 496-5440

Robert J. Rhead
Courthouse Square
Suite 1000
240 W. Main Street
Midland, MI 48640   (517) 832-8826

John H. Roberts
Dow Chemical Co.
1776 Building
Patent Dept.
Midland, MI 48674   (517) 636-0404

Joan H. Rogers#
Dow Chemical Co.
1776 Building
Patent Dept.
Midland, MI 48674   (517) 636-9125

Paula Sanders Ruhr
Dow Chemical Company
1776 Bldg.
Washington Street
Midland, MI 48674   (517) 636-1402

H. David Russell#
Dow Chemical Co.
1320 Waldo Road
Suite 225
Midland, MI 48640   (517) 636-2795

Bernd W. Sandt
Dow Chemical Co.
P.O. Box 1967
Midland, MI 48641   (517) 636-2274

Michael P. Santorsa
Dow Chemical Co.
Washington Street
1776 Building
Midland, MI 48640   (516) 636-2890

John R. Schenian
Dow Chemical Company
Patent Dept.
1776 Bldg.
Midland, MI 48674   (517) 636-0218

Robert W. Selby
Dow Chem. Co.
Pat. Dept.
1776 Building
Washington Street
Midland, MI 48674   (517) 636-1718

Sharon K. Severance#
Dow Corning Corp.
Mail No. Co1232
Midland, MI 48686   (517) 496-8120

Philip D. Shepherd
Dow Chem. Co.
Pat. Dept. - 1776 Bldg.
P.O. Box 1967
Midland, MI 48674   (517) 636-9815

Robert Spector#
Dow Corning Corporation
Patent Dept.
Mail 1232
Midland, MI 48640   (517) 496-5523

Timothy S. Stevens#
Dow Chemical Co.
1776 Bldg.
Midland, MI 48640   (513) 948-6490

John D. Thallemer
Dow Corning Corp.
CO 1232
Midland, MI 48686   (517) 496-8262

Duane C. Ulmer
Dow Chemical Co.
1776 Bldg.
Patent Dept.
Midland, MI 48674   (517) 636-8104

James B. Vanderkelen
Snow Machines, Inc.
1512 N. Rockwell
Midland, MI 48640   (517) 631-6091

Andrew H. Ward#
Dow Corning Corporation
S. Saginaw Road
Midland, MI 48640   (517) 496-4443

Noreen D. Warrick
Dow Chemical Co.
1776 Bldg.
Patent Dept.
Midland, MI 48674   (517) 636-9373

Richard G. Waterman
Dow Chemical Co.
Pat. Dept.
1776 Bldg.
Midland, MI 48674   (517) 636-0902

Alexander Weitz#
Dow Corning Corporation
Patent Dept.
Mail C 01232
Midland, MI 48686   (517) 496-5808

Michael L. Winkelman
Dow Chemical Company
Pat. Dept.
P.O. 1967
Midland, MI 48641   (517) 636-8449

John L. Wood
Dow Chemical Company
Pat. Dept.
1776 Bldg.
Midland, MI 48674   (517) 636-0891

William Miller Yates
27 Lexington Court
Midland, MI 48640   (517) 835-5493

Carl A. Yorimoto#
Dow Corning Corporation
Mail Stop CO 1232
Midland, MI 48646   (517) 631-1296

Lisha Simmonds Zindrick
Dow Chemical Company
Patent Dept.
1776 Building
Midland, MI 48674   (517) 636-2644

Thomas D. Zindrick
Dow Chemical Company
Patent Dept.
1776 Washington St.
Midland, MI 48674   (517) 636-1869

Ann M. Mueting#
Merchant, Gould, Smith,
Welter & Schmidt P.A.
3100 Norwest Center
90 S. 7th St.
Minneapolis, MI 55402   (612) 332-5300

David B. Kelley
Braunlich, Russow & Braunlich
111 S. Macomb St.
Monroe, MI 48161   (313) 241-8300

Ronald S. Courtney
17747 N. Nunneley Rd.
Mt. Clemens, MI 48043   (313) 286-3649

Robert K. Wallor
Waterfront Centre, Suite 315
1050 W. Western Avenue
Muskegon, MI 49441   (616) 728-0500

Ernest U. Lang
1510 Platt Street
Niles, MI 49120   (616) 683-1952

Andrew J. Haliw, III
Sullivan & Leavitt, P.C.
22375 Haggerty Road
P.O. Box 400
Northville, MI 48167   (313) 349-3980

Robert George Mentag
41000 Seven Mile Road
Northville, MI 48167   (313) 349-8833

Pamela S. Burt#
I. Weiner & Associates, P.C.
42400 Grand River Ave.
Suite 207
Novi, MI 48375   (313) 344-4422

Joseph P. Carrier#
I. Weiner & Associates, P.C.
42400 Grand River Ave., Ste. 207
Novi, MI 48375   (313) 344-4422

Robert M. Petrik#
I. Weiner & Associates, P.C.
42400 Grand River Ave.
Suite 207
Novi, MI 48375   (313) 344-4422

Irving M. Weiner
42400 Grand River Ave.
Suite 207
Novi, MI 48375   (313) 344-4422

Ian Campbell Mc Leod
2190 Commons Parkway
Okemos, MI 48864   (517) 347-4100

Michael F. Scalise
2190 Commons Parkway
Okemos, MI 48864   (517) 347-4100

Everett R. Casey
5845 Old Orchard Trail
Orchard Lake, MI 48033   (313) 682-0400

Arthur E. Kluegel
3111 Reese Road
Ortonville, MI 48462   (313) 625-0111

John F. Rohe
438 East Lake Street
Petoskey, MI 49770   (616) 347-7327

Robert H. Elliott, Sr.#
15688 Northville Forest Drive
Plymouth, MI 48170   (313) 420-2465

Domenica N.S. Hartman
15071 Bradner Road
Plymouth, MI 48170   (313) 420-0193

Gary M. Hartman#
15071 Bradner Road
Plymouth, MI 48170   (313) 420-0193

John J. Cantarella
1004 Joslyn Avenue
Pontiac, MI 48055   (313) 858-8871

Robert J. Madden
5292 Rosamond Lane
Pontiac, MI 48054   (313) 681-3354

Martha A. Gammill
6704 Pleasantview Dr.
Portage, MI 49002   (616) 385-7829

Donald E. Overbeek
210 E. Centre Ave.
Portage, MI 49002   (616) 327-8041

Daniel R. Edelbrock#
15019 Fenton
Redford, MI 48239   (313) 532-6857

William L. Fisher
2821 John R. Road
Rochester, MI 48307   (313) 852-8597

James P. De Clercq
17144 Dawn Street
Roseville, MI 48066   (313) 772-4687

Donald P. Bush
1608 Vinsetta Blvd.
Royal Oak, MI 48067   (313) 542-3669

Judith L. Olds#
3241 Merrill
Royal Oak, MI 48073   (313) 288-1347

Edmund C. Ross, Jr.
1415 N. Vermont
Royal Oak, MI 48067   (313) 545-1446

John F. Learman
Learman & Mc Culloch
5291 Colony Dr. North
Saginaw, MI 48603   (517) 799-5300

John K. Mc Culloch
Learman & Mcculloch
5291 Colony Dr., North
Saginaw, MI 48603   (517) 799-5300

John J. Swartz
Le Fevre, Swartz & Wilson
908 Court St.
Saginaw, MI 48602   (517) 793-8540

Gerald F. Stibitz
312 Lyon Boulevard
South Lyon, MI 48178   (313) 437-1058

William G. Abbatt
Brooks & Kushman
1000 Town Center
Twenty-Second Floor
Southfield, MI 48075   (313) 358-4400

Luis Miguel Acosta
Brooks & Kushman
1000 Town Center
Twenty-Second Floor
Southfield, MI 48075   (313) 358-4400

John A. Artz
Brooks & Kushman
1000 Town Center
Twenty-Second Floor
Southfield, MI 48075   (313) 358-4400

Robert C. Brandenburg
Brooks & Kushman
1000 Town Center
22nd Floor
Southfield, MI 48075   (313) 358-4400

Ernie L. Brooks
Brooks & Kushman
1000 Town Center
Twenty-Second Floor
Southfield, MI 48075   (313) 358-4400

Ralph M. Burton
Brooks & Kushman
1000 Town Center
Suite 2200
Southfield, MI 48075   (313) 358-4400

Mark A. Cantor
Brooks & Kushman
1000 Town Center
Twenty-Second Floor
Southfield, MI 48075   (313) 358-4400

Frank H. Cullen
Master Data Center, Inc.
29100 Northwestern Hwy.
Suite 300
Southfield, MI 48034   (313) 352-5810

Christopher J. Fildes
Brooks & Kushman
1000 Town Center
Twenty-Second Floor
Southfield, MI 48075   (313) 358-4400

Donald Joseph Harrington
Brooks & Kushman
1000 Town Center
22nd Floor
Southfield, MI 48075   (313) 358-4400

Kevin J. Heinl
Brooks & Kushman
1000 Town Center
Twenty-Second Floor
Southfield, MI 48075   (313) 358-4400

Robert F. Hess
Brooks & Kushman
1000 Town Center
Twenty-Second Floor
Southfield, MI 48075   (313) 358-4400

James A. Kushman
Brooks & Kushman
1000 Town Center
Twenty-Second Floor
Southfield, MI 48075   (313) 358-4400

Earl J. La Fontaine
Brooks & Kushman
2000 Town Center
Suite 2000
Southfield, MI 48075   (313) 358-4400

Thomas A. Lewry
Brooks & Kushman
1000 Town Center
Twenty-Second Floor
Southfield, MI 48075   (313) 358-4400

Paul L. Marshall
BASF Corp.
26701 Telegraph Rd.
Southfield, MI 48086   (313) 948-2020

Ronald M. Nabozny
Brooks & Kushman
1000 Town Center
Twenty-Second Floor
Southfield, MI 48075   (313) 385-4400

John E. Nemazi
Brooks & Kushman
1000 Town Center
Twenty-Second Floor
Southfield, MI 48075   (313) 358-4400

Timothy G. Newman
Brooks & Kushman
1000 Town Center
22nd Floor
Southfield, MI 48075   (313) 358-4400

Floyd K. Reynolds
24915 Thorndyke
Southfield, MI 48034   (313) 352-4460

Frederick M. Ritchie
Brooks & Kushman
1000 Town Center
Twenty Second Floor
Southfield, MI 48075   (313) 358-4400

Paul S. Rulon#
Eaton Corporation
Corp. Res. & Dev. Detroit Ctr.
26201 Northwestern Hwy.
P.O. Box 766
Southfield, MI 48037   (313) 354-5057

William J. Schramm
Brooks & Kushman
1000 Town Center
Twenty-Second Floor
Southfield, MI 48075   (313) 358-3351

Paul M. Schwartz
Brooks & Kushman
1000 Town Floor
Twenty-Second Floor
Southfield, MI 48075   (313) 358-4400

Markell Seitzman
Allied Corporation
20650 Civic Center Drive
Southfield, MI 48076   (313) 827-6280

David R. Syrowik
Brooks & Kushman
1000 Town Center
Twenty Second Floor
Southfield, MI 48075   (313) 358-4400

Robert C.J. Tuttle
Brooks & Kushman
1000 Town Center
22nd Floor
Southfield, MI 48075   (313) 358-4400

Loren H. Uthoff, Jr.#
Eaton Corporation
26201 Northwestern Hwy.
P.O. Box 766
Southfield, MI 48037   (313) 354-2871

Burton Howard Baker
2218 Mt. Curve
St. Joseph, MI 49085   (616) 983-7263

William Houseal
Kinney, Cook, Lindenfeld
& Kelley
Law & Title Bldg.
811 Ship Street
St. Joseph, MI 49085   (616) 983-0103

Gerald R. Black
Borg-Warner Automotive, Inc.
6700 18-1/2 Mile Road
P.O. Box 8022
Sterling Heights, MI 48311   (313) 726-4431

Gregory Dziegielewski
Borg-Warner Automotive, Inc.
6700 18-1/2 Mile Road
P.O. Box 8022
Sterling Heights, MI 48311   (313) 726-4431

E. Dennis O Connor
Masco Corp.
26855 Trolley Industrial Dr.
Taylor, MI 48180   (313) 291-3500

Leon E. Redman
Masco Corporation
21001 Van Born Rd.
Taylor, MI 48180   (313) 274-7400

Malcolm L. Sutherland
Masco Corporation
21001 Van Born Road
Taylor, MI 48180   (313) 274-7400

Edgar A. Zarins
Masco Corporation
21001 Van Born Road
Taylor, MI 48180   (313) 274-7400

Bradly D. Beams
Hook Drugs
1113 W. Michigan Avenue
Suite 611
Three Rivers, MI 49093   (616) 279-9702

Bill C. Panagos
GTE Valenite Corp.
750 Stephenson Hwy.
Tory, MI 48084   (313) 589-6287

Douglas S. Bishop
Elhart, Bishop & Thomas, P.C.
329 S. Union At Lake Street
Traverse City, MI 49685   (616) 946-4100

Richard P. Barnard
Reising, Ethington, Barnard
Perry & Milton
P.O. Box 4390
Troy, MI 48099   (313) 689-3554

Andrew Raymond Basile
Basile & Hanlon, P.C.
1650 W. Big Beaver Road
Suite 210
Troy, MI 48084   (313) 649-0990

John G. Batchelder
Basile & Hanlon, P.C.
1650 W. Big Beaver Road
Suite 210
Troy, MI 48084   (313) 649-0990

Ernest A. Beutler, Jr.
Harness, Dickey and Pierce
5445 Corporate Drive
Troy, MI 48098   (313) 641-1600

William D. Blackman
Weintraub, Du Ross & Brady, P.C.
3001 W. Big Beaver Road
Suite 504
Troy, MI 48084   (313) 649-3850

Charles H. Blair
Harness, Dickey and Pierce
5445 Corporate Drive
Troy, MI 48098   (313) 641-1600

Daniel H. Bliss
Dinnin & Dunn, P.C.
Top of Troy Building
755 West Big Beaver Road
Troy, MI 48084   (313) 362-2800

Wilfred S. Bobier#
Basile & Hanlon, P.C.
1650 W. Big Beaver Road
Suite 210
Troy, MI 48084   (313) 649-0990

Robert L. Boynton
Harness, Dickey & Pierce
5445 Corporate Drive
Troy, MI 48098   (313) 641-1600

Michael P. Brennan
Harness, Dickey & Pierce
5445 Coprorate Drive
Troy, MI 48098   (313) 641-1600

Christopher M. Brock
Harness, Dickey & Pierce
5445 Corporate Drive
Suite 400
Troy, MI 48098   (313) 641-1600

Joseph G. Burgess
Krass & Young
3001 W. Big Beaver
Suite 624
Troy, MI 48084   (313) 649-3333

Bernard J. Cantor
Harness, Dickey & Pierce
5445 Corporate Drive
Troy, MI 48098   (313) 641-1600

Richard L. Carlson
Harness, Dickey & Pierce
5445 Corporate Drive
Suite 400
Troy, MI 48098   (313) 641-1600

Edward R. Casselman
Harness, Dickey and Pierce
5445 Corporate Drive
Troy, MI 48098   (313) 641-1600

E. W. Christen
Reising, Ethington, Barnard,
Perry & Milton
P.O. Box 4390
Troy, MI 48099   (313) 689-3554

Julia F. Church
Basile & Hanlon, P.C.
1650 W. Big Beaver Rd.
Suite 210
Troy, MI 48084   (313) 649-0990

Ronald W. Citkowski
Krass & Young
3001 W. Big Beaver
Suite 624
Troy, MI 48084   (313) 649-3333

William J. Coughlin
Harness, Dickey & Pierce
5445 Corporate Dr.
Troy, MI 48098   (313) 641-1600

Paul S. Czarnota
Weintraub, Du Ross &
Brady, P.C.
3001 W. Big Beaver Rd.
Suite 504
Troy, MI 48084   (313) 64?-3850

Hugo A. Delevie#
Lyon & Mc Kinnon
Suite 2219
755 W. Big Beaver Road
Troy, MI 48084   (313) 362-2600

Michael R. Dinnin, Jr.
Dinnin & Dunn, P.C.
Top of Troy Building
755 West Big Beaver Rd.
Troy, MI 48084   (313) 362-2800

Eric M. Dobrusin
Harness, Dickey & Pierce
5445 Corporate Drive
Suite 400
Troy, MI 48098   (313) 641-1600

Garrett C. Donley#
1869 Kirts Blvd.
Apt. 203
Troy, MI 48084   (313) 649-6211

Robert A. Dunn
Dinnin & Dunn, P.C.
Top of Troy Building
755 West Big Beaver Road
Troy, MI 48084   (313) 362-2800

Laurie A. Ebling
Reising, Ethington, Barnard,
Perry & Milton
P.O. Box 4390
Troy, MI 48099   (313) 689-3554

David Bowerman Ehrlinger
Krass & Young
3001 W. Big Beaver
Suite 624
Troy, MI 48084   (313) 662-0270

Mark D. Elchuk
Harness, Dickey & Pierce
5445 Corporate Drive
4th Floor
Troy, MI 48098   (313) 641-1600

Paul J. Ethington
Reising, Ethington, Barnard,
Perry & Milton
P.O. Box 4390
Troy, MI 48099   (313) 689-3554

John Charles Evans
Reising, Ethington, Barnard,
Perry & Milton
P.O. Box 4390
Troy, MI 48099   (313) 689-3554

Lloyd Mason Forster
Suite 2218
755 W. Big Beaver Rd.
Troy, MI 48084   (313) 362-1115

Stephen J. Foss
Harness, Dickey & Pierce
5445 Corporate Drive
Troy, MI 48098   (313) 641-1600

Herman Foster
Budd Company
3155 W. Big Beaver Rd.
Troy, MI 48084   (313) 643-3530

Denise M. Glassmeyer
Basile & Hanlon, P.C.
1650 W. Big Beaver Road
Suite 210
Troy, MI 48084   (313) 649-0990

David A. Greenlee
Reising, Ethington, Barnard,
Perry & Milton
P.O. Box 4390
201 W. Big Beaver Rd., Ste. 400
Troy, MI 48099   (313) 689-3554

Thomas A. Hallin
Harness, Dickey & Pierce
5445 Corporate Drive
Troy, MI 48098   (313) 641-1600

William M. Hanlon, Jr.
Basile & Hanlon, P.C.
1650 W. Big Beaver Road
Suite 210
Troy, MI 48084   (312) 649-0990

Don K. Harness
Harness, Dickey and Pierce
5445 Corporate Dr.
Troy, MI 48098   (313) 641-1600

Gordon K. Harris, Jr.
Harness, Dickey & Pearce
5445 Corporate Drive
Troy, MI 48098   (313) 641-1600

Thomas D. Helmholdt
Basile & Hanlon, P.C.
1650 W. Big Beaver Road
Suite 210
Troy, MI 48084   (313) 649-0990

Warren D. Hill
Reising, Ethington, Barnard,
Perry & Milton
3290 W. Big Beaver
Suite 510
Troy, MI 48084   (313) 693-2345

Ronald L. Hofer
Harness, Dickey & Pierce
5445 Corporate Drive
Suite 400
Troy, MI 48098   (313) 641-1600

Richard W. Hoffmann
Reising, Ethington, Barnard,
Perry & Milton
P.O. Box 4390
Troy, MI 48099   (313) 689-3554

James R. Ignatowski
Gifford Van Ophem & Sprinkle
755 West Big Beaver
Suite 1313
Troy, MI 48084   (313) 362-1210

Steven G. Jonas
Newcor, Inc.
3270 W. Big Beaver Road
Troy, MI 48084   (313) 643-7730

Edward G. Jones
6738 Fredmoor St.
Troy, MI 48098   (313) 828-8032

Paul A. Keller
Harness, Dickey & Pierce
5445 Corporate Drive
Troy, MI 48098   (313) 641-1600

Kenneth I. Kohn
Reising, Ethington, Barnard,
Perry & Brooks
P.O. Box 4390
Troy, MI 48099   (313) 689-3554

Allen M. Krass
Krass & Young
3001 W. Big Beaver
Suite 624
Troy, MI 48084   (313) 649-3323

Anna M. Lewak
Harness, Dickey & Pierce
5445 Corporate Drive
Suite 400
Troy, MI 48098   (313) 641-1600

Lyman R. Lyon
Lyon & Delevie, P.C.
Suite 2224
755 W. Big Beaver Road
Troy, MI 48084   (313) 362-2600

Marshall G. Mac Farlane
Krass & Young
3001 W. Big Beaver
Suite 624
Troy, MI 48084   (313) 662-0270

Robert D. Marshall, Jr.
Krass & Young
3001 W. Big Beaver
Suite 624
Troy, MI 48084   (313) 662-0270

Kenneth M. Massaroni
OIS Optical Imaging Sys., Inc.
Pat. Dept.
1896 Barrett St.
Troy, MI 48084   (313) 362-2738

Gerald E. Mc Glynn, III
Reising, Ethington, Barnard,
Perry & Milton
P.O. Box 4390
Troy, MI 48099   (313) 689-3554

Gerald Edward Mc Glynn, Jr.
P.O. Box 4390
Troy, MI 48099   (313) 689-3554

Malcolm Robert Mc Kinnon
Mc Kinnon & Mc Kinnon
755 W. Big Beaver Road
Suite 2219
Troy, MI 48084   (313) 362-2600

H. Keith Miller
Harness, Dickey and Pierce
5445 Corporate Drive
Suite 400
Troy, MI 48098   (313) 641-1600

John A. Miller#
Harness, Dickey & Pierce
5445 Corporate Dr.
Troy, MI 48098   (313) 641-1600

Harold William Milton, Jr.
Reising, Ethington, Barnard,
Perry & Milton
P.O. Box 4390
Troy, MI 48099   (313) 689-3554

John P. Moran
Reising, Ethington, Barnard,
Perry & Milton
P.O. Box 4390
Troy, MI 48099   (313) 689-3554

Robert S. Nolan
Harness, Dickey & Pierce
5445 Corporate Drive
Troy, MI 48098   (313) 641-1600

Donald Alvin Panek
GTE Valenite Corp.
750 Stephenson Hwy.
P.O. Box 3950
Troy, MI 48007   (313) 589-6030

Joseph Richard Papp
Harness, Dickey & Pierce
5445 Corporate Drive
Suite 400
Troy, MI 48098   (313) 641-1600

Steven L. Permut
Krass & Young
3001 W. Big Beaver Road
Troy, MI 48084   (313) 649-3333

Owen E. Perry
Reising, Ethington, Barnard,
Perry & Milton
P.O. Box 4390
Troy, MI 48099   (313) 689-3554

Andrew E. Pierce
755 West Big Beaver
Suite 2216
Troy, MI 48084   (313) 362-3622

Paul J. Reising
Reising, Ethington, Barnard,
Perry & Milton
P.O. Box 4390
Troy, MI 48099   (313) 689-3554

Philip E. Rettig
Harness, Dickey & Pierce
5445 Corporate Drive
Troy, MI 48098   (313) 641-1600

Judith M. Riley
Krass & Young
3001 W. Big Beaver
Suite 624
Troy, MI 48084   (313) 649-3333

Jeffrey A. Sadowski
Harness, Dickey & Pierce
5445 Corporate Drive
Suite 400
Troy, MI 48098   (313) 641-1600

G. Gregory Schivley
Harness, Dickey & Pierce
5445 Corporate Drive
Suite 400
Troy, MI 48098   (313) 641-1600

David W. Schumaker#
Energy Conversion Devices, Inc.
1675 W. Maple
Troy, MI 48084   (313) 280-1900

John A. Sinclair
Harness, Dickey & Pierce
5445 Corporate Drive
Suite 400
Troy, MI 48098   (313) 641-1600

Marvin S. Siskind
Energy Conversion Devices, Inc.
1675 West Maple Road
Troy, MI 48084   (313) 280-1900

John V. Sobesky
Harness, Dickey and Pierce
5445 Corporate Drive
Troy, MI 48098   (313) 641-1600

Gregory A. Stobbs
Harness, Dickey & Pierce
5445 Corporate Drive
Troy, MI 48098   (313) 641-1600

W. R. Duke Taylor
Harness, Dickey & Pierce
5445 Corporate Drive
Suite 400
Troy, MI 48098   (313) 641-1600

Stanley C. Thorpe
Reising, Ethington, Barnard,
Perry & Milton
P.O. Box 4390
201 W. Big Beaver, Ste. 400
Troy, MI 48099   (313) 689-3554

Edward J. Timmer
Reising, Ethington, Barnard,
Perry & Milton
201 West Big Weaver Rd.
Suite 400
Troy, MI 48084   (313) 689-3554

Remy J. Van Ophem
755 W. Big Beaver Rd.
Suite 1313
Troy, MI 48084   (313) 362-1210

Richard P. Vitek
Harness, Dickey & Pierce
5445 Corporate Drive
Suite 400
Troy, MI 48098   (313) 641-1600

Neal A. Waldrop
755 West Big Beaver Road
Suite 2216
Troy, MI 48084   (313) 362-3620

Ronald W. Wangerow
Harness, Dickey & Pierce
5445 Corporate Drive
Troy, MI 48098   (313) 641-1600

Philip R. Warn
Harness, Dickey & Pierce
5445 Corporate Drive
Suite 400
Troy, MI 48098   (313) 641-1600

Arnold S. Weintraub
Weintraub, Du Ross & Brady, P.C.
3001 W. Big Beaver Road
Suite 504
Troy, MI 48084   (313) 649-3850

Donald L. Wenskay
Harness, Dickey & Pierce
5445 Corporate Drive
Troy, MI 48098   (313) 641-1600

Charles Richard White
Reising, Ethington, Barnard,
Perry & Milton & Learman & Mc Clulloch
Columbia Center
201 Big Beaver, Suite 400
Troy, MI 48084   (313) 689-3554

Gerald J. Woloson
5118 Berwyck Drive
Troy, MI 48098   (313) 828-8047

Donald L. Wood
Krass & Young
3001 W. Big Beaver
Suite 624
Troy, MI 48084   (313) 649-3323

Thomas N. Young
Krass & Young
3001 W. Big Beaver
Suite 624
Troy, MI 48084   (313) 649-3323

Frank D. Risko#
Pure-Pak, Inc.
850 Ladd Road
Bldg. D
Walled Lake, MI 48088   (313) 669-5800

David L. Kuhn*
U.S. Army
Tank-Automotive Command
Warren, MI 48397   (313) 574-8682

Richard Paul Mueller
Omi International Corp.
Legal Dept.
21441 Hoover Road
Warren, MI 48089   (313) 497-6892

Jess P. Santo
11161 Hanover
Warren, MI 48093   (313) 977-1712

Gail S. Soderling*
US Army Tank-Automotive Command
Attn: AMSTA-LP
Warren, MI 48397   (313) 574-8682

Peter Arthur Taucher*
U. S. Army Tank Automotive Comm.
Attn. AMSTA-LP
12 Mile & Van Dyke
Warren, MI 48090   (313) 573-2552

Gregory T. Zalecki
Zalecki & Bair, P.C.
30500 Van Dyke Avenue
Suite 704
Warren, MI 48093   (313) 573-2600

Jerry Gunther Beck
5328 Fairway Court
West Bloomfield, MI 48033   (313) 851-3228

Stanley J. Ference
Ference, Ference & Cicirelli
8623 N. Wayne Road
Suite 255
Westland, MI 48185   (313) 422-4666

Dennis V. Carmen
BASF Corporation
1419 Biddle Ave.
Wyandotte, MI 48192   (313) 246-6193

Martin P. Connaughton#
BASF Corp.
1419 Biddle Ave.
Wyandotte, MI 48192   (313) 246-6205

Norbert M. Lisicki
BASF Wyandotte Corp.
1419 Biddle Ave.
Wyandotte, MI 48192   (313) 246-6191

Minnesota

Frederick J.B. Wall#
13670 Tomahawk Drive South
Afton, MN 55001   (612) 436-8668

Sten E. Hakanson
IMI Cornellius Inc.
One Cornelius Place
Anoka, MN 55303   (612) 421-6120

Henry C. Kovar
P.O. Box 571
Anoka, MN 55303   (612) 421-6120

James R. Cwayna
8905 Woodcliff Road
Bloomington, MN 55438   (612) 944-0909

Roger W. Jensen
8127 Pennsylvania Circle
Bloomington, MN 55438   (612) 944-7525

Anthony A. Juettner
8430 Pennsylvania Road
Bloomington, MN 55438   (612) 942-8252

Joan S. Keps
8540 Irwin Road
Bloomington, MN 55437   (612) 897-1166

Oliver A. Ossanna, Jr.
2100 Overlook Dr.
Bloomington, MN 55431   (612) 884-2848

Robert C. Baker
R.C. Baker & Associates, Ltd.
200 TCF Bank Bldg.
12751 Nicollet Ave.
P.O. Box 1465
Burnsville, MN 55337   (612) 882-7777

Miles C. Huffstutler, Jr.#
1608 West 155th Street
Burnsville, MN 55337   (612) 831-8

Donald A. Jacobson
151 W. 126th St.
Burnsville, MN 55337   (612) 894-1055

Mary P. Bauman
Sanofi Diagnostics Pasteur
1000 Lake Hazeltine Drive
Chaska, MN 55318   (612) 347-7108

James H. Wills
Kalina, Wills & Woods
4111 Central Avenue N.E.
Suite 102 South
Columbia Heights, MN 55421
          (612) 789-9000

William C. Flynn
18202 Minnetonka Blvd.
Suite 200
Deephaven, MN 55391   (612) 473-2847

Joseph A. Speldrich#
Unisys Corporation
3333 Pilot Knob Rd.
Eagan, MN 55121   (612) 456-2936

David C. Bohn
Rosemount Inc.
Mail Stop W05
12001 Technology Drive
Eden Prairie, MN 55344   (612) 828-7804

Kathryn M. Howard#
Rosemount Inc.
12001 Technology Drive W05
Eden Prairie, MN 55344   (612) 828-7825

Robert C. Klinger#
Rosemount Inc.
12001 Technology Dr.
Eden Prairie, MN 55344   (612) 828-7806

Robert R. Kooiman
Rosemount Inc.
12001 Technology Drive
Eden Prairie, MN 55344   (612) 828-7802

Michael J. Schneider, Jr.
Rosemount Inc.
12001 Technology Drive
Eden Prairie, MN 55344   (612) 828-7803

Jane H. Arrett
Alliant Techsystems Inc.
5901 Lincoln Drive
Edina, MN 55436   (612) 939-2000

Henry L. Hanson
Chiefton Enterprises Inc.
5125 Blake Road
Edina, MN 55436   (612) 935-0564

Thomas Anthony Lennon
7101 York Ave. S.
Edina, MN 55435   (612) 921-3361

David A. Lingbeck
6450 York Avenue South
Suite 304
Edina, MN 55435   (612) 921-3361

William T. Udseth
Alliant Techsystems Inc.
5901 Llincoln Drive
Edina, MN 55436   (612) 931-7596

Jerold M. Forsberg
6 Old Still Road
Grand Rapids, MN 55744   (218) 327-1427

David A. Teicher
P.O. Box 170
Hamel, MN 55340   (612) 559-9504

William O. Ney
18 Williams Woods
Mahtomedi, MN 55115   (612) 426-2504

Salim A. Kassatly
7788 Ranchview Lane North
Maple Grove, MN 55369   (612) 420-4939

Augustus Jeter Hipp
Gnb Incorporated
1110 Highway 110
Mendota Heights, MN 55118   (612) 681-5417

John W. Adams
520 Norwest Midland Bldg.
401 2nd Avenue South
Minneapolis, MN 55401   (612) 339-4861

Philip G. Alden
Moore & Hansen
300 Norwest Center
90 South Seventh Street
Minneapolis, MN 55402   (612) 332-8200

Richard D. Allison
Palmatier & Sjoquist
2000 Northwestern Financial Center
7900 Xerxes Avenue S.
Minneapolis, MN 55431   (612) 831-5454

Edmund P. Anderson
Honeywell, Inc.
Honeywell Plaza MN 12-8251
Minneapolis, MN 55408   (612) 870-6494

Robert M. Angus
Control Data Corp.
P.O. Box O
Minneapolis, MN 55440   (612) 853-3266

Oliver F. Arrett
Vidas & Arrett, P.A.
1904 Plaza VII
45 S. 7th Street.
Minneapolis, MN 55402   (612) 339-8801

Richard A. Arrett
Vidas & Arrett, P.A.
1904 Plaza Vii
45 South 7th Street
Minneapolis, MN 55402   (612) 339-8801

Michael B. Atlass
Honeywell Inc.
Honeywell Plaza
Minneapolis, MN 55408   (612) 870-5857

William C. Babcock
Dorsey & Whitney
2200 First Bank Place East
Minneapolis, MN 55402   (612) 340-2635

Herman Hershel Bains
Williamson, Bains, Moore
& Hansen
608 Building, Suite 668
608 Second Avenue South
Minneapolis, MN 55402   (612) 332-2587

Richard Otto Bartz
Burd, Bartz & Gutenkauf
1300 Foshay Tower
821 Marquette Ave.
Minneapolis, MN 55402   (612) 332-6581

Robert C. Beck
Merchant, Gould, Smith, Edell,
Welter & Schmidt, P.A.
3100 Norwest Center
90 South 7th St.
Minneapolis, MN 55042   (612) 332-5300

Jennifer F. Becker
Kinney & Lange, P.A.
625 4th Ave., South
Suite 1500
Minneapolis, MN 55415   (612) 339-1863

Jerry F. Best
Onan Corp.
1400 73rd Ave., N.E.
Minneapolis, MN 55432   (612) 574-5802

Clyde C. Blinn
7250 Lewis Ridge Pkwy., #210
Minneapolis, MN 55439   (612) 941-8449

Douglas J. Bornemann
Faegre & Benson
2200 Norwest Center
90 South Seventh St.
Minneapolis, MN 55402   (612) 336-3407

Felicia J. Boyd
Faegre & Berson
90 South Seventh Street
Norwest Center
Minneapolis, MN 55402   (612) 336-3347

William A. Braddock
P.O. Box 14928
Minneapolis, MN 55414   (612) 331-2718

Joseph F. Breimayer
Medtronic, Inc.
7000 Central Avenue
Minneapolis, MN 55432   (612) 574-3278

Ronald J. Brown
Dorsey & Whitney
2200 First Bank Place E.
Minneapolis, MN 55402   (612) 340-2879

Steven C. Bruess
Merchant, Gould, Smith, Edell,
Welter & Schmidt, P.A.
3100 Norwest Center
90 S. 7th Street
Minneapolis, MN 55402　　　(612) 332-5300

Gregory A. Bruns
Honeywell, Inc.
Honeywell Plaza
Mn 12-8251
Minneapolis, MN 55408　　　(612) 870-2136

David D. Brush#
Kinney & Lange
625 Fourth Ave., South
Suite 1500
Minneapolis, MN 55415　　　(612) 339-1863

Robert L. Buckley, Jr.#
Toro Company
8111 Lyndale Avenue South
Minneapolis, MN 55420　　　(612) 888-8801

John W. Bunch
200 Southdale Place
3400 W. 66th Street
Minneapolis, MN 55435　　　(612) 922-2511

L. Paul Burd
Burd, Bartz & Gutenkauf
1300 Foshay Tower
821 Marquette Avenue
Minneapolis, MN 55402　　　(612) 332-6581

Alan G. Carlson
Merchant, Gould, Smith, Edell,
Welter & Schmidt, P.C.
3100 Norwest Center
90 South 7th Street
Minneapolis, MN 55402　　　(612) 332-5300

Philip P. Caspers
Merchant , Gould, Smith, Edell,
Welter & Schmidt, P.A.
3100 Norwest Center
90 S. 7th Street
Minneapolis, MN 55402　　　(612) 332-5300

Ronald E. Champion
Honeywell Inc.
Honeywell Plaza
Mn 12-8251
Minneapolis, MN 55408　　　(612) 870-6192

Judson K. Champlin
Kinney & Lange, P.A.
625 Fourth Ave., South
Suite 1500
Minneapolis, MN 55415　　　(612) 339-1863

John A. Clifford
Merchant, Gould, Smith, Edell
Welter & Schmidt, P.A.
3100 Norwest Center
90 South 7th Street
Minneapolis, MN 55402　　　(612) 332-5300

William J. Connors
Pillsbury Company
Pillsbury Center
Minneapolis, MN 55402　　　(612) 330-8709

Timothy R. Conrad
Merchant, Gould, Smith, Edell,
Welter & Schmidt, P.A.
3100 Norwest Center
90 South 7th Street
Minneapolis, MN 55402　　　(612) 332-5300

Robert W. Doyle
Doyle International Law Offs., Ltd.
4530 IDS Center
Minneapolis, MN 55402　　　(612) 338-7511

Reed A. Duthler
Medtronic Inc.
7000 Central Avenue
Minneapolis, MN 55432　　　(612) 574-3351

Wayne B. Easton
510 Plymouth Bldg.
Minneapolis, MN 55402　　　(612) 333-8723

Robert T. Edell
Merchant, Gould, Smith, Edell,
Welter & Schmidt, P.A.
3100 Norwest Center
90 South 7th Street
Minneapolis, MN 55402　　　(612) 332-5300

Mark S. Ellinger
Patterson & Keough, P.A.
615 Peavey Building
730 Second Ave., South
Minneapolis, MN 55402　　　(612) 342-2270

Michael D. Ellwein
Medtronic, Inc.
7000 Central Ave. N.E.
Minneapolis, MN 55432　　　(612) 574-3203

Robert A. Elwell
Oppenheimer Wolff & Donnelly
3400 Plaza VII Building
45 South Seventh Street
Minneapolis, MN 55402　　　(612) 344-9300

Gene O. Enockson
1600 Quebec Avenue N.
Minneapolis, MN 55427　　　(612) 545-2963

Allen H. Erickson#
Moore & Hansen
3000 Norwest Center
90 South Seventh St.
Minneapolis, MN 55402　　　(612) 332-8200

David R. Fairbairn
Kinney & Lange, P.A.
625 Fourth Avenue South
Suite 1500
Minneapolis, MN 55415　　　(612) 339-1863

Douglas B. Farrow
Graco Inc.
P.O. Box 1441
Minneapolis, MN 55440　　　(612) 623-6769

Alfred N. Feldman
9311 Franklin Avenue West
Minneapolis, MN 55426　　　(612) 545-3651

Harold D. Field, Jr.
Leonard, Street & Deinard
100 S. Fifth Street
Suite 1500
Minneapolis, MN 55402　　　(612) 337-1533

Robert C. Freed
Merchant, Gould, Smith, Edell,
Welter & Schmidt, P.A.
3100 Norwest Center
90 South 7th Street
Minneapolis, MN 55402　　　(612) 332-5300

Grady J. Frenchick
3433 Broadway St., N.E.
Suite 401
Broadway Place East
Minneapolis, MN 55413　　　(612) 331-1464

Norman P. Friederichs, Jr.
Merchant, Gould, Smith, Edell,
Welter & Schmidt, P.A.
3100 Norwest Center
90 South 7th Street
Minneapolis, MN 55402　　　(612) 332-5300

David N. Fronek
Dorsey & Whitney
2200 First Bank Place East
Minneapolis, MN 55402　　　(612) 340-2629

Keith J. Goar
10800 Lyndale Avenue
Apt. 250
Minneapolis, MN 55420　　　(612) 881-1601

John D. Gould
Merchant, Gould, Smith, Edell,
Welter & Schmidt, P.A.
3100 Norwest Center
90 South 7th Street
Minneapolis, MN 55402　　　(612) 332-5300

Alan G. Greenberg
701 Fourth Avenue South
Suite 1810
Minneapolis, MN 55415　　　(612) 333-7191

Craig M. Gregersen
Moore & Hansen
3000 Norwest Center
90 S. 7th St.
Minneapolis, MN 55402　　　(612) 332-8200

Leo Gregory
Reif and Gregory
1500 Dain Tower
527 Marquette Ave.
Minneapolis, MN 55402　　　(612) 333-7522

Robert W. Gutenkauf
Burd, Bartz & Gutenkauf
1300 Foshay Tower
821 Marquette Avenue
Minneapolis, MN 55402　　　(612) 332-6581

James R. Haller
Fredrikson & Byron, P.C.
1100 International Centre
900 Second Avenue South
Minneapolis, MN 55402　　　(612) 347-7017

Curtis B. Hamre
Merchant, Gould, Smith, Edell,
Welter & Schmidt, P.A.
3100 Norwest Center
90 South 7th Street
Minneapolis, MN 55402   (612) 332-5300

Conrad A. Hansen
Moore & Hansen
300 Norwest Center
90 South Seventh Street
Minneapolis, MN 55402   (612) 332-8200

James V. Harmon
1750 Northstar Center West
625 Marquette Avenue
Minneapolis, MN 55402   (612) 339-1400

Orrin M. Haugen
Haugen & Nikolai, P.A.
A Minnesota Corporation
900 Second Avenue South
Suite 820
Minneapolis, MN 55402   (612) 339-7461

John M. Haurykiewicz
Faegre & Benson
2200 Norwest Center
90 South Seventh Street
Minneapolis, MN 55402   (612) 336-3328

Gerald E. Helget
Palmatier & Sjoquist, P.A.
7900 Xerxes Avenue South
Suite 2000
Minneapolis, MN 55431   (612) 831-5454

Stuart R. Hemphill
Dorsey & Whitney
2200 First Bank Place East
Minneapolis, MN 55402   (612) 340-2734

Randall A. Hillson
Merchant, Gould, Smith, Edell,
Welter & Schmidt, P.A.
3100 Norwest Center
90 South 7th Street
Minneapolis, MN 55402   (612) 332-5300

Edward S. Hotchkiss
Fredrikson & Bryon, P.A.
1100 International Centre
900 2nd Avenue South
Minneapolis, MN 55402   (612) 347-7144

Keith R. Hughes
Robins, Kaplan, Miller & Ciresi
2800 La Salle Plaza
800 La Salle Avenue
Minneapolis, MN 55402   (612) 349-8500

Robert J. Jacobson
Palmatier & Sjoquist
2000 Northwestern Financial Ctr.
7900 Xeres Avenue South
Minneapolis, MN 55431   (612) 831-5454

Hugh D. Jaeger
Suite 207
3209 W. 76th Street
Minneapolis, MN 55435   (612) 830-1197

Cecilia M. Jaisle#
2728 Chowen Ave., S.
Minneapolis, MN 55416   (612) 920-7452

Harold D. Jastram
Oppenheimer, Wolff & Donnelly
Plaza Vii
45 S. 7th Street
Suite 3400
Minneapolis, MN 55402   (612) 344-9281

Larry M. Jennings
Bruce A. Rasmussen & Associates
2116 Second Ave., S.
Minneapolis, MN 55404   (612) 874-7071

Bruce A. Johnson
Medtronic, Inc.
7000 Central Avenue
Minneapolis, MN 55432   (612) 574-3275

Clayton Russell Johnson
3121 Dakota Avenue
Minneapolis, MN 55416   (612) 926-6939

Eugene L. Johnson
Dorsey & Whitney
2200 First Bank Place East
Minneapolis, MN 55402   (612) 340-2625

Gregory P. Kaihoi
Fredrikson & Byron, P.A.
1100 International Centre
900 Second Avenue South
Minneapolis, MN 55402   (612) 347-7049

Alan D. Kamrath
Peterson, Wicks, Nemer
& Kamrath, P.A.
1407 Soo Line Bldg.
Minneapolis, MN 55402   (612) 339-8501

Joseph R. Kelly
Kinney & Lange, P.A.
625 4th Ave., South
Suite 1500
Minneapolis, MN 55415   (612) 339-1863

Steven J. Keough
Patterson & Keough
615 Peavey Bldg.
730 Second Avenue South
Minneapolis, MN 55402   (612) 342-2270

Robert J. Klepinski
Medtronic, Inc.
7000 Central Avenue N.E.
Minneapolis, MN 55432   (612) 574-3234

Chad A. Klingbeil
Moore & Hansen
3000 Norwest Center
90 South Seventh Street
Minneapolis, MN 55402   (612) 332-8200

Daniel J. Kluth
Merchant, Gould, Smith, Edell,
Welter & Schmidt, P.A.
3100 Norwest Center
90 South 7th Street
Minneapolis, MN 55402   (612) 332-5300

John L. Knoble
Merchant, Gould, Smith, Edell,
Welter & Schmidt
3100 Norwest Center
90 South 7th Street
Minneapolis, MN 55402   (612) 336-4713

Franklin J. Knoll
C-1556 Government Center
Minneapolis, MN 55487   (612) 348-6971

Alan W. Kowalchyk
Merchant, Gould, Smith, Edell
Welter & Schmidt, P.A.
3100 Norwest Center
90 South 7th Street
Minneapolis, MN 55402   (612) 332-5300

Marc G. Kurzman
Kurzman Grant Ojala
2445 Park Avenue
Minneapolis, MN 55404   (612) 871-9004

Michael B. Lasky
Merchant, Gould, Smith,
Edell, Welter, Schmidt, P.A.
3100 Norwest Center
90 South 7th Street
Minneapolis, MN 55402   (612) 332-5300

Daniel W. Latham
Medtronic, Inc.
7000 Central Ave., N.E.
Minneapolis, MN 55432   (612) 574-3278

Ronald L. Laumbach
Cargill, Incorporated
P.O. Box 9300
Minneapolis, MN 55440   (612) 475-6366

Eric M. Lee
Dorsey & Whitney
2200 First Bank Place East
Minneapolis, MN 55402   (612) 340-2702

Donald J. Lenkszus
Honeywell Inc.
Honeywell Plaza
Mn12-8251
Minneapolis, MN 55408   (612) 870-2877

Robert B. Leonard
Honeywell Inc.
Honeywell Plaza
Mn12-8251
Minneapolis, MN 55408   (612) 870-5069

George A. Leone, Sr.
Leone & Moffa, P.A.
6500 Broadway Pl., West
1300 Godward St. N.E.
Minneapolis, MN 55413   (612) 378-1001

L. Meroy Lillehaugen
General Mills, Inc.
Number One General Mills Blvd.
Minneapolis, MN 55426   (612) 540-2283

Walter C. Linder
Kinney & Lange, P.A.
625 4th Avenue, South
Suite 1500
Minneapolis, MN 55415   (612) 339-1863

Ronald E. Lund
Medtronic Inc.
700 Central Avenue N.E.
Minneapolis, MN 55432   (612) 574-3013

Steven W. Lundberg
Merchant, Gould, Smith,
Edell, Welter & Schmidt, P.A.
3100 Norwest Center
90 South 7th Street
Minneapolis, MN 55402   (612) 332-5300

Ian D. Mac Kinnon#
Honeywell Inc.
Off. of Gen. Coun.
Honeywell Plaza
MN 12 - 8251
Minneapolis, MN 55408   (612) 870-6877

Joseph P. Martin
Toro Co.
8111 Lyndale Ave., S.
Minneapolis, MN 55420   (612) 887-8950

Daniel W. Mc Donald
Merchant, Gould, Smith, Edell,
Welter & Schmidt, P.A.
3100 Norwest Center
90 South 7th Street
Minneapolis, MN 55402   (612) 332-5300

Wendy M. Mc Donald
Merchant, Gould, Smith, Edell,
Welter & Schmidt, P.A.
3100 Norwest Center
90 South Seventh Street
Minneapolis, MN 55402   (612) 332-5300

Paul H. Mc Dowall
University of Minnesota
ORTTA- Off. of Pats.
& Licensing
1100 Washington Ave., South
Minneapolis, MN 55415   (612) 624-0550

Albin Medved
Honeywell Inc.
Office of Gen. Counsel
Honeywell Plaza
Minneapolis, MN 55408   (612) 870-6459

Charles G. Mersereau
Honeywell Inc.
Honeywell Plaza
Minneapolis, MN 55408   (612) 870-2875

Michelle M. Michel
Merchant, Gould, Smith, Edell,
Welter & Schmidt, P.A.
3100 Norwest Center
90 S. 7th Street
Minneapolis, MN 55402   (612) 336-4608

James W. Miller
8111 Lyndale Avenue South
Minneapolis, MN 55420   (612) 887-8903

Malcolm L. Moore
Moore & Hansen
3000 Norwest Center
90 South Seventh Street
Minneapolis, MN 55402   (612) 332-8200

Robert C. Moy
Merchant, Gould, Smith, Edell,
Welter & Schmidt, P.A.
3100 Norwest Center
90 South 7th Street
Minneapolis, MN 55402   (612) 332-5300

Lawrence M. Nawrocki
Suite 401, Broadway Pl., E.
3433 Broadway St. N.E.
Minneapolis, MN 55413   (612) 331-1464

Theodore F. Neils
Kinney & Lange
625 Fourth Avenue South
Suite 1500
Minneapolis, MN 55415   (612) 339-1863

Albin J. Nelson
Merchant, Gould, Smith, Edell,
Welter & Schmidt, P.A.
3100 Norwest Center
90 S. Seventh Street
Minneapolis, MN 55402   (612) 332-5300

Bruce A. Nemer
Wicks and Nemer, P.A.
1407 Soo Line Bldg.
105 S. 5th Street
Minneapolis, MN 55402   (612) 339-8501

Frederick W. Niebuhr
Haugen & Nikolai, P.A.
820 International Centre
900 Second Avenue South
Minneapolis, MN 55402   (612) 339-7461

Thomas J. Nikolai
Haugen & Nikolai, P.A.
820 International Centre
900 Second Avenue South
Minneapolis, MN 55402   (612) 339-7461

Angela N. Nwaneri
Oppenheimer Wolff & Donnelly
45 S. 7th St., Suite 3400
Minneapolis, MN 55402-2   (612) 344-9300

John A. O Toole
General Mills, Inc.
Number One General Mills Blvd.
Minneapolis, MN 55426   (612) 540-2422

Michael J. O'Loughlin
1012 Grain Exchange
400 South 4th Street
Minneapolis, MN 55415   (612) 332-0351

Kenneth D. Ohm
General Mills, Inc.
Number One General Mills Blvd.
Minneapolis, MN 55426   (612) 540-2284

James Edward Olds
10800 Lyndale Avenue South
Minneapolis, MN 55420   (612) 881-1601

Robert A. Pajak#
Honeywell Inc.
Honeywell Plaza
Minneapolis, MN 55408   (612) 870-2723

H. Dale Palmatier
Palmatier & Sjoquist, P.A.
7900 Xerxes Avenue South
Suite 2000
Minneapolis, MN 55431   (612) 831-5454

James H. Patterson
Patterson & Keough
615 Peavey Building
730 Second Ave., South
Minneapolis, MN 55402   (612) 342-2270

Harold R. Patton
Medtronic Inc.
7000 Central Ave., N.E.
Minneapolis, MN 55432   (612) 694-3156

Bradley D. Pedersen
Patterson & Keough, P.A.
615 Peavey Building
730 Second Avenue South
Minneapolis, MN 55402   (612) 342-2270

Stuart R. Peterson
Peterson, Wicks, Nemer
& Kamrath, P.A.
Suite 1407 Soo Line Bldg.
Minneapolis, MN 55402   (612) 339-8501

Thomas E. Popovich
Dorsey & Whitney
2200 First Bank Place East
Minneapolis, MN 55402   (612) 340-2964

Malcolm D. Reid
Network Systems Corp.
7600 Boone Ave. N.
Minneapolis, MN 55428   (612) 424-1582

Thomas A. Rendos#
Kinney & Lange
625 Fourth Avenue South
Suite 1500
Minneapolis, MN 55415   (612) 339-1863

John A. Rissman
Medtronic, Inc.
7000 Central Avenue N.E.
Law Dept.
Minneapolis, MN 55432   (612) 574-3279

John L. Rooney
Suite 401, Broadway Pl., East
3433 Broadway St. N.E.
Minneapolis, MN 55413   (612) 331-1464

Charles L. Rubow
Honeywell Inc.
Honeywell Plaza
Minneapolis, MN 55408   (612) 870-6461

Christopher John Rudy
4541 31st Avenue South
Minneapolis, MN 55406   (612) 721-5793

Zbigniew P. Sawicki
Kinney & Lange, P.A.
625 Fourth Avenue South
Suite 1500
Minneapolis, MN 55415   (612) 339-1863

Brian F. Schroeder
Schroeder & Siegfried, P.A.
234O IDS Center
Minneapolis, MN 55402   (612) 339-0120

Everett J. Schroeder
Schroeder & Siegfried, P.A.
2340 IDS Center
Minneapolis, MN 55402   (612) 339-0120

Mark D. Schuman
Merchants, Gould, Smith, Edell,
Welter & Schmidt, P.A.
3100 Norwest Center
90 South Seventh St.
Minneapolis, MN 55402   (612) 332-5300

Michael D. Schumann
Merchant, Gould, Smith,
Edell, Welter & Schmidt
3100 Norwest Center
90 S. 7th St.
Minneapolis, MN 55402   (612) 332-5300

Ronald J. Schutz
Robins, Kaplan, Miller
& Ciresi
1800 International Centre
900 Second Avenue South
Minneapolis, MN 55402   (612) 349-8500

Lew Schwartz
121 Washington Ave., South
Suite 504
Minneapolis, MN 55401   (612) 332-3023

Edward L. Schwarz
Honeywell Inc.
Honeywell Plaza
Minneapolis, MN 55408   (612) 870-6132

Micheal L. Schwegman
Merchant, Gould, Smith, Edell,
Welter & Schmidt, P.A.
3100 Norwest Center
90 South 7th Street
Minneapolis, MN 55402   (612) 332-5300

Gregory A. Sebald
Merchant, Gould, Smith, Edell,
Welter & Schmidt, P.A.
3100 Norwest Center
90 S. 7th Street
Minneapolis, MN 55402   (612) 336-4728

Michael S. Sherrill
Merchant, Gould, Smith, Edell,
Welter & Schmidt, P.A.
3100 Norwest Center
90 South 7th Street
Minneapolis, MN 55402   (612) 332-5300

John G. Shudy, Jr.
Honeywell Inc.
Honeywell Plaza
Minneapolis, MN 55408   (612) 870-6419

Wayne A. Sivertson
Kinney & Lange, P.A.
625 Fourth Avenue South
Suite 1500
Minneapolis, MN 55415   (612) 339-1863

Paul L. Sjoquist
Palmatier & Sjoquist
2000 Northwestern Financial Ctr.
7900 Xerxes Avenue S.
Minneapolis, MN 55431   (612) 831-5454

Phillip H. Smith
Merchant, Gould, Smith, Edell,
Welter & Schmidt, P.A.
3100 Norwest Center
Minneapolis, MN 55402   (612) 336-4614

Stephanie J. Smith#
Merchant, Gould, Smith, Edell,
Welter & Schmidt
90 S. Seventh St.
3100 Norwest Center
Minneapolis, MN 55402   (612) 336-4607

Charles E. Steffey
Faegre & Benson
2200 Norwest Center
Minneapolis, MN 55402   (612) 336-3322

Walter J. Steinkraus
Vidas & Arrett, P.A.
1904 Plaza Vii,
45 South 7th Street
Minneapolis, MN 55402   (612) 339-8801

Douglas A. Strawbridge
Merchant, Gould, Smith,
Edell, Welter & Schmidt, P.A.
3100 Norwest Center
90 South 7th Strret
Minneapolis, MN 55402   (612) 332-5300

Kristine M. Strodthoff
Merchant & Gould
3100 Norwest Center
90 S. Seventh St.
Minneapolis, MN 55402   (612) 336-4752

Warren A. Sturm
3989 Central Avenue N. E.
Suite 605
P.O. Box 21369
Minneapolis, MN 55421   (612) 781-3319

John P. Sumner
Merchant & Gould
3100 Norwest Center
Minneapolis, MN 55402   (612) 336-4624

John S. Sumners
Merchant, Gould, Smith, Edell,
Welter & Schmidt, P.A.
3100 Norwest Center
90 South 7th Street
Minneapolis, MN 55402   (612) 332-5300

David K. Tellekson
Merchant & Gould
3100 Norwest Center
Minneapolis, MN 55402   (612) 336-4612

Emery Lloyd Tracy
Kinney & Lange, P.A.
625 4th Avenue, South
Suite 1500
Minneapolis, MN 55415   (612) 339-1863

Albert L. Underhill#
Merchant & Gould
3100 Norwest Center
Minneapolis, MN 55402   (612) 332-5300

Robert O. Vidas
Vidas & Arrett, P.A.
1904 Plaza Vii
45 South 7th Street
Minneapolis, MN 55402   (612) 339-8801

Scott Q. Vidas
Vidas & Arrett, P.A.
1904 Plaza Vii
45 South 7th Street
Minneapolis, MN 55402   (612) 339-8801

John A. S. Webster#
2213 Nicollet Ave.
Minneapolis, MN 55404   (612) 871-9191

Paul A. Welter
Merchant, Gould, Smith, Edell,
Welter & Schmidt, P.A.
3100 Norwest Center
90 South 7th Street
Minneapolis, MN 55402   (612) 332-5300

Nickolas E. Westman
Kinney & Lange, P.A.
625 Fourth Avenue South
Suite 1500
Minneapolis, MN 55415   (612) 339-1863

Douglas J. Williams
Merchant, Gould, Smith, Edell,
Welter & Schmidt, P.A.
3100 Norwest Center
Minneapolis, MN 55402   (612) 336-4632

Warren D. Woessner
Merchant, Gould, Smith, Edell,
Welter & Schmidt, P.A.
3100 Norwest Center
90 South 7th Street
Minneapolis, MN 55402   (612) 332-5300

James L. Young
Kinney & Lange, P.A.
625 Fourth Avenue South
Suite 1500
Minneapolis, MN 55415   (612) 339-1863

Steven Bucher
3715 Huntingdon Dr.
Minnetonka, MN 55343   (612) 935-6647

Thomas W. Kenyon
4816 Sparrow Road
Minnetonka, MN 55345   (612) 474-9347

Jean Seaburg
4647 Baker Road S.
Minnetonka, MN 55343   (612) 936-7742

Patrick John Span
3832 Susan Lane
Minnetonka, MN 55345   (612) 473-2730

Janet P. Schafer
651 Driftwood Ct.
New Brighton, MN 55112   (612) 639-1683

S. Linda Ruffing#
577 Grospoint Lane
Oakdale, MN 55128   (612) 731-1714

Robert B. Moffatt
615 West Third Avenue
Pine City, MN 55063   (612) 629-6066

Thomas C. Naber
Schneider Division
Pfizer Inc.
5905 Nathan Lane
Plymouth, MN 55442   (612) 550-5500

Omund R. Dahle
7632 13th Avenue South
Richfield, MN 55423   (612) 550-5500

James Michael Anglin
IBM Corp.
3605 Highway 52 North
Department 917
Rochester, MN 55901   (507) 253-4661

Richard E. Billion
IBM Corporation
Dept. 917
3605 Highway 52 North
Rochester, MN 55901   (507) 253-7974

Keith T. Bleuer
1663 Wilshire Dr., N.E.
Rochester, MN 55901   (507) 288-4978

Bradley A. Forrest
IBM Corporation
Department 917
3605 Highway 52 North
Rochester, MN 55901   (507) 253-5331

Frederick W. Kellogg
Mayo Clinic
200 S.W. 1st Street
Rochester, MN 55905   (507) 284-4916

Robert Wyman Lahtinen
IBM Corporation
Department 917
3605 Highway 52 North
Rochester, MN 55901   (507) 253-4660

Curtis G. Rose
IBM Corporation
Department 917
3605 Highway 52 North
Rochester, MN 55901   (507) 253-2557

William J. Ryan
Michaels Seeger Rosenblad
& Arnold
550 Norwest Center
21 First Street S.W.
Rochester, MN 55902   (507) 288-7755

Roy W. Truelson
IBM Corporation
Dept. 917
3605 Highway 52 North
Rochester, MN 55901   (507) 253-8618

Donald F. Voss
2541 12th Ave., N.W.
Rochester, MN 55901   (507) 282-0396

Charles A. Johnson
Unisys Corporation
2276 Highcrest Road
Roseville, MN 55113   (612) 635-7702

Alan Maclean Staubly
17081 Sunset Avenue
Shakopee, MN 55379   (612) 445-4900

Thomas B. Tate
708-13th Ave., No.
P.O. Box 41
South St. Paul, MN 55075   (612) 457-6750

Charles J. Ungemach
3216 Rankin Road
St. Anthony, MN 55418   (612) 781-1661

Robert L. Kaner
1410 Colorado Ave., South
Suite 205
St. Louis Park, MN 55416   (612) 546-8861

Cruzan Alexander
P.O. Box 33427
St. Paul, MN 55133   (612) 733-1511

David W. Anderson
3M Company
Off. of Pat. Coun.
P.O. Box 33427
St. Paul, MN 55133   (612) 733-2221

Donna W. Bange#
Minnesota Mining & Manufacturing
3M Center 251-2A-09
St. Paul, MN 55144   (612) 733-5291

John C. Barnes
3M Company
Off. of Pat. Coun.
P.O. Box 33427
St. Paul, MN 55133   (612) 733-1519

William B. Barte
3M Company
Off. of Pat. Coun.
P.O. Box 33427
St. Paul, MN 55133   (612) 733-4395

Carolyn A. Bates
3M Company
Off. of Pat. Coun.
PO Box 33427
St. Paul, MN 55133   (612) 733-1535

Brian Hughes Batzli
Merchant, Gould, Smith, Edell,
Welter & Schmidt, P.A.
1000 Norwest Center
St. Paul, MN 55101   (612) 298-1055

Stephen Walter Bauer
Minn. Mining & Mfg. Company
Off. of Pat. Coun.
P.O. Box 33427
St. Paul, MN 55133   (612) 736-4533

William D. Bauer
3M Co.
3M Center
P.O. Box 33427
St. Paul, MN 55133   (612) 733-1532

Mark W. Binder
3M Office of Intellectual
Property Counsel
3M Center
PO Box 33427
St. Paul, MN 55133   (612) 737-5231

Dale A. Bjorkman#
3M Co.
Off. of Pat. Coun.
3M Center
P.O. Box 33427
St. Paul, MN 55133   (612) 733-1509

Jennie G. Boeder
3M Company
P.O. Box 33427
St. Paul, MN 55133   (612) 733-3084

William J. Bond
3M Co.
P.O. Box 33427
St. Paul, MN 55133   (612) 736-4790

Warren R. Bovee
3M Company
Off. of Pat. Coun.
P.O. Box 33427
St. Paul, MN 55133   (612) 733-1513

Glenn William Bowen, Sr.
Unisys Corporation
P.O. Box 64525
St. Paul, MN 55164   (612) 456-2682

Richard E. Brink
3M Company
Off. of Pat. Coun.
P.O. Box 33427
St. Paul, MN 55133   (612) 733-1517

Stephen W. Buckingham
3M Company
Off. of Pat. Coun.
P.O. Box 33427
2501 Hudson Road
St. Paul, MN 55133   (612) 733-3379

Robert Withy Burns
3M Company
Off. of Pat. Coun.
P.O. Box 33427
3M Center
St. Paul, MN 55144   (612) 733-1555

Linda M. Byrne
Merchant, Gould, Smith, Edell,
Welter & Schmidt, P.A.
1000 Norwest Center
St. Paul, MN 55101   (612) 298-1055

Gerald Frank Chernivec
Minn. Mining & Mfg. Co.
2501 Hudson
P.O. Box 33427
St. Paul, MN 55133   (612) 733-8398

James D. Christoff
3M Company
Off. of Pat. Coun.
3M Center
St. Paul, MN 55144   (612) 733-1512

David R. Cleveland
3M Company
Off. of Patent Coun.
P.O. Box 33427
St. Paul, MN 55133   (612) 733-1539

Stanley G. De La Hunt
3M Company
3M Center
St. Paul, MN 55133   (612) 733-1508

Mark J. Di Pietro
Merchant, Gould, Smith, Edell,
Welter & Schmidt, P.A.
Suite 1000
Norwest Center
St. Paul, MN 55101   (612) 298-1055

Janice L. Dowdall
3M Company
Off. of Pat. Coun.
3M Center
P.O. Box 33427
St. Paul, MN 55133   (612) 733-4995

Anthony G. Eggink
3100 First National Bank Bldg.
332 Minnesota Street
St. Paul, MN 55101   (612) 298-1171

Gregory A. Evearitt
167 N. Mc Knight Road
Apt. 208
St. Paul, MN 55119   (612) 739-3681

William G. Ewert
3M Company
Off. of Pat. Coun.
P.O. Box 33427
St. Paul, MN 55133   (612) 733-1533

Richard E. Fayling#
3M Company
3M Center
206-6b-05
St. Paul, MN 55144   (612) 733-5475

Frederick A. Fleming
1860 Highland Pkwy.
St. Paul, MN 55116   (612) 690-4656

James M. Foley#
3M Company
3M Center
Bldg., 260-6B-05
St. Paul, MN 55144   (612) 733-6629

Peter Forrest
3M Company
Office of Patent Counsel
P.O. Box 33427
St. Paul, MN 55133   (612) 733-3155

Richard Francis
3M Company
Office of Patent Counsel
P.O. Box 33427
St. Paul, MN 55133   (612) 733-7519

Thomas E. Fredeen#
3M Company
3M Center
Bldg. 236-GB-03
St. Paul, MN 55144   (612) 733-3975

Mark William Gehan
757 Fairmount Avenue
St. Paul, MN 55105   (612) 733-1509

Philip M. Goldman
3M Company
Office of Patent Counsel
P.O. Box 33427
St. Paul, MN 55133   (612) 733-4247

Charles E. Golla
Merchant, Gould, Smith, Edell,
Welter & Schmidt, P.A.
1000 Norwest Center
St. Paul, MN 55101   (612) 298-1055

Robert E. Granrud
1809 Colvin Avenue
St. Paul, MN 55116   (612) 698-7621

Edward P. Gray
Cardiac Pacemakers, Inc.
4100 Hamline Avenue North
St. Paul, MN 55112   (612) 638-4000

John Joseph Gresens
Merchant, Gould, Smith, Edell,
Welter & Schmidt, P.A.
1000 Norwest Center
St. Paul, MN 55101   (612) 298-1055

Karl G.A.L. Hanson
3M Company
Office of Patent Counsel
3M Center
P.O. Box 33427
St. Paul, MN 55133   (612) 736-7776

Jeffrey J. Hohenshell
Minnesota Mining & Mfg. Co.
3M Center
P.O. Box 33427
St. Paul, MN 55133   (612) 733-1555

Robert W. Hoke, II
3M Company
3M Center
P.O. Box 33427
St. Paul, MN 55133   (612) 736-9155

John H. Hornickel
3M Company
Office of Patent Counsel
P.O. Box 33427
St. Paul, MN 55133   (612) 733-4998

Susan M. Howard#
3M Company
3M Center
Bldg. 270-2S-06
St. Paul, MN 55144   (612) 733-6394

Allen J. Hubin#
3M Company
3M Center
St. Paul, MN 55144   (612) 733-2423

William L. Huebsch
3M Company
P.O. Box 33427
2501 Hudson Road
St. Paul, MN 55144   (612) 733-2835

Harold Hughesdon#
3M Company
3M Center
Bldg. 220-4W-01
St. Paul, MN 55144   (612) 733-2973

Dale E. Hulse
Minnesota Mining & Mfg. Co.
Office of Patent Counsel
P.O. Box 33427
St. Paul, MN 55133   (612) 736-9631

Mary M. Hurlocker#
3M Company
3M Center
Bldg. 260-6B-05
St. Paul, MN 55144   (612) 733-7731

Marvin Jacobson
Jacobson & Johnson
1 West Water St.
Suite 202
St. Paul, MN 55107   (612) 222-3775

Carl Lowell Johnson
Jacobson & Johnson
1 West Water Street
Suite 202
St. Paul, MN 55107   (612) 222-3775

David G. Johnson
Merchant, Gould, Smith, Edell,
Welter, Schmidt, P.A.
55 East Fifth Street
St. Paul, MN 55101   (612) 298-1055

Robert H. Jordan
Minnesota Mining & Mfg. Co.
P.O. Box 33427
2501 Hudson Road
St. Paul, MN 55133   (612) 733-6866

Barry D. Josephs
3M Company
Office of Patent Counsel
3M Center
P.O. Box 33427
St. Paul, MN 55133   (612) 736-5971

Thomas E. Jurgensen
3M Office of Intellectual
Property Counsel
P.O. Box 33427
St. Paul, MN 55133   (612) 733-1500

David B. Kagan
3M Company
3M Center
P.O. Box 33427
St. Paul, MN 55133   (612) 733-4879

Walter N. Kirn, Jr.
3M Company
Office of Patent Counsel
P.O. Box 33427
St. Paul, MN 55133   (612) 733-1523

Kent S. Kokko#
3M Company
3M Center
Bldg. 220-1S-11
St. Paul, MN 55144   (612) 733-3597

Mark A. Krull
Merchant & Gould
55 E. Fifth Street
St. Paul, MN 55101   (612) 298-1055

Bruce A. Langager
3M Company
3M Center
St. Paul, MN 55144   (612) 733-4028

Charles Houlton Lauder
Minnesota Mining & Mfg. Co.
P.O. Box 33427
St. Paul, MN 55133   (612) 733-1510

Louis M. Leichter#
3M Corporate Research Laboratories
Patent Liaison Function
3M Center
Bldg. 201-1S-13
St. Paul, MN 55144   (612) 733-9899

Charles D. Levine
Minnesota Mining &
Mfg. Co.
P.O. Box 33427
St. Paul, MN 55133   (612) 736-1255

James V. Lilly
3M Company
Off. of Pat. Coun.
P.O. Box 33427
St. Paul, MN 55133   (612) 733-1543

Mark A. Litman
3M Company
Off. of Pat. Coun.
P.O. Box 33427
St. Paul, MN 55133   (612) 733-1515

Douglas B. Little
Minnesota Mining & Mfg. Co.
P.O. Box 33427
St. Paul, MN 55133   (612) 733-1501

Gary F. Lyons
3M Center
P.O. Box 33428
St. Paul, MN 55133   (612) 733-1780

Eloise J. Maki
Minnesota Mining & Mfg. Co.
P.O. Box 33427
St. Paul, MN 55133   (612) 736-3373

Michael L. Mau
Merchant, Gould, Smith, Edell,
Welter & Schmidt, P.A.
1000 Norwest Center
St. Paul, MN 55101   (612) 298-1055

John A. Miller
3M Company
230-3F-06
3M Center
St. Paul, MN 55144   (612) 736-0800

Joan M. Mullins
15 W. Annapolis
St. Paul, MN 55118   (612) 222-7087

Darla P. Neaveill
Minnesota Mining & Mfg. Co.
Office of Pat. Coun.
3M Company
Box 3427
St. Paul, MN 55133   (612) 736-4986

Paul W. O Malley, Jr.
2189 Sargent Avenue
St. Paul, MN 55105   (612) 699-9589

Edward T. Okubo
3M Company
Off. of Pat. Coun.
P.O. Box 33427
St. Paul, MN 55133   (612) 733-1534

Sharon E. Pecore#
3M Company
3M Center
Bldg. 260-6B-05
St. Paul, MN 55144   (612) 736-1474

Carolyn V. Peters
3M
3M Center
P.O. Box 33427
St. Paul, MN 55133   (612) 733-1500

Terryl K. Qualey
3M Company
P.O. Box 33427
St. Paul, MN 55133   (612) 733-1940

Douglas E. Reedich#
3M Company
Off. of Pat. Coun.
P.O. Box 33427
St. Paul, MN 55133   (612) 736-2704

David L. Robbins#
3M Company
3M Center
Bldg. 260-6B-05
St. Paul, MN 55144   (612) 733-7913

Fred H. Roberts#
3M Company
3M Center
201-1S-13
St. Paul, MN 55144   (612) 733-2392

Joel A. Rothfus
Merchant, Gould, Smith,
Edell, Welter & Schmidt
55 East Fifth Street
Norwest Center, Ste. 1000
St. Paul, MN 55101   (612) 298-1055

Cecil C. Schmidt
Merchant, Gould, Smith, Edell,
Welter & Schmidt, P.A.
1000 Norwest Center
St. Paul, MN 55101   (612) 298-1055

Leland D. Schultz
3M Company
P.O. Box 33427
St. Paul, MN 55133   (612) 736-7170

Donald M. Sell
3M Company
Off. of Pat. Coun.
220-12W-01
P.O. Box 33427
St. Paul, MN 55133   (612) 733-1514

Lorraine R. Sherman
Minnesota Mining &
Manufacturing Co.
Off. of Pat. Coun.
P.O. Box 33427
St. Paul, MN 55133   (612) 733-1507

Miriam G. Simmons#
3M Company
3M Center
Bldg. 201-1S-13
St. Paul, MN 55144   (612) 736-2320

Joel D. Skinner, Jr.
Eggink & Associates
3100 First National Bank
St. Paul, MN 55101   (612) 298-1171

Steven E. Skolnick
3M Company
Off. Pat. Coun.
P.O. Box 33427
St. Paul, MN 55133   (612) 736-7796

James A. Smith
1609 Garden Ave.
St. Paul, MN 55113   (612) 646-5516

Andrew D. Sorensen
Merchant & Gould
55 East 5th St.
St. Paul, MN 55101   (612) 225-2005

Robert W. Sprague
3M Company
P.O. Box 33427
St. Paul, MN 55016   (612) 733-0052

Roger R. Tamte
Minn. Mining & Mfg. Company
2501 Hudson Road
P.O. Box 33427
St. Paul, MN 55133   (612) 733-1520

Gregory M. Taylor
Merchant, Gould, Smith,
Edell, Welter & Schmidt, P.A.
1000 Norwest Center
St. Paul, MN 55101   (612) 298-1055

John F. Thuente
University of Minnesota
Patents & Licensing
1919 University Avenue
St. Paul, MN 55104   (612) 624-2816

Carole Truesdale
3M Company
3M Center
Off. of Pat. Counsel
P.O. Box 33427
St. Paul, MN 55133   (612) 736-4151

Douglas L. Tschida
2819 Hamline Avenue North
Hamline Center
Suite 123
St. Paul, MN 55113   (612) 636-3727

J. Wade Van Valkenburg, Jr.
Accessible Technologies Inc.
494 Curfew St.
St. Paul, MN 55104   (612) 659-0569

Amy L. Watson#
3M Company
3M Center
Bldg. 230-1S-11
St. Paul, MN 55144   (612) 733-4933

David L. Weinstein
Minn. Mining & Mfg. Company
3M Center
Bldg. 220-12W-01
St. Paul, MN 55144   (612) 736-2681

Lucy Cline Weiss#
3M Company
P.O. Box 33427
St. Paul, MN 55133   (612) 733-9547

Jeffrey L. Wendt
3M Office of Intellectual
Property Counsel
3M Center
P.O. Box 33427
St. Paul, MN 55133   (612) 733-9547

Janet R. Westrom
Merchant, Gould, Smith, Edell,
Welter & Schmidt, P.A.
1000 Norwest Bank Bldg.
St. Paul, MN 55101   (612) 298-1055

Terry L. Wiles
Unisys Corp.
P.O. Box 64942
M.S. 4772
St. Paul, MN 55164   (612) 635-5082

William F. Wittman#
3M Company
3M Center
260-6B-05
St. Paul, MN 55144   (612) 733-4698

Bradford B. Wright#
3M Company
201-3S-01
3M Center
St. Paul, MN 55144   (612) 736-4172

Michael T. Koller
9225 84th Street
Stillwater, MN 55082   (612) 426-6451

Glen Ervin Schumann
Holsten & Schumann, P.A.
124 South Second Street
P.O. Box 206
Stillwater, MN 55082   (612) 439-0700

Louis B. Oberhauser, Jr.
1421 East Wayzata Blvd.
Wayzata, MN 55391   (612) 473-2521

Edmund Charles Meisinger
60 E. Marie
West St. Paul, MN 55118   (612) 457-2827

John M. Zangs, Jr.
1156 Allen Avenue
West St. Paul, MN 55118   (612) 457-5059

Gilbert B. Gehrenbeck#
5482 E. Bald Eagle Blvd.
White Bear Lake, MN 55110   (612) 429-9713

Arthur Mendel
4525 Oak Leaf Drive
White Bear Lake, MN 55127   (612) 429-1029

Richard J. Renk
768 Terrace Ln.
Winona, MN 55987   (507) 452-2461

Robert Leslie Marben
7101 Windgate Road
Woodbury, MN 55125   (612) 735-5481

Leigh E. Wood#
6548 Chamberlain Road
Woodbury, MN 55125   (612) 738-6148

Mississippi

Benjamin E. Long
United Technologies Electro Sys.
P.O. Box 2228
Mc Crary Road
Columbus, MS 39701   (601) 245-4301

Alexander F. Norcross, Sr.
P.O. Box 776
Gulfport, MS 39502   (601) 863-2944

Richard J. Hammond
Institute for Technology Dev.
700 N. State Street
Suite 500
Jackson, MS 39202   (601) 960-3600

G. Dempsey Ladner
1605 Kent Avenue
Jackson, MS 39211   (601) 969-2438

Lewis H. Wilson
803 South Forest Avenue
Long Beach, MS 39560   (601) 863-4470

Dewitt L. Fortenberry, Jr.
215 W. Jackson Avenue
P.O. Box 339
Mendenhall, MS 39114   (601) 847-3727

Leonard Weiss
527 Front Beach Dr., #25
Ocean Springs, MS 39564   (601) 875-6069

Emmette F. Hale, III
U.S. District Court for the
Northern District of Mississippi
P.O. Box 727
Oxford, MS 38655   (601) 234-1971

Harry Eugene Aine
R.R. 4, Box 767
Sumrall, MS 39482   (601) 268-6373

Missouri

Wayne B. Stone, Jr.
Colton & Stone, Inc.
R.R. 1, Box 122
Birch Tree, MO 65438   (314) 292-3624

Elmer J. Fischer
56 York Dr.
Brentwood, MO 63144   (314) 997-0251

Paul C. Krizov
360 Chateaugay Lane
Chesterfield, MO 63017   (314) 275-2641

Richard J. Sher
18734 Wild Horse Creek Rd.
Chesterfield, MO 63005   (314) 537-9525

Peter Nelson Davis
University of Missouri-Columbia
Columbia, MO 65211   (314) 882-2624

Donald Edward Gillihan
315 N. Washington
Farmington, MO 63640   (314) 756-6465

Charles R. Landholt
769 Waterfall Drive
Florissant, MO 63034   (314) 831-7343

William R. O Meara
1065 Jefferson Street
Florissant, MO 63031   (314) 837-9050

Robert J. Owens
1930 Pyrenees Drive
Florissant, MO 63033   (314) 837-1081

Glen R. Simmons
R.R. 2, Box 22
Green City, MO 63545   (816) 874-4332

Richard J. Grundstrom
P.O. Box 26
Holts Summit, MO 65043   (314) 634-8385

Harley R. Ball
US Sprint
Law Dept.
8140 Ward Parkway
MOKCMPO 54
Kansas City, MO 64114   (816) 276-6668

Charles N. Blitzer
Marion Laboratories
9300 Ward Parkway
Kansas City, MO 64114   (816) 966-4086

Joseph B. Bowman
Kokjer, Kircher, Bowman & Johnson
2414 Commerce Tower
911 Main St.
Kansas City, MO 64105   (816) 474-5300

Don M. Bradley
Kokjer, Kircher, Bradley, Wharton,
Bowman & Johnson
2414 Commerce Tower
911 Main St.
Kansas City, MO 64105   (816) 474-5300

Mark E. Brown
Litman, Mc Mahon & Brown
1600 One Kansas City Place
1200 Main Street
Kansas City, MO 64105 (816) 842-1587

John M. Collins
Schmidt, Johnson, Hovey
& Williams
1400 Mercantile Tower
1101 Walnut Street
Kansas City, MO 64106 (816) 474-9050

Dennis A. Crawford#
Litman, Day & Mc Mahon
1215 Commerce Trust Bldg.
922 Walnut Street
Kansas City, MO 64106 (816) 842-1587

William B. Day
Wirken & King
4740 Grand Avenue
Third Floor
Kansas City, MO 64112 (816) 753-6666

Steven R. Dickey
Schmidt, Johnson, Hovey
& Williams
1400 Mercantile Bank Tower
1101 Walnut Street
Kansas City, MO 64106 (816) 474-9050

Douglas J. Edmonds
Spencer, Fane, Britt & Browne
1400 Commerce Bank Building
1000 Walnut Street
Kansas City, MO 64106 (816) 474-8100

John P. Hazzard
Fermenta Animal Health Co.
7410 N.W. Tiffany Springs Pkwy.
P.O. Box 901350
Kansas City, MO 64190 (816) 891-5514

Robert D. Hovey
Hovey, Williams, Timmons
& Collins
1400 Mercantile Bank Tower
Kansas City, MO 64106 (816) 474-9050

Michael Bryan Hurd
Marion Laboratories, Inc.
10236 Marion Park Drive
P.O. Box 9627 - Park A
Kansas City, MO 64134 (816) 966-5000

Donald E. Johnson
1400 Mercantile Bank Tower
1101 Walnut Street
Kansas City, MO 64106 (816) 474-9050

Richard R. Johnson
Kokjer, Kircher, Bowman
& Johnson
2414 Commerce Tower
911 Main Street
Kansas City, MO 64105 (816) 474-5300

William Blaine Kircher
Kokjer, Kircher, Bowman
& Johnson
2414 Commerce Tower
911 Main Street
Kansas City, MO 64105 (816) 474-5300

Carter H. Kokjer
Kokjer, Kircher, Bowman
& Johnson
2414 Commerce Tower
911 Main St.
Kansas City, MO 64105 (816) 474-5300

Gerald M. Kraai
Husch & Eppenberger
1200 Main
Suite 1700
Kansas City, MO 64105 (816) 421-4800

Malcolm A. Litman
Litman Mc Mahon & Brown
1600 One Kansas City Place
1200 Main Street
Kansas City, MO 64105 (816) 842-1587

Robert E. Marsh
Blackwell, Sanders, Matheny,
Weary & Lombardi
2300 Main
Suite 1100
Kansas City, MO 64108 (816) 274-6847

Thomas P. Mc Bride
Kokjer, Kircher, Bradley, Wharton
Bowman & Johnson
2414 Commerce Tower
911 Main Street
Kansas City, MO 64105 (816) 474-5300

John C. Mc Mahon
Litman Mc Mahon & Brown
1600 One Kansas City Place
1200 Main Street
Kansas City, MO 64105 (816) 842-1587

Robert E. Mulloy, Jr.
H&R Block, Inc.
4410 Main Street
Kansas City, MO 64111 (816) 753-6900

Marcia J. Rodgers
Hovey, Williams, Timmons &
Collins
1101 Walnut Street
Suite 1400
Kansas City, MO 64106 (816) 474-9050

Taylor J. Ross
Kokjer, Kircher, Bowman
& Johnson
2414 Commerce Tower
911 Main Street
Kansas City, MO 64105 (816) 474-5300

William A. Rudy
Litman, Mc Mahon & Brown
1600 One Kansas City Pl.
1200 Main Street
Kansas City, MO 64105 (816) 842-1587

Cynthia L. Schaller
Litman, Mc Mahon & Brown
One Kansas City Place
1200 Main St., Ste. 1600
Kansas City, MO 64105 (816) 842-1590

Donald R. Schoonover
Litman, Mc Mahon
& Brown
1200 Main
Suite 1600
Kansas City, MO 64105 (816) 842-1587

Penny R. Slicer
Kokjer, Kircher, Bowman
& Johnson
2414 Commerce Tower
911 Main Street
Kansas City, MO 64105 (816) 474-5300

Stephen D. Timmons
Hovey, Williams, Timmons & Collins
1400 Mercantile Tower
1101 Walnut St.
Kansas City, MO 64106 (816) 474-9050

Thomas H. Van Hoozer
Hovey, Williams, Timmons
& Collins
1101 Walnut Street
Suite 1400
Kansas City, MO 64106 (816) 474-9050

John A. Weresh
Hovey, Williams, Timmons
& Collins
1101 Walnut Street, Ste. 1400
Kansas City, MO 64106 (816) 474-9050

J. David Wharton
Kokjer, Kircher, Bradley, Wharton,
Bowman & Johnson
2414 Commerce Tower
911 Main Street
Kansas City, MO 64105 (816) 474-5300

Warren N. Williams
Hovey, Williams, Timmons
& Coolins
1101 Walnut Street
Suite 1400
Kansas City, MO 64106 (816) 474-9050

Arvid V. Zuber#
Sherman, Wickens & Lysaught, P.C.
12th & Baltimore
P.O. Box 26530
Kansas City, MO 64196 (816) 471-6900

William Thomas Black
1549 Southlin Dr.
Kirkwood, MO 63122 (314) 966-8988

F. Travers Burgess
14 Taylor Woods
Kirkwood, MO 63122 (314) 822-0849

Norman Grant Steanson, Jr.
514 East 26th Avenue
North Kansas City, MO 64116
(816) 472-1116

Philip E. Hodur#
607 Seib Drive
O'Fallon, MO 63366 (314) 272-1881

John R. Miller
R.R. 1, Box 63
Owensville, MO 65066 (314) 437-2776

Chester Leslie Davis, Sr.
Perry, MO 63462   (314) 565-3570

Donald Myers
University of Missouri-Rolla
211 Parker Hall
Rolla, MO 65401   (314) 341-4135

Richard L. Marsh
Dayco Corporation
Dayco Technical Center
2601 W. Battlefield Road
Springfield, MO 65808   (417) 888-5426

Lawrence Edward Miller
King & Temple
The Hammons Tower
901 St. Louis St.
Suite 1501
Springfield, MO 65806   (417) 862-6100

Henry Wayne Cummings
124 North Main
St. Charles, MO 63301   (314) 946-0076

Frank R. Agovino
Senniger, Powers, Leavitt
& Roedel
One Metropolitan Square
16th Floor
St. Louis, MO 63102   (314) 231-5400

William I. Andress
Monsanto Co.
800 N. Lindbergh Blvd.
St. Louis, MO 63167   (314) 694-3165

Drew C. Baebler
Hullverson, Hullverson & Frank
1010 Market Street
Suite 1550
St. Louis, MO 63101   (314) 421-2313

Andrew Joseph Beck
Sherwood Medical Company
1831 Olive Street
St. Louis, MO 63103   (314) 621-7788

George R. Beck
Monsanto Co.
Patent Dept.
800 N. Lindbergh Blvd.
St. Louis, MO 63167   (314) 694-3187

Paul A. Becker, Sr.#
Emerson Electric Co.
White-Rodgers Div.
9797 Reavis Rd.
St. Louis, MO 63123   (314) 577-1513

Dennis A. Bennett
Monsanto Company
Bldg. A3 SD
800 N. Lindbergh Blvd.
St. Louis, MO 63167   (314) 694-1000

Jon H. Beusen
Monsanto Company
800 N. Lindbergh Blvd.
St. Louis, MO 63167   (314) 694-3191

George H. Blosser
Senniger, Powers, Leavitt
& Roedel
One Metropolitan Sq.
16th Floor
St. Louis, MO 63102   (314) 231-5400

Edward A. Boeschenstein
Gravely, Lieder and Woodruff
705 Olive St.
St. Louis, MO 63101   (314) 621-1457

James C. Bolding
Monsanto Company
800 N. Lindbergh Blvd.
St. Louis, MO 63166   (314) 694-3484

Grace L. Bonner
Monsanto Company
Mail Zone A3 SB
800 N. Lindbergh Blvd.
St. Louis, MO 63167   (314) 694-1000

Rebecca J. Brandau
Kalish & Gilster
818 Olive
Suite 1600 Paul Brown Bldg.
St. Louis, MO 63101   (314) 436-1331

Wendell W. Brooks
Monsanto Co.
800 N. Lindbergh Blvd.
St. Louis, MO 63167   (314) 694-3181

William George Bruns
Gravely, Lieder & Woodruff
705 Olive
St. Louis, MO 63101   (314) 621-1457

Robert M. Burton#
Burton International
Biomed
P.O. Box 13135
St. Louis, MO 63119   (314) 644-7332

Joseph C. Carr, Jr.
MI Tek Industries, Inc.
11710 Old Ballas Road
St. Louis, MO 63141   (314) 567-7127

James H. Casey*
U.S. Army Aviation System Comm.
ATTN: AMSAV-JP
Legal Office
4300 Goodfellow Blvd.
St. Louis, MO 63120   (314) 263-3591

John M. Charnecki
6401 West Court
St. Louis, MO 63116   (314) 752-0244

David H. Chervitz
Haverstock, Garrett & Roberts
611 Olive St.
Suite 1610
St. Louis, MO 63101   (314) 231-4427

Carol H. Clayman
Monsanto Company
800 North Lindbergh Blvd.
A3SA
St. Louis, MO 63167   (314) 694-3537

Arnold Harvey Cole
Monsanto Co.
800 N. Lindbergh Blvd.
St. Louis, MO 63167   (314) 694-3131

Ronald N. Compton
Summers, Compton, Wells
& Hamburg, P.C.
8909 Ladue Road
St. Louis, MO 63124   (314) 991-4999

Timothy H. Courson
Mc Donnell Douglas Corporation
P.O. Box 516
St. Louis, MO 63166   (314) 234-6052

William B. Cunningham, Jr.
Polster, Polster & Lucchesi
763 S. New Ballas Rd.
St. Louis, MO 63141   (314) 872-8118

Paul Michael Denk
763 S. New Ballas Rd.
St. Louis, MO 63141   (314) 872-8136

Samuel Digirolamo
Haverstock, Garrett & Roberts
611 Olive Street
St. Louis, MO 63101   (314) 241-4427

Rita E. Downard
Mallinckrodt, Inc.
675 Mc Donnell Blvd.
P.O. Box 5840
St. Louis, MO 63134   (314) 895-2909

William Harry Duffey
Monsanto Co.
Patent Dept.
800 N. Lindbergh Blvd.
St. Louis, MO 63167   (314) 694-3128

Wayne R. Eberhardt
Monsanto Agricultural Co.
800 N. Lindbergh Blvd.
St. Louis, MO 63167   (314) 694-3412

Rey Eilers
7419 Somerset
St. Louis, MO 63105   (314) 727-1217

Lawrence E. Evans, Jr.
Caruthers, Herzog, Crebs
& Mc Ghee
555 Washington Avenue
6th Floor
St. Louis, MO 63101   (314) 231-6700

Grace J. Fishel
Security Pl za Bldg., Ste. 100
929 Fee Road
Suite 100
St. Louis, MO 63043   (314) 878-0440

Donald J. Fitzpatrick
Interco Incorporated
101 S. Hanley Road
St. Louis, MO 63105   (314) 863-1100

Stanley Nelson Garber
Sherwood Medical Company
1831 Olive St.
St. Louis, MO 63103   (314) 621-7788

Robert M. Garrett
Haverstock, Garrett & Roberts
611 Olive Street
Suite 1610
St. Louis, MO 63101   (314) 241-4427

Peter S. Gilster
Kalish & Gilster
1614 Paul Brown Bldg.
818 Olive St.
St. Louis, MO 63101   (314) 436-1331

Michael E. Godar
Senniger, Powers, Leavitt
& Roedel
One Metropolitan Square
16th Floor
St. Louis, MO 63102   (314) 231-5400

Lynden Neal Goodwin
Guilfoil Petzall & Shoemake
100 North Broadway
St. Louis, MO 63102   (314) 241-6890

Guy Randall Gosnell#
Mc Donnell Douglas Corp.
P.O. Box 516
M.C. 1001240
St. Louis, MO 63166   (314) 234-2759

Edward P. Grattan
Monsanto Co.
800 N. Lindbergh Blvd.
St. Louis, MO 63167   (314) 694-3337

Jerome Arthur Gross
818 Olive Street
Suite 1610
St. Louis, MO 63101   (314) 241-7678

Richard E. Haferkamp
Rogers, Howell & Haferkamp
7777 Bonhomme
Suite 1700
St. Louis, MO 63105   (314) 727-5188

Charles Barker Haverstock
Haverstock, Garrett & Roberts
611 Olive St.
Suite 1610
St. Louis, MO 63101   (314) 241-4427

Edward J. Hejlek
Senniger, Powers, Leavitt
& Roedel
611 Olive Street
Suite 2050
St. Louis, MO 63101   (314) 231-0109

William H. Hellwege, Jr.#
9324 White Avenue
St. Louis, MO 63144   (314) 962-0057

Richard G. Heywood
Senninger, Powers, Leavitt
& Roedel
611 Olive Street
Suite 2050
St. Louis, MO 63101   (314) 231-0109

Virgil B. Hill
Ralston Purina Co.
Checkerboard Square
St. Louis, MO 63164   (314) 982-2164

Ronald W. Hind
Cohn, Powell & Hind, P.C.
7700 Clayton Rd.
Suite 103
St. Louis, MO 63117   (314) 645-2442

Dennis R. Hoerner, Jr.
Monsanto Company - BB 4F
700 Chesterfield Village Pkwy.
St. Louis, MO 63198   (314) 537-6099

Arthur Eugene Hoffman
Monsanto Co.
800 N. Lindbergh Blvd.
St. Louis, MO 63166   (314) 694-2714

John M. Howell
Rogers, Howell & Haferkamp
7777 Bonhomme
Suite 1700
St. Louis, MO 63105   (314) 727-5188

Lawrence J. Hurst
Ralston Purina Co.
835 S. 8th St.
St. Louis, MO 63188   (314) 982-2307

Kurt F. James
Senniger, Powers, Leavitt
& Roedel
One Metropolitan Sq.
16th Floor
St. Louis, MO 63102   (314) 231-5400

Frank B. Janoski
Coburn, Croft & Putzell
One Mercantile Center
Suite 2900
St. Louis, MO 63101   (314) 621-8575

Ralph W. Kalish
Kalish & Gilster
818 Olive Street
Suite 1614
St. Louis, MO 63101   (314) 436-1331

Neal Kalishman
221 Brooktrail Court
St. Louis, MO 63141   (314) 997-3236

Gene B. Kartchner
Sherwood Medical Company
1915 Olive Street
St. Louis, MO 63103   (314) 621-7788

J. Timothy Keane
G.D. Searle & Company
Monsanto Company
700 Chesterfield Village Pkwy.
St. Louis, MO 63198   (314) 537-6983

Thomas E. Kelley
Monsanto Company - G3NV
800 N. Lindbergh Blvd.
St. Louis, MO 63167   (314) 694-8494

Patrick D. Kelly
Haverstock, Garret & Roberts
611 Olive
Suite 1610
St. Louis, MO 63101   (314) 241-4427

John W. Kepler, III
Heller & Kepler
721 Emerson
Suite 569
St. Louis, MO 63141   (314) 993-4625

Curtis D. Kinghorn
Sherwood Medical Company
1915 Olive Street
St. Louis, MO 63103   (314) 241-5700

Roy J. Klostermann
Mallinckrodt, Inc.
675 Mcdonnell Blvd.
St. Louis, MO 63134   (314) 895-2915

Michael Kovac
Polster Polster & Lucchesi
763 South New Ballas Road
St. Louis, MO 63141   (314) 872-8118

William E. Lahey
Senniger, Powers, Leavitt
& Roedel
One Metropolitan Square
16th Floor
St. Louis, MO 63102   (314) 231-5400

John H. Lamming*
U.S. Dept of Army
4300 Goodfellow Blvd.
Attn: AMSAV-JP
St. Louis, MO 63120   (314) 263-2353

Donald G. Leavitt
Senniger, Powers, Leavitt
& Roedel
One Metropolitan Square
16th Floor
St. Louis, MO 63102   (314) 231-5400

Thomas B. Leslie
Haverstock, Garrett & Roberts
611 Olive Street
Suite 1610
St. Louis, MO 63101   (314) 241-4427

Linda L. Lewis
Monsanto Company
800 N. Lindbergh Avenue
St. Louis, MO 63167   (314) 694-1000

Roy A. Lieder
Gravely, Lieder and Woodruff
705 Olive Street
Suite 712
St. Louis, MO 63101   (314) 621-1457

Lawrence L. Limpus
Monsanto Co.
800 N. Lindbergh Blvd.
St. Louis, MO 63167   (314) 694-3145

James Charles Logomasini
1171 Big Bend Road
St. Louis, MO 63021   (314) 225-9408

Raymond C. Loyer, Sr.
Monsanto Company
800 N. Lindbergh Blvd.
St. Louis, MO 63167   (314) 694-3194

Lionel L. Lucchesi
Polster, Polster & Lucchesi
763 S. New Ballas Road
St. Louis, MO 63141   (314) 872-8118

Charles Emery Markham
7123 Stafford Court
St. Louis, MO 63123   (314) 843-3494

Stephen R. Matthews
Haverstock, Garrett & Roberts
611 Olive St., Suite 1610
St. Louis, MO 63101   (314) 241-4427

John E. Maurer
Senniger, Powers, Leavitt
& Roedel
One Metropolitan Sq.
16th Floor
St. Louis, MO 63102   (314) 231-5400

Richard C. Messnick
9174B Breeds Hill Lane
St. Louis, MO 63123   (314) 638-6160

Scott J. Meyer
Monsanto Co.
Bldg. A3SD
800 N. Lindbergh Blvd.
St. Louis, MO 63167   (314) 694-3117

Martha A. Michaels
Permea, Inc.
11444 Lackland Road
St. Louis, MO 63146   (314) 694-0212

Mc Pherson D. Moore
Armstrong, Teasdale, Schlafly,
Davis & Dicus
One Metropolitan Square
Suite 2600
St. Louis, MO 63102   (314) 621-5070

John J. Muller, II
Polster, Polster & Lucchusi
763 South New Ballas Road
St. Louis, MO 63141   (314) 872-8118

Keith F. Noe#
Senniger, Powers, Leavitt
& Roedel
One Metropolitan Sq.
16th Floor
St. Louis, MO 63102   (314) 231-5400

Alan H. Norman
Senniger, Powers, Leavitt
& Roedel
One Metropolitan Sq.
16th Floor
St. Louis, MO 63102   (314) 231-5400

Paul Leonard Passley
Monsanto Company
800 N. Lindbergh Blvd.
St. Louis, MO 63167   (314) 694-3192

Veo Peoples, Jr.
Peoples, Hale & Coleman
1221 Locust St.
Suite 200
St. Louis, MO 63103   (314) 231-9775

Donald Wilton Peterson, Sr.
Leydig, Voit & Mayer
One Mercantile Center
Suite 2900
St. Louis, MO 63101   (314) 621-9225

J. Philip Polster
Polster, Polster & Lucchesi
763 S. New Ballas Rd.
St. Louis, MO 63141   (314) 872-8118

Philip B. Polster
Polster, Polster & Lucchesi
763 S. New Ballas Rd.
Suite 160
St. Louis, MO 63141   (314) 872-8118

Irving Powers
Senniger, Powers, Leavitt
& Roedel
One Metropolitan Square
16th Floor
St. Louis, MO 63102   (314) 231-5400

Rudyard Kent Rapp
7 Villa Coublay
St. Louis, MO 63131   (314) 567-5703

Edward H. Renner
7700 Clayton Road
Suite 103
St. Louis, MO 63117   (314) 645-2442

Sidney B. Ring
9391 Roosevelt Dr.
St. Louis, MO 63127   (314) 843-0937

Glenn K. Robbins
Robbins & Robbins
314 N. Broadway
St. Louis, MO 63102   (314) 421-5712

Glenn K. Robbins, II
Robbins & Robbins
314 N. Broadway
St. Louis, MO 63102   (314) 421-6010

Herbert Barton Roberts
Haverstock, Garrett & Roberts
611 Olive St., #1610
St. Louis, MO 63101   (314) 241-4427

John K. Roedel, Jr.
Senniger, Powers,
Leavitt & Roedel
One Metropolitan Square
16th Floor
St. Louis, MO 63102   (314) 231-5400

Edmund C. Rogers
Rogers, Howell
& Haferkamp
7777 Bonhomme Avenue
Suite 1700
St. Louis, MO 63105   (314) 727-5188

John T. Rogers
Rogers, Howell & Haferkamp
7777 Bonhomme Avenue
Suite 1700
St. Louis, MO 63105   (314) 727-5188

Joseph M. Rolnicki
Rogers, Howell & Haferkamp
7777 Bonhomme
Suite 1700
St. Louis, MO 63105   (314) 727-5188

Stuart N. Senniger
Senniger, Powers, Leavitt
& Roedel
One Metropolitan Sq.
16th Floor
St. Louis, MO 63102   (314) 231-5400

Richard H. Shear
Monsanto Co.
800 N. Lindbergh Blvd.
G3NV
St. Louis, MO 63167   (314) 694-3361

Frank D. Shearin
Monsanto Co.
800 N. Lindbergh Blvd.
A3SA
St. Louis, MO 63167   (314) 694-5656

Gordon F. Sieckmann
Monsanto Co.
800 N. Lindbergh Blvd.
St. Louis, MO 63017   (314) 694-8694

Charles E. Smith
Monsanto Company
Patent Dept.
800 N. Lindenbergh Blvd.
St. Louis, MO 63167   (314) 694-3497

Charlie Smith
Sherwood Medical Co.
1831 Olive Street
St. Louis, MO 63103   (314) 241-5700

Montgomery W. Smith
Sherwood Medical Co.
Law Dept.
1831 Olive Street
St. Louis, MO 63103   (314) 241-5700

Jonathan P. Soifer
Polster, Polster & Lucchesi
763 S. New Ballas Rd.
St. Louis, MO 63141   (341) 872-8118

Kenneth Solomon
Rogers, Howell & Haferkamp
7777 Bonhomme
Suite 1700
St. Louis, MO 63105   (314) 727-5188

Howard Cromwell Stanley
Monsanto Co.
800 N. Lindbergh Blvd.
A3SB
St. Louis, MO 63167   (314) 694-3291

Larry R. Swaney
Monsanto Co.
700 Chesterfield Village Pkwy.
St. Louis, MO 63198   (314) 537-6264

Stanley M. Tarter
Rogers, Howell & Haferkamp
7777 Bonhomme
Suite 1700
St. Louis, MO 63105   (314) 727-5188

Arthur H. Tischer*
Dept. of the Army
U.S Army Aviation Sys. Comm.
Attn: AMSAV-JP
4300 Goodfellow Blvd.
St. Louis, MO 63120   (314) 263-2353

Gregory E. Upchurch
Polster, Polster & Lucchesi
763 South New Ballas
Suite 160
St. Louis, MO 63141   (314) 872-8118

Sarah P. Vaz#
39 Villawood Lane
St. Louis, MO 63119   (314) 961-0477

Mark F. Wachter
Monsanto Company
800 N. Lindbergh
A3SE
St. Louis, MO 63167   (314) 694-8651

Edward R. Weber
Pope & Weber, P.C.
915 Olive Street
Suite 1017
St. Louis, MO 63101   (314) 241-8465

Robert W. Welsh
Ralston Purina Co.
Checkerboard Sq.
St. Louis, MO 63164   (314) 982-3065

Bryan K. Wheelock
Senniger, Powers, Leavitt
& Roedel
One Metropolitan Sq.
16th Floor
St. Louis, MO 63102   (314) 621-5070

Jack W. White*
U.S. Dept. of Health &
Human Services
8706 Manchester Rd.
St. Louis, MO 63144   (314) 961-6602

Norman L. Wilson, Jr.
9121 Pennent Lane
St. Louis, MO 63126   (314) 645-2442

Evan R. Witt
Mallinckrodt Medical, Inc.
675 Mc Donnell Blvd.
P.O. Box 5840
St. Louis, MO 63134   (314) 895-2000

Frederick M. Woodruff
Gravely, Lieder and Woodruff
705 Olive St.
Suite 712
St. Louis, MO 63101   (314) 671-1457

Terry A. Witthaus
Route 1, Box 245
Washington, MO 63090   (314) 239-2113

Montana

David A. Veeder
Veeder & Broeder, P.C.
303 N. Broadway. Ste. 805
First Bank Bldg.
Billings, MT 59103   (406) 248-9156

John C. Conover#
4580 Bannock Drive
Bozeman, MT 59715   (406) 586-0532

Richard C. Conover
404 First National Bank Bldg.
P.O. Box 1329
Bozeman, MT 59715   (406) 587-4240

Jerry D. Guenther
311 Third Street
P.O. Box 2328
Havre, MT 59501   (406) 265-2257

Paul J. Van Tricht
804 North Hoback
Helena, MT 59601   (406) 443-7821

William D. West
805 Franklin Mine Rd.
Helena, MT 59601   (406) 449-8941

Arthur L. Urban
Box 4045
Red Lodge, MT 59068   (406) 446-1585

Nebraska

Gene D. Watson
Nebraska Public Power Dist.
P.O. Box 499
1414 - 15th Street
Columbus, NE 68601   (402) 563-5566

Vincent Laurence Carney
P.O. Box 80836
125 S. 52nd St.
Lincoln, NE 68501   (402) 489-0377

Austin P. Dodge
225 N. Brown Ave.
Minden, NE 68959   (308) 832-1121

John A. Beehner
502 Scoular Bldg.
2027 Dodge St.
Omaha, NE 68102   (402) 341-2433

Kevin Lynn Copple
1225 North 53rd Street
Omaha, NE 68132   (402) 551-3579

David D. Jensen#
15752 Jackson Dr.
Omaha, NE 68118   (402) 334-1862

William A. Klumper
4903 California
Suite 7
Omaha, NE 68132   (402) 554-8484

Philip J. Lee
Regency One Building
Suite 401
10050 Regency Circle
Omaha, NE 68114   (402) 397-3869

George R. Nimmer
1613 Farnam Street
Suite 514
Omaha, NE 68102   (402) 342-3077

Hiram A. Sturges#
#3 Park Plaza, Suite B
105 North 31st Ave.
Omaha, NE 68131   (402) 345-8932

Sean Patrick Suiter
7101 Mercy Road
Suite 400
Omaha, NE 68106   (402) 393-2093

Dennis L. Thomte
Zarley, Mc Kee, Thomte,
Voorhees & Sease
1111 Comm. Fed. Tower
Omaha, NE 68124   (402) 392-2280

James D. Welch
10328 Pinehurst Avenue
Omaha, NE 68124   (402) 391-4448

Bernard G. Fehringer
R.R. 2, Box 76
Sidney, NE 69162   (308) 254-2028

Nevada

Archie M. Cooke#
96 Arrowhead Drive
Carson City, NV 89706   (311) 150-7331

David H.T. Wayment
Nevada Supreme Court
Central Staff
Capitol Complex
Carson City, NV 89710   (702) 687-5179

Ian F. Burns#
P.O. Box 210
Glenbrook, NV 89413   (702) 749-5303

Herbert C. Schulze
P.O. Box 6070
Incline Village, NV 89450   (702) 831-3700

Robert W. Bass#
P.O. Box 85035
Las Vegas, NV 89185   (702) 733-3834

George Vincent Eltgroth
3601 Cambridge Street
Suite 265
Las Vegas, NV 89109   (702) 796-3569

Philip D. Junkins#
3101 Plaza De Rosa
Las Vegas, NV 89102   (702) 367-7960

Edward Polosky
2655 S. Decatur Blvd.
Unit No. 1036
Las Vegas, NV 89102   (702) 871-3591

Edward John Quirk
Quirk, Tratos & Roethel
550 E. Charleston Blvd.
Suite D
Las Vegas, NV 89104   (702) 386-1778

Chrispin M. Rivera
Jimmerson & Davis, P.C.
701 E. Bridger Ave.
Suite 600
Las Vegas, NV 89101   (702) 388-7171

John E. Roethel
Quirk, Tratos & Roethel
550 E. Charleston Blvd.
Suite D
Las Vegas, NV 89104   (702) 386-1778

John Joseph Roethel
550 E. Charleston Blvd.
Suite D
Las Vegas, NV 89104   (702) 386-1778

Richard A. Bjur
University of Nevada, Reno
School of Medicine
Dept. of Pharmacology
Howard Medical Science Bldg.
Reno, NV 89557   (702) 784-4116

Brian C. Kelly
1 East Liberty Street
Suite 416
P.O. Box 750
Reno, NV 89504   (702) 786-4646

New Hampshire

Christopher E. Blank
21 Stowell
Bedford, NH 03110   (603) 472-2479

Chris A. Caseiro#
Frankliln Pierce Law Center
2 White Street
Concord, NH 03301   (603) 228-1541

Raymond J. Di Lucci
4 Bicentennial Square
Concord, NH 03301   (603) 224-2100

Thomas G. Field, Jr.
Franklin Pierce Law Center
2 White St.
Concord, NH 03301   (603) 228-1541

David C. Goldman#
5 1/2 Hanover St.
Concord, NH 03301   (603) 226-4938

Karl Francis Jorda
Franklin Pierce Law Center
2 White Street
Concord, NH 03301   (603) 228-1541

Paul D. Parnass#
21 Chesterfield Drive
Concord, NH 03301   (603) 225-3442

Robert Harvey Rines
Rines and Rines
81 North State St.
Concord, NH 03301   (603) 228-0121

William B. Ritchie
43 Jackson Street
Concord, NH 03301   (603) 225-5212

Robert Shaw
209 East Side Drive
Concord, NH 03301   (603) 228-8601

Frank C. Henry
Red Hill Pond Road
Ctr. Sandwich, NH 03227   (603) 284-7091

Robert Gladden Crooks
1 Hamel Drive
Durham, NH 03824   (603) 868-5451

David Malcolm Keay
P.O. Box 69
Georges Mills, NH 03751   (603) 763-9711

Rochelle S. Ferber#
P.O. Box 186
Gilmanton, NH 03237   (603) 267-7326

H. Art Turner#
368 High Street
Hampton, NH 03842   (603) 926-8047

Philip E. Anderson#
5 Taylor River Road
Hampton Falls, NH 03844   (603) 778-1896

Harold L. Greenwald#
80 Lyme Road
Apt. 324
Hanover, NH 03755   (603) 926-8047

John Adams Thierry
Murray Hill Road
Hill, NH 03243   (603) 744-3540

George W. Dishong
Bryant Road
Jaffrey, NH 03452   (603) 532-7206

Janine J. Weins#
160 Mechanic Street
Lebanon, NH 03766   (603) 448-0123

Michael J. Weins
31 Bank Street
Lebanon, NH 03766   (603) 448-1922

Lee A. Strimbeck
42 Cottage Street
Littleton, NH 03561   (603) 444-2919

Joseph E. Funk, Sr.
82 Chase Road
Londonderry, NH 03053   (603) 432-4242

George Ward Whitney
The House By the Lake
R.R. 2, Box 278A
Lyme, NH 03768   (603) 795-2572

Michael J. Bujold
Davis, Bujold & Streck
175 Canal Street
Manchester, NH 03101   (603) 624-9220

Anthony G. M. Davis
Davis, Bujold & Streck
175 Canal Street
Manchester, NH 03101   (603) 624-9220

Steven J. Grossman
Hayes, Soloway, Hennessey
& Hage, P.C.
175 Canal Street
Manchester, NH 03101   (603) 668-1400

Susan H. Hage
Hayes, Soloway, Hennessey
& Hage, P.C.
175 Canal Street
Manchester, NH 03101   (603) 668-1400

Oliver W. Hayes
Hayes, Soloway, Hennessey
& Hage, P.C.
175 Canal Street
Manchester, NH 03101   (603) 668-1400

William O. Hennessey
Hayes, Soloway, Hennessey
& Hage, P.C.
175 Canal Street
Manchester, NH 03101   (603) 668-1400

Stephen J. Sand
177 Boutwell St.
Manchester, NH 03102   (603) 622-8503

Norman Peter Soloway
Hayes, Soloway, Hennessey
& Hage, P.C.
175 Canal Street
Manchester, NH 03101   (603) 668-1400

Edgar O. Rost
Canal Street
R.R. 4, Box 267
Meredith, NH 03253   (603) 279-4970

Linda M. Chinn#
28 Teak Drive
Nashua, NH 03062   (603) 595-2509

Louis Etlinger
15 Apache Road
Nashua, NH 03063   (603) 883-3884

David W. Gomes
Lockheed Sanders Inc.
P.O. Box 868
Mail Stop NHQ 1-719B
Nashua, NH 03061   (603) 885-2643

Larry B. Dufault#
Dufault & Dufault
Main Street
P.O. Box 306
New London, NH 03257   (603) 526-4472

Marc R.K. Bungeroth
Battles & Bungeroth
Main St. Prof. Bldg.
P.O. Box 769
North Conway, NH 03860   (603) 356-6966

C. Yardley Chittick
R.R. 1, Box 390
Ossipee, NH 03864   (603) 522-3275

Vincent W. Youmatz
365 A Pembroke St.
Pembroke, NH 03275   (603) 485-9010

John S. Beulick
229 Broad Street
Portsmouth, NH 03801   (603) 749-8371

Theodore C. Virgil
RFD1 Box 147 E. Rumney Rd.
Rumney, NH 03266   (603) 786-9401

Robert E. Brunson#
General Electric Co.
130 Main Street
Somersworth, NH 03878   (603) 749-8371

New Jersey

W. Patrick Quast
1 De Mercurio Drive
Allendale, NJ 07401   (201) 327-0006

Adel A. A. Ahmed
Cedar Grove Road
Box 68
Annandale, NJ 08801   (201) 735-4791

Kenneth R. Walton#
Exxon Research & Engineering Co.
Route 22 East
Annandale, NJ 08801   (201) 730-3071

Albert J. Mrozik, Jr.
1207 Bond Street
Asbury Park, NJ 07712   (201) 774-7987

Richard D. Foggio#
6328 Ocean Drive
Avalon, NJ 08202   (609) 967-4042

Lucian C. Canepa
15 Fieldstone Drive
Basking Ridge, NJ 07920   (201) 766-1726

Seymour E. Hollander
AT&T
131 Morristown Road
Basking Ridge, NJ 07920   (908) 204-8520

William T. Hough
312 South Finley Avenue
Basking Ridge, NJ 07920   (201) 766-3871

Joseph Louis Lazaroff
AT&T
131 Morristown Rd
Room B2090
Basking Ridge, NJ 07920   (908) 204-8410

Donald G. Marion
209 Madisonville Road
Basking Ridge, NJ 07920   (201) 766-5600

John Patrick Mc Donnell
AT&T
Rm. A2022
131 Morristown Rd.
Basking Ridge, NJ 07920   (908) 204-8400

Deborah L. Mellott
247 Penns Way
Basking Ridge, NJ 07920   (201) 580-0823

Rosemary A. Ryan
AT&T
131 Morristown Rd.
Room B2070
Basking Ridge, NJ 07920   (908) 204-8413

Donnie E. Snedeker
AT&T
131 Morristown Road
Room B2008
Basking Ridge, NJ 07920   (908) 204-8552

Lionel Norman White
35 Crest Drive
Basking Ridge, NJ 07920   (908) 647-4096

Donald F. Wohlers
30 Archgate Road
Basking Ridge, NJ 07920   (201) 647-4424

Gloria K. Koenig
867 Broadway
P.O. Box 226
Bayonne, NJ 07002   (201) 436-2247

Michael Y. Epstein
359 Griggstown Road
Belle Mead, NJ 08502   (201) 359-8453

Richard S. Roberts
24 Camden Road
Belle Mead, NJ 08502   (201) 359-2980

Frank Cozzarelli, Jr.
Cozzarelli, Mautone, Nardachone
& Mc Tigue
286 Union Avenue
Belleville, NJ 07109   (201) 751-4100

Charles F. Gunderson#
1622 N. Marconi Rd.
Belmar, NJ 07719   (201) 681-0464

Glen Erin Books#
AT&T
1 Oak Way
Berkeley Heights, NJ 07922   (201) 771-2244

William L. Keefauver
AT&T Bell Laboratories
1 Oak Way
Room 4ed106
Berkeley Heights, NJ 07922   (201) 771-2233

John Leary
AT&T
1 Oak Way
Berkeley Heights, NJ 07922   (201) 771-3232

Thomas Arthur Lennox
P.O. Box 127
100 State Hwy 73
Berlin, NJ 08009   (609) 767-6767

David A. Frank
12 Charlotte Hill Drive
P.O. Box 899
Bernardsville, NJ 07924   (609) 767-6767

Mary E. Porter
99 Mine Mount Rd.
Bernardsville, NJ 07924   (908) 766-9655

Sandra Gusciora Field
137 Raab Avenue
Bloomfield, NJ 07003   (908) 766-9655

Paul R. Gauer
56 Watsessing Avenue
Bloomfield, NJ 07003   (201) 743-7050

Richard N. Miller
188 Garner Avenue
Bloomfield, NJ 07003   (201) 338-4660

Sidney Shaievitz
Shaievitz & Berowitz
554 Bloomfield Avenue
Bloomfield, NJ 07003   (201) 743-7753

Clifford G. Frayne
44 Princeton Avenue
Brick Town, NJ 08723   (201) 840-9595

Salvatore J. Abbruzzese
Thomas & Betts Corporation
1001 Frontier Road
Bridgewater, NJ 08807   (201) 707-2367

Charles H. Davis, Sr.
936 Brown Road
Bridgewater, NJ 08807   (201) 722-7176

Jane Eileen Gennaro
National Starch & Chemical Co.
Finderne Ave.
P.O. Box 6500
Bridgewater, NJ 08807   (908) 685-5000

John J. Kissane
AT&T
55 Corporate Drive
Bridgewater, NJ 08807   (201) 658-6850

Robert M. Rodrick
Thomas & Betts Corp.
1001 Frontier Road
Bridgewater, NJ 08807   (201) 707-2364

Herbert M. Shapiro
92 Chelsea Way
Bridgewater, NJ 08807   (201) 725-2584

Louis J. Virelli, Jr.
National Starch & Chem. Co.
10 Finderne Avenue
Bridgewater, NJ 08807   (908) 685-5197

Walter Weick
283 Farmer Road
Bridgewater, NJ 08807   (201) 526-1329

Jesse Woldman
Thomas and Betts Corp.
1001 Frontier Rd.
Bridgewater, NJ 08807   (908) 707-2365

Eugene Zagarella, Jr.
National Starch & Chemical Co.
10 Finderne Ave.
P.O. Box 6500
Bridgewater, NJ 08807   (908) 685-5433

Edgar W. Adams, Jr.
7 Woodland Road
Brookside, NJ 07926   (201) 543-4606

J. Llewellyn Mathews
Apell & Mathews, P.C.
P.O. Box 95
57 Lakehurst Rd.
Browns Mills, NJ 08015   (609) 893-3122

Joseph F. Flayer
389 Route 46
Budd Lake, NJ 07828   (201) 691-9000

Ellen T. Dec#
RD 3, Box 179
Marigold Lane
Califon, NJ 07830   (201) 691-9000

John James Gallagher, Sr.
Hudson Maritime Services, Ltd.
800 Cooper Street
Camden, NJ 08102   (609) 342-7500

Ralph Thomas Lilore
163 Eileen Drive
Cedar Grove, NJ 07009   (201) 777-8876

Elizabeth Anne Bellamy#
Berlex Laboratories, Inc.
110 E. Hanover Ave.
Cedar Knolls, NJ 07927   (201) 540-8700

George W. Johnston, Jr.
100 Poplar Drive
Cedar Knolls, NJ 07927   (201) 539-7835

John Joseph Archer
57 Elmwood Ave.
Chatham, NJ 07928   (201) 635-8921

Richard A. Craig
28 Mountain View Road
Chatham, NJ 07928   (201) 635-7761

David L. Davis
27 Pembrooke Rd.
Chatham, NJ 07928   (201) 635-1490

Fong S. Lin
22 Edgehill Avenue
Chatham, NJ 07928   (201) 635-8142

Patrick J. Osinski
17 Clark Street
Chatham, NJ 07928   (201) 635-8274

Arthur Joseph Torsiglieri
2 Linden Lane
Chatham, NJ 07928   (201) 377-4321

Nicholas J. De Benedictis
1910 Morris Dr.
Cherry Hill, NJ 08003   (609) 795-1436

Charles Fredric Duffield
Suite 218
409 Route 70 East
Cherry Hill, NJ 08034   (609) 428-5338

Arnold Golden
108 Kingsdale Ave.
Cherry Hill, NJ 08003   (609) 424-2016

Harry Warren Hargis, III#
116 Fenwick Road
Cherry Hill, NJ 08034   (609) 428-6808

Norman E. Lehrer
1205 North Kings Highway
Cherry Hill, NJ 08034   (609) 429-4100

Philip O. Post
25 Appley Court
Cherry Hill, NJ 08002   (609) 667-7673

John H. Scarborough
Montano, Summers, Mullen, Manuel,
Owens & Gregorio
2 Executive Campus, Ste. 400
Route 70 & Cuthbert Blvd.
Cherry Hill, NJ 08002   (609) 665-9400

Robert D. Thompson
Pennington & Thompson, P.C.
135 Woodcrest Road
Cherry Hill, NJ 08003   (609) 795-0882

Jacob Trachtman
114 Rue Du Bois
Cherry Hill, NJ 08003   (609) 354-0967

Arthur M. Suga
107 Monticello Drive
Cinnaminson, NJ 08077   (609) 829-4693

Gary F. Danis#
69 Acorn Drive
Clark, NJ 07066   (201) 388-4872

Philip R. Arvidson
BASF Corporation
1255 Broad St.
Clifton, NJ 07015   (201) 365-3692

Cynthia Berlow#
414 Dwas Line Road
Clifton, NJ 07012   (201) 773-9568

Evelyn Berlow#
414 Dwas Line Road
Clifton, NJ 07012   (201) 773-9568

Anthony Frank Cuoco
66 Mount Prospect Avenue
Clifton, NJ 07013   (201) 779-0833

Joseph A. Giampapa
1054 Clifton Avenue
Clifton, NJ 07013   (201) 778-3203

Siegmar Silber
Silber & Cuoco
66 Mount Prospect Avenue
Clifton, NJ 07013   (201) 779-2580

Robert F. Tavares
Givaudan Corporation
100 Delawanna Ave.
Clifton, NJ 07015   (201) 363-8281

Linda A. Vag#
Givaudan Corporation
100 Delawanna Ave.
Clifton, NJ 07015   (201) 365-8165

John J. Herguth
Foster Wheeler Corporation
Perryville Corporate Park
Clinton, NJ 08809   (201) 730-4240

Marvin A. Naigur
Foster Wheeler Corporation
Perryville Corporate Park
Clinton, NJ 08809   (201) 730-4059

Robert J. Seman
Foster Wheeler Corporation
Law Dept. 275
Perryville Corporate Park
Clinton, NJ 08809   (201) 535-2269

Charles L. Willis
Foster Wheeler Corp.
Perryvile Corporate Park
Clinton, NJ 08809   (201) 730-4066

Carl P. Steinhauser#
11 Tallwood Court
Columbus, NJ 08022   (609) 298-3464

Rudolph J. Jurick
1 Canfield Terrace
Convent Station, NJ 07961   (201) 538-9248

C. Lance Marshall, Jr.
19 Piedmont Drive Road
Cranbury, NJ 08512   (609) 799-1306

Ellen K. Park
9 Kehoe Ct.
Cranbury, NJ 08512   (609) 799-1306

Homer J. Hall#
Rutgers University
SCILS
Cranford, NJ 07016   (201) 276-4311

Raphael A. Monsanto
Rohm & Monsanto
512 Springfield Avenue
Cranford, NJ 07016   (201) 276-3344

Benita J. Rohm
Rohm & Monsanto
512 Springfield Avenue
Cranford, NJ 07016   (201) 276-3344

Boris M. Pismenny#
14 Park Avenue
Cresskill, NJ 07626   (201) 567-7954

Frank A. Jones#
Chicopee Mfg. Co.
Research Div.
2351 U.S. Route 130
P.O. Box 940
Dayton, NJ 08810   (201) 274-3175

Edward F. Costigan*
U.S. Dept. of Army
H Q, Arnament, Munitions & Chem. Com.
AMSMC-GCL(D), Bldg. 3
Dover, NJ 07801   (201) 724-6594

Michael C. Sachs*
U.S. Army AMCCOM
Attn: AMSMC-GCL(D)
Picatinny Arsenal
Dover, NJ 07806   (201) 724-6595

Aurora A. Legarda
145 Lenox Avenue
Dumont, NJ 07628   (201) 384-9599

Morton Chirnomas
336 Second St.
Dunellen, NJ 08812   (908) 752-3598

Lawrence C. Edelman
3 Buffalo Run
East Brunswick, NJ 08816   (201) 251-3809

Joel F. Spivak
36 Yorktown Road
East Brunswick, NJ 08816   (201) 257-6635

Joseph J. Borovian, Sr.#
Sandoz Corporation
59 Route 10
East Hanover, NJ 07936   (201) 386-8532

Thomas C. Doyle#
Sandoz, Inc.
59 Route 10
East Hanover, NJ 07936   (201) 386-8177

Diane E. Furman
Sandoz Corporation
59 Route 10
Bldg. 418
East Hanover, NJ 07936   (201) 503-7332

Robert S. Honor
Sandoz Corporation
59 Route 10
East Hanover, NJ 07936   (201) 503-8474

Melvyn M. Kassenoff
Sandoz Pharmaceuticals Corp.
Pat. & Tdmk. Dept.
59 Route 10
East Hanover, NJ 07936   (201) 503-8477

Richard Kornutik
Nabisco Brands, Inc.
200 De Forest Avenue, TC/02
East Hanover, NJ 07936   (201) 503-4777

Thomas O. Mc Govern, Sr.
Sandoz, Inc.
59 Route 10
East Hanover, NJ 07936   (201) 503-8480

Gerald D. Sharkin
Sandoz, Inc.
59 Route 10
East Hanover, NJ 07936   (201) 503-8483

Richard E. Vila
Sandoz Corporation
59 Route 10
East Hanover, NJ 07936   (201) 503-7852

James H. Callwood
250 Prospect Street
East Orange, NJ 07017   (201) 675-1535

Hamlet E. Goore, Jr.
15 Prospect Street
East Orange, NJ 07017   (201) 674-5555

Martin Sachs
Sachs & Sachs, P.A.
614 U.S. #130
P.O. Box 968
East Windsor, NJ 08520   (609) 448-2700

Samuel L. Sachs
Sachs & Sachs, P.A.
614 U.S. #130
P.O. Box 968
East Windsor, NJ 08520   (609) 448-2700

Bill C. Giallovrakis
40 South Street
Eatontown, NJ 07724   (201) 542-2700

James J. Farrell
Lever Brothers Co.
Research Center
45 River Road
Edgewater, NJ 07020   (201) 943-7100

Milton L. Honig
Lever Brothers Company
45 River Road
Edgewater, NJ 07020   (201) 943-7100

A. Kate Huffman
Unilever United States, Inc.
45 River Road
Edgewater, NJ 07020   (201) 943-7100

Ronald A. Koatz
Unilever United States, Inc.
45 River Road
Edgewater, NJ 07020   (201) 943-7100

Matthew J. Mc Donald
Lever Brothers Company
Res. & Dev. Center
45 River Road
Edgewater, NJ 07020   (201) 943-7100

Gerard J. Mc Gowan, Jr.
Lever Brothers Company
45 River Road
Edgewater, NJ 07020   (201) 943-7100

Rimma Mitelman#
Unilever United States, Inc.
45 River Road
Edgewater, NJ 07020   (201) 943-7100

Mary A. Appollina#
Plevy & Selitto
146 Route 1, North
Edison, NJ 08817   (201) 572-5858

Omri M. Behr
Behr & Adams
325 Pierson Avenue
Edison, NJ 08837   (201) 494-5240

Joshua L. Cohen
Plevy & Selitto
146 Route 1, North
Edison, NJ 08817   (908) 572-5858

Robert A. Green
11 Perry Road
Edison, NJ 08817   (201) 572-0368

Markus Matzner#
23 Marshall Drive
Edison, NJ 08817   (908) 563-5398

Inez L. Moselle
Engelhard Corporation
33 Wood Avenue South
Edison, NJ 08818   (201) 321-5120

Arthur L. Plevy
146 Route 1, North
Edison, NJ 08817   (201) 572-5858

Ralph W. Selitto, Jr.
Plevy & Selitto
146 Route 1, North
Edison, NJ 08817   (201) 572-5858

Paul F. Swift
Plevy & Selitto
146 Route 1, North
Edison, NJ 08817   (201) 572-5858

James M. Nickels
280 Overlook Ave.
Elberon, NJ 07740   (201) 222-6457

Richard L. Cannaday
47 W. Grand St.
Elizabeth, NJ 07202   (201) 355-0499

Jack Gerber
60 Watson Avenue
Apt. 1-H
Elizabeth, NJ 07202   (201) 351-9098

Martin M. Glazer
Glazer & Kamel
40 Parker Road
Elizabeth, NJ 07208   (908) 354-2400

Alan M. Kamel
Glazer & Kamel
40 Parker Road
Elizabeth, NJ 07208   (908) 354-2400

Reuben Miller
531 Trotters Lane
Elizabeth, NJ 07208   (908) 354-2400

Barbara T. D'Avanzo
CPC International Inc.
P.O. Box 8000
International Plaza
Englewood Cliffs, NJ 07632   (201) 894-2126

Ellen P. Trevors
CPC International Inc.
International Plaza
P.O. Box 8000
Englewood Cliffs, NJ 07632   (201) 894-2716

Norbert Ederer
1-17 35th St
Fair Lawn, NJ 07411   (201) 797-5815

Frederick I. Levine
38-27 Wilson Street
Fair Lawn, NJ 07410   (201) 797-8014

Nicholas Anthony Gallo, III
FCS Laboratories, Inc.
Rd 3 Box 95
Darts Mill Vlg.
Flemington, NJ 08822   (201) 782-3353

Kenneth P. Glynn
Glynn, Byrnes & Schaul
6 Park Avenue
Flemington, NJ 08822 · (908) 788-0077

Merle V. Hoover#
174 Thatchers Hill Road
Flemington, NJ 08822 · (201) 782-7624

Joseph J. Allocca
Exxon Res. & Eng. Co.
P.O. Box 390
Florham Park, NJ 07932 · (201) 765-1446

John W. Ditsler
Exxon Res. & Eng. Co.
P.O. Box 390
Florham Park, NJ 07932 · (201) 765-1595

Joseph J. Dvorak
Exxon Res. & Eng. Co.
P.O. Box 390
180 Park Avenue
Florham Park, NJ 07932 · (201) 765-2255

Ronald David Hantman
Exxon Res. & Eng. Co.
P.O. Box 390
Florham Park, NJ 07932 · (201) 765-3647

Jerome Edward Luecke
Exxon Res. & Eng. Company
180 Park Avenue
Florham Park, NJ 07932 · (201) 765-2321

Richard E. Nanfeldt
Exxon Res. & Eng. Co.
P.O. Box 390
Florham Park, NJ 07932 · (201) 765-1534

Roy John Ott
Exxon Res. & Engineering Co.
P.O. Box 390
Florham Park, NJ 07932 · (201) 765-1549

F. Donald Paris
Exxon Res. & Eng. Co.
P.O. Box 390
180 Park Ave.
Florham Park, NJ 07932 · (201) 765-1550

Paul E. Purwin
Exxon Res. & Eng. Co.
180 Park Avenue
Florham Park, NJ 07932 · (201) 765-3643

Jay Simon
Exxon Res. and Eng. Co.
P.O. Box 390
180 Park Avenue
Florham Park, NJ 07932 · (201) 765-1580

Arthur F. Whitley
24 Puddingstone Way
Florham Park, NJ 07932 · (201) 377-7433

Maurice Leander Williams
Exxon Res. & Eng. Co.
180 Park Ave.
P.O. Box 390
Florham Park, NJ 07932 · (201) 765-1601

Erwin W. Pfeifle
39 Jackson Court
Fords, NJ 08863 · (201) 225-1920

Eugene M. Whitacre
1231 Orlando Drive
Forked River, NJ 08731 · (609) 971-0161

Richard P. Dyer
1530 Palisade Avenue
Apt. 23R
Fort Lee, NJ 07024 · (201) 592-6019

Lester Horwitz
6 Horizon Road
Fort Lee, NJ 07024 · (201) 224-0244

Jack Matalon
Sun Chemical Corp.
P.O. Box 1302
222 Bridge Plaza South
Fort Lee, NJ 07024 · (201) 224-4600

Constantine A. Michalos
330 New York Avenue
Fort Lee, NJ 07024 · (201) 944-9330

William H. Anderson*
U.S. Dept of Army
Communications-Electronics Command
AMSEL-LG-L
Fort Monmouth, NJ 07703 · (908) 532-4112

John Roger Drayer#*
U.S. Army Communications-Electronics
    Command
Attn: AMSEL-ED-CC-1
Fort Monmouth, NJ 07703 · (201) 532-1424

James J. Drew*
U.S. Army Communications-Electronics
    Command
Attn: AMSEL-LG-LP
Fort Monmouth, NJ 07703 · (201) 532-3384

Shelley R. Econom*
U.S. Army Communications
Electronics Command
Attn: AMSEL-LG-JAA
CECOM Office Bldg.
Fort Monmouth, NJ 07703 · (201) 532-5885

Victor J. Ferlise*
U.S. Army Communications
Electroncs Command
Attn: AMSEL-LG
Fort Monmouth, NJ 07703 · (201) 532-3045

Roy E Gordon#*
U.S. Army Communications
Electronics Command
Attn: DRSEL-LG-LS
Fort Monmouth, NJ 07703 · (201) 532-3187

Sheldon Kanars*
U.S. Army Electronics Comm.
AMSEL-LF-L
Fort Monmouth, NJ 07703 · (201) 532-4112

Ann M. Knab
U.S. Dept. of Army
Attn: AMSEL-LG-LS
Fort Monmouth, NJ 07703 · (201) 532-3373

Robert Allen Maikis*
Dept. of Army
U.S. Army Communications
Electronics Command
Attn: AMSEL-LG-LS
Fort Monmouth, NJ 07703 · (201) 532-4112

Judith A. Michal#*
U.S. Army
AMSEL-LG-JA
Fort Monmouth, NJ 07703 · (201) 532-3187

Kenneth J. Murphy*
U.S. Army Electronic Comm.
CECOM-AMSEL-LP-LG
Fort Monmouth, NJ 07703 · (201) 532-3062

John M. O Meara*
U.S. Army
Communications-Electronics Command
Attn: AMSEL-LG-LS
Fort Monmouth, NJ 07703 · (201) 532-3062

Kalman S. Pollen*
U.S. Army
CECOM
AMSEL-PC-C-E-SP-3
Fort Monmouth, NJ 07703 · (201) 532-5691

Michael J. Zelenka#*
U.S. Dept. Army
USACECOM Legal Office
Intel. Prop. Law Divison
Fort Monmouth, NJ 07703 · (201) 532-4112

Paul F. Koch, II
49 Main Street
Franklin, NJ 07416 · (201) 827-3020

Richard E. Brown#
Becton Dickinson & Company
One Becton Drive
Franklin Lakes, NJ 07417 · (201) 848-7110

Robert Joseph Dockery
Becton Dickinson & Co.
One Becton Drive
Franklin Lakes, NJ 07417 · (201) 848-7104

Robert Paul Grindle
Becton Dickinson & Co.
One Becton Drive
Franklin Lakes, NJ 07417 · (201) 848-7115

Robet M. Hallenbeck
Becton, Dickinson & Company
One Becton Drive
Franklin Lakes, NJ 07417 · (201) 848-7114

Gunter W. Koch
Becton, Dickinson & Co.
One Becton Drive
Franklin Lakes, NJ 07417 · (201) 848-7112

Richard J. Rodrick
Becton, Dickinson & Co.
One Becton Drive
Franklin Lakes, NJ 07417 · (201) 848-7116

Brian K. Stierwalt
Becton Dickinson & Co.
One Becton Drive
Franklin Lakes, NJ 07417 · (314) 694-1000

Nanette Semrau Thomas
Becton, Dickinson & Co.
One Becton Drive
Franklin Lakes, NJ 07417   (201) 848-7049

John L. Voellmicke
Becton, Dickinson & Co.
One Becton Dr.
Franklin Lakes, NJ 07417   (201) 848-7111

Ira M. Adler
Wemrock Professional Mall
Suite E-8
509 Stillwells Corner Rd.
Freehold, NJ 07728   (201) 577-9090

Patrick J. Pinto#
37 West Main Street
Freehold, NJ 07728   (201) 431-7662

Sylvia Jean Chin
470 Long Hill Road
Gillette, NJ 07933   (201) 647-2283

Ronald G. Goebel
17 Lorraine St.
Glen Ridge, NJ 07028   (201) 743-3136

Levonna Herzog#
16 Evergreen Court
Glen Ridge, NJ 07028   (201) 748-4329

Howard E. Thompson, Jr.
51 Ridgewood Ave.
Glen Ridge, NJ 07028   (201) 743-6364

Victor D. Behn
41 Beech Rd.
Glen Rock, NJ 07452   (201) 444-4853

Roger L. Fidler
400 Grove Street
Glen Rock, NJ 07452   (201) 445-8862

Richard C. Billups
Klauber & Jackson
Continental Plaza
411 Hackensack Ave.
Hackensack, NJ 07601   (201) 487-5800

Mitchell D. Bittman
Sequa Corporation
Three University Plaza
Hackensack, NJ 07601   (201) 343-1122

Richard M. Goldberg
25 East Salem Street
Fourth Floor
Hackensack, NJ 07602   (201) 343-7775

David A. Jackson
Klauber & Jackson
Continental Plaza
411 Hackensack Ave.
Hackensack, NJ 07601   (201) 487-5800

Arthur Jacob
Samuelson and Jacob
25 East Salem Street
P.O. Box 686
Hackensack, NJ 07602   (201) 488-8700

George M. Kachmar
Ostrager, Fieldman, Zucker,
Kachmar & Nirenberg
401 Hackensack Avenue
Hackensack, NJ 07601   (201) 488-7771

Stefan J. Klauber
Klauber & Jackson
Continental Plaza
411 Hackensack Avenue
Hackensack, NJ 07601   (201) 487-5800

Barbara L. Renda
Klauber & Jackson
One University Plaza
Hackensack, NJ 07601   (201) 487-5800

Eugene G. Reynolds
Herten, Burstein, Sheridan
& Cevasco
Court Plaza North
25 Main Street
Hackensack, NJ 07601   (201) 342-6000

Cyrus D. Samuelson
Samuelson and Jacob
25 East Salem Street
P.O. Box 686
Hackensack, NJ 07602   (201) 488-8700

Kenneth J. Bossong
401 Briarwood Avenue
Haddonfield, NJ 08033   (609) 858-7071

Cary A. Levitt
Archer & Greiner, P.C.
One Centennial Square
Haddonfield, NJ 08033   (609) 354-3016

Stanley Howard Zeyher
660 Clinton Avenue
Haddonfield, NJ 08033   (609) 429-1063

Edward Perley Barthel
Chrysler Motors Corp.
12000 Chrysler Drive
Highland Park, NJ 48288   (313) 956-4644

John J. Jones
121 Orchard Avenue
Hightstown, NJ 08520   (609) 448-7061

Martin G. Meder#
121 Park Avenue
Hightstown, NJ 08520   (609) 448-6469

Robert J. Molnar
63 Large Avenue
Hillsdale, NJ 07642   (201) 666-1915

Oleg Edward Alber
AT&T Bell Laboratories
P.O. Box 679
Holmdel, NJ 07733   (201) 949-3158

Gerard A. De Blasi
AT&T Bell Laboratories
101 Crawfords Corner Road
P.O. Box 3030
Holmdel, NJ 07733   (908) 949-3578

Barry H. Freedman
5 White Rock Terrace
Holmdel, NJ 07733   (908) 888-5580

Eugene S. Indyk
AT&T Bell Laboratories
Room 3K-206
101 Crawfords Corner Road
P.O. Box 3030
Holmdel, NJ 07733   (908) 949-5857

Arthur B. Larsen
AT&T Bell Laboratories
Room 2E-320
Crawfords Corner Road
Holmdel, NJ 07733   (201) 949-7467

Frederick B. Luludis
AT&T Bell Laboratories
Crawfords Corner Road
Holmdel, NJ 07733   (201) 948-6008

John F. Moran
29 Seven Oaks Circle
Holmdel, NJ 07733   (201) 946-3339

Gregory C. Ranieri
AT&T Bell Laboratories
Crawfords Corner Rd.
Room 3K-230
Holmdel, NJ 07733   (201) 949-6559

Thomas Stafford
AT&T Bell Laboratories
3K234, Crawfords Corner Rd.
Holmdel, NJ 07733   (201) 949-5780

Eli Weiss
AT&T
P.O. Box 679
Crawfords Corner Rd.
Holmdel, NJ 07733   (201) 949-3147

Stephen W. White#
892 Holmdel Road
Holmdel, NJ 07733   (201) 946-8664

Joseph D. Lazar
395 Province Line Road
Hopewell, NJ 08525   (609) 466-3480

Stanley Dubroff
12 Cooper Drive
Howell, NJ 07731   (201) 363-3991

Jack H. Stanley#
726 Fernmere Ave.
Interlaken, NJ 07712   (201) 531-0472

Sidney S. Kanter
1064 Clinton Ave.
Irvington, NJ 07111   (201) 371-3030

Raymond F. Keller
Engelhard Corp.
101 Wood Ave.
Iselin, NJ 08830   (908) 205-5000

Peter A. Luccarelli, Jr.
Siemens Corporation
Intellectual Property Dept.
186 Wood Avenue South
Iselin, NJ 08830   (201) 321-3006

Jeffrey P. Morris
Siemens Corporation
186 Wood Avenue South
Iselin, NJ 08830   (201) 321-3930

Edward J. Whitfield
Engelhard Corp.
101 Wood Avenue
Iselin, NJ 08838   (908) 205-5917

Ira L. Zebrak
Siemens Corp.
186 Wood Avenue S.
Iselin, NJ 08830   (201) 321-3020

Thomas A. Gallagher
30-605 Newport Parkway
Jersey City, NJ 07310   (201) 653-4269

Emil Scheller
Schering-Plough Corp.
2000 Galloping Hill Rd.
P.O. Box 520
Kenilworth, NJ 07033   (201) 709-2850

Lawrence Paul Benjamin
123 Fairfield Road
Kingston, NJ 08528   (609) 924-4292

Alfred Charles Hill#
9 Shirley Terrace
Kinnelon, NJ 07405   (201) 838-2731

N. Paul Klaas
51 Hoot Owl Terrace
Smoke Rise
Kinnelon, NJ 07405   (201) 838-8546

Mary U. O Brien
15 Derrygally Circle
Kinnelon, NJ 07405   (201) 492-9503

Neil D. Edwards
7 Turnbridge Row
Lakehurst, NJ 08733   (201) 657-2699

Stephen M. Hoffman
Advance Developments Company
480 Oberlin Ave., S.
Lakewood, NJ 08701   (201) 364-8855

Suzanne E. Babajko#
9 Highfield Court
Lawrenceville, NJ 08648   (609) 844-0094

Harold Christoffersen
53 Merritt Drive
Lawrenceville, NJ 08648   (609) 882-9123

Morris J. Cohen
1 Ivy Glen Lane
Lawrenceville, NJ 08648   (609) 896-9036

Clarence C. Richard, Jr.
20 Sunset Road
Lawrenceville, NJ 08648   (609) 896-1690

David A. Saar
8 Bennington Drive
Lawrenceville, NJ 08648   (609) 896-4831

John Edward Wilson
466 Lyons Road
Liberty Corner, NJ 07938   (908) 647-4174

Mitchell E. Alter
Exxon Chemical Co.
1900 E. Linden Avenue
Linden, NJ 07036   (201) 474-2537

Davis B. Dwinell#
American Flange & Mfg. Co., Inc.
Company, Inc.
1100 W. Blancke Street
Linden, NJ 07036   (201) 862-5000

Harold Einhorn
Exxon Chemical Co.
1900 E. Linden Avenue
Linden, NJ 07036   (201) 474-2256

Myron B. Kapustij
Exxon Chemical Company
1900 E. Linden Avenue
Linden, NJ 07036   (201) 474-2537

Robert A. Maggio
Exxon Chemical Company
P.O. Box 710
1900 E. Linden Avenue
Linden, NJ 07036   (201) 474-2297

John J. Mahon, Jr.
Exxon Chemical Co.
P.O. Box 710
Linden, NJ 07036   (201) 474-2518

Jack B. Murray, Jr.
Exxon Chem. Co.
P.O. Box 710
1900 E. Linden Avenue
Linden, NJ 07036   (201) 474-2271

John E. Schneider
Exxon Chemical Co.
P.O. Box 710
1900 E. Linden Ave.
Linden, NJ 07036   (908) 474-2330

Emil Richard Skula
Exxon Chemical Co.
1900 East Linden Avenue
Linden, NJ 07036   (201) 474-2418

James H. Takemoto
Exxon Company
1400 Park Ave.
Linden, NJ 07036   (908) 474-7757

Vivienne T. White
Exxon Chemical
1900 E. Linden Ave.
P.O. Box 710
Linden, NJ 07036   (201) 474-3656

Morris Irwin Pollack
Litton Industries Corp.
275 Paterson Ave.
P.O. Box 444
2nd Flr. Rear
Little Falls, NJ 07424   (201) 256-5550

Robert F. Rotella
Litton Industries
275 Paterson Avenue
Box 444
Little Falls, NJ 07424   (201) 256-5550

Herbert A. Stern
Litton Inds., Inc.
275 Patterson Avenue
P.O. Box 444
Little Falls, NJ 07424   (201) 256-5550

H. Gordon Dyke
134 Point Rd.
Little Silver, NJ 07739   (201) 842-7156

Robert Lawrence Lehman
192 Winding Way
Little Silver, NJ 07739   (201) 741-1537

Stephen B. Coan#
72 Sykes Avenue
Livingston, NJ 07039   (201) 992-3133

Edward M. Fink
Bell Communications Research, Inc.
290 W. Mount Pleasant Ave.
Livingston, NJ 07039   (201) 740-6420

Allen N. Friedman
Bell Communications Research
290 W. Mount Pleasant Ave.
Livingston, NJ 07039   (201) 740-6160

Stephen M. Gurey
34 Scarsdale Dr.
Livingston, NJ 07039   (201) 533-0821

John T. Peoples
Bell Communications Res., Inc.
290 W. Mt. Pleasant Ave.
Livingston, NJ 07039   (201) 740-6155

Leonard C. Suchyta
Bellcore
290 West Mt. Pleasant Avenue
Room Lcc 2E-301
Livingston, NJ 07039   (201) 740-4132

Max Yarmovsky
64 Bryant Drive
Livingston, NJ 07039   (201) 992-9515

Mark L. Hopkins
17 Cambridge Drive
Long Valley, NJ 07853   (201) 876-9252

John H.C. Blasdale#
Schering-Plough Corporation
One Giralda Farms
Madison, NJ 07940   (201) 822-7398

Matthew Boxer
Schering-Plough Corp.
One Giralda Farms
Madison, NJ 07940   (201) 822-7362

Eric S. Dicker
Schering-Plough Corporation
One Giralda Farms
Madison, NJ 07940   (201) 822-7383

Norman C. Dulak
Schering-Plough Corp.
Pat. Law Dept.
One Giralda Farms
Madison, NJ 07940   (201) 822-7375

Robert A. Franks
Schering-Plough Inc.
Patents & Licensing
One Giralda Farms
Madison, NJ 07940   (201) 822-7000

Thomas D. Hoffman
Schering-Plough Corporation
Pat. Dept.
One Giralda Farms
Madison, NJ 07940   (201) 822-7379

Henry C. Jeanette
Schering-Plough Corp.
Pat. Dept.
One Giralda Farms
Madison, NJ 07940   (201) 822-7378

Steinar V. Kanstad
Schering - Plough Corporation
One Giralda Farms
Madison, NJ 07940   (201) 822-7373

Walter E. Kupper#
65 Barnsdale Road
Madison, NJ 07940   (800) 638-8537

Warrick Edward Lee, Jr.
Schering-Plough Corp.
Pat. Dept., M3W
One Giralda Farms
Madison, NJ 07940   (201) 822-7382

Anita W. Magatti
Schering-Plough Corp,
Pat. Dept. M-3W
One Giralda Farms
Madison, NJ 07940   (201) 822-7389

John J. Maitner
Schering-Plough Corporation
One Giralda Farms
Madison, NJ 07940   (201) 822-7358

Joseph T. Majka
Schering-Plough Corp.
One Giralda Farms
P.O. Box 1000
Madison, NJ 07940   (201) 822-7397

Edward H. Mazer
Schering-Plough Corporation
One Giralda Farms
Madison, NJ 07940   (201) 822-7303

James R. Nelson
Schering - Plough Corporation
Patent Dept., M3W
One Giralda Farms
Madison, NJ 07940   (201) 822-7376

Robert O. Nimtz
28 Stafford Drive
Madison, NJ 07940   (201) 377-4309

Robert Ira Pearlman
6 Coursen Way
Madison, NJ 07940   (201) 771-6433

Leonard R. Fellen
26 Plaza Nine
Manalapan, NJ 07726   (201) 431-0473

James A. Curley
10 Hoffman Street
Maplewood, NJ 07040   (201) 761-5966

Michael H. Wallach
36 New England Road
Maplewood, NJ 07040   (201) 762-4316

Charles Ira Brodsky
9 S. Main St.
Marlboro, NJ 07746   (609) 431-1333

Eric B. Janofsky
23 Marigold Lane
Marlboro, NJ 07746   (201) 577-8205

Ilene Lapidus Janofsy
23 Marigold Lane
Marlboro, NJ 07746   (201) 577-8205

Theodore J. Criares
1997 Wash. Valley Rd.
Martinsville, NJ 08836   (201) 356-1978

William J. Barber
Weingram & Zall
197 W. Spring Valley Avenue
Maywood, NJ 07607   (201) 843-6300

Donald M. Boles
Weingram & Zall
197 W. Spring Valley Avenue
Maywood, NJ 07607   (201) 843-6300

Michael R. Friscia
Weingram & Zall
197 W. Spring Valley Ave.
Maywood, NJ 07604   (201) 843-6300

Edward R. Weingram
Weingram & Zall
197 W. Spring Valley Avenue
Maywood, NJ 07607   (201) 843-6300

Michael E. Zall
Weingram & Zall
197 W. Spring Valley Avenue
Maywood, NJ 07607   (201) 843-6300

John A. Caruso
5 Franklin Road
Mendham, NJ 07945   (201) 543-7272

Edward K. Kaprelian
15 Lowery Lane
Mendham, NJ 07945   (201) 543-7011

William E. Ringle
Box 335
Metuchen, NJ 08840   (609) 395-8592

John Abraham Caccuro
Bell Telephone Labs.
200 Laurel Avenue
Middletown, NJ 07748   (201) 957-3284

Alan Huang#
4 Burdge Drive
Middletown, NJ 07748   (201) 957-3284

David R. Padnes
AT&T Bell Laboratories
200 Laurel Ave., Room 4P-348
Middletown, NJ 07748   (201) 957-3285

Martin Pfeffer
100 Dwight Road
Middletown, NJ 07748   (201) 671-4528

Ronald D. Slusky
AT&T Bell Laboratories
200 Laurel Avenue
Room 4P358A
Middletown, NJ 07748   (201) 957-3282

Samuel R. Williamson
AT&T Bell Laboratories
200 Laurel Ave.
Middletown, NJ 07748   (908) 957-6608

Henry Freeman
825 Ridgewood Road
Millburn, NJ 07041   (201) 376-0213

Joseph Patrick Kearns, Jr.
14 Circle Drive
Millington, NJ 07946   (201) 647-3171

Charles S. Phelan
51 Circle Drive
Millington, NJ 07946   (908) 647-9226

Amirali Y. Haidri
460 Bloomfield Avenue
Suite 200
Montclair, NJ 07042   (201) 744-5757

William V. Ebs
9 Westmorland Ave.
Montvale, NJ 07645   (201) 391-9154

Albert Lewis Gazzola#
22 Main Street
Montvale, NJ 07645   (201) 843-6300

Howard J. Newby
11 Sunnyside Dr.
Montvale, NJ 07645   (201) 391-4686

Donald C. Simpson
Simpson & Simpson, P.C.
126 Borton Landing Road
Moorestown, NJ 08057   (609) 234-9590

Mark D. Simpson
Simpson & Simpson, P.C.
126 Borton Landing Road
Moorestown, NJ 08057   (609) 234-9590

Michael B. Einschlag
48 Vista Drive
Morganville, NJ 07751   (201) 972-7795

Stephen I. Miller
9 Vista Drive
Morganville, NJ 07751   (201) 536-6990

William S. Wolfe
General Electric Co.
GESD
Marne Hwy., Bldg. 108-108
Morrestown, NJ 08057   (609) 722-4906

Richard S. Bullitt*
Warner-Lambert Co.
201 Tabor Road
Morris Plains, NJ 07950   (201) 540-4790

Charles A. Gaglia, Jr.
Warner-Lambert Company
201 Tabor Road
Morris Plains, NJ 07950   (201) 540-4401

Anne M. Kelly
Warner Lambert Co.
201 Tabor Rd.
Morris Plains, NJ 07950   (201) 540-2602

Barbara E. Kurys#
2467 Route 10 E., #39-6B
Morris Plains, NJ 07950   (201) 292-9754

Charles E. Lents
Warner-Lambert Co.
201 Tabor Rd.
Morris Plains, NJ 07950   (201) 540-2151

Thomas M. Marshall
Tmmnet Ltd.
Powder Mill Village
89 Patriots Road
Morris Plains, NJ 07950   (201) 326-1892

Maxwell A. Pollack#
121 Glenbrook Rd.
Morris Plains, NJ 07950   (201) 539-1724

Royal N. Ronning, Jr.
100 Flintlock Rd.
Morris Plains, NJ 07950   (201) 898-1701

Daniel A. Scola, Jr.
Warner - Lambert Company
Corporate Pat. Dept.
201 Tabor Road
Morris Plains, NJ 07950   (201) 540-5960

Thomas L. Adams
Behr & Adams
10 Park Place
Suite 307
Morristown, NJ 07853   (201) 538-4112

Melanie L. Brown
Allied-Signal Inc.
Columbia Rd. & Park Ave.
P.O. Box 2245R
Morristown, NJ 07960   (201) 455-4851

Ernest D. Buff
Allied Corporation
Columbia Rd. & Park Ave.
Morristown, NJ 07960   (201) 455-3445

Salvatore R. Conte
6 Quaker Ridge Road
Morristown, NJ 07960   (203) 267-9785

David P. Cooke
Allied-Signal Inc.
P.O. Box 2245R
Morristown, NJ 07960   (201) 455-2817

Roger H. Criss
Allied-Signal Inc.
Law Dept.
P.O. Box 2245R
Morristown, NJ 07960   (201) 455-4796

Julius J. Denzler
Schenck, Price, Smith
& King
10 Washington Street
Morristown, NJ 07960   (201) 539-1000

Glen M. Diehl
Allied-Signal Inc.
P. O. Box 2245R
Morristown, NJ 07960   (201) 455-3548

Edward D. Dreyfus
AT&T
100 Southgate Parkway
Morristown, NJ 07960   (201) 898-8547

Jay Philip Friedenson
Allied Signal Inc.
Law Dept.
P.O. Box 2245R
Morristown, NJ 07960   (201) 455-2037

Gerhard Helmut Fuchs
Allied Signal Inc.
P.O. Box 2245R
Morristown, NJ 07960   (201) 455-3451

Ronald Gould
Shanley & Fisher
131 Madison Avenue
Morristown, NJ 07960   (201) 285-1000

Gus Theodore Hampilos
Allied-Signal, Inc.
Columbia Road & Park Avenue
Morristown, NJ 07960   (201) 455-3453

Karen A. Harding
Allied - Signal Inc.
Law Dept.
P.O. Box 2245
Morristown, NJ 07962   (201) 455-3679

Eileen Heldmann
Riker, Danzig, Scherer, Hyland
& Perretti
One Speedwell Ave.
Headquarters Plaza
Morristown, NJ 07962   (201) 538-0800

Francis Bradford Henry
Mennen Co.
Hanover Ave.
Morristown, NJ 07960   (201) 631-9361

Patrick L. Henry
Allied-Signal Inc.
P.O. Box 2245R
Morristown, NJ 07960   (201) 455-4705

Erwin Koppel
Allied-Signal Inc.
P.O. Box 2245R
Morristown, NJ 07960   (201) 455-6186

Anthony Lagani, Jr.
Laughlin Markensohn Lagani
& Pegg
129 Headquarters Plaza
Morristown, NJ 07960   (201) 539-0080

Richard T. Laughlin
Ribis, Graham & Curtin
4 Headquarters Plaza
CN-1991
Morristown, NJ 07960   (201) 292-1700

William C. Long
26 De Hart St.
Morristown, NJ 07960   (201) 540-8470

Christopher N. Malvone
Allied-Signal Inc.
PO Box 1057R
Bldg. AB/2
Morristown, NJ 07960   (201) 455-2041

Roy H. Massengill
Allied-Signal Inc.
P.O. Box 2245R
Morristown, NJ 07960   (201) 455-5127

Steven Mc Cann#
Cuyler, Burk & Matthews
P.O. Box 1947
Morristown, NJ 07962   (201) 765-9500

Andrew N. Parfomak
Allied-Signal Inc.
Morristown, NJ 07960   (201) 455-3790

Donald B. Paschburg
Allied-Signal Inc.
P.O. Box 2245R
Morristown, NJ 07962   (201) 455-3415

Arthur J. Plantamura
10 Butterworth Dr.
Morristown, NJ 07960   (201) 267-8671

Jonathan Plaut
Allied Corporation
P.O. Box 1013R
Morristown, NJ 07960   (201) 455-6570

Gerard P. Rooney
Allied-Signal Inc.
P.O. Box 2245R
Columbia Rd. & Park Ave.
Morristown, NJ 07962   (201) 455-2502

Anthony J. Stewart
Allied Corporation
P.O. Box 2245R
Morristown, NJ 07960   (201) 455-4033

Richard C. Stewart#
Allied Corporation
P.O. Box 2245R
Columbia Rd. and Park Ave.
Morristown, NJ 07962   (201) 455-3766

Colleen D. Szuch
Allied-Signal Inc.
P.O. Box 2245R
Morristown, NJ 07960-5000   (201) 455-2857

Darryl L. Webster
Allied-Signal, Inc.
Columbia Road & Park Ave.
A B-2
Morristown, NJ 07960   (201) 455-2945

Carol T. Wortmann
Riker, Danzig, Scherer, Hyland
& Perretti
One Speedwell Avenue
Morristown, NJ 07962   (201) 538-0800

Edith R. T. Grill
7 Raynor Road
Morristownship, NJ 07960   (201) 267-9153

Bernard John Murphy
10 Rockaway Terrace
Mountain Lakes, NJ 07046   (201) 334-9267

William Frank Pinsak
74 Tower Hill Road
Mountain Lakes, NJ 07046   (201) 335-2530

Herbert Smith Sylvester
16 Hillcrest Road
Mountain Lakes, NJ 07046   (201) 334-2606

F. W. Wyman#
34 Pollard Road
Mountain Lakes, NJ 07046   (201) 334-6331

Robert B. Ardis
1038 Ledgewood Road
Mountainside, NJ 07092   (201) 232-0865

Erwin Klingsberg
1597 Deer Path
Mountainside, NJ 07092   (201) 232-1108

Herbert I. Sherman
1492 Deer Path
Mountainside, NJ 07092   (201) 232-1063

Riggs T. Stewart
1170 Foothill Way
Mountainside, NJ 07092   (201) 654-3067

Louis Anthony Vespasiano
1059 Sunny Slope Drive
Mountainside, NJ 07092   (908) 233-8474

Joan E. Switzer
110A Hastings Way
Mt. Laurel, NJ 08054   (609) 866-8920

John C. Bonacci#
156 Gallinson Drive
Murray Hill, NJ 07974   (201) 464-8295

Henry T. Brendzel
AT&T Bell Laboratories
600 Mountain Avenue
Murray Hill, NJ 07974   (201) 582-4110

David I. Caplan
AT&T Bell Laboratores
600 Mountain Avenue
Room 3B-521
Murray Hill, NJ 07974   (201) 582-4937

Larry R. Cassett
BOC Group, Inc.
100 Mountain Avenue
Murray Hill, NJ 07974   (201) 771-6434

David A. Draegert
BOC Group, Inc.
Patent Trademark & Licensing Dept.
100 Mountain Avenue
Murray Hill, NJ 07974   (201) 771-6402

Samuel H. Dworetsky
AT&T Bell Laboratories
600 Mountain Ave.
Murray Hill, NJ 07974   (201) 771-2236

Geoffrey D. Green
AT&T Bell Laboratories
600 Mountain Avenue
Room 3C-509
Murray Hill, NJ 07974   (201) 582-4530

Jerry W. Herndon
AT&T Bell Laboratories
600 Mountain Avenue
Murray Hill, NJ 07974   (201) 582-4888

George S. Indig
At&T Bell Laboratories
600 Mountain Avenue
Room 3B-533
Murray Hill, NJ 07974   (201) 582-6117

Carol A. Nemetz
BOC Group, Inc.
100 Mountain Avenue
Murray Hill, NJ 07974   (201) 771-6446

Walter G. Nilsen, Sr.
AT&T Bell Laboratories
600 Mountain Avenue
Murray Hill, NJ 07974   (201) 582-3329

Irwin Ostroff
3 Lackawanna Blvd.
Murray Hill, NJ 07974   (201) 464-0248

Eugen E. Pacher
AT&T Bell Laboratories
600 Mountain Ave.
Murray Hill, NJ 07974   (201) 582-5337

David Lawrence Rae
BOC Group, Inc.
575 Mountain Avenue
Murray Hill, NJ 07974   (908) 771-4835

Coleman Robert Reap
BOC Group, Inc.
100 Mountain Avenue
Murray Hill, NJ 07974   (908) 771-6385

Thomas A. Restaino
AT&T Bell Laboratories
Room 3B-527
600 Mountain Ave.
Murray Hill, NJ 07974   (201) 582-3000

David M. Rosenblum
BOC Group, Inc.
100 Mountain Avenue
Murray Hill, NJ 07074   (201) 771-6167

William Ryan
135 Roland Road
Murray Hill, NJ 07974   (201) 464-2123

Bruce S. Schneider
AT&T Bell Laboratories
600 Mountain Ave.
Murray Hill, NJ 07974   (201) 582-6358

Charles W. Seabury#
AT&T Bell Laboratories
600 Mountain Avenue
Murray Hill, NJ 07974   (201) 581-3937

Peter V.D. Wilde
AT&T Bell Laboratories
600 Mountain Ave.
P.O. Box 636
Room 3A-515
Murray Hill, NJ 07974   (201) 582-3556

Mary M. Allen
Johnson & Johnson
One Johnson & Johnson Plaza
New Brunswick, NJ 08933   (201) 524-2796

Steven Paul Berman
Johnson & Johnson
Off. of Gen. Coun.
One Johnson & Johnson Plaza
New Brunswick, NJ 08933   (201) 524-2805

Joseph J. Brindisi
Johnson & Johnson
One Johnson & Johnson Plaza
New Brunswick, NJ 08993   (201) 524-2826

Audley A. Ciamporcero, Jr.
Johnson & Johnson
One Johnson & Johnson Plaza
New Brunswick, NJ 08933   (201) 524-2803

Andrea L. Colby
Johnson & Johnson
One Johnson & Johnson Plaza
New Brunswick, NJ 08933   (201) 524-2792

Paul A. Coletti
Johnson & Johnson
One Johnson & Johnson Plaza
New Brunswick, NJ 08933   (201) 524-2815

Geoffrey G. Dellenbaugh
Johnson & Johnson
One Johnson & Johnson Plaza
New Brunswick, NJ 08933   (201) 524-2818

Matthew S. Goodwin
Johnson & Johnson
One Johnson & Johnson Plaza
New Brunswick, NJ 08933   (201) 524-0400

Richard J. Grochala
Johnson & Johnson
One Johnson & Johnson Plaza
New Brunswick, NJ 08933   (201) 524-2819

Howard E. Heller#
3 Lansing Place
New Brunswick, NJ 08901   (201) 249-6169

Michael A. Kaufman
Johnson & Johnson
One Johnson & Johnson Plaza
New Brunswick, NJ 08933   (201) 524-2457

Benjamin Franklin Lambert
Johnson & Johnson
One Johnson & Johnson Plaza
New Brunswick, NJ 08933   (201) 524-2824

Jason Lipow
Johnson & Johnson
One Johnson & Johnson Plaza
New Brunswick, NJ 08933   (201) 524-2825

Gale F. Matthews
Johnson & Johnson
One Johnson & Johnson Plaza
New Brunswick, NJ 08933   (201) 524-2802

W. Scott Mc Nees
Johnson & Johnson
One Johnson & Johnson Plaza
New Brunswick, NJ 08933   (201) 524-3592

Charles Joseph Metz
Johnson & Johnson
One Johnson & Johnson Plaza
New Brunswick, NJ 08933   (201) 524-2814

Robert L. Minier
Johnson and Johnson
1 Johnson & Johnson Plaza
New Brunswick, NJ 08933   (201) 524-2817

Ralph R. Palo#
Johnson & Johnson
One Johnson & Johnson Plaza
New Brunswick, NJ 08933   (201) 524-2818

Raymond S. Parker, III
Johnson & Johnson
One Johnson & Johnson Plaza
New Brunswick, NJ 08933   (201) 524-2976

Joel R. Petrow
Johnson & Johnson
1 Johnson & Johnson Plaza, WH-4222
New Brunswick, NJ 08933   (201) 524-2825

Bernard F. Plantz
Johnson & Johnson
One Johnson & Johnson Plaza
New Brunswick, NJ 08933   (908) 524-2771

James Riesenfeld
Johnson & Johnson
One Johnson & Johnson Plaza
New Brunswick, NJ 08933   (201) 524-2641

Lawrence D. Schuler
Johnson & Johnson
One Johnson & Johnson Plaza
New Brunswick, NJ 08933   (201) 524-2811

Joseph F. Shirtz
Johnson & Johnson
One Johnson & Johnson Plaza
New Brunswick, NJ 08933   (201) 524-2812

Michael Stark
Johnson & Johnson
One Johnson & Johnson Plaza
New Brunswick, NJ 08933   (201) 524-2797

Lewis Stein
Johnson & Johnson
One Johnson & Johnson Plaza
New Brunswick, NJ 08933   (201) 524-2489

Michael Q. Tatlow
Johnson & Johnson
One Johnson & Johnson Plaza
New Brunswick, NJ 08933   (201) 524-2801

Hal B. Woodrow
Johnson & Johnson
Off. of Gen. Coun.
One Jonhson & Johnson Plaza
New Brunswick, NJ 08933   (201) 524-2801

John A. Casper
120 Passaic St.
New Providence, NJ 07974   (201) 464-2339

Edwin Blauvelt Cave
25 Alden Rd.
New Providence, NJ 07974   (201) 665-0780

Roger M. Rathbun
BOC Group, Inc.
100 Mouontain Avenue
New Providence, NJ 07974   (201) 771-6477

Sylvan Sherman
280 Woodbine Circle
New Providence, NJ 07974   (201) 665-1815

R. Hain Swope
BOC Group, Inc.
100 Mountain Avenue
Murray Hill
New Providence, NJ 07974   (201) 771-6292

Richard C. Winter
PCT International, Inc.
P.O. Box 573
New Vernon, NJ 07976   (201) 766-9458

Daniel H. Bobis
Popper, Bobis & Jackson
17 Academy St.
Newark, NJ 07102   (201) 623-1000

Silvio J. De Carli#
Carenter, Bennett & Morrissey
Three Gateway Center
Newark, NJ 07102   (201) 622-7711

Steven W. Grill
Immunomedics
5 Bruce Street
Newark, NJ 07103   (201) 456-4779

Ira J. Hammer
Crummy, Del Deo, Dolan,
Griffinger & Vecchione
One Riverside Plaza
Newark, NJ 07102   (201) 596-4500

Jacob Klapper
N.J. Institute of Tech.
323 King Blvd.
Newark, NJ 07102   (201) 546-3516

Rene Oliveras
5 Commerce Street
4th Floor
Newark, NJ 07102   (201) 622-1881

Henry Joseph Walsh
Crummy, Del Deo, Dolan,
Griffinger & Vecchione
One Riverfront Plaza
Newark, NJ 07102   (201) 596-4500

Brian Lowell Wamsley
Sills, Cummis, Zuckerman, Radin,
Tischman, Epstein & Gross, P.A.
Legal Center
One Riverfront Plaza
Newark, NJ 07102   (201) 643-5303

Harries A. Mumma, Jr.
24 Spruce Road
No. Caldwell, NJ 07006   (201) 228-1832

Burton E. Levin
7855 Boulevard East
North Bergen, NJ 07047   (201) 854-2458

James K. Mc Neal, III
915 Shore Road
Northfield, NJ 08225   (609) 641-8909

Ellen G. Ciambrone Coletti
Hoffmann-La Roche, Inc.
340 Kingsland St.
Nutley, NJ 07110   (609) 641-8909

William H. Epstein
Hoffmann-La Roche, Inc.
340 Kingsland St.
Nutley, NJ 07110   (201) 235-3723

Richard A. Gaither
Hoffmann-La Roche, Inc.
340 Kingsland Street
Nutley, NJ 07110   (201) 235-2147

George M. Gould
Hoffmann-La Roche, Inc.
340 Kingsland Street
Nutley, NJ 07110   (201) 235-3741

William G. Isgro
Hoffmann-La Roche, Inc.
340 Kingsland St.
Nutley, NJ 07110   (201) 235-4393

Alan P. Kass
Hoffmann-La Roche, Inc.
Patent Law Dept.
340 Kingsland Street
Nutley, NJ 07110   (201) 235-4205

William Krovatin
Hoffmann-La Roche, Inc.
340 Kingsland Street
Nutley, NJ 07110   (201) 235-4387

Bernard S. Leon
Hoffmann-La Roche, Inc.
340 Kingsland St.
Nutley, NJ 07011   (201) 235-4378

Bruce A. Pokras
Hoffmann-La Roche, Inc.
340 Kingsland Street
Nutley, NJ 07110   (201) 235-5000

Patricia S. Rocha
Hoffmann-La Roche, Inc.
340 Kingsland Street
Bldg. 85, 7th Fl.
Nutley, NJ 07110   (201) 235-2441

Catherine R. Roseman
Hoffmann-La Roche, Inc.
Patent Law Dept.
Bldg. 85, Room 721
340 Kingsland Street
Nutley, NJ 07110   (201) 235-6208

John J. Schlager
Hoffmann-La Roche, Inc.
340 Kingsland Street
Nutley, NJ 07110   (201) 235-2863

Dennis P. Tramaloni#
Hoffmann-La Roche, Inc.
340 Kingsland St.
Nutley, NJ 07110   (201) 235-4475

Bart J. Zoltan#
152 De Wolf Road
Old Tappan, NJ 07675   (201) 768-3580

Leroy G. Sinn
P.O. Box 559
Oldwick, NJ 08858   (201) 439-3551

Richard A. Joel
466 Kinderkamack Road
Oradell, NJ 07649   (201) 599-0588

Arnold D. Litt
Rusch & Litt
East 80 Route 4
Suite 310
Paramus, NJ 07652   (201) 845-7070

Richard D. Goldstein
Sony Corporation of America
Sony Drive
M.D. 3-76
Park Ridge, NJ 07656   (201) 930-7315

Shawn L. Mc Clintock
Sony Corp. of America
Sony Drive
Mail Drop 3-76
Park Ridge, NJ 07656   (201) 930-7445

Michael Robert Chipaloski#
BASF Corporation
100 Cherry Hill Road
Parsippany, NJ 07054   (201) 316-3000

Bruce M. Collins
Mathews Woodbridge Goebel
Pugh & Collins, P.A.
P.O. Box 5910
Parsippany, NJ 07054   (201) 299-1144

John W. Fisher
AT&T
5 Wood Hollow Road
Parsippany, NJ 07054   (201) 581-4939

Stephen L. Malaska
Mathews, Woodbridge &
Collins, P.A.
P.O. Box 5910
Parsippany, NJ 07054   (201) 299-1144

Louis E. Marn
400 Lanidex Center
Parsippany, NJ 07054   (201) 884-2122

Joyce L. Morrison
BASF
100 Cherry Hill Rd.
Parsippany, NJ 07054   (201) 316-3000

Richard A. Negin
Mathews, Woodbridge, Goebel,
Pugh & Collins, P.A.
3799 Route 46, Suite 309
P.O. Box 5910
Parsippany, NJ 07054   (201) 299-1144

Harry L. Newman
5 Wood Hollow Road
3J45
Parsippany, NJ 07054   (201) 581-7204

Robert M. Shaw
BASF Corporation
100 Cherry Hill Road
Parsippany, NJ 07054   (201) 316-3329

Stephen B. Shear
Nabisco Brands, Inc.
Law Dept.
7 Campus Drive
P.O. Box 311
Parsippany, NJ 07054   (201) 682-5000

Fran S. Wasserman
Mathews, Woodbridge & Collins
3799 Route 46
Suite 309
Parsippany, NJ 07054   (201) 299-1144

Harold F. Bennett#
RD 2 Box 55W
Pennington, NJ 08534   (609) 737-8789

Lawrence S. Levinson
4 Weidel Drive
Pennington, NJ 08534   (609) 737-1820

Eugene George Seems
28 Dublin Road
Pennington, NJ 08534   (609) 737-2372

Donald N. Timbie, Sr.
35 E. Curlis Avenue
Pennington, NJ 08534   (609) 737-0633

Frank Samuel Troidl
4 Mallard Drive
Pennington, NJ 08534   (201) 737-9180

Diamond C. Ascani#
1013 Carroll Avenue
Pennsville, NJ 08070   (609) 678-3446

Glenn B. Foster
Ingersoll-Rand Company
942 Memorial Parkway
Phillipsburg, NJ 08865   (201) 859-7656

Robert F. Palermo
Ingersoll Rand Corp.
Patent Dept.
942 Memorial Pkwy.
Phillipsburg, NJ 08865   (201) 859-7686

Betty B. Tibbott#
410 Ohio Ave.
Phillipsburg, NJ 08865   (201) 454-8726

David W. Tibbott
410 Ohio Ave.
Phillipsburg, NJ 08865   (201) 454-8726

Walter C. Vliet
Ingersoll-Rand Co.
942 Memorial Parkway
Phillipsburg, NJ 08865   (201) 859-7728

Michael J. Flynn#*
U.S. Army Armament
Res. & Eng. Center
SMCAR-FSI
Bldg. 3124
Picatinny Arsenal, NJ 78060   (201) 724-2665

Edward Goldberg*
U.S. Army AMCCOM
AMSMC-GCL D
Bldg. 3
Picatinny Arsenal, NJ 07806   (201) 724-6590

Anton B. Weber#*
Dept of the Army
AMCCOM-RD&E Ctr.
Pyro Systems Branch
Picatinny Arsenal, Bldg. 1515
Picatinny Arsenal, NJ 07806   (201) 724-2290

Gerald Stuart Rosen
12 Lancaster Ave.
Pine Brook, NJ 07058   (201) 227-4764

Francis H. Deef
38 Madison Ave.
Piscataway, NJ 08854   (201) 572-5853

Marta E. Delsignore#
Colgate-Palmolive Company
909 River Road
Piscataway, NJ 08854   (201) 878-7397

Max Fogiel#
61 Ethel Road West
Piscataway, NJ 08854   (201) 819-8880

Robert S. Mac Wright
Rutgers University
Off. of Corp. & Liaison &
Technology Transfer
377 Hoes Lane Cn 1179
Piscataway, NJ 08855   (201) 932-2074

Michael J. Mc Greal
Colgate-Palmolive Co.
909 River Road
P.O. Box 1343
Piscataway, NJ 08855   (908) 878-7152

Henry J. Nix
44 Stelton Road
Suite 230
Piscataway, NJ 08854   (201) 968-9595

James P. Scullin#
Huls America Inc.
80 Centennial Ave.
P.O. Box 456
Piscataway, NJ 08855   (201) 980-6867

Paul Shapiro
Colgate-Palmolive Co.
909 River Road
Piscataway, NJ 08854   (908) 878-7547

Robert L. Stone
Colgate-Palmolive Company
909 River Ross
Piscataway, NJ 08854   (201) 878-7417

Abraham Wilson
216 Stelton Road
B-1
Piscataway, NJ 08854   (908) 968-5900

John E. Callaghan
1120 Park Avenue
Plainfield, NJ 07060   (201) 757-0606

George S. Seltzer
181 Thoreau Drive
Plainsboro, NJ 08536   (609) 282-3133

Roderick Bruce Anderson
AT&T International
P.O. Box 900
Princeton, NJ 08540   (609) 639-2307

Arthur E. Bahr
GE & RCA Licensing
Management Operation, Inc.
P.O. Box 2023
Princeton, NJ 08540   (609) 734-9896

Donald Jay Barrack
Squibb Corporation
P.O. Box 4000
Princeton, NJ 08543   (609) 921-4328

Clement A. Berard, Jr.
General Electric Co.
Aerospace Pats. & Licensing
One Independence Way
P.O. Box 2023
Princeton, NJ 08540   (609) 734-9812

Allen Bloom
Liposome Company, Inc.
One Research Way
Princeton Forrestal Center
Princeton, NJ 08540   (609) 452-7060

Glenn Huber Bruestle
RCA Corporation
CN 5312
Princeton, NJ 08543   (609) 734-2356

William J. Burke
David Sarnoff
Research Center, Inc.
CN 5300
Princeton, NJ 08543   (609) 734-2560

Dean W. Chace
GE and RCA Licensing
Management Operation
Box 2023
2 Independence Way
Princeton, NJ 08540   (609) 734-9434

Richard G. Coalter
GE and RCA Licensing Management
Operation, Inc.
Two Independence Way
P.O. Box 2023
Princeton, NJ 08540   (609) 734-9782

Samuel Cohen
24 Littlebrook Road North
Princeton, NJ 08540   (609) 924-4561

Stephen B. Davis
Squibb Corporation
P.O. Box 4000
Princeton, NJ 08540   (609) 921-4338

Richard E. Elden
FMC Corp.
P.O. Box 8
Princeton, NJ 08540   (609) 452-2300

Peter Max Emanuel
GE and RCA Licensing Management
Operation, Inc.
Two Independence Way
P.O. Box 2023
Princeton, NJ 08540   (609) 734-9586

Nathan Feldstein#
Surface Technology, Inc.
P.O. Box 2027
Princeton, NJ 08540   (609) 452-2929

Anthony J. Franze
387 Gallup Road
Princeton, NJ 08540   (609) 683-9733

Harvey D. Fried
GE and RCA Licensing Management
Operation, Inc.
2 Independence Way
P.O. Box 2023
Princeton, NJ 08540   (609) 734-9839

Theodore R. Furman
Squibb Corporation
P.O. Box 4000
Princeton, NJ 08543   (609) 921-5735

Bernard Gerb
127 Meadowbrook Drive
Princeton, NJ 08540   (609) 921-9078

Dominic J. Giancola
12 Cameron Court
Princeton, NJ 08540   (609) 924-6686

Lester L. Hallacher
GE and RCA Licensing Management
Operation, Inc.
Two Independence Way
P.O. Box 2023
Princeton, NJ 08540   (609) 734-9629

Rita V. Hauck
Squibb Corp.
Lawrence-Princeton Road
Princeton, NJ 08543   (609) 921-5539

James B. Hayes
GE and RCA Licensing
Management
Two Independence Way
P.O. Box 2023
Princeton, NJ 08540   (609) 734-9570

Shabtay S. Henig
GE and RCA Licensing
Management Operation, Inc.
Two Independence Way
P.O. Box 2023
Princeton, NJ 08540   (609) 734-9751

Eric P. Herrmann
GE and RCA Licensing
Management Operation, Inc.
P.O. Box 2023
Two Independence Way
Princeton, NJ 08540   (609) 734-9754

Frank Ianno
FMC Corp.
P.O. Box 8
Princeton, NJ 08542   (609) 452-2300

Dennis H. Irlbeck, Sr.
GE and RCA Licensing
Management Operation, Inc.
Two Independence Way
P.O. Box 2023
Princeton, NJ 08540   (609) 734-9763

Robert D. Jackson#
FMC Corp.
Princeton, NJ 08542   (609) 452-2300

Herbert Leonard Jacobson
General Electric Company
P.O. Box 2023
Princeton, NJ 08540   (609) 734-9422

George J. Koeser
46 Carter Road
Princeton, NJ 08540   (609) 896-2926

Irwin M. Krittman
GE and RCA Licensing Management
Operation, Inc.
Two Independence Way
P.O. Box 2023
Princeton, NJ 08540   (609) 734-9556

Ronald H. Kurdyla
GE and RCA Licensing Management
Operation, Inc.
Two Independence Way
P.O. Box 2023
Princeton, NJ 08540   (609) 734-9701

Catherine L. Kurtz#
Liposome Co., Inc.
One Research Way
Princeton, NJ 08540   (609) 452-7060

Joseph J. Laks
GE and RCA Licensing Management
Operation, Inc.
Two Independence Way
P.O. Box 2023
Princeton, NJ 08540   (609) 734-9813

Thomas F. Lenihan
GE and RCA Licensing
Management Operation, Inc.
Two Independence Way
P.O. Box 2023
Princeton, NJ 08540   (609) 734-9826

Robert B. Levy
AT&T Bell Laboratories
Engineering Res. Center
P.O. Box 900
Princeton, NJ 08540   (609) 639-2305

Nicholas P. Malatestinic
Squibb Corporation
P.O. Box 4000
Princeton, NJ 08543   (609) 921-4301

Emery Marton
43 Woodland Drive
Princeton, NJ 08540   (609) 921-4301

Ronald J. Mc Caully
Wyeth-Ayerst Research
CN 8000
Princeton, NJ 08543   (201) 274-4414

William H. Meagher, Jr.
GE and RCA Licensing and Management
   Operation, Inc.
Two Independence Way
PO Box 2023
Princeton, NJ 08540   (609) 734-9837

William H. Meise
General Electric Company
Aerospace Patents & Licensing
One Independence Way
P.O. Box 2023
Princeton, NJ 08540   (609) 734-9816

Albert J. Molinari#
1506 Jonathan Court
Princeton, NJ 08540-00   (609) 924-3333

Birgit E. Morris
5 Tall Timbers Drive
Princeton, NJ 08540   (609) 921-1695

Richard R. Muccino
P.O. Box 1267
Princeton, NJ 08542   (609) 466-3407

Carlos Nieves
General Electric Company
Two Independence Way
Princeton, NJ 08540   (609) 734-9626

James L. O Brien
1 Constitution Hill
Princeton, NJ 08540   (609) 924-5918

Carl V. Olson
23 Broadripple Drive
Princeton, NJ 08540   (609) 924-3341

Ronald G. Ort
Liposome Co.
One Research Way
Princeton, NJ 08540   (609) 452-7060

Paul Jorgen Rasmussen
GE and RCA Licensing Management
Operation, Inc.
P.O. Box 2023
Two Independence Way
Princeton, NJ 08540   (609) 734-9883

Harry S. Reichard
Princeton Measurements Corp.
31 Air Park Road
Princeton, NJ 08540   (609) 924-7885

Burton Rodney
Bristol-Myers Squibb Co.
P.O. Box 4000
Princeton, NJ 08543   (609) 921-4336

Jerald E. Roehling
GE & RCA Licensing
Management Operatin, Inc.
P.O. Box 2023
Two Independence Way
Princeton, NJ 08540   (609) 734-9823

Timothy G. Rothwell
Squibb Pharmaceutical Group
P. O. Box 4500
Princeton, NJ 08540   (609) 243-6562

Albert Russinoff
119 Heather Lane
Princeton, NJ 08540   (609) 924-3473

David S. Saari#
American Cyanamid Co.
Agricultural Research Division
P.O. Box 400
Princeton, NJ 08543   (609) 799-0400

Franklyn Schoenberg
241 Dodds Lane
Princeton, NJ 08540   (609) 921-3474

Michael G. Schwarz
Mathews, Woodbridge & Collins
100 Thanet Circle
Suite 306
Princeton, NJ 08540   (609) 924-3773

Daniel Ethan Sragow
GE and RCA Licensing Management
Operation, Inc.
Two Independence Way
P.O. Box 2023
Princeton, NJ 08540   (609) 734-9891

Norman St. Landau
105 College Road East
Princeton, NJ 08540   (609) 924-0800

Richard H. Steen
P.O. Box 2178
Princeton, NJ 08543   (609) 871-5800

Stanley Allen Strober
Mobil Oil Corp.
P.O. Box 1031
Princeton, NJ 08543   (609) 737-5484

Joseph S. Tripoli
GE and RCA Licensing Management
Operation, Inc.
Two Independence Way
P.O. Box 2023
Princeton, NJ 08540   (609) 734-9443

Frederick A. Wein
GE and RCA Licensing
Management Operation, Inc.
Two Independence Way
P.O. Box 2023
Princeton, NJ 08543   (609) 734-9518

Nord F. Winnan
American Re-Insurance Co.
555 College Road East
Princeton, NJ 08543   (609) 243-4465

William K. Wissing
Union Camp Corp.
P.O. Box 3301
Princeton, NJ 08543   (609) 896-1200

Richard C. Woodbridge
Mathews, Woodbridge
& Collins, P.A.
100 Thanet Circle
Suite 306
Princeton, NJ 08540   (609) 924-3773

Kenneth Watov
Klauber & Jackson
P.O. Box 247
Princeton Junction, NJ 08550
                              (609) 243-0330

Charles M. Caruso
Merck & Company, Inc.
Patent Dept.
P.O. Box 2000
Rahway, NJ 07065   (201) 594-4830

Theresa Y. Cheng
Merck & Co., Inc.
Pat. Dept.
126 E. Lincoln Avenue
Rahway, NJ 07065   (201) 594-4982

Mark R. Daniel
Merck & Co., Inc.
Patent Department
P.O. Box 2000, RY 60-30
Rahway, NJ 07065   (201) 594-6609

Joseph F. Di Prima
Merck & Co., Inc.
Pat. Dept. RY 60-30
P.O. Box 2000
126 E. Lincoln Avenue
Rahway, NJ 07065   (201) 594-4365

Meredith C. Findlay
M&T Chemicals, Inc.
Pat. Dept.
P.O. Box 1104
Rahway, NJ 07065   (201) 499-2153

Frank P. Grassler
Merck & Co., Inc.
Patent Dept. RY 60-30
P.O. Box 2000
Rahway, NJ 07065   (908) 594-3462

John W. Harbour
Merck & Company
P.O. Box 2000
Rahway, NJ 07065   (201) 594-3507

549

Julian S. Levitt
Merck and Co., Inc.
126 E. Lincoln Avenue
P.O. Box 2000
Rahway, NJ 07065   (201) 594-4118

Menotti J. Lombardi, Jr.
2367 Jowett Pl.
Rahway, NJ 07065   (201) 382-0578

Gabriel Lopez
Merck & Co., Inc.
Patent Dept.
P.O. Box 2000
Rahway, NJ 07065   (201) 594-4417

Kevin J. Mc Gough
Merck & Company, Inc.
P.O. Box 2000
Rahway, NJ 07065   (201) 594-7042

Roy D. Meredith
Merck & Co., Inc.
Patent Dept.
P.O. Box 2000
Rahway, NJ 07065   (201) 594-4678

Salvatore C. Mitri
Merck & Co., Inc.
P.O. Box 2000
Rahway, NJ 07065   (201) 594-4454

Mario Anthony Monaco
Merck & Company, Inc.
P.O. Box 2000
Rahway, NJ 07065   (201) 594-5295

Edward W. Murray
Merck & Company, Inc.
126 E. Lincoln Avenue
Rahway, NJ 07065   (201) 594-5301

William H. Nicholson
Merck & Co., Inc.
Box 2000
Rahway, NJ 07065   (201) 594-5315

Robert J. North
Merck & Co., Inc.
P.O. Box 2000
Rahway, NJ 07065   (201) 594-7262

Richard B. Olson
Merck & Company, Inc.
Pat. Dept.
P.O. Box 2000
Rahway, NJ 07065   (201) 574-4057

Curtis C. Panzer
Merck & Co., Inc.
P.O. Box 2000
RY 60-30
Rahway, NJ 07065   (201) 594-3199

Richard S. Parr
Merck & Co., Inc.
P.O. Box 2000
Rahway, NJ 07065   (201) 594-4958

Donald J. Perrella
Merck & Co., Inc.
Pat. Dept.
126 E. Lincoln Ave.
Rahway, NJ 07065   (201) 594-5593

Hesna J. Pfeiffer
Merck and Co., Inc.
P.O. Box 2000
Rahway, NJ 07065   (201) 594-4251

Manfred Polk
Merck & Company, Inc.
P.O. Box 2000
Rahway, NJ 07065   (201) 594-4285

Alice Ota Robertson
Merck & Co., Inc.
P.O. Box 2000
Rahway, NJ 07065   (201) 594-4372

David L. Rose
Merck & Co., Inc.
Patent Dept., RY 60-30
Box 2000
126 E. Lincoln Avenue
Rahway, NJ 07065   (201) 594-4777

Roger G. Smith#
Merck & Co., Inc.
P.O. Box 2000
Rahway, NJ 07065   (201) 594-4817

Raymond M. Speer
Merck & Co., Inc.
P.O. Box 2000
126 E. Lincoln Ave., RY 60-30
Rahway, NJ 07065   (201) 594-4481

Michael C. Sudol, Jr.
Merck and Co., Inc.
126 E. Lincoln Avenue
Rahway, NJ 07065   (201) 594-5159

Richard A. Thompson
Merck & Co., Inc.
126 E. Lincoln Avenue
RY 60-30
Rahway, NJ 07065   (201) 574-4331

Jack L. Tribble
Merck & Co., Inc.
126 E. Lincoln Avenue
Rahway, NJ 07065   (908) 594-5321

Melvin Winokur
Merck & Company, Inc.
126 E. Lincoln Avenue
RY 60-30
Rahway, NJ 07065   (908) 594-7234

John F. Levis
266 Momar Dr.
Ramsey, NJ 07446   (908) 594-7234

Walter Francis Jewell#
23 Berry Lane
Randolph, NJ 07869   (201) 328-0371

James P. Demers#
R.W. Johnson Pharmaceutical
Research Institute
P.O. Box 300
Raritan, NJ 08869   (201) 218-7230

Ruby T. Hope#
Robert Woods Johnson
Pharmaceutical Res. Inst.
Route 202 S.
Raritan, NJ 08869   (201) 218-7261

Roger A. Clapp
70 Highway 35
Red Bank, NJ 07701   (201) 758-0800

Charles S. Guenzer
Bell Communications
Research, Inc.
Patent Law Dept.
P.O. Box 7040
Red Bank, NJ 07701   (908) 758-3205

Peter L. Michaelson
Stanger, Michaelson,
Spivak & Wallace, P.C.
328 Newman Springs Road
P.O. Box 8489
Red Bank, NJ 07701   (908) 530-6671

Revis Gale Rhodes, Jr.
Evans, Burgess, Osborne
& Kreitzman
P.O. Box BB
Red Bank, NJ 07701   (201) 741-9550

Charles L. Thomason
One Executive Drive
P.O. Box 8160
Red Bank, NJ 07701   (908) 530-9404

Donald R. Heiner
245 Main St.
Ridgefield Park, NJ 07660   (201) 440-8040

Herbert Lewis Davis, Jr.
136 Washington Place
Ridgewood, NJ 07450   (201) 444-6365

Lewis Messulam
152 South Irving Street
Ridgewood, NJ 07450   (201) 445-1879

Stanley I. Rosen
42 Kira Lane
Ridgewood, NJ 07450   (201) 445-1473

Douglas J. Kirk
R.D. 3, Box 376
Ringoes, NJ 08551   (201) 782-0578

Tobias Lewenstein
653 Woodside Ave.
River Vale, NJ 07675   (201) 391-1014

John V. Regan
4 Lemore Circle
Rocky Hill, NJ 08553   (609) 924-5924

Susan A. Capello
Carella, Byrne, Bain, Gilfillan,
Cecchi & Stewart
6 Becker Farm Road
Roseland, NJ 07068   (201) 935-6900

John G. Gilfillan, III
Carella, Byrne, Bain
& Gilfillan
6 Becker Farm Road
Roseland, NJ 07068   (201) 994-1700

Louis S. Gillow
Carella, Byrne, Bain
& Gilfillan
6 Becker Farm Road
Roseland, NJ 07068   (201) 994-1700

John Dennis Kaufmann
Carella, Byrne, Bain &
Gilfillan
6 Becker Farm Road
Roseland, NJ 07068   (201) 994-1700

Raymond J. Lillie
Carella, Byrne, Bain
& Gilfillan
6 Becker Farm Road
Roseland, NJ 07068   (201) 994-1700

George J. Minish
18 Thacke Ray Drive
Roseland, NJ 07068   (201) 288-9585

Elliot Michael Olstein
Carella, Byrne, Bain
& Gilfillan
7 Becker Farm Road
Roseland, NJ 07068   (201) 994-1700

Richard S. Serbin
61 Monroe Avenue
Roseland, NJ 07068   (201) 228-5182

Robert James Sanders, Jr.
39 Buena Vista Ave.
Rumson, NJ 07760   (201) 842-6269

Raymond W. Barclay
95 Raymond Avenue
Rutherford, NJ 07070   (201) 939-1295

Charles C. Marshall
248 Washington Avenue
Rutherford, NJ 07070   (201) 933-4759

Blake T. Biederman
Inco Patents & Licensing
Park 80 West - Plaza 2
Saddle Brook, NJ 07662   (201) 368-4847

Anthony D. Cipollone
49 Market Street
P.O. Box 542
Saddle Brook, NJ 07662   (201) 845-6626

Raymond J. Kenny
Inco Patent & Licensing
Park 80 West-Plaza 2
Saddle Brook, NJ 07662   (201) 368-4853

Francis John Mulligan, Jr.
Inco Patents & Licensing
Park 80 West - Plaza 2
Saddle Brook, NJ 07662   (201) 368-4849

Edward A. Steen
Inco Patents & Licensing
Park 80 West - Plaza 2
Saddle Brook, NJ 07662   (201) 368-4848

Peter A. Businger
344 Valley Scent Avenue
Scotch Plains, NJ 07076   (201) 322-4835

Robert D. Farkas
1776 Martine Ave.
Scotch Plains, NJ 07076   (201) 889-5200

Marthe L. Gibbons#
1485 Golf St.
Scotch Plains, NJ 07076   (201) 889-5200

Richard I. Samuel
1271 Cooper Road
Scotch Plains, NJ 07076   (201) 756-7406

Frank A. Sinnock
23 Black Birch Rd.
Scotch Plains, NJ 07076   (908) 889-6661

William F. Smith#
IG Glazing Systems
1202 Ocean Avenue
Sea Bright, NJ 07760   (201) 741-0517

Jeremiah Gerard Murray
P.O. Box 142
413 Philadelphia Blvd.
Sea Girt, NJ 08750   (908) 449-5751

Edward M. Farrell
108 36th Street
Sea Isle City, NJ 08243   (609) 263-2954

Peter A. Abruzzese
ITT Corporation
100 Plaza Drive
Secaucus, NJ 07096   (201) 601-4305

S. Michael Bender
Suite 127
Harmon Cove Towers
Secaucus, NJ 07094   (201) 867-6807

Julius F. Harms, III
300 Harmon Meadow Blvd.
Secaucus, NJ 07096   (201) 866-2500

Thomas N. Twomey
ITT Corporation
100 Plaza Drive
Secaucus, NJ 07096   (212) 258-1017

Michael Anthony Caputo
Hoechst Celanese Corp.
150 John F. Kennedy Pkwy.
Short Hills, NJ 07078   (201) 912-4892

Thomas Cifelli, Jr.
22 Cayuga Way
Short Hills, NJ 07078   (201) 379-4187

Clayton S. Gates
287 Taylor Rd., S.
Short Hills, NJ 07078   (201) 467-9866

John M. Genova
Hoechst Celanese Corp.
150 John F. Kennedy Parkway
Mack Center Building
Short Hills, NJ 07078   (201) 912-4898

Karen E. Klumas
Hoechst Celanese Corporation
Mack Center Building
150 John F. Kennedy Pkwy.
Short Hills, NJ 07078   (201) 912-4800

Jonathan E. Myers
19 Canterbury Lane
Short Hills, NJ 07078   (201) 379-4354

Daniel E. Nester
AT&T Bell Laboratories
101 J.F. Kennedy Parkway
Short Hills, NJ 07078   (201) 771-2208

Don Houghton Phillips
Hoechst Celanese Corp.
51 John F. Kennedy Pkwy.
Short Hills, NJ 07078   (201) 912-4912

Andrew F. Sayko, Jr.
Hoechst Celanese Corp.
150 John F. Kennedy Pkwy.
Mack Center Bldg.
Short Hills, NJ 07078   (201) 912-4891

Patrick E. Roberts
42 Dorchester Way
Shrewsbury, NJ 07702   (201) 912-4891

Edwin M. Szala#
68 Platz Drive
Skillman, NJ 08558   (201) 359-0410

Donna R. Fugit#
538 New Brunswick Road
Somerset, NJ 08873   (201) 846-5950

Stanley C. Corwin
GE Semiconductor Business
Route 202
Somerville, NJ 08876   (201) 685-6992

Hugh C. Crall
American Hoechst Corp.
Route 202-206 North
Somerville, NJ 08876   (201) 231-2842

Robert J. Ferb
15 W. High St.
P.O. Box 8109
Somerville, NJ 08876   (201) 722-4033

Kenneth A. Genoni
Hoechst Celanese Corporation
Route 202-206 North
Somerville, NJ 08876   (201) 231-4413

Kenneth R. Glick
Harris Semiconductor
724 Route 202
P.O. Box 591
Somerville, NJ 08876   (201) 685-6397

Tatsuya Ikeda
Hoechst-Celanese Corp.
Route 202-206 North
Somerville, NJ 08876   (201) 231-3341

Elliott Korsen
Hoechst Celanese Corporation
P.O. Box 2500
Route 202-206 North
Somerville, NJ 08876   (201) 231-2449

Stanley A. Marcus
M&T Chemicals Inc.
620 Old York Road
P.O. Box 1295
Somerville, NJ 08876   (201) 704-2301

Barbara V. Maurer
Hoechst Celanese Corp.
Route 202-206 North
P.O. Box 2500
Somerville, NJ 08876   (908) 231-4079

Rosemary M. Miano
Hoechst Celanese Corporation
Route 202-206 North
Somerville, NJ 08876   (201) 231-2000

Jerome Rosenstock
Hoechst Celanese Corporation
Route 202-206 North
P.O. Box 2500
Somerville, NJ 08876   (201) 231-2125

Kenneth Robert Schaefer
Norris, Mc Laughlin & Marcus
721 Rt. 202/206
P.O. Box 1018
Somerville, NJ 08876   (201) 722-0700

Henry I. Schanzer
Harris Corp.
Route 202
Somerville, NJ 08876   (201) 685-6830

Linda L. Setescak#
Hoechst - Roussel
Pharmaceuticals, Inc.
Route 202-206 North
Somerville, NJ 08876   (201) 231-3945

Raymond R. Wittekind
Hoechst Celanese Corp.
Route 202-206 North
P.O. Box 2500
Bldg. B-317
Somerville, NJ 08876   (201) 231-3391

Richard R. Gipson
11-B Parkwood Drive
South Amboy, NJ 08879   (201) 727-1891

David P. Alan
81 Second Street
South Orange, NJ 07079   (201) 762-6543

Neil O. Eriksen
Cohn & Cohn
14 South Orange Avenue
South Orange, NJ 07079   (201) 762-6444

Neal T. Levin
623 Hamilton Road
South Orange, NJ 07079   (201) 762-8037

Frank A. Santoro
300 Maple Avenue
South Plainfield, NJ 07080   (201) 561-6868

Arthur L. Lessler
540 Old Bridge Turnpike
South River, NJ 08882   (201) 254-5155

Vincent H. Gifford
106 Tuttle Avenue
Spring Lake, NJ 07762   (201) 449-8927

John J. Hart
214 Remsen Avenue
Spring Lake, NJ 07762   (201) 449-5881

Leo Fornero
P.O. Box 244
Springfield, NJ 07081   (201) 273-2306

Tennes I. Erstad
P.O. Box 100
Stanton, NJ 08885   (201) 735-9456

Chester A. Williams, Jr.
22 Sanford Road
Stockton, NJ 08559   (609) 397-0117

Herbert Joseph Winegar
440 Rosemont-Ringoes Rd.
Stockton, NJ 08559   (609) 397-4504

Charles B. Barris
Celanese Research Company
86 Morris Avenue
Summit, NJ 07901   (201) 522-7260

Henry Carpenter Dearborn
89 Canoe Brook Parkway
Summit, NJ 07901   (201) 273-9319

Michael W. Ferrell
Hoechst Celanese Corporation
86 Morris Avenue
Summit, NJ 07901   (201) 522-7260

Irving M. Fishman
CIBA-GEIGY Corporation
556 Morris Avenue
Summit, NJ 07901   (201) 277-4832

Kenneth B. Hamlin
60 Dale Drive
Summit, NJ 07901   (201) 273-9584

James M. Hunter, Jr.
Hoechst Celanese Corp.
86 Morris Ave.
Summit, NJ 07901   (908) 522-7747

Karen G. Kaiser#
CIBA-GEIGY Corporation
556 Morris Avenue
Building A-3024
Summit, NJ 07901   (908) 277-3318

Palaiyur S. Kalyanaraman#
Hoechst Celanese Corp.
86 Morris Ave.
Summit, NJ 07901   (201) 522-7568

Martin S. Landis
11 Garden Road
Summit, NJ 07901   (201) 522-7568

Joseph M. Mazzarese
Robert L. Mitchell Tech. Center
Hoechst Celanese Corp.
86 Morris Avenue
Summit, NJ 07901   (908) 522-7707

James L. Mc Ginnis#
Hoechst Celanese Corp.
86 Morris Avenue
Summit, NJ 07901   (201) 522-7586

Francis E. Morris
140 Rotary Drive
Summit, NJ 07901   (201) 277-2225

Everett Joseph Olinder
54 Rotary Drive
Summit, NJ 07901   (201) 273-7338

Walter R. Pfluger, Jr.
10 Myrtle Avenue
Summit, NJ 07901   (201) 273-0223

Howard R. Popper
Kerby, Cooper, English,
Danis, Popper & Garvin
480 Morris Avenue
Summit, NJ 07901   (201) 273-1212

Leonard P. Prusak
P.O. Box 339
Summit, NJ 07901   (201) 233-7092

Martha Greenewald Pugh
11 New Providence Avenue
Summit, NJ 07901   (201) 273-3487

Leo Stanger
Stanger, Michaelson, Spivak
& Wallace, P.C.
382 Springfield Ave.
Summit, NJ 07901   (212) 277-8588

Peter J. Tribulski, Jr.
183 Springfield Ave.
Summit, NJ 07901   (201) 273-6289

William Squire
North American Philips Corp.
580 White Plains Road
4th Fl.
Tarrytown, NJ 10591   (914) 332-0222

Brian J. Wieghaus
North American Philips Corp.
580 White Plains Road
Tarrytown, NJ 10591   (914) 332-0222

Arthur I. Degenholtz#
32 Vandelinda Ave.
Teaneck, NJ 07666   (201) 692-1292

Reno A. Del Ben
25 George St.
Tenafly, NJ 07670   (201) 567-9400

J. Gary Mohr
Concurrent Computer Corporation
197 Hance Avenue
Tinton Falls, NJ 07724   (201) 758-7819

William R. Rohrbach#
57 Waughaw Road
Towaco, NJ 07082   (201) 263-9097

Susan C. Gieser
Picco, Mack, Kennedy, Jaffe,
Perrella & Yoskin
One State Street Square
50 W. State St., Suite 1000
Trenton, NJ 08608   (609) 393-2400

John J. Kane
Sperry, Zoda & Kane
1 Highgate Drive, Suite D
Trenton, NJ 08618   (609) 882-7575

Nathan Levin
416 Highgate Dr.
Trenton, NJ 08618   (609) 883-6033

William L. Muckelroy
1901 N. Olden Avenue, Ste. 3A
Trenton, NJ 08618   (609) 882-2111

Frederick A. Zoda
Sperry, Zoda & Kane
1 Highgate Dr., Ste. D
Trenton, NJ 08618   (609) 882-7575

Nicholas L. Pollis, Jr.
Black Horse Pike & Evergreen Ave.
P.O. Box 704
Turnersville, NJ 08012   (609) 629-5000

Marian F. Kadlubowski
334 Meade Terrace
Union, NJ 07083   (201) 688-4390

Thomas W. Kennedy
134 Wildwood Avenue
Upper Montclair, NJ 07043   (201) 744-7914

Ira Meislik
Rabner, Allcorn & Widmark, P.C.
52 Upper Montclair Plaza
Upper Montclair, NJ 07043   (201) 744-0288

Mitchell P. Novick
52 Upper Montclair Plaza
P.O. Box 876
Upper Montclair, NJ 07043   (201) 744-5150

Carl C. Mueller
426-B Canterbury Court
Ventnor, NJ 08406   (201) 744-5150

James N. Blauvelt
60 Fells Rd.
Verona, NJ 07044   (201) 239-5839

David H. Klein
21 Valhalla Way
Verona, NJ 07044   (201) 857-0667

Jane Massey Licata
5 Red Oak Drive
Voorhees, NJ 08043   (609) 772-745

Edward A. Petko
617 Palmer Avenue
W. Allenhurst, NJ 07711   (201) 531-4993

Michael Bard
AT&T
Rm. 3A-E14
PO Box 4911
Warren, NJ 07060   (201) 580-5969

Maurice M. De Picciotto
AT&T
10 Independence Blvd.
P.O. Box 4911
Warren, NJ 07060   (201) 580-5966

John J. Halak
513 Warrenville Road
Warren, NJ 07060   (201) 992-9777

Alfred E. Hirsch, Jr.
6 Mountain Trail
Warren, NJ 07059   (201) 992-9777

William C. Hosford#
97 Stirling Rd.
Warren, NJ 07060   (201) 647-5564

Horst M. Kasper
13 Forest Drive
Warren, NJ 07060   (201) 757-2839

D. Laurence Padilla
AT&T Communications, Inc.
10 Independence Blvd.
Warren, NJ 07060   (201) 580-5972

Arthur S. Rosen
AT&T
10 Independence Blvd.
P.O. Box 4911
Warren, NJ 07060   (201) 580-5964

Harvey Zeller
AT&T
10 Independence Blvd.
Warren, NJ 07059   (908) 580-5970

Marilyn J. Maue
GAF Corp.
1361 Alps Rd.
Wayne, NJ 07470   (201) 628-3544

Edward P. Schmidt
124 Lake Drive W.
Wayne, NJ 07470   (201) 694-1118

Stanley J. Silverberg
American Cyanamid Co.
One Cyanamid Plaza
Wayne, NJ 07470   (201) 831-2120

Joshua J. Ward
GAF Corp.
1361 Alps Rd.
Bldg. 10
Wayne, NJ 07470   (201) 628-3529

Robert B. Green
5 Bonn Place
Weehawken, NJ 07087   (201) 863-3983

Bernard Olcott
Olcott & Blair
62 Hackensack Plank Rd.
Weehawken, NJ 07087   (201) 863-4201

Arthur I. Spechler
16 Alluvium Lakes Drive
West Berlin, NJ 08091   (609) 435-4609

Robert D. Polucki
RICOH Corporation
5 Dedrick Place
West Caldwell, NJ 07006   (201) 882-2128

Robert M. Skolnik
353 Monmouth Road
West Long Branch, NJ 07764
          (201) 222-1027

Mitchell G. Condos
307 Northfield Avenue
West Orange, NJ 07052   (201) 731-4375

Elijah H. Gold#
10 Roosevelt Avenue
West Orange, NJ 07052   (201) 736-3021

Clay Holland, Jr.
51 Brookside Road
West Orange, NJ 07052   (201) 736-4689

J. Russell Juten
7 Undercliff Terr.
West Orange, NJ 07052   (201) 731-6474

Burton I. Levine
71 Burnett Terrace
West Orange, NJ 07052   (201) 898-6957

John T. O Halloran
45 Fran Ave.
West Trenton, NJ 08628   (609) 882-8251

Thomas E. Arther#
408 Everson Place
Westfield, NJ 07040   (201) 283-0238

Peter J. Butch, III
Lerner, David, Littenberg,
Krumholz & Mentlik
600 South Avenue, West
Westfield, NJ 07090   (201) 654-5000

Robert B. Cohen
Lerner, David, Littenberg,
Krumholz & Mentlik
600 South Avenue, West
Westfield, NJ 07090   (201) 654-5000

Sidney David
Lerner, David, Littenberg,
Krumholz & Mentlik
600 South Avenue, West
Westfield, NJ 07090   (201) 654-5000

Arnold B. Dompieri
Lerner, David, Littenberg,
Krumholz & Mentlik
600 South Avenue, West
Westfield, NJ 07090   (201) 654-5000

Keith E. Gilman
Lerner, David, Littenberg,
Krumholz & Mentlik
600 South Avenue, West
Westfield, NJ 07090   (201) 654-5000

Stephen B. Goldman
Lerner, David, Littenberg,
Krumholz & Mentlik
600 South Avenue, West
Westfield, NJ 07090   (201) 654-5000

Charles A. Harris
8 Stoneleigh Park
Westfield, NJ 07090   (201) 232-6043

Walter Katz
116 Summit Ct.
Westfield, NJ 07090   (201) 233-5651

Paul H. Kochanski
Lerner, David, Littenberg,
Krumholz & Mentlik
600 South Avenue, West
Westfield, NJ 07090   (201) 654-5000

Marianne M. Kriman
Technitas International Trading
837 Carleton Road
Westfield, NJ 07090   (201) 233-6519

Arnold H. Krumholz
Lerner, David, Littenberg,
Krumholz & Mentlik
600 South Avenue, West
Westfield, NJ 07090   (201) 654-5000

Lawrence Irwin Lerner
Lerner, David, Littenberg,
Krumholz & Mentlik
600 South Avenue, West
Westfield, NJ 07090   (201) 654-5000

Joseph Saul Littenberg
Lerner, David, Littenberg,
Krumholz & Mentlik
600 South Avenue, West
Westfield, NJ 07090   (201) 654-5000

William L. Mentlik
Lerner, David, Littenberg,
Krumholz & Mentlik
600 South Avenue, West
Westfield, NJ 07090   (201) 654-5000

Marcus J. Millet
Lerner, David, Littenberg,
Krumholz & Mentlik
600 South Avenue, West
Westfield, NJ 07090   (201) 654-5000

Theodore Moss
611 Embree Crescent
Westfield, NJ 07090   (201) 654-4318

John R. Nelson
Lerner, David, Littenberg,
Krumholz & Mentlik
600 South Avenue, West
Westfield, NJ 07090   (201) 654-5000

Richard Manford Rabkin
245 Delaware Street
Westfield, NJ 07090   (201) 233-2467

Bruce H. Sales
Lerner, David, Littenberg,
Krumholz & Mentlik
600 South Avenue, West
Westfield, NJ 07090   (201) 654-5000

Elwood Joseph Schaffer
636 Prospect St.
Westfield, NJ 07090   (201) 232-1840

William Hays Smyers
229 Sylvania Pl.
Westfield, NJ 07090   (201) 233-2284

David Teschner
Lerner, David, Littenberg,
Krumholz & Mentlik
600 South Avenue, West
Westfield, NJ 07090   (201) 654-5000

Michael H. Teschner
Lerner, David, Littenberg,
Krumholz & Mentlik
600 S. Avenue, West
Suite 300
Westfield, NJ 07090   (201) 654-5000

Roy H. Wepner
Lerner, David, Littenberg,
Krumholz & Mentlik
600 South Avenue, West
Westfield, NJ 07090   (201) 654-5000

Craig H. Evans
56 Addison Avenue
Westmont, NJ 08108   (609) 858-4341

Richard P. Moon
739 Upper Way
Wharton, NJ 07885   (201) 366-3450

Lester H. Birnbaum
AT&T Bell Laboratories
1 Whippany Road
Whippany, NJ 07981   (201) 386-6377

Charles Edgar Graves
AT&T Bell Laboratories
1 Whippany Road
Whippany, NJ 07981   (201) 386-4475

Ruloff F. Kip, Jr.
AT&T Bell Laboratories
1 Whippany Road
Room 6B-107
Whippany, NJ 07981   (201) 386-6834

Alfred George Steinmetz
AT&T Bell Laboratories
1 Whippany Rd.
Whippany, NJ 07981   (201) 386-2718

James J. Trainor, Sr.
AT&T Bell Laboratories
1 Whippany Rd., Rm. 6B-102
Whippany, NJ 07981   (201) 386-4125

Eric A. La Morte
18 42nd Street
Whitehouse Station, NJ 08889
                              (201) 386-4125

Charlotte Howell Copperthite
Amerada Hess Corporation
1 Hess Plaza
Woodbridge, NJ 07095   (908) 750-6535

Ezra Sutton
Plaza 9
900 Route 9
Woodbridge, NJ 07095   (201) 634-3520

Alfred E. Riccardo
168 Pascack Road
Woodcliff Lake, NJ 07675   (201) 391-0079

Eugene C. Rzucidlo
48 Shaw Road
Woodcliff Lake, NJ 07675   (201) 573-0227

New Mexico

Edward L. Amonette#
5715 El Prado N.W.
Albuquerque, NM 87107   (505) 344-4143

Richard A. Bachand
5345 Wyoming Blvd. N.E.
Suite 107
Albuquerque, NM 87109   (505) 828-0411

James H. Chafin*
U.S. Dept. of Energy
Texas & K Street
Albuquerque, NM 87115   (505) 844-8231

Anne D. Daniel*
U.S. Dept. of Energy
P.O. Box 5400
Albuquerque, NM 87115   (505) 846-3123

Donovan F. Duggan
Montgomery & Andrews
201 Third St. N.W.
Suite 1300
P.O. Box 26927
Albuquerque, NM 87125   (505) 242-9677

Robert W. Harris
Poole, Tinnin & Martin, P.C.
P.O. Box 1769
Albuquerque, NM 87103   (505) 842-8155

Milton R. Kestenbaum
P.O. Box 5359
Albuquerque, NM 87185   (505) 844-8015

Dudley W. King
2510 Gen. Arnold N.E.
Albuquerque, NM 87112   (505) 299-0043

John R. Lansdowne
Singer, Smith & Williams, P.A.
300 Central SW, Suite 2000 West
P.O. Box 25565
Albuquerque, NM 87125   (505) 247-9532

George H. Libman
Sandia National Laboratories
Dept. 4050
P.O. Box 5800
Albuquerque, NM 87185   (505) 844-2125

Armand Mc Millan*
Dept. of Energy
Albuq. Operations Off.
Box 5400 Kirkland AFB
Albuquerque, NM 87115   (505) 844-8231

De Witt M. Morgan
Rodey, Dickason, Sloan, Akin
& Robb, P.A.
201 3rd St., N.W., Ste. 2200
P.O. Box 1888
Albuquerque, NM 87103   (505) 765-5900

Karla J. Ojanen
621 Georgia, Southeast
Albuquerque, NM 87108   (505) 266-8880

Kurt C. Olsen
Sandia National Laboratories
Pat. Dept. 4050
P.O. Box 5800
Albuquerque, NM 87185   (505) 845-9177

Deborah A. Peacock
Montgomery & Andrews
201 Third St. N.W.
Suite 1300
P.O. Box 26927
Albuquerque, NM 87125   (505) 242-9677

Frances Richey#
528-B Cardens St. S.E.
Albuquerque, NM 87108   (505) 255-6493

Donald P. Smith
Singer, Smith & Williams
P.O. Box 25565
Albuquerque, NM 87125   (505) 247-3911

Albert Sopp*
Univ. of New Mexico
Roma, NE
Scholes Hall, Rm. 102
Albuquerque, NM 87131   (505) 277-7646

Robert W. Weig
4701 Cutler N.E.
Albuquerque, NM 87110   (505) 884-3066

Charles C. Wells, Jr.
4741 Danube Drive
Albuquerque, NM 87111   (505) 296-2465

Hiram B. Gilson#
601 W. Orchard
Apt. 45
Carlsbad, NM 88220   (505) 885-3296

Dennis F. Armijo
3202 Aspen Dr.
Farmington, NM 87401   (505) 327-3543

Roger F. Phillips*
Department of the Air Force
Air Force Contract Mgmt. Div.
AFCMD/JAT
Kirtland A F B, NM 87117   (505) 844-0633

John A. Darden, III
Darden, Valentine & Driggers
200 W. Las Cruces Ave.
Las Cruces, NM 88001   (505) 526-6655

John H. Muetterties
2265 Durango Court
Las Cruces, NM 88001   (505) 521-1711

Bruce H. Cottrell
Los Alamos National Laboratory
MS D412 LC-IP
Los Alamos, NM 87505   (505) 667-3302

William A. Eklund
P.O. Box 822
Los Alamos, NM 87544   (505) 667-3766

Samuel M. Freund
11 Timber Ridge Road
Los Alamos, NM 87544   (505) 667-9701

Paul D. Gaetjens
Los Alamos National Lab.
MS-D412
Los Alamos, NM 87545   (505) 667-9701

Edward C. Walterscheid
Univ. of California
Los Alamos Natl. Lab.
P.O. Box 1663
Los Alamos, NM 87545   (505) 667-3970

Ray G. Wilson
Los Alamos National Laboratory
ADLC/PL, MS M326
P.O. Box 1663
Los Alamos, NM 87545   (505) 667-3302

Milton D. Wyrick
Los Alamos National Laboratory
P.O. Box 1663
LC/IP, MS D412
Los Alamos, NM 87545   (505) 667-3302

Richard J. Cordovano
735 Descanso Road
Santa Fe, NM 87501   (505) 988-7626

Robert B. Frailey
28 Camerada Road
Santa Fe, NM 87505   (505) 988-4281

Henry Heyman
RR 1, Box 201-B
Santa Fe, NM 87501   (505) 455-2473

George W. Price
2243 Calle Cacique
Santa Fe, NM 87505   (505) 982-9167

James Elbert Snead, III
Jones, Snead, Wertheim,
Rodriquez & Wentworth
215 Lincoln Avenue
P.O. Box 2228
Santa Fe, NM 87504   (505) 982-0011

Thomas A. Wilson
1117 North Luna
Santa Fe, NM 87501   (505) 982-6139

Edward C. Jason
P.O. Box 311
Williamsburg, NM 87942   (505) 894-6929

New York

Charles L. Guettel#
420 Sand Creek Road
Apt. 1-229
Albany, NY 12205   (518) 438-8078

Susan F. Gullotti
Heslin & Rothenberg, P.C.
450 New Karner Road
P.O. Box 12695
Albany, NY 12212   (518) 452-5600

Robert E. Heslin
Heslin & Rothenberg, P.C.
Wall Street Center
450 New Karner Road
P.O. Box 12695
Albany, NY 12212   (518) 452-5600

James J. Lichiello
Heslin & Rothenberg, P.C.
450 New Karner Road
Wall Street Center
P.O. Box 12695
Albany, NY 12212   (518) 785-5507

Nicholas Mesiti
Heslin & Rothenberg, P.C.
Wall Street Center
450 New Karner Road
P.O. Box 12695
Albany, NY 12695   (518) 452-5600

Allen C. Miller, Jr.
36 Willet Street
Box 7011
Albany, NY 12225   (518) 462-2382

Fredric T. Morelle
Schmeiser, Morelle & Watts
24 Computer Drive West
Albany, NY 12205   (518) 458-1850

Kevin P. Radigan
Heslin & Rothenberg, P.C.
Wall Street Center
450 New Karner Road
P.O. Box 12695
Albany, NY 12212   (518) 452-5600

Jeffrey Rothenberg
Heslin & Rothenberg, P.C.
450 New Karner Road
Albany, NY 12205   (518) 452-5600

Albert L. Schmeiser
Schmeiser, Morelle & Watts
24 Computer Drive West
Albany, NY 12205   (518) 458-1850

Lila T. Secrist#
Schmeiser, Morelle & Watts
24 Computer Drive West
Albany, NY 12205   (518) 458-1850

Susan J. Timian
Heslin & Rothenberg, P.C.
450 New Karner Road
P.O. Box 12695
Albany, NY 12212   (518) 452-5600

Charles T. Watts
Schmeiser, Morelle & Watts
24 Computer Drive West
Albany, NY 12205   (518) 458-1850

Jay R. Yablon
New York State Legislative
Commission on Science & Tech.
Agency Bldg. 4, 12th Floor
Albany, NY 12248   (518) 455-3919

David R. Schuster#
85 Manor Oak Drive
Amherst, NY 14228   (518) 455-3919

Jayme A. Huleatt
34th General Hospital
P.O. Box 111
Apo, NY 09178   (498) 215-1673

Roger S. Benjamin
25 Concord Road
Ardsley, NY 10502   (914) 693-6788

Harry Falber
CIBA-GEIGY Corp.
444 Saw Mill River Rd.
Ardsley, NY 10502   (914) 347-4700

Norbert Gruenfeld#
CIBA-GEIGY Corp.
444 Saw Mill River Rd.
Ardsley, NY 10502   (914) 347-4700

Paul John Juettner
598 Ashford Avenue
Ardsley, NY 10502   (914) 693-2969

Kevin T. Mansfield
CIBA-GEIGY Corp.
Patent Dept.
444 Saw Mill River Road
Ardsley, NY 10502   (914) 347-4700

Stephen V. O Brien
CIBA-GEIGY Corporation
444 Saw Mill River Road
Ardsley, NY 10502   (800) 431-1900

Norman N. Spain
31 Mt. View Avenue
Ardsley, NY 10502   (800) 431-1900

Leon E. Tenenbaum
67 Prospect Avenue
Ardsley, NY 10502   (914) 693-1315

Ernest F. Weinberger
51 Eastern Drive
Ardsley, NY 10502   (914) 693-1629

Carl C. Kling
22 Annandale St.
Armonk, NY 10504   (914) 273-8009

Kenneth J. Stempler
P.O. Box 523
Armonk, NY 10583   (914) 273-5494

David J. Zobkiw
1621 Hoover Rd.
R.R. 3
Attica, NY 14011   (716) 591-2854

Salvatore A. Alamia
42 Fire Island Avenue
Babylon, NY 11702   (516) 669-1730

R. Glenn Schroeder#
30 Totten Place
Babylon, NY 11702   (516) 669-0574

Philip M. Weiss#
3245 Bertha Drive
Baldwin, NY 11510   (516) 223-7664

Donald M. Winegar
2 Hillcrest Drive
Ballston Lake, NY 12189   (518) 877-5451

Michael T. Frimer
23-45 Bell Blvd.
Bayside, NY 11360   (718) 428-4705

Mary A. Whiting
18-15 215th St.
Bayside, NY 11361   (718) 423-9196

Marvin D. Genzer
New Indian Hill Road
Bedford, NY 10506   (914) 234-7741

Edward S. Drake#
Box 104
Bedford Hills, NY 10507   (914) 232-3051

George J. Brandt, Jr.
40 Ontario Road
Bellerose Village, NY 11001   (516) 437-8753

Harmon S. Potter
15 Rogers Ave.
Bellport, NY 11713   (516) 286-0096

Walter S. Zebrowski
3018 Wynnwood Drive
Big Flats, NY 14814   (607) 562-8770

Douglas M. Clarkson
48 Helen St.
Binghamton, NY 13905   (607) 797-6619

Mark E. Levy
Salzman & Levy
Press Bldg.
19 Chenango Street
Suite 606
Binghamton, NY 13901   (607) 722-6600

Robert S. Salzman
Salzman & Levy
19 Chenango St.
Suite 606
Binghamton, NY 13901   (607) 722-6600

Richard Glenn Stephens
Two Court St.
4th Floor
Binghamton, NY 13901   (607) 723-8295

Andrew Taras#
25 Clifton Blvd.
Binghamton, NY 13903   (607) 722-3915

John W. Young
22 Riverside Dr.
Binghamton, NY 13905   (607) 722-3426

Saul R. Bresch
28 Cypress Lane
Briarcliff Manor, NY 10510   (914) 762-4193

George E. Ham#
284 Pine Road
Briarcliff Manor, NY 10510   (914) 762-5682

Murray Nanes
63 Orchard Road
Briarcliff Manor, NY 10510   (914) 762-2342

S. C. Yuter
407 Cedar Drive West
Briarcliff Manor, NY 10510   (914) 762-0111

Ferdinand F. E. Kopecky
3356 Bronx Blvd.
Bronx, NY 10467   (212) 231-3659

Daniel J. Muccio
738 Calhoun Avenue
Bronx, NY 10465   (212) 518-0826

Alvin S. Rohssler
1790 Bruckner Blvd.
Bronx, NY 10473   (212) 991-8259

Leroy Eason
72 Pondfield Road West
Bronxville, NY 10708   (914) 793-5983

Janet C. Lentz
21 Sycamore Street
Bronxville, NY 10708   (914) 793-6049

Adrian T. Calderone
949 Bay Ridge Avenue
Brooklyn, NY 11219   (718) 745-2898

Gary Cohen
141 Joralemon Street
Brooklyn, NY 11201   (212) 624-6669

Stanley G. Harvey
5502 Kings Hwy.
Brooklyn, NY 11203   (212) 451-0756

Mark H. Jay#
P.O. Box 020083
General Post Office
Brooklyn, NY 11202   (718) 625-0399

Otto S. Kauder
Witco Corporation
Argus Division
633 Court Street
Brooklyn, NY 11231   (718) 858-5678

Ronald Lianides
350 65th Street
Brooklyn, NY 11220   (212) 884-6600

Bruce S. Londa
1111 Eighth Avenue
Brooklyn, NY 11215   (212) 971-9172

James J. Long#
6808 Tenth Avenue
Brooklyn, NY 11219   (718) 745-2185

Daniel R. Mc Glynn
266-76th Street
Brooklyn, NY 11209   (718) 238-6720

Nikolay Parada
1066 E. 13th St.
Brooklyn, NY 11230   (718) 258-8766

Gasper P. Quartararo#
Witco Corp.
633 Court Street
Brooklyn, NY 11231   (212) 858-5678

Lowell M. Rubin
141 Argyle Road
Brooklyn, NY 11218   (212) 282-4377

Robert Sherman
415 Beverly Rd., #5C
Brooklyn, NY 11218   (718) 436-0605

Perry Teitelbaum
Goodman & Teitelbaum
26 Court St.
Suite 1400
Brooklyn, NY 11242   (718) 643-0400

Richard T. Treacy#
34 Plaza St.
Apt. 1004
Brooklyn, NY 11238   (718) 802-4612

William Paul Vafakos
967 East 17th St.
Brooklyn, NY 11230   (212) 253-5014

Guy J. Agostinelli, III#
Kavinoky & Cook
120 Delaware Avenue
Buffalo, NY 14202   (716) 856-9234

Edwin T. Bean, Jr.
Hodgson, Russ, Andrews, Woods
& Goodyear
1800 One M&T Plaza
Buffalo, NY 14203   (716) 856-4000

John B. Bean
Bean, Kaufman & Bean
1313 Liberty Bk. Bldg.
Buffalo, NY 14202   (716) 852-7405

Thomas F. Daley
LTV Aerospace & Defense
Sierra Res. Division
247 Cayuga Road
Buffalo, NY 14225   (716) 681-6206

Joseph P. Gastel
722 Ellicott Square Bldg.
Buffalo, NY 14203   (716) 854-6285

Arthur F. Holz
Saperston & Day
1100 Goldome Center
Buffalo, NY 14203   (716) 856-5400

R. Craig Kauffman
Bean, Kauffman & Bean
1313 Liberty Bk. Bldg.
420 Main St.
Buffalo, NY 14202   (716) 852-7405

Alan S. Korman#
Hodgson, Russ, Andrews,
Woods & Goodyear
1800 One M&T Plaza
Buffalo, NY 14203   (716) 853-7778

Martin Gates Linihan, Jr.
Hodgson, Russ, Andrews,
Woods & Goodyear
1800 One M&T Plaza
Buffalo, NY 14203   (716) 856-4000

Laird F. Miller
National Gypsum Company
1650 Military Road
Buffalo, NY 14217   (716) 873-9751

Kellie Marie Muffoletto#
Saperston & Day, P.C.
Goldome Center
One Fountain Plaza
Buffalo, NY 14203   (716) 856-5400

James S. Nowak#
Saperston & Day
One Fountain Plaza
Goldome Center
Buffalo, NY 14203   (716) 856-5400

Daniel C. Oliverio
Sommer, Oliverio & Sommer
920 Liberty Bldg.
420 Main Street
Buffalo, NY 14202   (716) 853-7761

James C. Simmons
Hodgson, Russ, Andrews
Woods & Goodyear
1800 One M&T Plaza
Buffalo, NY 14203   (716) 856-4000

Peter K. Sommer
Sommer, Oliverio & Sommer
920 Liberty Bldg.
420 Main Street
Buffalo, NY 14202   (716) 853-7761

Alan H. Spencer
Bean, Kauffman & Spencer
1313 Liberty Bldg.
420 Main Street
Buffalo, NY 14202   (716) 852-7405

Kathleen R. Terry
State University of N.Y
At Buffalo
516 Capen Hall
Buffalo, NY 14260   (716) 636-3321

Ivan A. Mc Corkendale
190 Parrish St., Ext 51
Canandaigua, NY 14424   (716) 394-2192

Alfred B. Engelberg
Sedgewood Club
R.D. 12
Carmel, NY 10512   (914) 225-7099

Robert K. Bair#
3 Brashear Place
Castleton, NY 12033   (518) 732-7132

David M. Warren
655 Oakland Avenue
Cedarhurst, NY 11516   (516) 295-2054

John Finnie
152 Little Neck Road
Centerport, NY 11721   (516) 757-3085

Daniel H. Kane
22 Spring Hollow Road
Centerport, NY 11721   (516) 261-5712

William G. Valance#
P.O. Box 113
Central Islip, NY 11722   (516) 234-7698

Anthony Amaral, Jr.
575 Quaker Road
Chappaqua, NY 10514   (914) 238-8164

Howard G. Rath#
6 Valley View Road
Chappaqua, NY 10514   (914) 238-9304

James C. Jangarathis
366 Veterans Memorial Hwy.
Second Floor North
Commack, NY 11725   (516) 543-5563

Steven L. Krantz
Migliore & Infranco, P.C.
358 Veterans Memorial Hwy.
Commack, NY 11725   (516) 543-3663

Edwin D. Schindler
Five Hirsch Avenue
Coram, NY 11727   (516) 474-5373

Charles Q. Buckwalter, Jr.
Corning Glass Works
SP-FR-02-12
Corning, NY 14831   (607) 974-3455

Stephen R. Christian
Corning Glass Works
SP-FR-02-12
Corning, NY 14831   (607) 974-3956

Reginald J. Falkowski
Dresser-Rand Company
Barton Steuben Place
Corning, NY 14830   (607) 937-6444

Alexander R. Herzfeld
Corning Glass Works
Pat. Operations
SP-FR-02-12
Corning, NY 14831   (607) 974-3381

Clinton S. Janes, Jr.
Corning Glass Works
SP-FR-02-12
Corning, NY 14831   (607) 974-3323

Alfred L. Michaelsen
Corning Glass Works
Sullivan Park
SP-FR-02-12
Corning, NY 14831   (607) 974-3054

Milton M. Peterson
Corning Glass Works
Patent Dept.
SP-FR-02-12
Corning, NY 14831   (607) 974-3378

Timothy M. Schaeberle#
Corning Glass Works
SP-FR-02-12
Corning, NY 14830   (607) 974-3054

William J. Simmons, Jr.
Corning Glass Works
SP-FR-02-12
Corning, NY 14831   (607) 974-3322

Kenneth M. Taylor, Jr.
Corning Glass Works
SP-FR-02-12
Corning, NY 14831   (607) 974-3986

Kees Van Der Sterre
Corning Glass Works
Sullivan Park
SP-FR-02-12
Corning, NY 14831   (607) 974-3294

Richard N. Wardell
Corning Glass Works
SP-FR-02-12
Corning, NY 14831   (607) 974-3321

Kenneth William Greb#
Smith Corona Corporation
839 Route 13 South
Cortland, NY 13045   (607) 753-6011

Anthony J. Barbaro#
Technology Resource Group
P.O. Box 32
Cross River, NY 10518   (914) 763-8346

Paul S. Hoffman
139 Grand Street
P.O. Box 40
Croton-On-Hudson, NY 10520
                         (914) 271-5191

Jonathan B. Schafrann
17 Arlington Drive
Croton-On-Hudson, NY 10520
(212) 687-6625

Joseph Martin Weigman
9N Dove Court
P.O. Box 343
Croton-On-Hudson, NY 10520
(914) 271-6346

Michael F. Brown
Georgetown Manor, Inc.
150 W. Industry Court
Deer Park, NY 11729   (516) 242-0700

Ralph Cohen
58 Murray Ave.
Delmar, NY 12054   (518) 439-4685

Anne M. Rosenblum
163 Delaware Avenue
Suite 212
Delmar, NY 12054   (518) 475-0611

Annette M. Sansone
31 Brookedge Road
Depew, NY 14043   (716) 668-4860

Salvatore J. Levanti
10 Parkside Drive
Dix Hills, NY 11746   (516) 499-5376

Richard L. Miller#
12 Parkside Drive
Dix Hills, NY 11746   (516) 499-4343

Louis A. Morris
Akzo America Inc.
Livingston Avenue
Dobbs Ferry, NY 10522   (914) 693-1200

James K. Poole
Akzo American, Inc.
Patent & Trademark Dept.
1 Livingstone Ave.
Dobbs Ferry, NY 10522   (914) 693-1200

David H. Vickrey
Akzo America, Inc.
Pat. & Tdmk. Dept.
Livingstone Avenue
Dobbs Ferry, NY 10522   (914) 693-1200

John Edward Dumaresq
214 Manor Road
Douglaston, NY 11363   (718) 229-8373

Kevin E. Kavanagh
241-21 Rushmore Avenue
Douglaston, NY 11362   (718) 224-4660

Eric J. Sheets#
240-01 43rd Avenue
Douglaston, NY 11363   (212) 428-8721

Bernard Stimler
53-16 244th Street
Dc uglaston, NY 11362   (718) 229-3507

Robert C. Woodbury
87 East Fourth Street
P.O. Box 800
Dunkirk, NY 14048   (716) 366-8050

Elmer J. Lawson
Box 137
Best Road
East Greenbush, NY 12061   (518) 286-2893

William Gatewood Webb#
4 Timber Ledge
East Greenbush, NY 12061   (518) 477-7221

Gilbert P. Weiner
Pall Corp.
2200 Northern Blvd.
East Hills, NY 11548   (516) 484-5400

Harvey Lunenfeld
8 Patrician Drive
East Northport, NY 11731   (516) 754-1000

William E. Mear, III
300 Manor Drive
East Syracuse, NY 13057   (315) 656-8270

Theodore J. Koss, Jr.
81-11 Pettit Avenue
Elmhurst, NY 11373   (718) 898-8002

Lynn Lawrence Augspurger
IBM
Intellectual Property Law
N50/251-2
1701 N. Street
Endicott, NY 13760   (607) 755-0123

Shelley M. Beckstrand
IBM Corp.
1701 North Street
Endicott, NY 13760   (607) 757-6760

Lawrence R. Fraley
IBM Corporation
Intellectual Property Law Dept.
N50/251-2
1701 North Street
Endicott, NY 13760   (607) 757-6755

Pryor A. Garnett
IBM Corporation
Intellectual Property Law Dept.
N50/251-2
1701 North Street
Endicott, NY 13760   (607) 757-6743

John S. Gasper
IBM Corp.
1701 North Street
Endicott, NY 13760   (607) 757-6757

Richard M. Goldman
IBM Corporation
Intellectual Property Law Dept.
N50/251-2
1701 North Street
Endicott, NY 13760   (607) 757-6771

Judith D. Olsen
IBM Corp.
Bldg. 251-2, Dept. N50
1701 North St
Endicott, NY 13760   (607) 757-6745

Arthur J. Samodovit#
IBM Corporation
1701 North Street
Dept. N50/251-2
Endicott, NY 13760   (607) 757-6755

William H. Steinberg
IBM Corp.
1701 N. St.
Dept. N50/251-2
Endicott, NY 13760   (607) 757-6753

Norman R. Bardales
2735 Alexander Street
Endwell, NY 13760   (607) 754-3281

Paul M. Brannen#
609 Lacey Drive
Endwell, NY 13760   (607) 785-3128

Francis V.B. Giolma
3704 Frazier Rd.
Endwell, NY 13760   (607) 748-9242

George Weldon Killian
17 Charing Cross
Fairport, NY 14450   (716) 224-4839

J. Addison Mathews
415 Thayer Road
Fairport, NY 14450   (716) 253-6188

Carl W. Baker
7 The Gazebo
Fayetteville, NY 13066   (315) 637-4732

Frank N. Decker, Jr.
6954 Old Quarry Road
Fayetteville, NY 13066   (315) 637-4732

Lawrence P. Trapani
6883 East Genesee St.
P.O. Box 469
Fayetteville, NY 13066   (315) 445-8856

John P. Zacharias
145 Main Street
P.O. Box H
Fishkill, NY 12524   (914) 896-8643

John Maier, III
Main Street
Box C-1
Fleischmanns, NY 12430   (914) 254-5322

Nicholas J. Garofalo
259-09 81st Ave.
Floral Park, NY 11004   (212) 347-1619

Alfred W. Barber
32-44 Francis Lewis Blvd.
Flushing, NY 11358   (212) 463-3306

Seymour Gerald Bekelnitzky
147-09 72nd Dr.
Flushing, NY 11367   (718) 261-6475

George M. Kaplan
58-04 190 Street
Flushing, NY 11365   (212) 357-6230

Paul S. Martin
189-54 43rd Rd.
Flushing, NY 11358   (212) 358-5465

Nathaniel Altman#
68-37 Yellowstone Blvd.
Forest Hills, NY 11375   (212) 268-4541

J. Philip Anderegg
50 Exeter St.
Forest Hills, NY 11375   (212) 268-0206

Alexander Mencher
69-42 Ingram Street
Forest Hills, NY 11375   (212) 268-3807

Paul Sandler
108-50 71st Avenue
Forest Hills, NY 11375   (718) 268-2199

Julius Balogh
1046 Lorraine Drive
Franklin Square, NY 11010   (516) 775-0077

Paul W. Garbo#
48 Lester Ave.
Freeport, NY 11520   (516) 378-0393

Diane R. Bentley#
Scully, Scott, Murphy
& Presser, P.C.
400 Garden City Plaza
Garden City, NY 11530   (516) 742-4343

Donald T. Black
Scully, Scott, Murphy & Presser, P.C.
400 Garden City Plaza
Garden City, NY 11530   (516) 742-4343

Marvin Bressler
Scully, Scott, Murphy
& Presser, P.C.
400 Garden City Plaza
Garden City, NY 11530   (516) 742-4343

Ralph E. Bucknam
Bucknam & Archer
600 Old Country Road
Garden City, NY 11530   (516) 549-0956

Richard L. Catania
Scully, Scott, Murphy
& Presser, P.C.
400 Garden City Plaza
Garden City, NY 11530   (516) 742-4343

Mark J. Cohen
Scully, Scott, Murphy
& Presser, P.C.
400 Garden City Plaza
Garden City, NY 11530   (516) 742-4343

Frank S. Di Giglio
Scully, Scott, Murphy
& Presser, P.C.
400 Garden City Plaza
Garden City, NY 11530   (516) 742-4343

Robert J. Eichelburg
Scully, Scott, Murphy
& Presser, P.C.
400 Garden City Plaza
Garden City, NY 11530   (516) 742-4343

Paul J. Esatto, Jr.
Scully, Scott, Murphy
& Presser, P.C.
400 Garden City Plaza
Garden City, NY 11530   (516) 742-4343

Carl J. Evens#
Scully, Scott, Murphy
& Presser, P.C.
400 Garden City Plaza
Garden City, NY 11530   (516) 742-4343

Paul J. Farrell#
Scully, Scott, Murphy
& Presser, P.C.
400 Garden City Plaza
Garden City, NY 11530   (516) 742-4343

Fernanda Misani Fiordalisi
Bucknam & Archer
600 Old Country Road
Garden City, NY 11530   (516) 222-8885

Steven Fischman#
Scully, Scott, Murphy
& Presser, P.C.
400 Garden City Plaza
Garden City, NY 11530   (516) 742-4343

Richard Gail Geib
64 Whitehall Blvd.
Garden City, NY 11530   (516) 248-7443

Edward W. Grolz
Scully, Scott, Murphy
& Presser, P.C.
400 Garden City Plaza
Garden City, NY 11530   (516) 742-4343

Kenneth Lee King
Scully, Scott, Murphy & Presser, P.C.
400 Garden City Plaza
Garden City, NY 11530   (516) 742-4343

Ralph J. Mancini
Scullyl, Scott, Murphy
& Presser, P.C.
400 Garden City Plaza
Garden City, NY 11530   (516) 742-4343

Robert Mauer
1205 Franklin Avenue
Garden City, NY 11530   (516) 248-5803

William E. Mc Nulty
Scully, Scott, Murphy
& Presser, P.C.
400 Garden City Plaza
Garden City, NY 11530   (516) 742-4343

Jules M. Mencher
550 Stewart Avenue
Garden City, NY 11530   (516) 227-2221

Keith J. Murphy#
Scully, Scott, Murphy
& Presser, P.C.
400 Garden City Plaza
Garden City, NY 11530   (516) 742-4343

Stephen D. Murphy
Scully, Scott, Murphy
& Presser, P.C.
400 Garden City Plaza
Garden City, NY 11530   (516) 742-4343

James J. O'Connell
Scully, Scott, Murphy
& Presser, P.C.
400 Garden City Plaza
Garden City, NY 11530   (516) 742-4343

Gennaro L. Pasquale
600 Old Country Road
Suite 338
Garden City, NY 11530   (516) 222-2727

Leopold Presser
Scully, Scott, Murphy
& Presser, P.C.
400 Garden City Plaza
Garden City, NY 11530   (516) 742-4343

William C. Roch
Scully, Scott, Murphy
& Presser, P.C.
400 Garden City Plaza
Suite 300
Garden City, NY 11530   (516) 742-4343

Philip Sands
600 Old Country Rd.
Suite 529
Garden City, NY 11530   (516) 745-6677

Anthony C. Scott
Scully, Scott, Murphy
& Presser, P.C.
400 Garden City Plaza
Garden City, NY 11530   (516) 742-4343

John Francis Scully
Scully, Scott, Murphy
& Presser, P.C.
400 Garden City Plaza
Garden City, NY 11530   (516) 742-4343

John S. Sensny
Scully, Scott, Murphy
& Presser, P.C.
400 Garden City Plaza
Garden City, NY 11530   (516) 742-4343

Bruno P. Struzzi
Scully, Scott, Murphy & Presser, P.C.
400 Garden City Plaza
Garden City, NY 11530   (516) 742-4343

Bruce S. Weintraub
Scully, Scott, Murphy
& Presser, P.C.
400 Garden City Plaza
Garden City, NY 11530   (516) 742-4343

Patricia A. Wilczynski
Scully, Scott, Murphy
& Presser, P.C.
400 Garden City Plaza
Garden City, NY 11530   (516) 742-4343

M. Lisa Wilson#
Scully, Scott, Murphy &
Presser, P.C.
400 Garden City Plaza
Garden City, NY 11530   (516) 742-4343

Frances L. Olmsted
Route 9
Garrison, NY 10524   (914) 424-4077

Edgar N. Jay
Raynham Road
Glen Cove, NY 11542   (516) 671-0634

Stewart F. Moore
5 Oakwood Drive
Glens Falls, NY 12804   (518) 798-3844

Thomas J. Cione#
Norton & Christensen
60 Erie Street
Box 308
Goshen, NY 10924   (914) 294-7949

Alan D. Akers#
Moore Business Forms, Inc.
300 Lang Blvd.
Grand Island, NY 14072   (716) 773-0326

Robert F. Hause
3663 West River Pkwy.
Grand Island, NY 14072   (716) 773-7038

James F. Tao
Occidental Chemical Corporation
2801 Long Road
Grand Island, NY 14072   (716) 773-8400

Samson B. Leavitt
66 Essex Rd.
Great Neck, NY 11023   (516) 487-8417

Albert Levine
37 Ruxton Road
Great Neck, NY 11023   (516) 466-6663

Abner Sheffer
7 Piccadilly Rd.
Great Neck, NY 11023   (516) 482-5423

Howard A. Taishoff
105 Oxford Blvd.
Great Neck, NY 11023   (516) 482-4455

Philip Young
45 Cedar Drive
Great Neck, NY 11021   (516) 487-1238

James Anthony Kane, Jr.
34 O'Connell Court
Great River, NY 11739   (516) 222-1800

James F. Young
15 Duncan Lane
Halesite, NY 11743   (516) 421-4130

Elliott E. Haymovitz
225 Parsons Street
Harrison, NY 10528   (914) 835-1707

Alfred E. Page, Jr.
480 Mamaroneck Ave.
Harrison, NY 10528   (914) 698-8200

Jack Oisher
22 Kenneth Rd.
Hartsdale, NY 10530   (914) 693-0916

Robert John Patterson
107 Harvard Dr.
Hartsdale, NY 10530   (914) 948-6190

Edward D. Weil#
6 Amherst Drive
Hastings-On-Hudson, NY 10706
          (914) 478-3370

George Wesley Houseweart
Standard Microsystems Corporation
35 Marcus Blvd.
Hauppauge, NY 11788   (516) 273-3100

Alfred M. Walker
742 Veterans Memorial Highway
Hauppauge, NY 11788   (516) 361-8737

Michael W. Glynn
CIBA-GEIGY Corporation
Seven Skyline Drive
Hawhorne, NY 10532   (914) 785-7138

Luther A. R. Hall#
CIBA-GEIGY Corp.
7 Skyline Drive
Hawthorne, NY 10532   (914) 785-7123

Edward Mccreery Roberts
CIBA-GEIGY Corp.
Pat. Dept.
7 Skyline Drive
Hawthorne, NY 10532   (914) 347-4700

Joann L. Villamizar
CIBA-GEIGY Corporation
7 Skyline Drive
Hawthorne, NY 10532   (914) 785-7121

Lawrence S. Lawrence
Lawrence, Walsh & Kolins, P.C.
215 Hilton Avenue
Hempstead, NY 11550   (516) 538-2400

William Lars Ericson
933 Westwood Drive
Herkimer, NY 13350   (315) 866-0831

Robert R. Strack
10 Sutherland Road
Hicksville, NY 11801   (516) 586-6364

Michael B. Zapantis
86-53 Dunton Street
Holliswood, NY 11423   (718) 479-6968

Aziz M. Ahsan
26 Marges Way
Hopewell Junction, NY 12533
          (914) 226-1971

Ira D. Blecker
IBM Corporation
Intellectual Property Law Dept.
Dept. 901/Bldg. 300-482
Route 52
Hopewell Junction, NY 12533
          (914) 894-2580

Jeffrey L. Brandt
IBM Corporation
Dept. 901, Bldg. 300-482
Route 52
Hopewell Junction, NY 12533
          (914) 894-6213

Edward William Brown
IBM Corp.
Dept. 901, Bldg. 300-482
Route 52
Hopewell Junction, NY 12533
          (914) 894-3665

Wesley De Bruin
IBM Corp.
Route 52 D901
B/300-482
Hopewell Junction, NY 12533
          (914) 894-4713

Anne V. Dougherty
IBM Corporation
Route 52 - Bldg. 300-482
Hopewell Junction, NY 12593
          (914) 894-6919

John W. Henderson, Jr.
IBM Corporation
Intellectual Property Law
Dept. 901, Bldg. 300-482
Hopewell Junction, NY 12533
          (914) 894-2121

Harold Huberfeld
IBM Corp.
Dept. 34H, Bldg. 300-482
East Fishkill Facility
Route 52
Hopewell Junction, NY 12533
          (914) 894-2121

Graham S. Jones, II
IBM Corp.
Dept. 901, Bldg. 300-482
Route 52
Hopewell Junction, NY 12533
          (914) 894-3668

Richard A. Romanchik
IBM Corp.
Intellectual Property Law Dept.
Dept. 18G, Bldg. 300-482
Hopewell Junction, NY 12533
          (914) 894-6919

John A. Stemwedel
IBM Corporation
Dept. 901, Bldg. 300-482
Route #52
Hopewell Junction, NY 12533
          (914) 894-3664

Yen Sung Yee
IBM Corp.
Route 52
Bldg. 300-482
Hopewell Junction, NY 12533
          (914) 894-2210

Raymond J. Keogh
19 Whitehall Drive
Huntington, NY 11743   (516) 421-0176

Joseph Michael Maguire
Box 1245
Huntington, NY 11743   (516) 271-1026

Ralph Roger Barnard
200 E. Buffalo Street
Ithaca, NY 14850   (607) 273-1711

George M. Dentes
Hines & Dentes
417 N. Cayuga Street
Ithaca, NY 14850   (607) 273-6111

Joseph A. Edminister
Cornell University
College of Engineering
221 Carpenter Hall
Ithaca, NY 14853   (607) 255-8972

A. Joseph Gibbons
Cornell Research Foundation
Pat. & Licensing Dept.
East Hill Plaza
Ithaca, NY 14850   (607) 255-7367

H. Walter Haeussler
Cornell Research Foundation
East Hill Plaza
Ithaca, NY 14850   (607) 255-7367

Cynthia M. Hannah-White#
200 East Buffalo St.
Suite 102A
Ithaca, NY 14850   (607) 273-1711

Christopher A. Michaels#
200 East Buffalo St.
Suite 102A
Ithaca, NY 14850   (607) 273-1711

Marianne R. Rich
184-52 Radnor Road
Jamaica, NY 11432   (718) 380-0576

Ronald J. Baron
Hoffmann & Baron
350 Jericho Turnpike
Jericho, NY 11753   (516) 822-3550

Gerald T. Bodner
Hoffmann & Baron
350 Jericho Turnpike
Jericho, NY 11753   (516) 822-3550

Herbert L. Boettcher
Nolte, Nolte & Hunter, P.C.
350 Jericho Turnpike
Jericho, NY 11753   (516) 935-0180

Daniel P. Burke
Hoffmann & Baron
350 Jericho Turnpike
Jericho, NY 11753   (516) 822-3550

James Albert Eisenman
Hoffmann & Baron
350 Jericho Turnpike
Jericho, NY 11753   (516) 822-3550

Christopher B. Garvey
Nolte, Nolte & Hunter
350 Jericho Turnpike
Jericho, NY 11753   (516) 822-3550

Burton S. Heiko
15 Steuben Dr.
Jericho, NY 11753   (516) 938-1929

Charles R. Hoffmann
Hoffmann & Baron
350 Jericho Turnpike
Jericho, NY 11753   (516) 822-3550

Edward B. Hunter
Nolte, Nolte & Hunter, P.C.
350 Jericho Turnpike
Jericho, NY 11753   (516) 935-0180

A. Thomas Kammer
Hoffman & Baron
350 Jericho Turnpike
Jericho, NY 11753   (516) 822-3550

Michael N. Mercanti
Hoffmann & Baron
350 Jericho Turnpike
Jericho, NY 11753   (516) 822-3550

Arlene D. Morris
Hoffmann & Baron
350 Jericho Turnpike
Jericho, NY 11753   (516) 822-3550

Albert Charles Nolte, Jr.
Nolte, Nolte & Hunter, P.C.
350 Jericho Turnpike
Jericho, NY 11753   (516) 935-0180

Alan M. Sack
Hoffmann & Baron
350 Jericho Turnpike
Jericho, NY 11753   (516) 822-3550

James A. Stauber
Hoffmann & Baron
350 Jericho Turnpike
Jericho, NY 11753   (516) 822-3550

Leonard William Suroff
12 Tompkins Ave.
Jericho, NY 11753   (516) 681-3900

Juan C. Villar
Nolte, Nolte & Hunter, P.C.
350 Jericho Turnpike
Jericho, NY 11753   (516) 935-0180

Bernard Albert Chiama
P.O. Box 477
Keuka Park, NY 14478   (315) 536-8486

Fred I. Nathanson
123-33 83rd Avenue
Apt. 403
Kew Gardens, NY 11415   (718) 793-1986

David L. Adour
223 Sky Top Drive
Apartment 4
Kingston, NY 12401   (718) 793-1986

George Edmund Clark
IBM Corporation
Intellectual Property Law
Neighborhood Road
Kingston, NY 12401   (914) 385-4833

William A. Kinnaman, Jr.
IBM Corporation
Neighborhood Road
31 BA /058
Kingston, NY 12401   (914) 385-7055

Frederick D. Poag
38 Griffin Dr., Rd 8
Kingston, NY 12401   (914) 331-1934

Robert Lance Troike
IBM Corporation
Intellectual Property Law
Neighborhood Road
Dept. 31 BA-058
Kingston, NY 12401   (914) 385-4833

Mark S. Walker
IBM Corporation
Intellectual Property Law
Neighborhood Rd.
Dept. 31 BA-058
Kingston, NY 12401   (914) 385-7057

Russell G. Pelton
3 Oxford Road
Larchmont, NY 10538   (914) 834-7749

Benjamin Theodore Sporn
49 Lismore Road
Lawrence, NY 11559   (516) 239-6932

Peter F. Casella#
987 Elliott Drive
Lewiston, NY 14092   (716) 297-3658

James F. Mudd
610 Onondaga Street
Lewiston, NY 14092   (716) 754-4052

Mario D Arrigo
7515 Morgan Road
Liverpool, NY 13090   (315) 451-2383

Karl W. Brownell#
6232 Jacques Road
Lockport, NY 14094   (716) 778-5992

Robert P. Simpson
P.O. Box 96
Lockport, NY 14108   (716) 433-1661

Robert J. Jarvis
12 Woodridge Drive
Loudonville, NY 12211   (518) 482-7368

B. Woodrow Wyatt#
317 Loudonville Road
Loudonville, NY 12211   (518) 436-4934

Nicholas Noviello, Jr.
Noviello & Noviello, P.C.
Route 6
P.O. Box 863
Mahopac, NY 10541   (914) 628-4400

Robert Black
127 Eagles Crescent
Manhasset, NY 11030   (516) 484-3423

Charles J. Knuth
24 Bournedale Road South
Manhasset, NY 11030   (516) 365-8183

Melvin E. Libby
78 Parkwoods Road
Manhasset, NY 11030   (516) 627-7638

William V. Pesce
428 Broadway
Massapequa Park, NY 11762 (516) 799-4077

Joseph P. Hammond
Kollmorgen Corp.
PCK Technology Div.
322 S. Service Road
Melville, NY 11747   (516) 454-4410

Seymour Levine
33 Clafford Lane
Melville, NY 11747   (516) 549-8748

Edward Herbert Loveman
150 Broad Hollow Road
Melville, NY 11747   (516) 421-1122

Jacob Frederick Murbach
18 Alfred Road W.
Merrick, NY 11566   (516) 379-2776

Delavan Palmer Smith
Pine Ridge Road
P.O. Box H
Millerton, NY 12546   (518) 789-3679

Myron Amer
114 Old Country Road
Suite 310
Mineola, NY 11501   (516) 742-5290

Jerome Bauer
Bauer & Schaffer
114 Old Country Rd.
Suite 344
Mineola, NY 11501   (516) 248-1050

William D. Denson
190 Willis Avenue
Mineola, NY 11501   (516) 747-0300

Joseph L. Grosso
Grosso, Cordaro & Petrilli
114 Old Country Road
Mineola, NY 11501   (516) 248-1150

Betty A. Maier
153 Andrews Road
Mineola, NY 11501   (516) 746-6235

Allen Roosevelt Morganstern
98 Willis Avenue
Mineola, NY 11501   (516) 742-9000

Murray Schaffer
Bauer & Schaffer
114 Old Country Road
Suite 344
Mineola, NY 11501   (516) 248-1050

Richard P. Fennelly
1386 Quarry Drive
Mohegan Lake, NY 10547   (914) 962-5026

Albert F. Kronman
142 N. Columbus Avenue
Mt Vernon, NY 10553   (914) 667-6755

Harold L. Burstyn
Morrison Law Firm
142 N. Columbus Ave.
Mt. Vernon, NY 10553   (914) 667-6755

Thomas R. Morrison
142 North Columbus Avenue
Mt. Vernon, NY 10553   (914) 667-6755

Sam D. Walker
100 E. 1st St.
8th Fl.
Mt. Vernon, NY 10550   (914) 664-1116

Thomas H. Whaley
559 Gramatan Ave.
Suite 206
Mt. Vernon, NY 10552   (914) 699-8705

Daniel J. Roock#
216 Patricia Drive
N. Syracuse, NY 13212   (315) 457-3167

Arthur Dresner
16 Sundbury Drive
New City, NY 10956   (914) 634-6286

Joseph Hirshfeld
25 Ardsley Drive
New City, NY 10956   (914) 634-7418

Robert Frederick Dropkin
Special Metals Corporation
Middle Settlement Road
New Hartford, NY 13413   (315) 798-2023

Lawrence D. Cutter
11 Deer Path Drive
New Paltz, NY 12561   (914) 255-3276

Howard I. Podell#
28 Beachfront Lane
New Rochelle, NY 10805   (914) 632-1245

William Harold Saltzman
2 Trenor Dr.
New Rochelle, NY 10804   (914) 738-3247

Mark J. Abate#
Morgan & Finnegan
345 Park Ave.
New York, NY 10154   (212) 758-4800

Iman 'Al-Amin Abdallah
Patent Services
1186 Broadway
Suite 711
New York, NY 10001   (212) 889-7420

Alan Michael Abrams
Witco Corp.
520 Madison Ave.
New York, NY 10022   (212) 605-3841

Samuel B. Abrams
Pennie & Edmonds
1155 Avenue of the Americas
New York, NY 10036   (212) 790-9090

Ronald Abramson
Hughes, Hubbard & Reed
One Battery Park Plaza
New York, NY 10004   (212) 837-6000

Bruce L. Adams
Adams & Wilks
84 William Street
Suite 1000
New York, NY 10038   (212) 797-5610

Andrew P. Adler
Fish & Neave
875 Third Avenue
New York, NY 10022   (212) 715-0600

Sarah B. Adriano
Cooper & Dunham
30 Rockefeller Plaza
New York, NY 10112   (212) 977-9550

Cheryl H. Agris#
Pennie & Edmonds
1155 Avenue of the Americas
New York, NY 10036   (212) 790-9090

Lawrence C. Akers
Pfizer, Inc.
235 E. 42nd St.
New York, NY 10017   (212) 573-7743

Lawrence Alaburda
Fitzpatrick, Cella, Harper
& Scinto
277 Park Avenue
42nd Floor
New York, NY 10172   (212) 758-2400

Robert F. Alario
Carter-Wallace, Inc.
767 Fifth Avenue
New York, NY 10153   (212) 758-4500

Ronald W. Alice
ITT Corporation
320 Park Avenue
New York, NY 10022   (212) 940-1622

Prabodh I. Almaula
Bristol Myers Company
345 Park Avenue
New York, NY 10154   (212) 546-3649

Stanley Louis Amberg
Davis, Hoxie, Faithfull
& Hapgood
45 Rockefeller Plaza
New York, NY 10111   (212) 757-2200

Morton Amster
Amster, Rothstein & Engelberg
90 Park Ave.
New York, NY 10016   (212) 697-5995

Richard J. Ancel
Colgate-Palmolive Company
300 Park Avenue
New York, NY 10022   (212) 310-2959

William F. Andes, Jr.
Amster, Rothstein & Ebenstein
90 Park Avenue
New York, NY 10016   (212) 697-5995

Donna Louise Angotti#
Kenyon & Kenyon
One Broadway
New York, NY 10004   (212) 425-7200

Adriane M. Antler
Pennie & Edmonds
1155 Avenue of the Americas
New York, NY 10036   (212) 790-9090

Lawrence E. Apolzon
Weiss, Dawid, Fross, Zelnick
& Lehrman, P.C.
633 Third Avenue
New York, NY 10017   (212) 953-9090

Henry W. Archer
5 Tudor City Place
New York, NY 10017   (212) 687-8341

Steven John Armstrong
Kenyon & Kenyon
One Broadway
New York, NY 10004   (212) 425-7200

Craig James Arnold
Cooper & Dunham
30 Rockefeller Plaza
New York, NY 10112   (212) 977-9550

Seth J. Atlas
Morgan & Finnegan
345 Park Avenue
New York, NY 10154   (212) 758-4800

Peter D. Aufrichtig
300 East 42nd Street
New York, NY 10017   (212) 557-5040

Raymond W. Augustin
Pfizer, Inc.
Patent Dept., 20th Floor
235 East 42nd Street
New York, NY 10017   (201) 507-7588

M. Arthur Auslander
Auslander and Thomas
505 Eighth Avenue
New York, NY 10018   (212) 594-6900

Jack Babchik
Bergadano, Zichello & Babchik
420 Lexington Avenue
New York, NY 10176   (212) 972-5560

George E. Badenoch
Kenyon & Kenyon
One Broadway
New York, NY 10004   (212) 425-7200

James William Badie
Wyatt, Gerber, Burke
& Badie
645 Madison Ave.
New York, NY 10022   (212) 826-0171

Bradford J. Badke
1301 Avenue of the Americas
New York, NY 10019   (212) 259-8000

Robert Louis Baechtold
Fitzpatrick, Cella, Harper
& Scinto
277 Park Avenue
New York, NY 10172   (212) 758-2400

W. Edward Bailey
Fish & Neave
875 Third Avenue
New York, NY 10022   (212) 715-0600

Charles P. Baker
Fitzpatrick, Cella, Harper
& Scinto
277 Park Ave.
New York, NY 10172   (212) 758-2400

Victor N. Balancia
Pennie & Edmonds
1155 Avenue of the Americas
New York, NY 10036   (212) 790-6541

Geraldine F. Baldwin
Pennie & Edmonds
1155 Avenue of the Americas
New York, NY 10036   (212) 790-9090

Edmond R. Bannon
Pennie & Edmonds
1155 Avenue of the Americas
New York, NY 10036   (212) 790-9090

Jean B. Barish
Kenyon & Kenyon
1 Broadway
New York, NY 10028   (212) 425-7200

Bruce J. Barker
Pennie & Edmonds
1155 Avenue of the Americas
New York, NY 10036   (212) 790-9090

Richard M. Barnes
Fish & Neave
875 Third Avenue
New York, NY 10022   (212) 715-0600

Elizabeth Marion Barnhard
Bryan, Cave, Mc Pheeters
& Mc Roberts
245 Park Avenue
New York, NY 10167   (212) 692-1800

Steven J. Baron
685 Third Avenue
New York, NY 10017   (212) 878-6231

David K. Barr
Weil, Gotshal & Manges
767 Fifth Avenue
New York, NY 10153   (212) 310-8000

Richard S. Barth
Frishauf, Holtz, Goodman
& Woodward
600 Third Avenue
New York, NY 10016   (212) 972-1400

Mark T. Basseches
Colvin, Miskin, Basseches
& Mandelbaum
420 Lexington Ave.
New York, NY 10170   (212) 682-7280

Paula T. Basseches#
Colvin, Miskin, Basseches
& Mandelbaum
420 Lexington Ave.
New York, NY 10170   (212) 682-7280

Laura Anne Bauer
Fitzpatrick, Cella, Harper
& Scinto
277 Park Ave.
New York, NY 10172   (212) 758-2400

Walter Jacob Baum
Frishauf, Holtz, Goodman
& Woodward, P.C.
600 Third Avenue
30th Flr.
New York, NY 10016   (212) 972-1400

Steven C. Bauman
Fitzpatrick, Cella, Harper
& Scinto
277 Park Avenue
New York, NY 10172   (212) 758-2400

Charles Elliott Baxley
Hart, Baxley, Daniels
& Holtol
84 Williams St.
New York, NY 10038   (212) 668-0380

Steven H. Bazerman
Graham, Campaign & Mc Carthy
36 West 44th St.
New York, NY 10036   (212) 354-5650

Norman H. Beamer
Fish & Neave
875 Third Avenue
New York, NY 10022   (212) 715-0627

Thomas Augustus Beck
Felfe & Lynch
805 Third Avenue
New York, NY 10022   (212) 688-9200

Peter H. Behrendt
100 West 12th St.
New York, NY 10011   (212) 989-0174

Bernard Belkin
180 E. 79th Street
New York, NY 10021   (212) 988-8058

Vito Victor Bellino
American Home Products Corp.
685 Third Ave.
New York, NY 10017   (212) 878-6225

Leora Ben-Ami
Shea & Gould
1251 Ave. of the Americas
New York, NY 10020   (212) 827-3000

Egon E. Berg
American Home Products Co.
685 3rd Ave.
New York, NY 10017   (212) 878-6230

Michael J. Berger
Amster, Rothstein & Ebenstein
90 Park Ave.
New York, NY 10016   (212) 697-5995

Peter Lewis Berger
Levisohn, Lerner & Berger
757 Third Ave.
Suite 2400
New York, NY 10017   (212) 486-7272

James P. Bergin
Fish & Neave
875 Third Ave.
New York, NY 10022   (212) 715-0754

George B. Berka#
Striker, Striker & Stenby
360 Lexington Ave.
New York, NY 10017   (212) 687-5068

Richard G. Berkley
Brumbaugh, Graves, Donohue
& Raymond
30 Rockefeller Plaza
New York, NY 10112   (212) 408-2554

Jerome M. Berliner
Ostrolenk, Faber, Gerb
& Soffen
1180 Avenue of the Americas
New York, NY 10036   (212) 382-0700

Richard Kent Bernstein
555 Madison Avenue
New York, NY 10022   (212) 750-0544

E. Janet Berry
274 Madison Avenue
Room 401
New York, NY 10016   (212) 679-0581

William I. Bertsche
11 Broadway
New York, NY 10004   (212) 344-2930

Vincent F. Bick, Jr.
Exxon Corp.
1251 Ave. of Americas
Room 4804
New York, NY 10020   (212) 333-6416

Jordan B. Bierman
Bierman & Muserlian
757 Third Avenue
New York, NY 10017   (212) 752-7550

Linda G. Bierman
Synerflex Group, Inc.
215 East 59th Street
New York, NY 10022   (212) 486-5300

Nancy A. Bird
304 W. 91st Street
New York, NY 10024   (212) 724-6915

Patrick J. Birde
Kenyon & Kenyon
One Broadway
New York, NY 10004   (212) 425-7200

Julie Mae Blackburn
Revlon
625 Madison Avenue
New York, NY 10022   (212) 425-7200

Charles A. Blank
Felfe & Lynch
805 Third Avenue
New York, NY 10022   (212) 688-9200

Paul H. Blaustein
Hopgood, Calimafde, Kalil,
Blaustein & Judlowe
60 East 42nd Street
40th Floor
New York, NY 10165   (212) 986-2480

Herbert Blecker
Robin, Blecker & Daley
330 Madison Avenue
New York, NY 10017   (212) 682-9640

Ronald A. Bleker
W.R. Grace & Co.
1114 Ave. of the Americas
New York, NY 10036   (212) 819-5580

Edward M. Blocker
Blum, Kaplan, Friedman,
Silberman & Beran
1120 Avenue of the Americas
New York, NY 10036   (212) 704-0400

Mark H. Bloomberg
Fish & Neave
875 Third Avenue
New York, NY 10022   (212) 715-0600

Israel Blum
Morgan & Finnegan
345 Park Avenue
New York, NY 10154   (212) 758-4800

Norman Blumenkopf
35 East 85th Street
New York, NY 10028   (212) 988-0319

James M. Bogden
Morgan & Finnegan
345 Park Avenue
New York, NY 10154   (212) 758-4800

William Tilden Boland, Jr.
Kenyon & Kenyon
One Broadway
New York, NY 10004   (212) HA5-7200

James M. Bollinger
Hopgood, Calimafde, Kalil
& Blaustein
60 East 42nd Street
New York, NY 10165   (212) 986-2480

Jay A. Bondell
Schweitzer, Cornman & Gross
230 Park Avenue
New York, NY 10169   (212) 986-3377

Stevan J. Bosses
Fitzpatrick, Cella, Harper
& Scinto
277 Park Ave.
New York, NY 10172   (212) 758-2400

William L. Botjer
Hopgood, Calimafe, Kalil,
Blaustein & Judlowe
60 East 42nd Street
New York, NY 10165   (212) 986-2480

Alan T. Bowes
Kenyon & Kenyon
One Broadway
New York, NY 10004   (212) 425-7200

Livia S. Boyadjian
Dvorak & Traub
Wall Street Tower
20 Exchange, 37th Flr.
New York, NY 10005   (212) 968-1300

Douglas G. Brace
Kenyon & Kenyon
One Broadway
New York, NY 10004   (212) 425-7200

Charles W. Bradley, Jr.
Davis, Hoxie, Faithful
& Hapgood
45 Rockefeller Plaza
New York, NY 10111   (212) 757-2200

Abraham M. Bragin
504 Grand Street
Apt. A33
New York, NY 10002   (212) 982-0131

Charles R. Brainard
Kenyon & Kenyon
One Broadway
New York, NY 10004   (212) 425-7200

Brian S. Brandt
Fish & Neave
875 Third Ave.
New York, NY 10022   (212) 715-0600

Bradford S. Breen
Davis, Hoxie, Faithful
& Hapgood
45 Rockefeller Plaza
New York, NY 10111   (212) 757-2200

Albert J. Breneisen
Kenyon & Kenyon
One Broadway
New York, NY 10004   (212) 425-7200

Elaine P. Brenner
251 East 32nd Street
Apt. 14E
New York, NY 10016   (212) 889-9704

Sidney R. Bresnick
Fitzpatrick, Cella, Harper
& Scinto
277 Park Avenue
New York, NY 10172   (212) 758-2400

Gerow D. Brill#
Meller & Associates
50 East 42nd St.
New York, NY 10017   (212) 953-3350

Jeffrey K. Brinck
Milbank, Tweed, Hadlay
& Mc Cloy
1 Chase Manhattan Plaza
46th Floor
New York, NY 10005   (212) 530-5280

Richard Harold Brink
Bristol-Myers Co.
345 Park Ave.
New York, NY 10154   (212) 546-3637

Kurt G. Briscoe
Sprung, Horn, Kramer
& Woods
600 Third Avenue
New York, NY 10016   (212) 661-0520

Marilyn Brogan
Curtis, Morris & Safford, P.C.
530 Fifth Avenue
New York, NY 10036   (212) 840-3333

Mitchell P. Brook
Fish & Neave
875 Third Avenue
New York, NY 10022   (212) 715-0600

Adam L. Brookman#
Curtis, Morris & Safford
530 Fifth Avenue
New York, NY 10036   (212) 840-3333

Lorimer P. Brooks
Brooks, Haidt, Haffner
& Delahunty
99 Park Ave.
New York, NY 10016   (212) 697-3355

Daniel H. Brown
Rheem Manufacturing Company
405 Lexington Avenue
22nd Floor
New York, NY 10174   (212) 916-8150

Kathryn M. Brown#
Morgan & Finnegan
345 Park Ave.
23rd Floor
New York, NY 10154   (212) 415-9646

Ronald Erik Brown
Kane, Dalsimer, Sullivan, Kuracz,
Levy, Eisele & Richard
711 Third Avenue
New York, NY 10017   (212) 687-6000

G. M. Brumbaugh, Sr.
Brumbaugh, Graves, Donohue
& Raymond
30 Rockefeller Plaza
New York, NY 10112   (212) 408-2510

Granville M. Brumbaugh, Jr.
Brumbaugh, Graves, Donohue
& Raymond
30 Rockefeller Plaza
New York, NY 10112   (802) 496-3549

William John Brunet
Fitzpatrick, Cella, Harper
& Scinto
277 Park Avenue
New York, NY 10172   (212) 758-2400

Charles E. Bruzga
227 West 11 Street, #33
New York, NY 10014   (212) 645-1856

James F. Bryan
Cluett, Peabody & Co., Inc.
530 5th Ave.
New York, NY 10036   (212) 930-3025

Peter H. Bucci
Davis, Hoxie, Faithful
& Hapgood
45 Rockefeller Plaza
New York, NY 10111   (212) 757-2200

Harvey E. Bumgardner, Jr.
Suite 303
80 Eighth Avenue
New York, NY 10011   (212) 243-7763

Henry T. Burke
Wyatt, Gerber, Shoup,
Scobey & Badie
261 Madison Ave.
New York, NY 10016   (212) 687-0911

Jacob B. Burke
25 Central Park W.-11F
New York, NY 10023   (212) 245-8657

Gary M. Butter
Brumbaugh, Graves, Donohue
& Raymond
30 Rockefeller Plaza
New York, NY 10112   (212) 408-2500

Daniel N. Calder
Amster, Rothstein & Ebenstein
90 Park Avenue
New York, NY 10016   (212) 697-5995

John Michael Calimafde
Hopgood, Calimafde, Kalil,
Blaustein & Lieberman
Lincoln Bldg., Suite 4004
60 E. 42nd St.
New York, NY 10165   (212) 986-2480

Edward T. Callahan
Suite 1010
225 Central Pk. West
New York, NY 10024   (212) 595-5315

Joseph A. Calvaruso
Morgan & Finnegan
345 Park Avenue
New York, NY 10154   (212) 758-4800

Ronald J. Campbell#
Kenyon & Kenyon
One Broadway
New York, NY 10004   (212) 425-7200

Francis T. Carr
Kenyon & Kenyon
One Broadway
New York, NY 10004   (212) 425-7200

John J. Carrara
Westvaco Corp.
299 Park Ave.
New York, NY 10171   (212) 688-5000

Thomas G. Carulli
Cooper & Dunham
30 Rockefeller Plaza
New York, NY 10112   (212) 977-9550

Frederick C. Carver
Brumbaugh, Graves, Donohue
& Raymond
30 Rockefeller Plaza
New York, NY 10112   (212) 408-2528

Anthony J. Casella
Casella & Hespos
274 Madison Avenue
New York, NY 10016   (212) 725-2450

John J. Cassingham
Fish & Neave
875 Third Ave.
New York, NY 10022   (212) 715-0600

Vincent A. Castiglione
Morgan & Finnegan
345 Park Ave.
New York, NY 10154   (212) 758-4800

John Thomas Cella
Fitzpatrick, Cella,
Harper & Scinto
277 Park Ave.
New York, NY 10172   (212) 758-2400

Jules Louis Chaboty#
260 Seaman Ave., Apt. 7-4
New York, NY 10034   (212) 567-8594

Michael I. Chakansky
Brown, Raysman & Millstein
120 West 45th Street
New York, NY 10036   (212) 827-4789

Christopher E. Chalsen
Morgan & Finnegan
345 Park Avenue
New York, NY 10154   (212) 758-4800

Hugh A. Chapin
Kenyon & Kenyon
One Broadway
New York, NY 10004   (212) 425-7200

Frank Chau
Morgan & Finnegan
345 Park Ave.
New York, NY 10154   (212) 758-4800

Kent H. Cheng
Pennie & Edmonds
1155 Avenue of the Americas
New York, NY 10036   (212) 790-9090

Marshall J. Chick
Frishauf, Holtz, Goodman
& Woodward P.C.
600 Third Avenue
30th Floor
New York, NY 10016   (212) 972-1400

Edward G.H. Chin
Hertzog, Calamari & Gleason
100 Park Avenue
New York, NY 10017   (212) 481-9500

Leighton K.M. Chong
Ostrager & Chong
300 East 42nd Street
18th Floor
New York, NY 10017   (212) 697-8080

Joseph H. Church#
552 West 184th Street
New York, NY 10033   (212) 928-0828

Jay Saul Cinamon
Abelman, Frayne, Rezac
& Schwab
708 Third Avenue
New York, NY 10017   (212) 949-9022

Lester W. Clark
Cooper, Dunham, Griffin
& Moran
30 Rockefeller Plaza
New York, NY 10112   (212) 977-9550

Richard S. Clark
Brumbaugh, Graves,
Donohue & Raymond
30 Rockefeller Plaza
New York, NY 10112   (212) 408-2558

Kevin Barry Clarke
Carter-Wallace, Inc.
1345 Avenue of the Americas
New York, NY 10105   (212) 339-5000

Ronald A. Clayton
Fitzpatrick, Cella, Harper
& Scinto
277 Park Ave.
New York, NY 10172   (212) 758-2400

Alan B. Clement#
Hedman, Gibson, Costigan
& Hoare, P.C.
1185 Avenue of Americas
New York, NY 10036   (212) 302-8989

Peggy A. Climenson
Kenyon & Kenyon
1 Broadway
New York, NY 10004   (212) 425-7200

Peter Timothy Cobrin
Cobrin & Godsberg, P.C.
366 Madison Avenue
New York, NY 10017   (212) 687-6090

Nicholas Lazaros Coch
Shea & Gould
1251 Ave. of the Americas
New York, NY 10020   (212) 827-3000

Joseph S. Codispoti
Bonfiglio & Codispoti
71 Broadway
Suite 1201
New York, NY 10006   (212) 785-3600

Brian D. Coggio
Pennie & Edmonds
1155 Sixth Avenue
New York, NY 10036   (212) 790-9090

Julian Harris Cohen
Ladas & Parry
26 West 61st Street
New York, NY 10023   (212) 708-1888

Martin J. Cohen
Cohen & Silverman
666 Third Avenue
20th Floor
New York, NY 10017   (212) 986-8282

Myron Cohen
Cohen, Pontani & Lieberman
551 Fifth Avenue
Suite 1210
New York, NY 10176   (212) 687-2770

J. Bradley Cohn
314 E. 41st Street
New York, NY 10017   (212) 697-5088

Christine Cole
80 Wall Street
Suite 415
New York, NY 10005   (212) 425-3158

Henry D. Coleman
De Rosa, Vandenberg & Glenman
71 Broadway
New York, NY 11201   (212) 269-3020

Anthony C. Coles
Darby & Darby, P.C.
805 Third Ave., Floor 27
New York, NY 10020   (212) 697-7660

Arthur B. Colvin
420 Lexington Ave.
New York, NY 10170   (212) 986-4056

Dominick A. Conde#
Fitzpatrick, Cella, Harper
& Scinto
277 Park Ave.
New York, NY 10172   (212) 758-2400

William C. Conner*
U.S. District Court
Southern District of NY
U.S. Courthouse
40 Foley Square, Rm. 1902
New York, NY 10007   (212) 791-0934

Albert B. Cooper
20 East 74th Street, #14-C
New York, NY 10021   (212) 249-4435

Barry Arnold Cooper
Gottlieb, Rackman & Reisman
1430 Broadway
New York, NY 10018   (212) 869-2890

Leonard Cooper
Darby & Darby, P.C.
805 Third Avenue
New York, NY 10022   (212) 697-7660

Gordon D. Coplein
Darby and Darby, P.C.
805 Third Avenue
New York, NY 10022   (212) 692-7660

Michael Alexander Cornman
Mandeville and Schweitzer
230 Park Ave.
New York, NY 10169   (212) 986-3377

Thomas Johnson Corum
Colgate-Palmolive Co.
Patent Section
300 Park Ave.
New York, NY 10022   (212) 310-3145

Laura A. Coruzzi
Pennie & Edmonds
1155 Avenue of the Americas
New York, NY 10036   (212) 790-6431

James V. Costigan
Hedman, Gibson, Costigan
& Hoare, P.C.
1185 Avenue of the Americas
New York, NY 10036   (212) 302-8989

Robert A. Cote, Jr.
Davis, Hoxie, Faithful
& Hapgood
45 Rockefeller Plaza
New York, NY 10111   (212) 757-2200

Abigail F. Cousins#
Eslinger & Pelton, P.C.
600 Third Avenue
New York, NY 10016   (212) 972-4433

Gary L. Creason#
Fish & Neave
875 Third Ave
New York, NY 10022   (212) 715-0767

Thomas L. Creel
Kenyon & Kenyon
One Broadway
New York, NY 10004   (212) 425-7200

Richard L. Crisona
Duker & Barrett
90 Broad Street
New York, NY 10004   (212) 809-7700

Jack Saul Cubert
Fitzpatrick, Cella, Harper
& Scinto
277 Park Avenue
New York, NY 10172   (212) 758-2400

Andrew T. D'Amico, Jr.
Davis, Hoxie, Faithful
& Hapgood
45 Rockefeller Plaza
Suite 2800
New York, NY 10111   (212) 757-2200

Alfred A. D'Andrea, Jr.
Jordan & Hamburg
122 East 42nd St.
Suite 3303
New York, NY 10168   (212) 986-2340

Felix L. D'Arienzo, Jr.
Mandeville & Schweitzer
230 Park Ave.
Room 2200
New York, NY 10169   (212) 986-3377

J. Robert Dailey
Morgan & Finnegan
345 Park Avenue
New York, NY 10154   (212) 758-4800

J. David Dainow
Rosen, Dainow & Jacobs
489 5th Ave.
New York, NY 10017   (212) 692-7000

James J. Daley
Robin, Blecker & Daley
330 Madison Avenue
New York, NY 10017   (212) 682-9460

John J. Daniels
Adams & Wilks
500 Fifth Ave.
Suite 3200
New York, NY 10110   (212) 575-2600

Richard J. Danyko, Jr.#
Kane, Dalsimer, Sullivan, Kurucz,
Levy, Eisele & Richard
711 Third Avenue
20th Floor
New York, NY 10017   (212) 687-6000

Lynne Darcy
Kenyon & Kenyon
One Broadway
New York, NY 10004   (212) 425-7200

Clifford M. Davidson
Steinberg & Raskin
1140 Avenue of the Americas
New York, NY 10036   (212) 768-3800

Gerald B. Davis#*
U.S. Internal Revenue Service
120 Church Street
New York, NY 10008   (212) 264-3174

Karen De Benedictis
Pfizer, Inc.
Legal Dept.
219 East 42nd Street
5th Floor
New York, NY 10017   (212) 573-2323

Robert De Berardine
Jones, Day, Reavis
& Pogue
599 Lexington Avenue
32nd Floor
New York, NY 10022   (212) 326-3402

Joseph A. De Girolamo
Morgan & Finnegan
345 Park Avenue
New York, NY 10154   (212) 758-4800

John G. De La Rosa#
Pennie & Edmonds
1155 Avenue of the Americas
New York, NY 10036   (212) 790-9090

Jose R. De La Rosa
Cohen, Pontani & Lieberman
551 Fifth Avenue
New York, NY 10176   (212) 687-2770

Bruce D. De Renzi
Morgan & Finnegan
345 Park Ave.
New York, NY 10154   (212) 758-4800

Frank J. De Rosa
Rosen, Dainow & Jacobs
489 Fifth Avenue
New York, NY 10017   (212) 692-7000

Daniel A. De Vito#
Weil, Gotshal & Manges
767 Fifth Avenue
New York, NY 10153   (212) 310-8866

Aaron C. Deditch
Fitzpatrick, Cella, Harper
& Scinto
277 Park Avenue
42nd Floor
New York, NY 10172   (212) 758-2400

Donald E. Degling
Fish & Neave
875 Third Avenue
New York, NY 10022   (212) 715-0600

Marguerite Del Valle
Hopgood, Calimafde, Kalil
Blaustein & Judlowe
60 E. 42nd St.
New York, NY 10165   (212) 986-2480

G. Thomas Delahunty
Brooks, Haidt, Haffner
& Delahunty
99 Park Ave.
New York, NY 10016   (212) 697-3355

Richard L. Delucia
Kenyon & Kenyon
One Broadway
New York, NY 10004   (212) 425-7200

Manette Dennis
Ostrolenk, Faber, Gerb
& Soffen
1180 Avenue of the Americas
New York, NY 10036   (212) 382-0700

Leonard P. Diana
Fitzpatrick, Cella, Harper
& Scinto
277 Park Avenue
42nd Floor
New York, NY 10172   (212) 758-2400

John Andrew Diaz
Morgan & Finnegan
345 Park Avenue
New York, NY 10154   (212) 758-4800

Gerard F. Diebner
Kaye, Scholer, Fierman,
Hays & Handler
425 Park Avenue
New York, NY 10022   (212) 836-8802

John M. Dimatteo#
Egli International
599 Lexington Ave.
New York, NY 10022   (212) 836-4736

William H. Dippert
Kane, Dalsimer, Sullivan, Kurucz,
Levy, Eisele & Richard
711 Third Avenue
New York, NY 10170   (212) 687-6000

Lorraine M. Donaldson
Pfizer, Inc.
235 E. 42nd Street
New York, NY 10017   (212) 573-2858

Frederick J. Dorchak
Wyatt, Gerber, Burke
& Badie
645 Madison Avenue
New York, NY 10022   (212) 826-0171

Michael P. Dougherty
Morgan & Finnegan
345 Park Avenue
New York, NY 10154   (212) 758-4800

Earl M. Douglas
Morgan & Finnegan
345 Park Avenue
New York, NY 10154   (212) 758-4800

Donald Smith Dowden
Cooper & Dunham
30 Rockefeller Plaza
New York, NY 10112   (212) 977-9550

Thomas P. Dowling
Morgan & Finnegan
345 Park Avenue
New York, NY 10154   (212) 758-4800

Joel K. Dranove
401 Broadway
New York, NY 10013   (212) 431-7660

Dimitros T. Drivas
White & Case
1155 Avenue of the Americas
New York, NY 10036   (212) 819-8286

Mark Dryer
Pfizer Inc.
235 East 42nd Street
New York, NY 10017   (212) 573-7482

Samuel J. Du Boff
Bristol-Myers Co.
345 Park Ave.
New York, NY 10154   (212) 546-3632

William F. Dudine, Jr.
Darby and Darby, P.C.
805 Third Avenue
New York, NY 10022   (212) 697-7660

Louis C. Dujmich
Kenyon & Kenyon
One Broadway
New York, NY 10004   (212) 425-7200

Christopher Cooper Dunham
Cooper & Dunham
30 Rockefeller Plaza
New York, NY 10112   (212) 977-9550

Robert Secrest Dunham
Cooper, Dunham, Clark,
Griffin & Moran
30 Rockefeller Plaza
New York, NY 10112   (212) 977-9550

Diane F. Dunn
Darby & Darby P.C.
805 Third Ave.
New York, NY 10022   (212) 697-7660

Gerard F. Dunne
Wyatt, Gerber, Shoup,
Scobey & Badie
261 Madison Ave.
New York, NY 10016   (212) 687-0911

Jean M. Duvall
Morgan & Finnegan
345 Park Avenue
New York, NY 10154   (212) 758-4800

Daniel Simon Ebenstein
Amster, Rothstein & Ebenstein
90 Park Ave.
New York, NY 10016   (212) 697-5995

William F. Eberle
Brumbaugh, Graves, Donohue
& Raymond
30 Rockefeller Plaza
New York, NY 10112   (212) 408-2358

Michael Ebert
Hopgood, Calimafde, Kalil,
Blaustein & Judlowe
60 E. 42nd St.
New York, NY 10165   (212) 986-2480

Alan L. Edwards
Davis, Hoxie, Faithful
& Hapgood
45 Rockefeller Plaza - 28th Flr.
New York, NY 10111   (212) 757-2200

Joseph T. Eisele
Kane, Dalsimer, Sullivan, Kurucz,
Levy, Eisele & Richard
711 Third Avenue
New York, NY 10017   (212) 687-6000

Richard A. Elder
American Home Products Corp.
Law Dept., Pat. Section
685 Third Avenue
New York, NY 10017   (212) 878-6218

Robert L. Epstein
James & Franklin, P.C.
60 East 42nd Street
Suite 1217
New York, NY 10165   (212) 867-7260

Lewis H. Eslinger
Cooper & Dunham
30 Rockefeller Plaza
New York, NY 10112   (212) 977-9550

Barry Leonard Evans
Curtis, Morris & Safford, P.C.
530 5th Ave.
New York, NY 10036   (212) 840-3333

Emily A. Evans#
Fish & Neave
875 Third Avenue
29th Floor
New York, NY 10022   (212) 715-0600

William R. Evans
Ladas & Parry
26 West 61st Street
New York, NY 10023   (212) 708-1945

Alfred P. Ewert
Morgan & Finnegan
345 Park Avenue
New York, NY 10154   (212) 758-4800

Robert Charles Faber
Ostrolenk, Faber, Gerb
& Soffen
1180 Avenue of the Americas
New York, NY 10036   (212) 382-0700

Sidney G. Faber
Ostrolenk, Faber, Gerb
& Soffen
1180 Avenue of the Americas
New York, NY 10036   (212) 382-0700

Gary J. Falce#
American Standard Inc.
40 West 40th Street
New York, NY 10017   (212) 382-0700

James Warren Falk
Darby & Darby, P.C.
805 Third Ave.
New York, NY 10022   (212) 697-7660

Debra A. Fallek
Rosenman & Colin
575 Madison Ave.
New York, NY 10022   (212) 940-6433

F. Brice Faller
Felfe & Lynch
805 Third Avenue
New York, NY 10022   (212) 688-9200

Allan A. Fanucci
Pennie & Edmonds
1155 Avenue of the Americas
New York, NY 10036   (212) 790-6537

Martin Allen Farber
866 United Nations Plaza
New York, NY 10017   (212) 758-2878

Mark A. Farley
Pennie & Edmonds
1155 Ave. of the Americas
New York, NY 10036   (212) 790-9090

Vincent M. Fazzari
Felfe & Lynch
805 Third Avenue
New York, NY 10022   (212) 688-9200

Valerie M. Fedowich
Pfizer Inc.
235 E. 42nd Street
New York, NY 10017   (212) 573-3953

Ronald C. Fedus
Enzo Biochem, Inc.
345 Hudson Street
13th Floor
New York, NY 10014   (212) 337-3355

Henry M. Feiereisen#
500 Fifth Avenue
Suite 1205
New York, NY 10110   (212) 354-8262

William S. Feiler
Morgan, Finnegan, Pine,
Foley & Lee
345 Park Ave.
New York, NY 10154   (212) 758-4800

Irving N. Feit
IM Clone Systems Inc.
180 Varick Street
New York, NY 10014   (212) 645-1405

Marvin Feldman
Teitel & Feldman
6 E. 45th Street
Penthouse Suite
New York, NY 10017   (212) 286-0260

Stephen Edward Feldman
12 East 41st Street
New York, NY 10017   (212) 532-8585

Peter F. Felfe
Felfe & Lynch
805 Third Avenue
25th Floor
New York, NY 10022   (212) 688-9200

Stanley D. Ference, III
Brumbaugh, Graves, Donohue
& Raymond
30 Rockefeller Plaza
New York, NY 10112   (212) 408-2500

Richard P. Ferrara
Davis, Hoxie, Faithful
& Hapgood
45 Rockefeller Plaza
New York, NY 10111   (212) 757-2200

Albert E. Fey
Fish & Neave
875 Third Avenue
New York, NY 10022   (212) 715-0600

Robert W. Fiddler
Fiddler & Levine
7814 Empire State Bldg.
350 5th Ave.
New York, NY 10118   (212) 279-6088

Alan W. Fiedler
Fitzpatrick, Cella, Harper
& Scinto
277 Park Ave.
New York, NY 10172   (212) 758-2400

Paul Fields
Darby & Darby
805 Third Avenue
27th Floor
New York, NY 10022   (212) 697-7660

Robert D. Fier
Kenyon & Kenyon
One Broadway
New York, NY 10004   (212) 425-7200

Edward V. Filardi
White & Case
1155 Ave. of the Americas
New York, NY 10036   (212) 819-8200

Stephan J. Filipek
Davis, Hoxie, Faithful
& Hapgood
45 Rockefeller Plaza
28th Floor
New York, NY 10111   (212) 757-2200

James A. Finder
Ostrolenk, Faber, Gerb
& Soffen
1180 Avenue of the Americas
New York, NY 10036   (212) 382-0700

George B. Finnegan, Jr.
Morgan & Finnegan
345 Park Avenue
New York, NY 10154   (212) 758-4800

Robert H. Fischer
Fitzpatrick, Cella, Harper
& Scinto
277 Park Avenue
New York, NY 10172   (212) 758-2400

Julius Fisher
Mc Aulay, Fields, Fisher,
Goldstein & Nissen
The Chrysler Bldg.
405 Lexington Ave.
New York, NY 10174   (212) 986-4090

Barry H. Fishkin
Phillips, Nizer, Benjamin,
Krim & Ballon
40 West 57th Street
New York, NY 10019   (212) 977-9700

Joseph M. Fitzpatrick
Fitzpatrick, Cella, Harper
& Scinto
277 Park Ave.
New York, NY 10172   (212) 758-2400

Dennis M. Flaherty#
Rosen, Dainow & Jacobs
489 Fifth Avenue
New York, NY 10017   (212) 692-7058

Eugene L. Flanagan, III
Curtis, Morris & Safford, P.C.
530 Fifth Avenue
New York, NY 10036   (212) 840-3333

Porter F. Fleming#
Hopgood, Calimafde, Kalil,
Blaustein & Judlowe
Lincoln Bldg.
60 East 42nd Street
New York, NY 10165   (212) 986-2480

Gerald James Flintoft
Pennie & Edmonds
1155 Avenue of the Americas
New York, NY 10036   (212) 790-9090

Kenneth F. Florek
Hedman, Gibson, Costigan
& Hoare, P.C.
1185 Avenue of the Americas
New York, NY 10036   (212) 302-8989

John Aloysius Fogarty, Jr.
Kenyon & Kenyon
One Broadway
New York, NY 10004   (212) 425-7200

John Dennis Foley
Morgan & Finnegan
345 Park Avenue
New York, NY 10154   (212) 758-4800

Frank W. Ford, Jr.
Brumbaugh, Graves, Donohue
& Raymond
30 Rockefeller Plaza
New York, NY 10112   (212) 408-2524

James D. Fornari
Jarblum, Solomon & Fornari
650 Fifth Avenue
New York, NY 10019   (212) 265-1200

Mavis Kathleen Fowler
156-20 Riverside Dr., West
Apt. 8J
New York, NY 10032   (212) 928-2156

David R. Francescani
Kenny Group, Inc.
65 Broadway
New York, NY 10006   (212) 770-4905

Peter James Franco
6 Peter Cooper Road
New York, NY 10010   (212) 982-0999

Irene J. Frangos
National Pat. Dev. Corporation
9 West 57th Street
New York, NY 10019   (212) 230-9521

Howard M. Frankfort#
Darby & Darby, P.C.
805 Third Avenue
New York, NY 10022   (212) 697-7660

Monica Valencia Franklin#
Kenyon & Kenyon
One Broadway
New York, NY 10004   (212) 425-7200

Edward R. Freedman
Hopgood, Calimafde, Kalil,
Blaustein & Judlowe
Lincoln Bldg.
60 East 42nd Street
New York, NY 10165   (212) 986-2480

Robert M. Freeman
Brooks, Haidt, Haffner
& Delahunty
99 Park Avenue
New York, NY 10016   (212) 697-3355

Mark S. Frey
Dublirer, Haydon, Straci
& Victor
17 Battery Place
New York, NY 10004   (212) 943-0880

Thomas E. Friebel
Pennie & Edmonds
1155 Avenue of the Americas
New York, NY 10036   (212) 790-9090

Stewart Jay Fried
Abelman, Frayne, Rezac
& Schwab
708 Third Avenue
New York, NY 10017   (212) 949-9022

Jacob Friedlander
Leboeuf, Lamb, Leiby & Mac Rae
520 Madison Avenue
New York, NY 10022   (212) 715-8000

Alex Friedman
Blum, Kaplan, Friedman
Silberman & Beran
1120 Avenue of the Americas
New York, NY 10036   (212) 704-0400

Morton Friedman
Witco Corporation
520 Madison Avenue
New York, NY 10022   (212) 605-3848

Stephen H. Frishauf
Frishauf, Holtz, Goodman
& Woodward, P.C.
600 Third Avenue
30th Floor
New York, NY 10016   (212) 972-1400

William S. Frommer
Curtis, Morris & Safford, P.C.
530 5th Ave.
New York, NY 10036   (212) 840-3333

Albert Edward Frost#
Pfizer Inc.
Patent Dept., 20th Floor
235 E. 42nd Street
New York, NY 10017   (212) 840-3333

Grover F. Fuller, Jr.
Pfizer, Inc.
235 E. 42nd Street
New York, NY 10017   (212) 573-1390

Richard Guerard Fuller, Jr.
Brumbaugh, Graves, Donohue
& Raymond
30 Rockefeller Plaza
New York, NY 10112   (212) 408-2520

James W. Galbraith
Kenyon & Kenyon
One Broadway
New York, NY 10004   (212) 425-7200

Peter D. Galloway
Ladas & Parry
26 West 61st Street
New York, NY 10023   (212) 708-1905

Melvin C. Garner
Darby & Darby, P.C.
805 Third Avenue
New York, NY 10022   (212) 697-7660

Joseph D. Garon
Brumbaugh, Graves, Donohue
& Raymond
30 Rockefeller Plaza
New York, NY 10112   (212) 408-2540

Bradley B. Geist
Brumbaugh, Graves, Donohue
& Raymond
30 Rockefeller Plaza
New York, NY 10112   (212) 408-2562

Kenneth P. George
Amster, Rothstein & Ebenstein
90 Park Avenue
New York, NY 10016   (212) 697-5995

Eliot S. Gerber
Wyatt, Gerber, Burke,
& Badie
645 Madison Ave.
New York, NY 10022   (212) 826-0171

Stanley J. Gewirtz#
Solid State Systems Inc.
435 W. 119th St.
New York, NY 10027   (212) 222-2058

Thomas L. Giannetti
Fish & Neave
29th Floor
875 Third Avenue
New York, NY 10022   (212) 715-0600

Mark D. Giarrantana#
Kenyon & Kenyon
One Broadway
New York, NY 10004   (212) 425-7200

Thomas Martin Gibson
Hedman, Gibson, Costigan
& Hoare, P.C.
1185 Avenue of the Americas
New York, NY 10036   (212) 302-8989

Douglas J. Gilbert
Fish & Neave
875 Third Avenue
New York, NY 10022   (212) 715-0652

Stephen Paul Gilbert
Bryan, Cave, Mc Pheeters
& Mc Roberts
245 Park Avenue
New York, NY 10167   (212) 692-1800

William J. Gilbreth
Fish & Neave
875 Third Avenue
New York, NY 10022   (212) 715-0600

Alan D. Gilliland
Pennie & Edmonds
1155 Avenue of the Americas
New York, NY 10036   (212) 790-9090

Theresa M. Gillis
Jones, Day, Reavis & Pogue
599 Lexington Avenue
New York, NY 10022   (212) 326-3939

Michael R. Gilman
Gottlieb, Rackman
& Reisman, P.C.
1430 Broadway
New York, NY 10018   (212) 869-2890

Paul H. Ginsburg
Pfizer Inc.
Pat. Dept.
235 East 42nd Street
New York, NY 10017   (212) 573-2369

Ann L. Gisolfi
Pennie & Edmonds
1155 Avenue of the Americas
New York, NY 10036   (212) 790-9090

Howard M. Gitten
Blum, Kaplan, Friedman
Silberman & Beran
1120 Avenue of the Americas
New York, NY 10036   (212) 873-8633

Franklin M. Gittes
Skadden, Arps, Slate, Meagher
& Flom
919 Third Avenue
New York, NY 10022   (212) 735-3760

Marvin S. Gittes
366 Madison Avenue
New York, NY 10017   (212) 949-8787

Steven D. Glazer
Weil, Gotshal & Manges
33rd Floor
767 Fifth Avenue
New York, NY 10153   (212) 310-8000

Christopher P. Godziela
Fish & Neave
875 Third Avenue
New York, NY 10022   (212) 715-0600

Adda C. Gogoris
Darby & Darby, P.C.
805 Third Avenue
New York, NY 10022   (212) 697-7660

Jules Edward Goldberg
Mc Aulay, Fields, Fisher,
Goldstein & Nissen
Chrysler Bldg.
405 Lexington Avenue
New York, NY 10174   (212) 986-4090

Ron Goldman
Robin, Blecker, Daley
& Driscoll
330 Madison Ave.
New York, NY 10017   (212) 682-9640

Amy B. Goldsmith
Gottlieb, Rackman & Reisman, P.C.
1430 Broadway
New York, NY 10018   (212) 869-2890

S. Delvalle Goldsmith
Ladas & Parry
26 West 61st Street
New York, NY 10023   (212) 708-1910

Martin E. Goldstein
Darby & Darby
805 Third Avenue
New York, NY 10022   (212) 697-7660

Ronald B. Goldstein
Davis Hoxie Faithful
& Hapgood
45 Rockefeller Plaza
New York, NY 10111   (212) 757-2200

Donald J. Goodell
Pennie & Edmonds
1155 Avenue of the Americas
New York, NY 10036   (212) 790-6307

Herbert H. Goodman
Frishauf, Holtz, Goodman
& Woodward, P.C.
600 Third Ave., 30th Flr.
New York, NY 10016   (212) 972-1400

Beverly B. Goodwin
Darby & Darby, P.C.
805 Third Avenue
New York, NY 10022   (212) 697-7660

Lawrence B. Goodwin
Davis Hoxie Faithful
& Hapgood
45 Rockefeller Plaza
New York, NY 10111   (212) 757-2200

Jennifer Gordon
Pennie & Edmonds
1155 Avenue of the Americas
New York, NY 10036   (212) 790-9090

Marvin Norman Gordon
Hopgood, Calimafde, Kalil,
Blaustein & Lowe
60 E. 42nd St.
New York, NY 10165   (212) 986-2480

Philip H. Gottfried
Amster, Rothstein & Ebenstein
90 Park Ave.
New York, NY 10016   (212) 697-5995

George Gottlieb
Gottlieb, Rackman & Reisman, P.C.
1430 Broadway
New York, NY 10018   (212) 869-2890

James W. Gould
Morgan & Finnegan
345 Park Avenue
New York, NY 10154   (212) 758-4800

Eben M. Graves
Brumbaugh, Graves, Donohue
& Raymond
30 Rockefeller Plaza
New York, NY 10112   (212) 408-1500

Arthur D. Gray
Kenyon & Kenyon
One Broadway
New York, NY 10004   (212) 425-7200

William O. Gray, III
Ostrolenk, Faber, Gerb
& Soffen
1180 Avenue of the Americas
New York, NY 10036   (212) 382-0700

Edward W. Greason
Kenyon & Kenyon
One Broadway
New York, NY 10004   (212) 425-7200

Orville N. Greene
Greene and Durr
Rm. 4410
10 E. 40th St.
New York, NY 10016   (212) 686-9009

Richard S. Gresalfi
Kenyon & Kenyon
One Broadway
New York, NY 10004   (212) 425-7200

Gerald W. Griffin
Cooper & Dunham
30 Rockefeller Plaza
New York, NY 10112   (212) 977-9550

Murray M. Grill
Colgate-Palmolive Co.
300 Park Ave.
New York, NY 10022   (212) 310-3135

Marc S. Gross
Bryan, Cave, Mc Pheeters
& Mc Roberts
245 Park Avenue
New York, NY 10167   (212) 692-1800

Marta E. Gross
Fish & Neave
875 Third Avenue
New York, NY 10022   (212) 715-0600

Meyer A. Gross
Wolder, Gross & Bondell
230 Park Avenue
New York, NY 10169   (212) 986-3377

John F. Gulbin
Kane, Dalsimer, Sullivan, Kurucz,
Levy, Eisele & Richard
711 Third Avenue
New York, NY 10017   (212) 687-6000

Joseph H. Guth#
Natural Resources Defense Coun.
40 West 20th St.
New York, NY 10011   (212) 727-2700

Charles Guttman
Marmorek, Guttman & Rubenstein
330 Seventh Avenue
New York, NY 10001   (212) 594-9680

Bruce C. Haas
Fitzpatrick, Cella, Harper
& Scinto
277 Park Avenue
New York, NY 10172   (212) 758-2400

Gaylord Paul Haas, Jr.
150 E. 52nd Street
25th Floor
New York, NY 10022   (212) 906-3067

Alfred L. Haffner, Jr.
Brooks, Haidt, Haffner
& Delahunty
99 Park Ave.
New York, NY 10016   (212) 697-3355

Harold Haidt
Brooks, Haidt, Haffner
& Delahunty
99 Park Ave.
New York, NY 10016   (212) 697-3355

James F. Haley, Jr.
Fish & Neave
875 Third Avenue
New York, NY 10022   (212) 715-0600

C. Bruce Hamburg
Jordan & Hamburg
122 E. 42nd St.
New York, NY 10168   (212) 986-2340

Scot G. Hamilton
Memorial Sloan-Kettering
Cancer Center
Office of Industrial Affairs
1275 York Avenue
New York, NY 10021   (212) 639-3620

Thomas M. Hammond
Morgan & Finnegan
345 Park Avenue
New York, NY 10154   (212) 758-4800

Francis C. Hand
Kenyon & Kenyon
One Broadway
New York, NY 10004   (212) 425-7200

Joseph H. Handelman
Ladas & Parry
26 West 61st Street
New York, NY 10023   (212) 708-1880

Edward J. Handler, III
Kenyon & Kenyon
One Broadway
New York, NY 10004   (212) 425-7200

Walter E. Hanley, Jr.
Kenyon & Kenyon
One Bdwy.
New York, NY 10004   (212) 425-7200

Norman D. Hanson
Felfe & Lynch
805 Third Avenue
New York, NY 10022   (212) 688-9200

Cyrus S. Hapgood
Davis, Hoxie, Faithfull
& Hapgood
45 Rockefeller Plaza
New York, NY 10111   (212) 757-2200

Stephen M. Haracz
Jacobs & Jacobs, P.C.
521 Fifth Avenue
New York, NY 10175   (212) 687-1636

Stephen J. Harbulak
Pennie & Edmonds
1155 Avenue of the Americas
New York, NY 10036   (212) 790-9090

Robert M. Haroun
Brumbaugh, Graves, Donohue
& Raymond
30 Rockefeller Plaza
New York, NY 10112   (212) 408-2550

Edgar H. Haug
Curtis, Morris & Safford, P.C.
530 Fifth Avenue
New York, NY 10036   (212) 840-3333

Gunter A. Hauptman
Kaye, Scholer, Fierman,
Hays & Handler
425 Park Avenue
New York, NY 10022   (212) 836-8000

Edward Alan Hedman
Hedman, Gibson, Costigan
& Hoare, P.C.
1185 Avenue of the Americas
New York, NY 10036   (212) 302-8989

Samson Helfgott
Helfgott & Karas, P.C.
60th Floor
Empire State Bldg.
New York, NY 10118   (212) 643-5000

Paul Harold Heller
Kenyon & Kenyon
One Broadway
New York, NY 10004   (212) 425-7200

Kenneth B. Herman
Fish & Neave
875 Third Avenue
New York, NY 10022   (212) 715-0600

Gerald E. Hespos
Casella & Hespos
274 Madison Avenue
New York, NY 10016   (212) 725-2450

Robert J. Hess
Darby & Darby, P.C.
805 Third Avenue
New York, NY 10022   (212) 527-7751

Thomas V. Heyman
Jones, Day, Reavis & Pogue
599 Lexington Ave.
New York, NY 10022   (212) 326-3939

Christa Hildebrand
Egli International
599 Lexington Ave.
New York, NY 10022   (212) 836-4736

Ronald Budd Hildreth
Brumbaugh, Graves, Donohue
& Raymond
30 Rockefeller Plaza
New York, NY 10112   (212) 408-2544

John M. Hintz
Fish & Neave
875 Third Avenue
29th Floor
New York, NY 10022   (212) 715-0744

Joel Hirschel
253 West 72nd Street
New York, NY 10023   (212) 799-7767

George Philip Hoare, Jr.
Hedman, Gibson, Costigan
& Hoare, P.C.
1185 Avenue of the Americas
New York, NY 10036   (212) 302-8989

Steven M. Hoffberg
Cohen, Pontani & Lieberman
551 Fifth Avenue
New York, NY 10176　　　(212) 687-2770

Frank P. Hoffman
Bristol Myers Inc.
345 Park Avenue
New York, NY 10154　　　(212) 546-5698

Lawrence A. Hoffman
Townley & Updike
405 Lexington Ave.
New York, NY 10174　　　.(212) 973-6857

Norbert P. Holler
Gottlieb, Rackman & Reisman, P.C.
1430 Broadway
New York, NY 10018　　　(212) 869-2890

Gezina Holtrust
Pfizer Inc.
235 East 42nd Street
20th Floor
New York, NY 10017　　　(212) 573-7793

Leonard Holtz
Frishauf, Holtz, Goodman
& Woodward, P.C.
600 Third Avenue
30th Floor
New York, NY 10016　　　(212) 972-1400

Irving Holtzman
Bristol - Myers Co.
345 Park Ave.
New York, NY 10154　　　(212) 546-3666

Francis J. Hone
Brumbaugh, Graves, Donohue
& Raymond
30 Rockefeller Plaza
New York, NY 10112　　　(212) 408-2534

William J. Hone
Davis Hoxie Faithful
& Hapgood
45 Rockefeller Plaza
New York, NY 10111　　　(212) 757-2200

Roy C. Hopgood, Jr.
Hopgood, Calimafde, Kalil,
Blaustein & Judlowe
60 East 42nd Street
New York, NY 10165　　　(212) 986-2480

Leonard Horn
Sprung Horn Kramer
& Woods
1140 Avenue of the Americas
New York, NY 10036　　　(212) 391-0520

Steven Horowitz
Bass & Ullman
747 Third Avenue
New York, NY 10017　　　(212) 751-9494

Ethan Horwitz
Darby & Darby, P.C.
805 Third Avenue
New York, NY 10022　　　(212) 697-7660

Donald J. Howard
Mobil Oil Corp.
150 E. 42nd St.
New York, NY 10017　　　(212) 883-2742

Ben C. Hsing#
Fitzpatrick, Cella, Harper
& Scinto
277 Park Avenue
New York, NY 10172　　　(212) 758-2400

E. R. Hubbard
Fish & Neave
875 Third Avenue
New York, NY 10022　　　(212) 715-0600

Richard A. Huettner
Kenyon & Kenyon
One Broadway
New York, NY 10004　　　(212) 425-7200

Christopher A. Hughes
Morgan & Finnegan
345 Park Avenue
New York, NY 10154　　　(212) 758-4800

Julia A. Hull#
Lucas & Just
205 East 42nd Street
New York, NY 10017　　　(212) 682-4980

Jeffrey H. Ingerman
Fish & Neave
875 Third Avenue
New York, NY 10022　　　(212) 715-0600

Richard A. Inz
Fish & Neave
875 Third Avenue
New York, NY 10022　　　(212) 715-0600

Robert M. Isackson
Davis Hoxie Faithful
& Hapgood
45 Rockefeller Plaza
New York, NY 10111　　　(212) 757-2200

Shahan Islam
Morgan & Finnegan
345 Park Avenue
New York, NY 10154　　　(212) 758-4800

Robert E. Isner
Nims Howes Collison
& Isner
500 Fifth Avenue
Suite 3200
New York, NY 10110　　　(212) 382-1400

Alan Israel
Kirsthstein, Kirschstein,
Ottinger & Israel, P.C.
551 5th Avenue
New York, NY 10176　　　(212) 657-3750

Arthur E. Jackson#
Darby & Darby
805 Third Ave.
New York, NY 10022　　　(212) 351-2980

Robert R. Jackson
Fish & Neave
875 Third Avenue
New York, NY 10022　　　(212) 715-0600

Albert L. Jacobs
Jacobs & Jacobs, P.C.
521 5th Ave.
New York, NY 10175　　　(212) 687-1636

Albert L. Jacobs, Jr.
Rosenman & Colin
575 Madison Ave.
New York, NY 10022　　　(212) 940-8800

James David Jacobs
Rosen, Dainow & Jacobs
489 Fifth Ave.
New York, NY 10017　　　(212) 692-7000

Seth H. Jacobs
Davis, Hoxie, Faithfull
& Hapgood
45 Rockefeller Plaza
New York, NY 10111　　　(212) 757-2200

Howard R. Jaeger
Pfizer, Inc.
Patent Dept.
235 E. 42nd Street
New York, NY 10017　　　(212) 573-1229

Harold James
James & Franklin
60 East 42nd Street
Suite 1217
New York, NY 10165　　　(212) 867-7206

Isaac Jarkovsky
Bristol-Myers Co.
345 Park Ave.
New York, NY 10154　　　(212) 546-3653

Thomas L. Jarvis
Kenyon & Kenyon
One Broadway
New York, NY 10004　　　(212) 425-7200

Marius J. Jason
Felfe & Lynch
805 Third Avenue
New York, NY 10022　　　(212) 688-9200

Richard F. Jaworski#
Fitzpatrick, Cella,
Harper & Scinto
277 Park Avenue
New York, NY 10172　　　(212) 758-2400

Saul Jecies
605 Third Avenue
39th & 40th Streets
New York, NY 10158　　　(212) 972-1100

Jesse J. Jenner
Fish & Neave
875 Third Avenue
New York, NY 10022　　　(212) 715-0600

John E. Johnnidis
1010 5th Ave.
New York, NY 10028　　　(212) 628-0058

Harry Chapman Jones, III
Pennie & Edmonds
1155 Avenue of the Americas
New York, NY 10036　　　(212) 790-9090

Frank J. Jordan
Jordan & Hamburg
122 E. 42nd St.
New York, NY 10168   (212) 986-2340

John J. Jordan
Nynex Materiel Enterprises Co.
441 9th Avenue
7th Floor
New York, NY 10001   (212) 502-7192

Karen T. Judlowe
Fish & Neave
875 Third Ave.
New York, NY 10022   (212) 715-0600

Stephen Barry Judlowe
Hopgood, Calimafde, Kalil,
Blaustein & Judlowe
60 E. 42nd St.
New York, NY 10165   (212) 986-2480

David L. Just
Lucas & Just
205 East 42nd Street
New York, NY 10017   (212) 682-4980

Jeffrey M. Kaden#
Gottlieb, Rackman &
Reisman
1430 Broadway
New York, NY 10018   (212) 869-2890

Stephen D. Kahn
Weil, Gotshal & Manges
767 Fifth Avenue
New York, NY 10153   (212) 310-8820

Eugene J. Kalil
Hopgood, Calimafde, Kalil,
Blaustein & Judlowe
60 East 42nd Street
New York, NY 10165   (212) 986-2480

Nicholas N. Kallas
Fitzpatrick, Cella, Harper
& Scinto
277 Park Avenue
New York, NY 10172   (212) 758-2400

David A. Kalow
Lieberman, Rudolph & Nowak
292 Madison Avenue
New York, NY 10017   (212) 532-4447

Joseph S. Kaming
Kaming & Kaming
156 East 65th Street
New York, NY 10021   (212) 535-0245

Harold I. Kaplan
Blum Kaplan
1120 Avenue of the Americas
New York, NY 10036   (212) 704-0400

Jeffrey I. Kaplan#
Gottlieb, Rackman & Reisman, P.C.
1430 Broadway
New York, NY 10018   (212) 869-2890

Aaron B. Karas
Helfgott & Karas, P.C.
Empire State Bldg.
60th Floor
New York, NY 10118   (212) 643-5000

Pamela D. Kasa
7 Charles Street
New York, NY 10014   (212) 924-8987

Abraham Kasdan
Amster, Rothstein & Ebenstein
90 Park Avenue
New York, NY 10016   (212) 697-5995

David S. Kashman
Gottlieb, Rackman & Reisman
1430 Broadway
New York, NY 10018   (212) 869-2890

Yuri B. Kateshov#
5676 Riverdale Ave.
P.O. Box 900
New York, NY 10471   (212) 884-6600

Gabriel P. Katona
Schweitzer & Cornman
230 Park Avenue
New York, NY 10169   (212) 986-3377

Arthur A. Katz
Warshaw, Burstein, Cohen,
Schlesinger & Kuh
555 5th Ave.
New York, NY 10017   (212) 984-7764

Robert D. Katz
Cooper & Dunham
30 Rockefeller Plaza
New York, NY 10112   (212) 977-9550

Ivan S. Kavrukov
Cooper & Dunham
30 Rockefeller Plaza
New York, NY 10112   (212) 977-9550

Fred A. Keire
Curtis, Morris & Safford, P.C.
530 Fifth Ave.
New York, NY 10036   (212) 840-3333

Margaret B. Kelley
Morgan & Finnegan
345 Park Avenue
New York, NY 10154   (212) 758-4800

John J. Kelly, Jr.
Kenyon & Kenyon
One Broadway
New York, NY 10004   (212) 425-7200

Gail M. Kempler
Kenyon & Kenyon
1 Broadway
New York, NY 10004   (212) 425-7200

William B. Kempler
Ladas & Parry
26 West 61st Street
New York, NY 10023   (212) 708-3465

Wayne M. Kennard
Fish & Neave
875 Third Avenue
New York, NY 10022   (212) 715-0600

Mark P. Kesslen
Blum Kaplan
1120 Ave. of the Americas
New York, NY 10036   (212) 704-0400

John E. Kidd
Shea & Gould
1251 Ave. of the Americas
New York, NY 10020   (212) 827-3000

Gerald H. Kiel
Mc Aulay, Fisher, Nissen &
Goldberg
261 Madison Avenue
New York, NY 10016   (212) 986-4090

John M. Kilcoyne
Curtis, Morris & Safford, P.C.
530 Fifth Avenue
New York, NY 10036   (212) 840-3333

William F. Kilgannon
Davis, Hoxie, Faithfull
& Hapgood
45 Rockefeller Plaza
Suite 2800
New York, NY 10111   (212) 757-2200

Allen R. Kipnes
Hedman, Gibson, Costigan
& Hoare, P.C.
1185 Avenue of the Americas
New York, NY 10036   (212) 302-8989

Robert F. Kirchner
Curtis, Morris & Safford, P.C.
530 Fifth Avenue
New York, NY 10036   (212) 840-3333

Jules P. Kirsch
Cooper & Dunham
30 Rockefeller Plaza
New York, NY 10112   (212) 977-9550

David B. Kirschstein
Kirschstein, Kirschstein,
Ottinger & Israel, P.C.
551 Fifth Avenue
New York, NY 10176   (212) 697-3750

Richard B. Klar
Dvorak & Traub
Wall Street Tower
20 Exchange Place
37th Floor
New York, NY 10005   (212) 968-1300

Christopher A. Klein#
Morgan & Finnegan
345 Park Ave.
New York, NY 10154   (212) 758-4800

Milton Klein
150 West End Avenue
Apt. 24-S
New York, NY 10023   (212) 799-8608

573

Michael Klotz
151 East 83rd Street
New York, NY 10028   (201) 861-3971

Kenneth A. Koch
Asarco Inc.
Patent Dept.
180 Maiden Lane
New York, NY 10038   (212) 510-1942

Claire A. Koegler
Dewey, Ballantine, Bushby,
Palmer & Wood
140 Broadway
New York, NY 10005   (212) 820-1314

Thomas D. Kohler
Pennie & Edmonds
1155 Avenue of the Americas
New York, NY 10036   (212) 790-9090

Joseph G. Kolodny
Sprung, Horn, Kramer
& Woods
600 Third Avenue
New York, NY 10016   (212) 661-0520

Richard C. Komson
Morgan & Finnegan
345 Park Avenue
New York, NY 10154   (212) 758-4800

Gary L. Kosdan
Pennie & Edmonds
1155 Avenue of the Americaa
New York, NY 10036   (212) 790-9090

Melissa J. Koval#
362 Broome Street, Apt. 51
New York, NY 10013   (212) 713-5982

Thomas J. Kowalski
Curtis, Morris & Safford
530 Fifth Avenue
New York, NY 10036   (212) 840-3333

Gabriel P. Kralik
Morgan & Finnegan
345 Park Ave.
New York, NY 10154   (212) 758-4800

Jack Frank Kramer
Balogh, Osann, Kramer,
Dvorak, Genova & Traub
39 Broadway
New York, NY 10006   (212) 968-1300

Karen J. Kramer
Kenyon & Kenyon
One Broadway
New York, NY 10004   (212) 424-7200

Ronald Alan Krasnow
Fish & Neave
875 Third Ave.
New York, NY 10022   (212) 715-0600

John A. Krause
Fitzpatrick, Cella, Harper
& Scinto
277 Park Avenue
New York, NY 10172   (212) 758-2400

Stuart E. Krieger
Bristol-Myers Company
345 Park Ave.
New York, NY 10154   (212) 546-3646

Friedrich Kueffner
Toren, Mc Geady & Associates, P.C.
521 Fifth Avenue
New York, NY 10175   (212) 867-2912

Eve Kunen
Hopgood, Calimafde, Kalil,
Blaustein & Judlowe
Lincoln Bldg.
60 East 42nd Street
New York, NY 10165   (212) 986-2480

Lawrence Gerald Kurland
Bryan, Cave, Mc Pheeters
& Mc Roberts
245 Park Avenue
New York, NY 10167   (212) 692-1800

John Kurucz
Kane, Dalsimer, Sullivan, Kurucz,
Levy, Eisele & Richard
711 Third Avenue
New York, NY 10017   (212) 687-6000

John L. La Pierre
Pfizer, Inc.
235 East 42nd Street
New York, NY 10017   (212) 573-1594

Elias J. Lambiris
Novo Nordisk of North
America, Inc.
405 Lexington Ave.
Suite 6200
New York, NY 10017   (212) 573-1594

Thomas Langer
Frishauf, Holtz, Goodman
& Woodward, P.C.
600 Third Avenue
30th Floor
New York, NY 10016   (212) 972-1400

Andrew S. Langsam
Levisohn, Lerner & Berger
757 Third Avenue
Suite 2400
New York, NY 10017   (212) 486-7272

Marina T. Larson
Brumbaugh, Graves, Donohue
& Raymond
30 Rockefeller Plaza
New York, NY 10112   (212) 408-2580

Charles J. Laughon, II
Fish & Neave
875 Third Avenue
New York, NY 10022   (212) 715-0600

John J. Lauter
Pennie & Edmonds
1155 Avenue of the Americas
New York, NY 10036   (212) 790-9090

Stanton T. Lawrence, III
Pennie & Edmonds
1155 Avenue of the Americas
New York, NY 10036   (212) 790-9090

William F. Lawrence
Curtis, Morris & Safford, P.C.
530 5th Ave.
New York, NY 10036   (212) 840-3333

Jerome G. Lee
Morgan & Finnegan
345 Park Avenue
New York, NY 10154   (212) 758-4800

Steven J. Lee
Kenyon & Kenyon
One Broadway
New York, NY 10004   (212) 425-7200

Susan Lee
Skadden, Arps, Slate,
Meagher & Flom
919 Third Ave.
Room 3360
New York, NY 10022   (212) 735-3202

Susan K. Lehnhardt#
Brumbaugh, Graves, Donohue
& Raymond
30 Rockefeller Plaza
New York, NY 10112   (212) 408-2532

Paul Lempel
Kenyon & Kenyon
1 Broadway
New York, NY 10004   (212) 425-7200

Michael J. Lennon
Kenyon & Kenyon
One Broadway
New York, NY 10004   (212) 425-7200

Joseph B. Lerch
Darby & Darby, P.C.
805 Third Avenue
New York, NY 10022   (212) 697-7660

Henry R. Lerner
Levisohn, Niner & Lerner
757 Third Avenue
Suite 2400
New York, NY 10017   (212) 486-0296

Hallie R. Levie
Felfe & Lynch
805 Third Avenue
New York, NY 10022   (212) 688-9200

Alan H. Levine
Fiddler and Levine
350 5th Ave.
New York, NY 10118   (212) 239-4162

Jacob M. Levine
Bierman & Muserlian
757 Third Avenue
New York, NY 10017   (212) 752-7550

Martin A. Levitin
Bryan, Levitin, Franzino
& Rosenberg
330 Madison Avenue
33rd Floor
New York, NY 10017   (212) 972-8600

Gerald Levy
Kane, Dalsimer, Sullivan, Kurucz,
Levy, Richard & Eisele
711 Third Avenue
New York, NY 10017   (212) 687-6000

Bert J. Lewen
Rivkin, Radler, Dunne
& Bayh
805 Third Avenue
New York, NY 10022   (212) 418-5301

Dinah H. Lewitan
Rosen, Dainow & Jacobs
489 Fifth Avenue
New York, NY 10017   (212) 692-7000

Arthur L. Liberman
International Flavors
& Fragrances, Inc.
521 W. 57th Street
New York, NY 10019   (212) 708-7294

Arthur M. Lieberman
Lieberman, Rudolph & Nowak
292 Madison Avenue
New York, NY 10017   (212) 532-4447

Bernard Lieberman
Colgate Palmolive Company
300 Park Avenue
New York, NY 10022   (212) 310-3120

Lance J. Lieberman
Cohen, Pontani & Lieberman
551 Fifth Avenue
Suite 1210
New York, NY 10176   (212) 687-2770

Stanley H. Lieberstein
Ostrolenk, Faber, Gerb
& Soffen
1180 Avenue of the Americas
New York, NY 10036   (212) 382-0700

Maria C.H. Lin
Morgan & Finnegan
345 Park Avenue
New York, NY 10154   (212) 415-8520

Nels T. Lippert
White & Case
1155 Avenue of the Americas
New York, NY 10036   (212) 819-8200

Randy Lipsitz
Blum, Kaplan, Friedman,
Silberman & Beran
1120 Avenue of the Americas
18th Floor
New York, NY 10036   (212) 704-0400

Gordon K. Lister
Curtis, Morris & Safford, P.C.
530 Fifth Ave.
New York, NY 10036   (212) 840-3333

Nannellyn W. Lloyd
Pfizer Inc.
Legal Division
235 E. 42nd Street
New York, NY 10017   (212) 573-2100

Anthony F. Lo Cicero
Amster, Rothstein & Ebenstein
90 Park Ave.
New York, NY 10016   (212) 697-5995

Cecilia O Brien Lofters
Fish & Neave
875 Third Avenue
New York, NY 10022   (212) 715-0600

Denise L. Loring
Fish & Neave
875 Third Avenue
29th Floor
New York, NY 10022   (212) 715-0600

Michael D. Loughnane
Kenyon & Kenyon
One Broadway
New York, NY 10004   (212) 425-7200

Regina A. Loughran
151 Lexington Ave.
Suite 10D
New York, NY 10016   (212) 679-7841

Donald C. Lucas
Lucas & Just
205 East 42nd Street
New York, NY 10017   (212) 682-4980

William D. Lucas
Lucas & Just
205 E. 42nd St.
New York, NY 10017   (212) 682-4980

S. Peter Ludwig
Darby & Darby, P.C.
805 Third Avenue
New York, NY 10022   (212) 697-7660

John P. Luther
Felfe & Lynch
805 Third Avenue
New York, NY 10022   (212) 688-9200

Joel E. Lutzker
Amster, Rothstein & Ebenstein
90 Park Avenue
New York, NY 10016   (212) 697-5995

John E. D. Lynch
Felfe & Lynch
805 Third Avenue
New York, NY 10022   (212) 688-9200

Thomas D. Mac Blain
Brumbaugh, Graves, Donohue
& Raymond
30 Rockefeller Plaza
44th Floor
New York, NY 10112   (212) 408-2560

George W. Mac Donald, Jr.
Winston & Strawn, Cole & Deitz
175 Water Street
New York, NY 10038   (212) 269-2500

Leonard B. Mackey
ITT Corporation
320 Park Ave.
New York, NY 10022   (212) 940-1626

Kenneth Edward Madsen
Kenyon & Kenyon
One Broadway
New York, NY 10004   (212) 425-7200

Barry G. Magidoff
Sutton, Magidoff & Amaral
420 Lexington Avenue
New York, NY 10170   (212) 490-8533

Jay H. Maioli
Cooper & Dunham
30 Rockefeller Plaza
New York, NY 10112   (212) 977-9550

Bernard Malina
Malina & Wolfson
60 E. 42nd St.
New York, NY 10165   (212) 986-7410

Joseph M. Manak
Kaye, Scholer, Fierman,
Hays & Handler
425 Park Ave.
New York, NY 10022   (212) 836-8000

Adley F. Mandel
American Home Products Corp.
685 Third Avenue
New York, NY 10017   (212) 878-6223

Howard F. Mandelbaum
Miskin & Mandelbaum
350 Fifth Avenue
60th Floor
New York, NY 10118   (212) 268-0900

Hubert Turner Mandeville
230 Park Avenue
Room 2211
New York, NY 10169   (212) 697-4785

Raymond R. Mandra
Fitzpatrick, Cella, Harper
& Scinto
277 Park Ave., 42nd Fl.
New York, NY 10172   (212) 758-2400

Brian I. Marcus
Morgan & Finnegan
345 Park Avenue
New York, NY 10154   (212) 758-4800

Harry C. Marcus
Morgan & Finnegan
345 Park Avenue
New York, NY 10154   (212) 758-4800

James M. Markarian
Ladas & Parry
26 W. 61st Street
New York, NY 10023   (212) 708-1895

James G. Markey
Pennie & Edmonds
1155 Avenue of the Americas
New York, NY 10036   (212) 790-9090

Andrew S. Marks
Fish & Neave
875 3rd Avenue
New York, NY 10014   (212) 715-0600

Robert P. Marley
Fish & Neave
875 Third Avenue
New York, NY 10022   (212) 715-0600

Ernest Figdor Marmorek
Marmorek, Guttman & Rubenstein
330 Seventh Avenue
New York, NY 10001   (212) 594-9680

Walter G. Marple, Jr.
Anderson, Russell, Kill
& Olick, P.C.
666 Third Avenue
New York, NY 10017   (212) 850-0711

Jonathan A. Marshall
Pennie & Edmonds
1155 Avenue of the Americas
New York, NY 10036   (212) 790-9090

Larry H. Martin
601 5th Avenue
6th Floor
New York, NY 10017   (212) 888-1193

Patricia A. Martone
Fish & Neave
875 Third Avenue
New York, NY 10022   (212) 715-0600

Andrew D. Maslow
110 West End Avenue
New York, NY 10023   (212) 305-4026

Dennis A. Mason
Abelman, Frayne, Rezac
& Schwab
708 Third Avenue
New York, NY 10017   (212) 949-9022

Clifford J. Mass
Ladas & Parry
26 West 61 Street
New York, NY 10023   (212) 708-1800

Jane A. Massaro
Fish & Neave
875 Third Ave., 29th Flr.
New York, NY 10022   (212) 715-0600

Marla J. Mathias
Blum Kaplan
1120 Avenue of the Americas
New York, NY 10036   (212) 704-0400

James J. Maune
Brumbaugh, Graves,
Donohue & Raymond
30 Rockefeller Plaza
New York, NY 10112   (212) 408-2566

Richard L. Mayer
Kenyon & Kenyon
One Broadway
New York, NY 10004   (212) 425-7200

Lloyd Mc Aulay
Mc Aulay, Fields, Fisher,
Goldstein & Nissen
405 Lexington Ave.
New York, NY 10174   (212) 986-4091

Philip J. Mc Cabe
Kenyon & Kenyon
1 Broadway
New York, NY 10004   (212) 425-7200

David Martin Mc Conoughey
Curtis, Morris & Safford, P.C.
530 Fifth Avenue
New York, NY 10036   (212) 840-3333

Leslie A. Mc Donell
Fish & Neave
875 Third Ave.
29th Floor
New York, NY 10220   (212) 715-0600

Bernard X. Mc Geady
Toren, Mc Geady & Associates, P.C.
521 Fifth Avenue
New York, NY 10175   (212) 867-2912

Charles E. Mc Kenney
Pennie & Edmonds
1155 Avenue of the Americas
New York, NY 10036   (212) 790-9090

Kevin C. Mc Mahon
Weil, Gotshal & Manges
767 5th Avenue
New York, NY 10153   (212) 310-8448

John Q. Mc Quillan
Kenyon & Kenyon
One Broadway
New York, NY 10004   (212) 425-7200

James E. Mc Shane*
Dept. of Health & Human Ser.
Room 2909
26 Federal Plaza
New York, NY 10278   (212) 264-8834

Thomas F. Meagher
Kenyon & Kenyon
One Broadway
New York, NY 10004   (212) 425-7200

Edward A. Meilman
Ostrolenk, Faber, Gerb
& Soffen
1180 Avenue of the Americas
New York, NY 10036   (212) 382-0700

Michael Nicholas Meller
50 East 42nd Street
Suite 407
New York, NY 10017   (212) 953-3350

Thomas J. Meloro#
Kenyon & Kenyon
One Broadway
New York, NY 10004   (212) 425-7200

George A. Mentis
Bristol - Myers Co.
345 Park Avenue
New York, NY 10154   (212) 546-3656

James J. Merek
Fitzpatrick, Cella, Harper
& Scinto
277 Park
New York, NY 10172   (212) 758-2400

Michael A. Messina
Fish & Neave
875 Third Avenue
29th Floor
New York, NY 10022   (212) 715-0600

Dominic M. Mezzapelle
Bristol-Myers Co.
345 Park Ave.
New York, NY 10154   (212) 546-3651

Martin P. Michael
Rubin, Baum, Levin, Constant
& Friedman
30 Rockefeller Plaza
29th Floor
New York, NY 10112   (212) 698-7770

Peter C. Michalos
Notaro & Michalos, P.C.
350 Fifth Avenue
Suite 6902
New York, NY 10118   (212) 564-0200

Karl F. Milde, Jr.
Sprung, Horn, Kramer & Woods
1140 Avenue of the Americas
New York, NY 10036   (212) 391-0520

Charles E. Miller
Pennie & Edmonds
1155 Avenue of the Americas
New York, NY 10036   (212) 790-6544

Cynthia R. Miller#
Ladas & Parry
26 West 61st St.
New York, NY 10023   (212) 708-1914

Joel Miller
Weil, Gotshal & Manges
767 Fifth Avenue
New York, NY 10153   (212) 310-8067

Kerry P.L. Miller#
Brumbaugh, Graves, Donohue
& Raymond
30 Rockefeller Plaza
New York, NY 10012-2   (212) 408-2594

Loretta A. Miraglia
Fish & Neave
875 Third Ave.
New York, NY 10022   (212) 715-0600

Douglas A. Miro
Ostrolenk, Faber, Gerb
& Soffen
1180 Avenue of the Americas
New York, NY 10036   (212) 382-0700

Howard Charles Miskin
Miskin & Mandelbaum
350 Fifth Avenue
60th Floor
New York, NY 10118   (212) 268-0900

S. Leslie Misrock
Pennie & Edmonds
1155 Avenue of the Americas
New York, NY 10036   (212) 790-9090

John A. Mitchell
Fitzpatrick, Cella, Harper
& Scinto
277 Park Avenue
New York, NY 10172   (212) 758-2400

Marvin B. Mitzner
Stein, Davidoff & Malito
100 East 42nd Street
New York, NY 10017   (212) 557-7200

Gary R. Molnar
American Home Products Corp.
685 Third Avenue
New York, NY 10017   (212) 878-6225

Dennis J. Mondolino
Hopgood, Calimafde, Kalil,
Blaustein & Judlowe
60 E. 42nd St.
New York, NY 10165   (212) 986-2480

Thomas Francis Moran
Cooper & Dunham
30 Rockefeller Plaza
New York, NY 10112   (212) 977-9550

William R. Moran
333 East 43rd Street
Apt. 909
New York, NY 10017   (212) 986-5801

Robert C. Morgan
Fish & Neave
875 Third Avenue
New York, NY 10022   (212) 715-0600

Dolores A. Moro
Curtis, Morris & Safford, P.C.
530 Fifth Avenue
New York, NY 10036   (212) 840-3333

Eugene Moroz
Morgan & Finnegan
345 Park Avenue
New York, NY 10154   (212) 758-4800

Michael P. Morris
Fish & Neave
875 Third Avenue
New York, NY 10022   (212) 715-0600

Roberta J. Morris
Fish & Neave
875 Third Avenue
New York, NY 10022   (212) 715-0600

Kelly L. Morrow
Dewey, Ballantine, Bushby,
Palmer & Wood
140 Broadway
New York, NY 10005   (212) 820-1100

Mary Josephine Morry#
Morgan & Finnegan
345 Park Avenue
New York, NY 10154   (212) 758-4800

Robert H. Morse
303 West 66th Street
New York, NY 10023   (212) 787-8578

Max Moskowitz
Ostrolenk, Faber, Gerb
& Soffen
1180 Avenue of the Americas
New York, NY 10036   (212) 382-0700

Serle I. Mosoff
Enzo Biochem, Inc.
345 Hudson Street
New York, NY 10014   (212) 337-3355

Charles Gilmore Mueller
Brooks, Haidt, Haffner
& Delahunty
99 Park Ave.
New York, NY 10016   (212) 697-3355

David John Mugford
Bower & Gardner
110 E. 59th Street
New York, NY 10022   (212) 303-7021

John R. Mugno
Gottlieb, Rackman & Reisman
1430 Broadway
New York, NY 10018   (212) 869-2890

Mary Ann G. Mullen
Curtis, Morris & Safford, P.C.
530 Fifth Avenue
New York, NY 10036   (212) 840-3333

Edward F. Mullowney
Fish & Neave
875 Third Avenue
New York, NY 10022   (212) 715-0600

John D. Murnane
Brumbaugh, Graves, Donohue
& Raymond
30 Rockefeller Plaza
New York, NY 10112   (212) 408-2574

Brian P. Murphy
Hopgood, Calimafde, Kalil,
Blaustein & Judlowe
60 East 42nd St.
New York, NY 10165   (212) 986-2480

Francis J. Murphy
Hopgood, Calimafde, Kalil,
Blaustein & Judlowe
60 East 42nd Street
New York, NY 10165   (212) 986-2480

Michael M. Murray
Shea & Gould
1251 Ave. of the Americas
New York, NY 10020   (212) 827-3000

Charles A. Muserlian
Bierman & Muserlian
600 Third Avenue
New York, NY 10016   (212) 661-8000

Basam E. Nabulsi
Bryan, Cave, Mc Pheeters
& Mc Roberts
350 Park Avenue
New York, NY 10022   (212) 888-1199

George Nalaboff
Refac Technology Dev. Corp.
100 E. 42nd Street
New York, NY 10017   (212) 687-4741

John E. Nathan
Fish & Neave
875 Third Avenue
New York, NY 10022   (212) 715-0600

Howard Natter
Natter and Natter
25 W. 43rd St.
New York, NY 10036   (212) 840-8300

Seth Natter
Natter and Natter
25 W. 43rd St.
New York, NY 10036   (212) 840-8300

Peter J. Neckles
Skadden, Arps, Slate,
Meagher & Flom
919 Third Avenue
32nd Floor
New York, NY 10002-2   (212) 735-2466

Gregor N. Neff
Curtis, Morris & Safford, P.C.
530 Fifth Avenue
New York, NY 10036   (212) 840-3333

Lawrence W. Nelson#
Fitzpatrick, Cella, Harper
& Scinto
277 Park Avenue
New York, NY 10172   (212) 758-2400

Thomas R. Nesbitt, Jr.
Brumbaugh, Graves, Donohue
& Raymond
30 Rockefeller Plaza
New York, NY 10112   (212) 408-2550

Robert Neuner
Brumbaugh, Graves, Donohue
& Raymond
30 Rockefeller Plaza
New York, NY 10112   (212) 408-2552

Michael A. Nicodema
Morgan & Finnegan
345 Park Avenue
New York, NY 10154   (212) 758-4800

J. Harold Nissen
Mc Aulay, Fisher, Nissen &
Goldberg
261 Madisn Avenue
New York, NY 10016   (212) 986-4090

Kelsey I. Nix
Fish & Neave
875 Third Avenue
New York, NY 10022   (212) 715-0600

Sandra M. Nolan
Bristol-Myers Company
345 Park Avenue
6th Floor, Room 58
New York, NY 10154   (212) 546-3655

Angelo Notaro
Nataro & Michalos P.C.
350 Fifth Avenue
Suite 6902
New York, NY 10118   (212) 564-0200

Keith D. Nowak
Lieberman, Rudolph & Nowak
292 Madison Avenue
New York, NY 10017   (212) 532-4447

John A. O Brien
Fitzpatrick, Cella, Harper
& Scinto
277 Park Avenue
New York, NY 10172   (212) 758-2400

Robert D. O Brien
Bristol-Myers Company
345 Park Ave.
New York, NY 10154   (212) 546-3667

Francis M. O Connor
Shenier & O'Connor
122 East 42nd Street
New York, NY 10168   (212) 682-1986

Thomas P. O Hare
Steinberg & Raskin
551 Fifth Avenue
New York, NY 10175   (212) 628-2324

Thomas A. O Rourke
Wyatt, Gerber, Burke
& Badie
645 Madison Avenue
New York, NY 10022   (212) 826-0171

John H. Olding
28 West 44th Street
New York, NY 10036   (212) 840-1385

Milton M. Oliver
Frishauf, Holtz, Goodman
& Woodward, P.C.
600 Third Avenue
30th Floor
New York, NY 10016   (212) 972-1400

Paul J. Olivo
300 E. 40th Street
New York, NY 10016   (212) 867-4727

Stephen Martin Olko
Olko Engineering
136 West 21st Street
New York, NY 10011   (212) 645-9898

Kenneth Olsen
Schlumberger Limited
277 Park Avenue
New York, NY 10172   (212) 350-9425

Dara L. Onofrio
Ostrager & Chong
300 E. 42nd St.
New York, NY 10017   (212) 697-8080

Carl Oppedahl
Brumbaugh, Graves, Donohue
& Raymond
30 Rockefeller Plaza
New York, NY 10112   (212) 408-2500

Jerry Oppenheim
477 Madison Avenue
Suite 701
New York, NY 10022   (212) 935-8770

Daniel S. Ortiz
Burgess, Ryan & Wayne
370 Lexington Avenue
New York, NY 10017   (212) 683-8150

Glenn F. Ostrager
Ostrager & Chong
300 East 42nd Street
New York, NY 10017   (212) 697-8080

Glenn A. Ousterhout
Fish & Neave
875 Third Avenue
New York, NY 10022   (212) 715-0600

James N. Palik
Pennie & Edmonds
1155 Avenue of the Americas
New York, NY 10036   (212) 790-6219

Sheldon Palmer
Galvin & Palmer
425 Park Avenue
29th Floor
New York, NY 10022   (212) 421-4600

Maria Luisa Palmese#
Kenyon & Kenyon
1 Broadway
New York, NY 10004   (212) 425-7200

Brenda J. Panichi
Fish & Neave
875 Third Ave.
New York, NY 10022   (212) 715-0600

Michael J. Pantuliano
Pfizer Inc.
Pat. Dept.
235 E. 42nd Street
New York, NY 10017   (212) 573-2521

John P. Parise#
Lieberman, Rudolph & Nowak
292 Madison Avenue
New York, NY 10017   (212) 532-4447

Patricia A. Pasqualini
Amster, Rothstein & Ebenstein
90 Park Avenue
New York, NY 10016   (212) 697-5995

Walter Patton
American Home Products Corp.
685 Third Avenue
New York, NY 10017   (201) 274-4270

Robert E. Paulson
Morgan & Finnegan
345 Park Avenue
New York, NY 10154   (212) 758-4800

Martin B. Pavane
Schechter, Brucker & Pavane, P.C.
350 Fifth Avenue
Suite 4510
New York, NY 10118   (212) 244-6600

John B. Pegram
Davis, Hoxie, Faithfull
& Hapgood
45 Rockefeller Plaza
New York, NY 10111   (212) 757-2200

William E. Pelton
Cooper & Dunham
30 Rockefeller Plaza
New York, NY 10112   (212) 977-9550

Thomas J. Perkowski
Hopgood, Calimafde, Kalil,
Blaustein & Judlowe
60 East 42nd Street
New York, NY 10165   (212) 986-2480

Lawrence S. Perry
Fitzpatrick, Cella, Harper
& Scinto
277 Park Avenue
New York, NY 10172   (212) 758-2400

Robert F. Perry
Kenyon & Kenyon
One Broadway
New York, NY 10004   (224) 257-200

David H. Pfeffer
Morgan & Finnegan
345 Park Ave.
New York, NY 10154   (212) 758-4800

Peter J. Phillips
Cooper & Dunham
30 Rockefeller Plaza
Suite 3720
New York, NY 10112   (212) 977-9550

Margaret A. Pierri
Fish & Neave
875 Third Avenue
New York, NY 10022   (212) 715-0600

Granville Martin Pine
Morgan & Finnegan
345 Park Ave.
New York, NY 10154   (212) 758-4800

Nicola A. Pisano
Fish & Neave
875 Third Ave.
New York, NY 10022   (212) 715-0600

David C. Plache
Fish & Neave
875 Third Avenue
New York, NY 10022   (212) 715-0600

David W. Plant
Fish & Neave
875 Third Avenue
29th Floor
New York, NY 10022   (212) 715-0600

Roland Plottel
30 Rockefeller Plaza
New York, NY 10112   (212) 489-7073

Brian M. Poissant
Pennie & Edmonds
1155 Avenue of the Americas
New York, NY 10036   (212) 790-9090

Ann R. Pokalsky#
Weil, Gotshal & Manges
767 Fifth Ave.
New York, NY 10153   (212) 310-8908

Steven B. Pokotilow
1120 Avenue of the Americas
New York, NY 10036   (212) 704-0400

Grant E. Pollack
Hopgood, Calimafde, Kalil,
Blaustein & Judlowe
60 E. 42nd St.
Lincoln Bldg., Ste. 4000
New York, NY 10165   (212) 986-2480

Thomas C. Pontani
Cohen, Pontani & Lieberman
551 Fifth Avenue
Suite 1210
New York, NY 10176   (212) 687-2770

Jack Posin
340 East 64th Street
Apt. 6l
New York, NY 10021   (212) 319-4127

Donna M. Praiss
Kenyon & Kenyon
One Broadway
New York, NY 10004   (212) 425-7200

Mauro Premutico
Brumbaugh, Graves, Donohue
& Raymond
30 Rockefeller Plaza
New York, NY 10112   (212) 408-2628

Joseph J. Previto
Stoll, Previto & Hoffman
Empire State Building
350 Fifth Avenue
New York, NY 10118   (212) 736-2080

Peter H. Priest
Davis, Hoxie, Faithfull
& Hapgood
45 Rockefeller Plaza
New York, NY 10111   (212) 757-2200

David M. Quinlan
Fitzpatrick, Cella, Harper
& Scinto
277 Park Avenue
New York, NY 10172   (212) 758-2400

James A. Quinton
Frisenda, Quinton & Nicholson
605 Third Ave., 11th Flr.
New York, NY 10158   (212) 297-0015

Richard Raab#
Grow Tunneling Corp.
71 W. 23rd St.
Third Floor
New York, NY 10010   (212) 727-1760

Frederick H. Rabin
Felfe & Lynch
805 Third Avenue
New York, NY 10022   (212) 688-9200

Samuel S. Rabkin
211 East 53rd Street
New York, NY 10022   (212) 935-3739

Michael Irwin Rackman
Gottlieb, Rackman & Reisman, P.C.
1430 Broadway
New York, NY 10018   (212) 869-2890

Rory J. Radding
Pennie & Edmonds
1155 Avenue of the Americas
New York, NY 10036   (212) 790-9090

Bruce D. Radin
Morgan & Finnegan
345 Park Ave.
New York, NY 10154   (212) 758-4800

Arnold I. Rady
Morgan & Finnegan
345 Park Avenue
New York, NY 10154   (212) 758-4800

Martin G. Raskin
Steinberg & Raskin
1140 Avenue of the Americas
New York, NY 10036   (212) 768-3800

Susan H. Rauch
345 8th Avenue
New York, NY 10001   (212) 243-2623

Patricia L. Ray
Weil, Gotshal & Mages
767 5th Ave., Room 3344
New York, NY 10153   (212) 310-8532

Dana M. Raymond
Brumbaugh, Graves, Donohue
& Raymond
30 Rockefeller Plaza
New York, NY 10112   (212) 408-2518

Pasquale A. Razzano
Curtis, Morris & Safford, P.C.
530 Fifth Ave.
New York, NY 10036   (212) 840-3333

Scott K. Reed
Fitzpatrick, Cella, Harper
& Scinto
277 Park Avenue
New York, NY 10172   (212) 758-2400

John A. Reilly
Curtis, Morris & Safford, P.C.
530 5th Ave.
New York, NY 10036   (212) 840-3333

Barry D. Rein
Pennie & Edmonds
1155 Avenue of the Americas
New York, NY 10036   (212) 790-9090

Jesse David Reingold
Rosenman & Colin
575 Madison Ave.
New York, NY 10022   (212) 940-8800

James Reisman
Gottlieb, Rackman & Reisman, P.C.
1430 Broadway
9th Floor
New York, NY 10018   (212) 869-2890

Morris Relson
Darby and Darby, P.C.
805 Third Avenue
New York, NY 10022   (212) 697-7660

Henry J. Renk
Fitzpatrick, Cella, Harper
& Scinto
277 Park Avenue
New York, NY 10172   (212) 758-2400

Donald L. Rhoads
Fish & Neave
875 Third Ave.
New York, NY 10022   (212) 715-0600

James Madison Rhodes, Jr.
Hopgood, Calimafde, Kalil,
Blaustein & Judlowe
60 East 42nd Street
Suite 4000
New York, NY 10165   (212) 986-2480

John Richards
Ladas & Parry
26 West 61st Street
New York, NY 10023   (212) 708-1915

Peter C. Richardson
Pfizer Inc.
235 E. 42nd St.
New York, NY 10017   (212) 573-7805

Kurt Eugene Richter
Morgan & Finnegan
345 Park Ave.
New York, NY 10154   (212) 758-4800

Alan K. Roberts
Ladas & Parry
26 West 61st St.
New York, NY 10023   (212) 708-1800

Joseph R. Robinson
Darby & Darby, P.C.
805 Third Avenue
New York, NY 10022   (212) 697-7660

Kenneth P. Robinson
Brumbaugh, Graves, Donohue
& Raymond
30 Rockefeller Plaza
New York, NY 10112   (212) 408-2622

William R. Robinson
Brooks, Haidt, Haffner
& Delahunty
99 Park Avenue
New York, NY 10016   (212) 697-3355

Stephan A. Roen
420 East 64th Street
Suite 172
New York, NY 10021   (212) 355-1825

Laurence S. Rogers
Fish & Neave
875 Third Avenue
New York, NY 10022   (212) 715-0600

James J. Romano, Jr.
605 Third Avenue
18th Floor
New York, NY 10158   (212) 370-4044

Richard M. Rosati
Kenyon & Kenyon
One Broadway
New York, NY 10004   (212) 425-7200

Daniel M. Rosen
Rosen, Dainow & Jacobs
489 5th Ave.
New York, NY 10017   (212) 692-7000

Jeffrey A. Rosen
Sargoy, Stein, Rosen
& Shapiro
1790 Broadway
New York, NY 10019   (212) 581-2222

Lawrence Rosen
175 West 12th Street
Apt. 10B
New York, NY 10011   (212) 989-1390

Neal Lewis Rosenberg
Amster, Rothstein & Ebenstein
90 Park Avenue
New York, NY 10016   (212) 697-5995

Arthur Henry Rosenstein
Sterling Drug Inc.
90 Park Avenue
New York, NY 10016   (212) 907-3035

Lawrence Rosenthal
Blum Kaplan
1120 Avenue of the Americas
18th Floor
New York, NY 10036   (212) 704-0400

James E. Rosini
Kenyon & Kenyon
One Broadway
New York, NY 10004   (212) 425-7200

Warren H. Rotert
Morgan & Finnegan
345 Park Avenue
New York, NY 10154   (212) 415-8541

Jesse Aaron Rothstein
Amster, Rothstein & Ebenstein
90 Park Ave.
New York, NY 10016   (212) 697-5995

Philip E. Roux
Hedman, Gibson, Costigan
& Hoare, P.C.
1185 Avenue of the Americas
20th Floor
New York, NY 10036   (212) 302-8989

Thomas G. Rowan
Pennie & Edmonds
1155 Avenue of the Americas
New York, NY 10036   (212) 790-9090

Mark D. Rowland
Fish & Neave
875 Third Avenue
New York, NY 10022   (212) 715-0600

Bradley N. Ruben
Hopgood, Calimafde, Kalil,
Blaustein & Judlowe
60 East 42nd St., Ste. 4000
New York, NY 10165   (212) 715-0600

Allen I. Rubenstein
Gottlieb, Rackman & Reisman, P.C.
1430 Broadway
9th Floor
New York, NY 10018   (212) 869-2890

Kenneth Rubenstein
Marmorek, Guttman & Rubenstein
330 Seventh Avenue
New York, NY 10001   (212) 594-9680

John W. Ryan
White & Case
1155 Ave. of the Americas
New York, NY 10036   (212) 819-8200

Mary A. Ryan
White & Case
1155 Ave. of the Americas
New York, NY 10036   (212) 819-8539

Matthew K. Ryan
Curtis, Morris & Safford, P.C.
530 Fifth Ave.
New York, NY 10036   (212) 840-3333

A. Thomas S. Safford
Curtis, Morris & Safford, P.C.
530 Fifth Ave.
New York, NY 10036   (212) 840-3333

Martin I. Samuels
1725 York Avenue
Suite 16G
New York, NY 10128   (212) 427-2924

John J. Santalone
Lieberman, Rudolph,
& Nowak
292 Madison Ave.
New York, NY 10017   (212) 532-4447

Anthony M. Santini#
Bristol-Myers Squibb Co.
345 Park Ave.
Room 6-58
New York, NY 10154   (212) 546-4000

Leonard J. Santisi
Curtis, Morris & Safford, P.C.
530 5th Ave.
New York, NY 10036   (212) 840-3333

Ronald R. Santucci#
Kane, Dalsimer, Sullivan, Kurucz,
Levy, Eisele & Richard
711 Third Avenue
New York, NY 10017   (212) 687-6000

William J. Sapone
Mc Aulay, Fisher, Nissen
& Goldberg
261 Madison Avenue
New York, NY 10016   (212) 986-4090

Maria A. Savio
Kuhn & Muller
405 Lexington Avenue
New York, NY 10174   (212) 862-0864

Peter Saxon
Fitzpatrick, Cella, Harper
& Scinto
277 Park Avenue
New York, NY 10172   (212) 758-2400

David L. Schaeffer
Fitzpatrick, Cella, Harper
& Scinto
277 Park Avenue
New York, NY 10172   (212) 758-4800

Robert Schaffer
Darby & Darby
805 Third Avenue
New York, NY 10022   (212) 697-7660

Robert D. Schaffer
Hedman, Gibson, Costigan
& Hoare
1185 Ave. of the Americas
20th Floor
New York, NY 10036   (212) 302-8989

Daniel R. Schechter#
Darby & Darby, P.C.
805 Third Ave.
27th Floor
New York, NY 10022   (212) 351-2900

Peter C. Schechter
Darby & Darby, P.C.
805 Third Avenue
New York, NY 10022   (212) 351-2900

Frank F. Scheck
Pennie & Edmonds
1155 Avenue of the Americas
New York, NY 10036   (212) 790-9090

Robert C. Scheinfeld
Brumbaugh, Graves, Donohue
& Raymond
30 Rockefeller Plaza
44th Floor
New York, NY 10012   (212) 408-2512

Martin W. Schiffmiller
Kirschstein, Ottinger, Israel
& Schiffmiller, P.C.
551 Fifth Avenue
New York, NY 10176   (212) 697-3750

Barry J. Schindler
Morgan & Finnegan
345 Park Avenue
New York, NY 10154   (212) 758-4800

Richard L. Schmalz, Sr.
Westvaco Corp.
299 Park Avenue
New York, NY 10171   (212) 688-5000

Susan M. Schmitt
Kenyon & Kenyon
One Broadway
New York, NY 10004   (212) 425-7200

Caspar Carl Schneider, Jr.
Davis, Hoxie, Faithfull
& Hapgood
45 Rockefeller Plaza
New York, NY 10111   (212) 757-2200

Howard I. Schuldenfrei
441 Lexington Avenue
Suite 409
New York, NY 10017   (212) 286-9460

Jeffrey A. Schwab
Abelman, Frayne, Rezac
& Schwab
708 Third Avenue
New York, NY 10017   (212) 949-9022

Janet I. Schwadron
Ladas & Parry
26 W. 61st Street
New York, NY 10023   (212) 708-1800

Herbert Frederick Schwartz
Fish & Neave
875 Third Avenue
New York, NY 10022   (212) 715-0653

Fritz L. Schweitzer, Jr.
Schweitzer & Cornman
230 Park Ave.
New York, NY 10169   (212) 986-3377

Howard Myles Schwinger
225 Broadway
Suite 1806
New York, NY 10007   (212) 233-8820

Paul C. Scifo
233 Broadway
Suite 4703
New York, NY 10279   (212) 513-1122

Lawrence F. Scinto
Fitzpatrick, Cella,
Harper & Scinto
277 Park Ave.
New York, NY 10172   (212) 758-2400

Earl L. Scott
Wood, Williams, Rafalsky
& Harris
11 Hanover Square
New York, NY 10005   (212) 809-2900

Walter Scott
Kenyon & Kenyon
One Broadway
New York, NY 10004   (212) 425-7200

Thomas L. Secrest
Fish & Neave
875 Third Avenue
29th Floor
New York, NY 10022   (212) 715-0600

Rochelle K. Seide
Weil, Gotshal & Manges
767 Fifth Avenue
New York, NY 10153   (212) 310-8588

Robert J. Seligman
State of New York Mortgage Agency
260 Madison Avenue
New York, NY 10016   (212) 340-4266

Stuart D. Sender
Kenyon & Kenyon
One Broadway
New York, NY 10004   (212) 425-7200

Philip Thomas Shannon
Pennie & Edmonds
1155 6th Avenue
New York, NY 10036   (212) 790-6533

Henry L. Shenier
Shenier & O'Connor
380 Lexington Avenue
New York, NY 10168   (212) 682-1986

Richard S. Shenier
Shenier & O'Conner
380 Lexington Avenue
New York, NY 10168   (212) 682-1986

Robert F. Sheyka
Pfizer Inc.
Patent Department
235 East 42nd Street
New York, NY 10017   (212) 573-1189

Theodore F. Shiells
Curtis, Morris & Safford, P.C.
530 Fifth Avenue
New York, NY 10036   (212) 840-3333

Emma Shleifer#
Helfgott & Karas, P.C.
60th Floor
Empire State Bldg.
New York, NY 10118   (212) 643-5000

Doreen F. Shulman
Fish & Neave
875 Third Avenue
New York, NY 10022   (212) 715-0600

John J. Sideris#
Morgan & Finnegan
345 Park Ave.
New York, NY 10154   (212) 758-4800

Matthew W. Siegal
Blum, Kaplan, Friedman,
Silberman & Beran
1120 Avenue of the America
New York, NY 10036   (212) 704-0400

Ira E. Silfin
Amster, Rothstein &
Ebenstein
90 Park Avenue
New York, NY 10016   (212) 697-5995

Morton S. Simon
Bristol-Myers Squibb Co.
345 Park Avenue
Suite 6/54
New York, NY 10154   (212) 546-3645

Philip Y. Simons
Freeman, Wasserman & Schneider
90 John Street
New York, NY 10038   (212) 619-1770

Stuart J. Sinder
Kenyon & Kenyon
One Broadway
New York, NY 10004   (212) 425-7200

Alvin Sinderbrand
Curtis, Morris & Safford, P.C.
530 Fifth Avenue
New York, NY 10036   (212) 840-3333

John Patrick Sinnott
American Standard Inc.
1114 Avenue of the Americas
New York, NY 10036   (212) 703-5402

Brandon N. Sklar
Davis, Hoxie, Faithfull
& Hapgood
45 Rockefeller Plaza
28th Floor
New York, NY 10111   (212) 757-2200

Elizabeth O. Slade
Pfizer Inc.
235 E. 42nd St., 20th Flr.
Pat. Dept.
New York, NY 10017   (212) 573-1713

Robert L. Slater, Jr.
11 Broadway
New York, NY 10004   (212) 425-3158

Dennis M. Smid
Curtis, Morris & Safford, P.C.
530 Fifth Ave.
New York, NY 10036   (212) 840-3333

Alan D. Smith
Fish & Neave
875 Third Avenue
New York, NY 10022   (212) 715-0600

Charles B. Smith
Fish & Neave
875 Third Avenue
New York, NY 10022   (212) 715-0600

Keith G.W. Smith#
Schlumberger Limited
277 Park Avenue
New York, NY 10172   (212) 350-9554

Robert B. Smith
White & Case
1155 Ave. of the Americas
New York, NY 10036   (212) 819-8200

Stephen R. Smith
Morgan & Finnegan
345 Park Ave.
New York, NY 10154   (212) 758-4800

George B. Snyder
Curtis, Morris & Safford, P.C.
530 5th Ave.
New York, NY 10036   (212) 840-3333

Gerald Sobel
Kaye, Scholer, Fierman,
Hays & Handler
425 Park Ave
New York, NY 10022   (212) 836-8000

Joseph Sofer
Morgan & Finnegan
345 Park Ave., 22nd Flr.
New York, NY 10154   (212) 758-4800

Marvin C. Soffen
Ostrolenk, Faber,
Gerb & Soffen
1180 Avenue of the Americas
7th Floor
New York, NY 10036   (212) 382-0700

L. Teresa Solomon
Fish & Neave
875 Third Avenue
New York, NY 10022   (212) 715-0629

Kenneth H. Sonnenfeld#
Morgan & Finnegan
345 Park Avenue
New York, NY 10154   (212) 758-4800

Louis S. Sorell
Brumbaugh, Graves, Donohue
& Raymond
30 Rockefeller Plaza
New York, NY 10112   (212) 408-2620

Leonard S. Sorgi
Amster, Rothstein & Ebenstein
90 Park Avenue
New York, NY 10016   (212) 697-5995

Victor F. Souto
Kenyon & Kenyon
One Broadway
New York, NY 10004   (212) 425-7200

Mark H. Sparrow
Rosenman & Colin
575 Madison Ave.
New York, NY 10022   (212) 940-8800

Thomas E. Spath
Davis, Hoxie, Faithfull
& Hapgood
45 Rockefeller Plaza
New York, NY 10111   (212) 757-2200

William J. Spatz
Curtis, Morris & Safford, P.C.
530 Fifth Avenue
New York, NY 10036   (212) 840-3333

Charles B. Spencer
Kenyon & Kenyon
One Broadway
New York, NY 10004   (212) 425-7200

Reuben Spencer
8 Peter Cooper Road
New York, NY 10010   (212) 254-6690

Camil Peter Spiecens
11 Fifth Ave., #2E
New York, NY 10003   (212) 473-6697

Allen J. Spiegel#
Pfizer Inc.
235 E. 42nd St.
New York, NY 10017   (212) 573-2841

Seymour L. Spira#
1123 Broadway
New York, NY 10010   (212) 255-3346

Milton Springut
Lieberman, Rudolph &
Nowak
292 Madison Avenue
New York, NY 10017   (212) 532-4447

Walter E. Stalzer
Pennie & Edmonds
1155 Avenue of the Americas
New York, NY 10036   (212) 790-9090

Willard A. Stanback
Morgan & Finnegan
345 Park Ave.
New York, NY 10154   (212) 758-4800

Jon R. Stark
Pennie & Edmonds
1155 Avenue of the Americas
New York, NY 10036   (212) 790-9090

Daniel H. Steidl
Davis, Hoxie,
Faithfull & Hapgood
45 Rockefeller Plaza
New York, NY 10111   (212) 757-2200

Mitchell A. Stein
Lieberman, Rudolph & Nowak
292 Madison Avenue
New York, NY 10017   (212) 532-4447

Harold D. Steinberg
Steinberg & Raskin
1140 Avenue of the Americas
New York, NY 10036   (212) 768-3800

Gidon D. Stern
Pennie & Edmonds
1155 Avenue of the Americas
New York, NY 10036   (212) 790-9090

Henry Sternberg
Darby & Darby, P.C.
805 Third Ave.
27th Floor
New York, NY 10022   (212) 351-2994

Maurice B. Stiefel
Bryan, Cave, Mc Pheeters
& Mc Roberts
245 Park Avenue
New York, NY 10167   (212) 692-1838

Scott D. Stimpson
Pennie & Edmonds
1155 Avenue of the Americas
New York, NY 10036   (212) 970-9090

Klaus P. Stoffel
Toren, Mc Geady & Assoc.
521 Fifth Ave.
New York, NY 10175   (212) 867-2912

Robert S. Stoll
Stoll, Previto & Hoffman
5200 Empire State Bldg.
New York, NY 10118   (212) 736-0290

Samuel J. Stoll
Stoll, Previto & Hoffman
5200 Empire State Bldg.
New York, NY 10118   (212) 736-0290

John R. Storella
Fish & Neave
875 Third Avenue
29th Floor
New York, NY 10022   (212) 755-0600

Philip C. Strassburger
Bryan, Cave, Mc Pheeters
& Mc Roberts
245 Park Avenue
New York, NY 10167   (212) 692-1875

Richard L. Strauss
Pennie & Edmonds
1155 Ave. of the Americas
New York, NY 10036   (516) 764-2285

Michael John Striker
Striker, Striker & Stenby
360 Lexington Avenue
New York, NY 10017   (212) 687-5068

R. Neil Sudol
Coleman & Sudol
71 Broadway
Suite 1201
New York, NY 10006   (212) 269-3020

Joseph C. Sullivan
Kane, Dalsimer, Sullivan, Kurucz,
Levy, Eisle & Richards
711 Third Avenue
20th Floor
New York, NY 10017   (212) 687-6000

Robert C. Sullivan, Jr.
Darby & Darby P.C.
805 Third Avenue
New York, NY 10022   (212) 697-7660

Robert Cook Sullivan
Colgate-Palmolive Company
300 Park Avenue
New York, NY 10021   (212) 310-2972

Scott L. Sullivan#
Darby & Darby, P.C.
805 Third Avenue
New York, NY 10022   (212) 697-7660

Howard J. Susser
Morgan & Finnegan
345 Park Avenue
New York, NY 10154   (212) 758-4800

Paul J. Sutton
Sutton, Basseches, Magidoff
& Amaral
Graybar Bldg.
420 Lexington Ave.
New York, NY 10170   (212) 490-7900

Michael Jon Sweedler
Darby and Darby, P.C.
805 Third Avenue
New York, NY 10022   (212) 697-7660

John F. Sweeney
Morgan & Finnegan
345 Park Avenue
22nd Floor
New York, NY 10154   (212) 758-4800

Lawrence J. Swire
ITT Corporation
320 Park Ave.
New York, NY 10022   (212) 940-1632

Judith L. Sykes
370 East 76th St.
Apt. A1808
New York, NY 10021   (212) 628-8708

Henry Y. S. Tang
Brumbaugh, Graves, Donohue
& Raymond
30 Rockefeller Plaza
New York, NY 10112   (212) 408-2586

David G. Taylor
Murray & Hollander
400 Park Avenue
15th Floor
New York, NY 10022   (212) 755-6900

John Alton Taylor
307 W. 79th Street
Room 1043
New York, NY 10024   (212) 787-6600

Charles E. Temko
Temko & Temko
19 W. 44th St.
Suite 1109
New York, NY 10036   (212) 840-2178

Alan Tenenbaum
Morgan & Finnegan
345 Park Ave.
New York, NY 10154   (212) 758-4800

Arthur S. Tenser
Brumbaugh, Graves,
Donohue & Raymond
30 Rockefeller Plaza
44th Flr.
New York, NY 10112   (212) 408-2542

Berj A. Terzian
Pennie & Edmonds
1155 Avenue of the Americas
New York, NY 10036   (212) 790-6505

William J. Thomashower
Kaplan, Thomashower & Landau
747 Third Avenue
New York, NY 10017   (212) 593-1700

Roger S. Thompson
116 Pinehurst Avenue
Apt. D-14
New York, NY 10033   (212) 923-5145

Andrew L. Tiajoloff
Felfe & Lynch
805 Thrid Ave.
New York, NY 10022   (212) 688-9200

Daniel E. Tierney
Sullivan & Cromwell
125 Broad Street
New York, NY 10004   (212) 558-4848

Robert T. Tobin
Kenyon & Kenyon
One Broadway
New York, NY 10004   (212) 425-7200

Thomas W. Tobin
Wilson, Elser, Moskowitz,
Edelman & Dicker
150 East 42nd Street
New York, NY 10017   (212) 490-3000

Leonard M. Todd#
424 W. 119th St.
New York, NY 10027   (212) 865-3435

William G. Todd
Hopgood, Calimafde, Kalil,
Blaustein & Judlowe
60 E. 42nd St.
Lincoln Bldg.
New York, NY 10165   (212) 986-2480

John J. Tomaszewski
Asarco Incorporated
180 Maiden Lane
New York, NY 10038   (212) 510-1943

David Toren
Toren, Mc Geady
& Associates, P.C.
521 Fifth Avenue
New York, NY 10175   (212) 867-2912

John J. Torrente
Robin, Blecker, Deley & Driscoll
330 Madison Avenue
New York, NY 10017   (212) 682-9640

John O. Tramontine
Fish & Neave
875 Third Avenue
New York, NY 10022   (212) 715-0600

Wilma F. Triebwasser
Pennie & Edmonds
1155 Avenue of the Americas
New York, NY 10036   (212) 790-2126

Christine H. Tsai
Kenyon & Kenyon
One Broadway
New York, NY 10004   (212) 425-7200

John E. Tsavaris, II
Kenyon & Kenyon
One Broadway
New York, NY 10004   (212) 425-7200

Robert L. Tucker
160 East 84th Street
Suite 5-E
New York, NY 10028   (212) 472-6262

Marvin Turken
Jordan & Hamburg
122 East 42nd Street
Suite 3303
New York, NY 10168   (212) 986-2340

Helen Tzagoloff
152 E. 94th Street
Apt. 4J
New York, NY 10128   (212) 289-5902

Thomas J. Ungerland
Crane Company
757 Third Avenue
New York, NY 10017   (212) 415-7243

John D. Upham
165 E. 32nd, #16G
New York, NY 10016   (212) 532-1316

Robert W. J. Usher#
1133 Broadway
Suite 1515
New York, NY 10010   (212) 633-1076

Raymond Van Dyke
Morgan & Finnegan
345 Park Ave.
New York, NY 10154   (212) 758-4800

Edward E. Vassallo
Fitzpatrick, Cella, Harper
& Scinto
277 Park Ave.
New York, NY 10172   (212) 758-2400

John Charles Vassil
Morgan & Finnegan
345 Park Ave.
New York, NY 10154   (212) 758-4800

Vicki S. Veenker
Fish & Neave
875 Third Ave.
New York, NY 10022   (212) 715-0600

Anthony P. Venturino
Morgan & Finnegan
345 Park Avenue
New York, NY 10154   (212) 758-4800

Bartholomew Verdirame
Morgan & Finnegan
345 Park Ave.
New York, NY 10154   (212) 758-4800

Thomas J. Vetter
Fish & Neave
875 Third Avenue
New York, NY 10022   (212) 715-0600

Gilberto M. Villacorta#
Pennie & Edmonds
1155 Avenue of the Americas
New York, NY 10036   (202) 790-9090

Herbert H. Waddell
Memorial Sloan-Kettering
Cancer Center
1275 York Ave.
New York, NY 10021   (212) 639-3620

Mark E. Waddell
Bryan, Cave, Mc Pheeters
& Mc Roberts
245 Park Avenue
New York, NY 10167   (212) 692-1873

Linda A. Wadler
Fish & Neave
875 Third Avenue
New York, NY 10022   (212) 715-0600

Maxim H. Waldbaum
Jones, Day, Reavis
& Pogue
599 Lexington Avenue
New York, NY 10022   (212) 326-3607

George H. Wang
Boyle, Vogeler & Haimes
1270 Avenue of the Americas
New York, NY 10020   (212) 265-5100

John F. Ward
Fish & Neave
875 Third Ave.
29th Floor
New York, NY 10022   (212) 715-0600

Patrick D. Ward
720 West End Avenue
Room 408-C
New York, NY 10025   (212) 316-6000

James R. Warnot, Jr.
Shearman & Sterling
153 East 53rd Street
New York, NY 10022   (212) 848-4815

Gene Warzecha
Bristol-Myers Squibb Co.
345 Park Avenue
Suite 6-57
New York, NY 10154   (212) 546-3108

Milton J. Wayne
Burgess, Ryan & Wayne
370 Lexington Ave.
New York, NY 10017   (212) 683-8150

David Weild, III
Pennie & Edmonds
1155 Avenue of the Americas
New York, NY 10036   (212) 790-9090

Samuel Henry Weiner
Ostrolenk, Faber,
Gerb & Soffen
1180 Avenue of the Americas
New York, NY 10036   (212) 382-0700

Steven I. Weisburd
Ostrolenk, Faber, Gerb
& Soffen
1180 Avenue of the Americas
New York, NY 10036   (215) 382-0700

Morris L. Weiser
Vistatech Enterprises, Ltd.
935 Broadway
New York, NY 10010   (212) 254-9851

Laura S. Weiss
Cooper & Dunham
30 Rockefeller Plaza
37th Floor
New York, NY 10112   (212) 977-9550

Walter G. Weissenberger#
Felfe & Lynch
805 Third Ave.
New York, NY 10022   (212) 688-9200

Tiberiu Weisz
Kane, Dalsimer, Sullivan, Kurucz,
Levy, Eisele & Richard
711 Third Avenue
New York, NY 10017   (212) 687-6000

Mary C. Werner#
Kenyon & Kenyon
One Broadway
New York, NY 10004   (212) 425-7200

Paul B. West
Ladas & Parry
26 West 61st Street
New York, NY 10023   (212) 708-1980

John P. White
Cooper & Dunham
30 Rockefeller Plaza
New York, NY 10112   (212) 977-9550

Andrew M. Wilford
5676 Riverdale Ave.
P.O. Box 900
New York, NY 10471   (212) 884-6600

Van C. Wilks
Adams & Wilks
500 Fifth Ave.
Suite 3200
New York, NY 10110   (212) 575-2600

James Talbot Williams
Egli International
599 Lexington Avenue
New York, NY 10022   (212) 836-4736

Ira B. Winkler#
Hopgood, Calimafde, Kalil,
Blaustein & Judlowe
60 East 42nd Street
New York, NY 10165   (212) 986-2480

Drew M. Wintringham
Kenyon & Kenyon
One Broadway
New York, NY 10004   (212) 425-7200

Scott A. Wisser
Kenyon & Kenyon
One Broadway
New York, NY 10004   (212) 425-2200

Eric C. Woglom
Fish & Neave
875 Third Avenue
New York, NY 10022   (212) 715-0600

Michael I. Wolfson
Blum Kaplan
1120 Avenue of the Americas
New York, NY 10036   (212) 704-0400

Penina Wollman
Fitzpatrick, Cella, Harper
& Scinto
277 Park Ave.
New York, NY 10172   (212) 758-2400

Milton M. Wolson
Malina & Wolson
60 E. 42nd St.
New York, NY 10165   (212) 986-7410

William Redin Woodward
Frishauf, Holtz, Goodman
& Woodward, P.C.
600 Third Avenue
30th Floor
New York, NY 10016   (212) 972-1400

Douglas William Wyatt
Wyatt, Gerber, Burke
& Badie
645 Madison Avenue
New York, NY 10022   (212) 826-0171

Stanley J. Yavner
120 N. Main St.
Suite 203
New York, NY 10956   (914) 638-3310

Alex L. Yip#
Brumbaugh, Graves,
Donohue & Raymond
30 Rockefeller Plaza
New York, NY 10112   (212) 408-2500

Jeffrey T. Zachmann
IBM Corp.
590 Madison Ave.
New York, NY 10022   (212) 745-5836

Ilya Zborovsky#
Striker, Striker & Stenby
360 Lexington Ave.
New York, NY 10017   (212) 687-5068

Paul J. Zegger
Pennie & Edmonds
1155 Avenue of the Americas
New York, NY 10036   (212) 790-6352

Charles J. Zeller
Bristol-Myers Squibb Co.
345 Park Ave.
New York, NY 10154   (212) 546-3648

Cindy M. Zelson
Amster, Rothstein & Ebenstein
90 Park Avenue
New York, NY 10016   (212) 697-5995

Steve T. Zelson
Novo-Nordisk of North
America, Inc.
405 Lexington Avenue
Suite 6200
New York, NY 10017   (212) 867-0123

Bryan C. Zielinski
Davis, Hoxie, Faithfull
& Hapgood
45 Rockefeller Plaza
New York, NY 10111   (212) 757-2200

Alexander Zinchuk#
Darby & Darby P.C.
805 Third Ave.
27th Floor
New York, NY 10022   (212) 527-7700

Neil M. Zipkin
Amster, Rothstein & Ebenstein
90 Park Ave.
New York, NY 10016   (212) 697-5995

Norman H. Zivin
Cooper & Dunham
30 Rockefeller Plaza
New York, NY 10112   (212) 977-9550

Fredrick M. Zullow
Fitzpatrick, Cella, Harper
& Scinto
277 Park Avenue
New York, NY 10172   (212) 758-2400

Anthony M. Zupcic
Fitzpatrick, Cella,
Harper & Scinto
277 Park Avenue
New York, NY 10172   (212) 758-2400

Richard S. Milner
Cooper & Dunham
30 Rockefeller Plaza
New York, NY 10112   (212) 977-9550

William J. Ungvarsky
Kenyon & Kenyon
One Broadway
New York, NY 10004   (212) 425-7200

William J. Crossetta, Jr.
Dunn & Associates
P.O. Box 96
Newfane, NY 14108   (716) 433-1661

Michael L. Dunn
P.O. Box 96
Newfane, NY 14108   (716) 433-1661

Howard M. Ellis
Dunn & Associates
P.O. Box 96
Newfane, NY 14108   (716) 433-1661

John H. Engelmann
Occidental Chemical Corporation
P.O. Box 189
Niagara Falls, NY 14302   (716) 773-8400

Richard Dauster Fuerle
Occidental Chemical Corp.
P.O. Box 189
Niagara Falls, NY 14302   (716) 773-8459

Stanley J. Herowski, Jr.#
4037 Lewiston Road
Niagara Falls, NY 14305   (716) 284-0361

Wallace F. Neyerlin
521 College Avenue
Niagara Falls, NY 14305   (716) 284-6181

Stanley I. Laughlin
11 Frankie Lane
North Babylon, NY 11703   (516) 669-1999

William J. Mc Cabe
21 Fremont Road
North Tarrytown, NY 10591   (516) 669-1999

Arnold L. Albin
11 Robert Lennox Drive
Northport, NY 11768   (516) 757-1766

Mellor Alfred Gill
1 Sea Cove Road
Northport, NY 11768   (516) 261-9028

Edward A. Onders
146 Waterside Road
Northport, NY 11768   (516) 261-7785

Jane C. Mc Gregor
Norwich Eaton Pharmaceuticals, Inc.
17 Eaton Avenue
Norwich, NY 13815   (607) 335-2366

Randall L. Reed
6 West Park Place
Norwich, NY 13815   (607) 336-1800

Bernard J. Schulte
P.O. Box 42
Norwich, NY 13815   (607) 336-1800

David L. Suter
Norwich Eaton Pharmaceuticals, Inc.
P.O. Box 191
Norwich, NY 13815   (607) 335-2283

Martin C. Parkinson#
6 North Delaware Dr.
Nyack, NY 10960   (914) 358-3123

Douglas E. Holtz#
418 Jordan Street
Oceansisde, NY 11572   (516) 766-8778

Mark J. Egyed#
Morningside Drive
Ossining, NY 10562   (914) 762-4907

William L. Luc
204 Barnes Street
Ossining, NY 10562   (914) 941-6422

Frank R. Trifari#
8 Justamere Drive
Ossining, NY 10562   (914) 941-8347

Richard W. Watson#
Garlock Inc.
1666 Division St.
Palmyra, NY 14522   (315) 597-4811

Vern G. De Vries#
American Cyanamid Co.
Lederle Labs.
Pearl River, NY 10965   (914) 735-5000

Adma A. Ross#
Lederle Laboratories/
American Cyanamid
N. Middletown Rd.
Pearl River, NY 10965   (914) 732-3054

Sergei S. Brozski#
219 Walnut Street
Peekskill, NY 10566   (914) 739-9522

Beth K. Fields
22 Birch Brook Road
Peekskill, NY 10566   (914) 736-0095

Robert Joseph Bird
2070 Five Mile Line Road
Penfield, NY 14526   (716) 381-8920

David A. Howley,,
17 Feathery Circle
Penfield, NY 14526   (716) 264-0248

Robert Maurice Phipps
1118 Whalen Road
Penfield, NY 14526   (716) 377-7185

Ewan Campbell Mac Queen
886 Piermont Avenue
Piermont, NY 10968   (914) 359-0417

James Roy Frederick
44 Creek Ridge
Pittsford, NY 14534   (716) 381-7699

William Thomas French
4 Knobb Hill Drive
Pittsford, NY 14534   (716) 586-3649

Armin B. Pagel
24 Callingham Road
Pittsford, NY 14534   (716) 248-8432

Donald D. Schaper
14 Stonegate Lane
Pittsford, NY 14534   (716) 381-0589

John R. Schovee
10 Burr Oak Drive
Pittsford, NY 14534   (716) 248-8216

Bernard Donald Wiese
4 Northfield Gate
Pittsford, NY 14534   (716) 586-8944

Henry W. Koster
P.O. Box 890
Plandome, NY 11030   (516) 627-5124

Henry G. Mc Comb
3 Champlain Drive
Plattsburgh, NY 12901   (518) 563-0701

Israel Nissenbaum
Mechanical Plastics Corp.
Castleton Street
Pleasantville, NY 10570   (914) 769-8450

Richard R. Lloyd
7523 Cherry Valley Tpk.
Pompey, NY 13138   (315) 677-3660

Reginald Vincent Craddock
50 Bogart Ave.
Port Washington, NY 11050   (516) 944-8408

Martin Smolowitz
57 Driftwood Drive
Port Washington, NY 11050   (212) 244-3100

William Nelson Barret, Jr.
3 Brentwood Drive
Poughkeepsie, NY 12603   (914) 462-6286

Robert William Berray, Sr.
IBM Corp.
Dept. 447/Bldg. 414
South Road
Poughkeepsie, NY 12602   (914) 433-1161

Joseph A. Biela
IBM Corporation
Dept. 447 / Bldg. 414
P.O. Box 950
Poughkeepsie, NY 12602   (914) 433-1172

Carlton B. Fitchett#
28 S. White St.
Poughkeepsie, NY 12601   (914) 452-7894

Edward S. Gershuny
12 Round Hill Rd.
Poughkeepsie, NY 12603   (914) 462-3609

Bernard M. Goldman
IBM Corp.
Dept. 447/Bldg. 414
Box 390
Poughkeepsie, NY 12602   (914) 433-1162

Floyd A. Gonzalez
IBM Corporation
Intellectual Property Law Dept.
Dept. 447/Bldg. 414
P.O. Box 950
Poughkeepsie, NY 12602   (914) 433-1156

Robert J. Haase
16 Mark Vincent Drive
Poughkeepsie, NY 12603   (914) 471-7227

John F. Hanifin
17 Thornwood Drive
Poughkeepsie, NY 12603   (914) 454-9409

Edwin Lester
IBM Corp.
P.O. Box 950
Poughkeepsie, NY 12602   (914) 433-1176

Richard M. Ludwin
IBM Corp.
Dept. 447/Bldg. 414
P.O. Box 950
Poughkeepsie, NY 12602   (914) 433-1174

William B. Porter, III
IBM Corporation
P.O. Box 950
Poughkeepsie, NY 12602   (914) 433-1175

William S. Robertson, Jr.
55 Timberline Drive
Poughkeepsie, NY 12603   (914) 462-3018

George Oscar Saile, Jr.
20 Mc Intosh Drive
Poughkeepsie, NY 12603   (914) 452-5863

Joseph L. Spiegel
Spiegel, Pergament, Brown
& Basso
272 Mill Street
P.O. Box 831
Poughkeepsie, NY 12602   (914) 452-7400

Joseph Bernard Taphorn
8 Scenic Drive
Hagan Farms
Poughkeepsie, NY 12603   (914) 462-3262

Paul Denney Carmichael
IBM Corp.
2000 Purchase Street
Purchase, NY 10577   (914) 697-7233

E. Ronald Coffman
IBM Corp.
2000 Purchase Street
Purchase, NY 10577   (914) 697-7252

Colm J. Dobbyn
Pesico, Inc.
Law Dept., 4/2 M.D. 419
700 Anderson Hill Road
Purchase, NY 10577   (914) 253-3664

Ronald L. Drumheller
IBM Corp.
2000 Purchase Street
Purchase, NY 10577   (914) 697-6781

Terry J. Ilardi
IBM Corporation
Intellectual Property Law Dept.
2000 Purchase Street
Purchase, NY 10577   (914) 697-7308

Homer L. Knearl
IBM Corporation
Intellectual Property Law
2000 Purchase Street
Room 1F-40
Purchase, NY 10577   (914) 697-7252

William J. Mc Ginnis, Jr.
IBM Corporation
Intellectual Property Law
2000 Purchase Street
Purchase, NY 10577   (914) 697-7312

Steven J. Meyers
IBM Corporation
2000 Purchase Street
Purchase, NY 10577   (914) 697-7312

James Edward Murray, Jr.
IBM Corp.
2000 Purchase Street
Purchase, NY 10577   (914) 697-7380

Otho B. Ross, III
IBM Corp.
2000 Purchase Street 1F - 33
Purchase, NY 10577   (914) 697-6781

Saul A. Seinberg
IBM Corp.
2000 Purchase St.
Purchase, NY 10577   (814) 697-7367

Victor Siber
IBM Corporation
2000 Purchase Street
Mail Drop 1G-35
Purchase, NY 10577   (914) 697-7385

George A. Skoler
3010 Westchester Ave.
Purchase, NY 10577   (914) 694-1980

Marilyn D. Smith
IBM Corporation
2000 Purchase St.
Purchase, NY 10577   (914) 697-7710

Roger S. Smith
IBM Corp.
2000 Purchase Street
Purchase, NY 10577   (914) 697-7244

George Tacticos
IBM Corporation
Intellectual Property Law Dept.
1F-26
2000 Purchase Street
Purchase, NY 10577   (914) 697-7092

Saverio P. Tedesco
IBM Corporation
2000 Purchase Street
Purchase, NY 10577   (914) 697-7375

Alexander Tognino
IBM Corporation
2000 Purchase Street
1G-13
Purchase, NY 10577   (914) 697-7380

Gustave Goldstein#
22 James Drive
Putnam Valley, NY 10579   (914) 526-3588

Irving Karmin
32-22 92nd St.
Queens, NY 11369   (212) 779-3576

Douglas H. Tulley, Jr.#
21 Sylvan Ave.
Queensbury, NY 12804   (518) 793-4706

Paul E. Dupont#
Sterling-Winthrop Res. Inst.
Columbia Turnpike
Rensselaer, NY 12144   (518) 445-8292

Philip E. Hansen#
Sterling-Winthrop Res. Institute
Pat. Dept.
81 Columbia Turnpike
Rensselaer, NY 12144   (518) 445-8293

Thomas Lynn Johnson#
Sterling-Winthrop Res. Inst.
Columbia Turnpike
Rensselaer, NY 12144   (518) 445-8290

Theodore C. Miller
Sterling Drug Inc.
Sterling Res. Group
Pat. Dept.
Columbia Turnpike
Rensselaer, NY 12144   (518) 445-8284

Frederik W. Stonner#
Sterling Drug Inc.
81 Columbia Turnpike
Rensselaer, NY 12144   (518) 445-8291

Herbert Dubno
5676 Riverdale Avenue
Riverdale, NY 10471   (212) 884-6600

Andrew J. Anderson#
Eastman Kodak Co.
343 State St.
Rochester, NY 14650   (716) 722-9662

Dennis R. Arndt
Eastman Kodak Co.
343 State St.
Rochester, NY 14650   (716) 726-3896

Michael J. Balconi-Lamica
Eastman Kodak Co.
Pat. Dept.
Rochester, NY 14650   (716) 726-6828

Gerald E. Battist
Eastman Kodak Company
343 State Street
Rochester, NY 14650   (716) 724-4969

John Edward Beck
Xerox Corp.
Pat. Dept.
Xerox Square - 020
Rochester, NY 14644   (716) 423-3868

Francis H. Boos, Jr.
Eastman Kodak Company
343 State Street
Rochester, NY 14650   (716) 423-3868

Robert Francis Brothers
Eastman Kodak Co.
Patent Dept.
343 State St.
Rochester, NY 14650   (716) 724-4792

Judith L. Byorick
Xerox Corporation
Xerox Square
Rochester, NY 14644   (716) 423-4564

William A. Cammett
Xerox Corporation
Xerox Square 20A
Rochester, NY 14644   (716) 423-4132

Henry Merritt Chapin
239 Avalon Dr.
Rochester, NY 14618   (716) 442-1593

Ronald F. Chapuran
Xerox Corp.
Pat. Dept.
Xerox Square - 020
Rochester, NY 14644   (716) 423-4445

George Herman Childress
Eastman Kodak Co.
343 State St.
Rochester, NY 14650   (716) 722-7256

Robert A. Chittum
Xerox Corp.
Xerox Square 020
Rochester, NY 14644   (716) 423-4636

Thomas H. Close
Eastman Kodak Co.
343 State St.
Rochester, NY 14650   (716) 477-5272

Robert F. Cody
Eastman Kodak Co.
343 State St.
Rochester, NY 14650   (716) 726-3087

Gary B. Cohen
Xerox Corp.
Xerox Square 20A
Rochester, NY 14644   (716) 423-6612

Harold E. Cole
Eastman Kodak Co.
343 State St.
Rochester, NY 14650   (716) 722-9225

Mark Costello
Xerox Corporation
Xerox Square 20
Rochester, NY 14644   (716) 423-5006

Torger N. Dahl
Eastman Kodak Company
343 State St.
Rochester, NY 14650   (716) 724-4899

Thomas A. Davidson#
Fisons Corp.
755 Jefferson Rd.
Rochester, NY 14623   (716) 274-5329

Betty J. Deaton
Eastman Kodak Company
343 State Street
Rochester, NY 14650   (716) 477-0553

William F. Delaney, Jr.
Eastman Kodak Co.
343 State St.
Rochester, NY 14650   (716) 724-4960

Dennis M. Deleo
Eastman Kodak Co.
343 State St.
Rochester, NY 14650   (716) 724-7804

William C. Dixon, III
Eastman Kodak Co.
343 State St.
Pat. Dept.
Rochester, NY 14650   (716) 477-7418

Mark Z. Dudley
Eastman Kodak Company
Pat. Dept.
343 State Street
Rochester, NY 14650   (716) 724-4000

Edward Dugas
Eastman Kodak Company
Pat. Dept.
343 State Street
Rochester, NY 14650   (716) 477-9606

John R. Everett
Eastman Kodak Co.
Bldg. 83
343 State St.
Rochester, NY 14650   (716) 722-2776

Roger A. Fields
Eastman Kodak Co.
343 State St.
Rochester, NY 14650   (716) 726-2995

Philip Karnes Fitzsimmons
Shlesinger, Fitzsimmons & Shlesinger
183 East Main Street
Suite 1323
Rochester, NY 14604   (716) 325-4618

Henry Fleischer
Xerox Corp.
Xerox Square - 020
Rochester, NY 14644   (716) 423-4225

K. Donald Fosnaught
Eastman Kodak Co.
343 State Street
Rochester, NY 14650   (716) 724-3167

Norman H. Geil
Eastman Kodak Company
Legal Dept.
343 State Street
Rochester, NY 14650   (716) 724-5129

Samuel Richard Genca
2990 Culver Road
Rochester, NY 14622   (716) 266-4480

Robert A. Gerlach
Eastman Kodak Company
Pat. Dept.
343 State Street
Rochester, NY 14650   (716) 722-9430

Michael L. Goldman
Nixon, Hargrave, Devans
& Doyle
Clinton Square
P.O. Box 1051
Rochester, NY 14603   (716) 546-8000

Frank R. Gollon
133 Danbury Circle
Rochester, NY 14618   (716) 244-8814

Howard J. Greenwald
330 Executive Office Bldg.
36 Main Street West
Rochester, NY 14614   (716) 454-1200

Steve W. Gremban
2005 Westfall Road
Rochester, NY 14618   (716) 244-9711

Douglas Ian Hague
Eastman Kodak Company
343 State Street
Rochester, NY 14650   (716) 724-4181

Ralph E. Harper
Gleason Corporation
1000 University Ave.
P.O. Box 22970
Rochester, NY 14692   (716) 256-8750

Joseph J. Hawley
Eastman Kodak Co.
343 State St.
Rochester, NY 14650   (716) 722-9271

William A. Henry, II
Xerox Corp.
Xerox Square - 20A
Rochester, NY 14644   (716) 423-3086

Tom Hiatt
123 Scotch Lane
Rochester, NY 14617   (716) 467-4237

Ronald P. Hilst
Eastman Kodak Co.
343 State St.
Rochester, NY 14650   (716) 724-3391

Paul R. Holmes
87 Farm Brook Drive
Rochester, NY 14625   (716) 381-2946

Dwight J. Holter
Eastman Kodak Co.
343 State Street
Rochester, NY 14650   (716) 724-2883

John David Husser
Eastman Kodak Co.
343 State St.
Rochester, NY 14650   (716) 477-5256

Robert Hutter
Xerox Corporation
Xerox Square 020
Rochester, NY 14644   (716) 423-3811

John L. James
Eastman Kodak Co.
P.O. Box 15608
Rochester, NY 14615   (716) 722-9021

David F. Janci
Eastman Kodak Co.
Pat. Dept.
343 State St.
Rochester, NY 14650   (716) 722-9139

Ronald S. Kareken
Eastman Kodak Co.
Legal Dept.
343 State Street
Rochester, NY 14650   (716) 724-4669

Stephen C. Kaufman
Eastman Kodak Company
343 State Street
Rochester, NY 14650   (716) 726-3168

Lawrence P. Kessler
Eastman Kodak Co.
343 State St.
Rochester, NY 14650   (716) 722-7297

Thomas F. Kirchoff
Eastman Kodak Co.
343 State St.
Rochester, NY 14650   (716) 722-9349

William H. J. Kline
25 Indian Spring Lane
Rochester, NY 14618   (716) 244-6627

Richard Elliott Knapp
Eastman Kodak Co.
Patent Dept.
343 State St.
Rochester, NY 14650   (716) 722-9424

Peter H. Kondo
Xerox Corp.
Xerox Square 020
Rochester, NY 14644   (716) 423-4308

Antoinette F. Konski
Bausch & Lomb Inc.
Patent Law Dept.
1 Lincoln First Sq.
P.O. Box 54
Rochester, NY 14601   (716) 338-5573

David M. Krasnow
Bausch & Lomb Inc.
Patent Law Dept.
P.O. Box 54
Rochester, NY 14601   (716) 338-5573

Warren W. Kurz
Eastman Kodak Co.
343 State St.
Rochester, NY 14650   (716) 722-2396

Craig E. Larson
Bausch & Lomb
One Lincoln First Square
P.O. Box 54
Rochester, NY 14601   (716) 338-6613

Lawrence George Legg
Eastman Kodak Co.
343 State Street
Rochester, NY 14650   (716) 477-0908

Paul A. Leipold
Eastman Kodak Company
Pat. Dept.
343 State Street
Rochester, NY 14650   (716) 722-5023

Joshua Gerald Levitt
Eastman Kodak Co.
343 State St.
Rochester, NY 14650   (716) 722-9426

James Lord Lewis
Eastman Kodak Co.
343 State St.
Rochester, NY 14650   (716) 724-5721

Robert Allen Linn
Eastman Kodak Company
343 State Street
Bldg. 83
Rochester, NY 14650   (716) 722-5029

Alfred Paul Lorenzo
Eastman Kodak Co.
343 State St.
Rochester, NY 14650   (716) 477-3413

Martin Lukacher
Clinton Square
Suite 900
Rochester, NY 14604   (716) 263-1253

Joseph C. Mac Kenzie
Marjama & Pincelli, P.C.
488 White Spruce Blvd.
Rochester, NY 14623   (716) 272-8230

Paul J. Maginot
Xerox Corporation
Xerox Square - 20A
Rochester, NY 14644   (716) 423-6456

Seymour Manello
265 Warren Ave.
Rochester, NY 14618   (716) 473-2904

Owen D. Marjama
Marjama & Pincelli, P.C.
488 White Spruce Blvd.
Rochester, NY 14623   (716) 272-8230

Thomas R. Marton
102 Southland Drive
Rochester, NY 14623   (716) 424-3815

Norman Dean Mc Claskey
Eastman Kodak Co.
343 State St.
Rochester, NY 14650   (716) 724-2720

Robert L. Mc Dowell#
Gleason Works
1000 University Ave.
Rochester, NY 14692   (716) 473-1000

Frederick E. Mc Mullen
Xerox Corp.
Xerox Square 020
Rochester, NY 14644   (716) 423-3715

Stuart L. Melton
Eastman Kodak Company
343 State Street
Rochester, NY 14650   (716) 724-2048

Howard Anthony Miller
3156 Elmwood Avenue
Rochester, NY 14618   (716) 442-6411

Dennis P. Monteith
Eastman Kodak Co.
343 State St.
Rochester, NY 14650   (716) 726-3537

Paul F. Morgan
Xerox Corp.
Pat. Dept.
Xerox Square 20A
Rochester, NY 14644   (716) 423-3015

John A. Morrow
Eastman Kodak Co.
343 State St.
Rochester, NY 14650   (716) 724-2740

Samuel Elmore Mott, III
Xerox Corp.
Pat. Dept.
Xerox Square - 020
Rochester, NY 14644   (716) 423-3980

Jan A. Muddle
Eastman Kodak Co.
343 State St.
Rochester, NY 14650   (716) 477-5595

Robert Constantine Najjar
105 Lawson Road
Rochester, NY 14616   (716) 663-1811

Tallam I. Nguti
Eastman Kodak Company
343 State Street
Rochester, NY 14650   (716) 722-7643

John S. Norton
Bausch & Lomb Inc.
Intellectual Property Dept.
1 Lincoln First Square
Rochester, NY 14604   (716) 338-6611

William F. Noval
Eastman Kodak Company
343 State Street
Rochester, NY 14653   (716) 726-3299

Raymond L. Owens
Eastman Kodak Co.
343 State St.
Rochester, NY 14650   (716) 477-4653

Salvatore P. Pace
Bausch & Lomb
One Lincoln First Square
P.O. Box 54
Rochester, NY 14601   (716) 338-6001

Eugene Onofrio Palazzo
Xerox Corp.
100 South Clinton Ave.
Rochester, NY 14644   (716) 423-4687

Frank Pincelli
Marjama & Pincelli, P.C.
488 White Spruce Blvd.
Rochester, NY 14623   (716) 272-8230

Morton Arnold Polster
Eugene Stephens & Associates
56 Windsor Street
Rochester, NY 14605   (716) 232-7700

Denis A. Polyn
Bausch & Lomb Inc.
Asst. Gen. Coun. - Pat. Law
P.O. Box 54
Rochester, NY 14601   (716) 338-8417

Cecil D. Quillen, Jr.
Eastman Kodak Co.
Legal Dept.
343 State Street
Rochester, NY 14650   (716) 724-4839

Robert Lloyd Randall
Eastman Kodak Co.
Patent Dept.
343 State St.
Rochester, NY 14650   (716) 726-2132

Ronald Reichman
Eastman Kodak Co.
343 State Street
Rochester, NY 14650   (716) 253-3127

Sarah M. Roberts
Eastman Kodak Co.
Patent Legal Staff
343 State Street
Rochester, NY 14650   (716) 588-7488

Denis A. Robitaille
Xerox Corporation
Xerox Square - 20A
Rochester, NY 14644   (716) 423-6917

Judith A. Roesler
Eastman Kodak Co.
Rochester, NY 14650   (716) 477-0553

Norman Rushefsky
Eastman Kodak Co.
Pat. Dept.
343 State St.
Rochester, NY 14650   (716) 253-0125

Thomas B. Ryan#
Eugen Stephens & Associates
56 Windsor Street
Rochester, NY 14605   (716) 232-7700

Joseph R. Sakmyster
Xerox Corp.
Pat. Dept.
020 Xerox Square
Rochester, NY 14644   (716) 423-4708

Stephen B. Salai
Cumpston & Shaw, P.C.
850 Crossroads Bldg.
Rochester, NY 14614   (716) 325-5553

Milton Saunders Sales
Eastman Kodak Co.
343 State St.
Rochester, NY 14650   (716) 253-0128

Dana Murray Schmidt
Eastman Kodak Co.
343 State St.
Rochester, NY 14650   (716) 722-9151

Brian B. Shaw
Cumpston & Shaw
850 Crossroads Bldg.
Rochester, NY 14614   (716) 325-5553

George W. Shaw
Cumpston & Shaw, P.C.
850 Crossroads Office Bldg.
Rochester, NY 14614   (716) 325-5553

Svetlana Z. Short#
Eastman Kodak Co.
343 State St.
Rochester, NY 14650   (716) 253-0743

Benjamin B. Sklar, Jr.#
Xerox Corp.
Xerox Square
100 Clinton Ave.
Rochester, NY 14644   (716) 423-4554

Charles E. Snee, 111
Eastman Kodak Company
343 State Street
Rochester, NY 14650   (716) 477-7644

Eugene S. Stephens
56 Windsor Street
Rochester, NY 14605   (716) 232-7700

Elliott Stern
Eastman Kodak Co.
343 State St.
Rochester, NY 14650   (716) 724-5107

Hoffman Stone
1600 Midtown Tower
Rochester, NY 14604   (716) 325-7400

Donald W. Strickland
Eastman Kodak Co.
343 State St.
Rochester, NY 14650   (716) 588-0498

Herman J. Strnisha
53 Sansharon Dr.
Rochester, NY 14617   (716) 544-1493

Carl Otis Thomas
Eastman Kodak Co.
343 State St.
Rochester, NY 14650   (716) 722-9127

Leonard W. Treash, Jr.
Eastman Kodak Co.
343 State St.
Rochester, NY 14650   (716) 253-0152

Richard L. Troutman#
563 El Mar Drive
Rochester, NY 14616   (716) 663-0041

James L. Tucker
Eastman Kodak Co.
343 State St.
Rochester, NY 14650   (716) 722-9332

Marianne J. Twait#
Eastman Kodak Co.
343 State Street
Rochester, NY 14650   (716) 726-9818

S. C. Van Houten
141 Glen View Ln.
Rochester, NY 14609   (716) 482-9518

Robert L. Walker
Eastman Kodak Co.
343 State St.
Patent Dept.
Rochester, NY 14650   (716) 477-7419

Robert M. Wallace
Stanger, Michaelson, Spivak
& Wallace, P.C.
1151 Pittsford-Victor Rd.
Suite 105
Rochester, NY 14534   (716) 586-1730

Ogden H. Webster
Eastman Kodak Co.
343 State St.
Rochester, NY 14650   (716) 724-4437

Gerald T. Welch
Eastman Kodak Company
343 State Street
Rochester, NY 14650   (716) 724-7797

Doreen M. Wells
Eastman Kodak Co.
Patent Dept.
343 State St.
Rochester, NY 14650   (716) 477-0554

Roger Thomas Wolfe
3 Kingsbury Ct.
Rochester, NY 14618   (716) 586-3693

David M. Woods
Eastman Kodak Company
343 State Street
Rochester, NY 14653   (716) 726-2180

Harold S. Wynn#
320 N. Washington St.
Rochester, NY 14625   (716) 381-5374

Sandra M. Kotin
91 Rock Hill Dr.
P.O. Box 552
Rock Hill, NY 12775   (914) 791-6141

William P. Keegan
P.O. Box 293
Rockaway Park, NY 11694   (212) 634-3080

James P. Malone
1 Hillside Avenue
Rockville Centre, NY 11571   (516) 766-3810

Ike Aruti
257-37 149th Road
Rosedale, NY 11422   (718) 949-0840

Jack W. Benjamin#
257-27 149th Road
Rosedale, NY 11422   (718) 723-1001

Allison Charles Collard
Klein & Vibber, P.C.
1077 Northern Blvd.
Roslyn, NY 11576   (516) 365-9802

Thomas M. Galgano
Collard, Roe & Galgano, P.C.
1077 Northern Blvd.
Roslyn, NY 11576   (516) 365-9802

Kurt Kelman
1077 Northern Blvd.
Roslyn, NY 11576   (516) 627-9104

Edwin H. Keusey
Collard, Roe & Galgano, P.C.
1077 Northern Blvd.
Roslyn, NY 11576   (516) 365-9802

Chou H. Li
379 Elm Dr.
Roslyn, NY 11576   (516) 484-1719

Joseph J. Orlando
Collard, Roe & Galgano, P.C.
1077 Northern Blvd.
Roslyn, NY 11576   (516) 365-9802

Erwin S. Teltscher
69 Diana S Trail
Roslyn, NY 11576   (516) 484-2192

John F. Mc Cormack
116 Milburn Lane
Roslyn Heights, NY 11577   (516) 621-7830

Jackson B. Browning
51 Island Drive
Rye, NY 10580   (914) 939-6697

James Martin Heilman
Heilman & Heilman
10 Hix & Oakland Beach Aves.
Rye, NY 10580   (914) 967-2095

Robert Edward Kosinski
9 Philips Lane
Rye, NY 10580   (914) 967-4451

Theodore C. Jay
160 Brush Hollow Crescent
Rye Brook, NY 10573   (914) 939-6697

James W. Fitzsimmons
58 Washington Street
P.O. Box 414
Saratoga Springs, NY 12866   (518) 587-9656

Clario Ceccon#
Mc Glew & Tuttle, P.C
Scarborough Station
Scarborough, NY 10510   (914) 941-5600

Theobald J. Dengler#
Mc Glew & Tuttle, P.C.
Scarborough Station
Scarborough, NY 10510   (914) 941-5600

Christopher D. Goodman#
Mc Glew & Tuttle, P.C.
Scarborough Station
Scarborough, NY 10510   (914) 941-5600

Hilda Mc Glew
Mc Glew & Tuttle, P.C.
Scarborough Station
Scarborough, NY 10510   (914) 941-5600

John J. Mc Glew
Mc Glew & Tuttle, P.C.
Scarborough Station
Scarborough, NY 10510   (914) 941-5600

John James Mc Glew
Mc Glew & Tuttle, P.C.
Scarborough Station
Scarborough, NY 10510   (914) 941-5600

Philip D. Amins
Lackenbach, Siegel, Marzullo
& Aronson
One Chase Road
Scarsdale, NY 10583   (914) 723-4300

Howard N. Aronson
Lackenbach, Siegel, Marzullo
& Aronson
Penthouse Suite
One Chase Rd.
Scarsdale, NY 10583   (914) 723-4300

Robert E. Burns
Burns & Lobato, P.C.
Penthouse Suite
One Chase Road
Scarsdale, NY 10583   (914) 723-4300

Lawrence G. Fridman
Lackenbach, Siegel, Marzullo
& Aronson, P.C.
One Chase Road
Scarsdale, NY 10583   (914) 723-4300

David Barry Koss
164 White Rd.
Scarsdale, NY 10583   (914) 725-0542

Melvin H. Kurtz
93 Walworth Ave.
Scarsdale, NY 10583   (914) 723-7029

Richard Lau
Lackenbach, Siegel, Marzullo
& Aronson
One Chase Rd., Penthouse Suite
Scarsdale, NY 10583   (914) 723-4300

Emmanuel J. Lobato
Burns & Lobato, P.C.
Penthouse Suite
One Chase Road
Scarsdale, NY 10583   (914) 723-4300

Henry Anthony Marzullo, Jr.
Lackenbach, Siegel, Marzullo
& Aronson
One Chase Road
Scarsdale, NY 10583   (914) 723-4300

Irwin Pronin
P.O. Box 38
Scarsdale, NY 10583   (914) 725-1670

James E. Siegel
Lackenbach, Siegel, Marzullo
& Aronson, P.C.
One Chase Road
Scarsdale, NY 10583   (914) 723-4300

Gerhard K. Adam
823 State Street
Schenectady, NY 12307   (518) 346-7085

Jane M. Binkowski
General Electric Co.
P.O. Box 8
Corporate Res. & Dev.
Schenectady, NY 12308   (518) 387-6289

Jill M. Breedlove
General Electric Company
Corporate Res. & Dev.
P.O. Box 8 K1-3A63
Schenectady, NY 12301   (518) 387-6276

Donald R. Campbell
2475 Brookshire Drive
Apt. 22
Schenectady, NY 12309   (518) 346-1710

Joseph Vincent Claeys
2280 Pinehaven Drive
Schenectady, NY 12309   (518) 377-2029

Joseph T. Cohen
1320 Lexington Avenue
Schenectady, NY 12309   (518) 372-9481

James Clark Davis, Jr.
General Electric Co.
P.O. Box 8
Schenectady, NY 12301   (518) 387-6480

Sudhir G. Deshmukh
General Electric Co.
Corporate Res. & Dev.
P.O. Box 8, Bldg. K-1, 3A62
Schenectady, NY 12301   (518) 387-6275

Donald S. Ingraham
General Electric Co.
Res. & Dev. Center
Building K1, Room 3A68
P.O. Box 8
Schenectady, NY 12301   (518) 387-5073

Peter D. Johnson#
1100 Merlin Drive
Schenectady, NY 12309   (518) 785-5035

Allen L. Limberg
General Electric Company
Corp. Res. & Dev., K1-4A71
P.O. Box 8
Schenectady, NY 12345   (518) 387-7713

James Magee, Jr.
General Electric Company
Corp. Res. & Dev. Center
P.O. Box 8
Schenectady, NY 12301   (518) 387-6304

James R. Mc Daniel
GE Corporate Research
& Development
Bldg. K1, Rm. 3A61
P.O. Box 8
Schenectady, NY 12301   (518) 387-6289

James E. Mc Ginness
General Electric Company
Bldg. K-1, Room 3A64
Schenectady, NY 12301   (518) 387-6648

Mary A. Montebello
General Electric Company
Corporate Res. & Dev.
1 River Road
Schenectady, NY 12301   (518) 387-6283

Robert Ochis
P.O. Box 9106
Schenectady, NY 12309   (518) 864-5718

William Howard Pittman
General Electric Company
Corp. Res. & Dev. Center
P.O. Box 8, K1-4A64
Schenectady, NY 12301   (518) 387-5285

Paul Edward Rochford
General Electric Co.
Bldg. K1
P.O. Box 8
Schenectady, NY 12301   (518) 387-5927

Ralph M. Savage
General Electric Co.
1 River Road
Bldg. 2-406
Schenectady, NY 12345   (518) 385-4644

Patrick R. Scanlon
General Electric Company
P.O. Box 8
Bldg. K-1, Room 3A67
Schenectady, NY 12301   (518) 387-5286

Marvin Snyder
General Electric Co.
P.O. Box 8
Bldg. K-1, Room 3A58
Schenectady, NY 12301   (518) 387-6189

William A. Teoli
General Electric Co.
Corp. Research & Development
P.O. Box 8
Schenectady, NY 12301   (518) 387-5872

Wayne O. Traynham
General Electric Co.
1 River Road
Legal Operation
Bldg. 59E, Rm. 133
Schenectady, NY 12345   (518) 385-5834

Paul Richard Webb, II
General Electric Co.
Corp. Res. & Dev.
P.O. Box 8
Schenectady, NY 12301   (518) 387-5892

Lawrence P. Zale
GE Corp.
Research & Development
Bldg. K-1, Rm. 3A69
P.O. Box 8
Schenectady, NY 12301   (518) 387-7067

Kevin R. Kepner
92A Broad Street
P.O. Box 5
Schuylerville, NY 12871   (518) 695-6866

Leo I. Malossi
10 Park Lane
Scotia, NY 12302   (518) 399-5569

James W. Underwood#
9 Daphne Dr.
Scotia, NY 12302   (518) 399-4445

Francis T. Coppa
GE Plastics
1 Noryl Avenue
Selkirk, NY 12158   (518) 475-5204

Frederick H. Rinn#
39 Maple Street
Seneca Falls, NY 13148   (315) 568-6926

Anthony R. Barkume#
5 Baylor Drive
Smithtown, NY 11787-2000   (516) 724-3839

Leonard Belkin
11 Route 111
Suite 201
Smithtown, NY 11787   (516) 360-3235

Henry Hall Bassford, Jr.#
158 Douglas Rd.
Staten Island, NY 10304   (212) 273-0370

Frank V. Ponterio#
766 Pelton Avenue
Staten Island, NY 10310   (718) 273-8724

Maurice W. Ryan
23 Leslie Avenue
Staten Island, NY 10305   (718) 447-8885

Leo C. Krazinski
P.O. Box 556
Suffern, NY 10901   (212) 529-2272

George J. Darsa
15 Deer Path Lane
Syosset, NY 11791   (516) 921-1948

Morris Krapes
8 Jackson Ave.
Syosset, NY 11791   (516) 496-8466

Michael I. Kroll
171 Stillwell Lane
Syosset, NY 11791   (516) 367-7777

Marilyn L. Olshansky
5 Belmont Circle
Syosset, NY 11791   (516) 921-8060

Steven J. Winick
Ademco
165 Eileen Way
Syosset, NY 11791   (516) 921-6704

Charles E. Adams
Carrier Corporation
Patent Department
Carrier Parkway
P.O. Box 4800
Syracuse, NY 13221   (315) 432-6540

Vicki H. Audia#
Chemical Development Dept.
Bristol-Myers Squibb Co.
P. O. Box 4755
Syracuse, NY 13221   (315) 432-2502

Arthur S. Bickel
Wall & Roehrig
710 Hills Bldg.
217 Montgomery St.
Syracuse, NY 13202-2000   (315) 422-7383

Riva W. Bickel
Wall & Roehrig
Hills Bldg.
217 Montgomery St.
Seventh Floor
Syracuse, NY 13202   (315) 422-7383

Dana F. Bigelow
Carrier Corporation
P.O. Box 4800
Carrier Parkway
Syracuse, NY 13221   (315) 433-4642

Richard V. K. Bruns
Bruns & Wall
217 Montgomery Street
Syracuse, NY 13202   (315) 422-7383

Arthur A. Chalenski, Jr
Mackenzie Smith Lewis
Michell & Hughes
600 Onondaga Savings Bk. Bldg.
Syracuse, NY 13202   (315) 474-7571

Paul Checkovich
General Electric Co.
Electronics Park 6-102
Syracuse, NY 13221   (315) 456-3682

Donald Francis Daley
Carrier Corp.
6304 Carrier Parkway
Syracuse, NY 13221   (315) 433-4819

Frederick August Goettel, Jr.
Carrier Corporation
Patent Department
Carrier Pkwy., P.O. Box 4800
Syracuse, NY 13221   (315) 432-7454

Marvin A. Goldenberg
811 State Tower Bldg.
Syracuse, NY 13202   (312) 422-1191

Robert H. Kelly
Carrier Corporation
Patent Dept.
P.O. Box 4800
Carrier Parkway
Syracuse, NY 13221   (315) 433-4609

Richard V. Lang
General Electric Co.
Electronics Park, 6-102
P.O. Box 4840
Syracuse, NY 13221   (315) 456-2519

Charles Stevens Mc Guire
840 James Street
Syracuse, NY 13203   (315) 471-0361

Katherine H. Mc Guire
840 James Street
Syracuse, NY 13203   (315) 471-0361

Stephen T. Mc Mahon
128 Windemere Rd.
Syracuse, NY 13219   (315) 468-5267

Bernhard P. Molldrem, Jr.
Wall & Roehrig
710 Hills Bldg.
217 Montgomery Street
Syracuse, NY 13202   (315) 422-7383

Robert James Mooney
306 Brookford Road
Syracuse, NY 13224   (315) 446-6526

August Edward Roehrig, Jr.
Wall & Roehrig
710 Hills Bldg.
217 Montgomery Street
Syracuse, NY 13202   (315) 422-7383

Thomas J. Wall
Wall & Roehrig
217 Montgomery St.
Suite 710
Syracuse, NY 13202   (315) 422-7383

Robert O. Wright
Wall & Roehrig
710 Hills Bldg.
Syracuse, NY 13202   (315) 422-7383

Anne E. Barschall
U.S. Philips
580 White Plains Road
Tarrytown, NY 10591   (914) 332-0222

Steven R. Biren
U.S. Philips Corp.
580 White Plains Rd.
Tarrytown, NY 10591   (914) 332-0222

Thomas Allen Briody
U.S. Philips Corp.
580 White Plains Rd.
Tarrytown, NY 10591   (914) 332-0222

John C. Fox
U.S. Philips Corp.
580 White Plains Road
Tarrytown, NY 10591   (914) 332-0222

Bernard Franzblau
U.S. Philips Corp.
580 White Plains Rd.
Tarrytown, NY 10591   (914) 332-0222

Edward W. Goodman
U.S. Philips Corp.
580 White Plains Road
Tarrytown, NY 10591   (914) 332-0222

Alan J. Grant
Sprung, Horn, Kramer
& Woods
120 White Plains Road
Tarrytown, NY 10591   (914) 332-5289

Jeffrey M. Greenman
Technicon Instruments Corp.
511 Benedict Avenue
Tarrytown, NY 10591   (914) 333-6093

Jack E. Haken
U.S. Philips Corp.
580 White Plains Rd.
Tarrytown, NY 10591   (914) 332-0222

Nathaniel D. Kramer
Sprung, Horn, Kramer
& Woods
120 White Plains Road
Tarrytown, NY 10591   (914) 332-5065

Robert J. Kraus
U.S. Philips Corp.
580 White Plains Rd.
Tarrytown, NY 10591   (914) 332-0222

Emmanuel J. Lobato
U.S. Philips Corp.
580 White Plains Road
Tarrytown, NY 10591   (914) 352-0222

Michael E. Marion
U.S. Philips Corp.
580 White Plains Road
Tarrytown, NY 10591   (914) 332-0222

Robert T. Mayer
U.S. Philips Corp.
580 White Plains Rd.
Tarrytown, NY 10591   (914) 332-0222

Paul R. Miller
U.S. Philips Corp.
580 White Plains Rd.
Tarrytown, NY 10591   (914) 332-0222

Lesley Ann Rhyne
270 Carrollwood Drive
Tarrytown, NY 10591   (914) 332-0222

Stephen E. Rockwell
177 White Plains Road
Apt. 63-X
Tarrytown, NY 10591   (914) 631-0820

Ira J. Schaefer
Sprung, Horn, Kramer & Woods
120 White Plains Road
Tarrytown, NY 10591   (914) 332-5065

Jack D. Slobod
U.S. Philips Corp.
580 White Plains Road
Tarrytown, NY 10591   (914) 332-0222

Arnold Sprung
Sprung, Horn, Kramer & Woods
120 White Plains Road
Tarrytown, NY 10591   (914) 332-5065

Algy Tamoshunas
U.S. Philips Corp.
580 White Plains Rd.
Tarrytown, NY 10591   (914) 332-0222

Bernard Tiegerman
U.S. Philips Corp.
580 White Plains Road
Tarrytown, NY 10591   (914) 332-0222

David Robertson Treacy
U.S. Philips Corp.
580 White Plains Rd.
Tarrytown, NY 10591   (914) 332-0222

Arthur S. Cookfair#
201 Glenalby Road
Tonawanda, NY 14150   (716) 832-4088

Walter J. Olszewski
Union Carbide Corporation
P.O. Box 44
Tonawanda, NY 14151   (716) 879-2722

John Charles Thompson
69 Grayton Road
Tonawanda, NY 14150   (716) 832-9447

Michael J. Doyle
International Paper Company
Long Meadow Road
Tuxedo, NY 10987   (914) 577-7253

Walt Thomas Zielinski
International Paper Co.
Corporate Research Center
Long Meadow Road
Tuxedo, NY 10987   (914) 577-7283

Rocco S. Barrese
Dilworth & Barrese
50 Charles Lindbergh Blvd.
Uniondale, NY 11553   (516) 228-8484

David Michael Carter
Dilworth & Barrese
50 Charles Lindbergh Blvd.
Uniondale, NY 11553   (516) 228-8484

Joseph J. Catanzaro
Dilworth & Barrese
50 Charles Lindbergh Blvd.
Uniondale, NY 11553   (516) 228-8484

Peter De Luca
Dilworth & Barrese
333 Earle Ovington Blvd.
Uniondale, NY 11553   (516) 228-8484

Peter G. Dilworth
Dilworth & Barrese
50 Charles Lindbergh Blvd.
Uniondale, NY 11553   (516) 228-8484

Jeffrey S. Steen#
Dilworth & Barrese
333 Earle Ovington Blvd.
Uniondale, NY 11553   (516) 228-8484

Charles V. Grudzinskas#
501 N. Broadway
Upper Nyack, NY 10960   (914) 358-9232

Margaret C. Bogosian
Brookhaven National Lab.
Bldg. 355
Upton, NY 11973   (516) 282-7338

Vale P. Myles
Brookhaven National Laboratory
Bldg. 355
Upton, NY 11973   (518) 282-3312

Robert J. Speidel
Evans, Severn, Bankert
& Peet
231 Genesee Street
Utica, NY 13501   (315) 724-4151

Harold D. Berger
30 Eastwood Lane
Valley Stream, NY 11581   (516) 791-7179

Bernard S. Hoffman#
63 South Drive
Valley Stream, NY 11581   (516) 791-2488

Leonard H. King
P.O. Box 67
Valley Stream, NY 11582   (516) 997-7050

Kenneth P. Johnson
808 Sequoia Lane
Vestal, NY 13850   (607) 748-8752

Daniel Monroe Schaeffer
R.R. 2, 11 Fawn Meadow Path
Wading River, NY 11792   (516) 929-6478

Gregg D. Slepian#
2816 Riverside Drive
Wantagh, NY 11793   (516) 463-2101

Wolmar J. Stoffel
18 Shale Drive
Wappingers Falls, NY 12590   (914) 297-8234

John L. Young
General Electric Company
260 Hudson River Road
Waterford, NY 12188   (518) 233-2471

Fred L. Denson
14 E. Main Street
P.O. Box 801
Webster, NY 14580   (716) 265-2710

Gary G. Henry#
335 Brooksboro Drive
Webster, NY 14580   (716) 265-2710

Joseph A. Osofsky
300 Walton Street
West Hempstead, NY 11552   (516) 483-3235

Daniel Jay Tick
463 Dunster Court
West Hempstead, NY 11552   (516) 485-0481

William J. Eppig
1175 Montauk Hwy.
West Islip, NY 11795   (516) 587-8778

Thomas T. Kashiwabara
Champion International Corp.
West Nyack Road
West Nyack, NY 10994   (914) 578-7180

Philip Furgang
Centerock East
2 Crosfield Ave.
Suite 210
West Nyak, NY 10994   (914) 353-1818

Herbert S. Ingham
Perkin-Elmer Corporation
1101 Prospect Avenue
Westbury, NY 11590   (516) 683-2216

Leonard R. Kahn
Kahn Communications, Inc.
425 Merrick Avenue
Westbury, NY 11590   (516) 222-2221

Eugene Sheek Lovette
460 Canterbury Street
Westbury, NY 11590   (516) 334-6870

Terry S. Zisowitz
E-Z-E-M, Inc.
717 Main Street
Westbury, NY 11590   (516) 333-8230

James N. Hulme
Kelly & Hulme
277 Mill Road
Westhampton Beach, NY 11978
                                       (516) 288-2876

Robert A. Kelly
Kelly & Hulme
277 Mill Road
Westhampton Beach, NY 11978
                                       (516) 288-2876

Robert Augustin Kelly
Kelly & Hulme
277 Mill Road
Westhampton Beach, NY 11978
                                       (516) 288-2876

Curtis Ailes
Ailes & Ohlandt
175 Main Street
White Plains, NY 10601   (914) 949-7677

I. Walton Bader
Bader & Bader
65 Court St.
White Plains, NY 10601   (914) 682-0072

Charles L. Bauer
Texaco Dev. Corp.
2000 Westchester Ave.
White Plains, NY 10650   (914) 253-4041

Albert Brent
Texaco Dev. Corp.
2000 Westchester Ave.
White Plains, NY 10650   (914) 253-4541

Robert B. Burns
Texaco Dev. Corp.
2000 Westchester Ave.
White Plains, NY 10650   (914) 253-4542

Mark A. Campbell#
175 Main Street
8th Floor
White Plains, NY 10601   (914) 683-8223

George A. Coury#
Lilling & Lilling
123 Main Street
White Plains, NY 10601   (914) 684-0600

Daniel James Donovan
General Foods Co.
250 North St.
White Plains, NY 10625   (914) 335-9228

Henry H. Gibson
Texaco Development Corp.
2000 Westchester Ave.
White Plains, NY 10650   (914) 253-4538

Ronald G. Gillespie
Texaco Development Corp.
2000 Westchester Ave.
White Plains, NY 10650   (914) 253-4537

Myron Greenspan
Lilling & Greenspan
123 Main Street
White Plains, NY 10601   (914) 684-0600

Linn I. Grim
General Foods Corporation
250 North Street
White Plains, NY 10625   (914) 335-7806

Joseph T. Harcarik
General Foods Corp.
250 North St.
White Plains, NY 10625   (914) 335-9219

Charles J. Herron
175 Main Street
Suite 508
White Plains, NY 10601   (914) 761-6564

Joyce P. Hill
General Foods Corp.
250 North Street
White Plains, NY 10625   (914) 335-9188

C. Garman Hubbard
25 Hazelton Dr.
White Plains, NY 10605   (914) 949-9488

Robert A. Kulason
Texaco Inc.
2000 Westchester Ave.
White Plains, NY 10650   (914) 253-4042

Bruce E. Lilling
Lilling & Lilling
123 Main Street
Suite 936
White Plains, NY 10601   (914) 684-0600

Burton Lawrence Lilling
Lilling & Lilling
123 Main Street
Suite 936
White Plains, NY 10601   (914) 684-0600

Kenneth E. Macklin
180 South Bdwy.
White Plains, NY 10605   (914) 949-6550

Vincent A. Mallare
Texaco, Inc.
2000 Westchester Avenue
White Plains, NY 10650   (914) 253-4045

Thomas A. Marcoux
General Foods Corp.
250 North Street
White Plains, NY 10625   (914) 335-9220

Christopher Nicastri
Texaco Development Corp.
2000 Westchester Ave.
White Plains, NY 10650   (914) 253-6291

James J. O Loughlin
Texaco Dev. Corp.
2000 Westchester Ave.
White Plains, NY 10650   (914) 253-7943

John F. Ohlandt, Jr.
175 Main Street
Suite 614
White Plains, NY 10601   (914) 949-3389

David R. Plautz#
Vogt & O'Donnell
707 Westchester Ave.
White Plains, NY 10604   (914) 328-0055

Charles B. Rodman
Rodman & Rodman
7-11 South Broadway
White Plains, NY 10601   (914) 949-7210

Philip L. Rodman
Rodman & Rodman
7-11 South Broadway
Suite 400
White Plains, NY 10601   (914) 949-7210

Thomas Richard Savoie
Kraft General Foods, Inc.
250 North Street
White Plains, NY 10625   (914) 335-9222

Rolf E. Schneider
125 Lake Street
Apt. 7D-S
White Plains, NY 10604   (914) 997-1913

Carl G. Seutter
15 Idlewood Road
White Plains, NY 10605   (914) 997-1685

Martin J. Spellman, Jr.
34 South Broadway
Suite 333
White Plains, NY 10601   (914) 948-3500

Lawrence K. Stephens
4R Martine Ave., #1606
White Plains, NY 10606   (914) 422-3160

Loren C. Swingle
Nynex Corporation
500 Westchester Avenue
White Plains, NY 10604   (914) 683-2366

Dominick G. Vicari
Texaco Inc.
2000 Westchester Avenue
White Plains, NY 10650   (914) 253-7907

William H. Vogt, III
Vogt & O'Donnell
707 Westchester Ave.
White Plains, NY 10604   (914) 328-0055

Leo Zucker
50 Main Street
8th Floor
White Plains, NY 10606   (914) 761-7799

E. Herbert Liss
64 Segsbury Drive
Williamsville, NY 14221   (716) 857-3434

James J. Ralabate
5792 Main Street
Williamsville, NY 14221   (716) 634-2280

David A. Stein#
91 Exeter Road
Williamsville, NY 14221   (716) 634-9836

Charles J. Brown, Jr.
Brown, Kelleher, Zwickel
& Wilhelm
Main Street
P.O. Box 489
Windham, NY 12496   (518) 734-3800

Victor M. Serby#
VMS Consulting Engineers
255 Hewlett Neck Road
Woodmere, NY 11598   (516) 374-2455

Dale A. Bauer
10 Mayflower Dr.
Yonkers, NY 10710   (914) 779-7599

Thomas K. Landry
28 Birch Brook Road
Yonkers, NY 10708   (914) 779-5784

Louis C. Smith, Jr.
1200 Warburton Avenue
Apt. 22
Yonkers, NY 10701   (914) 423-6721

George P. Ziehmer
243 Scarsdale Road
Yonkers, NY 10707   (914) 779-6178

David Aker
IBM Corp.
T.J. Watson Research Center
P.O. Box 218
Yorktown Heights, NY 10598 (914) 241-4033

Jack M. Arnold
IBM Corp.
P.O. Box 218
Yorktown Heights, NY 10598 (914) 241-4044

Douglas W. Cameron
IBM Corporation
P.O. Box 218
Yorktown Heights, NY 10598 (914) 241-4288

Frank Chadurjian
IBM Corp.
P.O. Box 218
Yorktown Heights, NY 10598 (914) 241-4042

Michael J. De Sha
443 Granite Springs Rd.
Yorktown Heights, NY 10598 (914) 245-1640

Thomas P. Dowd
IBM Corp.
P.O. Box 218 - D/48-74
Yorktown Heights, NY 10598 (914) 241-4062

J. David Ellett, Jr.
IBM Corporation
Thomas J. Watson Res. Center
Bldg. 74, Room D/34
P.O. Box 218
Yorktown Heights, NY 10598 (914) 241-4060

Philip J. Feig
IBM Corporation
Intellectual Property Law Dept.
T.J. Watson Research Center
P.O. Box 218
Yorktown Heights, NY 10598 (914) 241-4055

William C. Gerstenzang
115 Timberlane Court
Yorktown Heights, NY 10598 (914) 245-2600

John J. Goodwin
IBM Corp.
Intellectual Property Law Dept.
P.O. Box 218
Yorktown Heights, NY 10598 (914) 241-4045

John A. Jordan
IBM Corp.
Intellectual Property Law Dept.
74-E50
P.O. Box 218
Yorktown Heights, NY 10598 (914) 241-4058

Thomas J. Kilgannon, Jr.
IBM Corp.
Thomas J. Watson Research Center
P. O. Box 218
Yorktown Heights, NY 10598 (914) 241-4040

Joe L. Koerber
1161 Baldwin Road
Yorktown Heights, NY 10598 (914) 962-3689

Daniel P. Morris
IBM Corporation
P.O. Box 218
Yorktown Heights, NY 10598 (914) 241-4041

Michael J. Quillinan
990 Lester Road
Yorktown Heights, NY 10598 (914) 245-7536

Marc D. Schechter
IBM Corporation
Intell. Property Law
T.J. Watson Res. Center
P.O. Box 218
Yorktown Heights, NY 10598 (914) 241-4278

Roy Ramon Schlemmer, Jr.
IBM Corp.
Box 218 74-DO6
Yorktown Heights, NY 10598 (914) 241-4057

J. E. Stanland#
IBM Corporation
Box 218
Yorktown Heights, NY 10598 (914) 241-4059

Robert M. Trepp
IBM Corp.
Thomas J. Watson Res. Ctr.
P.O. Box 218
Yorktown Heights, NY 10598 (914) 241-4093

Philip R. Wadsworth
IBM Corporation
T.J. Watson Research Ctr.
P.O. Box 218
Yorktown Heights, NY 10598   (914) 241-4268

North Carolina

David M. Carter
Patla, Straus, Robinson
& Moore, P.A.
P.O. Box 7625
29 North Market
Asheville, NC 28807   (704) 255-7641

Robert L. Daub
1100F Kensington Place
Asheville, NC 28803   (704) 667-7752

Bruce E. Harang
370 Sondley Drive East
Asheville, NC 28805   (704) 299-3530

Steven C. Schnedler
Roberts Stevens & Cogburn, P.A.
901 BB&T Building
P.O. Box 7647
Asheville, NC 28802   (704) 252-6600

Michael E. Smith
33 Page Avenue
Suite 200
Asheville, NC 28801   (704) 254-4778

Tommy R. Vestal
244 Bent Creek Ranch Rd.
Asheville, NC 28806   (704) 667-9739

Francis W. Young
175 Robinhood Road
Asheville, NC 28804   (704) 258-8568

J. Bowen Ross, Jr.
Davis & Newton, P.A.
412 Front Street
Beaufort, NC 28516   (919) 728-4080

James W. Williams, Jr.
2211 Front St.
Beaufort, NC 28516   (919) 728-4080

Gary J. Sertich#
201 NC 54 Bypass
#410
Carrboro, NC 27510   (919) 929-2904

William G. Buie, IV
Lord Corporation
405 Gregson Drive
P.O. Box 8225
Cary, NC 27512   (919) 469-3443

Irving M. Freedman
33 Wedgewood Road
Chapel Hill, NC 27514   (919) 942-4091

Lawrence A. Nielsen
416 Ridgecrest Drive
Chapel Hill, NC 27514   (919) 967-3572

W. Thad Adams, III
2180 First Union Plaza
301 South Tryon Street
Charlotte, NC 28282   (704) 375-9249

Herbert M. Adrian, Jr.
P.O. Box 220214
Charlotte, NC 28222   (704) 536-8651

Blas P. Arroyo#
Bell, Seltzer, Park
& Gibson
P.O. Drawer 34009
1211 East Morehead Street
Charlotte, NC 28234   (704) 377-1561

William M. Atkinson
Bell, Seltzer, Park & Gibson
P.O. 34009
Charlotte, NC 28234   (704) 377-1561

John J. Barnhardt, III
Bell, Seltzer, Park
& Gibson
1211 E. Moorehead Street
Charlotte, NC 28204   (704) 377-1561

Herman O. Bauermeister
7413 Traelight Ch. Road
Charlotte, NC 28227   (704) 545-4645

Paul B. Bell
Bell, Seltzer, Park
& Gibson
P.O. Drawer 34009
1211 E. Moorhead St.
Charlotte, NC 28234   (704) 377-1561

Mitchell S. Bigel
IBM Corp.
1001 W.T. Harris Blvd., W.
Charlotte, NC 28257   (704) 594-8300

Robert John Blanke
Celanese Corp.
Box 32414
Charlotte, NC 28232   (704) 554-2686

Stephen M. Bodenheimer, Jr.
Bell, Seltzer, Park & Gibson
P.O. Drawer 34009
Charlotte, NC 28234   (704) 377-1561

Jay M. Brown
Hoechst Celanese Corporation
P.O. Box 32414
Charlotte, NC 28232   (704) 554-2686

Barbara K. Caldwell
Bell, Seltzer, Park
& Gibson
1211 E. Morehead St.
P.O. Drawer 34009
Charlotte, NC 28234   (704) 377-1561

Julian E. Carnes, Jr.
Bell, Seltzer, Park
& Gibson
P.O. Drawer 34009
Charlotte, NC 28234   (704) 377-1561

Gregory N. Clements
Hoechst Celanese Corp.
P.O. Box 32414
Charlotte, NC 28232   (704) 554-2423

Michael S. Connor
Bell, Seltzer, Park
& Gibson, P.A.
P.O. Drawer 34009
1211 E. Morehead Street
Charlotte, NC 28234   (704) 377-1561

Ralph H. Dougherty
1515 Mockingbird Lane
Suite 410
Charlotte, NC 28209   (704) 527-7734

Charles B. Elderkin
Bell, Seltzer, Park
& Gibson, P.A.
1211 E. Morehead Street
P.O. Drawer 34009
Charlotte, NC 28234   (704) 377-1561

Shawn P. Foley#
Bell, Seltzer, Park
& Gibson
P.O. Drawer 34009
Charlotte, NC 28226   (704) 377-1561

Theodore E. Galanthay
1116 Milton Hall Place
Charlotte, NC 28226   (704) 377-1561

Floyd A. Gibson
Bell, Seltzer, Park
& Gibson, P.A.
1211 E. Moorhead St.
P.O. Drawer 34009
Charlotte, NC 28234   (704) 377-1561

Thomas H. Griswold
Homelite Divisio of Textron Inc.
P.O. Box 7047
Charlotte, NC 28241   (704) 588-3200

Robert H. Hammer, III
Hoechst Celanese Corporation
P.O. Box 32414
Charlotte, NC 28232   (704) 554-2490

John R. Hanway
5533 Birchhill Rd.
Charlotte, NC 28227   (704) 594-8305

Karl O. Hesse
IBM Corp.
1001 W.T. Harris Blvd.
Charlotte, NC 28257   (704) 594-8302

James B. Hinson
3126 Milton Road
Suite 222-B
Charlotte, NC 28215   (704) 536-6594

Clifton Tredway Hunt, Jr.
P.O. Box 15039
Charlotte, NC 28211   (704) 365-2844

Timothy R. Kroboth
5501 Providence Country Club Dr.
Charlotte, NC 28226   (704) 846-3105

Samuel Gilliland Layton, Jr.
Bell, Seltzer, Park
& Gibson, P.A.
P.O. Drawer 34009
1211 E. Morehead St.
Charlotte, NC 28234   (704) 377-1561

Ronald T. Lindsay
Collins & Aikman Corporation
701 Mc Cullough Drive
Charlotte, NC 28213   (704) 547-8500

Raymond O. Linker, Jr.
Bell, Seltzer, Park
& Gibson, P.A.
1211 East Morehead
P.O. Drawer 34009
Charlotte, NC 28234   (704) 377-1561

Ernest B. Lipscomb, III
Bell, Seltzer, Park
& Gibson, P.A.
1211 East Morehead Street
P.O. Drawer 34009
Charlotte, NC 28234   (704) 377-1561

Richard A. Lucey
725 East Trade Street
Suite 110
Charlotte, NC 28202   (704) 334-4137

Dickson M. Lupo
Bell, Seltzer, Park & Gibson
P.O. Drawer 34009
Charlotte, NC 28234   (704) 377-1561

Philip P. Mc Cann
Hoechst Celanese Corp.
P.O. Box 32414
Charlotte, NC 28232   (704) 554-2000

Michael D. Mc Coy
Bell, Seltzer, Park
& Gibson
1211 East Morehead Street
Charlotte, NC 28204   (704) 377-1561

Fritz Y. Mercer, Jr.
Carson & Mercer
730 East Trade Street
Charlotte, NC 28202   (704) 374-1005

Brian P. O Shaughnessy
Bell, Seltzer, Park
& Gibson
1211 E. Morehead Street
P.O. Drawer 34009
Charlotte, NC 28234   (704) 377-1561

Charles B. Park, III
Bell, Seltzer,
Park & Gibson, P.A.
P.O. Drawer 34009
Charlotte, NC 28234   (704) 377-1561

Paul F. Pedigo
Bell, Seltzer, Park
& Gibson, P.A.
1211 E. Morehead Street
Charlotte, NC 28204   (704) 377-1561

Francis M. Pinckney
Shefte, Pinckney & Sawyer
3740 One First Union Center
301 S. College St.
Charlotte, NC 28202   (704) 375-9181

Harris Emerson Potter#
Bell, Seltzer, Park
& Gibson, P.A.
1211 East Morehead St.
P.O. Drawer 34009
Charlotte, NC 28234   (704) 377-1561

Wilton Rankin
Sandoz Chemicals Corp.
4000 Monroe Road
Charlotte, NC 28205   (704) 331-7086

Christopher F. Regan
Bell, Seltzer, Park &
Gibson
1211 E. Morehead St.
Charlotte, NC 28234   (704) 377-1561

F. Michael Sajovec
Bell, Seltzer, Park
& Gibson
1211 East Morehead Street
Charlotte, NC 28204   (704) 377-1561

Karl S. Sawyer, Jr.
Shefte, Pinckney & Sawyer
3740 One First Union Center
301 S. College St.
Charlotte, NC 28202   (704) 375-9181

Kenneth A. Seaman
1001 W. T. Harris Blvd., West
Charlotte, NC 28257   (704) 594-8303

Donald Miller Seltzer
Bell, Seltzer, Park
& Gibson
1211 E. Morehead St.
Charlotte, NC 28201   (704) 377-1561

Dalbert Uhrig Shefte
Shefte, Pinckney & Sawyer
3740 One First Union Center
301 S. College Street
Charlotte, NC 28202   (704) 375-9181

Forrest Dale Stine
Hoechst Celanese Corp.
6000 Carnegie Blvd.
Charlotte, NC 28232   (704) 554-2573

John L. Sullivan, Jr.
Bell, Seltzer, Park
& Gibson
P.O. Drawer 34009
Charlotte, NC 28234   (704) 377-1561

Philip Summa
Bell, Seltzer, Park
& Gibson
1211 E. Morehead St.
P.O. Drawer 34009
Charlotte, NC 28234   (704) 377-1561

John H. Thomas
Bell, Seltzer, Park & Gibson
1211 East Morehead Street
P.O. Drawer 34009
Charlotte, NC 28204   (704) 377-1561

Joell T. Turner
Bell, Seltzer,
Park & Gibson, P.A.
P.O. Drawer 34009
Charlotte, NC 28234   (704) 377-1561

Russell W. Warnock
Shefte, Pinckney & Sawyer
3740 One First Union Center
301 South College St.
Charlotte, NC 28202   (704) 375-9181

Richard K. Warther
Bell, Seltzer, Park
& Gibson
1211 East Morehead Street
P.O. Drawer 34009
Charlotte, NC 28234   (704) 377-1561

Frank B. Wyatt, II
Bell, Seltzer, Park &
Gibson, P.A.
P.O. Drawer 34009
Charlotte, NC 28234   (704) 377-1561

William S. Burden#
6734 Amberley Lane
Clemmons, NC 27012   (704) 377-1561

Lynn E. Barber
Olive & Olive, P.A.
P.O. Box 2049
Durham, NC 27702   (919) 683-5514

Franklin H. Cocks#
Duke Univ.
Dept. of Mech. Eng.
Research Drive
Durham, NC 27706   (919) 684-2832

Steven J. Hultquist
Harlow, Reilly, Derr & Stark
1000 Park Forty Plaza
P.O. Drawer 13448
Research Triangle Park
Durham, NC 27702   (919) 683-5514

Richard E. Jenkins
Suite 200 S. Square Corp. Centre II
3708 Mayfair Street
Durham, NC 27707   (919) 688-8216

Michael E. Mauney
2003 Chapel Hill Road
Durham, NC 27707   (919) 683-2120

B. B. Olive
Olive & Olive
P.O. Box 2049
Durham, NC 27702   (919) 683-5514

Michael R. Philips
Olive & Olive, P.A.
P.O. Box 2049
Durham, NC 27702   (919) 683-5514

Charles E. Smith
North Carolina Central Univ.
School of Law
1801 Fayetteville St.
Turner Law Bldg.
Durham, NC 27707   (919) 560-6348

Karen M. Dellerman
BASF Corporation
Sand Hill Road
Enka, NC 28728   (704) 667-7685

Clarence R. Patty, Jr.
104 Paula Place
Flat Rock, NC 28731   (704) 697-2481

William D. Zahrt, II
Raychem Corp.
P.O. Box 3000
Fuquay-Varina, NC 27526   (919) 557-8639

Walter L. Beavers
338 N. Elm Street
Greensboro, NC 27401   (919) 275-7601

William Glenn Dosse
5307 Mockingbird Road
Greensboro, NC 27406   (919) 674-3743

Judith E. Garmon
501-B Forum VI
3200 Northline Ave.
Greensboro, NC 27408   (919) 294-0003

Howard A. Maccord, Jr.
Rhodes, Coats & Bennett
1600 First Union Tower
P.O. Box 2974
Greensboro, NC 27402   (919) 273-4422

John B. Maier
Burlington Inds., Inc.
3330 West Friendly Avenue
Greensboro, NC 27410   (919) 379-2134

Robert Yaeger Peters
AT&T
P.O. Box 25000
I-85 & Mt. Hope Church Road
Greensboro, NC 27420   (919) 279-3090

Charles R. Rhodes
Rhodes, Coats & Bennett
1600 First Union Tower
Greensboro, NC 27401   (919) 273-4422

Edward W. Rilee, Jr.
Rhodes, Coats & Bennett
1600 1st Union Tower
300 N. Greene St.
Greensboro, NC 27402   (919) 273-4422

Robert P. Miller
112 Radcliff Ct.
Hendersonville, NC 28739   (704) 693-0386

J. David Abernethy
Siecor Corporation
489 Siecor Park
Hickory, NC 28603   (704) 327-5354

Roy Bratton Moffitt
Siecor Corp.
800 17th St., N.W.
Hickory, NC 28603   (704) 327-5131

Robert R. Teall
Hickory Springs Mfg. Company
235 Second Avenue N.W.
P.O. Box 128
Hickory, NC 28601   (704) 328-2201

Hugh C. Bennett, Jr.
907 English Rd.
P.O. Box 660
High Point, NC 27261   (919) 883-2111

Gerald E. Smallwood
P.O. Box 1131
Kitty Hawk, NC 27949   (919) 261-1991

John F. Hohmann
Route 1, Box 1265
Mother Vineyard Road
Manteo, NC 27954   (919) 473-2254

Manford R. Haxton
2893 West Pine Street
Mount Airy, NC 27030   (919) 789-5034

Terry B. Mc Daniel
Westvaco Corp.
P.O. Box 2941105
5600 Virginia Ave.
North Charleston, NC 29411   (803) 745-3723

William J. Stellman
P.O. Box 786 CCNC
Pinehurst, NC 28374   (919) 692-7088

John F. Verhoeven
P.O. Box 1882
Pinehurst, NC 28374   (919) 295-3638

Robert W. Habel
P.H. Glatfelter Co.
Ecusta Division
P.O. Box 200
1 Ecusta Road
Pisgah Forest, NC 28768   (704) 877-2140

Paul M. Enlow
149 Ferrington Post
Pittsboro, NC 27312   (919) 542-3666

Henry S. Huff
402 Oakwood Drive
Pittsboro, NC 27312   (919) 542-5839

Reynold S. Allen
400 Oberlin Rd.
Suite 380
Raleigh, NC 27605   (919) 828-9508

Arthur D. Begun
8124 Brookwood Court
Raleigh, NC 27612   (919) 787-2748

David E. Bennett
Mills & Coats, P.A.
P.O. Box 5
909 Glenwood Avenue
Raleigh, NC 27602   (919) 832-3946

Freddie K. Carr#
University of North Carolina
820 Clay Street
Raleigh, NC 27605   (919) 733-4643

Larry L. Coats
Rhodes & Coats
P.O. Box 5
Raleigh, NC 27602   (919) 832-3946

Richard S. Faust
3739 National Drive
Cumberland Bldg.
Suite 225
Raleigh, NC 27612   (919) 787-6979

John B. Frisone
8828 Woody Hill Road
Raleigh, NC 27612   (919) 848-6834

Joseph H. Heard
Bell, Seltzer, Park & Gibson
1030 Washington Street
P.O. Box 10867
Raleigh, NC 27605   (919) 829-0616

James M. Kuszaj
Ogletree, Deakins, Nash,
Smoak & Stewart
P.O. Box 31608
Raleigh, NC 27622   (919) 787-9700

William Charles Lawton
P.O. Box 182
Raleigh, NC 27601   (919) 832-6861

David G. Matthews#
4500 Keswick Dr.
Raleigh, NC 27609   (919) 787-6382

James D. Myers
Bell, Seltzer, Park
& Gibson
310 UCB Plaza
P.O. Box 31107
Raleigh, NC 27622   (919) 881-3140

Robert G. Rosenthal
5856 Faringdon Place
Raleigh, NC 27609   (919) 872-4911

William E. Schadel
Carolina Power & Light
333 Fayetteville Street Mall
BB&T Building
Suite 513
Raleigh, NC 27602   (919) 546-4331

Kenneth D. Sibley
Bell, Seltzer, Park
& Gibson
3605 Glenwood Avenue
Suite 310, P.O. Drawer 31107
Raleigh, NC 27622   (919) 881-3140

Robert M. Wolters
8840 Reigate Lane
Raleigh, NC 27603   (919) 779-5138

Joscelyn G. Cockburn
IBM Corporation
972/002
Cornwallis Road
P.O. Box 12195
Research Triangle Park, NC 27709
(919) 543-9036

Gregory M. Doudnikoff
IBM Corporation
Dept. 972/B205
3039 Cornwallis Road
P.O. Box 12195
Research Triangle Park, NC 27709
(919) 543-9036

Edward H. Duffield
IBM Corp.
Dept. 972
P.O. Box 12195
Research Triangle Park, NC 27709
(919) 543-4710

Richard W. Evans
Harlow, Reilly, Derr
& Stark
P.O. Box 13448
Research Triangle Park, NC 27709
(919) 544-5555

Thomas F. Galvin
IBM Corporation
972/002, PO Box 12195
Cornwallis Road
Research Triangle Park, NC 27709
(919) 543-4184

Linda L. Gardner#
Semiconductor Research Corp.
P.O. Box 12053
4501 Alexander Drive
Suite 301
Research Triangle Park, NC 27709
(919) 541-9447

Hannah O. Green#
Burroughs Wellcome Company
3030 Cornwallis Road
Research Triangle Park, NC 27709
(919) 248-4832

Charles T. Joyner
Glaxo Inc.
5 Moore Drive
Research Triangle Park, NC 27709
(919) 248-2192

James R. Kersey
Murray & Kersey
P.O. Box 12719
Research Triangle Park, NC 27709
(919) 544-1764

Steven R. Lazar
CIBA-GEIGY Corporation
3054 Cornwallis Road
Research Triangle Park, NC 27709
(919) 541-8500

David J. Levy
Glaxo Inc.
Five Moore Drive
Research Triangle Park, NC 27709
(919) 248-2723

Robert J. Theissen#
Rhone-Poulenc Ag Co.
2 Alexander Drive
P.O. Box 12014
Research Triangle Park, NC 27709
(919) 549-2254

Gerald Ray Woods
IBM Corporation
P.O. Box 12195
972/B205
Research Triangle Park, NC 27709
(919) 543-7204

Herbert J. Zeh, Jr.
Reichhold Chemicals, Inc.
P.O. Box 13582
Research Triangle Park, NC 27709
(919) 361-7862

Donald L. Weinhold
318 North Main Street
Salisbury, NC 28144
(704) 637-2235

Davis T. Moorhead
P.O. Box 990
Saluda, NC 28773
(704) 749-9879

Philip L. Schlamp
587 Spyglass Lane
Sanford, NC 27330
(919) 499-6050

John Calvin Wiessler
6014 Cypress Pt.
Sanford, NC 27330
(919) 499-2605

Randall A. Davis
Lucas & Bryant, P.A.
P.O. Box 55
Selma, NC 27576
(919) 965-8184

George J. Cannon
423 Star Hill Drive
Swansboro, NC 28584
(919) 393-7699

Abner Milton Cornwell, Jr.
P.O. Box 887
Wadesboro, NC 28170
(704) 694-9550

John G. Mills, III
614 Durham Road
P.O. Box 587
Wake Forest, NC 27587
(919) 556-3838

Fletcher C. Eddens#
1730 Fairway Drive
Wilmington, NC 28403
(919) 762-0600

William H. Edgerton#
Formulations Development Co.
1200 N. 23rd Street
Unit 102
Wilmington, NC 28405
(919) 762-3120

Herbert J. Bluhm#
4101 Tangle Lane
Winston-Salem, NC 27106
(919) 723-7412

August J. Borschke
R.J. Reynolds Tobacco Co.
R.J. Reynolds Bldg.
Winston-Salem, NC 27102
(919) 773-5491

John M. Harrington
Petree, Stockton & Robinson
1001 West Fourth Street
Winston-Salem, NC 27101
(919) 725-2351

Jack B. Hicks
Womble, Carlyle, Sandridge & Rice
1600 One Triad Park
200 W. Second St.
Winston-Salem, NC 27101
(919) 721-3709

Charles Yount Lackey
3540 Buena Vista Rd.
Winston-Salem, NC 27106
(919) 724-4623

Grover M. Myers
R.J. Reynolds Tobacco Co.
4th & Main Streets
Winston-Salem, NC 27102
(919) 741-2694

Robert W. Pitts
AMP Inc.
3700 Reidsville Road
Winston-Salem, NC 27102
(919) 727-5051

North Dakota

Mack L. Thomas
Melroe Company
112 North University Drive
Fargo, ND 58102
(701) 241-8717

David G. Adams#
Devil S Lake Sioux Mfg.
Highway 57
Ft. Totten, ND 58335
(701) 766-4211

Robert Elick Kleve
1103 24th Avenue South
Grand Forks, ND 58201
(701) 772-4311

Ohio

Donald J. Bobak
Renner, Kenner, Greive,
Bobak & Taylor
1610 First National Tower
Akron, OH 44308
(216) 376-1242

Robert Walter Brown
Goodyear Tire & Rubber Co.
Pat. & Tdmk. Dept. 823
1144 E. Market Street
Akron, OH 44316
(216) 796-6389

Ford Whitman Brunner, Jr.
270 Stratford Road
Akron, OH 44313
(216) 867-5993

Richard Harvey Childress
Goodyear Tire & Rubber Co.
Dir. of Pats. & Tdmks.
1144 E. Market St.
Akron, OH 44316
(216) 796-4786

Mack Dickson Cook, II
Cook & Cook
900 1st Natl. Tower
Akron, OH 44308   (216) 376-1005

Alan A. Csontos
Uniroyal Goodrich Tire Co.
Pat. Law Dept., D/0530, UGB-6
600 S. Main Street
Akron, OH 44397   (216) 374-2951

Robert T. Cunningham, Jr.
Goodyear Tire & Rubber Co.
1144 E. Market Street
Akron, OH 44316   (216) 796-7880

Theodore Joseph Dettling
574 Castle Blvd.
Akron, OH 44313   (216) 836-4742

Marc R. Dion, Sr.
Goodyear Tire & Rubber Co.
1144 East Market Street
Akron, OH 44316   (216) 796-8251

David P. Dureska
Daniel J. Hudak Co.
156 S. Main St.
Transohio Bldg.
Suite 808
Akron, OH 44308   (216) 535-2220

Lee A. Germain#
Loral Systems Group
Div. of Loral Corporation
1210 Massillon Road
Akron, OH 44315   (216) 796-2070

Stephen L. Grant
Oldham & Oldham Co., L.P.A
1225 W. Market St.
Akron, OH 44313   (216) 864-5550

Edward G. Greive
Renner, Kenner, Greive,
Bobak & Taylor
1610 First National Tower
Akron, OH 44308   (216) 376-1242

Daniel N. Hall
Firestone Tire & Rubber Co.
Law Dept.
1200 Firestone Pkwy.
Akron, OH 44317   (216) 379-7543

Everett R. Hamilton
Renner, Kenner, Greive,
Bobak & Taylor
1610 First National Tower
Akron, OH 44308   (216) 376-1242

William R. Holland
807 Citi Center
146 South High Street
Akron, OH 44308   (216) 535-4114

Byron John Hook
1940 Revere Road
Akron, OH 44313   (216) 666-9269

Daniel J. Hudak
Transohio Building
156 S. Main St.
Suite 808
Akron, OH 44308   (216) 535-2220

Douglas J. Hura
Renner, Kenner, Greive, Bobak,
Taylor & Weber
Sixteenth Floor
First National Tower
Akron, OH 44308   (216) 376-1242

Thomas A. Kayuha
Ohio Edison Company
76 S. Main St.
Akron, OH 44308   (216) 384-5600

Phillip Lee Kenner
Renner, Kenner, Greive, Bobak,
Taylor & Weber
1610 First National Tower
Akron, OH 44308   (216) 376-1242

David L. King
Goodyear Tire & Rubber Co.
1144 E. Market St.
Akron, OH 44316   (216) 796-6389

Louis F. Kreek, Jr.
Oldham & Oldham Co., L.P.A.
1225 West Market Street
Akron, OH 44313   (216) 864-5550

Frederick K. Lacher
7 West Bowery Street
Suite 908
Akron, OH 44308   (216) 535-5522

Samuel B. Laferty
107 N. Pershing
Akron, OH 44313   (216) 867-0998

Robin Sue Levine#
Daniel J. Hudak Co., L.P.A.
Transohio Bldg., Suite 808
156 S. Main Street
Akron, OH 44308   (216) 535-2220

Thomas P. Lewandowski
Goodyear Tire & Rubber Co.
1144 E. Market St.
Akron, OH 44316   (216) 796-4219

James Robert Lindsay, Sr.
B.F. Goodrich Co.
3925 Embassy Parkway
Akron, OH 44313   (216) 374-2167

David M. Lowry
Tramonte, Kot, Davis
& Lowry
411 Wolf Ledges
Suite 100
Akron, OH 44311   (216) 434-1112

Frank C. Manak, III
General Tire, Inc.
One General Street
Akron, OH 44329   (216) 798-2941

Eric Marich
583 Ventura Blvd.
Akron, OH 44319   (216) 882-9752

Olaf Nielsen
679 N. Revere Road
Akron, OH 44313   (216) 867-6790

Nils E. Nilsson
77 Fir Hill
No. 4-B-12
Akron, OH 44304   (216) 384-5624

Edwin W. Oldham
Oldham & Oldham, L.P.A.
Twin Oaks Estate
1225 W. Market Street
Akron, OH 44313   (216) 864-5550

Scott M. Oldham
Oldham & Oldham, L.P.A.
1225 W. Market Street
Akron, OH 44313   (216) 864-5550

John G. Pere
2143 Pressler Rd.
Akron, OH 44312   (216) 733-6095

Jack L. Renner
Renner, Kenner, Greive, Bobak,
Taylor & Weber Co., L.P.A.
1610 First National Tower
Akron, OH 44308   (216) 376-1242

Howard S. Robbins
Renner, Kenner, Greive, Bobak,
Taylor & Weber Co., L.P.A.
1610 First National Tower
Akron, OH 44308   (216) 376-1242

Alvin T. Rockhill, III
Goodyear Tire & Rubber Co.
Dept. 823
1144 E. Market Street
Akron, OH 44316   (216) 796-8252

George W. Rooney, Jr.
Roetzel & Andress
75 E. Market Street
Akron, OH 44308   (216) 376-2700

James A. Rozmajzl
Goodyear Tire and Rubber Co.
1144 E. Market Street
Akron, OH 44316   (216) 796-7417

Ernst H. Ruf
Bridgestone/Firestone, Inc.
Legal Dept.
1200 Firestone Pkwy.
Akron, OH 44317   (216) 379-6851

Gordon B. Seward
Monsanto Company
260 Springside Drive
Akron, OH 44313   (216) 668-8257

Laura F. Shunk
Transohio Building
156 S. Main Street
Suite 808
Akron, OH 44308   (216) 535-2220

Robert W. Stachowiak
1685 Far View Road
Akron, OH 44312   (216) 644-2940

Robert W. Strozier#
Oldham & Oldham, L.P.A.
1225 W. Market St.
Akron, OH 44313   (216) 864-5550

Greg Strugalski
Uniroyal Goodrich Tire Company
Pat. Law Dept.
Dept. 0530, UGB-6
600 S. Main Street
Akron, OH 44397   (216) 374-4185

Perry Reese Taylor, Jr.
Renner, Kenner, Grieve,
Bobak & Taylor
1610 First National Tower
Akron, OH 44308   (216) 376-1034

David A. Thomas
Bridgestone/Firestone, Inc.
1200 Firestone Parkway
Akron, OH 44317   (216) 379-6850

Frank J. Troy, Sr.
Bridgestone/Firestone, Inc.
1200 Firestone Pkwy.
Akron, OH 44317   (216) 379-6178

Mark A. Watkins
Oldham & Oldham Co., L.P.A.
1225 W. Market Street
Akron, OH 44313   (216) 864-5550

Ray L. Weber
Renner, Kenner, Grieve,
Bobak, Taylor & Weber Co.
1610 First National Tower
Akron, OH 44308   (216) 376-1242

David E. Wheeler
Goodyear Tire & Rubber Co.
1144 E. Market Street
Akron, OH 44316   (216) 796-6364

Bruce H. Wilson
Tuccillo & Wilson
1000 Society Bldg.
159 S. Main St.
Akron, OH 44308   (216) 253-1900

Ronald P. Yaist
Goodyear Tire and Rubber Co.
1144 Market St.
Akron, OH 44316   (216) 796-4409

Henry Claude Young, Jr.
Goodyear Tire & Rubber Co.
1144 East Market
Akron, OH 44313   (216) 796-2956

Ralph L. Humphrey
1258 Prospect Rd.
Ashtabula, OH 44004   (216) 998-1112

Robert A. Sturges
3497 Nautilus Trail
Aurora, OH 44202   (216) 562-9668

J. Herman Yount, Jr.
31408 Adrich Dr.
Bay Village, OH 44140   (216) 871-1746

Charles Clyde Allshouse, Jr.
176 Debs Drive
Beavercreek, OH 45385   (513) 429-1268

Stanley M. Clark
8049 Robin Ln.
Brecksville, OH 44141   (216) 526-8809

Thoburn T. Dunlap#
B.F. Goodrich Company
9921 Brecksville Rd.
Brecksville, OH 44141   (216) 374-2555

Joseph Januszkiewicz
B.F. Goodrich Co.
9921 Brecksville Rd.
Brecksville, OH 44141   (216) 374-3082

Konrad H. Kaeding
B.F. Goodrich Co.
9921 Brecksville Rd.
Brecksville, OH 44141   (216) 374-3229

George A. Kap
B.F. Goodrich Company
9921 Brecksville Rd.
Brecksville, OH 44141   (216) 374-3237

George W. Moxon, II
B.F. Goodrich Co.
Patent Law Dept.
9921 Brecksville Rd.
Brecksville, OH 44141   (216) 447-5912

Debra L. Pawl
B.F. Goodrich Company
9921 Brecksville Rd.
Brecksville, OH 44141   (216) 374-2339

Joe A. Powell
B.F. Goodrich Co.
9921 Brecksville Rd.
Brecksville, OH 44141   (216) 374-4014

William F. Prout#
B.F. Goodrich Co.
9921 Brecksville Rd.
Brecksville, OH 44141   (216) 374-4014

David M. Ronyak
B.F. Goodrich Company
9921 Brecksville Road
Brecksville, OH 44141   (216) 447-2923

Nestor W. Shust
B.F. Goodrich Co.
9921 Brecksville Rd.
Brecksville, OH 44141   (216) 374-4014

Milton Lawrence Simmons
6572 Thorntree Dr.
Brecksville, OH 44141   (216) 526-2067

William A. Skinner
B.F. Goodrich Company
9921 Brecksville Road
Brecksville, OH 44141   (216) 374-2441

Mary A. Tucker
B.F. Goodrich Co.
9921 Brecksville Rd.
Brecksville, OH 44141   (216) 447-5221

John B. Frease
Frease & Bishop
519 National City Bldg
315 Tuscarawas St., W.
Canton, OH 44702   (216) 455-0331

Joseph Frease
Frease and Bishop
Suite 519 National City Bldg.
315 W. Tuscarawas Street
Canton, OH 44702   (216) 455-0331

Michael Sand
4450 Belden Village Ave., N.W.
Suite 201
Canton, OH 44718   (216) 492-1925

Dorothy I. Becker
7853 Betsy Ross Circle
Centerville, OH 45459   (216) 492-1925

Melvin Wiviott
1599 Ambridge Rd.
Centerville, OH 45459   (513) 434-1379

Lawrence R. Kempton
5224 Maple Spring Dr.
Chagrin Falls, OH 44022   (216) 338-1729

Richard G. Smith
115 Locust Lane
Chagrin Falls, OH 44022   (216) 247-8279

Joseph R. Teagno
219 Manor Brook Dr.
Chagrin Falls, OH 44022   (216) 338-3060

Arthur S. Collins#
12039 Fowlers Mill Road
Chardon, OH 44024   (216) 286-3140

Russell E. Baumann
7080 Hillcrest Lane
Chesterland, OH 44026   (216) 729-9503

Forrest L. Collins
7515 Avon Lane
Chesterland, OH 44026   (216) 729-0277

Emil F. Sos, Jr.
12690 Opalocka Dr.
Chesterland, OH 44026   (216) 729-1252

Robert Raymond Yurich
7862 Lake Rd.
Chippewa Lake, OH 44215   (216) 769-3047

Gregory J. Lunn
Wood, Herron & Evans
2700 Carew Tower
Cincinatti, OH 45202   (513) 241-2324

Edwin R. Acheson, Jr.
Frost & Jacobs
2500 Central Trust Center
201 E. Fifth Street
Cincinnati, OH 45202   (513) 651-6708

Gregory F. Ahrens
Wood, Herron & Evans
2700 Carew Tower
Cincinnati, OH 45202   (513) 241-2324

George W. Allen
Procter & Gamble Company
Pat. Division
P.O. Box 39175
Cincinnati, OH 45247   (513) 245-2912

Robert B. Aylor
Procter and Gamble Co.
Ivorydale Tech. Ctr.
Cincinnati, OH 45217   (513) 627-5144

Mark W. Bailey
SENCORP
8485 Broadwell Rd.
Cincinnati, OH 45244   (513) 388-2913

Jeffrey V. Bamber
Procter & Gamble Company
Winton Hill Technical Center
6100 Center Hill Road
Cincinnati, OH 45224   (513) 634-3003

Gerald A. Baracka#
National Distillers
& Chemical Corporation
11500 Northlake Drive
Cincinnati, OH 45249   (513) 530-6561

Bruce P. Bardes#
General Electric Co.
GE Aircraft Engines
Mail Drop H-85
P.O. Box 156301
Cincinnati, OH 45215   (513) 243-4268

Fernando A. Borrego
Procter & Gamble Co.
5299 Spring Grove Ave.
Cincinnati, OH 45217   (513) 627-5947

George P. Brandenburg
Third Floor
601 Main Street
Cincinnati, OH 45202   (513) 621-8700

Frederick H. Braun
Proctor and Gamble Co.
6100 Center Hill Road
Cincinnati, OH 45224   (513) 659-6332

Herbert C. Brinkman, Jr.
Wood, Herron & Evans
2700 Carew Tower
Cincinnati, OH 45202   (513) 241-2324

Thomas J. Burger
Wood, Herron & Evans
2700 Carew Tower
Cincinnati, OH 45202   (513) 241-2324

Charles Lyman Burgoyne#
7410 N. Mingo Lane
Cincinnati, OH 45243   (513) 793-5869

David F. Chalmers#
Procter & Gamble Co.
Ivorydale Tech. Ctr.
Spring Grove & June Sts.
Cincinnati, OH 45217   (513) 627-5515

J. Robert Chambers#
Wood, Herron & Evans
2700 Carew Tower
411 Vine St.
Cincinnati, OH 45244   (513) 241-2324

Karen F. Clark
Procter & Gamble Company
6071 Center Hill Road
Room F3A16
Cincinnati, OH 45224   (513) 634-5270

Kenneth J. Collier#
Marion Merrell Dow Inc.
2110 E. Galbraith Road
P.O. Box 156300
Cincinnati, OH 45215   (513) 948-7834

Brahm J. Corstanje
Procter & Gamble Co.
Miami Valley Laboratories
P.O. Box 398707
Cincinnati, OH 45239   (513) 245-2858

C. Thomas Cross
The Madison House, Ste. 903
2324 Madison Road
Cincinnati, OH 45208   (513) 475-2993

Rose Ann Dabek
Procter & Gamble Co.
Winton Hill Tech. Ctr.
6071 Center Hill Rd.
Cincinnati, OH 45224   (513) 659-5593

James P. Davidson
Frost & Jacobs
2500 Central Trust Center
Cincinnati, OH 45202   (513) 651-6993

J. Michael Dixon
Merrell Dow Pharmaceuticals Inc.
Pat. Dept.
2110 E. Galbraith Road
P.O. Box 156300
Cincinnati, OH 45215   (513) 948-7960

Donald Dunn#
1845 Greenbriar Place
Cincinnati, OH 45237   (513) 351-5022

Jack J. Earl
7618 Carriage Lane
Cincinnati, OH 45242   (513) 351-5022

C. Richard Eby
Thompson, Hine & Flory
312 Walnut Street
Suite 1400
Cincinnati, OH 45202   (513) 352-6700

Douglas E. Erickson
GE Aircraft Engines
1 Neuman Way
Mail Drop H-17
Cincinnati, OH 45215   (513) 243-9840

Richard Henry Evans
Wood, Herron & Evans
2700 Carew Tower
Cincinnati, OH 45202   (513) 241-2324

Thomas M. Farrell
Cincinnati Milacron Inc.
4701 Marburg Avenue
Cincinnati, OH 45209   (513) 841-8536

Julius P. Filcik
Procter and Gamble Co.
Winton Hill Tech. Ctr.
6090 Center Hill Rd.
Cincinnati, OH 45224   (513) 977-7503

Stanley H. Foster
Frost & Jacobs
250o Central Trust Center
201 E. 5th St.
Cincinnati, OH 45202   (513) 651-6975

Donald Francis Frei, Sr.
Wood, Herron & Evans
2700 Carew Tower
441 Vine Street
Cincinnati, OH 45202   (513) 241-2324

George Galanes#
Procter & Gamble Co.
Winton Hill Tech. Ctr.
6060 Center Hill Rd.
Cincinnati, OH 45224   (513) 659-4173

Edmund Frederick Gebhardt
2959 Annwood Street
Cincinnati, OH 45206   (513) 751-2717

Roger A. Gilcrest
Frost & Jacobs
2500 Central Trust Center
201 E. 5th Street
Cincinnati, OH 45202   (216) 651-6746

Steven J. Goldstein
Procter & Gamble Co.
Miami Valley Labs.
P.O. Box 398707
Cincinnati, OH 45239   (513) 245-2701

John V. Gorman
Procter and Gamble Co.
6090 Center Hill Rd.
Cincinnati, OH 45224   (513) 977-6148

William H. Gould
8601 Long Lane
Cincinnati, OH 45231   (513) 521-2378

Milton B. Graff
Procter & Gamble Company
Miami Valley Labs.
P.O. Box 398707
Cincinnati, OH 45239   (513) 245-2659

Edward A. Grannen, Jr.#
11186 Lincolnshire Drive
Cincinnati, OH 45240   (513) 851-5938

John W. Gregg
Cincinnati Milacron, Inc.
Pat. Dept.
4701 Marburg Ave.
Cincinnati, OH 45209   (513) 841-8344

Gerry S. Gressel#
7324 Timberknoll Dr.
Cincinnati, OH 45242   (513) 745-0749

Frederick H. Gribbell
Frost & Jacobs
2500 Central Trust Center
P.O. Box 5715
201 E. Fifth St.
Cincinnati, OH 45202   (513) 651-6416

Kurt L. Grossman
Wood, Herron & Evans
2700 Carew Tower
Cincinnati, OH 45202   (513) 241-2324

Eric W. Guttag
Procter & Gamble Co.
Winton Hill Tech. Center
6210 Center Hill Road
Cincinnati, OH 45224   (513) 659-2736

Kathleen M. Harleston
Procter & Gamble Company
Patent Division
Ivorydale Technical Center
Cincinnati, OH 45217   (513) 627-5946

Donald E. Hasse
Procter & Gamble Co.
Ivorydale Tech. Ctr.
5299 Spring Grove Avenue
Cincinnati, OH 45217   (513) 627-5144

Gretchen R. Hatfield
Procter & Gamble Company
Ivorydale Technical Center
Cincinnati, OH 45217   (513) 627-5947

James Harry Hayes
Frost & Jacobs
2500 Central Trust Center
201 E. 5th St.
Cincinnati, OH 45202   (513) 651-6800

William A. Heidrich, III
Quantum Chemical Corp.
Legal Dept.
11500 Northlake Drive
Cincinnati, OH 45249   (513) 530-6552

Ronald L. Hemingway
Procter & Gamble Co.
6071 Center Hill Road
Cincinnati, OH 45224   (513) 659-5593

Bart S. Hersko
Procter & Gamble Company
Miami Valley Labs.
P.O. Box 398707
Cincinnati, OH 45239   (513) 245-2889

Michael E. Hilton
Procter & Gamble Company
6090 Center Hill Road
Cincinnati, OH 45224   (513) 634-6748

Joseph V. Hoffman
Frost & Jacobs
2500 Central Trust Center
201 East 5th St.
Cincinnati, OH 45202   (513) 651-6800

Charles Marshall Hogan
8071 Village Drive
Cincinnati, OH 45142   (513) 489-2377

John M. Howell
Procter & Gamble Co.
Sharon Woods Tech. Center
11511 Reed Hartman Hwy.
Cincinnati, OH 45241   (513) 626-4416

Eileen L. Hughett#
Procter & Gamble Co.
Winton Hill Technical Center
Room A2M50G
6090 Center Hill Rd.
Cincinnati, OH 45224   (513) 634-4650

Larry L. Huston
Procter & Gamble Company
Winton Hill Tech. Center
6100 Center Hill Road
Cincinnati, OH 45224   (513) 634-2964

Richard G. Jackson
Quantum Chemical Corp.
11500 Northlake Drive
Cincinnati, OH 45249   (513) 530-6550

Maynard R. Johnson
Merrell Dow Pharmaceuticals, Inc.
Pat. Law Dept.
2110 E. Galbraith Road
Cincinnati, OH 45215   (513) 948-7967

Joseph R. Jordan
2800 Carew Tower
441 Vine
Cincinnati, OH 45202   (513) 891-4455

David J. Josephic
Wood, Herron & Evans
2700 Carew Tower
Cincinnati, OH 45202   (513) 241-2324

Thomas L. Kautz
Wood, Herron & Evans
2700 Carew Tower
Cincinnati, OH 45202   (513) 241-2324

James M. Kipling
Kenner Products
1014 Vine St.
Cincinnati, OH 45202   (513) 579-4808

John J. Kolano
Merrell Dow Pharmaceuticals Inc.
2110 E. Galbraith Rd.
Cincinnati, OH 45215   (513) 948-7964

William G. Konold
Wood, Herron & Evans
2700 Carew Tower
Cincinnati, OH 45202   (513) 241-2324

E.S. Lee, III
Kenney & Schenk
105 E. Fourth St.
Suite 720
Cincinnati, OH 45202   (513) 621-3220

Leonard W. Lewis
Procter & Gamble Company
Miami Valley Laboratories
P.O. Box 398707
Cincinnati, OH 45239   (513) 245-2924

James D. Liles
Frost & Jacobs
2500 Central Trust Center
Cincinnati, OH 45202   (513) 651-6707

Elmer K. Linman
Procter & Gamble Company
Pat. Division
6300 Center Hill Road
Cincinnati, OH 45224   (513) 634-6327

Jerrold J. Litzinger
Sencorp
8485 Broadwell Road
Cincinnati, OH 45244   (513) 388-2912

David M. Lockman
Wood, Herron & Evans
2700 Carew Tower
Cincinnati, OH 45202   (513) 241-2324

Clement H. Luken, Jr.
Wood, Herron & Evans
2700 Carew Tower
Cincinnati, OH 45202   (513) 241-2324

James D. Lykins
886 Pinewell Drive
Cincinnati, OH 45255   (513) 474-4729

Peter J. Manso
Wood, Herron & Evans
2700 Carew Tower
Cincinnati, OH 45202   (513) 241-2324

Raymond A. Mc Donald
Merrell Dow Pharmacenticals Inc.
Patent Dept.
2110 E. Galbraith Road
Cincinnati, OH 45215   (513) 948-7960

Mary P. Mc Mahon#
Procter & Gamble Co.
Ivorydale Technical Center
Cincinnati, OH 45217   (513) 948-7960

Terrence E. Miesle
Hilton Davis Company
2235 Langdon Farm Road
Cincinnati, OH 45237   (513) 841-4843

Steven W. Miller
Procter & Gamble Company
6100 Center Hill Road
Cincinnati, OH 45224   (513) 659-4490

Douglas C. Mohl
Procter & Gamble Company
11511 Reed Hartman Highway
Cincinnati, OH 45241   (513) 530-3991

Carolyn Davis Moon
Marion Merrell Dow Inc.
2110 East Galbraith Road
Cincinnati, OH 45215   (513) 948-7785

Charles L. Moore, Jr.
GE Aircraft Engines
One Neumann Way
M/D H17
Cincinnati, OH 45215   (513) 243-6393

Jacob F. Moskowitz
3300 Beredith Place
Cincinnati, OH 45213   (513) 272-0131

Walter S. Murray
6050 Stirrup Rd.
Cincinnati, OH 45244   (513) 231-3093

A. Ralph Navaro, Jr.
Wood, Herron & Evans
2700 Carew Tower
441 Vine Street
Cincinnati, OH 45202   (513) 241-2324

Daniel F. Nesbitt
Procter & Gamble Co.
6060 Center Hill Road
Cincinnati, OH 45224   (513) 634-3375

Stephen L. Nesbitt
Merrell Dow Pharmaceuticals, Inc.
Patent Dept.
2110 E. Galbraith Road
Cincinnati, OH 45215   (513) 948-7965

Carol J. Ney
Procter & Gamble Company
Legal Division
P.O. Box 599
Cincinnati, OH 45202   (513) 983-4517

Thomas H. O Flaherty
Procter and Gamble Co.
5299 Spring Grove Avenue
Cincinnati, OH 45217   (513) 627-6911

Daniel H. Owings
713 Locust Corner Road
Cincinnati, OH 45245   (513) 752-9289

James W. Pearce
1326 Michigan Avenue
Cincinnati, OH 45208   (513) 871-6551

Burton Perlman*
U.S. Courts
719 U.S.P.O. & Courthouse Bldg.
Cincinnati, OH 45202   (513) 684-2342

Walter A. Petersen#
8222 Monte Drive
Cincinnati, OH 45242   (513) 984-2637

John D. Poffenberger
Wood, Herron & Evans
2700 Carew Tower
Cincinnati, OH 45202   (513) 241-2324

John M. Pollaro
Procter & Gamble Co.
Winton Hill Tech. Ctr.
6100 Center Hill Rd.
Cincinnati, OH 45224   (513) 634-7324

John R. Rafter, II
General Electric Co.
One Neuman Way
P.O. Box 156301
Cincinnati, OH 45215   (513) 243-3701

William C. Rambo
Porter, Wright, Morris
& Arthur
250 E. Fifth Street
Suite 2200
Cincinnati, OH 45202   (513) 369-4250

T. David Reed#
Procter & Gamble Company
5299 Spring Grove Avenue
Cincinnati, OH 45217   (513) 983-1100

John David Rice
Quantum Chemical Corp.
11500 Northlake Drive
Cincinnati, OH 45249   (513) 530-6563

Lynda E. Roesch
Dinsmore & Shohl
1900 Chemed Center
255 East Fifth Street
Cincinnati, OH 45202   (513) 977-8139

Steven J. Rosen
4729 Cornell Road
Cincinnati, OH 45241   (513) 489-5383

Lee H. Sachs
15 Muirfield Lane
Cincinnati, OH 45241   (513) 793-9023

Carmen S. Santa Maria
General Electric Co.
One Neumann Way
Mail Drop H17
Cincinnati, OH 45215   (513) 243-6610

Michael J. Sayles
Merrell Dow Pharmaceuticals Inc.
2110 E. Galbraith Road
P.O. Box 156300
Cincinnati, OH 45215   (513) 948-6479

Jack D. Schaeffer
Procter & Gamble Co.
Ivorydale Technical Ctr.
5299 Spring Grove Avenue
Cincinnati, OH 45217   (513) 627-6868

Roy F. Schaeperklaus
Pearce and Schaeperklaus
36 E. 4th Street
Suite 1234
Cincinnati, OH 45202   (513) 241-1021

John G. Schenk
Kinney and Schenk
Suite 1306
105 E. 4th St.
Cincinnati, OH 45202   (513) 721-3440

David E. Schmit
Frost & Jacobs
2500 Central Trust Center
201 E. 5th St.
Cincinnati, OH 45202   (513) 651-6985

Bernard E. Shay
General Electric Co.
One Neumann Way
Mail Drop H17
Cincinnati, OH 45215   (513) 243-9835

Donald S. Showalter
Wood, Herron & Evans
2700 Carew Tower
Cincinnati, OH 45202   (513) 241-2324

Edlyn S. Simmons#
Merrell Dow Pharmaceuticals Inc.
2110 East Galbraith Road
Cincinnati, OH 45215   (513) 948-7829

Thomas J. Slone, Sr.
1257 Jeremy Ct.
Cincinnati, OH 45240   (513) 851-4549

Ronald J. Snyder
Frost & Jacobs
2500 Central Trust Center
201 East Fifth Street
Cincinnati, OH 45202   (513) 651-6992

Jerome Carmen Squillaro
General Electric Co.
1 Neumann Way
Mail Drop H-17
Cincinnati, OH 45215   (513) 243-9869

David S. Stallard
Wood, Herron & Evans
2700 Carew Tower
Cincinnati, OH 45202   (513) 241-2324

William J. Stein
Marion Merrell Dow
Pharmaceuticals Inc.
2110 E. Galbraith Road
Cincinnati, OH 45215   (513) 948-7492

Albert E. Strasser
Frost & Jacobs
2500 Central Trust Ctr.
201 E. 5th St.
Cincinnati, OH 45202   (513) 651-6977

Gary D. Street
Marion Merrell Dow Inc.
2110 East Galbraith Road
Cincinnati, OH 45215   (513) 948-7695

Stephen S. Strunck
General Electric Co.
1 Neumann Way, MD H17
Cincinnati, OH 45215   (513) 243-4903

Gary M. Sutter
Procter & Gamble Company
6110 Center Hill Road
FB0E08
Cincinnati, OH 45224   (513) 634-7939

Bruce Tittel
Wood, Herron & Evans
2700 Carew Tower
Cincinnati, OH 45202   (513) 241-2324

Chase D. Tonne#
1547 Cohasset Drive
Cincinnati, OH 45255   (513) 232-0091

Kenneth Dale Tremain
Quantum Chemical Corp.
11500 Northlake Drive
Cincinnati, OH 45249   (513) 530-6550

Edward J. Utz
36 E. Fourth St.
Bartlett Bldg.
Suite 1306
Cincinnati, OH 45202   (513) 241-8829

Louis J. Wille
Marion Merrell Dow, Inc.
2110 E. Galbraith Road
Cincinnati, OH 45215   (513) 948-6354

Leonard Williamson
Procter & Gamble Co.
SWTC-B1S08
11520 Reed Hartman Hwy.
Cincinnati, OH 45241   (513) 626-3387

Neal O. Willmann
Marion Merrell Dow Inc.
2110 E. Galbraith Road
Cincinnati, OH 45215   (513) 948-6451

Charles R. Wilson
4729 Cornell Road
Cincinnati, OH 45241   (513) 489-7484

Monte D. Witte
Procter & Gamble Co.
Ivorydale Technical Center
5299 Spring Grove Avenue
Cincinnati, OH 45217   (513) 627-6925

Richard C. Witte
Procter and Gamble Co.
Ivorydale Technical Center
5299 Spring Grove Avenue
Cincinnati, OH 45217   (513) 627-5666

Daniel A. Yaeger
1391 Devils Back Bone Road
Cincinnati, OH 45238   (513) 941-2725

Jerry J. Yetter
Procter & Gamble Co.
Miami Valley Laboratories
P.O. Box 398707
Cincinnati, OH 45239   (513) 941-2725

Gibson R. Yungblut
Frost & Jacobs
250o Central Trust Center
201 E. 5th St.
Cincinnati, OH 45202   (513) 651-6980

Kim William Zerby
Procter & Gamble Company
Sharon Woods Technical Center
11511 Reed Hartman Hwy.
Cincinnati, OH 45241   (513) 626-3993

Kenneth R. Adamo
Jones, Day, Reavis & Pogue
North Point
1901 Lakeside Avenue
Cleveland, OH 44114   (216) 586-7120

Clyde E. Bailey, Sr.*
NASA Lewis Research Center
21000 Brookpark Road
Mail Stop: LE-LAW
Cleveland, OH 44135   (216) 433-5755

James A. Baker
Parker Hannifin Corp.
17325 Euclid Ave.
Cleveland, OH 44112   (216) 531-3000

James E. Barlow
Watts, Hoffmann, Fisher
& Heinke
100 Erieview Plaza
Suite 2850
Cleveland, OH 44114   (216) 526-3056

Gordon P. Becker
Alcan Aluminum Corp.
100 Erieview Plaza
Cleveland, OH 44114   (216) 523-8250

Alfred Carpenter Body
Body, Vickers & Daniels
2000 Terminal Tower
Cleveland, OH 44113   (216) 623-0040

Heidi A. Boehlefeld#
B.P. America
200 Public Square
36-F-3454
Cleveland, OH 44114   (216) 586-4326

Armand Paul Boisselle, Sr.
Renner, Otto, Boisselle
& Sklar
1621 Euclid Ave., 19th Fl.
Cleveland, OH 44115   (216) 621-1113

Don W. Bulson
Renner, Otto, Boisselle
& Sklar
1621 Euclid Avenue
19th Floor
Cleveland, OH 44115   (216) 621-1113

David A. Burge
P.O. Box 22975
Cleveland, OH 44122   (216) 921-8900

Jay R. Campbell
Renner, Otto, Boisselle
& Sklar
1621 Euclid Avenue
19th Floor
Cleveland, OH 44115   (216) 621-1113

James E. Carson
Carson, Smith & Chandler
Suite 2020 Superior Bldg.
815 Superior Ave., N.E.
Cleveland, OH 44114   (216) 771-5818

Michael Chan
Taroll, Sundheim & Covell
1111 Leader Bldg.
Cleveland, OH 44114   (216) 621-2234

Albert E. Chrow
Eaton Corp.
Eaton Center
1111 Superior Avenue
Cleveland, OH 44114   (216) 523-4131

Kenneth A. Clark
Renner, Otto, Boisselle
& Sklar
1621 Euclid Ave., 19th Flr.
Cleveland, OH 44115   (216) 621-1113

George J. Coghill
10211 Lakeshore Blvd.
Cleveland, OH 44108   (216) 451-2323

Howard M. Cohn
Body, Vickers & Daniels
2000 Terminal Tower
Cleveland, OH 44113   (216) 623-0040

Hal D. Cooper
Jones, Day, Reavis & Pogue
1700 Huntington Bldg.
Cleveland, OH 44115   (216) 348-3939

Edward M. Corcoran
General Electric Company
Lighting Business Group
Nela Park - Noble Road
Cleveland, OH 44112   (216) 266-3640

Joseph J. Corso
Pearne, Gordon, Mc Coy
& Granger
1200 Leader Bldg.
Cleveland, OH 44114   (216) 579-1700

Calvin G. Covell
Tarolli, Sundheim & Covell
1111 Leader Bldg.
Cleveland, OH 44114   (216) 621-2234

Joseph G. Curatolo
Standard Oil Co.
200 Public Square
36-F
Cleveland, OH 44114   (216) 586-8460

Edward Kent Daniels, Jr.
Body, Vickers & Daniels
2000 Terminal Tower
Cleveland, OH 44113   (216) 623-0040

Walter C. Danison, Jr.#
Renner, Otto, Boisselle & Sklar
1621 Euclid Ave.
19th Fl.
Cleveland, OH 44115   (216) 621-1113

David B. Deioma
Pearne, Gordon, Mc Coy
& Granger
1200 Leader Bldg.
Cleveland, OH 44114   (216) 579-1700

Jayadeep R. Deshmukh#
D. Peter Hochberg Co., L.P.A.
Baker Bldg.
1940 East 6th Street
Cleveland, OH 44114   (216) 771-3800

Richard H. Dickinson, Jr.
Pearne, Gordon, Mc Coy
& Granger
6th & Superior Avenue
Cleveland, OH 44114   (216) 579-1700

James D. Donohoe
LTV Steel Company, Inc.
P.O. Box 6778
Cleveland, OH 44101   (216) 622-5624

Merton H. Douthitt
Watts, Hoffmann, Fisher
& Heinke Company
Suite 2850
100 Erieview Plaza
Cleveland, OH 44114   (216) 623-0775

James J. Drake
Weston, Hurd, Fallon,
Paisley & Howley
2500 Terminal Tower
Cleveland, OH 44113   (216) 241-6602

Neil A. Du Chez
Renner, Otto, Boiselle
& Sklar
1621 Euclid Avenue
19th Floor
Cleveland, OH 44115   (216) 621-1113

Richard Joseph Egan
Baldwin, Egan & Fetzer
816 Hanna Bldg.
Cleveland, OH 44115   (216) 621-2956

Roger D. Emerson
Fay, Sharpe, Beall, Fagan,
Minnich & Mc Kee
1100 Superior Avenue
Suite 700
Cleveland, OH 44114   (216) 861-5582

Michael F. Esposito
Standard Oil Company
200 Public Square
36-3454-F
Cleveland, OH 44114   (216) 586-8022

Larry W. Evans
B.P. America
200 Public Square
36-3556F
Cleveland, OH 44114   (216) 586-8450

Christopher Brendan Fagan
Fay, Sharpe, Beall, Fagan,
Minnich & Mc Kee
1100 Superior Avenue
Suite 700
Cleveland, OH 44114   (216) 861-5582

Regan J. Fay
Jones, Day, Reavis
& Pogue
North Point
901 Lakeside Avenue
Cleveland, OH 44114   (216) 586-7327

Robert Joseph Fetzer
Baldwin, Egan & Feetzer
816 Hanna Bldg.
Cleveland, OH 44115   (216) 621-2956

Thomas E. Fisher
Watts, Hoffman, Fisher
& Heinke Co., L.P.A.
100 Erieview Plaza
Cleveland, OH 44114   (216) 623-0775

Laurence D. Fogel#
B.P. America R&D
4440 Warrensville Center Road
Cleveland, OH 44128   (216) 581-5982

William Albert Gail
Pearne, Gordon, Mc Coy,
& Granger
1200 Leader Bldg.
Cleveland, OH 44114   (216) 579-1700

John X. Garred
Fay, Sharpe, Beall, Fagan,
Minnich & Mc Kee
1100 Superior Avenue
Suite 700
Cleveland, OH 44114   (216) 861-5582

Teresan W. Gilbert
Standard Oil Co.
200 Public Square
Cleveland, OH 44114   (216) 586-8475

Mary E. Golrick
Calfee, Halter & Griswold
1800 Society Bldg.
East 9th & Superior
Cleveland, OH 44114   (216) 622-8200

Charles Byron Gordon
Pearne, Gordon, Mc Coy
& Granger
1200 Leader Bldg.
Cleveland, OH 44114   (216) 579-1700

Howard D. Gordon
Eaton Corporation
Eaton Center
1111 Superior Avenue
Cleveland, OH 44114   (216) 523-4134

Charles H. Grace
Eaton Corporation
Eaton Center
1111 Superior Avenue
Cleveland, OH 44114   (216) 523-4127

Louis V. Granger
Pearne, Gordon, Sessions,
Mc Coy & Granger
1200 Leader Bldg.
Cleveland, OH 44114   (216) 579-1700

Thomas J. Gray
4515 St. Clair Avenue
Cleveland, OH 44103   (216) 391-7070

Vincent A. Greene
P.O. Box 14072
Cleveland, OH 44114   (216) 423-3511

Calvin P. Griffith
Jones, Day, Reavis & Pogue
North Point
901 Lakeside Ave.
Cleveland, OH 44114   (216) 586-7050

Kevin P. Hallquist
Calfee, Halter & Griswold
1800 Society Building
Cleveland, OH 44114   (216) 622-8505

M. Elaine Harmon
Woodling, Krost & Rust
530 National City East
Sixth Bldg.
Cleveland, OH 44114   (216) 241-4150

George E. Hawranko
General Electric Co.
Nela Park
Suite 1200
Cleveland, OH 44112   (216) 266-8649

Edward A. Hayman
Squire, Sanders & Dempsey
1800 Huntington Bldg.
Cleveland, OH 44115   (216) 687-8500

Lowell L. Heinke
Watts, Hoffmann, Fisher
& Heinke
Suite 2850
100 Erieview Plaza
Cleveland, OH 44114   (216) 623-0775

Kevin J. Heyd
Watts, Hoffmann, Fisher
& Heinke Co., L.P.A.
100 Erieview Plaza
Suite 2850
Cleveland, OH 44114   (216) 623-0775

Stephen A. Hill
Pearne, Gordon, Mc Coy
& Granger
1200 Leader Bldg.
Cleveland, OH 44114   (216) 579-1700

John R. Hlavka
Watts, Hoffmann, Fisher
& Heinke Co., L.P.A.
100 Erieview Plaza
Suite 2850
Cleveland, OH 44114   (216) 623-0775

D. Peter Hochberg
1510 Ohio Savings Plaza
1801 E. 9th Street
Cleveland, OH 44114   (216) 771-3800

James T. Hoffmann
Watts, Hoffman, Fisher
& Heinke
100 Erieview Plaza
Cleveland, OH 44114   (216) 623-0778

William Neill Hogg
23200 Chagrin Blvd.
Suite 605
Bldg. 3
Cleveland, OH 44122   (216) 765-8890

Roy F. Hollander
Jones, Day, Reavis & Pogue
North Point
901 Lakeside Avenue
Cleveland, OH 44114   (216) 765-8890

James A. Hudak
Hanna Bldg.
Suite 818
1422 Euclid Avenue
Cleveland, OH 44115   (216) 566-9700

Michael E. Hudzinski
Fay, Sharpe, Beall, Fagan,
Minnich & Mc Kee
1100 Superior Ave.
Suite 700
Cleveland, OH 44114   (216) 861-5582

Christopher H. Hunter
Calfee, Halter & Griswold
1800 Society Bldg.
Cleveland, OH 44114   (216) 622-8529

Robert R. Hussey
1610 Euclid Avenue
Second Floor
Cleveland, OH 44115   (216) 687-1111

Roger A. Johnston
Eaton Corp.
Eaton Center
Rm. 2505
Cleveland, OH 44114   (216) 523-4132

Ronald M. Kachmarik#
Tarolli, Sundheim & Covell
1111 Leader Bldg.
Cleveland, OH 44114   (216) 621-2234

Leslie J. Kasper
Eaton Corporation
Eaton Center
1111 Superior Avenue
Cleveland, OH 44114   (216) 523-4138

John A. Kastelic
Calfee, Halter & Griswold
1800 Society Bldg.
East 9th & Superior Ave.
Cleveland, OH 44114   (216) 622-8856

S. I. Khayat
Callaghan & Khayat
1040 Leader Bldg.
Cleveland, OH 44114   (216) 623-1040

Gordon D. Kinder
Renner, Otto, Boiselle
& Sklar
1621 Euclid Avenue
19th Floor
Cleveland, OH 44115   (216) 621-1113

Richard M. Klein
Fay, Sharpe, Beall, Fagan,
Minnich & Mc Kee
1100 Superior Avenue
Suite 700
Cleveland, OH 44114   (216) 861-5582

Thomas E. Kocovsky, Jr.
Fay & Sharpe
Suite 700
1100 Superior Avenue
Cleveland, OH 44114   (216) 861-5582

Sandra M. Koenig
Fay, Sharpe, Beall, Fagan,
Minnich & Mc Kee
1100 Superior Avenue
Suite 700
Cleveland, OH 44114   (216) 861-5582

Bruce B. Krost
Woodling, Krost & Rust
530 National City-East 6th Bldg.
Cleveland, OH 44114   (216) 241-4150

Mark M. Kusner
D. Peter Hochberg Co., L.P.A.
Baker Building
1940 E. Sixth St.
Sixth Floor
Cleveland, OH 44114   (216) 771-3800

Michael A. Lamanna
General Electric Company
Nela Park
Suite 1200
Cleveland, OH 44112   (216) 266-3026

James John Lazna
Body, Vickers & Daniels
2000 Terminal Tower
Cleveland, OH 44113   (216) 623-0040

Robert S. Lelkes
Renner, Otto, Boisselle &
Sklar
1621 Euclid Ave., 19th Flr.
Cleveland, OH 44115   (216) 621-1113

Leonard L. Lewis
Parker Hannifin Corporation
17325 Euclid Avenue
Cleveland, OH 44112   (216) 531-3000

William S. Lightbody
Woodling, Krost & Rust
530 National City East
Sixth Bldg.
1965 E. Sixth Street
Cleveland, OH 44114   (216) 241-4150

Alfred D. Lobo
Lobo & Co., L.P.A.
933 Leader Bldg.
526 Superior Avenue
Cleveland, OH 44114   (216) 566-1661

Jeanne E. Longmuir
Calfee, Halter & Griswold
1800 Society Bldg.
Cleveland, OH 44114   (216) 781-2166

James Andre Lucas
Mc Gean-Rohco, Inc.
50 Public Square, #1250
Cleveland, OH 44113   (216) 441-4900

John F. Luhrs
29925 Fairmount Blvd.
Cleveland, OH 44124   (216) 449-1375

Charles Stafford Lynch
B.P. America, Inc.
200 Public Square
36 F 3154
Cleveland, OH 44114   (216) 586-8469

Charles Bahlmann Lyon
Calfee, Halter & Griswold
1800 Society Bldg.
East 9th & Superior
Cleveland, OH 44114   (216) 781-2166

James A. Mackin*
NASA Lewis Research Ctr.
MS 301-6
21000 Brookpark Rd.
Cleveland, OH 44135   (216) 433-5752

Walter Maky
Renner, Otto, Boiselle
& Sklar
One Public Square
12th Floor
Cleveland, OH 44113   (216) 621-1113

Jane M. Marciniszyn
2675 Scarborough Road
Cleveland, OH 44106   (216) 621-1113

Vytas R. Matas
Mc Dermott Incorporated
26250 Euclid Avenue
Suite 927
Cleveland, OH 44132   (216) 261-9531

John P. Maxey
Renner, Otto, Boisselle
& Sklar
1621 Euclid Ave.
19th Floor
Cleveland, OH 44115   (216) 621-1113

Steven D.A. Mc Carthy
Watts, Hoffmann, Fisher
& Heinke Co., L.P.A.
100 Erieview Plaza
Suite 2850
Cleveland, OH 44114   (216) 623-0775

Robert John Mc Closkey
Eaton Corporation
9919 Clinton Road
Cleveland, OH 44144   (216) 281-2211

Scott A. Mc Collister
BP America
200 Public Square
36-F-3454
Cleveland, OH 44114   (216) 586-3136

William Charles Mc Coy, Jr.
Pearne, Gordon Mc Coy
& Granger
1200 Leader Bldg.
Cleveland, OH 44114   (216) 579-1700

Robert E. Mc Donald
Sherwin-Williams Co.
101 Prospect Ave., N.W.
Cleveland, OH 44115   (216) 566-2432

James Wm. Mc Kee
Fay, Sharpe, Beall, Fagan,
Minnich & Mc Kee
1100 Superior Avenue
Suite 700
Cleveland, OH 44114   (216) 861-5582

John P. Mc Mahon
General Electric Company
Nela Park
Cleveland, OH 44112   (216) 266-8649

Charles B. Meyer
Jones, Day, Reavis
& Pogue
North Point
901 Lakeside Avenue
Cleveland, OH 44114   (216) 586-7091

John E. Miller, Jr.
BP Chemicals, America
31-B4655
200 Public Square
Cleveland, OH 44124   (216) 586-5144

Stanley Ross Miller
White Consolidated Inds., Inc.
11770 Berea Rd.
Cleveland, OH 44111   (216) 252-3700

Richard J. Minnich
Fay, Sharpe, Beall, Fagan,
Minnich & Mc Kee
1100 Superior Avenue
Suite 700
Cleveland, OH 44114   (216) 861-5582

Jay F. Moldovanyi
Fay, Sharpe, Beall, Fagan,
Minnich & Mc Kee
1100 Superior Avenue
Suite 700
Cleveland, OH 44114   (216) 861-5582

James M. Moore
Pearne, Gordon, Mc Coy
& Granger
1200 Leader Bldg.
Cleveland, OH 44114   (216) 579-1700

Christopher H. Morgan
Parker Hannifin Corporation
17325 Euclid Avenue
Cleveland, OH 44112   (216) 531-3000

Philip J. Moy, Jr.
Fay, Sharpe, Beall, Fagan,
Minnich & Mc Kee
1100 Superior Avenue
Suite 700
Cleveland, OH 44114   (216) 861-5582

Cynthia S. Murphy
Calfee, Halter & Griswold
1800 Society Building
Cleveland, OH 44114   (216) 622-8200

John P. Murtaugh
Pearne, Gordon, Mc Coy
& Granger
1200 Leader Bldg.
Cleveland, OH 44114   (216) 579-1700

Timothy E. Nauman
Fay, Sharpe, Beall, Fagan,
Minnich & Mc Kee
1100 Superior Avenue
Suite 700
Cleveland, OH 44114   (216) 861-5582

Frank J. Nawalanic
Body, Vickers & Daniels
Suite 2000
Terminal Tower
Cleveland, OH 44113   (216) 623-0040

Norman A. Nixon
Pearne, Gordon, Mc Coy
& Granger
1200 Leader Bldg.
Cleveland, OH 44114   (216) 579-1700

Timothy J. O Hearn
Jones, Day, Reavis
& Pogue
North Point
901 Lakeside
Cleveland, OH 44114   (216) 586-1080

Lawrence R. Oremland
Calfee, Halter & Griswold
1800 Society Bldg.
Cleveland, OH 44114   (216) 781-2166

Stanley E. Ornstein
Tarolli, Sundheim & Covell
1111 Leader Bldg.
Cleveland, OH 44114   (216) 621-2234

Donald L. Otto
Renner, Otto, Boisselle
& Sklar
1621 Euclid Avenue
19th Fl.
Cleveland, OH 44115   (216) 621-1113

David J. Pasz
Kraig & Pasz
623 St. Clair Avenue N.W.
Cleveland, OH 44113   (216) 696-4009

David R. Percio
Cleveland Elec. Illuminating Co.
55 Public Square
Room 406
Cleveland, OH 44113   (216) 479-6522

Sue E. Phillips
BP America, Inc.
200 Public Sq., 36-F
Cleveland, OH 44114   (216) 586-8614

Wayne D. Porter, Jr.
Weston, Hurd, Fallon,
Paisley & Howley
2500 Terminal Tower
Cleveland, OH 44113   (216) 241-6602

Linn J. Raney
Watts, Hoffmann, Fisher
& Heinke Co., L.P.A.
Suite 2850
100 Erieview Plaza
Cleveland, OH 44114   (216) 623-0775

Carl A. Rankin
Pearne, Gordon, Mc Coy
& Granger
1200 Leader Bldg.
Cleveland, OH 44114   (216) 579-1700

John William Renner
Renner, Otto, Boisselle
& Sklar
1621 Euclid Avenue
19th Floor
Cleveland, OH 44115   (216) 621-1113

James A. Rich
Alcan Aluminum Corp.
100 Erieview Plaza
Cleveland, OH 44114   (216) 523-6991

Patrick R. Roche
Fay, Sharpe, Beall, Fagan,
Minnich & Mc Kee
1100 Superior Avenue
Suite 700
Cleveland, OH 44114   (216) 861-5582

Deborah S. Rodewig
Jones Day Reavis & Pogue
North Point
901 Lakeside Ave.
Cleveland, OH 44114   (216) 586-3939

Alan J. Ross
Fay, Sharpe, Beall, Fagan,
Minnich & Mc Kee
1100 Superior Avenue
Suite 700
Cleveland, OH 44114   (216) 861-5582

Donald A. Rowe
Eaton Corporation
Eaton Center
1111 Superior Avenue
Cleveland, OH 44114   (216) 523-4555

Charles R. Rust
Woodling, Krost & Rust
530 National City East
Sixth Bldg.
Cleveland, OH 44114   (216) 241-4150

Peter D. Sachtjen
545 Hanna Bldg.
Cleveland, OH 44115   (216) 696-6663

Frank M. Sajovec, Jr.
Eaton Corporation
Eaton Center
Cleveland, OH 44114   (216) 523-4136

Maurice R. Salada
TRW, Inc.
Automotive Sector
1900 Richmond Road
Cleveland, OH 44124   (216) 291-7392

Dan J. Sammon
Watts, Hoffmann, Fisher
& Heinke Co., L.P.A.
100 Erieview Plaza
Suite 2850
Cleveland, OH 44114   (216) 623-0775

Mark D. Saralino
Renner, Otto, Boisselle & Sklar
1621 Euclid Ave., 19th Fl.
Cleveland, OH 44115   (216) 621-1113

Stephen D. Scanlon
Tarolli, Sundheim & Covell
Leader Building
Suite 1111
526 Superior Avenue
Cleveland, OH 44114   (216) 621-2234

Thomas Paul Schiller
Pearne, Gordon, Mc Coy
& Granger
1200 Leader Bldg.
Cleveland, OH 44114   (216) 579-1700

Robert N. Schmidt
1721 Fulton Road
Cleveland, OH 44113   (216) 781-4096

Thomas M. Schmitz
Glidden Company
900 Huntington Bldg.
925 Euclid Ave.
Cleveland, OH 44115   (216) 344-8401

Frederic B. Schramm
20600 Chagrin Blvd.
Suite 503
Cleveland, OH 44122   (216) 283-0075

Stephen J. Schultz
Watts, Hoffmann, Fisher
& Heinke Co., L.P.A.
Suite 2850
100 Erieview Plaza
Cleveland, OH 44114   (216) 623-0775

Albert P. Sharpe, III
Fay, Sharpe, Beall, Fagan,
Minnich & Mc Kee
1100 Superior Avenue
Suite 700
Cleveland, OH 44114   (216) 861-5582

Richard A. Sharpe#
Watts, Hoffman, Fisher
& Heinke
100 Erieview Plaza
Suite 2850
Cleveland, OH 44114   (216) 623-0775

Howard G. Shimola
Pearne, Gordon, Mc Coy
& Granger
1200 Leader Bldg.
Cleveland, OH 44114   (216) 579-1700

Gene Edwin Shook, Sr.*
NASA Lewis Research Center
Off. of Chief Counsel
Mail Stop: LE-LAW
21000 Brookpark Rd.
Cleveland, OH 44135   (216) 433-4000

Warren A. Sklar
Renner, Otto, Boisselle & Sklar
1621 Euclid Avenue
19th Floor
Cleveland, OH 44115   (216) 621-1113

J. Helen Slough
200 Westview Towers
21010 Center Ridge Rd.
Suite 803
Cleveland, OH 44116   (216) 331-4892

Jeffrey J. Sopko
Pearne, Gordon, Mc Coy
& Granger
1200 Leader Bldg.
Cleveland, OH 44114   (216) 579-1700

David E. Spaw#
Pearne, Gordon, Mc Coy
& Granger
1200 Leader Bldg.
Cleveland, OH 44114   (216) 579-1700

Barry L. Springel
Jones, Day, Reavis & Pogue
North Point
901 Lakeside Avenue
Cleveland, OH 44114   (216) 586-7236

Kent N. Stone*
NASA Lewis Research Center
21000 Brookpark Road
Cleveland, OH 44135   (216) 433-2313

Robert Brandt Sundheim
Tarolli, Sundheim & Covell
1111 Leader Bldg.
Cleveland, OH 44114   (216) 621-2234

Mark S. Svat
Fay, Sharpe, Beall, Fagan,
Minnich & Mc Kee
1100 Superior Ave., Ste. 700
Cleveland, OH 44114   (216) 861-5582

Harold Duane Switzer
Jones, Day, Reavis & Pogue
North Point
901 Lakeside Avenue
Cleveland, OH 44114   (216) 586-7283

Paul E. Szabo
Tarolli, Sundheim & Covell
1111 Leader Bldg.
Cleveland, OH 44114   (216) 621-2234

Joseph H. Taddeo
Oldham & Oldham Co.
816 Hanna Bldg.
Cleveland, OH 44115   (216) 621-2956

Steven W. Tan
Sherwin-Williams Company
Corp. Legal Dept.
101 Prospect Avenue, N.W.
Cleveland, OH 44115   (216) 566-2487

Thomas L. Tarolli
Tarolli, Sundheim & Covell
1111 Leader Bldg.
Cleveland, OH 44114   (216) 621-2234

William A. Teoli, Jr.
Renner, Otto, Boisselle
& Sklar
1621 Euclid Avenue
19th Floor
Cleveland, OH 44115   (216) 621-1113

Richard Hadley Thomas
Tarolli, Sundheim & Covell
1111 Leader Bldg.
526 Superior Ave.
Cleveland, OH 44114   (216) 621-2234

John C. Tiernan
Fay, Sharpe, Beall,
Fagan, Minnich & Mc Kee
1100 Superior Ave.
Suite 700
Cleveland, OH 44114   (216) 861-5582

James H. Tilberry
3200 Terminal Tower
Cleveland, OH 44113   (216) 575-1000

William C. Tritt
Renner, Otto, Boisselle
& Sklar
1621 Euclid Ave., 19th Flr.
Cleveland, OH 44115   (216) 621-1113

Barry L. Tummino
Tarolli, Sundheim & Covell
1111 the Leader Bldg.
Cleveland, OH 44114   (216) 621-2234

David J. Untener
BP America
200 Public Square
36th Floor
Room 3753
Cleveland, OH 44114   (216) 586-8472

Michael W. Vary
Jones, Day, Reavis
& Pogue
North Point
901 Lakeside Avenue
Cleveland, OH 44114   (216) 586-1241

Robert Vernon Vickers
Body, Vickers & Daniels
2000 Terminal Tower
Cleveland, OH 44113   (216) 623-0040

Allan William Vogele
TRW, Inc.
1900 Richmond Road
Cleveland, OH 44124   (216) 291-7393

Richard R. Walling
530 National City East 6th Bldg.
1965 E. 6th Street
Cleveland, OH 44114   (216) 241-4150

James L. Wamsley, III
Jones, Day, Reavis
& Pogue
North Point
901 Lakeside Ave.
Cleveland, OH 44114   (216) 586-7251

James G. Watterson
Watts, Hoffmann, Fisher
& Heinke Co., L.P.A.
100 Erieview Plaza
Suite 2850
Cleveland, OH 44114   (216) 623-0775

Howard Lee Weinshenker
American Greetings Corp.
Tdmk. & Lic. Coun.
10500 American Rd.
Cleveland, OH 44144   (216) 252-7300

Louis Jerome Weisz
D. Peter Hochberg Co., L.P.A.
Baker Building
1940 E. 6th St., 6th Flr.
Cleveland, OH 44114   (216) 771-3800

Robert Philip Wright
Watts, Hoffmann, Fisher
& Heinke Co., L.P.A.
100 Erieview Plaza
Suite 2850
Cleveland, OH 44114   (216) 623-0775

Barbara A. Wrigley
Jones, Day, Reavis & Pogue
North Point
901 Lakeside Ave.
Cleveland, OH 44114   (216) 586-3939

Thomas E. Young
Body, Vickers & Daniels
2000 Terminal Tower
Cleveland, OH 44113   (216) 623-0040

Vincent Edward Young
B.P. America Inc.
200 Public Square
36-F-3655
Cleveland, OH 44114   (216) 586-8453

David P. Yusko
B.P. America
200 Public Square
36-F-3006
Cleveland, OH 44114   (216) 586-8461

Thaddeus Arthur Zalenski
LTV Steel Company, Inc.
25 W. Prospect Avenue
Cleveland, OH 44115   (216) 622-5626

Robert Emil Bielek
3298 Rumson Rd.
Cleveland Heights, OH 44118
                              (216) 566-2482

William E. Currie
3945 Northampton Road
Cleveland Heights, OH 44121
                              (216) 382-4706

I. Monica Olszewski
2558 Euclid Heights Blvd.
Cleveland Heights, OH 44106
                              (216) 932-5007

Thomas S. Baker, Jr.
2941 Kenny Road
Suite 240
Columbus, OH 43221   (614) 451-5122

Edwin Baranowski
Porter, Wright, Morris
& Arthur
41 South High Street
Columbus, OH 43215   (614) 227-2188

Dennis C. Belli
844 South Front Street
Columbus, OH 43206   (614) 444-6556

Barry S. Bissell
Battelle Dev. Corp.
505 King Ave.
Columbus, OH 43201   (614) 424-7798

John D. Boos
4642 Burbank Dr.
Columbus, OH 43220   (614) 457-7327

Richard A. Crane
2850 Columbus Avenue
Columbus, OH 43209   (614) 231-8717

Lonnie R. Drayer
Ross Laboratories
Department 104075-S1
Patent & Trademark Dept.
625 Cleveland Ave.
Columbus, OH 43215   (614) 227-3333

Daniel H. Dunbar
Kremblas, Foster, Millard
& Watkins
50 West Broad Street
Columbus, OH 43215   (614) 464-2700

Philip M. Dunson
1446 Friar Lane
Columbus, OH 43221   (614) 457-4314

Edward Paul Forgrave
Biebel, French & Nauman
50 West Broad Street
Suite 620
Columbus, OH 43215   (614) 464-2902

Frank H. Foster, III
7632 Slate Ridge Blvd.
Columbus, OH 43068   (614) 575-2100

Palmer Fultz
4386 Stinson Drive
Columbus, OH 43214   (614) 459-7356

John L. Gray
Emens, Hurd, Kegler
& Ritter
Capitol Square
65 East State St.
Columbus, OH 43215   (614) 462-5438

David L. Hedden
Ashland Chemical Company
P.O. Box 2219
Columbus, OH 43216   (614) 889-4265

Eric S. Lucas
5433 Coachman Rd.
Apt. G
Columbus, OH 43220   (614) 227-2000

Dwight A. Marshall
AT&T Bell Laboratories
6200 East Broad Street
Columbus, OH 43213   (614) 860-2127

William J. Mase
210 E. Cooke Road
Columbus, OH 43214   (614) 263-5728

George P. Maskas
3361 Stonehenge Court
Columbus, OH 43221   (614) 457-6549

Sidney Wayne Millard#
Kremblas, Foster & Millard
7632 Slate Ridge Boulevard
Columbus, OH 43068   (614) 575-2100

William Vernon Miller
Kremblas, Foster & Millard
7632 Slate Ridge Blvd.
Columbus, OH 43068   (614) 575-2100

Jerry K. Mueller, Jr.
Mueller & Smith, L.P.A.
7700 Rivers Edge Drive
Columbus, OH 43085   (614) 436-0600

Donald O. Nickey
Ross Laboratories
Div. of Abbott Labs.
625 Cleveland Avenue
Columbus, OH 43215   (614) 229-7080

Clarence Henry Peterson#
1491 Kirkley Rd.
Columbus, OH 43221   (614) 451-2243

Patrick P. Phillips
Biebel, French & Nauman
620 Le Veque Tower
50 West Broad Street
Columbus, OH 43215   (614) 464-2902

Paul C. Pink#
2449 Dale Avenue
Columbus, OH 43209   (614) 237-0140

Philip J. Pollick
Watkins, Dunbar & Pollick
2941 Kenny Rd., Ste. 260
Columbus, OH 43221   (614) 457-5700

Dennis H. Rainear
Borden Inc.
180 East Broad Street
Columbus, OH 43215   (614) 225-7069

Priscilla N. Ratliff#
1965 Glenn Avenue
Columbus, OH 43212   (614) 488-1622

Sheldon R. Schulte
1017 Conestoga Drive
Columbus, OH 43213   (614) 863-6120

Kenneth Earl Shaweker
Battelle Memorial Institute
505 King Ave.
Columbus, OH 43201   (614) 424-7453

Gerald Lee Smith
Mueller & Smith, L.P.A.
Mueller & Smith Bldg.
7700 Rivers Edge Drive
Columbus, OH 43235   (614) 436-0600

Jeffrey S. Standley
Porter, Wright, Morris
& Arthur
41 S. High Street
Columbus, OH 43215   (614) 227-2000

Robert E. Stebens
50 West Broad Street
Le Veque Tower
Suite 1930
Columbus, OH 43215   (614) 228-6359

Thomas L. Sweeney
Ohio State University
208 Bricker Hall
Columbus, OH 43210   (614) 292-1582

Kenneth P. Van Wyck
Borden, Inc.
180 E. Broad St.
Law Dept., 27th Floor
Columbus, OH 43215   (614) 225-3369

Vernon F. Venne
Ashland Chemical Company
P.O. Box 2219
Columbus, OH 43216   (614) 889-3975

Kenneth R. Warburton
7632 Slate Ridge Blvd.
Columbus, OH 43068   (614) 575-2100

Robert B. Watkins
Watkins, Dunbar & Pollick
2941 Kenny Road
Suite 260
Columbus, OH 43221   (614) 457-5700

Klaus H. Wiesmann
Battelle Memorial Institute
505 King Avenue
Columbus, OH 43201   (614) 424-6589

James B. Wilkens
Ohio State University
1960 Kenny Road
Columbus, OH 43210   (614) 292-6079

Jura Christine Zibas
1026 Deacon Circle
Columbus, OH 43214   (614) 457-6285

Robert L. Zieg
Battelle
505 King Ave.
Columbus, OH 43201   (614) 424-4243

Richard H. Haas
3071 Hariett Road
Cuyahoga Falls, OH 44224   (216) 688-4235

John Frank Jones
2724 Cedar Hill Rd.
Cuyahoga Falls, OH 44223   (216) 923-2952

Denbigh S. Matthews
1240 Lincoln Avenue
Cuyahoga Falls, OH 44223   (216) 923-8562

Robert W. Becker
Becker & Becker, Inc.
211 S. Main Street
Dayton, OH 45402   (513) 228-7801

Jerome P. Bloom
5751 N. Webster Street
P.O. Box 14553
Dayton, OH 45414   (513) 890-2079

James T. Candor
Candor, Candor & Tassone
P. O. Box 292305
Dayton, OH 45429   (513) 298-8606

J. T. Cavender
NCR Corp.
1700 S. Patterson Road
Dayton, OH 45479   (513) 445-2970

Robert L. Clark
NCR Corporation
1700 S. Patterson Blvd.
Dayton, OH 45479   (513) 445-2913

Robert L. Deddens
55 Park Avenue
Dayton, OH 45419   (413) 293-9696

John William Donahue
Biebel, French & Nauman
2500 Kettering Tower
Dayton, OH 45423   (513) 461-4543

Robert Kern Duncan*
4390 Baker Road
Dayton, OH 45424   (513) 233-3073

Roger S. Dybvig
Dybvig and Dybvig
22 Green Street
Dayton, OH 45402   (513) 461-1142

Thomas William Flynn
Biebel, French & Nauman
2500 Kettering Tower
Dayton, OH 45423   (513) 461-4543

Michael D. Folkerts
Biebel, French & Nauman
2500 Kettering Tower
Dayton, OH 45423   (513) 461-4543

Douglas S. Foote
NCR Corporation
Law Dept.
World Headquarters
Dayton, OH 45479   (513) 445-2968

Nathaniel R. French
Biebel, French & Nauman
2500 Kettering Tower
Dayton, OH 45423   (513) 461-4543

Gregory P. Gadson
NCR Corporation
World Headquarters, Law Dept.
1700 S. Patterson Blvd.
Dayton, OH 45479   (513) 445-2968

Charles Michael Gegenheimer
Killworth, Gottman, Hogan
& Schaeff
1400 One First Natonal Plaza
Dayton, OH 45402   (513) 223-2050

Donald P. Gillette
55 Westpark Road
Dayton, OH 45459   (513) 433-6563

James F. Gottman
Killworth, Gottman, Hagan
& Schaeff
1400 One First National Plaza
Dayton, OH 45402   (513) 223-2050

Joseph John Grass
Monarch Marking Sys., Inc.
P.O. Box 608
Dayton, OH 45401   (513) 865-2012

Timothy W. Hagan
Killworth, Gottman, Hagan
& Schaeff
1400 One First National Plaza
Dayton, OH 45402   (513) 223-2050

Wilbert Hawk, Jr.
NCR Corp.
1700 S. Patterson Blvd.
Dayton, OH 45479   (513) 445-2960

Louis E. Hay
847 Woodhill Rd.
Dayton, OH 45431   (513) 253-1645

Gilbert N. Henderson
Biebel, French & Nauman
2500 Kettering Tower
Dayton, OH 45423   (513) 461-4543

Ernest T. Hix
53 E. Thruston Blvd.
Dayton, OH 45409   (513) 298-1594

Michael A. Jacobs
NCR Corporation
Law Dept.
WHQ5
1700 S. Patterson Blvd.
Dayton, OH 45479   (513) 445-2928

William R. Jacox
Jacox and Meckstroth
2310 Far Hills Bldg.
2310 Far Hills Ave.
Dayton, OH 45419   (513) 298-2811

Matthew Richard Jenkins
729 Murrell Drive
Dayton, OH 45429   (513) 461-4543

Stephen F. Jewett
NCR Corp.
World Headquarters
1700 S. Patterson Blvd.
Dayton, OH 45479   (513) 445-2972

Richard A. Killworth
Killworth, Gottman, Hagan
& Schaeff
1400 One First National Plaza
Dayton, OH 45402   (513) 223-2050

Scott V. Kissinger
NCR Corporation
1700 S. Patterson Blvd.
Dayton, OH 45479   (513) 445-7415

Thomas C. Lagaly#
Killworth, Gottman, Hagan
& Schaeff
1400 One First National Plaza
Dayton, OH 45402   (513) 223-2050

Richard William Lavin
NCR Corporation
World Headquarters
Dayton, OH 45479   (513) 445-2914

Mark P. Levy
Thompson, Hine & Flory
2000 Courthouse Plaza N.E.
P.O. Box 8801
Dayton, OH 45401   (513) 443-6949

Theodore David Lienesch
Thompson, Hine & Flory
2000 Courthouse Plaza N.E.
P.O. Box 8801
Dayton, OH 45401   (513) 443-6958

Paul W. Martin
NCR Corporation
Law Dept.
1700 S. Patterson Blvd.
Dayton, OH 45479   (513) 445-2972

Ralph L. Marzocco
1434-B Miamisburg-Centerville Rd.
Dayton, OH 45458   (513) 435-5460

Alan F. Meckstroth
Jacox and Meckstroth
2310 Far Hills Bldg.
Dayton, OH 45419   (513) 298-2811

H. Stanley Muir, III
NCR Corporation
1700 South Patterson Blvd.
Dayton, OH 45479   (513) 445-2990

Joseph G. Nauman
2102 West Dorothy Lane
Dayton, OH 45439   (513) 294-8537

Ken K. Patel
Killworth, Gottman, Hagan
& Schaeff
One First National Plaza
Suite 1400
Dayton, OH 45402   (513) 223-2050

Bruce E. Peacock
Biebel, French & Nauman
2500 Kettering Tower
Dayton, OH 45423   (513) 461-4543

Jack R. Penrod
NCR Corporation
1700 S. Patterson Blvd.
Dayton, OH 45479   (513) 445-6740

Patricia L. Prior
Killworth, Gottman, Hagan
& Schaeff
1400 One First National Plaza
Dayton, OH 45402   (513) 223-2050

Albert L. Sessler, Jr.
NCR Corp.
1700 S. Patterson Blvd.
WHQ5
Dayton, OH 45479   (513) 445-2965

Charles N. Shane, Jr.
Mead Corp.
Courthouse Plaza N.E.
Dayton, OH 45462   (513) 495-3438

Robert L. Showalter, Jr.
Killworth, Gottman, Hagan
& Schaeff
1400 One First National Plaza
Dayton, OH 45402   (513) 223-2050

Michael W. Starkweather
NCR Corporation
Law Dept.
World Headquarters
Dayton, OH 45479   (513) 445-2995

Daniel J. Staudt
Biebel & French
2500 Kettering Tower
Dayton, OH 45423   (513) 461-4543

Lloyd B. Stevens, Jr.
1359 Tattersall Road
Dayton, OH 45459   (513) 433-0946

Richard C. Stevens
Killworth, Gottman, Hagan
& Schaeff
1400 One First Natl. Plaza
Dayton, OH 45402   (513) 223-2050

James M. Stover
NCR Corporation
Law Department
World Headquarters
Dayton, OH 45479   (513) 445-7663

Joseph V. Tassone
Dayco Products, Inc.
1 Prestige Place
Dayton, OH 45401   (513) 226-5725

Elmer Wargo
NCR Corporation
1700 S. Patterson Blvd.
Dayton, OH 45479   (513) 445-2969

Gregory A. Welte
NCR Corp.
1700 S. Patterson Blvd.
WHQ-5
Dayton, OH 45479   (513) 445-6240

Norman R. Wissinger
3103 Winding Way
Dayton, OH 45419   (513) 299-2003

Reuben Wolk
3849 Seiber Ave.
Dayton, OH 45405   (513) 278-1258

Daniel D. Mast
3520 Cackler Road
Delaware, OH 43015   (614) 362-3575

John Dabney Haney
13939 Doylestown Road
Doylestown, OH 44230   (216) 658-6580

Patricia A. Coburn
Adria Laboratories
Div. of Erbamont Inc.
5000 Post Road
Dublin, OH 43017   (614) 764-8121

Edward B. Dunning
Sherex Chem. Co., Inc.
P.O. Box 646
Dublin, OH 43017   (614) 764-6601

William Kammerer
8839 Birgham Ct. N.
Dublin, OH 43017   (614) 889-2711

Marc C. Pawl
Ashland Chemical Co.
5200 Blazer Memorial Pkwy.
Dublin, OH 43017   (614) 889-3105

Mary E. Picken
Ashland Chem. Inc.
5200 Blazer Parkway
Dublin, OH 43017   (614) 889-4694

Walter H. Schneider
P.O. Box 917
Dublin, OH 43017   (614) 889-5747

Michael A. Centanni
Gould Inc.
35129 Curtis Blvd.
Eastlake, OH 44095   (216) 953-5130

Clyde H. Haynes
1823 West River Road, N.
Elyria, OH 44035   (216) 324-4706

Richard J. Killoren
169 Locust Dr.
Fairborn, OH 45324   (513) 878-1272

Frank C. Rote, Jr.
Gencorp Inc.
175 Ghent Rd.
Fairlawn, OH 44333   (216) 869-4266

John Jay Freer
Eltech Systems Corporation
625 East Street
Fairport Harbor, OH 44077   (216) 357-4055

Girard R. Jetton, Jr.
P.O. Box 1003
Findlay, OH 45839   (419) 423-4680

Stan C. Kaiman
Cooper Tire & Rubber Co.
Lima & Western Avenue
Findlay, OH 45840   (419) 424-4233

James H. Sutton
6975 Scenic View Lane
Findlay, OH 45840   (419) 293-2423

Kevin M. Foley
Hepar Industries, Inc.
160 Industrial Drive
P.O. Box 338
Franklin, OH 45005   (513) 746-3603

Richard Charles Darr
915 Croghan Street
Fremont, OH 43420   (419) 332-4106

Jerry W. Semer
617 Croghan Street
Fremont, OH 43420   (419) 332-2221

Richard G. Stahr
1087 Arcaro Drive
Gahanna, OH 43230   (419) 332-2221

Harlan E. Hummer
P.O. Box 608
Gates Mills, OH 44040   (216) 449-6901

Charles E. Moore
55 Beechtree Lane
Granville, OH 43023   (614) 587-3846

Jack C. Mc Gowan
Baden, Jones, Scheper
& Crehan Co., L.P.A.
222 High Street
Hamilton, OH 45011   (513) 868-2731

John W. Teare
26 Hollytree Ct.
Hamilton, OH 45011   (513) 868-2564

Richard L. Kelly
918 Lakeshore Drive East
Hebron, OH 43025   (614) 928-4256

William Preston Hickey
1970 CR 32
Helena, OH 43435   (419) 457-5525

Timothy B. Gurin#
Picker International, Inc.
595 Miner Road
Highland Heights, OH 44143 (216) 473-3570

Ernest W. Legree
025 Eastlawn Drive
Highland Heights, OH 44143 (216) 449-4129

Ronald B. Brietkrenz#
1888 Bellus Road
Hinckley, OH 44233   (216) 225-2583

Woodrow W. Ban
7363 Winsted
Hudson, OH 44236   (216) 656-2953

William Earl Cleaver
5 Hudson Commons II
Hudson, OH 44236   (216) 656-2386

Albert C. Doxsey
200 Laurel Lake Drive
Apt. A-125
Hudson, OH 44236   (216) 656-2386

Robert L. Sahr
P.O. Box 597
219 N. Main Street
Hudson, OH 44236   (216) 653-3866

John M. Romanchik, Jr.
6107 Elmwood Ave.
Independence, OH 44131   (216) 524-8855

Joseph P. Burke
4050 Benfield Drive
Kettering, OH 45429   (513) 293-4998

Robert W. Graham
10 Southmoor Circle, N.W.
Suite 14
Kettering, OH 45429   (513) 293-8481

George J. Muckenthaler
1330 Tall Timber Trail
Kettering, OH 45409   (513) 293-2745

Albert H. Reuther
4489 Mapleridge Place
Kettering, OH 45429   (513) 293-0110

John W. Ball, Jr.
12030 Lake Avenue
Lakewood, OH 44107   (216) 228-5064

Robert Jesse Fay
12700 Lake Ave., #407
Lakewood, OH 44107   (216) 228-5064

Gustalo Nunez
4463 Oberlin Ave.
Lorain, OH 44053   (216) 282-9109

Edward J. Holler, Jr.
RR #2 Hilltop Drive
Magnolia, OH 44643   (216) 866-2289

Paul E. Milliken
9061 Wall Street, N.W.
Massillon, OH 44646   (216) 833-2740

Daniel S. Kalka
2289 Wilbur Road
Medina, OH 44256   (216) 239-2460

Ronald R. Stanley
P.O. Box 571
Medina, OH 44258   (216) 631-1020

Arthur L. Wolfe#
International Science Service
P.O. Box 181
Mentor, OH 44061   (216) 951-4369

Barbara Joan Haushalter
2440 Royal Ridge Dr.
Miamisburg, OH 45342   (513) 859-0838

Edward L. Brown, Sr.#
3011 Central Avenue
Middletown, OH 45044   (513) 422-7721

Robert J. Bunyard#
Armco Inc.
703 Curtis Street
Middletown, OH 45043   (513) 425-5973

Larry A. Fillnow#
Armco Inc.
703 Curtis
Middletown, OH 45043   (513) 425-2494

Robert H. Johnson
Armco Inc.
703 Curtis Street
Middletown, OH 45043   (513) 425-2432

Bruce M. Thomas
4201 Fisher Avenue
Middletown, OH 45042   (513) 425-0777

Thomas F. Mc Gann
30-D South Terrace Avenue
Newark, OH 43055   (614) 522-3958

Edward E. Schilling, Sr.
1816 Carlyle Court
Newark, OH 43055   (614) 344-4610

Francis D. Thomson#
735 Towne Krier Court
Newark, OH 43055   (614) 366-5520

Richardson Blackburn Farley
Hoover Corporation
101 E. Maple Ave.
North Canton, OH 44720   (216) 499-9200

Stephen H. Friskney
Hoover Co.
101 E. Maple St.
North Canton, OH 44720   (216) 499-9200

A. Burgess Lowe
Hoover Company
101 E. Maple Street
North Canton, OH 44720   (216) 499-9200

James D. Wolfe
1202 Lake Breeze Drive
North Canton, OH 44720   (216) 896-1881

Roy Davis
5916 Chapel Road
North Madison, OH 44057   (216) 428-2628

Ralph E. Jocke
6029 Barton Road
North Olmsted, OH 44070   (216) 777-3671

Robert August Wiedemann
Hiltz, Wiedemann & Allton Co.
401 Citizens Natl. Bank Bldg.
East Main Street
Norwalk, OH 44857   (419) 668-8211

Marvin L. Union
13530 Heath Rd.
Novelty, OH 44072   (216) 338-8185

John L. Shailer#
1720 Hiner Road
Orient, OH 43146   (614) 871-1972

Timothy E. Tinkler
Ricerca, Inc.
7528 Auburn Road
P.O. Box 1000
Painesville, OH 44077   (216) 357-3428

Eugene Nebesh
1949 W. Pleasant Valley Rd.
Parma, OH 44134   (216) 886-6112

Henry Kozak
30779 Shaker Blvd.
Pepper Pike, OH 44124   (216) 831-8394

Richard Roy Drown#
1041 Louisiana Avenue
Perrysburg, OH 43551   (419) 874-4418

John H. Miller
29666 Chatham Way
Perrysburg, OH 43551   (419) 878-1105

William E. Nobbe
10542 Fremont Pike
Apt. 105
Perrysburg, OH 43551   (419) 874-8950

Paul Joseph Rose, Jr.
126 Pheasant Drive
Perrysburg, OH 43551   (419) 874-4529

Kenneth H. Wetmore
Glasstech, Inc.
Ampoint Industrial Park
995 Fourth Street
Perrysburg, OH 43552   (419) 661-9500

David H. Wilson, Jr.
424 E. Front Street
Perrysburg, OH 43551   (419) 874-7661

Francis Thomas Kremblas, Jr.
Kremblas, Foster & Millard
7632 Slate Ridge Blvd.
Reynoldsburg, OH 43068   (614) 575-2100

Leslie Hamilton Blair
21111 W. Wagar Circle
Rocky River, OH 44116   (216) 331-0298

Albert L. Ely, Jr.
18951 Inglewood Road
Rocky River, OH 44116   (216) 333-5467

Frederic E. Naragon
248 East State Street
Salem, OH 44460   (216) 337-9578

Daniel Glenn Blackhurst
3353 Lansmere Rd.
Shaker Heights, OH 44122   (216) 283-8129

Sheldon B. Greenbaum#
Goal Oriented Strategies, Inc.
24139 Shelburne Road
Shaker Heights, OH 44122   (216) 831-2079

Robert Donovan Hart
3264 Kenmore Rd.
Shaker Heights, OH 44122   (216) 751-3864

P. Adrian Medert
20201 Shelburne Road
Shaker Heights, OH 44118   (216) 932-1310

Michael M. Rickin
3436 Colton Rd.
Shaker Heights, OH 44122   (216) 283-6541

Ramon Doyle Foltz
6841 Green Ridge Ave.
Solon, OH 44139   (216) 283-6541

Woodrow W. Portz
Portz & Portz
5530 S.O.M. Center Road
Solon, OH 44139   (216) 248-4500

Kathryn Wilson Grant#
4408 Forest Hill Road
Stow, OH 44224   (216) 678-1275

Donald A. Bergquist#
Patent Services Company
17145 Misty Lake Drive
Strongsville, OH 44136   (216) 238-1210

Andrew Beckerman-Rodau
3914 Barleyton Circle
Sylvania, OH 43560   (419) 772-2207

John B. Molnar
6832 S. Fredericksburg Drive
Sylvania, OH 43560   (419) 249-5734

Douglas R. Mc Kechnie
1 Westwood Court
Tiffin, OH 44883   (419) 447-0837

Vincent L. Barker, Jr.
William, Brinks, Olds, Hofer,
Gilson & Lione
1130 Edison Plaza
Toledo, OH 43604   (419) 244-6578

David Robert Birchall
3700 Heathesdowns Blvd.
Toledo, OH 43614   (419) 385-1627

Howard G. Bruss, Jr.
Owens-Illinois, Inc.
One Sea-Gate
Toledo, OH 43666   (419) 247-2036

Kenneth F. Cherry#
1533 East Gate
Toledo, OH 43614   (419) 385-9165

William J. Clemens
Marshall & Melhorn
4 Seagate, 8th Fl.
Toledo, OH 43604   (419) 249-7100

Richard B. Dence*
U.S. Defense Logistics Agency
DCMRCLE- Residency Off.
1330 Laskey Road
Toledo, OH 43612   (419) 259-7578

Richard Donovan Emch
Emch, Schaffer, Schaub
& Porcello Co., L.P.A.
One Sea Gate, Suite 1980
P.O. Box 916
Toledo, OH 43692   (419) 243-1294

Donald R. Fraser
Marshall & Melhorn
4 Seagate, 8th Fl.
Toledo, OH 43604   (419) 249-7100

R. La Mar Frederick
Owens-Corning Fiberglas Corp.
Fiberglas Tower
Toledo, OH 43659   (419) 248-8650

Ted C. Gillespie
Owens-Corning Fiberglas Corp.
Fiberglas Tower
Toledo, OH 43659   (419) 248-8461

Thomas H. Grafton
3333 Christie Blvd.
Toledo, OH 43606   (419) 531-6614

Allen D. Gutchess, Jr.
Suite 408 - 1806 Madison
Toledo, OH 43624   (419) 243-4353

Richard D. Heberling
Marshall & Melhorn
4 Seagate, 8th Fl.
Toledo, OH 43604   (419) 249-7100

David T. Innis
5003 Rudgate Blvd.
Toledo, OH 43623   (419) 885-1867

Hugh Adam Kirk
4120 Tantara Drive
Toledo, OH 43623   (419) 882-5995

Robert M. Leonardi
Dana Corporation
4500 Dorr Street
Toledo, OH 43615   (419) 535-4791

Richard S. Mac Millan
Mac Millan, Sobanski & Todd
905 First Federal Plaza
Toledo, OH 43624   (419) 255-5900

Catherine B. Martineau
Emch, Schaffer, Schaub
& Porcello Co. L.P.A.
One Sea Gate, Suite 1980
Toledo, OH 43604   (419) 243-1294

Frank B. Mc Donald
Dana Corporation
P.O. Box 1000
Toledo, OH 43697   (419) 535-4655

Thomas A. Meehan
William, Brinks, Olds, Hofer,
Gilson & Lione
1130 Edison Plaza
Toledo, OH 43604   (419) 244-6578

David D. Murray
William, Brinks, Olds, Hofer,
Gilson & Lione
1130 Edison Plaza
Toledo, OH 43604   (419) 244-6578

Phillip S. Oberlin
Marshall & Melhorn
4 Seagate, 8th Fl.
Toledo, OH 43604   (419) 249-7149

Patrick P. Pacella
Owens-Corning Fiberglas Corp.
Fiberglas Tower 26
Toledo, OH 43659   (419) 248-8230

Robert E. Pollock
Dana Corp.
P.O. Box 1000
Toledo, OH 43697   (419) 535-4653

James F. Porcello, Jr.
Emch, Schaffer, Schaub
& Porcello Co., L.P.A.
One Sea Gate, Suite 1980
P.O. Box 916
Toledo, OH 43692   (419) 243-1294

David C. Purdue
Suite B-2
2735 North Holland - Sylvania Rd.
Toledo, OH 43615   (419) 531-0599

John C. Purdue
Suite B-2
2735 North Holland - Sylvania Rd.
Toledo, OH 43615   (419) 531-0599

Philip M. Rice
Emch, Schaffer, Schaub
& Porcello Co., L.P.A.
One Sea Gate, Suite 1980
P.O. Box 916
Toledo, OH 43692   (419) 243-1294

George R. Royer
316 North Michigan St.
Suite 416
Toledo, OH 43624   (419) 241-6612

Robert F. Rywalski
William, Brinks, Olds, Hofer,
Gilson & Lione
1130 Edison Plaza
Toledo, OH 43604   (419) 244-6578

Charles R. Schaub
Emch, Schaffer, Schaub
& Porcello Co., L.P.A.
One Sea Gate, Suite 1980
P.O. Box 916
Toledo, OH 43692   (419) 243-1294

Charles F. Schroeder
2317 Valley Brook Drive
Toledo, OH 43615   (419) 244-3344

Donald A. Schurr
Marshall & Melhorn
4 Sea Gate, 8th Fl.
Toledo, OH 43604   (419) 249-7138

Ralph J. Skinkiss
2954 Shetland Rd.
Toledo, OH 43617   (419) 841-2158

Mark J. Sobanski
Mac Millan, Sobanski & Todd
905 First Federal Plaza
701 Adams Street
Toledo, OH 43624   (419) 255-5900

Paul F. Stutz
964 Spitzer Bldg.
Toledo, OH 43604   (419) 241-4211

Oliver E. Todd, Jr.
Mac Millan, Sobanski & Todd
905 First Federal Plaza
Toledo, OH 43624   (419) 255-5900

Edward A. Van Gunten
3619 Brookside
Toledo, OH 43606   (419) 537-0584

Donald K. Wedding
University of Toledo
2801 W. Bancroft St.
Toledo, OH 43606   (419) 537-2268

Robert E. Witt
316 N. Michigan Street
Suite 312
Toledo, OH 43624   (419) 241-3251

William Weigl
1805 Conwood Drive
Troy, OH 45373   (513) 339-1180

Robert L. Hearn
8367 Meadowlark Drive
West Chester, OH 45069   (513) 779-6680

Stewart A. Fraser#
Eveready Battery Co., Inc.
25225 Detroit Road
Westlake, OH 44145   (216) 835-7632

Michael Leo Gill
Nordson Corporation
28601 Clemens Road
Westlake, OH 44145   (216) 892-1580

Norman T. Musial
Musial & Musial Co., L.P.A.
30400 Detroit Road
Suite 207
Westlake, OH 44145   (216) 892-2040

Raymond J. Slattery, III
Nordson Corp.
28601 Clemens Rd.
Westlake, OH 44145   (216) 892-1580

Edmund J. Wasp
Nordson Corporation
28601 Clemens Rd.
Westlake, OH 44145   (216) 892-1580

Paul R. Edgar
360 South Yearling Road
Whitehall, OH 43213   (614) 237-8611

James Walter Adams, Jr.
Lubrizol Corp.
29400 Lakeland Blvd.
Wickliffe, OH 44092   (216) 943-4200

James L. Cordek#
Lubrizol Corporation
29400 Lakeland Blvd.
Wickliffe, OH 44092   (216) 943-4200

Charles A. Crehore
Lubrizol Corp.
29400 Lakeland Blvd.
Wickliffe, OH 44092   (216) 943-4200

Joseph P. Fischer#
Lubrizol Corporation
29400 Lakeland Blvd.
Wickliffe, OH 44092   (216) 943-4200

Roger Y. K. Hsu
Lubrizol Corp.
29400 Lakeland Blvd.
Wickliffe, OH 44092   (216) 943-4200

Gordon L. Vyrostek#
Lubrizol Corp.
29400 Lakeland Blvd.
Wickliffe, OH 44092   (216) 943-4200

Joseph B. Balazs
Figgie International Inc.
4420 Sherwin Road
Willoughby, OH 44094   (216) 953-2855

Frank B. Robb
Robb & Robb
37750 Euclid Ave.
Willoughby, OH 44094   (216) 951-2211

John F. Mc Devitt
2255 Par Lane
Willoughby Hills, OH 44094   (216) 585-0919

Richard B. O Planick
Rubbermaid Inc.
1147 Akron Road
Wooster, OH 44691   (216) 264-6464

Gary L. Loser
General Electric Company
GE Superabrasives
6325 Huntley Road
P.O. Box 568
Worthington, OH 43085   (614) 438-2438

Gregory P. Sturtz
6866 Mc Greegor St.
Worthington, OH 43085   (614) 846-9256

George Wolken, Jr.
6602 Hawthorne Street
Worthington, OH 43085   (614) 885-1411

William Scott Andes#*
U.S. Air Force
2750 Civil Engineering
Squadron/Deep
Wright-Patterson A.F.B., OH 45433
   (513) 257-3329

Charles E. Bricker*
U.S. Air Force
AF/JACPD
Area B, Bldg. 11, Room 100
Wright-Patterson A.F.B., OH 45433
   (513) 255-5052

Bernard E. Franz*
U.S. Air Force
AF/JACPD
Wright-Patterson A.F.B., OH 45433
   (513) 255-2838

Gerald B. Hollins*
U.S. Air Force
AF/JACPD
Wright-Patterson A.F.B., OH 45433
   (513) 255-2833

Thomas Louis Kundert*
U.S. Air Force
AF/JACPD
Wright-Patterson A.F.B., OH 45433
   (513) 255-2838

Richard Alan Lambert*
U.S. Air Force
Office of Judge Advocate Gen.
AF/JACPD
Wright-Patterson A.F.B., OH 45433
   (513) 255-5672

Edward W. Nypaver*
U.S. Air Force
Pat. Infringement Investigations
Off. Judge Adv. Gen.
Wright-Patterson A.F.B., OH 45433
   (513) 255-2872

Bobby D. Scearce*
U.S. Air Force
AF/JACPD
Area B, Bldg. 11
Room 100
Wright-Patterson A.F.B., OH 45433
   (513) 255-2838

Fredric L. Sinder*
U.S. Air Force
AF/JACPD
Area B, Bldg. 11, Room 100
Wright-Patterson A.F.B., OH 45433
(513) 255-2838

James D. Thesing*
U.S. Air Force
ASD/YSEA
Wright-Patterson A.F.B., OH 45433
(513) 255-9693

Ronald W. Kock#
577 Abilene Trail
Wyoming, OH 45215   (513) 255-9693

Richard C. Harpman#
Harpman & Harpman
400 City Centre One
Youngstown, OH 44503   (216) 747-1484

Webster B. Harpman
400 City Centre One
Federal Plaza East
Youngstown, OH 44503   (216) 747-1484

John A. Tomich
Commercial Intertech Corp.
1775 Logan Avenue
P.O. Box 239
Youngstown, OH 44501   (216) 746-8011

Oklahoma

G. Morrison Bennett
Phillips Petroleum Company
Research & Development
Bartlesville, OK 74004   (918) 661-6614

George E. Bogatie#
Phillips Petroleum Co.
287 Patent & Library Bldg.
Bartlesville, OK 74004   (918) 661-0560

Howard W. Bost#
1334 Quail Drive
Bartlesville, OK 74006   (918) 333-0699

Edward L. Bowman
Phillips Petroleum Company
Patent Library Bldg.
Bartlesville, OK 74004   (918) 661-0526

Karlheinz K. Brandes#
Phillips Petroleum Company
224 Patent Library Bldg.
Bartlesville, OK 74004   (918) 661-0563

James D. Brown#
Phillips Petroleum Company
240 Patent Library Bldg.
Bartlesville, OK 74004   (918) 661-0536

Kenneth A. Cannon#
Phillips Petroleum Company
263 Patent Library Bldg.
Bartlesville, OK 74004   (918) 661-0649

Lyell Henry Carver
Phillips Petroleum Company
216 Patent Library Bldg.
Bartlesville, OK 74004   (918) 661-0522

Paul S. Chirgott
Phillips Petroleum Company
204 Patent Library Bldg.
Bartlesville, OK 74006   (918) 661-0526

Carl D. Corvin
Phillips Petroleum Co.
Phillips Research Ctr.
280 Patent Library Bldg.
Bartlesville, OK 74004   (918) 661-9593

Ryan N. Cross
Phillips Petroleum Co.
Research Center
212 Patent Library Bldg.
Bartlesville, OK 74004   (918) 616-3934

Charles G. Cruzan#
918 S.E. King Drive
Bartlesville, OK 74006   (918) 335-0285

Beverly M. Dollar
Phillps Petroleum Company
234 Patent Library Bldg.
Bartlesville, OK 74004   (918) 661-0530

Richard C. Doss#
Phillips Petroleum Company
250 Patent Library Bldg.
Bartlesville, OK 74004   (918) 661-0671

John M. Fish, Jr.
Phillips Petroleum Co.
259 Patent Library Bldg.
Bartlesville, OK 74004   (918) 661-0524

Kenneth D. Goetz
Phillips Petroleum Company
226 Patent Library Bldg.
Bartlesville, OK 74004   (918) 661-0647

Bion E. Hitchcock
Phillips Petroleum Co.
236 Patent Library Bldg.
Bartlesville, OK 74004   (918) 661-0534

James H. Hughes
Fractionation Research, Inc.
P.O. Drawer F
Bartlesville, OK 74005   (918) 336-7140

Lynda S. Jolly
Phillips Petroleum Company
232 Patent Library Bldg.
Bartlesville, OK 74004   (918) 661-0611

David L. Kinsinger
Phillips Petroleum Co.
289 Patent Library Bldg.
Bartlesville, OK 74004   (918) 661-8971

Edmund T. Kittleman#
Phillips Petroleum Company
J.D. Brown
Phillips Research Center
240 Patent Library Bldg.
Bartlesville, OK 74003   (918) 661-8751

Robert C. Lutton
Phillips Petroleum Company
244 Patent Library Bldg.
Bartlesville, OK 74004   (918) 661-0675

Robert Scott Meece#
Phillips Petroleum Co.
269 Patent Library Bldg.
Bartlesville, OK 74004   (918) 661-6450

John W. Miller
Phillips Petroleum Company
246 Patent Library Bldg.
Bartlesville, OK 74004   (918) 661-0565

John D. Olivier
Phillips Petroleum Co.
270 Patent Library Bldg.
Bartlesville, OK 74004   (918) 661-0521

Jack Ewart Phillips
Phillips Petroleum Co.
214 Patent Library Bldg.
Bartlesville, OK 74004   (918) 661-0520

Allen W. Richmond
Phillips Petroleum Co.
202 Patent Library Bldg.
Bartlesville, OK 74004   (918) 661-0512

Archie Lew Robbins
Phillips Petroleum Co.
218 Patent Library Bldg.
Bartlesville, OK 74004   (918) 661-0546

William G. Roberts#
1912 Crestview
Bartlesville, OK 74003   (918) 336-3741

William R. Sharp
Phillips Petroleum Company
Phillips Research Center
228 Patent Library Bldg.
Bartlesville, OK 74004   (918) 661-0519

Lucas K. Shay#
Phillips Petroleum Co.
Phillips Research Center
295 Patent Library Bldg.
Bartlesville, OK 74005   (918) 661-0553

Ronald Brian Sherer
Phillips Petroleum Company
Phillips Research Center
208 Patent Library Bldg.
Bartlesville, OK 74004   (918) 661-0528

J. Michael Simpson
Brewer, Worten, Robinett,
Johnson, Worten & King
P.O. Box 1066
Bartlesville, OK 74005   (918) 336-4132

Charles W. Stewart
Phillips Petroleum Company
Phillips Research Center
R230 Patent Library Bldg.
Bartlesville, OK 74004   (918) 661-0533

Janelle D. Strode#
Phillips Petroleum Co.
285 Patent Library Bldg.
Bartlesville, OK 74004   (918) 661-6549

John Arthur Young
1525 Hillcrest Drive
Bartlesville, OK 74003   (918) 336-9249

C. Dean Domingue
Halliburton Services
1015 Bois D'Arc
Duncan, OK 73536   (401) 251-3641

James R. Duzan
Halliburton Services
1015 Bois D'Arc
Duncan, OK 73536   (405) 251-3487

Robert A. Kent
Halliburton Services
P.O. Box 1431
Duncan, OK 73536   (405) 251-3125

Thomas R. Weaver
P.O. Box 1405
Duncan, OK 73534   (405) 252-2710

Paul W. Hemminger
P.O. Box 1466
Eufaula, OK 74432   (918) 689-5383

James T. Robinson
303 South Peters
Norman, OK 73069   (405) 364-5444

William G. Addison
Kerr-Mc Gee Corp.
P.O. Box 25861
135 Robert S. Kerr Avenue
Oklahoma City, OK 73125   (405) 270-2821

David J. Alexander
Dunlap, Codding & Peterson
9400 N. Broadway
Suite 420
Oklahoma City, OK 73114   (405) 478-5344

Lucian W. Beavers
Laney, Dougherty, Hessin
& Beavers
2 Leadership Sq., Suite 1400
211 North Robinson
Oklahoma City, OK 73102   (405) 232-5586

Glen M. Burdick
Mc Carthy & Associates, Inc.
101 Park Avenue
Suite 250
Oklahoma City, OK 73102   (405) 232-5600

Charles Alan Codding
Dunlap, Codding & Peterson
9400 N. Broadway
Suite 420
Oklahoma City, OK 73114   (405) 478-5344

Clifford C. Dougherty, III
Laney, Dougherty, Hessin
& Beavers
2 Leadership Sq., Suite 1400
211 North Robinson
Oklahoma City, OK 73102   (405) 232-5586

Clifford Clark Dougherty, Jr.
Laney, Dougherty, Hessin
& Beavers
2 Leadership Sq., Suite 1400
211 North Robinson
Oklahoma City, OK 73102   (405) 232-5586

Jerry J. Dunlap
Dunlap, Codding & Peterson
9400 North Broadway
Suite 420
Oklahoma City, OK 73114   (405) 478-5344

Michael C. Felty
Fenton, Fenton, Smith,
Reneau & Moon
211 North Robinson
1 Leadership Square, Suite 800
Oklahoma City, OK 73102   (405) 235-4671

E. Harrison Gilbert, III
Laney, Dougherty, Hessin
& Beavers
2 Leadership Sq., Suite 1400
211 North Robinson
Oklahoma City, OK 73102   (405) 232-5586

Lynndon Michael Guard
Pate & Payne, P.C.
401 N. Hudson
Oklahoma City, OK 73102   (405) 235-4211

Herbert M. Hanegan
Kerr-Mc Gee Corporation
P.O. Box 25861
Oklahoma City, OK 73125   (405) 364-5444

Robert Marion Hessin
Laney, Dougherty, Hessin
& Beavers
2 Leadership Sq., Suite 1400
211 North Robinson
Oklahoma City, OK 73102   (405) 232-5586

Neal R. Kennedy
Laney, Dougherty, Hessin
& Beavers
2 Leadership Sq., Suite 1400
211 North Robinson
Oklahoma City, OK 73102   (405) 232-5586

William R. Laney
Laney, Dougherty, Hessin
& Beavers
2 Leadership Sq., Suite 1400
211 North Robinson
Oklahoma City, OK 73102   (405) 232-5586

Mary M. Lee
Dunlap, Codding & Peterson
9400 North Broadway
Suite 420
Oklahoma City, OK 73114   (405) 478-5344

Florence F. Mc Cann#
Laney, Dougherty, Hessin
& Beavers
2 Leadership Square
211 N. Robinson, Ste. 1400
Oklahoma City, OK 73102   (405) 232-5586

Billy D. Mc Carthy
101 Park Avenue
Suite 250
Oklahoma City, OK 73102   (405) 232-5600

Gary S. Peterson
Dunlap, Codding, Peterson & Lee
9400 N. Broadway
Suite 420
Oklahoma City, OK 73114   (405) 478-5344

Robert K. Rhea#
5350 S. Western, #305
Oklahoma City, OK 73109   (405) 634-1915

Ruel C. Terry#
2235 N.W. 55th St.
Oklahoma City, OK 73112   (405) 840-9586

John Teselle
U.S. Bankruptcy Judge
W.D. of Oklahoma
201 Dean A. Mcgee Ave.
Room 820
Oklahoma City, OK 73102   (405) 231-5925

Louis W. Watson
7 S.E. 78th Circle
Oklahoma City, OK 73149   (405) 634-9495

Roger R. Mc Fadden
P.O. Box 160
Okmulgee, OK 74447   (918) 756-3838

Ronald J. Carlson
Conoco Inc.
Patent & Licensing
P.O. Box 1267
1000 South Pine
Ponca City, OK 74603   (405) 767-2657

Robert B. Coleman, Jr.
2513 Mockingbird Lane
Ponca City, OK 74604   (405) 762-1371

Richard W. Collins
Conoco Inc.
P.O. Box 1267
Ponca City, OK 74603   (405) 767-4768

John E. Holder
Conoco Inc.
Patent & Licensing
P.O. Box 1267
4391 Rdw
Ponca City, OK 74603   (405) 767-2715

Henry H. Huth
1704 Monument Road
Ponca City, OK 74604   (405) 765-5169

William L. Martin, Sr.#
124 Elmwood Ave.
Ponca City, OK 74601   (405) 762-4140

William James Miller
P.O. Box 547
120 North 2nd
Ponca City, OK 74602   (405) 765-6697

A. Joe Reinert
Conoco, Inc.
P.O. Box 1267
1000 S. Pine
Ponca City, OK 74603   (405) 767-4724

Cortlan R. Schupbach, Jr.
Conoco, Inc.
P.O. Box 1267
Ponca City, OK 74603   (405) 767-3109

M. Kathryn Braquet Tsirigotis
Conoco, Inc.
P.O. Box 1267
Ponca City, OK 74603   (405) 767-5647

David W. Westphal
Conoco, Inc.
1000 South Pine Street
Ponca City, OK 74603   (405) 767-3254

Walter M. Benjamin
2620 North Boston Place
P.O. Box 6099
Tulsa, OK 74148   (918) 582-7257

Gary M. Bond
Amoco Corporation
P.O. Box 591
Tulsa, OK 74102   (918) 660-3625

Dennis D. Brown
Laney, Dougherty, Hessin
& Beavers
Williams Center Tower I, Ste. 1110
One West Third Street
Tulsa, OK 74103   (918) 592-6970

Frank J. Catalano
Head & Johnson, P.A.
Moore Manor
228 West 17th Place
Tulsa, OK 74136   (918) 584-4187

Robert R. Cochran
3532 East 71st Place
Tulsa, OK 74136   (918) 492-2505

William S. Dorman
Dorman & Kachigian, Inc.
1146 East 64th Street
Tulsa, OK 74136   (918) 747-1080

Mildred K. Flowers#
3913 E. 32nd Place
Tulsa, OK 74135   (918) 742-4316

John Dean Gassett
228 West 17th Place
Tulsa, OK 74119   (918) 584-4187

James R. Head
Head & Johnson, P.A.
228 W. 17th Place
Tulsa, OK 74119   (918) 584-4187

Fred E. Hook
Amoco Corporation
P.O. Box 591
Tulsa, OK 74102   (918) 660-3548

Paul H. Johnson
Head & Johnson
228 W. 17th Place
Tulsa, OK 74119   (918) 584-4187

Mark G. Kachigian
Head & Johnson
228 West 17th Place
Tulsa, OK 74119   (918) 587-2000

Stephen A. Littlefield
Dowell Schlumberger Inc.
5051 S. 129th East Avenue
Tulsa, OK 74134   (918) 250-4368

Marcy M. Lyles#
Amoco Corporation
4200 East Skelly Drive
P.O. Box 591
Suite 560
Tulsa, OK 74102   (918) 660-4410

Robert E. Massa
102 Brookcrest Square Off. Bldg.
1535 S. Memorial Dr.
Tulsa, OK 74112   (918) 664-2525

Serge Novovich
Nemorex Telex Corporation
P.O. Box 1526
Tulsa, OK 74101   (918) 627-2333

George L. Rushton#
OXY USA, Inc.
110 W. 7th St.
Tulsa, OK 74102   (918) 561-8540

Timothy D. Stanley
Amoco Corporation
P.O. Box 591
Tulsa, OK 74102   (918) 660-4403

Arthur L. Wade
15 W. 6th St.
1220 Fourth National Bank Bldg.
Tulsa, OK 74119   (918) 582-6151

Lawrence R. Watson
Laney, Dougherty, Hessin
& Beavers
Williams Center Tower One
One West 3rd St., Ste. 1110
Tulsa, OK 74103   (918) 592-6970

Oregon

William S. Lovell
17630 S.W. Butternut Drive
Aloha, OR 97007   (503) 642-4395

William K. Bucher#
Tektronix, Inc.
Pats., Tdmks. & Licensing
P.O. Box 500 50-Pat
Beaverton, OR 97077   (503) 627-7062

Francis I. Gray
Tektronix, Inc.
Pats., Tdmks. & Licensing
P.O. Box 500 50-Pat
Beaverton, OR 97077   (503) 627-7068

Robert S. Hulse
Tektronix, Inc.
Pats., Tdmks, & Licensing
P.O. Box 500 50-Pat
Beaverton, OR 97077   (503) 627-7074

Russell D. Mickiewicz
Tektronix, Inc.
P.O. Box 500
M/S 19-155
Beaverton, OR 97077   (503) 627-3455

Thomas J. Spence
Tektronix, Inc.
P.O. Box 500
Del. Sta. 50-Law
Beaverton, OR 97077   (503) 627-7016

John D. Winkelman
Tektronix, Inc.
Pats., Tdmks. & Licensing
P.O. Box 500 50-Pat
Beaverton, OR 97077   (503) 627-7050

Kenneth L. Brinich
Deschutes County Courthouse
Bend, OR 97701   (503) 388-6520

Dudley B. Smith
17328 La Bonte Lane
P.O. Box 938
Brookings, OR 97415   (503) 469-4595

Allegra J. Helfenstein
Molecular Probes, Inc.
4849 Pitchford Ave.
Eugene, OR 97402   (503) 344-3007

John F. Ingman
P.O. Box 5535
Eugene, OR 97405   (503) 342-8184

Thomas Edward Kristofferson
275 Foxtail Drive
Eugene, OR 97405   (503) 485-3635

Kenneth M. Durk#
2070 NE Laura Ct.
Hillsboro, OR 97123   (503) 648-4322

Sean P. Fitzgerald
Intel Corp.
Mail Stop: HF3-03
5200 N.E. Elam Young Pkwy.
Hillsboro, OR 97124   (503) 696-2833

Charles F. Roberts#
11105 Island Avenue
Island City, OR 97850   (503) 963-9765

Elmer W. Galbi
13314 Vermeer Drive
Lake Oswego, OR 97035   (503) 697-7844

Robert E. Howard
Entek Manufacturing Inc.
250 North Hansard Avenue
Lebanon, OR 97355   (503) 259-3901

Robert S. Thompson
P.O. Box 753
Mc Minnville, OR 97128   (503) 472-4721

Neil J. Driscoll
813 Mason Way Unit 5
Medford, OR 97501   (503) 772-7106

Henry George O Donohoe
13755 S.E. Maple Lane
Milwaukie, OR 97222   (503) 654-9675

William D. Haffner#
310 Walnut Drive
Monmouth, OR 97361   (503) 838-5219

Edward B. Watters#
18850 N.E. Ribbon Ridge Road
Box 2
Newberg, OR 97132   (503) 538-5687

Alan K. Aldous
Stoel Rives Boley
Jones & Grey
900 SW Fifth Avenue
Portland, OR 97204   (503) 224-9677

Paul S. Angello
Stoel, Rives, Boley,
Fraser & Wyse
900 SW Fifth Avenue
Suite 2300
Portland, OR 97204   (503) 294-9314

Mark L. Becker
Klarquist, Sparkman, Campbell,
Leigh & Whinston
1620 Willamette Center
121 S.W. Salmon St.
Portland, OR 97204   (503) 226-7391

Daniel J. Bedell
Dellett, Smith-Hill & Bedell
1425 One Main Place
101 S.W. Main
Portland, OR 97204   (503) 224-0115

William A. Birdwell
600 One Main Place
101 S.W. Main Street
Portland, OR 97204   (503) 228-1841

Glenn C. Brown#
Marger Johnson Mc Collom
& Stolowitz, Inc.
621 S.W. Morrison
Suite 650
Portland, OR 97205   (503) 222-3613

James Campbell, Jr.
Klarquist, Sparkman, Campbell,
Leigh & Whinston
121 S.W. Salmon Street
Suite 1620
Portland, OR 97204   (503) 226-7391

Daniel P. Chernoff
Chernoff, Vilhauer, Mc Clung
& Stenzel
600 Benjamin Franklin Plaza
One S.W. Columbia
Portland, OR 97258   (503) 227-5631

William Y. Conwell
Klarquist, Sparkman, Campbell,
Leigh, Whinston & Dellett
121 S.W. Salmon Street
Portland, OR 97204   (503) 226-7391

David P. Cooper
Kolisch Hartwell & Dickinson
200 Pacific Building
520 S.W. Yamhill St.
Portland, OR 97204   (503) 224-6655

Keith A. Cushing
1515 Building
1515 S.W. Fifth, Suite 1022
Portland, OR 97201   (503) 248-9149

Jack E. Day#
P.O. Box 25230
Portland, OR 97225   (503) 292-2113

John Philip Dellitt
Dellett, Smith-Hill & Bedell
101 S.W. Main
Portland, OR 97204   (503) 224-0115

Jon Macleod Dickinson
Kolisch, Hartwell & Dickinson
520 S.W. Yamhill Street
Portland, OR 97204   (503) 224-6655

Eugene M. Eckelman
Farley Bldg.
2400 S.W. 4th Avenue
Portland, OR 97201   (503) 222-1321

David A. Fanning
Kolisch, Hartwell & Dickinson
200 Pacific Bldg.
520 S.W. Yamhill Street
Portland, OR 97204   (503) 224-6655

Eugene D. Farley
Suite 100, Farley Bldg.
2400 S.W. 4th Ave.
Portland, OR 97201   (503) 222-1321

Douglas B. Ferguson
Chernoff, Vilhauer, Mc Clung
& Stenzel
1 S.W. Columbia, Suite 600
Benjamin Franklin Plaza
Portland, OR 97258   (503) 227-5631

Bruce J. Ffitch#
Suite 100, Farley Bldg.
2400 S.W. Fourth Avenue
Portland, OR 97201   (503) 222-1321

William O. Geny
Chernoff, Vilhauer, Mc Clung
& Stenzel
600 Benjamin Franklin Plaza
One S.W. Columbia
Portland, OR 97258   (503) 227-5631

James D. Givnan, Jr.#
209 Sylvan Westgate Bldg.
5319 S.W. Westgate Dr.
Portland, OR 97221   (503) 292-5758

Francine H. Gray
Kolisch, Hartwell & Dickinson
520 S.W. Yamhill Street
Portland, OR 97204   (503) 224-6655

Boulden G. Griffith
Dellett, Smith-Hill & Bedell
1070 One Main Place
101 S.W. Main
Portland, OR 97204   (503) 224-0115

Robert L. Harrington
1515 S.W. Fifth
Suite 1022
Portland, OR 97201   (503) 248-9149

Mortimer H. Hartwell, Jr.
Kolisch, Hartwell & Dickinson, P.C.
200 Pacific Bldg.
520 S.W. Yamhill
Portland, OR 97204   (503) 224-6655

Donald B. Haslett
Chernoff, Vilhauer, Mc Clung
& Stenzel
600 Benjamin Franklin Plaza
1 S.W. Columbia
Portland, OR 97258   (503) 227-5631

Peter E. Heuser
Kolisch, Hartwell & Dickinson
520 S.W. Yamhill Street
Suite 200
Portland, OR 97204   (503) 224-6655

Patrick W. Hughey
Stoel, Rives, Boley,
Jones & Grey
900 SW Fifth Avenue
Suite 2300
Portland, OR 97204   (503) 294-9222

James W. Jandacek
Stoel, Rives, Boley, Jones
& Grey
900 SW Fifth Ave., Ste. 2300
Portland, OR 97204   (503) 294-9119

Craig S. Jepson
Stoel, Rives, Boley,
Jones & Grey
900 SW Fifth Avenue
Portland, OR 97201   (213) 489-1600

Alexander C. Johnson, Jr.
Marger & Johnson, P.C.
621 S.W. Morrison Street
Suite 650
Portland, OR 97205   (503) 222-3613

Kenneth S. Klarquist
Klarquist, Sparkman, Campbell,
Leigh & Whinston
1 World Trade Center, Ste. 1600
121 S.W. Salmon St.
Portland, OR 97204   (503) 226-7391

Ramon A. Klitzke, III
Klarquist, Sparkman, Campbell,
Leigh & Whinston
121 S.W. Salmon
Portland, OR 97204   (503) 226-7391

J. Pierre Kolisch
Kolisch, Hartwell & Dickinson
520 SW Yamhill Street
200 Pacific Bldg.
Portland, OR 97204   (503) 224-6655

James S. Leigh
Klarquist, Sparkman, Campbell
Leigh & Whinston
121 S.W. Salmon Street
Suite 1600
Portland, OR 97204   (503) 226-7391

Michael L. Levine#
Stoel, Rives, Boley, Jones & Grey
900 SW Fifth Avenue
Standard Insurance Center
Suite 2300
Portland, OR 97204   (503) 220-2480

Timothy A. Long
Hyster Company
P.O. Box 2902
Portland, OR 97210   (503) 721-6063

Jay K. Malkin
Klarquist, Sparkman, Campbell,
Leigh & Whinston
121 S.W. Salmon Street
1 World Trade Center, Suite 1600
Portland, OR 97204   (503) 226-7391

Jerome S. Marger
621 S.W. Morrison Street
650 American Bank Bldg.
Portland, OR 97205   (503) 222-3613

Flory L. Martin
Klarquist, Sparkman, Campbell,
Leigh & Whinston
1 World Trade Center, Suite 1600
121 S.W. Salmon Street
Portland, OR 97204   (503) 226-7391

Charles D. Mc Clung
Chernoff, Vilhauer, Mc Clung
& Stenzel
600 Benjamin Franklin Plaza
Portland, OR 97258   (503) 227-5631

Alan T. Mc Collom
Marger & Johnson, P.C.
650 American Bank Bldg.
621 S.W. Morrison Street
Portland, OR 97205   (503) 222-3613

John M. Mc Cormack
Kolisch, Hartwell & Dickinson
520 S.W. Yamhill Street
Portland, OR 97204   (503) 224-6655

Mark M. Meininger
Stoel, Rives, Boley,
Jones & Grey
900 SW 5th Avenue
Suite 2300
Portland, OR 97204   (503) 294-9656

Erich W. Merrill, Jr.
Miller, Nash, Wiener,
Hager & Carlsen
111 S.W. Fifth Avenue
Portland, OR 97204   (503) 224-5858

Peter J. Meza
Marger, Johnson, Mc Collom
& Stolowitz, Inc.
Suite 650 American Bank Bldg.
621 S.W. Morrison St.
Portland, OR 97205   (503) 222-3613

William D. Noonan
Klarquist, Sparkman, Campbell,
Leigh & Whinston
1600 World Trade Center
121 S.W. Salmon Street
Portland, OR 97204   (503) 226-7391

Mark D. Olson
Olson & Olson
2400 S.W. 4th Avenue
Portland, OR 97201   (503) 222-1321

Oliver D. Olson
Olson & Olson
2400 S.W. 4th Avenue
Portland, OR 97201   (503) 222-1321

David P. Petersen
Klarquist, Sparkman, Campbell,
Leigh & Whinston
121 S.W. Salmon St.
Suite 1600
Portland, OR 97204   (503) 226-7391

Richard J. Polley
Klarquist, Sparkman, Campbell,
Leigh & Whinston
1 World Trade Center
121 S.W. Salmon, Ste. 1600
Portland, OR 97204   (503) 226-7391

Patrick J. Reynolds#
Oregon Health Sciences University
611 SW Campus Dr.
Portland, OR 97201   (503) 279-8958

David C. Ripma
Kolisch, Hartwell & Dickinson
520 S.W. Yamhill Street
Suite 200
Portland, OR 97204   (503) 224-6655

Howard P. Russell
Dellett, Smith-Hill
& Bedell
1070 One Main Pl.
101 S.W. Main
Portland, OR 97204   (503) 224-0115

Lee R. Schermerhorn
Farley Bldg., Suite 100
2400 S.W. 4th Avenue
Portland, OR 97201   (503) 222-1321

Carole Shlaes
9135 S.W. 35th Avenue
Portland, OR 97219   (503) 244-8837

John Smith - Hill
Dellett, Smith-Hill & Bedell
1070 One Main Place
101 S.W. Main
Portland, OR 97204   (503) 224-0115

Diane H. Sprunt
Klarquist, Sparkman, Campbell,
Leigh & Whinston
121 S.W. Salmon, Ste. 1600
Portland, OR 97204   (503) 226-7391

John P. Staples
Chernoff, Vilhauer, Mc Clung
& Stenzel
600 Benjamin Franklin Plaza
1 S.W. Columbia
Portland, OR 97258   (503) 227-5631

Dennis E. Stenzel
Chernoff, Vilhauer, Mc Clung
& Stenzel
600 Benjamin Franklin Plaza
1 S.W. Columbia
Portland, OR 97258   (503) 227-5631

Donald L. Stephens, Jr.
Klarquist, Sparkman, Campbell,
Leigh & Whinston
1 World Trade Center
121 SW Salmon St., Ste. 1600
Portland, OR 97204   (503) 226-7391

James G. Stewart
Kolisch, Hartwell & Dickinson
200 Pacific Building
520 S.W. Yamhill Street
Portland, OR 97204   (503) 224-6655

Micah D. Stolowitz
Marger & Johnson, Inc.
621 S.W. Morrison St.
Suite 650
Portland, OR 97205   (503) 222-3613

John W. Stuart
Blount, Inc.
4909 S.E. International Way
Portland, OR 97222   (503) 653-4335

John D. Vandenberg
Klarquist, Sparkman, Campbell,
Leigh & Whinston
121 S.W. Salmon, Suite 1600
Portland, OR 97204   (503) 226-7391

Robert D. Varitz
Kolisch, Hartwell, Dickinson,
Mc Cormack & Heuser, P.C.
200 Pacific Bldg.
520 S.W. Yamhill
Portland, OR 97204   (503) 224-6655

Jacob Ernest Vilhauer, Jr.
Chernoff, Vilhauer, Mc Clung
& Stenzel
600 Benjamin Franklin Plaza
1 S.W. Columbia
Portland, OR 97258   (503) 227-5631

Arthur Lewis Whinston
Klarquist, Sparkman, Campbell,
Leigh & Whinston
1 World Trade Center
121 S.W. Salmon St., Ste. 1600
Portland, OR 97204   (503) 226-7391

Edward B. Anderson
4079 Flournoy Valley Road
Roseburg, OR 97470   (503) 672-4719

Charles H. Hilke
Friel & Hilke
Suite 205 Equitable Ctr.
530 Center St., N.E.
Salem, OR 97301   (503) 362-1322

Kenneth John Isakson
P.O. Box 917
Wilsonville, OR 97070   (503) 685-9029

Pennsylvania

Arthur H. Swanson
1836 Harding Avenue
Abington, PA 19001   (215) 659-9388

Andrew Alexander
Aluminum Co. of America
Alcoa Center, PA 15069   (412) 337-2771

David W. Brownlee
Aluminum Co. of America
Patent Division
Alcoa Center, PA 15069   (412) 337-2773

Thomas J. Connelly
Aluminum Co. of America
Legal Dept.
Patent Division
Route 780
Alcoa Center, PA 15069   (412) 337-2759

Glenn E. Klepac
Aluminum Co. of America
Patent Div.
Alcoa Center, PA 15069   (412) 337-2770

William J. O'Rourke, Jr.
Aluminum Co. of America
Patent Division
Alcoa Center, PA 15069   (412) 337-2759

David W. Pearce-Smith
Aluminum Co. of America
Alcoa Center, PA 15068   (412) 337-2768

Elroy Strickland
Alcoa Tech. Corp.
Alcoa Tech. Center
Alcoa Center, PA 15069   (412) 337-2758

Gary P. Topolosky
Aluminum Co. of America
Alcoa Technical Center
Alcoa Center, PA 15069   (412) 337-2772

Nicholas F. Coates#
595 Golf Course Road
Aliquippa, PA 15001   (412) 375-5881

Russell Lee Brewer
Air Products & Chems., Inc.
P. O. Box 538
Allentown, PA 18105   (215) 481-7289

Geoffery L. Chase
Air Products & Chems., Inc.
P.O. Box 538
Allentown, PA 18105   (214) 481-7265

Richard Arthur Dannells, Jr.
Air Products & Chems., Inc.
P.O. Box 538
Allentown, PA 18105   (215) 481-8820

John M. Fernbacher#
Air Products & Chems., Inc.
Allentown, PA 18195   (215) 481-6560

James H. Fox
4465 Farm Drive
Allentown, PA 18104   (215) 439-8601

Keith D. Gourley
Air Products & Chemicals, Inc.
Pat. Dept.
Allentown, PA 18195   (215) 481-6566

E. Eugene Innis
Neuman, Williams, Anderson
& Olson
1929 Brookhaven Drive E.
Allentown, PA 18103   (215) 434-7915

Willard Jones, III
Air Products & Chems., Inc.
Allentown, PA 18195   (215) 481-4587

Lucie H. Laudenslager
1391 Springhouse Road
Allentown, PA 18104   (215) 398-0091

Richard D. Laumann
AT&T Bell Laboratories
555 Union Blvd.
Room 2a114
Allentown, PA 18103   (215) 439-7945

Michael Leach
Air Products & Chems., Inc.
Allentown, PA 18195   (215) 481-8519

William F. Marsh
Air Products & Chemicals, Inc.
Allentown, PA 18195   (215) 481-8660

Veronica O'Keefe
624 Ridge Avenue
Allentown, PA 18102   (215) 434-6900

Sanford J. Piltch
The Atrium - Suite 204
2895 Hamilton Blvd.
Allentown, PA 18104   (215) 433-6266

John T. Rehberg
AT&T Bell Laboratories
555 Union Blvd.
Room 2a114
Allentown, PA 18103   (215) 439-8601

Mark L. Rodgers
Air Products & Chemicals, Inc.
7201 Hamilton Blvd.
Allentown, PA 18195   (215) 481-8817

Thomas G. Ryder
Air Products & Chems., Inc.
7201 Hamilton Blvd.
Allentown, PA 18195   (215) 481-7851

James Charles Simmons
Air Products & Chems., Inc.
7201 Hamilton Blvd.
Allentown, PA 18195   (215) 481-8651

Seymour Traub
Traub, Butz & Fogerty, P.C.
1620 Pond Road
Suite 200
Allentown, PA 18104   (215) 395-1010

Michael J. Urbano
AT&T Bell Laboratories
555 Union Boulevard
Allentown, PA 18103   (215) 439-7925

Robert J. Wolff
Air Products & Chemicals, Inc.
Patent Group
7201 Hamilton Blvd.
Allentown, PA 18195   (215) 481-6150

E. Kears Pollock
2447 Trotter Drive
Allison Park, PA 15101   (412) 434-3242

George L. Kensinger
2803 Maple Avenue
2nd Floor
Altoona, PA 16601   (814) 944-1832

Gary W. Granzow
Henkel Corporation
300 Brookside Avenue
Ambler, PA 19002   (215) 628-1199

Gilbert W. Rudman
1408 Crystal Valley Drive
Ambler, PA 19002   (215) 646-8729

Max L. Williamson
643 Adams Court
Apollo, PA 15613   (412) 327-2786

James H. Rich, Jr.
122 Mill Creek Road
Ardmore, PA 19003   (215) 649-5238

John B. Sowell
182 Midfield Road
Ardmore, PA 19003   (215) 649-4815

Randolph J. Huis#
33 Bishops Drive
Aston, PA 19014   (215) 649-4815

Eugene Chovanes
Jackson & Chovanes
One Bala Plaza
Suite 319
Bala-Cynwyd, PA 19004   (215) 667-4392

Michael H. Minns
7504 Franks Drive
Bath, PA 18014   (908) 859-7725

Tracey G. Benson
Miller, Kistler & Campbell, Inc.
124 North Allegheny Street
Bellefonte, PA 16823   (814) 355-5474

Robert J. Mooney
Hosiery Corporation of
America, Inc.
3369 Progress Dr.
Bensalem, PA 19020   (215) 244-0997

Robert Clifford Nicander
575 Bair Road
Berwyn, PA 19312   (215) 644-2729

Francis A. Varallo#
241 Country Rd.
Berwyn, PA 19312   (215) 644-6066

Donald Stephen Ferito
6065 Murray Ave.
Bethel Park, PA 15102   (412) 835-2805

Nickolas C. Kotow
6390 Churchill Road
Bethel Park, PA 15102   (412) 831-7362

James L. Sherman
5860 Dashwood Drive
Bethel Park, PA 15102   (412) 831-8932

Anson W. Biggs#
329 Carver Dr.
Bethlehem, PA 18017   (215) 868-5920

George A. Heitczman
O'Hare & Heitczman
18 East Market Street
Bethlehem, PA 18018   (215) 691-5500

John I. Iverson
Bethlehem Steel Corp.
701 E. 3rd St.
Bethlehem, PA 18016   (215) 694-6401

John F. Lushis, Jr.
Bethlehem Steel Corp.
Martin Tower
Bethlehem, PA 18016   (215) 694-7312

Harold I. Masteller, Jr.#
Bethlehem Steel Corp.
Bethlehem, PA 18016   (215) 694-6415

Joseph J. O'Keefe
O'Keefe & Wilkinson
68 E. Broad Street
P.O. Box 1426
Bethlehem, PA 18016   (215) 867-9700

John S. Simitz#
2446 Greencrest Drive
Bethlehem, PA 18017   (215) 691-2167

Charles Alexander Wilkinson
O'Keefe & Wilkinson
68 E. Broad Street
P.O. Box 1426
Bethlehem, PA 18016   (215) 867-9700

Edward James Dwyer
3 Tally Ho Lane
Blue Bell, PA 19422   (215) 643-7288

Roger W. Robinson
Cyma Corp.
593 Skippack Pike
Blue Bell, PA 19422   (215) 542-8080

Anthony J. Rossi
ABB Power T&D Company Inc.
630 Sentry Park
Blue Bell, PA 19422   (215) 834-7420

Thomas J. Scott
Unisys Corporation
P.O. Box 500
Union Meeting
& Townshipline Rds.
Blue Bell, PA 19424   (215) 986-4116

Mark T. Starr
Unisys Corporation
Township Line & Union Meeting Rds.
P.O. Box 500
Msc1sw19
Blue Bell, PA 19424   (215) 986-4411

John Shaw Stevenson#
879 Crestline Dr.
Blue Bell, PA 19422   (215) 279-6274

Frank Joseph Vinci, Jr.
Unisys Corporation
P.O. Box 500
Blue Bell, PA 19424   (215) 542-4921

Wendy W. Koba
AT&T Bell Laboratories
Solid State Technology Center
Route 222
Breinigsville, PA 18031   (215) 391-2160

Scott W. Mc Lellan
AT&T Bell Laboratories
Solid State Technology Center
Route 222
Breinigsville, PA 18031   (215) 391-2161

J. Wesley. Haubner
2891 Gradyville Road
Broomall, PA 19008   (215) 353-4316

W. Melville Van Sciver
250 N. Central Blvd.
Broomall, PA 19008   (215) 356-7284

Robert S. Barton
714 Old Lancaster Rd.
Bryn Mawr, PA 19010   (215) 525-3784

George J. Harding, 3rd
921 Mt. Pleasant Road
Bryn Mawr, PA 19010   (215) 525-2039

George C. Atwell
Atwell & Morrow
P.O. Box 829
421 N. Main St.
Butler, PA 16003   (412) 283-9333

William Hintze
228 Allendale Way
Camp Hill, PA 17011   (717) 761-1316

William J. Keating
Dickinson School of Law
150 South College Street
Carlisle, PA 17013   (717) 243-4611

Arthur A. Murphy
Dickinson School of Law
S. College St.
Carlisle, PA 17013   (717) 243-4611

Francis K. Richwine
1213 Stratford Drive
Carlisle, PA 17013   (717) 245-2810

Arba G. Williamson, Jr.#
40 Swallow Hill Road
Carnegie, PA 15106   (412) 279-5036

Joseph C. Harkins, Jr.
1 Hunters Lane
Chadds Ford, PA 19317   (215) 558-1177

Edward J. Newitt
12 Ringfield
Chadds Ford, PA 19317   (215) 388-7190

Rene A. Kuypers
Molasky & Associates
Chalfont Centre
4 Limekiln Pike
Chalfont, PA 18914   (215) 822-3324

Joseph W. Molasky
4 S. Limekiln Pike
Chalfont, PA 18914   (215) 822-3324

Edward Franklin Possessky
Westinghouse Electric
Science & Tech. Center, Law Dept.
Churchill, PA 15235   (412) 256-5249

Walter G. Sutcliff
Westinghouse Elec. Corp.
Science & Tech. Center
1310 Beulah Rd.
Bldg. 801 - Rm. 4c65
Churchill, PA 15235   (412) 256-5242

Donald M. Mac Kay
2604 Leona Lane
Coraopolis, PA 15108   (412) 262-4999

Howard G. Massung
412 Fifth Avenue
Coraopolis, PA 15108   (412) 262-9199

Thomas R. Shaffer
Glassmire & Shaffer, P.C.
5 East Third Street
P.O. Box 509
Coudersport, PA 16915   (814) 274-7292

William E. Denk#
81 Steeplechase Rd.
Devon, PA 19333   (215) 688-2976

Edward J. Feeney, Jr.
151 Steeplechase Road
Devon, PA 19333   (215) 687-6583

Walter Scott Johns, III
P.O. Box 58
573 Tory Hill Road
Devon, PA 19333   (215) 688-5426

Gregory J. Gore
70 W. Oakland Avenue
Suite 316
Doylestown, PA 18901   (215) 348-1442

Armand M. Vozzo, Jr.
350 South Main Street
Suite 117
Doylestown, PA 18901   (215) 348-4733

George E. Bodenstein#
3255 Pebblewood Ln.
Dresher, PA 19025   (215) 576-0189

Frank J. Earnheart
Selas Corp. of America
2034 Limekiln Pike
P.O. Box 200
Dresher, PA 19025   (215) 283-8368

Charles E. Bartsch#
1001 Foss Ave.
Drexel Hill, PA 19026   (215) 589-5188

Robert J. Mc Donnell
Mc Donnell & Mc Donnell P.A.
Township Line Road
Drexel Hill, PA 19026   (215) 446-3290

Herbert L. Jones
P.O. Box 723
Easton, PA 18044   (215) 559-0871

Nathan D. Field#
373 Linden Drive
Elkins Park, PA 19117   (215) 887-2402

Aileen C. Addessi
604 East Line Ave.
Ellwood City, PA 16117   (412) 758-5859

John M. Fray
603 Crescent Avenue
Ellwood City, PA 16117   (412) 752-3729

Edward W. Goebel, Jr.
Mac Donald, Illig, Jones & Britton
600 1st Natl. Bank Bldg.
Erie, PA 16501   (814) 453-7611

William D. Gregory#
Gannon University
Dean of Science & Eng.
University Square
Erie, PA 16541   (814) 871-7618

Ralph Hammar
103 W. 10th St.
Erie, PA 16501   (814) 452-3494

Charles Lester Lovercheck
Lovercheck & Lovercheck
931 State Street
Erie, PA 16501   (814) 454-5218

Wayne L. Lovercheck
Lovercheck & Lovercheck
931 State St.
Erie, PA 16501   (814) 454-5218

Robert T. Payne
General Electric Co.
2901 East Lake Road
Erie, PA 16531   (814) 875-2724

Albert S. Richardson, Jr.
830 Washington Place
Erie, PA 16502   (814) 455-9730

Richard K. Thomson
Lord Corporation
2000 W. Grandview Blvd.
P.O. Box 10038
Erie, PA 16514   (814) 868-0924

Kenneth W. Wargo
Quinn, Gent, Buseck &
Leemhuis, Inc.
2222 W. Grandview Blvd.
Erie, PA 16506   (814) 833-2222

James W. Wright
Lord Corp.
2000 West Grandview Blvd.
Erie, PA 16514   (814) 868-0924

Heinrich Goretzky#
2913 Seneca Ct.
Export, PA 15632   (412) 327-5362

Donald Raymond Lackey
5035 Sequoia Dr.
Export, PA 15632   (412) 327-1196

Albert G. Marriott
5817 Terrance Drive
Export, PA 15632   (412) 327-1539

Clement L. Mc Hale
2915 Seneca Court
Export, PA 15632   (412) 327-3799

Donald J. Smith*
R.D. 3
Export, PA 15632   (412) 327-0077

Paul T. Teacher
5819 Washington Ave.
P.O. Box 487
Export, PA 15632   (412) 327-3891

Donald D. Joye#
226 Llandovery Drive
Exton, PA 19341   (215) 524-0296

Paul Lipsitz
205 Suffolk Rd.
Flourtown, PA 19031   (215) 233-4620

Frank A. Wolfe#
930 Fifth Avenue
Ford City, PA 16226   (412) 763-7125

Lawrence Thomas Fleming#
7 Mountain Laurel Drive
Forest City, PA 18421   (717) 785-5929

Imre Balogh
Rorer Group Inc.
500 Virginia Drive
Fort Washington, PA 19034   (215) 962-3309

Paul R. Darkes
Rhone-Poulenc Rorer Inc.
500 Virginia Drive, #3A
Fort Washington, PA 19034   (215) 962-4072

James A. Nicholson
Rorer Group, Inc.
500 Virginia Dr.
No. 2
Fort Washington, PA 19034   (215) 962-3310

Martin F. Savitzky
Rorer Group, Inc.
500 Virginia Drive
Fort Washington, PA 19034   (215) 628-6888

Albert F. Maier, Jr.
Maier & Maier
820 Main St.
Freeland, PA 18224   (717) 636-1140

John F. Daniels, III
101 Bruner Ave.
Glenolden, PA 19036   (215) 532-7558

Harrison H. Young, Jr.
529 Custis Road
Glenside, PA 19038   (215) 572-6362

Jon M. Lewis
205 Coulter Bldg.
Greensburg, PA 15601   (412) 836-4730

Nils H. Ljungman, Jr.
229 South Maple Avenue
P.O. Box 130
Greensburg, PA 15601   (412) 836-2305

Thomas N. Ljungman
229 S. Maple Avenue
P.O. Box 130
Greensburg, PA 15601   (412) 836-2305

Robert A. Shack
229 South Maple Avenue
Greensburg, PA 15601   (412) 836-2305

Ernest G. Szoke
Henkel Corp.
Triad .
2200 Renaissance Blvd.
Suite 200
Gulph Mills, PA 19406   (215) 270-8124

Annette M. Tobia
British Technology Group
2200 Renaissance Blvd.
Renaissance Bus. Park
Gulph Mills, PA 19406   (215) 278-1660

Leonard R. Hecker#
212 Tanglewood Way
Harleysville, PA 19438   (215) 628-5528

Philip D. Freedman
1101 North Front Street
Harrisburg, PA 17102   (717) 232-7000

Thomas Hooker
101 N. Front St.
Harrisburg, PA 17101   (717) 232-8771

Gerald K. Kita
AMP, Inc.
Mail Stop 140-62
P.O. Box 3608
Harrisburg, PA 17105   (717) 986-5464

Adrian John Larue
AMP, Inc.
Pat. Div. 140-62
P.O. Box 3608
Harrisburg, PA 17105   (717) 986-5875

Lisa Mumma Morgan
Penny Supply, Inc.
1001 Paxton Street
P.O. Box 3331
Harrisburg, PA 17105   (717) 233-4511

Katherine A.O. Nelson
AMP, Inc.
470 Friendship Road
P.O. Box 3608
Harrisburg, PA 17105   (717) 986-5470

Anton P. Ness
AMP, Inc.
Pat. Div. (140-62)
2901 Fulling Mill Road
Harrisburg, PA 17105   (717) 986-5477

William B. Noll, Sr.
AMP, Inc.
Mail Stop 140-62
P.O. Box 3608
Harrisburg, PA 17105   (717) 986-5468

Allan B. Osborne
AMP, Inc.
P.O. Box 3608
Harrisburg, PA 17105   (717) 986-5459

Gerald Post#
4108 Beechwood Lane
Harrisburg, PA 17112   (717) 652-3960

Frederick W. Raring#
4305 Fritchey Street
Harrisburg, PA 17109   (717) 653-8309

Jay Louis Seitchik
AMP, Inc.
P.O. Box 3608
Harrisburg, PA 17105   (717) 986-5461

David L. Smith
AMP, Inc.
P.O. Box 3608
Harrisburg, PA 17105   (717) 986-5458

James M. Trygg
AMP, Inc.
P.O. Box 3608
Harrisburg, PA 17105   (717) 986-5472

Bruce J. Wolstoncroft
AMP, Inc.
P.O. Box 3608
Mail Stop 140-062
Harrisburg, PA 17105   (717) 986-5471

Gerald D. Ames#
113 Franklin Ave.
Hatboro, PA 19040   (215) 672-4348

Jeffrey K. Rucker#
908 Edgewood Road
Havertown, PA 19083   (215) 853-1467

Daniel L. De Joseph
Ninth Floor
Northeastern Bldg.
Hazleton, PA 18201   (717) 455-6308

Bernard J. Burns#
2100 Winthrop Road
Huntingdon Valley, PA 19006
                              (215) 947-1518

Howard I. Forman
Albidale-Windmill Circle
P.O. Box 66
Huntingdon Valley, PA 19006
                              (215) 947-4154

Harold K. Hauger
Norwin Medical Center
28 Fairwood Drive
Irwin, PA 15642   (412) 864-6050

William K. Cox
Elliott Turbomachinery Co., Inc.
North Fourth Street
Jeannette, PA 15644   (412) 527-8720

William H. Eilberg
820 Homestead Road
Jenkintown, PA 19046   (215) 885-4600

Michael L. Ozalas
Ozalas & Mc Kinley
41 Broadway
Jim Thorpe, PA 18229   (717) 325-3616

M. Delcina M. Esser#
415 Greenwood Drive
Kennett Square, PA 19348   (215) 388-6526

Roger R. Horton
82 Kendal
Kennett Square, PA 19348   (215) 388-2077

Charles G. Rudershausen#
109 Taylor Lane
Kennett Square, PA 19348   (215) 444-3773

Robert L. Siegel#
Du Pont Company
811 William Thomas Drive
Kennett Square, PA 19348   (215) 347-1874

Michael F. Beausang, Jr.
Butera, Beausang, Moyer
& Cohen
700 Valley Forge Plaza
King of Prussia, PA 19406   (215) 265-0800

Stanley Bilker
2000 Valley Forge Circle
Apt. 927
King of Prussia, PA 19406   (215) 783-2678

Linda E. Hall
Smith, Kline, Beecham Corp.
Mail Code UW 2220
P.O. Box 1539
King of Prussia, PA 19406   (215) 270-5016

Herbert H. Jervis
Smith, Kline, Beecham
Pharmaceuticals
709 Swedeland Road
P.O. Box 1539
King of Prussia, PA 19406   (215) 270-5019

James M. Kanagy
Smith, Kline, Beecham Corp.
Corporate Patents-U.S., UW 2220
P.O. Box 1539
King of Prussia, PA 19406   (215) 270-5014

Benjamin E. Leace
Ratner & Prestia
500 N. Gulph Road
Suite 412
King of Prussia, PA 19406   (215) 256-6666

John J. Mc Aleese, Jr.
Mc Aleese, Mc Goldrick
& Susanin, P.C.
Suite 240 - Executive Terrace
455 S. Gulph Road
King of Prussia, PA 19406   (215) 337-4510

Henry J. Policinski
General Electric Co.
Pats. & Assigned Components
Suite 300
150 South Warner Road
King of Prussia, PA 19406   (215) 964-7655

Stephen Venetianer
Smith, Kline, Beecham Corp.
709 Swedeland Road
King of Prussia, PA 19406   (215) 270-5040

Janice E. Williams
Smith, Kline, Beecham Corp.
Mail Code UW 2220
P.O. Box 1539
King of Prussia, PA 19406   (215) 270-5012

Thomas J. Edgington,#
140 Westmont Drive
Kittanning, PA 16201   (412) 545-6922

John S. Stephen Bobb#
PQ Corporation
P.O. Box 258
Lafayette Hill, PA 19444   (215) 825-5000

James R. Bell
136 Crosswick Lane
Lancaster, PA 17601   (717) 397-6698

Vincent J. Coughlin, Jr.
379 Buch Avenue
Lancaster, PA 17601   (717) 560-9194

Robin M. Davis
Armstrong
Patent Dept.
150 North Queen Street
Lancaster, PA 17604   (717) 396-4122

Martin Fruitman
311 E. Orange Street
Lancaster, PA 17602   (717) 397-2314

Barry E. Haverstick
2916 Columbia Ave.
Lancaster, PA 17603   (717) 392-0446

George L. Herr
1340 Hunter Dr.
Lancaster, PA 17601   (717) 393-1195

Darrell F. Marquette
595 Laurel Lane
Lancaster, PA 17601   (717) 569-6320

Clifford B. Price, Jr.
43 Tennyson Dr.
Lancaster, PA 17602   (717) 397-6728

Theodore L. Thomas
Blakenger, Byler & Thomas P.C.
28 Penn Square
Lancaster, PA 17603   (717) 299-1100

Douglas E. Winters
Armstrong World Industries, Inc.
P.O. Box 3001
Lancaster, PA 17604   (717) 396-4070

Sylvia A. Gosztonyi
R.D. 2, Box 48A
Landenberg, PA 19350   (215) 274-8829

Christopher Egolf
Ecogen Incorporated
2005 Cabot Blvd. West
Langhorne, PA 19047   (215) 757-1590

Robert G. Danehower
17 Church Road
Lansdale, PA 19446   (215) 855-5924

Frank E. Manson
857 Sunnylea Road
Lansdale, PA 19446   (215) 855-1842

Lawrence R. Burns
16 Romar Avenue
Latrobe, PA 15650   (412) 539-2271

Larry R. Meenan
Kennametal, Inc.
P.O. Box 231
Latrobe, PA 15650   (412) 539-5485

James G. Porcelli
Kennametal, Inc.
P.O. Box 231
Latrobe, PA 15650   (412) 539-5838

John J. Prizzi
Kennametal, Inc.
P.O. Box 231
Latrobe, PA 15650   (412) 539-5331

Thomas R. Trempus
Kennametal, Inc.
P.O. Box 231
Latrobe, PA 15650   (412) 539-5337

John J. Selko
P.O. Box 2071
Lehigh Valley, PA 18001   (215) 264-9775

John S. Friderichs
Heatex of Philadelphia
100 Stevens
Suite 290
Lester, PA 19113   (215) 595-1052

Murray J. Ellman
39 Four Leaf Road
Levittown, PA 19056   (215) 946-4083

Florence U. Reynolds#
60 Yellowood Drive
Levittown, PA 19057   (215) 945-3336

Alan N. Mc Cartney
Consolidation Coal Company
4000 Brownsville Road
Library, PA 15129   (412) 854-6631

William E. Meason#
412 Franklin Street Extension
Ligonier, PA 15658   (412) 854-6631

James Irwin
3058 Leechburg Rd.
Suites 10 & 11
Lower Burrell, PA 15068   (412) 339-2225

Alfred R. Brady
Ecolaire Inc.
Two Country View Rd.
Malvern, PA 19355   (215) 648-8630

William J. Davis
Sterling Drug Inc.
9 Great Valley Parkway
Malvern, PA 19355   (215) 889-6612

George D. Hobbs, II
Centocor
244 Grant Valley Parkway
Malvern, PA 19355   (215) 296-4488

Irving Newman
Sterling Drug Inc.
9 Great Valley Parkway
Malvern, PA 19355   (215) 640-8824

Marshall M. Truex
117 Davis Road
R.D. #2
Malvern, PA 19355   (215) 647-0167

Donald L. Rose
P.O. Box 825
Mars, PA 16046   (412) 625-1917

William A. Behare
576 River Ridge Drive
Liberty Manor
Mc Keesport, PA 15133   (412) 672-8083

Nicholas A. Vonneuman
1396 Lindsay Lane
P.O. Box 3097
Meadowbrook, PA 19046   (215) 886-6244

Stuart S. Bowie
Box 55
23 West Second Street
Media, PA 19063   (215) 565-9160

Gerry J. Elman
Elman & Wilf
20 West Third Street
P.O. Box 703
Media, PA 19063   (215) 892-9580

Robert B. Famiglio
201 N. Jackson St.
P.O. Box 546
Media, PA 19063   (215) 565-4730

Han-Jolyon Lammers
431 Kirk Lane
Media, PA 19063   (215) 566-2528

Kenneth P. Lauria
60/40 Ware, Inc.
11 Blackhorse Lane
Media, PA 19063   (215) 891-1786

Robert S. Lipton
Lipton, Famiglio & Elman
201 N. Jackson Street
P.O. Box 546
Media, PA 19063   (215) 565-4730

Delbert E. Mc Caslin
12 Wyncroft Drive
Media, PA 19063   (215) 566-1964

Anthony J. Mc Nulty
115 N. Monroe St.
P.O. Box 605
Media, PA 19063   (215) 565-6700

Anthony Potts, Jr.
6 Well Fleet Dr.
Media, PA 19063   (215) 566-8587

Eugene E. Renz, Jr.
205 N. Monroe St.
Media, PA 19063   (215) 565-6090

Ashok Tankha
Elman & Wilf
20 W. 3rd St.
P.O. Box 703
Media, PA 19063   (215) 892-9580

Warren B. Gilbert#
6527 Edwards Drive
Mercersburg, PA 17236   (717) 328-3270

John A. Carroll#
Milford Town Green
Route 209
H.C. 395-23
Milford, PA 18337   (717) 328-3270

Oscar B. Brumback
1340 Towerlawn Drive
Monroeville, PA 15146   (412) 372-1162

Hymen Diamond
4409 Ruth Dr.
Monroeville, PA 15146   (412) 372-3555

William David Palmer
2317 Haymaker Road
Monroeville, PA 15146   (412) 373-0717

Mitchell John Halista
108 Lions Drive
Morrisville, PA 19067   (412) 373-0717

Philip W. Humer#
8 Glenwood South Gate
Morrisville, PA 19067   (215) 295-0479

Alphonso Henry Caser
20 S. Maple St.
Mount Carmel, PA 17851   (717) 339-3464

Daniel S. Buleza#
4412 West Run Road
Munhall, PA 15120   (412) 462-1452

Ronald S. Lombard
4430 Twin Oaks Lane
Murrysville, PA 15668   (412) 733-1999

David G. Maire
3908 Bridgewood Drive
Murrysville, PA 15668   (412) 327-7982

Elliott V. Nagle
3404 Oakdale Dr.
Murrysville, PA 15668   (412) 825-3448

Anthony J. Santantonio
5053 Northlawn Drive
Murrysville, PA 15668   (412) 327-1015

Michael J. Delaney
510 E. Main Street
Nanticoke, PA 18634   (717) 735-3950

Daniel David Dubosky
Box 128 Great Oaks Dr.
RR #1
Nesquehoning, PA 18240   (717) 645-3593

Stephen Ross Green
Gamble, Verterano, Mojock,
Piccione & Green
Suite 500
First Federal Plaza
New Castle, PA 16101   (412) 658-2000

Joseph A. Brown
New Holland, Inc.
500 Diller Avenue
New Holland, PA 17557   (717) 354-1439

James J. Kennedy
Ford New Holland, Inc.
500 Diller Avenue
New Holland, PA 17557   (717) 354-1439

Larry W. Miller
Ford New Holland, Inc.
500 Diller Avenue
New Holland, PA 17557   (717) 354-1353

John B. Mitchell
Ford New Holland, Inc.
500 Diller Avenue
New Holland, PA 17557   (717) 354-1447

Frank Allyn Seemar
Ford New Holland, Inc.
500 Diller Avenue
M.S. 641, P.O. Box 1895
New Holland, PA 17557   (717) 355-1341

Adolfo Arturo Mangieri
1170 Seventh St.
New Kensington, PA 15068   (412) 337-7795

Daniel A. Sullivan, Jr.
150 Chaney Court
New Kensington, PA 15068   (412) 335-8121

Fu Chen#
19 Meadow View Dr.
Newtown, PA 18940   (215) 355-3300

James D. Dee
SPS Technologies, Inc.
P.O. Box 1000
Newtown-Yardley Road
Newtown, PA 18901   (215) 860-3072

Timothy J. Gaul
11 Redwood Drive
Newtown, PA 18940   (215) 860-3072

Aaron Nerenberg
SPS Technologies, Inc.
Route 332
Newtown, PA 18940   (215) 860-3044

John J. Simkanich
Simkanich & Green
P.O. Box 671
Newtown, PA 18940   (215) 860-2040

Allen E. Amgott
315 Earls Lane
Newtown Square, PA 19073   (215) 353-5857

Joseph M. Corr
63 Charter Oak Drive
Newtown Square, PA 19073   (215) 353-9040

Stephen D. Harper#
Arco Chemical Co.
3801 W. Chester Pike
Newtown Square, PA 19073   (215) 359-2111

Michael S. Jarosz
Arco Chemical Co.
Legal Dept.
3801 W. Chester Pike
Newtown Square, PA 19073   (215) 359-2106

Dennis M. Kozak
Arco Chemical Co.
3801 W. Chester Pike
Newtown Square, PA 19073   (215) 359-2067

John Critchley Martin, Jr.
Arco Chemical Co.
3801 W. Chester Pike
Newtown Square, PA 19073   (215) 359-2109

Jonathan L. Schuchardt#
Arco Chemical Co.
3801 W. Chester Pike
Newtown Square, PA 19073   (215) 359-2276

Robert De Majistre
Seton Company
2500 Monroe Blvd.
Norristown, PA 19403   (215) 666-9600

Nicholas Montalto
Mc Brien & Montalto
514 Swede Street
Norristown, PA 19401   (215) 272-5300

John Bernard Sotak#
12191 Church Dr.
North Huntingdon, PA 15642   (412) 863-5277

Albert C. Martin
Leeds & Northrup Co.
Sumneytown Pike
North Wales, PA 19454   (215) 699-2000

William Henry Deitch
666 10th St.
Oakmont, PA 15139   (412) 828-8547

William G. Miller, Jr.
882 Hauck Road
P.O. Box 46
Perkiomenville, PA 18074   (215) 234-0571

Michael P. Abbott
Panitch,Schwarze,Jacobs
& Nadel
5 Penn Center Plaza, 36th Fl.
1601 Market Street
Philadelphia, PA 19103   (215) 567-2020

Marc S. Adler
Rohm & Haas Company
Independence Mall West
Philadelphia, PA 19105   (215) 592-3416

Robert L. Andersen
FMC Corporation
2000 Market Street
Philadelphia, PA 19103   (215) 299-6967

Raymond G. Arner
Schnader, Harrrison, Segal
& Lewis
1600 Market Street
Suite 3600
Philadelphia, PA 19103   (215) 751-2350

Stanford M. Back
2221 Panama St.
Philadelphia, PA 19103   (215) 732-3497

Patrick C. Baker, II
FMC Corporation
2000 Market Street
Philadelphia, PA 19103   (215) 299-6977

Alexis Barron
Synnestvedt & Lechner
2600 One Reading Center
1101 Market Street
Philadelphia, PA 19107   (215) 923-4466

Carl W. Battle
Rohm & Haas Company
Patent Dept. 7th Floor
Independence Mall West
Philadelphia, PA 19105   (215) 592-3052

Kirk Baumeister#
Panitch, Schwarze, Jacobs & Nadel
5 Penn Center Plaza, 36th Floor
1601 Market Street
Philadelphia, PA 19103   (215) 567-2020

Robert C. Beam
Paul & Paul
2900 Two Thousand Market Street
Philadelphia, PA 19103   (215) 568-4900

Stuart E. Beck
Weinstein, Trachtman, Beck
& Kimmelman
2410 Two Mellon Bank Center
Suite 2410
Philadelphia, PA 19102   (212) 569-9800

Martin G. Belisario#
Panitch, Schwarze, Jacobs
& Nadel
5 Penn Center Plaza, 36th Fl.
1601 Market Street
Philadelphia, PA 19103   (215) 567-2020

Philip E. Berens
Lewis Tower Bldg., Ste. 1401
Philadelphia, PA 19102   (215) 735-2425

Alan H. Bernstein
Caesar, Rivise, Bernstein,
Cohen & Pokotilow, Ltd.
7 Seven Penn Center Plaza, 12th Fl.
1635 Market Street
Philadelphia, PA 19103   (215) 567-2010

Thomas Joseph Bird, Jr.
General Electric Co.
Valley Forge Space Center
P.O. Box 8555
Philadelphia, PA 19101   (215) 354-4940

Mark G. Bocchetti
Scott Paper Co.
Scott Plaza II
Philadelphia, PA 19113   (215) 522-5804

Stephen J. Bor
Master, Donsky, Soffian
& Allen
230 S. Broad Street
Philadelphia, PA 19102   (215) 546-9800

Robert Sherman Bramson
Schnader, Harrison, Segal
& Lewis
Suite 3600
1600 Market Street
Philadelphia, PA 19103   (215) 751-2066

Royal E. Bright
Atochem North Amer., Inc.
Patent Dept.
Three Parkway
Philadelphia, PA 19102   (215) 587-7000

James R. Burdett
Woodcock, Washburn, Kurtz,
Mackiewicz & Norris
1 Liberty Place, 46th Fl.
Philadelphia, PA 19103   (215) 568-3100

Jean A. Buttmi
128 East Chestnut Hill Avenue
Philadelphia, PA 19118   (215) 247-6422

Abraham D. Caesar
Caesar, Rivise, Bernstein
Cohen & Pokotilow, Ltd.
21 S. 12th St.
Suite 800
Philadelphia, PA 19107   (215) 567-2010

John W. Caldwell
Woodcock, Washburn, Kurtz
Mackiewicz & Norris
1 Liberty Place, 46th Fl.
Philadelphia, PA 19103   (215) 568-3100

Carol Grobman Canter
Smith, Kline, Beckman Corp.
Corporate Pats. & Tdmks. N-160
1 Franklin Plaza
P.O. Box 7929
Philadelphia, PA 19101   (215) 751-6148

Charles Mclean Carter
Reading Company
1101 Market Street
Philadelphia, PA 19107   (215) 922-3303

John S. Child, Jr.
Dann, Dorfman, Herrell
& Skillman
3 Mellon Bank Center, No. 900
15th St. & S. Penn Square
Philadelphia, PA 19102   (215) 563-4100

Scott J. Childress#
2202 Hopkinson House
Philadelphia, PA 19106   (215) 627-3471

Joseph E. Chovanes
Paul & Paul
2900 Two Thousand Market St.
Philadelphia, PA 19103   (215) 568-4900

Thomas D. Christenbury
Miller & Quinn
1125 Land Title Bldg.
Broad & Chestnut Streets
Philadelphia, PA 19110   (215) 563-1810

Donald S. Cohen
Woodcock, Washburn, Kurtz,
Mackiewicz & Norris
1 Liberty Place, 46th Fl.
Philadelphia, PA 19103   (215) 568-3100

Gary M. Cohen
Weiser & Stapler
230 S. 15th Street
Suite 500
Philadelphia, PA 19102   (215) 875-8383

Lester H. Cohen
613 Avon Street
Philadelphia, PA 19116   (215) 676-5078

Stanley H. Cohen
Caesar, Rivise, Bernstein,
Cohen & Pokotilow, Ltd.
7 Penn Center Plaza, 12th Fl.
1635 Market Street
Philadelphia, PA 19103   (215) 567-2010

Norman L. Craig
FMC Corporation
2000 Market Street
Philadelphia, PA 19103   (215) 299-6934

Peter J. Cronk
Synnestvedt & Lechner
2600 One Reading Center
1101 Market Street
Philadelphia, PA 19107   (215) 923-4466

C. Marshall Dann
Dann, Dorfman, Herrell
& Skillman, P.C.
Three Mellon Bk. Ctr. - Suite 900
15th Street & South Penn Sq.
Philadelphia, PA 19102   (215) 563-4100

Q. Todd Dickinson
Sun Refining & Marketing Co.
1801 Market Street
Philadelphia, PA 19103   (215) 977-3142

Dara L. Dinner
Smith, Kline, Beckman Corp.
1 Franklin Plaza
P.O. Box 7929
Philadelphia, PA 19101   (215) 751-5182

John P. Donohue, Jr.
Woodcock, Washburn, Kurtz,
Mackiewicz & Norris
1 Liberty Place, 46th Fl.
Philadelphia, PA 19103   (215) 568-3100

John C. Dorfman
Dann, Dorfman, Herrell
& Skillman, P.C.
Three Mellon Bk. Ctr. - Suite 900
15th Street & South Penn Square
Philadelphia, PA 19102   (215) 563-4100

Jordan Joseph Driks
Rohm & Haas Co.
Independence Mall W.
6th & Market Sts.
Philadelphia, PA 19105   (215) 592-2478

James Albert Drobile
Schnader, Harrison, Segal
& Lewis
1600 Market Street
Suite 3600
Philadelphia, PA 19103   (215) 751-2242

Michael P. Dunnam
Woodcock, Washburn, Kurtz,
Mackiewicz & Norris
1 Liberty Place, 46th Fl.
Philadelphia, PA 19103   (215) 568-3100

Thomas J. Durling
Seidel, Gonda, Goldhammer
& Abbott P.C.
1800 Two Penn Center
Philadelphia, PA 19102   (215) 568-8383

Wayne J. Dustman
Smith, Kline, Beckman Corp.
P.O. Box 7929
One Franklin Plaza, N-160
Philadelphia, PA 19101   (215) 751-5125

Dianne B. Elderkin
Woodcock, Washburn, Kurtz,
Mackiewicz & Norris
1 Liberty Place, 46th Fl.
Philadelphia, PA 19103   (215) 568-3100

William H. Elliott, Jr.
Synnestvedt & Lechner
2600 One Reading Center
1101 Market Street
Philadelphia, PA 19107   (215) 823-4466

Susan B. Evans
Jackson & Evans
2043 Walnut Street
Philadelphia, PA 19103   (215) 557-8101

Martin L. Faigus
Caesar, Rivise, Bernstein,
Cohen & Pokotilow, Ltd.
12th Floor - Seven Penn Center
1635 Market Street
Philadelphia, PA 19103   (215) 567-2010

Michael B. Fein
Rohm and Haas Company
Independence Mall West
Philadelphia, PA 19105   (215) 592-3595

Charles C. Fellows
FMC Corp.
Chemical Pat. & Licensing Dept.
2000 Market St.
Philadelphia, PA 19103   (215) 299-6970

Scott Jason Fields
Synnestvedt & Lechner
2600 One Reading Center
1101 Market Street
Philadelphia, PA 19107   (215) 923-4466

Albert L. Free
Synnestvedt & Lechner
2600 One Reading Center
1101 Market Street
Philadelphia, PA 19107   (215) 923-4466

Darryl P. Frickey
Rohm & Haas Co.
Independence Mall West
Philadelphia, PA 19105   (215) 592-6745

Abram M. Goldfinger#
Thomas Jefferson University
1020 Locust St.
Room M-6
Philadelphia, PA 19107   (215) 955-6862

Joel S. Goldhammer
Panitch, Schwarze, Jacobs
& Nadel
36th Floor, Five Penn Center Plaza
1601 Market Street
Philadelphia, PA 19103   (215) 567-2020

Max Goldman
Caesar, Rivise, Bernstein,
Cohen & Pokotilow, Ltd.
12th Floor - Seven Penn Center
1635 Market Street
Philadelphia, PA 19103   (215) 567-2010

John W. Goldschmidt, Jr.
Reed, Smith, Shaw & Mc Clay
2500 One Liberty Place
1650 Market St.
Philadelphia, PA 19103   (215) 851-8100

Stuart M. Goldstein
Clark, Ladner, Fortenbaugh
& Young
32nd Floor
1818 Market St.
Philadelphia, PA 19103   (215) 241-1885

Daniel H. Golub
1600 Market Street
Suite 3600
Philadelphia, PA 19103   (215) 751-2332

Lewis F. Gould, Jr.
Eckert, Seamans, Cherin & Mellott
Room 3232
1700 Market St.
Philadelphia, PA 19103   (215) 563-8020

Roger K. Graham#
Rohm & Haas Co.
Patent Depart., 7th Flr.
6th & Market Sts.
Philadelphia, PA 19105   (215) 592-3000

Mark A. Greenfield
128 East Chestnut Hill Ave.
Philadelphia, PA 19118   (215) 247-6422

Stephan P. Gribok
Eckert, Seamans, Cherin & Mellott
1700 Market Street
Suite 3232
Philadelphia, PA 19103   (215) 563-8020

Patrick J. Hagan
Dann, Dorfman, Herrell
& Skillman, P.C.
Suite 900 - Three Mellon Bank Ctr.
15th St. & So. Penn Square
Philadelphia, PA 19102   (215) 563-4100

Carl A. Hechmer, Jr.
Pennwalt Corp.
Pat. Dept.
3 Pkwy.
Philadelphia, PA 19102   (215) 587-7700

William E. Hedges
The Warwick
17th & Locust Sts.
Philadelphia, PA 19103   (215) 546-4834

Louis M. Heidelberger
Reed Smith Shaw
& Mc Clay
1600 Avenue of the Arts Bldg.
Broad & Chestnut Streets
Philadelphia, PA 19107   (215) 875-4300

Dale M. Heist
Woodcock, Washburn, Kurtz,
Mackiewicz & Norris
One Liberty Place
46th Floor
Philadelphia, PA 19103   (215) 568-3100

Robert B. Henn
Atochem North America, Inc.
3 Parkway
Philadelphia, PA 19102   (215) 587-7792

Roger Wayne Herrell
Dann, Dorfman, Herrell
& Skillman, P.C.
Three Mellon Bk. Ctr. - Suite 900
15th Street & South Penn Sq.
Philadelphia, PA 19102   (215) 563-4100

James Edward Hess
Sun Refining & Marketing Co.
Patents & Licenses
Ten Penn Center
1801 Market Street
Philadelphia, PA 19103   (215) 977-3075

Robert G. Hoffmann
Atochem North Amer., Inc.
Patent Dept.
Three Parkway
Philadelphia, PA 19102   (215) 587-7694

Liza D. Hohenschutz
Woodcock, Washburn, Kurtz
Mackiewicz & Norris
One Liberty Place
46th Floor
Philadelphia, PA 19103   (215) 568-3100

Marianne Huseman#
Seidel, Gonda, Lavorgna
& Monaco, P.C.
Two Penn Center
Suite 1800
Philadelphia, PA 19102   (215) 568-8383

Lawrence A. Husick
Dilworth, Paxson, Kalish
& Kauffman
2600 Fidelity Bldg.
Philadelphia, PA 19109   (215) 875-8580

Richard Kay Jackson
Wyeth-Ayerst Labs.
P.O. Box 8299
Philadelphia, PA 19101   (215) 341-2310

Morton C. Jacobs
Millman & Jacobs
2940 PSFS Bldg.
12 South 12th Street
Philadelphia, PA 19107   (215) 592-6565

John Jamieson, Jr.
Panitch, Schwarze, Jacobs
& Nadel
5 Penn Center Plaza, 36th Fl.
1601 Market Street
Philadelphia, PA 19103   (215) 567-2020

Donald R. Johnson
Sun Refining & Marketing Co.
Pats. & Licenses
Ten Penn Center
Philadelphia, PA 19103   (215) 977-3074

Philip S. Johnson
Woodcock, Washburn, Kurtz,
Mackiewicz & Norris
One Liberty Place
46th Floor
Philadelphia, PA 19103   (215) 568-3100

John William Kane, Jr.
Scott Paper Co.
Scott Plaza II
Philadelphia, PA 19113   (215) 522-5810

Robert J. Kapalka
Eckert, Seamans, Cherin & Mellott
1700 Market Street
Suite 3232
Philadelphia, PA 19103   (215) 563-8020

Sheldon Kapustin
1011 Prospect Ave.
Philadelphia, PA 19126   (215) 635-2925

Leslie L. Kasten, Jr.
Panitch, Schwarze, Jacobs
& Nadel
5 Penn Center Plaza, 36th Fl.
1601 Market Street
Philadelphia, PA 19103   (215) 567-2020

Robert W. Kell
The Warwick, Suite 408
17th & Locust Sts.
Philadelphia, PA 19103   (215) 546-4834

Patrick J. Kelly#
Synnestvedt & Lechner
2600 One Reading Center
1101 Market St.
Philadelphia, PA 19107-215-   (215) 923-4466

Robert M. Kennedy
FMC Corp.
Chem. Pat. & Licensing Dept.
2000 Market Street
Philadelphia, PA 19103   (215) 299-6966

Albert T. Keyak
Woodcock, Washburn, Kurtz,
Mackiewicz & Norris
1 Liberty Place, 46th Fl.
Philadelphia, PA 19103   (215) 568-3100

William T. King
Smith, Kline, Beckman Corp.
1 Franklin Plaza
Philadelphia, PA 19101   (215) 751-5161

Charles M. Kinzig
Smith, Kline, Beckman Corp.
Corp. Pats. & Tdmks.
Box 7929, N-160
1 Franklin Plaza
Philadelphia, PA 19101   (215) 751-6137

Bruce H. Kleinstein
Information Ventures, Inc.
1500 Locust Street
Suite 3216
Philadelphia, PA 19102   (215) 732-9083

Louis Frank Kline, Jr.
Rohm & Haas Co.
Independence Mall W.
Philadelphia, PA 19105   (215) 592-2992

C. Frederick Koenig, III
Volpe & Koenig, P.C.
Suite 206
Ben Franklin Business Ctr.
Ninth & Chestnut Sts.
Philadelphia, PA 19107   (215) 238-0088

Robert A. Koons, Jr.
Schnader, Harrison, Segal
& Lewis
1600 Market Street
Suite 3600
Philadelphia, PA 19103   (215) 751-2180

Geoffrey H. Krauss
Aerospace Pat. Operations
General Electric Co.
Bldg. 100 - Room M3110
P.O. Box 8555
Philadelphia, PA 19101   (215) 354-5915

Richard E. Kurtz
Woodcock, Washburn, Kurtz,
Mackiewicz & Norris
1 Liberty Place, 46th Fl.
Philadelphia, PA 19103   (215) 568-3100

Jerome H. Lacheen
2400 Lewis Tower Bldg.
225 S. 15th St.
Philadelphia, PA 19102   (215) 545-6300

William E. Lambert, III
Rohm & Haas Company
Independence Mall West
Philadelphia, PA 19105   (215) 592-3025

Gregory J. Lavorgna
Seidel, Gonda, Lavorgna
& Monaco, P.C.
1800 Two Penn Center Plaza
Philadelphia, PA 19102   (215) 568-8383

Edward T. Lentz
Smith, Kline, Beckman Corp.
1 Franklin Plaza
P.O. Box 7929
Philadelphia, PA 19101   (215) 751-7012

Gary H. Levin
Woodcock, Washburn, Kurtz,
Mackiewicz & Norris
1 Liberty Place, 46th Fl.
Philadelphia, PA 19103   (215) 568-3100

John Lezdey
Steele, Gould & Fried
Suite 3232
1700 Market Street
Philadelphia, PA 19103   (215) 563-8020

Charles H. Lindrooth
Synnestvedt & Lechner
Suite 2600
1101 Market Street
Philadelphia, PA 19107   (215) 923-4466

Francis M. Linguiti
Schnader, Harrison, Segal
& Lewis
Suite 3600
1600 Market Street
Philadelphia, PA 19103   (215) 751-2387

Joseph Lucci
Woodcock, Washburn, Kurtz,
Mackiewicz & Norris
1 Liberty Place, 46th Fl.
Philadelphia, PA 19103   (215) 568-3100

John Jacob Mackiewicz
Woodcock, Washburn, Kurtz,
Mackiewicz & Norris
One Liberty Place, 46th Fl.
Philadelphia, PA 19103   (215) 568-3100

Paul Maleson
Maleson, Udell, Rosenberg,
Bilker & Farrell
1407 Lewis Tower Bldg.
Philadelphia, PA 19102   (215) 735-2678

Albert J. Marcellino
Woodcock, Washburn, Kurtz,
Mackiewicz & Norris
1 Liberty Place, 46th Fl.
Philadelphia, PA 19103   (215) 568-3100

Joseph A. Marlino, Sr.
Smith, Kline, Beckman Corp.
1 Franklin Plaza
Philadelphia, PA 19101   (215) 751-5183

Eric S. Marzluf
Caesar, Rivise, Bernstein,
Cohen & Pokotilow, Ltd.
7 Seven Penn Center, 12th Fl.
1635 Market Street
Philadelphia, PA 19103   (215) 567-2010

Mary E. Mc Carthy
Smith, Kline, Beckman Corp.
1 Franklin Plaza
P.O. Box 7929, N160
Philadelphia, PA 19101   (215) 751-3391

James C. Mc Connon
Paul and Paul
2000 Market Street
Suite 2900
Philadelphia, PA 19103   (215) 568-4900

Charles M. Mc Cuen
Montgomery, Mc Cracken,
Walker & Rhoads
Three Parkway
Philadelphia, PA 19102   (215) 563-0650

John F. Mc Nulty
Paul and Paul
2900 Two Thousand Market St.
Philadelphia, PA 19103   (215) 568-4900

Thomas F. Mccaffery
Smith, Kline, Beckman Corp.
Box 7929
Philadelphia, PA 19101   (215) 751-5060

Thomas E. Merchant
Pennwalt Corporation
3 Parkway
Philadelphia, PA 19102   (215) 587-7080

Otto O. Meyers, III
Seidel, Gonda, Lavorgna
& Monaco
Two Penn Center Plaza
Suite 1800
Philadelphia, PA 19102   (215) 568-8383

Austin R. Miller
Miller & Quinn
1125 Land Title Bldg.
Philadelphia, PA 19110   (215) 563-1810

Suzanne E. Miller
Woodcock, Washburn, Kertz,
Mackiewicz & Norris
1 Liberty Place, 46th Fl.
Philadelphia, PA 19103   (215) 568-3100

Max R. Millman
1919 Chestnut Street
Philadelphia, PA 19103   (215) 561-5250

William D. Mitchell
Atochem North Amer., Inc.
Patent Dept.
Three Parkway
Philadelphia, PA 19102   (215) 587-7445

Daniel A. Monaco
Seidel, Gonda, Lavorgna
& Monaco, P.C.
Two Penn Center Plaza
Suite 1800
Philadelphia, PA 19102   (215) 568-8383

Charles W. Morck, Jr.
3468 St. Vincent St.
Philadelphia, PA 19149   (215) 624-3678

William H. Murray
Benasutti & Murray
Suite 2701 One Reading Center
1101 Market Street
Philadelphia, PA 19107   (215) 923-6100

Alan S. Nadel
Panitch, Schwarze,Jacobs
& Nadel
5 Five Penn Center Plaza, 36th Fl.
1601 Market Street
Philadelphia, PA 19103   (215) 567-2020

Michael R. Nagy
Panitch, Schwarze, Jacobs
& Nadel
5 Penn Center Plaza, 36th Fl.
1601 Market St.
Philadelphia, PA 19102   (215) 567-2020

Jeffrey M. Navon
Synnestvedt & Lechner
2600 One Reading Center
1101 Market Street
Philadelphia, PA 19107   (215) 923-4466

Wallace D. Newcomb
Panitch, Schwarze, Jacobs
& Nadel
5 Penn Center Plaza, 36th Fl.
1601 Market Street
Philadelphia, PA 19103   (215) 567-2020

Driscoll A. Nina#
University of Pennsylvania
3400 Chestnut Street
Philadelphia, PA 19104   (215) 495-6481

Norman Leon Norris
Woodcock, Washburn, Kurtz,
Mackiewicz & Norris
1 Liberty Place, 46th Fl.
Philadelphia, PA 19103   (215) 568-3100

Ross J. Oehler
Panitch, Schwarze, Jacobs
& Nadel
5 Penn Center Plaza, 36th Fl.
1601 Market Street
Philadelphia, PA 19103   (215) 567-2020

Joel E. Oshtry
2751 Pennsylvania Ave., B-104
Philadelphia, PA 19130   (215) 232-5729

Vincent T. Pace
Dann, Dorfman, Herrell
& Skillman, P.C.
One Meridian Plaza, Suite 900
1414 South Penn Square
Philadelphia, PA 19102   (215) 563-4100

Francis Arthur Paintin
Woodcock, Washburn, Kurtz,
Mackiewicz & Norris
1 Liberty Place, 46th Fl.
Philadelphia, PA 19103   (215) 568-3100

Ronald L. Panitch
Panitch, Schwarze, Jacobs
& Nadel
1 Penn Center Plaza, 36th Flr.
1601 Market St.
Philadelphia, PA 19103   (215) 567-2020

Henrik D. Parker
Woodcock, Washburn, Kurtz,
Mackiewicz & Norris
1 Liberty Place, 46th Fl.
Philadelphia, PA 19103   (215) 568-3100

Steven R. Petersen
Reed, Smith, Shaw
& Mc Clay
2500 One Liberty Place
Philadelphia, PA 19103   (215) 851-8264

Marcia D. Pintzuk
FMC Corp.
2000 Market St.
Philadelphia, PA 19103   (215) 299-6965

Donald R. Piper, Jr.
Dann, Dorfman, Herrell
& Skillman, P.C.
One Meridian Plaza, Suite 900
1414 South Penn Square
Philadelphia, PA 19102   (215) 563-4100

Harris A. Platt
Duane, Morris & Heckscher
4200 One Liberty Place
Philadelphia, PA 19103   (215) 979-1000

Robert Charles Podwil
Reed, Smith, Shay & Mc Clay
2500 One Liberty Place
Philadelphia, PA 19103   (215) 851-8260

Manny Pokotilow
Caesar, Rivise, Bernstein,
Cohen & Pokotilow, Ltd.
7 Penn Center, 12th Fl.
1635 Market Street
Philadelphia, PA 19103   (215) 567-2010

Joseph F. Posillico
Synnestvedt & Lechner
2600 One Reading Center
1101 Market Street
Philadelphia, PA 19107   (215) 923-4466

Albert W. Preston, Jr.
Woodcock, Washburn, Kurtz,
Mackiewicz & Norris
1 Liberty Place, 46th Fl.
Philadelphia, PA 19103   (215) 568-3100

Jack D. Puffer
Boeing Defense & Space Group
Helicopters Division
P.O. Box 16858
Philadelphia, PA 19142   (215) 591-4188

Charles N. Quinn
Miller & Quinn
1125 Land Title Bldg.
Philadelphia, PA 19110   (215) 563-1810

Polly E. Ramstad
Rohm & Haas Company
Independence Mall West
Philadelphia, PA 19105   (215) 592-2423

Steven J. Rocci
Woodcock, Washburn, Kurtz,
Mackiewicz & Norris
1 Liberty Place, 46th Fl.
Philadelphia, PA 19103   (215) 568-3100

Robert E. Rosenthal
Schnader, Harrison, Segal
& Lewis
1600 Market Street, Suite 3600
Philadelphia, PA 19103   (215) 751-2680

Evelyn M. Sabino#
3722 Country Club Rd.
Philadelphia, PA 19131   (215) 877-1333

Edward A. Sager
Atochem North Amer., Inc
Patent Dept.
Three Parkway
Philadelphia, PA 19102   (215) 587-7688

J. Walter Schilpp
Paul & Paul
Suite 2900
2000 Market St.
Philadelphia, PA 19103   (215) 568-4900

William W. Schwarze
Panitch, Schwarze, Jacobs
& Nadel
1 Penn Center Plaza, 36th Fl.
1601 Market Street
Philadelphia, PA 19103   (215) 567-2020

Arthur Harris Seidel
Seidel, Gonda, Lavorgna & Monaco
2 Penn Center Plaza, Ste. 1800
Philadelphia, PA 19102   (215) 568-8383

Arthur G. Seifert
Wyeth-Ayerst Laboratories
P.O. Box 8299
Philadelphia, PA 19101   (215) 341-2314

John V. Silverio
Eckert, Seamans, Cherin & Mellott
1700 Market Street
Suite 3232
Philadelphia, PA 19103   (215) 563-8020

Michele L. Simons,#
Panitch, Schwarze, Jacobs
& Nadel
5 Penn Center Plaza, 36th Fl.
1601 Market St.
Philadelphia, PA 19103-0   (215) 567-2020

Henry H. Skillman
Dann, Dorfman, Herrell
& Skillman, P.C.
One Meridian Plaza, Suite 900
1414 South Penn Square
Philadelphia, PA 19102   (215) 563-4100

Alex R. Sluzas
Paul & Paul
2900 Two Thousand Market Street
Philadelphia, PA 19103   (215) 568-4900

Karl L. Spivak
Eckert, Seamans, Cherin & Mellott
1700 Market Street
Suite 3232
Philadelphia, PA 19103   (215) 563-8020

Alfred Stapler
7334 Rural Lane
Philadelphia, PA 19119   (215) 242-2698

Brian W. Stegman
Rohm & Haas Company
Pat. Dept.
Independence Mall West
Philadelphia, PA 19105   (215) 592-6818

Barry Allen Stein
Caesar, Rivise, Bernstein,
Cohen & Pokotilow, Ltd.
7 Penn Center Plaza, 12th Fl.
1635 Market Street
Philadelphia, PA 19103   (215) 567-2010

Michael David Stein
Woodcock, Washburn, Kurtz,
Mackiewicz & Norris
1 Liberty Place, 46th Fl.
Philadelphia, PA 19103   (215) 568-3100

Yuriy P. Stercho
Dann, Dorfman, Herrell
& Skillman, P.C.
Three Mellon Bank Ctr., Ste. 900
15th Street and S. Penn Square
Philadelphia, PA 19102   (215) 563-4100

Terence P. Strobaugh
Rohn & Haas Co.
Independence Mall W.
Philadelphia, PA 19105   (215) 592-3677

Stuart R. Suter
Smith, Kline, Beckman Corp.
1 Franklin Plaza
N160, P.O. Box 7929
Philadelphia, PA 19101   (215) 751-5186

Jeffrey A. Sutton#
Smith, Kline, Beckman Corp.
P.O. Box 7929
N-160
Philadelphia, PA 19101   (215) 751-3364

John T. Synnestvedt
Synnestvedt & Lechner
2600 One Reading Center
1101 Market Street
Philadelphia, PA 19107   (215) 923-4466

George Tarnowski
Wyeth Laboratories
P.O. Box 8299
Philadelphia, PA 19101   (215) 341-2312

John E. Taylor, III#
Rohm & Haas Co.
Pat. Dept.
Independence Mall W.
Philadelphia, PA 19105   (215) 592-3294

Frederick A. Tecce#*
U.S. Attorneys Office
Eastern District of Penn.
601 Market Street
Philadelphia, PA 19106   (215) 597-4087

Joseph A. Tessari
Paul & Paul
2900 Two Thousand Market St.
Philadelphia, PA 19103   (215) 568-4900

E. Arthur Thompson
Paul & Paul
Suite 2900
2000 Market St.
Philadelphia, PA 19103   (215) 568-4900

R. Duke Vickrey
Scott Paper Co.
Scott Plaza II
Philadelphia, PA 19113   (215) 521-5809

Anthony S. Volpe
Volpe & Koenig, P.C.
Benjamin Franklin Bus. Center
Ninth & Chestnut Streets
Suite 206
Philadelphia, PA 19107   (215) 238-0088

James G. Vouros#
Rohm & Haas Co.
Independence Mall West
Philadelphia, PA 19105   (215) 592-2564

Robert B. Washburn
Woodcock, Washburn, Kurtz,
Mackiewicz & Norris
1 Liberty Place, 46th Fl.
Philadelphia, PA 19103   (215) 568-3100

Richard D. Weber
Synnestvedt & Lechner
2600 One Reading Center
1101 Market Street
Philadelphia, PA 19107   (215) 923-4466

Louis Weinstein
1400 S. Penn Square
Suite 2410
Two Mellon Bank Center
Philadelphia, PA 19102   (215) 557-9797

Gerard J. Weiser
Weiser & Stapler
230 S. 15th Street
Suite 500
Philadelphia, PA 19102   (215) 875-8383

Bernard M. Weiss#
108-110 Almatt Place
Philadelphia, PA 19115   (215) 676-2280

John A. Weygandt
Scott Paper Co.
Scott Plaza II
Law Division
Philadelphia, PA 19113   (215) 522-5815

Robert Wiser
Wyeth-Ayerst Laboratories Div.
American Home Products Corp.
P.O. Box 8299
Philadelphia, PA 19101   (215) 341-2308

Zachary T. Wobensmith, III
707 Avenue of the Arts Bldg.
Broad & Chestnut Sts.
Philadelphia, PA 19107   (215) 735-3090

Joseph H. Yamaoka
1731 Lombard Street
Philadelphia, PA 19146   (215) 735-4381

Stephen A. Young
General Electric Co.
P.O. Box 8555
Bldg. 100 - Room M3110
Philadelphia, PA 19101   (215) 354-1190

William E. Zeiter
Morgan, Lewis & Bockius
2000 One Logan Square
Philadelphia, PA 19103   (215) 963-5367

Robert F. Zielinski
Dilworth, Paxson, Kalish
& Kauffman
123 S. Broad St., Ste. 2600
Philadelphia, PA 19109   (215) 875-8512

Robert J. Zinn
2011 Spring Garden St.
Philadelphia, PA 19130   (215) 896-5150

Glenn E.J. Murphy#
1285 East Evergreen Drive
Phoenixville, PA 19460   (215) 933-0924

Gary J. Connell#
PPG Industries, Inc.
One PPG Place
Pittsburg, PA 15272   (412) 434-2469

Daniel C. Abeles
Westinghouse Elec. Corp.
Law Dept., IPS
1310 Beulah Rd.
Pittsburgh, PA 15235   (412) 256-5227

John M. Adams
Buchanan Ingersoll
57th Floor
600 Grant Street
Pittsburgh, PA 15219   (412) 562-1067

Godfried R. Akorli#
Mobay Corp.
Patent Dept.
Mobay Road
Pittsburgh, PA 15205   (412) 777-2340

Lynn J. Alstadt
Buchanan Ingersoll, P.C.
56th Floor
600 Grant St.
Pittsburgh, PA 15219   (412) 562-1632

Edward Charles Arenz
Westinghouse Electric Corp.
Bldg. 801
R&D Center
Pittsburgh, PA 15235   (412) 256-5244

Brian W. Ashbaugh
Rose, Schmidt, Chapman,
Duff & Hasley
900 Oliver Bldg.
Pittsburgh, PA 15222   (412) 434-8859

Klaus J. Bach#
Westinghouse Electric Corp.
1310 Beulah Road
R&D Center
Pittsburgh, PA 15235   (412) 256-5257

Fred J. Baehr, Jr.
Westinghouse Elec. Corp.
Law Dept., IPS
1310 Beulah Rd.
Pittsburgh, PA 15235   (412) 256-5267

George P. Baier
Buchanan Ingersoll, P.C.
56th Floor
600 Grant Street
Pittsburgh, PA 15219   (412) 562-1635

Kent E. Baldauf
Webb, Burden, Ziesenheim
& Webb, P.C.
515 Oliver Bldg.
535 Smithfield Street
Pittsburgh, PA 15222   (412) 471-8815

Henry E. Bartony, Jr.
Reed, Smith, Shaw & Mc Clay
P.O. Box 2009
Pittsburgh, PA 15230   (412) 288-8594

Paul A. Beck
Buchanan Ingersoll, P.C.
56th Floor
600 Grant Street
Pittsburgh, PA 15219   (412) 562-1631

Richard M. Bies
217 Park Entrance
Pittsburgh, PA 15228   (412) 343-7421

Byron A. Bilicki
Kirkpatrick & Lockhart
1500 Oliver Bldg.
Pittsburgh, PA 15222   (412) 355-8653

Harry Donald Bishop#
8360 Remington Drive
Pittsburgh, PA 15237   (412) 364-6713

Walter J. Blenko, Jr.
Eckert, Seamans, Cherin
& Mellott
600 Grant Street
42nd Floor
Pittsburgh, PA 15219   (412) 566-6189

Paul Bogdon
Cauley & Conflenti
1212 Manor Complex
564 Forbes Ave.
Pittsburgh, PA 15219   (412) 391-7133

George John Bohrer#
188 Oak Park Place
Pittsburgh, PA 15243   (412) 561-6563

Stanley R. Bramham#
Westinghouse Electric Corp.
Law Dept.
Intellectual Property Section
1310 Beulah Road
Pittsburgh, PA 15235   (412) 256-5260

John S. Brams#
Carothers & Carothers
445 Fort Pitt Blvd.
Suite 500
Pittsburgh, PA 15219   (412) 471-3575

Edgar Wallace Breisch, Jr.
524 Olive Street
Pittsburgh, PA 15237   (412) 366-1621

John Reeder Bronaugh
Rockwell International Corp.
600 Grant St.
Pittsburgh, PA 15219   (412) 565-2902

Francis E. Browder
3931 Greensburg Pike
Pittsburgh, PA 15221   (412) 351-0942

David C. Bruening
Webb, Burden, Ziesenheim
& Webb, P.C.
515 Oliver Bldg.
535 Smithfield Street
Pittsburgh, PA 15222   (412) 471-8815

Joseph L. Brzuszek
Westinghouse Elec. Corp.
R&D Center, Bldg. 801
Law Dept.
1310 Beulah Road
Pittsburgh, PA 15235   (412) 256-5259

Richard L. Byrne
Webb, Burden, Ziesenheim
& Webb, P.C.
515 Oliver Bldg.
515 Smithfield Street
Pittsburgh, PA 15222   (412) 471-8815

Joseph John Carducci
708 Scurbgrass Road
Pittsburgh, PA 15243   (412) 563-6732

Floyd Barber Carothers
Carothers & Carothers
445 Fort Pitt Blvd., Suite 500
Pittsburgh, PA 15219   (412) 471-3575

Ernest A. Carpenter
Tippins Incorporated
435 Butler Street
Pittsburgh, PA 15223   (412) 782-7760

Raymond S. Chisholm
276 Trotwood Dr.
Pittsburgh, PA 15241   (412) 835-0418

Daniel P. Cillo
Westinghouse Electric Corp.
Law Dept., Intellectual Property Sec.
R&D Center
Churchill Borough
Pittsburgh, PA 15235   (412) 256-5264

Craig G. Cochenour
Eckert, Seamans, Cherin
& Mellott
600 Grant Street
42nd Floor
Pittsburgh, PA 15219   (412) 566-6083

Frederick H. Colen
Reed, Smith, Shaw
& Mc Clay
435 Sixth Avenue
Pittsburgh, PA 15219   (412) 288-4164

Andrew J. Cornelius
Klett, Lieber, Rooney & Schorling
1 Oxford Centre, 40th Fl.
Pittsburgh, PA 15219   (412) 392-2000

John E. Curley
459 Goldsmith Rd.
Pittsburgh, PA 15237   (412) 486-2754

Louis A. De Paul
Westinghouse Elec. Corp.
Law Dept., Intellectual Property Section
1310 Beulah Road
Pittsburgh, PA 15235   (412) 256-5228

Zigmund L. Dermer
Westinghouse Electric Corp.
R&D Center
1310 Beulah Rd.
Pittsburgh, PA 15235   (412) 256-5247

Michael L. Dever
Buchanan Ingersoll, P.C.
56th Floor
600 Grant Street
Pittsburgh, PA 15219   (412) 562-1637

George D. Dickos
Kirkpatrick & Lockhart
1500 Oliver Bldg.
Pittsburgh, PA 15222   (412) 355-6785

Alan M. Doernberg
Fisher Scientific Company
Law Dept.
711 Forbes Avenue
Pittsburgh, PA 15219   (412) 562-8451

Marvin R. Dunlap
Dickie, Mc Camey & Chilcote, P.C.
2 PPG Place
Suite 400
Pittsburgh, PA 15222   (412) 281-7272

Eugene Francis Dwyer
7 Creighton Ave.
Pittsburgh, PA 15205   (412) 922-6579

William A. Elchik
Westinghouse Electric Corp.
1300 Beulah Road
Pittsburgh, PA 15235   (412) 256-5215

Daniel W. Ernsberger
Behrend & Ernsberger
2400 Grant Bldg.
Pittsburgh, PA 15219   (412) 391-2515

Christine R. Ethridge
Kirkpatrick & Lockhart
1500 Oliver Bldg.
Pittsburgh, PA 15222   (412) 355-8619

Julian Falk
2131 5th Ave.
Pittsburgh, PA 15219   (412) 471-0774

Barry I. Friedman
Dickie, Mc Camey & Chilcote, P.C.
Suite 400
Two PPG Place
Pittsburgh, PA 15222   (412) 392-5479

Paula E. Ganz
Jones, Day, Reavis & Pogue
500 Grant St.
Pittsburgh, PA 15219   (412) 391-3939

Kevin J. Garber
Reed, Smith, Shaw & Mc Clay
435 Sixth Ave.
Pittsburgh, PA 15219   (412) 288-3170

Christopher H. Gebhardt
Kirkpatrick & Lockhart
1500 Oliver Bldg.
Pittsburgh, PA 15222   (412) 355-8641

Joseph C. Gil
Mobay Corporation
Mobay Road
Pittsburgh, PA 15205   (412) 777-2342

Douglas Gene Glantz
Alcoa
138 Alleyne Drive
Pittsburgh, PA 15215   (412) 784-1216

Lee R. Golden
Armstrong, Nikaido, Marmelstein,
Kubovcik & Murray
500 Wood Street, Ste. 721
Pittsburgh, PA 15222   (412) 281-2931

Kenneth P. Gournic#*
U.S. Postal Service
770 Trumbull
Pittsburgh, PA 15220   (412) 279-4861

David C. Hanson
Webb, Burden, Ziesenheim
& Webb, P.C.
515 Oliver Bldg.
535 Smithfield Street
Pittsburgh, PA 15222   (412) 471-8815

Raymond J. Harmuth
Webb, Burden, Ziesenheim
& Webb, P.C.
515 Oliver Bldg.
535 Smithfield St.
Pittsburgh, PA 15222   (412) 471-8815

Gordon R. Harris
Buchanan Ingersoll, P.C.
56th Floor
600 Grant Street
Pittsburgh, PA 15219   (412) 562-1636

Gene Harsh
Mobay Corporation
Mobay Road
Pittsburgh, PA 15205   (412) 777-2340

Richard E.L. Henderson
Mobay Corporation
Patent Dept.
Mobay Road
Pittsburgh, PA 15205   (412) 777-2341

Lee P. Johns
Westinghouse Elec. Co.
1310 Beulah Rd.
Pittsburgh, PA 15235   (412) 256-5213

Barbara E. Johnson
Webb, Burden, Ziesenheim
& Webb, P.C.
515 Oliver Bldg.
535 Smithfield Street
Pittsburgh, PA 15222   (412) 471-8815

Chester Arthur Johnston, Jr.
PPG Industries, Inc.
1 PPG Place
Pittsburgh, PA 15272   (412) 434-2931

Edward H. Jones, Jr.
Pennsylvania Dept. of
Environmental Resources
1303 Highland Bldg.
121 S. Highland Ave.
Pittsburgh, PA 15206   (412) 645-7454

John W. Jordan, IV
Gigsby, Gaca & Davies, P.C.
One Gateway Center
Tenth Floor
Pittsburgh, PA 15222   (412) 281-0737

William S. Karn
Karn Plastics Company
518 Dickson Avenue
Pittsburgh, PA 15202   (412) 766-1635

Harry B. Keck
Zurawsky & Keck
415 Lawyers Bldg.
Pittsburgh, PA 15219   (412) 281-7766

Suzanne Kikel
Eckert Seamans Cherin
& Mellott
600 Grant Street
42nd Floor
Pittsburgh, PA 15219   (412) 566-6130

Michael J. Kline#
Reed, Smith, Shaw & Mc Clay
435 Sixth Avenue
Pittsburgh, PA 15219   (412) 288-3338

William G. Kratz, Jr.
Armstrong, Nikaido, Marmelstein,
Kubovcik & Murray
500 Wood St., Suite 721
Pittsburgh, PA 15222   (412) 281-2931

William L. Krayer
Aristech Chemical Corp.
Room 2114
600 Grant St.
Pittsburgh, PA 15230   (412) 433-7538

Ronald L. Kuis
Kirkpatrick & Lockhart
1500 Oliver Bldg.
Pittsburgh, PA 15222   (412) 355-8614

James R. Kyper
Kirkpatrick & Lockhart
1500 Oliver Bldg.
Pittsburgh, PA 15222   (412) 355-6542

Charles P. Lawton
260 Fort Pitt Commons
Pittsburgh, PA 15219   (412) 355-5857

Edward B. Lee, III
Lindsay, Mc Cabe & Lee
200 Standard Life Bldg.
345 Fourth Avenue
Pittsburgh, PA 15222   (412) 471-2420

Robert P. Lenart
Westinghouse Electric Corp.
1310 Beulah Road
Pittsburgh, PA 15235   (412) 256-5205

Donald C. Lepiane
PPG Industries, Inc.
Patent Dept.
1 PPG Place
Pittsburgh, PA 15272   (412) 434-2936

John F. Letchford
Reed, Smith, Shaw & Mc Clay
435 Sixth Avenue
P.O. Box 2009
Pittsburgh, PA 15230   (412) 288-4058

Mark Levin
PPG Industries, Inc.
1 PPG Place
Pittsburgh, PA 15272   (412) 434-3792

Carl R. Lippert
513 Guyasuta Road
Pittsburgh, PA 15215   (412) 781-5355

William Henry Logsdon
Webb, Burden, Ziesenheim
& Webb, P.C.
515 Oliver Bldg.
535 Smithfield Street
Pittsburgh, PA 15222   (412) 471-8815

Daniel J. Long
PPG Industries, Inc.
1 PPG Place
Pittsburgh, PA 15272   (412) 434-2882

Charles M. Lorin
Westinghouse Electric Corp.
R&D Center
1310 Beulah Road
Pittsburgh, PA 15235   (412) 256-5235

Michael Patrick Lynch
Westinghouse Electric Corp.
1310 Beulah Road
Pittsburgh, PA 15235   (412) 256-5241

Richard E. Maebius
219 Mc Monagle Road
Pittsburgh, PA 15220   (412) 531-3489

George E. Manias
H.H. Robertson Co.
Two Gateway Center
Pittsburgh, PA 15222   (412) 281-3200

Michael I. Markowitz
5863 Nicholson St., Apt. 2
Pittsburgh, PA 15217   (412) 422-5241

Thomas H. Martin#
207 Shadowlawn Circle
Pittsburgh, PA 15236   (412) 882-1536

Edward I. Mates
1632 Sillview Drive
Pittsburgh, PA 15243   (412) 279-1936

Curtis W. Mc Bride*
U.S. Dept. of Energy
Pittsburgh Energy Technology Center
P.O. Box 10940
Pittsburgh, PA 15236   (412) 892-6161

David C. Mc Candless
135 Arden Road
Pittsburgh, PA 15216   (412) 561-7314

Douglas K. Mc Claine
Mine Safety Appliances Co.
P.O. Box 426
Pittsburgh, PA 15230   (412) 967-3316

John W. Mc Ilvaine, III
Webb, Burden, Ziesenheim
& Webb, P.C.
515 Oliver Bldg.
535 Smithfield St.
Pittsburgh, PA 15222-2000   (412) 471-8815

Thomas F. Mc Knight
Neville Chemical Co.
2800 Neville Road
Pittsburgh, PA 15225   (412) 331-4200

Charles Leroy Menzemer
Westinghouse Electric Corp.
R&D Center
Pittsburgh, PA 15235   (412) 256-5243

Jon F. Merz
Westinghouse Electric Corp.
1310 Beulah Road
Pittsburgh, PA 15235   (412) 256-5252

Alex Mich, Jr.
Westinghouse Electric Corp.
Law Dept., Intellectual Prop. Sec.
R&D Center
1310 Beulah Road
Pittsburgh, PA 15235   (412) 256-5263

Thomas G. Michalek
4 West Manilla Avenue
Pittsburgh, PA 15220   (412) 922-3331

Julian Kennedy Miller
4747 Bayard St.
Pittsburgh, PA 15213   (412) 682-2469

Thomas G. Miller
Armstrong, Nikaido, Marmelstein,
Kubovcik & Murray
500 Wood St., Ste. 721
Pittsburgh, PA 15222   (412) 281-2931

Dennis G. Millman
PPG Industries Inc.
1 PPG Place
Pittsburgh, PA 15272   (412) 434-2939

William C. Mitchell
Merck & Co., Inc.
Calgon Center
P.O. Box 1346
Pittsburgh, PA 15230   (412) 777-8922

Martin J. Moran
Westinghouse Electric Corp.
Law Dept., Intellectual Property Sec.
1310 Beulah Road
Pittsburgh, PA 15235   (412) 256-5225

George Douglas Morris
PPG Industries, Inc.
1 PPG Place
Pittsburgh, PA 15272   (412) 434-3797

Lee D. Moses
Lawyers Bldg., Ste. 70
Pittsburgh, PA 15219   (412) 471-8020

Thomas Henry Murray
2230 Koppers Bldg.
436 Seventh Avenue
Pittsburgh, PA 15219   (412) 765-1580

Bidyut K. Niyogi
3823 Henley Drive
Pittsburgh, PA 15235   (412) 243-8959

Russell D. Orkin
Webb, Burden, Ziesenheim
& Webb, P.C.
515 Oliver Bldg.
535 Smithfield Street
Pittsburgh, PA 15222   (412) 471-8815

William E. Otto
Rose, Schmidt, Hasley
& Di Salle
900 Oliver Building
Pittsburgh, PA 15222   (412) 434-8682

Michael G. Panian
Westinghouse Elec. Corp.
Science & Technology Ctr.
Law Dept., Intellectual Prop. Sec.
Bldg. 801
Pittsburgh, PA 15235   (412) 256-5233

Lia M. Pappas
1248 Old Meadow Rd.
Pittsburgh, PA 15241   (412) 221-3607

Daniel Patch
Koppers Bldg.
436 Seventh Ave.
Suite 2230
Pittsburgh, PA 15219   (412) 765-1580

John R. Pegan
Suite 721
500 Wood Street
Pittsburgh, PA 15222   (412) 281-2931

Edward L. Pencoske
Kirkpatrick & Lockhart
1500 Oliver Bldg.
Pittsburgh, PA 15222   (412) 355-8645

John H. Perkins
Perkins & Schildnecht
One Oliver Plaza
Pittsburgh, PA 15222   (412) 566-4714

Linda Pingitore
PPG Industries, Inc.
One PPG Place
Pittsburgh, PA 15272   (412) 434-3704

Clifford A. Poff#
2230 Koppers Bldg.
436 Seventh Avenue
Pittsburgh, PA 15219   (412) 765-1580

Aron Preis
Mobay Corp.
Mobay Road
Pittsburgh, PA 15205   (412) 777-2343

Stanley J. Price
4135 Brownsville Road
P.O. Box 98127
Pittsburgh, PA 15227   (412) 882-7170

David V. Radack
Eckert, Seamans, Cherin
& Mellott
600 Grant Street, 42nd Fl.
Pittsburgh, PA 15219   (412) 566-6183

James O. Ray, Jr.#
Armstrong, Nikaido, Marmelstein,
Kubovcik & Murray
500 Wood St., Suite 721
Pittsburgh, PA 15222   (412) 281-2931

George Raynovich, Jr.
Buchnan Ingersoll, P.C.
56th Floor
600 Grant St.
Pittsburgh, PA 15219   (412) 562-1633

Stanley J. Reisman
Law & Finance Bldg.
Pittsburgh, PA 15219   (412) 232-0433

Paul M. Reznick
Webb, Burden, Ziesenheim
& Webb, P.C.
515 Oliver Bldg.
535 Smithfield Street
Pittsburgh, PA 15222   (412) 471-8815

William F. Riesmeyer, III
USX Corporation
600 Grant St.
Room 1569
Pittsburgh, PA 15219   (412) 433-2842

Alvin E. Ring
Buchanan Ingersoll, P.C.
56th Floor
600 Grant Street
Pittsburgh, PA 15219   (412) 562-1634

Rita M. Rooney
Eckert, Seamans, Cherin
& Mellott
600 Grant Street
Pittsburgh, PA 15219   (412) 566-6091

Thomas W. Roy
Mobay Corporation
Mobay Road
Pittsburgh, PA 15220   (412) 777-2345

William J. Ruano
402 St. Clair Bldg.
Pittsburgh, PA 15241   (412) 835-3111

Donald M. Satina
USX Corporation
1551 USX Tower
600 Grant Street
Pittsburgh, PA 15219   (412) 433-2994

Nancy M. Scalise
Cauley & Conflenti
1212 Manor Complex
564 Forbes Avenue
Pittsburgh, PA 15219   (412) 433-2994

Leland P. Schermer
Dickie, Mc Camey & Chilcote, P.C.
2 PPG Place, Ste. 400
Pittsburgh, PA 15222   (412) 281-7272

C. William Schildnecht
Perkins & Schildnecht
1 Oliver Plaza, 18th Fl.
Pittsburgh, PA 15222   (412) 566-3225

Dean Schron
Westinghouse Electric Corp.
Science & Technology Ctr.
Law Dept., Intellectual Property Sec.
1310 Beulah Rd.
Pittsburgh, PA 15235   (412) 256-5237

Ansel M. Schwartz
425 N. Craig St.
Suite 301
Pittsburgh, PA 15213   (412) 621-9222

Frederick L. Segal
Suite 700 Manor Complex
564 Forbes Avenue
Pittsburgh, PA 15219   (412) 391-2263

Donna L. Seidel#
PPG Industries, Inc.
1 PPG Place
Pittsburgh, PA 15272   (412) 434-3798

Carl T. Severini
PPG Industries, Inc.
1 PPG Place
Pittsburgh, PA 15272   (412) 434-2938

Mark S. Shaffer
Dick Corporation
P.O. Box 10896
Pittsburgh, PA 15236   (412) 384-1287

Ronald H. Shakely
Mine Safety Appliances Co.
121 Gamma Drive
Pittsburgh, PA 15238   (412) 967-3215

Michael I. Shamos
Webb, Burden, Ziesenheim
& Webb, P.C.
515 Oliver Bldg.
535 Smithfield Street
Pittsburgh, PA 15222   (412) 471-8815

Thomas F. Shanahan
225 Foxcroft Road
Pittsburgh, PA 15220   (412) 279-3431

Samuel Shipkovitz
5829 Nicholson Street
Pittsburgh, PA 15217   (412) 521-3234

Arnold Barry Silverman
Eckert, Seamans, Cherin
& Mellott
600 Grant Street
42nd Floor
Pittsburgh, PA 15219   (412) 566-2077

Andrew C. Siminerio
PPG Industries, Inc.
1 PPG Place
Pittsburgh, PA 15272   (412) 434-4645

William R. Sittig, Jr.
909 Frick Building
437 Grant Street
Pittsburgh, PA 15219   (412) 391-3334

Neal L. Slifkin
Kirkpatrick & Lockhart
1500 Oliver Building
Pittsburgh, PA 15222   (412) 355-6216

William P. Smith
P.O. Box 6007
229 Shiloh St.
Pittsburgh, PA 15211   (412) 381-8088

Melvin C. Snyder, III
Kirkpatrick & Lockhart
1500 Oliver Bldg.
Pittsburgh, PA 15222   (412) 355-8374

Joseph C. Spadacene
Westinghouse Electric Corp.
Law. Dept., Intellectual Property Sec.
1310 Beulah Road
Pittsburgh, PA 15235   (412) 256-5374

Gay Ann Spahn
PPG Industries, Inc.
1 PPG Place, 39E
Pittsburgh, PA 15272   (412) 434-2923

Richard A. Speer
Rockwell International Corp.
600 Grant St.
Pittsburgh, PA 15219   (412) 565-7107

Kenneth J. Stachel
PPG Industries, Inc.
One PPG Place
Pittsburgh, PA 15272   (412) 434-3186

Patricia K. Staub
443 Morrison Drive
Pittsburgh, PA 15216   (412) 561-8929

Arland T. Stein
Reed, Smith, Shaw & Mc Clay
435 Sixth Avenue
Mellon Square
Pittsburgh, PA 15219   (412) 288-3100

Irwin M. Stein
PPG Industries, Inc.
One PPG Place
Pittsburgh, PA 15272   (412) 434-3799

Walter S.E. Stevens
Westinghouse Electric Corp.
1310 Beulah Road
Bldg. 801, Rm. 4N47
Pittsburgh, PA 15235   (412) 256-5226

Richard A. Stoltz
Westinghouse Electric Corp.
1310 Beulah Road
Pittsburgh, PA 15235   (412) 256-5202

Blair Ross Studebaker
Westinghouse Electric Corp.
Science & Tech. Ctr.
Law Dept., Intellectual Property Sec.
1310 Beulah Rd.
Pittsburgh, PA 15235   (412) 256-5211

Michael R. Swartz
205 Royal Oak Avenue
Pittsburgh, PA 15235   (412) 244-9205

James J. Tedjeske
Westinghouse Electric Corp.
Resource Energy Systems Div.
2400 Armore Blvd.
Pittsburgh, PA 15221   (412) 825-6534

Gordon Howard Telfer
Westinghouse Electric Corp.
1310 Beulah Rd.
Pittsburgh, PA 15235   (412) 256-5208

Frederick L. Tolhurst
Cohen & Grigsby
2900 CNG Tower
625 Liberty Avenue
Pittsburgh, PA 15222   (412) 394-4960

Frederic C. Trenor, II
Zimmer, Kunz, Loughren, Hart,
Lazaroff, Trenor, Banyas
& Conaway, P.C.
One Oxford Center, 43rd Floor
Pittsburgh, PA 15219   (412) 281-8000

James G. Uber
Reed, Smith, Shaw & Mc Clay
P.O. Box 2009
435 Sixth Ave.
Pittsburgh, PA 15230   (412) 288-4134

William J. Uhl#
PPG Industries, Inc.
One PPG Place
Pittsburgh, PA 15272   (412) 434-2881

David S. Urey
USX Corp.
600 Grant St.
Ste. 1551
Pittsburgh, PA 15219   (412) 433-2873

James C. Valentine
130 Morrison Drive
Pittsburgh, PA 15216   (412) 563-6513

Patrick J. Viccaro
Allegheny Ludlum Corp.
1000 Six PPG Place
Pittsburgh, PA 15222   (412) 394-2839

Robert F. Wagner
Dickie, Mc Camey & Chilcote, P.C.
2 PPG Place
Pittsburgh, PA 15222   (412) 281-7272

John M. Webb
Webb, Burden, Ziesenheim
& Webb, P.C.
515 Oliver Bldg.
535 Smithfield Street
Pittsburgh, PA 15222   (412) 471-8815

William Hess Webb
Webb, Burden, Ziesenheim
& Webb, P.C.
515 Oliver Bldg.
535 Smithfield Street
Pittsburgh, PA 15222   (412) 471-8815

Don L. Webber
Reed, Smith, Shaw & Mc Clay
435 Sixth Ave.
Pittsburgh, PA 15219   (412) 288-8596

Edward F. Welsh
Armstrong, Nikaido, Marmelstein,
Kubovcik & Murray
500 Wood St., Suite 721
Pittsburgh, PA 15222   (412) 281-2931

Richard V. Westerhoff
Eckert, Seamans, Cherin
& Mellott
42nd Floor
600 Grant Street
Pittsburgh, PA 15219   (412) 566-6090

Thomas C. Wettach
Reed, Smith, Shaw & Mc Clay
435 Sixth Avenue
Pittsburgh, PA 15219   (412) 288-3102

Lyndanne M. Whalen
Mobay Corporation
Patent Dept.
Mobay Road, Bldg. 4
Pittsburgh, PA 15205   (412) 777-2347

John K. Williamson
Westinghouse Electric Corp.
Law Dept -IPS
1310 Beulah Road
Pittsburgh, PA 15235   (412) 256-5219

Robert D. Yeager
Kirkpatrick & Lockhart
1500 Oliver Bldg.
Pittsburgh, PA 15222   (412) 355-8605

Frederick B. Ziesenheim
Webb, Burden, Ziesenheim
& Webb, P.C.
515 Oliver Bldg.
535 Smithfield Street
Pittsburgh, PA 15222   (412) 471-8815

Lawrence G. Zurawsky
Zurawsky & Keck
415 Lawyers Bldg.
428 Forbes Avenue
Pittsburgh, PA 15219   (412) 281-7766

John E. Drach#
Henkel Corp.
Law Dept.
140 Germantown Pike
Suite 150
Plymouth Meeting, PA 19462   (215) 832-2215

Real J. Grandmaison
Henkel Corporation
Law Department
140 Germantown Pike
Suite 150
Plymouth Meeting, PA 19462   (215) 832-2219

Wayne C. Jaeschke
Henkel Corporation
140 Germantown Pike
Suite 150
Plymouth Meeting, PA 19462   (215) 832-2222

Norvell E. Wisdom, Jr.
Henkel Corporation
Law Dept.
140 Germantown Pike
Suite 150
Plymouth Meeting, PA 19462   (215) 832-2226

Fred C. Battles#
516 Morgantown Street
Point Marion, PA 15474   (412) 725-5414

George R. Dohmann
604 West Market St.
Pottsville, PA 17901   (717) 622-3483

Arthur R. Eglington
121 Queen Street
Pottsville, PA 17901   (717) 622-0137

Walter B. Udell
Maleson, Udell & Rosenberg
P.O. Box 1010
Quakertown, PA 18951   (215) 536-4500

Henry Robinson Ertelt
184 Berwind Circle
Radnor, PA 19087   (215) 688-4315

Jack L. Foltz
Sun Company, Inc.
100 Matsonford Rd.
Radnor, PA 19087   (215) 293-6392

Richard O. Church
P.O. Box 15146
Reading, PA 19612   (215) 777-6487

Leonard M. Quittner
532 Court Street
Reading, PA 19601   (215) 376-7151

John G. Schwartz#
17F Congressional Circle
Reading, PA 19607   (215) 775-9980

Donald C. Watson
111 Montrose Blvd.
Reading, PA 19607   (215) 777-4660

Everett H. Murray, Jr.
141 Browning Lane
Rosemont, PA 19010   (215) 525-4497

Lucretia R. Quatrini#
409 South Wilbur Avenue
Sayre, PA 18840   (717) 888-7022

Harry D. Anspon#
29 Beaver Street
Sewickley, PA 15143   (412) 741-6029

Thomas W. Brennan#
132 Apache Drive
Shickshinny, PA 18655   (717) 256-3016

John R. Ewbank
1150 Woods Rd.
Southampton, PA 18966   (215) 357-3977

Cathy A. Musco
1206 Whitney Road
Southampton, PA 18966   (215) 355-1141

Charles M. Allen
Howson & Howson
Spring House Corp. Center
P.O. Box 457
Spring House, PA 19477   (215) 540-9200

Mary E. Bak
Howson & Howson
Spring House Corporate Center
P.O. Box 457
Spring House, PA 19477   (215) 540-9200

Ronald D. Bakule#
Rohm & Haas Company
727 Norristown Road
Spring House, PA 19477   (215) 641-7822

Henry Hansen
Howson & Howson
Spring House Corp. Center
P.O. Box 457
Spring House, PA 19477   (215) 540-9200

Thomas J. Howell#
Rohm & Haas Co.
727 Norristown Rd.
Spring House, PA 19477   (215) 641-7275

Stanley B. Kita
Howson & Howson
Spring House Corporate Center
P.O. Box 457
Spring House, PA 19477   (215) 540-9200

Wilson Oberdorfer
Howson & Howson
Spring House Corporate Center
Box 457
Spring House, PA 19477   (215) 540-9200

Karl F. Ockert
Rohm & Haas Company
Research Laboratories
727 Norristown Road
Spring House, PA 19477   (215) 641-7478

George A. Smith, Jr.
Howson & Howson
Spring House Corporate Center
P.O. Box 457
Spring House, PA 19477   (215) 540-9204

William Freedman
723 Rhoads Drive
Springfield, PA 19064   (215) 544-4330

Karl E. Geci
Twin Springs
962 Million Dollar Highway
St. Marys, PA 15857   (814) 781-3409

Virgil P. Quirk#
828 Johnsonburg Road
St. Marys, PA 15857   (814) 834-3610

Robert F. Custard
460 Glenn Road
State College, PA 16803   (814) 238-7323

Richard L. Hansen
Calder Square
P.O. Box 10361
State College, PA 16805   (814) 863-1160

Thomas E. Sterling
P.O. Box 14
234 E. College Ave.
State College, PA 16801   (814) 238-9455

Arthur W. Collins
514 School Ln.
Swarthmore, PA 19081   (215) 543-1620

Dante J. Picciano
31 Center Street
Tamaqua, PA 18252   (717) 668-4774

Jonathan D. Holmes#
R.D. 2, Box 859K
Thomasville, PA 17364   (717) 757-2600

Donald Russell Motsko
Box 294, Rt. 2
Thompson, PA 18465   (717) 727-2928

Steven D. Boyd
Betz Laboratories, Inc.
4636 Somerton Road
Trevose, PA 19047   (215) 355-3300

Gregory M. Hill
Betz Laboratories, Inc.
4636 Somerton Road
Trevose, PA 19047   (215) 355-3300

Richard A. Paikoff
Betz Laboratories, Inc.
4636 Somerton Road
Trevose, PA 19053   (215) 953-2489

Alexander D. Ricci
Betz Laboratories, Inc.
4636 Somerton Rd.
Trevose, PA 19047   (215) 953-2338

Philip H. Von Neida
Betz Laboratories, Inc.
4636 Somerton Road
Trevose, PA 19053   (215) 953-2418

Thomas J. Monahan
Intellectual Property Office
Pennsylvania State University
306 W. College Ave.
University Park, PA 16801   (814) 865-6277

James A. Spady
612 Beverly Blvd.
Upper Darby, PA 19082   (215) 352-7144

Lawrence E. Ashery#
Ratner & Prestia
P.O. Box 980
Valley Forge, PA 19482   (215) 265-6666

Lori Yanisko Beardell#
Ratner & Prestia
500 N. Gulph Rd.
Leighton Bldg., Ste. 412
P.O. Box 980
Valley Forge, PA 19482   (215) 265-6666

Kevin R. Casey
Ratner & Prestia
P.O. Box 980
500 North Gulph Rd.
Valley Forge, PA 19482   (215) 265-6666

Guy T. Donatiello
Ratner & Prestia
Box 980
Valley Forge, PA 19482   (215) 265-6666

John F. A. Earley
Harding, Earley, Follmer
& Frailey
86 The Commons At Valley Forge E.
1288 Valley Forge Rd., P.O. Box 750
Valley Forge, PA 19482   (215) 935-2300

John F.A. Earley, III
Harding, Earley, Follmer
& Frailey
86 The Commons At Valley Forge E.
1288 Valley Forge Rd., P.O. Box 750
Valley Forge, PA 19482   (215) 935-2300

Frank A. Follmer
Harding, Earley, Follmer
& Frailey
86 The Commons At Valley Forge E.
1288 Valley Forge Rd., P.O. Box 750
Valley Forge, PA 19482   (215) 935-2300

Andrew Louis Ney
Ratner & Prestia
P.O. Box 980
Valley Forge, PA 19482   (215) 265-6666

Kenneth N. Nigon
Ratner & Prestia
500 N. Gulph Road
P.O. Box 980
Valley Forge, PA 19482   (215) 265-6666

Michael F. Petock
46 The Commons At Valley Forge
1220 Valley Forge Road
P.O. Box 856
Valley Forge, PA 19481   (215) 935-8600

Ernest G. Posner
PQ Corporation
P.O. Box 840
500 E. Swedesford Road
Valley Forge, PA 19482   (215) 293-7354

Paul F. Prestia
Ratner & Prestia
Leighton Bldg.
500 N. Gulph Rd., Ste 412
P.O. Box 980
Valley Forge, PA 19482   (215) 265-6666

Allan Ratner
Ratner & Prestia
Box 980
Valley Forge, PA 19482   (215) 265-6666

William Schmonsees
Ratner & Prestia
500 N. Gulph Road
Suite 412
P.O. Box 980
Valley Forge, PA 19482   (215) 265-6666

Paul A. Serbinowski#
Harding, Earley, Follmer & Frailey
86 The Commons At Valley Forge E.
1288 Valley Forge Road
P.O. Box 750
Valley Forge, PA 19482   (717) 733-8719

James B. Bechtel*
U.S. Navy
Naval Air Development Center
Code 095
Warminster, PA 18974   (215) 411-3000

Marvin C. Gaer*
U.S. Navy
Naval Air Dev. Center
Code 4033
Warminster, PA 18970   (215) 441-2031

James Vincent Tura*
U.S. Dept. of Navy
Naval Air Development Center
Street & Jacksonville Rds.
Warminster, PA 18974   (215) 441-3000

Susan E. Verona*
U.S. Navy
Naval Air Dev. Center
Code 095
Warminster, PA 18974   (215) 441-3000

Allan H. Fried
121 N. Wayne Avenue
Wayne, PA 19087   (215) 254-0120

Frederick J. Olsson
Suite 216, Bldg. No. 8
Valley Forge Executive Mall
Wayne, PA 19087   (215) 687-1676

Francis S. Husar#
320 Antietam Drive
Waynesboro, PA 17268   (717) 762-3131

Thomas V. Flanagan
Novitiate of St. Isaac Jogues
Wernersville, PA 19565   (215) 678-8085

Randall J. Gort
Commodore Intl. Ltd.
1200 Wilson Drive
West Chester, PA 19380   (215) 431-9131

A. Newton Huff
407 Eaton Way
West Chester, PA 19380   (215) 431-0167

Marvin J. Powell
Green & Powell
105 Evans Street East
Evans Bldg., Suite C
P.O. Box 654
West Chester, PA 19381   (215) 436-4486

Laurence A. Weinberger
1414 Morstein Road
West Chester, PA 19380   (215) 431-1703

Lewis J. Young
1202 Clearbrook Road
West Chester, PA 19380   (215) 696-1076

Anthony J. Dixon
120 East Broad Street
West Hazleton, PA 18201   (717) 455-7112

Floyd S. Scheier,
391 Mc Kinney Road
Wexford, PA 15090   (412) 935-6746

John P. Blasko
Ferrill & Logan
Suite C-13, Executive MEWS
2300 Computer Avenue
Willow Grove, PA 19090   (215) 657-6850

Thomas M. Ferrill, Jr.
Ferrill & Logan
C-13 Executive MEWS
2300 Computer Avenue
Willow Grove, PA 19090   (215) 657-6850

David J. Johns
Ferrill & Logan
C-13 Executive MEWS
2300 Computer Avenue
Willow Grove, PA 19090   (215) 657-6850

John W. Logan, Jr.
Ferrill & Logan
C-13 Executive MEWS
2300 Computer Avenue
Willow Grove, PA 19090   (215) 657-6850

Michael S. L. Lovitz
Ferrill & Logan
C-13 Executive MEWS
2300 Computer Ave.
Willow Grove, PA 19090   (215) 657-6850

S. Debra Miller
27 Carriage House Drive
Willow Street, PA 17584   (717) 464-9125

B. Max Klevit
B-221 Cedarbrook Hill Apts.
Wyncote, PA 19095   (215) 886-9146

Frank Joseph Benasutti
616 Clothier Rd.
Wynnewood, PA 19096   (215) 896-6860

James Fay Hall, Jr.
926 Remington Road
Wynnewood, PA 19096   (215) 642-1610

Joel Y. Loewenberg
1364 Indian Creek Drive
Wynnewood, PA 19096   (215) 896-8144

Ervin B. Steinberg#
1219 W. Wynnewood Road
No. 410
Wynnewood, PA 19096   (215) 649-6833

E. Barron Batchelder
P.O. Box 6842
Wyomissing, PA 19610   (215) 678-6731

J. Spencer Overholser
58 Downing Drive
Wyomissing, PA 19610   (215) 670-0508

Donald D. Denton
2200 Yardley Road
Yardley, PA 19067   (215) 493-1532

Daniel E. Kramer
2009 Woodland Dr.
Yardley, PA 19067   (215) 493-4830

Edward J. Sites
Union Camp Corporation
816 Sumter Drive
Yardley, PA 19067   (215) 493-8526

Paul B. Weisz#
3 Delaware Rim Drive
Yardley, PA 19067   (215) 493-3551

Edward J. Hanson, Jr.
Dentsply International Inc.
570 W. College Avenue
York, PA 17405   (717) 845-7511

Charles H. Just
1654 Wyntre Brooke Drive, N.
York, PA 17403   (717) 741-2242

Samuel M. Learned, Jr.
149 East Market Street
York, PA 17401   (717) 846-9290

Charles J. Long
Smith & Le Cates
124 E. Market St.
York, PA 17401   (717) 845-9641

Dale R. Lovercheck
Dentsply International Inc.
570 West College Avenue
York, PA 17405   (717) 845-7511

Daniel J. O Connor
52 S. Duke Street
York, PA 17401   (717) 845-6593

Richard G. Weber
Precision Components Corporation
500 Lincoln Street
York, PA 17404   (717) 848-1126

Rhode Island

Daniel A. Curran
P.O. Box 432
Adamsville, RI 02801   (401) 635-8519

Louis S. Coppolino#
21 Houghton Street
Barrington, RI 02806   (401) 246-1788

Robert J. Doherty
10 George Street
Barrington, RI 02806   (401) 431-1320

Herbert B. Barlow, Jr.
Barlow & Barlow, Ltd.
1150 New London Ave.
Cranston, RI 02920   (401) 463-6830

John Edward Flanagan
37 Shore Drive
Middletown, RI 02840   (401) 847-7259

Peter C. Lauro
9 Brook Road
Narragansett, RI 02882   (401) 847-7259

Louis B. Applebaum
24 Toppa Blvd.
Newport, RI 02840   (401) 846-1241

Prithvi C. Lall*
U.S. Dept. of Navy
Off. of Gen. Coun.
Bldg. 112T
Newport, RI 02841   (401) 841-4736

Michael J. Mc Gowan*
Office of Naval Research
Off. of Gen. Counsel
Bldg. 112T
Newport, RI 02841   (401) 841-4736

Michael F. Oglo*
U.S. Dept of the Navy
Off. of Pat. Coun., Code OOOC
Naval Underwater Systems Ctr.
Newport, RI 02841   (401) 841-4736

Gordon Gale Menzies
210 W. Main St.
North Kingstown, RI 02852   (401) 294-4321

Arthur A. Mc Gill
7 Westwood Drive
Portsmouth, RI 02871   (401) 683-3532

Kurt R. Benson
Salter & Michaelson
Heritage Bldg.
321 South Main Street
Providence, RI 02903   (401) 421-3141

Ralph Douglas Gelling
Textron Inc.
40 Westminster Street
Providence, RI 02903   (401) 457-2320

Leonard Michaelson
Salter and Michaelson
321 South Main Street
Providence, RI 02903   (401) 421-3141

William W. Rymer, Jr.
Fish & Richardson
30 Kennedy Plaza
5th Floor
Providence, RI 02903   (401) 331-0181

Elliot A. Salter
Salter and Michaelson
321 S. Main Street
Providence, RI 02903   (401) 421-3141

Albert P. Davis
2080 Boston Neck Road
Saunderstown, RI 02874   (401) 294-3410

Burnett W. Norton
John Brown Inc.
1600 Division Road
West Warwick, RI 02893   (401) 884-9920

South Carolina

William A. Callahan#
Route 1, Box A-142
Aiken, SC 29801   (803) 652-3454

Harold M. Dixon*
U.S. Dept. of Energy
Office of Chief Counsel
P.O. Box A
Aiken, SC 29802   (803) 725-2497

Alfons G. Hutter
337 Lake Ridge Lane
Anderson, SC 29624   (803) 225-9828

William S. Rambo
204 Providence Villas
R.R. 4
Anderson, SC 29624   (803) 224-9667

William A. Dallis, Jr.
Dallis & Dreyfoos
124 Meeting Street
P.O. Box 1840
Charleston, SC 29402   (803) 577-9425

Billy C. Killough
120 Meeting Street
Charleston, SC 29401   (803) 577-9800

John P. Kozma
1333 S. Edgewater Drive
Charleston, SC 29407   (803) 766-2531

James Spool
Siebe North, Inc.
4090 Azalea Drive
Charleston, SC 29405   (803) 554-0660

Donald H. Feldman#
210-1 Cochran Road
Clemson, SC 29631   (803) 654-5483

Edwin B. Brading*
U.S. Dept. of Housing
& Urban Development
Strom Thurmond Fed. Bldg.
1835-45 Assembly St. Ste. 1155f
Columbia, SC 29201   (803) 765-5761

F. Rhett Brockington#
4016 Mac Gregor Drive
Columbia, SC 29206   (803) 787-7922

Timothy D. Harbeson
1205 Pendleton Street
Columbia, SC 29201   (802) 734-0457

Michael A. Mann
Suite 506
1401 Main Street
Columbia, SC 29201   (803) 254-8472

Joan R. Owen
432 Joshua St.
Columbia, SC 29205   (803) 254-1555

Benoni O. Reynolds
P.O. Drawer 6924
5019 Forest Lake Place
Columbia, SC 29260   (803) 782-9449

Leigh P. Gregory
W.R. Grace & Co.
P. O. Box 464
100 Rogers Bridge Rd.
Duncan, SC 29334   (803) 433-2332

William D. Lee, Jr.
W.R. Grace and Co.
Cryovac Div.
P.O. Box 464
Duncan, SC 29334   (803) 433-2334

Mark B. Quatt
W.R. Grace & Company
Cryovac Division
P.O. Box 464
Duncan, SC 29334   (803) 433-2817

Jennifer L. Skord
W.R. Grace & Co.
Cryovac Division
P.O. Box 464
Duncan, SC 29334   (803) 433-2496

John J. Toney
W.R. Grace & Co.
Cryovac Div.
100 Rogers Bridge Rd.
Bldg. A
Duncan, SC 29334   (803) 433-2332

Wesley L. Brown
Saint-Amand, Thompson & Brown
P.O. Box 936
210 S. Limestone Street
Gaffney, SC 29340   (803) 489-6052

James M. Bagarazzi
Dority & Manning
700 E. North Street
Suite 15
Greenville, SC 29601   (803) 271-1592

Ralph Bailey
Bailey & Hardaway
125 Broadus Ave.
Greenville, SC 29601   (803) 233-1338

Michael A. Cicero
Bailey & Hardaway
125 Broadus Ave.
Greenville, SC 29601   (803) 233-1338

Julian W. Dority
Dority & Manning
700 East North Street
Greenville, SC 29601   (803) 271-1592

Mark C. Dukes
Dority & Manning
700 E. North Street
Suite 15
Greenville, SC 26019   (803) 271-1592

Thomas W. Epting
Leatherwood, Walker, Todd
& Mann, P.C.
100 East Coffee St.
P.O. Box 87
Greenville, SC 29602   (803) 242-6440

Cort R. Flint, Jr.
P.O. Box 10827
Federal Station
501 E. Mc Bee Avenue
Greenville, SC 29601   (803) 232-4261

John B. Hardaway, III
Bailey & Hardaway
125 Broadus Ave.
Greenville, SC 29601   (803) 233-1338

Bryan F. Hickey
Haynsworth, Marion, Mckay
& Guerard
75 Beattie Place
C&S Tower, 11th Floor
Greenville, SC 29601   (803) 240-3246

Martin K. Lindemann
102 Independence Drive
Greenville, SC 29615   (803) 288-4799

Wellington M. Manning, Jr.
Dority & Manning
700 E. North Street
Suite 15
Greenville, SC 29601   (803) 271-1592

Ralph M. Mellom
Ogletree, Deakins, Nash,
Smoak & Stewart
P.O. Box 2757
1000 E. North St.
Greenville, SC 29602   (803) 242-1410

Richard M. Moose
Dority & Manning, P.A.
700 E. North Street
Suite 15
Greenville, SC 29601   (803) 271-1592

Robert R. Reed#
Michelin Americas R&D Corp.
515 Michelin Road
Greenville, SC 29605   (803) 277-8780

James E. Reynolds, Jr.
Fluor Daniel, Inc.
100 Fluor Daniel Drive
C303f
Greenville, SC 29607   (803) 281-8095

Roger F. Bley
103 Amhurst Dr.
Greenwood, SC 29646   (803) 223-0565

Robert F. Burnett
Route 1, Box 385
Greenwood, SC 29646   (803) 227-1332

C. Gordon Mc Bride
P.O. Box 2555
644 South Fourth Street
Hartsville, SC 29550   (803) 332-0193

James K. Everhart, Jr.
43 N. Port Royal Dr.
Hilton Head, SC 29928   (803) 681-2189

Mathew P. Mc Dermitt, Sr.
23 Rusty Rail Lane
Hilton Head, SC 29926   (803) 681-8808

Willard M. Hanger
Pope/Greenwood Office
Park Building 7
Suite 723
P.O. Box 3078
Hilton Head Island, SC 29928
(803) 785-7882

Milford W. Mac Donald
28 Angel Wing Drive
Hilton Head Island, SC 29926
(803) 681-5291

Raymond F. Mac Kay
82 Crosstree Drive
Hilton Head Island, SC 29928
(803) 681-8175

Wilbur C. Tupman
71 Tabby Trail
Hilton Head Island, SC 29926
(803) 837-4580

Francis W. Crotty
2 Pine Point Road
Lake Wylie, SC 29710
(803) 681-8808

John H. Gallagher
858 Sovereign Terrace
Mt. Pleasant, SC 29464
(802) 884-3540

H. Hume Mathews
East 411
5001 Little River Road
Myrtle Beach, SC 29577
(803) 449-7612

William J. Smith#
P.O. Box 814
Newberry, SC 29108
(803) 276-4946

Daniel B. Reece, IV
Westvaco Corp.
5600 Virginia Ave.
P.O. Box 2941105
North Charleston, SC 29411
(803) 745-3700

William R. Hovis
Hovis & Duncan
P.O. Box 10970
Rock Hill, SC 29731
(803) 324-1122

Stanley C. Dalton
23 Skipper
Salem, SC 29676
(803) 324-1122

Edward L. Bailey
180 Library Street
Spartanburg, SC 29301
(803) 582-3733

George M. Fisher
Milliken Research Corp.
Interstate 85 At I-585, S.E.
Spartanburg, SC 29304
(803) 573-1598

Kevin M. Kercherr
Milliken & Company
Legal Dept. (M-495)
P.O. Box 1926
920 Milliken Road
Spartanburg, SC 29302
(803) 573-2266

Earle Rollins Marden
Milliken Research Corp.
P.O. Box 1927
Spartanburg, SC 29304
(803) 573-1599

Timothy J. Monahan
Milliken & Company
Legal Dept. (M-495)
P.O. Box 1926
920 Milliken Road
Spartanburg, SC 29302
(803) 573-2266

Terry T. Moyer
Milliken Res. Corp.
M-405
P.O. Box 1927
Spartanburg, SC 29302
(803) 573-1600

Henry William Petry
Milliken Research Corp.
920 Milliken Road
Spartanburg, SC 29303
(803) 573-2266

Luke John Wilburn, Jr.
364 S. Pine Street
P.O. Box 5445
Spartanburg, SC 29304
(803) 585-6688

John C. Bigler
P.O. Box 177
Sullivan's Island, SC 29482
(908) 859-7741

South Dakota

Ivar M. Kaardal
805 N. Elmwood Ave.
Sioux Falls, SD 57104
(605) 336-9446

Tennessee

L. Aubrey Goodson
6005 Old Jonesboro Road
Bristol, TN 37620
(615) 878-6356

David J. Hill
6400 Lee Highway
P.O. Box 23307
Chattanooga, TN 37422
(615) 892-7120

Douglas T. Johnson
Miller & Martin
Suite 1000
Georgia Avenue
Chattanooga, TN 37402
(615) 756-6600

Micheline K. Johnson
Spears, Moore, Rebman
& Williams
8th Floor
Blue Cross Building
Chattanooga, TN 37402
(615) 756-7000

James L. Johnston
811 Chattanooga Bank Bldg.
Chattanooga, TN 37402
(615) 266-5531

Alan Ruderman
806 Maclellan Bldg.
721 Broad Street
Chattanooga, TN 37402
(615) 267-6980

John W. Tucker#
6 Brockhaven Road
Chattanooga, TN 37404
(615) 622-4696

Louis Milton Deckelmann#
303 Woodhaven Lane
Clinton, TN 37716
(615) 457-2491

Abram Wooldridge Hatcher
Nathan Smith Road
P.O. Box 112
College Grove, TN 37046
(615) 368-7256

Adrian J. Good
Great Lakes Research Division
P.O. Box 1031
Elizabethton, TN 37643
(615) 542-1703

Carl F. Peters
415 East K Street
Elizabethton, TN 37643
(615) 543-5022

Robert L. Taylor#
1433 West G St.
P.O. Box 681
Elizabethton, TN 37644
(615) 542-5475

Remega G. Hyder#
Route 1 Box 228
Hampton, TN 37658
(615) 725-3482

David Schwendinger#
Route 2 Box 1620
Pine Ridge Drive
Heiskel, TN 37754
(615) 457-8642

John P. Dority
104 Cherry Hill Drive
Hendersonville, TN 37075
(615) 824-8068

William J. Hite, III
4234 Sweden Drive
Hermitage, TN 37076
(615) 871-0950

Donald W. Spurrell
515 E. Watauga Ave.
Johnson City, TN 37601
(615) 929-1700

Greer C. Tidwell, Jr.
Baker, Worthington, Crossley,
Stansberry & Woolf
207 Mockingbird Lane
P.O. Box 3038 CRS
Johnson City, TN 37602
(615) 928-0181

Gary C. Bailey
Tennessee Eastman Company
P.O. Box 511
Kingsport, TN 37662
(615) 229-4941

George P. Chandler
Tennessee Eastman Co.
Eastman Road
P.O. Box 511
Kingsport, TN 37662
(615) 229-3620

Malcolm Graeme Dunn
Suite 122
1201 N. Eastman Road
Kingsport, TN 37664
(615) 378-6322

Bernard J. Graves, Jr.
Eastman Chemical Co.
P.O. Box 511
Kingsport, TN 37662   (615) 229-1365

William P. Heath, Jr.
Tennessee Eastman Co.
P.O. Box 511
Kingsport, TN 37662   (615) 229-3634

Charles R. Martin
Eastman Chemicals Division
P.O. Box 511
Kingsport, TN 37662   (615) 229-4863

Mark A. Montgomery
Eastman Chemicals Division
P.O. Box 511
Kingsport, TN 37662   (615) 229-8862

Daniel B. Reece, III
Tennessee Chemical Co.
P.O. Box 511
Kingsport, TN 37662   (615) 229-2097

Thomas R. Savitsky
Eastman Chemical Company
Legal Dept., Bldg. 75
P.O. Box 511
Kingsport, TN 37662   (615) 229-4305

John Frederick Stevens
Eastman Chemical Co.
Div. of Eastman Kodak Co.
P.O. Box 511
Kingsport, TN 37662   (615) 229-3618

John Frederick Thomsen
Tennessee Eastman Co.
Eastman Road
P.O. Box 511
Legal Dept., Bldg. 75
Kingsport, TN 37662   (615) 229-2282

Clyde L. Tootle
222 1/2 E. Center St.
P.O. Box 770
Kingsport, TN 37662   (615) 378-5771

Joseph A. Marasco
P.O. Box 1034
Kingston, TN 37763   (615) 376-9839

Martin J. Skinner#
836 Nelson Dr.
Kingston, TN 37763   (615) 376-6894

Monroe A. Brown
Suite 227 Digital Off. Plaza
9040 Executive Park Drive
Knoxville, TN 37923   (615) 531-3430

Judy W. Goans
900 S. Gay Street
Riverview Tower
Suite 1850
Knoxville, TN 37902   (615) 546-7770

Mark S. Graham
Luedeka, Hodges & Neely, P.C.
1030 First American Center
Knoxville, TN 37902   (615) 546-4305

Paul E. Hodges
Luedeka, Hodges & Neely, P.C.
1030 First American Center
Knoxville, TN 37902   (615) 546-4305

Geoffrey D. Kressin
Norton & Luhn, P.C.
Suite 2600
800 South Gay Street
Knoxville, TN 37902   (615) 522-2662

Edwin M. Luedeka
Luedeka, Hodges & Neely, P.C.
1030 First American Center
505 S. Gay Street
Knoxville, TN 37902   (615) 546-4305

Michael E. Mckee
Luedeka, Hodges & Neely, P.C.
1030 First American Center
Knoxville, TN 37902   (615) 546-4305

Andrew S. Neely
Luedeka, Hodges & Neely, P.C.
1030 First American Center
Knoxville, TN 37902   (615) 546-4305

Robert E. Pitts
Pitts & Brittian
P.O. Box 51295
1116 Weisgarber Building
Knoxville, TN 37950   (615) 584-0105

William T. Dixson, Jr.#
111 Caldwell Street
Mc Minnville, TN 37110   (615) 546-7770

Howard Roy Berkenstock, Jr.
Richards Medical Company
1450 Brooks Road
Memphis, TN 38116   (901) 396-2121

William B. Clemmons, Jr.
Heiskell, Donelson, Bearman,
Adams, Williams & Kirsch
20th Floor
First Tennessee Bldg.
Memphis, TN 38103   (901) 526-2000

Larry W. Mc Kenzie
Walker & Mc Kenzie, P.C.
6363 Poplar,
Suite 434
Memphis, TN 38119   (901) 685-7428

John J. Mulrooney
Suite 360
6055 Primacy Pkwy.
Memphis, TN 38119   (901) 763-3336

Loyal W. Murphy, III
5050 Poplar
Suite 1214
Memphis, TN 38157   (901) 767-4701

John Russell Walker, III
Walker & Mc Kenzie, P.C.
6363 Poplar Avenue
Suite 434
Memphis, TN 38119   (901) 685-7428

Johnny O. Younghanse#
4405 Ross Road
Memphis, TN 38141   (901) 362-8976

Leon Daniel Wofford, Jr.
Route 1 Box 54
Mulberry, TN 37359   (615) 759-7803

Stephen T. Belsheim
2014 Broadway
Suite 250
Nashville, TN 37203   (615) 321-3001

Edward D. Lanquist, Jr.
Manier, Herod, Hollabaugh,
& Smith
2200 One Nashville Place
150 Fourth Ave. N.
Nashville, TN 37219   (615) 244-0030

Lawrence C. Maxwell
Trabue, Sturdivant & Dewitt
25th Floor
Nashville City Center
511 Union Street
Nashville, TN 37219   (615) 244-9270

Gary V. Pack
Service Merchandise Co., Inc.
P.O. Box 24600
Nashville, TN 37202   (615) 366-3215

Mark J. Patterson
Manier, Herod, Hollabaugh
& Smith
150 4th Avenue N.
Suite 2200
Nashville, TN 37219   (615) 244-0030

Ira C. Waddey, Jr.
213 Fifth Avenue
Nashville, TN 37219   (615) 244-7545

Casey F. Wilson
Boult, Cummings, Conners
& Berry
222 Third Avenue, North
Nashville, TN 37201   (615) 252-2342

Irving Barrack*
U.S. Dept. of Energy
Federal Office Bldg.
P.O. Box E
Oak Ridge, TN 37831   (615) 576-1072

David E. Breeden#*
U.S. Dept. of Energy
Oak Ridge Oper. Off.
Administration Road
P.O. Box E
Oak Ridge, TN 37831   (615) 576-1082

George L. Craig
Martin Marietta Energy
Systems, Inc.
M.S. 8218
Oak Ridge, TN 37830   (615) 576-9676

Ivan L. Ericson
Martin Marietta Energy Sys., Inc.
P.O. Box 2009
Oak Ridge, TN 37831   (615) 574-2222

James D. Griffin#
Martin Marietta Energy Systems, Inc.
Systems, Inc.
P.O. Box Y
Oak Ridge, TN 37849   (615) 574-4178

Stephen David Hamel*
U.S. Dept. of Energy
P.O. Box E
Oak Ridge, TN 37830   (615) 576-1073

Herman L. Holsopple, Jr.#
101 Case Lane
Oak Ridge, TN 37830   (615) 482-1623

Earl L. Larcher*
U.S. Dept. of Energy
P.O. Box 2001
Oak Ridge, TN 37831   (615) 576-1080

Katherine Parks Lovingood*
U.S. Dept. of Energy
Office of Patent Counsel
P.O. Box 2001
Oak Ridge, TN 37831   (615) 576-1076

Robert Maxwell Poteat*
U.S. Dept. of Energy
P.O. Box 2000
Oak Ridge, TN 37830   (615) 576-1070

James M. Spicer#
Martin Marietta Energy Systems, Inc.
P.O. Box 2008
Building 4500 N, MS 6258
Oak Ridge, TN 37831   (615) 574-4180

David S. Zachry, Jr.
874 West Outer Drive
Oak Ridge, TN 37830   (615) 483-6725

Vivian G. Hyder#
P.O. Box 147
Rogersville, TN 37857   (615) 345-3950

Max L. Harwell#
Rt. 1, Box 280
Sardis, TN 38371   (901) 687-3854

Texas

F. Vern Lahart
DLM, Inc.
P.O. Box 3000
Allen, TX 75002   (214) 248-6300

John R. Merkling
Intermedics, Inc.
4000 Technology Drive
Angleton, TX 77515   (409) 848-4196

Richard L. Robinson
Intermedics, Inc.
4000 Technology Dr.
Angleton, TX 77515   (409) 848-4083

Frederick S. Frei
Snider & Moore
505 Ryan Plaza Drive
Suite 337
Arlington, TX 76011   (817) 861-1122

Charles William Mc Hugh
1010 Milby Road
Arlington, TX 76013   (817) 461-3113

Peggy L. Smith#
4814 Tamanaco Court
Arlington, TX 76017   (817) 467-9512

Michael P. Adams
Fulbright & Jaworski
600 Congress
2400 One American Center
Austin, TX 78701   (512) 474-5201

David D. Bahler
Arnold, White & Durkee
2300 One American Center
600 Congress Avenue
Austin, TX 78701   (512) 320-7200

Wayne P. Bailey#
IBM Corp.
11400 Burnet Road, #4054
Austin, TX 78758   (512) 823-1012

William G. Barber
Arnold, White & Durkee
2300 One American Center
Austin, TX 78701   (512) 320-7200

Thomas M. Blasey#
1840 Burton Drive, Apt. 137
Austin, TX 78741   (512) 443-9128

Thomas J. Bonk#
3M
P.O. Box 2963
Austin, TX 78769   (512) 984-3964

Andrea P. Bryant
IBM Corporation
Dept. 932
11400 Burnet Rd.
Austin, TX 78758   (512) 838-1003

Robert M. Carwell
IBM Corporation
Intellectual Property Law 932/815
11400 Burnet Road
Austin, TX 75758   (512) 823-1017

James L. Clingan, Jr.
Motorola, Inc.
3501 Ed Bluestein Blvd.
Austin, TX 78721   (512) 928-7004

Russell D. Culbertson
Shaffer & Culbertson
1250 Capital of Texas Hwy., S.
Bldg. Two, Suite 560
Austin, TX 78746   (512) 327-8932

Kevin L. Daffer
Arnold, White & Durkee
2300 One American Center
Austin, TX 78701   (512) 320-7200

Thomas George Devine
Dell Computer Corporation
9505 Arboretum Blvd.
Austin, TX 78759   (512) 338-8553

Dudley R. Dobie, Jr.
University of Texas System
Off. of Gen. Counsel
201 W. 7th Street
Austin, TX 78701   (512) 499-4462

Jasper W. Dockrey#
Motorola, Inc.
3501 Ed Bluestein Blvd.
V-Bldg., F-4
Austin, TX 78721   (512) 928-6539

Paul S. Drake
IBM Corp.
Intellectual Property Law
11400 Burnet Road - 4054
Austin, TX 78758   (512) 823-1005

Heinz D. Grether
Thompson & Knight
98 San Jacinto
Suite 1200
Austin, TX 78701   (512) 474-8211

Richard J. Groos
Arnold, White & Durkee
2300 One American Center
600 Congress Avenue
Austin, TX 78701   (512) 474-2583

Curtis A. Henschen*
Federal Bureau of Investigation
300 East 8th
Austin, TX 78767   (512) 478-8501

Daniel S. Hodgins
Arnold, White & Durkee
2300 One American Center
Austin, TX 78701   (512) 474-2583

Charles D. Huston
Schlumberger
8311 North, R.R. 620
P.O. Box 200015
Austin, TX 78720   (512) 331-3700

John Lionel Jackson
IBM Corp.
Intellectual Property Law
Dept. 932
11400 Burnet Road
Austin, TX 78758   (512) 823-1000

Prentiss Wayne Johnson#
3502 Denwood Drive
Austin, TX 78759   (512) 346-0912

Maurice J. Jones, Jr.
Motorola, Inc.
Intellectual Property Dept.
1 Texas Ctr., 505 Barton Springs Rd.
Suite 500, Mail Drop F4
Austin, TX 78704   (512) 322-8910

Jerry M. Keys
Thompson & Knight
98 San Jacinto Blvd.
Suite 1200
Austin, TX 78701   (512) 474-8211

William W. Kidd
Sematech, Inc.
2706 Montopolis Drive
Austin, TX 78741   (512) 356-3752

John R. King#
2018 W. Rundberg Lane
Apt. 12-C
Austin, TX 78758   (512) 356-3752

Robert L. King
Motorola, Inc.
One Texas Center
505 Barton Springs Rd.
Suite 500, Mail Drop F4
Austin, TX 78704   (512) 322-8910

Thomas E. Kirkland
5907 Ivy Hills Drive
Austin, TX 78759   (512) 322-8910

Barbara S. Kitchell
Arnold, White & Durkee
600 Congress Ave.
2300 One American Center
Austin, TX 78701   (512) 320-7200

Julius B. Kraft
IBM Corp.
11400 Burnet Road
Austin, TX 78759   (512) 823-1006

Dale A. Kubly
Fisher Controls International, Inc.
1712 Centre Creek Drive
Austin, TX 78754   (512) 834-7291

Douglas H. Lefeve
IBM Corp.
11400 Burnet Road
Dept. 932, Zip 4054
Austin, TX 78758   (512) 823-1013

Joseph F. Long
1411 West Ave.
Suite 101B
Austin, TX 78701   (512) 476-4066

James B. Marshall, Jr.
3M Co.
Bldg. A130-2N-32
P.O. Box 2963
Austin, TX 78769   (512) 984-3134

Denise L. Mayfield
Arnold, White & Durkee
2300 One American Center
600 Congress Avenue
Austin, TX 78701   (512) 320-7200

Mark E. Mc Burney
IBM Corporation
Intellectual Property Law Dept.
932/815 - 4054
11400 Burnet Road
Austin, TX 78758   (512) 823-0000

Michael S. Metteauer
Arnold, White & Durkee
2300 One American Center
600 Congress Ave.
Austin, TX 78701   (512) 320-7200

Jonathan P. Meyer
Motorola, Inc.
Patent Dept.
3501 Ed Bluestein Blvd.
Mail Drop F4
Austin, TX 78721   (512) 928-7006

Eric B. Meyertons
Arnold, White & Durkee
2300 One American Center
600 Congress Ave.
Austin, TX 78701   (512) 320-7200

Jack V. Musgrove
3M Austin Center
P.O. Box 2963
Austin, TX 78769   (512) 984-6224

Jeffrey V. Myers
Motorola, Inc.
6501 William Cannon West
Austin, TX 78735   (512) 440-2393

Floyd R. Nation
Arnold, White & Durkee
600 Congress Avenue
2300 One American Center
Austin, TX 78701   (512) 320-7200

Harry D. Nelson
Advanced Micro Devices Inc.
5204 E. Ben White Blvd.
Austin, TX 78741   (512) 462-4269

David L. Parker
Arnold, White & Durkee
2300 One American Center
6th & Congress
Austin, TX 78701   (512) 320-7200

Louis T. Pirkey
Arnold, White & Durkee
2300 One American Center
600 Congress Avenue
Austin, TX 78701   (512) 320-7200

Paul J. Polansky#
Motorola, Inc.
3501 Ed Bluestein Blvd.
V-Bldg. F4
Austin, TX 78721   (512) 928-7460

Kenneth R. Priem
10305 Mourning Dove Circle
Austin, TX 78750   (512) 258-9031

William D. Raman
Arnold, White & Durkee
2300 One American Center
600 Congress Avenue
Austin, TX 78701   (512) 320-7200

David A. Roth
1225 Westheimer Dr.
Suite 33
Austin, TX 78752   (512) 452-7121

Casimer K. Salys
IBM Corporation
Intellectual Property Law Dept.
Zip 4054
11400 Burnet Road
Austin, TX 78758   (512) 823-0092

Neil B. Schulte
3909 Peak Lookout Drive
Austin, TX 78738   (512) 263-2794

John N. Shaffer, Jr.
Shaffer & Culbertson
1250 Capital of Texas Hwy. S.
Bldg. 2, Suite 560
Austin, TX 78746   (512) 327-8930

Michael L. Sheldon
Fisher Controls
International, Inc.
1712 Centre Creek Dr.
Austin, TX 78754   (512) 834-7346

David M. Sigmond
MCC
3500 W. Balcones Center Dr.
Austin, TX 78759   (512) 338-3232

Russell R. Stolle
Texaco Development Corp.
P.O. Box 15730
Austin, TX 78761   (512) 483-0170

Blucher Stanley Tharp, Jr.
5930 Lookout Mountain Dr.
Austin, TX 78731   (512) 458-8037

Thomas E. Tyson
IBM Corporation
Pat. Operations 932/815
11400 Burnet Road
Austin, TX 78758   (512) 823-1004

Wayne E. Webb, Jr.
Fulbright & Jaworski
600 Congress Ave.
Suite 2400
Austin, TX 78701   (512) 474-5201

Charlotte B. Whitaker
Motorola, Inc.
3501 Ed Bluestein Blvd.
F-4
Austin, TX 78721   (512) 928-6570

Evan K. Butts
Exxon Chemical Co.
Law Technology Division
P.O. Box 5200
Baytown, TX 77522   (713) 425-2342

Ben C. Cadenhead
Exxon Chemical Co.
P.O. Box 5200
Baytown, TX 77522   (713) 425-5401

Wayne Hoover
P.O. Box 1463
Baytown, TX 77522   (713) 420-1358

John F. Hunt
Exxon Chemical Company
P.O. Box 5200
Baytown, TX 77522   (713) 425-2460

Myron Bernard Kurtzman
Exxon Chemical Co.
5200 Bayway Drive
Baytown, TX 77520   (713) 425-5964

Terry B. Morris
Exxon Chemical Company
5200 Bayway Drive
P.O. Box 5200
Baytown, TX 77522   (713) 425-2325

Linda K. Russell
Exxon Chemical Law Technology
Lone Star Bank Bldg.
1501 I-10 East
Suite 232
Baytown, TX 77522   (713) 425-1978

Jaimes Sher
Exxon Chemical Co.
Polymers Group
Law Technology Div.
P.O. Box 5200
Baytown, TX 75222   (713) 425-1832

Edward F. Sherer
Exxon Chemical Company
P.O. Box 5200
Baytown, TX 77522   (713) 425-5933

Thomas Dean Simmons
Exxon Chemical Company
P.O. Box 5200
Baytown, TX 77522   (713) 425-1950

M. Susan Spiering
Exxon Chemical Company
Polymer Group
Law Technology Div.
P.O. Box 5200
Baytown, TX 77522   (713) 425-1950

Alan J. Atkinson
4409 Acacia
Bellaire, TX 77401   (713) 661-2177

James L. Bailey
Texaco Dev. Corp.
P.O. Box 430
4800 Fournace Place
Bellaire, TX 77401   (713) 432-2629

Bill B. Berryhill
6300 W. Loop South
Suite 280
Bellaire, TX 77401   (713) 661-3305

Harold J. Delhommer
Texaco Development Corp.
P.O. Box 430
Bellaire, TX 77401   (713) 752-3294

Cynthia L. Kendrick
Texaco Development Corp.
P.O. Box 430
Bellaire, TX 77401   (713) 432-2623

Richard Albert Morgan
Texaco Development Corp.
P.O. Box 430
4800 Fournace Place
Bellaire, TX 77401   (713) 752-3292

Jack H. Park
Texaco Development Corp.
P.O. Box 430
Bellaire, TX 77402   (713) 520-3631

Harold Herbert Flanders
J.M. Huber Corp.
P.O. Box 2831
1100 Penn
Borger, TX 79007   (806) 274-6331

Alec H. Horn
J.M. Huber Corp.
1100 Penn Avenue
Borger, TX 79007   (806) 274-6331

Glwynn Robinson Baker
5410 C.R. 510
Route 5
Brazoria, TX 77422   (409) 798-4839

Melvin W. Barrow
304 West Texas Street
Brazoria, TX 77422   (409) 798-7580

Elizabeth G. Gammon
West, Adams, Webb
& Allbritton, P.C.
3000 Briarcrest
Suite 502
Bryan, TX 77802   (409) 776-2282

Lisa K. Jorgenson
SGS-Thomson
Microelectronics, Inc.
1310 Electronics Dr.
Carrollton, TX 75006   (214) 466-7414

Richard K. Robinson
SGS-Thomson Microelectronics, Inc.
1310 Electronics Drive
Carrollton, TX 75006   (214) 466-7674

Sidney J. Walker
35 Robinhood Lane
Clute, TX 77531   (409) 265-2901

Donald Raymond Cassady
Celanese Corporation
P.O. Box 9077
Corpus Christi, TX 78469   (512) 241-2343

G. Turner Moller
American Bank Plaza
Corpus Christi, TX 78401   (512) 883-7257

Ralph M. Pritchett#
Hoechst Celanese Corp.
P.O. Box 9077
Corpus Christi, TX 78469   (512) 242-4250

Stewart N. Rice
Wood, Boykin & Wolter, P.C.
1100 First City
Texas Tower
Corpus Christi, TX 78477   (512) 888-9201

Daniel Vera
P.O. Box 3782
Corpus Christi, TX 78463   (512) 851-1173

John K. Abokhair
Cosden Technology, Inc.
P.O. Box 410
Dallas, TX 75221   (214) 750-2585

Richard E. Aduddell
Richards, Medlock & Andrews
4500 Renaissance Tower
1201 Elm Street
Dallas, TX 75270   (214) 939-4500

Daniel R. Alexander
Dresser Industries, Inc.
Patent Dept.
P.O. Box 718
1600 Pacific Avenue
Dallas, TX 75221   (214) 740-6919

Rodney M. Anderson
Vinson & Elkins
3700 Trammell Crow Center
2001 Ross Avenue
Dallas, TX 75201   (214) 220-7749

Garland Paul Andrews
Richards, Harris, Medlock & Andrews
Suite 4500
Renaissance Tower
1201 Elm Street
Dallas, TX 75270   (214) 939-4500

Alva Harlan Bandy
Hubbard, Thurman, Turner
& Tucker
2100 One Galleria Tower
Dallas, TX 75240   (214) 233-5712

B. Peter Barndt
Texas Instruments, Inc.
P.O. Box 655474, M/S 219
Dallas, TX 75265   (214) 995-1373

Robert M. Betz
12660 Hillcrest Road.
Apt. 4101
Dallas, TX 75230   (214) 991-2813

Terry M. Blackwood
Rockwell International Corp.
Pat. Dept., M/S 407-111
P.O. Box 568842
Dallas, TX 75356   (214) 996-6494

Arthur C. Boos
Johnson, Bromberg & Leeds
2600 Lincoln Plaza
500 N. Akard
Dallas, TX 75201   (214) 740-2782

John F. Booth, Sr.
Crutsinger & Booth
1000 Thanksgiving Tower
1601 Elm Street
Dallas, TX 75201   (214) 741-4484

Robert G. Boydston
1030 Frito-Lay Tower
Dallas, TX 75235   (214) 350-7741

Stanton Connell Braden
Texas Instruments Inc.
Legal Dept.
P.O. Box 655474, M/S 219
Dallas, TX 75265   (214) 995-1375

James P. Bradley
Richards, Harris, Medlock & Andrews
Suite 4500
Renaissance Tower
1201 Elm Street
Dallas, TX 75270   (214) 939-4500

Randall C. Brown
Haynes & Boone
3100 First Republicbank Plaza
901 Main Street
Dallas, TX 75202   (214) 670-0050

John F. Bryan#
4250 West Lovers Lane
Dallas, TX 75209   (214) 902-0213

Michael S. Bush
Johnson & Swanson
900 Jackson Street
Suite 100
Dallas, TX 75202   (214) 977-9663

Michael J. Caddell
Fina Oil & Chemical Company
8350 North Central Expwy.
Dallas, TX 75206   (214) 750-2888

Mason M. Campbell#
Hubbard, Thurman, Turner,
Tucker & Harris
2100 One Galleria Tower
Dallas, TX 75240   (214) 750-2888

Michael Rocco Cannatti
Richards, Medlock & Andrews
4500 Renaissance Tower
1201 Elm Street
Dallas, TX 75270   (214) 939-4500

Thomas Lee Cantrell
Schley, Catrell, Kice,
Garland & Moore
5001 LBJ Freeway
Suite 705
Dallas, TX 75244   (214) 387-3804

Gregory W. Carr
Gardere & Wynne
1601 Elm Street
Suite 3000
Dallas, TX 75201   (214) 999-4542

Albert W. Carroll#
3008 Primrose Lane
Dallas, TX 75234   (214) 247-2738

David W. Carstens
Richards, Medlock & Andrews
4500 Renaissance Tower
1201 Elm Street
Dallas, TX 75270   (214) 939-4500

Roger N. Chauza
Baker, Smith & Mills, P.C.
2001 Ross Avenue
500 LTV Center
Dallas, TX 75201   (214) 220-8287

M. Norwood Cheairs
Amer. Petrofina Co. of Tex.
8350 North Central Expressway
Dallas, TX 75221   (214) 750-2532

Robert M. Chiaviello
Baker & Botts
800 Trammell Crow Center
2001 Ross Avenue
Dallas, TX 75201   (214) 953-6500

Tek Ling Chwang
Johnson & Gibbs
100 Founders Square
900 Jackson Street
Dallas, TX 75202   (214) 977-9617

Robert D. Clamon#
Dresser Industries, Inc.
P.O. Box 718
1600 Pacific
Dallas, TX 75221   (214) 740-6508

James T. Comfort
Texas Instruments, Inc.
P.O. Box 655,474
M/S 219
13510 N. Central Expressway
Dallas, TX 75243   (214) 995-4400

John M. Cone
Strasburger & Price
4300 Inter First Plaza
901 Main Street
Dallas, TX 75202   (214) 651-4655

Charles S. Cotropia
Richards, Harris, Medlock & Andrews
Suite 4500
Renaissance Tower
1201 Elm Street
Dallas, TX 75270   (214) 939-4500

Roland O. Cox#
Otis Engineering Corp.
P.O. Box 819052
Dallas, TX 75381   (214) 323-3883

Thomas Lynn Crisman
Johnson & Johnson
100 Founders Square
900 Jackson Street
Dallas, TX 75202   (214) 977-9614

Henry Croskell
Jones, Day Reavis
& Pogue
2300 Trammell Crow Center
2001 Ross Ave.
Dallas, TX 75201   (214) 220-3939

Morgan L. Crow#
2010 Elder Oaks Lane
Dallas, TX 75232   (214) 337-8120

Gerald G. Crutsinger, Sr.
Crutsinger & Booth
1000 Thanksgiving Tower
Dallas, TX 75201   (214) 741-4484

Jennifer R. Daunis
Hubbard, Thurman, Turner,
& Tucker
2100 One Galleria Tower
Dallas, TX 75240   (214) 233-5712

James Owen Dixon
Crutsinger & Booth
1000 Thanksgiving Tower
Dallas, TX 75201   (214) 741-4484

Richard L. Donaldson
Texas Instruments, Inc.
P.O. Box 225474
Dallas, TX 75265   (214) 995-5921

John A. Dondrea
Richards, Medlock & Andrews
4500 Renaissance Tower
1201 Elm Street
Dallas, TX 75270   (214) 939-4517

Terrence Dean Dreyer
Frito-Lay Inc.
P.O. Box 660634
Dallas, TX 75266   (214) 353-3814

Larry B. Dwight
4560 Belt Line Road
Suite 432
Dallas, TX 75244   (214) 701-8197

Elisabeth A. Evert
Richards, Medlock & Andrews
4500 Renaissance Tower
1201 Elm Street
Dallas, TX 75270   (214) 939-4519

Robert H. Falk
Hubbard, Thurman, Turner
& Tucker
2100 One Galleria Tower
Dallas, TX 75240   (214) 233-5712

Thomas R. Felger
Otis Eng. Corp.
P.O. Box 819052
Dallas, TX 75381   (214) 418-3882

Edward Gerald Fiorito
Dresser Industries, Inc.
1600 Pacific Avenue
Dallas, TX 75201   (214) 740-6901

Thomas R. Fitzgerald
Texas Instruments
P.O. Box 655474
MS 219
Dallas, TX 75265   (214) 995-1370

Michael David Folzenlogen
6774 Inverness
Dallas, TX 75214   (214) 823-5599

Gene Wilgus Francis, Jr.
Brice & Mankoff
300 Crescent Court
7th Floor
Dallas, TX 75201   (214) 855-3700

H. Mathews Garland
Johnson & Swanson
100 Founders Square
900 Jackson Street
Dallas, TX 75202   (214) 977-9615

Charles C. Garner
Ray Trotti Hemphill
& Finfrock, P.C.
1401 Elm St.
Suite 1800
Dallas, TX 75202   (214) 741-7447

Kenneth Roy Glaser
Glaser, Griggs & Schwartz
5430 LBJ Freeway
Suite 1540
Dallas, TX 75240   (214) 770-2400

Patrick J. Glynn
Sigalos, Levine & Montgomery
12750 Merit Drive
Suite 1000
Dallas, TX 75251   (214) 770-5444

David C. Godbey
Hughes & Luce
2800 Momentum Place
1717 Main Street
Dallas, TX 75201   (214) 939-5581

Howard R. Greenberg
Rockwell International Corp.
Pat. Dept. M/S 407-111
P.O. Box 568842
Dallas, TX 75356   (214) 996-5470

Lori J. Griebenow
Texas Instruments Inc.
13510 N. Central Expressway
Mail Station 219
Dallas, TX 75243   (214) 995-1365

Dennis T. Griggs
Glaser, Griggs & Schwartz
5430 LBJ Freeway
Suite 1540
Dallas, TX 75240   (214) 770-2400

Stephen A. Grimmer
Haynes & Boone
3100 First Republic Bank Plaza
901 Main
Dallas, TX 75202   (214) 670-0550

Robert Groover, III
Worsham, Forsythe, Sampels
& Wooldridge
Thirty-Two Hundred
2001 Bryan Tower
Dallas, TX 75201   (214) 979-3000

Rene E. Grossman
Texas Instruments Inc.
P.O. Box 655474
M/S 219
Dallas, TX 75265   (214) 995-1345

Norman L. Gundel
Crutsinger & Booth
1000 Thanksgiving Tower
1601 Elm Street
Dallas, TX 75201   (214) 220-0444

William R. Gustavson
Richards, Harris, Medlock
& Andrews
4500 Renaissance Tower
1201 Elm Street
Dallas, TX 75270   (214) 939-4510

Alfred E. Hall
Jones, Day, Reavis
& Pogue
2300 Trammell Crow Center
2001 Ross Avenue
Dallas, TX 75201   (214) 969-2975

Herbert J Hammond
Gardere & Wynne
1601 Elm St., Ste. 3000
Dallas, TX 75201   (214) 999-4612

Eugenia S. Hansen
Richards, Harris, Medlock
& Andrews
4500 Renaissance Tower
1201 Elm Street
Dallas, TX 75270   (214) 939-4500

Roy W. Hardin
Richards, Harris, Medlock
& Andrews
4500 Renaissance Tower
1201 Elm Street
Dallas, TX 75270   (214) 939-4508

William David Harris, Jr.
Richards, Harris, Medlock
& Andrews
4500 Renaissance Tower
1201 Elm Street
Dallas, TX 75270   (214) 939-4500

Bryan G. Harston
Johnson & Gibbs, P.C.
900 Jackson Street
Dallas, TX 75202   (214) 977-9636

Andrew M. Hassell
Park Central VII
12750 Merit Dr.
Suite 1000
Dallas, TX 75251   (214) 770-5444

Richard Blake Havill
Texas Instruments Inc.
P.O. Box 655474
Mail Stop 219
Dallas, TX 75265   (214) 995-2011

John N. Hazelwood
M.S. 76
P.O. Box 795549
Dallas, TX 75379   (214) 995-2011

Leo N. Heiting#
Texas Instruments Inc.
P.O. Box 655474
M.S. 219
Dallas, TX 75265   (214) 995-5493

William Eugene Hiller
Texas Instruments, Inc.
M.S. 219
13500 N. Central Expwy.
Dallas, TX 75231   (214) 995-1364

David L. Hitchcock
Richards, Harris, Medlock
& Andrews
4500 Renaissance Tower
1201 Elm Street
Dallas, TX 75270   (214) 939-4511

Carlton H. Hoel
Worsham, Forsythe, Sampels
& Wooldridge
2001 Bryan Tower, Suite 3200
Dallas, TX 75201   (214) 979-3000

John M. Holland
Hubbard, Thurman, Turner
& Tucker
2100 One Galleria Tower
Dallas, TX 75240   (214) 233-5712

Robby T. Holland
Texas Instruments, Inc.
P.O. Box 655474
M.S. 219
Dallas, TX 75265   (214) 995-5316

James F. Hollander
Texas Instruments, Inc.
P.O. Box 655474
M.S. 219
Dallas, TX 75265   (214) 995-5317

Gary C. Honeycutt
Texas Instruments, Inc.
P.O. Box 655474
M.S. 219
Dallas, TX 75265   (214) 995-1363

Marc A. Hubbard
Hubbard, Thurman, Tucker
& Harris
2100 One Galleria Tower
13355 Noel Road
Dallas, TX 75240   (214) 995-1363

Carl D. Hughes, Jr.
P.O. Box 610326
Dallas, TX 75261   (214) 761-9342

William N. Hulsey, III
Baker & Botts
800 Trammell Crow Center
2001 Ross Avenue
Dallas, TX 75201   (214) 953-6500

William D. Jackson
Richards, Harris, Medlock
& Andrews
4500 Renaissance Tower
Dallas, TX 75270   (214) 939-4500

David L. Joers
Crutsinger & Booth
1000 Thanksgiving Tower
1601 Elm Street
Suite 1000
Dallas, TX 75201   (214) 220-0444

Edward I. Jorgenson
Gardere & Wynne
1601 Elm Street
Suite 3000
Dallas, TX 75201   (214) 999-4531

David H. Judson
Hughes & Luce
2800 Momentum Place
Dallas, TX 75201   (214) 939-5672

Jack A. Kanz
Richards, Medlock & Andrews
4500 Renaissance Tower
1201 Elm Street
Dallas, TX 75270   (214) 939-4513

H. Dennis Kelly
Timmons & Kelly
North Tower, Suite 1720
Plaza of the Americas
Dallas, TX 75201   (214) 220-0310

Warren B. Kice
Haynes & Boone
3100 First Republic Bank Plaza
901 Main Street
Dallas, TX 75202 (214) 670-0634

John D. Kling
Texas Instruments Inc.
13510 N. Central Expressway
M.S. 219
Dallas, TX 75243 (214) 995-1348

J. Richard Konneker
Hubbard, Thurman, Turner
& Tucker
2100 One Galleria Tower
Dallas, TX 75240 (214) 233-5712

David N. Leonard
Leonard & Lott
5430 LBJ Freeway
Suite 875
Dallas, TX 75240 (214) 960-7477

Harold Levine
Park Central VII
12750 Merit Drive
Suite 1000
Dallas, TX 75251 (214) 770-5444

Stephen L. Levine
Baker & Botts
800 Trammell Crow Center
2001 Ross Avenue
Dallas, TX 75201 (214) 953-6687

Alan W. Lintel
Baker & Botts
800 Trammell Crow Center
2001 Ross Avenue
Dallas, TX 75201 (214) 220-8285

Ann Livingston
Baker & Botts
800 Trammell Crow Center
2001 Ross Avenue
Dallas, TX 75201 (214) 953-6500

Robert D. Lott
Leonard & Lott
5430 LBJ Freeway
Suite 875
Dallas, TX 75240 (214) 960-7477

George R. Love
Mc Cauley, Mac Donald, Love,
Devin & Brinker
3800 Renaissance Tower
1201 Elm Street
Dallas, TX 75270 (214) 744-3300

Seth Thomas Low*
U.S. Environmental Protection
Agency
1445 Ross Avenue
Dallas, TX 75202 (214) 655-2120

Bruce C. Lutz
Rockwell International Corp.
Pat. Dept. M.S. 07-111
P.O. Box 568842
Dallas, TX 75356 (214) 996-5770

Hulit L. Madinger
4207 Cobblers Lane
Dallas, TX 75287 (214) 380-5877

Thomas V. Malorzo
Otis Engineering Corp.
P.O. Box 819052
Dallas, TX 75381 (214) 418-3000

Roger L. Maxwell
Johnson & Swanson
100 Founders Square
900 Jackson Street
Dallas, TX 75202 (214) 977-9000

David L. Mc Combs
Haynes & Boone
3100 First Republic Bank Plaza
901 Main Street
Dallas, TX 75202 (214) 670-0050

William K. Mc Cord
Glaser, Griggs & Schwartz
3 Lincoln Centre
5430 LBJ Freeway
Suite 1540
Dallas, TX 75240 (214) 770-2400

Virgil Bryan Medlock, Jr.
Richards, Harris, Medlock
& Andrews
Suite 4500 Renaissance Tower
1201 Elm Street
Dallas, TX 75270 (214) 939-4502

Kevin J. Meek
Baker & Botts
800 Trammell Crow Center
2001 Ross Avenue
Dallas, TX 75201 (214) 953-6500

Harold Eugene Meier
Gardere & Wynne
Thanksgiving Tower
1601 Elm St., Ste. 3000
Dallas, TX 75201 (214) 999-4662

Michael E. Melton
Texas Instruments, Inc.
Pat. Dept.
P.O. Box 655474
M.S. 219
Dallas, TX 75265 (214) 995-6908

Norman R. Merrett
Texas Instruments, Inc.
13500 N. Central Expressway
M.S. 219
Dallas, TX 75265 (214) 995-1360

Jerry Woodrow Mills
Baker, Mills & Glast, P.C.
2001 Ross Avenue
Suite 500
Dallas, TX 75201 (214) 220-8283

Samuel M. Mims, Jr.
Texas Instruments, Inc.
13510 N. Central Expwy.
M.S. 241
Dallas, TX 75243 (214) 995-4294

John W. Montgomery
Sigalos, Levine & Montgomery, P.C.
2700 Tower II
NCNB Center
Dallas, TX 75201 (214) 953-1420

Stanley R. Moore
Johnson & Swanson
900 Jackson Street
Dallas, TX 75202 (214) 977-9616

David L. Mossman
5724 Everglade Road
Dallas, TX 75227 (214) 275-4206

James J. Murphy
Baker & Botts
800 Trammell Crow Center
2001 Ross Avenue
Dallas, TX 75201 (214) 953-6688

P. Weston Musselman, Jr.
Jenkens & Gilchrist, P.C.
1445 Ross Avenue
Suite 3200
Dallas, TX 75202 (214) 855-4764

Ronald Owen Neerings
Texas Instruments Inc.
P.O. Box 655474
M.S. 219
Dallas, TX 75524-3000 (214) 995-1804

Dale B. Nixon
Richards, Harris, Medlock
& Andrews
1201 Elm Street
Suite 4500
Dallas, TX 75270 (214) 939-4500

Michael A. O Neil
Gardere & Wynne
1601 Elm St.
Suite 3000
Dallas, TX 75201 (214) 999-4681

Ronald L. Palmer
Baker & Botts
900 Trammell Crow Center
2001 Ross Avenue
Dallas, TX 75201 (214) 953-6603

William R. Peoples
Dresser, Industries, Inc.
1600 Pacific Avenue
Dallas, TX 75201 (214) 740-6910

Daniel F. Perez
Jones, Day, Reavis & Pogue
2300 Trammell Crow Center
2001 Ross Ave.
Dallas, TX 75201 (214) 969-3607

Jefferson F. Perkins
Baker & Botts
800 Trammell Crow Center
2001 Ross Avenue
Dallas, TX 75201 (214) 953-6673

Jean M. Perron
Jones, Day, Reavis
& Pogue
2001 Ross Avenue
2300 Trammell Crow Center
Dallas, TX 75201   (214) 969-5290

John P. Pinkerton
Hubbard, Thurman, Turner
Tucker & Harris
13355 Noel Road
2100 One Galleria Tower
Dallas, TX 75240   (214) 233-5712

Harry C. Post, III
2001 Bryan Tower
Suite 815
Dallas, TX 75201   (214) 747-8511

J. Hughes Powell, Jr.
3315 Hanover Street
Dallas, TX 75225   (214) 361-6242

D. Carl Richards
Richards, Medlock & Andrews
4500 Renaissance Tower
Dallas, TX 75270   (214) 939-4500

Jonathan W. Richards
Thompson & Knight
3300 First City Center
1700 Pacific Avenue
Dallas, TX 75201   (214) 969-1259

Mark Alan Rogers
Haynes & Boone
3100 NCNB Plaza
901 Main St.
Dallas, TX 75202   (214) 670-0682

Daniel Rubin
North Tower, Suite 1720
Plaza of the Americas
Dallas, TX 75201   (214) 220-0310

Stephen S. Sadacca
LTV Corporation
P.O. Box 650003
MS-OWT-O5
Dallas, TX 75265   (214) 266-1908

Robert A. Samra
Johnson & Gibbs, P.C.
900 Jackson St., Ste. 100
Dallas, TX 75202   (214) 977-9317

Joseph A. Schaper
Fina Oil & Chemical Company
P.O. Box 2159
Dallas, TX 75221   (214) 750-2345

Larry C. Schroeder
Texas Instruments, Inc.
13500 N. Central Expressway
P.O. Box 655474
M.S. 241
Dallas, TX 75265   (214) 995-1357

Richard L. Schwartz
Glaser, Griggs & Schwartz
Three Lincoln Centre
5430 LBJ Freeway, Suite 1540
Dallas, TX 75240   (214) 770-2400

Jerry R. Selinger
Baker, Mills & Glast, P.C.
2001 Ross Avenue
500 Trammell Crow Center
Dallas, TX 75201   (214) 220-8300

V. Lawrence Sewell
Rockwell Internatl. Corp.
Pat. Dept.
P.O. Box 568842
M.S. 407-111
Dallas, TX 75356   (214) 996-2656

Melvin Sharp
Texas Instruments, Inc.
13500 N. Central Expressway
M.S. 219
Dallas, TX 75265   (214) 995-5865

John J. Sheedy
Goins, Underkofler, Crawford
& Langdon
3300 Thanksgiving Tower
1601 Elm Street
Dallas, TX 75201   (214) 969-5454

Jacob W Sietsema#
15409 Cypress Hills Drive
Dallas, TX 75248   (214) 661-0296

John Louis Sigalos
12750 Merit Drive
Suite 1000
Dallas, TX 75251   (214) 770-5444

Harry E. Simpson#
1409 Traymore
Dallas, TX 75217   (214) 391-5438

Bruce W. Slayden, II
Richards, Medlock & Andrews
1201 Elm Street
Suite 4500
Dallas, TX 75270   (214) 939-4500

Arthur M. Sloan
4020 Cedarbrush Drive
Dallas, TX 75229   (214) 350-0847

Douglas A. Sorensen
Texas Instruments Inc.
P.O. Box 655474
M.S. 219
Dallas, TX 75265   (214) 995-3329

Mark W. Stockman
Locke, Purnell, Rain & Harrell
2200 Rose Avenue
Suite 2200
Dallas, TX 75201   (214) 740-8000

David Harry Tannenbaum
Winstead, Sechrest
& Minick, P.C.
5400 Renaissance Tower
1201 Elm Street
Dallas, TX 75270   (214) 245-5354

Frederick J. Telecky, Jr.
Texas Instruments, Inc.
13510 N. Central Expressway
M.S. 219, North Bldg.
Dallas, TX 75243   (214) 995-1348

Peter J. Thoma
Richards, Medlock & Andrews
4500 Renaissance Tower
1201 Elm Street
Dallas, TX 75270   (214) 939-4526

Daniel V. Thompson
9223 Loma Vista
Dallas, TX 75243   (214) 349-2044

Ronald V. Thurman
Hubbard, Thurman, Tucker
& Harris
13355 Noel Rd.
One Galleria Tower, Ste. 2100
Dallas, TX 75240   (214) 716-3510

W. Thomas Timmons
Timmons & Kelly
North Tower, Suite 1720
Plaza of the Americas
Dallas, TX 75201   (214) 220-0310

Louis Touton
Jones, Day, Reavis
& Pogue
2001 Ross Avenue
2300 Trammell Crow Center
Dallas, TX 75201   (214) 969-3727

Laurey Dan Tucker
Hubbard, Thurman,
Tucker & Harris
13355 Noel Rd.
One Galleria Tower, Ste. 2100
Dallas, TX 75240   (412) 233-5712

Robert W. Turner
Jones, Day, Reavis & Pogue
2001 Ross Avenue
Suite 2300
Dallas, TX 75201   (214) 969-2984

Roy L. Van Winkle
Baker & Botts
800 Trammell Crow Center
2001 Ross Ave.
Dallas, TX 75201   (214) 953-6679

John E. Vick, Jr.
Hubbard, Thurman, Tucker
& Harris
2100 One Galleria Tower
Dallas, TX 75240   (214) 233-5712

Andrew S. Viger
Baker & Botts
2001 Ross Ave., Ste. 800
Dallas, TX 75201   (214) 953-6500

Sanford E. Warren, Jr.
Jones, Day, Reavis & Pogue
2300 Trammell Crow Center
2001 Ross Avenue
Dallas, TX 75201   (214) 969-4538

Harry J. Watson
Hubbard, Thurman Tucker
& Harris
One Galleria Tower
13355 Noel Rd., Ste. 2100
Dallas, TX 75240   (214) 233-5712

Stuart L. Watt
Richards, Medlock & Andrews
4500 Renaissance Tower
1201 Elm Street
Dallas, TX 75270   (214) 939-4516

Jimmy D. Wheelington*
Fina Oil & Chemical Co.
8350 N. Central Expressway
P.O. Box 2159
Dallas, TX 75221   (214) 750-2971

James William Williams
Richards, Harris, Medlock
& Andrews
4500 Renaissance Tower
1201 Elm Street
Dallas, TX 75270   (214) 939-4528

Robert Daniel Winn
5232 Moneta Lane
Dallas, TX 75236   (214) 296-8757

Brian R. Woodworth
Richards, Medlock & Andrews
4500 Renaissance Tower
1201 Elm Street
Dallas, TX 75270   (214) 939-4520

William John Scherback
402 W. Wheatland Road
Suite 111
Duncanville, TX 75116   (214) 780-9120

Mack J. Casner
P.O. Box 24073
El Paso, TX 79914   (915) 533-2681

R. Wayne Pritchard
Ginnings, Birkelbach,
Keith & Delgado
416 N. Stanton
Suite 700
El Paso, TX 79901   (915) 532-5929

Robert L. Eschenburg, II
P.O. Box 236
Floresville, TX 78114   (512) 393-6800

James A. Arno
Alcon Laboratories, Inc.
6201 S. Freeway
Fort Worth, TX 76134   (817) 551-8260

James E. Bradley
Felsman, Bradley, Gunter
& Kelly
2850 Continental Plaza
777 Main Street
Fort Worth, TX 76102   (817) 332-8143

Gregg C. Brown
Alcon Laboratories, Inc.
Pat. Dept.
6201 South Freeway
Fort Worth, TX 76134   (817) 551-8663

Julie J.L. Cheng
6037 Arbor Bend, #921
Fort Worth, TX 76132   (817) 551-8663

Barry L. Copeland
Alcon Laboratories, Inc.
6201 South Freeway
Fort Worth, TX 76134   (817) 551-4322

Andrew J. Dillon
Gearhart Industries, Inc.
P.O. Box 1936
Fort Worth, TX 76101   (817) 293-1300

James C. Fails
Wofford, Fails, Zobal & Mantooth
110 W. 7th Street
Suite 500
Fort Worth, TX 76102   (817) 332-1233

Robert A. Felsman
Felsman, Bradley & Gunter
2850 Continental Plaza
777 Main Street
Fort Worth, TX 76102   (817) 332-8143

George Galerstein
Bell Helicopter Textron, Inc.
P.O. Box 482
Fort Worth, TX 76101   (817) 280-2834

Elton F. Gunn#
5612 Oakmont Lane
Fort Worth, TX 76112   (817) 390-8640

Charles D. Gunter, Jr.
Felsman, Bradley, Gunter
& Dillon
2600 Continental Plaza
777 Main Street
Fort Worth, TX 76102   (817) 332-8143

Kenneth C. Hill
Felsman, Bradley, Gunter & Dillon
2600 Continental Plaza
777 Main St.
Fort Worth, TX 76102   (817) 332-8143

Melvin A. Hunn
Felsman, Bradley, Gunter & Dillon
777 Main Street
Suite 2850
Continental Plaza
Fort Worth, TX 76102   (817) 332-8143

William A. Linnell
Tandy Corp.
1800 One Tandy Center
Fort Worth, TX 76102   (817) 390-3700

Guy V. Manning
Felsman, Bradley, Gunter
& Dillon
2600 Continental Plaza
777 Main Street
Fort Worth, TX 76102   (817) 332-8143

Geoffrey A. Mantooth
Wofford, Fails, Zobal & Mantooth
110 West Seventh
Suite 500
Fort Worth, TX 76102   (817) 332-1233

Stephen S. Mosher#
Tandy Electronics Div. of
Tandy Corp.
900 Two Tandy Center
Fort Worth, TX 76102   (817) 390-3840

Frederick W. Padden
Tandy Corp.
1800 One Tandy Center
P.O. Box 17180
Fort Worth, TX 76102   (817) 390-3840

Robert E. Roehrs#
3729 Hulen Park
Fort Worth, TX 76109   (817) 924-8446

Charles E. Schurman
General Dynamics Corp.
General Dynamics Blvd.
Fort Worth, TX 76108   (817) 777-1711

Sally Yeager
Alcon Laboratories, Inc.
6201 South Freeway
Fort Worth, TX 76134   (817) 551-4031

Duke W. Yee
Felsman, Bradley, Gunter
& Dillon
2600 Continental Plaza
777 Main Street
Fort Worth, TX 76102   (817) 332-8143

Arthur Fred Zobal
Wofford, Fails, Zobal & Mantooth
110 West Seventh
Suite 500
Fort Worth, TX 76102   (817) 332-1233

Jeffrey S. Schira
Alcon Laboratories, Inc.
Patent Dept., Q-148
6201 South Freeway
Fort Worth, TX 76134   (817) 551-3063

A. Cooper Ancona#
Dow Chemical Co.
Bldg. B-1210
Pat. Dept.
Freeport, TX 77541   (409) 238-7259

James G. Carter#
Dow Chemical Co.
Bldg. B-1210
Pat. Dept.
Freeport, TX 77541   (713) 238-7250

Carol J. Cavender#
Dow Chemical Co.
Bldg. B-1210
Pat. Dept.
Freeport, TX 77541   (409) 238-3581

Benjamin G. Colley#
Dow Chemical Co.
Bldg. B-1210
Pat. Dept.
Freeport, TX 77541   (713) 238-7239

James H. Dickerson, Jr.
Dow Chemical Co.
Bldg. B-1210
Pat. Dept.
Freeport, TX 77541   (409) 238-7084

Glenn H. Korfhage
Dow Chemical Co.
Hwy. 227, B-1210 Bldg.
Freeport, TX 77541   (409) 238-2266

Stephen P. Krupp#
Dow Chemical Co.
Bldg. B-1210
Pat. Dept.
Freeport, TX 77541   (409) 238-2889

John A. Langworthy
Dow Chemical Co.
Bldg. B-1210
Pat. Dept.
Freeport, TX 77541   (409) 238-7222

Barbara J. Tribble#
Dow Chemical Co.
Bldg. B-1210
Pat. Dept.
Freeport, TX 77541   (409) 238-1638

L. Wayne White
Dow Chemical Company
Patent Dept.
Bldg. B-1210
Freeport, TX 77541   (409) 238-2149

C. Ray Holbrook, Jr.
County of Galveston
County Courthouse
Galveston, TX 77550   (713) 766-2244

Elizabeth F. Sporar#
4128 Avenue T
Galveston, TX 77550   (409) 762-0239

Richard M. Byron
3808 Shady Creek Dr.
Garland, TX 75042   (214) 276-8072

Thomas D. Copeland, Jr.
1900 Melody Lane
Garland, TX 75042   (214) 278-8012

Evert Allen Autrey
2007 South Main Street
Georgetown, TX 78626   (512) 863-3166

James M. Cate
LTV Corporation
LTV Aerospace & Defense Co.
1902 West Freeway
Grand Prairie, TX 75051   (214) 266-1908

Joseph P. Kulik, Jr.
E-Systems, Inc.
Division Counsel
Greenville Div.
P.O. Box 1056
Greenville, TX 75401   (214) 457-4633

C. M. Kucera
Route 3 Box 5683
Hallettsville, TX 77964   (512) 798-5597

Albert Julius Adamcik
Shell Dev. Co.
P.O. Box 2463
Houston, TX 77001   (713) 241-4729

Margaret E. Anderson
Browning, Bushman, Zamecki
& Anderson
Suite 1800
5718 Westheimer
Houston, TX 77057   (713) 266-5593

Gordon T. Arnold
Kirk & Lindsay, P.C.
3555 Timmons Lane
Suite 700
Houston, TX 77027   (713) 621-2021

Steven P. Arnold
Arnold, White & Durkee
P.O. Box 4433
Houston, TX 77210   (713) 787-1400

Tom Arnold
Arnold, White & Durkee
P.O. Box 4433
Houston, TX 77210   (713) 787-1400

Richard C. Auchterlonie
Arnold, White & Durkee
750 Bering Drive
P.O. Box 4433
Houston, TX 77210   (713) 787-1400

Martin S. Baer
5746 Valkeith Drive
Houston, TX 77096   (714) 729-0100

Douglas Baldwin, Jr.
Shell Dev. Co.
P.O. Box 2463
One Shell Plaza
Houston, TX 77001   (713) 241-3716

James Allen Bargfrede
2323 S. Voss
Suite 123
Houston, TX 77057   (713) 781-0422

Hardie R. Barr*
NASA
Johnson Space Center
Mail Code AL3
Houston, TX 77058   (713) 483-4871

Kenneth D. Baugh
4005 San Jacinto
Houston, TX 77004   (713) 529-2901

Robert E. Bayes
14403 Broadgreen Drive
Houston, TX 77079   (713) 497-7184

William James Beard
Halliburton Services
P.O. Box 42800
2135 Hwy. 6 South
Houston, TX 77242   (713) 496-8331

Emil J. Bednar
6131 Paisley
Houston, TX 77096   (713) 771-2941

Keith A. Bell
Exxon Production Res. Co.
P.O. Box 2189
Houston, TX 77001   (713) 965-7994

Christopher R. Benson
Arnold, White & Durkee
750 Bering Drive
P.O. Box 4433
Houston, TX 77210   (713) 789-1400

Peter A. Bielinski
Shell Development Company
P.O. Box 2463
Houston, TX 77001   (713) 241-4991

Johan A. Bjorksten#
9117 Almeda-Genoa Road
Houston, TX 77075   (713) 241-4991

Nelson A. Blish
Cooper Industries, Inc.
1001 Fannin
Suite 4000
Houston, TX 77002   (713) 739-5858

Ronald G. Bliss
Fulbright & Jaworski
1301 Mc Kinney Street
51st Floor
Houston, TX 77010   (713) 651-5151

Steven R. Borgman
Vinson & Elkins
2900 First City Tower
1001 Fannin Street
Houston, TX 77002   (713) 651-2002

Kirby L. Boston
Merichem Co.
4800 Texas Comm. Tower
Houston, TX 77002   (713) 224-3030

John H. Bouchard
Schlumberger Technology Corp.
5000 Gulf Freeway
P.O. Box 2175
Houston, TX 77252   (713) 928-4337

Margaret Anne Boulware
Vaden, Eickenroht, Thompson
& Boulware
One Riverway
Suite 1100
Houston, TX 77056   (713) 961-3525

Raymond H. Bradley#
850 Regal
Houston, TX 77034   (713) 944-1973

Margaret C. Bradshaw
6811 Greenyard Drive
Houston, TX 77086   (713) 893-9766

Patricia N. Brantley
Arnold, White & Durkee
P.O. Box 4433
Houston, TX 77210   (713) 787-1449

Michael P. Breston
2600 South Gessner
Suite 312
Houston, TX 77063   (713) 953-2990

Sylvester W. Brock, Jr.
Exxon Company, U.S.A.
P.O. Box 2180
Houston, TX 77252   (713) 656-4864

Anne E. Brookes
Arnold, White & Durkee
P.O. Box 4433
Houston, TX 77210   (713) 787-1400

Walter R. Brookhart
Browning, Bushman, Zamecki
& Anderson
Suite 1800
5718 Westheimer
Houston, TX 77057   (713) 266-5593

Scott H. Brown
Camco International Inc.
Patent Counsel
P.O. Box 14484
Houston, TX 77221   (713) 747-4000

W. Scott Brown
Vinson & Elkins
1001 Fannin Street
Houston, TX 77002   (713) 750-1105

Kevin A. Buford
Arnold, White & Durkee
750 Bering Dr.
Houston, TX 77057   (713) 787-1622

Timothy L. Burgess
Rosenblatt & Associates, P.C.
One Greenway Plaza, Suite 500
Houston, TX 77046   (713) 552-9900

Karen T. Burleson
Arnold, White & Durkee
P.O. Box 4433
Houston, TX 77210   (713) 787-1411

Gary L. Bush
Dodge, Bush & Moseley
950 Echo Lane
Suite 180
Houston, TX 77024   (713) 827-1054

C. James Bushman
Browning, Bushman, Zamecki
& Anderson
Suite 1800
5718 Westheimer
Houston, TX 77057   (713) 266-5593

Wendy K.B. Buskop
Lyondell Petrochemical
Legal Dept.
One Houston Center
1221 Mc Kinney
Houston, TX 77010   (713) 652-4588

Jesus D. Cabello
Compaq Computer Corporation
20555 F.M. 149
Houston, TX 77070   (713) 374-2634

Stephen H. Cagle
Arnold, White & Durkee
400 One Bering Park
750 Bering Drive
Houston, TX 77057   (713) 787-1400

Rodney K. Caldwell
Arnold, White & Durkee
P.O. Box 4433
Houston, TX 77210   (713) 787-1441

Norman R. Carlson#
13707 Tosca Lane
Houston, TX 77079   (713) 464-0269

Salvatore J. Casamassima
Exxon Company, U.S.A.
Pat. Dept.
P.O. Box 2180
Houston, TX 77252   (713) 656-3437

Del S. Christensen
Shell Oil Company
P.O. Box 2463
Houston, TX 77001   (713) 241-3997

Leslie J. Clark
Vinson & Elkins
3300 First City Tower
1001 Fannin
Houston, TX 77002   (713) 758-2951

Ronald L. Clendenen
Shell Dev. Co.
P.O. Box 2463
Houston, TX 77001   (713) 241-4738

Ned L. Conley
Butler & Binion
1600 First Interstate Bank Plaza
Houston, TX 77002   (713) 237-3195

Robert T. Cook
Arnold, White & Durkee
P.O. Box 4433
Houston, TX 77210   (713) 787-1452

Claude E. Cooke, Jr.
Pravel, Gambrell, Hewitt,
Kimball & Krieger
1177 W. Loop South, 10th Floor
Houston, TX 77027   (713) 850-0909

Harold Wade Coryell#
Shell Development Co.
P.O. Box 2463
One Shell Plaza
Houston, TX 77001   (713) 241-4708

Charles M. Cox
Pravel, Gambrell, Hewitt,
Kimball & Krieger
1177 West Loop South
10th Floor
Houston, TX 77027   (713) 850-0909

Hubert E. Cox, Jr.
Exxon Production Res. Co.
P.O. Box 2189
Houston, TX 77001   (713) 965-7956

Charles H. De La Garza
Arnold, White & Durkee
P.O. Box 4433
Houston, TX 77210   (713) 789-7600

Paul L. De Verter, II
Fulbright & Jaworski
1301 Mc Kinney Street
Suite 5100
Houston, TX 77010   (713) 651-5151

Albert B. Deaver, Jr.
Arnold, White & Durkee
750 Bering Dr., Ste. 400
Houston, TX 77057   (713) 787-1432

Marc L. Delflache
Pravel, Gambrell, Hewitt,
Kimball & Krieger
1177 W. Loop South
Suite 1010
Houston, TX 77027   (713) 850-0909

Stephen D. Dellett
Arnold, White & Durkee
P.O. Box 4433
Houston, TX 77210   (713) 787-1423

Harold Louis Denkler
823 Patchester
Houston, TX 77079   (713) 462-2996

Robert W. B. Dickerson
6200 Savoy
Suite 260
Houston, TX 77036   (713) 627-0252

James George Dieter
Evans & Kosut
Three Riverway
Suite 1776
Houston, TX 77056   (713) 850-8388

John H. Dodge, II
Dodge, Bush & Moseley
950 Echo Lane
Suite 180
Houston, TX 77024   (713) 827-1054

Michael Scott Dowler#
14531 Cindywood
Houston, TX 77079   (713) 497-3983

N. Elton Dry
Pravel, Gambrell, Hewitt,
Kimball & Krieger
1177 W. Loop S.
Suite 1010
Houston, TX 77027   (713) 850-0909

Arthur M. Dula
6900 Texas Commerce Tower
Houston, TX 77002   (713) 227-9000

George H. Dunn, III
Vinson & Elkins
2900 First City Tower
1001 Fannin
Houston, TX 77002   (713) 750-1073

William D. Durkee
Arnold, White and Durkee
P.O. Box 4433
Houston, TX 77210   (713) 787-1444

Lewis Hamilton Eatherton, III
1728 North Blvd.
Houston, TX 77098   (713) 529-1921

Russell J. Egan
Texaco Development Corp.
3336 Richmond Ave.
Suite 165 Box 20
Houston, TX 77098   (713) 520-3624

John S. Egbert
Harrison & Egbert
1018 Preston
Suite 100
Houston, TX 77002   (713) 223-4034

Marvin B. Eickenroht
Vaden, Eickenroht, Thompson,
& Boulware
One Riverway
Suite 1100
Houston, TX 77056   (713) 961-3525

James J. Elacqua
Arnold, White & Durkee
P.O. Box 4433
Houston, TX 77210   (713) 787-1465

Frank M. Elam*
NASA Johnson Space Center
Mail Code EH2
NASA Road 1
Houston, TX 77058   (713) 483-8280

Douglas H. Elliott
Arnold, White & Durkee
750 Bering Drive
Suite 400
Houston, TX 77057   (713) 787-1400

Clarence E. Eriksen
Arnold, White & Durkee
P.O. Box 4433
Houston, TX 77210   (713) 787-1400

A. H. Evans
Vinson & Elkins
1001 Fannin
2927 First City Tower
Houston, TX 77002   (713) 651-2356

William B. Farney
Arnold, White & Durkee
750 Bering
Suite 200
Houston, TX 77057   (713) 787-1400

John B. Farr
1409 Upland Drive
Houston, TX 77043   (713) 468-4205

Edward K. Fein*
NASA Johnson Space Center
Mail Code AL3
Houston, TX 77058   (713) 483-4871

Carl M. Fick
Exxon Production Research Co.
P.O. Box 2189
3120 Buffalo Speedway
Houston, TX 77252   (713) 965-7070

Donald H. Fidler
Fidler & Associates, P.C.
8955 Katy Freeway
Suite 201
Houston, TX 77024   (713) 468-3997

Albert M.T. Finch
Shell Oil Company
Pats. & Licensing
P.O. Box 2463
Houston, TX 77252   (713) 241-0401

Melvin F. Fincke
2210 Pelham Dr.
Houston, TX 77019   (713) 523-0855

Richard D. Fladung
Pravel, Gambrell, Hewitt,
Kimball & Krieger
1177 West Loop South
Suite 1010
Houston, TX 77027   (713) 850-0909

Fred Floersheimer
NL Industries, Inc.
3000 North Belt East
Houston, TX 77032   (713) 987-5152

Raymond C. Floyd
Exxon Chemical Company
13501 Katy Freeway
Houston, TX 77079   (713) 870-6714

Ernest A. Forzano
Shell Oil Co.
One Shell Plaza
P.O. Box 2463
Houston, TX 77001   (504) 588-4753

Mark M. Friedman
Pravel, Gambrell, Hewitt,
Kimball & Krieger
1177 W. Loop South
Houston, TX 77027   (713) 850-0909

James B. Gambrell
Pravel, Gambrell, Hewitt,
Kimball & Krieger
1177 W. Loop South
Suite 1010
Houston, TX 77027   (713) 850-0909

Bradley M. Ganz
Arnold, White & Durkee
750 Bering Drive
Suite 400
Houston, TX 77057   (713) 787-1400

Henry N. Garrana
Schlumberger Well Services
5000 Gulf Freeway
Houston, TX 77023   (713) 928-4071

M. H. Gay
Vinson & Elkins
First City Tower
1001 Fannin
Houston, TX 77002   (713) 651-2350

Henry C. Geller#
Shell Oil Company
P.O. Box 2463
Houston, TX 77001   (713) 241-3760

Jimmy Mark Gilbreth
Pravel, Gambrell, Hewitt,
Kimball & Krieger
10th Floor
1177 West Loop South
Houston, TX 77027   (713) 850-0909

Jefferson D. Giller
Fulbright & Jaworski
1301 Mc Kinney Street
51st Floor
Houston, TX 77010   (713) 651-5151

George E. Glober, Jr.
Exxon Chemical Co.
580 Westlake Park Blvd.
Houston, TX 77079   (713) 584-7666

Edward William Goldstein
Arnold, White & Durkee
750 Bering Drive
Houston, TX 77057   (713) 787-1499

Jack C. Goldstein
Arnold, White & Durkee
P.O. Box 4433
Houston, TX 77210   (713) 787-1400

Kenneth D. Goodman
Arnold, White & Durkee
750 Bering Drive
Suite 400
Houston, TX 77057   (713) 787-1400

W. Gary Goodson
Marathon Oil Company
P.O. Box 3128
Houston, TX 77253   (713) 629-6600

Alan H. Gordon
Arnold, White & Durkee
P.O. Box 4433
Houston, TX 77057   (713) 787-1492

John G. Graham
Texas Instruments Inc.
Box 1443
Houston, TX 77001   (713) 274-3657

Robert L. Graham
15603 Kuykendahl
Suite 115
Houston, TX 77090   (713) 444-7093

C. Donald Gunn
Gunn, Lee & Miller
11 Greenway Plaza
Suite 1616
Houston, TX 77046   (713) 850-9922

Donald F. Haas
Shell Oil Company
One Shell Plaza
P.O. Box 2463
Houston, TX 77001   (713) 241-3356

Mary P. Haddican
Shell Oil Company
700 Louisiana
Houston, TX 77002   (713) 241-5315

Peter K. Hahn
Gunn, Lee & Miller
11 Greenway Plaza
Suite 1616
Houston, TX 77046   (713) 850-9922

Wayne M. Harding
Arnold, White & Durkee
P.O. Box 4433
Houston, TX 77210   (713) 787-1400

Al Harrison
Harrison & Egbert
1018 Preston St.
Suite 100
Houston, TX 77002   (713) 223-4034

Howard Wayne Haworth
9400 Doliver
Apt. 23
Houston, TX 77063   (713) 974-4028

Jack W. Hayden
P.O. Box 22766
Houston, TX 77277   (713) 877-8893

Gerald D. Haynes
Chevron Corporation
1301 Mc Kinney Street
Suite 2108
Houston, TX 77010   (713) 754-3309

John T. Headley
905 Magdalene
Houston, TX 77024   (713) 754-3309

Michael F. Heim
Butler & Binion
1600 Allied Bank Plaza
1000 Louisiana
Houston, TX 77002   (713) 237-3646

Loren G. Helmreich
Browning, Bushman, Zamecki
& Anderson
5718 Westheimer
Suite 1800
Houston, TX 77057   (713) 266-5593

Lester L. Hewitt
Pravel, Gambrell, Hewitt,
Kimball & Krieger
1177 W. Loop South
Suite 1010
Houston, TX 77027   (713) 850-0909

Alden D. Holford
2450 Fondren Road
Suite 312
Houston, TX 77063   (713) 266-0050

Henry Welcker Hope
Fulbright & Jaworski
1301 Mc Kinney Street
51st Floor
Houston, TX 77010   (713) 651-5151

Earl W. Horne
Arnold, White & Durkee
P.O. Box 4433
Houston, TX 77210   (713) 787-1477

Roy F. House#
5726 Ettrick Street
Houston, TX 77035   (713) 721-2117

Eugene Y. Hsiao
15622 Stoney Fork Drive
Houston, TX 77084   (713) 774-4888

Dan L. Hubert#
GTE Mobile Communications Group
616 FM 1960 West
Suite 400
Houston, TX 77090   (713) 586-3660

Walter D. Hunter#
12118 Attlee Drive
Houston, TX 77077   (713) 558-4265

Rita M. Irani
Pravel, Gambrell, Hewitt,
Kimball & Krieger
1177 West Loop South
Suite 1010
Houston, TX 77027   (713) 850-0909

James Lonnie Jackson, Sr.
2929 Briarpark, Ste. 405
Houston, TX 77042   (713) 975-0010

John A. Jacobi
Tenneco Gas Pipeline Group
P.O. Box 2511
Houston, TX 77001   (713) 757-3670

George Byron Jamison, II
Jamison & Mc Gregor
4600 Republic Bank Center
Houston, TX 77002   (713) 224-1212

Paul M. Janicke
Arnold, White and Durkee
P.O. Box 4433
Houston, TX 77057   (713) 787-1455

Catherine K. Jen
7322 Southwest Freeway
Suite 468
Arena Tower I
Houston, TX 77074   (713) 778-1188

Edward L. Jensen
6363 Woodway
Suite 310
Houston, TX 77057   (713) 782-1311

Kenneth C. Johnson
Exxon Company U.S.A.
P.O. Box 2180
Houston, TX 77252   (713) 656-1719

Kenneth H. Johnson
P.O. Box 42415
Houston, TX 77242   (713) 780-7047

William Erby Johnson, Jr.
Browning, Bushman, Zamecki
& Anderson
5718 Westheimer
Suite 1800
Houston, TX 77057   (713) 266-5593

John W. Jones
Pravel, Gambrell, Hewitt,
Kimball & Krieger
1177 West Loop South
10th Floor
Houston, TX 77027   (713) 850-0909

Larry C. Jones
Fulbright & Jaworski
1301 Mc Kinney
51st Floor
Houston, TX 77010   (713) 651-5151

Patricia Ann Kammerer
Arnold, White & Durkee
P.O. Box 4433
Houston, TX 77210   (713) 787-1438

Barry C. Kane
Western Atlas International, Inc.
P.O. Box 1407
Houston, TX 77251   (713) 972-4930

Shaukat A. Karjeker
Pravel, Gambrell, Hewitt,
Kimball & Krieger
1177 W. Loop South
10th Floor
Houston, TX 77027   (713) 850-0909

Kenneth A. Keeling
2916 West T.C. Jester Blvd.
Houston, TX 77018   (713) 680-1447

Christopher D. Keirs
Compaq Computer Corporation
Legal Dept. - M123
P.O. Box 692000
Houston, TX 77269   (713) 374-7676

Denise M. Kettelberger#
Pravel, Gambrell, Hewitt,
Kimball & Krieger
1177 West Loop South
Tenth Floor
Houston, TX 77027   (713) 850-0909

Albert B. Kimball, Jr.
Pravel, Gambrell, Hewitt,
Kimball & Krieger
1177 West Loop S.
10th Floor
Houston, TX 77027   (713) 850-0909

John R. Kirk, Jr.
Baker, Kirk & Bissex, P.C.
3555 Timmons
Suite 700
Houston, TX 77027   (713) 621-2021

Susan K. Knoll
Arnold, White & Durkee
750 Bering Drive
Houston, TX 77057   (713) 787-1400

William A. Knox
8310 Ashcroft Drive
Houston, TX 77096   (713) 774-5414

Irene Kosturakis
Compaq Computer Corp.
20555 FM 149
Houston, TX 77070   (713) 370-0670

Michael P. Kovich
1385 Country Place Dr.
Houston, TX 77079   (713) 493-1906

Paul E. Krieger
Pravel, Gambrell, Hewitt,
Kimball & Krieger
1177 W. Loop South
10th Floor
Houston, TX 77027   (713) 850-0909

Kenneth E. Kuffner
Arnold, White & Durkee
P.O. Box 4433
Houston, TX 77210   (713) 787-1400

William L. Lafuze
Vinson & Elkins
2900 First City Tower
1001 Fannin
Houston, TX 77002   (712) 651-2595

Lee R. Larkin
Andrews & Kurth
4200 Texas Commerce Tower
600 Travis
Houston, TX 77002   (713) 220-4054

Charlene A. Launer
Fulbright & Jaworski
1301 Mc Kinney
Suite 5100
Houston, TX 77010   (713) 651-3634

Gary D. Lawson
Exxon Production Res. Company
P.O. Box 2189
Houston, TX 77252   (713) 965-4846

Sydney M. Leach
Arnold, White & Durkee
P.O. Box 4433
Houston, TX 77210   (713) 787-1511

Joseph D. Lechtenberger
Arnold, White & Durkee
P.O. Box 4433
Houston, TX 77210   (713) 787-1630

Richard F. Lemuth
Shell Oil Company
One Shell Plaza
P.O. Box 2463
Houston, TX 77252   (713) 241-3554

Robert E. Lowe
Western Atlas International, Inc.
P.O. Box 1407
Houston, TX 77251   (713) 972-4928

Gregory M. Luck
Arnold, White & Durkee
P.O. Box 4433
Houston, TX 77210   (713) 787-1400

Sheila M. Luck
Exxon Production Res. Co.
3120 Buffalo Speedway
P.O. Box 2189
Houston, TX 77252   (713) 965-4067

Mitchell D. Lukin
Baker & Botts
910 Louisiana
Houston, TX 77002   (713) 229-1733

Daniel N. Lundeen
4710 Bellaire
Suite 230
Houston, TX 77401   (713) 666-6911

Craig M. Lundell
Arnold, White & Durkee
P.O. Box 4433
Houston, TX 77210   (713) 787-1415

Keith E. Lutsch
Pravel, Gambrell, Hewitt
Kimball & Krieger
1177 West Loop South
10th Floor
Houston, TX 77027   (713) 850-0909

John F. Lynch
Arnold, White & Durkee
750 Bering Drive
Suite 400
Houston, TX 77057   (713) 787-1400

Michael L. Lynch
Arnold, White & Durkee
750 Bering Drive
Houston, TX 77057   (713) 787-1400

Gregory L. Maag
Butler & Binion
1600 First Interstate Bank Plaza
Houston, TX 77002   (713) 237-3130

Michael E. Macklin
Arnold, White & Durkee
P.O. Box 4433
Houston, TX 77210   (713) 787-1416

Paul S. Madan
Western Atlas International, Inc.
10205 Westheimer Road
Suite 1135
Houston, TX 77042   (713) 972-4925

Robert B. Mahley
M.W. Kellogg Company
3 Greenway Plaza
Houston, TX 77046   (713) 960-2147

Robert J. Marett
Marrett & Marrett
1800 West Loop South
Suite 800
Houston, TX 77027   (713) 622-2800

Rodney L. Marett
7703 Braes Meadow
Houston, TX 77071   (713) 622-2800

Fredrik Marlowe
Shell Oil Company
Pats. & Licensing
P.O. Box 2463
Houston, TX 77001   (713) 241-4746

Marvin J. Marnock
Fidler, Marnock & Ross
8955 Katy Freeway
Suite 201
Houston, TX 77024   (713) 468-3997

Thomas F. Marsteller, Jr.
P.O. Box 56265
Houston, TX 77256   (713) 784-6655

Julian C. Martin
Vinson & Elkins
First City Tower
1001 Fannin
Houston, TX 77002   (713) 651-2490

Danita J.M. Maseles
Baker & Botts
3000 One Shell Plaza
910 Louisiana
Houston, TX 77002   (713) 229-1225

Guy E. Matthews
1800 Augusta, Ste. 300
Houston, TX 77057   (713) 781-9595

Douglas H. May, Jr.
Baroid Corporation
P.O. Box 60087
Houston, TX 77205   (713) 987-5156

Carl O. Mc Clenny
6154 Willers Way
Houston, TX 77057   (713) 782-3620

Guy L. Mc Clung, III
Pravel, Gambrell, Hewitt,
Kimball & Krieger
1177 W. Loop South
10th Floor
Houston, TX 77027   (713) 850-0909

Cecil A. Mc Clure
Shell Oil Company
One Shell Plaza
P.O. Box 2463
Houston, TX 77252   (713) 241-3085

Pamela J. Mc Collough
Shell Oil Company
Legal Organ. - Pats. & Lic.
900 Louisiana
P.O. Box 2463
Houston, TX 77252   (713) 241-4091

Patrick H. Mc Collum
Atlas Wireline Services
P.O. Box 1407 0B8
10205 Westheimer
Houston, TX 77251   (713) 972-6024

Claude S. Mc Daniel#
Arnold, White & Durkee
750 Bering Drive
Houston, TX 77057   (713) 787-1400

Martin L. Mc Gregor, Jr.
Baker & Botts
One Shell Plaza
910 Louisiana
Houston, TX 77002   (713) 229-1874

Susan A. Mc Lean
Exxon Production Research Co.
P.O. Box 2189
Houston, TX 77252   (713) 965-4067

Michael T. Mc Lemore
Arnold, White & Durkee
750 Bering Drive
Suite 400
Houston, TX 77057   (713) 787-1454

Leonard P. Miller
Shell Oil Company
P.O. Box 2463
Houston, TX 77252   (713) 241-2538

Thomas A. Miller
Arnold, White & Durkee
P.O. Box 4433
Houston, TX 77210   (713) 787-1483

Peter E. Mims
Vinson & Elkins
2913 First City Tower
1001 Fannin
Houston, TX 77002   (713) 651-2732

Eric-Paul Mirabel
Butler & Binion
1500 First Interstate Bank Plaza
Houston, TX 77002   (713) 237-3149

Eugene R. Montalvo
Pravel, Gambrell, Hewitt,
Kimball & Krieger
1177 West Loop South
Suite 1010
Houston, TX 77027   (713) 850-0909

Raul R. Montes
Exxon Production Res. Company
P.O. Box 2189
3120 Buffalo Speedway
Houston, TX 77252   (713) 965-4064

William H. Montgomery
Vetco Gray Inc.
10777 Northwest Freeway
Houston, TX 77092   (713) 683-2429

Terry D. Morgan
Arnold, White & Durkee
750 Bering Drive
Houston, TX 77057   (713) 787-1400

Fay E. Morisseau, III
Vinson & Elkins
2900 First City Tower
1001 Fannin
Houston, TX 77002   (713) 651-2740

Paula D. Morris
Arnold, White & Durkee
P.O. Box 4433
Houston, TX 77210   (713) 787-1400

David L. Moseley
Dodge, Bush & Moseley
950 Echo Lane
Suite 180
Houston, TX 77024   (713) 827-1059

Kenneth R. Moseley
2047 Sheridan
Houston, TX 77030   (713) 660-0388

Richard L. Moseley
P.O. Box 42415
Houston, TX 77042   (713) 780-7047

Neal J. Mosely
2916 West T.C. Jester Blvd.
Suite 108
Houston, TX 77018   (713) 680-9676

Robert W. Mulcahy
14815 Tumbling Falls Court
Houston, TX 77062   (713) 480-0337

James J. Mullen
Shell Oil Co.
Legal Dept.
P.O. Box 2463
Houston, TX 77001   (713) 241-5765

Kimbley L. Muller
Shell Oil Company
1 Shell Plaza, Rm. 1148
Houston, TX 77001   (713) 241-2698

Kurt Sheridan Myers
7634 Braesdale
Houston, TX 77071   (713) 774-7152

Michael A. Nametz
Exxon Production Res. Co.
P.O. Box 2189
Houston, TX 77001   (713) 965-4554

Kenneth L. Nash
Browning, Bushman, Anderson
& Brookhart
5718 Westheimer
Suite 1800
Houston, TX 77057   (713) 266-5593

Francis D. Neruda
Baeder Neruda Interests, Inc.
5847 San Felipe
Suite 2949
Houston, TX 77057   (713) 784-3008

Nick A. Nichols, Jr.
Gunn, Lee & Miller
11 Greenway Plaza
Suite 1616
Houston, TX 77046   (713) 850-9922

John D. Norris
Arnold, White & Durkee
750 Bering Drive
Houston, TX 77057   (713) 787-1400

William C. Norvell, Jr.
Jackson & Walker
1100 Louisiana, Suite 4200
P.O. Box 4771
Houston, TX 77210   (713) 652-5100

Mark Alva Oathout
Kirk, Bissex & Lindsay, P.C.
3555 Timmons, Suite 700
Houston, TX 77027   (713) 621-2021

James O. Okorafor
Shell Oil Company
1110 One Shell Plaza
P.O. Box 2463
Houston, TX 77252   (713) 241-7236

David M. Ostfeld
Chamberlain, Hrdlicka, White,
Johnson & Williams
1200 Smith Street
1400 Citicorp Center
Houston, TX 77002   (713) 658-1818

Michael L. Parks
Parks & Associates, P.C.
440 Louisiana
Suite 1440
Houston, TX 77002   (713) 227-3050

Joe P. Parris#
Oil Tool Division
Cooper Industries, Inc.
P.O. Box 1212
Houston, TX 77251   (713) 939-2131

Melinda L. Patterson
Arnold, White & Durkee
750 Bering Drive
Suite 400
Houston, TX 77057   (713) 787-1400

William B. Patterson
Cooper Industries
First City Tower
Suite 4000
Houston, TX 77210   (713) 739-5559

Thomas D. Paul
Fulbright & Jaworski
1301 Mc Kinney Street
Houston, TX 77010   (713) 651-5325

Alton W. Payne, Jr.
Sroufe, Zamieck, Payne
& Lundeen
4710 Bellaire Blvd.
Suite 230
Houston, TX 77401   (713) 666-2288

Larry B. Phillips, III
Andrews & Kurth
4200 Texas Commerce Tower
Houston, TX 77002   (713) 220-4090

Richard F. Phillips
Exxon Company, U.S.A.
800 Bell St.
Houston, TX 77252   (713) 656-4864

Bernarr R. Pravel
Pravel, Gambrell, Hewitt,
Kimball & Krieger
1177 W. Loop South
Tenth Floor
Houston, TX 77027   (713) 850-0909

Andrew S. Pryzant#
Sroufe, Zamecki, Payne
& Lundeen
1700 W. Loop South
Suite 1230
Houston, TX 77027   (713) 840-8008

Julia A. Pryzant#
5211 Willowbend
Houston, TX 77096   (713) 729-4941

Frank Burruss Pugsley
Baker & Botts
3000 One Shell Plaza
Houston, TX 77002   (713) 229-1577

Jeffrey A. Pyle
Vaden, Eickenroht, Thompson
& Boulware
One Riverway
Suite 1100
Houston, TX 77056   (713) 961-3525

Ronald Ray Randall
Camco International Inc.
7030 Ardmore
Houston, TX 77054   (713) 749-5690

Richard T. Redano
Rosenblatt & Associates, P.C.
One Greenway Plaza, Suite 500
Houston, TX 77046   (713) 552-9900

James A. Reilly
Arnold, White & Durkee
P.O. Box 4433
720 Bering Drive
Houston, TX 77057   (713) 787-1400

Bernard A. Reiter
1800 Augusta Drive
Suite 300
Houston, TX 77057   (713) 266-6000

Stanley L. Renneker
3200 Southwest Freeway
Suite 2200
Houston, TX 77027   (713) 850-7400

James W. Repass
Fulbright & Jaworski
1301 Mc Kinney Street
51st Floor
Houston, TX 77010   (713) 651-5151

Ronald R. Reper
Shell Oil
P.O. Box 2463
Houston, TX 77252   (713) 241-3247

Fred S. Reynolds, Jr.
Shell Oil Company
One Shell Plaza
P.O. Box 2463
Houston, TX 77252   (713) 241-4177

Glenn W. Rhodes
Arnold, White & Durkee
P.O. Box 4433
Houston, TX 77210   (713) 787-1400

Eugene N. Riddle
Dodge Bush & Moseley
950 Echo Lane
Suite 180
Houston, TX 77024   (713) 827-1054

J. Albert Riddle
1800 Bering, Suite 900
Houston, TX 77057   (713) 781-9595

Carl G. Ries
13911 Kingsride
Houston, TX 77079   (713) 781-9595

James H. Riley
Pravel, Gambrell, Hewitt,
Kimball & Krieger
1177 West Loop South
10th Floor
Houston, TX 77027   (713) 850-0909

W. Ronald Robins
Vinson & Elkins
2918 First City Tower
1001 Fannin
Houston, TX 77002   (713) 651-2452

James B. Robinson
P.O. Box 891562
Houston, TX 77289   (713) 324-2921

Murray Robinson
Butler & Binion
1600 First Interstate Bank Plaza
Houston, TX 77002   (713) 237-3111

Kenneth A. Roddy#
Suite 108
2916 West T.C. Jester Blvd.
Houston, TX 77018   (713) 680-9676

Douglas W. Rommelmann
Pravel, Gambrell, Hewitt,
Kimball & Krieger
1177 W. Loop South
10th Floor
Houston, TX 77027   (713) 850-0909

David Alan Rose
Conley, Rose & Tayon, P.C.
1850 Texas Commerce Tower
600 Travis Street
Houston, TX 77002   (713) 228-8880

Steve Rosenblatt
One Greenway Plaza
Suite 500
Houston, TX 77046   (713) 552-0109

Alan D. Rosenthal
Baker & Botts
3000 One Shell Plaza
Houston, TX 77027   (713) 229-1584

Kent A. Rowald
Vaden, Eickenroht, Thompson
& Boulware
One Riverway
Suite 1100
Houston, TX 77056   (713) 961-3525

Carl A. Rowold
Baker Hughes Inc.
3900 Essex Lane
Suite 1200
Houston, TX 77027   (713) 439-8717

Robert E. Sandfield
16 West Shady Lane
Houston, TX 77063   (713) 789-3016

Ezra L. Schacht#
1620 W. Main Street
Houston, TX 77006   (713) 523-0515

Wilbur Allison Schaich
Hubbard, Thurman, Turner,
Tucker & Harris
6363 Woodway
Suite 275
Houston, TX 77057   (713) 266-1914

Russell E. Schlorff*
NASA Johnson Space Center
Mail Code A13
Houston, TX 77058   (713) 483-1001

John S. Schneider
P.O. Box 270241
Houston, TX 77277   (713) 667-0775

Michael B. Schroeder
Compaq Computer Corp.
20555 S.H. 249
Houston, TX 77070   (713) 374-6830

Willem G. Schuurman
Arnold, White & Durkee
P.O. Box 4433
Houston, TX 77210   (512) 320-7200

Eddie E. Scott
Cooper Industries, Inc.
1001 Fannin, Suite 4000
P.O. Box 4446
Houston, TX 77002   (713) 739-5534

Mark V. Seeley
Arnold, White & Durkee
750 Bering Drive
Suite 400
Houston, TX 77057   (713) 787-1502

Robert C. Shaddox
3050 Post Oak
Suite 1776
Houston, TX 77056   (713) 871-1776

Sue Z. Shaper
Pravel, Gambrell, Hewitt,
Kimball & Krieger
1177 W. Loop South
Tenth Floor
Houston, TX 77027   (713) 850-0909

William E. Shull
Butler & Binion
1500 First Interstate Bank Plaza
Houston, TX 77002   (713) 237-3645

Rand N. Shulman
Shell Oil Company
900 Louisiana
One Shell Plaza
Room 1130
Houston, TX 77001   (713) 241-5147

Peter J. Shurn, III
Arnold, White & Durkee
750 Bering Drive
Suite 400
Houston, TX 77056   (713) 787-1400

Charles D. Simmons
Schlumberger Technology Corp.
P. O. Box 2175
Houston, TX 77252   (713) 928-4811

Jan K. Simpson#
University of Texas
Health Science Center
6901 Bertner
P.O. Box 20334
Houston, TX 77225   (713) 792-4609

Paul F Simpson
Butler & Binion
1600 First Interstate Bank Plaza
Houston, TX 77002   (713) 237-2074

Thomas L. Sivak
Daniel Industries, Inc.
9753 Pine Lake Dr.
Houston, TX 77055   (713) 827-3863

Malcolm H. Skolnick
Neurophysiology Research Center
Univ. of Texas Health Science Ctr.
1343 Moursund Ave.
Houston, TX 77030   (713) 792-4542

Henry L. Smith, Jr.
Tenneco Inc.
1010 Milam
P.O. Box 2511
Houston, TX 77252　　　(713) 757-4116

James D. Smith
Arnold, White & Durkee
750 Bering Drive
Suite 400
Houston, TX 77057　　　(713) 787-1400

Mark A. Smith
Shell Oil Company
1178 One Shell Plaza
P.O. Box 2463
Houston, TX 77252　　　(713) 241-2094

Roy H. Smith, Jr.
5005 West 34th Street
Apt. A-120
Houston, TX 77092　　　(713) 957-2222

Gerald Warren Spinks
Baker Hughes Inc.
3900 Essex Lane
Suite 1200
Houston, TX 77027　　　(713) 439-8600

Jack R. Springgate
Vinson & Elkins
1001 Fannin
2924 First City Tower
Houston, TX 77002　　　(713) 651-2150

Darryl M. Springs
Western Atlas International, Inc.
10205 Westheimer Road
Houston, TX 77042　　　(713) 972-4926

Delmar L. Sroufe
Sroufe, Zamecki, Payne &
Lundeen
1700 W. Loop South
Suite 1230
Houston, TX 77027　　　(713) 840-8008

William Arnold Stout
Fulbright & Jaworski
1301 Mc Kinney Street
51st Floor
Houston, TX 77010　　　(713) 651-3738

Walter L. Stumpf, Jr.
Chevron Chemical Co.
P.O. Box 3725
Houston, TX 77253　　　(713) 754-3302

Michael O. Sutton
Arnold, White & Durkee
750 Bering Drive
Suite 400
Houston, TX 77056　　　(713) 787-1400

Keith M. Tackett
Shell Oil Company
One Shell Plaza, 11th Floor
P.O. Box 2463
Houston, TX 77252　　　(713) 241-2976

Jeffrey W. Tayon
Butler & Binion
1000 Louisianna Street
1600 First Interstate Bank Plaza
Houston, TX 77002　　　(713) 237-3238

Alan R. Thiele
Cooper Industries, Inc.
P.O. Box 4446
Houston, TX 77002　　　(713) 739-5855

E. Eugene Thigpen
Western Atlas International, Inc.
10205 Westheimer Road, No. 1135
P.O. Box 1407
Houston, TX 77251　　　(713) 972-4928

Klaus D. Thoma
Hollrah, Lange & Thoma
1331 Lamar
Suite 1570
Houston, TX 77010　　　(713) 650-1500

Jennings B. Thompson
Vaden, Eickenroht, Thompson
& Boulware
One Riverway
Suite 1100
Houston, TX 77056　　　(713) 961-3525

Marcus L. Thompson, Jr.
Andrews & Kurth
4200 Texas Commerce Tower
Houston, TX 77002　　　(713) 220-4180

Ben D. Tobor
Tudzin & Tobor
11757 Katy Freeway
Suite 1400
Houston, TX 77079　　　(713) 870-1173

D.C. Toedt, III
Arnold, White & Durkee
P.O. Box 4433
Houston, TX 77210　　　(713) 787-1400

Anastassios Triantaphyllis
Butler & Binion
1600 First Interstate Bank Plaza
Houston, TX 77002　　　(713) 237-3285

Timothy N. Trop
Arnold, White & Durkee
750 Bering Drive
Bering Park One
Suite 400
Houston, TX 77057　　　(713) 787-1400

Yung-Yi G. Tsang#
Shell Oil Co.
Pats. & Licensing Dept.
P.O. Box 2463
Houston, TX 77252　　　(713) 241-0956

Robert G. Upton#
Smith International, Inc.
16740 Hardy Street
P.O. Box 60068
Houston, TX 77205　　　(713) 233-5944

Frank Samuel Vaden, III
Vaden, Eickenroht, Thompson
& Boulware
One Riverway
Suite 1100
Houston, TX 77056　　　(713) 961-3525

Derek R. Van Gilder
2222 Summit Tower
11 Greenway Plaza
Houston, TX 77046　　　(713) 622-8344

Paul Van Slyke
Pravel, Gambrell, Hewitt,
Kimball & Krieger
1177 West Loop South
10th Floor
Houston, TX 77027　　　(713) 850-0909

Dean F. Vance
Shell Oil Company
One Shell Plaza
P.O. Box 2463
Houston, TX 77252　　　(713) 241-5648

Robert Michael Vargo
Smith International, Inc.
16740 Hardy Street
Houston, TX 77032　　　(713) 233-5945

Donald J. Verplancken
Conley, Rose & Tayon, P.C.
P.O. Box 3267
Houston, TX 77253　　　(713) 238-8000

Gordon G. Waggett
Arnold, White & Durkee
750 Bering Drive
Suite 400
P.O. Box 4433
Houston, TX 77057　　　(713) 787-1489

Craig W. Walford
M. W. Kellogg Company
601 Jefferson Ave.
P.O. Box 4557
Houston, TX 77210　　　(713) 753-2150

Darcell Walker
Exxon Production Res. Co.
3120 Buffalo Speedway
P.O. Box 2189
Houston, TX 77252　　　(713) 965-4197

Charles F. Walter
9131 Timberside Drive
Houston, TX 77025　　　(713) 661-0259

John P. Ward
M.W. Kellogg Company
601 Jefferson Ave.
P.O. Box 4557
Houston, TX 77210　　　(713) 753-2147

John D. Watts#
Patent Engineers
P.O. Box 79466
Houston, TX 77279　　　(713) 690-8722

Russell D. Weaver
Hollrah, Lange & Thoma
1331 Lamar
Suite 1570
Houston, TX 77010　　　(713) 650-1500

James F. Weiler
Clayton Foundation for Research
1 Riverway, Ste. 1560
Houston, TX 77056   (713) 626-8646

Walter T. Weller
5406 Theall Road
Houston, TX 77066   (713) 626-8646

W. David Westergard
Arnold, White & Durkee
750 Bering, Suite 400
Houston, TX 77057   (713) 787-1400

Albert S. Weycer
Weycer, Kaplan, Pulaski
& Zuber
11 Greenway Plaza
Suite 1414
Houston, TX 77046   (713) 961-9045

Karen T. White
Vinson & Elkins
3300 First City Tower
1001 Fannin
Houston, TX 77002   (713) 758-2388

Travis G. White
Arnold, White & Durkee
750 Bering Drive
Suite 400
Houston, TX 77057   (713) 787-1451

Robert V. Wilder
Compaq Computer Corporation
20555 SH 249
Houston, TX 77070   (713) 374-2124

Danny L. Williams
Arnold, White & Durkee
P.O. Box 4433
Houston, TX 77210   (713) 787-1493

Horace C. Wilson, Jr.
Pravel, Gambrell, Hewitt,
Kimball & Krieger
1177 W. Loop South
Suite 1010
Houston, TX 77027   (713) 850-0909

Michael E. Wilson
Baker & Botts
3000 One Shell Plaza
910 Louisiana
Houston, TX 77002   (713) 229-1819

Pamela L. Wilson
Exxon Production Res. Company
3120 Buffalo Speedway
Houston, TX 77098   (713) 965-4701

David S. Wise
Butler & Binion
1600 First Interstate Bank Plaza
Houston, TX 77002   (713) 237-3169

Mark R. Wisner
Vaden, Eickenroht, Thompson
& Boulware
One Riverway
Suite 1100
Houston, TX 77056   (713) 961-3525

Denise Y. Wolfs
Shell Oil Company
Legal Dept., Pats. & Licensing
P.O. Box 2463
Houston, TX 77252   (713) 241-4798

Russell T. Wong
Arnold, White & Durkee
750 Bering Drive
Suite 400
Houston, TX 77057   (713) 787-1400

E. Richard Zamecki
Sroufe, Zamecki, Payne
& Lundeen
1700 W. Loop South
Suite 1230
Houston, TX 77027   (713) 840-8008

Rene D. Zentner
University of Houston
4800 Calhoun Road
Houston, TX 77204   (713) 749-4662

Roland A. Dexter
8319 Laurel Leaf
Humble, TX 77346   (713) 852-2677

James Harold Barksdale, Jr.
IBM Corp.
Intellectual Property Law Dept.
220 Las Colinas Blvd.
Irving, TX 75039   (214) 556-5202

William R. Cohrs
Exxon Corporation
225 E. John W. Carpenter Frwy.
Irving, TX 75062   (214) 444-1412

Henry C. Goldwire
1405 Ben Drive
Irving, TX 75061   (214) 438-4013

William Lloyd Jones
2413 Crestview Circle
Irving, TX 75062   (214) 255-2361

John A. Odozynski
GTE Telephone Operations
5205 North O'Connor Blvd.
Irving, TX 75039   (214) 718-4441

Marcus S. Rasco#
P.O. Box 153729
Irving, TX 75015   (214) 570-7993

Brian M. Sakima
GTE Telephone Operations
P.O. Box 152092
Mail Code W11I23
Irving, TX 75015   (214) 718-6970

Peter Xiarhos
GTE Service Corporation
5205 N. O'Connor Blvd.
Irving, TX 75062   (214) 718-7810

C. W. Crady, Jr.
22334 Wetherburn Lane
Katy, TX 77449   (713) 574-4939

Walter Joe Lee
301 Ligustrum St.
Lake Jackson, TX 77566   (713) 297-6253

William Douglas Miller
128 Catalpa
Lake Jackson, TX 77566   (409) 297-6755

John E. Vandigriff
190 N. Stemmons Fwy.
Lewisville, TX 75067   (214) 436-0184

Glen F. Gallinger
1209 Phelps Ave.
Littlefield, TX 79339   (806) 385-3719

David E. Cotey
Texas Eastman Company
P.O. Box 7444
Longview, TX 75607   (214) 236-5454

Robert S. Nisbett
311 Anniversary Drive
Longview, TX 75604   (903) 297-3031

Dorsey L. Baker
4603 Eleventh Street
Lubbock, TX 79416   (806) 792-5868

Wendell Coffee
P.O. Box 3726
Lubbock, TX 79452   (806) 763-9252

Raywood H. Blanchard
Adobe Wells C. Club
Box 149
Mission, TX 78572   (312) 687-8050

John M. Duncan
3027 La Quinta Drive
Missouri City, TX 77459   (713) 437-0141

Clarence E. Keys
P.O. Box 1370
201 E. 4th Street
Monahans, TX 79756   (915) 943-4531

Theodore E. Bieber
270 Shoreline Drive
Nacogdoches, TX 75961   (409) 560-4135

Sherman D. Pernia#
1431 San Sebastian Lane
Nassau Bay, TX 77058   (713) 335-1651

Frank Streightoff
413 Canterbury Drive
New Braunfels, TX 78132   (512) 629-6320

Marcus L. Bates, Sr.#
111 W. 10th St.
Odessa, TX 79760   (915) 333-2121

Robert Christian Peterson
Milburn, Hudman & Perterson, P.C.
620 North Grant
Suite 507
P.O. Box 2626
Odessa, TX 79761   (915) 332-0463

Jackie Lee Duke
402 Handell
Pasadena, TX 77502   (713) 943-7017

John Paul Robinson, Jr.
801 W. Ellaine
Pasadena, TX 77506   (713) 477-0240

Wade J. Brady, III
7504 Hamner Lane
Plano, TX 75024   (214) 618-4837

John R. Casperson
Atlantic Richfield Company
Pat. & Tdmk Dept., Legal Div.
2300 West Plano Parkway
PHO 400V
Plano, TX 75075   (214) 754-3000

Dennis O. Kraft
2612 Chadbourne Drive
Plano, TX 75023   (214) 996-5335

Ronald R. Kranzow
Frito-Lay, Inc.
7701 Legacy Drive
Plano, TX 75024   (214) 353-3820

Robert Elsworth Lee, Jr.
Atlantic Richfield Company
Pat. & Tdmk Dept.
2300 W. Plano Parkway
Plano, TX 75075   (214) 754-3397

Roderick W. Mac Donald
Atlantic Richfield Co.
2300 W. Plano Pkwy.
Plano, TX 75075   (214) 754-3337

Michael E. Martin
Atlantic Richfield Company
2300 W. Plano Parkway
Plano, TX 75075   (713) 754-3104

Albert C. Metrailer
Arco Pat. Dept.
PHO 400
2300 W. Plano Parkway
Plano, TX 75075   (214) 754-3340

James D. Olsen
1901 Hillcrest
Plano, TX 75074   (214) 423-5278

John L. Palmer, Sr.
2525 N. Preston Rd., #611
Plano, TX 75075   (214) 964-3117

Tom F. Pruitt
Atlantic Richfield Co.
2300 W. Plano Pkwy.
PRC J2502
Plano, TX 75075   (214) 754-6179

Joseph E. Rogers
Arco
Patent Dept.
2300 W. Plano Parkway
Suite J2502
Plano, TX 75075   (214) 754-3386

Jasper C. Rowe
2605 Cielo Drive
Plano, TX 75074   (214) 424-7193

Thomas P. Schur
Frito-Lay, Inc.
Law Dept.
7701 Legacy Drive
Plano, TX 75024   (214) 353-3822

F. Lindsey Scott
Atlantic Richfield Company
2300 W. Plano Parkway
Plano, TX 75075   (214) 754-3136

Erving A. Trunk
P.O. Box 865043
Plano, TX 75023   (214) 754-3136

Joseph T. Yao
Atlantic Richfield Co.
2300 West Plano Pkwy.
Plano, TX 75075   (214) 754-6678

Calvin A. Rising, Sr.
4121 Everglades
Port Arthur, TX 77642   (409) 982-7159

Richard Warren Anderson#
417 Ridgewood Drive
Richardson, TX 75080   (214) 231-3951

Lawrence J. Bassuk
2907 Wyndham Lane
Richardson, TX 75081   (214) 231-6046

Roger C. Clapp
Ross, Howtson, Clapp
& Korn
740 E. Campbell Rd.
Suite 900
Richardson, TX 75081   (214) 231-9510

Robert J. Crawford
Rockwell International Corp.
1200 N. Alma Road
Richardson, TX 75081   (214) 996-5492

Drude Faulconer
Aetna Tower
2350 Lakeside Blvd.
Suite 850
Richardson, TX 75081   (214) 437-1204

Gregory M. Howison
Ross & Howison
740 E. Campbell Road
900 Park Pacific One
Richardson, TX 75081   (214) 231-9510

Thomas P. Hubbard, Jr.#
1321 Apache
Richardson, TX 75080   (214) 231-2871

Warren Henry Kintzinger
777 S. Central Expwy.
Suite 3-E
Richardson, TX 75080   (214) 234-3914

Martin Korn
Ross, Howison, Clapp
& Korn
Suite 900
740 E. Campbell Road
Richardson, TX 75081   (214) 231-9510

William L. Martin, Jr.
Tandy Corporation
2554 Buttercup
Richardson, TX 75082   (214) 235-0881

Monty L. Ross
Ross, Howison, Clapp & Korn
900 Park Pacific One
740 E. Campbell Road
Richardson, TX 75081   (214) 231-9510

Michael Anthony Sileo, Jr.
Electrospace Systems Inc.
1301 E. Collins Blvd.
Richardson, TX 75083   (214) 470-2517

Ronnie D. Wilson
710 First City Bank Center
Richardson, TX 75080   (214) 699-0041

Jonathan E. Jobe, Jr.
IBM Corporation
5 W. Kirkwood Blvd.
03-05-80
Roanoke, TX 76299   (817) 962-4085

David A. Mims, Jr.
IBM Corporation
5 W. Kirkwood Blvd.
03-05-80
Roanoke, TX 76299   (817) 962-4515

Ernest R. Archambeau, Jr.
2001 Christoval Road
San Angelo, TX 76903   (915) 653-2723

Homer O. Blair
902 N. Main #11
San Angelo, TX 76903   (915) 653-6502

Louis E. Barnett
Datapoint Corp.
9725 Datapoint Drive
M.S. #39
San Antonio, TX 78229   (512) 699-7501

Guy W. Caldwell
Cox & Smith Incorporated
600 National Bank of Comm. Bldg.
San Antonio, TX 78205   (512) 226-7000

Daniel D. Chapman
Gunn, Lee & Miller, P.C.
300 Convent
Suite 1650
San Antonio, TX 78205   (512) 222-2336

Donald R. Comuzzi
1631 Milam Bldg.
115 East Travis St.
San Antonio, TX 78205   (512) 225-0007

James M. Ezzell
Groce, Locke & Hebdon
2000 Frost Bk. Tower
San Antonio, TX 78205   (512) 231-6691

E. Manning Giles
7926 Broadway
Apt. 108
San Antonio, TX 78209   (512) 826-7539

Rosanne Goodman
Fulbright & Jaworski
2200 Interfirst Plaza
300 Convent Street
San Antonio, TX 78205-5000   (512) 224-5574

Gary W. Hamilton
Matthews & Branscomb
One Alamo Center
106 S. St. Mary's Street
Suite 800
San Antonio, TX 78205　　(512) 226-4211

Charles W. Hanor
Cox & Smith, Inc.
2000 NBC Bank Plaza
112 East Pecan Street
San Antonio, TX 78205　　(512) 554-5329

David G. Henry
Gunn, Lee & Jackson, P.C.
300 Convent, Suite 1650
San Antonio, TX 78205　　(512) 222-2336

Mark A. Kammer
Gunn, Lee & Jackson, P.C.
300 Convent, Suite 1650
San Antonio, TX 78205　　(512) 222-2336

Ted D. Lee
Gunn, Lee & Jackson, P.C.
300 Convent
Suite 1650
San Antonio, TX 78205　　(512) 222-2336

Christopher L. Makay#
Cox & Smith, Inc.
2000 NBC Bank Plaza
112 E. Pecan St.
San Antonio, TX 78205　　(512) 554-5500

Mark H. Miller
Gunn, Lee & Jackson, P.C.
300 Convent
Suite 1650
San Antonio, TX 78205　　(512) 222-2336

Martha F. Mims
11603 Sandman
San Antonio, TX 78216　　(512) 341-9310

William B. Nash
Gunn, Lee & Miller, P.C.
1650 NCNB Plaza
300 Convent Street
San Antonio, TX 78205　　(512) 222-2336

E. Suzanne Parr#
6842 Cerro Bajo
San Antonio, TX 78239　　(512) 655-4280

Gale R. Peterson
Cox & Smith
2000 NBC Bank Plaza
San Antonio, TX 78205　　(512) 554-5327

William H. Quirk, IV
KCI
4958 Stout Drive
San Antonio, TX 78219　　(512) 662-9191

Leland A. Sebastian
8030 Misty Canyon
San Antonio, TX 78250　　(512) 522-0534

Thomas E. Sisson
300 Convent
Suite 1650
San Antonio, TX 78205　　(512) 222-2336

Richard J. Smith
Matthews & Branscomb
One Alamo Center
106 S. St. Mary's Street
Suite 800
San Antonio, TX 78205　　(512) 226-4211

John C. Stahl
P.O. Box 13236
San Antonio, TX 78213　　(512) 344-7479

Robert P. Stecher
117 El Prado Drive West
San Antonio, TX 78212　　(512) 829-8900

Gustav N. Van Steenberg
Southwest Research Institute
6220 Culebra Road
P.O. Drawer 28510
San Antonio, TX 78228　　(512) 522-2211

Ronald H. Evans
6607 Saffron Hills Drive
Spring, TX 77379　　(713) 251-8101

Michael G. Fletcher#
8427 Burwood Park Drive
Spring, TX 77379　　(713) 251-8101

Gary H. Lantner
3627 Coltwood
Spring, TX 77388　　(713) 350-2889

W. Wayne Liauh
207 W. Sutton Sq.
Stafford, TX 77477　　(713) 491-1462

Theodore D. Lindgren
Texas Instruments Inc.
12201 Southwest Freeway
M.S. 676
Stafford, TX 77477　　(713) 274-3654

Philip M. Geren
3031 W. Hickory Park Circle
Sugar Land, TX 77479　　(713) 980-4112

John Jeffrey Ryberg
Schlumberger, Ltd.
200 Macco Blvd.
Sugar Land, TX 77478　　(713) 274-8331

Frederick D. Hamilton
205 Harris Drive
Sunnyvale, TX 75182　　(214) 226-4337

Mary Jane Gaskin
2170 Buckthorne Place
Suite 490
The Woodlands, TX 77380　　(713) 363-9121

Margareta Le Maire
70 North Misty Morning Trace
The Woodlands, TX 77381　　(713) 363-9121

John W. Adee#
P.O. Box 131505
Tyler, TX 75713　　(214) 566-4707

Charles W. Alworth#
502 Cumberland Road
Tyler, TX 75703　　(903) 581-7963

Ronald B. Sefrna
505 S. Bois D'Arc
P.O. Box 567
Tyler, TX 75710　　(214) 592-5965

James Duke Willborn
Intratech Corporation
3431 Harwood St.
Tyler, TX 75701　　(903) 595-6456

Charles C. M. Woodward
Howe-Baker Engineers, Inc.
3102 East Fifth Street
Tyler, TX 75701　　(903) 597-0311

Lawrence E. Johnson#
410 Mbank Tower
Waco, TX 76701　　(817) 756-7041

Utah

Edward E. Mc Cullough
P.O. Box 46
Brigham City, UT 84302　　(801) 723-6749

James W. Young
P.O. Box 1088
Centerville, UT 84014　　(801) 538-0900

Lucy L. Sonntag#
Dept. of the Air Force
USAF Hospital Hill
Hill Air Force Base
Clearfield, UT 84056　　(801) 777-5463

Richard B. Oberer*
Dept. of the Air Force
OO-ALC/TISFB
Hill Air Force Base, UT 84056
　　(801) 777-0263

Thompson E. Fehr
Olson & Hoggan, P.C.
56 West Center
P.O. Box 525
Logan, UT 84321　　(801) 752-1551

Robert B. Crouch
7050 Union Park Center
Suite 190
Midvale, UT 84047　　(801) 566-1684

J. David Nelson
Maddox, Nelson & Snuffer
488 East 6400 South
Suite 120
Murray, UT 84107　　(801) 263-2600

Ronald L. Lyons
Thiokol Corp.
2475 Washington Blvd.
M.S. Y01
Ogden, UT 84401　　(801) 629-2080

A. Ray Osburn
Mallinckrodt & Mallinckrodt
317 First Security Bank Bldg.
Ogden, UT 84401　　(801) 393-0331

Craig S. Barrus
1172 West 1300 North
Provo, UT 84604   (801) 373-1232

Joseph William Brown#
J.W. Brown & Associates
698 East 2320 North
Provo, UT 84604   (801) 373-1367

H. Tracy Hall#
1711 N. Lambert Lane
Provo, UT 84604   (801) 374-0300

George H. Mortimer
3687 N. Little Rock Drive
Provo, UT 84604   (801) 224-5647

Delbert R. Phillips
1280 West 1600 North
Provo, UT 84604   (801) 377-1097

Robert A. Bingham
IRECO Incorporated
Eleventh Floor
Crossroads Tower
Salt Lake City, UT 84144   (801) 364-4800

David R. Black
Van Cott, Bagley, Cornwall
& Mc Carthy
50 South Main Street
Suite 1600
Salt Lake City, UT 84144   (801) 532-3333

Richard F. Bojanowski
Suite 735, Judge Bldg.
8 East Broadway
Salt Lake City, UT 84111   (801) 533-0727

Laurence Blair Bond
Trask, Britt & Rossa
P.O. Box 2550
Salt Lake City, UT 84111   (801) 532-1922

William S. Britt
Trask, Britt & Rossa
525 South 300 East
P.O. Box 2550
Salt Lake City, UT 84110   (801) 532-1922

Berne S. Broadbent
1200 Beneficial Life Tower
36 S. State Street
Salt Lake City, UT 84111   (801) 363-1800

Kent S. Burningham
Workman, Nydegger & Jensen
1000 Eagle Gate Tower
60 East South Temple
Salt Lake City, UT 84111   (801) 533-9800

Jon C. Christiansen
Van Cott, Bagley, Cornwall
& Mc Carthy
50 S. Main Street
Suite 1600
Salt Lake City, UT 84144   (801) 532-3333

Kay S. Cornaby
Jones, Waldo, Holbrook
& Mc Donough
1500 First Interstate Plaza
170 South Main Street
Salt Lake City, UT 84101   (801) 521-3200

H. Brian Davis
Trask, Britt & Rossa
P.O. Box 2550
Salt Lake City, UT 84110   (801) 532-1922

Robert R. Finch
4322 Vallejo Drive
Salt Lake City, UT 84124   (801) 278-8184

Lynn Grant Foster
602 East 3rd South
Salt Lake City, UT 84102   (801) 364-5633

Bryan A. Geurts
36 South State
Suite 1200
Salt Lake City, UT 84111   (801) 596-3538

Andrew C. Hess
Callister, Duncan & Nebeker
Suite 800 Kennecott Bldg.
Salt Lake City, UT 84133   (801) 530-7318

Lee A. Hollaar#
1367 E. 100 South
Salt Lake City, UT 84102   (801) 363-8086

Allen R. Jensen
Workman, Nydegger & Jensen
1000 Eagle Gate Tower
60 East South Temple
Salt Lake City, UT 84111   (801) 533-9800

Mathew D. Madsen
123 Second Ave.
Apt. P-105
Salt Lake City, UT 84103   (801) 521-6769

Craig J. Madson
Madson & Metcalf
1200 Beneficial Life Tower
36 S. State Street
Salt Lake City, UT 84111   (801) 537-1700

Philip A. Mallinckrodt
Mallinckrodt & Mallinckrodt
10 Exchange Place
Suite 510
Salt Lake City, UT 84111   (801) 328-1624

Robert R. Mallinckrodt
Mallinckrodt & Mallinckrodt
10 Exchange Place
Suite 510
Salt Lake City, UT 84111   (801) 328-1624

Dennis L. Mangrum
7110 Highland Dr.
Salt Lake City, UT 84121   (801) 943-8107

Michael D. Mc Cully
2533 Catalina Drive
Salt Lake City, UT 84121   (801) 942-1883

Lloyd C. Metcalf
Madson & Metcalf
1200 Beneficial Life Tower
36 S. State St.
Salt Lake City, UT 84111   (801) 537-1700

Julie Kathryn Morriss
Trask, Britt & Rossa
525 South 300 East
Salt Lake City, UT 84111   (801) 532-1922

Rick D. Nydegger
Workman, Nydegger & Jensen
1000 Eagle Gate Tower
60 East South Temple
Salt Lake City, UT 84111   (801) 533-9800

Peter M. Peer, II
Mallinckrodt & Mallinckrodt
10 Exchange Pl.
Suite 510
Salt Lake City, UT 84111   (801) 328-1624

Michael Polacek
Baker Hughes Inc.
257 East 200 South
Suite 1025
Salt Lake City, UT 84111   (801) 366-7070

Thomas J. Rossa
Trask, Britt & Rossa
P.O. Box 2550
Salt Lake City, UT 84110   (801) 532-1922

M. Reid Russell
261 East 300 South
Suite 300
Salt Lake City, UT 84111   (801) 532-1601

David O. Seeley
Workman, Nydegger & Jensen
1000 Eagle Gate Tower
60 East South Temple
Salt Lake City, UT 84111   (801) 533-9800

Marlin Ralph Shaffer, Jr.
Eagle Gate Towers, Ste. 700-38
60 East South Temple St.
Salt Lake City, UT 84111   (801) 321-7800

James L. Sonntag
420 East South Temple
Suite 300
Salt Lake City, UT 84111   (801) 359-3762

Marcus G. Theodore
466 South 500 Easst
Salt Lake City, UT 84102   (801) 359-8622

Gale H. Thorne#
602 East Third South
Salt Lake City, UT 84102   (801) 364-5633

David V. Trask
Trask, Britt & Rossa
525 South 300 East
Salt Lake City, UT 84111   (801) 532-1922

Joseph A. Walkowski, Jr.
Trask, Britt & Rossa
525 South 300 East
Salt Lake City, UT 84111   (801) 532-1922

David A. Westerby
Pacificorp Electric Operations
1407 W. North Temple
Suite 330
Salt Lake City, UT 84140   (801) 220-4265

H. Ross Workman
Workman, Nydegger & Jensen
1000 Eagle Gate Tower
60 East South Temple
Salt Lake City, UT 84111   (801) 533-9800

Todd E. Zenger
Workman, Nydegger & Jensen
60 East South Temple St.
Suite 1000
Salt Lake City, UT 84111   (801) 533-9800

Scott L. Brown#
1009e Buchnell Drive
Sandy, UT 84070   (801) 533-9800

Grant R. Clayton
Thorpe, North & Western
9035 South 700 East
Suite 200
Sandy, UT 84070   (801) 566-6633

Terry M. Crellin
Thorpe, North & Western
9662 South State Street
Sandy, UT 84070   (801) 566-6633

Vaughn W. North
Thorpe, North & Western
9035 South 700 East
Suite 200
Sandy, UT 84070   (801) 566-6633

Calvin E. Thorpe
Thorpe, North & Western
9035 South 700 East
Suite 200
Sandy, UT 84070   (801) 566-6633

Marion Wayne Western
Thorpe, North & Western
9035 South 700 East
Suite 200
Sandy, UT 84070   (801) 566-6633

David Ferber#
P.O. Box 99
Springdale, UT 84767   (801) 772-3237

Craig O. Malin
155 E. Spring Drive
Unit 410
Woodland Hills, UT 84653   (801) 772-3237

Vermont

Mary Ann Mento
Rural Route 1
Box 1945
Arlington, VT 05250   (802) 375-6795

Bailin L. Kuch
General Electric Co.
Armanent Systems Dept.
Lakeside Ave.
Burlington, VT 05401   (802) 657-6592

Stephen John Limanek
75 De Forest Heights
Burlington, VT 05401   (802) 862-1200

Donald T. Steward, Sr.#
P.O. Box 39
Cambridgeport, VT 05141   (802) 869-2754

Francis Joseph Thornton
RD No. 1
Box 1105
Charlotte, VT 05445   (802) 425-2410

Robert Edward Smith
RFD #1
Box 247-106
Danby, VT 05739   (802) 293-5557

Mark F. Chadurjian
IBM Corporation
Intellectual Property Law
Dept. 915, Bldg. 972-2
1000 River Street
Essex Junction, VT 05452   (802) 769-8832

Douglas A. Lashmit
IBM Corp.
Intellectual Property Law
1000 River Street
Dept. 915/Bldg. 972-2
Essex Junction, VT 05452   (802) 769-8843

John Dennis Moore
IBM Corporation
Dept. 915/Bldg. 972-2
1000 River Street
Essex Junction, VT 05452   (802) 769-0111

William D. Sabo
IBM Corporation
Intellectual Property Law Dept.
Dept. 915, Bldg. 972-2
1000 River Road
Essex Junction, VT 05452   (802) 769-9454

Howard J. Walter, Jr.
IBM Corp.
1000 River Street
Essex Junction, VT 05452   (802) 769-9555

Kenneth F. Dusyn
P.O. Box 56 Prior Drive
Killington, VT 05751   (802) 422-3295

Clarence L. Carlson#
R.F.D. 2
Box 301
Lyndonville, VT 05851   (802) 626-8583

Thomas N. Neiman
R.D. 3
Meadow Ridge Lane
Milton, VT 05468   (802) 893-2342

John J. Welch, Jr.
56 West Street
Rutland, VT 05701   (802) 773-3384

Robert E. Dunn
205 Harbor Road
Shelburne, VT 05482   (802) 985-2120

J. Franklin Jones, Jr.
P.O. Box 737
103 Summer Street
Springfield, VT 05156   (802) 885-5127

William O. Moeser
Brownell & Moeser
7 Wall Street
Springfield, VT 05156   (802) 885-4591

Rudolph J. Anderson, Jr.
R.R. 3 Box 6480
Logging Hill Road
Stowe, VT 05672   (802) 253-9827

George E. Kersey
Brook Road
P.O. Box 64
Strafford, VT 05072   (802) 763-8502

Lawrence Shaper
Box 153B
Cream Street
Thetford Center, VT 05075   (802) 785-4403

Virginia

William Earl Fears
Fears & Agor
Box 210
Accomac, VA 23301   (804) 787-1560

Wescott B. Northam
P.O. Box 55
Accomac, VA 23301   (804) 787-1225

Viviana Amzel
6350 Pima Street
Alexandria, VA 22312   (703) 750-1441

Duane L. Antton#
Bacon & Thomas
625 Slaters Ln.
Suite 400
Alexandria, VA 22314   (703) 683-0500

M. Katherine Baumeister#
Burns, Doane, Swecker &
Mathis
Washington & Prince Streets
P.O. Box 1404
Alexandria, VA 22313   (703) 836-6620

Thomas E. Beall, Jr.
Fay, Sharpe, Beall, Fagan,
Minnich & Mc Kee
104 East Hume Avenue
Alexandria, VA 22301   (703) 684-1120

Stephen A. Becker
Lowe, Price, Le Blanc,
Becker & Shur
Suite 300
99 Canal Center Plaza
Alexandria, VA 22314   (703) 684-1111

Stephen A. Bent#
Schwartz, Jeffery, Schwaab, Mack,
Blumenthal & Evans, P.C.
P.O. Box 299
Alexandria, VA 22314   (703) 836-9300

Eugene Berman
5903 Mt. Eagle Drive
Apt. 714
Alexandria, VA 22303   (703) 960-0078

William P. Berridge
Oliff & Berridge
P.O. Box 19928
Alexandria, VA 22320   (703) 836-6400

Stanley M. Blacker
MATEC
8310 Centerbrook Pl.
Alexandria, VA 22308   (703) 360-9378

David A. Blumenthal
Schwartz, Jeffery, Schwaab, Mack,
Blumenthal & Evans, P.C.
1800 Diagonal Road
Suite 510
Alexandria, VA 22314   (703) 836-9300

Evon C. Blunk, Sr.
8701 Highgate Road
Alexandria, VA 22308   (703) 780-8186

Bruce J. Boggs, Jr.#
Burns, Doane, Swecker
& Mathis
699 Prince Street
Alexandria, VA 22314   (703) 836-6620

Craig E. Bolton#
Lowe, Price, La Blanc,
Becker & Shur
99 Canal Center Plaza, Ste. 300
Alexandria, VA 22314   (703) 684-1111

Eugene Michael Bond
Box 1251
Alexandria, VA 22313   (202) 842-0010

Joseph A. Boska#
7802 Strathdon Court
Alexandria, VA 22310   (703) 971-8366

Leonard D. Bowersox#
Oliff & Berridge
277 S. Washington Street
Alexandria, VA 22314   (703) 836-6400

Alfred W. Breiner
Breiner & Breiner
115 North Henry Street
P.O. Box 19290
Alexandria, VA 22314   (703) 684-6885

Mary J. Breiner
Breiner & Breiner
115 North Henry Street
Alexandria, VA 22314   (703) 684-5688

Theodore A. Breiner
Breiner & Breiner
115 North Henry Street
Alexandria, VA 22314   (703) 684-6885

Christopher W. Brody
Lowe, Price, Le Blanc,
Becker & Shur
99 Canal Center Plaza, Ste. 300
Alexandria, VA 22314   (703) 684-1111

John C. Brosky
Burns, Doane, Swecker
& Mathis
Wash. & Prince Sts.
P.O. Box 1404
Alexandria, VA 22313   (703) 836-6620

Kevin C. Brown#
Oliff & Berridge
P.O. Box 19928
Alexandria, VA 22320   (703) 836-6400

Curtis B. Brueske#
Burns, Doane, Swecker
& Mathis
Washington & Prince Sts.
P.O. Box 1404
Alexandria, VA 22313   (703) 836-6620

Stephen P. Burr#
Parkhurst, Wendel & Rossi
1421 Prince Street
Suite 210
Alexandria, VA 22314   (703) 739-0220

Robert Warren Carlson
6406 May Blvd.
Alexandria, VA 22310   (703) 971-4839

Herbert T. Carter, Sr.#
2115 Shiver Drive
Alexandria, VA 22307   (703) 768-0185

Michael J. Carvellas#
8417 Alyce Place
Alexandria, VA 22308   (703) 360-0656

Perry Carvellas
Sherman & Shalloway
413 N. Wash. St.
Alexandria, VA 22314   (703) 549-2282

Donald Clarke Casey
Lowe, Price, Le Blanc,
Becker & Shur
99 Canal Center Plaza, Ste. 300
Alexandria, VA 22314   (703) 684-1111

Lance W. Chandler#
Burns, Doane, Swecker
& Mathis
Washington & Prince Streets
P.O. Box 1404
Alexandria, VA 22313   (703) 836-6620

Chung - Chin Chen#
Bacon & Thomas
625 Slaters Lane
4th Floor
Alexandria, VA 22314   (703) 683-0500

Joseph A. Cooke
5903 Mount Eagle Drive
Suite 518
Alexandria, VA 22303   (703) 960-9327

Clifton B. Cosby#
710 Parkway Terrace
Alexandria, VA 22302   (703) 549-8862

Mario A. Costantino#
Oliff & Berridge
P.O. Box 19928
Alexandria, VA 22320   (038) 366-400

Ronald B. Cox
Cox & O'Connor
1213 Prince Street
Alexandria, VA 22314   (703) 548-2558

Gene P. Crosby#
2201 Scroggins Road
Alexandria, VA 22302   (703) 998-9125

Carol Lynn Cseh
Oliff & Berridge
700 S. Washington St.
Suite 300
Alexandria, VA 22314   (703) 836-6400

Thomas J. D Amico
Stevens, Davis, Miller
& Mosher
515 N. Washington Street
Box 1427
Alexandria, VA 22314   (703) 549-7200

Joseph De Benedictis
Bacon & Thomas
625 Slaters Lane, 4th Fl.
Alexandria, VA 22314   (703) 683-0500

Everett G. Diederiks, Jr.#
Bacon & Thomas
625 Slaters Lane, 4th Fl.
Alexandria, VA 22314   (703) 683-0500

David E. Dougherty
Bacon & Thomas
625 Slaters Lane, 4th Fl.
Alexandria, VA 22314   (703) 683-0500

Steven M. Du Bois#
Burns, Doane, Swecker & Mathis
George Mason Bldg.
Washington & Prince Streets
P.O. Box 1404
Alexandria, VA 22313   (703) 836-6620

Benton S. Duffett, Jr.
Burns, Doane, Swecker & Mathis
George Mason Bldg.
699 Prince Street
Alexandria, VA 22313   (703) 836-6620

John M. England, Jr.
P.O. Box 1674
Alexandria, VA 22313   (703) 548-2952

Joseph D. Evans
Schwartz, Jeffery, Schwaab,
Mack, Blumenthal & Evans, P.C.
1800 Diagonal Road
Suite 510
Alexandria, VA 22313   (703) 836-9300

John J. Feldhaus#
Schwartz, Jeffery, Schwaab,
Mack, Blumenthal & Evans, P.C.
1800 Diagonal Road
Suite 510
Alexandria, VA 22313   (703) 836-9300

Richard E. Fichter
Bacon & Thomas
625 Slaters Lane, 4th Fl.
Alexandria, VA 22314   (703) 683-0500

Milton M. Field
108 S. Columbus St.
Alexandria, VA 22314   (703) 683-4700

John Prince Floyd
704 N. Armistead St.
Alexandria, VA 22312   (703) 354-4235

Donald D. Forrer
Foley & Lardner, Schwartz,
Jeffery, Schwaab, Mack,
Blumenthal & Evans
1800 Diagonal Road, Suite 510
Alexandria, VA 22313   (703) 836-9300

Joel Mark Freed
Burns, Doane, Swecker & Mathis
699 Prince Street
Alexandria, VA 22313   (703) 836-6620

Keith E. George
Lowe, Price, Leblanc,
Becker & Shur
99 Canal Center Plaza
Suite 300
Alexandria, VA 22314   (703) 684-1111

Peter J. Georges
Breneman & Georges
3150 Commonwealth Ave.
Alexandria, VA 22305   (703) 683-8006

Michael M. Gerardi
Foley & Lardner, Schwartz, Jeffery,
Schwaab, Mack, Blumenthal & Evans
Suite 510
1800 Diagonal Rd., P.O. Box 299
Alexandria, VA 22313   (703) 836-9500

Erich J. Gess
Burns, Doane, Swecker
& Mathis
699 Prince Street
Alexandria, VA 22313   (703) 836-6620

Philip Goodman#
307 Yoakum Pkwy.
Alexandria, VA 22304   (703) 751-3631

Israel Gopstein
Lowe, Price, Le Blanc,
Becker & Shur
Suite 300
99 Canal Center Plaza
Alexandria, VA 22314   (703) 684-1111

Rosemary M. Graniewski#
7806 Ridgecrest Dr.
Alexandria, VA 22308   (703) 765-5595

Mary B. Grant
Burns, Doane, Swecker & Mathis
George Mason Bldg.
Washington & Prince Streets
P.O. Box 1404
Alexandria, VA 22313   (703) 838-6612

William J. Griffin#
Burns, Doane, Swecker
& Mathis
P.O. Box 1404
Alexandria, VA 22313   (703) 836-6620

Ronald L. Grudziecki
Burns, Doane, Swecker
& Mathis
699 Prince Street
Alexandria, VA 22313   (703) 836-6620

Michael S. Gzybowski#
Lowe, Price, Le Blanc,
Becker & Shur
99 Canal Center Plaza
Suite 300
Alexandria, VA 22314   (703) 684-1111

Richard C. Harris
Stevens, Davis, Miller
& Mosher
515 N. Washington St.
Alexandria, VA 22314   (703) 549-7200

Benjamin J. Hauptman
Lowe, Price, Le Blanc,
Becker & Shur
Suite 300
99 Canal Center Plaza
Alexandria, VA 22314   (703) 684-1111

Patricia W. Heenan#
2220 Martha's Road
Alexandria, VA 22307   (703) 660-9832

Karl F. Hoback#
Sherman & Shalloway
413 N. Washington St.
Alexandria, VA 22314   (703) 549-2282

Alan Holler
Sherman & Shalloway
413 N. Washington Street
Alexandria, VA 22314   (703) 549-2282

Carl B. Horton#
Burns, Doane, Swecker
& Mathis
699 Prince St.
P.O. Box 1404
Alexandria, VA 22313   (703) 836-6620

George A. Hovanec, Jr.
Burns, Doane, Swecker
& Mathis
699 Prince Street
Alexandria, VA 22313   (703) 836-6620

Kirk M. Hudson
Oliff & Berridge
P.O. Box 19928
Alexandria, VA 22320   (703) 836-6400

Deborah S. Humble
6000 Fort Hunt Road
Alexandria, VA 22307   (703) 329-1236

Horace H. Hunter#
8401 Bound Brook Lane
Alexandria, VA 22309   (703) 780-4308

Robert Danny Huntington
Burns, Doane, Swecker
& Mathis
699 Prince Street
Alexandria, VA 22313   (703) 836-6620

James W. Innskeep#
Burns, Doane, Swecker & Mathis
Washington & Prince St.
George Mason Bldg.
P.O. Box 1404
Alexandria, VA 22313-0   (703) 836-6620

Donald D. Jeffery
Schwartz, Jeffery, Schwaab, Mack,
Blumenthal & Evans, P.C.
Suite 510, 1800 Diagonal Road
P.O. Box 299
Alexandria, VA 22313   (703) 836-9300

William Ray Johnson#
Burns, Doane, Swecker
& Mathis
P.O. Box 1404
Wash. & Prince Sts.
Alexandria, VA 22313   (703) 836-6620

Patrick C. Keane
Burns, Doane, Swecker
& Mathis
P.O. Box 1404
699 Prince Street
Alexandria, VA 22313   (703) 836-6620

J. Ernest Kenney
Bacon & Thomas
625 Slaters Lane, 4th Fl.
Alexandria, VA 22314   (703) 683-0500

Jeffrey M. Ketchum
Fay, Sharpe, Beall, Fagan,
Minnich & Mc Kee
104 E. Hume Avenue
Alexandria, VA 22301   (703) 684-1120

Joseph M. Killeen
Rogers & Killeen
510 King Street
Suite 408
Alexandria, VA 22314   (703) 836-0400

Paul C. Kimball#
Foley & Lardner
1800 Diagonal Rd.
Suite 510
Alexandria, VA 22313   (703) 836-9300

Richard H. Kjeldgaard
Burns, Doane, Swecker
& Mathis
George Mason Bldg.
Washington & Prince Sts.
Alexandria, VA 22313   (703) 836-6620

William A. Knoeller
Stevens, Davis, Miller & Mosher
515 N. Washington Stteet
Alexandria, VA 22314   (703) 549-7200

John Kominski
1402 Key Drive
Alexandria, VA 22302   (703) 751-3026

Alan Edward Kopecki
Burns, Doane, Swecker
& Mathis
699 Prince Street
P.O. Box 1404
Alexandria, VA 22313   (703) 836-6620

Holly D. Kozlowski
Lowe, Price, Le Blanc,
Becker & Shur
Suite 300
99 Canal Center Plaza
Alexandria, VA 22314   (703) 684-1111

Charlotte M. Kraebel
340 Commerce Street
Alexandria, VA 22314   (703) 683-6226

Kenneth E. Krosin
Lowe, Price, Le Blanc,
Becker & Shur
Suite 300
99 Canal Center Plaza
Alexandria, VA 22314   (703) 684-1111

Richard R. Kucia
1707 Crestwood Drive
Alexandria, VA 22302   (703) 998-6937

Johnny A. Kumar#
Foley & Lardner
1800 Diagonal Road
Suite 500
Alexandria, VA 22313   (703) 836-9300

James A. La Barre
Burns, Doane, Swecker
& Mathis
699 Prince Street
P.O. Box 1404
Alexandria, VA 22314   (703) 836-6620

Joseph Labow
5417 Echols Ave.
Alexandria, VA 22311   (703) 820-7032

Glenn Law#
Foley & Lardner
1800 Diagonal Rd.
Suite 510
Alexandria, VA 22313   (703) 836-9300

Robert E. Le Blanc
Lowe, Price, Le Blanc,
Becker & Shur
Suite 300
99 Canal Center Plaza
Alexandria, VA 22314   (703) 684-1111

James E. Ledbetter
515 N. Washington Street
Alexandria, VA 22314   (703) 549-7200

Eugene M. Lee
Foley & Lardner
P.O. Box 299
1800 Diagonal Road
Suite 510
Alexandria, VA 22313   (703) 836-9300

Joseph R. Lentz
509 Duke St.
Alexandria, VA 22314   (703) 549-4453

Robert G. Lev#
Lowe, Price, Le Blanc,
Becker & Shur
99 Canal Center Plaza, Ste. 300
Alexandria, VA 22314   (703) 684-1111

Allan M. Lowe
Lowe, Price, Le Blanc,
Becker & Shur
99 Canal Center Plaza, Ste. 300
Alexandria, VA 22314   (703) 684-1111

Marc J. Luddy
211 North Union Street
Suite 260
Alexandria, VA 22314   (703) 836-6070

Peter G. Mack
Foley & Lardner
Suite 510
1800 Diagonal Road
Alexandria, VA 22313   (703) 836-9300

Robert R. Mackey#
6323 Phyllis Lane
Alexandria, VA 22312   (703) 354-9550

Joseph R. Magnone
Burns, Doane, Swecker & Mathis
George Mason Bldg.
Washington and Prince Sts.
P.O. Box 1404
Alexandria, VA 22313   (703) 836-6620

Ronni S. Malamud
Burns, Doane, Swecker & Mathis
George Mason Building
Washington & Prince Sts.
P.O. Box 1404
Alexandria, VA 22313   (703) 836-6548

Pamela H. Malech
Steven, Davis, Miller
& Mosher
515 N. Washington Street
Alexandria, VA 22314   (703) 549-7200

Shrinath Malur#
Fay, Sharpe, Beall, Fagan,
Minnich & Mc Kee
104 E. Hume Ave.
Alexandria, VA 22301   (703) 684-1120

Platon Nick Mandros
Burns, Doane, Swecker
& Mathis
699 Prince Street
P.O. Box 1404
Alexandria, VA 22313   (703) 836-6620

Eugene Mar
Bacon & Thomas
625 Slaters Lane, 4th Fl.
Alexandria, VA 22314   (703) 683-0500

William Lowrey Mathis
Burns, Doane, Swecker
& Mathis
699 Prince Street
Alexandria, VA 22314   (703) 836-6620

John R. Mattingly
Fay, Sharpe, Beall, Fagan,
Minnich & Mc Kee
104 E. Hume Avenue
Alexandria, VA 22301   (703) 684-1120

Isabelle Rodriguez Mc Andrews
Oliff & Berridge
700 S. Washington Street
Suite 300
Alexandria, VA 22314   (703) 836-6400

Barbara A. Mc Dowell
Foley, Lardner, Schwartz, Jeffery,
Schwaab, Mack, Blumenthal & Evans
King Street Station, Suite 510
1800 Diagonal Road
Alexandria, VA 22313   (703) 836-9300

Richard J. Mc Grath
Burns, Doane, Swecker
& Mathis
Washington & Prince Sts.
P.O. Box 1404
Alexandria, VA 22313   (703) 836-6620

Brian J. Mc Namara
Foley & Lardner, Schwartz, Jeffery,
Schwaab, Mack, Blumenthal & Evans
P.O. Box 299
1800 Diagonal Road, Ste. 510
Alexandria, VA 22313   (703) 836-9300

Frederick G. Michaud, Jr.
Burns, Doane, Swecker & Mathis
George Mason Bldg., Ste. 100
699 Prince Street
Alexandria, VA 22313   (703) 836-6620

Marc A. Miller
10 Ashby Street
Box D
Alexandria, VA 22305   (703) 836-7546

Robert A. Miller
Oliff & Berridge
P.O. Box 19928
Alexandria, VA 22320   (703) 836-6400

Samuel C. Miller
Burn, Doane, Swecker
& Mathis
699 Prince Street
Alexandria, VA 22313   (703) 836-6620

Demetra J. Mills
Oliff & Berridge
700 S. Washington Street
Alexandria, VA 22314   (703) 836-6400

Larry S. Millstein#
Foley & Lardner
1800 Diagonal Road
Suite 510
Alexandria, VA 22313   (703) 836-9300

Thomas J. Moore
Bacon & Thomas
625 Slaters Lane, 4th Fl.
Alexandria, VA 22314   (703) 683-0500

Ellsworth H. Mosher
Stevens, Davis, Miller
& Mosher
515 North Washington Street
Alexandria, VA 22314   (703) 549-7200

Robert G. Mukai
Burns, Doane, Swecker
& Mathis
Washington & Prince Sts.
P.O. Box 1404
Alexandria, VA 22313   (703) 836-6620

J.G. Mullins
Lowe, Price, Le Blanc,
Becker & Shur
99 Canal Plaza, Ste. 300
Alexandria, VA 22314   (703) 684-1111

David R. Murphy
Quaintance & Murphy
1213 Prince St.
Alexandria, VA 22314   (703) 549-1475

Earl A. Nielsen
9330 Mount Vernon Circle
Alexandria, VA 22309   (703) 360-2863

Chittaranjan N. Nirmel
Lowe, Price, Le Blanc,
Becker & Shur
99 Canal Center Plaza, Ste. 300
Alexandria, VA 22314   (703) 684-1111

Joseph M. Noto
Burns, Doane, Swecker
& Mathis
699 Prince Street
Alexandria, VA 22313   (703) 836-6620

James A. Oliff
Oliff & Berridge
P.O. Box 19928
Alexandria, VA 22320   (703) 836-6400

Glenn S. Ovrevik
7912 Telegraph Road
Alexandria, VA 22310   (703) 971-1824

M. Richard Page
Oliff & Berridge
P.O. Box 19928
Alexandria, VA 22320   (703) 836-6400

Thomas J. Pardini
Oliff & Berridge
P.O. Box 19928
Alexandria, VA 22320   (703) 836-6400

Roger W. Parkhurst
Parkhurst, Wendel & Rossi
1421 Prince Street, Ste. 201
Alexandria, VA 22314   (703) 739-0220

Thomas P. Pavelko
Stevens, Davis, Miller
& Mosher
515 North Washington St.
Alexandria, VA 22314   (703) 549-7200

Thomas W. Perkins
Rogers & Killeen
510 King Street
Suite 408
Alexandria, VA 22314   (703) 836-0400

Fred C. Philpitt
Lowe, Price, Le Blanc,
Becker & Shur
99 Canal Center Plaza, Ste. 300
Alexandria, VA 22314   (703) 684-1111

James A. Poulos, III
Stevens, Davis, Miller
& Mosher
515 North Washington St.
Alexandria, VA 22314   (703) 549-7200

Raymond H.J. Powell, Jr.
Oliff & Berridge
277 S. Washington St.
Alexandria, VA 22314   (703) 836-6400

Robert Lee Price
Lowe, Price, Le Blanc,
Becker & Shur
99 Canal Center Plaza, Ste. 300
Alexandria, VA 22314   (703) 684-1111

George E. Quillin
Foley & Lardner
1800 Diagonal Rd., Suite 500
P.O. Box 299
Alexandria, VA 22313   (703) 836-9300

Teresa Stanek Rea
Burns, Doane, Swecker
& Mathis
George Mason Bldg.
699 Prince Street
Alexandria, VA 22314   (703) 836-6620

David D. Reynolds
Burns, Doane, Swecker
& Mathis
699 Prince Street
Alexandria, VA 22314   (703) 836-6620

L. Lawton Rogers, III
510 King Street
Suite 408
Alexandria, VA 22314   (703) 836-0400

Simone A. Rose
Foley & Lardner
1800 Diagonal Rd.
P.O. Box 299
Suite 510
Alexandria, VA 22313   (703) 836-9300

Marc A. Rossi
Parkhurst, Wendel & Rossi
1421 Prince St.
Suite 210
Alexandria, VA 22314   (703) 739-0220

William C. Rowland
Burns, Doane, Swecker
& Mathis
Washington & Prince Sts.
P.O. Box 1404
Alexandria, VA 22313   (703) 836-6620

Gene Z. Rubinson
Lowe, Price, Le Blanc,
Becker, Shur
99 Canal Center Plaza, Ste. 300
Alexandria, VA 22314   (703) 684-1111

Colin G. Sandercock
Foley & Lardner Schwartz,
Jeffery, Schwaab, Mack,
Blumenthal & Evans
1800 Diagonal Rd., Suite 510
Alexandria, VA 22313   (703) 836-9300

Bernhard D. Saxe
Foley & Lardner, Schwartz, Jeffery,
Schwaab, Mack, Blumenthal & Evans
1800 Diagonal Road, Ste. 510
P.O. Box 299
Alexandria, VA 22313   (703) 836-9300

Stanley D. Schlosser
Schwartz, Jeffery, Schwaab, Mack,
Blumenthal & Evans, P.C.
Suite 510
1800 Diagonal Road
Alexandria, VA 22313   (703) 836-9300

Matthew L. Schneider
Burns, Doane, Swecker
& Mathis
George Mason Building
Prince & Washington Streets
Alexandria, VA 22313   (703) 836-6620

Robert M. Schulman
Burns, Doane, Swecker
& Mathis
699 Prince Street
Alexandria, VA 22313   (703) 836-6620

Richard L. Schwaab
Foley, Lardner, Schwartz,
Jeffery & Schwaab, P.C.
Suite 510, 1800 Diagonal Road
P.O. Box 299
Alexandria, VA 22313   (703) 836-9300

Arthur Schwartz
Schwartz, Jeffery, Schwaab,
Mack, Blumenthal & Evans, P.C.
Suite 510, 1800 Diagonal Road
Alexandria, VA 22313   (703) 836-9300

David J. Serbin
Burns, Doane, Swecker & Mathis
699 Prince Street
Suite 100
Alexandria, VA 22314   (703) 836-6620

Gary L. Shaffer
Foley & Lardner, Schwartz, Jeffery,
Schwaab, Mack, Blumenthal & Evans
1800 Diagonal Road
Alexandria, VA 22313   (703) 836-9300

Edwin A. Shalloway
Sherman and Shalloway
413 N. Wash. St.
Alexandria, VA 22314   (703) 549-2282

Robert H. Shapiro#
536 N. Imboden St., #304
Alexandria, VA 22304   (703) 370-2960

Anthony W. Shaw
Burns, Doane, Swecker
& Mathis
Suite 100
699 Prince Street
Alexandria, VA 22313   (703) 836-6620

Fred W. Sherling
1233 N. Picket Street
Alexandria, VA 22304   (703) 370-2445

Leonard W. Sherman
Sherman & Shalloway
413 N. Washington St.
Alexandria, VA 22314   (703) 549-2282

Chandrakant C. Shroff#
600 N. Pickett St.
Alexandria, VA 22304   (703) 823-2024

Henry Shur
Lowe, Price, Le Blanc,
Becker & Shur
Suite 300
99 Canal Center Plaza
Alexandria, VA 22314   (703) 684-1111

Peter K. Skiff
Burns, Doane, Swecker
& Mathis
Washington & Prince Sts.
George Mason Bldg., P.O. Box 1404
Alexandria, VA 22309   (703) 836-6620

Regis E. Slutter
Burns, Doane, Swecker & Mathis
George Mason Bldg.
Washington & Prince Sts.
P.O. Box 1404
Alexandria, VA 22313   (703) 836-6620

Peter H. Smolka
Burns, Doane, Swecker & Mathis
P.O. Box 1404
699 Prince Street
Alexandria, VA 22313   (703) 836-6620

H. Jay Spiegel
703 King Street
Alexandria, VA 22314   (703) 836-1507

Mary C. O. Stauss#
7701 Tauxemont Road
Alexandria, VA 22308   (703) 521-9360

Richard A. Steinberg#
Sherman & Shalloway
413 N. Washington St.
Alexandria, VA 22314   (703) 549-2282

Norman H. Stepno
Burns, Doane, Swecker
& Mathis
George Mason Bldg.
699 Prince St.
Alexandria, VA 22313   (703) 836-6620

William D. Stokes
707 Prince Street
Alexandria, VA 22314   (703) 548-0210

Edward L. Stolarun*
U.S. Army Materiel Command
Attn: Office of Command Counsel
5001 Eisenhower Ave.
Alexandria, VA 22333   (202) 274-8051

Michael J. Strauss
Lowe, Price, Le Blanc,
Becker & Shur
Suite 300
99 Canal Center Plaza
Alexandria, VA 22314   (703) 684-1111

Sharon E. Stroup
Burns, Doane, Swecker & Mathis
George Mason Bldg.
699 Prince Street
Alexandria, VA 22313   (703) 836-6620

Brereton Sturtevant
1227 Morningside Lane
Alexandria, VA 22308   (703) 765-8598

Robert S. Swecker
Burns, Doane, Swecker
& Mathis
699 Prince Street
Suite 100
Alexandria, VA 22314   (703) 836-6620

Angela D. Sykes
Oliff & Berridge
P.O. Box 19928
Alexandria, VA 22320   (703) 836-6400

Stanley H. Tollberg#
P.O. Box 22103
Alexandria, VA 22304   (703) 892-4825

Bruce H. Troxell
Bacon & Thomas
625 Slaters Lane, 4th Fl.
Alexandria, VA 22314   (703) 683-0500

Benjamin E. Urcia
Bacon & Thomas
625 Slaters Lane, 4th Fl.
Alexandria, VA 22314   (703) 683-0500

Michael J. Ure
Burns, Doane, Swecker
& Mathis
George Mason Bldg.
Washington & Prince Sts.
Alexandria, VA 22313   (703) 836-6580

Edward Philip Walker
Oliff & Berridge
P. O. Box 19928
Alexandria, VA 22320   (703) 836-6400

Todd R. Walters#
Burns, Doane, Swecker & Mathis
Washington & Prince Streets
P.O. Box 1404
Alexandria, VA 22313   (703) 836-6620

Robert S. Ward, Jr.#
4221 Shannon Hill Road
Alexandria, VA 22310   (703) 960-8986

Mark P. Watson
3207 Campbell Drive
Alexandria, VA 22303   (703) 960-7449

Eric H. Weisblatt
Burns, Doane, Swecker,
& Mathis
George Mason Bldg.
Washington & Prince Sts.
Alexandria, VA 22313   (703) 836-6620

Charles A. Wendel
Parkhurst, Wendel & Rossi
1421 Prince Street
Suite 210
Alexandria, VA 22314   (703) 739-0220

Farrell Roy Werbow
Stevens, Davis, Miller
& Mosher
515 N. Washington Street
Alexandria, VA 22314   (703) 549-7200

Robert F. White#
303 Princeton Blvd.
Alexandria, VA 22314   (703) 751-4542

Charles F. Wieland, III
Burns, Doane, Swecker
& Mathis
699 Prince St.
Alexandria, VA 22313   (703) 838-6604

Milton S. Winters
205 Yoakum Pkwy.
Apt. 621
Alexandria, VA 22304   (703) 370-1466

Edward J. Wise
Lowe, Price, Leblanc,
Becker & Shur
99 Canal Center Plaza, Ste. 300
Alexandria, VA 22314   (703) 684-1111

Charles R. Wolfe, Jr.
Bacon & Thomas
625 Slaters Lane, 4th Fl.
Alexandria, VA 22314   (703) 683-0500

David W. Woodward#
Foley & Lardner
Suite 510
1800 Diagonal Road
P.O. Box 299
Alexandria, VA 22313   (703) 836-9300

Jim Zegeer
Suite 108
801 N. Pitt St.
Alexandria, VA 22314   (703) 684-8333

Julia C. Ziurys
Burns, Doane, Swecker
& Mathis
P.O. Box 1404
Prince & Washington St.
Alexandria, VA 23313   (703) 836-6620

John R. Spielman
P.O. Box 871
Amherst, VA 24521   (804) 946-5266

Gladden L. Brilhart#
8206 Galahad Court
Annandale, VA 22003   (703) 560-9577

James E. Bryan
8209 Briar Creek Drive
Annandale, VA 22003   (703) 560-9577

Ernest F. Chapman
4840 Kingston Drive
Annandale, VA 22003   (703) 256-3808

Harold P. Deeley, Jr.
4921 S. Centaurs Court
Annandale, VA 22003   (703) 978-7531

Ronald H. Lazarus
Eskovitz, Lazarus & Pitrelli
7023 Little River Turnpike
Suite 202
Annandale, VA 22003   (703) 354-0561

Andrew Matthew Lesniak
4308 Gifford Pinchot Drive
Suite 101
Annandale, VA 22003   (703) 978-8212

Charles J. Myhre#
8106 Chivalry Road
Annandale, VA 22003   (703) 560-2530

Kevin F. O Brien
4544 Little River Run Drive
Annandale, VA 22003   (703) 354-5195

John F. Pitrelli
Eskovitz, Lazarus & Pitrelli
7010 Little River Turnpike
Suite 200
Annandale, VA 22003   (703) 941-3475

Carl D. Quarforth#
3902 Oliver Avenue
Annandale, VA 22003   (703) 256-2695

Vincent L. Ramik
Diller, Ramik & Wight, P.C.
Suite 101, Merrion Square
7345 Mc Whorter Place
Annandale, VA 22003   (703) 642-5705

Daniel J. Stanger#
4603 John Tyler Ct., #2
Annandale, VA 22003   (703) 642-5705

James D. Stokes, Jr.
8211 Toll House Road
Annandale, VA 22003   (703) 978-3012

Richard E. Aegerter
P.O. Box 7136
North Station
Arlington, VA 22207   (703) 536-2597

Dinesh Agarwal
Shlesinger, Arkwright & Garvey
3000 South Eads Street
Arlington, VA 22206   (703) 684-5600

Matthew L. Ajeman#
2101 Crystal Plaza Arcade
Suite 264
Arlington, VA 22202   (703) 684-5600

Penrose Lucas Albright
Pravel, Gambrell, Hewitt,
Kimball & Krieger
P.O. Box 2246
Arlington, VA 22202   (703) 979-3242

Richard L. Andrews#
1400 S. Joyce Street
Apt. C1205
Arlington, VA 22202   (703) 920-3838

Nicholas J. Aquilino
2001 Jefferson Davis Hwy.
Suite 802
Arlington, VA 22202   (703) 979-3315

George Alfred Arkwright, Jr.
Shlesinger, Arkwright,
Garvey & Fado
3000 S. Eads Street
Arlington, VA 22202   (703) 684-5600

John B. Armentrout
1600 S. Eads St., Ste. 1007N
Arlington, VA 22202   (703) 521-1143

James S. Bailey
6007 Williamsburg Blvd
Arlington, VA 22207   (703) 536-8298

Joseph Jay Baker
775 South 23rd Street
Arlington, VA 22202   (703) 979-5700

Stephen G. Baxter#
Oblon, Fisher, Spivak,
Mc Clelland & Maier, P.C.
4th Floor
1755 S. Jeff. Davis Hwy.
Arlington, VA 22202   (703) 521-5980

William E. Beaumont
Oblon, Fisher, Spivak,
Mc Clelland & Maier
Crystal Square Five - Suite 400
1755 S. Jeff. Davis Hwy.
Arlington, VA 22202   (703) 521-5940

John E. Benoit
Benoit, Smith & Laughlin
Suite 503
2001 Jefferson Davis Hwy.
Arlington, VA 22202   (703) 521-1677

Frederick L. Bergert
Nies, Kurz, Bergert
& Tamburro
2121 Crystal Drive
Suite 706
Arlington, VA 22202   (703) 521-6590

William C. Bergmann
705 N. Wayne St., #102
Arlington, VA 22201   (703) 521-6590

Jerry William Berkstresser
Shoemaker & Mattare, Ltd.
2001 Jeff. Davis Hwy.
Suite 1203, CP-1
Arlington, VA 22202   (703) 521-5210

Bruce H. Bernstein
Sandler & Greenblum
2920 South Glebe Road
Arlington, VA 22206   (703) 739-0333

Lois P. Besanko#
Dressler, Goldsmith, Shore,
Sutker & Milnamow, Ltd.
2001 Jeff. Davis Hwy.
Suite 810
Arlington, VA 22202   (703) 415-0880

Richard George Besha
Nixon & Vanderhye, P.C.
2000 North 15th Street
Suite 409
Arlington, VA 22201   (703) 875-0400

George Ladow Black#
IBM Corp.
1755 S. Jeff. Davis Hwy.
Suite 605
Arlington, VA 22202   (703) 769-2421

William A. Blake
Jones, Tullar & Cooper, P.C.
P.O. Box 2266
Eads Station
Arlington, VA 22202   (703) 521-5200

Samuel H. Blech
Oblon, Fisher, Spivak,
Mc Clelland & Maier, P.C.
1755 S. Jeff. Davis Hwy.
Arlington, VA 22202   (703) 521-5940

Peter A. Borsari#
2001 Jeff. Davis Hwy.
Suite 603
Arlington, VA 22202   (703) 920-2377

Charles Paul Boukus, Jr.
2001 Jeff. Davis Hwy., Ste. 202
Arlington, VA 22202   (703) 920-6120

Alan Edward Joseph Branigan
Griffin, Branigan & Butler
775 S. 23rd Street
Arlington, VA 22202   (703) 979-5700

Scott W. Brickner#
Jones, Tullar & Cooper, P.C.
2001 Jefferson Davis Hwy.
Suite 1002
Arlington, VA 22202   (703) 415-1500

Charles A. Brown
727 23rd St. South
Arlington, VA 22202   (703) 521-4536

Charles E. Brown
727 23rd St., S.
Arlington, VA 22202   (703) 892-2791

Laurence Ray Brown
2001 Jefferson Davis Hwy.
Suite 408
Arlington, VA 22202   (703) 521-7200

William S. Brown
2001 Columbia Pike
Apt. 404
Arlington, VA 22204   (703) 920-5916

Harold W. Burnam, Jr.
Griffin, Branigan & Butler
775 S. 23rd St.
P.O. Box 2326
Arlington, VA 22202   (703) 979-5700

James T. Busch*
U.S. Navy
Off. of Chief of Naval Res.
800 N. Quincy Street
Arlington, VA 22317   (202) 696-4008

F. Prince Butler
Griffin, Brannigan & Butler
775 S. 23rd St.
Arlington, VA 22202   (703) 979-5700

Duane M. Byers
Nixon & Vanderhye, P.C.
2200 Clarendon Blvd.
14th Floor
Arlington, VA 22201   (703) 875-0409

Robert Allen Cahill
General Electric Co.
2001 Jeff. Davis Hwy.
Arlington, VA 22202   (202) 637-4525

James A. Cairns
Sandler & Greenblum
2920 South Glebe Road
Arlington, VA 22206   (703) 739-0333

Mary K. Cameron
Sandler, Greenblum
& Bernstein
2920 S. Glebe
Arlington, VA 22206   (703) 739-0333

Alan W. Cannon
Sandler, Greenblum &
Bernstein
2920 S. Glebe Road
Arlington, VA 22206   (703) 739-0333

Marvin A. Champion
1600 S. Eads Street
Apt. 836-N
Arlington, VA 22202   (703) 521-1947

Richard L. Chinn#
Oblon, Spivak, Mc Clelland,
Maier & Neustadt, P.C.
1755 Jefferson Davis Hwy., 4th Fl.
Arlington, VA 22202   (703) 521-5940

Kathleen H. Claffy#
2301 Jeff. Davis Hwy.
Apt. 716
Arlington, VA 22202   (703) 979-6230

John H. O. Clarke
1755 Jefferson Davis Hwy., 4th Fl.
Arlington, VA 22202   (703) 521-4500

Herbert Cohen
Wigman & Cohen, P.C.
Crystal Plaza 3, Suite 200
1735 Jeff. Davis Hwy.
Arlington, VA 22202   (703) 892-4300

Sheldon I. Cohen
2009 N. 14th Street
Suite 708
Arlington, VA 22201   (703) 522-1200

George M. Cooper
Jones, Tullar & Cooper, P.C.
Suite 1002
2001 Jeff. Davis Hwy.
Arlington, VA 22202   (703) 521-5200

James F. Cottone#
2001 Jeff. Davis Hwy.
Crystal Plaza One
Suite 1008
Arlington, VA 22202   (703) 920-6772

John D. Crane
IBM Corporation
1755 S. Jeff. Davis Hwy.
Suite 605
Arlington, VA 22202   (703) 769-2411

Melvin L. Crane#
318 S. Cleveland Street
Arlington, VA 22204   (703) 892-0300

Arthur R. Crawford
Nixon & Vanderhye, P.C.
2000 N. 15th Street, Ste. 409
Arlington, VA 22201   (703) 875-0406

Robert Thompson Crawford
3233 North Pershing Drive
Arlington, VA 22201   (703) 524-9781

Felix J. D Ambrosio
Jones, Tullar & Cooper, P. C.
P.O. Box 2266, Eads Station
2001 Jeff. Davis Hwy.
Arlington, VA 22202   (703) 521-5200

Dennis R. Daley#
Oblon, Spivak, Mc Clelland,
Maier & Neustadt, P.C.
1755 Jeff. Davis Hwy.
Fourth Floor
Arlington, VA 22202   (703) 521-5940

William J. Daniel
2009 N. 14th St., Ste. 701
Arlington, VA 22201   (703) 527-0068

Kenneth E. Darnell#
2301 Jeff. Davis Hwy.
Suite 333
Arlington, VA 22202   (703) 892-0462

Bryan H. Davidson
Nixon & Vanderhye, P.C.
2000 North 15th Street
Suite 409
Arlington, VA 22201   (703) 875-0421

Anthony J. De Laurentis
2001 Jeff. Davis Hwy.
Suite 603
Arlington, VA 22202   (703) 920-2377

Josefino P. De Leon
Shlesinger & Myers
3000 S. Eads Street
Arlington, VA 22202   (703) 684-5600

Paul V. Del Giudice#
Shoemaker & Mattare, Ltd.
2001 Jeff. Davis Hwy.
Suite 1203
Arlington, VA 22202   (703) 521-5210

Donald L. Dennison
Dennison, Meserole, Pollack
& Scheiner
612 Crystal Square 4
1745 Jeff. Davis Hwy.
Arlington, VA 22202   (703) 521-1155

Robert I. Dennison
Dennison, Meserole,
Pollack & Scheiner
1911 Jefferson Davis Hwy.
Arlington, VA 22202   (703) 521-1155

George A. Depaoli
Depaoli & O'Brien
1911 Jefferson Davis Hwy.
Arlington, VA 22202   (703) 521-2110

John Bernard Dickman, III#
2001 Jeff. Davis Hwy., Ste. 1203
Arlington, VA 22202   (703) 521-1320

Michael H. Dickman
2525 North 10th Street
Apt. 303
Arlington, VA 22201   (703) 525-2410

Daniel K. Dorsey
Dennison, Meserole, Pollack
& Scheiner
612 Crystal Sq. 4
1745 Jefferson Davis Hwy.
Arlington, VA 22202   (703) 521-1155

A. Yates Dowell, Jr.
Dowell & Dowell
2001 Jeff. Davis Hwy.
Suite 705
Arlington, VA 22202   (703) 521-5550

A. Yates Dowell, III
Dowell & Dowell
2001 Jeff. Davis Hwy.
Suite 705
Arlington, VA 22202   (703) 521-5550

Ralph A. Dowell
Dowell & Dowell
2001 Jeff. Davis Hwy.
Suite 705
Arlington, VA 22202   (703) 521-5550

Gary R. Drew#
1400 S. Joyce St.
Apt. A-510
Arlington, VA 22202   (703) 521-5550

Frederic C. Dreyer#
2301 Jeff. Davis Hwy.
Apt. 1029
Arlington, VA 22202   (703) 920-1295

Stuart D. Dwork
Oblon, Fisher, Spivak,
Mc Clelland & Maier, P.C.
1755 S. Jefferson Davis Hwy.
Crystal Square Five-Suite 400
Arlington, VA 22202   (703) 521-5940

William T. Ellis*
IBM Corporation
Intellectual Property Law Dept.
1755 S. Jeff. Davis Hwy.
Suite 605
Arlington, VA 22202   (703) 769-2400

Charles R. Engle
George Mason Univ. Law School
3401 N. Fairfax Drive
Arlington, VA 22201   (703) 841-2614

Roger J. Erickson*
U.S. Dept. of Navy
Off. of Naval Research
800 N. Quincy Street
Arlington, VA 22217   (202) 696-4001

Anna P. Fagelson
2301 S. Jefferson Davis Hwy.
Arlington, VA 22202   (703) 892-4449

Charles W. Fallow
Shoemaker & Mattare, Ltd.
2001 Jefferson Davis Hwy.
Suite 1203
Arlington, VA 22202   (703) 521-5210

Robert W. Faris
Nixon & Vanderhye P.C.
2000 N. 15th Street
Suite 409
Arlington, VA 22201   (703) 875-0400

Walter C. Farley
1100 Wilson Blvd.
Suite 1701
Arlington, VA 22209   (703) 528-5282

Michael J. Fink
Sandler & Greenblum
2920 S. Glebe Road
Arlington, VA 22206   (703) 739-0333

Stanley Paul Fisher
Oblon, Fisher, Spivak,
Mc Clelland & Maier, P.C.
1755 S. Jeff. Davis Hwy.
Suite 400
Arlington, VA 22202   (703) 521-5940

Rodger H. Flagg#
2101 Crystal Plaza Arcade
Box 135
Arlington, VA 22202   (703) 553-0501

Howard N. Flaxman#
Hoffman, Wasson & Gitler
2361 Jefferson Davis Hwy.
Suite 522
Arlington, VA 22202   (703) 415-0100

Michael J. Flibbert
Roper & Quigg
Two Crystal Park
2121 Crystal Dr.
Suite 509
Arlington, VA 22202   (703) 920-8910

Andrew D. Fortney#
Oblon, Spivak, Mc Clelland,
Maier & Neustadt, P.C.
1755 Jeff. Davis Hwy., Ste. 400
Crystal Square 5
Arlington, VA 22202   (703) 521-5940

John P. Foryt
Jones, Tullar & Cooper, P.C.
P.O. Box 2266 Eads Station
Arlington, VA 22202   (703) 521-5200

Cynthia L. Foulke
900 N. Stafford St., No. 1618
Arlington, VA 22203   (703) 521-5200

William F. Frank#
2001 Jeff. Davis Hwy.
Suite 1008
Arlington, VA 22202   (703) 415-1785

Stuart D. Frenkel
Depaoli & O'Brien, P.C.
1911 Jeff. Davis Hwy.
Suite 1005
Arlington, VA 22202   (703) 521-2110

James D. Frew*
U.S. Dept. of Navy
Office of Naval Research
800 N. Quincy Street
Arlington, VA 22217   (202) 696-4002

Raymond L. Gable
Neuman, Williams, Anderson
& Olson
2001 Jeff. Davis Hwy.
Suite 903
Arlington, VA 22202   (703) 892-8787

William George Gapcynski
3833 N. Military Rd.
Arlington, VA 22207   (703) 525-6809

William C. Garvert*
Dept. of the Navy
Off. of the Chief of Naval Res.
Deputy Counsel Pats.
800 N. Quincy Street
Arlington, VA 22217   (202) 696-4000

George A. Garvey
Shlesinger, Arkwright & Garvey
3000 S. Eads Street
Arlington, VA 22202   (703) 684-5600

Charles L. Gholz
Oblon, Fisher, Spivak,
Mc Clelland & Maier
1755 S. Jeff. Davis Hwy.
Suite 400
Arlington, VA 22202   (703) 521-5940

Walter C. Gillis, Jr.
Larson & Taylor
727 23rd St., S.
Arlington, VA 22202   (703) 920-7200

Stewart L. Gitler
Hoffman, Wasson & Fallow
2361 Jeff. Davis Hwy.
Suite 522
Arlington, VA 22202   (703) 920-1434

Roger P. Glass
Sandler & Greenblum
2920 South Glebe Road
Arlington, VA 22206   (703) 739-0333

Robert F. Gnuse
Oblon, Fisher, Spivak,
Mc Clelland & Maier, P.C.
1755 Jeff. Davis Hwy.
Crystal Square Five, Suite 400
Arlington, VA 22202   (703) 521-5940

Douglas H. Goldhush
2627 S. Grant St.
Arlington, VA 22202   (703) 549-1918

Elmer Ellsworth Goshorn*
U.S. Dept. of Navy
P.O. Box 2304
Arlington, VA 22202   (202) 692-7136

Jonathan E. Grant
Longacre & White
1919 S. Eads St.
Suite 401
Arlington, VA 22202   (703) 521-1827

Neil F. Greenblum
Sandler & Greenblum
2920 South Glebe Road
Arlington, VA 22206   (703) 739-0333

Donald R. Greene
Leitner, Greene &
& Christensen, P.C.
1735 Jeff. Davis Hwy.
Crystal Sq. 3, Suite 203
Arlington, VA 22202   (703) 486-8100

Thomas J. Greer, Jr.
727 23rd St., S.
Arlington, VA 22202   (703) 892-2410

Edwin E. Greigg
Suite 220
727 23rd Street
Arlington, VA 22202   (703) 892-0300

Ronald E. Greigg
Suite 220
727 23rd Street South
Arlington, VA 22202   (703) 892-0300

Benjamin F. Griffin, Jr.
Griffin, Branigan and Butler
P.O. Box 2326
775 S. 23rd St.
Arlington, VA 22202   (703) 979-5700

Sarojini B. Grigsby#
1311 S. Norwood St.
Arlington, VA 22204   (703) 979-6315

Arnold G. Gulko
Dressler, Goldsmith, Shore,
Sutker & Milnamow
2001 Jefferson Davis Hwy.
Suite 810
Arlington, VA 22202   (703) 521-1880

Paul Anthony Guss#
1211 S. Eads Street
No. 1502
Arlington, VA 22202   (703) 920-6516

Robert W. Hahl#
Oblon, Spivak, Mc Clelland,
Maier & Neustadt
1755 S. Jeff. Davis Hwy.
Suite 400
Arlington, VA 22202   (703) 521-5940

George D. Hall
612-20th St. South
Arlington, VA 22202   (703) 979-3399

James D. Hamilton
Oblon, Fisher, Spivak,
Mc Clelland & Maier, P.C.
Crystal Square 5-Suite 400
1755 S. Jeff. Davis Hwy.
Arlington, VA 22202   (703) 521-5940

Diana Hamlet-King#
Millen, White & Zelano, P.C.
Arlington Courthouse Plaza 1
Suite 1201
2200 Clarendon Blvd.
Arlington, VA 22201   (703) 243-6333

Frederick R. Handren#
Suite 1110
2301 Jeff. Davis Hwy.
Arlington, VA 22202   (703) 418-1736

Douglas R. Hanscom
Jones, Tullar & Cooper, P.C.
P. O. Box 2266
Eads Station
Arlington, VA 22202   (703) 521-5200

Robert F. Hargest, III
Witherspoon & Hargest
745 S. 23rd Street
Suite 204
Arlington, VA 22202   (703) 521-0511

Blaney Harper#
IBM Corp.
1755 S. Jeff. Davis Hwy.
Suite 605
Arlington, VA 22202   (703) 769-2407

Boris Haskell
Paris & Haskell
2316 S. Eads St.
Arlington, VA 22202   (703) 979-4870

Brion P. Heaney#
Millen, White & Zelano, P.C.
Arlington Courthouse Plaza 1
Suite 1201
2200 Clarendon Blvd.
Arlington, VA 22201   (703) 243-6333

James W. Hellwege
Young & Thompson
745 South 23rd Street
Suite 200
Arlington, VA 22202   (703) 521-2297

Paul J. Henon
Nixon & Vanderhye, P.C.
2200 Clarendon Blvd.
14th Floor
Arlington, VA 22201   (703) 875-0417

Heinrich Wolfgang Herzfeld#
P.O. Box 2445
Arlington, VA 22202   (703) 230-2836

David Harmon Hill
B-1205 Riverhouse II
1400 S. Joyce Street
Arlington, VA 22202   (703) 920-7001

William R. Hinds
727 23rd St., S.
Arlington, VA 22202   (703) 920-7500

Robert V. Hines#
Three H Search Systems, Ltd.
2101 Crystal Plaza Arcade
Suite 333
Arlington, VA 22202   (703) 255-0461

Martin Paul Hoffman
Hoffman, Wasson & Fallow
522 Hayes Bldg.
2361 Jeff. Davis Hwy.
Arlington, VA 22202   (703) 920-1434

Patrick M. Hogan
ITT Defense
1000 Wilson Blvd.
Arlington, VA 22209   (703) 247-2990

William Harry Holt
727 23rd Street South
Suite 218
Arlington, VA 22202   (703) 553-0030

James T. Hosmer
Nixon & Vanderhye, P.C.
2000 N. 15th Street
Arlington, VA 22201   (703) 875-0400

Ross Franklin Hunt, Jr.
Larson & Taylor
727 23rd St., S.
Arlington, VA 22202   (703) 920-7200

Rupert B. Hurley, Jr.#
Sandler, Greenblum &
Bernstein
2920 South Glebe Rd.
Arlington, VA 22206   (703) 739-0333

Douglas E. Jackson
Larson & Taylor
727 23rd St., S.
Arlington, VA 22202   (703) 920-7200

William E. Jackson
Larson & Taylor
727 23rd St., S.
Arlington, VA 22202   (703) 920-7200

Fred Jacob
General Electric Company
2001 Jeff. Davis Hwy.
Arlington, VA 22202   (202) 637-4314

Kenneth E. Jacobs
1600 S. Eads Street
Apt. 311-S
Arlington, VA 22202   (703) 892-5167

Julius Jancin, Jr.
IBM Corp.
1755 S. Jeff. Davis Hwy.
Suite 605
Arlington, VA 22202   (703) 769-2401

David Leonard Johnson, Jr.
1600 S. Eads Street
708S
Arlington, VA 22202   (703) 920-3489

Richard J. Johnson#
5612 N. 18th Street
Arlington, VA 22205   (703) 536-6081

James L. Jones, Jr.#
727 23rd St., S.
Arlington, VA 22202   (703) 979-4586

Vivian P. Kafalenos
2023 N. 21st Street
Apt. 26
Arlington, VA 22201   (703) 841-9729

Manabu Kanesaka
Kanesaka & Takeuchi
727 23rd St., S.
Arlington, VA 22202   (703) 521-3810

A. Matt Kasap#
2101 Crystal Plaza Arcade
Suite 187
Arlington, VA 22202   (703) 684-1597

Michael J. Keenan
Nixon & Vanderhye, P.C.
2200 Clarendon Blvd., 14th Fl.
Arlington, VA 22201   (703) 875-0412

Steven B. Kelber
Oblon, Fisher, Spivak,
Mc Clelland & Maier, P.C.
Crystal Square Five - Suite 400
1755 S. Jeff. Davis Hwy.
Arlington, VA 22202   (703) 521-5940

Richard D. Kelly
Oblon, Fisher, Spivak,
Mc Clelland & Maier, P.C.
Suite 400, Crystal Square Five
1755 S. Jeff. Davis Hwy.
Arlington, VA 22202   (703) 521-5940

Solon B. Kemon
Shlesinger, Arkwright & Garvey
3000 S. Eads Street
Arlington, VA 22202   (703) 684-5600

Richard Benedict Kirk
Kirk & Smith
Suite 309
2001 Jefferson Davis Hwy.
Arlington, VA 22202   (703) 521-1820

Harry E. Kitchen#
2101 Crystal Plaza Arcade
Suite 379
Arlington, VA 22202   (703) 549-1733

Werner Warren Kleeman
Sandler, Greenblum
& Bernstein
2920 S. Glebe Rd.
Arlington, VA 22206   (703) 739-0333

William L. Klima
2001 Jefferson Davis Hwy.
Crystal Plaza One
Suite 903
Arlington, VA 22202   (703) 415-4777

Maurice H. Klitzman
IBM Corp.
1755 S. Jeff. Davis Hwy.
Suite 605
Arlington, VA 22202   (703) 769-2415

Steven Kreiss
Wigman & Cohen, P.C.
Crystal Square 3, Suite 200
1735 Jeff. Davis Hwy.
Arlington, VA 22202   (703) 892-4300

Dennis L. Kreps
2001 Jeff. Davis Hwy.
Suite 200
Arlington, VA 22202   (703) 920-2377

Eckhard H. Kuesters
Oblon, Fisher, Spivak,
Mc Clelland & Maier, P.C.
Suite 400
1755 Jeff. Davis Hwy.
Arlington, VA 22202   (703) 521-5940

George C. Kurtossy
Griffin, Branigan & Butler
775 South 23rd Street
Arlington, VA 22202   (703) 979-5700

Philip Elledge Kurz
Nies, Webner, Kurz
& Bergert
1911 Jeff. Davis Hwy.
Suite 700
Arlington, VA 22202   (703) 521-6590

John C. La Mont
323 S. Veitch Street
Apt. 8
Arlington, VA 22204   (703) 521-6090

Marc R. Labgold#
Oblon, Spivak, Mc Clelland,
Maier & Neustadt, P.C.
1755 Jefferson Davis Hwy.
Ste. 400
Arlington, VA 22202   (703) 521-5940

Randy W. Lacasse#
IBM Corp.
1755 S. Jefferson Davis Hwy.
Suite 605
Arlington, VA 22202   (703) 769-2400

Dennis H. Lambert
Neuman, Williams, Anderson
& Olson
2001 Jeff. Davis Hwy.
Crystal Plaza One, Ste. 903
Arlington, VA 22202   (703) 415-4777

Walter J. Landry
1817 N. Quinn Street
Suite 205
Arlington, VA 22209   (703) 528-3139

Munson H. Lane, Jr.
1515 N. Courthouse Road
Suite 402
Arlington, VA 22201   (703) 522-6762

Lawrence E. Laubscher, Jr.
Laubscher, Presta & Laubscher
745 South 23rd Street
Suite 300
Arlington, VA 22202   (703) 521-2660

Lawrence Edwin Laubscher
Laubscher, Presta & Laubscher
745 S. 23rd Street
Arlington, VA 22202   (703) 521-2660

James H. Laughlin, Jr.
Benoit, Smith & Laughlin
2001 Jeff. Davis Hwy.
Suite 501
Arlington, VA 22202   (703) 521-1677

Jean-Paul P.M. Lavalleye
Oblon, Fisher, Spivak,
Mc Clelland & Maier, P.C.
1755 S. Jeff. Davis Hwy.
Suite 400
Arlington, VA 22202   (703) 521-5940

Henry S. Layton
Suite 1114 South
2111 Jeff. Davis Hwy.
Arlington, VA 22202   (703) 979-5441

Saul Leitner
Leitner, Greene & Christensen, P.C.
1735 Jeff. Davis Hwy.
Arlington, VA 22202   (703) 486-8100

James A. Lisehora#
Oblon, Spivak, Mc Clelland,
Maier & Neustadt, P.C.
1755 Jefferson Davis Hwy., 4th Flr.
Arlington, VA 22202   (703) 521-5940

Richard C. Litman
1725 S. Jeff. Davis Hwy.
Suite 801
Arlington, VA 22202   (703) 920-6000

George A. Loud
Lorusso & Loud
2001 Jeff. Davis Hwy.
Suite 1003
Arlington, VA 22202   (703) 979-1960

Warren N. Low
Low & Low
2316 S. Eads St.
Arlington, VA 22202   (703) 979-4870

Robert Charles Lucke
2111 Jefferson Davis Hwy., N201
Arlington, VA 22202   (703) 920-0612

Jerry C. Lyell
2009 N. 14th Street
Suite 203
Arlington, VA 22201   (703) 243-3133

Gregory J. Maier
Oblon, Fisher, Spivak,
Mc Clelland & Maier, P.C.
1755 S. Jeff. Davis Hwy.
Suite 400
Arlington, VA 22202   (703) 521-5940

Alfred J. Mangels
Nies, Kurz, Bergert
& Tamburro
2121 Crystal Drive
Suite 706
Arlington, VA 22202   (703) 521-6590

Gary L. Manuse
P.O. Box 2227
Arlington, VA 22202   (703) 378-7496

Donald William Marks
542 South 23rd St.
Arlington, VA 22202   (703) 920-1688

Terrence Martin
Wigman & Cohen, P.C.
Crystal Square 3, Suite 200
1735 Jeff. Davis Hwy.
Arlington, VA 22202   (703) 892-4300

Brian J. Marton
Shlesinger, Arkwright & Garvey
3000 S. Eads Street
Arlington, VA 22202   (703) 684-5600

William B. Mason
Mason, Mason & Albright
2306 S. Eads St.
Arlington, VA 22202   (703) 979-3242

James T. Mc Call#
4083 S. Four Mile Run Drive
Arlington, VA 22204   (703) 486-8592

William F. Mc Carthy, Jr.*
U.S. Dept. of Navy
Office of Naval Research
800 N. Quincy St.
Arlington, VA 22217   (202) 696-4003

C. Irvin Mc Clelland
Oblon, Fisher, Spivak,
Mc Clelland & Maier, P.C.
1755 Jeff. Davis Hwy.
4th Floor
Arlington, VA 22202   (703) 521-5940

J. Andrew Mc Kinney, Jr.#
Jones, Tullar & Cooper, P.C.
2001 Jefferson Davis Hwy.
Suite 1002
Arlington, VA 22202   (703) 415-1500

Charles J. Merek
Research Publications
Suite 1821-D
1921 Jeff. Davis Hwy.
Arlington, VA 22202   (703) 920-5050

William H. Meserole, Jr.
Dennison, Meserole, Pollack
& Scheiner
1745 Jeff. Davis Hwy.
Suite 612
Arlington, VA 22202   (703) 521-1155

Anthony L. Miele#
Sandler, Greenblum & Bernstein
2920 S. Glebe Road
Arlington, VA 22206   (703) 739-0333

I. William Millen
Millen, White & Zelano, P.C.
Arlington Courthouse Plaza 1
Suite 1201
2200 Clarendon Blvd.
Arlington, VA 22201   (703) 243-6333

Anthony D. Miller
Hoffman, Wasson, Fallow
& Gitler, P.C.
2361 Jefferson Davis Hwy.
Suite 522
Arlington, VA 22202   (703) 415-0100

Davidson Church Miller
2538 23rd Road North
Arlington, VA 22207   (703) 525-1563

Robert C. Miller
Oblon, Fisher, Spivak,
Mc Clelland & Maier, P.C.
Suite 400
1755 S. Jeff. Davis Hwy.
Arlington, VA 22202   (703) 521-5940

Leonard C. Mitchard
Nixon & Vanderhye
14th Floor
2200 Clarendon Blvd.
Arlington, VA 22201   (703) 875-0400

James C. Mitchell
Jones, Tullar & Cooper, P.C.
2001 Jeff. Davis Hwy.
Suite 1002
Arlington, VA 22202   (703) 521-5200

A. Louis Monacell
6613 N. 29th St.
Arlington, VA 22213   (703) 241-8421

John R. Moses
Millen, White & Zelano, P.C.
Arlington Courthouse Plaza 1
Suite 1201
Arlington, VA 22201   (703) 243-6333

Janice M. Mueller
2250 Clarendon Blvd.
No. 318
Arlington, VA 22201   (703) 525-6249

Martin G. Mullen
Sandler & Greenblum
2920 South Glebe Road
Arlington, VA 22206   (703) 739-0333

George C. Myers, Jr.
Wigman & Cohen, P.C.
Crystal Square 3, Ste. 200
1735 Jeff. Davis Hwy.
Arlington, VA 22202   (703) 892-4300

Gary M. Nath
Roper & Quigg
Two Crystal Park
2121 Crystal Drive
Suite 509
Arlington, VA 22202   (703) 920-8910

Jeffry H. Nelson
Nixon & Vanderhye
2200 Clarendon Blvd., 14th
Arlington, VA 22201   (703) 875-0423

Arthur I. Neustadt
Oblon, Fisher, Spivak,
Mc Clelland & Maier, P.C.
Suite 400 - Crystal Square Five
1755 S. Jeff. Davis Hwy.
Arlington, VA 22202   (703) 521-5940

Frank G. Nieman*
U.S. Dept. of Navy
Office of Naval Research
800 N. Quincy St.
Arlington, VA 22217   (202) 696-4007

John Dirk Nies
Nies, Webner, Kurz
& Bergert
1911 Jeff. Davis Hwy.
Suite 700
Arlington, VA 22202   (703) 521-6590

William G. Niessen
5122 N. 16th Street
Arlington, VA 22205   (703) 528-4734

Joseph P. Nigon
2201 Jeff. Davis Hwy.
Arlington, VA 22202   (703) 521-1666

Larry S. Nixon
Nixon & Vanderhye, P.C.
2200 Clarendon Blvd., 14th Fl.
Arlington, VA 22201   (703) 875-0400

Jerome J. Norris
2001 Jeff. Davis Hwy.
Arlington, VA 22202   (703) 521-1666

David A. Novais#
Oblon, Fisher, Spivak,
Mc Clelland & Maier, P.C.
1755 S. Jeff. Davis Hwy.
Crystal Square Suite 400
Arlington, VA 22202   (703) 521-5940

Harold L. Novick
Larson & Taylor
727 23rd Street, South
Arlington, VA 22202   (703) 920-7200

Mark E. Nusbaum
Nixon & Vanderhye, P.C.
2000 N. 15th Street
14th Floor
Arlington, VA 22201   (703) 875-0414

William E. O'Brien
Depaoli & O'Brien, P.C.
1911 Jeff. Davis Hwy.
Arlington, VA 22202   (703) 521-2110

Paul T. O'Neil
O'Neil & Bean
1601 N. Kent Street
Suite 910
Arlington, VA 22209   (703) 525-3131

Kevin D. O'Shea#
2102 Crystal Plaza Arcade
Suite 406
Arlington, VA 22202   (703) 644-4852

Norman F. Oblon
Oblon, Fisher, Spivak,
Mc Clelland & Maier, P.C.
Crystal Square Five, Ste. 400
1755 S. Jeff. Davis Hwy.
Arlington, VA 22202   (703) 521-5940

Robert A. Ostmann
2001 Jeff. Davis Hwy.
Arlington, VA 22202   (703) 521-1221

Magdalene J. Palumbo#
1204 S. Oakcrest Rd.
Arlington, VA 22202   (703) 521-4089

Leslie J. Paperner
Sandler, Greenblum & Bernstein
2920 S. Glebe Road
Arlington, VA 22206   (703) 739-0333

Gayle Parker
727 23rd Street, South
Suite 300
Arlington, VA 22202   (703) 920-8831

Lutrelle F. Parker, Sr.
2016 S. Fillmore Street
Arlington, VA 22204   (703) 979-7696

Andrew J. Patch#
Young & Thompson
745 S. 23rd St.
Arlington, VA 22202   (703) 521-2297

Robert J. Patch
Young & Thompson
745 S. 23rd Street
Arlington, VA 22202   (703) 521-2297

Eugene J. Pawlikowski
Dennison, Meserole, Pollack
& Scheiner
612 Crystal Square 4
1745 Jeff. Davis Hwy.
Arlington, VA 22202   (703) 521-1155

Louis J. Percello
IBM Corp.
1755 S. Jefferson Davis Hwy., Ste. 605
Arlington, VA 22202   (703) 769-2417

Charles W. Peterson, Jr.#
IBM Corp.
1755 S. Jefferson Davis Hwy., Ste. 605
Arlington, VA 22202   (703) 769-2400

Marvin Petry
Larson & Taylor
727 23rd St., South
Suite 300
Arlington, VA 22202   (703) 920-7200

William Pieprz
Sandler & Greenblum
2920 S. Glebe Rd.
Arlington, VA 22206   (703) 739-0333

Carmen B. Pili Curtis#
Millen, White & Zelano
2200 Clarendon Blvd.
Suite 1201
Arlington, VA 22201   (703) 243-6333

David Pollack
Dennison, Meserole,
Pollack & Scheiner
Ste. 612, Crystal Square 4
1745 Jeff. Davis Hwy.
Arlington, VA 22202   (703) 521-1155

Robert T. Pous
Oblon, Spivak, Mc Clelland,
Maier, & Neustadt P.C.
1755 S. Jeff. Davis Hwy.
Suite 400
Arlington, VA 22202   (703) 521-5940

Frank Paul Presta
Laubscher, Presta & Laubscher
745 South 23rd St.
Suite 300
Arlington, VA 22202   (703) 521-2660

Donald J. Quigg
Roper & Quigg
Two Crystal Park
2121 Crystal Drive
Suite 509
Arlington, VA 22202   (703) 920-8910

Inna Reichstein#
Woolcott & Company, Inc.
1745 Jefferson Davis Highway
Suite 505
Arlington, VA 22202   (703) 998-5505

Louis K. Rimrodt#
SMX, Inc.
P.O. Box 15438
Arlington, VA 22215   (703) 548-3544

John Summerfield Roberts, Jr.
2111 Jeff. Davis Hwy., Ste. 609S
Arlington, VA 22202   (703) 418-6277

Alton D. Rollins
Oblon, Spivak, Mc Clelland,
Maier & Neustadt, P.C.
1755 S. Jefferson Davis Hwy.
Arlington, VA 22202   (703) 521-5940

Allen P. Rosenberg
Shoemaker & Mattare Ltd.
Crystal Plaza 1, Suite 1203
2001 Jeff. Davis Hwy.
Arlington, VA 22202   (703) 521-5210

Leo Ross
6340 12th Place, North
Arlington, VA 22205   (703) 532-7145

James L. Rowland
Sandler, Greenblum & Bernstein
2920 S. Glebe Road
Arlington, VA 22206   (703) 739-0333

Rene S. Rutkowski
Low & Low
2316 S. Eads Street
Arlington, VA 22202   (703) 979-4870

Surinder Sachar#
Oblon, Spivak, Mc Clelalnd,
Maier & Neustadt, P.C.
1755 Jefferson Davis Hwy.
Arlington, VA 22202   (703) 521-5940

Donald M. Sandler
Sandler & Greenblum
2920 South Glebe Road
Arlington, VA 22206   (703) 739-0333

Thomas P. Sarro
Larson & Taylor
727 23rd St., S.
Arlington, VA 22202   (703) 920-7200

Paul E. Sauberer#
2527 N. Lexington St.
Arlington, VA 22207   (703) 538-4449

Joseph Scafetta, Jr.
745 South 23rd Street
Suite 200
Arlington, VA 22202   (703) 521-1804

Michael J. Scheer#
IBM Corp.
1755 S. Jefferson Davis Hwy.
Suite 605
Arlington, VA 22202   (703) 769-2400

Burton S. Scheiner
Dennison, Meserole, Pollack
& Scheiner
1745 Jeff. Davis Hwy.
Suite 612, Crystal Sq. 4
Arlington, VA 22202   (703) 521-1155

Eric Paul Schellin, Sr.
2121 Crystal Drive
Suite 704, Two Crystal Park
Arlington, VA 22202   (703) 521-1666

Carl E. Schlier#
Oblon, Spivak, Mc Clelland,
Maier & Neustadt, P.C.
1755 Jefferson Davis Hwy.
Suite 400
Arlington, VA 22202   (703) 521-5940

Billy A. Schulman
Larson & Taylor
727 S. 23rd Street
Arlington, VA 22202   (703) 920-7200

Ira J. Schultz
Dennison, Meserole, Pollack
& Scheiner
1745 Jeff. Davis Hwy.
Suite 612
Arlington, VA 22202   (703) 521-1155

Stanley D. Schwartz
Schwartz & Weinrieb
2001 Jeff. Davis Hwy., Ste. 1109
Arlington, VA 22202   (703) 521-5250

Timothy R. Schwartz#
Oblon, Fisher, Spivak,
Mcclelland, Maier & Neustadt
1755 S. Jeff. Davis Hwy.
Suite 400
Arlington, VA 22202   (703) 521-5940

Mitchell W. Shapiro
Shapiro & Shapiro
1100 Wilson Blvd.
Suite 1701
Arlington, VA 22209   (703) 276-0700

Nelson Hirsh Shapiro
Shapiro & Shapiro
1100 Wilson Blvd.
Suite 1701
Arlington, VA 22209   (703) 276-0700

B. Edward Shlesinger, Jr.
Shlesinger & Myers
3000 S. Eads Street
Arlington, VA 22202   (703) 684-5600

Harry B. Shubin
Millen, White & Zelano, P.C.
Arlington Courthouse Plaza 1
Suite 1201
2200 Clarendon Blvd.
Arlington, VA 22201   (703) 243-6333

Thomas J. Sloyan#
1600 S. Eads Street
Apt. 629N
Arlington, VA 22202   (703) 521-5883

John Coventry Smith, Jr.
Benoit, Smith & Laughlin
2001 Jefferson Davis Hwy.
Suite 501
Arlington, VA 22202   (703) 521-8622

Richard Darwin Smith#
Kirk & Smith
2001 Jeff. Davis Hwy.
Suite 309
Arlington, VA 22202   (703) 521-1820

Yvonne H. Smith#
Russell, Georges & Breneman
745 S. 23rd Street
Suite 304
Arlington, VA 22202   (703) 521-1760

Eric S. Spector
Jones, Tullar & Cooper, P.C.
P.O. Box 2266 Eads Station
Arlington, VA 22202   (703) 521-5200

Marvin Jay Spivak
Oblon, Spivak, Mc Clelland,
Maier & Neustadt
1755 Jeff. Davis Hwy.
Fourth Floor
Arlington, VA 22202   (703) 521-5940

Stanley C. Spooner
Nixon & Vanderhye, P.C.
2200 Clarendon Blvd.
Arlington, VA 22201   (703) 875-0428

Clarence F. Stanback, Jr.
2009 N. 14th Street
Suite 307
Arlington, VA 22201   (703) 524-3800

Milton Sterman
Oblon, Fisher, Spivak,
Mc Clelland, Maier & Neustadt, P.C.
1755 S. Jeff. Davis Hwy.
4th Floor
Arlington, VA 22202   (703) 521-5940

Donley J. Stocking#
1106 N. Quantico St.
Arlington, VA 22205   (703) 533-0923

Robert C. Sullivan
P.O. Box 2402
Eads Station
Arlington, VA 22202   (703) 521-9361

Vincent J. Sunderdick
Oblon, Fisher, Spivak,
Mc Clelland & Maier
Suite 400 Crystal Square 5
1755 S. Jeff. Davis Hwy.
Arlington, VA 22202   (703) 521-5940

Hoge Tyler Sutherland
E.I. Du Pont De Nemours & Co.
2001 Jeff. Davis Hwy.
Suite 1001
Arlington, VA 22202   (703) 521-1800

Anton H. Sutto#
Three H Search Systems
2101 Crystal Plaza Arcade
Suite 333
Arlington, VA 22202   (703) 255-0461

Yusuke R. Takeuchi#
Kanesaka & Takeuchi
727 S. 23rd St.
Arlington, VA 22202   (703) 521-3810

David A. Tamburro
Nies, Kurz, Bergert
& Tamburro
2121 Crystal Dr., Ste. 706
Arlington, VA 22202   (703) 521-6590

Alvin E. Tanenholtz#
Oblon, Spivak, Mc Clelland,
Maier & Neustadt, P.C.
1755 Jeff. Davis Hwy.
Arlington, VA 22202   (703) 521-5940

John Paul Tarlano*
U.S. Dept. of the Navy
Strategic Sys. Prog. Off.
P.O. Box 15187
Arlington, VA 22215   (703) 695-4308

Andrew E. Taylor
Larson & Taylor
727 23rd St., South
Arlington, VA 22202   (703) 920-7200

A. Robert Theibault, Sr.
Wilkinson, Mawhinney & Theibault
2001 Jeff. Davis Hwy.
Crystal Plaza One
Suite 409
Arlington, VA 22202   (703) 521-2100

James R. Thein
2001 Jeff. Davis Hwy.
Suite 310
Arlington, VA 22202   (703) 415-0561

Murray Tillman#
Oblon, Fisher, Spivak,
Mc Clelland & Maier
Suite 400 - Crystal Sq. 5
1755 S. Jeff. Davis Hwy.
Arlington, VA 22202   (703) 521-5940

Richard J. Traverso
Millen, White & Zelano, P.C.
2200 Clarendon Blvd.
Suite 1201
Arlington, VA 22201   (703) 243-6333

John O. Tresansky
Oblon, Spivak, Mc Clelland,
Maier & Neustad
1755 S. Jefferson Davis Hwy.
Suite 400
Arlington, VA 22202   (703) 521-5940

Arnold Turk
Sandler, Greenblum & Bernstein
2920 S. Glebe Road
Arlington, VA 22206   (703) 739-0333

Eric J. Vacchio#
2001 Jefferson Davis Hwy.
Suite 810
Arlington, VA 22202   (301) 929-8517

L. S. Van Landingham, Jr.
Suite 507
2001 Jeff. Davis Hwy.
Arlington, VA 22202   (703) 979-4244

Robert A. Vanderhye
Nixon & Vanderhye, P.C.
2200 Clarendon Blvd., 14th Fl.
Arlington, VA 22201   (703) 875-0404

Ralph E. Varndell, Jr.
Varndell Legal Group
745 S. 23rd Street
Suite 100
Arlington, VA 22202   (703) 920-0460

Frederick D. Vastine#
Oblon, Spivak, Mc Clelland,
Maier & Neustadt
1755 S. Jefferson Davis Hwy.
Crystal Sq. 5, Ste. 400
Arlington, VA 22202   (703) 521-5940

David H. Voorhees
Shlesinger, Arkwright & Garrey
3000 South Eads St.
Arlington, VA 22202   (703) 684-5600

Gene Wan#
Pat Pro
2011 Crystal Drive
Suite 310
One Crystal Park
Arlington, VA 22202   (703) 271-8353

Mitchell B. Wasson
Hoffman, Wasson & Gitler, P.C.
2361 Jefferson Davis Hwy.
Suite 522
Arlington, VA 22202   (703) 415-0100

Harold W. Weakley#
515 S. Lexington Street
Arlington, VA 22204   (703) 578-4420

Steven W. Weinrieb
Schwartz & Weinrieb
2001 Jeff. Davis Hwy.
Crystal Plaza One
Suite 1109
Arlington, VA 22202   (703) 415-1250

John L. Welsh
2001 Jefferson Davis Hwy.
Suite 802
Arlington, VA 22202   (703) 415-0515

John Remon Wenzel
Litman Law Offices, Ltd.
Suite 800-801
Crystal Square II
1725 S. Jefferson Davis Hwy.
Arlington, VA 22202   (703) 902-6000

Walter F. Wessendorf, Jr.
P.O. Box 15846
Crystal City Station
Arlington, VA 22215   (703) 415-0550

Fred Smith Whisenhunt, Jr.
Griffin, Branigan & Butler
775 South 23rd Street
P.O. Box 2326
Arlington, VA 22202   (703) 979-5700

John L. White
Millen, White & Zelano, P.C.
Arlington Courthouse Plaza 1
Suite 1201
2200 Clarendon Blvd.
Arlington, VA 22201   (703) 243-6333

John M. White
Longacre & White
1919 South Eads Street
Suite 401
Arlington, VA 22202   (703) 521-1827

Michael D. White
Wigman & Cohen, P.C.
Crystal Square 3, Suite 200
1735 Jefferson Davis Hwy.
Arlington, VA 22202   (703) 892-4300

Victor M. Wigman
Wigman & Cohen, P.C.
Suite 200 - Crystal Square 3
1735 Jeff. Davis Hwy.
Arlington, VA 22202   (703) 892-4300

Howard S. Williams#
P.O. Box 15093
Arlington, VA 22215   (301) 299-4849

J. Paul Williamson
Arnold, White & Durkee
2001 Jeff. Davis Hwy.
Suite 401
Arlington, VA 22202   (703) 415-1720

Robert M. Wohlfarth, Sr.*
Dept. of Navy
Stategic Sys. Prog.
P.O. Box 15187
Arlington, VA 22217   (202) 695-4308

Mark H. Woolsey
1600 S. Joyce St., B-510
Arlington, VA 22202   (703) 521-5558

Daniel E. Wyman#
Oblon, Spivak, Mc Clelland,
Maier & Neustadt, P.C.
1755 S. Jefferson Davis Hwy.
Arlington, VA 22202   (703) 521-5940

John G. Wynn*
Office of Naval Research
800 N. Quincy Street
Arlington, VA 22217   (202) 696-4004

Michael M. Zadrozny#
Shlesinger & Myers
3000 S. Eads Street
Arlington, VA 22202   (703) 684-5600

Albert M. Zalkind
727 South 23rd St.
Suite 300
Arlington, VA 22202   (703) 920-2170

Anthony J. Zelano
Millen, White & Zelano, P.C.
Arlington Courthouse Plaza 1
Suite 1201
2200 Clarendon Blvd.
Arlington, VA 22201   (703) 243-6333

William C. Tyrrell#
P.O. Box 112
146 Ellen Circle
Bryce Resort
Basye, VA 22810   (703) 856-8283

Jere W. Sears
102 Brunswick Avenue
Blackstone, VA 23824   (804) 292-3828

Willie G. Abercrombie#
R.R. 1, Box 18-B
Broad Run, VA 22014   (703) 342-4937

Charlie T. Moon#
R.R. 1, Box 57D
Broad Run, VA 22014   (703) 347-0593

Gerald L. Brigance
R.R. 1, Box 5640
Bumpass, VA 23024   (703) 872-5532

Howard A. Birmiel
9314-C Old Keene Mill Road
Burke, VA 22015   (703) 451-4506

Martin H. Edlow#
9368 Tucker Woods Court
Burke, VA 22015   (703) 455-3639

Gary Steven Pisner
6018 Liberty Bell Court
Burke, VA 22015   (703) 455-3639

Morris Sussman
P.O. Box 1057
Callao, VA 22435   (804) 529-7155

Richard Murray
Route 1, Box 281
Castleton, VA 22716   (703) 987-8513

Keith F. Goodenough
P.O. Box 5231
Charlottesville, VA 22905   (804) 971-7100

Peter E. Rosden
1505 London Road
Charlottesville, VA 22901   (804) 971-3662

George M. J. Sarofeen
12501 Brook Lane
Chester, VA 23831   (804) 748-5483

Lawrence W. Langley
Vatell Corporation
P.O. Box 66
Christiansburg, VA 24073   (703) 961-2001

Robert A. Halvorsen
Box 182
Clarksville, VA 23927   (804) 374-2507

Lawrence R. Franklin#
P.O. Box 249
Clifton, VA 22024   (703) 323-6711

William Allen Marcontell
Westvaco Corp.
Law Dept. - Res. Center
Covington, VA 24426   (703) 969-5520

Sheldon H. Parker
Parker & Destefano
Route 3, Box 233B
Culpeper, VA 22701   (703) 825-2921

Wallace F. Poore#
P.O. Box 191
Culpeper, VA 22701   (703) 825-2411

Everette A. Powell, Jr.#
P.O. Box 489
Culpeper, VA 22701   (703) 825-1081

Harold D. Whitehead#
Route 2, Box 52
Culpeper, VA 22701   (703) 937-4433

Marguerite O. Dineen*
Office of Naval Res.
Code C73
Naval Surface Weapons Ctr.
Dahlgren, VA 22448   (703) 663-7121

John D. Lewis*
U.S. Dept. of Navy
Naval Surface Warfare Center
Code C72D
Dahlgren, VA 22448   (703) 663-8061

Robert L. Griffin#
4606 Spalding Drive
Dumfries, VA 22026   (703) 670-4926

Bennett A. Brown
3905 Railroad Avenue
Suite 200
Fairfax, VA 22030   (703) 591-3500

Walter L. Carlson#
3519 Kirkwood Drive
Fairfax, VA 22031   (703) 280-2358

Gene A. Church#
Church Associates, Inc.
4037 Autumn Court
Fairfax, VA 22030   (703) 644-5260

Ronald J. Cier
Mobil Oil Corporation
3225 Gallows Road
Fairfax, VA 22037   (703) 849-7792

I. J. Crickenberger
Fairfax Commons, Suite 71A
3921 Old Lee Highway
Fairfax, VA 22030   (703) 691-8900

Alan B. Croft
Hazel, Beckhorn & Hanes
4084 University Drive
Fairfax, VA 22030   (703) 273-6644

Kenneth L. Crosson
3900 University Drive
Suite 320
Fairfax, VA 22030   (703) 385-1010

Lori F. Cuomo#
Mobil Oil Corporation
3225 Gallows Road
Fairfax, VA 22037   (703) 849-7743

Ben W. Delos Reyes#
13136 Morning Spring Lane
Fairfax, VA 22033   (703) 968-6706

Benjamin Dobeck#
4628 Tara Drive
Fairfax, VA 22032   (703) 323-8517

Henry Louis Ehrlich
Mobil Oil Corp.
3225 Gallows Rd.
Fairfax, VA 22037   (703) 849-7790

Joseph P. Flanagan
Mohasco Corp.
4401 Fair Lakes Court
Fairfax, VA 22033   (703) 968-8041

Vincent J. Frilette#
3183 Readsborough Court
Fairfax, VA 22031   (703) 280-0444

Robert B. Furr, Jr.
Mobil Oil Corporation
OPC-8N208
3225 Gallows Road
Fairfax, VA 22037   (703) 849-7762

Robert Paul Gibson
5009 Prestwick Drive
Fairfax, VA 22030   (703) 273-7911

Michael G. Gilman
Mobil Oil Corp.
3225 Gallows Rd.
Fairfax, VA 22037   (703) 849-7763

Felix D. Gruber#
9524 Baccarat Drive
Fairfax, VA 22032   (703) 323-0854

Glenna M. Hendricks
9669-A Main St.
P.O. Box 2509
Fairfax, VA 22031   (703) 425-4250

Laurence P. Hobbes
Mobil Oil Corp.
3225 Gallows Rd.
Fairfax, VA 22037   (703) 849-7753

Malcolm Keen
Mobil Oil Corp.
Off. of Pat. Coun.
3225 Gallows Road
Fairfax, VA 22037   (703) 849-7795

Edward F. Kenehan
Mobil Oil Corp.
3225 Gallows Rd.
Fairfax, VA 22037   (703) 849-7768

James H. Knebel
4615 Tapestry Drive
Fairfax, VA 22032   (703) 425-3525

Alfons F. Kwitnieski*
5265 Pumphrey Drive
Fairfax, VA 22032   (703) 425-3525

James A. Leppink
9619 Jomar Drive
Fairfax, VA 22032   (703) 323-6325

Charles A. Malone
Mobil Oil Corp.
3225 Gallows Rd.
Fairfax, VA 22037   (703) 849-7742

James B. Marbert
10110 Farmington Drive
Fairfax, VA 22030　　(703) 273-5396

Giedre M. Mc Candless
9031 Pixie Court
Fairfax, VA 22031　　(703) 978-2661

Alexander J. Mc Killop
Mobil Oil Corp.
3225 Gallows Rd.
Fairfax, VA 22037　　(703) 849-7770

Lawrence O. Miller
Mobil Oil Corp.
3225 Gallows Rd.
Fairfax, VA 22037　　(703) 849-7794

Michael J. Mlotkowski
Mobil Oil Corporation
Off. of Pat. Coun.
3225 Gallows Road
Fairfax, VA 22037　　(703) 849-7767

William E. Mouzavires
Judicial Court
10615 Judicial Drive
Suite 703
Fairfax, VA 22030　　(703) 273-7300

James P. O Sullivan, Sr.
Mobil Oil Corp.
3225 Gallows Rd.
Fairfax, VA 22037　　(703) 849-7758

Alverna M. Paulan
Mobil Oil Corp.
3225 Gallows Rd.
Fairfax, VA 22037　　(703) 849-7746

James F. Powers, Jr.
P.O. Box 7187
Fairfax, VA 22039　　(703) 250-7874

Ernest Roy Purser, Sr.#
4915 Gadsen Drive
Fairfax, VA 22032　　(703) 323-5248

Peter W. Roberts#
Mobil Oil Corp.
3225 Gallows Rd.
Fairfax, VA 22037　　(703) 849-7732

Dennis P. Santini
Mobil Oil Corp.
3225 Gallows Rd.
Fairfax, VA 22037　　(703) 849-7760

Marina V. Schneller
Mobil Oil Corp.
3225 Gallows Road
Fairfax, VA 22037　　(703) 849-7765

Terrance L. Siemens#
Patentech
P.O. Box 2832
Fairfax, VA 22031　　(703) 978-4883

Jessica M. Sinnott
Mobil Oil Corp.
3225 Gallows Road
Fairfax, VA 22037　　(703) 849-7752

Charles J. Speciale
Mobil Oil Corp.
3225 Gallows Rd.
Fairfax, VA 22037　　(703) 849-7797

Richard D. Stone
Mobil Oil Corp.
3225 Gallows Rd.
Fairfax, VA 22037　　(703) 849-7745

Hastings S. Trigg
3707 John Barnes Lane
Fairfax, VA 22033　　(703) 620-8869

Edward Hatch Valance
Mobil Oil Corp.
3225 Gallows Rd.
Fair Oaks Room 12-101
Fairfax, VA 22037　　(703) 846-7748

Lowell G. Wise
Mobil Oil Corp.
3225 Gallows Rd.
Fairfax, VA 22037　　(703) 849-7756

Channing L. Pace#
2900 Hideaway Rd.
Fairfax County, VA 22031　　(703) 280-2118

Walter R. Baylor*
U.S. Army Laboratory Command
7701 Willowbrook Road
Fairfax Station, VA 22039　　(703) 250-9284

Peter P. Mitrano
P.O. Box 190
Fairfax Station, VA 22039　　(703) 591-7250

Robert F. Altherr, Jr.*
U.S. Army
Off. of Judge Advocate Gen.
Attn: JALS-PC
5600 Columbia Pike
Falls Church, VA 22041　　(202) 756-2623

Ralph J. Alvey
3709 S. George Mason Drive
Apt. 708E
Falls Church, VA 22041　　(703) 998-5910

James C. Arvantes
6539 Cedarwood Court
Falls Church, VA 22041　　(703) 256-0515

Johnny W. Bailey#
Birch, Stewart, Kolasch
& Birch
301 N. Washington Street
P.O. Box 747
Falls Church, VA 22046　　(703) 241-1300

Murray B. Baxter*
U.S. Army
5611 Columbia Pike
Falls Church, VA 22041　　(703) 756-2622

James L. Bean
Kerkam, Stowell, Kondracki
& Clarke, P.C.
Two Skyline Place
5203 Leesburg Pike, Ste. 600
Falls Church, VA 22041　　(703) 998-3302

John P. Beauchamp, Jr.#
6629 Kirby Court
Falls Church, VA 22043　　(703) 533-0987

Werten F.W. Bellamy#*
Dept. of the Army
Off. of Judge Adv. Gen.
5611 Columbia Pike
Falls Church, VA 22041　　(202) 756-2434

Anthony L. Birch
Birch, Stewart, Kolasch
& Birch
P.O. Box 209
301 N. Wash. St.
Falls Church, VA 22046　　(703) 241-1300

Herbert M. Birch
Birch, Stewart, Kolasch
& Birch
301 N. Wash. St.
Falls Church, VA 22046　　(703) 241-1300

Terrell C. Birch
Birch, Stewart, Kolasch
& Birch
301 N. Washington St.
Box 209
Falls Church, VA 22046　　(703) 241-1300

Robert Brown, Jr.
3329 Wilkins Dr.
Falls Church, VA 22041　　(703) 820-8017

William R. Browne#
3245 Rio Drive
Condo 406
Falls Church, VA 22041　　(703) 671-3608

Mark R. Buscher#
Birch, Stewart, Kolasch
& Birch
301 North Washington St.
P.O. Box 747
Falls Church, VA 22040　　(703) 241-1300

Frank Cacciapaglia, Jr.#
P.O. Box 5050
Falls Church, VA 22044　　(703) 536-2323

Terry L. Clark
Birch, Stewart, Kolesch
& Birch
301 North Washington St.
Falls Church, VA 22046　　(703) 241-1300

Dennis Philip Clarke
Kerkam, Stowell, Kondracki
& Clarke, P.C.
Two Skyline Place
5203 Leesburg Pike, Ste. 600
Falls Church, VA 22041　　(703) 998-3302

Charles E. Cohen#
Birch, Stewart, Kolasch
& Birch
301 N. Washington St.
P.O. Box 747
Falls Church, VA 22046　　(703) 241-1300

Donald J. Daley#
Birch, Stewart, Kolasch
& Birch
301 N. Washington St.
P.O. Box 747
Falls Church, VA 22040   (703) 241-1300

C. Joseph Faraci
Birch, Stewart, Kolasch
& Birch
301 N. Washington Street
Falls Church, VA 22046   (703) 241-1300

William L. Feeney
Kerkam, Stowell, Kondracki
& Clarke, P.C.
Two Skyline Place
5203 Leesburg Pike, Ste. 600
Falls Church, VA 22041   (703) 998-3302

Barbara A. Fisher#
Birch, Stewart, Kolasch
& Birch
301 N. Washington
Falls Church, VA 22046   (703) 241-1300

Michael J. Foycik, Jr.
One Skyline Place
Suite 310
5205 Leesburg Pike
Falls Church, VA 22041   (703) 820-4500

Abraham Frankel
3401 Glen Carlyn Drive
Falls Church, VA 22041   (703) 820-7140

Richard Gerard#
220 Midvale Street
Falls Church, VA 22046   (703) 534-3094

Raymond D. Gilbert#*
Dept. of Defense
Systems & Program Office-DSPO
VI Skyline Place, Suite 310
5109 Leesburg Pike
Falls Church, VA 22304   (703) 756-8990

Elliot A. Goldberg
Birch, Stewart, Kolasch
& Birch
301 N. Washington St.
P.O. Box 747
Falls Church, VA 22046   (703) 241-1300

Charles Gorenstein
Birch, Stewart, Kolasch
& Birch
301 N. Washington St.
P.O. Box 747
Falls Church, VA 22046   (703) 241-1300

James Walter Hiney, Jr.
6538 Bay Tree Court
Falls Church, VA 22041   (703) 941-3275

Edward J. Kelly
2921 Rosemary Ln.
Falls Church, VA 22042   (703) 534-1238

John C. Kerins
Kerkam, Stowell, Kondracki
& Clarke, P.C.
Two Skyline Place
5203 Leesburg Pike, Ste. 600
Falls Church, VA 22041   (703) 998-3302

Donald Craig Kolasch
Birch, Stewart, Kolasch
& Birch
301 N. Washington St.
P.O. Box 747
Falls Church, VA 22046   (703) 241-1300

Joseph Arlen Kolasch
Birch, Stewart, Kolasch
& Birch
301 N. Washington St.
P.O. Box 209
Falls Church, VA 22046   (703) 241-1300

Edward J. Kondracki
Kerkam, Stowell, Kondracki
& Clarke, P.C.
Two Skyline Place
5203 Leesburg Pike, Ste. 600
Falls Church, VA 22041   (703) 998-3302

Anthony Thomas Lane*
Dept. of the Army
Off. of Judge Advocate Gen.
Pats., Copyrights & Tdmks. Div.
5611 Columbia Pike
Falls Church, VA 22041   (703) 756-2617

Marion P. Lelong
1308 Seaton Lane
Falls Church, VA 22046   (703) 533-8827

Fred C. Mattern, Jr.
Birch, Stewart, Kolasch
& Birch
301 N. Washington St.
Falls Church, VA 22046   (703) 241-1300

Andrew D. Meikle
Birch, Stewart, Kolasch
& Birch
301 N. Washington St.
Falls Church, VA 22046   (703) 241-1300

Joe M. Muncy, Jr.
Birch, Stewart, Kolasch
& Birch
301 N. Washington St.
P.O. Box 747
Falls Church, VA 22046   (703) 241-1300

Gerald M. Murphy, Jr.
Birch, Stewart, Kolasch
& Birch
301 N. Washington St.
P.O. Box 747
Falls Church, VA 22046   (703) 241-1300

Michael K. Mutter
Birch, Stewart, Kolasch
& Birch
301 N. Washington St.
P.O. Box 747
Falls Church, VA 22047   (703) 241-1300

Timothy E. Newholm#
6814 Jackson Ave.
Falls Church, VA 22042   (703) 241-1300

Michael J. Nickerson
Birch, Stewart, Kolasch
& Birch
301 N. Washington St.
P.O. Box 747
Falls Church, VA 22046   (703) 241-1300

Meyer Perlin#
2016 Dexter Drive
Falls Church, VA 22043   (703) 893-9556

William W. Randolph*
U.S. Dept. of Army
5611 Columbia Pike
Nassif Building
Falls Church, VA 22041   (202) 756-2619

Earl T. Reichert*
Dept. of the Army
Off. of Judge Advocate Gen.
Intell. Prop. Law Division
5611 Columbia Pike, JALS-IP
Falls Church, VA 22041   (703) 756-2623

Andrew F. Reish
Birch, Stewart, Kolasch
& Birch
301 N. Washington St.
P.O. Box 747
Falls Church, VA 22046   (703) 241-1300

James M. Slattery
Birch, Stewart, Kolasch
& Birch
301 N. Washington St.
P.O. Box 747
Falls Church, VA 22046   (703) 241-1300

Raymond C. Stewart
Birch, Stewart, Kolasch
& Birch
301 N. Washington St.
P.O. Box 747
Falls Church, VA 22040   (703) 241-1300

Harold L. Stowell
Kerkham, Stowell, Kondracki
& Clarke, P.C.
Two Skyline Place
5203 Leesburg Pike, Ste. 600
Falls Church, VA 22041   (703) 534-6600

Mark A. Superko#
Birch, Stewart, Kolasch
& Birch
301 N. Washington St.
Falls Church, VA 22046   (703) 241-1300

Leonard R. Svensson
Birch, Stewart, Kolasch
& Birch
301 N. Washington St.
P.O. Box 747
Falls Church, VA 22046   (703) 241-1300

Christian L. Swartz
2354 Dunbar Lane
Falls Church, VA 22046   (703) 534-1505

677

Bernard L. Sweeney
Birch, Stewart, Kolasch
& Birch
301 N. Washington St.
P.O. Box 747
Falls Church, VA 22046   (703) 241-1300

Andrew J. Telesz#
Birch, Stewart, Kolasch
& Birch
301 N. Washington St.
P.O. Box 747
Falls Church, VA 22046   (703) 241-1300

David A. Testardi#
Polytechnic Patent Research
7230 Pimmit Court
Falls Church, VA 22043   (703) 821-8218

Marc S. Weiner
Birch, Stewart, Kolasch
& Birch
301 N. Washington St.
P.O. Box 747
Falls Church, VA 22046   (703) 241-1300

Harry Wong, Jr.#
Birch, Stewart, Kolasch
& Birch
301 N. Washington St.
P.O. Box 747
Falls Church, VA 22046   (703) 241-1300

Aubrey J. Dunn#*
U.S. Army CECOM
CNVEO
Attn: AMSEL-LG-P-NVEO
Fort Belvoir, VA 22060   (703) 664-5513

Darrell E. Hollis*
U.S. Army Corps of Eng.
Kingman Bldg.
CEHEC-OC
Fort Belvoir, VA 22060   (202) 355-3671

Gary W. Hudiburgh*
U.S. Army
Corps. of Engineers
Kingman Bldg.
Fort Belvoir, VA 22060   (202) 355-2160

Milton Wayne Lee*
U.S. Dept. of Army
Night Vision &
Electro-Optics Center
Fort Belvoir, VA 22060   (703) 664-2223

Charles D. Miller*
U.S. Army Belvoir Research
& Development Center
Attn: STRBE-L
Fort Belvoir, VA 22060   (703) 664-5411

Anthony Knight#
5718 Gladden Court
Franconia, VA 22303   (703) 960-6728

Edwin D. Grant
P.O. Box 225
Fredericksburg, VA 22404   (703) 960-6728

Lewis T. Jacobs#
P.O. Box 569
Front Royal, VA 22630   (703) 635-4383

Donavon Lee Favre
Virginia Manor
Route 1, Box 18
Glasgow, VA 24555   (703) 258-1292

Richard M. Foard
P.O. Box 356
Gloucester, VA 23061   (804) 693-5665

John R. Janes
Goode, VA 24556   (703) 586-4244

Robert Mistrot Meith
605 Clear Spring Road
Great Falls, VA 22066   (703) 444-4908

Joseph G. Seeber
P.O. Box 750
Great Falls, VA 22066   (703) 430-1702

George Franke Helfrich*
NASA
Langley Research Center
Mail Stop 279
Hampton, VA 23665   (804) 865-3725

Wallace J. Nelson
34 Salt Pond Road
Hampton, VA 23664   (804) 851-1667

Kevin B. Osborne
NASA
Langley Research Center
Mail Stop 279
Hampton, VA 23665   (804) 864-3523

Roger W. Bailey#
Perot Systems Corp.
13100 Worldgate Dr.
Suite 500
Herndon, VA 22070   (703) 709-3650

Cornelius J. Husar
12901 Pinecrest Road
Herndon, VA 22071   (703) 391-8939

Karen Stephan Young
3420 Doe Run Court
Herndon, VA 22071   (703) 904-0150

Raymond L. Greene
Route 658
James Store, VA 23080   (804) 725-4047

Americus Mitchell
P.O. Box 1335
Kilmarnock, VA 22482   (804) 435-3489

Bonnie L. Deppenbrock
706 Wage Drive, SW
Leesburg, VA 22075   (703) 777-3452

Richard Salvatore Sciascia
Route 1
Box 398-S
Linden, VA 22642   (703) 635-8560

Archie R. Borchelt#
P.O. Box 655
Locust Grove, VA 22508   (703) 399-1204

Robert C. Lampe, Jr.
General Electric Co.
GE Mobile Communications
Mountain View Road
Lynchburg, VA 24502   (804) 528-7400

Michael Masnik
International Intellectual
Property Associates, Inc.
1631 Belfield Place
Lynchburg, VA 24503   (804) 384-5970

Harold H. Dutton, Jr.
8711 Plantation Lane
Suite 301
P.O. Box 3110
Manassas, VA 22110   (703) 369-1922

Shelley Krasnow
Georator Corporation
9617 Center Street
P.O. Box 70
Manassas, VA 22110   (703) 368-2101

Francis L. Masselle
12683 Cobblestone Court
Manassas, VA 22111   (703) 368-2101

Gerald H. Bjorge
1300 Alps Drive
Mc Lean, VA 22102   (703) 893-7890

Francis X. Bradley, Jr.
7909 Falstaff Rd.
Mc Lean, VA 22101   (703) 356-0319

Charles Martin Bredehoft
6107 Woodland Terrace
Mc Lean, VA 22101   (703) 534-2189

James J. Brown
Gipple & Hall
6667-B Old Dominion Drive
Mc Lean, VA 22101   (703) 448-1770

Thomas W. Cole
Sixbey, Friedman, Leedom
& Ferguson
Suite 600
2010 Corporate Ridge
Mc Lean, VA 22102   (703) 790-9110

John B. Farmakides
5835 Upton Street
Mc Lean, VA 22101   (703) 241-0006

Gerald Joseph Ferguson, Jr.
Sixbey, Friedman, Leedom
& Ferguson
Suite 600
2010 Corporate Ridge
Mc Lean, VA 22102   (703) 790-9110

J. Howard Flint, Jr.#
1114 Dead Run Drive
Mc Lean, VA 22101   (703) 356-7453

Stuart Jay Friedman
Sixbey, Friedman, Leedom
& Ferguson
Suite 600
2010 Corporate Ridge
Mc Lean, VA 22102   (703) 790-9110

Terry M. Gernstein
1015 Salt Meadow Lane
Mc Lean, VA 22101   (703) 790-5945

John S. Hale
Gipple & Hale
6667-B Old Dominion Drive
Mc Lean, VA 22101   (703) 448-1770

William F. Hamrock
6862 Elm Street
Suite 300
Mc Lean, VA 22101   (703) 893-6928

Harrison L. Hinson#
935 Douglass Drive
Mc Lean, VA 22101   (703) 356-3041

Herman J. Hohauser
7926 Jones Branch Dr.
Suite 1000
Mc Lean, VA 22102   (703) 883-0399

Gerald L. Kaplan#
1231 Stoneham Court
Mc Lean, VA 22101   (703) 356-3788

Joan K. Lawrence
Sixbey, Friedman, Leedom
& Ferguson
2010 Corporate Ridge
Mc Lean, VA 22102   (703) 790-9110

Charles M. Leedom, Jr.
Sixbey, Friedman, Leedom
& Ferguson
Suite 600
2010 Corporate Ridge
Mc Lean, VA 22102   (703) 790-9110

Garo A. Partoyan
Mars, Inc.
6885 Elm Street
Mc Lean, VA 22101   (703) 821-4900

Joseph Edward Rusz
1102 Roberta Court
Mc Lean, VA 22101   (703) 356-5696

David S. Safran
Sixbey, Friedman, Leedom
& Ferguson
2010 Corporate Ridge
Suite 600
Mc Lean, VA 22102   (703) 790-9110

Theodore C. Salindong#
8380 Greensboro Dr., #416
Mc Lean, VA 22102   (703) 893-5491

Robert K. Schaefer
7735 Falstaff Road
Mc Lean, VA 22102   (703) 790-9486

Daniel W. Sixbey
Sixbey, Friedman, Leedom
& Ferguson
Suite 600
2010 Corporate Ridge
Mc Lean, VA 22102   (703) 790-9110

Donald R. Studebaker
Sixbey, Friedman, Leedom
& Ferguson
Suite 600
2010 Corporate Ridge
Mc Lean, VA 22102   (703) 790-9110

Steven P. Weihrouch#
Sixbey, Friedman, Leedom
& Ferguson
Suite 600
2010 Corporate Ridge
Mc Lean, VA 22102   (703) 790-9110

Glenn E. Wise#
6450 Georgetown Pike
Mc Lean, VA 22101   (703) 415-0579

James Creighton Wray
1493 Chain Bridge Road
Apt. 300
Mc Lean, VA 22101   (703) 442-4800

Lawrence D. Bush, Jr.
1120 Hanover Green Drive
P.O. Box 788
Mechanicsville, VA 23111   (804) 730-2200

Thomas N. Tarrant
P.O. Box 404
U. S. Route 13
Melfa, VA 23410   (804) 787-4995

Bradley K. De Sandro
Titus & Brynteson, P.C.
5104 W. Village Green Drive
Suite 100
Midlothian, VA 23112   (804) 744-6373

Claude E. Setliff
14105 Waters Edge Court
Midlothian, VA 23112   (804) 739-4384

David G. Mc Connell
Route 1, Box 372K
Mineral, VA 23117   (703) 894-5142

Robert S. Auten
Route 1, Box 302
Moneta, VA 24121   (703) 297-7562

Nelson W. Edgerton#
Route 4, Box 127
Moneta, VA 24112   (703) 721-3325

Clifford N. Rosen#
Newport News Shipbuilding
& Dry Dock Company
4101 Washington Ave.
E77 600-1
Newport News, VA 23607   (804) 380-2307

Stephen E. Clark#
916 W. 25th Street
Norfolk, VA 23517   (804) 625-1140

Adolph Charles Hugin
7602 Boulder St.
North Springfield, VA 22151   (703) 569-2233

Clyde R. Christofferson
2915 Hunter Mill Road
Suite 18
Oakton, VA 22124   (703) 281-1775

Jane Y. Sasai#
10308 Hickory Forest Dr.
Oakton, VA 22124   (703) 255-2033

Hanna S. Burke#
Allied-Signal Corporation
Fibers Div., Tech. Center
P.O. Box 31
Petersburg, VA 23803   (804) 520-3619

William H. Thrower, Jr.
Allied-Signal Corporation
P.O. Box 31
Petersburg, VA 23804-4000   (804) 520-3622

Robert James Buttermark#
6000 Derwent Road
Powhatan, VA 23139   (804) 375-9131

Marshall M. Curtis
Whitham & Marhoefer
11800 Sunrise Valley Dr.
Suite 220
Reston, VA 22091   (703) 391-2510

H. M. Hougen
Dyncorp
2000 Edmund Halley Dr.
Reston, VA 22091   (703) 264-9108

Laurence Joseph Marhoefer
Whitham & Marhoefer
11800 Sunrise Valley Dr.
Suite 220
Reston, VA 22091   (703) 391-2510

Mark J. Regen
Sprint International
Communications Corp.
12490 Sunrise Valley Drive
Reston, VA 22096   (703) 689-6905

David R. Schultz
Mobil Land Development Corp.
Suite 1400
11800 Sunrise Valley Drive
Reston, VA 22091   (703) 620-4780

Roland H. Shubert
P.O. Box 2339
Reston, VA 22090   (703) 435-4141

Alfred E. Smith#
1933 Red Lion Court
Reston, VA 22091   (703) 860-1709

Charles Lamont Whitham
Whitham & Marhoefer, P.C.
11800 Sunrise Valley Dr.
Suite 220
Reston, VA 22091   (703) 391-2510

Michael E. Whitham
Whitham & Marhoefer, P.C.
11800 Sunrise Valley Dr.
Suite 220
Reston, VA 22091   (703) 391-2510

Virginia S. Andrews
9302 Belfort Rd.
Richmond, VA 23229   (804) 740-2319

Alan M. Biddison
Reynolds Metals Company
Law Dept.
P.O. Box 27003
Richmond, VA 23261   (804) 281-2410

Ivan Christoffel#
A. H. Robins Co., Inc.
1407 Cummings Dr.
Richmond, VA 23220   (804) 257-2947

William B. Cridlin, Jr.
A. H. Robins Co., Inc.
1407 Cummings Dr.
Richmond, VA 23220   (804) 257-2184

John L. Dewey#
7012 Hunt Club Lane
Apt. 1834
Richmond, VA 23228   (804) 269-6869

Walter M. Dotts, Jr.#
2210 M Street
Richmond, VA 23223   (804) 644-3515

John Willard Gibbs, Jr.
204 Dryden Lane
Richmond, VA 23229   (804) 741-5855

Donald Edward Gillespie
A. H. Robins Co., Inc.
1407 Cummings Dr.
Richmond, VA 23220   (804) 257-2193

Arthur Lawrence Girard
8310 Poplar Hollow Tr.
Richmond, VA 23235   (804) 320-2964

Charles E. B. Glenn
Philip Morris Management Corp.
P.O. Box 26583
Richmond, VA 23261   (804) 274-2822

John F. C. Glenn
8915 Tolman Rd.
Richmond, VA 23229   (804) 740-8409

Edward H. Gorman, Jr.
A. H. Robins Co., Inc.
1407 Cummings Drive
Richmond, VA 23261   (804) 257-2125

Susan Addington Hutcheson#
Philip Morris Inc.
Res. & Dev.
P.O. Box 26583
Richmond, VA 23261   (804) 274-2162

Auzville Jackson, Jr.
Staas & Halsey
8652 Rio Grande Road
Richmond, VA 23229   (804) 740-6828

George Kapsalas#
224 Redmead Ln.
Richmond, VA 23236   (804) 320-1622

George William King
A. H. Robins Co., Inc.
1407 Cummings Dr.
Richmond, VA 23220   (804) 257-2191

Louis R. Lawson, Jr.
6416 Roselawn Rd.
Richmond, VA 23226   (804) 288-3991

Robert Chamberlayne Lyne, Jr.
Reynolds Metals Co.
6601 W. Broad St.
Richmond, VA 23230   (804) 281-2837

Alan T. Mc Donald
Reynolds Metals Co.
6601 W. Broad St.
Richmond, VA 23261   (804) 281-3791

R. Webb Moore
Hirschler, Fleischer, Weinberg,
Cox & Allen
629 East Main Street
P.O. Box 1Q
Richmond, VA 23202   (804) 771-9561

Arthur I. Palmer, Jr.
James River Corporation
P.O. Box 2218
Richmond, VA 23217   (804) 649-4200

Norman B. Rainer#
2008 Fondulac Road
Richmond, VA 23229   (804) 282-7109

Wayne W. Rupert
2500 E. Cary St.
Apt. 522
Richmond, VA 23223   (804) 282-7109

James Eric Schardt
Philip Morris Corp.
P.O. Box 26583
Richmond, VA 23261   (804) 274-2863

Myles T. Hylton
Parvin, Wilson, Barnett
& Hopper, P.C.
Dominion Bank Bldg.
213 S. Jefferson
Roanoke, VA 24011   (703) 343-1400

Arnold E. Renner
General Electric Co.
1501 Roanoke Blvd.
Room 104
Salem, VA 24153   (703) 387-7144

Jesse B. Grove, Jr.
P.O. Box 207
Scottsville, VA 24590   (804) 286-3134

Howard J. Osborn
3601 Seaford Road
Seaford, VA 23696   (804) 898-7677

Robert P. Auber
8231 Taunton Place
Springfield, VA 22152   (703) 451-6432

Thomas F. Callaghan#
7203 Countrywood Court
Springfield, VA 22151   (703) 256-9432

Kathleen E. Crotty
7817 Ravenel Court
Springfield, VA 22151   (703) 321-8051

Thomas De Benedictis, Sr.#
7123 Kerr Drive
Springfield, VA 22150   (703) 569-1184

Papan Devnani
National Tech. Information Ser.
5285 Port Royal Road
Springfield, VA 22161   (703) 487-4739

Christina M. Eakman#
Wiens Search Service
6214 Old Keene Mill Ct.
Springfield, VA 22152   (703) 451-1438

Stephen Gates*
U.S. Dept. of Commerce
NTIS
Room 304 Forbes
5285 Port Royal Road
Springfield, VA 22161   (703) 487-4838

Earl Levy
6130 Crozet Court
Springfield, VA 22150   (703) 971-6130

Neil F. Markva
8322-A Traford Lane
Springfield, VA 22152   (703) 644-5000

Richard P. Matthews
8410 Terra Woods Drive
Springfield, VA 22153   (703) 455-7087

Robert B. Reeves#
8523 Salisbury Court
Springfield, VA 22151   (703) 978-8139

Herman Karl Saalbach
6019 Brunswick Street
Springfield, VA 22150   (703) 451-3875

Susanne M. Hopkins#
Klima & Hopkins
26 Settler's Way
Stafford, VA 22554   (703) 720-0288

Rodney A. Corl#
3262 Catawba Rd.
Troutville, VA 24175   (703) 992-5475

Danita R. Byrd#*
U.S. General Services Agency
7980 Boeing Court
Vienna, VA 22182   (703) 760-7766

Hilary S. Cairnie
Dickstein, Shapiro & Morin
8300 Boone Blvd., Ste. 800
Vienna, VA 22182   (703) 847-9190

Robert M. Clark
STAC
111 Center Street S.
Suite B
Vienna, VA 22180   (703) 281-6351

Murriel E. Crawford
Dickstein, Shapiro & Morin
8300 Boone Blvd.
Suite 800
Vienna, VA 22180   (703) 847-9190

Michael C. Greenbaum
Dickstein, Shapiro & Morin
8300 Boone Blvd.
Suite 800
Vienna, VA 22180   (703) 847-9190

Donald A. Gregory
Dickstein, Shapiro & Morin
8300 Boone Blvd.
Suite 800
Vienna, VA 22180   (703) 847-9190

Daniel R. Gropper
9908 Dale Ridge Ct.
Vienna, VA 22181   (703) 281-7457

Jon D. Grossman
Dickstein, Shapiro & Morin
8300 Boone Blvd.
Suite 800
Vienna, VA 22180   (703) 847-1904

Gary M. Hoffman
Dickstein, Shapiro & Morin
8300 Boone Blvd.
Suite 800
Vienna, VA 22180   (703) 847-9190

Geoffrey M. Karny
Dickstein, Shapiro & Morin
8300 Boone Blvd.
Suite 800
Vienna, VA 22182   (703) 847-9190

George R. King#
801 Marjorie Lane
Vienna, VA 22180   (703) 938-6377

John R. Lastova
Dickstein, Shapiro & Morin
8300 Boone Blvd.
Suite 800
Vienna, VA 22180   (703) 847-1972

Donn Mc Giehan
8318 Mc Neil Street
Vienna, VA 22180   (703) 560-4562

Malcolm R. Uffelman#
1808 Horseback Trail
Vienna, VA 22182   (703) 938-6184

John Kenneth Mullarney
5556 Mac Goffie Street
Virginia Beach, VA 23464   (703) 938-6184

Craig S. Clarke*
U.S. Dept. of the Army
P.O. Box 1647, VHFS
Warrenton, VA 22186   (703) 349-4602

George W. Hager, Jr.
397 Woodstone Court
Warrenton, VA 22186   (703) 849-4086

John E. Leonarz
177 Old Orchard Lane
Warrenton, VA 22186   (703) 349-4740

Woolridge Brown Morton, Jr.
Route 1, Box 586
Warsaw, VA 22572   (804) 333-3311

L. Dewayne Rutledge#
109 Thorpes Parish
Williamsburg, VA 23185   (804) 220-5645

Peter J. Van Bergen
402 W. Duke of Gloucester St.
Suite 215
Williamsburg, VA 23185   (804) 220-2649

Arthur Livingston Branning
4 West Monmouth Street
Winchester, VA 22601   (703) 662-3540

Clyde I. Coughenour
16607 Sutton Place
Woodbridge, VA 22191   (703) 221-8677

Robert G. Hilton
12342 Oakwood Drive
Woodbridge, VA 22192   (703) 221-8677

Richard J. Scanlan, Jr.#
12028 Wm. & Mary Circle
Woodbridge, VA 22192   (703) 494-2718

Washington

Keith D. Gehr#
35820 57th Avenue South
Auburn, WA 98002   (206) 939-5997

John S. Sundsmo#
P.O. Box 11303
Bainbridge Island, WA 98110   (206) 842-2979

Dean A. Craine
800 Bellevue Way N.E.
Suite 300
Bellevue, WA 98004   (206) 637-3035

William G. Forster
13415 S.E. 30th
Bellevue, WA 98005   (206) 641-9000

Jeffrey T. Haley
Graybeal, Jackson,
Richardson & Haley
705 Key Bank Bldg.
10655 N.E. 4th St.
Bellevue, WA 98004   (206) 455-5575

Lawrence Adams Jackson
Jackson & Richardson
Key Bank Building
10655 NE 4th Street
Suite 705
Bellevue, WA 98004   (206) 455-5575

Robert W. Jenny#
77 Cascade Key
Bellevue, WA 98006   (206) 747-2936

John M. Johnson
Jackson & Richardson
Key Bank Building
10655 NE 4th Street
Suite 705
Bellevue, WA 98004   (206) 455-5575

Roy Edwin Mattern, Jr.
13415 S.E. 30th
Bellevue, WA 98005   (206) 641-9000

Don R. Mollick*
Mollick & Moravan
Suite 302
11671 S.E. 1st Street
Bellevue, WA 98005   (206) 454-2700

Gregory W. Moravan
Mollick & Moravan
11671 S.E. First
Suite 302
Bellevue, WA 98005   (206) 454-2700

David G. Pursel
Digital Equipment Corporation
14475 N.E. 24th Street
Bellevue, WA 98007   (206) 865-8844

Harry A. Richardson, Jr.
Jackson & Richardson
705 Key Bank Building
10655 N.E. 4th Street
Bellevue, WA 98004   (206) 455-5575

Gordon M. Stewart
Williams, Kastner & Gibbs
10900 N.E. 4th St., #2000
Bellevue, WA 98004   (206) 462-4700

J. Robert Cassidy
1112 Finnegan Way
Bellingham, WA 98225   (206) 647-9210

R. Reams Goodloe, Jr.
Hughes, Cassidy & Multer, P.S.
1720 Iowa Street
Bellingham, WA 98226   (206) 647-1296

Todd Nelson Hathaway
Hughes & Multer, P.S.
1720 Iowa St.
Bellingham, WA 98226   (206) 647-1296

Robert Bruce Hughes
Hughes & Multer, P.S.
1720 Iowa St.
Bellingham, WA 98226   (206) 647-1296

Michael R. Schacht
Hughes & Multer, P.S.
1720 Iowa St.
Bellingham, WA 98226   (206) 647-1296

Debra K. Leith
Micro Probe Corp.
1725 220th St. S.E., #104
Bothell, WA 98021   (206) 485-8566

W. Brinton Yorks, Jr.
Advanced Technology
Laboratories, Inc.
P.O. Box 3003
22100 Bothell Hwy. S.E.
Bothell, WA 98041   (206) 487-7152

Brent F. Logan
Hewlett-Packard Co.
18110 S.E. 34th Street
Camas, WA 98607   (206) 944-3232

Kent B. Roberts
375 Corbett Creek Road
Colville, WA 99114   (509) 684-8722

Mikio Ishimaru
John Fluke Mfg. Co., Inc.
P.O. Box C9090
Everett, WA 98206   (206) 356-5819

Richard A. Koske
John Fluke Mfg. Co., Inc.
P.O. Box C9090
M/S 203A
Everett, WA 98206   (206) 356-5819

George T. Noe
John Fluke Mfg. Co., Inc.
P.O. Box C9090
M/S 203A
Everett, WA 98206   (206) 356-6172

John D. Mc Auliffe
Box 3288
Federal Way, WA 98070   (206) 924-2539

Mark J. Zovko, Jr.
36504 28th Avenue South
Federal Way, WA 98003   (206) 838-1909

Karen M. Casto
1625 Main, #5
Ferndale, WA 98248   (206) 384-2823

Elizabeth F. Harasek
4415 Holly Lane
Gig Harbor, WA 98335   (206) 851-9969

Robert Iannucci
16813 S.E. Newport Way
Issaquah, WA 98027   (206) 649-8836

Donna J. Thies
790 Idylwood Drive, S.W.
Issaquah, WA 98027   (206) 392-9172

Robert Keith Sharp
906 N. Ledbetter Street
Kennewick, WA 99336   (509) 783-4796

William R. Bachand#
13532 S.E. 256th St.
Kent, WA 98042   (206) 630-1672

Marshall F. Gehring
25825 104th Avenue S.E.
Suite 375
Kent, WA 98031   (206) 631-4453

Earl R. Tarleton
Cassidy, Vance & Tarleton
303 Parkplace
Suite 132
Kirkland, WA 98033   (206) 889-8000

James R. Vance
Cassidy, Vance &
Tarleton, P.S.
303 Parkplace
Suite 132
Kirkland, WA 98033   (206) 889-8000

Fred R. Ahlers
3978 Holladay Park Dr. S.E.
Lacey, WA 98503   (206) 438-5976

Robert L. Jepsen#
2216 Beta Street
Lacey, WA 98503   (206) 438-5166

George W. F. Simmons
RR 2, Box 3377
Lopez, WA 98261   (206) 468-2068

Louis N. French
2010 - 151 St., S.E.
Mill Creek, WA 98012   (206) 468-2068

Charles A. Kingsford-Smith#
P.O. Box 894
Mukilteo, WA 98275   (206) 347-0569

Norris E. Faringer
140 Orchard Drive
Naches, WA 98937   (509) 965-3094

William I. Beach#
232 W. 10th St.
Port Angeles, WA 98362   (206) 452-8628

John B. Mason
Microsoft Corp.
One Microsoft Way
Redmond, WA 98052   (206) 882-8080

Frederick J. Mc Kinnon, Jr.
Westmark International
15220 NE 40th Street
Redmond, WA 98073   (206) 867-2077

Richard D. Multer
Hughes & Multer
15042 N.E. 40th
No. 205
Redmond, WA 98052   (206) 453-5701

David L. Tingey
1100 Maple Avenue, S.W.
Renton, WA 98055   (206) 271-7690

David J. Brown#
55 Jadwin Avenue
Apt. 110
Richland, WA 99352   (509) 946-0541

Joseph James Hauth#
Batelle Memorial Institute
P.O. Box 999
Richland, WA 99352   (509) 375-2226

Edward V. Hiskes*
U.S. Dept. of Energy
Federal Bldg.
Richland, WA 99352   (206) 376-3450

Stephen R. May
Battelle
Pacific N.W. Laboratories
Battelle Blvd.
P.O. Box 999
Richland, WA 99352   (509) 375-2387

Robert Southworth, III*
U.S. Dept. of Energy
825 Jadwin Ave.
Richland, WA 99352   (509) 376-7225

Paul W. Zimmerman#
Battelle Memorial Institute
Pacific Northwest Division
902 Battelle Blvd.
Richland, WA 99352   (509) 375-2981

Daniel T. Anderson
Boeing Comm. Airplane Co.
Box 3707
Mail Sta. 7E-25
Seattle, WA 98124   (206) 251-0229

Ronald M. Anderson
Christensen, O'Connor, Johnson
& Kindness
2800 Pacific First Centre
1420 Fifth Avenue
Seattle, WA 98101   (206) 682-8100

William C. Anderson
Boeing Commercial Airplanes
P.O. Box 3707
M/S 7E-25
Seattle, WA 98124   (206) 251-0237

Delbert J. Barnard
Barnard, Pauly & Kaser, P.S.
6000 Southcenter Blvd.
Suite 240
Seattle, WA 98188   (206) 246-0568

Robert J. Baynham
Seed & Berry
6300 Columbia Center
Seattle, WA 98104   (206) 622-4900

Robert Willis Beach
Beach & Brown
3107 Eastlake Ave. East
Seattle, WA 98102   (206) 325-6789

Glenn D. Bellamy
Barnard, Pauly & Kaser, P.S.
6000 Southcenter Blvd.
Suite 240
Seattle, WA 98188   (206) 246-0568

Robert M. Bellomy
Seed & Berry
6300 Columbia Center
Seattle, WA 98104   (206) 622-4900

Benjamin F. Berry
Seed & Berry
6300 Columbia Center
Seattle, WA 98104   (206) 622-4900

Michael W. Bocianowski
Christensen, O'Connor, Johnson
& Kindness
2800 Pacific First Centre
1420 Fifth Avenue
Seattle, WA 98101   (206) 682-8100

Thomas F. Broderick
Christensen, O'Connor,
Johnson & Kindness
2800 Pacific First Centre
1420 Fifth Avenue
Seattle, WA 98101   (206) 682-8100

Ward Brown
Beach & Brown
3107 Eastlake Ave., East
Seattle, WA 98102   (206) 325-6789

Carroll L. Bryan, II#
Bryan, Schiffrin & Mc Monagle
2701 First Avenue
Seattle, WA 98121   (206) 448-8100

Edward W. Bulchis
Seed & Berry
6300 Columbia Center
Seattle, WA 98104   (206) 622-4900

David V. Carlson
Seed & Berry
6300 Columbia Center
Seattle, WA 98104   (206) 622-4900

Morris A. Case
921 S.W. 152nd St.
Seattle, WA 98166   (202) 243-1240

Joseph O. Chalverus
12553 39th N.E.
Seattle, WA 98125   (206) 367-7816

Thomas D. Cohen#
3613 N.E. 43rd
Seattle, WA 98105   (206) 523-9342

George M. Cole
401 2nd Avenue S., #630
Seattle, WA 98104   (206) 622-3740

Robert R. Cook#
Christensen, O'Connor,
Johnson & Kindness
2800 Pacific First Center
1420 Fifth Ave.
Seattle, WA 98101   (206) 224-0723

Kenneth J. Cooper
Boeing Company
P.O. Box 3999
M.S. 80-PA
Seattle, WA 98124   (206) 773-2540

Harry Maybury Cross, Jr.
Dowrey & Cross
1254 Bank of Calif. Ctr.
Seattle, WA 98164   (206) 624-6535

Daniel D. Crouse
Christensen, O'Connor, Johnson
& Kindness
2800 Pacific First Centre
1420 Fifth Avenue
Seattle, WA 98101   (206) 682-8100

Paul C. Cullom, Jr.
Boeing Co.
P.O. Box 3707
M/S 6Y-25
Seattle, WA 98124   (206) 393-8056

Nicolaas De Vogel#
Boeing Company
P.O. Box 3707
M/S 7E-25
Seattle, WA 98124   (206) 251-0398

David H. Deits
Seed & Berry
6300 Columbia Center
Seattle, WA 98104   (206) 622-4900

Bernard A. Donahue
Boeing Commercial Airplane Co.
M.S. 7Y-25
P.O.Box 3707
Seattle, WA 98124   (206) 251-0443

Carl Gordon Dowrey, Sr.
Dowrey, Cross & Cole
401 Second Avenue South
Suite 630
Seattle, WA 98104   (206) 624-6535

Christopher Ogden Duffy
3031 The Bank of Calif. Center
Seattle, WA 98164   (206) 623-8088

Patrick M. Dwyer
220 West Mercer
Suite 500
Seattle, WA 98119   (206) 285-8480

William O. Ferron, Jr.
Seed & Berry
6300 Columbia Center
Seattle, WA 98104   (206) 622-4900

Michael J. Folise
Seed & Berry
6300 Columbia Center
Seattle, WA 98104   (206) 622-4900

L. Grant Foster
Seed & Berry
6300 Columbia Center
701 Fifth Avenue
Seattle, WA 98104   (206) 622-4900

George B. Fox
Seed & Berry
6300 Columbia Center
701 Fifth Avenue
Seattle, WA 98104   (206) 622-4900

Conrad Oliver Gardner
Boeing Commerical Airplane Co.
P.O. Box 3707
M.S. 7E-25
Seattle, WA 98124   (206) 251-0384

David Louis Garrison
3300 Westin Bldg.
2001 Sixth Ave.
Seattle, WA 98121   (206) 441-3440

David S. Goldenberg
Christensen, O'Connor, Johnson
& Kindness
2800 Pacific First Centre
1420 Fifth Avenue
Seattle, WA 98101   (206) 682-8100

Melvin M. Goldenberg
1555 28th Avenue West
Seattle, WA 98199   (206) 283-1019

John O. Graybeal
Graybeal, Jensen & Puntigam
1020 United Airlines Bldg.
2033 Sixth Avenue
Seattle, WA 98121   (206) 448-3200

Paul L. Griffiths
3300 Westin Building
2001 Sixth Avenue
Seattle, WA 98121   (206) 441-3440

Robert L. Gullette
Boeing Aerospace Co.
P.O. Box 24346
Mail Stop 7A-52
Seattle, WA 98124   (206) 865-5245

Henry William Haigh#
18982 Marine View Dr., S.W.
Seattle, WA 98166   (206) 243-6291

M. Henry Halle
Boeing Aerospace
P.O. Box 3999
M/S 85-74
Seattle, WA 98124   (206) 773-2572

Scott G. Hallquist
Immunex Corporation
51 University Street
Seattle, WA 98101   (206) 462-0362

James P. Hamley
Boeing Commercial Airplane Co.
Pats. & Licensing Staff
P.O. Box 3707
Mail Stop 7E-25
Seattle, WA 98124   (206) 251-0262

John C. Hammar
Boeing Defense & Space Group
P.O. Box 3999
M.S. 80-PA
Seattle, WA 98124   (206) 773-2540

H. Gus Hartmann#
Boeing Comm. Airplane Co.
P.O. Box 3707
M.S. 7E-25
Seattle, WA 98124   (206) 251-0237

Gregory W.J. Hauth#
University of Washington
Off. of Technology Transfer
4225 Roosevelt Way N.E.
Suite 301
Seattle, WA 98105   (206) 543-3970

Eugene O. Heberer
Barnard, Pauly & Kaser, P.S.
6000 Southcenter Blvd.
Suite 240
Seattle, WA 98188   (206) 246-0568

Thomas Waldo Hennen
Boeing Aerospace Company
P.O. Box 3999
M.S. 85-74
Seattle, WA 98124   (206) 773-2572

Karl R. Hermanns
Seed & Berry
6300 Columbia Center
Seattle, WA 98104   (206) 622-4900

Lynn H. Hess
Boeing Commercial
Airplane Co.
P.O. Box 3707
M.S. 7E-25
Seattle, WA 98124   (206) 251-0342

Ronald D. Hochnadel
234 27th Avenue, East
Seattle, WA 98112   (206) 322-8085

Julie A. Holly#
Zymo Genetics, Inc.
4225 Roosevelt Way, N.E.
Seattle, WA 98105   (206) 547-8080

Richard L. Hughes
Townsend & Townsend
1201 Third Avenue
Suite 2600
Seattle, WA 98101   (206) 467-9600

Russell W. Illich#
743 N. 35th St.
Seattle, WA 98103   (206) 954-7751

Robert A. Jensen
Graybeal, Jensen & Puntigan
2033 6th Avenue
Suite 1020
Seattle, WA 98121   (206) 448-3200

Lee E. Johnson
Christensen, O'Connor, Johnson
& Kindness
2800 Pacific First Centre
1420 Fifth Avenue
Seattle, WA 98101   (206) 682-8100

Bruce A. Kaser
Barnard, Pauly & Kaser, P.S.
6000 Southcenter Blvd.
Suite 240
Seattle, WA 98188   (206) 246-0568

Marcia S. Kelbon
Christensen, O'Connor,
Johnson & Kindness
2800 Pacific First Centre
1420 Fifth Ave.
Seattle, WA 98101   (206) 682-8100

John M. Kelly
Seed & Berry
6300 Columbia Center
Seattle, WA 98104   (206) 622-4900

Gary Stanley Kindness
Christensen, O'Connor,
Johnson & Kindness
2800 Pacific First Centre
1420 Fifth Avenue
Seattle, WA 98101   (206) 682-8100

Walter Louis Larsen
P.O. Box 18302
Seattle, WA 98118   (206) 723-0870

Kenneth M. Mac Intosh
710 2nd Avenue
Suite 1200
Seattle, WA 98104   (206) 464-6509

David J. Maki
Seed & Berry
6300 Columbia Center
701 Fifth Avenue
Seattle, WA 98104   (206) 622-4900

David D. Mc Masters
Seed & Berry
6300 Columbia Ctr.
701 Fifth Ave.
Seattle, WA 98104   (206) 622-4900

Lawrence H. Meier
Christensen, O'Connor, Johnson
& Kindness
2800 Pacific First Centre
1420 Fifth Avenue
Seattle, WA 98101   (206) 682-8100

Paul T. Meiklejohn
Seed & Berry
701 5th Avenue
Suite 6300
Seattle, WA 98104   (206) 622-4900

Jeffrey J. Miller
Neorx Corporation
410 W. Harrison St.
Seattle, WA 98119   (206) 286-2518

Landon C.G. Miller
Peat Marwick Main & Co.
2030 - 1st Avenue
Seattle, WA 98121   (206) 292-0410

J. Peter Mohn
Boeing Aerospace Co.
P.O. Box 3999
M.S. 85-74
Seattle, WA 98124   (206) 773-3763

Jerald E. Nagae
Christensen, O'Connor, Johnson
& Kindness
2800 Pacific First Centre
1420 Fifth Avenue
Seattle, WA 98101   (206) 682-8100

John M. Neary
Boeing Company
P.O. Box 3999
Seattle, WA 98124   (206) 773-2572

Adonis A. Neblett
Heller, Ehrman, White
& Mc Auliffe
6100 Columbia Center
701 Fifth Ave.
Seattle, WA 98104   (206) 447-0900

Bruce E. O'Connor
Christensen, O'Connor, Johnson
& Kindness
2800 Pacific First Centre
1420 Fifth Avenue
Seattle, WA 98101   (206) 682-8100

Jeffrey B. Oster
Immunex Corp.
Legal Affairs Dept.
51 University Street
Seattle, WA 98101   (206) 587-0430

Gary E. Parker#
Zymo Genetics, Inc.
4225 Roosevelt Way N.E.
Seattle, WA 98105   (206) 548-2322

Steven W. Parmelee
Townsend & Townsend
1201 Third Avenue
Suite 2600
Seattle, WA 98101   (206) 467-9600

Joan H. Pauly
Barnard, Pauly & Kaser, P.S.
6000 Southcenter Blvd.
Suite 240
Seattle, WA 98188   (206) 246-0568

Patricia Anne Perkins#
Legal Affairs Dept.
Immunex Corp.
51 University St.
Seattle, WA 98101   (206) 587-0430

Edward A. Peters#
US West Communications
1600 Bell Plaza
Room 1401
Seattle, WA 98191   (206) 345-3934

Roberta A. Picard#
Stoel, Rives, Boley Jones & Grey
600 University St.
Seattle, WA 98101   (206) 386-7581

Maurice J. Pirio
Seed & Berry
6300 Columbia Center
701 5th Ave.
Seattle, WA 98104   (206) 622-4900

Brian W. Poor#
Genetic Systems Corporation
3005 First Avenue
Seattle, WA 98121   (206) 728-4800

Clark A. Puntigam
Graybeal, Jensen & Puntigam
2033 6th Avenue
Suite 1020
Seattle, WA 98121   (206) 448-3200

Stacy Quan
Christensen, O'Connor, Johnson
& Kindness
2800 Pacific First Centre
1420 Fifth Ave.
Seattle, WA 98101   (206) 682-8100

Mary Y. Redman
Boeing Defense & Space Group
P.O. Box 3999
Mail Stop 80-PA
Seattle, WA 98124   (206) 773-2540

Jeffrey W. Reis
Christensen, O'Connor,
Johnson & Kindness
2800 Pacific First Centre
1420 Fifth Ave.
Seattle, WA 98101   (206) 682-8100

George C. Rondeau, Jr.
Seed & Berry
6300 Columbia Center
701 Fifth Avenue
Seattle, WA 98104   (206) 622-4900

Katie E. Sako
Christensen, O'Connor, Johnson
& Kindness
1420 Fifth Avenue
2800 Pacific First Centre
Seattle, WA 98101   (206) 682-8100

Jeffrey M. Sakoi
Christensen, O'Connor,
Johnson & Kindness
2800 Pacific First Centre
1420 Fifth Avenue
Seattle, WA 98101   (206) 682-8100

Laurence Arthur Savage
Boeing Associated Products
P.O. Box 3707
M.S. 6Y-14
Seattle, WA 98124   (206) 822-0227

Joe M. Scott#
5202 37th Ave., S.W.
Seattle, WA 98126   (206) 935-5595

Thomas W. Secrest, III
1023 N.E. 62nd
Seattle, WA 98115   (206) 523-2464

Richard W. Seed
Seed & Berry
6300 Columbia Center
701 Fifth Avenue
Seattle, WA 98104   (206) 622-4900

Kathryn Ann Seese#
Immunex Corporation
51 University Street
Seattle, WA 98101   (206) 587-0430

Richard G. Sharkey
Seed & Berry
6300 Columbia Center
701 Fifth Avenue
Seattle, WA 98104   (206) 622-4900

Dennis K. Shelton
Christensen, O'Connor, Johnson
& Kindness
2800 Pacific First Centre
1420 Fifth Avenue
Seattle, WA 98101   (206) 682-8100

Suzanne M. Shema
NEORX Corp.
410 West Harrison
Seattle, WA 98119   (206) 281-7001

Joseph M. Sorrentino
Oncogen
3005 First Ave.
Seattle, WA 98121   (206) 728-4800

Ann W. Speckman
Stoel, Rives, Boley,
Jones & Grey
3600 One Union Square
600 University St.
Seattle, WA 98101   (206) 624-0900

Robert H. Sproule
Boeing Commercial Airplanes
M.S. 6Y-25
P.O. Box 3707
Seattle, WA 98124   (206) 393-8061

Robert M. Storwick
Seed & Berry
6300 Columbia Center
Seattle, WA 98104   (206) 622-4900

Ronald E. Suter
Boeing Company
P.O. Box 3707
M.S. 76-52
Seattle, WA 98124   (206) 237-8755

Michael G. Toner
Christensen, O'Connor, Johnson
& Kindness
2800 Pacific First Centre
1420 Fifth Avenue
Seattle, WA 98101   (206) 682-8100

Rodney C. Tullett
Christensen, O'Connor,
Johnson & Kindness
2800 Pacific First Centre
1420 Fifth Ave.
Seattle, WA 98101   (206) 682-8100

James R. Uhlir
Christensen, O'Connor, Johnson
& Kindness
2800 Pacific First Centre
1420 Fifth Avenue
Seattle, WA 98101   (206) 682-8100

Christopher L. Wight
Immunex Corporation
51 University Street
Seattle, WA 98101   (206) 587-0430

Edward M. Yoshida
Imre Corporation
130 Fifth Avenue, N.
Seattle, WA 98109   (206) 448-1000

Kenneth William Thomas
20024 80th Ave. S.E.
Snohomish, WA 98290   (206) 481-6561

Keith S. Bergman
418 Symons Bldg.
S. 7 Howard Street
Spokane, WA 99204   (509) 838-2851

Randy A. Gregory
Wells, St. John & Roberts, P.S.
Suite 815
W. 601 Main Avenue
Spokane, WA 99201   (509) 624-4276

Daniel L. Hayes#
Wells, St. John
& Roberts, P.S.
W. 601 Main, Suite 815
Spokane, WA 99201   (509) 624-4276

Mark W. Hendricksen
Layman, Loft, Arpin
& White
820 Lincoln Bldg.
Spokane, WA 99201   (509) 455-8883

Lewis C. Lee#
Wells, St. John & Roberts
Suite 815
W. 601 Main Ave.
Spokane, WA 99201   (509) 455-8883

J. Christopher Lynch
Paine, Hamblen, Coffin,
Brooke & Miller
1200 Washington Trust Financial Ctr.
Spokane, WA 99204   (509) 455-6000

Mark S. Matkin
Wells, St. John & Roberts, P.S.
W. 601 Main
Suite 815
Spokane, WA 99201   (509) 624-4276

James L. Price#
Wells, St. John & Roberts
W. 601 Main Avenue
Suite 815
Spokane, WA 99201   (509) 624-4276

David P. Roberts
Wells, St. John & Roberts
W. 601 Main Avenue
Suite 815
Spokane, WA 99201   (509) 624-4276

Richard J. St. John
Wells, St. John & Roberts
W. 601 Main Avenue
Suite 815
Spokane, WA 99201   (509) 624-4276

Patrick Donlan Coogan
Weyerhaeuser Co.
Tacoma, WA 98477   (206) 924-2061

John M. Crawford
Weyerhaeuser Co.
Patent Dept.
Tacoma, WA 98477   (206) 924-5611

Kenneth S. Kessler
543 Broadway
Tacoma, WA 98402   (206) 383-1751

James F. Leggett
Leggett & Kram
1901 South I Street
Tacoma, WA 98405   (206) 272-7929

Bryan C. Ogden
Weyerhaeuser Co.
Pat. Dept.
Tacoma, WA 98477   (206) 924-2062

Bloor Redding, Jr.
Hewlett-Packard Company
P.O. Box 8906
Vancouver, WA 98668   (206) 896-2626

Susan L. Preston
16626 - 160th Place N.E.
Woodinvile, WA 98072   (206) 485-5000

George A. Cashman
4407 Carriage Hill Drive
Yakima, WA 98908   (206) 485-5000

West Virginia

James S. Lovell
136 Oakwood Rd.
Charleston, WV 25314   (304) 343-0989

Spencer D. Conard
General Electric Co.
Fifth & Avery Streets
Parkersburg, WV 26102   (304) 424-5422

Thomas Braden Hunter
Borg-Warner Chemicals, Inc.
International Center
Parkersburg, WV 26102   (304) 424-5564

George J. Neilan
2324 Woodland Ave.
South Charleston, WV 25303   (304) 744-8702

Mark S. Morrison
Weirton Steel Corporation
400 Three Springs Drive
Weirton, WV 26062   (304) 797-2040

Charles D. Bell
Bell, Mc Mullen & Cross
67 Seventh Street
Wellsburg, WV 26070   (304) 737-0771

Wisconsin

E. Frank Mc Kinney
Appleton Papers Inc.
P.O. Box 359
825 E. Wisconsin Avenue
Appleton, WI 54912   (414) 735-8660

Benjamin Mieliulis
Appleton Papers Inc.
P.O. Box 359
Appleton, WI 54912   (414) 735-8661

Paul S. Phillips, Jr.#
Appleton Papers Inc.
P.O. Box 359
Appleton, WI 54912   (414) 749-8885

Thomas D. Wilhelm#
1303 Palisades Dr.
Appleton, WI 54915   (414) 730-8575

Robert E. Emery
P.O. Box 519
Bayfield, WI 54814   (715) 779-5354

David J. Archer#
Beloit Corporation
1 St. Lawrence Avenue
Beloit, WI 53511   (608) 364-7018

Raymond W. Campbell
Beloit Corporation
1 St. Lawrence Avenue
Beloit, WI 53511   (608) 364-7042

Gerald Albert Mathews
Beloit Corporation
1 St. Lawrence Ave.
Beloit, WI 53511   (608) 364-7044

Dirk J. Veneman
Beloit Corporation
1 St. Lawrence Avenue
Beloit, WI 53511   (608) 364-7040

James L. Kirschnik
12845 West Burleigh Road
Brookfield, WI 53005   (414) 786-0200

William H. Nehrkorn
205 Bishops Way
Suite 201
Brookfield, WI 53005   (414) 789-5505

Richard C. Ruppin, Sr.
Harnischfeger Industries, Inc.
13400 Bishops Lane
Brookfield, WI 53005   (414) 797-6434

Willis B. Swartwout, III
21001 Wartertown Road
Suite 302
P.O. Box 1068
Brookfield, WI 53008   (414) 786-8614

John Lawrence Beales
1128 2nd Street
P.O. Box 726
Chetek, WI 54728   (715) 924-4801

Francis J. Bouda
13319 Centerville Road
Cleveland, WI 53015   (414) 693-8202

Carl William Laumann, Jr.
Box 1186
Cumberland, WI 54829   (414) 693-8202

William Arthur Denny
Denny & Yanisch
13500 Watertown Plank Road
Elm Grove, WI 53122   (414) 797-8777

Joseph S. Heino
Denny & Yanisch
13500 Watertown Plank Road
Elm Grove, WI 53122   (414) 797-8777

Donald Edwin Porter
15025 Westover Road
Elm Grove, WI 53122   (414) 782-3324

Donald Cayen
104 S. Main Street
Fond Du Lac, WI 54935   (414) 921-2288

Robert C. Curfiss
Mercury Marine
Div. of Brunswick Corp.
1939 Pioneer Road
Fond Du Lac, WI 54935   (414) 929-5419

John A. Reiter, Jr.#
247 North Street
Route 2 Box 9
Fountain City, WI 54629   (608) 687-3401

T. Lloyd Lafave
Weber, Raithel, Malm
& La Fave
Suite 210
5900 N. Port Washington Rd.
Glendale, WI 53217   (414) 964-5250

Gordon Paul Ralph
6642 N. Atwahl Drive
Glendale, WI 53209   (414) 352-7477

Peter Paul Kozak
1008 S. Van Buren
Green Bay, WI 54301   (414) 437-9235

Joseph M. Recka
Recka, Joannes & Faller S.C.
211 South Monroe Ave.
Green Bay, WI 54301   (414) 435-8159

Dennis J. Verhaagh
P.O. Box 995
Green Bay, WI 54305   (414) 435-6423

James T. Barr
P. O. Box 153
125 Main Street
Hancock, WI 54943   (715) 249-5182

Lewis L. Lloyd
U.S. Marine Corporation
& Bayliner Marine Corporation
105 Marine Drive
Hartford, WI 53027   (414) 673-2200

M. Paul Hendrickson
403 Main Street
P.O. Box 508
Holmen, WI 54636   (608) 526-4422

Gary L. Griswold
Cove Road
Route No. 3
Hudson, WI 54016   (612) 733-8904

Howard M. Herriot
Consigny, Andrews, Hemming
& Grant, S.C.
303 East Court Street
P.O. Box 1449
Janesville, WI 53547   (608) 755-5050

Edward R. Antaramian
Antaramian, Easton & Antaramian
2221 - 63rd Street
Kenosha, WI 53140   (414) 654-8669

Neil E. Hamilton
8306 42nd Avenue
Kenosha, WI 53142   (414) 694-6283

David J. Richter
Snap-On Tools Corporation
2801 80th Street
Kenosha, WI 53141   (414) 656-5322

Richard T. Naruo
1124 Woodland Road
Kohler, WI 53044   (414) 452-4215

William J. Beres
Trane Company
Patent Dept.
3600 Pammel Creek Road
La Crosse, WI 54601   (608) 787-4177

Peter D. Ferguson
Trane Co.
3600 Pammell Creek Rd.
La Crosse, WI 54601   (608) 787-3405

Robert J. Harter#
Trane Company
Pat. Dept.-12
3600 Pammell Creek Road
La Crosse, WI 54601   (608) 787-3860

William O'Driscoll
Trane Company
Patent Department-12
3600 Pammel Creek Road
La Crosse, WI 54601   (608) 787-2538

Raymond Lynn Balfour
Rayovac Corporation
601 Rayovac Drive
Madison, WI 53711   (608) 275-4584

Howard W. Bremer
Wisconsin Alumni Research Foundation
P.O. Box 7365
614 N. Walnut St.
Madison, WI 53707   (608) 263-2831

Barry U. Buchbinder#
Agrigenetics Advanced Science Co.
5649 E. Buckeye Road
Madison, WI 53716   (608) 221-5000

Tracy J. Dunn#
7022 Harvest Hill Road
Madison, WI 53717   (608) 231-1387

Harry C. Engstrom, Jr.
Foley & Lardner
One South Pinckney Street
P.O. Box 1497
Madison, WI 53701   (608) 258-4207

David T. Flanagan
Wisconsin Dept. of Justice
State Capitol
Madison, WI 53704   (608) 266-7971

Carl E. Gulbrandsen
25 West Main Street
Suite 300
Madison, WI 53703   (608) 257-2281

David J. Houser
Wisconsin Alumni Res. Foundation
614 N. Walnut Street
Madison, WI 53705   (608) 263-9395

Deborah Drayna Johnson
Lathrop & Clark
122 West Washington Avenue
Suite 1000
P.O. Box 1507
Madison, WI 53701   (608) 257-7766

James A. Kemmeter
Lathrop & Clark
P.O. Box 1507
122 W. Washington Ave.
Madison, WI 53701   (608) 257-7766

Trayton L. Lathrop
Lathrop & Clark
P.O. Box 1507
Madison, WI 53701   (608) 257-7766

Margaret M. Liss
Lathrop & Clark
122 West Washington Ave.
Suite 1000
P.O. Box 1507
Madison, WI 53701   (608) 257-7766

Theodore J. Long
Lathrop & Clark
P.O. Box 1507
Madison, WI 53701   (608) 257-7766

Charles J. Meyerson
6506 Offshore Drive
Madison, WI 52705   (608) 833-5733

Terry F. Peppard
4814 Marathon Dr.
Madison, WI 53705   (608) 233-8238

Costa Perchem
25 W. Main Street
Suite 711
Madison, WI 53703   (608) 256-0681

Douglas E. Pfrang#
Nicolet Instrument Corp.
5225 Verona Road
Madison, WI 53711   (608) 271-3333

Charles S. Sara
Andrus, Sceales, Starke & Sawall
3 S. Pinckney St., Suite 202
Madison, WI 53703   (608) 255-2022

William J. Scanlon
Foley & Lardner
One South Pinckney St.
P.O. Box 1497
Madison, WI 53701   (608) 258-4284

H. Keith Schoff
2257 E. Washington Ave.
Madison, WI 53704   (608) 241-2616

Nicholas J. Seay
Quarles & Brady
P.O. Box 2113
First Wisconsin Plaza
Madison, WI 53701   (608) 251-5000

David R.J. Stiennon
Lathrop & Clark
122 W. Washington Ave.
Suite 1000
P.O. Box 1507
Madison, WI 53701   (608) 257-7766

Patrick J.G. Stiennon
Lathrop & Clark
122 W. Washington Ave.
Suite 1000
Madison, WI 53703   (608) 257-7766

Janet Irmtraud Stockhausen*
USDA Forest Services
Forest Products Laboratory
1 Gifford Pinchot Drive
Madison, WI 53705   (608) 231-9502

Teresa J. Welch
Stroud, Stroud, Willink,
Thompson & Howard
25 W. Main St., Ste. 300
P.O. Box 2236
Madison, WI 53701   (608) 257-2281

Raymond Joseph Miller
326 Winnebago Avenue
Menasha, WI 54952   (414) 722-7804

William Arthur Autio
N85 W17231 Lee Place
Menomonee Falls, WI 53051   (414) 251-2863

David E. Stewart
Waste Management of
North America, Inc.
W124 N8925 Boundary Road
Menomonee Falls, WI 53051   (414) 251-4000

John Kay Crump
945 W. Heritage Court
N. 207
Meqoon, WI 53092   (414) 241-9279

Karl William Marquardt
3905 Sumac Circle
Middleton, WI 53562   (608) 836-8526

Frank S. Andrus
Andrus, Sceales, Starke
& Sawall
735 N. Water St.
Milwaukee, WI 53202   (414) 271-7590

Russell J. Barron
Foley & Lardner
777 E. Wisconsin Avenue
Milwaukee, WI 53202   (414) 289-3752

Ronald E. Barry
Foley & Lardner
777 East Wisconsin Avenue
Milwaukee, WI 53202   (414) 271-2400

Keith M. Baxter
General Electric Co.
Medical Systems Group
P.O. Box 414
Milwaukee, WI 53201   (414) 271-2400

John L. Beard
Godfrey & Kahn, S.C.
780 N. Water Street
Milwaukee, WI 53202   (414) 273-5585

Robert B. Benson
Allis-Chalmers Mfg. Co.
Patent Dept.
P.O. Box 512
Milwaukee, WI 53201   (414) 475-4038

James F. Boyle
Nilles & Nilles, S.C.
777 E. Wisconsin Ave.
Suite 3070
Milwaukee, WI 53202   (414) 276-0977

Glenn A. Buse
Michael, Best & Friedrich
100 East Wisconsin Avenue
Milwaukee, WI 53202   (414) 271-6560

Donald G. Casser
Quarles & Brady
411 E. Wisconsin Avenue
Milwaukee, WI 53202   (414) 277-5000

Robert E. Clemency
Michael, Best & Friedrich
100 East Wisconsin Avenue
Milwaukee, WI 53202   (414) 271-6560

John C. Cooper, III
Foley & Lardner
777 East Wisconsin Avenue
35th Floor
Milwaukee, WI 53202   (414) 289-3552

George R. Corrigan
Foley & Lardner
777 E. Wisconsin Ave.
Milwaukee, WI 53202   (414) 289-7037

Jeffrey N. Costakos
Foley & Lardner
777 E. Wisconsin Ave.
Milwaukee, WI 53202   (414) 271-2400

James R. Custin
Nilles & Nilles, S.C.
777 E. Wisconsin Avenue
Suite 3070
Milwaukee, WI 53202   (414) 276-0977

Thomas William Ehrmann
Quarles & Brady
411 E. Wisconsin Avenue
Milwaukee, WI 53202   (414) 277-5000

Arnold J. Ericsen
Whyte & Hirschboeck S.C.
111 East Wisconsin Avenue
Suite 2100
Milwaukee, WI 53202   (414) 271-8210

Gary A. Essmann
Andrus, Sceales, Starke
& Sawall
100 E. Wisconsin Ave.
Suite 1100
Milwaukee, WI 53202   (414) 271-7590

George A. Evans
735 N. Water St.
Milwaukee, WI 53202   (414) 271-7590

William L. Falk#
Foley & Lardner
777 East Wisconsin Avenue
Milwaukee, WI 53202   (414) 289-3682

Daniel Dawson Fetterley
Andrus, Sceales, Starke
& Sawall
100 East Wisconsin Ave.
Suite 1100
Milwaukee, WI 53202   (414) 271-7590

John D. Franzini
Quarles & Brady
411 E. Wisconsin Avenue
Milwaukee, WI 53202   (415) 277-5000

John F. Friedl
9102 W. Dixon
Apt. 105
Milwaukee, WI 53214   (414) 258-4835

Raymond E. Fritz, Jr.
Thomas Mc Kinnon Securities, Inc.
731 N. Water Street
Milwaukee, WI 53202   (414) 271-5670

Henry C. Fuller, Jr.
Fuller, Puerner & Hohenfeldt, S.C.
633 W. Wisconsin Avenue
Milwaukee, WI 53203   (414) 271-6555

Joseph A. Gemignani
Michael, Best & Friedrich
100 East Wisconsin Avenue
Milwaukee, WI 53202   (414) 271-6560

Victor M. Genco, Jr.
Godfrey & Kahn, S.C.
780 North Water St.
Milwaukee, WI 53202   (414) 273-3500

George G. Grigel
Godfrey & Kahn, S.C.
780 N. Water Street
Milwaukee, WI 53202   (414) 273-3500

George E. Haas
Quarles & Brady
411 E. Wisconsin Avenue
Milwaukee, WI 53202   (414) 277-5000

Aaron Lee Hardt, Sr.
Rexnord Corp.
Legal Dept.
4695 W. Greenfield Ave.
Milwaukee, WI 53214   (414) 643-2505

Charles F. Hauff, Jr.
Foley & Lardner
777 E. Wisconsin Avenue
Milwaukee, WI 53202   (414) 271-2400

Ralph G. Hohenfeldt
Fuller, Puerner & Hohenfeldt
633 W. Wisconsin Avenue
Milwaukee, WI 53203   (414) 271-6555

Guenther W. Holtz
Andrus, Sceales, Starke
& Sawall
100 East Wisconsin Ave.
Suite 1100
Milwaukee, WI 53202   (414) 271-7590

Joseph J. Jochman, Jr.
Andrus, Sceales, Starke
& Sawall
100 E. Wisconsin Ave.
Suite 1100
Milwaukee, WI 53202   (414) 271-7590

Ira Milton Jones
1037 N. Astor St.
Milwaukee, WI 53202   (414) 276-4210

Nicholas A. Kees
Fuller, Puerner & Hohenfeldt
633 W. Wisconsin Avenue
Milwaukee, WI 53203   (414) 271-6555

Timothy M. Kelley
Michael, Best & Friedrich
100 East Wisconsin Ave.
Milwaukee, WI 53202   (414) 271-6560

Thomas F. Kirby
Nilles, Custin & Kirby, S.C.
First Wisconsin Center
777 E. Wisconsin Avenue
Milwaukee, WI 53202   (414) 276-0977

Ramon A. Klitzke
Marquette Univ. Law School
1103 W. Wisconsin Ave.
Milwaukee, WI 53233   (414) 224-5366

Thomas O. Kloehn
Quarles & Brady
411 E. Wisconsin Avenue
Milwaukee, WI 53202   (414) 277-5773

Joseph A. Kromholz
Suite 450, Atrium Building
10400 W. North Ave.
Milwaukee, WI 53226   (414) 258-1545

Thad F. Kryshak
Quarles & Brady
411 E. Wisconsin Avenue
Milwaukee, WI 53202   (414) 277-5000

Casimir F. Laska
Andrus, Sceales, Starke
& Sawall
100 East Wisconsin Ave.
Suite 1100
Milwaukee, WI 53202   (414) 271-7590

Allan W. Leiser
Quarles & Brady
411 E. Wisconsin Avenue
Milwaukee, WI 53202   (414) 277-5000

Edward L. Levine
Johnson Controls, Inc.
5757 North Green Bay Avenue
P.O. Box 591
Milwaukee, WI 53201   (414) 228-2767

James E. Lowe, Jr
A. O. Smith Corp.
11270 W. Park Place
Milwaukee, WI 53224   (414) 359-4106

Allan O. Maki
Fuller, Ryan & Hohenfeldt
633 West Wisconsin Ave.
Milwaukee, WI 53203   (414) 271-6555

Stephen P. Malak
2934 E. Hartford Ave.
Milwaukee, WI 53211   (414) 332-2340

Deepak Malhotra#
Michael, Best & Friedrich
100 East Wisconsin Avenue
Milwaukee, WI 53202   (414) 271-6560

Philip P. Mann
Reinhart, Boerner, Van Deuren,
Norris & Rieselbach, S.C.
111 E. Wisconsin Ave.
Suite 1800
Milwaukee, WI 53202   (414) 271-1190

Andrew S. Mc Connell
Andrus, Sceales, Starke
& Sawall
100 E. Wisconsin Ave.
Suite 1100
Milwaukee, WI 53202   (414) 271-759O

Donald C. Mc Gaughey
Nilles & Nilles, S.C.
777 E. Wisconsin Avenue
Suite 3070
Milwaukee, WI 53202   (414) 276-0977

Michael J. Mc Govern
Quarles & Brady
411 E. Wisconsin Avenue
Milwaukee, WI 53202   (414) 277-5000

Philip G. Meyers
Foley & Lardner
777 E. Wisconsin Avenue
Milwaukee, WI 53202   (414) 289-3761

Bayard H. Michael
Michael, Best & Friedrich
100 East Wisconsin Avenue
Milwaukee, WI 53202   (414) 271-6560

Andrew J. Nilles
Nilles & Nilles, S.C.
777 E. Wisconsin Ave.
Suite 3070
Milwaukee, WI 53202   (414) 276-0977

James E. Nilles
Nilles & Nilles, S.C.
777 E. Wisconsin Avenue
Suite 3070
Milwaukee, WI 53202   (414) 276-0977

James P. O Shaughnessy
Foley & Lardner
777 East Wisconsin Avenue
Milwaukee, WI 53202   (414) 271-2400

Mark W. Pfeiffer#
Quarles & Brady
411 E. Wisconsin Avenue
Milwaukee, WI 53202   (414) 277-5000

John H. Pilarski
Godfrey & Kahn, S.C.
780 N. Water Street
Milwaukee, WI 53202   (414) 273-3500

Gary R. Plotecher
Whyte & Hirschboeck S.C.
Suite 2100
111 East Wisconsin Avenue
Milwaukee, WI 53202   (414) 271-8210

Paul R. Puerner
633 W. Wisconsin Avenue
Suite 1305
Milwaukee, WI 53203   (414) 223-4281

Joseph D. Radtke
Kotecki & Radtke, S.C.
606 W. Wisconsin Avenue
Milwaukee, WI 53203   (414) 273-6363

Michael D. Rechtin
Reinhart, Boerner, Van Deuren,
Norris & Rieselbach, S.C.
111 E. Wisconsin Avenue
Suite 1800
Milwaukee, WI 53202   (414) 271-1190

Andrew O. Riteris
Michael, Best & Friedrich
100 East Wisconsin Avenue
Suite 3300
Milwaukee, WI 53202   (414) 271-6560

Joseph E. Root, III
Johnson Controls, Inc.
Intellectual Property Law X-73
5757 North Green Bay Avenue
P. O. Box 591
Milwaukee, WI 53201   (414) 228-1200

Daniel D. Ryan, III
Fuller, Ryan & Hohenfeldt
633 W. Wisconsin Ave.
Milwaukee, WI 53203   (414) 271-6555

John Phillip Ryan
1610 N. Prospect Ave.
Milwaukee, WI 53202   (414) 273-1352

Barry E. Sammons
Quarles & Brady
411 E. Wisconsin Avenue
Milwaukee, WI 53202   (414) 277-5000

Eugene R. Sawall
Andrus, Sceales, Starke
& Sawall
100 E. Wisconsin Ave.
Suite 1100
Milwaukee, WI 53202   (414) 271-7590

Carl R. Schwartz
Quarles & Brady
411 E. Wisconsin Avenue
Milwaukee, WI 53202   (414) 277-5715

David B. Smith
Michael, Best & Friedrich
100 East Wisconsin Avenue
Milwaukee, WI 53202   (414) 271-6560

George H. Solveson
Andrus, Sceales, Starke
& Sawall
100 East Wisconsin Ave.
Milwaukee, WI 53202   (414) 271-7590

Warren T. Sommer#
3137 N. Cramer St.
Milwaukee, WI 53211   (414) 962-7076

Glen O. Starke
Andrus, Sceales, Starke
& Sawall
735 N. Water St.
Suite 1102
Milwaukee, WI 53202   (414) 271-7590

Robert J. Steininger
825 N. Jefferson St.
Milwaukee, WI 53202   (414) 272-0530

Richard C. Steinmetz, Jr.
Allen-Bradley Co.
1201 S. 2nd St.
Milwaukee, WI 53204   (414) 382-2134

Douglas E. Stoner
General Elec. Co.
P.O. Box 414
Milwaukee, WI 53201   (414) 544-3439

C. Thomas Sylke
Whyte & Hirschboeck S.C.
111 East Wisconsin Avenue
Suite 2100
Milwaukee, WI 53202   (414) 271-8210

Michael E. Taken
Andrus, Sceales, Starke
& Sawall
100 E. Wisconsin Avenue
Suite 1100
Milwaukee, WI 53202   (414) 375-4780

Richard Paul Ulrich
4524A W. Oklahoma Ave.
Milwaukee, WI 53219   (414) 543-0215

Thomas E. Valentyn
Johnson Controls, Inc.
5757 North Green Bay Ave.
P.O. Box 591
Milwaukee, WI 53201   (414) 228-2278

Larry G. Vande Zande#
Eaton Corporation
4201 N. 27th Street
Milwaukee, WI 53216   (414) 449-6264

Kenneth D. Wahlin
Michael, Best & Friedrich
100 East Wisconsin Ave.
Milwaukee, WI 53202   (414) 271-6560

Allan B. Wheeler
10400 W. North Ave.
Suite 450 Atrium Bldg.
Milwaukee, WI 53226   (414) 258-1213

Fred Wiviott
Michael, Best & Friedrich
100 East Wisconsin Avenue
Suite 3300
Milwaukee, WI 53202   (414) 271-6560

Thomas M. Wozny
Andrus, Sceales, Starke
& Sawall
100 East Wisconsin Ave.
Suite 1100
Milwaukee, WI 53202   (414) 271-7590

Elroy J. Wutschel
Kirst, Surges & Wutschel
522 N. Water Street
Milwaukee, WI 53202   (414) 276-0220

Elwin J. Zarwell
Quarles & Brady
411 E. Wisconsin Avenue
Milwaukee, WI 53202   (414) 277-5000

David R. Price
Michael, Best & Friedrich
100 East Wisconsin Avenue
Milwuakee, WI 53202   (414) 271-6560

James A. Wilke
S76 W12620 Mc Shane Drive
Muskego, WI 53150   (414) 529-1575

Robert S. Alexander
James River Corp.
1915 Marathon Ave.
P.O. Box 899
Neenah, WI 54957   (414) 729-8360

Gregory E. Croft
Kimberly-Clark Corporation
401 N. Lake Street
Neenah, WI 54956   (414) 721-3616

Thomas M. Gage
Kimberly-Clark Corp.
Legal Dept.
401 N. Lake St.
Neenah, WI 54956   (414) 721-2000

John P. Kirby, Jr.
Kimberly-Clark Corp.
401 N. Lake Street
Neenah, WI 54957   (414) 721-2985

Thomas Joe Mielke
Kimberly-Clark Corporation
401 N. Lake Street
Neenah, WI 54956   (414) 721-2000

Douglas L. Miller
Kimberly-Clark Corporation
401 North Lake Street
Neenah, WI 54956   (414) 721-2343

Eckhard C.A. Schwarz#
Biax-Fiberfilm Corp.
884 Chapman Avenue
Neenah, WI 54956   (414) 722-3180

Donald L. Traut
Kimberly-Clark Corporation
401 North Lake Street
Neenah, WI 54956   (414) 721-2433

Paul Y. P. Yee
Kimberly-Clark Corp.
401 North Lake Street
Neenah, WI 54956   (414) 721-2435

Carl M. Lewis
642 Winter Street
Onalaska, WI 54650   (608) 783-3924

David L. Polsley
2102 Esther Dr.
Onalaska, WI 54650   (608) 783-7427

James J. Getchius
711 Jackson Street
Oshkosh, WI 54901   (414) 235-0601

Chester J. Giuliani
Box 53
Phelps, WI 54554   (906) 548-2138

Robert T. Johnson#
603 Collins Street
Plymouth, WI 53073   (414) 892-8556

Ronald P. Brockman#
Thompson & Coates, Ltd.
840 Lake Avenue
P.O. Box 516
Racine, WI 53403   (414) 632-7541

Jerome Donald Drabiak
S.C. Johnson & Son, Inc.
1525 Howe Street
M.S. 077
Racine, WI 53403   (414) 631-2000

J. William Frank, III
S.C. Johnson & Sons, Inc.
Pat. Section
1525 Howe St.
Racine, WI 53403   (414) 631-2673

Alexander M. Gerasimow
JI Case Company
700 State Street
Racine, WI 53404   (414) 636-6873

Robert D. Godard#
Dykema Gossett
5719 Cambridge Lane, Unit #4
Racine, WI 53406   (414) 886-3705

Arthur John Hansmann
5200 Washington Avenue
Racine, WI 53406   (414) 632-2818

Peter N. Jansson
245 Main Street
Suite M
Racine, WI 53403   (414) 632-6900

Joseph T. Kivlin, Jr.
Hartig, Bjelajac, Michelson
& Kivlin
601 Lake Avenue
Racine, WI 53403   (414) 633-9800

Richard E. Rakoczy
S.C. Johnson & Son, Inc.
1525 Howe Street
Racine, WI 53403   (414) 631-2909

Dorothy L. Sander
S.C. Johnson & Son, Inc.
1525 Howe Street
Racine, WI 53403   (414) 631-2206

Larry L. Shupe
245 Main Street
Suite M
Racine, WI 53403   (414) 632-6900

Stanley E. Sutherland
S.C. Johnson & Sons, Inc.
Law Dept.
1525 Howe St.
Racine, WI 53403   (414) 631-2995

Tipton L. Randall#
Zimpro/Passavant
301 W. Military Road
Rothschild, WI 54474   (715) 359-7211

Fabian A. Brusok
3519 High Cliff Circle
Sheboygan, WI 58083   (414) 458-6833

Ray Gunnar Olander
Bucyrus-Erie Co.
1100 Milwaukee Ave.
South Milwaukee, WI 53172   (414) 768-4099

Arnold J. De Angelis
527 Oakwood Drive
Thiensville, WI 53092   (414) 242-5893

Harold Warner Grothman
621 Grand Ave.
Thiensville, WI 53092   (414) 242-1574

Ronald W. O'Keefe
GE Medical Systems
3000 N. Grandview Blvd.
Waukesha, WI 53188   (414) 544-3155

James O. Skarsten
General Electric Company
3000 North Grandview Blvd.
Waukesha, WI 53188   (414) 544-3175

James R. Sommers
Hunter & Sommers
259 South Street
Waukesha, WI 53186   (414) 547-7788

Russell E. Weinkauf
1403 Steuban Street
Wausau, WI 54401   (715) 842-7066

Robert C. Jones
3860 North 102 Street
Wauwatosa, WI 53222   (414) 461-2366

James G. Morrow
639 North 76th Street
Wauwatosa, WI 53213   (414) 258-9335

Arthur Lowell Nelson
669 N. 78th Street
Wauwatosa, WI 53213   (414) 771-9672

Cyril M. Hajewski
12209 W. Holt Avenue
West Allis, WI 53227   (414) 543-7102

Russell L. Johnson#
P.O. Box 161
Weyauwega, WI 54983   (414) 867-3482

Neal Seegert
5158 N. Lake Drive
Whitefish Bay, WI 53217   (414) 332-0642

Wyoming

James E. Wolber
2200 N. Lake Creek
Casper, WY 82604   (414) 332-0642

Ronald W. Redo#
P.O. Box 15641
Cheyenne, WY 82003   (307) 778-4831

Part Four

Appendices and Master Index

Appendix A

U.S. Patent Classifications

The *Index to the U.S. Patent Classification* is an alphabetical list of the subject headings referring to specific classes and subclasses of the U.S. patent classification system. The classifications are intended as an initial means of entry into the Patent and Trademark Office's (PTO's) classification system and should be particularly useful to those who lack experience in using the classification system or who are unfamiliar with the technology under consideration.

The classifications are to searching a patent what the card catalog is to looking for a library book. It is the only way to discover the existing prior art. The classifications are a star to steer by, without which no meaningful patent search can be conducted.

Although continual changes are made in the classification system, the following pages will serve as an invaluable tool to 1) understand how inventions are classified by the PTO, and 2) prepare for a patent search.

Before you begin your patent search, use the classifications list that follows to clarify your direction. First, look for the term which you feel best represents your invention. If a match cannot be found, look for terms of approximately the same meaning, for example, the essential function or the effect of use or application of the object.

The classifications are arranged with subheadings that can extend to four levels of indentation. A complete reading of a subheading includes the title of the most adjacent higher heading, and so on until there are no higher headings. Some headings will reference other related or preferred entries with a "(see...)" phrase.

Once you have identified the numbers of possibly pertinent classes and subclasses for your idea, go to a Patent and Trademark Depository Library (PTDL). Here you may look at the complete *Manual of Classification*, a loose-leaf PTO volume listing the numbers and descriptive titles of more than 300 classes and 95,000 subclasses used in the subject classification of patents, along with the same index to classifications.

New classes and subclasses are continuously based upon breaking developments in science and technology. Old classes and subclasses are rendered obsolete by technological advance. If you have

suggestions for future revisions of the classifications, or find omissions or errors, you are encouraged to alert the PTO. Send your suggestions to Editor, *Index to the U.S. Patent Classification*, Office of Documentation, U.S. Patent and Trademark Office, Washington, DC 20231.

Symbols and abbreviations used in the Index are as follows:

A. . .Z   An Alpha designation following the official numeric subclass identifies an unofficial subclass, i.e., a grouping of patents selected out from an official subclass by an examiner and then made an indented subclass under the official subclass, usually for purposes of further breaking down a concept. For example, in Class 273 Amusement Devices, Games, Subclass 58A collects balls which are solid and which are covered. Topics not specifically selected out are left in a residual subclass, designated R. For example, 58R is generic to balls not provided for elsewhere. Unofficial subclass groupings are used in the Examiners' Files; they are not available in the Public Search Room.

Ctg.   "Containing" may be abbreviated "ctg" in long chemical phrases.

D   Design classes are preceded by the letter D, as in D2.

DIG.   A digest, indicated by DIG. and a subclass number, is an unofficial collection of patents based on a concept which relates to a class but not to any particular subclass of that class. In this Index 123 DIG. 12 is a collection of Hydrogen fueled engines in Class 123, Internal Combustion Engines. Digests have been created over the years by examiners to facilitate their searches within the arts under their jurisdiction. They are unofficial collections and are not available in the Public Search Room.

PLT   Plant patents are identified as class PLT.

TM   A few trademark names have been included where a particular field of search could best be suggested by such. These are identified by the letters TM.

+   The plus sign following the subclass indicates that the entry includes that subclass and all subclasses indented thereunder. (See the Manual of Classification for class schedules with subclasses.)

X-art   Cross reference art collections may be described in the text as "X-art" collections. These are official collections of patents based on a concept which relates to a class but not to any particular subclass of that class. For example, Class 425 contains molding apparatus described by a functional structure and cross-reference art collection 801 collects button molding patents into one group.

*   The asterisk following a subclass designates a cross-reference art collection.

Users are encouraged to make suggestions for future revisions of this Index. Identification of errors and omissions are welcomed. When recommending new entries, please include the classification with the descriptor, if possible. Send suggestions to *Index to U.S. Patent Classification,* Office of Documentation, U.S. Patent and Trademark Office, Washington, D.C. 20231.

The Index is available from the Superintendent of Documents, U.S. Government Printing Office, Washington, D.C. 20402. Each subscription to the *Manual of Classification* includes one copy of the Index and any new edition issuing during the life of the subscription. Responsive to the many requests for multiple copies, the Government Printing Office now sells the Index as a separate stand-alone document.

CLASSES ARRANGED IN ALPHABETICAL ORDER

CLASS	Title of Class
51	Abrading
181	Acoustics
357	Active Solid State Devices, e.g., Transistors, Solid State Diodes
156	Adhesive Bonding and Miscellaneous Chemical Manufacture
226	Advancing Material of Indeterminate-Length
244	Aeronautics
366	Agitating
420	Alloys or Metallic Compositions
102	Ammunition and Explosives
86	Ammunition and Explosive-Charge Making
330	Amplifiers
272	Amusement and Exercising Devices
273	Amusement Devices, Games
446	Amusement Devices, Toys
119	Animal Husbandry
2	Apparel
223	Apparel Apparatus
221	Article Dispensing
236	Automatic Temperature and Humidity Regulation
4	Baths, Closets, Sinks and Spit.oons
136	Batteries, Thermoelectric and Photoelectric
384	Bearings
5	Beds
449	Bee Culture
402	Binder Device Releasably Engaging Aperture or Notch of Sheet
8	Bleaching and Dyeing; Fluid Treatment and Chemical Modification of Textiles and Fibers
10	Bolt, Nail, Nut, Rivet and Screw Making
412	Bookbinding: Process and Apparatus
281	Books, Strips and Leaves
12	Boot and Shoe Making
36	Boots, Shoes and Leggings
175	Boring or Penetrating the Earth

CLASS	Title of Class
215	Bottles and Jars
188	Brakes
14	Bridges
300	Brush, Broom and Mop Making
15	Brushing, Scrubbing and General Cleaning
24	Buckles, Buttons, Clasps, Etc.
441	Buoys, Rafts, and Aquatic Devices
17	Butchering
79	Button Making
40	Card, Picture and Sign Exhibiting
502	Catalyst, Solid Sorbent, or Support Therefor, Product or Process of Making
59	Chain, Staple and Horseshoe Making
297	Chairs and Seats
194	Check-Actuated Control Mechanisms
436	Chemistry: Analytical and Immunological Testing
260	Chemistry, Carbon Compounds
204	Chemistry, Electrical and Wave Energy
429	Chemistry, Electrical Current Producing Apparatus, Product and Process
71	Chemistry, Fertilizers
518	Chemistry Fischer-Tropsch Processes; or Purification or Recovery of Products Thereof
585	Chemistry, Hydrocarbons
423	Chemistry, Inorganic
435	Chemistry: Molecular Biology and Microbiology
530	Chemistry, Natural Resins or Derivatives Peptides or Proteins; Lignins or Reaction Products Thereof
23	Chemistry, Physical Processes
279	Chucks or Sockets
209	Classifying, Separating and Assorting Solids
134	Cleaning and Liquid Contact with Solids
292	Closure Fasteners
160	Closures, Partitions and Panels, Flexible and Portable
192	Clutches and Power-Stop Control
118	Coating Apparatus

697

CLASS	Title of Class
401	Coating Implements with Material Supply
427	Coating Processes
341	Coded Data Generation or Conversion
372	Coherent Light Generators
453	Coin Handling
431	Combustion
342	Communications, Directive Radio Wave Systems and Devices (e.g., Radar, Radio Navigation)
340	Communications, Electrical
367	Communications, Electrical: Acoustic Wave System and Devices
343	Communications, Radio Wave Antennas
252	Compositions
501	Compositions: Ceramic
106	Compositions, Coating or Plastic
7	Compound Tools
159	Concentrating Evaporators
193	Conveyers, Chutes, Skids, Guides and Ways
198	Conveyers, Power-Driven
406	Conveyors, Fluid Current
147	Coopering
460	Crop Threshing or Separating
380	Cryptography
30	Cutlery
407	Cutters, For Shaping
83	Cutting
408	Cutting by Use of Rotating Axially Moving Tool
329	Demodulators
433	Dentistry
232	Deposit and Collection Receptacles
222	Dispensing
202	Distillation: Apparatus
203	Distillation: Processes, Separatory
201	Distillation: Processes, Thermolytic
424	Drug, Bio-Affecting and Body Treating Compositions
514	Drug, Bio-Affecting and Body Treating Compositions

CLASS	Title of Class
34	Drying and Gas or Vapor Contact with Solids
369	Dynamic Information Storage or Retrieval
360	Dynamic Magnetic Information Storage or Retrieval
172	Earth Working
434	Education and Demonstration
219	Electric Heating
313	Electric Lamp and Discharge Devices
314	Electric Lamp and Discharge Devices, Consumable Electrodes
315	Electric Lamp and Discharge Devices, Systems
445	Electric Lamp or Space Discharge Component or Device Manufacturing
363	Electric Power Conversion Systems
392	Electric Resistance Heating Devices
381	Electrical Audio Signal Processing Systems, and Devices
364	Electrical Computers and Data Processing Systems
439	Electrical Connectors
310	Electrical Generator or Motor Structure
377	Electrical Pulse Counters, Pulse Dividers or Shift Registers: Circuits and Systems
338	Electrical Resistors
307	Electrical Transmission or Interconnection Systems
320	Electricity, Battery and Condenser Charging and Discharging
200	Electricity, Circuit Makers and Breakers
174	Electricity, Conductors and Insulators
361	Electricity, Electrical Systems and Devices
337	Electricity, Electrothermally or Thermally Actuated Switches
335	Electricity, Magnetically Operated Switches, Magnets and Electro-Magnets
324	Electricity, Measuring and Testing
318	Electricity, Motive Power Systems
388	Electricity, Motor Control Systems
323	Electricity, Power Supply, or Regulation Systems
322	Electricity, Single Generator Systems
191	Electricity, Transmission to Vehicles

CLASSES ARRANGED IN ALPHABETICAL ORDER—Continued

CLASS	Title of Class
902	Electronic Funds Transfer
187	Elevators
227	Elongated-Member-Driving Apparatus
474	Endless Belt Power Transmission Systems and Components
229	Envelopes, Wrappers and Paperboard Boxes
371	Error Detection/Correction and Fault Detection/Recovery
37	Excavating
411	Expanded, Threaded, Driven, Headed, Tool-Deformed, or Locked-Threaded Fastener
92	Expansible Chamber Devices
149	Explosive and Thermic Compositions or Charges
168	Farriery
256	Fences
42	Firearms
182	Fire Escapes, Ladders, Scaffolds
169	Fire Extinguishers
43	Fishing, Trapping and Vermin Destroying
383	Flexible Bags
141	Fluent Material Handling, with Receiver or Receiver Coacting Means
137	Fluid Handling
303	Fluid-Pressure Brake and Analogous Systems
416	Fluid Reaction Surfaces (i.e., Impellers)
239	Fluid Sprinkling, Spraying and Diffusing
426	Food or Edible Material: Processes, Compositions and Products
99	Foods and Beverages: Apparatus
450	Foundation Garments
410	Freight Accommodation on Freight Carrier
44	Fuel and Related Compositions
110	Furnaces
261	Gas and Liquid Contact Apparatus
48	Gas, Heating and Illuminating
55	Gas Separation
409	Gear Cutting, Milling, or Planing

CLASS	Title of Class
935	Genetic Engineering: Recombinant DNA Technology, Hybrid or Fused Cell Technology and Related Manipulations of Nucleic Acids
33	Geometrical Instruments
65	Glass Manufacturing
294	Handling, Hand and Hoist-Line Implements
54	Harness
56	Harvesters
165	Heat Exchange
432	Heating
237	Heating Systems
108	Horizontally Supported Planar Surfaces
368	Horology: Time Measuring Systems or Devices
405	Hydraulic and Earth Engineering
362	Illumination
382	Image Analysis
494	Imperforate Bowl, Centrifugal Separators
254	Implements or Apparatus for Applying Pushing or Pulling Force
376	Induced Nuclear Reactions, Systems and Elements
336	Inductor Devices
373	Industrial Electric Heating Furnaces
123	Internal-Combustion Engines
63	Jewelry
277	Joint Packing
403	Joints and Connections
289	Knots and Knot Tying
280	Land Vehicles
278	Land Vehicles, Animal Draft Appliances
296	Land Vehicles, Bodies and Tops
298	Land Vehicles, Dumping
301	Land Vehicles, Wheels and Axles
69	Leather Manufactures
122	Liquid Heaters and Vaporizers
210	Liquid Purification or Separation
70	Locks

CLASSES ARRANGED IN ALPHABETICAL ORDER—Continued

CLASSES ARRANGED IN ALPHABETICAL ORDER—Continued

CLASS	Title of Class
355	Photocopying
354	Photography
358	Pictorial Communication; Television
285	Pipe Joints or Couplings
138	Pipes and Tubular Conduits
475	Planetary Gear Transmission Systems and Components
47	Plant Husbandry
111	Planting
PLT	Plants
264	Plastic and Nonmetallic Article Shaping or Treating: Processes
425	Plastic Article or Earthenware Shaping or Treating: Apparatus
419	Powder Metallurgy—Processes
60	Power Plants
100	Presses
290	Prime-Mover Dynamo Plants
283	Printed Matter
101	Printing
422	Process Disinfecting, Deodorizing, Preserving or Sterilizing, and Chemical Apparatus
623	Prosthesis (i.e., Artificial Body Members), Parts Thereof or Aids and Accessories Therefor
375	Pulse or Digital Communications
417	Pumps
150	Purses, Wallets, and Protective Covers
250	Radiant Energy
430	Radiation Imagery Chemistry—Process, Composition or Product
213	Railway Draft Appliances
258	Railway Mail Delivery
105	Railway Rolling Stock
246	Railway Switches and Signals
295	Railway Wheels and Axles
104	Railways
238	Railways, Surface Track

CLASS	Title of Class
220	Receptacles
503	Record Receiver Having Plural Leaves or a Colorless Color Former, Method of Use or Developer Therefor
346	Recorders
62	Refrigeration
235	Registers
152	Resilient Tires and Wheels
404	Road Structure, Process and Apparatus
901	Robots
418	Rotary Expansible Chamber Devices
415	Rotary Kinetic Fluid Motors or Pumps
464	Rotary Shafts, Gudgeons, Housings and Flexible Couplings for Rotary Shafts
109	Safes, Bank Protection and Related Devices
234	Selective Cutting (e.g., Punching)
437	Semiconductor Device Manufacturing: Process
225	Severing by Tearing or Breaking
112	Sewing
271	Sheet Feeding or Delivering
270	Sheet-Material Associating
413	Sheet Metal Container Making
114	Ships
116	Signals and Indicators
241	Solid Material Comminution or Disintegration
206	Special Receptacle or Package
75	Specialized Metallurgical Processes, Compositions for Use Therein, Consolidated Metal Powder Compositions,
267	Spring Devices
365	Static Information Storage and Retrieval
249	Static Molds
52	Static Structures, e.g., Buildings
428	Stock Material or Miscellaneous Articles
125	Stone Working
126	Stoves and Furnaces
127	Sugar, Starch and Carbohydrates

CLASS	Title of Class	CLASS	Title of Class
505	Superconductor Technology-Apparatus, Material, Process	57	Textiles, Spinning, Twisting and Twining
248	Supports	139	Textiles, Weaving
312	Supports, Cabinet Structures	374	Thermal Measuring and Testing
211	Supports, Racks	131	Tobacco
128	Surgery	132	Toilet
600	Surgery	173	Tool Driving or Impacting
604	Surgery	81	Tools
606	Surgery	291	Track Sanders
520	Synthetic Resins or Natural Rubbers—Part of the Class 520 Series	212	Traversing Hoists
521	Synthetic Resins or Natural Rubbers—Part of the Class 520 Series	190	Trunks and Hand Carried Luggage
522	Synthetic Resins or Natural Rubbers—Part of the Class 520 Series	334	Tuners
523	Synthetic Resins or Natural Rubbers—Part of the Class 520 Series	82	Turning
524	Synthetic Resins or Natural Rubbers—Part of the Class 520 Series	199	Type Casting
525	Synthetic Resins or Natural Rubbers—Part of the Class 520 Series	276	Type Setting
526	Synthetic Resins or Natural Rubbers—Part of the Class 520 Series	400	Typewriting Machine
527	Synthetic Resins or Natural Rubbers—Part of the Class 520 Series	27	Undertaking
528	Synthetic Resins or Natural Rubbers—Part of the Class 520 Series	171	Unearthing Plants or Buried Objects
455	Telecommunications	251	Valves and Valve Actuation
178	Telegraphy	293	Vehicle Fenders
379	Telephonic Communications	98	Ventilation
135	Tents, Canopies, Umbrellas and Canes	333	Wave Transmission Lines and Networks
87	Textiles, Braiding, Netting and Lace Making	177	Weighing Scales
26	Textiles, Cloth Finishing	166	Wells
19	Textiles, Fiber Preparation	305	Wheel Substitutes for Land Vehicles
68	Textiles, Fluid Treating Apparatus	157	Wheelwright Machines
38	Textile, Ironing or Smoothing	231	Whips and Whip Apparatus
66	Textiles, Knitting	242	Winding and Reeling
28	Textiles, Manufacturing	245	Wire Fabrics and Structure
		140	Wireworking
		142	Wood Turning
		217	Wooden Receptacles
		144	Woodworking
		269	Work Holders
		378	X-Ray or Gamma Ray Systems or Devices

DESIGN CLASSES ARRANGED IN ALPHABETICAL ORDER

CLASS	Title of Class
D30	Animal Husbandry
D2	Apparel and Haberdashery
D22	Arms, Pyrotechnics, Hunting, Fishing, and Trapping Equipment
D4	Brushware
D25	Building Units and Construction Elements
D29	Devices and Equipment Against Fire Hazards, for Accident Prevention and for Rescue
D1	Edible Products
D23	Environmental Heating and Cooling, Fluid Handling and Sanitary Equipment
D7	Equipment for Preparing or Serving Food or Drink not Elsewhere Specified
D13	Equipment for Production, Distribution or Transformation of Energy
D6	Furnishings
D21	Games, Toys, and Sports Goods
D11	Jewelry, Symbolic Insignia and Ornaments
D26	Lighting
D15	Machines, Not Elsewhere Specified
D34	Material or Article Handling Equipment

CLASS	Title of Class
D10	Measuring, Testing or Signalling Instruments
D24	Medical and Laboratory Equipment
D99	Miscellaneous
D17	Musical Instruments
D19	Office Supplies, Artists' and Teachers' Materials
D9	Packages and Containers for Goods
D28	Pharmaceutical, Cosmetic Products, and Toilet Articles
D16	Photography and Optical Equipment
D18	Printing and Office Machinery
D14	Recording, Communication or Information Retrieval Equipment
D20	Sales and Advertising Equipment
D5	Textile, or Paper Yard Goods; Sheet Material
D27	Tobacco and Smokers' Supplies
D8	Tools and Hardware
D12	Transportation
D3	Travel Goods and Personal Belongings
D32	Washing, Cleaning, or Drying Machine

	Class	Subclass

	Class	Subclass
Catalyst regeneration with	502	27
Organic	502	28
Containers	206	524.1+
Detergents containing	252	142+
Electrolytic synthesis of	204	103+
Organic	562	
Carboxylic halides	562	840
Halides	562	800+
Higher fatty electric discharge treatment	204	167
Higher fatty electromagnetic wave treatment	204	157.6
Hydrocarbon purification using	585	866
Phosphate fertilizer	71	33+
Acidophilus Milk, Cream, Buttermilk	426	71
Acorn Type Electronic Tube	313	
Acoustics (See Sound)	181	
Building construction	52	144+
Cells		
Altering optical element	350	358
Isolation of motion picture housing and supports	352	35
Coupler		
Combined telephone-phonograph	379	78
Design	D14	242
Oscillators	331	155
Electrical, wave systems & devices	367	
Airborne shock-wave detection	367	906*
Collision avoidance	367	909*
Coordinate determination	367	907*
Doppler compensation systems	367	904*
Material level detection	367	908*
Noise reduction in nonseismic receiving system	367	901*
Particular well-logging apparatus	367	911*
Portable sonar devices	367	910*
Side lobe reduction or shading	367	905*
Sonar time varied gain control	367	900*
Speed of sound compensation	367	902*
Transmit-receive circuitry	367	903*
Energy measuring	73	861.25+
Fluid pressure leak detector	73	40.5 A
Holography, vibration measuring	73	603
Measuring instruments	73	570+
Stock material	181	284+
Target for golf	273	181C
Transducer making	29	594
Transducer making	29	602.1
TV		
Mechanical-optical scanning by acoustic wave	358	201
Picture projection device	358	235
Wall construction	52	144+
Acridine	546	102+
Anthrone or anthraquinone nuclei	546	
Azo compounds	534	753+
Acridone Dye	8	662
Acriflavine	546	104+
Acrilan T M (polyacrylonitrile Copolymers)	526	342
Acrolein		
Polymers (See also synthetic resin or natural rubber)	526	315
Acroncines	546	62+
In drug	514	285
Acrylon T M (polyacrylonitrile Methyl Methacrylate)	526	329.4
Acrylonitrile (See also Synthetic Resin or Natural Rubber)	526	341
Actinic Glass	501	900*
Developing by heat	65	30.1+
Actinide Series Metals (See Radioactive)		
Alloys	420	1+
Compositions	252	625
Electrolysis	204	1.5
Heat treatment	148	132
Metallurgy	75	393+
Nuclear fuels	376	409+
Organic compounds	534	11+
Actinometer	356	213+
Actinomycetales	435	169
X-art collections	435	822*
Actinospectacin	549	361
Activated Ergosterol	552	653
Activating Carbon	502	180+
Activation		
Lamp filaments	445	6
Of catalyst before use in hydrocarbon synthesis	585	906*
Vitamins in foods	426	72+

	Class	Subclass
Active Solid State Devices (See Transistor)	357	
Actuated and Actuation (See Type of Device Actuated as Door Gate Valve, etc or Means Employed to Actuate as Cam, Gear, Motor Pawl, Speed Responsive Device, Spring Thermostat)		
Acupuncture Test, Electric	128	735
Acupuncture, X-art	128	907*
Acyclic Hydrocarbon (See Diolefin; Olefin; Parafin; Triple Bond)		
Acyloins	568	303+
Adamantane Synthesis	585	352
Adapter		
Electrical connector socket	439	638+
Pipe joint	285	176+
Adding		
Dispenser cycle totalizer	222	36+
Machine	235	
Design	D18	3+
With recorder	235	58R+
Pencil	235	64
Addressing Machine	101	47+
Address printers	D18	13+
Plates	101	369+
Adducts		
Stabilizing an enzyme by	435	188
Urea	564	1.5+
Adenine	544	277
Adenoid Removal Devices	606	110
Adhesive		
Applying		
Stacking apparatus with means to apply adhesive to articles	414	904*
To pluck fowls	17	11.1 A
With laminating	156	
Bonding (See extensive digest list)	156	
Apparatus	156	346+
Methods	156	60+
Brackets attached by	248	205.3+
Carrier component having adherent surface	224	901*
Compositions		
Alkali metal silicate	106	600+
Biocide with	424	
Carbohydrate gum	106	205+
Cellulose liberation liquor	106	123.1
Core oils	106	38.2+
Protein	106	124+
Rubber ctg (See synthetic resin or natural rubber)	520	1+
Starch	106	210+
Synthetic resin ctg (See synthetic resin or natural rubber)	520	1+
Anerobic	523	176
Dental	523	116
Optical cement	523	168
Surgical	523	118
Vermin catching, trapping	424	77
Digest	24	DIG. 11
Fastener for diaper	604	389+
Insect catching and destroying	424	77
Insect trap	43	114+
Jewelry	63	DIG. 1
Manifolding digest	282	DIG. 2
Metal working involving	29	DIG. 1
Moisteners	118	
Envelope sealing combined	156	441.5+
Process of forming	427	207.1+
Separator for sheet feeding	271	33
Tape	428	261+
Coated on both sides	427	208
Coated on outermost layer	428	343
Coating process	427	207.1
Holder with edge for tearing	225	6+
Laminated	428	343+
Removable layer	428	40+
Rolls	206	411
Testing	73	150A
Toys, detachably adhesive	446	85+
X-art, special receptacle or package	206	813*
Adipic Acid	562	590
Adjuncts		
Static mold	249	205+
Sink head or hot top	249	202
Adjustable (See Device to be Adjusted)		
Adrenalin T M	564	365
Medicine containing	424	520+
Adrenochrome	548	512
Adrenocorticotrophin Acth	530	306

	Class	Subclass
Adrenosterone	552	621
Adsorbent (See Sorbent)		
Advancing		
Device for tool	173	141+
Automatic	173	4+
Cyclic self acting	173	19
Impacting device combined	173	112+
Material of indeterminate-length	226	
Advertising		
Aerial	40	212+
Calendars	40	107
Card support	D6	512
Design	D20	10+
Display cards	40	124.1
Mirror or reflector devices	350	600+
Mounting bracket for sign	D8	354+
Printed matter	283	56
Show boxes and cards	D20	40
Sign exhibiting	40	
Skywriting	40	213
Telegraphophones & annunciators	369	
Vehicle body	296	21
Adzes	30	308.1
Design	D8	76
Aeration		
Aerator	D23	213+
Agitation mixers	366	101+
Bait container	43	57
Dispensing with gas agitating	222	195
Distillation separatory		
Distillate	202	203
Vapor	202	201
Faucet attachment for	239	428.5
Digest	261	DIG. 22
Food, cereal puffing	99	323.4
Liquid purification		
Contact surface means	210	150+
Process, biological	210	620+
Oxidation	210	758+
With flotation	210	703+
Submerged fluid inlet	210	220+
Oxygen transfer technique	435	818*
Pumping by aerating liquid column	417	108+
Sewage process	210	620+
Diffusers	261	DIG. 7
Rotating	261	DIG. 71
Aerial (See Antenna)		
Bombs	102	382+
Cableway	104	112+
Camera	354	65+
Films, radiation imagery	430	928*
Fire fighting apparatus	244	136
Photo instruments	33	1A
Sighting	33	229+
Aerogram Design	D19	2
Aerometer	73	30.1+
Aeronautics	244	
Aeroplane (See Aircraft)	244	
Aerosol T M	252	305
Aerostatic Exhibiting Devices	40	212
Skywriting	40	213
Aerosteam Engines	60	200.1+
Aesculin	536	18.1
Affixing Apparatus		
Label pasting and paper hanging	156	
Sheet metal container making	413	
Afghan Blanket	D6	603
African Violets	PLT	69
Afterburner, Reaction Motor with	60	261
Aftertreatment of Polymers		
Chemically from ethylenic monomers only	525	326.1+
Physically	528	480+
Agar	536	3
Age Hardening		
Ferrous alloys	148	12.3
Ferrous alloys	148	142
Nonferrous alloys	148	12.7 R
Nonferrous alloys	148	158+
Permanent magnet material	148	102
Agglomerating		
Apparatus	425	222
Inorganic compounds	23	313R+
Metal powder or scrap	425	222
Ores	75	746+
Preventing		
Agents	252	381+
Comminution combined	241	16
Aggregate and Pellet	425	DIG. 101
Agitating (See Compacting, Jarring, Jogging, Mixer, Stirrer, Vibrators)	366	

	Class	Subclass
Explosive	149	
Organic compounds	260	
Oxides	423	635
Peroxides	423	583
Phosphates	423	305+
Pigments fillers or aggregates	106	461+
Pyrometallurgy	75	605+
Refractory compositions ctg	501	123+
Sulphates	423	554+
Hydroxide preparation	423	641
By electrolysis	204	98+
Metal		
Alloys	420	400
Lead	420	564
Carbonates	423	419R+
Chemical agents containing	252	192
Chemical agents for absorbing	252	193
Crystallization of compounds	23	302R+
Electrolytic synthesis	204	68
Electrolytic synthesis of	204	98
Electrolytic synthesis of hydroxides and oxides	204	98+
Electrolytic treatment of hydroxides and oxides	204	153
Halides	423	462+
Hydroxides and oxides	423	592
Inorganic compound recovery	423	179
Mercury alloys by electrolysis	204	125
Misc. chemical manufacture	156	DIG. 71
Nitrates	149	
Explosive containing	401	118+
Organic compounds	260	
Oxy-halogen salt	423	472
Explosive	149	77
Phosphates	423	304+
Preparation	75	745
Pyrometallurgy	75	589
Solvents, sodium chloride & potassium chloride	422	902*
Sulphates	423	551
Nitrates	423	
Organic	260	
Oxids		
Packing	206	524.1+
Secondary battery		
Chromium cobalt or cadmium	429	
Iron	429	
Nickel	429	
Alkaline Earth		
Carbonates	423	430+
Metals & cds	156	DIG. 78
Oxides and hyclroxides	423	635
Pyrometallurgy	75	605+
Silicates	423	326
Sulfates	423	554+
Sulfides	423	554
Alkaloids		
Biocides containing	424	
Cinchona	546	134+
Ergot	546	67+
Medicines containing	424	
Opium	546	44+
Solanum	546	124
Tobacco	546	282
Undetermined constitution	424	195.1
Alkamine		
Acyclic	564	503
Alkanoic esters	560	253
Monocyclic carboxylic acid esters	560	110
Monocyclic carboxylic acid esters	560	106+
Nitro carboxylic acid esters	560	20+
Nitro carboxylic acid esters	560	156
Alkane (See Hydrocarbon; Parafin)	585	700+
Alkenes (See Olefines)		
Alkines (See Acetylene; Triple Bond Hydrocarbon)		
Alkyd Resins (See Synthetic Resin or Natural Rubber)		
Alkylate Detergent Synthesis	585	455+
Alkylation		
Acyclic hydrocarbons	585	709+
Alicyclic hydrocarbons	585	375
Aromatic hydrocarbons	585	446+
Mineral oil	208	46+
Phenols	568	780+
Allantoin	548	308
In drug	514	390
Allochiral Arrangements		
Bulb or tube special package	206	422
Closure actuators	49	299
Closure link system	49	111+

	Class	Subclass
Tool with series of teeth	83	852
Tool-face, metal deforming	72	198
Alloy		
Age or precipitation hardened	148	405
Amorphous	148	403
Bearing compositions	252	12+
Compositions	420	
Making when casting	164	55.1+
Continuously casting	164	473
Deformation of	72	700
Dispersion strengthened	75	951*
Electrodes for electrolytic apparatus	204	293
Electrolytic		
Coating from aqueous bath	204	43.1+
Synthesis from aqueous bath	204	123+
Synthesis from fused bath	204	71
Treatment	204	140+
Electron emissive cathode of	420	
Ferrous (See ferrous alloys)	420	
Iron	420	8+
Loose metal particles	75	255+
Mechanical memory	148	402
Metal treatment	148	
Metallurgy	420	
Powder metallurgy		
Processes	419	
Products	75	228+
Pyrophorous	420	416
Railway surface track	238	150
Shape memory	148	402
Alloying		
Fusing dopant with substrate	437	134+
Alpha		
Beta gamma survey meter	250	336.1
Scintillation type	250	361R+
Hand counter	250	374
Proportional counter	376	153+
Similar counter tubes	313	93
System	250	374
Scintillation detector		
Crystal mounting	250	361R
Electric	250	361R
Non-electric	250	483.1
With sample holder	250	361R+
Survey meters	250	374
Electric	250	361R
Nonelectric	250	483.1
Scintillation type	250	361R+
Alpha Globulins	530	392
Alphabet		
Block	434	172
Design, type	D18	24+
Letter training	434	159+
Playing cards	273	299
Specialized font	382	
Stencil	101	127+
Alphanumeric		
Display		
By matrix	340	748+
Cathode ray tube	340	756+
Altars	81	
Design	D 6	329+
Alternating Current (See Type of Equipment Using))		
Converters to direct current	363	
Electromagnet with armature	335	243+
Electromagnetic switches	335	99+
Generator	310	159+
Prime mover driven	290	5
Lamp or electronic tube	315	246+
Meter	324	137+
Motor	310	159+
Polyphase to arc lamp	314	31
Polyphase to lamp or electronic tube	315	137+
System		
Telautograph	178	19
Telegraph	178	66.1+
Transmission	307	
Alternating Mechanical Motions (see Gearing)	74	
Alternator Structure	310	
Altimeter (See Altitude Instrument)	D10	70
Electrical echo systems	367	87+
Radio wave systems	342	120
Altitude Instrument		
Barometric	73	384+
Design	D10	70+
Directional radio	342	462
Light ray type	33	282
Aerial bomb sight	33	229
Optical	356	3+

	Class	Subclass
Solar position locator	33	268+
Telescopes	350	537+
Vertical and horizontal angle	33	281
Vertical angle	33	282+
Passive sonic or supersonic object detection type	367	118+
Radar type	51	229+
Radiant energy	342	120+
Radiosonde	340	870.1+
System	340	870.1
Sonic or supersonic echo system, ie active sonar	367	87
With position indicating	342	462
Alto		
Wind instrument (See instrument)		
Alum	423	544+
Alumina	423	625
Refractory containing	501	127+
Magnesium compound combined	501	118+
Aluminates	423	592+
Aluminosilicate		
Alicyclic compound synthesis	585	475
By isomerization	585	481+
Compounds with plural metals	423	118
Saturated compound synthesis	585	722
Saturated compound synthesis	585	739
Silicon and silicon compounds	423	328+
Unsaturated compound synthesis	585	533
Unsaturated compound synthesis	585	666
Aluminum		
Alloys	420	528
Copper	420	529
Compounds		
Aluminates	423	592
Azo compounds containing	534	692+
Azoles containing	548	101+
Carbocyclic or acyclic compounds containing	556	170+
Diarylene ortho diketones containing	552	209
Diazine containing	544	225
Diazo compounds containing	534	562+
Halides	423	462
Higher fatty acids containing	260	414
Inorganic	423	
Misc. chemical manufacture	156	DIG. 61
With 3 &/or 4 compounds	156	DIG. 104
Nitrides	423	409
Organic	556	170+
Chelates	556	175
Clay or zeolite containing	556	173
Salicylates	556	175
Sesquihalides	556	180
Oxazine containing	544	64
Oxides	423	625
Proteins containing	530	400+
Pyridine nucleus compounds containing	546	2
Sulphates	423	544+
Thiazine containing	544	4
Triarylmethanes containing	552	103
Electrolytic coating of	204	58
Cleaning or etching combined	204	33
Electrolytic synthesis	204	67
Electrothermic processes	75	10.1+
Heat treatment	148	159
Metal working using	29	DIG. 2
Pigments	106	404
Plate	428	650+
Pyrometallurgy	75	671+
Working & heat treatment	148	11.5 A
Including age hardening	148	12.7 A
Amalgam Mixer, eg Dental Filling	366	602*
Amalgamator (See Mercury)		
Coated surface adhesion	209	50+
Mercury suspension	209	174+
Other separators combined	209	12+
Selective differentiation	209	48
Amantadine, in Drug	514	656
Amberlite T M (ion Exchange Resins)	521	25+
Ambulance	296	19+
Ambulatory Pneumatic Splints	128	83.5
Amides	564	
Higher fatty acid	260	404+
Sulphonated	260	401
Amidines	564	225+
Aromatic	564	244+
N prime-aryl formamidines	564	245
Polyamidines	564	243
Amidino Hydrazines or Hydrazones	564	226+
Amidoximes	564	229

	Class	Subclass
Antikink Devices		
Appliance cords		
Extensible cords	174	69
Springs	267	74
Antiknock		
Cyclomatic compounds	556	1+
Engine accessories	123	198A
Exhaust gas treatment	422	168+
Fuel combined with antiknock agent	44	300+
Gaseous compositions	252	373
Measuring knock	73	35
Tetraalkyl lead	556	95
With preservative or stabilizer	556	3+
Antileak Joint	105	424
Antimagnetic Timepiece	368	293
Antimicrobial Activity		
Fermentative analysis or testing	435	32
Antimony		
Alloys	420	576
Antimonates	423	592+
Carbocyclic or acyclic compounds containing	556	64+
Halides	423	462+
Misc. chemical manufacture	156	DIG. 94
Oxides	568	477
Pigments fillers or aggregates	106	455
Pyrometallurgy	75	703+
Sulphides inorganic	423	511
Antinoise		
Bed	5	309
Closure buffer	16	82+
Cranks and wrist pins	74	604
Ear shields	128	864+
Flywheels and rotors	74	574
Gears	74	443
Land vehicle		
Animal draft gear	278	61+
Axle connection	280	138
Tongue	280	108
Levers and linkage	74	490
Lock mufflers	70	463
Machinery support		
Bases	248	678+
Bracket	248	674+
Suspension	248	637+
Resilient	248	610+
Motor support vehicle	248	560+
Mufflers	181	212+
Pawl and rachet	74	576
Railway		
Rolling stock	105	452
Track	238	382
Wheels	295	7
Reaction motor nozzle	239	265.13
Rotary motors or pump	415	119
Supports	248	562+
Typewriter		
Muffler	400	689+
Platen	400	661+
Antioronze Agent or Process	430	929*
Antioxidant		
Compositions	252	397+
Edible oils and fats	426	601+
Fats, fatty oils, fatty acids, ester type waxes	260	398.5
Polymers treated (See class 523, 524)		
Solid polymers containing (See synthetic resin compositions)		
Antipanic Door Bolts	292	92+
Antipecking Poultry Restraint	119	97.3
Antiphone Ear Sound Stoppers	128	864+
Antipick Locks	70	419+
Antipicking Poultry Restraint	119	97.3
Antipyrine	548	368
Antirachitic Vitamin		
D compound	552	653
Foods and processes	426	
Medicines	514	167+
Antirattle	16	DIG. 6
Lock	70	463
Keeper	292	341.12+
Window	292	76
Car	49	428+
Antireflection Films	350	164+
Antiricochet Projectiles	102	398
Antiroosting Poultry Restraint	119	903*
Antiscorbutic Vitamin		
C compound	549	315
Foods and processes	426	72+
Medicines	514	474

	Class	Subclass
Antiseptics (See Disinfection;		
Preserving)	422	
Compositions	424	
Detergents combined	252	106+
Medicators	604	19+
Water softener combined	252	175+
Antiside Draft Device (See Equalizer)		
Animal draft	278	3+
Harvesters	56	321
Antisiphon for Fluid Handling Systems	137	215+
Antislip or Antiskid		
Compositions, coating or plastic	106	36
Horseshoe calks	168	29+
Ladder terminals and feet	182	107+
Overshoes	36	7.6+
Pavements	404	19+
Block surface	404	19+
Shoe		
From foot	36	58.5
Tread	36	59C
Supports		
Illumination	362	389
Tires	152	208+
Armored	152	167+
Track sanders	291	
Traction mats	238	14
Antismoking Product or Device	131	270
Antismut		
Compositions, coating repellent	106	2
Printing combined	101	416.1+
Slip sheet		
Apparatus	34	94
Processes	34	6
Antisnoring		
Body restraints	128	871
Chin supporters	128	164
Mouth restraints	128	859+
Antispasmodics	424	
Antisplash		
Vehicle mud guards	280	847+
Antistatic		
Electrically conductive coating compositions	252	500+
Polymers treated (See class 523, 524)		
Process for synthetic fibers	57	901*
Antitheft Devices, eg Locks	70	
Alarms	340	500+
For vehicle	340	426+
Vehicle ignition disabling	307	10.3+
Antithumb Sucking	128	857+
Antitoxins	424	85.1+
Syringes	604	187+
Antivermin	43	124+
Animal treatment	119	156+
Perches fowl	119	25
Antivibration (See Antinoise)		
Anvil	D 8	46
Assembling and disassembling	29	283
Blacksmith	72	476+
Bolt and rivet making	10	22
Hammer attached	72	407+
Hammer or press	72	462+
Hollow rivet setting	227	61+
Machine combined	29	560+
Nail extractor	254	19
Pile or post	173	128
Saw setting	76	73
Saw swaging	76	57
Tool driving	173	128
Vises combined	72	457+
Apa-6	540	312
Apartments	52	234+
Rolling partitions	160	351
Hanging or drape type	160	350
Aperture Card	493	944*
Apiaries	449	
Apparel (See Type of Article Worn;		
Clothes)	2	
Adornment, attachable	D11	
Apparatus	223	
Boot and shoe making	12	
Knitting	66	
Patterns	33	2R+
Sewing	112	
Weaving	139	
Blouse, shirt	D 2	208+
Boots shoes and leggings	36	
Design	D 2	264+
Making	12	
Cape, stole, shawl	D 2	179+

	Class	Subclass
Catamenial and diaper	604	358+
Design	D24	50
Ceremonial robe	D 2	79
Cleaning		
Brushing devices	15	
Ironing or pressing	38	
Methods	8	137+
Washing devices	68	
Coat, jacket, vest	D 2	183+
Collar making	83	901*
Containers for	206	278+
Making	493	938*
Making garment bag	493	935*
Making with support	493	939*
Costume	D 2	79
Design	D 2	1
Hosiery	D 2	329+
Negligees	D 2	12+
Dress or suit	D 2	49+
Female support undergarments	450	1+
Fire fighters	169	48+
Footwear	D 2	264+
Hand coverings	D 2	610+
Headwear	D 2	509+
Knitted	66	169R+
Layout and measurement means	33	403+
Footwear	33	4+
Garments	33	11+
Leather manufactures	69	
Light thermal electrical treatment	128	379+
Lingerie	D 2	1+
Locks for	70	59
Marking & measuring instruments	33	2R+
Neckwear	D 2	600+
Respirators	128	200.24+
Skirt	D 2	223+
Stockings, socks	D 2	329+
Swimming suit	D 2	40+
Travel bags for	206	278+
Undergarments	D 2	1+
Vestment	D 2	79
Appendage	425	DIG. 3
Apple Trees	PLT	34
Appliances		
Cooking	D 7	323+
Food preparation, household	D 7	
Major household (See types)	D15	
Applicator		
Brushing	15	
Material supply combined	401	268+
Coating by absorbent	427	429
Coating by brush	427	429
Coating or cleaning implement	15	104R+
Material supply combined	401	
Daubing	15	
Dispenser combined	141	110+
Electrical	128	783+
Electrolytic	204	224R
Hair dye	D28	7+
Kinesitherapy	128	24R+
Light	128	395+
Loaders	15	257.5+
Material supply combined	401	
Medicators	604	19+
Medicinal	D24	63
Polishing	15	
Material supply combined	401	
Powder compact	132	293+
Pad overlying material including	401	130
Separable from combined supply container	401	118+
Thermal	128	399+
Wiping	15	
Appliers (See Applicator)		
Abrading sheet	51	275
Antiskid device	152	213R+
Boot and shoe cement	118	410+
Boot and shoe lining	12	39
Box bail	493	88
Box end enclosure	493	102+
Cigar and cigarette making	131	88+
Hose clamp	81	9.3+
Label pasting and paper hanging	156	
Metallic leaf	156	540+
Receptacle closure	53	287+
Sealing wax	401	1+
Dispenser with heating means	222	146.2
Dispenser with illuminator or burner	222	113
Shaped cosmetic	132	320
Sheet metal container making components	413	

	Class	Subclass
Caps and cartridges	86	12
Chain links	59	25
Chains	59	1+
Chains sprocket	59	7
Cleaning combined	134	59
Electrostatic type	29	900*
Liquid purification	210	232
Matrices type casting	199	9+
Nailing and stapling combined	227	19+
Press type filter	210	230
Supporters garment	223	49
Tire and rim	157	
Welding combined	228	4.1+
Bolts nuts and washers	10	155R+
Boxes		
Closure cap and liner	413	56+
Laminating	156	
Electric discharge devices	445	23+
Hinges	29	11
Lamps and vacuum tubes	445	23
Apparatus	445	60+
Processes	445	1+
Mold parts	164	339+
Processes		
Batteries	29	623.1+
Electrostatic type	29	900*
Receptacle with closure	413	2+
Apparatus	413	26+
Sheet material	270	
Shoe heels	12	50+
Shoe uppers	12	52
Textile fibers	19	144+
Well	166	373+
Assembly Means		
Fluid handling	137	315
Pipe joint	285	18
Associating Sheet Material	270	
Tobacco leaf	131	327
Assorter (See Associated Process, Article, or Machine)		
Brush making	300	18
Cleaning machines combined	15	3+
Coin	453	3+
Livestock	119	155
Printing machines combined	101	
Solid material	209	
Astrolabe	33	1R
Astrology	434	106
Astronomical Instrument		
Clock	368	15+
Globes	33	268+
Optical	356	138+
Solar position locator	33	268+
Telescopes	350	537+
Tellurions	33	268+
Astronomy Teaching	434	284+
Astrophysical Instruments		
Angle measurement	33	281+
Solar	33	268+
Spectroscope	356	300+
Telescope	350	537+
Astygmatism		
Ophtahlmic lenses for correction	351	176
Test chart	351	241
Athletic		
Equipment (See type)		
Mens shoes	D 2	309
Protective hand covering	D29	20+
Shoes, making	12	142P
Shorts	D 2	42
Atmometer	374	39+
Atmosphere		
Special atmosphere	156	DIG. 89
Utilizing special		
Glass making	65	32.1+
Refrigeration	62	78
Atmospheric Electricity	307	149
Chemically inert or reactive	425	815*
Collecting	310	308+
With condenser	320	1
Discharging	361	212
Grounding	174	6+
Lightning rods	174	2+
Atomic Energy	376	
Fission reactions	376	349
Fusion reactions	376	100
Power utilization	376	336
Power utilization	376	402
Power utilization	376	370
Power utilization	376	391
Transmutation reactions	376	156

	Class	Subclass
Well processes	166	247
Atomic Hydrogen Welding	219	75
Atomizer	239	
Container or bottle with	D28	91.1
Electrically heated	392	386+
Electrically heated	392	324+
Fumigator	422	305+
Gas liquid contact	261	78.1
Non-carburetor	261	78.2
Internal combustion engine supply	123	434+
Liquid fuel burner	239	
Medical	D24	62
Medicator	128	200.14+
Perfume	239	355+
Sprayer	239	337+
Temperature responsive control	236	8
Atropine	546	131
In drug	514	304
Attack Defeating Locks		
Mechanism	70	1.5+
Key lock	70	416+
Attempering		
Canned food processors	99	367+
Conveyor combined	99	361
Attenuator	333	81R
Light conductor with photocell	250	227.11
Pads	333	24R+
Resistance structure	338	
Voltage magnitude control resistors	323	369
Attic		
Fans	416	
Inlet	98	39.1
Inlet and outlet	98	33.1
Outlet	98	42.2
Ventilators	98	29+
Attitude or Pitch Indicator	33	328+
Attrition Comminutor		
Apparatus	241	284
Suspended particles	241	39
Processes	241	26
Suspended particles	241	5
Auction		
Livestock tags	40	300+
Audience		
Radio survey	455	2
Reaction apparatus		
Loudness meter	73	646
Voting machines	235	51+
Television survey	358	84
Audio Amplifier	330	
Audio-visual Teaching Machine	D19	60
Audiometer	73	585
Design	D24	21+
Audion T M		
Amplifier, electronic tube type	330	
Cathode ray tube structure	313	364+
Miscellaneous systems	328	
Oscillation generators	331	
Structure	313	
Audiphone	381	68+
Amplifier for	330	
Auditorium	52	6
Acoustics	181	30
Auger (See Bit; Drill)		
Boring machines wood	408	199
Earth boring	175	394+
Land anchors	52	157
Manufacture		
Dies	72	470+
Machines	76	2+
Processes	76	102
Twisting stock	72	299
Mortising machines wood	144	69+
Post base	52	157
Wood	408	199+
Auramine	564	269
Auricle	181	129+
Aurines	552	115
Fuchsone	552	115
Autoclaves	422	242
Cooker (See steaming food)		
Autocollimator	33	297+
With photocell	356	152+
Autofrettage	29	1.11
Autogenous Bonding		
Of particles	264	123+
Of running length fibers	156	180+
Of self-sustaining laminae	156	308.2+
Autogenous Welding	228	196+
Electric	219	78.1+
Autogiro	244	17.11+

	Class	Subclass
Aerial toy	446	36+
Airplane combined	244	8
Autograph Register	D18	21
Autoharp T M	84	8+
Automatic (See Controller & Control)		
Automated machine		
Broaching	409	245+
Gear cutting	409	2+
Milling	409	80+
Planing	409	289+
Camera		
Exposure control	354	400+
Focusing	354	410+
Cutouts for electric circuit	200	
Electromagnetic	335	
For lamp circuit	315	
Dispenser	222	52+
With cutoff operator	222	14+
With recorder register or indicator	222	23+
Drying and gas or vapor contact with solids	34	43+
Earth boring means	175	24+
Fire alarms	116	101+
Electric	340	500+
Flame	340	577+
Heat	340	584+
Smoke	340	628+
With indicator	116	5
Fish butchering	17	54
Flexible partitions and panels eg curtains	160	1+
Food treating machine	99	486+
Glass making apparatus	65	160+
Grain hullers and scourers	99	488
Grapples	294	110.1+
Gun (See gun)		
Heat exchange	165	13
Heating and cooling	165	14
Heating systems	237	2R+
Metal deforming		
Mining control	299	1
Apparatus	72	6+
Musical instruments	84	2+
Comb type	84	94.1+
Electrical tone generation	84	600+
Stringed eg pianos	84	7+
Two different instruments	84	2+
Wind eg organs	84	83+
Nuclear reactions	376	207
Press control	100	43
Refrigeration	62	132+
Solid material comminution or disintegration	241	33+
Teller machine		
Using cryptographic code	380	24
Temperature or humidity regulation	236	
Cereal puffing apparatus	99	323.4+
Cooking apparatus	99	325+
Distillation separatory processes	203	2
Distillation separatory systems	202	160
Electronic tube or lamp	315	117
Humidor	312	31+
Hygrostats	73	335+
Infuser, eg percolator	99	281+
Liquid heaters and vaporizers	122	446+
Mineral oil distillation condensers	196	141
Mineral oil distillation vaporizors	196	132
Thermostats	60	527+
Water heater vessels	126	374
Tool driving or impacting device	173	2+
Volume control		
Amplifier coupling impedance	330	144+
Amplifier having of bias or power supply	330	127+
Amplifier thermal coupling impedance	330	143
In radio receiver	455	234+
Wave transmission lines & networks	333	17.1+
Weigher	177	60+
Cigar and cigarette making	131	280
Automation (See Robot)	901	1*
Automobile (See Land Vehicle; Motor Vehicle)		
Air bag passenger restraints	280	728+
Beds	5	118+
Cribs	5	94
Body crushers	100	901*
Carburetor (See carburetor)	261	
Compound repair tools	7	100+
Cooler	62	243

	Class	Subclass
Door handle	D 8	300+
Electric distribution systems	307	10.1+
Electric motors	D13	112
Fender covers	280	762+
Frame straightener	72	705
Headlight	362	61+
Design	D26	28+
Dimming system	315	82+
Electric system for	315	82+
Mutually responsive automatic dimmers	315	82+
Retractable	362	65
Sealed beam	313	113+
Heater	237	12.3 R
Heater and cooler	165	42+
Hood light	362	80
Horn	D10	116+
Insignia	D11	95+
License tags	40	200
Changeable year section	40	207+
Lift or hoist	269	58+
Lights combined with part	362	61+
Ceiling or roof light	362	74
Mirrors	350	600+
Plural	350	612+
Plural reflections	350	618+
Miscellaneous equipment design	D12	155+
Name plate fasteners	40	662+
Radio	455	345
Portable	455	346
Portable transmitter	455	99
Registration plates	D20	13+
Roll bars	280	756
Safety belt or harness	297	464+
Motor vehicle system responsive	180	268+
Passive	280	802+
Safety promoting means,	180	271+
Seats	297	
Crash	297	216
Design	D 6	334+
With body modification	296	63+
Shipped as freight	410	4+
Group of bodies shipped	410	43
Signals	116	28R+
Electric	340	425.5+
Lamp supply systems	315	77+
Stop electric	340	479+
Simulators	434	65+
Steering post or wheel		
Fluid handling	137	352
Lock for pilot wheel control or linkage	70	253
Suspension systems	280	688+
Stub axle mounts	280	660+
Tailpipe muffler	181	227
With exhaust pipe	181	228
Tape storage for cars	D 3	35
Toy	446	431+
Sounding	446	409+
Transmission jack	254	DIG. 16
Trunk light	362	61+
Switch	200	61.62+
Turntables	104	44
Ventilation	98	2+
Wheels and axles for land vehicles	301	
Aligning tools, hand	29	273
Lock for spare or mounted wheel	70	259
Window defrosting (See window)	52	171+
Autophon	84	83+
Autosyn (See Selsyn)		
Averaging, Aromatic Hydrocarbon Synthesis	585	474
Aviaries (See Bird)		
Awakener		
Bedstead related	5	131+
Clock	368	12
Alarm	368	244+
Person contacting	116	205
Time operated	368	12
Time alarm except clock	111	91+
Electric	340	309.15+
Awl	D 8	47
Leather	30	366
Needles	223	104
Punch type	30	366
Sewing		
Implement	112	169
Machine	112	48
Shoe making	12	103
Surgery	606	222+
Awning	160	45+

	Class	Subclass
Design, fabric type	D25	57
Head rod bracket	248	273
Non rigid structure	160	45+
Storage or shield combined	160	22
Rigid structure	52	74+
Design	D25	57
Window closing	49	71
Ax	30	308.1+
Design	D 8	76
Diaphragm	92	96+
Holder	224	234
Making		
Blanks	76	103
Dies	72	470+
Machines	76	7+
Processes	76	103
Rack	211	13+
Type cutlery	30	318
Axle		
Box		
Manufacture dies forging	72	356
Manufacture dies forging railway car	72	343+
Railway car	29	168
Railway car mounting	105	218.1+
Wheel combined	301	109+
Gauge	33	193
Housings	74	607
Lubricating jack	254	32
Manufacture		
Assembly or disassembly tools	29	700+
By metal deforming process	72	
Dies forging railway car	72	343+
Dies forging vehicle	72	356+
Lathes turning	82	104+
Puller	29	244+
Railway	295	36.1+
Removing device	29	277
Rethreading dies	408	20+
Rethreading dies	408	22+
Shafts	301	124R
Skein	301	134+
Spindle	301	131+
Vehicle	301	124R+
Design	D12	160
Vehicle wheel combined	301	1+
Axminster		
Carpet	139	399+
Making apparatus	139	7+
Azaporphyrins	540	121+
Azeotropes	203	50+
Azetididiones	540	356
Azetidines	548	950+
Azetidinones	540	200+
Azides	423	
Inorganic metal azides	149	35
Explosive or thermic containing	149	35
Organic radical containing	552	1+
Azidocillin	540	331
Azimuth Instrument		
Ammunition	102	372
Horizontal and vertical angle	33	281
Horizontal angle	33	285
Optical	356	138+
Solar locating	33	268+
Telescope	350	537+
Azines	544	1+
Azines	564	249
Azo compounds	534	751+
Heavy metal containing	534	701+
Aziridines	548	954+
Azo Compounds	534	573+
Anthraquinone containing	534	654+
Dye compositions containing	8	662+
Several dyes	8	639+
Textile printing	8	445+
Fiber-reactive	534	617+
Indane containing	534	659
Quaternary ammonium ctg.	534	603+
Resorcinol containing	534	682+
Salicylic acid containing	534	660+
Stilbene containing	534	689+
Azoles	548	100+
Acridine nucleus containing	546	26+
Aluminum containing	548	101+
Arsenic containing	548	102
Azo compounds	534	769+
Heavy metal containing	534	710+
Boron containing	548	110
Heavy metal containing	548	101+
Phosphorus containing	548	111+

	Class	Subclass
Silicon containing	548	110
Azomethine	564	271+
Benzoselenazoles	548	121
Benzothiazoles	548	152
Benzoxazoles	548	219
Heavy metal containing	556	32+
Pyridines	546	268
Quinolines	546	152
Azoxy Compound	534	566+
B		
B Naphthol	568	735+
Monoazo compounds	534	840+
Preparation from aryl halides	568	739
Baby		
Bath	D23	278
Bottle warmer	126	261+
Chemical reaction	126	263
Electric	219	429+
Carriage	280	47.38+
Convertible to cradle or crib	280	31
Folding	280	639+
Mosquito nets and canopies	5	416
Steering	280	47.11
Chairs	297	
Combined with table	297	136+
Convertible high and low	297	345+
Cradle	5	101+
Cribs	5	93.1+
Diapers	604	358+
Design	D24	50
Doorway safety guard	49	50+
Removable	49	463+
Fences	256	24+
Flatware for	D 7	642+
Food or bottle warmer	D 7	326
Harness (See definition notes)	119	96+
Highchair	D 6	339
Tray	D 6	509
Incubators	600	22
Jumpers	297	274+
Pacifier	606	234+
Design	D24	45
Pen	256	25
With floor	5	99.1
Playpen	D 6	331
Rattles	446	419
Design	D21	65
Rubber pants	2	400+
Absorbent pad holders	604	393+
Design	D 2	10
Safety garments	128	869+
Shoes		
Bronzing	204	20
Teething device	D24	45
Training toilet	D23	296+
Walker	280	87.51+
Vehicle	280	87.2
With seat	297	5+
Bacillus	435	832*
Hydrolase from	435	221+
Bacitracins	530	320
Back		
Pad harness	54	66
Rest	297	452+
Bed	5	70
Boat	114	363
Design	D 6	502
Scratcher	128	62R
Backband		
Harness	54	4
Backfire	261	DIG. 6
Backfire Preventer	48	192
Backgammon	273	248
Backgrounds Photographic	354	291
Backing Dental		
Instrument	433	141+
Making	29	160.6
Metalware shaping	72	54+
Backlash Take up		
Between meshing gears	74	409
Planetary gearing	475	346+
Sectional gear	74	440+
Milling work feeds	409	146
Backpack	224	153+
Backrest	D 6	502
Backspace on Typewriter	400	308+
Backstay on Car Tops	296	144
Backstop		
Baseball	273	26A
Projectile	273	410
Target	273	404+

	Class	Subclass
Backup Auto Lights	362	257+
Backwash		
Filter cleaning	210	108
Filter with	210	333.1
Fluid cleaning airpump	210	411
In multi-way valve	210	425
Sand bed with rehabilitation means	210	275+
With additional cleaner	210	393
Bacon	426	645
Packaging	53	DIG. 1
Preservation	426	332
Bacteria	435	
Fertilizer preparation with	71	6+
Liquid purification by	210	601+
Virus culture on	435	235+
Bactericidal		
Compositions	424	
Bacteriophage	435	235
Measuring or testing	435	5
Bacteriostatic		
Compositions	424	
Badge	40	1.5+
Design	D11	95+
Film radioactivity	250	475.2
Film radioactivity	250	472.1
Badminton (See Tennis)	273	411
Rackets	273	67R
Baffle (See Process or Machine with which Associated)	416	
Acoustical muffler structure per se	181	264+
Acoustical muffler with	181	264+
Amalgamator agitator	209	187
Animal muzzle	119	131
Cleaning and liquid contact with solids	134	182
Decanter	210	513+
Diverse separators	210	294+
Drop water	122	188
Fire tube boilers	122	44.1+
Fuel	122	503
Furnace	110	322+
Loud speaker	181	175+
Mercury coated for amalgamating	209	54
Particulate material separators	210	285+
Pipe and tube	138	37+
Railway dumping car	105	279
Tank gas and liquid contact	261	123
Tower gas and liquid contact	261	108+
Turbine stator	415	148+
Vane	415	208.1+
Vehicle body top board	296	33
Water tube boiler	122	235.17+
Bag (See Pouch; Receptacle; Sack)		
Abrading material filled	51	294
Airship gas cell construction and arrangement	244	128
Balloons	244	31+
Inflation	244	98
Belt or body attached	224	205
Straps crisscross shoulders	224	209+
Theft or loss resistant	150	102
Blood	604	403
Brief bag or case	190	900*
Closures	383	42+
Clothespin		
Design	D32	36
Package	D 9	305
Colostomy	604	338+
Explosive powder	102	282
Fabric	139	389+
Fastener	24	30.5 R+
Shirring cord	383	72
Traveling bag handle	16	110R
Feed	119	65+
Flexible	383	
Closures	383	42
Fountain syringe with tubing	206	69
Garment	206	278
Gas separator	55	361+
Golf	206	315.3
Holder	248	95+
Sorting racks as for mail bags	211	12
Hot water	383	901*
Ice	383	901*
Inflatable for raising vessels	114	54
Locks for	70	64+
Mouth support	141	391
Paper		
Closures	383	42+
Making of bag	493	186+
Opening	206	601+

	Class	Subclass
Pocketbook	150	100+
Closures	150	118+
Saddlebag	224	191+
Sewing machine	112	10
Filled bags	112	11
Filling and closing	53	139
Sleeping	2	69.5
With bed structure	5	413
Striking or punching	272	77+
Tea or coffee	426	77
Traveling	206	278+
Design	D 3	30.1+
Turner or reverser	223	39+
Wheeled traveling	280	37
Bagasse Furnace	110	235+
Baggage	190	
Cars	105	355+
Design	D 3	30.1+
Identification tags	40	6
Rack on vehicle, illumination	362	73
Sample case with terraced trays	190	16
Supports	190	18R
Trunks	190	19+
Vehicle attached carrier	224	273+
Wheeled	280	29+
With picnic or lunch unit	190	12R
Bail		
Applying to paper bag	493	88
Ear		
Metallic receptacle	220	91+
Wooden bucket	217	126
Handle		
Basket	217	125
Lantern	362	399
Metallic receptacle	220	94R+
Making or forming wire	140	75
Railway coupling link	213	75R+
Bailer		
Cistern	294	68.22+
Hoisting bucket type	294	68.22+
Well	166	162+
Bait		
Animal food and preparation	426	1
Food		
With poison	424	410
Glass drawing	65	352+
Holder	43	55
Poison	424	84
Baked Products	426	549+
Design	D 1	
Bakelite T M	528	129
Baker and Bakery		
Ovens	432	120+
Peels	294	49
Products	D 1	
Baking		
Molds	99	372+
Ovens domestic	126	273R+
Pans	220	
Balanced to Unbalanced		
Coupling network with frequency characteristic	333	25
In amplifier stage	330	116
Powder	426	562+
With phase inverter	330	116
Balancer	D10	82
Balancing method	29	901*
Beam scales	177	246+
Equal arm weigher	177	190+
Horology	368	169+
Making	29	178
Sash	16	193+
Balancing Machine Parts (See Counterbalance)		
Comminuting elements combined	241	292
Determining balance by computer	364	463
Flywheels and rotors combined	74	573R
Grindstone weights	51	169
Instrument calibrating	73	1R+
Rotor unbalance testing	73	66
Weights for land wheel	301	5R
Wheels vehicle	301	5R
Balata (See Rubber)		
Balconies	52	73
Window connected	182	53+
Bale		
Band tightener and twister	140	93.2+
Portable & detachable tensioner	254	199+
Tightener and sealer	140	150+
Grappling hooks	294	26
Round hay bale handling	414	24.5

	Class	Subclass
Tie	24	16R+
Design	D 8	394+
Making	140	73
Baling	100	
Ball		
Amusement devices	273	58R+
Design	D21	204+
And socket joints (See joints)		
Bearing for typewriter type bar	400	447
Dental tool support	433	64
Vehicle steering mechanism	180	258
Baseball curvers	273	28
Bearing		
Hinge	16	275+
Inking mechanism	400	270.2
Making	29	898.6+
Mounting for ball tool	401	212
Radial	384	452+
Retainer design	D15	143
Thrust	384	590+
Billiard ball spotting racks	273	22
Bowling		
Ball gauge	33	509+
Ball return	273	47+
Grip testing	33	510
Calculator	235	68
Carriers	224	919*
By hand	294	137+
Cash register check	235	18
Chaser grinder		
Frictionally driven	241	103+
Loose ball	241	173+
Clock pendulum	368	179
Cocks toilet tank	137	409+
Games	273	118R+
Golf ball making	156	146
Design	D21	204+
Making	29	899+
Abrading processes	51	289S
Glass	65	21.1+
Hollow metal spheres	72	348
Laminating	156	
Spheroid winding	242	3
Winding	242	2
Wooden balls	142	1
Mill	241	170+
Point pen	401	209+
Point stylus	81	9.2
Racks	211	14+
Pool table attached	273	10+
Sport or game balls	273	58R+
Tea	426	77+
Receptacle type	99	323
Tethered bowling ball	273	40
Time balls	116	200
Ball Cock	137	409
Ballast		
Aircraft	244	93+
Cleaning		
Excavating combined	171	16
Loose material separation	209	
Railway grading combined	37	104
Resistor	338	20+
Ship	114	121+
Tamper	104	10+
Ballers	425	332+
Kitchen hand tool	D 7	681
Balloon		
Aircraft	244	31+
Barrage, antiaircraft	89	36.16
Aerial mine carrying	102	409
Montgolfier type	244	31+
Pyrotechnic device combined	102	356
Railway car attached	104	22
Rocket combined	102	347
Support with antenna	343	706
Toy	446	220+
Design	D21	84
Ballot	283	5
Box	232	2+
Registering	235	57
Counting or voting machine	235	51+
Marking, teaching	434	306
Baluster	52	720
Balustrade	256	59+
Bamboo Seat	D 6	369
Banana		
Banana plugs electric	439	825+
Insulated	439	625+
Band		
Endless band cutting	83	935*

	Class	Subclass
Instruments	84	1+
Lessons	84	470R+
Saw	83	788+
Saw	30	380
Blades	83	835+
Spreading in radio tuner	334	
Filter	333	167+
Using distributed impedance only	333	219+
Band Width		
Reduction system		
Facsimile	358	426+
Multiplex	370	118
Pulse or digital communications	375	122
Radio	455	72
Recording	369	60
Magnetic	360	8+
Speech	381	29+
Television	358	133+
Regulator		
Amplitude modulation	332	159
Frequency modulation	332	123
Pulse modulation	332	107
Bandage	D24	49
Body treating	604	304+
Controlled release of medication	604	890.1+
Electric	128	82.1
Package for coiled type	206	389+
Scissors	30	286
Splints	128	87R
Surgical	128	155+
Application	128	82+
Tire patches	152	371+
Webbing	128	156
Winding	242	60
Banding		
Bale	100	1+
Box covering	493	111+
Brush broom and mop making	300	15
Hat making	223	22
Label pasting and paper hanging	156	
Paper bunch	493	386+
Bandoleer	224	203
Bandsaw	30	380
Banister	256	59+
Banjo	84	269+
Clocks	368	229
Banking, Aircraft Control	244	75R+
Banks & Banking		
Burglar traps	43	59
Checks and deposit slips	283	57+
Coin handling	453	
Deposit apparatus	D99	34+
Depository	D99	43
Depository with verification means	109	24.1
Photographic or microfilming	346	22
Time record-marking	346	22
Indicator for degree of	33	328+
Piggy bank	D99	35+
Protection devices	109	2+
Safes	109	
Toy	446	8+
Banners		
Flag type	116	173+
Design	D11	165+
Rod	D11	181+
Sign type	40	
Baptismal Font	D99	25
Bar (See Rod)		
Boring	408	199+
Counters	312	140.1+
Crow	254	120
Horseshoe blank	59	62+
Lingual	433	190
Manufacture (See particular operation)		
Compound metal bars	228	4.1+
Drawing	72	274+
Extruding metal	72	253.1+
Indirectly	72	273.5
Horseshoe blank rolling	59	63+
Juxtapose and bond metal bars	228	101+
Rolling	72	199+
Masonry reinforcing	52	720+
Metal shape	428	544+
Design	D25	119+
Metallic plural layers	428	615+
Mining bar cutter	299	79+
Needle for sewing machine	112	222
Sash bar	52	777+
Splice rail joint	238	243+
Insulated	238	159+

	Class	Subclass
Towel	211	105.1+
Type printing	101	401.1
Forming machines	101	401.6
Processes of forming	101	401.4
Wrecking	254	120
Barb		
Applying to wire, barbing	140	58+
Cutting fence barbs	29	7.1
Fences	256	2+
Hayfork type of harpoon having	294	127+
Nails having barbs	411	456
Staple barbing	59	73+
Tag fastener	40	669+
Wire	140	59
Barbecues	99	419+
Cooker	D 7	332+
Fork	30	322+
Turners	99	419+
Barber		
Cabinets	312	209
Chairs	297	68+
Design	D 6	334
Implement with light	362	115
Poles	40	538
Design	D20	16
Tweezers	606	133
Barbital	544	307
Barbituric Acids	544	299+
In drug	514	270
Barges	114	26+
Cranes	212	190+
Barium		
Alkali earth metal	23	304+
Naphthenate	562	511
Bark		
Debarking	144	340
Extract	560	68+
Descaling agent ctg	252	83+
Osier peelers	144	207
Rossing	144	208R
Stripping machine	144	208R
Branches	144	207
Barkhausen Kurz Oscillator	331	92+
Barn	119	16
Barometer	73	384+
Design	D10	55
Barometric Material Feed		
Animal watering trough	119	77
Dispenser trap chamber	222	457
Gas liquid contact device	261	73
Inkstand or inkwell	222	585+
Liquid level miscellaneous	137	453+
Lubricator	184	84
Supply container and independent applicator	401	120
Barrel (See Cask; Container)		
Bungs	217	98+
Charring	432	224
Cleaning		
Brushing implements	15	164+
Brushing machines	15	57+
Methods	134	22.1+
Washing apparatus	134	43+
Wipers	15	211+
Closures	217	76+
Design, package simulates shape	D 9	325
Design, shipping	D34	39
Drying	34	104+
Methods	34	21
Electrolytic apparatus	204	213+
Filling	141	
Firearm	42	76.1+
Breech hinged	42	8
Forward sliding	42	10+
Moving and recoiling	89	160+
Muzzle loading	42	51
Recoiling with revolving chambers	89	157
Revolver hinged	42	63+
Side swinging	42	12+
Stock fastenings	42	75.1+
Swinging barrel locks	42	44+
Heads	217	76+
Interior illuminators	362	154
Laterally directed		
Drop bomb	102	383
Shell	102	383
Liquid level indicators	73	290R+
Making	147	
Sheet metal container	413	1+
Metal receptacle	220	
Ordnance	89	14.5+

	Class	Subclass
Moving and recoiling	89	160+
Practice	89	29
Recoil checks	89	42.1+
Organs	84	86
Paper receptacle	102	480
Periscope for examining interior	350	540+
Plugs	217	110
Collapsible bulb	604	212+
Removing bungs	137	324+
Syringe pump hand held	604	187+
Tapping	137	317+
With cutter or punch	222	81+
Watch	368	142+
Wooden receptacle	217	72+
Insulated	217	131
Barrette	132	278+
Design	D28	39+
Barrier (See Ceiling; Floor; Gate; Wall)		
Building construction	52	
Static mold or form for making	249	18+
Flight deck	244	110R
Neutron	376	287
Neutron	250	518.1
Neutron	376	347
Floating	405	63+
Impact absorbing closure	49	9
Race track	119	15.2
Radiation	250	505.1
Process	55	16
Solid, for gas separation	55	158
Subterranean moisture	405	38
Barrier Layer		
Coating	437	
By electrolysis	204	14.1+
Dry rectifier	357	
Composition	252	62.3 R
Manufacture	29	25.2+
Electrolytic condenser or rectifie	361	500+
Electrolyte	252	62.2
Manufacture	29	25.1+
Material P N type	148	33
Making P N type	437	
Barrow		
Dumping	298	2+
Hand	294	15+
Wheel	280	47.31
Bascule Bridges	14	
Base Exchange		
Compositions	252	193
Water treating	423	328+
Silicates	423	
Water purification	210	660+
Apparatus	210	263+
Baseball	273	25+
Balls	273	60R
Bats	273	72R
Card or title games	273	298
Cover sewing machine	112	121.28
Game board	D21	28
Glove	2	19
Design	D29	21
Projector	124	
Simulated game	273	88+
Bases (See Foundation)	248	346
Apparel apparatus	223	120
Bed plate	52	292+
Compositions	252	193
Curved wall	52	247
Dispenser supports	222	173+
Electric lamp	439	611+
Electric switch	200	293+
Flatiron	248	117.2+
Inorganic	423	
Lanterns and lamps	362	190
Air preheating type	362	172
Chimney	362	314
Design	D26	93+
Shade	362	441+
Machinery	248	678
Organic	260	
Pole	52	292+
Concrete type	52	294+
Prop and brace	248	357+
Robot	901	
Staff	248	519+
Stand	248	188.1+
Swinging for movable receptacle	248	144
Vehicle runner	280	28
Basin (See Bowl)		
Auxiliary in ship lifting locks	405	4+
Bath	4	619+

	Class	Subclass
Design	D23	284+
Bath fittings	4	191+
Cabinet combined	312	228
Canal locks	405	85+
Dispenser drip	222	108+
Drinking fountain catch	239	28+
Model	73	148
Receptacle type	220	
Sewerage catch	210	532.1+
Grated inlet	210	163+
With strainer	210	299+
Surgical receptors	604	317+
Basket (See Container; Creel)		
Closures	217	124
Design	D11	143+
Creel	D 3	38
Electrolytic cell electrode	204	259
Flaccid material on frame	220	9.1+
Forming	147	48
Ski pole	280	824
Wire	220	485
Wooden	217	122+
Basketball Devices	273	1.5 R
Balls	273	65R
Teaching	434	248
Bass		
Bars for violin	84	276
Drums	84	411R+
Finders	84	470R+
Violin	84	274+
Viols	84	274+
Bassinet	D 6	390
Bassoons	84	380R+
Basting	99	345+
Brush	D 4	130
Bat (See Batt)		
Forming apparatus for hats	19	148
Game and amusement device	273	67R+
Baseball	273	72R
Design	D21	211+
Harvesting reel	56	219+
Batch Charger		
Furnace type	414	167+
Glass furnace combined	65	335
Batch Mixer		
Bread pastry and confection	366	69+
Batch feeder and furnace	65	335
Mortar	366	1+
With heat exchanger	165	109.1
Bath	4	
Bird	119	1
Design	D30	123
Brush	D 4	132
Cloth	D28	63
Design	D23	277+
Fused salt	266	120
Heater	D23	318+
Material supply coated or		
impregnated	15	104.94
Mitt	15	227
Design	D28	63
Photographic (See photography)		
Quench	266	130+
Sponges	15	209R+
With soap	401	201
Room accessories	D 6	524
Sponges	15	209R
Design	D28	63
With soap	401	201
Steam or sauna	D24	37
Therapeutic	128	365+
Tub or footbath, therapic	D24	38
Bathing		
Garments	2	67+
Buoyant	441	88+
Hat	D 2	510+
Suit	D 2	36+
Trunks	D 2	42+
Pools	4	488+
Swimming pool purification	210	169
Shoes	36	8.1
Sandals	36	11.5
Shower	4	596+
Shower nozzles	239	548+
Tubs	4	538
Bathometer	73	300
Bathroom		
Disinfection	4	222+
Fixture liners	4	DIG. 18
Prison bathrooms	4	DIG. 15
Ventilation	4	209R+

	Class	Subclass
Bathtub (See Receptacles)	4	538+
Antislip mat	4	582
Bath and basin fittings	4	191+
Design	D23	277+
Fittings	4	191+
Seat	4	578
Soap dishes with soap handling		
means	206	77.1
Support	4	571+
Therapeutic	128	369+
Water cutoff device	4	191+
Bating	8	94.17
Fermentative	435	265
Batiste	139	426R
Baton	D21	100
Illuminated	362	102
Music conductors type	84	477B
Batt		
Bonding with	156	62.6+
Forming articles from	264	116
Impregnation of workpiece by	264	136
Mechanical molding of article	264	257
Multilayer metal receptacle with	220	452
Special package for reel or roll	206	417
Battery (See Electret)	136	
Applications	136	291*
Assembling components	29	730+
Methods	29	623.1+
Bonding apparatus	228	58*
Bonding process	228	901*
Cable terminals	429	179+
Carrier	224	902*
By hand	294	149+
By hand	294	903*
Cases	429	164+
Charge indicator	324	427+
Battery attached	429	90+
Charger	320	2+
Design	D13	107+
Material electrolyte	429	189+
Material regenerating	429	17+
Circuit applications	136	293*
Clamps for terminals	439	754+
Deferred action	429	110+
Depolarizer	252	
Design	D13	+
Electro chemical	429	
Electrode receptacle		
Conducting	429	239+
Nonconducting	429	235+
Electrolytic cell internal battery	204	248+
Electronic tube having	315	55
Filler	429	233
Flash lights using	362	157+
Grid	429	233+
Filling	141	32+
Making	29	2
Pasting	141	32+
Handling hand and hoist line		
implements	294	
Holddowns, vehicle mounted	248	503+
Holders in automobiles	180	68.5
Hydrometers	429	90+
Internal or electrolytic		
Cell	204	248+
Cleaning process	204	144
Protecting apparatus	204	197
Protecting process	204	148
Water treatment	204	150
Jar	429	163+
Metal platemaking	29	623.1
Motor fed systems	318	139
Nuclear	310	301+
Photoelectric	136	243+
Plate forming	204	6
Plate making	29	2
Plugs	429	89
Prime mover dynamo plant with	290	50
Radioisotope-powered	136	202
Radioisotope-powered	376	317
Receptacle acidproof	206	524.1+
Regenerating material	429	19+
Separator	429	247+
Solar cell	136	243+
Space satellite applications	136	292*
Stoppers	429	89
Storage	429	149+
Surgical application	128	391+
Switches	320	2+
Terminal	429	179+
Applying or removing apparatus	29	246

	Class	Subclass
Protector	429	65
Tester	D10	77
Testing	324	426+
Accessories	429	90+
Calibrating, treating, testing	136	290*
Charging combined	320	47
Hydrometer	73	441+
Integral with battery	429	90+
Thermoelectric	136	200+
Vent caps	429	89+
Batting	19	296
Package for	206	389+
Battle Lanterns		
Flashlight type	362	157+
Bay Window	52	201
Bayonet	30	
Gun combined	42	86
Holder	224	232+
Holder	224	232
Joint (See joints)	362	35
Beacon		
Aircraft	362	262
Radio	342	385+
Directive radio	342	350+
Rotating	342	398+
Floating marine	441	1+
Illuminating lights	362	35
Radar transponder	342	42+
I F F	342	45
Signal lights	340	
Air craft	340	981+
Bead and Beading		
Adsorbent or catalytic	502	527
Bead chain making	59	2
Garment supporting	2	300
Glass compositions	501	33+
Glass manufacturing	65	21.1+
Apparatus	65	142
Lathing corner beads	52	255+
Metal beading	72	67+
Metal molding of	164	179
Curving and	72	298+
Tube forming and	72	149+
Metal tube corrugating	72	367
Necklace	63	2
Design, strung	D11	11
Rosary	235	123
Design	D99	26
Sewing on with sewing machine	112	104+
Shoe		
Heel beading machines	12	49
Upper beading machines	12	57.1
Spreaders for tires	254	50.1+
Spread holders	81	15.3
Tire mounting combined	157	1.17+
Stringing	223	48
Apparatus	29	241
Method	29	433
Tire beads	152	539
Window bead fasteners	16	220
Beakers (See Receptacles)	422	102
Dispensing	222	566+
Molten metal	222	591+
Measuring	73	426+
Beam		
Brake	188	219.1+
Design, architectural	D25	126+
I-shaped beam cutter	83	DIG. 2
Locomotive truck buffer	105	173
Masonry	52	723
Mortar and block	52	433
Method	72	365.2
Plow	172	681+
Power amplifier	313	299
Rectifier	313	298
With beam forming electrodes	313	299+
With two cathodes	313	5
Railroad ties I beam	238	65+
Two part	238	57
Scales	177	246+
Static mold	249	50
Support for brackets	248	228
Pipe or cable type	248	72
Beam Lead Frame or Device Structure	357	69+
Making	29	827
Beam-trammel Distance Device	33	810+
Beaming	28	190+
Beanie	D 2	256
Beans	426	629
Bearing (See Journal)	384	
Brasses or linings	384	276+

	Class	Subclass
Clutch throwout	192	110B
Connecting rod adjustable	74	594
Design	D15	143
Fishing reel	242	320
Hydrostatic bearing with gearing	409	904*
Lathe headstock	82	148
Manufacture		
Assembly apparatus roller and bearing ball	29	724+
Processes	29	898+
Fluid	29	898.2
Linear	29	898.3
Rotary	29	898.4
Rotary anti-friction	29	898.6+
Rotary self adjusting	29	898.43+
Rotary sleeves and bushings	29	898.54+
Rotary thrust	29	898.41
Processes of grinding	51	289R+
Materials	252	12+
Piano pedal	84	228
Pullers	29	244+
Rudder post	114	169
Separator, imperforate bowl centrifugal	494	83
Testing frictional resistance	73	9+
Turntable railway	104	46
Typewriter bar	400	448+
Typewriter carriage	400	354+
Watch and clock	368	324
Wringers	68	269R
Beat Note in Detector		
Autodyne detector	455	321+
Heterodyne detector	455	313+
Homodyne	455	324
Beater		
Agitator	366	343+
Cleaning	15	89+
Implements	15	141.1+
Rotary	15	141.2
Cloth finishing	26	25+
Comminuting processes	241	27
Drum and cymbal	84	422.1
Fiber liberation and preparing	19	85+
Brakes and beaters	19	30
Decorticating	19	33
Heat exchanging	165	92
Meat tenderer	17	25+
Paper fiber engine	241	97
Pastry and confection	366	69+
Rotary beater mill	241	185R+
Perforate casing	241	86+
Screen	209	299
Clearer	209	383
Shoe welt	12	67.2
Textile braider and	87	36
Wire twister and	140	32
Portable machine	140	42
Beauty Parlor Equipment	D28	9+
Bed (See Associated Machine or Tool)	5	
Activated sludge	210	623+
Animal	119	1
Design	D30	118
Attachments and accessories	5	508+
Backpack convertible to	224	156
Baggage convertible to	190	2
Bedding	D 6	382
Bedstead	5	131+
Camping land vehicle	296	174
Clothes holder	24	72.5
Clothing	5	482+
Bed attached	5	498
Weight supporting	5	505
Combination furniture	5	2.1+
Cooling	5	284
Heater combined	165	46
Mattress	5	421+
Refrigerating	62	261
Crib	5	93.1+
Design	D 6	382
Doll or toy	446	482
Exercising device	272	144+
Fan combined	416	146R
Filter (See filter)		
Frame component	D 6	503+
Garment		
Gown type	2	114
Pajama type	2	83
Glass molding		
Planar platen	65	256+
Heater	126	205
Cooling combined	165	46

	Class	Subclass
Electric	219	217
Mattress	5	421+
Mattress electric	219	217
Medicator combined	604	113+
Surgical	128	376+
Horticultural	47	18
Hotbed	47	19
Invalids	5	60+
Jacket, apparel	D 2	12+
Light combined	362	130
Light thermal electrical application	128	376+
Machine or tool part		
Addressing machines	101	57+
Apparel plaiting, fluting, shirring	223	31
Bed and platen printing	101	287+
Clothes pressing platen	38	17+
Clothes pressing roll	38	44+
Clothes washing scrubbing	68	63+
Clothes washing squeezing	68	94+
Coopering stave jointing	147	26
Glass molding	65	361
Inkers for printing devices	101	335+
Intaglio printing bed and cylinder	101	158+
Lathe	82	150
Metal bending	72	214+
Multicolor printing bed and cylinder	101	186+
Numbering and printing	101	72+
Planer reciprocating bed	409	321+
Planers woodworking endless	144	128
Planographic copying	101	131+
Planographic printing bed and cylinder	101	146
Printing member and inker	101	104+
Rolling contact printing bed and cylinder	101	214+
Selective or progressive bed and platen printing	101	93+
Solid material stratifier	209	422+
Ticket printing	101	66+
Mattress or cushion	5	448+
Design	D 6	596+
Midrib and center strip	5	192
Modular bedroom lighting	362	801*
Mosquito nets and canopies	5	414+
Pan	4	450+
Design	D24	57
Particulate material separator	210	263+
Pavement or road	404	27+
Person restrainer for	128	869+
Quilts	D 6	603+
Railway car with	105	316+
Ships	114	188+
Sleeping bags	5	413
Sofa	5	12.1+
Spread or cover	D 6	596+
Supported horizontal surface	108	49
Tray	D 6	406
Trunks convertible	190	2
Waterbed	5	451
Window ventilator combined	98	89
Wire bottom frame attaching	140	110
Bedbug Trap	43	123
Bedpan	4	450+
Design	D24	57
Rinser in flush toilet	4	300.2
Bedplate (See Bases)		
Bedroom Lighting, eg Modular Combinations	362	801*
Bedspread	D 6	596
Bedsteads		
Design	D 6	382
Folding	5	174+
Bee Culture	449	
Beeswax purification	260	420+
Feeder	449	5+
Hive	449	3+
Honeycomb type receptacle	449	17+
Smokers	43	127+
Beef Jack		
Closure operators	49	69
Trolley transfer	104	97
Beer	426	592
Apparatus for making	99	275+
Beeswax Purification	260	420+
Beet		
Harvesting devices	171	
Digging then topping	171	26+
Lifters	171	50+
Topping then digging	171	26+
Toppers	56	

	Class	Subclass
Washers	15	3.1+
Beetling	26	26
Beheader Fowl	17	12
Belaying Pin	114	221R
Bell		
And hopper charger	414	204
Combined with blast furnace	266	184
Animal and sleigh	116	170
Automatic musical instruments	84	103
Bicycle	116	166
Church	116	150
Design	D10	116
Combined with escutcheon	D 8	350+
Electric	340	392+
Electrically actuated signals	340	392+
Electrically simulated church bell	340	398
Gas holder		
Inverted bell and tank	48	176
Moving bell	48	179
Sectional telescoping bell	48	177
Glasses for protection of plants	47	26+
Highway crossing type	246	296
Mechanical	116	148+
Musical bell	84	406
Design	D17	22
Musical bell with striker	84	407
Ornamental	428	11
Railway crossing protection	246	111+
Signals & indicators	116	148+
Sound producing	116	148+
Swinging eg church	116	150
Typewriter margin signal	400	712+
Bellows	D23	384
Accordions	84	376R
Cameras	354	187+
Dispenser	222	206+
Fluid pressure responsive	92	3+
Making	29	454
Metal tube corrugating	72	54+
X-art collection	493	940*
Meter	73	262+
Organs	84	355+
Pump	417	472+
Fumigator combustion air	422	305
Belt (See Machine with which Associated)		
Apparel	2	311+
Compressor type	450	154
Design	D 2	627+
Design, garter or sanitary	D 2	625
Elements of supporters	2	338+
Buckle attached	D 2	627+
Carrier mounted onto	224	163
Cartridge feeding	89	35.1+
Cigar and cigarette type machine	131	55
Comminutor type	241	200
Conveyors (See conveyor, endless)		
For dredges	37	69
Couplers	24	31B+
Hook	24	35
Lacing	24	34
Design	D 2	627+
Drive train design	D15	148
Drives	474	
Endless for vehicle fender	293	20
Fastener	24	31R+
Structure for nuclear reactor moderator	376	302
Gearing	474	
Coating to prevent slippage	106	36
Hanger	D 6	315+
Holder for cartridges, fishing lures	D 3	100
Holder or rack	D 6	315+
Holster combined	D 3	101
Hook inserters	29	243.51
Lifting tool	474	130
Making	156	137+
Of leather	69	
Special sewing machine	112	121.27
Mechanisms	474	
Belt and sprockets	474	
Guards	474	144
Money	224	229
Paper making		
Endless drying	34	243R
Fourdrinier	162	348
Pulleys	474	
Registers	235	125
Reinforced (See reinforcement, fabric)	474	268+
Safety	182	3+

	Class	Subclass
Aircraft	244	122B
Vehicle seat	297	464+
Hand vehicle	280	801+
Motor vehicle	180	268+
Seat belt safety buckle	D11	200+
Shifter	474	101+
For variable speed drive	474	80+
Starting gas pumps and fans	417	223
Stock material	428	
Conveyor type	198	
Drive type	474	237+
Knitted	66	169R+
Laminated	428	
Woven	139	383R+
Support in leather working machine	69	41
Tightener (See notes under)	474	101+
Abrasive belt	51	148
Band type twisting apparatus	57	105
Belt type drying apparatus	34	118
Railway wheel and axle drive	105	105+
Tool holder carried by	224	904*
Tree trunk guards	47	24
Trusses surgical	128	99.1+
V-type	474	237+
Bench		
Cabinet combined	312	235.2+
Greenhouse	47	18
Ladder combined	182	33
Shoemakers	12	122
Table	108	
Wash tub or machine	68	236
Woodworking	144	286R+
Clamps	269	
Design	D 6	396+
Dogs	144	306+
Support for	144	286R
Tool chest combined	144	285
Bench Mark	52	103
Bed plate	52	292+
Bending (See Crimping; Folding)		
Brake	72	310
Chain making machines	59	27
Sheet metal	59	15
Hoof and shoe expanders	168	47
Horseshoe making machines	59	36+
Lock washer making machines	10	73
Metal		
Machines and processes for	72	
Machines combining with other metal shaping	29	34R+
Wire	140	
Working, with bending	29	DIG. 3
Nut making machines	10	74
Other metal shaping	72	
Scale removal by flexing	29	81.3+
Sheet metal seaming by shaping	228	144
Sheet metal seaming machines	72	48+
Sheet metal seaming processes	228	137
Staple making and setting apparatus	227	82+
Staple making machines	59	71
Strength of materials testing by	73	849+
Sweep arm bender	72	217
Tubes	72	367+
Wood	144	254+
Beneficiating Ores	423	1+
Apparatus	266	168+
Bentonite	106	DIG. 4
Benz C Fluoran	549	224+
Benzaldehyde	568	425
Benzanthrones	552	286+
Benzene (See Aromatic Hydrocarbon)		
Benzidine	534	822+
Disazo compounds from	534	822+
Disazo pyrazoles from	534	760
Tetrakisazo compounds from	534	808
Trisazo compounds from	534	813+
Benzoates		
Ammonia	562	493
Benzyl	560	106
Bismuth	556	78
Caffeine	544	274
Lithium	562	493
Magnesium	562	493
Mercury	556	132
Benzocaine	560	19
In drug	514	535
Benzodianthrones	552	282+
Benzofuranes	549	462+
Benzoic		
Acid	562	493
Anhydride	562	887

	Class	Subclass
Benzoin	568	331
Benzol (See Aromatic Hydrocarbon)		
Motor spirits	585	14
Benzomorphans, in Drug	514	295
Benzophenazines	544	343
Benzophenone	568	332
Benzopyrenequinones	552	284+
Benzoquinoline	546	101
Benzoquinolzines	546	95
Benzoquinone	552	293
Benzoselenazoles	548	121
Benzothiazines	544	49+
Benzothiazoles	548	152+
Benzothiophenes	549	49+
Benzotrichloride	570	185
Preparation	570	191+
Benzoxazoles	548	217+
Benzyl		
Alcohol	568	715
Butyrate	560	254
Disulphide	568	25
Ether	568	659
Salicylate	560	71
Thiocyanate	558	10
Benzylamine	564	391
Bergamot Oils	512	5
Berry		
Clipper		
Catcher combined	56	331
Holder combined	30	124+
Nipper type	30	175+
Crates	217	40
Berth		
Bunk beds	5	9.1+
Self leveling	114	192+
Sleeping car	105	316+
Beryllium		
Alloy compositions	420	401
Electrolytic synthesis from fused bath	204	65
Pyrometallurgy	75	593
Bessemer Converter	266	243+
Bessemerizing	75	
Iron or steel	75	528+
Beta		
Crystal mounting	250	361R
Gamma survey meter	250	336.1
With sample holder	250	336.1
Beta Alanine	562	576
Beta Globulins	530	394
Betaines	562	
Bethamethasone	552	574
Bevel		
Gear	74	640+
Intercontrolled blades	33	455
Planetary gearing	475	336
Protractor	33	455
Square	33	474
Beverages	426	
Apparatus	99	275+
Beer or pop can	206	139
Can cover	220	903
Carbonated	426	590+
Carbonater	261	
Carbonating and flavoring	99	323.1+
Coaster or mat	D 7	624
Cooler	62	389+
Foam forming or inhibiting	137	170.1+
With selection from plural materials	222	144.5
Electrolytic treatment	204	138+
Mixer blender, household	D 7	376+
Mixing		
By mechanical agitating	366	
Milk shake type	366	197+
By nozzle shape	239	549
Preparation or dispensing machines	D 7	300+
Reusable infusion receptable for preparing beverages	99	323
Storage receptacle	D 7	602+
Vending machines	D20	5
Bezel		
Design	D26	139
Test instrument	73	431
Watch	368	295
Bias Voltage Control in an Amplifier	330	129+
Bib	2	49R+
Dental patient	433	137
Design	D 2	226+
Bicarbonate	423	419R
Drugs or bio-affecting	424	717

	Class	Subclass
Bichromate Cells	429	
Bicycle	280	200+
Attached carrier	224	30R+
Bell	116	166
Brake	188	24.11+
Coaster	192	6R
Convertible	280	7.1+
Cranks and pedals	74	594.1+
Cyclometers (See odometer)		
Design	D12	111
Dust and mud guards	280	152.1+
Exercising devices	272	73
Generator	310	
System	322	1
Handle bars	74	551.1+
Handle or grip design	D 8	DIG. 8
Horn	116	137R+
Hub ball bearing	384	545
Lights	362	72+
Electric signal	340	432
Generator bulb system	315	76
Generator control	322	1
Generator per se	310	
Supports	362	382+
With wheel driven generator	362	193+
Locks	70	233+
Wheel	70	225+
Making		
Frame assembling	29	700+
Methods	29	428+
Pedal crank bearings	384	431
Propelled marine pedomotors	440	30
Racks	211	17+
Rack or holder, design	D12	115
Reflector	350	97+
Seats	297	195+
Simulations	280	828+
Umbrella for	135	88
Wheel	301	5R+
Guards	280	160.1
Scrapers and cleaners	280	158.1
Bidet	4	443+
Design	D23	295+
Bier	27	27
Wheeled	280	47.34+
Extensible	280	640
Tiltable	280	47.17+
Bifocal		
Contact lens	351	161
Spectacle lens	351	168+
Bile Acids	552	548+
Bilge Discharge	114	183R+
Bilirubin	436	97*
Billboard Type Signs	D20	39+
Billets	428	577+
Fault removal	29	526.2+
Harness	24	182
Hame strap	54	28
Hame tugs	54	32+
Piercing	72	325+
Billfold	150	132+
Leather, design	D 3	56
Billiards	273	2+
Balls	273	59R
Chalk	273	17
Cues	273	68+
Tips	273	70
Design, cues and accessories	D21	210
Register operating device	235	91B
Stick or rest	273	23
Tables		
Beds	273	6
Covers	273	13
Design	D21	232
Tops	273	6
Timing device	368	3
Billing Machines (See Printers)		
Bills in Knotting	289	11
Billy Club	273	84R
Design	D22	117
Guns	42	1.16
Biltmore Sticks	33	483
Bimetallic Stock (See Thermostats)	428	615+
Bonded layers	228	101+
Casting	164	91+
Juxtapose and bond layers	228	101+
Making by extrusion	72	258
Shape or structure	428	577+
Ingots	428	585
Bin (See Box; Receivers)		
Charging or discharging	414	288+

	Class	Subclass
Deforming	72	700
Discharging (See dispensing)	222	
Charging and	414	288
Granaries		
Hopper with port	52	192+
Sifter associated	209	370+
Ventilated grain bins	98	55
Binary Compounds, Inorganic	423	509+
Uranium containing	423	253+
Binaural	381	1+
Binder and Binding		
Baling press combined	100	8+
Binder device releasably engaging aperture or notch of sheet	402	
Depository	402	73+
Expander	402	80R
Filler	402	80R
Fly-strip fly-leaf	402	80R
Marker position holder	402	80R
Books	412	
Design	D19	26+
Cooking apparatus combined	99	350
Covering with metal	29	33.2
Severing base combined	29	33.5
Edge binding guide for sewing	112	137
Handle or grip design	D 8	DIG. 8
Harvesting	56	432+
Cutter combined	56	67+
Hat making	223	22
Post	439	775+
Press	100	
Resilient sectional tire	152	300+
Ski	280	611+
Snow shoe	36	126
Table top edge	52	783+
Laminated top	52	783+
Tobacco product	131	365
Vehicle load	410	96+
Water ski	441	70
Wire or band tensioners		
Combined with sealers	140	93.2
Portable & detachable	254	199+
Binding		
With flexible filament band or strand	100	1+
Binding Assays	436	500+
Binding Proteins	530	387
Bingo Games	273	269
Binoculars	350	145
Carrier	224	909*
Cases	206	316.3
Design	D16	133+
Microscope	350	514
Telescope	350	545
Bio-affecting Compositions (See Drug, Bio-affecting and Body Treating Compositions)	514	
Biochemical Oxygen Demand Tests, Bod	436	62*
Biocidally Protected Polymers	523	122
Biocides	424	
Coating or plastic composition	106	15.5+
Detergent with	252	106+
Mineral oil derived	208	2
Water softening or purifying agent	252	175+
Biodegradable		
Plant receptacle	47	74+
Biology		
Coating processes	427	4
Preservation	47	
Teaching	434	295+
Biopsy	128	739
Biopsy device	128	751+
Biosynthesis	435	
Biotin	548	303
In drug	514	387
Biphenyl (See Aromatic Hydrocarbon)		
Biphosphates, Inorganic	423	
Bird		
Antiroosting structure	52	101
Artificial	428	16
Feather	428	6
Baths	119	1
Design	D30	123
Cage	119	17+
Design	D30	114+
Perches	119	26
Perches, design	D30	119+
Calls	446	204+
Feeders or waterers	D30	121+
Food	D 1	
Houses	119	23

	Class	Subclass
Design	D30	110+
Perches or cage attachments	D30	119+
Repellents		
Electric fence	256	10
Electric prod	231	7
Scarecrows	40	
Training	119	
Traps	43	
Birefringent Element		
Color televison	358	61
Polarized light examination	356	365
Birth Control (See Contraceptives)		
Biscuit Package	206	830*
Biscuit Shredded	426	560
Apparatus	425	289+
Bismuth		
Alloys containing	420	577
Misc. chemical manufacture	156	DIG. 79
Organic compounds containing	556	64+
Pyrometallurgy	75	705
Bisnorcholenic Acids	552	553+
Bisphthalimides	548	461+
Bisulphates, Inorganic	423	520+
Bisulphides	423	511
Bisulphites	423	520
Reducing composition	252	188.21
Bit (See Cutter; Drill)		
Bench plane	30	493
Bit clamps	30	492
Earth boring	175	327+
Frame	408	53
Gauges	33	201
Harness bridle	54	7+
Design	D30	136
Weaning type	119	134
Ice boring	175	18
Ice cutting in situ	299	24+
Key bitting	70	409+
Mining cutter	299	79
Stock	81	28+
Wood boring double ended bit stock	408	36
Wood boring movable work machines	408	62+
Stationary bitstock	144	98
Wood boring stationary work machines	408	72R+
Inclined bitstock	144	100
Tobacco pipe	131	227+
Wood auger	408	199+
Bitts	114	218
Bituminous		
Abrasives containing	51	305
Compositions	106	273.1+
Emulsions	252	311.5
Mineral oil recovery from	208	400+
Polymers containing (See asphalt, polymers containing)		
Roads	404	17+
Bivalve Opener, Marine Animal	17	74+
Black Liquor	106	123.1
Blackboard	248	441.1+
Cabinet combined	312	230
Compasses	33	27.2+
Compositions	106	32.5
Design	D19	52+
Easels	248	441.1+
Erasers	15	208+
Writing surface	106	32.5
Blacking (See Coating)		
Box	15	258
Applicator combined	401	118+
Holder	15	259
Stand	15	265+
Blackjack	273	84R
Carrier	224	914*
Bladder Instrument	606	170
Blade (See Associated Machine)		
Electric switch	200	271+
Hand manipulated cutlery	30	
Holder for blades	30	329+
Making	76	104.1+
Saws	76	25.1+
Metalworking cutter	407	
Propeller	416	
Adjustable	416	147+
Stoneworking saw		
Reciprocating	125	18
Rotary	125	15
Blanching		
Fruits and vegetables	426	506+

	Class	Subclass
Grain apparatus	99	600+
Grain process	426	506+
Nuts	99	623+
Blankets		
Animal	D30	145
Bed	5	482+
Electrically heated	219	212
Horse	54	79
Design	D30	145
Household linen	D 6	603
Nuclear fertile material	376	172
Printers	428	909*
Retaining devices	24	72.5
Blanking		
Cathode ray tube circuit	315	384
Metal workpiece for deformation	72	324+
Blanks (See Stock)	428	542.8
Cartridges	102	530+
Cigar	131	364
Horseshoe	59	62+
Metal	428	577+
Double blanks	29	DIG. 2
Holder for during deformation	72	293+
Metallic receptacle	220	62
Nut	411	427
Paper envelope	229	75
Design	D19	3+
Phonograph record	428	64+
Printed	283	
Processes and		
Chain	59	35.1
Garment collar	2	143
Printing members	101	401.1
Sprocket chain	59	8
Staple	59	77
Tools and implements	76	101.1+
Weldless chain	59	12
Shelf bracket of single blank sheet material	248	248
Shoe sole prepared for attachment	36	22R
With upper structure	36	12+
Shoe upper	36	47+
Spike	10	62
Blast		
Abrading by sand	51	410+
Boiler cleaning by sand	122	395
Cleaning	15	300.1+
Beater and or brush combined	15	363+
Suction combined	15	345+
Decorticating fibrous material	19	9
Filter cleaning	210	407+
Fluid flow dispensing	222	630+
Forges	110	195
Furnace (See shaft, furnace)	266	197+
Furnace reduction of iron compounds	75	458+
Gas agitation	222	195
Insect powder dusters	43	132.1+
Metal treating	266	
Pipes for locomotives	60	685+
Jet pump type	417	155+
Pneumatic conveyors	406	
Track sanders	291	3+
Blasting	102	301+
Caps	102	275.12
Cartridges	102	314+
Well	102	301+
Blaugas	48	211
Bleachers	52	8+
Design	361	2+
Bleaching	8	101+
Apparatus textiles	68	
Compositions		
Cake, tablet or powder form	8	524+
Detergents combined	252	94+
Oxidative	252	186.1+
Reductive	252	188.1+
Dyeing and	8	931*
Electrolytic	204	133
Foods and beverages	426	253+
Apparatus	99	467+
Flour	426	253+
Paper stock	162	1+
Photographic	430	430+
Resin bleach	8	DIG. 6
Silicones	8	DIG. 1
Sugar	127	64
Textile bleaching	8	101+
Tobacco	131	290
Electrical or radiant energy	131	299
Vinyl sulfones & precursors thereof	8	DIG. 2
Waste paper	162	4+

	Class	Subclass
Detachable from neck cup smoke treating type	131	214
Feeder	131	180
Lined or coated	131	220
Lined or coated material traps	131	204
Reversible	131	221
Spaced inner bowl	131	196
Storage means	131	180
Toilet		
Closures	4	253
Couplings and supports	4	252R+
Covers	4	234+
Disinfection	4	222+
Drip catchers	4	252A
Obstruction removers	4	257
Ventilation	4	216
Water closet		
Bowl	4	420+
Design	D23	295+
Plunger	4	420+
Seat combined	4	234+
Side receptacle	4	341+
Siphon	4	421+
Tank combined	4	300+
Urinal	4	311+
Valved	4	434+
Washout	4	420+
Bowling	273	37+
Ball	273	63R+
Case	D 3	36
Grip test	33	510
Bowling alley equipment	D21	233
Bowling game tables	D21	233
Games	273	37+
Pins	273	82R
Digest	76	DIG. 1
Shoes	36	130
Teaching	434	249
Box (See Bin; Chest; Crate; Receptacle; Safe)		
Annealing	266	262+
Axle process of manufacture of		
Railway car irons	29	168+
Ballot	232	2+
Registering	235	57
Beam	52	731
Blacking box (See blacking box)	15	258
Closure		
Paper	229	124+
Paperboard box	229	124+
Wooden	217	56+
Cooled commodity containing	62	371+
Core	164	228
Couch	297	192+
Covering	493	111+
Deposit and collection	232	
Design	D 9	414+
Egg candling	356	64+
Electric battery	429	96+
Fare	232	7+
File	206	425+
Fire (See firebox)		
Flower attachable to window	47	68+
For transport, goods handling	D34	40+
Hat	206	8+
Housings for electrical conductors	174	50+
Vacuum or fluid containing	174	17LF+
Ice	62	459+
Air controller combined	62	420+
Joint clamp	269	111+
Land vehicle axle	301	109+
Letter boxes	232	22
Lubricating railway car journal	384	160+
Making machine		
Assembling and nailing	227	19+
Grooving handhold	144	136D
Nail driving	227	
Paper box	493	52
Sheet metal bending	72	394+
Sheet metal stamping	72	343+
Staple forming and setting	227	82+
Staple setting	227	
Metallic	220	
Outlet or junction type	220	3.2+
Sheet metal container making methods	413	1+
Meter	73	201
Circuit protectors	361	364+
Miter	83	746+
Money	446	8+
Opener		

	Class	Subclass
Cutting	30	2
Prying	254	18+
Paperboard	229	100+
Collapsible	229	117.1+
Joints	229	198.1
With handle	229	117.9+
Pipe and box joints	285	128+
Presses		
Expressing	100	127
Movable bale	100	221+
Railway axle box mounting	105	218.1+
Registering	235	100
Ballot	235	57
Rheostat plug	338	77
Savings	229	8.5
Sewing	223	107
Special articles and packages	206	
Splints	128	93
Extension appliances	128	86
Switch	200	293+
Ties	229	125.22
Toilet kits	132	286+
Powder	132	293+
Toy money	446	8+
Design	D99	34+
Traps for animals	43	60+
Wooden	217	5+
Egg cells	217	18+
Ice	217	130
Insulated	217	128+
Boxing Gloves	2	18
Brace (See Suspenders)		
Body		
Bandages	128	155+
Combined with corset	450	96
Fracture apparatus	128	83+
Orthopedic	128	68+
Shoe attached skates	280	11.36
Shoulder and back	2	44+
Trusses	128	95.1+
Walking irons	128	83.5
Boilers	122	493
Closure fasteners	292	338+
Design		
Dispenser spouts	222	573
Handle combined	222	475
Door	254	39
Drill	81	28+
Freight on carrier	410	121+
Bar wall-to-wall	410	143+
Panel wall-to-wall	410	129+
Yieldable	410	117+
Making	29	897+
Orthodontic	433	2+
Instruments	433	3+
Pole or wall	52	146+
Wing type	52	153+
Rail		
Tie plates	238	292+
Track fastenings	238	336+
Supports	248	351+
Clothesline	248	353
Vehicle spring	267	66+
Bracelet	63	3+
Design	D11	3+
Identification	40	625+
Bracket	248	200+
Antenna on	343	892
Article carried for storage support	248	682
Article support	248	200+
Specially mounted or attached	248	205.1+
Bicycle carrier	224	39+
Book or music holder support	248	441.1+
Clamped to support	248	225.31+
Dental engine	433	103+
Design	D 8	354+
Dispenser casing	222	180+
Eaves trough to wall	248	48.2
Guides for sliding panels and doors	16	90
Ladder rung	182	220
Light supports	362	432
Design	D26	138+
Machinery supports	248	674+
Making and assembly	29	897+
Mine car axle	295	42
Nursing bottle type	248	103+
Orthodontic	433	8+
Pipe or cable	248	65+
Ray generation machine support	250	522.1
Receptacle support	248	311.2
Rotary shaft to wall	384	442+

	Class	Subclass
Staff	248	534+
Stand alternative	248	126
Machinery supports	248	676
Stand combined	248	121+
Static mold form	249	219.1
Stove shelf	126	333
Thermometer supports	374	208
Track combined for sliding panels and doors	16	94R+
Vehicle attached	224	42.45 R+
Watch and clock	248	115
Whip socket	280	175
Brad	411	439+
Braiding	87	
Guide for sewing	112	139
Packages	206	389+
Sewing machine	112	23
Shaft packing material	277	227+
Tire carcass material	152	548
Trimming design	D 5	7+
Braille	434	113+
Typewriter	400	122+
Brake (See Machine or Device Combined)	188	
Airplane	244	110A+
Skids	244	108
Back pedaling	192	5
Beam	188	219.1+
Fulcrum	188	231+
Guide	188	233.3
Beams and bars for vehicles	188	219.1+
Bending brake	72	457
Bicycle	188	24.11+
Caliper type	188	24.12
Coaster	192	6R
Disc type	188	26
Block	188	250R+
Centrifugal casting machine combined	164	294
Closure safety	49	322
Clutch combined	192	12R+
Cooling	301	6CS
Deformable impact absorber	188	371+
Device with fluid	92	8+
Closed loop system	188	297+
Combined with spring	267	
Fluid spring	267	64.11+
Mechanical spring	267	217+
Open loop system	92	8
Disc type	188	71.1+
Door checks	16	82+
Electric		
Electrodynamic torque type	310	92+
Locomotive	105	61
Motor braking	318	362+
Motor with brake	310	77
Electrodynamic	310	92+
Elevator	187	73+
Expansible chamber	92	15+
Flexible panel or closure		
Automatic control	160	8
Roll type	160	291+
Fluid		
Internal resistance	188	266+
Internal resistance mechanical combined	188	271
Systems	303	
Testing fluid pressure systems	73	39
Inertial vibration damper	188	378+
Internal resistance motion retarder	188	266+
Master cylinder	60	533+
Motor control combined	192	1.1+
Power stop mechanisms	192	116.5+
Reel type	242	99
Fishing reel	242	283+
Spinning reel	242	243+
Ship propeller shaft	440	74
Shoes	188	250R+
Composition ctg synthetic resin or natural rubber	523	149+
Spinning or twisting element	57	113
Strand tension	242	156
System control	291	15
Testing	73	121+
Fluid pressure systems	73	39
Track sander control combined	291	14
Transmission control combined	192	4R+
Vehicle fender combined	293	2+
Wheel lock of brake type	70	228
Wheeled skate	280	11.2
Bran	426	618+

Entry	Class	Subclass
Comminution processes	241	6+
Removal from grain		
Apparatus	99	600+
Processes	426	482+
Separation	209	
Comminution combined	241	7
Brandboards	68	224
Branding		
Electrically heated stamp	219	228
Food cooking apparatus combined	99	430
Slice toaster or broiler type	99	388
Ink	106	20+
Instruments (See pyrographic)		
Iron		
Electric	219	228
External heat	101	31
Fuel burner	126	402+
Stamp	219	228
Brass		
Compositions	420	477
Knuckles	273	84R
Plating	204	44
Brasses		
Bearings with brasses	384	191+
Lubricating means	384	162+
Lubricating reservoir	384	163+
Musical instruments	84	387R+
Brassieres	450	1+
Combined with corset or girdle	450	7+
Design	D 2	24
Design combined with girdle	D 2	3
Brazing (See Soldering)		
Electric heating	219	85.1
Metal working with	29	DIG. 4
Methods	228	101+
Bread		
Box	D 7	609
Compositions and processes	99	
Cutters	83	761
Design	D 1	129+
Kneading board, household	D 7	698
Loaf design	D 1	129
Testing	73	169
Toaster	99	385+
Breaker		
Circuit (See circuit)		
Cornstalk harvester	56	52
Fiber preparation	19	35+
Ice breaker ships	114	40+
Metal	225	93+
Sheet, strip, rod, strand	225	93+
Solid disintegrator	241	
Stalk choppers	56	500+
Strips for tires	152	542
Wood chip	144	243+
Disintegrator	241	
Breaking Emulsion		
Agents	252	358
Mineral oil	252	328+
Combined processes	208	187+
Processes	252	319+
Breaking or Tearing	225	
Container opening	206	601+
Device	119	29
Harness	54	71+
Breaking Strength Testing Machines	73	788+
Breakwater	405	21+
Breast		
Artificial body member	623	7+
Board		
Plow moldboard	172	754+
Washboard	68	225
Drills	81	28+
Harness	54	20
Protectors	2	2
Pump	604	73+
Shields	128	890
Strap	54	58+
Design	D30	134+
Breasting Shoe Heels	12	47+
Nailing combined	12	43
Breath Alcohol	436	900*
Breath Testing	73	23.22
Breathers		
Caps	220	367+
Crank case	92	78+
Internal combustion engine	123	41.86
Plugs	220	367+
Roof		
Tanks	220	85R
Tank	220	367+
Valves, safety	137	455+
Breech	42	16+
Block	89	17+
Breeching	54	5
Holdbacks	278	128+
Breeder Reactor Nuclear	376	171
Briar Pipe	131	230
Brick (See Block; Panel)		
Carrier	294	62+
Checker	165	9.1+
Cleaning	125	26
Coking ovens	202	267.1
Composition	501	141+
Cutting after firing	125	23.1
Double course wall	52	561+
Veneer	52	434+
Drying apparatus		
Combined with kilns	432	128
Furnace, solid fuel	110	338+
Glazing	427	376.2
By tie	52	582+
On plural face	52	589+
Mold	425	398
Mold	425	412+
Static	249	117+
Molding machines	425	130+
Molding machines	425	218+
Pavements	404	34+
Presses	425	253+
Presses	425	363+
Presses	425	383+
Presses	425	413
Cutting while green	425	289
Cutting while green	83	
Road modules	404	34+
Sanding apparatus	118	308+
Through passage	52	606+
Brickkiln	432	120+
Bricklaying Machine	52	749
Brickmaking	425	
Brickset Cookstove	126	8
Brickwork		
Imitation		
Static mold	249	16
Brides Dress	D 2	50+
Bridge	14	
Arch	14	24+
Cable traction railways at swing bridges	104	188
Closure shiftable to bridge pit	49	33
Covering	14	74
Cranes horizontally swinging	212	205+
Horizontally	212	226
Deck	14	73
Dental	433	167+
Draw bridge	14	31+
Duplex bridge telegraph system	370	28
Electrical		
Amplifier (See amplifier, bridge)		
Bridge networks	323	365
Testing bridge	324	+
Distributive parameters	324	648
Eye glass bridges	351	124+
Fire hose bridges for spanning railroads	104	275+
Floating	14	27+
Floor bridges for vaults & safes	109	87
Gangways	14	69.5+
Girder	14	17
Glass furnace barrier	65	342+
Glass furnace in	65	342
Guitar bridges	84	298
Tailpiece combined	84	299
Irons, process of making	72	
Masonry or concrete	52	263
Arch	52	86+
Musical bridge details	84	307+
Piano bridges	84	209+
Piers	14	75+
Portable	14	2.4+
Railway bridge warnings or telltales	246	486
Railway safety bridges between cars	105	458+
Railway track tread bridges	238	218+
Steel or wood	14	
Stringed instruments (See music)	84	307+
Design	D17	21
Guitar	84	298
Guitar tailpiece	84	299
Piano	84	209
Structural		
Steel or wood	14	
Suspension	14	18+
Truss	14	3+
Wheatstone (See electrical)		
Bridgeman Type Crystallization	156	616.4
Bridle		
Bit		
Design	D30	136+
Brush and broom	15	168+
Harness	54	6R+
Design	D30	134+
Pipe or cable supporting bracket	248	69
Brief Case	190	900*
Design	D 3	76+
Molded integral handle	D 3	73
Zipper	D 3	71
Locks	70	67+
Bright Polishing by Etching	156	664+
X-art	156	903*
Brightening (See Burnishing; Polishing)		
Coatings	427	158
Electrolytic	204	36
Electrolytic	204	140
Radiation imaging, brightener ctg	430	933*
Brim		
Curler	223	14
Trimmer	223	16
Wirer	223	17
Brinell Testers	73	78+
Brines		
Purifying	210	
Briquette (See Block)		
Metal for furnace charging	75	303+
Solid fuel	44	530+
Briquetting	425	579
Distillation combined	201	5+
Fuel	44	593+
Nuclear	264	.5
Meat	17	32
Metal eg powder scrap		
Apparatus	425	78+
Apparatus	100	
Processes	419	61+
Processes sintering	419	
Ore	75	746+
Bristle	15	159R+
Brushware, design	D 4	130+
Dressing and assorting	300	18
Fastening in brooms and brushes	300	
Trimming	300	17
Broach or Broaching	409	243+
Bolt heads	10	20
Buttons	79	1
Cutters for	407	13+
Dental instruments for	433	102
Gear cutting machines	409	58+
Nuts	10	81
Broadcasting		
Container & scattering means for non-fluid material	239	650+
Insect powder dusters or sprayers	239	
Planting	111	130+
Planting with drilling	111	8+
Radiant energy systems eg radio	455	
Broadcloth	139	383R+
Wool	26	19+
Brocade	139	416
Broiler		
Attachments	126	14
Drip collecting	99	444+
Gas stove combined	126	41R
Griddles	99	422+
Grids	99	450
Slice	99	385+
Bromacil	544	313
Bromate	423	475
Bromides Inorganic	423	462+
Bromine	423	500+
Halocarbons	570	101+
Bromural	564	45
Bronchoscopes	128	4+
Bronze	420	470
Aluminum	420	471
Pigment	106	403
Brooch	D11	40+
Brooder	119	31+
Automatically controlled heater	236	6
Heater, design	D23	334
Heating system	237	14+
Automatic control	237	3+
Incubator combined	119	30

Broom U.S. Patent Classifications **Buna T M**

	Class	Subclass
Broom	15	159R+
Cabinet	312	206+
Cleaning material combined	401	268+
Design	D 4	135+
Non-bristle	D32	40+
Holder	15	146
Lawn or grass	56	400.17+
Making	300	
Machines	300	12+
Sewing machines	112	6
Rack	211	65+
Support	248	110+
Whisk broom	D 4	135
Brucine	546	35
Brush or Brushing (See Associated Process, Machine or Device)	15	
Attachments	15	246+
Bottle or jar closure supporting brush	215	208+
Container closure supporting brush	206	15.2
Drip cups and shields	15	248R+
For moving surface cleaning	15	256.5+
Bristle brush with scraper	D 4	118
Broom	D 4	135
Brushes and brooms	15	159R+
Cabinet	312	206+
Cleaner for	15	38
Cleaning implements	15	104R+
Making containers, X-art	493	942*
Cleaning machines	15	21.1+
Cleaning or coating	15	
Material supply combined	401	268+
Cleaning processes	134	6+
Clothes and textiles washing machine scrubber	68	
Design, bristled	D 4	130+
Clothes, hat, fingernail	D 4	
Dispenser clearing or striking	222	352
Dispensing, design	D 4	114+
Dustpan combined	15	104.8
Fly fan	416	501
For vacuum or floor polisher	D32	15+
Ginning saw picker	19	60
Hardware	16	2+
Implement	15	104R+
Light thermal and electrical application to body	128	393
Machine	15	3+
Air blast or suction combined	15	363+
Brush cleaner combined	15	48
Fruit, vegetable, meat or egg	15	3.1+
Receptacle cleaner	15	56+
Shoe	15	36+
Shoe blacking and shining	15	30+
Street cleaner	15	78+
Textile cloth finishing	26	29R+
Thread finishing	28	217+
Magnetoelectric	310	248+
Holder	310	239+
Making	300	
Individual wire bristle or tooth setting	227	79
Machines	300	2+
Massage	128	44+
Multiple-tip multiple-discharge	401	28
Non-bristle	D32	40+
Paint	15	159R+
Design	D 4	
Reservoir attached	401	268+
Pastery or basting	D 4	
Plural	D 4	119+
Powered	D 4	100
Rack	211	65+
Receptacle	206	361+
Roughening by wire brush	29	76.1
Scraper can attachment	220	90
Sifter cleaner combined	209	385+
Simulative	D 4	124+
Solid material assorting	209	615
Support	248	110+
Tobacco		
Feeding combined	131	109.1
Leaf	131	324+
Leaf stemming with smoothing or cleaning	131	315+
Toilet bowl	D 4	
Toilet kit	132	313
Material supply combined	401	118+
Shaving brush type including toilet article, eg mirror	132	289+
Soap and brush combined	401	123+
Tooth brush type including	132	308+
Tooth	15	167.1+
Tooth paste supply combined	401	268+
With massage tool	15	110
Washboard surface	68	227
Wick trimmers	431	120
Work supports	15	268
Shoe	15	36+
X-art collection	425	805*
Bubble		
Amusement devices	446	15+
Bubble domain caculating	364	714
Magnetic bubbles	365	1+
Solid separation processes	209	164+
Specific gravity test	73	439
Tower (See dephlegmators)		
Porous mass	261	94+
Wet baffle	261	108+
Bucket (See Pail, Tub)	220	
Conveyer (See type conveyer)		
Design	D32	53
Coal scuttle	D23	410
Minnow	D22	136
Excavating	37	
Clamshell	37	183R+
Orange peel	37	182
Fire extinguisher	169	34
Flaccid material	220	904+
Heater for dinner bucket	126	266+
Hoist line	414	564+
Vertically swinging shovel	414	565
Vertically swinging support	414	680+
Minnow	43	56+
Paperboard	229	910+
Rotary kinetic fluid motor	415	
Sap	47	50+
Staved wooden	217	72+
Tank type meter with rotary bucket	73	217+
Turpentine and rubber collecting	47	11
Type conveyer	198	701+
Dispensing	222	369
Dispensing endless	222	371
Wind motor collapsible	416	142+
Wooden	217	
Bucking Bar	72	457
Buckle	24	163R+
Belt attached buckled, clasp, slide	D 2	627+
Design	D 2	3
Harness	D30	139
Design apparel	D11	200+
Making	29	3
Seat belt	D11	200+
Shoe	D11	200+
Turnbuckle	403	43+
Buckstay Construction	52	86+
Bucky Grid	378	154
Budding and Grafting	47	7
Buddles	209	458+
Buffers (See Springs)		
Amplifier	330	
Radio transmitter with	455	91+
Bedstead	5	309
Bridge	14	48
Loom picker and picker stick	139	167+
Manicuring	132	76.4+
Compound tool	132	75.6
Polishing wheels	51	358+
Railway		
Draft gear	213	220+
Dump car door	105	285
Freight car interior end	105	374
Locomotive buffer beams	105	173
Store service	186	24+
Tire combined	152	158
Track	104	249
Trunk and baggage	190	37
Typewriter	400	686+
Warship	114	13
Water closet combined	4	248
Bug Type Telegraph Key	178	79+
Bugle	84	387R
Buhrstone or Burrstone	241	296+
Mills	241	244+
Builder Mechanism for Bobbins and Cops	242	26.1+
Builders		
Detergents	252	89.1+
Elevators	182	141+
Traveling scaffold	182	12+
Hardware	16	
Design	D 8	
Buildings	52	
Amusement	272	2+
Animal barns and sheds	119	16
Apparatus for moving material to a position for erection or repair	414	10+
Assembly or disassembly	52	127.1+
Auditorium or stadium feature	52	6+
Bay window	52	201
Burial vault	52	128+
Lifting or handling	52	124.1+
Construction elements	D25	
Convertible	52	64+
Cupola	52	200+
Drip deflector	52	97
Foundation	52	292+
Heated and cooled	165	48.1
Ventilated	165	16
Jail type structure	52	106
Lifting or handling of primary component	52	122.1+
Position adjusting	52	126.1+
Molding in situ	264	31+
Multi-room or multi-level	52	234+
Non-rectangular	52	236.1+
Panel or facer	52	474+
Prefabricated	52	
Subenclosure	52	79.1+
Railroad car roof construction	52	45+
Refrigerated	62	259.1+
Shipbuilding	114	65R+
Static mold	249	13+
Stepped or stair	52	182+
Synthetic resin component	52	309.1+
Tents and canopies	135	
Toy	446	476+
Construction	446	108+
Design	D21	114
Ventilation	98	29+
With terranean relationship	52	169.1+
Work holder positions in installed location	269	904*
Bulb		
Discharge tube	220	2.1 R+
Electronic device envelope	220	2.1 R+
Making of glass	65	
Expansible chamber device	92	92
Fluorescent lamp	313	484+
Gas or vapor lamp	220	2.1 R+
Incandescent lamp	220	2.1 R+
Reflector	313	113+
Planters	111	909*
Radio tube	220	2.1 R+
Bulkhead		
Compartment and	114	78
Door and	114	116+
Bulldozer		
Blade	172	701.1+
With vehicle mount	172	811+
Cab	180	89.12
Design	D15	23+
Bullet	D22	116
Bullet (See Ammunition; Projectiles)	102	501+
Making	29	1.22+
Resistant		
Tanks	220	415+
Setting	86	43
Speed electrically measured	324	160
Timing electrically	324	178+
Timing recorder	346	38
Bulletin Board	D19	52+
Frame	D19	54
Bulletproof	428	911*
Apparel, guards and protectors	2	2.5
Plural layers	428	
Walls and panels	109	78+
Bulls Eye Liquid Level Gauge	73	331
Bumper		
Automobile	293	102+
Guards	293	142+
Design	D12	167+
Fluid	293	108
Horizontal bar	293	120+
Lights	362	82
Closure checks	16	82+
Railway car end	213	220+
Railway car stops	104	254+
Release of closure latch	49	364
Vehicle, design	D12	163+
Buna T M (See also Synthetic Resin or Natural Rubber)	526	339

	Class	Subclass
Bunching (See Bundling)	493	386
Vegetable end cutting combined	99	635
Bundle		
Fuel	44	541
Bundle Formers		
Hay	56	401+
Bundling	100	
Compacted trash	428	2
Compression	53	523+
Plug tobacco	131	112+
Raking and	56	341+
Wood splitting and	144	192
Bung		
Extractors	81	3.7+
Plug type closure for dispenser	222	544+
Wooden barrel	217	98+
Bunk		
Bed	5	9.1+
Log bunks for logging truck	105	160
Buns	426	556+
Bunsen Burner	431	354
Buoy	441	1+
Design	D10	107
Signal	116	107+
Submarine escape	114	323+
Submarine marker	114	326+
Buoyancy Motors	60	495+
Tide or wave	60	497+
Buoyant Devices		
Convertible, inflatable	441	125+
Life preservers	441	80+
Garment type	441	88+
Seats	441	126+
Bur		
Dental	433	165
Pipe reamer	408	227
Removers		
Horseshoe making	59	59
Textile fiber	19	84
Bureau	312	
Baggage convertible	190	3+
Plural		
Burette, Chemical Apparatus	422	100
Chemical analysis, apparatus	422	82.5
Graduated container	222	158
Measuring vessel	73	426+
Burgee	116	173+
Burglar		
Alarm		
Combined with other function	116	6+
Electric	340	541+
For vehicle	340	426+
Electric approach operated	340	541+
Electric switch for	200	
Indicator combined	116	5
Mechanical	116	75+
Trap	43	59
Burglarproof Structures	109	
Burial		
Body preparation	27	21.1+
Coffins	27	2+
Casing portable	27	35
Garment	2	64
Shoes	36	8.2
Vault	52	128+
Vault design	D99	1+
Burner (See Furnace; Heat)		
Bunsen	431	351
Cap	431	129+
Cap	431	144+
Carbon black apparatus	422	150+
Control	431	18+
Design	D26	114
Disinfecting material preparation	43	127+
Dispensers having	222	113
Extinguisher	431	129+
Extinguisher	431	144+
Automatic control	431	33+
Flame protector with	362	376
Flashback prevention	431	22
Flashback prevention	431	346
Fluid distributor for	239	
Fluid fuel	431	
Distributor for	239	
Stoves	126	
Insect destroying	43	144
Light distributor with	362	93
Lime light	431	347
Distributor for	239	
Making	29	890.2
Nozzle	239	

	Class	Subclass
Operation process	431	2+
Pilot with	431	278+
Pot type	431	331+
Pyrotechnic single use	102	335+
Rotary	239	214+
Solid fuel	110	
Stoves	126	
Tobacco smoking devices	131	330+
Toilet closets having	4	111.4+
Torch	239	398+
Torch	141	345
Capillary mass	431	327
Welding	239	398+
Welding	228	
Wick	431	298+
Burning		
Electric heater for	219	227+
Insulation from wire	81	9.4
Lime, cement, etc. kilns	432	
Burning in	29	89.5
Burnishing (See Polishing)		
Electrolytic coating combined	204	36
Implement		
Mutilating eraser combined	7	124+
Shoe	12	104+
Machines	29	90.1+
Paper tube	493	467
Photograph	354	351+
Shoe	12	70+
Wearing in	29	89.5
Burr (See Bur)		
Bus (See Land Vehicles)		
Baggage rack ladders	182	127
Bar (See conductor, electric)	174	
Attachment means for switches	337	191
Distribution or control panels	361	355
Duct	174	50
Electrical housings for	361	341
Electronic service distribution	361	361
Printed circuit board	361	407
Design	D12	84
Destination signs	40	446+
Motor changed	40	470+
Bushing	16	2+
Abrading processes	51	290+
Applying and removing		
Couplings to conduits	29	237
Grommets	227	55+
Processes	29	428+
Pulling or pushing type	29	244+
Resilient bushings	29	235+
Tube and coextensive core	29	234
Bearing type	384	276+
Bung wooden barrel	217	113
Glass dispensing orifice	65	1+
Glass filament intrusion	65	325+
Insulating	174	152R+
Arcing or stress distributing	174	142+
Fluid or vacuum containing	174	31R
For antenna	174	151+
In line insulator	174	167
Railway wheel	295	35
Wooden barrel	285	130
Business		
Computer controlled, monitored	364	401+
Forms		
Carbon copy sets	282	28R
Carbonless copy sets	503	200+
Printed forms	283	
Teaching	434	107+
Bust Support (See Brassiere)	450	1+
Bustle	2	210
Design	D 2	699
Butadiene (See Diolefin Hydrocarbon)		
Liquid polymer	585	507+
Solid polymer (See synthetic resin or natural rubber)		
Synthesis of	585	601+
By cracking	585	615
Butchering	17	
Fowl	17	11+
Hog scalders	17	15
Hog scrapers	17	16+
Marine animals	17	53+
Meat tenderers	17	25+
Processes	17	45+
Sausage stuffers	17	35+
Supports & shackles	17	44+
Butter		
Apparatus	99	452+
Compositions	426	

	Class	Subclass
Cutters	83	
Design	D 1	
Dish		
Cooled	62	457.6
Design, general	D 7	502
Design, refrigerating	D15	89
Fruit	426	616
Mold	249	
Cutter combined	425	289+
Packaging	426	392+
Peanut	426	633
Preservation	426	317
Processes manipulative	426	
Worker	99	452+
Butterfly		
Resonator	331	95
Tuner	334	67
Valve, fluid handling, unbalanced	137	484
Valve, rotary	251	305
Button	24	572+
Abrading cleaning shields	51	265
Apparel, design	D11	222+
Art collection	425	801*
Attaching		
Processes	2	265
Sewing machine	112	104+
Bell buttons mechanical	116	172
Design	D11	222+
Lapel	D11	95+
Eyelet and rivet setting	227	55+
Fabric	24	92
Fastener making	29	4
Fasteners	24	90R+
Feeder	227	31
Sewing machine	112	113
Fluorescent and phosphorescent	250	462.1
French cuff button	24	102FC
Hinged leaf	24	97+
Hole (See buttonhole)	24	659+
Cutters	30	118+
Marking guides	33	575+
Hook	D 2	643
Loop	24	660
Making	79	
Making, metal	29	4
Metallic	79	3+
Pearl & composition	79	6+
Setting machine	227	55
Shank buttons	79	2
Strap or cable button	24	114.5
Switch	200	329+
Telegraph key	178	110
Whip	231	6
Work supports	79	18
Buttoner	24	40
Design, shoe	D 2	643
Machine for shoes	12	58.3
Buttonhole	24	659+
Cutting	83	905*
Hand tool	30	118
Marker	33	575+
Sewing machines	112	65+
Processes and seams	112	264.1
Shoe closures	36	52
Stitch machine	112	157
Workholder in making	269	
Buttress, Vertical Wall	405	112
Butyl Alcohol	568	840
Fermentative preparation	435	160+
Preparation by olefine hydration	568	895+
Preparation from alkyl halides	568	891+
Preparation from alkyl sulfates	568	888
Butyn T M	560	49
Buzzers		
Electric	340	392+
Toys		
Noisemaking tops	446	213+
Sounding	446	397+
Spinning and whirling	446	236+
BX TM Cable	174	109
Fitting	174	50+
Staples	411	457+
Driving implement	227	140+
Insulated	174	159+
By Product Recovery (See Source Material or Material Recovered)		
Byte	370	83
Assembly and formatting	370	99

C

	Class	Subclass
C-clamp	D 8	73
C-vinyl Aromatic Hydrocarbon		
Synthesis	585	435

	Class	Subclass
Cab		
Closure operators		
Door and window	49	72
Window	49	324+
Locomotive		
Dust guards	98	28
Ventilation	98	3
Window condensation preventers	98	91
Motor vehicle body	180	89.1+
Railway	105	456
Signal or train control	246	167R+
Block signal systems	246	20+
Drawbridge protection	246	119
Grade crossing track protection	246	117
Automatic	246	115
Switch stands	246	394
Train dispatching central signal	246	4
Train dispatching train order	246	6
Squirts	239	174
Cabin and Stateroom		
Pressurized	98	1.5
Aircraft structure combined	244	59
Heated and cooled	165	15
Ship	114	71
Furniture arrangement	114	189
Toy construction logs	446	106
Cabinet	312	
Bath tubs	4	538+
Bed combined	5	2.1
Card or sheet magazine	312	50
Card or sheet retainer	312	183+
Cooking stove	126	37R
Dark cabinet	354	307+
Dispensing (See dispensing)		
Escutcheons	70	452
File	312	183+
Instrument sterilizing	422	
Ironing table support	108	33+
Joint of sectional cabinets	312	111
Light combined	362	133
Light etc application to body	128	371+
Locks for	70	78+
Metallic receptacle spaced wall	220	445+
Music	84	DIG. 17
Office machines	D18	
Outside players for music instruments	84	108
Phonograph type	312	8
Console type	D14	175+
Record holders	312	9+
Photography fluid treating	354	297+
Radio type	312	7.1
Design	D14	188+
Receiver	455	347+
Transceiver	455	90
Transmitter	455	128
With antenna	343	702
Safes	109	
Combined	109	23+
Convertible	109	22
Incidentally movable	109	45+
Plural compartment	109	53+
Shields and protectors	109	49.5
Supports and mountings	109	50+
Wall and panel structure	109	58+
Show case	312	114+
Shower bath	4	596+
Design	D23	283
Sign exhibiting reel	40	515
Sound or video recording etc instrument	369	75.1+
Slot for disc insertion	369	77.1+
Structure	312	
Television	358	254
Toy	446	482
Type case accessory	276	44
Wall or ceiling mountable	312	242
Cable and Cable Making	57	
Advancing material of indeterminate length	226	
Block and tackle	254	389+
Bridge suspension	14	22
Burglar alarm cable controlled	116	94
Button		
Strap chain cable attachment	24	114.5
Supports pipe or cable	248	49+
Supports suspended	248	317+
Car (See traction below)		
Clamp		
Bridge suspension	14	22
Cord and rope	24	115R+

	Class	Subclass
Pipe or cable	248	49+
Clip (See cable clamp)	104	200+
Coaxial	333	239+
Composite plastic cable molding	425	500+
Compound die expressed plastic	425	113+
Concentric conductors	174	102R+
Connector	403	
Container for	206	389+
Conveyor		
Hoist linear traversing	212	76+
Ship coaling type	212	72
Material or article handling	414	
Power driven	198	
Endless	198	804+
Endless single cable	198	643
Store service systems	186	14+
Cutting	30	
Fishing line	43	43.12
Ships anchor cable	114	221R
Well cable	166	54.5+
Well torpedo cable	89	1.14
Drum and cable mechanism	74	506
Antifriction	254	901*
Slip clutch	254	903*
Specific material	254	902*
Wave motion responsive actuator	254	900*
Electric	174	
Furnace, cooling of	174	15.7
Loaded	178	45+
Telegraphy	178	63R
Thermal responsive circuit breaker incorporated	337	415
Welding, cooling of	174	15.7
Electric conductor making and or joining	156	47+
Electrical assembly	29	755
Emergency elevator	187	71
Fairing	114	243
Gear drum and cable mechanism	74	505
Grip (See clamp above)		
Cable driving stops railway	104	239
Car control vehicle fender	293	3
Motor placement railway rolling stock	105	134
Guides (See guides, cable line)		
Hauling or strand placing	254	134.3 R+
Hoisting		
Apparatus for hauling or hoisting load	254	264
Crane overhead travelling cable driven	212	205+
Crane rotary drive	212	246
Elevator car brakes and catches	187	81+
Grapples pivoted jaws separate cable actuator	294	111
Hook tackle	294	82.11+
Inclined elevator brake	187	13+
Ladders extension	182	207+
Ladders track type	182	63+
Lever actuator pushing and pulling	254	127+
Linear traverse	212	76+
Material or article handling		
Combined carriers endless	414	564+
Elevator	414	592+
Traversing type	414	560+
Pushing and pulling implements or apparatus	254	264
Ship coaling type linear traverse	212	72
Testing physical	73	158
Trucks	254	4R
Vehicle body lifter	254	47+
Impregnating	118	
Apparatus	118	
Indefinite length electrical conductor	156	47+
Metal casting, continuous	164	462+
Joining wire	140	111+
Joints and couplings (See joints, pipes, shafts, rods, plates)		
Laying apparatus	405	154+
Protector from abrasion	174	136
Rail	104	112+
Railway	104	240
Switch cable and pully connection	246	332
Releaser	187	72
Remote control	74	500.5+
Stop		
Drum automatic control slack cable	254	272
Ship	114	199+
Traction railway	104	239

	Class	Subclass
Submarine		
Loaded	178	45
Support for pipe or cable	248	49+
Bracket	248	65+
Suspended	248	58+
Table equalizers	108	87
Tension device	254	199+
Tension equalizer	187	1A
Tension reliever	114	215
Tightener belt	474	101+
Towing		
Land vehicle train	280	480
Marine propulsion	440	34+
Ship	114	242+
Traction		
Closure operator	49	347
Hauling apparatus	254	264+
Lever single throw	254	127+
Linear hoists	212	76+
Linear hoists ship coaling type	212	72
Overhead cranes	212	214+
Railway rails	104	112+
Railway rails combined with rigid	104	87
Railway rolling stock suspended	105	151
Railway traction	104	173.1+
Railway turntable actuator	104	39
Rotary cranes	212	246
Store service systems	186	14+
Tube or rod splicing	425	802*
Twisting wire	140	149
Underground conduits for pipes or cables	138	105
Winding & reeling	242	
Wireworking article making or forming	140	71R+
Cable Tool Rig	173	81+
Cacao	426	631+
Cachets	604	403+
Cacodyl Compounds	556	70+
Cadmium		
Alloys	420	525
Batteries	429	
Compounds	23	
Electrolytic coating with	204	50.1
Electrolytic synthesis	204	114+
Fluorescent or phosphorescent compositions containing	252	301.6 R
Misc. chemical manufacture	156	DIG. 92
Organic compound containing	556	118+
Pigments fillers or aggregates	106	452
Pyrometallurgy	75	668+
Combined with hydrometallurgy	75	416+
Cafeteria		
Coffee service unit	D34	14
Tray dispenser unit	D34	14
Caffeine	544	274
Extraction from coffee	426	427+
Cage		
Animal mouth guard	119	133
Bank protection	109	10+
Bearing		
Ball	384	523+
Making	29	898.64
Radial	384	512+
Thrust	384	623
Bird and animal confining	119	17+
Design	D30	114+
Fish and animal traps	43	58+
Design	D22	121
Material handling		
Hand type	294	26.5
Hoist line type	294	68.1+
Mills	241	188R
Plant supports	47	45
Queen bee	449	28
Shower bath spray	4	601+
Solid fuel grate	110	294
Water cooled	122	373
Type antenna	343	896
Washing apparatus having	68	
Caisson		
Box or open	405	11+
Diving bell type	405	185+
Ordnance	89	40.7
Pressurized	405	8+
Shaft lining	405	133
Ship raising	114	49
Ship raising salvage	114	53
Ship repair	405	12
Shoring, temporary	405	272
Cake		
Compositions	426	552+

	Class	Subclass		Class	Subclass		Class	Subclass
Cover	312	283+	Structure	313	359.1	Stock material	428	919*
Cutter or server	D 7	673	Cam (See Apparatus Using)	74	567+	Tents	135	87+
Design	D 1	129+	Actuated or operated	74		Warship	114	15
Dish or cover	D 7	610	In bolt to nut lock	411	272	**Camp and Camping**		
Enfolding with enfolding with cloth			**Camber**			Beds	5	112+
before	100	101	Correction			Bodies land vehicles	296	156+
Inedible	D11	184	Bending for	72	704	Chairs folding	297	16+
Pans	249	DIG. 1	Bending three point jacks	72	386+	Ground mat	5	417+
Stripping cloth from cake after			Forging for	72	704	Kit combined or convertible to table.	190	12A
pressing	100	298	Vehicle running gear spring	280	688+	Lunch kits	206	541+
Trimmers	100	94+	Vehicle running gear stub axle	280	661	Sleeping bag	2	69.5
Calciferol	552	653	Test wheel gauge	33	203	With bed structure	5	413
Calcimines (See Kalsomines)			**Cambric**	139	426R	Stools folding	297	166
Calcining			**Camel**			Stoves cooking	126	25R+
Furnaces	432		Ship raising	114	49	Stoves heating	126	59
Portland cement processes	106	739+	Salvage	114	53	Tables convertible baggage	190	12R
Calcium (See Alkali, Earth Metal)	75		**Cameo Molds**	249	104	Tent	135	87+
Carbide	423		**Camera**	354	288	Heater	126	59
Acetylene generator carbide feed.	48	38+	Animated cartoon	352	87	Trailer design	D12	101+
Cartridges	48	59+	Carrier or holder	224	908*	Vehicle design	D12	100
Compounds inorganic	423		Hand	294	139	**Camphene (See Alicyclic Hydrocarbon)**	585	350
Calculator (See Computer)	235	61R+	Cases	206	316.2	Terpene isomerization	585	355
Design	D18	6+	Cinematographic	D16	205+	**Camphor**	568	339+
Instruments simulating	33	1SB	Compound lens system with	350	502	**Camphoric Acid**	562	504
Specific gravity test	73	443	Copy making, eg mechanical			**Camping (See Camp)**		
Vehicle controlling	364	424.1+	negative, eg mechanical	430	951*	**Camshafts**	74	567
Calendar	40	107+	Copying	355	18+	Drive means internal combustion		
Clocks	368	28+	Design	D16	200+	engine	123	90.31
Coin controlled	40	107+	Electronic still	358	909	Overhead location	123	90.27
Design	D19	20+	Scanned semiconductor matrix	358	213.11+	**Can (See Container)**		
Desk article combined	40	358	Exposure control, automatic	354	410+	Beer, pop, or food	206	139
Holder design	D19	20+	Film winding spools	242	71	Carrier by hand	220	
Pads	40	121	Motion picture	242	210	Crushers	100	902*
Printed matter type	283	2+	Flashlight	354	126+	Presses in general	100	
Watches	368	28	Focusing, automatic	354	400+	Cutting to scrap	83	923
Calender			Lens mount	354	286	Design	D 9	
Roll	29	110+	Angularly adjustable camera front			Filling and closing	53	266R+
Paper manufacture	100	161+	or lens mount	354	189	Insulating jacket	220	903
Textile	38	63+	Lucida	350	121	Labeling machines	156	
Calendering			Mammography study	128	915	Leakage tester	73	40+
Cloth finishing	38	52	Microscope	350	511	Discarding type	73	45+
Laundry	38	44+	Motion picture	352		Making		
Paper	100	161+	Film winding	242	179+	Body straightening	72	367+
Plastics	425	363+	Multiple exposure on single plate	354	120+	Fusion-bonding apparatus	228	
Calesthenic Wands	272	93+	Obscura	350	122	Metal cap preparing machine	413	56+
Calibrating			Panoramic	354	94+	Sheet metal container making	413	
Battery calibration	136	290*	Periscope	350	541	Metal and miscellaneous	220	
Electrophotographic copier	355	203	Photographing internal body organs			Milk	220	
Instrument	73	1R+	ultrasonically	128	660.1+	Milking machine	119	14.46
Measuring and test instrument	73	1+	Reflex	354	152	Nuclear fuel containing	376	409
Radar systems	342	165	Reproducing	355	18+	Openers	30	400+
Radiant energy device	250	252.1	Schmitt			Closure combined	220	260
Caliper	33	783+	Lens and mirror	350	443	Cutter type combined	7	152+
Assorter combined	209	602+	Motion picture	352	69	Design	D 8	33+
Brake	188		Television	358	209+	Electric	30	401
Bicycle	188	24.11+	Shutter	354	226+	Key holder combined	D 3	64
Disc type	188	71.1+	Automatic control	354	456+	Other tool combined	7	151
Design	D10	73	Between lens	354	233	Receptacle puncturing and closing	220	278
Gauge	33	501.7+	Mechanical test	73	5	With dispensing features	222	81+
Calk			Optical test	73	5	Packing in portable receptacles	53	531+
Forming on horseshoes	59	65	Timer	354	267.1+	Processors	99	359+
Combined machines	59	39	Stereoscopic	354	112+	Resilient wall	222	206+
Horseshoe calk making	59	66+	Submarine motion picture	352	132	Testers leak detector	73	52
Horseshoe type	168	29+	Supports	354	293+	Vacuumizing & sealing methods	53	403+
Sharpener	51	172	Tripods	248	163.1+	Valves for conveying		
Tool	168	46	Telescope	350	557	Bin charge or discharge	414	288+
Shoe device	36	67R	Television	D16	202	Chute retarders	193	32
Tire antiskid device	152	229+	Time lapse	352	84	Chute switches	193	31R
Calking			Tripod	354	293+	Cooker inlet or outlet	99	366
Compositions	106		Twin lens reflex	354	151	Horizontal axis rotary conveyor	198	670
Dispenser	222		Ultrasonic scanning	128	660.1	Horizontal axis spiral	198	778
Gun	222	326+	Blood	128	660.2	Rotary conveyor	198	803.1
Edge sealing laminated glass	156	107	Flow	128	661.8	Thrower	198	642
Implements	81	8.1	Breast	128	915	Washers and cleaners		
Metal packing	72	462+	Mammography	128	915	Brushing machine	15	70+
Pointers for masonry	15	235.3	Pressure	128	660.2	Liquid contact	134	
Ship calking	114	224	Temperature	128	660.2	With valved outlet	222	
Metal working	72	462+	Therapy	128	660.3	**Canal**	405	84+
Methods	114	86	View finders	354	219+	Amusement park	104	73
Packing fibrous material	81	8.1	Geometrical	33	290+	Boats		
Putting devices	425	87	Motion picture	354	219+	Form	114	60
Putting devices	425	458	Perpendicular to objective lens	D16	217	Structure	114	70
Seaming ships combined	114	86	**Camisole**	D 2	23	Locks	405	85+
Calling Card Receiver	D 6	449	**Camouflage**			**Canceler**		
Calliope	84	330	Canopies	135	87+	Fare box ticket puncher combined	232	8
Callus Culture	435	240.48+	Coat hunters	2	94	Stamp		
Calomel	423	491	Coating composition	106		Barrel bung	217	114
Medicines containing	424	645	Coating non-uniform	427	256+	Hand	101	371
Calorimeter	374	31+	Illusion	273		Machine	101	233+
Calutron			Including open mesh	428	255+	**Cancer Treatment**		
Mass spectrometer	250	281+	Ordnance	89	36.1	5-flurouracil	544	313

	Class	Subclass
Candelabra	431	295
Electric	362	410+
Design	D26	24
Candle	431	288
Candlestick	D26	9+
Compositions for making	44	275
Design	D26	6
Electric lamp	362	157+
Electric socket simulating	200	51.1
Extinguishers	431	144+
Forming	425	803*
Fumigants	424	40
Holders	431	289+
Christmas tree	362	249+
Design	D26	9
Imitation	362	810*
Lanterns using	362	161
Lighters (See igniter and ignition)		
Lighting imitation of	362	810*
Making		
Casting or molding apparatus	425	117+
Casting processes	264	271.1+
Dipping processes	427	442+
Physical and ornamental structure	431	126
Shade or bowl supports for	362	447
Simulated	431	125
Snuffers	431	144+
Design	D29	2
Miners	7	104+
Candling	356	52+
Candling, instrument	D10	48
Candy		
Coating apparatus	118	13+
Compositions	426	
Cutting machine	83	
Design	D 1	127+
Kneading-pulling equipment	366	70
Making, apparatus	99	450.1
Making, processes	426	
Cotton candy	425	9
Preservation	426	321+
Stick, container	493	943*
Cane	135	65+
Design	D 3	5+
Detonating toy	446	402+
Fabrics	139	424
Gun combined	42	52
Light combined	362	102
Locks for cane racks	70	59
Racks	211	62+
Simulated umbrella case	135	18
Stand convertible	248	155
Stand or stool converting to	248	155+
Therapeutic appliers combined	128	394
Umbrellas combined	135	17
Whip convertible	231	3
Canister		
Design	D 7	612+
Gas separation	55	
Projectile	102	507+
Canning		
Cooking and subsequent	99	356
Cooking filled receptacles	99	359+
Filling and closing	53	266R+
Food preserving apparatus	99	467+
Food preserving processes	426	392+
Hermetic	426	392+
Cannon	89	
Toy	42	55
Cannula	604	239
Attaching means	604	240+
Medicator combined	604	187+
Structure	604	239
Canoe	114	347+
Chair	114	363
Design	D12	302
Paddles	440	101
Sails	114	103+
Canopy	135	87+
Bed	5	414+
Camp	5	113
Hammock and canopy with		
common support	5	128
Hammocks	5	121
Body or belt attached	224	186
Building light type	350	258+
Cabinet type	312	3+
Coffin	27	9
Design	D25	56
Flexible panel combined	160	19+
Frames	135	101+

	Class	Subclass
Light shade	362	351+
Light support	362	404+
Padlock shields	70	56
Reflector	362	341+
Rigid type	52	74+
Support type	248	345
Tent-type design	D21	253+
Cant Hook	294	17
Canteen		
Design	D 3	30.1+
Flow controller or closure	222	544+
Jacket or spaced wall	220	415+
Jacketed dispensing type	222	131
Kit with	206	547
Separable cup or funnel	141	379+
Cantilever		
Bridge	14	7+
Building	52	73+
Design	D25	61
Spring	267	41
Canting Element in Bolt to Nut Lock	411	274+
Canvas		
Awning stripe	139	417
Fabric	139	426R
Shoes rubber sole	36	9R
Caoutchouc (See Rubber)		
Cap		
Ammunition	102	204
Making	86	10+
Apparel	2	195+
Bathing	D 2	68
Bathing design	D 2	510+
Design (See headwear)	D 2	509+
Protective against blows	2	410+
With eye shield	2	10
With light support	362	106
Beading and crimping	72	102+
Blasting	102	275.12
Bottle seal	425	809*
Brush and broom type	15	175
Burner		
Extinguishing	431	146
Chimney		
Coping	52	244
Cowls	98	61+
With valve	98	59
Closure		
Auto radiator or gas tank	D12	197+
Bottle	215	200+
Depressible oil cup cap	184	89
Design	D 9	435+
Flexible bag	383	80
Inflation stem air & dust cap	138	89.1+
Inflation stem air cap	138	89.3
Inflation stem dust cap	138	89.4
Inflation stem dust cap and tool		
combined	152	431
Lining apparatus	156	
Lining combined with metal		
working	413	
Lining process	156	262
Making from sheet metal	413	8+
Apparatus	413	56+
Removers	81	3.7+
Screw type	220	288
Slip type	220	352
Soldering to can	228	
Coating implement with material		
supply combined	401	
Column	52	301
Detonating toys	446	398+
Mechanical projector combined	124	2
Pistol simulating	42	54+
Electric fuse terminal	337	248
Fastenings for saw teeth	83	840+
Fountain pen bifurcate nib-type	401	243+
Gas or radiator	D12	197+
Insulator	174	188+
Lock	70	158+
Making		
Ammunition	86	10
Die shaping sheet metal	72	343+
Spirally grooving metal	72	105+
Mast	212	266
Nail screw and nut		
Nut caps	411	429+
Pile driving	173	128+
Pile protector	405	255
Pipe vented	98	122
Radiator		
Ornament	428	31

	Class	Subclass
Ornament bird or insect	40	411+
Removing	29	245
With thermometer	209	684
Spinning machines	57	74
Strand twisting and laying	57	127
Surfacing	52	465+
Tower	52	649
Typewriter key	400	490+
Ventilating	98	122
Wheel hub	301	108R
Whip	231	6
Capacitance Interelectrode Testing		
Electronic Tube	324	409+
Cathode ray tube	324	404
Capacitor (See Condenser; Electret)		
Air dielectric	361	326+
Design	D13	125
Neutralizing	361	212+
Capacity Measuring Bridge	324	680+
Cape Body Garment	2	88+
Design	D 2	179+
Combined	D 2	180+
Combined with dress	D 2	89+
Combined with evening dress	D 2	53+
Combined with suit	D 2	87+
Capillary		
Active agents (See wetting)	252	351
Heat exchange pipe	165	104.26
Tube, refrigeration using	62	511
Capotasto	84	318
Capping		
Bottles	53	287+
Nails and screws	10	156
Screw closure applying		
Soldering iron having guide	228	25+
Capric Acid	260	413+
Caprylic Acid	260	413+
Capsaicin, in Drug	514	627
Capstans	254	266+
Capsule (See Container)		
Composite stock material with liquid	428	321.5
Filling and closing	53	266R+
Making by dipping	425	269+
Making, X-art	425	804*
Medicine vehicle	424	451+
Microcapsule (See microcapsule)		
Nonmetallic collapsible wall	222	107
Sheet metal making	72	347+
Stock with liquid	428	321.5
Car (See Vehicle)		
Airship	244	30
Bodies and tops	296	
Control		
Electric railway external	104	295+
Elevator	187	28+
Elevator brakes and catches	187	73+
Elevator fluid governors	187	68+
Railway switches and signals	246	
Vehicle fenders combined	293	2+
Derailer railway	246	163
Car attached	104	261
Elevator or hoist	414	227+
Dumb waiter type	187	5
Superposed	187	16
Fire escape reciprocating moving		
rope	182	12+
Fire escape reciprocating stationary		
rope	182	189+
Fire escape vertically movable	182	40
Hand	105	162
Heated and cooled	165	42
Automatic control	165	25
Land vehicle	280	
Loading or unloading	414	373+
Railway unloading using tractor-		
trailer	414	333
Self	414	467+
Train of unloading vehicles	414	339
Motor	180	
Propulsion (See electric)		
Elevator	187	17+
Railway	105	
Railway cable crossing conduit	104	187
Railway cable curves	104	190
Railway systems	104	287+
Road vehicle	180	
Store service	186	
Pusher		
Implement	254	35+
Railway hand crank	105	90.1
Railway	105	

	Class	Subclass
Cell (continued)		
Jail	52	106
Leeching or diffusing	127	7
Photo electric	136	243+
Queen bee	449	8
Storage battery	429	149+
Support for removable cell	429	96+
Cellar	52	169.1+
Cello (See Violin)	84	274
Cellophane		
Compositions	106	168
Making article from	493	
Cellosolve T M	568	678
Cellular (See Porous)		
Glass body making	65	22
Glass composition	501	39
Laminate	428	304.4+
Forming by spaced sealing	156	290+
Magazine type		
Cabinet	312	97.1
Dispenser	221	69
Plant or animal material, polymer ctg or derived from (See synthetic resin or natural rubber)		
Polymer (See also synthetic resin or natural rubber)	521	134
Spring mattresses	5	475+
Celluloid T M	106	169+
Cellulose and Derivatives	536	56+
Compositions	106	163.1+
Esters	536	58+
Compositions	106	169+
Ethers	536	84+
Compositions	106	169+
Fermentation of	435	179
Fermentative treatment	435	277+
Hydrolysis to sugars	127	37
Liberation	162	1+
Nitrogen containing	536	30+
Cement (See Adhesive)		
Aluminous cement	106	692
Clinker	106	739+
Clinker	106	739+
Compositions containing	106	638+
Refractory	501	124
Cycle tire	106	33
Earth treatment with	405	266+
Injector	405	269
Magnesia sorel	106	685+
Manufacturing processes	106	739+
Portland roman	106	713+
Roads	404	17+
Making	404	72+
Sheet structure	404	17+
Shoe sole attached by	36	19.5
Cementing		
Implements	15	235.4+
Lining and coating pipes	118	105
Lining and coating pipes	118	125
Lining and coating pipes	118	404+
Lining and coating pipes	425	110+
Lining and coating pipes	425	460
Shoe making	118	410+
Well cementing processes	166	285+
Center		
Drilling work support	408	72R+
Gauge	33	670+
Lathe centers	82	151+
Making in work for mounting	408	199+
Milling tail stock	409	242
Woodworking lathe	142	53
Center of Gravity Measuring	73	65
Center Punch	30	366
Locating	33	670+
Sheets or bars	30	366
Centerboard	114	127+
Centerers	82	170
Centering Devices		
Aligning tool	29	271+
Circular saw on work holder	76	79
Locating	82	170
Mold for arch or tunnel		
Repositioning means for progressive molding	425	63
Using subterranean feature	425	59
With vehicle to move apparatus	425	62
Slide in cylinder	175	325
Well casing type	166	241
Centerless Grinding	51	103R
Propelling work through	51	74R+
Work rest for	51	238R
Central Heating Systems	237	13

	Class	Subclass
Centrifugal Force		
Assorting	209	642
Brake systems fluid	384	168
Brakes	188	180+
Fishing reel	242	287
Centrifugally operated devices		
Abrading tools	51	332+
Bearings	384	135+
Bobbin winding	242	46.5
Cable drum drive or brake	254	267
Fuse arming	102	237+
Governor (See governor)		
Light support	362	384
Lubricator force feed	184	43
Lubricator gravity feed	184	77
Pipe and tube cleaners	15	104.14
Switch	200	80R
Classifying by fluid suspension	209	148
Clutches	192	103R+
Comminution apparatus	241	275
Comminution process	241	5
Conveyers and feeders		
Feeder	209	915*
Pneumatic	406	100+
Structure, rotary	198	803.1
Structure, thrower	198	638
Dispensing	222	410+
Drying apparatus	34	58+
Drying process	34	8
Film formers or spreaders		
Coating machine	118	56
Coating process	427	240
Evaporators	159	6.1
Fluid treatment cleaning and washing textile fibers yarns	68	147+
Forming or casting		
Liquid comminuting	425	6+
Liquid comminuting	425	425
Metal apparatus	164	286+
Metal process	164	114+
Molding apparatus	425	425
Process	264	310+
Pulp	162	384+
Sugar crystals	127	19
Grain hullers	99	519
Gun	124	4+
Lubricators	184	70
Meter fluid flow	73	861.29
Mufflers and sound filters	181	274
Nut crackers	99	571
Pumps		
Gas rotary	415	
Gas screw combined	415	143
Liquid rotary	415	203+
Material agitating	366	263+
Scatterers or sprayers		
Fluid sprayer	239	214+
Fuel burners	239	214+
Gas and liquid contact	261	83+
Rotary	239	681+
Spray fluid actuated	239	381
Separating colloids	252	349
Separator (See separators and separating, centrifugal)		
Spinning apparatus	57	76+
Centrifuge	D24	22
Centripetal Apparatus		
Dispenser rotary central discharge	222	411
Extractors with inward flow	210	360.2+
Pumps rotary	415	120+
Cephaeline	546	96
Cephalosporin	540	215+
In drug	514	200+
Cepharanthine	546	31
Ceramics		
Compositions	501	1+
Fluxes	106	313
Firing apparatus	432	
Firing processes	264	56+
Forming and molding	425	
Uniting combined	264	241
Cerates	424	
Cereals	426	618+
Apparatus	99	
Comminution	241	6+
Design	D 1	
Preservation	426	321+
Puffing, processing	99	323.4+
Ceremonial Robe	D 2	79
Ceresin	208	24+
Cerium		
Alloys	420	416

	Class	Subclass
Inorganic compounds	423	
Recovery	423	21.1+
Pyrometallurgy	75	610
Cermet	75	230+
Certificate		
Design	D19	9
Of deposit	D19	9
Cesium		
Compounds	23	
Preparation	75	745
Pyrometallurgy	75	589
Radioactive compositions	252	625
Cesspool	210	542
Sewage purifying	210	532.1+
Cevadine	546	34
Chafe		
Guard positioned between artice and bearer	224	907*
Prevention on trawling net	43	9.9
Chafe Iron Vehicle	280	161+
Chafing Dish	D 7	355
Chain	59	78+
Bracelets	63	4
Cable stoppers	114	200
Chain saw machines	83	788+
Mining	299	29+
Saw element	83	830+
Design	D 8	499+
Ornamental	D11	13
Door fastener	292	264
End holders	24	116R
Endless chain coupling	464	49
Endless chain pumps	415	5
Endless earth boring tool	175	89+
Fabrics	245	4
Flexible shafts	464	
General purpose	D 8	499+
Grates	110	269+
Jewelry attachments	63	21+
Design	D11	13
Key ring	70	457
Link formation by soldering	72	192
Lubricators	184	15.1+
Making	59	1+
Molds for chains	249	57
Ornamental	59	80+
Design	D11	13
Propellers	416	7+
Marine type	440	95+
Reaction, nuclear	376	
Skid chains	152	231+
Applying tool	81	15.8
Design	D12	154
Sprocket	474	152+
Bicycle	D12	123
Design	D15	148+
Guards	474	144+
Stirrer, chain-type	366	607*
Stitch sewing machine	112	197
Tire	152	208+
Design	D12	154
Traction railways	104	172.1+
Warp threads	28	175+
Watch chain snap hooks	24	905
Welding	59	31+
Electric	219	51
Wheel combined for traction	301	42
Wood mortising cutter	144	72+
Chain Saw	30	381+
Chair	297	
Baggage convertible to	190	8
Barbers	297	68+
Design	D 6	334+
Bath	4	578
Cabinet combined	312	235.2+
Chair lift	D12	52
Chaisse lounge	D 6	360
Design, seat	D 6	334+
Doll or toy	446	482
Dry closet	4	465
Elevator landing	187	75+
Exercising	272	144
Fan combined	416	146R
Folding	297	16+
Folding vehicle occupant seated	280	650
Kinesitherapy	128	33+
Ladders	182	33
Light thermal or electrical application	128	376+
Locks	70	261
Milking stool type	297	175+
Pedestal design	D 6	364

	Class	Subclass
Cereal puffer	99	323.8
Dispensers	222	2
Lock releasing slots	194	247+
Motion picture machine	352	104
Self photograph	354	76+
Telephone systems	379	143+
Voting machines	235	53
Weighers	177	125
Workmens time recorder	346	
Door	16	82+
Closer combined	16	49+
Design	D 8	330+
Gun breach gas	89	26
Gun recoil	89	42.1+
Harness underchecks	54	57
Labels tags and	283	81
Labels, tags, design	D20	22+
Pay check with statement of		
deductions	282	DIG. 1
Protectors		
Embossing	101	24
Fraud preventing coating	427	7
Writer machines	D18	5
Check-out Counter	186	59+
Checkbook Cover Design	D19	26+
Checker Brick	165	9.1+
Checkers	D21	53
Game board design	D21	24
Checkhooks	54	61+
Checking and Unchecking Device	54	70
Underchecks	54	57
Checking Watermarks	356	432+
Checkout Counter	D 6	396+
Checkreins	54	16+
Cheek Plate Railway Draft	213	54+
Cheese		
Compositions and processes	426	582+
Cutter	83	
Cutter and expresser	425	308
Design	D 1	
Making apparatus	99	452+
Packaging	426	392+
Preservation	426	330.2
Presses	100	
Textile		
Fluid treatment	8	155.2
Packages	242	159+
Treating apparatus	99	452+
Cheesecloth	139	426R
Chelate		
Aluminum containing	556	175
Chelation ion exchange	423	DIG. 14
Detergents having agent	252	DIG. 11
Drug compositions	424	DIG. 6
Food or edible material	426	271
Heavy metals containing	556	1+
Printing dye with metal chelating		
group	8	452
Tobacco smoke separator or treater	131	334
Chelidamic Acid	546	299
Chemical Modification, Protein	530	402
Chemical Reaction Heat Producers		
Irreversible reaction	44	250+
Readily reversible reaction	252	70
Chemiluminescent	252	700
Lighting	362	34
Tests involving	436	172*
Chemistry & Chemical (See Particular		
Compound, Composition, or Art)	422	
Analytical and analytical control		
methods	436	43+
Apparatus	422	44+
Carbon compounds	260	
Carbon compounds	560	
Chemical composition		
Crosslinking	252	183.11+
Curing	252	182.11+
Hardening	252	182.11+
Interactive	252	182.11+
Non-interactive	252	182.11+
Vulcanizing	252	182.11+
Chemical manufacture, misc.	156	
Chemical oxygen demand tests, cod.	436	62*
Chemical symbol character	400	900*
Classifying, separating, & assorting		
solids	209	
Cleaning & liquid contact with solids	134	1+
Apparatus	134	42+
Coating apparatus	118	
Coating processes	427	
Printing	101	

	Class	Subclass
Cooling by chemical reaction	62	4
Distillation		
Apparatus	202	
Separatory processes	203	
Thermolytic processes	201	
Dyestuffs	8	
Electrical current producing		
apparatus, product & process	429	
Electrical radiant or wave energy	204	
Fermentation	435	
Fertilizers	71	
Foodstuffs	99	
Inorganic	423	
Measuring instrument	D10	81
Metallurgical	75	
Reactions	376	
Modification of synthetic resins or		
natural rubbers	525	
Modification of textiles & fibers	8	115.51+
Fluid treating apparatus	68	
Manipulative fluid treatment	8	147+
Photographic chemistry, processes &		
products	430	
Physical processes	23	293R+
Refrigeration	62	4
Sterilizers, refrigeration combined	62	78
Synthetic resins (See synthetic resin		
or natural rubber)	520	1+
Teaching	434	298
Chemotherapy Drugs		
5-flurouracil	544	313
Chendeoxycholic Acid	552	551
Chenille	223	45
Misc. manufacture	28	144
Spinning twisting twining		
Apparatus	57	24
Strand structure	57	203
Tufting sewing machine	112	80.1+
Woven pile fabrics	139	393+
Strands	139	395
Cherry Seeder	30	113.1+
Chess Pieces	D21	52
Chest (See Box)		
Cedar	206	278.1
Medicine	206	828*
Protector	2	92
Tool	206	349+
Wind	84	54
Woodworking tool and workbench		
combined	144	285
Chest of Drawers	D 6	432
Cheviot	26	19+
Chevron	2	246
Badge	40	1.5+
Chewing Gum	426	3+
Package	206	800*
Tobacco combined	131	347+
Chicken (See Fowl)		
Chiffon, Foamable Edible Material	426	571
Chiffoniers, Trunks	190	19
Childproof Bottle Closures	215	201+
Chilean Type Mill	241	107+
Chill		
Brake shoes	188	260
Fastener bedstead	5	289
Metal founding	164	371+
Chilling iron processes	164	127
Consumable	164	357
Patterns chill supporting	164	230
Sand mold or core having	164	352+
Static mold having	249	111
Chime		
Bells	84	406+
Clocks	368	273
Electric	340	392+
Musical	84	402+
Design	D17	22
Electric tone type	84	600+
Signal	116	141
Chimney		
Building flue heaters	98	46
Cap, masonry	52	244
Cap, ventilating	98	67
Cleaning	134	1+
Sweep or brush	15	162+
Lamp chimney lantern	362	180
Chimney element	362	312+
Chimney element design	D26	118
Masonry smoke flue	52	245+
Static mold for making	249	17
Ventilation	98	58+

	Class	Subclass
Design	D23	374
Chin		
Rest		
Firearm stocks	42	71.1+
Violin	84	279
Violin combined with support	84	278
Strap		
Hats	132	58
Surgical	128	164
Support		
Corpse	27	25.1
Surgical	128	164
China Oil	260	398+
Compositions	106	244+
Synthetic resin or natural rubber		
ctg (See synthetic resin or		
natural rubber)		
Chinaware	D 7	500+
Display stand	D 6	512+
Vase	D11	143
Chintz	139	426R
Chip		
Breaker	144	243+
Cutting tool with	407	2+
Cutting	83	906*
Making		
Comminuting type	241	83+
Slicer type	144	162R+
Poker	40	27.5
Design	D21	53
Chiropody		
Bandages	128	157
Kinesitherapy	128	33
Pedicuring instrument	132	73+
Shields	128	882
Chisel		
Cold	30	168
Design	D 8	47
Dovetail	144	88
Mortising	144	67+
Sawtooth cutting machine	76	28
Stoneworking	125	41
Tenoning	144	202
Wood turning	142	42+
Pattern guide combined	142	21
Woodworking implement	30	167+
Chitin	423	413
Chloracetic Acid	562	602
Chloral	568	495+
Hydrate	568	844
Chloramine	423	383
Chloramphenicol	564	213
Chloranil	552	308
Chlorate	423	475
Electrolytic synthesis	204	95
Per compounds	204	82+
Oxidizing compositions	252	186.1+
Chlorazene Chloramine T	562	837
Chlorhydrins	568	841+
Hydrolysis of	568	859
Chloride		
Binary halides inorganic	423	462
Biocides containing	424	
Dissolver, sodium chloride &		
potassium chloride	422	902*
Halocarbons	570	101+
Hydrochloric acid	423	481+
Organic acid halides	562	800+
Oxidizing compositions	252	186.1
Salts inorganic	423	462+
Sulfur	423	462
Chlorination		
Hydrocarbon	570	101+
Water purification	210	198.1+
Processes	210	754+
Chlorine	423	500
Biocides containing	424	661
Bleaching with	8	101+
Electrolytic	204	133
Electrolytic synthesis	204	128
Feeder	137	87+
Chlorine Di, -monoxide Oxidizing		
Composition	252	187.21
Chlorocarbonic Acid Esters	558	280+
Chloroform	570	181
Preparation from acyclic halides	570	101+
Chlorophenols	568	774+
Chlorophyll	540	145
Chloroprene	570	189
Preparation	570	216
Solid polymers (See synthetic resin		

	Class	Subclass
Annunciators	340	815.28
Arc lamp cutout or shunt	314	10+
Arc lamp system	314	131
Check control telephone	379	146+
Electromagnetic switching systems	361	139+
Key telegraph	178	101+
Lamp or electrode tube system	315	362
Motor system	318	
Switchboard details	361	332+
Telegraph manipulating	178	75
Telephone	379	414+
Printed	428	901*
Manufacture including etching	156	901*
On organic base	156	902*
Protector		
Lightning arrester	361	117+
Meter	324	110
Retainer automatic telegraph repeater	178	72
Work holder for electrical circuit	269	903*
Circular		
Knitting machine		
Independent needles	66	8+
United needles	66	79+
Looms	139	457+
Saw		
Dressing and jointing	76	48
Metal sawing	83	594
Wood sawing	83	469+
Circumcision Devices	606	118
Cistern	52	169.1+
Cleaners	15	246.5
Hoisting bucket type	294	68.22+
Molds for	425	59+
Sewage purification	210	532.1+
Citazinic Acid	546	296
Citral	568	448
Citric Acid	562	584
Citrus Fruit		
Cutters	30	123.5+
Machines	99	539+
Plants	PLT	45
Civetone	568	375
Clack Valve	251	298+
Clamp & Clamping Device (See Clasp; Fastener)	24	455+
Adjustable	403	DIG. 1
Apparel apparatus	269	
Applier		
Hose clamp	81	9.3
Surgical clip	29	243.56
Binder device releasably engaging aperture or notch of sheet	402	
Book or leaf holder	281	45+
Bracket clamped to support	248	225.31+
Bracket with article holding clamp	248	316.1+
Building components	52	584
Anchor or tie	52	698+
Clasp, clip, or support clamp	24	455+
Design	D19	65
Clip board	24	67.3+
Closure fastener	292	256+
Rod clamp type	292	305+
Cord rope and strand anchors	24	115R+
Bridge cable	14	22
Sash cord	16	205
Dental implement	433	153+
Dam	433	138+
Design	D 8	72
Hose	24	19
Appliers	81	9.3
With support	248	75+
Jar lid	D 9	435+
Locked	70	19
Metal deforming apparatus	72	293+
Moderator structure for nuclear reactor	376	458
Paper fastener	24	67R+
Portable	269	
Railway track fastening	238	338+
Saw setting	76	78.1+
Scaffold binders	24	16R+
Shoemakers	12	103
Static mold	249	219.1
Strand tensioning	242	149+
Surgical implement	606	151+
Design	D24	27
Testing strength of specimen with	73	856+
Woodworking	269	
Clamped Work (See Holders)		
Abrading machine	51	217R+

	Class	Subclass
Cutting & punching sheets & bars	83	
Cutting tables	83	451+
Drawcut	83	642+
Drawcut	83	644+
Pivoted cutter	83	597+
Punching or pricking	83	
Reciprocating cutter	83	613+
Table with	83	
Daguerreotype plates	354	2
Engraving	81	4
Fish dressing	17	70
Fowl supporting	17	44.3
Sewing machine presser	112	235+
Shoemaking	12	
Specimen testing by stress or strain application	73	856+
Vegetable and meat cutting	269	
Vises	269	
Pipe or cable	269	
Wire fabric making	140	51
Wood bending	144	269
Wood turning	142	
Woodworking	269	
Machine	144	278R
Clamshell Bucket	294	68.23+
Clappers for Bells (See Bells)		
Clarinet	84	382+
Clasp	24	455+
Clothespin (See clothespin)		
Design	D32	61+
Cuff holder	24	43+
Design apparel	D 2	200+
Fastener combined	24	326+
For nuclear reactor moderator structure	376	304
General purpose	D 8	382+
Jewelry	D11	86+
Pen or pocket	24	11R+
Pen or pencil	D19	41+
Project-retract means associated	401	104+
Utilitarian, garment use	D 2	200+
Classifying Solids	209	
Cutting or comminuting combined		
Apparatus	241	68+
Cereal processes	241	9+
Fluid applying apparatus	241	38+
Gas applying processes	241	19
Liquid applying processes	241	20
Processes	241	24
Vegetable and meat apparatus	99	510
Vegetable and meat processes	426	518+
Dorr type	209	462
Clathrates (See Definition Note (3))	428	402.2
Clavier	84	DIG. 25
Clavulanic Acid	540	349
Claw Bar Lever Nail Extractor	254	25+
Clay		
Acid activated	502	81
Burning in kilns	264	56+
Cleaner	209	
Comminuting combined	241	
Compositions	501	141+
Abrasive	51	308
Adsorbents	502	407+
Catalyst	502	80+
Catalyst carrier	502	439+
Porcelain	501	141+
Refractory	501	94+
Molding and casting	425	
Pigeon	273	362+
Projector	124	6+
Pipe or tube	138	175+
Shaping processes	264	
Firing ware combined	264	60+
Treatment	501	145+
Chemical	501	146+
Cleaner and Cleaning (See Clearer)		
Abrading file	51	414
Antismut device combined	101	423+
Applicator	15	104R+
Material supply combined (See implements below)	401	
Ash tray combined with cleaning device	131	232
Attachment	15	246+
Curtain or shade	160	11
Electric insulator	174	211
Electrode tip	314	23+
Hand rake	56	400.8+
Harrow teeth	172	90
Liquid level sight glass	73	324

	Class	Subclass
Printing apparatus	101	425
Razor	30	41
Solid material comminuter member	241	166+
Textile knitting apparatus	66	168
Tobacco smoking device	131	184.1
Type casting apparatus	199	62
Typewriting machine	400	701+
Band saw pulley	83	168
Boiler with means for	122	379+
Boot and shoe scrapers	15	237+
Bottle washing machine attachment	15	104.9
Bread and pastry	15	3.11
Bread and pastry	15	3.16
Brushes for, design	D 4	
Brushing scrubbing wiping general cleaning	15	
Burner	431	121
Burner tip	362	458
Button cleaning	51	171
Garment shields	51	265
Carpet	15	3+
Sweeper type	15	41.1+
Cistern	15	246.5
Comb		
Brush machine	15	39
Combined	132	119
Implement	15	104.5+
Combustion means	110	236+
Composition		
Abrasive	51	293+
Detergent	252	89.1+
Flue	44	640
Lubricant combined	252	11
Metal fluxing	148	23+
Soap	252	369+
Soap products	252	368
Conveyor	198	493+
Currycomb	119	88
Design	D30	159
Design	D32	1+
Dispenser	222	148+
Receiver inlet cleaning attachment	141	90
Distillation apparatus carbon removers	202	241
Processes	201	2
Drive belt and chain	474	92
Earth boring means with	175	84
Earth boring tool with	175	313
Electric trolley head	191	62
Electrolysis		
Cleaning or etching	204	129.1+
Coating with cleaning or etching	204	32.1+
Electrophotographic copier	355	296+
Electrostatic	15	1.51+
Engine cylinder	15	104.11
Filter	210	407+
Flue		
Brush and broom	15	162+
Flue attached	15	249
Scraper	15	242+
Fluid handling devices	137	237+
Fluid nozzle	239	106+
Foundry apparatus with	164	158
Fruits		
Processes	426	478+
Fur treatment	69	23+
Gas separator	55	282+
By liquid	55	242+
Electrical precipitator	55	108+
Process	55	96
Ginning saw	19	64
Glass mfg apparatus	65	27
Gun bore	15	104.5+
Cartridge contained	102	442
Gun combined	42	95
Harvesting binder needle cleaner	56	448
Hat cleaning	223	23
Heat exchange apparatus	165	95
Hoof	168	48.1+
Implement	15	104R+
Brush broom or mop with material supply	401	268+
Coated or impregnated with cleaning material	15	104.93+
Container, srapper for	493	942*
Erasing (See eraser)	15	424+
Material supply combined	401	
Padlike, bladelike or apertured with material supply	401	261+
Porous pad with material supply	401	196+

	Class	Subclass
Stationary and with material supply	15	104.92
Stranded fabric mop or mop head	15	229.1+
With fixed handle	15	229.2+
With pivoted handle	15	229.6+
Intestine	17	43
Liquid contact		
Solids apparatus	134	
Textile apparatus	68	
Textile processes	8	137+
Liquid heaters and vaporizers	122	379+
Water tube heater	122	361
Loom cleaning	139	1C
Machines	15	3+
Lights on	362	91
Marine hull	114	222
Metal working with (See metal debris)	29	DIG. 7
Cutter engaging cleaner	29	DIG. 97
Rotating, reciprocating, or oscillating cleaner	29	DIG. 98
Nozzle	239	106+
Oil vaporizing apparatus carbon removers	196	122
Processes	208	48R
Paper making felt & wire combined	162	274+
Pen	15	423
Piano action combined	84	64
Pipe and tube (See apparatus or process using)		
Brushing machines	15	88
Implements	15	104.3+
Liquid contacting apparatus	134	166R+
Processes	134	1+
Tobacco smokers	131	243+
Piston ring groove	15	104.12
Railway		
Snowplows	37	214+
Switch point	246	444
Track	104	279+
Track sweepers	15	54+
Receptacle		
Brush and broom implement	15	164+
Brushing machines	15	56+
Liquid contacting	134	
Wiper implement	15	211+
Refrigeration apparatus	62	303
Sand blast nozzle combined	291	12
Scale removing from metal	29	
Scouring with abradant	51	16+
Solids cleaning apparatus cleaner combined	34	85
Sound record	15	
Spark plug	15	104.11
Steam boilers	122	379+
Stove flue	126	16
Stratifier	209	487
Street		
Air blast or suction	15	300.1+
Flushers	239	159+
Sweepers	15	78+
Teeth	433	216
Textile fiber working combined	19	262
Roll cleaning	15	256.51
Tobacco smoking device	131	243+
Tobacco treatment		
Fluids or fluent material	131	300+
Leaf	131	325
Stemming and cleaning	131	315+
Tool driving or impacting combined	173	57+
Automatic control	173	3
Track sander delivery pipe	291	42+
Vacuum	15	300.1+
Nozzles	15	415.1+
Processes	134	21
Tool driving motor fluid caused	173	60
Valve	137	244+
Vegetables		
Processes	426	478+
Washing-drying machines	D32	6+
Welding apparatus combined	228	18
Well	166	170+
Processes	166	311+
Screen with washing point or shoe	166	157+
Wheel scraper vehicle attached	280	855+
Wood saw	83	169
Clearance Reduction		
Compressor	92	60.5
Regulator	417	274+
Internal combustion engine	123	48R

	Class	Subclass
Clearer (See Cleaner)		
Diggers potato	171	114+
Fork	294	50+
Harvesters	56	314+
Hat making	223	11
Plows	172	606+
Propeller screw	440	73
Railway track	104	279+
Rakes		
Hand	56	400.8+
Horse drawn	56	395
Roadway snow excavators	37	196+
Sifters	209	379+
Textile machinery		
Drawing roll	19	258+
Loom shuttle	139	261+
Spinning	57	300+
Warp beaming	28	173
Cleat		
Insulator electrical	174	157
Bracket for	248	67.5
Ship	114	218
Shoe sole edge gripping	36	7.7
Shoe spike or	D 2	317
Sliding car door	49	472
Traction for wheels	301	43+
Cleaver	30	308
Clerestory	52	18
Clevis	403	157
Ball socket	403	142
Coupler	403	302
Hoist line		
Insulators	174	207
Links	59	86+
Thill couplings	278	52+
Clew	114	115
Clicker	242	304
Climber		
Mountain climbing aids	248	925
Plants		
Rose	PLT	2+
Vine	PLT	54+
Playground apparatus	D21	242+
Pole post or rope	182	133+
Tire attached surface rail climber	301	90
Toy figure	446	314+
Clincher Rim	152	382+
Clinometer	33	365+
Clip		
Applier	29	243.5+
Surgical	29	243.56
Bottle attachment	215	101
Design		
Apparel	D11	200+
Hair	D28	39+
Jewelry	D11	40+
Paper, pen, pencil	D19	56
Inserters	606	139+
Leaf spring	267	53
Making and joining		
Fabric wire	29	243.56
Spring head	29	172
Wire	140	82+
Nose animal	119	135
Paper	24	455+
Paper fastener	D19	65
Pedal crank combined with shoe type	74	594.6
Pen and pencil	24	10R+
Design	D19	56
Project-retract means associated	401	104+
Rafters and	33	DIG. 16
Rail fastenings	238	378
Structural unit fasteners	403	
Surgical	606	151+
Wire fabric	245	3
Clippers		
Clip applying	29	243.5+
Cloth finishing cutters	26	7+
Fur hair removing	69	25
Hair	D28	52+
Hand manipulated cutters	30	
Harvesters	56	
Hat making	223	19
Nails	D28	60+
With hair catching	30	124+
Cloak	2	
Clock	368	62+
Atomic	331	3
Atomic frequency control for	331	3
Cases	224	164+

	Class	Subclass
Clock movement casing	206	18
Design	D10	1+
Cordless electric	310	191
Crystal mounting	29	807
Design	D10	1+
Digital exhibitor	D10	15
Electric motor for	310	162+
Floor standing model	D10	16
Illumination combined	362	23+
Keys	81	122+
Power substitution for electric	307	64
Processes for making	29	177+
Radio	455	344
Radio encased	455	231
Radio or phonograph combined	D14	170+
Radio station selector	334	
Spring assembly	29	228
Staking	29	231+
Supports	248	114+
Testing	73	6
Time clock	D10	41
Transistor operated	318	127
Winding key	D 8	347+
Work control (See time operated and controlled)	368	1
Fuses	102	276
Switch operation	200	35R+
Time locks	70	272+
Clocking Device (See Timers)		
Clod Crusher	172	
Cultivator and roller	172	170+
Harrow and roller	172	170+
Land roller	404	122+
Plow and roller	172	170+
Soil elevator and treater	172	33
With separating after earth working	172	32
Clog, Footwear	D 2	292
Platform heel	D 2	324
Platform sole	D 2	322
Close (See Closer; Closing; Closure)		
Closed Circuit Grinding		
Apparatus	241	80
Fluid applying	241	61
Gas swept grinder	241	52+
Processes	241	24
Cereal	241	10
Gas applying	241	19
Liquid applying	241	20
Closed-die Metal Shaping	72	343+
Forging	72	352+
Closer (See Circuit; Closure; Fly; Operator)		
Door	16	71+
Check combined	16	49+
Design	D 8	330
Fly closer lasting tool	12	113
Paper		
Bag closing	493	186
Box closing	493	52
Slot		
Cable railway	104	194
Lever	74	566
Pneumatic tube	104	161
Closet		
Sanitary	4	300+
Basin combined	4	664+
Bed combined	5	90
Bowl couplings supports and drip collectors	4	252R+
Bowl outlet obstruction remover	4	257
Bowls	4	420+
Cleaning devices	4	257
Closure	4	253
Covers for seats	4	242+
Design	D23	295+
Disinfection	4	222+
Dry closet and furnace combined	4	111.1+
Dry type	4	449+
Dry type with drier or burner	4	111.1+
Flushing tanks	4	353+
Head foot and body rests	4	254
Seats	4	237+
Static mold	249	58
Tub and basin combined	4	663
Ventilation	4	209R+
Warming	126	18
Stove base combined	126	55
Stove pipe heated	126	17
Closing		
Ammunition loading and shell closing	86	25+
Apparel pocket attachment	2	252

	Class	Subclass
Bottom opening hopper door equalizer	298	37
Conductor conduit	191	25
Third rail	191	31
Inkstand (See inkwell)		
Pen, actuated for opening	15	257.75
Metal tube ends		
Welding apparatus	228	60
Paper box or tube		
Folding setting up end closing	493	183+
Receptacle filling and closing	53	266R+
Cooking and filling with cooked food	99	356
Multifolded closure	53	378+
Closure (See also Article Having Closure eg Window)	160	
Adaptable for various sizes	220	287
Apparel		
Buttons and fasteners	24	90R+
Design	D 2	222+
Coat	2	96
Collar	2	141R+
Design	D 2	200+
Flies	2	234
Lacing devices	24	712+
Pocket	2	252
Separable fasteners	24	572+
Shirt	2	128
Shoe	36	50+
Skirt placket	2	218
Skirt waistband	2	219
Sleeves or legs	2	270
Slit	24	437+
Trouser waistband	2	235
Applicator attached	401	126+
Bathtub	4	293+
Bayonet		
Metallic receptacle	220	293+
Wooden barrel bung	217	107
Boiler header	122	360+
Tube	122	364+
Bottles & jars	215	33+
Designs	D 9	367+
Box		
Designs	D 9	414+
Breech	89	17+
Building construction with	52	204+
Bung hole	217	98+
Cabinet		
Foldaway device support combined	312	22+
Inner and outer	312	291
Sectional unit	312	109+
Check	16	82+
Door closer combined	16	49+
Photographic plate holder	354	281
Closet bowl	4	253
Obstruction remover combined	4	255+
Coffin	27	14+
Cooking ovens	126	190+
Disc manufacture of	493	379+
Dispensing	222	544+
Distilling apparatus	202	242+
Fastener	292	
Barrel	217	89
Bottle and jar	215	273+
Breakable neck bottle and jar	215	35+
Burglar alarm combined	116	12+
Cork retaining wire	140	85
Envelope	229	77+
House letter box	232	23
Keepers for	292	340+
Lantern reflector combined	362	374
Letter sheet	229	92.5+
Marine hatch	114	203
Metallic receptacles	220	315
Paperboard box	229	125.19+
Pivoted type closure bottle and jar	215	237+
Railway track joint	238	312
Safes	109	59R+
Sectional metallic receptacle	220	7
Fences	256	
Flexible and portable	160	
Flexible bag closure, misc.	383	42+
Flexible, eg sliding desk top type	206	816*
Floating roof type	220	216+
Foot pedal actuated	220	262
Friction held	220	352
Gas retorts	48	124
Grate		

	Class	Subclass
Fireplace summer front	126	540
Stove draft control	126	160
Hinged metal receptacle	220	334
Horizontal building entry	52	19+
House paint can	220	354
Humidifier	312	31.1
Ice bunker	98	7
Lock	70	77+
Magnetic	220	230
Mail box	232	45+
Mailing sheet	229	92.5+
Manhole	404	25+
Mesh sifter element	209	391
Motor vehicle	49	
Having vehicle system responsive to its position	180	286
Related to theft prevention	180	289
Responsive to vehicle movement	180	282+
With vehicle feature	296	146+
Movable or removable	49	
Oil cup lubricators	184	90+
Operator (See closer)	49	324+
Alarm	116	100
Burglar alarm	116	85+
Cabinet	312	319
Car dumping tilting track	414	354+
Curtain shade screen (See appropriate subclasses in)	160	
Elevator door	187	51+
Furnace door	110	176+
Gate	49	324+
Pen-operated inkwell	15	257.75
Power plant X-art	60	903*
Railway dumping car	105	286+
Railway dumping car control	105	311.1+
Railway passenger car	105	341+
Receptacle and stand	248	147
Receptacle pivotally mounted on rack	211	83
Seat and cover	4	251
Sectional cabinet equalizers	312	110
Showcase	312	139
Stand combined	248	147
Stove door	126	192
Tilting receptacle and stand	248	134
Ornamental, metal receptacle	220	376
Partitions & panels, flexible & portable	160	
Permanent	220	600+
Pipe (tobacco)	131	176+
Pipe or conduit	138	89+
Railway tie end closure	238	104
Protective grill or safety guard	49	50+
Purses	150	118+
Railway rail joint	238	312
Receptacle	220	200+
Bag, flexible	383	42+
Barrels	217	76+
Baskets	217	124
Batteries	429	163+
Bottles and jars	215	200+
Bucket paper	229	910+
Bucket staved wooden	217	76+
Cigar cigarette smoking device holder	131	242
Collapsible material guide combined	222	528
Compartmental letter box	232	25+
Deposit and collection receptacle	232	44+
Dispenser	222	544+
Electrical boxes and housings	174	65R
Electrical outlet and junction box	220	3.8
Envelope	229	76+
Flexible closure	206	816*
Folded blank paper bucket	229	910+
Lantern	362	375
Lubricator valve combined	184	80
Metallic receptacles	220	200+
Milk pail	220	200+
Mixing chamber	366	347
Movable material guide combined	222	531+
Nuclear fuel	376	434
Oil cup	184	88.1+
Paper or web container	493	962*
Paperboard box	229	124+
Purses	150	118+
Rack and closure	211	71+
Sectional for dispenser	222	502
Sifter	209	372
Sliding apertured cap type	222	522+
Spittoon	4	267+

	Class	Subclass
Surface condenser	165	73+
Tube	229	93
Usable as cup or funnel	141	381
Water heater overflow combined	126	384
Wooden boxes	217	56+
Safes	109	64+
Screw-type	220	288
Ship opening	114	201R+
Showcase	312	138.1+
Rack interconnected	312	129+
Tray interconnected	312	127
Sink	4	293+
Soldered or sealed	220	361
Sorter for container closure	209	928*
Static mold	249	204
Stopper type	220	307
Tab tops	220	271
Tie string	493	962*
Time controlled	49	29+
Tire casing	152	515+
Transparent	220	377
Boiler type sight glass	73	330+
Door peep	49	171
Sight glass	73	334
Stove oven door	126	200
Window	49	
Valve	493	962*
Vault pavement type	404	25+
Weatherstrip seal	49	475+
Well	166	325+
Zipper, ie receptacle securement	206	810*
Cloth (See Fabric; Textile)		
Abrasive filled	51	294
Bags	383	
Buttons covered	24	92
Making	79	5
Cleaning or polishing	15	208+
Cutting cloth stock	83	936+
Finishing	26	
Beating	26	25+
Cutting	26	7+
Expanding device for webs	26	71+
Inspecting	26	70
Napping	26	29R+
Pile fabrics	26	2R+
Rubbing	26	27+
Sheet stretching frame	38	102+
Shrinking	26	18.5+
Singeing	26	51+
Stretching or spreading & working	26	51+
Tenter for webs	26	89+
Web spreader	26	87+
Web stretcher	26	71+
Biaxial	26	72+
Fireproof	427	223+
Packages or reels of	242	222+
Purses	150	100+
Reeling and unreeling	242	62
Clothes, Clothing (See Apparel)		
Bed clothing	5	482
Boots and shoes	36	
Design	D 2	264+
Brushes, bristle	D 4	130+
Brushes, non-bristle	D28	9+
Card clothing	19	114
Applying	140	97+
Cleaning machines	D32	1+
Devices for putting on & removing	223	111+
Dryers (See driers drying)		
Handling devices		
Laundry sticks	294	23.5
Tongs	294	8.5
Hanger	223	85+
Holder on chairs	297	190
Knitted	66	169R+
Design	D 2	43
Design	D 2	44+
Label	D 5	63+
Patterns	33	2R+
Pounder	294	23.5
Rack	211	
Bed attachment	5	504+
Sewing methods	112	262.1+
Thermal and electrical treatment	128	379+
Clothesline		
Cleaners	15	256.6
Guide pulley	254	389+
Isolated supports	211	119.1+
Props	248	353
Reels	242	100+

	Class	Subclass
Reels	D 8	358+
Support or attachment	D32	60+
Clothespin	24	455+
Bag design	D32	36
Clasps	24	455+
Design	D32	61+
One piece rigid	24	570+
Pivoted	24	489+
Resilient	24	530+
Track or way guided	24	522+
Wire working making	140	83
Woodworking making	144	9
Clotting (See Coagulating)		
Analysis by coagulometer	422	73
Test with blood clotting factor	435	13
Cloud Chamber	250	335
Clover Huller	460	15
Cloxacillin	540	327
Club		
Carrier for policemans weapon	224	914*
Dumbbell and exerciser	272	122+
Game	273	67R+
Design	D21	210+
Golf	D21	214+
Clutch (See Pawl; Rachet)	192	
Brake and	192	12R+
Coaster brakes	192	6R
Steering by driving	180	6.2+
Crane	212	170+
Eddy current	310	105+
Electric clutch control system	310	92+
Automatic	310	94+
Electromagnetic	310	103+
Fishing reel	242	255+
Freewheeling	192	41R+
Gear combined	192	20+
Lathe carriage feeds	82	132+
Lathe transmissions	82	905
Mechanical movements	74	
Metal rolling mill drives	72	249
Transmission and clutch controls	192	3.51+
Impact delivering	173	93.5
Intermittent grip type movement	74	111+
Exhibiting device		
Electrical control	40	463+
Mechanical control	40	446+
Magnetic		
Automatic electric friction	192	40
Operator electric friction	192	84R
Motor combined	192	
Railway wheel and axle drive	105	130
Reversing mechanisms		
Planers	409	336
Tapping	408	123
Side, in bolt to nut lock	411	296+
Take up	192	110R
Exhibiting device		
Electrical control	40	463+
Mechanical control	40	446+
Transmission combined	192	3.51
Crane	212	170+
Coagulating	252	315.1+
Blood		
Composition for	424	
Properties of fluids determining	73	54+
Surgical	606	111+
Electrical	606	40
Synthetic resin or natural rubber	528	480+
Coal		
Bag or carrier	294	149+
Bin		
Building construction	52	192+
Charging or discharging	414	288
Locomotive tender	105	237
Briquette	44	550+
Buckets or hods	220	2
Cleaner or separator	209	
Coking	201	
Apparatus	202	
Comminutor	241	
Compositions	44	
Conversion to hydrocarbon	585	943*
Cutter	299	79+
Endless flexible	299	82
Rotary	299	89
Feed		
Boiler control	122	449
Boiler water grate combined	122	376
Furnace structure	110	101R+
Locomotive tender combined	105	232+
Metallurgical furnace combined	266	176+

	Class	Subclass
Stokers	110	267+
Fertilizer containing	71	24
Gas making	48	
Liquefaction	208	400+
Locomotives (See locomotive)		
Mineral oil recovery	208	400+
Mining	299	
Ships	114	26+
Shovels	294	49+
Stove	126	
Treating process	44	620+
Coanda and Coanda Effect	239	DIG. 7
Flow affected by	137	803
Coaptation Splints	128	85
Coaptator	606	213+
Coast Artillery (See Gun)		
Coaster		
Brake	192	5+
Drink mat or holder	D 7	624
Glass or tumbler	248	346.1
Wagon	280	87.1+
Coat		
Apparel	2	93+
Boat mast	114	93
Fluid control device using		
Design	D 2	183
Design, dress &	D 2	91+
Design, evening dress or suit	D 2	53+
Design, three piece ensemble	D 2	87+
Forms	223	68+
Hangers	223	92+
Design	D 6	315
Coated Article (See Type of Article)	428	
Electrolytically produced	204	14.1+
Explosive or thermic stock	149	3+
Fuel	44	542
Metal particle	428	570
Metal stock	428	615+
Coating		
Anticorrosive	106	14.5+
Apparatus	118	
Deterring	118	639
Electromagnetic	118	623
Electrostatic	118	621+
Lubrication	184	
Match dipping	144	50+
Mold metal casting	118	
Mold plastics	118	
Paper hanging combined	156	574+
Pipes	118	708+
Pipes	118	105
Pipes	118	125
Pipes	118	404+
Pipes	425	110+
Pipes	425	460
Printing	101	
Typecasting	199	
Typesetting	276	
Typewriting machines	400	
Semiconductor vapor doping	118	900*
Xerographic transfer	118	644*
Applicator or applier	15	
Material supply combined	401	
Separable from supply	401	118+
Classifying separating assorting		
solids by	209	47+
Coated surface or mass	209	49+
Cleaning by	134	4
Combined with other manufacturing		
operations (See particular art)		
Composition (See composition		
coating)		
Window glass to prevent deposit		
and freezing of moisture	106	13
Electrical barrier layer	437	
Composition	252	62.3 R
Electro less	427	304
Implement	15	104R+
Material supply combined (See		
brush, pens, etc.)	401	
Metal founding combined	164	
Mold making with	164	14
Mold surface	164	72+
Workpiece	164	75
Paper making combined		
Apparatus	162	265+
Process	162	135+
Paper undried	162	158+
Plastic	106	
Printing	101	
Processes	427	

	Class	Subclass
Abrasive tool	51	295
Catalysts	502	100+
Cathode sputtering	204	192.1+
Dyeing combined	8	495+
Electrically conductive	427	58+
Electro less	427	304
Electrolytic	204	14.1+
Electron emissive	427	77+
Electrophoretic or electro osmotic	204	180.2+
Electrostatic	427	13
Foods	426	289+
Metals by chemical action	148	240+
Molds	427	133+
Textiles chemical modification		
combined	8	115.6
Recovery	427	345
Removal		
By electric spark or arc	219	383+
Repellent	106	3
Manufacture	65	24+
Robot	901	43*
Synthetic resin or natural rubber		
(See class 523, 524)		
Testing	73	150R
Textile operation combined		
Braiding, netting or lace making	87	23
Covering or wrapping	57	7
Spinning twisting or twining	57	295+
Warp preparation	28	178+
With glass mfg	65	60.1
Coaxial Cables	174	102R+
Making	29	828
Switch for	200	
Wave guide type	333	239+
Cobalamin, in Drug	514	52
Cobalt		
Alloys	420	435
Electrolytic		
Coating	204	48
Synthesis	204	112+
Hydrometallurgy	75	711+
Metal stock	428	668+
Organic compounds containing	556	138+
Pyrometallurgy	75	626
Combined with hydrometallurgy	75	416+
Radiation imagery	430	936*
Cocaine	546	130
Cochineal Solution	252	408.1
Cock (See Valve)		
Abrading machines	51	28
Barrel bung	217	99+
With cutter or punch	222	81+
Carbide feed generator	48	40
Design	D23	233+
Dispensing	222	544+
Nozzle terminal	239	445
Pinch	251	4+
Sea	114	198
Support	248	67
Test liquid level	73	297
Time actuated	137	624.11+
With time indicator	137	552.7
Cocker	56	401+
Firearms	42	40+
Cockpit	244	119+
Cocktail Shaker	220	568
Agitating feature	366	
Design	D 7	300.1
Cocktail Stick	D 7	683
Cocoa Preparation	426	631
Apparatus	99	467
Cocoanut	426	617
Comminuting	241	
Shelling	99	568+
Cod Liver Oil	260	398+
Extraction	260	412.1
Medicine containing	424	555
Cod, Chemical Oxygen Demand Tests	436	62*
Code and Code Signaling		
Card or tape punching	83	
Code conversion	234	69+
Finger print as code for lock	70	277
Punch selection	234	94+
Punch selector	234	94+
Selective	234	
Synthetic speech from	381	51+
Code conversion	234	69
Coded record		
Magnetic record medium	360	131+
Record controlled calculators	235	419+
Records per se	235	487+

	Class	Subclass
Insertion / removal tool	29	213.1
Collimator		
Image projector having	353	102
Light beam projector having	362	257+
Faceted reflector	362	297+
Optical system	356	138+
Signal reflector having	350	102+
Testing device having	356	153
Collisions (See Protection)		
Avoidance	367	909*
Aircraft systems	342	29
Ship systems	342	41
Collocating Gauge	33	613+
Collodion	106	169+
Colloids	252	302+
Radioactive	252	634+
Cologne	D28	5
Colophony (See Rosin)		
Color & Coloring (See Dyeing; Pigment)		
Application training	434	81+
Assorting by	209	580
Charts		
Color display	434	98+
Unknown color test	356	421+
Electrophotographic copier	355	326+
Filter	350	311+
Photographic chemically defined	430	6+
Flowers by absorption of dye	427	4
Food	426	540+
Preserving or modifying color	426	262
Fresh flowers	47	58
Glass heat developed	65	111+
Lanterns	362	166+
Signal	362	168
Luminous, inorganic	252	301.6 R+
Luminous, organic	252	301.16
Mineral oil	208	12
Musical instrument accessory	84	464R
Paper	8	
Photography (See photography)	430	357+
Separation record making	430	356
Seams	112	17
Colorimeter	356	402+
Chemical analysis, apparatus	422	82.9
Photoelectric	250	226
Visual or photoelectric	356	402+
With computer	364	526
With light conductor	250	227.23
Colorimetric Radiation Dosimeter	250	474.1
Colostomy		
Bags	604	338+
Receptacle engaging stoma	604	332+
Colter	172	
Bearings	384	157
Plow combined	172	165+
Rolling	172	518+
Column (See Mast; Pole; Post)	52	720+
Aerated column pump	417	108+
Design	D25	126+
Fractionating	261	75+
Combined	202	158+
Mineral oil	196	139
Still combined	202	153+
Making	29	897.33
Static mold	249	48
Comb and Combing	D28	21+
Cleaner		
Brush machine	15	39
Comb combined	132	119
Implement	15	104.5+
Cornstalk huskers	56	114+
Cotton harvester	56	34+
Curry	119	86+
Design	D30	159
Fiber liberating	19	29
Fiber working	19	115R+
Grids combined	15	142
Hair	132	219+
Design	D28	21
Hat fasteners	132	61
Honey foundation	449	44+
Honey frames	449	42+
Kits toilet	132	121
Leather manufactures	69	27
Making wire	140	100
Making wood	144	26
Making, X-art	83	908*
Making, X-art	425	805*
Music	84	94.1+
Scissors combined	7	136+

	Class	Subclass
Seed gatherers	56	127+
Surgical	128	393
Textile		
Fiber preparation	19	115R+
Thread finishing	28	222+
Warp preparing	28	212+
Comber Board	139	86
Combinational Code		
Conversion, for punching	234	69+
Punching, selective	234	
Combustion (See Thermal)	431	
Attenuating sound	431	114
Burner control	431	18+
Chamber	431	350+
Exhaust pump	431	157
Fluid fuel burner	431	308+
Fluid fuel burner	431	350+
Outlet forms jet nozzle	431	158
Pre-chamber	123	253+
Control of nitrous oxides	60	900*
Earth boring by	175	12
Earth in situ	166	256
Engine	123	
Exhaust treatment	60	272+
Combustion products with	60	39.5+
Fluid injector	239	584+
Fluid motor combined	60	616+
Free piston device	60	596
Vacuum generated	60	397
Waste heat driven motor	60	597+
Pump drive	417	364
Pump regulator	417	34
Reaction motor	60	200.1+
Reaction motor with aircraft	244	74
Rotary piston	123	200+
Temperature measurement	374	144+
Turbine combined	60	598+
Turbine combined, product	60	39.34+
Failure responsive fuel safety cutoff	137	65+
Fumigant	424	40+
Improver for solid fuel	44	641
Processes	431	2+
Product		
Exhaust treatment	60	272+
Fire extinguishing agent	169	12
Generator	60	722+
Power plant	60	39.1+
Pulsating	431	1
Pump	417	73+
Tube		
Comforters	5	502+
Bedding	D 6	596
Comminuting	241	
Liquid metal	75	331+
Refrigeration combined	62	320
Refuse treatment means	110	222
Rock in situ	299	
Shredding metal	29	4.51+
Shredding metal and sintering	419	33
Strainer combined	210	173+
Combined	201	7+
Tobacco	131	311+
Feeding combined	131	109.1
Vegetables and meat	241	
Comminutor Element	241	293+
With cooperating surface	241	261.1
Commode		
Dry closet	4	449+
Communication Equipment (See Type of Equipment Desired)		
Aids for handicapped	434	112+
Communications		
Demodulators	329	
Electrical	340	
Alarms	340	500+
Audible, eg simulated noise in trainers	340	384R+
Character recognition systems	382	
Code converters	341	
Code transmitters	341	173+
Operator controlled	341	20+
Compressional wave	367	
Continuously variable indicating, eg telemetering	340	870.1+
Digital comparator systems	340	146.2+
Echo systems	367	87+
Elevator	187	130+
Error checking systems	371	
Image analysis	382	
Selective, eg remote control	340	825+
Signals, visual	340	815.1+

	Class	Subclass
Tactual signals	340	407
Traffic control signal	340	907+
Underwater	367	131+
Vehicle mounted signal	340	425.5+
Well bore, compressional wave	367	81+
Well bore, electrical	340	853+
Light wave	455	600+
Modulated carrier wave communications systems	455	
Modulators	332	
Multiplexing	370	
Non-electric miscellaneous	116	
Pulse or digital	375	
Radio	455	
Recorders	346	
Recording registers & calculators	235	
Sound recorders	369	
Telecommunications	455	
Telegraphic carrier wave	375	
Telegraphy	178	
Telephony	379	
Television, facsimile	358	
Commutation Book and Ticket	283	26
Commutator		
Bars	310	233+
Brushes	310	248+
Holders	310	239
Flanging tube to hold segments	29	243.517+
Grinders	51	244
Making	29	597
Assembling apparatus	29	732+
Structure	310	233+
Turning portable lathe	82	128
Type motors		
Alternating current	310	173+
Universal	310	158
Compacting (See Jarring)		
Agitating	366	
Beverage infuser combined	99	287
Burnishing	29	90.1+
Cooking apparatus combined	99	349+
Earth boring by	175	19+
Foundry mold sand	164	37+
Glass manufacturing	65	111+
Leather hammering	69	1
Machinery design	D15	20
Metals	419	61
With sintering	419	
Nuclear fuel	264	.5
Pipe making combined	425	110+
Pipe making combined	425	262
Powder metallurgy	419	38
Slip or slurry	419	40
Presses and pressing	100	
Foundry mold material	164	207+
Receptacle filling combined	141	71+
Bag	53	523+
Refuse treatment means	110	223
Textile thread	26	18.6
Trash bundle	428	2
Compactor	D15	20
Compacts	132	293+
Absent special toilet article	401	126+
Design	D28	78
Illuminated	362	135+
Compander	333	14
Compressor amplifier		
Thru bias control	330	129+
Thru impedance control	330	144+
Thru thermal impedance control	330	143
Expander amplifier		
Thru bias control	330	129+
Thru impedance control	330	144+
Thru thermal impedance control	330	143
In electrical recording or reproducing	369	174
Comparators		
Configuration	356	388+
Microscope subcombination	350	507+
Optical	356	
Projection type	356	391+
Compartment		
Aircraft		
Freight	244	118.1+
Passenger	244	118.5+
Safety lowering device	244	140+
Bed element	5	308
Boiler	122	37+
Building	52	234+
Cigar cigarette smoking device holder	131	242
Land vehicle bodies	296	24.1+

	Class	Subclass
Luggage	296	37.1+
Marine	114	78
Motormans	105	342
Pillow	5	442
Receptacle		
Apparel pocket	2	253
Bag, flexible	383	38
Bag, multiple pocket	383	38+
Baggage convertible to furniture		
type bureau	190	3+
Barrel	217	75
Beverage infuser	99	316+
Bottles and jars	215	6
Cabinet	312	
Collapsible	222	94
Cover for barrel	217	82
Dispenser	222	129+
Drying apparatus rotary drum	34	109
Envelope	229	72
Fire extinguishers	169	71+
House type letter box	232	21
Ice in wooden receptacle	217	130
Kilns and driers	34	209+
Letter box	232	24+
Material holder for heating	432	213+
Medication containing	604	403+
Metallic	220	500+
Nested drum driers	34	128
Oil vaporizer	196	111
Paperboard box	229	120.2+
Pitcher	222	129+
Purse	150	112+
Showcase	312	117+
Sifter	209	373
Sliding tray type paper box	229	10
Spaced wall or jacket	220	445+
Special receptacles	206	
Traveling bags	190	109+
Type cases	276	44+
Washing machine tumbling drum	68	139+
Wooden box	217	7+
Safes	109	53+
Movable	109	45+
Secret cabinet with	312	204
Sofa bed element	5	58
Tub	4	514
Outlet fittings	4	208
Supply fittings	4	193
Water closet flush tank	4	363+
Compass		
Course recorder	346	8
Design	D19	38
Divider	33	558.1+
Gyromagnetic	33	316+
Gyroscopic	33	324+
Magnetic	33	355R+
Repeater stations combined	340	870.15+
Saws	83	835+
Scriber	33	27.2
Beam type	33	27.3
Scriber compound	33	26
Compensating		
Abrading tool wear	51	155
Alternating current phase	323	212
Backlash	409	146
Gear cutter	409	5
Brake wear	188	214+
Clock and watch balances	368	171
Regulating	368	170
Clock pendulums	368	182
Clutch	192	110R
Compass magnetic	33	356+
Conductor electric	174	12R+
Flow meter	73	254
Fluid pressure	138	26+
Gearing	475	220+
Gearless differential	74	650
Gun sight	33	237
Internal combustion engine	123	192R
Lamp electric	314	13+
Musical instrument winding	84	131
Pump pressure	417	540+
Railway switch interlocking	246	152+
Thermometer	374	197
For thermocouple circuit reference		
junction	374	181
Transmission electric	333	18
Valve	137	625.34+
Vehicle	180	76
Vehicle steering	180	6.24+
Weighing scale temperature	177	226+

	Class	Subclass
Compensator		
Expansion or contraction		
Static mold	249	82
Compliance Device	33	832
Robot	901	45*
Composing		
Typesetting	276	
Sticks	276	38+
Composite Article (See Laminate)		
All metal or with adjacent metals	428	615+
Making by powder metallurgy	419	5
Making core and winding	242	7.1
Multipart article shaping	264	45.1+
Static mold for making	249	83+
Composition (See Type)	252	
Abrasive	51	293+
Absorbent	502	400+
Algicides	71	67
Alloy	420	
Antibiotic mixture	424	114
Antifog	106	287.1+
Antihemorrhagic	424	
Aquatic plant destruction	71	66+
Asphalt, tar, pitch	106	
Mineral oil only	208	22
Biocidal	424	
Coating or plastic combined	106	15.5+
Mineral oil only	208	2
Candles	44	275
Catalytic	502	100+
Ceramic	501	1+
Chemiluminescent	252	700
Combined light source	362	34
Coating	106	
Match splint	149	3+
Metal base reactive	148	240+
Mold	106	38.22+
Colloidal	252	302+
Core		
Metal casting	106	38.2+
Plastic molding	106	38.2+
Crayons	106	19
Crosslinking	252	182.11+
Curing	252	182.11+
Defoliant	71	69+
Desuckering	71	78
Detergent	252	89.1+
Diagnostic biological	424	2+
Dielectric fluid	252	570+
Hydrocarbon	585	6.3+
Distilling apparatus	202	267.1
Dye	8	
Electric insulation fluid	252	570+
Hydrocarbon	585	6.3
Electrically conductive	252	500+
Electrode (See electrodes)		
Electrolyte	204	
Battery	429	188+
Condenser or rectifier	252	62.2
Electron emissive	252	500+
Embalming	424	75
Emulsifying	252	351+
Explosive	149	
Fertilizer	71	
Fire extinguishing	252	2+
Flux	148	23+
Food and beverage	99	
Friction for matches	149	
Fuel	44	
Gaseous	48	197R+
Jet	149	1
Mineral oil only	208	15+
Growth regulating for plants	71	65+
Hardening	252	182.11+
Herbicides	71	65+
High boiling	252	71
Humidistatic	252	194
Hydrocarbon	585	1+
Inorganic materials only	106	286.1
Insulators		
Fluent electric	252	570+
Interactive	252	183.11+
Light transmission modifying	252	582
Low pour point	252	71
Lubricant	252	9+
Mineral oil only	208	18+
Luminous, inorganic	252	301.6 R+
Luminous, organic	252	301.16
Mantle	252	492
Match	149	
Medicinal	424	

	Class	Subclass
Mold forming	106	38.2+
Molten metal treatment	75	303+
Monomers	252	182.11+
Non-interactive	252	182.11+
Optical filter	252	582+
Photo conductive	430	56+
Photographic		
Developing	430	464+
Sensitizing	430	570+
Sensitizing and developing	430	486+
Plastic	106	
Mineral oil only	208	22+
Polymer precursors	252	182.11+
Pyrotechnic	149	37+
Radioactive	252	625
Medicinal	424	1.1
Production	376	171
Rubber containing (See synthetic		
resin or natural rubber)		
Tanning	8	94.19 R+
Textile treating	252	8.6+
Thermoelectric	136	236.1+
Undertaking	424	75
Utilizing wave energy	522	71+
Vulcanizing	252	182.11+
Water softening	252	175+
Wax	106	
Mineral oil only	208	20+
Welding flux	148	23+
Wick	502	400
X-ray contrast	424	4
Compound (See Type)		
Casting metal	164	91
Electrolytic production of	204	59R+
Hydrocarbon	585	16+
Lumber	52	782+
Mail receiving	52	364+
Metallic stock	428	615+
Organic	260	
Electrolytic production of	204	72+
Electromagnetic production of	204	157.15+
Electrostatic or electric discharge		
production of	204	165+
Fermentative production of	435	
Mineral oils	208	
Sugar and starch	127	
Refrigeration	62	332+
Automatic	62	175
Steam engines	91	152+
Temperature sensing member	428	616
In thermometer	374	205+
Tools	7	
Compressed Air (See Pneumatic and Fluid Operated)		
Vehicle brakes	303	
Compressing Apparatus (See Compressor, Molding, Pressing, Shaping)	417	
Barrel	147	4+
Harvesters		
Binding combined	56	432+
Shocker combined	56	430
Refrigeration	62	498+
Automatic	62	190+
Rims	157	2
Seams		
Die	413	31+
Folding combined	413	27+
Roller	413	31+
Tube side seam	72	48+
Snow	37	225+
Tobacco plug making	131	111+
Compression Member (See Column)		
Spring	267	70+
Compressional Wave		
Generators	116	137R
Submarine signal	367	131+
Compressor		
Body		
Brassiere		
Brassiere design	D 2	24
Brassiere or chest bandage	450	1+
Corset or girdle	450	94+
Corset or girdle design	D 2	2+
For upper leg or thigh	450	101
Design of machine	D15	7+
Gas pumps	417	
Refrigerator compressor muffler	181	403*
Tube pinching valve	137	384.2+
Comptometer	235	82C
Computer		
Applications & systems	364	400+

	Class	Subclass
Computer		
Automatic control of recorder mechanism	360	69+
Cassette	360	131+
Cassette or cartridge holder	206	387
Character recognition	382	
Coded document sorting	209	547+
Computerized switching in telephone systems	379	284
Controlled display	364	518+
Converting analog signal to digital	360	32
Data processing equipment	D14	100+
Disc	360	135
Drum	360	136
Dynamic magnetic information storage or retrieval	360	
Electrical	364	
Analog	364	800+
Applications		
Computational systems	364	400+
Data processors, general	364	200
Data processors, special	364	900
Measuring system, electrical	364	481+
Measuring systems, physical	364	550+
Calculators		
Analog	364	800+
Digital	364	700+
Control systems	364	130+
General purpose digital data processor	364	200
Hybrid type	364	600+
Special purpose digital data processor	364	900
Systems controlled by data bearing records	235	375+
Equipment	D14	100+
Error checking system in computer	371	
General processing of digital signal	360	39+
General processing of TV signal	360	33.1+
General recording or reproducing	360	55+
Geophysical system with digital computer	367	60
Head	360	110+
Mounting	360	104+
Transport	360	101+
Holder for magnetic disc assembly	206	444
Memory system subcombinations	365	
Data processing	364	900
Microprocessor	364	200
Camera photographic operation	354	412
Modulating & demodulating	360	29+
Monitoring or testing progress of recording	360	31
Non-electrical	235	
Fluidic	235	200R
Mechanical	235	61R
Record controlled	235	419
Ordnance	235	400+
Programmed data processors	364	200
Radar system combined	342	195
Record controlled, electronic	235	375+
Record copying	360	15+
Record editing	360	13+
Record medium	360	131+
Record transport	360	81+
Restrictive access over telephone line	379	95
Signal splitting	360	22
Special purpose television system application	358	93+
Tape	360	134
Teaching computer science	434	118
TV application	358	903*
Computing		
Distance rolling contact	33	733
Nuclear reaction control	376	217
Registers	235	
Scales	177	25.11+
Weighing	177	25.11+
Solar	33	268
Tape length	33	763
Workmens labor and wage recorder	346	
Concealed Data		
Printed matter	283	72+
Concealment (See Camouflage)		
Warship	114	15
Concentrating		
Beverage	426	425+
Cathode ray beams	313	441+
Evaporators	159	
Ionized gas for thermo nuclear reaction	376	100

	Class	Subclass
Liquids in liquids		
Apparatus	422	256+
Processes	23	306+
Milk	426	587+
Ores	423	1+
Ores	209	
Rare element, inorganic	423	
Separating solids	209	
Rubber		
Natural	528	937*
Synthetic or natural	523	335
Sugar crystallization combined	127	16
Concentric Tube or Cylinder		
Burner gas	431	195+
Electric cable flexible	174	102R+
Fluid vacuum or air	174	28+
With insulation spacer	174	99R+
Expansible chamber device	92	51+
Flue	122	160+
Header water tube	122	362
Heat exchanger	165	154+
Pipe joint	285	133.1+
Pipe joint flexible pipe	285	149
Transmission line	174	103+
Involving line parameters	333	236+
With nonsolid insulation	174	28+
Concertina	84	375
Keyboard type	84	376R
Concrete		
Beam or joist	52	723+
Block machine	249	117+
Compositions	106	713+
Filling subterraneah cavity	405	267
Freight car	105	405
Gas content testing	73	19.2
Making concrete blocks	264	333
Marine structure cast in situ	405	222+
Mixers	366	1+
Mold	249	
Adjuncts	249	205
Elements	249	187.1
In situ construction type	249	207
Clamps or brackets	249	219.1
Water type	249	219.2
Mold	425	62
Mold	425	63
Mold	425	87
Mold	425	458
Tie wire puller	254	18+
Piles	405	256+
Casting in situ	405	233+
Sheet	405	275
Pipe machine	425	110+
Pipe machine	425	262
Pipes	138	175+
Composite	138	153
Polymer containing (See synthetic resin or natural rubber)		
Prestressed making	264	228
Radiation barrier nuclear	376	293
Radiation barrier nuclear	376	287
Radiation barrier nuclear	250	515.1
Railway surface track	238	5
Center way concrete	238	7
Pedestals	238	115+
Stringers	238	25
Rammers	425	424
Rammers	425	425
Rammers	425	456
Rammers	425	469
Reinforcement	52	720+
Reinforcing	264	271.1
Reinforcing bar	D25	164
Ribbed construction	52	319+
Structures	52	
Arch	52	88
Beam or joist	52	723
Corner	52	250+
Faced	52	378
Hollow	52	380+
Lift slab	52	125.1
Moisture remover	52	310
Position adjusting	52	126.1+
Reinforcement	52	720+
Ribbed	52	319+
Stair	52	189
Void former	52	577
Concretion Removal	606	127
Condensation (See Alkylation; Polymerization)		
Preventer		

	Class	Subclass
Electric heaters	219	203
Windows and windshields	52	171+
Windows by ventilation	98	90+
Condensed		
Milk	426	587
Condensed Billing		
Typewriter		
Line spacing	400	546+
Paper feeding	400	595+
Condenser		
Atmospheric condensate	62	272+
Automatic	62	150+
Process	62	80+
Electric		
Amplifying device system	330	7
Anti induction telephone or system	379	414+
Bushing insulating	174	143
Charging and discharging	320	1
Circuit element	361	271+
Electrolytic type	361	500+
Induction apparatus combined	361	268+
Laminated making	D22	115+
Lamp and electronic tube supply circuit combined with	315	227R+
Making	29	25.41+
Making electrolytic	29	25.3
Motor system	318	781+
Nonelectrolytic type	361	271+
Phase regulation	323	212
Reactance tube system	333	213
Synchronous motors	310	162+
Testing	324	673
Treatment living body	128	783+
Variable type	361	277+
With variable resistor	323	370
X ray system combined with	378	103
Furnace electric	373	56+
Mechanical		
Making	29	890.7
Microscopes	350	414+
Lens construction	350	374+
Mounting, vehicle type	180	68.4+
Press	100	
Refrigeration including	62	
Steam engines	60	643+
Steam heaters	261	DIG. 1
Textile fiber preparation		
Article screen	19	148
Feed with	19	89
Multiple rotor feed with	19	88
Web forming screen	19	304+
Vapor	165	110+
Arc lamp fumes	314	26+
Beverage infuser	99	293+
Cooking apparatus basting	99	347
Drying combined	34	73+
Heat exchanger	165	110+
Jet gas and liquid contact apparatus	261	76+
Jet gas pump combined	417	173+
Jet heat exchanger	165	112+
Jet power plant exhaust	60	264+
Jet power plant exhaust	60	266+
Mercury vapor lamp	313	34
Mineral oil distillation	208	347+
Mineral oil distillation apparatus	196	138+
Mineral oil distillation with vaporizing apparatus	196	98+
Refrigerant	62	506+
Steam	165	110+
Still separatory	202	164
Still separatory preheater	202	180
Still separatory vapor treating	202	185.1+
Water heater combined with	126	381+
Water heater with liquid supply	126	380
Zinc pyrometallurgy	75	665+
Condiment	426	650+
Holder and dispenser		
Design	D 7	590+
General types	222	
Hand manipulable shaker type	222	142.1+
Conductor		
Electric	174	
Coated	428	375+
Coated	427	58+
Coated or covered	174	110R
Coated welding electrodes	427	59+
Coated with metal	428	615+
Coated with plural coatings	427	118+
Coating	427	117+

	Class	Subclass
Fryer or egg boiler control	99	336
Hydraulic conveyors	406	
Pigment and vehicle in pen	401	40+
Preserving disinfecting and sterilizing	422	
Soap and water in cleaning implement	401	40+
Solid metal treating apparatus	266	114+
Textile treating apparatus	68	
Making clocks	368	164
Period electric switch	200	19R+
Point electric switch	200	275+
Preventing		
Drying apparatus	34	94
Drying processes	34	6
Printing antismut devices	101	420+
Stenciling machine internal inker	101	120
Rails	191	22DM+
Rectifier	357	
Composition	252	62.3 R
Manufacture	29	25.2
Process	437	
Rolls electrical transmission	191	45R+
Sterilization by metal	426	322
Type microphone	381	178+
Wheels electrical transmission	191	45R+
Contact Lens for Eyes	351	160R+
Applicators	294	1.2
Case or cover	206	5+
Fitting instruments	351	247
Synthetic resin compositions	523	106+
Synthetic resin or natural rubber	525	937*
Container (See Ampoule; Barrel; Basket; Can; Capsule; Holder; Receptacle; or Under Item Contained)		
Acetylene generator carbide	48	29
Acid-proof	206	524.5
Adjustable bottom	206	45.16
Metallic	220	93
Wooden	217	64
Baggage	190	
Baked cereal product container type	426	138+
Bottles and jars	215	
Design	D 9	367+
Breaking	206	601+
Cabinet structure	312	
Coupon is detachable	206	831
Dispensing	222	
Drying houses and	34	201+
Edible design	D 1	
Electrophotographic developer	355	260
Envelopes	229	
Flexible bags	383	
Food sterilizing in	426	521
Freight	220	1.5
Hermetic food packaging	426	392+
Insulating jacket	220	903
Latching-type	206	405
Loose material filter	210	263+
Manufacturing from paper	493	
Decoration article	493	955+
Pallet	493	964+
Pliable	493	916+
Rigid	493	901+
Textile	493	937+
Material	428	34.1+
Medical syringes bulb combined	604	212+
Medicator	604	232+
Metallic	220	
Opening	206	601+
Opening	206	601+
Ornamental	428	34.1+
Paperboard boxes	229	100+
Plastic metal working	72	272
Poison alarms	116	72
Separation	206	602
Sheet metal container making	413	1+
Shock protecting	206	521
Snap-opening actions	206	404
Special packaging	206	
Of coiled items	206	389+
Specialty carriers & boxes	D 3	30.1+
Open top bins, trays	D34	40+
Sterilizer apparatus	422	
Storage convertible to display	206	44R
Tearing	206	601+
Toilet kits	132	286+
Absent special toilet article, eg mirror	401	118+
Bristled brush design	D 4	130+

	Class	Subclass
Toothpicks with thread supply	132	324
Toys	446	71+
Type static mold	249	117
Wooden	217	
Contamination, Reduction of	435	800*
Contour		
Non-condition responsive sorting by	209	940*
Contraceptives		
Condom	604	349
I U D	128	833
Pessaries	128	834+
Spermacides	424	DIG. 14
Control or Controller (See Operator; Pattern Control; Program; Regulator; Time, Operated & Controlled)		
Aircraft	244	75R+
Airship	244	96+
Bulldozer positioning	172	812
Car controlled fenders	293	2+
Circuit		
Clock operator	368	184
Motor system	318	
Railway signal	246	28R+
Sectional conductor vehicle transmission	191	14+
Switch	200	
Clutches and power stop control	192	
Power stop control	192	116.5+
Condition responsive metal deforming machine	72	6+
Driftage fence	256	12.5
Drill press	408	8
Electric		
Battery charging & discharging	320	2+
Car steps railway	105	444
Electric furnace	373	
Electrochemical apparatus	204	193+
Fluid pressure brake	303	20
Fluid pressure brake with fluid control	303	15+
Gear cutting	409	290+
Generator	322	
Limit stop	192	142A
Locomotive railway	105	61
Motor	318	
Motor and brake	192	1.1+
Motor and clutch	192	
Music tone generation	84	600+
Musical instrument	84	113
Musical instrument	84	171
Phonograph turntable stop mechanism	369	237
Prime mover dynamo plants	290	7+
Radiant energy	250	
Light wave	455	603
Railway block signal systems	246	20
Robot	318	568.11+
Stop mechanism	192	127
Switch	200	
Telegraphy remote control	178	4.1 R
Telephone systems	379	
Textile making	87	15
Transmission and brake	192	9
Tuner remote control	334	8+
Turntable railway	104	38
Vehicle brake	188	137+
Vehicle brake operator	188	182
Elevator	187	100+
Fluid		
Feed in solid fuel furnace	110	188+
Pressure brake systems	303	
Pressure regulators	137	505+
Valve actuation	251	
Humidity	236	44R
Lock		
Control and machine elements	70	174+
Time	70	267+
Machine elements		
Lever and linkage	74	469+
Trips	74	2+
Motor or pump (See regulator)		
Brake combined	192	1.1+
Clutch combined	192	
Pump regulators	417	279+
Motor vehicle	180	315+
Mechanism	180	315+
Radiated wave responsive means	180	167+
Velosity responsive means	180	170+
Nuclear reactor	376	207
Nuclear reactor	376	220

	Class	Subclass
Nuclear reactor	376	339
Nuclear, control component	376	327
Parachute	244	152
Pipe		
Fluid pressure brake	303	86
Fluid pressure brake automatic	303	29
Fluid pressure brake automatic and direct	303	25+
Fluid pressure brake charging	303	66+
Fluid pressure brake releasing	303	81+
Power stop	192	116.5+
Press	100	43+
Robot	901	2+
Electric	318	568.11+
Speed (See speed)		
Temperature (See cooling; heating)		
Automatic	236	
Controlled Release of Medicines	604	890.1+
Conversion of Power, Electrical	363	
Conversion Reaction, Nuclear	376	156
Conversion Systems, Electrical	363	
Converter		
Bessemer	266	243+
Electric current	363	
Cryogenic	363	14
Frequency	310	160
Apparatus	310	160
System	363	157+
Heptode pentagrid	313	300
Metallurgical	266	243+
Octode	313	300
Pentagrid	313	300
Triode heptode	313	298
With two cathodes	313	6
Triode hexode	313	298
With two cathodes	313	6
Convertible (See Type of Device)		
Armchair to bed	297	115+
Auto bodies	296	
Chair	297	118+
To cane	297	118
To stacked bunks	297	62
To table	297	119+
Cutlery	30	122
Garment supporter	2	301
Horizontally supported		
Planar surfaces	108	11+
Plural seats to single bed	297	63+
Racks	211	2+
Refrigeration related	62	326+
Stand to cane	248	155+
Vehicle to chair or seat	280	30+
Walker to vehicle	280	7.1+
Converting, Mechanical Process of	29	401.1
Conveyor (See Carrier; Chute; Feeder)	193	
Air blast &/or vacumm	29	DIG. 78
Article support for label affixing	156	556+
Bearing support for roller	384	418+
Belt (See endless below)		
Acetylene generator carbide feed	48	39
Automatic gauging apertured	209	681
Automatic gauging pocketed	209	685
Automatic gauging slotted	209	665+
Coated surface separator receiver with	209	63
Magazine dispenser	221	76+
Magazine dispensing slender article	221	76+
Manual assorting	209	705
Manure spreader type	239	650+
Mattress filling	53	255+
Paper pasting strip cutter feed	156	510
Power driven	198	
Sifters	209	307+
Sifting liquid treatment	209	272
Stratifiers liquid treatment	209	428+
Textile smoothing transfer	38	8+
Butchering	17	24
Centrifugal fluid distributor feed	239	215
Chain or belt	29	DIG. 73
Chute	193	2R+
Apertures automatic gauging	209	682
Drying apparatus	34	165+
Gaseous suspension assorting	209	149
Gauging automatic by rolling	209	696
Liquid suspension assorting	209	196
Reciprocating	209	196
Stationary	209	202
Manual assorting	209	703+
Movable bed stratifiers	209	479+
Liquid treatment, spiral	209	434

	Class	Subclass		Class	Subclass		Class	Subclass

	Class	Subclass
Box	164	228+
Planes	30	479
Sections hinged	164	233
Brush and broom	15	204+
Casting		
Boxes	164	228+
Cement pipe mold	425	59+
Cement pipe mold	425	406+
Chaplets	164	398+
Materials	106	38.2+
Means to expand or contract	249	178+
Metal founding sand	164	369+
Molds metal separable	249	142+
Molds sand combined	164	365+
Of plural sections	249	184+
Of resilient material	249	183
Plastic molds	425	468
Removing	164	132
Removing apparatus	164	345+
Static mold	249	175+
Cutters auger type	408	199+
Cutting		
Annular drills	408	204+
Button making	79	16
Earth	175	403+
Tubular wood saw	408	204+
Wood disk	408	204+
Dam	405	109
Deforming metal		
Destructible core	72	57+
Flexible core	72	466
Distillation retort	202	225
Electromagnet	335	297
Circuit breaker armature	335	281+
Circuit breaker multiple	335	180+
Flexible for tube bending	72	466
Magnetic		
Electromagnet with armature	335	220+
Reactor	336	233+
Transformer	336	233+
Materials	106	38.2+
Molding machines for foundries	164	228+
Nuclear reactor	376	347
Nuclear reactor	376	409
Pile driving type	405	245+
Pillow	5	439
Plates for storage batteries	429	209+
Railway pedestal block	238	117
Reeling fabrics	242	68+
Screw threading taps		
Cam	408	158
Wedge	408	179
Static mold having		
Removable	249	142
Strand structure covered	57	210+
Supports	164	397+
Textile covering processes	57	3+
Textiles covering	57	3+
Tires casing enclosed	152	310+
Corer		
Hand manipulated	30	113.1+
Plural cooperating blades	30	174
Machine	99	547
Doffer parer combined	99	541+
Doffer parer segmenter combined	99	543
Parer combined	99	541
Parer slicer combined	99	543
Parer slicer combined	99	592
Segmenter combined	99	545
Slicer combined	99	552
Cork		
Board making	264	124
Bottle stoppers	215	355+
Coating or plastic compositions		
containing	106	204.1
Fastener making	140	85
Hoof covering	168	26+
Paving	404	17+
Polymers ctg (See synthetic resin or natural rubber)		
Press	144	284
Pullers	81	3.7+
Removing implement	D 8	42
Wire applying to	140	94+
Corkscrew	81	3.45
Hayfork manipulator	294	121
Making	140	86
Corn		
Cob splitter	225	93+
Ear holder	294	5
Files	132	76.4
Foods	426	627+
Flakes	426	621
Grinding	241	
Harvesting	56	51+
Holder for eating	D 7	683
Husking		
Machines	460	25
Means to butter ear corn	401	12
Medicines for	424	
Pads for	128	882
Planter	111	
Clutches	192	23
Popping apparatus	99	323.5+
Shellers	460	45+
Green corn	99	567
Stringers weaving	139	19
Stripper	30	121.5
Corncob		
Pipe	131	230
Splitter	225	93+
Corneal Cutting Device	606	166
Corner		
Building component	52	272+
Concrete or plaster	52	250+
Embedded protector	52	254+
Panel edge binder	52	264+
Trim or shield	52	288
Window frame post	52	282
Cabinets	312	238
Connector		
Folded wall extension	229	190+
Folded web	229	186+
Structure		
Bedstead fastenings	5	288+
Block	52	284+
Bolted table leg	108	156
Bracket type support	248	220.1
Building component	52	272
Coffins	27	10
Curbs and gutters	404	8
Elastic flat bedbottom	5	205+
Land vehicle body	296	29+
Panel type switchboard	361	362
Razors	30	76
Resilient tires sectional	152	183
Trunk shields and buffers	190	37
Cornet	84	388+
Design	D17	11
Drainage	84	397
Mouthpieces	84	398+
Mutes	84	400
Tremolos	84	401
Cornice		
Curtain shade or screen combined	160	19+
Design, fabric	D 6	575
Design, rigid	D25	55
Design, structural	D25	55
Metal sheathing	52	287
Trim or shield	52	288
Corona		
Discharge photography	354	3
Discharge process	430	937*
Chemical apparatus	422	907*
Electrophotographic copier	355	221
Electrophotographic copier	355	274+
Prevention		
Conductors	174	127
Dynamos	310	196
Corpse		
Carriers	27	28
Coolers		
Display type	62	246+
Open support	62	458
Cremating furnace	110	194
Undertaking	27	
Corrosion Preventing		
Agents	252	387+
Coating composition	106	14.5+
Distilling combined	203	6+
In hydrocarbon synthesis	585	950*
In nuclear reactors	376	305
Mineral oil conversion combined	208	47
Processes	422	7
Refrigeration with	62	85
Corrugated Structure		
Building panel	52	630
Metal	428	603+
Pipes		
Flexible	138	121+
Reinforced	138	173
Railway ties	238	82
Receptacle wooden straps for	217	67
Reflector lantern combined	362	297+
Sheet metal	428	603+
Ship	114	80
Stock material	428	179
Making apparatus	156	205+
Making process	264	286+
Wrappers for bottles	229	90
Corrugating (See Crimping)		
Land furrowing implement	172	
Metal	72	190+
Curving and	72	177
Slitting and	29	6.1
Tube	72	102+
Web or sheet	72	196
Paper		
Crepe paper making	162	111+
Processes for	264	286+
Plastic & earthenware shaping	425	396
Ruffler for sewing machine	112	132+
Corsage	D11	117+
Corset	450	94
Combined with brassiere	450	7+
Design	D 2	3
Covers	2	110
Design	D 2	2
Making	223	53
Reinforcing	223	4
Corticosterone	552	588
Cortisone	552	577
Cortisone, in Drug	514	179
Corundum	423	
Abrasive composition	51	309
Refractory composition	501	127+
Magnesium compound containing	501	119
Corynanthine	546	53
Cosmetic Product	D28	4+
Applicator, bristled brush	D 4	
Applicator, general	D28	7+
Box	206	542
Brush	D 4	
Cold cream	514	772+
Compacts	401	126+
Design	D28	78
Facial	514	844*
Including toilet article,	132	286+
Eg mirror	132	316
Compact	132	301+
With light	362	135+
Lipsticks	424	64
Container	206	385
Lipstick in container	D28	76+
Package for	206	803*
Powder applicator	D28	8
Powder in container	D28	7+
Shaped appliers	132	320
Design	D28	76+
Cosmic Ray Detection	250	336.1
Cloud chamber	250	
Detector tubes	376	153+
Detector tubes	313	93
Photographic film	250	475.2
Scintillation	250	361R*
Costume	D 2	79
Garment prior to 1900	D 2	26
Cots	5	110+
Canopies	5	414+
Cotter	411	513+
Nut locks		
Design	D 8	382
In bolt to nut lock	411	213
In bolt to nut lock	411	320
Making	29	5
Pullers	29	247+
Railway car wheel	295	50
Shaft joint type	403	378+
Slot punching	10	14
Cotton		
Chopper type plows	172	
Fiber preparation	19	
Harvesting	56	28+
Ripening	47	5
Cotton Cady Maker	425	9
Couch		
Box type	297	192+
Furniture design	D 6	334
Hammock	5	124+
Kinestherapy	128	33
Light thermal and electrical applying	128	376+
Orthopedic	128	70+
Roll	162	314

	Class	Subclass
Piano	84	177
Pillow	5	490
Pipe end (conduit type)	138	96R+
Plant	47	28.1+
Cut or artificial plant	206	423
Platen	355	72+
Electrophotographic copier	355	230+
Playing field (baseball)	273	27
Pool table	273	13
Protective for cigarette pack	D27	186+
Protective for matches	D27	173+
Purse (decorative cover)	150	105
Racquet	150	163
Including press	273	74
Radiator	237	79
Railcar step	105	450
Receptacle	150	154
Roller skate	206	315.1
Boot attached cover	280	811
Scabbard	280	825
Sap bucket	47	54
Saw	150	161
Rigid guard type	30	504
Sewing machine	150	164
Design of cover	D15	75
Housing attached cover	312	208
Shelf (covering)	428	904.4
Shoe (during manufacture)	36	72R+
Bag type	383	
Special receptacle type	206	278+
Skate (roller or ice)	206	315.1
Boot attached	280	811
Scabbard	280	825
Ski	206	315.1
Binding	150	154
Body supported	224	191+
Clamp or tie attached to ski	280	814+
Vehicle supported	224	309+
Soap (porous cover)	401	201
Sponge	15	244.1
Spoon	30	326
Stair	52	177+
Steering wheel	74	558
Stirrer	366	349
Suitcase	190	26
Surgical drape type	128	849+
Thermal probe	374	209
Swimming pool	4	
Building type	52	
Switch actuator	200	333
Table	108	90
Tarpaulin	52	3+
Telephone	379	451+
Tent or canopy	135	115
Thermal probe	374	209
Tire or inflated inner tube	206	304+
Design of cover	D12	202
Vehicle attached tire	224	42.2
Toilet (tank or seat)	4	661
Design of cover	D23	311
Seat	4	242+
Trailer hitch ball	280	507
Tree	47	20+
Christmas tree bag type	206	423
Trunk guard type	47	23+
Typewriter	150	165
Design of cover	D18	12
Housing attached cover	312	208
Umbrella	135	33R+
Vault pavement type	404	25+
Design	D25	36
Vehicle	150	166
Body (while in operation; including a 'bra')	280	770
Engine attached drip catcher type	180	69.1
Load	296	100+
Design of cover	D12	156
Motorcycle or bicycle	150	167
Roll up type	296	98
Tire	206	304+
Design of cover	D12	202
Vehicle attached tire	224	42.2
Wall (wallpaper)	428	904.4
Washing machine	68	196
Water heater		
Boiler structure	122	494
Electrical	219	
Tank	220	
Window	150	168
Rigid cover sections	160	370.2
Vehicle combined	296	95.1

	Class	Subclass
Windshield	150	168
Rigid cover sections	160	370.2
Vehicle combined	296	95.1
Wire coil	206	303
Wooden container	217	3CV
Barrel	217	81
Coffin	27	19
Covering (See Apparel; Clothes; Coating; Garment)		
Applying by laminating	156	
Bridge	14	74
Floor		
Carpet	428	85+
Laminated	428	54
Slatted	52	664+
Insulating for electric conductors	174	
Making and applying	156	
Button covering	79	4+
By laminating	156	
Extruding metal sheath on wire	72	268
Extruding plastic on core	425	113
Indefinite length electrical conductor	156	47+
Metal covering	29	33.2
Paper	493	
Sheath to tube or rod	228	126+
Shoe heel	12	49.1
Strap	12	59
Textile spinning	57	3+
Wire	29	428+
With sheet metal	29	243.57
Shelf	428	904.4
Stair	52	177+
Wall	428	904.4
Covert	139	416
Cowl		
Ventilating	98	61
Design	D23	374+
Intake and outlet	98	35
Crack-off	225	
Glass	225	94+
Cracker		
Firecracker	102	361
Food	99	
Design	D1	128+
Cracking		
Hydrocarbon synthesis		
Aromatic	585	483+
Diolefin	585	613+
Olefin	585	648+
Paraffin	585	752
Triple bond	585	534+
Oils	208	46+
Electrostatic or electric discharge	204	172
Crackled Coating	427	257
Crackled Glass Making	65	111
Cradle		
Battery	429	96+
Bed	5	101+
Design	D6	382
Grain	56	324
Land vehicle convertible	280	31
Receptacle stand	248	139
Squeezer type washing machine	68	114
Cramps		
Block tie	52	712+
Crane (See Derrick)		
Automatic control	212	153+
Cut-off	212	149+
Cyclic	212	161
Sway	212	147+
Cab	212	165
Travelling bridge	212	206
Collapsible	212	182+
Cornstalk cutter with discharger	56	72+
Counterweight	212	195+
Removable	212	178+
Davit	114	368
Design	D34	33
Electric control	212	124+
Excavator with cable operated boom	37	116
Floating	212	190+
Invalid lift	5	87
Locomotive with turntable	105	28
Material or article handling	414	561+
Remote or dual control	212	160
Rotary	212	223+
Self-erecting	212	176
Climbing	212	199+
Ship mounted	212	190
Transmission of electricity to	191	

	Class	Subclass
Traveling	212	71+
Bridge	212	205+
Trolley type railway with transfer	104	98
Trucks for overhead	105	163.1+
Vehicle mountable	212	180
Window	49	148
Crank		
Foot operated	74	594.1
Hand lever	74	545+
Mechanical movement including	74	
Operated (See device operated)		
Paddle		
Impeller	416	78
Marine	440	26+
Water motor	416	78
Pedals combined	74	594.1+
Velocipede design	D12	123
Wrist pin combined	74	595+
Crankcase		
Breathers	92	78+
Drainers	184	1.5
Housings	74	606R
Internal combustion engine	123	196CP
Crankshaft		
Bearing		
Antifriction	384	457
Plain	384	429+
Thrust	384	250
Lapping	51	73R
Making		
Apparatus	29	6.1
By turning	82	106+
Method	29	888.8
Straightening	72	389+
Twisting	72	299
Methods	72	371
Crash		
Fabric	26	26
Seat	297	216
Signal for airplane		
Radio	342	385+
Water dye	116	211
Crate	D9	414+
Animal neck stocks combined	119	99
Animal shipping, design	D30	109
Fermentation apparatus	435	287+
Freight car combined	105	373
Shooks	217	43R
Wooden receptacle	217	36+
Cravats	2	144+
Fasteners	24	49R+
Crayon	401	49+
Compositions	106	19
Holder	401	88+
Advancing means combined	401	55+
Modifying thermometers	374	183*
Molding device	425	117
Molding device	425	425
Cream		
Cold	514	772+
Separating and gas	55	189+
Separating decanter	210	514+
Separators	D7	369+
Treating processes	426	334+
Whipping	261	DIG. 16
Cream of Tartar	562	585
Baking powder	426	562+
Creaming Latex		
Natural rubber	528	937*
Synthetic or natural rubber	523	335
Creaser (See Folder)		
Boot and shoe making		
Loose upper shaping	12	54.1
Box, cut and crease	493	59+
Envelope	493	186+
Horseshoe	59	36+
Plaiting fluting shirring and cross creasing	223	29
Sewing machine	112	131
Leather	112	51
Creche	D11	122+
Credit Card Systems	235	380+
Creel (See Basket)		
Bobbin supporter and holder	242	131
Fishing	43	55
Design	D3	38
Creeper		
Antislip shoe attachment	36	59R+
Baby	297	5+
Electric wire placing	254	134.5+
Horseshoe calk	168	30

	Class	Subclass
Cyclic Thiocarbonates	549	30+
Dithiocarbonates, monothiocarbonate	549	30
Trithiocarbonates	549	36
Cyclization		
Mineral oil	208	46+

3

	Class	Subclass
3,5-cycloandrostanes	552	500+
Cyclobutyl Carboxylic Acid Esters	560	123
3,5-cyclocholesterols	552	500+
3,5-cyclogonanes	552	500+
Cyclohexamide	546	
In drug	514	328
Cyclohexanol	568	835
Cyclohexanone	568	376
Cyclohexylamine	564	462
Cyclometers	235	95R+
Cyclone Separator	406	173
Gas separation	55	337
Liquid seoaratiom	210	512.2
Cycloolefines	585	23
Synthesis	585	350+
By dehydrogenation	585	379+
Cycloparaffins	585	20+
Synthesis	585	350+
Cyclopentadiene Synthesis	585	350+
By alkylation	585	446
Cyclopentanohydrophenanthrene		
Compounds	552	502+
Heterocyclic	540	2+
Cyclopentyl Carboxylic Acid Esters	560	121+
Cyclopropane	585	350+
Cyclopropyl Carboxylic Acid Esters	560	124
Cycloserine	548	244
3,5-cyclosteroids	552	500+
Cyclotrons		
Energizing systems	328	234
Nuclear fusion reactions use in	376	100
Structure	313	62
Systems including	328	234
Tubes	313	62
Cyclovitamin D	552	653
3,5-cyclovitamin D	552	653
Cylinder		
Airplane with sustaining rotor	244	10
Bracket support	248	230+
Caster	16	22+
Compressor		
Movable cylinder with		
reciprocating piston	417	460+
Multiple cylinder	417	521+
Rotary cylinder	417	462
Door closing and liquid checking	16	57
Earthworking	172	554
Escapement	368	129
Expansible chamber device	92	169.1+
Concentric cylinders	92	51+
Lubricating means	92	153+
Movable cylinder	91	196+
Moving cylinder multple	92	117R+
Gauges	33	542+
Glass cylinder making	65	
Glass handling	414	24
Grinding ie internal grinding	51	290
Handles, hollow cylinder	414	910*
Head packing	277	
Heat exchange means	165	181
Holding member	292	261
Internal combustion engine		
Construction	123	193R
Cooling	123	41.72+
Multiple	123	52R+
Oscillating	123	42
Reciprocating	123	50R
Rotary	123	43R+
Joint for rotating cylinder	285	134+
Lagging	165	135+
Lock	70	357+
Housing	70	449
Key	70	406
Lubricator	184	18+
Machine gun	89	13.5+
Material treating		
Abrading floors	51	176
Abrading with rotary flexible tool	51	358+
Abrading with rotary rigid tool	51	206R+
Acetylene generator	48	
Baffle	210	512.1
Brushing	15	3+
Comminuting	241	293+
Corn sheller	460	45
Drying and cooling with rotary		
cylinder	34	63

	Class	Subclass
Fiber carding	19	112
Fiber combing	19	126+
Filter element	210	497.1
Huller	99	600+
Leather treatment	69	37
Movable filter	210	402+
Paper stuff straining	210	402+
Photographic printing	355	104+
Potato digger with screen	171	111+
Reactor compartment	210	207+
Scouring	51	22+
Textile treatment with fluid	68	5R+
Threshing	460	59
Web forming	162	323+
Weighted land roller	404	111+
Winding	242	7.21
Meter having expansible chamber	73	232+
Music boxes automatic playing	84	95.1+
Comb actuator	84	95.1+
Organ pin cylinder	84	86
Separable from keyboard	84	106
Panel hanger	16	89
Percussive tool rammer	173	125
Pistol or rifle revolving cylinder	42	59+
Pump (See pump fluid)		
Multiple cylinder	417	521+
Rotary cylinder	417	462
Rotary cylinder with reciprocating		
piston	417	460+
Rotor sustaining airplane	244	21
Sheet handling		
Associating	270	60
Folding and associating	270	32+
Folding, associating & printing	270	1.1+
Solid separation devices	209	
Steam locomotive	105	42
Syringe	604	218+
Telescoping	92	51+
Toothed cylinder manufacture		
Apparatus	29	23.1
Method	29	895.31
Vehicle support resilient	152	
Wood and metal treating		
Barrel stave cutter	147	35
Corrugating	72	
Cutter	144	221+
Planer with rotary cutter	144	116+
Slicer with rotary cutter	144	172+
Cylindrical		
Photographic printing apparatus	84	422.1
Cymbal Beater	84	422.1
Cymel T M (aminoplast)	528	254
Cymene (See Aromatic Hydrocarbon)	585	350+
Cyproheptadine, in Drug	514	325
Cysteine Ester, in Drug	514	550
Cystine	562	557
Preparation from protein	562	516
Cystoscope	128	7
Cytosine	544	317
Czuchralski Type Crystallization	156	617.1
D D T, in Drug	514	748
D D V P, in Drug	514	136
Dacron Insulated Cable	174	100
Dacron T M (See also Synthetic Resin		
or Natural Rubber)	528	308.1
Dado		
Cutter	144	222
Lapped multiplanar surfacing	52	536
Machine	144	133R
Daggers	30	295
Daguerreotypy	354	2+
Analysis and analytical control		
Apparatus	422	50+
Apparatus design	D16	
Dairy Type Bottles	D 9	
Dairy, Food Treatment	99	452+
Dam	405	107+
Dental	433	136+
Mortar	52	421
Power extraction from water	405	78
Dampening (See Damper; Moistener;		
Steaming)	68	5R
Consumable electrode electric lamp		
Electromagnet feeding with		
damping	314	128
Feed with damping	314	99+
Instrument mechanism damping	73	430
Bourdon tube	73	739+
Musical instrument damping	84	
Overhead electric conductor with		
vibration damping	174	42

	Class	Subclass
Paper making calender	100	73+
Planographic printing	101	147+
Copying apparatus	101	132.5
Pressure compensator		
Pipes and tubular conduits	138	26+
Pumps combined	417	540+
Vibration damping support	248	562+
Nonresilient	248	636
Resilient	248	562+
Suspension	248	610+
Damper		
Ear sound stoppers	128	864+
Electric measuring instruments for	324	162+
Electric musical instrument for	84	600
Envelope shaping (i. e., attack,		
decay, sustain, or release)	84	627
Envelope shaping (i. e., attack,		
decay, sustain, or release)	84	663
Envelope shaping (i. e., attack,		
decay, sustain, or release)	84	702
Envelope shaping (i. e., attack,		
decay, sustain, or release)	84	738
Volume control	84	633
Volume control	84	665
Volume control	84	711
Volume control	84	741
Electric winding for alternating or		
direct current dynamos	310	183+
Fluid damper for planetary gear		
element	475	92
Flywheels and rotors having	74	574
Furnace draft	110	163
Heat exchanger	165	69
Impact absorption	49	9
By flexible barrier	49	9
By plastic deformation	188	371+
By resilient deformation	188	268
Inertial vibration damper	188	378+
Microphone or diaphragm	381	158
Piano damper head and stems	84	255
Harmonic	84	234
Held	84	218
Lifted	84	217
Railway truck	105	193
Stove or furnace flue	126	285R+
Open evaporating pans having	159	41
Time controlled	126	285.5
Weighing scale	177	184
Zither	84	287+
Damping Diode	313	317
With emissive cathode	313	310
With thermionic cathode	313	310
With two cathodes and anodes	313	1
Dancing		
Figure toys	446	330+
Shoes	36	8.3
Teaching	434	250
Dandy Roll	162	314
Dark		
Cabinet photography	354	307+
Lantern with shutter or screens	362	167+
Room	354	307+
Illuminators	362	293
Ventilators	98	29+
Darning		
Knitting	66	2
Last	223	100
Sewing machines	112	121
Design	D15	66+
Elements	112	236
Weaving	139	33.5
Dart	273	416+
Design	D22	115
Amusement devices	D21	49
Projector	D22	107
Projector	124	22
Dashboard		
Land vehicle	296	70+
Lighting		
Automobile	362	23+
Trolley car or rail car	362	77
Motor vehicle	180	90
Dasher		
Food	366	69+
Gas and liquid contact	261	32+
Dashpot		
Check valve	137	514+
Closure check	16	82+
Closure check brake type	188	266+
Door check and closer	16	49+
Electrode feed retarder	314	99+

	Class	Subclass
Degradability Enhanced Synthetic Resin		
or Natural Rubber Composition	523	124+
Cellular product	521	916
Degreasing		
Distillation		
Still and extractor	202	168+
Still extractor	202	170
Solids	134	
Textiles	8	139+
Degumming	8	138
Dyeing combined	8	
Dehalogenation		
Preparation of epoxy compounds	549	518+
Preparation of olefines	585	641+
Preparation of unsaturated esters	560	213
Dehorners		
Electric heater type	606	164
Hand tool	30	
Dehumidifier	D23	359+
Dehydrating Oil		
Fatty oils	260	405.5
Mineral oil	208	187+
Electrical separation	204	188+
Electrophoretic or electroosmotic	204	181.8+
Dehydration (See Drier)		
Acid anhydrides by	562	869+
Acyclic ketones from alcohols	568	403+
Aldehydes from alcohols	568	485+
Compositions	252	194
Concentrating evaporating	159	
Distillation	203	12+
Fatty oils or acids	260	405.5
Foods	426	443+
Apparatus	99	467
Gas or vapor contact with solids	34	
Gas separation	55	
Refrigeration combined	62	271
Refrigeration combined process	62	93+
Mineral oils	208	187+
Electrical	204	188+
Electrophoretic or electroosmotic	204	181.8+
Olefine production by	585	638
Refrigerant	62	474+
Automatic	62	195
Process	62	85
Dehydroabietic Acid	562	404
Dehydroandrosterone	552	636

7

	Class	Subclass
7-dehydrocholesterol	552	547
Dehydrogenation		
Acyclic ketones from alcohols	568	403+
Aldehydes from alcohols	568	485+
Aromatic hydrocarbon synthesis	585	440
Diolefin hydrocarbon synthesis	585	616+
Esters from alcohols	560	239
Mineral oil	208	46+
Olefines from hydrocarbons	585	654+
Synthesis aromatic from alicyclic	585	430
Dehydrohalogenation		
Alicyclic hydrocarbon synthesis	585	359
Diolefin hydrocarbon synthesis	585	612
Olefin hydrocarbon synthesis	585	641+
Dehydrothiotoluidine Azo Compounds	534	800
Deicer (See Thawing)		
Aircraft	244	134C
Antenna	343	704
Carburetors	261	DIG. 2
Combustion product as motive fluid	60	39.93
Trolley collector head	191	62
Deinking		
Paper stock	162	4+
Textiles	8	137
Delasting	12	15.1
Delay Networks		
Delay lines including		
A lumped parameter	333	138+
Elastic bulkwave propagation		
means	333	141+
Elastic surface wave propagation		
means	333	150+
Long line elements	333	156+
Electronic tube system	328	55
Delinter	19	40+
Carbonizing processes	8	140
Delinting Cotton Seed	47	1.1
Processes	47	58
Fluid treatment	8	140
Delivery		
Collection receptacle	232	
Coins	232	64+
Letter box chute	232	53

	Class	Subclass
Conveyer	198	
Carrier controlled selective	198	570+
Dumping vehicle with chute	298	7
Letter box	232	17+
Mail from aircraft	258	1.2+
Planting chute	111	76
Railway mail	258	
Railway selective	104	88+
Record holder	312	15+
Sheet	271	278+
Store service	186	3
Delrin T M (polyoxymethylene)	528	270
Delustering		
Cellulose ether or ester compositions	106	192
Viscose	106	166
Coating or impregnating processes	427	170
Demagnetizing	361	267
Demagnetizer	D13	183
Erase, head	360	118
General recording biasing, erasing	360	66
Sorting treatment	209	8
Demijohns (See Carboy)	D 9	367+
Demodulator	329	
Demonstrating Apparatus or Product	434	365+
Demonstrator Radioactivity (See		
Radioactivity Demonstrator)		
Demountable Rim	301	10R+
Demulsifying Compositions	252	358
Denatured Compositions	252	365
Alcohol	252	366
Densitometer		
Emulsion exposing printer	354	20
Emulsion opaqueness measuring	356	436+
Density		
Analysis gas	73	30.1+
Analysis liquids	73	32R+
Dent Remover Sheet Metal	72	457
Dental and Dentistry	433	
Amalgams and alloys	420	526
Amalgam mixer, eg dental filling	366	602*
Apparatus, fixed	D24	4+
Apparatus, portable	D24	10+
Cabinet	312	209
Cassettes	378	168
Chairs	297	68+
Design	D 6	334
Compositions	106	35
Containers for dental use	206	63.5
Dentures	433	167+
Die shaping	72	
Engines	433	103+
Equipment stands	433	25+
Fillings	433	226+
Floss	132	321
Holders	132	323+
Design	D28	64
Impression devices	433	34+
Impression material	106	38.2+
Instruments	433	25+
Medicines	424	
Molding devices	425	2
Molding devices	425	175+
Molding devices	249	54
Flasks	425	175+
Flasks for metal casting	164	376
Molds	425	175+
Molding processes	264	16+
Orthodontic devices	433	2+
Practice	433	
Processes manufacturing	29	160.6
Spittoons	4	263+
Spotlights	362	257+
X-art collection	362	804*
Supply packages	206	63.5
Teaching devices & methods	434	263
Teeth	433	167+
Tool container	206	368+
Waste receptacles	206	63.5
Dentifrices	49	
Dentiphone		
Acoustic hearing aid	181	127
Electrical hearing aid	381	68.3
Dentistry (See Dental)	433	
Design	D24	
Deodorant		
Body	424	65+
Cosmetics containing	424	65+
Fertilizer containing	71	3
Milk	426	488
Non body	424	76.1+
Preserving, disinfecting, and		
sterilizing	422	5

	Class	Subclass
Deodorizer or Ozonizer	D23	366
Deoxidant Compositions	252	188.1+
Descaling compositions containing	252	81
Water softening or scale inhibiting		
compositions ctg	252	178
Dephlegmators		
Mineral oil	196	139
Still combined	202	
Depilating		
Animal carcass	17	47
Process	17	47
Compositions		
Electric needle	606	44
Needle supports	606	44
Fermentative	435	265
Hides and skins	8	94.16
Mechanical	19	2+
Depolarizer	252	
Depolarizing		
Compositions	252	
Depolymerization to Obtain		
Hydrocarbon Mixture	585	241
Deposit		
Certificates	283	59
Change gates	232	14
Collection and deposit receptacle	232	
Cabinet combined	312	211+
Depository design	D99	28+
Letter boxes	232	22
Milk bottle collection	232	42
Receptacles for	232	
Registering or receipt printing		
combined	109	24.1+
Photograph or microfilm record	346	22
Safes or depositories	109	
Design	D99	28+
Receipting means included	109	24.1
Depositor		
Bread etc making depositor	425	447+
Implement	222	
External body surface	604	289+
Medicine	604	47+
Solid materials	604	57+
Planting	111	
Dibbling	111	89+
Hill	111	34+
Toy figure for coins	446	10+
Depressor		
Necktie	24	53+
Railway crossing cable	104	186
Tongue	128	15+
Atomizer combined	128	200.15
Depth Bomb	102	390+
Depth Gauge		
Compressional wave	367	99+
Depth measuring instrument	D10	46+
Fluid level	33	
Geometrical	33	713+
Geophysical exploration	324	323+
Hydrophone	367	141+
Liquid	73	290R+
Radar	342	120+
Other than air, eg, underwater	342	118
Sound	181	
Derailer		
Actuated cab signal or train control	246	170+
Cycle or motorcycle	D12	124
Enclosure or guard	D12	127
Guards		
Railway	104	242+
Store service	186	30
Lever and cable operator	74	502.2
Railway		
Car attached	104	261
Interlocking	246	163
Derailleur (See Derailer)		
Derectifiers Electric	363	
Dereverberator	381	66
Dermatological Device		
Light application	606	9
Needle surgical	606	44
Support	606	44
Surgical instrument	606	131+
Dermatome	606	132
Derrick		
Extensible or movable	52	111+
Pushing and pulling elements		
Cable hoists	254	283+
Screw	254	99
Skeleton tower	52	648+
Desalting Sea Water	210	642

	Class	Subclass		Class	Subclass		Class	Subclass
Chain welding	59	33	Processes	127	43+	Germanium	357	
Coiling spiral	72	141	**Diffusing (See Extracting)**			Glow discharge	313	567+
Deforming	72	462+	Air sterilizing	422	120+	High mu triode	313	303
Design	D15	136+	Aircraft structure	244	136	With two cathodes	313	5
Forging anvil adjustable	72	418	Cabinets	312	31+	Light emitting	362	800*
Forging presses enclosed	72	343+	Fans	416		Pentode	313	298
Horseshoe	59	60	Fumigators	422	305+	With two cathodes	313	5
Injecting	164	303+	Gases for separation	55	158	Power amplifier pentode	313	298
Process for making	76	107.1+	Process	55	16	With two cathodes	313	5
Screw threading platen	72	88	Thermal diffusion	55	209	R F	313	317+
Through-die	72	467+	Thermal diffusion process	55	81	With emissive cathode	313	310
Wire cutting	83		Liquids through membranes	210	634+	With thermionic cathode	313	310
Wire joining	140	116	Process in semiconductor device			Sharp cutoff pentode	313	298
Work-orbiting screw threader	72	191+	making	437	141	With two cathodes	313	5
Paper box	493	167+	Sacchariferous material	127	3+	Triode pentode	313	298
Punches			Sprinkling and spraying	239		With plural cathodes	313	6
Sheets and bars	83		**Digallic Acid**	560	70	Triple	313	1
Wood match making	144	53	**Digester & Digestion (See Extracting)**			Triple high mu triode	313	303
Woodworking	144	197	Apparatus	422	307+	With plural cathodes	313	5
Punching dies	83	651+	Fermentative	435		Triple triode	313	303
Screw threads	72	88+	Liquid purification	210	601+	With plural cathodes	313	5
Shaping			Mineral oils	208	46+	Twin	313	306
Closure applying	53	341+	Pulp apparatus	162	233+	With two cathodes	313	1
Embossing members	101	16+	Pulp processes	162	1+	**Diolefin Hydrocarbon**	585	16+
Metal can seaming	413	31+	Still combined	202	107	Purification	585	800+
Metal sheet shaping	72	462+	**Digger**	171		Synthesis of	585	601+
Printing embossing hot	101	8+	Design	D15	10+	**Dioramas**	272	9+
Stretch press	72	302	Hand tool	D 8		**Dioxane**	549	377
Wire fabric	140	107	Ditcher	37	80R+	**Dioxazine**	544	63+
Shoe sole	12	38	Elevator	414	596+	**Dioxazoles**	548	124
Textile	57	138	Dredger	37	54+	**Dioxirane**	549	200
Dielectric			Endless excavator	37	191R+	**Dip Pipe**	202	255+
Capacitors in	361	301+	Fire arm combined	42	93	**Dip-cup-provided Inkwell**	15	257.7+
Fluent compositions	252	570+	Posthole			Dispenser type	222	576+
Hydrocarbon	585	6.3	Implement	294	50.6+	**Diphenoquinone**	552	304
Heating apparatus	219	6.5+	Machine	175		**Dipole Antenna**	343	793+
Heating methods	219	10.41+	Potato	171		**Dipper**		
Hygrometer	324	666	Rotary excavator	37	189+	Excavators	414	685+
Lens	333	248+	Tunnel excavating	299	29+	**Dipping (See Coating)**		
Diels Alder Synthesis (See Diene Synthesis)			**Digital**			Animals	119	158
Hydrocarbon synthesis	585	361	Clock	D10	15	Apparatus for molding	425	269+
Diene Synthesis			Motor art digest	91	DIG. 1	Molds or forms	425	275
Higher unsaturated fatty acid	260	404.8	Readouts	73	901*	Channel pumps	415	88+
Maleic acid	562	498+	**Digital Data**			Hat	223	10
Maleic acid anhydride	549	262	Communications	375		Match making	144	60
Carbocylic compound	549	234+	Computer systems, general	364	200	Cutting and framing combined	144	52+
Polymerization	585	507+	Error correction	371	30+	Frames	144	62+
Rosin	530	214	Processing machines	364	900	Framing combined	144	58
Differential			Synchronous transmission	375	106+	Processes for forming by	264	301+
Gearing			Transmission of coded intelligence	178	2B	Trap chamber dispensing	222	356+
Cash register key operated	235	14+	**Digital Memory**	365		Endless belt	222	371
Change speed and	475	198+	Speech reproduction from	381	51+	Rotary	222	369
Conveyers having different speed			**Digitalis Glycosides**	536	6.1	**Dips Animal**	424	
zones	198	792	**Dihydroabietic Acid**	562	404	**Direct Current Distribution**	307	
Plural conveyer sections	198	577	**Dihydronovobiocin**	536	13	Pulsating telegraph systems	178	66.1+
Plural sections with differing			**Dihydrostreptomycin**	536	15	Transmission to vehicles	191	
speeds	198	579	**Dihydrotachysterol**	552	653	**Direction**		
Elevator rope drive sheave	187	23	**Diketo Purines**	544	267+	Radio signaling	342	350+
Multiple driving motors	475	1+	**Diketopyrimidines**	544	309+	**Direction Indicator**		
Planetary	475	220+	**Dilatometry**	374	55	Design	D10	65+
Register single axis	235	119	**Dilator**			Earth boring combined	175	45
Register transfer mechanism	235	136	Surgical instruments	606	191+	Fluid flow velocity combined	73	188
Sectional rotary bodies	74	444	Surgical instruments combined	606		Geographical	33	320
Speed responsive device	73	507	**Diluents Coating Compositions**	106	311	Navigation	73	178R
Gearless	74	650	**Dimer (See Polymerization)**			Radar	342	147
Material treatment			**Dimmer for Lamps**	362	257+	Register fare	235	48
Comminution of mixed solids	241	14	Headlight systems	315	82+	Ships course	116	19
Sorting pretreatment	209	4+	Lamp systems	315	291+	**Traffic**		
Mechanism			Structure	323	905*	Barrier	404	6+
Brake operator movement	188	134	**Dimpling Sheet Metal**	72	414+	Director	404	9+
Chemical feed pressure control	137	100+	Testing ductility by	73	87	Vehicle	116	35R+
Fire extinguisher automatic valve	169	22	**Dinas Brick**	501	141	Design	D26	28+
Flow meter pressure type	73	861.42+	**Dining Car**	105	327	Electrically operated system	340	465
Railway draft springs	213	26+	**Dining Room Store Service**	186	38+	**Director Antenna**	343	912+
Sewing machine stitch forming			**Dinitroanthraquinones**	552	254	**Dirigible**	244	24+
feed	112	312+	**Diode**	313		**Disappearing Instrument Cabinet**	312	21+
Ships logs pressure type	73	182+	Beam power amplifier	313	298	**Disassembly (See Assembling)**		
Telegraph system duplex	370	27	With two cathodes	313	5	Apparatus	29	700+
Telephone transmitter electrodes	381	168+	Damping	313	317	Processes	29	426.1+
Motor	91	415+	With emissive cathode	313	310	Battery	29	
Motor vehicle with	180	76	With thermionic cathode	313	310	Repairing combined	29	402.3+
Differential Amplifier	330	69	With two cathodes and anodes	313	1	**Disc Recording or Reproducing**		
Differentiating Circuit	333	19	Duo pentode	313	298	Television	358	342+
Calculators having	235	61R+	Triode	313	303	Color	358	322
Electronic tube type	328	127+	Duplex hi mu triode	313	303	**Disconnecting Devices (See Stop Mechanism)**		
Diffraction	350	162.11+	With two cathodes	313	5	**Dish (See Cup; Tray)**	215	
Grating	350	162.17+	Duplex pentode	313	298	Butter refrigerating	62	457.6
Diffuser			With plural cathodes	313	5	Design, general	D 7	502
Slow diffuser holder	239	34+	Duplex triode	313	303	Design, refrigerating	D15	89
Sugar making	127	3+	With plural cathodes	313	5	Cleaner		
			Duplex twin	313	307			

	Class	Subclass
Brushing	15	74
Liquid contacting	134	
Compartmented	D 7	555
Cover	215	200+
Covered	D 7	538
Design	D 7	500+
Design, simulative	D 7	571+
Drainer		
Compartmented	220	572+
Rack	211	41
Wire receptacle	220	487
Heater	126	246
Making		
Clay	425	263+
Clay	425	459
Paper and paper board	162	387+
Sheet metal container	413	
Stand	248	128+
Design	D 6	310+
Washer	134	
Design	D32	2+
Disher		
Ice cream cone	425	118
Ice cream scoops	425	221
Ice cream scoops	425	276+
Dishwasher	D32	2+
Dishwashing racks	D32	55
Disinfectants & Disinfecting	422	
Apparatus for treating air	422	120+
Baths closets sinks and spittoons	4	222+
Sewer	4	220
Spittoon	4	261
Strainer or stopper cover	4	294
Urinal	4	309+
Cabinet combined	312	31
Chemical holders for flush toilet	4	228+
Disinfectants	424	
Electrolytic	204	130+
Ozonizer	422	186.7+
Preparation and distribution	43	127
Animal treating	119	156+
Dusters or sprayers	239	
Fumigators vermin destroying	43	125+
Intermittent discharge type	422	123+
Receptacle attachment	220	87.1+
Telephone attachment	379	452
Thermometer case	206	306
Tobacco	131	290
Electrical or radiant energy	131	299
Vapor and fume generator	422	305+
Disintegrator (See Comminuting; Crusher)	241	
Etching with electric arc	219	68+
Machining with electron beam	219	68
Paper fiber	241	
Peat	44	633
Rock in situ	299	
Sugar treatment combined	127	4
Textile fiber preparation	19	82+
Tobacco leaf	131	311+
Feeding combined	131	109.1
Disk		
Agricultural implement	172	518+
Bearing for	384	460
Bearing support for	384	157
Harvester cutter	56	255+
Harvester cutter with conveyor	56	157+
Hedge trimmer	56	235
Planting by drilling	111	163+
Plow moldboard	172	167
Scraper for	172	558+
Sharpener combined	172	437
Amusement roundabout	272	46+
Animal powered motor	185	18
Brake	188	71.1+
For railway vehicle	188	58+
For velocipede	188	26
Brush rotary	15	180
Boot clean black and polish	15	34+
Handle mount	15	28+
Street sweeper	15	87
Calendar	40	113+
Carbon black making collector	422	150+
Card exhibitor	40	495+
Cathode ray tube stream concentrator	313	441+
Clock geographical	368	27
Closure seal	292	308+
Comminuting and grinding		
Abrading tool	51	358+
Cherry stoner	99	566

	Class	Subclass
Comminutor	241	
Corn shellers	460	45
Disc grinding	51	
Floor surfacing machine	51	177
Grain huller	99	600+
Scouring machine	51	24+
Counting mechanism	235	98R
Coin	453	58+
Cutting		
Annular drills	408	204
Barrel stave jointing cutter	147	29
Button cutting	79	16
Can opener	30	435
Earth boring tool	175	373
Pipe cutter external	30	101+
Portable auger	408	204
Roller hand tool	30	307
Saw sharpening	76	45
Stone sawing	125	20
Sweep auger	408	199+
Tool sharpener	76	82
Tubular saw wood	408	204+
Wood	408	204
Wood dovetailing	144	89
Wood planer rotary cutter	144	118+
Wood sawing knife disc	83	469+
Wood turning chisel feed pattern	142	44
Educational elements	434	402
Electric conductor insulation	174	111
Electric generator armature	310	268+
Electrode consumable	314	44
Evaporator moving film support	159	9.1+
Feed for wood lathe	142	19
Feed for wood saw mill carriage	83	403.1
Feeder for nails	10	169
Flowmeter variable restriction	73	861.55+
Fluid distributor rotary	239	224
Gas and liquid contact impeller	261	84+
Gear sectional body	74	444+
Gearing frictional wheel and	74	194+
Laminated fabric disk making	156	
Magazines for target	124	46+
Making or working by rolling metal	72	67+
Metal rolling apparatus with disk platen	72	80+
Motion picture	352	102+
Paper closure assembling with container	493	108
Projectile holders	124	42+
Railway wheels	295	1+
Recorder record receiver	346	137
Recording and reproducing		
Compact disk	369	275.1+
Player	369	100+
Floppy disk	360	97.1
Player	360	99.1+
Optical disk	369	275.1+
Player	369	100+
Shaft coupling yielding element	464	98+
Non metallic	464	92+
Sound record	369	272+
Apparatus for molding with label applying	425	500+
Stacking racks for	211	49.1
Vehicle land wheel	301	63R+
Making	29	894.32+
Dislodging of Anchors	114	297
Dispatch Pneumatic	406	
Dispatching of Trains	246	2R+
Dispensing (See Notes to Main Class)	222	
Articles	221	
Beverage and food machines	D 7	300+
Cabinet		
Article	221	
Mixed drinks	222	129.1+
Spool holder and thread cutter	83	
Tearing	225	
Twine holder with cutter	83	
Coin handling	453	
Confetti	446	475
Container opening	206	601+
Dental apparatus	433	80+
Deposit & collection receptacles	232	
Device with wheels, skids, etc.	222	608+
Filling portable receptacles with fluent material	141	
Fire extinguishers	169	
Gas dispensing article	222	3+
Inkwell	222	576+
Molten metal	222	591+
Continuous casting	164	437+

	Class	Subclass
Planting apparatus	111	25+
Refrigerated liquids	62	389+
Sprinkling, spraying, & diffusing fluids	239	
Store dispensing service	186	
Tape		
Gummed	83	
Length measuring	33	732+
Length measuring with cutting	83	522.11+
Magazine type dispenser	312	39
Roll or spool	206	389+
With fixed severing edge	225	
Track sanders	291	
Trap chambers	222	424.5+
Movable or conveyer type	222	344+
Unwinding and cutting fabrics	242	55+
Vending machines	D20	1+
Web or strand feeding	226	
Dispersing (See Colloids; Emulsifying)		
Agents	252	351+
Solids combined	252	363.5
Agitating	366	
Colloids	252	302+
Radioactive	252	634+
Compositions including agents for		
Bituminous material	106	278
Carbohydrate gum	106	208
Casein	106	146
Cellulose	106	203
Glue or gelatine	106	135
Natural resin	106	236+
Pigments	106	499+
Prolamine	106	153
Protein	106	161
Starch	106	213
Synthetic resin (See class 523, 524)		
Wax	106	271
Dyes including agents for	8	
Fermentation apparatus	435	287+
Foam control	435	812*
Medium for coating or plastic compositions	106	311
Solid disintegrating	241	
Display		
Animated	340	724+
Bank counters	109	10+
Boxes	217	11
Closures hinged	217	58
Compartmented	217	10
Design	D 9	414+
Folding	217	9
Sliding closure	217	63
Cabinet	312	114+
Frozen food, produce or meat	D 6	432+
With display opening	312	234
Card picture and sign	40	
Design	D20	10+
Design misc	D20	10+
Cathode ray tube	340	720+
Changeable exhibitor	40	446+
Container convertible for storage	206	44R
Device with gas or liquid movement	40	406+
Drying apparatus combined	34	88
Easels	248	441.1+
Educational devices	434	365
Electro-optic	350	330+
Electroluminescent	315	169.3
Electrophotographic copier	355	271
Envelopes	229	71
Forms	D20	10+
Furniture		
Design	D 6	
Garment forms	223	66+
Gas panel	315	169.4
Heads up	340	705
Integrated with circuit	340	718+
Letters sheets	229	92.3
Light systems electric control	340	286.1+
Liquid crystal	350	330+
Matrix	340	752+
Monogram	340	756+
Optical projection screen	350	117+
Packages	206	44R+
Photographic	355	18
Printed advertising	283	56
Racks	211	
For picture or business card	40	124
Receptacles	206	44R+
Refrigerators	62	246+
Open access	62	458

	Class	Subclass
Show cases	312	114+
Stands	261	DIG. 14
T V channel	358	192.1
Three dimensional movable figure	40	411+
Timepiece	368	223+
Trays	206	557+
Vehicle body feature	296	21
Wooden containers	217	9+
Wrappers	229	92.3
Disposal		
Closets sinks spittoons	4	
Crematory	110	194
Cutlery combined	30	124+
Incinerator	110	235+
Incinerator garbage	110	235+
Nuclear radioactive waste	252	626+
Razor combined	30	41
Receptacles collection	232	
Sewage treatment	210	
Sewerage	137	
Disproportionation		
Alicyclic hydrocarbon synthesis	585	375+
Aromatic synthesis	585	470+
Hydrogen exchange (See dehydrogenation; hydrogenation)		
Olefin synthesis	585	643+
Paraffin synthesis	585	708
Dissemination (See Scattering)		
Dissolves in Motion Pictures	352	91R
Dissolving (See Solvent)	423	658.5
Actinide series compounds	423	249+
Inorganic compounds	423	210+
Solids to produce cooling	62	4
Distance Measuring Devices	33	700+
Speed integrator	73	490
Distillation		
Apparatus	202	
Azeotropic	203	50+
Beverages alcoholic	426	
Bromine or iodine	423	500
Chemical apparatus combined	422	189+
Convertive	203	49
Distillate treatment (See vapor)		
Destructive distillation	201	29
Distillation combined	203	
Mineral oils	208	349
Extractive	203	50+
Filming	203	89
Filming	203	72
Flash	203	88
Fluidized bed	201	31
Heating and illuminating gas	48	
Mineral oils	208	347+
Conversion combined	208	46+
Rectification	208	350+
Separatory	203	
Stills	202	81+
Mineral oil	196	104+
Thermolytic	201	
Distillery		
Waste as fertilizer	71	26
Stock feed	426	635
Distortion Bucking in an Amplifier	330	149
Distortion Control		
Amplification	330	
Amplitude modulation	332	159+
Frequency modulation	332	123+
Metal deforming	72	701
Pulse modulation	332	107
Telegraphy	178	69A
Telephone circuit	379	414
Wave transmission lines and networks	333	
Distributing Material		
Ammunition filamentary material	102	504+
Bolt nail nut rivet screw making	10	162R+
Coating implement with material supply	401	
Conveyer power driven thrower	198	638+
Conveyer thrower powerdriven	198	638+
Distilland in retort	201	40
Drying		
Agitators	34	241+
Drum rotary	34	130+
Kilns	34	218+
Kilns plural	34	210+
Processes	34	11
Stationary receptacle	34	24
Ensilage	406	164+
Fluid centrifugal	239	214+
Gear friction pressure balancing	74	196

	Class	Subclass
Gear worm	74	427
Gearing pressure	74	410
Hay	414	25
Letter boxes compartment	232	26
Mobile orchard type	239	77+
Pamphlets from aircraft	40	216
Pastry machine combined	425	289+
Railway track layers	104	5+
Road material	404	101+
Scattering unloader	239	650+
Sifters	209	254
Stratifiers	209	498
Type cases	276	44+
Type casting machine	199	33+
Type casting machine combined	199	14+
Type setting machine	276	22+
Type setting machine combined	276	2+
Wire nail making	10	44
Distribution		
Boxes for electric conductors	174	50+
Electric current	307	
Fluid (See ventilation)		
Gas or mist	102	367+
Handling	137	
Motors expansible chamber	91	
Motors valve actuation	91	
Nozzles	239	
Gas in gas mains	48	190+
Distribution Amplifier	330	54
Distributor Electric		
Auto	200	6R+
Ignition circuit	315	210
Ignition system	123	146.5 R+
Conductors		
Combined	174	73.1
Overhead	174	43
Underground	174	38
Ignition		
Structure	200	23+
System	123	143R+
System	315	211+
Insulators combined	174	140R
Signalling transmitters	341	192
Switch multiple circuit	200	1R+
Switch periodic	200	19DR+
Systems miscellaneous	307	
Telegraphy multiplex	370	46+
Telephone switchboards	379	319+
Transmission to vehicle	191	2+
X ray tubes potential stress	378	139
Ditch Filler	37	142.5
Ditcher	37	80R+
Elevator	414	596+
Filler	37	142.5
Dithiazoles	548	123
Dithionites		
Reducing composition	252	188.22
Dithiosulfurous Acid Esters	560	310
Diversity Receiver	455	132+
Divider		
Assorter discharging	209	493
Dough, severing apparatus general	225	93
Fleece	19	151
Fluid flow porportional	137	118+
Gauges point markers	33	665
Center	33	670+
Proportional	33	663+
Harvester track clearing	56	314
Opposed contact	33	558.1+
Road	404	6+
Scribing	33	18.1+
Diving Apparatus	405	185+
Artificial gill	128	200.25
Board	D21	236
Helmets	2	2.1 R
Suits	2	2.1 R
Air or oxygen supply	128	201.29+
Buoyant or swimming feature	441	88+
Submarine working device	405	186+
Divining Rods	324	800*
Dobby Loom	139	66R+
Griff vibrating	139	67
Dock		
Drydocks	405	4+
Floating	114	45+
Doctor Blade (See Device or Apparatus with which Associated))		
Electrophotographic copier	355	299
Document	283	
Handling		
Electrophotographic copier	355	308+

	Class	Subclass
Holder		
Electrophotographic copier	355	230+
Sales	283	60.1+
Documentation Computer Application	364	419
Doffing		
Bobbin and cop winding combined	242	41
Carding combined	19	106A
Conveyer section	198	622
Ginning combined		
Delinters	19	42+
Saw	19	58+
Magnetic solids separator or classifier combined	209	229+
Spinning, twisting, or twining combined	57	266+
Preparation for	57	276+
Vegetable cutter or comminutor combined		
Parer and corer	99	542
Parer corer and segmenter	99	543
Parer corer and slicer	99	543
Peeler or parer	99	588+
Dog		
Collar	119	106
Design	D30	152
With leash or tether	119	118
Driven ratchet bar combined	74	169
Foodstuff	D 1	
For bolt	70	467+
Furnace fire	126	298
Harness	54	
House	D30	108+
Kennels	119	19
Muzzle	119	30+
Sawmill	83	721
Set works end	83	730
Sewing machine feeding	112	324
Turning lathe	82	166+
Woodworking bench	144	306+
Doily	428	81+
Embroidery, design	D 6	613+
Furniture protector	D 6	613+
Dollies		
Article supporting	280	47.34+
Design	D34	23
Hoisting trucks	254	2R+
Rivet	227	
Washing machine	68	138
Dolls	446	268+
Aquatic	446	156+
Assembling and disassembling	29	805
Design	D21	148+
Eating drinking nursing	446	304+
Sleeping	446	345+
Talking crying	446	297+
Wheeled	446	270+
Voices	446	297+
Pneumatic	446	188+
Wheeled	446	269+
Inflatable	446	226
Dolomite Refractory Compositions	501	113
Dome		
Arcuate design	D25	19
Building structures	52	80+
Fire tube	122	116
Geodesic design	D25	13
Lantern	362	182
Tubular	362	173
Lights vehicle	362	61
Steam	122	508
Separator	122	492
Superheater	122	486
Track sander hopper	291	40
Dominoes	273	292+
Donning		
Empty spinning bobbins	57	266+
Preparatory	57	276+
Door (See Closure)		
Bar for jamming closed	D 8	331+
Bells electric	340	392+
Braces	52	291
Brakes, track or guideway	16	DIG. 2
Cabinet		
Sectional unit	312	109+
Showcase	312	138.1+
Channel for sliding door	D25	119+
Checks	16	82+
Checks and closures	16	49+
Chimes	D10	118
Coin controlled	194	
Curved cast archway	52	85

	Class	Subclass
Door		
Design	D25	48+
Electric		
Contacts for burglar alarms	200	61.93
Elevator	187	51+
Flexible roll	160	238+
Folding	160	229.1+
Sliding together	160	222+
Frame	49	504
In situ structure	52	204+
Guards	49	50+
Handles (See --, knob)	16	110R+
Latch	16	DIG. 32
Opening apparatus	16	DIG. 7
Hanger or track	16	87R+
Hinges	16	221+
Jail	49	15+
Keyhole		
Fluorescent	250	466.1
Illuminator	362	100+
Knob	16	121+
Design	D 8	
With spindle attachment	292	347+
Knockers	116	148
Design	D 8	401+
Land vehicle	296	146+
Latch (See closure; fastener)		
Latch bolts, biased	70	144
Dog for bolt	70	467+
Lock	70	91+
Keepers	292	340+
Railway	105	395
Mat	15	215+
Motor vehicle (See closure)		
Operators	49	324+
Ornamental panel	52	311+
Overhead door	16	DIG. 1
Panel warp correction	52	291
Plate or sign for	D20	43+
Plural panels	52	455+
Pneumatic-type closer	D 8	330
Railway		
Dumping car	105	280+
Emergency	105	348+
Platform trap	105	426+
Vehicle restrained	246	304+
Releasers	292	341.16
Removers	414	684.3
Traversing hoist	212	166
Sectional panel	52	455+
Ships	114	116+
Silencer	D 8	400
Sliding	49	404+
Stop	D 8	402
Stove, furnace or oven		
Casings furnace	110	181
Cooling furnace	110	180
Furnace	110	173R+
Stove	126	190+
Switches	200	61.62+
Lamp or electronic tube system	315	84
Track	D 8	377
Trim	52	211+
Troughs	49	408
Ventilating	98	87
Stove	126	198
Work holder for door and frame	269	905*
Dop Jewel Holder	51	229
Doping		
To form semiconductor P-N type		
junction	437	16+
Alloying	437	134+
Diffusing	437	141+
Fusing dopant with substrate	437	134+
Using energy beam	437	16+
While depositing material	437	81+
Doping Agent Source Material	252	950*
For vapor transport	252	951*
Doppler Compensation Systems	367	904*
Dorr Classifier	209	462
Dosage-related Cabinet	312	234+
Dosing Device		
Dispensing	222	
Indicator for medicine	116	308
Medicators	604	
Dot		
Dot type mosaic screen cathode ray		
tube	313	472
Printer	400	124
Double Antibody Test	436	540
Double Bond Shift (See Isomerization)		
Doubler	202	199

	Class	Subclass
Doublet Antenna	343	793+
Doubling		
Twisting strands		
Covering or wrapping	57	14
Delivery twist	57	59+
Ends or hanks	57	26+
Receiving twist	57	66+
Winding bobbins and cops	242	42
Fault detecting	242	38
Fault detecting load	242	40
Douche	604	36+
Hand held	604	212
Nozzle with separate ingress &		
egress	604	39
Treating material introduced	604	54
Dough		
Compositions	426	549+
Crimping devices	425	293
Cutting machines	83	
Dividing machines	425	289+
Forming, molding and working	425	197+
Forming, molding and working	425	200+
Kneading machines	366	69+
Kneading machines	425	197+
Kneading machines	425	200+
Mixing machines	366	69+
Packaging or wrapping	426	392+
Presses	425	
Raisers	126	281+
Rollers	425	294
Rollers	425	329
Testing	73	169
Doughnut	426	496
Cookers		
Deep fat fryer type	99	403+
Forming or shaping combined	99	354
Opposed heated surface type	99	382
Support combined	99	442
Deep fry process	426	439
Design	D 1	120
Dough fermentation	426	18
Doup Heddle	139	51+
Doup Heddle		
Dovetailing		
Bedstead corner fastening	5	300
Design	D 6	503+
Calks	168	35
Woodworking	144	85+
Dowel		
Jigs	408	72R+
Pins	403	292+
Making	144	12
Module connector	52	585
Road joint	404	47+
Downdraft		
Furnaces	110	315
Hot air furnace	126	103
Liquid heaters and vaporizers		
Fire tube	122	97
Water tube	122	286
Stove	126	76
Downspout	52	16
Doxycycline	552	203
Draft		
Air current		
Back draft preventer	98	119
Boiler	122	
Control to combustion chamber	431	20
Downdraft furnace	110	315
Forced draft fumigant burner	43	125+
Heating stove downdraft	126	76
Hot air furnace downdraft	126	103
Intermediate in superimposed fire		
box with air or steam	110	267+
Stove damper	126	290
Superimposed fire box and		
intermediate draft	110	317
Appliance		
Animal	278	
Articulated vehicle train	280	400
Cable hoist	212	76+
Cable railway	104	193
Horse drawn sweep	185	23
Monorail animal draft	105	143
Railway	213	
Wheeled horse rake draft		
dumping	56	386+
Equalizer		
Animal draft appliance	278	3+
Horse drawn sweep	185	22
Gauges	73	700+
Regulator		

	Class	Subclass
Boiler	122	38
Damper	126	285R+
Damper automatic	236	45
Furnace	110	147+
Spark arrester combined	110	123
Drafting (See Drawing)		
Board		
Cabinet combined	312	231
Implement	D19	35+
Box for	206	371
Curved ruler type	33	561.2+
Scriber type	33	18.1+
Straightedge type	33	403+
Table	D 6	420
Textile fibers	19	236+
Drag		
Animal restraining	119	107
Classifier	209	462
Conveyer chains	198	717+
Conveyor chains for viscous fluids	198	643
Drilling planting machine	111	84
Earth working	172	
Fishing reel	242	283+
Land vehicle type of	280	19
Line scraper	172	26.5
Metal casting sand	164	349+
Oil distributor	114	234
Reciprocating saw machine	83	746+
Saw guide	83	821
Sea anchor	114	311
Design	D12	215
Ships log	73	184+
Spinning reel	242	243+
Textile spinning and twisting	57	113
Bobbin	57	72
Flier type bobbin	57	70
Drain and Drainage		
Building construction	52	302+
Cabinets with	312	229
Conduit electrical	191	26
Dispenser	141	364
Footwear with	36	3R
Fryer deep fat	99	410+
Machine fluid treating	68	208
Photographic wet plates	354	280
Pipe cleaner	15	104.31+
Pump	417	437
Road or pavement	404	2+
Roll type clothes wringer	68	271
Roofing interior	52	553
Gutter	52	11+
Safes or banks	109	28
Sand	405	50
Sewerage	210	163+
Shield	D23	261
Sink	4	650+
Soil	405	36+
Static mold having fluid	249	141
Sugar centrifugals	127	19
Surgical	604	264+
Table surgical	108	24
Washing machines	68	208
Wringer roll	68	271
Drain Board (See Dish, Drainer)		
Sink accessory	D32	56
Tub	4	578
Attachment	4	656
Design	D23	277+
Draperies	160	330+
Hook	16	93D
Surgical	128	
Tie back, fabric	D 6	578
Tie back, rigid	D 8	368
Draw		
Press	72	343+
Drawbar		
Articulated vehicles combined	280	400+
Bedstead corner fastening	5	298
Elastic extension device inclosing	267	72
Railway	213	62R+
Car forging dies	72	343+
Car making	72	
Drawbridge	14	31+
Electrical transmission to	191	9
Floating	14	29+
Protection switch or signal	246	118+
Drawer		
Bed combined	5	308
Cabinet	312	
Horizontal sliding	312	330.1+
Cash register operation	235	22

	Class	Subclass
Element of sofa bed	5	58
Key set cash register operated by	235	10
Knob labels and tags	40	331+
Locks	70	85+
Plate or sign for	D20	43+
Pull labels and tags	40	325
Refrigerated	62	382
Air blocking when open	62	266
Drawing (See Drafting)		
Abrading reel	51	75
Board		
Cabinet combined	312	231
Design	D19	52
Straight edge mounting	33	430+
Die	72	467
Easels	248	441.1+
Etching combined with	156	658
Glass	65	193+
Process	65	66+
Implement	D19	35+
Manufacturing glass		
Product by	65	
Material, design of	D19	35+
Mechanical instruments	33	
Mercury vapor lamp starting by	313	170+
Metal	72	
Cup or shell	72	347+
Method & apparatus	29	DIG. 11
Push drawing	72	343
Rolling combined	72	206
Sheet die shaping	72	380+
Spinning	72	82+
Wire or tube	72	274+
Pen	401	267
Design	D19	41+
Perspective instruments	33	1K
Reverse drawing on metal	72	708*
Textile fiber	19	236
Textile spinning, twisting, twining		
Combined	57	315+
Rollers	57	97
Triangles	33	474
Visual arts	434	85+
Drawknives	30	313
Drawplate		
Diamond setting	76	107.1+
Drawstring	24	712
Dredge		
Excavating	37	54+
Submerged vessel raising	114	55
Dredge Top Dispenser	222	480
Dress	D 2	49+
Dressmaker	D 2	195
Garment	2	71+
Forms	223	68+
Hangers	223	92+
Protector	2	46+
Shield pressing	223	54
Weights	2	273
Maternity	2	76
Sifting screen	209	392+
Wheel guards	280	160.1
Dresser, Fish Preparation Tool	D 7	693+
Dresser, Furniture	D 6	432+
Dressing		
Brushes and brooms	300	18
Meat and fowl	17	
Saw making		
Jointing and gaging combined	76	46
Jointing combined	76	50.4+
Separating screen	209	392+
Separating solids	209	
Stone working	125	2+
Diamond tool for	125	39
Grindstones	125	11.11
Millstone	125	27+
Surgical (See bandaging)		
Tobacco		
Fluid or fluent material	131	300+
Smoothing brushing rolling	131	324+
Stemming combined	131	315+
Warps	28	178+
Dressmaker	D 2	195
Dried Fruits and Vegetables	426	640+
Drier and Drying (See Dehydration)	34	
Blotter or towel type	34	95+
Non-drying device combined	34	89.1+
Cigar and cigarette making		
Tip or mouthpiece applying	131	92
Wrapper sealing	131	68
Clothes	34	
Automatic	34	43+
Clothes drying rack	D32	58
Machine design	D32	8
Coating combined with drying	427	372.2+
Collar cuff and bosom making	223	3
Composition	106	310
Drying oils combined	106	264
For coating	106	310
Dress coat or skirt forming and stretching	223	69
Dry closet	4	111.1+
Fishing line	242	104
Fruits and vegetables dehydrating	426	471+
Garbage and sewerage furnace	110	224+
Gas and vapor contact with solids	34	
Hair dryer	D28	12+
Hand dryer	D28	54.1
Methods	34	3+
Milk dehydrating	426	471+
Mortar mixer combined	366	22+
Paper article making		
Box making	493	141+
Envelope making	493	265
Rack	211	
Refuse treatment means	110	224+
Shoe and boot	34	104
Solids apparatus	34	43+
Textiles		
Thread finishing, heating or drying	28	219
Washing machine combined	68	20
Web spreader combined	26	92
Web stretching combined	26	106
Tobacco	131	300+
Wood bending	144	254
Drift		
Indicator (See direction indicator)		
Pin	29	275
Drill (See Bit)		
Bit	D15	138+
Dental	433	165+
Chucks	279	
Cigar tip perforators	131	254
Dentists	433	165
Design	D24	12
Drilling machines	408	
Drills	408	199+
Design	D15	139
Design hand tool	D 8	59
Earth boring	175	327+
Expansible	175	263+
Grain planting	111	14+
Grinding processes	51	288
Making and sharpening machines	76	5.1+
Making blanks and processes	76	108.1+
Press	408	72R+
Press type combined metal working machine	29	26R+
Pyramidal end	125	1
Rack for	211	69
Rail portable	408	77
Receptacles special for	206	379
Rock	175	327+
Trepan	606	176+
Twist for metalworking	408	230
Twisting stock to make	72	299
Vise attached	29	560.1
Well	175	
Woodworking	408	199+
Drilling (See Boring)		
Borehole and drilling study	73	151
Brush making tuft setting and	300	3
Button making	79	11+
Pearl button surfacing and	79	6
Earth	175	
Grain drill chute	193	9
Machines		
Dental	433	103+
Driving or impacting	173	
Radial	408	236+
Rock	175	
Mining implement	299	79+
Muds		
Analysis	73	153
Compositions	252	8.51
Earth boring with	175	65+
Planting	111	14+
Broadcasting and	111	8+
Liquid or gas	111	118+
Plant setting	111	109+
Solid material	408	
Electric	310	
Well	175	
Marine platform	405	195+
Drink		
Preparation or dispensing machines	D 7	300+
Registers	235	94R
Drinking Fountain	239	24+
Animal watering devices	119	72+
Design	D30	121+
Design	D 7	304
Tube	239	33
Filter combined	210	251
Drinking Straw	239	33
Drinking Vessel		
Container for collapsible type	206	218+
Drinking Water		
Chemical purification	210	198.1+
Process	210	601+
Filtering	210	348+
Decanting combined	210	294+
Drip		
Building attached deflector	52	97
Catcher (See drip pan)		
Closet bowl	4	252A
Coating implement adjunct	401	15
Cooking apparatus	99	444+
Dispenser	222	108+
Fluid sprayer or sprinkler	239	120+
Gas separation apparatus	55	280
Griddle	99	425
Inverted container support with	141	364
Refrigerator defrost water	62	285+
Refrigerator ice melt	62	459+
Refrigerator ice melt filter	62	318
Toaster	99	400
Waffle type cooker	99	375
Water heating vessels	126	383+
Coffee maker	99	306
Collector	431	119
Cooled		
Ice melt gas contactor	62	312+
Ice melt heat exchanger	62	460+
Wet wall type refrigerator	62	278
Cup		
Cleaning attachment	15	248R+
Umbrella	135	48
Liquid diffusers	239	38
Meters	604	251
Drip sensor	604	253
Filter	604	252
Flow control	604	246
Pan (See drip catcher)		
Liquid fuel cooking stove	126	51
Lubricators	184	106
Vehicle body	296	38
Plate		
Boiler feed heater	122	417
Water heater	126	355
Drive in Theatres	52	6+
Anchor	52	155
Supported building component	52	292+
Earth piercer	52	155+
Terranean relationship	52	169.1+
Drive or Driver		
Barrel hoop	147	7+
Barrel tap	81	27
With dispensing	222	81+
Bolt or nut	81	52+
Centrifugal separators	210	360.1+
Conveyor endless	198	832+
Earth boring tool combined	175	170+
Impact	175	135
Elongated-member driving apparatus	227	
Golf club	273	77R
Design	D21	214+
Hammer	173	90+
Design	D 8	75+
Hill planting machine belt feed	111	19
Locomotive	105	26.5+
Mallets	81	19
Nail		
Implements	227	140+
Machine	227	
Pile	173	90+
Railway car wheel or axle	105	96+
Railway turntable actuator	104	41
Screw driver implement	81	436+
Screw driving machine	81	54
Sewing machine	112	220+
Shoe lasting and nailing	12	13.1
Tap driving ratchet	408	120+

	Class	Subclass
Tool	173	
Track spike	104	17.1
Trackmans car	105	86+
Turning work	82	166+
Well point	175	
Driven Headed and Screw Threaded		
Devices	411	
Railway spikes	238	366+
Drop		
Annunciator systems	340	286.11+
Annunciators	340	815.29+
Bomb	102	382+
Forging	72	435+
Fluid operated	72	453.1+
Hammer	72	435+
Pick up for overhead railway car	104	122
Store service	186	22+
Telephone switchboard restorers	379	315+
Dropper		
Animal waterer	119	72.5
Cash register indicator tablet	235	25
Dispenser	222	420+
Medicine dropper type	141	24
Medicine	604	295+
Combined with bottle, design	D 9	338
Combined with closure, design	D 9	447
Droppings Catcher		
Dispenser	222	108+
Railway	104	133
Drugs	424	
Bio-affecting & body treating		
compositions	514	
Acronycines	514	285
Allantoin	514	390
Amantadine	514	656
Amphetamine	514	654
Ampicillin	514	198
Amprotropine	514	534
Aspirin	514	165
Atropine	514	304
Barbituric acid	514	270
Benzocaine	514	535
Benzomorphans	514	295
Biotin	514	387
Capsaicin	514	627
Cephalosporins	514	200+
Chlorpheniramine	514	357
Cholecalciferol	514	167
Chrysanthemic acid	514	572
Cobalamin	514	52
Codeine	514	282
Colchicine	514	629
Cortisone	514	179
Cupreine	514	305
Cycloheximide	514	328
Cyproheptadine	514	325
Cysteine ester	514	550
D D T	514	748
D D V P	514	136
Dextromethorphan	514	289
Dyphylline	514	263
Ephedrine	514	653
Estradiol	514	182
Fluspirilene	514	278
Glaucine	514	284
Glaumine	514	669
Griseofulvin	514	462
Hexachlorophene	514	735
Hydrocortisone	514	179
Isoniazid	514	354
Malathion	514	122
Melatonin	514	415
Meperidine	514	330
Methadone	514	648
Methapyrilene	514	336
Methomyl	514	477
Morphinans	514	289
Morphine	514	282
Nandrolone	514	178+
Niacinamide	514	355
Nicotinamide	514	356
Nicotine acid	514	356
Nortestosterone	514	178
Novocaine	514	535
Oxolinic acid	514	291
Parathion	514	132
Penicillin G	514	199
Perdnisolone	514	179
Perimidines	514	269
Phenylephrine	514	653
Phenyltoloxamine	514	651

	Class	Subclass
Pilocarpine	514	397
Pimozide	514	323
Piromidic acid	514	303
Procaine	514	535
Progesterone	514	177
Psoralen	514	455
Pteridine	514	249
Purines	514	261
Quinicine	514	314
Quinidine	514	305
Quinine	514	305
Quinoxaline	514	249
Riboflavines	514	251
Salinomycin	514	460
Scopolamine	514	291
Tartaric acid	514	574
Tetracycline	514	152
Tetramisole	514	368
Theophilline	514	263
Thiamines	514	276
Tocopherols	514	458
Tripelennamine	514	352
Tryptophan	514	419
Tyrosine	514	567
Uracil	514	274
Vinblastine	514	283
Vincamine	514	283
Viquidil	514	314
Childproof bottle closures	215	201+
Drugs of abuse	436	901*
Radioactive	424	1.1+
Drums (See Cylinders; Rollers)		
Armature (See armature)	310	265
Brake	188	218R
Cask or	D34	39
Centrifugal (See centrifugal force, separators)		
Container type	220	
Earthworking or smoothing (See rolls & rollers, land)		
Evaporator moving film support	159	9.1
Treated material inside drum	159	9.2
Material separating (See separator)		
Musical	84	411R+
Automatic	84	104
Rotary furnace	432	103+
Metallurgical	266	173
Rotary heat exchange	165	89+
Rotary or tilting support	384	549
Stand	248	128+
Tumbling device for abrading	51	164.1
Winding (See coiling winding)		
Drumstick	84	422.4
Dry Cell Battery	429	156
Recharging system	320	4+
Thermoelectric & photoelectric	136	
Dry Cleaning	8	142
Cleaner compositions	252	89.1+
Machinery design	D32	10
Dry Closet	4	449+
Furnace combined	4	111.1+
Dry Disk Rectifier	357	
Dry Dock	405	4+
Floating	114	45+
Dry Ice T M Solid Carbon Dioxide		
Making	62	8+
Process	62	10
Shaping combined	62	35
Refrigeration by	62	384+
Automatic	62	165+
Dryers (See Drier & Drying)		
Drying Japans	106	310
Drying Oil	260	398+
Composition containing	106	252+
Natural resin containing	106	222+
Fatty oil with preservative	106	263
Hydrocarbon	585	945*
Mineral oil	208	1
Polymers ctg (See synthetic resin or natural rubber)		
Drywall Construction	52	344+
Strip	52	459+
Dual-in-line Electronic Package	174	52.4
With semiconductor	357	74
Duck		
Calls	446	207+
Decoys	43	3
Duct (See Conduit)		
Humidifier	261	DIG. 15
Ductility Testing	73	87
Dulcimer	84	284

	Class	Subclass
Dumb Waiter	187	3+
Dumbbell	272	122+
Dumdum Bullet	102	507+
Dummy		
Ammunition	102	444+
Bomb	102	395
Shell	102	498
Clothes and other display	40	538
Figure dispenser	222	78
Figure toys	446	268+
Aquatic	446	156+
Inflatable	446	226
Wheeled	446	269+
Pyrotechnics	102	355
Tackling	273	55R
Dumping Mechanism		
Cable hoist	212	79+
Clothes washing machines	68	210
Egg candling trays	356	65
Furnace ash pan	110	167+
Hand shovel	414	722+
Mortar mixer	366	45+
Nuclear reactor	376	261
Portable receptacle	414	403+
Safes	109	46
Shelf to shelf flow drying apparatus	34	172
Stove grates	126	162
Rocking bar	126	177
Shaking combined	126	158
Vertical axis oscillatory	126	171
Vehicle	298	
Horse rake	56	386+
Railway	105	282
Moving car	105	241.1
Rake and tedder	56	368
Roadway with external	414	354+
Scoop	37	140
Scow	114	27+
Wheeled toy	446	428
Water closet	4	365+
Dune Buggy Design	D12	87
Dunnage		
Element	267	136+
For container	220	429
For freight	410	121
Edge-around	410	155
Honeycomb	410	154
Duplex Copy		
Electrophotographic copier	355	319+
Duplex Diplex System	370	36
Duplex Telegraph System	370	24+
Duplicate Whist and Other Card Hand		
Holding Apparatus	273	151
Duplicating (See Manifolding)		
Cutting machine	409	93
Keys	409	81+
Machines	D18	13+
Pattern controlled		
Milling machine	409	79+
Sheet or card punching	234	59+
Plural ribbon typewriter	400	204+
Printing machine	101	113
Selecting or progressive	101	90
Ticket printing machine	101	67+
Durez T M (phenoplast)	528	129
Duroquinone	552	310
Dust		
Carburetor	406	
Collectors		
Abrading machines	51	273
Cabinet with	312	229
Dental engine	433	116
Design	D32	15+
Earth boring cuttings	175	207+
Gas separator with	55	429
Textile spinning apparatus	57	300+
With air blast or suction cleaner	15	347+
Woodworking machines	144	252R
Conveyor		
Threshing machine	460	117
Woodworking	144	252R
Cover		
Billiard and pool table	273	13
Inflation stem type	138	89.4
Inflation stem type air and	138	89.1+
Vehicle top	296	136
Wheel and valve stem with	152	428
Fixation	404	76
Guard		
Car ventilation	98	25+
Locomotive cab	98	28

	Class	Subclass
Frequency conversion	363	157+
Phase conversion	363	148+
Cooking by current through food	99	358
Cooling produced by	62	3.1+
Cord		
Storing	191	12R+
Course control	244	175+
Current distribution	307	
Current producing apparatus, product, process	429	
Demagnetizers	361	267
Depilatories	606	43
Design	D13	
Air conditioning	D23	351+
Clocks	D10	1+
Fan	D23	370+
Heating	D23	314+
Lamps	D26	
Measuring testers	D10	75+
Medical xray	D24	2
Photo chemical etching	430	313+
Soldering iron	D 8	30
Diagnostics medical	128	630
Distribution	D13	
Distributor (See distributor, electric)		
Door or gate	49	59
Drill	30	500*
Electrical device making processes	29	592.1
Electrical power conversion systems	363	
Electrical systems & devices	361	
Electrocardiographs	128	697
Electrolytic device	361	500+
Batteries	136	
Chemistry, electrical & wave energy	204	
Electromagnetic operation	335	
Electronic tube (See electric space discharge devices)		
Electrothermally or thermally actuated switches	337	
Electrothermic metallurgical processes	75	10.1+
Elevators, electric control	187	100+
Engine starting motor systems	290	
Motor per se	310	
Equipment, design	D13	
Eye (See photoelectric)		
Fan	310	40.5
Guards	416	247R
With body motion	310	40.5
Fault isolating safety systems	361	1+
Fault location	324	512+
Fault testing	324	500+
Fence	256	10
Insulator	174	137R+
Filter (See filter, electric)		
Fire starting device	D 7	416+
Flashlight	362	208
Fluid heater	392	311+
Fluid motor system incorporating electrical system	60	911*
Furnace	373	
Fuses	337	142+
For explosives	102	
In arc lamp	314	10+
In lamp or electronic tube	315	74+
Gas & liquid contact apparatus	261	DIG. 8
Geiger muller counter	250	374+
Geiger muller counter	313	93
Generation	310	
In lamp or electronic tube	315	55
System	322	
Generator (See generators)	310	
Bicycle driven with lamp	362	193
Nonelectric prime mover driven	290	
System	322	
Generator control	322	44+
Polarity control	322	5+
Power transmitting mechanism control	322	40+
Generator systems	322	
Automatic control	322	11
Automatic control	322	17+
Excitation control	322	59+
Piezoelectric	322	2R
Gyroscopic compass	33	324+
Heating	219	
Distillation processes		
Paratory	203	
Thermolytic	201	19
Electrothermic metallurgy	75	10.1+

	Class	Subclass
Ferrous metal local hardening	148	150
Ferrous metal treatment	148	154
Gas generator retort	48	103
Incubator type heater	236	3
Knife	30	140
Metal working with	29	DIG. 13
Static mold	249	78
Thawing pipes and freeze protection	138	33
Track sanders	291	20
High frequency medical	128	804+
Igniter with burner (See igniter)	431	258+
Ignition, internal combustion engine	123	143R+
Impulse generator		
Condenser discharge	320	1
Dynamoelectric	310	10+
Electromagnetic periodic switch	335	87+
Electronic tube system	328	59+
Gas tube relaxation generator	331	129
Nondynamoelectric	310	300
Periodic switch	200	19R+
Radar system	342	175+
Relaxation oscillation generator	331	143
Signal transmitters	341	173+
Supplied to lamp or electronic tube	315	289
Telegraph system	178	
Telephone call transmitters	379	
Transistor relaxation generator	331	111
Inductor devices	336	
Insulation, electrical	493	949*
Insulator	D13	129+
Interrupter (See interrupter)		
Ionization chamber	313	93
For nuclear fusion reaction	376	100
System	250	335
Junction box	D13	152
Key (See key, electric)		
Lamp (See lamps, electric)	313	
Consumable electrode arc lamp	314	
Display systems, matrix	340	780
Display systems, monogram	340	760+
Flash lights	362	208+
Heaters using	219	552+
Lamp systems	315	
Making	445	
Making glassworking machine	65	152+
Making glassworking processes	65	36+
Making miscellaneous	445	
Repair apparatus	445	61
Repair methods	445	2
Signs, multiple function	40	553+
Lead frame stock	428	572
Line (See line, electric)		
Liquid purifier insulating or discharging	210	243
Magnetic surgery	600	9+
Magnetically operated switches, magnets, & electromagnets	335	
Massage	128	24.1+
Mattress, flexible heater element	219	549
Measuring & testing	324	
Instruments, design	D10	46+
Medical and surgical treatment	128	362+
Medicators	604	20+
Melting furnace	373	
Glass	373	27
Meter (See meter, electric)	324	76R+
Coin controlled	194	
Recording	346	
Miscellaneous	361	
Miscellaneous electron space discharge device systems	328	
Miscellaneous non-linear reactor systems	307	401+
Miscellaneous non-linear solid solid device circuits	307	200.1
Motive power (See motor, electric)	310	
Aircraft control actuator	244	175+
Arc electrode actuator	314	69+
Arc electrode actuator	314	94+
Automatic miscellaneous	318	445+
Generating electric locomotives	105	35+
Induction motors	310	166+
Induction motors systems	318	727+
Interchangeably locked gearing operator	74	365
Ironing or smoothing machine	38	38+
Lock operating mechanism	70	277+
Locomotives	105	49+
Motor acceleration	388	842+

	Class	Subclass
Motor braking	318	362+
Motor making processes	29	596+
Motor reversing	318	280+
Motor vehicle	180	65.1+
Piezoelectric	310	311+
Planetary gearing combined	475	149+
Plural motor control	318	255+
Plural motors	318	34+
Potentiometer controller	D13	125
Reciprocating motor systems	318	119+
Speed control	388	800+
Steering gear operator	180	79.1
Synchronous motor systems	318	700+
Systems	318	
Textile twisting apparatus	57	100
Time controlled locks	70	271
Toy machinery	446	
Toys	446	484
Vehicle	446	457
Vehicle	D13	112+
Motor generators	310	113
System	322	
Motors alternating current	310	159+
Music instrument	84	600+
Nail hammers	227	131
Needle	606	44
Oil well tubing dewaxer	392	301+
Organ (See organ, electric)	84	600+
Orthopedics	128	68.1
Oscillators	331	
Outlet box support	248	906*
Oven	219	391+
Packaging for components in roll form	206	328+
Pencil sharpener	144	28.1+
Phase control systems	323	212
Photography, electric	430	31+
Plant culture	47	1.3
Plasma control	376	143
Plug (See plug, electric)		
Power conversion systems	363	
Power supply or regulation systems	323	
Inrush current limiters	323	908*
Lamp dimmer structure	323	905*
Medical electronics	323	911*
Optical coupling to semiconductor	323	902*
Precipitators	323	903*
Remote sensing	323	909*
Solar cell systems	323	906*
Starting circuits	323	901*
Temperature compensation of semiconductor	323	907*
Touch systems	323	904*
Two of three phases regulated	323	910*
Precipitator	55	101+
Distillation processes &	201	19
Distillation processes &	201	40
Processes	55	2
Preserving, disinfecting, sterilizing methods	422	22+
Food or beverage	426	237
Apparatus	99	451
Fumigator with electric fan	422	124
Prime-mover dynamo plants	290	
Printed circuits	361	380+
Prod	231	7
Production	D13	
Protection (See protection, electric)		
Public utility meter	D10	100
Radio tube (See electric space discharge devices)		
Radioactive surgery	600	1+
Rail bonds	238	14.1+
Recorder	346	
Electrochemical	346	165
Lock indicator	70	434
Photographic oscillograph	346	109
Spark perforating	346	163
Rectifier		
Electrolytic	361	436+
Electronic tube type structure	313	
Supplying electronic tube or lamp	315	200R+
Supplying radio	328	262+
Systems	363	13+
Relay (See relay, electric)		
Resistor devices (See resistance)	338	
Rivet heaters	219	157
Safety devices (See safety devices)		
Score boards	235	1B
Timing type	340	309.15
Shock apparatus (See shocker)		

	Class	Subclass
Heat exchanging	165	51+
Hot air	60	508+
Expansion of medium	60	516+
Externally applied heat	60	643+
Internal combustion engine	123	
Cooling	123	41.1
Design	D15	1+
Excess air to assist exhaust	60	900*
Exhaust treatment of rotaries	60	901*
Lubricators	123	196R
Nozzle	239	584+
Oil by-pass heat exchange	165	35
Pump drive	417	364
Refrigeration utilizing	62	323.1
Rotary piston	123	200+
Rotary reactor, separator or		
treater of exhaust	60	902*
Speed responsive throttle control	60	906*
Temperature measurement	374	144+
Locomotive	105	26.5+
Microwave	60	203.1
Muffler for hobby craft engine	181	404*
Plural motors system	60	698+
Power plant	60	
Prime mover dynamo plant	290	
Pulp refining	241	
Reaction motor	60	200.1+
Signal systems ships	116	21
Solar	60	641.8+
Starter gearing	74	6+
Steam	91	
Design	D15	2
Steam traction	180	36+
Strap starter	74	139+
Testing	73	116+
Toy	446	
Valve gear (See valve, actuation)		
Washing machine system	60	908*
Engineering Hydraulic and Earth	405	
Engraving		
Equipment design	D18	13
Photographic process	430	269+
Plate cutting	409	79+
Printing	101	150+
Stippling	72	76
Tools	30	164.9
Enlargers Photographic	355	18+
Ensilage Compaction	100	65+
Enteric Coating of Medicinal Tablets	424	475+
Entrenching Earthworking Apparatus	105	161
Envelope (See Bulb; Casing)		
Battery plate separator	429	136+
Closures	229	76+
Cutting blank form	83	911*
Feeding or delivering	271	2
Making	493	186+
Sealing and stamping combined	156	442
Tear strip opener	493	923*
Window	493	919*
X-art collection	493	917*
Manifolding		
Leaf combined	282	25
Strip connected	282	11.5 R
Moistening and sealing	156	441.5+
Opener	83	912*
Design	D 8	102+
Paper	229	68R+
Design	D19	3+
Environmental Impact or Decreasing		
Pollution	422	900*
Survival computerized, monitored	364	413.3+
Enzyme	435	183
Enzyme or microbe electrode	435	817*
Fruit juice treatment	426	51
Medicinal compositions	424	94.1+
Separation or purification	435	814*
By sorption	435	815*
Ignition, internal combustion	435	816*
Tobacco treatment	131	308
Epaulet	2	246
Badge type	40	1.5+
Epdm (See Synthetic Resin or Natural Rubber)		
Ephedrine	564	364
In drug	514	653
Epicycle		
Land vehicle occupant propelled	280	207
Railway		
Drive wheel	105	100
Wheel	295	3
Epicyclic Gearing	475	

	Class	Subclass
Episcope	353	65
Episcotister		
Light valve	350	484+
Photometer	356	217
Epon T M (See Synthetic Resin or Natural Rubber)	528	87+
Epoxide Resin (See Synthetic Resin or Natural Rubber)		
Epoxy Compounds	549	512+
Equalized and Equalizing		
Animal draft	278	3+
Horses abreast	278	5+
Animal powered sweep	185	22
Bed bottom	5	278
Brake position adjusters	188	204R
Dumping vehicle hopper door closing	298	37
Extension table	108	87
Hot air furnace	126	105R
Locomotive	105	82
Railway brake series	188	46
Railway motor torque	105	135
Railway truck	105	209
Bogie	105	194
Sectional cabinet closure	312	110
Serial valves dependent motion	137	613+
Direct response	137	512+
Stop	137	629+
Trolleys for single rail suspended		
car	105	152
Vehicle fender car and truck	293	14
Vehicle running gear frame	280	104
Equatorial Telescope	350	568
Equilin	552	625
Eraser	15	424+
Abradant type mutilating	51	
Blackboard	15	208+
Blade type	30	
Design	D19	53+
Disintegrable type	15	424+
Brush combined	15	105.52
Cutter e. g., pencil sharpener		
combined	15	105.53
Mechanical	15	3.53
Ruler combined	15	105.51
Sharpener pencil combined	15	105.53
Mutilating type and burnisher	7	124+
Scraper type mutilating	30	169+
Shield	51	262.1
Typewriter attachment	400	695+
Ergolines	546	67+
Ergometer with Feedback	272	DIG. 6
Ergosterol	552	547+
Ergot Alkaloids	546	67+
Erosion Preventing	405	15+
Mineral oil conversion combined	208	47
Error		
Checking, electrical	371	
Arithmetic operations	364	737+
Communications synchronization	371	47.1
By error rate or count	371	5.4
Computers	371	16.1+
Data processing systems, general	364	200
Memory devices	371	21.1+
Program debugging	371	19
Correcting circuit	371	30+
Correcting storage on typewriter	400	6
Detecting	371	
Positional servo systems	318	638
Prevention means on communication		
keyboards	341	24+
Erysodine	546	72
Erysonine	546	72
Erysothiopine	546	72
Erysothiovine	546	72
Erysotrine	546	72
Erysovine	546	72
Erythraline	546	48
Erythramine	546	48
Erythratine	546	48
Erythritol	568	853+
Erythroidine	546	66
Erythromycin	536	7.2+
Escalator	198	321+
Design	D34	30
Escape		
Acetylene generator automatic		
safety	48	56
Fire	182	
Endless conveyor	182	42+
Single strand stile or pole	182	189+
Tower or chute	182	48+

	Class	Subclass
Gas holder high pressure safety	48	175
Submarine	114	323+
Escapement		
Clock and watch structure	368	124+
Clock electric driving type	368	131
Electrode feed	314	82
Gong striker	116	161+
Meter electric	324	139
Motor	74	1.5
Motor composite with	185	5
Motor spring with	185	38
Motor weight with	185	31
Music leaf turner	84	498+
Type casting machine	199	23
Typewriter feed mechanism	400	319+
Typewriter line locks	400	672+
Escutcheon	D 8	350+
Bushings or lining thimbles	16	2+
Covers or face plates	220	241
Design	D 8	350+
Electrical apparatus	174	66+
Fluorescent or phosphorescent		
apparatus	250	466.1
Design	D 8	
Door knob handle or rose plate	292	357
Fluid handling	137	377+
Key operated lock	70	452
Pipe joint	285	46
Radio plate for	D14	257
Television, plate for	D14	239
Wall outlet with	D13	137+
Espalier	D25	100
Esr, Electron Spin Resonance	436	173*
Essential Oil	512	5
Biocides containing	424	195.1+
Esters	260	
Carboxylic acid	560	1+
Hydrophenanthrene nucleus		
containing	560	5+
Cellulose	536	58+
Nitrogen containing	536	30+
Higher fatty acid	260	410+
Resin acid	530	200+
Free acid	560	5+
Rosin acids	530	215+
Starch	536	107+
Nitrogen containing	536	48+
Sulfate	558	20+
Tall oil acids	530	232
Thiocyanic acid	558	10+
Estradiol	552	625
Estradiol, in Drug	514	182
Estrenes	552	642+
Estriol	552	617
Estrone	552	625
Estrous Cycle Monitoring	128	738
Etching	156	625+
Apparatus	134	
Bright polishing	156	903*
Composition	252	79.1+
Differential apparatus	156	345
Direction indicator combined	33	305+
Electrolytic	204	129.1+
Electrolytic coating combined	204	32.1+
Glass	156	663
Glass with glass working	65	31
Metal working with	29	DIG. 16
Misc. chemical manufacture	156	DIG. 111
Photo	430	299+
Multicolor printing surface	430	301
Printing plate making	156	905*
Resist compositions	156	904*
Semiconductor making	156	625+
Semiconductor making	156	901*
Photoetching	430	313
Ethanol (See Alcohol)	568	840+
Beverages containing	426	592+
Dehydrating	203	19
Ethers	568	
Cellulose	536	84+
Nitrogen containing	536	43+
Cyclic hetero-o-type	549	200+
Oxy	568	579+
Phenol	568	630+
Thio	568	38+
Vinyl	568	687+
Ethers, Crown	549	347+
Ethyl		
Acetate	560	265
Cellulose not plastic	536	100+
Formate	560	265

	Class	Subclass
Nitrite nitrous ether	558	488
Ethylene (See Olefin Hydrocarbon)		
Ethylene diamine tetraacetic acid	562	566
Glycol	568	852
Polymer	526	352
Synthesis	585	500+
By cracking	585	648+
Urea	548	300+
Ethylenic Monomers		
Solid polymer solely therefrom	526	72+
Etiocholanic Acids	552	610+
Eucalyptol	549	397
Eudiometers	73	23.2+
Eugenol	568	652
Euphoniums	84	402
Eurhodines	544	347+
Eutectic, Refrigeration Utilizing	62	430+
Automatic	62	139
Process	62	59
Evacuated Chamber		
Means to introduce material into	414	217+
Means to seal chamber	414	292
Evacuating Receptacles	141	65
Air displacement in filling		
receptacles	141	59
Successive receptacles	141	65
Apparatus	141	65
Electric lamp etc manufacture	445	38+
Apparatus	445	73
Combined operation methods	445	56+
Methods	445	53+
Food preserving	426	404+
Storage receptacle	99	472
Gas filling combined	141	66
Hermetic closing of filled receptacle,		
method	53	403+
Radiant energy tubes	313	553+
Having heating means	313	549
Evaginated Shoe	12	142A
Sole attaching means	36	17A
Evaporator		
Boiler	122	
Cabinet combined	312	31.1+
Concentrating	159	
Spray type	159	3
Gaseous current	159	4.1+
Gas and liquid contact apparatus		
Porous mass	261	94+
Porous sheet	261	100+
Making	29	890.7
Oil combustion engine	123	522
Refrigerator	62	515+
Automatic	62	216+
Defrost water by heater	62	275+
Defrost water by refrigerator		
heat	62	279
Making	29	890.35
Sugar manufacture	127	16
Ventilator		
Building	98	30
Car	98	12
Car inlet	98	17
Evaporometer	374	39
Evener		
Animal draft vehicle equalizer	278	13
Horse detacher	278	32
Grain harvesters, head or butt	56	468
Textile loom, dobby	139	74
Excavation (See Digger)	37	
Caisson	405	272
Pressurized	405	8+
Clamshell buckets	37	183R+
Ditch filler	37	142.5+
Ditchers	37	80R+
Dredgers	37	54+
Mine ventilation	98	50
Mining or quarrying with separating	299	7
Orange-peel buckets	37	182+
Railway graders	37	104+
Railway snow excavators	37	198+
Road grader type	37	108R+
Roadway snow excavators	37	196+
Self-loading vehicles	37	4+
Shoring or bracing	405	272+
Snow excavators & melters	37	227+
Stump and removers	37	2R+
Tunnel excavating	299	
Tunnel ventilation	98	49
Excavator		
Dipper	414	690+
Force pump	37	75+

	Class	Subclass
Pipe and cable laying combined	405	154+
Screw	175	394+
Suction pump	37	58+
Excelsior		
Making	144	185+
Package filling	493	967*
Exchange		
Heat	165	
Mortar mixers having	366	22+
Processes	366	4
Temperature regulation	236	
Tobacco smoking device	131	194+
Telephone	379	
Execution Devices		
Electric chairs	128	377+
Medicators	604	19+
Poisons	424	
Surgical instruments	606	1+
Trap animal		
Choking or squeezing	43	85+
Electrocuting	43	98+
Everset	43	64+
Explosive	43	84
Impaling or smiting	43	77+
Self reset	43	73+
Trap insect		
Adhesive	43	114+
Electrocuting	43	112
Exercising		
Arm wrestling	272	901*
Devices	272	93+
Breathing improvement	272	99
Field sports	272	100
Isometric	272	125
Weights	272	117+
Kinesitherapy	128	25R+
Light thermal and electrical		
application combined	128	363+
Music	84	465+
Opponent-supplied resistance	272	902*
Poultry feeding combined	119	70
Exfoliation		
Cereals	426	449+
Compositions	252	378R
Decorticating	19	5R+
Seed or seed parts	241	6+
Stone	125	23.1+
Exhaust (See Evacuating Receptacles)		
Mufflers	181	212+
Design	D12	194
Operated		
Brakes	188	154
Pipe making	29	890.8
Horns and whistles	116	138
Recirculation	123	568
Silencers	181	212+
For internal combustion engines	60	272+
Steam engine	60	681+
Treatment		
Combustion device feed in	60	683
Combustion engine	60	272+
Excess air to internal		
combustion engine	60	900+
Serially connected motors	60	679+
Furnace	432	67+
Furnace, metallurgical	266	144+
Furnace, solid fuel	110	203
Motors	60	685+
Rotary engines	60	901*
Exhauster (See Evacuating		
Receptacles)		
Filled receptacle cooking apparatus	99	359+
Heating etc gas making apparatus	48	173
Water closet tank outlet siphon	4	368+
Exhibiting		
Cards pictures and signs	40	
Design	D20	10+
Design display card	D20	40
Check controlled	194	
Dispensing combined	222	23+
Gem setting feature	63	30
Indicator combined	116	307
Exhibitor		
Cabinet with	312	234
Exit		
Bee hive	449	20+
Door lock	70	92
Fire escape chute	182	48+
Insect	160	12+
Railway car		
Control	105	341+

	Class	Subclass
Emergency	105	348+
Stratifier discharge	209	494+
Expanded Metal	52	670+
Making	29	6.1
Expander and Expanding		
Anchor	411	15+
Expanding or piercing earth	52	155+
Boiler formers	72	393
Boring head cutter	408	199+
Building component	52	67
Cereals	426	449+
Apparatus	99	323.4+
Collapsible taps	408	147+
Exfoliated composition	252	378R
Face plate railway car	105	10+
Felly	157	9+
Gas liquefying and separating	62	36+
Processes	62	9+
Gate valve faces	251	193+
Hoof and shoe	168	47
Joints (See device incorporating)		
Liquefied gas fuel as refrigerant	62	7
Mandrels	269	48.1
Chuck	279	2R
Lathe work mount	82	169
Winding reel	242	63
Winding reel core	242	110
Metal making	29	6.1
Article	428	596+
Orthodontic	433	7
Pipe expanders	72	393
Methods of joint expanding	29	890.44
Piston ring applier	29	222+
Reamer	408	227+
Earth boring type	175	263+
Refrigeration by	62	
Rod or piston packings	277	138
Sheet metal shaping die	72	393
Tube cutter	30	103+
Wheel felly	157	9
Expansible Cabinets	312	205
Expansible Chamber Devices	92	
Expansion (See Expander)		
Engine, refrigeration producer	62	402+
Automatic	62	172
Liquefied gas producing	62	38+
Process	62	86
Joints		
Between articulated members	403	52+
Bridge	14	16.1
Building	52	573
Distillation apparatus	202	268
Flexible diaphragm or bellows	403	50+
Fluid distributor nozzle	239	397.5
Heat exchanger	165	81+
Pipe	285	298+
Resilient building	52	313
Static mold forming	249	9
Plug		
Metallic receptacle	220	233+
Pipe	138	89+
Reeds warp preparing	28	213
Valve, refrigeration utilizing	62	527+
Automatic	62	222+
Exploring		
Borehole and drilling study	73	151+
Depth sounding	73	290R+
Electro-acoustic	367	99+
Sound	181	
Diagnostics medical	128	630+
Earth boring	175	50
Electrical	324	
Electrical ore testing	324	376
Meteorology	73	170R+
Radar	342	191+
Radiosonde	340	870.1+
System	340	870.1+
Sound	181	
Explosion Control		
Acetylene generator	48	56
Blast or cupola furnace	266	174
Blowout plugs, frangible or fusible	220	89.1
Containers of inflammable		
substances	220	88.1+
Davy lamp or wire gauze protected		
Dry cleaning and laundry apparatus	68	209
Drying apparatus	34	51
Gas cooking stoves	126	42
Gas distribution	48	192
Gas holder high pressure safety		
escape	48	175

	Class	Subclass
Treatment	204	140+
Electrothermic processes	75	10.1+
Heat treatment	148	134+
Special composition with	148	14
Working combined	148	12R+
Pyrometallurgy	75	507+
Ferrule	16	108+
Assembly and dissassembly	29	282
Hand tools	29	280+
Percussive or explosive operator	29	255
Design	D 4	199
Digest, brushes & cleaning tools	15	DIG. 4
Making		
Metalworking methods	29	428+
Wireworking	140	76
Ring ferrules	16	109
Ferry		
Boats	114	70
Slips	114	231
Suspended car or load carrier	104	90
Fertile Nuclear Conversion Material		
(See Nuclear Reactions, Reactors)	376	172
Fertility, Pregnancy	436	510*
Fertility tests	435	806*
Fertilizers	71	
Adjuvant	71	DIG. 1
Apparatus	422	193+
Chelating agent	71	DIG. 2
Defluorination	71	DIG. 3
Distributing machines	222	
Application to plant	47	48.5
Scattering unloaders	239	650+
Distributing machines below ground	111	
In irrigating systems	47	1.1
Festoon	428	10
Collapsible	428	9
Paper, making	493	957*
Festooning (See Loopers)		
Fetters		
Animal husbandry	119	127
Butchering	17	44+
Design	D30	151+
Portable locks	70	15+
Fez	D 2	256+
Fiber (See Filament)		
Bicomponent	428	373
Blend	162	181.1
Blend	162	141
Bonding apparatus	156	433
Bonding methods	156	166+
Bristles	15	159A+
Brushes design	D 4	
Chemical modification of	8	115.51
Cleaning or laundering	8	137+
Coating	427	212
Coating with	427	180+
Composite	428	373+
Crimped	428	370
Composite metal	428	607
Composition (See type)		
Crimped	428	369+
Staple length	428	362
Dyeing & bleaching	8	
Fiber optics	358	901+
Fluid treatment of	8	147+
Glass compositions	501	35+
Glass manufacturing	65	1+
High modulus	428	902*
Insulated electric conductor	174	124R
Plural layers	174	122R
Plural layers impregnated	174	121R
Isocyanate & carbonate fiber		
modification	8	DIG. 11
Layers needled	428	300
Liberation & paper making	162	
Making		
Powder metallurgy	419	4
Micro fiber	428	903*
Optics	350	96.1+
Sign	40	547
Polyvinyl halide esters or alcohol		
fiber modification	8	DIG. 1
Preparation (See textile prep)	19	
Refractory compositions	501	95
Retting		
Chemical	162	1+
Fermentation	435	279+
Rods	350	96.1+
Fiber optics	358	901*
Photocopy device use	355	1
Spinning mechanical	57	315+

	Class	Subclass
Staple length	428	359+
Strand structure	57	200+
Swelling & stretching	8	DIG. 3
Testing	73	159+
Textile (See type)		
Coated or impregnated	428	245+
Treatment		
Bleaching	8	101+
Carroting	8	112
Chemical modification	8	115.51+
Cleaning or laundering	8	137+
Coating	427	412
Coating, general processes	427	
Comminution	241	
Dyeing	8	
Electrolytic treatment	204	132+
Fluid apparatus	68	
Fluid processes manipulative	8	147+
Liberation chemical	162	1+
Liberation physical	19	1+
Swelling or plasticizing	8	130.1+
With structure or coating	428	364+
Fiber Board Making	156	62.2+
Fiber Board Making	264	109+
Fiber Optic Lamp	D26	27
Fiber-reactive Azo Dyes	534	617+
Fiberglass		
Coating	427	389.8
Fishing rods	43	18.5
Metallic receptacles digest	220	DIG. 23
Fibrillation		
Cellular materials	264	DIG. 8
Diagnostic testing for	128	705
Textiles fibrillated	57	907*
Fibroin	106	161
Fids	114	221R
Field		
Artillery	89	
Mounts	89	40.1+
Glasses	350	545+
Design	D16	133+
Holders	350	547+
Optics	350	481
Magnet for electric machine	310	180+
Field Sport Devices	272	100+
Fifth Wheel		
Land vehicle		
Articulated	280	433+
Running gear swinging axle	280	80.1+
Figure		
Decorative with illumination	362	808*
Toys	446	268+
Air actuated	446	199+
Aquatic	446	156+
Design	D21	148+
Knockdown	446	97+
Spinning or rotating	446	236+
Toy money box with coin deposit		
figure release	446	10+
Velocipede type	280	1.13+
Wheeled	446	269+
Filament (See Strand; Wire)		
Bonding apparatus	156	441
Bonding methods	156	60+
Filament formation	156	167
Filaments only	156	180+
Filaments winding	156	169+
Stressed filaments	156	161
Cereal	426	560
Container for non-spooled	206	388
Container or wrapper for coiled	206	389+
Cutting filament-to-staple fiber	83	913*
Distributing explosive device	102	504
Electric lamp	313	341+
Compositions	252	500+
Mounting methods	445	29+
Electronic tube	313	341+
Replaceable filament	313	237
In electronic tube	315	46+
In electronic tube	315	49
Laminating	156	166
Lantern projector combined	362	211+
Making		
Electric conductor	156	47+
Electroforming	204	12+
Electrolytic apparatus	204	206+
Electrophoretic	204	180.4
Glass apparatus	65	1+
Glass processes	65	2+
Liquid comminuting and solidifying		
apparatus	425	6+

	Class	Subclass
Liquid comminuting and solidifying		
processes	264	5+
Metal casting	164	423+
Metal drawing	72	274+
Metal working plastic	72	253.1+
Molding apparatus	425	66
Molding apparatus	425	67
Molding apparatus	425	76
Molding apparatus	425	382.2
Molding processes	264	165+
Powder metallurgy	419	4+
Spinning twisting or twining	57	
Textile misc processes	28	299
Metal	428	606
Coating of	427	216+
Compound	428	607
Multiple filament lamp and electronic		
tube	315	64+
Spun twisted or twined	57	200+
Design yarn cord	D 5	7+
Stock material	428	364+
Supply circuit for electronic tube	315	94+
Testing	73	160
Tires combined	152	168
Transformer	336	
With anode supply winding	336	170+
With anode supply	315	94+
With grid bias	328	226
With grid bias power packs	307	150
File or Filing		
Binder device releasably engaging		
aperture or notch of sheet	402	
Cabinet, design	D 6	432+
Cabinet with material retainer	312	183+
Card spacer	283	36+
Dental	433	141+
Directory	40	371+
Folder	493	947*
Handles	29	80
Hoof trimming	168	48.1+
Leather	69	1
Implement	69	20
Manicuring	132	76.4+
Combined in compound tool	132	75.6
Metal and metal working files	29	76.1+
Cleaning and resharpening by		
electrolysis	204	141.5
Cleaning and resharpening by		
sand blast	51	414
Making	76	12+
Processes of abrading	51	313+
Pencil sharpening (See pencil		
sharpener)	29	78
Rasps and	29	78+
Receptacle file partition	220	529+
Riffle file	40	372
Rotary or flip card	D19	75
Sharpening tools by	76	82+
Saw sharpening	76	31+
Sorting type desk	211	11
Fillers		
Barometric control of dispenser	222	479
Bars for beds	5	283
Cables conduits with fluid or vacuum	174	8+
Cables conduits with insulation	174	68.3+
Composition	106	400+
Shoe	106	38
Curtain shade or screen seals	160	40+
Dispenser and receiver combined	141	
Portable receptacle	141	173+
Dispenser with refilling of supply	141	18+
Ditch	37	142.5
Floor mats pivoted link	15	240
Fountain pen combined	401	118+
Force means in filler	141	20.5
Gas generator	48	74
Insulating heat or sound	252	62
Mattress	5	448+
Foam materials	5	481
Pastry	99	
Pastry	425	230+
Printing member	101	402+
Rail joint containing	238	163
Safe wall structure combined	109	84
Sheet metal cup die shaping	72	54+
Sound mufflers having	181	282
Synthetic resin or natural rubber		
(See class 523, 524)		
Wheel web combined	295	23
Filling		
Abrasive bag making	51	294

	Class	Subclass
Straps leather	69	17
Thread	28	217+
Type casting combined	199	81+
Fins		
Heat exchanger	165	185
Nuclear fuel having	376	435
Stabilizing ships	114	126
Steering ships	114	152
Submarine diving	114	331+
Vertical on aircraft	244	91
Fire (See Burner; Fireproof)		
Alarms		
Electric detectors	340	577+
Electric system automatic	340	577+
Electric systems	340	287+
Smoke bells	362	379
Thermal control	116	101+
Thermal electric switch	337	298
Thermometer systems	374	141+
Electrical	374	166+
With indicator	116	5
Brick	501	94+
Design	D23	422
Cigarette lighters	D27	141+
Detector (See alarms)		
Doors furnace	110	173R+
Electric starter device	D 7	416+
Engine	169	24
Boiler feed heater	122	418+
Heating system combined	237	12.2
Hose and ladder type	280	4
Ladder or escape	182	63+
Escape	182	
Rope and brake	188	65.1+
Extinguisher	169	
Boiler fires	122	504+
Compositions	252	2+
Fire retarding	169	45
Design	D29	2+
Gas agent	169	11+
Toy vehicle	446	432
Extinguishing	169	
Aerial fire apparatus	244	136
Methods of spraying fluids	239	1+
Nozzles automatic	169	37+
Processes chemical	169	43+
Shipboard	169	5
Fighting	D29	
Hose	D23	266
Hose nozzles	239	
Design	D23	213+
Hydrant	D23	234
Ladder	D25	62+
Mask	D29	7+
Vehicle design	D12	13
Kindling	431	6+
Composition	44	542+
Design solid fuel	D13	100
Matches design	D27	139+
Plate	126	148
Pots and linings	126	144+
Thermostats	236	104
Preventing		
Devices, design	D29	
Receptacle attachments	220	88.1+
Textile apparatus	68	209
Retarding composition	252	601+
Biocidal composition with	424	
Containing gas	252	605
Dispersion or collodian systems	252	611
Fiber chemically modified	8	115.51+
For living matter	252	603
For synthetic polymer	252	609
Intumescent	252	606
Paper-making or fiber liberated	162	159
Stock material mechanically		
interengaged strands	428	276
Fire or flame proof features	428	921*
Synthetic polymer use	252	609
Trees, grass use	252	603
Woven material use	252	608
Screen	126	544
Starting devices	D 7	416+
Stop	52	317
Telegraphs	340	287+
Tong	294	11
Design	D 8	52+
Tube	122	
Fire-polish Glass	65	
Firearm (See Gun)	42	
Automatic	89	125+

	Class	Subclass
Barrels	42	76.1
Bayonets	42	86
Bolt making	29	903*
Breech loading	42	2+
Cane guns	42	52
Converted to spring gun	124	28
Design	D22	100+
Firing mechanism	42	69.1+
Imitation, light flashing	362	112
Light combined	362	110+
Magazine chargers	42	87+
Making	29	1.1
Muzzle loaders	42	51
Pistol swords	42	53
Plural triggers	42	42.2
Revolvers	42	59+
Sights	33	233+
Silencers or mufflers	181	223+
Simulating		
Dispenser	222	79
Light	362	157+
Stocks	42	71.1+
Toy	42	54+
Design	D21	145+
Firebox (See Furnace)		
Feeding steam air or water	110	297+
Feeding water to	110	297+
Fire tube	122	44.1+
Flue	122	135.1+
Fuel disperser installed in	431	159
Liquid heater or boiler combined	122	
Locomotive		
Structure combined	105	37+
Type combustion features	110	
Type water heating	122	
Smoke and gas return to	110	203+
Solid fuel	110	317+
Straw and wet fuel burning	110	196
Water		
Tube	122	235.34+
Wall	122	294
Firebrick		
Carbide containing	501	87+
Design	D23	422
Refractory	501	94+
Firecracker	102	361
Firedamp		
Analyzing and testing apparatus	422	50+
Indicators	73	23.2+
Firedogs	126	298
Design	D23	407
Firefighting (See Fire)		
Fireless Cookers		
Heat accumulators	126	400
Ovens	126	273.5
Vessels	126	375
Fireplace	126	500
Arches	52	219
Dampers	126	536
Design	D23	343+
Fenders	126	201+
Design	D23	403+
Grates	126	540+
Design	D23	398+
Heating systems having	126	509
Logs		
Electric	272	8R
Gas	431	125+
Gas design	D23	409
Mantel design	D23	404
Screen	126	544
Digest	16	DIG. 11
Shelves	126	505
Fireproof (See Fire, Retarding; Insulating, Heat)		
Miscellaneous compositions	252	601+
Plastic or coating compositions	106	15.5+
Reactive building component	52	232
Stock material	428	920*
Synthetic resin or natural rubber (See synthetic resin or natural rubber)		
Cellular product	521	50+
Fireworks	102	335+
Loading	86	20.1+
Firing		
Clayware processes	264	56+
Devices or mechanisms		
Ammunition & explosive devices	102	200+
Automatic firearms and guns	89	132+
Blasting	102	200+

	Class	Subclass
Drop bomb	102	396
Firearm nonautomatic	42	69.1+
Gun cartridge	102	470+
Mechanical guns	124	31+
Mine	102	424+
Mines marine	102	416+
Ordnance nonautomatic	89	27.11+
Electric	89	28.5+
First Aid		
Cabinets	312	209
Kits	206	570+
X-art collection	206	803*
Fischer-tropsch Crude Processing	208	950*
Fischer-tropsch Synthesis	518	700+
Fish (See Fishing)		
Butchering apparatus	17	53+
Fish bowl design	D30	101+
Hooks	43	43.16
Knife for kitchen use	D 7	649+
Liver extraction		
Fats etc	260	412.1
Medicine containing	424	554+
Vitamins	260	
Net	43	7+
Design	D22	135
Fabric and processes for making	87	12
Making apparatus	87	53
Plates	238	243
Special, bolt or nut to substructure lock	411	94
Special culture	119	3
Fisherman's Tool	D22	149+
Fishing	43	4+
Amusement fish ponds	273	140
Automatic hookers	43	15+
Bait, tackle, and catch container	43	54.1+
Bait or lure carrier	224	920*
Minnow buckets	43	56
Design	D22	136
Baskets and creels	43	55
Belt with lure holder	D 3	100+
Containers & carrying cases	D 3	38
Design	D22	134+
Equipment	D22	134+
Files, work holder for	269	907*
Fish traps	43	100+
Hooks	43	43.16
Design	D22	144
Making	29	9
Line	43	44.98
Reel	242	223+
Attachments	242	322
Design	D22	137+
Rod	43	18.1+
Cover or container for disjointed	43	26
Fly tieing	242	7.19
Holder	224	922*
Method	242	7.2
Snelling	242	7.19
Stringers	224	103
Tackle and accessories	43	4+
Tool		
Compound	7	106+
Electricians	254	134.3 R
Traps	43	100+
Design	D22	121
Trotlines	43	27.4
Holder	43	57.3
Vessel	114	255+
Weighing of fish	D10	87+
Fishing Rod	D22	137+
Fishing Rod Holder	D22	147
Fission Counter Scintillations	250	361R+
Fission Sustaining Nuclear Reactors	376	
Fitting Boots and Shoes	36	8.4
Fittings		
Bath basin and receptacle	4	191+
Design	D23	271+
Electrical		
Boxes housings	174	50+
Conductor insulator combined	174	169+
Making		
Gas and water	29	890.14+
Pipe	285	120+
Fluid distribution design	D23	200
Gaslight design	D26	113+
Scaffolding design	D25	66
Suspended support	248	342+
Fixatives		
Artists	106	
Electrophotographic copier	355	282+

	Class	Subclass
Perfume	512	2
Photographic	430	455+
Flag	116	173+
Badge	40	1.5
Design	D11	165+
Design, insignia or jewelry	D11	165+
Pole or post structure	52	720+
Socket	52	298
Staff	116	173+
Decorative element	52	301
Staff type support	248	511+
Flagpole	52	720
Design	D11	181
Holder	D11	182+
Socket	52	298
Flail		
Cotton harvester	56	29
Threshing machine	460	61
Flails Flexible	51	334
Flake Ice Making	62	320+
Flakes		
Coating	427	212+
Electroforming	204	10
Flame (See Fire; Fireproof)		
Arc lamp	313	231.1+
Movable electrode	314	
Arrester	48	192
Dispenser having	222	189
Receptacle having	220	88.1+
Respirator	128	202.24
Bonding by	228	902*
Colorant composition for fuel	44	642
Deflector		
Deflector in furnace	431	171+
Resistance, coated	427	393.3
Retardant, textile collection	57	904*
Thrower	431	91
Flanger		
Pipe or tube	72	316+
Sheet	72	101
Flannel and Flannelette Napping	26	29R+
Flap		
Bag closures	383	84
Envelope moistening & sealing	156	441.5
Head coverings	2	172
Paper receptacles		
Closure for paperboard box	229	126+
Letter sheet closure	229	92.7
Pneumatic tire	152	514
Shoe sole channel machines		
Layers	12	29
Turners	12	30
Flare		
Burners		
Magnesium strip	431	99
Solid fuel	102	336+
Dropping	89	1.51+
Guns	42	1.15
Package for	206	803*
Pyrotechnic	102	335+
Composition for	149	
Design	D22	112
Parachute	102	337+
Torches	102	336+
Flaring Tool (See Pipe Making)		
Flash		
Camera attachment	D16	239+
Electric weld	219	97+
Flash bulb mount	D16	239+
Flash removal	425	806*
Spotlight	D26	24+
Trimmers	83	914*
Flash Bulb		
Combustion type	431	358+
Mount design	D16	239+
Flashing	52	408+
Exterior type, eg roof	52	58+
Glass	65	60.1
Lamps		
Sign	40	541+
Sign systems	340	791+
Switch in lamp	315	72
Systems	315	
Light as signal	40	902*
Pipe and plate joint roof flashing type	285	42+
Signals	340	331
Flashlight		
Combined with structure of another device	362	157+
Canes	362	102

	Class	Subclass
Pen or pencil	362	118
Electric	362	208
Bicycle operated generator	362	193
Design	D26	37+
Hand generator per se	310	50
Hand operated generator	362	192
Electric generator combined	362	192+
Bicycle	362	193
Flash powder and photoflash	431	357+
Flask		
Adjustable	164	377
Pattern with	164	237+
Plastic molding	425	175+
Flat Pack Electronic Device	174	52.4
With semiconductor	357	74
Flat Tire Alarm		
Electric	340	442
Switch	200	61.22+
Nonelectric	116	34R
Flatiron	38	74+
Design	D32	68+
Heating		
Automatic temperature regulators	236	7
Electric	219	245+
Liquid and gaseous fuel	126	230
Self heating liquid and gaseous fuel	126	411+
Stove and furnace	126	227+
Stands combined	38	142
Steam	38	77.1+
Support	248	117.1+
Ironing table combined	38	107
Flats		
Bed bottoms	5	186.1+
Carding	19	113
Grinding	51	243
Stationary	19	104
Stripping	19	110+
Traveling	19	102
Flattening (See Crusher; Straightening Devices)		
Glass	65	67
Sheet material rolling or wrapping combined	53	116+
Textile ironing and smoothing	38	
Tobacco leaf	131	324+
Stemming combined	131	315
Flatware-tableware	D7	642+
Disposable	D7	642+
Flavanthrone	546	31
Flavones	549	403
Flavors	426	534+
Artificial	426	533+
Flavylium, Basic Dye	8	657+
Animal-derived natural fiber	8	426
Printing permanently	8	454
Flaw Detection		
Magnetic	324	238+
Induced voltage-type sensor	324	240+
Oscillator type	324	237
Utilizing radiant energy	250	
Flax (See Fiber)		
Retting	435	279+
Thresher	460	24
Flechettes	244	3.24+
Ammunition	102	703*
Fleece		
Dividers in fiber preparation	19	151
Knitting		
Fabric or articles	66	194
Needle cooperating wheels	66	94
Unknitted materials	66	191
Fletching Jig	269	38
Flexible		
Concentric electric cable	174	102R+
Fluid vacuum or air	174	28+
With insulation spacer	174	99R+
Flail tools	51	334
Remote control cable	74	500.5+
Shafts	464	
Tubing	138	118+
Flicker Beam Photometer	356	217
Flier		
Spinning machine	57	67+
Flier elements	57	115+
Flies		
Design fishing bait	D22	125+
Sheet delivering	271	83
Trouser	2	234
Flight Training Simulator	D25	1
Flint		
Composition	420	416

	Class	Subclass
Firearm firing device	42	69.1+
Lighter	431	273+
Special packaging for	206	528+
Flip Flop Oscillator	331	144
Driven type	328	193+
Gas tube type	331	113R
Flippers for Swimming	D21	239
Floating Compasses	33	364
Floats	169	DIG. 1
Ballasting	114	123
Marine pipe or cable laying	405	158+
Submergible floating matter barrier	405	64
Submergible marine structure	405	205
Submergible wave or flow dissipator	405	23
Buoy	441	1+
Closure	220	216+
Controlled		
Alarm	116	110+
Alarm electric	340	623+
Boiler safety device	122	505
Bottle nonrefillable	215	20
Burner liquid fuel	431	64
Closure	49	21+
Dispenser fluid pressure discharge	222	62
Dispenser indicator	222	51
Dispenser material level	222	67+
Flood gate	49	11
Funnel valve	141	199+
Funnel valve air displacement	141	199+
Gas liquid contact	261	70
Liquid feeder traps	137	165+
Liquid level indicator	73	305+
Liquid purifier	210	121+
Liquid purifier level indicator	210	86
Pump electric	417	36+
Pump jet	417	182.5
Pump pneumatic displacement	417	126+
Pump starter	417	36+
Pump throttle	417	36+
Trap thermal	236	53
Valve	137	409+
Valve steam trap	137	192+
Water closet outlet	4	395+
Water closet plural flusher	4	324+
Water closet valved bowl	4	437+
Watering trough	119	78+
Decoy ducks	43	3
Fishing	43	43.1+
Design	D22	146
Fluid pressure gauge	73	713+
Impeller		
Endless chain	416	7
Paddle wheel	416	84
Masonry implement	15	235.4+
Mortar-joint finisher	15	235.3
Oil distributor	114	234
Operated		
Acetylene generator	48	53.4
Beverage infuser	99	320
Brake	188	179
Dispenser indicator	222	51
Heater fluid	236	31
Liquid level gauge	73	305+
Motor fluid	60	495+
Pump tide motor	417	333+
Specific gravity indicator	73	451+
Switch electric	200	84R
Thermal air relief valve	236	62+
Thermal regulator	236	52
Thermometer combined	374	156
Watergate or weir	405	96+
Plural connected, bath	4	DIG. 1
Regulator		
Fluid pressure	137	505.19
Pump starter	417	36+
Pump throttle	417	36+
Skimmer glass furnace	65	206
Structure	73	322.5
Specific gravity tester	73	448+
Supported liquid separator	210	242.1
Traversing hoist	212	190+
Valve	137	409+
Vessel raising and docking	114	44+
Warship battery	114	4
Water closet design	D23	303
Water motor	60	495+
Flocculating and Deflocculating	209	5
Prevention of	210	696+
With precipitation	210	702+

	Class	Subclass
Flocculation		
Liquid separation	210	702+
Solid separation	209	5
Flocking		
Apparatus	118	308+
Coating combined	427	180+
Laminating combined	156	279
Floodgate	49	10+
Floodlights	362	
Design	D26	61+
Dirigible support	362	418+
Floor or Flooring	52	
Animal stall	119	28
Bridge	14	73
Bridges for safes	109	87
Building		
Friction or wear surface	52	177
Joists to wall	52	289
Masonry	53	391+
Mat	52	660+
Nailing beam	52	364+
Parquet	52	390+
Rotatable	52	65
Spacing sleeper	52	480
Static mold for making	249	18+
Supporting wall	52	264
Vertically adjustable	52	126.1+
Yielding sub-structure	52	403
Carpet	428	85+
Design	D 6	582+
Cleaners	D32	15+
Cleaning		
Compositions	252	88
Machinery design	D32	15+
Machines and devices	15	
Disintegrating machine	299	36+
Drying apparatus	34	237+
Endless excavating vehicle	37	7
Heat exchanger	165	56
Heated vehicle	237	43
Jack	254	11+
Lamp	D26	93+
Lighting	362	153
Mat, portable	15	215+
Mop, sponge, non-bristle broom	D32	40+
Outlet or junction box	220	3.3+
Packaging for coiled floor covering	206	389+
Parquet	428	50
Pavement	404	17+
Polishing		
Compositions	106	3+
Railway car	105	422
Auxiliary	105	375
Registers ventilation	98	102+
Scrubbing or brushing machine	D32	15+
Sectional	428	44+
Ship	114	76
Slatted covering	52	660+
Surfacing machines	51	174+
Vehicle floor mat	D12	203
Flora	428	17+
Framework for	211	
Floral		
Holders	47	41.1+
Supports	248	27.8
Design, apparel attached	D11	1+
Floss, Dental		
Holder	D28	64
Toilet kit with	132	323+
Flotation		
Liquid purifying	210	703+
Gas-liquid surface contactor	210	150
Liquid surface outlet	210	221.1
Physical separation agents	252	60+
Separating solids	209	162+
Sugar manufacturing	127	57
Flouncing Lace Goods	87	10
Flour	426	622+
Bleaching	426	253+
Manufacture by comminution	241	6+
Testing	73	169
Flouring or Flour Dusting	425	101
Flow (See Proportional, Feeding)		
Control (See valve, actuation)		
Dryer	34	54
Electrical purification of liquids	204	306
Electrolytic apparatus	204	229+
Gas filtration particulate material	55	474
Gas filtration reticulated		
Gas liquid contact porous sheet	261	100+
Gas liquid contact porous mass	261	94+

	Class	Subclass
Gas liquid contact wet baffle	261	108+
Gas separator combined	55	
Heat exchange	165	32+
Pump reversible	417	315
Radiator heat exchange	165	96+
Counter nuclear		
Determine characteristic of fluid	250	336.1
Through counter tube	250	374+
Indicators		
Dispenser combined	222	40
Electric	340	606+
Fluid	116	273+
Refrigerant	62	125
Lines making, metal working	29	DIG. 17
Meters		
Electric current	324	76R+
Fluid	73	861+
Thermal	73	204.11+
Fluid direction combined	73	189
Proportional	137	98+
Recorders	346	72
Regulator	138	37+
Earth boring tool combined	175	232+
General utility type	138	37+
Impact activated valves	251	76
Responsive to element		
deformation	137	67
Sprinkling and spraying	239	533.1+
Valve combined	251	118+
Responsive		
Alarm	116	112
Alarm electric	340	606+
Dispenser cutoff	222	59+
Dispenser preset cutoff	222	14
Heater closed fluid	236	25R
Pump throttle valve	417	43
Switch electric	200	81.9 R
Valve	137	455+
Retarder		
Brake fluid pressure	303	84.1
Fluid in conduit	138	37+
Railway switch indication	246	109
Flower	PLT	
Curbing for flower bed	D25	38+
Holder	47	41.1+
Apparel attached	D11	1+
Corsage boutonniere personal		
wear	24	5+
For plant shrub vine etc	47	44+
Furniture type	D 6	403+
Vases	D11	143+
Wall mounted	D 6	556+
Wreaths and sprays	248	27.8
Imitation	428	17+
Plants	PLT	
Pots	47	66+
Design	D11	143+
Support, depended	D 6	513+
Support, shelf	D 6	556+
Preserving		
Coating or impregnating	427	4
Plant catalyst or stimulant	71	68
Flowmeter (fluid)	73	861+
Thermal	73	204.11
Flowmeter for Heat	374	29+
Flowrate	73	19.4
Determination gas content	73	19.4
Floxacillin	540	327
Flue		
Boiler	122	
Building attached	52	219
Cleaner		
Brush and broom	15	162+
Brush and broom design	D 4	
Scraper	15	242+
Collar adjustable	126	315
Connection making	29	890.145
Cover		
Design		
Expanders (See expander)		
Gas heating	432	219+
Masonry	52	245+
Mineral oil still	196	116+
Regenerative	431	170
Checker bricks	165	9.1+
Stopper	126	319
Thermolytic retorts having	202	124+
Horizontal	202	138+
Sprinkling and spraying	239	533.1+
Fluent Material Handling	141	
Fluent Tool Metal Shaping	72	54+

	Class	Subclass
Fluid (See Gas; Liquid)		
Amplifier	137	803+
Bearing	384	100+
Breakwater	405	22
Calculator	235	200R+
Cleansing tool drive combined	173	57+
Clutch	192	58R
Collection from body	604	317+
Conveyer (See conveyer)		
Coupling	60	325+
Gearing combined	74	730.1+
Planetary gearing combined	475	31+
Shaft connection	464	24
Delivery		
Check controlled device	194	
Sprinkling, spraying, diffusing	239	
Track sander	291	3+
Distribution	137	
Conduit and electrical conductor		
combined	174	47
Cooling and or insulating electrical		
conductors	174	8+
Expansible chamber motor	91	
Gas separator combined	55	
Outlet nozzles	239	
Reaction motor discharge	239	265.11+
Reaction motor discharge cooling	239	127.1+
Sound muffler	181	212
Drive	60	325+
Earth boring with	175	65+
Electrifying	361	225
Floating matter barrier	405	62
Flow direction measurement	73	188+
Flow indicator	116	273+
Fluidic device		
Making	29	890.9
Fuel		
Efficiency test	73	113+
Gas analysis	73	23.2+
Illuminating testing	73	36
Liquid	44	300+
Liquid analysis	73	53+
Fuel burner		
Combustion feature	431	
Dispersing feature	239	
Fuel for heating		
Material or bodies	432	
Structures	237	
Fuel in nuclear reactors	376	151
Fuel in nuclear reactors	376	172
Fuel in nuclear reactors	376	356
Fuel in nuclear reactors	376	356
Governor	73	521+
Elevator	187	68+
Handling equipment	137	
Mounted on vehicular support	137	899+
Systems	137	561R+
Branched passage flow control	137	861+
Heat exchange	165	
Packaged	D 9	
Heater (See heat heater and heating liquid)		
Electric	392	311+
Infrared treatment of	250	432R
Jet texturizer	28	271+
Level gauges	73	290R+
Mattress	5	449+
Measuring vessels	73	426+
Mixture sensing	236	12.1+
Motor (See propeller)		
Brake	303	
Expansible chamber	91	
Manual displacement energized	49	32
Manufacture	29	888+
Meter	73	861+
Power plant	60	
Rotary kinetic	415	
Dental hardpiece	433	132
Turbine	415	
Operated		
Aircraft control	244	78
Alarm	116	109+
Apparel or garment turning	223	43
Belt and pulley control	474	
Body vibrators or massagers	128	24R+
Boiler fire extinguisher	122	506
Brake	188	151R+
Camera shutter	354	257
Changeable exhibitor	40	477+
Chuck	279	4
Closure by ambient fluid	49	21+

	Class	Subclass
Clutch operator	192	82R+
Depth gauge	73	299+
Die shaping	72	54+
Dumping vehicles	298	22R
Earthenware apparatus	425	
Explosive arming devices	102	222
Fluid scatterer or sprayer	239	214+
Governor	73	521+
Governor modifier	73	521+
Hat shaping	223	13
Impacting device	173	90+
Indicator	116	264+
Linear hoist	212	138
Lubricators	184	14+
Machine elements	74	
Planetary gearing controls	475	31+
Milling machine work feed	409	170
Mortar mixers	366	10+
Process	366	3+
Motors temperature controlled	236	79+
Overhead hoist	212	213
Photographic printing	355	91+
Pipe and tube cleaner	15	104.3
Pipe cleaners	131	244
Piston packing	277	
Pneumatic tire alarm	116	34R
Power hammer and presses	72	407
Press	100	269R+
Pressure gauge	73	700+
Puller or pusher assembler or disassembler	29	252
Pushing and pulling implements	254	93R
Railway car stop	104	256
Railway switches and signals	246	
Railway turntable	104	37
Rate of climb meter	73	179+
Reciprocating cutter	83	639.1
Rod or piston packing	277	
Rotary hoist	212	250
Sausage stuffer	17	39
Sawmill set works	83	726
Seal with bearing	384	131
Ship steering	114	150+
Ships log	73	181+
Sound record molding device	425	385
Sound record molding device	425	406+
Sprayer	239	237+
Steering gear	180	79+
Switch	200	81R+
Syringe	604	131+
Testing device	73	37+
Thermometer	374	201+
Thermostat	60	530+
Tool turret	29	42
Toy	446	176+
Track sander dispenser	291	24
Oscillator	239	589.1
Trolley retriever	191	85+
Trolley stand	191	67
Valve responsive to flow	137	455+
Valve responsive to pressure	362	384
Vehicle head lamp	362	37+
Well tubing cutter	166	55+
Wire placing apparatus	254	134.4
Work holder for grinder	51	233
Pillow	5	441
Pressure gauges	73	700+
Pressure regulator (See pressure)	137	505+
Proportioning	236	12.1+
Pumps	417	
Ratio maintenance	236	12.1+
Reaction surfaces, ie impellers	416	
Regulator and control		
Volume or rate of flow meters combined	73	199
Samplers	73	863+
Screen	110	179
Shark screen	405	22
Sprinkling, spraying, diffusing	239	
Storage in earth cavity	405	53+
Transmission (See transmission)		
Treatment		
Bleaching and dyeing	8	
Coating implement with material supply	401	
Drying and gas contact with solid	34	
Food apparatus	99	516+
Food, butchering process	17	51
Gas and liquid contact	261	
Gas separation	55	
Metal	148	

	Class	Subclass
Shoe soles	12	41.3
Textile fabric apparatus	68	
Textile fibers	19	66R+
Thread finishing with	28	217+
Ultraviolet treatment of	250	432R
Vortex amplifier	137	812+
Vortex generator	137	808+
Fluke		
Anchor	114	301+
Anchor trippers	114	210
Flume	405	119+
Gates	405	87+
Screen	210	154+
Comminuting	210	173+
Fluoranthene (See Aromatic Hydrocarbon)	585	400+
Fluorene (See Aromatic Hydrocarbon)	585	400+
Fluorenone	568	326
Fluorescein	549	223
Fluorescent		
Cathode ray screen	313	467
Cathode ray tube	313	467+
Coating	427	157
Composition		
Inorganic	252	301.6 R
Mineral oil	208	12
Organic	252	301.16
Radioactive	252	625+
Devices and applications	250	462.1
Electric light	313	484+
Ballast	336	155+
System	315	
Electric lights and electronic tubes	313	495+
Illuminating devices with	362	217+
Screens	250	483.1
Signs	40	542+
Fluorides Inorganic	423	462+
Binary	423	489
Fluorimeter		
Chemical analysis, apparatus	422	82.8
Fluorine	423	500
Acids	423	481
Halocarbon containing	570	123+
Fluorocarbons, Eg, Teflon, Kel-f		
And memory plastics, pipe joints	285	909
Insulated cable and conductors	174	110FC
Plastic article forming, resin	264	127
Fluoroscope and Fluorescent	250	458.1
Article inspection	250	458.1
Fluorescent applications	250	458.1
Fluoroscopic equipment	D24	2
Tables	378	38
Fluosilicates	423	342+
Fluosilicic Acid	423	341
Flush Tank		
Liquid level control	137	409+
Water closet	4	353+
Bowl combined	4	300+
Urinal	4	353+
Flusher		
Nozzle cleaner	239	106+
Sewerage	137	240
Siphon including automatic	137	132+
Street	239	146+
Valve	251	15+
Water closet	4	300+
Design	D23	295+
Fluspirilene, in Drug	514	278
Flute	84	384
Fluter	223	28+
Flux		
Bituminous material	106	279+
Ceramic	106	313
Earth boring combined	175	13
Metallurgical	75	303+
Natural resin	106	240
Protein containing	106	161
Casein	106	147
Gelatine or glue	106	136
Prolamine	106	153
Soldering feeders for	228	33+
Solid metal treating	148	23+
Synthetic resin or natural rubber (See class 523, 524)		
Fluxmeter		
Magnetic flux	324	244+
Fly		
Brush	416	501
Closer lasting tool	12	113
Fishing	43	42.24+
Design	D22	125+

	Class	Subclass
Holders for fishhooks, flies	43	57.1+
Net	54	81
Paper box making	493	52+
Swatter	43	137
Design	D22	124
Textile spinning	57	96
Trap	43	122
Fly Ash	106	DIG. 1
Processes of using	264	DIG. 49
Stock material	428	406
Flying (See Aircraft)		
Blind		
Aircraft controls	244	75R+
Traffic control systems	340	947+
Machine	244	
Design	D12	319+
Design toy	273	362+
Propeller toys	446	36+
Target	273	362+
Design	D22	114+
Drone	D12	16.1
Trainers	434	30+
Flying Tool		
Cutting	83	284+
Wire tool	83	307.1+
Metal deforming	72	184+
Flypaper	43	114+
Catching implements	43	136
Packages	206	447
Flytraps	43	107+
Design	D22	122
Flywheels		
Fan combined	416	60
Fluid	60	330+
Impellers combined	416	60
Machine elements	74	572+
Occupant propelled vehicle power storing	280	217
Sewing machine abrading attachment	51	257
Vehicle drive combined	180	165
With planetary gearing	475	267+
Foam	252	307
Breaking		
Distillation apparatus	202	264
Distillation process combined	203	20
Processes	252	321
Coating implement with material supply	401	44+
Control	435	812*
Fire extinguishing	252	3
Stabilizing agent containing	252	8.5
Froth flotation agents	252	61
Gas & liquid contact apparatus	261	DIG. 26
Glass making	65	22
Laminate	428	304.4+
In situ foaming	156	77+
Metallic receptacles collection	220	902*
Plastic molding	264	41+
Producing compositions	252	350
Protective coating or zone	252	382
Separating gases from	55	178
Processes	55	87
Fob	D11	79+
Focusing		
Camera	354	195.1+
Automatic	354	400+
Hood	354	287
Flashlight	362	187
Focus coil and cathode ray tube	313	442
Focus coil per se	335	213+
Photocell controlled	250	201.2+
In dynamic information storage and retrieval	369	44.23
Projector	362	317+
Adjustable light	362	285+
X ray tube combined	378	138
Fodder	426	636
Fog	252	305
Combustion	126	59.5
Dispelling appliances	98	1
By spraying	239	2.1
Horns	116	137R+
Preventing compositions	106	13
Foil (See Sheet)		
Dental filling	433	227+
Fencing	272	98
Laminate		
Manufacture	29	17.1+
Metal	428	606+
Folder (See Creaser)		
Apparel apparatus	223	37+

	Class	Subclass
Collar cuff bosom making	223	3
Collar, cuff or neckband ironers	223	52.1+
Plaiting etc with fold guides	223	34
Plaiting fluting shirring with cross creasing	223	29
Boot and shoe making		
Lasting	12	12+
Uppers	12	55
Cigar or cigarette tip or mouthpiece forming	131	89+
Folder carried indicia	40	359
Paper article making		
Bags, envelopes	493	243+
Boxes	493	162+
Sheet material associating or folding		
Associating and folding	270	32+
Associating folding and rotary printing combined	270	4+
Folding and rotary printing	493	321+
Wrapping folding and or printing combined	53	116+
Sheet metal can seam	413	69+
Sheet metal seaming	29	243.5+
Folding	493	405+
Boats	114	353
Chairs	297	16+
Tables	108	115+
Foliage		
Artificial	428	17+
Design	D11	117+
Folic Acid	544	261
Follower		
Ammunition dispenser (See magazine)	42	49.1+
Cabinet		
Phonograph record holder	312	9+
Cam	74	569
Cooking apparatus combined	99	349+
Dispenser combined		
Container	222	386+
Insertable cartridge	222	326+
Fountain pen	401	141+
Lubricator pump	184	37
Mattress filling apparatus	53	255+
Metallic receptacle	220	93
Followers and partitions	220	529+
Partition follower	220	559+
Pipe joint piston type	285	302
Rack card or sheet	211	51+
Self loading material handler	414	506
Sheet retainer with releasable keeper	402	60+
Smoking device	131	181
Trunk	190	36
Wooden		
Barrel	217	86
Box	217	64
Follow-up Mechanism		
Aircraft control	244	76R+
Controlled gearing	74	388R+
Fluid pressure brake valve	303	54+
Gearing	74	388R+
Recorder drive	346	31+
Servometer fluid	91	358R+
Ship steering	114	144R+
Font		
Sorting	199	40+
Typewriter embossing	400	127+
Typewriter, penetrating	400	135+
Food (See Type)	426	
Animal foodstuffs	D 1	
Animal mastication aids	D30	199
Apparatus	99	
Canning	426	392+
Filling receptacles	141	
Cereal pupping	99	323.4
Conductivity control	99	DIG. 11
Cooking apparatus	99	324+
Dairy food treatment	99	452
Dehydration	426	443+
Design	D 1	
Edible containers	D 1	
Edible laminated product making	99	450.1
Enclosed modified atmosphere treating	99	467
Food mixer-grinder-blenders	D 7	368+
Fruit and vegetable peelers	D 7	693
Hand tools for food preraration	D 7	669+
Induction heating	99	DIG. 14
Inedible stick or holder &	D 1	105
Live stock	426	635+

	Class	Subclass
Means to treat food	99	485+
Non-cooking heat treatment	99	483
Non-protein nitrogen	426	69
Preparation or dispensing machines	D 7	
Preparing and treating (See type)		
Apparatus	99	
Preserving	426	
Processes, compositions & products	426	
Equipment design	D 7	368+
Storage receptacle, household	D 7	601+
Strip or stick form	D 1	
Temperature responsive	99	DIG. 1
Vinegar making	99	323.12
Foot		
Bandages	128	165+
Bath	D24	38
Closure by foot pedal	220	262
Controlled dibbling apparatus	111	98
Electric applicator	128	795
Flipper	441	64
Guard railway track	238	379+
Levers pedals	74	594.4
Measuring devices	33	3R
Monitor radioactivity	378	38
Miscellaneous	378	38
Scintillation type	250	458.1
Operated		
Anvil vises forging	D21	1+
Closure bolt releasers	292	255
Control lever and linkage systems	74	469+
Cutting tool	30	275
Cutting tool plural blade	30	297
Dispenser	222	179
Earthworking tool lift	172	495+
Gang punching machine	83	620+
Runner vehicle	280	12.1+
Sheet punching machine	83	613+
Ship steering apparatus	114	153
Vehicle steering apparatus	280	87.1
Wheeled vehicle	280	200+
Orthopedics	128	80R+
Piece		
Dustpan	15	257.5
Fork and shovel hand	294	60
Posts for skates	280	11.17
Resilient machinery support	248	615+
Rest (See --, support)	297	423+
Chair reversible back	297	73+
Chair with folding	297	30
Closet	4	254
Control lever system type	74	564
Design	D 6	501
Land vehicle body	296	75
Land vehicle combined	280	291
Piano	84	232
Rocker with	297	271
Substitute	297	428
Support		
Boat	114	363
Life preserver with	441	80+
Skate joined runner & support	280	11.15
Skate resiliently mounted	280	11.14
Surgical table	269	
Supporter	36	71
Warmer		
Heat radiator	237	77
Stove	126	204
Surgical body wear	128	382+
Football	273	65R
Electric scoreboard	340	323R
Game	273	55R
Game board	D21	29
Scoreboard	116	222+
Shoes	36	128
Simulated game	273	94
Tackling dummy	273	55R
Teaching	434	251
Footboard		
Bed	5	53.1+
Foot or leg support	5	443+
Motor vehicle body	180	90.6
Railway rolling stock	105	460
Footholds		
Building with	52	184+
Friction surface	52	177+
Cane	135	70
Crutch	135	70
Ladder terminals	182	108+
Pavement	404	19+
Stair covers	428	85+
Walking aids with tips	135	77+

	Class	Subclass
Footing (See Bases; Foundation)		
Footlights		
Cleaning type	15	238+
Footwear	36	
Design, general	D 2	264+
Mens	D 2	308+
Stockings & socks	D 2	329+
Womens	D 2	282+
Foraminous		
Cabinet wall structure	312	213
Force Measuring	73	
Coin controlled devices	194	
Electric	324	
Force Pumps	4	255+
Forceps		
Dental	433	159
Railway traction cable	104	217
Surgical	606	205+
Design	D24	27
Forcer		
Closure applying die	53	341+
Sheet metal die shaping	72	54+
Forehearth	266	166
Forestry	47	
Hand tools design	D 8	1+
Forge	110	195
Tuyeres	110	182.6
Water cooled	122	6.7
Forging	72	
Machines with other tools	29	34R
Metal working with	29	DIG. 18
Nuts and washers	10	76R+
Piston rings	29	888.7+
Processes	72	372+
Wrought nails	10	55+
Fork (See Shovel)		
Bicycle type	280	276+
Compound tool with culinary fork	7	112
Compound tool with pitch fork	7	115
Cutlery	30	322+
Design	D 7	653+
Guard	30	323
Hand	294	55.5
Handles	D 8	DIG. 4
Material handling		
Blanks and processes for making	76	111
Cooking spit or impaler	99	419
Excavating scoops	37	120+
Hand	294	55.5+
Hayfork	294	120+
Hayfork type grapple	294	86.4+
Inclined elevator or hoist	414	595+
Tilting	414	598+
Making dies	72	470+
Railway mail delivery	258	21+
Railway mail delivery support	258	11+
Single throw lever fulcrumed	254	131.5
Tedder	56	374
Vertically swinging	414	685+
Table (See cutlery)	30	322+
Blanks & processes for making	76	105
Combined tool	7	112+
Combined with knife	30	148
Dies for making	72	470+
With ejector	30	129
With material holder	30	137
Tuning or musical		
Music instrument	84	409
Tuning	84	457
Wheel occupant propelled vehicle		
Front	280	279+
Rear	280	288
Rear yielding	280	284+
Forklift Truck	414	529
Ramp design	D12	
Formaldehyde	568	448
Dehydrating	203	17
Synthetic resin (See synthetic resin or natural rubber resin or natural rubber)		
Formamide	564	215
Formates		
Alkali forming metal	562	609
Cellulose	536	67
Esters	560	231+
Formazans	534	652
Former (See Die; Molding; Shaping)		
Formic Acid	562	609
Forming		
Metal by plastic shaping	72	
Fluent tool or energy field	72	54+

	Class	Subclass
Forms		
Display	223	66+
Design	D20	10+
Dress	223	68
In situ construction engineering or building type	249	1+
Forte Devices (See Piano)		
Fortune Telling Devices	273	161
Coin controlled	194	
Playing cards	D21	42+
Toys	D21	240
Foundation (See Bases)		
Bee comb	449	44+
Building	52	292+
Hair	132	54+
Marine	405	195+
Ballasted	405	207+
Piles	405	231+
Road bed	404	27+
Founding	164	
Foundry Apparatus	164	
Foundry Mold or Mold Core Binder	523	139+
Fountain		
Beverage infuser	99	313+
Gravity feed	99	310+
Comb	132	112+
Design	D23	201
Drinking	D 7	304
Stock watering	D30	121+
Drinking	239	24+
Design	D 7	304
Inkwell (See inkwell)	222	577
Machine		
Bed and platen oscillating printing	101	315
Bed and platen reciprocating printing	101	321
Dampener planographic	101	148
Inker	101	364+
Inker multicolor	101	210
Inker roller	101	363
Printing members and inker	101	330+
Pen (See fountain pen)		
Sprinkling	239	17+
Stencil addressing	101	48+
Stock watering	D30	121+
Stock watering	119	74+
Syringe	604	131+
Gravity feed from plural reservoirs	604	80
Fountain Pen (See Pens)	401	
Ball point	401	209+
Follower fluid	401	142
Bifurcate nib type	401	221+
Ink retainer attachment at tool	401	252+
Filling	141	21+
Filling	141	18+
Projectable and retractable tool	401	99+
Rupturable cartridge	401	132+
Sign or picture carrying	40	334+
Stylographic	401	258+
Fourdrinier Machines	162	348+
Fowl	17	11+
Brooders	119	31+
Butchering	17	11+
Care and handling design	D30	
Confining and housing devices	119	21+
Design	D30	108+
Feeding devices	119	51.1+
Foods	426	635
Medicines	424	
Supports and shackles	17	44.1
Watering devices	119	72+
Design	D30	121+
Fowler Flap for Aircraft	244	216
Fracture Apparatus Surgical	128	83+
Frame	52	
Abrading machine	51	166R+
Agricultural implement	172	776
Aircraft		
Airfoil	244	123+
Airship hull	244	125
Fuselage	244	119+
Apparel		
Armpit shield	2	57
Hat	2	180
Apparel apparatus		
Dress coat or skirt stretching	223	69
Hat frame making	223	8
Tie	223	65
Trousers stretching	223	63
Bag holder	248	99+

	Class	Subclass
Bed		5
Bee		
Comb	449	35+
Hive	449	3+
Camera	354	161
Canopy	135	101+
Car roof	52	45+
Caster	16	31R+
Clock	368	88
Closures partitions panels	160	371+
Awning	160	76
Mount or support	160	369
Nonrigid	160	354
Plural	160	353
Roll type	160	239+
Door	49	504+
Metal	52	782+
Drill press	408	234+
Drying		
Curtain	38	102.1+
Electric lamp	314	130+
Arc	314	130+
Fabric cleaner	15	233
Fabric stretching	38	102.1+
Fastening savings box	232	6
Fireplace	126	500
Fishing reel	242	309+
For building	52	648
For expansible chamber device	92	161
Hair foundation	132	54
Hand and hoistline implements	294	67.1+
Brick carriers	294	63.1+
Grapple	294	113+
Vacuum	294	65
Harvester	56	
Headwear receptacle	206	9
Internal combustion engine	123	195R
Land vehicle		
Body structure	296	187+
Body supporting	280	781+
Design	D12	159
Drop frame convertible to	280	7.11
Dropped body	280	2
Endless track carried by	305	16+
Equalized	280	104+
Roller bearing	280	105
Velocipede	280	281.1+
Velocipede steered	280	274+
Leather making work holder	69	19
Light shade or bowl	362	433+
Luggage	190	121+
Metallic receptacle wall	220	668
Mirror or reflector	52	
Mounted motor	180	58+
Music		
Hammer action	84	250
Key	84	432
Piano	84	184
Open work	52	633+
Panel	52	474+
Edging	428	21+
Sectional	52	455+
Pavement light or lens	404	22+
Picture	40	152+
Picture frame molding	D25	119+
Planting drill	111	52+
Portal with building structure	52	204+
Printing	355	122+
Quilting	112	119
Radiator	165	149
Railway		
Bogie side	105	206.1+
Car	105	396+
Locomotive	105	172+
Scroll saw	83	776+
Sewing	112	258+
Sheet drying apparatus	34	163
Sifting machine	209	404+
Sign or bulletin board	D19	52
Slate	434	422
Spreader	294	81.1
Table	248	163.1+
Tent	135	101+
Textile		
Hand loom	139	34
Pile tufting tubes	139	10
Textile thread	28	149+
Trolley pole	191	65
Trunk	190	24
Truss surgical	128	99.1+
Typewriter	400	691+

	Class	Subclass
Ventilating register	98	114
Wall decoration	D 6	300+
Window	49	504+
Wireworking	140	92.1
Woodworking		
Bit	408	53
Clamps	269	
Handsaw	30	166.3+
Match dipping	144	65
Frangible		
Attachment for metallic receptacle	220	89.1+
Barrel bung	217	111
Bottle or jar cap	215	250+
Building structures	52	98+
Cartridge fountain pen	401	132+
Container opening	206	601+
Dispenser outlet	222	542
Electric signal box element	340	303
Element in pipe joint	285	3+
Feature		
Static mold	249	61
Fluid handling means	137	797
Locks		
Key bit	70	410
Key operating mechanism	70	422
Material used in glass process	65	23
Metallic receptacle closure	220	265+
Multiple shoe brake connection	188	241
Railway switch element	246	289
Railway train pipe closure	246	199
Derailment operated type	246	172
Valve controlling element	137	68.1+
Frankfurter (See Sausage)		
Packaging	53	DIG. 1
Fraternal Insignia	D11	95+
Fraud Preventor or Detector		
Check controlled apparatus	194	302+
Combined	194	202+
Coating or impregnating	427	7+
Electrophotographic copier	355	201
Printed matter	283	72+
Stock material	428	916*
Transfer or decalcomania	428	915*
Free Fall Item Feeding	209	908*
Free Radical, Hydrocarbon Synthesis		
Involving	585	942*
Free Wheeling Clutches	192	41R+
Freeze-out Trap	62	55.5
Freezer and Freezing (See Congelation)		
Flowable material	62	340+
Automatic	62	135+
Process	62	66+
Ice cream machine design	D15	82
Pressure compensators	138	27+
For internal combustion engines	123	41.5
Refrigerator-freezer design	D15	81+
Safety systems for fluid handling	137	59+
Work holder using	269	7
Freight Accomodation on Freight		
Carrier	410	
Lashing	410	96+
Anchor	410	101+
Load brace	410	121+
Bar wall-to-wall	410	143+
Panel wall-to-lading	410	127+
Panel wall-to-wall	410	129+
Retainer	410	77+
Yieldable	410	117+
Particular article	410	2+
Automobile shipment	410	4+
Grouped	410	31+
Massive	410	44
Freight Carrier		
Freight accomodating construction	410	
Land vehicle	280	
Body	296	
Rail car	105	355+
Framing	105	404+
French Curves	33	561.2+
French Dressing	426	615+
French Fryers	99	403+
Freon T M	570	134
Frequency		
Amplitude study mechanical	73	579+
Analyzer	324	77R
Automatic control		
Gain, level, volume	455	234
Plural receivers	455	136
System radio	455	71
Television receiver	358	194.1
Transceiver	455	119

	Class	Subclass
Transmitter	455	313+
Changers	310	127+
Conversion systems	363	157+
Converters	310	
Systems	363	157+
Demodulators	329	315+
Divider	363	157+
Electronic tube type	328	15
Electronic tube type	328	25
Dividers oscillator	331	51
Group frequency radio signaling	455	
Measuring	324	78R+
Absorption type	324	81
Heterodyne	324	79R
Wave meter	250	250
Modulators	332	117+
Multipliers oscillator	331	53
Radio signaling	455	
Response modification	358	904*
Speech range compression or expansion in telephone system	381	29+
Synthesis	328	14
Fret Saws	30	513+
Frets for Fingerboards of Stringed Instruments	84	314R
Control	84	DIG. 3
Friction		
Clutch	192	
Brake position adjuster combined	188	199
Gearing	74	190+
Planetary	475	183+
Single speed	74	206
Generators		
Boiler heating	122	26
Electric	310	310
Heater	126	247
Igniting mass or surface combined with fuel	44	506+
Producing compositions	106	36
Removable closure retained by	220	352
Tape	428	343+
Coated on both sides	427	208
Package	206	411
Test		
Between solid bodies	73	9
Liquid viscosity	73	55
Torque	73	862.12
Welding apparatus	228	2
Nonmetallic	156	580.1
Methods	156	73.1+
Sonic or ultrasonic	156	580.1
Welding process, nonmetallic	156	73.5
Sonic or ultrasonic	156	73.1
Fringe		
Apparel trimming	2	244
Making on looms	139	118+
Misc. manufacture	28	145+
Of sewn strands	112	409
Design	D 5	7+
On web or sheet	428	115+
Sewing machine forming	112	64
Trimming design	D 5	
Turning guide for sewing machines	112	149
Woven fabric	139	385
Frisbee T M	446	46+
Frisket	101	421
Frits, Glass Compositions	501	14+
Frog		
Flower stem holder	47	41.13
Design	D11	147
Railway	246	454+
Continuous rail	246	275+
Frog obviating switches	246	417+
Switch connected	246	382+
Switch connected vehicle actuated	246	274
Trolley transfer	104	105
Textile weaving	139	346
Front		
Boards	5	54
Boiler	122	497+
Fireplace	126	544+
Design	D23	343+
Furnace	110	172
Upending bedstead case		
Panel	5	160
Wing	5	161
Winter or storm		
Land vehicle	296	77.1+
Railway car	105	353
Frost Indicator for Refrigerator	62	128
Frost Preventing		
Composition	252	70

	Class	Subclass
Coating	106	13
Plants and fruit	47	2
Refrigerator	62	272+
Automatic	62	150+
Process	62	80+
Frosting Lamp Bulbs	156	625+
Froth	252	307
Flotation	209	163+
Agents	252	61
Copper	209	901*
Phosphate	209	902*
Preventing	252	321
Fruit	426	616+
Apparatus for treating	99	467
Cleaning		
Brushing or wiping apparatus	15	3.1+
Fluid treatment apparatus	134	
Processes	426	478+
Clipper	30	123.5+
Coffee substitutes from	426	594+
Corer	99	547+
Design	D 7	693
Gatherer	56	328.1+
Design	D 8	13
Halving cutters	83	
Juice extractor		
Comminuting citrus fruits	99	495
Design	D 7	665+
Interfitting cup type press	100	213
Presses	100	104+
Peelers or parers	99	588+
Design	D 7	693
Plants	PLT	33+
Preservation	426	321+
Canning	426	392
Coating	426	289+
Dehydration	426	471+
Refrigeration	426	524
Seeding and stoning	99	547+
Sorting machines	209	509+
Stemmer	99	635+
Frying Pan		
Deep fat	99	403+
Heat transfer feature	126	373+
Structure	220	
Surface griddles	99	422+
Ftorafur	544	313
Fuchsin Acid	552	112
Fuchsine	552	114+
Fuel	44	
Bound animal blood or animal manure containing	44	552
Bound sewage containing	44	552
Boxes	126	283
Can	D23	202+
Cell	429	12
Active material electrode	429	27+
Automatic control means	429	22+
Catalytic electrode structure	429	40+
Cell electrical generator	429	12+
Chemically specified electrode	429	46
Coal slurries	44	280+
Design solid	D13	100
Mineral oil	208	15
Feed and supply systems	137	
Engine combined generally	60	
Engine internal combustion combined	123	495+
Fuel injection	123	445
Rotary piston	123	200+
Liquid fuel nozzle generally	239	
Solid fuel burners combined	110	101R+
With combustion feature	431	
Gas		
Distribution	48	190+
Generator	48	61+
Gaseous	48	
Gelled liquid	44	265+
Heat exchange feature	429	26+
High energy compounds	149	120+
Igniting devices & fuel	44	
Injector combined with rotary piston engine	123	200+
Liquid	44	300+
Acyclic peroxy compound containing	44	322
Liquid hydrocarbon monomer mixtures	585	1+
Aldehyde or ketone containing	44	437+
Azo compound containing	44	328
Carbohydrate containing	44	313

	Class	Subclass
Carboxamide containing	44	418+
Explosive or thermic containing	149	1
Free metal or alloy containing	44	321
Halohydrocarbon containing	44	456
Heavy metal containing except lead	44	354+
Heterocyclic compound containing	44	329+
Inorganic compound containing	44	457+
Material, actinide or radioactive inorganic compound	423	249+
Nitrate or nitrite ester containing	44	323+
Nitrile or isonitrile containing	44	384
Nitro compound containing	44	413+
Organic amine containing	44	412+
Organic carboxylate containing	44	388+
Organic oxygen compounds containing	44	436+
Organic sulfate containing	44	369
Organic sulfonate containing	44	370+
Organic sulfur or organic phosphorus containing	44	435+
Organo lead compounds containing	44	454+
Phosphorus organic acids or esters containing	44	375+
Quinone containing	44	312
Silicon containing	44	320
Solid hydrocarbon polymer containing	44	459
Solidifying	44	265+
Urea or thiourea containing	44	417
Log	44	535
Mineral oil	208	15+
Nuclear	376	
Alloys for actinide groups	420	1+
Blanket arrangements	376	172
Bundle or pack	376	434
By-product treatment	376	189
Canned	376	412
Carrier, insertable	376	342
Cladding	376	457
Coated or impregnated	376	414
Complementary segments	376	429
Compositions	252	626+
Concentric layers	376	431
Coolant included	376	424
Core arrangements	376	348
Electrolytic production	204	1.5
In fission plenum	376	412
Particles	376	411
Reprocessing	423	249+
Handling	376	260
Heat treatment	148	132
Layered	376	416
Leak detecting	376	250
Warning structure	376	450
Loading and unloading	376	264
Manufacturing		
Chemical milling	156	625+
Chemical, misc	156	67
Coating	427	6+
Etching	156	625+
Irradiation or conversion	376	156
Laminating	156	67
Metal stock heat treated	148	401+
Metallurgical production	75	393+
Metallurgical production	420	1+
Powder metallurgy	419	
Material, actinide or radioactive		
X-ref patents	376	901*
Alloy	376	422
Organic compounds	534	10+
Mechanical spacing means or securing	376	438
Molten	376	359
Projections, prongs fins	376	454
Metal alloys	420	1+
Seal for	376	451
Segmented	376	426
Non-metal	264	21+
X-art	376	903
Sintered process	419	
Sintered stock	75	228+
Slurry, associated components	376	356
Slurry, conversion process	376	171
Spacing means or securing	376	438
Waste disposal	252	626+
Nuclear device making	29	DIG. 9
Plural cells	429	18+
Process of operating	429	13+
Pumps	415	

	Class	Subclass
Connections	169	42
Metal, in bolt to nut lock	411	257
Plugs for automatic fire sprinkler		
heads	169	37+
Receptacle attachment	220	89.3
Sensor signal system	340	590+
Stove damper release	126	287.5
Switch	337	401+
Thermal current element	200	
Thermal current heater	337	182+
Thermometer	374	106
Valve operator	137	72+
Valves and controllers	222	54
Fire extinguisher	169	19+
Ventilator cowl release	98	86
Fusion (See Melting)		
Bonding apparatus	228	
Coating combined	427	375
Separation combined	422	285+
Sulphur	423	567R
Fusion Bonding		
Glass preforms of	65	36+
Apparatus	65	152+
Fusion Nuclear Reactors	376	100
G Acid	562	80
Gable, Roof	52	90+
Roof end	52	94+
Gaff		
Fishing	43	5
Gamecock	30	297
Grappling	294	19.3
Ship spars	114	97+
Gag		
Fishing tackle	43	53.5
Runner for bridles	54	14
Side cutting harvester lever	56	271+
Specula	128	12+
Gage (See Gauge)		
Gagger	164	411
Flask with	164	382
Gain Control, Automatic	73	900*
Gaining	144	133R
Gaiters	36	2R
Design	D 2	267
Fastenings	24	
Making	12	142W
Galley		
Ships stove	126	24+
Type setting	276	40+
Justifying machine	276	33
Gallic Acid	562	476
Gallium		
Pyrometallurgy	75	688
Gallocyanines	544	99+
Galloons		
Lace	87	10
Gallows (See Execution Devices)		
Galvanized Sheets	428	659
Galvanizing		
Metals	427	433+
Galvanometers	324	76R+
Galvanoplastic Processes	204	6
Gambrels	294	79+
Butchers	17	44+
Roof design	D25	24
Game		
Amusement railways	104	53+
Annunciator systems	340	323R
Board games	273	236
Carrier	224	103
Other than stringer	224	921*
Check or coin control	194	
Computer controlled, monitored	364	410+
Design	D21	1+
Game apparatus	273	
Peg tally board	235	90
Pieces	273	288+
Design, misc	D21	51+
Score boards	116	222+
Spinners & indicators for	D21	22
Tables	D21	121+
Tally sheet	283	49+
Tv, input control	273	
Input control per se	340	709+
Gamecock		
Boxing muff	119	143
Gaff	30	297
Gamma		
Geiger survey meter	250	374+
Radiation reflector	376	220
Radiation reflector	376	350

	Class	Subclass
Radiation reflector	376	458
Radiation shield	376	287
Radiation shield	376	347
Radiation survey meter	250	336.1
Scintillation type	250	361R+
Ray detector	250	336.1
Gang		
Button, eyelet, rivet setting machine	227	51+
Circular saw roller feed	83	425.2+
Earthworking disk	172	599+
Plural	172	579+
Scrapers	172	558+
Harvesters	56	6+
Metal working rotary cutter	407	31
Nail feeding delivery	227	
Plows		
Wheeled	172	314
Punching machine	83	620+
Woodworking		
Rotary disc cutter	144	237
Slivering saw	144	189
Gang Bar Type Locking or Latching		
Means		
Cabinet	312	216+
Sectional unit type	312	107.5
Ganged Radio Tuner	334	
Operator only for tuner	74	10.45
Gangway		
Animal traffic	119	82
Car loading side	105	436
Conveyer skids	193	41
Conveyer skidways	193	38+
Endless conveyer type	14	70
One end attached	14	71.1
Ship attached one end floating	114	258
Unattached	14	69.5
Gantry	212	218+
Gap Jumper Amusement Railway	104	54
Garage	52	
Vehicle handler	414	227+
With ramp	52	174
Garbage		
Boiler plants for destroying	122	2
Carbonization of	201	21
Dump & trash truck design	D12	15
Fat etc recovery from		
Rendering	260	412.6+
Solvent extraction	260	412.8
Fertilizer from	71	14
Grinding and disposal		
Comminuting solids	241	
Design of machine	D 7	375
Grinder	4	DIG. 4
Sinks combined	4	629
Incinerator	110	235+
Insect trap for can	43	120
Receptacle		
Household design	D34	1+
Stand with closure operator	248	147
Truck, compactor type	100	100
Utilization for food	426	655
Wet fuel furnace	110	238
Garden (See Plant Husbandry)		
Garland	428	10
Design	D11	119
Paper, making	493	958*
Garment		
Absorbent pad holding	604	393+
Apparel	2	
Apparatus	223	
Design	D 2	26+
Design costumes	D 2	79
Design costumes bifurcated	D 2	30+
Design lingerie or undergarments	D 2	1+
Design surgical	D 2	
Bag, protector	493	935*
Bandaging type	128	82+
Respirator mask	128	206.12+
Suspensory	128	159
Body discharge shield	604	356+
Buttons buckles clasps etc	24	
Article holders	24	3R+
Shielded buckles	24	184
Cleaning		
Fluid treatment	8	
Fluid treatment apparatus	68	
Demonstration	434	395+
Fastener bonding	156	66
Fastener making	140	81
Forms	223	66+
Foundation type	450	7+

	Class	Subclass
Hangers	223	85+
Making	140	81.5
Knitted	66	171+
Design	D 2	44+
Design dresses, suits, skirts	D 2	43
Life preservers	441	88+
Light etc application to body	128	379+
Parachutes attached to	244	143
Protectors boot and shoe	36	70R
Receptacles for	206	278
Hat and headwear	206	8+
Seat attached holders for	297	190
Stay making	140	91
Stiffener making	223	28+
Supporters		
Blouses	2	107
Brassiere with	450	28+
Combined	2	300
Connected spaced for plural	2	304+
Convertible or reversible	2	301
Corset or girdle with	450	110+
Elements	2	336+
Foundation garment with	450	26
Mens outer shirts	2	117
Partially encircling	2	309
Plural encircling type	2	308
Rigid vertical type	2	302+
Shoulder suspension and		
encircling	2	310
Strip connected spaced	2	323+
Torso or limb encircling	2	311+
Underwear	2	112
Garnet Paper	51	394+
Garnetting Machines	19	98+
Garters		
Apparel design	D 2	
Corset or girdle with	D 2	2+
Fasteners	D 2	625+
Brassiere with	450	28+
Combined	2	300
Connected spaced for plural	2	304+
Corset or girdle with	450	110+
Elements	2	336+
Encircling	2	311+
Foundation garment with	450	26
Partially encircling	2	309
Plural encircling type	2	308
Rigid vertical type	2	302+
Shoulder suspension and encircling	2	310
Strip connected spaced	2	323+
Gas (See Fluid)		
Ammunition and explosive devices	102	367+
Drop bomb	102	369
Grenade	102	368
Gun cartridge	102	370
Shell	102	367
Analysis		
Chemical apparatus	422	50+
Chemical process	436	153+
Content	73	19.1
Electrical	324	
Physical	73	23.2+
Spectrometry	250	281+
Apparatus, gas & liquid contact	261	
Batteries	429	12+
Blood gas analysis	436	68*
Burners	431	
Burners	239	
Nozzles per se	239	
Cap for automobile	D12	197+
Cigarette lighter	431	150
Igniter correlated with feed	431	130+
Igniter correlated with feed	431	254
Circulated in circuit	261	DIG. 27
Coal	48	210
Coating contacting	427	248.1+
Compositions	252	372+
Concentrating evaporators	159	
Content of nongaseous material	73	19.1
Cooler	62	404+
Automatic	62	186+
Liquid contact with gas	62	304+
Process	62	89+
Deflector for separation	55	434+
Detectors		
Alarms	116	67R
Analysis chemical apparatus	422	50+
Analysis chemical methods	436	153+
Analysis miscellaneous	436	153
Apparatus	435	807*
Chromatographic	73	23.4

	Class	Subclass
Density	73	30.4
Electric alarms	340	632+
Moisture content	73	29.5
Nonchemical gas analysis	73	31.5+
Thermal property	73	25.5
Vapor content	73	29.5
Vibration testing	73	24.6
Discharge display panels	313	582+
Discharging nozzles	239	
Dispensing	222	3+
Packaging combined, method	53	403+
Distribution	137	
Airship control	244	96+
Cooling and or insulating electrical conductors	174	16.1
Explosion prevention	48	192
Heating and illuminating	48	190+
Leak detectors	137	455+
Leakage detection & prevention	48	193+
Musical instruments	84	330+
Outlet nozzles	239	
Electrochemical processes	204	124+
Engines	123	
Extinguishers	169	11
Foam forming	169	14+
Filling lamp or radio tube	141	66
Filters	55	
Fittings, design	D26	113+
Furnace		
Coking	110	230+
Domestic hot air	126	99R+
Industrial	432	
Power plant combined	60	669
Producer	110	229+
Solid fuel furnace combined	110	260+
Getter (See getters)		
Guns	42	
Bank protection	109	20
Bank protection combined	109	29+
Heaters and stoves	126	
Boilers	122	
Carburetor	261	127+
Cooking stoves	126	39R+
Design	D 7	339+
Design	D23	314+
Heating stoves	126	58+
Hot air furnace	126	116R
Metallurgical apparatus combined	266	138+
Stove burners	431	
Nozzle	239	398+
Stove structure	126	
Heating and illuminating	48	
Acetylene	48	1+
Carburetors	48	144+
Distribution	48	61+
Generators	48	180.1+
Mixers	48	198.1+
Natural	48	197R+
Processes	48	128+
Purifiers	48	119+
Retorts	48	
Holders and packages	206	.6+
Acetylene generator combined	48	
Heating or illuminating	48	174+
High pressure	220	3
Inhaler combined	128	203.12
Medicator	604	403+
Structures	220	
Telescopic	48	174
Illuminating	48	
Burner nozzles	239	
Inert	55	66
Separating	55	66
Injection		
Bathtub combined	4	542+
Convective distillation	203	49
Iron steel (molten)	75	528+
Liquid contact apparatus	261	
Medicators	604	19+
Mortar mixer	366	10+
Process	366	3+
Ionization in thermo nuclear reaction	376	100
Thermolytic distillation	201	31
Thermolytic distillation	201	36+
Lamp brackets, fittings	D26	113+
Light incandescent burners	431	100+
Design	D26	1+
Liquefaction	62	36+
Processes	62	9+
Liquefied handling apparatus	62	45.1+
Receptacles	220	901*

	Class	Subclass
Liquefiers and solidifiers		
Acetylene generator combined	48	1
Refrigeration	62	
Separating gases	55	209
Separating gases processes	55	81
Liquid contact with	261	
Automatic refrigeration	62	171
Concentrating evaporators	159	
Distillation	203	49
Gas separation combined	55	
Heat exchanger	261	127+
Mineral oil	208	
Percolation control	261	DIG. 81
Refrigerating process	62	121
Refrigeration	62	304+
Sugar treating	127	12
Log heater-fireplace	D23	409
Making		
Coking furnace	110	230+
Compositions	252	4+
Furnace	110	229+
Furnace and boiler combined	122	5
Furnace using liquid fuel	431	161+
Generators chemical	422	164
Heating and illuminating	48	
Life preservers	441	98+
Oil	48	211+
Mantles	431	100
Compositions	252	492
Mask	128	206.12+
Canisters for	55	DIG. 33
Fillers	502	400+
Processes of making corrugated tubes for	264	DIG. 52
Separating gases	55	
X-art collection	425	815*
Meters	73	861+
Coin controlled	194	
Mixers		
Acetylene generator combined	48	3R
Fuel burner	431	354
Heating and illuminating	48	180.1+
Internal combustion engine charge forming device	123	434+
Monitor	73	23.2+
Photoelectric	250	573
Radioactivity	250	336.1
Scintillation type	250	336.1
Nuclear fusion fuel	376	151
Oil	48	211
Pump	D15	9.1
Water	48	214R+
Operated		
Firearms and guns	89	125+
Rotary kinetic motors	415	
Rotary kinetic pumps	415	
Permeable coated fabrics	427	245
Poison	424	
Pressure storage	137	206+
Pressurized gas treatment of textiles	8	DIG. 15
Producers		
Chemical	422	126
Heating and illuminating	48	
Protecting joint during bonding	228	214
Public utility meter	D10	99+
Pumps	415	
Impellers	416	
Oil or gasoline pump	D15	9.1
Purifying	55	
Chemical	423	210+
Compositions, absorbent	502	400
Compositions, chemical agents	252	182.11+
Heating and illuminating	48	128
Ranges	126	1R+
Rapid producers of	149	
Rare	423	262
Electric furnace atmosphere	373	
Arc furnace	373	60
Resistance furnace	373	110+
Electric lamps	313	567+
Electronic tube	313	567+
Food preservation with	426	512+
Lantern	362	263+
Liquefying and separating	62	22
Purification chemical	423	210+
Separating	55	66
Rectifier full wave	313	581
Rectifier half wave	313	567
With hot cathode	313	629
Refrigeration produced by	62	401+

	Class	Subclass
Automatic	62	172
Heat generation type	62	6
Process	62	86+
Samplers	73	863+
Separation	55	
Chemical	423	210+
Electrostatic precipitators	55	101+
Flowmeter combined	73	200
Flue	110	148+
Liquid contacting means	55	220+
Liquid from	55	159+
Non-liquid cleaning means	55	282+
Recycle means	55	338+
Refrigeration apparatus	62	36+
Refrigeration process	62	11+
Solid	73	19.1
Nonchemical gas content detecting	73	19.1
Solid contact with	34	
Comminuting	241	
Distillation	201	
Gas separation	55	
Textile treating with	68	
Thyratron	313	591+
Triode	313	567+
Tube amplifier system	330	41
Ultraviolet treating		
Apparatus	250	432R+
Chemical change in	204	124+
Sterilizing	422	121+
Water heaters	126	344+
Gaseous Fuel Burners	431	
Discharge nozzle	239	
Igniters	431	
Discharge nozzle	239	
Heating attachments	126	249+
Tool heating attachments	126	231+
Synthetic	585	14
Gasket	277	
Applying		
Apparatus laminating	156	
To metal cap or closure	413	8+
Apparatus	413	58+
Barrel bung	217	109
Design	D23	269
Making by laminating	156	
Pipe joint combined	285	335+
Gasoline	208	
Carrier truck	D12	95
Explosion preventing appliances in containers	220	88.1+
Fuel composition containing	44	300+
Pick up truck mounted	D12	156
Pump	222	
Design	D15	9.1+
Storage tank systems	220	85S
Gasometer (See Gas Holders)		
Gastroscopes	128	8
Gate	49	
Design	D25	50+
Hinges	D 8	323+
Drawbridge	14	50+
End for vehicle bodies	296	50+
Standing top	296	106
Hanger	16	86.1+
Hinge	D 8	323+
Latch	D 8	331+
Motion picture	352	221+
Operators	49	324+
Railway highway crossing		
Automatic	246	292+
Automatic electric	246	125+
Fluid motor	246	261
Vehicle actuated	246	272
Register operated by	235	93
Sorting animal	119	155
Umbrella joint	135	32
Valve (See valve)		
Water gates	405	87+
With flow meter	73	215
Gated Beam Discriminator	313	300
Gatherer		
Brush and broom making		
Tuft gathering	300	7
Tuft gathering and setting	300	5+
Conveyor type loading machine	198	506+
Fish with conveyor to boat	43	6.5
Glass	65	
Ladling and gathering	65	125
Tank furnace gathering pool	65	336+
Harvesters	56	
Cornstalk type	56	119

	Class	Subclass
Acetylene	48	3R
Acetylene	48	4+
Chemical	422	198+
Chemical pressure in fire extinguisher	169	6+
Combustible mixture for internal combustion engine	123	3
Electric	310	
Battery charging	320	61+
Bicycle	362	193
Chair operated therapeutic	128	378
Combined in electronic tube	315	55
Compressional wave actuated	367	140+
Correlated with burner feed	431	255
Driven by compressional wave	367	178
Electric meter having	324	76R+
Electrically driven	322	
Electrostatic	310	309
For supplying engine ignition	123	149R
Infrared heat	392	407+
Locomotive	105	35+
Motor generator	310	113
Musical	84	600+
Nonelectric prime mover plant	290	
Nonmagnetic	310	300+
Nuclear	310	301+
Reactor	376	100
Reactor	376	347
Oscillator	331	
Piezoelectric	310	311+
Piezoelectric systems	322	2R
Plural generator in system	307	43+
Rotary	310	40R+
Sound wave actuated	367	140+
Superimposed different currents	307	1+
Supplying different circuits	307	18+
Synchronous inductive	310	171
System of control	322	
Systems	322	
Temperature measuring plural	374	110+
Thermoelectric	136	200+
Voltage controlled	322	
Fumigators sterilizing	422	305+
Gas	48	61+
Power calculation for	364	492
Refrigerator	62	
Sinusoidal		
Electronic tube type oscillator	331	
Generator structure	310	
Generator system	322	
Steam	122	
Steam combined with stove	126	5
Vapor pressure thermostat	374	201+
Genetic		
Expression	935	33*
Transfection	935	52*
Transformation	935	52*
Vectors	935	72*
Genetic Engineering	435	172.1+
Apparatus	935	85*
Assay	935	76*
Recombination	435	172.3
X-art collection	935	*
Genetically Engineered Cells	435	240.1+
Genetically Engineered Cells	435	243+
Utilization	435	
X-art collection	935	59*
X-art collection	935	66*
Geneva		
Drive	414	905*
Gear	74	436
Gearing	74	84R
Planetary combined	475	14+
Genital		
Body wear bands combining	128	386
Catheter	604	264+
Curette	606	160
Cutter		
Castrating	606	137
Embryotome	606	126
Urethrotome	606	170+
Dilator with light and electrical application	606	13+
Kinesitherapic appliance	128	31
Male electric applicator	128	794
Obstetric instruments	606	119
Forceps	606	205+
Orthopedic appliance	128	79
Pessary and inserter	128	834+
Receptor		
Body protruding member	604	346+

	Class	Subclass
Catamenial and diaper	604	358+
Obstetric	604	317+
Urethrorrhoeal	604	327+
Sexual restrainer	128	883+
Stallion shield	119	145
Suspensory bandage	128	158+
Tampon		
Inserters	604	11+
Structure	604	358+
X-art collection	604	904*
Vaginal		
Douche (See douche)	604	36+
Electric applicator	128	788
Geodisic Dome	52	81
Design	D25	13
Geographical		
Clock	368	21+
Teaching	434	130+
Geology, Teaching	434	299
Geometrical Instruments (See Type of Instrument)	33	
Blocks educational design	D21	108
Design	D10	61+
Teaching geometry	434	211
Geophysical Prospecting (See Prospecting)		
Acoustic means	181	101
Digital computer with system	367	60+
Exploration in situ	324	323+
Geochemical exploration	436	25
Nuclear	250	253+
Underwater seismometer spreads	367	15+
Vibration transducers	367	140+
Geothermal Power Plant	60	641.2
Geraniol	568	875
Masonry	52	723
German Silver	420	481
Germanium, Misc, Chemical Manufacture	156	DIG. 67
Alloys	420	556
Organic compounds	556	81+
Pyrometallurgy	75	689
Germicides	424	
Germicidal lamp (See ultraviolet; X ray)		
Water softening compositions containing	252	175+
Germination		
Retarding & accelerating compositions	71	65+
Seed germinator	47	61
Germine	546	28
Gerotor Motor	418	61.3
Gettering	437	10+
Getters (See Absorbent)		
Compositions	252	181.1+
Gettering	156	DIG. 66
Processes of use	445	55
Pumps	417	48+
Gib (See Key)		
Slide bearings	384	39
Gill Net	43	10
Gilsonite (See Asphalt; Pitch)	208	22+
Gimbal	248	183
Gimlets	408	230
Gin		
Beverage	426	
Cotton	19	39+
Saw sharpening and gumming	76	32+
Ginger Ale	426	590+
Girder (See Beam)	52	721+
Bridge	14	17
Composite	52	730+
Dissimilar material	52	722+
Design	D25	126+
I-beam	52	729
Making	29	897.35
Mortar bonded	52	433+
Static mold	249	50
Girdles	450	94+
Combined with bra	450	7+
Design	D 2	2
Design bifurcated	D 2	4
Design combined with bra	D 2	3+
For nuclear reactor	376	302
Moderator structure	376	304
Girths		
Harness	54	23
Packing	277	
Gland		
Extracts		

	Class	Subclass
Medicines containing	424	520+
Glare		
Reduction	350	276R+
Automobile screen	296	97.1+
Filters	350	311+
Glass antireflection coating	350	164+
Polarizers	350	399
Rearview mirror	350	277+
Glass	65	
Actinic composition	501	900*
Developing by heat	65	111+
Bead		
Reflector	350	104+
Reflector making	427	163+
Block construction	52	596+
Translucent component	52	306
Blowing		
Charging glass melt furnace	414	165+
Compositions	501	11+
Cutters (See search notes)	83	879+
Drinking glass	D 7	509+
Electric furnaces	373	27+
Etching	156	663
Fiber, coated	428	357+
Glass-to-metal seal	428	432
Metallic	428	636
Gratings optical	350	162.11+
Grinding processes	51	283R+
Handling cylinders of	414	24
Household glassware	D 7	509+
Laminated and safety	428	426+
Wire glass	428	38+
Manufacturing	65	
Filament or fiber making	65	1+
Process	65	17+
Processes of uniting glass	65	36+
With electric lamp making	445	22
With metal founding	164	
With metal working	29	
Ornamenting		
By abrasive blasting method	51	317+
By grinding	51	
Photochromic	501	13
Plate		
Grinding and polishing	51	
Pot furnaces	432	156+
Pots	432	262+
Product manufacturing computerized	364	473
Scribers	33	18.1+
Stand, rack or tray for	D 7	616
Stemware	D 7	524+
Structural	428	268
Wire mesh	428	38
Treating hard glass	65	
Apparatus	65	348+
Process	65	111+
Window	428	38
Glasses (See Eye Glasses)		
Eye and spectacles	351	41+
Design	D16	102+
Metal frame making	29	20
Optics	350	481
Ground for cameras	354	161
Sight for liquid level gauge	73	323+
Glassware	D 7	509+
Glaucine, in Drug	514	284
Glaze		
Compositions	501	14+
Glaziers Points	411	477+
Setters	227	
Glazing		
Earthenware	427	376.2
Firing	264	60+
Fruits and vegetables	426	289+
Glider		
Aerial toy	446	61+
Aircraft	244	16
Furniture design	D 6	344+
Porch swing	297	282
Hammock couch type	5	124+
Glides	16	42R+
Furniture design	D 8	374
Globe		
Clock	368	23
Design	D10	10
Lamp	362	809*
Design	D26	118+
Lantern	362	363
Design	D26	37
Operators for	362	174+
Manipulating pole	81	53.11+

	Class	Subclass
Operator tube lantern	362	174+
Railway amusement	104	68
Teaching	434	131+
Terrestrial globe	362	809*
Globin	530	385
Globulins	530	386
Glove	2	159+
Baseball	2	19
Boxing	2	18
Buttoners	24	40
Design	D 2	643
Design	D 2	617+
Fingerless	D 2	610+
Forms	223	78+
Knitted	66	174
Circular machine	66	45
Straight machine	66	65
Sewing machine	112	16+
Glow		
Discharge, chemical apparatus	422	907*
Discharge diode	313	567+
Transfer counter	377	103+
Glower (See Electric; Electrode; Filament; Lamp)		
Glucamines	564	507
In drug	514	669
Gluconic Acid	562	587
Glucosamines	536	55.2
Glucose	127	30
Making	127	36+
Tests for	436	95*
Glucosides	536	4.1+
Glue (See Adhesive)		
Compositions	106	125+
Containing	106	125+
Gluing	156	
Applicator	D19	70+
Clamps	269	
Glue heating pots	126	284
Press	100	
Using particular adhesive	156	325+
Glutamates	562	573
Foods ctg	426	656
Preparation by protein hydrolysis	562	516
Glutamic Acid	562	573
Glutamine	562	563
Glutathione	530	332
Glyceric Acid	562	587
Glycerides (See Fats)	260	398+
Rosin	530	216+
Glycerine	568	852+
Fermentative production	435	159+
Glycerol (See Glycerine)		
Glycidol	549	554+
Glycine	562	575
Glycocholic Acid	552	550
Glycols	568	852+
Ethers	568	672+
Oleate	260	410.6
Silicate	556	482+
Stearate	260	410.6
Glycopeptides	530	322
Glycoproteins	530	395
Glycosides	536	4.1+
Cardiac	536	5+
Glyoxal	568	494
Treatment of textiles, polyaldehyde combined	8	DIG. 17
Glyoxaline	548	300+
Glyptal T M (See Alkyd Resins)		
Good	231	2.1
Design animal	D30	156+
Goal, Game Element	D21	48+
Goblet	D 7	524
Goddard Patent Collection	60	915*
Godet		
Heated by induction	219	10.492
Strand feeding wheel	226	168+
Goggles	2	426+
Design	D16	102+
Combined with head covering	D 2	509+
Gold		
Alloys	420	507
Beating	72	420+
Carbon compounds	556	110+
Electrolysis		
Coating	204	47.5+
Synthesis	204	109+
Foil	428	606
Hydrometallurgy	75	711+
Laminate	428	457+

	Class	Subclass
Metal-to-metal	428	672
With glass	428	433+
Misc. chemical manufacture	156	DIG. 101
Plate	428	672
Pyrometallurgy	75	637
Golf		
Accessory	D21	234
Bag	206	315.3
Bag cart	D34	15
Holder	248	96
On wheeled carrier	280	47.26
Digest	280	DIG. 6
Ball	273	62
Design	D21	204+
Making	156	146
Means to print on balls	101	DIG. 4
Cart	224	274
Cart motorized	280	DIG. 5
Club	273	77R+
Design	D21	214
Rack or support for	211	70.2
Woodworking machines for making	144	2XA
Courses	273	32R
Miniature	273	176R
Covers, flaccid	150	160
Game board	273	245
Pocket or capture feature	D21	18
Simulation	D21	27
Surface projectile game	D21	11
Gauge	33	508
Glove	2	161R
Design	D29	22
Handles or grips	273	81R
Design	D 8	DIG. 6
Holes	273	32R
Putting	273	34R
Powered passenger cart	D12	16
Practice devices	273	35R+
Score registers	273	87R
Shoes	36	127
Simulated game	273	87R
Targets	273	181R
Miniature golf	273	176R
Putting	273	34R
With ball return	273	182R
Teaching devices	434	252
Tees	273	33+
Carriers	224	191+
Design	D21	234
Device for setting	273	32.5
Gondola		
Railway		
Drop bottom	105	244+
Freight	105	406.1+
Inclined bottom	105	256
Side door	105	258+
Gong (See Bell)		
Goniometer		
Crystal testing	356	31
Light ray type		
Horizontal angle	33	285
Vertical and horizontal angle	33	281
Vertical angle	33	282+
Photoelectric		
Measure of light scattered by particles	356	340
With light conductor	250	227.3
Radio	342	428+
Goniometer device	342	441
Straight edge type	33	403+
Goniophotometer (See Goniometer)		
Governor (See Brake; Control; Photoelectric Regulator)		
Elevator	187	68+
Fire escape	182	191+
Gearing control	74	336.5
Reversing means	74	404.5
Internal combustion engine	123	319+
Pump liquid	417	279+
Speed recorder	346	73
Speed responsive	73	488+
Switch electric	200	80R
Valve	137	47+
Vehicle motion and direction indication	116	38
Wind motor		
Impeller deflection control	416	9
Impeller speed control	416	44
Rotary wheel deflection	416	9+

	Class	Subclass
Gr-n Rubber (See Synthetic Resin or Natural Rubber)		
Gr-s Rubber (See also Synthetic Resin or Natural Rubber)	526	340
Grab (See Grapple)		
Cable traversing hoist		
Load dumping draft rope	212	81
Load suspension	212	84
Clam shell buckets	37	183R+
Dredges	37	55
Furnace charging	414	186
Hoistline grab hooks	294	82.1+
Irons railway rolling stock	105	461
Magnetic	294	65.5
Material handling elevator	414	618+
Magnet and	414	606
Orange peel buckets	37	182
Rotary crane	212	243
Self propelled traversing hoist electric	212	127
Guide bar	212	129
Vertically swinging load support	414	680+
Graders and Grading (See Separator)	209	
Design	D15	23+
Grinding abradant supplier	51	264
Railway	37	104+
Road	37	108R+
Scraper type	172	
Gradometers	33	365+
Graduates	73	426+
Graft Copolymer (See Synthetic Resin or Natural Rubber)	525	50+
Grafting	47	6+
Textile fiber grafting	8	DIG. 18
Grain		
Adjusters	56	466+
Car		
Doors	49	404+
Temporary closures	160	370.1
Casse-grain type antenna	343	781CA
Comminuting processes	241	6+
Cradles	56	324+
Drill		
Chutes	193	9
Hoppers	222	
Planters	111	
Drying & gas or vapor contact with solids	34	
Feeder or dispenser	222	
Automatic control of flow	99	488
Food		
Apparatus	99	467
Design	D 1	
Preservation	426	321+
Processes and products	426	549+
Harvesters	56	
Hulling	99	600
By applying fluid	99	518
Land vehicle tank	296	15
Paper grain, digest	229	DIG. 5
Pattern and size		
Grain boundaries in metal stock	148	33.2
Metallurgical linings having specific grain size	266	284
Optical grain sensitizing in photography	430	571
Paper grain digest	229	DIG. 5
Stock, angular grain	428	105
Stock, artifical wood or leather surface	428	151
Stock, parallel grain	428	114
Superimposed grain as color photographic technique	430	365
Wood grain	428	106
Scourers		
Apparatus	99	600+
Processes	426	483+
Scouring by brushing	15	3.1+
Shelled-grain catchers	56	207
Standing-grain gatherers	56	219+
Storage units	D25	14
Bin heaters	D23	334
Ventilation	98	52+
Threshing	460	
Flying grain arrestors	460	59
Recleaners	460	903
Savers	460	905
Separators	460	79+
Wheels and casters	56	322
Wood grain, stock material	428	106
Artificial woodsurface	428	151

	Class	Subclass
Spool holders combined	242	140
Textile spinning	57	352+
Textile spinning centrifugal pot	57	76+
Textile spinning rail or support	57	136+
Textile spinning receiving twist flier	57	71
Textile strand feeding	226	196
Textile warp	28	172+
Thread stripping guide	28	222+
Winding and reeling	242	157R+
Camera		
Focusing	354	191
Magazine type quarter turned plate	354	183
Sliding plate magazine type	354	180
Closure		
Automobile glass	49	404+
Door and window frames	49	404+
Hanging or drape type	160	330+
Parallelogram type	160	159
Plural panel type	160	181+
Roll panel type	160	238+
Sliding car door	49	404+
Sliding door or panel	16	87R+
Venetian blind	160	172
Compartmented letter box	232	26
Conveyer belt	198	837+
Automatic shew	198	837
Load retainer	198	836.1+
Conveyers	193	
Coopers stave jointing saw	147	34
Copyholder with static	226	196+
Cutting tool	83	821+
Deflector for fluid motor	415	208.1+
Derailment preventing	104	245+
Dispenser with material guide	222	
Movable guide type	222	526+
Drawer	384	20+
Ball bearing	384	18
Cabinet combined	312	350
Combined with drawer	312	330.1+
Roller bearing	384	19
Earth boring tool combined	175	408
Electric arc furnace electrode	373	
Arc furnace devices	373	60+
Electrode support	373	94+
With electrode holder or guide guide	373	69
Elevator	187	95+
Fluid or fluent material (See distribution, fluid; feeder; nozzle; pipe)		
Acoustical mufflers	181	233+
Chutes skids guides ways	193	
Conveyer fluid type	406	191+
Metal founding furnace or crucible	266	275+
Nozzle, spout, pouring devices	222	566+
Molten metal	222	591+
Receptacle filling	141	365+
Receptacle filling	141	391
Water control	405	80+
Forging hammer or press head	72	428
Foundry flask	164	385+
Gate hanger		
Bracket	16	90
Roller	16	91
Wheel mounts	16	106
Hair cutting and dressing	132	213+
Hand (See hand grips guards guides rests and straps)		
Hat fastener	132	65.1+
Line on fishing rod	43	24
Liquid level or depth gauge	73	320
Lock or latch part		
Bolt casing	292	337
Combination setting	70	331
Internal key guides	70	453
Key insertion	70	454
Machine or instrument		
Bottle cleaner	15	66
Cultivator lister furrow	172	26
Cultivator tree or stake guided	172	24
Drill	408	72R+
Drill planter check correcting	111	16+
Peripheral face abrading machine	51	102
Radial face abrading machine	51	128
Radial face opposed abrading machine	51	116
Rotary cylinder abrading tool	51	208
Rotary disk abrading tool	51	210

	Class	Subclass
Shoe blacking polisher	15	266
Sole and heel edge trimmer	12	90
Stationary abrading tool	51	214
Manual cutting tool (See machine or instrument guide)	30	286+
Coring or gage paring knives	30	282
Ice pick	30	164.7
Miter box	83	761+
Nippers	30	179
Razor and sharpener	30	37
Shears	30	233
Tool sharpener	76	88+
Metal can head applying	413	26+
Metal coiling	72	142+
Metal rolling	72	250+
Metal turning template	82	11.1
Miter box	83	761+
Monorail system	104	118+
Rolling stock	105	141+
Paper calender doctor combined	100	173+
Pattern		
Gear cutting	409	2
Wood sawing	83	747
Wood shaping	144	137+
Wood turning	142	
Pencil sharpening implement	30	451+
Pencil sharpening machine	144	28.1+
Photography film developing	354	339
Piano key	84	436
Piano key fulcrum combined	84	434
Plain bearings	384	26+
Rail clamp lateral adjusting	238	347
Rail splice bar	238	246+
Railway automatic switch	246	348
Railway rolling stock truck element		
Axle box	105	218.1+
Bogie bolster	105	207
Swinging bolster column mounted	105	191
Saw making		
Band saw type	76	74
Filing	76	36
Screw driver combined	81	451+
Sewing machine		
Glove	112	20
Hat	112	15
Leather channel	112	50
Leather edge and crease	112	51
Leather welt	112	52
Needle	112	227
Work	112	136+
Shaft bearing	384	29
Stage scenery	272	23
Tap metal screw thread type	408	72R+
Tool sharpener	76	88+
Traffic (See traffic lights and guides)		
Type casting	199	3+
Typewriter part		
Keyboard hand guide	400	715+
Ribbon	400	248+
Type bar	400	456+
Upending bedstead shifting fulcrum	5	137
Violin bow	84	283
Washing machine wringer	68	264+
Wave electromagnetic	333	239+
Weaving pattern	139	332
Web or strand	226	
Weigher loading	177	161+
Wheelwright spoke setter type	157	4
Winding and reeling fabrics	242	76
Wood saw	83	821+
Woodworking		
Beveling slicer	144	169
Curved work shaping	144	152
Tool	408	72R+
Work	144	253R
Wrench element		
Pivoted outer jaw traveling fulcrum	81	109
Writing or scribing (See masking; stencils and stenciling)		
Educational	434	164
Scribers	33	18.1+
Straightedge	33	403+
Guided Missile		
Control	244	75R+
By radio (See type of servomotor system)		
Design	D12	16.1
Manned	244	75R+
Radar not within missile	342	62
Space ship	244	158R+

	Class	Subclass
Telemetering	340	870.1+
Torpedo type	114	20.1+
Tracking by antenna	342	75+
Aiming a gun	342	67
Unmanned	244	3.1+
Guillotine (See Execution Devices)	83	613+
Shears	83	613+
Guitar	84	267
Bridges	84	298
With tailpiece	84	299
Electric amplifier	84	723+
Gum (See Synthetic Resin or Natural Rubber)	520	1+
Chewing gum package	206	800*
Gum ball vending machines	D20	7
Gums and derivatives	536	114
Nitrogen containing	536	52
Resins (See resins)		
Rubber	520	
Tablet form		
Design	D 1	
Gumming		
Envelope manufacture	493	220+
Saw aperture	83	850
Saw making	76	30+
Gun (See Firearm)	42	
Airgun	124	56+
Antiaircraft	89	
Automatic	89	125+
Blowback	89	194+
Firing devices	89	132+
Gas piston type	89	191.1+
Movable barrel	89	160+
Movable chamber	89	155+
Band type clasps	24	2.5
Barrel		
Firearms	42	76.1+
Materials or coatings	42	76.2
Ordnance	89	14.5+
Bayonet combined	42	86
Billy club	42	1.16
Blowgun	124	62
Bore inspection	356	241
Breakdown type	42	40
Cane gun	42	52
Cattle slaughter type	42	1.12
Control calculators	235	400+
Gun training mechanism	89	41.1+
Motor operated	89	41.2+
Cotton	536	35+
Composition containing	149	94+
Over 10%	149	96+
Design	D22	100+
Grease	D 8	14.1
Pistol	D22	104+
Racks	D 6	552+
Sights	D22	109
Toy	D21	146+
Dummy	42	106
Ejectors	42	25
Electrically operated		
Firearms	42	84
Lighting devices	362	110+
Ordnance	89	135
Electron	313	441+
Extractors	42	16+
Firing mechanisms	42	69.1
Revolver	42	65
Upward tilting breech	42	41+
Flare	42	1.15
Fluid pressure adapter	124	58
Foob, ie fire out of battery	89	42.3
Gatling type	89	12
Grenade launchers	42	105
Gun engaging means	102	483+
Handles	89	1.42
Heaters	89	1.12
Howitzer	89	
Implement combined	42	90+
Indicators	42	1.1
K gun	89	1.1
Knife combined	42	53
Loading	89	45
Lubricating or caulking type	D 8	14.1
Machine gun	89	9+
Toy	D21	146+
Magazine	42	87+
Making	29	1.1+
Mechanical	124	
Mount	89	37.1+
Training mechanism	89	41.1+

	Class	Subclass
Mounted	89	
Movable chambers		
Firearms	42	39.5
Ordnance	89	155
Multiple barrel	89	1.41
Nonrecoil	89	1.7
Pen and knife	42	1.9
Port ship	114	173+
Stopper	114	175
Portable	42	
Powder		
Ammunition loading with	86	
Bags	102	282
Engine starters	123	183
Engines	123	24R
Forms	102	283+
Racks	211	64
Rapid fire	124	72+
Recoil operated	89	162
Recoilless	89	1.7
Rests	42	94
Revolver	42	59+
Safety mechanism	42	70.1+
Automatic guns	89	137+
Revolvers	42	66
Semiautomatic	89	4.5+
Shields	89	36.1+
Deflected ray tube	356	253+
Shotguns	42	
Sidearms	D22	
Sights	33	233+
Design	D22	109
Design, telescopic	D16	132
Optical system	356	247+
Stocks	42	71.1+
Teargas	42	1.8
Telescopic gunsight	D16	132
Toy simulating	42	54+
Ammunition	102	281
Machine gun or projector	124	29
Non-detonating	446	473
With sound	446	405+
Training in gunnery	434	16+
Trigger protectors	42	70.7
Underwater	42	1.14
Walking cane combined	4	515+
Water gun	124	56+
Toy	D21	146+
Water pistol	222	79
Well tubing perforator	175	2
Y gun	89	1.1
Gussets		
Garment	2	275
Gut or Gut Treatment	8	94.11
Splitter	83	932*
Guttapercha	525	331.9+
Gutter	405	119+
Eaves trough	52	11+
Electric conductor underground		
structure	174	39
Road and pavement	404	2+
Support design	D 8	363
Guy	52	146+
Bed spring and frame	52	272
Gymnastic Devices	272	109+
Coin controlled apparatus	194	
Gypsum	423	554
Calcining	106	722+
Coating or plastic compositions		
containing	106	772+
Alkali metal silicate	106	611
Gyrating		
Reciprocating sifter		
Actuating means	209	366+
Horizontal and vertical shake	209	326
Horizontal shake	209	332
Gyratory Crusher		
Jaw crushers rotary component	241	207+
Parallel flow through plural zones	241	140
Series flow through plural zones	241	156
Gyro Stabilized		
Article support	248	183
Furniture for ships	114	119
Gyroplane (See Aircraft)	244	17.11+
Gyroscope	74	5R+
Aerial camera combined	354	70
Aircraft control	244	79
Direction indicator	33	318+
Gimbals	248	182
Gun sight combined	33	236
Gyroscopic compass	33	324+

	Class	Subclass
Telemetric system combined	340	870.7+
Gyroscopic light valve, photoelectric	250	231.12
Monorail rolling stock	105	141+
Suspended	105	150+
Rotors	74	572
Rotors and flywheels	74	
Ship antiroll	114	112
Ship stabilizer	114	122
Ship steering	114	144R
Speed responsive devices	73	504
Torpedo	114	24
Torpedo steering	114	24
Toy	446	233+
Transmission	74	64
H Acid	562	71
Habitat, Submarine	114	314
Hack Saw	30	507+
Combined	7	149+
Design	D 8	96
Hanging	83	783+
Hackling		
Combing	19	115R+
Decorticating	19	5R+
Hacks Tree	30	121
Haemocytometer	356	39
Testing lenses	356	124+
Hair		
Artificial furs	428	85+
Artificial structure	132	201
Beauty parlor equipment	D28	10+
Brush	15	160+
Carried hat fasteners	132	60
Clippers	D28	52+
Coating compositions	106	155+
Curlers		
Curling iron	D28	38
Electrically heated	219	222+
Fluid fuel heated	126	408+
Cutters	30	
Design	D 8	57
Design clippers	D28	52+
For inside ear or nose	30	29.5
Hair planers	30	30+
Drying on head		
Apparatus	34	96+
Processes	34	3
Supports for	34	101
Dye applicator	D28	7+
Dyeing and dyes	8	405+
Fasteners	132	273+
Design	D28	39+
Fertilizer from	71	18
Hairpiece	D28	92+
Inserters	606	187
Jeweled fastener	63	2
Net	132	274
Pins (See hairpins)	132	276+
Planers	30	30+
Removing (See notes)	30	32
Burial preparation	27	21.1+
Butchering	17	1D
Coarse or water hair from fur	69	24+
Cutters for inside ear or nose	30	29.5
Depilating untanned skins	8	94.16
Depilatories	8	94.16+
Electric needle	606	44
Electric needle supports	606	44
Fiber liberating	19	2
Fur treatment	69	24+
Process	17	47
Razors	30	32+
Surgical instruments	606	222+
Tweezers	606	133
Shampooing apparatus	4	515+
Shearing, fur finishing	26	15R+
Thinning shears	30	195
Springs	368	175+
Strand making		
Covering by spinning etc	57	4
Spinning etc	57	28+
Synthetic resin or natural rubber		
containing compositions	524	12
Textile spinning etc	57	29
Thinners	30	195
Design	D28	52+
Toilet preparations	424	70+
Treating process	132	200+
Tufting hair in doll or wig	112	80.2
Waving	132	200+
Hairpin	132	276+
Design	D28	39+

	Class	Subclass
Dispenser	132	332
Hat fastener cord or loop and	132	61
Making	140	87
Packaging	227	25
Half Belts	2	309
Half Wave		
Gas rectifier	313	567+
With hot cathode	313	629
High voltage rectifier	313	317+
With emissive cathode	313	310
With thermionic cathode	313	310
Rectifier system	363	13+
Circuit interrupter for	200	
Dynamoelectric machine	310	10+
Electronic tube for	313	317+
Gas tube type	363	114+
Power packs	307	150
Unidirectional impedance for	357	
Vacuum tube type	363	114+
With filter	363	39+
With voltage regulator	363	84+
Halftone		
Blanks and processes printing	101	401.1
Etching	156	654+
X-art	156	905*
Photographic process	430	396
Photographic screens	350	322+
Chemically defined	430	6+
Printing plates	101	395
Halides (See Material Halogenated)		
Hydrocarbon	570	101+
As azeotropes	203	67
Electromagnetic wave synthesis	204	157.15+
Electrostatic field or electrical		
discharge synthesis	204	169
Metal	423	462+
Electrolytic synthesis	204	94
Nitroaromatic	568	927+
Nonmetal inorganic	423	462
Organic acid	562	800+
Rubber hydrohalide	525	332.3
Hall Effect Means in an Amplifier	330	6
Haloamines	564	114+
Acyclic	564	118+
Hydroxy or ether containing	564	119
Plural difluoramine groups	564	121+
Unsaturated	564	120
Alicyclic	564	117
Amidines	564	116
Halogen Compounds (See Material		
Halogenated)		
Halogenated Carboxylic Acid Esters	560	1+
Acyclic acid esters	560	226+
Of phenals	560	145
Acyclic amino acid esters	560	172
Acyclic carbamic acid esters	560	161
Acyclic oxy acid esters	560	184
Acyclic polycarboxylic acid esters	560	192
Acyclic unsaturated acid esters	560	219
Alicyclic	560	125
Aromatic amino acid esters	560	47
Aromatic carbamic acid esters	560	30
Aromatic polycarboxylic acid esters	560	23
Oxybenzoic acid esters	560	65
Phenoxyacetic acid esters	560	62+
Halohydrin	568	841+
Halothiocarbonate Esters	558	249
Halowax	570	181
Halter		
Brassiere type garment	450	1+
Feed bags supported on	119	66
Harness	54	24+
Design	D30	134+
Poke with bar and	119	141
Snap releasers	119	114
Hamburger		
Cookers	99	422
Grinders	241	
Molding and shaping (See		
briquetting meat)		
Hames	54	25+
Collar combined	54	18R
Design	D30	137
Traces and connectors	54	30+
Tugs	54	32+
Hammer	81	20+
Automobile fender straightening	72	705
Burglar alarm	116	88+
Claw	254	26R
Combined with additional tools	7	143+
Design	D 8	75

	Class	Subclass
Manicuring tool	132	75
Razor	30	86
Retractable	16	115
Sporting goods, misc	D21	
Switch	200	329+
Sword	D 8	DIG. 5
Tool and implement	81	489+
Cleaning tool	15	143R+
Curry comb	119	94
Dispenser combined	222	191
Drive control	173	170
Drive fluid passage	173	168+
Flexible abradant	51	392+
Forks and shovels	294	57+
Golf club	273	81R+
Light support	362	399
Manipulating for drive or hammer	173	
Multiple tool	7	167+
Printing stamp	101	405+
Razor	30	85+
Rigid abradant	51	205R
Sadiron	38	90+
Saw	30	166.3+
Shears detachable blade and handle	30	260
Wrench	81	177.1+
Valve		
Detachable actuator	251	291+
Disabling means	251	89+
Stops	251	284+
Vehicle seat back grips	297	183
With indicia	40	642
Handling (See Conveyor)		
Castings with casting device	164	269
Completed work, of sewing machine	112	121.29
Earth boring drilling fluid or cuttings	175	207+
Fluid handling		
Baths, closets, sinks, & spittoons	4	
Pipes or tubular conduits	138	
Valves & valve actuation	251	
Ventilation	98	
Handicapped person	414	921*
Handler-type toys	414	915*
Handlers using parallel links	414	917*
Handlers with spring devices	414	913*
Implements		
Bakers peel	294	49
Combined or convertible	414	912*
Hand and hoist line implements	294	
Hand tools	81	
Sorting manually	209	614
Mechanism		
Coin	453	
Dryer for hollow articles with	34	105+
Dryer rotary drum type with	34	108+
Dryer rotary drum with	34	108+
Dryer shelf or tray	34	194
Dryer with for dried material	34	236
Fish net	43	8
Furnace charging	414	147+
Target	273	406
Trap shooting targets	273	406
Metal forging work	72	419+
Nuclear reactor components	376	146
Nuclear reactor components	376	260
Nuclear reactor components	376	260
Nuclear reactor components	376	308
Charging or discharging	414	146
Photographic fluid treating	354	340
Pipe or cable	414	745.1+
Submerged	405	158+
Rolling mill work shifter	72	200
Shiftable handler, transmission line for	414	918*
Solid material handling		
Chutes, skids, guides, & ways	193	
Delivery to or from moving vehicle	258	
Elevators	187	
Fluid current conveyers	406	
Food	99	
Furnace ash	110	165R+
Hand & hoist-line implements	294	
Hoists	414	592+
Marine loading or unloading	414	137.1+
Material or article	414	
Package & article carriers	224	
Pneumatic dispatch	406	
Power-driven conveyers	198	
Stacking	414	788.1+
Store service	186	

	Class	Subclass
Traversing hoists	212	
Unstacking	414	795.4+
Toy, material handling	446	424+
Traversing hoists	212	
Vehicles with overhead guard for operator	414	914*
Handpiece (See Hand, Manipulated)		
Dental	433	114+
Handrail	D25	119+
Support	D 8	363
Handsaw	30	166.3+
Combined with diverse tool	7	148+
Handwheel	74	552+
Hanger (See Suspension Devices)	224	232+
Bed bottom		
Frame	5	207+
Slat	5	238
Broom type	248	110+
Butchering	17	44.2+
Eaves trough	248	48.1
Garment	223	85+
Attached to garment	2	271
Design	D 6	315+
Making	140	81.5
Gate	16	86.1+
Illuminating light support	362	432+
Movable component	52	64+
Panel movable	16	87R+
Pipe	248	58+
Rack	211	
Railway		
Cable rail	104	115+
Traction cable	104	182
Trolley	104	111
Rod for sheet or leaf handling	294	5.5
Scabbard	224	232
Shaft	384	215+
Sifter support	209	415
Slender article type, including broom	211	60.1
Suspended roof	52	83
Tent	135	119
Train vestibule face plate rod	105	22
Hanging		
Cable traversing hoist	212	133
Canopies	135	905*
Grindstone	51	168+
Paper	156	574+
Design	D 5	
Plant receptacles	47	67+
Ships rudder	114	165
Sign exhibiting	40	617
Wood saw	83	469+
Hanks		
Finishing	28	287+
Sails and rigging	114	114
Spinning twisting and twining	57	25+
Winding	242	53
Harbor		
Channels	405	84+
Floating	114	258
Wharves	405	284+
Hard-hat	2	411+
Hardening		
Aging	148	158+
Iron	148	142
Iron with working	148	12.3
Cast iron	148	141
Electric hardening		
Apparatus	219	50+
Inductive apparatus	219	6.5+
Processes	148	13+
Zone hardening of ferrous metal	148	150
Electrophotographic fixation	355	282+
Fat	260	409
Glass	65	111+
Quenching iron	148	143+
Localized heat treated stock	148	902+
Localized treatment	148	145+
Surface treatment	148	152
Working combined	148	12.4
Steel	148	
Hardness Test	73	78+
Electric space discharge devices vacuum	324	413
Hardware Miscellaneous	16	
Brushings or lining thimbles	16	2+
Carpet fasteners	16	4+
Casters	16	18R+
Checks & closers	16	49+
Design of	D 8	
Ferrules, rings & thimbles	16	108+

	Class	Subclass
Gate hangers	16	86.1+
Handles	16	110R+
Hinges	16	221+
Nesting hinge leaves	16	DIG. 29
Sash balances	16	193+
Sash weights	16	216+
Sash-cord fasteners	16	202+
Sash-cord guides	16	210+
Tracks, travelers, panel hangers	16	87R+
Window-bead fasteners	16	220+
Harmidine	546	86
Harmonic		
Electric		
Filtering	333	175+
Generation by oscillators	331	53
Harmonic or reed telegraphs	178	47+
Intensifier for loudspeaker or earphone	381	161
Party line ringing	379	179
Semiautomatic telephone ringing	379	252
Telephone ringing	379	372+
Wave analysis	324	77R
Music		
Tunings or arrangements of string	84	312R
Musical		
Electric organ selective control of tone partials	84	659+
Electric organ with tuned generator control	84	622+
Piano harmonic dampers	84	234
Relays	335	94
Circuits	361	182+
Selective system	340	825.39
Relays	335	101
Structure	335	243+
Harmonicas	84	377+
Design	D17	12
Harmonica digest	84	DIG. 14
Harmoniums (See Organ, Reed)	84	351+
Harness (See Body, Harness)	54	
Animal stocks	119	101
Body		
Fire escape	182	3+
Land vehicle occupant	280	290
Parachute	244	151R
Bridles	54	6R+
Buckles	24	164+
Design	D30	139
Checking devices	54	70
Collars	54	19R+
Design	D30	134+
Feed bags supported by	119	67
Hames	54	25+
Hand package carrier	294	157+
Loom	139	82+
Drawing in warp	28	203+
Hand pushed	139	30+
Motions for dobby looms	139	66R+
Pads	54	65+
Saddles	54	37+
Stirrups	54	47+
Trunk	190	27
Protecting	190	26
Harp		
Aeolian ie wind driven	84	330
Autoharp or zither	84	286+
Jews harp	84	375
Toy	446	397+
Musical stringed instrument	84	264+
Design	D17	16
Piano	84	258
Support for lamp	362	417
Harpoons		
Bomb lances	102	371
Fishing	43	6
Gun	89	1.7
Mechanical	124	
Handling implement	294	125+
Line carrying shells	102	504
Harrow	172	
Cultivator combined	172	133+
Design	D15	10+
Disc sharpener	76	85
Abrading attachment	51	246+
Plow combined	172	133+
Roller combined	172	173
Scraper or drag combined	172	197
Harvester		
Agriculture	56	
Bundle discharging carrier	56	474+
Carrier	56	473.5

	Class	Subclass
Compressing and driving	56	432+
Guards	56	DIG. 24
Motorized	56	10.1+
Rake and rakers		
Bundling and	56	341+
Hand rakes	56	400.1+
Horse rake	56	375+
Loading and	56	344+
Tedders and	56	365+
Shocker	56	401+
Stalk choppers	56	500+
Catching and	56	194+
Conveying and	56	153+
Conveying binding, and	56	131+
Raking and	56	193+
Windrowing	56	192+
Design	D15	26+
Knife holder	51	222+
Knife sharpener	51	36
Marine	56	8+
Mining	299	
Potato or beet digger	171	
Tree felling	144	34R
Hasp		
Closure fastener	292	281+
Lock	70	2+
Hassocks	297	462
Design	D 6	349+
Hat		
Attached fasteners	132	57.1+
Band	2	179
Box	206	8+
Brush	D 4	
Design	D 2	509+
Eye shields attached to	2	10
Forms	223	24+
Labels and tags	40	329
Making apparatus	223	7+
Making apparatus	D15	122+
Ornaments		
Design	D 2	263
Pins	132	57.1+
Design	D11	209
Stickpin	D11	47
Protective packing	206	9+
Racks	211	30+
Design	D 6	315+
Receptacles	206	8+
Safety helmet	D29	12+
Sewing	112	12+
Methods and seams	112	263.1
Shower cap	D 2	510+
Try on linings	2	63
Wires making	140	77
Hatch		
Building	52	20
Closure operators for	49	324+
Elevator	187	62+
Fasteners	292	256.5
Freight car	105	377
Ship	114	201R+
Hatchets	30	308.1+
Carrier	224	234
Design	D 8	76
Making		
Dies	72	470+
Processes and blanks	76	103
Hatpin	132	57.1+
Non ornamental	D11	207
Protective packing	206	9+
Stickpin	D11	47
Hauling (See Pull)		
Cables drum type	254	266+
Hawks, Masonry and Concrete	294	3.5
Hawse Pipes	114	179+
Hawsers (See Rope)		
Hay		
Handling		
Bale accumulator, vehicle carried	414	111
Curing and preserving	426	321+
Distributors	414	25
Forks	294	107+
Grapples expanding pivoted	294	98
Grapples fixed and movable jaw	294	105
Grapples pivoted jaws	294	107+
Harpoon	294	126+
Hay retainer for vertically swinging fork	414	721
Hoist line fork	294	120+
Load binders	280	180
Pitch forks	294	55.5+

	Class	Subclass
Round bale	414	24.5
Stack shapers	414	132
Hayracks		
Brakes for	188	14
Self-loading movable car	414	522
Vehicle body	296	6+
Convertible box and	296	11+
Pelleters	100	
Hcg	436	818*
Head		
Brake		
Beam adjustable head	188	219.6+
Beam combined	188	219.1+
Beam fixed head	188	222.1+
Shoe fastener	188	236
Shoe fastener interlocking	188	242+
Coupling		
Animal draft D type	278	72+
Animal draft L type	278	67
Animal draft T type	278	68+
Animal draft whiffle tree	278	105+
Railway draft cushioned	213	18+
Railway draft link or bar	213	182+
Rod to base plate or head	403	187+
Through intermediate member	403	187+
Tool handle fastening plate	403	
Tool handle fastening split	403	302
Covering with light	362	105+
Drum musical	84	414
Gear telephone	379	430
Implement and tool		
Brush	15	171+
Hand fork and shovel pivoted and adjustable	294	53.5
Mop	15	228+
Plow rotary chopper	172	518+
Razor handle	30	89
Internal combustion engine cooling	123	41.72+
Multiple for bolt		
Hooked end	292	116+
Sliding	292	156+
Sliding and rotary	292	59
Sliding and swinging	292	68
Spring arm	292	91
Swinging	292	213+
Swinging and hooked end	292	56
Nail, spike or tack	411	439+
Design	D 8	388+
Nozzle		
Sprinkler, fire	169	37+
Prop	248	357
Puller carcass	17	1R
Rail		
Detachable	238	143+
Joint	238	221+
Reversible	238	133
Rudder	114	169
Screw machines for working on	10	5+
Sewing machine	112	259
Shampooing apparatus	4	515+
Spar ship	114	94
Stock lathe	82	142+
Toy	446	397+
Tripod	248	177+
Trolley electric	191	59+
Valve		
Cooperating seat	251	333+
Cooperating stem	251	84+
Rotary plug	251	309+
Structure	251	356+
Swiveled	251	88
Well casing	166	75.1+
Headboards and Sections		
Bedstead extension	5	183
Bedstead folding	5	178+
Bedstead upending	5	148
Invalid bed	5	67+
Sofa	5	53.1+
Headed Fastenings	411	500+
Making	10	
Header		
Boiler	122	
Harvester	56	
Manufacture	29	890.52
Radiator	165	153
Heading (See Forging)		
Barrel	147	6
Bars by upsetting	72	470+
Bolts	10	11R+
Cartridge cases	29	1.31
Filling combined	141	73+

	Class	Subclass
Filling combined	141	115+
Nails	10	47+
Cut	10	28+
Wire	10	43+
Pins	163	6
Glass headed	65	142+
Rivets	10	11R+
Screws making	10	2+
Sheet metal cans	413	26+
Apparatus for making	413	69+
Soldering apparatus	228	
Headlight		
Dirigible	362	62+
Sealed beam	313	113+
System electric	315	82+
Testing	356	121+
Vehicle	362	61+
Headphones		
Earpieces	381	187
Electrical	381	183+
Supports	381	188+
In or under pillow	381	158
Mechanical	181	
Pillow type	381	158+
Radio	455	344
Support for phone	381	205+
With radio	455	149
Headrest		
Bath tubs	4	578
Chair	297	391+
Chair reclining	297	61
Coffin	27	13
Design	D 6	501
(See also)	D 6	601
Shampooing apparatus	4	523+
Sofa	5	52
Water closet	4	254
Headset	D14	205+
Headwear		
Attached hat fastener	132	57.1+
Decorative	2	171+
Design	D 2	509+
Guard and protector	2	410+
Helmet	2	410+
Protective	2	410+
Receptacle	206	8+
Sports	2	425+
Structure	2	171+
Surgical	128	380
Health		
Computer controlled, monitored	364	413.1+
Lamps	128	395
Heap Still	202	210
Hearing Aid	D24	35
Design	D24	35
Ear trumpets	181	129+
Electrical	381	68
Non electrical	181	129+
Dentiphone	181	127
Ear trumpet	181	129+
Stethoscope	181	126+
Stethoscopes	181	126+
Hearing Therapy		
Voice reflector	D24	35
Hearses	296	16+
Design	D12	82+
Heart		
Artifical	623	3
Artery	623	1
Valve	623	2
Energized magnet actuator	251	65
Motor operated by motivating mass	60	516
Pivoted, line condition change responsive	137	527
Cardiac assist device	600	16+
Expansible chamber pump	417	472
Heart cam, register	235	144HC
Heart-lung digest	128	DIG. 3
Pacemaker	128	419P+
Shaped packaging	D 9	315
Hearth		
Cooking	126	12+
Electric furnace		
Arc	373	60
Electroslag	373	42
Glass	373	27
Induction	373	138
Resistance	373	109
Fireplace	126	500
Fenders	126	544

	Class	Subclass
Material heating	432	
Thermolytic still	202	102+
Heat, Heater & Heating (See Furnace; Radiator)	432	
Absorption meters	374	39
Accumulators	165	
By change of state	165	104.15
Ovens having	126	273.5
Regenerators	165	4
Vessels having	126	375
Air		
Automatic temperature regulation	236	10+
Cookstove combined	126	6
Distributing system combined	237	
Fireplace	126	500+
Furnaces	126	99R+
Gas separation combined	55	
Heat exchange	165	
Heated ventilating flue	98	45+
Illuminating burner attachments	126	248+
Nuclear reactors	376	383
Regenerative	432	214+
Stoves	126	58+
Animal or plant husbandry		
Brooders and incubators	119	30+
Electrically heated branding irons	219	245+
Fuel heated branding irons	126	402+
Greenhouses	47	17+
Hot beds	47	19
Orchard heaters	126	59.5
Preventing plant frosting	47	2
Seed testers	47	16
Sterilizing gathered soil	47	1.44
Tree covers	47	22
Waterers	119	73
Apparel forms	223	
Automatic temperature regulation	236	
Heating and cooling combined	165	14+
Heating systems	237	2A+
Automobile	237	28+
Design	D23	324+
Bathtub	4	545
Bedsteads	5	284
Bending metal by heating	72	54
Body	604	113+
Treating material	604	291+
Brooder and incubator		
Automatic temperature control	236	2+
Automatically controlled heating systems	237	3+
Brooder	119	31+
Combined	119	30
Heating systems	237	14+
Incubator	119	35+
Buildings	237	
Burner feed	431	
Cabinet combined	312	236
Carburetors	261	127+
Internal combustion engine	123	543
Chemical	126	263
Closets		
Dry closets	4	111.1+
Collectors, solar	4	461+
Combustion	431	
Computerized or monitored in product manufacturing	364	477
Cutting (See 4 note to class 30)	30	
Distillation and cracking	201	
Apparatus	202	
Mineral oils	208	46+
Solid carbonaceous material	201	
Drier	34	
Textile web spreader combined	26	92
Textile web stretcher combined	26	106
Earth	405	131
Earth boring by	175	11+
Earth boring combined	175	17
Electric fuses having	337	182+
Electric railway conduit having	191	27
Electrical	219	
Earth boring by	175	16
Element	338	
Furnaces	373	
Heating gas generator	48	103
Rivet heaters	219	157
Embossing hot die	101	27
Exchange (See radiator)	165	
Agitating with	366	144+
Mortar mixer	366	22+
Cabinet combined	312	236
Combustion products generator	60	730
Distillation apparatus	202	
Distillation processes	201	
Distillation processes	203	
Gas separators having	55	
Gas washers having	261	127+
Gun barrel cooling	89	14.1
Induction heating	219	10.491+
Internal combustion engine	123	41.1+
Mineral oil process	208	347+
Mineral oil stills having	196	134
Nuclear reactor with	376	369
Nuclear reactor with	376	378
Nuclear reactor with	376	391
Nuclear reactor with	376	404
Nuclear reactor with	376	405
Particulate	165	920
Static mold having	249	79+
Temperature control of	236	17+
Tobacco smoking device	131	194+
Fans		
Electric	392	360+
Fire screen or guard		
Fluid screen	110	179
Stove and fireplace	126	544
Fluid fuel burner	431	
Nozzle per se	239	
Fluid handling systems or devices with	137	334+
For glass manufacturing		
Apparatus	65	355
Discharge nozzle	239	128+
Freight car heaters	105	451
Fume		
Generators	422	305+
Furnace, solid or combined fuels	110	
Garments & bodywear (See body, heaters and warmers)		
Boot or shoe	36	2.6
Electrically heated	219	211
Heat exchanger	165	46
Therapeutic body member inclosing	128	402
Therapeutic body wear	128	379+
Gas & liquid contact apparatus	261	DIG. 31
Heaters and condensers	261	DIG. 32
Heaters, spray	261	DIG. 33
Gas manufacture		
Chemical purification	423	210+
Heating and illuminating	48	
Gun	89	1.12
Hair curling straightening or treating		
Drying apparatus	34	96+
Drying processes	34	3
Heated brushes	15	160
Heated combs	132	118
Process	132	211
Heat or cooling, conveyer act	198	952*
Heat pump, reversible	62	324.1+
Automatic	62	160
With supplemental heat source	165	29
Heat storage control	165	18
Heater submerged in liquid	261	DIG. 29
Heating equipment	D23	330+
Heating systems	237	
Humidity regulation	236	
Immersion type	126	367+
Fluid fuel	126	360R
Insulating and screening	52	
Alkali metal silicate compositions	106	600+
Bifurcated garments	2	81
Bottles	215	12.1+
Cement compositions	106	672+
Ceramic compositions	501	94+
Compositions miscellaneous	252	62
Flatiron	38	89
Head protectors	2	7+
Laminated fabric making	156	
Miscellaneous plastic or coating compositions	106	122
Safe doors	109	65
Safe walls	109	78+
Stock material	428	920*
Toaster or broiler	99	401
Internal combustion engine charge	123	543
Leather sewing machine	112	41
Liquid	122	
Automatic temperature control	236	20R+
Automatically controlled heating systems	237	2R+
Concentrators	159	
Cookstove combined	126	38+
Discharge nozzle	239	128+
Distillation	203	
Distributing system combined	237	
Electric	392	441+
Fireplace	126	513+
Heat exchange	165	
Heated street oilers	239	128+
Heating system combined	237	
Hot air furnace combined	126	101
Liquid separator heated	210	175+
Open type	126	344+
Pressure generators	122	
Radiator and boiler combined	237	16+
Range fluid fuel water back	126	53+
Range water back	126	34+
Solar	126	417+
Textile washing apparatus combined	68	15+
Liquid separators or purifiers	210	175+
Dispensers with heater	222	146.2+
Dispensers with heating jacket	222	131
Dispensers with illuminator or burner	222	113
Distillation apparatus containing	202	
Mortar mixers	562	625+
Portable receptacle fillers	53	127
Solid material comminution combined	241	
Lubricators having	184	104.1
Material heating	432	
Material treating (See material treated)		
Coating, baking or drying	427	372.2+
Driers	34	
Electrolyte	204	
Electrolytic coating	204	37.1+
Fertilizer	71	44+
Foods	99	
Ozonizers	422	186.8+
Peat fuel	44	492
Road or pavement material	404	93+
Solids separators	209	
Tobacco	131	290+
Water purifiers	210	177+
Mattresses	5	421+
Measuring and responsive instruments		
Calorimeters	374	31+
Electrothermal switches	337	14+
Heat responsive metal stock	428	616
Hygrostats	73	335+
Thermal batteries	136	200+
Thermometers	374	147
Thermostatic switches	337	298+
Metal		
Agglomerating processes	75	746+
Casting apparatus with	164	338.1
Deforming with heating or cooling	72	
Electric heating	219	50+
Electric heating working and welding	219	50+
Electrolytic coating heating	204	37.1+
Founding	164	
Heat treatment	148	
Metallurgical apparatus	266	
Miscellaneous heating	432	
Pyrometallurgy processes	75	414+
Rolling and heating	72	200
Rolling and heating	72	252
Soldering apparatus	228	
Working with	29	DIG. 21
Working with electric heating	29	DIG. 13
Nozzle combined	239	128+
Pile installation combined	405	234
Pipe	165	104.15
Electrical apparatus combined	174	15.2
Pipe joint	285	41
Power plant combined	60	
Engine heated vehicles	237	12.1+
Processes	432	1+
Producing compositions	252	70
Fuel	44	
Pump	62	324.1+
Register or diffuser	D23	308+
Resolving colloids by	252	346+
Sealing wax applier	401	1+
Searing		
Leather	69	7.5
Seat combined	297	180
Separator, imperforate bowl centrifugal	494	13+

	Class	Subclass
Shields for aircraft	244	117R+
Shields, general	244	158R
Shoes having heater	36	2.6
Single room heater	D23	330+
Sink	165	
Design	D13	179
Ventilation	174	16.3
Snow excavator and melter	37	227+
Solid fuel burners	110	
Special ray generator combined	250	495.1
Stabilized (See synthetic resin or natural rubber, class 523, 524)		
Sterilizing	422	
Fume generators	422	305+
Stoves	126	
Design	D23	342+
Surgical	604	113+
Systems	237	
Electric	219	482+
Temperature & humidity regulation	236	
Tents having	135	92
Textiles or strands		
Heating or drying threads	28	217+
Singeing textiles	26	3+
Web spreader combined	26	92
Web stretching combined	26	92
Tools and instruments		
Apparel forms	223	
Branding iron	126	402+
Brushes	15	160
Coating implement combined	401	1+
Cutlery	30	140
Dress or coat forms	223	70
Electrically	219	221+
Flatirons	38	82+
Glove forms	223	79
Hair combs	132	118
Hat forms	223	26
Ice cutter	62	320+
Rolls	126	410
Shoe formers	12	129.4
Soldering irons	228	51+
Sterilizer for	422	
Stocking forms	223	76
Tool heating stoves	126	226+
Tools self heated by fluid fuel	126	401
Tools self heated electrically	219	221+
Trouser or sleeve forms	223	73
Windshield cleaners	15	250.5
Track sanders having	291	19+
Transfer mediums	252	70
Expanding or vaporizing type	252	67+
Low freezing or high boiling point	252	71+
Tray heaters	261	DIG. 3
Treating by induction	219	10.41+
Apparatus	219	6.5+
Treatment of metals	148	
Vaporizers	122	
Vehicle systems	237	28+
Power plant combined	237	12.1+
Vulcanizing combined	425	28.1+
Vulcanizing combined	425	340+
Vulcanizing combined	425	363
Welding with heat	228	
With heat and pressure	228	228
Wells		
Electric heaters	392	301+
Having heaters	166	57+
Processes using heat	166	302+
Solid material recovering	299	
Heat Producer		
Flameless composition	44	250+
Hecogenin	540	19
Hectograph	101	131+
Compositions	106	14.5
Flies, work holder for	101	463.1
Coating processes	427	144
Transfer sheet	428	914*
Paper	428	488.1
Processes	101	463.1
Heddle		
Making		
Cord	29	4.6
Wire	140	72
Needle winding	242	51
Structure	139	93+
Doup	139	52
Hedge		
Fence	256	20
Training	47	4
Trimmer	56	233+

	Class	Subclass
Heel		
Clamping work supports	12	125
Cushioned horseshoes	168	15
Cutting machines		
Die cutting	83	
Heel lifts for shoes	D 2	323
Interfitted sole and	36	24.5
Lasts		
Plates and sockets	12	139
Separate heel block	12	135R+
Punching machines	83	
Shoe	36	34R+
Design	D 2	323+
Heel engaging shoe retainers	36	58.6
Protectors	36	73+
Supporters	36	69
Shoemaking machines	12	42R+
Burnishing	12	70+
Edge trimmers	12	85+
Heel seat forming	12	31.5
Loading	227	125
Nailing	227	140+
Plate attaching	227	
Stiffener	12	61R+
Shoemaking processes	12	147R
Skate attaching clamps	280	11.31+
Spring with soles	36	27
Straps for horseshoes	168	22
Heelless		
Overshoes	36	7.2
Rubber overshoes	36	7.4
Sandals	36	7.7
Helical (See Gearing)		
Bed spring connectors	5	269
Brass music instrument valve movement	84	391
Course		
Amusement ride	104	56+
Filled receptacle cooker	99	365
Textile liquid treatment	68	176
Nuclear fuel structures	376	362
Pump interengaging rotary impellers	418	201.1+
Rotary pump dipping channel	415	88
Springs	267	166+
Torsion	267	155+
Switch	200	500
Track sander feeder vertical plunger guide	291	37
Traction railway	104	167
Helicopter	244	17.11+
Airplane combined or convertible	244	6+
Airship combined	244	26
Rotor	416	
Blade positioning means	416	147+
Blade structure	416	223R+
Motor combined	416	20R+
Toy	446	230+
Flying	446	37+
Heliographic Code Signaling	116	20
Heliostats	353	3
Heliotropine	549	436
Helium (See Gas, Rare)		
Isolation by physical processes	55	66
Apparatus for	55	
Liquefaction	62	11+
Refrigeration apparatus for	62	36+
Speech	381	54
Helix (See Screw; Thread)		
Electrical resistance	338	296+
Rheostat	338	143+
Type product		
Static mold	249	59
Helmets	2	410+
Aviators	2	6
Design	D29	12+
Diving with air or oxygen	128	201.27+
Firemans	2	5
Hood type	2	205
Illumination combined	362	105
Miners hat	362	164
Military	2	6
Space suit	2	2.1 A
Sports headgear	2	425+
Helmitol	544	185
Hemacytometers		
Colorimeter type	356	39
Microscope type	356	39+
Hematin	540	145
Hemmers Sewing Machine	112	141+
Hemocytometer (See Haemocytometer)		
Hemoglobin	540	145

	Class	Subclass
Tests for	436	66*
Hemoglobinometer	356	40+
Hemostatic Devices	606	
Artery or vein	606	157
Umbilical	606	120
Hemstitch		
Machine	112	81+
False	112	179
Heparin	536	21
Medicines containing	514	56
Heptode		
Mixer	313	300
Pentagrid converter	313	300
Triode	313	298
With two cathodes	313	6
Herbicides	71	65+
Hermetic Sealing		
Closing portable receptacles	53	285+
Gas filling of portable receptacle, method	53	403+
Glass uniting processes	65	36+
Metallic receptacle closures	220	200+
Soldering apparatus	228	
Hernia Pads	128	112.1
Heroin	546	44
Hertzian Wave (See Radio)		
Railway signal systems	246	30
Transmission line systems	333	
Hetaamoxicillin	540	325
Hetacillin	540	325
Heterocyclic Compounds	540	1+
Azo compounds	534	751+
Heavy metal ctg	534	701+
Silver sensitized formation in color photography development	430	376+
Bio-affecting & body treating composition	424	
Wave energy preparation	204	157.69+
Heterodyne Frequency Measurement	324	79R
Heterodyne Receiver	455	130+
Hevea (See Synthetic Resin or Natural Rubber)		
Hexachlorocyclohexane	570	212
Hexachlorophene, in Drug	514	735
Hexamethylene Tetramines	544	185+
Hexestrol	568	729
Hexode Triode	313	298
Hexylresorcinol	568	766
Hides Skins and Leather		
Biocidal saturant for	424	
Chemical or fluid treatment	8	94.1 R+
Manipulative	8	150.5
Coating	427	
Compositions for treating	252	8.57
Deliming	8	94.17
Dyeing	8	436+
Electrolytic treatment	204	135
Manufacture	69	
Removing fat	8	139+
Tanning	8	94.19 R+
Treatment of	8	94.1 R+
High Altitude Compartments		
Aircraft cabin	98	1.5
Aircraft power plant	244	59
High Energy Metal Forming	72	56
High Frequency		
Coaxial cable	333	239+
Triode	313	293+
High Frequency Heater (See Induction, Heating)		
High mu Triode	313	293+
Highchair		
Childs furniture	D 6	329+
Tray	D 6	396+
Highway	404	
Crossing railway signal automatic	246	293+
Electric	246	125+
Guard		
Fence	256	13.1
Pipe or cable bracket	248	66
Guide or barrier	404	6+
Track combined	238	8
Track for nonflanged wheel	238	3+
Traffic director	404	9+
Hilling and Hill Planting	111	
Ridging and covering plows	172	642
Hinge or Hinged Joint (See Device Hinged)	16	221+
Adjustable	16	235+
Belt fastener	24	33R+
Breakaway	16	222

	Class	Subclass
Check and closer combined	16	50
For nuclear reactor moderator structure	376	285
Liquid check	16	54+
Pneumatic check	16	68
Design	D 8	323+
Flaccid closure mount	49	147
Flask sections	164	391+
Headlight hinge	16	DIG. 26
Magnetic	16	DIG. 14
Making assembly or mounting		
Metal	29	11
Wood seat cutting	144	27
Nesting hinge leaves	16	DIG. 29
Pipe joints	285	273
Detachable	285	283+
Pivotal sheet retainer	402	26+
Plastic	16	DIG. 13
Rod	403	119+
Adjustable angle	403	83+
Rubber sleeve bearings & hinges	16	DIG. 33
Separable	16	254+
Snap-hinge	16	227
Spring	16	277+
Stepladder	182	174+
Stop	16	374+
Stove door	126	194
Toilet seat	4	240
Cover combined	4	236
Tool handle fastenings		
Adjustable angle	403	84+
Freely swinging	403	52+
Watchcase	368	313
Windshield	296	92
Hinged		
Closures	49	381+
Barrel	217	83
Bottom opening hopper vehicle	298	29+
Building shutters	49	381+
Dispenser	222	556+
Drawbridge gates	14	50+
Fence panels	256	26
Flexible and portable	160	
Gates	49	
Lockets	63	19
Metallic receptacle	220	334
Oil cup	184	90+
Operators for	49	324+
Paperboard box	229	125.8+
Purse	150	119
Railway car dumping	105	280+
Ship port	114	178
Spectacle case	206	6
Vault road or pavement	404	25
Vehicle windows and doors	296	146+
Vehicle windshields	296	86+
Vehicle windshields hinges per se	296	92
Windows sliding and swinging sash	49	208+
Windows swinging sash	49	381+
Wooden box	217	57+
Devices		
Base for folding camera	354	192+
Bed combined	5	2.1+
Bellows for roll type camera	354	194
Binder device releasably engaging aperture or notch of sheet	402	
Bracelet multiple	63	9
Bracelet single	63	7+
Buttons and fasteners hinged leaf	24	97+
Calendars hinged leaves	40	119
Card holding clamp	40	647+
Changing exhibitor	40	446
Display items	40	530
Endless type signs	40	524
Cultivator center	172	617
Cutter frame for combine	56	125
Demountable wheel rims sectional	301	32
Drawbridge bascule	14	36+
Extension ladder	182	163+
Extension table	108	65+
Fluid fuel water heater	126	357
Folding bedsteads	5	174+
Folding table	108	115+
Freight car sectional deck	105	372
Garment stays	2	262
Horseshoes sectional	168	7+
Invalid beds	5	60+
Ironing table with supports	108	115+
Lamp shade etc drop support	362	451
Lister type cultivator laterally hinged	172	640

	Class	Subclass
Lockets	63	19
Movable on horizontal axis	211	169.1
Plow moldboard	172	736
Pneumatic tire rims sectional	152	414
Propped ladder eg stepladder	182	165+
Revolver barrel	42	63+
Scoop type vehicle fender	293	44+
Screw threading die stocks	408	180+
Sectional board ironing	38	139
Sectional mattress	5	465
Sectional scows	114	29
Shoe lasts	12	136R
Sleeping car berths	105	321
Slice toaster or broiler grids	99	402
Sofa beds	5	32.1+
Surgical splints	128	88
Upending bedsteads	5	133+
Waffle iron type cooker	99	380+
Wall supported table	108	134+
Weather strips	49	475+
Hip artifical	623	22+
Hippuric Acid	562	450
Histidine	548	344
History, Teaching	434	154
Hitching		
Animal restraining devices	119	109+
Posts design	D30	154
Harness type straps	54	34
Holders	54	64
Hitching post	D30	154
Vehicle attachments	280	186+
Hive		
Bee	449	3+
Connectors	449	27+
Hmx	540	475
Hobbing	409	11
Hobs	407	23+
Hobby Horses	272	52+
Hockey Stick	273	67R
Hod		
Household bin sifter	209	376
Masons	224	44.5
Endless elevator for	198	803.3+
Receptacle with pouring lip	222	572
Hoe	172	371+
Design	D 8	11
Making		
Dies	72	358+
Processes and blanks	76	109
Sausage stuffers	17	35+
Scraper	30	171
Weeder combined	172	375
Hog		
Cholera antigens and sera	424	88+
Oiler	119	157
Scalders	17	15
Scraper combined	17	13+
Scrapers	17	16+
Singers	17	20
Hoist & Hoisting (See Elevator; Lift)		
Alarms for	116	68
Arc lamps	248	317+
Automobile lift	269	58+
Building with	52	122.1+
Cable drum type	254	266+
Derrick	254	283+
Design	D34	33+
Endless or rotary carrier combined	414	564
Fire escape	182	141+
Lateral traverse combined	182	12+
Forklift truck ramp design	D12	
Hoist line implements	294	
Inclined for blast furnaces	414	208
Receptacle moved back & forth along incline	414	168
Invalid lift	5	83+
Beds with	5	61
Load handling	414	592+
Load handling type traversing	414	560+
Measuring and testing cables and ropes	73	158
Nuclear reactor control components	376	219
Fuel charging & discharging	376	268
Ordnance loading	89	46
Railway car type	414	391+
Transports from one vehicle to another	414	342
Ramp design	D12	
Skidway combined	414	571+
Sleeping berth vertically movable	105	318
Slings	294	74+

	Class	Subclass
Submerged vessel	114	51
Tilting railway car body	105	273
Traversing	212	
Truck ladder	182	63+
Trucks	254	2R+
Vehicle hoist design	D12	
Vertically adjustable vehicle	280	43+
Vertically swinging load support combined	414	569
Weighing	177	147
Wheel brake one way for	188	82.1+
With weighing scale	177	147
Hoist Line Implements	294	
Hold		
Backs		
Animal draft	278	126+
Breeching	54	5
Hill hold transmission control	192	4R+
Hill hold vehicle	188	82.1+
Hill hold wheel	188	30
Hold-downs	248	500+
Electrophotographic copier	355	230
Food cooking	99	349+
Food cooking filled receptacle	99	369
Freight retaining on freight carrier	410	
Container twist-lock	410	82+
Photocopying of book and sheet	355	72+
Punching machines	83	374+
Punching machines	83	438+
Punching machines	83	451+
Supports for machines combined	248	680
Holder (see Container; Receptacles)		
Animal catching and holding	119	151+
Arc furnace electrode	373	94+
Arc lamps electrode	403	
Article locking racks	211	4
Assembly jig	269	37
Bag	248	95+
Golf	248	96
Bait fishing	43	54.1+
Live bait inclosing	43	41
Design	D22	136
Bicycle article carrier spring	224	37+
Billiard cue chalkers	273	18+
Blind slat	292	345
Body and belt attached cartridge	224	239+
Body or clothing attached article	24	3R+
Book and sheet		
Book leaf	281	42
Book or sheet	281	45+
Camera plate	354	276+
Copy (See copyholder)		
Depository	402	73+
Guides	402	24
Label pasting etc	156	
Leaf	281	45+
Manifolding leaf	282	29R
Music and	248	441.1+
Printing flexible sheet	101	415.1
Sheet feeding	271	8.1+
Stencil	101	127.1+
Typewriter card	400	521+
Boot tree	12	123.5
Brackets	248	200+
Article	248	309.1+
Shelf type	248	250
Brush for dynamo	310	239+
Camera roll	354	212+
Candle	431	289+
Cane	211	62
Card	40	642+
Chain	24	116R+
Cigar and cigarette	131	187+
Design	D27	183+
Wind guard or ash receiver combined	131	175
Cigar tip cutter with	131	250+
Clasp, clip or support clamp	24	455+
Shirt collar	24	900
Cleaner cover	15	247
Closure fastener portable securing bar	292	289+
Coin	206	.8+
Copy (See copyholder)	248	441.1+
Movable copy or line guide	40	341+
Cord rope thread and wire	24	115R+
Fabric delivery roll	242	55.2+
Fishing floats with	43	44.92+
Harness hitching	54	64
Harness reins	54	74
Insulator and conductor embracing	174	154+

Description	Class	Subclass
Insulator with conductor	174	168+
Measuring and dispensing	33	732+
Piano bridge	84	214
Piano bridge on sounding board with	84	213
Sash cord slack	16	208
Spooled strand material	242	134+
Textile spinning apparatus	57	353
Textile spinning receiver with	57	131
Twine unwinding	242	141+
Tying cords or strands	289	
Cutlery and combined material	30	124+
Dental dam	433	138+
Dental floss	132	323+
Design garment or belt attached	D 3	100+
Hunting or fishing	D 2	14
Dispenser handle and spout	222	475
Ear corn	294	5
Electric lamp etc consumable electrode	314	130+
Fluid diffusers	239	34+
Fluid supply and combined hose	137	355.16+
With nozzle	239	195+
Flypaper	43	115
Garment and fabric		
Article supporting	24	3R+
Bed attached bedclothes	5	498
Bed attached pillow sham	5	498
Bedclothes	24	72.5
Chair attachments	297	190
Closure panel elongated element	160	383+
Closure panel frame type	160	371+
Clothespins (See clothespin)		
Connected substantially spaced plural garment	2	304+
Cuff	24	41+
Drying continuous strips edge	34	158
Drying sheets etc movable	34	163
Garment type combined	2	300
Garment type convertible or reversible	2	301
Garment type elements	2	336+
Garment type partially encircling limb or torso	2	309
Garment type plural encircling	2	308
Garment type strip connected spaced	2	323+
Garment type suspension and encircling	2	310
Glove and fur sewing work	112	18+
Looped fabric sewing work	112	27
Napkin	138	177
Necktie	24	49R+
Rack for paper or textile sheets	211	45+
Rigid vertical type	2	302+
Skirt lifter and	2	217
Sorting rack bag	211	12
Stacked article type rack	211	50+
Torso or limb encircling	2	311+
Towel service cabinet	312	37
Vehicle lap robe	296	77
Work manipulating sewing machine	112	148
Gear cutting, milling, planing	409	903*
Hair		
Combs	132	219+
Mustache	132	215
Handling type weigher	177	148+
Hatpin point guard	132	72.1
With guides	132	69.1
Hollow wound package liquid treating	68	198
Hot dish lifting	294	27.1+
Illumination lens	362	455+
Implement and machine parts (See mandrel)		
Apparel making	223	106+
Axial metalworking tools and movable slide rest	29	52+
Bobbin winding or unwinding	242	130
Bolt	81	13
Brush and broom	15	146
Brush bristle	15	206
Can head seaming roller	413	31+
Chucks or sockets	279	
Cigar making machine with wrapper feed or holder	131	36
Cigar wrapper	131	105
Coil winding or unwinding	242	129
Compound tools	7	167+
Cutlery blade detachable	30	329+
Drill	408	238+
Electric generator and motor brush	310	239+
Engine starter handle	74	551
Fabric reeling or unreeling	242	68+
Illuminating lens	362	455+
Machine tool cutter mount	407	66+
Machine type wrench with bolt	81	55+
Main tapping multiple tool	408	31+
Manicuring file etc with separable	132	76.5
Mechanical gun projectile	124	41.1+
Metal deforming die	124	41.1+
Metal file and rasp	29	80
Metalworking turret with rotary	29	40
Metalworking turret with sliding	29	41
Mirror	52	
Mop	15	147R+
Multiple metalworking tools oscillating	29	56
Pivoted clasp handle engaging	24	489+
Razor with sharpener	30	37+
Reciprocating abrading machine	51	68+
Reciprocating wood saw	83	783+
Rectilinearly movable multiple metalworking tools	29	54+
Rotary reciprocating abrading machine	51	47
Saw making file	76	36
Screw driver attachments	81	451+
Sewing machine attachment	112	257
Sewing machine bobbin	112	231
Sewing machine shuttle	112	196
Sheet punching machine die	83	651
Skein winding or unwinding	242	127
Socket wrench with nut	81	125
Sound box stylus	369	160
Sponge	15	244.1+
Having pivoted handle	15	244.2
Spool winding or unwinding	242	134+
Spring drum type article	242	107+
Stationary metalworking tools and movable work	29	51
Tool handle fastenings	403	
Type casting mechanism matrix	199	49
Wiper dauber and polisher sheet	15	231+
Insect poison	43	131
Key	70	456R+
Music keyboard	84	441
Light guard and	362	376
Live bait inclosing	43	41
Design	D22	136
Lock tumbler	70	328
Permutation compound tumbler	70	317
Nuclear fuel	376	434
Packet	24	17R+
Pencil	401	88+
Penholders	15	435+
Phonograph record	312	9+
Photographic plate	354	276+
Roll film	354	212+
Picture	40	152+
Support type	248	488+
Support type	248	490+
Piles arrangements spacers for	206	
Pocket and personal use	206	37+
Ticket	206	39
Racks	211	
Bridles, saddles, whips, design	D30	143
Radioactive body treatment	600	7+
Receptacle		
Acetylene carbide feed	48	38+
Acetylene generator and	48	2
Acetylene water feed	48	4+
Animal cage food and water	119	18
Article carrying tray	294	172
Bait or catch	43	55
Blacking box	15	259
Blood	604	403+
Boiler or deep fat fryer	99	403+
Brush and blacking box	15	258
Camera with storage for plate	354	174+
Closet bowl disinfectant	4	222+
Closure remover with receptacle	81	3.8
Collapsible wall dispenser with	222	105
Cover containing a	220	521+
Dispenser with cutter and	222	81+
Dispenser with support or	222	173+
Extracting or leaching	422	261+
Fishing hook and tackle	43	54.1+
Flower	47	41.1+
Handle	220	96
Handle containing a	16	110.5
Heating and illuminating gas	48	174+
High pressure gas	220	3
Inkstand	248	127+
Dispenser type	222	577
Insect poison	43	131
Inverted container drainage	141	364
Match and toothpick	206	102+
Matchbox	206	121+
Mattress filling bat holder	53	255+
Medicating inhaler gas	128	203.12+
Medicator	604	403
Motor vehicle battery	180	68.5
Plural source dispenser stacking	222	143
Racks for	211	71+
Rotatable assembly of dispensers	222	144
Spittoon flexible material and	4	285
Spray fluid	239	302+
Rein	54	74
Design	D30	141
Whip socket combined	280	177
Scale load	177	253+
Handling type weigher	177	148+
Sleeping car article	105	325
Stringed music instrument with article	84	329
Tail	54	78
Cow	119	105
Telephone	379	453+
Tobacco ash receiver with		
Cigar cigarette or smoking device	131	240.1+
Match	131	239
Tooth	433	141+
Tops with spinning devices	446	262+
Train dispatching safety system signal	246	16
Vacuum-operated load engagers	294	64.1+
Vacuum-type holding means	269	21
Valve head	251	89+
Friction detent	251	297
Vehicle	280	182+
Hand lantern	362	61+
Rein	280	182
Vehicle attachment	280	182+
Washboard soap	68	224
Washing machine transferable clothes	68	10
Welding electrode	219	145.1+
Wheel	157	14
Work (See mandrel)	269	
Abrading	51	216R+
Abrading work rests	51	238R+
Abrading work tables	51	240R
Ammunition shell loading	86	44
Apparel apparatus	269	
Book typewriter	400	24+
Boring machines	408	
Brush boring	408	89+
Brushing scrubbing cleaning	15	268
Brushmaking	300	10+
Butchering	17	44+
Can opener	30	400+
Capping nail and screw making	10	158
Chain welding	59	34+
Circular saw	83	409+
Cleaning and liquid contact with solids	134	137+
Collapsible structure	269	901*
Door	269	133
Drilling machines	408	
Electric heating	219	158+
Electrolytic apparatus	204	297R
Feeding stock combined	29	56.6
Forging machines	269	
Gear cutting milling and planing	269	
Glove and fur sewing	112	18+
Hair dryers	34	101
Hair shampooing apparatus	4	515+
Hat sewing machines	112	13+
Horseshoe nail finishing	10	39
Immersion scouring rotary	51	19
Ironing	38	64
Ironing table	38	103+
Leather manufacturing	69	19
Looped fabric sewing	112	27
Magnetic	335	285+
Manual sewing	269	
Meat blocks	269	289R+
Metal working	269	
Metalworking multiple tool laterally moving stock	29	51

	Class	Subclass
Discharge to gravity section	198	550.1+
Gravity discharge holder	198	540+
Dispenser.......	222	
Dumping land vehicles	298	24+
Dumping railway car	105	247
Flat convertible to	105	243
Feeding and feeder		
Animal	119	52.1+
Animal, design	D30	121+
Article	221	
Bag filling	53	570+
Cigarette tube	131	74
Filler tobacco or bunch......	131	44
Furnace charging bell and ...	414	204
Furnace fuel	110	108+
Furnace fuel blower	110	105
Match dipping frames	144	64
Rotary drying drum external type	34	112
Sifter......	209	244+
Static mold having	249	108
Tobacco for cigar etc making	131	108+
Track sander	291	38+
Track sander with jet in	291	18
Track sander with oscillating	291	30
Wood slicing	144	180
Planting combined		
Auxiliary frame hill type......	111	24
Dibble with revolving	111	90+
Floating auxiliary frame	111	63+
Revolving	111	74
Vibrating	111	75
Power driven conveyor combined		
Bottom forming......	198	550.1
Chute and	198	568+
Discharge for	198	523+
Feeder	198	523+
Plural conveyors	198	570+
Hopples	119	126+
Slings combined	119	100
Hops and Hopping......	426	11+
Apparatus	99	278+
Plants	PLT	89
Stirrer......	366	343+
Vine stripper	460	123+
Horizon Artificial......	356	248+
Hormones......	552	502+
Animal extracts		
Medicines containing	424	520+
Heterocyclic	540	2+
Sex	552	502+
Hormones, Protein......	530	399
Horn......	D10	120
Antenna	343	786
Ear trumpet	181	129+
Electric	340	388+
Fertilizer	71	18
Fog	116	137R+
Lamp combined	116	3
Loud speaker	381	156
Making by deforming sheet metal ...	72	
Megaphone......	181	177+
Musical instrument	84	387R+
Design	D17	11
Periodic sounding	116	24
Phonograph	369	163
Shield	119	144
Shoe......	223	118+
Design	D 2	642
Shoe inseam trimmer	12	4.5
Stringed instrument resonator.......	84	294
Structure......	116	137R+
Toy	446	209
Vehicle......	116	59
Design	D10	120+
Horology......	368	
Antimagnetic devices	368	293
Balances	368	169+
Clocks	368	62+
Design	D10	30+
Dial trains	368	220+
Dials	368	232+
Dials hands	368	238+
Escapements	368	124+
Invertible, hour glass type	368	93
Parking meter type	368	90+
Pendulums	368	134+
Safety wheels	368	181+
Watches	368	62+
Horse		
Blankets	54	79
Design	D30	145

	Class	Subclass
Boots	54	82
Design	D30	146+
Collars	54	19R
Design	D30	137
Hames combined	54	18R
Shaping	69	3
Stuffing	69	4
Confining and housing devices ...	119	15+
Detachers	278	21+
Four legged support	182	181+
Unitary foldable	182	153+
Grooming and currying devices ...	119	83+
With air draft......	15	363+
Harness	54	
Powered motor	185	15+
Composite or multiple	185	3
Rake horse drawn	56	375+
Training apparatus	54	71
Horseshoe	D30	147+
Calk	D30	150
Farriery	168	
Making	59	36+
Nail making		
Cut type	10	36+
Wrought type	10	64+
Nails	411	439+
Horticulture (See Plant Husbandry)		
Horticultural tools	D 8	1+
Illumination	362	805*
Hose (See Stocking)		
Alarm system via fluid hose	340	320
Applying and removing......	29	235+
Bridge	104	275
Brush or mop combined	401	289
Carrier reel	242	86+
Clamp applier	81	9.3
Clamps	24	19
Cutoff	251	4+
Connections	285	238+
Dispensed material guide	222	527+
Dispensing guide interlock	222	74+
Fire fighting vehicle	280	4
Holder and fluid supply	137	355.16+
With nozzle	239	195+
Liquid level gauge combined	73	294
Making	156	143+
Nozzle	239	589+
Pantyhose	2	409
Protectors	138	110
Structure......	138	177
Design	D23	266
Support	248	75+
Supporters		
Fluid conduit or nozzle	248	75+
Plural garment	2	306
Strip connected spaced holder	2	335
Torso or limb encircling	2	311+
Water hose	D23	266
Wheeled hose reels	242	86.2+
Hospital Furniture and Equipment (See Type)		
Hot Air Furnace	126	99R
Hot Bearing Indicator	116	214
Train hotbox	246	169A
Hot Bed	47	19
Hot Blast Stoves	432	214+
Hot Dog		
Cooker	D 7	323+
Packaging	53	DIG. 1
Hot Houses	47	17+
Engine cooling vents	180	68.3
Hot Pad or Mitt	D29	20+
Hot Plate	219	443+
Design	D 7	362+
Hot Topping	164	122+
Hot Tops	249	106
Hot Water Bottle	383	901*
Design	D24	43
Hot Water Furnace	122	
Hot Wire Meters	324	106
Hounds		
Vehicle	280	139
Draft pole combined......	278	43
Hour Glass	368	93
House	52	
Amusement	272	2
Animal	119	15+
Annunciators	340	815.1+
Compound maintenance tools	7	105+
Drying ie kilns	34	201+
Heating and cooling system	165	48.1

	Class	Subclass
Heating system	237	
House trailer	D12	103
Moving	52	143
Portable track	238	13
Numbers		
Fluorescent	250	462.1
Luminous	40	542
Plate	D20	10+
Plumbing (See plumbing)	4	211+
Toy	446	476+
Construction	446	108+
Furnishings	446	479+
Ventilation	98	
Toilet	4	209R+
Houseboat		
Design	D12	315
Housecoat or Robe	D 2	12+
Household Articles	D 7	
Linen	D 6	595+
Measuring implement	D10	46.2
Housing		
Abrading apparatus	51	268+
Animal confining and	119	15+
Brush and broom	15	184+
Closure seals	292	317
Combination lock	70	443
Fence supported by	70	322
Electricity conductors	174	50
Fluid or vacuum type......	174	17R+
Wall mounted	174	48+
For an electrical device	174	50+
For different electrical devices	361	331+
For specific electrical device (See the type of device)		
Key operated locks	70	448+
Lock on rotary shaft	70	184+
Lock on valve	70	179
Material treating drum with gas flow	34	139
Heating	34	133
Metal rolling apparatus	72	237+
Padlock	70	52
Nonshackle type	70	33
Radio receiver	455	347+
Vehicle	455	343+
Rotary shafting	464	170+
Shaft combined	464	52+
Swinging door locks		
Interfitting keeper and	70	102+
Swinging deadbolt type	70	136
Switchboard	361	331+
Television	358	254
Thermometer	374	208+
Transceiver	455	90
Transmitter	455	128
Tuner with	334	85
Hovering Vehicles	180	116+
Howitzers (See Gun)		
Hub		
Automobile light combined	362	78
Bearing		
Ball	384	544
Roller	384	589
Brake		
Operator	188	114
Transmission	192	6R
Vehicle	188	17+
Velocipede	188	26
Cap	301	108R
Lock	70	259+
Groove railway	295	45+
Impeller	416	204R+
Land vehicle	301	105R+
Compression plural spoke series ..	301	74+
Compression single spoke series ..	301	80+
Tension	301	59+
Wrench	301	116
Manufacture and assembly	29	894.36+
Boring	408	72R
Special woodworking machines ..	144	16
Runner vehicle	280	14
Wrench	81	75+
Hub Odometers	235	95B+
Hull		
Airship	244	125
Cleaning implements	114	222
Rotary for propulsion	440	99
Ship	114	56+
Small boat	114	355+
Huller (See Husker)		
Boll	19	35+

	Class	Subclass
Fermentative	435	30
Inorganic Compounds	23	
Conversion to hydrocarbon	585	943*
Wave energy preparation	204	157.4+
Inorganic Settable Material		
Compositions	106	638+
Building guard	52	101
Catching nets	43	143
Destroying	43	132.1+
Electrocuting	43	112
Exits in closures and screens	160	12+
Fumigators	43	125+
Netting for beds	5	414+
Preserving, teaching taxidermy	434	296+
Traps	43	107+
Design	D22	122+
Insect		
Insecticides	424	
Coated or impregnated vehicle	424	411+
Coating or plastic compositions containing	106	15.5+
Container	206	524.1
Detergents containing	252	106+
Fertilizer containing	71	3+
Garment hangers combined	223	86
Insecticide sprayer	D23	213+
Apparatus	239	
X-art collection	43	900*
Method	43	132.1
Water softening or purifying compositions containing	252	175
Insemination, Artificial	604	906
Inserters (See Nailing; Stapling)		
Assembling apparatus	29	700+
Belt hook	29	243.51
Box lining	493	93+
Sausage stuffers	17	35+
Saw tooth	76	80
Surgical implement	606	117+
Needle	606	222
Tape inserting apparel	223	50
Tire patch inserters	81	15.5+
Wick	431	120
Inserts		
Bearing surface	384	282+
Bedstead	5	280
Brake shoes cast	188	256
Composite metal casting	164	91+
Cooking molds	249	112+
Earth boring bit insert	175	410
Garment	2	400
Gearing	74	451
Holding		
Core	164	231
Mold	164	332+
Pattern	164	236
Nonmetallic composite casting	249	83+
Insignia		
Apparel	2	246
Badges	40	1.5+
Design	D11	95+
Emblem per se	D11	99
Insoles		
Making method	12	146B
Reinforcing machine	12	20
Shoe (See soles)	36	43+
Laminated	36	44
Prepared for securing	36	22R
Ventilated	36	3R
With body treating means	128	383
With local padding	36	71
With shank stiffener	36	76R
Therapeutic	128	383
Inspection & Inspecting (See Specific Subject; Testing)		
Cloth	26	70
Dispenser with means for	222	154+
Drying apparatus with means for	34	88
Gauges	33	501+
Lamp for	362	138
Measuring and testing	73	
Metal founding device with means for	164	150+
Robot	901	44*
Instrument (See Type of Instrument)		
Board	73	866.3
Illumination, automobile	362	61+
Switch boards	361	331+
Telephone switchboard	379	319+
Calibrating	73	1R+
Electrical instruments	324	74+

	Class	Subclass
Casings	73	431
Dental	433	25+
Electric switch combined with	200	56R
Fluorescent or phosphorescent indicating	250	483.1
Flying of aircraft	342	407+
Apparel	33	2R+
Area integrators	33	121+
Direction or force indicator	33	300+
Distance	33	700+
Landing	342	410+
Scribers	33	18.1+
Straight edge type	33	431+
Navigational	33	457
Straight line light ray type	33	227+
Household measuring	D10	46.2
Materials	33	DIG. 11
Measuring	D10	
Mechanical expedients	33	DIG. 12
Mechanism damping	73	430
Medical	D24	8+
Musical	84	1+
Design	D17	
Electrical	84	600+
Stringed	84	173+
Wind	84	330+
Panel in vehicle	D12	192
Panel mounted	248	27.1+
Signalling	D10	
Sterilizing, disinfecting apparatus	422	
Support heads for	248	177+
Testing	D10	
Transformers	336	
Insufflators	604	58
Insulation and Insulating		
Building wall	52	404+
By lamina	52	408+
Yieldable	52	393+
Electric	174	
Battery separator	429	247+
Bridged rail joint	246	48
Bushings	174	152R+
Ceramic composition	501	1+
Conductor cover burnishing	29	90.5
Fluent composition	252	570+
Handle	16	116R+
Indefinite length conductor covering	156	47+
Lamp sockets	439	
Magnet or coil tape	242	7.8
Making metal parts for	29	631
Meters and testers	324	555+
Pipe joints or couplings	285	47+
Rail joints	238	152+
Railway rolling stock parts	105	60
Railway tie tie plate and rail	238	107+
Ring winding tape	242	6
Shaft coupling	464	900*
Trolley conductor section	191	39
Trolley conductor support	191	42
X-art	493	949*
Heat		
Alkali metal silicate containing composition	106	601+
Boiler outlet flue	122	164
Bottle or jar	215	12.1+
Building	52	408+
Cabinets	312	214
Ceramic porous composition	501	80+
Compositions	252	62
Container	206	542
X-art	493	903*
Covering laminated type making	156	
Double-wall receptacle, wooden	217	128+
Flatiron body	38	89
Flatiron stand or base	248	117.3+
Fluid sprinkler, sprayer, diffuser	239	397.5
Gas separator	55	267
Handle	16	116R+
Inorganic ingredient containing composition	106	672
Metal receptacle spaced wall	220	445+
Muffler or sound filter	181	282
Oven	126	273.5
Pore forming composition	106	122
Stock material	428	920*
Tobacco smoking devices	131	194+
Water heater	126	375
Sound (See muffler)		
Aircraft body structure	244	119
Alkali metal silicate containing composition	106	601+

	Class	Subclass
Auditorium	181	30
Ceramic porous composition	501	80+
Compositions	252	62
Gas separator	55	276
Gearing	74	443
Inorganic ingredient containing composition	106	672+
Lock silencer or muffler	70	463
Motion picture camera	352	35
Muffler or sound filter	181	207+
Piano resonance device	84	191
Pore forming composition	106	122
Structure	181	
Tape	428	343+
Coating	427	207.1+
Coating opposite sides	427	208
Packaged roll	206	411
Insulation Strippers		
Tube or sheath splitter	30	90.1+
Wire	81	9.4+
Insulin, T.m	530	303
Medicine containing	514	3
Insurance Documents	283	54
Intaglio		
Electrolytic reproduction	204	4+
Ornamentation	428	156+
Photographic production of	430	307+
Printing	101	150+
Combined with another type	101	DIG. 43
Static mold for forming product having	249	104
Intake		
Channel	405	127
Filter for pipe	210	460+
Fluid current conveyor	406	108+
Adjustable	406	113+
Plural	406	117+
Pneumatic straw stacker	406	139+
Gravity positioned sprayer	239	334
Pump rotary center	415	206
Integrated Circuit (See Transistor)		
Integrated Optical Circuit	350	96.11+
Integrators		
Area	33	121+
Calculators	235	61C
Distance	73	490
Ships log combined	73	183
Velocity	73	503
Volume or rate of flow meters		
Area velocity	73	227
Pressure differential type	73	861.43+
Weigher combined	177	16
Intensifiers (See Amplifiers)		
Diaphragms acoustic	181	157+
Photographic	430	487
Telephone	379	156+
Intercalates (See Definition Note (3))	428	402.2
Intercommunication System		
Electric	379	167+
Multiplex	370	
Paging	340	311.1
Selective	340	825.44+
Intercom	D14	159
Telephone-type	D14	142+
Intercomparison		
Article or material	434	367
Printed matter	281	51
Interconnection Systems	307	
Interelectrode Capacitance Testing		
Electronic Tube	324	409+
Interference Film Antireflection	350	162.19
Interference Suppression (See Transmission, Electric)		
Interferometer	356	345+
Digest	33	DIG. 4
Holography	356	347
Interferon	435	811*
Spectroscopy	356	346
With light conductor	250	227.27
Interferon	530	351
Interleaving		
Electrophotographic copier	355	325
Interleukin	530	351
Interliners Tires	152	203+
Interlock and Interlocking		
Clutches and power stop control	192	
Electric switches	200	50R
Machine tool functions		
Cutting machine	83	399+
Selective cutting	234	52
Multiple controlling elements	74	483R+

	Class	Subclass		Class	Subclass		Class	Subclass
Jodhpurs	2	227+	Pipe dredging	37	72	Tube pneumatic piano	84	57
Jogger (See Agitating; Impact; Jarring)	271	221+	Piston rod	403	230+	**Junk Yard Crushers & Presses**	100	1+
Joining			Articulated with piston	403	52+	**Junket**	426	578
Adhesive for	428	343+	Railway rail	238	151+	**Jury Rudders**	114	168
Box corner staying	493	89	Reinforcing bar	52	719	**Justification**		
Labels	156		Rod	403		Counter or indicator for	234	4+
Stamps	156		Smoking device	131	225	Type casting	199	51
Wallpaper	156	574+	Thill	278	49	Typesetting	276	28+
Removable layer	428	40	Tool to handle	403		Circulating	276	8+
Wire	140	111+	Toy figure	446	376+	Typewriter letter space	400	1+
Fabric stay applying clip	140	10	Umbrella frame	135	29+	**K Acid**	562	71
Sleeve deforming machines	29	282	Valve combined	251	142+	**K Guns**	89	1.1
Sleeve deforming tools	29	282	Whip	231	6	**Kaleidoscopes**	350	4.1+
Tire applying with	157	1.14	Railway track	238	151+	Design	D21	60
Joint Packing	277		Riveted by hollow rivet	227	51+	**Kalsomines**	106	
Jointing			Riveted by solid rivet	227	51+	**Kanamycin**	536	13.7+
Indefinite length conductors	156	49	Robot	901	28*	**Kaolin (See Clay)**		
Saws	76	50.4	Safe wall	109	79	**Karaya Gum**	536	114
Gaging combined	76	46	Sectional cabinets joining means	312	111	**Kauri Gum**	530	203
Staves	147	25+	Showcase	312	140	**Kayaks**	114	347
Joints (See Bond; Connection; Hinge;			Snap fitting	29	453+	Design	D12	302
Seam)	403		Stirrup	403	232.1	**Keels**		
Artifical body member	623	18+	Stove	126	98	Blocks for ships in dry dock	405	7
Ball and socket			Toy blocks bars sheets	446	85+	Rudders mounted on or below keel	114	149
Lockable at fixed position	403	90	Universal (See joints ball and)	464	106+	Ship	114	140+
Pipe	285	261+	Vehicle			**Keenes Cement**	106	777
Pipe insulated	285	51	Airplane structure	244	131	**Keepers**		
Pipe multiple	285	135	Animal draft	278		Lock or latch	292	340+
Plural	403	56	Articulated land	280	400+	Casing interfitting for swinging		
Road joint, per se	404	47+	Body land vehicle	296	29+	door	70	102
Rod to base plate head	403	68	Body welt land vehicle	280	153.5	Extensible for sliding door	70	96+
Surgical truss	128	122.1	Bumper end connections	293	154	Movable checkhook	54	62
Thermal expansion sprayer			Bumper joints	293	155	**Kefir**	426	583
combined	239	397.5	Felly and rim	301	99+	**Keg (See Barrel)**		
Tripod head	248	181+	Freight car granular material seal	105	424	**Kelly**	464	163
Universal	403	122+	Ship	114	88	**Kelp**		
Vehicle hitch	280	511+	Tops land vehicle	296	29+	Alkali metal inorganic recovery from	423	179+
Vibrating mounts for rod packing	277	100	Wooden box	217	65	Iodine recovery from	423	500
With static joint	403	76+	**Joist**			**Kennels**	119	19
Barrel	217	96	Bridging floors	52	695	**Keratins**	530	357
Bayonet (See joints closure)			Connection to wall	52	289	Fermentative treatment	435	272+
Pipe joint	285	396	Hanger	52	702	**Kerf**		
To gun	42	86	Static mold	249	18+	Cutting		
Bridge truss	14	14	**Journal (See Bearing)**	384		Mining or disintegrating in situ	299	
Building block	52		Box			Woodworking	144	136R
Building facings or shingles	52	519	Dust guards	277	130+	Guide	83	102.1
Closure			Jacks	254	33+	Spreading	83	102.1
Coffin lid	27	17	Design	D15	143	**Ketchup (See Catsup)**		
Safe interfitting	109	74	Machine elements	384		**Ketenes**	568	301+
Safe seals	109	75	**Joystick (electrical Input Device)**			**Ketimines**	564	271+
Weather strips	49	475+	Code signal generating	341	20+	Acyclic	564	276+
Electrical	439		Computer	364	709.9	Aromatic	564	271
Conduit or cable	174	21R+	Display	340	709	**Ketol**	568	414+
Joining indefinite lengths	156	49	Game device	273	148R	**Ketole (See Indole)**		
Length	156	47+	Resistive	338	68+	**Ketones**	568	303+
Rail bonds	238	14.1+	Writing	178	18	Carboxylic acid esters		
Rail bonds block signal system	246	48	**Juice**			Acyclic	560	174+
Rail bonds block signal system	246	57	Comminution and extraction	99	510+	Alicyclic	560	126
Rail bonds block signal system	246	75	Dehydration	426	471+	Aromatic	560	51+
Switch with connector	200	51R+	Juice extractor	D 7	665+	Ketone bodies	436	128*
Fabric fastening (See notes to)	160	382+	Preservation	426	330.5	**Ketoximes**	564	253+
Furnace hot air	126	119	Presses	100	104+	**Kettle (See Pot)**	220	
Geometrical instruments			Sugar crystallizing	127	58	Drums	84	419
Pivot joints	33	495+	Evaporating and concentrating	127	61	Furnace	126	345+
Squares, pivoted	33	465+	Extracting	127	43+	Liquid heating	126	373+
Squares, separable	33	478	Purifying	127	46.1+	**Key**		
Making, general (See welding)	228	4.1+	Separating	127	53+	Board		
Caulking ships	114	86	Treatment	426	305+	Code generator	341	22+
Juxtapose and bonding	228	101+	**Juke Box**	369	30+	Controlled light valves with light		
Metal to nonmetal	228	903	Check controlled		194	conductors	250	227.22
Mortar	52	415+	Design	D14	173	Display control	340	711+
Riveted	227	51+	Selective or remote electric control	369	24+	Electric signal systems	341	22+
Sheet metal container making	413		**Jumpers**			Keyboard actvated display	340	711+
Static mold	249	98+	Baby chair	297	274+	Musical instrument transposer	84	445
Metal			Railway amusement park			Musical instruments	84	423R+
Receptacle	220	677	Gap	104	54	Design	D17	1+
To nonmetal bonded	403	265+	Passenger	104	82	Operation training	434	227+
Using thermal characteristic	403	28	**Jumping Jack**	446	322+	Punch machine, selective	234	123+
Metal casting	164	108+	**Jumping Shoes**	36	7.8	Registers	235	145R+
Mining tooth to head	299	91+	**Junction**			Registers key locks	235	27
Packing ends	277	220+	Boxes			Typewriter	400	472+
Paperboard box	229	198.1+	Electrical	174	50+	Typewriter key levers	400	496+
Pavement	404	47+	Metallic receptacle	220	3.2+	Typewriter key locks	400	663+
Static mold for making	249	9	Pipe combined	285	128	Can opening	220	274+
Pipes shafts rods and plates (See			Switch	200	297	Clock winding	D 8	347+
joints ball and socket)			Semiconductor P-N type process of			Code instruction	434	222+
Bedstead	5	282.1+	forming	437	15+	Controlled closure	220	210
Fence rail	256	65	Semiconductor process of forming	437	15+	Controlled musical instruments	84	1+
Flexible shaft	464		Thermoelectric	136	200+	Automatic players	84	2+
Packings	277		Thermometer	374	179+	Cotter type	411	513+
Pipe	285		Thermometer zone	374	179+	Nail lock	411	356+

	Class	Subclass
Electric		
Signal box transmission	340	287+
Switch	200	
Telegraph	178	101+
Telephone	379	251+
Holders	70	456R+
Case	206	37.1+
Design, cases or rings	D 3	61+
With light	362	116
Hole		
Alarm	116	80
Cover	70	455
Escutcheons	70	452
Guards	70	423+
Illuminator	362	100
Horological apparatus	81	122+
Making	76	114
Winding attached	368	214
Identification tags	283	74+
Lock illuminator	362	100
Lock type (See key hole)	70	393+
Alarm	116	80
Cutting by milling	409	89+
Design	D 8	347+
Holders	70	456R+
Identification	70	460
Insertion guide	70	454
Processes and blanks	76	110
Locking bolt or nut to substructure		
Laterally introduced key	411	137
Longitudianal key	411	110
Locking bolt to nut		
Elongated key, disposed axially	411	216
Elongated key, disposed axially	411	321
Laterally movable key	411	199
Laterally movable key	411	209
Laterally movable key	411	315
Received in each of plural nuts	411	224
Swingable key	411	205
Used with thread lock		
Longitudinal key	411	292
Radial key	411	294
Tangential key	411	300
Mechanism		
Bowed stringed instruments	84	315+
Plucked stringed instruments	84	320+
Morse	178	101+
Operated (See device operated)		
Rail joints utilizing	238	151+
Railway wheel tire fastening	295	19
Ring or fob	D 3	61+
Seater	409	190
Telegraph	178	101+
Torque transmitting	403	355+
Pullers	29	250
Slidable for gearing	74	356+
Keyed Block (See Block)		
Keyer		
In transmitter	375	68+
Alternating current telegraphy	178	66.2
Vacuum tube	328	247
Keystone Blocks	52	575
Kick Plate	16	1R
Kick guard	16	DIG. 2
Kicker	83	707+
Kickers and Blowing Agent Moderators (See Synthetic Resin or Natural Rubber)		
Kier Treatment of Fibers	8	157
Kiln	432	120+
Carbonizing	202	
Drying	34	201+
Kiln	D15	144.1
Kimonos	2	74
Kindler		
Composition	44	542+
Kinematographic Apparatus (See Motion Picture)		
Kinescope		
Cathode ray tube	313	364+
Motion picture	352	129
Television system using	358	242+
Kinesitherapy	128	24R+
Design	D24	36
King Pin	403	
Kingbolts	72	472
Kitchen		
Cabinets	312	
With sifter	312	210.5
Utensils	D 7	368+
Kites	244	153R+
Design	D21	88
Kits		
Camp	206	541+
Convertible to table	190	12A
Diagnostic	206	569
Lunch	206	541+
Mess	206	541+
Photographic	354	278+
Pocket stove	126	38
Signal for trainmen	246	488
Toilet kit	132	286+
Absent special toilet articles eg, mirror	401	118+
Klystron	315	5+
Amplifier	330	45
Demodulator	329	
Modulator	332	
Oscillator	331	83
Reflex	315	5.18+
Knapsack	224	209+
Backpack convertible to other devise	224	153
With belt	224	215
Kneader (See Mixer)		
Bread board	269	302.1
Bread, pastry or confection	366	69+
Implements	416	70R
Surgical	128	60
Vulcanizable gums		
Apparatus	425	197+
Apparatus	425	200+
Knee		
Action		
Running gear springs	280	688+
Shaft transmission mechanism combined	180	348+
Stub axle mounts	280	96.1+
Vehicle spring devices	267	228+
Artificial	623	39+
Knee bone	623	20
Garment protectors	2	62
Length trousers or overalls	2	228
Lock for composing stick	276	39
Operated control lever or linkage	74	515R
Pads or rests	2	24
Ship	114	87
Support surgical table combined	269	
Vehicle runner	280	27
Kneehole Desk	312	194+
Knife		
Box or holder	D 7	637
Changing	83	954*
Cutlery	30	138+
Cutting	76	82+
Design	D 7	649+
Electric or battery driven	D 7	646
Hand manipulable cutting	30	
Additional tool combined	7	151+
Handles	D 8	DIG. 4
Harvester	51	36
Clamp	51	222
Holder, body or belt attached	224	232
Knife edge texturing	28	260
Machette or hunting knife	D22	118
Machine knife in situ	51	246+
Processes	51	285
Rest	248	37.3+
Sharpener attachment for food mixer	D 8	35
Sharpener, powered	D 8	63
Snap type	200	472
Solid material comminution	241	291+
Swaging	76	89.2
Switch, knife blade	200	554+
Knit Fabrics and Articles	66	169R+
Caps	2	201
Design	D 5	47+
Knit-deknit texturing	28	218
Neckties	2	147
Knitting		
Darning	66	2
Design	D 5	11+
Fabric manipulation	66	147+
Fabric or article	66	169R+
Feeding	66	125R+
Looper pin & hook	66	3
Machines		
Elements	66	90+
Needles (See needle & needle)	66	116+
Independent needle machines	66	7+
Needle beds	66	114+
Needle cooperating elements	66	90+
United needle machines	66	78+
Pattern mechanism	66	231+
Reknitting	66	1.5
Sheaths	30	151+
Stopping	66	157+
Additional tool combined	7	158+
Knobs	16	121+
Bed post ornament	5	281
Combination lock operator	70	445
Door	292	347
Attaching devices	292	348+
Bearings	292	356
Design	D 8	300+
Rose plates	292	357
Insulated	16	118
Label combined	40	331+
Lock combined	70	207+
Machine element	74	553
Making		
Deforming sheet metal	72	
Metal	29	161
Pull or	D 8	300+
Signal combined (See knob)	116	
Knockdown		
Beds and cots	5	
Box wooden	217	12R+
Bureaus	312	266+
Cabinets	312	257.1+
Cooking rack	99	449
Crate wooden	217	43R+
Desks kneehole type	312	195
Foundry core bar	164	400
Freight car	105	363
Hand rake	56	400.19+
Hold knife or razor	30	332
Ladder	182	194
Pedestal table	108	150
Picture frames	40	155
Razor	30	47
Reel	242	115
Sifter support	209	414
Stand	248	529
Plural leg	248	165
Receptacle	248	150
Stoves	126	9R
Toys	446	85+
Velocipede	280	278
Frames	280	287
Weighing scale	177	126+
Knockers		
Boll hulling combined	19	37
Breaker combined	19	36
Cable hoist	212	116
Door	D 8	401
Sifting element	209	382
Knockouts	220	265+
Knot and Knotting (See Tying)	289	
Book sewing machine	112	22
Detector	28	227
Harvester		
Compressing and binding	56	433+
Cutting conveying and binding	56	131+
Raking and bundling	56	343
Mesh fabric	87	53
Making	87	12
Necktie	2	148+
Sash cord	16	207
Sewing machine stitch former	112	156
Strand joint	403	
Wireworking	140	101
Knuckle	213	100R+
Knurling Metal	72	703
By rolling	72	191+
Methods & apparatus	29	DIG. 23
Kraton T M (See Synthetic Resin or Natural Rubber)		
Kraut	426	
Canned	426	131
Kyropoulos Type Crystallization	156	616.3
Label		
Clothing label	D 5	63+
Design	D20	22+
Exhibiting	283	81+
Gumming apparatus	118	
And applying	156	
Dynamic information storage and retrieval	360	
Identifying plug tobacco	131	368
Label or tape embossing tool	D18	19+
Labels	283	81+
Magnetic bubbles	365	1+

	Class	Subclass
Read only memories	365	97+
Static information storage and retrieval	365	
Obliterating stamp scarifier	81	9.21
X-art	493	961*
Labeling (See Digests 1-51)	156	
Cigarette packaging machine	131	281+
Fabrics or deformable surfaces	156	DIG. 19
Feeding	156	DIG. 28
Flat rigid surfaces	156	DIG. 1
Magazines	156	DIG. 29
Manual dispenser	156	DIG. 48
Non-flat surfaces	156	DIG. 5
Plug tobacco	131	113+
Tobacco products	131	106
Laboratory Apparatus	422	99+
Burettes	422	100
Cabinet	312	209
Elements	422	99+
Laboratory equipment	D24	
Monitor nuclear	250	472.1
Scintillation type	250	483.1
Labyrinth	272	19
Memory systems		
Dynamic electric information storage and retrieval	369	
Dynamic magnetic information storage and retrieval	360	
Packing	277	53+
Lace		
Digest art	D 5	47+
Table cloth	D 6	617+
Tipping	425	807*
Laces and Lacings		
Belt end connectors	24	34
Belt operated corset closure	450	119
Design, narrow	D 5	11+
Garment supports	2	326+
Knot or end holders	24	712.1+
Lacing and unlacing machines	12	58.5
Lacing device	24	712+
Making	87	
Fiber destruction by chemicals	8	114.6
Packages for	206	49
Sewing machine for lacing or whipping	112	121.2
Shoe lacing machine	12	58.5
Shoelace fasteners	24	712.2+
Shoelaces	24	715.4+
Design	D 2	316
Stud making	29	12
Stud setting machine	227	51+
Tipping	72	282
Lacquer (See Varnish)		
Cellulose ether or ester (See also synthetic resin or natural rubber compositions)	106	169
Synthetic resin or natural rubber (See class 523, 524)		
Lacrosse	273	326
Lactams		
Caprolactam	540	485+
Delta lactams eg alpha piperidone	546	243
Gamma lactams, eg alpha pyrrolidine	548	543+
Isatins	548	485
Naphthostyrils	548	437
Lactic Acid	562	589
Preparation from carbohydrates	562	515
Fermentation	435	139
Lactide	549	274
Lactometer	73	32R+
Lactones	549	263+
Poly hetero O		
5 membered ring	549	296
6 membered ring	549	274
7 membered ring	549	267
Lactose	127	31
Ladders	182	194
Attached supports	248	210
Bracket attachments	248	210+
Scaffold or shelf type	248	238
Chute or escape combined	182	49
Collapsible	182	156+
Convertible	182	21+
To chair	182	33+
To scaffold	182	27
Design	D25	62+
Door combined	182	77+
Elevating platform	182	101+
Extension	182	207+

	Class	Subclass
Fire trucks	280	4
Hose or nozzle support	248	76
Attached	248	77+
Land vehicle	182	127
Platform detachable	182	120+
Propped	182	165+
Safety devices for	182	107+
Scaffold support	182	116+
Self sustaining	182	180
Sleeping car	105	326
Wall or floor attached	182	93+
Ladles	141	110
Glass handling	65	324+
Hoist traversing	212	130
Kitchen	30	324+
Design	D 7	691
Metal dispensing type	222	
Molding device with	164	335+
Lagging		
Reels	206	398+
Lake		
Dye fixing with	8	
Mordanting metal	8	
Tannate combined	8	596
Pigments containing	106	402
Laminate (See Class 428 Glossary)	428	
Abrasive tool		
Flexible cylindrical	51	358+
Making	51	297
Rigid cylindrical	51	207
Belt transmission	474	237+
Brushes rotary	15	181
Building constructions	52	408+
Curvilinear	52	267
Building panel	52	662
Conductors insulated	174	120C+
Electrode for electrolysis	204	290R+
Glass	428	426+
Making	156	99+
Hinges	16	372
Insoles	36	44
Making	156	
Coating and stripping	264	
Coating apparatus	118	
Coating processes	427	
Metallic	428	615+
Nuclear fuel structures	376	416
Panel fences	256	24+
Pipe	138	140+
Flexible	138	137+
Pulleys	474	166+
Rail bonds	238	14.1+
Shoe parts	12	
Sole leveling	12	33+
Soles	36	30R
Stencils	101	128.21
Wipers, eg blackboard erasers	15	223+
Wood	144	
Lampblack	423	445+
Agglomerating	23	314
Apparatus	422	150+
Electrostatic field or electrical discharge preparation	204	173
Lamps (See also Electrode)		
Arc	313	
Barrel illumination	362	154
Bicycle	362	72
Burner		
Camera attached light	D16	239+
Candle in lantern	362	161
Candle simulating	362	157+
Carburetors for	48	144+
Chimney cleaners	15	211+
Chimneys	362	312+
Consumable electrode type	314	
Daylight illuminating fixture	362	1
Daylight lamp structure	313	110+
Fluorescent material type	313	486+
Design	D26	
Diagnostic	128	23
Dimmer structure	323	905*
Discharge device, energizing	315	
Disinfecting, ultraviolet	250	504R+
Dough raiser	126	282
Egg candling		
Box with lamp control	356	67
Lamp attachment	356	68
Electric	313	
Arc lamp	313	41
Arc lamp electrode feeding	314	
Arc lamp gas or vapor	313	567+

	Class	Subclass
Arc lamp mercury vapor	313	163+
Automatic substitution of filament	315	65
Circuit element in multiple filament	315	64+
Connectors for clusters of	439	638+
Consumable electrode	314	
Exhausting	141	65
Filament compositions	252	500+
Filament for gas or vapor lamps	313	341+
Filament shapes	313	341+
Gas filling	141	66
Gas or vapor	313	567+
Glassworking apparatus for making	65	152+
Glassworking processes for making	65	36+
Glow discharge	313	567+
Incandescent bulb manufacturing	445	27
Mercury vapor	313	567+
Multiple filament	313	316
Nernst	313	14
Portable self contained	362	157+
Reflector built in	313	113+
Repair apparatus	445	61
Repair processes	445	2
Sockets	439	
Sockets rheostat combined	338	70+
Space discharge device manufacturing	445	
Systems of supply for	315	
Electrophotographic copier	355	228+
Extinguishing devices	431	144+
Flashlight	362	208+
Camera shutter synchronized	354	126
Photographic type	431	358+
Globe or bowl	D26	118+
Heater attachment	126	255+
Tool	126	235
Heating with electric	219	552+
Illuminated		
Sign attachment	40	553+
Sign lamp box	40	564+
Incandescent	D26	2+
Inspection	362	138
Magnesium	431	99
Miners	362	164+
Nonwired supply		
High frequency field	315	248
Induction type	315	248
Radio	315	149+
Transformer in lamp	315	57
Ornamental bulb	220	2.1 R+
Fluorescent	313	493
Gas or vapor	313	567+
Reflector	313	113+
Sign	40	541+
Photographic lighting	362	3+
Photographic safe lamp	362	803*
Pole		
Decorative element	52	301
Post combined with	D26	67+
Shades	362	351+
Design	D26	118+
Making	493	950*
Rotating	362	35
Signal lamp, horn combined	116	3
Signaling devices having	340	
Design	D10	111+
Socket	D13	134
Stands	362	382+
Switchboard	379	315+
Testing of	379	20
Therapeutic	128	396+
Tool heater combined	126	241
Vehicle	362	61+
Water heater type	126	358
Wick trimmers	431	120
Wicks	431	325
Lampworking		
Glass manufacturing	65	
Lances		
Bomb	102	371
Fishing	43	6
Metallurgical	266	225+
Lancet	606	181+
Land		
Anchors	52	155+
Markers for plows	172	126+
Rollers	404	122+
Land Vehicle (See Specific Types)	280	
Air bags	280	728+

	Class	Subclass
Cathode ray tube		
Structure	313	317+
System	315	14+
Cleaner		
Brushing	15	
Composition	252	89.1+
Scrubbing	134	
Coated	350	165
Coating method	350	320
Coating per se	350	165
Combined with electric lamp	313	110+
Contact for eye	351	160R+
Contact lens fitting	351	247
Design	D16	134+
Echelon lens	350	451
Electromagnetic wave	343	909
Dielectric type	343	911R
With antenna	343	753+
Zoned type	343	910
Electron microscope	250	311
Eye glasses	351	159+
Finder	354	152+
Fog lamp	362	317+
Fresnel lens	350	452
Galilean telescope	350	453
Gauge	33	507
Grinding machines	51	284R
Design	D15	124+
Hoods	350	580+
Intraocular	623	4+
Lenticular elements	350	167
Light fixture	D26	120+
Luneberg type	343	911L
Making	65	37+
Grinding machines	51	
Grinding processes	51	284R
Shaping glass surfaces	65	61
Uniting glass	65	37+
Marker gage	33	200
Mold	425	808*
Moulding glass while hot	65	66+
Moulding plastic	425	
Mounts and holders		
Eye glasses and spectacles	351	41+
Light source combined	362	455+
Sidewalk light transmitting	350	409+
Optical	350	409+
Panoramic lens	350	441
Periscope	350	540+
Projector combined		
Antiglare	362	351+
Reflector combined	362	341+
Reflector signal combined	350	104+
Refracting combined	362	326+
Schmidt lens	350	443
Shades	350	580+
Sidewalk light transmitting	404	22+
Spectacles	351	159+
Lens lining or rims	351	154
Spectacle mountings	351	83+
Telephoto	350	454+
Testers	356	124+
Thermometer tube combined	374	194
Toroidal lens	350	434
Fuse or primer combines	102	200+
Variable focal length	350	423+
Zoom lens	350	423+
Lenticular	430	946*
Lepidine	546	181
Letoff (See Clutch; Ratchet)		
Warp	139	100+
Letter Box	232	17+
Closures and chutes	232	45+
Design	D99	29
Electric switch combined	200	61.63
Letter Opener	D 8	102
Lettering Guides, Stencils	D19	40
Letters Sign Interchangeable	40	618+
Lettuce		
Planters	111	914*
Levees	405	107+
Leveling		
Apparel	33	7
Clocks	248	188.4
Furniture	248	188.2+
Caster adjustment	16	19
Illuminated	33	348.2
Load levelers	267	64.16+
Metal straightening	72	160+
Wire fabric	140	107+
Railway trucks	105	164

	Class	Subclass
Self leveling supports	114	191+
Shaft	33	412
Sole machines		
Fastener inserting and laying combined	12	33.1
Laying combined	12	33+
Straightedge geometric instrument	33	451
Wall guide & plumb	33	404+
Suspended railway car	105	149
Track	33	287
Tripod or stand head	248	180
Self leveling	248	182
Vehicle body land type	280	840+
Body relative to wheels	180	41
Steam boiler body	180	39
Levels (See Leveling)	33	365+
Illuminated	33	348.2
Light ray type	33	290
Liquid		
Alarm electric	340	618+
Alarm mechanical	116	109+
Dispenser material level control	222	64+
Gauge	73	290R+
Indicator	116	227+
Responsive or maintaining systems	137	386+
Material level detection	367	908*
Plumb	33	392+
Spirit level	D10	69
Lever	74	519+
Design	D 8	88
Flexible material tensioning	254	243
Carpet	254	209
Floor jack	254	17
Handle, asymmetric	D 8	DIG. 1
Link systems	74	469+
Locks	70	192+
Switch combined	70	254
Switch combined	200	61.85+
Nail extractor	254	21+
Pushing or pulling implement	254	120+
Rail or tie shifters	254	44
Shaft coupling	403	
Treadle	74	561
Vehicle spring	267	228+
Levigation		
With electric heating	219	7.5
Levitation & Reduced Gravity	156	DIG. 62
Magnetic railway	104	281+
Levulinic Acid	562	577
Esters	560	174
Lewises	294	89
Lewisite	556	70
Lexan T M (See Synthetic Resin or Natural Rubber)		
Leyden Jars	361	301+
Liberating		
Fermentative	435	262+
Fibers chemical	162	1+
Fibers mechanical	19	1+
Library Security by Metal Detection	340	568+
License Plate		
Design	D20	13+
Frames	40	209
Frames	40	152+
Holder with light	D26	31
Jewel reflectors	350	97+
Licorice	426	638
Lid (See Closure; Cover)		
Applying	53	287+
With compacting of contents	100	54+
Burner	431	144+
Car journal	384	189+
Lifter	294	12+
Lock	70	158+
Manhole	404	25+
Purse	150	119
Railway ties tubular type	238	70+
Retort	48	124
Stove	126	211+
Lie Detectors		
Electrocardiograph	128	697
Life Boats	114	348+
Design	D12	316+
Life craft handling devices	114	365+
Life Preservers	441	88+
Aquatic lifesaving devices	441	80+
Design	D21	238
Life vests	441	106+
Racks	114	190
Life Rafts	441	35+

	Class	Subclass
Life Sciences, Computer Applications	364	413.1+
Life Signals		
Computer controlled, monitored	364	413.2+
Signalling from burial grave	27	31
Undertaking equipment	27	31
Lifesaving Apparatus		
Aquatic	441	80
Artificial respiration	128	28+
Buoyant devices	99	311
Drowning prevention	441	80
Fenders	293	
Fire escapes and body catchers	182	137+
Parachutes	244	142+
Submarine emergency equipment	114	323+
Supporting and launching boats Handling equipment	114	365+
Watercraft lowering drum	254	266+
Lift Modifiers for Aircraft	244	198+
By flap or spoiler	244	213+
Lift Trucks		
Elevator type	187	9R
Handling assistant	414	444
Jack type	254	2R+
Running gear lifts	280	43+
Lift-slab Construction	52	125.1
Lifts, Lifters, Lifting (See Conveyor; Dispensing; Elevator)		
Aircraft	244	34R
Apparel skirt	2	217
Beet	171	50+
Beet	171	103
Belt shifter	474	101+
Bridge lifts	14	42
Canal lock raising	405	86
Car coupling link	213	192+
Car journal box	254	33+
Cartridge	42	17
Clutch for plow	192	62
Copyholder platen	40	349+
Damper musical instrument	84	217
Dobby	139	71+
Door	414	684.3
Traversing hoist	212	166
Drawbar locomotive	213	5
Drum sifter	209	294+
Foot sewing machine	112	237+
Harvester cutter	56	273
Harvester cutter bar	56	283+
Harvester cutter bar finger	56	312+
Harvester cutter bar with rocking	56	276+
Harvester reels	56	364
Hinge combined	16	233
Hoist	254	266
Horse drawn rake	56	397+
Incline plane	254	88
Invalid beds	5	60+
Invalid from bed	5	81R+
Jacquard	139	65
Lantern globe	362	174+
Magnets		
Electric	335	291+
Electro combined	294	65.5
Multiple	254	89R+
Pallets fork lift	108	51.1+
Pawl driving	74	149+
Pawl holding	74	155
Pipe or rod	254	30+
Planer cutter	409	347
Planter tool bar	111	67+
Plate and receptacle	294	27.1+
Plows	172	
Clutch	192	62
Control	172	452+
Plant	172	517
Power take-off	172	492
Servomotor actuator	172	491
Pot lid	294	12
Rail or tie	254	121
Shelf loading type	414	246+
Horizontal & vertical movement	414	260
Ships and boats	405	3+
Spool in winder	242	46
Thread boards	57	360
Traction cable	104	199
Valve lift regulation	123	345+
Vehicle body	254	45+
Ligators	606	139+
Ligatures		
Packaged	206	63.3
Structure	606	228+
Light (See Electric; Electrooptical; Illumination)		
Arc lamp	314	

	Class	Subclass
Contact and	261	2+
Deaerating	55	159+
Deaerating processes	55	36+
Distillation	203	
Electrical	204	186+
Electrical apparatus for	204	302+
Emulsion breaking	252	319+
Gas separation combined	55	421
Gas separation process combined	55	45
Ice melt and impurities	62	318
Liquid meters	73	200
Liquid-solid	4	DIG. 19
Oil dehydration	208	187+
Oil electrophoresis	204	181.8+
Oil paraffin separation	208	28+
Oil residual from paraffin	208	30+
Selective freezing	62	123+
Selective freezing process	62	67
Sprinkling, spraying, diffusing	239	
Testing	73	53+
Treatment		
Amylaceous liquids	127	69+
Coating implement with material supply	401	
Combs employing	132	219+
Drying	34	
Electrolytic foods and beverages	426	
Electrolytic water and sewage	204	149+
Fermentation	435	
Food preserving by	426	330+
Food sterilizing and pasteurizing	99	467+
Grain huller apparatus	99	600+
Grain huller process	426	482+
Lighted tobacco extinguisher	131	236
Metal heat treating combined	266	121+
Metal quenching apparatus	266	114+
Continuous strip	266	111+
Metal quenching processes	148	13+
Sound muffling by	181	220
Disparate fluid mingling	181	259+
Sterilization with except foods	422	
Textile apparatus for	68	
Textile combined	28	167+
Textile fiber decorticating	19	7+
Textile ironing	38	77.1+
Textile processes for	8	
Thread finishing	28	217+
Vaporizing	122	
Warp preparing	28	178+
Liquid Crystals		
Article or stock	428	1
Cholesteremic	552	544
Composition	252	299.1
Display device	350	330+
Light control	350	
Display of message	340	784
Hand and dial	368	84
Monogram	340	765
Time indication	368	242
Optical filter composition	252	582
Liquors	426	592
Malt	426	64
Listers	172	642
Drilling	111	83
Walking cultivator or plow	172	351+
Listing Forms	283	66.1+
Lite Pipe Sign	40	547
Litharge	423	619
Lithium (See Alkali Metal)	423	179.5
Lithocholic Acid	552	552
Lithography	101	130+
Lithographic emulsion	430	949*
Lithopone	106	421+
Drying oil containing	106	255
Lithotrites	606	127
Litter	5	81R+
Litter bag	D34	1+
Portable	5	82R+
Wheeled	296	20
Livestock		
Care	119	
Design	D30	
Cars	119	7
Living Product	800	2+
Lixiviating (See Leaching)		
Load		
Braced on freight carrier	410	121+
Bridging vehicle	414	460+
Cell	177	
Hauling or hoisting	254	266
Lashed on freight carrier	410	96+

	Class	Subclass
Levelers	267	64.16+
Starters		
Animal draft appliances combined	278	2
Wheeled vehicle attachments	280	151
Loaders & Loading (See Gun; Magazine)		
Aircraft	244	127+
Ammunition	86	23+
Antenna	343	749+
Cartridge belts	86	48
Conveyor chutes	193	3
Dump truck	298	
End gate loading chute combined	296	61
Excavator	37	4+
Fireworks	86	20.1+
Glider pickups	244	2+
Harvester combined with raker	56	344+
Loaded		
Conductors for signaling	178	45+
Line	178	45
Feeding to amplifier	178	45
Phantom circuit having	178	45
With attenuator	178	45
Marine	414	137.1+
Nuclear reactors	376	260
Nuclear reactors	376	268
Railway car bodies dumping	105	239+
Railway mail delivery	258	
Self loading vehicles	414	467+
Sheaf	56	120
Shoe nailer	221	
Textile fluid treatment	68	210
Trains in motion	104	18+
Vertically swinging shovels	414	680+
Weighing scales	177	145+
Loafer, Footwear	D 2	269
Lobster Claw Clamp	17	1R
Localizer	342	413
Locating (See Detection; Testing)		
Foreign bodies	128	737+
Location determining devices		
Acoustic	181	125
Radar	342	126+
Railway track defect	73	146+
Train protecting combined	246	120+
Sound		
Mechanically	181	125
Submarine	367	87+
Lock (See Definition Notes, Associated Devices; Fastener; Detent; Latch; Seal; Securing Means)	70	
Air		
Caissons	405	8
Diving bell combined	405	192
Dryers	34	92
Dryers vacuum	34	242
Furnace charging	414	162+
Material moved between unlike pressure areas	414	217+
Pneumatic conveyor discharging	406	169
Submarine escape	114	334+
Article dispenser	221	154
Automobile	70	237+
Bicycle	70	233+
Book binder sheet retainers	402	2
Box and chest	70	63
Cabinet with		
Selector operator	312	215
Canal	405	85+
Cane, umbrella, apparel	70	59
Closure	70	77+
Barrel bung	217	106+
Bottles and jars	215	207
Burglar alarm combined	116	8+
Camera plate holders	354	281
Car dump	105	308.1+
Check controlled	194	
Drawers	312	333
Elevator door	187	61
Fasteners	292	
Freight car automatic lock	105	395
Railway car platform trap door	105	435
Safe alarm combined	109	44
Safe bolt work	109	59R+
Safe door poison gas releaser combined	109	30
Special receptacle	206	1.5
Till drawer alarm combined	116	76
Vault covers	404	25+
Combination	70	286+
Design	D 8	331+

	Class	Subclass
Dispensers with	222	153
Dog for bolt	70	467+
Door	70	91+
Door latch bolts, biased	70	144
Electric lamp	439	133+
Expansible chamber device with	92	15+
Fluid	92	8+
Firearm safety mechanism	42	70.1+
Fluid handling systems or devices with	137	383+
Gang bar type for cabinet	312	216+
Sectional unit type cabinet	312	107.5
Hasp	70	2+
Key removal preventing	70	389
Keyholders	70	456R+
Lap robe in car body	296	77
Machine elements valves etc	70	175+
Mechanical gun	124	40
Mounting devices	70	466
Nail	411	439+
Railway spike	238	375
Nut and bolt		
Bolt pivoted end	411	340+
Bolt threadless key type	411	340+
Bolt to nut	411	190+
Railway splice bar	238	262
Openers electric or magnetic	70	277+
Operating mechanism	70	266+
Pad	70	20+
Permutation	70	315+
Pipe joint	285	80
Portable	70	14+
Puzzle	70	289+
Rack combined	211	4+
Railway switches and signals	246	
Register	235	130R
Key locks	235	27
Safety chains	292	262+
Seal tamper detecting	70	440
Latch	292	307R+
Padlock	70	50
Shear line	70	496
Side lock	411	191
Sidebar	70	496
Slotted bolt, railway splice bar type	238	252
Special application	70	57+
Stitch making	112	
Switch		
Lock combined	200	61.64+
Operated	200	43.1+
Systems	70	262+
Telephone	379	445
Operated over telephone line	379	103
Time	70	267+
Trunk	70	69+
Typewriter	400	663+
Valve	70	175+
Automobile fuel	70	242+
Gravity feed lubricators	184	86
Lock operated	137	384.2+
Ward guard	70	420
Weigher	177	124
Wheel	70	225
Brake	188	69
Caster	16	35R+
Yale	70	357+
Lockets	63	18+
Design	D11	80
Lockout		
Key locks	70	337
Telephone	379	194+
Intercommunicating	379	168+
Polystation	379	184+
Locomotive		
Boilers	122	
Chain grate progressive feed	110	198+
Design	D12	38
Feeding air		
Firebox	110	198+
Feeding air and steam		
Firebox	110	199+
Heater	110	201
Undergrate	110	202
Firebox	110	198+
Fuel feeding stoker	110	105.5
Headlight and signal illuminators	362	61+
Journal	384	187
Light supports automatic	362	66+
Railway	105	26.5+
Safety bridges	105	459
Truck frames	105	172+

	Class	Subclass
Shed fixtures	104	51+
Smoke and gas return	110	198+
Stoker type comminutor	241	276
Tender scoops	417	232
Toy	446	467
Illuminated	446	438+
Smoking	446	25
Sounding	446	410
Valves, gears	91	218+
Ventilation of cab	98	3
Log		
Borehole testing	73	152
Building	52	233
Toy	446	106
Gas burner	431	125
Gas or electric, simulative	D23	409
Loading vehicle	414	460
Ship		
Course recorder	346	8
Navigation	73	181+
Splitter	144	193R+
Wood sawing		
Band saw	83	788+
Log deck block	83	708+
Log transfer	83	707+
Log turner	83	708+
Log turner and deck block	83	708+
Woodworking		
Stay log slicer	144	177+
Stay log veneer lathe	144	214
Logging		
Cable type traversing hoists	212	76+
Pole or tree handlers	414	23
Vehicle	414	460
Wood sawing	83	703+
Logic Gate Active Element Oscillator	331	DIG. 3
Logotype	400	95+
Looms	139	
Circular	139	457+
Lace	87	27
Special type	139	11
Stationary weft supply	139	429
Traveling wave shed	139	436
Loop		
Amusement railway with vertical	104	55
Animal catching lasso	119	153
Antenna	343	866+
High frequency type	343	741+
Belt		
Machine for sewing	112	121.27
Cartridge belt	224	223+
Clothes line endless	211	119.2+
Fastening device		
Billet loops	24	182
Button engaging	24	660+
Button engaging garter type	24	464+
Engaging with camming lever	24	69R+
Hat combined with comb or hairpin	132	61
Lace guides	24	714.6+
Pipe couplings	285	71
Tag attaching pins	40	668
Tie to collar button	24	58
Garment	2	271
Garment supporter	2	321+
Crossed loop	2	327+
Plural loops	2	331
Handle	16	125+
Insulated	16	119
Harness		
Bridle crown	54	13
Checkrein hook	54	17
Design	D30	140+
Strap connector	54	87
Lock		
Element on trunk	70	36
Rail holddown	238	43
Sewing machine	112	32
Railroad splice bar	238	256
Textile pile	26	2R+
Cutting	26	7+
Trap	43	86
Wire fabric interlocking	245	5
Loopers Festooner		
Carpet rag	28	148
Drier combined	34	157
Knitting	66	
Packet	227	77
Sewing	112	
Web or strand feeding	226	104
Wireworking	140	102+

	Class	Subclass
Loose Leaf		
Binder device releasably engaging aperture or notch of sheet	402	
Opening reinforcement	402	79
Lorgnettes	351	56
Design	D16	105
Supports	351	56
Lorol (See Lauryl Alcohol)	568	840
Lotions	424	
Loud Speaker (See Telephone)	381	150+
Baffle	181	155
Design	D14	221+
Public address system	381	82+
Loudness Meter	73	646
Lounges	52	473
Louvers		
Movable	49	74+
Shutters		
Unit design	D25	152+
Ventilating	98	121.1
Low mu Triode	313	293+
Lowering		
Beds	5	65
Coffins	27	32+
Cable	254	266+
Hoist hook	212	89
Life craft	114	365+
Load with automatic return	414	594
Plow	172	464
Support releasable	248	320+
Vehicle tops	296	112+
Lozenges		
Design	D 1	127
Medicinal	D28	1+
Presses	99	353+
Lubrication and Lubricators (See Oiler and Oiling)	184	
Air line lubricators	261	DIG. 35
Ammunition	102	511
Axle jacks	254	32
Compositions (See lubricants)		
Earth boring drilling fluid	175	65+
Engines, internal combustion	123	196R
Poppet valve operating mechanism	123	188A+
Fishing reel	242	319
Fitting	285	169+
Valved	251	142+
Fluid handling	137	246+
Gun	D 8	14.1
Lubricants (See oil)	252	9+
Mineral oil only containing	208	18+
Purifying used	208	179
Testing	73	64
Frictional characteristics	73	10
Gas content	73	19.11
Lubricating apparatus	184	
Gun	D 8	14.1
Oil dispensers	222	
Machine parts		
Band saw pulley	83	169
Bearings	384	
Belt and sprocket	474	91
Brakes	188	264B
Casters	16	36
Clutches	192	113B
Connecting rods	74	587
Crank and wrist pins	74	605
Earth boring tool combined	175	227
Endless flexible track pivot	305	14
Gears	74	467+
Glass making apparatus	65	170
Hinges	16	274
Metal founding apparatus	164	149
Pistons	92	153+
Piston rings	277	
Planes wood	30	483
Planetary gearing	475	159+
Sewing machines	112	256
Sewing machines leather	112	43
Shaft coupling	464	7+
Spinning rings and travelers	57	120
Spinning whirls	57	133
Springs leaf	81	3.7
Springs leaf	267	50
Springs leaf covered	267	37.4+
Sprocket chain pivot	474	91
Textile drawing saddles	19	284
Valve actuator	251	355
Valve interface	137	246+
Vehicle wheels spring type	152	2+

	Class	Subclass
Material working		
Ammunition making	86	19
Cutters	407	11
Die expressing	425	197+
Die expressing	425	200+
Dies wire drawing	72	41+
Screw threading	408	
Pipe joint	285	94
Refrigeration apparatus	62	468+
Automatic	62	192+
Process	62	84
Reservoir for packing assembly	277	17+
Separator, imperforate bowl centrifugal	494	15
Textile compositions	252	8.6+
Textile processes	28	169+
Braid	87	1
Coating or impregnating	427	
Strands	57	295+
Strands covered	57	7
Lucite T M (See Synthetic Resin or Natural Rubber)		
Luggage (See Baggage)	190	
Design	D 3	30.1+
Latch	D 8	331+
Travel bag	206	278+
Lumber		
Building component	52	
Compound	52	782+
Denailing	254	18
Impregnated	428	541
Laminated	428	18
Luminal	544	305
Luminaries	362	296+
Luminous (See Fluorescent; Illuminated; Light)	362	
Cathode ray screens	313	461+
Photo process of making	430	23+
Compositions fluorescent or phosphorescent	252	301.16
Inorganic	252	301.6 P+
Organic	252	301.16
Compositions radioactive	252	625+
Fluorescent and phosphorescent	250	458.1
Heads for figure toys	446	391+
House number	250	462.1
Sign	40	541+
Illuminating devices with material	362	84
Lamps	313	
Push button	250	465.1
Toys	446	485
Chemiluminescent	446	219
Lunch		
Heaters	126	261+
Kit or box	206	1+
Design	D 7	626+
With heater	126	266
Wagons	296	22
Luneberg Lens Type	343	911L
Lungs		
Heat-lung digest	128	DIG. 3
Lungmotor respirators	128	28+
Lures		
Fishing	43	4+
Artificial	43	42+
Design	D22	126+
Food bait	426	1
Lustering		
Fabrics	26	27+
Lutidine	546	348
Lymphokines	530	351
Lysergic Acid	546	69
Lysergic Acid Diethylamide Lsd	546	69
Lysine	562	562
Lythrine	546	43
Macaroni	426	557
Apparatus		
Die expressing	425	288
Die expressing	425	289+
Die expressing	425	331
Die expressing	425	381.2
Design	D 1	126+
Drying digest	34	
Machete	30	
Design	D22	118
Machine Elements	74	
Automatic operation or control	74	2+
Bearings	384	
Design	D15	143
Belt & pulley drive system	474	
Brakes	188	

	Class	Subclass
Fluid pressure brake & analogous systems	303	
Clutches & power-stop control	192	
Control lever & linkage systems	74	469+
Cylinders and pistons	92	
Design, general	D15	
Engine starters	74	6+
Gearing	74	640+
Planetary	475	
Gyroscopes	74	5R+
Intermittent grip type movements	74	111+
Lighting	362	89+
Mechanical movements	74	20+
Power take-off	74	11+
Pushing & pulling implements, force multipliers	254	1+
Rotary member or shaft indexing	74	813R+
Shaft operators	74	10R+
Shafting and flexible shaft couplings	464	
Lost motion couplings	464	160+
Rotary shafting	464	179+
Tool driving or impacting	173	
Work holders	269	
Machine Gun	89	9+
Explosive propellant type	89	9+
Charger	89	1.4
Self loading explosion operated	89	125+
Mechanical	124	
Centrifugal	124	4+
Machine Supports	248	637+
Machinery, Toy	446	
Mackinaws	2	93+
Macrame	289	16.5
Magazine (See Handle Receptacle Type)		
Ammunition and explosive		
Body or belt attached	224	223+
Bandoleer	224	203
Covered flap	224	239
Chargers	42	87+
Gun combined	42	2+
Loading	86	47
Mechanical gun or projector	124	45+
Ordnance	89	33.1+
Railway torpedo	246	212+
Revolver combined	42	60
Structure	42	49.1+
Article dispenser (See class defs)	221	
Boiler		
Liquid heater or vaporizer	122	
Camera	354	174+
Coin return, empty dispensing	194	200
Dry closets	4	464+
Electrode	314	5
Fire pots combined with magazine section	D23	386+
Holder as furniture	D 6	396+
Implements and machines with		
Assembling apparatus	29	809+
Check controlled machine	194	
Detonating cane	446	403
Label pasting machine	156	
Digest, apparatus	156	DIG. 29
Nail drivers	227	
Pencil	401	89+
Pencil selectively replenishing feed guide	401	56
Pencil sequentially replenishing feed guide	401	57
Razor	30	40
Riveter	227	
Riveter eyelet	227	51+
Sound recording tapes	D14	120+
Stapling	227	
Type cases	276	44+
Type casting machine	199	1+
Type setter	276	15
Type setter and distributor	276	3+
Typewriter paper feed	400	624+
Wire and slat	140	27
Wire and slat portable	140	38
Wireworking	140	53+
Labeling	156	DIG. 29
Lock magazine cylinder & pin tumblers	70	DIG. 5
Motion picture film	352	78R
Paper cover removal	83	933*
Photographic	354	174+
Slide projector, design	D16	236
Razor blade holder	206	352+
With means of feeding blades into razor	30	40
Stove and furnace		
Furnace structure	110	293+
Stove	126	
Toaster	99	387
With article removing facilities	312	35
Magenta (See Fuchsine)		
Magic		
Amusement devices	272	8R+
Clock	368	81
Slides	40	152+
Magnesia	423	635
Cement	106	801
Magnesium		
Alloys	420	402
Aluminum	420	542
Aluminum copper	420	533
Electrolytic synthesis	204	70
Compound inorganic	423	
Carbonates	423	430+
Inorganic	423	
Oxide and hydroxide	423	635
Electrothermic processes	75	10.1+
Volatilization of magnesium	75	10.33
Heat treatment	148	161
Metal working with	29	DIG. 27
Oxychloride cement	106	685+
Peroxide	423	583
Plate	428	649+
Preparation		
Electrolytic	204	70
Pyrometallurgy	75	594+
Strip type solid fuel	431	99
Magnet and Magnetic (See Type of Magnet)	D13	183
Actuated motor for railway switches	246	227+
Actuated motor for railway switches	246	231+
Actuated switch	335	2+
Alloys	420	
Analog storage systems	365	48
Antimagnetic devices for watches and clocks	368	293
Arc or ray deflecting or focusing		
Arc lamp	313	153+
Arc lamp with discharge deflecting	314	20
Cathode ray tube	313	421+
Cathode ray tube circuits with plural	315	14+
Electron microscope	250	311
Electronic tube controlled by	315	267
Electronic tube controlled by	315	344+
X ray tube	378	137
Board games	273	239
Bracket	248	206.5
Holding article by	248	309.4
Bubbles, magnetic	365	1+
Chargers	335	284
Chuck	279	1R
Holding magnet	335	285+
Circuit and switch control		
Arc blowout	315	344
Arc blowout in switch	200	147R
Electromagnet circuits	361	139+
Electromagnetic switch	335	2+
Electromagnetic switching	361	139+
Operator for brake	188	161+
Railway cab signal	246	178+
Railway cab signal	246	193+
Railway cab signal	246	197
Railway cab signal	246	202
Railway point thrower	246	265
Railway switch actuation	246	225+
Railway train trip	246	360
Transmission to vehicles	191	16
Transmission to vehicles	191	17+
Circuit controller for railway section	191	17+
Arc lamp operator	314	104
Closures	220	230
Clutches	310	92+
Compasses	33	355R+
Compositions	252	62.51
Consumable electrode feed or control	314	
Control for frictional gearing	74	210
Conveyers and feeders		
Holding item by magnetic attraction	209	904*
Magnetic feeder	209	907*
Cooling by	62	3.1
Coupling	464	29
Demodulator for radio receiver	329	
Digital signal processing	360	39+
Disk	360	135
Drum	360	136
Transport systems	360	81+
Dynamic magnetic info storage and retrieval	360	
Automatic control system	360	69+
General recording or reproducing	360	55+
Head	360	110+
Physical elements	360	102+
Record medium	360	131+
Signal processing	360	18+
Special purpose devices	360	1+
Earth boring tool	175	328
Electro	335	209
Lifting	335	291
Removing splinters from the eye	600	11
Electrochemistry apparatus	422	186.1+
Explosive igniters		
Fuses	102	206+
Marine mines	102	417
Feeler mechanism loom	139	274
Fender securing device	293	DIG. 6
Fishing reel brake	242	286
Flip flops	307	406
Hoist line type	294	65.5
Induction current transmission to vehicles	191	10
Instruments	33	DIG. 1
Joints and connections digest	403	DIG. 1
Key lock	70	413
Levitation railway	104	281+
Lifting handling or holding	335	285
Bracket supports	248	206.5
Conveyors	198	690.1
Enhanced by magnetic hold	198	690.1+
Detachable electric connector	439	38+
Grapples	294	88
Hand and hoist line implements	294	65.5
Light supports	362	398
Material handling	414	606
Metal founding apparatus	164	146
Railway mail delivery apparatus	258	4
Lightning arrester	361	133+
Locomotive traction regulator	105	77+
Loom shuttle actuator	139	134
Magnetic bubbles	365	1
Magnetic memory	365	
Associative memories	365	50
Dynamic information storage &	360	
Formation by shaping	264	104
Laminated type	428	900*
Magnetic bubbles	365	1+
Read-only memories	365	97+
Static information storage & retrieval	365	
With coated metal base	437	
Magnetic recording digest	29	DIG. 28
Making	29	603+
Material treatment	148	100+
In magnetic field	148	108
Metal casting with	164	498+
Continuous casting metal with	164	466+
Nonmetallic	252	62.51
Particulate material	148	105
Permanent	148	101+
Permanent in magnetic field	148	103
Permanent with age hardening	148	102
Silicon steel	148	110+
Metal working digest	29	DIG. 95
Microphone	381	177+
Mines	102	401+
Motor art digest	91	DIG. 4
Motor operated lock	70	276
Motors with armatures	335	220
Motors with armatures	310	10+
Music leaf turner	84	521
Music tone generator	84	641+
Package or receptacle	206	818*
Permanent and electromagnet structure	335	209+
Making	29	607+
Press digest	100	917*
Pulse generator	307	419
Read only memories	365	97+
Record and recording	360	1+
Coating	427	128+
Copying	360	15+
Editing	360	13+
General reproducing	360	55+
Medium	360	131+

	Class	Subclass
Paper pulp dip moulding	162	411+
Plastic mold	425	298+
Printers	428	908*
Printing and process of making	101	401.2+
Forming machine	101	401.5+
Selective	340	825.79+
Static mold	249	
Toe and heel stiffener	12	66
Type casting	199	
Matte		
Preparation; copper	423	48
Pyrometallurgy; copper	75	643+
Mattock		
Design	D 8	11
Hoes	172	375
Making	76	109
Root cutters and grubbers	172	
Mattress		
Air or gas filled	5	449+
Bed type	5	448+
Covers for	5	499
Design	D 6	596+
Design	D 6	596+
Earth control revetment	405	19
Fasteners floppers turners	5	411
Filling and closing	53	524
Fluid filled	5	449+
Invalid	5	431+
Life raft	441	135
Protectors	5	499
Rack	211	28
Sewing machines	112	2.1
Springs	5	478+
Waterbed	5	451
Mausoleums		
Undertaking	27	1
Vaults, masonry or concrete	52	134
Mayonnaise	426	605
Maytansinol	540	462
Maze	272	19
Meal	426	622
Measuring (See Definition Notes)	73	
Acceleration	73	488+
Acetylene generator charge	48	46+
Apparel	33	2R+
Capacity measuring	324	658+
Winding coil apparatus in	242	7.12
Winding coil method in	242	7.3
Change of orientation	73	505
Coin controlled mechanism	194	
Coin size	194	339+
Conveyor combined	198	502.1+
Direction	33	300+
Dispensing combined	222	
Distance	33	700+
Distance by electromagnetic radiation	356	4+
Earth boring combined	175	40+
Electric	324	
Amplifier condition	330	2
Amplifier with measuring means	324	326+
Capacitance	324	658+
Voltage or current	324	76R+
Filling with stationary measure	141	
Filling with travelling measure	141	135+
Footwear	33	3R+
Fuel burner charge	431	287
Galvanometer	324	76R+
Geometric	33	
Lens testers	D10	61+
Household implement	D10	46.2
Instruments for	D10	
Lens testers	356	124+
Liquefied gas handling combined	62	50.1+
Lubricator	184	83
Medical	D24	8+
Meter electric	324	76R+
Nuclear reactor condition	376	245
Control of reactor, combined	376	207
Object by X-ray energy	378	
Optical	356	
Optical test	356	
Pyrometer	356	43+
Phase electric	324	83R+
Pressure	73	700+
Printing combined	101	73+
Radar, general	342	
Altitude	342	120
Distance determining	342	120+
Radiant energy	250	
Operational nuclear reactor	376	245

	Class	Subclass
Radio		
Receiver	455	226
Transmitter	455	95+
Range by electromagnetic radiation	356	4+
Refrigeration combined	62	125+
Signal energy in antenna	343	703
Speed	73	488+
Spoon, household	D10	46.2
Tapes, design	D10	71
Temperature	374	
Testing electric properties	324	
Testing physical properties	73	
Using ultrasonic	73	570+
Vessels	73	426+
Volume or rate of flow	73	861+
Wave meter electric	250	250
Weighing scales	177	
Meat	426	641+
Blocks	269	289R+
Resurfacing wood	144	2D
Briquetters	17	32
Butchering	17	
Choppers hand	30	
Cleaning	17	
Brushing or wiping machines	15	3.1+
Liquid contact apparatus	134	
Tenderer	17	25+
Comminutor	241	
Cutting boards	269	289R+
Cutting machine	83	
Foodstuffs design	D 1	
Hanging	17	44+
Preservation	426	321+
Canning	426	392+
Coating	426	289+
Dehydration	426	471+
Packaging	426	392+
Press	100	910*
Refrigeration	426	524+
Products, edible	D 1	
Slicers	D 7	381+
Tenderers	17	25
Pounders	17	30
Mechanical Memory Alloy	148	402
Mechanical Movements	74	
Belt & pulley drive system	474	
Mechanical Removal Assistant	29	DIG. 96
Mechanics Teaching	434	302+
Medal		
Badge	40	1.5
Jewelry	63	20
Ornamental design	D11	95+
Medical & Surgical Equipment		
Acupuncture	128	907*
Applicator	128	362+
Aspirators	604	35
Bandaging	128	82.1
Biological signal amplifier	128	902*
Blood pressure recorder	128	900*
Cabinets	312	209
Centrifuge (e. g., for blood)	494	16+
Colostomy bags	604	338+
Electric shock protection for patient	128	908*
Electrocardiograph	128	697
Electronics	323	911*
Feedback to patient of biological signal other than brain electric signal	128	905*
Fixed apparatus	D24	1.1+
Hypodermics	604	187+
Piston structure	604	218+
Injection devices	604	187+
Instrument	606	1+
Kinesiatric	128	24.1+
Magnetic	600	9+
Measuring instruments	D24	21+
Medicinal	604	19+
Multiphasic diagnostic clinic	128	906*
Noise suppression in electric signal	128	901*
Nursing	D24	34+
Orthopedic	128	68.1
Portable apparatus	D24	8+
Powder dispensers	604	58
Pressure infusion	604	131+
Radio telemetry	128	903*
Snake or insect venom extractors	604	314
Syringes (See syringe)		
Telephone telemetry	128	904*
Vacuum removal of material	604	131
From external body surface	604	313+
Venom extractors	604	314

	Class	Subclass
Medicated Devices		
Animal		
Antivermin	119	156+
Overshoes	168	2
Sole pads	168	27
Coated or impregnated base	424	443+
Medicinal applicator	D24	63
Nest eggs	119	46
Papers and fabrics	424	443+
Truss pads	128	114.1
Medicated Food	424	439+
Medicated Matrix	424	484+
Medicator	604	890.1+
Capsule	604	403+
Controlled release	604	890.1+
Medicine	424	
Anesthetic	424	
Cabinets	312	209
Bathroom	D 6	559+
Capsules	424	451+
Chest	206	438+
Coated or impregnated base	424	443+
Compositions	424	
Computer controlled, monitored	364	413.2+
Controlled release means	604	890.1+
Design, pharmaceutical product	D28	1+
Dose		
Administering devices	604	187+
Containers	206	528
Dispenser with removable doser	141	22+
Pump doser	141	23+
Indicator	116	308
Package	206	828*
Structured dosage unit	424	400+
Tear-off packaging	206	532
Special package	206	828*
Sulfa drug ctg	514	155+
Medium mu Twin Triode	313	301
With two cathodes	313	6
Megaphone	181	177+
Electric	381	75
Megohmmeters	324	691+
Melamine (See Synthetic Resin or Natural Rubber)	544	200
Aldehyde resins (See synthetic resin or natural rubber)		
Melatonin, in Drug	514	415
Melodeon	84	351+
Melting		
Cleaning by	134	5
Electric furnace	373	
Glass and slag	373	27+
Fat rendering	260	412.6+
Furnaces	432	161
Heaters	126	343.5 R
Iron processes	75	571+
Purifying combined	75	507+
Metal casting combined	164	47+
Metals furnaces	266	200+
Metals furnaces, X-art	266	900*
Nonferrous metals	75	585+
Partial melting	156	DIG. 105
Pots for glass	432	262+
Product manufacturing computerized	364	477
Separating combined	422	285+
Processes	23	308R+
Snow	37	227+
Spraying combined	239	79+
Sugar starch and carbohydrates	127	22
Sulphur	423	567R
Melton Fabric	26	19+
Membrane		
Dental dam	433	136+
Diaphragm	92	196+
Acoustical	181	157+
Collapsible	137	843+
Valve	251	331
Valve biased	137	540+
Pipe organ valve	84	336
Switches	200	512+
Vibrant surgical	128	64
Memorial Tablets	40	124.5
Memory Systems		
Computer storage unit design	D14	107
Dynamic electric information storage & retrieval	369	
Dynamic magnetic information storage & retrieval	360	
Malfunction detection, correction	371	
Plural diverse storage	364	900
Static information storage and retrieval	365	

	Class	Subclass
Cutter		
Air current generated by	29	DIG. 82
Ambulatory with fluent conduit	29	DIG. 64
Complete immersion	29	DIG. 71
Contained supply reservoir	29	DIG. 65
Drive	29	DIG. 55
Engaging cleaner	29	DIG. 97
Fan coaxial with	29	DIG. 83
Feed	29	DIG. 57
Fluid channel in	29	DIG. 92
Fluid spreader contacts	29	DIG. 69
Hood encased	29	DIG. 86
Partial immersion	29	DIG. 74
Rotating	29	DIG. 67
Debris		
Chutes	29	DIG. 102
Control	29	DIG. 5
Receptacle, remover	29	DIG. 79
Remover, catcher, deflector	29	DIG. 94
Removers, plural type	29	DIG. 61
Separators from workpiece	29	DIG. 53
Design	D15	122+
Digest	29	DIG. 11
Dip or splash supply	29	DIG. 72
Double blank	29	DIG. 2
Drawing	72	274+
Electric heating	29	DIG. 13
Electroplating	29	DIG. 12
Electrostatic	29	DIG. 95
Elevated tank supply	29	DIG. 9
Enameling	29	DIG. 14
Etching	29	DIG. 16
Extruding	29	DIG. 47
Fan coaxial with cutter	29	DIG. 83
Filter or separator	29	DIG. 77
Flexible conveyor	29	DIG. 99
Flow line & crystal making	29	DIG. 17
Fluid applicator, rotary or oscillating	29	DIG. 7
Cutting fluid application & debris control	29	DIG. 5
Fluid channel in cutter	29	DIG. 92
Fluid control interrelated with machine tool mechanism	29	DIG. 54
Fluid control valve	29	DIG. 85
Fluid conveyor or applicator	29	DIG. 63
Fluid flow to channeled cutter	29	DIG. 66
Fluid paths, multiple	29	DIG. 87
Fluid spreader contacts cutter	29	DIG. 69
Forging	29	DIG. 18
Forging and welding	228	
Forming article on end of long stock then cutting off	29	DIG. 15
Fusion-bonding apparatus	228	
Gas entrained liquid supply	29	DIG. 89
Gear cutting, milling, planing	409	1+
Grinding	29	DIG. 19
Heat treatment combined	148	11.5 R+
Heating	29	DIG. 21
Combined	219	149+
Local heating	29	DIG. 24
Hollow spindle contains	29	DIG. 68
Hood		
Actuated by tool or work approach	29	DIG. 59
Encased cutter	29	DIG. 86
Progressive displacement of	29	DIG. 6
Shield or hood movement	29	DIG. 56
Indium	29	DIG. 22
Knurling	29	DIG. 23
Machining	29	DIG. 26
Magnesium	29	DIG. 27
Magnetic	29	DIG. 95
Magnetic recording digest	29	DIG. 28
Mechanical manufacture	29	400.1
Mechanical removal assistant	29	DIG. 96
Molding combined	29	527.1
Molding with other steps	29	DIG. 29
Movable liquid-carrying trough	29	DIG. 76
Ornamenting	29	DIG. 3
Pans, beds & tables	29	DIG. 101
Particular materials	29	DIG. 25
Plastic state	72	
Plastically shaping	72	
Plural discharge loci	29	DIG. 93
Plural discharge openings	29	DIG. 91
Power stop control for movable element	29	DIG. 51
Pressing powder	29	DIG. 31
Reels	242	78+
Rolling	72	199+

	Class	Subclass
Rubbing transfer of solid coating to rotating element	29	DIG. 62
Digest	29	DIG. 32
Scaling	29	DIG. 34
Sheet and foil making	29	17.1+
Sheet metal container making	413	
Shot blasting	29	DIG. 36
Shrink fitting	29	DIG. 35
Soldering	219	85.1
Spot welding	29	DIG. 38
Spraying	29	DIG. 39
Stamping	29	DIG. 37
Staple making	59	72
Stock & blanks	428	544+
Strips of repeated articles cut up later	29	DIG. 4
Taper tube making	29	DIG. 41
Tension applied during working	29	DIG. 42
Titanium	29	DIG. 45
Tools and implements making	76	
Assembly, magnetic or indicia tools	29	DIG. 105
Tool or work attached	29	DIG. 104
Upsetting & flanging tube end	29	DIG. 43
Vacuum	29	DIG. 44
Vacuum exhaust type	29	DIG. 84
Vibration	29	DIG. 46
Welding	29	DIG. 48
Wire	140	
Work hardening	29	DIG. 49
Workpiece & debris separators	29	DIG. 53
Metaldehyde	549	353
Metalizing	427	
Electrolytic	204	14.1+
Metallic		
Metallurgy (See Metal & Metallurgy)	75	
Metampicillin	540	331
Metanilic Acid	562	58
Metaphosphates Inorganic	423	
Metaphosphoric Acid	423	307
Metasilicates	423	
Metatarsal Pads	128	69
Meteorology	73	170R+
Teaching	434	217
Meter and Metering	73	
Design	D10	46+
Dispenser combined	222	23+
Dispenser cutoff	222	14+
Automatic	222	59+
Dispenser volume or rate of flow	222	71+
Electric	324	76R
Amplifier combined with meter	324	123R+
Amplifier condition measuring system	330	2
Amplitude modulaton	332	150
Circuit protection	324	110
Coin controlled	194	
Electrolytic	324	94
Frequency	324	78R+
Frequency modulation	332	118
Galvanometer	324	76R+
Leakage	324	555+
Phase	324	83R+
Polyphase current	324	107+
Radiant energy	250	
Radio frequency	455	155
Recording	346	
Resistance	324	691+
Slip	324	
Switchboard	361	364+
Time interval of energization	324	139
Time lapse	368	114+
Triphase current	324	107+
Volt	324	76R+
Wave	250	250
Exposure photographic	356	213+
Flowmeters	73	861+
Gas separator combined	55	270
Hardness for roentgen tubes	324	413+
Illumination	356	213+
Noise exposure	73	572+
Noise exposure alarms	340	540
Parking		
Alarm electric	340	932.2
Coin controlled	194	
Time controlled	368	90+
Photo	356	443+
Photography		
Measuring shutter speed	73	5
Volume or rate of flow	73	861+
Dual pipe connections	285	30

	Class	Subclass
Methacrylate Resins (See Synthetic Resin or Natural Rubber)		
Azeotropic distillation with	203	66
Methadone, in Drug	514	648
Methanol (See Alcohol)		
Distillation destructive	201	
Fischer-tropsch synthesis	518	700
Low freezing point compositions	252	73+
Purification or recovery	568	913+
Dehydration	203	18
Methapyrilene, in Drug	514	336
Methicillin	540	339
Methionine	562	559
Methomyl	514	477
Methyl		
Methacrylate	560	205
Synthetic resins (See synthetic resin or natural rubber)		
Orange	534	845
Salicylate	560	71
Violet	552	114
Methylenecitric Acid	549	296
Metol T M	564	443
Metronome	84	484
Metronome design	D10	43
Mica	106	DIG. 3
Chemical treatment of	423	
Comminution	241	4
Exfoliation of	252	378R
Flake	428	363
Laminate	428	454+
Laminate non-structural	428	446+
Layer	428	363
Mass, structurally defined	428	363
Pigment containing	106	415+
Splitting	125	24
Micellar Solution	252	302+
Michlers Ketone	564	328
Microbiology (See Molecular Biology & Microbiology)	435	
Microcapsules	428	402.2+
Gaseous core	521	76
Liquid core	428	402.2
Solid core	428	402.24
Microcard		
Photocopying	355	54
Selective data retrieval	353	27R
Microemulsions (See Microcapsule)		
Lubricants	252	49.5
Microfiber	428	903*
Microfiche		
Camera for producing	355	54
Card rack with pockets	40	124.2
Cartridge	353	27A
Slide transfer mechanism	353	103+
Film mounts	40	159.2
Photocopying	355	78+
Plural image recording on single plate	354	120
Projectors, picture carrier	353	120
Readers-printer design	D16	223
Reading machines	350	241
Transparent film viewers	40	361
Microfilm (See Microcard, Microfiche)		
Frame mats & mounts	40	159
Image transferred from document	355	64
Means to attach film to spool	242	74+
Photocopying	355	271+
Vieners	40	361+
Audio-visual equipment	D16	225+
Projector	353	74+
Reader-printer design	D16	223
Selective data retrieval	353	26R
With detailed optics	350	239+
With image projection on screen	353	
Micrometer	33	813+
Design	D10	73
Magnifiers	350	114+
Microorganism	435	820+
Microphone	D14	225+
Dynamic	381	177+
Granular	381	180+
Mechanical or acoustical	181	
Muffler at mouthpiece	181	242
Musical instrument attached	84	723+
Ribbon	381	176+
Moving coil type	381	177
Telephone	381	168+
Transformer	336	
Microphotography	355	18+
With fiber optics	355	1

	Class	Subclass
Microphotometer	356	213+
Microporous (See Pore Forming)		
Microprocessor	364	200
Camera photographic operation control	354	412
Microprojector	353	39
Microscope	350	507+
Container for slides	206	456
Design	D16	131
Electron	250	311
Generally	350	507+
Illuminator	350	253
Interference	350	509+
Objectives	350	414+
Phase contrast	350	509+
Slide	350	529+
Stage	350	529
Box	206	456
Microtome	83	915.5*
Microwave		
Coating by direct application of microwave	427	45.1
Concentrating evaporators electric field digest	159	DIG. 26
Containers, special for cooking	D 7	354+
Cooking food	426	241
Drying & gas or vapor contact	34	4
Food specially packaged or wrapped	426	107
Induction heating digest	99	DIG. 14
Material subjected to electrical energy	34	1
Oven	219	10.55 R
Design	D 7	351
Thermometer	374	149
Piercing probe	374	155
Radio &microwave absorption wavemeters	250	250
Thermal analysis	374	21
Thermometer	374	122
Used in reaction to prepare inorganic or organic material	204	157.15+
Used in reaction to prepare or treat synthetic resin or natural rubber	522	1+
Mildew Proofing	428	907*
Process	424	405+
Proofed stock	428	907*
Mileage Transportation Indicia	283	24
Military		
Computer controlled, monitored	364	423
Insignia	D11	95+
Milk		
Artificial	426	580+
Bottle		
Box for door or window	232	42+
Brushes	15	164+
Caps	215	200+
Easy removal disk caps	215	298
Bottles baby feeding	215	11.1+
Breast pumps	604	36+
Cans	220	
Design	D 9	367+
With light	362	154
Cartons ,paper	206	621.1+
Holder or rack for	D 7	616+
Liquid proofed	229	3.1
Chocolate	426	584+
Condensing	159	
Dairy food treatment	99	452+
Evaporating	159	
Filter	210	348+
Milking devices	119	14.1+
Claw	119	14.54+
Releasers	119	14.5+
Stools	297	175+
With cooling	119	14.9
Milking machines	119	14.1+
Methods of milking	119	14.2+
Releasers	119	14.5+
Modification	426	317+
Preservation	426	330.2
Canning	426	392+
Dehydration	426	471+
Sterilization & pasturization	426	521+
Protein foods from	426	656+
Receptacles deposit and collection	232	41R+
Receptacles for dairy use	220	
Combined with stool	297	175+
Leg or lap supported	220	17.1+
With strainer	210	464+
Shaped bottle	D 9	367+
Testing	422	74

	Class	Subclass
Processes	436	20
Sediment	73	61R
Miller Hook	213	105+
Millinery		
Hat linings	2	63
Hats caps etc	2	171+
Head protectors	2	2.5+
Manufacturing apparatus	223	7+
Milling (See Comminuting)		
Chemical etching	156	625+
Cutting machine	409	64+
Chemical	156	345
Cutters for	407	30+
Electro chemical		
Nut	10	82
Saw shaping	76	44
Gear cutting	409	1+
Test by	73	7+
Mills		
Coffee mill or grinder	D 7	372+
Condiment grinder, household	D 7	679
Gristmill alarms	116	71
Metal rolling	72	199+
Solid material comminutors	241	
Vegetable and meat comminutors	241	
Millstone		
Dress	241	296
Dressing	125	27+
Gauges	33	196
Picks	125	42
Mimeograph		
Plates or surfaces	101	463.1
Processes	101	463.1
Rotary	101	116+
Mincing		
Meat	241	
Hand cutters	30	
Mine & Mining		
Car		
Axles	295	41+
Bodies	105	364
Brakes four wheel spreading	188	55
Lubrication	184	2
Trucks	105	161
Door electric signal railway	246	125+
Elevator crosshead	187	96
Explosive	102	401+
Aerial	102	405
Airplane sustained	244	14
Laying submarines	114	316+
Laying vessels	114	18
Ship protection	114	240R+
Sweeping implements	114	221R
Towed sweeping devices	114	242+
Machine trucks for	105	161
Miners lamp	D26	37+
Combined with tool	7	104+
Safety	362	164+
Testing of	73	164
Picks	125	43
Carrier	224	
Hand	294	
Prop	405	288+
Ventilation	98	50
Mineral		
Acids	423	
Boron compound	423	277+
Carbon	423	414+
Halogen	423	462+
Nitrogen	423	385+
Phosphorus compound	423	304+
Silicon	423	324+
Sulfur	423	511+
Detectors		
Electric prospecting	324	323+
Electric testing	324	
Glass		
Ceramic compositions	501	11+
Coating	427	216+
Laminating	156	62.2+
Layer or laminated	428	426
Making apparatus	425	6+
Making from slag	65	1+
Oil emulsion		
Biocidal	514	938*
Oils	208	
Apparatus	196	
Compositions lubricating and miscellaneous	252	9+
Electrical separation or purification	204	188+

	Class	Subclass
Electrolytic treatment	204	136
Electrophoretic or electro osmotic treatment	204	181.8+
Electrostatic field or electrical discharge treatment	204	168+
Extracting	208	311+
Fischer-tropsch crude	208	950*
Process & products	208	
Recovering or disintegrating in situ	299	
Substances chemical purification	23	
Fermentative, preexisting cd	435	262+
Fermentative, synthesizing cd	435	168
Mineral Block	D 1	100
Mineral Wool Compositions	501	36
Miniature Electronic Tubes	313	
Mining (See Mine & Mining)	299	
Minnow		
Buckets	43	56+
Design	D22	136
Nets	43	11+
Design	D22	135
Mirrors and Reflectors	350	600+
Beds having	5	280
Bracket support for	248	475.1+
Cabinet combined with	312	224+
Cavity inspecting hand	350	640
Changing effects in mirror	40	900*
Comparing or demonstrating	434	371
Correct answer revealed	434	331
Dental	433	3+
Design	D24	13
Design	D 6	300+
Diagnostic	128	3+
Fireplace heat reflectors	126	552
Fluorescent or phosphorescent devices with reflector	250	467.1
Hand-held	D28	64.1
Hats having	2	185R+
Hook or support for	D 8	354+
Support hardware	D 8	373+
Illuminating devices and electric lamps having	362	296
Lamp structure combined	313	113+
Lights combined with mirrors	362	135
Furniture combined	362	128
Inspection facilitation	362	138
Liquid level gage glass having	73	327
Magnetic, use in thermo nuclear reactions	376	140
Coating apparatus	118	
Coating methods	427	162
Electroforming	204	7
Electroplating	204	19
Molding or blowing glass	65	
Making	350	320
Maze or labyrinth illusion	272	19
Observation	350	600+
Examining inner surface of hollow body	350	640
Examining inner walls of cars	350	640
Firearm sighting	33	233+
Reflex camera	354	152+
One way transparent	350	259
Railway track reflectors	246	474
Rear view	D12	187
Remote control	74	502.1
Signs having	362	253
Stage having	272	13
Support	248	466+
Frames or holders combined	52	
Surface configuration	428	409
Electroformed	204	7
Electroplated	204	19
Metal	428	687+
Surgeons specula	128	3+
Surgical	128	21+
Light combined	362	138
Oral	128	10+
Telescope attachments	350	574+
Television scanners	358	206+
Toilet kits having	132	316+
Brush type	132	102+
Compact	132	301+
Powder box and applicator	132	301+
Triple	350	102+
Typewriters having	400	716+
X ray reflectors	378	70
Missiles	D12	16.1
Aerial projectile amusement devices	273	317+
Bow	124	23.1+
Control	244	3.1+

	Class	Subclass
Explosive gun ejected	102	501+
Guided	244	3.1+
Mortar	102	372+
Projectile making	29	1.2+
Projectiles, mechanical	124	
Rocket projected	102	374+
Simulated projectile game	273	313+
Slingshot	124	17+
Stabilized	244	3.1+
Mist Treating		
Coating	427	428+
Textile machines	68	5R+
Miter		
Box	83	746
Wood sawing apparatus	83	761+
Mitering Devices		
Metal sawing	269	
Printers leads cutting	83	559
Wood sawing		
Boxes	83	746+
Machines	83	471.2+
Woodworking		
Clamps	269	
Cutters	144	216+
Mitomycins	548	422
Mitt		
Apparel	2	158+
Bath mitt or cloth	D28	63
Fingerless glove	D 2	610
Kitchen hot pad	D29	20
Mitten	D 2	622+
Apparel, knitted hand covers	66	174+
Fabric wipers and polishers	15	227
Knitting machine circular	66	45
Knitting machine straight	66	65
Mixer (See Agitating; Air, Mixer; Kneader)		
Agitating	366	69+
Animal food mixer	366	603*
Auxiliary air inlet for combustible gas mixture	48	189.3
Board for dentists	433	49
Bread manufacturing implement	416	70R
Bread manufacturing machine	366	69+
Carbohydrate dryer mixing tank	127	14
Carbohydrate granulation and mixing	127	21
Combustible mixture supply		
Engine charge forming	123	434+
For starting	123	180R
Gas and liquid	261	
Gas generator	48	180.1+
Acetylene combined	48	3R
Oil retort combined	48	104
Liquid or gaseous fuel burner	431	
Concrete	D15	19
Truck mounted	D12	14
Dental amalgam mixer	366	602*
Design	D 7	376+
Dough machines	366	69+
Electronic tube	313	300
Electronic tube type	328	158+
First detector type, heterodine	455	313+
Faucet aerator	239	428.5
Fluid or vapor		
Agitating device	366	101+
Aspiration	137	888
Beverage mixing nozzles	239	549
Carbohydrate granulation and mixing tank	127	14
Chemical feeders, water purification	210	198.1+
Dispenser with plural sources	222	129
Drying apparatus	34	
Faucets gas charged liquid	137	896+
Fire extinguisher chemical tank		
Automatic receptacles	169	27+
Gas pressure type	169	71+
Portable	169	78+
Fire extinguisher fluid system	169	14+
Fluid distributer	137	597+
Inhaler	128	203.12+
Inhaler vaporizer	128	203.26+
Oronasal	128	203.29+
Plural diverse fluids	137	896+
Sprayers dissolvers and mixers	239	310+
Temperature control mixing valves	236	12.1+
Valve multiway unit	128	203.29
Sequential	137	625.12+
Valves plural	137	602+
Sequential	137	628+

	Class	Subclass
For radio signals	455	313+
Glass manufacture	65	178+
Heptode	313	300
Household mixer-grinders	D 7	376+
In line between carburetor and internal combustions engine	48	189.4+
Mixing slab		
Dental	269	302.1
Pill	269	302.1
Mortar	366	10+
Process	366	3+
Tractor mounted mixing chamber	366	606*
Paint	366	605*
Mobile Equipment		
Medical and laboratory	D24	1.1
Radio	455	154+
Repeater or replay	455	11+
Vechile and tuning	455	152
With vehicle	455	297
Robot	901	1*
Transceiver	455	89
Transmitter	455	95+
Mobile Home Design	D12	101+
Moccasin	36	11
Design	D 2	268+
Making	12	142MC
Model (See Toys)		
Airplanes	244	190
Demonstrating	434	365+
Dental	433	213
Photographic	354	292
Modular Construction	52	
Sub-enclosure section	52	79.1+
Modulator	332	
Bedroom combinations with lighting	362	801*
Carrier type electric wave	332	
Amplitude modulators	332	149+
Combined with radio transmitter	455	91+
Frequency modulators	332	117+
Keyers	375	68+
Phase modulators	332	144+
Pulse modulators	332	106+
Cathode ray intensity	313	441
Demodulator amplifier per se	330	10+
With meter	324	118+
Modulation meter		
Amplitude	332	150
Frequency	332	118
Pressure modulating relays or followers	137	82+
Radar type pulse modulators	342	194+
General	342	
With pulse frequency	342	201+
With phase	342	200
Module (See Block; Brick; Panel; Slab; Tile)		
Road or pavement, per se	404	34+
Static mold forming		
Barrier having plural simulated	249	16
Having hollow portions through	249	144
Intaglio or cameo areas on	249	140
Joined by tie	249	85
Modulus		
Fiber or filament, high	428	902*
Moistener and Moistening		
Abrading combined	51	213
Air	239	34+
For hot air heating systems	237	78R+
Generally	261	
Barrel	217	97
Dispenser with	222	190
Envelope sealer combined	156	441.5+
Fireplace combined	126	508
Furnace hot air combined	126	113
Gas and liquid contact	261	
Heat exchanger combined	165	110+
Implement with material combined	401	
Label pasting or paper hanging machine combined	156	
Radiator combined	237	78R
Sewing machine combined	112	43
Stovepipe combined	126	313
Surgical electric applicator	128	803
Textile smoothing combined	38	3
Flatiron	38	77.1+
Thread finishing	28	217+
Tobacco	131	300+
Ventilator combined		
Floor	98	105
Wall	98	109
Moisture		
Curtains shades and screens operated	160	5

	Class	Subclass
Indicator		
Content in gas	73	29.1
Thermal testing	73	25.5
Vibration testing	73	24.4
Electric alarms	340	604
Meteorological	340	602
Electric testing capacitive	324	658+
Electric testing resistive	324	691+
Humidity controller	236	44R
Hygrometer	73	335+
Meteorology	73	170R+
Rain alarm	116	69
Rain gauges	73	171
Testing content of	73	73+
Molasses	127	
Electrolytic treatment	204	138
Foods	426	658+
Animal	426	635
Waste as fertilizer	71	26
Moldboard	172	754+
Bulldozer	172	701.1+
Adjustable	172	811+
Supplemental or attachment type	172	817
Shiftable for hillside	172	204+
Supplemental or attachment type	172	754+
Molding (See Casting; Molds)		
Apparatus and appliances		
Blow molding	425	522+
Bread, pastry, confection	99	426+
Butterworker combined	425	200+
Casting mold making	164	159+
Ceramic materials	425	
Concrete	425	
Dairy cutters combined	425	289+
Dental impression	433	34+
Design	D15	135
Foam	425	4R+
Glass	65	
Heel and toe stiffener	12	64+
Heels	12	48
Injection molding	425	542+
Isostatic mold	425	405.2
Metal founding	164	139+
Plastics	425	
Press	425	406+
Soap	425	289+
Soap	425	290+
Sole blank	12	21+
Sugar draining combined	127	18+
Apparel	223	52+
Building element	52	716+
Article supporting	52	27
Baseboard	52	242
Corner attached	52	282
Corner type	52	288
Design	D25	136+
Embedded corner	52	164+
Filler strip	52	312
Veneer applying press	156	349
Carpet fasteners	16	7
Composition, mold forming	106	38.2
Electrical device in insulation	264	272.11
Metal working with other steps	29	DIG. 29
Picture frame	D25	119+
Plastic		
Apparatus	425	
Composition	523	1+
Processes		
Battery electrodes	264	
Battery grids	164	
Bread, pastry, confections	426	660+
Casting mold making	164	6+
Clay, concrete block, earthen and mineral	264	
Computer controlled, monitored	364	476
Glass	65	66+
Glass blowing combined	65	78+
Laminating combined	156	
Metal founding	164	1+
Metal working combined	29	527.1
Particulate material	264	109
Plastic	264	
Elastic memory	264	230
Sugar starch carbohydrates	127	59
Pulp	162	382+
Pulp	162	218+
Apparatus	162	382+
Tire mold	409	902*
Tobacco	131	
Molds		
Edibles, plastic deformation		

	Class	Subclass
Fan driven by	416	170R
Fluid type	310	11
Linear	310	12+
Railway	104	290+
Magnetic conveyor	198	690.1
Motor and gearing	310	80
Motor control system	318	9+
Nonmagnetic	310	300+
Oscillating	310	36+
Prime mover dynamo plants	290	
Pump driven by	417	410+
Pyromagnetic	310	306+
Railway mail delivery having	258	4
Expansible chamber (See device associated therewith)	91	
Devices	92	
Digital	91	DIG. 1
Electrical control	91	459+
Electrical control	91	361+
Exhaust throttled motor control	91	DIG. 2
Large area valve	91	DIG. 3
Magnets	91	DIG. 4
One shot explosion actuated	60	632+
Plural expansible chamber	91	165+
Plural expansible chamber	91	508+
Fluid (See associated device; fluid, motor)	415	
Internal combustion engine	60	597+
Free piston device	60	596
Vacuum generated	60	397
Waste heat driven motor	60	597
Making	29	888+
Fluid reaction surfaces	416	
Fuels (See fuels; fluid, fuel)		
Liquid	44	300+
Generators	310	113
Heat		
Valve actuator	251	11
Hydraulic	60	325+
Hydro-pneumatic	91	4R
Impellers	416	
Induction systems	318	727+
Braking	318	757+
With reversing	318	741+
Starting	318	778+
Internal combustion	123	
Fluid motor combined	60	597+
Free piston device	60	596
Vacuum generated	60	397
Waste heat driven motor	60	597
Induction type governor valve	137	479+
Power plant	60	632+
Turbine combined	60	598+
Ion	60	202
Marine	440	84+
Mechanical motors	185	
Meter lights	362	25
Meters	324	76R+
Meters	324	139
Multiple	60	698+
One shot explosion actuated	60	632+
Outboard	440	53+
Container or package for	206	319
X-art collection	440	900*
Polyphase current	310	159+
Power plants	405	
Pumps	417	
Reaction (See jet; motor)	60	200.1+
Rotary expansible chamber devices	418	
Rotary kinetic fluid motor or pump	415	
Single phase current	310	159+
Spring	185	37+
Steering actuating	244	175+
Support	248	637+
System of diverse motors	60	698+
Punch card or tape control	60	703
Testing	73	116+
Tide or wave (See associated device)	60	497+
Pulsating flow drive	416	6
Utilizing natural heat	60	641.1+
Vehicle (See motor vehicle)	180	
Ward leonard systems	318	140+
Weight	185	27+
Motor Home		
Trailer design	D12	103
Vehicle design	D12	100
Motor Vehicle (See Vehicle)	180	
Airstream reactive	180	903*
Attracted to surface by fluid force	180	164
Baby seats	297	250+

	Class	Subclass
Body	180	89.1+
Boiler plants	122	3
With vehicle structure	180	310
Brake	188	2R+
Carriage or crib, infant	180	166
Collision, protecting occupant or motor from	180	232
Control means		
Radiated wave responsive	180	167+
Velocity reponsive	180	170+
Electrical servo mechanism	180	178+
Fluid servo mechanism	180	175+
Control mechanism	180	314+
Crank and gear case lubrication	184	1.5
Demonstrating	434	373+
Differential mechanism	180	76
Door	49	
Energy accumulation by fluid or mechanical menas	180	165
Four wheels driven	180	233+
And steerable	180	234+
Differentially	180	248+
Including electric or fluid motor	180	242+
Frame	180	311+
Ground effect	180	116+
Harvester	56	10.1+
Heat and power	237	12.3 R
Land	280	
Laterally driven	180	199+
Leveling device	180	41
Locomotive	105	26.5+
Moped	180	205+
Operation instruction	434	62+
Parking, driven laterally	180	199+
Parking, programmable, steerable wheels	180	204
Plow	172	
Power, eg, traction motor	180	54.1+
Battery mount	180	68.5
Electric motor	180	65.1+
Four wheel driven	180	242+
Three-wheeled, coaxial wheels	180	214
Three-wheeled, steerable	180	216
Two-wheeled	180	220
Exhaust handling	180	309
Hood	180	69.2+
Motor driven by expansible gas	180	302+
Steam	180	303+
Motor driven by noncompressible fluid	180	305+
Motor powered from external source	180	2.1+
Force of nature, eg, sun, wind	180	2.2
Motor to frame relationship	180	291+
Radiator mounting	180	68.4+
Turbine, combustion product driven	180	301
Propelled by other than wheel	180	7.1+
Jet	180	7.3
Propeller	180	7.4
Screw	180	7.2
Stepper	180	8.1+
Track, portable	180	9+
Vehicle mounted winch	180	7.5
Radiator	165	41
Railway car propulsion	104	287+
Restraining belt or harness, system related	180	268+
Ignition or starter circuits	180	270
Transmission	180	269
Runner supported, eg, snowmobile	180	182+
Substitutable for wheel	180	183+
Endless track	180	184+
Surface engaging propulsion element	180	186+
Safety promoting means	180	271+
Absent, inattentive or unqualified operator	180	272
Acceleration, deceleration or tilt responsive	180	282+
Perimeter engagement responsive	180	274+
Unauthorized use prevention	180	287+
Vehicle load weight responsive	180	290
Seats with body modification	296	63+
Ship	114	
Ship propulsive systems	440	
Ski or skate type	180	180+
Snow excavator	37	231+
Steam powered tractive vehicle	180	36+
Steerable wheel, driven	180	252+
Five or more wheels	180	23+

	Class	Subclass
Four wheels driven, all steerable	180	234+
Motor-carrying attachment	180	12+
Steam tractive vehicle	180	37+
Three-wheeled vehicle	180	211+
Two-wheeled vehicle	180	223+
Rotary-element driven wheel	180	222
Steering	180	79+
By differential traction	180	6.2
Endless track	180	6.7
By walking attendant	180	19.1+
Electrical power	180	79.1
Fluid power	180	132+
Four wheels driven, articulated frame	180	235
Gear	180	79+
Mechanical power	180	79.3
Shaft	180	78
Utilizing terrestrial guide	180	131
Suction effect	180	164
Supported by, & driving, other vehicle	180	198
Surface effect	180	116+
Traction motor powers external device	180	53.1+
Trailer combined	280	400+
Trains of vehicles	180	14.1+
Transmission mechanism	180	337+
Vertical surface transversing device	180	901
Wheel base, special	180	21+
Changeable location or number of wheels	180	209
Collapsible or knockdown	180	208
Five or more wheels	180	22+
Three-wheeled	180	210+
Two-wheeled	180	218+
Tandem	180	219+
Wheel chair, motorized	180	907
Wheel slip during acceleration, detecting	180	197
Motorcycle	180	
Design	D12	110
Frames	280	281.1+
Lights	362	72
Props	280	293+
Safety devices for motor vehicles	180	271+
Seat support	267	132
Steering	280	263+
Tandem design	180	219+
Design	D12	109
Three-wheeled evhicle	180	210+
Tracked	180	9.25
Two-wheeled vehicle	180	218+
Mount (See Supports)		
Acoustic diaphragm	181	198+
Aircraft		
Airfoils resilient	244	38
Power plant	244	54
Battery	180	68.5
Bearing		
Ball	384	537+
Roller	384	584+
Bracket special	248	205.1+
Electricity		
Conduit or housing	174	48+
Electric device	174	52.1+
Insulator	174	158R+
Eye glass lens	351	41+
Brace arm	351	103
Continuous rim	351	83
Pince-nez type	351	65
Rimless	351	110
Split rim	351	90
Fishing reel	43	22
Gun	89	37.1+
Spring	124	29
Outlet or junction box	220	3.3+
Photograph	40	152+
Picture frame	40	158.1+
Design	D 6	300+
Printing stencil	101	127.1+
Safes bank protection devices etc	109	50+
Sound box	369	157
Tool and machine part		
Brush bristle	15	186+
Brushing machine handle	15	22.1+
Comminutor	241	285R+
Earthworking	172	
Bulldozer	172	811+
Frame mount abrading	51	166R+
Gripper lever	74	156+
Gripper slide	74	160+

	Class	Subclass
Hand control lever	74	511R
Printing plate	101	382.1+
Snowplow	37	231+
Spring for spring motor	185	45
Squeezing machine textile	68	125
Work abrading	51	277
Wringer roller textile	68	272
Tripod	354	293
Camera support	354	293
Ladder prop	182	169
Pivoted ladders	182	104
Induction furnace	373	140+
Support stand	248	163.1+
Trolley wheel	191	63.3
Vehicle for conveyer	198	300+
Wheel and axle	301	
Cultivator, adjustable	172	395+
Land marker, slidable	172	126+
Land vehicle	301	
Panel hangers travelers	16	97+
Plow, adjustable	172	395+
Railway	295	
Railway motor axle mounting	105	136+
Railway rolling stock combined	105	
Railway truck axle box mounting	105	218.1+
Skate wheel	280	11.27+
Vehicle combined	280	29+
Mountain Climbing	248	925
Mounting		
Tires on wheels	157	
Mouse		
Guard		
Piano pedal openings	84	233
Traps	43	58+
Mouth		
Bag holders	248	99+
Boiler	122	499+
Doll part	446	395
Electric applicator	128	787
Electrical input device (see joystick)		
Guard animal type	119	129+
Guard human type	128	859+
Harmonica	84	377+
Harp	84	375
Organ	84	93
Organs	84	377+
Pieces		
Acoustical muffler	181	242
Animal bits	54	8+
Applying or forming cigar or cigarette	131	88+
Automatic wind instrument	84	93
Brass musical instrument	84	398+
Cigar or cigarette	131	361+
Clarinet	84	383R
Ear piece combined speaking tube type	181	20
Flexible bag closure	383	80+
For wind instruments	84	398+
Speaking tube	181	21
Telephone	381	157+
Antiseptic	379	452
Woodwind instrument	84	383R
Wash	424	
Movement Mechanical	74	
Belt & pulley drive system	474	
Step by step		
Annunciator systems	340	316+
Annunciators	340	815.8
Boring	408	70+
Copy holders movable copy	40	342+
Copy holders movable marker	40	352+
Electric motor operated temperature controls	236	76
Electromagnetic switch	335	138+
Endless conveyors	198	750+
Pusher	198	736
Floor jacks	254	14
Garment supporters and mortising cutters	144	71
Pencil lead advancing	401	65+
Separable fasteners	24	580+
Shingle sawing machines	83	704+
Strap tighteners	24	70R+
Telephone call systems	379	299+
Watch	368	76
Receptacles	206	18
Ring	368	299
Mover		
Bed	5	510+
Trunk tray	190	29+

	Class	Subclass
Movie Apparatus	D16	205+
Moving Picture (See Motion Picture)	D16	205+
Mower		
Harvesting	56	
Sharpener	76	82.1
Mucic Acid	562	582
Mucilage Holder	15	
Design	D 9	
Muck and Scrap Bar	428	576+
Mud		
Bath	128	365
Drum cleaning	122	393
Gas content testing	73	19.9
Guard	280	847+
Design	D12	184+
Gun	266	273
Lifters for wells	166	162+
Ring	122	369
Cleaning	122	385
Traps		
Washers	134	109
Muff		
Apparel design, earmuff	D29	19
Apparel design, handmuff	D 2	611+
Ear and throat	2	208+
Fur	2	66
Hanger	223	97
Muffle & Muffler (See Insulating; Noise, Muffler)	181	212
Acoustical	181	175+
And heat exchanger combined	165	135
Making	29	890.8
Apparel	2	91
Design	D29	19
Dry closet with	4	481+
Exhaust treatment	60	272+
For distilling zinc	266	148+
Hobby craft engine	181	404*
Hood combined	2	203
Lock with silencer	70	463
Mouthpiece, ie telephone	181	242
Refrigerator compressor	181	403*
Screen or fence absorbing jet noise	181	210
Silencer for firearms	181	223
Sound absorbing material combined	181	256+
Jet engine muffler devices	181	222
Straight through continous passage	181	248
Through passage	181	252
Mulch and Mulching	47	9
Planters	111	144+
Transplanting through	111	102
Mule	57	320+
Womans footwear	D 2	283
Muller		
Comminutors	241	
Mullion	52	777+
Sash component	52	821+
Multicellular Living Organisms and Unmodified Parts Thereof	800	
Multicolor		
Ink block	101	134
Pen or pencil	401	34+
Box	206	371
Photography processes	430	357+
Printing	101	171+
Intalgio	101	151+
Planographic	101	135+
Stenciling	101	115
Weaving	139	453+
Multilevel Buildings	52	234+
Multiple		
Effect evaporators	159	17.1+
Expansion engine	91	152+
Semi-compound	91	8+
Filament lamps	313	316
Plant receptacles	47	85+
Telephone switchboard	379	310+
Tip multiple discharge coating implement	401	28
Valves	137	
Actuation	137	
Gas and liquid contact	261	42+
Wire systems for continuous current distribution	307	
Multiplex		
Diplexing	370	37
Duplexing	370	24
Facsimile	358	425
Frequency division	370	69.1
General purpose programmable computer system	364	200

	Class	Subclass
Teletypewtiter	370	41
Time division	370	77+
General purpose system	370	
Motion picture	352	81
Nippers with plural cutting blades	30	176
Optical	370	1+
Radio	370	
Shears	30	226
Multiplex shearing position	30	197
Special purpose digital data processors	364	900
Switching	370	53
Telegraph	370	41+
Telemetering	340	870.11
Seismic prospecting	367	78+
Telephone	370	
Multiplier		
Electronic tube	313	103R+
In amplifier circuit	330	42+
Frequency	363	157+
Electronic tube type	328	15+
Electronic tube type	328	20+
Multipliers Pattern Mechanisms	139	326
Multiplying		
Machines	235	61F+
Multipolar Instruments	33	1MP
Multiroom Builders	52	234+
Muntin (See Mullion)	52	766+
Murexide	544	296
Muscular Strength Measuring	73	379+
Coin controlled	194	
Mushroom		
Anchors	114	300
Culture	47	1.1
Fertilizing	71	5
Plant	PLT	89
Music (See Musical Instrument)	84	
Accessories	84	453+
Boxes such as swiss music boxes	84	94.1+
Design	D17	24
Cabinets	84	DIG. 17
Cases for music roll	206	308
Cases for musical instruments	206	314+
Cathode ray tube	84	DIG. 6
Chiff	84	DIG. 5
Chime	D17	22
Chimes for clocks	368	273
Chorus, ensemble, celeste	84	DIG. 4
Combined in another device		
Toys figure	446	297+
Toys figure wheeled	446	270+
Toys wheeled	446	409+
Crescendo	84	DIG. 16
Electric key switch feature	84	DIG. 7
Electronic gates for tones	84	DIG. 23
Exercising devices	84	465+
Feedback	84	DIG. 1
Filtering	84	DIG. 9
Foldable, detachable, collapsible	84	DIG. 3
Frequency dividers	84	DIG. 11
Gas discharge tube	84	DIG. 13
Holder		
Design	D 6	417
Music stand	248	441.1+
Piano music desks or racks	84	180
Piano music receptacles	84	181
Instruments	84	
Keyed oscillators	84	DIG. 8+
Leaf turners	40	530
Leaf turners for music	84	486+
Light sensitive resistor	84	DIG. 19
Note sheets for automatic players	84	161+
Page turners	84	486+
Piezoelectrical transducers	84	DIG. 24
Pitch (See tuning)		
Players and talking machines combined	84	4+
Plural speakers	84	DIG. 1
Preference networks	84	DIG. 2
Printed matter	84	483.2
Reverberation	84	DIG. 26
Service manuals	84	DIG. 28
Sheets	84	483.2
For mechanical players	84	161+
Side rhythm and percussion devices	84	DIG. 12
Stereo	84	DIG. 27
Striking trains for clocks	368	272
Tape	84	DIG. 29
Teaching devices	84	470R+
Toys	446	397+
Tuning	84	DIG. 18

	Class	Subclass
Tuning devices	84	454+
Typewriting machines	400	117+
Musical Instruments	84	1+
Accessories	84	453+
Automatic or self playing	84	2+
Bars	84	102
Bells	84	103
Combined	84	3+
Combs	84	94.1
Cylinder	84	106
Drums	84	104
Organ	84	84+
Piano	84	13+
Rolls for automatic players	84	161+
Selectors	84	115+
Stringed instrument	84	7+
Wind instrument	84	83+
Bowed		
Automatic device	84	10
Bowed piano	84	256+
Violin bow guides	84	283
Violin, cello, and bass	84	274+
Brasses	84	387R+
Trombone	84	395+
Valve type	84	388+
Cases for	206	314+
Hand carriers	224	910*
Violin case	206	14
Design	D17	
Accordians	D17	3
Instrument cases	D 3	30.1+
Organ cabinet	D17	5+
Percussion instruments	D17	22+
Pianos	D17	7
String instruments	D17	14+
Wind instruments	D17	10+
Drums	84	411R
Electric	84	600+
Fret control	84	DIG. 3
Keyboards	84	423R+
Keyed oscillators	84	DIG. 8
Manufacture of musical instrument	29	169.5
Mechanical resonator	84	DIG. 21
Monophonic	84	DIG. 2
Picking or plucking devices	84	320
Recording movement of keys	84	461+
Punched record	234	49+
Punched record	234	109+
Rigid vibrators	84	402+
Bells	84	406+
Service manuals	84	DIG. 28
Side rhythm and percussion	84	DIG. 12
Stringed	84	173+
Banjo	84	269+
Dulcimer	84	284
Guitars	84	267
Harp	84	264+
Mandolin	84	268
Pedal clavier	84	DIG. 25
Piano	84	173+
Violin	84	274+
Zither	84	285
Teaching devices	84	470R+
Top	446	258
Transposers	84	445+
Tuning devices	84	454+
Wind	84	330
Accordian	84	376R+
Accordian digest	84	DIG. 15+
Chord organ	84	DIG. 22
Clarinet	84	382
Flute	84	384
Harmonica	84	377+
Harmonica	84	DIG. 14
Pipe organ	84	331+
Actions and valves	84	332+
Valve tremolos	84	348
Reed organ	84	351+
Actions	84	365+
Stop actions and valves	84	369+
Saxophone	84	385R
Woodwinds	84	380R+
Musk Artificial	568	932+
Muskets	42	
Mustache Holder	132	215
Mustard	426	638
Gas	568	56
Mutation	435	172.1
Mutes		
Banjo	84	273
Brass wind instruments	84	400
Violin	84	310+
Mutoscopes	352	99
Muzzle	119	130+
Dog	D30	152
Mylar T M (See also Synthetic Resin or Natural Rubber)	528	308.1
X-art collection	310	800*
N-cl Compound	8	DIG. 2
Oxidizing compositions	252	186.35+
Nacelles		
Engine	123	41.7
Aircraft combined	244	53R+
Nafcillin	540	339
Nail	411	439+
Brush	15	167.3
Buffer	132	76.4+
Compound manicure tool with	132	75.6
Capping nail	10	156+
Design	D 8	391+
Digest	10	DIG. 4+
Extractor	254	18+
Feeding and distributing	227	107
Sorting	10	162A
Files	132	76.4
Design	D28	59
Fluorescent or phosphorescent	250	462.1
Hammer	227	133
Hole plug	52	514
Horseshoe	168	23
Calk	168	38
Insulated, electrically	174	159
Making	10	28+
And driving	227	82+
Placer hammer	81	23+
Polish applicator	D 4	
Sets	81	44
Stripper	193	38
Nailing	227	
Broom making	300	13
Receiving strip	52	364
Reel	242	103
Shoe heel working and	12	43+
Shoe lasting and	12	13.1
Tool	12	108
Nainsook	139	426R
Name Plate	232	46
Design	D20	15
Holders	40	642+
Name Tag	D20	22+
Nandrolone, in Drug	514	178+
Naphthacetols	564	222
Naphthalene	585	400+
Hydrogenation of	585	266+
Sulfonic acids	562	89
Naphthalic Acid	562	488
Decarboxylation	562	479
Imides	546	
Naphthazarine	552	298
Naphthenic Acid	562	511
Naphthionic Acid	562	68
Naphthisatins	548	451
Naphthols	568	735+
Acylamino	564	123+
Acylaminosulfo	562	47
Amino	564	428+
Aminosulfo	562	70
Azo compounds ctg	534	573+
Carboxylic acids	562	466+
Amides	564	173
Esters	560	56
Nitro	568	707
Sulfo acids	562	79
Naphthothiophenes	549	43+
Naphthylamine	564	428+
Naphthyridine	546	122
Napkin	D 6	595
Design	D 6	595+
Design, paper	D 6	595
Holder		
Body supported	24	7+
Design	D 7	631
Stands	248	127+
Sanitary	604	358+
Design	D24	49
Materials for	604	367+
Napping, Cloth Finishing	26	29R+
Hat making	223	18
Narceine	549	444
Natural Rubber (See Synthetic Resin or Natural Rubber)		
Naval (See Marine)		
Navigation		
Calculators	235	61NV+
Celestial navigation training	434	111
Computer controlled, monitored	364	443+
Course recorders	346	8
Decca	342	397
Instruments (See notes thereto)	73	178R+
Geometric	33	
Loran	342	388+
Omega	342	396
Radionavigation training	434	239
Tacan	342	399
Underwater signal for position finding	367	137
Nebulizers		
Sprayer structure	239	338
Surgical	128	200.14
Neck		
Appliance for thermal surgical treatment	128	380
Bottle	215	31+
Nonrefillable separate	215	30
Cleaning brush for bottle	15	67
Glass reshaping to form	65	108+
Labelling machine for bottle body &	156	476+
Musical stringed instrument	84	293
Pad for harness	54	67
Stock for animal restraining	119	98
Support bracket engaging receptacle	248	312
Supported shower bath	4	618
Tobacco pipe bowl detachable from	131	222
Trap	131	214
Yoke for animal	D30	137
Draft appliance	278	119+
Restraining poke	119	136+
Necklace	D11	3+
Neckwear		
Collars	2	129
Design	D 2	600+
Ironer engaging neckband for clamping	38	13
Ironers for collars and neckbands	223	52.1+
Ironing table with neckband shaper	38	109+
Napkin holder inclosing	24	9
Shaping forms	223	81+
Shirt neckbands	2	127
Ties	2	144+
Design	D 2	605+
Fasteners	24	49R+
Design	D11	202+
Hanger	D 6	315+
Machines for sewing	112	121.22
Rack or holder	D 6	315+
Needle	369	173
Blasting fuse setting	86	21
Compass	33	355R+
Cord knotter	289	16
Harvester needle cleaner or guard	56	448
Crochet	66	118
Gun	42	16+
Hypodermic	604	187+
Pocket carrier for	206	365+
Injecting for pickling	99	532+
Injector liquid applying to textile	68	201
Knitting	66	116+
Design	D 3	28
Machine		
Buttonhole sewing	112	65+
Embroidering	112	78+
Knitting textiles	66	
Independent	66	7+
United	66	78+
Leather sewing	112	28+
Sewing	112	
Stitch forming	112	154+
Textile needling	28	107+
Fluid needles	28	104+
Weaving	139	440+
Making	163	
Abrading process	51	281R+
Crochet	163	2
Eye polishers	51	152
Straightening or bending	72	319
Packages or cases	206	380+
Design	D 9	303
Drill holder type	206	379
Phonograph	369	173
Design	D14	264
Radioactive seed holder	600	7+
Seine making	87	60
Sewing	223	102+
Holder for	223	109R

	Class	Subclass
Orthopedics		
Ankle supports and braces	128	166
Skates with	280	11.36
Appliance	D24	64
Arch supports	36	71
Boot or shoe	128	581+
Pressure distribution testing	73	172
Shoe making	12	142N
Miscellaneous parts	12	146M
Shoe pads	36	71
Surgical	128	68+
Orthoquinones	552	291
Oscillation Circuit		
Generator	331	
Therapeutic treatment type	128	422
Oscillator	331	
Audio	331	
Coil electric	336	
Electronic	331	
Hartley type	331	170
Transistor	331	108R
Multi phase	331	45
Tuned circuit for	334	
Feed conveyer, oscillating-type	209	921*
Laser beam applications		
Logic gate active element	331	DIG. 3
Phase locked loop having locd		
indicating or detecting means	331	DIG. 2
Tetrode	331	
Oscillographs (See Oscilloscope)		
Cathode ray tube circuits	315	1+
Cathode ray tube structure	313	364+
Design	D10	76+
Electrical quantity measurement	324	97
Cathode ray type	324	121R
Facsmile systems	358	296+
Photographic recorder	346	108+
Plural function recorder	346	45+
Steam engine indicator	346	3+
Oscilloscope (See Oscillograph)	D10	76
Cathode ray tube system	315	1+
Chemical analysis, apparatus	422	82.8
Expanded scale	324	131
Cathode ray	324	121R
Meter	324	97
Cathode ray type	324	121R
Television	358	242+
Tube structure	313	364+
Osmium		
Alloys	420	461
Hydrometallurgy	75	711+
Pyrometallurgy	75	631
Osmosis (See Desalting Seawater)		
Osones or Osozones	536	1.1+
Osteal Adjuster	128	69+
Otec, Ocean Thermal Energy		
Conversion	60	641.7
Otoscope	128	9
Otterboard, Trawl Net Type	43	9.7
Ottoman	297	462
Outboard Motor	440	53+
Container or package for	206	319
X-art collection	440	900*
Outdrives	440	900*
Outer Garment	2	115+
Outlet		
Bath basin fitting	4	195+
Compartment tub	4	208
Box		
Digest	33	DIG. 1
Electric conduit	174	50+
Electric connector	439	892+
Electric junction box	220	3.2+
Support	248	906*
Conduit wiring	439	207+
Support	248	27.1+
Cover or escutcheon plate	D 8	350+
Decanter	210	109+
Distillation vapor	196	136
Drinking fountain		
Catch basin	239	28+
Faucet combined	239	25+
Muffler	181	238
Nozzle	239	
Outlet plug	D13	137+
Pneumatic stacker tube	406	154+
Pressure control		
Fluid pressure brake	303	59+
Fluid pressure regulator	137	505+
Smoking device stem	131	229
Sprinkler pipe	239	266+

	Class	Subclass
Stratifier	209	494
Ventilation		
Building	98	42.1+
Car	98	19+
Cowl	98	66.1+
Wall mounted electric conduit	174	49
Water closet valve		
Siphon	4	368+
Valve	4	378+
Water closet ventilation	4	218
Outriggers		
Awning type		
Multipart	160	78+
Rigid	160	83.1
Swinging	160	81+
Vehicle	280	763.1+
Crane	212	189
Outsoles	432	120+
Ovens and Accessories (See Heat)	432	120+
Coke	202	81+
Cooking	126	
Design, household	D 7	348+
Design, special use	D15	144.1
Distilling	202	
Drying	34	
Electric	219	391+
Lights	362	92
Overalls		
Nether type	2	227+
Union type	2	79
Overcoat	2	85+
Overcoming Dead Center	74	36
Overflow		
Bath and basin fitting	4	198+
Conneyer art	198	954+
Decanter and filter	210	299+
Filling portable receptacle	141	115
Gas and liquid contact baffle	261	114.1
Textile machine	68	208
Trough	261	DIG. 44
Water closet		
Outlet valve	4	390+
Outlet valve float	4	395+
Siphon bowl	4	427+
Water control gate	405	101
Water heater	126	359
Receptor	126	383+
Overlaps	2	121
Overload		
Automatic weighers	177	82
Fluid receptacle filling devices	141	115+
Relay		
Electric trips	335	21+
Thermal current trips	337	70
Thermal trips	337	356+
Release		
Comminution apparatus	241	32
Earthworking implements	172	261+
Bulldozer	172	816
Hydraulic control	172	260.5
Elastic couplings ships	114	217
Flexible shaft couplings	464	30+
Motor vehicle trains	180	14.5
Pitmans and connecting rods	74	584+
Planetary gearing	475	263+
Plows	172	261+
Snow plows	37	232+
Tools	81	467+
Torque responsive clutches	192	56R
Trailer couplings	280	449+
Stop	192	150
Spring wheels	152	18
Support movable or disengageable		
on impact or overload	248	900*
Overshoes		
Horse	168	1+
Design	D30	146+
Design	D30	146+
Mens	36	7.1 R+
Design	D 2	271+
Overthrow Preventer	235	131R
Overwinding Preventer		
Spring motor combined	185	43
Plural motors	185	13
Watch or clock key combined	81	467
Watches combined	368	209
Weight motor combined	185	35
Ox		
Shoe	168	5
Making	59	70
Yoke	54	77

	Class	Subclass
Oxacillin	540	327
Oxadiazoles, 1,2, 3-	548	125
Oxadiazoles, 1,2, 4-	548	131+
Oxadiazoles, 1,2, 5-	548	125
Oxadiazoles, 1,3, 4-	548	143+
Oxalic Acid	562	597
Oxazines	544	63+
Bio-affecting & body treating		
composition	514	228.8+
Oxazoles	548	215+
Oxazoles, 1, 2-	548	240+
Oxazoles, 1, 3-	548	215+
Oxazolidones	548	225+
Oxidants		
Bound solid fuel	44	558
Oxidation (See Substance Produced)		
Absorbent regeneration	502	20+
Alcohols		
Acyclic ketone production	568	402
Aldehyde production	568	471+
Carboxylic acid production	562	418+
Aldehydes to produce acids	562	418+
Aldehydes to produce acids	562	531+
Bleaching (See bleaching)		
Carbon monoxide	423	415R
Compositions	252	186.1+
Detergents	252	95+
Compounds for	149	119
Dyeing	8	649
Oxidizable dyes	436	904*
Electrolytic	204	
Fermentative	435	
Glyceride oils	260	406
Purification by	260	423
Hydrocarbons		
Acyclic acid production	562	542+
Anthracene	552	208
Aromatic acid production	562	408+
Aromatic to acid anhydrides	549	231+
Electrolytic	204	80+
Electromagnetic	204	157.15+
Electrostatic	204	169
Mineral oil	208	3+
Nonaromatic mixtures	568	950+
Preparing acyclic aldehydes	568	469.9
Preparing acyclic carboxylic		
acids esters	560	241.1
Preparing acyclic carboxylic		
acids	562	512.2+
Preparing acyclic hydroxy ctg		
compounds	568	910.5
Preparing acyclic ketones	568	398.8
Preparing fats, fatty oils,		
ester-type waxes or higher		
fatty acids	260	398.6
Olefines to epoxy compounds	549	523+
Liquid purification	210	758+
Sewage treatment	210	601+
Metallurgy	75	
Metals	23	
Surface	148	284+
Nonmetals	23	
Resins natural	530	200+
Oxides		
Heat producing compositions	149	37+
Iron	423	632
Magnesium	423	635
Metal	423	579
Alkali	423	592
Alkali earth	423	583
Nitrogen	423	400+
Combustion control	60	900*
Electrostatic	204	179
Phosphorus	423	304
Sulfur	423	512R+
Titanium	423	608
Electrosynthesis	204	96+
Oxidized Hydrocarbons of		
Undetermined Structure	568	959
Oximes	564	253+
Acyclic	564	268
Alpha oxy or oxo	564	258
Aromatic	564	265+
Cyclohexanone	564	267
O-carbamoyl	564	255
O-esters	564	254+
O-ethers	564	256+
Preparation	564	259+
Oximeters	356	41
Ear attached	128	633
Oxindole	548	484+

	Class	Subclass
Oxirane	8	DIG. 8+
Oxolinic Acid, in Drug	514	291
Oxonium	568	557
Oxy Containing Carboxylic Acid Esters		
Acyclic	560	179+
Alicyclic	560	126
Aromatic	560	55+
Oxyacetylene Gas Torches	239	398+
Oxychlorid	423	462
Cement	106	685+
Oxygen	423	579+
Blood oxygenators	261	DIG. 28
Electrolytic synthesis	204	129
Generative composition	252	186.1+
Liquefaction	62	8+
Mask	128	205.25
Rings	549	200+
Azo compound containing	534	751+
Oxazine	544	63+
Oxazole	548	215+
Sulfurized	549	200+
Tent	128	205.26
Tests for determining		
Biochemical oxygen demand, bod.	436	62*
Chemical oxygen demand, cod	436	62*
Total oxygen demand, tod	436	62*
Oxyhalogenation of Hydrocarbons	570	224
Oxysulfate	106	683+
Oxytocin	530	315
Oyster		
Culture	119	4
Dredger	37	55
Opener	17	74+
Opening process	17	48
Scoop	37	119
Ozone	423	581
Electrostatic field or electrical		
discharge synthesis	204	176
Generators	422	186.7+
Food treatment	426	236
Liquid purification	210	760
Preliminary manufacture	210	192
Ozonides	549	430+
Ozonizer		
Electromagnetic wave energy or	422	186.7+
Gas & liquid contact apparatus	261	DIG. 42
Inhaler	128	202.25
Process	62	78
Refrigeration combined	62	264
Pacemaker	128	419P+
Pacifiers	606	234+
Design	D24	45
Pack		
Advancing and feeding sheets from	271	8.1+
Pack holders	271	145+
Animal		
Load attaching means	224	905*
Camera	354	174+
Firearm	42	88+
Forming by receiving sheets	271	278+
Jogger	271	221+
Metal foil formation	29	17.4+
Nuclear fuel components	376	434
Packet		
Clasping or holding devices	24	17R+
Inserting binding loops	227	77
Making and wrapping	53	206
Postal card type	229	92.8
Parachute	244	148
Saddle	54	37
Package	206	
Adhesive	206	813*
Automatic filling and weighing	177	52+
Blister	206	461+
Button eyelet or rivet	206	348
Chewing gum	206	800*
Coins	206	.8+
Card type	229	92.9
Paper wrapper	229	87.2
Cosmetic article	206	823*
Design	D 9	
Detergent containing	252	90+
Emergency packs	206	803*
Fastenings or ties	24	16R+
Filler for space	206	814*
Finger opening	206	815*
Fire kindling	44	541
First aid packs	206	803*
Flares	206	803*
Flexible closure, eg sliding desk top		
type	206	816*

	Class	Subclass
Follower	206	817*
Land vehicle compartment	296	37.1+
Living fungi ferment containing	435	
Lubricant containing	252	10
Magnet	206	818*
Mail box clips or pockets	232	33
Making	53	
Material	206	819*
Medicinal content	206	828*
Nail and staple	206	338
Package and article carriers	224	
Sewing machine case	312	208
Plant	47	84+
Powder puff	206	823*
Recoil-type retainer	206	825*
Registers operated by	235	98R
Rubber band	206	805*
Rupture means	206	601+
Sausage shirred casing	206	802*
Separable, strip-like plural items	206	820*
Sheet material folding	270	39+
Special shape	206	822*
Stacking member	206	821*
Suction cup adjunct	206	829*
Suspension	206	806*
Tamper-proof	206	807*
Tear apertures for roll, strip, or		
sheet	206	824*
Textile yarn or cloth		
Laundering or cleaning methods	8	154
Liquid treating for cleaning	68	
Thread finishing	28	217+
Ties	24	16R+
Toilet article	206	823*
Towel, packaged	206	812*
Vacuum	206	829*
Water treating agent containing	252	176
Waterproof	206	811*
Wound strand	242	159+
Wrapper	229	87.1+
Gift wrap	229	87.19
Wrapper, packaging bag	D 9	305
X-type	206	809*
Zipper, ie receptacle securement	206	810*
Packaging Methods		
Automatic or triggered control	53	52+
Bags, wrappers, tubes	D 9	305
Baling	100	1+
Binding	100	1+
Cigar and cigarette making	131	283+
Closing packages	53	285+
Collar, cuff and bosom making		
combined	53	117
Containers	D 9	414+
Contents material treating	53	111R+
Depositing articles into receptacles	53	235+
Filling and closing portable		
receptacles	53	266R+
Foods	426	392+
In liquids	426	392+
Group forming into unit	53	147+
Needles and pins	227	25
Plug tobacco making combined	131	112+
Receptacle filling & closing	53	467+
Receptacle opening	53	381R+
Textile strands	28	289+
Wrapping machines	53	203+
Packer		
Well	166	179+
Closing port	166	141
Handling	166	373+
Preset	166	114
Preset conduit carried	166	116
With expanding anchor	166	118+
With outward pressure means	166	101
With passage control	166	142+
With pump or plunger	166	106
Packing Method (See Stack)		
Bags & wrappers	D 9	305
Cartridge	86	47
Cases and materials		
Design	D 9	
For cartridges	206	3
For chemicals	206	524.1+
For cigars and cigarettes	206	242+
For cycles	217	37+
For metal leaf	206	71
For photographic papers	206	455
Cigar bundling	131	107
Dispensing devices	222	143
Felting roll pack	28	125+

	Class	Subclass
Closure with gasket or packing	220	378+
Foods	426	392+
Joints	277	
Bottle cap	215	341+
Closure	220	200+
Cup or flange type	277	212R
Dispenser outlet	222	512
Dispenser outlet	222	542
Pipe joint	285	335+
Prefabricated unit	277	35
Receptacle seam	220	681
Valve	251	356+
Materials		
For sensitive plates	206	454+
Molded separator sheets	D 9	456
Piston	277	
Ring making	29	888.7+
Portable receptacles		
Bags and sacks	53	570+
Cigarette	53	148+
Uniform density	141	71+
Receptacle filling & closing	53	467+
Sand founding molds	164	37+
Tobacco plug	131	112+
Tools	81	8.1
Weather strips	49	475+
Wrappers	229	87.1+
Packs		
Change for plates and films	354	276+
Photographic plate	354	276+
Pad		
Abrasive sheet holding	51	358+
Handled	51	392+
Absorbent	604	358
Animal hampering pads	119	108
Body	2	267+
Button covering	24	114
Cleaning material supply combined	401	261+
Closet seat	4	244+
Desk or typewriter	D19	99
Diaper	604	358
Electrically heated	219	528+
Foot	36	71
Furniture	248	188.8
Ground mat	5	420
Hand	2	20
Harness	54	65+
Design	D30	134+
Horseshoe	168	26+
Design	D30	147+
Modified shoe	168	12+
Hot pot holders	2	20
Hand manipulable	294	25
Handle type	16	116R+
Knitted base	66	190+
Lifters	294	27.1+
Of textile material	428	224+
Inking and copying	118	264+
Printing machine having	101	
Knee	2	24
Label pasting	156	
Ladder foot	182	108+
Lock		
Shackles	70	53
Mattress	5	500
Pedal	74	563
Receptor		
Catamenial and diaper	604	358+
Shoe	36	71
Heel cushioning	36	37
Skirt combined	2	215
Sleeping bag with pad	5	413
Stair	428	
Conforming to tread nose	52	179
Table	428	53+
Table, cover or casing for	428	68+
Table, scuff plate	248	346.1
Table, sectional layer for	428	44+
Truss	128	112.1+
Padded		
Bottle wrapper	229	91
Container with shock protection	206	521+
Crate	217	53
Paddle and Paddle Wheel	416	78+
Aircraft propulsion	244	70+
Aircraft sustension		
Heavier than air	244	4R+
Lighter than air	244	27+
Boat propulsion manual	440	21
Design	D12	306
Game apparatus	D21	48

	Class	Subclass
Pumps	417	436
Ship and boat propulsion	440	90+
Retractable blades	440	92+
Ship steering	114	147+
Ships	114	58
Water motors	416	78
Padlock	70	20+
Condition indicator combined		
Recorder	70	435
Register	70	437
Design	D 8	334+
Latch combined	292	
Design	D 8	334+
Pail (See Bucket; Receptacle)		
Milk	220	9.1
Making	413	69+
With strainer	210	464+
Scrub	15	264
Winding material applying	242	7.22
Method	242	7.2
Paint	106	
Anticorrosive	106	14.5+
Antifouling	106	15.5+
Artists box	206	1.7
Brushes	15	159R+
Design	D 4	
Material supply combined	401	268+
Drier	106	310
Luminous		
Inorganic	252	301.6 R
Organic	252	301.16
Radioactive	252	625+
Signs	40	542
Mill	241	
Mixer	366	605*
Painters equipment		
Artists boxes	206	1.7+
Brushes	15	159R+
Design	D 4	
Graining tool	D 8	
Roller type	D 4	122+
Easels	248	441.1+
Fabric stretching frames	160	374.1
Stencils (See stencil)		
Teaching	434	84
Removing	134	38
Removing-burning tool	D 8	29.1
Roller tray	D32	53.1
Rollers & trays for	D 4	122+
Scraper	D32	46+
Pajamas	2	83
Design	D 2	8+
Palladium (See Platinum)		
Alloys	420	463
Pallet (See Pawl)		
Bed clothing	5	482
Container with	206	386
Dental mixing slab	433	49
Hawk	294	3.5
Horological drive	368	131
Mold	425	253+
Concrete or ceramic type	425	253
Organ valve	84	368
Synthetic plastic industrial platform	108	901*
Made with deformable integral		
fastening elements	108	902*
Type for lift trucks	108	51.1+
Palmitic Acid	260	413+
Pamphlet		
Covering	412	4+
Distribution from aircraft	40	216
Pan (See Pot)		
Ash		
Boiler feed preheater combined	122	413
Furnace	110	166+
Stove	126	243+
Bed	4	450+
Boiler	122	394
In steam space	122	435+
With blowoff	122	386
Carbonless	503	200+
Closet bowl combined	4	429+
Closet bowl valve	4	442+
Cooking container	D 7	354+
Drip (See drip pan)		
Drying	34	237+
Housed or supported	34	192+
Dust	15	257.1+
Evaporating and concentrating		
Closed	159	22+
Combined open and closed	159	21

	Class	Subclass
Open	159	32+
Land vehicle drip	296	38
Liquid fuel burner	431	331+
Lubricator drip catching	184	106
Mill	241	107+
Motor vehicle splash	180	69.1
Pan shaped metal bending dies	72	394+
Railway tender coal feed	105	235
Scale	177	262+
Stove drip	126	51
Street sweeper elevator	15	85
Water heating	126	373
Pancake		
Fryer	99	422+
Turner	294	7+
Pancreas Extracts	530	845
Medicine containing	514	3+
Pane, Window (See Panel)		
Fasteners	52	764+
Hermetically sealed	428	34
Laminated glass	428	426+
Parallel panes	52	304
Drier	52	171+
Resiliently spaced	52	398
Vented	52	304
Ventilated	98	96
Panel		
Bed	5	280
Board switchboards	361	331+
Telephone	335	108+
Building components (See block)	52	
Adhered tile	52	390+
Attached handle	52	125.2+
Corrugated component	52	795+
Dimpled	52	792
Edged laminated	52	783+
Edged laminated and separate	52	821+
Edged resilient	52	393
Edgewise connected	52	578+
Embossed	52	792
Hollow	52	785
Integral reinforced	52	630
Internal fastener received	52	787
Multicellular core	52	806+
Ornamental	52	311+
Partition	52	238.1+
Perforate sheet	52	794
Perforate sheet, edgeing separate	52	815
Sandwich type	52	785+
Shaft attached	52	474+
Shield	52	202+
Stonelike	52	596+
Sub-structure attached	52	506+
Casket sliding panel	27	15
Closure fastener operator	292	93
Decorative sectional layer	428	44+
Decorative sheet cover or casing	428	68+
Disc	428	64+
Edge lighted	362	31
Fasteners	24	573.2
Fences	256	24+
Panel element	256	73
Flexible	160	
For floor, wall, roof	D25	138+
Hangers	16	87R+
Heat exchanger	165	168+
Heating stove	126	64
Door transparent	126	200
Joints, readily disengageable	403	DIG. 1
Mirror with backing	52	786
Mounting for instrument	248	27.1+
Movable or removable closures	49	
Partitions &, flexible & portable	160	
Phonograph blanks	428	64+
Portable	160	351
Radiant heating wall	165	49
Railway passenger car		
Convertible open to closed	105	332+
End panel open car	105	347
Static mold form	249	189
Travelers	16	87R+
Wall panel, molding continuous	264	46.4+
Windshield	296	90
Sash convertible to screen	160	128
Panic Bar Latches	292	92+
With lock	70	92
Panification		
Processes	426	549+
Panoramic Cameras	354	94+
Pantographic Scribers	33	23.1+
Curved surface	33	22

	Class	Subclass
Cutting machine for shaping	409	86+
Multiplane	33	24.1
Pantolactone	549	319
Pantothenic Acid	562	569
Pants	2	227+
Design	D 2	28+
Design, lingerie or undergarments		
Pantyhose	2	409+
Papaverine	546	149
Paper		
Article design	D 5	
Article making	493	
Computer controlled, monitored	364	471
Bag making	493	186+
Bag opening	493	309+
Bags	383	
Barrels	229	67
Binder device releasably engaging		
aperture or notch of sheet	402	
Blueprint	430	540
Bobbins and spools	242	118+
Box	229	100+
Gift wrapped	229	923
Box making	493	52+
Carbon	428	488.1+
Compositions	106	20+
Paper, design	D19	1+
Cartridge shells	86	11+
Clip or clamp fasteners	D19	65
Clips	24	455+
Coating	427	411+
Stripping or edging	427	284+
Collars	2	136
Combined into book	D19	26+
Cutter	D18	34
Cutter, machine-type	D15	127
Disc closure making	493	962*
Disinfecting or sterilizing	422	243+
Driers	34	148+
Processes	34	23+
Dyeing	8	
Electroprinting	204	2
Envelope making	493	186+
Envelopes	229	+
Fasteners	24	67R+
Making	29	13
Penetrating-type	24	901
Festoon making	493	957*
Flypaper holder	43	115
Flypaper package	206	447
Folding or plaiting	493	394+
Heat seal	493	375+
Scraps	100	914*
Tag applying	493	269+
Tube making	493	958*
Garland making	493	958*
Hanging	156	574
Design	D 5	
Interlocked sheets	493	390
Lined metal receptacle	220	460+
Making & fiber liberation	162	
Matchmaking	144	51
Medicated	424	414+
Moistener	134	122R+
Money sorting	209	534
Opening reinforcements	402	79
Other than in book	D19	33
Package or article carrier	224	906*
Parchmentized or linen finish	8	119
Photographic	430	270+
Photographic paper packages	206	455
Photoprinting	430	270+
Punch	D19	72
Punching	83	
Roll type dispenser packages	206	389+
With severing edge	225	39+
Self toning	430	559
Sheet dispensing packages	206	449+
With tearing edge	225	27+
Sheet manipulating devices		
Flipper, inserter, lifter,	402	80R
Remover, transfer fork	402	80R
Transfer post	402	47
Sheet stock	428	
Shelf paper	D 5	
Shifting plate	29	270
Stationery	D19	
Strand twisting or twining	57	31+
Stripping photographic	430	259+
Tablet	D19	26
Tearing straight edge	225	91+

	Class	Subclass
Pawl carried by washer	411	138
Pawl carried movably	411	145+
Resilient pawl with side lock	411	125+
Bolt to nut lock and ratchet	411	326
Used with jam nut	411	227+
Paying Devices	114	224
Pay check or envelope with statement of deductions	282	DIG. 1
Peach Stoner	30	113.1+
Peanut	426	632+
Butter	426	633
Oil	260	398+
Pearl Button Making	79	6+
Pearl Essence	544	276
Peat	44	490+
Carbonizing	201	25
Excavator	37	3
Fertilizers from	71	24
Mineral oil from	208	400+
Peavy	294	17
Pebble Mill	241	170+
Pectin	536	2
Fermentative production	435	275
Jellies containing	426	577+
Pedal	74	560+
Clavier	84	DIG. 25
Closure actuator	49	263+
Crank	74	594.4+
Bearing	384	429+
Bearing antifriction	384	458
Combined	74	594.1+
Design	D12	125
Hand control combined	74	481
Link and lever system control	74	512+
Music keyboard		
Auxilliary	84	426
Chord player	84	444
Piano	84	229+
Automatic	84	72+
Reed organ	84	353
Key action	84	366
Wind control	84	357+
Valve actuator	251	295
Pedestal		
Bearing support	384	440
Spring type	248	162.1
Vertically adjustable	248	161+
Horizontal planar surface	108	88
Supporting	108	150
Railway surface track	238	109+
Stand supporting	248	158+
Extension	108	88
Pedicure Implements	D28	56
Pedomotors (See Velocipedes)	235	105
Boat	440	21+
Peeler or Parer (See Cutter; Stripper)		
Corer combined	99	542+
Corer segmenter combined	99	545
Food	99	588
Fruit & vegetable	99	588+
Design	D 7	693
Osier	144	207
Rossing bark	144	208R
Slicer combined	99	592
Corer combined	99	543
Wire stripper	81	9.4+
Peeling		
Bark	144	208R
Bark branch	144	207
Fruit and vegetable	99	584+
Peen and Peening	72	76
Shot blasting	72	53
Thread texturizer	28	277
Peepholes	220	662+
Door mounted	49	171
Stove	126	200
Drying apparatus (See notes to)	34	88
Measuring sight glass plate type	73	334
Shield mounted	109	58.5
Peg (See Pin)		
Cutters	12	47.1
Implements	12	104.5
Cutting and driving	227	93+
Indicators	116	325+
Receivers lasts	12	140
Sole fastening means	36	24
String fastener musical	84	304+
Tally boards	235	90
Tent	135	118
Violin	84	304+
Peg-board Type Article Supports	248	216.1+

	Class	Subclass
Pelican Hook	294	82.24+
Pellet (See Pills)		
Depositors	604	57+
Design	D 1	
Medicinal	D28	1+
Hay pellet makers in general	100	
Hypodermic injectors	604	187+
Machine	425	222
Process for food	426	512+
Receptacles for pocket	206	528+
Pellet	D22	116
Pellicles for Optical Instruments	350	171+
Peltier Effect (See Thermocouple)		
Pelts	69	22
Carroting	8	112
Chemical or fluid treatment	8	94.14
Cleaning elements made of	15	235
Cutting	83	
Dyeing	8	404+
Fiber liberation from	19	2+
Shearing machines	26	15R+
Stretching frames	38	19+
Treatment	69	23+
Pen		
Advertizing on	40	905*
Animal	119	20
Design	D30	108+
Box for	206	371
Caps	401	243+
Cleaners	15	423
Desig	D19	98
Extractors	29	808
Facsimile	358	296+
Facsimile design	D19	41+
Fountain (See fountain pen)	401	
Filling devices	141	21+
Tattooing	81	9.22
Holder (See penholders)	15	435+
Design	D19	81+
Inkstand combined	211	69.2+
Light, photoelectric		
With light conductor	250	227.12
Pill carrier	206	537
Play	256	25
With bottom	5	99.1
Points (See penpoints)	15	435+
Rack	211	69.1+
Design	D19	81+
Ruling	401	221
Adjustable gap	401	256+
Adjustable gap fountain	401	233
Type flashlight	362	118
Penamecillin	540	342
Pencil	401	49+
Advertizing on	40	905*
Book cover combined	281	30
Box or case	D 3	30.1+
Box sectioned	206	371
Calculator combined	235	64
Calendar combined	40	335
Clip	D19	56
Compositions	106	19
Cosmetic eg lipstick or eyebrow	401	49+
Shaped	132	320
Indicia carrying	40	334+
Calendar	40	335
Mechanical	401	49+
Package	206	214
Pocket fastener	24	10R+
Design	D19	56+
Rack	211	69.1+
Sharpeners (See pencil sharpener)	30	451+
Electric	144	28.1+
Pencil combined	401	50+
Type electronic tubes	313	
Wood making	144	28
Pencil Sharpener	D19	73+
Implement	30	451+
Blade	30	165+
Debris receiver combined	30	453+
Eraser combined	15	105.53
Pencil combined	401	50+
Tool combined	7	160+
Machine	144	
Edge bevelling	144	28.2
Hand manipulable	144	28.11
Pendant		
Earrings	63	13
Jewelry	D11	79+
Watch pendant gearing	368	190+
Watchcase	368	301+

	Class	Subclass
Watchcase making	29	179
Pendulum		
Clock	368	134+
Design	D10	130
Electric	368	165+
Escapement	368	134+
Facsimile synchronization using	358	415
Level and plumb	33	353+
Bubble combined	33	353
Lubricator		
Force feed follower	184	44
Force feed pump	184	30
Gravity feed reciprocating valve	184	78
Gravity feed rotatable valve	184	73
Mill	241	129+
Operated changeable exhibitor	40	485
Scale	177	216+
Testing		
Dynamometer	73	862.38
Tensile stress	73	836
Tide or wave motor	60	497+
Penetrating (See Punch; Puncturing)		
Nailing and stapling	227	
Nozzles	239	271
Perforating drilling	83	
Pole climbers	182	221
Printing by	101	3.1+
Sorting	209	688+
Testing by	73	81+
Typewriter	400	135+
Penetration Resistant Stock	428	911*
Penetrometers	73	59
Penholders	15	435+
Desk type for drafting	206	371
Indicia carrying	40	334+
Calendar	40	335
Picture combined	40	334+
Penicillamine	562	558
Penicillin F	540	314
Penicillin G	540	342
Penicillin G, in Drug	514	199
Penicillin K	540	314
Penicillin N	540	335
Penicillin O	540	314
Penicillin S	540	314
Penicillin V	540	341
Penicillin X	540	342
Penicillins	540	304+
By fermentation	435	43
Medicines containing	514	192+
Penmanship Instruction	434	162+
Pennant	116	173+
Penpoints	15	435+
Cleaner	15	423
Extractors or inserters	29	808
Ink retainer attached	401	252+
Ink retainer integral	401	221
Ruling	401	256+
Reservoir	401	233
Stylus ink retaining	401	292
Wiper	15	423
Pentaerythritol	568	853
Nitrated compositions	149	93
Pentaerythritol Tetranitrate	558	485
Explosives containing	149	93
Pentagrid Heptode Converter	313	300
Pentanediol	568	853
Pentode		
Amplifier	313	300
Cutoff	313	300+
Diode	313	298
With two cathodes	313	5
Diode triode	313	298
Tube structure	313	297+
Video	313	300
Pepper		
Design, shaker or mill	D 7	591+
Hand mill	241	168+
Shaker	222	142.1+
Peptide Synthesis	530	333
Peptides	530	300
Production by fermentation	435	68.1+
Including recombinant dna technique	435	69.1+
Including tissue culture use	435	70.1+
Per Acids	562	1+
Electrolytic production	204	82+
Perambulator (See Baby Carriage)	280	47.38+
Convertible to walker	280	7.1+
Springs	267	5
Perborate Esters	560	305

	Class	Subclass
Perborates	423	
Electrolytic production	204	85
Perbunan T M (See Synthetic Resin or Natural Rubber)		
Percale	139	426R
Percarbonates	423	419R
Percarboxylate Esters	560	302
Perch for Fowl	119	24+
Design	D30	119+
Perchlorates	423	476
Compositions		
Chemical agent type	252	186.1+
Detergent type	252	95+
Explosive	149	42
Explosive	149	75+
Percolator Beverage	99	307+
Electrically heated	219	438+
Stand combined	219	429+
Pitcher type	222	
Pumps	417	208+
Strainer	210	473+
Top transparent	220	377.1
Percussion		
Electric weld	219	95+
Musical instruments	D17	22+
Percussive		
Appliance kinestherapy	128	54+
Fuses	102	272+
Signs	40	579
Perdnisolone, in Drug	514	179
Perforating (See Penetrating; Punch)		
By chemical dissolution	156	644+
Devices		
Cigars	131	253+
Design	D19	72+
Earthenware multitubular pipe	425	290+
Electric heater	219	229+
Fountain pen combined	81	9.22
Glass	65	166
Inker combined, eg tattooer	81	9.22
Plastic block expressing and	425	290+
Sheets or bars	83	
Well tubing in situ	166	55
With glass mfg means	65	166
Electric discharge	219	384
Recorder	346	153.1
Handling perforated article	414	908*
Metal	83	
Billet piercing	72	345
Expanded	29	6.1
Multiperforated making	29	6.1+
Perforated devices		
Control members (See type)		
Cards	235	487
Musical instrument	84	161+
Tape	178	112
Typewriter tape	400	73+
Drive device for tool	173	90+
Recording printing or typing		
Printing	101	3.1+
Recorder	346	78+
Telegraph code recorder	178	92
Typewriter	400	80+
Telegraph code tape		
By telegraph signals	178	70R
Selective, by keyboard	234	
Tobacco products	131	253+
With electric spark or arc	219	384
Perfume Compositions	512	
Design	D28	5+
Encapsulated	512	4
Essential oils	512	5
Fixative for	512	2
Garment hanger combined	223	86
Head coverings with therapeutic device, medicament or perfume	2	171.2
Perfumed jewelry	63	DIG. 2
Preserver, stabilizer for	512	2
Solids, gels and shaped bodies	512	4
Perhalate Esters	560	306
Perhydroxamate Esters	560	311
Perhypohalite Esters	560	300
Perimidine	544	249
In drug	514	269
Periodic Switch		
Electric circuit makers & breakers	200	19R+
Electromagnetic	335	87+
Electric discharge device supply control	315	209R+
Periscopes	350	540+
Compound lens system	350	541+

	Class	Subclass
Submarine combined	114	340
View simulation	434	26
Perlite	501	28
Digest	106	DIG. 2
Permanent Magnet		
Alloy	420	
Permanent Press		
Chemically modified fiber	8	115.51
Coated	427	393.2+
Compositions	252	8.6+
Permanent Wave Machine for Hair	D28	11
Permeability Tuned Transformer	336	130
Permutation		
Locks		
Combination	70	315+
Key	70	382+
Key bitting	70	411
Switch combined	200	42.1+
Switch	200	42.1+
Typewriter	400	100+
Valves	137	269
Permutit T M	423	328
Peroxides (See Oxides)		
Alkali metal	423	582
Bleaching with	8	101+
Electrolytic coating with	204	57
Electrolytic synthesis	204	83+
Electrostatic field or electrical discharge	204	175
Heat producing compositions	149	37+
Hydrogen	423	584+
Organic	568	558+
Acids or salts	562	1+
Oxidizing	252	186.22+
Peroxynitrate Esters	560	304
Perspective Plotters	33	432+
Persulfonate Esters	560	318
Persulphides (See Polysulfide)		
Perthioborate Esters	560	305
Perthiocarboxylate Esters	560	302
Pessaries	128	834+
Pesticides (See Vermin Destroying)	D22	120
Pestle	241	291
Grain hulling	99	600+
Mortar & pestle	241	199+
Petrolatum	208	20+
Treatment or recovery	208	24+
Petroleum	208	
Emulsion breaking	252	319+
Fuels	44	
Mineral oil only	208	15+
Fuels containing	44	
Lubricants	252	9+
Mineral oil only	208	18+
Production		
From tunnels	299	2
Thermal or solvent eg secondary recovery	166	
Wells and related processes	166	
Receptacles and storage with explosion preventing devices (See explosion control)		
Sulfonates	562	33
Petrothene T M (polyethylene)	526	352
Pew	D 6	334+
PH Electrode or PH Sensor		
Chemical analysis, apparatus	422	82.4
PH Measurements Electrolytic	324	438
Phantom Circuit		
With loading coil	178	45
Pharmaceutical Product	D28	1+
Pharyngoscopes	128	10+
Phase		
Changer		
Same current in input & output	323	212
Using controlled electronic tubes	328	24
Control electrical	323	212
Conversion, electrical	363	148+
Indication electrical	324	83R+
Measuring and indicating	324	83R+
Meters	324	83R+
Phase inverter amplifiers		
Balanced to unbalanced coupling	330	116
Transistor amplifiers	330	301
Unbalanced to balanced coupling	330	117
Phase locked loop having lock indicating or detecting means	331	DIG. 2
Phase locked loop in demodulator circuit	329	325
Regulators for alternating current generators	322	20+

	Class	Subclass
Shift keying	375	52
Shifter		
Electronic tube type	328	155
Shifting and power factor control	323	205
Electronic tube circuit having factor control	315	247
Electronic tube circuit having shifting	315	194+
Meter adjustment	324	138
Wave transmission type	333	138+
Phenacetin	564	223
Phenanthraquinone	552	292+
Phenanthrene Purification (See Aromatic Hydrocarbon)	585	800+
Phenanthridine	546	108
Phenanthroline	546	88+
Phenates (See Phenols)	568	716+
Phenazines	544	347+
Phenethicillin	540	341
Phenetidine	564	443
Phenobarbital	544	305
Phenolphthalein	549	308
Phenols (See Naphthols)	568	716+
Alkylation	568	780+
Amino	564	443+
Addition salts	564	280
Sulfo	562	58
Carboxylic acid esters	560	67+
Monoazo compounds	534	660+
Halo	568	774+
Mineral oil containing	208	2
Nitro	568	706+
Explosive compositions	149	55+
Explosive compositions	149	68
Explosive compositions	149	80
Explosive compositions	149	99
Explosive compositions	149	103
Explosive compositions	149	105+
Polymers (See synthetic resin or natural rubber)		
Sulfo acids	562	74
Phenones	568	335
Phenoplast (See Synthetic Resin or Natural Rubber)		
Phenothiazines	544	35+
Phenoxazines	544	102+
Phenoxy Resin (See Synthetic Resin or Natural Rubber)		
Phenylarsonic Acid	556	73
Phenylazonaphthyl	534	839+
Heavy metal containing	534	713+
Phenylazonaphthylazonaphthyl	534	832+
Phenylazophenyl	534	843+
Phenylazophenylazonaphthyl	534	832+
Phenyldimethylpyrazolone	544	356+
Phenylene		
Diamin	564	305
Phenylephrine, in Drug	514	653
Phenylglycine	562	443
Indole preparation from	548	489
Phenylhydrazine	564	310
Phenylquinolin	546	173
Carboxylic acid	546	170
Phenyltoloxamine, in Drug	514	651
Phenylurethane	560	75
Phloroglucin	568	763+
Phones		
Sound power	381	177+
Phonograph	369	
Automatic disk changer	369	178+
Cabinets	312	8+
Check operated switch	194	239
Coin controlled	194	
Design	D14	173
Design	D14	199+
Electrical reproducing or recording	369	128+
Figure toy combined	446	297+
Illuminator	362	87
Indexing devices	369	30
Magnetic	360	86
Motion picture apparatus combined	352	32+
Musical instrument combined		
Automatic musical instrument	84	4+
Comb organ	84	100
Electric tone generation eg organ	84	639
Needle	369	173
Photographic	369	125
Radio combined	369	6
Reciprocating tone arm motions	74	28
Record and composition		
Cabinets for	312	8+

	Class	Subclass
Heat exchanger	165	177+
Hickeys (See bending)		
Holder	D27	180+
Hydraulic ram	417	226+
Irrigation, vehicle supported	239	723+
Jack for pipes or rods	254	29R+
Joints	285	
Making	29	428+
Making	29	890.43
Valved	251	142+
Laying pipe and cable	405	154+
Submarine	405	158+
Making		
Bending	72	367+
Clay smoking	425	233+
Earthenware machine	425	376.1+
Earthenware mold	249	136+
Encircling with wire on tape	242	7.22+
Expander	72	317
Fittings	29	890.14+
Flanging	72	317
Foundry mold making	164	176+
Glass	65	
Laminated	156	
Filament winding by	169	
Metal casting	164	
Apparatus	164	421+
Metal fusion-bonding apparatus	228	
Metal joint mold	249	89
Reamer	408	227+
Riveting	227	52
Soldering clamp	269	
Swaging	72	399+
Trimming	408	28+
Organ	84	331+
Automatic	84	91
Pattern sprayer	239	548+
Pocket case for	206	244
Rack	211	70.3+
Smoking device	131	260
Rotary shaft	464	183
Joints	464	18+
Sand delivery	291	41+
Shearing machines	30	92+
Smokers pipe literature	131	DIG. 1
Soap bubble	446	19+
Sprinkler	239	266+
Stand (See tank)		
Flue boiler	122	184+
Superheater	122	475
Water tube boiler	122	321+
Water tube boiler coil or loop	122	244+
Water tube boiler loop	122	281
Water tube boiler spur	122	307+
Stems	131	227+
Stove	126	307R+
Damper	126	292+
Oven	126	17
Water heater	126	364+
Strainers	4	DIG. 13
Structure	138	177
End combined	138	109
Supports	248	49+
Tamper tobacco smokers	131	247+
Thread	285	390+
Protectors	138	96R
Threading		
Cutting	408	
Milling	409	65+
Rolling	72	126
Tobacco	131	330+
Cartridge	131	348
Toy	446	24
Train	239	723+
Waste	4	191+
Well		
Cleaner combined	166	173+
Cutter or perforator	166	55
Destroying	166	373+
Expansible	166	207
Joining	166	373+
Plural concentric	285	138+
Screen combined	166	227+
Structure	166	242
Weakening	166	55
Wrenches	81	52+
Pipeline		
Cleaner	15	104.3
Filters	210	435+
Intake	210	460+
Piperazines	544	358+

	Class	Subclass
Piperidine	546	184+
Piperonal	549	436
Pipette	73	864.1+
Pipette		
Absorption for gas analysis	422	100
Chemical analysis, apparatus	422	83+
Body contacting, eg, hypodermic		
needle, syringe	128	
Design	D24	55
Design, chicken baster	D24	60+
Chemical analysis, apparatus	422	82.5
Chemical apparatus	422	
Chemical methods	23	
Container cap combined	141	29
Container or container closure	141	22+
Bulb type, eg eye dropper	141	24
Dispensing	222	
Drop formers	222	420
Medicinal	604	295+
Design	D24	55
Successive receivers filled from		
continuously flowing source	141	130
Pipings	36	57
Pipings, Upholstery Trim	428	123
Piromidic Acid, in Drug	514	303
Pistol		
Grip	42	71.2+
Magazine	42	7
Holsters	224	193
Design	D 3	101
For either side of torso	224	192
Leather or flexible fabric	224	911*
Metal or rigid material	224	912*
Rotatably or swingably attached	224	198*
Shoulder	224	206*
With closure flap	224	238*
With gun positioner	224	243*
Revolvers	42	59+
Design	D22	104*
Swords	42	53
Design	D22	118
Water	222	79
Piston (See Associated Device)		
Cleaner for plugs and pistons	15	104.11+
Door check	16	49+
Expansible chamber device	92	
Annular	92	107+
Multiple	92	61+
Oscillating	92	120+
Filler for fountain pens	401	171+
Fire extinguisher	169	33
Free piston	60	910*
Hammer of tool drive	173	134+
Hypodermic	604	218+
Internal combustion engine	123	193R
Manufacture	29	888.4+
Grinding	51	289R
Meter	73	232+
Packing	277	
Digest	33	DIG. 15
Pressure gauge	73	744+
Ring	277	
Assembling and disassembling on		
piston	29	222+
Gauge, ring & bearing race	33	DIG. 17
Group contracting for working	29	269
Lubrication combined with		
Making	29	888.7+
Testing	73	120
Testing for leakage	73	47
Rod (See rod connecting)	74	579R+
Lubricators	184	24+
Valved piston compressor or		
pump combined	417	545+
Rotary engine	123	200+
Rotary for motors	418	68
Sausage stuffer	17	36+
Syringe	604	218+
Valve	251	324+
Check type	137	538
Valved		
Pump liquid	417	545+
Pit Remover	30	113.1+
Pitch (See Tar)		
Changing		
Conveyor screw	198	659
Impeller	416	25+
Musical instrument	84	
Pump impeller	415	129+
Ship propeller	440	50
Fuel containing	44	
Ionized gas		

	Class	Subclass
Fluent material supplied space		
discharge device	313	231.1+
Making, treating and recovery	208	39+
Mineral oil only	208	22+
Music device for giving pitch	84	454+
Plastic compositions containing	106	
Production by distillation	201	
Pitcher	222	
Dispensing type with handle	222	465.1+
Handle, asymmetric	D 8	DIG. 9
With flow controller or closure	222	556+
Pitchfork	294	55.5+
Cutter combined	7	115+
Making		
Blanks and processes	76	111
Dies	72	470+
Pitching Casks	118	
Pitchometers	33	530
Pitman (See Connecting Rod)	74	579R+
Bearing	384	266+
Making	29	888.9
Piton	248	216.1+
Mountain climbing aids	248	925
Pitot Meter	73	861.65+
Pitter	99	547+
Pivampicillin	540	336
Pivot (See Hinge; Joints)		
Bearing	384	192+
Knife-edge fulcrum	384	4+
Rotary crane	212	253
Teeth	433	220+
Turntable	104	46
Pku	436	108*
Placard Frames	40	152+
Placemat		
Furniture protector	D 6	613+
Paper	D 6	613+
Placers (See Arranging; Piling)		
Electrical wiring	254	134.3 R
Hammer with nail placer	81	23+
Track ties	104	6
Track torpedo	246	216
Placket	2	218
Plait and Plaiting		
Apparel making	223	28+
Box making	493	162+
Knitting	66	14
Laminated pile fabric making	156	
Tucking	112	145
Planes		
Woodworking	30	478+
Planetarium	434	286
Planetary (See Gear and Gearing)		
Planimeter	33	122
Planing Devices		
Bench planes for wood	30	478+
Hair planer	30	30+
Planer type combined machine	29	30
Planing machine	409	288+
Stave jointing	147	33
Stoneworking	125	9
Wood saw teeth	83	835+
Wood slivering scoring	144	186+
Woodworking	144	114R+
Woodworking combined machine	144	36+
Plank and Planking (See Panel)		
Clamp portable for	269	91+
Pavement	404	17+
Railway truck plank	105	208
Scaffold	182	222+
Ship	114	84
Planographic		
Printing	101	130+
Electrolytic coating	204	17
Multicolor surface making		
photographic	430	301
Surface making photographic	430	300+
Plans		
Building	52	234+
Store	52	33
Traffic guide	52	174
Plant	800	200+
Bud	800	200
Carnation	800	DIG. 1
Chrysanthemum	800	DIG. 12
Corn	800	DIG. 56
Derived from embryogenesis	800	205+
Derived from mutagenesis	800	230
Derived from protoplast cell fusion;		
somatic cell fusion	800	220
Derived from recombinant or		
transgenic processes;		
transformation	800	205

	Class	Subclass
Derived from spontaneous mutation	800	230
Derived from tissue culture regeneration	800	205
Derived from traditional breeding techniques	800	200
Dicotyledon seed (dicot)	800	255
Embryo	800	200+
Embryoid	800	200+
Flower	800	200
Flowering plant	800	DIG. 67+
Fruit	800	200
Fruit plant	800	DIG. 64+
Fusion product	800	220
Graft or plant with grafted part	800	240
Gymnosperm	800	DIG. 49+
Leaf	800	200
Meristem	800	200
Mutant plant	800	230+
Monocot (e. g. maize)	800	235
Other plant parts	800	200
Pear	800	DIG. 35
Plantlet	800	200+
Recombinant plant	800	205
Rice	800	DIG. 57
Root	800	200
Rose	800	DIG. 36
Seedling	800	200
Seeds, genetically modified	800	250
Shoot	800	200+
Stem	800	200
Tuber	800	200
Vegetable plant	800	DIG. 71
Wheat	800	DIG. 58
Plant Cell Culture	435	240.4+
Plant Cell Growth Media	435	240.54+
Plant Cell Per se	435	240.4+
Planters and Planting		
Agricultural	111	
Broadcasting	111	130+
Dibbling	111	89+
Drilling	111	14+
Liquid or gas	111	118+
Plant setting	111	100+
Corn planter clutches	192	23
Plants		
Attack resistant stock	428	907*
Catalysts and stimulants	71	65+
Coating of	427	4
Cutting and comminuting		
Discriminate	99	
Indiscriminate comminuting	241	
Indiscriminate cutting	83	
Extracts (See extracts)		
Food	99	
Gas making	48	
Heat or power		
Aircraft	244	53R+
Boiler	122	1R+
Dynamoelectric	290	
Heat and power combined	237	12.1+
Heating systems	237	13
Power	60	
Husbandry (See plant class)	47	
Compound tools	7	114+
Covers, shades and screens	47	26+
Diggers	171	
Fertilizers	71	
Fertilizers	47	48.5
Fruit cutters and corers	30	113.3
Garden hose	138	118+
Garden rake	56	400.1
Greenhouses	47	17+
Harrows	172	
Harvesters	56	
Hotbeds and coldframes	47	19
Nutrient supply	47	62+
Ornamental beds	47	33
Planting	111	
Plows and cultivators	172	
Pollination	47	1.41
Pots and receptacles	47	66+
Powder dusters or sprayers	239	
Presses for balers	100	
Sprayers mixing and dissolving	239	310+
Terrarium	47	69+
Threshing	460	
Tree fumigators	43	126
Tree hacks	30	121
Water culture	47	59+
Lights	362	805*
Living	800	200+

	Class	Subclass
Metallurgical	266	142+
Molding	425	
Brickmaking	425	88
Nameplates or cards for	40	645+
Nuclear	376	
Patented varieties of	PLT	
Planting	111	
Dispensing seed	222	
Grain drill chutes	193	9
Preserving	47	58
Coating	427	4
Cut flower holder	47	41.1+
Stalks retting	162	1+
Retting bacteriological	435	279+
Stands	47	39+
Thinners	172	534+
Earth working function	172	900*
Plant sensor activates power	47	1.43
Plaque	428	913.3
Holder	248	488+
Holder	248	490+
Layer non-planar	428	174+
Layer varying thickness	428	156+
Panel type, sectional layer	428	44+
Wall decoration	D11	132+
Plasma	424	530
Blood component	424	530
Drying and freezing	34	5
Gas		
Chemical process employing	204	164+
Injecting fuel into body of plasma	376	127
Ion producer	361	230
Magnetohydrodynamic generator	310	11
Moving or shifting plasma	376	107
Nuclear fusion reactions	376	100
Containment by magnetic field	376	121
Systems	315	111.21+
Plasmid	935	27*
Plaster of Paris (see Plaster)	106	772+
Plaster or Plastered (See Plastering; Plastic)		
Board	52	443+
Attached to furring	52	344+
Edge joint	52	448+
Laminate	428	411.1+
Making	156	39
Corner	52	250+
Cutters		
Hand manipulable	30	165+
Machines	83	
Pointers	15	235.3
Medicated composition	424	
On brick	52	444
Panel strip	52	459
Plaster of paris	106	772+
Alkali metal silicate composition	106	611+
Compositions	106	772+
Processes of burning	106	765+
Splint	128	91R+
Screed	52	364+
Slatted lath	52	341+
Surgical body wear	128	390
Trimmers	433	144
Plastering		
Applicator-material supplied	15	104.93
Fountain-trowel	15	235.4
Machines	425	62
Machines	425	87
Mortar-joint finisher	15	235.3
Combined	15	105.5
Trowel	15	235.4+
Plastic (See Plaster)		
Block and earthenware apparatus	425	
Chain, staple, horseshoe making	59	
Coating	106	
Compositions	106	
Cutters	30	115
Deformation, pressing, or treating		
Metal deformation	72	
Metal fusion bonding	228	
Presses	100	
Shaping or treating apparatus	425	
Sheet metal container making	413	
Static molds	249	
Wireworking	140	
Dyeing of	8	
Jewelry	63	DIG. 3
Metalworking	72	253.1+
Superplastic material	72	709*
Plastics	425	
Screws, bolts, nails etc	411	904*

	Class	Subclass
Shaping or treating articles		
Apparatus	425	
Computer controlled, monitored	364	473
Processes	264	
Structures		
Brush or broom tuft socket	15	193
Coffins	27	7
Fence connections rail	256	66
Fence connections wire	256	50+
Fences	256	19
Hook	24	907
Lacing eyelet	24	713.8
Lacing hocks	24	714.3
Railway ties	238	84+
Seat covers	297	219+
Settable material masonry	52	
Splints	128	90+
Synthetic resin building component	52	309.1+
Testing of	73	150R
Plasticizing		
Casein ctg compositions	106	147
Gelatine ctg compositions	106	136
Polymer treating (See synthetic resin or natural rubber)		
Textiles	8	130.1+
Plate		
Antiskid for tires	152	225R+
Baking	99	372
Cleaners		
Brushing	15	77
Wiping	15	102
Compound metal	428	615+
Dental	433	167+
Design, household	D 7	500+
Distribution for electric wires	174	40R+
Electric		
Conductor	191	13
Consumable electrode	314	44
Face	220	241
Face switch	174	66+
Fluorescent	250	466.1
Switch	200	271
Switchboard	361	346+
Electrophotographic copier	355	211+
Escutcheons	70	452
Door knob	292	357
Face	220	241
Electric housing	174	66
Grinding and polishing	51	
Lock	70	450
Making	65	
Hinge	16	251
Illumination lens perforated	362	456
Joints and couplings (See joints)		
Pipe and plate	285	189+
Pipe and plural plate	285	128+
Pipe and plural plate spaced	285	19+
Rod and plate	403	230+
Through plate insulator	174	151+
With intermediate member	403	187+
Kick	16	1R
Kick guard	16	DIG. 2
Knob rose	292	357
License	40	200+
Lifters	294	27.1+
Lifting clamp	294	901*
Liquid heater or vaporizer	122	208
Nail plates	10	33
Distributors	10	171+
Horseshoe nails	10	41
Name	40	
Houses	40	
Letter box	232	46
Nuclear fuel	376	432
Packages	206	449+
Pattern foundry type	164	241
Photography		
Camera	354	
Coating apparatus	118	
Plate holders	354	276+
Plates	430	495+
X ray plate changers	378	172
X ray plate holders	378	167
Portable closure fastener	292	289+
Printing	101	395
Addressing	101	369+
Casting	164	2+
Coating electrolytically	204	17
Making electrolytically	204	6
Making stereotype	29	21
Mountings	101	382.1+

	Class	Subclass
Prism for overhead lighting	362	326+
Push	16	1R
Racks for	211	41
Railway		
Draft cheek	213	54+
Frog	246	462
Rail base plate	238	209+
Switch	246	453
Tie	238	287+
Rectifiers, dry	357	
Sounding for pianos	84	184+
Stencil	101	127+
Plate	101	112
Stock metal	428	615+
Storage battery	429	209+
Stove	126	81
Stovepipe thimble	126	317
Table article	220	
Design	D 7	500+
Design, simulative	D 7	571
Transformer	336	
Tumblers lock type	70	377
Turners	294	6
Watch	368	318
Design	D10	123+
Platen		
Copyholding (See copyholder)		
Electrophotographic copier	355	230+
Heat exchanger	165	168+
Liquid extracting press	100	295+
Memorandum pad type	281	6+
Metal rolling	72	88+
Roll and platen	72	88+
Screw thread	72	88+
Printing press bed and platen type	101	287+
Addressing machine	101	57+
Intaglio	101	163+
Multicolor	101	193+
Numbering	101	78+
Roller and plate inker	101	338+
Roller type inker	101	359+
Selective or progressive	101	93+
Special article	101	41+
Ticket machine	101	66+
Reciprocating press	100	214+
Smoothing and pressing textiles	38	17+
Conveyor belt combined	38	10
Implements	38	71+
Multiple cooperative	38	5
Structure of	38	66
With fluid flow	38	15+
Stamping recorder	346	78
Textile felting	28	139+
Typewriter	400	648+
Card holder combined	400	525+
Case vertical shift	400	264+
Divided	400	585+
Flat	400	23+
Index wheel	400	158+
Key wheel	400	158+
Rocker platen	400	649+
Shift lock	400	274+
Wear distributors	400	554+
Washing textiles		
Roll and platen scrubber type	68	86+
Scrubbing by carrier roll and platen	68	42
Platers Railway Tie	104	16
Platform		
Actuator		
Closure fastener	292	94
Door	49	263+
Register	235	99R
Dumping scow	114	31+
Floating landing	114	263
Gyrostabilized	248	183
Harvester platform adjustments	56	208+
Illuminators vehicle	362	76
Industrial	108	51.1+
Synthetic plastic	108	901*
Deformable integral fastening elements	108	902*
Knock-down	108	51.3
Ladder	182	116+
Land vehicle		
Swinging axle	280	114+
Connections	280	130+
Rocking connections	280	127+
Load-transfer	108	52.1
Marine drilling	405	195+
Pole or tree	182	187+

	Class	Subclass
Portable lift truck	108	51.1+
Railroad car	105	425+
End buffer	213	222
Truck platform	105	211+
Railroad station	104	30+
Scaffold	182	222+
Elevating	182	141+
Tunnel construction	405	148
Scale	177	253+
Shoes	36	19.5
Stove and furnace	126	277+
Synthetic plastic industrial platform	108	901+
Made with deformable integral fastening elements	108	902*
Take-it-or-leave-it	108	52.1
Vehicle illuminators	362	76
Window	182	53+
Plating (See Coating; Electrolytic Devices)		
Electrolytic coating processes	204	14.1+
Piano keys having	84	437
Warship armor	114	11+
Platinum		
Alloys	420	466
Compounds	23	
Organic	556	136+
Electrolysis	204	
Igniting devices	431	268
Misc. chemical manufacture	156	DIG. 101
Pyrometallurgy	75	633
Player		
Automatic music instrument	84	2+
Automatic musical player attachable to keyboard	84	105+
Automatic player piano	84	13+
Musical cord played by single key	84	443+
Page turner operated by player	84	502+
Playground Climber	272	113
Design	D21	242+
Playing Cards	273	292+
Design	D21	42+
Playpen	256	25
Childs furniture		
Design	D 6	331
With floor	5	99.1
Pleat		
Apparatus	223	28
Method	223	28
Pleated		
Web or sheet	428	181
Plectrum		
Piano	84	258
Picking device	84	320
Plenum		
Fluid mixer thermal	236	13
Surgical receptor	604	313+
Plexiglass T M (See Synthetic Resin or Natural Rubber)		
Pliabilizing		
Shoe soles	12	41.2+
In shoe	12	41.1
Pliers	81	300+
Assembling tool	29	268
Bending	72	409+
Clip applying	72	409+
Nose ring	72	409+
Compound tool	7	125+
Tong type	294	3
Cotter removers	29	248
Dental	433	159+
Orthodontic	433	4
Design	D 8	52+
Ear tag	72	409+
Eyelet setting	227	144
Hollow rivet setting	227	144
Implements	12	110+
Lever pulling or pushing	29	268
Making		
Dies	72	470+
Pivoted	76	64
Sliding	76	69
Piston ring tools	29	223+
Punches	30	363
Spectacle	81	3.6
Staking	29	232
Tongs	294	
Valve tools engine	29	221
Wire twisters	140	121
Plinth (See Base)		
Pliophen T M (phenoplast)	528	129
Plotters Perspective	33	432+

	Class	Subclass
Plow	172	
Cable hauled	104	169
Conduit or trolley		
Electric current collector	191	48
Railway block signal actuation	246	84
Railway switch actuation	246	252
Trap	104	145
Design	D15	11+
Disk bearing	384	157
Radial antifriction	384	460
Ditcher type	37	98+
Forging dies	72	474+
Gondola car unloading	105	259
Hillside	172	204+
Cable hauled	104	169
Iron making	29	14
Lifting clutch	192	62
Making	29	14
Mole	37	193
Cable laying combined	405	184
Pipe or cable laying	405	180+
Submerged	405	164+
Potato digger	171	71+
Potato digger	171	111+
Road grader type	37	110+
Snow		
Excavator and melter type	37	227+
Railway type	37	214+
Roadway type	37	266+
Subsoil	172	699+
Plucks		
Fowl plucking apparatus	17	11.1 R
Fur plucking apparatus	69	26
Musical instrument (See music)		
Plug		
Barrel bung	217	110+
Blasting	102	333
Blasting cartridge combined with	102	304
Bottle and jar closures	215	200+
Dispensing flow control		
Axial discharge stationary	222	521
Cap carried axial	222	546
Closure	222	563
Rotary	222	554
Rotary screw	222	552
Electric connector or switch		
Boxes and housings plug receptacle type	174	53
Cleaner ie spark plug	15	104.11+
Conductor reel combined with	191	12.4
Electric fuse	337	254+
Electric fuse and socket	337	213+
Electric switch combined with	200	51R
Electric switch type	200	253.1+
Interposed two part	439	625+
Outlet box and wall plug	439	638+
Quick detachable	439	180
Resistance plug boxes	338	77
Screw	439	662+
Telephone multiple switchboard plug terminal	379	312+
Telephone multiple switchboard test	379	20+
Telephone switchboard	379	319+
Flow meter with orifice and tapered plug	73	861.58
Incandescent lamp seal	220	2.1 R+
Lock cylinder type		
Cylinder and plug assembly	70	367+
Element	70	375
Rotary	70	490+
Sliding	70	361
Sliding and rotary	70	360
Lubricator choke	184	52
Metal receptacle closure expansible type	220	233+
Metallurgical furnace tap hole	266	271
Nail hole	52	514
Nuclear fuel closure	376	451
Pipe and conduit	138	89+
Pipe coupling solid type	285	242
Plugging instruments dental	433	25+
Rail spike receiving	238	370
Spark (See spark plugs)		
Swivel pipe joint with	285	121+
Telephone jack and plug	439	676+
In central office switchboard	379	319+
Tire patch combined with	152	370
Tire plug inserting tool	81	15.5+
Tobacco		
Attaching cigar and cigarette	131	94

	Class	Subclass
Combined	131	366+
Compositions	131	352
Design	D27	101
Inserting cigar and cigarette	131	72
Making	131	111+
Shapes	131	111+
Wrappers or binders	131	365
Valve		
Bath or basin outlet type	4	204
Float operated	4	205+
With inside operator	4	204+
Combined with strainer	4	287
Head	251	356+
Hydrant type	137	272+
Liquid filter with screened type	210	429+
Movable liquid filter with screened type	210	390
Multiple outlet nozzle rotary	239	392+
Nonstop reciprocating	137	528+
Plug cock and seat abrading machine	51	28
Reciprocating	251	318+
Rotary	251	309+
Ship bottom type	114	197
Stopper type	4	295
Well		
Cementing or plugging	166	285+
Closing port	166	141
Handling	166	373+
Preset	166	114
Preset conduit carried	166	116
With expanding anchor	166	118+
With outward pressure means	166	101
With pump or plunger	166	106
Plug in Units		
Contact and socket structure	439	626+
Structural combinations	361	392+
Plumb Bob	33	392+
Design	D10	61+
Plumbago (See Graphite)		
Plumbanes	556	95
Hexaorganodiplumbanes	556	82
Plumbates	423	592+
Plumbers Friend	4	255+
Plumbing		
Baths and closets	4	
Design	D23	200
House fittings	4	DIG. 7
Sewerage	137	356+
Snake	15	104.33
Trap making	29	890.146
Water flow control	137	
Water pressure regulator	137	505+
Plumbs		
Structure	33	391+
Wall construction guides	33	404+
Wireworking	140	41
Plumes	428	6
Plume abatement	261	DIG. 77
Plunger		
Air pressure guns having	124	65+
Conveyer elements	198	717+
Die or cutter		
Box making	493	167+
Pastry cutter	83	
Plastic metal shaping	72	273
Sheet or bar die cutter	83	
Stationary vegetable or meat cutter	83	401+
Wood slivering	144	190
Drilling or dibbling planters having	111	
Expansion chambers having	138	31
Feed (See type of machine)		
Coal spreader for furnace	110	114
Locomotive tender	105	233
Manual dibblers	111	96
Railway sand feeders	291	36+
Rotary peripheral face abraders	51	82R
Gongs having	116	152+
Mechanical gun plunger release mechanism	124	33+
Press		
Block forming plungers	425	398+
Block forming plungers	425	344+
Block forming plungers	425	412+
Block forming presses		
Portable	425	
Reciprocating mold	425	352+
Reciprocating mold	425	406+
Stationary mold	425	352+
Stationary mold	425	406+

	Class	Subclass
Die expressing metal	72	273
Earthenware die expressers	425	406
Glass mold plungers	65	362
Metal mold sand packers	164	207+
Miscellaneous die expressers presses	425	406+
Pump having	417	437+
Textile fluid treating apparatus having	68	
Vegetable comminutors having	241	
Water closet having	4	300+
Plural Amplifier Channel Systems	330	124R+
Transistor type	330	295
Plutonium Compounds		
Inorganic	423	251
Organic	534	11+
Ply		
Delaminating	156	344
Laminate	428	
Wood	428	106
Wood	428	533+
Laminated	428	411.1+
Sewing machine trimmer combined	112	127
Tires resilient	152	151+
Woven fabrics	139	409+
Plywood (See Veneering)	52	
Making		
Glue appliers and presses	156	
Presses	156	580+
Processes	156	
Processes combined	144	346+
Pneumatic (See Hydraulic)		
Apparel turning	223	43
Burner extinguisher	431	145
Carding stripping	19	109
Chuck	279	4
Cleaners	15	300.1+
Clock	368	54
Combined automatically played musical instruments	84	6
Conveyers and feeders		
Agitation and	222	195
Article dispensing	221	278
Conveyer	406	
Pneumatic or liquid stream feeding item	209	906*
Sheet or web feeding or delivery	271	
Stacker	406	
Cotton harvester	56	30+
Motor driven	56	13.1+
Dispatch	406	
Tube or system	D18	35
Endless flexible track for land vehicle	305	34
Face plate expander	105	11
Folder	493	450
Gas separator cleaning	55	282+
Grooming device	15	401+
Gun	124	56+
Harvester	56	12.9+
Jet for marine propulsion	440	38
Mattresses	5	449+
Metal casting injector	164	306+
Musical instrument player attachable to keyboard	84	107
Musical instrument tracker	84	151+
Musical instrument tracker combined with sheet	84	146+
Nail driving device	227	9
Piano automatically played	84	24+
Pipe organ	84	331+
Power plant	60	407+
Pump	417	118+
Air jet combined	417	86
Jet type pump	417	151+
Railway cable gripper	104	203
Railway propulsion system	104	155
Locomotives	105	63+
Selector for automatically played musical instrument	84	115+
Sheet turner for sheet music	84	519
Shoe forms	12	114.4
Shoe sole cushion	36	29
Tamper for track laying	104	10+
Tire (See tire, pneumatic)	152	151+
Toy	446	176+
Typewriter	400	176+
Carriage return	400	182+
Velocipede seat	297	199+
Wheel	152	1
Wind instrument automatically played	84	83+

	Class	Subclass
Pneumothorax Apparatus	604	26
Pocket		
Alarm watch or clock	368	244
Apparel element	2	247+
In garment supporter	2	300
Shoe with	36	1
Book covers	281	31
Calendars combining card pockets	40	122
Carried		
Brush with recess for comb	132	121
Lighter	431	129+
Lighter	431	146+
Match safe	206	96+
Special receptacle carried in pocket	206	37+
Spittoon carried in pocket	4	259
Ticket case	206	39
Cartridge belt with pocket holder	224	224+
Cartridge belt with tilting	224	224
Compound tool	7	118+
Pocketknife	30	155+
Dispenser with		
Barometric or angle of repose trap	222	457
Movable trap chamber	222	344+
Tiltable container trap only	222	454+
Trap chamber cutoff	222	425+
Game table with		
Ball game	273	123R+
Moving surface type	273	113+
Pool	273	12
Pool with chutes connected	273	11R
Lamps	362	157+
With writing pads	362	99
Letter box with	232	33
Letter box with pivoted trap	232	47+
Machine		
Article dispenser ejector	221	208+
Bevel machine table	144	125
Cutting window stile pockets	144	19
Dryer rotary drums	34	109
Dryer shelves	34	172
Endless conveyor with flexible pocket	198	715
Felting machine roller	28	126+
Locomotive tender draft device	213	6
Loom pushed shed	139	28
Pneumatic conveyer feeder	406	63+
Manifolding strip form pockets for sheets	282	11.5 R
Match safe	206	96+
Design	D27	173+
Paper wrappers with	229	87.5
Picture mounts pocketed	40	159
Railway car post holder	105	408
Railway flat car stake holder	105	390+
Releasable	105	383
Sliding window sash	49	372+
Solid separator employing	209	684+
Stove grates with	126	159
Stoves	126	38
Support with	211	55+
Telescopes	350	546
Watches	368	62+
Wooden receptacle with	217	28
Woven	139	389+
Pocketbook	150	100+
Closure	150	118+
Design	D 3	42+
Pocketknife	30	155+
Compound tool	7	118+
Pod		
Augers		
Wood	408	199+
Pogo Stick	272	114
Design	D21	67
Walking	272	70+
Point & Points (See Impaling; Prong; Punch)		
Bands wheel hub	301	107
Glaziers	411	477+
Guards or holders		
Hatpin	132	72.1
With hatpin guide	132	69.1
Marker guide	33	574+
With gauge	33	666+
Nails spikes tacks	411	493+
Needles (See notes to)	223	102
Parallelizing divider	33	558.3
Pen (See penpoints)	15	445+
Plow	172	681+

	Class	Subclass
Printers for the blind	400	122+
Protectors pencil sharpener	30	460
Railway switch	246	435R+
Thrower	246	264+
Setter glazier	227	
Spreaders nail spike tack	411	359+
Switch		
Foot guard	238	380
Structure	246	435R+
Thrower railway	246	264+
Trolley transfer	104	104+
Ties and braces mattresses	5	458
Trolley transfer	104	104
Vaccine supply	206	367
Well casing	175	19+
Pointer		
Chance device	273	141R
Indicator type	116	
Arm structure	116	327+
Masonry implement	15	235.3
Pointing (See Calking)		
Barrel hoop lapping and	147	45
Bolts	10	21
Nail cutting and	10	31
Pencil (See pencil sharpeners)	30	451+
Screw threading and	10	4
Screws	10	9
Spike	10	59
Staple cutting bending and	59	74
Wooden pin	144	30
Poise		
Scale		
Automatic	177	235+
Beam	177	246+
Beam automatic	177	212+
Computing	177	35
Poison	424	
Holder or container		
Alarm combined	116	72
Indicating contents	215	367
Vermin destroying	43	131
Poker	294	14
Poker Chips	D21	53
Rack	211	49.1
Pokes	119	136+
Polarimeters	356	367+
Polariscopes	356	366
Polarity Testers	324	133
Polarized Devices and Polarizers		
Electromagnet with armature	335	230+
Electromagnetic wave	343	909
In wave guide	333	21R
With antenna	343	756
Examination devices generally	356	356
Gem and crystal examining	356	30+
Glass	350	370+
Headlight	362	19
Illuminated signs	40	548
Image projectors	353	20
Light polarizer	350	370+
Combined with illuminator	362	19
Light valve	350	374+
Making		
Structural orientation	264	1.3
Radio wave polarizations	343	909+
Spectro-photometers	356	326
Stage illusions using	273	9+
Stereoscope, stereo systems	350	132
Storage and retrieval of information	365	121
Switch electromagnetic	335	78+
Telegraph code recorder	178	93
With antenna	343	756
Testing with polarizers	356	364+
Vision testing means	351	49
Polarographic Chemical Analysis	204	400+
Apparatus	422	82.1+
Pole (See Mast; Post)		
Barber	40	607
Boat	440	36
Changer	361	245+
Changing in alternating current motors		
Induction	318	773
Synchronous	318	704
Climbers	182	133+
Leg or shoe attachable	182	221
Collectors trolley	191	64
Couplings	191	60.5
Curtain design	D 8	376+
Electrical conductor	174	45R
For electric wire	52	4

	Class	Subclass
Fire escape	182	189+
Wall attached	182	100
Flagstaffs	116	173+
Hammock ridgepole	5	130
Handlers	414	23
Land vehicle		
Supports	278	86+
Thills	278	33+
Tow	280	493+
Light support	362	431
Adjustable light	362	418+
Magnetoelectric armatures	310	264
Manipulating	294	19.1+
Plate or piece		
Switch	200	271
Telephone magnet	381	199+
Rack type	211	105.1+
Rossing bark	144	208R
Scaffolds for	182	187+
Ski	280	819+
Spray	239	532
Switch electric	200	49
Tent	135	114
Window operating	49	460+
Poler		
Bearing plate for poling railway cars	105	462
Marine propulsion device	440	36
Side track push rods	213	224
Police or Fire Alarms and Signals		
Signal systems	340	287+
Telephone circuit combined	379	37+
Telegraphs	340	287+
Whistle	116	137R+
Policemens Clubs or Batons	273	84R
Carriers for	224	914*
Polishing and Polishing Appliances (See Burnishing)		
Abrading		51
Boot and shoe machines		
Cleaning and blacking combined	15	30+
Cleaning combined	15	36+
Component parts in making	12	79.5
Coating treating process	427	355
Compositions	106	3+
Dental grinders combined	433	142
Motor driven	433	125
Fabric type	15	209R+
Hat pouncing combined	223	20
Machinery design	D15	124+
Metal burnishing	29	90.1+
Needle eye working	51	152
Paper calendering	29	90.1
Scouring machines combined	51	4
Planing combined	144	38
Turning combined	144	47
Shoe stands with polisher guide	15	266
Shoebrush and dauber	401	39
Woodworking machines		
Pollinating	47	1.41
Pollution		
Decreasing, & environmental impact	422	900*
Polo		
Ball	273	58R
Mallets	273	67R
Pony boots	36	111
Polyacrylyl (See Synthetic Resin or Natural Rubber)		
Fiber dyeing	8	
Safety glass employing	428	442
Polyamides (See Synthetic Resin or Natural Rubber)		
Fiber dyeing	8	
Processes of molding	264	DIG. 61
Polyarylmethyl Compounds	552	101+
Amines	564	315+
Ketones	568	332
Carboxylic acids	562	460
Polybenzanthronyl	552	273
Polycarbodiimide (See Synthetic Resin or Natural Rubber)		
Polycarboxylic (See Type Compound)	260	
Acyclic	562	590+
Acyclic esters	560	190+
Alicyclic esters	560	127
Anhydrides	549	231+
Aromatic esters	560	76+
Polymers derived from (See synthetic resin or natural rubber)		
Polychlorinated Biphenyl(pcb) Removal	208	262.5
Polycycles	280	282
Polycyclic Hydrocarbons		
Alicyclic synthesis	585	360

	Class	Subclass
Aromatic synthesis	585	400
Polyelectrolytes	252	62.2
Polyelectrolytes	429	188+
Electrolytic chemical processes	204	
Polyepoxide		
Polymers from (See synthetic resin or natural rubber)		
Polyester		
Fiber dyeing	8	
Fibers	8	DIG. 4
Polymers from (See synthetic resin or natural rubber)		
Polygon	33	482
Polygonal Turning and Cutting		
Broaching	409	243+
Drilling and boring	82	1.2+
Wood square hole	408	30
Milling	409	79+
Planing	409	289+
Turning	82	18
Wood	142	2+
Wood		
Polygonal cutters	144	224+
Shaping	144	138
Polyhydric Alcohol	568	852+
Carbohydrates	536	1.1+
Ethers	568	672+
Fermentative production	435	158+
Polyimide		
Polymers from (See synthetic resin or natural rubber)		
Polymer		
Dissolver	422	901*
Liquid hydrocarbon	585	17+
Synthesis	585	502+
Low molecular weight hydrocarbon polymer containing composition	106	901*
Replacing hard animal tissue	528	950*
Solid synthetic resin or natural rubber (See synthetic resin or natural rubber)		
Polymerization (See Compound Produce)		
Addition (See also synthetic resin or natural rubber)	526	59+
Apparatus for	526	920*
Aftertreating	252	182.11+
Crosslinking	252	182.11+
Curing	252	182.11+
Fatty oils	260	407
Hardening	252	182.11+
Hydrocarbon synthesis, cyclic		
Alicyclic	585	360+
Aromatic	585	422+
Aromatic cyclic	585	415+
Monomer polymerized in presence of transition metal containing catalyst (See catalyst)		
Olefin, fiber dyeing	8	
Photo image	430	269+
Precursors	252	182.11+
Utilizing electric, radiant		
Within nuclear reactor	376	
Utilizing wave energy	522	1+
Vulcanizing	252	182.11+
Polymixins	530	319
Polyolefin	8	DIG. 9
Polyolefin-type foam		
Injection molding	264	DIG. 83
Polyp Removal Devices	606	110
Polyphase		
Current meters	324	107+
Polyphenylene Oxide (See Synthetic Resin or Natural Rubber)		
Polypod		
Breakwater	405	29
Polypropylene		
Molding foamed articles	264	DIG. 16
Polysaccharides	536	1.1+
Heteropolysaccharides	536	123
Polysulfide	423	511+
Fatty oils	260	399
Organic	568	21+
Resins (See synthetic resin or natural rubber)		
Polysulfone (See Synthetic Resin or Natural Rubber)		
Polyurethane		
Cellular product	521	159+
Filters in gas separation	55	DIG. 13
High resiliency or cold cure	521	904*

	Class	Subclass
Nonstructural laminate	428	423.1
Nontextile dyeing	8	515
Radiation sensitive binder	430	906*
Stock material	428	160
Synthetic fiber dyeing	8	926*
Synthetic resins with solvent		
Alcohols	524	379+
Hydrocarbons	524	474+
Phenols	524	323+
Terminal ethylenic unsaturation	525	920*
Tire or tube tread or carcass	156	128.1
Wave energy preparation or treatment	522	1+
Polyvinyl Acetals	525	61
Safety glass employing	428	437
Pom-pom Gun	89	126
Pompom (See Tassel)	428	4+
Pontoons		
Buoys	441	3
Landing platforms	114	258
Salvage	114	44+
Pool		
Billiard game		
Ball	273	59R
Cues	273	68+
Tables and accessories	273	2+
Glass heating furnace	65	336+
Swimming, purification		
Process	210	749+
Purification	210	169
Solar heated	126	415+
Swimming-type nuclear reactor	376	174
Swimming-type nuclear reactor	376	361
Swimming-type nuclear reactor	376	403
Popcorn		
Machine	D 7	325
Poppers	99	323.5
Presses	425	406+
Sheller	460	
Popgun	124	63+
Design	D21	146+
Popsicle T M Forming Machine	425	DIG. 219
Porcelain		
Compositions	501	141+
Electric insulating	501	141
Shaping processes	264	
Pore Forming or Porous		
Compositions	106	122
Abrasive	51	296
Alkali metal silicate ctg	106	601+
Cement hydraulic setting	106	672+
Ceramic	501	80+
Glass	501	39
Metal casting	164	79
Molding	264	41+
Mortar mixer combined	366	10+
Processes	366	2+
Synthetic resin ctg (See cellular under synthetic resin or natural rubber)		
Wells		
Gravel placing means	166	51
Gravelling	166	278
Porous cementitious filter	166	276
Porous material screens	166	228
Pork	D 1	
Porosity Test	73	38
Porous Devices (See Pore Forming & Type of Device)		
Stock material, cellular layer layer	428	304.4
Porous Forming (See Cellular Under Synthetic Resin or Natural Rubber)		
Porphyrins	540	145
Port		
Boiler	122	6.5
Lifting or handling	52	122.1+
Shield mounted	109	58.5
Ship	114	173+
Valve seat	251	359+
Internal combustion engine	123	188S+
Rotary	251	314+
Portable Circular Saw	30	360+
Portable Houses	52	143+
Portfolio	190	900*
Design	D 3	30.1+
Locks	70	67+
Portland Cement		
Compositions containing	106	713+
Alkali metal silicate containing	106	601+
Refractory	501	124
Making processes	106	739+

	Class	Subclass
Position Detectors Indicators and Markers	116	209+
Elevator	116	226
Electric	187	130+
Indexed printed matter	283	37+
Liquid level gauge	73	314
Optical measurement of	356	375
Global	250	206.1+
Local, of shaft position or	250	231.13+
Speed	250	231.13+
Railway switch stand	246	401+
Steering wheel	116	31
Train	246	122R+
Weaving shuttle	139	341+
Positive Feedback Amplifier (See Regenerative, Amplifier)		
Post (See Column; Mast; Pole)	52	720+
Architectural, design	D25	126+
Bearing	114	169
Binding electrical	439	775+
Drilling machine	408	236
Drivers	175	
Continuous push	254	29R+
Impact	173	90+
Electrical conductor	174	45R
Hitching	119	122+
Design	D30	154
Jacks	254	30+
Kiln furniture	432	253+
Lamp or light combined	D26	67+
Light support	362	431
Railway	104	125
Rotatable machine element	74	504+
Scaffold	248	351+
Sign attachments	40	607
Static mold	249	51
Steering	74	492+
Tool	82	162
Top	52	301
Bed	5	281
Tower	52	648
Transmission tower	D25	126+
Velocipede	280	281.1+
Violin sound	84	277
Window corner	52	282
Postal Card or Packet	229	92.8
Design	D19	1+
Posthole Digger		
Implement	294	50.6+
Machine	175	
Pot (See Bowl; Jar; Kettle; Pan)		
Burner	431	331+
Chamber	4	479+
Coffee	99	279+
Coffee or teapot strainers	210	473+
Receptacle spout attached	210	466+
Cooking container	D 7	354+
Dash (See dashpot)		
Fire pot thermostats	236	104
Flower	47	66+
Design	D11	143+
Furnace and metallurgical		
Fire pot	126	144+
Fire pot thermostats	236	104
Glue pot heater	126	284
Material heating furnaces	432	156+
Material heating holders	432	252+
Metal melting and casting	164	335+
Solder pot heater	126	240
Glue heating	126	284
Holder	16	116R+
Liquid heating	126	373+
Railway rail supports	238	110+
Railway signals	246	475
Vulcanizing	425	283
Potassium (See Alkali Metal)		
Chloride dissolver	422	902*
Potato		
Chip machine	99	403+
Slicing combined	99	353
Chips	426	637
Digger	171	
Masher	241	
Implements	241	169.2
Picker and hopper	221	213+
Planters	111	908*
Preservation	426	321+
Canning	426	321+
Coating	426	289+
Compositions	426	637+
Dehydration	426	471+

	Class	Subclass
Refrigeration	426	524
Washing machines	15	3.1+
Potential		
Regulators for dynamos	322	28
Potentiometer	338	68
Potentiometric Electrode		
Assembling apparatus for making	29	729
Chemical analysis, apparatus	422	82.1
Current generating thermometer with	374	179
Galvanometer with	324	98+
Structure	338	68+
Systems	323	364
Testing	324	714
Pothead	174	74R+
Fluid or vacuum type	174	19+
Potted Antenna	343	873
Potted Electrical Device	361	331
Pottery		
Extruders	425	289+
Molds	425	406+
Potters wheels	425	263+
Potters wheels	425	459
Potting Electrical Device	264	272.11
Pouch (See Bag; Satchel)		
Animal attachment for manure	119	95
Colostomy	604	332+
Diaper having	604	348
Key containing	206	37.1+
Tobacco	D27	185
Poultry		
Butchering	17	11+
Husbandry	119	
Design	D30	
Litter	119	1
Pouncing Hat Machines	223	20
Pounder		
Meat	17	30+
Textiles fluid treatment	68	215+
Pouring		
Articles with pouring spout	D 7	312+
Automatic weigher discharge	177	115
Dispenser	222	
Receptacle filler	141	
Powder (See Coating; Dispensing; Distributing)		
Ammunition powder bags	102	282
Ammunition powder forms	102	283+
Baking	425	526+
Cans	220	
Dispensing	222	
Coating composition ctg synthetic resin or natural rubber (See also 524-904)	525	934*
Comminuted metal	75	331+
Cosmetic	424	69
Compact	D28	78
Design	D28	8
Dispenser	604	58
Duster aircraft borne	244	136
Duster insect killing	239	
Dye	8	524+
Engine driven by	123	24R
Engine starting with	123	183
Explosive compositions	149	
Explosive thawer	126	269
Fluent solid applier pocketed, eg powder puff	401	200+
Formed by electrolysis	204	10
Fusing and spraying	239	79+
Insect	424	
Marking ground with	111	26
Medicator for applying	604	58
Metallic powder containing trunks	106	403+
Metallurgy		
Briquetting apparatus	425	78+
Comminuted metal	419	
In blank	428	546+
In laminated blank	428	548+
Heat treating, other than sintering	419	31
Making powders	75	343+
Pressing powder with other steps	29	DIG. 31
Sintered stock materials	75	228+
Sintering	419	
Sintering then working	419	28
Paper folder for containing	493	948*
Photographic developer	430	499
Powdered fuel	44	500+
Puff		
Box, cosmetic	206	823*
Design	D28	91

	Class	Subclass
Sheet retainer type	402	25
Surgical instrument	606	185+
Punts	114	343+
Pupin Coil	336	
With condenser	333	
With long line	178	45+
Puppets	446	363
Hand	446	327+
Purger, Refrigerant	62	474+
Automatic	62	195
Burner	431	29+
Burner	431	121+
Process	62	85
Purging Sugar Crystals	127	18+
Apparatus	127	20
Purification of Synthetic Resin or Natural Rubber	528	480+
Purification, Proteins	530	412+
Purifiers & Purifying (See Bleaching; Cleaner; Dehydration; Disinfection; Distillation; Evaporator; Extracting; Fermenting; Recovery; Sterilizing)		
Beneficiating ores	423	1+
Apparatus	266	168+
Gas	55	
Hydrocarbon purification	585	800+
Liquid	210	
Used oil	208	179+
Purines	544	264+
In drug	514	261
Purl Knit Fabrics	66	197
Purling Sewing Machine Stitch	112	161
Purpurin	552	261
Purses	150	100+
Coin	150	150+
Design	D 3	42
French purse	D 3	42+
Latch	D 8	331+
Push (See Pull)		
Button		
Electric switch	200	341+
Fluorescent	250	465.1
Mechanical bell	116	172
Carts	280	47.26+
Cutter	30	
Plate	16	1R
Pushing devices	254	
Rods or bars		
Closure operator	292	
Side track car	213	224
Push Pull Amplifiers	330	118+
Cascade type	330	71+
With distributed parameter coupling	330	55
With feedback	330	81+
Pusher (See Conveyor, Reciprocating)		
Blocks for saw logs	83	710+
Car pushing implement	254	35+
Check controlled article delivery		
Check connector	194	293+
Gravity forwardly turning part	194	262+
Gravity reciprocating part	194	260+
Thrust type	194	249+
Feed		
Check controlled article delivery	194	
Circular wood saw	83	707+
Combined with other types of conveyer sections	198	570+
Conveyer	198	717+
Drying kiln material	34	206
Furnace fuel	110	109
Label pasting strip server and cutter	156	510
Saw log	83	710+
Load discharging		
Fork tine or shovel clearers	294	50+
Furnace	414	214
Self loading vehicle	414	511
Vehicle, external unloader	414	395+
Locomotive track pusher	105	31
Plastic block press portable mold	425	453+
Railway car propulsion system	104	162+
Roadway snow plow V type	37	272+
Sled	280	14.2+
Stock to be machined	414	14+
Truckmans car drive	105	90.1+
Putlogs or Spacer	182	229
Ladder combine	182	214
Scaffold wall embedded	182	87
Putting on		
Garments	223	111+
...t accessory	2	101

	Class	Subclass
Garment supporters	2	300
Putty		
Devices	425	458
Including putty supply	425	87
Removing		
Attachment, moving	15	256.5
Implement scraper	15	236.1+
Puzzle	273	153R+
Design	D21	104
Jigsaw type	428	33
Locks	70	289
Pvc-isocyanate Resins		
Molding cellular	264	DIG. 3
Pvdf	310	800*
Pycnometers	73	32R
Pyramidon	548	366
Pyrans	549	356+
Pyranthrones	552	282
Pyrazines	544	336+
Pyrazoles	548	373+
Acridine nucleus containing	546	26
Azo compounds	534	751+
Heavy metal containing	534	710+
Barbiturate addition compounds	544	300
Pyrazolone (See Pyrazoles)		
Pyrene (See Aromatic Hydrocarbon)	585	400+
Pyrethrum	560	124
Insecticides containing	514	65+
Pyridine	546	348+
Azo compounds	534	770+
Heavy metal containing	546	2+
Pyridium	534	605+
Pyridoanthraquinones	546	78
Pyridone	546	290
Pyridoxine	546	301
Pyrimidines	544	242+
Pyrocatechin	568	763+
Pyroelectric Lamps		
Structure	313	14
System of supply	315	359
Electric heater for	315	115+
Plural loads	315	114
Pyrogallol	568	763+
Pyrographic		
Burning tools	126	401
Electric	219	227+
Fuel	368	248+
Electric tools	219	227+
Hand tool spark	219	384
Machine spark	219	68+
Pens	219	229
Recorders	346	76R
Telegraph receiver	178	94
Textile finishing	26	3+
Pyroligneous Acid	562	607
Production of	201	
Pyrolysis (See Cracking)		
Acetylene production by	585	534+
Aromatic hydrocarbons by	585	476
Carbon preparation by	423	449+
Natural resins	530	200+
Rosin	530	226+
Pyromagnetoelectric Devices	310	306+
Pyrometallurgy	75	414+
Pyrometers	374	100+
Optical	356	43+
Radiation	374	121+
Thermoelectric	136	213+
Relay temperature regulation	236	69+
Pyromucic Acid	549	484
Pyrones	549	416+
Pyronine	549	388+
Pyrophoric Igniters	431	273+
Alloys	420	416
Correlated with burner feed	431	254
Tobacco product combined	131	351
With burner cover	431	146+
Correlated actuation	431	129+
Pyrophosphate Esters	558	152
Formation of	558	127+
Pyrotechnic		
Compositions	149	37+
Devices	102	335+
Flare	D22	112
Illusion amusement devices	272	20
Signals	102	335+
Detonating burglar alarms	116	83
Door fastener	116	11
Railroad torpedoes	246	486
Pyroxylin	536	35+
Apparatus	425	90+

	Class	Subclass
Compositions	106	169+
Pyrrocolines	546	73
Pyrrolidine	548	579
Pyrroline	548	565
Pyrrols	548	400+
Quadrant		
Angle measurement		
Horizontal and vertical angles	33	281
Horizontal angle	33	285
Vertical angle	33	282+
Spinning mule	57	322+
Quadrasonic	381	19+
Quadruplex Telegraphy	370	34+
Quarrying	299	15
Quarter Wave Transmission Line	333	27
Quarters Shoe Blank	36	48+
Quartz (See Piezocrystals)		
Glass manufacturing	65	
Compositions	501	27+
Lamps	313	112
Mercury vapor	313	112
Quaternary Ammonium	564	281
Acyclic	564	291+
Aromatic	564	282+
Azo compound ctg.	534	603+
Quebracho Extract	560	68+
Descaling agent containing	252	83+
Tanning with	8	94.32
Aluminum compound containing	8	94.3
Queen Bee Cells and Cages	449	28
Quenchers and Quenching		
Coating cooling	427	398.3
Coke quenchers	239	750+
Distillation residues	201	39
Autothermic systems	202	95
Thermolytic apparatus	202	227+
Metal heat treatment processes	148	143+
Metal heating and quenching	266	114+
Apparatus	266	121+
Metal quenching apparatus	266	114+
Still bottom closures	202	253
Question and Answer	434	322+
Correct answer indicated	434	327+
Electrical recording means	434	362
Examination card or sheet	434	363
Grading of response form	434	353+
Responses electrically communicate to monitor or recorder	434	350+
Quick Acting		
Fluid pressure brake systems		
Automatic synchronizing	303	37+
Motor releasing	303	69
Guns	89	125+
Portable wood working clamps		
Parallel screw	269	140+
U beam	269	
Quick setting portable rail drills	408	72R+
Quick traverse bobbin and cop	242	31
Cone	242	31
Winding	242	43R
Quick wheel release trolley heads	191	63.1
Quick Detachable		
Electric connectors	439	180
Plow elements	172	749+
Shaft coupling	464	901*
Quicksilver (See Mercury)		
Quillstock Spinning	57	4
Quilting	112	117+
Frames with cloth winding means	242	67.1 R+
Quilts	5	482+
Design	D 6	603+
Bedding per se	D 6	595
Quinaldine	546	181
Quinazolines	544	283+
Quinhydrones	552	293
Quinic Acid	562	508
Quinicine, in Drug	514	314
Quinidine, in Drug	514	305
Quinine	546	134
In drug	514	305
Quinizarin	552	262
Quinoline	546	152+
Azo compounds	534	770+
Quinolinium Compounds	546	182
Quinolizines	546	138
Quinonediamine	552	301+
Quinonimines	552	302
Quinophthalone	546	173
Quinoxalines	544	353+
In drug	514	249
Quinuclidines	546	133

	Class	Subclass
Quiver	D 3	36+
Quoins	254	40+
Quonset Hut	52	86
R Acid	562	80
Race		
Bearing		
Assembly apparatus	29	724+
Ball	384	513+
Divided	384	199+
Fastening means	384	541
Manufacturing process	29	898+
Roller	384	569+
Take up	384	500
Thrust	384	622
Starting barriers	119	15.2
Track	272	4+
Water wheels	405	118+
Raceway		
Bearing raceway grinding	51	291
Grinding bearings in	51	130
Sewing machine shuttle	112	196
Racing		
Amusement railway	104	60
Arena	272	4+
Simulated games	273	86R
Rack (See Ratchet; Support)	211	
Adjustable	211	175
Animal husbandry	119	20
Feeding	119	60
Article (See also supports for		
specific articles)	211	
Ball billiard table attached	273	10+
Battery	429	96+
Card	40	124
Check label and tag	40	657
Clothes bed attached	5	504
Gun	211	64
Hat	211	30+
Hose or nozzle	248	89+
Ironing table combined		
Life preserver	114	190
Music piano attached	84	180
Pen or pencil	211	69.1+
Portable	294	143
Radiator	165	67+
Cabinet combined	312	
Canning		
Food cooking	99	359+
Receptacle	211	71+
Receptacle stationary	248	146+
Collapsible	211	195+
Disinfecting, preserving	422	297
Knockdown framework	211	189+
Mechanism	74	422
Oven	126	337R+
Pinion and (See device in which		
used)	74	422
Mechanical movements	74	20+
Intermittent grip type	74	111+
Pushing and pulling mechanism	254	95+
Rail	238	123
Elevator	187	19
Locomotives	105	29.1+
Switches	104	132
Wheel	295	4
Releasable keeper for sheet retainer	402	60+
Showcase combined	312	128+
Sorting-type	211	10
Spool	211	59.1
Stackable	211	194
Terraced trays	206	211
Vehicle body	296	3+
Racket or Racquet	273	67R+
Croquet	273	83
Design	D21	212
Restringing tension meter	73	862.43
Tennis	273	73R
Racking Hoops	147	44
Radar	342	147+
Design	D10	104
Digital computer combined system	342	195
Light ray radar	356	3+
Simulators	434	2+
Sound or supersonic	367	87+
Radiant Energy (See Heat; Illuminated;		
Music; Optics; Photography; Radio;		
Sound; Space Induction; Solar;		
Television)	250	
Barrier	250	505.1
Detection (See radio)		
Computer controlled, monitored	364	413.26

	Class	Subclass
Infrared	250	338.1
Invisible light	250	336.1
Neutrons	250	390.1+
Nuclear reactions control	376	207
Photoelectric	250	200+
X ray	250	336.1
Devices, railway switches	246	DIG. 1
Electric heater	392	407+
Imagery chemistry (See X-art list)	430	
Binder ctg	430	905*
Binder-free emulsion	430	932*
Initiator ctg	430	913*
Radiation-activated cross linking		
agent ctg	430	927*
Radiation-chromic compound	430	962*
Spectral sensitizer ctg	430	926*
Material treatment		
Article or object apparatus	250	492.1
Chemical and other apparatus	422	186+
Chemical apparatus	204	193
Coating	427	12+
Coating apparatus	118	620+
Coating with fibers or	427	58+
Particulate material	427	12+
Drying processes etc radiant		
heating	34	39+
Drying processes etc selected		
range	34	4
Dyeing apparatus	8	103
Dyeing processes	8	444
Electric discharge apparatus	422	186.21+
Electrical energy drying and gas		
or vapor contact with solids	34	1
Electrons	250	306+
Fluent material apparatus	250	428+
Food and beverage apparatus	99	451
Refrigeration combined	62	264
Solar drying process and gas or		
vapor contact with solids	34	93
Sterilizing and pasteurizing		
processes	426	521+
Sterilizing and pasteurizing		
processes	426	237+
Tobacco treatment	131	299
Vitamin activation processes	426	72
Medical and surgical applications	128	362+
Application to body	604	20
Bandaging	128	82.1
Computer controlled, monitored	364	413.26
Injection means	604	21
Instruments	606	2+
Kinesitherapy eg massage	128	24.1+
Magnetic	600	9+
Medicators	604	20+
Nuclear reaction	376	100
Nuclear reaction	376	156
Orthopedics	128	68.1
Processes	376	156
Nuclear reaction produced energy	376	
Radiation measurer	D10	47
Computer controlled, monitored	364	413.26
Radiation resistant hydrocarbon		
composition	585	944*
Responsive		
Battery	136	243
Cathode ray tube	313	365+
Cathode ray tube circuit	315	10+
Circuits miscellaneous	250	200+
Closure operator	49	25
Explosive igniters fuses	102	213
Explosive igniters marine mines	102	416+
Explosive igniters mines	102	427
Photocell electronic tube type	313	523+
Resistor	338	15+
To invisible light X ray etc	250	336.1
Secondary emission		
Cathode ray tube circuit	315	11
Cathode ray tube plural cathodes	313	375
Cathode ray tube with photo		
sensitive electrode	313	387
Electronic tube	313	103R+
Secondary radiation minimizing		
X ray screen	378	154
X ray tube	378	140
Shielding		
Against invisible rays	250	515.1
Nuclear reactor combined	376	287
Nuclear reactor combined	376	347
Simulation of radiation	434	218
Thermometers	374	121
Toaster	D 7	324+

	Class	Subclass
Radiation Imagery--chemical	430	
Achromatic image from chromatic	430	356
By sound or nondigital compressive		
forces	430	3
Color imaging process	430	351+
Control feature responsive to test or		
measurement	430	30
Diazo reproduction	430	141+
Effecting frontal radiation during		
exposure, eg screening etc	430	396
Electric or magnetic imagery	430	31+
Exposure step or specified pre-		
exposure step perfecting		
exposure	430	494
Imaged product chemically defined	430	9+
Imaging affecting physical property		
of radiation sensitive material, or		
producing non-planar or printing		
surface	430	269
Liquid crystal	430	20
Microcapsule	430	138
Micrography	430	8
Non-radiation sensitive image	430	449+
Plural exposure steps	430	394
Post imaging process	430	404
Producing cathode ray tube element	430	23
Product having sound record or		
process of making	430	140
Radiation sensitive product and		
composition	430	495+
Regenerating image processing		
composition	430	398
Registration or layout process	430	22
Silver halide colloid tanning	430	264+
Stripping process or element	430	256+
Thermographic process	430	348+
Transfer procedures between image		
layer & image receiving layer	430	199
Using reflected radiation	430	395
Visible imaging	430	198
Firing or sintering step	430	198
Radiation only, other than heating		
by surface contact or		
convection	430	346
Radiator (See Heat)		
Automobile radiator and light	362	79
Automobile radiator leak testing	73	49.7
Automobile type	165	148+
Design	D12	163+
Brushes	15	159R+
Cap ornaments	428	31
Thermometer type	374	146
Caps	220	200+
Covers		
Automobile	165	98
Incased	165	129
Shields	237	79
Design	D23	330+
Electrical	392	375+
Radiating panel	392	435
Radiating plate	392	435
Fluid heated radiators	237	70+
Gas burning	126	91R
Incandescent fire grate	126	92R
Heat exchangers	165	148+
Heating systems having	237	
Making	29	890.3+
Assembling	29	726+
Headers	29	890.52
Tube joint	29	890.44
Work stands	269	
Relief valves and vents	137	455+
Temperature controlled	236	61+
Sectional	165	130
Supported foot rests	248	218.4+
Supports		
Bases	248	348
Brackets	248	232+
Window mounting brackets	248	209
Work holders	269	
Temperature controlled	236	22
Cooling radiator	236	34+
Heating radiator	236	36+
Vehicle, design	D12	166+
Vehicle mountings	180	68.4+
Aircraft	244	57
Viscosity controlling	165	35+
Wall panel heating and cooling	165	49
Radio		
Aircraft control	244	75R
Aircraft control automatic	244	175+

	Class	Subclass
Amplifiers	330	
Automatic volume control	455	234+
Amplifier stage only with bias control	330	129+
Amplifier stage only with impedance control automatic	330	144+
Amplifier stage only with thermal impedance control automatic	330	143
Band changing	334	47+
Block signal systems hertzian wave	246	30
Block signal systems no line wire	246	38
Block signal systems no line wire	246	44+
Cabinets	312	7.1
Design	D14	188+
With antenna	343	702
Chassis	455	347+
Housed	174	50+
Prewired sub assembly	361	422
Clock combined	D14	170+
Condenser	361	271+
Adjustable	361	277+
Electrolytic	361	500+
Console with phonograph	D14	169
Control of devices		
Aircraft	244	175+
Camera shutter	354	131
Frequency responsive	340	825.72
Motors	318	16
Pulse responsive	340	825.69
Transmitter for	341	176
Demodulators	329	
Direction		
Finder receiver	342	417+
Directive	342	350+
Reflected wave systems, ie radar	342	
Electronic tube structure	313	238+
Headphones	455	350+
Indicators	116	241+
Magnetic induction transmission to vehicle	191	10
Miscellaneous	455	
Oscillators	331	
Phonograph combined	369	6+
Pocket	455	351
Transmitter	455	100
Power transmission		
Electric motor	318	16
Miscellaneous	307	149
Vehicle	191	10
Program indicator	40	901*
Pulse or digital communications	375	
Receiver	375	75+
Transmitter	375	59+
Radio and phonograph	369	6+
Receivers	455	130+
Design	D14	188+
Pulse or digital	375	
Regenerative	455	336
Stereo	381	2+
Superheterodyne	455	313+
Superregenerative	455	336
Testing of	455	226
Tuned radio frequency	455	150+
Remote control by radio waves	340	825.72+
Pulse responsive	340	825.69
Remote control of receiver	455	352+
Tuning	455	151+
Remote control of transmitter	455	92
Remote signal course control	244	189
Responsive explosive igniters		
Ammunition & explosive devices	102	214
Mines	102	427
Mines marine type	102	418
Second detector	329	
Shielded	330	68
Static eliminator	455	296+
Supersonic remote control for	455	352
Teledynamic		
Motor control	318	16
Telemetric	340	870.28+
Telegraph space induction	178	43
Onto signal line	178	49
Telemetry	128	903*
Telephone superaudible composite system	379	64+
Torpedo control	114	21.1
Train dispatching inductive telegraphy or telephony	246	8
Transmitter	455	91+
Tubes	313	
Envelope	313	248

	Class	Subclass
Manufacture	445	
Manufacture and repair	445	
Part of tube	313	313
Shield	174	35TS
Testing	324	405+
Tuners	334	
Band spreading	334	
Capacitors variable	361	277+
Dial operator	74	10R+
Electric motor operated	334	20+
Electric wave filters	333	167+
Ganged	334	
Limit stop mechanism	192	138+
Operator only for tuner	74	10.45
Push button	74	10.1+
Using distributed impedance only	333	219+
Wave hf transmission	333	
Wave meter	250	250
With gain control by		
Automatic impedance control	330	144+
Automatic thermal impedance control	330	143
Bias control	330	129+
Radioactive & Radioactivity	250	493.1+
Actinide series metals		
Alloys	420	1+
Compositions	252	636+
Electrolysis	204	1.5
Heat treatment	148	132
Inorganic compounds	423	249+
Metallurgy	75	393+
Organic compounds	534	11+
Alloys	420	1+
Battery	310	301+
Carbocyclic or acyclic compounds	534	10+
Nitro	568	924+
Compositions	252	625+
Molding	264	21
Contamination detector	250	336.1
Demonstrator	250	336.1
Cloud chamber	250	
Electrically	250	336.1
Fluorescent	250	458.1
Drug, bio-affecting & body treating compositions	424	1.1+
Electrical & wave energy, chemistry	204	1.5
Energy measurement	250	336.1
Material	52	
Apparatus	422	903*
Compositions	252	625
Containing electrically conductive or emissive compounds	252	517
For influencing fluent	250	493.1
Material incinerator	110	237
Production	376	156
Treating	264	
Geiger type system	250	374+
Measurement of	436	178+
Medicine	424	1.1
Metallurgy alloys cerium and	420	1+
Nuclear reactions & systems	376	
Neutron type system	250	390.1+
Photography	250	475.2
Producing in material	376	156
Scintillation	250	361R+
Plastic shaping of radioactive material	264	.5
Powder metallurgy	419	
Radioactive isotope of another element	423	249+
Technetium radioactive metal	432	2+
Organic compounds	534	14
Test method containing	424	1.1
Antigen-antibody	436	500+
Therapeutic application	128	11+
Treating mixture to obtain	423	2+
Waste disposal storing	252	626+
X-ray or neutralized shield	252	478
Radiochemical (See Radiant Energy)		
Apparatus chemical change	204	193
Compositions radioactive	252	625+
Food treatment		
Preservation	426	237+
Vitamin activation	426	72+
Medicine radioactive	424	1.1
Nuclear reactions	376	
Sound recording and reproduction	369	100+
Tobacco treatment	131	299
Radiography	250	
Radioimmunoassay	436	545*
Radioisotope Powered Generator	310	301+

	Class	Subclass
Radioisotopic Iodine for Testing	436	804*
Radiometers		
Light meters	356	213+
Pyrometers	356	43+
Radio wave meter	250	250
Rotating vane type	356	216
Thermometers	374	121
Ultraviolet	250	372+
Radiosonde	340	870.1+
System	340	870.1+
Radiotelegraphy	375	
Earth transmission systems	375	6
Radiotelephony	455	
Earth transmission systems	455	40
Radium (See Radioactive)		
Radius & Spiral Measurement	33	1SP
Radius Rod		
Brake	188	191
Spring braces	267	66+
Rafter	52	92+
Clips and	33	DIG. 16
Rafts and Rafting		
Life rafts	441	35+
Rafting and booming	441	48+
Rail (See Railway)	104	23.1+
Airplane	104	23.1+
Amusement	104	53+
Anchors	238	315+
Balloon	104	22
Bed	5	
Camp	5	117
Elements and details	5	286+
Extension	5	184
Folding	5	177
Rotating	5	302
Benders	72	210
Portable	72	210
Billiard table	273	8+
Bonds for electric railways	238	14.5+
By electric weld	219	53+
Manufacture and installation	228	4.1+
Manufacture and installation	228	101+
Cable rails	104	112+
Car propulsion systems	104	287+
Car replacer	104	262+
Car stops	104	249
Car yards	104	26.1+
Chalk and eraser	434	417
Circuit		
Power	191	
Signal	246	
Cleaner	15	54+
Clearer	104	279+
Snow	37	198+
Clip	238	378
Derail guards	104	242+
Drills, portable	408	
Elevators	104	127+
Fasteners	238	310+
Fence	256	59+
Fissure detector	73	146
By abrasion	73	8
Magnetic	324	217+
Grinders	51	178
Guard		
Railway surface track	238	17+
Track	256	14+
Hose bridges	104	275+
Joints	238	151+
Land vehicle		
Dashboard	296	71
Top	296	123
Manufacture	72	
Curving	72	210
Punching	83	
Monorail	104	118+
Moving train	104	18+
Razor		
Form	30	83
Guard	30	80
Seat-guard control	104	241
Seats	238	264+
Shapes	238	122+
Shifters	254	43+
Ship belaying pin	114	218
Sledways	104	134+
Slotted conduit	104	140+
Suspended	104	89+
Swimming pool grab rail	D25	41
Switches	104	130+
Terminals & stations	104	27+

	Class	Subclass
Textile making apparatus	57	136+
Toy railroad	104	DIG. 1
Track clearers	104	279+
Track layers	104	2+
Trackmans-car drive	105	86+
Traction	104	165+
Trains	105	1.4+
Trolley rails	104	106+
Trolley transfer	104	96+
Truck changers	104	32.1+
Trucks	105	157.1+
Tunnel ventilation	98	49
Turn tables	104	35+
Railways (See Rail)	104	
Antileak joints	105	424
Brakes	188	33+
Cabs	105	456
Car design	D12	39+
Car framing and structure	105	396+
Car locks for elevators	187	97
Chair making	29	16
Draft appliances	213	
Electric vehicle, streetcar type	D12	36+
Appliance	D13	128
Floors	105	422
Geometrical instruments	33	1Q
Graders	37	104+
Linings	105	423
Locomotives	105	26.5+
Cab ventilation	98	3
Car ventilation	98	4+
Dust guards	98	28
Mail delivery	258	
Marine	405	2
Monorail	105	141+
Motor placement	105	133+
Pit scale	177	134
Platforms	105	425+
Rail or tie shifters	254	43
Rolling stock	105	
Design	D12	36+
Sounding toy	446	410
Safety bridges	105	458+
Ship	405	2
Side guards	105	439+
Signals & switches	246	
Block system	246	20+
Snow excavators	37	198+
Sound deadeners	105	452+
Special car bodies	105	238.1+
Steps	105	443+
Steps with gate	105	437+
Surface track	238	
Electrical connections	238	14.1+
Fence or guard rails	238	17+
Juxtapose and bonding rails	228	101+
Leveling and spacing gauge	33	338
Rail bond making	228	4.1+
Spike pullers	254	18
Suspended	105	148+
Switches and shifters	246	
Tie clamp	D12	51
Toy railroad	104	DIG. 1
Toy simulation	446	444+
Design	D21	129
Electric	104	288+
Wheels and axles	295	
Axle making	72	
Wheel making	29	168
Wheel or axle drive	105	96+
Rain		
Gauges	73	171
Producers	239	14.1
Raincoats	2	87
Raisers (See Hoist)		
Corpse in coffin	27	12
Dough		
Baking powder	426	562+
Heater	126	281+
Jacks	254	
Track	104	7.1+
Wick	431	304+
Wick	431	315+
Raising Ships	114	44+
Rakes		
Agricultural implements	56	
Blanks and processes for making	76	111
Bundling combined	56	341+
Cutter and ground rake	56	193
Cutter with conveying rake	56	158+
Cutter with detachable rake	56	4

	Class	Subclass
Design	D 8	13
Dies for making	72	470+
Grain separator	460	
Hand	56	400.1+
Horse drawn	56	375+
Loading combined	56	344+
Potato digger conveyor	171	13
Side delivery	56	376
Flume screens	210	154+
Furnace grate	110	285
Hand fork combined	294	52
Making	83	908*
Pan type evaporator	159	33
Scoop type	37	120+
Stove grate	126	173
Vegetable or meat comminutor		
feeder	241	
Vehicle fender	293	53
Rammers (See Tamper)		
Ammunition loading	86	29+
Earth compacting	404	133
Road construction type	404	133
Foundry mold press type compactor	164	207+
Plural rammers	164	172+
Machine for concrete	425	425
Machine for concrete	425	456
Ordnance loading	89	47
Road and pavement	404	133
Ramp, Loading (See Gangway)		
Ramrod		
Firearm	42	90
Rams		
Hydraulic	417	226+
Ice breaker	114	41
Warship	114	2+
Rand for Shoe	36	78
Making machine	12	67R+
Random Control		
Cutting machine	83	58+
Selective	234	22+
Range		
Shooting	273	348.1+
Range Finders		
By straight line light ray	33	284+
Camera focusing, automatic	354	400+
Radar	342	147+
Distance determining	342	118
Sound wave type	181	
Submarine type	367	99+
Using wave radiation	356	4+
Ranges	126	1R+
Electric		
Heating and cooking	219	392
Ranque Tube	62	5
Rare Earth Metal		
Pyrometallurgy	75	610
Rare Earths	156	DIG. 63
Organic compounds	534	15+
Rare Gases (See Gas, Rare)		
Raschig Rings	261	94
Surface contact	261	95
Rasp	29	78+
Making	76	13
Wood	29	78+
Rat		
Guards for mooring lines	114	221R
Hair grooming device	132	55
Trap	43	58+
Ratchet		
Drills	408	120+
Combined	408	120+
Pawl and	74	575+
Bit stock	81	28+
Fishing reel brake	242	296+
Fishing reel drive	242	254
Intermittent grip movements	74	111+
Wrenches	81	60+
Rate of Climb		
Gyro controlled	73	504
Pressure type	73	179
Ratine		
Twisted strands	57	210+
Yarn making	57	3+
Apparatus	57	3+
Rattan Seat	D 6	369
Rattles		
Toy	446	419
Design	D21	65
Vehicle shakers	73	669+
Raunitidine	546	48
Ravelers	28	171

	Class	Subclass
Ray (See Radiant Energy)		
Applications and devices	250	
Cathode ray		
Circuit for tube	315	1+
Structure of tube	313	364+
Television receiver	358	242+
Television scanning	358	199+
Television system	358	83+
Television transmitter	358	217+
Light ray geometrical instrument	33	227+
Light ray projector type game	273	310+
Photocell special ray sensitive	313	523+
Television optical ray	358	209+
Rayon (See Artificial, Silk)		
Razor	30	32+
Blade	30	346.5+
Abrading machines for	51	34BR
Rotary tool, one way traverse	51	80B
Blade design	D28	45+
Manufacturing	76	DIG. 8
Package	206	352+
Sharpeners	76	DIG. 9
Cleaner	15	218
Combined with razor	30	41
Design	D28	45+
Dry shaver	30	43
Electric	30	45
Holder	D 6	526
Hones for	51	211H
Illuminated	362	115
Package or container for	206	208
Design	D 9	342
Kit with razor	206	228
Special package for powered		
razor	206	351
Safety	30	51+
Sharpeners and straps	30	35+
Design	D 8	92
Sharpening	51	
Processes	51	285
Razor combined	30	35+
Strops	76	81+
Reaches		
Land vehicle	280	140+
Crossed	280	102
Pivoted and sliding	280	101
Reaction Motor (See Jet, Motor)	60	200.1+
Discharge nozzle	239	265.11+
Reactivation (See Revivification)		
Reactive Power Control	323	205
Reactors	336	
Lamp or electronic tube system		
having	315	289+
Nuclear	376	
Reader		
Cabinet structures	312	
Coded record & reader, invisible		
radient energy type	250	556
Document verification or graph		
reader	250	271
Reader-printer Machines	D16	223
Data processing equipment	D14	111
Reading		
Aids		
Blind or sightless user	434	113+
Bookholder	248	441.1+
Bookmark	116	234+
Copyholder (See copyholder)		
Line indicator	116	240
Magnifier	350	245+
Forms	283	
Teaching	434	178+
Rapid reading	434	179+
Reamers	408	199+
Cleaning implement	15	104.2
Cutlery blade type	30	352
Design	D15	138
Earth boring	175	406+
Expansible	175	263+
Grinding	51	288
Rotary fruit juicer	99	495+
Screw threading, reaming & tapping	408	118
Smoking pipe bowl	131	246
Wood	408	227+
Reapers	56	
Rearrangement (See Isomerization)		
Reatomizers	261	DIG. 55
Rebar	52	737+
Rebar chair	52	677+
Recapping	156	95+
Receipts		
Transportation cash fare	283	30+

	Class	Subclass
Page printing	178	28+
Photographic	178	15
Printing selector	178	34
Receiver	178	89+
Reed printer	178	48
Siphon	178	91
Transmitter	178	17R+
Type wheel printer	178	35
Telephone		
Calling number	379	142
Conversation time	379	114+
Telegraphophone	369	132+
Television	358	335+
Camera with recorder	358	906*
Color	358	310+
Magnetic	360	33.1+
Pause control	358	908*
Track skipper	358	907*
Television system combined	358	296+
Thermoplastic, storage retrieval of information	365	126
Thermoplastic, visual record	346	151
Television	358	344
Color	358	310
Facsimile	358	300
Time		
Plural recordings	346	45+
Printing or punching	346	80+
Transparency	365	127
Verifying	73	156
Voting machine	235	50R
Wire	346	150
Sound type	360	89
X ray photography	378	
Recovery (See Material Recovered)		
Coating excess	427	345
Dyes	8	440
Fats and oils	260	412+
Inorganic actinide compounds	423	249+
Nuclear materials	376	189
Nuclear materials	376	201
Nuclear materials	376	308
Scrap metal salvaging	29	403.1+
Separation of solids	209	
Solvent apparatus	68	18R
Rectangular Proportioner	33	DIG. 9
Rectifiers and Rectifying	D13	110+
Beam power amplifier	313	298
With two cathodes	313	5
Distillation	203	
Apparatus	196	
Apparatus	202	
Liquids	203	
Mineral oil	208	350+
Doubler high vacuum	313	306+
With two cathodes	313	1
Electrical	363	13+
Barrier layer coating	437	
Barrier layer compositions	252	62.3 R
Electrolytes	252	62.2
Electrolytic devices	361	436
Full wave high vacuum	313	306
Lamp or electronic tube supply	315	200R+
Manufacture	29	25.1+
Process	437	
X ray circuit supply	378	101
Gas and liquid contact apparatus	261	
Gas liquefaction	62	42+
Process	62	32+
Process, plural separation	62	24+
Gas separation processes	55	36+
Half wave system	363	13+
Circuit interrupter for	200	
Dynamoelectric machine	310	10+
Electronic tube for	313	317+
Gas tube type	363	114+
Power packs	307	150
Unidirectional impedance for	357	
Vacuum tube type	363	114+
With filter	363	39+
With voltage regulator	363	84+
Mercury vapor system	363	114+
Rectifier control only	315	246+
Mineral oils	208	350+
Vacuum or gas tube	313	
Recuperator		
Distillation retort combined	202	111
Gas bench	202	148
Horizontal	202	140
Inclined	202	130
Vertical	202	122

	Class	Subclass
Furnace		
Fuel burned in permeable mass	431	170
Gas heating furnaces	432	179
Red Lead	423	619
Redox Catalyst	526	915*
Reducer		
Commutation sparking	310	220+
Machines		
Cabinets	128	371+
Exercising	272	93+
Massage	128	24R+
Pipe joint	285	177
Screw	285	392
Pressure (See pressure, regulator)		
Socket wrench	81	185
Reduction (See Hydrogenation)		
Amines primary production by	564	415+
Apparatus for ores	266	168+
Catalysts treatment	502	100+
Chemical agents	252	188.1
Detergent	252	105
Hydrogen production by	423	648.1
Inorganic sulfide production by	423	
Iron and alloys electrothermic	75	10.1+
Of metal ores and metal compounds	75	
Separation distillation	203	32
Sulphur production by	423	567R
Thermit type	75	959
Reed		
Musical		
Accordian	84	376R
Harmonica	84	377+
Nonorgan reed instruments	84	375+
Organ reed pipes	84	350
Reed organs	84	351+
Toy	446	207+
Tuning reed for furnishing pitch	84	456
Wood winds	84	380R+
Station selective signalling	340	825.39
Telephone call type	379	360+
Telegraph	178	47+
Telephone call transmitter	379	360+
Textile reed loom	139	192
Movement thereof separate from raceway motion	139	191
Terry reed	139	26
Valve	137	855+
Vibration sensing	73	651
Reeding	28	201+
Reef, Artificial	405	21+
Reefing Sails	114	104+
Reel		
Agricultural implements		
Cutting and gathering	56	
Line wire driven planter	111	44
Potato digger	171	111+
Thresher straw carrier	460	85
Antenna	343	877
Antismut printing device	101	418
Baking oven combined	432	141
Body supported	224	162
Cable drum type, driven	254	266+
Carriers	242	85+
Changeable exhibitor	40	514+
Calendar type	40	116+
Electric cord	191	12.2 R+
Suspended support	248	328+
Fishing rod combined	43	20+
Design	D22	137+
Reel seat	43	22
Fishing type	242	223+
Attachments	242	322
Flexible ladder	182	73+
Holder	D 3	30.1+
Hose type with spray device	239	195+
Portable	239	722+
Measuring tape	33	761+
Midline tightener	24	71.2
Motion picture film winding	242	179+
With projector	352	124
Packages	206	389
Plumb bob	33	393+
Resistors	338	79
With motion picture projector	352	124
Rewind motion picture film	242	210
Sheet web or strand drying apparatus	34	153
Sound recording or reproducing tape drive	242	179+
Sounding	33	715+
Strand type fire escapes	182	236+

	Class	Subclass
Strop combined	76	81.4
Tape measure	D10	71
Winding and reeling type	242	77+
Automatically contracting	242	63
Belt type	242	67
Revoluble carrier	242	64
Trundle type	242	94
Winding drum or sand reel type	242	117
Reeling	242	
Information-bearing web machine convertible	242	179+
Textile warp preparing	28	190+
Winding &	60	905*
Reeves Drive	474	8+
Refilling (See Fillers; Filling)		
By dispensing means	141	21
Fountain pen filling mechanisms	401	143+
Prevention of		
Bottles	215	14+
Dispensers	222	147+
Specially related dispenser and receiver	141	348+
Refining (See Purifiers)		
Alcoholic	99	277.1+
Apparatus		
Mineral oil	196	
Carbohydrate manufacture	127	34+
Starch manufacture	127	65+
Sugar manufacture	127	42+
Electrolysis	204	59R+
Fats oils and waxes	260	420+
Fatty oils	260	420+
Recovering	260	412+
Glass making in	65	
Metal casting apparatus combined	164	266
Metals	75	
Mineral oils	208	177+
Conversion combined	208	46+
Deasphalting	208	309
Gum or gum former removal	208	255+
Metal contaminant removal	208	251R+
Nitrogen removal	208	254R
Organic acid removal	208	263
Phenol removal	208	263
Sulfur removal	208	208R+
Sweetening	208	189+
Paper making	241	
Smelting combined	266	161+
Sugar	127	
Reflectors (See Mirrors & Reflectors)		
Antenna	343	912+
Collapsible	343	915
Design	D26	118+
Glass bead type	350	104+
Nuclear reactor including for neutrons	376	220
Nuclear reactor including	376	347
For neutrons	376	346
Radar	342	5
Reflection reducing films on optical surfaces	350	164+
Reflectometers	356	445+
Chemical analysis, apparatus	422	82.5
Solar heat collecting	126	438+
Warning reflector triangle for highway use	40	903*
Design	D10	114
Reflex Klystron	315	5.18+
Reflux, Vapor Barrier	156	DIG. 9
Refractometer	356	128+
Refractors		
Electric lamp having	313	110+
Fluorescent device having	250	485.1
Illuminating device type	362	326+
Illuminating reflector combined	362	327
Lenses	350	409+
Prisms	350	286+
Refractory Material		
Building block	52	596
Checker brick furnace structure	165	9.1+
Compositions	501	94+
Carbide containing	501	87+
Gas generator including	48	74
Metal heat treatment	148	133
Molding processes	264	56+
Pyrometallurgy	75	611+
Refrigerators (See Cooler and Cooling)		
Cabinet structure combined	312	236
Showcase type	312	116
Car	62	239+
Body structure	105	355+

	Class	Subclass
Processes	264	271.1
Radiating from column	52	260
Railway tie	238	85+
Ribbed	52	319+
Ribbed slab	52	602
Roadway	404	134+
Rods	52	720+
Safe or vault wall	109	83
Packing joint	277	
Pipe conduit	138	172+
Receptacle		
Flexible bag	383	119
Metallic	220	639+
Metallic edge	220	640+
Safe	109	83
Trunk	190	25
Rings	16	108+
Shoe insole	12	20
Spoons	30	328
Switch point railway	246	437
Textile, reinforcing or tire cords	57	902*
Thimbles	16	108
Tires		
Inner tube	152	511
Tire	152	393+
Tire and valve stem	152	430
Valve head	251	358
Vehicle frame	296	30
Wheels		
Hubs	301	106+
Spokes	301	83
Spokes plural series	301	77
Reins	54	36
Check	54	16+
Guards	54	73
Design	D30	141
Holders		
Design	D30	141
Vehicle attachment	280	182+
Whip socket combined	280	177
Holds	54	74
Supports and guards vehicle type	280	181
Reject Catcher or Deflector		
Glass making apparatus	65	165
Refrigeration producer		
Impurity removal	62	195
Reknitting	66	1.5
Relaxation Oscillator	331	
Driven type	328	193+
Relay		
Auto		
Headlight systems	315	77+
Horn systems	340	425.5+
Electric		
Amplifier	330	
Casing	200	302.1
Directional safety system	361	1+
Electric thermal responsive system	236	78R
Electromagnetic switch	335	2+
Electromagnetic switching systems	361	139+
Frequency division repeaters	370	75
High speed	335	2+
Light wave repeaters	455	601
Motor operated	335	68+
Pulse or digital repeaters	375	3
Railway signalling system	246	
Systems	361	139+
Systems safety	361	1+
Telegraph repeaters	178	70A+
Telephone central line-signal control	379	317+
Thermal current	337	14+
Thermal responsive system	236	68R
Time division repeaters	370	97
Pneumatic dispatch	406	181+
Pressure modulating relays or followers	137	82+
Structure electromagnetic		
Electromagnetic switch	335	
Electrothermal switch	337	14+
Temperature and humidity control by	236	
Time lag	200	33R+
Release & Releasers (See Uncoupling)		
Article	221	289+
Brakes and clutches		
Brakes	188	
Check controlled lock release	194	
Clutch and brake	192	12R
Clutch operators	192	82R+
Clutches	192	30R+

	Class	Subclass
Drive release and brake	192	144+
Fluid pressure brake motor releasing	303	68+
Fluid pressure system synchronizing	303	36
Overload release	192	150
Phonograph stop mechanism	369	268
Speed responsive clutch operator	192	103R+
Transmission control and automatic brake	192	7+
Typewriter feed mechansim	400	334+
Coating		
Compositions	106	2
Methods	427	
Load-responsive release	403	DIG. 3
Overload (See clutch; crank; latch; lever; overload; trip)		
Automatic trip mechanisms	74	2
Elevator cable	187	72
Hand wheels	74	556
Harpoon type hay fork	294	126+
Hoist line sling	294	75+
Intermittent grip mechanical movements	74	111+
Key holding catch for locks	70	459
Lever carried pawl	74	535+
Lever detent and release	74	529
Lock emergency release	70	465
Locks	70	
Mechanical gun triggers	124	31+
Music leaf turners	84	486+
Pitman automatic trips	74	584
Railway switches and signals	246	
Rotatable platform trap	43	72
Store service cable propulsion	186	16
Tiltable platform trap	43	70
Tilting hay fork	294	124
Time lock emergency release	70	274
Time switch latch trip	200	40+
Toy money box figure or mechanism	446	9
Traction railway cable grippers	104	207
Typewriter line spacing	400	565+
Valve	251	66+
Well torpedo lowering	102	313
Quick release	403	DIG. 4
Relief		
Electrochemical process	204	4+
Maps	434	152
Milling	409	69
Non-planar layer ornamented	428	187
Photographic process	430	322+
Picture	40	160
Relievers		
Electric cord strain	439	449+
Internal combustion engine pressure	123	182
Metalworking roll adjustment	72	237+
Photographs in relief making	430	269+
Radiator air valves	236	61+
Receptacle filling displaced air pressure	141	
Safe explosive pressure	109	27
Scale platform gear	177	151+
Ship tension	114	213+
Religion Teaching	434	245+
Religious		
Altar	D 6	396+
Apparel	2	
Article not elsewhere specified	D99	25+
Artifact cruciform type	428	3
Baptismal font	D99	25
Cabinet	312	
Creche	D11	122+
Educational device	434	245+
Incense burner	D11	131.1
Jewelry	63	
Medal	D11	95+
Monument	52	103+
Receptacle	206	19
Tabernacle	D99	25
Relishes	426	615+
Relishing	144	6
Remote Control (See Type of Control or Device Controlled)		
By flexible cable	74	500.5+
By light waves	455	600
By radio	340	825.72
Pulse responsive	340	825.69
By transmitter	341	176
Handlers	414	909*
Mining machine	299	30

	Class	Subclass
Remote Cutoff		
RF pentode	313	300
Tetrode	313	297+
Remover (See Cleaner; Pull; Push; Scraper)		
Antiskid means on wheels	152	213R+
Bur removing		
Chains	59	59
Pipe	408	199+
Closure	81	3.7+
Can opener cutter combined	7	152+
Other tool combined	7	151+
Plier tool combined	7	126+
Dental crown	433	141+
Disassembly	29	700+
Garment	223	111
Heel shoe	12	50.5
Ingot strippers	164	405
Insulation from wire	81	9.4+
Obstruction conduit	4	255+
Residue from filter	210	407+
Rind	99	584+
Scale (See cleaner and cleaning)	29	81.1
Scum or sediment	210	523+
Distinct separators	210	299
Shell nut and vegetable	99	568+
Skin fruit and vegetable	99	584+
Sludge		
Acetylene generator	48	57
Sewage	210	523+
Stem	99	635+
Stump excavators	37	2R
Threshing	460	
Tire	157	11
Rubber tire	157	1.1
Welding flash planers	409	288+
Removing		
Caffein or tannin	426	427+
Cast article from mold	164	131+
Demountable rim	301	14
Disassembling processes	29	426.1+
Tools and appliances	29	700+
Drying	34	
Electric insulator moisture or dirt	174	211
Emulsifying agent from mineral oil	208	177+
Excess coating material	427	235
Fermenting processes		
Fermentate removal	435	262+
Fowl sinew	17	11.3
Garments	223	111+
Growth in water courses	56	8+
Rendering	260	412.6+
Liquid separation combined	210	767+
Apparatus	210	175+
Melting separators for chemicals	422	285+
Press	100	104+
Rennet or Rennin		
Cheese preparation with	426	582
Renovating		
Butter	426	663
Apparatus	99	452+
Coatings	427	140+
Dress		
Dry cleaning	8	142
Dyeing combined	8	441
Feathers	34	2
Repairing (See Darning; Patch)	29	402.1+
Access to fluid handling systems	137	583+
Adhesive bonding	156	94+
Building structure	52	514
Assembly or disassembly	52	127.1+
Coating processes	427	140+
Electrolytic	204	16
Disassembly steps included	29	402.3+
Electric		
Conductor	156	49
Discharge device	445	
Apparatus	445	61
Process	445	2
Lamp	445	
Apparatus	445	61
Process	445	2
Fluid handling systems or devices	137	315+
Furnace wall	264	30
Glass making apparatus	65	27
Liquid purification or separation apparatus	210	232+
Sectional chamber press type	210	230
Metal casting	164	92.1
Patching	29	402.9+
Pile	405	250

	Class	Subclass
Rice		
Bran	426	618+
Cleaning and polishing	426	481+
Apparatus	99	600+
Planters	111	911*
Ricinoleic Acid	260	413+
Ricks		
Hay straw	414	132
Rickshaw	224	159
Riddles	209	331+
Ridge		
Bed pole	5	130
Gondola car drop bottom		
Central	105	245
Rising	105	246
Ridging		
Plow	172	701
Riding		
Boots	D 2	271+
Equipment	D30	134+
Rifamycins	540	458+
Riffles		
Liquid suspension separator having	209	206
Stratifier having	209	506+
Rifles (See Gun)		
Automatic	89	125+
Carrier	224	913*
Receiver holding butt end	224	149
Sling carrier	224	150
Design	D22	103+
Firearm	42	
Ranges	273	348.1+
Recoilless	89	1.7+
Targets	273	348.1+
Rifling		
Cutting	409	306
Firearm barrel	42	78
Rigging		
Brake	188	2R+
Sails and rigging	114	102+
Ships implements	114	223
Rigs		
Earth boring	175	
Rim		
Breaker	157	1.35
Compressor	157	2
Demountable	301	10R+
Gear		
Segmental	74	448
Separate	74	446+
Handwheel grip or cover	74	558
Integral for tires	152	378R
Locks for	70	260
Making	29	894.35+
Pulley	474	166+
Expansible	474	47+
Cone pulley	474	8+
Separable for tires	152	396+
Tire rim bands	152	513
Wheel	301	95
Ring (See Band; Collar)	16	108+
Container for annular article	206	303
Fasteners connectors or supports		
Article holding bracket	248	315
Bolts closure fastener	292	299
Bottle and jar closure	215	274+
Demountable rim	301	26+
Draft pole coupling	278	125
Filter medium	210	495
Gimbal shaft coupling	464	125+
Handled filter	210	471
Insulator cap type	174	192
Insulator pin type	174	198
Metal receptacle closure	292	256.6+
Panel to track attaching	16	87.2+
Pipe conduit bracket	248	69
Pipe conduit bracket	248	74.1
Pipe conduit suspended	248	62
Shaft joint	403	
Spring lamp shade support	362	440
Tire to rim	152	409+
Tobacco users finger	131	258
Handle	16	127
Jewelry		
Earrings	63	12+
Design	D11	40+
Finger	63	15+
Design	D11	26+
Trays for	206	566
Key	70	456R+
Design	D 3	61+

	Class	Subclass
Lock cylinder	70	381
Locking, bolt or nut to substructure		
lock	411	109
Magnetoelectric apparatus	228	179
Armature retaining	310	271
Electric heating	219	51+
Electroforming process	204	9
Grinding process	51	290+
Jewelry forming and sizing	29	8
Packings	29	888.7
Rolling	72	67+
Slip	310	232
Spinning, twisting, twining		
Apparatus	57	21
Process	57	21
Welding process	228	179+
Winding ring wound armature	242	4R+
Wire	140	88
Making by deforming metal	72	
Gem and jewel setting	29	10
Napkin holder	D 7	633
Nose animal	119	135
Piston	277	
Slip for electric current machine	310	232
Snap ring	24	16R
Spinning frames	57	75
Textile spinning element	57	119+
Tire bead grommets	245	1.5
Trapeze	272	61
Trim	301	37R
Watchcase movement	368	299
Welding backup	228	50
Joint integratable	285	21
Ringers and Ringing		
Telephone		
Automatic	379	232+
Party line	379	179+
Rinks, Ice	62	235
Grooming apparatus	37	219+
Rinsing		
Solid work pieces	134	
Textiles and fibers		
Machines	68	
Methods	8	147+
Rip Strips		
For metal containers	220	266
For paper boxes	206	601+
Riprap	405	32+
Risers		
Bed bottom		
Corner riser	5	205+
Frame shiftable on riser	5	209
Railway car folding step	105	448
Railway switch	246	432
Nonderailing	246	421
Sand mold	164	359+
Stair riser carpet fastenings	16	11
Rivet	411	500+
Design	D 8	386
Making	10	11R+
Seal type	292	314
Riveting		
Blind	227	51+
Button riveting machine	227	51+
Design	D15	122+
Driving and heading	227	51+
Electric heating and	219	150R
Hollow rivets	227	51+
Solid rivets	29	243.53+
Riving	144	182+
Shaving combined	144	40
Roads	404	
Cement structure	404	71
Chair	404	136
Curb	404	7
Expansion joints	404	47+
Finishing by compacting	404	113+
Grader	37	108R+
Gravelling means	404	101+
Gutter or drain	404	2+
Land roller	404	122+
Manhole cover	404	25
Pavement	404	17+
Apparatus for making	404	83+
Crushing in situ	404	90
Marking	404	93+
Paving machinery	404	83+
Process for making	404	72+
Railroad	238	2+
Railway graders	37	104+
Reinforcement	404	134

	Class	Subclass
Roadbed	404	27+
Screed or drag	404	118+
Street curb sewage inlet	404	4
Surface marker, signaller	D10	113
Tamper	404	133
Testing	73	146
By abrasion etc	73	8
Traffic barrier	404	6+
Traffic directors in pavement	404	9+
Mirrors and reflectors	404	14
Roaster		
Coffee	99	286
Domestic ovens	126	273R+
Gas ovens and broilers	126	39R+
Metallurgical	266	171+
Miscellaneous materials	432	
Rotary driers	34	108+
Spits	99	419+
Stove broiling attachments	126	14
Vegetable and fruit peelers	99	483
Processes	426	479+
Roasting		
Coffee	426	523
Metallurgical apparatus	266	171+
Smelting combined	266	175
Robes		
Apparel	2	69
Design, ceremonial	D 2	79
Design, housecoat	D 2	12+
Locks for	70	60
Vehicle lap robe holders	296	77+
Robot	901	
Arm movement	901	14+
Control, electric	318	568.11+
Coating	901	43*
Compliance device	901	45*
Control, electric	318	568.11+
Counterbalance	901	48*
Drive systems	901	19+
Electric	318	568.11+
Gearing	901	25*
Grip	901	31+
Industrial		
Grappling functions	294	86.4+
Magnet or piston controlled	294	88
Pivoted jaws	294	106
Gripper functions, body	623	64
Fluid actuated	623	57+
Material or article handling	414	4+
Horizontal linear motion	414	753
Horizontal swing motion	414	744.1+
Vertical swing motion	414	730
Motion and control, electrical	318	568.11+
Spray painting	118	323
Structure, control lever and		
linkage systems	74	469+
Welding functions	219	125.1
Inspection	901	44+
Mechanically actuated	901	11+
Mobile	901	1*
Control, electric	318	568.12
Programmable		
Article handling	318	568.11+
Artificial intelligence	364	513
Speech signal processing	364	513.5
Self organizing controls	364	513
Safety device	901	49*
Sensor	901	46*
Spray painting	901	43*
System	901	6+
Teaching system	318	568.13+
Track-guided	901	14+
Welding	901	42*
Wrist	901	29*
Rochelle Salt	562	585
Rock		
Bits	175	
Blasting	102	301+
Methods	102	301+
Drilling machines	175	
Rocker		
Arms machine elements		
Poppet valve operating, internal		
combustion engine	123	90.39+
Bearings		
Pivot	384	154+
Shaft	384	154+
Spinning spindle	384	238
Bogie truck	105	187
Chair	297	258+
Design	D 6	344+

	Class	Subclass
Saponin	536	4.1+
Saran T M	526	343
Sarcophagi Molds	425	
Sarcosine	562	575
Sash		
Accessories		
Balances	16	193+
Burglar alarms and fasteners	116	16+
Cord fasteners	16	202+
Cord guides	16	210+
Electric switches	200	61.62+
Weights	16	216+
Garment supporter	2	300
Making	144	5+
Panel	52	782+
Sectioned	52	455+
Window		
Locks for	70	89+
Ventilating	98	97
Satchel (See Pouch)		
Satellite (See Space Ship)	244	158R+
Aplications, photoelectric cell	136	292*
Saturable Core Reactor		
External or operator controlled	323	329
Input responsive	323	302
Linear amplifier systems	330	8
Non-linear reactor systems	307	401
Output responsive	323	249
Self regulating	323	310
Transformer systems	323	250
Voltage magnitude control	323	253
Saturating		
Fireproofing compositions	252	601
Preserving	422	
Transformer	336	155
External or operator controlled	323	329
Input responsive	323	302
Output responsive	323	249
Regulating system	323	
Self regulating	323	310
Sauces	426	589+
Sauerkraut (See Kraut)		
Sauna		
Bath	D24	37
Heater	D23	315
Sausage	426	129
Casing	426	140+
Inedible	428	34.1+
Shirred casing	206	802*
Wall structure	138	
Cookers		
Electrode type	99	358
Opposed heated surface type	99	380+
Spit or impaling type	99	419+
Support combined	99	441
Grinders	241	
Linkers	17	34
Stuffers combined	17	33
Stuffers	17	35+
Savings Boxes	232	4R+
Paper type	229	8.5
Toy	446	8+
With register	235	100
Saw	30	166.3+
Band	30	380
Band saw	83	788+
Barrel head making	147	38
Bone	606	82
Trephine	606	172+
Bow	30	507+
Butchering	17	23
Button blank sawing	79	15+
Chain	30	381+
Chain saw	83	788+
Cleaners and oilers	83	169
Dado	144	222
Drag saw	83	771+
Frame	83	859
Hand	30	166.3+
Fret	30	513
Gauge	33	202
Gin	19	55R+
Gummers	76	25.1+
Wheels	76	45
Handsaws	30	166.3
Combined with other cutter	30	144
Combined with other tools	7	148+
Hole saw	408	204
Making	76	25.1+
Methods	76	112
Metal sawing machines	83	

	Class	Subclass
Saws	83	835+
Rotary	30	388+
Scroll	83	783+
Hand	30	513+
Sets	76	58+
Stave jointing	147	28+
Stone sawing	125	12
Surgical	606	176
Table saw	83	477+
Tables	269	289R
Gauges	83	522.16+
Modified for blade	269	83+
Teeth	83	835+
Fastenings	83	840+
Vegetable and meat combined with cutter	83	
Wood sawing devices	83	
Design hand saw	D 8	95+
Design, power saws	D 8	64+
Design, sawing machinery	D15	133
Guards	83	440.2
Guides	83	821+
Wood slivering gang	144	189
Woodworking rotary cutter	144	222+
Sawbucks and Sawhorses		
Buck, ie holding or clamping	269	296+
Horse, ie scaffold or plank supporting	182	181+
Sawdust		
Briquette	44	590+
Burners (See furnaces solid fuel burning sawdust refuse and wet fuel)		
Fuel	44	506
Fur cleaning by	69	23+
Making processes	241	28
Sawhorse	D25	67
Sawing (See Saw)		
Button blanks	79	15+
Dog	269	
Feed	83	401+
Metal	83	
Sawmill carriage wheel guards	37	194
Sawmill dogs with carriage	83	721
Stone	125	12+
Wood	83	
Saw table gauges	83	522.16+
Sawtooth Oscillator		
Driven	328	193+
Free running	331	
Saxophones	84	385R
Scabbards (See Sheath)	224	232+
Design	D 3	102
Ice skate	280	825
Sword	224	232+
Design	D 3	102
Scaffolds	182	179
Bracket supports	248	235+
Combined	182	129
Convertible	182	20
Ladder	182	27
Design	D25	62+
Design, element or coupling	D25	68+
Elevating means	182	141+
Foldable	182	152+
Platform	182	222+
Shaft supported	182	128
External	182	187
Scalder		
Cleaning	134	
Fowl	17	11.2
Hog	17	13+
Peeler or parer combined	99	483
Scale (See Weighers)		
Armored tire	152	206
Bathroom	D10	87
Lamellate composition	106	415+
Metal scaling	29	DIG. 34
Music		
Automatic note selector	84	152+
Enharmonic	84	451
Keyboard touch plate	84	428
Optical element combined	350	110+
Optical pointers	350	110+
Pans	177	262+
Price	177	25.11+
Register	235	124
Remover and preventer		
Composition	252	80+
Dentistry	433	143
Electrolytic apparatus	204	196+

	Class	Subclass
Electrolytic process	204	147+
Fish	17	64
Hammer	29	81.15
Magnetic apparatus	422	186.2+
Metalworking	29	81.1+
Water softener	252	175+
Rule or	D10	71+
Ruler	33	494
Square	33	474
Typewriter	400	705+
Line spacing	400	548+
Weighing instrument	177	
Design	D10	87
Scalpels	606	167+
Scandium	156	DIG. 114
Scanning		
Antenna	343	757+
Facsimile	358	474+
Optical	350	
Electrophotographic copier	355	233+
Periodically moving optical element	350	6.1+
Periscope	350	540+
Photoelectric	250	200+
Television	358	199+
Telescope	350	537+
Scarecrow	40	
Noise maker		
Bells	116	148+
Horns	116	137R+
Scarf		
Apparel	2	207
Design fur	D 2	601
Design haberdashery item	D 2	500+
Design shaped	D 2	500+
Design shaped ascot	D 2	605+
Design with evening dress	D 2	53+
Hanger for fur	223	97
Joint	238	230+
Slide holder	D 2	202+
Scarfing (See Skiving)		
Apparatus for metal	266	51+
Blowpipes	239	398+
Processes for metal	148	9R
Scarifier		
Earthworking	172	
With scraper	37	108R+
Stamp	81	9.21
Surgical		
Lancet	606	181+
Other instruments	606	186
Receptor combined	604	22
Scattering (See Sprayer; Sprinkling)		
Drying apparatus centrifugal	34	59
From aircraft	244	136
Insect powder dusters	239	
Non-fluid material scattering	239	650+
Planting broadcasting	111	8+
Pneumatic conveyor discharge	406	157+
Road treating material	404	101+
Track sanders	291	
Scavenger & Scavenging		
Anesthesia gas scavenging system	128	910*
Oxygen scavinging	252	186.1+
School		
Bus signal lights	340	433
Desks and seats	297	135+
Design	D 6	335+
Design, with drawer	D 6	335+
Schreinerized Fabrics	26	69R
Science Teaching	434	276+
Scillin	536	4.1+
Scissors (See Nipper; Shearing)	30	194+
Buttonhole	30	120
Can opening	30	428
Design	D 8	57
Knife combined	30	146
Manicuring	30	29
Compound tool	132	75.4+
Material holder combined	30	131+
Nipper combined	30	145
Plier combined	7	135+
Sharpening		
Abrading design	D 8	93
Apparatus	76	82.2
Combined with can opener	D 8	35
Power sharpener	D 8	63
Process	51	285
Scleroscope	73	79
Scoop	D 7	691
Bag filling	141	108+
Ditchers	37	103

	Class	Subclass
Pens	15	435+
Shaft	464	179
Aligning and leveling	33	412
Angle transducer	250	231.13
Arrows	273	416+
Axle land vehicle	301	126
Bearings	384	
Clutch	192	110R
Couplings	403	
Flexible	464	
Crank		
Bearing antifriction	384	457+
Bearing plain	384	250
Lathes	82	106+
Making		
Apparatus	29	6.1
Method	29	888.8
Earth boring means combined	175	320+
Earth, eg mine	405	133+
Flexible	464	
Portable drill	408	127
Furnace	432	95+
Fume arrester with	266	144+
Metallurgical	266	197+
Hanger	384	442+
Housing	464	
Flexible shaft combined	464	52+
Hub connections	403	
Joint packings	277	
Joints (See joints)		
Locks	70	182+
Metallurgical furnace	266	197+
Mucking machines	414	916*
Process abrading	51	289R
Speed recorders	346	
Wagon	278	33+
Wall retainer	405	272+
Shaker		
Shaking		
Apparatus		
Grates	126	152R+
Automobile	73	669+
Devices		
Cleaning machine	15	94
Filled cooking receptacle in oven	99	371
Gas filter cleaning	55	304+
Heat exchanger	165	84
Portable receptacle unloading	414	415
Separator receiver	209	64
Shotting combined for cleaning	15	96
Sifter horizontal or inclined	209	309+
Sifters	209	233+
Sprinkler or sprayer	239	374
Vehicle unloader	414	375
Vehicle unloader tilting track	414	357
Pivotable movement	414	362
Receptacle	366	
Clothes washing machine	68	171+
Table		
Grain separator straw carrier	460	90+
Separator receiver	209	64
Sifter	209	309+
Shale		
Distillation	201	
Mineral oil from	208	400+
Shale Oil		
Recovery	208	400+
Sham Holders Pillow	5	492
Shampoo	252	89.1+
Shank		
Buttons	24	92
Cloth	24	92
Making	79	2
Sewing machine attaching	112	108+
Key	70	402
Extensible	70	397
Foldable	70	396
Plural	70	401
Manufacture of metal knobs	29	161
Padlock nonshackle type	70	34
Railway spike shank brace	238	309
Railway track clamp anchor	238	345
Shoe	36	76R
Loose upper shaping	12	54.1+
Machines	12	40.5
Tool and machine parts		
Handle fastenings	403	
Slidably adjustable wrenches	81	129+
Wrench attachments	81	184
Wrench handles and	81	177.1+
Shape Memory Alloy	148	402

	Class	Subclass
Shaping (See Product Itself; Molding)	29	
Abrading or grinding	51	
Boring	408	
Comminution combined	241	3
Cone wound bobbin or cop shaper	242	34
Dental instruments for	433	25+
Die (See die)		
Directly applied fluid pressure differential	264	500+
Drilling	408	
Food	99	
Bread pastry and confection making, food apparatus	99	
Bread pastry and confection making, shaping apparatus	425	
Cooking and	99	324+
Dairy molds	249	
Meat briquetting apparatus	17	32
Hay stack shapers	414	132
Metal		
Bending to angle	72	176+
Casting combined	29	527.5+
Coating combined	29	527.2+
Corrugating	72	180
Deforming	72	
Drawing sheet metal	72	347+
Drawing through die	72	274+
Electric heating and	219	149+
Extruding	72	253.1+
Indirectly	72	273.5
Molten metal	164	
Plastic metal	72	
Powder apparatus	425	78+
Powder processes	419	61+
Pressure or impact forging	72	
Rolling	72	199+
Sheet metal by dies	72	
Spinning	72	82+
Wire fabric	140	107
Wire into articles and fabrics	140	
Milling	409	64+
Nonmetallic		
Briquetting fuel	44	593+
Coopering	147	
Glass	65	
Ironing and smoothing textiles	38	
Leather	69	
Paper	493	395+
Plastic block and earthenware	425	
Plastics	425	
Sand molds for metal	164	
Stone	125	
Tobacco plug	131	111+
Wood sawing	83	
Wood turning	142	
Woodworking	144	
Pencil sharpeners (See pencil sharpener)		
Planing	409	288+
Screw threading	408	
Stack or pile shaping	414	788.9+
Turning	82	
Shares for Plows	172	681+
Forging dies for making	72	425
Making machines and processes	29	14
Sharpener	83	174+
Abrading or grinding	51	
Design, knife or scissors	D 8	93
Button surfacing with tool	79	8
Combined with cutting machine tool	83	174+
Cutlery with	30	138+
Disk plow	172	439
Drill forming and	76	5.1+
Harvester cutting reel type	56	250
Horseshoe calk	168	46
Making and sharpening	59	65
Knife or scissors		
Combined with can opener	D 8	35
Powered tool	D 8	63
Metal etching	204	141.5
Pencil (See pencil sharpener)	30	451+
Design of	D19	73
Machine	144	28.1+
Razor with	30	35+
Saw	76	30+
Tool	76	82+
Shaving and Shaver		
Brushes	D 4	135
Design	D 4	114
With drip cup or shield	15	248R+
With soap supply	401	268+

	Class	Subclass
Comminutor type		
Ice	30	136+
Mill	241	83+
Vegetable and meat	241	
Cream	424	
Cups	220	
Electric	30	45
Gear	409	37
Hair planers	30	30+
Design	D28	44
Hat	223	19
Kits	132	289+
Brush or soap retained	401	118+
Pencil (See pencil sharpener)	30	451+
Razor	30	32+
Blades	30	346+
Dry shaver	30	43
Screw head	10	8
Spoke	30	281
Woodworking	144	155+
Shavings Feeding to Furnace	110	102
Shawls	2	91
Design	D 2	179+
Sheaf Handling		
Binding	56	432+
Carriers	56	474+
Composite construction	52	334
Cornstalk loader	56	120
Carrier	56	121
Crop threshing or separating	460	19
Shocker	56	401+
Shear Line, Cylinder Lock	70	496
Shear Pin	464	32+
Shearing & Shears (See Cutter)		
Bolt and rivet making	10	25
Buttonhole	30	120
Can opener	30	428
Cutlery combined		
Knife	30	146
Nipper	30	145
Design	D 8	93
Gauges for shears	33	631
Ground supported hand operated harvester	56	241
Handle and blade connections	30	341
Holder or disposal combined	30	131+
Manicure	30	29
Plier combined	7	125+
Modified handle	7	131+
Receptacle combined		
Cigar receptacle	206	238+
Dispensing sheet receptacle	206	216
Ticket holder receptacle	30	124+
Shears hand manipulable	30	194+
Stone shearing	125	23.1+
Surgical	606	174
Textile		
Cloth finishing	26	15R+
Thread finishing	28	226+
Sheath (See Scabbard; Sheathing)	224	191+
Awl or prick punch	30	368
Cable	74	502.5
Cutlery	30	151+
Two or more cutters	30	143
Design	D 3	102
Hair comb	132	143
With brush	132	121
Manicuring tool	132	76.2
Padlock	70	55
Splitters	30	90.1+
Surface supporting	108	150+
Wire strippers	81	9.4+
Sheathing (See Facer Sheathing (See Facer Construction;)		
Earth control	405	272+
Tunnel or shaft lining	405	132+
Electrical conductor	174	68.1+
Making		
Indefinite length electrical conductor	156	47+
Metal extrusion	72	268
Packing	277	229+
Railway car		
Freight	105	409+
Passenger	105	401
Ship	114	84
Sheave (See Pulley)		
Reeling storage	242	47.5
Shed		
Animal husbandry	119	16
Buildings	52	

	Class	Subclass
For jacquard looms	139	64
Locomotive	104	51+
Shedding		
Loom	139	55.1
Cam	139	79
Dobby	139	66R
Jacquard	139	59
Sheet and Sheet Material (see Composition)		
Accounting and listing form	283	66.1+
Advancing material of indeterminate length	226	
Apertured	428	131+
Associating and folding	270	
Spacer for drying	34	6
Bedding	5	482
Contour bedding	5	497
Restraining	5	498
Binder device releasably engaging aperture or notch of sheet	402	
Boiler		
Crown	122	496
Tube	122	512
Bracket shelf of	248	247+
Cloth	D 5	
Coated	428	411.1+
Sheets	446	108+
Composite	428	411.1+
Composite with cellular layer	428	304.4+
Composite with foamed layer	428	304.4
Construction toys of	446	108+
Container for coiled form	206	389+
Control (See pattern control)		
Electric circuit maker	200	46
Stop mechanism	192	126+
Cookie	99	422+
Corner structure	428	81+
Counting	235	89R
Cover with	428	68+
Creped	428	152+
Discontinous coating with	428	195+
Easel of	248	441.1+
Edge feature	428	192+
Edge fold	428	121+
Edge spliced	428	57+
Expanded metal making	29	6.1
Feeding and delivery	271	
Dispensing cabinet magazine	312	37
Electrophotographic	355	308+
Collating	355	323
Imbricating	355	322
Interleaving	355	325
Separators, pneumatic	355	312
Sorting	355	323
Stacking	355	322
Strippers	355	315
Magnetic oferations	271	901+
Manifolding	282	
Reverse direction of sheet movement	271	902
Separators, pneumatic	271	90+
Separators, rotary	271	109+
Strippers	271	900
Support rack for paper	211	45+
Support rack stacked	211	50+
Towel service locking rack	211	6
Towel service rack	211	16
Typewriter paper on edge	271	903
Typewriter paper	400	624+
Winding and reeling	242	
Figure toys of	446	387+
Filter	210	348+
Fire escape body catcher	182	138+
Flanged pipe joint packing of	285	363+
Garment hanger	223	87
Gas and liquid contact porous	261	100+
Grooved	428	161+
Holders and securing devices		
Billiard table cloth	273	7
Cleaning fabric	15	231+
Cylindrical for sandpaper	51	358+
Endless conveyor with gripper	198	688.1+
Handled for sandpaper	51	392+
Manifolding pockets	282	11.5 R
Music stands	248	441.1+
Pack holders	271	145+
Printing	101	415.1
Rack	211	45+
Sandpaper disc	51	358+
Shoe last with retainers	12	141
Stack rack	211	50+

	Class	Subclass
Tray	206	557+
Wall mounted tray	211	88
Joined	428	57+
Laminate	428	411.1+
Making		
Coating and uniting	156	
Drying	34	
Electroforming	204	12+
Electrolytic coating metal	204	27+
Glass	65	
Metal and foil	29	17.1+
Metal by rolling	72	199+
Pastry rolling	425	363+
Pastry rolling	425	335+
Plastic rolling or pressing	425	
Pulp molding compressor	162	396+
Pulp molding winder	162	283+
Metal		
Can making machines	413	26+
Cap preparing	413	8+
Apparatus	413	56+
Pattern and texture design	D 5	1+
Metal laminate	428	615+
Metal shaping (See particular operation)		
Chain making	59	13+
Container making	413	
Sprocket chain making	59	6
Staple making	59	72
Wire barbing	140	66
Metal structures and articles	405	276+
Bale tie band	24	23R+
Cabinets	312	
Can closures	220	200+
Chain link	59	91
Closure seal	292	310+
Connecting rods	74	588
Containers	220	
Display packages	206	44B+
Drive chain	474	202+
Electroplating	204	27+
Flumes	405	119+
Gearing sectional	74	449
Heat radiator	165	170
Hook	24	698.2
Jar closures	215	200+
Joint packing reinforcement	277	
Nails spikes tacks	411	439+
Piling	405	274+
Pipe joint	285	424
Pulley	474	166+
Rack support for receptacle	211	72+
Railway ties	238	63+
Railway wheel web	295	22+
Resilient clasps	24	530+
Sash cord guide casing	16	212
Separable-fasteners	24	572+
Shaft hangers	384	443
Shelf type with support	108	
Diverse support for articles	108	28
Spools	242	118.8
Stove for heating	126	65+
Stove legs	126	306
Nonrectangular	428	80+
Nonstructural	428	411.1+
Note musical		
Music teaching	84	483.1
Musical automatic instrument	84	146+
Opening cutter	402	1
Pack holders	271	145+
Package of coiled sheets	206	449+
Package of flat sheets	206	449+
Paper dispensing package	206	449+
Paper wrapping	229	87.1+
Mailing	229	92
Pavement	404	17+
Combined block and	404	18
Perimeter structure	428	81+
Picture support of	248	441.1+
Piling	405	274+
Printing smut preventing	101	419
Railway tie		
Folded sheet	238	77
Reinforcing	238	93
Receptacle stand of	248	152
Receptacle support of	211	72+
Ribbed	428	161+
Roofing	427	147+
Sorting type rack	211	10+
Spacer	402	80R
Spliced	428	57+

	Class	Subclass
Spot bonded	428	198+
Stand of	248	174
Stencil	101	127
Structurally defined	428	98+
Structurally defined element therein	428	221+
Tearing or breaking	225	
Testing sheet material	73	159+
Bursting strength	73	838+
Flatness, optically	356	371
Transparency, optically	356	432+
Thread for pin type insulator	174	204
Treating		
Abrasive applying	51	275
Brushing machines	15	77
Chemically cleaning	134	
Coating	118	
Coating electrolytically	204	22+
Cutting	83	
Drying apparatus	34	148+
Drying method	34	18
Gas contacting	34	23+
Heat	108	28
Ironing shaking out	38	143
Radiant heating	34	41
Stencil making	101	128.4
Wiping machines	15	102
Winding & reeling	242	
Wrapper for coiled form	206	389+
Wrinkled	428	152+
Shelf and Shelving	108	
Attached laterally of support	108	152
Baggage convertible	190	9+
Bookcase shelving		
Design	D 6	567+
Enclosed	312	
Open	108	59+
Cabinet with racks or shelves	312	
Showcase type	312	128+
Covering	428	68
Apertured	428	131+
Edge fold	428	121+
Design	D 6	567+
Display receptacle	206	45
Drying apparatus		
Removable	34	192+
Shelf to shelf material flow	34	165+
Endless conveyer with coacting support	198	860.1
Plural conveying sections	198	600
Heat exchange	165	168+
Ironing board with folding support	108	115+
Loading or unloading	414	266+
Loading or unloading	414	227+
Main or coin chute	232	63
Paper design	D 5	
Rack, plural shelf	108	59+
Shoe or boot rack	211	36
Stove	126	332+
Supports	248	235+
Terraced shelf rack convertible to table	108	17
With structural installation	108	42+
With upward extending support	108	149
Shell		
Ammunition	102	
Cartridge	102	464+
Design	D22	116+
Explosive	102	473+
Extracting implements	81	3.5
Feeding devices (See gun loading)	86	46
Loading	86	23+
Making	86	10+
Making metal working	29	1.3+
Brake shoe	188	254
Edible		
Cereal	426	138+
Design	D 1	
Electrical connector socket	439	753+
Foundry molds	164	361
Shell openers	30	120.1
Shellac	530	201
Sheller	460	45
Corn	460	45+
Feeders for	460	16+
Green	99	567
Cutlery	30	120.1+
Nutcracker	30	120.2
Pea	460	131
Vegetable cutter combined	99	
Shellfish		
Apparatus	17	53

	Class	Subclass
Sifter and Sifting (See Screen)		
Cabinet with	312	210.5
Kitchen utensil	D 7	667
Pneumatic conveyor intake	406	121
Potato diggers	171	71+
Potato diggers	171	111+
Separating solids		
Combined operations	209	12+
Comminuting surface apertured for	241	83+
Comminution and gas borne material	241	49+
Comminution combined	241	68+
Stove ashpan	126	244
Sight		
Bow, arrow	33	265
Dispenser inspection	222	154+
Drift	33	320
Feed lubricator	184	96+
Glass liquid level gauge	73	323+
Gun	33	233+
Design	D22	109
Design, telescopic	D16	132
Fiducial telesopes	356	247+
Illuminated sights	362	110
Letter box signal	232	34+
Line controlled scriber	33	20.1+
Opening in umbrella	135	35S
Projector	362	383
Telescopes	350	567+
Sign and Sign Exhibiting	40	584+
Circuit control flashing light etc.	40	902*
Clerical desk article combined	40	358
Directional design	D20	29+
Electric signal display system	340	700+
Fiber optics, eg light pipe	40	547
Frame	D19	52
Illuminated	40	541+
Illumination itself	362	812*
Mirror having changing effect	40	900*
Mounting bracket	D 8	354+
Out-to-lunch type signs	40	907*
Pen and pencil X-art collection	40	905*
Reflecting triangle for highway use	40	903*
Roadway warning	D20	10+
Sound X-art collection	40	906*
Street	40	612
Supports	248	466+
Towed through sky	40	215
TV or radio program indicator	40	901*
Vehicle motion and direction	116	42+
Warning reflector or flasher	D10	114+
Window shade	160	10
Sign, Illuminated	362	812*
Signal	116	
Acoustic	116	67R+
Bells	116	148+
Horns	116	137R+
Sirens	116	147+
Whistles	116	137R+
Alarms	116	67R+
Amplifier condition	330	2
Annunciators	340	815.1+
Arc lamp	314	9
Automobile horns	116	137R+
Awakening device (See awakener)		
Balloons	116	DIG. 9
Bank protection	109	21
Beverage infuser	99	285
Bicycle bells	116	166
Boiler combined	122	504.2
Brake wear	188	1.11
Buoy	116	26+
Burglar electric	340	541+
Burglar mechanical	116	6+
Cabinet with	312	234
Calculator or register combined	235	128
Check controlled apparatus		
Check in circuit	194	304+
Check operated switch	194	239+
Clock alarm	368	244+
Electrically operated	368	250+
Closure position	49	13
Code signalling	116	18+
Coin operated	194	
Consumable electrode	314	9
Conveyer power driven	198	502.1+
Cooking apparatus	99	342+
Design, instruments	D10	104+
Detonating (see detonators, signal and indicators)		

	Class	Subclass
Dispenser combined	222	23+
Distance measurement combined	33	740+
Cutting combined	83	522.11+
Tearing combined	225	18
Drawbridge	14	49
Dropping, ordnance	89	1.51+
Earth boring combined	175	40+
Electric	340	
Alarms	340	500+
Arc lamp	314	9
Extinguisher combined	340	289
Float gauge combined	73	308
Lamp system	315	129+
Noise or unwanted signal reduction	367	901*
Radio analog	455	
Radio directive	343	
Radio pulse or digital	375	
Telegraphy	178	
Telephony	379	
Vehicle mounted	340	425.5+
Elevator		
Electric	187	100+
Electric annunciators	187	121+
Mechanical	116	64
Mechanical alarm	116	68
Expansible chamber motor combined	91	1
Fire alarm		
Extinguisher combined	340	289
Flame	340	577+
Smoke	340	628+
Systems	340	287+
Thermal control	340	584+
Fire escape	182	18
Fire extinguisher combined	169	23
Fishing	43	16
Automatic hooker	43	17
Flags & flagstaffs	116	173+
Flexible panel combined	160	10
Float gauge combined	73	307+
Foghorn	116	137R
Electric actuation	340	384R+
Folder carried indicia	40	359
Foot operated linkage	74	514
Gas separator combined	55	274+
Illuminated by locomotive headlight	362	61+
Index card	283	36+
Lamp for vehicle	D26	28
Lamp or reflector	D10	111+
Lantern	362	166+
Letter box	232	34+
Life in undertaking	27	31
Liquid purification	210	85+
Lock condition	70	441
Masts and flagstaffs	116	173+
Mechanical	116	
Mirror combined	350	97
Nautical	116	26+
Ordnance	89	1.51+
Periodic	116	22R+
Pneumatic dispatch	406	34+
Press combined	100	99
Pyrotechnic	102	335+
Cartridge	102	346
Design	D22	112+
Parachute flares	102	337+
Railway torpedos	246	487
Skyrocket	102	347+
Torches	102	336+
Railway	246	
Audible and other signal	246	217
Audible for switch	246	217
Block signal systems	246	20+
Bridge warning	246	486
Cab signal	246	167R+
Crossing signal on train	246	208
Dispatching combined	246	2R+
Drawbridge protection	246	118+
Electric actuation	246	218+
Fluid motor actuation	246	257+
Grade crossing	246	111+
Highway crossing	246	292+
Highway crossing electric	246	125+
Interlocking	246	131
Mail delivery	258	2
Mechanical motor actuation	246	262+
Passenger at station	246	209
Pilot car	246	166
Roadway defect	246	120+
Signal car	246	166
Torpedo	246	487

	Class	Subclass
Torpedo placing	246	210+
Traction cable	104	179
Train dispatching	246	2R
Train position	246	122R+
Tunnel warning	246	486
Vehicle energy actuation	246	270R+
Whistle	116	137R+
Recorder	346	
Recorder combined	346	17+
Reeling tension device	242	128
Rudder position	116	303
Safes combined	109	38+
Security instrument	D10	104+
Sheet carried indicia	40	360
Ships telegraph	116	21
Speaking tubes	181	19
Street traffic	116	63R+
Support structure	248	347
Telephony		
Central system	379	242+
Divided central	379	310+
Thermostatic regulator combined	236	94
Time	368	243+
Tire deflation	116	34R
Electric	340	442
Torpedo for railway	246	210+
Placing mechanism	246	487
Track sander combined	291	17
Traffic	116	63R
Electric	340	907+
Typewriter ribbon	400	249+
Underwater sound waves	367	131+
Vehicle	116	28R+
Watches alarm	368	244
Water heater	126	388
Water heater vessel combined	126	388
Signaling Teaching	434	223
Signature Gatherers and Conveyors		
Conveyor power driven	198	644
Machine for writing several	33	23.6
Sheet associating	270	54+
Signing machine	D18	21
Silastic T M (See Synthetic Resin or Natural Rubber)		
Silazanes	568	410+
Silencer (See Insulating, Sound; Muffle; Noise Deadening)		
Firearm and gun	89	14.4
Muffler for firearm	181	223
Silent Butler	D32	74
Silica	423	335
Abrasive compositions containing	51	308
Absorbents containing	502	407+
Aerogel	106	287.34
Catalysts containing	502	232+
Cement containing	106	737
Gel	423	338
Dispersing or stabilizing agent combined	252	315.1
Glass compositions containing	501	53+
Glass manufacturing	65	
Lime compositions containing	106	796+
Lubricants containing	252	9+
Silicane	556	465+
Silicates & Silicon Compounds,		
Inorganic	423	324+
Abrasive compositions containing	51	308
Absorbents containing	502	407+
Catalysts containing	502	232+
Coating or plastic compositions containing		
Alkali metal	106	286.7+
Carbide	501	87+
Glass	501	11+
Lime combined	106	792+
Opacifiers for enamels	106	312
Pigments fillers aggregates	106	400+
Portland cement type	106	713+
Portland cement type making	106	739+
Refractories	501	94+
Slag	106	789+
Detergents containing	252	89.1+
Dyeing	8	523+
Fertilizers containing	71	
Hydrofluosilicic acid	423	341
Lubricants containing	252	9+
Organic compounds	556	400+
Esters	556	482+
Silica	423	335
Silicofluorides	423	324
Silicic Acid	423	325

	Class	Subclass
Silicofluoric Acid	423	341
Silicofluorides	423	324+
Silicol	556	463
Silicon	423	348
Alloys		
Aluminum	420	548
Aluminum copper	420	537
Copper	420	490
Metallic	420	578
Carbide	423	345
Ferro producing	420	578
Misc. chemical manufacture	156	DIG. 64
Organic compounds in compositions	106	287.1+
Polymers (See synthetic resin or natural rubber)		
Room-temperature curable	528	901*
Silicon Steel	148	110+
Silicone	556	465+
Compositions	106	287.13
Sililation of Polymers	525	100+
Silk		
Artificial making		
Apparatus	425	66
Apparatus	425	67+
Apparatus	425	76
Apparatus	425	382.2
Processes	264	165+
Chemical treatment	8	128.1+
Degumming	8	138
Fermentative treatment	435	268
Liberating	19	3
Screen printing	101	114+
Process	101	129
Thread finishing	28	217+
Silkworm Culture	119	6
Sill	52	204
Railway car	105	396+
Weatherstrips	49	303+
Sillimanite		
Ceramic containing	501	141+
Porcelain containing	501	141+
Refractory containing	501	128+
Silos		
Circular	52	245+
Compacting ensilage	100	65+
Portable, & means for erecting	414	919*
Prestressed	52	224
With port	52	192+
Siloxanes	556	450+
Resinous solids (See also synthetic resin or natural rubber)	528	10+
Silver		
Alloys	420	501
Chest or box for	D 3	75
Composite metal stock	428	673
Electrodeposition	204	46.1
Electrolytic synthesis	204	109+
Mercury alloy	204	126+
Hydrometallurgy	75	711+
Misc. chemical manufacture	156	DIG. 101
Organic compounds	556	110+
Proteins	530	400
Pyrometallurgy	75	634+
Silverware		
Chest or box for	D 3	75
Sorter	209	926*
Tray or container	206	557
Simulations		
Article dispensing	221	24
Cabinets with	312	204
Display	40	538+
Material dispensing combined	222	80+
Pyrotechnics	102	335
Supports digest	248	908*
Toys	446	
Velocipedes	280	827+
Sine Bar	33	536+
Sinew Removing	17	11.3
Singeing		
Cloth finishing	26	3+
Electric burner	219	223+
Hair combs for	132	118
Hogs	17	20
Thread finishing	28	239
Warps	28	174
Single Grid Twin Plate Control Tube	313	303
Single Roll Crusher	241	221+
Singletree		
Horse detacher		
From thill or evener	278	32
Traces from singletree	278	24+

	Class	Subclass
Sink (See Receptacles)	4	
Cabinet combined	312	228
Design	D23	284+
Disappearing bowl	4	DIG. 2
Drain board or tray	D32	56
Lining mat	D32	57
Strainers and stoppers	4	286+
Structure	4	619+
Traps	4	191+
Tub combined	4	553+
Sinker		
Fishing type	43	43.1+
Design	D22	145
Knitting machine	66	104+
Traps	43	68
Sinkhead	249	106
Sintering		
Apparatus, ore	266	176
Stationary grate	266	185
Cellular synthetic resin or natural rubber product	521	919*
Ceramic composition	501	1
Clay	264	56+
Glass granules	65	18.1
Apparatus	65	144
Ores	75	746+
Portland type cement making	106	739+
Powder metallurgy		
Processes	419	
Liquid phase	419	47
Products	75	228+
Processes	432	1+
Lime gypsum cement	106	765+
Sinusoidal Generator		
Electronic tube type oscillator	331	
Generator structure	310	
Generator system	322	
Siphon		
Acetylene generator	48	17
Boiler	122	68
Bottle	215	4+
Design	D 7	300.1
Filling	141	14+
Bowl	4	421+
Code recorder	178	91
Dispensers	222	416
Discharge assistant combined	222	204
Fluid flow meter	73	222
Fluid handling	137	123+
Automatic	137	132+
Lubricator	184	85
Plumbing	137	247.11+
Syringes	604	131+
Water closet		
Siphon bowl	4	421+
Tank	4	368+
Siren	116	147
Design	D10	120
Electric	340	405
Toy	446	205
Sirup (See Syrups)		
Sizing (See Assorter)		
Bread, pastry and confections	99	352+
Chain making	59	29
Coating compositions		
Natural resin containing	106	238
Protein containing	106	124+
Starch containing	106	210+
Coating processes	427	
Dyeing processes combined	8	495+
Finger ring forming and	29	8
Hat making	223	10
Paper tube making	493	269
Textiles chemical modification of	8	115.6
Warp threads	28	178
Skate and Skaters Appliances	280	841+
Clamp for sharpening by abrading	51	228
Design	D21	224+
Ice, powered	180	180+
Roller, powered	180	180+
Sharpening by cutting	76	83
Skaters supports	272	70+
Snow	280	600
Design	D21	224+
Strap	280	11.3
Water	441	65+
Wheels for	301	5.3+
Skeeball Game	273	352
Skein		
Axles	301	134+
Fluid treating package	8	155.2

	Class	Subclass
Reeling	441	26+
Holders	242	127
Skeining	28	291
Skeleton		
Figure toys	446	373+
Keys	70	394
Plug valve	251	310+
Towers	52	648+
Skelping Metal	72	176+
Skewer Pins	17	1S
Ski (See Aquaplane; Sled)	280	601+
Apparel design	D 2	31
Bindings	280	611+
Clamp type having plural diverse axes clampsps	280	DIG. 12
Controlled by skier postion or by remote control	280	DIG. 13
Brake	280	604+
Carriers	224	917*
Carriers, hand	294	147
Case	206	315.1
Clamp	280	814
Design	D21	229+
Handle or grip design	D 8	DIG. 6
Poles and sticks	280	819+
Presses	280	815
Simulators	280	842
Snow, ie, non-water, powered	180	180+
Shoe fasteners	280	611+
Trail groomer	37	219+
Teaching skiing	434	253
Water skis	441	68+
Tow line	114	253+
Work holder for	269	906*
Skid	193	41
Aircraft landing gear	244	108
Anti		
Armored resilient tire	152	167+
Boots and shoes	36	59R+
Devices for resilient tires	152	208+
Chain applier	81	15.8
Inverted vehicle	280	32
Platform	108	51.1+
Railway		
Car stop	104	260
Truck	105	216
Skidding		
Traversing hoist	212	167
Skidway	193	38+
Hoisting rope combined	414	571
Railway	104	134+
Vehicle loading conveyor	414	537+
Skillet	126	373+
Skimmer		
Concentrating evaporator	159	42
Fermentation apparatus combined	435	287+
Floating separator	210	242.1
Froth flotation	210	221.1
Gravitational separator	210	513+
Stratifiers	209	493
Skimobile	180	182
Skin Diving (See Diving Apparatus)		
Skinner		
Animal	17	21
Fish	17	62
Nut	99	584+
Process	17	50
Skins and Leather	69	
Carroting of fur	8	112
Cutting for grafting	606	132
Dyeing	8	436+
Electrolytic treatment processes	204	135
Fluid or chemical treatment	8	94.1 R+
Manipulative	8	150.5
Forming integral skin, plastic and non-metallic	264	45.5+
Skip		
Chute for rotary drum	193	10
Furnace charging	414	178
Accumulating means before release	414	168
Loading or unloading	414	639+
Mortar mixer feeding	366	39
Liquid and	366	36
Vertically swinging load handler	414	680+
Skipping		
Exercising devices	272	74+
Figure toys	446	311+
Rope skipping	446	307
Skirt		
Christmas tree base	D11	130

	Class	Subclass
Rotatable rod in	403	165
Whip	280	170+
Wrenches	81	121.1+
Socks	D 2	329+
Sod Cutters	172	19+
Soda and Potash	423	
Carbon compound	423	421
Oxygen compound	423	641
Recovery	423	179
Soda Fountain		
Cabinet structure, ie syrup compartments	312	
With dispensing	222	173+
Soda water apparatus (See carbonation; dispensing; foam)		
Design	D 7	300.1+
Soderberg Electrode Arc Furnace	373	60
Electrolytic apparatus	204	225+
Sodium (See Alkali Metal)		
Amalgam	420	526
Electrolytic synthesis	204	125
Chloride dissolver	422	902*
Organic compounds	260	665R
Preparation	75	745
Pyrometallurgy	75	589+
Valve cooled by	123	41.16
Making	29	888.45
Vapor lamps	313	567+
Sofa		
Bed	5	12.1+
Construction elements	5	52
Chairs and seats	297	232+
Design	D 6	334
Swing type	297	282
Exercise	272	144+
Table bed type	5	7
Soft Drinks	426	590+
Carbonated	426	590+
Softening		
Leather	69	
Protection		
By data encryption	380	4
Water	210	
Ion exchange process	210	660+
Soil		
Gas content testing	73	19.9
Handlers or conveyors	172	33
Mixer, machinery design	D15	19
Stabilizing compositions	106	900*
Treatment machinery, design	D15	21
Sol (See Colloids; Dispersing; Suspensoids)	252	302
Radioactive	252	634+
Solanum Alkaloids	546	124
Solar		
Bleaching	8	103
Celestial observation instruments	33	268
Cells	136	243
Systems	323	906*
Concentrating evaporators	159	903+
Cooker or warmer	D 7	324
Crystallization process	23	295S
Distillation of water	203	10+
Driers	34	93
Food treatment	99	
Heat accumulator structures	126	400
Heat collectors	126	417+
Heat exchange		
Automatic control of heat	165	18
Heating and cooling	165	48.1
Heating systems	237	1R
Kinestherapy	D24	39
Living body treatment	128	372+
Panel	136	243
Photovoltaic or photoelectric generator, ie solar cell	136	243
Power plants	60	641.8+
Energizing & storage	60	659
Reflectors, curved surface	350	629+
Refrigeration utilizing	62	235.1
Spacecraft altitude control by solar pressure	244	168
Sterilizing	422	22+
Apparatus	250	428+
Stills	202	234
Distillation digest	203	DIG. 1
Stoves and furnaces	126	417+
Thermoelectric generator	136	206
Solder & Soldering (See Alloy)	228	56.3
Apparatus	228	
Design	D 8	30

	Class	Subclass
Blow torches	431	344
Can body forming and side seam	228	17.5+
Can head or body seam	228	
Electric	219	129
Fuse self soldering	337	152
Heat	219	129
Metal heating	219	85.1
Soldering iron	219	227+
With pressure	219	85.1
Iron	228	51+
Design	D10	46+
Electrically heated	219	227+
Heater portable furnace	126	236+
Self heating with burner	126	413+
Permanent closure	220	600+
Stick	428	558
Temporary closure	220	200
Solderless		
Terminals	439	
Flexible wire joint	403	206+
Wire joining machines	140	113+
Sole		
Forming and attaching		
Attaching means	36	12+
Machines	12	17R+
Sewing machines	112	29+
Straighteners	12	120.5
Trimmers	12	85+
Pad		
Cleaning devices	15	227
Cushioned horseshoe	168	14
Foot supporters and	36	71
Horseshoe	168	26+
Horseshoe, design	D30	149+
Shoe	36	25R+
Antislipping devices	36	59R+
Design	D 2	317
Attaching means	36	12+
Design	D 2	319+
Design, insoles	D 2	318
Design, protectors	D 2	271
Filling compositions	106	38
Insoles	36	43+
Interfitted heel and	36	24.5
Light etc application to body	128	382+
Outsoles	36	25R+
Protectors	36	73+
Solenoid (See Magnet)		
Solid (See Liquid)		
Detectors	422	68.1+
Solid State Circuits and Devices (See Transistor)		
Solidifying (See Crystallization)		
Borehole inclinometer		
Earth control by	405	263+
Gas	62	8
Liquid by cooling	62	340+
Automatic	62	135+
Process	62	66+
Liquid comminuting combined		
Apparatus	425	222+
Processes	264	5+
Liquid fuel	44	265+
Sugar processes	127	58+
Solids Dissolving for Cooling	62	4
Solids Separating and Stratifying	209	
Solvent (See Extracting)		
Coating or plastic composition for	106	311
Compositions	252	364
Detergent having	252	89.1+
Extraction to recover hydrocarbon	585	833+
Gas package or receptacle having	206	.7
Polymer dissolver	422	901*
Recovery by drying or distillation		
Concentrating evaporators	159	
Distillation	203	
Distillation apparatus	202	
Drying combined	34	72+
Mineral oils	208	179+
Refrigerant combined	252	69
Sodium chloride & potassium chloride dissolver	422	902*
Treatment of coating	427	335
Well		
Cleaning with	166	304
Solid material recovering	299	5
Sonar and Sonics		
Flow	261	DIG. 78
Gas separation	55	277
Process	55	15
Portable sonar devices	367	910*

	Class	Subclass
Simulators, sonar	434	6+
Sonar design	D10	46+
Sonar time varied gain control systems	367	900*
Sonic tests		
Compressional wave in fluid	367	87+
Gas analysis	73	24.1
Gas content	73	24.1
Seismic	367	14+
Vibration	73	570+
Vibrators	261	DIG. 48
Wave bonding, nonmetallic		
Apparatus	156	580.1+
Process	156	73.1+
Wave measurers	33	1P
Sonic		
Used in reaction to prepare inorganic or organic material	204	157.15+
Used in reaction to prepare or treat synthetic resin or natural rubber	522	1+
Soot		
Blower		
Boiler installed	122	390+
Heater installed	165	95
Catcher	126	280
Sorbent (See Absorbents)		
Air contacting with refrigeration	62	271
Process	62	94
Gas adsorption thermometer	374	159
Gas separation	55	387+
Processes	55	74+
Liquefied gas		
Producing	62	17+
Storing	62	46.1+
Refrigeration utilizing	62	476+
Automatic	62	141+
Process	62	101+
Sorbic Acid	562	601
Sorbitol	568	852+
Sorel Cement	106	685+
Sorghum	426	658+
Sorption to Recover Hydrocarbon	585	820+
Sorters and Sorting		
Automatic machines	209	509+
Belts sifters	209	307+
Coin	453	3+
Data processing equipment	D14	110
Electrostatic	209	127.1+
Fluid suspension	209	132
For paper sheets	271	287
Grates or sifters	209	233+
Machines for statistical records		
Coded edges	209	608
Coded holes	209	613
Magnetic	209	212+
Racks	211	10+
Refuse treatment means	110	220
Sheet feeding or delivery	271	
By items following different trajectories	209	638
Thrower	209	642
Sorting & classifying solid items	209	
Bottles	209	522+
By sensing radiant energy	209	576+
Bottles, jars, etc	209	524+
Cigarettes, etc	209	536
Laser	209	579
X-ray	209	589
By size	209	659+
By susceptibility to deform		
Condition responsive means	209	599
Separating means	209	699
Cigarettes	209	535+
Eggs	209	510+
Electrostatic	209	127.1+
Fluid suspension	209	132+
Lumber	209	517+
Magnetic	209	212+
Mail	209	900*
Paper money	209	534
Preparing items for sorting	209	3.1+
Sifting	209	233+
Special items	209	509+
Stratifiers	209	422+
Stratifiers	209	422
Weight operated	209	645
Condition responsive controls	209	592+
Sound (See Insulating; Recording; Reproducers; Acoustics)		
Bell	116	148+
Electric	340	392+

	Class	Subclass
Railway spring planks	105	208
Rocking running gear	280	112.1
Roller skate trucks	280	11.28
Runner type	280	25
Running gear	280	688+
Skates with foot supports	280	11.14
Spring wheels	152	1+
Steam traction engine mounting	180	40
Stub axle mounts	280	660+
Swinging and rocking axle	280	110
Swinging axle	280	113+
Swinging axle connections	280	125+
Thill supports	278	81+
Tires spring and pneumatic	152	156
Traction increasing lugs for wheels	301	51
Trailer draft device	280	483+
Two wheel vehicle	280	65+
Velocipede seat	297	195+
Wheeled vehicle	33	418
Watch etc hairsprings	368	175
Sprinkler	D23	214+
Sprinkling	239	
Carts	239	146+
Embedded	239	201+
Flow regulator or modifier	239	533.1+
Head fire extinguishing	169	37+
Rack combined	211	127
Showcase combined	312	115
Textile treating	68	205R+
Sprocket	474	152+
Chain	474	202+
Making	29	893
Chain	59	5+
Railway cable	104	236
Chain cable	104	237
Spud Anchor	37	73+
Spur		
Attached shoe protectors	36	74
Gearing (See gearing spur)		
Horseshoe spur fastening	168	21
Riding	54	83R
Design	D30	157
Stirrups combined with	54	49.5
Tag fasteners	40	669+
Tube		
Boiler	122	
Heat exchangers	165	42
Mineral oil distillation	196	129
Wheel traction increasing	301	43+
Sput	285	201+
Sputtering (See Cathode Sputtering)		
Sputum Cup	4	259
Spyglasses	350	537+
Square	33	474+
Combined		
Fixed straightedge	33	429
Hypotenuse	33	420
Nonpivoted straightedge sliding	33	427
Pivoted straightedge sliding	33	425
Leveling feature	33	451
Squeegee (See Wiper and Wiping)		
Cleaning device	15	245
Brush combined	15	117
Design	D32	41
Material supply combined	401	261+
Wiper combined	15	121
Squeezer		
Animal traps	43	85+
Butter worker	99	452+
Citrus fruit	99	495+
Cotton gin	19	49
Expressing	100	104+
Paste tube	222	95+
Raisin seeders	99	547+
Textile fluid treating apparatus	68	94+
Squeezing		
Liquids from solids	100	35+
Apparatus for, with drain	100	104+
Squirt		
Locomotive cab	239	174
Preventer		
Type casting integral line	199	53
Type casting separate type	199	90
Stabilizing (See Preserving)		
Aircraft		
Propellers	244	92
Weights	244	93+
Colloids	252	351+
Gyroscope	74	5.22
Mineral oil	208	351

	Class	Subclass
Missiles, manned	244	75R+
Missiles, unmanned	244	3.1+
Projectiles	244	3.1+
Ships	114	121+
Space ships	244	158R+
Synthetic resin or natural rubber (See class 523, 524)		
Latex	528	935*
Stable		
Animal	119	15+
Stalls	119	27+
Stalls, design	D30	108+
Stanchions	119	147.1
Stack and Stacker (See Packing)		
Beds stacked	5	8
Bottle and jar	215	10
Calender rolls	100	162R
Dispensing	221	
Cabinet	312	42+
Rack	211	59.2+
Electrical insulator	174	137R+
Forming apparatus	414	788.1+
Coal storage type	414	133
Furnace and flue		
Draft regulator	110	159+
Feed water heater	122	439+
Marine	114	187
Masonry or concrete	52	245
Smoke stacks	110	184
Spark arrester	110	121
Hay stack shaping	414	132
Horizontally attached receptacle set	220	23.2+
Nuclear fuel components	376	433
Pneumatic	406	
Receptacles	206	501+
Special package, stacked item	206	821*
Rotary rolling contact printer	101	232+
Sheet		
Cutting combined	83	86+
Delivering	271	278+
Electrophotographic copier	355	322
Feeding	271	8.1+
Material folder	83	83+
Oscillating blade	271	264+
Stacked article support	211	49.1+
Adjusts for different sizes	414	900*
Card or sheet	211	50+
With impaling means, follower	211	54.1
With impaling means, terraced	211	57.1
With article counter	414	901*
With impaling means	211	59.1
With load-supporting fluid cushion	414	903*
With means to apply adhesive to articles	414	904*
With operational sequence control for stack pattern forming	414	902*
Stacking for dispensing	221	175+
Stacking Apparatus	414	788.1+
Stadium (See Theatre)	52	6
Construction	D25	12
Staff		
Flag	116	173+
Horological	368	324
Planting	111	97
Railway signal system	246	19
Support	248	511+
Tobacco ash receiver support	131	259
Stage (See Theater)	52	6+
Appliances	272	21+
Scenery shifting	272	22+
Illusions	272	9+
Lighting	315	312+
Motion picture set	352	88+
Set design	D25	58+
Stagger Tuned Amplifiers	330	154
Stained		
Glass mosaics	428	38+
Stainless Steel	420	8+
Austenitic	420	43
Heat treated stock	148	325+
Heat treating	148	135+
Pyrometallurgy	75	507+
Stains Paint	106	34
Dyeing only	8	402
Stair (See Ladder)	52	182+
Amusement	272	2
Design	D25	62+
Escalators	198	326+
Lighting	362	146
Mobile for boarding aircraft	D34	30
Pads and covers		

	Class	Subclass
Conforming to tread nose	52	179
Rug fasteners	16	10+
Slatted	52	180
Scribers for stair curves	33	29
Static mold for making	249	14
Tread	D25	69
Stake		
Animal hitching	119	121
Posts, design	D30	154
Guided cultivator	172	24
Land vehicle body	296	43
Railway freight car	105	380+
Tent	135	118
Staking		
Hide	69	34
Watches and clocks	29	231+
Stalk and Stalk Handling		
Choppers	56	500+
Fiber liberating	19	10
Harvester		
Cornstalk type	56	51+
Woven	139	424
Stalls		
Animal	119	27+
Thermolytic distillation	202	97+
Stamp (See Branding)		
Affixers	156	
Alteration preventing or detecting	283	71
Brand burning	219	228
Brand burning	101	27
Design	D19	9+
Dispensers	83	242+
Coin controlled	194	
Hand stamp collocator	33	614+
Handles	101	405+
Holder, household type	D19	67
Imprinting or embossing stamp	D18	15+
Key, workmans time recorder	346	53
Mail cancelling	101	371
Mill	241	270+
Pad	118	264+
Plural parallel blade cutlery	30	305
Pocket receptacle	206	39+
With tearing edge	225	39+
Postage roll dispenser	D19	67+
Printed labels	283	81
Purse with holder	150	106
Racks for hand stamps	211	39
Registers	235	101
Rubber	101	368+
Endless band	101	111
Scarifiers	81	9.21
Time	346	80+
Time with automatic recording mechanism	346	80+
Vending machine for	D20	2
Stamping and Stamped Devices		
Bed bottom risers	5	206
Cigar	131	106
Cigarette package	131	283
Envelope		
Printing and stamping	493	320+
Sealing and stamping	156	
Hand	D18	14+
Metal working	29	DIG. 37
Mill	241	270+
Picture frames	40	154
Plug tobacco	131	368
Sheet metal chain	59	15
Stanchions	119	147.1
Sure stop	119	904*
Stand (See Holder; Support)	248	127
Apparel apparatus	223	120
Bag holder	248	97+
Beehive	449	26
Bicycle		
Kick stands	280	293+
Locked	70	235
Racks	211	17+
Book and music	248	441.1+
Bracket		
Alternative	248	126
Combined	248	121+
Cabinet structures	312	
Calendar	40	120
Christmas tree	248	519+
Copyholder	248	441.1+
Dental equipment	433	25+
Dish	248	128+
Flatiron	248	117.2+
Accessory combined	38	142

	Class	Subclass
Brakes	188	
Cable drum control	254	267+
Drill feed	408	5
Drying automatic control	34	55
Elevator	187	28+
Fan regulator	416	170R
Hoist traversing cable	212	86+
Lathe carriage	82	132+
Metal working	72	1+
Motor vehicle safety	180	271+
Planer tool feed	409	326+
Power driven conveyors	198	832+
Power stop control	192	116.5+
Pump regulator	417	1+
Punching machine shaft drive	83	58+
Punching machine shaft drive	83	69
Punching machine shaft drive	83	203+
Register control	235	132R
Rotary crane	212	149+
Telegraph code recorder	178	95
Tool turret carriage	29	65
Type casting	199	52+
Type distributing	199	43
Typewriter carriage	400	298+
Typewriter case shifts	400	277+
Typewriter line spacing	400	547+
Typewriter tabulator	400	294+
Windmill	416	14
Wire feeding for working	72	461
Stop Watches	368	101
Stopcock Boxes	220	3.2+
Stopper		
Bath and sink	4	286
Bottle and jar	215	355+
Breakable neck	215	33+
Design	D 9	439
Closure for metal receptacle	220	307
Flexible bag	383	96
Flue	126	319
Ingot or pig molds	249	204
Ordnance	89	30
Ship or vessel		
Cable	114	199+
Hawse pipe	114	180
Leak	114	227+
Port	114	174+
With measuring devices	222	424.5+
Tiltable	222	454+
Stopwatch	D10	30
Storage		
Battery	429	
Charging and discharging	320	2+
Connections	439	754+
Electrolyte	429	189+
Half lead	429	
Lead	429	
Plates and grids	429	209+
Building & devices (See garage; silo)	52	
Ballast storage for aircraft	244	94
Bed with stowage	5	503
Bin with power driven conveyor	198	550.1
Cordage	242	47.5
Crib with stowage	5	96
Fire escapes	182	82+
Food storage receptacle	99	467+
Magazine for label pasting	156	
Material handling	414	589
Positioning material for building erection or repair	414	10+
Parachute storage	244	147+
Photographic plate holder	354	174+
Portable storage & means for erecting	414	919*
Ramp	52	175
Storage provision on passenger car	105	334+
Table leaf	108	86+
Table leaf terraced	108	93
Vehicle top	296	140+
Ventilation of storage structure	98	52+
X ray film holders	378	167
Cryogen	62	45.1+
Floating vessel	114	256
Hydrogen, alloys for	420	900*
Radio active waste material	252	633
Submerged vessel	114	257
Tank	D23	202
Underground fluid	405	53+
Store Service	186	
Trolley type ladders	182	36+
Using punch cards	186	56

	Class	Subclass
Storefront Design	D25	59
Stores	52	28+
Bay windows	52	201
Self service	52	33
Show window panel	52	777+
Show window panel	52	764+
Toy	446	476
Storm Fronts		
Railway passenger cars	105	353
Vehicle body	296	77.1+
Window	52	202
Screen combination	160	90+
Storm Windows	49	61+
Stove	126	
Absorber coating	126	901*
Air preheating	266	138+
Dampers	126	285R+
Automatic	236	45
Combined with furnace features	110	163
Design	D 7	339+
Handle or knob	D 8	300+
Heating		
Electric	392	418+
Illumination	362	92+
Implements	294	9+
Compound	7	109+
Polish	106	3+
Still combined	202	165
Ventilators	D23	371+
Working fluid compositions	126	900*
Stovepipe	126	307R+
Design	D23	394
Elbow bending	72	307
Hood combined	126	301+
Thimbles	126	314+
Water heaters	126	364+
Straight Jackets	128	869+
Straight Line Motions	74	103
Straightedge	33	403+
Design	D10	61+
Leveling	33	341+
Rod joint	403	
Tearing or breaking	225	91
Straightening Devices		
Decurling coated flap of envelope	493	395+
Grain harvester	56	473
Metal		
Automobile frame	72	705
Leveler	72	160+
Stretcher	72	302
Paper web	162	270+
Shoe sole	12	120.5
Tobacco treatment	131	324+
Stemming	131	315+
Trouser leg	2	233
Wire	140	147
Cutting combined	140	139+
Strain		
Connector electrical	439	296+
Gauge	33	DIG. 13
Relief		
Pipe joint	285	114+
Reliever electrical cord terminal	439	449+
Strainer (See Filter; Screen; Sifter)		
Baths closets etc	4	286+
Sinks	4	652
Concentrating evaporator	159	42
Dispenser combined	222	189
Dispensing type sprinkler etc	222	565
Eaves, troughs	52	11
Kitchen utensil	D 7	667
Liquid separation	210	348+
Detachable	210	473+
Faucet attached	210	449
Mold for metal having	164	358
Paper stuff	210	
Press	100	130
Separating yeasts ferments etc	435	287+
Separatory stills	202	178
Sewer inlet	210	163+
Sewerage trap	210	435+
Support	248	94
Textile fluid treating apparatus	68	
Vegetable etc comminutor type	241	
Well casing	166	227+
Strand (See Filament; Wire)		
Advancing material of indeterminate length	226	
Brakes	188	65.1+
Drying apparatus	34	148+
Drying processes		

	Class	Subclass
Diverse	34	18
Gas contact	34	23+
Heat	34	41
Vacuum or pressure	34	16
Feeder	226	
Finishing apparatus	28	217+
Fire escape	182	189+
Floor or wall attached	182	100
Joining (See joints)		
Electric conductor	156	49
Rope splices	57	202
Rope splicing devices	57	22+
Tying	289	
Joint packing	277	227+
Laminating	156	166
Making	28	217+
Braiding	87	
Casting	164	418+
Knitting	66	3+
Sliver assembling	19	157+
Spinning or twisting	57	
Waxed end	57	20
Package	206	389+
Wound	242	159+
Packaging	28	289+
Placing or cable hauling	254	134.3 R+
Structure		
Barbed wire	256	6+
Conductors	174	126.1+
Conductors insulated	174	110R+
Conductors sheathed	174	102R+
Fence wire	256	46
Textile	428	364+
Textile braided	87	8+
Textile twisted	57	200+
Tearing or breaking	225	
Tire bead grommets	245	1.5
Winding & reeling	242	
Strangler		
Bridle	54	15
Strap		
Animal shoe fastening	168	22
Antiskid for wheels	152	221+
Apparel supporting	2	338+
Spaced holders	2	323+
Auto folding top	296	130
Boot	36	56
Button loops	24	660
Closure holding		
Wooden boxes	217	66+
Clutch	192	80+
End attaching devices	24	265R+
Hand		
Casket lowering	27	33+
Package and article carriers	294	149+
Railway car	105	354
Harness		
Breast	54	58+
Breech	278	128
Hame	54	28
Hitching	54	34
Loops	24	182
Hat	132	58
Machine element	74	570+
Making		
Box straps	140	74
Strap finishing	69	17
Nuclear reactor moderator support	376	304
Safety	182	3+
Tighteners	24	68R+
Bale and package ties	24	19
Portable, detachable tensioners	254	199
Trouser	24	72
Weaving machine		
Harnessing connecting	139	88
Shuttle fly lug connecting	139	153
Wheel brake	188	77R
Wrist watch	224	164
End attaching devices	24	265R+
Stratifying (See Separator, Solids)	209	
Straw		
Drinking	239	33
Drinking device	D 7	300.2
Holder	D 7	635
Mattresses	5	448+
Stacks forming	414	132
Streamlined		
Aircraft	244	117R+
Railway trains	105	1.4
Street (See Road)	404	
Cleaners and sweepers		

	Class	Subclass
Brushing	15	78+
Machinery	D32	15+
With air blast or suction	15	300.1+
Flushers	239	146+
Lantern	362	257+
Oilers	239	130
Pavements	404	17+
Signal		
Marker	404	9+
Mirrors and reflectors	404	14
Traffic	116	63R
Signs	40	612
Street light	D26	67+
Design	D20	17
Underground electric conductors	174	39
Streetcar	D12	36+
Appliance	D13	128
Strength Measuring		
By acoustic emission	73	801
By impact or shock	73	12+
By testing model of structure	73	804
Hardness	73	78+
Muscular	73	379
Coin controlled	194	
Optically sensed	73	800
Periodic	73	808+
Alternating tension & compression	73	797
Stress strain	73	788+
To cracking of specimen	73	799
Streptomycin	536	14
Stress Distributing (See Equalizing)		
Conduit electric	174	73.1
Insulator electric	174	140R+
Testing	73	760+
X ray tube	378	139
Stretchers & Stretching (See Definition Notes; Tension; Tighteners)	254	199+
Boot and shoemaking		
Loose upper shaping	12	54.1+
With diverse operations	12	52.5
Coating process	427	171+
Coating process combined	427	401
Design	D12	128+
Fabric		
Apparel	223	52+
Carpet	254	200+
Carpet	294	8.6
Carpet fastener combined	16	5
Fastener applier combined	227	12
Frame, eg curtain stretcher	38	102.1+
Framed panel	160	378
Hat shaping	223	15
Hung or draped	160	328+
Ironing table combined	38	108
Smoothing	38	102
Smoothing implements combined	38	70
Smoothing machine combined	38	12+
Wire	140	108+
Working combined	26	51+
Fence incorporated	256	37+
Frames		
Tack puller for	254	18
Textile sheets	38	102+
Horseshoe	168	9
Invalid lift		
Bottoms and slings	5	89
Field type	5	82R+
Wheeled	296	20
Leather		
Belt	69	1.5
Frame	69	19.1+
Lasting machine	12	7+
Lasting tool	12	108+
Shoe spot stretchers	12	115.2+
Translating tool	69	46
Work holder	69	19
Metal curving or straightening	72	302
Metal slitting and	29	6.1
Necktie	223	65
Pelt	69	19.1+
Pushing and pulling implements or apparatus	254	199
Resilient article assembling	29	235
Saw making	76	26+
Screws for ropes and cables	28	240+
Twisting combined	57	310
Ships sail	114	109+
Shoe former	12	128V+
Skein washing combined	68	159
Spinning and twisting combined	57	310
Strap	24	68D

	Class	Subclass
Stretch forming	72	295+
Textile web running lengths	26	71+
Biaxial	26	72+
Heater-dryer and spreader	26	92
Heater-dryer combined	26	106
Roller spreader	26	99+
Spreader	26	87+
Tubular fabric spreader	26	80+
Web condition responsive control	26	74+
Thread finishing	28	240+
With liquid treating	28	246
With stopping	28	242
Umbrella tent	135	20R
Wire (See notes to)	140	123.5
Striated Effects Surface Type	428	30
Striker and Striking		
Ammunition fuse etc	102	272+
Animal trap	43	77+
Clocks	368	243+
Electric	368	250+
Comminuting		
Hammer mills	241	86+
Hammer mills	241	185R+
Hammer mills plural parallel zones	241	138
Hammer mills plural series zones	241	154
Impact processes	241	27
Diaphragm horn with rotary	116	143+
Fixed bell with pivoted	116	167
Gong with pivoted	116	155+
Lock	292	340+
Mechanism for music players	84	323+
Piano	84	236+
Movable figure toys	446	336
Music instruments		
Automatic outside players	84	111
Piano auxiliary	84	224
Piano hammer actions	84	236+
Pneumatic pianos	84	54
Rigid vibrators with	84	404
Stringed with	84	12
Punching bags	272	76+
Railway draft block	213	58+
Surfaces for matches	44	643
Compositions	149	
Typewriter		
Index wheel	400	165.1+
Key wheel	400	174+
String (See Cord; Strand)	57	
Catgut	84	297S
Covering with copper or silver wire	57	3
Draw	24	712
Gut	8	94.11
Insulators	174	150
Stress distributing combined	174	141R
Musical instrument	84	297S+
Covering with wire	57	3
Design	D17	14+
Piano	84	199
Tuning devices	84	455
Nail manufacture	10	45
Nailers	227	95+
Plate piano	84	188
Frame and sounding board combined	84	184
Frame combined	84	185
Sounding board combined	84	187
Shoe	24	715.4+
Tag	40	662+
Band type	40	665
Tear	206	616+
Stringed Instruments	84	
Automatic player pianos	84	7+
Electric pickup	84	723+
Stringer (See Girder)		
Railway	238	24+
Tie ends	238	42
Seed corn	139	19
Stringing		
Assembling by (See notes to)	29	241
Beads	29	700+
Apparel trimming	223	48
Sewing machine for	112	121.19
Tennis racquet	273	73R
Tobacco	414	26
Striping		
Brush	15	166
With material supply	401	193
Coating		
Implement	401	48

	Class	Subclass
Machine	118	
Processes	427	286+
Fabric weaving	139	417
Stripper and Stripping		
Berry	56	330
Bobbin	28	292+
Coating	118	100+
Coil	242	81
Color stripping	8	102
Cotton	56	33+
Cutlery combined	30	128+
Electrolytic	204	146
Apparatus combined	204	193+
Expressing press cake	100	218
Cloth removing	100	298
Fiber and thread		
Carding	19	108+
Guides	28	222+
File blank	29	77
Filling tube	141	265
Harvesting	56	126+
Insulation from wire	81	9.4+
Molds and castings	164	
Ingot strippers	164	405+
Nail strippers	193	38
Of electrophotographic copy sheets	355	315
Of electrophotographic copy sheets	430	256+
Photographic paper and films	430	256+
Prick punch	30	368
Punch press	83	111
Tube or sheath splitter	30	90.1+
Vine and seed	460	123
Strips (See Tape)		
Adhesively affixing	156	
With cutting	156	510+
Apparel	2	338+
Breaker strip	152	526
Building joint covering	52	459+
Buttonhole	2	266
Carpet fastening	16	16
Continuous binding	281	23
Dispensing	221	33+
Continuous strip	312	39
Towel	312	37
Feeding and delivering	226	
Form pockets	282	11.5 R
Guide sewing machine	112	152
Handling (See feeder; sheet)		
Manifolding	282	12R+
Metallic	428	544+
Compound	428	615+
Strips of repeated articles, cut up later	29	DIG. 4
Paper	281	5+
Leaves books combined	282	2+
Leaves combined	281	2
Printed	283	62
Pasters	156	
Perforated telegraph tape	178	112
Recorder record receiver	346	136
Recording paper	346	136
Servers (See notes & definitions)	226	
Spiral cutting	83	680
Supports	211	45+
Stacked	211	50+
Switchboard	361	331
Telephone	379	325
Tack	411	439+
Tape appliers shoe making	12	59.5
Tearing or breaking	225	
Container opening	206	605+
Telegraph tape	178	111+
Towel supports	211	16
Locked	211	6
Vehicle body wear	296	41
Weather	49	475+
Windshield	296	93
Winding and reeling	242	
Woodcutters for strips		
Bench planes	30	478+
Slicers	144	164+
Veneer lathe	144	215
Stroboscopes		
Electric		
Frequency	324	78R+
General	324	
Phase	324	83R+
Electric lamp systems	315	
Condenser discharge	315	227R+
Optical	356	23+
Stroboscopic Light Synchonizer	356	23+

	Class	Subclass
Sulfenate Esters	558	62
Sulfides		
Binary inorganic	423	511
Inorganic	423	511+
Containing	424	706+
Medicine or poison containing	424	706+
Organic	568	38+
Amine	564	100+
Containing	514	706
Sulfinamides	564	101
Sulfinate Esters	558	61
Sulfinic Acid	562	125
Sulfinylamines	560	317
Sulfite		
Inorganic	423	519
Liquor		
Carboxylic acids from	562	513
Coating or plastic compositions containing	106	123.1
Compositions	162	83+
Fermentation by yeasts	435	251
Lignins from	530	500+
Paper stock treatment	162	83+
Synthetic resins containing	527	400+
Tanning agents containing	8	94.31
Organic	260	
Reducing composition	252	188.21
Sulfite Esters	558	59
Sulfo Acids	562	30+
Sulfoacetic Acid	562	109
Sulfohydroxamate Esters	560	303
Sulfonamides	564	80+
Acyclic	564	95+
Aromatic	564	84+
Hydrazine containing	564	81
Plural	564	82+
Sulfa type	564	86
Sulfonate Esters	558	44+
Sulfonation of Fatty Oils	260	400
Fatty oils	260	400
Sulfones	568	28
Sulfonic Acids	562	30+
Aromatic preservation	562	41
Fat cleavage with	260	416
Hydrophenanthrene nucleus containing	562	31
Mineral oil sludge source	208	13
Triphenylmethane	552	112+
Wetting agent containing	252	353+
Sulfonium	568	18
Sulfophthalic Acid	562	54
Sulfoxides	568	27
Sulfur		
Acids and acid anhydrides	423	512R+
Electrolytic synthesis	204	104
Binding agents	44	589+
Combined with fuel	44	604
Burner	422	160+
Compounds organic		
Carboxylic acids		
Acyclic	562	512+
Acyclic acid moiety	560	147+
Alicyclic esters	560	125
Aromatic	560	9+
Fats fatty oils ester type waxes higher fatty acids	260	399+
Heterocyclic	549	1+
Dye compositions	8	
Food fumigation with	426	319
Medicine or poison containing	424	
Mining	299	3+
Misc. chemical manufacture	156	DIG. 72
Preparation inorganic	423	
Electrolytic	204	128
Removed from coal	44	622+
Rendering harmless in fuel	208	177+
Salts inorganic	423	
Electrolytic synthesis	204	92+
Sulfureted hydrogen	423	563
Sulfuric acid	423	522+
Apparatus		
Chambers	422	
Concentrating	422	160+
Trains	422	160
Electrolytic synthesis	204	104
Esters	558	20+
Mineral oil sludge source	208	13
Sulfurization		
Fatty oils	260	399
Rubber molding combined	264	
Synthetic resin or natural rubber	525	343+

	Class	Subclass
Synthetic resin from	528	389
Synthetic resin or natural rubber compositions containing	524	1+
Treating agent containing	526	917*
Sulfurous Acid	423	521
Apparatus trains	422	160
Paper stock treatment	162	83+
Sulfuryl Chlorid	423	462
Sulky Vehicle	280	63+
Go cart	280	47.24
Sultones	549	10+
Naphthosultone, seven membered hetero ring	549	10
Five members	549	33
Four members	549	89
Six members	549	15
Sun (See Solar)		
Sun Baths	128	372
Sun Roof, Automobile	296	
Sundial	33	270
Design	D10	45
Sunglasses		
Clip on type	D16	102
Supercharger		
Centrifugal pump	415	
Internal combustion engine	123	434+
Combined	123	559.1+
Fluid pressure motor combined	60	598+
Turbine combined	60	598+
Turbine combined	60	597+
Superconductive		
Alloys	420	901*
Cable	174	125.1
With cooling	174	15.5
Electromagnet	335	216
Inductor	336	DIG. 1
In power system	323	360
Magnetometer	324	248
Oscillator	331	107S
Process of making conductor	29	599
By coating	427	62+
Resistor	338	32S
Solid state system	307	200.1+
Wave transmission	333	99S
Superconductor		
High temperature (above 30 degrees k)	505	1
Apparatus	505	1
Material	505	1
Process	505	1
Low temperature (at or below 30 degrees k)	505	800+
Alloy or metallic composition	420	901
Coating	427	62
Control circuit for electromagnetic device	361	141
Electric conductor structure	174	125.1
Cooling of	174	125.1+
Electric pulse counter, pulse divider, or shift register	377	93
Electrical computer or data processing system	364	775
Housing or mounting assembly with plural diverse electrical components	361	331+
Impact ionization or thin film device	307	277
Inductor	336	DIG. 1
Josephson junction device	357	5
Magnet or electromagnet	335	216
Magnetic field shield	307	91
Magnetic lens	250	396ML
Magnetic measuring device	324	248
Making by coating, forming, etching, or sputtering	204	192.24
Mechanical manufacture	29	599
Nonlinear solid state device and circuit	307	200.1+
Digital logic	307	462
Digital logic function (and, or, nand, or not)	307	476
For attenuation	307	541
For gating	307	245
With josephson junction	307	306
Power supply or regulation system including transformer or inductor	323	360
Protective circuit	361	19
Resistor responsive to magnetic field	338	32S
Rotary dynamoelectric motor	310	40R+

	Class	Subclass
With cooling or fluid feeding	310	52+
Solid state active element oscillator	331	107S
Solid state device with housing and cryogenic cooling	357	83
Squid	357	5
Static information storage device	365	160
Josephson junction type	365	162
Thin film type	365	161
Stock material	428	930
Temperature measurement	374	176
Wave transmission line or network	333	99S
Superfinish Grinder	51	57+
Superheater		
Concentrating evaporator	366	127
Cupola combined	48	64
Steam	122	459+
Locomotive combined	105	45
Power plant combined	60	643+
Superheterodyne Receiver	455	130+
Superphosphates	423	
Fertilizers	71	33+
Superplastic	420	902*
Superregenerative Radio	455	336+
Superscript Character	400	904*
Supersonic Wave	367	
Agitation	366	
Bonding non-metallic		
Apparatus	156	580.1+
Process	156	73.1+
Chemical apparatus	204	193
Chemical change causing		
Apparatus	204	193
Code signaling	116	18+
Controls direction of travel	318	460
Airplane	244	76R+
Electric motor actuated	318	460
Torpedo	114	23
Gas separation	55	277
Process	55	15
Generator	116	137A
Piezoelectric, resonator combined	310	322
Piezoelectric with mechanical energy coupling means	310	334
Laminating processes	156	73.1+
Measuring using	73	570+
Metal casting using	164	501
Nautical signaling	116	26+
Seismic surveying	181	
Submarine signalling	367	131+
Testing miscellaneous	73	570+
Welding apparatus using	228	1.1
Supervisory Control	340	825.6+
Support (See Article or Device Supported)	248	
Airplane motor	248	554+
Antenna combined	343	878+
Apparel	2	300+
Apparatus	223	120
Design	D 2	624+
Arch	128	166.5
Armrest	248	118
Article by building	52	27+
Articles		
Or object being X rayed	378	38
Plural	211	
Plural locked	70	62
Single	248	
Ash tray smokers	131	240.1+
Athletic	128	157+
Bag	248	95+
Baggage	190	18R
Ball rack	211	14+
Basin	4	643+
Bathtub	4	592+
Bather	4	571+
Bearing	384	428+
Bed linens	24	72.5
Bed attached	5	504+
Beds	5	
Blotter, ie desk pad	248	346+
Boat		
Foot supports	114	363
Seat	114	363
Bobbin	242	130+
Body catchers	182	137+
Body harness	182	3+
Baby jumper	297	275
Body in bed	5	431+
Boiler	122	510

Sweep **Inventing & Patenting Sourcebook, 2nd Edition** Synthetic Resin or Natural Rubber

	Class	Subclass
Wood disk cutting machine	408	72R+
Sweeper		
Brush broom vacuum carpet street	15	
Rake eg lawn broom	56	400.17+
Sweeping Compound	252	88
Sweet Potato Musical Instrument	84	380R
Sweetening		
Mineral oil	208	189+
Swelling		
Artificial fibers	8	130.1+
Dyeing combined	8	
Fiber swelling & stretching	8	DIG. 3
Swells		
Electric tone generation with expression device	84	633
Piano	84	182
Pipe organ	84	346+
Reed organ	84	372
Swimming		
Aids	441	55+
Buoyant	D21	237+
Foot attached	441	61+
Foot flipper	441	64
Exercising devices for simulating	272	71
Mask	D29	8+
Pool		
Design	D25	2
Grab rail	D25	41
Inground	52	169.7+
Nuclear reactor	376	174
Nuclear reactor	376	361
Nuclear reactor	376	403
Portable design	D21	252
Sea baths	4	487+
Tank baths	4	488+
Water purification	210	169+
Shoes	36	8.1
With fins	441	76+
Structure of pool	52	261
Concrete	52	250
Curved	52	245
Suit	D 2	40+
Teaching	434	254
Swinging Door	49	381+
Swings		
Amusement	D21	246
Chairs and seats	272	85+
With fan	416	146R
Roundabout	272	28R
Switch		
Apparatus for elevators	187	
Automatic		
Electromagnetic	335	6+
Thermal	337	298+
Automatic for acetylene generator	48	7
Battery	320	
Box	220	3.2+
Burglar alarm	200	61.58 R+
System	340	541+
Charging for storage batteries	320	
Combination	200	1R+
Conveyer type		
Chute	193	31R
Rollerway	193	36
Skidway	193	39
Spiral chute	193	13
Zigzag tube	193	28
Door actuated	200	61.62+
Electric (See controller)	200	
Arc lamp system having	314	131
Arc preventing or extinguishing	200	144R+
Automotive vehicle combined with lock	70	237+
Box or housing	174	53+
Check operated	194	211+
Coin	194	239+
Combined with distribution circuit	307	
Design	D13	158+
Display system control	340	806+
Door or window	200	61.62+
Electric illuminating fixture	362	394
Electric lamp having	315	32
Electromagnetic switching systems	361	139+
Electronic tube having	315	56
Electrothermally actuated	337	
Fire actuated	337	298+
Lamp or electronic tube system having	315	362
Light projector combined with	362	253
Magnetically operated	335	
Making	29	622
Mechanically actuated, general	200	
Multiple filament lamp having	315	64
Plate or socket lighting	362	95
Position or condition responsive	362	802*
Post supported electric light combined with	362	431
Supported electric light combined with	362	382+
Telegraph circuit manipulator	178	75
Telephone	379	422+
Telephone check operated	379	146+
Telephone switchboard	379	319+
Telephone system automatic selective	335	108+
Testing of automatic telephone type	379	17
Thermal control of alarm	340	593+
Thermally actuated	335	
Trolley conductor type	191	38
Tuner with	334	47+
Electronic	328	76
Heat controlled	337	298+
Inductor combined with switch	336	105
To neutralize capacitance	307	134
Interference from elimination	307	134+
Knife	200	
Luminous		
Faceplate	250	465.1
Handle or key	250	465.1
Push button	250	465.1
Movable carrier of textile braider with pattern mechanism	87	16
Movable for carrier of textile braiding apparatus	87	37+
Photoelectrically controlled	361	173+
Plates (See face, plates)		
Pneumatic dispatch system	406	181+
Railway	246	
Cable	104	184
Cable slot type	104	195
Foot guard	238	380
Rail joint	238	165
Track	104	130+
Trolley conductor	191	38
Stands	246	393+
Automatic face setting	246	278+
Manual operation with signal actuation	246	144+
Signal	246	476
Stoplight		
Auto	200	61.59
Hydraulic	200	81R+
Store service system	186	19+
Thermal responsive	337	298+
Heated by electric current	337	14+
Tilting track section type	414	354+
Vehicle unloading	414	357
Travelling crane system	414	562
Vertical axis for cutting and conveying harvester	56	170+
Switchboard		
Electric	361	331+
Telephone	379	319+
Divided central	379	310+
Through ringing	379	256
Switches		
Membrane type	200	512
Rolamite	200	503
Switching Systems	307	
Lamp systems	315	313+
Relay systems	361	160+
Signalling	340	286.1+
Selective	340	825+
Transmitters	341	
Telegraph	178	
Telephone	379	242
Automatic	379	258+
Calling	379	350+
Central	379	242+
Intercommunicating	379	167+
Party line	379	177+
Plural exchanges	379	219+
Switchboard	379	310+
Swivels	403	164+
Bracelets	63	10
Chain	59	95
Making	59	9
Chute horizontal swinging	193	22
Endless chain pumps	198	643
Fish bait artificial	43	42+
Fishhooks	43	44.83+
Panel hanger wheel mount	16	104
Pipe joint	285	272+
Multiple	285	168
Punching bag	272	77+
Railway car suspending	105	156
Rod joint	403	164+
Sleds multiple	280	17
Socket combined	24	136L+
Tool handle	403	52+
Trolley head electric collector	191	60+
Turnbuckle	403	60
Valve disc	251	264+
Valve gate	251	193+
Watch chain snap hook	24	905
Wheeled vehicle running gear	280	86
Wood saw	83	747
Swizzle Stick	D 7	300.2
Swords	30	165+
Cane	135	66
Carrier	224	232+
Design	D22	118+
Handles	D 8	DIG. 5
Pistol	42	53
Sylphon T M (See Bellows)		
Flowmeter	73	262
Fluid pressure guage	73	729+
Fluid pressure regulator	137	505.36
Making	29	454
Corrugating	72	54+
Pump	417	472+
Synchronizing		
Camera and flash powder	354	129+
Camera and photoflash testing	356	72
Camera shutter release and light	354	129+
Clock	368	52
Clutches	192	52+
Dynamoelectric machine interconnection	307	43+
Electric motor	318	41+
Electric motor	318	85
Fluid pressure brake systems	303	35+
Gearing	74	339+
Governor	73	507
Impeller combined	416	34
Motion picture and sound	352	12+
Multiple motor	60	700+
Multiple motor	60	706+
Musical instrument	84	119
Plural radio stations	235	58M
Telegraphy	375	106+
Telegraphy multiplex	370	47+
Television	358	148+
Recording or reproducing color signal	358	319
Synchronous Motor	310	162+
Systems	318	700+
Synchroscope	324	91
Synthesis		
Electro	204	59R+
Inorganic compounds	23	
Organic compounds	260	
Polymers (See synthetic resin or natural rubber)		
Synthetic		
Alumina	423	625
Diamond and making	423	446+
Gem	501	86
Resin (See synthetic resin or natural rubber)		
Rubber (See rubber; synthetic resin or natural rubber)		
Synthetic Resin or Natural Rubber		
Abs (See also prepared from, below)	526	338
Addition polymerization	526	72+
Manipulative process features	526	59+
Adhesive material	520	1+
Compositions (See class 523, 524)		
Admixtures of resins or rubbers	525	50+
Aftertreatments		
Chemical reaction involved	525	50+
To produce ion exchange resin	521	25+
Utilizing wave energy	522	1+
Vulcanization of rubber (products)	525	331.9+
Vulcanization of rubber with sulfur or sulfur compound	525	343+
With asphalt or bituminous material	525	54.5+
With carbohydrate	525	54.2+

	Class	Subclass
Packages	206	529
Suprarenal Extracts		
Medicine containing	424	563
Surface Active Agents	252	351+
Surfaces		
Active agent	252	351+
Aircraft sustenation	244	200
Bearing		
Inserts	384	276+
Treating	384	625
Bed bottom	5	186.1+
Carbonizing of iron	148	16.5
Chair or seat		
Changeable supporting surface	297	283
Headrest with changeable surface	297	220+
Spring supported upholstery	267	80+
Cooking mold	99	372+
Sheet or griddle	99	422+
Demonstration	434	365+
Electric insulator	174	212
Electrically charged for safes	109	35
Erasable, eg chalkboard	434	408+
Leaves	281	39
Evaporators sugar making	127	16
Fluid reaction ie impellers	416	
Friction gearing	74	214+
Gauge plates	33	567
Heater	126	271.1+
Water vessel	126	390+
Horizontally supported planar	108	
Laminated type photographic	101	178
Stencil	101	128.21
Pavement	404	17+
Antislip	404	19+
Sheet type	404	18
Surface treating	404	83+
Planographic printing	101	453+
Platen or pressure	100	295
Printing member	101	401
Railway rail tread surface	238	148
Tie plate top surface	238	304
Receptacle stand for sloping surface	248	148
Roll or rollers	29	121.1+
Roughness testing	73	105
Smoothing or pressing textiles	38	
Street mirror or reflector T	404	14
Surgical electric applicator	128	783+
Textile drawing roll	29	110+
Washboard	68	226+
Washtub with scrub surface	68	233
Yielding surface printing member	101	379+
Rolling contact	101	376
Surfacing		
Button	79	7+
Drilling combined	79	6
Grinding	51	
Roads	404	17+
Stone	125	
Surfboard	441	74+
Design	D21	228+
Surgery and Surgical (See Classes 128 and 604)		
Acupuncture	128	907*
Adhesive fastener for absorbent pad	604	389+
Article package	206	438+
Basin	D24	56+
Biological signal amplifier	128	902*
Feedback to patient	128	905*
Blood coagulation	128	DIG. 22
Breathing apparatus without patient cross-contamination	128	909*
Cannula supporters	128	DIG. 26
Catheters (See catheter)		
Cervical collars	128	DIG. 23
Clip appliers	29	243.56
Collagen	128	DIG. 8
Computer controlled, monitored	364	413.1+
Cyrogenic	128	DIG. 27
Design	D24	
Electric shock, patient protection	128	908*
Fluid amplifiers	128	DIG. 1
Heart-lung	128	DIG. 3
Heat shrinkable film	128	DIG. 18
Infusion monitoring	128	DIG. 13
Intravenous injection support	128	DIG. 6
Invalid beds	5	60+
Mask	D29	7
Medical-surgical bags	128	DIG. 24
Medicament receptacle	604	403+
Mixing receptacles	604	416
Multiphasic diagnostic clinic	128	906*

	Class	Subclass
Pressure infusion	128	DIG. 12
Radio telemetry	128	903*
Servo-systems	128	DIG. 7
Silicone	128	DIG. 21
Sphineters artificial	128	DIG. 25
Splint, clavicle	128	DIG. 19
Splint, inflatable	128	DIG. 2
Spotlights	362	804*
Stapler	227	175
Supplies packaged	206	438+
Suppression of noise in electric signal	128	901*
Syringes (See syringe)		
Table	269	
Teaching	434	262
Teflon T M	128	DIG. 14
Tree	47	8
Velcro T M	128	DIG. 15
Surlyn D T M (polyethylene Methacrylic Acid)	526	317.1
Survey Meter Nuclear Energy	250	472.1+
Scintillation type	250	361R
Surveying Instruments	33	290+
Bent line of sight type	356	253+
Angle measuring	356	138+
Range or height finders	356	3+
Optical features	356	247+
Design	D10	61+
Survival in Hostile Environment		
Computer controlled, monitored	364	413.3+
Suspended		
Roof	52	83
Swings	272	85+
Suspender		
Artificial leg	623	32
Fasteners design	D 2	640+
Garment		
Design	D 2	626
End elements	2	340
Fasteners for	D11	200+
Plural	2	304+
Saddle type cord retainers	2	341+
Strip connected spaced	2	323+
Trouser attached	2	230
Suspension Devices (See Hangers)		
Battery active material, faure electrodes	429	208
Clothesline single run	211	119.12
Drying		
Hollow articles	34	106
Sheets webs or strands	34	151+
Filter household	210	105
Fluid (See suspensoids)		
Current drying method	34	10
Separating solids	209	132+
Folding bed fulcrum	5	141
Insulator support	174	160
Knob flexible	16	122
Light bowl	362	453+
Picture support	248	489+
Pipe or cable	248	58+
Rack for broom and brush	211	66
Radiator bracket	248	233
Roundabout amusement vehicle	272	40+
Special receptacles or packages	206	806*
Stage appliances	272	24
Structures & vehicles (See hanger)		
Boiler regulation	122	450
Bridge cantilever truss	14	8
Bridges	14	18+
Dumping railway car	105	242
Fences	256	23
Railway car body	105	453
Railway car suspended	105	148+
Railways	104	89+
Scaffolds	182	142+
Track	104	123
Unloader for suspended vehicle	414	377+
Vehicle body spring	280	788
Suspended supports	248	317+
Swinging seat	272	85+
Tea ball type beverage infuser	99	322
Telephone supports	379	454+
Tool and machine		
Brush flue cleaner	15	163
Cable drum	254	380+
Car journal box lifting	254	34
Conveyor, load suspending	198	678.1+
Conveyor loop	198	678.1+
Cultivator wheeled	172	
Grapple pivoted saw operator	294	112

	Class	Subclass
Harvester knife grinding clamp	51	223
Machinery support	248	637+
Resilient	248	610
Plow wheeled	172	
Scraper flue cleaner	15	243
Sheet delivering conveyors	271	198+
Strop	76	81+
Vehicle springs	267	
Stub axle mounts	280	660+
Vehicle structure combined	280	788
Vehicle systems	280	688+
Suspensoids		
Colloidal	252	302+
Comminuting or grinding	241	
Electrical purification	204	186+
Electrolytic		
Forming processes	204	10
Treatment processes	204	130+
Electrophoretic or electro osmosis using	204	180.2+
Suspensories	128	158
Sustained or Differential Release		
Capsules	424	457+
Implants	424	422+
Tablets	424	468+
Sustained Release Implants	424	422+
Sutureless Closures	606	213+
Sutures		
Appliers surgical	29	243.56
Surgical	606	228+
Textile knitting	66	43
Hosiery seam	66	179
Swab	15	208+
Container for	206	209
Culture medium combined	435	292+
Diagnostic	128	759
Lubricator	184	102
Making	19	145+
Medicated	604	1+
Sampling	73	854
Valved pump piston	417	545+
Swaging		
Bolt side	10	18
Chain making shaping and	59	30
File cutting by	76	14
Horseshoe nail side	10	68
Horseshoe shaping and	59	58
Metal forging	72	
Nut forging side	10	79+
Pipe joint	285	382
Pipe to plate joint	285	222
Saw making	76	51+
Saw setting	76	72
Screw threading by	10	152R+
Tool sharpening by	76	89.2
Swarm Appliances	449	27+
Swash Plates	74	60
Expansible chamber device	92	71
Flow meter	73	244
Internal combustion engine		
Multiple parallel cylinder	123	58R
Rotating cylinder	123	43R
Motor	91	472+
Pump	417	269+
Swathing Attachments	56	189
Swatters	43	137
Sweater		
Animal	D30	145
Apparel	2	90
Design	D 2	44+
Knit fabric	66	176
Sweats or Sweat Band	2	181+
Hat making	223	22
Sweep		
Animal powered motor	185	19+
Cable drum actuation	254	381+
Circuit		
Cathode ray tube having	315	378
Electronic tube type oscillator	331	
Triggered	328	193+
Cultivator shovels	172	681+
Cutter	30	310+
Can opener	30	435+
Plural blade	30	310+
Foundry mold making device	164	161+
Pivoted arm bending machine	72	217
Plow trailing		
Separately attached wings	172	722+
Winged	172	730
Sheet cutting machine	82	
Wood auger	408	199+

	Class	Subclass
Perbunan T M type rubber		
(copolymer butadiene-		
acrylonitrile)	526	338
Phenolic resin	528	129+
Phenoplast	528	129
Phenoxy resin	528	86+
Plexiglas T M type resin		
(polymethyl methacrylate)	526	329.7
Polyamide resin (See also 528-		
332+)	528	310+
Polycarbodiimide	528	44+
Polyesters	528	272+
Polyimide	528	310+
Polyphenylene oxide	525	905*
Polysulfide	528	373+
Prepared from		
Acetylene	526	285
Acrolein	526	315
Acrylates	526	328+
Acrylic acid	526	317.1+
Acrylonitrile	526	341+
Aldehyde e. g. homopolymer of		
saturated etc.	528	230
Ethylenically unsaturated	526	315
Hydrocarbon	528	247
Melamine	528	254*
Urea	528	259+
With amine e. g. aniline etc.	528	266+
With ketone	528	227
With phenol	528	129+
Asphalt or bituminous material	527	500
Unsaturated	526	290
Butadiene	526	335+
Carbohydrates e. g. sugar, starch		
etc.	527	300+
Unsaturated carbohydrate		
derivative	526	238.2+
Carboxylic acid or derivative	528	271+
Ethylenically unsaturated	526	317.1+
Cashew nut liquor	528	2+
Cellular plant or animal material	527	100+
Chloroprene	526	295
Coal derived material (See		
asphalt or bituminous material,		
above)		
Cork	527	100+
Coumarone	526	266+
Diolefin hydrocarbon	526	335+
Cycloaliphatic	526	308
Fused or bridged ring	526	280+
Epoxide compound	528	403+
Diglycidyl ether of bisphenol		
compound (See also 525-		
523+; 525-107+;525-		
403+)	528	87+
Ethylenically unsaturated	526	273+
Ethylene (homopolymer)	526	352+
Copolymer	526	72+
Ethylenic monomer (solely		
therefrom)	526	72+
With non-ethylenic monomer		
(See also class 527)	528	
Formaldehyde (See aldehyde,		
above)		
Furfural (See also aldehyde,		
above)	526	270
With non-ethylenic monomer	528	230+
Hydrocarbons	528	396+
Ethylenically unsaturated (See		
also 526-335;526-346)	526	348+
Indene	526	280+
With coumarone	526	266+
Isobutylene	526	348.7
With butydiene	526	339
Isocyanate compound	528	44+
Isoprene (note: for purposes of		
the class 520 series natural		
rubber is considered to be		
polyisoprene)	526	340.2
Lignin	527	400+
Melamine (See aldehyde, above)		
Metal compounds	528	395
Ethylenically unsaturated		
monomer	526	240
Organometallic	528	9
Methacrylate	526	319+
Methacrylic acid	526	317.1+
Methylmethacrylate	526	319+
Monoepoxides (See epoxide,		
above)		
Olefin	526	72+

	Class	Subclass
Oxirane compounds (See epoxide,		
above)		
Phosphorous compounds	528	398+
Ethylenically unsaturated	526	274+
Plant extract of undetermined		
constitution	528	1+
Polyepoxide compound (See		
epoxide, above)		
Propylene (homopolymer)	526	351
Copolymer	526	72+
Protein or biologically active		
polypeptide	527	200+
Ethylenically unsaturated		
derivative	526	238.1
Silicon compound	528	
Ethylenically unsaturated	526	279
Si- H or si- C	528	10+
Tannin	527	400+
Ethylenically unsaturated		
monomer	526	238.2+
Tar or pitch	527	500+
Ethylenically unsaturated		
derivative	526	290
Tetra fluoroethylene	526	242+
Urea (See aldehyde, above)		
Vinyl chloride	526	344+
Vinyl compounds	526	72+
Vinylidene compounds	526	72+
Purification of	528	480+
Composition	524	310
Reclaiming	521	40+
Rubber	520	1+
Diene rubber	526	72+
Spandex T M type	528	44+
Thiokol T M type	528	374+
Vulcanized	525	332.5+
E P D M	525	331.8
Rubber hydrocholride	525	332.3
Silastic (organosilicon polymer)	528	10+
Silicone resin	528	10+
Compositions	524	858+
Room temperature vulcanizing		
(rtv rubber)	528	901*
Synthetic rubber (for purposes of		
class 520, is considered to be a		
synthetic resin)		
Teflon T M (polytetrafluoro-		
ethylene)	526	255
Wave energy utilized	522	1+
Synthol	518	700+
Syringe		
Dental	D24	14
Dispenser	222	206+
With follower	222	386+
Hand held	604	187+
Pump	417	437+
Specific gravity tester	73	441+
Sprayer		
Liquid pump feed	239	329+
Piston feed	239	320+
Surgical	604	187+
Air	604	26
Design	D24	60+
Ear	604	187+
Eye	604	295+
Fluid operated	604	131+
Fountain	604	131+
Hand held	604	187+
Hydrant	604	150
Indication of defect or use	604	111
Irrigation	604	187+
Motorized	128	DIG. 1
Nozzle	604	275
Piston	604	218+
Pocket container for	206	364+
Preventing reuse	604	110
Projectile	604	130
Receptacle for	206	364+
Siphon	604	131+
Thermal medicator	604	113+
With light and electrical		
application	604	20+
Syrups		
Condiments and flavors	426	658+
Electrolytic treatment	204	138
Malt	426	44
Sugars only	127	30+
Aircraft control	244	175+
Stoves and furnaces	126	
T Squares	33	474+
Design	D10	65+

	Class	Subclass
Tab		
Bolt or nut, side lock, tab		
deformable in situ	411	122+
Exhibitor type	40	662+
Protrusion to receive identification	40	359+
Suspender ends	2	340
Tabernacle	D99	25
Tables	108	17
Animal restraining	119	103
Bed	5	3+
Bed combined	5	507
Billiard and pool	273	3R+
Design	D21	232
Cabinet convertible to	312	241
Cabinet with separable table	312	277
Chair combined	297	135+
Chair convertible	297	119+
Cooling	62	458
Cover or cloth	D 6	617
Croquet	273	57
Disappearing instrument or device	312	21+
Examining or surgical	D24	3
Furniture	108	
Design	D 6	396+
Game	273	309
Design	D21	1
Game boards	273	287
Collapsible	273	285+
Interchangeable pattern	273	284
Rotatably mounted	273	280
With separable pattern	273	283
Gauging eg grizzlies	209	680
Gravity assorting in liquid		
suspension	209	172+
Invalid back rest combined	5	70
Ironing	38	103+
Combined with sewing machines	112	217.1+
Plural surface	38	135+
Single surface	38	137+
Light table, illumination for	362	97
Luggage convertible	190	11+
Machine part		
Box nailing machine	227	154
Cigar and cigarette making	131	41+
Cutting machine	83	
Grain separator straw carrier	460	90
Metal can making and soldering	413	
Metal flanging	72	102
Plastic material cutting	83	
Plastic material cutting	425	289+
Powered table for operating		
attachments and auxiliaries	74	16
Shingle machine	83	704+
Slicer woodworking	144	171
Typewriter paper feeding	400	647+
Woodworking saw table	269	289R
Modified for blade	269	83
Manual assorting	209	703+
Pads	248	346.1
Phonograph or sound recording and		
reproducing		
Flat tablet traveling table	369	213
Mandrels	369	260
Tablet	369	267
Rack convertible to table	108	16
Refrigerator combined	62	258
Self levelling	114	195
Steam cooking	126	33
Stratifier	209	422+
Starch making	127	26
Surgical	269	323+
Telegraph instrument	178	114
Tennis	273	30
Toy or doll	446	482
Transfer	104	48+
Electricity transmission to vehicle	191	9
Trunk	190	11+
Turntables	104	35
Automobile	104	44
Plural related surfaces	108	103+
Self service eg lazy susan	108	94+
Self service power driven	108	20+
Single surface	108	139+
Wall	108	152
With structural installation	108	72+
Work (see also table machine part)		
Abrading	51	240R
Butter working	99	452+
Cloth inspecting	26	70
Illuminated operating table	362	33
Ironing textiles	38	103+

	Class	Subclass
With fatty acid or fatty acid glycerol ester	525	50+
With natural resin	525	54.4+
With protein or biologically active polypeptide	525	54.1+
Physical treatment	528	480+
Compositions	523	300+
Compositions: creaming, concentrating or agglomerating latex	523	335
Natural rubber e. g. extraction of rubber from plant source	528	931*
Natural rubber latex	528	934*
Natural rubber latex coagulating	528	936*
Natural rubber latex concentrating	528	937*
Natural rubber latex preserving	528	935*
Alkyd resin (solid polymer)	528	272+
Compositions (See class 523, 524)		
Containing chemically combined fatty acid or fatty acid glyceride	528	295.5
Containing chemically combined natural resin	527	604
Aminoplast resin (See also prepared from aldehydes, below)	528	266
Bakelite T M type resin	528	129+
Balata (See natural rubber, below)		
Blends of solid polymers	525	50+
Block and block type copolymers	524	88+
Buna T M type rubber	526	339
Buna-n (butadiene nitrile rubber)	526	338
Catalysts		
Aldehyde polymerization	528	232+
With phenol	528	137+
Chalcogen ring opening	528	406+
Epoxy resin	528	88+
Ethylenic polymerization	526	89+
Graft polymerization	525	244+
Redox	526	915*
Transition metal	526	90+
Polyester	528	274+
With ethylenic monomer	525	11+
Silicone	528	12+
Urethane	528	48+
Cellular or porous	521	50+
Blends of synthetic resin or natural rubber	521	134+
Blowing agent i. e. foaming agent	521	82+
Resin formation in the presence of	521	99+
Carbodiimide polymer	521	901*
Degradability enhanced	521	916*
Dye or pigment containing	521	920*
Electrical or wave energy used during cell forming	521	50.5*
Ethylene polymer	521	143+
Fireproofed	521	82+
Non-urethane resin	521	907*
Resin formation in presence of fireproofing agent	521	99+
Urethane resin	521	906*
Hydraulic cement ctg (See water settable inorganic composition ctg, below)		
Integral skin foam	521	51
Isocyanurate polymer	521	902*
Kicker or blowing agent moderator	521	909*
Latex rubber	521	65+
Natural resin containing	521	84.1
Synthetic resin formation in the presence of	521	109.1
Natural rubber containing	521	150
Latex	521	65+
Nucleating agents for	521	908*
Phenolic resin	521	180+
Pigment containing	521	920*
Polyester resin	521	182+
With unsaturated monomer	521	128
Protein containing	521	84.1
Resin formation in the presence of	521	109.1
Reduced smoke formation	521	903*
Reticulated foam	521	52
Separated reactive materials e. g. :two package system:	521	912*
Sintering process for formation of porus product	521	919*
Styrene polymer	521	146+
Foamed beads	521	56+
Syntactic foam	521	53+
Non resinous hollow particles	523	218+
Urethane polymer	521	155+
Vinyl chloride polymer	521	145
Compositions (See class 523, 524)		
Adhesive (See class 523, 524)		
Anerobic	523	176
Glass substrate	523	168
Antioxidant containing (See class 523, 524)		
Antistatic	524	910*
Asphalt containing	524	59+
Resin produced in presence of	524	705
With epoxy resin	523	450
With polyester-unsaturated monomer	523	518
Biocidally protected	523	122
Bitumen ctg (See asphalt ctg, above)		
Body contact or implantable (intended use)	523	105+
Brake shoe (intended use)	523	149+
Carbohydrate containing	524	27+
Resin produced in presence of	524	732+
With epoxy resin	523	447+
With polyester-unsaturated monomer	523	509
Carbon black containing (See class 523, 524)		
Resin produced in presence of	524	847
Specified dimension	524	495+
With epoxy resin	523	440
With polyester-unsaturated monomer	523	512
Castor oil containing	524	310
Resin produced in presence of	524	769
Cellular plant or animal material containing	524	9+
Resin produced in presence of	524	702
With epoxy resin	523	446
With polyester-unsaturated monomer	523	500+
Cellulose ether or ester ctg (See carbohydrate ctg, above)		
With polyester-unsaturated monomer	523	509
Chinawood oil containing (See fatty acid triglyceride, below)		
Coating (See class 523, 524)		
Antifogging	523	169
Glass enamel	523	170
Ink for glass substrate	523	160
Concrete ctg (See inorganic water-settable, below)		
Contact lens, used for (See also class 527-937)	523	106+
Cork ctg (See cellular plant or animal material, above)		
Degradability enhanced i. e. pro-degradant containing	523	124+
Dispersions (See emulsions, below)		
Drying oil ctg (See fatty acid triglyceride, below)		
Dye ctg (See class 523, 524)		
Emulsions (See class 523, 524)		
Aqueous (See also 524-501)	524	800+
Epoxide resin containing	523	400+
Fatty acid triglyceride ctg	524	313
Resin produced in presence of	524	773+
With epoxy resin	523	455
With polyester-unsaturated monomer	523	511
Feather ctg (See cellular plant or animal material, above)		
Fireproofed (See class 523, 524)		
Flux ctg (See class 523, 524)		
Foundry mold or mold core binder	523	139+
Hair ctg (See cellular plant or animal material, above)		
Heat stabilized (See class 523, 524)		
Inorganic water settable material ctg e. g. hydraulic cement	524	2+
Resin produced in presence of	524	650+
With epoxy resin	523	402+
With polyester-unsaturated monomer	523	501+
Lacquer (See class 523, 524)		
Latex (See class 523, 524)		
Coagulating	528	936*
Concentrating	528	937*
Creaming, agglomerating or coalescing	523	335
Physical treatment of natural rubber latex	528	934*
Preserving or stabilizing	528	935*
Resin produced in presence of water	524	800
Leather ctg (See cellular plant or animal material, above)		
Lignin or tannin containing	524	76
Resin produced in presence of	524	735
With epoxy resin	523	400+
With polyester-unsaturated monomer	523	500+
Linseed oil ctg (See fatty acid triglyceride, above)		
Molding (See class 523, 524)		
Natural resin ctg (See also rosin, below)	524	77
Synthetic resin produced in the presence of	524	764
With epoxy resin	523	400+
With polyester-unsaturated monomer	523	500+
Natural resin ctg (see class 523, 524 treated as polyisoprene)		
Guayule	528	930
Phonograph record (intended use)	523	174
Pigment ctg (See class 523, 524)		
Plasticizer ctg (See class 523, 524)		
Protein or biologically active polypeptide ctg	524	17+
Resin produced in presence of	524	704
With epoxy resin	523	449
With polyester-ethylenic monomer	523	500+
Rosin containing	524	270+
Synthetic resin produced in the presence of	524	764
Tannin ctg (See lignin, above)		
Tar ctg (See asphalt, above)		
Varnish (See class 523, 524)		
Wax containing	524	487+
Oxygen containing wax	524	275+
Oxygen ctg wax with epoxy resin	523	455+
Oxygen ctg wax with polyester unsaturated monomer	523	511
Resin produced in presence of oxygen ctg wax	524	763+
Condensation polymerization	528	
Manipulative process features	526	59+
Coumarone resin	526	266+
Dacron T M type resin	528	308.1+
Ebonite T M type resin	525	332.6+
Elastomer (See rubber, below)		
E P (ethylene-propylene)	526	348+
E P D M (ethylene-propylene-diene monomer)	526	
Cured or vulcanized	525	331.7+
Epoxide resin (See also classes 525, 528)	528	87+
Foams	521	50+
G R - N (acrylonitrile-butadiene rubber)	526	338
G R - S (styrene-butadiene rubber)	526	340
Graft or graft type	525	50+
Guayule (See also natural rubber)	528	930*
Hevea (See natural rubber)		
Hypalon T M type resin	525	333.9
Ion exchange resin	521	25+
Ionomer resin systems	525	919*
Carboxyl bearing	526	317.1+
Kraton T M type resin (styrene-butadiene rubber)	526	340
Lexan T M type resin	528	196+
Lucite T M type resin	526	329.7
Melamine resin	528	254+
Natural rubber (note: for purposes of the class 520 series natural rubber is considered to be polyisoprene)		
Neoprene	526	295
Vulcanized	525	330.9+
Nylon	528	310+

	Class	Subclass
Ironing with work catcher	38	111
Leather working	69	19
Milling planing or gear cutting	409	219+
X ray	378	209
Tablets		
Calendar tear off type	40	121
Coffee	426	594
Dye compositions	8	526
Food design	D 1	
Medicinal	424	464+
Design	D28	1+
Memorial	40	124.5
Molding presses for	425	406+
Phonograph	369	272+
Making by turning	82	1.11
Railway signalling system	246	19
Tabulators		
Calculator	235	85+
Teletypewriter	178	25+
Typewriter	400	284+
Indicating	400	705.4+
Tachometer	73	488+
Design	D10	98
Electric	324	160+
Rotation counter	235	104
Rotation counter, with electrical speed indicator	324	166+
Tack	411	439+
Pullers	254	18+
Shoe lasting machine	12	16
Shoe sole sewing machine	112	30
Strips		
Making	10	29
Nailing implement	227	136
Tackle		
Cable and hook	294	82.11+
Fishing	43	4+
Design	D22	134+
Pushing and pulling implements		
Block and	254	390+
Single throw lever	254	128
Tensioners	254	242
Traversing hoist cable type	212	76+
Tag		
Affixing tags	156	DIG. 23
Collision released identification	116	32
Envelopes	229	74
Exhibitor type	283	74+
Fastenings	40	662+
Inserters	606	117+
Machines		
Cigar	131	106
Cigar etc packaging combined	131	283
Label pasting	156	
Paper tag applying	493	375+
X-art collection	493	961*
Paper wrapping machine	53	128+
Plug tobacco	131	113+
Printed matter	283	74+
Taillights	362	61+
Tailpieces	84	299+
Stringed instrument	84	297R+
Tailpipe		
With muffler	181	227+
With exhaust pipe	181	228
Tailstocks		
Lathe	82	148
Milling machine	409	242
Take up		
Bed bottom cord tightener	5	212
Braiding netting and lace making	87	31
Brake position adjusters	188	196R+
Brake wear	188	79.51+
Electric cable extensible	174	69
Electricity flexible extension to vehicle	191	12R+
Embroidering machine thread	112	96
Harvesting binder cord knotter	56	450
Hitching post	119	124
Leather sewing machine thread	112	57
Railway freight car shrinkage	105	412
Railway traction cable	104	196
Sewing machine	112	241+
Textile knitting fabric	66	149R+
Trolley rope slack	191	93
Twine holder	242	129.1+
Weaving fabric	139	304+
Weaving warp feeding	139	99
Winding and reeling	242	129.1+
Recording tape or motion picture film	242	179+

	Class	Subclass
Talcum Powders (See Phonograph)	424	69
Talking Machine (See Phonograph)		
Book	434	317
Tall Oil	530	205+
Derivatives	530	230+
Pulp preparation combined	162	14+
Tally (See Counter)		
Game	116	222+
Printed forms	283	48.1+
Register	235	
Strip feeding	400	586+
Tam-o-shanter	D 2	255
Tambourines	84	418
Tamper (See Rammer)		
Blasting charge	86	20.11+
Earthenware pipe	425	456
Female	425	425+
Male	425	469
Railway track	104	10+
Road or earth	404	133
Tobacco		
Pipe	131	248
Pipe combined	131	184.1
Pipe cover combined	131	177
Well torpedo combined	102	304
Tamper-proof Packages	206	807*
Fasteners with anti-tamper means	269	910*
Tampions (See Tompions)		
Tampon	604	358+
Depositor	604	11+
Making	28	118+
Sanitary	604	358+
Materials	604	367+
X-art collection	604	904*
Tandem		
Cycles	280	200+
Design	D12	109
Motor	180	219+
Three tandem wheels	280	239
Cylinder internal combustion engine	123	57R
Horse hitch	278	3+
Power plants	60	698+
Two wheeled skates	280	11.23
Tang		
Drill	408	213
Socket		
Barbed or pronged tang	279	104
Molded or cast in tang	279	105
Offset tang	279	93+
Shouldered tang	279	89+
Tool handle fastenings	403	
Tank (See Vessel)	220	
Cars, trucks, ships		
Ballasting	114	125
Grain tank body	296	15
Marine tanker	114	74R
Railway car	105	358+
Railway tender	105	236
Self levelling vehicle	280	7
Tank truck	280	830+
Vessel raising or docking air tank	114	52+
Water wagon	280	830+
Chemical reaction	422	
Cooled heat transfer, liquid		
containing	62	430+
Commodity contacting	62	373+
Liquid congealing	62	356
Depth gauges	73	290R+
Filling or emptying with fluids	137	561R+
Fire extinguishers	169	
Fluid distribution	137	561R+
Fowl and hog scalding		
Fowl	17	11.2
Hog	15	15
Hog scalder with scraper	17	13+
Fuel with burner	431	344
Heated	431	203+
Gas and liquid contact		
Contacting	261	119.1+
Liquid supply	261	72.1+
Gas holding	220	3
Heating and illuminating gas	48	174+
Package	206	.6+
With dispensing	222	3+
Gasoline holding for motor vehicle	280	830+
Glass melting	432	195
Heat exchange		
Trickler	165	118
With agitator or conveyer	165	109.1
Heating system expansion	237	66
Ice mold	249	117+

	Class	Subclass
Lined metallic	220	415+
Liquefied gas storing	62	45.1+
Liquid heating	122	
Cooking vessels	126	344+
Lock valve combined	137	384.2+
Making		
Outlet nozzle attaching	29	890.142
Sheet metal	413	1+
Masonry and concrete	52	264
Curved	52	245
Material or article charging or discharging	414	288
Material separating		
Liquid separation	210	
Solids	209	
Military	89	36.8
Endless track drive	180	9.1+
Five or more wheels	180	22+
Gun mounts	89	40.3
Steering by driving	180	6.7
Toy	446	433
Model basin or testing tank	73	148
Mold	249	144+
Nuclear reactor combined	376	403
Oil tank with fire extinguisher	169	66+
Photographic developing	354	331+
Rack for	211	71+
Sand blast equipment	51	436
Self healing	428	912*
Settling (See settling tanks and chambers)		
Shower bath	4	602+
Spring wheel air tank	152	10
Stage or theatrical	272	26
Stock material	428	34.1+
Sugar starch and carbohydrate treating	127	
Supply for ambulant discharge sprayer	239	302+
Swimming pool	4	488+
Water purification	210	169
Tank-type carrier truck	D12	95
Tower	52	40
Underground	405	53+
Volume or rate of flow meter	73	217+
Washing		
Brushing scrubbing general cleaning	15	
Clothes washing machines & tubs	68	
Dish and solids washing machine	134	
Water closet and urinal flushing	4	353+
Disinfection	4	222+
Wheelwright tire setting	157	7
Wooden	217	4
Tanker	114	74R
Tannic Acid	560	68+
Tanning	8	94.19 R+
Apparatus	69	29+
Descaling agent ctg	252	83+
Electrolytic	204	135
Fish skins	8	94.12
Hides	8	94.19 R+
Processes manipulative	8	150.5
Tannins	560	68
Glucosidal	536	4.1+
Tanning with	8	94.19 R+
Tantalate Ceramic Compositions	501	134
Tantalum		
Alloys	420	427
Compound stock	428	662
Electrolysis	204	
Inorganic compound recovery	423	62
Misc. chemical manufacture	156	DIG. 87
Tap		
Barrel	137	317+
Bung	217	98+
Changer transformers	336	137
Transformer systems	323	
Connector electric	439	638+
Dancing shoe	36	8.3
Gas taps illumination	362	381
Piercing electrical connector	439	387+
Pipe	408	215
Screw threading	408	215+
Machine	408	72R+
Wrenches ratchet	408	120+
Wheel bolt	301	117
Tape (See Adhesive; Strips; Taping)		
Adhesive	428	343+
Processes of coating	427	207.1+
Box making	493	116+

	Class	Subclass
Cutting to sever	83	
Inserting in apparel	223	50
Needles	223	103
Record	360	134
Magnetic		
Coated	360	134+
Recorded, moving head	360	83
Recorded, stationary head	360	90
Measuring	33	755+
Design	D10	71+
Reel	242	84.8+
Package	206	389+
Tape cartridge package	206	387
Record medium	360	134
Recorders sound transport system	360	90+
Recording head	360	110+
Seed	47	56
Machine	53	545+
Planting	111	199
Sound reproducing, reel drive means	242	179+
Spinning twisting apparatus	57	31+
Spinning twisting process	57	31
Coating or impregnating	57	295+
Spool holders	D19	67+
Strand structure spun or twisted	57	260
Coated or impregnated	57	259
Covered or wrapped	57	235
Tape deck	D14	161+
Tearing or breaking	225	
Container opening	206	605+
Telegraphic	178	111+
Ticker	178	34
Taper Forming (See Pointing)		
Gauges	33	531+
Needle and pin making	163	
Paper tube making	493	296
Pencil (See pencil sharpener)	30	451+
Machine	144	28.1+
Taper tube making	29	DIG. 41
Tapering resistor	338	217+
Rheostat	338	138+
Thread milling	409	65+
Turning	82	15+
Wood	142	
Tapestries	139	408
Taping	242	7.1+
Hollow object	242	7.23
Pail or pipe	242	7.22
Laminating applying	156	
Object making	242	7.2+
Ring winding combined	242	6
Shoe upper making	12	59.5
Strip pasting and applying	156	
Tappers Electric Line	439	387+
Tappet		
Plow	171	71
Poppet valve operating internal combustion engine	123	90.48+
Tapping		
Container with cutter or punch	222	81+
Drilling and	408	
Fluid handling system	137	317+
Mains under pressure	137	317
Metallurgical plugging and	266	271+
Power plant exhaust re expansion	60	677+
Screw threading internally	408	222
Tar (See Pitch)		
Compounds containing		
Abrasive	51	305
Biocidal	424	
Coating or plastic	106	
Fuel liquid	44	311
Polymer ctg (See synthetic resin or natural rubber)		
Distillation	208	42+
Destructive	202	
Production by	201	
Mineral oil	208	22+
Making, treating and recovery	208	39+
Tar Sand		
Mineral oil from	208	400+
Treatment with liquid	208	390+
Target		
Aerial projectile type	273	348.1+
Projector combined	273	317+
Tethered projectile combined	273	331+
Thermal	273	348.1
Animal holder and releaser	119	15.6
Bowling	273	41
Cathode ray	313	461+
Design, hunting or shooting	D22	113+

	Class	Subclass
Disk		
Holder and releaser	124	43
Magazine	124	46+
Following, with photocell control	250	203.1+
For nuclear energy treatment	376	156
Global position determination	250	206.1+
Golf (See also golf)	273	181R
Moving	273	359+
Surface	273	127R
Surveying	33	293
X ray	378	143
Tartaric Acid	562	585
In drug	514	574
Tassel (See Product)	428	28
Garment trimming	D 5	7+
Making and attaching	223	46
Making	28	147
Ornamented handle	16	121+
Shoe lace tip	24	715.5+
Tatting		
Apparatus	87	52
Shuttles	87	58+
Shuttle winding	242	52
Tattooing Pen	81	9.22
Taurine	562	104
Taurocholic Acid	552	550
Tawing	8	94.29+
Taxidermy, Teaching	434	296
Exhibition panel	40	160
Taximeter		
Computer controlled, monitored	364	467
Fare recorder	346	
Register combined	346	15+
Fare register	235	33+
Tea	426	592+
Apparatus general	99	485+
Bag	426	77
Design	D 1	199
Balls	426	77+
Extracts	426	597
Infusers	99	279+
Ddesign	D 1	199
Teaching Devices	434	
Audio recording & visual means	434	308+
Audio recording combined	434	319+
Audio visual machines	D19	60
Blocks & cards	434	403
Dental	434	263+
Design	D19	
Graphic materials	434	314+
Design	D19	62+
Image projector combined	434	324+
Keyboard operation	434	227+
Music	84	470R+
Panel, chart or graph	434	330+
Science demonstrator	D19	62+
Telegraphy	178	115
Toy	D21	59+
Vehicle dual control		
Aircraft	244	229
Automobile	180	322
Teapot		
Handle, asymmetric	D 8	DIG. 9
Tear Gas		
Gun	42	1.8
Tearing	225	
Breaking or	225	
Container opening	206	601+
Container seam	206	631
Label pasting paper hanging	156	510
Machines	225	93+
Paper gauge	225	91+
Paper straight edge	225	91
Sheets, strands or webs	225	
Tear strip		
Cigarette package opening	206	264
Envelope	493	923*
Pliable container	493	930*
Roll, strip, or sheet having	206	824*
Tensile test	73	835
Tobacco stemming	131	319+
Weakened line	206	620+
Teat Cups	119	14.47+
Pulsator combined	119	14.38
Technetium		
Organic compounds	534	14
Tedders	56	370+
Rake combined	56	365+
Teeming (See Casting)		
Teepee	135	100
Tees, Golf	273	33

	Class	Subclass
Device for setting	273	32.5
Teeth (See Tooth)		
Dentistry	433	
Selection guide	433	26
Display package with	206	83
Gearing	74	457+
Lubrication	74	468
Razor guard	30	82
Static mold	249	54
Teething Devices	606	235+
Design	D24	45
Teflon T M	526	255+
Electrical cable insulation	174	110FC
Pipe & tubular conduit digest	138	DIG. 3
Surgery digest	128	DIG. 14
Telautograph	178	18+
Telegraphophones	369	
Telegraphs	178	
Combined with telephone		
Alternative	379	108+
Simultaneous	370	125
Earth transmission	375	6
Optical light transmission	455	600+
Photo telegraphs	358	400+
Radio transmission	375	
Ships telegraphs	116	21
Systems	178	2R+
Pulse	375	
Train telegraphs	246	7+
Writing on telautograph	178	18+
Telemetering	340	870.1+
Radio	128	903*
Telephone	128	904*
Telephone	379	
Answering device		
Answering-recording system	D14	141
Calling number recorder	379	142
Remote inquiry	379	76
Sound recorder or reproducer	379	70+
Attachment	379	441+
Base pad	248	
Index	40	336+
Pad	281	44
Roll type pad	281	11
Automatic systems	379	258+
Call	379	350+
With recorded message	379	69
Card	40	336+
Common control	379	268
Party line	379	182+
Booths	52	27+
Design	D25	16
Movable wall	52	71
Calling	379	352+
Card attached to telephone	40	336+
Coin collectors for pay stations	194	
Coin operated	379	143+
Computerized switching	379	284
Cordless	379	61
Design	D14	142+
Dial		
Dial structure	379	362+
Illuminated	362	24
Locking	379	445
Pulse transmitter	379	362+
Self luminous	40	337
Telephone system	379	258
Directory	40	371
Earth transmission	455	40
Handset	379	433
Headgear support	379	430
Key systems	379	156+
Light wave telephony	455	600+
Lights for telephones	362	88
Mechanical telephones	181	138
Message counter	379	139
Muffler for mouthpiece	181	242
Over composite line used for other services	379	90+
Pad or book holder combined	248	441.1+
Party line	379	177+
Plural phone systems	D14	241
Push button call transmitters	379	368+
Radio transmission	455	
Receiver and transmitter combined	379	433+
Repeaters	379	338+
Dial pulse	379	341+
Conversion	379	339
Voice frequency	379	338+
Repertory dialers	379	355+
Sets	379	419+

	Class	Subclass		Class	Subclass		Class	Subclass
Dyeing combined	8	495+	In drug	514	263	Thermometer	374	179+
Textile operation combined	28	167+	**Therapeutic**			Combined	374	141
Manufacturing	28	100+	Lamps	250	504R+	**Thermoelectric**		
Computer controlled, monitored	364	470+	Vitamins	424		Battery	136	200+
Molten metal, bleaching digest	8	DIG. 19	**Thermal**			Controls burner	431	80
Mordanting textiles	8		Activating adhesion in labeling	156	DIG. 21	Motors	310	306+
Nylon	8	DIG. 21	Activating closure	220	201	Expansion and contraction type	60	516+
Organic titanium compounds on	8	DIG. 5	Activator for bank protector	109	33	Systems	318	117
Polyester fibers	8	DIG. 4	Battery and cells	136	200+	**Thermoforming**		
Polyolefin	8	DIG. 9	Compositions or charges	149		Directly applied fluid pressure		
Polyvinyl halide esters or alchol			Controlled			differential	264	544
fiber modification	8	DIG. 1	Boiler feeders	122	451.1+	**Thermolysis (See Cracking)**		
Pressurized & high temperature			Closure	49	1+	**Thermometers**		
liquid treatment	8	DIG. 16	Dampers	126	287.5	Bimetallic stock	428	616+
Pressurized gas treatment	8	DIG. 15	Electrical alarms	340	584+	Case for	206	306
Printing			Electrical signals	340	584+	Design	D10	57+
Dyeing type	8	445	Flexible panels	160	1+	Electrically heated	60	516+
Printing type	101		Freeze responsive drain valves	137	59+	Expanding fluid	374	201+
Reinforcing or tire cords	57	902*	Mechanical alarms	116	101+	Expanding solid	374	187+
Resin bleach	8	DIG. 6	Mining	299	3+	Bi-metallic	374	204+
Sewing threads	57	903*	Non-fusible closure release	49	2	Heated	60	516+
Silicones	8	DIG. 1	Refrigeration, heat transmission	62	383	Hydrometer combined	73	442+
Spinning twisting twining	57		Temperature regulation	236		With float	73	449
Apparatus	57	1R+	Timers	60	516+	Liquid gauge combined	73	292
Processes	57	362	Vertically sliding sash			Making by glass working	65	62
Strand structure	57	200+	Weighing scale compensator	177	226	And calibrating	374	1+
Stock material	428	224+	Electric motors	310	306+	Structure	374	100+
Coated or impregnated	428	245+	**Electrical devices**			Testing of	374	1+
Swelling & stretching of fibers	8	DIG. 3	Batteries	136	200+	Thermostat	D10	50
Treating compositions	252	8.6+	Electromagnet and thermal			**Thermomigration**	437	14
Biocide containing	424		current switch	335	141+	**Thermonuclear Reaction Fusions**	376	100
Vinyl sulfones & precursors thereof	8	DIG. 2	Lightning arresters	361	117+	**Thermophore**		
Wave energy treatment	8	DIG. 12	Meters	324	105+	Compresses	128	399+
Weaving	139		Multiple circuit switch	200	3	**Thermophoric Compositions**	252	70
Weighting or mordanting	8	443	Pyromagneto electric	310	306+	**Thermopiles**	136	224+
Wire fabrics & structure	245		Thermal current switch	337	14+	**Thermoregulators**	236	
Texturing Thread	28	247+	Thermal switch	337	298+	**Thermos T M Bottle**	215	12.1+
Thatch Roof	52	750	Expansive instruments	33	DIG. 19	Ceramic	215	12.1+
Thaumatrope	446	244	Gas separation	55	209	Design	D 7	608+
Thawing			Processes	55	81	Metallic	220	420+
Antenna	343	704	Heat exchange devices	165		Vacuumizing	220	420+
Compositions	252	70	I-aging composition	430	964*	Prefilled	220	411+
Defrosters	62	272+	Impedance in amplifier system path	330	143	**Thermostat (See Thermometers)**		
Automatic	62	151	Insulated rigid containers	493	903*	**Thermostatic**		
By ventilation	98	2	Liquid level or depth gages	73	295	Control and regulation		
Frost indicator combined	62	128	Measurement	374		Arc lamps	314	89+
Frosted window panes	52	171+	Motor	60	516+	Burner	431	18+
Process	62	80+	Neutrons	376	158	Refrigeration	62	132+
Earth	405	131	Neutrons	376	347	Temperature and humidity	236	
With excavating	299	3+	Reactive building component	52	232	Valve cutoff	137	457
Engines	123	41.1+	Recording	346	76R	Materials		
Ice tank	62	349+	Surgical applications	128	362+	Composition	252	70
Process	62	73	Inhalers	128	204.17	Compound metal	428	616+
Pipes	138	32+	Instruments	606	27+	Switches	337	298+
Radiators by own steam	165	73	Kinesitherapy	128	24.1+	Electrically heated	337	14+
Theater			Medicators	604	20+	Electromagnet	335	141+
Arena buildings auditoriums	52	6+	Orthopedics	128	68.1	**Thetine**	549	89
Toy	446	82+	Switches	337	298+	**Thiadiazines, 1,2, 3-**	544	8+
Car drive in	52	6+	Electrically heated	337	14+	**Thiadiazoles, 1,2, 3-**	548	127
Chairs indicating or illuminating			Tests	374		**Thiadiazoles, 1,2, 4-**	548	128+
Electric indicator	340	667	Borehole formation logging	73	154	**Thiadiazoles, 1,2, 5-**	548	134+
Furniture with light	362	131	Calorimeters	374	31+	**Thiadiazoles, 1,3, 4-**	548	136+
Indicators	116	200+	Gas analysis	73	25.1+	**Thialdehydes**	568	20
Curtains	160		Volume or rate of fluid flow meters	73	204.11	**Thiamines**	544	327
Derrick hoist	254	266+	**Thermionic Devices**	313		In drug	514	276
Illusion devices	272	8R+	Circuit element combined structurally	315	32+	**Thiazines**	544	3+
Lights	362	227+	Electric lamps	313		Bio-affecting and body treating		
Control systems for	315		**Thermistor**	338	22R	compositions	514	222.2+
Dimming systems	315	291+	Systems	323	369	**Thiazoles**	548	146+
Switching systems	315	313+	**Thermit T M**			Azo compounds	534	751+
Seat	D 6	334+	Casting	164	53	Heavy metal ctg	534	701+
Stage appliances	272	21+	Compositions	149	37	Synthetic rubber or natural rubber		
Design	D25	58	Explosive or thermic	149	37+	accelerator	525	349
Stages	52	6+	Composition containing	149	37+	Thiazoles, 1, 2-	548	206+
Movable	52	29+	Track joint welder	104	15	Thiazoles, 1, 3-	548	146+
Tickets	283	53	Type reduction	75	959	**Thiazolidine**	548	146+
Theelin	552	625	**Thermocouple**			**Thiazoline**	548	146+
Theft			Body treating electric elements			**Thief Devices (See Sampler)**		
Alarms			Body wear	128	391+	**Thienamycin**	540	351
Electrical	340	568+	Electrical applicators	128	783+	**Thills**		
Mechanical	116	75	Chemical analysis, apparatus	422	82.12	Coupling jacks	254	28.5
Prevention, motor vehicle	180	287+	Electric motor	310	306+	Land vehicle animal draft	278	33+
Theine	544	274	Motive power system	318	117	Couplings	278	52+
Theobromine	544	274	Electric relay circuit	361	160+	Supports	278	81+
Theodolites	33	290+	Electricity to heat to electricity	322	2R	Tugs	54	50+
Design	D10	66	Generator systems	322	2R	**Thimbles**	D 3	29
Horizontal and vertical angle	33	281	Making (process)	437		Bushing	16	2+
Horizontal angle	33	285	Refrigerator	62	3.2+	Clews	114	115
Optical	356	138+	Peltier effect battery	136	203+	Combined with cutting devices	7	121+
Optical readout	33	1T	Structure	136	200+	Conduit end lining	174	83
Theophyllines	544	267+	Temperature control system	236	69+	Cutter tool combined with	7	121+

	Class	Subclass
Hardware	16	108+
Insulator pin type socket	174	200
Making	72	356
Packed shaft joint	277	112
Sewing apparatus	223	101
Design	D 3	29
Stovepipe	126	314+
Ventilation valved	98	118
Thinner		
Composition coating or plastic	106	311
Hair	30	195
Plant	172	534
Cross-art collection	172	900*
Thioacetic Acid	562	26
Esters	558	250
Thioacetol	568	63
Thioamides	564	74+
Thiobarituric Acids	544	299+
Thiobiurets	564	22
Thioborate Esters	558	285
Thiocarbamate Esters	558	232+
Thiocarbamic	562	27
Thiocarbanilide	564	26+
Thiocarbazides or Semicarbazides	564	18+
Thiocarbazones or Semicarbazones	564	19+
Thiocarbonic Acid	423	415R
Esters	558	243+
Cellulose	536	60+
Thiocarboxamides	564	74
Thiocarboxylic Acid	562	26
Esters	558	230+
Cellulose	536	60+
Thiocarboxylic Acid Anhydrides	562	886
Thiocyanate	423	366
Esters	558	10+
Thioethers	568	38+
Thioglycollic Acid	562	512
Esters	560	147
Thiohydroxamate Esters	560	312+
Thioimidic Acid Esters	558	1+
Imidothiocarbonates	558	2
Pseudothioureas	558	4
Thiohydroximidic acid esters	558	3
Thioindigo	548	457+
Bicyclo ring system	549	52
Indole nucleus ctg	548	464
Thioketones	568	20
Thiokol T M (polyalkylenepolysulfide)	528	388
Thiolsulfinate Esters	560	310
Thiolsulfonate Esters	560	307
Thiolsulfonic Acids	568	21
Thiomorpholines	544	59+
Thionitrate Esters	558	480+
Thionitrite Esters	558	488
Thionyl Chlorid	423	462
Thiooxamides	564	77
Thiophene	549	29+
Indole nucleus combined	548	464
Phosphorus attached	549	6
Thiophenols	568	67
Addition salts	564	280
Thiophosgene	562	839
Thiosulfate Esters	560	308+
Thiosulfates, Inorganic	423	514+
Thiosulfenamides	564	100
Thioureas	564	17
Stabilized or preserved	564	3
Synthetic rubber or natural rubber accelerator	525	352
Thioxanthenes		
Chalcogen or nitrogen attached	549	27
Third Rail		
Conductor	191	29R+
Shoe	191	49
Snow excavator	37	204
Switch	246	419
Thiuram Disulphide	564	76
Thorium		
Alloys	420	1+
Composition radio active	252	625+
Compound recovery	423	
Electrolytic production	204	1.5
Metallurgy	75	394+
Nuclear fuel material	376	181
Nuclear fuel material	376	412
Nuclear fuel material	376	901
Organic compounds	534	11+
Thread		
Catchers		
Spinning and twisting frames	57	353
Counters of threads in fabrics	356	242

	Class	Subclass
Cutters on sewing machines	112	285+
Dental floss holders	132	323+
Envelope affixing	493	376
Frames	28	149+
Generating helix or thread	409	65+
Guides		
Knitting machines	66	125R+
Spinning and twisting frames	57	352+
In bolt to nut lock		
Gripper	411	246+
Jam nut oppositely threaded nut	411	244
Intersecting thread, on bolt	411	245
Jam nut reversed threads on one	411	243
Thread lock by pitch difference	411	307
Thread lock differential thread	411	263
Thread lock dissimilar threads	411	308+
One deforms other	411	309
One deforms other near crest	411	311
Thread lock interrupted threads	411	304
Thread lock non-circular thread	411	282
Thread lock plural threads, oppositely biased	411	312
Thread lock rocking thread section, on bolt	411	264
Lock, bolt or nut to substructure	411	106
Lock, bolt to nut	411	259+
Including elastic gripping	411	301+
Packages	206	389+
Protectors for spinning and twisting frames	57	352+
Sensing elements on sewing machine	112	278
Sewing	57	903*
Sewing machine handling of	112	
Spooling machines	242	16+
Textile manufacture		
Coating or dyeing combined	28	217+
Finishing	28	217+
Spinning	57	
Winding and reeling	242	
Threaded Device (See Bolt, Nut, Screw)		
Bottle or jar cap	215	329+
Bottle or jar stopper	215	356
Fastening	215	283+
Interrupted thread fastening	215	281
Chuck or socket adjusting	279	
Fastenings	411	
Bolt or nut substructure lock	411	81
Bolt to nut lock	411	190+
Chuck grip	279	7
Element, deformed	411	333+
Making	10	2+
Nuts	411	427+
Wire nail	411	439+
Making		
Die cutting	408	215+
Drilling and tapping tool	408	215
Glass molding presses	65	309+
Grinding apparatus	51	95R
Grinding methods	51	288
Metal molding apparatus	164	216
Milling	409	65+
Platen rolling	72	88+
Rolling	72	104
Spirally corrugating tubes	72	103+
Static mold	249	59
Swaging	10	152R+
Tapping	408	215+
Turning	82	110+
Pipe joint	285	390+
Thimble	285	386+
Protectors		
Disassembly impact receiving	29	277
Pipe	138	96R
Safe closure	109	72
Tool adjusting	81	
Threader and Threading		
Buttons and staples	227	32
Setting combined	227	34
Curtain rod	223	105
Loom shuttle	139	221+
Implements	139	381+
Replenishing loom	139	259
Needle		
Hand	223	99
Sewing machine	112	224
Sewing machine needle setting combined	112	225
Stuffer box crimpers	28	268+
Textile twisting	57	279+
Three Dimensional Object		
Imitation or treated natural	428	15

	Class	Subclass
Three Dimensional Photography (See Stereoscopy)		
Three Phase		
Current meters	324	107+
Motors	310	159+
Threonine	562	570
Threshers	460	
Combines	56	122+
Motorized	56	14.6
Design	D15	10+
Thresholds for Buildings	49	467+
Throstle Frames	57	
Throttle (See Valve)		
Engine speed responsive	60	906*
Governors for internal combustion combustion engines	123	319+
Nozzle	239	546
Pneumatic piano	84	41
Pump regulator	417	1+
Train control	246	186+
Thrower (See Projector)		
Belt	474	101
Conveyor	198	638+
Fluid sprinkling, spraying and diffusing	239	
Or splasher	239	214+
Gaseous suspension classifier	209	153
Grenade		
Firearm attachment	42	105
Hand	102	486
Ordnance	89	1.1
Mechanical	124	
Pneumatic conveyor	406	71
Point switch railways	246	264+
Scattering non-fluid material	239	650+
Water guns	222	79
Throwoff		
Cradle rocker	5	106
Printing machine		
Bed and cylinder	101	283+
Multicolor bed and cylinder	101	191+
Thrum Mechanism	112	69
Thrust		
Antifriction bearing	384	590+
Radial thrust bearing combined	384	452+
Balanced pistons	92	126
Bar		
Typewriter	400	390+
Bearings plain	384	420+
Railway car journal	384	188
Compositions for developing	149	
Thumbtack	411	439+
Applier	81	44
Remover	254	18+
Thymol	568	781
Thyratron (See Space Discharge Device Space Discharge Device)		
Thyroid Extracts		
Medicine containing	424	568
Thyroxin	562	447
Ticket	283	
Booth	D25	16
Cases for pocket	206	39
For card	150	147+
With cutter	30	124
Dispensing package	206	39.4+
Fare boxes	232	7+
Holders	206	39+
Railway cars	40	642+
Letter box deposit or collection record	232	18
Machine	221	
Magazine	312	50
Making		
Pin ticket tag applying	493	375+
Printing	101	66+
Registering device	235	32
Theatre	283	53
Ticklers, Carburetor Primers	261	DIG. 73
Ticktack	446	417
Tiddlywink Game	273	353
Tide or Wave Motors	60	497+
Channel	405	76+
Combined with other	60	495+
Gravity	60	639+
Weight	185	30
Fluid current motors	416	
Gas pump actuator	417	330+
Liquid piston type	417	100
Generator drive	290	53
Electric control combined	290	42

	Class	Subclass
Prison bathrooms	4	DIG. 15
Seat	4	234+
Pads and covers	4	242+
Seat or cover	D23	311+
Water additive or substitute	4	DIG. 1
Toilet Water, Perfume	D28	5
Tokens		
Checks labels and tags	283	74+
Coin type	40	27.5
Coin	D11	95+
Tolidine (See Benzidine)		
Tollers	73	863+
Toluene (See Aromatic Hydrocarbon)	585	400+
Oxidation of	562	409+
Toluidin	564	305
Tombstones (See Monument)	52	103+
Design	D99	17+
Name plate for	40	124.5
Picture attachment for	40	124.5
Tompions	89	31
Tone		
Arms of phonographs		
Electrical	369	135+
Mechanical	369	158
Control means	333	24R+
Equalizer network	333	28R
With amplifier	330	
Automatic impedance control	330	144+
Automatic thermal impedance control	330	143+
Cathode feedback	330	94+
Feedback	330	109+
Input coupling impedance	330	157+
Interstage coupling impedance	330	185+
Output coupling impedance	330	195+
Musical pitch (See music)		
Toner	430	105+
Concentration control	118	689+
Radiation imaging	430	965*
Tong (See Lazy Tongs)		
Agriculture or forestry	D 8	4
Animal catching	119	154
Extension ladder	182	157+
Fire escapes	182	69
Fireplace equipment	D 8	52+
Fireplace tong stand	D 6	416
Food preparation & serving	D 7	686
Handling implement	294	3
Clothes	294	8.5
Design	D 7	686
Fire	294	11
Grapple	294	119
Harvester self raking	56	165
Making	76	
Metal drawing appliance	72	290
Roofers sheet metal seaming	29	243.5+
Tools	81	300+
Tongue		
Depressors	128	15+
Atomizers with	128	200.15
Design	D24	19
Envelope closure	229	84
Sealing	229	82
Harvester type	56	218
Tilting platform central cutter	56	188
Knitting needle making	163	3
Land vehicle		
Animal draft	278	33+
Antivibrators	280	108
Attachments	280	155
Extension actuators	280	46
Trailer, adjustable	280	462
Truck	280	82+
Lock type shoe sole sewing machine	112	33
Punching tie band	29	21.1
Railway switch point	246	442+
Rigid vibrator music instruments	84	408
Shoe	36	54
Single throw lever actuated by	414	436
Straddle row cultivator without	172	338+
Vehicle brake operator		
Auxiliary	188	115+
Movable	188	119+
Tonometers		
Hardness testing	128	645+
Medical diagnostics	128	645+
Musical pitch	84	454
Sphygmomanometer	128	677+
Tonsillotome	606	110+
Hemostasis	606	111+
Other treatment combined	606	34+

	Class	Subclass
Tools (See Implements; Instruments)	81	
Abrading and polishing	51	
Advancing means	173	141+
Assembling	29	700+
Belt lifters	474	130
Boxes	206	349+
Chest	312	DIG. 33
Cleaning	15	104R+
Cleansing means	173	57+
Fish	17	66+
Hog carcass scrapers	17	19
Ship hull	114	222
Closure applying and removing		
Ammunition capping	86	37+
Ammunition shell closing	86	28
Bag with opening device	206	610
Bottle and jar closing	53	287+
Bottle cap with opener	215	228+
Box openers	254	18+
Can head with opener	220	260+
Can openers	30	400+
Cap removal facilitated	215	295+
Closure removing	81	3.7+
Complex for deforming metal	72	394+
Compound	7	
Magnetic feature, collection	7	900*
Resilient pivot, collection	7	901*
Cooler	62	293
Cutlery	30	
Dental	433	25+
Orthodontic	433	3+
Design of	D 8	
Detachable, portable, power-driven	30	500*
Diamond	125	39
Drill having diamond edge	408	145
Holders for abrading and polishing	51	229
Making blank or process	76	101.1+
Making dies	76	107.1+
Making drills	76	108.1+
Precious stone working	125	30.1
Driving means	173	
Metal deforming machine	72	429+
Vibration isolation	173	162.1+
Earthworking	172	
Boring or drilling	175	
Hand forks and shovels	294	49+
Hand manipulated cultivating	172	371+
Hand rakes	56	400.1+
Handling hand and hoist line	294	2
Mining and disintegrating in situ	299	79+
Spade or shovel with cutter	7	116+
Subsoil irrigators and weed killers	405	35+
Electrically heated	219	221+
Elongated handles	D 8	DIG. 7+
Farriery	168	45
Feed	173	141+
Earth boring type	175	162
Food handling	D 7	669
Cherry seeding	30	113.1+
Corn husking	460	26
Corn shellers	460	45
Cutlery	30	
Meat and vegetable preparing	30	
Oyster openers	17	74+
Peach stoner	30	113.1+
Raisin seeding	30	113.1+
Scoring	30	164.9
Hair dressing or manicuring	D28	9
Hand	81	
Assembly or disassembly	29	270+
Cooler	62	293
Handle	81	489+
Handles	7	167+
Handling	294	
Holders		
Carried by belt	224	904*
For dental machines	433	77+
For metal lathes	82	159+
Impacting device	173	90+
Leather	69	20
Boot and shoe making	12	103
Making	76	
By die rolling	72	184+
Cutting spirally grooved	409	65+
Grinding to shape or sharpen	51	
Sharpening electrolytically	204	129.1+
Manicuring	132	73+
Medical or surgical	D24	
Metal working		
Assembly, magnetic, or indicia tools	29	DIG. 105

	Class	Subclass
Bolt nail nut rivet and screw making	10	
Cutting & punching sheets & bars	83	
Deforming	72	462+
Design	D15	122+
Horseshoe making	59	60
Metal tool and implement making	76	
Soldering	228	349+
Tool or work attached	29	DIG. 104
Wire working	140	
Motors		
Multipurpose	7	
Nut lock tool	81	10
Pliers and tongs	81	300+
Pushing and pulling	254	
Rack-type holder	D 8	71
Rack-type holder	D 8	349+
Self heating with burner	126	401+
Sharpener	83	174
Shaving or hair clipping	D28	44+
Shoe making	12	103
Stone working	125	36+
Telephone dialing	379	456
Tire		
Bead spreading	254	50.1+
Breakers for split rims	157	1.35+
Repairing	81	15.2+
Repairing with vulcanizing	425	12
Setting and removing	157	1.1+
Valve stem cap with tool	152	431
Top and dies	408	215+
Valve removal tools	29	221.6
Vehicle-mounted	83	928*
Vises	269	
Barrel chamfering	147	42
Design, hardware	D 8	
Design, machinery	D15	122+
Turning	142	56
Wrenches	81	52+
Impact	81	463+
Tooth (See Teeth)		
Agricultural and earthworking implements		
Cultivator tooth structure	172	713
Diggers	171	
Excavating tooth	37	142R
Hand cultivating tool	172	378+
Hand rakes	56	400.21
Harrows	172	
Horse drawn rakes	56	400
Planting drill	111	154+
Raking machine	56	344+
Brushes	15	167.1
Design	D 4	104+
Dispensing	132	311
Holder	D 6	524+
Kit	132	308
Material supply combined	401	268+
Receptacle for	206	361+
With massage tool	15	110
Comb teeth cutting	144	26
Combs	132	152+
Dentifrices	424	49+
Dentistry	433	167+
Design	D24	10+
Pulp tester	433	32
Gear	74	457+
Lubrication	74	468
Testing	73	162
Inserted tooth metalworking cutter	407	33+
Kit		
Brush or paste retained	401	118+
Making		
Apparatus	29	23.1
Forging	72	376
Gear cutting	409	1+
Method	29	895.31
Saw tooth	76	25.1+
Toothed article	83	908*
Toothed cylinder		
Paste	424	49+
Holder-squeezer	D 6	541
Tube	222	92+
Picks	132	321+
Design	D28	64
Making	144	185+
Packaging	53	236
Receptacle	206	380+
Stone sawing	125	22
Tooth article making	83	908*
Vehicle fender	293	53

	Class	Subclass
Power plant	60	
Pump with motor	417	379+
Railway car axle drive	105	96.2
Scales	177	208+
Guide for a shiftable handler	414	918*
Heat		
Heat exchangers	165	185
Liquid heaters and vaporizers	122	367.1
Internal combustion engine	123	197R
Lamp electrode power control	314	79+
Latch operated clutch	192	24+
Line	174	103+
Concentric	174	103+
Involving line parameters	333	236+
Resonant	333	236+
With nonsolid insulation	174	28+
Machine element	74	
Belt & pulley drive system	474	
Planetary gearing	475	
Motor vehicle	180	337+
Having two wheels in tandem	180	230
Safety belt responsive to transmission thereof	180	269
Railway car axle box mounted	105	96.1
Rotary crane	212	247+
Tower or post	D25	126+
Transmutation of Elements Fission	376	156
By fusion	376	100
Transom (See Window)		
Operators	49	356
Railway truck	105	208+
Bolster combined	105	202+
Transparency		
Product with abnormal	428	918*
Testing and inspecting sheets	356	432+
Viewer	40	361+
Walled receptacle	220	662+
Transplanters	111	100+
Biodegradable receptacle	47	74+
Dibbling implements	111	92+
Receptacles	47	73+
Big ball tree	47	76
Seedling	47	77+
Tree	111	101
Transponders		
Radar	342	42
Transportation (See Vehicle Handling)		
Advancing material of indeterminate length	226	
Printed matter	283	23+
Vehicle design	D12	
Transporting Attachments Harvesters	56	228
Transposers		
Electric conductors	174	33+
Music		
Indicators with transposing dial	84	480
Keyboard transposers	84	445+
Teaching device with transposing dial	84	474
Teaching device with transposing slide	84	473
Tracker and music sheet shifted to change key	84	145
Transuranium Compounds		
Inorganic	423	250+
Organic	534	11+
Trap & Trapping		
Animal etc	43	58+
Bank protection devices	109	3+
Burglar	43	59
Chamber dispensing	222	
Agitator combined with movable	222	226+
Agitator or ejector for movable	222	216+
Barometric type stationary	222	457
Fluid flow discharge from movable	222	636
Into conduit	406	63+
Inkwell dip well	222	576+
Jarring movable	222	197
Movable or conveyor type	222	251+
Removable from container for discharge	141	110+
Rotatable container with	222	170
Stationary	222	424.5+
Stationary with cutoffs	222	425+
Supply movable relatively to	222	162
Tiltable assembly	222	454+
Comminutor projected material	241	82
Currycomb combined	119	87
Deposit receptacle coin	232	55+
Design	D22	119+

	Class	Subclass
Doors		
Ladder combined	182	77+
Manhole cover	404	25+
Railway car platforms	105	426+
Vertically swinging outside		
False or picking key	70	390
Fish traps	43	100+
Fish, game, vermin traps	D22	119+
Design	D22	119+
Fish locating instrument	D10	46+
Hooks	43	34+
Design	D22	144
Lures	D22	125+
Nets, landing spears, etc	D22	135+
Gas & liquid (See seal, liquid)		
Automatic heating radiator with	236	41
Automatic temperature etc regulation	236	53+
Bath etc	4	191+
Beverage infusers	99	301
Boiler	122	
Cigars etc	131	331
Concentrating evaporators	159	31
Decanters with	210	513+
Dispenser inlet or outlet	222	188
Filters with sediment	210	299+
Fluid current conveyor	406	62+
Fluid delivery track sanders	291	11.1+
Fluid distribution	137	171+
Gas and liquid contact apparatus	261	86
Gas separator combined	55	355
Manufacturing processes	29	890.146
Nonrefillable bottle	215	15
Sewerage	137	247.11+
Smoking pipes etc	131	201+
Spittoon	4	283
Steam	137	171+
Water closet	4	300+
Ion	335	210+
With cathode ray tube	313	424+
With TV picture tube	313	424+
Nests	119	47+
Railway plow	104	145
Safety pockets for garments	2	254
Steam	137	171+
Thermostatic	236	53+
Target		
Centrifugal projectors	124	4+
Disk holders and carriers	124	42+
Disk magazines	124	46+
Projectors	124	
Trap shooting target handling	273	406
Vehicle fender	293	15+
Trapezes	272	61
Trapping	43	
Trash (See Refuse; Waste)		
Bundle	428	2
Metal	428	576
Burners	126	222+
Furnace structure for garbage and sewage	110	235+
Furnace structure for refuse	110	235+
Can liners	220	404+
Can support	248	907*
Clearers for plows	172	606+
Travelers		
Animal hitching	119	120
Ship	114	204+
Ship mast	114	112
Textile spinning rings and	57	119+
Track hardware	16	87R+
Traveling		
Bags	206	278
Design	D 3	30.1+
Grates	110	267+
Special sewing machine	112	121.14
Wave loom	139	436
Wave tube		
Demodulator	329	
Modulator	332	
With delay line	315	3.5+
Traverse Rod	D 8	377
Trawl	43	9.1+
Lines	43	27.4
Tray (See Dish; Receptacle)		
Animal stock	119	103
Automobile supported table	108	44+
Bank protection deal	109	19
Cabinet showcase type	312	126+
Cafeteria dispenser unit	D34	14
Circular household article	D 7	552+

	Class	Subclass
Coin pickup or delivering		
Coin deliverers	232	64+
Special receptacles for coins	206	.8+
Compartmented	D 7	553
Dental equipment	433	77+
Designed for glasses	D 7	553
Drying etc	34	237+
Removable tray type	34	192+
Egg candling	356	61
Gas filter tray	55	494
Handle supported	294	172
Special article	206	557+
Heaters	261	DIG. 3
Ice cube	249	117+
Incubator	119	43+
Making wood	144	33
Paper receptacle	229	190+
Photographic fluid treating	354	331+
Rack	211	126+
Wall or window mounted	211	88
Sample case	190	17
Slide projector	40	361+
Special receptacle	206	557+
Tobacco ash receiving	131	231+
Trunk	190	35
Warming	D 7	363
Tread		
Burglar alarm operated by	116	98
Electric circuit controller	200	86R
Burglar alarm	340	565
Railway car wheel	295	31.1+
Railway rail with	238	122+
Joint with bridge	238	218+
Surface per se	238	148
Resilient tire		
Antiskid	152	208+
Armored	152	167+
Shoe antislip	36	59C+
Testing	73	146
Abrasion or rubbing	73	8
Vehicle step	280	169
Wheel guard	293	58
Treadle		
Closure actuator	49	263+
Grinding wheel mount	51	167
Levers	74	512+
Machine element	74	561
Sewing machine conversion	112	217.4
Trackmans car drive	105	93
Treadmill		
Closure operating	49	262
Exercising devices	272	69
Motors	185	
Occupant propelled vehicle	280	228
Treasure Locator (See Prospecting)		
Tree		
Covers	47	20+
Felling	144	34R
Portable circular saw	30	360+
Hacks	30	121
Handlers for trees & poles	414	23
Husbandry	47	
Imitation	428	18+
Tree- like ornament lighted	362	123
Impregnation	47	57.5
Plants		
Broadleaf	PLT	51+
Conifer	PLT	50
Fruit	PLT	33+
Nut	PLT	30+
Ship masts		
Cross and trestle	114	92
Shoe	12	114.2+
Stand for supporting	248	519+
Supports and props	47	42+
Surgery	47	8
Trunk		
Guards	47	23+
Insect traps	43	108
Trellis	47	44+
Design	D25	100
Tremolos	381	62
Brass wind instrument tremolo	84	401
Electric oscillator having	331	182
Electric tone generator tremolo	84	629
Fan tremolo for reed organs	84	374
Stringed instrument	84	313
Valve tremolo for pipe organs	84	348
Trepan	606	176+
Trephine	606	172+
Trestle		
Scaffold or saw horse	182	181+

Entry	Class	Subclass
Unitary foldable	182	153+
Tire setter	157	7
Trees ship mast	114	92
Type propped extension ladder	182	105
Triamcinolone	552	566
Triangle		
Geometrical instrument	33	474
Musical instrument	84	402+
Reflective triangle for highway use	40	903*
Triarylmethyl	552	101+
Triazines	544	3+
Azo compounds	534	751+
Stilbene ctg	544	193.1
Triazoles, 1,2, 3-	548	255+
Triazoles, 1,2, 4-	548	262.2+
Triazolobenzodiazepinone	540	499
Tricarballylic Acid	562	590
Esters	560	190+
Trichloracetic Acid	562	602
Esters	560	226+
Trickler		
Coffee pots	99	306
Heat exchanger	165	115+
Ice melt cooler	62	312+
Process	62	64
Textile treating	68	205R+
Trickling Filter	261	94+
Liquid purification	210	150+
Processes	210	601+
Tricot Knitting Machine	66	87
Tricycle (See Velocipede)		
Design	D12	112+
Triethanolamine	564	506
Triethylene Diamines	544	351
Trigger (See Trip and Trigger)		
Circuits		
Electronic tube type	328	191+
Gas tube type	315	
Firearms	42	
Triglycerides	436	71*
Trigonometry, Teaching	434	211
Trimelletic Anhydride	549	245
Trimmers (See Cutters)		
Billiard cue	30	494
Brick	425	289+
Brick	83	
Burner wick	431	120
Capacitor	361	271+
Cigar and cigarette	131	
Condenser	361	271+
With inductor	334	78+
Cooking mold	99	430
Embroidery	83	910
Farriery hoof	168	48.1+
Hat	223	24
Hedgerow	56	233+
Lawn	172	13+
Photograph	83	
Pie	425	293
Plastic block press	425	289+
Sewing machine	112	122+
Attachment	112	130
Sheet cutting	83	
Scrap cutting	83	923
Shoe inseam machine	12	4.3
Shoe sole and heel edge	12	85+
Textile selvage	139	302+
Waffle iron drip	99	375
Trimming		
Apparel	223	44+
Design	D 5	7+
Battery grid	83	903*
Brush and broom	300	17
Cartridge case	29	1.32
Chain sizing and	59	29
Dental plaster	433	144
Design	D 5	7+
Garment	2	244+
Making and attaching	223	44+
Gun	42	85
Harness	54	75+
Design	D30	134+
Hat	2	186
Banding	223	22
Brim	223	16
Horseshoe	59	59
Leather skiving and splitting	69	9.3
Packages	206	389+
Pearl button surfacing	79	10
Screw threading dies	408	215+
Shoe sole and heel machines	12	
Design	D15	122+
Stock material	428	
Woodworking box shaping	144	135
Trinitrotoluene	568	935
Containing	149	
Explosives containing	149	69
Explosives containing	149	107
Triode	313	293+
Amplifier	313	293+
Detector amplifier	313	293+
Diode high mu	313	303
With two cathodes	313	5
Diode pentode	313	298
With plural cathodes	313	6
Direct coupled twin	313	3
Double	313	301
With two cathodes	313	6
Dual grid	313	297
Duo diode	313	303
Duplex diode	313	303
With plural cathodes	313	5
Duplex diode high mu	313	303
With two cathodes	313	5
Gas	313	581+
Hot cathode type	313	592
Germanium	357	
Heptode converter	313	298
With two cathodes	313	6
Hexode converter	313	298
With two cathodes	313	6
High frequency	313	293+
High frequency twin	313	301
With two cathodes	313	6
High mu	313	293+
High mu twin	313	301
With two cathodes	313	6
Low mu	313	293+
Low mu uhf	313	293+
Medium mu	313	293+
Medium mu twin	313	301
With two cathodes	313	6
Power amplifier	313	293+
Power amplifier twin	313	301
With two cathodes	313	6
Triple diode	313	303
With plural cathodes	313	5
Triple diode high mu	313	303
With plural cathodes	313	5
Twin	313	301
With two cathodes	313	6
Twin power amplifier	313	301
With two cathodes	313	6
Two grid	313	297
U H F	313	293+
Trioxane	549	367
Trip and Trigger		
Animal trap	43	58+
Belt conveyor unloader	198	633+
Cable hoist	212	116
Closure fastener	292	332+
Electric switch latch trip		
Clock train	200	39R+
Double snap, blocked	200	411+
Double snap, restrained	200	415+
Electromagnetic	335	174+
Reciprocating	200	434+
Single snap, blocked	200	470
Single snap, restrained	200	471+
Thermal current	337	70+
Expansible chamber motor		
Trip gear	91	338
Firearm	42	70.1+
Guns and ordnance	89	27.11+
Automatic gun	89	132+
Electric	89	28.5+
Harvester		
Shocker	56	401+
Mechanical gun	124	31+
Odometer	235	97
Operated file cutter	76	15+
Planting		
Depositing mechanisms	111	34+
Railway		
Cable gripper	104	205+
Railway switches and signals		
Derailment contact trips	246	171+
Track trip automatic electric	246	76
Track trips	246	201+
Train trips	246	359+
Released valve	251	66+
Rotary printing member	101	234+
Saw setting		
Pivoted set	76	62
Sliding set	76	67
Ship anchor	114	210
Swinging door lock	70	157
Trip mechanisms	169	DIG. 3
Trips	74	2+
Compressing and binding	56	436+
Vehicle brake operator	188	111
Vehicle fenders		
Dash and wheel	293	33
Drop fender	293	34+
Tripelennamine, in Drug	514	352
Triphenylmethanes	552	101+
Triple		
Bond hydrocarbon		
Purification	585	800+
Synthesis	585	534+
Diode	313	1
Mirrors	350	102+
Tripod	248	163.1+
Camera supports	354	293
Design	D16	244+
Ladder prop	182	169
Support stand	248	163.1+
Heads	248	177+
Trirams	114	61
Trituration	241	
Trocars	604	264+
Trolley		
Crane	212	142.1
Swing boom carries	212	250+
Vertically	212	257
Excavating		
Cable operated trolley supported	37	117
Ladders	182	36+
Linear traversing hoists		
Rope carrier	212	119
Railway		
Collectors	191	50+
Conductors	191	33R+
Nonretracting trolley stops	191	95
Rails	104	106+
Retrievers	191	85+
Suspended single rail	105	148+
Transfer	104	96+
Railway switches	246	419
Trolley actuated controller automatic	246	84
Trolley actuated controllers	246	252
Trolley completed circuits	246	254
Wire support	191	40+
Trombones	84	395+
Electric transmission line	333	219+
Mutes	84	400
Section in high frequency transmission lines	333	33+
Tropanes	546	124+
Trophy	428	542.4
Design	D11	131
Tropine	546	127
Trotlines	43	27.4
Trough		
Animal feeding	119	61+
Design	D30	121+
Belt conveyor	198	818+
Burners	431	52+
Chute unloading	193	6
Closure	49	408
Dryer gravity flow	34	166
Flume	405	119+
Holder for soldering	269	
Railway track rail	238	129
Roof eaves	52	11+
Support	248	48.1
Sprinkler supply	239	724+
Vibrating conveyor	198	771
Wood sawing log transfer	83	707+
Trousers	2	227+
Design	D 2	202+
Undergarment	D 2	6+
Forms	223	72+
Guards and straps	24	72
Hangers	223	95+
Knee protectors	2	62
Leg guards attached to	2	23
Stretching	223	63
Union type	2	79
Trowels	15	235.4
Design	D 8	10
Handle fastening	403	

927

	Class	Subclass
Multiway unit	137	625+
Bypass	137	599+
Gate	137	625+
Plural valves sequential opening	137	630.16
Plural valves single outlet	137	602+
Unit	137	625+
Music (See music)		
Needle valves	261	DIG. 38
Optical element displaced or rotated	350	484
Oscillating		
Fluid flow meter combined	73	268
Outlet exhaust and discharge		
Bath or basin	4	203+
Bath or basin fitting	4	194
Bath or basin overflow	4	198+
Bath or basin trap	4	197
Filter	210	418+
Filter, faucet attached	210	449
Four cycle engine	123	84+
Furnace ashpan	110	169
Furnace spark arrester	110	129
Gas separation	55	417
Grain huller	99	611
Internal combustion engine	123	188R+
Sewerage catch basin	210	109+
Sewerage trap	137	247.13+
Spittoon	4	266
Spittoon floor orifice	4	276
Water closet	4	378+
Pipe	251	
System	137	
Piston for pumps	417	481+
Pivoted	251	298+
Check	137	527+
Closet bowl	4	421+
Cowls	98	74
Register	98	102
Position indicator	116	277
Pressure reducing and regulating	137	505+
Pressure relief	137	455+
Puppet or poppet	251	82
Abrading machine	51	29
Acetylene generator	48	53+
Coolant containing	123	41.16
Four cycle engine	123	79R
Internal combustion engine	123	188A
Making	29	888.4+
Radiator vent	137	455+
Thermostatic	236	61+
Reciprocating	251	318+
Animal watering	119	80
Biased check	137	528+
Engine nozzle	239	584+
Furnace ashpan	110	169
Gate	251	326+
Lubricator horizontal	184	79
Lubricator vertical	184	74+
Reversing	137	309+
Refrigeration expansion	62	527+
Automatic	62	222+
Regulator		
Fluid pressure	137	505+
Pump	417	279+
Removal tool	29	213.1
Reversing	137	309+
Reciprocating regenerating furnace type	137	309+
Rotary regenerating furnace type	137	311
Rocking and oscillating		
Animal watering	119	79
Four cycle engine	123	81R
Lubricator	184	34
Ventilator cowl	98	74
Wall register	98	106
Rotary or oscillating		
Disk	251	304+
Floor register	98	102
Fluid flow meter	73	265+
Internal combustion engine	123	190R
Lubricator force feed	184	35
Lubricator gravity feed	184	71+
Plug	251	309+
Safety		
Acetylene generator	48	56
Boiler	122	507
Boiler feed heater	122	437
Fluid pressure relief	137	455+
High temperature cutoff	137	457
Muffler	181	237
Pipe break cutoff	137	456+
Seat grinders		

	Class	Subclass
Abrading	51	27+
Milling	409	64+
Seat making	29	890.122
Fluid motor	29	888.44
Slide		
Abrading machine	51	30
Acetylene generator	48	55
Soda water type		
Combined with plural sources	222	129+
Multiple valve	137	170.1+
Nozzles	239	445+
Valve	251	
With agitating	366	195+
With deflector nozzles	239	499+
With fizzing stream	239	445
With multiple inlet nozzles	239	407
With variable discharge	239	435
Stem cap	138	89+
Combined tool	152	431
Packing	277	
Switch combined	200	61.86
Thermostat and temperature	236	
Burner combined	431	83+
Cutoff	137	457
Freeze responsive drain	137	59+
Throttle and engineers		
Fluid brake control	303	50+
Train control	246	189+
Tire	137	223+
Caps and cores	137	232+
Combined gauge	137	227+
Combined pressure regulator	152	431
Train control pipe	246	190
Well closures	166	316+
Valved Device		
Acetylene generator		
Flap	48	43
Bag closure	383	44
Filling	141	68
Filling	141	315
Barrel bung	217	99+
Bath basin	4	191+
Battery cell, weight actuated	429	85
Battery vent	429	53+
Blow-out type	429	56
Bottle		
Closure	215	311+
Nonrefillable	215	17+
Chute	193	21
Counterweight	193	20
Closet vent	4	216
Coating implement	401	
Cooking stove	126	52
Dispenser	222	
Drinking fountain	239	29
Bubble cup	239	31
Dry closet	4	466
Bowl	4	471+
Elastic	429	54
Excavating scoop	37	136
Filter	210	418+
Float	429	76
Fluid flow meter		
Diaphragm	73	265
Piston	73	248
Proportional	73	203
Fluid flow path	251	
Interlocking	48	83
Plate	48	52
Pop	48	53
Slide	48	58
System	137	
Fountain pen	401	
Funnel	141	344
Air displacement	141	297+
Gas and liquid contact apparatus	261	38+
Gas separator inflow control	55	418+
Gas separator outflow control	55	417
Heating gas purifier	48	170
Hoist bucket	414	657+
Illuminating fixture	362	394+
Inhaler	128	203.12+
Nasal	128	203.12+
Internal combustion engine		
Cooling	123	41.1+
Igniter	123	146
Liquid level gauge	73	332+
Lubricator		
Force feed pump	184	26+
Gravity feed	184	65+
Music brasses	84	388+

	Class	Subclass
Pastry depositor	222	
Pastry depositor	425	381.2
Pastry depositor	425	461+
Piano pneumatic	84	60+
Bleeds	84	59
Pipe	251	
System	137	
Pipe organ		
Action	84	336
Key	84	342
Tremolo	84	348
Pipe taps	137	317
Piston		
Flow meter	73	248
Fluid motor	91	422
Internal combustion engine	123	47R
Pump	417	545+
Pump oscillating	417	484
Pump reciprocating	417	545+
Pneumatic tire		
Tire and valve stem	152	429+
Tool and valve stem cap	152	431
Wheel and valve stem	152	427+
Preserving apparatus	422	
Pump	417	559
Diaphragm	417	480
Regulator	417	279+
Receptacle filler	141	
Refrigeration producer	62	
Refrigerator, air controlling	62	408+
Automatic	62	186+
Blocking air at access opening	62	265+
Sand blast machine	51	436
Ship		
Plug	114	197
Sea cock	114	198
Ventilator	114	212
Sifter		
Discharging	209	258
Feeding hopper	209	246
Sink strainer	4	287
Smoking device	131	223
Sound muffler	181	254
Spray nozzle	239	569+
Deflector	239	506
Multiple outlet	239	562+
Whirler	239	476+
Strainer	210	156
In pipeline	210	418+
Multiple in pipeline	210	294+
Textile pounder	68	218
With plunger	68	217
Track sander	291	6
Traps		
Sewerage	137	247.13+
Steam	137	171+
Ventilation		
Air pump	98	116
Building inlet	98	41.1
Chimney cap	98	59
Thimbles	98	118
Water closet		
Bowl	4	434+
Siphon bowl	4	421+
Tank	4	378+
Vamp and Quarters	36	48+
Trimming machines	12	57.6
Vanadium		
Alloys, ferrous	420	127
Alloys, non-ferrous	420	424
Carbon compounds	556	42+
Compound stock	428	662
Electrolytic preparation	204	64R
Inorganic compound recovery	423	62
Misc. chemical manufacture	156	DIG. 106
Pyrometallurgy	75	622
Vanes		
Aircraft control	244	82
Ammunition and explosive device combined	244	3.24+
Drop bomb combined	102	384
Fuse arming	102	225+
Parachute flare	102	339
Rocket combined	102	348
Fluid flow direction measuring	73	188+
Forming working member for expansible chamber device	92	121+
Gate mounted wind balance	49	135
Meter	73	861.74+
Expansible chamber	73	232+
Pump	418	

	Class	Subclass
Ships log....................	73	186+
Signaling (See semaphores)		
Turbine guide vane	415	148+
Ventilating cowl	98	61+
Windmill guide vane	416	12+
Vanillin	568	442
Vanity		
Cases or compacts............	132	293+
Absent non-coating article eg,		
mirror	401	118+
Illuminated mirror	362	136
Vapor (See Gas)		
Baths	4	524+
Coating.....................	427	248.1+
Condenser	165	110+
Contact with solids.........	34	
Cookstove	126	44
Design	D 7	339+
Deposition	118	715+
Process to form semiconductor P-		
N type junction...........	437	
Doping, semiconductor	118	900*
Process to form junction....	437	
Electric lamp...............	313	567+
Arc	313	567+
Lanterns	362	263+
Electric space discharge tubes	313	567+
Generator for disinfecting..	422	305+
Heating burners	431	161+
Heating burners	431	207+
Heating stoves	126	95
Liquefied or solidified.....	62	8
Medicators	604	23+
Testing for explosive		
Catalytic	73	25.1+
Miners lamp	362	164+
Therapeutic baths	128	367+
Thermometer.................	374	201+
Treating	34	
Cabinets	312	31+
Coating processes	427	335+
Distilled vapors treated .	203	
Distilled vapors treated .	201	29
Gases with separation ...	55	
Mineral oil vapors treated	208	3
Textile apparatus	68	5R+
Textile processes	8	149.2+
Tube amplifying system	330	41
Vaporizer Making	29	890.7
Vaporizing and Vaporization	122	
Concentrators	159	
Design	D23	360+
Distilling	201	
Distilling	203	
Drying	34	
Electrical heating for......	392	386+
Fumigators	43	129+
Heat motors	60	531
Heat or energy exchange agents	252	67+
Internal combustion engines	123	250+
External vaporizing	123	522+
Jet gas pumps	417	152+
Liquids	122	
Medical inhalers	128	203.12+
Mineral oils	208	
Nuclear reactors involved in	376	370
Nuclear reactors involved in	376	371
Nuclear reactors involved in	376	378
Pumps.......................	417	208+
Recovery of actinide compounds by.	423	249
Refrigeration	62	
Resolving colloids by	252	346+
Vaporizers	261	DIG. 65
Variable Capacitor	361	277
With variable resistor	323	354
Variable Denier		
Extruding apparatus	425	76
Stretching	28	243
Variometer	336	115+
Radio tuning system	334	62
Varnish		
Asphalt	106	273.1+
Fatty oil	106	246+
Natural resin		
Fatty oil	106	220+
Polymer ctg (See synthetic resin		
or natural rubber)		
Solvent	106	236
Removing	134	38
Materials	252	89.1+
Varnishing	427	

	Class	Subclass
Shells	86	17
Vase	D11	143+
Vasodilators	424	
Vasopressin	530	315
Vats (See Vessel)		
Cheese	99	452+
Metallic and miscellaneous .	220	
Textile fluid treating apparatus		
Movable carrier in vat....	68	157
Separate centrifuge and vat	68	26
Vat or sulfur dyeing	8	650+
Several dyes including ...	8	642
Textile printing	8	453
Textile printing	8	461
Wooden	217	4
Vaulted Roof Design	D25	18+
Vaults		
Building lights	350	258+
Burial	52	128+
Lifting or handling	52	124.1+
Design	D99	1+
Dry closet	4	474+
Road or pavement vault cover	404	25+
Design	D25	36
Vcr (See Recording and Recorders)	360	33.1
Vegetable		
Cleaning		
Apparatus brushing or wiping	15	3.1+
Apparatus liquid treatment	134	
Processes	426	478+
Coffee substitutes from ...	426	594
Comminuting machines	241	
Cutting, comminuting, peeling,		
paring	99	588+
Fluid or chemical	426	287+
Fertilizers from	71	23+
Fibers		
Liberation chemical	162	1+
Liberation mechanical ...	19	
Gathering	56	327.1
Digger type	171	
Grating	241	95+
Jellies	426	573+
Juices	426	599+
Peeling	99	584+
Pit stem or core removal		
Corers	99	544+
Processes	426	484+
Seeders and stoners	99	547+
Stemmers	99	635+
Planters	111	913*
Preservation	426	321+
Canning	426	392+
Coating	426	289+
Dehydration	426	471+
Refrigeration	426	524
Protein foods	426	656
Stalks		
Retting	162	1+
Retting bacteriological..	435	277+
Synthetic resins from extracts	528	1+
Vegetation		
Fuel	44	605+
Briquette	44	589+
Hydrocarbon recovery from ..	585	240+
Imitation	428	17+
Treated	428	17+
Vehicle (See Specific Type)		
Air	244	
Air bag passenger restraints	280	728+
Air conditioner	D23	351+
Air cushion	180	116
Cushion design	D12	5
Animal draft appliances	278	
Attachments		
Abrading	51	258
Antennas	343	700R+
Automobile carried	40	591
Body lifters	254	48
Cutting tool	83	928
Electric lamps	315	77+
Fluid sprinkling etc.....	239	172+
Jack	254	418+
Land	280	727+
Land wheel ornament	301	37R
Load accomodating on freight		
carrier	410	
Package and article carrier	224	273+
Receptacles	224	273+
Sifter support	209	421
Steps	280	163+

	Class	Subclass
Weigher	177	136+
Automobile (See automobile)		
Bevel gear	74	424
Bicycle	280	200+
Bodies and tops	296	
On & off freight load bearer	410	52+
Body ornament	428	31
Brake operator		
Automatic	188	110+
Fluid pressure	188	152+
Multiple	188	106R+
Speed responsive	188	181R
Spring	188	167+
Vehicle step	188	108
Weight	188	176+
Brake position adjusters		
Supports	188	208
Vehicle body movement ...	188	190+
Brakes	188	2R+
Clutch combined	192	13R
Cab or body		
On & off freight bearer..	410	52
Car (See car)		
Carriers, attachable	224	273+
Carriers, attachable	294	904*
Cleaning	15	53.1+
Apparatus	134	123
Railway car brushing machines	15	54
Collision avoidance	367	909*
Compositions coating or plastic	106	311
Computer controlled, monitored	364	424.1+
Control calculator	364	424.1+
Convertible	280	30+
Cycle	280	200+
Dials	D12	192
Draft devices	280	400+
Animal	278	
Draft appliances	213	
Dumping	298	
External cooperating means	414	373+
Railway	105	239+
Electric power		
Locomotive	104	35+
Locomotive	104	49+
Motor vehicle, traction motor	180	65.1+
Motors	D13	112
Electricity transmission to	191	
Excavators		
Self loading	37	4+
Vehicle actuated scoops .	37	132
Fender	293	
Design	D12	186
Floor mat design	D12	203
Fluid cushion	180	116+
Fluid handling device supported by	137	899+
Freight accomodating	410	
On & off freight container	410	52+
Gas pumps and fans		
Combined	417	331+
Fans combined	416	146R
Operating devices vehicle		
actuated	417	231+
Ground effect	180	116+
Headlights and spotlights .	362	61+
Design	D26	28+
Headrest....................	D 6	501
Heat and power plant	237	12.3 R
Motor vehicle boiler plants	122	3
Heating	237	28+
Heating and cooling........	165	41
Inflatable passenger restraint	280	728+
Instrument panel	D12	192
Land (See land vehicles) ...	296	
Bodies & tops	296	
Dumping	298	
Freight accomodating	410	
Motor	180	
License plates	40	200+
Lighting	362	61+
Circuits	315	77+
Design	D26	
Loading and unloading		
Chutes	193	5
Dumping	298	
External cooperating means	414	373+
Self loading or unloading vehicles	414	467+
Locks	70	237+
Medicine		
Inorganic	514	769+
Organic	514	772+
X-art collections	514	800+

	Class	Subclass
Mirrors	350	600+
Mirror frames	52	
Plural	350	612+
Plural combined	350	612+
Signal	350	97
Moped	180	205+
Motion picture	352	132
Motor	180	
Boiler plants	122	3
Supports	248	637+
Motorcycle	180	210+
Two-wheeled	180	218+
Tandem	180	219+
Movement display	434	305
Operation teaching	434	30+
Periscopes	350	540+
Railways	104	
Freight accomodating	410	
Rolling stock	105	
Making	29	168+
Refrigerated	62	239+
Automatic	62	133+
Process	62	61
Root rack	D12	157
Running gear		
Land vehicle	280	80.1+
Railway trucks	105	157.1+
Suspension	280	688+
Stub axle mounts	280	660+
Safety belt or harness	280	801+
Passive	280	802+
System responsive	180	268+
With seat structure	297	468+
Safety guard	280	748+
Safety promoting means	180	271+
Seat belt passenger restraints	280	801+
Seats	297	195+
Shelter support	135	88+
Signals	340	425.5+
Simulators	434	29+
Snowmobile	180	182+
Special purpose	D12	1+
Springs	267	2+
Steering mechanism moves pivoted lamp	362	37
Steering switches light system	362	36
Suction effect	180	164
Surface effect	180	116+
Suspension	280	688+
Springs	267	2+
Tires, resilient	152	
Tool driving or impacting device combined	173	22+
Tool, mounted	83	928*
Toy	446	431+
Design	D21	128+
Trains		
Draft appliances	213	
Electricity transmission to	191	11
Land	280	400+
Motor	180	142
Braking	188	112R+
Railway	105	1.4+
Braking	188	124
Tricycle	280	200+
Washer	134	123+
Head lamp	239	284.2
Windshield	239	284.1
Water, ie ships	114	
Buoys, rafts, aquatic devices	441	
Marine propulsion	440	
Weigher combined	177	136+
Wheel propelled	280	3
Wheel substitutes	305	
Wheelchair (See wheelchair)	280	250.1
Wheels & axles (See axle; wheel)		
Axle making	72	
Land	301	
Railway	295	
Resilient wheels	152	
Wheel making	29	894+
Veils	2	207
Design	D 5	47+
Veining		
Textile dyeing	8	478+
Velcro T M	24	306
Belt fasteners	24	31V
Brackets attached by	248	205.2
Buildings, static structures digest	52	DIG. 13
Chairs & seats digest	297	DIG. 6
Separable fastener	428	100

	Class	Subclass
Surgery digest	128	DIG. 15+
Velocipede		
Brake	188	24.11+
Canopies	135	88
Design	D12	107+
Figure toys	446	440
Land vehicle		
Coasters	280	87.1+
Convertible	280	7.1+
Dust and mud guards	280	152.1+
Occupant propelled	280	200+
Occupant steered	280	263+
Simulations	280	827+
Wheel guards	280	160.1
Windshields	296	77.1+
Racks	211	17+
Locking	211	5
Seats for	297	195+
Walkers nonsteered	280	87.2
Vending Machine	D20	1+
Veneer and Veneering		
Laminate	428	411.1+
Metal-to-metal	428	615+
Static mold on product or existing building structure	249	15
Tile adhered	52	390
Tile backer	52	384
Woodworking		
Lathe	144	209R+
Press	156	349+
Venetian Blind (See Blind)	160	166.1+
Cleaning machines	15	77
Wiping	15	102
Design	D 6	577
Making	160	405
Assembly	29	24.5+
Method	160	405
Tape cutting	26	11.4
Woven float cutting	26	7
Wood slat slotting	144	136R
Venom Extractors	604	314
Ventilators & Ventilated Items	98	
Abrasive tool or support	51	356
Aircraft cabin having	98	1.5
Automobile having	98	2+
Barrel	217	74
Bath closets sinks spittoons	4	209R+
Bed mattresses	5	468+
Fluid filled	5	453
Forced air heating, cooling	5	423
Blowers	415	
Boot or shoe	36	3R
Building	98	29+
Cabinets having	312	213
Car	98	4+
Chimneys	98	58+
Cleaning or polishing tool or support	15	230.1
Consumable electrode arc lamp having	314	26+
Cooking oven	126	21R
Cowls	98	61+
Crankcase	92	78+
Crates	217	42
Door	98	87
Dry closet		
Inclosed receptacle	4	477
Receptacle	4	482
Valved bowl	4	472
Vault	4	475
Electric batteries	362	74
Electric batteries	429	82+
Cap type	429	89
Gang type	429	87
Non-spill type	429	84
Plug type	429	89
Reactive, absorbable type	429	86
Separate inlet and exhaust	429	83
Stopper	429	89
Weight actuated valve type	429	83
Electric cables conduits boxes	174	16.1
Electric insulators	174	187
Equipment	D23	370+
Fans	416	
Fire escapes	182	47
Hats	2	171.3
Sweats	2	181.6
Heating and cooling systems	165	59
Heating systems	237	46+
Hoods and offtakes	98	115.1+
Light projectors	362	294
Locomotive cab	98	3

	Class	Subclass
Dust guards	98	28
Louvers	98	121.1+
Mask	128	205.25+
Mine	98	50
Oven doors	126	198
Pavement vault cover	404	25+
Photographic dark cabinet	354	309
Pressure maintenance	98	1.5
Provision safes	98	51
Railway cars	98	4+
Railway tunnels and subways	98	49
Registers	98	101+
Roof eaves	98	DIG. 6
Ships	114	211+
Shoes, making	12	142V
Solar power	98	900*
Storage	98	52+
Stove attachments	126	80
Fluid fuel stoves	126	84
Platforms	126	279
Stovepipe combined	126	312
Damper combined	126	293
Thimble combined	126	316
Temperature regulated	236	49.1+
Tents having	135	93
Umbrella cover	135	35V
Water closet		
Siphon bowl	4	351+
Urinal	4	306+
Valved bowl	4	348+
Washout bowl	4	348+
Window	98	88.1+
Vents		
Automobile no-draft	49	390+
Ball point pen	401	217
Building with	52	198+
Fountain pen	401	242
Molding devices		
Chills	164	372
Metal founding	164	410
Metal founding core	164	234
Metal injection die having	164	305
Static molds	249	117+
Ordnance	89	30
Pump priming and venting	417	435
Rotary	415	11
Receptacle		
Batteries	429	82+
Bottle closure	215	307+
Check valves for radiator	137	511+
Closures	220	367+
Cutter for vent and dispensing openings	222	85+
Dispenser	222	478+
Dispenser handle with	222	468
Dispenser movable trap chamber with	222	332
Dispenser trap chamber with	222	442
Dispenser with fluid trap seal for	222	188
Dispenser with follower and	222	387
Float closure for dispenser vent	222	69
Fluid operated lubricator	184	59
Fluid pressure brake systems	303	70
Tube bottle closure	215	307+
Water heating	126	389
Wick burner fuel supply	431	321
Rubber or heavy plastic mixer	366	75
Sifting bins with	209	378
Valve		
Closet etc ventilation	4	216
Independent multiple air	137	583+
Multiway rotary plug	137	625.22+
Serial dependent drainage	137	512+
Direct response	137	512+
Selective motion actuation	137	636+
Separate actuators	137	637+
Sequential	137	628*
Venting	425	812+
Wall with	52	302+
Venturi		
Carburetor	261	DIG. 12
Flow meter	73	861.63+
Restrictor type	138	44
Scrubbers	261	DIG. 54
Variable venturi	261	DIG. 56
Veratramine	546	195
Veratridine	546	34
Veratrol	568	648
Verel T M (polyacrylonitrile Vinylidine Chloride)	526	342
Verifier		
Card or tape punch	234	34

	Class	Subclass
D	552	653
Design	D28	1+
E	549	408
Electrostatic treatment	204	166
Foods containing	426	72+
G (See vitamin b2)		
K	552	299
Preparation by wave energy	204	157.67
Vitreous (See Glass)		
Vitrification or Devitrification		
Color processes photo	430	198
Glass composition	501	2+
Glass preform	65	33
Voice		
Artificial electric	381	
Doll	446	297+
Air operated	446	188+
Operated circuits relay	361	160+
Reflector for therapy	D24	35
Void Former	52	577
Within module	52	576
Volleyball	273	411
Voltage		
Differentiator electric	333	19
Electronic tube type	328	127+
Plural different voltages	307	15
Integrator electric	333	19
Electronic tube type	328	127+
Magnitude & phase control systems	323	
Regulation electronic	323	
Regulator (See regulator; electricity)		
Stabilization (see regulator)		
Voltage Magnitude Control		
Audio signal	381	104+
Facsimile	358	400+
Pulse or digital communications	375	98
Radio	455	232+
Telecommunications	455	232+
Television	358	174+
Color	358	27
Voltmeter (See Galvanometer)		
Amplifier tube for	313	293+
Volume Control Automatic		
Amplifier type with		
Bias control	330	129+
Impedance control	330	144+
Thermal impedance control	330	143
Public address type	381	
Radio receiver type	455	234+
Input coupling impedance adjustment	330	185+
Interstage coupling impedance adjustment	330	157+
Output coupling impedance adjustment	330	192+
Volume Control Manual		
Amplifier type by		
Bias adjustment	330	129
Cathode self bias impedance adjustment	330	142
Volume Meters		
Calibration of	73	3
Dispenser combined	222	71+
Check controlled	194	
Preset cutoff	222	14+
Fluid	73	861+
Liquid level	73	290R+
Measurers	33	1V
Measuring vessel	73	426+
Proportional flow control	137	87+
Receptacle filling		
Stationary measure	141	
Traveling measure	141	135+
Timing fluid flow	137	552.7
Volumetric content measuring	73	149
Vortex Tube for Cooling	62	5
Vortical Separator	209	
Gas separation	55	447+
Liquid contact combined	55	235+
Voting		
Booths	52	36
Flexible wall	52	63
Computer controlled, monitored	364	409
Machines	235	51+
Teaching	434	306
Vulcanizable Gums		
Apparatus for	425	28.1+
Apparatus for	425	113
Apparatus for	425	381.2
Dyeing of	8	
Electrophoresis involving	204	180.3+

	Class	Subclass
Synthetic resin (See synthetic resin or natural rubber)		
Vulcanizing		
Machinery design	D15	199
Molding apparatus for composite articles	425	500+
Molding devices	425	363+
Molding devices	425	383+
Presses	425	335+
Processes		
Molding combined	264	
Synthetic resin or natural rubber	525	50+
Wabble Mounted Circular Saw	144	238
Wabbler Plate	74	60
Gearing	475	163+
Wadding or Filling for Containers	493	967*
Wads	102	532
Columns	102	532
With cartridge	102	448+
Waffle Irons	99	372+
Batter feeding combined	99	373
Design	D 7	359
Waffle Type Ceiling	52	337
Wagering, Computer Controlled	364	412
Waging War	89	1.11
Wagons	280	
Bodies	296	
Design	D12	17+
Dumping	298	
Ice	62	239+
Lunch		
Heater	126	268
Ovens	126	276
Bakers	126	276
Toy	D21	134+
Wainscoting	52	506+
Waistbands		
Skirts	2	220+
Closures	2	219
Special sewing machine	112	121.27
Trousers	2	236+
Union type skirted	2	76
Waists		
Body garments	2	104+
Wakers (See Alarms)		
Walkers		
Baby	272	70+
Design	D12	130
Water	441	76+
Walkie Talkie Radio	455	73+
Separate transmitter and receiver	455	39+
Walking		
Aids	135	68+
Seats	297	5+
Cane or stick	135	65+
Irons	128	83.5
Mechanism for dolls	446	377+
Motor driven	446	355
Walkways	404	17+
Wall		
Absorbent filtering gas	34	81
Aperture thermometer support	374	208
Board cutting	83	
Bracket attached to wall anchor	248	231.91
Building (See facer construction)	52	
Article support	52	27
Coated	52	515+
Heat exchanging	165	47+
Hollow block	52	503+
Hollow block static mold	249	144+
Hollow ventilating	98	31
Lateral block courses	52	561+
Plasterboard	52	344+
Static mold for making	249	33
Surfacing abrading machine	51	180
Ventilating register	98	106+
Cabinet	312	245
In wall or panel recess	312	242
Chute	193	33+
Cleaning machines	15	49.1+
Decoration or plaque	D11	132+
Electric furnace		
Resistance	373	109+
Resistance in	373	
Water cooled		
Arc	373	75
Induction	373	138
Resistance	373	113
Exterior elevator	187	6+
Freight car vacuum	105	357
Heating and cooling panel	165	49

	Class	Subclass
Insulator through wall	174	151+
Ironing board attached laterally of support	108	152
Folding	108	134+
Ladder with platforms	182	83+
Mirror plural	350	612+
Mounted conduits	174	48+
Outlets	174	48+
Wiring	174	50+
Panels, continuous molding of	264	46.4+
Plug electrical	439	638+
Protector for furniture	248	346.1
Racks	211	
Receptacle		
Bottle or jar spaced	215	12.1+
Electrical	174	53+
Insulated, wooden	217	131
Metallic	220	660+
Metallic spaced	220	415+
Metallic transparent	220	662+
Outlet wall mounted	220	3.3+
Paper	229	4.5
Pitcher spaced	222	131
Purse	150	127+
Trunk	190	23
Wooden box	217	17
Insulated	217	128+
Retaining	405	284+
Cribbing	405	273
Sheet piling	405	274+
Wave or flow dissapating	405	30+
Retort	202	223
Safe and bank	109	78+
Table attached laterally of support	108	152
Folding	108	134+
Water for liquid heater	122	106
Bridge	122	192
Wallets	150	132+
Wallpaper	428	346+
Cleaning device	15	219
Design	D 5	
Steamer	126	271.1
Walnuts		
Novelties, jewelry motif	D11	117
Plants	PLT	32
Wankel Engine	418	61.2
Ward		
Distributing	276	26
Setting combined	276	6
Wardrobe	312	
Trunk type	190	13R+
Warfare		
Electronic	342	13
Gases	424	
Teaching	434	11
Warmers		
Bed mattress	5	421+
Bedstead	5	284
Body	126	204+
Heat exchanger	165	46
Heated shoe	36	2.6
Bottle	99	359
Food	126	
Household appliances	D 7	362+
Foot warming radiators	237	77
Receptacles	220	3.1
Solar	D 7	324
Surgical devices	128	362+
Bandaging	128	82.1
Instruments	606	1+
Kinesitherapy	128	24.1+
Medicators	604	113+
Orthopedic	128	68.1
Receptors	604	358+
Warnings (See Indicators; Signal)		
Warning or flashing signs	D10	114+
Warp		
Feeding	139	97
Frames	139	34
Irregular feed	139	24+
Knitting		
Circular independent needles	66	10
Circular united needles	66	81
Fabrics	66	192
Fabrics having unknit materials	66	195
Straight united needles	66	203+
Straight united needles unknit material incorporating	66	84R+
Knotting machines	28	209+
Panel straightener	52	291
Pile weaving	139	37+

	Class	Subclass
Compressional wave signaling	367	
Electric musical instrument	84	723
Electric wave transmission	333	
Explosive mine firing	102	427
Fuses, primers, and igniting devices	102	211
Guides electrical	333	
Interference type electric musical instrument	84	600
Marine mine firing	102	416+
Preserving disinfecting and sterilizing processes	422	20+
Microwave	422	21
Radio	455	
Pulse or digital	375	37+
Shaping	328	34+
Sound and supersonic generators	116	137A
Surgical or medical applications (See radiant energy)		
Textile treatment	8	DIG. 12
Tuners	334	
Use in nuclear plasma reactions	376	100
Used in reaction to prepare inorganic or organic material	204	157.15+
Used in reaction to prepare or treat synthetic resin or natural rubber	522	1+
Explosive devices	102	366
Fluid breakwaters	405	22
Form analysis		
Electrical	324	77R
Generator		
Earth boring combined	175	56
Water	405	79
Guide electric	333	239+
Antenna	343	772+
Conduit only	138	118+
In cathode ray tube	315	4+
In electronic tube	315	39
Making	29	600
Standing wave indicator	324	633+
With frequency characteristic	333	239+
With insulation	138	137+
Meters	250	250
Motors	60	495+
Resonator	333	219+
In cathode ray tube	315	4+
In electronic tube	315	32+
Shaping		
Electronic tube type system	328	34+
Transmission, wireless	455	
Pulse or digital	375	37+
Waveguides		
Chemical analysis apparatus	422	82.11
Railway rail joints	238	151+
Railway rails	238	122+
Railway wheels	295	21+
Railway wheels integral hollow	295	28
Textile fiber forming	19	296+
Feeding	226	
Stepping-motor drive for web	400	902*
Machine part		
Calendars double reel and	40	117
Calendars single reel and	40	116
Changeable exhibitors	40	446+
Drier drum external	34	123
Drier plural drums external	34	116+
Drier rotary drum external	34	111
Drier spacers	34	94
Printing	101	
Sheet feeding or delivery	271	
Supplied machines		
Automatic drier with control by	34	49
Box making, folding	493	162+
Drier for	34	148+
Drier spacer web	34	94
Drying processes diverse types	34	18
Drying processes radiant etc	34	41
Drying processes with vacuum	34	16
Drying processes with vapor etc	34	23+
Drying processes with web spacer	34	6
Envelope making	493	186+
Fabric coating and uniting	156	
Fabric reeling web fasteners	242	74
Fabric reeling web roll supply	242	58
Folding or associating	270	
Printing	101	
Severing	83	
Textile fluid treating	68	177+
Typewriter paper feed	400	611+

	Class	Subclass
Tacky web cutting	83	922*
Winding & reeling	242	
Waving Hair		
Methods	132	210
Wax		
Centrifugal honey extractor filter	210	361+
Coating with	427	
Compositions		
Abrasive	51	305
Biocidal	424	
Coating or plastic	106	
Detergent	252	89.1+
Ester type	260	398+
Inks	106	31
Mold	106	38.8
Mold coating	106	38.25
Polishes	106	10
Synthetic resin or natural rubber ctg (See synthetic resin or natural rubber)		
Dyeing	8	521
Electromagnetic wave treatment	204	157.15+
Electrostatic field or electrical discharge treatment	204	167
Ester type	260	398+
Fermentative treatment	435	271
Matches	44	507+
Melting apparatus for honey separation	210	175+
Modeling dental	433	213
Paper	428	486
Paraffin	208	20+
Waxing		
Bobbin winding combined	242	18R+
Coating pads and forms	118	264+
Leather sewing and	112	42
Shoe making	12	79.5
Waxy hydrocarbon polymer production	585	946*
Ways		
Conveyor	193	
Marine	405	1
Railway		
Freight cars for tubular	105	365
Suspended	104	123
Tubular	104	138.1+
Weaning Devices	119	134
Weapons (See Ammunition; Firearms; Ordnance)		
Wear		
Building surface	52	177+
Compensating brake beam	188	214+
Compensating radial bearings	384	247+
Compensating wheel bearings	384	261+
Compensators for clutches	192	110R
Distributor for typewriter platen	400	554+
Earth boring bit indicating	175	39
Element removable from pipe fitting	285	16+
Indicator for pipe	138	36
Pavement	404	17+
Plates for railway truck	105	225
Plates hub thrust	384	247+
Resistant surface	428	908.8
Strips vehicle body	296	41
Surface rod joint packing	277	
Take up for wheel brake	188	79.51+
Wearing in	29	89.5
Weather (See Humidity; Moisture)		
Barometers	73	384+
Computer controlled, monitored	364	420
Control of	239	2.1
Operated curtain shade screen	160	5
Proofing buildings (See water proof)		
Radar	342	26
Rain actuated window closers	49	21+
Strips	49	475+
Land vehicle windshield	296	93
Vanes	73	188+
Design	D10	59
Weatherstrip, metal	428	595
Weavers Implements and Irons	139	380+
Design	D 8	
Dies for forging	72	462+
Slat and wire fabric	140	35
Portable machines	140	48
Warp manipulation	139	35
Web (See Sheet, Fabric)		
Article part		
Coating or forming by electrophoresis etc	204	180.4
Electroforming	204	12+

	Class	Subclass
Electrolytic coating	204	27+
Formed web assembling	19	161.1+
Paper bag making	493	186+
Paper type forming apparatus	162	289+
Paper type forming processes	162	202+
Wedge		
Bolt to nut lock		
Thread gripper with tapered section	411	253
Thread lock with tapered surface	411	265+
Coupling and securing		
Bale tie	24	25
Buckle	24	194+
Buckle design	D11	200+
Clasp	24	526
Also pivots	24	491
Clasp design	D11	87
Closure fasteners	292	342+
Cord holder	24	136R+
Harness clamp	24	171
Rail anchor	238	324
Rail clamp	238	353+
Rod joint sleeve	403	274+
Tool socket fastening	403	367+
Wire fence clamp	256	56
Design	D 8	47
Printers quoins	254	42
Pushing and pulling	254	104
Rock splitting in situ	299	20+
Wood splitting	144	192+
Weed Killers	71	65+
Weeders	30	
Electric	47	1.3
Flame or burner type	126	271.1+
Plant cultivators	47	1.44
Fluid introduction into soil	111	118+
Grappling fork or shovel	294	50.6+
Harrows	172	705+
Process of introducing fluid into soil	47	58
Root puller	254	132
Subsurface blade	172	720
Turners	172	514+
Weft		
Inactive		
Restrainers for	139	170.3+
Measuring and storing	139	452+
Pile fabrics	139	392+
Selecting	139	453+
Stop motions	139	370.1+
Thread cutting devices	139	302+
Thread replenishing	139	224R+
Thread tensioning devices	139	194
Winding	242	26.4+
Weighers (See Scales)		
Automatic supply control	177	60+
Cigar and cigarette making	131	280
Coin or check controlled	194	
Conveyer pneumatic combined	406	10+
Conveyer power driven combined	198	502.1+
Design	D10	87+
Dispensing	222	77
Gas separation combined	55	270
Illuminated	177	177+
Liquid level gauge	73	296
Material and handling	414	21
Packaging machine combined	141	129+
Receptacle filling combined	177	59
Recording	177	2+
Printing or perforating	346	9+
Scales	177	
With diverse exhibitors	40	458
Sifting	209	239
Sorting	209	645
Condition responsive controls	209	592+
Specific gravity	73	433+
Traversing hoist	212	158
Weighing Machines	D10	87+
Weight (See Notes to Def of)	16	DIG. 8
Aircraft stabilizing	244	93+
Animal restraining	119	107
Bed counterbalance	5	166.1+
Bottle valve	215	22+
Brake	188	174+
Cigar making	131	347
Closure fastener	292	344
Clutch	192	89W+
Coin tester	194	339+
Computer controlled, monitored	364	466
Controlled operated and regulated		
Outlet valve	4	386+
Outlet valve latched open	4	386+

	Class	Subclass
Metal molds	249	56
Panel hanger	16	107
Mounts	16	97+
Plow		
Guide	172	286+
Potato diggers		
Sifter	171	71+
Sifter	171	111+
Potters	425	263+
Potters	425	459+
Power take off	74	13+
Pullers	29	244+
Pulleys		
Friction drive	474	166+
Positive drive	474	152+
Pumps		
Water motor	417	334+
Rack	211	23+
Velocipede type	211	20+
Railway	366	
Railway cable gripper		
Sprocket	104	236
Whelp	104	235
Resilient	152	
Pneumatic	152	53+
Pneumatic and coil	152	35
Pneumatic and leaf	152	34
Pneumatic and rubber	152	30
Rolling	72	67+
Rotary crane		
Bull wheel	212	246
Rotary fender		
Horizontal	293	19
Sash cord guides		
Casing combined	16	215
Multiple wheel	16	213
Sash weight attached	16	219
Ship steering		
Paddle wheel	114	147
Wheel and drum	114	160
Skates	301	5.3
Slip		
Acceleration induced, detecting in motor vehicle	180	197
During braking, regulating	188	2A
Spring	152	1+
Substitutes	305	
Brakes	305	9
Traction (See traction)		
Training for cycle	D12	122+
Trolley		
Guard or finder	191	76
Head	191	63+
Typewriter		
Index	400	165.1+
Key	400	174+
Vehicle		
Bearings	384	416
Land vehicle	301	5R+
Land vehicle design	D12	204+
Locks	70	259+
Marine towing	440	36
Pivoted carrier	280	38+
Railway	295	
Railway trucks	105	157.1+
Resilient	152	
Runner attachment	280	13
Runner combined	280	8+
Vertically adjustable	280	43+
Wheelbarrow	280	47.31
Water motor		
Wheeled bag	248	98
Wheeled hose reels	242	86.2
Wheelwright machines	157	
Wheel holding stands	157	14
Wind motor		
Woodworking		
Hub boring	144	97+
Special work machines	144	15+
Spoke turning	144	206
Workholding stand for	157	14
Wheelbarrow	280	47.31
Design	D34	16
Wheelchair	280	250.1+
Attachments	280	304.1
Design	D12	131
Folding	280	647+
Motorized	180	907
Seating aspect	297	DIG. 4
Wheeled (See Wheel)		
Aircraft landing gear	244	103R+

	Class	Subclass
Ambulance stretchers	296	20
Beds		
Cable drum carriers	254	279
Coasters steered	280	87.1+
Earth working tools	172	
Fire engines	169	24+
Harrows	172	
Harvester	56	322+
Horse rakes	56	396+
Rear delivery	56	384+
Rear delivery revolving	56	380+
Ladders	182	12+
Plows	172	669+
Frames	172	669+
Land markers	172	126+
Pneumatic dispatch		
Carriers	406	185
Receptacles or trucks	280	
Roofing scaffolds	182	36+
Scoops	37	124+
Fork or rake	37	121
Skates	280	11.19+
Stands		
Bag holder	248	98
Movable receptacle	248	129
Store service		
Carriers	186	27+
Toys		
Aircraft	446	230+
Figure	446	269+
Knockdown	446	93+
Miscellaneous	446	431+
Rotating	446	237
Sounding	446	409+
Vehicles	280	
Vehicle	280	
Convertible to chair or seat	280	30+
Convertible to cradle or crib	280	31
Extensible	280	638+
Bier or casket type	280	640
Folding	280	639+
Folding carriage or stroller	280	642
Folding wheel chair	280	647
Fenders	293	43
Hand	280	47.17+
Motor	180	
Pivoted wheel carrier	280	38+
Railway rolling stock	105	
Shipment on freight carrier	410	3+
Shipping receptacle for	206	335
Walkers	280	87.2
Convertible to steered vehicles	280	7.1+
Steered	280	87.1+
Winding and reeling		
Hose reel carriers	242	85+
Reel carriers	242	86.5 R+
Wheelwright Machines (See Tire)	157	
Whetstone		
Compositions	51	293+
Structures	51	211R+
Whey	426	583
Whiffletree	278	90+
Harness connectors	54	53
Whip	231	2.1+
Animal prod	D30	156+
Cotton harvesters	56	29
Design	D30	156
Kitchen beater	D 7	380
Machines	231	1
Rack	211	67
Design	D30	143
Sifter clearers	209	383
Sockets	280	170+
Treadmill etc combined	185	15+
Whipping Devices		
Agitating or mixing	366	
Animal powered motor combined	185	24
Sweep	185	20
Goad	231	2.1+
Design	D30	156+
Electric	231	7
Initiating devices	272	27R
Sewing machine	112	121.2
Vehicle combined	280	178
Whipstock	166	117.5+
Earth boring combined	175	79+
Whirlers and Whirling		
Gas separation by deflection	55	447+
Nozzles	239	463+
Pneumatic conveyors	406	92
Rockets		

	Class	Subclass
Rotating	102	350
Whirlers with	102	359
Toys	446	236+
Whirligigs (See Whirlers)		
Noise making toys	446	215
Spinning toys	446	236+
Flying propellers	446	36+
Whisk Brooms (See Brooms)	D 4	130+
Whisker Growth	156	DIG. 112
Whiskey	426	
Whistles		
Buoy	116	108
Electric	340	406
Liquid level	137	213+
Musical instrument	84	330+
Periodic	116	24
Security or signal instrument	D10	119
Signal	116	137R+
Steam	116	137R+
Structure	116	137R+
Toy	446	204+
Design	D21	64
White Lead	423	592+
Whiting	423	592
Wick		
Atmosphere	239	34+
Boiler	122	366
Burners	431	298+
Blast lamps having	431	252
Blast lamps having	431	300
Cookstoves having	126	45+
Heating stoves having	431	102+
Mantle or incandescent element combined	431	102
Perforated combustion tube type	431	195+
Porous block type	431	326
Coated	431	325
Combined dispenser	222	187
Combined with fuel	44	519
Compositions	502	400
Heat exchange, capillary heat pipe	165	104.26
Liquid purification or separation, capillary	210	294
Stops	431	301
Structure	431	325
Trimmer or handler	431	120
Wickets Doors	49	169+
Wideband Cable	333	239+
Wiener Roaster	99	324+
Electric	99	358
Elongated type	99	441
Wigner Effect	376	350
Wigner Effect	376	351
Wigner Effect	376	358
Wigner Effect	376	458
Wigs	132	53+
Accessory	D28	93
Design	D28	92+
Rooting hair by tufting	112	80.2
Toy	446	394
Willowers		
Textile fiber	19	85+
Wilton		
Carpet	139	391+
Making apparatus	139	37+
Winches (See Windlass)		
Wind		
Guards and shields		
Couch hammock combined	5	125
Harvester combined	56	190
Land vehicle combined	296	77.1+
Match receptacles combined	206	102+
Smoking devices combined	131	174+
Instruments, musical	84	330
Design	D17	10+
Motors (See current fluid; motor; windmills)	416	
Control for windmill driven generator	290	44
Generator drive system	290	55
Marine propulsion device	440	8
Ventilator drive device	98	
Musical instruments	84	330+
Automatic	84	83+
Automatic piano	84	24+
Pressure meters	73	861+
Propelled revolving sign	40	479
Illuminated	40	480
Screens for vehicles	296	84.1+
Socks and vanes	73	188+
Tunnels	73	147

	Class	Subclass		Class	Subclass		Class	Subclass
Wooden box stays	217	71	Bending	144	254+	Power unit efficiency	73	112+
Wooden box straps	217	68	Boring	408		Work table lighting system	362	33
Cutter	30		Buildings	52		**Worm (See Gearing)**		
Cutting coiled wire	83	907*	Carbonizing	201		Culture	119	6
Cutting wires	83	950*	Carrier	294	149+	Electrical expeller	47	1.3
Dental arch	433	20	Coating	427		Gearing	74	425+
Electric			By immersion	427	440	**Worship**		
Conductors	174		Comminution	241	28	Cabinets	312	33
Magnetic body treating coils	600	13+	Distillation	201		Teaching	434	245+
Meter hot wire measuring	324	106	Drying and vapor treatment	34		**Wort**		
Telephony	379		Dyeing	8	402	Making	426	16
Tower support	52	40	Fastener made of	269	905*	Apparatus	99	278
Electroplating	204	27+	Fences	256	19	**Wrap-forming Metal**	72	296+
Fabric	139	425R	Fuel	44	606	**Wrapper (See Package)**		
Bed bottoms	5	186.1+	Briquette	44	590	Bicycle article carrier	224	34
Coating with rubber etc	204	180.4	Gas generation	48	209	Coin type	229	87.2
Design	D 5	47+	Apparatus	48	111+	Label pasting	156	
Design for floor or wall	D 5	47+	Hydrocarbon recovery from	585	242	Newspaper	229	92
Electroforming	204	11	**Impregnating**			Paper	229	87.1+
Electrolytic coating	204	24	Processes for	427		Paperboard boxes	229	40
Pile textile weaving	139	37+	Insulated compartment for box	217	131	Tobacco	131	365
Wireworking	140		Moldings	52	716+	Cigar etc tube feeding	131	74+
Gauge, strain &	33	DIG. 13	Nails	411	439+	Holders and carriers	131	105
Glass, safety glass	428	255+	**Ornamenting**			**Wrapping and Wrapped**		
Manufacturing	65		Panel track combined	16	95W	Chain links	59	92
Heat treated	148	400+	Flexible panel attachment	16	87.4 W+	Chain making	59	19
Coated	428		**Pulp making**			Cigar and cigarette making	131	27.1+
Handling running length work	148	155+	Comminution processes	241	21	Packaging combined	131	283
Making	148	4+	Pulp mill	241	280+	Wrapping materials and devices	131	58+
Instruments, ribbon &	33	DIG. 7	**Receptacles**	217		Containers	D 9	305
Insulation cutter	30	90.1+	**Saturating or indurating**			Food preservation	426	392+
And remover	81	9.4+	Compositions	106	12	Machines	53	203+
Making			Compositions mineral oil	208	2	Receptacle forming & filling	53	558+
Casting	164	423+	Fireproofing	252	607	Paper	D 5	
Coating with rubber etc	204	180.4	Screws metal	411	378+	**Sheet material**		
Covered or wrapped	57	3+	Making	10	2+	Folding	53	116+
Drawing	72	274+	Shavings making	241		Printing and folding	53	116+
Drawing dies	72	467	Ships	114	82	Printing and rolling	53	116+
Electroforming	204	12+	Shoe soles	36	33	Rolling	270	32+
Electrolytic coating apparatus	204	206+	Attaching means	36	13	Spinning etc apparatus	57	3+
Electrolytic coating processes	204	27+	Slicing	144	162R+	Spinning etc processes	57	3+
Metal rolling processes	72	365.2+	Splitting	144	193R+	Strand structure	57	210+
Screw thread rolling	72	95+	Staining	8	402	Textile smoothing implements	38	73
Wire glass	65		**Structural**			Warp preparing	28	176
Metal miscellaneous			Composite columns	52	720+	Wire fabric stay applying	140	10+
By electrolysis	204	12+	Compound lumber	52	782+	Portable machines	140	16+
From powders	419	4+	Joist connections	403	230+	**Wreath**		
Making miscellaneous	72		Through intermediate member	403	187+	Artificial	428	10
Packaging for	206	389+	Nailing beam	52	376	Design	D11	120
Photo systems (See facsimile)			Panels	52	782+	Illuminated	362	122
Recorders sound	360	89	Rail pedestal	238	118	Natural flora sustaining	47	41.1
Recording machines	360	89	Rail seat cushions	238	285	**Wrenches**	81	52+
Wire per se	360	131	Splices and joints	403		Adapted to turn eye screw	81	901*
Rope making	57		Wheel making	144	15	Compound	7	138+
Server/cooler	62	457.8	Sugar production	127	37	Constructed from specific material	81	900*
Sheet attached lathing	52	454	Apparatus	127	1	Design	D 8	21+
Splicing			Turning lathes	142		Impact	81	463+
Bale or package tie	24	16R+	Wind musical instruments	84	380R+	Impact delivery clutch	173	93.5+
Electric conductor	156	49	**Working**	144		**Making**		
Joint	403	206+	Design	D15	122+	Forging dies for	72	470+
Textile analogue	57	22+	Sawing	83		Machines	76	10
Trolley wire	191	44.1	Tools (See tools)			Methods	76	114
Stitching stapling	227		Turning	142		Monkey, combined with other tools	7	139+
Stretchers	254	199+	**Woodbury Type**	101		Pump and oiler combined with	81	2
Strippers	81	9.4+	With photographic step	430	300+	Tap, ratchet	408	120+
Working and using machines			**Woodwinds**	84	380R+	**Wrigglers**	239	229
Bobbin and cop winding	242	25R	**Wool**			**Wringers**	68	241+
Broom banding	300	15	Chemical modification	8	128.1	Bevel gearing for	74	387
Coiling	72	135+	Cleaning or laundering	8	137+	Drier combined	34	70
Electric wire placing apparatus	254	134.3 R+	Devices and machines	68		Mechanisms for washer and	74	17
Etching to section	134		Waste cleaning	68	1	**Mop**	15	260+
Grain etc compressing and binding	56	451+	Dyeing	8		Brush combined	15	116.1+
Harvester binder type	56	132	Mixed textiles	8	529+	Combined	15	119R+
Hat frame	223	8	Vat or sulphur dyes	8	650+	**Supports**	68	239
Hatbrim wire shaping	223	17	Fertilizers containing	71	18	Wash bench combined	68	236
Metal forging and welding feeder	72	419+	Liberation mechanical	19	2+	Wash tub etc attachments	68	238
Nail making	10	43+	Mercerizing	8	128.1	**Wrinkling**		
Nailing and stapling	227		Removing grease from	8	139+	Coatings	427	257
Reels	242	78	Steel wool making	29	4.51+	Corrugating metal	72	191+
Stapling	227		**Waste**			Paper apparatus	162	280+
Textile spinning twisting etc	57		Cleaning and reconditioning	68	1	Paper processes	162	111+
Working	140		**Work**			**Wrist Pins**	74	595+
Wired Music			Carrier and support for sewing			Bearing	384	266+
Distributing system	379	101	machine work	112	121.15	For piston rod connection	384	268+
With phonograph	379	87	Holder	269		**Wrist Watch (See Watch)**		
Electric organs	84	600+	Dental	433	49+	Design	D10	30+
Electric phonograph	379	87	Electric welding	219	158+	Fastener	63	3
Wireless (See Radio)			For testing by stress or strain			**Wristlets**	2	170
Wood (See Plywood)			application to specimen	73	856+	**Writing**		
Alcohol (See methanol)			**Measuring**			Forms	283	45
Ball making	142	1	Dynamometers	73	862+	Guides	434	164

Appendix B

Patent and Trademark Office Telephone Directory

I have found that the best way to avoid making mistakes in PTO applications, procedures, and fees and thereby save myself time, money, and mental torment, is to pick up the phone and, as AT&T suggests, reach out and touch someone. Devitalizing as the federal bureaucracy can be, its official representatives, when approached one-on-one, can be warm, helpful, and sympathetic to your plight.

To assist you in locating the appropriate officials (who are scattered throughout some 14 buildings in Crystal City, Virginia), the complete organizational telephone directory for the PTO is included here. While people may change positions through career moves, the job slots remain constant, and so do the office telephone numbers.

If you find yourself in a game of telephone tag with any of the PTO officials listed on the following pages, call the PTO's Office of General Services at (703) 557-0813. General Services is responsible for updating and publishing the PTO's telephone directory and furnishing accurate personnel locator information to the public. As a last resort, do not hesitate to call the Commissioner's Office at (703) 557-3071. An administrative secretary or special assistant will be glad to help you.

If you would like to write to a PTO official, refer to the listings on the following pages and send your correspondence to: U.S. Patent and Trademark Office, Washington, DC 20231.

Note: The area code for all the phone numbers that follow is (703).

OFFICE OF THE ASSISTANT SECRETARY AND COMMISSIONER OF PATENTS AND TRADEMARKS

Assistant Secretary and Commissioner Harry F. Manbeck Jr. rm 906 PK2 — 305-8600
    Administrative Secretary Norma M. Rose rm 906 PK2 — 305-8600
Executive Assistant to the Commissioner and Director of Interdisciplinary Programs
Edward R. Kazenske rm 906 PK2 — 305-8600
    Secretary Georgia A. Maddox rm 906 PK2 — 305-8600
    Program Analyst Ann Farson rm 906 PK2 — 305-8600
Deputy Assistant Secretary and Deputy Commissioner Douglas B. Comer rm 904 PK2 — 305-8700
    Secretary Kathy Kennedy rm 904 PK2 — 305-8700
Assistant Commissioner for Patents James E. Denny (Designate) rm 919 PK2 — 305-8800
    Secretary Patricia R. Appelle rm 919 PK2 — 305-8800
Assistant Commissioner for Trademarks Jeffrey M. Samuels rm 910 PK2 — 305-8900
    Secretary Sheila G. Pellman rm 910 PK2 — 305-8900
Assistant Commissioner for Administration Theresa A. Brelsford rm 908 PK2 — 305-9100
    Secretary Karon Hricik rm 908 PK2 — 305-9100
Assistant Commissioner for Finance & Planning Bradford R. Huther rm 904 PK2 — 305-9200
    Secretary Vickie T. Bryant rm 904 PK2 — 305-9200
Assistant Commissioner for External Affairs Michael K. Kirk rm 902 PK2 — 305-9300
    Secretary Johnell M. Bersano rm 902 PK2 — 305-9300
Assistant Commissioner for Information Systems Thomas P. Giammo rm 916 PK2 — 305-9400
Secretary Helen White rm 916 PK2 — 305-9400

OFFICE OF THE SOLICITOR

Solicitor Fred E. McKelvey rm 918 PK2 — 305-9035
    Secretary Olga M. Suarez rm 918 PK2 — 305-9035
Deputy Solicitor Albin F. Drost rm 918 PK2 — 305-9035
    Secretary rm 918 PK2 — 305-9035
Associate and Assistant Solicitors:
    Lee E. Barrett rm 918 PK2 — 305-9035
    Muriel C. Crawford rm 918 PK2 — 305-9035
    John W. Dewhirst rm 918 PK2 — 305-9035
    Robert D. Edmonds rm 918 PK2 — 305-9035
    Teddy S. Gron rm 918 PK2 — 305-9035
    Jamerson Lee rm 918 PK2 — 305-9035
    Harris A. Pitlick rm 918 PK2 — 305-9035
    John H. Raubitschek rm 918 PK2 — 305-9035
    Richard E. Schafer rm 918 PK2 — 305-9035
    Linda N. Skoro rm 918 PK2 — 305-9035
    Nancy C. Slutter rm 918 PK2 — 305-9035
Paralegal Specialists:
    Teresa M. Byerley rm 918 PK2 — 557-4046
    Patricia D. McDermott rm 918 PK2 — 557-4031
    Maryann B. Volkmar rm 918 PK2 — 557-4022
Solicitor's Library
    Theresa Trierweiler-Cappo rm 918 PK2 — 557-4052

OFFICE OF ENROLLMENT AND DISCIPLINE

Director Cameron Weiffenbach rm 4830-ST — 308-9618
    Secretary Louwilda Turner rm 810 PK1 — 308-9618
    Harry I. Moatz rm 810 PK1 — 308-9618
    Shirley Rasheed rm 810 PK1 — 308-9618
    Patricia M. Jordan rm 810 PK1 — 308-9618
    Steve Morrison rm 810 PK1 — 308-9618
    Roster Information rm 810 PK1 — 557-1728

BOARD OF PATENT APPEALS AND INTERFERENCES

Chairman Saul I. Serota rm 12C12 CG2 — 557-4072
    Secretary Twanna Hawkins rm 12C12 CG2 — 557-4072
Vice Chairman Ian A. Calvert rm 10D10 CG2 — 557-4000
    Secretary Wanda G. Banks rm 10D10 CG2 — 557-4000
General Information:
    Ex parte Appeals rm 12C08 CG2 — 557-4101
    Interferences rm 10C01 CG2 — 557-4007

Examiners-in-Chief:

Neal E. Abrams rm 12D04 CG2	557-4057
James R. Boler rm 10C12 CG2	557-4009
Raymond F. Cardillo Jr. rm 10A04 CG2	557-0782
Marc L. Caroff rm 10C04 CG2	557-4009
Irwin C. Cohen rm 12B18 CG2	557-4703
Jerry D. Craig rm 10C22 CG2	557-4058
Mary F. Downey rm 10B14 CG2	557-4003
Stephen J. Emery rm 12D12 CG2	557-4326
Charles E. Frankfort rm 12D10 CG2	557-4057
Bradley R. Garris rm 12D06 CG2	557-4057
Melvin Goldstein rm 10D08 CG2	557-4068
John T. Goolkasian rm 10A10 CG2	557-4068
Kenneth W. Hairston rm 10A02 CG2	557-4069
Edward C. Kimlin rm 10D02 CG2	557-4003
Errol A. Krass rm 10B12 CG2	557-7517
William F. Lindquist rm 10C18 CG2	557-4061
Charles N. Lovell rm 10D06 CG2	557-4070
William E. Lyddane rm 12C04 CG2	557-4703
Thomas E. Lynch rm 10B04 CG2	557-7517
John C. Martin rm 10A12 CG2	557-4063
Harrison E. McCandlish rm 12B14 CG2	557-4703
John P. McQuade rm 10A20 CG2	557-7645
James M. Meister rm 10B06 CG2	557-4063
Edward J. Meros rm 10C20 CG2	557-4009
Andrew H. Metz rm 12B10 CG2	557-4326
Marion Parsons, Jr. rm 12C02 CG2	557-4025
William F. Pate, III rm 10B10 CG2	557-4069
Irving R. Pellman rm 10A16 CG2	557-4064
Verlin R. Pendegrass rm 10A08 CG2	557-4067
James A. Seidleck rm 10A22 CG2	557-4070
John D. Smith rm 12B16 CG2	557-4393
Ronald H. Smith rm 12D02 CG2	557-4057
William F. Smith rm 10A06 CG2	557-4069
Michael Sofocleous rm 10C06 CG2	557-4066
Lawrence J. Staab rm 10A18 CG2	557-7645
Robert F. Stahl rm 12B02 CG2	557-4025
Arthur J. Steiner rm 10A14 CG2	557-4062
David L. Stewart rm 10C10 CG2	577-4063
Bruce H. Stoner, Jr. rm 12B12 CG2	557-4394
Henry W. Tarring, III rm 10B02 CG2	557-4001
James D. Thomas rm 10C02 CG2	557-4007
Norman G. Torchin rm 10B16 CG2	557-4009
Stanley M. Urynowicz, Jr. rm 10C14 CG2	557-4066
Sherman D. Winters rm 10D04 CG2	557-4001
Thomas G. Wiseman rm 12D08 CG2	557-4326

Programs and Resources Administrator

Craig R. Feinberg rm 12C10 CG2	557-7169

Service Branch:

Chief Clerk of Board T. Maxine Duvall rm 12C06 CG2	557-4101
Deputy Clerk Nannie B. Henry rm 10C01A CG2	557-4007
Deputy Clerk Shirley A. Jefferys rm 12C08 CG2	557-4101
Deputy Clerk Eunice I. Price rm 10C07 CG2	557-4101
Ex parte Legal Clerk Group 120	
Eleanor R. Green rm 10C08 CG2	557-4108
Ex parte Legal Clerk Groups 130-180	
Paula Goldring rm 12C08 CG2	557-3100
Ex parte Legal Clerk Groups 110-150 & A.U. 223	
Karen Sweeney rm 10C09 CG2	557-4109
Ex parte Legal Clerk Groups 210-230-250-260-290 & A.U. 222	
Wanda Tigner rm 10C09 CG2	557-4107
Ex parte Legal Clerk Groups 330-340-350	
Mabel A. Neal rm 12C08 CG2	557-4106
Ex parte Legal Clerk Groups 240-310-320 & A.U. 221	
Leslie R. Chase rm 12C08 CG2	557-7189
Inter partes Legal Clerk	
Olivia M. Duvall rm 10C01 CG2	557-4006

Inter partes Legal Clerk
Carrie Evans rm 10C01 CG2   557-4004
Inter partes Evidence Clerk
Margaret E. Branson rm 10C01 CG2   557-4011

OFFICE OF QUALITY REVIEW

Director James D. Trammell rm 1100 CP6   603-0525
Secretary Carolyn D. Ballard rm 1100 CP6   603-0525

OFFICE OF ASSISTANT COMMISSIONER FOR PATENTS

Assistant Commissioner James Denny rm 917 PK2   305-8800
Secretary Patricia R. Appelle rm 917 PK2   305-8800
Patent Policy and Projects Administrator Charles Van Horn rm 919 PK2   557-3054
Secretary Donna E. Ellis rm 919 PK2   557-3054
Paralegal Specialist Sherry D. Brinkley rm 919 PK2   557-3054
Manual of Patent Examining Procedure Editor Louis O. Maassel rm 919 PK2   557-3070

Special Program Examination Unit
Supervisory Special Program Examiner Manuel A. Antonakas rm 923 PK2   557-8384
Special Program Examination Unit Information   557-8384
Deputy Assistant Commissioner for Patents Steven G. Kunin (Acting) rm 919 PK2   305-8850
Secretary Donna E. Ellis (Acting) rm 919 PK2   305-8850
Patent Programs Administrator Michael J. Lynch rm 917 PK2   305-8800
Patent Policy and Resources Director   305-8800
Trudy O. Ketler rm 917 PK2   305-8800
Supervisory Petitions Examiner
Jeffrey V. Nase rm 913 PK2   305-9282
Petitions Information rm 913 PK2   305-9282

Office of Patent Program and Documentation Control
Director Richard H. Rouck rm 925 PK2   305-9222
Receptionist Leonay Wynn rm 925 PK2   557-9182
Budget Coordinator Carolyn Arrington rm 925 PK2   557-9184
Planning and Evaluation Coordinator John Mielcarek rm 925 PK2   557-4214
Palm Coordinator Pamela Reinhart-Ganous rm 925 PK2   557-5142
Patent Academy Richard McGarr rm 501 PK1   557-2086

PATENT DOCUMENTATION ORGANIZATION

Administrator for Documentation William S. Lawson rm 703 PK2   557-0400
Secretary Jim Doyle rm 703 PK2   557-0400
Data Base Administrator Philip K. Olson rm 703 PK2   557-0400

Deputy Administrator for Documentation
Edward J. Earls rm 300 CM2   557-0400
Secretary   557-0400
Office of Classification Support Staff
Director Sally Middleton rm 300 CM2   557-0400
Secretary   557-0400
Contract Monitoring and Reclassification Division Janice Burse rm 967 CM2 557-8877
Secretary Michelle K. Smith rm 967 CM2   557-8877
Project Monitoring Unit Inez Roberts rm 967 CM2   557-5164
Processing Unit I - Janice Burse rm 964 CM2   557-2590
Processing Unit II - Linda McDowell rm 968 CM2   557-7467
Processing Unit III - Pat Walker rm 965 CM2   557-7455
Special Projects Unit Nan Galloway (Acting) rm 969 CM2   557-3396
Weekly Issue (new U.S. patents) Pat Harris rm 1240 CP6   603-0621
Temporary Search Room rm 316 CM2   557-9579
New Document Processing Division (Foreign Patents)
Chiquita Clark (Acting) Lobby CP6   603-0572
Preprocessing Branch Jerry Redmond Lobby CP6   603-0576
Final Processing Branch Daisy Turner Lobby CP6   603-0574
Special Processing Natalie Jackson Lobby CP6   557-5108
Editorial Division Mary Louise McAskill rm 318 CM2   557-2101
Patent Index rm 304 CM2   557-3951

Data Control Technician Division Marcia Smith rm 305	557-0400
Chemical-Electrical Classification Group Director Donald J. Hoffman rm 901CM2	557-2825
Secretary Sandra P. Crawford rm 903 CM2	557-2820
Receptionist Chenette Foster rm 900 CM2	557-3814
Unit I Diane B. Russell rm 971 CM2	557-2755
Unit II Eugene B. Woodruff rm 935 CM2	557-2826
Unit III Gary Solyst rm 923 CM2	557-3505
Unit IV Kendall J. Dood rm 912 CM2	557-0151
Mechanical-General Classification Group Director John W. Will rm 310 CM2	557-0107
Secretary Christina Boska rm 310 CM2	557-0107
Receptionist Debbie Hawkins rm 310 CM2	557-0109
Unit I Donald P. Rooney rm 310 CM2	557-0134
Unit II Robert Craig rm 310 CM2	557-0136
Unit III Harold P. Smith rm 310 CM2	557-2446
Unit IV John Leonard rm 912 CM2	557-0151
Special Projects Classification Group SPC Leslie Wolf rm 982 CM2	557-2781
Secretary Melvina Jarrett rm 980 CM2	557-0173
Office of International Patent Documentation Director J. Russell Goudeau rm 300 CM2	557-0667
Secretary Sharon K. Fink rm 300 CM2	557-3756
International Patent Classifiers: Doris Funderburk-Penn rm 300 CM2	557-3756
Palmer W. Sullivan rm 300 CM2	557-3756
Exchange Agreements Frank L. Rytlewski rm 300 CM2	557-0400
Office of Documentation Planning and Support Director George Chadwick rm 300 CM2	557-0431
Scientific Library And Technical Information Center Program Manager Henry Rosicky rm 2C08 CP34	308-0808
Secretary Gail Owens rm 2C08 CP34	308-0808
Administrative Librarian Irene Heisig rm 2C08 CP34	308-0808
Foreign Patents Division Barry Balthrop rm 2C01 CP34	308-0817
Bindery Unit Ronald Knickerbocker FERN	557-1530
Document Retrieval and Copy Branch Lendoria Roberson rm 2C01 CP34	308-1076
Receipts and Records Branch Eunice Foster Suite 1821D CM2 Concourse	557-0186
Reference Service Bernard Hamilton rm 2C01 CP34	308-1076
Scientific Literature Division Kay Melvin rm 2C06 CP34	308-0810
Commercial Database Contracts Gay Posey rm 2C15 CP34	308-3244
Technical Services Branch Jesse Gibson rm 2C06A CP3	308-0813
Collections Development Dale Ram rm 2C01 CP34	308-0853
User Services Branch Dora Weinstein rm 2C04CP34	308-0810
Circulation rm 2C01 CP34	308-0810
Computer Searching rm 2C01 CP34	308-0810
Interlibrary Loans rm 2C01 CP34	308-0810
Reference Service rm 2C01 CP3	308-0810
Translations Division Dean Thorne rm 2C15 CP34	308-0881
Receptionist Carol Releford rm 2C15 CP34	308-0881

CHEMICAL EXAMINING GROUPS

110 General, metallurgical, Inorganic, Petroleum and Electrical Chemistry and Engineering rm 9Cl7 CP3	308-0661
Director Dennis E. Talbert rm 9D19 CP3	308-3729
Secretary Constance L. Morgan rm 9D19 CP3	308-3729
General Information/Receptionist rm 9C17 CP3	308-0661
SAC Dorothy Dawkins rm 9C17 CP3	308-0661
111 Metallurgical methods and apparatus, alloys, hydrometallurgy	
Richard Dean rm 10E06 CP3	308-2548
112 Electro-chemistry, process and apparatus	
John F. Niebling rm 10B04 CP3	308-3325
113 Inorganic compounds and non-metallic elements (except radioactive), chemical gas purification processes, beneficiating ores, metal stock, fertilizers (chemical and biological)	
Michael Lewis rm 10A15 CP3	308-2535
114 Methods for semiconductor treating and manufacturing batteries	
Brian E. Hearn rm 10D35 CP3	308-2552
115 Chemical compositions, dying and pigments	
Paul Leiberman rm 10A15 CP3	308-3316
116 Catalytic compositions, chemistry of hydrocarbons	
Helen Sneed rm 9D35 CP3	308-0840
117 Semiconductor treating and manufacturing, single crystals, disinfecting	

Olik Chaudhuri rm 9D27 CP3	308-0434
118 Refractory glass and cement compositions, abrasives, coating compositions, and pigments	
William R. Dixon, Jr. rm 10B02 CP3	308-3324
II9A Mineral oils processes and products, cleaning, carbohydrates, coating compositions, and batteries	
Theodore Morris rm 10E02 CP3	308-2546
119B Fuel and lubricating compositions, detergents, miscellaneous, chemical compositions, bleaching and dyeing	
Prince Willis rm 10D07 CP3	308-3313
120 Organic Chemistry Drug, Bio-Affecting and Body Treating	
Composition rm 8C13 CP2	308-0210
Director John F. Terapane rm 8A07 CP2	308-0193
Secretary Robin Brown (Acting) rm 8A07 CP2	308-0193
General Information/Receptionist rm 8C13 CP2	308-1235
SAC Helen Childs rm 8C13 CP2	308-0210
121 Five, four or three membered ring, certain acids, esters, helerocyclic chemistry, nitriles, and azo chemistry	
Mary C. Lee rm 8A19 CP2	308-0409
122 Nitrogen containing heterocyclic compounds, six, seven or more ring members	
Mukund Shah rm 8B32 CP2	308-1256
123 Six membered hetrocyclics and oxygen containing hetrocyclic compounds	
Frederick Waddell rm 8B10 CP2	308-1235
124 Organic carboxylic acids, esters and organometallics	
Jose Dees rm 8A09 CP2	308-0092
125 Pharmaceutical methods and compounds	
Stanley Friedman rm 8B02 CP2	308-0119
126 Carbo, oxy, aldehyde, keto and halo compounds	
Donald Moyer rm 7D01 CP2	308-0252
129 Pesticides, amines and cosmetics	
Glennon H. Hollrah rm 8D31 CP2	308-0322
130 Specialized Chemical Industries and Chemical Engineering rm 8C17 CP3	308-0651
Director Barry S. Richman rm 8D19 CP3	308-1193
Secretary Vickie Beach rm 8D19 CP3	308-1193
General Information/Receptionist rm 8C17 CP3	308-0651
SAC Ola Sims rm 8C17 CP3	308-3835
131 Adhesive bonding and miscellaneous chemical manufacture	
Michael W. Ball rm 9E02 CP3	308-0651
132 Food or edible material, processes, compositions and products	
Donald E. Czaja rm 8D01 CP3	308-0651
133 Paper making and fiber liberation, glass manufacture, gas, heating and illumination apparatus and processes, minerals, oils apparatus, and coating apparatus	
Richard Fisher rm 8B36 CP3	308-0651
134 Compositions, methods and apparatus for etching in chemical manufacture, coating apparatus, metal treating, and adhesive bonding	
David Simmons rm 9E16 CP3	308-0651
135 General molding or treating apparatus, static molds gas, separation, gas and liquid contact	
Jay H. Woo rm 8D35 CP3	308-0651
136 Liquid purification or separation and processes of plastic and nonmetallic article shaping or treating	
Richard Dawson rm 7A09 CP3	308-0651
137 Processes of plastic and nonmetallic article shaping or treating	
Jan H. Silbaugh rm 8A01 CP3	308-0651
138 Liquid purification or separation, concentrating apparatus and processes and separating and assorting solids - froth flotation	
Stanley S. Silverman rm 7D11 CP3	308-0651
139 Coating processes	
Norman Morgenstern rm 7D17 CP3	308-0651
150 High Polymer Chemistry, Plastics, Coating, Photography, Stock Materials and Compositions rm 11C19 CP2	308-2351
Director James O. Thomas, Jr. rm 11A04 CP2	308-2359
Secretary Cheryl P. Gibson rm 11A04 CP2	308-2359
Deputy Director John E. Kittle rm 11B16 CP2	308-2428
General Information/Receptionist rm 11C19 CP2	308-2351
SAC Kathryn Perry rm 11B14 CP2	308-2365
151 Mixed synthetic resin compositions, block and graft copolymers	
John Bleutge rm 11B02 CP2	308-2368
152 Drug, Bio-affecting and body treating compositions	
Merrell Cashion rm 10A10 CP2	308-2367
153 Foams, condensation polymers of cellulose, phenols, isocyanates, polyesters, polyepoxides, stabilization of polymers	
John Kight, III rm 10A22 CP2	308-2453
154 Stock materials or miscellaneous articles of manufacture	

George Lesmes rm 11E16 CP2	308-2362
155 Polymer compositions, addition polymers, and carbohydrates	
Joseph L. Schofer rm 10E02 CP2	308-2452
156 Radiation imagery chemistry - silver halide and diazophotosentitive compositions and processes	
Paul Michl rm 10D02 CP2	308-2451
157 Radiation imagery chemistry - photopolymerization, irradiation of polymers and electrophotography	
Marion McCamish rm 10D32 CP2	308-3961
158 Stock materials or miscellaneous articles and record receivers	
Ellis Robinson rm 11E02 CP2	308-2364
180 Biotechnology rm 9C13 CP2	308-1123
Director Edward E. Kubasiewicz rm 9A09 CP2	308-0196
Secretary Betty Kaminsky rm 9A09 CP2	308-1123
Deputy Director Charles F. Warren rm 9D01 CP2	308-2928
General Information/Receptionist rm 9C13 CP2	308-0196
SAC Ellen Scott rm 9C13 CP2	308-1175
181 Chemical apparatus such as analyzers, reactors and sterilizers; processes of chemical and clinical analysis, sterilizing and preserving; liquid purifications or separation by living organisms	
Robert Warden rm 5A15 CP2	308-2920
182 Clinical chemistry, microbiology, immunology and enzymology, purification and chemical engineering	
Esther Kepplinger rm 6A01 CP2	308-1219
183 Peptide and carbohydrate chemistry, and drug, bioaffecting and body treating compositions containing peptides, carbohydrates, and undetermined constitution	
Johnnie R. Brown rm 6D01 CP2	308-1179
184 Multicellular organisms (animal/plant), molecular genetics, cell culture, hybridoma, plant and animal molecular biology, microbiology, protozoa and algae	
Elizabeth Weimar rm 9B32 CP2	308-0254
185 Molecular genetics and molecular biology of bacteria and fungi genetic engineering and vectors, enzyme production	
Richard Schwartz rm 9D31 CP2	308-1083
186 Proteins, antibody, antigen chemistry and drug, bioaffecting and body treating compositions containing proteins	
Margaret Moskowitz rm 9A01 CP2	308-0452
187 Nucleic acid assays, AIDS assays, immunology, chemical assay standards, fermentation apparatus	
Robert Wax rm 5D33 CP2	308-2454
188 Enzymes, fermentation, microbiology, drug, bioaffecting and body treating compositions containing enzymes or animal extracts of undetermined compositions	
(Vacant) rm 5D33 CP2	308-2454

ELECTRICAL EXAMINING GROUPS

210 Industrial Electronics, Physics and Related Elements rm 9C17 CP4	308-1782
Director Donald G. Kelly rm 9D19 CP4	308-0658
Secretary Danita L. Ingram rm 9D19 CP4	308-0658
General Information/Receptionist rm 9C17 CP4	308-1782
SAC Romaine D. Bowling rm 9C17 CP4	308-2930
211 Photography, photocopying, motion pictures, electric and magnetographic recorders, optics, capacitors	
L. Thomas Hix rm 9B40 CP4	308-2969
212 Electrical motor-generator structure, piezoelectric elements and devices, generator systems, battery and condenser charging and discharging, power supply regulation, conversion systems, and music	
Steven L. Stephans rm 9A01 CP4	308-3778
213 Conductors, insulators, inductors, electromagnets, magnetic and thermal switches, electric lamp and discharge devices, industrial electric furnaces, acoustics and register	
Leo P. Picard rm 10D01 CP4	308-3304
214 Electrical switches and arc suppression, heating by induction, plasma and electric discharge machining, protection of electrical systems and devices, electromagnetic control systems, electric charge devices and systems, electrical component housing and mounting and, electrical transmission systems, electrical elevator controls, and prime mover dynamo plants	
A. David Pellinen rm 10D17 CP4	308-3301
215 Electric photocopying and coating apparatus	
Arthur T. Grimley rm 9E16 CP4	308-2968
216 Recorders, scales, electric heating, electric welding, resistors and pictorial communications	
Bruce A. Reynolds rm 10B02 CP4	308-3305
217 Motor control systems, coded data generation or conversion and horology	
William M. Shoop Jr rm 9D01 CP4	308-0905
220/290 Special Laws Administration & Designs rm 11D17 CP3	308-0766
Director Robert E. Garrett rm 11D17 CP3	308-0753

Secretary Tereta Gilchrist rm 11D19 CP3	308-0765
General Information/Receptionist Group 220 rm 11C17 CP3	308-0766
General Information/Receptionist Group 290 rm 1106 CP6	308-0744
SAC Joanne Hodge rm 11D11 CP3	308-3518
Licensing and Review Hilda Grimes rm 11D09 CP3	308-1714
Mildred Scott rm 11D08 CP3	308-1715

221 Weapons (firearms, ordnance, ammunition, explosive devices, aeronautics and ships as well as all classified mechanical applications

Deborah L. Kyle rm 11A15 CP3	308-0918

222 Radio, optic, acoustic, wave communications systems and all classified electrical applications

Thomas H. Tarcza rm 11D35 CP3	308-1689

223 Chemical specials use compositional including five extinguishing and retarding earth boring and well treating, liquid crystals, colloids and dispersants, preservative agents, and microcapsule and body treating radioactive as well as classified chemical applications

Robert L. Stoll rm 11E02 CP3	308-1701

224 Chemical engineering including radioactive materials, powder metallurgy rocket fuels, explosives, thermal and photoelectric batteries, and nuclear reactors systems and related technologies, and classified chemical applications

Brooks H. Hunt rm 11A01 CP3	308-1682

291 Ornamental designs in the area of industrial arts

Wallace R. Burke rm 11E02 CP6	603-0569

292 Ornamental designs for fine arts

Bernard Ansher rm 11A02 CP6	603-0558

293 Ornamental designs, industrial/fine arts

Bruce Dunkins rm 1215-1 CP6	603-0555

294 Ornamental designs, industrial/fine arts

A. Hugo Word rm 11B04 CP6	603-0571

230 Information Processing, Storage and Retrieval rm 11C17 CP4	308-0754
Director Gerald Goldberg rm 11D37 CP4	308-1785
Secretary Teresa E. Dugan rm 11D37 CP4	308-1785
General Information/Receptionist rm 11C17 CP4	308-0754
SAC Katherine A. Nelson rm 11C17 CP4	308-1322

231 Computer data presentation systems, digital arithmetic circuitry, speech analysis and synthesis systems

Gary V. Harkcom rm 10E16 CP4	308-2826

232 General and special purpose digital data processing systems

Archie E. Williams Jr. rm 11B02 CP4	308-0538

233 Static information storage and retrieval and elements of dynamic magnetic information storage

Stuart N. Hecker rm 11D01 CP4	308-0079

234 Ordnance or weapon system computers and special applications of computers including vehicle control, navigation, measuring, testing and monitoring

Parshotam S. Lall rm 11D17 CP4	308-1373

235 Electrical dynamic information storage and retrieval and record controlled systems

(Vacant) rm 11D11 CP4	308-3126

236 Computer control systems, computers in business, computers in medicine, computer aided product manufacturing, analog and hybrid computers, and error correction and detection systems

Jerry Smith rm 11E10 CP4	308-3101

237 General and special purpose digital data processing systems

Gareth D. Shaw rm 11B40 CP4	308-3103

238 Artificial intelligence, general purpose programmable digital computer systems, computers in earth science, games and social sciences, and miscellaneous digital data processing systems

Michael R. Fleming rm 10E06 CP4	308-1691

239 Systems controlled by data bearing records, coded record sensors, record controlled calculators, dynamic information storage or retrieval and telephone answering machines

Stuart S. Levy rm 11D35 CP4	308-0725

240 Packages, Cleaning, Textiles and Geometrical Instruments rm 6C17 CP4	308-0771
Director Trygve Blix rm 6D37 CP4	308-0777
Secretary Donna P. Magaha rm 6D37 CP4	308-0777
General Information/Receptionist rm 6C17 CP4	308-0771
SAC Doretha A. Marcelli rm 6D15 CP4	308-0771

241 Packaging art including glass, fabric, metal, wood, paper and plastic receptacles plus closures

Stephen Marcus rm 6B02 CP4	308-1499

242 Fluid treating, presses, food apparatus, cleaning, agitating, centrifuges, and web feeding

Harvey C. Hornsby rm 7A15 CP4	308-1374

243 Conduits, bathroom facilities, cleaning apparatus, filling apparatus, switches, and article carriers

Henry Recla rm 6B30 CP4	308-1382

244 Special receptacles or packages, shoes and shoe making

Paul T. Sewell rm 7D19 CP4	308-1543

245 Textiles, winding and reeling, pushing and pulling, bearings, and flexible torque transmitters
    Daniel P. Stodola rm 7E16 CP4   308-2673
246 Measuring and testing, dynamic information storage or retrieval, optical image projectors and joint packing
    William Cuchlinski rm 6E14 CP4   308-3873
247 Textile and leather manufacture, apparel, and textiles
    Werner Schroeder rm 6D01 CP4   308-0949
250 Electronic and Optical Systems and Devices rm 8D17 CP4   308-0956
    Director Joe Rolla rm 8D19 CP4   308-0530
        Secretary Deborah P. Leeper rm 8D19 CP4   308-0530
        General Information/Receptionist rm 8C17 CP4   308-0956
        SAC JoAnn Davis rm 8D21 CP4   308-1578
251 Lasers, fiber optic devices and antennas
    William L. Sikes rm 8E02 CP4   308-0700
252 Electronic modulators, demodulators, oscillators, amplifiers, tuners and wave transmission lines and networks

    Eugene R. Laroche rm 8B34 CP4   308-2878
253 Semiconductor devices
    Andrew J. James rm 7B02 CP4   308-1610
254 Semiconductor and vacuum tube circuits and systems and electronic and electromechanical counting circuits
and systems
    Stanley D. Miller rm 7E02 CP4   308-1532
255 Optical measuring and testing systems and photocell circuits
    Davis L. Willis rm 8A15 CP4   308-1393
256 Radiant energy systems
    Janice A. Howell rm 8B02 CP4   308-0612
257 Optical systems and elements and vision testing and correcting
    Bruce Arnold rm 8E16 CP4   308-0758
258 Semiconductor devices
    Rolf Hille rm 7B12 CP4   308-1598
260 Communications, Measuring, Testing and Lamp/Discharge
Group rm 5D19 CP4   308-0962
    Director Stewart J. Levy (Acting) rm 5D19 CP4   308-2802
        Secretary Iyone L. Miles-Brooks rm 5D19 CP4   308-2802
        General Information/Receptionist rm 5C14 CP4   308-0962
        SAC Vivian C. Harris rm 5D21 CP4   308-0525
261 Telegraphy, telephony and audio systems Jin F. Ng rm 4D01 CP4   308-2991
262 Television and television facsimile
    James J. Groody rm 4B02 CP4   308-2912
263 Multiplex communications, digital communications and telecommunications
    Douglas W. Olms rm 5E02 CP4   308-0305
264 Electrical communications and acoustic wave systems
    Donald J. Yusko rm 6E02 CP4   308-2126
265 Measuring and testing of non-electrical phenomenon
    Hezron E. Williams (Acting) rm 5B24 CP4   308-2674
266 Image analysis
    David K. Moore rm 5B02 CP4   308-2686
267 Measuring and testing of electrical phenomenon
    Reinhard J. Eisenzopf rm 5D01 CP4   308-2696
268 Condition responsive communications, measuring and testing
    Joseph A. Orsino rm 5D01 CP4   308-0979
269 Selective Visual Display/Facsimile
    Alvin E. Oberley rm 4B16 CP4   308-3255

MECHANICAL EXAMINING GROUPS

310 Handling and Transporting Media rm 5D19 CP3   308-1113
    Director Bobby R. Gray rm 5D17 CP3   308-1134
        Secretary Annette L. Pray rm 5D19 CP3   308-1134
        General Information/Receptionist rm 5C17 CP3   308-1113
        SAC Margaret Stevens rm 5D14 CP3   308-1113
311 Conveyors, article dispensing, elevators, and sheet feeding or delivering devices
    Robert Olszewski rm 5A01 CP3   308-2588
312 Motor vehicle wheels and bodies, fluid conveying, fire extinguishers and handling implements
    Margaret Focarino rm 4D35 CP3   308-0885
313 Brakes, fluid pressure brake systems, spring devices railways, and railway equipment
    Robert Oberleitner rm 5B18 CP3   308-2569
314 Spraying devices and land and motor vehicles

Andres Kashnikow rm 5D01 CP3	308-1137
315 Aeronautics and marine arts	
Joseph F. Peters, Jr. rm 5E02 CP3	308-2561
316 Land and motor vehicles	
Charles Marmor rm 6A01 CP3	308-0361
317 Article handling, check-controlled apparatus, railway track and switches, coin handling, dumping vehicles, merchandizing and freight accomodation	
Robert J. Spar rm 5D35 CP3	308-2555
320 Material Shaping, Article Manufacturing, Tools rm 6C17 CP3	308-1148
Director Nicholas P. Godici rm 6D19 CP3	308-1078
Secretary Laura Dorsey rm 6D19 CP3	308-1078
General Information/Receptionist rm 6C17 CP3	308-1148
SAC Vera Thomas rm 6C17 CP3	308-2326
321 Metal deforming, packaging machinery, butchering and woodworking	
Robert Spruill rm 7E02 CP3	308-3837
322 Electrical connectors, gear cutting, milling and chucks	
(Vacant) rm 6D35 CP3	308-1152
323 Abrading, workholders, tools and paper manufactures	
Frederick Schmidt rm 7D35 CP3	308-2333
324 Cutting, cutlery, tools, bookmaking, and printed matter	
Frank T. Yost rm 6E02 CP3	308-2058
325 Metal founding, metal turning, miscellaneous hardware, fishing, vermin trapping and welding	
Richard K. Seidel rm 6D01 CP3	308-2319
326 Metal working, comminution and wire working	
Howard Goldberg rm 7A01 CP3	308-3852
330 Surgery, Plant and Animal Husbandry, Amusement and Exercise Devices and Printing rm 4C17 CP4	308-2192
Director John J. Love rm 4D19 CP4	308-0873
Secretary Norma L. Watson rm 4D19 CP4	308-0873
Gerneral Information/Receptionist rm 4C17 CP4	308-0858
SAC Clara S. Desmukes rm 4D21 CP4	308-2192
PTO Facsimile Center rm 3D56 CP34	308-1353
Paper Correlating Center rm 3C04 CP4	308-2645
331 Surgical instruments, toys, coating implements	
Robert Hafer rm 3A01 CP3	308-2121
332 Exercising and therapy devices, tobacco, artificial bodyparts, orthopedics, and educational devices	
Richard Apley rm 4E16 CP4	308-1065
333 Dentistry, animal husbandry, and toiletries	
Gene Mancene rm 3E02 CP3	308-2156
334 Amusement games and medical therapy	
Edward Coven rm 3E02 CP3	308-0871
335 Surgery diagnostics	
Kyle Howell rm 4E02 CP4	308-3256
336 Surgery instruments, medicators or receptors	
C. Fred Rosenbaum rm 3D39 CP4	308-0495
337 Printing, typewriting, surgery-respiratory	
Edgar Burr rm 3D01 CP3	308-3144
338 Prosthetic devices, medical apparatus	
Randall Green rm 4D25 CP4	308-1006
339 Surgical Instruments and medicators	
Stephen Pellegrino rm 3D35 CP4	308-1015
340 Solar, Heat, Power and Fluid Engineering Devices rm 3C17 CP4	308-0861
Director Carlton R. Croyle (Acting) rm 3D19 CP4	308-0975
Secretary Carol A. Laroda rm 3D19 CP4	308-0975
General Information/Receptionist rm 3C17 CP4	308-0861
SAC Verlene D. Green rm 3C17 CP4	308-0861
341 Expansible chamber motors and fluid power systems, impellers and fluid reaction devices	
Edward K. Look rm 3B22 CP4	308-1044
342 Internal combustion engines including charge forming and ignition systems	
Charles J. Myhre rm 2D01 CP4	308-1946
343 Jet engines of the air-breathing type, aerospace propulsion systems including solid and liquid fueled rockets, electric and electromagnetic types, gas turbine power plants for vehicles and stationary power generation, turbocharged, supercharged and rotary internal combustion engines, rotary expanisible chamber devices, and pumps including expansible chamber, fluid entrainment, and motor-driven types	
Carlton J. Croyle rm 3B02 CP4	308-0102
344 Environmental control, including heating, drying air conditioning, refrigeration, ventilation, and controls	
Albert Makay rm 3D01 CP4	308-0101

346 Devices and methods for illumination, exhaust gas treatment of internal combustion engines, power plants of the type using natural heat, lubrication, stoves and furnaces including heating the environment and extracting heat from the sun and combustion

Ira Lazarus rm 2B02 CP4	308-1935

347 Fluid handling which includes valves, pressure regulators and flow controllers for liquids and gases, heat exchange, and methods for transferring heat from one material to another

Martin Schwadron rm 2B42 CP4	308-2597

350 General Construction, Petroleum and Mining Engineering

rm 3C17 CP3	308-2168
Director Al Lawrence Smith rm 3D19 CP3	308-1020
Secretary Fran E. Lynah rm 3D19 CP3	308-1020
General Information/Receptionist rm 3C17 CP3	308-2168
SAC Joyce G. Hill rm 3D13 CP3	308-3765

351 Joints and connections, pipe couplings, fences, earth and hydraulic engineering

Randolph A. Reese rm 4D01 CP3	308-2498

352 Gearing, power transmissions, clutches, machine elements

Leslie A. Braun rm 2B36 CP3	308-0830

354 Building structures and components

David A. Scherbel rm 2D35 CP3	308-2144

355 Supports, racks, fire escapes, ladders, scaffolds, flexible partitions

Carl D. Friedman rm 4E02 CP3	308-0331

356 Petroleum, mining, highway and bridge engineering, well drilling, endless belts

Ramon S. Britts rm 4A01 CP3	308-0839

357 Tables, chairs, windows, doors, cabinets, buckles, buttons, clasps,

Kenneth J. Dorner rm 3E16 CP3	308-0866

358 Fasteners, safes, locks, closure fasteners, beds, control levers and linkages

Gary L. Smith rm 4D17 CP3	308-0827

OFFICE OF THE ASSISTANT COMMISSIONER FOR EXTERNAL AFFAIRS

Assistant Commissioner Michael K. Kirk rm 902 PK2	305-9300
Secretary Johnell M. Bersano rm 902 PK2	305-9300
Director of Congressional Affairs Patricia Callahan rm 902 PK2	557-1310
Congressional Liaison Janie F. Cooksey rm 902 PK2	557-1310
Office of Public Affairs	
Director Gil Weidenfeld rm 208B PK1	305-8341
Public Information Specialist Oscar G. Mastin rm 208B PK1	305-8341
Office of Legislation and International Affairs rm 902 PK2	557-3065
Legislative and International Intellectual Property	
Specialists: Denise W. DeFranco rm 902 PK2	557-3065
H. Dieter Hoinkes rm 902 PK2	557-3065
Paul Salmon rm 902 PK2	557-3065
Lee J. Schroeder rm 902 PK2	557-3065
G. Lee Skillington rm 902 PK2	557-3065
Attorney Advisors:	
Rosemarie G. Bowie rm 902 PK2	557-3065
Michael S. Keplinger rm 902 PK2	557-3065
Richard C. Owens rm 902 PK2	557-3065

OFFICE OF THE ASSISTANT COMMISSIONER FOR TRADEMARKS

Assistant Commissioner Jeffrey M. Samuels rm 910 PK2	305-8900
Secretary Sheila G. Pellman rm 910 PK2	305-8900
Deputy Assistant Commissioner for Trademarks Robert M. Anderson rm 910 PK2	305-8910
Trademark Legal Administrator Lynn G. Beresford rm 910 PK2	305-9464
Secretary Carol P. Smith rm 910 PK2	557-3061
Staff Attorney Gerald Rogers rm 910 PK2	557-2521
Trademark Program Analyst Karen Strohecker rm 910 PK2	557-2221

Office of Trademark Program Analysis

Director Kimberly Krehely rm 3C06 CP2	308-0928
Secretary Percy L. Turner III rm 3C06 CP2	308-0928
Program Analyst Betty Andrews rm 3C06 CP2	308-0928
Program Analyst Blake Pearl rm 3C06 CP2	308-0928
Paralegal Assistant Kathy Dixon rm 3C06 CP2	308-0928
Data Base Maintenance Yvonne Evans rm 4C16 CP2	308-2850
Quality Control Staff Charles Phucas rm 411 CP1	603-0188

Office of Trademark Quality Review
    Director Charles J. Condro rm 1205 CP6   603-0657
        Secretary Lisa Y. Jones rm 1205 CP6   603-0657
        Attorney Advisor Donald J. Fingeret rm 1205 CP6   603-0657
        Attorney Advisor David M. Soroka rm 1205 CP6   603-0657
        Attorney Advisor Joseph H. Webb rm 1205 CP6   603-0657

TRADEMARK EXAMINING OPERATION

Director David E. Bucher rm 3C06 CP2   308-0928
        Secretary rm 3C06 CP2   308-0928
        Administrator for Trademark Operations Patricia M. Davis rm 3C06 CP2   308-0928
        Secretary Beth Acker rm 3C06 CP2   308-0928
        Administrator for Trademark Policy and Procedures James T. Walsh rm 3C06 CP2   308-0928
        Secretary rm 3C06 CP2   308-0928
        Administrator for Petitions and Classification Jessie N. Marshall rm 3C06 CP2   308-0928
        Petitions Assistant Deborah Mackall rm 3C06 CP2   308-0928
        Administrator for Trademark Procedures and Special Projects Mary C. Coyle rm 3C06 CP2   308-0928
        Secretary Everett J. Henson rm 3C06 CP2   308-0928
        Trademark Program Analyst Nancy P. Miller rm 3C06 CP2   308-0928
        Paralegal Assistant Melanie Meon rm 3C06 CP2   308-0928
        Paralegal Assistant Darlene Bullock rm 3C06 CP2   308-0928
        Paralegal Assistant Sarah Hyde rm 3C06 CP2   308-0928
        Trademark Law Offices
    Managing Attorney Law Office I Deborah Cohn rm 3C28 CP2   308-0933
    Senior Attorney Kathryn Dobbs, Nancy Hankins, Henry Zak
    Supervisory Applications Examiner Johnson Anderson III
    Managing Attorney Law Office II Ron Williams rm 2C24 CP2   308-0936
    Senior Attorney Craig Morris, Mary Hannon
    Supervisory Applications Examiner Sylvia Hammett
    Managing Attorney Law Office III Myra Kurzbard rm 2C22 CP2   308-3714
    Senior Attorney Jody Drake, Robert Freely, Patrick Hines
    Supervisory Applications Examiner Jackie Perry
    Managing Attorney Law Office IV Thomas Lamone rm 3C13 CP2   308-3709
    Senior Attorney John Demos, Terry Ellen Holtzman
    Supervisory Applications Examiner Sharon Wilson
    Managing Attorney Law Office V Paul Fahrenkoph rm 2C11 CP2   308-1875
    Senior Attorney Sharon Marsh, Terry Rupp, Ronald Sussman
    Supervisory Applications Examiner Thurmond Streater
    Managing Attorney Law Office VI Mary Sparrow rm 3C27 CP2   603-0700
    Senior Attorney Robert Crowe, Mary Kay McDonald
    Supervisory Application Examiner Janice Martin
    Managing Attorney Law Office VII Thomas Howell rm 4C13 CP2   308-1807
    Senior Attorney Michael Hamilton, David Shallant, Michelle Wiseman
    Supervisory Applications Examiner Karen McCray
    Managing Attorney Law Office VIII Sidney Moskowitz rm 4C11 CP2   308-1812
    Senior Attorney Michael Bodson, David Stine
    Supervisory Applications Examiner Ada Rollins
    Managing Attorney Law Office IX Ronald E. Wolfington rm 4C27 CP2   308-4069
    Senior Attorney Ira Goodsaid, Jean Logan
    Supervisory Applications Examiner Deborah Mays
    Post Registration Law Office Managing Attorney Jackie Cole rm 6C24 CP2   308-0483
Trademark Services Division
    Director Doreane Poteat rm 6C27 CP2   308-1819
        Secretary Sophia Brock rm 6C27 CP2   308-1819
    Quality Review Clerk Deborah Ahmed rm 6C27 CP2   308-1819
    Publication & Issue Supervisor Valarie Bean rm 6C27 CP2   308-1817
    Affidavit Examiners rm 6C27 CP2   308-0485
    Renewal Examiners rm 6C27 CP2   308-0485
    Mail Reader/Messenger rm 6C27 CP2   308-1827
    Microfilm Section rm 6C27 CP2   308-1825
    Intent-To-Use Managing Attorney Frank Hellwig rm 4C27 CP2   308-4061

TRADEMARK TRIAL AND APPEAL BOARD

Members of the Board:
Chairman J. David Sams rm 1008 CS5   557-3551

Ellen Seeherman rm 1008 CS5	557-3551
Robert F. Cissel rm 1008 CS5	557-3551
Louise E. Rooney rm 1008 CS5	557-3551
G. Douglas Hohein rm 1008 CS5	557-3551
Janet E. Rice rm 1008 CS5	557-3551
Rany L. Simms rm 1008 CS5	557-3551
Elmer W. Hanak, III rm 1008 CS5	557-3551
T. Jeffrey Quinn rm 1008 CS5	557-3551
Attorney-Examiners:	
Paula T. Hairston rm 1008 CS5	557-3551
Beth A. Chapman rm 1008 CS5	557-3551
Carla C. Calcagno rm 1008 CS5	557-7049
Marc A. Bergsman rm 1008 CS5	557-3551
Helen R. Wendel rm 1008 CS5	557-3551
Paralegal Specialist Gladys R. Springer rm 1008 CS5	557-3551
Administrator Jean Brown rm 1008 CS5	557-3551
Legal Technician Sheila H. Veney rm 1008 CS5	557-3551

OFFICE OF THE ASSISTANT COMMISSIONER FOR ADMINISTRATION

Assistant Commissioner Theresa A. Brelsford rm 908 PK2	305-9100
Secretary Karon Hricik rm 908 PK2	305-9100
Deputy Assistant Commissioner for Administration Wesley H. Gewehr rm 908 PK2	305-9110
Secretary Dee Dee Walker rm 908 PK2	305-9110
Program Analyst Joan S. Griffey rm 908 PK2	557-2290

Office of General Services

Director John D. Hassett rm 803 PK1	305-8183
Secretary Peggy Fewell rm 803 PK1	305-8183
Deputy Director G. William Richardson rm 803 PK1	305-8183
Security/Safety Officer Thomas J Killcullen rm 803 PK1	305-8183
Correspondence and Mail Division Sallye Rayford rm 1A03 CP2	308-0792
Deputy (Vacant) rm 1A03 CP2	308-0792
Incoming-Outgoing Mail Branch	
William Rivers rm 1A01 CP2	308-0907
Initial Review and Serializing Branch	
Shirley Steele rm 1B03 CP2	308-0906
Ethel Dillard rm 1B03 CP2	308-1056
Correspondence Branch rm 1A03 CP2	308-0901
Facilities Management Division	
Robert Randolph rm 802C PK1	305-8442
Records and Property Management Branch Florence Stanmore rm 802C PK1	305-8410
Space and Telecommunications Branch William Morris rm 802C PK1	305-8331
Office Services Division Constant G. Fearing rm 803 PK1	305-8183
Travel Arrangements rm 803 PK1	305-8183
Support Services Branch Robert Fenwick FERN	308-7000
Transportation Unit John Holmes FERN	557-1531
File Information Unit William Satterwhite rm 1DO1 CP3	308-2733
Official Search Unit Emily Hambrick FERN	557-9690

Office of Public Services

Director Patrick Rowe rm 503 PK1	305-8543
Secretary Teresa Knight rm 503 PK1	305-8543
Deputy Director Mary E. Turowski rm 503 PK1	305-8543
Public Search Services Division Cheryl Davis (Acting) rm 1A01 CP3	308-0595
Patent Search Branch Robin Roark rm 1A01 CP3	308-0595
Secretary Vickie Woods rm 1A01 CP3	308-0595
Trademark Assignment Search Branch Doris Kahn (Acting) rm 2C08 CP2	308-0940
Supv. Conveyance Exam. Shirley Royall rm 5C22 CP2	308-1200
Supv. TM Search Asst. Fontella Gray rm 2C08 CP2	308-2794
Program Control Division Sharon Carver Lobby CP1	603-0696
Secretary Doris Rodgers Lobby CP1	603-0696
Public Service Center Sharon Furbush rm 208A PK1	305-8155
Secretary Trina Parks rm 208A PK1	305-8155
PTO Help Line	557-4357
Assignment/Certification Services Division Cathy Kern rm 5C11 CP2	308-3008
Secretary Sissy Bassford rm 800 PK1	308-3008

Assignment Branch rm 6C30 CP2	308-0017
Secretary Connie Green rm 7D19 CP2	308-0017
Examination Section Audrey Britt rm 6C14 CP2	308-2600
Digest & Recording Section Diane Russele rm 5C16 CP2	308-0914
Supv. Legal Tech. Valita Barbour rm 5C14 CP2	308-0924
Supv. Legal Tech. Francis Morris rm 5C14 CP2	308-0924
Supv. Legal Tech. Pat Ragland rm 5C14 CP2	308-1941
Quality Control Section Joyce Johnson (Acting) rm 6C30 CP2	308-0049
Certification Branch Lannie C. Anderson rm 800 PK1	557-1552
Secretary Joycelyn Gaskins rm 800 PK1	557-1552
Input Records and Control Section Mary Gartrell rm 800 PK1	557-1587
Microfiche and Printing Section Helen Phillips rm 800 PK1	557-1603
Certification Section Linda Smith rm 800 PK1	557-1564
Office of National and International Application Review	
Director Anne Kelly rm 7D25 CP2	308-0910
Secretary Donna Bonne rm 7D25 CP2	308-0910
Clerk Typist David G. Lancaster rm 7D25 CP2	308-0910
Application Processing Division Willie Bowman, Jr. rm 7D19 CP2	308-0921
Secretary Bernadette Peterson rm 7D19 CP2	308-0921
Application Clerk Mary Frances Diggs rm 7D19 CP2	308-0921
Classification and Routing Branch Norma White rm 7C19 CP2	308-1214
Special Processing/Administrative Branch Mose Montgomery rm 7C10 CP2	308-1202
Supervisor Jeanette Gatling rm 7C10 CP2	308-0377
Application Processing Division Team I Everette Oliver rm 7C10 CP2	308-1901
Application Processing Division Team II Delora Dillard rm 7C10 CP2	308-1157
Application Processing Division Team III Macia Fletcher rm 7C10 CP2	308-0919
File maintanence branch	
Jeanette Gatling rm 7C10 CP2	308-0377
Night shift Gordon Toney rm 7D25 CP2	308-0910
Micrographics Division Ronald Adams rm 7D25 CP2	308-0910
Application Filming Branch Calvin Pullen 6C22 CP2	308-0849
Quality Control Branch Mary Smith 1627D HCHB	377-5501
Patent Filming Branch Thomas Hawkins 1627D HCHB	377-4968
PCT International Division	
Administrator Vince Turner rm 1248 CP6	603-0465
Secretary Tonya Pritchett rm 1248 CP6	603-0465
Program Coordinator Mary Reed rm 1248 CP6	603-0465
Receiving Office Branch Terry Johnson rm 1245 CP6	603-0465
DO/EO/IPEA Branch (Vacant) rm 1245 CP6	603-0465
Legal Review Branch Richard Lazarus rm Lobby CP6	603-0465

Office of Publications

Director Richard A. Bawcombe rm 513 PK1	305-8594
Secretary Karna Cooper rm 513 PK1	557-0698
Deputy Director Michael Stellabotte rm 513 PK1	305-8237
Publishing Division Sylvia F. Martin rm 512 PK1	557-3283
Deputy Manager Martina Thompson rm 512 PK1	557-6388
Allowed Files and Assembly Branch rm 512 PK1	557-6395
Production Control Branch Willard D. Ireland rm 512	557-6412
Editorial Branch James Alexander rm 512 PK1	557-6393
Data Base Query Section Kenneth Folks rm 512 PK1	557-6392
Patent Copy Inspection Section Patricia Smalls rm 512 PK1	557-6393
Drafting Branch Martin Baum rm 510 PK1	557-6404
Statistical Analysis Division rm 513 PK1	557-1963
Data Base Inspection Branch Melvinia Gary rm 513 PK1	557-6414
Certificates of Corrections Branch Mary H. Allen rm 809 PK1	557-0709
Technical Development Division Edwin P. Hall rm 513 PK1	557-1992
Patent Maintenance Division C.H. Griffen rm 811 PK1	557-6945

Office of Civil Rights

Director R. Jacqueline Dees rm 600A PK1	305-8292
Supervisory Secretary Tracy Welch rm 600A PK1	305-8292
Supervisory Equal Opportunity Specialist (Vacant) rm 600A PK1	205-8292
Equal Opportunity Specialists:	
Kay Adams rm 600A PK1	305-8292
Kathy Broadbent rm 600A PK1	305-8292
Jessica Hamilton rm 600A PK1	305-8292

Anita O'Neal rm 600A PK1	305-8292
Paula Reid rm 600A PK1	305-8292

Office of Management and Organization

Director Sara E. Bjorge rm 505 PK1	305-8325
Secretary (Vacant) rm 505 PK1	305-8325
Project Managers:	
Alvin Dorsey rm 505 PK1	305-8325
Greg P. Mullen rm 505 PK1	305-8325
Management Analysts:	
Susan Hadley rm 505 PK1	305-8325
Joseph Jones rm 505 PK1	305-8325

Office of Procurement

Director Stanley H. Livingstone rm 806 PK1	305-8014
Secretary Cristina M. Moran rm 806 PK1	305-8014
Contract Division Page A. Etzel rm 806 PK1	305-8014
Small Purchases Division Muriel Brown rm 806 PK1	305-8014

Office of Patent Depository Library Programs

Director Carole A. Shores rm 306 CM2	577-9686
Secretary Susan Fink Jenkins rm 306 CM2	577-9686
Technical Information Specialists:	
James A. Arshem rm 306 CM2	557-9686
Martha L. Crockett rm 306 CM2	557-9686
Amanda Putnam rm 306 CM2	557-9686
Fellowship Librarians:	
Neil Massong rm 306 CM2	557-9686
Chris Marhenke rm 306 CM2	557-9686

OFFICE OF THE ASSISTANT COMMISSIONER FOR FINANCE AND PLANNING

Assistant Commissioner Bradford R. Huther rm 904 PK2	305-9200
Secretary Vickie T. Bryant rm 904 PK2	305-9200

Office of Budget

Director James R. Lynch rm 805 PK1	305-8175
Deputy Director Miguel B. Perez rm 805 PK1	305-8175
Program and Budget Division Nancy J. Wine rm 805 PK1	305-8175
Execution and Control Division Miguel B. Perez (Acting) rm 805 PK1	305-8175

Office of Finance

Director W. B. Erwin rm 802A PK1	305-8051
Secretary Virginia R. Clark rm 802A PK1	305-8051
Financial Accounting Division Douglas M. Grady rm 802B PK1	305-8051
General Accounting Branch Charles J. Yaple rm 802 PK1	557-3423
Receipts Division John L. Oliff rm 1B01 CP2	308-4024
Financial Systems Division Robert M. Kopson rm 802A PK1	557-3051

Office of Long-Range Planning and Evaluation

Director Frances Michalkewicz rm 507 PK1	305-8510
Secretary Dawana Penn rm 507 PK1	305-8510
Project XL Administrator Ruth Nyblod rm 507 PK1	557-2345

Office of Personnel

Personnel Officer Carolyn P. Acree rm 700 PK1	305-8062
Assistant to the Personnel Officer Larry Tabachnick rm 700 PK1	305-8062
Secretary Mildred Newman rm 700 PK1	305-8062
Classification and Employment Division Cynthia Nelson rm 700 PK1	305-8231
Employee and Labor Relations Division Richard Haisch rm 601 PK1	305-8121
Workforce Effectiveness Division Robert Ramig rm 700 PK1	305-8431
Personnel Payroll Processing Branch Steve Burke rm 700 PK1	557-1208
Office of Labor Law Counsel Jimmy Lawence rm 601 PK1	557-9684

OFFICE OF THE ASSISTANT COMMISSIONER FOR INFORMATION SYSTEMS

Assistant Commissioner for Information Systems Thomas P. Giammo rm 916 PK2	305-9400
Secretary Helen White rm 916 PK2	305-9400
Deputy Assistant Commissioner for Information Systems Boyd Alexander rm 916 PK2	305-9400
Secretary Michele Helms rm 916 PK2	305-9400
Program Management Support Services Director L Liddle rm 1002 PK2	557-6000
Secretary Judy Barbour rm 1002 PK2	557-6000
Contracting Staff Director James Murphy rm 509 CP1	603-0711
Secretary rm 509 CP1	603-0711

Office of Documentation Information

Director Jane S. Myers rm 304 CM2	557-5652
Secretary Carolyn Johnson rm 304 CM2	557-5652
D-ROM Projet Manager Jim Peterson rm 304 CM2	557-6115
Technology Assessment and Forecast (Vacant) rm 304 CM2	557-5652
Information Services Lynn Smith rm 308 CM2	557-5655

Directorate for Automated Patent Systems

Director Boyd Alexander rm 916 PK2	557-9093
Secretary Michelle Helms rm 916 PK2	557-9093
Deputy Director Donald LeCrone rm 1000 PK2	557-6156

APS Program Management Office

Director Gerald Findley rm 1000 PK2	557-6156
Secretary Audrey Jackson rm 1000 PK2	557-6156

Office of Electronic Data Conversion and Distribution

Director David Grooms rm 1100B PK2	557-6154
Secretary Beverly Allen rm 1100B PK2	557-6154

Directorate for Automated Trademark Systems

Director Raymond Rahn (Acting) rm 1001 PK2	557-3544
Secretary April Irondi rm 1001 PK2	557-3544

Office of Administrative Systems and Microcomputer Applications

Director Doug Hines rm 1001 PK2	557-6330
Secretary Sharrie Ruppel rm 1001 PK2	557-6330

Directorate for Systems Engineering and Evaluation

Director Jim Oberthaler rm 1004 PK2	557-7862
Secretary Ruth Harrison rm 1004 PK2	557-7862

Office of Systems Engineering and Data Communications

Director Stephen Jacobsen rm 1004 PK2	557-7862
Program Assistant Eliza Davis rm 1004 PK2	557-7862

Office of Technical Review and Evaluation

Director Larry Despain rm 1004 PK2	557-7862
Secretary Linda Bilbo rm 1004 PK2	557-7862

Office of System Test and Evaluation

Director Stephenie Rolle rm 1050 PK2	557-4114

Directorate for Central Computer Operations

Director Bob Mason rm 1100A PK2	557-3646
Secretary Virginia Richardson rm 1100A PK2	557-3646
Deputy Director William Maykrantz rm 1100A PK2	557-3646

Office of User Support

Director Mike Harrigan rm 1100A PK2	557-3646
Deputy Director Laura Ghaffari rm 1100A PK2	557-3646

Office of Operating Systems Support

Director A.K. Borough rm 1100A PK2	557-3646

Office of Computer Operations

Director John Fancovic 11th Floor Computer Room PK2	557-1225

Appendix C

Copyright Office Telephone Directory

If you would like to write to a copyright official, refer to the listings on the following pages and send your correspondence to: Copyright Office, Library of Congress, Washington, DC 20559.

Note: The area code for all the phone numbers that follow is (202).

General Information	707-5000
Office of the Register of Copyrights and Assistant Librarian of Congress for Copyright Services	
Register of Copyrights Ralph Owens LM	707-8350
Executive Assistant to the Register Sherry A. Baden LM 403	707-8350
Staff Assistant Henrietta Terry LM 403	707-8350
Associate Register of Copyrights for Management Michael R. Pew LM 403	707-8370
Secretary Mamie Muse LM 403	707-8370
Associate Register for Legal Affairs (General Counsel) Dorothy M. Schrader LM 403	707-8380
Secretary Mary E. Grant LM 403	707-8370
Assistant General Counsel LM 403	707-8380
Assistant Register Anthony P. Harrison LM 403	707-8350
Policy Planning Adviser William Patry LM 403	707-8350
Policy Planning Adviser Eric Schwartz LM 403	707-8350
Policy Planning Adviser Marybeth Peters	707-8350
Policy Planning Adviser Louis I. Flacks	707-8350
Senior Administrative Officer Eric Reid LM 403	707-8370
Senior Administrative Officer Donette S. Vandell LM 403	707-8370
Receiving and Processing Division	
Chief Orlando L. Campos LM 435	707-7700
Secretary Velma Wigglesworth LM 435	707-7700
Assistant Chief Richard Neldon LM 435	707-7700
Head Fiscal Control Section Joseph C. Bogert LM 437	707-6883
Head Mail and Correspondence Control Section LM 424	707-8002
Head Materials Control Section Victor A. Holmes LM 432	707-8239
Cataloging Division	
Chief LM 513	707-8040
Secretary LM 513	707-8040
Assistant Chief Raoul leMat LM 513	707-8040
Head Visual Arts Section Edward A. Kapusciarz LM 504	707-8070
Head Audio Visual Section Virginia L. Kass LM 502	707-8060
Head Literary Section Jacquelyn R. Watts LM 517	707-8080
Head Performing Arts Section Joanne Roussis LM 516	707-8090

959

Head Serials Section Rodney Hall LM 524 707-8100
Head Technical Support Gloria Y. Ayers LM 513 707-8040
Examining Division
Chief Harriet L. Oler LM 445 707-8200
    Secretary Carol Lowe Pickett LM 445 707-8200
Assistant Chief Jodi Rush LM 445 707-8200
Head Visual Arts Section Frank J. Vitalos LM 447 707-8202
Head Literary Section I Nancy H. Lawrence LM 441 707-8250
Head Literary Section II Phillip L. Gill LM 441 707-8250
Head Performing Arts Section Julia Huff LM 449 707-6040
Head Renewals Section Bernard F. Dietz LM 444 707-8180
Information and Reference Division
Chief Joan A. Doherty LM 453 707-6800
    Secretary Ann Tayman LM 453 707-6800
Assistant Chief James P. Cole LM 453 707-6800
Head Certifications and Documents Section Maxine Marshall LM 402 707-6787
Head Information Section Victor Marton LM 401 707-8700
Head Copyright Publications Section Joseph G. Ross Jr. LM 455 707-6804
Head Records Management Section Stephen G. Soderberg LM 438A 707-6891
Head Reference and Bibliography Section James C. Roberts LM 450 707-6850
Licensing Division
Chief Walter D. Sampson LM 454 707-8130
    Secretary Michele Murphy LM 454 707-8130
Assistant Chief John E. Martin Jr. LM 454 707-8130
Head Licensing Information Section Dorothy P. Harden LM 454 707-8150
Head Examining Section Willie A. Adams LM 454 707-8160
Head Fiscal Section Vincent M. Murzinski LM 454 707-8150

Appendix D

Patent and Trademark Office Fee Schedule

The U.S. Patent and Trademark Office has amended its rules of practice in patent and trademark cases, Parts 1 and 2 of Title 37, Code of Federal Regulations to adjust patent and trademark fee amounts.

Any fee payment due and paid on or after December 16, 1991, must be paid in revised amount. The date of mailing indicated on a proper Certificate of Mailing under either 37 CRF 1.8 or 37 CFR 1.10 will be considered to be the date of receipt and payment in the Office.

The fees which are subject to reduction for small entities who have established status (37 CFR 1.27) are shown in a separate column.

For additional information, please contact the PTO Public Service Center at (703) 305-HELP.

Patent Fees

Fee Code 37 CFR	Description	Fee	Small Entity Fee
Filing Fees			
101/201 1.16(a)	Basic filing fee, utility	690.00	345.00
102/202 1.16(b)	Independent claims in excess of three	72.00	36.00
103/203 1.16(c)	Claims in excess of twenty	20.00	10.00
104/204 1.16(d)	Multiple dependent claim	220.00	110.00
105/205 1.16(e)	Surcharge—Late filing fee or oath or declaration	130.00	65.00
106/206 1.16(f)	Design filing fee	280.00	140.00
107/207 1.16(g)	Plant filing fee	460.00	230.00
108/208 1.16(h)	Reissue filing fee	690.00	345.00
109/209 1.16(i)	Reissue independent claims over original patent	72.00	36.00
110/210 1.16(j)	Reissue claims in excess of 20 and over original patent	20.00	10.00
139 1.17(k)	Non-English specification	130.00	
Extension Fees			
115/215 1.17(a)	Extension for response within first month	110.00	55.00
116/216 1.17(b)	Extension for response within second month	350.00	175.00
117/217 1.17(c)	Extension for response within third month	810.00	405.00
118/218 1.17(d)	Extension for response within fourth month	1,280.00	640.00
Appeals/Interference Fees			
119/219 1.17(e)	Notice of appeal	260.00	130.00
120/220 1.17(f)	Filing a brief in support of an appeal	260.00	130.00
121/221 1.17(g)	Request for an oral hearing	260.00	110.00
Issue Fees			
142/242 1.18(a)	Utility issue fee	1,130.00	565.00
143/243 1.18(b)	Design issue fee	400.00	200.00
144/244 1.18(c)	Plant issue fee	570.00	285.00
Miscellaneous Fees			
111 1.20(j)	Extension of term of patent	1,000.00	
112 1.17(n)	Requesting publication of SIR—Prior to examiner's action	790.00	
113 1.17(o)	Requesting publication of SIR—After examiner's action	1,580.00	
145 1.20(a)	Certificate of correction	70.00	
147 1.20(c)	For filing a request for reexamination	2,180.00	
148/248 1.20(d)	Statutory Disclaimer	110.00	55.00
Patent Petition Fees			
122	Petitions to the Commissioner, unless otherwise specified	130.00	
138 1.17(j)	Petition to institute a public use proceeding	1,310.00	
140/240 1.17(l)	Petition to revive unavoidably abandoned application	110.00	55.00
141/241 1.17(m)	Petition to revive unintentionally abandoned application	1,130.00	565.00
Maintenance Fees	**Applications Filed on or after December 12, 1980**		
183/283 1.20(e)	Due at 3.5 years	900.00	450.00
184/284 1.20(f)	Due at 7.5 years	1,810.00	905.00
185/285 1.20(g)	Due at 11.5 years	2,730.00	1,365.00
186/286 1.20(h)	Surcharge—Late payment within 6 months	130.00	65.00
187 1.20(i)	Surcharge after expiration	600.00	
PCT Fees—National Stage			
154/254 1.492(e)	Surcharge—Late filing fee or oath or declaration	130.00	65.00
156 1.492(f)	English translation—after 20 months	130.00	
956/957 1.492(a)(1)	IPEA—US	620.00	310.00
958/959 1.492(a)(2)	ISA—US	690.00	345.00
960/961 1.492(a)(3)	PTO not ISA or IPEA	920.00	460.00
962/963 1.492(a)(4)	Claims meet PCT Article 33(1)-(4)-IPEA-US	90.00	45.00
964/965 1.492(b)	Claims—extra independent (over 3)	72.00	36.00

Fee Code	37 CFR	Description	Fee	Small Entity Fee
PCT Fees—National Stage (cont.)				
966/967	1.492(c)	Claims—extra independent (over 20)	20.00	10.00
968/969	1.492(d)	Claims—multiple dependent	220.00	110.00
970/971	1.492(a)(5)	For filing with EPO or JPO search report	800.00	400.00
PCT Fees—International Stage				
150	1.445(a)(1)	Transmittal letter	190.00	
151	1.445(a)(2)	PCT search fee—no U.S. application	600.00	
152	1.445(a)(3)	Supplemental search per additional invention	160.00	
153	1.445(a)(2)	PCT search fee—prior U.S. application	400.00	
190	1.482(a)(1)	Preliminary examination fee—ISA was the U.S.	440.00	
191	1.482(a)(1)	Preliminary examination fee—ISA was not the U.S.	650.00	
192	1.482(a)(2)	Additional invention—ISA was the U.S.	140.00	
193	1.482(a)(2)	Additional invention—ISA was not the U.S.	220.00	
PCT Fees to WIPO				
800		Basic fee (first 30 pages)	490.00	
801		Basic supplemental fee (for each page over 30)	10.00	
803		Handling fee	150.00	
803-896		Designation fee per country	119.00	
PCT Fee to EPO				
802		International search	1,320.00	
Patent Service Fees				
561	1.19(a)(1)(i)	Printed copy of patent w/o color, regular service	3.00	
562	1.19(a)(1)(ii)	Printed copy of patent w/o color, expedited local service	6.00	
563	1.19(a)(1)(iii)	Printed copy of patent w/o color, ordered via EOS, expedited service	25.00	
564	1.19(a)(2)	Printed copy of plant patent, in color	12.00	
565	1.19(a)(3)	Copy of utility patent or SIR, with color drawings	24.00	
566	1.19(b)(1)(i)	Certified or uncertified copy of patent application as filed, regular service	12.00	
567	1.19(b)(1)(ii)	Certified or uncertified copy of patent application, expedited local service	24.00	
568	1.19(b)(2)	Certified or uncertified copy of patent-related file wrapper and contents	150.00	
569	1.19(b)(3)	Certified or uncertified copy of document, unless otherwise provided	25.00	
570	1.19(b)(4)	For assignment records, abstract of title and certification, per patent	20.00	
571	1.19(c)	Library Service	50.00	
572	1.19(d)	List of U.S. patents and SIRs in subclass	3.00	
573	1.19(e)	Uncertified statement re status of maintenance fee payments	10.00	
574	1.19(f)	Copy of non-U.S. document	12.00	
575	1.19(g)	Comparing and Certifying Copies, Per Document, Per Copy	25.00	
576	1.19(h)	Additional filing receipt, duplication or corrected due to applicant error	20.00	
577	1.21(c)	Filing a Disclosure Document	10.00	
578	1.21(d)	Local delivery box rental, per annum	50.00	
579	1.21(e)	International type search report	35.00	
580	1.21(g)	Self-Service copy charge, per page	0.25	
581	1.21(h)	Recording each patent assignment, agreement or other paper, per property	40.00	
583	1.21(i)	Publication in Official Gazette	20.00	
584	1.21(j)	Labor charges for services, per hour or fraction thereof	30.00	
585	1.21(k)	Unspecified other services	At cost	
586	1.21(l)	Retaining abandoned application	130.00	
587	1.21(n)	Handling fee for incomplete or improper application	130.00	
588	1.21(o)	Automated Patent System (APS—text) terminal session time, per hour	40.00	
589	1.296	Handling fee for withdrawal of SIR	130.00	
590	1.24	Patent coupons	3.00	

Fee Code	37 CFR	Description	Fee

Patent Enrollment Fees

609	1.21(a)(1)	Admission to examination	290.00
610	1.21(a)(2)	Registration to practice	100.00
611	1.21(a)(3)	Reinstatement to practice	15.00
612	1.21(a)(4)	Copy of certificate of good standing	10.00
613	1.21(a)(4)	Certificate of good standing (suitable for framing)	20.00
615	1.21(a)(5)	Review of decision of Director, Office of Enrollment and Discipline	120.00
616	1.21(a)(6)	Regrading of Examination	120.00

Trademark Fees

Trademark Processing Fees

361	2.6(a)(1)	Application for registration, per class	200.00
362	2.6(a)(2)	Filing an Amendment to Allege Use under § 1(c) per class	100.00
363	2.6(a)(3)	Filing a Statement of Use under § 1(d)(1), per class	100.00
364	2.6(a)(4)	Filing a Request for 6-month Extension of Time for Filing a Statement of Use under § 1(d)(1), per class	100.00
365	2.6(a)(5)	Application for renewal, per class	300.00
366	2.6(a)(6)	Additional fee for late renewal, per class	100.00
367	2.6(a)(7)	Publication of mark under § 12(c), per class	100.00
368	2.6(a)(8)	Issuing new certificate of registration	100.00
369	2.6(a)(9)	Certificate of Correction, registrant's error	100.00
370	2.6(a)(10)	Filing disclaimer to registration	100.00
371	2.6(a)(11)	Filing amendment to registration	100.00
372	2.6(a)(12)	Filing § 8 affidavit, per class	100.00
373	2.6(a)(13)	Filing § 15 affidavit, per class	100.00
374	2.6(a)(14)	Filing combined § § 8 & 15 affidavit, per class	200.00
375	2.6(a)(15)	Petition to the Commissioner	100.00
376	2.6(a)(16)	Petition for cancellation, per class	200.00
377	2.6(a)(17)	Notice of opposition, per class	200.00
378	2.6(a)(18)	Ex parte appeal, per class	100.00

Trademark Service Fees

461	2.6(b)(1)(i)	Printed copy of each registered mark, regular service	3.00
462	2.6(b)(1)(ii)	Printed copy of each registered mark, expedited local service	6.00
463	2.6(b)(1)(iii)	Printed copy of each registered mark ordered via EOS, expedited service	25.00
464	2.6(b)(4)(i)	Certified copy of registered mark, with title and/or status, regular service	10.00
465	2.6(b)(4)(ii)	Certified copy of registered mark, with title and/or status, expedited local service	20.00
466	2.6(b)(2)(i)	Certified or uncertified copy of trademark application as filed, regular service	12.00
467	2.6(b)(2)(ii)	Certified or uncertified copy of trademark application as filed, expedited local service	24.00
468	2.6(b)(3)	Certified or uncertified copy of trademark-related file wrapper and contents	50.00
469	2.6(b)(5)	Certified or uncertified copy of trademark document, unless otherwise provided	25.00
470	2.6(b)(7)	For assignment records, abstracts of title and certification per registration	20.00
475	1.19(g)	Comparing and certifying copies, per document, per copy	25.00
480	2.6(b)(9)	Self-service copy charge, per page	0.25
481	2.6(b)(6)	Recording trademark assignment, agreement or other paper, first mark per document	40.00
482		For second and subsequent marks in the same document	25.00
484	2.6(b)(10)	Labor charges for services, per hour or fraction thereof	30.00
485	2.6(b)(11)	Unspecified other services	At cost
488	2.6(b)(8)	Each hour of T-SEARCH terminal session time	40.00
490	1.24	Trademark coupons	3.00

General Fees

Fee Code	37 CFR	Description	Fee
Finance Service Fees			
607	1.21(b-1)	Establish deposit account	10.00
608	1.21(b)(2)	Service charge for below minimum balance	20.00
608	1.21(b)(3)	Service charge for below minimum balance restricted subscription deposit account	20.00
617	1.21(m)	Processing returned checks	50.00
Computer Service Fees			
618		Computer records	At cost

Appendix E

Significant Inventors and Inventions in U.S. History

The U.S. patent system has played an important role in the development of our nation. As President and patentee Abraham Lincoln said, it gives "the fuel of interest to the fire of genius." President McKinley told Americans that the country must turn to invention "as one of the most powerful aids...in the enlargement and advance of science, industry, and commerce."

"The patent system is a cornerstone of America's free enterprise economy," wrote C. Marshall Dann, Commissioner of Patents and Trademarks, at the time of our Bicentennial celebration. "The individuals who have been named as inventors in U.S. patents have made great contributions to the nation's technological strength."

The Patent and Trademark Office compiled the following lists of patents to draw attention to at least some of the more significant patents during our first 200 years. These lists are by no means comprehensive, but interesting and inspiring reading nevertheless.

Inventors. Many of the inventors listed in this section are better known for their social, political, or other endeavors, but have also been granted U.S. patents. Others on the list, whose names are not commonly known, merit recognition as the first members of a minority and other groups to obtain patents. Still others have been selected by the National Council of Intellectual Property Law Associations as especially significant and have been inducted into the National Inventors Hall of Fame.

Utility Patents. This section cites a sampling of important patents that have had an impact on our nation. Some of the patentees are well-known; others have received little publicity, but their contributions have been substantial.

Design Patents. This section features a few design patents whose subjects you will instantly recognize.

Plant Patents. This section highlights a few plant patents for flora you are sure to know.

Inventors

There are 9,957 patents that were issued prior to July 13, 1836, at which time patent numbering began. Inventors are listed in alphabetic order by last name.

Bardeen, John Walter H. Brattain. Transistor; "Three-Electrode Circuit Element Utilizing Semiconductive Materials" (1950). Patent Number: 2,524,035. Members of National Inventors Hall of Fame.

Bell, Alexander Graham. Telephone; "Improvement in Telegraphy" (1876). Patent Number: 174,465. Member of National Inventors Hall of Fame.

Blair, Henry. "Corn Planter" (October 14, 1834). First black patentee.

Boone, Sarah. "Ironing-Board" (1892). Patent Number: 473,653. One of the first black female inventors.

Brattain, Walter H.; John Bardeen. Transistor; "Three-Electrode Circuit Element Utilizing Semiconductive Materials" (1950). Patent Number: 2,524,035. Members of National Inventors Hall of Fame.

Carver, George Washington. Cosmetics; "Cosmetic and Process of Producing the Same" (1925). Patent Number: 1,522,176. Black educator.

Clemens, Samuel L. (Mark Twain). Suspenders; "Improvement in Adjustable and Detachable Straps for Garments" (1871). Patent Number: 121,992. Author, humorist, and steamboat pilot.

Coolidge, William D. Tungsten lamp filament; "Tungsten and Method of Making Same for Use as Filaments of Incandescent Electric Lamps and for Other Purposes" (1913). Patent Number: 1,082,933. "X-Ray Tube" (1933). Patent Number: 1,917,099. "X-Ray Tube" (1934). Patent Number: 1,946,312. "Cathode Ray Tube" (1935). Patent Number: 2,010,712. Member of National Inventors Hall of Fame.

Diesel, Rudolf. Diesel engine; "Internal-Combustion Engine" (1898). Patent Number: 608,845. Member of National Inventors Hall of Fame.

Edison, Thomas A. Quadruplex telegraph; "Improvement in Printing-Telegraphs" (1873). Patent Number: 140,488. Phonograph; "Improvement in Phonograph or Speaking Machines" (1878). Patent Number: 200,521. Incandescent lamp; "Electric lamp" (1880). Patent Number: 223,898. "Speaking Telegraph" (1892). Patent Number: 474,230. Motion picture projector; "Apparatus for Exhibiting Photographs of Moving Objects" (1893). Patent Number: 493,426. "Kinetographic Camera" (1897). Patent Number: 589,168. Member of National Inventors Hall of Fame; 1,093 patents were granted to him.

Fermi, Enrico; Leo Szilard. "Neutronic Reactor" (1955). Patent Number: 2,708,656. Member of National Inventors Hall of Fame.

Goodyear, Charles. Vulcanized rubber; "Improvement in India-Rubber Fabrics" (1844). Patent Number: 3,633. Member of National Inventors Hall of Fame.

Goodwin, Hannibal. Celluloid photographic film; Photographic Pellicle and Process of Producing Same" (1898). Patent Number: 610,861. Clergyman patentee.

Hall, Charles M. Process for the "Manufacture of Aluminum" (1889). Patent Number: 400,665. Member of National Inventors Hall of Fame.

Hollerith, Herman. Punch card accounting; "Art of Compiling Statistics" (1889). Patent Number: 395,782. Former Patent Office employee.

Hopkins, Samuel. Process of making potash and pearl ash; "Improvement, not known before such Discovery, in the making of Pot ash and Pearl ash by a new Apparatus and Process" (July 31, 1780). First U.S. patent.

Kies, Mary. "Weaving Straw With Silk or Thread" (May 5, 1809). First woman patentee.

Lincoln, Abraham. A device for "Bouying Vessels Over Shoals" (1849). Patent Number: 6,469. U.S. President patentee.

Marconi, Guglielmo. Wireless telegraphy; "Transmitting Electrical Signals" (1897). Patent Number: 586,193. Member of National Inventors Hall of Fame.

McCormick, Cyrus H. Reaper; "Improvement in Machines for Reaping Small Grain" (June 21, 1834). Member of National Inventors Hall of Fame.

Morse, Samuel F.B. Telegraph signs; "Improving in the Mode of Communicating Information by Signals by the Application of Electro-Magnetism" (1840). Patent Number: 1,647. Member of National Inventors Hall of Fame.

Mullikin, Samuel. Four machine patents; "Machine for Threshing Grain and Corn"; Machine for Breaking and Swinging Hemp"; "Machine for Cutting and Polishing Marble"; and "Machine for Raising a Nap on Cloths" (March 11, (1791). First patentee to obtain more than one patent.

Ruggles, John. Locomotive steam engine; "Locomotive Steam-Engine for Rail and Other Roads" (1836). Patent Number: 1. First chairman of the Senate Committee on Patents, whose report culminated in the 1836 law which is the "basis of the patent system as it functions today."

Russell, Lillian. "Dresser-Trunk" (1912). Patent Number: 1,014,853. Movie star patentee.

Schawlow, Arthur L; Charles H. Townes. Laser (optical maser); "Masers and Maser Communications System" (1960). Patent Number: 2,929,922.

Shockley, William. Transistor; "Circuit Element Utilizing Semiconductive Material" (1951). Patent Number: 2,569,347. Member of National Inventors Hall of Fame.

Stuart, James E.B. "Method of Attaching Sabers to Belts" (1859). Patent Number: 25,684. Well-known Confederate General.

Szilard, Leo; Enrico Fermi. "Neutronic Reactor" (1955). Patent Number: 2,708,656.

Tesla, Nikola. Induction type of electric motor; "Electrical Transmission of Power" (1888). Patent Number: 382,280. "Pyromagneto-Electric Generator" (1890). Patent Number: 428,057. Member of National Inventors Hall of Fame.

Townes, Charles H. Maser; "Production of Electromagnetic Energy" (1959). Patent Number: 2,879,439. Laser (optical maser) with Arthur L. Schawlow• "Masers and Maser Communications System"' (1960). Patent Number: 2,929,922. Member of National Inventors Hall of Fame.

Whitney, Eli. Cotton gin; "Description of New Invented Cotton Gin, or Machine for Separating Cotton From Its Seed" (March 14,1794). Member of National Inventors Hall of Fame.

Winslow, Samuel. Salt extracting method (1641). First patent granted in North America, by Massachusetts Bay Colony.

Wright, Orville and Wilbur. Airplane with motor; "Flying-Machine" (1906). Patent Number: 821,393. Members of National Inventors Hall of Fame.

Utility Patents

There are 9,957 patents that were issued prior to July 13, 1836, at which time patent numbering began. Inventions are listed in chronological order. The title of the invention is surrounded by quotation marks.

NAIL-MAKING MACHINE; "a machine for making nails." Patented by Samuel Briggs, Sr. and Samuel Briggs, Jr. on August 2,1791.

CARDING MACHINE; "an improvement in manufacturing cards." Patented by Amos Whittemore on June 5, 1797.

CAST IRON PLOW; "an improvement in the art of plow making." Patented by Charles Newbold on June 26,1797.

HIGH-PRESSURE STEAM ENGINE. Patented by Oliver Evans on February 14, 1804.

STEAMBOAT, EXPERIMENTAL. Patented by Robert Fulton on February 11, 1809.

STEAMBOAT, PRACTICAL. Patented by Robert Fulton on February 9, 1811.

WASHING MACHINE. Patented by Chester Stone on February 17, 1827.

FOUNTAIN PEN. Patented by D. Hyde on May 20, 1830.

MOWING MACHINE; "improvement in the machine for cutting grass and grain." Patented by W. Manning on May 3, 1831.

STEAM BOILER; "improvement in the construction of steam boilers or Cooper's Steam Boiler." Patented by Peter Cooper on October 13, 1831.

REVOLVER; "improvement in fire arms." Patented by Samuel Colt on February 25, 1836.

MATCH; "manufacture of friction-Matches." Patented by Alonzo D. Phillips in 1836. Patent Number: 68.

SOWING MACHINE; "machine for sowing plaster, ashes, seed, and other separable substances. Patented by Julius Hatch in 1838. Patent Number: Re. 1. First reissued patent.

WATER WHEEL; "spiral-bucket water-wheel." Patented by Lorenzo Dow Adkins in 1839. Patent Number: 1,154.

COLLAPSIBLE TUBE (as for toothpaste); "improvement in the construction of vessels or apparatus for preserving paint, etc." Patented by John Rand in 1841. Patent Number: 2,252.

CYLINDER LOCK; "door-Lock." Patented by Linus Yale in 1844. Patent Number: 3,630.

SEWING MACHINE; "improvement in sewing machines." Patented by Elias Howe, Jr. in 1846. Patent Number: 4,750.

RUBBER TIRE; "improvement in carriage wheels, etc." Patented by Robert W. Thomson in 1847. Patent Number: 5,104.

ROTARY PRINTING PRESS; "improvement in printing-presses." Patented by Richard M. Hoe in 1847. Patent Number: 5,188.

VALVE GEAR, STEAM ENGINE; "cut-off and working the valves of steam-engines." Patented by George H. Corliss in 1849. Patent Number: 6,162.

SAFETY PIN; "dress-pin." Patented by Walter Hunt in 1849. Patent Number: 6,281.

ELECTRO-MAGNETIC ENGINE; "improvement in electro-magnetic engines." Patented by Jacob Neff in 1851. Patent Number: 7,889.

ICE-MAKING MACHINE; "improved process for the artificial production of ice." Patented by John Gorrie in 1851. Patent Number: 8,080.

RAILROAD-CAR, TRUCK BRAKE; "railroad-car truck and brake." Patented by E.G. Otis in 1852. Patent Number: 8,973.

"IMPROVEMENT IN FIRE-ARMS." Patented by Horace Smith and Daniel B. Wesson in 1854. Patent Number: 10,535.

DECOMPOSING FAT; "improvement in process for purifying fatty bodies." Patented by Richard A. Tilghman in 1854. Patent Number: 11,766.

"IMPROVEMENT IN SEWING-MACHINES." Patented by Isaac M. Singer in 1855. Patent Number: 13,661.

STEEL PROCESS; "improvement in the manufacture of iron and steel." Patented by Henry Bessemer in 1856. Patent Number: 16,082.

MASON JAR; "improvement in screw-neck bottles." Patented by John L. Mason in 1858. Patent Number: 22,186.

RAILROAD AIR BRAKE; "improvement in railroad brakes." Patented by Nehemiah Hodge in 1860. Patent Number: 28,670.

BREECH LOADING GUN. Patented by Van Houten in 1861.

MACHINE GUN; "improvement in revolving battery guns." Patented by Richard J. Gatling in 1862. Patent Number: 36,836.

WEB PRINTING PRESS; "printing-machine." Patented by William Bullock in 1863. Patent Number: 38,200.

TYPEWRITER; "improvement in typewriting machine." Patented by C. Latham Sholes, Carlos Glidden, and Samuel W. Soule in 1868. Patent Number: 79,265.

TAPE MEASURE; "improvement in tape measures." Patented by Alvin J. Fellows in 1868. Patent Number: 79,965.

AIR BRAKE; "improvement in steam-power-brake devices." Patented by George Westinghouse, Jr. in 1869. Patent Number: 88,929.

VACUUM CLEANER; "improved sweeping-machine." Patented by Ives W. McGaffey in 1869. Patent Number: 91,145.

GUN CARRIAGE; "improvement in gun carriages." Patented by James B. Eads. Patent Number: 93,691.

"IMPROVEMENT IN POWER-LOOM FOR WEAVING INGRAIN CARPET." Patented by William and John W. Murkland in 1869. Patent Number: 97,106.

"IMPROVEMENT IN TREATING AND MOLDING PYROXYLINE" (resulted in the development of celluloid). Patented by John W. Hyatt, Jr. and Isaiah S. Hyatt in 1870. Patent Number: 105,338.

GOODYEAR WELT; "improvement in machines for sewing boots and shoes." Patented by Charles Goodyear, Jr. in 1871. Patent Number: 111,197.

VASELINE rgt; "improvement in products from petroleum." Patented by Robert A. Chesebrough in 1872. Patent Number: 127,568.

ELECTRIC TRAIN SIGNALING APPARATUS; "improvement for electric-signaling apparatus for railroads." Patented by William Robinson in 1872. Patent Number: 130,661.

PASTEURIZATION; "improvement in brewing beer and ale." Patented by Louis Pasteur in 1873. Patent Number: 135,245.

AUTOMATIC CAR COUPLINGS; "improvement in car-couplings." Patented by Eli H. Janney in 1873. Patent Number: 138,405.

BARBED WIRE; "improvement in wire fences." Patented by Joseph F. Glidden in 1874. Patent Number: 157,124.

"IMPROVEMENT IN CARPET-SWEEPERS." Patented by Melville R. Bissell in 1876. Patent Number: 182,346.

INTERNAL COMBUSTION ENGINE; "improvement in gas-motor engines." Patented by Nicolaus A. Otto in 1877. Patent Number: 194,047.

REINFORCED CONCRETE; "improvement in composition floors, roofs, pavements, etc." Patented by Thaddeus Hyatt in 1878. Patent Number: 206,112.

"IMPROVEMENT IN VEGETABLE-ASSORTERS." Patented by John H. Heinz in 1879. Patent Number: 212,000.

CARBON ARC LAMP; "electric lamp." Patented by Charles F. Brush in 1879. Patent Number: 219,208.

"MICROPHONE." Patented by Emile Berliner in 1880. Patent Number: 224,573.

"AIR-ENGINE." Patented by John Ericsson in 1880. Patent Number: 226,052.

CREAMER; "centrifugal creamer." Patented by Edwin J. Houston and Elihu Thomson in 1881. Patent Number: 239,659.

"BUTTON-HOLE SEWING-MACHINE." Patented by John Reece in 1881. Patent Number: 240,546.

"LIFE-RAFT." Patented by Frederick S. Allen in 1881. Patent Number: 240,634.

PLAYER PIANO; "mechanical musical instrument." Patented by John McTammany, Jr. in 1881. Patent Number: 242,786.

"SPEAKING-TELEPHONE." Patented by Francis Blake in 1881. Patent Number: 250,126.

"STETHOSCOPE." Patented by William F. Ford in 1882. Patent Number: 257,487.

"ELECTRIC FLAT-IRON." Patented by Henry W. Seeley in 1882. Patent Number: 259,054.

"CASH REGISTER AND INDICATOR." Patented by James Ritty and John Birch in 1883. Patent Number: 271,363.

TRANSPARENT PHOTOGRAPHIC PAPER STRIP FILM; "photographic film." Patented by George Eastman in 1884. Patent Number: 306,954.

ELECTRICAL WELDING; "apparatus for electric welding." Patented by Elihu Thomson in 1886. Patent Number: 347,140.

RECORD, DISC; "gramophone." Patented by Emile Berliner in 1887. Patent Number: 372,786.

ELECTRIC METER; "meter for alternating electric currents." Patented by Oliver B. Shallenberger in 1888. Patent Number: 388,003.

ROLL FILM CAMERA (KODAK™); "camera." Patented by George Eastman in 1888. Patent Number: 388,850.

BALLPOINT PEN; "pen." Patented by John J. Loud in 1888. Patent Number: 392,046.

LINOTYPE™; "machine for producing linotypes, type-matrices, etc." Patented by Ottmar Mergenthaler in 1890. Patent Number: 436,532.

"PHOTOGRAPHIC FILM." Patented by George Eastman in 1890. Patent Number: 441,831.

"AUTOMATIC TELEPHONE-EXCHANGE." Patented by Almon B. Strowgerin 1891. Patent Number: 447,918.

AUTOMATIC TELEPHONE; "automatic telephone or other electrical exchange." Patented by Almon B. Stowger in 1892. Patent Number: 486,909.

HALF-TONE PRINTING; "photogravure-printing plate." Patented by Frederick E. Ives in 1893. Patent Number: 495,341.

ELECTRIC TROLLEY CAR; "travelling contact for electric railways." Patented by Charles J. Van Depoele in 1893. Patent Number: 495,443.

ZIPPER™; "clasp locker or unlocker for shoes." Patented by Whitcomb L. Judson in 1893. Patent Number: 504,038.

GASOLINE AUTOMOBILE; "road-vehicle." Patented by Charles E. Duryea in 1895. 540,648.

ROAD CARRIAGE; "road-engine." Patented by George B. Selden in 1895. Patent Number: 549,160.

"ROTARY-DISC PLOW." Patented by Clement A. Hardy in 1896. Patent Number: 556,972.

MONOTYPE™; "machine for making justified lines of type. Patented by Tolbert Lanston in 1896. Patent Number: 557,994.

PRODUCTION OF CARBORUNDUM™; "electrical furnace." Patented by Edward Goodrich Acheson in 1896. Patent Number: 560,291.

EVEN KEEL SUBMARINE; "submarine vessel." Patented by Simon Lake in 1897. Patent Number: 581,213.

"CARBURETOR." Patented by Henry Ford in 1898. Patent Number: 610,040.

"NAVIGABLE BALLOON." Patented by Ferdinand Graf Zeppelin in 1899. Patent Number: 621,195.

MOTOR-DRIVEN VACUUM CLEANER; "pneumatic carpet-renovator." Patented by John S. Thurman in 1899. Patent Number: 634,042.

GAS TURBINE; "apparatus for generating mechanical power. Patented by Charles G. Curtis in 1899. Patent Number: 635,919.

ASPIRIN; "acetyl salicylic acid." Patented by Felix Hoffmann in 1900. Patent Number: 644,077.

STANLEY STEAMER; "motor-vehicle." Patented by Francis E. and Freelan O. Stanley in 1900. Patent Number: 657,711.

MAGNETIC TAPE RECORDER; "method of recording and reproducing sounds or signals." Patented by Valdemar Poulsen in 1900. Patent Number: 661,619.

HIGH SPEED STEEL TOOLS; "metal cutting tool and method of making same." Patented by Frederick W. Taylor and Maunsel White in 1901. Patent Number: 668,269.

"MOTOR CARRIAGE." Patented by Henry Ford in 1901. Patent Number: 686,046.

"SUBMARINE BOAT OR VESSEL." Patented by John P. Holland in 1902. Patent Number: 694,154.

AUTOMATIC STEREOTYPE-PRINTING; "automatic stereotype-printing-plate casting and finishing apparatus." Patented by Henry A. Wood in 1903. Patent Number: 721,117.

STARTING MOTOR FOR AUTO; "means for operating motor-vehicles." Patented by Clyde J. Cole-man in 1903. Patent Number: 745,157.

AUTOMOBILE TIRE MAKING MACHINE; "machine for making outer casings for double-tube tires." Patented by Frank A. Seiberling and William C. Stevens in 1904. Patent Number: 762,561.

GLASS SHAPING BOTTLE MACHINE; "glass shaping machine." Patented by Michael J. Owens in 1904. Patent Number: 766,768.

SAFETY RAZOR; "razor" (title for both patents). Patented by King C. Gillette in 1904. Patent Numbers: 775,134 and 775,135.

TWO-ELEMENT VACUUM TUBE; "instrument for converting alternating electric currents into continuous currents." Patented by John Ambrose Fleming in 1905. Patent Number: 803,684.

RADIO TUBE DETECTOR; "oscillation-responsive device." Patented by Lee De Forest in 1906. Patent Number: 836,070.

RADIO AMPLIFIER TUBE; "device for amplifying feeble electrical currents." Patented by Lee De Forest in 1907. Patent Number: 841,387.

THREE-ELEMENT VACUUM TUBE (AUDION); "space-telegraphy." Patented by Lee De Forest in 1908. Patent Number: 879,532.

BAKELITE™; "condensation product and method of making same." Patented by Leo H. Baekeland in 1909. Patent Number: 942,809.

ELECTRICAL INSULATORS; "insulating material." Patented by James C. Dow in 1910. Patent Number: 952,513.

SLEEVE VALVE ENGINE; "internal-combustion engine." Patented by Charles Y. Knight in 1910. Patent Number: 968,166.

SYNTHETIC AMMONIA; "production of ammonia." Patented by Fritz Habor and Robert Le Rossignol in 1910. Patent Number: 971,501.

MERCURY VAPOR LAMP; "apparatus for the electrical production of light." Patented by Peter Cooper Hewitt in 1912. Patent Number: 1,030,178.

CRACKING PROCESS; "manufacture of gasoline." Patented by William M. Burton in 1913. Patent Number: 1,049,667.

ROCKET ENGINE; "rocket apparatus." Patented by Robert H. Goddard in 1914. Patent Number: 1,103,503.

"WIRELESS RECEIVING SYSTEM." Patented by Edwin H. Armstrong in 1914. Patent Number: 1,113,149.

"BRASSIERE." Patented by Mary Phelps Jacob (Caresse Crosby) in 1914. Patent Number: 1,115,674.

HYDRO AIRPLANE; "heavier-than-air flying-machine." Patented by Glenn H. Curtiss in 1915. Patent Number: 1,156,215.

SELECTIVE RADIO TUNING SYSTEM; "selective tuning system." Patented by Ernst F.W. Alexanderson in 1916. Patent Number: 1,173,079.

INCANDESCENT GAS LAMP; "incandescent electric lamp." Patented by Irving Langmuir in 1916. Patent Number: 1,180,159.

"RADIOTELEPHONY." Patented by Carl R. Englund in 1917. Patent Number: 1,245,446.

GYROCOMPASS; "gyroscopic compass." Patented by Elmer A. Sperry in 1918. Patent Number: 1,279,471.

"DIVER'S SUIT" (permitting escape). Patented by Harry Houdini in 1921. Patent Number: 1,370,376.

HIGH VACUUM RADIO TUBE; "electrical discharge apparatus and process of preparing and using the same." Patented by Irving Langmuir in 1925. Patent Number: 1,558,436.

LEAD ETHYL GASOLINE; "method and means for using motor fuels." Patented by Thomas Midgley, Jr. in 1926. Patent Number: 1,573,846.

ELECTRIC RAZOR; "shaving implement." Patented by Jacob Schick in 1928. Patent Number: 1,721,530.

PACKAGED FROZEN FOODS; "method of preparing food products." Patented by Clarence Birdseye in 1930. Patent Number: 1,773,079.

"TELEVISION SYSTEM." Patented by Philo T. Farnsworth in 1930. Patent Number: 1,773,980.

"TELEVISION RECEIVING SYSTEM." Patented by Philo T. Farnsworth in 1930. Patent Number: 1,773,981.

REFRIGERATION APPARATUS; "refrigeration." Patented by Albert Einstein and Leo Szilard in 1930. Patent Number: 1,781,541.

MICROFILMING CAMERA; "photographing apparatus. Patented by George Lewis McCarthy in 1931. Patent Number: 1,806,763.

REFRIGERANTS, LOW-BOILING FLUORINE COMPOUND (FREON™); "heat transfer." Patented by Thomas Midgley, Jr., Albert R. Henne, and Robert R. McNary in 1931. Patent Number: 1,833,847.

"POLARIZING REFRACTING BODIES." Patented by Edwin H. Land and Joseph S. Friedman in 1933. Patent Number: 1,918,848.

TWO-PATH FM RADIO; "radio signalling system." Patented by Edwin H. Armstrong in 1933. Patent Number: 1,941,066.

CYCLOTRON; "method and apparatus for the acceleration of ions." Patented by Ernest O. Lawrence in 1934. Patent Number: 1,948,384.

MONOPOLY™; "board game apparatus." Patented by Charles B. Darrow in 1935. Patent Number: 2,026,082.

NYLON; "linear condensation polymers," and "Fiber and Method of Producing It." Patented by Wallace H. Carothers in 1937. Patent Numbers: 2,071,250 and 2,071,251.

CATHODE RAY TUBE, TELEVISION; "cathode ray tube." Patented by Vladimir K. Zworykin in 1938. Patent Number: 2,139,296.

XEROGRAPHY; "electron photography." Patented by Chester F. Carlson in 1940. Patent Number: 2,221,776.

D.D.T. (dichlorodiphenyltrichloroethane); "devitalizing composition of matter." Patented by Paul Muller in 1943. Patent Number: 2,329,074.

WIRE RECORDER, METHOD AND MEANS OF MAGNETIC RECORDING. Patented by Marvin Camras in 1944: "method and means of magnetic recording" Patent number: 2,351,004.

SULFONAMIDE; "diazine compounds." Patented by James M. Sprague in 1946. Patent Number: 2,407,966.

ANTIBIOTIC; "process and culture media for producing new penicillins." Patented by Otto K. Behrens, Joseph W. Corse, Reuben G. Jones, and Quentin F. Soper in 1949. Patent Number: 2,479,295.

DUCTILE CAST IRON; "cast ferrous alloy." Patented by Keith Dwight Millis, Albert Paul Gagnebin, and Norman Baden Pilling in 1949. Patent Number: 2,485,760.

DIURETIC; "heterocyclic sulfonamides and methods of preparation thereof." Patented by James W. Clapp and Richard O. Robin in 1951. Patent Number: 2,554,816.

CORTICOSTEROID; "oxygenation of steroids by mucorales fungi." Patented by Herbert C. Murray and Durey H. Peterson in 1952. Patent Number: 2,602,769.

TRANQUILIZER; "anti-excitatory compositions." Patented by Joseph Seifter, Anthony L.

Monaco, and Franklin Judson Hoover in 1957. Patent Number: 2,799,619.

"SATELLITE STRUCTURE." Patented by Robert C. Baumann in 1958. Patent Number: 2,835,548.

VIDEOTAPE RECORDER; "broad band magnetic tape system and method." Patented by Charles P. Ginsburg, Shelby F. Henderson, Jr., Ray M. Dolby, and Charles E. Anderson in 1960. Patent Number: 2,956,114.

INTEGRATED CIRCUIT; "semiconductor device-and-lead structure." Patented by Robert N. Noyce in 1961. Patent Number: 2,981,877.

GOEDESIC DOME; "suspension building." Patented by Richard Buckminster Fuller in 1964. Patent Number: 3,139,957.

"APPARATUS FOR COOLING AND SOLAR HEATING A HOUSE." Patented by Harry E. Thomason in 1967. Patent Number: 3,295,591.

"COMBINATION SMOKE AND HEAT DETECTOR ALARM." Patented by Sidney Jacoby in 1976. Patent Number: 3,938,115.

Design Patents

TYPE FACE; "printing type." Patented by George Bruce in 1842. Patent Number: Des. 1.

STATUE OF LIBERTY; "design for a statue." Patented by Auguste Bartholdi in 1879. Patent Number: Des. 11,023.

CONGRESSIONAL MEDAL OF HONOR; "design for a badge." Patented by George L. Gillespie in 1904. Patent Number: Des. 37,236.

LADY'S STOCKING; "design for a lady's stocking." Patented by William G. Bley in 1948. Patent Number: Des. 151,732.

Plant Patents

"CLIMBING OR TRAILING ROSE." Patented by Henry F. Bosenberg in 1931. Patent Number: Plant Pat. 1.

"GRAPEVINE." Patented by Chester A. Sanderson in 1948. Patent Number: Plant Pat. 782.

"STRAWBERRY PLANT." Patented by Frank J. Keplinger in 1953. Patent Number: Plant Pat. 1,183.

"APPLE TREE." Patented by Ralph Banta in 1971. Patent Number: Plant Pat. 3,045.

"WALNUT TREE." Patented by Louis Rodhouse in 1972. Patent Number: Plant Pat. 3,159.

"GRAPEFRUIT TREE." Patented by Richard A. Hensz in 1972. Patent Number: Plant Pat. 3,222.

Appendix F

Top 200 Corporations Receiving U.S. Patents in 1990

Intellectual Property Owners, Inc. (IPO) compiled this list from data provided by the U.S. Patent and Trademark Office. The numbers are patents issued during 1990 for which an assignment of title was of record in the Office when the patent was issued. Patents in the names of subsidiaries, other related companies, or divisions are not combined with patents in the name of the parent.

Rank	Company	Number	Rank	Company	Number
1.	Hitachi	908	38.	Hewlett—Packard Co.	218
2.	Toshiba Corp.	891	39.	Robert Bosch GmbH	204
3.	Canon K.K.	868	40.	Boeing Co.	192
4.	Mitsubishi Denki K.K.	862	41.	Amoco Corp.	187
5.	General Electric Co.	785	42.	Hoechst Celanese Corp.	182
6.	Fuji Photo Film Co., Ltd.	767	43.	Olympus Optical Co., Ltd.	179
7.	Eastman Kodak Co.	720	44.	Brother Kogyo K.K.	178
8.	U.S. Philips Corp.	637	45.	Imperial Chemical Industries PLC	176
9.	IBM Corp.	608	46.	Honeywell Inc.	167
10.	Siemens A.G.	506	47.	Sunstrand	155
11.	Bayer A.G.	499	48.	Ford Motor Co.	154
12.	E.I. du Pont de Nemours & Co.	481		Pioneer Electronic Corp.	154
13.	NEC Corp.	436	50.	United Technologies Corp.	151
14.	Westinghouse Electric Co.	435	51.	Toyota Jidosha K.K.	149
15.	AT&T Co.	429	52.	Konica Corp.	147
16.	Ciba—Geigy Corp.	409	53	Eaton Corp.	143
17.	Dow Chemical Co.	400	54.	American Cyanamid Co.	136
18.	BASF A.G.	394	55.	Merck & Co., Inc.	134
	Motorola Inc.	394	56.	Mazda Motor Corp.	132
20.	General Motors Corp.	379	57.	Aisin Seiki K.K.	131
21.	Minolta Camera Co., Ltd.	376		Phillips Petroleum Co.	131
22.	Nissan Motor Co., Ltd.	372	59.	Procter & Gamble Co.	127
23.	Honda Motor Co., Ltd.	363	60.	Exxon Research & Engineering Co.	126
24.	Mobil Oil Corp.	353	61.	Digital Equipment Corp.	124
25.	Sharp Corp.	349	62.	Fancu Ltd.	123
26.	Matsushita Electric Industrial Co.	340		Sumitomo Electric Industries	123
27.	3M Corp.	323	64.	Dow Corning Corp.	122
28.	Ricoh Co., Ltd.	292		GTE Products Corp.	122
29.	Texas Instruments, Inc.	284	66.	Henkel KGAA	117
30.	Hoechst A.G.	278	67.	Fuji Jukogyo K.K.	116
31.	Shell Oil Co.	263	68.	Rockwell International Corp.	115
32.	Fujitsu Ltd.	260	69.	Sumitomo Chemical Co., Ltd.	114
33.	Xerox Corp.	252		Texaco Inc.	114
34.	Hughes Aircraft Co.	248		Thomson—CSF	114
35.	Allied Signal Inc.	243	72.	Chrysler Motors Corp.	112
36.	Sony Corp.	238	73.	Sanyo Electric Co., Ltd.	111
37.	AMP Inc.	223	74.	Massachusetts Inst. of Tech.	110

Rank	Company	Number	Rank	Company	Number
75.	Ethyl Corp.	108		University of California	62
	Warner—Lambert Co.	108	142.	E.R. Squibb & Sons, Inc.	61
77.	W.R. Grace & Co.	105		Messerschmitt-Bolkow-Blohm GmbH	61
78.	Yamaha Corp.	104		Nippon Seiko K.K.	61
79.	Asahi Kogaku Kogyo K.K.	103		Samsung Electronics Co., Ltd.	61
	Daimler—Benz A.G.	103	146.	Colgate—Palmolive Co.	60
81.	PPG Industries Inc.	100	147.	Aluminum Co. of America	59
82.	Yazaki Corp.	99		Baxter International Inc.	59
83.	Air Products & Chemicals Inc.	98		RCA Licensing Corp.	59
84.	Mitsui Toatsu Chemicals Inc.	94	150.	Lever Brothers Co.	58
85.	Tektronic Inc.	90		National Research Development Corp.	58
86.	Alps Electric Co., Ltd.	89		Sulzer Brothers, Ltd.	58
	Deere & Co.	89	153.	B.F. Goodrich Co.	57
88.	Agency of Indus. Science & Tech.	87		Casio Computer Co., Ltd.	57
89.	North American Philips Corp.	86		Chevron Research Co.	57
	Pitney—Bowes, Inc.	86	153.	ITT Corp.	57
91.	Takeda Chemical Industries Ltd.	85		National Semiconductor Corp.	57
	UOP	85		Nippon Steel Corp.	57
93.	Brunswick Corp.	84		Raychem Corp.	57
	Mitsui Petrochemical Indus. Ltd.	84	160.	British Telecommunications	56
	Monsanto Co.	84		Diesel Kiki Co., Ltd.	56
	NCR Corp.	84		Emerson Electric Co.	56
	Union Carbide Chem & Plastics Co.	84		Fuji Xerox Co., Ltd.	56
98.	Corning Inc.	83		Leybold A.G.	56
	Goodyear Tire & Rubber Co.	83		Yoshida Kogyo K.K.	56
	NGK Insulators, Ltd.	83	166.	Union Oil Co. of California	55
101.	Advanced Micro Devices, Inc.	82		Zenith Electronics Corp.	55
	Bridgestone Corp.	82	168.	Becton, Dickinson & Co.	54
	OKI Electric Industry Co., Ltd.	82		Nalco Chemical Co.	54
	Pfizer Inc.	82		Seiko Epson Corp.	54
	Unisys Corp.	82	171.	Fujisawa Pharmaceutical Co., Ltd.	53
106.	Raytheon Co.	80		GTE Laboratories, Inc.	53
107.	Hoffmann—La Roche Inc.	78		Kanegafuchi Chemical Indus. Co.	53
108.	Nippondenso Co., Ltd.	76		Sanden Corp.	53
109.	FMC Corp.	75		Victor Co. of Japan, Ltd.	53
	Northern Telecom, Ltd.	75	176.	Caterpillar Inc.	52
	Rhone—Poulenc Chimie De Base	75		Conoco, Inc.	52
	Schering A.G.	75		Degussa A.G.	52
113.	Olin Corp.	74		KAO Corp.	52
114.	Eli Lilly & Co.	73		Mitsubishi Rayon Co., Ltd.	52
	Exxon Chemicals Patents Inc.	73		Semiconductor Energy Lab. Co.	52
	Harris Corp.	73		Square D Co.	52
117.	Institut Francais Du Petrole	72		TRW Inc.	52
	Nikon Corp.	72		University of Texas	52
119.	Sanshin Kogyo K.K.	71	183.	Boehringer Manheeim GmbH	51
120.	Fuji Electric Co., Ltd.	70		Emhart Industries, Inc.	51
	Grumman Aerospace Corp.	70		Halliburton Co.	51
122.	Commissariat A L'Energie Atomique	69		Hamamatsu Photonics K.K.	51
123.	Kimberly—Clark Corp.	68		Polaroid Corp.	51
	Murata Manufacturing Co., Ltd.	68		Toyoda Gosei K.K.	51
	Spectra—Physics, Inc.	68	191.	Alfred Teves GmbH	50
126.	Cooper Industries Inc.	67		Daikin Kogyo Co., Ltd.	50
	Mitsubishi Kasei Corp.	67		Kawasaki Steel Corp.	50
	Salomon S.A.	67		Kureha Chemical Industry Co.	50
129.	American Home Products Corp.	66		Nestec, S.A.	50
130.	Lucas Industries Public Ltd. Co.	65	196.	AGFA—Gevaert, A.G.	49
	Mead Co.	65		Akzo N.V.	49
	Mitsubishi Jukogyo K.K.	65		Dainippon Screen Mfg. Co., Ltd.	49
133.	Abbott Laboratories	64		Ford New Holland, Inc.	49
	Atlantic Richfield Co.	64		Hoechst—Roussel Pharm. Inc.	49
	ICI Americas Inc.	64		Huels A.G.	49
	Yamaha Motor Co., Ltd.	64		Mita Industrial Co., Ltd..	49
137.	Konishiroku Photo Industry Co., Ltd.	63		SGS—Thomson Microelectronics	49
	Omron Teteisi Electronics Co.	63		Societe Nationale Indus. Aerosp.	49
139.	Schlumberger Technology Corp.	62			
	Shin Etsu Chemical Co., Ltd.	62			

Appendix G

Glossary

This glossary contains some of the terms used in discussing commercialization and innovation processes. It was prepared by the U.S. Department of Energy and Argonne National Laboratory.

art-technology: An art-technology is one whose invention (or use) follows from know-how, craft skill, or experience, rather than from formal scientific and engineering knowledge. Inventions based on art-technology occur in virtually all fields. In many applications, such inventions are readily accepted. When they occur in industries based on formal scientific and engineering principles, however, art-technologies can face formidable market barriers. Roentgen's X-ray photography, which preceded scientific knowledge of radiation, was an art-technology that led to new scientific knowledge. In the computer industry, by way of contrast, the tantalizing possibilities inherent in art-technology, in the form of new software, or modifications to existing software packages, have created a market for art-technology that often results in the fragmentation of supposedly standardized technology into locally distinct usages dependent on know-how.

best-available-technology: In some highly regulated industries (such as hazardous waste disposal) government regulations mandate purchasing equipment or processes under a best-available-technology standard. Thus, if testing can establish that quality, a technology has a clear-cut marketing strategy. When the technology fails to prove itself "best," however, or when testing criteria work against

innovation, best-available-technology regulations erect virtually insurmountable *market barriers*. Common mythology also ascribes variants of best-available-technology standards to other, non-regulated industries: "When they see it they'll have to buy it," says the inventor. In fact, counting on best-available-technology standards seldom, if ever, constitutes a viable marketing strategy outside those few closely regulated industries. Even there, marketing strategies relying on best-available-technology standards are likely to become time-consuming, frustrating, and very risky. (See **20/30 rule; market barrier.**)

boiler plate: Those standard, legal sounding paragraphs appearing in all contracts, such as licensing agreements, and in most venture capital and investment documents.

bootstrapping: See **financing.**

business plan: A standard business document, the business plan (typically 25-35 pages long) is a written statement intended to crystallize business objectives, inform readers about the business, and provide a guidebook for managing the company. Often used as a prospectus when seeking financing, the standard business plan will contain a brief executive summary, a history and description

979

of the business, and sections detailing the company's market analysis, marketing strategy, financial projections, organization, and capitalization. A typical business plan may also contain appendices detailing such things as patents, financial projections, explanations of special problems or capabilities, and resumes for the company's key personnel.

capital: The total money and property a business owns or has at its disposal. It is important to recognize how the various specific types and sources of capital typically correlate with the technical, marketing, and business development steps in the commercialization process for small businesses:

> **sweat equity**—The unpaid effort and labor the owner of an intellectual property brings to the commercialization process. Actually a form of capital, sweat equity (along with personal and family savings) will usually suffice to move from concept to working model and to make the first serious passes at market analysis and business planning. In some cases sweat equity and personal savings will take a technical development program through the engineering prototype.

> **seed capital**—Early stage, limited capital (typically in the 25,000 to 100,000 range for the very earliest stages, 100K to 500K later). Usually raised locally through networks of friends and informal investors, seed capital will probably bring a technical development program to production prototyping while market analysis and business planning become formalized.

> **pre-venture capital**—Typically in the 500K to 1M range, pre-venture capital often brings more active involvement from investors. This is the capital that commonly produces product qualification models, limited production, and the first introduction of the product or process into the market. Market strategy and business planning must be set, even as they still require fine tuning.

> **venture capital**—Formal (or institutional) venture capital is almost always the last form of equity capital to appear in the commercialization process (other than an SEC regulated stock offering). Usually 1M and up, venture capital is most often available to businesses that already have achieved market penetration and are headed toward the break-even point. Formal venture capitalists are only interested in businesses that have potential for rapid growth. Anyone seeking venture capital must recognize the implications of the 10/5 rule (as a basic standard, formal venture capitalists expect start-ups to produce a 10 times return on investment in 5 years). Full production capability, a real market and defined marketing strategy, and a working business structure—these are the things that attract venture capital.

captive inventor: Inventorship and ownership of an invention are actually separate issues. Ownership, which by definition involves "property," can become a contractual matter. A captive inventor is one who works under an arrangement that assigns ownership to someone else (usually a situation specified among the terms of employment). Determining ownership of an invention can become a complex legal matter, and some states have enacted laws governing the circumstances under which ownership of an invention is assigned to an employer, rather than to the employee-inventor.

cash flow: One of the most important financial measures for any business. Cash flow is the difference between the amount of money coming in during a given time and the amount going out over the same time (usually the short term—calculated in months, or even in weeks or days). When the money coming in is greater, there is positive cash flow. When expenses exceed income, a business has a negative cash flow. The importance of a positive cash flow is seen in the plight of any small company with few cash reserves, a large backlog of new orders and a negative cash flow. At best, such a business will need substantial new credit or loans to meet short-term expenses; at worst, negative cash flow will spell disaster for an otherwise healthy firm.

cross-licensing: In many industries—such as automobiles, equipment manufacturing, and petroleum, communications—individual companies commonly exchange technology through cross-licensing agreements. Under such agreements, firms typically grant royalty-free licenses to other participants, in exchange for reciprocal rights to their competitors' technologies. In effect, such cross-licensing agreements create industry-wide technology pools. (See licensing; royalty-free license.)

due-diligence: A legal term, due-diligence refers to the formal investigative procedures a business must undergo when entering into certain regulated financial arrangements, such as making a public stock offering. More generally, inventors and small businesspersons might be well advised to pursue their own due diligence investigations when negotiating with investors or prospective licensees.

end user: The actual user of a technological product or those products derived from technological processes. The significance of this term appears when the end user is distinguished from the customer. Frequently the customer (the person who actually buys) and the end user are different individuals, as is almost always the case with industrial tools, supplies, or products—always the case with sales to an OEM (original equipment manufacturer). The customer and the end user do not always share the same incentives to buy a new technology, and the difference in their willingness to employ innovations often forms a critical market barrier. Distinguishing the end user from the customer can be the most crucial step in developing an effective marketing strategy.

engineering prototype: See prototype.

entrepreneur: A person who undertakes to start and operate a business, usually assuming the greater part of the financial risks involved—and consequently reaping a large part of any rewards earned. In the commercialization of new technologies, the entrepreneur is frequently someone other than the inventor.

equity: Normally describes the total value of the preferred and common stock of a business. The term equity is also used frequently in describing the percentage of ownership a person or group holds in a business.

exclusive license: See license.

exit: The sale of equity (ownership) in a business.

exit strategy: The plan or method those holding shares of ownership intend to use when liquidating equity.

financing: A general term used to describe the ways to acquire capital necessary for establishing, operating, or expanding a business. While financing strategies vary considerably in complexity, those most small businesses can use for sustainable commercialization fall into just three types:

> bootstrapping—Self-generated financing from current income (requires a reliable positive cash flow).

> debt financing—Borrowed money.

> equity financing—Sale of a share in ownership to acquire capital.

intellectual property: A general term describing the legally protected ownership of copyrights, inventions, know-how, logos, patents, service marks, trademarks, trade names, or trade secrets.

invasionary technology: A technology or technological process whose commercialization requires competing directly with other technologies already dominating that particular market.

license: An agreement under which the owner of an intellectual property allows someone else to make, use, or sell things protected by ownership. With an exclusive license the licensee gains sole right to employ the intellectual property governed by the license, although such a license may carry limitations on territory, field of use, product, or time. Under a limited or nonexclusive license the person granting the license is free to grant other similar licenses on the same intellectual property. (Also see cross-licensing; royalty-free licensing.)

licensee: The person or company gaining rights to an intellectual property under a licensing agreement.

licensing: The general term describing the legal process in which a license is granted on an intellectual property. One of the two basic commercialization strategies available to individual inventors. (Also see **venturing**; **cross-licensing**.)

licensor: The person who grants use of an intellectual property under a licensing agreement.

limited license: See **license**.

linchpin technology: A technology for which commercialization increases the market potential for other supporting or ancillary technologies. In some cases, commercialization of a linchpin technology will actually call for the invention of new technologies, just as inventing the light bulb called for new electrical generating, transmission, and distribution technologies. In other cases, the linchpin technology will reorder or revitalize existing technology, as the automobile did to the petroleum refining industry. Generally, linchpin inventions face formidable market barriers.

market barrier: Those obstacles other than the needs for technical development, market analysis, and business planning that must be overcome in commercializing a technology. Indeed, the normal commercialization activities (technical development, market analysis, and business planning) will expose market barriers, which can be things like extraordinary capital costs, user acceptance problems, the need to establish extensive advertising, sales, distribution, user education, or maintenance capabilities, the NIH syndrome, linchpinning, or an inability to meet the 20/30 rule. Obviously, no list of market barriers can be exhaustive, but all such barriers must be identified and addressed before sustainable commercialization is really possible.

market channels: The step-by-step paths along which technologies move from producer to the end user. Writing these out (or diagraming them) is one of the basic first steps in market analysis.

marketing: Those activities involved in analyzing the sales potential of a product or process, as well as those activities involved in customer service, advertising, distribution, and selling. In the commercialization process marketing actually breaks down into three vital parts:

> **market research and planning**—Analysis and evaluation of the market, which includes such tasks as identifying market barriers, channels of distribution, market size, and who will buy. Market research should begin at the concept development stage, and play a continuing role in technical development as well as in developing market strategy and business organization.

> **market management**—Advertising, promotion, and customer service. These critical service functions play a central role in sustaining the commercialization process.

> **sales and distribution**—Management of the channels of distribution and sales force. By definition, sales and distribution are the obvious goals of any commercialization effort. Less obviously, perhaps, these activities can also furnish important information leading to product improvements, the development of new applications, or even to new technologies.

model: See **prototype**.

negative cash flow: See **cash flow**.

NIH: Initials standing for "Not Invented Here," a phrase used to describe industry reluctance to adopt innovations originating outside that industry's normal R&D channels. The NIH syndrome can form a crucial market barrier, especially in some of the older, more established technology-based industries such as automobiles, steel, oil, metallurgy, or transportation.

OEM: Initials standing for Original Equipment Manufacturer. Such firms typically purchase various parts, supplies, or even sub-assemblies from other manufacturers. (See **end user**.)

paid-up license: See **royalty-free license**; **license**.

positive cash flow: See **cash flow**.

product qualification model: See prototype.

production prototype: See prototype.

prototype: A prototype can be a mock-up, model, or actual working version of a technological device or process. Prototypes are used to generate information that will help design or perfect the final product/process.

> **working model**—A reduction to practice, proof of concept. The working model is often less than full-scale, inexpensively and crudely constructed, and need not function optimally. Intended to test the most basic operating parameters and to aid in the design of an engineering prototype.

> **engineering prototype**—An actual working version of a product, apparatus, or process used to gather data on operation, performance, and production requirements. Most often one-of-a-kind and commonly fitted with special instrumentation, this model is usually hand-made, but always of sufficient technical quality to determine whether a production prototype can (or should) be built.

> **production prototype**—A full-scale, completely operational model designed to determine production and fabrication requirements for the production item. Also used to generate the final pre-production performance data on operation and durability. Usually hand-built, the production prototype must conform as closely as possible to the design standards for the final full-production product or process.

> **production qualification model**—A full-scale, fully operational model manufactured in an initial, limited production run under conditions as close as possible to final production. Used to ensure final production runs will produce a product meeting design standards. Product qualification prototypes are often subjected to independent third-party testing, especially if the product must meet industry or government regulatory standards.

Together, the sequential development of these various prototypes and models forms the core of a complete technical development program, one that will lead to a viable production item or process.

royalty-free license: A license requiring no further royalty payments. Also called a paid-up license. At times such licenses are granted with an up-front, one-time cash payment. Other times they are granted without any financial consideration involved; this is particularly the case under cross-licensing agreements and with government use of inventions developed under public funding.

seed capital: See capital.

sweat equity: See capital.

technology: Commonly thought of simply as mechanical or science-based ways of doing work, this word actually warrants careful attention. "Technology" comes in all varieties, and on all scales, from the smallest consumer item to vast industrial complexes. For the sake of clarity it is worthwhile to point out that all technologies, large or small, will fall into one of four categories:

> **product**—An actual thing to be manufactured, used, or consumed.

> **process**—A way of doing things, making things, or controlling a manufacturing activity.

> **tool**—Those things needed to make products or implement a process. (Something will be a tool to end users, even while those who manufacture and sell it consider it a "product.")

> **know-how**—Knowledge or experience allowing effective and economical use of technological products, processes, or tools. Often mistakenly considered intangible, or even of negligible commercial value, know-how actually constitutes one of the most marketable intellectual properties inventors can bring to the commercialization process. In some industries, electronics for example, know-how often furnishes the only basis for commercialization, whether through venturing or licensing.

All four of these technological entities can be protected as intellectual property, and any of the four can become the object of commercialization. Indeed, with some inventions commercialization may be possible through more than one of these four technology categories. In that case, deciding whether to commercialize the invention as product, process, tool, or know-how constitutes a crucial first step toward the market. When commercialization requires developing an invention through more than one of these forms, the invention is probably a linchpin technology.

10/5 rule: See **venture capital** under the glossary listing for **capital**.

20/30 rule: A very general rule of thumb for assessing market potential with an invasionary technology. Variously stated by different people, the 20/30 rule really just says that to succeed in the market a new technology must do its job 20 percent better and 30 percent cheaper (or vice-versa) than existing technology. (Also see **best-available-technology**.)

venture capital: See **capital**.

venturing: A general term to describe a commercialization strategy based on creating a new business. Sometimes the meaning of venturing is expanded to describe a commercialization involving significant expansion of an existing small business. One of the two basic commercialization strategies available to individual inventors. (See **entrepreneur; licensing**.)

working model: See **prototype**.

Master Index

Names of organizations, programs, publications, and institutions that appear in both the essay and directory sections are arranged alphabetically in this index. Essay citations appear with page numbers (p. or pp. precede the italicized page reference). Directory citations appear with an entry number; a star (★) before an entry number indicates a name mentioned within the text of an entry. Names may also be cited under the keywords they contain (e.g., "Creative Behavior, The Journal of").

Master Index

Master Index

Master Index

Master Index

J

K

L

Master Index

Master Index

P

Master Index

Q

R

Master Index

Weapons Laboratory; Air Force – U.S.
 Department of the Air Force.........1703, 2084
Weapons Laboratory; Benet – U.S.
 Department of the Army......................1730
Weapons Station Earle; Naval – U.S.
 Department of the Navy......................2175
Weapons Station; Naval – U.S. Department
 of the Navy...............................1764
Weapons Support Center; Naval – U.S.
 Department of the Navy......................1765
Weapons Systems Center; Naval – U.S.
 Department of the Navy......................1766
Weed Science Laboratory; Southern – U.S.
 Department of Agriculture..................1947
Weed Science Research Laboratory;
 Foreign Disease- – U.S. Department of
 Agriculture 1887
Weeds and Crops Research; Nematodes, –
 U.S. Department of Agriculture – Georgia
 Coastal Plain Experiment Station 1890
Weekly Reader National Invention
 Contest 2372
Weisfeld; Lewis B..............................412
Weiss, Peck and Greer Venture Partners
 L.P. (Boulder).............................963
Weiss, Peck and Greer Venture Partners
 L.P. (New York)...........................1293
Weiss, Peck and Greer Venture Partners
 L.P. (San Francisco) 956
Welsh, Carson, Anderson, and Stowe........1294
Wendt Library; Kurt F. – University of
 Wisconsin—Madison 2687
West Palm Beach Post of Duty – U.S. Small
 Business Administration 1565
West Virginia Capital Company Credit
 Program – West Virginia Economic
 Development Authority 1540
West Virginia Economic Development
 Authority – West Virginia Capital
 Company Credit Program 1540
West Virginia Small Business Development
 Center.....................................748
Western Cotton Research Laboratory – U.S.
 Department of Agriculture 1969
Western Financial Capital Corp............1412
Western Illinois University – Center for
 Business and Economic Research 553
Western Illinois University – Technology
 Commercialization Center 554
Western Michigan University – Institute for
 Technological Studies......................603
Western Michigan University – Technology
 Transfer Center............................604
Western New York Invention Program 2373
Western Regional Research Center – U.S.
 Department of Agriculture 1970
Westinghouse Credit Corp. – VenWest,
 Inc..1351
WESTLAW Intellectual Property Library.....2613
WESTLAW Texts and Periodicals
 Library....................................2614
Westman & Associates; Robert A. 395
Wetlands Research Center; National – U.S.
 Department of the Interior – Fish and
 Wildlife Service 2133
What to Do? 2502
What's Next................................2503
Wheeling Jesuit College – National
 Technology Transfer Center 749
White River Capital Corp. 1046
White, Robert M. ....*p. 234*
White Sands Missile Range – U.S.
 Department of the Army.....................1740
Who's Who in Venture Capital..............2337
Wichita District Office – U.S. Small Business
 Administration.............................1579

Wichita State University – Ablah Library.....2642
Wichita State University – Center for
 Entrepreneurship..........................568
Wichita State University – Kansas Small
 Business Development Center 569
Wiedenmann & Associates 151
Wildlife Health Research Center; National –
 U.S. Department of the Interior 2143
Wildlife Research Center—Anchorage;
 Alaska Fish and – U.S. Department of the
 Interior 2126
Wildlife Research Center; Denver – U.S.
 Department of the Interior 2130
Wildlife Research Center; Northern Prairie –
 U.S. Department of the Interior 2144
Wildlife Research Center; Patuxent – U.S.
 Department of the Interior 2145
Wilkes-Barre Branch Office – U.S. Small
 Business Administration 1625
William B. Hardy...........................320
William Blair and Co. (Chicago)...........1040
William Blair and Co. (Denver) 964
William G. Garner.........................152
William H. Bates Linear Accelerator Center
 – U.S. Department of Energy 2049
Willis S. Steinitz.........................387
Wilmington Branch Office – U.S. Small
 Business Administration 1560
Wind Point Partners (Chicago)............1041
Wind Point Partners (Racine)............1443
Wine Investor.............................153
Winfield Capital Corp.....................1295
Wisconsin Bureau of Development Finance
 – SBIR Bridge Financing Program 1541
Wisconsin Bureau of Development Finance
 – Wisconsin SBIR Bridge Financing
 Program...................................1822
Wisconsin for Research, Inc...............760
Wisconsin Intellectual Property Law
 Association 2738
Wisconsin SBIR Bridge Financing Program
 – Wisconsin Bureau of Development
 Finance....................................1822
Wisconsin Small Business Development
 Center – University of Wisconsin 753
Wise Owl.................................2504
Wolfensohn Associates, L.P. (New York)...1296
Wolfensohn Associates, L.P. (Palo Alto) ...957
Wolff Associates, Inc. 321
Wolff Consultants.........................292
*Women Who Invent or Want To; The Book
 for* 2386
Wood River Capital Corp...................1297
Worcester Area Inventors USA 37
Worcester Polytechnic Institute –
 Management of Advanced Automation
 Technology Center 592
World Bank of Technology 2615
*World Directory of Sources of Patent
 Information*..............................2338
World Electronic Developments...........2339
World Intellectual Property Report.......2340
World Patent Information.................2341
World Patents Index.......................2616
*World Technology/Patent Licensing
 Gazette*..................................2342
WPI/APIPAT................................2617
WPI (World Patents Index) 2616
Wrenn, Bob ....*p. 240*
Wright Aeronautical Laboratories; Air Force
 – U.S. Department of the Air Force 2085
Writing in Action; Creative.............2404
Wyoming Science, Technology and Energy
 Authority 1542
Wyoming Small Business Development
 Center....................................761

X

Xerox Venture Capital (Stamford) 986
Xpand Inc.....................................236

Y

Yakima Agricultural Research Laboratory –
 U.S. Department of Agriculture 1971
Yale University – Economic Growth
 Center.....................................504
Yale University – Office of Cooperative
 Research 505
Yankee Ingenuity Program.....................75
Ychem International Corporation..............154
Yellowstone Inventors Association.............56
Young Inventors Fair.......................2374
Young Inventors Program; New
 Hampshire 2361
Young Inventors Program; New York 2365
Youth Education Program; Inventors
 Association of New England 2353
Youth Programs; Inventors Association of
 St. Louis..................................2354
Youth Research (Division of Consumer
 Sciences Inc.) 184
Yuma Proving Ground – U.S. Department of
 the Army 1741
Yusa Capital Corp. 1298

Z

Z.J. Loussac Public Library – Anchorage
 Municipal Libraries.......................2620
Zanes Communications Consultants............200
Zero Stage Capital Co., Inc.
 (Cambridge)...............................1130
Zero Stage Capital Co., Inc. (State
 College) 1352
Zoological Park & Conservation Center;
 National – Smithsonian Institution.........1850